Elsevier Academic Press
525 B Street, Suite 1900, San Diego, California 92101-4495, USA
84 Theobald's Road, London WC1X 8RR, UK

Library of Congress Cataloging-in-Publication Data

Biomaterials science : an introduction to materials in medicine / edited by
Buddy D. Ratner ... [et al.].– 2nd ed.
    p. ; cm.
  Includes bibliographical references and index.
  ISBN 0-12-582463-7 (hardcover : alk. paper)
  1. Biomedical materials.
  [DNLM: 1. Biocompatible Materials. QT 37 B6145 1996] I. Ratner, B. D.
(Buddy D.), 1947-
  R857.M3B5735 2004
  610′.28–dc22

                             2003027823

British Library Cataloguing in Publication Data
A catalogue record for this book is available from the British Library

ISBN: 0-12-582463-7

For all information on all Academic Press publications
visit our Web site at www.academicpress.com

Printed in China
04 05 06 07 08   9 8 7 6 5 4 3 2 1

BIOMATERIALS SCIENCE

An Introduction to Materials in Medicine
2nd Edition

Edited by

Buddy D. Ratner, Ph.D.
Professor, Bioengineering and Chemical Engineering
Director of University of Washington Engineered Biomaterials (UWEB), an
NSF Engineering Research Center
University of Washington, Seattle, WA USA

Allan S. Hoffman, ScD.
Professor of Bioengineering and Chemical Engineering
UWEB Investigator
University of Washington, Seattle, WA USA

Frederick J. Schoen, M.D., Ph.D.
Professor of Pathology and Health Sciences and Technology (HST)
Harvard Medical School
Executive Vice Chairman, Department of Pathology
Brigham and Women's Hospital
Boston, MA USA

Jack E. Lemons, Ph.D.
Professor and Director of Biomaterials Laboratory Surgical Research
Departments of Prosthodontics and Biomaterials, Orthopaedic Surgery/Surgery and
Biomedical Engineering, Schools of Dentistry, Medicine and Engineering
University of Alabama at Birmingham, AL USA

ELSEVIER
ACADEMIC
PRESS

Amsterdam Boston Heidelberg London New York Oxford
Paris San Diego San Francisco Singapore Sydney Tokyo

CONTENTS

PART II
BIOLOGY, BIOCHEMISTRY, AND MEDICINE

CHAPTER 3   Some Background Concepts

CHAPTER 4   Host Reactions to Biomaterials and Their Evaluation

CHAPTER 5   Biological Testing of Biomaterials

CHAPTER 6   Degradation of Materials in the Biological Environment

EDITORS AND LEAD CONTRIBUTORS

Numbers in parentheses indicate the pages on which the authors' contributions begin.

Harold Alexander (180) Orthogen Corporation, Springfield, NJ 07081

James M. Anderson (296, 360, 771) Institute of Pathology, Case Western Reserve University, Cleveland, OH 44106

Richard W. Bianco (379) Division of Experimental Surgery, Department of Surgery, University of Minnesota, Minneapolis, MN 55455

John B. Brunski (137, 823) Department of Biomedical Engineering, Rensselaer Polytechnic Institute, Troy, NY 12180

Thomas M. S. Chang (507) Artificial Cells and Organ Research Centre, McGill University, Montreal, Quebec H3G 1Y6, Canada

Andér Colas (80, 697) Dow Corning Life Sciences, B-7180 Seneffe, Belgium

Francis W. Cooke (23) Orthopedic Research Institute, Wichita, KS 67214

Stuart L. Cooper (67) Ohio State University, Department of Chemical Engineering, Raleigh, Columbia, OH 43210

Joachim Kohn (115) Department of Chemistry, Rutgers, The State University of New Jersey, Piscataway, NJ 08854

Bill Costerton (345) Center for Biofilm Engineering, College of Engineering, Montana State University, Bozeman, MT 59717

Arthur J. Coury (411) Biomaterials Research, Genzyme Corporation, Cambridge, MA 02139

A. Norman Cranin (555) Brookdale University Hospital and Medical Center, The Dental Implant Group, Brooklyn, NY 11212

Jim Curtis (80, 697) Life Sciences Industry, Medical Device Operations, Dow Corning Corporation, Midland, MI 48686

Elaine Duncan (788) Paladin Medical, Stillwater, MN 55082

Nadim James Hallab (526) Department of Orthopedic Surgery, Rush Medical College, Chicago, IL 60612

Stephen R. Hanson (328, 367) Department of Biomedical Engineering, Oregon Health Sciences University, Beaverton, OR 97006

Kip D. Hauch (396) Department of Chemical Engineering, University of Washington, Seattle, WA 98195

Jorge Heller (628) A.P. Pharma, Department of Research, Redwood City, CA 94063

Larry L. Hench (153) Department of Materials, Imperial College of Science, Technology and Medicine, University of London, London SW7 2BP, United Kingdom

Arne Hensten-Pettersen (328) Scandinavian Institute of Dental Materials (NIOM), Haslum, Norway

Simon P. Hoerstrup (712) Clinic for Cardiovascular Surgery, University Hospital, CH8091, Zurich, Switzerland

Allan S. Hoffman (1, 67, 109, 197, 201, 225, 628, 760, 805) Department of Bioengineering, University of Washington, Seattle, WA 98195

Thomas A. Horbett (234) Center for Bioengineering and Department of Chemical Engineering, University of Washington, Seattle, WA 98195

John A. Jansen (218) Department of Biomaterials, Dental School, University of Nijmegen, 6500 HB, Nijmegen, The Netherlands

Richard J. Johnson (318) Exploratory Research, Baxter Healthcare Coporation, Round Lake, IL 60073

John B. Kowalski (754) Sterilization Science & Technology, Johnson & Johnson, New Brunswick, NJ 08906

Jack E. Lemons (1, 23, 455, 783, 805) Department of Biomaterials and Surgery, School of Dentistry and Medicine, University of Alabama, Birmingham, AL 35294

Michael J. Lysaght (728) Center for Biomedical Engineering, Brown University, Providence, RI 02912

Paul S. Malchesky (514) International Center for Artificial Organs and Transplantation, Painesville, OH 44077

Cristina L. Martins (819) INEB-Instituto de Engenharia Biomédica, Laboratório de Biomateriais, Universidade do Porto, 4150-180 Porto, Portugal

Jay P. Mayesh (797) Kaye, Scholer, LLP, New York, NY 10022

Larry V. McIntire (282) Department of Bioengineering, Institute of Bioscience & Bioengineering, Rice University, Houston, TX 77005

Claudio Migliaresi (181) Department of Materials Engineering and Industrial Technologies, University of Trento, 38050 Trento, Italy

Antonios G. Mikos (735) Department of Bioengineering, Rice University, Houston, TX 77251

Richard N. Mitchell (246, 260, 304) Department of Pathology, Brigham and Women's Hospital, Boston, MA 02115

Robert B. More (170) Medical Carbon Research Institute, Austin, TX 78754

Jeffrey R. Morgan (602) Department of Molecular Pharmacology, Physiology, and Biotechnology, Biomedical Center, Providence, RI 02912

Sharon J. Northup (356) Northup RTS, Highland Park, IL 60035

Robert F. Padera, Jr. (470) Department of Pathology, Brigham and Women's Hospital, Boston, MA 02115

Anil S. Patel (591) Alcon Labs, Seattle, WA 98115

Nicholas A. Peppas (100) Department of Chemical Engineering, The University of Texas at Austin, Austin, TX 78712

Buddy D. Ratner (1, 10, 40, 197, 201, 237, 355, 367, 411, 803) University of Washington Engineered Biomaterials, University of Washington, Seattle, WA 98195

Miguel F. Refojo (583) Department of Opthalmology, The Schepens Eye Research Institute, Harvard Medical School, Boston, MA 02114

Mark S. Roby (614) United States Surgical, North Haven, CT 06473

Subrata Saha (793) Biomedical Engineering Science Program, Alfred University, Alfred, NY 14802

Frederick J. Schoen (1, 246, 260, 293, 338, 360, 439, 455, 470, 709, 753, 760, 771, 805) Department of Pathology, Brigham and Women's Hospital, Harvard Medical School, Boston, MA 02115

Michael V. Sefton (456) Institute of Biomaterials and Biomedical, University of Toronto, Toronto, ON M53 3G9, Canada

Steven M. Slack (813) Department of Biomedical Engineering, University of Memphis, Memphis, TN 38152

Dennis C. Smith (572) Centre for Biomaterials, University of Toronto, Toronto, ON L9Y 3Y9, Canada

Francis A. Spelman (656) Advanced Cochlear Systems, Snoqualmie, WA. Department of Bioengineering, University of Washington, Seattle, WA 98195

Peter J. Tarcha (684) Abbott Laboratories, Department of Advanced Drug Delivery, Abbott Park, IL 60064

Ramakrishna Venugopalan (648) Codman and Shurtleff, A J&J Company, Raynham, MA 02767

Ivan Vesely (32) The Saban Research Institute of Children's Hospital, Los Angeles, Los Angeles, CA 90027

Erwin A. Vogler (59) Department of Materials Science and Engineering and Bioengineering, Materials Research Institute, Penn State University, University Park, PA 16802

William R. Wagner (454) Presbyterian University Hospital, University of Pittsburgh, Pittsburgh, PA 15219

Steven Weinberg (86) Biomedical Device Consultants and Laboratories, Inc., Webster, TX 77598

David F. Williams (430) Department of Clinical Engineering, Royal Liverpool University Hospital, The University of Liverpool, Liverpool, L69 3BX, United Kingdom

Paul Yager (669) Department of Bioengineering, University of Washington, Seattle, WA 98195

Ioannis V. Yannas (127) Department of Mechanical Engineering, Massachusetts Institute of Technology, Cambridge, MA 02139

PREFACE

The interest and excitement in the field of biomaterials has been validated by sales of the first edition of this textbook: more than 10,000 copies sold. Also, the first edition has been widely adopted for classroom use throughout the world. The concept behind the first edition was that a balanced textbook on the subject of biomaterials science was needed. As with the first edition, the intended audience is multidisciplinary: students of medicine, dentistry, veterinary science, engineering, materials science, chemistry, physics, and biology (not an all-inclusive list) can find essential introductory material to permit them a reasonably knowledgeable immersion into the key professional issues in biomaterials science.

Textbooks by single authors too strongly emphasize their own areas of expertise and ignore other important subjects. Articles from the literature are commonly used in the classroom setting, but these are difficult to weave into a cohesive curriculum. Handout materials from professors are often graphically unsophisticated, and again, slanted to the specific interests of the individual. In *Biomaterials Science: An Introduction to Materials in Medicine*, 2nd edition, we the editors (whose 140+ person-year expertise spans materials science, pathology, and hard- and soft-tissue applications), endeavor to present a balanced perspective on an evolving discipline by integrating the experience of many leaders in the biomaterials field. Balanced presentation means appropriate representation of hard biomaterials and soft biomaterials; of orthopedic ideas, cardiovascular concepts, ophthalmologic ideas, and dental issues; a balance of fundamental biological concepts, materials science background, medical/clinical concerns, and government/societal issues; and coverage of biomaterials past, present, and future. In this way, we hope that the reader can visualize the scope of the field, absorb the unifying principles common to all materials in contact with biological systems, and gain a solid appreciation for the special significance of the word *biomaterial* as well as the rapid and exciting evolution and expansion of biomaterials science and its applications in medicine.

More than 108 biomaterials professionals from academia, industry, and government have contributed to this work. Certainly, such a distinguished group of authors provides the needed balance and perspective. However, such a diverse group of authors also leads to unique complexities in a project of this type. Do the various writing styles clash? Does the presentation of material, particularly controversial material, result in one chapter contradicting another? Even with so many authors, all subjects relevant to biomaterials cannot be addressed—subjects should be included and which left out? How should such a project be refereed to ensure scientific quality, pedagogical effectiveness, and the balance we strive for?

These are some of the issues the editors grappled with over the years from conception of the second edition in 1998 to publication in 2004. From this complex editorial process, a unique volume has evolved that the editors feel can make an ongoing contribution to the development of the biomaterials field. An educational tool has been synthesized here directing those new to biomaterials, be they engineers, physicians, materials scientists, or biochemists, on a path to appreciating the scope, complexity, basic principles, and importance of this enterprise.

What's new in *Biomaterials Science: An Introduction to Materials in Medicine*, 2nd edition? All chapters have been updated and rewritten, most extensively. A large number of new chapters have been added. The curricular organization for teaching the fundamental cell biology, molecular biology, tissue organization, and histology, key subjects that support the modern biomaterials research endeavor, has been restructured. A new, three-chapter section on tissue engineering has been added. The total content and size of the book have been significantly increased. A Web site has been coupled to the book offering supplemental material including surgery movies and homework problems. The graphics design has been upgraded. You have in your hands a new book that can address biomaterials in the 21st century.

Acknowledgments and thanks are in order. First, let us address the Society For Biomaterials that served as sponsor and inspiration for this book. The Society For Biomaterials is a model of "scientific cultural diversity" with engineers, physicians, scientists, veterinarians, industrialists, inventors, regulators, attorneys, educators, and ethicists all participating in an endeavor that is intellectually exciting, humanitarian, and profitable. As with the first edition, all royalties from this volume are being returned to the Society For Biomaterials to further education and professional advancement related to biomaterials. For further information on the Society For Bio-materials, visit the SFB Web site (http://www.biomaterials.org/).

Next, we offer a special thanks to those who enthusiastically invested time, energy, experience, and intelligence to author the chapters that are this textbook. The many scientists, physicians, and engineers who contributed their expertise and perspectives are clearly the backbone of this work and they deserve high praise—their efforts will strongly affect the education of the next generation of biomaterials scientists. Also, some reviewers assisted the editors in carefully refereeing chapters. We thank Kip Hauch, Colleen Irvin, Gayle Winters, Tom Horbett, and Steven Slack.

The support, encouragement, organizational skills, and experience of the staff, first at Academic Press and now at Elsevier Publishers, have led this second edition from vision to volume. Thank you, Elsevier, for this contribution to the field of biomaterials.

Finally, a unique person at the University of Washington has contributed to the assembly and production aspects of this work. We offer special thanks to Elizabeth Sharpe for her superb editorial/organizational efforts. This volume, deep down, has Elizabeth's intelligent and quality-oriented stamp all over it. Clearly, she cares!

The biomaterial field has always been ripe with opportunities, stimulation, compassion, and intellectual ideas. As a field we look to the horizons where the new ideas from science, technology, and medicine arise. We aim to improve the quality of life for millions through biomaterials-based, improved medical devices and tissue engineering. We editors hope the biomaterials overview you now hold will stimulate you as much as it has us.

Buddy D. Ratner
Allan S. Hoffman
Jack E. Lemons
Frederick J. Schoen
May 2004

Biomaterials Science: A Multidisciplinary Endeavor

BUDDY D. RATNER, ALLAN S. HOFFMAN, FREDERICK J. SCHOEN, JACK E. LEMONS

BIOMATERIALS AND BIOMATERIALS SCIENCE

Biomaterials Science: An Introduction to Materials in Medicine addresses the properties and applications of materials (synthetic and natural) that are used in contact with biological systems. These materials are commonly called biomaterials. Biomaterials, an exciting field with steady, strong growth over its approximately half century of existence, encompasses aspects of medicine, biology, chemistry, and materials science. It sits on a foundation of engineering principles. There is also a compelling human side to the therapeutic and diagnostic application of biomaterials. This textbook aims to (1) introduce these diverse elements, particularly focusing on their interrelationships rather than differences and (2) systematize the subject into a cohesive curriculum.

We title this textbook *Biomaterials Science: An Introduction to Materials in Medicine* to reflect, first, that the book highlights the scientific and engineering fundamentals behind biomaterials and their applications, and second, that this volume contains sufficient background material to guide the reader to a fair appreciation of the field of biomaterials. Furthermore, every chapter in this textbook can serve as a portal to an extensive contemporary literature. The magnitude of the biomaterials endeavor, its interdisciplinary scope, and examples of biomaterials applications will be revealed in this introductory chapter and throughout the book.

Although biomaterials are primarily used for medical applications (the focus of this text), they are also used to grow cells in culture, to assay for blood proteins in the clinical laboratory, in equipment for processing biomolecules for biotechnological applications, for implants to regulate fertility in cattle, in diagnostic gene arrays, in the aquaculture of oysters, and for investigational cell-silicon "biochips." How do we reconcile these diverse uses of materials into one field? The common thread is the interaction between biological systems and synthetic or modified natural materials.

In medical applications, biomaterials are rarely used as isolated materials but are more commonly integrated into devices or implants. Although this is a text on materials, it will quickly become apparent that the subject cannot be explored without also considering biomedical devices and the biological response to them. Indeed, both the effect of the materials/device on the recipient and that of the host tissues on the device can lead to device failure. Furthermore, a biomaterial must always be considered in the context of its final fabricated, sterilized form. For example, when a polyurethane elastomer is cast from a solvent onto a mold to form the pump bladder of a heart assist device, it can elicit different blood reactions than when injection molding is used to form the same device. A hemodialysis system serving as an artificial kidney requires materials that must function in contact with a patient's blood and also exhibit appropriate membrane permeability and mass transport characteristics. It also must employ mechanical and electronic systems to pump blood and control flow rates.

Because of space limitations and the materials focus of this work, many aspects of device design are not addressed in this book. Consider the example of the hemodialysis system. The focus here is on membrane materials and their biocompatibility; there is little coverage of mass transport through membranes, the burst strength of membranes, flow systems, and monitoring electronics. Other books and articles cover these topics in detail.

The words "biomaterial" and "biocompatibility" have already been used in this introduction without formal definition. A few definitions and descriptions are in order and will be expanded upon in this and subsequent chapters.

A definition of "biomaterial" endorsed by a consensus of experts in the field, is:

> A biomaterial is a nonviable material used in a medical device, intended to interact with biological systems (Williams, 1987).

If the word "medical" is removed, this definition becomes broader and can encompass the wide range of applications suggested above.

If the word "nonviable" is removed, the definition becomes even more general and can address many new tissue-engineering and hybrid artificial organ applications where living cells are used.

"Biomaterials science" is the physical and biological study of materials and their interaction with the biological environment. Traditionally, the most intense development and investigation have been directed toward biomaterials synthesis, optimization, characterization, testing, and the biology of host–material interactions. Most biomaterials introduce a non–specific, stereotyped biological reaction. Considerable current effort is directed toward the development of engineered surfaces that could elicit rapid and highly precise reactions with cells and proteins, tailored to a specific application.

Indeed, a complementary definition essential for understanding the goal (i.e., specific end applications) of biomaterials science is that of "biocompatibility."

Biocompatibility is the ability of a material to perform with an appropriate host response in a specific application (Williams, 1987).

Examples of "appropriate host responses" include the resistance to blood clotting, resistance to bacterial colonization, and normal, uncomplicated healing. Examples of specific applications include a hemodialysis membrane, a urinary catheter, or a hip-joint replacement prosthesis. Note that the hemodialysis membrane might be in contact with the patient's blood for 3 hours, the catheter may be inserted for a week, and the hip joint may be in place for the life of the patient.

This general concept of biocompatilility has been extended recently in the broad approach called "tissue engineering" in which *in-vitro* and *in-vivo* pathophysiological processes are harnessed by careful selection of cells, materials, and metabolic and biomechanical conditions to regenerate functional tissues.

Thus, in these definitions and discussion, we are introduced to considerations that set biomaterials apart from most materials explored in materials science. Table 1 lists a few applications for synthetic materials in the body. It includes many materials that are often classified as "biomaterials." Note that metals, ceramics, polymers, glasses, carbons, and composite materials are listed. Such materials are used as molded or machined parts, coatings, fibers, films, foams and fabrics. Table 2 presents estimates of the numbers of medical devices containing biomaterials that are implanted in humans each year and the size of the commercial market for biomaterials and medical devices.

Five examples of applications of biomaterials now follow to illustrate important ideas. The specific devices discussed were chosen because they are widely used in humans with good success. However, key problems with these biomaterial devices are also highlighted. Each of these examples is discussed in detail in later chapters.

EXAMPLES OF BIOMATERIALS APPLICATIONS

Heart Valve Prostheses

Diseases of the heart valves often make surgical repair or replacement necessary. Heart valves open and close over 40 million times a year and they can accumulate damage sufficient to require replacement in many individuals. More than

TABLE 1 Some Applications of Synthetic Materials and Modified Natural Materials in Medicine

Application	Types of materials
Skeletal system	
Joint replacements (hip, knee)	Titanium, Ti–Al–V alloy, stainless steel, polyethylene
Bone plate for fracture fixation	Stainless steel, cobalt–chromium alloy
Bone cement	Poly(methyl methacrylate)
Bony defect repair	Hydroxylapatite
Artificial tendon and ligament	Teflon, Dacron
Dental implant for tooth fixation	Titanium, Ti–Al–V alloy, stainless steel, polyethylene
	Titanium, alumina, calcium phosphate
Cardiovascular system	
Blood vessel prosthesis	Dacron, Teflon, polyurethane
Heart valve	Reprocessed tissue, stainless steel, carbon
Catheter	Silicone rubber, Teflon, polyurethane
Organs	
Artificial heart	Polyurethane
Skin repair template	Silicone–collagen composite
Artificial kidney (hemodialyzer)	Cellulose, polyacrylonitrile
Heart–lung machine	Silicone rubber
Senses	
Cochlear replacement	Platinum electrodes
Intraocular lens	Poly(methyl methacrylate), silicone rubber, hydrogel
Contact lens	Silicone-acrylate, hydrogel
Corneal bandage	Collagen, hydrogel

80,000 replacement valves are implanted each year in the United States because of acquired damage to the natural valve and congenital heart anomalies. There are many types of heart valve prostheses and they are fabricated from carbons, metals, elastomers, plastics, fabrics, and animal or human tissues chemically pretreated to reduce their immunologic reactivity and to enhance durability. Figure 1 shows a bileaflet tilting-disk mechanical heart valve, one of the most widely used designs. Other types of heart valves are made of chemically treated pig valve or cow pericardial tissue. Generally, almost as soon as the valve is implanted, cardiac function is restored to near normal levels and the patient shows rapid improvement. In spite of the overall success seen with replacement heart valves, there are problems that may differ with different types of valves; they include induction of blood clots, degeneration of tissue, mechanical failure, and infection. Heart valve substitutes are discussed in Chapter 7.3.

Artificial Hip Joints

The human hip joint is subjected to high levels of mechanical stress and receives considerable abuse in the course of

TABLE 2 The Biomaterials and Healthcare Market—Facts and Figures (per year) (U.S. numbers—Global numbers are typically 2–3 times the U.S. number)

Total U.S. health care expenditures (2000)	$1,400,000,000,000
Total U.S. health research and development (2001)	$82,000,000,000
Number of employees in the medical device industry (2003)	300,000
Registered U.S. medical device manufacturers (2003)	13,000
Total U.S. medical device market (2002)	$77,000,000,000
U.S. market for disposable medical supplies (2003)	$48,600,000,000
U.S. market for biomaterials (2000)	$9,000,000,000
Individual medical device sales:	
Diabetes management products (1999)	$4,000,000,000
Cardiovascular Devices (2002)	$6,000,000,000
Orthopedic-Musculoskeletal Surgery U.S. market (1998)	$4,700,000,000
Wound care U.S. market (1998)	$3,700,000,000
In Vitro diagnostics (1998)	$10,000,000,000
Numbers of devices (U.S.):	
Intraocular lenses (2003)	2,500,000
Contact lenses (2000)	30,000,000
Vascular grafts	300,000
Heart valves	100,000
Pacemakers	400,000
Blood bags	40,000,000
Breast prostheses	250,000
Catheters	200,000,000
Heart-Lung (Oxygenators)	300,000
Coronary stents	1,500,000
Renal dialysis (number of patients, 2001)	320,000
Hip prostheses (2002)	250,000
Knee prostheses (2002)	250,000
Dental implants (2000)	910,000

FIG. 1. A replacement heart valve.

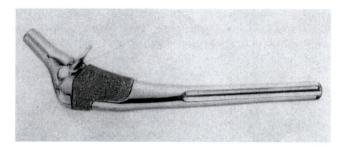

FIG. 2. A metalic hip joint. (Photograph courtesy of Zimmer, Inc.)

normal activity. It is not surprising that after 50 or more years of cyclic mechanical stress, or because of degenerative or rheumatological disease, the natural joint wears out, leading to considerable loss of mobility and often confinement to a wheelchair. Hip-joint prostheses are fabricated from titanium, stainless steel, special high-strength alloys, ceramics, composites, and ultrahigh-molecular-weight polyethylene. Replacement hip joints (Fig. 2) are implanted in more than 200,000 humans each year in the United States alone. With some types of replacement hip joints and surgical procedures that use a polymeric cement, ambulatory function is restored within days after surgery. For other types, a healing-in period is required for integration between bone and the implant before the joint can bear the full weight of the body. In most cases, good function is restored. Even athletic activities are possible, although they are generally not advised. After 10–15 years, the implant may loosen, necessitating another operation. Artificial hip joints are discussed in Chapter 7.7.

Dental Implants

The widespread introduction of titanium implants (Fig. 3) has revolutionized dental implantology. These devices form an implanted artificial tooth anchor upon which a crown is affixed and are implanted in approximately 300,000 people each year, with some individuals receiving more than 12 implants. A special requirement of a material in this application is the ability to form a tight seal against bacterial invasion where the implant traverses the gingiva (gum). One of the primary advantages originally cited for the titanium implant was its osseous integration with the bone of the jaw. In recent years, however, this attachment has been more accurately described as a tight apposition or mechanical fit and not true bonding. Loss of tissue support leading to loosening remains an occasional problem along with infection and issues associated with the mechanical properties of unalloyed titanium that is subjected to long-term cyclic loading. Dental implants are discussed in Chapter 7.8.

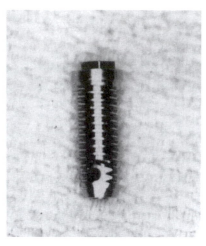

FIG. 3. A titanium dental implant. (Photograph courtesy of Dr. A. Norman Cranin, Brookdale Hospital Medical Center, Brooklyn, NY.)

Intraocular Lenses

A variety of intraocular lenses (IOLs) have been fabricated of poly(methyl methacrylate), silicone elastomer, soft acrylic polymers, or hydrogels and are used to replace a natural lens when it becomes cloudy due to cataract formation (Fig. 4). By the age of 75, more than 50% of the population suffers from cataracts severe enough to warrant IOL implantation.

FIG. 4. An intraocular lens. (Photograph courtesy of Alcon Laboratories, Inc.)

This translates to almost 4 million implantations in the United States alone each year, and double that number worldwide. Good vision is generally restored almost immediately after the lens is inserted and the success rate with this device is high. IOL surgical procedures are well developed and implantation is often performed on an outpatient basis. Recent observations of implanted lenses using a microscope to directly observe the implanted lens through the cornea show that inflammatory cells migrate to the surface of the lenses after implantation. Thus, the conventional healing pathway is seen with these devices, similar to that observed with materials implanted in other sites in the body. Outgrowth of cells from the posterior lens capsule stimulated by the IOL can cloud the vision, and this is a significant complication. IOLs are discussed in Chapter 7.11.

Left Ventricular Assist Device

With a large population of individuals with seriously failing hearts (estimated at as many as 50,000 per year) who need cardiac assist or replacement and an available pool of donor hearts for transplantation of approximately 3000 per year, effective and safe mechanical cardiac assist or replacement has been an attractive goal. Left ventricular assist devices (LVADs), that can be considered as one half of a total artificial heart, have evolved from a daring experimental concept to a life-prolonging tool. They are now used to maintain a patient with a failing heart while the patient awaits the availability of a transplant heart and some patients receive these LVADs as a permanent ("destination") therapy. An LVAD in an active adult is illustrated in Fig. 5. He is not confined to the hospital bed, although this pump system is totally supporting his circulatory needs. Patients have lived on LVAD support for more than 4 years. However, a patient with an LVAD is always at risk for infection and serious blood clots initiated within the device. These could break off (embolize) and possibly obstruct blood flow to a vital organ. LVADs are elaborated upon in Chapter 7.4.

These five cases, only a small fraction of the many important medical devices that could have been described here, spotlight a number of themes. Widespread application with good success is generally noted. A broad range of synthetic materials varying in chemical, physical, and mechanical properties are used in the body. Many anatomical sites are involved. The mechanisms by which the body responds to foreign bodies and heals wounds are observed in each case. Problems, concerns, or unexplained observations are noted for each device. Companies are manufacturing each of the devices and making a profit. Regulatory agencies are carefully looking at device performance and making policy intended to control the industry and protect the patient. Are there ethical or social issues that should be addressed? To set the stage for the formal introduction of biomaterials science, we will return to the five examples just discussed to examine the issues implicit to each case.

CHARACTERISTICS OF BIOMATERIALS SCIENCE

Now that we've defined some terms and reviewed a few specific examples, we can discern characteristics central to the

FIG. 5. A left ventricular assist device worn by a patient. (Photograph courtesy of Novacor.)

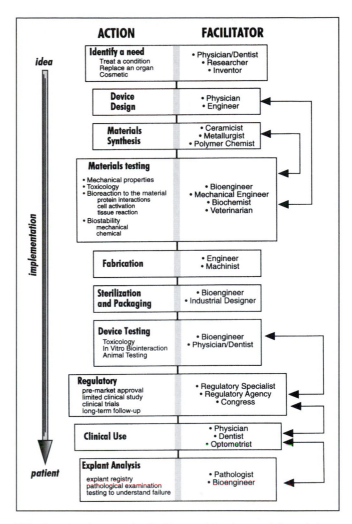

FIG. 6. Disciplines involved in biomaterials science and the path from a need to a manufactured medical device.

field of biomaterials. Here are a few considerations that are so central that it is hard to imagine biomaterials without them.

Multidisciplinary

More than any other field of contemporary technology, biomaterials science brings together researchers from diverse backgrounds who must communicate clearly. Figure 6 lists some of the disciplines that are encountered in the progression from identifying the need for a biomaterial or device to its manufacture, sale, and implantation.

Many Diverse Materials

The biomaterials scientist will have an appreciation of materials science. This may range from an impressive command of the theory and practice of the field demonstrated by the professional materials scientist to a general understanding of the properties of materials that might be demonstrated by the physician or biologist investigator involved in biomaterials-related research.

A wide range of materials is routinely used (Table 1), and no one researcher will be comfortable synthesizing, characterizing,

and designing with all these materials. Thus, specialization is common and appropriate. However, a broad appreciation of the properties and applications of these materials, the palette from which the biomaterials scientist creates, is a hallmark of professionals in the field.

There is a tendency to group biomaterials and researchers into the "hard-tissue replacement" camp, typically represented by those involved in orthopedic and dental materials, and the "soft-tissue replacement" camp, frequently associated with cardiovascular implants and general plastic-surgery materials. Hard-tissue biomaterials researchers are thought to focus on metals and ceramics while soft-tissue biomaterials researchers are considered polymer experts. In practice, this division is artificial: a heart valve may be fabricated from polymers, metals, and carbons. A hip joint will be composed of metals and polymers (and sometimes ceramics) and will be interfaced to the body via a polymeric bone cement. There is a need for a general understanding of all classes of materials and the common conceptual theme of their interaction with the biological milieu. This book provides a background to the important classes of materials, hard and soft.

Development of Biomaterials Devices

Thomas Edison once said that he would only invent things that people would buy. In an interesting way, this idea is central to biomaterials device development. The process of biomaterial/medical device innovation is driven by clinical need: a patient or a physician defines a need and then initiates an invention. Figure 6 illustrates multidisciplinary interactions in biomaterials and shows the progression in the development of a biomaterial or device. It provides a perspective on how different disciplines work together, starting from the identification of a need for a biomaterial through development, manufacture, implantation, and removal from the patient.

Magnitude of the Field

The magnitude of the medical device field expresses both a magnitude of need and a sizeable commercial market (Table 2). A conflict of interest can arise with pressures from both the commercial quarter and from patient needs. Consider four commonly used biomaterial devices: a contact lens, a hip joint, a hydrocephalus drainage shunt, and a heart valve. All fill medical needs. The contact lens offers improved vision and, some will argue, a cosmetic enhancement. The hip joint offers mobility to the patient who would otherwise need a cane or crutch or be confined to a bed or wheelchair. The hydrocephalus shunt will allow an infant to survive without brain damage. The heart valve offers a longer life with improved quality of life. The contact lens may sell for $100, and the hip joint, hydrocephalus shunt, and heart valve may sell for $1000–4000 each. Each year there will be 75 million contact lenses purchased worldwide, 275,000 heart valves, 5000 hydrocephalus shunts, and 500,000 total artificial hip and knee prostheses. Here are the issues for consideration: (1) the number of devices (an expression of both human needs and commercial markets), (2) medical significance (cosmetic to life saving), and (3) commercial potential (who will manufacture it and why—for example, what is the market for the hydrocephalus shunt?). Always, human needs and economic issues color this field we call "biomaterials science." Medical practice, market forces, and bioethics come into play most every day.

Lysaght and O'Laughlin (2000) have estimated that the magnitude and economic scope of the contemporary organ replacement enterprise are much larger than is generally recognized. In the year 2000, the lives of more than 20 million patients were sustained, supported, or significantly improved by functional organ replacement. The impacted population grows at over 10% per year. Worldwide, first-year and follow-up costs of organ replacement and prostheses exceeds $300 billion U.S. dollars per year and represents between 7% and 8% of total worldwide health-care spending. In the United States, the costs of therapies enabled by organ-replacement technology exceed 1% of the gross national product. The costs are also impressive when reduced to the needs of the individual patient. For example, the cost of a substitute heart valve is roughly $4000. The surgery to implant the device entails a hospital bill and first-year follow-up costs of approximately $60,000. Reoperation for replacing a failed valve will have

these same costs. Reoperations for failed valves now exceed 10% of all valve replacements.

Success and Failure

Most biomaterials and medical devices perform satisfactorily, improving the quality of life for the recipient or saving lives. However, no manmade construct is perfect. All manufactured devices have a failure rate. Also, all humans are different with differing genetics, gender, body chemistries, living environment, and degrees of physical activity. Furthermore, physicians implant or use these devices with varying degrees of skill. The other side to the medical device success story is that there are problems, compromises, and complications that occur with medical devices. Central issues for the biomaterials scientist, manufacturer, patient, physician, and attorney are, (1) what represents good design, (2) who should be responsible when devices perform "with an *in*appropriate host response," and (3) what are the cost/risk or cost/benefit ratios for the implant or therapy?

Some examples may clarify these issues. Clearly, heart valve disease is a serious medical problem. Patients with diseased aortic heart valves have a 50% chance of dying within 3 years. Surgical replacement of the diseased valve leads to an expected survival of 10 years in 70% of the cases. However, of these patients whose longevity and quality of life have clearly been enhanced, approximately 60% will suffer a serious valve-related complication within 10 years after the operation. Another example involves LVADs. A clinical trial called Randomized Evaluation of Mechanical Assistance for the Treatment of Congestive Heart Failure (REMATCH) led to the following important statistics (Rose *et al.*, 2001). Patients with an implanted Heartmate LVAD (Thoratec Laboratories) had a 52% chance of surviving for 1 year, compared with a 25% survival rate for patients who took medication. Survival for 2 years in patients with the Heartmate was 23% versus 8% in the medication group. Also, the LVAD enhanced the quality of life for the patients—they felt better, were less depressed, and were mobile. Importantly, patients participating in the REMATCH trial were not eligible for a heart transplant. In the cases of the heart valve and the LVAD, long-term clinical complications associated with imperfect performance of biomaterials do not preclude clinical success overall.

These five characteristics of biomaterials science: multidisciplinary, multimaterial, need-driven, substantial market, and risk–benefit, flavor all aspects the field. In addition, there are certain subjects that are particularly prominent in our field and help delineate biomaterials science as a unique endeavor. Let us review a few of these.

SUBJECTS INTEGRAL TO BIOMATERIALS SCIENCE

Toxicology

A biomaterial should not be toxic, unless it is specifically engineered for such a requirement (e.g., a "smart" drug delivery system that targets cancer cells and destroys them). Since the nontoxic requirement is the norm, toxicology for biomaterials has evolved into a sophisticated science. It deals with the substances that migrate out of biomaterials. For example, for polymers, many low-molecular-weight "leachables" exhibit some level of physiologic activity and cell toxicity. It is reasonable to say that a biomaterial should not give off anything from its mass unless it is specifically designed to do so. Toxicology also deals with methods to evaluate how well this design criterion is met when a new biomaterial is under development. Chapter 5.2 provides an overview of methods in biomaterials toxicology. Implications of toxicity are addressed in Chapters 4.2, 4.3 and 4.5.

Biocompatibility

The understanding and measurement of biocompatibility is unique to biomaterials science. Unfortunately, we do not have precise definitions or accurate measurements of biocompatibility. More often than not, biocompatibility is defined in terms of performance or success at a specific task. Thus, for a patient who is doing well with an implanted Dacron fabric vascular prosthesis, few would argue that this prosthesis is not "biocompatible." However, the prosthesis probably did not recellularize (though it was designed to do so) and also is embolic, though the emboli in this case usually have little clinical consequence. This operational definition of biocompatible ("the patient is alive so it must be biocompatible") offers us little insight in designing new or improved vascular prostheses. It is probable that biocompatibility may have to be specifically defined for applications in soft tissue, hard tissue, and the cardiovascular system (blood compatibility). In fact, biocompatibility may have to be uniquely defined for each application.

The problems and meanings of biocompatibility will be explored and expanded upon throughout this textbook, in particular, see Chapters 4 and 5.

Functional Tissue Structure and Pathobiology

Biomaterials incorporated into medical devices are implanted into tissues and organs. Therefore, the key principles governing the structure of normal and abnormal cells, tissues, and organs, the techniques by which the structure and function of normal and abnormal tissue are studied, and the fundamental mechanisms of disease processes are critical considerations to workers in the field.

Healing

Special processes are invoked when a material or device heals in the body. Injury to tissue will stimulate the well-defined inflammatory reaction sequence that leads to healing. Where a foreign body (e.g., an implant) is present in the wound site (surgical incision), the reaction sequence is referred to as the "foreign-body reaction" (Chapter 4.2). The normal response of the body will be modulated because of the solid implant. Furthermore, this reaction will differ in intensity and duration depending upon the anatomical site involved. An understanding of how a foreign object alters the normal inflammatory reaction sequence is an important concern for the biomaterials scientist.

Dependence on Specific Anatomical Sites of Implantation

Consideration of the anatomical site of an implant is essential. An intraocular lens may go into the lens capsule or the anterior chamber of the eye. A hip joint will be implanted in bone across an articulating joint space. A substitute heart valve will be sutured into cardiac muscle and will contact both soft tissue and blood. A catheter may be placed in an artery, a vein, or the urinary tract. Each of these sites challenges the biomedical device designer with special requirements for geometry, size, mechanical properties, and bioresponses. Chapter 3.4 introduces these ideas about special requirements to consider for specific anatomical sites.

Mechanical and Performance Requirements

Each biomaterial and device has mechanical and performance requirements that originate from the need to perform a physiological function consistent with the physical (bulk) properties of the material. These requirements can be divided into three categories: mechanical performance, mechanical durability, and physical properties. First, consider mechanical performance. A hip prosthesis must be strong and rigid. A tendon material must be strong and flexible. A tissue heart valve leaflet must be flexible and tough. A dialysis membrane must be strong and flexible, but not elastomeric. An articular cartilage substitute must be soft and elastomeric. Then, we must address mechanical durability. A catheter may only have to perform for 3 days. A bone plate may fulfill its function in 6 months or longer. A leaflet in a heart valve must flex 60 times per minute without tearing for the lifetime of the patient (realistically, at least for 10 or more years). A hip joint must not fail under heavy loads for more than 10 years. The bulk physical properties will also address other aspects of performance. The dialysis membrane has a specified permeability, the articular cup of the hip joint must have high lubricity, and the intraocular lens has clarity and refraction requirements. To meet these requirements, design principles are borrowed from physics, chemistry, mechanical engineering, chemical engineering, and materials science.

Industrial Involvement

A significant basic research effort is now under way to understand how biomaterials function and how to optimize them. At the same time, companies are producing implants for use in humans and, appropriate to the mission of a company, earning profits on the sale of medical devices. Thus, although we are now only learning about the fundamentals of biointeraction, we manufacture and implant millions of devices in humans. How is this dichotomy explained? Basically, as a result of considerable experience we now have a set of materials that performs satisfactorily in the body. The medical practitioner can use them with reasonable confidence, and the performance in the patient is largely acceptable. Though the devices and materials are far from perfect, the complications associated with the devices are less than the complications of the original diseases.

The complex balance between the desire to alleviate suffering and death, the excitement of new scientific ideas, the corporate imperative to turn a profit, the risk/benefit relationship, and the mandate of the regulatory agencies to protect the public forces us to consider the needs of many constituencies. Obviously, ethical concerns enter into the picture. Also, companies have large investments in the development, manufacture, quality control, clinical testing, regulatory clearance, and distribution of medical devices. How much of an advantage (for the company and the patient) will be realized in introducing an improved device? The improved device may indeed work better for the patient. However, the company will incur a large expense that will be perceived by the stockholders as reduced profits. Moreover, product liability issues are a major concern of manufacturers. The industrial side of the biomaterials field raises questions about the ethics of withholding improved devices from people who need them, the market share advantages of having a better product, and the gargantuan costs (possibly nonrecoverable) of introducing a new product into the medical marketplace. If companies did not have the profit incentive, would there be any medical devices, let alone improved ones, available for clinical application?

When the industrial segment of the biomaterials field is examined, we see other essential contributions to our field. Industry deals well with technologies such as packaging, sterilization, storage, distribution, and quality control and analysis. These subjects are grounded in specialized technologies, often ignored in academic communities, but have the potential to generate stimulating research questions. Also, many companies support in-house basic research laboratories and contribute in important ways to the fundamental study of biomaterials science.

Ethics

A wide range of ethical considerations impact biomaterials science. Some key ethical questions in biomaterials science are summarized in Table 3. Like most ethical questions, an absolute answer may be difficult to come by. Some articles have addressed ethical questions in biomaterials and debated the important points (Saha and Saha, 1987; Schiedermayer and Shapiro, 1989). Chapter 10.4 introduces ethics in biomaterials.

Regulation

The consumer (the patient) demands safe medical devices. To prevent inadequately tested devices and materials from coming on the market, and to screen out individuals clearly unqualified to produce biomaterials, the United States

---

TABLE 3 Ethical Concerns Relevant to Biomaterials Science

Is the use of animals justified? Specifically, is the experiment well designed and important so that the data obtained will justify the suffering and sacrifice of the life of a living creature?

How should research using humans be conducted to minimize risk to the patient and offer a reasonable risk-to-benefit ratio? How can we best ensure informed consent?

Companies fund much biomaterials research and own proprietary biomaterials. How can the needs of the patient be best balanced with the financial goals of a company? Consider that someone must manufacture devices—these would not be available if a company did not choose to manufacture them.

Since researchers often stand to benefit financially from a successful biomedical device and sometimes even have devices named after them, how can investigator bias be minimized in biomaterials research?

For life-sustaining devices, what is the trade-off between sustaining life and the quality of life with the device for the patient? Should the patient be permitted to "pull the plug" if the quality of life is not satisfactory?

With so many unanswered questions about the basic science of biomaterials, do government regulatory agencies have sufficient information to define adequate tests for materials and devices and to properly regulate biomaterials?

Should the government or other "third-party payors" of medical costs pay for the health care of patients receiving devices that have not yet been formally approved for general use by the FDA and other regulatory bodies?

Should the CEO of a successful multimillion dollar company that is the sole manufacturer a polymer material (that is a minor but crucial component of the sewing ring of nearly all heart valves) yield to the stockholders' demands that he/she terminate the sale of this material because of litigation concerning one model of heart valve with a large cohort of failures? The company sells 32 pounds of this material annually, yielding revenue of approximately $40,000?

Should an orthopedic appliance company manufacture two models of hip joint prostheses: one with an expected "lifetime" of 20 years (for young, active recipients) and another that costs one-fourth as much with an expected lifetime of 7 years (for elderly individuals), with the goal of saving resources so that more individuals can receive the appropriate care?

government has evolved a complex regulatory system administered by the U.S. Food and Drug Administration (FDA). Most nations of the world have similar medical device regulatory bodies. The International Standards Organization (ISO) has introduced international standards for the world community. Obviously, a substantial base of biomaterials knowledge went into establishing these standards. The costs to comply with the standards and to implement materials, biological, and clinical testing are enormous. Introducing a new biomedical device to the market requires a regulatory investment of tens of millions of dollars. Are the regulations and standards truly addressing the safety issues? Is the cost of regulation inflating the cost of health care and preventing improved devices from reaching those who need them? Under this regulation topic, we see the intersection of all the players in the biomaterials community: government, industry, ethics, and basic science. The answers are not simple, but the problems must be addressed every day. Chapters 10.2 and 10.3 expand on standards and regulatory concerns.

BIOMATERIALS LITERATURE

Over the past 50 years, the field of biomaterials has evolved from individual medical researchers innovating to save the lives of their patients into the sophisticated, regulatory/ethics-driven multidisciplinary endeavor we see today. Concurrent with the evolution of the discipline, a literature has also developed addressing basic science, applied science, engineering, and commercial issues. A bibliography is provided in Appendix D "The Biomaterials Literature" to highlight key reference works and technical journals in the biomaterials field.

BIOMATERIALS SOCIETIES

The evolution of the biomaterials field, from its roots with individual researchers and clinicians who intellectually associated their efforts with established disciplines such as medicine, chemistry, chemical engineering, or mechanical engineering, to a modern field called "biomaterials," parallels the formation of biomaterials societies. Probably the first biomaterials-related society was the American Society for Artificial Internal Organs (ASAIO). Founded in 1954, this group of visionaries established a platform to consider the development of devices such as the artificial kidney and the artificial heart. A Department of Bioengineering was established at Clemson University, Clemson, South Carolina, in 1963. In 1969, Clemson began organizing annual International Biomaterials Symposia. In 1974–1975, these symposia evolved into the Society For Biomaterials, the world's first biomaterials society.

Founding members, those who joined in 1975 and 1976, numbered about 50 and included clinicians, engineers, chemists, and biologists. Their common interest, biomaterials, was the engaging focus for the multidisciplinary participants. The European Society for Biomaterials was founded in 1975. Shortly after that, the Canadian Society For Biomaterials and the Japanese Society of Biomaterials were formed. The Controlled Release Society, a group strongly rooted in biomaterials for drug delivery, was founded in 1978. At this time there are many national biomaterials societies and related societies. The development of biomaterials professionalism and a sense of identity for the field called biomaterials can be attributed to these societies and the researchers who organized and led them.

SUMMARY

This chapter provides a broad overview of the biomaterials field. It provides a vantage point from which the reader can gain a perspective to see how the subthemes fit into the larger whole.

Biomaterials science may be the most multidisciplinary of all the sciences. Consequently, biomaterials scientists must master certain key material from many fields of science, technology, engineering, and medicine in order to be competent and conversant in this profession. The reward for mastering this volume of material is immersion in an intellectually stimulating endeavor that advances a new basic science of biointeraction and contributes to reducing human suffering.

Bibliography

Lysaght, M. J., and O'Laughlin, J. (2000). The demographic scope and economic magnitude of contemporary organ replacement therapies. *ASAIO J.* 46: 515–521.

Rose, E. A., Gelijns, A. C., Moskowitz, A. J., Heitjan, D. F., Stevenson, L. W., Dembitsky, W., Long, J. W., Ascheim, D. D., Tierney, A. R., Levitan, R. G., Watson, J. T., Ronan, N. S., Shapiro, P. A., Lazar, R. M., Miller, L. W., Gupta, L., Frazier, O. H., Desvigne-Nickens, P., Oz, M. C., Poirier, V. L., and Meier, P. (2001). Long-term use of a left ventricular assist device for end-stage heart failure. *N. Engl. J. Med.* 345: 1435–1443.

Saha, S., and Saha, P. (1987). Bioethics and applied biomaterials. *J. Biomed. Mater. Res. Appl. Biomater.* 21: 181–190.

Schiedermayer, D. L., and Shapiro, R. S. (1989). The artificial heart as a bridge to transplant: ethical and legal issues at the bedside. *J. Heart Transplant* 8: 471–473.

Society For Biomaterials Educational Directory (1992). Society For Biomaterials, Mt. Laurel, NJ.

Williams, D. F. (1987). *Definitions in Biomaterials. Proceedings of a Consensus Conference of the European Society for Biomaterials,* Chester, England, March 3–5, 1986, Vol. 4, Elsevier, New York.

A History of Biomaterials

BUDDY D. RATNER

At the dawn of the 21st century, biomaterials are widely used throughout medicine, dentistry and biotechnology. Just 50 years ago biomaterials as we think of them today did not exist. The word "biomaterial" was not used. There were no medical device manufacturers (except for external prosthetics such as limbs, fracture fixation devices, glass eyes, and dental devices), no formalized regulatory approval processes, no understanding of biocompatibility, and certainly no academic courses on biomaterials. Yet, crude biomaterials have been used, generally with poor to mixed results, throughout history. This chapter will broadly trace from the earliest days of human civilization to the dawn of the 21st century the history of biomaterials. It is convenient to organize the history of biomaterials into four eras: prehistory, the era of the surgeon hero, designed biomaterials/engineered devices, and the contemporary era leading into a new millennium. However, the emphasis of this chapter will be on the experiments and studies that set the foundation for the field we call biomaterials, largely between 1920 and 1980.

BIOMATERIALS BEFORE WORLD WAR II

Before Civilization

The introduction of nonbiological materials into the human body was noted far back in prehistory. The remains of a human found near Kennewick, Washington, USA (often referred to as the "Kennewick Man") was dated (with some controversy) to be 9000 years old. This individual, described by archeologists as a tall, healthy, active person, wandered through the region now know as southern Washington with a spear point embedded in his hip. It had apparently healed in and did not significantly impede his activity. This unintended implant illustrates the body's capacity to deal with implanted foreign materials. The spear point has little resemblance to modern biomaterials, but it was a "tolerated" foreign material implant, just the same.

Dental Implants in Early Civilizations

Unlike the spear point described above, dental implants were devised as implants and used early in history. The Mayan people fashioned nacre teeth from sea shells in roughly 600 A.D. and apparently achieved what we now refer to as bone integration (see Chapter 7.8), basically a seamless integration into the bone (Bobbio, 1972). Similarly, an iron dental implant in a corpse dated 200 A.D. was found in Europe

(Crubezy *et al.*, 1998). This implant, too, was described as properly bone integrated. There were no materials science, biological understanding, or medicine behind these procedures. Still, their success (and longevity) is impressive and highlights two points: the forgiving nature of the human body and the pressing drive, even in prehistoric times, to address the loss of physiologic/anatomic function with an implant.

Sutures for 32,000 Years

There is evidence that sutures may have been used as long as 32,000 years ago (NATNEWS, 1983, 20(5): 15–7). Large wounds were closed early in history by one of two methods—cautery or sutures. Linen sutures were used by the early Egyptians. Catgut was used in the Middle Ages in Europe.

Metallic sutures are first mentioned in early Greek literature. Galen of Pergamon (circa 130–200 A.D.) described ligatures of gold wire. In 1816, Philip Physick, University of Pennsylvania Professor of Surgery, suggested the use of lead wire sutures noting little reaction. In 1849, J. Marion Sims, of Alabama, had a jeweler fabricate sutures of silver wire and performed many successful operations with this metal.

Consider the problems that must have been experienced with sutures in eras with no knowledge of sterilization, toxicology, immunological reaction to extraneous biological materials, inflammation, and biodegradation. Yet sutures were a relatively common fabricated or manufactured biomaterial for thousands of years.

Artificial Hearts and Organ Perfusion

In the 4th century B.C., Aristotle called the heart the most important organ in the body. Galen proposed that veins connected the liver to the heart to circulate "vital spirits throughout the body via the arteries." English physician William Harvey in 1628 espoused a relatively modern view of heart function when he wrote, "The heart's one role is the transmission of the blood and its propulsion, by means of the arteries, to the extremities everywhere." With the appreciation of the heart as a pump, it was a logical idea to think of replacing the heart with an artificial pump. In 1812, the French physiologist Le Gallois expressed his idea that organs could be kept alive by pumping blood through them. A number of experiments on organ perfusion with pumps were performed from 1828–1868. In 1881, Étienne-Jules Marey, a brilliant scientist and thinker who published and invented in photography theory, motion

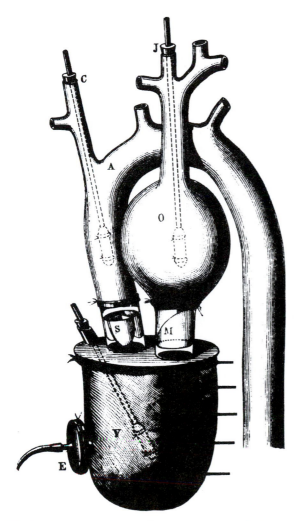

FIG. 1. An artificial heart by Étienne-Jules Marey, Paris, 1881.

studies and physiology, described an artificial heart device (Fig. 1), but probably never constructed such an apparatus.

In 1938, aviator (and engineer) Charles Lindbergh and surgeon (and Nobel prize winner) Alexis Carrel wrote a visionary book, *The Culture of Organs.* They addressed issues of pump design (referred to as the Lindbergh pump), sterility, blood damage, the nutritional needs of perfused organs and mechanics. This book must be considered a seminal document in the history of artificial organs. In the mid-1950s, Dr. Paul Winchell, better known as a ventriloquist, patented an artificial heart. In 1957, Dr. Willem Kolff and a team of scientists tested the artificial heart in animals. (The modern history of the artificial heart will be presented later in Chapter 7.4).

Contact Lenses

Leonardo DaVinci, in the year 1508, developed the contact lens concept. Rene Descartes is credited with the idea of the corneal contact lens (1632) and Sir John F. W. Herschel (1827) suggested that a glass lens could protect the eye. Adolf Fick, best known for his laws of diffusion, was an optometrist

by profession. One of his inventions (roughly 1860) was a glass contact lens, possibly the first contact lens offering real success. He experimented on both animals and humans with contact lenses. In a period from 1936 to 1948, plastic contact lenses were developed, primarily poly(methyl methacrylate).

Basic Concepts of Biocompatibility

Most implants prior to 1950 had a low probability of success because of a poor understanding of biocompatibility and sterilization. As will be elaborated upon throughout the textbook, factors that contribute to biocompatibility include the chemistry of the implant, leachables, shape, mechanics, and design. Early studies, especially with metals, explored primarily chemistry ideas to explain the observed bioreaction.

Possibly the first study assessing the *in vivo* bioreactivity of implant materials was performed by H. S. Levert (1829). Gold, silver, lead, and platinum specimens were studied in dogs and platinum, in particular, was found to be well tolerated. In 1886, bone fixation plates of nickel-plated sheet steel with nickel-plated screws were studied. In 1924, A. Zierold published a study on tissue reaction to various materials in dogs. Iron and steel were found to corrode rapidly leading to resorption of adjacent bone. Copper, magnesium, aluminum alloy, zinc, and nickel discolored the surrounding tissue while gold, silver, lead, and aluminum were tolerated but inadequate mechanically. Stellite, a Co–Cr–Mo alloy, was well tolerated and strong. In 1926, M. Large noted inertness displayed by 18-8 stainless steel containing molybdenum. By 1929 Vitallium alloy (65% Co–30% Cr–5% Mo) was developed and used with success in dentistry. In 1947, J. Cotton of the UK discussed the possible use for titanium and alloys for medical implants.

The history of plastics as implantation materials is not nearly as old as metals, simply because there were few plastics prior to the 1940s. What is possibly the first paper on the implantation of a modern synthetic polymer, nylon as a suture, appeared in 1941. Papers on the implantation of cellophane, a polymer made from plant sources, were published as early as 1939, where it was used as a wrapping for blood vessels. The response to this implant was described as a "marked fibrotic reaction." In the early 1940s papers appeared discussing the reaction to implanted poly(methyl methacrylate) and nylon. The first paper on polyethylene as a synthetic implant material was published in 1947 (Ingraham *et al.*). The paper pointed out that polyethylene production using a new high-pressure polymerization technique began in 1936. This process enabled the production of polyethylene free of initiator fragments and other additives. Ingraham *et al.* demonstrated good results on implantation (i.e., a mild foreign body reaction) and attributed these results to the high purity of the polymer they used. A 1949 paper commented on the fact that additives to many plastics had a tendency to "sweat out" and this may be responsible for the strong biological reaction to those plastics (LeVeen and Barberio, 1949). They found a vigorous foreign body reaction to cellophane, Lucite, and nylon but extremely mild reaction to "a new plastic," Teflon. The authors incisively concluded, "Whether the tissue reaction is due to the dissolution of traces of the unpolymerized chemical used in plastics manufacture or actually to the solution of an infinitesimal amount of the plastic itself cannot be determined." The possibility that cellulose might trigger the severe reaction by activating the complement system could not have been imagined because the complement system was not yet discovered.

POST WORLD WAR II—THE SURGEON/ PHYSICIAN HERO

At the end of World War II, high-performance metal, ceramic, and especially polymeric materials transitioned from wartime restricted to peacetime available. The possibilities for using these durable, novel, inert materials immediately intrigued surgeons with needs to replace diseased or damaged body parts. Materials originally manufactured for airplanes and automobiles were taken "off the shelf" by surgeons and applied to medical problems. These early biomaterials include silicones, polyurethanes, Teflon, nylon, methacrylates, titanium, and stainless steel.

A historical context helps us appreciate the contribution made primarily by medical and dental practitioners. After World War II, there was little precedent for surgeons to collaborate with scientists and engineers. Medical and dental practitioners of this era felt it was appropriate to invent (improvise) on their own where the life or functionality of their patient was at stake. Also, there was minimal government regulatory activity and minimal human subjects protections. The physician was implicitly entrusted with the life and health of the patient and had much more freedom than is seen today to take heroic action where other options were exhausted.[1] These medical practitioners had read about the post–World War II marvels of materials science. Looking at a patient open on the operating table, they could imagine replacements, bridges, conduits, and even organ systems based on such materials. Many materials were tried on the spur of the moment. Some fortuitously succeeded. These were high-risk trials, but usually they took place where other options were not available. The term "surgeon hero" seems justified since the surgeon often had a life (or a quality of life) at stake and was willing to take a huge technological and professional leap to repair the individual. This laissez faire biomaterials era quickly led to a new order characterized by scientific/engineering input, government quality controls, and a sharing of decisions prior to attempting high-risk, novel procedures. Still, a foundation of ideas and materials for the biomaterials field was built by courageous, fiercely committed, creative individuals and it is important to look at this foundation to understand many of the attitudes, trends, and materials common today.

---

[1] The regulatory climate in the United States in the 1950s was strikingly different from now. This can be appreciated in this recollection from Willem Kolff about a pump oxygenator he made and brought with him from Holland to the Cleveland Clinic (Kolff, 1998): "Before allowing Dr. Effler and Dr. Groves to apply the pump oxygenator clinically to human babies, I insisted they do 10 consecutive, successful operations in baby dogs. The chests were opened, the dogs were connected to a heart-lung machine to maintain the circulation, the right ventricles were opened, a cut was made in the interventricular septa, the septa holes were closed, the right ventricles were closed, the tubes were removed and the chests were closed. (I have a beautiful movie that shows these 10 puppies trying to crawl out of a basket)."

Intraocular Lenses

Sir Harold Ridley, M.D. (1906–2001) (Fig. 2), inventor of the plastic intraocular lens (IOL), made early, accurate observations of biological reaction to implants consistent with currently accepted ideas of biocompatibility. After World War II, he had the opportunity to examine aviators who were unintentionally implanted in their eyes with shards of plastic from shattered canopies in Spitfire and Hurricane fighter planes. Most of these flyers had plastic fragments in their eyes for years. The conventional wisdom at that time was that the human body would not tolerate implanted foreign objects, especially in the eye—the body's reaction to a splinter or a bullet was cited as examples of the difficulty of implanting materials in the body. The eye is an interesting implant site because you can look in through a transparent window to see what happened. When Ridley did so, he noted that the shards had healed in place with no further reaction. They were, by his standard, tolerated by the eye. Today, we would describe this type of stable healing without significant ongoing inflammation or irritation as "biocompatible." This is an early observation of "biocompatible" in humans, perhaps the first, using criteria similar to those accepted today. Based on this observation, Ridley traced down the source of the plastic domes, ICI Perspex poly(methyl methacrylate), and ordered sheets of the material. He used this material to fabricate implant lenses (intraocular lenses) that were found, after some experimentation, to function reasonably in humans as replacements for surgically removed natural lenses that had been clouded by cataracts. The first implantation in a human was November 29, 1949. For many years, Ridley was the center of fierce controversy because he challenged the dogma that spoke against implanting foreign materials in eyes—it hard to believe in the 21st century that the implantation of a biomaterial would provoke such an outcry. Because of this controversy, this industry did not spontaneously arise—it has to await the early 1980s before IOLs became a major force in the biomedical device market. Ridley's insightful observation, creativity, persistence, and surgical talent in the late 1940s evolved to an industry that presently puts more than 7,000,000 of these lenses annually in humans. Through all of human history, cataracts meant blindness, or a surgical procedure that left the recipient needing thick, unaesthetic eye glasses that poorly corrected the vision. Ridley's concept, using a plastic material found to be "biocompatible," changed the course of history and substantially improved the quality of life for millions of individuals with cataracts. Harold Ridley's story is elaborated upon in an obituary (Apple and Trivedi, 2002).

Hip and Knee Prostheses

The first hip replacement was probably performed in 1891 by a German surgeon, Theodore Gluck, using a cemented ivory ball. This procedure was not successful. Numerous attempts were made between 1920 and 1950 to develop a hip replacement prosthesis. Surgeon M. N. Smith-Petersen, in 1925, explored a glass hemisphere to fit over the ball of the hip joint. This failed because of poor durability. Chrome-based alloys

FIG. 2. Sir Harold Ridley, inventor of the intraocular lens.

FIG. 3. Sir John Charnley.

and stainless steel offered improvements in mechanical properties and many variants of these were explored. In 1938, the Judet Brothers of Paris, Robert and Jean, explored an acrylic surface for hip procedures, but it had a tendency to wear and loosen. The idea of using fast-setting dental acrylics to anchor prosthetics to bone was developed by Dr. Edward J. Haboush in 1953. In 1956, McKee and Watson-Farrar developed a "total" hip with an acetabular cup of metal that was cemented in place. Metal-on-metal wear products probably led to high complication rates. It was John Charnley (1911–1982) (Fig. 3), working at an isolated tuberculosis sanatorium in Wrightington, Manchester, England, who invented the first really successful hip joint prosthesis. The femoral stem, ball head, and plastic acetabular cup proved to be a reasonable solution to the problem of damaged joint replacement. In 1958, Dr. Charnley used a Teflon acetabular cup with poor outcomes due to wear debris. By 1961 he was using a high-molecular-weight polyethylene cup and was achieving much higher success rates. Interestingly, Charnley learned of high-molecular-weight polyethylene from a salesman selling novel plastic gears to one of his technicians. Dr. Dennis Smith contributed in an important way to the development of the hip prosthesis by introducing Dr. Charnley to poly(methyl methacrylate) cements, developed in the dental community, and optimizing those cements for hip replacement use. Total knee replacements borrowed elements of the hip prosthesis technology and successful results were obtained in the period 1968–1972 with surgeons Frank Gunston and John Insall leading the way.

Dental Implants

Some of the "prehistory" of dental implants was described earlier. In 1809, Maggiolo implanted a gold post anchor into fresh extraction sockets. After allowing this to heal, he fastened to it a tooth. This has remarkable similarity to modern dental implant procedures. In 1887, this procedure was used with a platinum post. Gold and platinum gave poor long-term results and so this procedure was never widely adopted. In 1937, Venable used surgical Vitallium and Co–Cr–Mo alloy for such implants. Also around 1937, Strock at Harvard used a screw-type implant of Vitallium and this may be the first successful dental implant. A number of developments in surgical procedure and implant design (for example, the endosteal blade implant) then took place. In 1952, a fortuitous discovery was made. Per Ingvar Branemark, an orthopedic surgeon at the University of Lund, Sweden, was implanting an experimental cage device in rabbit bone for observing healing reactions. The cage was a titanium cylinder that screwed into the bone. After completing the experiment that lasted several months, he tried to remove the titanium device and found it tightly integrated in the bone (Branemark *et al.*, 1964). Dr. Branemark named the phenomenon osseointegration and explored the application of titanium implants to surgical and dental procedures. He also developed low-impact surgical protocols for tooth implantation that reduced tissue necrosis and enhanced the probability of good outcomes. Most dental implants and many other orthopedic implants are now made of titanium and its alloys.

The Artificial Kidney

Kidney failure, through most of history, was a sentence to unpleasant death lasting over a period of about a month. In 1910, at Johns Hopkins University, the first attempts to remove toxins from blood were made by John Jacob Abel. The experiments were with rabbit blood and it was not possible to perform this procedure on humans. In 1943, in Nazi-occupied Holland, Willem Kolff (Fig. 4), a physician just beginning his career at that time, built a drum dialyzer system from a 100-liter tank, wood slats, and sausage-casing (cellulose) as the dialysis membrane. Some successes were seen in saving lives where prior to this there was only one unpleasant outcome to kidney failure. Kolff took his ideas to the United States and in 1960, at the Cleveland Clinic, developed a "washing machine artificial kidney" (Fig. 5). Major advances in kidney dialysis were made by Dr. Belding Scribner (1921–2003) at the University of Washington. Scribner devised a method to routinely access the bloodstream for dialysis treatments. Prior to this, after just a few treatments, access sites to the blood were used up and further dialysis was not possible. After seeing the potential of dialysis to help patients, but only acutely, Scribner tells the story of waking up in the middle of the night with an idea to gain easy access to the blood—a shunt implanted between an artery and vein that emerged through the skin as a "U." Through the exposed portion of the shunt, blood access could be readily achieved. When Dr. Scribner heard about this new plastic, Teflon, he envisioned how to get the blood out of and into the blood vessels. His device used Teflon tubes to access the vessels, a Dacron sewing cuff through the skin, and a silicone rubber tube for blood flow. The Scribner shunt made chronic dialysis possible and is said to be responsible for more than a million patients being alive today. Additional important contributions to the artificial kidney were made by Professor Les Babb of the University of Washington who, working with Scribner, improved dialysis performance and invented a proportioning mixer for the dialysate fluid.

The Artificial Heart

Willem Kolff was also a pioneer in the development of the artificial heart. He implanted the first artificial heart in the Western hemisphere in a dog in 1957 (a Russian artificial heart was implanted in a dog in the late 1930s). The Kolff artificial heart was made of a thermosetting poly(vinyl chloride) cast inside hollow molds to prevent seams. In 1953, the heart–lung machine was invented by John Gibbon, but this was useful only for acute treatment as during open heart surgery. After the National Heart and Lung Institute of the NIH in 1964 set a goal of a total artificial heart by 1970, Dr. Michael DeBakey implanted a left ventricular assist device in a human in 1966 and Dr. Denton Cooley implanted a polyurethane total artificial heart in 1969. In the period 1982–1985, Dr. William DeVries implanted a number of Jarvik hearts with patients living up to 620 days on the devices.

Breast Implants

The breast implant evolved to address the poor results achieved with direct injection of substances into the breast for augmentation. In fact, in the 1960s, California and Utah classified silicone injections as a criminal offense. In the 1950s,

FIG. 4. Dr. Willem Kolff at age 92. (Photo by B. Ratner.)

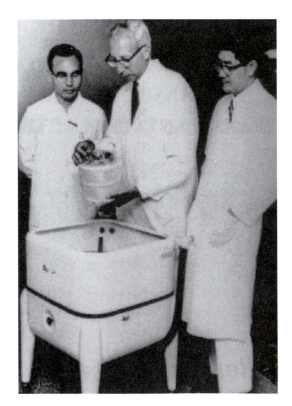

FIG. 5. Willem Kolff (center) and the washing machine artificial kidney.

poly(vinyl alcohol) sponges were implanted as breast prostheses, but results with these were also poor. University of Texas plastic surgeons Thomas Cronin and Frank Gerow invented the first silicone breast implant in the early 1960s, a silicone shell filled with silicone gel. Many variants of this device have been tried over the years, including cladding the device with polyurethane foam (the Natural Y implant). This variant of the breast implant was fraught with problems. However, the basic silicone rubber–silicone gel breast implant was generally acceptable in performance (Bondurant *et al.*, 1999).

Vascular Grafts

Surgeons have long needed methods and materials to repair damaged and diseased blood vessels. Early in the century, Dr. Alexis Carrel developed methods to anastomose (suture) blood vessels, an achievement for which he won the Nobel Prize in medicine in 1912. In 1942, Blackmore used Vitallium metal tubes to bridge arterial defects in war-wounded soldiers. Columbia University surgical intern Arthur Voorhees (1922–1992), in 1947, noticed during a post-mortem that tissue had grown around a silk suture left inside a lab animal. This observation stimulated the idea that a cloth tube might also heal by being populated by the tissues of the body. Perhaps such a healing reaction in a tube could be used to replace an artery? His first experimental vascular grafts were sewn from a silk handkerchief and then parachute fabric (Vinyon N), using his wife's sewing machine. The first human implant of a prosthetic vascular graft was in 1952. The patient lived many years after this procedure, inspiring many surgeons to copy the procedure. By 1954, another paper was published establishing the clear benefit of a porous (fabric) tube over a solid polyethylene tube (Egdahl *et al.*, 1954). In 1958, the following technique was described in a textbook on vascular surgery (Rob, 1958): "The Terylene, Orlon or nylon cloth is bought from a draper's shop and cut with pinking shears to the required shape. It is then sewn with thread of similar material into a tube and sterilized by autoclaving before use."

Stents

Partially occluded coronary arteries lead to angina, diminished heart functionality, and eventually, when the artery occludes (i.e., myocardial infarction), death of a section of the heart muscle. Bypass operations take a section of vein from another part of the body and replace the occluded coronary artery with a clean conduit—this is major surgery, hard on the patient and expensive. Synthetic vascular grafts in the 3-mm diameter appropriate to the human coronary artery anatomy will thrombose and thus cannot be used. Another option is percutaneous transluminal coronary angioplasty (PTCA). In this procedure, a balloon is threaded on a catheter into the coronary artery and then inflated to open the lumen of the occluding vessel. However, in many cases the coronary artery can spasm and close from the trauma of the procedure. The invention of the coronary artery stent, an expandable metal mesh that holds the lumen open after PTCA, was a major revolution in the treatment of coronary occlusive disease. In his own words,

Dr. Julio Palmaz (Fig. 6) describes the origins and history of the cardiovascular stent.

I was at a meeting of the Society of Cardiovascular and Interventional Radiology in February 1978, New Orleans when a visiting lecturer, Doctor Andreas Gruntzig from Switzerland, was presenting his preliminary experience with coronary balloon angioplasty. As you know, in 1978 the mainstay therapy of coronary heart disease was surgical bypass. Doctor Gruntzig showed his promising new technique to open up coronary atherosclerotic blockages without the need for open chest surgery, using his own plastic balloon catheters. During his presentation, he made it clear that in a third of the cases, the treated vessel closed back after initial opening with the angioplasty balloon because of elastic recoil or delamination of the vessel wall layers. This required standby surgery facilities and personnel, in case of acute closure after balloon angioplasty prompted emergency coronary bypass. Gruntzig's description of the problem of vessel reclosure elicited in my mind the idea of using some sort of support, such as used in mine tunnels or in oil well drilling. Since the coronary balloon goes in small (folded like an umbrella) and is inflated to about 3–4 times its initial diameter, my idealistic support device needed to go in small and expand at the site of blockage with the balloon. I thought one way to solve this was a malleable tubular criss-cross mesh. I went back home in the Bay Area and started making crude prototypes with copper wire and lead solder, which I first tested in rubber tubes mimicking arteries. I called the device a BEIS or balloon-expandable intravascular graft. However, the reviewers of my first submitted paper wanted to call it a stent. When I looked the word up, I found out that it derives from Charles Stent, a British dentist who died at turn of the century. Stent invented a wax material to make dental molds for dentures. This material was later used by plastic surgeons to keep tissues in place, while healing after surgery. The word "stent" was then generically used for any device intended to keep tissues in place while healing.

I made the early experimental device of stainless steel wire soldered with silver. These were materials I thought would be appropriate for initial laboratory animal testing. To carry on with my project I moved to the University of Texas Health Science Center in San Antonio (UTHSCSA) were I had a research laboratory and time for further development. From 1983–86 I performed mainly bench and animal testing. Dozens of ensuing projects showed the promise of the technique and the potential applications it had in many areas of vascular surgery and cardiology. With a UTHSCSA pathologist, Doctor Fermin Tio, we observed our first microscopic specimen of implanted stents in awe. After weeks to months after implantation by catheterization under X-ray guidance, the stent had remained open, carrying blood flow. The metal mesh was covered with translucent, glistening tissue similar to the lining of a normal vessel. The question remained whether the same would happen in atherosclerotic vessels. We tested this question in the atherosclerotic rabbit model and to our surprise, the new tissue free of atherosclerotic plaque encapsulated the stent wires, despite the fact that the animals were still on a high cholesterol diet. Eventually, a large sponsor (Johnson and Johnson) adopted the project and clinical trials were instituted under the scrutiny of the Food and Drug Administration, to compare stents to balloon angioplasty.

Coronary artery stenting is now performed in well over 1.5 million procedures per year.

FIG. 6. Dr. Julio Palmaz, inventor of the coronary artery stent.

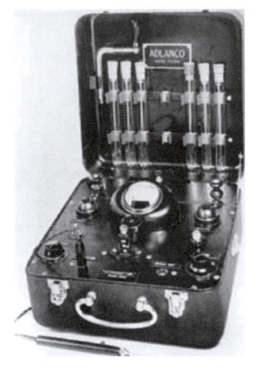

FIG. 7. The Albert Hyman Model II portable pacemaker, circa 1932–1933. (With permission of NASPE Heart Rhythm Society.)

Pacemakers

In London, in 1788, Charles Kite wrote "An Essay Upon the Recovery of the Apparently Dead" where he discussed electrical discharges to the chest for heart resuscitation. In the period 1820–1880, it was already known that electric shocks could modulate the heartbeat (and, of course, consider the Frankenstein story from that era). The invention of the portable pacemaker, hardly portable by modern standards, may have taken place almost simultaneously in two groups in 1930–31— Dr. Albert S. Hyman (USA) (Fig. 7) and Dr. Mark C. Lidwill (working in Australia with physicist Major Edgar Booth).

Canadian electrical engineer John Hopps, while conducting research on hypothermia in 1949, invented an early cardiac pacemaker. Hopps' discovery was that if a cooled heart stopped beating, it could be electrically restarted. This led to Hopps' invention of a vacuum tube cardiac pacemaker in 1950. Paul M. Zoll developed a pacemaker in conjunction with the Electrodyne Company in 1952. The device was about the size of a large table radio, was powered with external current, and stimulated the heart using electrodes placed on the chest—this therapy caused pain and burns, though it could pace the heart.

In the period 1957–58, Earl E. Bakken, founder of Medtronic, Inc., developed the first wearable transistorized (external) pacemaker at the request of heart surgeon, Dr. C. Walton Lillehei. Bakken quickly produced a prototype that Lillehei used on children with postsurgery heart block. Medtronic commercially produced this wearable, transistorized unit as the 5800 pacemaker.

In 1959, the first fully implantable pacemaker was developed by engineer Wilson Greatbatch and cardiologist W. M. Chardack. He used two Texas Instruments transitors, a technical innovation that permitted small size and low power drain. The pacemaker was encased in epoxy to inhibit body fluids from inactivating it.

Heart Valves

The development of the prosthetic heart valve paralleled developments in cardiac surgery. Until the heart could be stopped and blood flow diverted, the replacement of a valve would be challenging. Charles Hufnagel, in 1952, implanted a valve consisting of a poly(methyl methacrylate) tube and nylon ball in a beating heart. This was a heroic operation and basically unsuccessful, but an operation that inspired cardiac surgeons to consider that valve prostheses might be possible. The 1953 development of the heart–lung machine by Gibbon allowed the next stage in the evolution of the prosthetic heart valve to take place. In 1960, a mitral valve replacement was performed in a human by surgeon Albert Starr using a valve design consisting of a silicone ball and poly(methyl methacrylate) cage (later replaced by a stainless steel cage). The valve was invented by engineer Lowell Edwards. The heart valve was based on a design for a bottle stopper invented in 1858. Starr was quoted as saying, "Let's make a valve that works and not worry about its looks," referring to its design that was radically different from the leaflet valve that nature evolved in mammals. Prior to the Starr–Edwards valve, no human had lived with a prosthetic heart valve longer than 3 months. The Starr–Edwards valve was found to permit good patient survival. The major issues in valve development in that era were thrombosis and durability. Warren Hancock started the development of the first leaflet tissue heart valve in 1969 and his company and valve were acquired by Johnson & Johnson in 1979.

DESIGNED BIOMATERIALS

In contrast to the biomaterials of the surgeon-hero era, largely off-the-shelf materials used to fabricate medical devices, the 1960s on saw the development of materials designed specifically for biomaterials applications. Here are some key classes of materials and their evolution from commodity materials to engineered/synthesized biomaterials.

Silicones

Though the class of polymers known as silicones has been explored for many years, it was not until the early 1940s that Eugene Rochow of GE pioneered the scale-up and manufacture of commercial silicones via the reaction of methyl chloride with silicon in the presence of catalysts. In Rochow's 1946 book, *The Chemistry of Silicones* (John Wiley & Sons, Publishers), he comments anecdotally on the low toxicity of silicones but did not propose medical applications. The potential for medical uses of these materials was realized shortly after this. In a 1954 book on silicones, McGregor has a whole chapter titled "Physiological Response to Silicones." Toxicological studies were cited suggesting to McGregor that the quantities of silicones that humans might take into their bodies should be "entirely harmless." He mentions, without citation, the application of silicone rubber in artificial kidneys. Silicone-coated rubber grids were also used to support a dialysis membrane (Skeggs and Leonards, 1948). Many other early applications of silicones in medicine are cited in Chapter 2.3.

Polyurethanes

Polyurethanes, reaction products of diisocyanates and diamines, were invented by Otto Bayer and colleagues in Germany in 1937. The chemistry of polyurethanes intrinsically offered a wide range of synthetic options leading to hard plastics, flexible films, or elastomers (Chapter 2.2). Interestingly, this was the first class of polymers to exhibit rubber elasticity without covalent cross-linking. As early as 1959, polyurethanes were explored for biomedical applications, specifically heart valves (Akutsu *et al.*, 1959). In the mid-1960s a class of segmented polyurethanes was developed that showed both good biocompatibility and outstanding flex life in biological solutions at 37°C (Boretos and Pierce, 1967). Sold under the name Biomer, these segmented polyurethanes comprised the pump diaphragms of the Jarvik 7 hearts that were implanted in seven humans.

Teflon

DuPont chemist Roy Plunkett discovered a remarkably inert polymer, Teflon (polytetrafluoroethylene), in 1938. William L. Gore and his wife Vieve started a company in 1958 to apply Teflon for wire insulation. In 1969, their son Bob found that Teflon, if heated and stretched, forms a porous membrane with attractive physical and chemical properties. Bill Gore tells the story that, on a chairlift at a ski resort, he pulled from his parka pocket a piece of porous Teflon tubing to show to his fellow ski lift passenger. The skier was a physician and asked for a specimen to try as a vascular prosthesis. Now, Goretex porous Teflon is the leading synthetic vascular graft and has numerous applications in surgery and biotechnology.

Hydrogels

Hydrogels have been found in nature since life on earth evolved. Bacterial biofilms, hydrated living tissues, extracellular matrix components, and plant structures are ubiquitous, hydrated, swollen motifs in nature. Gelatin and agar were also explored early in human history. But, the modern history of hydrogels as a class of materials designed for medical applications can be accurately traced.

In 1936, DuPont scientists published a paper on recently synthesized methacrylic polymers. In this paper, poly(2-hydroxyethyl methacrylate) (polyHEMA) was mentioned. It was briefly described as a hard, brittle, glassy polymer and clearly not considered of importance. After that paper, this polymer was essentially forgotten until 1960. Wichterle and Lim published a paper in *Nature* describing the polymerization of HEMA monomer and a cross-linking agent in the presence of water and other solvents (Wichterle and Lim, 1960). Instead of a brittle polymer, they obtained a soft, water-swollen, elastic, clear gel. This innovation led to the soft contact lens industry and to the modern field of biomedical hydrogels as we know them today.

Interest and applications for hydrogels have steadily grown over the years and these are described in detail in Chapter 2.5. Important early applications included acrylamide gels for electrophoresis, poly(vinyl alcohol) porous sponges (Ivalon) as implants, many hydrogel formulations as soft contact lenses, and alginate gels for cell encapsulation.

Poly(ethylene glycol)

Poly(ethylene glycol) (PEG), also called poly(ethylene oxide) (PEO) in its high-molecular-weight form, can be categorized as a hydrogel, especially when the chains are cross-linked. However, PEG has many other applications and implementations. It is so widely used today that it is best discussed in its own section.

The low reactivity of PEG with living organisms has been known since at least 1944 where it was examined as a possible vehicle for intravenously administering fat-soluble hormones (Friedman, 1944). In the mid-1970s, Abuchowski and colleagues (Abuchowski *et al.*, 1977) discovered that if PEG chains were attached to enzymes and proteins, they would a have a much longer functional residence time *in vivo* than biomolecules that were not PEGylated. Professor Edward Merrill of MIT, based upon what he called "various bits of evidence" from the literature, concluded that surface-immobilized PEG would resist protein and cell pickup. The experimental results from his research group in the early 1980s bore this conclusion out (Merrill, 1992). The application of PEGs to wide range of biomedical problems has been significantly accelerated by the synthetic chemistry developments of Dr. Milton Harris while at the University of Alabama, Huntsville.

Poly(lactic–glycolic acid)

Though originally discovered in 1833, the anionic polymerization from the cyclic lactide monomer in the early 1960s made materials with mechanical properties comparable to Dacron possible. The first publication on the application of poly(lactic acid) in medicine may have been by Kulkarni *et al.* (1966). This group demonstrated that the polymer degraded slowly after implantation in guinea pigs or rats and was well tolerated by the organisms. Cutright *et al.* (1971) was the first to apply this polymer for orthopedic fixation. Poly(glycolic acid) and copolymers of lactic and glycolic acid were subsequently developed. Early clinical applications of polymers in this family were for sutures. The glycolic acid/lactic acid polymers have also been widely applied for controlled release of drugs and proteins. Professor Robert Langer's group was the leader in developing these polymers in the form of porous scaffolds for tissue engineering (Langer and Vacanti, 1993).

Hydroxyapatite

Hydroxyapatite is one of the most widely studied materials for healing in bone. It is both a natural component of bone (i.e., a material of ancient history) and a synthetic material with a modern history. Hydroxyapatite can be easily made as a powder. One of the first papers to apply this material for biomedical application was by Levitt *et al.* (1969), in which they hot-pressed the hydroxyapatite power into useful shapes for biological experimentation. From this early appreciation of the materials science aspect of a natural biomineral, a literature of thousands of papers has evolved. In fact, the nacre implant described in the prehistory section may owe its effectiveness to hydroxyapatite—recent data have shown that the calcium carbonate of nacre can transform in phosphate solutions to hydroxapatite (Ni and Ratner, 2003).

Titanium

In 1791, William Gregor, a Cornish amateur chemist, used a magnet to extract the ore that we now know as ilmenite from a local river. He then extracted the iron from this black powder with hydrochloric acid and was left with a residue that was the impure oxide of titanium. After 1932, a process developed by William Kroll permitted the commercial extraction of titanium from mineral sources. At the end of World War II, titanium metallurgy methods and titanium materials made their way from military application to peacetime uses. By 1940, satisfactory results had been achieved with titanium implants (Bothe *et al.*, 1940). The major breakthrough in the use of titanium for bony tissue implants was the Branemark discovery of osseointegration, described earlier in the section on dental implants.

Bioglass

Bioglass is important to biomaterials as one of the first completely synthetic materials that seamlessly bonds to bone. It was developed by Professor Larry Hench and colleagues. In 1967 Hench was an assistant professor at the University of Florida. At that time his work focused on glass materials and their interaction with nuclear radiation. In August of that year, he shared a bus ride to an Army Materials Conference in Sagamore, New York, with a U.S. Army Colonel who had just returned from Vietnam where he was in charge of supplies to 15 MASH units. He was not terribly interested in the radiation resistance of glass. Rather, he challenged Hench with the following: hundreds of limbs a week in Vietnam were being amputated because the body was found to reject the metals and polymer materials used to repair the body. "If you can make a material that will resist gamma rays, why not make a material the body won't resist?"

Hench returned from the conference and wrote a proposal to the U.S. Army Medical R and D Command. In October 1969 the project was funded to test the hypothesis that silicate-based glasses and glass-ceramics containing critical amounts of Ca and P ions would not be rejected by bone. In November 1969 Hench made small rectangles of what he called 45S5 glass (44.5 wt.% SiO_2) and Ted Greenlee, Assistant Professor of Orthopaedic Surgery at the University of Florida, implanted them in rat femurs at the VA Hospital in Gainesville. Six weeks later Greenlee called—"Larry, what are those samples you gave me? They will not come out of the bone. I have pulled on them, I have pushed on them, I have cracked the bone and they are still bonded in place." Bioglass was born, and with the first composition studied! Later studies by Hench using surface analysis equipment showed that the surface of the Bioglass, in biological fluids, transformed from a silicate-rich composition to a phosphate-rich structure, possibly with resemblance to hydroxyapatite (Clark *et al.*, 1976).

THE CONTEMPORARY ERA (MODERN BIOLOGY AND MODERN MATERIALS)

It is probable that the modern era in the history of biomaterials, biomaterials engineered to control specific biological reactions, was ushered in by rapid developments in modern biology. In the 1960s, when the field of biomaterials was laying down its foundation principles and ideas, concepts such as cell-surface receptors, growth factors, nuclear control of protein expression and phenotype, cell attachment proteins, and gene delivery were either controversial observations or undiscovered. Thus, pioneers in the field, even if so moved, could not have designed materials with these ideas in mind. It is to the credit of the biomaterials community that it has been quick to embrace and exploit new ideas from biology. Similarly, new ideas from materials science such as phase separation, anodization, self-assembly, surface modification, and surface analysis were quickly assimilated into the biomaterial scientists' toolbox and vocabulary. A few of the important ideas in the biomaterials literature that set the stage for the biomaterials science we see today are useful to list:

Protein adsorption
Biospecific biomaterials
Nonfouling materials
Healing and the foreign-body reaction
Controlled release
Tissue engineering
Regenerative medicine

Since these topics are well elaborated upon in *Biomaterials Science: An Introduction to Materials in Medicine,* 2nd edition, they will not be expanded upon in this history section. Still, it is important to appreciate the intellectual leadership of many researchers that promoted these ideas that make up modern biomaterials.

CONCLUSIONS

Biomaterials have progressed from surgeon-heroes, sometimes working with engineers, to a field dominated by engineers and scientists, to our modern era with the biologist as a critical player. As *Biomaterials Science: An Introduction to Materials in Medicine*, 2nd edition, is being published, many individuals who were biomaterials pioneers in the formative days of the field are well into their ninth decade. A number of leaders of biomaterials, pioneers who spearheaded the field with vision, creativity, and integrity, have passed away. Biomaterials is a field with a history modern enough so the first-hand accounts of its roots are available. I encourage readers of the textbook to document their conversations with pioneers of the field (many of whom still attend biomaterials conferences), so that the exciting stories that led to the successful and intellectually alive field we see today are not lost.

Bibliography

Abuchowski, A., McCoy, J. R., Palczuk, N. C., van Es, T., and Davis, F. F. (1977). Effect of covalent attachment of polyethylene glycol on immunogenicity and circulating life of bovine liver catalase. *J. Biol. Chem.* **252**(11): 3582–3586.

Akutsu, T., Dreyer, B., and Kolff, W. J. (1959). Polyurethane artificial heart valves in animals. *J. Appl. Physiol.* **14**: 1045–1048.

Apple, D. J., and Trivedi, R. H. (2002). Sir Nicholas Harold Ridley, Kt, MD, FRCS, FRS. *Arch. Ophthalmol.* **120**(9): 1198–1202.

Bobbio, A. (1972). The first endosseous alloplastic implant in the history of man. *Bull. Hist. Dent.* **20**: 1–6.

Bondurant, S., Ernster, V., and Herdman, R. (ed.) (1999). *Safety of Silicone Breast Implants*. National Academies Press, Washington, D. C.

Boretos, J. W., and Pierce, W. S. (1967). Segmented polyurethane: a new elastomer for biomedical applications. *Science* **158**: 1481–1482.

Bothe, R. T., Beaton, L. E., and Davenport, H. A. (1940). Reaction of bone to multiple metallic implants. *Surg., Gynec. & Obstet.* **71**: 598–602.

Branemark, P. I., Breine, U., Johansson, B., Roylance, P. J., Röckert, H., Yoffey, J. M. (1964). Regeneration of bone marrow. Acta Anat. **59**: 1–46.

Clark, A. E., Hench, L. L., and Paschall, H. A. (1976). The influence of surface chemistry on implant interface histology: a theoretical basis for implant materials selection. *J. Biomed. Mater. Res.* **10**: 161–177.

Crubezy, E., Murail, P., Girard, L., and Bernadou, J-P (1998). False teeth of the Roman world. *Nature* **391**: 29.

Cutright, D. E., Hunsuck, E. E., Beasley, J. D. (1971). Fracture reduction using a biodegradable materials, polylactic acid. *J. Oral Surg.* **29**, 393–397.

Egdahl, R. H., Hume, D. M., Schlang, H. A. (1954). Plastic venous prostheses. *Surg. Forum* **5**: 235–241.

Friedman, M. (1944). A vehicle for the intravenous administration of fat soluble hormones. *J. Lab. Clin. Med.* **29**: 530–531.

Ingraham, F. D., Alexander, E., Jr. and Matson, D. D. (1947). Polyethylene, a new synthetic plastic for use in surgery. *JAMA* **135**(2): 82–87.

Kolff, W. J. (1998). Early years of artificial organs at the Cleveland Clinic, Part II: Open heart surgery and artificial hearts. *ASAIO J.* **44**(3): 123–128.

Kulkarni, R. K., Pani, K. C., and Neuman, C., Leonard, F. (1966). Polylactic acid for surgical implants. *Arch. Surg.* **93**: 839–843.

Langer, R, and Vacanti, J. P. (1993). Tissue engineering. *Science* **260**: 920–926.

LeVeen, H. H., and Barberio, J. R., (1949). Tissue reaction to plastics used in surgery with special reference to Teflon. *Ann. Surg.* **129**(1): 74–84.

Levitt, S. R., Crayton, P. H., Monroe, E. A., and Condrate, R. A. (1969). Forming methods for apatite prostheses. *J. Biomed. Mater. Res.* **3**: 683–684.

McGregor, R. R. (1954). *Silicones and Their Uses.* McGraw-Hill, New York.

Merrill, E. W. (1992). Poly(ethylene oxide) and blood contact. in *Poly(ethylene glycol) Chemistry: Biotechnical and Biomedical Applications,* J. M. Harris (ed.). Plenum Press, New York, pp. 199–220.

Ni, M., and Ratner, B. D. (2003). Nacre surface transformation to hydroxyapatite in a phosphate buffer solution. *Biomaterials* **24**: 4323–4331.

Rob, C. (1958). Vascular surgery. in *Modern Trends in Surgical Materials,* L. Gillis (ed.). Butterworth & Co., London, pp. 175–185.

Scales, J. T. (1958). Biological and mechanical factors in prosthetic surgery. *in Modern Trends in Surgical Materials.* L. Gillis (ed.). Butterworth & Co., London, pp. 70–105.

Skeggs, L. T., and Leonards, J. R. (1948). Studies on an artificial kidney: preliminary results with a new type of continuous dialyzer. *Science* **108**: 212.

Wichterle, O., and Lim, D. (1960). Hydrophilic gels for biological use. *Nature* **185**: 117–118.

I

Materials Science and Engineering

CHAPTER

1

Properties of Materials

EVELYN OWEN CAREW, FRANCIS W. COOKE, JACK E. LEMONS, BUDDY D. RATNER,
IVAN VESELY, AND ERWIN VOGLER

1.1 INTRODUCTION

Jack E. Lemons

The bulk and surface properties of biomaterials used for medical implants have been shown to directly influence, and in some cases, control the dynamic interactions that take place at the tissue–implant interface. These interactions are included in the concept of compatibility, which should be viewed as a two-way process between the implanted materials and the host environment that is ongoing throughout the *in vivo* lifetime of the device.

It is critical to recognize that synthetic materials have specific bulk and surface properties or characteristics. These characteristics must be known prior to any medical application, but also must be known in terms of changes that may take place over time *in vivo*. That is, changes with time must be anticipated at the outset and accounted for through selection of biomaterials and/or design of the device.

Information related to basic properties is available from national and international standards, plus handbooks and professional journals of various types. However, this information must be evaluated within the context of the intended biomedical use, since applications and host tissue responses are quite specific for given areas, e.g., cardiovascular (flowing blood contact), orthopedic (functional load bearing), and dental (percutaneous).

The following chapters provide two chapters on basic information about bulk and surface properties of biomaterials based on metallic, polymeric, and ceramic substrates, a chapter on finite element modeling and analyses, and a chapter specific to the role(s) of water and surface interaction with biomaterials. Also included are details about how some of these characteristics have been determined. The content of these chapters is intended to be relatively basic and more in-depth information is provided in later chapters and in the references.

1.2 BULK PROPERTIES OF MATERIALS

Francis W. Cooke

INTRODUCTION: THE SOLID STATE

Solids are distinguished from the other states of matter (liquids and gases) by the fact that their constituent atoms are held together by strong interatomic forces (Pauling, 1960). The electronic and atomic structures, and almost all the physical properties, of solids depend on the nature and strength of the interatomic bonds. For a full account of the nature of these bonds one would have to resort to the modern theory of quantum mechanics. However, the mathematical complexities of this theory are much beyond the scope of this book and we will instead content ourselves with the earlier, classical model, which is still very adequate. According to the classical theory there are three different types of strong or primary interatomic bonds: ionic, covalent, and metallic.

Ionic Bonding

In the ionic bond, electron donor (metallic) atoms transfer one or more electrons to an electron acceptor (nonmetallic) atom. The two atoms then become a cation (e.g., metal) and an anion (e.g., nonmetal), which are strongly attracted by the electrostatic or Coulomb effect. This attraction of cations and anions constitutes the ionic bond (Hummel, 1997).

In ionic solids composed of many ions, the ions are arranged so that each cation is surrounded by as many anions as possible to reduce the strong mutual repulsion of cations. This packing further reduces the overall energy of the assembly and leads to a highly ordered arrangement called a crystal structure (Fig. 1). Note that in such a crystal no discrete molecules exist, but only an orderly collection of cations and anions. The loosely bound electrons of the atoms are now tightly held in the locality of the ionic bond. These bound electrons are no longer available to serve as charge carriers and ionic solids are poor electrical conductors. Finally, the low overall energy state of these substances endows them with relatively low chemical reactivity. Sodium chloride ($NaCl$) and magnesium oxide (MgO) are examples of ionic solids.

Covalent Bonding

Elements that fall along the boundary between metals and nonmetals, such as carbon and silicon, have atoms with four valence electrons and about equal tendencies to donate and accept electrons. For this reason, they do not form strong ionic bonds. Rather, stable electron structures are achieved by sharing valence electrons. For example, two carbon atoms can

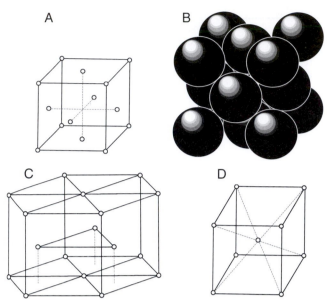

FIG. 1. Typical metal crystal structures (unit cells). (A) Face-centered cubic (FCC). (B) Full size atoms in FCC. (C) Hexagonal close-packed (HCP). (D) Body-centered cubic (BCC).

each contribute an electron to a shared pair. This shared pair of electrons constitutes the covalent bond $-\overset{|}{\underset{|}{C}}-\overset{|}{\underset{|}{C}}-$ (Hummel, 1997).

If a central carbon atom participates in four of these covalent bonds (two electrons per bond), it has achieved a stable outer shell of eight valence electrons. More carbon atoms can be added to the growing aggregate so that every atom has four nearest neighbors with which it shares one bond each. Thus, in a large grouping, every atom has a stable electron structure and four nearest neighbors. These neighbors often form a tetrahedron, and the tetrahedra in turn are assembled in an orderly repeating pattern (i.e., a crystal) (Fig. 2). This is the structure of both diamond and silicon. Diamond is the hardest of all materials, which shows that covalent bonds can be very strong. Once again, the bonding process results in a particular electronic structure (all valence electrons in pairs localized at the covalent bonds) and a particular atomic arrangement or crystal structure. As with ionic solids, localization of the valence electrons in the covalent bond renders these materials poor electrical conductors.

Metallic Bonding

The third the least understood of the strong bonds is the metallic bond. Metal atoms, being strong electron donors, do not bond by either ionic or covalent processes. Nevertheless, many metals are very strong (e.g., cobalt) and have high melting points (e.g., tungsten), suggesting that very strong interatomic bonds are at work here, too. The model that accounts for this bonding envisions the atoms arranged in an orderly, repeating, three-dimensional pattern, with the valence electrons migrating between the atoms like a gas.

It is helpful to imagine a metal crystal composed of positive ion cores, atoms without their valence electrons, about which the negative electrons circulate. On the average, all the electrical charges are neutralized throughout the crystal and bonding arises because the negative electrons act like a glue between the positive ion cores. This construct is called the free electron model of metallic bonding. Obviously, the bond strength increases as the ion cores and electron "gas" become more

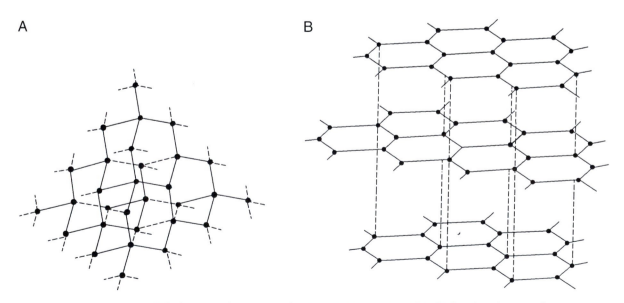

FIG. 2. Crystal structures of carbon. (A) Diamond (cubic). (B) Graphite (hexagonal).

tightly packed (until the inner electron orbits of the ions begin to overlap). This leads to a condition of lowest energy when the ion cores are as close together as possible.

Once again, the bonding leads to a closely packed (atomic) crystal structure and a unique electronic configuration. In particular, the nonlocalized bonds within metal crystals permit plastic deformation (which strictly speaking does not occur in any nonmetals), and the electron gas accounts for the chemical reactivity and high electrical and thermal conductivity of metallic systems (Hummel, 1997).

Weak Bonding

In addition to the three strong bonds, there are several weak secondary bonds that significantly influence the properties of some solids, especially polymers. The most important of these are van der Waals bonding and hydrogen bonding, which have strengths 3 to 10% that of the primary C–C covalent bond.

Atomic Structure

The three-dimensional arrangement of atoms or ions in a solid is one of the most important structural features that derives from the nature of the solid-state bond. In the majority of solids, this arrangement constitutes a crystal. A crystal is a solid whose atoms or ions are arranged in an orderly repeating pattern in three dimensions. These patterns allow the atoms to be closely packed [i.e., have the maximum possible number of near (contacting) neighbors] so that the number of primary bonds is maximized and the energy of the aggregate is minimized.

Crystal structures are often represented by repeating elements or subdivisions of the crystal called unit cells (Fig. 1). Unit cells have all the geometric properties of the whole crystal. A model of the whole crystal can be generated by simply stacking up unit cells like blocks or hexagonal tiles. Note that the representations of the unit cells in Fig. 1 are idealized in that atoms are shown as small circles located at the atomic centers. This is done so that the background of the structure can be understood. In fact, all nearest neighbors are in contact, as shown in Fig. 1B (Newey and Weaver, 1990).

MATERIALS

The technical materials used to build most structures are divided into three classes, metals, ceramics (including glasses), and polymers. These classes may be identified only roughly with the three types of interatomic bonding.

Metals

Materials that exhibit metallic bonding in the solid state are metals. Mixtures or solutions of different metals are alloys.

About 85% of all metals have one of the crystal structures shown in Fig. 1. In both face-centered cubic (FCC) and hexagonal close-packed (HCP) structures, every atom or ion is surrounded by twelve touching neighbors, which is the closest packing possible for spheres of uniform size. In any enclosure filled with close-packed spheres, 74% of the volume will be occupied by the spheres. In the body-centered cubic (BCC) structure, each atom or ion has eight touching neighbors or eightfold coordination. Surprisingly, the density of packing is only reduced to 68% so that the BCC structure is nearly as densely packed as the FCC and HCP structures (Hummel, 1997).

Ceramics

Ceramic materials are usually solid inorganic compounds with various combinations of ionic and covalent bonding. They also have tightly packed structures, but with special requirements for bonding such as fourfold coordination for covalent solids and charge neutrality for ionic solids (i.e., each unit cell must be electrically neutral). As might be expected, these additional requirements lead to more open and complex crystal structures. Aluminum oxide or alumina (Al_2O_3) is an example of a ceramic that has found some use as an orthopedic implant material. (Kingery, 1976).

Carbon is often included with ceramics because of its many ceramic-like properties, even though it is not a compound and conducts electrons in its graphitic form. Carbon is an interesting material since it occurs with two different crystal structures. In the diamond form, the four valence electrons of carbon lead to four nearest neighbors in tetrahedral coordination. This gives rise to the diamond cubic structure (Fig. 2A). An interesting variant on this structure occurs when the tetrahedral arrangement is distorted into a nearly flat sheet. The carbon atoms in the sheet have a hexagonal arrangement, and stacking of the sheets (Fig. 2B) gives rise to the graphite form of carbon. The (covalent) bonding within the sheets is much stronger than the bonding between sheets.

The existence of an element with two different crystal structures provides a striking opportunity to see how physical properties depend on atomic and electronic structure (Table 1).

Inorganic Glasses

Some ceramic materials can be melted and upon cooling do not develop a crystal structure. The individual atoms have

TABLE 1   Relative Physical Properties of Diamond and Graphite[a]

Property	Diamond	Graphite
Hardness	Highest known	Very low
Color	Colorless	Black
Electrical conductivity	Low	High
Density (g/cm^3)	3.51	2.25
Specific heat (cal/gm atm/deg.C)	1.44	1.98

[a] Adapted from D. L. Cocke and A. Clearfield, eds., *Design of New Materials*, Plenum Publ., New York, 1987, with permission.

nearly the ideal number of nearest neighbors, but an orderly repeating arrangement is not maintained over long distances throughout the three-dimensional aggregates of atoms. Such noncrystals are called glasses or, more accurately, inorganic glasses and are said to be in the amorphous state. Silicates and phosphates, the two most common glass formers, have random three-dimensional network structures.

Polymers

The third category of solid materials includes all the polymers. The constituent atoms of classic polymers are usually carbon and are joined in a linear chainlike structure by covalent bonds. The bonds within the chain require two of the valence electrons of each atom, leaving the other two bonds available for adding a great variety of atoms (e.g., hydrogen), molecules, functional groups, etc.

Based on the organization of these chains, there are two classes of polymers. In the first, the basic chains are all straight with little or no branching. Such "straight" chain or linear polymers can be melted and remelted without a basic change in structure (an advantage in fabrication) and are called thermoplastic polymers. If side chains are present and actually form (covalent) links between chains, a three-dimensional network structure is formed. Such structures are often strong, but once formed by heating will not melt uniformly on reheating. These are thermosetting polymers.

Usually both thermoplastic and thermosetting polymers have intertwined chains so that the resulting structures are quite random and are also said to be amorphous like glass, although only the thermoset polymers have sufficient cross linking to form a three-dimensional network with covalent bonds. In amorphous thermoplastic polymers, many atoms in a chain are in close proximity to the atoms of adjacent chains, and van der Waals and hydrogen bonding holds the chains together. It is these interchain bonds together with chain entanglement that are responsible for binding the substance together as a solid. Since these bonds are relatively weak, the resulting solid is relatively weak. Thermoplastic polymers generally have lower strengths and melting points than thermosetting polymers (Billmeyer, 1984).

Microstructure

Structure in solids occurs in a hierarchy of sizes. The internal or electronic structures of atoms occur at the finest scale, less than 10^{-4} μm (which is beyond the resolving power of the most powerful direct observational techniques), and are responsible for the interatomic bonds. At the next higher size level, around 10^{-4} μm (which is detectable by X-ray diffraction, field ion microscopy, scanning tunneling microscopy, etc.), the long-range, three-dimensional arrangement of atoms in crystals and glasses can be observed.

At even larger sizes, 10^{-3} to 10^2 μm (detectable by light and electron microscopy), another important type of structural organization exists. When the atoms of a molten sample are incorporated into crystals during freezing, many small crystals are formed initially and then grow until they impinge on each

other and all the liquid is used up. At that point the sample is completely solid. Thus, most crystalline solids (metals and ceramics) are composed of many small crystals or crystallites called grains that are tightly packed and firmly bound together. This is the microstructure of the material that is observed at magnifications where the resolution is between 1 and 100 μm.

In pure elemental materials, all the crystals have the same structure and differ from each other only by virtue of their different orientations. In general, these crystallites or grains are too small to be seen except with a light microscope. Most solids are opaque, however, so the common transmission (biological) microscope cannot be used. Instead, a metallographic or ceramographic reflecting microscope is used. Incident light is reflected from the polished metal or ceramic surface. The grain structure is revealed by etching the surface with a mildly corrosive medium that preferentially attacks the grain boundaries. When this surface is viewed through the reflecting microscope the size and shape of the grains, i.e., the microstructure, is revealed.

Grain size is one of the most important features that can be evaluated by this technique because fine-grained samples are generally stronger than coarse-grained specimens of a given material. Another important feature that can be identified is the coexistence of two or more phases in some solid materials. The grains of a given phase will all have the same chemical composition and crystal structure, but the grains of a second phase will be different in both these respects. This never occurs in samples of pure elements, but does occur in mixtures of different elements or compounds where the atoms or molecules can be dissolved in each other in the solid state just as they are in a liquid or gas solution.

For example, some chromium atoms can substitute for iron atoms in the FCC crystal lattice of iron to produce stainless steel, a solid solution alloy. Like liquid solutions, solid solutions exhibit solubility limits; when this limit is exceeded, a second phase precipitates. For example, if more Cr atoms are added to stainless steel than the FCC lattice of the iron can accommodate, a second phase that is chromium rich precipitates. Many important biological and implant materials are multiphase (Hummel, 1997). These include the cobalt-based and titanium-based orthopedic implant alloys and the mercury-based dental restorative alloys, i.e., amalgams.

MECHANICAL PROPERTIES OF MATERIALS

Solid materials possess many kinds of properties (e.g., mechanical, chemical, thermal, acoustical, optical, electrical, magnetic). For most (but not all) biomedical applications, the two properties of greatest importance are strength (mechanical) and reactivity (chemical). The chemical reactivity of biomaterials will be discussed in Chapters 1.4 and 6. The remainder of this section will, therefore, be devoted to mechanical properties, their measurement, and their dependence on structure. It is well to note that the dependence of mechanical properties on microstructure is so great that it is one of the fundamental objectives of materials science to control mechanical properties by modifying microstructure.

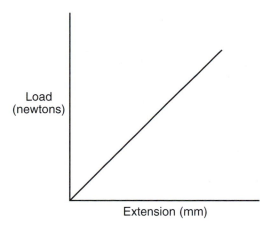

FIG. 3. Initial extension is proportional to load according to Hooke's law.

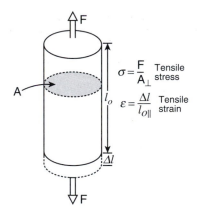

FIG. 4. Tensile stress and tensile strain.

Elastic Behavior

The basic experiment for determining mechanical properties is the tensile test. In 1678, Robert Hooke showed that a solid material subjected to a tensile (distraction) force would extend in the direction of traction by an amount that was proportional to the load (Fig. 3). This is known as Hooke's law and simply expresses the fact that most solids behave in an elastic manner (like a spring) if the loads are not too great.

Stress and Strain

The extension for a given load varies with the geometry of the specimen as well as its composition. It is, therefore, difficult to compare the relative stiffness of different materials or to predict the load-carrying capacity of structures with complex shapes. To resolve this confusion, the load and deformation can be normalized. To do this, the load is divided by the cross-sectional area available to support the load, and the extension is divided by the original length of the specimen. The load can then be reported as load per unit of cross-sectional area, and the deformation can be reported as the elongation per unit of the original length over which the elongation occurred. In this way, the effects of specimen geometry can be normalized.

The normalized load (force/area) is stress (σ) and the normalized deformation (change in length/original length) is strain (ε) (Fig. 4).

Tension and Compression

In tension and compression the area supporting the load is perpendicular to the loading direction (tensile stress), and the change in length is parallel to the original length (tensile strain).

If weights are used to provide the applied load, the stress is calculated by adding up the total number of pounds-force (lb) or newtons (N) used and dividing by the perpendicular cross-sectional area. For regular specimen geometries such as cyclindrical rods or rectangular bars, a measuring instrument, such as a micrometer, is used to determine the dimensions. The units of stress are pounds per inch squared (psi) or newtons

per meter squared (N/m^2). The N/m^2 unit is also known as the pascal (Pa).

The measurement of strain is achieved, in the simplest case, by applying reference marks to the specimen and measuring the distance between with calipers. This is the original length, l_o. A load is then applied, and the distance between marks is measured again to determine the final length, l_f. The strain, ε, is then calculated by:

$$\varepsilon = \frac{l_f - l_o}{l_o} = \frac{\Delta l}{l_o}. \tag{1}$$

This is essentially the technique used for flexible materials like rubbers, polymers, and soft tissues. For stiff materials like metals, ceramics, and bone, the deflections are so small that a more sensitive measuring method is needed (i.e., the electrical resistance strain gage).

Shear

For cases of shear, the applied load is parallel to the area supporting it (shear stress, τ), and the dimensional change is perpendicular to the reference dimension (shear strain, γ) (Fig. 5).

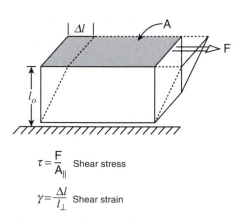

$$\tau = \frac{F}{A_{\parallel}} \quad \text{Shear stress}$$

$$\gamma = \frac{\Delta l}{l_{\perp}} \quad \text{Shear strain}$$

FIG. 5. Shear stress and shear strain.

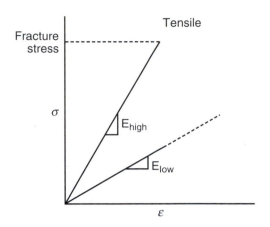
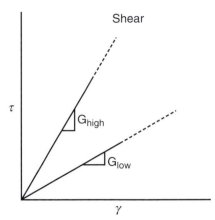

FIG. 6.   Stress versus strain for elastic solids.

Elastic Constants

By using these definitions of stress and strain, Hooke's law can be expressed in quantitative terms:

$$\sigma = E\,\varepsilon, \text{ tension or compression}, \tag{2a}$$

$$\tau = G\,\gamma, \text{ shear}. \tag{2b}$$

E and G are proportionality constants that may be likened to spring constants. The tensile constant, E, is the tensile (or (Young's) modulus and G is the shear modulus. These moduli are also the slopes of the elastic portion of the stress versus strain curve (Fig. 6). Since all geometric influences have been removed, E and G represent inherent properties of the material. These two moduli are direct macroscopic manifestations of the strengths of the interatomic bonds. Elastic strain is achieved by actually increasing the interatomic distances in the crystal (i.e., stretching the bonds). For materials with strong bonds (e.g., diamond, Al_2O_3, tungsten), the moduli are high and a given stress produces only a small strain. For materials with weaker bonds (e.g., polymers and gold), the moduli are lower (Hummel, 1997). The tensile elastic moduli for some important biomaterials are presented in Table 2.

Isotropy

The two constants, E and G, are all that are needed to fully characterize the stiffness of an isotropic material (i.e., a material whose properties are the same in all directions).

Single crystals are anisotropic (not isotropic) because the stiffness varies as the orientation of applied force changes relative to the interatomic bond directions in the crystal. In polycrystalline materials (e.g., most metallic and ceramic specimens), a great multitude of grains (crystallites) are aggregated with multiply distributed orientations. On the average, these aggregates exhibit isotropic behavior at the macroscopic level, and values of E and G are highly reproducible for all specimens of a given metal, alloy, or ceramic.

On the other hand, many polymeric materials and most tissue samples are anisotropic (not the same in all directions) even at the macroscopic level. Bone, ligament, and sutures are all stronger and stiffer in the fiber (longitudinal) direction than they are in the transverse direction. For such materials, more

TABLE 2   Mechanical Properties of Some Implant Materials and Tissues

	Elastic modulus (GPa)	Yield strength (MPa)	Tensile strength (MPa)	Elongation to failure (%)
Al_2O_3	350	—	1000 to 10,000	0
CoCr Alloy[a]	225	525	735	10
316 S.S.[b]	210	240 (800)[c]	600 (1000)[c]	55 (20)[c]
Ti–6Al–4V	120	830	900	18
Bone (cortical)	15 to 30	30 to 70	70 to 150	0–8
PMMA	3.0	—	35 to 50	0.5
Polyethylene[d]	0.6–1.8	—	23 to 40	200–400
Cartilage	[e]	—	7 to 15	20

[a] 28% Cr, 2% Ni, 7% Mo, 0.3% C (max), Co balance.
[b] Stainless steel, 18% Cr, 14% Ni, 2 to 4% Mo, 0.03 C (max), Fe balance.
[c] Values in parentheses are for the cold-worked state.
[d] High density polyethylene (HDPE) and ultrahigh molecular weight polyethylene (UHMWPE)
[e] Strongly viscoelastic.

than two elastic constants are required to relate stress and strain properties.

MECHANICAL TESTING

To conduct controlled load-deflection (stress–strain) tests, a load frame is used that is much stiffer and stronger than the specimen to be tested (Fig. 7). One cross-bar or cross-head is moved up and down by a screw or a hydraulic piston. Jaws that provide attachment to the specimen are connected to the frame and to the movable cross-head. In addition, a load cell to monitor the force being applied is placed in series with the specimen. The load cell functions somewhat like a very stiff spring scale to measure the applied loads.

Tensile specimens usually have a reduced gage section over which strains are measured. For a valid determination of fracture properties, failure must also occur in this reduced section and not in the grips. For compression testing, the direction of cross-head movement is reversed and cylindrical

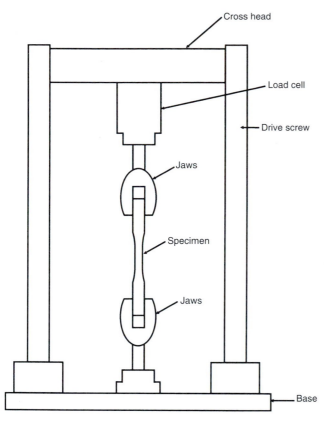

FIG. 7.   Mechanical testing machine.

or prismatic specimens are simply squeezed between flat anvils. Standardized specimens and procedures should be used for all mechanical testing to ensure reproducibility of results (see the publications of the American Society for Testing and Materials, 100 Barr Harbor Dr., West Conshohocken, PA 19428-2959).

Another useful test that can be conducted in a mechanical testing machine is the bend test. In bend testing, the outside of the bowed specimen is in tension and the inside in compression. The outer fiber stresses can be calculated from the load and the specimen geometry (see any standard text on strength of materials; Meriam, 1996). Bend tests are useful because no special specimen shapes are required and no special grips are necessary. Strain gages can also be used to determine the outer fiber strains. The available formulas for the calculation of stress states are only valid for elastic behavior. Therefore, they cannot be used to describe any nonelastic strain behavior.

Some mechanical testing machines are also equipped to apply torsional (rotational) loads, in which case torque versus angular deflection can be determined and used to calculate the torsional properties of materials. This is usually in important consideration when dealing with biological materials, especially under shear loading conditions (Hummel, 1997).

Elasticity

The tensile elastic modulus, E (for an isotropic material), can be determined by the use of strain gages, an accurate load cell,

and cyclic testing in a standard mechanical testing machine. To do so, Hooke's law is rearranged as follows:

$$E = \frac{\sigma}{\varepsilon}. \qquad (3)$$

Brittle Fracture

In real materials, elastic behavior does not persist indefinitely. If nothing else intervenes, microscopic defects, which are present in all real materials, will eventually begin to grow rapidly under the influence of the applied tensile or shear stress, and the specimen will fail suddenly by brittle fracture. Until this brittle failure occurs, the stress–strain diagram does not deviate from a straight line, and the stress at which failure occurs is called the fracture stress (Fig. 6). This behavior is typical of many materials, including glass, ceramics, graphite, very hard alloys (scalpel blades), and some polymers like polymethylmethacrylate (bone cement) and unmodified polyvinyl chloride (PVC). The number and size of defects, particularly pores, is the microstructural feature that most affects the strength of brittle materials.

Plastic Deformation

For some materials, notably metals, alloys, and some polymers, the process of plastic deformation sets in after a certain stress level is reached but before fracture occurs. During a tensile test, the stress at which 0.2% plastic strain occurs is called the 0.2% offset yield strength. Once plastic deformation starts, the strains produced are very much greater than those during elastic deformation (Fig. 8); they are no longer proportional to the stress and they are not recovered when the stress is removed. This happens because whole arrays of atoms under the influence of an applied stress are forced to move, irreversibly, to new locations in the crystal structure. This is the microstructural basis of plastic deformation. During elastic straining, on the other hand, the atoms are displaced only slightly by reversible stretching of the interatomic bonds.

Large scale displacement of atoms without complete rupture of the material, i.e., plastic deformation, is only possible in the presence of the metallic bond so only metals and alloys exhibit true plastic deformation. Since long-distance rearrangement of atoms under the influence of an applied stress cannot occur in ionic or convolutely bonded materials, ceramics and many polymers do not undergo plastic deformation.

Plastic deformation is very useful for shaping metals and alloys and is called ductility or malleability. The total permanent (i.e., plastic) strain exhibited up to fracture by a material is a quantitative measure of its ductility (Fig. 8). The strength, particularly the 0.2% offset yield strength, can be increased significantly by reducing the grain size as well as by prior plastic deformation or cold work. The introductions of alloying elements and multiphase microstructures are also potent strengthening mechanisms.

Other properties can be derived from the tensile stress–strain curve. The tensile strength or the ultimate tensile stress (UTS) is the stress that is calculated from the maximum load experienced during the tensile test (Fig. 8).

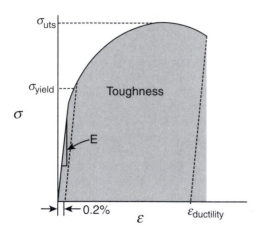

FIG. 8.   Stress versus strain for a ductile material.

TABLE 3   Mechanical Properties Derivable from a Tensile Test

| | Units | | |
Property	Fundamental[a]	International	English
1. Elastic modulus (E)	F/A	N/m^2 (Pa)	lbf[b]/in.^{2} (psi)
2. Yield strength (σ_{yield})	F/A	N/m^2 (Pa)	lbf/in.^{2} (psi)
3. Ultimate tensile strength (σ_{uts})	F/A	N/m^2 (Pa)	lbf/in.^{2} (psi)
4. Ductility ($\epsilon_{ductility}$)	%	%	%
5. Toughness (work to fracture per unit volume)	$F \times d/V$	J/m^3	in lbf/in.^{3}

[a] F, force; A, area; d, length; V, volume.
[b] lbf, pounds force.

The area under the tensile curve is proportional to the work required to deform a specimen until it fails. The area under the entire curve is proportional to the product of stress and strain, and has the units of energy (work) per unit volume of specimen. The work to fracture is a measure of toughness and reflects a material's resistance to crack propagation (Fig. 8) (Newey and Weaver, 1990). The important mechanical properties derived from a tensile test and their units are listed in Table 3. Representative values of these properties for some important biomaterials are listed in Table 2.

Creep and Viscous Flow

For all the mechanical behaviors considered to this point, it has been assumed that when a stress is applied, the strain response is instantaneous. For many important biomaterials,

including polymers and tissues, this is not a valid assumption. If a weight is suspended from an excised ligament, the ligament elongates essentially instantaneously when the weight is applied. This is an elastic response. Thereafter the ligament continues to elongate for a considerable time even though the load is constant (Fig. 9A). This continuous, time-dependent extension under load is called "creep."

Similarly, if the ligament is extended in a tensile machine to a fixed elongation and held constant while the load is monitored, the load drops continuously with time (Fig. 9B). The continuous drop in load at constant extension is called stress relaxation. Both these responses are the result of viscous flow in the material. The mechanical analog of viscous flow is a dashpot or cylinder and piston (Fig. 10A). Any small force is enough to keep the piston moving. If the load is increased, the rate of displacement will increase.

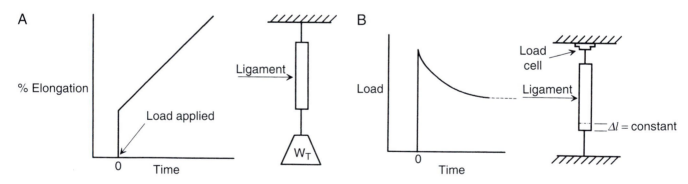

FIG. 9.   (A) Elongation versus time at constant load (creep) of ligament. (B) Load versus time at constant elongation (stress relaxation) for ligament.

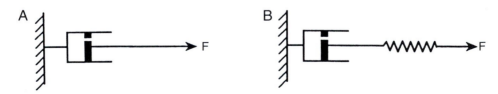

FIG. 10.   (A) Dash pot or cylinder and piston model of viscous flow. (B) Dash pot and spring model of a viscoelastic material.

Despite this liquid-like behavior, these materials are functionally solids. To produce such a combined effect, they act as though they are composed of a spring (elastic element) in series with a dashpot (viscous element) (Fig. 10B). Thus, in the creep test, instantaneous strain is produced when the weight is first applied (Fig. 9A). This is the equivalent of stretching the spring to its equilibrium length (for that load). Thereafter, the additional time-dependent strain is modeled by the movement of the dashpot. Complex arrangements of springs and dashpots are often needed to adequately model the actual behavior of polymers and tissues.

Materials that behave approximately like a spring and dashpot system are viscoelastic. One consequence of viscoelastic behavior can be seen in tensile testing where the load is applied at some finite rate. During the course of load application, there is time for some viscous flow to occur along with the elastic strain. Thus, the total strain will be greater than that due to the elastic response alone. If this total strain is used to estimate the Young's modulus of the material ($E = \sigma/\varepsilon$), the estimate will be low. If the test is conducted at a more rapid rate, there will be less time for viscous flow during the test and the apparent modulus will increase. If a series of such tests is conducted at ever higher loading rates, eventually a rate can be reached where no detectable viscous flow occurs and the modulus determined at this critical rate will be the true elastic modulus, i.e., the spring constant of the elastic component. Tests at even higher rates will produce no further increase in modulus. For all viscoelastic materials, moduli determined at rates less than the critical rate are "apparent" moduli and must be identified with the strain rate used. Failure to do this is one reason why values of tissue moduli reported in the literature may vary over wide ranges.

Finally, it should be noted that it may be difficult to distinguish between creep and plastic deformation in ordinary tensile tests of highly viscoelastic materials (e.g., tissues). For this reason, the total nonelastic deformation of tissues or polymers may at times be loosely referred to as plastic deformation even though viscous flow is involved.

OTHER IMPORTANT PROPERTIES OF MATERIALS

Fatigue

It is not uncommon for materials, including tough and ductile ones like 316L stainless steel, to fracture even though the service stresses imposed are well below the yield stress. This occurs when the loads are applied and removed for a great number of cycles, as happens to prosthetic heart valves and prosthetic joints. Such repetitive loading can produce microscopic cracks that then propagate by small steps at each load cycle.

The stresses at the tip of a crack, a surface scratch, or even a sharp corner are locally enhanced by the stress-raising effect. Under repetitive loading, these local high stresses actually exceed the strength of the material over a small region. This phenomenon is responsible for the stepwise propagation of the cracks. Eventually, the load-bearing cross-section becomes so small that the part finally fails completely.

Fatigue, then, is a process by which structures fail as a result of cyclic stresses that may be much less than the ultimate tensile stress. Fatigue failure plagues many dynamically loaded structures, from aircraft to bones (march- or stress-fractures) to cardiac pacemaker leads.

The susceptibility of specific materials to fatigue is determined by testing a group of identical specimens in cyclic tension or bending (Fig. 11A) at different maximum stresses. The number of cycles to failure is then plotted against the maximum applied stress (Fig. 11B). Since the number of cycles to failure is quite variable for a given stress level, the prediction of fatigue life is a matter of probabilities. For design purposes, the stress that will provide a low probability of failure after 10^6 to 10^8 cycles is often adopted as the fatigue strength or endurance limit of the material. This may be as little as one third or one fourth of the single-cycle yield strength. The fatigue strength is sensitive to environment, temperature, corrosion, deterioration (of tissue specimens), and cycle rate (especially for viscoelastic materials) (Newey and Weaver, 1990). Careful attention to these details is required if laboratory fatigue results are to be successfully transferred to biomedical applications.

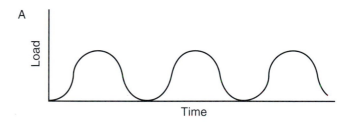

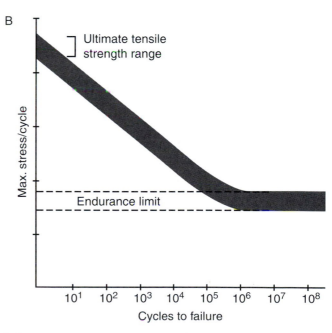

FIG. 11. (A) Stress versus time in a fatigue test. (B) Fatigue curve: fatigue stress versus cycles to failure.

Toughness

The ability of a material to plastically deform under the influence of the complex stress field that exists at the tip of a crack is a measure of its toughness. If plastic deformation does occur, it serves to blunt the crack and lower the locally enhanced stresses, thus hindering crack propagation. To design "failsafe" structures with brittle materials, it has become necessary to develop an entirely new system for evaluating service worthiness. This system is fracture toughness testing and requires the testing of specimens with sharp notches. The resulting fracture toughness parameter is a function of the apparent crack propagation stress and the crack depth and shape. It is called the critical stress intensity factor (K_{lc}) and has units of $Pa\sqrt{m}$ or $N \cdot m^{3/2}$ (Meyers and Chawla, 1984). For materials that exhibit extensive plastic deformation at the crack tip, an energy-based parameter, the J integral, can be used. The energy absorbed in impact fracture is also a measure of toughness, but at higher loading rates (Newey and Weaver, 1990).

Effect of Fabrication on Strength

A general concept to keep in mind when considering the strength of materials is that the process by which a material is produced has a major effect on its structure and hence its properties (Newey and Weaver, 1990). For example, plastic deformation of most metals at room temperature flattens the grains and produces strengthening while reducing ductility. Subsequent high-temperature treatment (annealing) can reverse this effect. Polymers drawn into fibers are much stronger in the drawing direction than are undrawn samples of the same material.

Because strength properties depend on fabrication history, it is important to realize that there is no unique set of strength properties of each generic material (e.g., 316L stainless steel, polyethylene, aluminum oxide). Rather, there is a range of properties that depends on the fabrication history and the microstructures produced.

CONCLUSION

The determination of mechanical properties is not only an exercise in basic materials science but is indispensable to the practical design and understanding of load-bearing structures. Designers must determine the service stresses in all structural members and be sure that at every point these stresses are safely below the yield strength of the material. If cyclic loads are involved (e.g., lower-limb prostheses, teeth, heart valves), the service stresses must be kept below the fatigue strength.

In subsequent chapters where the properties and behavior of materials are discussed in detail, it is well to keep in mind that this information is indispensable to understanding the mechanical performance (i.e., function) of both biological and manmade structures.

Bibliography

Billmeyer, F. W. (1984). *Textbook of Polymer Science*. John Wiley and Sons Inc., New York.

Hummel, R. E. (1997). *Understanding Materials Science*. Springer-Verlag, New York.

Kingery, W. D. (1976). *Introduction to Ceramics*. John Wiley and Sons Inc., New York.

Meriam, J. L. (1996). *Engineering Mechanics*, Vol. 1, *Statics*, 4th ed. John Wiley and Sons Inc., New York.

Meyers, M. A., and Chawla, K. K. (1984). *Mechanical Metallurgy*. Prentice-Hall Inc., Upper Saddle River, NJ.

Newey, C., and Weaver, G. (1990). *Materials Principals and Practice*. Butterworth-Heinemann Ltd., Oxford, UK.

Pauling, L. (1960). *The Nature of the Chemical Bond and the Structure of Molecules and Crystals*. Cornell Univ. Press, Ithaca, NY.

1.3 FINITE ELEMENT ANALYSIS

Ivan Vesely and Evelyn Owen Carew

INTRODUCTION

The previous chapter introduced the reader to the concepts of elasticity, stress, and strain. Estimations of material stress and strain are necessary during the course of device design to minimize the chance of device failure. For example, artificial hip joints need to be designed to withstand the loads that they are expected to bear without fracture or fatigue. Stress analysis is therefore required to ensure that all components of the device operate below the fatigue limit. For deformable structures such as diaphragms for artificial hearts, an estimate of strains or deformations is required to ensure that during maximal deformation, components do not contact other structures, potentially causing interference and unexpected failure modes such as abrasion.

For simple calculations, such as the sizing of a bolt to connect two components that bear load, simple analytical calculations usually suffice. Often, these calculations are augmented by reference to engineering tables that can be used to refine the stress estimates based on local geometry, such as the pitch of the threads. Such analytical methods are preferred because they are exact and can be supported by a wealth of engineering experience. Unfortunately, analytical solutions are usually limited to linear problems and simple geometries governed by simple boundary conditions. The boundary conditions can be considered input data or constraints on the solution that are applied at the boundaries of the system. Most practical engineering problems involve some combination of material or geometrical nonlinearity, complex geometry, and mixed boundary conditions. In particular, all biological materials have nonlinear elastic behavior and most experience large strains when deformed. As a result, nonlinearities of one form or the other are usually present in the formulation of problems in biomechanics. These nonlinearities are described by the equations relating stress to strain and strain to displacement. Applying analytical methods to such problems would require so many assumptions and simplifications that the results would have poor accuracy and would thus be of little engineering value. There is therefore no alternative but to resort to approximate or numerical methods. The most popular numerical method for solving problems in continuum mechanics is the finite element method (FEM), also referred to as finite element analysis (FEA).

FEA is a computational approach widely used in solid and fluid mechanics in which a complex structure is divided into a large number of smaller parts, or elements, with interconnecting nodes, each with geometry much simpler than that of the whole structure. The behavior of the unknown variable within the element and the shape of the element are represented by simple functions that are linked by parameters that are shared between the elements at the nodes. By linking these simple elements together, the complexity of the original structure can be duplicated with good fidelity. After boundary conditions are taken into account, a large system of equations for the unknown nodal parameters always results; these equations are solved simultaneously by a computer, using indirect or iterative means.

Finite element analysis is extremely versatile. The size and configuration of the elements can be adjusted to best suit the problem; complex geometries can be discretized and solutions can be stepped through time to analyze dynamic systems. Very often, simple analytical methods are used to make a first approximation to the design of the device, and FEA is subsequently used to further refine the design and identify potential stress concentrations. FEA can be applied to both solids and fluids or, with additional complexity, to systems containing both. FEA software is very mature and computing power is now sufficiently cheap to allow finite element methods to be applied to a wide range of problems. In fluid flow, FEA has been applied to weather forecasting and supersonic flow around aircraft and within engines, and in the medical field, to optimizing blood pumps and cannulas. In solids, FEA has been used to design, build, and crash automobiles, estimate the impact of earthquakes, and reconstruct crime scenes. In biomaterials, FEA has been applied to almost every implantable device, ranging from artificial joints to pacemaker leads. Although originally developed to help structural engineers analyze stress and strain, FEA has been adopted by basic scientists and biologists to study the dynamic environment within arteries, muscles and even cells.

In this chapter we hope to introduce the reader to finite element methods without digressing into detailed discussion of some of the more difficult concepts that are often required to properly define and execute a real-world problem. For that, the reader is referred to the many excellent texts in the field, some of which are included in the bibliography at the end of this chapter.

OVERVIEW OF THE FINITE ELEMENT METHOD

The essential steps in implementing the FEM follow:

(i) The region of interest (continuum) is discretized, that is, subdivided into a smaller number of regions called elements, interconnected at nodal points. Nodes may also be placed in the interior of an element. In one dimension, the elements are line segments; in two dimensions, they are usually triangles or quadrilaterals (Fig. 1); in three dimensions, they can be rectangular prisms (hexahedra) or triangular prisms (tetrahedra),

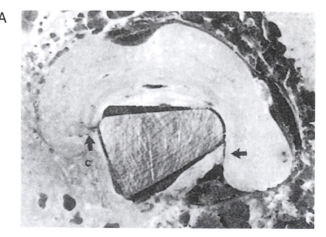

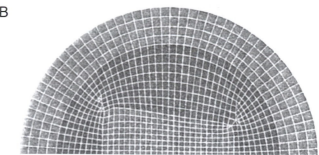

FIG. 1. (A) Cross-section of an autopsy-retrieved femur showing a cracked mantle (arrows). (B) Mixed planar quadrilateral/triangle FE representation of (A). (From Middleton *et al.*, 1996, p. 35. Reproduced with permission of Gordon and Breach Publishers, Overseas Publishers Assn., Amsterdam.)

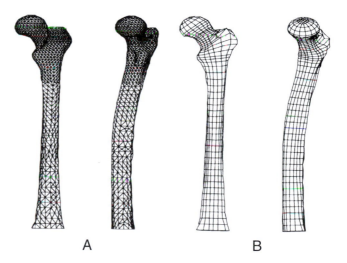

FIG. 2. 3D FE representations of the human femur. (A) Tetrahedral elements; (B) hexahedral elements. (From Middleton *et al.*, 1996, p. 125. Reproduced with permission of Gordon and Breach Publishers, Overseas Publishers Assn., Amsterdam.)

for example (Fig. 2). Elements may be quite general with the possibility of non-planar faces and curvilinear sides or edges (Desai, 1979; Zienkiewicz and Taylor, 1994).

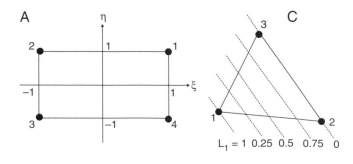

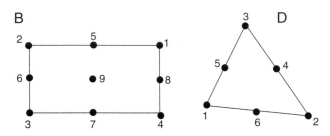

FIG. 3.   Examples of two-dimensional elements and their corresponding local coordinate systems [embedded in (A) and (C)]. For the rectangles, the local coordinates (ξ, η) are referred to a cartesian system with $-1 \leq \xi, \eta \leq 1$; for the triangles, the local coordinates (L_1, L_2, L_3) are area coordinates satisfying $0 \leq L_1, L_2, L_3 \leq 1$. Elements with linear interpolating functions (first order) are shown in (A) and (C). Quadratic elements (second-order interpolating functions) are shown in (B) and (D).

(ii)   The unknown variables within the continuum (e.g., displacement, stress, or velocity components) are defined within each element by suitable interpolating functions. Interpolating functions are traditionally piecewise polynomials and are also known as basis or shape functions. The order of the interpolating functions (i.e., first, second, or third order) is usually used to fix the number of nodes in the elements (Fig. 3).

(iii)   The equations that define the behavior of the unknown variable, such as the equations of motion or the relationships between stress and strain or strain and displacement, are formulated for each element in the form of matrices. These element matrices are then assembled into a global system of equations for the entire discretized domain. This system is defined by a coefficient matrix, an unknown vector of nodal values, and a known "right-hand side" (RHS) vector. Boundary conditions in derivative form would already be included in the RHS vector at this stage, but those that set the unknown function to a known value at the boundary have to be incorporated into the system matrix and RHS vector by overwriting relevant rows and columns. Since the RHS vector contains information about the boundary conditions, it is sometimes called the "external load vector."

(iv)   The final step in FEA involves solving the global system of equations for the unknown vector. In theory, this can be achieved by premultiplying the RHS vector by the inverse of the coefficient matrix. The result is

the discrete (pointwise) solution to the original problem. If the problem is linear and isotropic, the elements of the matrix are constants and the required matrix inversion can be done. If the defining equations are nonlinear or the material is anisotropic, the coefficient matrix itself will be a function of the unknown variables and matrix inversion is not straightforward. Some kind of linearization is necessary before the matrix can be inverted (e.g., successive approximation or Newton's methods; see, for example, Harris and Stöcker, 1998). In practice, the global system matrix, whether linear or nonlinear, is seldom inverted directly, usually because it is too large. Some indirect method of solving the system of equations is preferred [i.e., lower-upper (LU) decomposition, Gaussian elimination; see, for example, Harris and Stöcker, 1998].

The evaluation of element matrices, their assembly into the global system, and the possible linearization and eventual solution of the global system is a task that is always passed on to a high-speed computer. This usually requires complex computer programs written in a high-level language, such as Fortran. Indeed, it is the advent of high-speed computers and workstations and the continuous improvements in processor speed, memory management, and disk storage that have enabled large-scale FE problems to be tackled with relative ease.

The modern-day FEA toolbox also includes facilities for data pre- and postprocessing. Data preprocessing usually involves input formatting and grid definition, the latter of which may require some ingenuity, because mesh design may affect the convergence and accuracy of the numerical solution. Element size is governed by local geometry and the rate of change of the solution in different parts of the domain. Mesh refinement (a gradation of element size) in the vicinity of sharp corners, boundary layers, high solution gradients, stress concentrations or vortices is done routinely to enhance the accuracy and convergence of the solution. Adaptive procedures that allow the mesh to change with the solution according to some error criteria are usually incorporated into the FE process (George, 1991; Brebbia and Aliabadi, 1993; Zienkiewicz and Taylor, 1994). Typically, this means that the mesh is refined in areas where the solution gradient is high, and elements are removed from regions where the solution is changing slowly. The result is usually a dramatic improvement in convergence, accuracy, and computational efficiency. Postprocessing of data involves the evaluation of *ad hoc* variables such as strains, strain rates, stresses; generating plots such as simple xy-plots, contour plots, and particle paths; and solution visualization and animation. All of the additional information facilitates the understanding and interpretation of the results.

The importance of checking and validating FE solutions cannot be overemphasized. The most basic validation involves a "patch test" (Zienkiewicz and Taylor, 1994) in which a few elements (i.e., a patch of the material) are analyzed to verify the formulation of interpolating functions and the consistency of the code itself. Second, a very simple problem with known analytical solution is simulated with a coarse grid to verify that the code reproduces the known solution with acceptable accuracy. For example, parabolic flow in a tube can be simulated with a very coarse grid and the result quickly compared against the analytical solution. We caution, however, that reproducing the solution in a simpler problem does not guarantee that the code will work in more realistic and complicated cases. It is also recommended that numerical solutions be obtained from at least three meshes with increasing degrees of mesh refinement. Such solutions should converge with mesh refinement (*h*-convergence, Strang and Fix, 1973). Comparison of numerical results to experimental data should always be made where possible. Last, especially in the absence of analytical solutions or experimental data, numerical solutions should be compared across different numerical methods, or across different numerical codes if the same method is used. There is no gold standard for the number of validation tests that is required for any particular problem. The greater the variety of test problems and checks, the greater the degree of confidence one can have in the results of the finite element method.

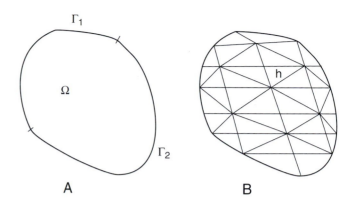

FIG. 4. (A) A continuum Ω enclosed by the boundary $\Gamma = \Gamma_1 \cup \Gamma_2$; the function itself is specified on Γ_1 and its derivative on Γ_2. (B) A finite element representation of the continuum. The domain has been discretized with general arbitrary triangles of size h, with the possibility of having curved sides.

THE CONTINUUM EQUATIONS

Whether we use FEA to compute the stress in a prosthetic limb or to simulate blood flow in bifurcating arteries, the first objective in setting up an FEA problem is to identify and specify the equations that define the behavior of unknown variables in the continuum. Such equations typically result from applying the universal laws of conservation mass, momentum, and energy, as well as the constitutive equations that define the stress–strain or other relationships within the material. The resulting differential or integral equations must then be closed by specifying the appropriate boundary conditions.

A "well-behaved" solution to the continuum problem is guaranteed if the differential or integral equations and boundary conditions systems are "well posed." This means that a solution to the continuum problem should exist, be unique, and only change by a small amount when the input data change by a small amount. Under these circumstances the numerical solution is guaranteed to converge to the true solution. Proving in advance that a general continuum problem is "well posed" is not a trivial exercise. Fortunately, consistency and convergence of the numerical solution can usually be monitored by other means, for example, the already mentioned "patch test" (Zienkiewicz and Taylor, 1994).

The equations governing the description of a continuum can be formulated via a differential or variational approach. In the former, differential equations are used to describe the problem; in the latter, integral equations are used. In some cases, both formulations can be applied to a problem. As an illustration we present a case for which both formulations apply and later show that these lead to the same FE equations.

The Differential Formulation

Consider the function $u(x, y)$, defined in some two-dimensional domain Ω bounded by the curve Γ (Fig. 4), which satisfies the differential equation

$$-\nabla^2 u + qu = f \text{ in } \Omega \tag{1a}$$

subject to the boundary conditions

$$u = g \text{ on } \Gamma_1 \tag{1b}$$

$$\frac{\partial u}{\partial n} = 0 \text{ on } \Gamma_2 \tag{1c}$$

where $\nabla^2 \equiv \partial^2/\partial x^2 + \partial^2/\partial y^2$ is the Laplacian operator in two dimensions, n is the unit outward normal to the boundary, and q, f, g are assumed to be constants for simplicity, with $q \geq 0$. Here, the boundary Γ is made up of two parts, Γ_1 and Γ_2, where different boundary conditions apply.

When $f = 0$, Eq. 1a means that the spatial change of the gradient of u at any point in the $x - y$ space is proportional to u. The boundary condition 1b sets u to have a fixed value at one part of the boundary. On another part of the boundary, the rate of change of u in the normal direction is set to zero (boundary condition 1c). The system represented by Eqs. 1a–c can be used to describe the transverse deflection of a membrane, torsion in a shaft, potential flows, steady-state heat conduction, or groundwater flow (Desai, 1979; Zienkiewicz and Taylor, 1994).

The Variational Formulation

A variational equation can arise, for example, from the physical requirement that the total potential energy (TPE) of a mechanical system must be a minimum. Thus the TPE will be a function of a displacement function, for example, itself a function of spatial variables. A "function of a function" is referred to as a functional. We consider, as an example, the functional $I(v)$ of the function $v(x, y)$ of the spatial variables x and y, defined by:

$$I(v) = \iint_\Omega \left\{ (\nabla v)^2 + qv^2 - 2vf \right\} d\Omega \tag{2}$$

(Strang and Fix, 1973; Zienkiewicz and Taylor, 1994). The relevant question is that of all the possible functions $v(x, y)$ that satisfy Eq. 2, what particular $v(x, y)$ minimizes $I(v)$? We get the answer by equating the first variation of $I(v)$, written $\delta I(v)$, to zero. To perform the variation of a functional, one uses the standard rules of differentiation. It can be shown that the variation of $I(v)$ over v results in Eq. 1a, provided Eqs. 1b and 1c hold and the variation of v is zero on Γ_1. Thus the function that minimizes the functional defined in Eq. 2 is the same function that solves the boundary value problem given by Eqs. 1a–c.

THE FINITE ELEMENT EQUATIONS

There are four basic methods of formulating the equations of finite element analysis. These are: (i) the direct or displacement method, (ii) the variational method, (iii) the weighted residual method, and (iv) the energy balance method. Only the more popular variational and weighted residual methods will be described here. The integral equation 2 will be used to illustrate the variational method, while the differential equation system 1a–c will be used to illustrate the weighted residual method.

The Variational Approach

The FEM is introduced in the following way. The region is divided into a finite number of elements of size h (Fig. 4). The h notation is to be interpreted as referring to the subdivided domain. Instead of seeking the function v that minimizes $I(v)$ in the continuous domain, i.e., the exact solution, we instead seek an approximate solution by looking for the function v^h that minimizes $I(v^h)$ in the discrete domain. The following trial functions are defined over the discretized domain:

$$v^h(x, y) = \sum_{i=1}^n v_i N_i(x, y) \qquad (3)$$

where N_i are global basis or shape functions and v_i are nodal parameters. The sum is over the total number of nodes n in the mesh. Using Eq. 3 in Eq. 2, the functional becomes

$$I(v^h) = \sum_{i,j} v_i v_j \iint_\Omega \nabla N_i \nabla N_j \, d\Omega$$
$$+ q \sum_{i,j} v_i v_j \iint_\Omega \nabla N_i N_j \, d\Omega - 2 \sum_{i,j} v_i \iint_\Omega f N_i \, d\Omega$$

$$\qquad (4)$$

which can be written in matrix notation as

$$I(v^h) = v^T K v + q v^T M v - 2 v^T F \qquad (5)$$

where

$$K = \iint_\Omega \nabla N_i \nabla N_j \, d\Omega, \quad M = \iint_\Omega N_i N_j \, d\Omega, \quad F = \iint_\Omega f N_i \, d\Omega$$

v^T represents the *transpose* of the vector v; K is known as the stiffness matrix, M as the mass matrix, and F as the local load vector. The function v^h that minimizes Eq. 5 should satisfy $\delta I(v^h) = 0$. This gives

$$(K + qM)v = F \qquad (6)$$

that is, a set of simultaneous equations for the nodal parameters v.

Weighted Residual Approach

The weighted residual approach can be applied directly to any system of differential equations such as 1a–c and even to those problems for which a variational principle may not exist. This approach is therefore more general. The method assumes an approximation $u^h(x, y)$ for the real solution $u(x, y)$. Because u^h is approximate, its substitution into Eq. 1a will result in an error or residual R^h:

$$R^h = -\nabla^2 u^h + q u^h - f \qquad (7)$$

The weighted residual approach requires that some weighted average of the error due to nonsatisfaction of the differential equation by the approximate solution u^h (Eq. 7) vanish over the domain of interest:

$$\iint_\Omega R^h w \, d\Omega = \iint_\Omega \left(-\nabla^2 u^h + q u^h - f \right) w \, d\Omega = 0 \quad (8)$$

where $w(x, y)$ is a weighting function. A function u^h that satisfies Eq. 8 for all possible w selected from a certain class of functions must necessarily satisfy the original differential equations 1a–c. It actually does so only in an average or "weak" sense. Equation 8 is therefore known as a "weak form" of the original equation 1a. The second-order derivatives of the ∇^2 term are usually reduced to first order derivatives by an integration by parts (Harris and Stocker, 1998). The result is another weak form:

$$\iint_\Omega \left\{ \nabla u^h \nabla w + q u^h w - f w \right\} d\Omega = 0 \qquad (9)$$

which has the advantage that approximating functions can now be chosen from a much larger space, a space where the function only needs be once-differentiable. Again, we divide the region into a finite number of elements and assume that the approximate solution can be represented by the sum of the product of unknown nodal values v_j and interpolating functions $N_j(x, y)$, defined at each node j of the mesh:

$$u^h = \sum_{j=1}^n v_j N_j \qquad (10)$$

When Eq. 10 is substituted into Eq. 9 with $w = N_i$, Eq. 6 results as before, proving the equivalence of the weighted residual and variational methods for this particular example. We note that the weighting function is required to be zero on those parts of the boundary where the unknown function is specified (Γ_1, in our example) and that there can be other choices of the weighting function w. Choosing weighting functions to be the same as interpolating functions defines the Galerkin finite element method (Strang and Fix, 1973; Zienkiewicz and Taylor, 1994).

Properties of Interpolating Functions

The process of discretizing the continuum into smaller regions means that the global shape functions $N_j(x, y)$ are replaced by local shape functions $N_j^e(\xi, \eta)$, defined within each element e, where ξ, η are the local coordinates within the element (Fig. 3). In the FEM, interpolating functions are usually piecewise polynomials that are required to have (a) the minimum degree of smoothness, (b) continuity between elements, and (c) "local support."

The minimum degree of smoothness is dictated by the highest derivative of the unknown function that occurs in the "weak" or variational form of the continuum problem. The requirement for continuity between elements can always be satisfied by an appropriate choice of the approximating polynomial and number of boundary nodes that define the element. The requirement for "local support" means that within an element,

$$N_i^e(\xi, \eta) = \begin{cases} 1 & \text{at node } i \\ 0 & \text{at all other nodes} \end{cases} \quad (11)$$

as shown in Fig. 5. This is the single most important property of the interpolating functions. This property makes it possible for the contributions of all the elements to be summed up to give the response of the whole domain.

The notation P^m is conventionally used to indicate the degree m of the interpolating polynomial. The notation C^n is used to indicate that all derivatives of the interpolating function, up to and including $n - 1$, exist and are continuous. By convention, the notation $P^m - C^n$ is therefore used to indicate the order and smoothness properties of the interpolating polynomials.

EXAMPLES FROM BIOMECHANICS

The following are examples of FEA applications in biomaterials science and biomechanics.

Analysis of Commonplace Maneuvers at Risk for Total Hip Dislocation

Dislocation is a frequent complication of total hip arthroplasty (THA). In this FE study (Nadzadi *et al.*, 2003), a motion tracking system and a recessed force plate were used to capture the kinematics and ground reaction forces from several trials of realistic dislocation-prone maneuvers performed by actual subjects. Kinematics and kinetic data associated with the experiments were imported into a FE model of THA dislocation. The FE model was used to compute stresses developed within the implant, given the observed angular motion of the hip and contact force inferred from inverse dynamics. The FE mesh (Fig. 6A) was created using PATRAN version 8.5 and the simulations were executed with ABAQUS version 5.8. In the FE analysis, the resultant resisting moment developed around the hip-cup center was tracked, as a function of hip angle. The peak of this resistive moment was a key outcome measure used to estimate the relative risk of dislocations from the motions. All seven maneuvers studied led to frequent instances of computationally predicted dislocation (Fig. 6B). The authors conclude that this library of dislocation-prone maneuvers appear to substantially extend the information base previously available to study this important complication of THA. Additionally the hope is that their results will contribute to improvements in implant design and surgical technique and reduce *in vivo* incidence.

A Finite Element Model for the Lower Cervical Spine

A parametric study was conducted to determine the variations in the biomechanical responses of the spinal components in the lower cervical spine (Yoganandan *et al.*, 1997). Axial compressive load was imposed uniformly on the superior surface of the C4-C6 unit. The various components were assumed to have linear isotropic and homogeneous elastic behavior and appropriate material parameters were taken from the literature. A detailed 3D finite element model was reconstructed from 1.0-mm CT scans of a human cadaver, resulting in a total of 10,371 elements (Fig. 7A). The results show that an increase in elastic moduli of the disks resulted in an increase in endplate stresses and that the middle C5 vertebral body produced the highest compressive stresses (Fig. 7B). The model appears to confirm clinical experience that cervical fractures are induced by external compressive forces.

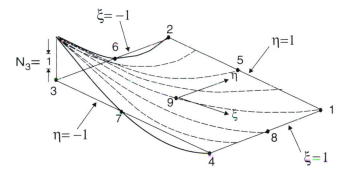

$$N_1^e(\xi, \eta) = \xi\eta\,(1+\xi)\,(1+\eta)/4$$
$$N_2^e(\xi, \eta) = -\xi\eta\,(1-\xi)\,(1+\eta)/4$$
$$N_3^e(\xi, \eta) = \xi\eta\,(1-\xi)\,(1-\eta)/4$$
$$N_4^e(\xi, \eta) = -\xi\eta\,(1+\xi)\,(1-\eta)/4$$
$$N_5^e(\xi, \eta) = \eta\,(1-\xi^2)\,(1+\eta)/2$$
$$N_6^e(\xi, \eta) = -\xi(1-\xi^2)\,(1-\eta^2)/2$$
$$N_7^e(\xi, \eta) = -\eta(1-\xi^2)\,(1-\eta)/2$$
$$N_8^e(\xi, \eta) = \xi(1+\xi^2)\,(1-\eta^2)/2$$
$$N_9^e(\xi, \eta) = \xi(1-\xi^2)\,(1-\eta^2)$$

FIG. 5.   Sample shape functions for a nine-noded rectangular element. Shape functions are defined in terms of local coordinates ξ and η where $-1 \le \xi, \eta \le 1$; $N_3^e(\xi, \eta)$ is shown in the plot. It can be checked that $N_i = 1$ at node i and zero at all other nodes (compact support) as required.

A

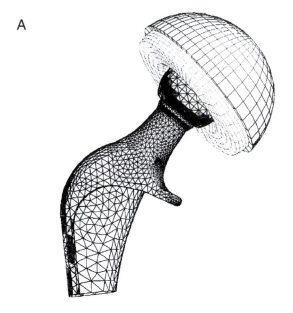

B

Maneuver	No. of trials	No. of dislocations	% of trials dislocating
Low sit-to-stand	47	41	87
Normal sit-to-stand	55	33	64
Tie	69	31	45
Leg cross	64	22	34
Stoop	42	6	14
Post. disloc. maneuvers	277	133	48
Pivot	58	23	40
Roll	19	12	63
Ant. disloc. maneuvers	77	35	45
Overall series	353	168	47

FIG. 6. (A) Finite element model of a contemporary 22-mm modular THA system. (B) Table of FE dislocation predictions of the seven challenge maneuvers simulated. (Reproduced with permission from Nadzadi *et al.*, 2003.)

Finite Element Analysis of Indentation Tests on Pyrolytic Carbon

Pyrolytic carbon (PyC) heart valves are known to fail through cracks initiated at the contact areas between leaflets and their housing. In Gilpin *et al.* (1996), this phenomenon is simulated with a 5.1-mm steel ball indenting a graphite sheet coated on each side with PyC, similar to the makeup of real heart valves. Two types of contacts were analyzed: when the surface material is thick (rigid backing) and when it is fairly thin (flexible backing). FEA was used to evaluate the stresses resulting from a range of loads. The geometry was taken to be axisymmetric, PyC was assumed to be an elastic material and quadrilateral solid elements were used. Figure 8A shows part of the FE mesh. Note that the mesh is refined in the contact areas but gets progressively coarser toward the noncontact areas. Figure 8B shows the maximum principal stress on the PyC surface adjacent to ball contact, as a function of the indentation load. "Flexible backing" is seen to greatly reduce the maximum principal stress in this area. The FE results were correlated with data from experiments and used to develop failure criteria for contact stresses. This in turn provided criteria for designing contact regions in pyrolytic heart valves.

Numerical Analysis of 3D Flow in an Aorta through an Artificial Heart Valve

Three-dimensional transient flow past a Björk-Shiley valve in the aorta is simulated by the FEM combined with a time-stepping algorithm (Shim and Chang, 1997). The FE mesh is shown in Fig. 9A, comprising some 32,880 elements and 36,110 nodes. The results indicate that the flow is split into two major jet flows by the valve, which later merge downstream. A 3D plot of velocity vectors show large velocities in the upper and lower jet flow regions in the sinus region, large

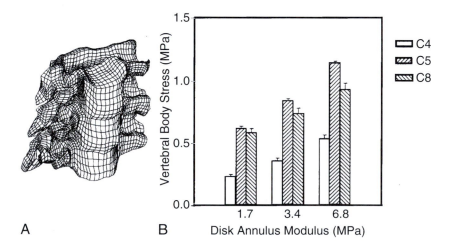

A   B

FIG. 7. Finite element model of the C4-C6 unit of the lower cervical spine: (A) mesh showing 3D solid elements and (B) plot of vertebral body stress as a function of disk annulus moduli. (Reproduced with permission from Yoganandan *et al.*, 1997.)

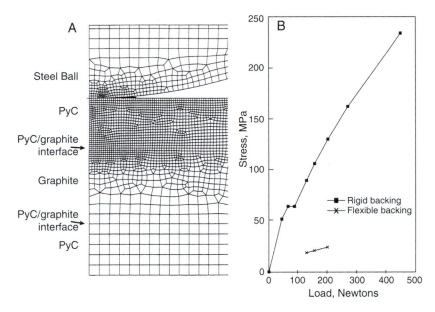

FIG. 8. Finite element analysis of indentation tests on pyrolytic carbon (PyC). (A) Part of the FE mesh showing a steel ball in contact with a PyC/graphite material. (B) Maximum principal stress on the PyC surface adjacent to ball contact radius. (Reproduced with permission from Gilpin *et al.*, 1996.)

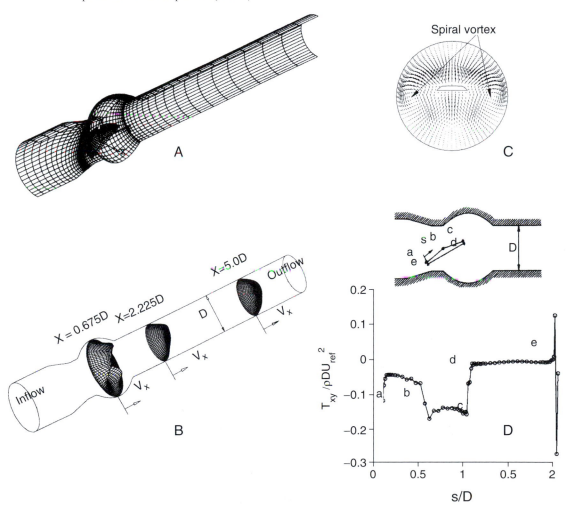

FIG. 9. FE analysis of transient 3D flow past a Bjork-Shiley valve in the aorta: (A) surface grid of aorta and fully opened Bjork-Shiley valve prosthesis, (B) carpet plot of axial velocity vectors, (C) secondary flow vector plot showing spiral vortices, and (D) shear stress along the valve surface in the symmetric mid-plane. (Reproduced with permission from Shim and Chang, 1997.)

velocities only in the upper part of the merged jet, and an almost uniform paraboloid distribution near the outflow region (Fig. 9B). Twin spiral vortices are generated immediately downstream of the valve, in the sinus region (Fig. 9C) and are convected downstream, where they quickly die away by diffusion. Shear stress along the surface of the valve is shown to be a maximum in the vicinity of its leading edge (Fig. 9D). A study such as this provides useful information on the function of the valve *in vivo*.

CONCLUSION

The FEM is an approximate, numerical method for solving boundary-value problems of continuum mechanics that are posed in differential or variational form. The main advantages of the FEM over other numerical methods lie in its generalization to three dimensions and the relative ease in which arbitrary geometries, boundary conditions, and material anisotropy can be incorporated into the solution process. The same FE code can be applied to solve a wide range of nonrelated problems. Its main disadvantage has been its complexity to implement. Fortunately, the abundance and availability of commercial codes in recent years and an emphasis on a "black box" approach with minimum user interaction have reduced the level of expertise required in the implementation of FEA to most engineering problems.

Bibliography

Brebbia, C. A., and Aliabadi, M. H., eds. (1993). *Adaptive Finite and Boundary Element Methods.* Elsevier, New York.

Desai, C. S. (1979). *Elementary Finite Element Method.* Prentice-Hall, Upper Saddle River, NJ.

George, P. L. (1991). *Automatic Mesh Generation: Application to Finite Element Methods.* Wiley, New York.

Gilpin, C. B., Haubold, A. D., and Ely, J. L. (1996). Finite element analysis of indentation tests on pyrolytic carbon. *J. Heart Valve Dis.* 5(**Suppl. 1**): S72.

Harris, J. W., and Stöcker, H. (1998). *Handbook of Mathematics and Computational Science.* Springer, New York.

Middleton, J., Jones, M. L., and Pande, G. N., eds. (1996). *Computer Methods in Biomechanics and Biomedical Engineering.* Gordon and Breach, Amsterdam.

Nadzadi, M. E., Pedersen, D. R., Yack, H. J., Callaghan, J. J., and Brown, T. D. (2003). Kinematics, kinetics, and finite elements analysis of commonplace maneuvers at risk for total hip dislocation. *J. Biomech.* 36: 577.

Shim, E. B., and Chang, K. S. (1997). Numerical analysis of three-dimensional Björk-Shiley valvular flow in an aorta. *J. Biomech. Eng.* 119: 45.

Strang, G., and Fix, G. J. (1973). *An Analysis of the Finite Element Method.* Prentice-Hall, Upper Saddle River, NJ.

Yoganandan, N., Kumaresan, S., Voo, L., and Pintar, F. A. (1997). Finite element model of the human lower cervical spine: parametric analysis of the C4-C6 unit. *J. Biomech. Eng.* 119: 87.

Zienkiewicz, O. C., and Taylor, R. L. (1991, 1994). *The Finite Element Method*, 4th ed., 2 vols. McGraw-Hill, London.

1.4 SURFACE PROPERTIES AND SURFACE CHARACTERIZATION OF MATERIALS

Buddy D. Ratner

INTRODUCTION

Consider the atoms that make up the outermost surface of a biomaterial. As we shall discuss in this section, these atoms that reside at the surface have a special organization and reactivity. They require special methods to characterize them and novel methods to tailor them, and they drive many of the biological reactions that occur in response to the biomaterial (protein adsorption, cell adhesion, cell growth, blood compatibility, etc.). The importance of surfaces for biomaterials science has been appreciated since the 1960s. Almost every biomaterials meeting will have sessions addressing surfaces and interfaces. In this chapter we focus on the special properties of surfaces, definitions of terms, methods to characterize surfaces, and some implications of surfaces for bioreaction to biomaterials.

In developing biomedical implant devices and materials, we are concerned with function, durability, and biocompatibility. In order to function, the implant must have appropriate properties such as mechanical strength, permeability, or elasticity, just to name a few. Well-developed methods typically exist to measure these bulk properties—often these are the classic methodologies of engineers and materials scientists. Durability, particularly in a biological environment, is less well understood. Still, the tests we need to evaluate durability have been developed over the past 20 years (see Chapters 1.2, 6.2, and 6.3). Biocompatibility represents a frontier of knowledge in this field, and its study is often assigned to the biochemist, biologist, and physician. However, an important question in biocompatibility is how the device or material "transduces" its structural makeup to direct or influence the response of proteins, cells, and the organism to it. For devices and materials that do not leach undesirable substances in sufficient quantities to influence cells and tissues (i.e., that have passed routine toxicological evaluation; see Chapter 5.2), this transduction occurs through the surface structure – the body "reads" the surface structure and responds. For this reason we must understand the surface structure of biomaterials. Chapter 9.4 elaborates on the biological implications of this idea.

General Surface Considerations and Definitions

This is the appropriate point in this chapter to highlight general ideas about surfaces, especially solid surfaces. First, the surface region of a material is known to be of unique reactivity (Fig. 1A). Catalysis (for example, as used in petrochemical processing) and microelectronics both capitalize on special surface reactivity—thus, it would be surprising if biology did not also use surfaces to do its work. This reactivity also leads to surface oxidation and other surface chemical reactions. Second, the surface of a material is inevitably different from the bulk. The traditional techniques used to analyze the bulk structure of materials are not suitable for surface determination because they typically do not have the sensitivity to observe

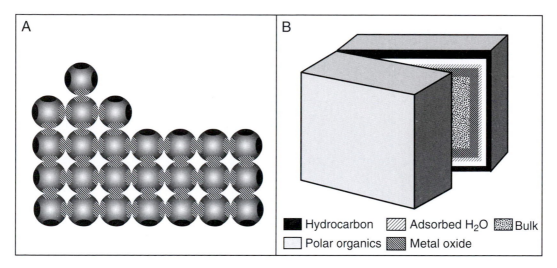

FIG. 1. (A) A two-dimensional representation of a crystal lattice illustrating bonding orbitals (black or crosshatched ovals). For atoms in the center (bulk) of the crystal (crosshatched ovals), all binding sites are associated. At planar exterior surfaces, one of the bonding sites is unfulfilled (black oval). At corners, two bonding sites are unfulfilled. The single atom on top of the crystal (an adatom) has three unfulfilled valencies. Energy is minimized where more of these unfulfilled valencies can interact. (B) In a "real world" material (e.g., a block of metal from an orthopedic device), if we cleave the block (under ultrahigh vacuum to prevent recontamination) we should find hydrocarbon on the outermost layer (perhaps 3 nm, surface energy $\sim$22 ergs/cm^2), polar organic molecules (>1 nm, surface energy $\sim$45 ergs/cm^2), adsorbed water (<1 nm, surface energy $\sim$72 ergs/cm^2), metal oxide (approximately 5 nm, surface energy $\sim$200 ergs/cm^2), and finally, the uniform bulk interior (surface energy $\sim$1000 ergs/cm^2). The interface between air and material has the lowest interfacial energy ($\sim$22 ergs/cm^2). The layers are not drawn to scale.

the small amount of material comprising the unique surface chemistry/structure. Third, there is not much total mass of material at a surface. An example may help us to appreciate this—on a 1 cm^3 cube of titanium, the 100 Å oxide surrounding the cube is in the same proportion as a 5-m wide beach on each coast of the United States is to the roughly 5,000,000 m distance from coast to coast. Fourth, surfaces readily contaminate with components from the vapor phase (some common examples are hydrocarbons, silicones, thiols, iodine). Under ultrahigh vacuum conditions (pressures <10^{-7} Pa) we can retard this contamination. However, in view of the atmospheric pressure conditions under which all biomedical devices are used, we must learn to live with some contamination. The key questions here are whether we can make devices with controlled and acceptable levels of contamination and also avoid undesirable contaminants. This is critical so that a laboratory experiment on a biomaterial generates the same results when repeated after 1 day, 1 week, or 1 year, and so that the biomedical device is dependable and has a reasonable shelf life. Finally, the surface structure of a material is often mobile. A modern view of what might be seen at the surface of a real-world material is illustrated in Fig. 1B.

The movement of atoms and molecules near the surface in response to the outside environment is often highly significant. In response to a hydrophobic environment (e.g., air), more hydrophobic (lower energy) components may migrate to the surface of a material—a process that reduces interfacial energy (Fig. 1B). Responding to an aqueous environment, the surface may reverse its structure and point polar (hydrophilic) groups outward to interact with the polar water molecules.

Again, energy minimization drives this process. An example of this is schematically illustrated in Fig. 2. For metal alloys, one metal tends to dominate the surface, for example, silver in a silver–gold alloy or chromium in stainless steel.

The nature of surfaces is complex and the subject of much independent investigation. The reader is referred to one of many excellent monographs on this important subject for a complete and rigorous introduction (see Somorjai, 1981, 1994; Adamson and Gast, 1997; Andrade, 1985). For overviews of the relationship between surface science, biology and biomaterials, see Castner and Ratner (2002), Tirrell *et al.* (2000), Ratner (1988).

When we say "surface," a question that immediately comes to mind is, "how deep into the material does it extend?" Although formal definitions are available, for all practical purposes, the surface is the zone where the structure and composition, influenced by the interface, differs from the average (bulk) composition and structure. This value often scales with the size of the molecules making up the surface. For an "atomic" material, for example gold, after penetration of about five atomic layers (0.5–1 nm), the composition becomes uniform from layer to layer (i.e., you are seeing the bulk structure). At the outermost atomic layer, the organization of the gold atoms at the surface (and their reactivity) can be substantially different from the organization in the averaged bulk. The gold, in air, will always have a contaminant overlayer, largely hydrocarbon, that may be roughly 2 nm thick. This is also a difference in composition between bulk and surface, but it is not the atomic/molecular rearrangements we are discussing here. For a polymer, the unique surface zone may extend from 10 nm

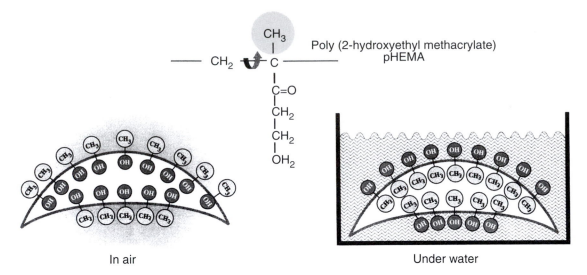

Poly (2-hydroxyethyl methacrylate)
pHEMA

In air

Under water

FIG. 2. Many materials can undergo a reversal of surface structure when transferred from air into a water environment. In this schematic illustration, a hydroxylated polymer (for example, a pHEMA contact lens) exhibits a surface rich in methyl groups (from the polymer chain backbone) in air, and a surface rich in hydroxyl groups under water. This has been observed experimentally (Ratner *et al.*, 1978, *J. Appl. Polym. Sci.* **22**: 643; Chen *et al.*, 1999, *J. Am. Chem. Soc.* **121**(2): 446).

to 100 nm (depending on the polymeric system and the chain molecular weight). Figure 1B addresses some of these issues about surface definitions. Two more definitions must be considered. An interface is the transition between two phases, in principle an infinitely thin separation plane. An interphase is the unique compositional zone between two phases. For the example, for gold, we might say that the interphase between gold and air is 3 nm thick (the structurally rearranged gold atoms + the contaminant layer).

Parameters to Be Measured

Many parameters describe a surface, as shown in Fig. 3. The more of these parameters we measure, the better we can piece together a complete description of the surface. A complete characterization requires the use of many techniques to compile all the information needed. Unfortunately, we cannot yet specify which parameters are most important for understanding biological responses to surfaces. Studies have been published on the importance of roughness, wettability, surface mobility, chemical composition, electrical charge, crystallinity, and heterogeneity to biological reaction. Since we cannot be certain which surface factors are predominant in each situation, the controlling variable or variables must be independently ascertained.

SURFACE ANALYSIS TECHNIQUES

General Principles

A number of general ideas can be applied to all surface analysis. They can be divided into the categories of sample preparation and analysis described in the following paragraphs.

Sample Preparation

A key consideration for sample preparation is that the sample should resemble, as closely as possible, the material or device being subjected to biological testing or implantation. Needless to say, fingerprints on the surface of the sample will cover up things that might be of interest. If the sample is placed in a package for shipping or storage prior to surface analysis, it is critical to know whether the packaging material can induce surface contamination. Plain paper in contact with most specimens will transfer material (often metal ions) to the surface of the material. Many plastics are processed with silicone oils or other additives that can be transferred to the specimen. The packaging material used should be examined by surface analysis methods to ascertain its purity. Samples can be surface analyzed prior to and after storage or shipping in containers to ensure that the surface composition measured is not due to the container. As a general rule, the polyethylene press-close bags used in electron microscopy and cell culture plasticware are clean storage containers. However, abrasive contact must be avoided and each brand must be evaluated so that a meticulous specimen preparation is not ruined by contamination. Many brands of aluminum foil are useful for packing specimens, but some are treated with a surface layer of stearic acid that can surface contaminate wrapped biomaterials, implants or medical devices. Aluminum foil should be checked for cleanliness by surface analysis methods before it is used to wrap important specimens.

Surface Analysis General Comments

Two general principles guide sample analysis. First, all methods used to analyze surfaces also have the potential to alter the surface. It is essential that the analyst be aware of the

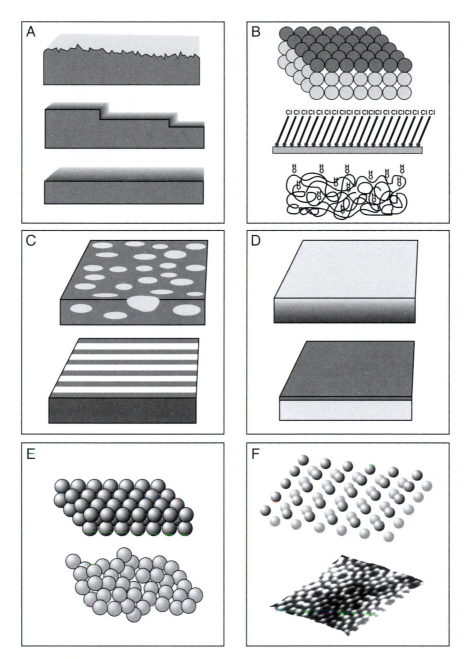

FIG. 3. What might be measured to define surface structure? (A) Surfaces can be rough, stepped or smooth. (B) Surfaces can be composed of different chemistries (atomic, supramolecular, macromolecular). (C) Surfaces may be structurally or compositionally inhomogeneous in the plane of the surface such as phase-separated domains or microcontact printed lanes. (D) Surfaces may be inhomogeneous with depth into the specimen or simply overlayered with a thin film. (E) Surfaces may be highly crystalline or disordered. (F) Crystalline surfaces are found with many organizations such as a silicon (100) unreconstructed surface or a silicon (111)(7 × 7) reconstructed surface.

damage potential of the method used. Second, because of the potential for artifacts and the need for many pieces of information to construct a complete picture of the surface (Fig. 3), more than one method should be used whenever possible. The data derived from two or more methods should always be corroborative. When data are contradictory, be suspicious and question why. A third or fourth method may then be necessary

to allow confident conclusions to be drawn about the nature of a surface.

These general principles are applicable to all materials. There are properties (only a few of which will be presented here) that are specific to specific classes of materials. Compared with metals, ceramics, glasses, and carbons, organic and polymeric materials are more easily damaged by surface

TABLE 1   Common Methods to Characterize Biomaterial Surfaces

Method	Principle	Depth analyzed	Spatial resolution	Analytical sensitivity	Cost[c]
Contact angles	Liquid wetting of surfaces is used to estimate the energy of surfaces	3–20 Å	1 mm	Low or high depending on the chemistry	$
ESCA (XPS)	X-rays induce the emission of electrons of characteristic energy	10–250 Å	10–150 μm	0.1 at%	$$$
Auger electron spectroscopy[a]	A focused electron beam stimulates the emission of Auger electrons	50–100 Å	100 Å	0.1 atom%	$$$
SIMS	Ion bombardment sputters secondary ions from the surface	10 Å–1 μm[b]	100 Å	Very high	$$$
FTIR-ATR	IR radiation is adsorbed and excites molecular vibrations	1–5 μm	10 μm	1 mol%	$$
STM	Measurement of the quantum tunneling current between a metal tip and a conductive surface	5 Å	1 Å	Single atoms	$$
SEM	Secondary electron emission induced by a focused electron beam is spatially imaged	5 Å	40 Å, typically	High, but not quantitative	$$

[a] Auger electron spectroscopy is damaging to organic materials and is best used for inorganics.
[b] Static SIMS ≈ 10 Å, dynamic SIMS to 1 μm
[c] $, up to $5000; $$, $5000–$100,000; $$$, >$100,000.

analysis methods. Polymeric systems also exhibit greater surface molecular mobility than inorganic systems. The surfaces of inorganic materials are contaminated more rapidly than polymeric materials because of their higher surface energy. Electrically conductive metals and carbons will often be easier to characterize than insulators using the electron, X-ray, and ion interaction methods. Insulators accumulate a surface electrical charge that requires special methods (e.g., a low-energy electron beam) to neutralize. To learn about other concerns in surface analysis that are specific to specific classes of materials, published papers become a valuable resource for understanding the pitfalls that can lead to artifact or inaccurate results.

Table 1 summarizes the characteristics of many common surface analysis methods, including their depth of analysis and their spatial resolution (spot size analyzed). A few of the more frequently used techniques are described in the next section. However, space limitations prevent a developed discussion of these methods. The reader is referred to many comprehensive books on the general subject of surface analysis (Andrade, 1985; Briggs and Seah, 1983; Feldman and Mayer, 1986; Vickerman, 1997). References on specific surface analysis methods will be presented in sections on each of the key methods.

Contact Angle Methods

A drop of liquid sitting on a solid surface represents a powerful, but simple, method to probe surface properties. Experience tells us that a drop of water on highly polished automobile body surfaces will stand up (bead up), whereas if that car has not been polished in a long time, the liquid will flow evenly over the surface. This observation, with some understanding of the method, tells us that the highly polished car probably has silicones or hydrocarbons at its surface, while the unpolished car surface consists of oxidized material. This type of observation, backed up with a quantitative measurement of the drop angle with the surface, has been used in biomaterials science to predict the performance of vascular grafts and the adhesion of cells to surfaces.

The phenomenon of the contact angle can be explained as a balance between the force with which the molecules of the liquid (in the drop) are being attracted to each other (a cohesive force) and the attraction of the liquid molecules for the molecules that make up the surface (an adhesive force). An equilibrium is established between these forces, the energy minimum. The force balance between the liquid–vapor surface tension (γ_{lv}) of a liquid drop and the interfacial tension between a solid and the drop (γ_{sl}), manifested through the contact angle (θ) of the drop with the surface, can be used to quantitatively characterize the energy of the surface (γ_{sv}). The basic relationship describing this force balance is:

$$\gamma_{sv} = \gamma_{sl} + \gamma_{lv}\cos\theta$$

The energy of the surface, which is directly related to its wettability, is a useful parameter that has often correlated strongly with biological interaction. Unfortunately, γ_{sv} cannot be directly obtained since this equation contains two unknowns, γ_{sl} and γ_{sv}. Therefore, the γ_{sv} is usually approximated by the Zisman method for obtaining the critical surface tension (Fig. 4), or calculated by solving simultaneous equations with data from liquids of different surface tensions. Some critical surface tensions for common materials are listed in Table 2.

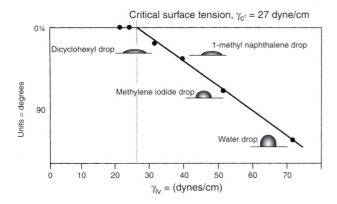

FIG. 4. The Zisman method permits a critical surface tension value, an approximation to the solid surface tension, to be measured. Drops of liquids of different surface tensions are placed on the solid, and the contact angles of the drops are measured. The plot of liquid surface tension versus angle is extrapolated to zero contact angle to give the critical surface tension value.

TABLE 2 Critical Surface Tension Values for Common Materials Calculated from Contact Angle Measurements

Material	Critical surface tension (dyn/cm)
Polytetrafluoroethylene	19
Poly(dimethyl siloxane)	24
Poly(vinylidine fluoride)	25
Poly(vinyl fluoride)	28
Polyethylene	31
Polystyrene	33
Poly(2-hydroxyethyl methacrylate)	37
Poly(vinyl alcohol)	37
Poly(methyl methacrylate)	39
Poly(vinyl chloride)	39
Polycaproamide (nylon 6)	42
Poly(ethylene oxide)-diol	43
Poly(ethylene terephthalate)	43
Polyacrylonitrile	50

Experimentally, there are a number of ways to measure the contact angle, and some of these are illustrated in Fig. 5. Contact angle methods are inexpensive, and, with some practice, easy to perform. They provide a "first line" characterization of materials and can be performed in any laboratory. Contact angle measurements provide unique insight into how the surface will interact with the external world. However, in performing such measurements, a number of concerns must be addressed to obtain meaningful data (Table 3). Review articles are available on contact angle measurement for surface characterization (Andrade, 1985; Good, 1993; Zisman, 1964; McIntire, *et al.*, 1985).

Electron Spectroscopy for Chemical Analysis

Electron spectroscopy for chemical analysis (ESCA) provides unique information about a surface that cannot be obtained by other means (Andrade, 1985; Ratner, 1988; Dilks, 1981; Ratner and McElroy, 1986; Ratner and Castner, 1997; Watts and Wolstenholme, 2003). In contrast to the contact angle technique, ESCA requires complex, expensive apparatus (Fig. 6A) and demands considerable training to perform the measurements. However, since ESCA is available from commercial laboratories, university analytical facilities, national centers (for example, NESAC/BIO at the University of Washington), and specialized research laboratories, most biomaterials scientists can get access to it to have their samples analyzed. The data can be interpreted in a simple but still useful fashion, or more rigorously. ESCA has contributed significantly to the development of biomaterials and medical devices, and to understanding the fundamentals of biointeraction.

The ESCA method (also called X-ray photoelectron spectroscopy, XPS) is based upon the photoelectric effect, properly described by Einstein in 1905. X-rays are focused upon a specimen. The interaction of the X-rays with the atoms in the specimen causes the emission of a core level (inner shell) electron. The energy of this electron is measured and its value provides information about the nature and environment of the atom from which it came. The basic energy balance describing this process is given by the simple relationship:

$$BE = h\nu - KE$$

where BE is the energy binding the electron to an atom (the value desired), KE is the kinetic energy of the emitted electron (the value measured in the ESCA spectrometer), and $h\nu$ is the energy of the X-rays, a known value. A simple schematic diagram illustrating an ESCA instrument is shown in Fig. 6B. Table 4 lists some of the types of information about the nature of a surface that can be obtained by using ESCA. The origin of the surface sensitivity of ESCA is described in Fig. 7.

ESCA has many advantages, and a few disadvantages, for studying biomaterials. The advantages include the high information content, the surface localization of the measurement, the speed of analysis, the low damage potential, and the ability to analyze most samples with no specimen preparation. The last advantage is particularly important because it means that many biomedical devices (or parts of devices) can be inserted, as fabricated and sterilized, directly in the analysis chamber for study. The disadvantages include the need for vacuum compatibility (i.e., no outgassing of volatile components), the vacuum environment (particularly for hydrated specimens), the possibility of sample damage by X-rays if long analysis times are used, the need for experienced operators, and the cost associated with this complex instrumentation. The vacuum limitations can be sidestepped by using an ESCA system with a cryogenic sample stage. At liquid nitrogen temperatures, samples with volatile components, or even wet, hydrated samples, can be analyzed.

The use of ESCA is best illustrated with a brief example. A poly(methyl methacrylate) (PMMA) ophthalmologic device is to be examined. Taking care not to touch or damage the surface of interest, the device is inserted into the ESCA instrument introduction chamber. The introduction chamber is then

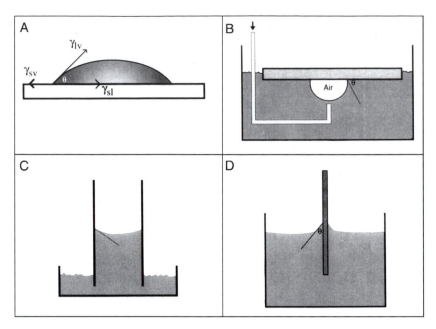

FIG. 5. Four possibilities for contact angle measurement: (A) sessile drop, (B) captive air bubble method, (C) capillary rise method, (D) Wilhelmy plate method.

TABLE 3   Concerns in Contact Angle Measurement

The measurement is operator dependent.

Surface roughness influences the results.

Surface heterogeneity influences the results.

The liquids used are easily contaminated (typically reducing their ν_{lv}).

The liquids used can reorient the surface structure.

The liquids used can absorb into the surface, leading to swelling.

The liquids used can dissolve the surface.

Few sample geometries can be used.

Information on surface structure must be inferred from the
  data obtained.

pumped down to 10^{-6} torr (1.33×10^{-4} Pa) pressure. A gate valve between the introduction chamber and the analytical chamber is opened and the specimen is moved into the analysis chamber. In the analysis chamber, at 10^{-9} torr (1.33×10^{-7} Pa) pressure, the specimen is positioned (on contemporary instruments, using a microscope or TV camera) and the X-ray source is turned on. The ranges of electron energies to be observed are controlled (by computer) with the retardation lens on the spectrometer. First, a wide scan is made in which the energies of all emitted electrons over, typically, a 1000 eV range are detected (Fig. 8). Then, narrow scans are made in which each of the elements detected in the wide scan is examined in higher resolution (Fig. 9).

From the wide scan, we learn that the specimen contains carbon, oxygen, nitrogen, and sulfur. The presence of sulfur and nitrogen is unexpected for PMMA. We can calculate elemental ratios from the wide scan. The sample surface contains 58.2% carbon, 27.7% oxygen, 9.5% nitrogen, and 4.5% sulfur. The narrow scan for the carbon region (C1s spectrum) suggests four species: hydrocarbon, carbons singly bonded to oxygen (the predominant species), carbons in amide-like molecular environments, and carbons in acid or ester environments. This is different from the spectrum expected for pure PMMA. An examination of the peak position in the narrow scan of the sulfur region (S2p spectrum) suggests sulfonate-type groups. The shape of the C1s spectrum, the position of the sulfur peak, and the presence of nitrogen all suggest that heparin was immobilized to the surface of the PMMA device. Since the stoichiometry of the lens surface does not match that for pure heparin, this suggests that we are seeing either some of the PMMA substrate through a <100 Å layer of heparin, or we are seeing some of the bonding links used to immobilize the heparin to the lens surface. Further ESCA analysis will permit the extraction of more detail about this surface-modified device, including an estimate of surface modification thickness, further confirmation that the coating is indeed heparin, and additional information about the nature of the immobilization chemistry.

Secondary Ion Mass Spectrometry

Secondary ion mass spectrometry (SIMS) is an important addition to the armamentarium of tools that the surface analyst can bring to bear on a biomedical problem. SIMS produces

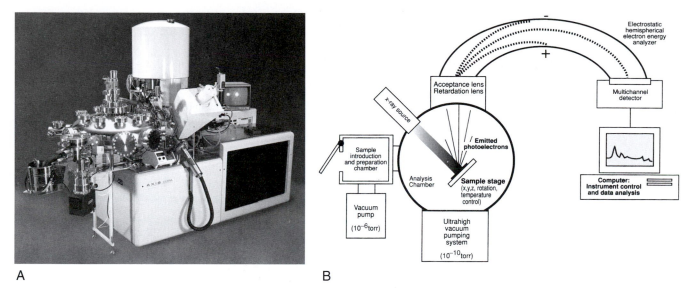

FIG. 6. (A) Photograph of a contemporary ESCA instrument (photo by Kratos Corp.). (B) Schematic diagram of a monochromatized ESCA instrument.

TABLE 4 Information Derived from an ESCA Experiment

In the outermost 100 Å of a surface, ESCA can provide:

Identification of all elements (except H and He) present at concentrations >0.1 at %

Semiquantitative determination of the approximate elemental surface composition (±10%)

Information about the molecular environment (oxidation state, bonding atoms, etc.)

Information about aromatic or unsaturated structures from shake-up $\pi^* \leftarrow \pi$ transitions

Identification of organic groups using derivatization reactions

Nondestructive elemental depth profiles 100 Å into the sample and surface heterogeneity assessment using angular-dependent ESCA studies and photoelectrons with differing escape depths

Destructive elemental depth profiles several thousand angstroms into the sample using argon etching (for inorganics)

Lateral variations in surface composition (spatial resolution 8–150 μm, depending upon the instrument)

"Fingerprinting" of materials using valence band spectra and identification of bonding orbitals

Studies on hydrated (frozen) surfaces

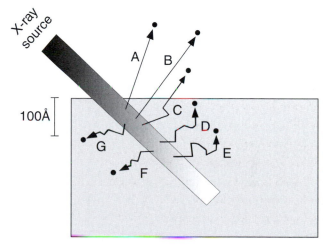

FIG. 7. ESCA is a surface-sensitive method. Although the X-ray beam can penetrate deeply into a specimen, electrons emitted deep in the specimen (D, E, F, G) will lose their energy in inelastic collisions and never emerge from the surface. Only those electrons emitted near the surface that lose no energy (A, B) will contribute to the ESCA signal used analytically. Electrons that lose some energy but still have sufficient energy to emerge from the surface (C) contribute to the background signal.

a mass spectrum of the outermost 10 Å of a surface. Like ESCA, it requires complex instrumentation and an ultrahigh vacuum chamber for the analysis. However, it provides unique information that is complementary to ESCA and greatly aids in understanding surface composition. Some of the analytical capabilities of SIMS are summarized in Table 5. Review articles on SIMS are available (Ratner, 1985; Scheutzle et al., 1984;

Briggs, 1986; Davies and Lynn, 1990; Vickerman et al., 1989; Benninghoven, 1983; Van Vaeck et al., 1999; Belu et al., 2003).

In SIMS analysis, a surface is bombarded with a beam of accelerated ions. The collision of these ions with the atoms and molecules in the surface zone can transfer enough energy to them so they sputter from the surface into the vacuum phase. The process is analogous to racked pool balls that are ejected from the cluster by the impact of the cue ball; the harder the cue ball hits the rack of balls, the more balls are emitted from the rack. In SIMS, the cue balls are ions (xenon, argon, cesium,

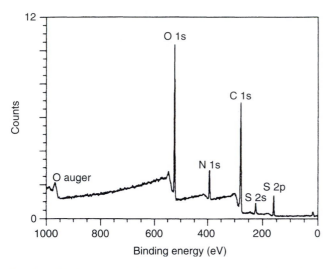

FIG. 8. ESCA wide scan of a surface-modified poly(methyl methacrylate) ophthalmologic device.

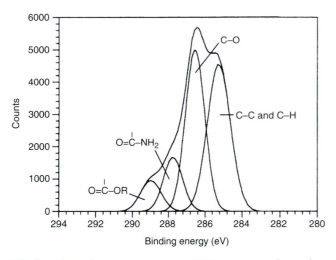

FIG. 9. The carbon 1s narrow scan ESCA spectrum of a surface-modified poly(methyl methacrylate) ophthalmologic device. Narrow scan spectra can be generated for each element seen in low-energy resolution mode in Fig. 8.

TABLE 5 Analytical Capabilities of SIMS

	Static SIMS	Dynamic SIMS
Identify hydrogen and deuterium	✓	✓
Identify other elements (often must be inferred from the data)	✓	✓
Suggest molecular structures (inferred from the data)	✓	-
Observe extremely high mass fragments (proteins, polymers)	✓	-
Detection of extremely low concentrations	✓	✓
Depth profile to 1 μm into the sample	*	✓
Observe the outermost 1–2 atomic layers	✓	-
High spatial resolution (features as small as ~400 Å)	✓	✓
Semiquantitative analysis (for limited sets of specimens)	✓	✓
Useful for polymers	✓	-
Useful for inorganics (metals, ceramics, etc.)	✓	✓
Useful for powders, films, fibers, etc.	✓	✓

*Cluster ion sources may allow depth profiling with static SIMS-like information content

and gallium ions are commonly used) that are accelerated to energies of 5000–20,000 eV. The particles ejected from the surface are positive and negative ions (secondary ions), radicals, excited states, and neutrals. Only the secondary ions are measured in SIMS. In ESCA, the energy of emitted particles (electrons) is measured. SIMS measures the mass of emitted ions (more rigorously, the ratio of mass to charge, m/z) using a time-of-flight (TOF) mass analyzer or a quadrupole mass analyzer.

There are two modes for SIMS analysis, depending on the ion flux: dynamic and static. Dynamic SIMS uses high ion doses in a given analysis time. The primary ion beam sputters so much material from the surface that the surface erodes at an appreciable rate. We can capitalize on this to do a depth profile into a specimen. The intensity of the m/z peak of a species of interest (e.g., sodium ion, $m/z = 23$) might be followed as a function of time. If the ion beam is well controlled and the sputtering rate is constant, the sodium ion signal intensity measured at any time will be directly related to its concentration at the erosion depth of the ion beam into the specimen. A concentration depth profile (sodium concentration versus depth) can be constructed over a range from the outermost atoms to a micron or more into the specimen. However, owing to the damaging nature of the high-flux ion beam, only atomic fragments can be detected. Also, as the beam erodes deeper into the specimen, more artifacts are introduced in the data by "knock-in" and scrambling of atoms.

Static SIMS, in comparison, induces minimal surface destruction. The ion dose is adjusted so that during the period of the analysis less than 10% of one monolayer of surface atoms is sputtered. Since there are typically 10^{13}–10^{15} atoms in 1 cm^2 of surface, a total ion dose of less than 10^{13} ions/cm^2 during the analysis period is appropriate. Under these conditions, extensive degradation and rearrangement of the chemistry at the surface does not take place, and large, relatively intact molecular fragments can be ejected into the vacuum for measurement. Examples of molecular fragments are shown in Fig. 10. This figure also introduces some of the ideas behind SIMS spectral interpretation. A more complete introduction to the concepts behind static SIMS spectral interpretation can be found in Van Vaeck et al., (1999) or in standard texts on mass spectrometry.

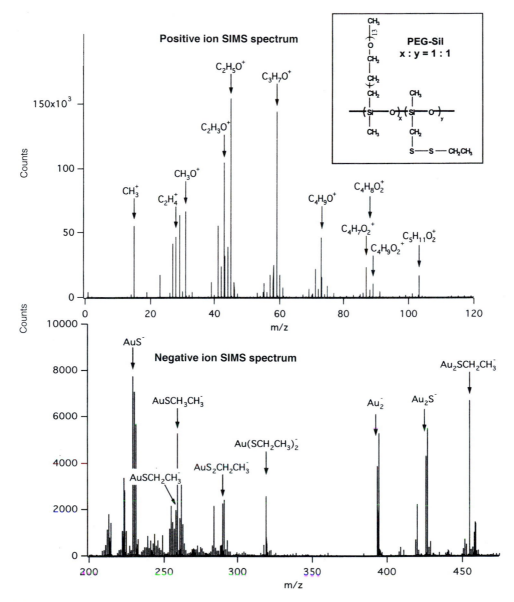

FIG. 10. Static positive and negative ion SIMS spectra of a poly(ethylene glycol)–poly(dimethyl siloxane) copolymer containing disulfide side groups on a gold surface. The primary peaks are identified. The low-mass region of the negative ion spectrum offers little insight into the polymer structure, but the high-mass region is rich in information. In this case, the low-mass positive spectrum is rich in information. Further details on this class of polymers can be found in *Macromolecules* **27**: 3053 (1994). (Figure supplied by D. Castner.)

Magnetically or electrostatically focusing the primary ion beam permits the SIMS technique to have high spatial resolution in the x,y plane. In fact, SIMS analysis can be performed in surface regions of 10 nm diameter or smaller. For static SIMS analysis, less than 10% of the atoms in any area are sampled. Thus, as the spot size gets smaller, the challenge to achieve high analytical sensitivity increases sharply. Still, static SIMS measurements have been performed in areas as small as 40 nm. Newly developed cluster ion sources (for example, using gold molecular clusters, Au_3, or C_{60} as the impacting primary particles) show high secondary ion yields and relatively low surface damage. These may improve spatial resolution and also permit depth profiling of organic surfaces by sputtering down into a surface while monitoring secondary ion emission as a function of time.

If the focused primary ion beam is rastered over the surface and the x,y position of the beam correlated with the signal emitted from a given spot, the SIMS data can be converted into an elemental image. Patterning and spatial control of chemistry is becoming increasingly important in biomaterials surface design. For example, microcontact printing allows patterned chemistry to be transferred to surfaces at the micron level using a relatively simple rubber stamp. Imaging SIMS is well suited to studying and monitoring such spatially defined chemistry.

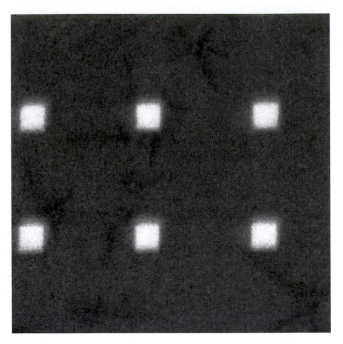

FIG. 11. Static SIMS image of protein islands on a poly(ethylene oxide) background. (For details, see Veiseh, M., Wickes, B. T., Castner, D. G., Zhang, M., "Guided cell patterning by surface molecular engineering." In press.)

An example is presented in Fig. 11. Imaging SIMS is also valuable for observing defects in thin films (pinholes), analyzing the chemistry of fine particulates or assessing causes of implant failure.

Scanning Electron Microscopy

Scanning electron microscopy (SEM) images of surfaces have great resolution and depth of field, with a three-dimensional quality that offers a visual perspective familiar to most users. SEM images are widely used and much has been written about the technique. The comments here are primarily oriented toward SEM as a surface analysis tool.

SEM functions by focusing and rastering a relatively high-energy electron beam (typically, 5–100 keV) on a specimen. Low-energy secondary electrons are emitted from each spot where the focused electron beam impacts. The measured intensity of the secondary electron emission is a function of the atomic composition of the sample and the geometry of the features under observation. SEM images surfaces by spatially reconstructing on a phosphor screen [or charged coupled device (CCD) detector] the intensity of the secondary electron emission. Because of the shallow penetration depth of the low-energy secondary electrons produced by the primary electron beam, only secondary electrons generated near the surface can escape from the bulk and be detected (this is analogous to the surface sensitivity described in Fig. 7). Consequently, SEM is a surface analysis method.

Nonconductive materials observed in the SEM are typically coated with a thin, electrically grounded layer of metal to minimize negative charge accumulation from the electron beam. However, this metal layer is always so thick (>200 Å) that the electrons emitted from the sample beneath cannot penetrate. Therefore, in SEM analysis of nonconductors, the surface of the metal coating is, in effect, being monitored. If the metal coat is truly conformal, a good representation of the surface geometry will be conveyed. However, the specimen surface chemistry no longer influences secondary electron emission. Also, at very high magnifications, the texture of the metal coat and not the surface may be under observation.

SEM, in spite of these limitations in providing true surface information, is an important corroborative method to use in conjunction with other surface analysis methods. Surface roughness and texture can have a profound influence on data from ESCA, SIMS, and contact angle determinations. Therefore, SEM provides important information in the interpretation of data from these methods.

The development of low-voltage SEM offers a technique to truly study the surface chemistry (and geometry) of nonconductors. With the electron accelerating voltage lowered to approximately 1 keV, charge accumulation is not as critical and metallization is not required. Low-voltage SEM has been used to study platelets and phase separation in polymers. Also, the environmental SEM (ESEM) permits wet, uncoated specimens to be studied.

The primary electron beam also results in the emission of X-rays. These X-rays are used to identify elements with the technique called energy-dispersive X-ray analysis (EDXA). However, the high-energy primary electron beam penetrates deeply into a specimen (a micron or more). The X-rays produced from the interaction of these electrons with atoms deep in the bulk of the specimen can penetrate through the material and be detected. Therefore, EDXA is not a surface analysis method.

The primary use of SEM is to image topography. SEM for this application is elaborated upon in the chapter on microscopy in biomaterials research (Chapter 5.6).

Infrared Spectroscopy

Infrared spectroscopy (IRS) provides information on the vibrations of atomic and molecular species. It is a standard analytical method that can reveal information on specific chemistries and the orientation of structures. Fourier transform infrared (FTIR) spectrometry offers outstanding signal-to-noise ratio (S/N) and spectral accuracy. However, even with this high S/N, the small absorption signal associated with the minute mass of material in a surface region can challenge the sensitivity of the spectrometer. Also, the problem of separating the vastly larger bulk absorption signal from the surface signal must be addressed.

Surface FTIR methods couple the infrared radiation to the sample surface to increase the intensity of the surface signal and reduce the bulk signal (Allara, 1982; Leyden and Murthy, 1987; Urban, 1993; Dumas *et al.*, 1999). Some of these sampling modes, and their characteristics, are illustrated in Fig. 12.

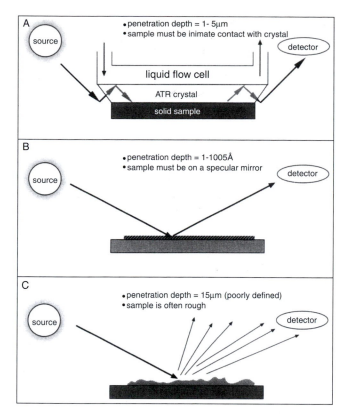

FIG. 12. Three surface-sensitive infrared sampling modes: (A) ATR-IR, (B) IRAS, (C) diffuse reflectance.

The attenuated total reflectance (ATR) mode of sampling has been used most often in biomaterials studies. The penetration depth into the sample is 1–5 μm. Therefore, ATR is not highly surface sensitive, but observes a broad region near the surface. However, it does offer the wealth of rich structural information common to infrared spectra. With extremely high S/N FTIR instruments, ATR studies of proteins and polymers under water have been performed. In these experiments, the water signal (which is typically 99% or more of the total signal) is subtracted from the spectrum to leave only the surface material (e.g., adsorbed protein) under observation.

Another infrared method that has proven immensely valuable for observing extremely thin films on reflective surfaces is infrared reflection absorption spectroscopy (IRAS), Fig. 12. This method has been widely applied to self-assembled monolayers (SAMs), but is applicable to many surface films that are less than 10 nm in thickness. The surface upon which the thin film resides must be highly reflective and metal surfaces work best, though a silicon wafer can be used. IRAS gives information about composition, crystallinity and molecular orientation. Infrared spectroscopy is one member of a family of methods called vibrational spectroscopies. Two other vibrational spectroscopies, sum frequency generation and Raman, will be mentioned later in the section on newer methods.

Scanning Tunneling Microscopy, Atomic Force Microscopy, and the Scanning Probe Microscopies

In the 10 years since the first edition of this book, scanning tunneling microscopy (STM) and atomic force microscopy (AFM) have developed from novel research tools to key methods for biomaterials characterization. AFM has become more widely used than STM because oxide-free, electrically conductive surfaces are not needed with AFM. General review articles (Binnig and Rohrer, 1986; Avouris, 1990; Albrecht *et al.*, 1988) and articles oriented toward biological studies with these methods (Hansma *et al.*, 1988; Miles *et al.*, 1990; Rugar and Hansma, 1990; Jandt, 2001) are available.

The STM was invented in 1981 and led to a Nobel Prize for Binnig and Rohrer in 1986. The STM uses quantum tunneling to generate an atom-scale, electron density image of a surface. A metal tip terminating in a single atom is brought within 5–10 Å of an electrically conducting surface. At these distances, the electron cloud of the atom at the "tip of the tip" will significantly overlap the electron cloud of an atom on the surface. If a potential is applied between the tip and the surface, an electron tunneling current will be established whose magnitude, J, follows the proportionality:

$$J \propto e^{(-Ak_0 S)}$$

where A is a constant, k_0 is an average inverse decay length (related to the electron affinity of the metals), and S is the separation distance in angstrom units. For most metals, a 1 Å change in the distance of the tip from the surface results in an order of magnitude change in tunneling current. Even though this current is small, it can be measured with good accuracy.

To image a surface, this quantum tunneling current is used in one of two ways. In constant current mode, a piezoelectric driver scans a tip over a surface. When the tip approaches an atom protruding above the plane of the surface, the current rapidly increases, and a feedback circuit moves the tip up to keep the current constant. Then, a plot is made of the tip height required to maintain constant current versus distance along the plane. In constant height mode, the tip is moved over the surface and the change in current with distance traveled along the plane of the surface is directly recorded. A schematic diagram of a scanning tunneling microscope is presented in Fig. 13. Two STM scanning modes are illustrated in Fig. 14.

The STM measures electrical current and therefore is well suited for conductive and semiconductive surfaces. However, biomolecules (even proteins) on conductive substrates appear amenable to imaging. It must be remembered that STM does not "see" atoms, but monitors electron density. The conductive and imaging mechanism for proteins is not well understood. Still, Fig. 15 suggests that valuable images of biomolecules on conductive surfaces can be obtained.

The AFM uses a similar piezo drive mechanism. However, instead of recording tunneling current, the deflection of a tip mounted on a flexible cantilever arm due to van der Waals and electrostatic repulsion and attraction between an atom at the tip and an atom on the surface is measured. Atomic-scale measurements of cantilever arm movements can be made by reflecting a laser beam off a mirror on the cantilever arm (an optical lever). A one-atom deflection of the cantilever arm can

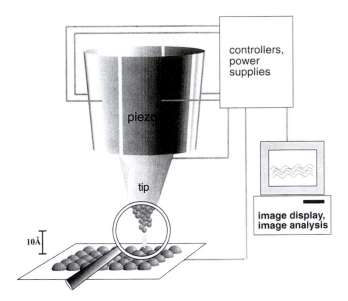

FIG. 13. Schematic diagram illustrating the principle of the scanning tunneling microscope—a tip terminating in a single atom permits localized quantum tunneling current from surface features (or atoms) to tip. This tunneling current can be spatially reconstructed to form an image.

Tips are important in AFM as the spatial resolution of the method is significantly associated with tip terminal diameter and shape. Tips are made from microlithographically fabricated silicon or silicon nitride. Also carbon whiskers, nanotubes, and a variety of nanospherical particles have been mounted on AFM tips to increase their sharpness or improve the ability to precisely define tip geometry. Tips are also surface-modified to alter the strength and types of interactions with surfaces (static SIMS can be used to image these surface modifications). Finally, cantilevers are sold in a range of stiffnesses so the analysis modes can be tuned to needs of the sample and the type of data being acquired. The forces associated with the interaction of an AFM tip with a surface as it approaches and is retracted are illustrated in Fig. 16. Since force is being measured and Hooke's law applies to the deformation of an elastic cantilever, the AFM can be used to quantify the forces between surface and tip. An exciting application of AFM is to measure the strength of interaction between two biomolecules (for example, biotin and streptavidin; see Chilkoti *et al.*, 1995).

easily be magnified by monitoring the position of the laser reflection on a spatially resolved photosensitive detector. Other principles are also used to measure the deflection of the tip. These include capacitance measurements and interferometry. A diagram of a typical AFM is presented in Fig. 16.

AFM instruments are commonly applied to surface problems using one of two modes, contact mode and tapping mode. In contact mode, the tip is in contact with the surface (or at least the electron clouds of tip and surface essentially overlap). The pressures resulting from the force of the cantilever delivered through the extremely small surface area of the tip can be damaging to soft specimens (proteins, polymers, etc). However, for more rigid specimens, excellent topographical imaging can be achieved in contact mode. In tapping mode, the tip is oscillated at a frequency near the resonant frequency

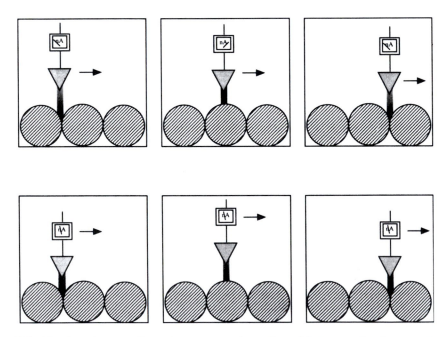

FIG. 14. Scanning tunneling microscopy can be performed in two modes. In constant height mode, the tip is scanned a constant distance from the surface (typically 5–10 Å) and the change in tunneling current is recorded. In constant current mode, the tip height is adjusted so that the tunneling current is always constant, and the tip distance from the surface is recorded as a function of distance traveled in the plane of the surface.

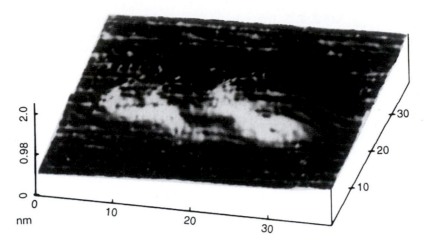

FIG. 15. Scanning tunneling micrograph image of a fibrinogen molecule on a gold surface, under buffer solution (image by Dr. K. Lewis).

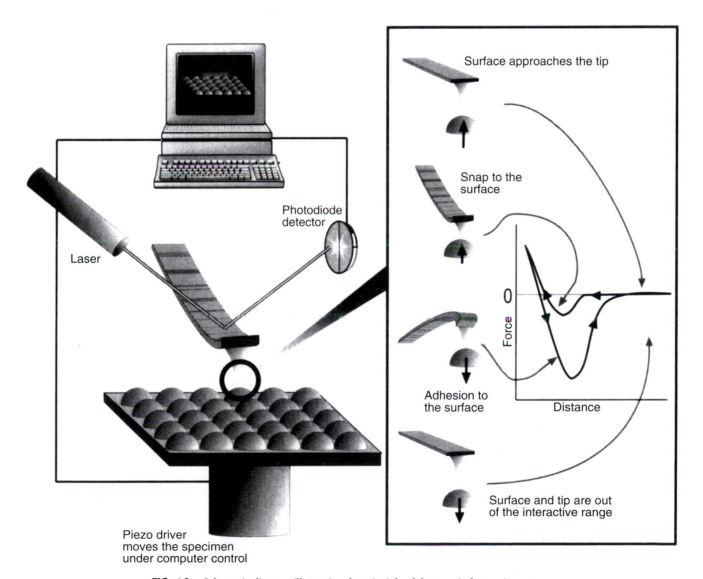

FIG. 16. Schematic diagram illustrating the principle of the atomic force microscope.

TABLE 6 Scanning Probe Microscopy (SPM) Modes

Name	Acronym	Use
Contact mode	CM-AFM	Topographic imaging of harder specimens
Tapping (intermittent force) mode	IF-AFM	Imaging softer specimens
Noncontact mode	NCM-AFM	Imaging soft structures
Force modulation (allows slope of force–distance curve to be measured)	FM-AFM	Enhances image contrast based on surface mechanics
Scanning surface potential microscopy (Kelvin probe microscopy)	SSPM, KPM	Measures the spatial distribution of surface potential
Magnetic force microscopy	MFM	Maps the surface magnetic forces
Scanning thermal microscopy	SThM	Maps the thermal conductivity characteristics of a surface
Recognition force microscopy	RFM	Uses a biomolecule on a tip to probe for regions of specific biorecognition on a surface
Chemical force microscopy	CFM	A tip derivatized with a given chemistry is scanned on a surface to spatially measure differences of interaction strength
Lateral force microscopy	LFM	Maps frictional force on a surface
Electrochemical force microscopy	EFM	The tip is scanned under water and the electrochemical potential between tip and surface is spatially measured
Nearfield scanning optical microscopy	NSOM	A sharp optical fiber is scanned over a surface allowing optical microscopy or spectroscopy at 100-nm resolution
Electrostatic force microscopy	EFM	Surface electrostatic potentials are mapped
Scanning capacitance microscopy	SCM	Surface capacitance is mapped
Conductive atomic force microscopy	CAFM	Surface conductivity is mapped with an AFM instrument
Nanolithographic AFM		An AFM tip etches, oxidizes, or reacts a space permitting pattern fabrication at 10 nm or better resolution
Dip-pen nanolithography	DPN	An AFM tip, inked with a thiol or other molecule, writes on a surface at the nanometer scale

of the cantilever. The tip barely grazes the surface. The force interaction of tip and surface can affect the amplitude of oscillation and the oscillating frequency of the tip. In standard tapping mode, the amplitude change is translated into topographic spatial information. Many variants of tapping mode have been developed allowing imaging under different conditions and using the phase shift between the applied oscillation to the tip and the actual tip oscillation in the force field of the surface to provide information of the mechanical properties of the surface (in essence, the viscoelasticity of the surface can be appreciated).

The potential of the AFM to explore surface problems has been greatly expanded by ingenious variants of the technique. In fact, the term "atomic force microscopy" has been generalized to "scanning probe microscopy (SPM)." Table 6 lists many of these creative applications of the AFM/STM idea.

Since the AFM measures force, it can be used with both conductive and nonconductive specimens. Force must be applied to bend a cantilever, so AFM is subject to artifacts caused by damage to fragile structures on the surface. Both AFM and STM can function well for specimens under water, in air, or under vacuum. For exploring biomolecules or mobile organic surfaces, the "pushing around" of structures by the tip

is a significant concern. This surface artifact can be capitalized upon to write and fabricate surface structures at the nanometer scale (Fig. 17) (Boland *et al.*, 1998; Quate, 1997; Wilson *et al.*, 2001).

Newer Methods

There are many other surface characterization methods that have the potential to become important in future years. Some of these are listed in Table 7. A few of these evolving techniques that will be specifically mentioned here include sum frequency generation (SFG), Raman, and synchrotron methods.

SFG uses two high-intensity, pulsed laser beams, one in the visible range (frequency $= \omega_{\text{visible}}$) and one in the infrared (frequency $= \omega_{\text{ir}}$), to illuminate a specimen. The light emitted from the specimen by a non-linear optical process, $\omega_{\text{sum}} = \omega_{\text{visible}} + \omega_{\text{ir}}$, is detected and quantified (Fig. 18). The intensity of the light at ω_{sum} is proportional to the square of the sample's second-order nonlinear susceptibility ($\chi^{(2)}$). The term susceptibility refers to the effect of the light field strength on the molecular polarizability. The ω_{sum} light intensity vanishes when a material has inversion symmetry, i.e., in the bulk of the material. At an interface, the inversion symmetry is broken and an SFG signal is generated. Thus, SFG is exquisitely sensitive

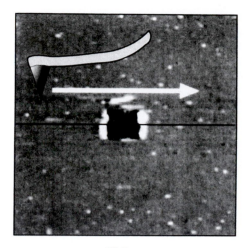

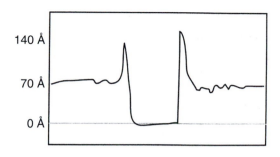

FIG. 17.   An AFM tip, using relatively high force, was used to scratch a rectangular feature into a thin (70 Å) plasma deposited film. The AFM could also characterize the feature created.

TABLE 7   Methods that may have Applicability for the Surface Characterization of Biomaterials

Method	Information obtained
Second-harmonic generation (SHG)	Detect submolayer amounts of adsorbate at any light-accessible interface (air–liquid, solid–liquid, solid–gas)
Surface-enhanced Raman spectroscopy (SERS)	High-sensitivity Raman at rough metal interfaces
Ion scattering spectroscopy (ISS)	Elastically reflected ions probe only the outermost atomic layer
Laser desorption mass spectrometry (LDMS)	Mass spectra of adsorbates at surfaces
Matrix assisted laser desorption ionization (MALDI)	Though generally a bulk mass spectrometry method, MALDI has been used to analyze large adsorbed proteins
IR photoacoustic spectroscopy (IR-PAS)	IR spectra of surfaces with no sample preparation based on wavelength-dependent thermal response
High-resolution electron energy loss spectroscopy (HREELS)	Vibrational spectroscopy of a highly surface-localized region, under ultrahigh vacuum
X-ray reflection	Structural information about order at surfaces and interfaces
Neutron reflection	Thickness and refractive index information about interfaces from scattered neutrons—where H and D are used, unique information on interface organization can be obtained
Extended X-ray absorption fine structure (EXAFS)	Atomic-level chemical and nearest-neighbor (morphological) information
Scanning Auger microprobe (SAM)	Spatially defined Auger analysis at the nanometer scale
Surface plasmon resonance (SPR)	Study aqueous adsorption events in real time by monitoring changes in surface refractive index
Rutherford backscattering spectroscopy (RBS)	Depth profiling of complex, multiplayer interfacial systems

to the plane of the interface. In practice, ω_{ir} is scanned over a vibrational frequency range—where vibrational interactions occur with interface molecules, then the SFG signal is resonantly enhanced and we see a vibrational spectrum. The advantages are the superb surface sensitivity, the cancellation of bulk spectral intensity (for example, this allows measurements at a water/solid interface), the richness of information from vibrational spectra, and the ability to study molecular orientation due to the polarization of the light. SFG is not yet a routine method. The lasers and optical components are expensive and require precision alignment. Also, the range in the infrared over which lasers can scan is limited (though it has slowly expanded with improved equipment). However, the power of SFG for biomaterials studies has already been

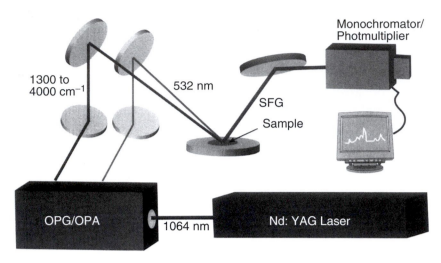

FIG. 18. Schematic diagram of a sum frequency generation (SFG) apparatus (based upon a diagram developed by Polymer Technology Group, Inc.).

proven with studies on hydrated hydrogels, polyurethanes, surface active polymer additives, and proteins (Shen, 1989; Chen *et al.*, 2002).

In Raman spectroscopy a bright light is shined on a specimen. Most of the light scatters back at the same frequency as the incident beam. However, a tiny fraction of this light excites vibrations in the specimen and thereby loses or gains energy. The frequency shift of the light corresponds to vibrational bands indicative of the molecular structure of the specimen. The Raman spectroscopic technique has been severely limited for surface studies because of its low signal level. However, in recent years, great strides in detector sensitivity have allowed Raman to be applied for studying the minute mass of material at a surface. Also, surface enhanced Raman spectroscopy (SERS), Raman spectra taken from molecules on a roughened metal surface, can enhance Raman signal intensity by 10^6 or more. Raman spectra will be valuable for biomedical surface studies because water, which absorbs radiation very strongly in the infrared range, has little effect on Raman spectra that are often acquired with visible light (Storey *et al.*, 1995).

Synchrotron sources of energetic radiation that can be used to probe matter were originally confined to the physics community for fundamental studies. However, there are now more synchrotron sources, better instrumentation, and improved data interpretation. Synchrotron sources are typically national facilities costing >\$100M and often occupying hundreds of acres (Fig. 19). By accelerating electrons to near the speed of light in a large, circular ring, energies covering a broad swath of the electromagnetic spectrum (IR to energetic X-rays) are emitted. A synchrotron source (and ancillary equipment) permits a desired energy of the probe beam to be "dialed in" or scanned through a frequency range. Other advantages include high source intensity (bright light) and polarized light. Some of the experimental methods that can be performed with great success at synchrotron sources include crystallography, scattering, spectroscopy, microimaging, and nanofabrication.

FIG. 19. The Advanced Photon Source, Argonne National Laboratories, a modern synchrotron source.

Specific surface spectroscopic methods include scanning photoemission microscopy (SPEM, 100 nm spatial resolution), ultraESCA (100 μm spatial resolution, high energy resolution), and near edge X-ray absorption spectrometry (NEXAFS).

STUDIES WITH SURFACE METHODS

Hundreds of studies have appeared in the literature in which surface methods have been used to enhance the understanding of biomaterial systems. A few studies that demonstrate the power of surface analytical methods for biomaterials science are briefly described here.

Platelet Consumption and Surface Composition

Using a baboon arteriovenous shunt model of platelet interaction with surfaces, a first-order rate constant of reaction of platelets with a series of polyurethanes was measured. This rate constant, the platelet consumption by the material, correlated in an inverse linear fashion with the fraction of hydrocarbon-type groups in the ESCA C1s spectra of the polyurethanes (Hanson *et al.*, 1982). Thus, surface analysis revealed a chemical parameter about the surface that could be used to predict long-term biological reactivity of materials in a complex *ex vivo* environment.

Contact-Angle Correlations

The adhesion of a number of cell types, including bacteria, granulocytes, and erythrocytes, has been shown, under certain conditions, to correlate with solid-vapor surface tension as determined from contact-angle measurements. In addition, immunoglobulin G adsorption is related to v_{sv} (Neumann *et al.*, 1983).

Contamination of Intraocular Lenses

Commercial intraocular lenses were examined by ESCA. The presence of sulfur, sodium, and excess hydrocarbon at their surfaces suggested contamination by sodium dodecyl sulfate (SDS) during the manufacture of the lenses (Ratner, 1983). A cleaning protocol was developed using ESCA to monitor results that produced a lens surface of clean PMMA.

Titanium

The discoloration sometimes observed on titanium implants after autoclaving was examined by ESCA and SIMS (Lausmaa *et al.*, 1985). The discoloration was found to be related to accelerated oxide growth, with oxide thicknesses to 650 Å. The oxide contained considerable fluorine, along with alkali metals and silicon. The source of the oxide was the cloth used to wrap the implant storage box during autoclaving. Since fluorine strongly affects oxide growth, and since the oxide layer has been associated with the biocompatibility of titanium implants, the authors advise avoiding fluorinated materials during sterilization of samples. A newer paper contains detailed surface characterization of titanium using a battery of surface methods and addresses surface preparation, contamination, and cleaning (Lausmaa, 1996).

SIMS for Adsorbed Protein Identification and Quantification

All proteins are made up of the same 20 amino acids and thus, on the average, are compositionally similar. Surface analysis methods have shown the ability to detect and quantify surface-bound protein, but biological tools have, until recently, been needed to identify specific proteins. Modern static SIMS instrumentation, using a multivariate statistical analysis of the data, has demonstrated the ability to distinguish between more than 13 different proteins adsorbed on surfaces (Wagner and Castner, 2001). Also, the limits of detection for adsorbed proteins on various surfaces were compared by ESCA and SIMS (Wagner *et al.*, 2002).

Poly(glycolic acid) Degradation Studied by SIMS

The degradation of an important polymer for tissue engineering, poly(glycolic acid), has been studied by static SIMS. As well as providing useful information on this degradation process, the study illustrates the power of SIMS for characterizing synthetic polymers and their molecular weight distributions (Chen *et al.*, 2000).

CONCLUSIONS

The instrumentation of surface analysis steadily advances and newer instruments and techniques can provide invaluable information about biomaterials and medical devices. The information obtained can be used to monitor contamination, ensure surface reproducibility, and explore fundamental aspects of the interaction of biological systems with living systems. Considering that biomedical experiments are typically expensive to perform, the costs for surface analysis are modest to ensure that the surface is as expected, stable and identical surface from experiment to experiment. Surface analysis can also contribute to the understanding of medical device failure (and success). Myriad applications for surface methods are found in device optimization, manufacture and quality control. Predicting biological reaction based on measured surface structure is a frontier area for surface analysis.

Acknowledgment

Support was received from the UWEB NSF Engineering Research Center and the NESAC/BIO National Resource, NIH grant EB-002027, during the preparation of this chapter and for some of the studies described herein.

QUESTIONS

1. Scan the table of contents and abstracts from the last three issues of the *Journal of Biomedical Materials Research* or *Biomaterials*. List all the surface analysis methods used in the articles therein and briefly describe what was learned by using them.
2. How is critical surface tension related to wettability? For the polymers in Table 2, draw the chemical formulas of the chain repeat units and attempt to relate the structures to the wettability. Where inconsistencies are noted, explain those inconsistencies using Table 3.
3. A titanium dental implant was manufactured by the Biomatter Company for the past 8 years. It performed well clinically. For economic reasons, manufacturing of the titanium device was outsourced to Metalsmed, Inc. Early clinical results on this Metalsmed implant, supposedly identical to the Biomatter implant, suggested increased inflammation. How would you compare the surface chemistry and structure of these two devices to see if a difference that might account for the difference in clinical performance could be identified?

Bibliography

Adamson, A. W., and Gast, A. (1997). *Physical Chemistry of Surfaces*, 6th ed. Wiley-Interscience, New York.

Albrecht, T. R., Dovek, M. M., Lang, C. A., Grutter, P., Quate, C. F., Kuan, S. W. J., Frank, C. W., and Pease, R. F. W. (1988). Imaging and modification of polymers by scanning tunneling and atomic force microscopy. *J. Appl. Phys.* **64**: 1178–1184.

Allara, D. L. (1982). Analysis of surfaces and thin films by IR, Raman, and optical spectroscopy. *ACS Symp. Ser.* **199**: 33–47.

Andrade, J. D. (1985). *Surface and Interfacial Aspects of Biomedical Polymers, Vol. 1: Surface Chemistry and Physics.* Plenum Publishers, New York.

Avouris, P. (1990). Atom-resolved surface chemistry using the scanning tunneling microscope. *J. Phys. Chem.* **94**: 2246–2256.

Belu, A. M., Graham, D. J., and Castner, D. G. (2003). Time-of-flight secondary ion mass spectrometry: techniques and applications for the characterization of biomaterial surfaces. *Biomaterials* **24**: 3635–3653

Benninghoven, A. (1983). Secondary ion mass spectrometry of organic compounds (review). in *Springer Series of Chemical Physics: Ion Formation from Organic Solids*, Vol. 25, A. Benninghoven, ed. Springer-Verlag, Berlin, pp. 64–89.

Binnig, G., and Rohrer, H. (1986). Scanning tunneling microscopy. *IBM J. Res. Develop.* **30**: 355–369.

Boland, T., Johnston, E. E., Huber, A., and Ratner, B. D. (1998). Recognition and nanolithography with the atomic force microscope. in *Scanning Probe Microscopy of Polymers*, Vol. 694, B. D. Ratner and V. V. Tsukruk, eds. American Chemical Society, Washington, D.C., pp. 342–350.

Briggs, D. (1986). SIMS for the study of polymer surfaces: a review. *Surf. Interface Anal.* **9**: 391–404.

Briggs, D., and Seah, M. P. (1983). *Practical Surface Analysis.* Wiley, Chichester, UK.

Castner, D. G., and Ratner, B. D. (2002). Biomedical surface science: foundations to frontiers. *Surf. Sci.* **500**: 28–60.

Chen, J., Lee, J.-W., Hernandez, N. L., Burkhardt, C. A., Hercules, D. M., and Gardella, J. A. (2000). Time-of-flight secondary ion mass spectrometry studies of hydrolytic degradation kinetics at the surface of poly(glycolic acid). *Macromolecules* **33**: 4726–4732.

Chen, Z., Ward, R., Tian, Y., Malizia, F., Gracias, D. H., Shen, Y. R., and Somorjai, G. A. (2002). Interaction of fibrinogen with surfaces of end-group-modified polyurethanes: A surface-specific sum-frequency-generation vibrational spectroscopy study. *J. Biomed. Mater. Res.* **62**: 254–264.

Chilkoti, A., Boland, T., Ratner, B. D., and Stayton, P. S. (1995). The relationship between ligand-binding thermodynamics and protein-ligand interaction forces measured by atomic force microscopy. *Biophys. J.* **69**: 2125–2130.

Davies, M. C., and Lynn, R. A. P. (1990). Static secondary ion mass spectrometry of polymeric biomaterials. *CRC Crit. Rev. Biocompat.* **5**: 297–341.

Dilks, A. (1981). X-ray photoelectron spectroscopy for the investigation of polymeric materials. in *Electron Spectroscopy: Theory, Techniques, and Applications*, Vol. 4, A. D. Baker and C. R. Brundle, eds. Academic Press, London, pp. 277–359.

Dumas, P., Weldon, M. K., Chabal, Y. J. and Williams, G. P. (1999). Molecules at surfaces and interfaces studied using vibrational spectroscopies and related techniques. *Surf. Rev. Lett.*, **6**(2): 225–255.

Feldman, L. C., and Mayer, J. W. (1986). *Fundamentals of Surface and Thin Film Analysis.* North-Holland, New York.

Good, R. J. (1993). Contact angle, wetting, and adhesion: a critical review. in *Contact Angle, Wettability and Adhesion*, K. L. Mittal, ed. VSP Publishers, The Netherlands.

Hansma, P. K., Elings, V. B., Marti, O., and Bracker, C. E. (1988). Scanning tunneling microscopy and atomic force microscopy: application to biology and technology. *Science* **242**: 209–216.

Hanson, S. R., Harker, L. A., Ratner, B. D., and Hoffman, A. S. (1982). Evaluation of artificial surfaces using baboon arteriovenous shunt model. in *Biomaterials 1980, Advances in Biomaterials*, G. D. Winter, D. F. Gibbons, and H. Plenk, eds., Vol. 3, Wiley, Chichester, UK, pp. 519–530.

Jandt, K. D. (2001). Atomic force microscopy of biomaterials surfaces and interfaces. *Surf. Sci.* **491**: 303–332.

Lausmaa, J. (1996). Surface spectroscopic characterization of titanium implant materials. *J. Electron Spectrosc. Related Phenom.* **81**: 343–361.

Lausmaa, J., Kasemo, B., and Hansson S. (1985). Accelerated oxide growth on titanium implants during autoclaving caused by fluorine contamination. *Biomaterials* **6**: 23–27.

Leyden, D. E., and Murthy, R. S. S. (1987). Surface-selective sampling techniques in Fourier transform infrared spectroscopy. *Spectroscopy* **2**: 28–36.

McIntire, L., Addonizio, V. P., Coleman, D. L., Eskin, S. G., Harker, L. A., Kardos, J. L., Ratner, B. D., Schoen, F. J., Sefton, M. V., and Pitlick, F. A. (1985). *Guidelines for Blood-Material Interactions—Devices and Technology Branch, Division of Heart and Vascular Diseases*, National Heart, Lung, and Blood Institute, NIH Publication No. 85–2185, revised July 1985, U.S. Department of Health and Human Services.

Miles, M. J., McMaster, T., Carr, H. J., Tatham, A. S., Shewry, P. R., Field, J. M., Belton, P. S., Jeenes, D., Hanley, B., Whittam, M., Cairns, P., Morris, V. J., and Lambert, N. (1990). Scanning tunneling microscopy of biomolecules. *J. Vac. Sci. Technol. A* **8**: 698–702.

Neumann, A. W., Absolom, D. R., Francis, D. W., Omenyi, S. N., Spelt, J. K., Policova, Z., Thomson, C., Zingg, W., and Van Oss, C. J. (1983). Measurement of surface tensions of blood cells and proteins. *Ann. N. Y. Acad. Sci.* **416**: 276–298.

Quate, C. F. (1997). Scanning probes as a lithography tool for nanostructures. *Surf. Sci.* **386**: 259–264.

Ratner, B. D. (1983). Analysis of surface contaminants on intraocular lenses. *Arch. Ophthal.* **101**: 1434–1438.

Ratner, B. D. (1988). *Surface Characterization of Biomaterials.* Elsevier, Amsterdam.

Ratner, B. D., and Castner, D. G. (1997). Electron spectroscopy for chemical analysis. in *Surface Analysis—The Principal Techniques.* J. C. Vickerman, ed. John Wiley and Sons, Ltd., Chichester, UK, pp. 43–98.

Ratner, B. D., and McElroy, B. J. (1986). Electron spectroscopy for chemical analysis: applications in the biomedical sciences. in *Spectroscopy in the Biomedical Sciences*, R. M. Gendreau, ed. CRC Press, Boca Raton, FL, pp. 107–140.

Rugar, D., and Hansma, P. (1990). Atomic force microscopy. *Physics Today* **43**: 23–30.

Scheutzle, D., Riley, T. L., deVries, J. E., and Prater, T. J. (1984). Applications of high-performance mass spectrometry to the surface analysis of materials. *Mass Spectrom.* **3**: 527–585.

Shen, Y. R. (1989). Surface properties probed by second-harmonic and sum-frequency generation. *Nature* **337**(6207): 519–525.

Somorjai, G. A. (1981). *Chemistry in Two Dimensions: Surfaces.* Cornell Univ. Press, Ithaca, NY.

Somorjai, G. A. (1994). *Introduction to Surface Chemistry and Catalysis.* John Wiley and Sons, New York.

Storey, J. M. E., Barber, T. E., Shelton, R. D., Wachter, E. A., Carron, K. T., and Jiang, Y. (1995). Applications of surface-enhanced Raman scattering (SERS) to chemical detection. *Spectroscopy* **10**(3): 20–25.

Tirrell, M., Kokkoli, E., and Biesalski, M. (2000). The role of surface science in bioengineered materials. *Surf. Sci.* **500:** 61–83.

Urban, M. W. (1993). *Vibrational Spectroscopy of Molecules and Macromolecules on Surfaces.* Wiley-Interscience, New York.

Van Vaeck, L., Adriaens, A., and Gijbels, R. (1999). Static secondary ion mass spectrometry (S-SIMS): part I. Methodology and structural interpretation. *Mass Spectrom. Rev.* **18:** 1–47.

Vickerman, J. C. (1997). *Surface Analysis: The Principal Techniques.* John Wiley and Sons, Chichester, UK.

Vickerman, J. C., Brown, A., and Reed, N. M. (1989). *Secondary Ion Mass Spectrometry, Principles and Applications.* Clarendon Press, Oxford.

Wagner, M.S., and Castner, D.G. (2001). Characterization of adsorbed protein films by time-of-flight secondary ion mass spectrometry with principal component analysis. *Langmuir* **17:** 4649–4660.

Wagner, M. S., McArthur, S. L., Shen, M., Horbett, T. A., and Castner, D. G. (2002). Limits of detection for time of flight secondary ion mass spectrometry (ToF-SIMS) and X-ray photoelectron spectroscopy (XPS): detection of low amounts of adsorbed protein. *J. Biomater. Sci. Polymer Ed.* **13**(4): 407–428.

Watts, J. F., and Wolstenholme, J. (2003). *An Introduction to Surface Analysis by XPS and AES.* John Wiley & Sons, Chichester, UK.

Wilson, D. L., Martin, R., Hong, S. I., Cronin-Golomb, M., Mirkin, C. A., and Kaplan, D. L. (2001). Surface organization and nanopatterning of collagen by dip-pen nanolithography. *Proc. Natl. Acad. Sci. USA* **98**(24): 13,360–13,664.

Zisman, W. A. (1964). Relation of the equilibrium contact angle to liquid and solid constitution. in *Contact Angle, Wettability and Adhesion, ACS Advances in Chemistry Series,* Vol. 43, F. M. Fowkes, ed. American Chemical Society, Washington, D.C., pp. 1–51.

1.5   ROLE OF WATER IN BIOMATERIALS

Erwin A. Vogler

The primary role water plays in biomaterials is as a solvent system. Water is the "universal ether" as it has been termed (Baier and Meyer, 1996), dissolving inorganic salts and large organic macromolecules such as proteins or carbohydrates (solutes) with nearly equal efficiency (Pain, 1982). Water suspends living cells, as in blood, for example, and is the principal constituent of the interstitial fluid that bathes tissues. Water is not just a bland, neutral carrier system for biochemical processes, however. Far from this, water is an active participant in biology, which simply could not and would not work the way it does without the special mediating properties of water. Moreover, it is widely believed that water is the first molecule to contact biomaterials in any clinical application (Andrade *et al.,* 1981; Baier, 1978). This is because water is the majority molecule in any biological mixture, constituting 70 wt % or more of most living organisms, and because water is such a small and agile molecule, only about 0.25 nm in the longest dimension. Consequently, behavior of water near surfaces and the role of water in biology are very important subjects in biomaterials science. Some of these important aspects of water are discussed in this chapter.

WATER SOLVENT PROPERTIES

Figures 1A–1D collect various diagrams of water illustrating the familiar atomic structure and how this arrangement leads to the ability to form a network of self-associated molecules through hydrogen bonding. Self-association confers unique properties on water, many of which are still active areas of scientific investigation even after more than 200 years of chemical and physical research applied to water (Franks, 1972).

Hydrogen bonds in water are relatively weak 3–5 kcal/mole associations with little covalent character (Iassacs *et al.,* 1999; Marshall, 1999). As it turns out, hydrogen bond strength is approximately the same as the energy transferred from one molecule to another by collisions at room temperature (Vinogradov and Linnell, 1971). So hydrogen bonds are quite transient in nature, persisting only for a few tens of picoseconds (Berendsen, 1967; Luzar and Chandler, 1996). Modern molecular simulations suggest, however, that more than 75% of liquid-water molecules are interconnected in a three-dimensional (3D) network of three or four nearest neighbors at any particular instant in time (Robinson *et al.,* 1996). This stabilizing network of self-associated water formed from repeat units illustrated in Fig. 1D is so extensive, in fact, that it is frequently termed "water structure," especially in the older literature (Narten and Levy, 1969). These somewhat dated water-structure concepts will not be discussed further here, other than to caution the reader that the transient nature of hydrogen bonding greatly weakens the notion of a "structure" as it might be practically applied by a chemist for example (Berendsen, 1967) and that reference to water structure near solutes and surfaces in terms of "icebergs" or "melting" should not be taken too literally, as will be discussed further subsequently.

A very important chemical outcome of this propensity of water to self-associate is the dramatic effect on water solvent properties. One view of self-association is from the standpoint of Lewis acidity and basicity. It may be recalled from general chemistry that a Lewis acid is a molecule that can accept electrons or, more generally, electron density from a molecular orbital of a donor molecule. An electron-density donor molecule is termed a Lewis base. Water is amphoteric in this sense because, as illustrated in Figs. 1A and 1D, it can simultaneously share and donate electron density. Hydrogen atoms (the Lewis acids) on one or more adjacent water molecules can accept electron density from the unshared electron pairs on the oxygen atom (the Lewis bases) of another water molecule. In this manner, water forms a 3D network through Lewis acid–base self-association reactions.

If the self-associated network is more complete than some arbitrary reference state, then there must be relatively fewer unmatched Lewis acid–base pairings than in this reference state. Conversely, in less-associated water, the network is relatively incomplete and there are more unmatched Lewis acid–base pairings than in the reference state. These unmatched pairings in less associated water are readily available to participate in other chemical reactions, such as dissolving a solute molecule or hydrating a water-contacting surface. Therefore, it can be generally concluded that less-associated water is a stronger solvent than more-associated water because it has a greater potential to engage in reactions other than self-association. In chemical terminology, less self-associated water has a greater chemical potential than more self-associated water. Interestingly, more self-associated water with

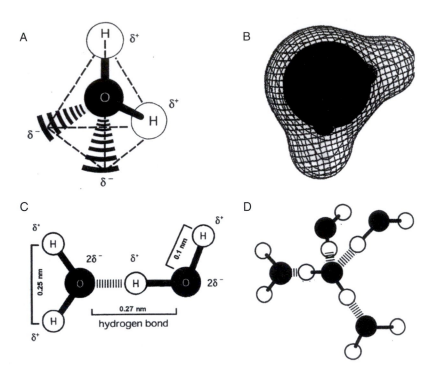

FIG. 1. Atomic structure of water illustrating (A) tetrahedral bonding arrangement wherein hydrogen atoms (H, light-colored spheres) are Lewis acid centers and the two lone-pair electrons on oxygen (O, dark-colored spheres) are Lewis base centers that permit water to hydrogen bond with four nearest-neighbor water molecules; (B) electron density map superimposed on an atomic-radius sphere model of water providing a more authentic representation of molecular water; (C) approximate molecular dimensions; and (D) five water molecules participating in a portion of a hydrogen-bond network.

a relatively more complete 3D network of hydrogen bonds must be less dense (greater partial molar volume) than less self-associated water because formation of linearly directed hydrogen bonds takes up space (Fig. 1C), increasing free volume in the liquid. This is why water ice with a complete crystalline network is less dense than liquid water and floats upon unfrozen water, a phenomenon with profound environmental impact. Thus, less associated water is not only more reactive but also more dense. These inferred relationships between water structure and reactivity are summarized in Table 1, which will be a useful aid to subsequent discussion.

A variety of lines of evidence ranging from molecular simulations (Lum *et al.*, 1999; Robinson *et al.*, 1996) to experimental studies of water solvent properties in porous media (Qi and Soka, 1998; Wiggins, 1988) suggest that water expands and contracts in density (molar volume) with commensurate changes in chemical potential to accommodate presence of imposed solutes and surfaces. The word "imposed" is specifically chosen here to emphasize that a solute (e.g., an ion or a macromolecule) or an extended surface (e.g., the outer region of a biomaterial) must in some way interfere with self-association. Simply stated, the solute or surface gets in the way and water molecules must reorient to maintain as many hydrogen bonds with neighbors as is possible in this imposed presence of solute or surface. Water may not be able to maintain an extensive hydrogen-bond network in certain cases and this has important and measurable effects on water solvency. The next sections will first consider "hydrophobic" and "hydrophilic" solute molecules and then extend the discussion to hydrophobic and hydrophilic biomaterial surfaces, at least to the extent possible within the current scientific knowledge base.

THE HYDROPHOBIC EFFECT

The hydrophobic effect is related to the insolubility of hydrocarbons in water and is fundamental to the organization of lipids into bilayers, the structural elements of life as

TABLE 1 Relationships among Water Structure and Solvent Properties

Extent of water self-association	Density	Partial molar volume	Chemical potential (number of available hydrogen bonds)
More	Less	More	Less
Less	More	Less	More

we know it (Tanford, 1973). Clearly then, the hydrophobic effect is among the more fundamental, life-giving phenomena attributable to water. Hydrocarbons are sparingly soluble in water because of the strong self-association of water, not the strong self-association of hydrocarbons as is sometimes thought. Thus water structure is seen to be directly related to solvent properties in this very well-known case.

The so-called "entropy of hydrophobic hydration" (ΔS) has received a great deal of research attention from the molecular-simulation community because it dominates the overall (positive) free energy of hydrophobic hydration (ΔG) at ambient temperatures and pressures. The rather highly negative entropy of hydration of small hydrocarbons ($\Delta S \approx -20$ e.u.; see Kauzmann, 1959, for discussion related to lipids and proteins) turns out to be substantially due to constraints imposed on water-molecule orientation and translation as water attempts to maintain hydrogen-bond neighbors near the solute molecule (Paulaitis *et al.*, 1996). Apparently, there are no structural "icebergs" with enhanced self-association around small hydrocarbons (Besseling and Lyklema, 1995) as has been invoked in the past to account for ΔS (Berendsen, 1967). Instead, water surrounding small solutes such as methane or ethane may be viewed as spatially constrained by a "solute-straddling" effect that maximizes as many hydrogen-bonded neighbors as possible at the expense of orientational flexibility. Interestingly, while these constraints on water-molecule orientation do not significantly promote local self-association (i.e., increase structure), this lack of flexibility does have the effect of reducing repulsive, non-hydrogen-bonding interactions between water-molecule neighbors, accounting for a somewhat surprisingly exothermic (≈ -2 kcal/mol) enthalpy of hydration (ΔH) of small hydrocarbons (Besseling and Lyklema, 1995). The strong temperature sensitivities of these entropic and enthalpic effects are nearly equal and opposite and compensate in a way that causes the overall free energy of hydration ($\Delta G = \Delta H - T\Delta S$) to be essentially temperature insensitive. Increasing temperature expands the self-associated network of water, creating more space for a hydrophobic solute to occupy, and ΔS becomes more positive ($-T\Delta S$ more negative). On the other hand, nonbonding (repulsive) contacts between water molecules increase with temperature, causing ΔH to become more positive.

As one might imagine, difficulties in maintaining a hydrogen-bonded network are exacerbated near very large hydrophobic solutes where no orientations can prevent separation of water-molecule neighbors. Another water-driven mechanism comes into play in some of these cases wherein hydrophobic patches or domains on a solute such as a protein aggregate, exclude water, and participate in what has been termed "hydrophobic bonding" (DeVoe, 1969; Dunhill, 1965; Kauzmann, 1959; Tanford, 1966). This aspect of the hydrophobic effect is very important in biomaterials because it controls the folding of proteins and is thus involved in protein reactions at surfaces, especially denaturation of proteins at biomaterial surfaces induced by unfolding reactions in the adsorbed state. Water near large hydrophobic patches will be further considered in relation to hydrophobic surfaces that present analogous physical circumstances to water.

THE HYDROPHILIC EFFECT

There is no broadly recognized "hydrophilic effect" in science that is the antithesis to the well-known hydrophobic effect just discussed. But generally speaking, the behavior of water near hydrophilic solutes is so substantially different from that occurring near hydrophobic solutes that hydrophilicity may well be granted a distinguishing title of its own. The terms hydrophilic and hydrophobic are poorly defined in biomaterials and surface science (Hoffman, 1986; Oss and Giese, 1995; Vogler, 1998), requiring some clarification at this juncture since a distinction between hydrophilic and hydrophobic needs to be made. For the current purposes, let the term hydrophilic be applied to those solutes that compete with water for hydrogen bonds. That is to say, hydrophilic solutes exhibit Lewis acid or base strength comparable to or exceeding that of water, so that it is energetically favorable for water to donate electron density to or accept electron density from hydrophilic solutes instead of, or at least in competition with, other water molecules. For the sake of clarity, let it be added that there is no chemistry or other energetic reason for water to hydrogen bond with a hydrophobic solute as defined herein. Generally speaking, free energies of hydrophilic hydration are greater than that of hydrophobic hydration since acid–base chemistry is more energetic than the non-bonding "hydrophobic" reactions previously considered, and this frequently manifests itself in large enthalpic contributions to ΔG.

Familiar examples of hydrophilic solutes with biomedical relevance would include cations such as Na^+, K^+, Ca^{2+}, and Mg^{2+} or anions such as Cl^-, HCO_3^{-1} and HPO_4^{-2}. These ions are surrounded by a hydration sphere of water directing oxygen atoms toward the cations or hydrogen atoms toward the anions. Water structuring near ions is induced by a strong electric field surrounding the ion that orients water dipole moments in a manner that depends on ionic size and extent of hydration (Marcus, 1985). Some ions are designated "structure promoting" and others "structure breaking." Structure-promoting ions are those that impose more local order in surrounding water than occurs distant from the ion whereas structure-breaking or "chaotropic" ions increase local disorder and mobility of adjacent water molecules (Wiggins, 1971). Another feature of ion solvation important in biomaterials is that certain ions such as Ca^{2+} and Mg^{2+} are more hydrated (surrounded by more water molecules) than K^+ and Na^+ in the order of the so-called Hofmeister or lyotropic series. This implies that highly hydrated ions will partition into less associated water with more available hydrogen bonds for solvation (see Table 1) preferentially over more associated water with fewer available hydrogen bonds (Christenson *et al.*, 1990; Vogler, 1998; Wiggins, 1990; Wiggins and Ryn, 1990). This ion partitioning can have dramatic consequences on biology near surfaces since Ca^{2+} and Mg^{2+} have strong allosteric effects on enzyme reactions, a point that will be raised again in the final section of this chapter.

As in the hydrophobic effect, size plays a big role in the solvation of hydrophilic ions. Small inorganic ions are completely ionized and lead to separately hydrated ions in the manner just discussed above. Hydration of a polyelectrolyte such as hyaluronic acid or a single strand of DNA is more complicated because a counterion "atmosphere" surrounds the dissolved polyelectrolyte. The countercharge distribution within this atmosphere is not uniform in space but instead diminishes in concentration with distance from the polyelectrolyte. This means that water in a hypothetical compartment near the polyelectrolyte is enriched in countercharges (higher ionic strength, lower water chemical potential) relative to that of an identical compartment distant from the polyelectrolyte (lower ionic strength, higher water chemical potential). Since concentration (chemical potential) gradients cannot persist at equilibrium, there must be some route to making chemical potentials uniform throughout solution. Wiggins has argued that the only means available to such a system of dissolved polyelectrolytes at constant temperature, pressure, and fixed composition (including water) is adjustment of water density or, more precisely, partial molar volume (Wiggins, 1990). That is to say, in order to increase water chemical potential in the near compartment relative to that of water in the distant compartment, water density must increase (see row 2 of Table 1, more molecules/unit volume available for chemical work). At the same time, in order to decrease water chemical potential in the distant compartment relative to that in the closer one; water density must decrease (see row 2 of Table 1, fewer molecules/unit volume available for chemical work). This thinking gives rise to the notion of contiguous regions of variable water density within a polyelectrolyte solution. Here again, it is evident that adjustment of water chemical potential to accommodate the presence of a large solute molecule appears to be a necessary mechanism to account for commonly observed hydration effects. The next section will explore how these same effects might account for surface wetting effects.

THE SURFACE WETTING EFFECT

It is a very common observation that water wets certain kinds of surfaces whereas water beads up on others, forming droplets with a finite "contact angle." This and related wetting phenomena have intrigued scientists for almost three centuries, and the molecular mechanisms of wetting are still an important area of research to this day. The reason for such continued interest is that wetting phenomena probe the various intermolecular forces and interactions responsible for much of the chemistry and physics of everyday life. Some of the remaining open questions are related to water structure and solvent properties near different kinds of surfaces.

Although surfaces on which water spreads are commonly termed hydrophilic and those on which water droplets form hydrophobic, the definitions employed in preceding sections based on presence or absence of Lewis acid/base groups that can hydrogen bond with water will continue to be used here, as this is a somewhat more precise way of categorizing biomaterials. Thus, hydrophobic surfaces are distinguished from hydrophilic by virtue of having no Lewis acid or base functional groups available for water interaction.

Water near hydrophobic surfaces finds itself in a predicament similar to that briefly mentioned in the preceding section on the hydration of large hydrophobic solutes in that there are no configurational options available to water molecules closest to the surface that allow maintenance of nearest-neighbor hydrogen bonds. These surface-contacting water molecules are consequently in a less self-associated state and, according to row 2 of Table 1, must temporarily be in a state of higher chemical potential than bulk water. The key word here is temporarily, because chemical potential gradients cannot exist at equilibrium. At constant temperature and pressure, the only recourse available to the system toward establishment of equilibrium is decreasing local water density by increasing the extent of water self-association (row 1, Table 1). Thus it is reasoned that water in direct contact with a hydrophobic surface is less dense than bulk water some distance away from the hydrophobic surface.

This reasoning has been recently corroborated theoretically through molecular simulations of water near hydrophobic surfaces (Besseling and Lyklema, 1995; Lum et al., 1999; Silverstein et al., 1998) and experimentally by application of sophisticated vibrational spectroscopies (Du et al., 1994; Gragson and Richmond, 1997). Although there is not precise uniformity among all investigators using a variety of different computational and experimental approaches, it appears that density variations propagate something of the order of 5 nm from a hydrophobic surface, or about 20 water layers.

There are at least two classes of hydrophilic surfaces that deserve separate mention here because these represent important categories of biomaterials as well (Hoffman, 1986). One class includes surfaces that *ad*sorb water through the interaction with surface-resident Lewis acid or base groups. These water–surface interactions are constrained to the outermost surface layer, say the upper 1 nm or so. Examples of these biomaterials might include polymers that have been surface treated by exposure to gas discharges, use of flames, or reaction with oxidative reagents as well as ceramics, metals, and glass. Another category of hydrophilic surfaces embraces those that significantly *ab*sorb water. Examples here are hydrogel polymers such as poly(vinyl alcohol) (PVA), poly(ethylene oxide) (PEO), or hydroxyethylmethacrylate (HEMA) that can visibly swell or even go into water solution, depending on the molecular weight and extent of crosslinking. Modern surface engineering can create materials that fall somewhere between water-adsorbent and -absorbent by depositing very thin films using self-assembly techniques (P.-Grosdemange et al., 1991; Prime and Whitesides, 1993), reactive gas plasma deposition (Lopez et al., 1992), or radiation grafting (Hoffman and Harris, 1972; Hoffman and Kraft, 1972; Ratner and Hoffman, 1980) as examples. Here, oligomers that would otherwise dissolve in water form a thin-film surface that cannot swell in the usual, macroscopic application of the word. In all of the mentioned cases, however, water hydrogen bonds with functional groups that may be characterized as either Lewis acid or base. In the limit of very strong (energetic) surface acidity or basicity, water can become ionized through proton or hydroxyl abstraction.

The subject of water structure near hydrophilic surfaces is considerably more complex than water structuring at hydrophobic surfaces just discussed, which itself is no trivial matter. This extra complexity is due to three related features of hydrophilic surfaces. First, each hydrophilic surface is a unique combination of both type and surface concentration of water-interactive Lewis acid or base functional groups (amine, carboxyl, ether, hydroxyl, etc.). Second, hydrophilic surfaces interact with water through both dispersion forces and Lewis acid–base interactions. This is to be contrasted to hydrophobic surfaces that interact with water only through dispersion forces (dispersion forces being a class of intermolecular interactions between the momentary dipoles in matter that arise from rapid fluctuations of electron density within molecular orbitals). Third, as a direct result of these two features, the number of possible water interactions and configurations is very large, especially if the hydrophilic surface is heterogeneous on a microscopic scale. These features make the problem of water behavior at hydrophilic surfaces both computationally and experimentally challenging.

In spite of this complexity, the reasoning and rationale applied to large polyelectrolytes in the preceding section should apply in an approximate way to extended hydrophilic surfaces, especially the more water-wettable types where acid–base interactions with water predominate over weaker dispersion interactions. This would suggest, then, that water near hydrophilic surfaces is more dense than bulk water, with a correspondingly less extensive self-associated water network (row 2, Table 1). There is some support for this general conclusion from simplified molecular models (Besseling, 1997; Silverstein et al., 1998).

Thickness of this putative denser-water layer must depend in some way on the surface concentration (number) of Lewis acid/base sites and on whether the surface is predominately acid or predominately basic, but these relationships are far from worked out in detail. One set of experimental results suggesting that hydration layers near water-wettable surfaces can be quite thick comes from the rather startling finding by Pashley and Kitchener (1979) of 150-nm-thick, free-standing water films formed on fully water-wettable quartz surfaces from water vapor. These so-called condensate water films would comprise some 600 water molecules organized in a layer through unknown mechanisms. Perhaps these condensate films are formed from water-molecule layers with alternating oriented dipoles similar to the water layers around ions briefly discussed in the previous section. Note that this hypothetical arrangement defeats water self-association throughout the condensate-film layer in a manner consistent with the inferred less self-associated, high-density nature of water near hydrophilic surfaces.

Stepping back and viewing the full range of surface wetting behaviors discussed herein, it is apparent that water solvent properties (structure) near surfaces can be thought of as a sort of continuum or spectrum. At one end of the spectrum lie perfectly hydrophobic surfaces with no surface-resident Lewis acid or base sites. Water interacts with these hydrophobic surfaces only through dispersion forces mentioned above. At the other end of the spectrum, surfaces bear a sufficient surface concentration of Lewis sites to completely disrupt bulk water structure through a competition for hydrogen bonds, leading to complete water wetting (0° contact angle). Structure and solvent properties of water in contact with surfaces between these extremes must then exhibit some kind of graded properties associated with the graded wettability observed with contact angles. If the

surface region is composed of molecules that hydrate to a significant degree, as in the case of hydrogel materials, then the surface can adsorb water and swell or dissolve. At the extreme of water–surface interactions, surface acid or base groups can abstract hydroxyls or protons from water, respectively, leading to water ionization at the surface.

Finally, in closing this section on water properties near surfaces, it is worthwhile to note that whereas insights gained from computational models employing hypothetical surfaces and experimental systems using atomically smooth mica and highly polished semiconductor-grade silicon wafers provide very important scientific insights, these results have limited direct biomedical relevance because practical biomaterial surfaces are generally quite rough relative to the dimensions of water (Fig. 1C). At the 0.25-nm scale, water structure near a hydrophobic polymer such as polyethylene, for example, might better be envisioned as a result of hydrating molecular-scale domains where methyl- and methylene-group protrusions from a "fractal" surface solvate in water rather than a sea of close-packed groups disposed erectly on an infinitely flat plane that one might construct in molecular modeling. Surfaces of functionalized polymers such as poly(ethylene terephthalate) (PET) would be even more complex. Both surface topography and composition will play a role in determining water structure near surfaces.

WATER AND THE BIOLOGICAL RESPONSE TO MATERIALS

It has long been assumed that the observed biological response to materials is initiated or catalyzed by interactions with material residing in the same thin surface region that affects water wettability, arguably no thicker than about 1 nm. In particular, it is frequently assumed that biological responses begin with protein adsorption. These assumptions are based on the observations that cells and proteins interact only at the aqueous interface of a material, that this interaction seemingly does not depend on the macroscopic thickness of a rigid material, and that water does not penetrate deeply into the bulk of many materials (excluding those that absorb water). Thus, one may conclude that biology does not "sense" or "see" bulk properties of a contacting material, only the outermost molecular groups protruding from a surface. Over the past decade or so, the validity of this assumption seems to have been confirmed through numerous studies employing self-assembled monolayers (SAMs) supported on glass, gold, and silicon in which variation of the outermost surface functional groups exposed to blood plasma, purified proteins, and cells indeed induces different outcomes (Fragneto et al., 1995; Liebmann-Vinson et al., 1996; Margel et al., 1993; Mooney et al., 1996; Owens et al., 1988; Petrash et al., 1997; Prime and Whitesides, 1993; Scotchford et al., 1998; Singhvi et al., 1994; Sukenik et al., 1990; Tidwell et al., 1997; Vogler et al., 1995a, b). But exactly how surfaces influence "biocompatibility" of a material is still not well understood.

Theories attempting to explain the role of surfaces in the biological response fall into two basic categories. One asserts that surface energy is the primary correlating surface property (Akers *et al.*, 1977; Baier, 1972; Baier *et al.*, 1969), the other that water solvent properties near surfaces are the primary causative agent (Andrade *et al.*, 1981; Andrade and Hlady, 1986; Vogler, 1998). The former attempts to correlate surface energy factors such as critical surface energy σ_c or various interfacial tension components while the latter attempts correlations with water contact angle θ or some variant thereof such as water adhesion tension $\tau = \sigma_{lv} \cos \theta$, where σ_{lv} is the interfacial tension of water (see Chapter 1.4). Both approaches attempt to infer structure–property relationships between surface energy/wetting and some measure of the biological response. These two ideas would be functionally equivalent if water structure and solvent properties were directly related to surface energy in a straightforward way (e.g., linear), but this appears not to be the case (Vogler, 1998) because of water structuring in response to surface (adsorption) energetics, as described in preceding sections.

Both surface energy and water theories acknowledge that the principle interfacial events surfaces can promote or catalyze are adsorption and adhesion. Adsorption of proteins and/or adhesion of cells/tissues is known (or at least strongly suspected) to be involved in the primary interactions of biology with materials. Therefore, it is reasonable to anticipate that surfaces induce a biological response through adsorption and/or adhesion mechanisms. The surface energy theory acknowledges this connection by noting that surface energy is the engine that drives adsorption and adhesion. The water theory recognizes the same but in a quite different way. Instead, water theory asserts that surface energetics is the engine that drives adsorption of water and then, in subsequent steps, proteins and cells interact with the resulting hydrated interface either through or by displacing a so-called vicinal water layer that is more or less bound to the surface, depending on the energetics of the original water–surface interaction. Furthermore, water theory suggests that the ionic composition of vicinal water may be quite different than that of bulk water, with highly hydrated ions such as Ca^{2+} and Mg^{2+} preferentially concentrating in water near hydrophilic surfaces and less hydrated ions such as Na^+ and K^+ preferentially concentrating in water near hydrophobic surfaces. It is possible that the ionic composition of vicinal water layers further accounts for differences in the biological response to hydrophilic and hydrophobic materials on the basis that divalent ions have allosteric effects on enzyme reactions and participate in adhesion through divalent ion bridging.

Water is a very small, but very special, molecule. Properties of this universal biological solvent, this essential medium of life as we understand it, remain more mysterious in this century of science than those of the very atoms that compose it. Self-association of water through hydrogen bonding is the essential mechanism behind water solvent properties, and understanding self-association effects near surfaces is a key to understanding water properties in contact with biomaterials. It seems safe to conclude that no theory explaining the biological response to materials can be complete without accounting for water properties near surfaces and that this remains an exciting topic in biomaterials surface science.

Acknowledgments

The author is indebted to the editors for helpful and detailed discussion of the manuscript and to Professor J. Kubicki for molecular models used in construction of figures. Mr. Brian J. Mulhollem is thanked for reading the manuscript for typographical errors.

Bibliography

Akers, C. K., Dardik, I., Dardik, H., and Wodka, M. (1977). Computational methods comparing the surface properties of the inner walls of isolated human veins and synthetic biomaterials. *J. Colloid Interface Sci.* **59**: 461–467.

Andrade, J. D., and Hlady, V. (1986). Protein adsorption and materials biocompatibility: a tutorial review and suggested mechanisms. *Adv. Polym. Sci.* **79**: 3–63.

Andrade, J. D., Gregonis, D. E., and Smith, L. M. (1981). Polymer–water interface dynamics. in *Physicochemical aspects of polymer surfaces*, K. L. Mittal, ed. Plenum Press, New York, pp. 911–922.

Baier, R. E. (1972). The role of surface energy in thrombogenesis. *Bull. N. Y. Acad. Med.* **48**: 257–272.

Baier, R. E. (1978). Key events in blood interactions at nonphysiologic interfaces—a personal primer. *Artificial Organs* **2**: 422–426.

Baier, R. E., and Meyer, A. E. (1996). Physics of solid surfaces. in *Interfacial Phenomena and Bioproducts*, J. L. Brash and P. W. Wojciechowski, eds. Marcel Dekker, New York, pp. 85–121.

Baier, R. E., Dutton, R. C., and Gott, V. L. (1969). Surface chemical features of blood vessel walls and of synthetic materials exhibiting thromboresistance. in *Surface Chemistry of Biological Systems*, M. Blank, ed. Plenum Publishers, New York, pp. 235–260.

Berendsen, H. J. C. (1967). Water structure. in *Theoretical and Experimental Biophysics*, A. Cole, ed. Marcel Dekker, New York, pp. 1–74.

Besseling, N. A. M. (1997). Theory of hydration forces between surfaces. *Langmuir* **13**: 2113–2122.

Besseling, N. A. M., and Lyklema, J. (1995). Hydrophobic hydration of small apolar molecules and extended surfaces: a molecular model. *Pure Appl. Chem.* **67**: 881–888.

Christenson, H. K., Fang, J., Ninham, B. W., and Parker, J. L. (1990). Effect of divalent electrolyte on the hydrophobic attraction. *J. Phys. Chem.* **94**: 8004–8006.

DeVoe, H. (1969). Theory of the conformations of biological macromolecules in solution. in *Structure and Stability of Biological Molecules*, S. N. Timasheff and G. D. Fasman, eds. Marcel Dekker, New York, pp. 1–59.

Du, Q., Freysz, E., and Shen, Y. R. (1994). Surface vibrational spectroscopic studies of hydrogen bonding and hydrophobicity. *Science* **264**: 826–828.

Dunhill, P. (1965). How proteins acquire their structure. *Sci. Progr.* **53**: 609–619.

Fragneto, G., Thomas, R. K., Rennie, A. R., and Penfold, J. (1995). Neutron reflection study of bovine casein adsorbed on OTS self-assembled monolayers. *Science* **267**: 657–660.

Franks, F. (1972). Introduction—water, the unique chemical. in *Water: A Comprehensive Treatise*, F. Franks, ed. Plenum Publishers, New York, pp. 1–17.

Gragson, D. E., and Richmond, G. L. (1997). Comparisons of the structure of water at near oil/water and air/water interfaces as determined by vibrational sum frequency generation. *Langmuir* **13**: 4804–4806.

Hoffman, A. S. (1986). A general classification scheme for "hydrophilic" and "hydrophobic" biomaterial surfaces. *J. Biomed. Mater. Res.* **20**: ix–xi.

Hoffman, A. S., and Harris, C. (1972). Radiation-grafted hydrogels on silicone rubber surfaces — a new biomaterial. *Polym. Preprints* **13**: 740–746.

Hoffman, A. S., and Kraft, W. G. (1972). Radiation-grafted hydrogels on polyurethane surfaces—a new biomaterial. *Polym. Preprints* **13**: 723–728.

Iassacs, E. D., Shukla, A., Platzman, P. M., Hamann, D. R., Bariellini, B., and Tulk, C. A. (1999). Covalency of the hydrogen bond in ice: a direct X-ray measurement. *Phys. Rev. Lett.* **82**: 600–603.

Kauzmann, W. (1959). Some factors in the interpretation of protein denaturation. *Adv. Protein Chem.* **14**: 1–63.

Liebmann-Vinson, A., Lander, L. M., Foster, M. D., Brittain, W. J., Vogler, E. A., Majkrak, C. F., and Satija, S. (1996). A neutron reflectometry study of human serum albumin adsorption *in Situ*. *Langmuir* **12**: 2256–2262.

Lopez, G. P., Ratner, B. D., Tidwell, C. D., Haycox, C. L., Rapoza, R. J., and Horbett, T. A. (1992). Glow discharge plasma deposition of tetraethylene glycol dimethyl ether for fouling-resistant biomaterial surfaces. *J. Biomed. Mater. Res.* **26**: 415–439.

Lum, K., Chandler, D., and Weeks, J. D. (1999). Hydrophobicity at small and large length scales. *J. Phys. Chem. B* **103**: 4570–4577.

Luzar, A., and Chandler, D. (1996). Hydrogen-bond kinetics in liquid water. *Nature* **379**: 55–57.

Marcus, Y. (1985). *Ion Solvation*. Wiley, New York.

Margel, S., Vogler, E. A., Firment, L., Watt, T., Haynie, S., and Sogah, D. Y. (1993). Peptide, protein, and cellular interactions with self-assembled monolayer model surfaces. *J. Biomed. Mater. Res.* **27**: 1463–1476.

Marshall, E. (1999). Getting to the bottom of water. *Science* **283**: 614–615.

Mooney, J. F., Hunt, A. J., McIntosh, J. R., Leberko, C. A., Walba, D. M., and Rogers, C. T. (1996). Patterning of functional antibodies and other proteins by photolithograpy of silane monolayers. *Proc. Natl. Acad. Sci. USA* **93**: 12287–12291.

Narten, A. H., and Levy, H. A. (1969). Observed diffraction pattern and proposed models of liquid water. *Science* **165**: 447–454.

Oss, C. J. v., and Giese, R. F. (1995). The hydrophilicity and hydrophobicity of clay minerals. *Clays Clay Miner.* **43**: 474–477.

Owens, N. F., Gingell, D., and Trommler, A. (1988). Cell adhesion to hydroxyl groups of a monolayer film. *J. Cell Sci.* **91**: 269–279.

P.-Grosdemange, C., Simon, E. S., Prime, K. L., and Whitesides, G. M. (1991). Formation of self-assembled monolayers by chemisorption of derivatives of oligo(ethylene glycol) on gold. *J. Am. Chem. Soc.* **113**: 13–20.

Pain, R. H. (1982). Molecular hydration and biological function. in *Biophysics of Water*, F. Franks and S. Mathias, eds. John Wiley and Sons, Chichester, UK, pp. 3–14.

Pashley, R. M., and Kitchener, J. A. (1979). Surface forces in adsorbed multilayers of water on quartz. *J. Colloid Interface Sci.* **71**: 491–500.

Paulaitis, M. E., Garde, S., and Ashbaugh, H. S. (1996). The hydrophobic effect. *Curr. Opin. Colloid Interface Sci.* **1**: 376–383.

Petrash, S., Sheller, N. B., Dando, W., and Foster, M. D. (1997). Variation in tenacity of protein adsorption on self-assembled monolayers with monolayer order as observed by X-ray reflectivity. *Langmuir* **13**: 1881–1883.

Prime, K. L., and Whitesides, G. M. (1993). Adsorption of proteins onto surfaces containing end-attached oligo ethylene oxide: a model system using self-assembled monolayers. *J. Am. Chem. Soc.* **115**: 10714–10721.

Qi, Z., and Soka, M. (1998). Dynamic properties of individual water molecules in a hydrophobic pore lined with acyl chains: a molecular dynamics study. *Biophys. Chem.* **71**: 35–50.

Ratner, B. D., and Hoffman, A. S. (1980). Surface characterization of hydrophilic–hydrophobic copolymer model systems I. A preliminary study. *Polymer. Sci. Technol.* **12B**: 691–706.

Robinson, G. W., Zhu, S.-B., Singh, S., and Evans, M. W. (1996). *Water in Biology, Chemistry, and Physics*. World Scientific, London.

Scotchford, C. A., Cooper, E., Leggett, G. J., and Downes, S. (1998). Growth of human osteoblast-like cells on alkanethiol on gold self-assembled monolayers: the effect of surface chemistry. *J. Biomed. Mater. Res.* **41**: 431–442.

Silverstein, K. A. T., Haymet, A. D. J., and Dill, K. A. (1998). A simple model of water and the hydrophobic effect. *J. Am. Chem. Soc.* **120**: 3166–3175.

Singhvi, R., Kumar, A., Lopez, G. P., Stephanopoulos, G. N., Wang, D. I. C., Whitesides, G. M., and Ingber, D. E. (1994). Engineering cell shape and function. *Science* **264**: 696–698.

Sukenik, C. N., Balachander, N., Culp, L. A., Lewandowska, K., and Merritt, K. (1990). Modulation of cell adhesion by modification of titanium surfaces with covalently attached self-assembled monolayers. *J. Biomed. Mater. Res.* **24**: 1307–1323.

Tanford, C. (1966). *Physical Chemistry of Macromolecules*. John Wiley and Sons, New York.

Tanford, C. (1973). *The Hydrophobic Effect: Formation of Micelles and Biological Membranes*. John Wiley and Sons, New York.

Tidwell, C. D., Ertel, S. I., Ratner, B. D., Tarasevich, B. J., Atre, S., and Allara, D. L. (1997). Endothelial cell growth and protein adsorption on terminally functionalized, self-assembled monolayers of alkanethiols on gold. *Langmuir* **13**: 3404–3413.

Vinogradov, S. N., and Linnell, R. H. (1971). *Hydrogen Bonding*. Van Nostrand Reinhold, New York.

Vogler, E. A. (1998). Structure and reactivity of water at biomaterial surfaces. *Adv. Colloid Interface Sci.* **74**: 69–117.

Vogler, E. A., Graper, J. C., Harper, G. R., Lander, L. M., and Brittain, W. J. (1995a). Contact activation of the plasma coagulation cascade. 1. Procoagulant surface energy and chemistry. *J. Biomed. Mater. Res.* **29**: 1005–1016.

Vogler, E. A., Graper, J. C., Sugg, H. W., Lander, L. M., and Brittain, W. J. (1995b). Contact activation of the plasma coagulation cascade. 2. Protein adsorption on procoagulant surfaces. *J. Biomed. Mater. Res.* **29**: 1017–1028.

Wiggins, P. M. (1971). Water structure as a determinant of ion distribution in living tissue. *J. Theor. Biol.* **32**: 131–146.

Wiggins, P. M. (1988). Water structure in polymer membranes. *Prog. Polym. Sci.* **13**: 1–35.

Wiggins, P. M. (1990). Role of water in some biological processes. *Microbiol. Rev.* **54**: 432–449.

Wiggins, P. M., and Ryn, R. T. v. (1990). Changes in ionic selectivity with changes in density of water in gels and cells. *Biophys. J.* **58**: 585–596.

2

Classes of Materials Used in Medicine

Sascha Abramson, Harold Alexander, Serena Best, J. C. Bokros, John B. Brunski,
André Colas, Stuart L. Cooper, Jim Curtis, Axel Haubold, Larry L. Hench,
Robert W. Hergenrother, Allan S. Hoffman, Jeffrey A. Hubbell, John A. Jansen,
Martin W. King, Joachim Kohn, Nina M. K. Lamba, Robert Langer, Claudio Migliaresi,
Robert B. More, Nicholas A. Peppas, Buddy D. Ratner, Susan A. Visser,
Andreas von Recum, Steven Weinberg, and Ioannis V. Yannas

2.1 INTRODUCTION

Allan S. Hoffman

Biomaterials can be divided into four major classes of materials: polymers, metals, ceramics (including carbons, glass-ceramics, and glasses), and natural materials (including those from both plants and animals). Sometimes two different classes of materials are combined together into a composite material, such as silica-reinforced silicone rubber or carbon fiber- or hydroxyapatite particle-reinforced poly (lactic acid). Such composites are a fifth class of biomaterials. What is the history behind the development and application of such diverse materials for implants and medical devices, what are the compositions and properties of these materials, and what are the principles governing their many uses as components of implants and medical devices? This chapter critically reviews this important literature of biomaterials.

The wide diversity and sophistication of materials currently used in medicine and biotechnology is testimony to the significant scientific and technological advances that have occurred over the past 50 years. From World War II to the early 1960s, relatively few pioneering surgeons were taking commercially available polymers and metals, fabricating implants and components of medical devices from them, and applying them clinically. There was little government regulation of this activity, and yet these earliest implants and devices had a remarkable success. However, there were also some dramatic failures. This led the surgeons to enlist the aid of physical, biological, and materials scientists and engineers, and the earliest interdisciplinary "bioengineering" collaborations were born.

These teams of physicians and scientists and engineers not only recognized the need to control the composition, purity, and physical properties of the materials they were using, but they also recognized the need for new materials with new and special properties. This stimulated the development of many new materials in the 1970s. New materials were designed *de novo* specifically for medical use, such as biodegradable polymers and bioactive ceramics. Some were derived from existing materials fabricated with new technologies, such as polyester fibers that were knit or woven in the form of tubes for use as vascular grafts, or cellulose acetate plastic that was processed as bundles of hollow fibers for use in artificial kidney dialysers. Some materials were "borrowed" from unexpected sources such as pyrolytic carbons or titanium alloys that had been developed for use in air and space technology. And other materials were modified to provide special biological properties, such as immobilization of heparin for anti-coagulant surfaces. More recently biomaterials scientists and engineers have developed a growing interest in natural tissues and polymers in combination with living cells. This is particularly evident in the field of tissue engineering, which focuses on the repair or regeneration of natural tissues and organs. This interest has stimulated the isolation, purification, and application of many different natural materials. The principles and applications of all of these biomaterials and modified biomaterials are critically reviewed in this chapter.

2.2 POLYMERS

Stuart L. Cooper, Susan A. Visser,
Robert W. Hergenrother, and Nina M. K. Lamba

Many types of polymers are widely used in biomedical devices that include orthopedic, dental, soft tissue, and cardiovascular implants. Polymers represent the largest class of biomaterials. In this section, we will consider the main types of polymers, their characterization, and common medical applications. Polymers may be derived from natural sources, or from synthetic organic processes.

The wide variety of natural polymers relevant to the field of biomaterials includes plant materials such as cellulose, sodium alginate, and natural rubber, animal materials such as tissue-based heart valves and sutures, collagen, glycosaminoglycans (GAGs), heparin, and hyaluronic acid, and other natural materials such as deoxyribonucleic acid (DNA), the genetic material of all living creatures. Although these polymers are undoubtedly important and have seen widespread use in numerous applications, they are sometimes eclipsed by the seemingly endless variety of synthetic polymers that are available today. Synthetic polymeric biomaterials range from hydrophobic, non-water-absorbing materials such as silicone rubber (SR), polyethylene (PE), polypropylene (PP), poly(ethylene terephthalate) (PET), polytetrafluoroethylene (PTFE), and poly(methyl methacrylate) (PMMA) to somewhat more polar materials such as poly(vinyl chloride) (PVC), copoly(lactic–glycolic acid) (PLGA), and nylons, to water-swelling materials such as poly(hydroxyethyl methacrylate) (PHEMA) and beyond, to water-soluble materials such as poly(ethylene glycol) (PEG or PEO). Some are hydrolytically unstable and degrade in the body while others may remain essentially unchanged for the lifetime of the patient.

Both natural and synthetic polymers are long-chain molecules that consist of a large number of small repeating units. In synthetic polymers, the chemistry of the repeat units differs from the small molecules (monomers) that were used in the original synthesis procedures, resulting from either a loss of unsaturation or the elimination of a small molecule such as water or HCl during polymerization. The exact difference between the monomer and the repeat unit depends on the mode of polymerization, as discussed later.

The task of the biomedical engineer is to select a biomaterial with properties that most closely match those required for a particular application. Because polymers are long-chain molecules, their properties tend to be more complex than those of their short-chain precursors. Thus, in order to choose a polymer type for a particular application, the unusual properties of polymers must be understood.

This chapter introduces the concepts of polymer synthesis, characterization, and property testing as they are relevant to the eventual application of a polymer as a biomaterial. Following this, examples of polymeric biomaterials currently used by the medical community are cited and their properties and uses are discussed.

MOLECULAR WEIGHT

In polymer synthesis, a polymer is usually produced with a distribution of molecular weights. To compare the molecular weights of two different batches of polymer, it is useful to define an average molecular weight. Two statistically useful definitions of molecular weight are the number average and weight average molecular weights. The number average molecular weight (M_n) is the first moment of the molecular weight distribution and is an average over the number of molecules. The weight average molecular weight (M_w) is the second moment of the molecular weight distribution and is an average over the weight of each polymer chain. Equations 1 and 2 define the two averages:

$$M_n = \frac{\sum N_i M_i}{\sum N_i} \qquad (1)$$

$$M_w = \frac{\sum N_i M_i^2}{\sum N_i M_i} \qquad (2)$$

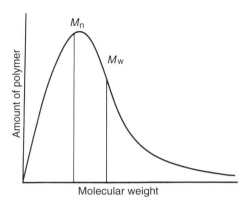

FIG. 1.   Typical molecular weight distribution of a polymer.

where N_i is the number of moles of species i, and M_i is the molecular weight of species i.

The ratio of M_w to M_n is known as the polydispersity index (PI) and is used as a measure of the breadth of the molecular weight distribution. Typical commercial polymers have polydispersity indices of 1.5–50, although polymers with polydispersity indices of less than 1.1 can be synthesized with special techniques. A molecular weight distribution for a typical polymer is shown in Fig. 1.

Linear polymers used for biomedical applications generally have M_n in the range of 25,000 to 100,000 and M_w from 50,000 to 300,000, and in exceptional cases, such as the PE used in the hip joint, the M_w may range up to a million. Higher or lower molecular weights may be necessary, depending on the ability of the polymer chains to crystallize or to exhibit secondary interactions such as hydrogen bonding. The crystallinity and secondary interactions can give polymers additional strength. In general, increasing molecular weight corresponds to increasing physical properties; however, since melt viscosity also increases with molecular weight, processability will decrease and an upper limit of useful molecular weights is usually reached. Mechanical properties of some polymeric biomaterials are presented in Table 1.

SYNTHESIS

Methods of synthetic polymer preparation fall into two categories: addition polymerization (chain reaction) and condensation polymerization (stepwise growth) (Fig. 2). (Ring opening is another type of polymerization and is discussed in more detail later in the section on degradable polymers.) In addition polymerization, unsaturated monomers react through the stages of initiation, propagation, and termination to give the final polymer product. The initiators can be free radicals, cations, anions, or stereospecific catalysts. The initiator opens the double bond of the monomer, presenting another "initiation" site on the opposite side of the monomer bond for continuing growth. Rapid chain growth ensues during the propagation step until the reaction is terminated by reaction with another radical, a solvent molecule, another polymer molecule, an initiator, or an added chain transfer agent. PVC, PE, and PMMA are

TABLE 1 Mechanical Properties of Biomedical Polymers

Polymer	Water absorption (%)	Bulk modulus (GPa)	Tensile Strength (MPa)	Elongation at break (%)	T_g (K)	T_m (K)
Polyethylene	0.001–0.02	0.8–2.2	30–40	130–500	160–170	398–408
Polypropylene	0.01–0.035	1.6–2.5	21–40	100–300	243–270	433–453
Polydimethyl-siloxane	0.08–0.1		3–10	50–800	148	233
Polyurethane	0.1–0.9	1.5–2	28–40	600–720	200–250	453–523*
Polytetrafluoro-ethylene	0.01–0.05	1–2	15–40	250–550	293–295	595–600
Polyvinyl-chloride	0.04–0.75	3–4	10–75	10–400	250–363	423*
Polyamides	0.25–3.5	2.4–3.3	44–90	40–250	293–365	493–540
Polymethyl-methacrylate	0.1–0.4	3–4.8	38–80	2.5–6	379–388	443*
Polycarbonate	0.15–0.7	2.8–4.6	56–75	8–130	418	498–523
Polyethylene-terephthalate	0.06–0.3	3–4.9	42–80	50–500	340–400	518–528

* = decomposition temperature

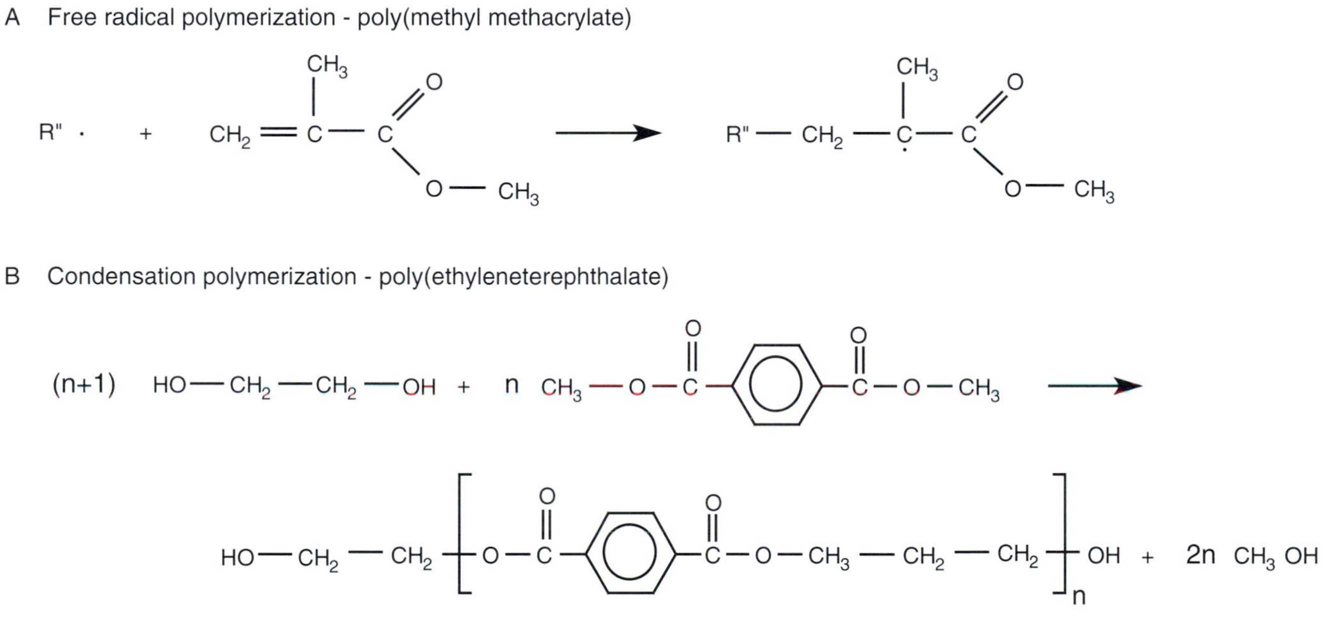

A   Free radical polymerization - poly(methyl methacrylate)

B   Condensation polymerization - poly(ethyleneterephthalate)

FIG. 2. (A) Polymerization of methyl methacrylate (addition polymerization). (B) Synthesis of poly(ethylene terephthalate) (condensation polymerization).

relevant examples of addition polymers used as biomaterials. The polymerization of MMA to form PMMA is shown in Fig. 2A.

Condensation polymerization is completely analogous to condensation reactions of low-molecular-weight molecules. Two monomers react to form a covalent bond, usually with elimination of a small molecule such as water, hydrochloric acid, methanol, or carbon dioxide. Nylon and PET (Fig. 2B) are typical condensation polymers and are used in fiber or fabric form as biomaterials. The reaction continues until almost all of one reactant is used up. There are also polymerizations that resemble the stepwise growth of condensation polymers, although no small molecule is eliminated. Polyurethane synthesis bears these characteristics, which

is sometimes referred to as polyaddition or rearrangement polymerization (Brydson, 1995).

The choice of polymerization method strongly affects the polymer obtained. In free radical polymerization, a type of addition polymerization, the molecular weights of the polymer chains are difficult to control with precision. Added chain transfer agents are used to control the average molecular weights, but molecular weight distributions are usually broad. In addition, chain transfer reactions with other polymer molecules can produce undesirable branched products (Fig. 3) that affect the ultimate properties of the polymeric material. In contrast, molecular architecture can be controlled very precisely in anionic polymerization. Regular linear chains with PI indices close to unity can be obtained. More recent methods

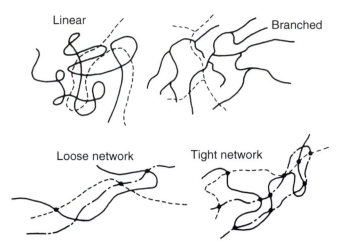

FIG. 3. Polymer arrangements. (From F. Rodriguez, *Principles of Polymer Systems*, Hemisphere Publ., 1982, p. 21, with permission.)

of living free radical polymerizations called ATRP and RAFT may also yield low PIs.

Polymers produced by addition polymerization can be homopolymers, i.e., polymers containing only one type of repeat unit, or copolymers with two or more types of repeat units. Depending on the reaction conditions and the reactivity of each monomer type, the copolymers can be random, alternating, graft, or block copolymers, as illustrated in Fig. 4. Random copolymers exhibit properties that approximate the weighted average of those of the two types of monomer units, whereas block copolymers tend to phase separate into a monomer-A-rich phase and a monomer-B-rich phase, displaying properties unique to each of the homopolymers. Figure 5 shows the repeat units of many of the homopolymers used in medicine.

Condensation polymerization can also result in copolymer formation. The properties of the condensation copolymer depend on three factors: the chemistry of monomer units; the molecular weight of the polymer product, which can be controlled by the ratio of one reactant to another and by the time

of polymerization; and the final distribution of the molecular weight of the copolymer chains. The use of bifunctional monomers gives rise to linear polymers, while multifunctional monomers may be used to form covalently cross-linked networks. Figure 6 shows the reactant monomers and polymer products of some biomedical copolymers.

Postpolymerization cross-linking of addition or condensation polymers is also possible. Natural rubber, for example, consists mostly of linear molecules that can be cross-linked to a loose network with 1–3% sulfur (vulcanization) or to a hard rubber with 40–50% sulfur (Fig. 3). In addition, physical, rather than chemical, cross-linking of polymers can occur in microcrystalline regions, that are present in nylon (Fig. 7A). Alternatively, physical cross-linking can be achieved through incorporation of ionic groups in the polymer (Fig. 7B). This is used in acrylic acid cement systems (e.g., for dental cements) where divalent cations such as zinc and calcium are incorporated into the formulation and interact with the carboxyl groups to produce a strong, hard material. The alginates, which are polysaccharides derived from brown seaweed, also contain anionic residues that will interact with cations and water to form a gel. The alginates are used successfully to dress deep wounds and are also being studied as tissue engineering matrices.

THE SOLID STATE

Tacticity

Polymers are long-chain molecules and, as such, are capable of assuming many conformations through rotation of valence bonds. The extended chain or planar zigzag conformation of PP is shown in Fig. 8. This figure illustrates the concept of tacticity. Tacticity refers to the arrangement of substituents (methyl groups in the case of polypropylene) around the extended polymer chain. Chains in which all substituents are located on the same side of the zigzag plane are isotactic, whereas syndiotactic chains have substituents alternating from side to side. In the atactic arrangement, the substituent groups appear at random on either side of the extended chain backbone.

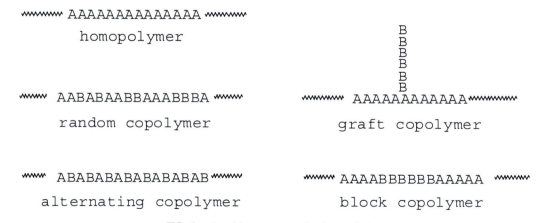

FIG. 4. Possible structures of polymer chains.

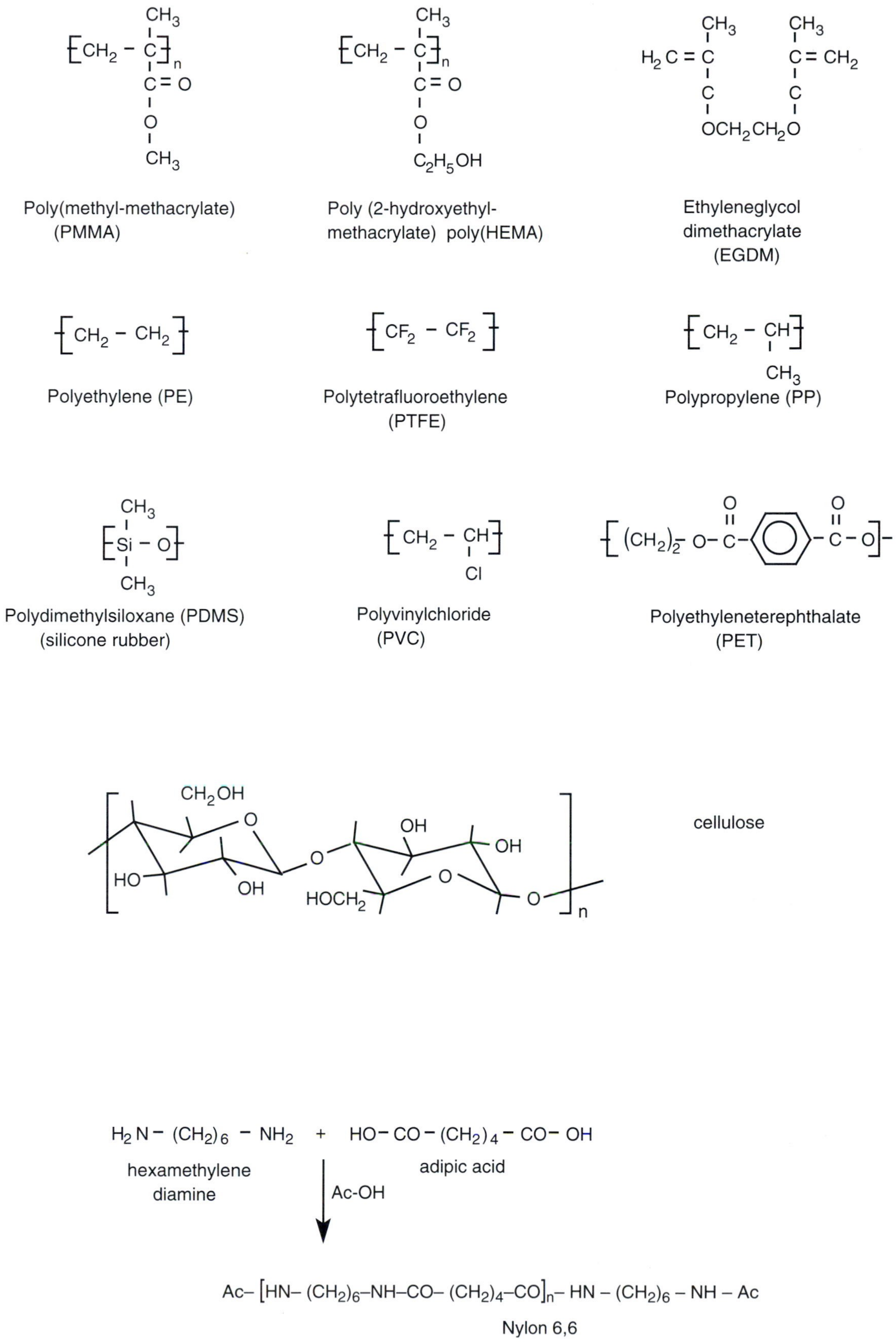

$$\left[CH_2 - \underset{\underset{\underset{CH_3}{|}}{\underset{|}{O}}}{\overset{\overset{CH_3}{|}}{\underset{|}{C}}}\right]_n$$

Poly(methyl-methacrylate)
(PMMA)

$$\left[CH_2 - \underset{\underset{\underset{C_2H_5OH}{|}}{\underset{|}{O}}}{\overset{\overset{CH_3}{|}}{\underset{|}{C}}}\right]_n$$

Poly (2-hydroxyethyl-
methacrylate) poly(HEMA)

$$H_2C = \underset{\underset{OCH_2CH_2O}{|}}{\overset{\overset{CH_3}{|}}{C}} \quad \underset{\underset{}{|}}{\overset{\overset{CH_3}{|}}{C}} = CH_2$$

Ethyleneglycol
dimethacrylate
(EGDM)

$$\left[CH_2 - CH_2\right]$$

Polyethylene (PE)

$$\left[CF_2 - CF_2\right]$$

Polytetrafluoroethylene
(PTFE)

$$\left[CH_2 - \underset{\underset{CH_3}{|}}{CH}\right]$$

Polypropylene (PP)

$$\left[\underset{\underset{CH_3}{|}}{\overset{\overset{CH_3}{|}}{Si}} - O\right]$$

Polydimethylsiloxane (PDMS)
(silicone rubber)

$$\left[CH_2 - \underset{\underset{Cl}{|}}{CH}\right]$$

Polyvinylchloride
(PVC)

$$\left[(CH_2)_2 - O - \overset{\overset{O}{||}}{C} - \underset{}{\bigcirc} - \overset{\overset{O}{||}}{C} - O\right]$$

Polyethyleneterephthalate
(PET)

cellulose

$H_2N - (CH_2)_6 - NH_2 \quad + \quad HO - CO - (CH_2)_4 - CO - OH$

hexamethylene   adipic acid
diamine

$\downarrow$ Ac-OH

$Ac - \left[HN - (CH_2)_6 - NH - CO - (CH_2)_4 - CO\right]_n - HN - (CH_2)_6 - NH - Ac$

Nylon 6,6

FIG. 5. Homopolymers used as biomaterials.

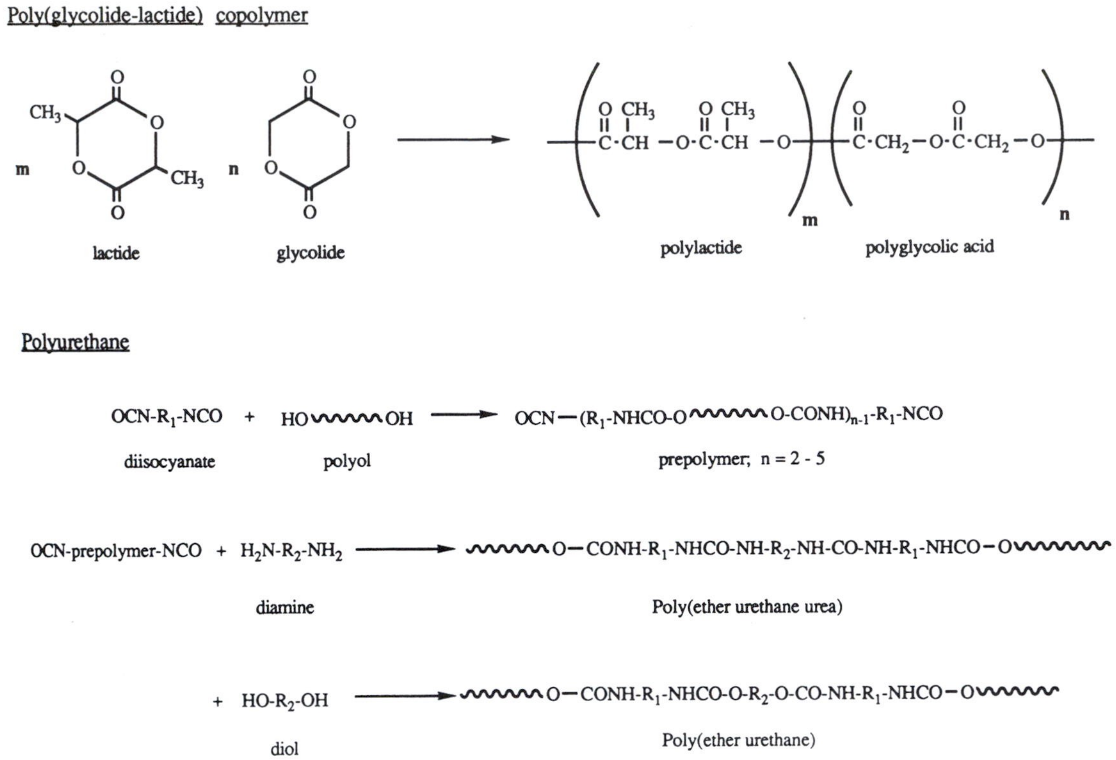

Poly(glycolide-lactide) copolymer

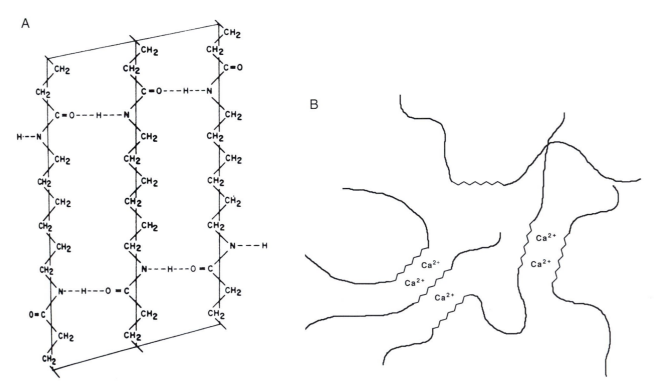

lactide   glycolide   polylactide   polyglycolic acid

Polyurethane

OCN-R₁-NCO   +   HO〰〰OH ⟶ OCN—(R₁-NHCO-O〰〰O-CONH)ₙ₋₁-R₁-NCO

diisocyanate   polyol   prepolymer; n = 2 - 5

OCN-prepolymer-NCO + H₂N-R₂-NH₂ ⟶ 〰〰O—CONH-R₁-NHCO-NH-R₂-NH-CO-NH-R₁-NHCO—O〰〰

diamine   Poly(ether urethane urea)

+ HO-R₂-OH ⟶ 〰〰O—CONH-R₁-NHCO-O-R₂-O-CO-NH-R₁-NHCO—O〰〰

diol   Poly(ether urethane)

FIG. 6.   Copolymers used in medicine and their base monomers.

FIG. 7.   (A) Hydrogen bonding in nylon-6,6 molecules in a triclinic unit cell: σ form. (From L. Mandelkern, *An Introduction to Macromolecules*, Springer-Verlag, 1983, p. 43, with permission.) (B) Ionic aggregation giving rise to physical cross-links in copolymers.

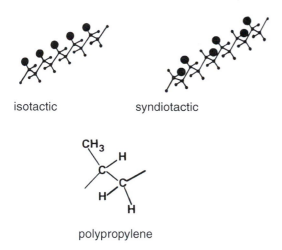

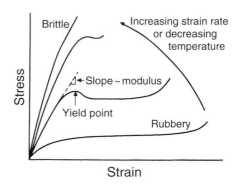

FIG. 9. Tensile properties of polymers.

FIG. 8. Schematic of stereoisomers of polypropylene. (From F. Rodriguez, *Principles of Polymer Systems*, Hemisphere Publ., 1982, p. 22, with permission.)

Atactic polymers usually cannot crystallize, and an amorphous polymer results. Isotactic and syndiotactic polymers may crystallize if conditions are favorable. PP is an isotactic crystalline polymer used as sutures. Crystalline polymers, such as PE, also possess a higher level of structure characterized by folded chain lamellar growth that results in the formation of spherulites. These structures can be visualized in a polarized light microscope.

Crystallinity

Polymers can be either amorphous or semicrystalline. They can never be completely crystalline owing to lattice defects that form disordered, amorphous regions. The tendency of a polymer to crystallize is enhanced by the small side groups and chain regularity. The presence of crystallites in the polymer usually leads to enhanced mechanical properties, unique thermal behavior, and increased fatigue strength. These properties make semicrystalline polymers (often referred to simply as crystalline polymers) desirable materials for biomedical applications. Examples of crystalline polymers used as biomaterials are PE, PP, PTFE, and PET.

Mechanical Properties

The tensile properties of polymers can be characterized by their deformation behavior (stress-strain response (Fig. 9). Amorphous, rubbery polymers are soft and reversibly extensible. The freedom of motion of the polymer chain is retained at a local level while a network structure resulting from chemical cross-links and chain entanglements prevents large-scale movement or flow. Thus, rubbery polymers tend to exhibit a lower modulus, or stiffness, and extensibilities of several hundred percent, as shown in Table 1. Rubbery materials may also exhibit an increase of stress prior to breakage as a result of strain-induced crystallization assisted by molecular orientation in the direction of stress. Glassy and semicrystalline polymers have higher moduli and lower extensibilities.

The ultimate mechanical properties of polymers at large deformations are important in selecting particular polymers for biomedical applications. The ultimate strength of polymers is the stress at or near failure. For most materials, failure is catastrophic (complete breakage). However, for some semicrystalline materials, the failure point may be defined by the stress point where large inelastic deformation starts (yielding). The toughness of a polymer is related to the energy absorbed at failure and is proportional to the area under the stress-strain curve.

The fatigue behavior of polymers is also important in evaluating materials for applications where dynamic strain is applied. For example, polymers that are used in the artificial heart must be able to withstand many cycles of pulsating motion. Samples that are subjected to repeated cycles of stress and release, as in a flexing test, fail (break) after a certain number of cycles. The number of cycles to failure decreases as the applied stress level is increased, as shown in Fig. 10 (see also Chapter 6.4). For some materials, a minimum stress exists below which failure does not occur in a measurable number of cycles.

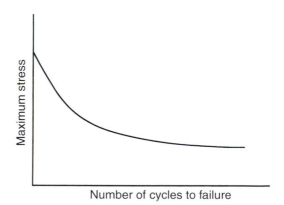

FIG. 10. Fatigue properties of polymers.

Thermal Properties

In the liquid or melt state, a noncrystalline polymer possesses enough thermal energy for long segments of each polymer to move randomly (Brownian motion). As the melt is cooled, a temperature is eventually reached at which all long-range segmental motions cease. This is the glass transition temperature (T_g), and it varies from polymer to polymer. Polymers used below their T_g, such as PMMA, tend to be hard and glassy, while polymers used above their T_g, such as SR, are rubbery. Polymers with any crystallinity will also exhibit a melting temperature (T_m) owing to melting of the crystalline phase. These polymers, such as PET, PP, and nylon, will be relatively hard and strong below T_g, and tough and strong above T_g. Thermal transitions in polymers can be measured by differential scanning calorimetry (DSC), as discussed in the section on characterization techniques. All polymers have a T_g, but only polymers with regular chain architecture can pack well, crystallize, and exhibit a T_m. The T_g is always below the T_m.

The viscoelastic responses of polymers can also be used to classify their thermal behavior. The modulus versus temperature curves shown in Fig. 11 illustrate behaviors typical of linear amorphous, cross-linked, and semicrystalline polymers. The response curves are characterized by a glassy modulus below T_g of approximately 3×10^9 Pa. For linear amorphous polymers, increasing temperature induces the onset of the glass transition region where, in a 5–10°C temperature span (depending on heating rate), the modulus drops by three orders of magnitude, and the polymer is transformed from a stiff glass to a leathery material. The relatively constant modulus region above T_g is the rubbery plateau region where long-range segmental motion is occurring but thermal energy is insufficient to overcome entanglement interactions that inhibit flow. This is the target region for many biomedical applications. Finally, at high enough temperatures, the polymer begins to flow, and a sharp decrease in modulus is seen over a narrow temperature range. This is the region where polymers are processed into various shapes, depending on their end use.

Crystalline polymers exhibit the same general features in modulus versus temperature curves as amorphous polymers; however, crystalline polymers possess a higher plateau modulus owing to the reinforcing effect of the crystallites. Crystalline polymers tend to be tough, ductile plastics whose properties are sensitive to processing history. When heated above their flow point, they can be melt processed and will crystallize and become rigid again upon cooling.

Chemically cross-linked polymers exhibit modulus versus temperature behavior analogous to that of linear amorphous polymers until the flow regime is approached. Unlike linear polymers, chemically cross-linked polymers do not display flow behavior; the cross links inhibit flow at all temperatures below the degradation temperature. Thus, chemically cross-linked polymers cannot be melt processed. Instead, these materials are processed as reactive liquids or high-molecular-weight amorphous gums that are cross-linked during molding to give the desired product. SR is an example of this type of polymer. Some cross-linked polymers are formed as networks during polymerization, and then must be machined to be formed into useful shapes. The soft contact lens, poly(hydroxyethyl methacrylate) or polyHEMA, is an example of this type of network polymer; it is shaped in the dry state, and used when swollen with water.

Copolymers

In contrast to the thermal behavior of homopolymers discussed earlier, copolymers can exhibit a number of additional thermal transitions. If the copolymer is random, it will exhibit a T_g that approximates the weighted average of the T_g values of the two homopolymers. Block copolymers of sufficient size and incompatible block types, such as the polyurethanes, will exhibit two individual transitions, each one characteristic of the homopolymer of one of the component blocks (in addition to other thermal transitions) but slightly shifted owing to incomplete phase separation.

CHARACTERIZATION TECHNIQUES

Determination of Molecular Weight

Gel permeation chromatography (GPC), a type of size exclusion chromatography, involves passage of a dilute polymer solution over a column of porous beads. High-molecular-weight polymers are excluded from the beads and elute first, whereas lower molecular-weight molecules pass through the pores of the bead, increasing their elution time. By monitoring the effluent of the column as a function of time using an ultraviolet or refractive index detector, the amount of polymer eluted during each time interval can be determined. Comparison of the elution time of the samples with those of monodisperse samples of known molecular weight allows the entire molecular weight distribution to be determined. A typical GPC trace is shown in Fig. 12.

Osmotic pressure measurements can be used to measure M_n. A semipermeable membrane is placed between two chambers.

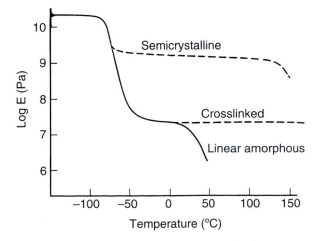

FIG. 11. Dynamic mechanical behavior of polymers.

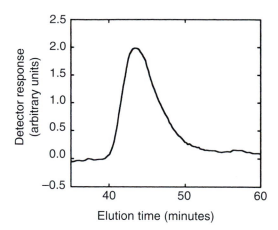

FIG. 12. A typical trace from a gel permeation chromatography run for a poly(tetramethylene oxide)/toluene diisocyanate-based polyurethane. The response of the ultraviolet detector is directly proportional to the amount of polymer eluted at each time point.

Only solvent molecules flow freely through the membrane. Pure solvent is placed in one chamber, and a dilute polymer solution of known concentration is placed in the other chamber. The lowering of the activity of the solvent in solution with respect to that of the pure solvent is compensated by applying a pressure π on the solution. π is the osmotic pressure and is related to M_n by:

$$\frac{\pi}{c} = RT \left[\frac{1}{M_n} + A_2 c + A_3 c^2 + \cdots \right] \quad (3)$$

where c is the concentration of the polymer in solution, R is the gas constant, T is temperature, and A_2 and A_3 are virial coefficients relating to pairwise and triplet interactions of the molecules in solution. In general, a number of polymer solutions of decreasing concentration are prepared, and the osmotic pressure is extrapolated to zero:

$$\lim_{c \to 0} \frac{\pi}{c} = \frac{RT}{M_n} \quad (4)$$

A plot of π/c versus c then gives as its intercept the number average molecular weight.

A number of other techniques, including vapor pressure osmometry, ebulliometry, cryoscopy, and end-group analysis, can be used to determine the M_n of polymers up to molecular weights of about 40,000.

Light-scattering techniques are used to determine M_w. In dilute solution, the scattering of light is directly proportional to the number of molecules. The scattered intensity is observed at a distance r and an angle θ from the incident beam I_o is characterized by Rayleigh's ratio R_θ:

$$R_\theta = \frac{i_o r^2}{I_o} \quad (5)$$

The Rayleigh ratio is related to M_w by:

$$\frac{Kc}{R_\theta} = \frac{1}{M_w} + 2 A_2 c + 3 A_2 c^2 + \cdots \quad (6)$$

A number of solutions of varying concentrations are measured, and the data are extrapolated to zero concentration to determine M_w.

Determination of Structure

Infrared (IR) spectroscopy is often used to characterize the chemical structure of polymers. Infrared spectra are obtained by passing infrared radiation through the sample of interest and observing the wavelength of the absorption peaks. These peaks are caused by the absorption of the radiation and its conversion into specific motions, such as C–H stretching The infrared spectrum of a polyurethane is shown in Fig. 13, with a few of the bands of interest marked.

Nuclear magnetic resonance (NMR), in which the magnetic spin energy levels of nuclei of spin 1/2 or greater are probed, may also be used to analyze chemical composition. ^{1}H and ^{13}C NMR are the most frequently studied isotopes. Polymer chemistry can be determined in solution or in the solid state. Figure 14 shows a ^{13}C NMR spectrum of a polyurethane with a table assigning the peaks to specific chemical groups. NMR is also used in a number of more specialized applications relating to local motions and intermolecular interactions of polymers.

Wide-angle X-ray scattering (WAXS) techniques are useful for probing the local structure of a semicrystalline polymeric solid. Under appropriate conditions, crystalline materials diffract X-rays, giving rise to spots or rings. According to Bragg's law, these can be interpreted as interplanar spacings. The interplanar spacings can be used without further manipulation or the data can be fit to a model such as a disordered helix or an extended chain. The crystalline chain conformation and atomic placements can then be accurately inferred.

Small-angle X-ray scattering (SAXS) is used in determining the structure of many multiphase materials. This technique requires an electron density difference to be present between two components in the solid and has been widely applied to morphological studies of copolymers and ionomers. It can probe features of 10–1000 Å in size. With appropriate modeling of the data, SAXS can give detailed structural information unavailable with other techniques.

Electron microscopy of thin sections of a polymeric solid can also give direct morphological data on a polymer of interest, assuming that (1) the polymer possesses sufficient electron density contrast or can be appropriately stained without changing the morphology and (2) the structures of interest are sufficiently large.

Mechanical and Thermal Property Studies

In stress-strain or tensile testing, a dog-bone-shaped polymer sample is subjected to a constant elongation, or strain, rate, and the force required to maintain the constant elongation rate is monitored. As discussed earlier, tensile testing gives information about modulus, yield point, and ultimate strength of the sample of interest.

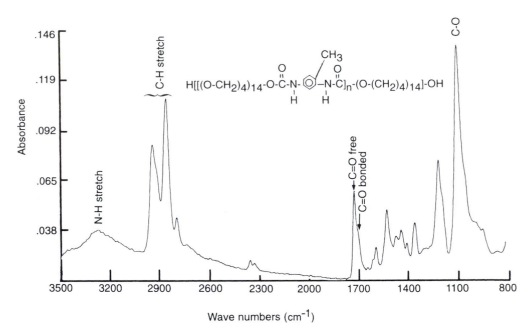

FIG. 13. Infrared spectrum of a poly(tetramethylene oxide)/toluene diisocyanate-based polyurethane.

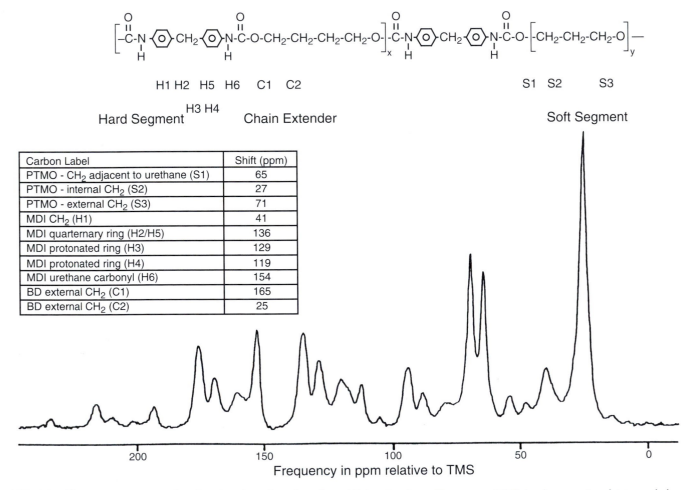

Carbon Label	Shift (ppm)
PTMO - CH$_2$ adjacent to urethane (S1)	65
PTMO - internal CH$_2$ (S2)	27
PTMO - external CH$_2$ (S3)	71
MDI CH$_2$ (H1)	41
MDI quarternary ring (H2/H5)	136
MDI protonated ring (H3)	129
MDI protonated ring (H4)	119
MDI urethane carbonyl (H6)	154
BD external CH$_2$ (C1)	165
BD external CH$_2$ (C2)	25

FIG. 14. ^{13}C NMR spectrum and peak assignation of a polyurethane [diphenylmethane diisocyanate (MDI, hard segment), polytetramethylene oxide (PTMO, soft segment), butanediol (BD, chain extender)]. Obtained by cross-polarization magic angle spinning of the solid polymer. (From Okamoto, D. T., Ph.D. thesis, University of Wisconsin, 1991. Reproduced with permission.)

Dynamic mechanical analysis (DMA) provides information about the small deformation behavior of polymers. Samples are subjected to cyclic deformation at a fixed frequency in the range of 1–1000 Hz. The stress response is measured while the cyclic strain is applied and the temperature is slowly increased (typically at 2–3 degrees/min). If the strain is a sinusoidal function of time given by:

$$\varepsilon(\omega) = \varepsilon_o \sin(\omega t) \qquad (7)$$

where ε is the time-dependent strain, ε_o is the strain amplitude, ω is the frequency of oscillation, and t is time, the resulting stress can be expressed by:

$$\sigma(\omega) = \sigma_o \sin(\omega t + \delta) \qquad (8)$$

where σ is the time-dependent stress, σ_o is the amplitude of stress response, and δ is the phase angle between stress and strain. For Hookean solids, the stress and strain are completely in phase ($\delta = 0$), while for purely viscous liquids, the stress response lags by 90°. Real materials demonstrate viscoelastic behavior where δ has a value between 0° and 90°.

A typical plot of tan δ versus temperature will display maxima at T_g and at lower temperatures where small-scale motions (secondary relaxations) can occur. Additional peaks above T_g, corresponding to motions in the crystalline phase and melting, are seen in semicrystalline materials. DMA is a sensitive tool for characterizing polymers of similar chemical composition or for detecting the presence of moderate quantities of additives.

Differential scanning calorimetry is another method for probing thermal transitions of polymers. A sample cell and a reference cell are supplied energy at varying rates so that the temperatures of the two cells remain equal. The temperature is increased, typically at a rate of 10–20 degrees/min over the range of interest, and the energy input required to maintain equality of temperature in the two cells is recorded. Plots of energy supplied versus average temperature allow determination of T_g, crystallization temperature (T_c), and T_m. T_g is taken as the temperature at which one half the change in heat capacity, ΔC_p, has occurred. The T_c and T_m are easily identified, as shown in Fig. 15. The areas under the peaks can be quantitatively related to enthalpic changes.

Surface Characterization

Surface characteristics of polymers for biomedical applications are critically important. The surface composition is inevitably different from the bulk, and the surface of the material is generally all that is contacted by the body. The main surface characterization techniques for polymers are X-ray photoelectron spectroscopy (XPS), contact angle measurements, attenuated total reflectance Fourier transform infrared (ATR-FTIR) spectroscopy, and scanning electron microscopy (SEM). The techniques are discussed in detail in Chapter 1.4.

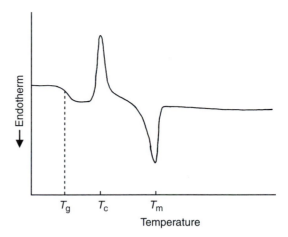

FIG. 15. Differential scanning calorimetry thermogram of a semicrystalline polymer, showing the glass transition temperature (T_g), the crystallization temperature (T_c), and the melting temperature (T_m) of the polymer sample.

FABRICATION AND PROCESSING

Before a polymer can be employed usefully in a medical device, the material must be manipulated physically, thermally, or mechanically into the desired shape. This can be achieved using the high-molecular-weight polymer at the start of the process and may require additives in the material to aid processing, or the end use. Such additives can include antioxidants, UV stabilizers, reinforcing fillers, lubricants, mold release agents, and plasticizers.

Alternatively, polymer products can be fabricated into end-use shapes starting from the monomers or low-molecular-weight prepolymers. In such processes, the final polymerization step is carried out once the precursors are in a casting or molding device, yielding a solid, shaped end product. A typical example is PMMA dental or bone cement, which is cured *in situ* in the body. Polymers can be fabricated into sheets, films, rods, tubes, and fibers, as coatings on another substrate, and into more complex geometries and foams.

It is important to realize that the presence of processing and functional aids can affect other properties of a polymer. For example, plasticisers are added to rigid PVC to produce a softer material, e.g., for use as dialysis tubing and blood storage bags. But additives such as plasticizers and mold release agents may alter the surface properties of the material, where the tissues come into contact with the polymer, and may also be extracted into body fluids.

Prior to use, materials must also be sterilized. Agents used to reduce the chances of clinical infection include, steam, dry heat, chemicals, and irradiation. Exposing polymers to heat or ionizing radiation may affect the properties of the polymer, by chain scission or creating cross-links. Chemical agents such as ethylene oxide may also be absorbed by a material and later could be released into the body. Therefore, devices sterilized with ethylene oxide require a period of time following sterilization for any residues to be released before use.

POLYMERIC BIOMATERIALS

PMMA is a hydrophobic, linear chain polymer that is transparent, amorphous, and glassy at room temperature and may be more easily recognized by such trade names as Lucite or Plexiglas. It is a major ingredient in bone cement for orthopedic implants. In addition to toughness and stability, it has excellent light transmittance, making it a good material for intraocular lenses (IOLs) and hard contact lenses. The monomers are polymerized in the shape of a rod from which buttons are cut. The button or disk is then mounted on a lathe, and the posterior and anterior surfaces machined to produce a lens with defined optical power. Lenses can also be fabricated by melt processing, compression molding, or casting, but lathe machining methods are most commonly used.

Soft contact lenses are made from the same methacrylate family of polymers. The substitution of the methyl ester group in methylmethacrylate with a hydroxyethyl group (2-hydroxyethyl methacrylate or HEMA) produces a very hydrophilic polymer. For soft contact lenses, the poly(HEMA) is slightly cross-linked with ethylene glycol dimethyacrylate (EGDMA) to retain dimensional stability for its use as a lens. Fully hydrated, it is a swollen hydrogel. PHEMA is glassy when dried, and therefore, soft lenses are manufactured in the same way as hard lenses; however, for the soft lens a swelling factor must be included when defining the optical specifications. This class of hydrogel polymers is discussed in more detail in Chapter 2.5.

Polyacrylamide is another hydrogel polymer that is used in biomedical separations (e.g., polyacrylamide gel electrophoresis, or PAGE). The mechanical properties and the degree of swelling can be controlled by cross-linking with methylene-bis-acrylamide (MBA). Poly(N-alkylacrylamides) are environmentally sensitive, and the degree of swelling can be altered by changes in temperature and acidity. These polymers are discussed in more detail in Chapters 2.6 and 7.14; see also Hoffman (1997).

Polyacrylic acids also have applications in medicine. They are used as dental cements, e.g., as glass ionomers. In this use, they are usually mixed with inorganic salts, where the cation interacts with the carboxyl groups of the acid to form physical cross-links. Polyacrylic acid is also used in a covalently cross-linked form as a mucoadhesive additive to mucosal drug delivery formulations (See Chapter 7.14). Polymethacrylic acid may also be incorporated in small quantities into contact lens polymer formulations to improve wettability.

PE is used in its high-density form in biomedical applications because low-density material cannot withstand sterilization temperatures. It is used as tubing for drains and catheters, and in ultrahigh-MW form as the acetabular component in artificial hips and other prosthetic joints. The material has good toughness and wear resistance and is also resistant to lipid absorption. Radiation sterilization in an inert atmosphere may also provide some covalent cross-linking that strengthens the PE.

PP is an isotactic crystalline polymer with high rigidity, good chemical resistance, and good tensile strength. Its stress cracking resistance is excellent, and it is used for sutures and hernia repair.

PTFE, also known as PTFE Teflon, has the same structure as PE, except that the four hydrogens in the repeat unit of PE are replaced by fluorines. PTFE is a very high melting polymer ($T_m = 327°C$) and as a result it is very difficult to process. It is very hydrophobic, has excellent lubricity, and is used to make catheters. In microporous form, known generically as e-PTFE or most commonly as the commercial product Gore-Tex, it is used in vascular grafts. Because of its low friction, it was the original choice by Dr. John Charnley for the acetabular component of the first hip joint prosthesis, but it failed because of its low wear resistance and the resultant inflammation caused by the PTFE wear particles.

PVC is used mainly as tubing and blood storage bags in biomedical applications. Typical tubing uses include blood transfusion, feeding, and dialysis. Pure PVC is a hard, brittle material, but with the addition of plasticizers, it can be made flexible and soft. PVC can pose problems for long-term applications because the plasticizers can be extracted by the body. While these plasticizers have low toxicities, their loss also makes the PVC less flexible.

Poly(dimethyl siloxane) (PDMS) or SR is an extremely versatile polymer, although its use is often limited by its relatively poor mechanical strength. It is unique in that it has a silicon–oxygen backbone instead of a carbon backbone. Its properties are less temperature sensitive than other rubbers because of its very low T_g. In order to improve mechanical properties, SR is usually formulated with reinforcing silica filler, and sometimes the polysiloxane backbone is also modified with aromatic rings that can toughen it. Because of its excellent flexibility and stability, SR is used in a variety of prostheses such as finger joints, heart valves, and breast implants, and in ear, chin, and nose reconstruction. It is also used as catheter and drainage tubing and in insulation for pacemaker leads. It has also been used in membrane oxygenators because of its high oxygen permeability, although porous polypropylene or polysulfone polymers have recently become more used as oxygenator membranes. Silicones are so important in medicine that details on their chemistry are provided in Chapter 2.3 and their medical applications are discussed in Chapter 7.19.

PET is one of the highest volume polymeric biomaterials. It is a polyester, containing rigid aromatic rings in a "regular" polymer backbone, which produces a high-melting ($T_m = 267°C$) crystalline polymer with very high tensile strength. It may be fabricated in the forms of knit, velour, or woven fabrics and fabric tubes, and also as nonwoven felts. Dacron is a common commercial form of PET used in large-diameter knit, velour, or woven arterial grafts. Other uses of PET fabrics are for the fixation of implants and hernia repair. PET can also be used in ligament reconstruction and as a reinforcing fabric for tissue reconstruction with soft polymers such as SR. It is used in a nonwoven felt coating on the peritoneal dialysis shunt (where it enters the body and traverses the skin) to enhance ingrowth and thereby reduce the possibility of infection.

PEG is used in drug delivery as conjugates with low solubility drugs and with immunogenic or fairly unstable protein drugs, to enhance the circulation times and stabilities of the drugs. It is also used as PEG–phospholipid conjugates to enhance the stability and circulation time of drug-containing liposomes. In both cases it serves to "hide" the circulating drug system from immune recognition, especially in the liver (See Chapter 7.14). PEG has also been immobilized on polymeric biomaterial surfaces to make them "nonfouling." PEGs usually exist in a highly hydrated state on the polymer surfaces, where they can exhibit steric repulsion based on an osmotic or entropic mechanism. This phenomenon contributes to the protein- and cell-resistant properties of surfaces containing PEGs (See Chapter 2.13).

Regenerated cellulose, for many years, was the most widely used dialysis membrane. Derivatives of cellulose, such as cellulose acetate (CA), are also used, since CA can be melt processed as hollow fibers for the hollow fiber kidney dialyser. CA is also used in osmotic drug delivery devices (See Chapter 7.14).

Polymerization of bisphenol A and phosgene produces polycarbonate, a clear, tough material. Its high impact strength dictates its use as lenses for eyeglasses and safety glasses, and housings for oxygenators and heart–lung bypass machine. Polycarbonate macrodiols have also been used to prepare copolymers such as polyurethanes. Polycarbonate segments may confer enhanced biological stability to a material.

Nylon is the name originally given by Du Pont to a family of polyamides; the name is now generic, and many other companies make nylons. Nylons are formed by the reaction of diamines with dibasic acids or by the ring opening polymerization of lactams. Nylons are used as surgical sutures (see also Chapter 2.4).

Biodegradable Polymers

PLGA is a random copolymer used in resorbable surgical sutures, drug delivery systems, and orthopedic appliances such as fixation devices. The degradation products are endogenous compounds (lactic and glycolic acids) and as such are nontoxic. PLGA polymerization occurs via a ring-opening reaction of a glycolide and a lactide, as illustrated in Fig. 6. The presence of ester linkages in the polymer backbone allows gradual hydrolytic degradation (resorption). The rate of degradation can be controlled by the ratio of polyglycolic acid to polylactic acid (See Chapter 7.14).

Copolymers

Copolymers are another important class of biomedical materials. A copolymer of tetrafluoroethylene with a small amount of hexafluoropropylene (FEP Teflon) is used as a tubing connector and catheter. FEP has a crystalline melting point near 265°C compared with 327°C for PTFE. This enhances the processability of FEP compared with PTFE while maintaining the excellent chemical inertness and low friction characteristic of PTFE.

Polyurethanes are block copolymers containing "hard" and "soft" blocks. The "hard" blocks, having T_g values above room temperature and acting as glassy or semicrystalline reinforcing blocks, are composed of a diisocyanate and a chain extender. The diisocyanates most commonly used are 2,4-toluene diisocyanate (TDI) and methylene di(4-phenyl isocyanate) (MDI), with MDI being used in most biomaterials. The chain extenders are usually shorter aliphatic glycol or diamine materials with two to six carbon atoms. The "soft" blocks in polyurethanes are typically polyether or polyester polyols whose T_g values are much lower than room temperature, allowing them to give a rubbery character to the materials. Polyether polyols are more commonly used for implantable devices because they are stable to hydrolysis. The polyol molecular weights tend to be on the order of 1000 to 2000.

Polyurethanes are tough elastomers with good fatigue and blood-containing properties. They are used in pacemaker lead insulation, catheters, vascular grafts, heart assist balloon pumps, artificial heart bladders, and wound dressings.

FINAL REMARKS

The chemistry, physics, and mechanics of polymeric materials are highly relevant to the performance of many devices employed in the clinic today. Polymers represent a broad, diverse family of materials, with mechanical properties that make them useful in applications relating to both soft and hard tissues. The presence of functional groups on the backbone or side chains of a polymer also means that they are readily modified chemically or biochemically, especially at their surfaces. Many researchers have successfully altered the chemical and biological properties of polymers, by immobilizing anticoagulants such as heparin, proteins such as albumin for passivation and fibronectin for cell adhesion, and cell-receptor peptide ligands to enhance cell adhesion, greatly expanding their range of applications (See Chapter 2.16).

Bibliography

Billmeyer, F. W., Jr. (1984). *Textbook of Polymer Science*, 3rd ed. Wiley-Interscience, New York.

Black, J., and Hastings, G. (1998). *Handbook of Biomaterial Properties*. Chapman and Hall, London.

Brydson, J. A. (1995). *Plastics Materials*, 3rd ed. Butterworth Scientific, London.

Flory, P. J. (1953). *Principles of Polymer Chemistry*. Cornell Univ. Press, Ithaca, NY.

Hoffman, A. S. (1997). Intelligent Polymers. in *Controlled Drug Delivery*, K. Park, ed. ACS Publications, ACS, Washington, D.C.

Lamba, N. M. K., Woodhouse, K. A. and Cooper, S. L. (1998). *Polyurethanes in Biomedical Applications*. CRC Press, Boca Raton, FL.

Mandelkern, L. (1983). *An Introduction to Macromolecules*. Springer-Verlag, New York.

Rodriguez, F. (1996). *Principles of Polymer Systems*, 4th ed. Hemisphere Publishing, New York.

Sperling, L. H. (1992). *Introduction to Physical Polymer Science*, 2nd ed. Wiley-Interscience, New York.

Stokes, K., McVenes, R., and Anderson, J. M. (1995). Polyurethane elastomer biostability. *J. Biomater. Appl.* 9: 321-354.

Szycher, M. (ed.) *High Performance Biomaterials*. Technomic, Lancaster, PA.

2.3 SILICONE BIOMATERIALS: HISTORY AND CHEMISTRY

André Colas and Jim Curtis

CHEMICAL STRUCTURE AND NOMENCLATURE

Silicones are a general category of synthetic polymers whose backbone is made of repeating silicon to oxygen bonds. In addition to their links to oxygen to form the polymeric chain, the silicon atoms are also bonded to organic groups, typically methyl groups. This is the basis for the name "silicones," which was assigned by Kipping based on their similarity with ketones, because in most cases, there is on average one silicone atom for one oxygen and two methyl groups (Kipping, 1904). Later, as these materials and their applications flourished, more specific nomenclature was developed. The basic repeating unit became known as "siloxane" and the most common silicone is polydimethylsiloxane, abbreviated as PDMS.

$$-\left(\begin{array}{c}R \\ | \\ Si-O- \\ | \\ R\end{array}\right) \quad \text{and if R is } CH_3, \quad -\left(\begin{array}{c}CH_3 \\ | \\ Si-O- \\ | \\ CH_3\end{array}\right)_n$$

"siloxane"   "polydimethylsiloxane"

Many other groups, e.g., phenyl, vinyl and trifluoropropyl, can be substituted for the methyl groups along the chain. The simultaneous presence of "organic" groups attached to an "inorganic" backbone gives silicones a combination of unique properties, making possible their use as fluids, emulsions, compounds, resins, and elastomers in numerous applications and diverse fields. For example, silicones are common in the aerospace industry, due principally to their low and high temperature performance. In the electronics field, silicones are used as electrical insulation, potting compounds and other applications specific to semiconductor manufacture. Their long-term durability has made silicone sealants, adhesives and waterproof coatings commonplace in the construction industry. Their excellent biocompatibility makes many silicones well suited for use in numerous personal care, pharmaceutical, and medical device applications.

Historical Milestones in Silicone Chemistry

Key milestones in the development of silicone chemistry—thoroughly described elsewhere by Lane and Burns (1996), Rochow (1987), and Noll (1968)—are summarized in Table 1.

Nomenclature

The most common silicones are the polydimethylsiloxanes trimethylsilyloxy terminated, with the following structure:

$$CH_3 - \underset{\underset{CH_3}{|}}{\overset{\overset{CH_3}{|}}{Si}} - O - \left(\underset{\underset{CH_3}{|}}{\overset{\overset{CH_3}{|}}{Si}} - O\right)_n - \underset{\underset{CH_3}{|}}{\overset{\overset{CH_3}{|}}{Si}} - CH_3,$$

$$(n = 0, 1, \dots)$$

These are linear polymers and liquids, even for large values of n. The main chain unit, $-(Si(CH_3)_2 O)_n-$, is often represented by the letter D because, as the silicon atom is connected with two oxygen atoms, this unit is capable of expanding within the polymer in two directions. M, T and Q units are defined in a similar manner, as shown in Table 2.

The system is sometimes modified by the use of superscript letters designating nonmethyl substituents, for example, $D^H = H(CH_3)SiO_{2/2}$ and M^ϕ or $M^{Ph} = (CH_3)_2(C_6H_5)SiO_{1/2}$ (Smith, 1991). Further examples are shown in Table 3.

Preparation

Silicone Polymers

The modern synthesis of silicone polymers is multifaceted. It usually involves the four basic steps described in Table 4. Only step 4 in this table will be elaborated upon here.

Polymerization and Polycondensation. The linear [4] and cyclic [5] oligomers resulting from dimethyldichlorosilane hydrolysis have chain lengths too short for most applications. The cyclics must be polymerized, and the linears condensed, to give macromolecules of sufficient length (Noll, 1968).

Catalyzed by acids or bases, cyclosiloxanes $(R_2SiO)_m$ are ring-opened and polymerized to form long linear chains.

TABLE 1 Key Milestones in the Development of Silicone Chemistry

1824	Berzelius discovers silicon by the reduction of potassium fluorosilicate with potassium: $4K + K_2SiF_6 \rightarrow Si + 6KF$. Reacting silicon with chlorine gives a volatile compound later identified as tetrachlorosilane, $SiCl_4$: $Si + 2Cl_2 \rightarrow SiCl_4$.
1863	Friedel and Craft synthesize the first silicon organic compound, tetraethylsilane: $2Zn(C_2H_5)_2 + SiCl_4 \rightarrow Si(C_2H_5)_4 + 2ZnCl_2$.
1871	Ladenburg observes that diethyldiethoxysilane, $(C_2H_5)_2Si(OC_2H_5)_2$, in the presence of a diluted acid gives an oil that decomposes only at a "very high temperature."
1901–1930s	Kipping lays the foundation of organosilicon chemistry with the preparation of various silanes by means of Grignard reactions and the hydrolysis of chlorosilanes to yield "large molecules." The polymeric nature of the silicones is confirmed by the work of Stock.
1940s	In the 1940s, silicones become commercial materials after Hyde of Dow Corning demonstrates the thermal stability and high electrical resistance of silicone resins, and Rochow of General Electric finds a direct method to prepare silicones from silicon and methylchloride.

TABLE 2 Shorthand Notation for Siloxane Polymer Units

CH_3 $\vert$ $CH_3-Si-O-$ $\vert$ CH_3	CH_3 $\vert$ $-O-Si-O-$ $\vert$ CH_3	O $\vert$ $-O-Si-O-$ $\vert$ CH_3	O $\vert$ $-O-Si-O-$ $\vert$ $O-$
M	D	T	Q
$(CH_3)_3 SiO_{1/2}$	$(CH_3)_2 SiO_{2/2}$	$CH_3 SiO_{3/2}$	$SiO_{4/2}$

TABLE 3 Examples of Silicone Shorthand Notation

At equilibrium, the reaction results in a mixture of cyclic oligomers plus a distribution of linear polymers. The proportion of cyclics will depend on the substituents along the Si—O chain, the temperature, and the presence of a solvent. Polymer chain length will depend on the presence and concentration of substances capable of giving chain ends. For example, in the KOH-catalyzed polymerization of the cyclic tetramer octamethylcyclotetrasiloxane $(Me_2SiO)_4$ (or D_4 in shorthand notation), the average length of the polymer chains will depend on the KOH concentration:

$$x(Me_2SiO)_4 + KOH \rightarrow (Me_2SiO)_y + KO(Me_2SiO)_zH$$

A stable hydroxy-terminated polymer, $HO(Me_2SiO)_zH$, can be isolated after neutralization and removal of the remaining cyclics by stripping the mixture under vacuum at elevated temperature. A distribution of chains with different lengths is obtained.

The reaction can also be made in the presence of $Me_3SiOSiMe_3$, which will act as a chain end blocker:

$$\cdot\cdot\ddot{\,}\cdot\cdot\ddot{\,}\cdot\cdot Me_2SiOK + Me_2SiOSiMe_3$$
$$\rightarrow \cdot\cdot\ddot{\,}\cdot\cdot\ddot{\,}\cdot\cdot Me_2SiOSiMe_3 + Me_3SiOK$$

where $\cdot\cdot\ddot{\,}\cdot\cdot\ddot{\,}\cdot\cdot$ represents the main chain.

The Me_3SiOK formed will attack another chain to reduce the average molecular weight of the linear polymer formed.

The copolymerization of $(Me_2SiO)_4$ in the presence of $Me_3SiOSiMe_3$ with Me_4NOH as catalyst displays a surprising viscosity change over time (Noll, 1968). First a peak or viscosity maximum is observed at the beginning of the reaction. The presence of two oxygen atoms on each silicon in the cyclics makes them more susceptible to a nucleophilic attack by the base catalyst than the silicon of the endblocker, which is substituted by only one oxygen atom. The cyclics are polymerized first into very long, viscous chains that are subsequently reduced in length by the addition of terminal groups provided by the endblocker, which is slower to react. This reaction can be described as follows:

$$Me_3SiOSiMe_3 + x(Me_2SiO)_4 \xrightarrow{cat} Me_3SiO(Me_2SiO)_nSiMe_3$$

or, in shorthand notation:

$$MM + x D_4 \xrightarrow{cat} MD_nM$$

where $n = 4x$ (theoretically).

The ratio between D and M units will define the average molecular weight of the polymer formed.

Catalyst removal (or neutralization) is always an important step in silicone preparation. Most catalysts used to prepare silicones can also catalyze the depolymerization (attack along the chain), particularly at elevated temperatures in the presence of traces of water.

$$\cdot\cdot\ddot{\,}\cdot\cdot\ddot{\,}\cdot\cdot(Me_2SiO)_n\cdot\cdot\ddot{\,}\cdot\cdot\ddot{\,}\cdot\cdot + H_2O$$
$$\xrightarrow{cat} \cdot\cdot\ddot{\,}\cdot\cdot\ddot{\,}\cdot\cdot(Me_2SiO)_y H + HO(Me_2SiO)_z\cdot\cdot\ddot{\,}\cdot\cdot\ddot{\,}\cdot\cdot$$

It is therefore essential to remove all remaining traces of the catalyst, providing the silicone optimal thermal stability. Labile catalysts have been developed. These decompose or are volatilized above the optimum polymerization temperature and consequently can be eliminated by a brief overheating.

TABLE 4   The Basic Steps in Silicone Polymer Synthesis

1. Silica reduction to silicon

$$SiO_2 + 2C \rightarrow Si + 2CO$$

2. Chlorosilanes synthesis

$$Si + 2CH_3Cl \rightarrow \underset{[1]}{(CH_3)_2SiCl_2} + \underset{[2]}{CH_3SiCl_3} + \underset{[3]}{(CH_3)_3SiCl} + CH_3HSiCl_2 + \cdots$$

3. Chlorosilanes hydrolysis

$$\underset{[1]}{Cl - \underset{\underset{CH_3}{|}}{\overset{\overset{CH_3}{|}}{Si}} - Cl} + 2H_2O \rightarrow \underset{linears \ [4]}{HO-\left(-\underset{\underset{CH_3}{|}}{\overset{\overset{CH_3}{|}}{Si}}-O-\right)_x - H} + \underset{cyclics \ [5]}{\left(\underset{\underset{CH_3}{|}}{\overset{\overset{CH_3}{|}}{Si}}-O\right)_{3,4,5}} + HCl$$

4. Polymerization and polycondensation

$$\underset{cyclics \ [5]}{\left(\underset{\underset{CH_3}{|}}{\overset{\overset{CH_3}{|}}{Si}}-O\right)_{3,4,5}} \rightarrow \underset{polymer}{-\left(-\underset{\underset{CH_3}{|}}{\overset{\overset{CH_3}{|}}{Si}}-O-\right)_y-}$$

$$\underset{linears \ [4]}{HO-\left(-\underset{\underset{CH_3}{|}}{\overset{\overset{CH_3}{|}}{Si}}-O-\right)_x-H} \rightarrow \underset{polymer}{-\left(-\underset{\underset{CH_3}{|}}{\overset{\overset{CH_3}{|}}{Si}}-O-\right)_z-} + zH_2O$$

In this way, catalyst neutralization or filtration can be avoided (Noll, 1968).

The cyclic trimer $(Me_2SiO)_3$ has an internal ring tension and can be polymerized without reequilibration of the resulting polymers. With this cyclic, polymers with narrow molecular-weight distribution can be prepared, but also polymers only carrying one terminal reactive function (living polymerization). Starting from a mixture of different "tense" cyclics also allows the preparation of block or sequential polymers (Noll, 1968).

Linears can combine when catalyzed by many acids or bases to give long chains by intermolecular condensation of silanol terminals (Noll, 1968; Stark *et al.*, 1982).

$$\cdots O-\underset{\underset{Me}{|}}{\overset{\overset{Me}{|}}{Si}}-OH + HO-\underset{\underset{Me}{|}}{\overset{\overset{Me}{|}}{Si}}-O\cdots$$

$$\underset{+H_2O}{\overset{-H_2O}{\rightleftharpoons}} \quad \cdots O-\underset{\underset{Me}{|}}{\overset{\overset{Me}{|}}{Si}}-O-\underset{\underset{Me}{|}}{\overset{\overset{Me}{|}}{Si}}-O\cdots$$

A distribution of chain lengths is obtained. Longer chains are favored when working under vacuum and/or at elevated temperatures to reduce the residual water concentration. In addition to the polymers described above, reactive polymers can also be prepared. This can be achieved when reequilibrating oligomers or existing polymers to obtain a polydimethyl-methylhydrogenosiloxane, $MD_zD_w^HM$.

$$Me_3SiOSiMe_3 + x(Me_2SiO)_4 + Me_3SiO(MeHSiO)_ySiMe_3$$
$$\overset{cat}{\rightarrow} cyclics + Me_3SiO(Me_2SiO)_z(MeHSiO)_wSiMe_3$$

$$[6]$$

Additional functional groups can be attached to this polymer using an addition reaction.

$$Me_3SiO(Me_2SiO)_z(MeHSiO)_wSiMe_3 + H_2C = CHR$$

$$[6]$$

$$\overset{Pt \ cat}{\longrightarrow} Me_3SiO(Me_2SiO)_z \underset{\underset{CH_2CH_2R}{|}}{(Me \ Si \ O)_w}SiMe_3$$

The polymers shown are all linear or cyclic, comprising difunctional units, D. In addition to these, branched polymers or resins can be prepared if, during hydrolysis, a certain amount of T or Q units are included, which will allow molecular expansion, in three or four directions, as opposed to just two. For example, consider the hydrolysis of methyltrichlorosilane in the presence of trimethylchlorosilane, which leads to a branched polymer as shown next:

$$x \ Me - \underset{\underset{Me}{|}}{\overset{\overset{Me}{|}}{Si}} - Cl + y \ Me - \underset{\underset{Cl}{|}}{\overset{\overset{Cl}{|}}{Si}} - Cl \underset{-HCl}{\overset{+H_2O}{\longrightarrow}} z$$
$$\qquad\qquad [3] \qquad\qquad\qquad [2]$$

$$Me - \underset{\underset{Me}{|}}{\overset{\overset{Me}{|}}{Si}} - O - \underset{\underset{O}{|}}{\overset{\overset{Me}{|}}{Si}} - O - \underset{\underset{OH}{|}}{\overset{\overset{Me}{|}}{Si}} - O \cdots$$

$$Me - \underset{\underset{O}{|}}{\overset{\overset{}{}}{Si}} - O \cdots$$

$$Me - \underset{\underset{Me}{|}}{\overset{\overset{}{}}{Si}} - Me$$

The resulting polymer can be described as $(Me_3SiO_{1/2})_x$ $(MeSiO_{3/2})_y$ or M_xT_y, using shorthand notation. The formation of three silanols on the $MeSiCl_3$ by hydrolysis yields a three-dimensional structure or resin, rather than a linear polymer. The average molecular weight depends upon the number of M units that come from the trimethylchlorosilane, which limits the growth of the resin molecule. Most of these resins are prepared in a solvent and usually contain some residual hydroxyl groups. These could subsequently be used to cross-link the resin and form a continuous network.

Silicone Elastomers

Silicone polymers can be easily transformed into a three-dimensional network by way of a cross-linking reaction, which allows the formation of chemical bonds between adjacent chains. The majority of silicone elastomers are cross-linked according to one of the following three reactions.

1. Cross-Linking with Radicals Efficient cross-linking with radicals is only achieved when some vinyl groups are present on the polymer chains. The following mechanism has been proposed for the cross-linking reaction associated with radicals generated from an organic peroxide (Stark, 1982):

$$R^{\cdot} + CH_2 = CH - Si \equiv \rightarrow R - CH_2 - CH^{\cdot} - Si \equiv$$

$$RCH_2 - CH^{\cdot} - Si \equiv + CH_3 - Si \equiv$$
$$\rightarrow RCH_2 - CH_2 - Si \equiv + \equiv Si - CH_2^{\cdot}$$

$$\equiv Si - CH_2^{\cdot} + CH_2 = CH - Si \equiv$$
$$\rightarrow \equiv Si - CH_2 - CH_2 - CH^{\cdot} - Si \equiv$$

$$\equiv Si - CH_2 - CH_2 - CH^{\cdot} - Si \equiv + \equiv Si - CH_3$$
$$\rightarrow \equiv Si - CH_2 - CH_2 - CH_2 - Si \equiv + \equiv Si - CH_2^{\cdot}$$

$$2 \equiv Si - CH_2^{\cdot} \rightarrow \equiv Si - CH_2 - CH_2 - Si \equiv$$

where $\equiv$ represents two methyl groups and the rest of the polymer chain.

This reaction has been used for high-consistency silicone rubbers (HCRs) such as those used in extrusion or injection molding, as well as those that are cross-linked at elevated temperatures. The peroxide is added before processing. During cure, some precautions are needed to avoid the formation of voids by the peroxide's volatile residues. Postcure may also be necessary to remove these volatiles, which can catalyze depolymerization at high temperatures.

2. Cross-Linking by Condensation Although mostly used in silicone caulks and sealants for the construction industry and do-it-yourselfer, this method has also found utility for medical devices as silicone adhesives facilitating the adherence of materials to silicone elastomers, as an encapsulant and as sealants such as around the connection of a pacemaker lead to the pulse generator (Fig. 1 shows Silastic Medical Adhesive, type A).

These products are ready to apply and require no mixing. Cross-linking starts when the product is squeezed from the cartridge or tube and comes into contact with moisture,

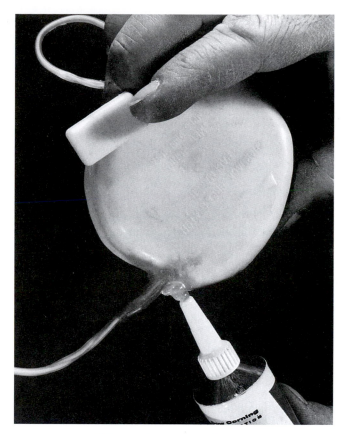

FIG. 1. RTV silicone adhesive.

typically from humidity in the ambient air. These materials are formulated from a reactive polymer prepared from a hydroxy end-blocked polydimethylsiloxane and a large excess of methyltriacetoxysilane.

$$HO - (Me_2SiO)_x - H + excess \; MeSi(OAc)_3$$

$$\xrightarrow[-2AcOH]{} (AcO)_2MeSiO(Me_2SiO)_xOSiMe(OAc)_2$$
$$[7]$$

where $Ac = -\overset{\displaystyle CH_3}{\underset{\displaystyle |}{C}} = O$

Because a large excess of silane is used, the probability of two different chains reacting with the same silane molecule is remote. Consequently, all the chains are end-blocked with two acetoxy functional groups. The resulting product is still liquid and can be packaged in sealed tubes and cartridges. Upon opening the acetoxy groups are hydrolyzed by the ambient moisture to give silanols, which allow further condensation to occur.

$$\cdots \overset{\displaystyle Me}{\underset{\displaystyle |}{O - Si - OAc}} \xrightarrow[-AcOH]{+H_2O} \cdots \overset{\displaystyle Me}{\underset{\displaystyle |}{O - Si - OH}}$$
$$\qquad\quad OAc \qquad\qquad\qquad\qquad OAc$$
$$\quad\;\; [7] \qquad\qquad\qquad\qquad\qquad [8]$$

$$
\cdots\cdots\cdots\cdots O-\underset{\underset{OAc}{|}}{\overset{\overset{Me}{|}}{Si}}-OH + AcO-\underset{\underset{OAc}{|}}{\overset{\overset{Me}{|}}{Si}}-O\cdots\cdots\cdots\cdots
$$

$$[8]\qquad\qquad[7]$$

$$
\xrightarrow[-AcOH]{}\;\cdots\cdots\cdots\cdots O-\underset{\underset{OAc}{|}}{\overset{\overset{Me}{|}}{Si}}-O-\underset{\underset{OAc}{|}}{\overset{\overset{Me}{|}}{Si}}-O\cdots\cdots\cdots\cdots
$$

In this way, two chains have been linked, and the reaction will proceed further from the remaining acetoxy groups. An organometallic tin catalyst is normally used. The cross-linking reaction requires moisture to diffuse into the material. Accordingly cure will proceed from the outside surface inward. These materials are called one-part RTV (room temperature vulcanization) sealants, but actually require moisture as a second component. Acetic acid is released as a by-product of the reaction. Problems resulting from the acid can be overcome using other cure (cross-linking) systems that have been developed by replacing the acetoxysilane $RSi(OAc)_3$ with oximosilane $RSi(ON=CR'_2)_3$ or alkoxysilane $RSi(OR')_3$.

Condensation curing is also used in some two-part systems where cross-linking starts upon mixing the two components, e.g., a hydroxy end-blocked polymer and an alkoxysilane such as tetra-n-propoxysilane (Noll, 1968):

$$
4\cdots\cdots\cdots\underset{\underset{Me}{|}}{\overset{\overset{Me}{|}}{Si}}-OH + nPrO-\underset{\underset{O}{|}}{\overset{\overset{nPr}{|}}{\underset{O}{Si}}}-OnPr
$$

$$
\xrightarrow[-4nPrOH]{cat}
$$

$$
\cdots\cdots\cdots Si-O-Si-O-Si\cdots\cdots\cdots
$$

Here, no atmospheric moisture is needed. Usually an organotin salt is used as catalyst, but it also limits the stability of the resulting elastomer at high temperatures. Alcohol is released as a by-product of the reaction, leading to a slight shrinkage upon cure (0.5 to 1% linear shrinkage). Silicones with this cure system are therefore not suitable for the fabrication of parts with precise tolerances.

3. Cross-linking by Addition Use of an addition-cure reaction for cross-linking can eliminate the shrinkage problem mentioned above. In addition cure, cross-linking is achieved by reacting vinyl endblocked polymers with Si–H groups carried by a functional oligomer such as described above [6]. A few polymers can be bonded to this functional oligomer [6], as follows (Stark, 1982):

$$
\cdots\cdots\cdots O-\underset{\underset{Me}{|}}{\overset{\overset{Me}{|}}{Si}}-CH=CH_2 + H-Si\equiv
$$

$$[5]$$

$$
\xrightarrow{cat}\;\cdots\cdots\cdots O-\underset{\underset{Me}{|}}{\overset{\overset{Me}{|}}{Si}}-CH_2-CH_2-Si\equiv
$$

where $\equiv$ represents the remaining valences of the Si in [6].

The addition occurs mainly on the terminal carbon and is catalyzed by Pt or Rh metal complexes, preferably as organometallic compounds to enhance their compatibility. The following mechanism has been proposed (oxidative addition of the $\equiv$Si to the Pt, H transfer to the double bond, and reductive elimination of the product):

$$\equiv Si-Pt-H \qquad a+\equiv Si-CH=CH_2$$
$$\rightleftarrows\;\equiv Si-CH_2-CH_2-Pt-Si\equiv$$
$$\xrightarrow[-Pt]{}\;\equiv Si-CH_2-CH_2-Si\equiv$$

where, to simplify, other Pt ligands and other Si substituents are omitted.

There are no by-products with this reaction. Molded pieces made with silicone using this addition-cure mechanism are very accurate (no shrinkage). However, handling these two-part products (i.e., polymer and Pt catalyst in one component, SiH oligomer in the other) requires some precautions. The Pt in the complex is easily bonded to electron-donating substances such as amine or organosulfur compounds to form stable complexes with these "poisons," rendering the catalyst inactive and inhibiting the cure.

The preferred cure system can vary by application. For example, silicone-to-silicone medical adhesives use acetoxy cure (condensation cross-linking), and platinum cure (cross-linking by addition) is used for precise silicone parts with no by-products.

4. Elastomer Filler In addition to the silicone polymers described above, the majority of silicone elastomers incorporate "filler." Besides acting as a material extender, as the name implies, filler acts to reinforce the cross-linked matrix. The strength of silicone polymers without filler is generally unsatisfactory for most applications (Noll, 1968). Like most other noncrystallizing synthetic elastomers, the addition of reinforcing fillers reduces silicone's stickiness, increases its hardness and enhances its mechanical strength. Fillers might also be employed to affect other properties; for example, carbon black is added for electrical conductivity, titanium dioxide improves the dielectric constant, and barium sulfate increases radiopacity. These and other materials are used to pigment the otherwise colorless elastomer; however, care must be taken to select only pigments suitable for the processing temperatures and end-use application.

Generally, the most favorable reinforcement is obtained using fumed silica, such as Cab–O–Sil, Aerosil, or Wacker HDK. Fumed silica is produced by the hydrolysis of silicon tetrachloride vapor in a hydrogen flame:

$$SiCl_4 + 2H_2 + O_2 \xrightarrow{1800°C} SiO_2 + 4HCl$$

Unlike many naturally occurring forms of crystalline silica, fumed silica is amorphous. The very small spheroid silica particles (on the order of 10 nm diameter) fuse irreversibly while still semimolten, creating aggregates. When cool, these aggregates become physically entangled to form agglomerates. Silica produced in this way possesses remarkably high surface area, 100 to 400 m²/g as measured by the BET method developed by Brunauer, Emmett, and Teller (Brunauer *et al.*, 1938; Noll, 1968; Cabot Corporation, 1990).

The incorporation of silica filler into silicone polymers is called "compounding." This is accomplished prior to cross-linking, by mixing the silica into the silicone polymers on a two-roll mill, in a twin-screw extruder, or in a Z-blade mixer capable of processing materials with this rheology.

Reinforcement occurs with polymer adsorption encouraged by the silica's large surface area and when hydroxyl groups on the filler's surface lead to hydrogen bonds between the filler and the silicone polymer, thereby contributing to the production of silicone rubbers with high tensile strength and elongation capability (Lynch, 1978). The addition of filler increases the polymer's already high viscosity. Chemical treatment of the silica filler with silanes further enhances its incorporation in, and reinforcement of, the silicone elastomer, resulting in increased material strength and tear resistance (Lane and Burns, 1996) (Fig. 2).

Silicone elastomers for medical applications normally utilize only fillers of fumed silica, and occasionally appropriate pigments or barium sulfate. Because of their low glass transition temperature, these compounded and cured silicone materials are elastomeric at room and body temperatures without the use of any plasticizers—unlike other medical materials such as PVC, which might contain phthalate additives.

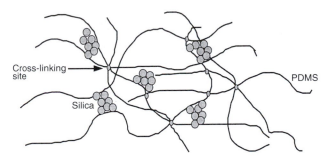

FIG. 2. Silicone elastomer matrix.

Physicochemical Properties

Silicon's position just under carbon in the periodic table led to a belief in the existence of analog compounds where silicon would replace carbon. Most of these analog compounds do not exist, or behave very differently. There are few similarities between Si–X bonds in silicones and C–X bonds (Corey, 1989; Hardman, 1989; Lane and Burns, 1996; Stark, 1982).

Between any given element and Si, bond lengths are longer than for C with this element. The lower electronegativity of silicon ($\chi^{Si} \approx 1.80$, $\chi^{C} \approx 2.55$) leads to more polar bonds compared to carbon. This bond polarity also contributes to strong silicon bonding; for example, the Si–O bond is highly ionic and has large bond energy. To some extent, these values explain the stability of silicones. The Si–O bond is highly resistant to homolytic scission. On the other hand, heterolytic scissions are easy, as demonstrated by the reequilibration reactions occurring during polymerizations catalyzed by acids or bases (see earlier discussion).

Silicones exhibit the unusual combination of an inorganic chain similar to silicates and often associated with high surface energy, but with side methyl groups that are very organic and often associated with low surface energy (Owen, 1981). The Si–O bonds are quite polar and without protection would lead to strong intermolecular interactions (Stark, 1982). Yet, the methyl groups, only weakly interacting with each other, shield the main chain (see Fig. 3).

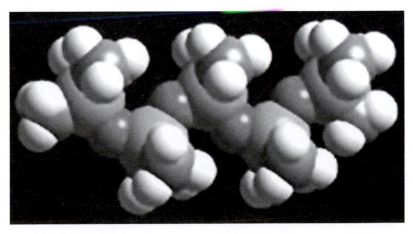

FIG. 3. Three-dimensional representation of dodecamethylpentasiloxane, Me_3SiO-$(SiMe_2O)_3SiMe_3$ or MD_3M. (Courtesy S. Grigoras, Dow Corning.)

This is made easier by the high flexibility of the siloxane chain. Barriers to rotation are low and the siloxane chain can adopt many configurations. Rotation energy around a H_2C-CH_2 bond in polyethylene is 13.8 kJ/mol but only 3.3 kJ/mol around a Me_2Si-O bond, corresponding to a nearly free rotation. In general, the siloxane chain adopts a configuration such that the chain exposes a maximum number of methyl groups to the outside, whereas in hydrocarbon polymers, the relative rigidity of the polymer backbone does not allow a "selective" exposure of the most organic or hydrophobic methyl groups. Chain-to-chain interactions are low, and the distance between adjacent chains is also greater in silicones. Despite a very polar chain, silicones can be compared to paraffin, with a low critical surface tension of wetting (Owen, 1981).

The surface activity of silicones is evident in many ways (Owen, 1981):

- The polydimethylsiloxanes have a low surface tension (20.4 mN/m) and are capable of wetting most surfaces. With the methyl groups pointing to the outside, this gives very hydrophobic films and a surface with good release properties, particularly if the film is cured after application. Silicone surface tension is also in the most promising range considered for biocompatible elastomers (20 to 30 mN/m).

- Silicones have a critical surface tension of wetting (24 mN/m) higher than their own surface tension. This means that silicones are capable of wetting themselves, which promotes good film formation and good surface covering.

- Silicone organic copolymers can be prepared with surfactant properties, with the silicone as the hydrophobic part, e.g., in silicone glycols copolymers.

The low intermolecular interactions in silicones have other consequences (Owen, 1981):

- Glass transition temperatures are very low, e.g., 146 K for a polydimethylsiloxane compared to 200 K for polyisobutylene, the analog hydrocarbon.

- The presence of a high free volume compared to hydrocarbons explains the high solubility and high diffusion coefficient of gas into silicones. Silicones have a high permeability to oxygen, nitrogen, or water vapor, even though liquid water is not capable of wetting a silicone surface. As expected, silicone compressibility is also high.

- The viscous movement activation energy is very low for silicones, and their viscosity is less dependent on temperature compared to hydrocarbon polymers. Furthermore, chain entanglements are involved at higher temperature and contribute to limit the viscosity reduction (Stark, 1982).

CONCLUSION

Polydimethylsiloxanes are often referred to as silicones. They are used in many applications because of their stability, low surface tension, and lack of toxicity. Methyl group substitution or introduction of tri- or tetra-functional siloxane units leads to a wide range of structures. Polymers are easily cross-linked at room or elevated temperature to elastomers, without loosing the above properties.

Acknowledgments

Part of this section (here revised) was originally published in *Chimie Nouvelle*, the journal of the Société Royale de Chimie (Belgium), Vol. 8 (30), 847 (1990) by A. Colas and are reproduced here with the permission of the editor. The authors thank S. Hoshaw and P. Klein, both from Dow Corning, for their contribution regarding breast implant epidemiology.

Bibliography

Brunauer, S., Emmett, P. H., and Teller, E. (1938). Adsorption of gases in multimolecular layers. *J. Am. Chem. Soc.* **60**: 309.

Cabot Corporation (1990). *CAB-O-SIL Fumed Silica Properties and Functions.* Tuscola, IL.

Corey, J. Y. (1989). Historical overview and comparison of silicone with carbon. in *The Chemistry of Organic Silicon Compounds*, Part 1, S. Patai and Z. Rappoport eds. John Wiley & Sons, New York.

Hardman, B. (1989). Silicones. *Encyclopedia of Polymer Science and Engineering.* John Wiley & Sons, New York, Vol. 15, p. 204.

Kipping, F. S. (1904). Organic derivative of silicon. Preparation of alkylsilicon chlorides. *Proc. Chem. Soc.* **20**: 15.

Lane, T. H., and Burns, S. A. (1996). Silica, silicon and silicones... unraveling the mystery. *Curr. Top. Microbiol. Immunol.* **210**: 3–12.

Lynch, W. (1978). *Handbook of Silicone Rubber Fabrication.* Van Nostrand Reinhold, New York.

Noll, W. (1968). *Chemistry and Technology of Silicones.* Academic Press, New York.

Owen, M. J. (1981). Why silicones behave funny. *Chemtech* **11**: 288.

Rochow, E. G. (1987). *Silicon and Silicones.* Springler-Verlag, New York.

Smith, A. L. (1991). Introduction to silicones. *The Analytical Chemistry of Silicones.* John Wiley & Sons, New York.

Stark, F. O., Falender, J. R., and Wright, A. P. (1982). Silicones. In Comprehensive Organometallic Chemistry, G. Wikinson, F. G. A. Sone, and E. W. Ebel, eds. Pergamon Press, Oxford, Vol. 2, pp. 288–297.

2.4 MEDICAL FIBERS AND BIOTEXTILES

Steven Weinberg and Martin W. King

The term "medical textiles" encompasses medical products and devices ranging from wound dressings and bandages to high-technology applications such as biotextiles, tissue engineered scaffolds, and vascular implants (King, 1991). The use of textiles in medicine goes back to the Egyptians and the Native Americans who used textiles as bandages to cover and draw wound edges together after injury (Shalaby, 1985). Over the past several decades, the use of fibers and textiles in medicine has grown dramatically as new and innovative fibers, structures, and therapies have been developed. Advances in fabrication techniques, fiber technology, and composition have led to numerous new concepts for both products and therapies, some of which are still in development or in clinical trials. In this chapter, an introduction to fiber and textile fabric technology will be presented along with discussion of

TABLE 1 Textile Structures and Applications (Ko, 1990)

Application	Material	Yarn structure	Fabric structure
Arteries	Dacron T56 Teflon	Textured Multifilament	Weft/warp knit Straight/ bifurcations Woven/non-woven
Tendons	Dacron T56 Dacron T55 Kevlar	Low-twist filament Multifilament	Coated woven tape
Hernia repair	Polypropylene	Monofilament	Tricot knit
Esophagus	Regenerated collagen	Monofilament	Plain weave Knit
Patches	Dacron T56	Monofilament Multifilament	Woven Knit/knit velour
Sutures	Polyester Nylon Regenerated collagen Silk	Monofilament Multifilament	Braid Woven tapes
Ligaments	Polyester Teflon Polyethylene	Monofilament Multifilament	Braid
Bones and joints	Carbon in thermoset or thermoplastic Matrix	Monofilament	Woven tapes Knits/braids

both old and new application areas. Traditional and nontraditional fiber and fabric constructions, processing issues, and fabric testing will be included in order to offer an overview of the technology associated with the use of textiles in medicine. Table 1 illustrates some of the more common application areas for textiles in medicine. As can be seen from this table, the products range from the simplest products (i.e., gauze bandages) to the most complex textile products such as vascular grafts and tissue scaffolds.

MEDICAL FIBERS

All textile-based medical devices are composed of structures fabricated from monofilament; multifilaments; or staple fibers formulated from synthetic polymers, natural polymers (biopolymers), or genetically engineered polymers. When choosing the appropriate fiber configuration and polymer for a specific application, consideration must be given to the device design requirements and the manner in which the fiber is to be used. For example, collagen-based implantable hemostatic wound dressings are available in multiple configurations including loose powder (Avitine), nonwoven mats (Helistat and Surgicel Fibrillar Hemostat), and knitted collagen fibrils (Surgicel Nu-Knit). In addition, other materials are also available for the same purpose (e.g., Surgicel Absorbable Hemostat is knitted from regenerated cellulose). Fibers can be fabricated from nonabsorbable synthetic polymers such as

poly(ethylene terephthalate) or polyester (e.g., Dacron) and polytetrafluoroethylene (e.g., Teflon), or absorbable synthetic materials such as polylactide (PLA) and polyglycolide (PGA) (Hoffman, 1977). Natural materials (biopolymers), such as collagen or polysaccharides like alginates, have also been used to fabricate medical devices (Keys, 1996). And there are recent reports that biomimetic polymers have been synthesized in experimental quantities by genetic engineering of peptide sequences from elastin, collagen, and spider dragline silk protein, and expressed in *Escherichia coli* and yeast using plasmid vectors (Huang, 2000; Teule, 2003).

Cotton was and still is commonly being used for bandages, surgical sponges, drapes, and surgical apparel, and in surgical gowns. In current practice, cotton has been replaced in many applications by coated nonwoven disposable fabrics, especially in cases when nonabsorbency is critical.

It is important to note that most synthetic polymers currently used in medicine were originally developed as commercial polymers for nonmedical applications and usually contain additives such as dyes, delustrants, stabilizers, antioxidants, and antistatic agents. Some of these chemicals may not be desirable for medical applications, and so must be removed prior to use. To illustrate this point, poly(ethylene terephthalate) (PET), formerly Dacron, which at present is the material of choice for most large-caliber textile vascular grafts, was originally developed for apparel use. A complex cleaning process is required before the material can be used in an implant application. Additional reading relating to this point can be found in Goswami *et al.* (1977) and Piller (1973).

Synthetic Fibers

Various synthetic fibers have been used to fabricate medical devices over the past 25 years. Starting in the 1950s, various materials were evaluated for use in vascular grafts, such as Vinyon (PVC copolymer), acrylic polymers, poly(vinyl alcohol), nylon, polytetrafluoroethylene, and polyester (PET) (King, 1983). Today, only PTFE and PET are still used for vascular graft applications since they are reasonably inert, flexible, resilient, durable, and resistant to biological degradation. They have withstood the test of time, whereas other materials have not proven to be durable when used in an implant application. Table 2 shows a partial list of synthetic polymers that have been prepared as fibers, their method of fabrication, and how they are used in the medical field.

Most synthetic fibers are formed either by a melt spinning or a wet spinning process.

Melt Spinning

With melt spinning the polymer resin is heated above its melting temperature and extruded through a spinneret. The number of holes in the spinneret defines the number of filaments in the fiber being produced. For example, a spinneret for a monofilament fiber contains one hole, whereas 54 holes are required to produce the 54-multifilament yarn that is commonly used in vascular graft construction. Once the monofilament or multifilament yarn is extruded, it is then drawn and cooled prior to being wound onto spools. The yarn can also be further processed to form the final configuration. For example,

TABLE 2 Synthetic Polymers (Shalaby, 1996)

Type	Chemical and physical aspects	Construction/useful forms	Comments/applications
Polyethylene (PE)	High-density PE (HDPE): melting temperature $T_m = 125°C$ Low-density PE (LDPE): $T_m = 110°C$, Linear low-density (LLDPE) Ultrahigh molecular weight PE (UHMWPE) ($T_m = 140–150°C$), exceptional tensile strength and modulus	Melt spun into continuous yarns for woven fabric and/or melt blown onto nonwoven fabric Converted to very high tenacity yarn by gel spinning	The HDPE, LDPE and LLDP are used in a broad range of health care products Used experimentally as reinforced fabrics in lightweight orthopedic casts, ligament prostheses, and load-bearing composites
Polypropylene (PP)	Predominantly isotactic, $T_m = 165–175°C$; higher fracture toughness than HDPE	Melt spun to monofilaments and melt blown to nonwoven fabrics Hollow fibers	Sutures, hernia repair meshes, surgical drapes, and gowns Plasma filtration
Poly(tetrafluoro-ethylene) (PTFE)	High melting ($T_m = 325°C$) and high crystallinity polymer (50–75% for processed material)	Melt extruded	Vascular fabrics, heart valve sewing rings, orthopedic ligaments
Nylon 6	$T_g = 45°C$, $T_m = 220°C$, thermoplastic, hydrophilic	Monofilaments, braids	Sutures
Nylon 66	$T_g = 50°C$, $T_m = 265°C$, thermoplastic, hydrophilic	Monofilaments, braids	Sutures
Poly(ethylene terephthalate) (PET)	Excellent fiber-forming properties, $T_m = 265°C$, $T_g = 65–105°C$	Multifilament yarn for weaving, knitting, and braiding	Sutures, hernia repair meshes, and vascular grafts

most yarns used for application in vascular grafts are texturized to improve the handling characterizes of the final product. In contrast to flat or untexturized yarn, texturization results in a yarn that imparts bulk to the fabric for improved "hand" or feel, flexibility, ease of handling and suturing, and more pores for tissue ingrowth. Melt spinning is typically used with thermoplastic polymers that are not affected by the elevated temperatures required in the melt spinning process. Figure 1 is a schematic representation of a melt spinning process.

In this process, the molten resin is extruded through the spinning head containing one (monofilament) or multiple

holes (multifilament). Air is typically used to cool and solidify the continuous threadline prior to lubricating, twisting, and winding up on a bobbin.

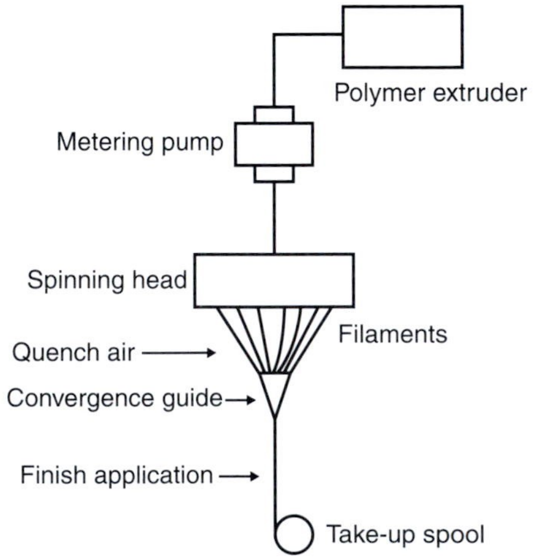

Wet Spinning

If the polymer system experiences thermal degradation at elevated temperatures, as is the case with a polymer containing a drug, a low-temperature wet solution spinning process can be used. In this process the polymer is dissolved in a solvent and then extruded through a spinneret into a nonsolvent in a spin bath. Because the solvent is soluble in the spin bath, but the polymer is not, the continuous polymer stream precipitates into a solid filament, which is then washed to remove all solvents and nonsolvents, drawn, and dried before winding up (Adanur, 1995). Figure 2 presents a schematic of a typical wet solution spinning process.

Electrospinning

The diameters of fibers spun by melt spinning and wet solution spinning are controlled by the size of the hole in the spinneret and the amount of draw or stretch applied to the filament prior to wind-up. So the diameters of conventional spun fibers fall in a range from about 10 μm for multifilament yarns to 500 μm or thicker for monofilaments. To obtain finer fiber diameters it is necessary to employ alternative spinning technologies such as the bicomponent fiber (BCF) approach (see later section entitled "Hybrid Bicomponent Fibers"), or an electrospinning technique. This method of manufacturing microfibers and nanofibers has been known since 1934 when the first patent was filed (Formhals, 1934). Since then

FIG. 1. Melt spinning process.

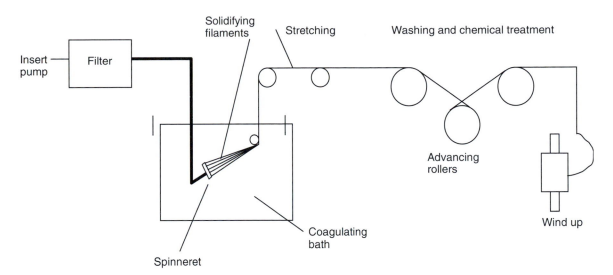

FIG. 2. Wet solution spinning process.

Freudenberg Inc. has used this process for the commercial production of ultrahigh-efficiency filters (Groitzsch, 1986).

Electrospinning occurs when a polymer solution or melt is exposed to an electrostatic field by the application of a high voltage (5–30 kV), which overcomes the surface tension of the polymer and accelerates fine jets of the liquid polymer towards a grounded target (Reneker *et al.*, 2000). As the polymer jets cool or lose solvent they are drawn in a series of unstable loops, solidified, and collected as an interconnected web of fine fibers on a grounded rotating drum or other specially shaped target (Fig. 3).

The fineness of the fibers produced depends on the polymer chemistry, its solution or melt viscosity, the strength and uniformity of the applied electric field, and the geometry and operating conditions of the spinning system. Fiber diameters in the range of 1 μm down to 100 nm or less have been reported.

In addition to being used to fabricate ultrathin filtration membranes, electrospinning techniques have also been applied to the production of nonwoven mats for wound dressings (Martin *et al.*, 1977), and there is currently much interest in making scaffolds for tissue engineering applications. Nonwoven scaffolds spun from Type I collagen and synthetic polymers such as poly(L-lactide), poly(lactide-*co*-glycolide), poly(vinyl alcohol), poly(ethylene-*co*-vinyl acetate), poly(ethylene oxide), polyurethanes, and polycarbonates have been reported (Stitzel *et al.*, 2001; Matthews *et al.*, 2002; Kenawy *et al.*, 2002; Theron *et al.*, 2001; Schreuder-Gibson *et al.*, 2002). In addition genetic engineering has been used to synthesize an elastin–biomimetic peptide polymer based on the elastomeric peptide sequence of elastin and expressed from recombinant plasmid pRAM1 in *Escherichia coli*. The protein has been electrospun into fibers with diameters varying between 3 nm and 200 nm (Huang *et al.*, 2000) (Fig. 4).

Polymer and Fiber Selection

When deciding on a polymer and fiber structure to be incorporated into the construction of a medical fabric, careful

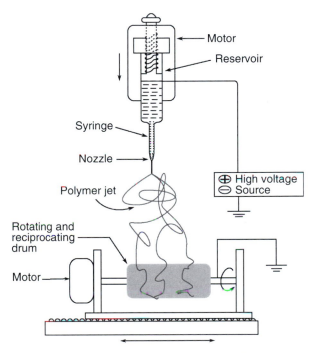

Schematic representation of laboratory electrospinning system

FIG. 3. Electrospinning system.

consideration of the end use is necessary. Issues such as the duration of body contact, device mechanical properties, fabrication restrictions, and sterilization methods must be considered. To illustrate this point, polypropylene has been successfully used in many implantable applications such as a support mesh for hernia repair. Experience has shown that polypropylene has excellent characteristics in terms of tissue compatibility and can be fabricated into a graft material with adequate mechanical strength. A critical question remaining

A

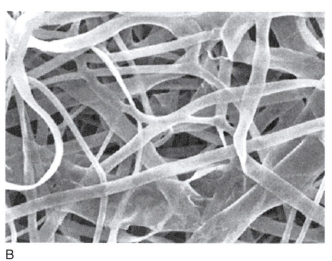

B

FIG. 4. Electrospun fibers from biomimetic-elastin peptide.

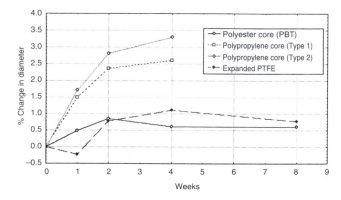

FIG. 5. Creep characteristics of various graft materials (Weinberg, 1998).

Absorbable Synthetic Fibers

Another series of synthetic fibers used in clinical applications are constructed from polymers that are designed to be absorbed over time when placed in the body. They classically have been used as sutures, but have also been used experimentally for neurological, vascular graft, and tissue scaffold applications. Table 3 is a list of bioabsorbable polymers that have been used in the past to fabricate medical devices. When in contact with the body, these polymers degrade either by hydrolysis or by enzymatic degradation into nontoxic by-products. They break down or degrade either through an erosion process that starts on the exterior surface of the fiber and continues until the fiber has been totally absorbed, or by a bulk erosion mechanism in which the process is autocatalytic and starts in the center of the fiber. Caution should be exercised when using these types of materials. In vascular applications, the risk of distal embolization to the microvasculature may occur if small pieces of the polymer break off during the erosion or absorption process.

Modified Natural Fibers

In addition to synthetic polymers, a class of fibers exists that is composed of natural biopolymer based materials. In contrast to synthetic fibers that have been adapted for medical use, natural fibers have evolved naturally and so can be particularly suited for medical applications. Cellulose, which is obtained from processed cotton or wood pulp, is one of the most common fiber-forming biopolymers. Because of the highly absorbent nature of cellulose fibers, they are commonly used in feminine hygiene products, diapers, and other absorbable applications, but typically are not used *in vivo* because of the highly inflammatory reactions associated with these materials. In certain cases, these properties can be used to advantage such as in the aforementioned hemostat Surgicel. In this application, the thrombogenicity and hydration characteristics of the regenerated cellulose are used in stopping internal bleeding from blood vessels and the surface of internal organs. Also of growing interest are fibers created from modified polysaccharides including alginates, xanthan gum, chitosan, dextran, and reticulated cellulose (Shalaby and Shah, 1991; Keys, 1996).

is whether the graft will remain stable and survive as a long-term implant. Figure 5 demonstrates the creep characteristics of grafts fabricated from expanded polytetrafluoroethylene (e-PTFE), polyester, and a bicomponent fiber (BCF) containing polypropylene yarns. In the case of the first BCF design (see later section), the polypropylene was used as the nonabsorbable core material and the main structural component of the fiber. Figure 5 represents the outer diameter of a series of pressurized graft materials as a function of time. Classical graft materials such as PET and e-PTFE show no creep over time, whereas the polypropylene-based materials continue to creep over time, making them unacceptable for long-term vascular implants. However, in other applications such as for hernia repair meshes and sutures, polypropylene has been used very successfully. It should be noted that in the second-generation BCF design, the core material was changed to poly(butylene terephthalate) (King *et al.*, 2000).

TABLE 3 Absorbable Synthetic Polymers

Type	Chemical and physical aspects	Construction/useful forms	Comments/applications
Poly(glycolide) (PGA)	Thermoplastic crystalline polymer ($T_m = 225°C$, $T_g = 40–45°C$)	Multifilament yarns, for weaving, knitting and braiding, sterilized by ethylene oxide	Absorbable sutures and meshes (for defect repairs and periodontal inserts)
10/90 Poly(L-lactide-co-glycolide) (Polyglactin 910)	Thermoplastic crystalline co-polymer ($T_m = 205°C$, $T_g = 43°C$)	Multifilament yarns, for weaving, knitting and braiding, sterilized by ethylene oxide	Absorbable sutures and meshes
Poly(p-dioxanone) (PDS)	Thermoplastic crystalline co-polymer, ($T_m = 110–115°C$, $T_g = 10°C$)	Melt spun to monofilament yarn	Sutures, intramedullary pins and ligating clips
Poly(alkylene oxalates)	A family of absorbable polymers with T_m between 64 and 104°C	Can be spun to monofilament and multifilament yarns	Experimental sutures
Isomorphic poly(hexamethylene-co-trans-1, 4-cyclohexane dimethylene oxalates)	A family of crystalline polymers with T_m between 64 and 225°C	Can be spun to monofilament and multifilament yarns	Experimental sutures

These materials are obtained from algae, crustacean shells, and through bacterial fermentation. A list of several forms of alginates and their proposed uses is presented in Table 4 (Keys, 1996). Another natural material, chitosan, has been used to fabricate surgical sutures and meshes, and it is currently under investigation for use as a substrate or scaffold for tissue-engineered materials (Skjak-Break and Sanford, 1989). Chitosan and alginate fibers are formed when the polymer is coagulated in a wet solution spinning process.

Silk and collagen are two natural fibers that have been widely used in medicine for multiple applications. Silk from the silkworm, *Bombyx mori*, has been used for decades as a suture. Because of the fineness of individual silk fibers, it is necessary to braid the individual fibers or brins together into thicker yarn bundles. Collagen has been used either in a reconstituted form or in its natural state. Reconstituted collagen is obtained from enzymatic chemical treatment of either bovine skin or tendon followed by reconstitution into fibrils. These fibrils can then be spun into fibers and fabricated into textile structures or can be left in their native fibrillar form for use in hemostatic mats and tissue-engineered substrates. "Catgut," a natural collagen-based suture material obtained from ovine intestine, which is cross-linked and cut into narrow strips, was one of the first bioabsorbable fibers used in surgery.

TABLE 4 Potential Uses of Alginates (Keys, 1996)

Type	Current use
Ca alginate (non-woven)	Absorbent wound dressings Pledgets Scaffold for cell culture Surgical hemostats
Ca alginate (particulate)	Acid-labile conjugates of alginate and doxirubicin Sequestration of 90Sr from ingested contaminated food and water
Na alginate (ultra pure)	Microencapsulation Bioreactors
Ca/Na alginate (hydrogel)	Wound management

Hybrid Bicomponent Fibers

Hybrid bicomponent fiber technology is a novel fiber concept that has been under development for a number of years for use in vascular grafts and other cardiovascular applications. One of the configurations of a bicomponent fiber is a sheath of an absorbable polymer around an inner core of a second nonabsorbable or less absorbable polymer. With a multifilament BCF yarn, each of the filaments of the yarn bundle is identical and contains an identical inner core and outer sheath. Prior to the development of the BCF yarn, when a bicomponent fabric was to be produced it was fabricated by weaving, braiding, or knitting together two (or more) homogenous yarns (e.g., a polyester yarn and a PLA yarn). With such constructions, the tissue or blood sees multiple polymers at the same time. In contrast, with BCF technology only one polymer in the sheath makes initial contact with the tissue. If the outer sheath of a BCF fiber is composed of a bioabsorbable material such as PGA, the inner core polymer is only exposed when the sheath is absorbed. The composition and molecular weight of the polymer and the thickness of the sheath regulate its absorption rate. The hypothesis relating to the BCF concept is that the healing process can be modulated by slowly exposing the less biocompatible inner core material. Preliminary data has shown that the absorption rate can be regulated and will affect the healing process (King *et al.*, 1999). By constructing the inner core from a nonabsorbable biostable polymer such as PET, or a slower absorbing polymer such as PLA, the strength of the fiber will be maintained even as the outer sheath dissolved.

Additionally, drugs can be incorporated into the outer absorbable sheath and delivered at predefined rates depending on the choice and thickness of the outer polymer. By using this BCF technology, both the material strength profile and the biological properties can be engineered into the fiber to meet specific medical requirements.

CONSTRUCTION

After a fiber or yarn is produced, it is then fabricated into a textile structure in order to obtain the desired mechanical and biological properties. Typical biotextile structures used for medical applications include nonwovens, wovens, knits, and braids. Within each of these configurations, many variations exist. Each type of construction has positive and negative attributes, and in most cases, the final choice represents a compromise between desired and actual fabricated properties. For example, woven fabrics typically are stronger and can be fabricated with lower porosities or water/blood permeability as compared to knits, but are stiffer, less flexible, and more difficult to handle and suture. Knits have higher permeability than woven designs and are easier to suture, but may dilate after implantation. Braids have great flexibility, but can be unstable except when subject to longitudinal load, as in the case of a suture. Multilayer braids are more stable, but are also thicker and less flexible than unidimensional braids. Each construction is a compromise.

Nonwovens

By definition, a nonwoven is a textile structure produced directly from fibers without the intermediate step of yarn production. The fibers are either bonded or interlocked together by means of mechanical or thermal action, or by using an adhesive or solvent or a combination these approaches. Figure 6 is a representation of both wet and dry nonwoven forming processes. The fibers may be oriented randomly or preferentially in one or more directions, and by combining multiple layers one can engineer the mechanical properties independently in the machine (lengthwise) and cross directions. The average pore size of a nonwoven web depends on the density of fibers, as well as the average fiber diameter, and falls under a single distribution (Krcma, 1971). For this reason some tissue-engineered substrates under development use nonwovens to form the underlying tissue scaffold (Chu, 2002).

Woven Fabrics

The term "woven" is used to describe a textile configuration where the primary structural yarns are oriented at 90° to each other. The machine direction is called the warp direction and the cross direction is identified as the filling or weft direction. Because of the orthogonal relationship between the warp and filling yarns, woven structures display low elongation and high breaking strength in both directions. There are many types of woven constructions including plain, twill, and satin weaves (Robinson, 1967). Figure 7 is a sketch showing several weave

Dry process

Dry forming (Air-Laid)

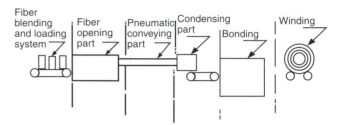

Wet process

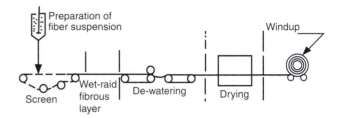

FIG. 6. Wet and dry nonwoven processes.

designs commonly used in vascular graft fabrications. Water permeability is one critical parameter used in the assessment of textile structures for vascular implants. Water permeability is a measure of the water flux through a fabric under controlled conditions and is given in units of $ml\,cm^{-2}\,min^{-1}$. It is measured by placing fabric into a test fixture having a fixed orifice size and applying a pressure of 120 mm Hg across the fabric. The water passing through the fabric is collected and measured over time and water permeability is calculated (ISO 7198, Section 8.2.2, Water Permeability). Surgeons use this parameter as a guide to determine if "pre-clotting" of a graft material is necessary prior to implantation. "Pre-clotting" is a process where a graft material is clotted with a patient's blood prior to implantation, rendering the fabric nonpermeable to blood after implantation. Fabric grafts with water permeability values less than $50\,ml\,cm^{-2}\,min^{-1}$ usually do not require pre-clotting prior to implantation. The water permeability of the woven graft fabrics can be controlled through the weaving and finishing process and can range from a low of $50\,ml\,cm^{-2}\,min^{-1}$ up to about $350\,ml\,cm^{-2}\,min^{-1}$. Above this range, a woven fabric starts becoming mechanically unstable.

Table 5 offers a list of a number of commercial woven graft designs with their respective mechanical properties. As can be seen, many variations in design are possible, presenting a difficult selection process for the surgeon. It is interesting to note that the choice of a graft by a surgeon is often based on the graft's "ease of handling" or "ease of suturing" rather than on its reported long-term performance. Plain weaves, in contrast to knits, can be made very thin (< 0.004 in.) and have thus become the material of choice for many endovascular graft designs.

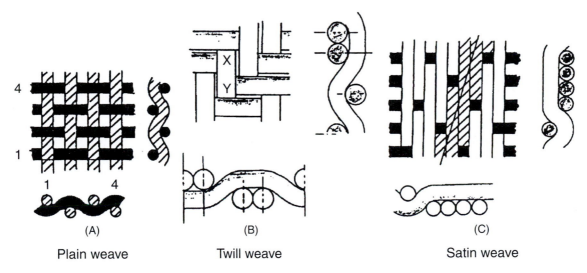

FIG. 7. Examples of woven graft designs.

TABLE 5 Woven Graft Properties and Construction (King, 1991)

Prosthesis	Type of weave	Ends per inch	Picks per inch	Bursting strength (N)	Water permeability	Suture retention strength (N)	Dilatation at 120 mm Hg (%)
Twill woven	1/1 Plain with float	42p22f	48	280	330	25	0
Debakey soft woven	1/1 Plain	52	32	366	220	35	0.2
Debaky extra low porosity	1/1 Plain	55	40	439	50	40	—
Vascutek woven	1/1 Plain	56	30	227	80	30	0.5
Meadox woven double velour	6/4 Satin+ 1/1 plain	36s36p	38	310	310	48	1.2
Meadox cooley verisoft	1/1 Plain	58	35	211	180	30	0.2
Intervascular oshner 200	1/1 Plain with leno	42p14L	21	268	250	22	0.5
Intervascular oshner 500	1/1 Plain with leno	42p14L	21	259	530	26	1.2

Knits

Knitted constructions are made by interloping yarns in horizontal rows and vertical columns of stitches. They are softer, more flexible and easily conformable, and have better handling characteristics than woven graft designs. Knit fabrics can be built with water permeability values as high as 5,000 ml cm^{-2} min^{-1} and still maintain structural stability. Currently, highly porous grafts materials are usually coated or impregnated with collagen or gelatin so that the surgeon does not have to perform the time consuming pre-clotting process at the time of surgery. The water permeability values for non-coated knitted grafts range from about 1200 ml cm^{-2} min^{-1} up to about 3500 ml cm^{-2} min^{-1}. When knits are produced, the fabric is typically very open and requires special processing to tighten the looped structure and lower its permeability. This compaction process is usually done using a chemical shrinking agent such as methylene chloride or by thermal shrinking. Because of their open structure, knits are typically easier to suture and have better handling characteristics; however, in vascular graft applications, some ultralightweight designs have been known to continuously dilate or expand when implanted in hypertensive patients. It is not uncommon to have lighter weight weft knitted grafts increase up to 20% in diameter shortly after implantation.

As is the case with woven structures, there are several variations in knits; the most common are the weft knit and warp knit constructions (see Fig. 8). Warp knitted structures have

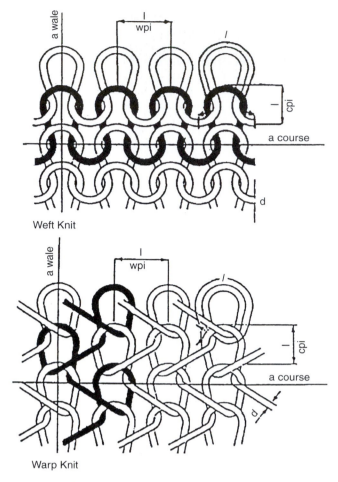

FIG. 8.   Types of knit fabrics (Spencer, 1983).

less stretch than weft knits, and therefore are inherently more dimensionally stable, being associated with less dilation *in vivo*. Warp knits do not run and ravel when cut at an angle (King, 1991). Warp knits can be further modified by the addition of an extra yarn in the structure, which adds thickness, bulk, and surface roughness to the fabric. This structure is commonly known as a velour knit. The addition of the velour yarn, while making the fabric feel softer, results in a more intense acute inflammatory reaction and increases the amount of tissue ingrowth into the fabric.

Figures 9A and 9B demonstrate the difference in the level of inflammatory response as seen with plain and velour knit designs, respectively. Figure 9A is a photomicrograph of a Golaski Microkit weft knit with high water permeability. This high porosity weft knit design utilized nontexturized yarns that resulted in a mild inflammatory response as seen at 4 weeks. In contrast, the Microvel fabric, which is a warp knit velour design using texturized yarns, shows an intense acute response at 3 days (Fig. 9B). This more intense acute reaction was designed intentionally so as to make the graft easier to pre-clot and to increase the extent of tissue incorporation into the graft wall.

Braids

Braids have found their way into medical use primarily in the manufacture of suture materials and anterior cruciate ligament (ACL) prostheses. Common braided structures involve the interlacing of an even number of yarns, leading to diamond, regular, and Hercules structures that can be either two- or three-dimensional (see Fig. 10). A myriad of structural forms can be achieved with 3D braiding, such as "I" beams, channels, and solid tubes. A sketch of a flat braiding machine is included in Fig. 11.

PROCESSING AND FINISHING

Once a fabric has been produced from yarn, the subsequent processing steps are known as finishing. As mentioned previously, the starting yarn may contain additives that can result in cytotoxicity and adverse reactions when in contact with tissue. Some of these additives, such as titanium dioxide, which is used as a delusterant to increase the amount of light reflected, are inside the spun fiber and cannot be removed in the finishing operation. Other surface finishes, on the other hand, such as yarn lubricants, can be removed with the proper cleaning and scouring operations. Typically such surface additives are mineral oil based and demand specially designed aqueous-based washing procedures or dry-cleaning techniques with organic solvents to ensure complete removal. In addition to such surface lubricants, the warp yarns may be coated with a sizing agent prior to weaving. This sizing protects the yarns from surface abrasion and filament breakage during weaving. Since each polymer and fabrication process is different, the finishing operation must be material and device specific. Finishing includes such steps as cleaning, heat setting, bleaching, shrinking (compaction), inspection, packaging, and sterilization and will influence the ultimate properties of the biotextile fabric. Figure 12 represents a schematic of a typical finishing operation used in vascular graft manufacturing. The chemicals used in the finishing operation may differ among manufacturers and are usually considered proprietary. If the cleaning process is properly designed, all surface finishes are removed during the finishing process. Testing of the finished product for cytotoxicity and residual extractables is typically used to ensure all the surface additives are removed from the product's surface prior to packaging and sterilization.

TESTING AND EVALUATION

Once the biotextile is in its final form, it must be tested and evaluated to confirm that it meets published standards and its intended end use. The testing will include component testing on each component including the textile as well as final functional testing of the entire device. When developing and implementing a testing program, various pieces of reference information may apply, including ASTM standards, AAMI/ISO standards, FDA documents, prior regulatory submissions, and the results of failure analyses. In setting up the test plan a fine balance is

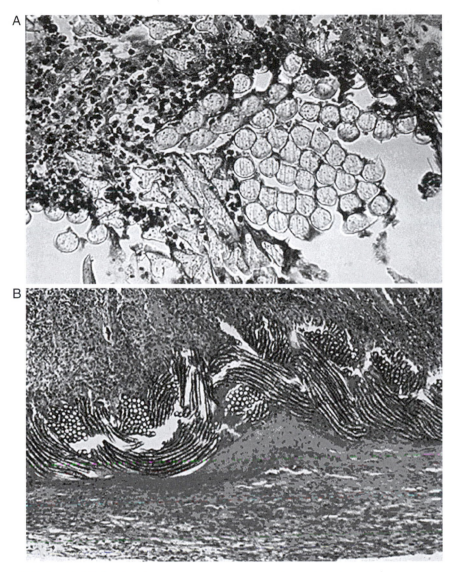

FIG. 9. (A) Weft knit inflammatory response at 4 weeks (Golaski Microkit); (B) Warp knit inflammatory response at 3 days (Microvel). (See color plate)

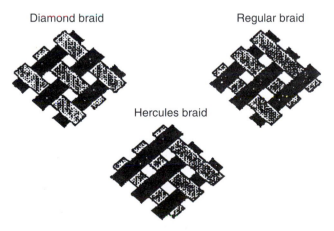

FIG. 10. Braided constructions.

needed so as to minimize the scope of the testing program while still ensuring that the polymer, textile, and final product will be safe and efficacious. Table 6 is a list of the suggested test methods used in the development of a textile-based vascular graft for large vessel replacement (ANSI/AAMI/ISO, 2001).

APPLICATIONS

The application of fibers and biotextiles as components for implantable devices is widespread and covers all aspects of medicine and health care. Textiles are used as basic care items such as drapes, protective apparel, wound dressings, and diapers and in complex devices such as heart valve sewing rings, vascular grafts, hernia repair meshes, and percutaneous access devices.

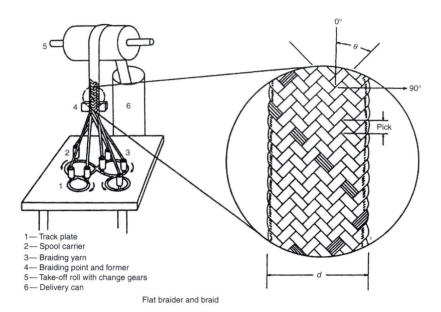

1— Track plate
2— Spool carrier
3— Braiding yarn
4— Braiding point and former
5— Take-off roll with change gears
6— Delivery can

Flat braider and braid

FIG. 11. Sketch of flat braider.

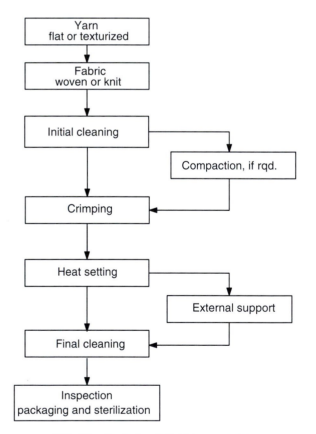

FIG. 12. Typical graft finishing operation.

TABLE 6   Sample Test Methods for Large-Diameter Textile Grafts

Test	Required regulatory testing	Routine quality testing
Visual inspection for defects	X	X
Water permeability	X	X
Longitudinal tensile strength	X	
Burst strength	X	X
Usable length	X	X
Relaxed internal diameter	X	X
Pressurized internal diameter	X	
Wall thickness	X	
Suture retention strength	X	
Kink diameter/radius	X	
Dynamic compliance	X	
Animal trials	X	
Shelf life	X	
Sterility	X	X
Biomaterials/toxicity and pyrogen testing	X	X

Drapes and Protective Apparel

The most common nonimplantable medical use of textiles is for protective surgical gowns, operating room drapes, masks, and shoe covers. Nonwovens and wovens are most frequently used for these applications, with nonwovens being the material of choice for single-use (disposable) products, and wovens for reusable items. Most of these barrier-type fabrics are made from cellulose (cotton, viscose rayon, and wood pulp), polyethylene, and polypropylene fibers. Many fabrics contain finishes that render them water repellent depending on the clinical need. Additionally, such fabrics must generally be fire retardant because of the risk of explosions due to exposure to flammable gases used for anesthesia. In applications such as facemasks, the fabric must minimize the passage of bacteria through the mask. This can be ensured by engineering the appropriate pore size distribution in the filtration fabric (Schreuder-Gibson, 2002). Antibacterial coatings are also placed on surgical drapes to minimize the risk of wound contamination. Drapes and protective apparel typically require some assembly that can be done either through conventional sewing or by ultrasonic seaming methods. The latter method is preferred for those products used in sterile fields since the holes created by conventional sewing needles can render the fabric permeable to liquids and liquid-borne pathogens. Drapes are usually constructed of a nonwoven fabric laminated to a plastic film to ensure that they are impervious to blood and other fluids. Another common use of textiles is in the fabrication of adhesive tapes. These tapes generally consist of an adhesive layer that is laminated onto a woven, knitted, or nonwoven fabric substrate.

Topical and Percutaneous Applications

Textiles have been used for many years as bandages, wound coverings, and diapers. Gauze, which is basically an open woven structure made from cotton fiber, is manufactured in many forms and sold by many companies worldwide. Elastic bandages are basically woven tapes where an expandable yarn, such as spandex polyurethane, is placed in the warp direction to allow for longitudinal stretch and recovery. Development continues to improve wound dressing products by the addition of antibiotics, barrier fabrics, growth factors, and modification of the basic underlining bandage construction. One example of the latter is the work of Karamuk *et al.* (2001), in which a three-layered laminate was formed from a nonwoven polyester/polypropylene/cotton outer layer, a monofilament polyester middle layer, and a three-dimensional embroidered polyester inner layer with large pores to promote angiogenesis.

Blood access devices are a class of medical devices where tubes, wires, or other components pass through the skin. These include percutaneous drug delivery devices, blood access shunts, air or power lines for heart and left ventricular assist devices, and many types of leads. All of these devices suffer from the same basic problem, a high risk of infection at the skin–device interface due to the migration of bacteria along the surface of the percutaneous lead. If a textile cuff is placed around the tube, at the point of entry through the skin, aggressive tissue ingrowth into the fabric reduces the risk of infection at the percutaneous site. These cuffs are usually made from knits, nonwoven felts, and velour materials. Once a device is infected, it must be removed to prevent further spreading of the infection. Surface additives, such as silver or antibiotics, are sometimes coated on the fabric to reduce infection rates (Butany, 2002; Takai, 2002).

In Vivo Applications

Cardiovascular Devices

Biotextiles developed for cardiovascular use include applications such as heart valve sewing rings, angioplasty rings,

vascular grafts, valved conduits, endovascular stent grafts, and the components of left ventricular assist devices. One of the most important uses of textile fabrics in medicine is in the fabrication of large diameter vascular grafts (10 mm to 40 mm in diameter). As previously noted, polyester [poly(ethylene terephthalate)] is the principal polymer used to fabricate vascular grafts. These grafts can either be woven or knitted and are produced in straight or bifurcated configurations. Within each type of construction, various properties can be incorporated into the product as illustrated in Table 5. Manufacturers recommend that all woven and knit grafts with water permeability rates over 50 ml cm^{-2} min^{-1} be pre-clotted to prevent blood loss through the fabric at the time of implantation. To eliminate the need for this pre-clotting procedure, textile-based vascular grafts are usually manufactured with a coating or sealant of collagen or gelatin.

Today a substantial amount of research activity is being directed toward the development of a small vessel prosthesis with diameters less than 6 mm for coronary artery bypass and tibial/popliteal artery replacement. Currently, no successful commercial products exist to meet this market need. The question still remains as to whether a biotextile will work as a small vessel prosthesis if it is fabricated to have the required compliance and mechanical properties and its surface is modified with surface coatings, growth factors, and other bioactive agents to prevent thrombosis and thrombo-embolic events. Current development activities are directed toward tissue-engineered grafts (Teebken, 2002; Huang, 2000), coated or surface-modified synthetic and textile grafts (Chinn, 1998), and biologically based grafts (Weinberg, 1995).

During the past 10 years, large amounts of financial and personnel resources have gone into the development of endovascular stent grafts (Makaroun, 2002). These grafts have been used for aortic aneurysm repair, occlusive disease, and vascular trauma. Endovascular prostheses or stent grafts are tubular grafts with an internal or external stent or rigid scaffold. The stent grafts range in size from about 20 mm up to 40 mm ID and are collapsed and folded into catheters and inserted through the femoral artery, thus avoiding the need for open surgery. The stents are typically made from nitinol, stainless steel, and Elgiloy wires and are similar to the coronary stents, however, much larger in diameter (e.g., 24 mm versus 4 mm, respectively). There are balloon expandable or self-expanding stents, which are manufactured in straight or bifurcated configurations. The stents are then covered in either ultrathin ePTFE (Cartes-Zumelzu, 2002) or woven polyester (Areydi, 2003). Most of the endovascular graft designs incorporate an ultrathin woven polyester tube. Most biotextile tubes are plain woven structures with water permeabilities ranging from 150 to 300 ml cm^{-2} min^{-1} depending on the manufacturer. They have been woven from 40 or 50 denier untexturized polyester yarn so as to minimize the overall wall thickness of the device.

General Surgery

Three key applications of biotextiles in general surgery are sutures, hemostatic devices, and hernia repair meshes. Commercial sutures are typically monofilament or braided; they can be constructed of natural materials such as silk

TABLE 7 Comparison between Commercial Hemostats (Ethicon, 1998)

	Surgicel Fibrillar Hemostat	Oxycel	Collagen power	Gelfoam
Bacterial activity	Inhibits bacterial growth	No antibacterial activity	No antibacterial activity	No antibacterial activity
Hemostasis time	3.5 to 4.5 minutes	2 to 8 minutes	2 to 4 minutes	Not specified
Bioresorbability	7 to 14 days	3 to 4 weeks	8 to 10 weeks	4 to 6 weeks
Packaging	Foil/Tyvek Sterile	Glass vials	Glass jars	Peel envelope
Preparation	Packaged for use	Packaged for use	Packaged for use	Must be cut/soaked

or collagen (catgut), or synthetic materials such as nylon, polypropylene, and polyester. Sutures can be further classified into absorbable and nonabsorbable types. For obvious reasons, when blood vessels are ligated, only nonabsorbable sutures are used, and these are typically constructed of either braided polyester or polypropylene monofilaments. On the other hand, when ligating soft tissue or closing wounds subcutaneously, absorbable sutures are preferred. Absorbable sutures do not create a chronic inflammatory response and do not require removal. These are typically made from poly(glycolic acid) (PGA) or poly(glycolide-co-lactide) copolymers.

Another common application of biotextiles and fiber technology in general surgery is the use of absorbable hemostatic agents, including those constructed of collagen and oxidized regenerated cellulose. As mentioned previously, these can be fabricated as nonwoven mats or woven and knitted fabrics, or they can be left in fibrillar form. Table 7 highlights some commercially available hemostatic agents and their representative properties. As can be seen in Table 7, collagen-based hemostatic devices are available in layered fibril, foam, and powdered forms. The regenerated cellulose pad is also available as a knitted fabric and is sold under the trade name of Surgicel. This material is commonly used to control suture line bleeding. The nonwoven and powdered forms are generally used to stop diffuse bleeding that results from trauma to the liver and spleen. Experience has shown that the loose fibril form is more difficult to use, so most surgeons prefer the more structured form of the product.

Various forms of open mesh fabrics are used as secondary support material in hernia repair. Traditional constructions are warp knitted from polypropylene monofilaments, and some forms of the mesh are preshaped for easy installation. More recently three-dimensional Raschel knits using polyester multifilament yarns have been found to be more flexible and therefore can be implanted endoscopically. As with other textile structures, various properties can be engineered into the mesh to meet design goals that may include added flexibility, increased strength, reduced thickness, improved handling, and better suture holding strength. Some designs include a protein or microporous PTFE layer on one side only, which reduces the risk of unwanted adhesions *in vivo*.

Orthopedics

Attempts have been made to construct replacement ligaments and tendons using woven and braided fabrications. One design, which has had some limited clinical success, is a prestretched knitted graft, material used to repair separated shoulder joints. A similar design, using a high-tenacity polyester woven web inside of a prestretched knitted graft, was evaluated for anterior cruciate ligament (ACL) repair in the knee joint with limited success. In general, biotextiles have had limited success in orthopedic ligament and tendon applications as a result of abrasion wear problems, inadequate strength, and poor bone attachment (Guidoin, 2000). An attempt was made to use a braided PTFE structure for ACL repair, but early failures occurred as a result of creep problems associated with the PTFE polymer. Roolker (2000) recently reported on using the e-PTFE ligament prosthesis on 52 patients. However, during the follow-up they experienced increasing knee instability over time indicating prosthesis failure. Copper (2000) and Lu (2001) have reported the development of a three-dimensional bioabsorbable braid using poly(glycolide-co-lactide) fibers for ligament replacement. They were able to modify the scaffold porosity, mechanical properties and matrix design using a three-dimensional braiding technique. A successful ACL ligament replacement would be a significant advance for orthopedic surgery, but at present, no biotextile or other type of prosthesis has shown clinical promise.

Tissue Engineering Scaffolds and The Future

One key area of research gaining significant attention over the past several years is tissue engineered scaffolds. This technology combines an engineered scaffold, or three-dimensional structure, with living cells. These scaffolds can be constructed of various materials and into various shapes depending on the desired application. One such concept is the use of the biodegradable hydrogel–textile substrate (Chu, 2002). Their concept uses a 3D porous biodegradable hydrogel on a nonwoven fabric structure. An alternate concept developed by Karamuk (2000, 2001) uses a 3D embroidered scaffold to form a tissue-engineered substrate. With this concept, polyester yarns were used to form a complex textile structure, which allowed for easy deformation that they believe will enhance cellular attachment and cell growth. Risbud (2002) reported on the development of 3D chitosan–collagen hydrogel coating for fabric meshes to support endothelial cell growth. They are directing their research toward the development of liver bioreactors.

Further in the future, various novel concepts will be undergoing development. Heim (2002) reported on the development of a textile-based tissue engineered heart valve.

Using microfiber woven technology, Heim *et al.* hypothesized that the filaments could be oriented along the stress lines and the fabric based leaflet structure would have good fatigue resistance with minimal bending stiffness. Significant development is required before this concept can be used *in vivo*. Coatings on textile based vascular grafts continue to be an area of interest. Coury (2000) reported on the use of a synthetic hydrogel coating based on poly(ethylene glycol) (PEG) to replace collagen. If successful, the use of a synthetic coating would be preferable to use of a collagen one since it will reduce manufacturing costs and graft-to-graft variability that typically occurs with naturally derived collagen materials. As mentioned earlier, even silk is undergoing modifications to enhance its biocompatibility for cardiovascular applications by sulfation and copolymerization with various monomers (Tamada, 2000). These concepts will provide new and novel implantable products for advancing medical treatments and therapies in the future.

SUMMARY

In summary, it can be stated that the use of biotextiles in medicine will continue to grow as new polymers, coatings, constructions, and finishing processes are introduced to meet the device needs of the future. In particular, advances in genetic engineering, fiber spinning, and surface modification technologies will provide a new generation of biopolymers and fibrous materials with unique chemical, mechanical, biological, and surface properties that will be responsible for achieving the previously unobtainable goal of tissue-engineered organs.

Acknowledgments

The authors thank Ruwan Sumansinghe and Henry Sun for their technical assistance in preparing this manuscript.

Bibliography

Adanur, S. (1995). *Wellington Sears Handbook of Industrial Textiles*. Technomic Publishing Company, Lancaster, PA, pp. 57–65.

Ayerdi, J., McLafferty, R. B., Markwell, S. J., Solis, M. M., Parra, J. R., Gruneiro, L. A., Ramsey, D. E., and Hodgson, K. J. (2003). Indications and outcomes of AneuRx phase III trial versus use of commercial AneuRx stent graft (In Process Citation). *J. Vascular Surgery* 37(4): 739–743.

ANSI/AAMI/ISO 7198: 1998/2001. *Cardiovascular Implants—Vascular Prostheses*, 2001. Association for the Advancement of Medical Instrumentation.

Butany, J., Scully, H. E., Van Arsdell, G., and Leask, R. (2002). Prosthetic heart valves with silver-coated sewing cuff fabric: Early morphological features in two patients. *Can. J. Cardiol.* 18(7): 733–738.

Cartes-Zumelzu, F., Lammer, J., Hoelzenbein, T., Cejna, M., Schoder, M., Thurnher, S., and Kreschmer, G. (2002). Endovascular placement of a nitinol-ePTFE stent-graft for abdominal aortic aneurysms: Initial and midterm results. *J. Vasc. Interv. Radiol.* 13(5): 465–473.

Chinn, J. A., Sauter, J. A., Phillips, R. E., Kao, W. J., Anderson, J. M., Hanson, S. R., and Ashton, T. R. (1998). Blood and tissue compatability of modified polyester: Thrombosis, inflammation, and healing. *J. Biomed. Mater. Res.* 39(1): 130–140.

Chu, C., Zhang, X. Z., and Van Buskirk, R. (2002). Biodegradable hydrogel-textile hybrid for tissue engineering. National Textile Center Research Briefs—Materials Competency: June 2002 (NTC Project: M01–B01).

Cooper, J. A., Lu, H. H., Ko, F. K., and Laurencin, C. T. (2000). Fiber-based tissue engineered scaffold for ligament replacement: Design considerations and in vitro evaluation, 208. Society for Biomaterials, Sixth World Biomaterials Congress Transactions.

Coury, A., Barrows, T., Azadeh, F., Roth, L., Poff, B., VanLue, S., Warnock, D., Jarrett, P., Bassett, M., and Doherty, E. (2000). Development of synthetic coatings for textile vascular prostheses, 1497. Society for Biomaterials, Sixth World Biomaterials Congress Transactions.

Ethicon, Inc. (1998). Surgicel Fibrillar, Absorbable Hemostat. Somerville, NJ.

Formhals A. (1934). Process and apparatus for preparing artificial threads. US Patent 1,975,504.

Goswami, B. C., Martindale, J. G., and Scardono, F. L. (1977). *Textile Yarns: Technology, Structure and Applications*. John Wiley and Sons, New York.

Groitzsch D., and Fahrbach, E. (1986). Microporous multiplayer nonwoven material for medical applications. US Patent 4,618,524.

Guidoin, M. F., Marois, Y., Bejui, J., Poddevin, N., King, M. W., and Guidoin, R. (2000). Analysis of retrieved polymer fiber based replacements for the ACL. *Biomaterials* 21(23): 2461–2474.

Heim, F., Chakfe, N., and Durand, B. (2002). A new concept of a flexible textile heart valve prosthesis, 665. Society for Biomaterials, 28th Annual Meeting Transactions.

Hoffman, A. S. (1977). Medical application of polymeric fibers. *J. Appl. Polym. Sci., Appl. Polym. Symp.* 31: 313.

Huang, L., McMillan, R. A., Apkarian, R. P., Pourdeyhimi, B., Conticello, V. P., and Chaikof, E. L. (2000). Generation of synthetic elastin-mimetic small diameter fibers and fiber networks. *Macromolecules* 33: 2989–2997.

Karamuk, E., Raeber, G., Mayer, J., Wagner, B., Bischoff, B., Billia, M., Seidl, R., and Wintermantel, E. (2000). Structual and mechanical aspects of embroidered textile scaffolds for tissue engineering, 4. Society for Biomaterials, Sixth World Biomaterials Congress Transactions.

Karamuk, E., Mayer, J., Selm, B., Bischoff, B., Ferrario, R., Heller, M., Billia, M., Seidel, R., Wanner, M., and Moser, R. (2001). Development of a structured wound dressing based on a textile composite funtionalised by embroidery technology. Tissupor, KTI. Projekt N–511.

Kenawy, E. R., Bowlin, G. L., Mansfield, K., Layman, J., Simpson, D. G., Sanders, E., and Wnek, G. E. (2002). Release of tetracycline hydrochloride from electrospun poly(ethylene-co-vinyl acetate), poly(l-lactic acid) and a blend. *J. Control Release* 81(1–2): 57–64.

Keys, A. F. (1996). Presentation to the Texticeutical Meeting, 16 January.

King, M. W. (1991). Designing fabrics for blood vessel replacement. *Canadian Textile Journal* 108(4): 24–30.

King, M. W., Guidoin, R. G., Gunasekera, K. R., and Gosselin, C. (1983). Designing polyester vascular prostheses for the future. Medical Progress Technology, Springer-Verlag.

King, M. W., Ornberg, R. L., Marois, Y., Marinov, G. R., Cadi, R., Roy, R., Cossette, F., Southern, J. H., Joardar, S. J., Weinberg, S. L., Shalaby, W., and Guidon, R. (1999). Healing response of partially bioresorbably bicomponent fibers: A subcutaneous rat study. Society for Biomaterials, 25th Annual Meeting Transactions, Providence R.I.

King, M. W. (1991). Designing fabrics for blood vessel replacement. *Canadian Textile Journal* 108(4): 24–30.

King, M. W., Ornberg, R. L., Marois, Y., Marinov, G. R., Cadi, R., Southern, J. H., Joardar, S. J., Weinberg, S. L., Shalaby, S. W., and Guidoin, R. (2000). Partially bioresorbable bicomponent fibers for tissue engineering: mechanical stability of core polymers, 533. Sixth World Biomaterials Congress, May 15–20, Kamuela, Hawaii.

Krcma, R. (1971). Manual of Nonwovens. Textile Trade Press, Manchester, England.

Ko, F. K. (1990). Presentation on fabrication, structure and properties of fibrous assemblies for medical applications, Drexel University and Medical Textiles, Inc. Philadelphia, PA. Workshop on Medical Textiles, Society for Biomaterials 16th Annual Meeting, Charleston, South Carolina, May 19.

Lu, H. H., Cooper, J. A., Ko, F. K, Attawia, M. A., and Laurencin, C. T. (2001). Effect of polymer scaffold composition on the morphology and growth of anterior cruciate ligament cells, 140. Society of Biomaterials, 27th Annual Meeting Transactions.

Makaroun, M. S., Chaikof, E., Naslund, T., and Matsumura, J. S. (2002). Efficacy of a bifurcated endograft versus open repair of abdominal aortic aneurysms: A reappraisal. J. Vascular Surg. 35: 203–210.

Martin, C. E., and Cockshott, I. D. (1977). US Patent 4,043,331.

Matthews, J. A., Wnek, G. E., Simpson, D. G., and Bowlin, G. L. (2002). Electrospinning of collagen nanofibers. Biomacromolecules 3: 232–239.

Piller, B. (1973). Bulked Yarns. SNTL/Textile Trade Press, Manchester, England.

Reneker, D. H., Yarin, A. L., Fong, H., and Koombhongse, S. (2000) Bending instability of electrically charged liquid jets of polymer solutions in electrospinning. J. Appl. Phys., Part 1 87: 4531.

Risbud, M. V., Karamuk, E., Moser, R., and Mayer, J. (2002). Hydrogen-coated textile scaffolds as three-dimensional growth support for human umbilical vein endothelial cells (HUVECs): Possibilities as coculture system in liver tissue engineering. Cell Transplant 11(4): 369–377.

Robinson, A. T. C., and Marks, R. (1967). Woven Cloth Construction. Plenum Press, New York.

Roolker, W., Patt, T. W., Van Dijk, C. N., Vegter, M., and Marti, R. K. (2000). The Gore-Tex Prosthetic Ligament as a Salvage Procedure in Deficient Knees. Knee. Surg. Sports Taumatol. Arthrosc. 8(1): 20–25.

Schreuder-Gibson, H., Gibson, P., Senecal, K., Sennett, M., Walker, J., Yeoman, W., Ziegler D., and Tsai, P. P. (2002). Protective textile materials based on electrospun nanofibers. J. Adv. Maters. 34(3): 44–55.

Shalaby, S. W. (1985). Fibrous materials for biomedical applications. in High Technology Fibers, Part A, M. Lewin and J. Preson, eds. Marcel Dekker, New York.

Shalaby, S. W. (1996). Fabrics. in Biomaterials Science: An Introduction to Materials in Medicine. Hoffman, Lemons, Ratner & Schoen, eds., 118–124. Academic Press, Boston.

Shalaby S. W., and Shah, K. R., (1991). Chemical modification of natural polymers and their technological relevance. in Water-Soluable Polymers: Chemisty and Applications, S. W. Shalaby, G. B. Butler, and C. L. McCormick, eds., 74. ACS Symposium Series, American Chemical Society, Washington, D.C.

Skjak-Braek, G., and Sanford, P. A. eds. (1989). Chitin and Chitosan: Sources, Chemistry, Biochemistry, Physical Properties, and Applications. Elsevier, New York.

Spencer, D. J. (1983). Knitting Technology. Pergamon Press, Oxford.

Stitzel, J. D., Pawlowski, K. J., Wnek, G. E., Simpson, D. G., and Bowlin, G. L. (2001). Arterial smooth muscle cell proliferation on a novel biomimiking, biodegradable vascular graft scaffold. J. Biomaterials Applications 15:1.

Takai, K., Ohtsuka, T., Senda, Y., Nakao, M., Yamamoto, K., Matsuoka, J., and Hiari, Y. (2002). Antibacterial properties of antimicrobial-finished textile products. Microbiol. Immunol. 46(2): 75–81.

Tamada, Y., Furuzono, T., Ishihara, K., and Nakabayashi, N. (2000). Chemical modification of silk to utilize as a new biomaterial. Society for Biomaterials, Sixth World Biomaterials Congress Transactions.

Teebken, O. E., and Haverich, A. (2002). Tissue engineering of small diameter vascular graft. Eur. J. Vasc. Endovasc. Surg. 23(6): 475–487.

Teule, F., Aube, C., Ellison, M., and Abbott, A. (2003). Biomimetic manufacturing of customized novel fiber proteins for specialized applications, 38–43. Proceedings 3rd Autex Conference, Gdansk, Poland.

Theron, A., Zussman, E., and Yarin, A. L. (2001). Electrostatic field assisted alignment of electrospun nanofibers. Nanotechnology 12: 384–390.

Weinberg, S. L. (1998). Biomedical Device Consultants Laboratory Data.

Weinberg, S., Abbott, W. M., (1995). Biological vascular grafts: Current and emerging technologies. in Vascular Surgery: Theory and Practice, A. D. Callow and C. B. Ernst, eds., 1213–1220. McGraw-Hill, New York.

2.5 HYDROGELS

Nicholas A. Peppas

Hydrogels are water-swollen, cross-linked polymeric structures containing either covalent bonds produced by the simple reaction of one or more comonomers, physical cross-links from entanglements, association bonds such as hydrogen bonds or strong van der Waals interactions between chains (Peppas, 1987), or crystallites bringing together two or more macromolecular chains (Hickey and Peppas, 1995). Hydrogels have received significant attention because of their exceptional promise in biomedical applications. The classic book by Andrade (1976) offers some of the best work that was available prior to 1975. The more recent book and other reviews by Peppas (1987, 2001) addresses the preparation, structure, and characterization of hydrogels.

Here, we concentrate on some features of the preparation of hydrogels, as well as characteristics of their structure and chemical and physical properties.

CLASSIFICATION AND BASIC STRUCTURE

Depending on their method of preparation, ionic charge, or physical structure features, hydrogels maybe classified in several categories. Based on the method of preparation, they may be (i) homopolymer hydrogels, (ii) copolymer hydrogels, (iii) multipolymer hydrogels, or (iv) interpenetrating polymeric hydrogels. Homopolymer hydrogels are cross-linked networks of one type of hydrophilic monomer unit, whereas copolymer hydrogels are produced by cross-linking of two comonomer units, at least one of which must be hydrophilic to render them swellable. Multipolymer hydrogels are produced from three or more comonomers reacting together (see e.g., Lowman and Peppas, 1997, 1999). Finally, interpenetrating polymeric hydrogels are produced by preparing a first network that is then swollen in a monomer. The latter reacts to form a second intermeshing network structure. Based on their ionic charges, hydrogels may be classified (Ratner and Hoffman, 1976; Brannon-Peppas and Harland, 1990) as (i) neutral hydrogels, (ii) anionic hydrogels, (iii) cationic hydrogels, or (iv) ampholytic hydrogels. Based on physical structural features of the system, they can be classified as (i) amorphous hydrogels, (ii) semicrystalline hydrogels, or (iii) hydrogen-bonded or complexation structures. In amorphous hydrogels, the macromolecular chains are arranged randomly. Semicrystalline hydrogels are characterized by dense regions of ordered macromolecular chains (crystallites). Finally, hydrogen bonds and complexation structures may be responsible for the three-dimensional structure formed.

Structural evaluation of hydrogels reveals that ideal networks are only rarely observed. Figure 1A shows an ideal macromolecular network (hydrogel) indicating tetrafunctional cross-links (junctions) produced by covalent bonds. However, in real networks it is possible to encounter multifunctional junctions (Fig. 1B) or physical molecular entanglements (Fig. 1C) playing the role of semipermanent junctions. Hydrogels with molecular defects are always possible. Figures 1D and 1E indicate two such effects: unreacted functionalities with partial entanglements (Fig. 1D) and chain loops (Fig. 1E). Neither of these effects contributes to the mechanical or physical properties of a polymer network.

The terms "cross-link," "junction," or "tie-point" (an open circle symbol in Fig. 1D) indicate the connection points of several chains. These junctions may be carbon atoms, but they are usually small chemical bridges [e.g., an acetal bridge in the case of cross-linked poly(vinyl alcohol)] with molecular weights much smaller than those of the cross-linked polymer chains. In other situations, a junction may be an association of macromolecular chains caused by van der Waals forces, as in the case of the glycoproteinic network structure of natural mucus, or an aggregate formed by hydrogen bonds, as in the case of aged microgels formed in polymer solutions.

Finally, the network structure may include effective junctions that can be either simple physical entanglements of permanent or semipermanent nature, or ordered chains forming crystallites. Thus, the junctions should never be considered as points without volume, which is the usual assumption made when developing structural models for analysis of the cross-linked structure of hydrogels (Flory, 1953). Instead, they have a finite size and contribute to the deformational distribution during biomedical applications.

PREPARATION

Hydrogels are prepared by swelling cross-linked structures in water or biological fluids. Water or aqueous solutions may be present during the initial preparation of the cross-linked structure. Methods of preparation of the initial networks include chemical cross-linking, photopolymerization, or irradiative cross-linking (Peppas *et al.*, 2000).

Chemical cross-linking calls for direct reaction of a linear or branched polymer with at least one difunctional,

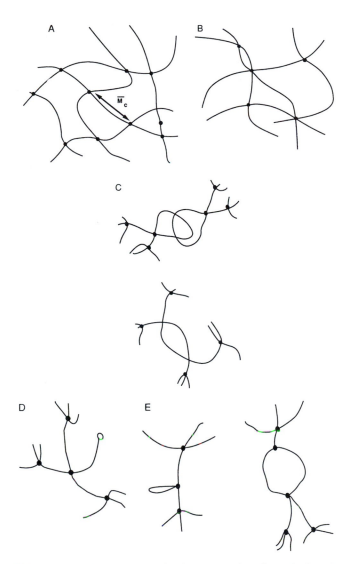

FIG. 1. (A) Ideal macromolecular network of a hydrogel. (B) Network with multifunctional junctions. (C) Physical entanglements in a hydrogel. (D) Unreacted functionality in a hydrogel. (E) Chain loops in a hydrogel.

small molecular weight, cross-linking agent. This agent usually links two longer molecular weight chains through its di- or multifunctional groups. A second method involves a copolymerisation-cross-linking reaction between one or more abundant monomers and one multifunctional monomer that is present in relatively small quantities. A third method involves using a combination of monomer and linear polymeric chains that are cross-linked by means of an interlinking agent, as in the production of polyurethanes. Several of these techniques can be performed in the presence of UV light leading to rapid formation of a three-dimensional network. Ionizing radiation cross-linking (Chapiro, 1962) utilizes electron beams, gamma rays, or X-rays to excite a polymer and produce a cross-linked structure via free radical reactions.

SWELLING BEHAVIOR

The physical behavior of hydrogels is dependent on their equilibrium and dynamic swelling behavior in water, since upon preparation they must be brought in contact with water to yield the final, swollen network structure. Figure 2 shows one of two possible processes of swelling. A dry, hydrophilic cross-linked network is placed in water. Then, the macromolecular chains interact with the solvent molecules owing to the relatively good thermodynamic compatibility. Thus, the network expands to the solvated state.

The Flory–Huggins theory can be used to calculate thermodynamic quantities related to that mixing process. Flory (1953) developed the initial theory of the swelling of cross-linked polymer gels using a Gaussian distribution of the polymer chains. His model describing the equilibrium degree of cross-linked polymers postulated that the degree to which a polymer network swelled was governed by the elastic retractive forces of the polymer chains and the thermodynamic compatibility of the polymer and the solvent molecules. In terms of the free energy of the system, the total free energy change upon swelling was written as:

$$\Delta G = \Delta G_{elastic} + \Delta G_{mix} \qquad (1)$$

Here, $\Delta G_{elastic}$ is the contribution due to the elastic retractive forces and ΔG_{mix} represents the thermodynamic compatibility of the polymer and the swelling agent (water).

Upon differentiation of Eq. 1 with respect to the water molecules in the system, an expression can be derived for the chemical potential change of water in terms of the elastic and mixing contributions due to swelling.

$$\mu_1 - \mu_{1,0} = \Delta \mu_{elastic} + \Delta \mu_{mix} \qquad (2)$$

Here, μ_1 is the chemical potential of water within the gel and $\mu_{1,0}$ is the chemical potential of pure water.

At equilibrium, the chemical potentials of water inside and outside of the gel must be equal. Therefore, the elastic and mixing contributions to the chemical potential will balance one another at equilibrium. The chemical potential change upon mixing can be determined from the heat of mixing and the entropy of mixing. Using the Flory–Huggins theory, the chemical potential of mixing can be expressed as:

$$\Delta \mu_{mix} = RT \left(\ln(1 - 2\upsilon_{2,s}) + \upsilon_{2,s} + \chi_1 \upsilon_{2,s}^2 \right) \qquad (3)$$

where χ_1 is the polymer–water interaction parameter, $\upsilon_{2,s}$ is the polymer volume fraction of the gel, T is absolute temperature, and R is the gas constant.

This thermodynamic swelling contribution is counterbalanced by the retractive elastic contribution of the cross-linked structure. The latter is usually described by the rubber elasticity theory and its variations (Peppas, 1987). Equilibrium is attained in a particular solvent at a particular temperature when the two forces become equal. The volume degree of swelling, Q (i.e., the ratio of the actual volume of a sample in the swollen state divided by its volume in the dry state), can then be determined from Eq. 4.

$$\upsilon_{2,s} = \frac{\text{Volume of polymer}}{\text{Volume of swollen gel}} = \frac{V_p}{V_{gel}} = 1/Q \qquad (4)$$

Researchers working with hydrogels for biomedical applications prefer to use other parameters in order to define the equilibrium-swelling behavior. For example, Yasuda et al. (1969) introduced the use of the so-called hydration ratio, H, which has been accepted by those researchers who use hydrogels for contact lens applications (Peppas and Yang, 1981). Another definition is that of the weight degree of swelling, q, which is the ratio of the weight of the swollen sample to that of the dry sample.

In general, highly swollen hydrogels include those of cellulose derivatives, poly(vinyl alcohol), poly(N-vinyl-2-pyrrolidone) (PNVP), and poly(ethylene glycol), among others. Moderately and poorly swollen hydrogels are those of poly(hydroxyethyl methacrylate) (PHEMA) and many of its derivatives. In general, a basic hydrophilic monomer can be copolymerized with other more or less hydrophilic monomers to achieve desired swelling properties. Such processes have led to a wide range of swellable hydrogels, as Gregonis et al. (1976), Peppas (1987, 1997), and others have pointed out. Knowledge of the swelling characteristics of a polymer is of utmost importance in biomedical and pharmaceutical applications since the equilibrium degree of swelling influences (i) the solute diffusion coefficient through these hydrogels, (ii) the surface properties and surface mobility, (iii) the optical properties,

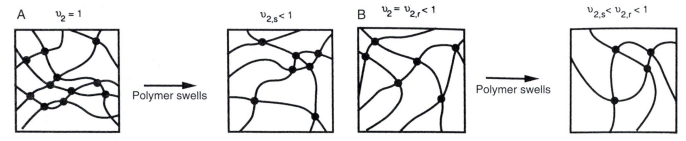

FIG. 2. (A) Swelling of a network prepared by cross-linking in dry state. (B) Swelling of a network prepared by cross-linking in solution.

especially in relation to contact lens applications, and (iv) the mechanical properties.

DETERMINATION OF STRUCTURAL CHARACTERISTICS

The parameter that describes the basic structure of the hydrogel is the molecular weight between cross-links, $\overline{M}_c$ (as shown in Fig. 1A). This parameter defines the average molecular size between two consecutive junctions regardless of the nature of those junctions and can be calculated by Eq. 5.

$$\frac{1}{\overline{M}_c} = \frac{2}{\overline{M}_c} - \frac{(v/V_1)\left[\ln(1 - v_{2,s}) + v_{2,s} + \chi_1 v_{2,s}^2\right]}{\left(v_{2,s}^{1/3} - v_{2,s}/2\right)} \quad (5)$$

An additional parameter of importance in structural analysis of hydrogels is the cross-linking density, ρ_x, which is defined by Eq. 6.

$$\rho_x = \frac{1}{v\overline{M}_c} \quad (6)$$

In these equations, v is the specific volume of the polymer (i.e., the reciprocal of the amorphous density of the polymer), and $\overline{M}_n$ is the initial molecular weight of the un-cross-linked polymer.

PROPERTIES OF IMPORTANT BIOMEDICAL HYDROGELS

The multitude of hydrogels available leaves numerous choices for polymeric formulations. The best approach for developing a hydrogel with the desired characteristics for biomedical application is to correlate the macromolecular structures of the polymers available with the swelling and mechanical characteristics desired (Peppas *et al.*, 2000; Peppas, 2001).

The most widely used hydrogel is water-swollen, cross-linked PHEMA, which was introduced as a biological material by Wichterle and Lim (1960). The hydrogel is inert to normal biological processes, shows resistance to degradation, is permeable to metabolites, is not absorbed by the body, is biocompatible, withstands heat sterilization without damage, and can be prepared in a variety of shapes and forms.

The swelling, mechanical, diffusional, and biomedical characteristics of PHEMA gels have been studied extensively. The properties of these hydrogels are dependent upon their method of preparation, polymer volume fraction, degree of cross-linking, temperature, and swelling agent.

Other hydrogels of biomedical interest include polyacrylamides. Tanaka (1979) has done extensive studies on the abrupt swelling and deswelling of partially hydrolyzed acrylamide gels with changes in swelling agent composition, curing time, degree of cross-linking, degree of hydrolysis, and temperature. These studies have shown that the ionic groups produced in an acrylamide gel upon hydrolysis give the gel a structure that shows a discrete transition in equilibrium-swollen volume with environmental changes.

Discontinuous swelling in partially hydrolyzed polyacrylamide gels has been studied by Gehrke *et al.* (1986). Besides HEMA and acrylamides, *N*-vinyl-2-pyrrolidone (NVP), methacrylic acid (MAA), methyl methacrylate (MMA), and maleic anhydride (MAH) have all been proven useful as monomers for hydrogels in biomedical applications. For instance, cross-linked PNVP is used in soft contact lenses. Small amounts of MAA as a comonomer have been shown to dramatically increase the swelling of PHEMA polymers. Owing to the hydrophobic nature of MMA, copolymers of MMA and HEMA have a lower degree of swelling than pure PHEMA (Brannon-Peppas and Peppas, 1991). All of these materials have potential use in advanced technology applications, including biomedical separations, and biomedical and pharmaceutical devices.

INTELLIGENT OR SMART HYDROGELS

Hydrogels may exhibit swelling behavior dependent on the external environment. Over the past 30 years there has been a significant interest in the development and analysis of environmentally or physiologically responsive hydrogels (Peppas, 1991). Environmentally responsive materials show drastic changes in their swelling ratio due to changes in their external pH, temperature, ionic strength, nature and composition of the swelling agent, enzymatic or chemical reaction, and electrical or magnetic stimuli (Peppas, 1993). In most responsive networks, a critical point exists at which this transition occurs.

An interesting characteristic of numerous responsive gels is that the mechanism causing the network structural changes can be entirely reversible in nature. The ability of pH- or temperature-responsive gels to exhibit rapid changes in their swelling behavior and pore structure in response to changes in environmental conditions lend these materials favorable characteristics as carriers for bioactive agents, including peptides and proteins. This type of behavior may allow these materials to serve as self-regulated, pulsatile drug delivery systems.

pH-Sensitive Hydrogels

One of the most widely studied types of physiologically responsive hydrogels is pH-responsive hydrogels. These hydrogels are swollen ionic networks containing either acidic or basic pendant groups. In aqueous media of appropriate pH and ionic strength, the pendant groups can ionize developing fixed charges on the gel. All ionic materials exhibit a pH and ionic strength sensitivity. The swelling forces developed in these systems are increased over those of nonionic materials. This increase in swelling force is due to the localization of fixed charges on the pendant groups. As a result, the mesh size of the polymeric networks can change significantly with small pH changes.

Temperature Sensitive Hydrogels

Another class of environmentally sensitive gels exhibits temperature-sensitive swelling behavior due to a change in the polymer/swelling agent compatibility over the temperature range of interest. Temperature-sensitive polymers typically exhibit a lower critical solution temperature (LCST), below which the polymer is soluble. Above this temperature, the polymers are typically hydrophobic and do not swell significantly in water (Kim, 1996). However, below the LCST, the cross-linked gel swells to significantly higher degrees because of the increased compatibility with water.

Complexing Hydrogels

Some hydrogels may exhibit environmental sensitivity due to the formation of polymer complexes. Polymer complexes are insoluble, macromolecular structures formed by the non-covalent association of polymers with affinity for one another. The complexes form as a result of the association of repeating units on different chains (interpolymer complexes) or on separate regions of the same chain (intrapolymer complexes). Polymer complexes are classified by the nature of the association as stereocomplexes, polyelectrolyte complexes, or hydrogen-bonded complexes. The stability of the associations is dependent on such factors as the nature of the swelling agent, temperature, type of dissolution medium, pH and ionic strength, network composition and structure, and length of the interacting polymer chains.

In this type of gel, complex formation results in the formation of physical cross-links in the gel. As the degree of effective cross-linking is increased, the network mesh size and degree of swelling is significantly reduced. As a result, if hydrogels are used as drug carriers, the rate of drug release will decrease dramatically upon the formation of interpolymer complexes.

APPLICATIONS

Biomedical Applications

The physical properties of hydrogels make them attractive for a variety of biomedical and pharmaceutical applications. Their biocompatibility allows them to be considered for medical applications, whereas their hydrophilicity can impart desirable release characteristics to controlled and sustained release formulations.

Hydrogels exhibit properties that make them desirable candidates for biocompatible and blood-compatible biomaterials (Merrill et al., 1987). Nonionic hydrogels for blood contact applications have been prepared from poly(vinyl alcohol), polyacrylamides, PNVP, PHEMA, and poly(ethylene oxide) (Peppas et al., 1999). Heparinized polymer hydrogels also show promise as materials for blood-compatible applications (Sefton, 1987).

One of the earliest biomedical applications of hydrogels was in contact lenses (Tighe, 1976; Peppas and Yang, 1981) because of their relatively good mechanical stability, favorable refractive index, and high oxygen permeability.

Other potential applications of hydrogels include (Peppas, 1987) artificial tendon materials, wound-healing bioadhesives, artificial kidney membranes, articular cartilage, artificial skin, maxillofacial and sexual organ reconstruction materials, and vocal cord replacement materials (Byrne et al., 2002).

Pharmaceutical Applications

Pharmaceutical hydrogel applications have become very popular in recent years. Pharmaceutical hydrogel systems include equilibrium-swollen hydrogels, i.e., matrices that have a drug incorporated in them and are swollen to equilibrium. The category of solvent-activated, matrix-type, controlled-release devices comprises two important types of systems: swellable and swelling-controlled devices. In general, a system prepared by incorporating a drug into a hydrophilic, glassy polymer can be swollen when brought in contact with water or a simulant of biological fluids. This swelling process may or may not be the controlling mechanism for diffusional release, depending on the relative rates of the macromolecular relaxation of the polymer and drug diffusion from the gel.

In swelling-controlled release systems, the bioactive agent is dispersed into the polymer to form nonporous films, disks, or spheres. Upon contact with an aqueous dissolution medium, a distinct front (interface) is observed that corresponds to the water penetration front into the polymer and separates the glassy from the rubbery (gel-like) state of the material. Under these conditions, the macromolecular relaxations of the polymer influence the diffusion mechanism of the drug through the rubbery state. This water uptake can lead to considerable swelling of the polymer with a thickness that depends on time. The swelling process proceeds toward equilibrium at a rate determined by the water activity in the system and the structure of the polymer. If the polymer is cross-linked or if it is of sufficiently high molecular weight (so that chain entanglements can maintain structural integrity), the equilibrium state is a water-swollen gel. The equilibrium water content of such hydrogels can vary from 30% to 90%. If the dry hydrogel contains a water-soluble drug, the drug is essentially immobile in the glassy matrix, but begins to diffuse out as the polymer swells with water. Drug release thus depends on two simultaneous rate processes: water migration into the device and drug diffusion outward through the swollen gel. Since some water uptake must occur before the drug can be released, the initial burst effect frequently observed in matrix devices is moderated, although it may still be present. The continued swelling of the matrix causes the drug to diffuse increasingly easily, ameliorating the slow tailing off of the release curve. The net effect of the swelling process is to prolong and linearize the release curve. Details of hydrogels for medical and pharmaceutical applications have been presented by Korsmeyer and Peppas (1987) for poly(vinyl alcohol) systems, and by Peppas (1981) for PHEMA systems and their copolymers. One of numerous examples of such swelling-controlled systems was reported by Franson and Peppas (1983), who prepared cross-linked copolymer gels of poly(HEMA-co-MAA) of varying compositions. Theophylline release was studied and it was found that near zero-order release could be achieved using copolymers containing 90% PHEMA.

Poly(vinyl alcohol)

Another hydrophilic polymer that has received attention is poly(vinyl alcohol) (PVA). This material holds tremendous promise as a biological drug delivery device because it is nontoxic, is hydrophilic, and exhibits good mucoadhesive properties. Two methods exist for the preparation of PVA gels. In the first method, linear PVA chains are cross-linked using glyoxal, glutaraldehyde, or borate. In the second method, Peppas and Hassan (2000), semicrystalline gels were prepared by exposing aqueous solutions of PVA to repeating freezing and thawing. The freezing and thawing induced crystal formation in the materials and allowed for the formation of a network structure cross-linked with the quasi-permanent crystallites. The latter method is the preferred method for preparation as it allows for the formation of an "ultrapure" network without the use of toxic cross-linking agents. Ficek and Peppas (1993) used PVA gels for the release of bovine serum albumin using novel PVA microparticles.

Poly(ethylene glycol)

Hydrogels of poly(ethylene oxide) (PEO) and poly(ethylene glycol) (PEG) have received significant attention for biomedical applications in the past few years (Graham, 1992). Three major preparation techniques exist for the preparation of cross-linked PEG networks: (i) chemical cross-linking between PEG chains, (ii) radiation cross-linking of PEG chains, and (iii) chemical reaction of mono- and difunctional PEGs. The advantage of using radiation-cross-linked PEO networks is that no toxic cross-linking agents are required. However, it is difficult to control the network structure of these materials. Stringer and Peppas (1996) have prepared PEO hydrogels by radiation cross-linking. In this work, they analyzed the network structure in detail. Additionally, they investigated the diffusional behavior of smaller molecular weight drugs, such as theophylline, in these gels. Kofinas et al. (1996) have prepared PEO hydrogels by a similar technique. In this work, they studied the diffusional behavior of various macromolecules in these gels. They noted an interesting, yet previously unreported dependence between the cross-link density and protein diffusion coefficient and the initial molecular weight of the linear PEGs.

Lowman et al. (1997) have presented an exciting new method for the preparation of PEG gels with controllable structures. In this work, highly cross-linked and tethered PEG gels were prepared from PEG dimethacrylates and PEG monomethacrylates. The diffusional behavior of diltiazem and theophylline in these networks was studied. The technique presented in this work is promising for the development of a new class of functionalized PEG-containing gels that may be of use in a wide variety of drug delivery applications.

pH-Sensitive Hydrogels

Hydrogels that have the ability to respond to pH changes have been studied extensively over the years. These gels typically contain side ionizable side groups such as carboxylic acids or amine groups. The most commonly studied ionic polymers include poly(acrylamide) (PAAm), poly(acrylic acid) (PAA), poly(methacrylic acid) (PMAA), poly(diethylaminoethyl methacrylate) (PDEAEMA), and poly(dimethylaminoethyl methacrylate) (PDMAEMA). The swelling and release characteristics of anionic copolymers of PMAA and PHEMA (PHEMA-co-MAA) have been investigated. In acidic media, the gels did not swell significantly; however, in neutral or basic media, the gels swelled to a high degree because of ionization of the pendant acid group. Brannon-Peppas and Peppas (1991) have also studied the oscillatory swelling behavior of these gels.

Temperature-Sensitive Hydrogels

Some of the earliest work with temperature-sensitive hydrogels was done by Hirotsu et al. (1987). They synthesized cross-linked poly(N-isopropyl acrylamide) (PNIPAAm) and determined that the LCST of the PNIPAAm gels was 34.3°C. Below this temperature, significant gel swelling occurred. The transition about this point was reversible. They discovered that the transition temperature was raised by copolymerizing PNIPAAm with small amounts of ionic monomers. Dong and Hoffman (1991) prepared heterogeneous gels containing PNIPAAm that collapsed at significantly faster rates than homopolymers of PNIPAAm. Yoshida et al. (1995) and Kaneko et al. (1996) developed an ingenious method to prepare comb-type graft hydrogels of PNIPAAm. The main chain of the cross-linked PNIPAAm contained small-molecular-weight grafts of PNIPAAm. Under conditions of gel collapse (above the LCST), hydrophobic regions were developed in the pores of the gel resulting in a rapid collapse. These materials had the ability to collapse from a fully swollen conformation in less than 20 minutes, whereas comparable gels that did not contain graft chains required up to a month to fully collapse. Such systems show major promise for rapid and abrupt or oscillatory release of drugs, peptides, or proteins.

Complexation Hydrogels

Another promising class of hydrogels that exhibit responsive behavior is complexing hydrogels. Bell and Peppas (1995) have discussed a class of graft copolymer gels of PMAA grafted with PEG, poly(MAA-g-EG). These gels exhibited pH-dependent swelling behavior due to the presence of acidic pendant groups and the formation of interpolymer complexes between the ether groups on the graft chains and protonated pendant groups. In these covalently cross-linked, complexing poly(MAA-g-EG) hydrogels, complexation resulted in the formation of temporary physical cross-links due to hydrogen bonding between the PEG grafts and the PMAA pendant groups. The physical cross-links were reversible in nature and dependent on the pH and ionic strength of the environment. As a result, these complexing hydrogels exhibit drastic changes in their mesh size in response to small changes of pH.

Promising new methods for the delivery of chemotherapeutic agents using hydrogels have been recently reported. Novel biorecognizable sugar-containing copolymers have been investigated for the use in targeted delivery of anti-cancer drugs. Peterson et al. (1996) have used poly(N-2-hydroxypropyl methacrylamide) carriers for the treatment of ovarian cancer.

Self-Assembled Structures

In the past few years there have been new, creative methods of preparation of novel hydrophilic polymers and hydrogels that may represent the future in drug delivery applications. The focus in these studies has been the development of polymeric structures with precise molecular architectures. Stupp *et al.* (1997) synthesized self-assembled triblock copolymer nanostructures that may have very promising biomedical applications.

Star Polymers

Dendrimers and star polymers (Dvornik and Tomalia, 1996) are exciting new materials because of the large number of functional groups available in a very small volume. Such systems could have tremendous promise in drug targeting applications. Merrill (1993) has offered an exceptional review of PEO star polymers and applications of such systems in the biomedical and pharmaceutical fields. Griffith and Lopina (1995) have prepared gels of controlled structure and large biological functionality by irradiation of PEO star polymers. Such new structures could have particularly promising drug delivery applications when combined with emerging new technologies such as molecular imprinting.

Bibliography

Andrade, J. D. (1976). *Hydrogels for Medical and Related Applications.* ACS Symposium Series, Vol. 31, American Chemical Society, Washington, D.C.

Bell, C. L., and Peppas, N. A. (1995). Biomedical membranes from hydrogels and interpolymer complexes. *Adv. Polym. Sci.* **122:** 125–175.

Brannon-Peppas, L., and Harland, R. S. (1990). *Absorbent Polymer Technology.* Elsevier, Amsterdam.

Brannon-Peppas, L., and Peppas, N. A. (1991). Equilibrium swelling behavior of dilute ionic hydrogels in electrolytic solutions. *J. Controlled Release* **16:** 319–330.

Brannon-Peppas, L., and Peppas, N. A. (1991). Time-dependent response of ionic polymer networks to pH and ionic strength changes. *Int. J. Pharm.* **70:** 53–57.

Byrne, M. E., Henthorn, D. B., Huang, Y., and Peppas, N. A. (2002). Micropatterning biomimetic materials for bioadhesion and drug delivery. in *Biomimetic Materials and Design: Biointerfacial Strategies Tissue Enginering and Targeted Drug Delivery*, A. K. Dillow and A. M. Lowman, eds. Dekker, New York, pp. 443–470.

Chapiro, A. (1962). *Radiation Chemistry of Polymeric Systems.* Interscience, New York.

Dong, L. C., and Hoffman, A. S. (1991). A novel approach for preparation of pH-sensitive hydrogels for enteric drug delivery. *J. Controlled Release* **15:** 141–152.

Dvornik, P. R., and Tomalia, D. A. (1996). Recent advances in dendritic polymers. *Curr. Opin. Colloid Interface Sci.* **1:** 221–235.

Ficek B. J., and Peppas, N. A. (1993). Novel preparation of poly(vinyl alcohol) microparticles without crosslinking agent. *J. Controlled Rel.* **27:** 259–264.

Flory, P. J. (1953). *Principles of Polymer Chemistry.* Cornell Univ. Press, Ithaca, NY.

Franson, N. M., and Peppas, N. A. (1983). Influence of copolymer composition on water transport through glassy copolymers. *J. Appl. Polym. Sci.* **28:** 1299–1310.

Gehrke, S. H., Andrews, G. P., and Cussler, E. L. (1986). Chemical aspects of gel extraction. *Chem. Eng. Sci.* **41:** 2153–2160.

Graham, N. B. (1992). Poly(ethylene glycol) gels and drug delivery. in *Poly(ethylene glycol) Chemistry, Biotechnical and Biomedical Applications*, J. M. Harris, ed. Plenum Press, New York, pp. 263–281.

Gregonis, D. E., Chen, C. M., and Andrade, J. D. (1976). The chemistry of some selected methacrylate hydrogels. in *Hydrogels for Medical and Related Applications*, J. D. Andrade, ed. ACS Symposium Series, Vol. 31. American Chemical Society, Washington, D.C., pp. 88–104.

Griffith, L., and Lopina, S. T. (1995). Network structures of radiation cross-linked star polymer gels. *Macromolecules* **28:** 6787–6794.

Hassan, C. M., and Peppas, N. A. (2000). Structure and morphology or freeze/thawed PVA hydrogels. *Macromolecules* **33:** 2472–2479.

Hickey, A. S., and Peppas, N. A. (1995). Mesh size and diffusive characteristics of semicrystalline poly(vinyl alcohol) membranes. *J. Membr. Sci.* **107:** 229–237.

Hirotsu, S., Hirokawa, Y., and Tanaka, T. (1987). Swelling of gels. *J. Chem. Phys.* **87:** 1392–1395.

Kaneko, Y., Saki, K., Kikuchi, A., Sakurai, Y., and Okano, T. (1996). Fast swelling/deswelling kinetics of comb-type grafted poly(*N*-isopropyl acrylamide) hydrogels. *Macromol. Symp.* **109:** 41–53.

Kim, S. W. (1996). Temperature sensitive polymers for delivery of macromolecular drugs. in *Advanced Biomaterials in Biomedical Engineering and Drug Delivery Systems*, N. Ogata, S. W. Kim, J. Feijen, and T. Okano, eds. Springer, Tokyo, pp. 125–133.

Kofinas, P., Athanassiou, V., and Merrill, E. W. (1996). Hydrogels prepared by electron beam irradiation of poly(ethylene oxide) in water solution: unexpected dependence of cross-link density and protein diffusion coefficients on initial PEO molecular weight. *Biomaterials* **17:** 1547–1550.

Korsmeyer, R. W., and Peppas, N. A. (1981). Effects of the morphology of hydrophilic polymeric matrices on the diffusion and release of water soluble drugs. *J. Membr. Sci.* **9:** 211–227.

Lowman, A. M., and Peppas, N. A. (1997). Analysis of the complexation/decomplexation phenomena in graft copolymer networks. *Macromolecules* **30:** 4959–4965.

Lowman, A. M., and Peppas, N. A. (1999). Hydrogels. in *Encyclopedia of Controlled Drug Delivery*, E. Mathiowitz, ed. Wiley, New York, pp. 397–418.

Lowman, A. M., Dziubla, T. D., and Peppas, N. A. (1997). Novel networks and gels containing increased amounts of grafted and crosslinked poly(ethylene glycol). *Polymer Preprints* **38:** 622–623.

Merrill, E. W. (1993). Poly(ethylene oxide) star molecules: synthesis, characterization, and applications in medicine and biology. *J. Biomater. Sci. Polym. Edn.* **5:** 1–11.

Merrill, E. W., Pekala, P. W., and Mahmud, N. A. (1987). Hydrogels for blood contact. in *Hydrogels in Medicine and Pharmacy*, N. A. Peppas, ed. CRC Press, Boca Raton, FL, Vol. 3, pp. 1–16.

Peppas, N. A. (1987). *Hydrogels in Medicine and Pharmacy.* CRC Press, Boca Raton, FL.

Peppas, N. A. (1991). Physiologically responsive hydrogels. *J. Bioact. Compat. Polym.* **6:** 241–246.

Peppas, N. A. (1993). Fundamentals of pH- and temperature-sensitive delivery systems. in *Pulsatile Drug Delivery*, R. Gurny, H. E. Juninger, and N. A. Peppas, eds. Wissenschaftliche Verlagsgesellschaft, Stuttgart, pp. 41–56.

Peppas, N. A. (1997). Hydrogels and drug delivery. *Crit. Opin. Colloid Interface Sci.* **2:** 531–537.

Peppas, N. A. (2001). Gels for drug delivery. in *Encyclopedia of Materials: Science and Technology.* Elsevier, Amsterdam, pp. 3492–3495.

Peppas, N. A., and Yang, W. H. M. (1981). Properties-based optimization of the structure of polymers for contact lens applications. *Contact Intraocular Lens Med. J.* **7**: 300–321.

Peppas, N. A., Huang, Y., Torres-Lugo, M., Ward, J. H., and Zhang, J. (2000). Physicochemical foundations and structural design of hydrogels in medicine and biology. *Ann. Rev. Biomed. Eng.* **2**: 9–29.

Peppas, N. A., Keys, K. B., Torres-Lugo, M., and Lowman, A. M. (1999). Poly(ethylene glycol)-Containing Hydrogels in Drug Delivery. *J. Controlled Release* **62**: 81–87.

Peterson, C. M., Lu, J. M., Sun, Y., Peterson, C. A., Shiah, J. G., Straight, R. C., and Kopecek, J. (1996). *Cancer Res.* **56**: 3980–3985.

Ratner, B. D., and Hoffman, A. S. (1976). Synthetic hydrogels for biomedical applications. in *Hydrogels for Medical and Related Applications*, J. D. Andrade, ed. ACS Symposium Series, American Chemical Society, Washington, D.C., Vol. 31, pp. 1–36.

Sefton, M. V. (1987). Heparinized hydrogels. in *Hydrogels in Medicine and Pharmacy*, N. A. Peppas, ed. CRC Press, Boca Raton, FL, Vol. 3, pp. 17–52.

Stringer, J. L., and Peppas, N. A. (1996). Diffusion in radiation-crosslinked poly(ethylene oxide) hydrogels. *J. Controlled Rel.* **42**: 195–202.

Stupp, S. I., LeBonheur, V., Walker, K., Li, L. S., Huggins, K. E., Keser M., and Amstutz, A. (1987). *Science* **276**: 384–389 (1997).

Tanaka, T. (1979). Phase transitions in gels and a single polymer. *Polymer* **20**: 1404–1412.

Tighe, B. J. (1976). The design of polymers for contact lens applications. *Brit. Polym. J.* **8**: 71–90.

Wichterle, O., and Lim, D. (1960). Hydrophilic gels for biological use. *Nature* **185**: 117–118.

Yasuda, H., Peterlin, A., Colton, C. K., Smith, K. A., and Merrill, E. W. (1969). Permeability of solutes through hydrated polymer membranes. III. Theoretical background for the selectivity of dialysis membranes. *Makromol. Chem.* **126**: 177–186.

Yoshida, R., Uchida, K., Kaneko, Y., Sakai, K., Kikcuhi, A., Sakurai, Y., and Okano, T. (1995). Comb-type grafted hydrogels with rapid deswelling response to temperature changes. *Nature* **374**: 240–242.

2.6 APPLICATIONS OF "SMART POLYMERS" AS BIOMATERIALS

Allan S. Hoffman

INTRODUCTION

Stimulus-responsive, "intelligent" polymers are polymers that respond with sharp, large property changes to small changes in physical or chemical conditions. They are also known as "smart" or "environmentally sensitive" polymers. These polymers can take many forms; they may be dissolved in aqueous solution, adsorbed or grafted on aqueous–solid interfaces, or cross-linked in the form of hydrogels.

Many different stimuli have been investigated, and they are listed in Table 1. Typically, when the polymer's critical response is stimulated, the behavior will be as follows (Fig. 1):

- The smart polymer that is dissolved in an aqueous solution will show a sudden onset of turbidity as it phase separates, and if its concentration is high enough, it will convert from a solution to a gel.

TABLE 1 Environmental Stimuli

Physical
  Temperature
  Ionic strength
  Solvents
  Radiation (UV, visible)
  Electric fields
  Mechanical stress
  High pressure
  Sonic radiation
  Magnetic fields

Chemical
  pH
  Specific ions
  Chemical agents

Biochemical
  Enzyme substrates
  Affinity ligands
  Other biological agents

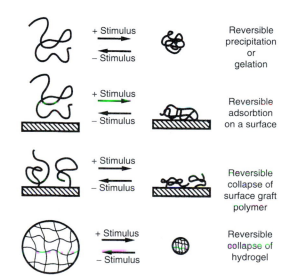

FIG. 1. Schematic illustration showing the different types of responses of "intelligent" polymer systems to environmental stimuli. Note that all systems are reversible when the stimulus is reversed (Hoffman *et al.*, *Journal of Biomedical Materials Research* © 2000).

- The smart polymer that is chemically grafted to a surface and is stimulated to phase separate will collapse, converting that interface from a hydrophilic to a hydrophobic interface. If the smart polymer is in solution and it is stimulated to phase separate, it may physically adsorb to a hydrophobic surface whose composition has a balance of hydrophobic and polar groups that is similar to the phase-separated smart polymer.
- The smart polymer that is cross-linked in the form of a hydrogel will exhibit a sharp collapse, and release much of its swelling solution.

These phenomena are reversed when the stimulus is reversed. Sometimes the rate of reversion is slower when the polymer has to redissolve or the gel has to reswell in aqueous media. The rate of collapse or reversal of smart polymer systems is sensitive to the dimensions of the smart polymer system, and it will be much more rapid for systems with nanoscale dimensions.

Smart polymers may be physically mixed with or chemically conjugated to biomolecules to yield a large and diverse family of polymer–biomolecule hybrid systems that can respond to biological as well as to physical and chemical stimuli. Biomolecules that may be combined with smart polymer systems include proteins and oligopeptides, sugars and polysaccharides, single and double-stranded oligonucleotides, RNA and DNA, simple lipids and phospholipids, and a wide spectrum of recognition ligands and synthetic drug molecules. In addition, polyethylene glycol (PEG, which is also a smart polymer) may be conjugated to the smart polymer backbone to provide it with "stealth" properties (Fig. 2).

Combining a smart polymer and a biomolecule produces a new, smart "biohybrid" system that can synergistically combine the individual properties of the two components to yield new and unusual properties. One could say that these biohybrids are "doubly smart." Among the most important of these systems are the smart polymer–biomolecule conjugates, especially the polymer–drug and polymer–protein conjugates. Such smart bioconjugates, and even a physical mixture of the individual smart polymers and biomolecules, may be physically adsorbed or chemically immobilized on solid surfaces. The biomolecule may also be physically or chemically entrapped in smart hydrogels. All of these hybrid systems have been extensively studied and this chapter reviews these studies. There have been a number of successful applications in both medicine and biotechnology for such smart polymer–biomolecule systems, and as such, they represent an important extension of polymeric biomaterials beyond their well-known uses in implants and medical devices. Several review articles are available on these interesting smart hybrid biomaterials (Hoffman, 1987, 1995, 1997; Hoffman *et al.*, 1999, 2000; Okano *et al.*, 2000).

SMART POLYMERS IN SOLUTION

There are many polymers that exhibit thermally induced precipitation (Table 2), and the polymer that has been studied most extensively is poly(N-isopropyl acrylamide), or PNIPAAm. This polymer is soluble in water below 32°C, and it precipitates sharply as temperature is raised above 32°C (Heskins and Guillet, 1968). The precipitation temperature is called the lower critical solution temperature, or LCST. If the solution contains buffer and salts the LCST will be

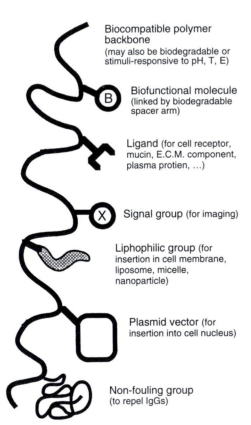

Biocompatible polymer backbone (may also be biodegradable or stimuli-responsive to pH, T, E)

Biofunctional molecule (linked by biodegradable spacer arm)

Ligand (for cell receptor, mucin, E.C.M. component, plasma protien, ...)

Signal group (for imaging)

Liphophilic group (for insertion in cell membrane, liposome, micelle, nanoparticle)

Plasmid vector (for insertion into cell nucleus)

Non-fouling group (to repel IgGs)

FIG. 2. Schematic illustration showing the variety of natural or synthetic biomolecules which may be conjugated to a smart polymer. In some cases, only one molecule may be conjugated, such as a recognition protein, which may be linked to the protein at a reactive terminal group of the polymer, or it may be linked at a reactive pendant group along the polymer backbone. In other cases more than one molecule may be conjugated along the polymer backbone, such as a targeting ligand along with many drug molecules (Hoffman *et al.*, *Journal of Biomedical Materials Research* © 2000).

TABLE 2 Some Polymers and Surfactants that Exhibit Thermally-Induced Phase Separation in Aqueous Solutions

Polymers/Surfactants with Ether Groups
    Poly(ethylene oxide) (PEO)
    Poly(ethylene oxide/propylene oxide) random copolymers
      [poly(EO/PO)]
    PEO-PPO-PEO triblock surfactants (Polyoxamers or Pluronics)
    PLGA-PEO-PLGA triblock polymers
    Alkyl-PEO block surfactants (Brij)
    Poly(vinyl methyl ether)

Polymers with Alcohol Groups
    Poly(hydropropyl acrylate)
    Hydroxypropyl cellulose
    Methylcellulose
    Hydroxypropyl methylcellulose
    Poly(vinyl alcohol) derivatives

Polymers with Substituted Amide Groups
    Poly(N-substituted acrylamides)
    Poly(N-acryloyl pyrrolidine)
    Poly(N-acryloyl piperidine)
    Poly(acryl-L-amino acid amides)

Others
    Poly(methacrylic acid)

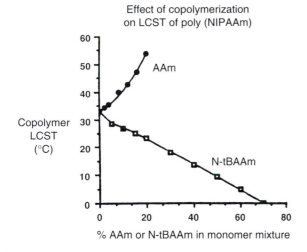

FIG. 3. Copolymerization of a thermally sensitive polymer, PNIPAAm, with a more hydrophilic comonomer, AAm, raises the LCST of the copolymer, whereas copolymerization with a more hydrophobic comonomer, N-tBAAm, lowers the LCST (Hoffman *et al.*, *Journal of Biomedical Materials Research* © 2000).

reduced several degrees. If NIPAAm monomer is copolymerized with more hydrophilic monomers such as acrylamide, the LCST increases and may even disappear. If NIPAAm monomer is copolymerized with more hydrophobic monomers, such as *n*-butylacrylamide, the LCST decreases (Fig. 3) (Priest *et al.*, 1987). NIPAAm may also be copolymerized with pH-sensitive monomers, leading to random copolymers with temperature- and pH-responsive components (Dong and Hoffman, 1987; Zareie *et al.*, 2000) (see also Chapter 7.14 on drug delivery systems). NIPAAm has been copolymerized with pH-responsive macromonomers, leading to graft copolymers that independently exhibit two separate stimulus-responsive behaviors (Chen and Hoffman, 1995).

A family of thermally gelling, biodegradable triblock copolymers has been developed for injectable drug delivery formulations (Vernon *et al.*, 2000; Lee *et al.*, 2001; Jeong *et al.*, 2002). They form a medium viscosity, injectable solution at room temperature and a solid hydrogel at 37°C. These polymers are based on compositions of hydrophobic, degradable polyesters combined with PEO. The copolymers are triblocks with varying MW segments of PLGA and PEO. Typical compositions are PEO-PLGA-PEO and PLGA-PEO-PLGA.

Tirrell (1987) and more recently, Stayton, Hoffman, and co-workers have studied the behavior of pH-sensitive alpha-alkylacrylic acid polymers in solution (Lackey *et al.*, 1999; Murthy *et al.*, 1999; Stayton *et al.*, 2000). As pH is lowered, these polymers become increasingly protonated and hydrophobic, and eventually phase separate; this transition can be sharp, resembling the phase transition at the LCST. If a polymer such as poly(ethylacrylic acid) or poly(propylacrylic acid) is in the vicinity of a lipid bilayer membrane as pH is lowered, the polymer will interact with the membrane and disrupt it. These polymers have been used in intracellular drug delivery to disrupt endosomal membranes as pH drops in the endosome, enhancing the cytosolic delivery of drugs, and avoiding

exposure to lysosomal enzymes. (See further discussion in Chapter 7.14 on drug delivery systems.)

SMART POLYMER–PROTEIN BIOCONJUGATES IN SOLUTION

Smart polymers may be conjugated randomly to proteins by binding the reactive end of the polymer or reactive pendant groups along the polymer backbone to reactive sites on the protein (Fig. 4). One may utilize chain transfer free radical polymerization to synthesize oligomers with one functional end group, which can then be derivatized to form a reactive group that can be conjugated to the protein. NIPAAm has also been copolymerized with reactive comonomers (e.g., N-hydroxysuccinimide acrylate, or NHS acrylate) to yield a random copolymer with reactive pendant groups, which have then been conjugated to the protein. Vinyl monomer groups have been conjugated to proteins to provide sites for copolymerization with free monomers such as NIPAAm. These synthesis methods are described in several publications (Cole *et al.*, 1987; Monji and Hoffman, 1987; Shoemaker *et al.*, 1987; Chen *et al.*, 1990; Chen and Hoffman, 1990, 1994; Yang *et al.*, 1990; Takei *et al.*, 1993a; Monji *et al.*, 1994; Ding *et al.*, 1996) (see also Chapter 2.16 on biologically functional materials).

Normally the lysine amino groups are the most reactive protein sites for random polymer conjugation to proteins, and N-hydroxysuccinimide (NHS) attachment chemistry is most often utilized. Other possible sites include –COOH groups of aspartic or glutamic acid, –OH groups of serine or tyrosine, and –SH groups of cysteine residues. The most likely attachment site will be determined by the reactive group on the polymer and the reaction conditions, especially the pH. Because these conjugations are generally carried out in a nonspecific way, the conjugated polymer can interfere sterically with the protein's active site or modify its microenvironment, typically reducing the bioactivity of the protein. On rare occasions the

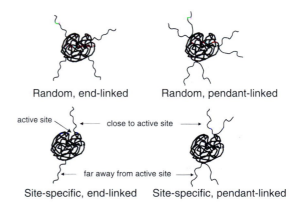

FIG. 4. Various types of random and site-specific smart polymer–protein conjugates. In the latter case, conjugation near the active site of the protein is intended to provide stimulus control of the recognition process of the protein for its ligand, whereas conjugation far away from the active site should avoid any interference of the polymer with the protein's natural activity (Hoffman *et al.*, *Journal of Biomedical Materials Research* © 2000).

conjugation of a polymer increases the activity of the protein. (e.g., Ding *et al.*, 1998).

Biomedical uses of smart polymers in solution have mainly been as conjugates with proteins. Random conjugation of temperature-sensitive (mainly) and pH-sensitive (occasionally) polymers to proteins has been extensively investigated, and applications of these conjugates have been focused on immunoassays, affinity separations, enzyme recovery, and drug delivery (Schneider *et al.*, 1981; Okamura *et al.*, 1984; Nguyen and Luong, 1989; Taniguchi *et al.*, 1989, 1992; Chen and Hoffman, 1990; Monji *et al.*, 1990; Pecs *et al.*, 1991; Park and Hoffman, 1992; Takei *et al.*, 1993b, 1994; Galaev and Mattiasson, 1993; Fong *et al.*, 1999; Anastase-Ravion *et al.*, 2001). In some cases the "smart" polymer is a polyligand, such as polybiotin or poly(glycosyl methacrylate), which is used to phase separate target molecules by complexation to multiple binding sites on target proteins, such as streptavidin and Concanavalin A, respectively (Larsson and Mosbach, 1979; Morris *et al.*, 1993; Nakamae *et al.*, 1994). Wu, Hoffman, and Yager (1992, 1993) have synthesized PNIPAAm–phospholipid conjugates for use in drug delivery formulations as components of thermally sensitive composites and liposomes.

SMART POLYMERS ON SURFACES

One may covalently graft a polymer to a surface by exposing the surface to ionizing radiation in the presence of the monomer (and in the absence of air), or by preirradiating the polymer surface in air, and later contacting the surface with the monomer solution and heating in the absence of air. (See also Chapter 1.4 on surface properties of materials.) These surfaces exhibit stimulus-responsive changes in wettability (Uenoyama and Hoffman, 1988; Takei *et al.*, 1994; Kidoaki *et al.*, 2001). Ratner and co-workers have used a gas plasma discharge to deposit temperature-responsive coatings from a NIPAAm monomer vapor plasma (Pan *et al.*, 2001). Okano and Yamato and co-workers have utilized the radiation grafting technique to form cell culture surfaces having a surface layer of grafted PNIPAAm. (Yamato and Okano, 2001; Shimizu *et al.*, 2003). They have cultured cells to confluent sheets on these surfaces at 37°C, which is above the LCST of the polymer. When the PNIPAAm collapses, the interface becomes hydrophobic and leads to adsorption of cell adhesion proteins, enhancing the cell culture process. Then when the temperature is lowered, the interface becomes hydrophilic as the PNIPAAm chains rehydrate, and the cell sheets release from the surface (along with the cell adhesion proteins). The cell sheet can be recovered and used in tissue engineering, e.g., for artificial cornea and other tissues. Patterned surfaces have also been prepared (Yamato *et al.*, 2001). Smart polymers may also be grafted to surfaces to provide surfaces of gradually varying hydrophilicity and hydrophobicity as a function of the polymer composition and conditions. This phenomenon has been applied by Okano, Kikuchi, and co-workers to prepare chromatographic column packing, leading to eluate-free ("green") chromatographic separations (Kobayashi *et al.*, 2001; Kikuchi and Okano, 2002). Ishihara *et al.* (1982, 1984b) developed photoresponsive coatings and membranes that reversibly changed surface wettability or swelling, respectively, due to the photoinduced isomerization of an azobenzene-containing polymer.

SITE-SPECIFIC SMART POLYMER BIOCONJUGATES ON SURFACES

Conjugation of a responsive polymer to a specific site near the ligand-binding pocket of a genetically engineered protein is a powerful new concept. Such site-specific protein–smart polymer conjugates can permit sensitive environmental control of the protein's recognition process, which controls all living systems. Stayton and Hoffman *et al.* (Stayton *et al.*, 2000) have designed and synthesized smart polymer–protein conjugates where the polymer is conjugated to a specific site on the protein, usually a reactive –SH thiol group from cysteine that has been inserted at the selected site (Fig. 5). This is accomplished by utilizing cassette mutagenesis to insert a site-specific mutation into the DNA sequence of the protein, and then cloning the mutant in cell culture. This method is applicable only to proteins whose complete peptide sequence is known. The preparation of the reactive smart polymer is similar to the method described above, but now the reactive end or pendant groups and the reaction conditions are specifically designed to favor conjugation to –SH groups rather than to –NH2 groups. Typical SH-reactive polymer end groups include maleimide and vinyl sulfone groups.

The specific site for polymer conjugation can be located far away from the active site (Chilkoti *et al.*, 1994), in order to avoid interference with the biological functioning of the protein, or nearby or even within the active site, in order to control the ligand–protein recognition process and the biological activity of the protein (Fig. 4) (Ding *et al.*, 1999, 2001; Bulmus *et al.*, 1999; Stayton *et al.*, 2000; Shimoboji *et al.*, 2001, 2002a, b, 2003). The latter has been most studied by

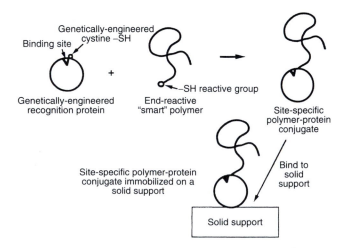

FIG. 5. Schematic illustration of the process for preparing an immobilized, site-specific conjugate of a smart polymer with a genetically-engineered, mutant protein (Hoffman *et al.*, *Journal of Biomedical Materials Research* © 2000).

the Stayton/Hoffman group. Temperature-, pH-, and light-sensitive smart polymers have been used to form such novel, "doubly smart" bioconjugates. Since the objective is to control the activity of the protein, and not to phase separate it, these smart polymer–engineered protein bioconjugates have usually been immobilized on the surfaces of microbeads or nanobeads. Stayton, Hoffman, and co-workers have used such beads in microfluidic devices for immunoassays (Malmstadt *et al.*, 2003). Earlier work by Hoffman and co-workers established the importance of matching the smart polymer composition with the surface composition in order to enhance the stimulus-driven adsorption of the smart polymer on the surface (Miura *et al.*, 1994). Others have also recently utilized this phenomenon in microfluidic devices (Huber *et al.*, 2003).

The proteins that have been most studied by the Stayton/Hoffman group to date include streptavidin and the enzyme cellulase. PNIPAAm–streptavidin site-specific bioconjugates have been used to control access of biotin to its binding site on streptavidin, and have enabled separation of biotinylated proteins according to the size of the protein (Ding *et al.*, 2001). Ding, Stayton, and Hoffman *et al.* (1999) also found that raising the temperature to thermally induce the collapse of the polymer "triggered" the release of the bound biotin molecules (Ding *et al.*, 1999). For the site-specific enzyme conjugates, a combined temperature- and light-sensitive polymer was conjugated to specific sites on an endocellulase, which provided on–off control of the enzyme activity with either light or temperature (Shimoboji *et al.*, 2001, 2002a, b, 2003).

Triggered release of bound ligands by the smart polymer–engineered protein bioconjugates could be used to release therapeutics, such as for topical drug delivery to the skin or mucosal surfaces of the body, and also for localized delivery of drugs within the body by stimulated release at pretargeted sites using noninvasive, focused stimuli, or delivery of stimuli from catheters. Triggered release could also be used to release and recover affinity-bound ligands from chromatographic and other supports in eluate-free conditions, including capture and release of specific cell populations to be used in stem cell and bone marrow transplantation. These processes could involve two different stimulus-responsive polymers with sensitivities to the same or different stimuli. For delicate target ligands such as peptides and proteins, recovery could be affected without the need for time-consuming and harsh elution conditions. Triggered release could also be used to remove inhibitors, toxins, or fouling agents from the recognition sites of immobilized or free enzymes and affinity molecules, such as those used in biosensors, diagnostic assays, or affinity separations. This could be used to "regenerate" such recognition proteins for extended process use. Light-controlled binding and release of site-specific protein conjugates may be utilized as a molecular switch for various applications in biotechnology, medicine, and bioelectronics, including hand-held diagnostic devices, biochips, and lab-on-a-chip devices.

Fong, Stayton, and Hoffman (Fong *et al.*, 1999) have developed an interesting construct to control the distance of the PNIPAAm from the active site. For this purpose, they conjugated one sequence of complementary nucleotides to a specific site near the binding pocket of streptavidin, and a second sequence to the end of a PNIPAAm chain. Then, by controlling the location and length of the complementary sequence, the self-assembly via hybridization of the two single-chain DNA sequences could be used to control the distance of the polymer from the streptavidin binding site.

SMART POLYMER HYDROGELS

When a smart polymer is cross-linked to form a gel, it will collapse and re-swell in water as a stimulus raises or lowers it through its critical condition. PNIPAAm gels have been extensively studied, starting with the pioneering work of Toyoichi Tanaka in 1981 (Tanaka, 1981). Since then, the properties of PNIPAAm hydrogels have been widely investigated in the form of beads, slabs, and multilamellar laminates (Park and Hoffman, 1992a, b, 1994; Hu *et al.*, 1995, 1998; Mitsumata *et al.*, 2001; Kaneko *et al.*, 2002; Gao and Hu, 2002). Okano and co-workers have developed smart gels that collapse very rapidly, by grafting PNIPAAm chains to the PNIPAAm backbone in a cross-linked PNIPAAm hydrogel (Yoshida *et al.*, 1995; Masahiko *et al.*, 2003). Smart hydrogel compositions have been developed that are both thermally gelling and biodegradable (Zhong *et al.*, 2002; Yoshida *et al.*, 2003). These sol-gel systems have been used to deliver drugs by *in vivo* injections and are discussed in the section on smart polymers in solution, and also in more detail in Chapter 7.14 on drug delivery systems.

Hoffman and co-workers were among the first to recognize the potential of PNIPAAm hydrogels as biomaterials; they showed that the smart gels could be used (a) to entrap enzymes and cells, and then turn them on and off by inducing cyclic collapse and swelling of the gel, and (b) to deliver or remove biomolecules, such as drugs or toxins, respectively, by stimulus-induced collapse or swelling (Dong and Hoffman, 1986, 1987, 1990; Park and Hoffman, 1988, 1990a, b, c) (Fig. 6). One unique hydrogel was developed by Dong and Hoffman (1991). This pH- and temperature-sensitive hydrogel was based on a random copolymer of NIPAAm and AAc, and it was shown to release a model drug linearly over a 4-hour period as the pH went from gastric to enteric conditions at 37°C. At body temperature the NIPAAm component was trying to maintain the gel in the collapsed state, while as the pH went from acidic to neutral conditions, the AAc component was becoming ionized, forcing the gel to swell and slowly release the drug (see Fig. 6B).

Kim, Bae, and co-workers have investigated smart gels containing entrapped cells that could be used as artificial organs (Vernon *et al.*, 2000). Matsuda and co-workers have incorporated PNIPAAm into physical mixtures with natural polymers such as hyaluronic acid and gelatin, for use as tissue engineering scaffolds (Ohya *et al.*, 2001a, b).

Peppas and co-workers (Robinson and Peppas, 2002) have studied pH-sensitive gels in the form of nanospheres. Nakamae, Hoffman, and co-workers developed novel compositions of smart gels containing phosphate groups that were used to bind cationic proteins as model drugs and then to release them by a combination of thermal stimuli and ion exchange (Nakamae *et al.*, 1992, 1997; Miyata *et al.*, 1994).

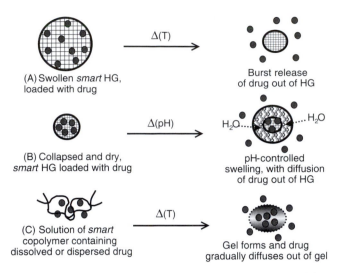

(A) Swollen *smart* HG, loaded with drug

$\Delta(T)$

Burst release of drug out of HG

(B) Collapsed and dry, *smart* HG loaded with drug

$\Delta(pH)$

H_2O .... H_2O

pH-controlled swelling, with diffusion of drug out of HG

(C) Solution of *smart* copolymer containing dissolved or dispersed drug

$\Delta(T)$

Gel forms and drug gradually diffuses out of gel

FIG. 6. Schematic illustration showing three ways that smart gel formulations may be stimulated to release bioactive agents: (A) thermally induced collapse, which is relevant to skin or mucosal drug delivery; (B) pH-induced swelling, which is relevant to oral drug delivery, where the swelling is induced by the increase in pH in going from the gastric to enteric regions; and (C) sol-to-gel formation, which is relevant to injectable or topical formulations of a triblock copolymer solution that are thermally gelled at body temperature. For *in vivo* uses, the block copolymer is designed to be degradable. The first two apply to cross-linked gels applied topically or orally, and the third is relevant to thermally induced formation of gels from polymer solutions that are delivered topically or by injection. (See also Chapter 7.14 on drug delivery systems.)

SMART GELS THAT RESPOND TO BIOLOGICAL STIMULI

A number of drug delivery devices have been designed to respond to biologic signals in a feedback manner. Most of these gels contain an immobilized enzyme. Heller and Trescony (1979) were among the first to work with smart enzyme gels. In this early example, urease was immobilized in a gel, and urea was metabolized to produce ammonia, which caused a local pH change, leading to a permeability change in the surrounding gel. Ishihara *et al.* (1985) also developed a urea-responsive gel containing immobilized urease. Smart enzyme gels containing glucose oxidase (GOD) were designed to respond to a more relevant signal, that of increasing glucose concentration. In a typical device, when glucose concentration increases, the entrapped GOD converts the glucose in the presence of oxygen to gluconic acid and hydrogen peroxide. The former lowers pH, and the latter is an oxidizing agent. Each of these byproduct signals has been used in various smart hydrogel systems to increase the permeability of the gel barrier to insulin delivery (Horbett *et al.*, 1984; Albin *et al.*, 1985; Ishihara *et al.*, 1983, 1984a; Ishihara and Matsui, 1986; Ito *et al.*, 1989; Iwata and Matsuda, 1988).

In one case, the lowered pH due to the GOD by-product, gluconic acid, accelerated hydrolytic erosion of the polymer matrix that also contained entrapped insulin, releasing the insulin (Heller *et al.*, 1990). Siegel and co-workers have used the glucose-stimulated swelling and collapse of hydrogels containing entrapped glucose oxidase to drive a hydrogel piston in an oscillating manner, for release of insulin in a glucose-driven, feedback manner (Dhanarajan *et al.*, 2002). Other smart enzyme gels for drug delivery have been developed based on activation of an inactivated enzyme by a biologic signal (Schneider *et al.*, 1973; Roskos *et al.*, 1993).

Smart gels have also been developed that are based on affinity recognition of a biologic signal. Makino *et al.* (1990) developed a smart system that contained glycosylated insulin bound by affinity of its glucose groups to an immobilized Concanavalin A in a gel. When glucose concentration increases, the free glucose competes off the insulin, which is then free to diffuse out of the gel. Nakamae *et al.* (1994) developed a gel based on a similar concept, using a cross-linked poly(glycosylethyl methacrylate) hydrogel containing physically or chemically entrapped Concanavalin A. In this case, the ConA is bound by affinity to the pendant glucose groups on the polymer backbone, acting as a cross-linker because of its four affinity binding sites for glucose; when free glucose concentration increases, the ConA is competed off the polymer backbone. This leads to swelling of the gel, which acts to increase permeation of insulin through the gel. Miyata and co-workers have designed and synthesized smart affinity hydrogels that are stimulated to swell or collapse by the binding of affinity biomolecules (Miyata *et al.*, 1999, 2002).

Chapter 7.14 covers smart bioresponsive gels as drug delivery systems in more detail.

CONCLUSIONS

Smart polymers in solution, on surfaces, and as hydrogels have been utilized in many interesting ways, especially in combination with biomolecules such as proteins and drugs. Important applications include affinity separations, enzyme processes, immunoassays, drug delivery, and toxin removal. These smart polymer–biomolecule systems represent an important extension of polymeric biomaterials beyond their well-known uses in implants and medical devices.

Bibliography

Albin, G., Horbett, T. A., and Ratner, B. D. (1985). Glucose sensitive membranes for controlled delivery of insulin: insulin transport studies. *J. Controlled Release* **2**: 153–164.

Anastase-Ravion, S., Ding, Z., Pelle, A., Hoffman, A. S., and Letourneur, D. (2001). New antibody purification procedure using a thermally-responsive polyNIPAAm-dextran derivative conjugate. *J. Chromatogr. B.* **761**: 247–254.

Bulmus, V., Ding, Z., Long, C. J., Stayton, P. S., and Hoffman, A. S. (1999). Design, synthesis and site-specific conjugation of a pH- and temperature-sensitive polymer to streptavidin for pH-controlled binding and triggered release of biotin. *Bioconj. Chem.* **11**: 78–83.

Chen, J. P., and Hoffman, A. S. (1990). Polymer–protein conjugates. II. Affinity precipitation of human IgG by poly(N-isopropyl acrylamide) – protein A conjugates. *Biomaterials* **11**: 631–634.

Chen, G., and Hoffman, A. S. (1994). Synthesis of carboxylated poly(NIPAAm) oligomers and their application to form thermo-reversible polymer-enzyme conjugates. *J. Biomater. Sci. Polymer Edn.* **5**: 371–382.

Chen, G., and Hoffman, A. S. (1995). Graft copolymer compositions that exhibit temperature-induced transitions over a wide range of pH. *Nature* **373**: 49–52.

Chen, J. P., Yang, H. J., and Hoffman, A. S. (1990). Polymer–protein conjugates. I. Effect of protein conjugation on the cloud point of poly(*N*-isopropyl acrylamide). *Biomaterials* **11**: 625–630.

Chilkoti, A., Chen, G., Stayton, P. S., and Hoffman, A. S. (1994). Site-specific conjugation of a temperature-sensitive polymer to a genetically-engineered protein. *Bioconj. Chem.* **5**: 504–507.

Cole, C. A., Schreiner, S. M., Priest, J. H., Monji, N. and Hoffman, A. S. (1987). *N*-Isopropyl acrylamide and *N*-acryl succinimide copolymers: a thermally reversible water soluble activated polymer for protein conjugation. *Reversible Polymeric Gels and Related Systems*, ACS Symposium Series, Vol. 350, P. Russo, ed. ACS, Washington, D.C., pp. 245–254.

Dhanarajan, A. P., Misra, G. P., and Siegel, R. A. (2002). Autonomous chemomechanical oscillations in a hydrogel/enzyme system driven by glucose. *J. Phys. Chem.* **106**: 8835–8838.

Ding, Z. L., Chen, G., and Hoffman, A. S. (1996). Synthesis and purification of thermally-sensitive oligomer–enzyme conjugates of poly(NIPAAm)–trypsin. *Bioconj. Chem.* **7**: 121–125.

Ding, Z. L., Chen, G., and Hoffman, A. S. (1998). Properties of polyNIPAAm–trypsin conjugates. *J. Biomed. Mater. Res.* **39**: 498–505.

Ding, Z. L., Long, C. J., Hayashi, Y., Bulmu, E. V., Hoffman, A. S., and Stayton, P. S. (1999). Temperature control of biotin binding and release with a streptavidin–polyNIPAAm site-specific conjugate. *J. Bioconj. Chem.* **10**: 395–400.

Ding, Z. L., Shimoboji, T., Stayton, P. S., and Hoffman, A. S. (2001). A smart polymer shield that controls the binding of different size biotinylated proteins to streptavidin. *Nature* **411**: 59–62.

Dong, L. C., and Hoffman, A. S. (1986). Thermally reversible hydrogels: III. Immobilization of enzymes for feedback reaction control. *J. Contr. Rel.* **4**: 223–227.

Dong, L. C., and Hoffman, A. S. (1987). Thermally reversible hydrogels: swelling characteristics and activities of copoly(NIPAAm-AAm) gels containing immobilized asparaginase. in *Reversible Polymeric Gels and Related Systems*, ACS Symposium Series, Vol 350, P. Russo, ed. ACS, Washington, D.C., pp. 236–244.

Dong, L. C., and Hoffman, A. S. (1990). Synthesis and application of thermally-reversible heterogels for drug delivery. *J. Contr. Release* **13**: 21–32.

Dong, L. C., and Hoffman, A. S. (1991). A novel approach for preparation of pH- and temperature-sensitive hydrogels for enteric drug delivery. *J. Contr. Release* **15**: 141–152.

Fong, R. B., Ding, Z. L., Long, C. J., Hoffman, A. S., and Stayton, P. S. (1999). Thermoprecipitation of streptavidin via oligonucleotide-mediated self-assembly with poly(NIPAAm). *Bioconj. Chem.* **10**: 720–725.

Galaev, I. Y., and Mattiasson, B. (1993). Affinity thermoprecipitation: Contribution of the efficiency and access of the ligand. *Biotechnol. Bioeng.* **41**: 1101–1106.

Gao, J., and Hu, Z. B. (2002). Optical properties of *N*-isopropylacrylamide microgel spheres in water. *Langmuir* **18**: 1360–1367.

Heller, J., and Trescony, P. V. (1979). Controlled drug release by polymer dissolution II. Enzyme-mediated delivery device. *J. Pharm. Sci.* **68**: 919–921.

Heller, J., Chang, A. C., Rodd, G., and Grodsky, G. M. (1990). Release of insulin from a pH-sensitive poly (ortho ester). *J. Controlled Release* **14**: 295–304.

Heskins, H., and Guillet, J. E. (1968). Solution properties or poly(*N*-isopropyl acrylamide). *J. Macromol. Sci. Chem. A2* **6**: 1209.

Hoffman, A. S. (1987). Applications of thermally reversible polymers and hydrogels in therapeutics and diagnostics. *J. Contr. Rel.* **6**: 297–305.

Hoffman, A. S. (1995). Intelligent polymers in medicine and biotechnology. *Macromol. Symp.* **98**: 645–664.

Hoffman, A. S. (1997). Intelligent polymers in medicine and biotechnology. in *Controlled Drug Delivery*, K. Park, ed. ACS Publications, Washington, D.C.

Hoffman, A. S., *et al.* (2000). Really smart bioconjugates of smart polymers and receptor proteins. *J. Biomed. Mater. Res.* **52**: 577–586.

Hoffman, A. S., Chen, G., Wu, X., Ding, Z., Matsuura, J. E., and Gombotz, W. R. (1999). Stimuli-responsive polymers grafted onto polyacrylic acid and chitosan backbones as bioadhesive carriers for mucosal drug delivery. in *Frontiers in Biomedical Polymer Applications*, R. M. Ottenbrite, ed. Technomic Publ., Lancaster, UK, pp. 17–29.

Horbett, T. A., Kost, J., and Ratner, B. D. (1984). Swelling behavior of glucose sensitive membranes. in *Polymers as Biomaterials*, S. Shalaby, A. S. Hoffman, B. D. Ratner, and T. A., Horbett, eds. Plenum Press, New York, pp. 193–207.

Hu, Z. B., Zhang, X. M., and Li, Y. (1995). Synthesis and application of modulated polymer gels. *Science* **269**: 525.

Hu, Z. B., Chen, Y. Y., Wang, C. J., Zheng, Y. Y. and Li, Y. (1998). Polymer gels with engineered environmentally responsive surface patterns. *Nature* **393**: 149.

Huber, D. L., Manginell, R. P., Samara, M. A., Kim, B. I., and Bunkar, B. C. (2003). Programmed adsorption and release of proteins in a microfluidic device. *Science* **301**: 352.

Ishihara, K., and Matsui, K. (1986). Glucose-responsive insulin release from polymer capsule. *J. Polymer Sci., Polymer Lett. Ed.* **24**: 413–417.

Ishihara, K., Okazaki, A., Negishi, N., Shinohara, I., Okano, T., Kataoka, K., and Sakurai, Y. (1982). Photo-induced change in wettability and binding ability of azoaromatic polymer. *J. Appl. Polymer Sci.* **27**: 239–245.

Ishihara, K., Kobayashi, M., and Shinohara, I. (1983). Control of insulin permeation through a polymer membrane with responsive function for glucose. *Makromol. Chem. Rapid Commun.* **4**: 327–331.

Ishihara, K., Kobayashi, M., Ishimaru, N., and Shinohara, I. (1984a). Glucose-induced permeation control of insulin through a complex membrane consisting of immobilized glucose oxidase and a poly(amine). *Polymer J.* **16**: 625–631.

Ishihara, K., Hamada, N., Kato, S., and Shinohara, I. (1984b). Photo-induced swelling control of amphiphilic azoaromatic polymer membrane. *Polymer Sci., Polymer Chem. Ed.* **22**: 21–128.

Ishihara, K., Muramoto, N., Fujii, H., and Shinohara, I. (1985). Preparation and permeability of urea-responsive polymer membrane consisting of immobilized urease and a poly(aromatic carboxylic acid). *J. Polymer Sci., Polymer Lett. Ed.* **23**: 531–535.

Ito, Y., Casolaro, M., Kono, K., and Imanishi, Y. (1989). An insulin-releasing system that is responsive to glucose. *J. Controlled Release* **10**: 195–203.

Iwata, H., and Matsuda, T. (1988). Preparation and properties of novel environment-sensitive membranes prepared by graft polymerization onto a porous substrate. *J. Membrane Sci.* **38**: 185–199.

Jeong, B., Kim, S. W., and Bae, Y. H. (2002). Thermosensitive sol–gel reversible hydrogels. *Adv. Drug Delivery Rev.* **54**: 37–51.

Kaneko, D., Gong, J. P., and Osada, Y. (2002). Polymer gels as soft and wet chemomechanical systems—an approach to artificial muscles. *J. Mater. Chem.* **12**: 2169–2177.

Kawaguchi H., Kisara K., Takahashi T., Achiha K., Yasui M., and Fujimoto K. (2000). Versatility of thermosensitive particles. *Macromol. Symp.* **151:** 591–598.

Kawaguchi H., Isono Y., and Tsuji S. (2002). Hairy particles prepared by living radical graft polymerization. *Macromol. Symp.* **179:** 191–206.

Kidoaki, S., Ohya, S., Nakayama, Y., and Matsuda T. (2001). Thermoresponsive structural change of a PNIPAAm graft layer measured with AFM. *Langmuir* **17:** 2402–2407.

Kikuchi, A., and Okano, T. (2002). Intelligent thermoresponsive polymeric stationary phases for aqueous chromatography of biological compounds. *Progr. Polymer Sci.* **27:** 1165–1193.

Kobayashi, J., Kikuchi, A., Sakai, K., and Okano, T. (2001). Aqueous chromatography utilizing pH-/temperature-responsive polymers as column matrix surfaces for separation of ionic bioactive compounds. *Anal. Chem.* **73(9):** 2027–2033.

Lackey, C. A., Murthy, N., Press, O. W., Tirrell, D. A., Hoffman, A. S., and Stayton, P. S. (1999). Hemolytic activity of pH-responsive polymer–streptavidin bioconjugates. *Bioconj. Chem.* **10:** 401–405.

Larsson, P. O., and Mosbach, K. (1979). Affinity precipitation of enzymes. *FEBS Lett.* **98:** 333–338.

Lee, D. S., Shim, M. S., Kim, S. W., Lee, H., Park, I., and Chang, T. (2001). Novel thermoreversible gelation of biodegradable PLGA-*block*-PEO-*block*-PLGA triblock copolymers in aqueous solution. *Macromol. Rapid Commun.* **22:** 587–592.

Makino, K., Mack, E. J., Okano, T., and Kim, S. W. (1990). A microcapsule self-regulating delivery system for insulin. *J. Controlled Release* **12:** 235–239.

Malmstadt, N., Yager, P., Hoffman, A. S., and Stayton, P. S. (2003). A smart microfluidic affinity chromatography matrix composed of poly(*N*-isopropylacrylamide)-coated beads. *Anal. Chem.* **75:** 2943–2949.

Masahiko, A., Matsuura, T., Kasai, M., Nakahira, T., Hara, Y., and Okano, T. (2003). Preparation of comb-type N-isopropylacrylamide hydrogel beads and their application for size-selective separation media. *Biomacromolecules* **4:** 395–403.

Mitsumata, T., Gong, J. P., and Osada, Y. (2001). Shape memory functions and motility of amphiphilic polymer gels. *Polymer Adv. Technol.* **12:** 136–150.

Miura, M., Cole, C. A., Monji, N., and Hoffman, A. S. (1994). Temperature-dependent adsorption/desorption behavior of LCST polymers on various substrates. *J. Biomater. Sci. Polymer Ed.* **5:** 555–568.

Miyata, T., Nakamae, K., Hoffman, A. S., and Kanzaki, Y. (1994). Stimuli-sensitivities of hydrogels containing phosphate groups. *Macromol. Chem. Phys.* **195:** 1111–1120.

Miyata, T., Asami, N., and Uragami, T. (1999). A reversibly antigen-responsive hydrogel. *Nature* **399:** 766–769.

Miyata, T., Uragami, T., and Nakamae, K. (2002). Biomolecule-sensitive hydrogels. *Adv. Drug Delivery Rev.* **54:** 79–98.

Monji, N. and Hoffman, A. S. (1987). A novel immunoassay system and bioseparation process based on thermal phase separating polymers. *Appl. Biochem. Biotechnol.* **14:** 107–120.

Monji, N., Cole, C. A., and Hoffman, A. S. (1994). Activated, N-substituted acrylamide polymers for antibody coupling: application to a novel membrane-based immunoassay. *J. Biomater. Sci. Polymer Ed.* **5:** 407–420.

Monji, N., Cole, C. A., Tam, M., Goldstein, L., Nowinski, R. C., and Hoffman, A. S. (1990). Application of a thermally-reversible polymer–antibody conjugate in a novel membrane-based immunoassay. *Biochem. Biophys. Res. Commun.* **172:** 652–660.

Morris, J. E., Hoffman, A. S., and Fisher, R. R. (1993). Affinity precipitation of proteins by polyligands. *Biotechnol. Bioeng.* **41:** 991–997.

Murthy, N., Stayton, P. S., and Hoffman, A. S. (1999). The design and synthesis of polymers for eukaryotic membrane disruption. *J. Controlled Release* **61:** 137–143.

Nakamae, K., Miyata, T., and Hoffman, A. S. (1992). Swelling behavior of hydrogels containing phosphate groups. *Macromol. Chem.* **193:** 983–990.

Nakamae, K., Miyata, T., Jikihara, A., and Hoffman, A. S. (1994). Formation of poly(glucosyloxyethyl methacrylate)–Concanavalin A complex and its glucose sensitivity. *J. Biomater. Sci. Polymer Ed.* **6:** 79–90.

Nakamae, K., Nizuka, T., Miyata, T., Furukawa, M., Nishino, T., Kato, K., Inoue, T., Hoffman, A. S., and Kanzaki, Y. (1997). Lysozyme loading and release from hydrogels carrying pendant phosphate groups. *J. Biomater. Sci., Polymer Ed.* **9:** 43–53.

Nguyen, A. L., and Luong, J. H. T. (1989). Syntheses and application of water soluble reactive polymers for purification and immobilization of biomolecules. *Biotechnol. Bioeng.* **34:** 1186–1190.

Ohya, S., Nakayama, Y., and Matsuda, T. (2001a). Thermoresponsive artificial extracellular matrix for tissue engineering: hyaluronic acid bioconjugated with poly(*N*-isopropylacrylamide) grafts. *Biomacromolecules* **2:** 856–863.

Ohya, S., Nakayama, Y., and Matsuda, T. (2001b). Material design for an artificial extracellular matrix: cell entrapment in poly(*N*-isopropylacrylamide) (PNIPAM)-grafted gelatin hydrogel. *J. Artif. Organs* **4:** 308–314.

Okamura, K., Ikura, K., Yoshikawa, M., Sakaki, R., and Chiba, H. (1984). Soluble–insoluble interconvertible enzymes. *Agric. Biol. Chem.* **48:** 2435–2440.

Okano, T., Kikuchi, A., and Yamato, M. (2000). Intelligent hydrogels and new biomedical applications. in *Biomaterials and Drug Delivery toward the New Millennium*. Han Rim Won Publishing Co., Seoul, Korea, pp. 77–86.

Pan, Y. V., Wesley, R. A., Luginbuhl, R., Denton, D. D., and Ratner, B. D. (2001). Plasma-polymerized *N*-isopropylacrylamide: synthesis and characterization of a smart thermally responsive coating. *Biomacromolecules* **2:** 32–36.

Park, T. G., and Hoffman, A. S. (1988). Effect of temperature cycling on the activity and productivity of immobilized β-galactosidase in a thermally reversible hydrogel bead reactor. *Appl. Biochem. Biotechnol.* **19:** 1–9.

Park, T. G., and Hoffman, A. S. (1990a). Immobilization and characterization of β-galactosidase in thermally reversible hydrogel beads. *J. Biomed. Mater. Res.* **24:** 21–38.

Park, T. G., and Hoffman, A. S. (1990b). Immobilization of *A. simplex* cells in a thermally-reversible hydrogel: effect of temperature cycling on steroid conversion. *Biotech. Bioeng.* **35:** 52–159.

Park, T. G., and Hoffman, A. S. (1990c). Immobilized biocatalysts in reversible hydrogels. in *Enzyme Engineering X*, A. Tanaka, ed. *Ann. N.Y. Acad. Sci.*, Vol. 613, pp. 588–593.

Park, T. G., and Hoffman, A. S. (1992a). Preparation of large, uniform size temperature-sensitive hydrogel beads. *J. Polymer Sci. A Polymer Chem.* **30:** 505–507.

Park, T. G., and Hoffman, A. S. (1992b). Synthesis and characterization of pH- and/or temperature-sensitive hydrogels. *J. Appl. Polymer Sci.* **46:** 659–671.

Park, T. G., and Hoffman, A. S. (1993). Synthesis and characterization of a soluble, temperature-sensitive polymer-conjugated enzyme. *J. Biomater. Sci. Polymer Ed.* **4:** 493–504.

Park, T. G., and Hoffman, A. S. (1994). Estimation of temperature-dependent pore sizes in poly(NIPAAm) hydrogel beads. *Biotechnol. Progr.* **10:** 82–86.

Pecs, M., Eggert, M., and Schügerl, K. (1991). Affinity precipitation of extracellular microbial enzymes. *J. Biotechnol.* **21:** 137–142.

Priest, J. H., Murray, S., Nelson, R. G., and Hoffman, A. S. (1987). LCSTs of aqueous copolymers of N-isopropyl acrylamide and other N-substituted acrylamides. *Reversible Polymeric Gels and Related Systems*, ACS Symposium Series, Vol. 350, P. Russo, ed. ACS, Washington, D.C., pp. 255–264.

Robinson, D. N., and Peppas, N. A. (2002). Preparation and characterization of pH-responsive poly(methacrylic acid-g-ethylene glycol) nanospheres. *Macromolecules* 35: 3668–3674.

Roskos, K. V., Tefft, J. A. and Heller, J. (1993). A morphine-triggered delivery system useful in the treatment of heroin addiction. *Clin. Mater.* 13: 109–119.

Schneider, M., Guillot, C., and Lamy, B. (1981). The affinity precipitation technique: application to the isolation and purification of trypsin from bovine pancreas. *Ann. N.Y. Acad. Sci.* 369: 257–263.

Schneider, R. S., Lidquist, P., Wong, E. T., Rubenstein, K. E., and Ullman, E. F. (1973). Homogeneous enzyme immunoassay for opiates in urine. *Clin. Chem.* 19: 821–825.

Shimizu, T., Yamato, M., Kikuchi, A., and Okano, T. (2003). Cell sheet engineering for myocardial tissue reconstruction. *Biomaterials* 24: 2309–2316.

Shimoboji, T., Ding, Z., Stayton, P. S., and Hoffman, A. S. (2001). Mechanistic investigation of smart polymer–protein conjugates. *Bioconj. Chem.* 12: 314–319.

Shimoboji, T., Ding, Z. L., Stayton, P. S., and Hoffman, A. S. (2002a). Photoswitching of ligand association with a photoresponsive polymer–protein conjugate. *Bioconj. Chem.* 13: 915–919.

Shimoboji, T., Larenas, E., Fowler, T., Kulkarni, S., Hoffman, A. S., and Stayton, P. S. (2002b). Photoresponsive polymer–enzyme switches. *Proc. Natl. Acad. Sci. USA* 99: 16592–16596.

Shimoboji, T., Larenas, E., Fowler, T., Hoffman, A. S., and Stayton, P. S.(2003). Temperature-induced switching of enzyme activity with smart polymer–enzyme conjugates. *Bioconj. Chem.* 14: 517–525.

Shoemaker, S., Hoffman, A. S., and Priest, J. H. (1987). Synthesis of vinyl monomer–enzyme conjugates. *Appl. Biochem. Biotechnol.* 15: 11.

Stayton, P. S., Hoffman, A. S., Murthy, N., Lackey, C., Cheung, C., Tan, P., Klumb, L. A., Chilkoti, A., Wilbur, F. S., and Press, O. W. (2000). Molecular engineering of proteins and polymers for targeting and intracellular delivery of therapeutics. *J. Contr. Rel.* 65: 203–220.

Stayton, P. S., Shimoboji, T., Long, C., Chilkoti, A., Chen, G., Harris, J. M., and Hoffman, A. S. (1995). Control of protein–ligand recognition using a stimuli-responsive polymer. *Nature* 378: 472–474.

Takei, Y. G., Aoki, T., Sanui, K., Ogata, N., Okano, T., and Sakurai, Y. (1993A). Temperature-responsive bioconjugates. 1. Synthesis of temperature-responsive oligomers with reactive end groups and their coupling to biomolecules. *Bioconj. Chem.* 4: 42–46.

Takei, Y. G., Aoki, T., Sanui, K., Ogata, N., Okano, T., and Sakurai, Y. (1993b). Temperature-responsive bioconjugates. 2. Molecular design for temperature-modulated bioseparations. *Bioconj. Chem.* 4: 341–346.

Takei, Y. G., Matsukata, M., Aoki, T., Sanui, K., Ogata, N., Kikuchi, A., Sakurai, Y., and Okano, T. (1994a). Temperature-responsive bioconjugates. 3. Antibody-poly(N-isopropylacrylamide) conjugates for temperature-modulated precipitations and affinity bioseparations. *Bioconj. Chem.* 5: 577–582.

Takei, Y. G., Aoki, T., Sanui, K., Ogata, N., Sakurai, Y., and Okano, T. (1994b). Dynamic contact angle measurements of temperature-responsive properties for PNIPAAm grafted surfaces. *Macromolecules* 27: 6163–6166.

Tanaka, T. (1981). Gels. *Sci. Am.* 244: 124.

Taniguchi, M., Kobayashi, M., and Fujii, M. (1989). Properties of a reversible soluble–insoluble cellulase and its application to repeated hydrolysis of crystalline cellulose. *Biotechnol. Bioeng.* 34: 1092–1097.

Taniguchi, M., Hoshino, K., Watanabe, K., Sugai, K., and Fujii, M. (1992). Production of soluble sugar from cellulosic materials by repeated use of a reversibly soluble-autoprecipitating cellulase. *Biotechnol. Bioeng.* 39: 287–292.

Tirrell, D. (1987). Macromolecular switches for bilayer membranes. *J. Contr. Rel.* 6: 15–21.

Uenoyama, S., and Hoffman, A. S. (1988). Synthesis and characterization of AAm/NIPAAm grafts on silicone rubber substrates. *Radiat. Phys. Chem.* 32: 605–608.

Vernon, B., Kim, S. W., and Bae, Y. H. (2000). Thermoreversible copolymer gels for extracellular matrix. *J. Biomed. Mater. Res.* 51: 69–79.

Wu. X. S., Hoffman, A. S., and Yager, P. (1992). Conjugation of phosphatidylethanolamine to poly(NIPAAm) for potential use in liposomal drug delivery systems. *Polymer* 33: 4659–4662.

Wu, X. S., Hoffman, A. S., and Yager, P. (1993). Synthesis of and insulin release from erodible polyNIPAAm-phospholipid composites. *J. Intell. Mater. Syst. Struct.* 4: 202–209.

Yamato, M., and Okano, T. (2001). Cell sheet engineering for regenerative medicine. *Macromol. Chem. Symp.* 14(2): 21–29.

Yamato, M., Kwon, O. H., Hirose, M., Kikuchi, A., and Okano, T. (2001). Novel patterned cell co-culture utilizing thermally responsive grafted polymer surfaces. *J. Biomed. Mater. Res.* 55: 137–140.

Yang, H. J., Cole, C. A., Monji, N., and Hoffman, A. S. (1990). Preparation of a thermally phase-separating copolymer with a controlled number of active ester groups per polymer chain. *J. Polymer Sci. A., Polymer Chem.* 28: 219–226.

Yoshida, R., Uchida, K., Kaneko, Y., Sakai, K., Kikuchi, A., Sakurai, Y., and Okano, T. (1995). Comb-type grafted hydrogels with rapid de-swelling response to temperature changes. *Nature* 374: 240–242.

Yoshida, T., Aoyagi, T., Kokufuta, E., and Okano, T. (2003). Newly designed hydrogel with both sensitive thermoresponse and biodegradability. *J. Polymer Sci. A: Polymer Chem.* 41: 779–787.

Zareie, H. M., Bulmus, V., Gunning, P. A., Hoffman, A. S., Piskin, E., and Morris, V. J. (2000). Investigation of a pH- and temperature-sensitive polymer by AFM. *Polymer* 41: 6723–6727.

Zhong, Z., Dijkstra, P. J., Feijen, J., Kwon, Y.-Mi., Bae, Y. H., and Kim, S. W. (2002). Synthesis and aqueous phase behavior of thermoresponsive biodegradable poly(D,L-3-methyl glycolide)-b-poly(ethylene glycol)-b-poly(D,L-3-methyl glycolide) triblock copolymers. *Macromol. Chem. Phys.* 203: 1797–1803.

2.7 BIORESORBABLE AND BIOERODIBLE MATERIALS

Joachim Kohn, Sascha Abramson, and Robert Langer

INTRODUCTION

Since a degradable implant does not have to be removed surgically once it is no longer needed, degradable polymers are of value in short-term applications that require only the temporary presence of a device. An additional advantage is that the use of degradable implants can circumvent some of the problems related to the long-term safety of permanently implanted devices. A potential concern relating to the use of degradable implants is the toxicity of the implant's degradation products. Since all of the implant's degradation products are released into the body of the patient, the design of a degradable implant requires careful attention to testing for potential toxicity of the degradation products. This chapter covers basic definitions relating to the process of degradation and/or erosion, the most important types of *synthetic*, degradable polymers available today, a classification of degradable medical implants, and a number of considerations specific for the design and use of degradable medical polymers (shelf life, sterilization, etc.).

DEFINITIONS RELATING TO THE PROCESS OF EROSION AND/OR DEGRADATION

Currently four different terms (biodegradation, bioerosion, bioabsorption, and bioresorption) are being used to indicate that a given material or device will eventually disappear after having been introduced into a living organism. However, when reviewing the literature, no clear distinctions in the meaning of these four terms are evident. Likewise, the meaning of the prefix "bio" is not well established, leading to the often-interchangeable use of the terms "degradation" and "biodegradation," or "erosion" and "bioerosion." Although efforts have been made to establish generally applicable and widely accepted definitions for all aspects of biomaterials research (Williams, 1987), there is still significant confusion even among experienced researchers in the field as to the correct terminology of various degradation processes.

Generally speaking, the term "degradation" refers to a chemical process resulting in the cleavage of covalent bonds. Hydrolysis is the most common chemical process by which polymers degrade, but degradation can also occur via oxidative and enzymatic mechanisms. In contrast, the term "erosion" refers often to physical changes in size, shape, or mass of a device, which could be the consequence of either degradation or simply dissolution. Thus, it is important to realize that erosion can occur in the absence of degradation, and degradation can occur in the absence of erosion. A sugar cube placed in water erodes, but the sugar does not chemically degrade. Likewise, the embrittlement of plastic when exposed to UV light is due to the degradation of the chemical structure of the polymer and takes place before any physical erosion occurs.

In the context of this chapter, we follow the usage suggested by the Consensus Conference of the European Society for Biomaterials (Williams, 1987) and refer to "biodegradation" only when we wish to emphasize that a biological agent (enzyme, cell, or microorganism) is causing the chemical degradation of the implanted device. After extensive discussion in the literature, it is now widely believed that the chemical degradation of the polymeric backbone of poly(lactic acid) is predominantly controlled by simple hydrolysis and occurs independently of any biological agent (Vert *et al.*, 1991). Consequently, the degradation of poly(lactic acid) to lactic acid should not be described as "biodegradation." In agreement with Heller's suggestion (Heller, 1987), we define a "bioerodible polymer" as a water-insoluble polymer that is converted under physiological conditions into water-soluble material(s) without regard to the specific mechanism involved in the erosion process. "Bioerosion" includes therefore both physical processes (such as dissolution) and chemical processes (such as backbone cleavage). Here the prefix "bio" indicates that the erosion occurs under physiological conditions, as opposed to other erosion processes, caused for example by high temperature, strong acids or bases, UV light, or weather conditions. The terms "bioresorption" and "bioabsorption" are used interchangeably and often imply that the polymer or its degradation products are removed by cellular activity (e.g., phagocytosis) in a biological environment. These terms are somewhat superfluous and have not been clearly defined.

OVERVIEW OF CURRENTLY AVAILABLE DEGRADABLE POLYMERS

From the beginnings of the material sciences, the development of highly stable materials has been a major research challenge. Today, many polymers are available that are virtually nondestructible in biological systems, e.g., Teflon, Kevlar, or poly(ether ether ketone) (PEEK). On the other hand, the development of degradable biomaterials is a relatively new area of research. The variety of available, degradable biomaterials is still too limited to cover a wide enough range of diverse material properties. Thus, the design and synthesis of new, degradable biomaterials is currently an important research challenge, in particular within the context of tissue engineering where the development of new biomaterials that can provide predetermined and controlled cellular responses is a critically needed component of most practical applications of tissue engineering (James and Kohn, 1996).

Degradable materials must fulfill more stringent requirements in terms of their biocompatibility than nondegradable materials. In addition to the potential problem of toxic contaminants leaching from the implant (residual monomers, stabilizers, polymerization initiators, emulsifiers, sterilization by-products), one must also consider the potential toxicity of the degradation products and subsequent metabolites. The practical consequence of this consideration is that only a limited number of nontoxic, monomeric starting materials have been successfully applied to the preparation of degradable biomaterials.

Over the past decade dozens of hydrolytically unstable polymers have been suggested as degradable biomaterials; however, in most cases no attempts have been made to develop these new materials for specific medical applications. Thus, detailed toxicological studies *in vivo*, investigations of degradation rate and mechanism, and careful evaluations of the physicomechanical properties have so far been published for only a very small fraction of those polymers. An even smaller number of synthetic, degradable polymers have so far been used in medical implants and devices that gained approval by the Food and Drug Administration (FDA) for use in patients. Note that the FDA does not approve polymers or materials per se, but only specific devices or implants. As of 1999, only five distinct synthetic, degradable polymers have been approved for use in a narrow range of clinical applications. These polymers are poly(lactic acid), poly(glycolic acid), polydioxanone, polycaprolactone, and a poly(PCPP-SA anhydride) (see later discussion). A variety of other synthetic, degradable biomaterials currently in clinical

TABLE 1 Degradable Polymers and Representative Applications under Investigation

Degradable polymer	Current major research applications
Synthetic degradable polyesters	
Poly(glycolic acid), poly(lactic acid), and copolymers	Barrier membranes, drug delivery, guided tissue regeneration (in dental applications), orthopedic applications , stents, staples, sutures, tissue engineering
Polyhydroxybutyrate (PHB), polyhydroxyvalerate (PHV), and copolymers thereof	Long-term drug delivery, orthopedic applications, stents, sutures
Polycaprolactone	Long-term drug delivery, orthopedic applications, staples, stents
Polydioxanone	Fracture fixation in non-load-bearing bones, sutures, wound clip
Other synthetic degradable polymers	
Polyanhydrides	Drug delivery
Polycyanoacrylates	Adhesives, drug delivery
Poly(amino acids) and "pseudo"-Poly(amino acids)	Drug delivery, tissue engineering, orthopedic applications
Poly(ortho ester)	Drug delivery, stents
Polyphosphazenes	Blood contacting devices, drug delivery, skeletal reconstruction
Poly(propylene fumarate)	Orthopedic applications
Some natural resorbable polymers	
Collagen	Artificial skin, coatings to improve cellular adhesion, drug delivery, guided tissue regeneration in dental applications, orthopedic applications, soft tissue augmentation, tissue engineering, scaffold for reconstruction of blood vessels, wound closure
Fibrinogen and fibrin	Tissue sealant
Gelatin	Capsule coating for oral drug delivery, hemorrhage arrester
Cellulose	Adhesion barrier, hemostat
Various polysaccharides such as chitosan, alginate	Drug delivery, encapsulation of cells, sutures, wound dressings
Starch and amylose	Drug delivery

use are blends or copolymers of these base materials such as a wide range of copolymers of lactic and glycolic acid. Note that this listing does not include polymers derived from animal sources such as collagen, gelatin, or hyaloronic acid.

Recent research has led to a number of well-established investigational polymers that may find practical applications as degradable implants within the next decade. It is beyond the scope of this chapter to fully introduce all of the polymers and their applications under investigation, thus only representative examples of these polymers are described here. This chapter will concern itself mostly with *synthetic* degradable polymers, as *natural* polymers (e.g., polymers derived from animal or plant sources) are described elsewhere in this book. Table 1 provides an overview of some representative degradable polymers. For completeness, some of the natural polymers have also been included here. Structural formulas of commonly investigated synthetic degradable polymers are provided in Fig. 1. It is an interesting observation that a large proportion of the currently investigated, *synthetic*, degradable polymers are polyesters. It remains to be seen whether some of the alternative backbone structures such as polyanhydrides, polyphosphazenes, polyphosphonates, polyamides, or polycarbonates will be able to challenge the predominant position of the polyesters in the future.

Polydioxanone (PDS) is a poly(ether ester) made by a ring-opening polymerization of *p*-dioxanone monomer. PDS has gained increasing interest in the medical field and pharmaceutical field due to its degradation to low-toxicity monomers *in vivo*. PDS has a lower modulus than PLA or PGA. It became the first degradable polymer to be used to make a monofilament suture. PDS has also been introduced to the market as a suture clip as well as a bone pin marketed under the name OrthoSorb in the USA and Ethipin in Europe.

Poly(hydroxybutyrate) (PHB), poly(hydroxyvalerate) (PHV), and their copolymers represent examples of bioresorbable polyesters that are derived from microorganisms. Although this class of polymers are examples of *natural* materials (as opposed to *synthetic* materials), they are included here because they have similar properties and similar areas of application as the widely investigated poly(lactic acid). PHB and its copolymers with up to 30% of 3-hydroxyvaleric acid are now commercially available under the trade name "Biopol" (Miller and Williams, 1987). PHB and PHV are intracellular storage polymers providing a reserve of carbon and energy to microorganisms similar to the role of starch in plant metabolism. The polymers can be degraded by soil bacteria (Senior *et al.*, 1972) but are relatively stable under physiological conditions (pH 7, 37°C). Within a relatively narrow window, the rate of degradation can be modified slightly by varying the copolymer composition; however, all members of this family of polymers require several years for complete resorption *in vivo*. *In vivo*, PHB degrades to D-3-hydroxybutyric acid, which is a normal constituent of human blood (Miller and Williams, 1987). The low toxicity of PHB may at least in part be due to this fact.

PHB homopolymer is very crystalline and brittle, whereas the copolymers of PHB with hydroxyvaleric acid are

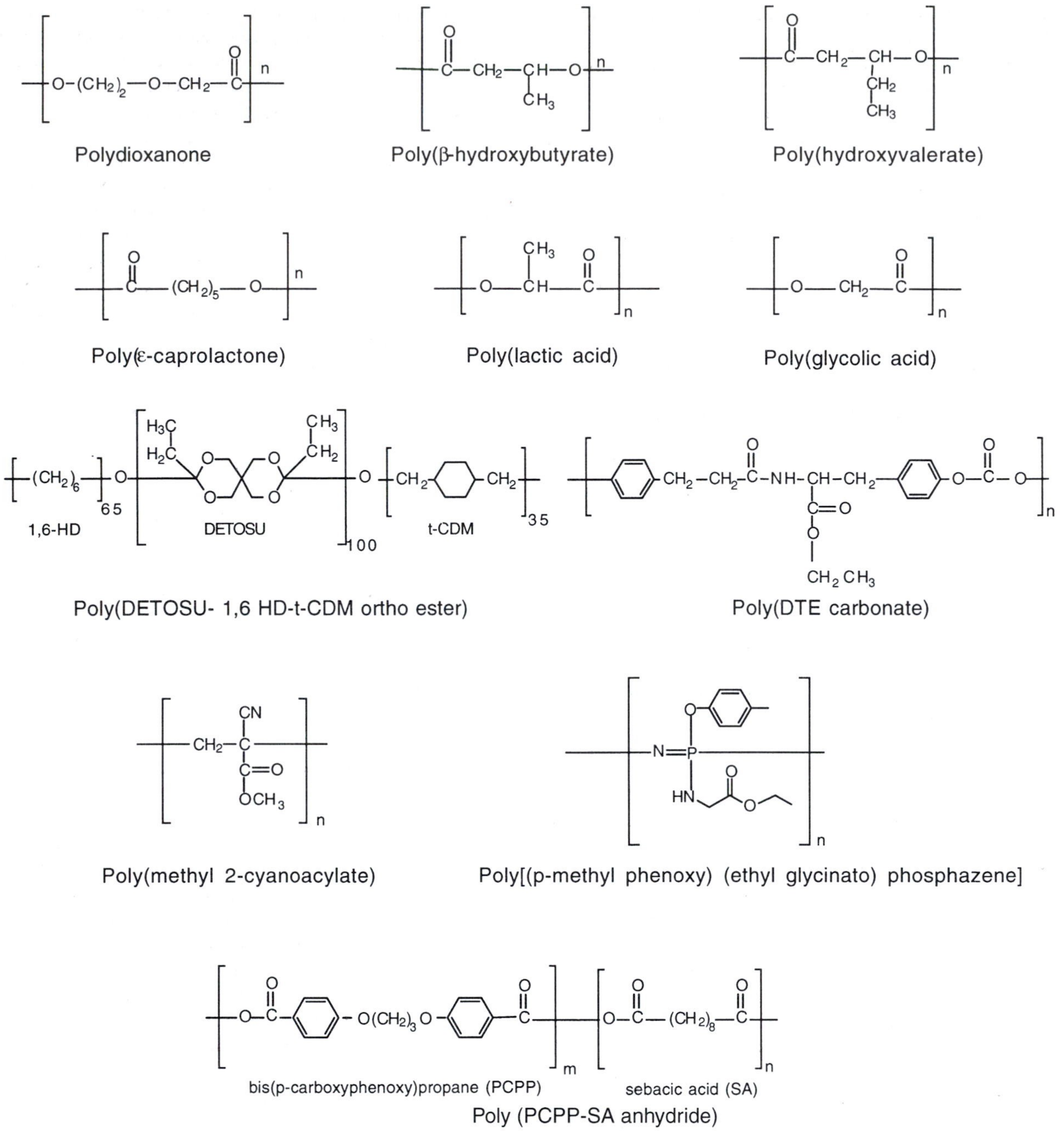

FIG. 1. Chemical structures of widely investigated degradable polymers.

less crystalline, more flexible, and more readily processible (Barham *et al.*, 1984). The polymers have been considered in several biomedical applications such as controlled drug release, sutures, artificial skin, and vascular grafts, as well as industrial applications such as medical disposables. PHB is especially attractive for orthopedic applications because of its slow degradation time. The polymer typically retained 80% of its original stiffness over 500 days on *in vivo* degradation (Knowles, 1993).

Polycaprolactone (PCL) became available commercially following efforts at Union Carbide to identify synthetic polymers that could be degraded by microorganisms (Huang, 1985). It is a semicrystalline polymer. The high solubility of polycaprolactone, its low melting point (59–64°C), and its exceptional ability to form blends has stimulated research on its application as a biomaterial. Polycaprolactone degrades at a slower pace than PLA and can therefore be used in drug delivery devices that remain active for over 1 year. The release characteristics of polycaprolactone have been investigated in detail by Pitt and his co-workers (Pitt *et al.*, 1979). The Capronor system, a 1-year implantable contraceptive device (Pitt, 1990), has become commercially available in Europe and the United States. The toxicology of polycaprolactone has been extensively studied as part of the evaluation of Capronor. Based on a large number of tests, ε-caprolactone and polycaprolactone are currently regarded as nontoxic and tissue-compatible materials. Polycaprolactone is currently being researched as part of wound dressings, and in Europe, it is already in clinical use as a degradable staple (for wound closure).

Polyanhydrides were explored as possible substitutes for polyesters in textile applications but failed ultimately because of their pronounced hydrolytic instability. It was this property that prompted Langer and his co-workers to explore polyanhydrides as degradable implant materials (Tamada and Langer, 1993). Aliphatic polyanhydrides degrade within days, whereas some aromatic polyanhydrides degrade over several years. Thus aliphatic–aromatic copolymers are usually employed which show intermediate rates of degradation depending on the monomer composition.

Polyanhydrides are among the most reactive and hydrolytically unstable polymers currently used as biomaterials. The high chemical reactivity is both an advantage and a limitation of polyanhydrides. Because of their high rate of degradation, many polyanhydrides degrade by surface erosion without the need to incorporate various catalysts or excipients into the device formulation. On the other hand, polyanhydrides will react with drugs containing free amino groups or other nucleophilic functional groups, especially during high-temperature processing (Leong *et al.*, 1986). The potential reactivity of the polymer matrix toward nucleophiles limits the type of drugs that can be successfully incorporated into a polyanhydride matrix by melt processing techniques. Along the same line of reasoning, it has been questioned whether amine-containing biomolecules present in the interstitial fluid around an implant could react with anhydride bonds present at the implant surface.

A comprehensive evaluation of the toxicity of the polyanhydrides showed that, in general, the polyanhydrides possess excellent *in vivo* biocompatibility (Attawia *et al.*, 1995). The most immediate applications for polyanhydrides are in the field of drug delivery. Drug-loaded devices made of polyanhydrides can be prepared by compression molding or microencapsulation (Chasin *et al.*, 1990). A wide variety of drugs and proteins including insulin, bovine growth factors, angiogenesis inhibitors (e.g., heparin and cortisone), enzymes (e.g., alkaline phosphatase and β-galactosidase), and anesthetics have been incorporated into polyanhydride matrices, and their *in vitro* and *in vivo* release characteristics have been evaluated (Park *et al.*, 1998; Chasin *et al.*, 1990). Additionally, polyanhydrides have been investigated for use as nonviral vectors of delivering DNA in gene therapy (Shea and Mooney, 2001). The first polyanhydride-based drug delivery system to enter clinical use is for the delivery of chemotherapeutic agents. An example of this application is the delivery of BCNU (bis-chloroethylnitrosourea) to the brain for the treatment of glioblastoma multiformae, a universally fatal brain cancer (Madrid *et al.*, 1991). For this application,

BCNU-loaded implants made of the polyanhydride derived from bis-*p*-carboxyphenoxypropane and sebacic acid received FDA regulatory approval in the fall of 1996 and are currently being marketed under the name Gliadel.

Poly(ortho esters) are a family of synthetic, degradable polymers that have been under development for a number of years (Heller *et al.*, 1990). Devices made of poly(ortho esters) can erode by "surface erosion" if appropriate excipients are incorporated into the polymeric matrix. Since surface eroding, slab-like devices tend to release drugs embedded within the polymer at a constant rate, poly(ortho esters) appear to be particularly useful for controlled-release drug delivery applications. For example, poly(ortho esters) have been used for the controlled delivery of cyclobenzaprine and steroids and a significant number of publications describe the use of poly(ortho esters) for various drug delivery applications (Heller, 1993). Poly(ortho esters) have also been investigated for the treatment of postsurgical pain, ostearthritis, and ophthalmic diseases (Heller *et al.*, 2002). Since the ortho ester linkage is far more stable in base than in acid, Heller and his co-workers controlled the rate of polymer degradation by incorporating acidic or basic excipients into the polymer matrix. One concern about the "surface erodability" of poly(ortho esters) is that the incorporation of highly water-soluble drugs into the polymeric matrix can result in swelling of the polymer matrix. The increased amount of water imbibed into the matrix can then cause the polymeric device to exhibit "bulk erosion" instead of "surface erosion" (see below for a more detailed explanation of these erosion mechanisms) (Okada and Toguchi, 1995).

By now, there are three major types of poly(ortho esters). First, Choi and Heller prepared the polymers by the transesterification of 2,2'-dimethoxyfuran with a diol. The next generation of poly(ortho esters) was based on an acid-catalyzed addition reaction of diols with diketeneacetals (Heller *et al.*, 1980). The properties of the polymers can be controlled to a large extent by the choice of the diols used in the synthesis. Recently, a third generation of poly(ortho esters) have been prepared. These materials are very soft and can even be viscous liquids at room temperature. Third-generation poly(ortho esters) can be used in the formulation of drug delivery systems that can be injected rather than implanted into the body.

Poly(amino acids) and "Pseudo"-Poly(amino acids) Since proteins are composed of amino acids, it is an obvious idea to explore the possible use of poly(amino acids) in biomedical applications (Anderson *et al.*, 1985). Poly(amino acids) were regarded as promising candidates since the amino acid side chains offer sites for the attachment of drugs, cross-linking agents, or pendent groups that can be used to modify the physicomechanical properties of the polymer. In addition, poly(amino acids) usually show a low level of systemic toxicity, due to their degradation to naturally occurring amino acids.

Early investigations of poly(amino acids) focused on their use as suture materials (Miyamae *et al.*, 1968), as artificial skin substitutes (Spira *et al.*, 1969), and as drug delivery systems (McCormick-Thomson and Duncan, 1989). Various drugs have been attached to the side chains of poly(amino acids), usually via a spacer unit that distances the drug from the backbone. Poly(amino acid)–drug combinations investigated include poly(L-lysine) with methotrexate and pepstatin (Campbell *et al.*, 1980), and poly(glutamic acid) with adriamycin, a widely used chemotherapeutic agent (van Heeswijk *et al.*, 1985).

Despite their apparent potential as biomaterials, poly(amino acids) have actually found few practical applications. Most poly(amino acids) are highly insoluble and nonprocessible materials. Since poly(amino acids) have a pronounced tendency to swell in aqueous media, it can be difficult to predict drug release rates. Furthermore, the antigenicity of polymers containing three or more amino acids limits their use in biomedical applications (Anderson *et al.*, 1985). Because of these difficulties, only a few poly(amino acids), usually derivatives of poly(glutamic acid) carrying various pendent chains at the γ-carboxylic acid group, have been investigated as implant materials (Lescure *et al.*, 1989). So far, no implantable devices made of a poly(amino acid) have been approved for clinical use in the United States.

In an attempt to circumvent the problems associated with conventional poly(amino acids), backbone-modified "pseudo"-poly(amino acids) were introduced in 1984 (Kohn and Langer, 1984, 1987). The first "pseudo"-poly(amino acids) investigated were a polyester from N-protected *trans*-4-hydroxy-L-proline, and a polyiminocarbonate derived from tyrosine dipeptide. The tyrosine-derived "pseudo"-poly(amino acids) are easily processed by solvent or heat methods and exhibit a high degree of biocompatibility. Recent studies indicate that the backbone modification of poly(amino acids) may be a generally applicable approach for the improvement of the physicomechanical properties of conventional poly(amino acids). For example, tyrosine-derived polycarbonates (Nathan and Kohn, 1996) are high-strength materials that may be useful in the formulation of degradable orthopedic implants. One of the tyrosine-derived pseudo-poly(amino acids), poly(DTE carbonate) exhibits a high degree of bone conductivity (e.g., bone tissue will grow directly along the polymeric implant) (Choueka *et al.*, 1996; James and Kohn, 1997).

The reason for the improved physicomechanical properties of "pseudo"-poly(amino acids) relative to conventional poly(amino acids) can be traced to the reduction in the number of interchain hydrogen bonds: In conventional poly(amino acids), individual amino acids are polymerized via repeated amide bonds leading to strong interchain hydrogen bonding. In natural peptides, hydrogen bonding is one of the interactions leading to the spontaneous formation of secondary structures such as α-helices or β-pleated sheets. Strong hydrogen bonding also results in high processing temperatures and low solubility in organic solvents which tends to lead to intractable polymers with limited applications. In "pseudo"-poly(amino acids), half on the amide bonds are replaced by other linkages (such as carbonate, ester, or iminocarbonate bonds) that have a much lower tendency to form interchain hydrogen bonds, leading to better processibility and, generally, a loss of crystallinity.

Polycyanoacrylates are used as bioadhesives. Methyl cyanoacrylates are more commonly used as general-purpose glues and are commercially available as "Crazy Glue."

Methyl cyanoacrylate was used during the Vietnam war as an emergency tissue adhesive, but is no longer used today. Butyl cyanoacrylate is approved in Canada and Europe as a dental adhesive. Cyanoacrylates undergo spontaneous polymerization at room temperature in the presence of water, and their toxicity and erosion rate after polymerization differ with the length of their alkyl chains (Gombotz and Pettit, 1995). All polycyanoacrylates have several limiting properties: First, the monomers (cyanoacrylates) are very reactive compounds that often have significant toxicity. Second, upon degradation polycyanoacrylates release formaldehyde resulting in intense inflammation in the surrounding tissue. In spite of these inherent limitations, polycyanoacrylates have been investigated as potential drug delivery matrices and have been suggested for use in ocular drug delivery (Deshpande *et al.*, 1998).

Polyphosphazenes are very unusual polymers, whose backbone consists of nitrogen–phosphorus bonds. These polymers are at the interface between inorganic and organic polymers and have unusual material properties. Polyphosphazenes have found industrial applications, mainly because of their high thermal stability. They have also been used in investigations for the formulation of controlled drug delivery systems (Allcock, 1990). Polyphosphazenes are interesting biomaterials, in many respects. They have been claimed to be biocompatible and their chemical structure provides a readily accessible "pendant chain" to which various drugs, peptides, or other biological compounds can be attached and later released via hydolysis. Polyphosphazenes have been examined for use in skeletal tissue regeneration (Laurencin *et al.*, 1993). Another novel use of polyphosphazenes is in the area of vaccine design where these materials were used as immunological adjuvants (Andrianov *et al.*, 1998).

Poly(glycolic acid) and poly(lactic acid) and their copolymers are currently the most widely investigated, and most commonly used synthetic, bioerodible polymers. In view of their importance in the field of biomaterials, their properties and applications will be described in more detail.

Poly(glycolic acid) (PGA) is the simplest linear, aliphatic polyester (Fig. 1). Since PGA is highly crystalline, it has a high melting point and low solubility in organic solvents. PGA was used in the development of the first totally synthetic, absorbable suture. PGA sutures have been commercially available under the trade name "Dexon" since 1970. A practical limitation of Dexon sutures is that they tend to lose their mechanical strength rapidly, typically over a period of 2 to 4 weeks after implantation. PGA has also been used in the design of internal bone fixation devices (bone pins). These pins have become commercially available under the trade name "Biofix."

In order to adapt the materials properties of PGA to a wider range of possible applications, copolymers of PGA with the more hydrophobic poly(lactic acid) (PLA) were intensively investigated (Gilding and Reed, 1979, 1981). The hydrophobicity of PLA limits the water uptake of thin films to about 2% and reduces the rate of backbone hydrolysis as compared to PGA. Copolymers of glycolic acid and lactic acid have been developed as alternative sutures (trade names "Vicryl" and "Polyglactin 910").

It is noteworthy that there is no linear relationship between the ratio of glycolic acid to lactic acid and the physicomechanical properties of the corresponding copolymers. Whereas PGA is highly crystalline, crystallinity is rapidly lost in copolymers of glycolic acid and lactic acid. These morphological changes lead to an increase in the rates of hydration and hydrolysis. Thus, 50:50 copolymers degrade more rapidly than either PGA or PLA.

Since lactic acid is a chiral molecule, it exists in two steroisomeric forms that give rise to four morphologically distinct polymers: the two stereoregular polymers, D-PLA and L-PLA, and the racemic form D, L-PLA. A fourth morphological form, *meso*-PLA, can be obtained from D, L-lactide but is rarely used in practice.

The polymers derived from the optically active D and L monomers are semicrystalline materials, while the optically inactive D,L-PLA is always amorphous. Generally, L-PLA is more frequently employed than D-PLA, since the hydrolysis of L-PLA yields L(+)-lactic acid, which is the naturally occurring stereoisomer of lactic acid.

The differences in the crystallinity of D, L-PLA and L-PLA have important practical ramifications: Since D,L-PLA is an amorphous polymer, it is usually considered for applications such as drug delivery, where it is important to have a homogeneous dispersion of the active species within the carrier matrix. On the other hand, the semicrystalline L-PLA is preferred in applications where high mechanical strength and toughness are required, such as sutures and orthopedic devices.

PLA and PGA and their copolymers have been investigated for more applications than any other degradable polymer. The high interest in these materials is based, not on their superior materials properties, but mostly on the fact that these polymers have already been used successfully in a number of approved medical implants and are considered safe, nontoxic, and biocompatible by regulatory agencies in virtually all developed countries. Therefore, implantable devices prepared from PLA, PGA, or their copolymers can be brought to market in less time and for a lower cost than similar devices prepared from novel polymers whose biocompatibility is still unproven.

Currently available and approved products include sutures, GTR membranes for dentistry, bone pins, and implantable drug delivery systems. The polymers are also being widely investigated in the design of vascular and urological stents and skin substitutes, and as scaffolds for tissue engineering and tissue reconstruction. In many of these applications, PLA, PGA, and their copolymers have performed with moderate to high degrees of success. However, there are still unresolved issues: First, in tissue culture experiments, most cells do not attach to PLA or PGA surfaces and do not grow as vigorously as on the surface of other materials, indicating that these polymers are actually poor substrates for cell growth *in vitro*. The significance of this finding for the use of PLA and PGA as tissue engineering scaffolds *in vivo* is currently a topic of debate. Second, the degradation products of PLA and PGA are relatively strong acids (lactic acid and glycolic acid). When these degradation products accumulate at the implant site, a delayed inflammatory response is often observed months to years after implantation (Bergsma *et al.*, 1995; Athanasiou *et al.*, 1998; Törmälä *et al.*, 1998).

APPLICATIONS OF SYNTHETIC, DEGRADABLE POLYMERS AS BIOMATERIALS

Classification of Degradable Medical Implants

Some typical short-term applications of biodegradable polymers are listed in Table 2. From a practical perspective, it is convenient to distinguish between five main types of degradable implants: the temporary support device, the temporary barrier, the drug delivery device, the tissue engineering scaffold, and the multifunctional implant.

A temporary support device is used in those circumstances in which the natural tissue bed has been weakened by disease, injury, or surgery and requires some artificial support. A healing wound, a broken bone, or a damaged blood vessel are examples of such situations. Sutures, bone fixation devices (e.g., bone nails, screws, or plates), and vascular grafts would be examples of the corresponding support devices. In all of these instances, the degradable implant would provide temporary, mechanical support until the natural tissue heals and regains its strength. In order for a temporary support device to work properly, a gradual stress transfer should occur: As the natural tissue heals, the degradable implant should gradually weaken. The need to adjust the degradation rate of the temporary support device to the healing of the surrounding tissue represents one of the major challenges in the design of such devices.

Currently, sutures represent the most successful example of a temporary support device. The first synthetic, degradable sutures were made of poly(glycolic acid) (PGA) and became available under the trade name "Dexon" in 1970. This represented the first routine use of a degradable polymer in a major clinical application (Frazza and Schmitt, 1971). Later copolymers of PGA and poly(lactic acid) (PLA) were developed. The widely used "Vicryl" suture, for example, is a 90:10 copolymer of PGA/PLA, introduced into the market in 1974.

TABLE 2 Some "Short-Term" Medical Applications of Degradable Polymeric Biomaterials

Application	Comments
Sutures	The earliest, successful application of synthetic degradable polymers in human medicine.
Drug delivery devices	One of the most widely investigated medical applications for degradable polymers.
Orthopedic fixation devices	Requires polymers of exceptionally high mechanical strength and stiffness.
Adhesion prevention	Requires polymers that can form soft membranes or films.
Temporary vascular grafts and stents made of degradable polymers	Only investigational devices are presently available. Blood compatibility is a major concern.
Tissue engineering or guided tissue regeneration scaffold	Attempts to recreate or improve native tissue function using degradable scaffolds.

Sutures made of polydioxanone (PDS) became available in the United States in 1981. In spite of extensive research efforts in many laboratories, no other degradable polymers are currently used to any significant extent in the formulation of degradable sutures.

A temporary barrier has its major medical use in adhesion prevention. Adhesions are formed between two tissue sections by clotting of blood in the extravascular tissue space followed by inflammation and fibrosis. If this natural healing process occurs between surfaces that were not meant to bond together, the resulting adhesion can cause pain, functional impairment, and problems during subsequent surgery. Surgical adhesions are a significant cause of morbidity and represent one of the most significant complications of a wide range of surgical procedures such as cardiac, spinal, and tendon surgery. A temporary barrier could take the form of a thin polymeric film or a meshlike device that would be placed between adhesion-prone tissues at the time of surgery. To be useful, such as temporary barrier would have to prevent the formation of scar tissue connecting adjacent tissue sections, followed by the slow resorption of the barrier material (Hill *et al.*, 1993). This sort of barrier has also been investigated for the sealing of breaches of the lung tissue that cause air leakage.

Another important example of a temporary barrier is in the field of skin reconstruction. Several products are available that are generally referred to as "artificial skin" (Beele, 2002). The first such product consists of an artificial, degradable collagen/glycosaminoglycan matrix that is placed on top of the skin lesion to stimulate the regrowth of a functional dermis. Another product consists of a degradable collagen matrix with preseeded human fibroblasts. Again, the goal is to stimulate the regrowth of a functional dermis. These products are used in the treatment of burns and other deep skin lesions and represent an important application for temporary barrier type devices.

An implantable drug delivery device is by necessity a temporary device, as the device will eventually run out of drug or the need for the delivery of a specific drug is eliminated once the disease is treated. The development of implantable drug delivery systems is probably the most widely investigated application of degradable polymers (Langer, 1990). One can expect that the future acceptance of implantable drug delivery devices by physicians and patients alike will depend on the availability of degradable systems that do not have to be explanted surgically.

Since poly(lactic acid) and poly(glycolic acid) have an extensive safety profile based on their use as sutures, these polymers have been very widely investigated in the formulation of implantable controlled release devices. Several implantable, controlled release formulations based on copolymers of lactic and glycolic acid have already become available. However, a very wide range of other degradable polymers have been investigated as well. Particularly noteworthy is the use of a type of polyanhydride in the formulation of an intracranial, implantable device for the administration of BCNU (a chemotherapeutic agent) to patients suffering from glioblastoma multiformae, a usually lethal form of brain cancer (Chasin *et al.*, 1990).

The term *tissue engineering scaffold* will be used in this chapter to describe a degradable implant that is designed to act as an artificial extracellular matrix by providing space for cells to grow into and to reorganize into functional tissue (James and Kohn, 1996).

It has become increasingly obvious that manmade implantable prostheses do not function as well as the native tissue or maintain the functionality of native tissue over long periods of time. Therefore, tissue engineering has emerged as an interdisciplinary field that utilizes degradable polymers, among other substrates and biologics, to develop treatments that will allow the body to heal itself without the need for permanently implanted, artificial prosthetic devices. In the ideal case, a tissue engineering scaffold is implanted to restore lost tissue function, maintain tissue function, or enhance existing tissue function (Langer and Vacanti, 1993). These scaffolds can take the form of a feltlike material obtained from knitted or woven fibers or from fiber meshes. Alternatively, various processing techniques can be used to obtain foams or sponges. For all tissue engineering scaffolds, pore interconnectivity is a key property, as cells need to be able to migrate and grow throughout the entire scaffold. Thus, industrial foaming techniques, used for example in the fabrication of furniture cushions, are not applicable to the fabrication of tissue engineering scaffolds, as these industrial foams are designed contain "closed pores," whereas tissue engineering scaffolds require an "open pore" structure. Tissue engineering scaffolds may be preseeded with cells *in vitro* prior to implantation. Alternatively, tissue engineering scaffolds may consist of a cell-free structure that is invaded and "colonized" by cells only after its implantation. In either case, the tissue engineering scaffold must allow the formation of functional tissue *in vivo*, followed by the safe resorption of the scaffold material.

There has been some debate in the literature as to the exact definition of the related term "guided tissue regeneration" (GTR). Guided tissue regeneration is a term traditionally used in dentistry. This term sometimes implies that the scaffold encourages the growth of specific types of tissue. For example, in the treatment of periodontal disease, periodontists use the term "guided tissue regeneration" when using implants that favor new bone growth in the periodontal pocket over soft-tissue ingrowth (scar formation).

One of the major challenges in the design of tissue engineering scaffolds is the need to adjust the rate of scaffold degradation to the rate of tissue healing. Depending upon the application the scaffold, the polymer may need to function on the order of days to months. Scaffolds intended for the reconstruction of bone illustrate this point: In most applications, the scaffold must maintain some mechanical strength to support the bone structure while new bone is formed. Premature degradation of the scaffold material can be as detrimental to the healing process as a scaffold that remains intact for excessive periods of time. The future use of tissue engineering scaffolds has the potential to revolutionize the way aging-, trauma-, and disease-related loss of tissue function can be treated.

Multifunctional devices, as the name implies, combine several of the functions just mentioned within one single device. Over the past few years, there has been a trend toward increasingly sophisticated applications for degradable biomaterials. Usually, these applications envision the combination of several functions within the same device and require the design of custom-made materials with a narrow range of predetermined materials properties. For example, the availability of biodegradable bone nails and bone screws made of ultrahigh-strength poly(lactic acid) opens the possibility of combining the "mechanical support" function of the device with a "site-specific drug delivery" function: a biodegradable bone nail that holds the fractured bone in place can simultaneously stimulate the growth of new bone tissue at the fracture site by slowly releasing bone growth factors (e.g., bone morphogenic protein or transforming growth factor β) throughout its degradation process.

Likewise, biodegradable stents for implantation into coronary arteries are currently being investigated (Agrawal *et al.*, 1992). The stents are designed to mechanically prevent the collapse and restenosis (reblocking) of arteries that have been opened by balloon angioplasty. Ultimately, the stents could deliver an antiinflammatory or antithrombogenic agent directly to the site of vascular injury. Again, it would potentially be possible to combine a mechanical support function with site specific drug delivery.

Various functional combinations involve the tissue engineering scaffold. Perhaps the most important multifunctional device for future applications is a tissue engineering scaffold that also serves as a drug delivery system for cytokines, growth hormones, or other agents that directly affect cells and tissue within the vicinity of the implanted scaffold. An excellent example for this concept is a bone regeneration scaffold that is placed within a bone defect to allow the regeneration of bone while releasing bone morphogenic protein (BMP) at the implant site. The release of BMP has been reported to stimulate bone growth and therefore has the potential to accelerate the healing rate. This is particularly important in older patients whose natural ability to regenerate tissues may have declined.

The Process of Bioerosion

One of the most important prerequisites for the successful use of a degradable polymer for any medical application is a thorough understanding of the way the device will degrade/erode and ultimately resorb from the implant site. Within the context of this chapter, we are limiting our discussion to the case of a solid, polymeric implant. The transformation of such an implant into water-soluble material(s) is best described by the term "bioerosion." The bioerosion process of a solid, polymeric implant is associated with macroscopic changes in the appearance of the device, changes in its physicomechanical properties and in physical processes such as swelling, deformation, or structural disintegration, weight loss, and the eventual depletion of drug or loss of function.

All of these phenomena represent distinct and often independent aspects of the complex bioerosion behavior of a specific polymeric device. It is important to note that the bioerosion of a solid device is not necessarily due to the chemical cleavage of the polymer backbone or the chemical cleavage of cross-links or side chains. Rather, simple solubilization of the intact polymer, for instance, due to changes in pH, may also lead to the erosion of a solid device.

Two distinct modes of bioerosion have been described in the literature. In "bulk erosion," the rate of water penetration into the solid device exceeds the rate at which the polymer is transformed into water-soluble material(s). Consequently, the uptake of water is followed by an erosion process that occurs throughout the entire volume of the solid device. Because of the rapid penetration of water into the matrix of hydrophilic polymers, most of the currently available polymers will give rise to bulk eroding devices. In a typical "bulk erosion" process, cracks and crevices will form throughout the device that may rapidly crumble into pieces. A good illustration for a typical bulk erosion process is the disintegration of an aspirin tablet that has been placed into water. Depending on the specific application, the often uncontrollable tendency of bulk eroding devices to crumble into little pieces can be a disadvantage.

Alternatively, in "surface erosion," the bioerosion process is limited to the surface of the device. Therefore, the device will become thinner with time, while maintaining its structural integrity throughout much of the erosion process. In order to observe surface erosion, the polymer must be hydrophobic to impede the rapid imbibition of water into the interior of the device. In addition, the rate at which the polymer is transformed into water-soluble material(s) has to be fast relative to the rate of water penetration into the device. Under these conditions, scanning electron microscopic evaluation of surface eroding devices has sometimes shown a sharp border between the eroding surface layer and the intact polymer in the core of the device (Mathiowitz *et al.*, 1990).

Surface eroding devices have so far been obtained only from a small number of hydrophobic polymers containing hydrolytically highly reactive linkages in the backbone. A possible exception to this general rule is enzymatic surface erosion. The inability of enzymes to penetrate into the interior of a solid, polymeric device may result in an enzyme-mediated surface erosion mechanism. Currently, polyanhydrides and poly(ortho esters) are the best known examples of polymers that can be fabricated into surface eroding devices.

Mechanisms of Chemical Degradation

Although bioerosion can be caused by the solubilization of an intact polymer, chemical degradation of the polymer is usually the underlying cause for the bioerosion of a solid, polymeric device. Several distinct types of chemical degradation mechanisms have been identified (Fig. 2) (Rosen *et al.*, 1988). Chemical reactions can lead to cleavage of cross-links between water-soluble polymer chains (mechanism I), to the cleavage of polymer side chains resulting in the formation of polar or charged groups (mechanism II), or to the cleavage of the polymer backbone (mechanism III). Obviously, combinations of these mechanisms are possible: for instance, a cross-linked polymer may first be partially solubilized by the cleavage of crosslinks (mechanism I), followed by the cleavage of the backbone itself (mechanism III). It should be noted that water is key to all of these degradation schemes. Even enzymatic degradation occurs in aqueous environment.

Since the chemical cleavage reactions described above can be mediated by water or by biological agents such as enzymes and microorganisms, it is possible to distinguish between hydrolytic degradation and biodegradation, respectively. It has often been

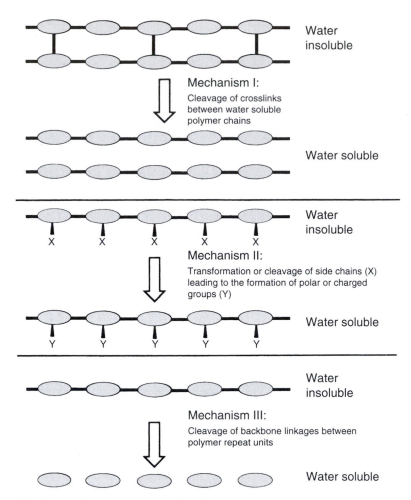

FIG. 2. Mechanisms of chemical degradation. Mechanism I involves the cleavage of degradable cross-links between water-soluble polymer chains. Mechanism II involves the cleavage or chemical transformation of polymer side chains, resulting in the formation of charged or polar groups. The presence of charged or polar groups leads then to the solubilization of the intact polymer chain. Mechanism III involves the cleavage of unstable linkages in the polymer backbone, followed by solubilization of the low-molecular-weight fragments.

stated that the availability of water is virtually constant in all soft tissues and varies little from patient to patient. On the other hand, the levels of enzymatic activity may vary widely not only from patient to patient but also between different tissue sites in the same patient. Thus polymers that undergo hydrolytic cleavage tend to have more predictable *in vivo* erosion rates than polymers whose degradation is mediated predominantly by enzymes. The latter polymers tend to be generally less useful as degradable medical implants.

Factors That Influence the Rate of Bioerosion

Although the solubilization of intact polymer as well as several distinct mechanisms of chemical degradation have been recognized as possible causes for the observed bioerosion of a solid, polymeric implant, virtually all currently available implant materials erode because of the hydrolytic cleavage of the polymer backbone (mechanism III in Fig. 2). We therefore limit the following discussion to solid devices that bioerode because of the hydrolytic cleavage of the polymer backbone.

In this case, the main factors that determine the overall rate of the erosion process are the chemical stability of the hydrolytically susceptible groups in the polymer backbone, the hydrophilic/hydrophobic character of the repeat units, the morphology of the polymer, the initial molecular weight and molecular weight distribution of the polymer, the device fabrication process used to prepare the device, the presence of catalysts, additives, or plasticizers, and the geometry (specifically the surface area to volume ratio) of the implanted device.

The susceptibility of the polymer backbone toward hydrolytic cleavage is probably the most fundamental parameter. Generally speaking, anhydrides tend to hydrolyze faster than ester bonds that in turn hydrolyze faster than amide bonds. Thus, polyanhydrides will tend to degrade faster than polyesters that in turn will have a higher tendency to bioerode than polyamides. Based on the known susceptibility of the polymer backbone structure toward hydrolysis, it is possible to make predictions about the bioerosion of a given polymer.

However, the actual erosion rate of a solid polymer cannot be predicted on the basis of the polymer backbone structure alone. The observed erosion rate is strongly dependent on the ability of water molecules to penetrate into the polymeric matrix. The hydrophilic versus hydrophobic character of the polymer, which is a function of the structure of the monomeric starting materials, can therefore have an overwhelming influence on the observed bioerosion rate. For instance, the erosion rate of polyanhydrides can be slowed by about three orders of magnitude when the less hydrophobic sebacic acid is replaced by the more hydrophobic bis(carboxy phenoxy)propane as the monomeric starting material. Likewise, devices made of poly(glycolic acid) erode faster than identical devices made of the more hydrophobic poly(lactic acid), although the ester bonds have about the same chemical reactivity toward water in both polymers.

The observed bioerosion rate is further influenced by the morphology of the polymer. Polymers can be classified as either semicrystalline or amorphous. At body temperature (37°C) amorphous polymers with T_g above 37°C will be in a glassy state, and polymers with a T_g below 37°C will in a rubbery state. In this discussion it is therefore necessary to consider three distinct morphological states: semicrystalline, amorphous–glassy, and amorphous–rubbery.

In the crystalline state, the polymer chains are densely packed and organized into crystalline domains that resist the penetration of water. Consequently, backbone hydrolysis tends to occur in the amorphous regions of a semicrystalline polymer and at the surface of the crystalline regions. This phenomenon is of particular importance to the erosion of devices made of poly(L-lactic acid) and poly(glycolic acid) which tend to have high degrees of crystallinity around 50%.

Another good illustration of the influence of the polymer morphology on the rate of bioerosion is provided by a comparison of poly(L-lactic acid) and poly(D, L-lactic acid): Although these two polymers have chemically identical backbone structures and an identical degree of hydrophobicity, devices made of poly(L-lactic acid) tend to degrade much more slowly than identical devices made of poly(D, L-lactic acid). The slower rate of bioerosion of poly poly(L-lactic acid) is due to the fact that this stereoregular polymer is semicrystalline, while the racemic poly(D, L-lactic acid) is an amorphous polymer.

Likewise, a polymer in its glassy state is less permeable to water than the same polymer when it is in its rubbery state. This observation could be of importance in cases where an amorphous polymer has a glass transition temperature that is not for above body temperature (37°C). In this situation, water sorption into the polymer could lower its T_g below 37°C, resulting in abrupt changes in the bioerosion rate.

The manufacturing process may also have a significant effect on the erosion profile. For example, Mathiowitz and co-workers (Mathiowitz *et al.*, 1990) showed that polyanhydride microspheres produced by melt encaspulation were very dense and eroded slowly, whereas when the same polymers were formed into microspheres by solvent evaporation, the microspheres were very porous (and therefore more water permeable) and eroded more rapidly.

The preceding examples illustrate an important technological principle in the design of bioeroding devices: The bioerosion rate of a given polymer is not an unchangeable property, but depends to a very large degree on readily controllable factors such as the presence of plasticizers or additives, the manufacturing process, the initial molecular weight of the polymer, and the geometry of the device.

Storage Stability, Sterilization, and Packaging

It is important to minimize premature polymer degradation during fabrication and storage. Traces of moisture can seriously degrade even relatively stable polymers such as poly(bisphenol A carbonate) during injection molding or extrusion. Degradable polymers are particularly sensitive to hydrolytic degradation during high-temperature processing. The industrial production of degradable implants therefore often requires the construction of "controlled atmosphere" facilities where the moisture content of the polymer and the ambient humidity can be strictly controlled.

After fabrication, γ-irradiation or exposure to ethylene oxide may be used for the sterilization of degradable implants. Both methods have disadvantages and as a general rule, the choice is between the lesser of two evils. γ-Irradiation at a dose of 2 to 3 Mrad can result in significant backbone degradation. Since the aliphatic polyesters PLA, PGA, and PDS are particularly sensitive to radiation damage, these materials are usually sterilized by exposure to ethylene oxide and not by γ-irradiation. Unfortunately, the use of the highly dangerous ethylene oxide gas represents a serious safety hazard as well as potentially leaving residual traces in the polymeric device. Polymers sterilized with ethylene oxide must be degassed for extended periods of time.

Additionally, for applications in tissue engineering, biodegradable scaffolds may be preseeded with viable cells or may be impregnated with growth factors or other biologics. There is currently no method that could be used to sterilize scaffolds that contain viable cells without damaging the cells. Therefore, such products must be manufactured under sterile conditions and must be used within a very short time after manufacture. Currently, a small number of products containing preseeded, living cells are in clinical use. These products are extremely expensive, are shipped in special containers, and have little or no shelf life.

Likewise, it has been shown that sterilization of scaffolds containing osteoinductive or chondroinductive agents leads to significant losses in bioactivity, depending on the sterilization method used (Athanasiou *et al.*, 1998). The challenge of producing tissue engineering scaffolds that are preseeded with viable cells or that contain sensitive biological agents has not yet been fully solved.

After sterilization, degradable implants are usually packaged in air-tight aluminum-backed plastic-foil pouches. In some cases, refrigeration may also be required to prevent backbone degradation during storage.

Bibliography

Agrawal, C. M., Hass, K. F., Leopold, D. A., and Clark, H. G. (1992). Evaluation of poly(L-lactic acid) as a material for intravascular polymeric stents. *Biomaterials* **13**: 176–182.

Allcock, H. R. (1990). In *Biodegradable Polymers as Drug Delivery Systems*, M. Chasin, and R. Langer, eds. Marcel Dekker, New York, pp. 163–193.

Anderson, J. M., Spilizewski, K. L., and Hiltner, A. (1985). In *Biocompatibility of Tissue Analogs*, Vol. 1, D. F. Williams, ed. CRC Press Inc., Boca Raton, FL, pp. 67–88.

Andrianov, A. K., Sargent, J. R., Sule, S. S., LeGolvan, M. P., Woods, A. L., Jenkins, S. A., and Payne, L. G. (1998). Synthesis, physico-chemical properties and immunoadjuvant activity of water-soluble polyphosphazene polyacids. *J. Bioactive Comp. Polym.* **13**: 243–256.

Athanasiou, K., Agrawal, M., Barber, A., and Burkhart, S. (1998). Orthopaedic applications for PLA–PGA biodegradable polymers. *Arthroscopy* **14**: 726–737.

Attawia, M. A., Uhrich, K. E., Botchwey, E., Fan, M., Langer, R., and Laurencin, C. T. (1995). Cytotoxicity testing of poly(anhydride-co-imides) for orthopedic applications. *J. Biomed. Mater. Res.* **29**: 1233–1240.

Barham, P. J., Keller, A., Otun, E. L., and Holmes, P. A. (1984). Crystallization and morphology of a bacterial thermoplastic: poly-3-hydroxybutyrate. *J. Mater. Sci.* **19**: 2781–2794.

Beele, H. (2002). Artificial skin: Past, present and future. *Int. J. Artificial Organs* **25**: 163–173.

Bergsma, J. E., de Bruijn, W. C., Rozema, F. R., Bos, R. R. M., and Boering, G. (1995). Late degradation tissue response to poly(L-lactic) bone plates and screws. *Biomaterials* **16**: 25–31.

Campbell, P., Glover, G. I., and Gunn, J. M. (1980). Inhibition of intracellular protein degradation by pepstatin, poly(L-lysine) and pepstatinyl-poly (L-lysine). *Arch. Biochem. Biophys.* **203**: 676–680.

Chasin, M., Domb, A., Ron, E., Mathiowitz, E., Langer, R., Leong, K., Laurencin, C., Brem, H., and Grossman, S. (1990). In *Biodegradable Polymers as Drug Delivery Systems*, M. Chasin and R. Langer, eds. Marcel Dekker, New York, pp. 43–70.

Choueka, J., Charvet, J. L., Koval, K. J., Alexander, H., James, K. S., Hooper, K. A., and Kohn, J. (1996). Canine bone response to tyrosine-derived polycarbonates and poly(L-lactic acid). *J. Biomed. Mater. Res.* **31**: 35–41.

Deshpande, A. A., Heller, J., and Gurny, R. (1998). Bioerodible polymers for ocular drug delivery. *Crit. Rev. Thera. Drug Carrier Syst.* **15**: 381–420.

Frazza, E. J., and Schmitt, E. E. (1971). A new absorbable suture. *J. Biomed. Mater. Res.* **1**: 43–58.

Gilding, D. K., and Reed, A. M. (1979). Biodegradable polymers for use in surgery—poly(glycolic)/poly(lactic acid) homo- and copolymers: 1. *Polymer* **20**: 1459–1464.

Gilding, D. K., and Reed, A. M. (1981). Biodegradable polymers for use in surgery—poly(glycolic)/poly(lactic acid) homo- and copolymers: 2: In vitro degradation. *Polymer* **22**: 494–498.

Gombotz, W. R., and Pettit, D. K. (1995). Biodegradable polymers for protein and peptide drug delivery. *Bioconjugate Chem.* **6**: 332–351.

Heller, J. (1987). in *Controlled Drug Delivery, Fundamentals and Applications*, 2nd ed. J. R. Robinson and V. H. L. Lee, eds. Marcel Dekker, New York, pp. 180–210.

Heller, J. (1993). Polymers for controlled parenteral delivery of peptides and proteins. *Adv. Drug Delivery Rev.* **10**: 163–204.

Heller, J., Barr, J., Ng, S. Y., Abdellauoi, K. S., and Gurny, R. (2002). Poly(ortho esters): synthesis, characterization, properties and uses. *Adv. Drug Delivery Rev.* **54**: 1015–1039.

Heller, J., Penhale, D. W. H., and Helwing, R. F. (1980). Preparation of poly(ortho esters) by the reaction of diketene acetals and polyols. *J. Polymer Sci. (Polymer Lett. Ed.)* **18**: 619–624.

Heller, J., Sparer, R. V., and Zentner, G. M. (1990). in *Biodegradable Polymers as Drug Delivery Systems*, M. Chasin and R. Langer, eds. Marcel Dekker, New York, pp. 121–162.

Hill, J. L., Sawhney, A. S., Pathak, C. P. and Hubbell, J. A. (1993). Prevention of post-operative adhesions using biodegradable hydrogels. 19th Annual Meeting of the Society of Biomaterials, Minneapolis, MN, p. 199.

Huang, S. (1985) in *Encyclopedia of Polymer Science and Engineering*, Vol. 2. F. H. Mark, N. M. Bikales, C. G. Overberger, G. Menges, and J. I. Kroshwitz, eds. John Wiley, New York, pp. 220–243.

James, K., and Kohn, J. (1996). New biomaterials for tissue engineering. *MRS Bull.* **21**: 22–26.

James, K., and Kohn, J. (1997). in *Controlled Drug Delivery: Challenges and Strategies*, K. Park, ed. American Chemical Society, Washington, D.C., pp. 389–403.

Knowles, J. C. (1993). Development of a natural degradable polymer for orthopedic use. *J. Med. Eng. Technol.* **17**: 129–137.

Kohn, J., and Langer, R. (1984). in *Polymeric Materials, Science and Engineering*, Vol. 51. American Chemical Society, Washington, D.C., pp. 119–121.

Kohn, J., and Langer, R. (1987). Polymerization reactions involving the side chains of α-L-amino acids. *J. Am. Chem. Soc.* **109**: 817–820.

Langer, R. (1990). New methods of drug delivery. *Science* **249**: 1527–1533.

Langer, R., and Vacanti, J. P. (1993). Tissue engineering. *Science* **260**: 920–926.

Laurencin, C. T., Norman, M. E., Elgendy, H. M., El-Amin, S. F., Allcock, H. R., Pucher, S. R., and Ambrosio, A. A. (1993). Use of polyphosphazenes for skeletal tissue regeneration. *J. Biomed. Mater. Res.* **27**: 963–973.

Leong, K. W., D' Amore, P. D., Marletta, M., and Langer, R. (1986). Bioerodible polyanhydrides as drug-carrier matrices. II: Biocompatibility and chemical reactivity. *J. Biomed. Mater. Res.* **20**: 51–64.

Lescure, F., Gurney, R., Doelker, E., Pelaprat, M. L., Bichon, D., and Anderson, J. M. (1989). Acute histopathological response to a new biodegradable, polypeptidic polymer for implantable drug delivery system. *J. Biomed. Mater. Res.* **23**: 1299–1313.

Madrid, Y., Langer, L. F., Brem, H., and Langer, R. (1991). New directions in the delivery of drugs and other substances to the central nervous system. *Adv. Pharmacol.* **22**: 299–324.

Mathiowitz, E., Kline, D., and Langer, R. (1990). Morphology of polyanhydride microsphere delivery systems. *J. Scanning Microsc.* **4**: 329–340.

McCormick-Thomson, L. A., and Duncan, R. (1989). Poly(amino acid) copolymers as a potential soluble drug delivery system. 1. Pinocytic uptake and lysosomal degradation measured *in vitro*. *J. Bioact. Biocompat. Polymer* **4**: 242–251.

Miller, N. D., and Williams, D. F. (1987). On the biodegradation of poly-β-hydroxybutyrate (PHB) homopolymer and poly-β-hydroxybutyrate-hydroxyvalerate copolymers. *Biomaterials* **8**: 129–137.

Miyamae, T., Mori, S. and Takeda, Y. (1968). Poly-L-glutamic acid surgical sutures. US Patent 3,371,069.

Nathan, A. and Kohn, J. (1996). in *Protein Engineering and Design*, P. Carey, ed. Academic Press, New York, pp. 265–287.

Okada, H., and Toguchi, H. (1995). Biodegradable microspheres in drug delivery. *Crit. Rev. Ther. Drug Carrier Syst.* **12**: 1–99.

Park, E.-S., Maniar, M., and Shah, J. (1998). Biodegradable polyanhydride devices of cefazolin sodium, bupivacaine, and taxol for local drug delivery: preparation, kinetics and mechanism of in vitro release. *J. Control Rel.* **52**: 179–189.

Pitt, C. G. (1990). in *Biodegradable Polymers as Drug Delivery Systems*, M. Chasin and R. Langer, eds. Marcel Dekker, New York, pp. 71–120.

Pitt, C. G., Gratzl, M. M., Jeffcoat, A. R., Zweidinger, R., and Schindler, A. (1979). Sustained drug delivery systems II: Factors affecting release rates from poly(ε-caprolactone) and related biodegradable polyesters. *J. Pharm. Sci.* **68:** 1534–1538.

Rosen, H., Kohn, J., Leong, K., and Langer, R. (1988). in *Controlled Release Systems: Fabrication Technology*, D. Hsieh, eds. CRC Press, Boca Raton, FL, pp. 83–110.

Senior, P. J., Beech, G. A., Ritchie, G. A. and Dawes, E. A. (1972). The role of oxygen limitation in the formation of poly-β-hydroxybutyrate during batch and continuous culture of *Azotobacter beijerinckii. Biochem. J.* **128:** 1193–1201.

Shea, L. D., and Mooney, D. J. (2001). Nonviral DNA delivery from polymeric systems *Methods Mol. Med.* **65:** 195–207.

Spira, M., Fissette, J., Hall, C. W., Hardy, S. B., and Gerow, F. J. (1969). Evaluation of synthetic fabrics as artificial skin grafts to experimental burn wounds. *J. Biomed. Mater. Res.* **3:** 213–234.

Tamada, J. A., and Langer, R. (1993). Erosion kinetics of hydrolytically degradable polymers. *Proc. Natl. Acad. Sci. USA* **90:** 552–556.

Törmälä, P., Pohjonen, T. and Rokkanen, P. (1998). Bioabsorbable polymers: materials technology and surgical applications. *Proc. Inst. Mech. Engr.* **212:** 101–111.

van Heeswijl, W. A. R., Hoes, C. J. T., Stoffer, T., Eenink, M. J. D., Potman, W., and Feijen, J. (1985). The synthesis and characterization of polypeptide–adriamycin conjugates and its complexes with adriamycin. Part 1. *J. Control Rel.* **1:** 301–315.

Vert, M., Li, S., and Garreau, H. (1991). More about the degradation of LA/GA-derived matrices in aqueous media. *J. Control Release* **16:** 15–26.

Williams, D. F. (1987). *Definitions in Biomaterials—Proceedings of a Consensus Conference of the European Society for Biomaterials.* Elsevier, New York.

2.8 NATURAL MATERIALS

Ioannis V. Yannas

Natural polymers offer the advantage of being very similar, often identical, to macromolecular substances which the biological environment is prepared to recognize and to deal with metabolically (Table 1). The problems of toxicity and stimulation of a chronic inflammatory reaction, as well as lack of recognition by cells, which are frequently provoked by many synthetic polymers, may thereby be suppressed. Furthermore, the similarity to naturally occurring substances introduces the interesting capability of designing biomaterials that function biologically at the molecular, rather than the macroscopic, level. On the other hand, natural polymers are frequently quite immunogenic. Furthermore, because they are structurally much more complex than most synthetic polymers, their technological manipulation is quite a bit more elaborate. On balance, however, these opposing factors have conspired to lead to a substantial number of biomaterials applications in which naturally occurring polymers, or their chemically modified versions, have provided unprecedented solutions.

An intriguing characteristic of natural polymers is their ability to be degraded by naturally occurring enzymes, a virtual guarantee that the implant will be eventually metabolized by physiological mechanisms. This property may, at first glance,

appear as a disadvantage since it detracts from the durability of the implant. However, it has been used to advantage in biomaterials applications in which it is desired to deliver a specific function for a temporary period of time, following which the implant is expected to degrade completely and to be disposed of by largely normal metabolic processes. Since, furthermore, it is possible to control the degradation rate of the implanted polymer by chemical cross-linking or other chemical modifications, the designer is offered the opportunity to control the lifetime of the implant.

A potential problem to be dealt with when proteins are used as biomaterials is their frequently significant immunogenicity, which, of course, derives precisely from their similarity to naturally occurring substances. The immunological reaction of the host to the implant is directed against selected sites (antigenic determinants) in the protein molecule. This reaction can be mediated by molecules in solution in body fluids (immunoglobulins). A single such molecule (antibody) binds to single or multiple determinants on an antigen. The immunological reaction can also be mediated by molecules that are held tightly to the surface of immune cells (lymphocytes). The implant is eventually degraded. The reaction can be virtually eliminated provided that the antigenic determinants have been previously modified chemically. The immunogenicity of polysaccharides is typically far lower than that of proteins. The collagens are generally weak immunogens relative to the majority of proteins.

Another potential problem in the use of natural polymers as biomaterials derives from the fact that these polymers typically decompose or undergo pyrolytic modification at temperatures below the melting point, thereby precluding the convenience of high-temperature thermoplastics processing methods, such as melt extrusion, during the manufacturing of the implant. However, processes for extruding these temperature-sensitive polymers at room temperature have been developed. Another serious disadvantage is the natural variability in structure of macromolecular substances which are derived from animal sources. Each of these polymers appears as a chemically distinct entity not only from one species to another (species specificity) but also from one tissue to the next (tissue specificity). This testimonial to the elegance of the naturally evolved design of the mammalian body becomes a problem for the manufacturer of implants, which are typically required to adhere to rigid specifications from one batch to the next. Consequently, relatively stringent methods of control of the raw material must be used.

Most of the natural polymers in use as biomaterials today are constituents of the extracellular matrix (ECM) of connective tissues such as tendons, ligaments, skin, blood vessels, and bone. These tissues are deformable, fiber-reinforced composite materials of organ shape as well as of the organism itself. In the relatively crude description of these tissues as if they were manmade composites, collagen and elastin fibers mechanically reinforce a "matrix" that primarily consists of protein polysaccharides (proteoglycans) highly swollen in water. Extensive chemical bonding connects these macromolecules to each other, rendering these tissues insoluble and, therefore, impossible to characterize with dilute solution methods unless the tissue is chemically and physically degraded. In the latter case, the solubilized components are subsequently extracted and

TABLE 1   General Properties of Certain Natural Polymers

	Polymer	Incidence	Physiological function
A. Proteins	Silk	Synthesized by arthropods	Protective cocoon
	Keratin	Hair	Thermal insulation
	Collagen	Connective tissues (tendon, skin, etc.)	Mechanical support
	Gelatin	Partly amorphous collagen	(Industrial product)
	Fibrinogen	Blood	Blood clotting
	Elastin	Neck ligament	Mechanical support
	Actin	Muscle	Contraction, motility
	Myosin	Muscle	Contraction, motility
B. Polysaccharides	Cellulose (cotton)	Plants	Mechanical support
	Amylose	Plants	Energy reservoir
	Dextran	Synthesized by bacteria	Matrix for growth of organism
	Chitin	Insects, crustaceans	Provides shape and form
	Glycosaminoglycans	Connective tissues	Contributes to mechanical support
C. Polynucleotides	Deoxyribonucleic acids (DNA)	Cell nucleus	Direct protein biosynthesis
	Ribonucleic acids (RNA)	Cell nucleus	Direct protein biosynthesis

characterized by biochemical and physicochemical methods. Of the various components of extracellular materials that have been used to fashion biomaterials, collagen is the one most frequently used. Other important components, to be discussed later, include the proteoglycans and elastin.

Almost inevitably, the physicochemical processes used to extract the individual polymer from tissues, as well as subsequent deliberate modifications, alter the native structure, sometimes significantly. The description in this section emphasizes the features of the naturally occurring, or native, macromolecular structures. Certain modified forms of these polymers are also described.

STRUCTURE OF NATIVE COLLAGEN

Structural order in collagen, as in other proteins, occurs at several discrete levels of the structural hierarchy. The collagen in the tissues of a vertebrate occurs in at least 10 different forms, each of these being predominant in a specific tissue. Structurally, these collagens share the characteristic triple helix, and variations among them are restricted to the length of the nonhelical fraction, as well as the length of the helix itself and the number and nature of carbohydrates attached on the triple helix. The collagen in skin, tendon, and bone is mostly type I collagen. Type II collagen is predominant in cartilage, while type III collagen is a major constituent of the blood vessel wall as well as being a minor contaminant of type I collagen in skin. In contrast to these collagens, all of which form fibrils with the distinct collagen periodicity, type IV collagen, a constituent of the basement membrane that separates epithelial tissues from mesodermal tissues, is largely nonhelical and does not form fibrils. We follow here the nomenclature that was proposed by W. Kauzmann (1959) to describe in a general way the structural order in proteins, and we specialize it to the case of type I collagen (Fig. 1).

The primary structure denotes the complete sequence of amino acids along each of three polypeptide chains as well as the location of interchain cross-links in relation to this sequence. Approximately one-third of the residues are glycine and another quarter or so are proline or hydroxyproline. The structure of the bifunctional interchain cross-link is the relatively complex condensation product of a reaction involving lysine and hydroxylysine residues; this reaction continues as the organism matures, thereby rendering the collagens of older animals more difficult to extract from tissues.

The secondary structure is the local configuration of a polypeptide chain that results from satisfaction of stereochemical angles and hydrogen-bonding potential of peptide residues. In collagen, the abundance of glycine residues (Gly) plays a key configurational role in the triplet Gly–X–Y, where X and Y are frequently proline or hydroxyproline, respectively, the two amino acids that control the chain configuration locally by the very rigidity of their ring structures. On the other hand, the absence of a side chain in glycine permits close approach of polypeptide chains in the collagen triple helix.

Tertiary structure refers to the global configuration of the polypeptide chains; it represents the pattern according to which the secondary structure is packed within the complete macromolecule and it constitutes the structural unit that can exist as a physicochemically stable entity in solution, namely, the triple helical collagen molecule.

In type I collagen, two of the three polypeptide chains have identical amino acid composition, consisting of 1056 residues and are termed a1(I) chains, while the third has a different composition, it consists of 1038 residues and is termed a2(I). The three polypeptide chains fold to produce a left-handed helix, whereas the three-chain supercoil is actually right-handed with an estimated pitch of about 100 nm (30–40 residues). The helical structure extends over 1014 of the residues in each of the three chains, leaving the remaining residues at the ends (telopeptides) in a nonhelical configuration. The residue

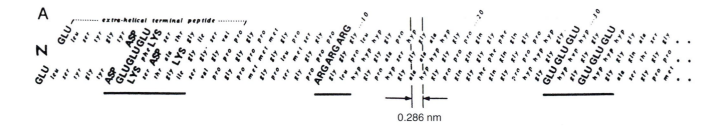

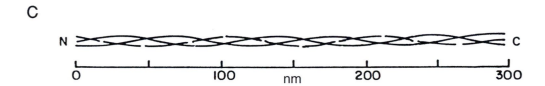

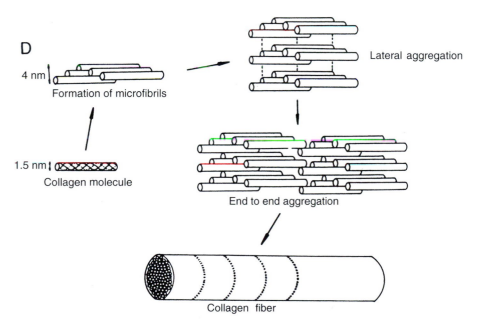

FIG. 1. For legend see opposite page.

spacing is 0.286 nm and the length of the helical portion of the molecule is, therefore, 1014×0.286 or 290 nm long.

The fourth-order or quaternary structure denotes the repeating supermolecular unit structure, comprising several molecules packed in a specific lattice, which constitutes the basic element of the solid state (microfibril). Collagen molecules are packed in a quasi-hexagonal lattice at an interchain distance of about 1.3 nm, which shrinks considerably when the microfibril is dehydrated. Adjacent molecules in the microfibril are approximately parallel to the fibril axis; they all point in the same direction along the fibril and are staggered regularly, giving rise to the well-known D-period of collagen, about 64 nm, which is visible in the electron microscope. Higher levels of order, eventually leading to gross anatomical features that can be readily seen with the naked eye, have been proposed, but there is no general agreement on their definition.

BIOLOGICAL EFFECTS OF PHYSICAL MODIFICATIONS OF THE NATIVE STRUCTURE OF COLLAGEN

Crystallinity in collagen can be detected at two discrete levels of structural order: the tertiary (triple helix) (Fig. 1C) and the quaternary (lattice of triple helices) (Fig. 1D). Each of these levels of order corresponds, interestingly enough, to a separate melting transformation. A solution of collagen triple helices is thus converted to the randomly coiled gelatin by heating above the helix–coil transition temperature, which is approximately 37°C for bovine collagen, or by exceeding a critical concentration of certain highly polarizable anions, e.g., bromide or thiocyanate, in the solution of collagen molecules. Infrared spectroscopic procedures, based on helical marker bands in the mid- and far infrared, have been developed to assay the gelatin content of collagen in the solid or semisolid states in which collagen is commonly used as an implant. Since implanted gelatin is much more rapidly degradable than collagen, a characteristic that can seriously affect implant performance, these assays are essential tools for quality control of collagen-based biomaterials. Frequently, such biomaterials have been processed under manufacturing conditions that may threaten the integrity of the triple helix.

Collagen fibers also exhibit a characteristic banding pattern with a period of about 65 nm (quaternary structure). This pattern is lost reversibly when the pH of a suspension of collagen fibers in acetic acid is lowered below 4.25 ± 0.30. Transmission electron microscopy or small-angle X-ray diffraction can be used to determine the fraction of fibrils that possess banding as the pH of the system is altered. During this transformation, which appears to be a first-order thermodynamic transition, the triple helical structure remains unchanged. Changes in pH can, therefore, be used to selectively abolish the quaternary structure while maintaining the tertiary structure intact.

This experimental strategy has made it possible to show that the well-known phenomenon of blood platelet aggregation by collagen fibers (the reason for use of collagen sponges as hemostatic devices) is a specific property of the quaternary rather than of the tertiary structure. Thus collagen that is thromboresistant *in vitro* has been prepared by selectively "melting out" the packing order of helices while preserving the triple helices themselves. Figure 2 illustrates the banding pattern of such collagen fibers. Notice that short segments of banded fibrils persist even after very long treatment at low pH, occasionally interrupting long segments of nonbanded fibrils (Fig. 2, inset).

The porosity of a collagenous implant normally makes an indispensable contribution to its performance. A porous structure provides an implant with two critical functions. First, pore channels are ports of entry for cells migrating from adjacent tissues into the bulk of the implant for tissue serum (exudate) that enters via capillary suction or of blood from a hemorrhaging blood vessel nearby. Second, pores endow a material with a frequently enormous specific surface that is made available either for specific interactions with invading cells (e.g., myofibroblasts bind extensively on the surface of porous collagen–glycosaminoglycan copolymer structures that induce regeneration of skin in burned patients) or for interaction with coagulation factors in blood flowing into the device (e.g., hemostatic sponges).

Pores can be incorporated by first freezing a dilute suspension of collagen fibers and then inducing sublimation of the ice crystals by exposing the suspension to low-temperature vacuum. The resulting pore structure is a negative replica of the network of ice crystals (primarily dendrites). It follows that

---

FIG. 1. Collagen, like other proteins, is distinguished by several levels of structural order. (A) Primary structure—the complete sequence of amino acids along each polypeptide chain. An example is the triple chain sequence of type I calf skin collagen at the N-end of the molecule. Roughly 5% of a complete molecule is shown above. No attempt has been made to indicate the coiling of the chains. Amino acid residues participating in the triple helix are numbered, and the residue-to-residue spacing (0.286 nm) is shown as a constant within the triple helical domain, but not outside it. Bold capitals indicate charged residues which occur in groups (underlined) (Reprinted from J. A. Chapman and D. J. S. Hulmes (1984). In *Ultrastructure of the Connective Tissue Matrix*, A. Ruggeri and P. M. Motta, eds. Martinus Nijhoff, Boston, Chap. 1, Fig. 1, p. 2, with permission.) (B) Secondary structure—the local configuration of a polypeptide chain. The triplet sequence Gly-Pro-Hyp illustrates elements of collagen triple-helix stabilization. The numbers identify peptide backbone atoms. The conformation is determined by trans peptide bonds (3-4, 6-7, and 9-1); fixed rotation angle of bond in proline ring (4-5); limited rotation of proline past the C=O group (bond 5-6); interchain hydrogen bonds (dots) involving the NH hydrogen at position 1 and the C=O at position 6 in adjacent chains; and the hydroxy group of hydroxyproline, possibly through water-bridged hydrogen bonds. (Reprinted from K. A. Piez and A. H. Reddi, editors (1984). *Extracellular Matrix Biochemistry*. Elsevier, New York, Chap. 1, Fig 1.6. p. 7, with permission.) (C) Tertiary structure—the global configuration of polypeptide chains, representing the pattern according to which the secondary structures are packed together within the unit substructure. A schematic view of the type I collagen molecule, a triple helix 300 nm long. (Reprinted from K. A. Piez and A. H. Reddi, editors (1984). *Extracellular Matrix Biochemistry*. Elsevier, New York, Chap. 1, Fig. 1.22, p. 29, with permission.) (D) Quaternary structure—the unit supermolecular structure. The most widely accepted unit is one involving five collagen molecules (microfibril). Several microfibrils aggregate end to end and also laterally to form a collagen fiber that exhibits a regular banding pattern in the electron microscope with a period of about 65 nm. (Reprinted from E. Nimni, editor (1988). *Collagen*, Vol. I, Biochemistry, CRC Press, Boca Raton, FL Chap. 1, Fig. 10, p. 14, with permission.)

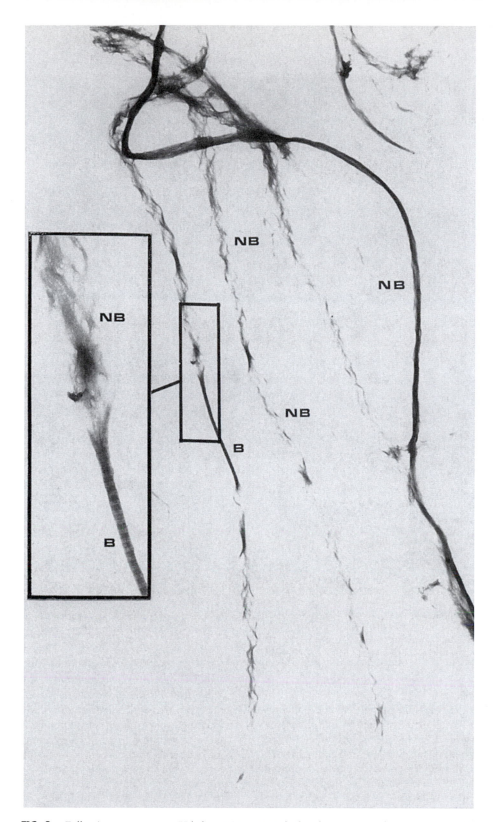

FIG. 2. Following exposure to pH below 4.25 ± 0.30 the banding pattern of type I bovine hide collagen practically disappears. Short lengths of banded collagen (B) do, however, persist next to very long lengths of nonbanded collagen (NB), which has tertiary but not quaternary structure. This preparation does not induce platelet aggregation provided that the fibers are prevented from recrystallizing to form banded structures when the pH is adjusted to neutral in order to perform the platelet assay. Stained with 0.5 wt.% phosphotungstic acid. Banded collagen period, about 65 nm. Original magnification: 15,000×. Inset original mag.: 75,000×. (Reprinted from M. J. Forbes, M. S. dissertation, Massachusetts Institute of Technology, 1980, courtesy of MIT.)

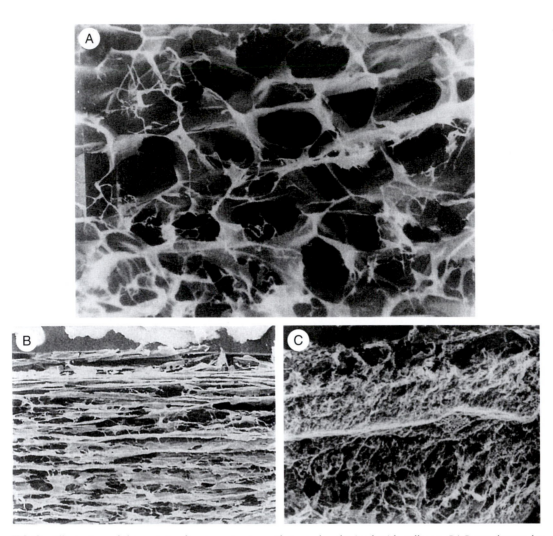

FIG. 3.   Illustration of the variety of porous structures that can be obtained with collagen–GAG copolymers by adjusting the kinetics of crystallizaton of ice to the appropriate magnitude and direction. Pores form when the ice dendrites are eventually sublimed. Scanning electron microscopy. (Courtesy of MIT.)

control of the conditions for ice nucleation and growth can lead to a large variety of pore structures (Fig. 3).

In practice, the average pore diameter decreases with decreasing temperature of freezing while the orientation of pore channel axes depends on the magnitude of the major heat flux vector during freezing. In experimental implants the mean pore diameter has ranged between about 1 and 800 μm; pore volume fractions have ranged up to 0.995; the specific surface has been varied between about 0.01 and 100 m²/g dry matrix; and the orientation of axes of pore channels has ranged from strongly uniaxial to almost random. The ability of collagen–glycosaminoglycan copolymers to induce regeneration of tissues such as skin, the conjunctiva and peripheral nerves depends critically, among other factors, on the adjustment of the pore structure to desired levels, e.g., a pore size range of about 20–125 μm for skin regeneration and less than 10 μm for sciatic nerve regeneration appear to be mandatory. Determination of pore structure is based on principles of stereology, the discipline which allows the quantitative statistical properties of three-dimensional structures of implants to be related to those of two-dimensional projections, e.g., sections used for histological analysis.

CHEMICAL MODIFICATION OF COLLAGEN AND ITS BIOLOGICAL CONSEQUENCES

The primary structure of collagen is made up of long sequences of some 20 different amino acids. Since each amino acid has its own chemical identity, there are 20 types of pendant side groups, each with its own chemical reactivity, attached to the polypeptide chain backbone. As examples, there are carboxylic side groups (from glutamic acid and aspartic acid residues), primary amino groups (lysine, hydroxylysine, and arginine residues), and hydroxylic groups (tyrosine and hydroxylysine). The collagen molecule is therefore subject to modification by a large variety of chemical reagents. Such versatility comes with a price: Even though the choice of reagents is large, it is important to ascertain that use of a given reagent has actually led to modification of a substantial fraction of the residues of an amino acid in the molecule. This is equivalent to proof that a reaction has proceeded to a desired "yield." Furthermore, proof that a given reagent has attacked only a specific type of amino acid, rather than all amino acid residue types carrying the same functional group, also requires chemical analysis.

Historically, the chemical modification of collagen has been practiced in the leather industry (since about 50% of the protein content of cowhide is collagen) and in the photographic gelatin industry. Today, the increasing use of collagen in biomaterials applications has provided renewed incentive for novel chemical modification, primarily in two areas. First, implanted collagen is subject to degradative attack by collagenases, and chemical cross-linking is a well-known means of decelerating the degradation rate. Second, collagen extracted from an animal source elicits production of antibodies (immunogenicity) and chemical modification of antigenic sites may potentially be a useful way to control the immunogenic response. Although it is widely accepted that implanted collagen elicits synthesis of antibodies at a far smaller concentration than is true of most other implanted proteins, treatment with specific reagents, including enzymatic treatment, or cross-linking, is occasionally used to reduce significantly the immunogenicity of collagen.

Collagen-based implants are normally degraded by mammalian collagenases, naturally occurring enzymes that attack the triple helical molecule at a specific location. Two characteristic products result, namely, the N-terminal fragment, which amounts to about two-thirds of the molecule, and the C-terminal fragment. Both of these fragments become spontaneously transformed (denatured) to gelatin at physiological temperatures via the helix–coil transition and the gelatinized fragments are then cleaved to oligopeptides by naturally occurring enzymes that degrade several other tissue proteins (nonspecific proteases).

Collagenases are naturally present in healing wounds and are credited with a major role in the degradation of collagen fibers at the site of trauma. At about the same time that degradation of collagen and of other ECM components proceeds in the wound bed, these components are being synthesized *de novo* by cells at the same anatomical site. Eventually, new architectural arrangements of collagen fibers, such as scar tissue, are synthesized. Although it is not a replica of the intact tissue, scar tissue forms a stable endpoint to the healing process and acts as a tissue barrier that allows the healed organ to continue functioning at a nearly physiological level. One of the frequent challenges in the design of collagen implants is to modify collagen chemically in a way that the rate of its degradation at the implantation site is either accelerated or slowed down to a desired level.

An effective method for reducing the rate of degradation of collagen by naturally occurring enzymes is by chemical cross-linking. A very simple self-cross-linking procedure, dehydrative cross-linking, is based on the fact that removal of water below ca. 1 wt.% insolubilizes collagen as well as gelatin by inducing formation of interchain peptide bonds. The nature of cross-links formed can be inferred from results of studies using

chemically modified gelatins. Gelatin that had been modified either by esterification of the carboxylic groups of aspartyl and glutamyl residues, or by acetylation of the ε-amino groups of lysyl residues, remained soluble in aqueous solvents after exposure of the solid protein to high temperature, while unmodified gelatins lost their solubility. Insolubilization of collagen and gelatin following severe dehydration has been, accordingly, interpreted as the result of drastic removal of the aqueous product of a condensation reaction that led to formation of interchain amide links. The proposed mechanism is consistent with results, obtained by titration, showing that the number of free carboxylic groups and free amino groups in collagen are both significantly decreased following high-temperature treatment.

Removal of water to the extent necessary to achieve a density of cross-links in excess of 10^{-5} mol cross-links/g dry protein, which corresponds to an average molecular weight between crosslinks, M_c, of about 70 kDa, can be achieved within hours by exposure to temperatures in excess of 105°C under atmospheric pressure. The possibility that cross-linking achieved under these conditions is caused by a pyrolytic reaction has been ruled out. Furthermore, chromatographic data have shown that the amino acid composition of collagen remains intact after exposure to 105°C for several days. In fact, it has been observed that gelatin can be cross-linked by exposure to temperatures as low as 25°C provided that a sufficiently high vacuum is present to achieve the drastic moisture removal that drives the cross-linking reaction.

Exposure of highly hydrated collagen to temperatures in excess of ca. 37°C is known to cause reversible melting of the triple helical structure, as described earlier. The melting point of the triple helix increases with the collagen–diluent ratio from 37°C, the helix–coil transition of the infinitely dilute solution, to about 120°C for collagen swollen with as little as 20 wt.% diluent and up to 210°C, the approximate melting point of anhydrous collagen. Accordingly, it is possible to cross-link collagen using the drastic dehydration procedure described above without loss of the triple helical structure. It is simply sufficient to adjust the moisture content of collagen to a low enough level prior to exposure to the high temperature levels required for rapid dehydration.

Dialdehydes have been long known in the leather industry as effective tanning agents and in histological laboratories as useful fixatives. Both of these applications are based on the reaction between the dialdehyde and the ε-amino group of lysyl residues in the protein, which induces formation of interchain cross-links. Glutaraldehyde cross-linking is a relatively widely used procedure in the preparation of implantable biomaterials. Free glutaraldehyde is a toxic substance for cells; it cross-links vital cell proteins. However, clinical studies and extensive clinical use of implants have shown that the toxicity of glutaraldehyde becomes effectively negligible after the unreacted glutaraldehyde has been carefully rinsed out following reaction with an implant, e.g., one based on collagen. The nature of the cross-link formed has been the subject of controversy, primarily due to the complex, apparently polymeric, character of this reagent. Considerable evidence supports a proposed anabilysine structure, which is derived from two lysine side chains and two molecules of glutaraldehyde.

Evidence for other mechanisms has been presented. By comparison with other aldehydes, glutaraldehyde has shown itself to be a particularly effective cross-linking agent, as judged, for example, by its ability to increase the crosslink density to very high levels. Values of the average molecule weight between cross-links (M_c) provide the experimenter with a series of collagens in which the enzymatic degradation rate can be studied over a wide range, thereby affording implants that effectively disappear from tissue between a few days and several months following implantation. The mechanism of the reaction between glutaraldehyde and collagen at neutral pH is understood in part; however, the reaction in acidic media has not been studied extensively. Evidence that covalent cross-linking is involved comes from measurements of the equilibrium tensile modulus of films that have been treated to induce cross-linking and have subsequently been gelatinized by treatment in 1 M NaCl at 70°C. Under such conditions, only gelatin films that have been converted into a three-dimensional network by cross-linking support an equilibrium tensile force; by contrast, un-cross-linked specimens dissolve readily in the hot medium.

Several other methods for cross-linking collagen have been studied, including hexamethylene diisocynate, acyl azide, and a carbodiimide, 1-ethyl-3-(3-dimethlyaminopropyl) carbodiimide (EDAC).

The immunogenicity of the collagen used in implants is not insignificant and has been studied assiduously using laboratory preparations. However, the clinical significance of such immunogenicity has been shown to be very low and is often considered to be negligible. The validity of this simple approach to using collagen as a biomaterial was long ago recognized by manufacturers of collagen-based sutures. The apparent reason for the low antigenicity of type I collagen mostly stems from the small species difference among type I collagens (e.g., cow versus human). Such similarity is, in turn, probably understandable in terms of the inability of the triple helical configuration to incorporate the substantial amino acid substitutions that characterize species differences with other proteins. The relative constancy of the structure of the triple helix among the various species is, in fact, the reason why collagen is sometimes referred to as a "successful" protein in terms of its evolution or, rather, the relative lack of it.

In order to reduce the immunogenicity of collagen it is useful to consider the location of its antigenic determinants, i.e., the specific chemical groups that are recognized as foreign by the immunological system of the host animal. The configurational (or conformational) determinants of collagen depend on the presence of the intact triple helix and, consequently, are abolished when collagen is denatured into gelatin; the latter event (see earlier discussion) occurs spontaneously after the triple helix is cleaved by a collagenase. Gelatinization exposes effectively the sequential determinants of collagen over the short period during which gelatin retains its macromolecular character, before it is cleared away following attack by one of several nonspecific proteases. Control of the stability of the triple helix during processing of collagen, therefore, partially prevents the display of the sequential determinants.

Sequential determinants also exist in the nonhelical end (telopeptide region) of the collagen molecule, and this region has been associated with most of the immunogenicity of

TABLE 2   Certain Applications of Collagen-Based Biomaterials

Applications	Physical state
Sutures	Extruded tape (Schmitt, 1985)
Hemostatic agents	Powder, sponge, fleece (Stenzel *et al.*, 1974; Chvapil, 1979)
Blood vessels	Extruded collagen tube, processed human or animal blood vessel (Nimni, 1988)
Heart valves	Processed porcine heart valve (Nimni, 1988)
Tendon, ligaments	Processed tendon (Piez, 1985)
Burn treatment (dermal regeneration)	Porous collagen–glycosaminoglycan (GAG) copolymers (Yannas *et al.*, 1981, 1982, 1989: Burke *et al.*, 1981; Heimbach *et al.*, 1988)
Peripheral nerve regeneration	Porous collagen–GAG copolymers (Chang and Yannas, 1992)
Meniscus regeneration	Porous collagen–GAG copolymers (Stone *et al.*, 1989, 1997)
Skin regeneration (plastic surgery)	Porous collagen–GAG copolymers
Intradermal augmentation	Injectable suspension of collagen particles (Piez, 1985)
Gynecological applications	Sponges (Chvapil, 1979)
Drug-delivery systems	Various forms (Stenzel *et al.*, 1974; Chvapil, 1979)

collagen-based implants. Several enzymatic treatments have been devised to cleave the telopeptide region without destroying the triple helix. Treatment of collagen with glutaraldehyde not only reduces its degradation rate by collagenase but also appears to reduce its antigenicity. The mechanism of this effect is not well understood.

Certain applications of collagen-based biomaterials are shown in Table 2

PROTEOGLYCANS AND GLYCOSAMINOGLYCANS (GAG)

Glycosaminoglycans (GAGs) occur naturally as polysaccharide branches of a protein chain, or protein core, to which they are covalently attached via a specific oligosaccharide linkage. The entire branched macromolecule, which has been described as having a "bottle brush" configuration, is known as a proteoglycan and typically has a molecular weight of about 10^3 kDa.

The structure of GAGs can be generically described as that of an alternating copolymer, the repeat unit consisting of a hexosamine (glucosamine or galactosamine) and of another sugar (galactose, glucuronic acid, or iduronic acid). Individual GAG chains are known to contain occasional substitutions

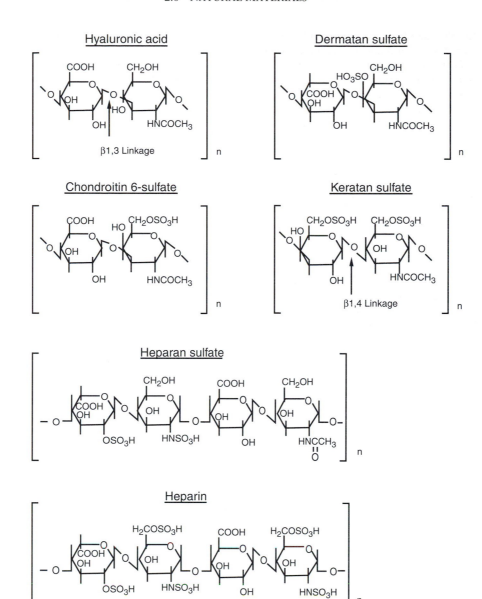

FIG. 4. Repeat units of glycosaminoglycans. (Reprinted from J. Uitto and A. J. Perejda, editors (1987). *Connective Tissue Disease, Molecular Pathology of the Extracellular Matrix*, Vol. 12 in the series The Biochemistry of Disease. Marcel Dekker, New York, Chapter 4, Figs. 1 and 2, p. 85, with permission.)

of one uronic acid for another; however, the nature of the hexosamine component remains invariant along the chain. There are other deviations from the model of a flawless alternating copolymer, such as variations in sulfate content along the chain. It is, nevertheless, useful for the purpose of getting acquainted with the GAGs to show their typical (rather, typified) repeat unit structure, as in Fig. 4. The molecular weight of many GAGs is in the range 5–60 kDa with the exception of hyaluronic acid, the only GAG which is not sulfated; it exhibits molecular weights in the range 50–500 kDa. Sugar units along GAG chains are linked by α or β glycosidic bonds that are 1,3 or 1,4 (Fig. 4). There are several naturally occurring enzymes which degrade specific GAGs, such as hyaluronidase and chondroitinase. These enzymes are primarily responsible

for the physiological turnover rate of GAGs, which is in the range 2–14 days.

The nature of the oligosaccharide linkage appears to be identical for the GAGs, except for keratan sulfate, and is a galactosyl–galactosyl–xylose, with the last glycosidically linked to the hydroxyl group of serine in the protein core.

The very high molecular weight of hyaluronic acid is the basis of most uses of this GAG as a biomaterial: Almost all make use of the exceptionally high viscosity and the facility to form gels that characterize this polysaccharide. Hyaluronic acid gels have found considerable use in ophthalmology because they facilitate cataract surgery as well as retinal reattachment. Other reported uses of GAGs are in the treatment of degenerative joint dysfunction in horses and in the treatment of certain orthopedic dysfunctions in humans. On the other hand, sulfated GAGs are anionically charged and can induce precipitation of collagen at acidic pH levels, a process that yields collagen–GAG coprecipitates that can be subsequently freeze-dried and covalently cross-linked to yield biomaterials that have been shown capable of inducing regeneration of skin (dermis), peripheral nerves, and the conjunctiva (Table 2).

ELASTIN

Elastin is one of the least soluble protein in the body, consisting as it does of a three-dimensional cross-linked network. It can be extracted from tissues by dissolving and degrading all adjacent substances. It appears to be highly amorphous and thus has eluded elucidation of its structure by crystallographic methods. Fortunately, it exhibits ideal rubber elasticity and it thus becomes possible to study certain features of the macromolecular network. For example, mechanical measurements have shown that the average number of amino acid units between cross-links is 71–84. Insoluble elastin preparations can be degraded by the enzyme elastase. The soluble preparations prepared thereby have not yet been applied extensively as biomaterials.

GRAFT COPOLYMERS OF COLLAGEN AND GLYCOSAMINOGLYCANS

The preceding discussion in this chapter has focused on the individual macromolecular components of ECMs. Naturally occurring ECMs are insoluble networks comprising several macromolecular components. Several types of ECMs are known to play critical roles during organ development. During the past several years certain analogs of ECMs have been synthesized and have been studied as implants. This section summarizes the evidence for the unusual biological activity of a small number of ECM analogs.

In the 1970s it was discovered that a highly porous graft copolymer of type I collagen and chondroitin 6-sulfate was capable of modifying dramatically the kinetics and mechanism of healing of full-thickness skin wounds in rodents. In the adult mammal, full-thickness skin wounds represent anatomical sites that are demonstrably devoid of both epidermis and dermis, the two main tissues that comprise skin, respectively. Such wounds normally close by contraction of wound edges and by synthesis of scar tissue. Previously, collagen and various glycosaminoglycans, each prepared in various forms such as powder and films, had been used to cover such deep wounds without observation of a significant modification in the outcome of the wound healing process.

Surprisingly, grafting of these wounds with the porous CG copolymer on guinea pig wounds blocked the onset of wound contraction by several days and led to synthesis of new connective tissue within about 3 weeks in the space occupied by the copolymer. The copolymer underwent substantial degradation during the 3-week period, at the end of which it had degraded completely at the wound site. Studies of the connective tissue synthesized in place of the degraded copolymer eventually showed that the new tissue was distinctly different from scar and was very similar, though not identical, to physiological dermis. In particular, new hair follicles and new sweat glands had not been synthesized. This marked the first instance where scar synthesis was blocked in a full-thickness skin wound of an adult mammal and, in its place, a nearly physiological dermis had been synthesized. That this result was not confined to guinea pigs was confirmed by grafting the same copolymer on full-thickness skin wounds in other adult mammals, including swine and, most importantly, human victims of massive burns as well as humans who underwent reconstructive surgery of the skin.

Although a large number of CG copolymers were synthesized and studied as grafts, it was observed that only one possessed the requisite activity to dramatically modify the wound healing process in skin. In view of the nature of its unique regenerative activity this biologically active macromolecular network has been referred to as dermis regeneration template (DRT). The structure of DRT required specification at two scales: At the nanoscale, the average molecular weight of the cross-linked network that was required to induce regeneration of the dermis was $12,500 \pm 5000$; at the microscale, the average pore diameter was between 20 and 120 μm. Relatively small deviations from these structural features led to loss of activity.

The regeneration of dermis was followed by regeneration of a quite different organ, the peripheral nerve. This was accomplished using a distinctly different ECM analog, termed nerve regeneration template (NRT). Although the chemical composition of the two templates was nearly identical, there were significant differences in other structural features. NRT degrades considerably more slowly than DRT (half-life of about 6 weeks for NRT compared to about 2 weeks for DRT) and is also characterized by a much smaller average pore diameter (about 5 μm compared to 20–120 μm for DRT). DRT was also shown capable of inducing regeneration of the conjunctiva, a specialized structure underneath the eyelid that provides for tearing and other functions that preserve normal vision. The mechanism of induced organ regeneration by templates appears to consist primarily of blocking of contraction of the injured site followed by synthesis of new physiological tissue at about the same rate that the tissue originally present is degraded (synchronous isomorphous replacement). These combined findings suggest that other ECM analogs, still to be discovered, could induce regeneration of organs such as a kidney or the pancreas.

Bibliography

Burke, J. F., Yannas, I. V., Quimby, W. C., Jr., Bondoc, C. C., and Jung, W. K. (1981). Successful use of a physiologically acceptable artificial skin in the treatment of extensive burn injury. *Ann. Surg.* **194:** 413–428.

Chamberlain, L. J., Yannas, I. V., Hsu, H-P., Strichartz, G., and Spector, M. (1998). Collagen-GAG substrate enhances the quality of nerve regeneration through collagen tubes up to level of autograft. *Exp. Neurol.* **154:** 315–329.

Chang, A. S., and Yannas, I. V. (1992). Peripheral nerve regeneration. in *Neuroscience Year* (Suppl. 2 to *The Encyclopedia of*

Neuroscience), B. Smith and G. Adelman, eds. Birkhauser, Boston, pp. 125–126.

Chvapil, M. (1979). Industrial uses of collagen. in *Fibrous Proteins: Scientific, Industrial and Medical Aspects*, D. A. D. Parry and L. K. Creamer, eds. Academic Press, London, Vol. 1, pp. 247–269.

Compton, C. C., Butler, C. E., Yannas, I. V., Warland, G., and Orgill, D. P. (1998). Organized skin structure is regenerated *in vivo* from collagen-GAG matrices seeded with autologous keratinocytes. *J. Invest. Dermatol.* **110**: 908–916.

Davidson, J. M. (1987). Elastin, structure and biology. in *Connective Tissue Disease*, J. Uitto and A. J. Perejda, eds. Marcel Dekker, New York, Chap. 2, pp. 29–54.

Heimbach, D., Luterman, A., Burke, J., Cram, A., Herndon, D., Hunt, J., Jordan, M., McManus, W., Solem, L., Warden, G., and Zawacki, B. (1988). Artificial dermis for major burns. *Ann. Surg.* **208**: 313–320.

Hsu, W.-C., Spilker M. H., Yannas I. V., and Rubin P. A. D. (2000). Inhibition of conjunctival scarring and contraction by a porous collagen-GAG implant. *Invest. Ophthalmol. Vis. Sci.* **41**: 2404–2411.

Kauzmann, W. (1959). Some factors in the interpretation of protein denaturation. *Adv. Protein Chem.* **14**: 1–63.

Li, S.-T. (1995). Biologic biomaterials: tissue-derived biomaterials (collagen). in *The Biomedical Engineering Handbook*, J. D. Bronzino, ed. CRC Press, Boca Raton, FL, Chap. 45, pp. 627–647.

Nimni, M. E., editor. (1988). *Collagen*, Vol. III, *Biotechnology*. CRC Press, Boca Raton, FL.

Piez, K. A. (1985). Collagen. in *Encyclopedia of Polymer Science and Technology*, Vol. 3, pp. 699–727.

Schmitt, F. O. (1985). Adventures in molecular biology. *Ann. Rev. Biophys. Biophys. Chem.* **14**: 1–22.

Shalaby, S. W. (1995). Non-blood-interfacing implants for soft tissues. in *The Biomedical Engineering Handbook*, J. D. Bronzino, ed. CRC Press, Boca Raton, FL, Chap. 46.2, pp. 665–671.

Silbert, J. E. (1987). Advances in the biochemistry of proteoglycans. in *Connective Tissue Disease*, J. Uitto and A. J. Perejda, eds. Marcel Dekker, New York, Chap. 4, pp. 83–98.

Stenzel, K. H., Miyata, T., and Rubin, A. L. (1974). Collagen as a biomaterial. in *Annual Review of Biophysics and Bioengineering*, L. J. Mullins, ed. Annual Reviews Inc., Palo Alto, CA, Vol. 3, pp. 231–252.

Stone, K. R., Steadman, R., Rodkey, W. G., and Li, S.-T. (1997). Regeneration of meniscal cartilage with use of a collagen scaffold. *J. Bone Joint Surg.* **79-A**: 1770–1777.

Yannas, I. V. (1972). Collagen and gelatin in the solid state. *J. Macromol. Sci.-Revs. Macromol. Chem.* C7(1): 49–104.

Yannas, I. V., Burke, J. F., Orgill, D. P., and Skrabut, E. M. (1982). Wound tissue can utilize a polymeric template to synthesize a functional extension of skin. *Science* **215**: 174–176.

Yannas, I. V., Lee, E., Orgill, D. P., Skrabut, E. M., and Murphy, G. F. (1989). Synthesis and characterization of a model extracellular matrix which induces partial regeneration of adult mammalian skin. *Proc. Natl. Acad. Sci. USA* **86**: 933–937.

Yannas, I. V. (1990). Biologically active analogs of the extracellular matrix. *Angew. Chem. Int. Ed.* **29**: 20–35.

Yannas, I. V. (1997). In vivo synthesis of tissue and organs. in *Principles of Tissue Engineering*, R. P. Lanza, R. Langer, and W. L. Chick, eds. R. G. Landes, Austin, Chap. 12, pp. 169–178.

Yannas, I. V. (2004). Synthesis of tissues and organs. *Chembiochem.* 5(1): 26–39.

Yannas, I. V. (2001). Tissue and Organ Regeneration in Adults. New York: Springer.

Yannas, I. V., and Hill, B. J. (2004). Selection of biomaterials for peripheral nerve regeneration using data from the nerve chamber model. *Biomaterials.* 25(9): 1593–1600.

2.9 METALS

John B. Brunski

INTRODUCTION

Implant materials in general, and metallic implant materials in particular, have a significant economic and clinical impact on the biomaterials field. The worldwide market for all types of biomaterials was estimated at over $5 billion in the late 1980s, but grew to about $20 billion in 2000 and is likely to exceed $23 billion by 2005. With the recent emergence of the field known as tissue engineering, including its strong biomaterials segment, the rate of market growth has been estimated at about 12 to 20% per year.

For the United States, the biomaterials market has been estimated at about $9 billion as of the year 2000, with a growth rate of about 20% per year. The division of this market into various submarkets is illustrated by older data: in 1991 the total orthopedic implant and instrument market was about $2 billion and was made up of joint prostheses made primarily of metallic materials ($1.4 billion), together with a wide variety of trauma products ($0.340 billion), instrumentation devices ($0.266 billion), bone cement accessories ($0.066 billion), and bone replacement materials ($0.029 billion). Estimates for other parts of the biomaterials market include $0.425 billion for oral and maxillofacial implants and $0.014 billion for periodontal treatments, and materials for alveolar ridge augmentation or maintenance.

Estimates of the size of the total global biomaterials market are substantiated by the statistics on clinical procedures. For example, of the approximately 3.6 million orthopedic operations per year in the United States, four of the 10 most frequent involve metallic implants: open reduction of a fracture and internal fixation (1 on the list); placement or removal of an internal fixation device without reduction of a fracture (6); arthroplasty of the knee or ankle (7), and total hip replacement or arthroplasty of the hip (8). Moreover, 1988 statistics show that although reduction of fractures was first on the list of inpatient procedures (631,000 procedures), second on the list was excision or destruction of an intervertebral disk (250,000 procedures). Since the latter often involves a bone graft of some kind (from the same patient of from a bone bank) and internal fixation with plates and screws, this represents yet another clinical procedure involving significant use of biomaterials. Overall, including all clinical specialties in 1988, statistics showed that about 11 million Americans (about 4.6% of the civilian population) had at least one implant (Moss *et al.*, 1990).

In view of this wide utilization of implants, many of which are metallic, the objective of this chapter is to describe the composition, structure, and properties of current metallic implant alloys. Major themes are the metallurgical principles underlying structure–property relationships, and the role that biomaterials play in the larger problem of design, production, and proper utilization of medical devices.

STEPS IN THE FABRICATION OF IMPLANTS

Understanding the structure and properties of metallic implant materials requires an appreciation of the metallurgical significance of the material's processing history. Since each metallic device will ordinarily differ in exactly how it is manufactured, generic processing steps are outlined in Fig. 1A.

Metal-Containing Ore to Raw Metal Product

With the exception of the noble metals (which do not represent a major fraction of implant metals), metals exist in the Earth's crust in mineral form wherein the metal is chemically combined with other elements, as in the case of metal oxides. These mineral deposits (ore) must be located and mined, and then separated and enriched to provide ore suitable for further processing into pure metal and/or various alloys.

For example, with titanium, certain mines in the southeastern United States yield sands containing primarily common quartz but also mineral deposits of zircon, titanium, iron, and rare earth elements. The sandy mixture can be concentrated by using water flow and gravity to separate out the metal-containing sands into titanium-containing compounds such as rutile (TiO_2) and ilmenite ($FeTiO_3$). To obtain rutile, which is particularly good for making metallic titanium, further processing typically involves electrostatic separations. Then, to extract titanium metal from the rutile, one method involves treating the ore with chlorine to make titanium tetrachloride liquid, which in turn is treated with magnesium or sodium to produce chlorides of the latter metals and bulk titanium "sponge" according to the Kroll process. At this stage, the titanium sponge is not of controlled purity. So, depending on the purity grade desired in the final titanium product, it is necessary to refine it further by using vacuum furnaces, remelting, and additional steps. All of this can be critical in producing titanium with the appropriate properties. For example, the four most common grades of commercially pure (CP) titanium differ in oxygen content by only tenths of a percent, but these small differences in oxygen content can make major differences

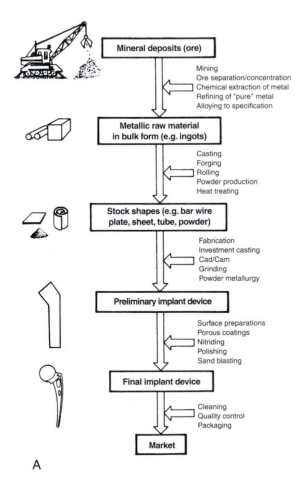

A   B

FIG. 1. (A) Generic processing history of a typical metallic implant device, in this case a hip implant. (B) Image of one step during the investment casting ("lost wax") process of manufacturing hip stems; a rack of hip stems can be seen attached to a system of sprues through which molten metal can flow. At this point, ceramic investment material composes the mold into which the molten metal will flow and solidify during casting, thereby replicating the intended shape of a hip stem.

in mechanical properties such as yield and tensile and fatigue strength of titanium, as discussed later in this chapter. In any case, from the preceding extraction steps, the resulting raw metal product eventually emerges in some type of bulk form, such as ingots, which can be supplied to raw materials vendors or metal manufacturers.

In the case of multicomponent metallic implant alloys, the raw metal product will usually have to be processed further both chemically and physically. Processing steps include remelting, the addition of alloying elements, and controlled solidification to produce an alloy that meets certain chemical and metallurgical specifications. For example, to make ASTM (American Society for Testing and Materials) F138 316L stainless steel, iron is alloyed with specific amounts of carbon, silicon, nickel, and chromium. To make ASTM F75 or F90 alloy, cobalt is alloyed with specific amounts of chromium, molybdenum, carbon, nickel, and other elements. Table 1 lists the chemical compositions of some metallic alloys for surgical implants.

Raw Metal Product to Stock Metal Shapes

A metal supplier further processes the bulk raw metal product (metal or alloy) into "stock" bulk shapes, such as bars, wire, sheet, rods, plates, tubes, or powders. These stock shapes may then be sold to specialty companies (e.g., implant manufacturers) who need stock metal that is closer to the final form of the implant. For example, a maker of screw-shaped dental implants might want to buy rods of the appropriate metal to simplify the machining of the screws from the rod stock.

The metal supplier might transform the metal product into stock shapes by a variety of processes, including remelting and continuous casting, hot rolling, forging, and cold drawing through dies. Depending on the metal, there may also be heat-treating steps (carefully controlled heating and cooling cycles) designed to facilitate further working or shaping of the stock; relieve the effects of prior plastic deformation (e.g., as in annealing); or produce a specific microstructure and properties in the stock material. Because of the high chemical reactivity of some metals at elevated temperatures, high-temperature processes may require vacuum conditions or inert atmospheres to prevent unwanted uptake of oxygen by the metal, all of which adds to cost. For instance, in the production of fine powders of ASTM F75 Co–Cr–Mo alloy, molten metal is often ejected through a small nozzle to produce a fine spray of atomized droplets that solidify while cooling in an inert argon atmosphere.

For metallic implant materials in general, stock shapes are often chemically and metallurgically tested at this early stage to ensure that the chemical composition and microstructure of the metal meet industry standards for surgical implants (ASTM Standards), as discussed later in this chapter. In other words, an implant manufacturer will want assurance that they are buying an appropriate grade of stock metal.

Stock Metal Shapes to Preliminary and Final Metal Devices

Typically, an implant manufacturer will buy stock material and then fabricate preliminary and final forms of the device from the stock material. Specific steps depend on a number of factors, including the final geometry of the implant, the forming and machining properties of the metal, and the costs of alternative fabrication methods.

TABLE 1 Chemical Compositions of Stainless Steels Used for Implants

Material	ASTM designation	Common/trade names	Composition (wt.%)	Notes
Stainless steel	F55 (bar, wire)	AISI 316 LVM	60–65 Fe	F55, F56 specify 0.03 max for P,S.
	F56 (sheet, strip)	316L	17.00–20.00 Cr	F138, F139 specify 0.025 max for
	F138 (bar, wire)	316L	12.00–14.00 Ni	P and 0.010 max for S.
	F139 (sheet, strip)	316L	2.00–3.00 Mo	LVM = low vacuum melt.
			max 2.0 Mn	
			max 0.5 Cu	
			max 0.03 C	
			max 0.1 N	
			max 0.025 P	
			max 0.75 Si	
			max 0.01 S	
Stainless steel	F745	Cast stainless steel cast 316L	60–69 Fe	
			17.00–20.00 Cr	
			11.00–14.00 Ni	
			2.00–3.00 Mo	
			max 0.06 C	
			max 2.0 Mn	
			max 0.045 P	
			max 1.00 Si	
			max 0.030 S	

Fabrication methods include investment casting (the "lost wax" process), conventional and computer-based machining (CAD/CAM), forging, powder metallurgical processes (e.g., hot isostatic pressing, or HIP), and a range of grinding and polishing steps. A variety of fabrication methods are required because not all implant alloys can be feasibly or economically made in the same way. For instance, cobalt-based alloys are extremely difficult to machine by conventional methods into the complicated shapes of some implants. Therefore, many cobalt-based alloys are frequently shaped into implant forms by investment casting (e.g., Fig. 1B) or powder metallurgy. On the other hand, titanium is relatively difficult to cast, and therefore is frequently machined even though titanium in general is not considered to be an easily machinable metal.

Another aspect of fabrication, which comes under the heading of surface treatment, involves the application of macro- or microporous coatings on implants, or the deliberate production of certain degrees of surface roughness. Such surface modifications have become popular in recent years as a means to improve fixation of implants in bone. The surface coating or roughening can take various forms and require different fabrication technologies. In some cases, this step of the processing history can contribute to metallurgical properties of the final implant device. For example, in the case of alloy beads or "fiber metal" coatings, the manufacturer applies the coating only over specific regions of the implant surface (e.g., on the proximal portion of a femoral stem), and the means by which such a coating is attached to the bulk substrate may involve a process such as high-temperature sintering. Generally, sintering involves heating the coating and substrate to about one-half or more of the alloy's melting temperature, which is meant to enable diffusive mechanisms to form necks that join the beads in the coating to one another and to the implant's surface (Fig. 2). Such temperatures can also modify the underlying metallic substrate.

An alternative surface treatment to sintering is plasma or flame spraying a metal onto an implant's surface. Hot, high-velocity gas plasma is charged with a metallic powder and directed at appropriate regions of an implant surface. The powder particles fully or partially melt and then fall onto the substrate surface, where they solidify rapidly to form a rough coating (Fig. 3).

Other surface treatments are also available, including ion implantation (to produce better surface properties), nitriding, and coating with a thin diamond film. In nitriding, a high-energy beam of nitrogen ions is directed at the implant under vacuum. Nitrogen atoms penetrate the surface and come to rest at sites in the substrate. Depending on the alloy, this process can produce enhanced properties. These treatments are commonly used to increase surface hardness and wear properties.

Finally, the manufacturer of a metallic implant device will normally perform a set of finishing steps. These vary with the metal and manufacturer, but typically include chemical cleaning and passivation (i.e., rendering the metal inactive) in appropriate acid, or electrolytically controlled treatments to remove machining chips or impurities that may have become embedded in the implant's surface. As a rule, these steps are conducted according to good manufacturing practice (GMP) and ASTM specifications for cleaning and finishing implants.

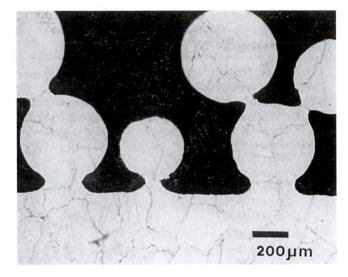

FIG. 2. Low-power view of the interface between a porous coating and solid substrate in the ASTM F75 Co–Cr–Mo alloy system. Note the structure and geometry of the necks joining the beads to one another and to the substrate. Metallographic cross section cut perpendicular to the interface; lightly etched to show the microstructure. (Photo courtesy of Smith & Nephew Richards, Inc. Memphis, TN.)

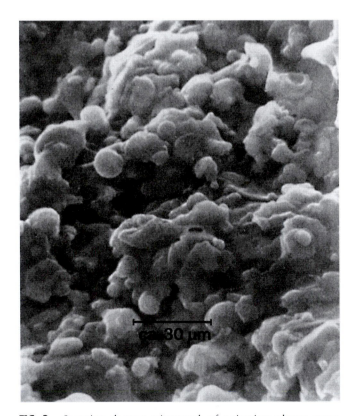

FIG. 3. Scanning electron micrograph of a titanium plasma spray coating on an oral implant. (Photo courtesy of A. Schroeder, E. Van der Zypen, H. Stich, and F. Sutter, *Int. J. Oral Maxillofacial Surg.* 9: 15, 1981.)

It is worth emphasizing that these steps can be extremely important to the overall biological performance of the implant because they can affect the surface properties of the medical device, which is the surface that comes in direct contact with the blood and other tissues at the implant site.

MICROSTRUCTURES AND PROPERTIES OF IMPLANT METALS

In order to understand the properties of each alloy system in terms of microstructure and processing history, it is essential to know (1) the chemical and crystallographic identities of the phases present in the microstructure; (2) the relative amounts, distribution, and orientation of these phases; and (3) the effects of the phases on properties. This section of the chapter emphasizes mechanical properties of metals used in implant devices even though other properties, such as surface properties and wear properties, must also be considered and may actually be more critical to control in certain medical device applications. (Surface properties of materials are reviewed in more depth in Chapter 1.4 of this book.) The following discussion of implant alloys is divided into the stainless steels, cobalt-based alloys, and titanium-based alloys, since these are the most commonly used metals in medical devices.

Stainless Steels

Composition

Although several types of stainless steels are available for implant use (Table 1), in practice the most common is 316L (ASTM F138, F139), grade 2. This steel has less than 0.030% (wt.%) carbon in order to reduce the possibility of *in vivo* corrosion. The "L" in the designation 316L denotes low carbon content. The 316L alloy is predominantly iron (60–65%) with significant alloying additions of chromium (17–20%) and nickel (12–14%), plus minor amounts of nitrogen, manganese, molybdenum, phosphorus, silicon, and sulfur.

With 316L, the main rationale for the alloying additions involves the metal's surface and bulk microstructure. The key function of chromium is to permit the development of corrosion-resistant steel by forming a strongly adherent surface oxide (Cr_2O_3). However, the downside to adding Cr is that it tends to stabilize the ferritic (BCC, body-centered cubic) phase of iron and steel, which is weaker than the austenitic (FCC, face-centered cubic) phase. Moreover, molybdenum and silicon are also ferrite stabilizers. So to counter this tendency to form weaker ferrite, nickel is added to stabilize the stronger austenitic phase.

The main reason for the low carbon content in 316L is to improve corrosion resistance. If the carbon content of the steel significantly exceeds 0.03%, there is increased danger of formation of carbides such as $Cr_{23}C_6$. Such carbides have the bad habit of tending to precipitate at grain boundaries when the carbon concentration and thermal history are favorable to the kinetics of carbide growth. The negative effect of carbide precipitation is that it depletes the adjacent grain boundary regions of chromium, which in turn has the effect of diminishing

formation of the protective, chromium-based oxide Cr_2O_3. Steels in which such grain-boundary carbides have formed are called "sensitized" and are prone to fail through corrosion-assisted fractures that originate at the sensitized (weakened) grain boundaries.

Microstructure and Mechanical Properties

Under ASTM specifications, the desirable form of 316L is single-phase austenite (FCC); there should be no free ferritic (BCC) or carbide phases in the microstructure. Also, the steel should be free of inclusions or impurity phases such as sulfide stringers, which can arise primarily from unclean steel-making practices and predispose the steel to pitting-type corrosion at the metal–inclusion interfaces.

The recommended grain size for 316L is ASTM #6 or finer. The ASTM grain size number n is defined by the formula:

$$N = 2^{n-1} \tag{1}$$

where N is the number of grains counted in 1 square inch at 100-times magnification (0.0645 mm^2 actual area). As an example, when $n = 6$, the grain size is about 100 microns or less. Furthermore, the grain size should be relatively uniform throughout (Fig. 4A). The emphasis on a fine grain size is explained by a Hall–Petch-type relationship (Hall, 1951; Petch, 1953) between mechanical yield stress and grain diameter:

$$t_y = t_i + kd^{-m} \tag{2}$$

Here t_y and t_i are the yield and friction stress, respectively; d is the grain diameter; k is a constant associated with propagation of deformation across grain boundaries; and m is approximately 0.5. From this equation it follows that higher yield stresses may be achieved by a metal with a smaller grain diameter d, all other things being equal. A key determinant of grain size is manufacturing history, including details on solidification conditions, cold-working, annealing cycles, and recrystallization.

Another notable microstructural feature of 316L as used in typical implants is plastic deformation within grains (Fig. 4B). The metal is often used in a 30% cold-worked state because cold-worked metal has a markedly increased yield, ultimate tensile, and fatigue strength relative to the annealed state (Table 2). The trade-off is decreased ductility, but ordinarily this is not a major concern in implant products.

In specific orthopedic devices such as bone screws made of 316L, texture may also be a notable feature in the microstructure. Texture means a preferred orientation of deformed grains. Stainless steel bone screws show elongated grains in metallographic sections taken parallel to the long axis of the screws (Fig. 5). Texture arises as a result of the cold drawing or similar cold-working operations inherent in the manufacture of bar rod stock from which screws are usually machined. In metallographic sections taken perpendicular to the screw's long axis, the grains appear more equiaxed. A summary of representative mechanical properties of 316L stainless is provided in Table 2, but this should only be taken as a general guide, given that final production steps specific to a given implant may often affect properties of the final device.

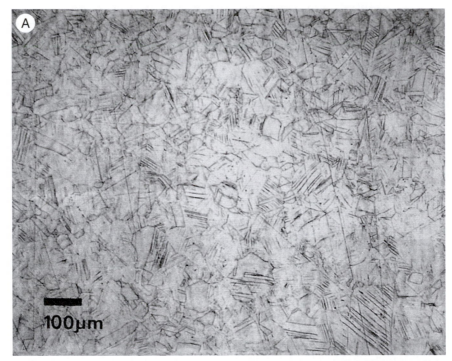

FIG. 4. (A) Typical microstructure of cold-worked 316L stainless steel, ASTM F138, in a transverse section taken through a spinal distraction rod. (B) Detail of grains in cold-worked 316L stainless steel showing evidence of plastic deformation. (Photo in B courtesy of Zimmer USA, Warsaw, IN.)

TABLE 2 Typical Mechanical Properties of Implant Metals[a]

Material	ASTM designation	Condition	Young's modulus (GPa)	Yield strength (MPa)	Tensile strength (MPa)	Fatigue endurance limit (at 10^7 cycles, $R = -1$[c]) (MPa)
Stainless steel	F745	Annealed	190	221	483	221–280
	F55, F56, F138, F139	Annealed	190	331	586	241–276
		30% Cold worked	190	792	930	310–448
		Cold forged	190	1213	1351	820
Co–Cr alloys	F75	As-cast/annealed	210	448–517	655–889	207–310
		P/M HIP[b]	253	841	1277	725–950
	F799	Hot forged	210	896–1200	1399–1586	600–896
	F90	Annealed	210	448–648	951–1220	Not available
		44% Cold worked	210	1606	1896	586
	F562	Hot forged	232	965–1000	1206	500
		Cold worked, aged	232	1500	1795	689–793 (axial tension $R = 0.05$, 30 Hz)
Ti alloys	F67	30% Cold-worked Grade 4	110	485	760	300
	F136	Forged annealed	116	896	965	620
		Forged, heat treated	116	1034	1103	620–689

[a] Data collected from references noted at the end of this chapter, especially Table 1 in Davidson and Georgette (1986).
[b] P/M HIP; Powder metallurgy product, hot-isostatically pressed.
[c] R is defined as $\sigma_{min}/\sigma_{max}$.

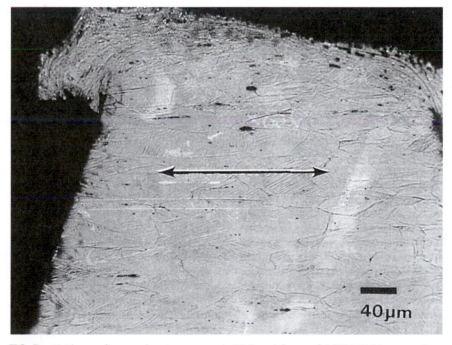

FIG. 5. Evidence of textured grain structure in 316L stainless steel ASTM F138, as seen in a longitudinal section through a cold-worked bone screw. The long axis of the screw is indicated by the arrow.

Cobalt-Based Alloys

Composition

Cobalt-based alloys include Haynes-Stellite 21 and 25 (ASTM F75 and F90, respectively), forged Co–Cr–Mo alloy (ASTM F799), and multiphase (MP) alloy MP35N (ASTM F562). The F75 and F799 alloys are virtually identical in composition (Table 3), each being about 58–70% Co and 26–30% Cr. The key difference is their processing history, as discussed later. The other two alloys, F90 and F562, have slightly less Co and Cr, but more Ni in the case of F562, and more tungsten in the case of F90.

Microstructures and Properties

ASTM F75 The main attribute of this alloy is corrosion resistance in chloride environments, which is related to its bulk composition and surface oxide (nominally Cr_2O_3). This alloy has a long history in both the aerospace and biomedical implant industries.

When F75 is cast into shape by investment casting ("lost wax" process), the alloy is melted at 1350–1450°C and then poured or pressurized into ceramic molds of the desired shape (e.g., femoral stems for artificial hips, oral implants, dental partial bridgework). The sometimes intricately shaped molds are made by fabricating a wax pattern to near-final dimensions of the implant and then coating (or investing) the pattern with a special ceramic, which then holds its shape after the wax is burned out prior to casting—hence the "lost wax" name of the process. Molten metal is poured into the ceramic mold through sprues, or pathways. Then, once the metal has solidified into the shape of the mold, the ceramic mold is cracked away and processing of the metal continues toward the final device.

TABLE 3 Chemical Compositions of Co-Based Alloys for Implants

Material	ASTM designation	Common trade names	Composition (wt.%)	Notes
Co–Cr–Mo	F75	Vitallium Haynes-Stellite 21 Protasul-2 Micrograin-Zimaloy	58.9–69.5 Co 27.0–30.0 Cr 5.0–7.0 Mo max 1.0 Mn max 1.0 Si max 2.5 Ni max 0.75 Fe max 0.35 C	Vitallium is a trade mark of Howmedica, Inc. Haynes-Stellite 21 (HS 21) is a trademark of Cabot Corp. Protasul-2 is a trademark of Sulzer AG, Switzerland. Zimaloy is a trademark of Zimmer USA.
Co–Cr–Mo	F799	Forged Co–Cr–Mo Thermomechanical Co–Cr–Mo FHS	58–59 Co 26.0–30.0 Cr 5.0–7.00 Mo max 1.00 Mn max 1.00 Si max 1.00 Ni max 1.5 Fe max 0.35 C max 0.25 N	FHS means, "forged high strength" and is a trademark of Howmedica, Inc.
Co–Cr–W–Ni	F90	Haynes-Stellite 25 Wrought Co–Cr	45.5–56.2 Co 19.0–21.0 Cr 14.0–16.0 W 9.0–11.0 Ni max 3.00 Fe 1.00–2.00 Mn 0.05–0.15 C max 0.04 P max 0.40 Si max 0.03 S	Haynes-Stellite 25 (HS25) is a trademark of Cabot Corp.
Co–Ni–Cr–Mo–Ti	F562	MP 35 N Biophase Protasul-1()	29–38.8 Co 33.0–37.0 Ni 19.0–21.0 Cr 9.0–10.5 Mo max 1.0 Ti max 0.15 Si max 0.010 S max 1.0 Fe max 0.15 Mn	MP35 N is a trademark of SPS Technologies, Inc. Biophase is a trademark of Richards Medical Co. Protasul-10 is a trademark of Sulzer AG, Switzerland.

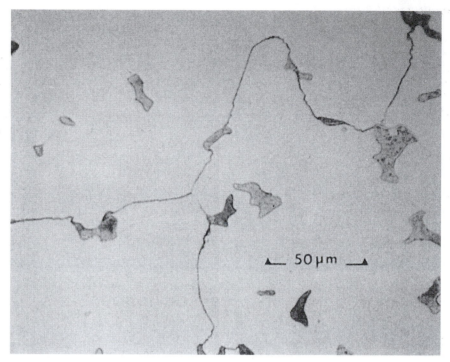

FIG. 6. Microstructure of as-cast Co–Cr–Mo ASTM F75 alloy, showing a large grain size plus grain boundary and matrix carbides. (Photo courtesy of Zimmer USA, Warsaw, IN.)

Depending on the exact casting details, this process can produce at least three microstructural features that can strongly influence implant properties, often negatively.

First, as-cast F75 alloy (Figs. 6 and 7A) typically consists of a Co-rich matrix (alpha phase) plus interdendritic and grain-boundary carbides (primarily $M_{23}C_6$, where M represents Co, Cr, or Mo).

There can also be interdendritic Co and Mo-rich sigma intermetallic, and Co-based gamma phases. Overall, the relative amounts of the alpha and carbide phases should be approximately 85% and 15%, respectively. However, because of nonequilibrium cooling, a "cored" microstructure can develop. In this situation, the interdendritic regions become solute (Cr, Mo, C) rich and contain carbides, while the dendrites become depleted in Cr and richer in Co. This is an unfavorable electrochemical situation, with the Cr-depleted regions being anodic with respect to the rest of the microstructure. (This is also an unfavorable situation if a porous coating will subsequently be applied by sintering to this bulk metal.) Subsequent solution-anneal heat treatments at 1225°C for 1 hour can help alleviate this situation.

Second, the solidification during the casting process results not only in dendrite formation, but also in a relatively large grain size. This is generally undesirable because it decreases the yield strength via a Hall–Petch relationship between yield strength and grain diameter (see Eq. 2 in the section on stainless steel). The dendritic growth patterns and large grain diameter (~4 mm) can be easily seen in Fig. 7A, which shows a hip stem manufactured by investment casting.

Third, casting defects may arise. Figure 7B shows an inclusion in the middle of a femoral hip stem. The inclusion was a particle of the ceramic mold (investment) material, which presumably broke off and became entrapped within the interior of the mold while the metal was solidifying. This contributed to a fatigue fracture of the implant device *in vivo*, most likely because of stress concentrations and crack initiation sites associated with the ceramic inclusion. For similar reasons, it is also desirable to avoid macro- and microporosity arising from metal shrinkage upon solidification of castings. Figures 7C and 7D exemplify a markedly dendritic microstructure, large grain size, and evidence of microporosity at the fracture surface of a ASTM F75 dental device fabricated by investment casting.

To avoid problems such as the above with cast F75, and to improve the alloy's microstructure and mechanical properties, powder metallurgical techniques have been used. For example, in hot isostatic pressing (HIP), a fine powder of F75 alloy is compacted and sintered together under appropriate pressure and temperature conditions (about 100 MPa at 1100°C for 1 hour) and then forged to final shape. The typical microstructure (Fig. 8) shows a much smaller grain size (~8 μm) than the as-cast material. Again, according to a Hall–Petch relationship, this microstructure gives the alloy higher yield strength and better ultimate and fatigue properties than the as-cast alloy (Table 2). Generally speaking, the improved properties of the HIP versus cast F75 result from both the finer grain size and a finer distribution of carbides, which has a hardening effect as well.

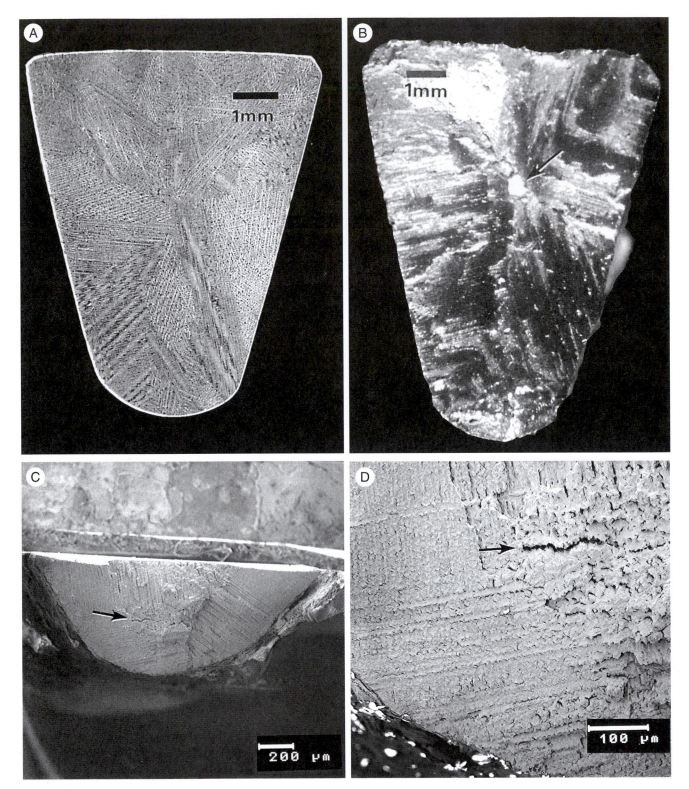

FIG. 7. (A) Macrophoto of a metallographically polished and etched cross section of a cast Co–Cr–Mo ASTM F75 femoral hip stem, showing dendritic structure and large grain size. (B) Macrophoto of the fracture surface of the same Co–Cr–Mo ASTM F75 hip stem as in (A). Arrow indicates large inclusion within the central region of the cross section. Fracture of this hip stem occurred *in vivo*. (C), (D) Scanning electron micrographs of the fracture surface from a cast F75 subperiosteal dental implant. Note the large grain size, dendritic microstructure, and interdendritic microporosity (arrows).

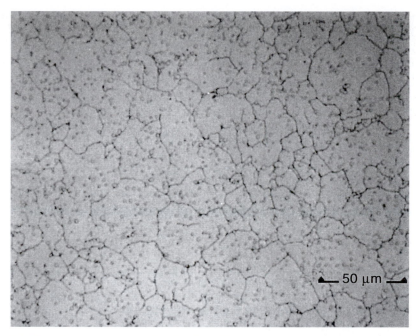

FIG. 8. Microstructure of the Co–Cr–Mo ASTM F75 alloy made via hot isostatic pressing (HIP), showing the much smaller grain size relative to that in Fig. 6. (Photo courtesy of Zimmer USA, Warsaw, IN.)

In porous-coated prosthetic devices based on F75 alloy, the microstructure will depend on the prior manufacturing history of the beads and substrate metal as well as on the sintering process used to join the beads together and to the underlying bulk substrate. With Co–Cr–Mo alloys, for instance, sintering can be difficult, requiring temperatures near the melting point (1225°C). Unfortunately, these high temperatures can decrease the fatigue strength of the substrate alloy. For example, cast-solution-treated F75 has a fatigue strength of about 200–250 MPa, but it can decrease to about 150 MPa after porous coating treatments. The reason for this decrease probably relates to further phase changes in the nonequilibrium cored microstructure in the original cast F75 alloy. However, it has been found that a modified sintering treatment can return the fatigue strength back up to about 200 MPa (Table 2).

Beyond these metallurgical issues, a related concern with porous-coated devices is the potential for decreased fatigue performance due to stress concentrations inherent in the geometrical features where particles are joined to the substrate (e.g., Fig. 2).

ASTM F799 The F799 alloy is basically a modified F75 alloy that has been mechanically processed by hot forging (at about 800°C) after casting. It is sometimes known as thermomechanical Co–Cr–Mo alloy and has a composition slightly different from that of ASTM F75. The microstructure reveals a more worked grain structure than as-cast F75 and a hexagonal close-packed (HCP) phase that forms via a shear-induced transformation of FCC matrix to HCP platelets. This microstructure is not unlike that which occurs in MP35N (see ASTM F562).

The fatigue, yield, and ultimate tensile strengths of this alloy are approximately twice those of as-cast F75 (Table 2).

ASTM F90 Also known as Haynes Stellite 25 (HS-25), F90 alloy is based on Co–Cr–W–Ni. Tungsten and nickel are added to improve machinability and fabrication. In the annealed state, its mechanical properties are about the same as those of F75 alloy, but when cold worked to 44%, the properties more than double (Table 2).

ASTM F562 Known as MP35N, F562 alloy is primarily Co (29–38.8%) and Ni (33–37%), with significant amounts of Cr and Mo. The "MP" in the name refers to the multiple phases in its microstructure. The alloy can be processed by thermal treatments and cold working to produce a controlled microstructure and a high-strength alloy, as follows.

To start with, under equilibrium conditions pure solid cobalt has an FCC Bravais lattice above 419°C and a HCP structure below 419°C. However, the solid-state transformation from FCC to HCP is sluggish and occurs by a martensitic-type shear reaction in which the HCP phase forms with its basal planes ⟨0001⟩ parallel to the close-packed ⟨111⟩ planes in FCC. The ease of this transformation is affected by the stability of the FCC phase, which in turn is affected by both plastic deformation and alloying additions. Now, when cobalt is alloyed to make MP35N, the processing includes 50% cold work, which increases the driving force for the transformation of the FCC to the HCP phase. The HCP phase emerges as fine platelets within FCC grains. Because the FCC grains are small (0.01–0.1 µm, Fig. 9) and the HCP platelets further

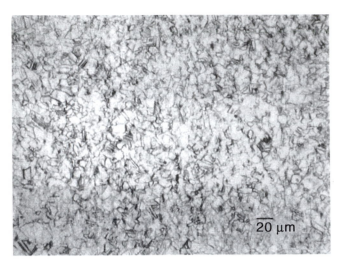

FIG. 9.   Microstructure of Co-based MP35N, ASTM F562, Biophase. (Photo courtesy of Smith & Nephew Richards, Inc., Memphis, TN.)

impede dislocation motion, the resulting structure is significantly strengthened (Table 2). It can be strengthened even further (as in the case of Richards Biophase) by an aging treatment at 430–650°C. This produces Co_3Mo precipitates on the HCP platelets. Hence, the alloy is truly multiphasic and derives strength from the combination of a cold-worked matrix phase, solid solution strengthening, and precipitation hardening. The resulting mechanical properties make the family of MP35N alloys among the strongest available for implant applications.

Titanium-Based Alloys

Composition

Commercially pure (CP) titanium (ASTM F67) and extra-low interstitial (ELI) Ti–6Al–4V alloy (ASTM F136) are the two most common titanium-based implant biomaterials. The F67 CP Ti is 98.9–99.6% Ti (Table 4). The oxygen content of

CP Ti (and other interstitial elements such as C and N) affects its yield and its tensile and fatigue strengths significantly, as discussed shortly.

With Ti–6Al–4V ELI alloy, the individual Ti–Al and Ti–V phase diagrams suggest the effects of the alloying additions in the ternary alloy. That is, since Al is an alpha (HCP) phase stabilizer while V is a beta (BCC) phase stabilizer, it turns out that the Ti–6Al–4V alloy used for implants is an alpha-beta alloy. The alloy's properties depend on prior treatments.

Microstructure and Properties

ASTM F67   For relatively pure titanium implants, as exemplified by many current dental implants, typical microstructures are single-phase alpha (HCP), showing evidence of mild (30%) cold work and grain diameters in the range of 10–150 μm (Fig. 10), depending on manufacturing. The nominal mechanical properties are listed in Table 2. Interstitial elements (O, C, N) in both pure titanium and the Ti–6Al–4V alloy strengthen the metal through interstitial solid solution strengthening mechanisms, with nitrogen having approximately twice the hardening effect (per atom) of either carbon or oxygen.

As noted, it is clear that the oxygen content of CP Ti (and the interstitial content generally) will affect its yield and its tensile and fatigue strengths significantly. For example, data available in the ASTM standard show that at 0.18% oxygen (grade 1), the yield strength is about 170 MPa, whereas at 0.40% (grade 4) the yield strength is about 485 MPa. Likewise, the ASTM standard shows that the tensile strength increases with oxygen content.

The literature establishes that the fatigue limit of unalloyed CP Ti is typically increased by interstitial content, in particular the oxygen content. For example, Fig. 11A shows data from Beevers and Robinson (1969), who tested vacuum-annealed CP Ti having a grain size in the range 200–300 μm in tension-compression at a mean stress of zero, at 100 cycles/sec. The 10^7 cycle endurance limit, or fatigue limit, for Ti 115 (0.085 wt.% O, grade 1), Ti 130 (0.125 wt.% O, grade 1), and Ti 160 (0.27 wt.% O, grade 3) was 88.3, 142, and

TABLE 4   Chemical Compositions of Ti-Based Alloys for Implants

Material	ASTM designation	Common/trade names	Composition (wt.%)	Notes
Pure Ti, grade 4	F67	CP Ti	Balance Ti max 0.10 C max 0.5 Fe max 0.0125–0.015 H max 0.05 N max 0.40 O	CP Ti comes in four grades according to oxygen content— Grade 1 has 0.18% max O Grade 2 has 0.25% max O Grade 3 has 0.35% max O Grade 4 has 0.40% max O
Ti–6Al–4V ELI*	F136	Ti–6Al–4V	88.3–90.8 Ti 5.5–6.5 Al 3.5–4.5 V max 0.08 C max 0.0125 H max 0.25 Fe max 0.05 N max 0.13 O	

*A more recent specification can be found from ASTM, the American Society for Testing and Materials, under *F136-98e1 Standard Specification for Wrought Titanium-6 Aluminium-4 ELI (Extra Low Interstitial) Alloy (R56401) for Surgical Implant Applications.*

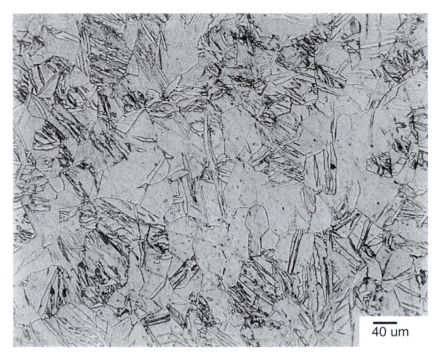

FIG. 10. Microstructure of moderately cold-worked commercial purity titanium, ASTM F67, used in an oral implant.

216 MPa, respectively. Figure 11B shows similar results from Turner and Roberts' (1968) fatigue study on CP Ti (tension–compression, 160 cycles/sec, mean stress = zero) having a grain size in the range 26–32 μm. Here the fatigue limit for "H. P. Ti" (0.072 wt.% O, grade 1), Ti 120 (0.087 wt.% O, grade 1), and Ti 160 (0.32 wt.% O, grade 3) was 142, 172, and 295 MPa, respectively—again increasing with increasing oxygen content. Also, for grade 4 Ti in the cold-worked state, Steinemann *et al.* (1993) reported a 10^7 endurance limit of 430 MPa.

Figure 11C, from Conrad *et al.* (1973), summarizes data from several fatigue studies on CP Ti at 300K. Note that the ratio of fatigue limit to yield stress is relatively constant at about 0.65, independent of interstitial content and grain size. Conrad *et al.* suggest that this provides evidence that "the high cycle fatigue strength is controlled by the same dislocation mechanisms as the flow [yield] stress" (p. 996). The work of Turner and Roberts also reported that the ratio *f* (fatigue limit/ultimate tensile strength)—which is also called the "fatigue ratio" in materials design textbooks (e.g., Charles and Crane, 1989, p. 106)—was 0.43 for the high-purity Ti (0.072 wt.% O), 0.50 for Ti 120 (0.087 wt.% O), and 0.53 for Ti 160 (0.32 wt.% O). It seems clear that interstitial content affects the yield and tensile and fatigue strengths in CP Ti.

Also, cold work appears to increase the fatigue properties of CP Ti. For example, Disegi (1990) quoted bending fatigue data from for annealed versus cold-worked CP Ti in the form of unnotched 1.0 mm-thick sheet (Table 5); there was a moderate increase in UTS and "plane bending fatigue strength" when comparing annealed versus cold-rolled Ti samples. In these data, the ratio of fatigue strength to ultimate tensile strength ("endurance ratio" or "fatigue ratio", see paragraph

above) varied between 0.45 and 0.66. Whereas the ASM *Metals Handbook* (Wagner, 1996) noted that the fatigue limit for high-purity Ti was only about 10% larger for cold-worked versus annealed material, Desegi's data shows that the fatigue strength increased by about 28%, on average.

In recent years there has been increasing interest in the chemical and physical nature of the oxide on the surface of titanium and its 6Al–4V alloy and its biological significance. The nominal composition of the oxide is TiO_2 for both metals, although there is some disagreement about exact oxide chemistry in pure versus alloyed Ti. Although there is no dispute that the oxide provides corrosion resistance, there is some controversy about exactly how it influences the biological performance of titanium at molecular and tissue levels, as suggested in literature on osseointegrated oral and maxillofacial implants by Brånemark and co-workers in Sweden (e.g., Kasemo and Lausmaa, 1988).

ASTM F136 This alloy is an alpha–beta alloy, the microstructure of which depends upon heat treating and mechanical working. If the alloy is heated into the beta phase field (e.g., above 1000°C, the region where only BCC beta is thermodynamically stable) and then cooled slowly to room temperature, a two-phase Widmanstätten structure is produced (Fig. 12). The HCP alpha phase (which is rich in Al and depleted in V) precipitates out as plates or needles having a specific crystallographic orientation within grains of the beta (BCC) matrix. Alternatively, if cooling from the beta phase field is very fast (as in oil quenching), a "basketweave" microstructure will develop, owing to martensitic or bainitic (nondiffusional shear) solid-state transformations. Most commonly, the F136 alloy

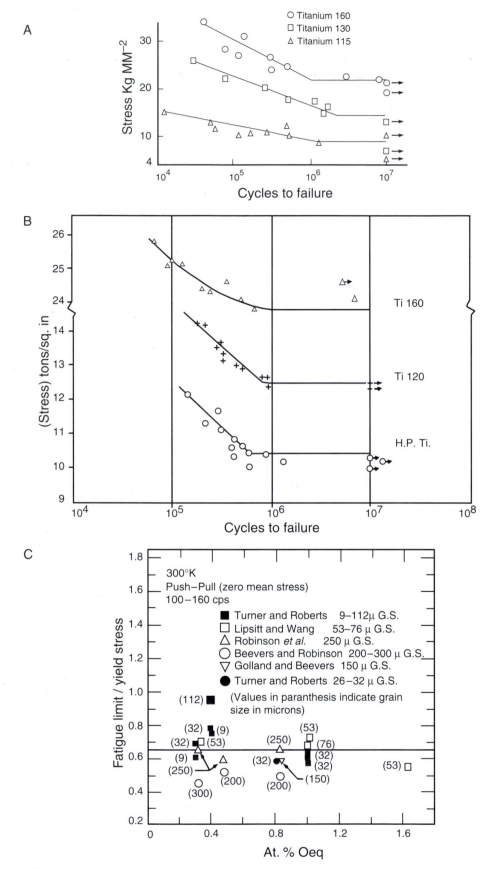

FIG. 11. (A) S–N curves (stress amplitude–number of cycles to failure) at room temperature for CP Ti with varying oxygen content (see text for O content of Ti 160, 130, and 115), from Beevers and Robinson (1969). (B) S–N curves at room temperature for CP Ti with varying oxygen content (see text), from Turner and Roberts (1968a). (C) Ratio of fatigue limit to yield stress in unalloyed Ti at 300 K as a function of at.% oxygen and grain size, from Conrad *et al.* (1973).

TABLE 5   Plane Bending Fatigue Data for Unnotched
1.0-mm-Thick Unalloyed Titanium Sheet, Tested at
58 cycles/sec in Air (from Disegi, 1990)

Ultimate tensile strength (MPa)	Sample condition	Plane bending fatigue strength (MPa)
371	Annealed	246
402	Annealed	235
432	Annealed	284
468	Annealed	284
510	Cold rolled	265
667	Cold rolled	314
667	Cold rolled	343
745	Cold rolled	334
766	Cold rolled	343
772	Cold rolled	383
820	Cold rolled	383

is heated and worked at temperatures near but not exceeding
the beta transus, and then annealed to give a microstructure
of fine-grained alpha with beta as isolated particles at grain
boundaries (mill annealed, Fig. 13).

Interestingly, all three of the just-noted microstructures in
Ti–6Al–4V alloy lead to about the same yield and ultimate
tensile strengths, but the mill-annealed condition is superior in high-cycle fatigue (Table 2), which is a significant
consideration.

Like the Co-based alloys, the above microstructural aspects
for the Ti systems need to be considered when evaluating the
structure–property relationships of porous-coated or plasma-
sprayed implants. Again, as in the case of the cobalt-based
alloys, there is the technical problem of successfully attaching
the coating onto the metallic substrate while maintaining adequate properties of both coating and substrate. Optimizing
the fatigue properties of Ti–6Al–4V porous-coated implants
becomes an interdisciplinary design problem involving not only
metallurgy but also surface properties and fracture mechanics.

CONCLUDING REMARKS

It should be evident that metallurgical principles guide
understanding of structure–property relationships and inform
judgments about implant design, just as they would in the
design process for any well-engineered product. Although this
chapter's emphasis has been on mechanical properties (for the
sake of specificity), other properties, in particular surface texture, are receiving increasing attention in relation to biological
performance of implants. Timely examples of this are (a) efforts

FIG. 12.   Widmanstätten structure in cast Ti–Al–4V, ASTM F136. Note prior beta grains (three large
grains are shown in the photo) and platelet alpha structure within grains. (Photo courtesy of Zimmer
USA, Warsaw, IN.)

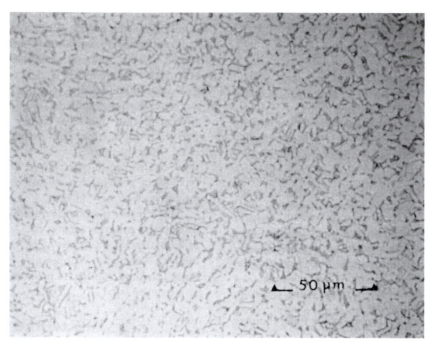

FIG. 13. Microstructure of wrought and mill-annealed Ti–6Al–4V, showing small grains of alpha (light) and beta (dark). (Photo courtesy of Zimmer USA, Warsaw, IN.)

to attach relevant biomolecules to metallic implant surfaces to promote certain desired interfacial activities; and (b) efforts to texture implant surfaces to optimize molecular and cellular reactions.

Another point to remember is that the intrinsic material properties of metallic implants—such as elastic modulus, yield strength, or fatigue strength—are not the sole determinant of implant performance and success. Certainly it is true that inadequate attention to material properties can doom a device to failure. However, it is also true that even with the best material, a device can fail because of faulty structural properties, inappropriate use of the implant, surgical error, or inadequate mechanical design of the implant in the first place. As an illustration of this point, Fig. 14 shows a plastically deformed 316L stainless steel Harrington spinal distraction rod that failed *in vivo* by metallurgical fatigue. An investigation of this case concluded that failure occurred not because 316L cold-worked stainless steel had poor fatigue properties per se, but rather due to a combination of factors: (a) the surgeon bent the rod to make it fit a bit better in the patent, but this increased the bending moment and bending stresses on the rod at the first ratchet junction, which was a known problem area; (b) the stress concentrations at the ratchet end of the rod were severe enough to significantly increase stresses at the first ratchet junction, which was indeed the eventual site of the fatigue fracture; and (c) spinal fusion did not occur in the patient, which contributed to relatively persistent loading of the rod over several months postimplantation. Here the point is that all three of these factors could have been anticipated and addressed during the initial design of the rod, during which both structural and material properties would be considered in various stress analyses related to possible failure modes. It must always be recalled that implant design is a multifaceted problem in which materials selection is only a part of the problem.

Bibliography

American Society for Testing and Materials (1978). ASTM Standards for Medical and Surgical Materials and Devices. Authorized Reprint from *Annual Book of ASTM Standards*, ASTM, Philadelphia, PA.

Beevers, C. J., and Robinson, J. L. (1969). Some observations on the influence of oxygen content on the fatigue behavior of α-titanium. *J. Less-Common Metals* 17: 345–352.

Brunski, J. B., Hill, D. C., and Moskowitz, A. (1983). Stresses in a Harrington distraction rod: their origin and relationship to fatigue fractures in vivo. *J. Biomech. Eng.* 105: 101–107.

Charles, J. A., and Crane, F. A. A. (1989). *Selection and Use of Engineering Materials*. 2nd ed. Butterworth–Heinemann Ltd., Halley Court, Oxford.

Compte, P. (1984). Metallurgical observations of biomaterials. in *Contemporary Biomaterials*, J. W. Boretos and M. Eden, eds. Noyes Publ., Park Ridge, NJ, pp. 66–91.

Conrad, H., Doner, M., and de Meester, B. (1973). Critical review: deformation and fracture. in *Titanium Science and Technology*, Vol. 2, R. I. Jaffee and H. M. Burte, eds. Plenum Press, New York, pp. 969–1005.

Cox, D. O. (1977). The fatigue and fracture behavior of a low stacking fault energy cobalt–chromium–molybdenum–carbon casting alloy used for prosthetic devices. Ph.D. dissertation, Engineering, University of California at Los Angeles.

Davidson, J. A., and Georgette, F. S. (1986). State-of-the-art materials for orthopaedic prosthetic devices. in *Implant Manufacturing and*

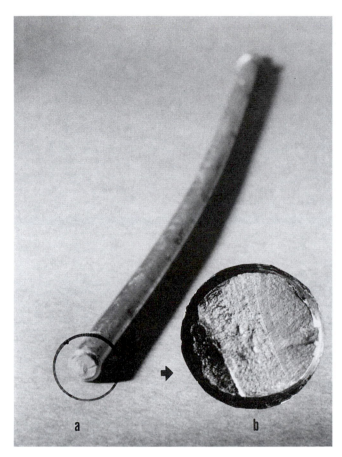

FIG. 14. The smooth part of a 316L stainless steel Harrington spinal distraction rod that fractured by fatigue *in vivo*. Note the bend in the rod (the rod was originally straight) and (insert) the relationship of the crack initiation zone of the fracture surface to the bend. The inserted photo shows the nature of the fatigue fracture surface, which is characterized by a region of "beach marks" and a region of sudden overload failure. (Photo courtesy of J. B. Brunski, D. C. Hill, and A. Moskowitz, 1983. Stresses in a Harrington distraction rod: their origin and relationship to fatigue fractures *in vivo. J. Biomech. Eng.* **105:** 101–107.)

Material Technology. Proc. Soc. of Manufacturing Engineering, Itasca, IL.

Disegi, J. (1990). AO/ASIF Unalloyed Titanium Implant Material. Technical Brochure available from Synthes (USA), P.O. Box 1766, 1690 Russell Road, Paoli, PA, 19301–1222.

Golland, D. I., and Beevers, C. J. (1971). Some effects of prior deformation and annealing on the fatigue response of α-titanium. *J. Less-Common Metals* **23:** 174.

Golland, D. I., and Beevers, C. J. (1971). The effect of temperature on the fatigue response of alpha-titanium. *Met. Sci. J.* **5:** 174.

Gomez, M., Mancha, H., Salinas, A., Rodríguez, J. L., Escobedo, J., Castro, M., and Méndez, M. (1997). Relationship between microstructure and ductility of investment cast ASTM F-75 implant alloy. *J. Biomed. Mater. Res.* **34:** 157–163.

Hall, E. O. (1951). The deformation and ageing of mild steel: Discussion of results. *Proc. Phys. Soc. (London)* **64B:** 747–753.

Hamman, G., and Bardos, D. I. (1980). Metallographic quality control of orthopaedic implants. in *Metallography as a Quality Control Tool,* J. L. McCall and P. M. French, eds. Plenum Publishers, New York, pp. 221–245.

Honeycombe, R. W. K. (1968). *The Plastic Deformation of Metals.* St. Martin's Press, New York, p. 234.

Kasemo, B., and Lausmaa, J. (1988). Biomaterials from a surface science perspective. in *Surface Characterization of Biomaterials,* B. D. Ratner, ed. Elsevier, New York, Ch. 1, pp. 1–12.

Lipsitt, H. A., and Wang, D. Y. (1961). The effects of interstitial solute atoms on the fatigue limit behavior of titanium. *Trans. AIME* **221:** 918.

Moss, A. J., Hamburger, S., Moore, R. M. Jr., Jeng, L. L., and Howie, L. J. (1990). Use of selected medical device implants in the United States, 1988. *Adv. Data* (191): 1–24.

Nanci, A., Wuest, J. D., Peru, L., Brunet, P., Sharma, V., Zalzal, S., and McKee, M. D. (1998). Chemical modification of titanium surfaces for covalent attachment of biological molecules. *J. Biomed. Mater. Res.* **40:** 324–335.

Petch, N. J. (1953). The cleavage strength of polycrystals. *J. Iron Steel Inst. (London)* **173:** 25.

Pilliar, R. M., and Weatherly, G. C. (1984). Developments in implant alloys. *CRC Crit. Rev. Biocompatibility* **1**(4): 371–403.

Richards Medical Company (1985). Medical Metals. Richards Medical Company Publication No. 3922, Richards Medical Co., Memphis, TN. [Note: This company is now known as Smith & Nephew Richards, Inc.]

Robinson, S. L., Warren, M. R., and Beevers, C. J. (1969). The influence of internal defects on the fatigue behavior of α-titanium. *J. Less-Common Metals* **19:** 73–82.

Steinemann, S. G., Mäusli, P.-A., Szmuckler-Moncler, S., Semlitsch, M., Pohler, O., Hintermann, H.-E., and Perren, S. M. (1993). Beta-titanium alloy for surgical implants. In *Titanium '92 Science and Technology,* F. H. Froes and I. Caplan, eds. The Minerals, Metals & Materials Society, pp. 2689–2698.

Turner, N. G., and Roberts, W. T. (1968a). Fatigue behavior of titanium. *Trans. Met. Soc. AIME* **242:** 1223–1230.

Turner, N. G., and Roberts, W. T. (1968b). Dynamic strain ageing in titanium. *J. Less-Common Metals* **16:** 37.

www.biomateria.com/media_briefing.htm

www.sric-bi.com/Explorer/BM.shtml

Wagner, L. (1996). Fatigue life behavior. in *ASM Handbook,* Vol. 19, *Fatigue and Fracture,* S. Lampman, G. M. Davidson, F. Reidenbach, R. L. Boring, A. Hammel, S. D. Henry, and W. W. Scott, Jr., eds., ASM International, pp. 837–853.

Zimmer USA (1984a). *Fatigue and Porous Coated Implants.* Zimmer Technical Monograph, Zimmer USA, Warsaw, IN.

Zimmer USA (1984b). *Metal Forming Techniques in Orthopaedics.* Zimmer Technical Monograph, Zimmer USA, Warsaw, IN.

Zimmer USA (1984c). *Physical and Mechanical Properties of Orthopaedic Alloys.* Zimmer Technical Monograph, Zimmer USA, Warsaw, IN.

Zimmer USA (1984d). *Physical Metallurgy of Titanium Alloy.* Zimmer Technical Monograph, Zimmer USA, Warsaw, IN.

2.10 CERAMICS, GLASSES, AND GLASS-CERAMICS

Larry L. Hench and Serena Best

Ceramics, glasses, and glass-ceramics include a broad range of inorganic/nonmetallic compositions. In the medical industry, these materials have been essential for eyeglasses, diagnostic instruments, chemical ware, thermometers, tissue culture flasks, and fiber optics for endoscopy. Insoluble porous glasses have been used as carriers for enzymes, antibodies, and antigens, offering the advantages of resistance to microbial attack, pH changes, solvent conditions, temperature, and packing under high pressure required for rapid flow (Hench and Ethridge, 1982).

Ceramics are also widely used in dentistry as restorative materials such as in gold–porcelain crowns, glass-filled ionomer cements, and dentures. These dental ceramics are discussed by Phillips (1991).

This chapter focuses on ceramics, glasses, and glass-ceramics used as implants. Although dozens of compositions have been explored in the past, relatively few have achieved clinical success. This chapter examines differences in processing and structure, describes the chemical and microstructural basis for their differences in physical properties, and relates properties and tissue response to particular clinical applications. For a historical review of these biomaterials, see Hulbert *et al.* (1987).

TYPES OF BIOCERAMICS—TISSUE ATTACHMENT

It is essential to recognize that no one material is suitable for all biomaterial applications. As a class of biomaterials, ceramics, glasses, and glass-ceramics are generally used to repair or replace skeletal hard connective tissues. Their success depends upon achieving a stable attachment to connective tissue.

The mechanism of tissue attachment is directly related to the type of tissue response at the implant–tissue interface. No material implanted in living tissue is inert because all materials elicit a response from living tissues. There are four types of tissue response (Table 1) and four different means of attaching prostheses to the skeletal system (Table 2).

A comparison of the relative chemical activity of the different types of bioceramics, glasses, and glass-ceramics is shown in Fig. 1. The relative reactivity shown in Fig. 1A correlates very closely with the rate of formation of an interfacial bond of ceramic, glass, or glass-ceramic implants with bone (Fig. 1B). Figure 1B is discussed in more detail in the section on bioactive glasses and glass-ceramics in this chapter.

The relative level of reactivity of an implant influences the thickness of the interfacial zone or layer between the material and tissue. Analyses of implant material failures during the past 20 years generally show failure originating at the biomaterial–tissue interface. When biomaterials are nearly inert (type 1 in Table 2 and Fig. 1) and the interface is not chemically or biologically bonded, there is relative movement and progressive development of a fibrous capsule in soft and hard tissues.

TABLE 1 Types of Implant–Tissue Response

If the material is toxic, the surrounding tissue dies.

If the material is nontoxic and biologically inactive (nearly inert), a fibrous tissue of variable thickness forms.

If the material is nontoxic and biologically active (bioactive), an interfacial bond forms.

If the material is nontoxic and dissolves, the surrounding tissue replaces it.

TABLE 2 Types of Bioceramic–Tissue Attachment and Their Classification

Type of attachment	Example
1. Dense, nonporous, nearly inert ceramics attach by bone growth into surface irregularities by cementing the device into the tissues or by press-fitting into a defect (termed "morphological fixation").	Al_2O_3 (Single crystal and polycrystalline)
2. For porous inert implants, bone ingrowth occurs that mechanically attaches the bone to the material (termed "biological fixation").	Al_2O_3 (Polycrystalline) Hydroxyapatite-coated porous metals
3. Dense, nonporous surface-reactive ceramics, glasses, and glass-ceramics attach directly by chemical bonding with the bone (termed "bioactive fixation").	Bioactive glasses Bioactive glass-ceramics Hydroxyapatite
4. Dense, nonporous (or porous) resorbable ceramics are designed to be slowly replaced by bone.	Calcium sulfate (Plaster of Paris) Tricalcium phosphate Calcium-phosphate salts

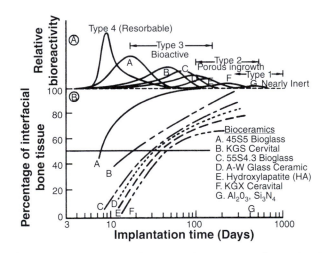

FIG. 1. Bioactivity spectra for various bioceramic implants: (A) relative rate of bioreactivity, (B) time dependence of formation of bone bonding at an implant interface.

The presence of movement at the biomaterial—tissue interface eventually leads to deterioration in function of the implant or the tissue at the interface, or both. The thickness of the nonadherent capsule varies, depending upon both material (Fig. 2) and extent of relative motion.

The fibrous tissue at the interface of dense Al_2O_3 (alumina) implants is very thin. Consequently, as discussed later, if alumina devices are implanted with a very tight mechanical fit and are loaded primarily in compression, they are very successful. In contrast, if a type 1 nearly inert implant is loaded so that interfacial movement can occur, the fibrous capsule can become several hundred micrometers thick, and the implant can loosen very quickly.

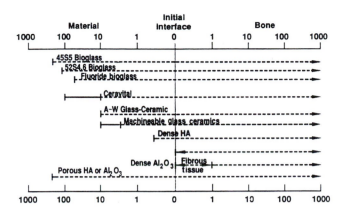

FIG. 2. Comparison of interfacial thickness (μm) of reaction layer of bioactive implants of fibrous tissue of inactive bioceramics in bone.

The mechanism behind the use of nearly inert microporous materials (type 2 in Table 2 and Fig. 1) is the ingrowth of tissue into pores on the surface or throughout the implant. The increased interfacial area between the implant and the tissues results in an increased resistance to movement of the device in the tissue. The interface is established by the living tissue in the pores. Consequently, this method of attachment is often termed "biological fixation." It is capable of withstanding more complex stress states than type 1 implants with "morphological fixation." The limitation with type 2 porous implants, however, is that for the tissue to remain viable and healthy, it is necessary for the pores to be greater than 50 to 150 μm (Fig. 2). The large interfacial area required for the porosity is due to the need to provide a blood supply to the ingrown connective tissue (vascular tissue does not appear in pore sizes less than 100 μm). Also, if micromovement occurs at the interface of a porous implant and tissue is damaged, the blood supply may be cut off, the tissues will die, inflammation will ensue, and the interfacial stability will be destroyed. When the material is a porous metal, the large increase in surface area can provide a focus for corrosion of the implant and loss of metal ions into the tissues. This can be mediated by using a bioactive ceramic material such as hydroxyapatite (HA) as a coating on the metal. The fraction of large porosity in any material also degrades the strength of the material proportional to the volume fraction of porosity. Consequently, this approach to solving interfacial stability works best when materials are used as coatings or as unloaded space fillers in tissues.

Resorbable biomaterials (type 4 in Table 2 and Fig. 1) are designed to degrade gradually over a period of time and be replaced by the natural host tissue. This leads to a very thin or nonexistent interfacial thickness (Fig. 2). This is the optimal biomaterial solution, if the requirements of strength and short-term performance can be met, since natural tissues can repair and replace themselves throughout life. Thus, resorbable biomaterials are based on biological principles of repair that have evolved over millions of years. Complications in the development of resorbable bioceramics are (1) maintenance of strength and the stability of the interface during the degradation period and replacement by the natural host tissue, and

(2) matching resorption rates to the repair rates of body tissues (Fig. 1A) (e.g., some materials dissolve too rapidly and some too slowly). Because large quantities of material may be replaced, it is also essential that a resorbable biomaterial consist only of metabolically acceptable substances. This criterion imposes considerable limitations on the compositional design of resorbable biomaterials. Successful examples of resorbable polymers include poly(lactic acid) and poly(glycolic acid) used for sutures, which are metabolized to CO_2 and H_2O and therefore are able to function for an appropriate time and then dissolve and disappear (see Chapters 2, 6, and 7 for other examples). Porous or particulate calcium phosphate ceramic materials such as tricalcium phosphate (TCP) have proved successful for resorbable hard tissue replacements when low loads are applied to the material.

Another approach to solving problems of interfacial attachment is the use of bioactive materials (type 3 in Table 2 and Fig. 1). Bioactive materials are intermediate between resorbable and bioinert. A bioactive material is one that elicits a specific biological response at the interface of the material, resulting in the formation of a bond between the tissues and the material. This concept has now been expanded to include a large number of bioactive materials with a wide range of rates of bonding and thicknesses of interfacial bonding layers (Figs. 1 and 2). They include bioactive glasses such as Bioglass; bioactive glass-ceramics such as Ceravital, A-W glass-ceramic, or machinable glass-ceramics; dense HA such as Durapatite or Calcitite; and bioactive composites such as HA-polyethylene, HA-Bioglass, Palavital, and stainless steel fiber–reinforced Bioglass. All of these materials form an interfacial bond with adjacent tissue. However, the time dependence of bonding, the strength of bond, the mechanism of bonding, and the thickness of the bonding zone differ for the various materials.

It is important to recognize that relatively small changes in the composition of a biomaterial can dramatically affect whether it is bioinert, resorbable, or bioactive. These compositional effects on surface reactions are discussed in the section on bioactive glasses and glass-ceramics.

CHARACTERISTICS AND PROCESSING OF BIOCERAMICS

The types of implants listed in Table 2 are made using different processing methods. The characteristics and properties of the materials, summarized in Table 3, differ greatly, depending upon the processing method used.

The primary methods of processing ceramics, glasses, and glass-ceramics are summarized in Fig. 3. These methods yield five categories of microstructures:

1. Glass
2. Cast or plasma-sprayed polycrystalline ceramic
3. Liquid-phase sintered (vitrified) ceramic
4. Solid-state sintered ceramic
5. Polycrystalline glass-ceramic

Differences in the microstructures of the five categories are primarily a result of the different thermal processing steps

TABLE 3   Bioceramic Material Characteristics and Properties

Composition

Microstructure
  Number of phases
  Percentage of phases
  Distribution of phases
  Size of phases
  Connectivity of phases

Phase state
  Crystal structure
  Defect structure
  Amorphous structure
  Pore structure

Surface
  Flatness
  Finish
  Composition
  Second phase
  Porosity

Shape

required to produce them. Alumina and calcium phosphate bioceramics are made by fabricating the product from fine-grained particulate solids. For example, a desired shape may be obtained by mixing the particulates with water and an organic binder, then pressing them in a mold. This is termed "forming." The formed piece is called green ware. Subsequently, the temperature is raised to evaporate the water (i.e., drying) and the binder is burned out, resulting in bisque ware. At a very much higher temperature, the part is densified during firing. After cooling to ambient temperature, one or more finishing steps may be applied, such as polishing. Porous ceramics are produced by adding a second phase that decomposes prior to densification, leaving behind holes or pores (Schors and Holmes, 1993), or transforming natural porous organisms, such as coral, to porous HA by hydrothermal processing (Roy and Linnehan, 1974).

The interrelation between microstructure and thermal processing of various bioceramics is shown in Fig. 3, which is a binary phase diagram consisting of a network-forming oxide such as SiO_2 (silica), and some arbitrary network modifier oxide (MO) such as CaO. When a powdered mixture of MO and SiO_2 is heated to the melting temperature T_m, the entire mass will become liquid (L). The liquid will become homogeneous when held at this temperature for a sufficient length of time. When the liquid is cast (paths 1B, 2, 5), forming the shape of the object during the casting, either a glass or a polycrystalline microstructure will result. Plasma spray coating follows path 1A. However, a network-forming oxide is not necessary to produce plasma-sprayed coatings such as hydroxyapatites, which are polycrystalline (Lacefield, 1993).

If the starting composition contains a sufficient quantity of network former (SiO_2), and the casting rate is sufficiently slow, a glass will result (path 1B). The viscosity of the melt increases greatly as it is cooled, until at approximately T_1, the glass transition point, the material is transformed into a solid.

If either of these conditions is not met, a polycrystalline microstructure will result. The crystals begin growing at T_L and complete growth at T_2. The final material consists of the equilibrium crystalline phases predicted by the phase diagram. This type of cast object is not often used commercially because the large shrinkage cavity and large grains produced during cooling make the material weak and subject to environmental attack.

If the MO and SiO_2 powders are first formed into the shape of the desired object and fired at a temperature T_3, a liquid-phase sintered structure will result (path 3). Before firing, the composition will contain approximately 10–40% porosity, depending upon the forming process used. A liquid will be formed first at grain boundaries at the eutectic temperature, T_2. The liquid will penetrate between the grains, filling the pores, and will draw the grains together by capillary attraction. These effects decrease the volume of the powdered compact.

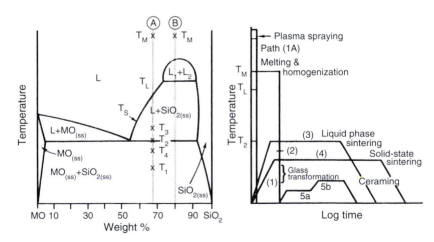

FIG. 3.   Relation of thermal processing schedules of various bioceramics to equilibrium phase diagram.

Since the mass remains unchanged and is only rearranged, an increased density results. Should the compact be heated for a sufficient length of time, the liquid content can be predicted from the phase diagram. However, in most ceramic processes, liquid formation does not usually proceed to equilibrium owing to the slowness of the reaction and the expense of long-term heat treatments.

The microstructure resulting from liquid-phase sintering, or vitrification as it is commonly called, will consist of small grains from the original powder compact surrounded by a liquid phase. As the compact is cooled from T_3 to T_2, the liquid phase will crystallize into a fine-grained matrix surrounding the original grains. If the liquid contains a sufficient concentration of network formers, it can be quenched into a glassy matrix surrounding the original grains.

A powder compact can be densified without the presence of a liquid phase by a process called solid-state sintering. This is the process usually used for manufacturing alumina and dense HA bioceramics. Under the driving force of surface energy gradients, atoms diffuse to areas of contact between particles. The material may be transported by either grain boundary diffusion, volume diffusion, creep, or any combination of these, depending upon the temperature or material involved. Because long-range migration of atoms is necessary, sintering temperatures are usually in excess of one-half of the melting point of the material: $T > T_L/2$ (path 4).

The atoms move so as to fill up the pores and open channels between the grains of the powder. As the pores and open channels are closed during the heat treatment, the crystals become tightly bonded together, and the density, strength, and fatigue resistance of the object improve greatly. The microstructure of a material that is prepared by sintering consists of crystals bonded together by ionic–covalent bonds with a very small amount of remaining porosity.

The relative rate of densification during solid-state sintering is slower than that of liquid-phase sintering because material transport is slower in a solid than in a liquid. However, it is possible to solid-state sinter individual component materials such as pure oxides since liquid development is not necessary. Consequently, when high purity and uniform fine-grained microstructures are required (e.g., for bioceramics) solid-state sintering is essential.

The fifth class of microstructures is called glass-ceramics because the object starts as a glass and ends up as a polycrystalline ceramic. This is accomplished by first quenching a melt to form the glass object. The glass is transformed into a glass-ceramic in two steps. First, the glass is heat treated at a temperature range of 500–700°C (path 5a) to produce a large concentration of nuclei from which crystals can grow. When sufficient nuclei are present to ensure that a fine-grained structure will be obtained, the temperature of the object is raised to a range of 600–900°C, which promotes crystal growth (path 5b). Crystals grow from the nuclei until they impinge and up to 100% crystallization is achieved. The resulting microstructure is nonporous and contains fine-grained, randomly oriented crystals that may or may not correspond to the equilibrium crystal phases predicted by the phase diagram. There may also be a residual glassy matrix, depending on the duration of the ceraming heat treatment. When phase separation occurs

(composition B in Fig. 3), a nonporous, phase-separated, glass-in-glass microstructure can be produced. Crystallization of phase-separated glasses results in very complex microstructures. Glass-ceramics can also be made by pressing powders and a grain boundary glassy phase (Kokubo, 1993). For additional details on the processing of ceramics, see Reed (1988) or Onoda and Hench (1978), and for processing of glass-ceramics, see McMillan (1979).

NEARLY INERT CRYSTALLINE CERAMICS

High-density, high-purity (>199.5%) alumina is used in load-bearing hip prostheses and dental implants because of its excellent corrosion resistance, good biocompatibility, high wear resistance, and high strength (Christel *et al.*, 1988; Hulbert, 1993; Hulbert *et al.*, 1987; Miller *et al.*, 1996). Although some dental implants are single-crystal sapphires (McKinney and Lemons, 1985), most Al_2O_3 devices are very fine-grained polycrystalline $< \alpha\text{-}Al_2O_3$ produced by pressing and sintering at $T = 1600–1700°C$. A very small amount of MgO ($< 0.5\%$) is used to aid sintering and limit grain growth during sintering.

Strength, fatigue resistance, and fracture toughness of polycrystalline $< \alpha\text{-}Al_2O_3$ are a function of grain size and percentage of sintering aid (i.e., purity). Al_2O_3 with an average grain size of $< 4\,\mu m$ and $> 99.7\%$ purity exhibits good flexural strength and excellent compressive strength. These and other physical properties are summarized in Table 4, along with the International Standards Organization (ISO) requirements for alumina implants. Extensive testing has shown that alumina implants that meet or exceed ISO standards have excellent resistance to dynamic and impact fatigue and also resist subcritical crack growth (Drre and Dawihl, 1980). An increase in

TABLE 4 Physical Characteristics of Al_2O_3 Bioceramics

	High alumina ceramics	ISO Standard 6474
Alumina content (% by weight)	>99.8	≥99.50
Density (g/cm³)	>3.93	≥3.90
Average grain size (μm)	3–6	<7
Ra (μm)[a]	0.02	
Hardness (Vickers hardness number, VHN)	2300	>2000
Compressive strength (MPa)	4500	
Bending strength (MPa) (after testing in Ringer's solution)	550	400
Young's modulus (GPa)	380	
Fracture toughness (K_1C) (MPa^{12})	5–6	
Slow crack growth	10–52	

[a] Surface roughness value.

average grain size to >17 μm can decrease mechanical properties by about 20%. High concentrations of sintering aids must be avoided because they remain in the grain boundaries and degrade fatigue resistance.

Methods exist for lifetime predictions and statistical design of proof tests for load-bearing ceramics. Applications of these techniques show that load limits for specific prostheses can be set for an Al_2O_3 device based upon the flexural strength of the material and its use environment (Ritter *et al.*, 1979). Load-bearing lifetimes of 30 years at 12,000-N loads have been predicted (Christel *et al.*, 1988). Results from aging and fatigue studies show that it is essential that Al_2O_3 implants be produced at the highest possible standards of quality assurance, especially if they are to be used as orthopedic prostheses in younger patients.

Alumina has been used in orthopedic surgery for nearly 20 years (Miller *et al.*, 1996). Its use has been motivated largely by two factors: its excellent type 1 biocompatibility and very thin capsule formation (Fig. 2), which permits cementless fixation of prostheses; and its exceptionally low coefficients of friction and wear rates.

The superb tribiologic properties (friction and wear) of alumina occur only when the grains are very small (<4 μm) and have a very narrow size distribution. These conditions lead to very low surface roughness values (Ra < 4 0.02 μm, Table 4). If large grains are present, they can pull out and lead to very rapid wear of bearing surfaces owing to local dry friction.

Alumina on load-bearing, wearing surfaces, such as in hip prostheses, must have a very high degree of sphericity, which is produced by grinding and polishing the two mating surfaces together. For example, the alumina ball and socket in a hip prosthesis are polished together and used as a pair. The long-term coefficient of friction of an alumina–alumina joint decreases with time and approaches the values of a normal joint. This leads to wear on alumina-articulating surfaces being nearly 10 times lower than metal–polyethylene surfaces (Fig. 4).

Low wear rates have led to widespread use in Europe of alumina noncemented cups press-fitted into the acetabulum of the hip. The cups are stabilized by the growth of bone into grooves or around pegs. The mating femoral ball surface is also made of alumina, which is bonded to a metallic stem. Long-term results in general are good, especially for younger patients. However, Christel *et al.* (1988) caution that stress shielding, owing to

the high elastic modulus of alumina, may be responsible for cancellous bone atrophy and loosening of the acetabular cup in old patients with senile osteoporosis or rheumatoid arthritis. Consequently, it is essential that the age of the patient, nature of the disease of the joint, and biomechanics of the repair be considered carefully before any prosthesis is used, including alumina ceramics.

Zirconia (ZrO_2) is also used as the articulating ball in total hip prostheses. The potential advantages of zirconia in load-bearing prostheses are its lower modulus of elasticity and higher strength (Hench and Wilson, 1993). There are insufficient data to determine whether these properties will result in higher clinical success rates over long times (>15 years).

Other clinical applications of alumina prostheses reviewed by Hulbert *et al.* (1987) include knee prostheses; bone screws; alveolar ridge and maxillofacial reconstruction; ossicular bone substitutes; keratoprostheses (corneal replacements); segmental bone replacements; and blade, screw, and post dental implants.

POROUS CERAMICS

The potential advantage offered by a porous ceramic implant (type 2, Table 2, Figs. 1 and 2) is its inertness combined with the mechanical stability of the highly convoluted interface that develops when bone grows into the pores of the ceramic. The mechanical requirements of prostheses, however, severely restrict the use of low-strength porous ceramics to nonload-bearing applications. Studies reviewed by Hench and Ethridge (1982), Hulbert *et al.* (1987), and Schors and Holmes (1993) have shown that when load-bearing is not a primary requirement, porous ceramics can provide a functional implant. When pore sizes exceed 100 μm, bone will grow within the interconnecting pore channels near the surface and maintain its vascularity and long-term viability. In this manner, the implant serves as a structural bridge or scaffold for bone formation.

Commercially available porous products originate from two sources: hydroxyapatite converted from coral (e.g., Pro Osteon) or animal bone (e.g., Endobon). Other production routes; e.g., burnout techniques (e.g., Fang *et al.*, 1991) and decomposition of hydrogen peroxide (Peelen *et al.*, 1977; Driessen *et al.*, 1982) are not yet used commercially. The optimal type of porosity is still uncertain. The degree of interconnectivity of pores may be more critical than the pore size. Eggli *et al.* (1988) demonstrated improved integration in interconnected 50–100 μm pores compared with less connected pores with a size of 200–400 μm. Similarly Kühne *et al.* (1994) compared two grades of 25–35% porous coralline apatite with average pore sizes of 200 and 500 μm and reported bone ingrowth to be improved in the 500 μm pore sized ceramic. Holmes (1979) suggests that porous coralline apatite when implanted in cortical bone requires interconnections of osteonic diameter for transport of nutrients to maintain bone ingrowth. The findings clearly indicate the importance of thorough characterisation of porous materials before implantation, and Hing (1999) has recommended a range of techniques that should be employed.

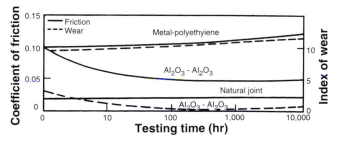

FIG. 4. Time dependence of coefficient of friction and wear of alumina–alumina versus metal–polyethylene hip joint (*in vitro* testing).

Porous materials are weaker than the equivalent bulk form in proportion to the percentage of porosity, so that as the porosity increases, the strength of the material decreases rapidly. Much surface area is also exposed, so that the effects of the environment on decreasing the strength become much more important than for dense, nonporous materials. The aging of porous ceramics, with their subsequent decrease in strength, requires bone ingrowth to stabilize the structure of the implant. Clinical results for non-load-bearing implants are good (Schors and Holmes, 1993).

BIOACTIVE GLASSES AND GLASS-CERAMICS

Certain compositions of glasses, ceramics, glass-ceramics, and composites have been shown to bond to bone (Hench and Ethridge, 1982; Gross *et al.*, 1988; Yamamuro *et al.*, 1990; Hench, 1991; Hench and Wilson, 1993). These materials have become known as bioactive ceramics. Some even more specialized compositions of bioactive glasses will bond to soft tissues as well as bone (Wilson *et al.*, 1981). A common characteristic of bioactive glasses and bioactive ceramics is a time-dependent, kinetic modification of the surface that occurs upon implantation. The surface forms a biologically active carbonated HA layer that provides the bonding interface with tissues.

Materials that are bioactive develop an adherent interface with tissues that resist substantial mechanical forces. In many cases, the interfacial strength of adhesion is equivalent to or greater than the cohesive strength of the implant material or the tissue bonded to the bioactive implant.

Bonding to bone was first demonstrated for a compositional range of bioactive glasses that contained SiO_2, Na_2O, CaO, and P_2O_5 in specific proportions (Hench *et al.*, 1972) (Table 5). There are three key compositional features to these bioactive glasses that distinguish them from traditional soda–lime–silica glasses: (1) less than 60 mol% SiO_2, (2) high Na_2O and CaO content, and (3) a high CaO/P_2O_5 ratio. These features make the surface highly reactive when it is exposed to an aqueous medium.

Many bioactive silica glasses are based upon the formula called 45S5, signifying 45 wt.% SiO_2 (S = the network former) and 5 : 1 ratio of CaO to P_2O_5. Glasses with lower ratios of CaO to P_2O_5 do not bond to bone. However, substitutions in the 45S5 formula of 5–15 wt.% B_2O_3 for SiO_2 or 12.5 wt.% CaF_2 for CaO or heat treating the bioactive glass compositions to form glass-ceramics has no measurable effect on the ability of the material to form a bone bond. However, adding as little as 3 wt.% Al_2O_3 to the 45S5 formula prevents bonding to bone.

The compositional dependence of bone and soft tissue bonding on the Na_2O–CaO–P_2O_5–SiO_2 glasses is illustrated in Fig. 5. All the glasses in Fig. 5 contain a constant 6 wt.% of P_2O_5. Compositions in the middle of the diagram (region A) form a bond with bone. Consequently, region A is termed the bioactive bone-bonding boundary. Silicate glasses within region B (e.g., window or bottle glass, or microscope slides) behave as nearly inert materials and elicit a fibrous capsule at the implant–tissue interface. Glasses within region C are resorbable and disappear within 10 to 30 days of implantation. Glasses within region D are not technically practical and therefore have not been tested as implants.

TABLE 5 Composition of Bioactive Glasses and Glass-Ceramics (in Weight Percent)

	45S5 Bioglass	45S5F Bioglass	45S5.4F Bioglass	40S5B5 Bioglass	52S4.6 Bioglass	55S4.3 Bioglass	KGC Ceravital	KGS Ceravital	KGy213 Ceravital	A-W GC	MB GC
SiO_2	45	45	45	40	52	55	46.2	46	38	34.2	19–52
P_2O_5	6	6	6	6	6	6				16.3	4–24
CaO	24.5	12.25	14.7	24.5	21	19.5	20.2	33	31	44.9	9–3
$Ca(PO_3)_2$							25.5	16	13.5		
CaF_2		12.25	9.8							0.5	
MgO							2.9			4.6	5–15
MgF_2											
Na_2O	24.5	24.5	24.5	24.5	21	19.5	4.8	5	4		3–5
K_2O							0.4				3–5
Al_2O_3									7		12–33
B_2O_3				5							
Ta_2O_5/TiO_2									6.5		
Structure	Glass	Glass	Glass	Glass	Glass		Glass-ceramic	Glass-ceramic		Glass-ceramic	Glass-ceramic
Reference	Hench *et al.* (1972)	Hench *et al.* (1972)	Hench *et al.* (1972)	Hench *et al.* (1972)	Hench *et al.* (1972)	Hench *et al.* (1972)	Gross *et al.* (1988)	Gross *et al.* (1988)		Nakamura *et al.* (1985)	Höhland and Vogel (1993)

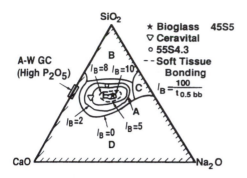

FIG. 5. Compositional dependence (in wt.%) of bone bonding and soft tissue bonding of bioactive glasses and glass-ceramics. All compositions in region A have a constant 6 wt.% of P_2O_5. A-W glass ceramic has higher P_2O_5 content (see Table 5 for details). I_B, Index of bioactivity.

The collagenous constituent of soft tissues can strongly adhere to the bioactive silicate glasses that lie within the dashed line region in Fig. 5. The interfacial thicknesses of the hard tissue–bioactive glasses are shown in Fig. 2 for several compositions. The thickness decreases as the bone-bonding boundary is approached.

Gross *et al.* (1988) and Gross and Strunz (1985) have shown that a range of low-alkali (0 to 5 wt.%) bioactive silica glass-ceramics (Ceravital) also bond to bone. They found that small additions of Al_2O_3, Ta_2O_5, TiO_2, Sb_2O_3, or ZrO_2 inhibit bone bonding (Table 5, Fig. 1). A two-phase silica–phosphate glass-ceramic composed of apatite $[Ca_{10}(PO_4)_6(OH_1F_2)]$ and wollastonite $[CaO, SiO_2]$ crystals and a residual silica glassy matrix, termed A-W glass-ceramic (A-WGC) (Nakamura *et al.*, 1985; Yamamuro *et al.*, 1990; Kokubo, 1993), also bonds with bone. The addition of Al_2O_3 or TiO_2 to the A-W glass-ceramic also inhibits bone bonding, whereas incorporation of a second phosphate phase, B-whitlockite $(3CaO, P_2O_5)$, does not.

Another multiphase bioactive phosphosilicate containing phlogopite $(Na, K) Mg_3[AlSi_3O_{10}]F_2$ and apatite crystals bonds to bone even though Al is present in the composition (Höhland and Vogel, 1993). However, the Al^{3+} ions are incorporated within the crystal phase and do not alter the surface reaction kinetics of the material. The compositions of these various bioactive glasses and glass-ceramics are compared in Table 5.

The surface chemistry of bioactive glass and glass-ceramic implants is best understood in terms of six possible types of surface reactions (Hench and Clark, 1978). A high-silica glass may react with its environment by developing only a surface hydration layer. This is called a type I response (Fig. 6). Vitreous silica (SiO_2) and some inert glasses at the apex of region B (Fig. 5) behave in this manner when exposed to a physiological environment.

When sufficient SiO_2 is present in the glass network, the surface layer that forms from alkali–proton exchange can repolymerize into a dense SiO_2-rich film that protects the glass from further attack. This type II surface (Fig. 6) is characteristic of most commercial silicate glasses, and their biological response of fibrous capsule formation is typical of many within region B in Fig. 5.

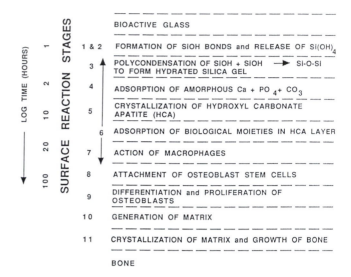

FIG. 6. Types of silicate glass interfaces with aqueous or physiological solutions.

At the other extreme of the reactivity range, a silicate glass or crystal may undergo rapid, selective ion exchange of alkali ions, with protons or hydronium ions leaving a thick but highly porous and nonprotective SiO_2-rich film on the surface (a type IV surface) (Fig. 6). Under static or slow flow conditions, the local pH becomes sufficiently alkaline (pH > 19) that the surface silica layer is dissolved at a high rate, leading to uniform bulk network or stoichiometric dissolution (a type V surface). Both type IV and V surfaces fall into region C of Fig. 5.

Type IIIA surfaces are characteristic of bioactive silicates (Fig. 6). A dual protective film rich in CaO and P_2O_5 forms on top of the alkali-depleted SiO_2-rich film. When multivalent cations such as Al^{3+}, Fe^{3+}, and Ti^{4+} are present in the glass or solution, multiple layers form on the glass as the saturation of each cationic complex is exceeded, resulting in a type IIIB surface (Fig. 6), which does not bond to tissue.

A general equation describes the overall rate of change of glass surfaces and gives rise to the interfacial reaction profiles shown in Fig. 6. The reaction rate (R) depends on at least four terms (for a single-phase glass). For glass-ceramics, which have several phases in their microstructures, each phase will have a characteristic reaction rate, R_i.

$$R = -k_1 t^{0.5} - k_2 t^{1.0} - k_3 t^{1.0} + k_4 t^y + k_n t^z \qquad (1)$$

The first term describes the rate of alkali extraction from the glass and is called a stage 1 reaction. A type II nonbonding glass surface (region B in Fig. 6) is primarily undergoing stage 1 attack. Stage 1, the initial or primary stage of attack, is a process that involves an exchange between alkali ions from the glass and hydrogen ions from the solution, during which the remaining constituents of the glass are not altered. During stage 1, the rate of alkali extraction from the glass is parabolic ($t^{1/2}$) in character.

The second term describes the rate of interfacial network dissolution that is associated with a stage 2 reaction. A type IV surface is a resorbable glass (region C in Fig. 5) and is experiencing a combination of stage 1 and stage 2 reactions. A type V surface is dominated by a stage 2 reaction. Stage 2, the second stage of attack, is a process by which the silica structure breaks down and the glass totally dissolves at the interface. Stage 2 kinetics are linear ($t^{1.0}$).

A glass surface with a dual protective film is designated type IIIA (Fig. 6). The thickness of the secondary films can vary considerably—from as little as 0.01 μm for Al_2O_3–SiO_2-rich layers on inactive glasses, to as much as 30 μm for CaO–P_2O_5-rich layers on bioactive glasses.

A type III surface forms as a result of the repolymerization of SiO_2 on the glass surface by the condensation of the silanols (Si<pisbOH) formed from the stage 1 reaction. For example:

$$Si-OH + OH-Si \rightarrow Si-O-Si + H_2O \qquad (2)$$

Stage 3 protects the glass surface. The SiO_2 polymerization reaction contributes to the enrichment of surface SiO_2 that is characteristic of type II, III, and IV surface profiles (Fig. 6). It is described by the third term in Eq. 1. This reaction is interface controlled with a time dependence of $+k_3t^{1.0}$. The interfacial thickness of the most reactive bioactive glasses shown in Fig. 2 is largely due to this reaction.

The fourth term in Eq. 1, $+k_4t^y$ (stage 4), describes the precipitation reactions that result in the multiple films characteristic of type III glasses. When only one secondary film forms, the surface is type IIIA. When several additional films form, the surface is type IIIB.

In stage 4, an amorphous calcium phosphate film precipitates on the silica-rich layer and is followed by crystallization to form carbonated HA crystals. The calcium and phosphate ions in the glass or glass-ceramic provide the nucleation sites for crystallization. Carbonate anions (Co_3^{2-}) substitute for OH^- in the apatite crystal structure to form a carbonate hydroxyapatite similar to that found in living bone. Incorporation of CaF_2 in the glass results in incorporation of fluoride ions in the apatite crystal lattice. Crystallization of carbonate HA occurs around collagen fibrils present at the implant interface and results in interfacial bonding.

In order for the material to be bioactive and form an interfacial bond, the kinetics of reaction in Eq. 1, and especially the rate of stage 4, must match the rate of biomineralization that normally occurs *in vivo*. If the rates in Eq. 1 are too rapid, the implant is resorbable, and if the rates are too slow, the implant is not bioactive.

By changing the compositionally controlled reaction kinetics (Eq. 1), the rates of formation of hard tissue at an implant interface can be altered, as shown in Fig. 1. Thus, the level of bioactivity of a material can be related to the time for more than 50% of the interface to be bonded ($t_{0.5bb}$) [e.g., I_B index of bioactivity: = $(100/t_{0.5bb})$] (Hench, 1988). It is necessary to impose a 50% bonding criterion for an IB since the interface between an implant and bone is irregular (Gross *et al.*, 1988). The initial concentration of macrophages, osteoblasts, chondroblasts, and fibroblasts varies as a function of the fit of the implant and the condition of the bony defect (see Chapters 3 and 4). Consequently, all bioactive implants require an incubation period before bone proliferates and bonds. The length of this incubation period varies widely, depending on the composition of the implant.

The compositional dependence of IB indicates that there are isoI_B contours within the bioactivity boundary, as shown in Fig. 5 (Hench, 1988). The change of I_B with the SiO_2/(Na_2O + CaO) ratio is very large as the bioactivity boundary is approached. The addition of multivalent ions to a bioactive glass or glass-ceramic shrinks the isoI_B contours, which will contract to zero as the percentage of Al_2O_3, Ta_2O_5, ZrO_2, or other multivalent cations increases in the material. Consequently, the isoI_B boundary shown in Fig. 5 indicates the contamination limit for bioactive glasses and glass-ceramics. If the composition of a starting implant is near the I_B boundary, it may take only a few parts per million of multivalent cations to shrink the I_B boundary to zero and eliminate bioactivity. Also, the sensitivity of fit of a bioactive implant and length of time of immobilization postoperatively depends on the I_B value and closeness to the $I_B = 0$ boundary. Implants near the I_B boundary require more precise surgical fit and longer fixation times before they can bear loads. In contrast, increasing the surface area of a bioactive implant by using them in particulate form for bone augmentation expands the bioactive boundary. Small (<200 μm) bioactive glass granules behave as a partially resorbable implant and stimulate new bone formation (Hench, 1994).

Bioactive implants with intermediate I_B values do not develop a stable soft tissue bond; instead, the fibrous interface progressively mineralizes to form bone. Consequently, there appears to be a critical isoI_B boundary beyond which bioactivity is restricted to stable bone bonding. Inside the critical isoI_B boundary, the bioactivity includes both stable bone and soft-tissue bonding, depending on the progenitor stem cells in contact with the implant. This soft tissue–critical isoI_B limit is shown by the dashed contour in Fig. 5.

The thickness of the bonding zone between a bioactive implant and bone is proportional to its I_B (compare Fig. 1 with Fig. 2). The failure strength of a bioactively fixed bond appears to be inversely dependent on the thickness of the bonding zone. For example, 45S5 Bioglass with a very high I_B develops a gel bonding layer of 200 μm, which has a relatively low shear strength. In contrast, A-W glass-ceramic, with an intermediate I_B value, has a bonding interface in the range of 10–20 μm and a very high resistance to shear. Thus, the interfacial bonding strength appears to be optimum for I_B values of ~4. However, it is important to recognize that the interfacial area for bonding is time dependent, as shown in Fig. 1. Therefore, interfacial strength is time dependent and is a function of such morphological factors as the change in interfacial area with time, progressive mineralization of the interfacial tissues, and resulting increase of elastic modulus of the interfacial bond, as well as shear strength per unit of bonded area. A comparison of the increase in interfacial bond strength of bioactive fixation of implants bonded to bone with other types of fixation is given in Fig. 7 (Hench, 1987).

Clinical applications of bioactive glasses and glass-ceramics are reviewed by Gross *et al.* (1988), Yamamuro *et al.* (1990), and Hench and Wilson (1993) (Table 6). The 8-year history of successful use of Ceravital glass-ceramics in middle ear surgery

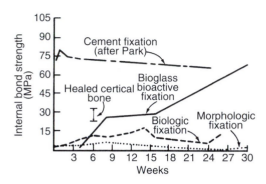

FIG. 7. Time dependence of interfacial bond strength of various fixation systems in bone. (After Hench, 1987.)

TABLE 6 Present Uses of Bioceramics

Orthopedic load-bearing applications Al_2O_3	Coatings for tissue ingrowth (Cardiovascular, orthopedic, dental and maxillofacial prosthetics) Al_2O_3
Coatings for chemical bonding (Orthopedic, dental and maxillofacial prosthetics) HA Bioactive glasses Bioactive glass-ceramics	Temporary bone space fillers Tricalcium phosphate Calcium and phosphate salts
Dental implants Al_2O_3 HA Bioactive glasses	Periodontal pocket obliteration HA HA–PLA composite Trisodium phosphate Calcium and phosphate salts Bioactive glasses
Alveolar ridge augmentations Al_2O_3 HA HA–autogenous bone composite HA–PLA composite Bioactive glasses	Maxillofacial reconstruction Al_2O HA HA–PLA composite Bioactive glasses
Otolaryngological Al_2O_3 HA Bioactive glasses Bioactive glass-ceramics	Percutaneous access devices Bioactive glasses Bioactive composites
Artifical tendon and ligament PLA–carbon fiber composite	Orthopedic fixation devices PLA–carbon fibers PLA–calcium/phosphorus–base glass fibers

(Reck *et al.*, 1988) is especially encouraging, as is the 10-year use of A-W glass-ceramic in vertebral surgery (Yamamuro *et al.*, 1990), the 10-year use of 45S5 Bioglass in endosseous ridge maintenance (Stanley *et al.*, 1996) and middle-ear replacement, and the 6-year success in repair of periodontal defects (Hench and Wilson, 1996; Wilson, 1994).

CALCIUM PHOSPHATE CERAMICS

Bone typically consists, by weight of 25% water, 15% organic materials and 60% mineral phases. The mineral phase consists primarily of calcium and phosphate ions, with traces of magnesium, carbonate, hydroxyl, chloride, fluoride, and citrate ions. Hence calcium phosphates occur naturally in the body, but they occur also within nature as mineral rocks, and certain compounds can be synthesized in the laboratory. Table 7 summarizes the mineral name, chemical name and composition of various phases of calcium phosphates.

Within the past 20–30 years interest has intensified in the use of calcium phosphates as biomaterials, but only certain compounds are useful for implantation in the body since both their solubility and speed of hydrolysis increase with a decreasing calcium-to-phosphorus ratio. Driessens (1983) stated that those compounds with a Ca/P ratio of less than 1 : 1 are not suitable for biological implantation.

The main crystalline component of the mineral phase of bone is a calcium-deficient carbonate hydroxyapatite, and various methods have been investigated to produce synthetic hydroxyapatite. The commercial routes are based on aqueous precipitation or conversion from other calcium compounds. Aqueous precipitation is most often performed in one of two ways: a reaction between a calcium salt and an alkaline phosphate (Collin, Hayek and Newsley, 1963; Eanes *et al.*, 1965; Bonel *et al.*, 1987; Young and Holcomb, 1982; Denissen *et al.*, 1980; Jarcho *et al.*, 1976; Kijima and Tsutsumi, 1979) or a reaction between calcium hydroxide or calcium carbonate and phosphoric acid (Mooney, 1961; Irvine, 1981; Nagai *et al.*, 1985; Rao and Boehm, 1974; McDowell *et al.*, 1977; Akao *et al.*, 1981).

Other routes include solid–state processing (Monma *et al.*, 1981; Fowler, 1974; Young and Holcomb, 1982; Rootare *et al.*, 1978; Lehr *et al.*, 1967); hydrolysis (Schleede *et al.*, 1932; Perloff *et al.*, 1956; Morancho *et al.*, 1981; Young and Holcomb, 1982); hydrothermal synthesis (Young and Holcomb, 1982; Fowler, 1974; Roy, 1971; Skinner, 1973; Arends *et al.*, 1979; Perloff and Posner, 1960).

The route and conditions under which synthetic HA is produced will greatly influence its physical and chemical characteristics. Factors that affect the rate of resorption of the implant include physical factors such as the physical features of the material (e.g., surface area, crystallite size), chemical factors such as atomic and ionic substitutions in the lattice, and biological factors such as the types of cells surrounding the implant and location, age, species, sex, and hormone levels.

The thermodynamic stability of the various calcium phosphates is summarized in the form of the phase diagram shown in Fig. 8. The binary equilibrium phase diagram between CaO and P_2O_5 gives an indication of the compounds formed between the two oxides, and by comparing this with Table 7 it is possible to identify the naturally occurring calcium phosphate minerals. The diagram does not indicate the phase boundaries of apatite due to the absence of hydroxyl groups.

TABLE 7   Calcium Phosphates

Ca : P	Mineral name	Formula	Chemical name
1.0	Monetite	$CaHPO_4$	Dicalcium phosphate (DCP)
1.0	Brushite	$CaHPO_4 \cdot 2H_2O$	Dicalcium phosphate Dihydrate (DCPD)
1.33	—	$Ca_8(HPO_4)_2(PO_4)_4 \cdot 5H_2O$	Octocalcium phosphate (OCP)
1.43	Whitlockite	$Ca_{10}(HPO_4)(PO_4)_6$	
1.5	—	$Ca_3(PO_4)_2$	Tricalcium phosphate (TCP)
1.67	Hydroxyapatite	$Ca_{10}(PO_4)_6(OH)_2$	
2.0		$Ca_4P_2O_9$	Tetracalcium phosphate

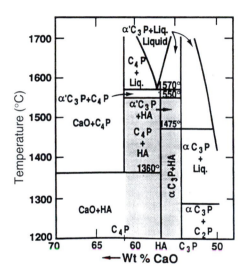

FIG. 8.   Phase equilibrium diagram of calcium phosphates in a water atmosphere. Shaded area is processing range to yield HA-containing implants. (After K. de Groot. 1988. *Ann. N. Y. Acad. Sci.* **523:** 227.)

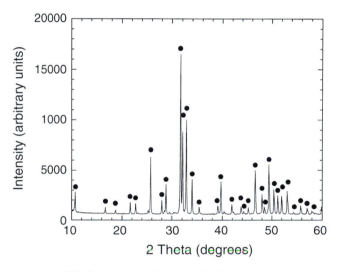

FIG. 9.   X-ray diffraction of hydroxyapatite.

However, from the binary diagram an indication may be obtained of the stability of other calcium phosphates with temperature.

The stoichiometry of HA is highly significant where thermal processing of the material is required. Slight imbalances in the stoichiometric ratio of calcium and phosphorus in HA (from the standard molar ratio of 1.67), can lead to the appearance of either α or β-tricalcium phosphate on heat treatment. Many early papers concerning the production and processing of HA powders reported problems in avoiding the formation of these extraneous phases (Jarcho *et al.*, 1976; De With *et al.*, 1981a, b; Peelen *et al.*, 1978). However, using stoichiometric hydroxyapatite it should be possible to sinter, without phase purity problems, at temperatures in excess of 1300°C.

X-ray diffraction (Fig. 9) and infra-red spectroscopy (Fig. 10) should be used to reveal the phase purity and level of hydroxylation of HA. Kijima and Tsutsumi (1979) used these techniques to study hydroxyapatite sintered at different temperatures and reported that after sintering at 900°C, the

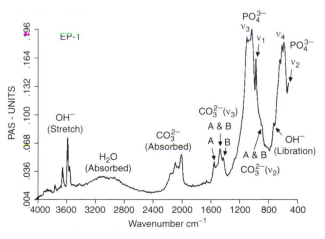

FIG. 10.   Typical FT-IR spectrum for HA.

material was fully hydroxylated, but after sintering at temperatures higher than this, dehydroxylation occurred. Dehydration of HA, produced by processes such as high temperature solid state reaction, result in the formation of oxyhydroxyapatite: $Ca_{10}(PO_4)_6(OH)_{2-2x}O_xV_x$ (where V is a hydroxyl vacancy). Hydroxyapatite has a $P6_3/m$ space group: This signifies that the lattice is primitive Bravais, there is a sixfold axis parallel to the c axis and a 1/2 (3/6) translation along the length of the c axis (a screw axis) with a mirror plane situated perpendicular to the screw axis and the c axis. The a and c parameters for hydroxyapatite are 0.9418 nm and 0.6884 nm, respectively.

The structure assumed by any solid is such that, on an atomic level, the configuration of the constituents is of the lowest possible energy. In phosphates, this energy requirement results in the formation of discrete subunits within the structure and the PO_4^{3-} group forms a regular tetrahedron with a central P^{5+} ion and O^{2-} ions at the four corners. In a similar manner, the $(OH)^-$ groups are also ionically bonded. In terms of the volume occupied, the oxygen ions exceed all other elements in phosphates. Any other elements present may therefore be considered as filling the interstices, with the exact position being determined by atomic radius and charge (See Fig. 11).

The hydroxyapatite lattice contains two kinds of calcium positions; columnar and hexagonal. There is a net total of four "columnar calcium" ions that occupy the [1/3, 2/3, 0] and [1/3, 2/3, 1/2] lattice points. The "hexagonal calcium" ions are located on planes parallel to the basal plane at $c = 1/4$ and $c = 3/4$ and the six PO_4^{3-} tetrahedra are also located on these planes. The $(OH)^-$ groups are located in columns parallel to the c axis, at the corners of the unit cell, which may be viewed as passing through the centers of the triangles formed by the "hexagonal calcium" ions. Successive hexagonal calcium triangles are rotated through 60° (See Fig. 12).

Defects and impurities in hydroxyapatite may be identified as either substitutional or as discrete, extraneous crystalline phases (as discussed above). Methods of detection of impurities include X-ray diffraction, infrared spectroscopy, and spectrochemical analysis. It is important to make a full spectrochemical analysis of hydroxyapatite since contact with any metal

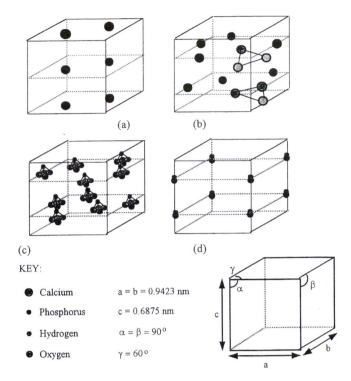

KEY:

- Calcium $a = b = 0.9423$ nm
- Phosphorus $c = 0.6875$ nm
- Hydrogen $\alpha = \beta = 90°$
- Oxygen $\gamma = 60°$

FIG. 12. Theoretical positions of the ionic species within the unit cell of hydroxyapatite (Hing, 1995).

ions during production can lead to high levels of impurities in the product. Typical data for one commercial hydroxyapatite powder are shown in Table 8.

Ions that may be incorporated into the HA structure, either intentionally or unintentionally, include carbonate ions (substituting for hydroxyl or phosphate groups), fluoride ions (substituting for hydroxyl groups), silicon, or silicate ions (substituting for phosphorus or phosphate groups) and magnesium ions substituting for calcium, e.g., Newsley, 1963; Le Geros, 1965; Barralet, 1997; Jha *et al.*, 1998; Gibson *et al.*, 1999).

The presence of carbonate may be observed directly, using infrared spectroscopy, in the form of weak peaks at between 870 and 880 cm^{-1} and a stronger doublet between 1460 and 1530 cm^{-1}, and also through alterations in the hydroxyapatite

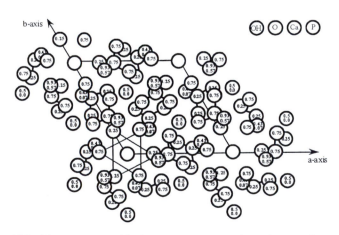

FIG. 11. Structure of hydroxyapatite projected on the x,y plane, adapted from Kay *et al.*, 1964 (Hing, 1995).

TABLE 8   Trace Elements in a Commercial Hydroxyapatite

Trace element	ppm
Al	600
Cu	1
Fe	1000
Ge	100
Mg	2000
Mn	300
Na	3000
Pb	4
Si	500
Ti	30

TABLE 9 Typical Mechanical Properties of Dense Hydroxyapatite Ceramics

Theoretical density	3.156 g cm^3
Hardness	500–800 HV, 2000–3500 Knoop
Tensile strength	40–100 MPa
Bend strength	20–80 MPa
Compressive strength	100–900 MPa
Fracture toughness	approx. 1 Mpa m$^{0.5}$
Young's modulus	70–120 GPa

lattice parameters from X-ray diffraction (Rootare and Craig, 1978; LeGeros, 1965; Barralet et al., 1997). The substitution of electronegative anions such as fluorine and chlorine for (OH)$^-$ has also been reported to alter the lattice parameters of the material (e.g., Young and Elliot, 1966; Kay et al., 1964; see also Elliot, 1994).

Hydroxyapatite may be processed as a ceramic using compaction (die pressing, isostatic pressing, slip casting, etc.) followed by solid-state sintering (discussed earlier in this chapter). When reporting methods for the production and sintering of hydroxyapatite powders it is very important to adequately characterize the morphology of the product including the surface area, particle size distribution, mean particle size, and physical appearance of the powders, since this will greatly influence the handling and processing characteristics of the material (Best and Bonfield, 1994). There is great deal of variation in the reported mechanical performance of dense hydroxyapatite ceramics, dependent on phase purity, density and grain size, but the properties cited generally fall in the range shown in Table 9.

CALCIUM PHOSPHATE COATINGS

The clinical application of calcium phosphate ceramics is largely limited to non-major-load-bearing parts of the skeleton because of their inferior mechanical properties, and it was partly for this reason that interest was directed toward the use of calcium phosphate coatings on metallic implant subtrates. A very good review of techniques for the production of calcium phosphates was given by Wolke et al. in 1998. Many techniques are available for the deposition of hydroxyapatite coatings, including electrophoresis, sol–gel routes, electrochemical routes, biomimetic routes, and sputter techniques, but the most popular commercial routes are those based on plasma spraying. In plasma spraying, an electric arc is struck between two electrodes and stream of gases is passed through the arc. The arc converts the gases into a plasma with a speed of up to 400 m/sec and a temperature within the arc of 20,000 K. The ceramic powder is suspended in the carrier gas and fed into the plasma where it can be fired at a substrate. There are many variables in the process including the gases used, the electrical settings, the nozzle/substrate separation and the morphology, particle size, and particle size distribution of the powder. Because of the very high temperatures but very short

times involved, the behavior of the hydroxyapatite powder particle is somewhat different than might be predicted in an equilibrium phase diagram. However, according to the particular conditions used, it is likely that at least a thin outer layer of the powder particle will be in a molten state and will undergo some form of phase transformation, but by careful control of the operating variables the transformed material should represent a relatively small volume fraction of the coating and the product should maintain the required phase purity and crystallinity (Cook et al., 1988; Mayer et al., 1986; Wolke et al., 1992).

A number of factors influence the properties of the resulting coating, including coating thickness (this will influence coating adhesion and fixation—the agreed optimum now seems to be 50–100 μm; Soballe et al., 1993 and de Groot et al., 1987), crystallinity (this affects the dissolution and biological behavior; Klein et al., 1994a, b; Clemens, 1995; Le Geros et al., 1992), biodegradation (affected by phase purity, chemical purity, porosity, crystallinity), and adhesion strength (these may range between 5 and 65 MPa (97)).

The mechanical behavior of calcium phosphate ceramics strongly influences their application as implants. Tensile and compressive strength and fatigue resistance depend on the total volume of porosity. Porosity can be in the form of micropores (< 1 μm diameter, due to incomplete sintering) or macropores (> 100 μm diameter, created to permit bone growth). The dependence of compressive strength (σ_c) and total pore volume (V_p) is described in de Groot et al. (1990) by:

$$\sigma_c = 700 \exp -5V_p \quad \text{(in MPa)}, \quad (3)$$

where V_p is in the range of 0–0.5.

Tensile strength depends greatly on the volume fraction of microporosity (V_m):

$$\sigma_t = 220 \exp -20V_m \text{(in MPa)}. \quad (4)$$

The Weibull factor (n) of HA implants is low in physiological solutions ($n = 12$), which indicates low reliability under tensile loads. Consequently, in clinical practice, calcium phosphate bioceramics should be used (1) as powders; (2) in small, unloaded implants such as in the middle ear; (3) with reinforcing metal posts, as in dental implants; (4) as coatings (e.g., composites); or (5) in low-loaded porous implants where bone growth acts as a reinforcing phase.

The bonding mechanism of dense HA implants appears to be very different from that described above for bioactive glasses. The bonding process for HA implants is described by Jarcho (1981). A cellular bone matrix from differentiated osteoblasts appears at the surface, producing a narrow amorphous electron-dense band only 3 to 5 μm wide. Between this area and the cells, collagen bundles are seen. Bone mineral crystals have been identified in this amorphous area. As the site matures, the bonding zone shrinks to a depth of only 0.05 to 0.2 μm (Fig. 2). The result is normal bone attached through a thin epitaxial bonding layer to the bulk implant. TEM image analysis of dense HA bone interfaces show an almost perfect epitaxial alignment of some of the growing bone crystallites with the apatite crystals in the implant. A consequence of this ultrathin bonding zone is a very high gradient in elastic modulus at the bonding interface between HA and bone. This is one of the major differences between the bioactive apatites and the bioactive glasses and glass-ceramics. The implications of this difference for the implant interfacial response to Wolff's law is discussed in Hench and Ethridge (1982, Chap. 14).

RESORBABLE CALCIUM PHOSPHATES

Resorption or biodegradation of calcium phosphate ceramics is caused by three factors:

1 Physiochemical dissolution, which depends on the solubility product of the material and local pH of its environment. New surface phases may be formed, such as amorphous calcium phosphate, dicalcium phosphate dihydrate, octacalcium phosphate, and anionic-substituted HA.
2 Physical disintegration into small particles as a result of preferential chemical attack of grain boundaries.
3 Biological factors, such as phagocytosis, which causes a decrease in local pH concentrations (de Groot and Le Geros, 1988).

Ideally, one would wish for a replacement material to be slowly resorbed by the body once its task of acting as a scaffold for new bone has been completed. Degradation or resorption of calcium phosphates *in vivo* occurs by a combination of phagocytosis of particles and the production of acids. However, when selecting a resorbable material for implantation, care must be taken to match the rate of resorption with that of the expected bone tissue regeneration. Where the solubility of calcium phosphates is higher than the rate of tissue regeneration, they will not be of use to fill bone defects. As mentioned previously, the rate of dissolution increases with decreasing calcium-to-phosphorus ratio, and consequently, tricalcium phosphate, with a Ca:P ratio of 1.5, is more rapidly resorbed than hydroxyapatite. Tricalcium phosphate has four polymorphs, α β γ and super α. The γ polymorph is a high pressure phase and the super α polymorph is observed at temperatures above approximately 1500°C (Nurse *et al.*, 1959). Therefore the most frequently observed polymorphs in bioceramics are α and β-TCP. X-ray diffraction studies indicate that the β polymorph transforms to the α polymorph at temperatures between 1120°C and 1290°C (Gibson *et al.*, 1996).

In the 1980s, the idea of a new bone susbtitute material was introduced and the materials were referred to as calcium phosphate bone cements. These materials offer the potential for *in situ* molding and injectability. There are a variety of different combinations of calcium compounds (e.g., α-tricalcium phosphate and dicalcium phosphate) that are used in the formulation of these bone cements, but the end product is normally based on a calcium-deficient hydroxyapatite (Fernandez *et al.*, 1998, 1999a, b).

All calcium phosphate ceramics biodegrade to varying degrees in the following order: increasing rate HA.

The rate of biodegradation increases as:

1. Surface area increases (powders > porous solid > dense solid)
2. Crystallinity decreases

3. Crystal perfection decreases
4. Crystal and grain size decrease
5. There are ionic substitutions of CO_2^{-3}, Mg^{2+}, and Sr^{2+} in HA

Factors that tend to decrease the rate of biodegradation include (1) F^- substitution in HA, (2) Mg^{2+} substitution in β-TCP, and (3) lower β-TCP/HA ratios in biphasic calcium phosphates.

CLINICAL APPLICATIONS OF HA

Calcium phosphate-based bioceramics have been used in medicine and dentistry for nearly 20 years (Hulbert *et al.*, 1987; de Groot, 1983, 1988; de Groot *et al.*, 1990; Jarcho, 1981; Le Geros, 1988; Le Geros and Le Geros, 1993). Applications include dental implants, periodontal treatment, alveolar ridge augmentation, orthopedics, maxillofacial surgery, and otolaryngology (Table 6).

Most authors agree that HA is bioactive, and it is generally agreed that the material is osseoconductive, where osseoconduction is the ability of a material to encourage bone growth along its surface when placed in the vicinity of viable bone or differentiated bone-forming cells. A good recent review of *in vitro* and *in vivo* data for calcium phosphates has been prepared by Hing *et al.* (1998), who observed that there are a large number of "experimental parameters," including specimen, host, and test parameters, which need to be carefully controlled in order to allow adequate interpretation of data.

Hydroxyapatite has been used clinically in a range of different forms and applications. It has been utilised as a dense, sintered ceramic for middle ear implant applications (van Blitterswijk, 1990) and alveolar ridge reconstruction and augmentation (Quin and Kent, 1984; Cranin *et al.*, 1987), in porous form (Smiler and Holmes, 1987; Bucholz *et al.*, 1987), as granules for filling bony defects in dental and orthopaedic surgery (Aoki, 1994; Fujishiro, 1997; Oonishi *et al.*, 1990; Froum *et al.*, 1986; Galgut *et al.*, 1990; Wilson and Low, 1992), and as a coating on metal implants (Cook *et al.*, 1992a, b; De Groot, 1987).

Another successful clinical application for hydroxyapatite has been in the form of a filler in a polymer matrix. The original concept of a bioceramic polymer composite was introduced by Bonfield *et al.* (1981) and the idea was based on the concept that cortical bone itself comprises an organic matrix reinforced with a mineral component. The material developed by Bonfield and co-workers contains up to 50 vol % hydroxyapatite in a polyethylene matrix, has a stiffness similar to that of cortical bone, has high toughness, and has been found to exhibit bone bonding *in vivo*. The material has been used as an orbit implant for orbital floor fractures and volume augmentation (Tanner *et al.*, 1994) and is now used in middle ear implants, commercialized under the trade name HAPEX (Bonfield, 1996).

Bibliography

Akao, M., Aoki, H., and Kato, K. (1981). Mechanical Properties of Sintered Hydroxyapatite for Prosthetic Applications. *J. Mat. Sci.* **16**: 809.

Aoki, H. (1994). *Medical Applications of Hydroxyapatite*. Ishiyaku EuroAmerica Inc. Tokyo.

Levin, E. M., Robbins, C. R., and McMurdie, H. F. (eds.) (1964). *American Ceramic Society, Phase Diagrams for Ceramists*. American Ceramics Society, Ohio, p. 107.

Arends, J. Schutof, J., van der Linden, W. H., Bennema, P., and van den Berg, P. J. (1979). Preparation of Pure Hydroxyapatite Single Crystals by Hydrothermal Recrystallisation. *J. Cryst Growth* **46**: 213.

Barralet, J. E., Best, S. M., and Bonfield, W. (1998). Carbonate substitution in precipitated hydroxyapatite: An investigation into the effects of reaction temperature and bicarbonate Ion concentration. *J. Biomed. Mater. Res.* **41**: 79–86.

Berndt, C. C., Haddad, G. N., and Gross, K. A. (1989). Thermal spraying for bioceramics application. in *Bioceramics 2*. Proceedings of the 2nd International Symposium on Ceramics in Medicine, Heidelberg, pp. 201–206.

Best, S. M., and Bonfield, W. (1994). Processing behaviour of hydroxyapatite powders of contrasting morphology. *J. Mater. Sci. Mat. Med.* **5**: 516.

Bocholz, R. W., Carlton, A., and Holme, R. E. (1987). Hydroxyapatite and tricalcium phosphate bone graft substitute. *Orthop. Clin. North Am.* **18**: 323–334.

Bonel, G., Heughebeart, J-C., Heughebaert, M., Lacout, J. L., and Lebugle, A. (1987). Apatitic calcium orthophosphates and related compounds for biomaterials preparation. *Annals of the New York Academy of Science* 115.

Bonfield, W., Grynpas, M. D., Tully, A. E., Bowman, J., and Abram, J. (1989). Hydroxyapatite reinforced polyethylene—A mechanically compatible implant. *Biomaterials* **2**: 185–186.

Bonfield, W. (1996). Composite biomaterials. in *Bioceramics 9*, T. Kukubo, T. Nakamura, and F. Miyaji, eds. Proceedings of the 9th International Symposium on Ceramics in Medicine, Pergamon.

Christel, P., Meunier, A., Dorlot, J. M., Crolet, J. M., Witvolet, J., Sedel, L., and Boritin, P. (1988). Biomechanical compatability and design of ceramic implants for orthopedic surgery. in *Bioceramics: Material Characteristics versus In-Vivo Behavior*, P. Ducheyne and J. Lemons, eds. *Ann. N. Y. Acad. Sci.* **523**: 234.

Clemens, J. A. M. (1995). Fluorapatite Coatings for the Osseointegration of Orthopaedic Implants, Thesis, University of Leiden, Leiden, The Netherlands.

Clemens J. A. M., Klein C. P. A. T., Vriesde R. C., de Groot K., and Rozing P. M. (1995). Large gaps around calcium phosphate coated bone implants: Deficient bone apposition despite use of allograft bone. in *Proceedings of 21st Annual Meeting*. Society for Biomaterials, San Francisco.

Collin, R. L. (1959). Strontium—Calcium hydroxyapatite solid solutions: Preparations and lattice constant measurements. *J. Am. Chem. Soc.* **81**: 5275.

Cook, S. D., Thomas, K. A., Kay, J. F., and Jarcho, M. (1998). Hydroxylapatite coated titanium for orthopaedic implant applications. *Clin. Orthop. Rel. Res.* **232**: 225–243.

Cook, S. D., Thomas, K. A., Dalton, J. E., Volkman, R. K., Whitecloud, T. S., and Kay, J. E. (1999). Hydroxylapatite coating of porous implants improves bone ingrowth and interface attachment strength. *J. Biomed. Mater. Res.* **26**: 989–1001.

Cranin, A. N., Tobin, G. P., and Gelbman, J. (1987). Applications of hydroxyapatite in oral and maxilofacial surgery, part II: Ridge augmentation and repair of major oral defects. *Compend. Contin. Educ. Dent.* **8**: 334–345.

DeGroot, K. (1984). Surface chemistry of sintered hydroxyapatite: On possible relations with biodegredation and slow crack propagation. in *Adsorption and Surface Chemistry of Hydroxyapatite*. Ed. Misra, D. N. Plenum Publishing, New York.

DeGroot, K. (1987). Hydroxylapatite coatings for implants in surgery. in *High Tech Ceramics*. Ed. Vincenzini, P. Elsevier, Amsterdam.

DeGroot, K., Geesink, R. G. T., Klein, C. P. A. T., and Serekian, P. (1987). Plasma sprayed coatings of hydroxylapatite. *J. Biomed. Mater. Res.* **21**: 1375–1381.

DeGroot, K., Wolke, J. G. C., and Jansen, J. A. (1998). Calcium phosphate coatings for medical implants. *Proc. Inst. Mech. Engrs.* **212** (part H): 437.

Denissen, H. W., DeGroot, K., Driessen, A. A., Wolke, J. G. C., Peelen, J. G. J., van Dijk, H. J. A., Gehring, A. P., and Klopper, P. J. (1980a). Hydroxylapatite implants: Preparation, properties and use in alveolar ridge preservation. *Sci Ceram.* **10**: 63.

Denissen, H. W., DeGroot, K., Klopper, P. J., van Dijk, H. J. A., Vermeiden, J. W. P., and Gehring, A. P. (1980b). Biological and mechanical evaluation of calcium hydroxyapatite made by hot pressing. in *Mechanical Properties of Biomaterials*, Chapter 40, Eds. Hastings, G. W., and Williams, D. F. John Wiley and Sons, New York.

De With, G., van Dijk, H. J. A., and Hattu, N. (1981a). Mechanical behaviour of biocompatible hydroxyapatite ceramics. *Proc. Brit. Ceram. Soc.* **31**: 181.

De With, G., van Dijk, H. J. A., Hattu, N., and Prijs, K. (1981b). Preparation, microstructure and mechanical properties of dense polycrystalline hydroxyapatite. *J. Mat. Sci.* **16**: 1592.

Dickens, B., Schroeder, L. W., and Brown, W. E. (1974). Crystallographic studies of the role of Mg as a stabilising impurity in β-tricalcium phosphate. *J. Solid State Chem.* **10**: 232.

Driessen, A. A., Klein, C. P. A. T., and de Groot, K. (1982). Preparation and some properties of sintered ß-Whitlockite. *Biomaterials* **3**: 113–116.

Driessens, F. C. M. (1983). Formation and stability of calcium phosphate in relation to the phase composition of the mineral in calcified tissue. in *Bioceramics of Calcium Phosphate*, Ed. DeGroot, K. CRC Press, Boca Raton, Florida.

de Groot, K. (1983). *Bioceramics of Calcium-Phosphate*. CRC Press, Boca Raton, FL.

de Groot, K. (1988). Effect of porosity and physicochemical properties on the stability, resorption, and strength of calcium phosphate ceramics. in *Bioceramics: Material Characterics versus In-Vivo Behavior*, *Ann. N. Y. Acad. Sci.* **523**: 227.

de Groot, K., and Le Geros, R. (1988). in *Position Papers in Bioceramics: Materials Characteristics versus In-Vivo Behavior*, P. Ducheyne and J. Lemons, eds. *Ann. N. Y. Acad. Sci.* **523**: 227, 268, 272.

de Groot, K., Klein, C. P. A. T., Wolke, J. G. C., and de Blieck-Hogervorst, J. (1990). Chemistry of calcium phosphate bioceramics. in *Handbook on Bioactive Ceramics*, T. Yamamuro, L. L. Hench, and J. Wilson, eds. CRC Press, Boca Raton, FL, Vol. II, Ch. 1.

Drre, E., and Dawihl, W. (1980). Ceramic hip endoprostheses. in *Mechanical Properties of Biomaterials*, G. W. Hastings and D. Williams, eds. Wiley, New York, pp. 113–127.

Eanes, E. D., Gillessen, J. H., and Posner, A.L. (1965). Intermediate states in the precipitation of hydroxyapatite. *Nature* **208**: 365.

Eggli, P.S., Muller, W., and Schenk, R.K. (1988). Porous hydroxyapatite and tricalcium phosphate cylinders with two different pore size ranges implanted in the cancellous bone of rabbits. *Clin. Orthop. Relat. Res.* **232**: 127–138.

Elliot, J. R. (1994). *Structure and Chemistry of Apatites and Other Calcium Orthophosphate*. Elsevier, Amsterdam.

Fang, Y., Agrawal, D. K., Roy, D. M., and Roy, R. (1992). Fabrication of Porous Hydroxyapatite by Microwave Processing. *J. Mater. Res.* **7**(2): 490–494.

Fernandez, E., Gil, F. X., Ginebra, M. P., Driessens, F. C. M., Planell, J. A, and Best, S. M. (1999a). Calcium Phosphate Bone Cements for Clinical Applications, Part I, Solution Chemistry. *J. Mater. Sci. Mater. in Med.* **10**: 169–176.

Fernandez, E., Gil, F. X., Ginebra, M. P., Driessens, F. C. M., Planell, J. A, and Best, S. M. (1999b). Calcium Phosphate Bone Cements for Clinical Applications, Part II, Precipitate Formation during Setting Reactions. *J. Mater. Sci. Mater. in Med.* **10**: 177–184.

Fernandez, E., Planell, J. A., Best, S. M., and Bonfield W. (1998). Synthesis of Dahllite Through a Cement Setting Reaction, *J. Mater. Sci. Mater. in Med.* **9**: 789–792.

Fowler, B.O. (1974). Infrared studies of apatites, part II: Preparation of normal and isotopically substituted calcium, strontium and barium hydroxyapatites and spectra-structure correlations. *Inorg. Chem.* **13**: 207.

Froum, S. J., Kushner, J., Scopp, L., and Stahl, S. S. (1986). Human clinical and histologic responses to durapatite implants in intraosseous lesions: Case reports. *J. Periodontol.* **53**: 719–725.

Fujishiro, Y., Hench, L. L., and Oonishi, H. (1997). Quantitative rates of in-vivo bone generation for bioglass and hydroxyapatite particels as bone graft substitute, *J. Mater. Sci. Mater. in Med.* **8**: 649–652.

Galgut, P. N., Waite, I. M., and Tinkler, S. M. B. (1990). Histological investigation of the tissue response to hydroxyapatite used as an implant material in periodontal treatment. *Clin. Mater.* **6**: 105–121.

Gibson, I. R., Best, S. M., and Bonfield, W. (1996). Phase transformations of tricalcium phosphates using high temperature x-ray diffraction, Bioceramics **9**, Eds. Kokubo, Nakamura, and Miyaji, Publ. Pergamon Press, Oxford, pp. 173–176.

Gibson, I. R., Best, S. M., and Bonfield, W. (1999). Chemical characterisation of silicon-substituted hydroxyapatite. *J. Biomed. Mater. Res.* **44**: 422–428.

Gross, V., and Strunz, V. (1985). The interface of various glasses and glass-ceramics with a bony implantation bed. *J. Biomed. Mater. Res.* **19**: 251.

Gross, V., Kinne, R., Schmitz, H. J., and Strunz, V. (1988). The response of bone to surface active glass/glass-ceramics. *CRC Crit. Rev. Biocompatibility* **4**: 2.

Hayek, E., and Newsley, H. (1963). Pentacalcium monohydroxy-orthophosphate. *Inorganic Synthesis* **7**: 63.

Hench, L. L., and Wilson, J. (1993). *An Introduction to Bioceramics.* World Scientific, Singapore.

Hing, K. A. (1995). Assessment of porous hydroxyapatite for bone replacement. PhD Thesis, University of London.

Hing, K. A., Best, S. M., and Bonfield, W. (1999). Characterisation of porous hydroxyapatite. *J. Mater. Sci. in Med.* **10**: 135–160.

Hing, K. A., Best, S. M., Tanner, K. E., Revell, P. A., and Bonfield, W. (1998). Histomorphological and biomechamical characterisation of calcium phosphates in the osseous environment. *Proc. Inst. Mech. Engrs.* **212** (part H): 437.

Holmes, R. E. (1979). Bone regeneration within a coralline hydroxyapatite implant. *Plast. Reconstr. Surg.* **63**: 626–633.

Hulbert, S. F., Young, F. A., Mathews, R. S., Klawitter, J. J., Talbert, C. D., and Stelling, F. H. (1970). Potential of ceramic materials as permanantly implantable skeletal prosthesis. *J. Biomed. Mater. Res.* **4**: 433–456.

Hench, L. L. (1987). Cementless fixation. in *Biomaterials and Clinical Applications*, A. Pizzoferrato, P. G. Marchetti, A. Ravaglioli, and A. J. C. Lee, eds. Elsevier, Amsterdam, p. 23.

Hench, L. L. (1988). Bioactive ceramics. in *Bioceramics: Materials Characteristics versus In-Vivo Behavior*, P. Ducheyne and J. Lemons, eds. *Ann. N. Y. Acad. Sci.* **523**: 54.

Hench, L. L. (1991). Bioceramics: From concept to clinic. *J. Am. Ceram. Soc.* **74**: 1487–1510.

Hench, L. L. (1994). *Bioactive Ceramics: Theory and Clinical Applica,* O. H. Anderson and A. Yli-Urpo, eds. Butterworth–Heinemann, Oxford, England, pp. 3–14.

Hench, L. L., and Clark, D. E. (1978). Physical chemistry of glass surfaces. *J. Non-Cryst. Solids* **28**(1): 83–105.

Hench, L. L., and Ethridge, E. C. (1982). *Biomaterials: An Interfacial Approach.* Academic Press, New York.

Hench, L. L., and Wilson, J. W. (1993). *An Introduction to Bioceramics.* World Scientific, Singapore.

Hench, L. L., and Wilson, J. W. (1996). *Clinical Performance of Skeletal Prostheses.* Chapman and Hall, London.

Hench, L. L., Splinter, R. J., Allen, W. C., and Greenlec, T. K., Jr. (1972). Bonding mechanisms at the interface of ceramic prosthetic materials. *J. Biomed. Res. Symp. No. 2.* Interscience, New York, p. 117.

Höhland, W., and Vogel, V. (1993). Machineable and phosphate glass-ceramics. in *An Introduction to Bioceramics*, L. L. Hench and J. Wilson, eds. World Scientific, Singapore, pp. 125–138.

Hulbert, S. (1993). The use of alumina and zirconia in surgical implants. in *An Introduction to Bioceramics*, L. L. Hench and J. Wilson, eds. World Scientific, Singapore, pp. 25–40.

Hulbert, S. F., Bokros, J. C., Hench, L. L., Wilson, J., and Heimke, G. (1987). Ceramics in clinical applications: past, present, and future. in *High Tech Ceramics*, P. Vincenzini, ed. Elsevier, Amsterdam, pp. 189–213.

Irvine, G. D. (1981). Synthetic bone ash. British Patent number 1 586 915.

Irwin, G. R. (1959). Analysis of stresses and strains near the end of a crack traversing a plate. *J. Appl. Mechanics* **24**: 361.

Jarcho, M., Bolen, C. H., Thomas, M. B., Bobick, J., Kay, J. F., and Doremus, R. H. (1976). Hydroxylapatite synthesis in dense polycrystalline form. *J. Mat. Sci.* **11**: 2027.

Jha, L., Best, S. M., Knowles, J., Rehman, I., Santos, J., and Bonfield, W. (1997). Preparation and characterisation of fluoride-substituted apatites. *J. Mater. Sci. Mater. in Med.* **8**: 185–191.

Johnson, P. D., Prener, J. S., and Kingsley, J. D. (1963). Apatite: Origin of blue color. *Science* **141**: 1179.

Jarcho, M. (1981). Calcium phosphate ceramics as hard tissue prosthetics. *Clin. Orthop. Relat. Res.* **157**: 259.

Kay, M. I., Young, R. A., and Posner, A. S. (1964). Crystal structure of hydroxyapatite. *Nature* **12**: 1050.

Kijima, T., and Tsutsumi, M. (1979). Preparation and thermal properties of dense polycrystalline oxyhydroxyapatite. *J. Am. Ceram. Soc.* **62**(9): 455.

Klawitter, J. J., Bagwell, J. G., Weinstein, A. M., Sauer, B. W., and Pruitt, J. R. (1976). An evaluation of bone growth into porous high density polyethylene. *J. Biomed. Mater. Res.* **10**: 311–323.

Klein, C. P. A. T., Driessen, A.A., and de Groot, K. (1983a). Biodegredation of calciumphosphate ceramics – ultrastructural geometry and dissolubility of different calcium phosphate ceramics. in Proceedings of the 1st International Symposium on Biomaterials in Otology, 84–92.

Klein, C. P. A. T., Driessen, A.A., de Groot, K., and van den Hooff, A. (1983b). Biodegration behaviour of various calcium phosphate materials in bone tissue. *J. Biomed. Mater. Res.* **17**: 769–784.

Klein, C. P. A. T., Wolke, J. G. C., de Blieck-Hogervorst, J. M. A., and DeGroot, K. (1994). Features of calcium phosphate coatings: an in-vitro study. *J. Biomed. Mater. Res.* **28**: 961–967.

Klein, C. P. A. T., Wolke, J. G. C., de Blieck-Hogervorst, J. M. A., and DeGroot, K. (1994). Features of calcium phosphate coatings: An in-vivo study. *J. Biomed. Mater. Res.* **28**: 909–917.

Kühne, J.H., Bartl, R., Frish, B., Hanmer, C., Jansson, V., Zimmer, M. (1994). Bone formation in coralline hydroxyapatite: Effects of pore size studied in rabbits. *Acta Orthop. Scand.* **65** (3): 246–252.

Kokubo, T. (1993). A/W glass-ceramics: Processing and properties. in *An Introduction to Bioceramics*, L. L. Hench and J. Wilson, eds. World Scientific, Singapore, pp. 75–88.

LeGeros, R. Z. (1965). Effect of carbonate on the lattice parameters of apatite. *Nature* **206**: 403.

LeGeros, R. Z. (1988). Calcium phosphate materials in restorative dentistry: A review. *Adv. Dent Res.* **2** (1): 164.

LeGeros, R. Z., Daculsi, G., Orly, I., Gregoire, M., Heughebeart, M., Gineste, M., and Kijkowska, R. (1992). Formation of carbonate apatite on calcium phosphate materials: Dissolution/precipitation processes. in *Bone Bonding Biomaterials*. Eds. Ducheynes, P., Kukubo, T., and van Blitterswijk, C. A.. Reed Healthcare Communications, Leiderdorp, The Netherlands, pp. 78088.

Lehr, J. R, Brown, E. T., Frazier, A.W., Smith, J. P., and Thrasher, R. D. (1967). *Crystallographic Properties of Fertilizer Compounds*. Tennessee Valley Authority Chemical Engineering Bulletin, 6.

Lacefield, W. R. (1993). Hydroxylapatite coatings. in *An Introduction to Bioceramics*, L. L. Hench and J. Wilson, eds. World Scientific, Singapore, pp. 223–238.

Le Geros, R. Z. (1988). Calcium phosphate materials in restorative densitry: a review. *Adv. Dent. Res.* **2**: 164–180.

Le Geros, R. Z., and Le Geros, J. P. (1993). Dense hydroxyapatite. in *An Introduction to Bioceramics*, L. L. Hench and J. Wilson, eds. World Scientific, Singapore, pp. 139–180.

Maxian, S. H., Zawaddsky, J. P., and Dunn, M. G. (1994). Effect of calcium phosphate coating resorption and surgical fit on the bone/implant interface. *J. Biomed. Mater. Res.* **28**: 1311–1319.

Maxian, S.H., Zawaddsky, J.P., and Dunn, M.G. (1993). Mechanical and histological evaluation of amorphous calcium phosphate and poorly crystalised hydroxyapatite coatings on titanium implants. *J. Biomed. Mater. Res.* **27**: 717–728.

McDowell, H., Gregory, T. M., and Brown, W. E. (1977). Solubility of $Ca_5(PO_4)_3OH$ in the system $Ca(OH)_2$ - H_3PO_4 - H_2O at 5, 15, 25 and 37°C. *J. Res. Natl. Bureau Standards* **81A**: 273.

Monma, H., Ueno, S., and Kanazawa, T. T. (1981). Properties of hydroxyapatite prepared by the hydrolysis of tricalcium phosphate. *J. Chem Tech. Biotechnol*. **31**: 15.

Mooney, R. W., and Aia, M. A. (1961). Alkaline earth phosphates. *Chem. Rev.* **61**: 433.

Morancho, R., Ghommidh, J., and Buttazoni, B.G. (1981). Constant, thin films of several calcium phosphates obtained by chemical spray of aque ous calcium hydrogen phosphate solution: A route to hydroxyapatite films. *Proceedings of the 8th International Conference on Chemical Vapour Deposition*. Electrochemical Society, New York.

Moroni, A., Caja, V. J., Egger, E. L., Trinchese, L., and Chao, E. Y. S. (1994). Histomorphometry of hydroxyapatite coated and uncoated porous titanium bone implants. *Biomaterials* **15** (11):926–930.

McKinney, Jr., R. V., and Lemons, J. (1985). *The Dental Implant*. PSG Publ., Littleton, MA.

McMillan, P. W. (1979). *Glass-Ceramics*. Academic Press, New York.

Miller, J. A., Talton, J. D., and Bhatia, S. (1996). in *Clinical Performance of Skeletal Prostheses*, L. L. Hench and J. Wilson, eds. Chapman and Hall, London, pp. 41–56.

Nagai, H., and Nishimura, Y. (1985). Hydroxyapatite ceramic material and process for the preparation thereof. US Patent no. 4,548,59.

Newsley, H. (1963). Crystallographic and morphological study of carbonate-apatite, M.h.f. *Chem.* **95**: 270.

Nakamura, T., Yamumuro, T., Higashi, S., Kokubo, T., and Itoo, S. (1985). A new glass-ceramic for bone replacement: Evaluation of its bonding to bone tissue. *J. Biomed. Mater. Res.* **19**: 685.

Oonishi, H., Tsuji, E., Ishimaru, H., Yamamoto, M., and Delecrin, J. (1990). Clinical sginificance of chemical bonds and bone in orthopaedic surgery. in *Bioceramics 2: Proceedings of the 2nd International Symposium on Ceramics in Medicine*. Ed, Heimke, G. German Ceramic Society, Cologne.

Onoda, G., and Hench, L. L. (1978). *Ceramic Processing before Firing*. Wiley, New York.

PDF card 9-432. ICDD, Newton Square, PA.

Peelen, J. G. J., Rejda, B. V., and DeGroot, K. (1978). Preparation and properties of sintered hydroxylapatite. *Ceramurgica International* **4** (2): 71.

Perloff, A., and Posner, A. S. (1956). Preparation of pure hydroxyapatite crystals. *Science* **124**: 583.

Phillips, R. W. (1991). *Skinners Science of Dental Materials*, 9th ed., Ralph W. Phillips, ed. Saunders, Philadelphia.

Quinn, J. H., and Kent, J .N. (1984). Alveolar ridge maintenance with solid non-porous hydroxylapatite root implants. *Oral Surg.* **58**: 511–516.

Rao, R. W., and Boehm, J. (1974). A study of sintered apatites. *J. Dent Res.* 1351.

Rahn, B. A., Neff, J., Leutenegger, A., Mathys, R., and Perren, S. M. (1986). Integration of synthetic apatite of various pore size and density in bone. in *Biological and Biomechanical Performance of Biomaterials*. Eds, Christel, P., Meunier, A., and Lee, A. J. C. Elsevier Science Publishers, Amsterdam.

Rootare, H., and Craig, R. G. (1978). Characterisation of hydroxyapatite powders and compacts at room temperature after sintering at 1200°C. *J. Oral Rehab.* **5**: 293.

Roy, D. M. (1971). Crystal growth of hydroxyapatite. *Mater. Res. Bull.* **6**: 1337.

Reck, R., Storkel, S., and Meyer, A. (1988). Bioactive glass-ceramics in middle ear surgery: an 8-year review. in *Bioceramics: Materials Characteristics versus In-Vivo Behavior*, P. Ducheyne and J. Lemons, eds. *Ann. N. Y. Acad. Sci.* **523**: 100.

Reed, J. S. (1988). *Introduction to Ceramic Processing*. Wiley, New York.

Ritter, J. E., Jr., Greenspan, D. C., Palmer, R. A., and Hench, L. L. (1979). Use of fracture mechanics theory in lifetime predictions for alumina and bioglass-coated alumina. *J. Biomed. Mater. Res.* **13**: 251–263.

Roy, D. M., and Linnehan, S. K. (1974). Hydroxyapatite formed from coral skeletal carbonate by hydrothermal exchange. *Nature* **247**: 220–222.

Schleede, A., Schmidt, W., and Kindt, H. (1932). Zu kenntnisder calciumphosphate und apatite. *Z. Elektrochem.* **38**: 633.

Skinner, H. C. W. (1973). Phase relations in the $CaO-P_2O_5-H_2O$ system from 300 to 600°C at 2kb H_2O pressure. *J. Am. Sci.* **273**: 545.

Smiler, D. G., and Holmes, R. E. (1987). Sinus life procedure using prous hydroxyapaitte: A preliminary clinical report. *J. Oral. Implantology* **13**: 17–32.

Soballe, K., Hansen, E. S., Brockstedt-Rasmussen, H. B., and Bunger, C. (1993). Hydroxyapatite coating converts fibrous tissue to bone around loaded implants. *J. Bone Jt Surgery* **75B**: 270–278.

Stephenson, P. K., Freeman, M. A. R. F., Revell, P. A., German, J., Tuke, M., Pirie, C. J. (1991). The effect of hydroxyapatite coating on ingrowth of bone into cavities in an implant. *J. Arthroplasty* **6** (1): 51–58.

Schors, E. C., and Holmes, R. E. (1993). Porous hydroxyapatite. in *An Introduction to Bioceramics*, L. L. Hench and J. Wilson, eds. World Scientific, Singapore, pp. 181–198.

Stanley, H. R., Clark, A. E., and Hench, L. L. (1996). Alveolar ridge maintenance implants. in *Clinical Performance of Skeletal Prostheses*. Chapman and Hall, London, pp. 237–254.

Tanner, K.E., Downes, R. N., and Bonfield, W. (1994). Clinical applications of hydroxyapatite reinforced materials. *Brit. Ceram. Trans.* **4** (93): 104–107.

van Blitterswijk, C. A., Hessling, S. C., Grote, J. J., Korerte, H. K., and DeGroot, K. (1990). The biocompatibility of hydroxyapatite ceramic: A study of retrieved human middle ear implants. *J. Biomed. Mater. Res.* **24:** 433–453

Wilson, J., and Low, S. B. (1992). Bioactive ceramics for periodontal treatment: Comparative study in Patus monkey. *J. App. Biomat.* **2:** 123–129.

Wolke, J. G. C., de Blieck-Hogervorst, J. M. A., Dhert, W. J. A., Klein, C. P. A. T., and DeGroot, K. (1992). Studies on thermal spraying of apatite bioceramics. *J. Thermal Spray Technology* **1:** 75–82.

White, E., and Schors, E. C. (1986). Biomaterials aspects of interpore-200 porous hydroxyapatite. *Dent. Clin. North Am.* **30:** 49–67.

Wilson, J. (1994). *Clinical applications of bioglass implants.* in *Bioceramics-7*, O. H. Andersson, ed. Butterworth–Heinemann, Oxford, England.

Wilson, J., Pigott, G. H., Schoen, F. J., and Hench, L. L. (1981). Toxicology and biocompatibility of bioglass. *J. Biomed. Mater. Res.* **15:** 805.

Yamamuro, T., Hench, L. L., Wilson, J. (1990). *Handbook on Bioactive Ceramics,* Vol. I: *Bioactive Glasses and Glass-Ceramics,* Vol. II: *Calcium-Phosphate Ceramics.* CRC Press, Boca Raton, FL.

Young, R. A., and Elliot, J. C. (1966). Atomic scale bases for several properties of apatites. *Arch. Oral Biol.* **11:** 699.

Young, R. A., and Holcomb, D. W. (1982). Variability of hydroxyapatite preparations. *Calcif. Tiss Int.* **34:** S17.

2.11 PYROLYTIC CARBON FOR LONG-TERM MEDICAL IMPLANTS

Robert B. More, Axel D. Haubold, and Jack C. Bokros

INTRODUCTION

Carbon materials are ubiquitous and of great interest because the majority of substances that make up living organisms are carbon compounds. Although many engineering materials and biomaterials are based on carbon or contain carbon in some form, elemental carbon itself is also an important and very successful biomaterial. Furthermore, there exists enough diversity in structure and properties for elemental carbons to be considered as a unique class of materials beyond the traditional molecular carbon focus of organic chemistry, polymer chemistry, and biochemistry. Through a serendipitous interaction between researchers during the late 1960s the outstanding blood compatibility of a special form of elemental pyrolytic carbon deposited at high temperature in a fluidized bed was discovered. The material was found to have not only remarkable blood compatibility but also the structural properties needed for long-term use in artificial heart valves (LaGrange *et al.,* 1969). The blood compatibility of pyrolytic carbon was recognized empirically using the Gott vena cava ring test. This test involved implanting a small tube made of a candidate material in a canine vena cava and observing the development of thrombosis within the tube in time. Prior to pyrolytic carbon, only surfaces coated with graphite, benzylalkonium chloride, and heparin would resist thrombus formation when exposed to blood for long periods. The incorporation of pyrolytic carbon in mechanical heart valve implants was declared "an exceptional event" (Sadeghi, 1987) because it

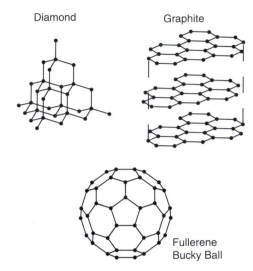

FIG. 1. Allotropic crystalline forms of carbon: diamond, graphite, and fullerene.

added the durability and stability needed for heart valve prostheses to endure for a patient's lifetime. The objective of this chapter is to present the elemental pyrolytic carbon materials currently is used in the fabrication of medical devices and to describe their manufacture, characterization, and properties.

ELEMENTAL CARBON

Elemental carbon is found in nature as two crystalline allotrophic forms: graphite and diamond. Elemental carbon also occurs as a spectrum of imperfect, turbostratic crystalline forms that range in degree of crystallinity from amorphous to the perfectly crystalline allotropes. Recently a third crystalline form of elemental carbon, the fullerene structure, has been discovered. The crystalline polymorphs of elemental carbon are shown in Fig. 1.

The properties of the elemental carbon crystalline forms vary widely according to their structure. Diamond with its tetrahedral sp^3 covalent bonding is one of the hardest materials known. In the diamond crystal structure, covalent bonds of length 1.54 Å connect each carbon atom with its four nearest neighbors. This tetrahedral symmetry repeats in three dimensions throughout the crystal (Pauling, 1964). In effect, the crystal is a giant isotropic covalently bonded molecule; therefore, diamond is very hard.

Graphite with its anisotropic layered in-plane hexagonal covalent bonding and interplane van der Waals bonding structure is a soft material. Within each planar layer, each carbon atom forms two single bonds and one double bond with its three nearest neighbors. This bonding repeats in-plane to form a giant molecular (graphene) sheet. The in-plane atomic bond length is 1.42 Å, which is a resonant intermediate (Pauling, 1964) between the single-bond length of 1.54 Å and the double-bond length of 1.33 Å. The planer layers are held together by relatively weak van der Waals bonding at a distance of 3.4 Å, which is more than twice the 1.42-Å bond length (Pauling, 1964). Graphite has low hardness and a lubricating property because the giant molecular sheets can readily slip past one another against the van der Waals bonding. Nevertheless, although large-crystallite-size natural graphite is used as a lubricant, some artificially produced graphites can be very abrasive if the crystallite sizes are small and randomly oriented.

Fullerenes have yet to be produced in bulk, but their properties on a microscale are entirely different from those of their crystalline counterparts. Fullerenes and nanotubes consist of a graphene layer that is rolled up or folded (Sattler, 1995) to form a tube or ball. These large molecules, C_{60} and C_{70} fullerenes and (C_{60+18j}) nanotubes, are often mentioned in the literature (Sattler, 1995) along with more complex multilayer "onion skin" structures.

There exist many possible forms of elemental carbon that are intermediate in structure and properties between those of the allotropes diamond and graphite. Such "turbostratic" carbons occur as a spectrum of amorphous through mixed amorphous, graphite-like and diamond-like to the perfectly crystalline allotropes (Bokros, 1969). Because of the dependence of properties upon structure, there can be considerable variability in properties for the turbostratic carbons. Glassy carbons and pyrolytic carbons, for example, are two turbostratic carbons with considerable differences in structure and properties. Consequently, it is not surprising that carbon materials are often misunderstood through oversimplification. Properties found in one type of carbon structure can be totally different in another type of structure. Therefore it is very important to specify the exact nature and structure when discussing carbon.

PYROLYTIC CARBON (PyC)

The biomaterial known as pyrolytic carbon is not found in nature: it is manmade. The successful pyrolytic carbon biomaterial was developed at General Atomic during the late 1960s using a fluidized-bed reactor (Bokros, 1969). In the original terminology, this material was considered a low-temperature isotropic carbon (LTI carbon). Since the initial clinical implant of a pyrolytic carbon component in the DeBakey–Surgitool mechanical valve in 1968, 95% of the mechanical heart valves implanted worldwide have at least one structural component made of pyrolytic carbon. On an annual basis this translates into approximately 500,000 components (Haubold, 1994). Pyrolytic carbon components have been used in more than 25 different prosthetic heart valve designs since the late 1960s and have accumulated a clinical experience on the order of 16 million patient-years. Clearly, pyrolytic carbon is one of the most successful, critical biomaterials both in function and application. Among the materials available for mechanical heart valve prostheses, pyrolytic carbon has the best combination of blood compatibility, physical and mechanical properties, and durability. However, the blood compatibility of pyrolytic carbon in heart-valve applications is not perfect; chronic anticoagulant therapy is needed for patients with mechanical heart valves. Whether the need for anticoagulant therapy arises from the biocompatible properties of the material itself or from the particular hydrodynamic interaction of a given device and the blood remains to be resolved.

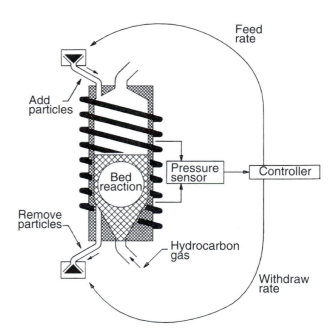

FIG. 2. Fluidized-bed reactor schematic.

The term "pyrolytic" is derived from "pyrolysis," which is thermal decomposition. Pyrolytic carbon is formed from the thermal decomposition of hydrocarbons such as propane, propylene, acetylene, and methane, in the absence of oxygen. Without oxygen the typical decomposition of the hydrocarbon to carbon dioxide and water cannot take place; instead a more complex cascade of decomposition products occurs that ultimately results in a "polymerization" of the individual carbon atoms into large macroatomic arrays.

Pyrolysis of the hydrocarbon is normally carried out in a fluidized-bed reactor such as the one shown in Fig. 2. A fluidized-bed reactor typically consists of a vertical tube furnace that may be induction or resistance heated to temperatures of 1000 to 2000°C (Bokros, 1969). Reactor diameters ranging from 2 cm to 25 cm have been used; however, the most common size used for medical devices has a diameter of about 10 cm. These high-temperature reactors are expensive to operate, and the reactor size limits the size of device components to be produced.

Small refractory ceramic particles are placed into the vertical tube furnace. When a gas is introduced into the bottom of the tube furnace, the gas causes the particle bed to expand: Interparticle spacing increases to allow for the flow of the gas. Particle mixing occurs and the bed of particles begins to "flow" like a fluid. Hence the term "fluidized bed." Depending upon the gas flow rate and volume, this expansion and mixing can be varied from a gentle bubbling bed to a violent spouting bed. An oxygen-free, inert gas such as nitrogen or helium is used to fluidize the bed, and the source hydrocarbon is added to the gas stream when needed.

At a sufficiently high temperature, pyrolysis or thermal decomposition of the hydrocarbon can take place. Pyrolysis products range from free carbon and gaseous hydrogen to a mixture of C_xH_y decomposition species. The pyrolysis reaction is complex and is affected by the gas flow rate, composition, temperature and bed surface area. Decomposition products, under the appropriate conditions, can form gas-phase nucleated droplets of carbon/hydrogen, which condense and deposit on the surfaces of the wall and bed particles within the reactor (Bokros, 1969). Indeed, the fluidized-bed process was originally developed to coat small (200–500 micrometer) diameter spherical particles of uranium/thorium carbide or oxide with pyrolytic carbon. These coated particles were used as the fuel in the high temperature gas-cooled nuclear reactor (Bokros, 1969).

Pyrolytic carbon coatings produced in vertical-tube reactors can have a variety of structures such as laminar or isotropic, granular, or columnar (Bokros, 1969). The structure of the coating is controlled by the gas flow rate (residence time in the bed), hydrocarbon species, temperature and bed surface area. For example, an inadequately fluidized or static bed will produce a highly anisotropic, laminar pyrolytic carbon (Bokros, 1969).

Control of the first three parameters (gas flow rate, hydrocarbon species, and temperature) is relatively easy. However, until recently, it was not possible to measure the bed surface area while the reactions were taking place. As carbon deposits on the particles in the fluidized bed, the diameter of the particles increases. Hence the surface area of the bed changes, which in turn influences the subsequent rate of carbon deposition. As surface area increases, the coating rate decreases since a larger surface area now has to be coated with the same amount of carbon available. Thus the process is not in equilibrium. The static-bed process was adequate to coat nuclear fuel particles without attempting to control the bed surface area, because such thin coatings (25 to 50 μm thick) did not appreciably affect the bed surface area.

It was later found that larger objects could be suspended within the fluidized bed of small ceramic particles and also become uniformly coated with carbon. This finding led to the demand for thicker, structural coatings, an order of magnitude thicker (250 to 500 μm). Bed surface area control and stabilization became an important factor (Akins and Bokros, 1974) in achieving the goal of thicker, structural coatings. In particular, with the discovery of the blood-compatible properties of pyrolytic carbon (LaGrange *et al.*, 1969), thicker structural coatings with consistent and uniform mechanical properties were needed to realize the application to mechanical heart-valve components. Quasi-steady-state conditions as needed to prolong the coating reaction were achieved empirically by removing coated particles and adding uncoated particles to the bed while the pyrolysis reaction was taking place (Akins and Bokros, 1974). However, the rates of particle addition and removal were based upon little more than good guesses.

Three of the four parameters that control carbon deposition could be accurately measured and controlled, but a method to measure and control bed surface area was lacking. Thus, the quasi-steady-state process was more of an art than a science. If too many coated particles were removed, the bed became too small to support the larger components within it and the bed collapsed. If too few particles were removed, the rate of deposition decreased, and the desired amount of coating was not achieved in the anticipated time. Furthermore, there were

considerable variations in the mechanical properties of the coating from batch to batch. It was found that in order to consistently achieve the hardness needed for wear resistance in prosthetic heart valve applications, it was necessary to add a small amount of β-silicon carbide to the carbon coating. The dispersed silicon carbide particles within the pyrolytic carbon matrix added sufficient hardness to compensate for potential variations in the properties of the pyrolytic carbon matrix. The β-silicon carbide was obtained from the pyrolysis of methyltrichlorosilane, CH_3SiCl_3. For each mole of silicon carbide produced, the pyrolysis of methyltrichlorosilane also produces 3 moles of hydrogen chloride gas. Handling and neutralization of this corrosive gas added substantial complexity and cost to an already complex process. Nevertheless, this process allowed consistency for the successful production of several million components for use in mechanical heart valves.

A process has been developed and patented that allows precise measurement and control of the bed surface area. A description of this process is given in the patent literature and elsewhere (Emken *et al.*, 1993, 1994; Ely *et al.*, 1998). With precise control of the bed surface area it is no longer necessary to include the silicon carbide. Elimination of the silicon carbide has produced a stronger, tougher, and more deformable pure pyrolytic carbon. Historically, pure carbon was the original objective of the development program because of the potential for superior biocompatibility (LaGrange *et al.*, 1969). Furthermore, the enhanced mechanical and physical properties of the pure pyrolytic carbon now possible with the improved process control allows prosthesis design improvements in the hemodynamic contribution to thromboresistance (Ely *et al.*, 1998).

Structure of Pyrolytic Carbons

X-ray diffraction patterns of the biomedical-grade fluidized-bed pyrolytic carbons are broad and diffuse because of the small crystallite size and imperfections. In silicon-alloyed pyrolytic carbon, a diffraction pattern characteristic of the β form of silicon carbide also appears in the diffraction pattern along with the carbon bands. The carbon diffraction pattern indicates a turbostatic structure (Kaae and Wall, 1996) in which there is order within carbon layer planes, as in graphite; but, unlike graphite, there is no order between planes. This type of turbostatic structure is shown in Fig. 3 compared to that of graphite. In the disordered crystalline structure, there may be lattice vacancies and the layer planes are curved or kinked. The ability of the graphite layer planes to slip is inhibited, which greatly increases the strength and hardness of the pyrolytic carbon relative to that of graphite. From the Bragg equation, the pyrolytic carbon layer spacing is reported to be 3.48 Å, which is larger than the 3.35 Å graphite layer spacing (Kaae and Wall, 1996). The increase in layer spacing relative to graphite is due to both the layer distortion and the small crystallite size, and is common feature for turbostatic carbons. From the Scherrer equation the crystallite size is typically 25–40 Å (Kaae and Wall, 1996).

During the coating reaction, gas-phase nucleated droplets of carbon/hydrogen form that condense and deposit on the surfaces of the reactor wall and bed particles within the reactor.

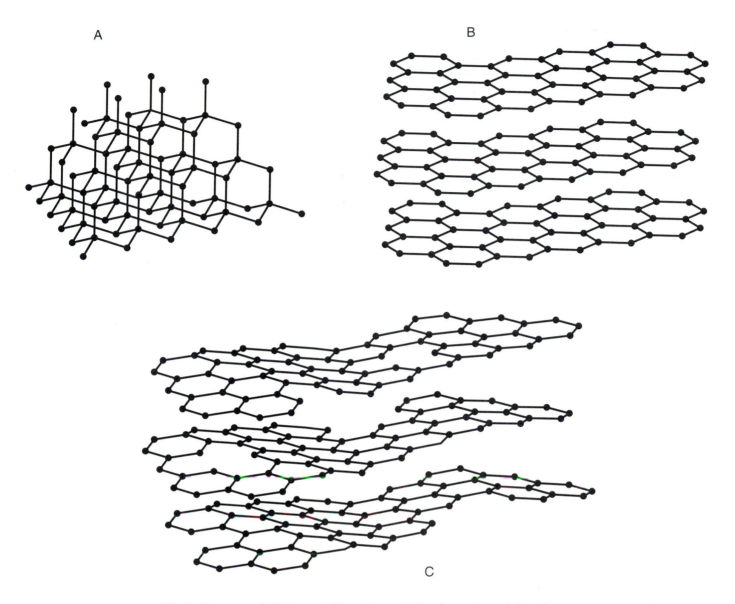

FIG. 3. Structures of (A) diamond, (B) graphite, and (C) turbostratic pyrolytic carbon.

These droplets aggregate, grow, and form the coating. When viewed with high-resolution transmission electron microscopy, a multitude of near-spherical polycrystalline growth features are evident as shown in Fig. 4 (Kaae and Wall, 1996). These growth features are considered to be the basic building blocks of the material, and the shape and size are related to the deposition mechanism. In the silicon-alloyed carbon small silicon carbide particles are present within the growth features as shown in Fig. 5. Based on a crystallite size of 33 Å, each growth feature contains about 3×10^9 crystallites. Although the material is quasi-crystalline on a fine level, the crystallites are very small and randomly oriented in the fluidized bed pyrolytic carbons so that overall the material exhibits isotropic behavior.

Glassy carbon, also known as vitreous carbon or polymeric carbon, is another turbostratic carbon form that has been proposed for use in long-term implants. However, its strength is low and the wear resistance is poor. Glassy carbons are quasi crystalline in structure and are named 'glassy' because the fracture surfaces closely resemble those of glass (Haubold *et al.*, 1981).

Vapor-deposited carbons are also used in heart-valve applications. Typically, the coatings are thin (< 1 μm) and may be applied to a variety of materials in order to confer the biochemical characteristics of turbostratic carbon. Some examples are vapor-deposited carbon coatings on heart-valve sewing cuffs and metallic orifice components (Haubold *et al.*, 1981).

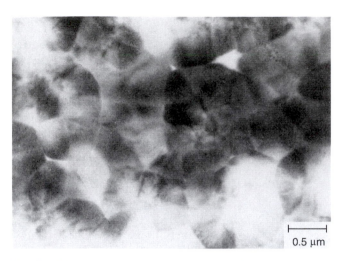

FIG. 4. Electron micrograph of pure pyrolytic carbon microstructure showing near-spherical polycrystalline growth features formed during deposition (Kaae and Wall, 1996).

TABLE 1 Mechanical Properties of Biomedical Carbons

Property	Pure PyC	Typical Si-alloyed PyC	Typical glassy carbon
Flexural strength (MPa)	493.7 ± 12	407.7 ± 14.1	175
Young's modulus (GPa)	29.4 ± 0.4	30.5 ± 0.65	21
Strain-to-failure (%)	1.58 ± 0.03	1.28 ± 0.03	
Fracture toughness (MPa $\sqrt{m}$)	1.68 ± 0.05	1.17 ± 0.17	0.5–0.7
Hardness (DPH, 500 g load)	235.9 ± 3.3	287 ± 10	150
Density (g/cm^3)	1.93 ± 0.01	2.12 ± 0.01	<1.54
CTE (10–6 cm/cm °C)	6.5	6.1	
Silicon content (%)	0	6.58 ± 0.32	0
Wear resistance	Excellent	Excellent	Poor

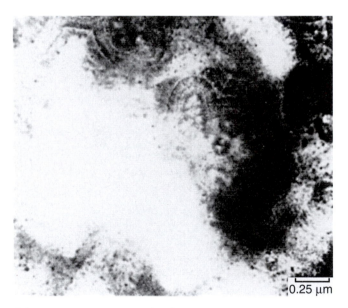

FIG. 5. Electron micrograph of silicon-alloyed pyrolytic carbon microstructure showing near-spherical polycrystalline growth features formed during deposition (Kaae and Wall, 1996). Small silicon carbide particles are shown in concentric rings in the growth features.

Mechanical Properties

Mechanical properties of pure pyrolytic carbon, silicon-alloyed pyrolytic carbons and glassy carbon are given in Table 1. Pyrolytic carbon flexural strength is high enough to provide the necessary structural stability for a variety of implant applications and the density is low enough to allow for components to move easily under the applied forces of circulating blood. With respect to orthopedic applications, Young's modulus is in the range reported for bone (Reilly and Burstein, 1974; Reilly et al., 1974), which allows for

compliance matching. Relative to metals and polymers, the pyrolytic carbon strain-to-failure is low; it is a nearly ideal linear elastic material and requires consideration of brittle material principles in component design. Strength levels vary with the effective stressed volume or stressed area as predicted by classical Weibull statistics (De Salvo, 1970). The flexural strengths cited in Table 1 are for specimens tested in four-point bending, third-point loading (More et al., 1993) with effective stressed volumes of 1.93 mm^3. The pyrolytic carbon material Weibull modulus is approximately 10 (More et al., 1993).

Fracture toughness levels reflect the brittle nature of the material, but the fluidized-bed isotropic pyrolytic carbons are remarkably fatigue resistant. Recent fatigue studies indicate the existence of a fatigue threshold that is very nearly the single-cycle fracture strength (Gilpin et al., 1993; Ma and Sines, 1996, 1999, 2000). Fatigue-crack propagation studies indicate very high Paris-Law fatigue exponents, on the order of 80, and display clear evidence of a fatigue-crack propagation threshold (Ritchie et al., 1990; Beavan et al., 1993; Cao, 1996).

Crystallographic mechanisms for fatigue-crack initiation, as occur in metals, do not exist in the pyrolytic carbons (Haubold et al., 1981). In properly designed and manufactured components, and in the absence of externally induced damage, fatigue does not occur in pyrolytic-carbon mechanical heart-valve components. In the 30 years of clinical experience, there have been no clear instances of fatigue failure. Few pyrolytic carbon component fractures have occurred, less than 60 out of more than 4 million implanted components (Haubold, 1994), and most are attributable to induced damage from handling or cavitation (Kelpetko et al., 1989; Kafesjian et al., 1994; Richard and Cao, 1996).

Wear resistance of the fluidized-bed pyrolytic carbons is excellent. The strength, stability, and durability of pyrolytic carbon are responsible for the extension of mechanical-valve lifetimes from less than 20 years to more than the recipient's expected lifetime (Wieting, 1996; More and Silver, 1990; Schoen, 1983; Schoen et al., 1982).

Pyrolytic carbon in heart-valve prostheses is often used in contact with metals, either as a carbon disk in a metallic valve orifice or as a carbon orifice stiffened with a metallic ring. Carbon falls with the noble metals in the galvanic series (Haubold *et al.*, 1981), the sequence being silver, titanium, graphite, gold, and platinum. Carbon can accelerate corrosion when coupled to less noble metals *in vivo*. However, testing using mixed potential corrosion theory and potentiostatic polarization has determined that no detrimental effects occur when carbon is coupled with titanium or cobalt–chrome alloys (Griffin *et al.*, 1983; Thompson *et al.*, 1979). Carbon couples with stainless steel alloys are not recommended.

STEPS IN THE FABRICATION OF PYROLYTIC CARBON COMPONENTS

To convert a gaseous hydrocarbon into a shiny, polished black component for use in the biological environment is not a trivial undertaking. Furthermore, because of the critical importance of long-term implants to a recipient's health, all manufacturing operations are performed to stringent levels of quality assurance under the auspices of U.S. Food and Drug Administration Good Manufacturing Practices and International Standards Organization ISO-9000 regulations. As in the case of fabrication of metallic implants, numerous steps are involved. Pyrolytic carbon is not machined from a block of material, as is the case with most metallic implants, nor is it injection or reactive molded, as are many polymeric devices. An overview of the processing steps leading to a finished pyrolytic carbon coated component for use in a medical device is shown in Fig. 6 and is further described in the following sections.

Substrate Material

Since pyrolytic carbon is a coating, it must be deposited on an appropriately shaped, preformed substrate (preform). Because the pyrolysis process takes place at high temperatures, the choice of substrates is severely limited. Only a few of the refractory materials such as tantalum or molybdenum/rhenium alloys and graphite can withstand the conditions at which the pyrolytic carbon coating is produced. Some refractory metals have been used in heart-valve components; for example, Mo/Re preforms were coated to make the struts for the Beall-Surgitool mitral valve. It is important for the thermal expansion characteristics of the substrate to closely match those of the applied coating. Otherwise, upon cooling of the coated part to room temperature the coating will be highly stressed and can spontaneously crack. For contemporary heart-valve applications, fine-grained isotropic graphite is the most commonly used substrate. This substrate graphite can be doped with tungsten in order to provide radiopacity for X-ray visualizations of the implants. The graphite substrate does not impart structural strength. Rather, it provides a dimensionally stable platform for the pyrolytic carbon coating both at the reaction temperature and at room temperature.

Preform

Once the appropriate substrate material has been selected and prior to making a preform, it must be inspected to ensure that the material meets the desired specifications. Typically, the strength and density of the starting material are measured. Thermal expansion is ordinarily validated and monitored through process control. The preform, which is an undersized replica of the finished component, is normally machined using conventional machining methods. Because the fine-grained isotropic graphite is very abrasive, standard machine tools have given way to diamond-plated or single-point diamond tools. In the case of heart valves, numerical control machining methods are often required to maintain critical component dimensional tolerances. After the preform is completed, it is inspected to ensure that its dimensions fall within the specified tolerances and that it contains no visible flaws or voids.

Coating

Generally numerous preforms are coated in one furnace run. A batch to be coated is made up of substrates from a single lot of preforms. Such batch processing by lot is required in order to maintain "forward and backward" traceability. In other words, ultimately it is necessary to know all of the components that were prepared using a specific material lot, given either the starting material lot number (forward) or given the specific component serial number (backward). The number of parts that can be coated in one furnace run is dictated by the

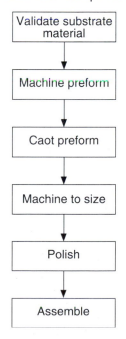

Steps in the fabrication of pyrolic carbon components

FIG. 6. Schematic of manufacturing processing steps.

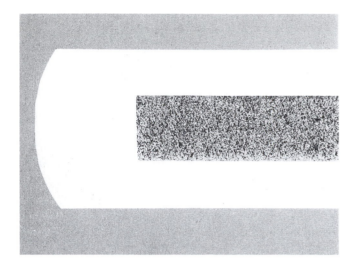

FIG. 7. Metallographic mount cross section of heat valve component. The light-colored pyrolytic carbon layer is coated over the interior, darker colored granular-appearing graphite substrate.

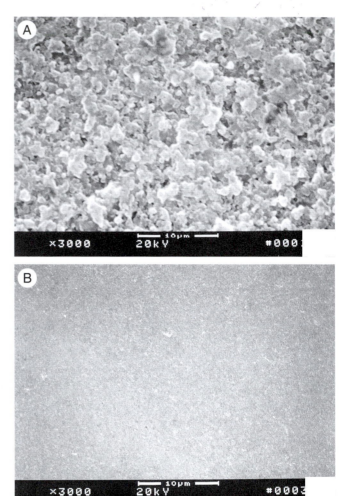

FIG. 8. Scanning electron microscope micrographs of (A) as-coated and (B) as-polished surfaces.

size of the furnace and the size and weight of the parts to be coated. The batch of substrates is placed within the fluidized bed in the vertical tube furnace and is coated to the desired thickness. Coating times are generally on the order of a few hours, but the entire cycle (heat-up, coating, and cool-down) may take as long as a full day.

A statistical sample from each coating lot is taken for analysis. At this point, typical measurements include coating thickness, microhardness and microstructure. The microhardness, and microstructure are determined from a metallographically prepared cross section of the coated component taken perpendicular to plane of deposition. Thus, this test is destructive. An example of a metallographically prepared cross section of a pyrolytic carbon component is shown in Fig. 7.

Machine to Size

The components used to manufacture medical devices have strict dimensional requirements. Because of the inability, until recently, to precisely measure and control bed size and indirectly coating thickness, the preforms were generally coated more thickly than necessary to ensure adequate pyrolytic carbon coating thickness on the finished part. The strict dimensional requirements were then achieved through precision grinding or other machining operations. Because pyrolytic carbon is very hard, conventional machine tools again cannot be used. Diamond-plated grinding wheels and other diamond tooling are required. The dimensions of final machined parts are again verified.

Polish

The surface of as-deposited, machined and polished components is shown in Fig. 8. It was found early on in experiments (LaGrange *et al.*, 1969; Haubold *et al.*, 1981; Sawyer *et al.*, 1975) that clean polished pyrolytic carbon surfaces

of tubes when placed within the vasculature of experimental animals accumulated minimal if any thrombus and certainly less than pyrolytic carbon tubes with the as-deposited surface. Consequently, the surfaces of pyrolytic carbon have historically been polished, either manually or mechanically, using fine diamond or aluminum oxide pastes and slurries. The surface finish achieved has roughness measured on the scale of nanometers. As can be seen from Table 2 (More and Haubold, 1996), the surfaces of polished pyrolytic carbon (30–50 nm) are an order of magnitude smoother than the as-deposited surfaces (300–500 nm).

Once the desired surface quality is achieved, components are again inspected. The final component inspection may include measurement of dimensions, X-ray inspection in two orientations to verify coating thickness, and visual inspection for surface quality and flaws. In many cases, automated inspection methods with computer-controlled coordinate measurement machines are used. X-ray inspection can be used to ensure that minimum coating thickness requirements are met. Two orthogonal views ensure that machining and grinding of

TABLE 2   Surface Finish (R_a, Average, and R_q, Root Mean Square) of Pyrolytic Carbon Heat Valve Components[a]

Specimen	R_a (nm)	R_q (nm)	Comments
Glass microscope slide	17.14	26.80	
On-X leaflet	33.95	42.12	Clinical
Sorin Bicarbon leaflet	40.12	50.63	Nonclinical
SJM leaflet	49.71	62.74	Clinical
CMI (SJM) leaflet	67.98	85.56	Nonclinical
Sorin Monoleaflet	99.59	128.10	Clinical
DeBakey-Surgitool ball	129.78	157.93	Nonclinical
As-coated slab	389.07	503.72	

[a]Components/prepared by: On-X/Medical Carbon Research Institute, Austin, TX; Sorin/Sorin Biomedica, Saluggia, Italy; SJM/Saint Jude Medical, Saint Paul, MN; CMI (SJM)/CarboMedics, Austin, TX; DeBakey-S/CarboMedics, San Diego (circa 1968). "Clinical" was from as-packaged valve; "nonclinical" lacks component traceability.

the coating was achieved uniformly and that the coating is symmetrical. The machining and grinding operation after coating is not without the risk of inducing cracks or flaws in the coating, which may subsequently affect the service life of the component. Such surface flaws are detected visually or with the aid of dye-penetrant techniques. Components may also be proof-tested to detect and eliminate components with subsurface flaws. With the advent of bed size control, which allows coating to exact final dimensions, the concerns about flaws introduced during the machining and grinding operation have been eliminated.

The polished and inspected components, thus prepared, are now ready for assembly into devices, or are packaged and sterilized in the case of stand-alone devices. Shown in Fig. 9 are the three pyrolytic carbon components for a bileaflet mechanical heart valve. The components were selected and matched for assembly using the data generated from the final dimensional inspection to achieve the dimensional requirements specified in the device design. In Fig. 10, the pyrolytic carbon components for a replacement metacarpophalangeal total joint prosthesis are shown.

Assembly

The multiple components of a mechanical heart must be assembled. The brittleness of pyrolytic carbon poses a significant assembly problem. Because the strain-to-failure is on the order of 1.28 to 1.58%, there is a limited range of deformation that can be applied in order to achieve a proper fit. Relative fit between the components defines the capture and the range of motion for components that move to actuate valve opening and closing. Furthermore, component obstructive bulk and tolerance gaps are critical to hemodynamic performance.

In designs that use a metallic orifice, the metallic components are typically deformed in order to insert the pyrolytic

FIG. 9.   Components for On-X bileaflet heart valve.

FIG. 10.   Replacement metacarpophalangeal total joint prosthesis components, Ascension Orthopedics, Austin TX.

carbon occluder disk. For the all-carbon bileaflet designs, the carbon orifice must be deflected in order to insert the leaflets. As the valve diameter decreases and as the section modulus of the orifice design increases, the orifice stiffness increases. The possibility of damage or fracture during assembly was a limiting factor in early orifice design. For this reason, the orifices in valve designs using silicon-alloyed pyrolytic carbon were simple cylindrical geometries, and the smallest sizes limited to the equivalent of a 19-mm-diameter tissue annulus. The simple cylindrical orifice designs are often reinforced with a metallic stiffening ring that is shrunk on after assembly. The stiffening ring ensures that physiological loading will not produce deflections that can inhibit valve action or result in leaflet escape.

The increased strain-to-failure of pure pyrolytic carbon, relative to the silicon-alloyed carbon, allows designs with more complex orifice section moduli. This allows designers to utilize hydrodynamically efficient shapes such as flared inlets and to incorporate external stiffening bands that eliminate the need for a metallic stiffening ring. The increased strain-to-failure of On-X carbon has been used to advantage in the On-X mechanical heart valve design (Ely *et al.*, 1998).

Cleaning and Surface Chemistry

Pyrolytic carbon surface chemistry is important because the manufacturing and cleaning operations to which a component is subjected can change and redefine the surface that is presented to the blood. Oxidation of carbon surfaces can produce surface contamination that detracts from blood compatibility (LaGrange *et al.*, 1969; Bokros *et al.*, 1969). Historically, the initial examinations of pyrolytic carbon biocompatibility assumed de facto that the surface needed to be treated with a thromboresistant agent such as heparin (Bokros *et al.*, 1969). It was found, however, that the non-heparin-coated surface was actually more blood compatible than the treated surface. Hence, the efforts toward surface coating with heparin were abandoned.

In general it is desired to minimize the surface oxygen and any other non-carbon surface contaminants. From X-ray photoelectron spectroscopy (XPS) analyses, a typical heart valve component surface has 76–86% C, 12–21% O, 0–2% Si, and 1–2% Al (More and Haubold, 1996; King *et al.*, 1981; Smith and Black, 1984). Polishing compounds tend to contain alumina and some alumina particles may become imbedded in the carbon surface. Other contaminants that may be introduced at low levels < 2% each are Na, B, Cl, S, Mg, Ca, Zn, and N. The XPS carbon 1s peak when scanned at high resolution can be deconvoluted to determine carbon oxidations states. The carbon 1s peak will typically consist primarily of hydrocarbon-like carbon (60–81%), ether alcohol/ester-like carbon (10–24%), ketone-like carbon (0–6%), and ester/acid-like carbon (1–12%) (More and Haubold, 1996). Each manufacturing, cleaning and sterilizing operation potentially redefines the surface. The effect of modified surface chemistry on blood compatibility is not well characterized, so this adds a level of uncontrolled variability when considering the blood compatibility of pyrolytic-carbon heart-valve materials from different manufacturing sources and different investigators. In general, the presence of oxygen and surface contaminants should be eliminated.

BIOCOMPATIBILITY OF PYROLYTIC CARBON

The suitability of a material for use in an implant is a complex issue. Biocompatibility testing is the focus of other chapters. In the case of pyrolytic carbon, its successful history interfacing with blood in mechanical heart valves attests to its suitability for this application. A note of caution, however, is in order. Until about a decade ago, the pyrolytic carbon used so successfully in mechanical heart valves was produced by a single manufacturer. The material, many applications in the biological environment, and the processes for producing the material were all patented. Since the expiration of the last of these patents in 1989, other sources for pyrolytic carbon have appeared that are copies of the original General Atomic material. When considering alternative carbon materials, it is important to recognize that the proper combination of physical, mechanical, and blood-compatible properties is required for the success of the implant application. Furthermore, because there are a number of different possible pyrolysis processes, it should be recognized that each can result in different microstructures with different properties. Just because a material is carbon, a turbostratic carbon, or a pyrolytic carbon does not qualify its use in a long-term human implant (Haubold and Ely, 1995). For example, pyrolytic carbons prepared by chemical vapor deposition processes, other than the fluidized-bed process, are known to exhibit anisotropy, nonhomogeneity, and considerable variability in mechanical properties (Agafonov *et al.*, 1999). Although these materials may exhibit biocompatibility, the potential for variability in structural stability and durability may lead to valve dysfunction.

The original General Atomic–type fluidized-bed pyrolytic carbons all demonstrate negligible reactions in the standard Tripartite and ISO 10993-1 type biocompatibility tests. Results from such tests are given in Table 3 (Ely *et al.*, 1998). Pure pyrolytic carbon is so non-reactive that it can serve as a negative control for these tests.

It is believed that pyrolytic carbon owes its demonstrated blood compatibility to its inertness and to its ability to quickly absorb proteins from blood without triggering a protein denaturing reaction. Ultimately, the blood compatibility is thought to be a result of the protein layer formed upon the carbon surface. Baier observed that pyrolytic carbon surfaces have a relatively high critical surface tension of 50 dyn/cm, which immediately drops to 28 to 30 dyn/cm following exposure to blood (Baier *et al.*, 1970). The quantity of sorbed protein was thought to be an important factor for blood compatibility. Lee and Kim (1974) quantified the amount of radiolabeled

TABLE 3   Biological Testing of Pure PyC

Test description	Protocol	Results
Klingman maximization	ISO/CD 10993–10	Grade 1; not significant
Rabbit pyrogen	ISO/ DIS 10993–11	Nonpyrogenic
Intracutaneous injection	ISO 10993–10	Negligible irritant
Systemic injection	ANSI/AAMI/ISO 10993–11	Negative—same as controls
Salmonella typhimurium reverse mutation assay	ISO 10993–3	Nonmutagenic
Physicochemical	USP XXIII, 1995	Exceeds standards
Hemolysis—rabbit blood	ISO 10993–4/NIH 77–1294	Nonhemolytic
Elution test (L929 mammalian cell culture)	ISO 10993–5, USP XXIII, 1995	Noncytotoxic

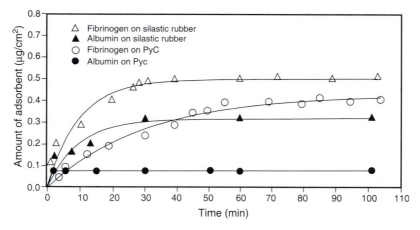

FIG. 11. Fibrinogen and albumin adsorption on pyrolytic carbon (PyC) and Silastic silicone rubber.

proteins sorbed from solutions of mixture proteins (albumin, fibrinogen, and gamma-globulin). While pyrolytic carbon does absorb albumin, it also absorbs considerable quantity of fibrinogen as shown in Fig. 11. As can be seen in Fig. 11, the amount of fibrinogen absorbed on pyrolytic carbon surfaces is far greater than the amount of albumin on these surfaces and is comparable to the amount of fibrinogen that sorbed on silicone rubber. The mode of albumin absorption, however, appears to be drastically different for these two materials. Albumin sorbs immediately on the pyrolytic carbon surfaces, whereas the buildup of fibrinogen is much slower. In the case of silicone rubber, both proteins sorb at a much slower rate. It appears that the mode of protein absorption is important and not the total amount sorbed.

Nyilas and Chiu (1978) studied the interaction of plasma proteins with foreign surfaces by measuring directly the heats of absorption of selected proteins onto such surfaces using microcalorimetric techniques. They found that the heats of absorption of fibrinogen, up to the completion of first monolayer coverage, are a factor of 8 smaller on pyrolytic carbon surfaces than on the known thrombogenic control (glass) surface as shown in Fig. 12. Furthermore, the measured net heats of absorption of gamma globulin on pyrolytic carbon were about 15 times smaller than those on glass. They concluded that low heats of absorption onto a foreign surface imply small interaction forces with no conformational changes of the proteins that might trigger the clotting cascade. It appears that a layer of continuously exchanging blood proteins in their unaltered state "masks" the pyrolytic carbon surfaces from appearing as a foreign body.

There is further evidence that the minimally altered sorbed protein layers on pyrolytic carbon condition blood compatibility. Salzman *et al.* (1977), for example, observed a significant difference in platelet reaction with pyrolytic carbon beads in packed columns prior to and after pretreatment with albumin. With no albumin preconditioning treatment, platelet retention by the columns was high, but the release of platelet constituents was low. However, with albumin pretreatment, platelet retention and the release of constituents was minimal.

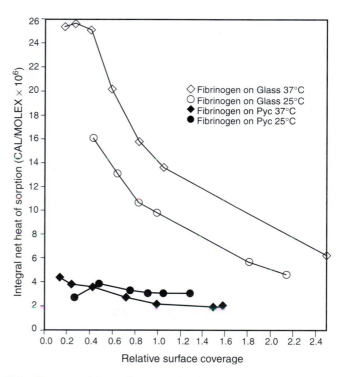

FIG. 12. Integral heat of sorption for fibrinogen on glass and fibrinogen on PyC at two different temperatures (Nyilas and Chiu, 1978).

The foregoing observations led to the view that pyrolytic carbon owes its demonstrated blood compatibility to its inertness and to its ability to quickly absorb proteins from blood without triggering a protein-denaturing reaction (Haubold *et al.*, 1981; Nyilas and Chiu, 1978). However, the assertion that pyrolytic carbon is an inert material and induces minimal conformational changes in adsorbed protein was reexamined by Feng and Andrade (1994). Using differential scanning calorimetry and a variety of proteins and buffers, they found that pyrolytic carbon surfaces denatured all of the proteins studied. They concluded that whether or not a surface denatures protein cannot be the sole criteria for blood compatibility. Their suggestion was that the specific proteins and the sequence in which they are denatured may be important. For example, it was suggested that pyrolytic carbon may first adsorb and denature albumin, which forms a layer that subsequently passivates the surface and inhibits thrombosis.

Chinn *et al.* (1994) reexamined the adsorption of albumin and fibrinogen on pyrolytic carbon surfaces and noted that relative large amounts of fibrinogen were adsorbed and speculated that the adsorbed fibrinogen was rapidly converted to a nonelutable form. If the elutable form is more reactive to platelets than the nonelutable form, then the nonelutable protein layer may contribute to the passivating effect.

Work on visualizing the carbon surface and platelet adhesion done by Goodman *et al.* (1995) using low-accelerating-voltage scanning electron microscopy, along with critical-point drying techniques, has discovered that the platelet spreading on pyrolytic carbon surfaces is more extensive than previously observed (Haubold *et al.*, 1981). However, platelet loading was in a static flow situation that does not model the physiological flow that a heart valve is subjected to. Hence, this approach cannot resolve kinetic effects on platelet adhesion. However, Okazaki, Tweden, and co-workers observed adherent platelets on valves following implantation in sheep that were not treated with anticoagulants (Okazaki *et al.*, 1997). There were no instances of valve thrombosis even though platelets were present on some of the valve surfaces. But, the relevance of this observation to clinical valve thromboses is not clear because human patients with mechanical heart valves undergo chronic anticoagulant therapy (Edmunds, 1987) and have a hemostatic system different from that of sheep.

A more contemporary version of the mechanism of pyrolytic carbon blood compatibility might be to reject the assumption that the surface is inert, as it is now thought by some that no material is totally inert in the body (Williaims, 1998), and to accept that the blood–material interaction is preceded by a complex, interdependent, and time-dependent series of interactions between the plasma proteins and the surface (Hanson, 1998) that is as yet poorly understood. To add to the confusion, it must also be recognized that much of the forementioned conjecture depends on the assumption that all of the carbon surfaces studied were in fact pure and comparable to one another.

CONCLUSION

Because the blood compatibility of pyrolytic carbon in mechanical heart valves is not perfect, anticoagulant therapy is required for mechanical heart valve patients. However, pyrolytic carbon has been the most successful material in heart valve applications because it offers excellent blood and tissue compatibility which, combined with the appropriate set of physical and mechanical properties and durability, allows for practical implant device design and manufacture. Improvements in biocompatibility are desired, of course, because when heart valves and other implants are used, a deadly or disabling disease is often treated by replacing it with a less pathological,

more manageable chronic condition. Ideally, an implant should not lead to a chronic condition.

It is important to recognize that the mechanism for the blood compatibility of pyrolytic carbons is not fully understood, nor is the interplay between the biomaterial itself, design-related hemodynamic stresses, and the ultimate biological reaction. The elucidation of the mechanism for blood and tissue compatibility of pyrolytic carbon remains a challenge.

It is also worth restating that the suitability of carbon materials from new sources for long-term implants is not assured simply because the material is carbon. Elemental carbon encompasses a broad spectrum of possible structures and mechanical properties. Each new candidate carbon material requires a specific assessment of biocompatibility based on its own merits and not by reference to the historically successful General Atomic–type pyrolytic carbons.

Bibliography

Agafonov, A., Kouznetsova, E., Kouznetsova, V., and Reif, T. (1999). TRI carbon strength and macroscopic isotropy of boron carbide alloyed pyrolytic carbon. *Artif. Organs.* 23(7): 80.

Akins, R. J., and Bokros, J. C. (1974). The deposition of pure and alloyed isotropic carbons and steady state fluidized beds. *Carbon* 12: 439–452.

Baier, R. E., Gott, V. L., and Feruse, A. (1970). Surface chemical evaluation of thromboresistant materials before and after venous implantation. *Trans. Am. Soc. Artif. Intern. Organs* 16: 50–57.

Beavan, L. A., James, D. W., and Kepner, J. L. (1993). Evaluation of fatigue in pyrolite carbon. in *Bioceramics*, Vol. 6, P. Ducheyne and D. Christiansen, eds. Butterworth–Heinemann, Oxford, pp. 205–210.

Bokros, J. C. (1969). Deposition, structure and properties of pyrolytic carbon. in *Chemistry and Physics of Carbon*, Vol. 5, P. L. Walker, ed. Marcel Dekker, New York, pp. 1–118.

Bokros, J. C., Gott, V. L., LaGrange, L. D., Fadall, A. M., Vos, K. D., and Ramos, M. D. (1969). Correlations between blood compatibility and heparin adsorptivity for an impermeable isotropic pyrolytic carbon. *J. Biomed. Mater. Res.* 3: 497–528.

Cao, H. (1996). Mechanical performance of pyrolytic carbon in prosthetic heart valve applications. *J. Heart Valve Dis.* 5(Suppl. I): S32–S49.

Chinn, J. A., Phillips, R. E., Lew, K. R., and Horbett, T. A. (1994). Fibrinogen and albumin adsorption to pyrolite carbon. *Trans. Soc. Biomater.* 17: 250.

De Salvo, G. (1970). *Theory and Structural Design Applications of Weibull Statistics*, Report WANL-TME-2688, Westinghouse Electric Corporation.

Dillard, J. G. (1995). X-ray photoelectron spectroscopy (XPS) and electron spectroscopy for chemical analysis (ESCA). in *Characterization of Composite Materials*, Vol. 1, H. Ishida and L. E. Fitzpatrick, eds. Butterworth–Heinemann, Boston, pp. 22.

Edmunds, L. H. (1987). Thrombotic and bleeding complications of prosthetic heart valves. *Ann. Thorac. Surg.* 44: 430–445.

Ely, J. L., Emken, M. R., Accuntius, J. A., Wilde, D. S., Haubold, A. D., More, R. B., and Bokros, J. C. (1998). Pure pyrolytic carbon: preparation and properties of a new material, on-X carbon for mechanical heart valve prostheses. *J. Heart Valve Dis.* 7: 626–632.

Emken, M. R., Bokros, J. C., Accuntius, J. A., and Wilde, D. S. (1993). Precise control of pyrolytic carbon coating. Presented at the 21st Biennial Conference on Carbon, Buffalo, New York, June 13–18, 1993, Extended Abstracts and Program Proceedings, pp. 531–532.

Emken, M. R., Bokros, J. C., Accuntius, J. A., and Wilde, D. S. (1994). U.S. Patent No. 5,284,676, Pyrolytic deposition in a fluidized bed, Feb. 8, 1994.

Feng, L., and Andrade, J. D. (1994). Protein adsorption on low-temperature isotropic carbon: I. Protein conformational change probed by differential scanning calorimetry. *J. Biomed. Mater. Res.* **28**: 735–743.

Gilpin, C. B., Haubold, A. D., and Ely, J. L. (1993). Fatigue crack growth and fracture of pyrolytic carbon composites. in *Bioceramics*, Vol. 6, P. Ducheyne and D. Christiansen, ed. Butterworth–Heinemann, Oxford, pp. 217–223.

Goodman, S. L., Tweden, K. S., and Albrecht, R. M. (1995). Three-dimensional morphology and platelet adhesion on pyrolytic carbon heart valve materials. *Cells Mater.* **5**(1): 15–30.

Griffin, C. D., Buchanan, R. A., and Lemons, J. E. (1983). In vitro electrochemical corrosion study of coupled surgical implant materials. *J. Biomed. Mater. Res.* **17**: 489–500.

Hanson, S. R. (1998). Blood–material interactions. in *Handbook of Biomaterial Properties*, J. Black and G. Hastings, eds. Chapman and Hall, London, pp. 545–555.

Haubold, A. D. (1994). On the durability of pyrolytic carbon in vivo. *Medi. Prog. Technol.* **20**: 201–208.

Haubold, A. D., and Ely, J. L. (1995). Carbons used in mechanical heart valves. *Transactions Society for Biomaterials, 21st Annual Meeting, San Francisco*, p. 275.

Haubold, A. D., Shim, H. S., and Bokros, J. C. (1981). Carbon in medical devices. in *Biocompatibility of Clinical Implant Materials*. David, P. Williams, ed. CRC Press, Boca Raton, Florida, pp. 3–42.

Kaae, J. L., and Wall, D. R. (1996). Microstructural characterization of pyrolytic carbon for heart valves. *Cells Mater.* **4**: 281–290.

Kafesjian, R., Howanec, M., Ward, G. D., Diep, L., Wagstaff, L., and Rhee, R. (1994). Cavitation damage of pyrolytic carbon in mechanical heart valves. *J. Heart Valve Dis.* **3**(Suppl. I): S2–S7.

Kelpetko, V., Moritz, A., Mlzoch, J., Schurawitzki, H., Domanig, E., and Wolner, E. (1989). Leaflet fracture in Edwards-Duromedics bileaflet valves. *J. Thorac. Cardiovasc. Surg.* **97**: 90–94.

King, R. N., Andrade, J. D., Haubold, A. D., and Shim, H. S. (1981). Surface analysis of silicon: alloyed and unalloyed LTI pyrolytic carbon. in *Photon, Electron and Ion Probes of Polymer Structure and Properties*, ACS Symposium Series 162, D. W. Dwight, T. J. Fabish, and H. R. Thomas, eds. American Chemical Society, Washington, D.C., pp. 383–404.

LaGrange, L. D., Gott, V. L., Bokros, J. C., and Ramos, M. D. (1969). Compatibility of carbon and blood. in *Artificial Heart Program Conference Proceedings*, R. J. Hegyeli, ed. U.S. Government Printing Office, Washington, D.C., pp. 47–58.

Lee, R. G., and Kim, S. W. (1974). Adsorption of proteins onto hydrophobic polymer surfaces: adsorption isotherms and kinetics. *J. Biomed Mater. Res.* **8**: 251.

Ma, L., and Sines, G. (1996). Fatigue of isotropic pyrolytic carbon used in mechanical heart valves. *J. Heart Valve Dis.* **5**(Suppl. I): S59–S64.

Ma, L., and Sines, G. (1999). Unalloyed pyrolytic carbon for implanted heart valves. *J. Heart Valve Dis.* **8**(5): 578–585.

Ma, L., and Sines, G. (2000). Fatigue behavior of pyrolytic carbon. *J. Biomed. Mater. Res.* **51**: 61–68.

More, R. B., and Haubold, A. D. (1996). Surface chemistry and surface roughness of clinical pyrocarbons. *Cells Mater.* **6**: 273–279.

More, R. B., and Silver, M. D. (1990). Pyrolytic carbon prosthetic heart valve occluder wear: in vivo vs. in vitro results for the Björk–Shiley prosthesis. *J. Appl. Biomater.* **1**: 267–278.

More, R. B., Kepner, J. L., and Strzepa, P. (1993). Hertzian fracture in pyrolite carbon. in *Bioceramics*, Vol. 6, P. Ducheyne and D. Christiansen, eds. Butterworth–Heinemann, Oxford, pp. 225–228.

Nyilas, E., and Chiu, T. H. (1978). Artificial surface/sorbed protein structure/hemocompatibility correlations. *Artif. Organs* **2**(Suppl): 56–62.

Okazaki, Y., Wika, K. E., Matsuyoshi, T., Fukamachi, K., Kunitomo, R., Tweeden, K. S., and Harasaki, H. (1997). Platelets were early postoperative depositions on the leaflet of a mechanical heart valve in sheep without postoperative anticoagulants or antiplatelet agents. *ASAIO J.* **42**: M750–M754.

Pauling, L. (1964). *College Chemistry*, 3rd ed. W. H. Freeman and Company, San Francisco.

Reilly, D. T., and Burstein, A. H. (1974). The mechanical properties of bone. *J. Bone Joint Surg. Am.* **56**: 1001.

Reilly, D. T., Burstein, A. H., and Frankel, V. H. (1974). The elastic modulus for bone. *J. Biomech.* **7**: 271.

Richard, G., and Cao, H. (1996). Structural failure of pyrolytic carbon heart valves. *J. Heart Valve Dis.* **5**(Suppl. I): S79–S85.

Ritchie, R. O., Dauskardt, R. H., Yu, W., and Brendzel, A. M. (1990). Cyclic fatigue-crack propagation, stress corrosion and fracture toughness behavior in pyrolite carbon coated graphite for prosthetic heart valve applications. *J. Biomed. Mat. Res.* **24**: 189–206.

Sadeghi, H. (1987). Dysfonctions des prostheses valvulaires cardaques et leur traitment chirgical. *Schwiez. Med. Wochenschr.* **117**: 1665–1670.

Salzman, E. W., Lindon, J., Baier, D., and Merril, E. W. (1977). Surface-induced platelet adhesion, aggregation and release. *Ann. N.Y. Acad. Sci.* **283**: 114.

Sattler, K. (1995). Scanning tunneling microscopy of carbon nanotubes and nanocones. *Carbon* **7**: 915–920.

Sawyer, P. N., Lucas, L., Stanczewski, B., Ramasamy, N., Kammlott, G. W., and Goodenough, S. H. (1975). Evaluation techniques for potential cardiovascular prosthetic alloys experience with titanium aluminum 6-4 ELI tubes. *Proceedings of the San Diego Biomedical Symposium*, Vol. 14, pp. 423–427.

Schoen, F. J. (1983). Carbons in heart valve prostheses: foundations and clinical performance. in *Biocompatible Polymers, Metals and Composites*, M. Zycher, ed. Technomic, Lancaster, PA, pp. 240–261.

Schoen, F. J., Titus, J. L., and Lawrie, G. M. (1982). Durability of pyrolytic carbon–containing heart valve prostheses. *J. Biomed. Mater. Res.* **16**: 559–570.

Smith, K. L., and Black, K. M. (1984). Characterization of the treated surfaces of silicon alloyed pyrolytic carbon and SiC. *J. Vac. Sci. Technol.* **A2**: 744–747.

Thompson, N. G., Buchanan, R. A., and Lemons, J. E. (1979). In vitro corrosion of Ti-6Al-4V and Type 316L stainless steel when galvanically coupled with carbon. *J. Biomed. Mater. Res.* **13**: 35–44.

Wieting, D. W. (1996). The Björk–Shiley Delrin tilting disc heart valve: historical perspective, design and need for scientific analyses after 25 years. *J. Heart Valve Dis.* **5**(Suppl. I): S157–S168.

Williams, D. F. (1998). General concepts of biocompatibility. in *Handbook of Biomaterial Properties*, J. Black, and G. Hastings, eds. Chapman and Hall, London, pp. 481–489.

2.12 COMPOSITES

Claudio Migliaresi and Harold Alexander

INTRODUCTION

The word *composite* means "consisting of two or more distinct parts." At the atomic level, materials such as metal alloys and polymeric materials could be called composite materials in that they consist of different and distinct atomic groupings. At the microstructural level (about 10^{-4} to 10^{-2} cm), a metal alloy such as a plain-carbon steel containing ferrite and pearlite could be called a composite material since the ferrite and pearlite are distinctly visible constituents as observed in the optical microscope. At the molecular and microstructural level, tissues such as bone and tendon are certainly composites with a number of levels of hierarchy.

In engineering design a composite material usually refers to a material consisting of constituents in the micro- to macro-size range, favoring the macrosize range. For the purpose of discussion in this chapter, composites can be considered materials consisting of two or more chemically distinct constituents, on a macroscale, having a distinct interface separating them. This definition encompasses the fiber and particulate composite materials of primary interest as biomaterials. Such composites consist of one or more discontinuous phases embedded within a continuous phase. The discontinuous phase is usually harder and stronger than the continuous phase and is called the *reinforcement* or *reinforcing material*, whereas the continuous phase is termed the *matrix*.

Properties of composites are strongly influenced by the properties of their constituent materials, their distribution and content, and the interaction among them. The composite properties may be the volume fraction sum of the properties of the constituents, or the constituents may interact in a synergistic way due to geometrical orientation so as to provide properties in the composite that are not accounted for by a simple volume fraction sum. Thus in describing a composite material, besides specifying the constituent materials and their properties, one needs to specify the geometry of the reinforcement, their concentration, distribution, and orientation.

Most composite materials are fabricated to provide desired mechanical properties such as strength, stiffness, toughness, and fatigue resistance. Therefore, it is natural to study together the composites that have a common strengthening mechanism. The strengthening mechanism strongly depends upon the geometry of the reinforcement. Therefore, it is quite convenient to classify composite materials on the basis of the geometry of a representative unit of reinforcement. Figure 1 shows a commonly accepted classification scheme.

With regard to this classification, the distinguishing characteristic of a particle is that it is nonfibrous in nature. It may be spherical, cubic, tetragonal, or of other regular or irregular shape, but it is approximately equiaxial. A fiber is characterized by its length being much greater than its cross-sectional dimensions. Particle-reinforced composites are sometimes referred to as *particulate composites*. Fiber-reinforced composites are, understandably, called *fibrous composites*. *Laminates* are composite structures made by stacking laminae of fiber composites oriented to produce a structural element. Characteristics, number, and orientation of laminae are such as to match specific design requirements.

Fibers are much more mechanically effective than particles, and polymer-fiber composites can reach stiffness and strength comparable to those of metals and even higher.

Moreover, whereas particle-reinforced composites are isotropic, fiber-reinforced composites are basically anisotropic. Properties in different directions can be in most cases designed to match specific requirements.

At the molecular and microstructural level, tissues such as bone and tendon or vessels are certainly composites with a number of levels of hierarchy. Their properties are highly anisotropic, and the only possibility to mimic them is to use composites.

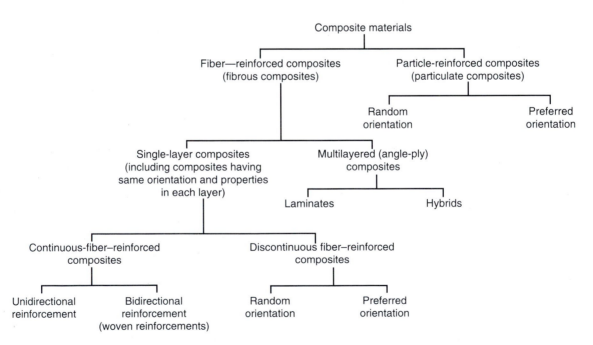

FIG. 1. Classification of composite materials. (From Agarwal and Broutman, 1980.)

Failure of a composite material implant can expose fibers or particles to the surrounding biological environment. In many cases failure in composites is preceded by the failure of the interface between filler and matrix, this being due to idrothermal aging or stresses exceeding the interface strength. Sterilization methods or conditions can play an important role.

As with all biomaterials, the question of biocompatibility (tissue response to the composite) is paramount. Being composed of two or more materials, composites provide enhanced probability of causing adverse tissue reactions. Also, the fact that one constituent (the reinforcement) usually has dimensions on the cellular scale always leaves open the possibility of cellular ingestion of particulate debris that can result in either the production of tissue-lysing enzymes or transport into the lymph system.

Although durability and biocompatibility can be considered major issues in a composite medical device, composites offer unique advantages in terms of design ability and fabrication. These advantages can be used to construct isocompliant arterial prostheses (Gershon et al., 1990, 1992), intervertebral disks duplicating the natural structure (Ambrosio et al., 1996), or fixation plates and nails with controlled stiffness (Veerabagu et al., 2003)

For some applications, moreover, radiolucency is considered to be a further potential advantage. An example is external or internal fracture fixation devices not shielding the bone fracture site from the X-ray radiography.

Design flexibility, strength, and lightweight have made polymeric composite materials, mostly carbon fiber reinforced, the ideal materials also for orthotic aids able to return walking and even athletic performances to impaired people (Dawson, 2000).

REINFORCING SYSTEMS

The main reinforcing materials that have been used in biomedical composites are carbon fibers, polymer fibers, ceramics, and glasses. Depending upon the application, the reinforcements have been either inert or absorbable.

Carbon Fiber

Carbon fiber is a lightweight, flexible, high-strength, high-tensile-modulus material produced by the pyrolysis of organic precursor fibers, such as rayon, polyacrylonitrile (PAN), and pitch in an inert environment. The term carbon is often indifferently interchanged with the term graphite, but carbon and graphite fibers differ in the temperature of fabrication, thermal treatment, and the content of carbon (93–95% for carbon fibers and more than 99% for graphite fibers). Because of their low density (depending on the precursor from 1.7 to 2.1 g/cm^3) and high mechanical properties (elastic modulus up to 900 GPa and strength up to 4.5 GPa, depending on the precursor and on the fabrication process—hence they can be much stiffer and stronger than steel!) these fibers are used in composites in a variety of applications demanding lightness and high mechanical properties. Their disadvantage is that carbon fibers have poor shear strength.

In medicine, several commercial products have used carbon fibers. Some of the first devices, however, have experienced severe negative effects and have been recalled from the market. Two examples are:

- Short carbon fiber reinforced UHMWPE for orthopedic applications. The assumption was that increase of strength and decrease of creep would increase the bearing longevity. The favorable indications of the laboratory wear tests contrasted with the *in vivo* results: Many patients presented with osteolysis and failure of the tibial inserts (Kurtz et al., 1999).
- In the 1980s carbon fibers have been used to develop a scaffolding device to induce tendon or ligament repair. The low shear strength of fibers caused fiber breakage and the formation of harmful debris. A resorbable polymeric coating was somewhat successful in preventing carbon fiber breakage and localizing debris. However, because of poor performance and permanent wear debris in the joint, the carbon fiber device was not approved by the FDA for ACL reconstruction (Dunn, 1998).

In spite of these early failures, however, carbon fibers display unique properties for the fabrication of load-bearing medical devices.

Polymer Fibers

Whereas carbon fibers have been used for their superior mechanical properties, polymer fibers are not comparably strong or stiff as reinforcements for other polymers, with the possible exceptions of aramid fibers or ultrahigh-molecular-weight polyethylene (UHMWPE) fibers. For biomedical applications, biocompatibility, of course, and high strength and fatigue resistance are compulsory, while stiffness is a design parameter to be adapted to the specific conditions. This is why for some applications PET fibers have been used. In addition, thanks to their absorbability, not to their mechanical superiority, certain absorbable fibers have been employed.

- *Aramid* is the generic name for aromatic polyamide fibers. The most well known aramids are Kevlar and Nomex (DuPont trademarks), and Twaron (made by Teijin/Twaron of Japan). Kevlar is produced by spinning a sulfuric acid/poly(p-phenylene terephthalamide) solution through an air layer into a coagulating water bath. Aramid fibers are light (density = 1.44 g/cm^3), stiff (the modulus can go up to 190 GPa), and strong (tensile strength about 3.6 GPa); moreover, they resist impact and abrasion damage. A negative point that can be relevant for biomedical applications is that aramid fibers absorb moisture, and worth noting is their poor compressive strength, about 1/8 of the tensile strength. Aramid fiber composites are used commercially where high tensile strength and stiffness, damage resistance, and resistance to fatigue and stress rupture are important. In medicine, these composites have not seen extensive use, due perhaps to some concerns about their biocompatibility or long-term fate. Main applications have been in dentistry

(Pourdeyhimi *et al.*, 1986; Vallittu, 1996) and ligament prostheses (Wening *et al.*, 1994).

- Commercially available high-strength, high-modulus polyethylene fibers include Spectra from Honeywell Performance Fibers (Colonial Heights, VA), Dyneema from DSM (Heerlen, The Netherlands), and Toyobo fibers from Toyobo (Shiga, Japan). UHMWPE fibers are produced by a gel-spinning technique starting from an approximately 2–8 wt.% solution of the ultrahigh-molecular-weight polymer ($M_w > 10^6$) in a common solvent, such as decalin. Spinning at 130–140°C and hot drawing at very high draw ratios produces fibers with the highest specific strength of all commercial fibers available to date. UHMWPE fibers possess high modulus and strength, besides displaying light weight (density about 0.97 g/cm^3) and high energy dissipation ability, compared to other fibers. In addition PE fibers resist abrasion and do not absorb water. However, the chemical properties of UHMWPE fibers are such that few resins bond well to the fiber surfaces, and so the structural properties expected from the fiber properties are often not fully realized in a composite. The low melting point of the fibers (about 147°C) impedes high-temperature fabrication. Bulk UHMWPE has extensive applications in medicine for the fabrication of bearings for joint prostheses, displaying excellent biocompatibility but with lifetime restricted by its wear resistance. Polyethylene fibers are used to reinforce acrylic resins for application in dentistry (Ladizesky *et al.*, 1994; Karaman *et al.*, 2002; Brown, 2000), or to make intervertebral disk prostheses (Kotani *et al.*, 2002). They have been also used for the fabrication of ligament augmentation devices (Guidoin *et al.*, 2000).
- Dacron is the name commonly used to indicate poly(ethylene terephthalate) fibers. These fibers have several biomedical uses, most in cardiovascular surgery for arterial grafts. Poly(ethylene terephthalate) fibers, however, have been proposed in orthopedics for the fabrication of artificial tendons or ligaments (Kolarik *et al.*, 1981) and ligament augmentation devices, as fibers or fabrics alone, or imbedded in different matrices in composites. Other proposed applications include soft-tissue prostheses, intervertebral disks (Ambrosio *et al.*, 1996), and plastic surgery applications.
- Polylactic and polyglycolic acid and their copolymers are the principal biodegradable polymers used for the fabrication of biodegradable fibers. These fibers have been used for a number of years in absorbable sutures. Properties of these fibers depend upon several factors, such as crystallinity degree, molecular weight, and purity (Migliaresi and Fambri, 1997). Fibers and tissues have been proposed for ligament reconstruction (Durselen *et al.*, 2001) or as scaffolds for tissue engineering applications (Lu and Mikos, 1996). They also have been employed in composites, in combination with parent biodegradable matrices. Examples are the intramedullary biodegradable pins and plates (Vert *et al.*, 1986, Middleton and Tipton, 2000) and biodegradable scaffolds for bone regeneration (Vacanti *et al.*, 1991, Kellomaki *et al.*, 2000).

Ceramics

A number of different ceramic materials have been used to reinforce biomedical composites. Since most biocompatible ceramics, when loaded in tension or shear, are relatively weak and brittle materials compared to metals, the preferred form for this reinforcement has usually been particulate. These reinforcements have included various calcium phosphates, aluminum- and zinc-based phosphates, glass and glass-ceramics, and bone mineral. Minerals in bone are numerous. In the past, bone has been defatted, ground, and calcined or heated to yield a relatively pure mix of the naturally occurring bone minerals. It was recognized early that this mixture of natural bone mineral was poorly defined and extremely variable. Consequently, its use as an implant material was limited.

The calcium phosphate ceramic system has been the most intensely studied ceramic system. Of particular interest are the calcium phosphates having calcium-to-phosphorus ratios of 1.5–1.67. Tricalcium phosphate and hydroxyapatite form the boundaries of this compositional range. At present, these two materials are used clinically for dental and orthopedic applications. Tricalcium phosphate has a nominal composition of $Ca_3(PO_4)_2$. The common mineral name for this material is whitlockite. It exists in two crystographic forms, α- and β-whitlockite. In general, it has been used in the β-form.

The ceramic hydroxyapatite has received a great deal of attention. Hydroxyapatite is, of course, the major mineral component of bone. The nominal composition of this material is $Ca_{10}(PO_4)_6(OH)_2$.

Tricalcium phosphates and hydroxyapatite are commonly referred as bioceramics, i.e., bioactive ceramics. The definition refers to their ability to elicit a specific biological response that results in the formation of bond between the tissues and material (Hench *et al.*, 1971). Hydroxyapatite ceramic and tricalcium phosphates are used in orthopedics and dentistry alone or in combination with other substances, or also as coating of metal implants. The rationale behind the use of bioceramics in combination with polymeric matrix for composites is in their ability to enhance the integration in bone, while improving the device mechanical properties. An example are the HA-PE composites developed by Bonfield (Bonfield, 1988; Bonfield *et al.*, 1998), and today commercialized with the name of HAPEX (Smith & Nephew ENT, Memphis, TN).

Glasses

Glass fibers are used to reinforce plastic matrices to form structural composites and molding compounds. Commercial glass fiber plastic composite materials have the following favorable characteristics: high strength-to-weight ratio; good dimensional stability; good resistance to heat, cold, moisture, and corrosion; good electrical insulation properties; ease of fabrication; and relatively low cost. De Santis *et al.* (2000) have stacked glass and carbon/PEI laminae to manufacture a hip prosthesis with constant tensile modulus but with bending modulus increasing in the tip–head direction. An isoelastic intramedullary nail made of PEEK and chopped glass fibers has been evaluated by Lin *et al.* (1997), and glass fibers have been used to increase the mechanical properties of acrylic resins for applications in dentistry (Chen *et al.*, 2001).

Zimmerman *et al.* (1991) and Lin (1986) introduced an absorbable polymer composite reinforced with an absorbable calcium phosphate glass fiber. This allowed for the fabrication of a completely absorbable composite implant material. Commercial glass fiber produced from a lime–aluminum–borosilicate glass typically has a tensile strength of about 3 GPa and a modulus of elasticity of 72 GPa. Lin (1986) estimates the absorbable glass fiber to have a modulus of 48 GPa, comparing favorably with the commercial fiber. The tensile strength, however, was significantly lower, approximately 500 MPa.

MATRIX SYSTEMS

Ceramic matrix or metal matrix composites have important technological applications, but their use is restricted to specific cases (e.g., cutting tools, power generation equipment, process industries, aerospace), with just a few examples for biomedical applications (e.g., calcium phosphate bone cements).

Most biomedical composites have polymeric matrices, mostly thermoplastic, bioabsorbable or not.

The most common matrices are synthetic nonabsorbable polymers. By far the largest literature exists for the use of polysulfone, poly(ether ether ketone) (PEEK), ultrahigh-molecular-weight polyethylene (UHMWPE), polytetrafluoroethylene (PTFE), poly(methyl methacrylate) (PMMA), and hydrogels. These matrices, reinforced with carbon fibers, polyethylene fibers, and ceramics, have been used as prosthetic hip stems, fracture fixation devices, artificial joint bearing surfaces, artificial tooth roots, and bone cements. Also, epoxy composite materials have been used. However, because of concerns about the toxicity of monomers (Morrison *et al.*, 1995) the research activity on epoxy composite for implantable devices gradually decreased.

Materials used and some examples of proposed applications are reported in Table 1.

Not all the proposed systems underwent clinical trial and only some of them are today regularly commercialized.

A review on biomedical applications of composites is in Ramakrishna *et al.* (2001).

Absorbable composite implants can be produced from absorbable α-polyester materials such as polylactic and polyglycolic polymers. Previous work has demonstrated that for most applications, it is necessary to reinforce these polymers to obtain adequate mechanical strength. Poly(glycolic acid) (PGA) was the first biodegradable polymer synthesized (Frazza and Schmitt, 1971). It was followed by poly(lactic acid) (PLA) and copolymers of the two (Gilding and Reed, 1979). These α-polyesters have been investigated for use as sutures and as implant materials for the repair of a variety of osseous and soft tissues. Important biodegradable polymers include poly(ortho esters), synthesized by Heller and co-workers (Heller *et al.*, 1980), and a class of bioerodable dimethyl-trimethylene carbonates (DMTMCs) (Tang *et al.*, 1990). A good review of absorbable polymers by Barrows (1986) included poly(lactic acid), poly(glycolic acid), poly(lactide-*co*-glycolide), polydioxanone, poly(glycolide-*co*-trimethylene

carbonate), poly(ethylene carbonate), poly(iminocarbonates), polycaprolactone, polyhydroxybutyrate, poly(amino acids), poly(ester amides), poly(ortho esters), poly(anhydrides), and cyanoacrylates. The more recent review by Middleton and Tipton (2000) focused on biodegradable polymers suited for orthopedic applications, mainly poly(glycolic acid) and poly(lactic acid). The authors examined chemistry, fabrication, mechanisms, degradation, and biocompatibility of different polymers and devices.

Natural-origin absorbable polymers have also been utilized in biomedical composites. Purified bovine collagen, because of its biocompatibility, resorbability, and availability in a well-characterized implant form, has been used as a composite matrix, mainly as a ceramic composite binder (Lemons *et al.*, 1984). A commercially available fibrin adhesive (Bochlogyros *et al.*, 1985) and calcium sulfate (Alexander *et al.*, 1987) have similarly been used for this purpose.

Reis *et al.* (1998) proposed alternative biodegradable systems to be used in temporary medical applications. These systems are blends of starch with various thermoplastic polymers. They were proposed for a large range of applications such as temporary hard-tissue replacement, bone fracture fixation, drug delivery devices, or tissue engineering scaffolds.

FABRICATION OF COMPOSITES

Composite materials can be fabricated with different technologies. Some of them are peculiar for the type of filler (particle, short or long fiber) and matrix (thermoplastic or thermosetting). Some make use of solvents whose residues could affect the material biocompatibility, hence not being applicable for the fabrication of biomedical composites. The selection of the most appropriate manufacturing technology is also influenced by the relatively low volumes of the production, compared to other applications, and by the relatively low dominance of the manufacturing cost over the overall cost of the device.

Some biomedical composites, moreover, are fabricated "*in situ.*" This is the case of composite bone cements.

The most common fabrication technologies for composites are:

1. Hand lay up
2. Spray up
3. Compression molding
4. Resin transfer molding
5. Injection molding
6. Filament winding
7. Pultrusion

In principle all of the listed technologies could be used for the fabrication of biomedical composites. Only some of them, however, have found practical use.

Fabrication of Particle-Reinforced Composites

Injection molding, compression molding, and extrusion are the most common fabrication technologies for biomedical particulate composites. In some applications composites

TABLE 1   Some Examples of Biomedical Composite Systems

Applications	Matrix/reinforcement	Reference
External fixator	Epoxy resin/CF	Migliaresi *et al.*, 2004; Baidya *et al.*, 2001
Bone fracture fixation plates, pins, screws	Epoxy resins/CF	Ali *et al.*, 1990; Veerabagu *et al.*, 2003
	PMMA/CF	Woo *et al.*, 1974
	PSU/CF	Claes *et al.*, 1997
	PP/CF	Christel *et al.*, 1980
	PE/CF	Rushton and Rae, 1984
	PBT/CF	Gillett *et al.*, 1986
	PEEK/CF	Fujihara *et al.*, 2001
	PEEK/GF	Lin *et al.*, 1997
	PLLA/HA	Furukawa *et al.*, 2000a
	PLLA/PLLA fibers	Tormala, 1992; Rokkanen *et al.*, 2000
	PGA/PGA fibers	Tormala, 1992; Rokkanen *et al.*, 2000
Spine surgery	PU/bioglass	Claes *et al.*, 1999
	PSU/bioglass	Marcolongo *et al.*, 1998
	PEEK/CF	Ciappetta *et al.*, 1997
	Hydrogels/PET fibers	Ambrosio *et al.*, 1996
Bone cement	PMMA/HA particles	Morita *et al.*, 1998
	PMMA/glass beads	Shinzato *et al.*, 2000
	Calcium phosphate/aramid fibers,CF,GF,PLGA fibers	Xu *et al.*, 2000
	PMMA/UHMWPE fibers	Yang *et al.*, 1997
Dental cements and other dental applications	Bis-GMA/inorganic particles	Moszner and Salz, 2001
	PMMA/KF	Pourdeyhimi *et al.*, 1986; Vallittu, 1996
Acetabular cups	PEEK/CF	Wang *et al.*, 1998
Hip prostheses stem	PEI/CF-GF	De Santis *et al.*, 2000
	PEEK/CF	Akay and Aslan, 1996; Kwarteng, 1990
Bone replacement, substitute	PE/ HA particles	Bonfield, 1988; Bonfield *et al.*, 1998
Bone filling, regeneration	Poly(propylene fumarate)/TCP	Yaszemski *et al.*, 1996
	PEG-PBT/HA	Qing *et al.*, 1997
	PLGA/HA fibers	Thomson *et al.*, 1998
	P(DLLA-CL)/HA particles	Ural *et al.*, 2000
	Starch/HA particles	Reis and Cunha, 2000; Leonor *et al.*, 2003
Tendons and ligaments	Hydrogels/PET	Kolarik *et al.*, 1981; Iannace *et al.*, 1995
	Polyolefins/UHMWPE fibers	Kazanci *et al.*, 2002
Vascular grafts	PELA /Polyurethane fibers	Gershon *et al.*, 1990; Gershon *et al.*, 1992
Prosthetic limbs	Epoxy resins/CF,GF,KF	Dawson, 2000

Legenda: PMMA, polymethylmethacrylate; PSU, polysulfone; PP, polypropylene; PE, polyethylene; PBT, poly(butylene terephthalate); PEEK, poly(ether ether ketone); PLLA, poly(L-lactic acid); PGA, poly(glycolic acid); PU, polyurethane; PET, poly(ethylene terephthalate); Bis-GMA, bis-glycidil dimethacrylate; PEI, poly(ether-imide); PEG, poly(ethylene glycol); PLGA, lactic acid–glycolic acid copolymer; PDLLA, poly(D,L-lactic acid); CL, poly(ε-caprolactone acid); PELA, ethylene oxide/lactic acid copolymer; CF, carbon fibers; GF, glass fibers; HA, hydroxyapatite; UHMWPE, ultrahigh-molecular-weight polyethylene; TCP, tricalcium phosphate; KF, Kevlar fibers.

are manufactured *in situ.* This is the case of dental restorative composites and particle-reinforced bone cements.

Fabrication of Fiber-Reinforced Composites

Fiber-reinforced composites are produced commercially by one of two classes of fabrication techniques: open or closed molding. Most of the open-molding techniques are not appropriate to biomedical composites because of the character of the matrices used (mainly thermoplastics) and the need to produce materials that are resistant to water intrusion.

Consequently, the simplest techniques, the hand lay-up and spray-up procedures, are seldom, if ever, used to produce biomedical composites. The two open-molding techniques that may find application in biomedical composites are the vacuum bag–autoclave process and the filament-winding process.

Vacuum Bag–Autoclave Process

This process is used to produce high-performance laminates, usually of fiber-reinforced epoxy. Composite materials produced by this method are currently used in aircraft and aerospace applications. The first step in this process, and indeed many other processes, is the production of a "prepreg." This basic structure is a thin sheet of matrix imbedded with uniaxially oriented reinforcing fibers. When the matrix is epoxy, it is prepared in the partially cured state. Pieces of the prepreg sheet are cut out and placed on top of each other on a shaped tool to form a laminate. The layers, or plies, may be placed in different directions to produce the desired strength and stiffness.

After the laminate is constructed, the tooling and attached laminate are vacuum-bagged, with a vacuum being applied to remove entrapped air from the laminated part. Finally, the vacuum bag enclosing the laminate and the tooling is put into an autoclave for the final curing of the epoxy resin. The conditions for curing vary depending upon the material, but the carbon fiber–epoxy composite material is usually heated at about 190°C at a pressure of about 700 kPa. After being removed from the autoclave, the composite part is stripped from its tooling and is ready for further finishing operations. This procedure is potentially useful for the production of fracture fixation devices and total hip stems.

Filament-Winding Process

Another important open-mold process to produce high-strength hollow cylinders is the filament-winding process. In this process, the fiber reinforcement is fed through a resin bath and then wound on a suitable mandrel (Fig. 2). When sufficient layers have been applied, the wound mandrel is cured. The molded part is then stripped from the mandrel. The high degree of fiber orientation and high fiber loading with this method produce extremely high tensile strengths. Biomedical applications

for this process include intramedullary rods for fracture fixation, prosthetic hip stems, ligament prostheses, intervertebral disks, and arterial grafts.

Closed-Mold Processes

There are many closed-mold methods used for producing fiber-reinforced plastic materials. The methods of most importance to biomedical composites are compression and injection molding and continuous pultrusion. In compression molding, the previously described prepregs are arranged in a two-piece mold that is then heated under pressure to produce the laminated part. This method is particularly useful for use with thermoplastic matrices. In injection molding the fiber–matrix mix is injected into a mold at elevated temperature and pressure. The finished part is removed after cooling. This is an extremely fast and inexpensive technique that has application to chopped fiber–reinforced thermoplastic composites. It offers the possiblity to produce composite devices, such as bone plates and screws, at much lower cost than comparable metallic devices.

Continuous pultrusion is a process used for the manufacture of fiber-reinforced plastics of constant cross section such as structural shapes, beams, channels, pipe, and tubing. In this process, continuous-strand fibers are impregnated in a resin bath and then are drawn through a heated die, which determines the shape of the finished stock (Fig. 3). Highly oriented parts cut from this stock can then be used in other structures or they can be used alone in such applications as intramedullary rodding or pin fixation of bone fragments.

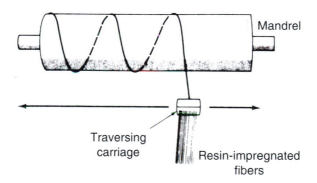

FIG. 2. Filament-winding process for producing fiber-reinforced composite materials.

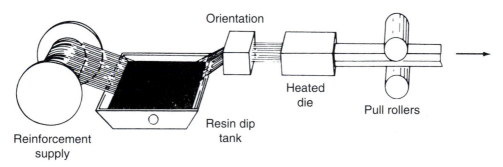

FIG. 3. The pultrusion process for producing fiber-reinforced polymer composite materials. Fibers impregnated with polymer are fed into a heated die and then are slowly drawn out as a cured composite material with a constant cross-sectional shape.

MECHANICAL AND PHYSICAL PROPERTIES OF COMPOSITES

Continuous Fiber Composites

Laminated continuous fiber-reinforced composites are described from either a micro- or macromechanical point of view. Micromechanics is the study of composite material behavior wherein the interaction of the constituent materials is examined on a local basis. Macromechanics is the study of composite material behavior wherein the material is presumed homogeneous and the effects of the constituent materials are detected only as averaged apparent properties of the composite. Both the micromechanics and macromechanics of experimental laminated composites will be discussed.

Micromechanics

There are two basic approaches to the micromechanics of composite materials: the mechanics of materials and the elasticity approach. The mechanics-of-materials approach embodies the concept of simplifying assumptions regarding the hypothesized behavior of the mechanical system. It is the simpler of the two and the traditional choice for micromechanical evaluation. The most prominent assumption made in the mechanics-of-materials approach is that strains in the fiber direction of a unidirectional fibrous composite are the same in the fibers and the matrix. This assumption allows the planes to remain parallel to the fiber direction. It also allows the longitudinal normal strain to vary linearly throughout the member with the distance from the neutral axis. Accordingly, the stress will also have a linear distribution.

Some other important assumptions are as follows:

1. The lamina is macroscopically homogeneous, linearly elastic, orthotropic, and initially stress-free.
2. The fibers are homogeneous, linearly elastic, isotropic, regularly spaced, and perfectly aligned.
3. The matrix is homogeneous, linearly elastic, and isotropic.

In addition, no voids are modeled in the fibers, the matrix or between them.

The mechanical properties of a lamina are determined by fiber orientation. The most often used laminate coordinate system has the length of the laminate in the x direction and the width in the y direction. The principal fiber direction is the 1 direction, and the 2 direction is normal to that. The angle between the x and 1 directions is ϕ. A counterclockwise rotation of the 1–2 system yields a positive ϕ.

The mechanical properties of the lamina are dependent on the material properties and the volume content of the constituent materials. The equations for the mechanical properties of a lamina in the 1–2 directions are:

$$E_1 = E_f V_f + E_m V_m \tag{1}$$

$$E_2 = \frac{E_f E_m}{V_m E_f} + V_f E_m \tag{2}$$

$$v_{12} = V_m v_m + V_f v_f \tag{3}$$

$$v_{21} E_1 = v_{12} E_2 \tag{4}$$

$$G_{12} = \frac{G_f G_m}{V_m G_f} + V_f G_m \tag{5}$$

$$V_m = 1 - V_f \tag{6}$$

where E is Young's modulus, G is the shear modulus, V is the volume fraction, v is Poisson's ratio, and subscripts f and m represent fiber and matrix properties, respectively. These equations are based on the law of mixtures for composite materials.

Macromechanics of a Lamina

The generalized Hooke's law relating stresses to strains is

$$\sigma_i = C_{ij}\varepsilon_j \qquad ij = 1, 2, \ldots, 6 \tag{7}$$

where s_i = stress components, C_{ij} = stiffness matrix, and ε_j = strain components. An alternative form of the stress–strain relationship is

$$\varepsilon_{ij} = S_{ij}\sigma_i \qquad ij = 1, 2, \ldots, 6 \tag{8}$$

where S_{ij} = compliance matrix.

Given that $C_{ij} = C_{ji}$, the stiffness matrix is symmetric, thus reducing its population of 36 elements to 21 independent constants. We can further reduce the matrix size by assuming the laminae are orthotropic. There are nine independent constants for orthotropic laminae. In order to reduce this three-dimensional situation to a two-dimensional situation for plane stress, we have

$$\tau_3 = 0 = \sigma_{23} = \sigma_{13} \tag{9}$$

thus reducing the stress–strain relationship to

$$\begin{vmatrix} \varepsilon_1 \\ \varepsilon_2 \\ \gamma_{12} \end{vmatrix} = \begin{vmatrix} S_{11} & S_{12} & 0 \\ S_{21} & S_{22} & 0 \\ 0 & 0 & S_{66} \end{vmatrix} \begin{vmatrix} \sigma_1 \\ \sigma_2 \\ \tau_{12} \end{vmatrix} \tag{10}$$

The stress–strain relation can be inverted to obtain

$$\begin{vmatrix} \sigma_1 \\ \sigma_2 \\ \tau_{12} \end{vmatrix} = \begin{vmatrix} Q_{11} & Q_{12} & 0 \\ Q_{21} & Q_{22} & 0 \\ 0 & 0 & Q_{66} \end{vmatrix} \begin{vmatrix} \varepsilon_1 \\ \varepsilon_2 \\ \gamma_{12} \end{vmatrix} \tag{11}$$

where Q_{ij} are the reduced stiffnesses. The equations for these stiffnesses are

$$Q_{11} = \frac{E_1}{1 - v_{21}v_{12}} \tag{12}$$

$$Q_{12} = \frac{v_{12}E_2}{1 - v_{12}v_{21}} = \frac{v_{21}E_l}{1 - v_{12}v_{21}} = Q_{21} \tag{13}$$

$$Q_{22} = \frac{E_2}{1 - v_{21}v_{21}} \tag{14}$$

$$Q_{66} = G_{12} \tag{15}$$

The material directions of the lamina may not coincide with the body coordinates. The equations for the transformation of stresses in the 1–2 direction to the x–y direction are

$$\begin{vmatrix} \sigma_x \\ \sigma_y \\ \tau_{xy} \end{vmatrix} = \begin{bmatrix} T^{-1} \end{bmatrix} \cdot \begin{vmatrix} \sigma_1 \\ \sigma_2 \\ \tau_{12} \end{vmatrix} \tag{16}$$

where $[T^{-1}]$ is

$$[T^{-1}] = \begin{vmatrix} \cos^2 \Phi & \sin^2 \Phi & -2\sin\Phi\cos\Phi \\ \sin^2 \Phi & \cos^2 \Phi & 2\sin\Phi\cos\Phi \\ \sin\Phi\cos\Phi & -\sin\Phi\cos\Phi & \cos^2 \Phi - \sin^2 \Phi \end{vmatrix} \tag{17}$$

The x and 1 axes form angle Φ. This matrix is also valid for the transformation of strains,

$$\begin{vmatrix} \varepsilon_x \\ \varepsilon_y \\ \frac{1}{2}\gamma_{xy} \end{vmatrix} = [T^{-1}] \cdot \begin{vmatrix} \varepsilon_1 \\ \varepsilon_2 \\ \frac{1}{2}\gamma_{12} \end{vmatrix} \tag{18}$$

Finally, it can be demonstrated that

$$\begin{vmatrix} \sigma_x \\ \sigma_y \\ \tau_{xy} \end{vmatrix} = [\overline{Q}_{ij}] \cdot \begin{vmatrix} \varepsilon_x \\ \varepsilon_y \\ \gamma_{xy} \end{vmatrix} \tag{19}$$

where $[\overline{Q}_{ij}]$ is the transformed reduced stiffness. The transformed reduced stiffness matrix is

$$[\overline{Q}_{ij}] = \begin{vmatrix} \overline{Q}_{11} & \overline{Q}_{12} & \overline{Q}_{16} \\ \overline{Q}_{21} & \overline{Q}_{22} & \overline{Q}_{26} \\ \overline{Q}_{16} & \overline{Q}_{26} & \overline{Q}_{66} \end{vmatrix} \tag{20}$$

where,

$$\overline{Q}_{11} = Q_{11}\cos^4 \Phi + Q_{22}\sin^4 \Phi$$
$$+ 2(Q_{12} + 2Q_{66})\sin^2 \Phi \cos^2 \Phi \tag{21}$$

$$\overline{Q}_{22} = Q_{11}\sin^4 \Phi + Q_{22}\cos^4 \Phi$$
$$+ 2(Q_{12} + 2Q_{66})\sin^2 \Phi \cos^2 \Phi \tag{22}$$

$$\overline{Q}_{12} = (Q_{11} + Q_{22} - 4Q_{66})\sin^2 \Phi\cos^2 \Phi$$
$$+ Q_{12}(\sin^4 \Phi + \cos^4 \Phi) \tag{23}$$

$$\overline{Q}_{66} = (Q_{11} + Q_{22} - 2Q_{12} - 2Q_{66})\sin^2 \Phi\cos^2 \Phi$$
$$+ Q_{66}(\sin^4 \Phi + \cos^4 \Phi) \tag{24}$$

$$\overline{Q}_{16} = (Q_{11} - Q_{12} - 2Q_{66})\sin\Phi\cos^3 \Phi$$
$$- (Q_{22} - Q_{12} - 2Q_{66})\sin^3 \Phi\cos\Phi \tag{25}$$

$$\overline{Q}_{26} = (Q_{11} - Q_{12} - 2Q_{66})\sin^3 \Phi\cos\Phi$$
$$- (Q_{22} - Q_{12} - 2Q_{66})\sin\Phi\cos^3 \Phi \tag{26}$$

$\overline{Q}_{16} = \overline{Q}_{26} = 0$ for a laminated symmetric composite.

The transformation matrix $[T^{-1}]$ and the transformed reduced stiffness matrix $[\overline{Q}_{ij}]$ are very important matrices in the macromechanical analysis of both laminae and laminates. These matrices play a key role in determining the effective in-plane and bending properties and how a laminate will perform when subjected to different combinations of forces and moments.

Macromechanics of a Laminate

The development of the A, B, and D matrices for laminate analysis is important for evaluating the forces and moments to which the laminate will be exposed and in determining the stresses and strains of the laminae. As given in Eq. (19),

$$(\sigma_k) = \lfloor \overline{Q}_{ij} \rfloor (\varepsilon_k) \tag{27}$$

where σ = normal stresses, ε = normal strains, and $[\overline{Q}_{ij}]$ = stiffness matrix. The A, B, and D matrices are equivalent to the following:

$$[A_{ij}] = \sum_{k=1}^{n} (\overline{Q}_{ij})_k (h_k - h_{k-1}) \tag{28}$$

$$[B_{ij}] = \frac{1}{2} \sum_{k=1}^{n} (\overline{Q}_{ij})_k (h_k^2 - h_{k-1}^2) \tag{29}$$

$$[D_{ij}] = \frac{1}{3} \sum_{k=1}^{n} (\overline{Q}_{ij})_k (h_k^3 - h_{k-1}^3) \tag{30}$$

The matrix $[A]$ is called the *extensional stiffness matrix* because it relates the resultant forces to the midplane strains, while matrix $[D]$ is called the *bending stiffness matrix* because it relates the resultant moments to the laminate curvature. The so called *coupling stiffness matrix*, $[B]$, accounts for coupling between bending and extension, which means that normal and shear forces acting at the laminate midplane are causing laminate curvature or that bending and twisting moments are accompanied by midplane strain.

The letter k denotes the number of laminae in the laminate with a maximum number (N). The letter h represents the distances from the neutral axis to the edges of the respective laminae. A standard procedure for numbering laminae is used where the 0 lamina is at the bottom of a plate and the Kth lamina is at the top.

The resultant laminate forces and moments are:

$$\begin{vmatrix} N_x \\ N_y \\ N_{xy} \end{vmatrix} = [A_{ij}] \cdot \begin{vmatrix} \varepsilon_x \\ \varepsilon_y \\ \gamma_{xy} \end{vmatrix} + [B_{ij}] \cdot \begin{vmatrix} k_x \\ k_y \\ k_{xy} \end{vmatrix} \tag{31}$$

$$\begin{vmatrix} M_x \\ M_y \\ M_{xy} \end{vmatrix} = [B_{ij}] \cdot \begin{vmatrix} \varepsilon_x \\ \varepsilon_y \\ \gamma_{xy} \end{vmatrix} + [D_{ij}] \cdot \begin{vmatrix} k_x \\ k_y \\ k_{xy} \end{vmatrix} \tag{32}$$

The k vector represents the respective curvatures of the various planes. The resultant forces and moments of a loaded composite can be analyzed given the *ABD* matrices. If the laminate is assumed symmetric, the force equation reduces to

$$\begin{vmatrix} N_x \\ N_y \\ N_{xy} \end{vmatrix} = [A_{ij}] \cdot \begin{vmatrix} \varepsilon_x \\ \varepsilon_y \\ \gamma_{xy} \end{vmatrix} \tag{33}$$

Once the laminate strains are determined, the stresses in the xy direction for each lamina can be calculated. The most useful information gained from the *ABD* matrices involves the determination of generalized in-plane and bending properties of the laminate.

In a generic laminate, normal stresses N_x and/or N_y (or thermal stresses or liquid sorption) will cause deformations in the directions x and/or y, but also shear strains, unless A_{16} and A_{26} of the extensional stiffness matrix are equal to 0. These coefficient become 0 if the laminate is balanced, i.e., has the same number of laminae oriented at Φ and $-\Phi$.

Moreover, in a generic laminate, normal or shear stresses will produce bending, and bending or twisting will cause midplane strains. The coupling between bending and extension can be eliminated if the coefficients of the B_{ij} matrix are equal to zero, that is, if the laminate is fabricated symmetric with respect to its midplane.

The equivalent elastic constants $(E_x, E_y, G_{xy}, \nu_{xy})$ of a symmetric and balanced laminate can be easily evaluated from the A_{ij} coefficients (Barbero, 1998):

$$E_x = \frac{1}{h}\frac{A_{11}A_{22} - A_{12}^2}{A_{22}} \tag{34}$$

$$E_y = \frac{1}{h}\frac{A_{11}A_{22} - A_{12}^2}{A_{11}} \tag{35}$$

$$\nu_{xy} = \frac{A_{12}}{A_{22}} \tag{36}$$

$$G_{xy} = \frac{1}{h}A_{66} \tag{37}$$

In the equations above h is the total thickness of the laminate.

Short-Fiber Composites

A distinguishing feature of the unidirectional laminated composites discussed above is that they have higher strength and modulus in the fiber direction, and thus their properties are amenable to alteration to produce specialized laminates. However, in some applications, unidirectional multiple-ply laminates may not be required. It may be advantageous to have isotropic laminae. An effective way of producing an isotropic lamina is to use randomly oriented short fibers as the reinforcement. Of course, molding compounds consisting of short fibers that can be easily molded by injection or compression molding may be used to produce generally isotropic composites. The theory of stress transfer between fibers and matrix in short-fiber composites goes beyond this text; it is covered in detail by Agarwal and Broutman (1980). However, the longitudinal and transverse moduli (E_L and E_T, respectively) for an aligned short-fiber lamina can be derived from the generalized Halpin-Tsai equations (Halpin and Kardos, 1976), as:

$$\frac{E_L}{E_m} = \frac{1 + ((2l/d)\eta_L V_f)}{1 - \eta_L V_f} \tag{38}$$

$$\frac{E_T}{E_m} = \frac{1 + 2\eta_T V_f}{1 - \eta_T V_f} \tag{39}$$

$$\eta_L = \frac{E_f/E_m - 1}{E_f/E_m + 2(l/d)} \tag{40}$$

$$\eta_T = \frac{E_f/E_m - 1}{E_f/E_m + 2} \tag{41}$$

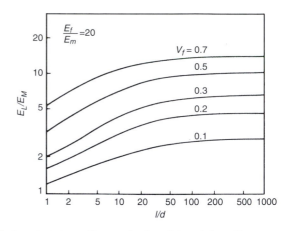

FIG. 4. Variations of longitudinal modulus of short-fiber composites against aspect ratio for different fiber volume fractions ($E_f/E_m = 20$).

In the previous equations E_m is the elastic modulus of the matrix, l and d are the fiber length and diameter respectively, and V_f is the fiber volume fraction.

For a ratio of fiber to matrix modulus of 20, the variation of longitudinal modulus of an *aligned* short-fiber lamina as a function of fiber aspect ratio, l/d, for different fiber volume fractions is shown in Fig. 4. It can be seen that approximately 85% of the modulus obtainable from a continuous fiber lamina is attainable with an aspect ratio of 20.

The problem of predicting properties of *randomly oriented* short-fiber composites is more complex. The following empirical equation can be used to predict the modulus of composites containing fibers that are randomly oriented in a plane:

$$E_{\text{random}} = \frac{3}{8}E_L + \frac{5}{8}E_T \tag{42}$$

where E_L and E_T are respectively the longitudinal and transverse moduli of an aligned short-fiber composite having the same fiber aspect ratio and fiber volume fraction as the composite under consideration. Moduli E_L and E_T can either be determined experimentally or calculated using Eqs. 38 and 39.

Particulate Composites

The reinforcing effect of particles on polymers was first recognized for rubbery matrices during studies of the effect of carbon black on the properties of natural rubber.

Several models have been introduced to predict the effect of the addition of particles to a polymeric matrix, starting from the equation developed by Einstein in 1956 to predict the viscosity of suspensions of rigid spherical inclusions. The paper by Ahmed and Jones (1990) well reviews theories developed to predict strength and modulus of particulate composites. One of the most versatile equation predicting the shear modulus of composites of polymers and spherical fillers is due to Kerner (1956):

$$G_c = G_m \left(1 + \frac{V_f}{V_m}\frac{15(1 - \nu_m)}{(8 - 10\nu_m)}\right) \tag{43}$$

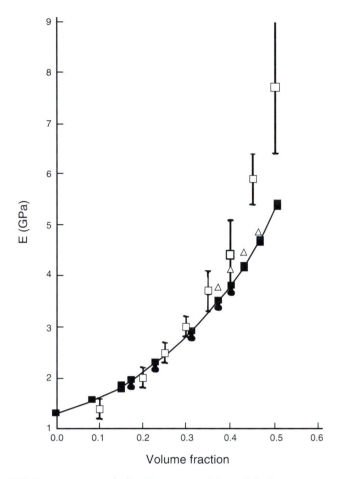

FIG. 5. Variation of the Young's modulus of hydroxyapatite–polyethylene composites modulus with volume fraction: experimental values, □, and predicted values before and after the application of the statistical model; ■, primary; △, equal strain; ▲, equal stress (from Guild and Bonfield, 1993).

A more generalized form was developed by Nielsen (1974),

$$M_c = M_m \frac{1 + ABV_f}{1 - B\psi V_f} \tag{44}$$

where M_c is any modulus—shear, Young's or bulk- of the composite, the constant A takes into account for the filler geometry and the Poisson's ratio of the matrix and the constant B depends on the relative moduli of the filler (M_f) and the matrix (M_m). The function Ψ depends on the particle packing fraction.

By using a finite element analysis method Guild and Bonfield (1993) predicted the elastic modulus of hydroxyapatite–polyethylene reinforced composites for various filler content. Their result (Fig. 5) indicated a good agreement between theoretical and experimental data, except at higher hydroxyapatite volume fraction.

While elastic modulus of a particulate composites increases with the filler content, strength decreases in tension and increases in compression. Size and shape of the inclusion play an important role, with a higher stress concentration cause

by irregularly shaped inclusions. For spherical particles, tensile strength can be predicted by the equation (Nicolais and Narkis, 1971):

$$\sigma_{cu} = \sigma_{mu} \left(1 - 1.21 V_f^{2/3}\right) \tag{45}$$

where σ_{cu} and σ_{mu} are tensile strength of composite and matrix, respectively.

ABSORBABLE MATRIX COMPOSITES

Absorbable matrix composites have been used in situations where absorption of the matrix is desired. Matrix absorption may be desired to expose surfaces to tissue or to release admixed materials such as antibiotics or growth factors (drug release) (Yasko *et al.*, 1992). However, the most common reasons for the use of this class of matrices for composites has been to accomplish time-varying mechanical properties and assure complete dissolution of the implant, eliminating long-term biocompatibility concerns. A typical clinical example is fracture fixation (Daniels *et al.*, 1990; Tormala, 1992).

Fracture Fixation

Rigid internal fixation of fractures has conventionally been accomplished with metallic plates, screws, and rods. During the early stages of fracture healing, rigid internal fixation maintains alignment and promotes primary osseous union by stabilization and compression. Unfortunately, as healing progresses, or after healing is complete, rigid fixation may cause bone to undergo stress protection atrophy. This can result in significant loss of bone mass and osteoporosis. Additionally, there may be a basic mechanical incompatibility between the metal implants and bone. The elastic modulus of cortical bone ranges from 17 to 24 GPa, depending upon age and location of the specimen, while the commonly used alloys have moduli ranging from 110 GPa (titanium alloys) to 210 GPa (316L steel). This large difference in stiffness can result in disproportionate load sharing, relative motion between the implant and bone upon loading, as well as high stress concentrations at bone–implant junctions.

Another potential problem is that the alloys currently used corrode to some degree. Ions so released have been reported to cause adverse local tissue reactions as well as allogenic responses, which in turn raises questions of adverse effects on bone mineralization as well as adverse systemic responses such as local tumor formation (Martin *et al.*, 1988). Consequently, it is usually recommended that a second operation be performed to remove hardware.

The advantages of absorbable devices are thus twofold. First, the devices degrade mechanically with time, reducing stress protection and the accompanying osteoporosis. Second, there is no need for secondary surgical procedures to remove absorbable devices. The state of stress at the fracture site gradually returns to normal, allowing normal bone remodeling.

Absorbable fracture fixation devices have been produced from poly(L-lactic acid) polymer, poly(glycolic acid) polymer, and polydioxanone. An excellent review of the mechanical properties of biodegradable polymers was prepared by Daniels and co-workers (Daniels *et al.*, 1990; see Figs. 6 and 7).

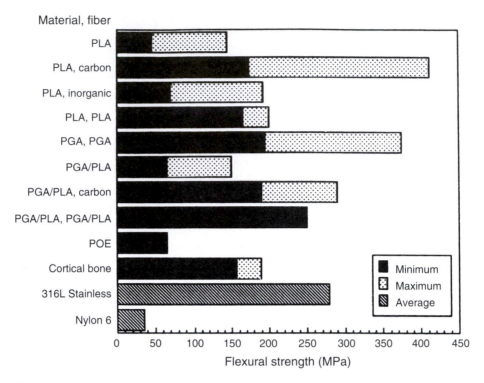

FIG. 6. Representative flexural strengths of absorbable polymer composites (from Daniels *et al.*, 1990).

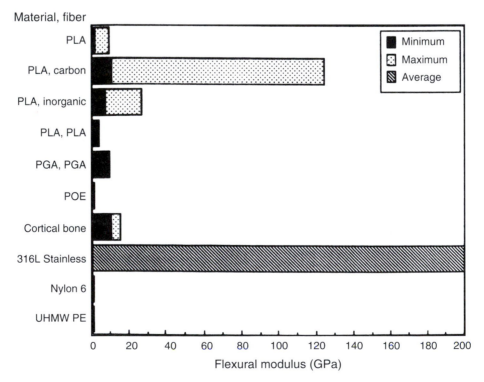

FIG. 7. Representative flexural moduli of absorbable polymer composites (from Daniels *et al.*, 1990).

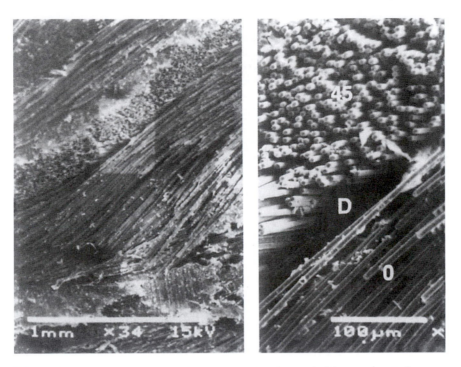

FIG. 8. Scanning electron micrograph of laminae buckling and delamination (D) between lamina in a carbon fiber-reinforced PLA fracture fixation plate (from Zimmerman *et al.*, 1987).

Their review revealed that unreinforced biodegradable polymers are initially 36% as strong in tension as annealed stainless steel, and 54% in bending, but only 3% as stiff in either test mode. With fiber reinforcement, highest initial strengths exceeded those of stainless steel. Stiffness reached 62% of stainless steel with nondegradable carbon fibers, 15% with degradable inorganic fibers, but only 5% with degradable polymeric fibers.

Most previous work on absorbable composite fracture fixation has been performed with PLLA polymer. PLLA possesses three major characteristics that make it a potentially attractive biomaterial:

1. It degrades in the body at a rate that can be controlled.
2. Its degradation products are nontoxic, biocompatible, easily excreted entities. PLA undergoes hydrolytic deesterification to lactic acid, which enters the lactic acid cycle of metabolites. Ultimately it is metabolized to carbon dioxide and water and is excreted.
3. Its rate of degradation can be controlled by mixing it with poly(glycolic acid) polymer.

Poly(L-lactic acid) polymer reinforced with randomly oriented chopped carbon fiber was used to produce partially degradable bone plates (Corcoran *et al.*, 1981). It was demonstrated that the plates, by virtue of the fiber reinforcement, exhibited mechanical properties superior to those of pure polymer plates. *In vivo*, the matrix degraded and the plates lost rigidity, gradually transferring load to the healing bone. However, the mechanical properties of such chopped fiber plates were relatively low; consequently, the plates were only adequate for low-load situations. Zimmerman *et al.* (1987) used composite theory to determine an optimum fiber layup for a long fiber composite bone plate. Composite analysis suggested the mechanical superiority of a $0°/\pm45°$ laminae layup. Although the $0°/\pm45°$ carbon/polylactic acid composite possessed adequate initial mechanical properties, water absorption and subsequent delamination degraded the properties rapidly in an aqueous environment (Fig. 8). The fibers did not chemically bond to the matrix.

In an attempt to develop a totally absorbable composite material, a calcium-phosphate-based glass fiber has been used to reinforce poly(lactic acid). Experiments were pursued to determine the biocompatibility and in vitro degradation properties of the composite (Zimmerman *et al.*, 1991). These studies showed that the glass fiber–PLA composite was biocompatible, but its degradation rate was too high for use as an orthopedic implant.

Shikinami and Okuno (2001), have produced miniplates, rods, and screws made of hydroxyapatite poly(L-lactide). These composites have been principally applied for indications such as repair of bone fracture in osteosynthesis and fixation of bony fragments in bone grafting and osteotomy, exhibiting total resorbability and osteological bioactivity while retaining sufficient stiffness high stiffness retainable for a long period of time to achieve bony union. These plates are commercialized with the name of Fixsorb-MX.

Furukawa *et al.* (2000b) have investigated the *in vivo* biodegradation behavior of hydroxyapatite/poly(L-lactide) composite rods implanted *sub cutem* and in the intramedullary cavities of rabbits, showing that after 25 weeks of *sub cutem* implantation rods maintained a bending strength higher than 200 MPa. Their conclusion was that such a strength was sufficient for application of the rods in the fixation of human bone fractures.

By using a sintering technique, Tormala *et al.* (1988) have produced self-reinforced PGA (SR-PGA) rods that have been used in the treatment of fractures and osteotomies. Afterwards, by using the same technique, self-reinforced PLLA (SR-PLLA) pins and screws have been produced. The higher initial mechanical properties of SR-PLGA are counterbalanced by their faster decrease with respect to the SR-PLLA material, which has a slower degradation rate and is reabsorbed in 12–16 months. These products are commercially available.

NONABSORBABLE MATRIX COMPOSITES

Nonabsorbable matrix composites are generally used as biomaterials to provide specific mechanical properties unattainable with homogeneous materials. Particulate and chopped-fiber reinforcement has been used in bone cements and bearing surfaces to stiffen and strengthen these structures.

For fracture fixation, reduced-stiffness carbon-fiber-reinforced epoxy bone plates to reduce stress-protection osteoporosis have been made. These plates have also been entered into clinical use, but were found to not be as reliable or biocompatible as stainless steel plates. Consequently, they have not generally been accepted in clinical use. By far the most studied, and potentially most valuable use of nonabsorable composites has been in total joint replacement.

Total Joint Replacement

Bone resorption in the proximal femur leading to aseptic loosening is an all-too-common occurrence associated with the implantation of metallic femoral hip replacement components. It has been suggested that proximal bone loss may be related to the state of stress and strain in the femoral cortex. It has long been recognized that bone adapts to functional stress by remodeling to reestablish a stable mechanical environment. When applied to the phenomenon of bone loss around implants, one can postulate that the relative stiffness of the metallic component is depriving bone of its accustomed load. Clinical and experimental results have shown the significant role that implant elastic characteristics play in allowing the femur to attain a physiologically acceptable stress state. Femoral stem stiffness has been indicated as an important determinant of cortical bone remodeling (Cheal *et al.*, 1992). Composite materials technology offers the ability to alter the elastic characteristics of an implant and provide a better mechanical match with the host bone, potentially leading to a more favorable bone remodeling response.

Using different polymer matrices reinforced with carbon fiber, a large range of mechanical properties is possible. St. John (1983) reported properties for ±15° laminated test specimens (Table 2) with moduli ranging from 18 to 76 GPa.

TABLE 2   Typical Mechanical Properties of Polymer–Carbon Composites (Three-Point Bending)

Polymer	Ultimate strength (MPa)	Modulus (GPa)
PMMA	772	55
Polysufone	938	76
Epoxy		
Stycast	535	30
Hysol	207	24
Polyurethane	289	18

However, the best reported study involved a novel press-fit device constructed of carbon fiber/polysulfone composite (Magee *et al.*, 1988). The femoral component designed and used in this study utilized composite materials with documented biologic profiles. These materials demonstrated strength commensurate with a totally unsupported implant region and elastic properties commensurate with a fully bone-supported implant region. These properties were designed to produce constructive bone remodeling. The component contained a core of unidirectional carbon/polysufone composite enveloped with a bidirectional braided layer composed of carbon/polysufone composite covering the core. These regions were encased in an outer coating of pure polysufone (Fig. 9). Finite-element stress analysis predicted that this construction would cause minimal disruption of the normal stresses in the intact cortical bone. Canine studies carried out to 4 years showed a favorable bone remodeling response. The authors proposed that implants fabricated from carbon/polysulfone composites should have the potential for use in load-bearing

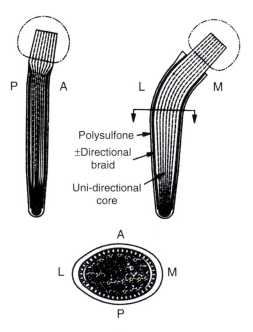

FIG. 9.   Construction details of a femoral stem of a composite total hip prosthesis. (From Magee *et al.*, 1988.)

applications. An implant with appropriate elastic properties provides the opportunity for the natural bone remodeling response to enhance implant stability.

Adam *et al.* (2002) reported on the revision of 51 epoxy resin/carbon fiber composite press fit-hip prostheses implanted in humans. Their result showed that within 6 years 92% of the prostheses displayed aseptic loosening, i.e., did not induce bone ongrowth. Authors attributed the failure to the smoothness of the stem surface. No osteolysis or wear or inflammatory reaction were, however, observed.

Different fibers matrices and fabrication technologies have been proposed for the fabrication of hip prostheses. Reviews of materials and methods are in Ramakrishna *et al.* (2001) and in de Oliveira Simopes and Marques (2001).

CONCLUSIONS

Biomedical composites have demanding properties that allow few, if any, "off the shelf" materials to be used. The designer must almost start from scratch. Consequently, few biomedical composites are yet in general clinical use. Those that have been developed to date have been fabricated from fairly primitive materials with simple designs. They are simple laminates, chopped fiber, or particulate reinforced systems with no attempts made to react or bond the phases together. Such bonding may be accomplished by altering the surface texture of the filler or by the introduction of coupling agents: molecules that can react with both filler and matrix. However, concerns about the biocompatibility of coupling agents and the high development costs of surface texture alteration procedures have curtailed major developments in this area. It is also possible to provide three-dimensional reinforcement with complex fiber weaving and impregnation procedures now regularly used in high-performance aerospace composites. Unfortunately, the high development costs associated with these techniques have restricted their application to biomedical composites.

Because of the high development costs and the small-volume market available, few biomedical materials have, to date, been designed specifically for biomedical use. Biomedical composites, because of their unique requirements, are probably be the first general class of materials developed exclusively for implantation purposes.

Bibliography

Adam, F., Hammer, D. S., Pfautsch, S., and Westermann, K. (2002). Early failure of a press-fit carbon fiber hip prosthesis with a smooth surface. *J. Arthroplasty* **17**: 217–223.

Agarwal, B. D., and Broutman, L. J. (1980). *Analysis and Performance of Fiber Composites.* Wiley-Interscience, New York.

Ahmed, S., and Jones F. R. (1990). A review of particulate reinforcement theories for polymer composites. *J. Mater. Sci.* **25**: 4933–42.

Akay, M., and Aslan, N. (1996). Numerical and experimental stress analysis of a polymeric composite hip joint prosthesis. *J. Biomed. Mater. Res.* **31**: 167–82.

Alexander, H., Parsons, J. R., Ricci, J. L., Bajpai, P. K., and Weiss, A. B. (1987). Calcium-based ceramics and composites in bone reconstruction. *CRC Crit. Rev. Biocompat.* **4**(1): 43–77.

Ali, M. S., Hastings, G. W., Rushton, N., Ross, E. R. S., and Wynn-Jones, C. H. (1990). Carbon fiber composite plates. *J. Bone Joint Surg.* **72-B**: 586–591.

Ambrosio, L., Netti, P., Iannace, S., Huang, S. J., and Nicolais, L. (1996). Composite hydrogels for intervertebral disc prostheses. *J. Mater. Sci. Mater. Med.* **7**: 251–4.

Baidya, K. P., Ramakrishna, S., Rahman, M., and Ritchie A. (2001). Advanced textile composite ring for Ilizarov external fixator system. *Proc. Inst. Mech. Eng. Part H, J. Eng. Med.* **215**: 11–23.

Barbero, E. J. (1998). *Introduction to Composite Materials Design.* Taylor and Francis, Philadelphia.

Barrows, T. H. (1986). Degradable implant materials: a review of synthetic absorbable polymers and their applications. *Clin. Mater.* **1**: 233.

Bochlogyros, P. M., Hensher, R., Becker, R., and Zimmerman, E. (1985). A modified hydroxyapatite implant material. *J. Maxillofac. Surg.* **13**(5): 213.

Bonfield, W. (1988). Composites for bone replacement. *J. Biomed. Eng.* **10**: 522.

Bonfield, W., Wang, M., and Tanner K. E. (1998). Interfaces in analogue biomaterials. *Acta Mater.* **7**: 2509–2518.

Brown, D. (2000). Fibre-reinforced materials. *Dent. Update* **27**(9): 442–448.

Cheal, E. J., Spector, M., and Hayes, W. C. (1992). Role of loads and prosthesis material properties on the mechanics of the proximal femur after total hip arthroplasty. *J. Orthop. Res.* **10**: 405–422.

Chen, S. Y., Liang, W. M., and Yen, P. S. (2001). Reinforcement of acrylic denture base resin by incorporation of various fibers. *J. Biomed. Mater. Res.* **58**(2): 203–208.

Christel, P., Leray, J., Sedel, L., and Morel, E. (1980). Mechanical evaluation and tissue compatibility of materials for composite bone plates. in *Mechanical Properties of Biomaterials*, G. Hasting and D. F. Williams, eds. Wiley, New York, pp. 367–377.

Ciappetta, P., Boriani, S., and Fava, G. P. (1997). A carbon fiber reinforced polymer cage for vertebral body replacement: technical note. *Neurosurgery* **4**(5): 1203–1206.

Claes, L., Hutter, W., and Weiss, R. (1997). Mechanical properties of carbon reinforced polysulfone plates for internal fixation. in *Biological and Biomechanical Performance of Biomaterials*, P. Christel, A. Meunier, and A. J. C. Lee, eds. Elsevier, Amsterdam, pp. 81–86.

Claes, L., Schultheiss, M., Wolf, S., Wilke, H. J., Arand, M., and Kinzl, L. (1999). A new radiolucent system for vertebral body replacement: its stability in comparison to other systems. *J. Biomed. Mater. Res. Appl. Biomater.* **48**(1): 82–89.

Corcoran, S., Koroluk, J., Parsons, J. R., Alexander, H., and Weiss, A. B. (1981). The development of a variable stiffness, absorbable composite bone plate. in *Current Concepts for Internal Fixation of Fractures*, H. K. Uhthoff, ed. Springer-Verlag, New York, pp. 136.

Daniels, A. U., Melissa, K. O., and Andriano, K. P. (1990). Mechanical properties of biodegradable polymers and composites proposed for internal fixation of bone. *J. Appl. Biomater.* **1**(1): 57–78.

Dawson, D. K. (2000). Medical devices. in *Comprehensive Composite Materials*, Vol. 6, A. Kelly, ed. Elsevier, pp. 1–32.

de Oliveira Simoes, J. A., and Marques, A. T. (2001). Determination of stiffness properties of braided composites for the design of hip prostheses. *Composites, Part A* **32**: 655–662.

De Santis, R., Ambrosio, L., and Nicolais, L. (2000). Polymer-based composite hip prosthesis. *J. Inorg. Biochem.* **79**: 97–102.

Dunn, M. G. (1998). Anterior cruciate ligament prostheses. in *Encyclopedia of Sports Medicine and Science*, T. D. Fahey, ed. Internal Society for Sports Science, http://sportsci.org.

Durselen, L., Dauner, M., Hierlemann, H., Planck, H., Claes, L. E., and Ignatius, A. (2001) Resorbable polymer fibers for ligament augmentation. *J. Biomed. Mater. Res.* **58**(6): 666–672.

Einstein, A. (1956). in *Investigation of Theory of Brownian Motion.* Dover, New York.

Frazza, E. J., and Schmitt, E. E. (1971). A new absorbable suture. *J. Biomed. Mater. Res.* **10**: 43.

Furukawa, T., Matsusue, Y., Yasunaga, T., Nakagawa, Y., Shikinami, Y., Okuno, M., and Nakamura, T. (2000a). Bone bonding ability of a new biodegradable composite for internal fixation of bone fractures. *Clin. Orthop.* **379**: 247–258.

Furukawa, T., Matsusue, Y., Yasunaga, T., Shikinami, Y., Okuno, M., and Nakamura, T. (2000b). Biodegradation behavior of ultra-high-strength hydroxyapatite/poly(L-lactide) composite rods for internal fixation of bone fractures. *Biomaterials* **21**: 889–898.

Gershon, B., Cohn, D., and Marom, G. (1992). Compliance and ultimate strength of composite arterial prostheses. *Biomaterials* **13**: 38–43.

Gershon, B., Cohn, D., and Marom, G. (1990). The utilization of composite laminate theory in the design of synthetic soft tissue for biomedical prostheses. *Biomaterials* **11**: 548–552.

Gilding, D. K., and Reed, A. M. (1979). Biodegradable polymers for use in surgery: PGA/PLA homo- and copolymers. 1. *Polymer* **20**: 1459.

Gillett, N., Brown, S. A., Dumbleton, J. H., and Pool, R. P. (1986). The use of short carbon fiber reinforced thermoplastic plates for fracture fixation. *Biomaterials* **6**: 113–21.

Guidoin, M. F., Marois, Y., Bejui, J., Poddevin, N., King, M. W., and Guidoin R. (2000). Analysis of retrieved polymer fiber based replacements for the ACL. *Biomaterials* **21**: 2461–2474.

Guild, F. J., and Bonfield, W. (1993). Predictive modeling of hydroxyapatite–polyethylene composite. *Biomaterials* **14**(13): 985–993.

Halpin J. C., and Kardos J. L. (1976). The Halpin-Tsai equations: a review. *Poly. Eng. Sci.* **16**(5): 344–335.

Heller, J., Penhale, D. W. H., and Helwing, R. F. (1980). Preparation of poly(ortho esters) by the reaction of diketene acetals and polyols. *J. Polymer Sci. Polymer Lett. Ed.* **18**: 619.

Hench, L. L., Splinter, R. J., Allen, W. C., and Greenlee, T. K. (1971). Bonding mechanisms at the interface of ceramic prosthetic materials. *J. Biomed. Mater. Res.* **72**: 117–141.

Iannace, S., Sabatini, G., Ambrosio, L., and Nicolais, L. (1995). Mechanical behavior of composite artificial tendons and ligaments. *Biomaterials* **16**: 675–680.

Jarcho, M., Kay, J. F., Gumaer, K. I., Domerus, R. H., and Droback, H. P. (1977). Tissue cellular and subcellular events at a bone-ceramic hydroxyapatite interface. *J. Bioeng.* **1**: 79–92.

Karaman, A. I., Kir, N., and Belli, S. (2002). Four applications of reinforced polyethylene fiber material in orthodontic practice. *Am. J. Orthod. Dentofacial. Orthop.* **121**(6): 650–654.

Kazanci, M., Cohn, D., Marom, G., Migliaresi, C., and Pegoretti, A. (2002). Fatigue characterization of polyethylene fiber reinforced polyolefin biomedical composites. *Composites, Part A* **33**: 453–458.

Kellomaki, M., Niiranen, H., Puumanen, K., Ashammakhi, N., Waris, T., and Tormala, P. (2000). Bioabsorbable scaffolds for guided bone regeneration and generation. *Biomaterials* **21**: 2495–2505.

Kerner, E. H. (1956). The elastic and thermo-elastic properties of composite media. *Proc. Phys. Soc. B* **69**: 808–13.

Kolarik, J., Migliaresi, C., Stol, M., and Nicolais, L. (1981). Mechanical properties of model synthetic tendons, *J. Biomed. Mater. Res.* **15**: 147.

Kotani, Y., Abumi, K., Shikinami, Y., Takada, T., Kadoya, K., Shimamoto, N., Ito, M., Kadosawa, T., Fujinaga, T., and

Kaneda, K. (2002). Artificial intervertebral disc replacement using bioactive three-dimensional fabric: design, development, and preliminary animal study. *Spine* **27**(9): 929–935.

Kurtz, S. M., Muratoglu, O. K., Evans, M., and Edidin, A. A. (1999). Advances in the processing, sterilization, and crosslinking of ultra-high molecular weight polyethylene for total joint arthroplasty. *Biomaterials* **20**: 1659–1688.

Kwarteng, K. B. (1990). Carbon fiber reinforced PEEK (APC-2/AS-4) composites for orthopaedic implants. *SAMPE Quart.* **21**(2): 10–14.

Ladizesky, N. H., Chow, T. W., and Cheng, Y. Y. (1994). Denture base reinforcement using woven polyethylene fiber. *Int. J. Prosthodont.* **7**(4): 307–314.

Lemons, J. E., Matukas, V. J., Nieman, K. M. W., Henson, P. G., and Harvey, W. K. (1984). Synthetic hydroxylapatite and collagen combinations for the surgical treatment of bone. in *Biomedical Engineering*, Vol. 3, Sheppard, L. C., ed. Pergamon, New York, p. 13.

Leonor, I. B., Ito, A., Kanzaki, N., and Reis R. L. (2003). In vitro bioactivity of starch thermoplastic/hydroxyapatite composite biomaterials: an in situ study using atomic force microscopy. *Biomaterials* **24**: 579–585.

Lin, T. C. (1986). Totally absorbable fiber reinforced composite from internal fracture fixation devices. *Trans. Soc. Biomater.* **9**: 166.

Lin, T. W., Corvelli, A. A., Frondoza, C. G., Roberts, J. C., and Hungerford D. S. (1997). Glass PEEK composite promotes proliferation and osteocalcin production of human osteoblastic cells. *J. Biomed. Mater. Res.* **36**(2): 137–144.

Lu, L., and Mikos A.G. (1996). The importance of new processing techniques in tissue engineering. *MRS Bull.* **21**(11): 28–32.

Magee, F. P., Weinstein, A. M., Longo, J. A., Koeneman, J. B., and Yapp, R. A. (1988). A canine composite femoral stem. *Clin. Orthop. Rel. Res.* **235**: 237.

Marcolongo, M., Ducheyne, P., Garino, J., and Schepers, E. (1998). Bioactive glass fiber/polymeric composites bond to bone tissue. *J. Biomed. Mater. Res.* **39**(1): 161–170.

Martin, A., Bauer, T. W., Manley, M. T., and Marks, K. E. (1988). Osteosarcoma at the site of total hip replacement. A case report. *J. Bone Joint Surg. Am.* **70**: 1561–1567.

Middleton, J. C., and Tipton, A. J. (2000). Synthetic biodegradable polymers as orthopedic devices. *Biomaterials* **21**: 2335–2346.

Migliaresi, C., and Fambri, L. (1997). Processing and degradation of poly(L-lactic acid) fibres. *Macromol. Symp.* **123**: 155–161.

Migliaresi, C., Nicoli, F., Rossi, S., and Pegoretti, A. (2004). Novel uses of carbon composites for the fabrication of external fixators. *Comp. Sci. Tech.* **64**(6): 873–883.

Morita, S., Furuya, K., Ishihara, K., and Nakabayashi, N. (1998). Performance of adhesive bone cement containing hydroxyapatite particles. *Biomaterials* **19**(17): 1601–1606.

Morrison, C., Macnair, R., MacDonald, C., Wykman, A., Goldie, I., and Grant, M. H. (1995). In vitro biocompatibility testing of polymers for orthopaedic implants using cultured fibroblasts and osteoblasts. *Biomaterials* **16**(3): 987–992.

Moszner, N., and Salz, U. (2001). New developments of polymeric dental composites. *Prog. Polymer Sci.* **26**(4): 535–576.

Nicolais, L., and Narkis, M. (1971). Stress-strain behavior of SAN/glass bead composites in the glassy region. *Polym. Eng. Sci.* 194–199.

Nielsen, L. E. (1974). Mechanical properties of polymers and composites, Marcel Dekker, Inc., New York.

Pourdeyhimi, B., Robinson, H. H., Schwartz, P., and Wagner, H. D. (1986). Fracture toughness of Kevlar 29/poly(methyl methacrylate) composite materials for surgical implantations. *Ann. Biomed. Eng.* **14**(3): 277–294.

Qing, L., de Wijn, J. R., and van Blitterswijk, C. A. (1997). Nano-apatite/polymer composites: mechanical and physicochemical characteristics. *Biomaterials* **18**: 1263–1270.

Ramakrishna, S., Mayer, J., Wintermantel, E., and Leong, K. W. (2001). Biomedical applications of polymer composite materials: a review. *Composites Sci. Technol.* **61**: 1189–1224.

Reis, R. L., and Cunha, A. M. (2000). New degradable load-bearing biomaterials based on reinforced thermoplastic starch incorporatin blends. *J. Appl. Med. Polymer.* **4**: 1–5.

Reis, R. L., Cunha, A. M., and Bevis, M. J. (1998). Shear controlled orientation injection molding of polymeric composites with enhanced properties. *SPE Proceedings*, 57th Annual Technical Conference, Atlanta, USA, pp. 487–493.

Rokkanen, P. U., Bostman, O., Hirvensalo, E., Makela, E. A., Partio, E. K., Patiala, H., Vainionpaa, S., Vihtonen, K., and Tormala, P. (2000). Bioabsorbable fixation in orthopaedic surgery and traumatology. *Biomaterials* **21**: 2607–2613.

Rushton, N., and Rae, T. (1984). The intra-articular response to particulate carbon fiber reinforced high density polyethylene and its constituents: an experimental study in mice. *Biomaterials* **5**: 352–356.

Shikinami, Y., and Okuno, M. (2001). Bioresorbable devices made of forged composites of hydroxyapatite (HA) particles and poly L-lactide (PLLA). Part II: practical properties of miniscrews and miniplates. *Biomaterials* **22**: 3197–3211.

Shinzato, S., Kobayashi, M., Farid Mousa, W., Kamimura, M., Neo, M., Kitamura, Y., Kokubo, T., and Nakamura, T. (2000). Bioactive polymethyl methacrylate–based bone cement: Comparison of glass beads, apatite- and wollastonite-containing glass-ceramic, and hydroxyapatite fillers on mechanical and biological properties. *J. Biomed. Mater. Res.* **51**(2): 258–272.

St. John, K. R. (1983). Applications of advanced composites in orthopaedic implants. in *Biocompatible Polymers, Metals, and Composites*, M. Szycher, ed. Technomic, Lancaster, PA, p. 861.

Tang, R., Boyle, Jr., W. J., Mares, F., and Chiu, T.-H. (1990). Novel bioresorbable polymers and medical devices. *Trans. 16th Ann. Mtg. Soc. Biomater.* **13**: 191.

Thomson, R. C., Yaszemski, M. J., Powers, J. M., and Mikos, A. G. (1998). Hydroxyapatite fiber reinforced poly(α-hydroxy ester) foams for bone regeneration. *Biomaterials* **19**: 1935–1943.

Tormala, P. (1992). Biodegradable self-reinforced composite materials: manufacturing structure and mechanical properties. *Clin. Mater.* **10**: 29–34.

Tormala, P., Rokkanen, P., Laiho, J., Tamminmaki, M., and Vainionpaa, S. (1988). Material for osteosynthesis devices. U.S. Patent No. 4,734,257.

Ural, E., Kesenci, K., Fambri, L., Migliaresi, C., and Piskin, E. (2000). Poly(D,L-lactide/ε-caprolactone)/hydroxyapatite composites. *Biomaterials* **21**: 2147–2154.

Vacanti, C. A., Langer, R, Schloo, B., and Vacanti, J. P. (1991). Synthetic polymers seeded with chondrocytes provide a template for new cartilage formation. *Plast. Reconstr. Surg.* **88**(5): 753–759.

Vallittu, P.K. (1996). A review of fiber-reinforced denture base resins. *J. Prosthodont.* **5**(4): 270–276.

Veerabagu, S., Fujihara, K., Dasari, G. R., and Ramakrishna, S. (2003). Strain distribution analysis of braided composite bone plates. *Composites Sci. Technol.* **61**: 427–435.

Vert, M., Christel, P., Garreau, H., Audion, M., Chanavax, M., and Chabot, F. (1986). Totally bioresorbable composites systems for internal fixation of bone fractures. in *Polymers in Medicine*, Vol. 2, C. Migliaresi and L. Nicolais, eds. Plenum, New York, pp. 263–275.

Wang, A., Lin, R., Polineni, V. K., Essner, A., Stark, C., and Dubleton, J. H. (1998). Carbon fiber reinforced polyether ether ketone composite as a bearing surface for total hip replacement. *Tribology Int.* **31**: 661–667.

Wening, J. V., Katzer, A., Nicolas, V., Hahn, M., Jungbluth, K. H., and Kratzer, A. (1994). Imaging of alloplastic ligament implant. An in vivo and in vitro study exemplified by Kevlar. *Unfallchirurgie* **20**(2): 61–65.

Woo, S. L. Y., Akeson, W. H., Levenetz, B., Coutts, R. D., Matthews, J. V., and Amiel, D. (1974). Potential application of graphite fiber and methylmethacrylate resin composites as internal fixation plates. *J. Biomed. Mater. Res.* **8**: 321–328.

Xu, H. K., Eichmiller, F. C., and Giuseppetti, A. A. (2000). Reinforcement of a self-setting calcium phosphate cement with different fibers. *J. Biomed. Mater. Res.* **52**(1): 107–114.

Yang, J. M., Huang, P. Y., Yang, M. C., and Lo, S. K. (1997). Effect of MMA-g-UHMWPE grafted fiber on mechanical properties of acrylic bone cement. *J. Biomed. Mater. Res.* **38**(4): 361–369.

Yasko, A., Fellinger, E., Waller, S., Tomin, A., Peterson, M., Wang, E., and Lane, J. (1992). Comparison of biological and synthetic carriers for recombinant human BMP induced bone formation. *Trans. Orth. Res. Soc.* **17**: 71.

Yaszemski, M. J., Paune, R. G., Hayes, W. C., Langer, R., and Mikos, A. G. (1996). In vitro degradation of a poly(propylene fumavate)-based composite materials. *Biomaterials* **17**: 2127–2130.

Zimmerman, M. C., Alexander, H., Parsons, J. R., and Bajpai, P. K. (1991). The design and analysis of laminated degradable composite bone plates for fracture fixation. in *Hi-Tech Textiles*, T. Vigo, ed. ACS Publications, Washington, D.C.

Zimmerman, M. C., Parsons, J. R., and Alexander, H. (1987). The design and analysis of a laminated partially degradable composite bone plate for fracture fixation. *J. Biomed. Mater. Res. Appl. Biomater.* **21A**(3): 345.

2.13 NONFOULING SURFACES

Allan S. Hoffman and Buddy D. Ratner

INTRODUCTION

"Nonfouling" surfaces (NFSs) refer to surfaces that resist the adsorption of proteins and/or adhesion of cells. They are also loosely referred to as protein-resistant surfaces and "stealth" surfaces. It is generally acknowledged that surfaces that strongly adsorb proteins will generally bind cells, and that surfaces that resist protein adsorption will also resist cell adhesion. It is also generally recognized that hydrophilic surfaces are more likely to resist protein adsorption, and that hydrophobic surfaces usually will adsorb a monolayer of tightly adsorbed protein. Exceptions to these generalizations exist, but, overall, they are accurate statements.

An important area for NFSs focuses on bacterial biofilms. Bacteria are believed to adhere to surfaces via a "conditioning film" of molecules (often proteins) that adsorbs first to the surface. The bacteria stick to this conditioning film and begin to exude a gelatinous slime layer (the biofilm) that aids in their protection from external agents (for example, antibiotics). Such layers are particularly troublesome in devices such as urinary catheters and endotracheal tubes. However, they also form on vascular grafts, hip joint prostheses, heart valves, and other long-term implants where they can stimulate significant inflammatory reaction to the infected device. If the conditioning film can be inhibited, bacterial adhesion and biofilm formation can also be reduced. NFSs offer this possibility.

NFSs have medical and biotechnology uses as blood-compatible materials (where they may resist fibrinogen adsorption and platelet attachment), implanted devices, urinary catheters, diagnostic assays, biosensors, affinity separations, microchannel flow devices, intravenous syringes and tubing, and nonmedical uses as biofouling-resistant heat exchangers and ship bottoms. It is important to note that many of these uses involve *in vivo* implants or extracorporeal devices, and many others involve *in vitro* diagnostic assays, sensors, and affinity separations. As well as having considerable medical and economic importance, nonfouling surfaces offer important experimental and theoretical insights into one of the important phenomena in biomaterials science, protein adsorption. Hence, they have been the subject of many investigations. Aspects of nonfouling surfaces are addressed in many other chapters of this textbook including the chapters on water at interfaces (Chapter 1.5), surface modification (Chapter 2.14) and protein adsorption (Chapter 3.2).

The majority of the literature on non-fouling surfaces focuses on surfaces containing the relatively simple polymer poly(ethylene glycol) or PEG:

$$(-CH_2CH_2O-)_n$$

When n is in the range of 15 to 3500 (molecular weights of approximately 400–100,000), the PEG designation is used. When molecular weights are greater than 100,000, the molecule is commonly referred to as poly(ethylene oxide) (PEO). Where n is in the range of 2–15, the term oligo(ethylene glycol) (oEG) is often used. An interesting article on the origins of the use of PEG to enhance the circulation time of proteins in the body has recently been published by Davis (2002). Other natural and synthetic polymers besides PEG show nonfouling behavior, and they will also be discussed in this chapter.

BACKGROUND

The published literature on protein and cell interactions with biomaterial surfaces has grown significantly in the past 30 years, and the following concepts have emerged:

- It is well established that hydrophobic surfaces have a strong tendency to adsorb proteins irreversibly (Horbett and Brash, 1987, 1995; Hoffman, 1986). The driving force for this action is most likely the unfolding of the protein on the surface, accompanied by release of many hydrophobically structured water molecules from the interface, leading to a large entropy gain for the system (Hoffman, 1999). Note that adsorbed proteins can be displaced from the surface by solution phase proteins (Brash *et al.*, 1974).

- It is also well known that at low ionic strengths cationic proteins bind to anionic surfaces and anionic proteins bind to cationic surfaces (Hoffman, 1999; Horbett and Hoffman, 1975). The major thermodynamic driving force for these actions is a combination of ion–ion

coulombic interactions, accompanied by an entropy gain due to the release of counterions along with their waters of hydration. However, these interactions are diminished at physiologic conditions by shielding of the protein ionic groups at the 0.15 N ionic strength (Horbett and Hoffman, 1975). Still, lysozyme, a highly charged cationic protein at physiologic pH, strongly binds to hydrogel contact lenses containing anionic monomers (see Bohnert *et al.*, 1988, and Chapter 7.10, for discussion of class IV contact lenses).

- It has been a common observation that proteins tend to adsorb in monolayers, i.e., proteins do not adsorb nonspecifically onto their own monolayers (Horbett, 1993). This is probably due to retention of hydration water by the adsorbed protein molecules, preventing close interactions of the protein molecules in solution with the adsorbed protein molecules. In fact, adsorbed protein films are, in themselves, reasonable nonfouling surfaces with regard to other proteins (but not necessarily to cells).

- Many studies have been carried out on surfaces coated with physically or chemically immobilized PEG, and a conclusion was reached that the PEG molecular weight should be above a minimum of ca.2000 in order to provide good protein repulsion (Mori *et al.*, 1983; Gombotz *et al.*, 1991; Merrill, 1992). This seems to be the case whether PEG is chemically bound as a side chain of a polymer that is grafted to the surface (Mori *et al.*, 1983), is bound by one end to the surface (Gombotz *et al.*, 1991; Merrill, 1992), or is incorporated as segments in a cross-linked network (Merrill, 1992). The minimum MW was found to be ca. 500–2000, depending on packing density (Mori *et al.*, 1983; Gombotz *et al.*, 1991; Merrill, 1992).

- The mechanism of protein resistance by the PEG surfaces may due to be a combination of factors, including the resistance of the polymer coil to compression due to its desire to retain the volume of a random coil (called "entropic repulsion" or "elastic network" resistance) plus the resistance of the PEG molecule to release both bound and free water from within the hydrated coil (called "osmotic repulsion") (Gombotz *et al.*, 1991; Antonsen and Hoffman, 1992). The size of the adsorbing protein and its resistance to unfolding may also be an important factor determining the extent of adsorption on any surface (Lim and Herron, 1992). The thermodynamic principles governing the adsorption of proteins onto surfaces involve a number of enthalpic and entropic terms favoring or resisting adsorption. These terms are summarized in Table 1. The major factors favoring adsorption will be the entropic gain of released water and the enthalpy loss due to cation–anion attractive interactions between ionic protein groups and surface groups. The major factors favoring resistance to protein adsorption will be the retention of bound water, plus, in the case of an immobilized hydrophilic polymer, entropic and osmotic repulsion of the polymer coils.

- In spite of the evidence for a PEG molecular weight effect, excellent protein resistance can be achieved with very

TABLE 1 Thermodynamics of Protein Adsorption

		Favoring adsorption
ΔH_{ads}	(−)	VdW interactions (short-range)
	(−)	ion–ion interactions (long-range)
ΔS_{ads}	(+)	desorption of many H_2Os
	(+)	unfolding of protein
		Opposing adsorption
ΔH_{ads}	(+)	dehydration (interface between surface and protein)
	(+)	unfolding of protein
	(+)	chain compression (PEO)
ΔS_{ads}	(−)	adsorption of protein
	(−)	protein hydrophobic exposure
	(−)	chain compression (PEO)
	(−)	osmotic repulsion (PEO)

short chain PEGs (OEGs) and PEG-like surfaces (Lopez *et al.*, 1992; Sheu *et al.*, 1993).

- Surface-assembled monolayers (SAMs) of lipid–oligoEG molecules have been studied, and it has been found that at least about 50% of the surface should be covered before significant resistance to protein adsorption is observed (Prime and Whitesides, 1993). This suggests that protein resistance by OEG-coated surfaces may be related to a "cooperativity" between the hydrated, short OEG chains in the "plane of the surface," wherein the OEG chains interact together to bind water to the surface, in a way that is similar to the hydrated coil and its osmotic repulsion, as described above. It has also been observed that a minimum of 3 EG units are needed for highly effective protein repulsion (Harder *et al.*, 1998). Based on all of these observations, one may describe the mechanism as being related to the conformation of the individual oligoEG chains, along with their packing density in the SAM. It has been proposed that helical or amorphous oligoEG conformations lead to stronger water–oligoEG interactions than an all-trans oligoEG conformation (Harder *et al.*, 1998).

Packing density of the nonfouling groups on the surface is difficult to measure and often overlooked as an important factor in preparing nonfouling surfaces. Nevertheless, one may conclude that the one common factor connecting all nonfouling surfaces is their resistance to release of bound water molecules from the surface. Water may be bound to surface groups by both hydrophobic (structured water) and hydrophilic (primarily via hydrogen bonds) interactions, and in the latter case, the water may be H-bonded to neutral polar groups, such as hydroxyl (–OH) or ether (–C–O–C–) groups, or it may be polarized by ionic groups, such as –COO⁻ or –NH₃⁺. *The overall conclusion from all of the above observations is that resistance to protein adsorption at biomaterial interfaces is directly related to resistance of interfacial groups to the release of their bound waters of hydration.*

Based on these conclusions, it is obvious why the most common approaches to reducing protein and cell binding to biomaterial surfaces have been to make them more hydrophilic.

This has been accomplished most often by chemical immobilization of a hydrophilic polymer (such as PEG) on the biomaterial surface by one of the following methods: (a) using UV or ionizing radiation to graft copolymerize a hydrophilic monomer onto surface groups; (b) depositing such a polymer from the vapor of a precursor monomer in a gas discharge process; or (c) directly immobilizing a preformed hydrophilic polymer on the surface using radiation or gas discharge processes. Other approaches to make surfaces more hydrophilic have included the physical adsorption of surfactants or chemical derivatization of surface groups with neutral polar groups such as hydroxyls, or with negatively charged groups (especially since most proteins and cells are negatively charged) such as carboxylic acids or their salts, or sulfonates. Gas discharge has been used to covalently bind nonfouling surfactants such as Pluronic polyols to polymer surfaces (Sheu *et al.*, 1993), and it has also been used to deposit an "oligoEG-like" coating from vapors of triglyme or tetraglyme (Lopez *et al.*, 1992). More recently, a hydrophilic polymer containing phosphorylcholine zwitterionic groups along its backbone has been extensively studied for its nonfouling properties (Iwasaki *et al.*, 1999). Coatings of many hydrogels including poly(2-hydroxyethyl methacrylate) and polyacrylamide show reasonable nonfouling behavior. There have also been a number of naturally occurring biomolecules such as albumin, casein, hyaluronic acid, and mucin that have been coated on surfaces and have exhibited resistance to nonspecific adsorption of proteins. Naturally occurring ganglioside lipid surfactants having saccharide head groups have been used to make "stealth" liposomes (Lasic and Needham, 1995). One paper even suggested that the protein resistance of PEGylated surfaces is related to the "partitioning" of albumin into the PEG layers, causing those surfaces to "look like native albumin" (Vert and Domurado, 2000).

Recently, SAMs presenting an interesting series of headgroup molecules that can act as H-bond acceptors but not as H-bond donors have been shown to yield surfaces with unexpected protein resistance (Chapman *et al.*, 2000; Ostuni *et al.*, 2001; Kane *et al.*, 2003). Interestingly, PEG also fits in this category of H-bond acceptors but not donors. However, this generalization does not explain all nonfouling surfaces, especially a report in which mannitol groups with H-bond donor –OH groups were found to be nonfouling (Luk *et al.*, 2000). Another hypothesis proposes that the functional groups that impart a nonfouling property are kosmotropes, order-inducing molecules (Kane *et al.*, 2003). Perhaps because of the ordered water surrounding these molecules, they cannot penetrate the ordered water shell surrounding proteins so strong intermolecular interactions between surface group and protein cannot occur. An interesting kosmotrope molecule with good nonfouling ability described in this paper is taurine, $H_3N^+(CH_2)_2SO_3^-$. Table 2 summarizes some of the different compositions that have been applied as nonfouling surfaces.

It is worthwhile to mention some computational papers (supported by some experiments) that offer new insights and ideas on NFSs (Lim and Herron, 1992; Pertsin *et al.*, 2002; Pertsin and Grunze, 2000). Also, many new experimental methods have been applied to study the mechanism of nonfouling surfaces including neutron reflectivity to measure the

TABLE 2 "Nonfouling" Surface Compositions

Synthetic Hydrophilic Surfaces
- PEG polymers and surfactants
- Neutral polymers
    Poly(2-hydroxyethyl methacrylate)
    Polyacrylamide
    Poly(N-vinyl-2-pyrrolidone)
    Poly(N-isopropyl acrylamide) (below 31°C)
- Anionic polymers
- Phosphoryl choline polymers
- Gas discharge-deposited coatings (especially from PEG-like monomers)
- Self-assembled n-alkyl molecules with oligo-PEG head groups
- Self-assembled n-alkyl molecules with other polar head groups

Natural Hydrophilic Surfaces
- Passivating proteins (e.g., albumin and casein)
- Polysaccharides (e.g., hyaluronic acid)
- Liposaccharides
- Phospholipid bilayers
- Glycoproteins (e.g., mucin)

water density in the interfacial region (Schwendel et al., 2003), scanning force microscopy (Feldman et al., 1999), and sum frequency generation (Zolk et al., 2000).

Finally, it should be noted that bacteria tend to adhere and colonize almost any type of surface, perhaps even many protein-resistant NFSs. However, the best NFSs can provide acute resistance to bacteria and biofilm build-up better than most surfaces (Johnston et al., 1997). Resistance to bacterial adhesion remains an unsolved problem in surface science. Also, it has been pointed out that susceptibility of PEGs to oxidative damage may reduce their utility as nonfouling surfaces in real-world situations (Kane et al., 2003).

CONCLUSIONS AND PERSPECTIVES

It is remarkable how many different surface compositions appear to be nonfouling. Although it is difficult to be sure about the existence of a unifying mechanism for this action, it appears that the major factor favoring resistance to protein adsorption will be the retention of bound water by the surface molecules, plus, in the case of an immobilized hydrophilic polymer, entropic and osmotic repulsion by the polymer coils. Little is known about how long a nonfouling surface will remain nonfouling in vivo. Longevity and stability for nonfouling biomaterials remains an uncharted frontier. Defects (e.g., pits, uncoated areas) in NFSs may provide "footholds" for bacteria and cells to begin colonization. Enhanced understanding of how to optimize the surface density and composition of NFSs will lead to improvements in quality and fewer microdefects. Finally, it is important to note that a clean, "nonfouled" surface may not always be desirable. In the case of cardiovascular implants or devices, emboli may be shed when such

a surface is exposed to flowing blood (Hoffman et al., 1982). This can lead to undesirable consequences, even though (or perhaps especially because) the surface is an effective nonfouling surface. In the case of contact lenses, a protein-free lens may seem desirable, but there are concerns that such a lens will not be comfortable. Although biomaterials scientists can presently create surfaces that are nonfouling for a period of time, applying such surfaces must take into account the specific application, the biological environment, and the intended service life.

Bibliography

Antonsen, K. P., and Hoffman, A. S. (1992). Water structure of PEG solutions by DSC measurements. in Polyethylene Glycol Chemistry: Biotechnical and Biomedical Applications, J. M. Harris, ed. Plenum Press, New York, pp. 15–28.

Bohnert, J. L., Horbett, T. A., Ratner, B. D., and Royce, F. H. (1988). Adsorption of proteins from artificial tear solutions to contact lens materials. Invest. Ophthalom. Vis. Sci. 29(3): 362–373.

Brash, J. L., Uniyal, S., and Samak, Q. (1974). Exchange of albumin adsorbed on polymer surfaces. Trans. Am. Soc. Artif. Int. Organs 20: 69–76.

Chapman, R. G., Ostuni, E., Takayama, S., Holmlin, R. E., Yan, L., and Whitesides, G. M. (2000). Surveying for surfaces that resist the adsorption of proteins. J. Am. Chem. Soc. 122: 8303–8304.

Davis, F. F. (2000). The origin of pegnology. Adv. Drug. Del. Revs. 54: 457–458.

Feldman, K., Hahner, G., Spencer, N. D., Harder, P., and Grunze, M. (1999). Probing resistance to protein adsorption of oligo(ethylene glycol)-terminated self-assembled monolayers by scanning force microscopy. J. Am. Chem. Soc. 121(43): 10134–10141.

Gombotz, W. R., Wang, G. H., Horbett, T. A., and Hoffman, A. S. (1991). Protein adsorption to PEO surfaces. J. Biomed. Mater. Res. 25: 1547–1562.

Harder, P., Grunze, M., Dahint, R., Whitesides, G. M., and Laibinis, P. E. (1998). Molecular conformation and defect density in oligo(ethylene glycol)-terminated self-assembled monolayers on gold and silver surfaces determine their ability to resist protein adsorption. J. Phys. Chem. B 102: 426–436.

Hoffman, A. S. (1986). A general classification scheme for hydrophilic and hydrophobic biomaterial surfaces. J. Biomed. Mater. Res. 20: ix.

Hoffman, A. S. (1999). Non-fouling surface technologies. J. Biomater. Sci., Polymer Ed. 10: 1011–1014.

Hoffman, A. S., Horbett, T. A., Ratner, B. D., Hanson, S. R., Harker, L. A., and Reynolds, L. O. (1982). Thrombotic events on grafted polyacrylamide–Silastic surfaces as studied in a baboon. ACS Adv. Chem. Ser. 199: 59–80.

Horbett, T. A. (1993). Principles underlying the role of adsorbed plasma proteins in blood interactions with foreign materials. Cardiovasc. Pathol. 2: 137S–148S.

Horbett, T. A., and Brash, J. L. (1987). Proteins at interfaces: current issues and future prospects. in Proteins at Interfaces, Physicochemical and Biochemical Studies, ACS Symposium Series, Vol. 343, T. A. Horbett and J. L. Brash, eds. American Chemical Society, Washington, D.C., pp. 1–33.

Horbett, T. A., and Brash, J. L. (1995). Proteins at interfaces: an overview. in Proteins at Interfaces II: Fundamentals and Applications, ACS Symposium Series, Vol. 602, T. A. Horbett and J. L. Brash, eds. American Chemical Society, Washington, D.C., pp. 1–25.

Horbett, T. A., and Hoffman, A. S. (1975). Bovine plasma protein adsorption to radiation grafted hydrogels based on hydroxyethyl-methacrylate and *N*-vinylpyrrolidone, Advances in Chemistry Series, Vol. 145, *Applied Chemistry at Protein Interfaces*, R. Baier, ed. American Chemical Society, Washington D.C., pp. 230–254.

Iwasaki, Y., *et al.* (1999). Competitive adsorption between phospholipid and plasma protein on a phospholipid polymer surface. *J. Biomater. Sci. Polymer Ed.* **10**: 513–529.

Johnston, E. E., Ratner, B. D., and Bryers, J. D. (1997). RF plasma deposited PEO-like films: Surface characterization and inhibition of *Pseudomonas aeruginosa* accumulation. in *Plasma Processing of Polymers*, R. d'Agostino, P. Favia and F. Fracassi, eds. Kluwer Academic, Dordrecht, The Netherlands, pp. 465–476.

Kane, R. S., Deschatelets, P., and Whitesides, G. M. (2003). Kosmotropes form the basis of protein-resistant surfaces. *Langmuir* **19**: 2388–2391.

Lasic, D. D., and Needham, D. (1995). The "stealth" liposome: A prototypical biomaterial. *Chem. Rev.* **95**(8): 2601–2628.

Lim, K., and Herron, J. N. (1992). Molecular simulation of protein–PEG interaction. in *Polyethylene Glycol Chemistry: Biotechnical and Biomedical Applications* J. M. Harris, ed. Plenum Press, New York, p. 29.

Lopez, G. P., Ratner, B. D., Tidwell, C. D., Haycox, C. L., Rapoza, R. J., and Horbett, T. A. (1992). Glow discharge plasma deposition of tetraethylene glycol dimethyl ether for fouling-resistant biomaterial surfaces. *J. Biomed. Mater. Res.* **26**(4): 415–439.

Luk, Y., Kato, M., and Mrksich, M. (2000). Self-assembled monolayers of alkanethiolates presenting mannitol groups are inert to protein adsorption and cell attachment. *Langmuir* **16**: 9605.

Merrill, E. W. (1992). Poly(ethylene oxide) and blood contact: a chronicle of one laboratory. in *Polyethylene Glycol Chemistry: Biotechnical and Biomedical Applications*, J. M. Harris, ed. Plenum Press, New York, pp. 199–220.

Mori, Y., *et al.* (1983). Interactions between hydrogels containing PEO chains and platelets. *Biomaterials* **4**: 825–830.

Ostuni, E., Chapman, R. G., Holmlin, R. E., Takayama, S., and Whitesides, G. M. (2001). A survey of structure–property relationships of surfaces that resist the adsorption of protein. *Langmuir* **17**: 5605–5620.

Pertsin, A. J., and Grunze, M. (2000). Computer simulation of water near the surface of oligo(ethylene glycol)-terminated alkanethiol self-assembled monolayers. *Langmuir* **16**(23): 8829–8841.

Pertsin, A. J., Hayashi, T., and Grunze, M. (2002). Grand canonical monte carlo simulations of the hydration interaction between oligo(ethylene glycol)-terminated alkanethiol self-assembled monolayers. *J. Phys. Chem. B.* **106**(47): 12274–12281.

Prime, K. L., and Whitesides, G. M. (1993). Adsorption of proteins onto surfaces containing end-attached oligo(ethylene oxide): a model system using self-assembled monolayers. *J. Am. Chem. Soc.* **115**: 10715.

Schwendel, D., Hayashi, T., Dahint, R., Pertsin, A., Grunze, M., Steitz, R., and Schreiber, F. (2003). Interaction of water with self-assembled monolayers: neutron reflectivity measurements of the water density in the interface region. *Langmuir* **19**(6): 2284–2293.

Sheu, M.-S., Hoffman, A. S., Terlingen, J. G. A., and Feijen, J. (1993). A new gas discharge process for preparation of non-fouling surfaces on biomaterials. *Clin. Mater.* **13**: 41–45.

Vert, M., and Domurado, D. (2000). PEG: Protein-repulsive or albumin-compatible? *J. Biomater. Sci., Polymer Ed.* **11**: 1307–1317.

Zolk, M., Eisert, F., Pipper, J., Herrwerth, S., Eck, W., Buck, M., and Grunze, M. (2000). Solvation of oligo(ethylene glycol)-terminated self-assembled monolayers studied by vibrational sum frequency spectroscopy. *Langmuir* **16**(14): 5849–5852.

2.14 PHYSICOCHEMICAL SURFACE MODIFICATION OF MATERIALS USED IN MEDICINE

Buddy D. Ratner and Allan S. Hoffman

INTRODUCTION

Much effort goes into the design, synthesis, and fabrication of biomaterials and devices to ensure that they have the appropriate mechanical properties, durability, and functionality. To cite a few examples, a hip joint should withstand high stresses, a hemodialyzer should have the requisite permeability characteristics, and the pumping bladder in an artificial heart should flex for millions of cycles without failure. The bulk structure of the materials governs these properties.

The biological response to biomaterials and devices, on the other hand, is controlled largely by their surface chemistry and structure (see Chapters 1.4 and 9.4). The rationale for the surface modification of biomaterials is therefore straightforward: to retain the key physical properties of a biomaterial while modifying only the outermost surface to influence the biointeraction. If such surface modification is properly effected, the mechanical properties and functionality of the device will be unaffected, but the bioresponse related to the tissue–device interface will be improved or modulated.

Materials can be surface-modified by using biological, mechanical, or physicochemical methods. Many biological surface modification schemes are covered in Chapter 2.16. Generalized examples of physicochemical surface modifications, the focus of this chapter, are illustrated schematically in Fig. 1. Surface modification with Langmuir–Blodgett (LB) films has elements of both biological modification and physicochemical modification. LB films will be discussed later in this chapter. Some applications for surface modified biomaterials are listed in Table 1. Physical and chemical surface modification methods, and the types of materials to which they can be applied, are listed in Table 2. Methods to modify or create surface texture or roughness will not be explicitly covered here, though chemical patterning of surfaces will be addressed.

GENERAL PRINCIPLES

Surface modifications fall into two categories: (1) chemically or physically altering the atoms, compounds, or molecules in the existing surface (chemical modification, etching, mechanically roughening), or (2) overcoating the existing surface with a material having a different composition (coating, grafting, thin film deposition) (Fig. 1). A few general principles provide guidance when undertaking surface modification:

Thin Surface Modifications

Thin surface modifications are desirable. The modified zone at the surface of the material should be as thin as possible. Modified surface layers that are too thick can change the mechanical and functional properties of the material.

Surface Modification Possibilities

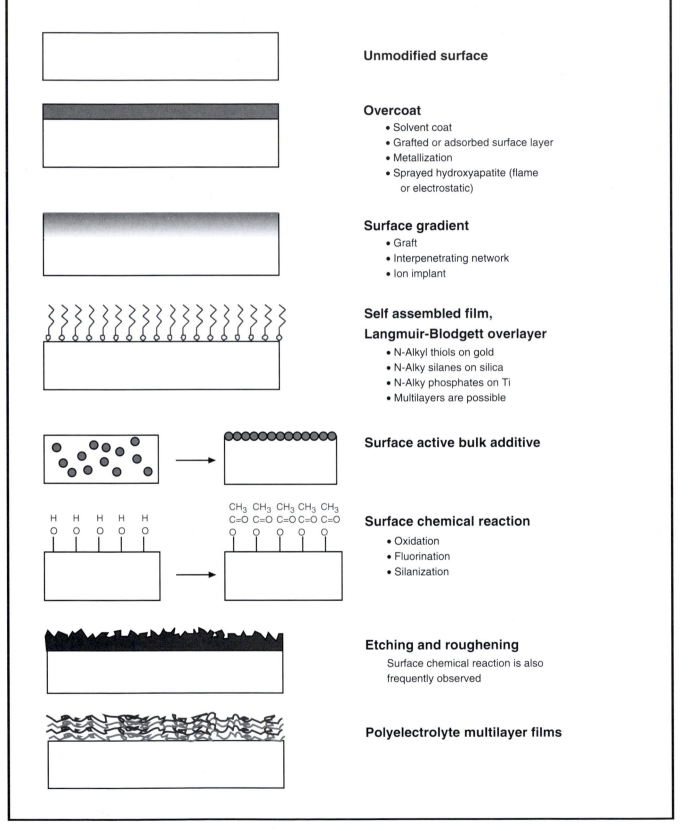

Unmodified surface

Overcoat
- Solvent coat
- Grafted or adsorbed surface layer
- Metallization
- Sprayed hydroxyapatite (flame or electrostatic)

Surface gradient
- Graft
- Interpenetrating network
- Ion implant

Self assembled film, Langmuir-Blodgett overlayer
- N-Alkyl thiols on gold
- N-Alky silanes on silica
- N-Alky phosphates on Ti
- Multilayers are possible

Surface active bulk additive

Surface chemical reaction
- Oxidation
- Fluorination
- Silanization

Etching and roughening
Surface chemical reaction is also frequently observed

Polyelectrolyte multilayer films

FIG. 1. Schematic representations of methods to modify surfaces.

TABLE 1 Some Physicochemically Surface-Modified Biomaterials

To modify blood compatibility
Octadecyl group attachment to surfaces (albumin affinity)
Silicone-containing block copolymer additive
Plasma fluoropolymer deposition
Plasma siloxane polymer deposition
Radiation grafted hydrogel
Chemically modified polystyrene for heparin-like activity

To influence cell adhesion and growth
Oxidized polystyrene surface
Ammonia plasma-treated surface
Plasma-deposited acetone or methanol film
Plasma fluoropolymer deposition (reduce endothelial adhesion to IOLs)

To control protein adsorption
Surface with immobilized poly(ethylene glycol) (reduce adsorption)
Treated ELISA dish surface (increase adsorption)
Affinity chromatography column
Surface cross-linked contact lens (reduce adsorption)

To improve lubricity
Plasma treatment
Radiation grafting (hydrogels)
Interpenetrating polymeric networks

To improve wear resistance and corrosion resistance
Ion implantation
Diamond deposition
Anodization

To alter transport properties
Polyelectrolyte grafting

To modify electrical characteristics
Polyelectrolyte grafting
Magnetron sputtering of titanium

Thick coatings are also more subject to delamination and cracking. How thin should a surface modification be? Ideally, alteration of only the outermost molecular layer (3–10 Å) should be sufficient. In practice, thicker films than this will be necessary since it is difficult to ensure that the original surface is uniformly covered when coatings and treatments are so thin. Also, extremely thin layers may be more subject to surface reversal (see later discussion) and mechanical erosion. Some coatings intrinsically have a specific thickness. For example, the thickness of LB films is related to the length of the amphiphilic molecules that form them (25–50 Å). Other coatings, such as poly(ethylene glycol) protein-resistant layers, may require a minimum thickness (a dimension related to the molecular weight of chains) to function. In general, surface modifications should be the minimum thickness needed for uniformity, durability, and functionality, but no thicker. This is often experimentally defined for each system.

Delamination Resistance

The surface-modified layer should be resistant to delamination and cracking. Resistance to delamination is achieved by covalently bonding the modified region to the substrate, intermixing the components of the substrate and the surface film at an interfacial zone (for example, an interpenetrating network), applying a compatibilizing ("primer") layer at the interface, or incorporating appropriate functional groups for strong intermolecular adhesion between a substrate and an overlayer (Wu, 1982).

Surface Rearrangement

Surface rearrangement can readily occur. It is driven by a thermodynamic minimization of interfacial energy and enhanced by molecular mobility. Surface chemistries and structures can "switch" because of diffusion or translation of surface atoms or molecules in response to the external environment (see Chapter 1.4 and Fig. 2 in that chapter). A newly formed surface chemistry can migrate from the surface into the bulk, or molecules from the bulk can diffuse to cover the surface. Such reversals occur in metallic and other inorganic systems, as well as in polymeric systems. Terms such as "reconstruction," "relaxation," and "surface segregation" are often used to describe mobility-related alterations in surface structure and chemistry (Ratner and Yoon, 1988; Garbassi *et al.*, 1989; Somorjai, 1990, 1991). The driving force for these surface changes is a minimization of the interfacial energy. However, sufficient atomic or molecular mobility must exist for the surface changes to occur in reasonable periods of time. For a modified surface to remain as it was designed, surface reversal must be prevented or inhibited. This can be done by cross-linking, sterically blocking the ability of surface structures to move, or by incorporating a rigid, impermeable layer between the substrate material and the surface modification.

Surface Analysis

Surface modification and surface analysis are complementary and sequential technologies. The surface-modified region is usually thin and consists of only minute amounts of material. Undesirable contamination can readily be introduced during modification reactions. The potential for surface reversal to occur during surface modification is also high. The surface reaction should be monitored to ensure that the intended surface is indeed being formed. Since conventional analytical methods are often insufficiently sensitive to detect surface modifications, special surface analytical tools are called for (Chapter 1.4).

Commercializability

The end products of biomaterials research are devices and materials that are manufactured to exacting specifications for use in humans. A surface modification that is too complex will be difficult and expensive to commercialize. It is best to minimize the number of steps in a surface modification process and to design each step to be relatively insensitive to small changes in reaction conditions.

METHODS FOR MODIFYING THE SURFACES OF MATERIALS

General methods to modify the surfaces of materials are illustrated in Fig. 1, with many examples listed in Table 2. A few of the more widely used of these methods will be briefly described. Some of the conceptually simpler methods such as solution coating of a polymer onto a substrate or metallization by sputtering or thermal evaporation will not be elaborated upon here.

Chemical Reaction

There are hundreds of chemical reactions that can be used to modify the chemistry of a surface. Chemical reactions, in the context of this article, are those performed with reagents that react with atoms or molecules at the surface, but do not overcoat those atoms or molecules with a new layer. Chemical reactions can be classified as nonspecific and specific.

Nonspecific reactions leave a distribution of different functional groups at the surface. An example of a nonspecific surface chemical modification is the chromic acid oxidation of polyethylene surfaces. Other examples include the corona discharge modification of materials in air; radio-frequency glow discharge (RFGD) treatment of materials in oxygen, argon, nitrogen, carbon dioxide, or water vapor plasmas; and the oxidation of metal surfaces to a mixture of suboxides.

Specific chemical surface reactions change only one functional group into another with a high yield and few side reactions. Examples of specific chemical surface modifications for polymers are presented in Fig. 2. Detailed chemistries of biomolecule immobilization are described in Chapter 2.16.

Radiation Grafting and Photografting

Radiation grafting and related methods have been widely applied for the surface modification of biomaterials starting in the late 1960s (Hoffman *et al.*, 1972), and comprehensive review articles are available (Ratner, 1980; Hoffman, 1981; Hoffman *et al.*, 1983; Stannett, 1990; Safrany, 1997). The earliest applications, particularly for biomedical applications, focused on attaching chemically reactable groups (–OH, –COOH, –NH$_2$, etc) to the surfaces of relatively inert hydrophobic polymers. Within this category, three types of

TABLE 2   Physical and Chemical Surface Modification Methods

	Polymer	Metal	Ceramic	Glass
Noncovalent coatings				
Solvent coating	✓	✓	✓	✓
Langmuir–Blodgett film deposition	✓	✓	✓	✓
Surface-active additives	✓	✓	✓	✓
Vapor deposition of carbons and metals[a]	✓	✓	✓	✓
Vapor deposition of parylene (*p*-xylylene)	✓	✓	✓	✓
Covalently attached coatings				
Radiation grafting (electron accelerator and gamma)	✓	—	—	—
Photografting (UV and visible sources)	✓	—	—	✓
Plasma (gas discharge) (RF, microwave, acoustic)	✓	✓	✓	✓
Gas-phase deposition				
• Ion beam sputtering	✓	✓	✓	✓
• Chemical vapor deposition (CVD)	—	✓	✓	✓
• Flame spray deposition	—	✓	✓	✓
Chemical grafting (e.g., ozonation + grafting)	✓	✓	✓	✓
Silanization	✓	✓	✓	✓
Biological modification (biomolecule immobilization)	✓	✓	✓	✓
Modifications of the original surface				
Ion beam etching (e.g., argon, xenon)	✓	✓	✓	✓
Ion beam implantation (e.g., nitrogen)	—	✓	✓	✓
Plasma etching (e.g., nitrogen, argon, oxygen, water vapor)	✓	✓	✓	✓
Corona discharge (in air)	✓	✓	✓	✓
Ion exchange	✓[b]	✓	✓	✓
UV irradiation	✓	✓	✓	✓
Chemical reaction				
• Nonspecific oxidation (e.g., ozone)	✓	✓	✓	✓
• Functional group modifications (oxidation, reduction)	✓	—	—	—
• Addition reactions (e.g., acetylation, chlorination)	✓	—	—	—
Conversion coatings (phosphating, anodization)	—	✓	—	—
Mechanical roughening and polishing	✓	✓	✓	✓

[a] Some covalent reaction may occur.
[b] For polymers with ionic groups.

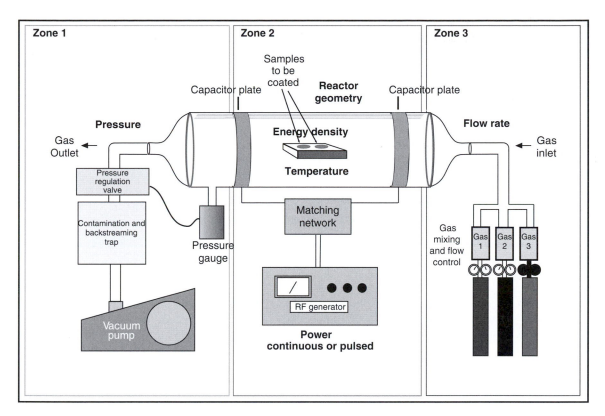

FIG. 2. A diagram of a capacitively coupled RF plasma reactor. Important experimental variables are indicated in bold type-face. Zone 1 shows gas storage and mixing. Zone 2 shows components that power the reactor. Zone 3 highlights components of the vacuum system.

reactions can be distinguished: grafting using ionizing radiation sources (most commonly, a cobalt-60 or cesium-137 gamma radiation source) (Dargaville *et al.*, 2003), grafting using UV radiation (photografting) (Srinivasan and Lazare, 1985; Matsuda and Inoue, 1990; Dunkirk *et al.*, 1991; Swanson, 1996), and grafting using high-energy electron beams (Singh and Silverman, 1992). In all cases, similar processes occur. The radiation breaks chemical bonds in the material to be grafted, forming free radicals, peroxides, or other reactive species. These reactive surface groups are then exposed to a monomer. The monomer reacts with the free radicals at the surface and propagates as a free radical chain reaction incorporating other monomers into a surface grafted polymer. Electron beams and gamma radiation sources are also used for biomedical device sterilization (see Chapter 9.2).

These high-energy surface modification technologies are strongly dependent on the source energy, the radiation dose rate, and the amount of the dose absorbed. Gamma sources have energies of roughly 1 MeV (1 eV = 23.06 kcal/mol). Typical energies for electron beam processing are 5–10 MeV. UV radiation sources are of much lower energy (<6 eV). Radiation does rates are low for UV and gamma and very high for electron beams. The amount of energy absorbed is measured in units of grays (Gy) where 1 kilogray (kGy) = 1000 joules/kilogram. Units of megarads (MR) are often used for gamma sources; $1 MR = 1 \times 10^6$ ergs/gram = 10 kGy.

Three distinct reaction modes can be described: (a) In the mutual irradiation method, the substrate material is immersed in an oxygen-free solution (monomer ± solvent) that is then exposed to the radiation source. (b) The substrate materials can also be exposed to the radiation under an inert atmosphere or at low temperatures (to stabilize free radicals). In this case, the materials are later contacted with a monomer solution to initiate the graft process. (c) Finally, the exposure to the radiation can take place in air or oxygen, leading to the formation of per-oxide groups on the surface. Heating the material to be grafted in the presence of monomer, or addition of a redox reactant (e.g., Fe^{2+}) that will decompose the peroxide groups to form free radicals, can initiate the graft polymerization (in O_2-free conditions).

Graft layers formed by energetic irradiation of the substrate are often thick (>1 μm) and composed of relatively high-molecular-weight polymer chains. However, they are typically well-bonded to the substrate material. Since many polymerizable monomers are available, a wide range of surface chemistries can be created. Mixtures of monomers can form unique graft copolymers (Ratner and Hoffman, 1980). For example, the hydrophilic/hydrophobic ratio of surfaces can be controlled by varying the ratio of a hydrophilic and a hydrophobic monomer in the grafting mixture (Ratner and Hoffman, 1980; Ratner *et al.*, 1979).

Photoinitiated grafting (usually with visible or UV light) represents a unique subcategory of surface modifications in which there is growing interest. There are many approaches to effect this photoinitiated covalent coupling. For example, a phenyl azide group can be converted to a highly reactive nitrene upon UV exposure. This nitrene will quickly react with many organic groups. If a synthetic polymer is prepared with phenyl azide side groups and this polymer is exposed simultaneously to UV light and a substrate polymer or polymeric medical device, the polymer containing the phenyl azide side groups will be immobilized to the substrate (Matsuda and Inoue, 1990). Another method involves the coupling of a benzophenone molecule to a hydrophilic polymer (Dunkirk et al., 1991). In the presence of UV irradiation, the benzophenone is excited to a reactive triplet state that can abstract a hydrogen leading to radical cross-linking.

Radiation, electron beam, and photografting have frequently been used to bond hydrogels to the surfaces of hydrophobic polymers (Matsuda and Inoue, 1990; Dunkirk et al., 1991). Electron beam grafting of N-isopropyl acrylamide to polystyrene has been used to create a new class of temperature-dependent surfaces for cell growth (Kwon et al., 2000) (also see Chapter 2.6). The protein interactions (Horbett and Hoffman, 1975), cell interactions (Ratner et al., 1975; Matsuda and Inoue, 1990), blood compatibility (Chapiro, 1983; Hoffman et al., 1983), and tissue reactions (Greer et al., 1979) of hydrogel graft surfaces have been investigated.

RFGD Plasma Deposition and Other Plasma Gas Processes

RFGD plasmas, as used for surface modification, are low-pressure ionized gas environments typically at ambient (or slightly above ambient) temperature. They are also referred to as glow discharge or gas discharge depositions or treatments. Plasmas can be used to modify existing surfaces by ablation or etching reactions or, in a deposition mode, to overcoat surfaces (Fig. 1). Good review articles on plasma deposition and its application to biomaterials are available (Yasuda and Gazicki, 1982; Hoffman, 1988; Ratner et al., 1990; Chu et al., 2002; Kitching et al., 2003). Some biomedical applications of plasma-modified biomaterials are listed in Table 3.

The application of RFGD plasma surface modification in biomaterials development is steadily increasing. Because such coatings and treatments have special promise for improved biomaterials, they will be emphasized in this chapter. The specific advantages of plasma-deposited films (and to some extent, plasma-treated surfaces) for biomedical applications are:

1. They are conformal. Because of the penetrating nature of a low-pressure gaseous environment in which mass transport is governed by both molecular (line-of-sight) diffusion and convective diffusion, complex geometric shapes can be treated.
2. They are free of voids and pinholes. This continuous barrier structure is suggested by transport studies and electrical property studies (Charlson et al., 1984).
3. Plasma-deposited polymeric films can be placed upon almost any solid substrate, including metals, ceramics, and semiconductors. Other surface-grafting or surface-modification technologies are highly dependent upon the chemical nature of the substrate.
4. They exhibit good adhesion to the substrate. The energetic nature of the gas-phase species in the plasma

TABLE 3   Biomedical Applications of Glow Discharge Plasma-Induced Surface Modification Processes

A. Plasma treatment (etching)
   1. Clean
   2. Sterilize
   3. Cross-link surface molecules

B. Plasma treatment (etching) and plasma deposition
   1. Form barrier films
      a. Protective coating
      b. Electrically insulating coating
      c. Reduce absorption of material from the environment
      d. Inhibit release of leachables
      e. Control drug delivery rate

   2. Modify cell and protein reactions
      a. Improve biocompatibility
      b. Promote selective protein adsorption
      c. Enhance cell adhesion
      d. Improve cell growth
      e. Form nonfouling surfaces
      f. Increase lubricity

   3. Provide reactive sites
      a. For grafting or polymerizing polymers
      b. For immobilizing biomolecules

reaction environment can induce mixing, implantation, penetration, and reaction between the overlayer film and the substrate.
5. Unique film chemistries can be produced. The chemical structure of the polymeric overlayer films generated from the plasma environment usually cannot be synthesized by conventional chemical methods.
6. They can serve as excellent barrier films because of their pinhole-free and dense, cross-linked nature.
7. Plasma-deposited layers generally show low levels of leachables. Because they are highly cross-linked, plasma-deposited films contain negligible amounts of low-molecular-weight components that might lead to an adverse biological reaction. They can also prevent leaching of low-molecular-weight material from the substrate.
8. These films are easily prepared. Once the apparatus is set up and optimized for a specific deposition, treatment of additional substrates is rapid and simple.
9. The production of plasma depositions is a mature technology. The microelectronics industry has made extensive use of inorganic plasma-deposited films for many years (Sawin and Reif, 1983; Nguyen, 1986).
10. Plasma surface modifications, although they are chemically complex, can be characterized by infrared (IR) (Inagaki et al., 1983; Haque and Ratner, 1988; Krishnamurthy et al., 1989), nuclear magnetic resonance (NMR) (Kaplan and Dilks, 1981), electron spectroscopy for chemical analysis (ESCA) (Chilkoti et al., 1991a), chemical derivatization studies (Everhart and Reilley, 1981; Gombotz and Hoffman, 1988; Griesser and Chatelier, 1990; Chilkoti et al., 1991a), and static secondary ion mass spectrometry (SIMS

(Chilkoti *et al.*, 1991b, 1992; Johnston and Ratner, 1996).

11. Plasma-treated surfaces are sterile when removed from the reactor, offering an additional advantage for cost-efficient production of medical devices.

It would be inappropriate to cite all these advantages without also discussing some of the disadvantages of plasma deposition and treatment for surface modification. First, the chemistry produced on a surface is often ill-defined. For example, if tetrafluoroethylene gas is introduced into the reactor, polytetrafluoroethylene will not be deposited on the surface. Rather, a complex, branched fluorocarbon polymer will be produced. This scrambling of monomer structure has been addressed in studies dealing with retention of monomer structure in the final film (Lopez and Ratner, 1991; Lopez *et al.*, 1993; Panchalingham *et al.*, 1993). Second, the apparatus used to produce plasma depositions can be expensive. A good laboratory-scale reactor will cost $10,000–30,000, and a production reactor can cost $100,000 or more. Third, uniform reaction within long, narrow pores can be difficult to achieve. Finally, contamination can be a problem and care must be exercised to prevent extraneous gases and pump oils from entering the reaction zone. However, the advantages of plasma reactions outweigh these potential disadvantages for many types of modifications that cannot be accomplished by other methods.

The Nature of the Plasma Environment

Plasmas are atomically and molecularly dissociated gaseous environments. A plasma environment contains positive ions, negative ions, free radicals, electrons, atoms, molecules, and photons (visible and UV). Typical conditions within the plasma include an electron energy of 1–10 eV, a gas temperature of 25–60°C, an electron density of 10^{-9} to 10^{-12}/cm^3, and an operating pressure of 0.025–1.0 torr.

A number of processes can occur on the substrate surface that lead to the observed surface modification or deposition. First, a competition takes place between deposition and etching by the high-energy gaseous species (ablation) (Yasuda, 1979). When ablation is more rapid than deposition, no deposition will be observed. Because of its energetic nature, the ablation or etching process can result in substantial chemical and morphological changes to the substrate.

A number of mechanisms have been postulated for the deposition process. The reactive gaseous environment and UV emission may create free radical and other reactive species on the substrate surface that react with and polymerize molecules from the gas phase. Alternately, reactive small molecules in the gas phase could combine to form higher-molecular-weight units or particulates that may settle or precipitate onto the surface. Most likely, the depositions observed are formed by some combination of these two processes.

Production of Plasma Environments for Deposition

Many experimental variables relating both to reaction conditions and to the substrate onto which the deposition is placed affect the final outcome of the plasma deposition process (Fig. 2). A diagram of a typical inductively coupled radio frequency plasma reactor is presented in Fig. 2. The major subsystems that make up this apparatus are a gas introduction system (control of gas mixing, flow rate, and mass of gas entering the reactor), a vacuum system (measurement and control of reactor pressure and inhibition of backstreaming of molecules from the pumps), an energizing system to efficiently couple energy into the gas phase within the reactor, and a reactor zone in which the samples are treated. Radio-frequency, acoustic, or microwave energy can be coupled to the gas phase. Devices for monitoring the molecular weight of the gas-phase species (mass spectrometers), the optical emission from the glowing plasma (spectrometers), and the deposited film thickness (ellipsometers, vibrating quartz crystal microbalances) are also commonly found on plasma reactors. Technology has been developed permitting atmospheric-pressure plasma deposition (Massines *et al.*, 2000; Klages *et al.*, 2000). Another important development is "reel-to-reel" (continuous) plasma processing, opening the way to low-cost treatment of films, fibers, and tubes.

RFGD Plasmas for the Immobilization of Molecules

Plasmas have often been used to introduce organic functional groups (e.g., amine, hydroxyl) on a surface that can be activated to attach biomolecules (see Chapter 2.16). Certain reactive gas environments can also be used for directly immobilizing organic molecules such as surfactants. For example, a poly(ethylene glycol)-*n*-alkyl surfactant will adsorb to polyethylene via the propylene glycol block. If the polyethylene surface with the adsorbed surfactant is briefly exposed to an argon plasma, the *n*-alkyl chain will be cross-linked, thereby leading to the covalent attachment of pendant poly(ethylene glycol) chains (Sheu *et al.*, 1992).

High-Temperature and High-Energy Plasma Treatments

The plasma environments described above are of relatively low energy and low temperature. Consequently, they can be used to deposit organic layers on polymeric or inorganic substrates. Under higher energy conditions, plasmas can effect unique and important inorganic surface modifications on inorganic substrates. For example, flame-spray deposition involves injecting a high-purity, relatively finely divided (~100 mesh) metal powder into a high-velocity plasma or flame. The melted or partially melted particles impact the surface and rapidly solidify (see Chapter 2.9). An example of thermal spray coating on titanium is seen in Gruner (2001).

Silanization

Silane treatments of surfaces involve a liquid-phase chemical reaction and are straightforward to perform and low cost. A typical silane surface modification reaction is illustrated in Fig. 4. Silane reactions are most often used to modify hydroxylated surfaces. Since glass, silicon, germanium, alumina, and quartz surfaces, as well as many metal oxide surfaces, are rich in

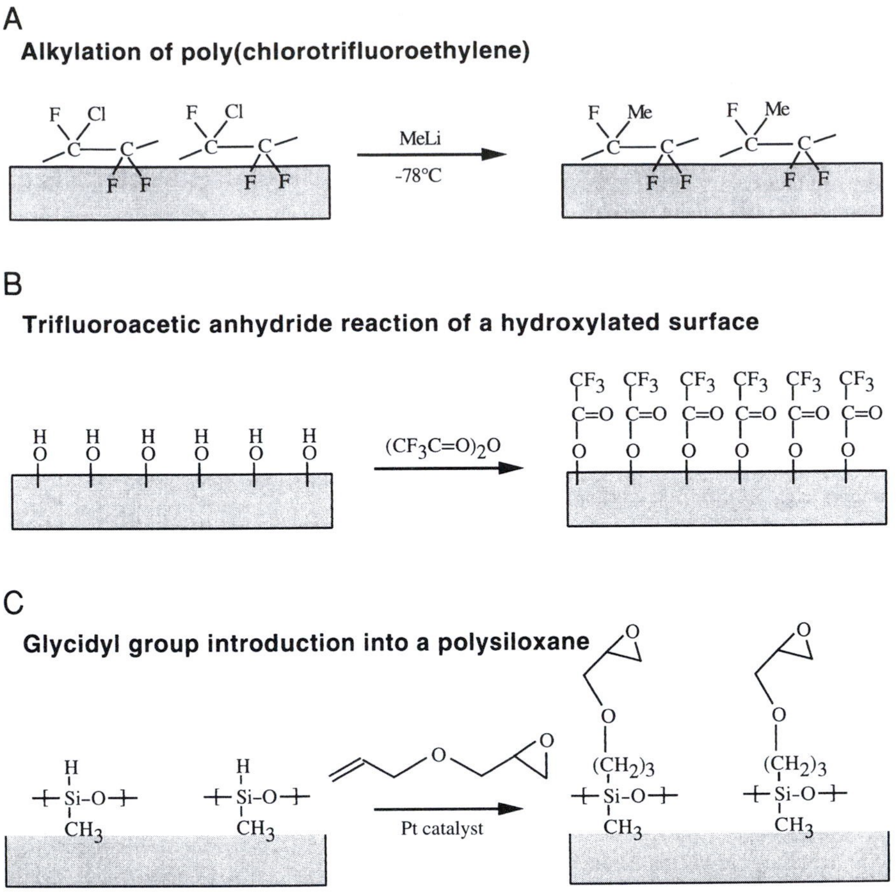

A

Alkylation of poly(chlorotrifluoroethylene)

B

Trifluoroacetic anhydride reaction of a hydroxylated surface

C

Glycidyl group introduction into a polysiloxane

FIG. 3. Some specific chemical reactions to modify surfaces.

hydroxyl groups, silanes are particularly useful for modifying these materials. Numerous silane compounds are commercially available, permitting a broad range of chemical functionalities to be incorporated on surfaces (Table 4). The advantages of silane reactions are their simplicity and stability, attributed to their covalent, cross-linked structure. However, the linkage between a silane and an hydroxyl group is also readily subject to basic hydrolysis, and film breakdown under some conditions must be considered (Wasserman et al., 1989).

Silanes can form two types of surface film structures. If only surface reaction occurs (perhaps catalyzed by traces of adsorbed surface water), a structure similar to that shown in Fig. 4 can be formed. However, if more water is present, a thicker silane layer can be formed consisting of both Si–O groups bonded to the surface and silane units participating in a "bulk," three-dimensional, polymerized network. The initial stages in the formation of a thicker silane film are suggested by the further reaction of the group at the right side of Fig. 4D with solution-phase silane molecules. Without careful control of silane liquid purity, water concentration, and reaction conditions, thicker silane films can be rough and inhomogeneous.

A new class of silane-modified surfaces based upon monolayer silane films and yielding self-assembled, highly ordered

structures is of particular interest in precision engineering of surfaces (Pomerantz et al., 1985; Maoz et al., 1988; Heid et al., 1996). These self-assembled monolayers are described in more detail later in this chapter.

Many general reviews and basic science studies on surface silanization are available (Arkles, 1977; Plueddemann, 1980; Rye et al., 1997). Applications for silanized surface-modified biomaterials are on the increase and include cell attachment (Matsuzawa et al., 1997; Hickman and Stenger, 1994), biomolecule and polymer immobilization (Xiao et al., 1997; Mao et al., 1997), nonfouling surfaces (Lee and Laibinis, 1998), surfaces for DNA studies (Hu et al., 1996), biomineralization (Archibald et al., 1996), and model surfaces for biointeraction studies (Jenney and Anderson, 1999).

Ion Beam Implantation

The ion-beam method injects accelerated ions with energies ranging from 10^1 to 10^6 eV (1 eV = 1.6×10^{-19} joules) into the surface zone of a material to alter its surface properties. It is largely, but not exclusively, used with metals and other inorganics such as ceramics, glasses, and semiconductors. Ions formed from most of the atoms in the periodic table

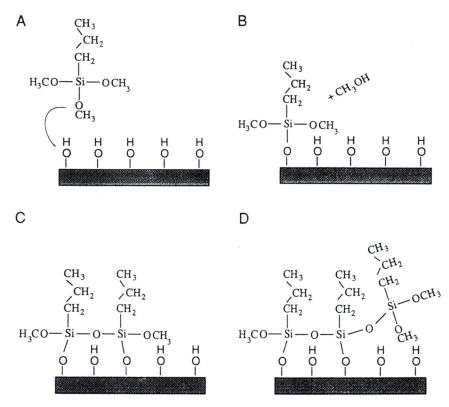

FIG. 4. The chemistry of a typical silane surface modification reaction. (A) A hydroxylated surface is immersed in a non-aqueous solution containing *n*-propyl trimethoxysilane (nPTMS). (B) One of the methoxy groups of the nPTMS couples with a hydroxyl group releasing methanol. (C) Two of the methoxy groups on another molecule of the nPTMS have reacted, one with a hydroxyl group and the other with a methoxy group from the first nPTMS molecule. (D) A third nPTMS molecule has reacted only with a methoxy group. This molecule is tied into the silane film network, but is not directly bound to the surface.

TABLE 4 Silanes for Surface Modification of Biomaterials

$$
\begin{array}{c}
X \\
| \\
X - Si - R \\
| \\
X
\end{array}
$$

X = leaving group	R = functional group
–Cl	–$(CH_2)_nCH_3$
–OCH_3	–$(CH_2)_3NH_2$
–OCH_2CH_3	–$(CH_2)_2(CF_2)_5CF_3$
	$\begin{array}{l} \quad\quad\quad CH_3 \\ -(CH_2)_3O\text{-}C\text{-}C\text{=}CH_2 \\ \quad\quad\quad\; \| \\ \quad\quad\quad\; O \end{array}$
	–CH_2CH_2–⬡

can be implanted, but not all provide useful modifications to the surface properties. Important potential applications for biomaterial surfaces include modification of hardness (wear), lubricity, toughness, corrosion, conductivity, and bioreactivity.

If an ion with kinetic energy greater than a few electron volts impacts a surface, the probability that it will enter the surface is high. The impact transfers much energy to a localized surface zone in a very short time interval. Some considerations for the ion implantation process are illustrated in Fig. 5. These surface changes must be understood quantitatively for engineering of modified surface characteristics. Many review articles and books are available on ion implantation processes and their application for tailoring surface properties (Picraux and Pope, 1984; Colligon, 1986; Sioshansi, 1987; Nastasi *et al.*, 1996).

Specific examples of biomaterials that have been surface altered by ion implantation processes are plentiful. Iridium was ion implanted in a Ti–6Al–4V alloy to improve corrosion resistance (Buchanan *et al.*, 1990). Nitrogen implanted into titanium greatly reduces wear (Sioshansi, 1987). The ion implantation of boron and carbon into type 316L stainless steel improves the high cycle fatigue life of these alloys (Sioshansi, 1987). Silver ions implanted into polystyrene permit cell attachment (Tsuji *et al.*, 1998).

Langmuir–Blodgett Deposition

The Langmuir–Blodgett (LB) deposition method overcoats a surface with one or more highly ordered layers of surfactant molecules. Each of the molecules that assemble into this

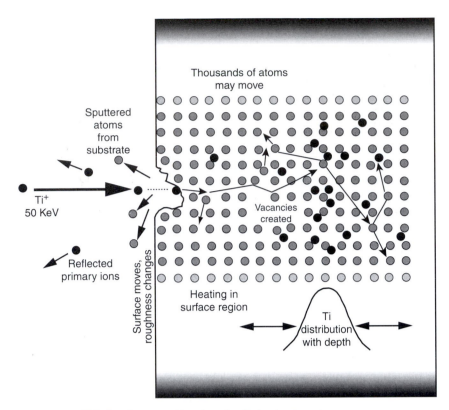

FIG. 5. Some considerations for the ion implantation process.

layer contains a polar "head" group and a nonpolar "tail" group. The deposition of an LB film using an LB trough is illustrated schematically in Fig. 6. By withdrawing the vertical plate through the air–water interface, and then pushing the plate down through the interface, keeping the surface film at the air–water interface compressed at all times (as illustrated in Fig. 6), multilayer structures can be created. Some compounds that form organized LB layers are shown in Fig. 7. The advantages of films deposited on surfaces by this method are their high degree of order and uniformity. Also, since a wide range of chemical structures can form LB films, there are many options for incorporating new chemistries at surfaces. The stability of LB films can be improved by cross-linking or internally polymerizing the molecules after film formation, often through double bonds in the alkyl portion of the chains (Meller *et al.*, 1989). A number of research groups have investigated LB films for biomedical applications (Hayward and Chapman, 1984; Bird *et al.*, 1989; Cho *et al.*, 1990; Heens *et al.*, 1991). A unique cross between silane thin films and LB layers has been developed for biomedical surface modification (Takahara *et al.*, 2000). Many general reviews on these surface structures are available (Knobler, 1990; Ulman, 1991).

Self-Assembled Monolayers

Self-assembled monolayers (SAMs) are surface films that spontaneously form as highly ordered structures (two-dimensional crystals) on specific substrates (Maoz *et al.*, 1988; Ulman, 1990, 1991; Whitesides *et al.*, 1991; Knoll, 1996). In some ways SAMs resemble LB films, but there are important differences, in particular their ease of formation. Examples of SAM films include *n*-alkyl silanes on hydroxylated surfaces (silica, glass, alumina), alkane thiols [e.g., $CH_3(CH_2)_n SH$] and disulfides on coinage metals (gold, silver, copper), amines and alcohols on platinum, carboxylic acids on aluminum oxide, and silver and phosphates (phosphoric acid or phosphonate groups) on titanium or tantalum surfaces. Silane SAMs and thiols on gold are the most commonly used types. Most molecules that form SAMs have the general characteristics illustrated in Fig. 8. Two processes are particularly important for the formation of SAMs (Ulman, 1991): a moderate to strong adsorption of an anchoring chemical group to the surface (typically 30–100 kcal/mol), and van der Waals interaction of the alkyl chains. The bonding to the substrate (chemisorption) provides a driving force to fill every site on the surface and to displace contaminants from the surface. This process is analogous to the compression to the LB film by the movable barrier in the trough. Once adsorption sites are filled on the surface, the chains will be in sufficiently close proximity so that the weaker van der Waals interactive forces between chains can exert their influence and lead to a crystallization of the alkyl groups. Fewer than nine CH_2 groups do not provide sufficient interactive force to stabilize the 2D quasicrystal and are difficult to assemble. More than 24 CH_2 groups have too many options for defects in the crystal and are also

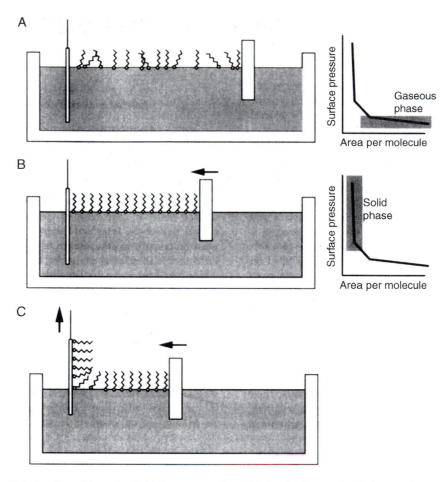

FIG. 6. Deposition of a lipid film onto a glass slide by the Langmuir–Blodgett technique. (A) The lipid film is floated on the water layer. (B) The lipid film is compressed by a moveable barrier. (C) The vertical glass slide is withdrawn while pressure is maintained on the floating lipid film with the moveable barrier.

difficult to assemble. Molecules with lengths between nine and 24 methylene groups will assemble well. Molecular mobility is an important consideration in this surface crystal formation process so that (1) the molecules have sufficient time to maneuver into position for tight packing of the binding end groups at the surface and (2) the chains can enter the quasicrystal.

The advantages of SAMs are their ease of formation, their chemical stability (often considerably higher than that of comparable LB films) and the many options for changing the outermost group that interfaces with the external environment. Many biomaterials applications have already been suggested for SAMs (Lewandowska *et al.*, 1989; Mrksich and Whitesides, 1996; Ferretti *et al.*, 2000). Useful SAMs for creating molecularly-engineered functional surfaces include headgroups of ethylene glycol oligomers, biotin, free radical initiators, N-hydroxysuccinimide esters, anhydrides, perfluoro groups, and amines, just to list a small sampling of the many possibilities. Though most SAMs are based on *n*-alkyl chain assembly, SAMs can form from other classes of molecules including proteins (Sara and Sleytr, 1996), porphyrins, nucleotide bases and aromatic ring hydrocarbons.

Multilayer Polyelectrolyte Absorption

A new strategy for the surface modification of biomaterials has been developed within the past few years (Decher, 1996) and has already found application in biomaterials devices. Multilayer polyelectrolyte absorption requires a surface with either a fixed positive or a fixed negative charge. Some surfaces are intrinsically charged (for example, mica) and others can be modified with methods already described in this chapter. If the surface is negatively charged, it is dipped into an aqueous solution of a positively charged polyelectrolyte (e.g., polyethyleneimine). It is then rinsed in water and dipped in an aqueous solution of a negatively charged polyelectrolyte. This process is repeated as many times as desired to build up a polyelectrolyte complex multilayer of the appropriate thickness for a given application. Once a thin layer of a charged component adsorbs, it will repel additional adsorption thus tightly controlling the layer thickness and uniformity. The outermost layer can be the positively charged or negatively charged component. This strategy works well with charged biomolecules, for

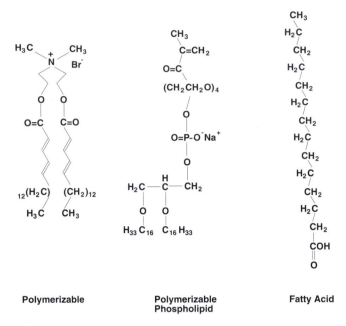

FIG. 7.   Three examples of molecules that form organized Langmuir–Blodgett films.

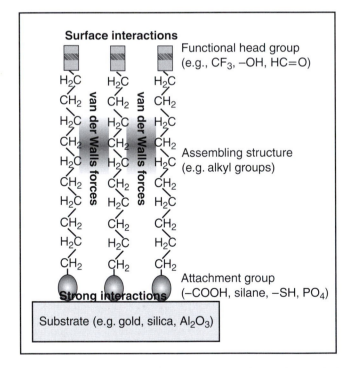

FIG. 8.   General characteristics of molecules that form self-assembled monolayers.

example hyaluronic acid (−) and chitosan (+). Layers formed are durable and assembly of these multiplayer structures is simple. The pH and ionic strength of polyelectrolyte solutions are important process variables. Such overlayer films are now being explored for application in contact lenses.

Surface-Modifying Additives

Specifically designed and synthesized surface-active compositions can be added in low concentrations to a material during fabrication and will spontaneously rise to and dominate the surface (Ward, 1989; Wen *et al.*, 1997). These surface-modifying additives (SMAs) are well known for both organic and inorganic systems. A driving force to minimize the interfacial energy causes the SMA to concentrate at the surface after blending homogeneously with a material. For efficient surface concentration, two factors must be taken into consideration. First, the magnitude of interfacial energy difference between the system without the additive and the same system with the SMA at the surface will determine the magnitude of the driving force leading to a SMA-dominated surface. Second, the molecular mobility of the bulk material and the SMA additive molecules within the bulk will determine the rate at which the SMA reaches the surface, or if it will get there at all. An additional concern is the durability and stability of the SMA at the surface.

A typical SMA designed to alter the surface properties of a polymeric material will be a relatively low molecular weight diblock or triblock copolymer (see Chapter 2.2). The "A" block will be soluble in, or compatible with, the bulk material into which the SMA is being added. The "B" block will be incompatible with the bulk material and have lower surface energy. Thus, the A block will anchor the B block into the material to be modified at the interface. This is suggested schematically in Fig. 9. During initial fabrication, the SMA might be distributed uniformly throughout the bulk. After a period for curing or an annealing step, the SMA will migrate to the surface. Low-molecular-weight end groups on polymer chains can also provide the driving force to bring the end group to the surface.

As an example, on SMA for a polyurethane might have a low-molecular-weight polyurethane A block and a poly(dimethyl siloxane) (PDMS) B block. The PDMS component on the surface may confer improved blood compatibility to the polyurethane. The A block will anchor the SMA in the polyurethane bulk (the polyurethane A block should be reasonably compatible with the bulk polyurethane), while the low-surface-energy, highly flexible silicone B block will be exposed at the air surface to lower the interfacial energy (note that air is "hydrophobic"). The A block anchor should confer stability to this system. However, consider that if the system is placed in an aqueous environment, a low-surface-energy polymer (the B block) is now in contact with water—a high interfacial energy situation. If the system, after fabrication, still exhibits sufficient chain mobility, it might phase-invert to bring the bulk polyurethane or the A block to the surface. Unless the system is specifically engineered to do such a surface phase reversal, this inversion is undesirable. Proper choice of the bulk polymer and the A block can impede surface phase inversion.

An example of a polymer additive that was developed by 3M specifically to take advantage of this surface chemical inversion phenomenon is a stain inhibitor for fabric. Though not a biomaterial, it illustrates design principles for this type of system. The compound has three "arms." A fluoropolymer arm, the lowest energy component, resides at the fabric surface in air. Fluoropolymers and hydrocarbons (typical stains) do not mix,

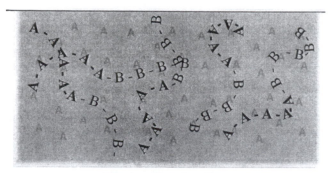

During fabrication

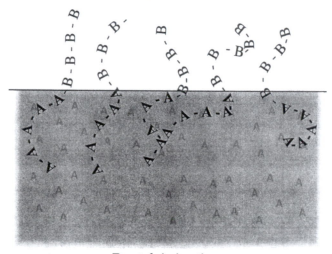

Post-fabrication

FIG. 9. A block copolymer surface-modifying additive comprising an A block and a B block is blended into a support polymer (the bulk) with a chemistry similar to the A block. During fabrication, the block copolymer is randomly distributed throughout the support polymer. After curing or annealing, the A block anchors the surface-modifying additive into the support, while the low-energy B block migrates to the air–polymer interface.

so hydrocarbons are repelled. A second arm of hydrophilic poly(ethylene oxide) will come to the surface in hot water and assist with the washing out of any material on the surface. Finally, a third arm of hydrocarbon anchors this additive into the fabric.

Many SMAs for inorganic systems are known. For example, very small quantities of nickel will completely alter the structure of a silicon (111) surface (Wilson and Chiang, 1987). Copper will accumulate at the surface of gold alloys (Tanaka *et al.*, 1988). Also, in stainless steels, chromium will concentrate (as the oxide) at the surface, imparting corrosion resistance.

There are a number of additives that spontaneously surface-concentrate, but are not necessarily designed as SMAs. A few examples for polymers include PDMS, some extrusion lubricants (Ratner, 1983), and some UV stabilizers (Tyler *et al.*, 1992). The presence of such additives at the surface of a polymer may be unplanned and they will not necessarily form stable, durable surface layers. However, they can significantly contribute (either positively or negatively) to the bioresponse to the surface.

Conversion Coatings

Conversion coatings modify the surface of a metal into a dense oxide-rich layer that imparts corrosion protection, enhanced adhesivity, altered appearance (e.g., color) and sometimes lubricity to the metal. For example, steel is frequently phosphated (treated with phosphoric acid) or chromated (with chromic acid). Aluminum is electrochemically anodized in chromic, oxalic, or sulfuric acid electrolytes. Electrochemical anodization may also be useful for surface-modifying titanium and Ti–Al alloys (Bardos, 1990; Kasemo and Lausmaa, 1985).

The conversion of metallic surfaces to "oxide-like," electrochemically passive states is a common practice for base-metal alloy systems used as biomaterials. Standard and recommended techniques have been published (e.g., ASTM F4-86) and are relevant for most musculoskeletal load-bearing surgical implant devices. The background literature supporting these types of surface passivation technologies has been summarized (von Recum, 1986).

Base-metal alloy systems, in general, are subject to electrochemical corrosion ($M \rightarrow M^+ + e^-$) within saline environments. The rate of this corrosion process is reduced 10^3–10^6 times by the presence of a dense, uniform, minimally conductive, relatively inert oxide surface. For many metallic devices, exposure to a mineral acid (e.g., nitric acid in water) for times up to 30 minutes will provide a passivated surface. Plasma-enhanced surface passivation of metals, laser surface treatments, and mechanical treatments (shot peening) can also impart many of these characteristics to metallic systems.

The reason that many of these surface modifications are called "oxide-like" is that the structure is complex, including OH, H, and subgroups that may, or may not, be crystalline. Since most passive surfaces are thin films (5–500 nm) and are transparent or metallic in color, the surface appears similar before and after passivation. Further details on surfaces of this type can be found in Chapters 1.4, 2.9, and 6.3.

Parylene Coating

Parylene (*para*-xylylene) coatings occupy a unique niche in the surface modification literature because of their wide application and the good quality of the thin film coatings formed (Loeb *et al.*, 1977a; Nichols *et al.*, 1984). The deposition method is also unique and involves the simultaneous evaporation, pyrolysis, deposition, and polymerization of the monomer, di-*para*-xylylene (DPX), according to the following reaction:

$$CH_2 \!-\!\!\bigcirc\!\!-\! CH_2 \qquad\qquad CH_2 \!=\!\!\bigcirc\!\!=\! CH_2$$

Di-para-xylylene   para-xylylene
1) vaporize   2) pyrolyze

$$\longrightarrow \ \Big[\!\!- CH_2 \!-\!\!\bigcirc\!\!-\! CH_2 \!-\!\!\Big]_n$$

Poly(para-xylylene)
3) deposit

The DPX monomer is vaporized at 175°C and 1 torr, pyrolyzed at 700°C and 0.5 torr, and finally deposited on a substrate at 25°C and 0.1 torr. The coating has excellent electrical insulation and moisture barrier properties and has been used for protection of implant electrodes (Loeb *et al.*, 1977b; Nichols *et al.*, 1984) and implanted electronic circuitry (Spivack and Ferrante, 1969). Recently, a parylene coating has been used on stainless steel cardiovascular stents between the metal and a drug-eluting polymer layer (see Chapters 7.3 and 7.14).

Laser Methods

Lasers can rapidly and specifically induce surface changes in organic and inorganic materials (Picraux and Pope, 1984; Dekumbis, 1987; Chrisey *et al.*, 2003). The advantages of using lasers for such modification are the precise control of the frequency of the light, the wide range of frequencies available, the high energy density, the ability to focus and raster the light, the possibilities for using both heat and specific excitation to effect change, and the ability to pulse the source and control reaction time. Lasers commonly used for surface modification include ruby, neodymium : yttrium aluminum garnet (Nd : YAG), argon, and CO_2. Treatments are pulsed (100 nsec to picoseconds pulse times) and continuous wave (CW), with interaction times often less than 1 msec. Laser-induced surface alterations include annealing, etching, deposition, and polymerization. Polymers, metals, ceramics, and even tooth dentin have been effectively surface modified using laser energy. The major considerations in designing a laser surface treatment include the absorption (coupling) between the laser energy and the material, the penetration depth of the laser energy into the material, the interfacial reflection and scattering, and heating induced by the laser.

PATTERNING

Essentially all of the surface modification methods described in this chapter can be applied to biomaterial surfaces as a uniform surface treatment, or as patterns on the surface with length scales of millimeters, microns or even nanometers. There is much interest in deposition of proteins and cells in surface patterns and textures in order to control bioreactions (Chapter 2.16). Furthermore, devices "on a chip" frequently require patterning. Such devices include microfluidic systems ("lab on a chip"), neuronal circuits on a chip, and DNA diagnostic arrays. An overview of surface patterning methods for bioengineering applications has been published (Folch and Toner, 2000).

Photolithographic techniques that were developed for microelectronics have been applied to patterning of biomaterial surfaces when used in conjunction with methods described in this chapter. For example, plasma-deposited films were patterned using a photoresist lift-off method (Goessl *et al.*, 2001).

Microcontact printing is a newer method permitting simple modification. Basically, a rubber stamp is made of the pattern that is desired on the biomaterial surface (Fig. 10). The stamp can be "inked" with thiols (to stamp gold), silanes (to stamp silicon), proteins (to stamp many types of surfaces) or

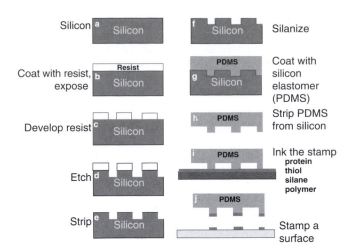

FIG. 10. Fabrication of a silicone elastomer stamp for microcontact printing. The sequence of steps is a-j.

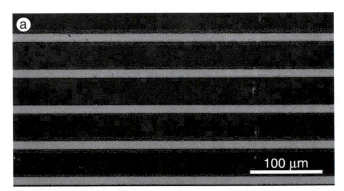

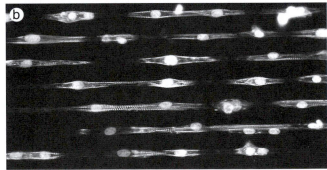

FIG. 11. (a) Microcontact printed lines of laminin protein (fluorescent labeled) on a cell-resistant background. (b) Cardiomyocyte cells adhering and aligning on the laminin printed lines (see *J. Biomed. Mater. Res.* **60**: 472 for details) (used with the permission of P. Stayton, C. Murry, S. Hauschka, J. Angello and T. McDevitt).

polymer solutions (again, to stamp many types of surfaces). Spatial resolution of pattern features in the nanometer range has been demonstrated, though most patterns are applied in the micron range. Methods have been developed to accurately stamp curved surfaces. An example of cells on laminin-stamped lines is shown in Fig. 11. These laminin lines were durable for at least 2 weeks of cell contact. Durability remains a major consideration with patterns on surface generated by this relatively simple method.

There are many other options to pattern biomaterial surfaces. These include ion-beam etching, electron-beam lithography, laser methods, inkjet printers, and stochastic patterns made by phase separation of two components (Takahara *et al.*, 2000).

CONCLUSIONS

Surface modifications are being widely explored to enhance the biocompatibility of biomedical devices and improve other aspects of performance. Since a given medical device may already have appropriate performance characteristics and physical properties and be well understood in the clinic, surface modification provides a means to alter only the biocompatibility of the device without the need for redesign, retooling for manufacture, and retraining of medical personnel.

Acknowledgment

The suggestions and assistance of Professor J. Lemons have enhanced this chapter and are gratefully appreciated.

QUESTIONS

1. You are assigned the task of designing a proteomics array for cancer diagnostics. Six hundred and twenty-five proteins must be attached to the surface of a standard, glass microscope slide in a 25×25 array. Design a scheme to make such a proteomic chip. What are the important surface issues? Which strategies might you apply to address each of the issues? You may find helpful ideas in Chapters 1.4, 2.13, and 2.16.

2. A hydrogel surface must be put on a silicone rubber medical device. A viscous solution of the hydrogel polymer is used to spray-coat the device. When it is placed in aqueous buffer solution the hydrogel layer quickly delaminates from the silicone. How might you permanently attach a hydrogel layer to a silicone device? Briefly describe the method you would use and the general steps needed to produce a reliable coating.

3. List the molecular and design factors that can contribute to increasing the durability of an *n*-alkyl thiol self-assembled monolayer on gold.

Bibliography

Archibald, D. D., Qadri, S. B., and Gaber, B. P. (1996). Modified calcite deposition due to ultrathin organic films on silicon substrates. *Langmuir* 12: 538–546.

Arkles, B. (1977). Tailoring surfaces with silanes. *Chemtech* 7: 766–778.

Bardos, D. I. (1990). Titanium and titanium alloys. in *Concise Encyclopedia of Medical and Dental Materials*, 1st ed., E. Williams, R. W. Cahn, and M. B. Bever, eds. Pergamon Press, Oxford, pp. 360–365.

Bird, R. R., Hall, B., Hobbs, K. E. F., and Chapman, D. (1989). New haemocompatible polymers assessed by thrombelastography. *J. Biomed. Eng.* 11: 231–234.

Buchanan, R. A., Lee, I. S., and Williams, J. M. (1990). Surface modification of biomaterials through noble metal ion implantation. *J. Biomed. Mater. Res.* 24: 309–318.

Chapiro, A. (1983). Radiation grafting of hydrogels to improve the thrombo-resistance of polymers. *Eur. Polym. J.* 19: 859–861.

Charlson, E. J., Charlson, E. M., Sharma, A. K., and Yasuda, H. K. (1984). Electrical properties of glow-discharge polymers, parylenes, and composite films. *J. Appl. Polymer Sci. Appl. Polymer Symp.* 38: 137–148.

Chilkoti, A., Ratner, B. D., and Briggs, D. (1991a). Plasma-deposited polymeric films prepared from carbonyl-containing volatile precursors: XPS chemical derivatization and static SIMS surface characterization. *Chem. Mater.* 3: 51–61.

Chilkoti, A., Ratner, B. D., and Briggs, D. (1991b). A static secondary ion mass spectrometric investigation of the surface structure of organic plasma-deposited films prepared from stable isotope-labeled precursors. Part I. Carbonyl precursors. *Anal. Chem.* 63: 1612–1620.

Chilkoti, A., Ratner, B. D., Briggs, D., and Reich, F. (1992). Static secondary ion mass spectrometry of organic plasma deposited films created from stable isotope-labeled precursors. Part II. Mixtures of acetone and oxygen. *J. Polymer Sci., Polymer Chem. Ed.* 30: 1261–1278.

Cho, C. S., Takayama, T., Kunou, M., and Akaike, T. (1990). Platelet adhesion onto the Langmuir–Blodgett film of poly-(gamma-benzyl L-glutamate)–poly(ethylene oxide)–poly(gamma-benzyl L-glutamate) block copolymer. *J. Biomed. Mater. Res.* 24: 1369–1375.

Chrisey, D. B., Piqué, A., McGill, R. A., Horowitz, J. S., Ringeisen, B. R., Bubb, D. M., and Wu, P. K. (2003). Laser deposition of polymer and biomaterial films. *Chem. Rev.* 103: 553–576.

Chu, P. K., Chen, J. Y., Wang, L. P., and Huang, N. (2002). Plasma surface modification of biomaterials. *Mater. Sci. Eng. Rep.* 36: 143–206.

Colligon, J. S. (1986). Surface modification by ion beams. *Vacuum* 36: 413–418.

Dargaville, T. R., George, G. A., Hill, D. J. T., and Whittaker, A. K. (2003). High energy radiation grafting of fluoropolymers. *Prog. Polymer Sci.* 28: 1355–1376.

Decher, G. (1996). Layered nanoarchitectures via directed assembly of anionic and cationic molecules. in *Comprehensive Supramolecular Chemistry*, Vol. 9, *Templating, Self-Assembly and Self-Organization*, J.-P. Sauvage and M. W. Hosseini, eds. Pergamon Press, Oxford, pp. 507–528.

Dekumbis, R. (1987). Surface treatment of materials by lasers. *Chem. Eng. Prog.* 83: 23–29.

Dunkirk, S. G., Gregg, S. L., Duran, L. W., Monfils, J. D., Haapala, J. E., Marcy, J. A., Clapper, D. L., Amos, R. A., and Guire, P. E. (1991). Photochemical coatings for the prevention of bacterial colonization. *J. Biomater. Appl.* 131–156.

Everhart, D. S., and Reilley, C. N. (1981). Chemical derivatization in electron spectroscopy for chemical analysis of surface functional groups introduced on low-density polyethylene film. *Anal. Chem.* 53: 665–676.

Ferretti, S., Paynter, S., Russell, D. A., and Sapsford, K. E. (2000). Self-assembled monolayers: a versatile tool for the formulation of bio-surfaces. *Trends Anal. Chem.* 19(9): 530–540.

Folch, A., and Toner, M. (2000). Microengineering of cellular interactions. *Ann. Rev. Bioeng.* 2: 227–256.

Garbassi, F., Morra, M., Occhiello, E., Barino, L., and Scordamaglia, R. (1989). Dynamics of macromolecules: A challenge for surface analysis. *Surf. Interface Anal.* 14: 585–589.

Goessl, A., Garrison, M. D., Lhoest, J., and Hoffman, A. (2001). Plasma lithography—thin-film patterning of polymeric biomaterials by RF plasma polymerization I: surface preparation and analysis. *J. Biomater. Sci. Polymer Ed.* **12**(7): 721–738.

Gombotz, W. R., and Hoffman, A. S. (1988). Functionalization of polymeric films by plasma polymerization of allyl alcohol and allylamine. *J. Appl. Polymer Sci. Appl. Polymer Symp.* **42**: 285–303.

Greer, R. T., Knoll, R. L., and Vale, B. H. (1979). Evaluation of tissue-response to hydrogel composite materials. *SEM* **2**: 871–878.

Griesser, H. J., and Chatelier, R. C. (1990). Surface characterization of plasma polymers from amine, amide and alcohol monomers. *J. Appl. Polymer Sci. Appl. Polymer Symp.* **46**: 361–384.

Gruner, H. (2001). Thermal spray coating on titanium. in *Titanium in Medicine*, D.M. Brunette, P. Tengvall, M. Textor and P. Thomsen, eds. Springer-Verlag, Berlin.

Haque, Y., and Ratner, B. D. (1988). Role of negative ions in the RF plasma deposition of fluoropolymer films from perfluoropropane. *J. Polymer Sci., Polymer Phys. Ed.* **26**: 1237–1249.

Hayward, J. A., and Chapman, D. (1984). Biomembrane surfaces as models for polymer design: The potential for haemocompatibility. *Biomaterials* **5**: 135–142.

Heens, B., Gregoire, C., Pireaux, J. J., Cornelio, P. A., and Gardella, J. A., Jr. (1991). On the stability and homogeneity of Langmuir–Blodgett films as models of polymers and biological materials for surface studies: An XPS study. *Appl. Surf. Sci.* **47**: 163–172.

Heid, S., Effenberger, F., Bierbaum, K., and Grunze, M. (1996). Self-assembled mono- and multilayers of terminally functionalized organosilyl compounds on silicon substrates. *Langmuir* **12**(8): 2118–2120.

Hickman, J. J., and Stenger, D. A. (1994). Interactions of cultured neurons with defined surfaces. in *Enabling Technologies for Cultured Neural Networks*, D. A. Stenger, T. N. McKenna, eds. Academic Press, San Diego, pp. 51–76.

Hoffman, A. S. (1981). A review of the use of radiation plus chemical and biochemical processing treatments to prepare novel biomaterials. *Radiat. Phys. Chem.* **18**: 323–342.

Hoffman, A. S. (1988). Biomedical applications of plasma gas discharge processes. *J. Appl. Polymer Sci. Appl. Polymer Symp.* **42**: 251–267.

Hoffman, A. S., Schmer, G., Harris, C., and Kraft, W. G. (1972). Covalent binding of biomolecules to radiation-grafted hydrogels on inert polymer surfaces. *Trans. Am. Soc. Artif. Int. Organs* **18**: 10–17.

Hoffman, A. S., Cohn, D. C., Hanson, S. R., Harker, L. A., Horbett, T. A., Ratner, B. D., and Reynolds, L. O. (1983). Application of radiation-grafted hydrogels as blood-contacting biomaterials. *Radiat. Phys. Chem.* **22**: 267–283.

Horbett, T. A., and Hoffman, A. S. (1975). Bovine plasma protein adsorption on radiation-grafted hydrogels based on hydroxyethyl methacrylate and N-vinyl-pyrrolidone. in *Applied Chemistry at Protein Interfaces*, Advances in Chemistry Series, R. E. Baier, ed. American Chemical Society, Washington, D.C., pp. 230–254.

Hu, J., Wang, M., Weier, U. G., Frantz, P., Kolbe, W., Ogletree, D. F., and Salmeron, M. (1996). Imaging of single extended DNA molecules on flat (aminopropyl)triethozysilane-mica by atomic force microscopy. *Langmuir* **12**(7): 1697–1700.

Inagaki, N., Nakanishi, T., and Katsuura, K. (1983). Glow discharge polymerizations of tetrafluoroethylene, perfluoromethylcyclohexane and perfluorotoluene investigated by infrared spectroscopy and ESCA. *Polymer Bull.* **9**: 502–506.

Jenney, C. R., and Anderson, J. M. (1999). Alkylsilane-modified surfaces: inhibition of human macrophage adhesion and foreign body giant cell formation. *J. Biomed. Mater. Res.* **46**: 11–21.

Johnston, E. E., and Ratner, B. D. (1996). XPS and SSIMS characterization of surfaces modified by plasma deposited oligo(glyme) films.

in *Surface Modification of Polymeric Biomaterials*, B. D. Ratner, D. G. Castner, eds. Plenum Press, New York, pp. 35–44.

Kaplan, S., and Dilks, A. (1981). A solid state nuclear magnetic resonance investigation of plasma-polymerized hydrocarbons. *Thin Solid Films* **84**: 419–424.

Kasemo, B., and Lausmaa, J. (1985). Metal selection and surface characteristics. in *Tissue-Integrated Prostheses*, P. I. Branemark, G. A. Zarb and T. Albrektsson, eds. Quintessence Publishing, Chicago, pp. 99–116.

Kitching, K. J., Pan, V., and Ratner, B. D. (2003). Biomedical applications of plasma-deposited thin films. in *Plasma Polymer Films*, H. Biederman, ed. Imperial College Press, London.

Klages, C. -P., Höpfner, K., and Kläke, N. (2000). Surface functionalization at atmospheric pressure by DBD-based pulsed plasma polymerization. *Plasmas Polymer* **5**: 79–89.

Knobler, C. M. (1990). Recent developments in the study of monolayers at the air–water interface. *Adv. Chem. Phys.* **77**: 397–449.

Knoll, W. (1996). Self-assembled microstructures at interfaces. *Curr. Opin. Colloid Interface Sci.* **1**: 137–143.

Krishnamurthy, V., Kamel, I. L., and Wei, Y. (1989). Analysis of plasma polymerization of allylamine by FTIR. *J. Polymer Sci., Polymer Chem. Ed.* **27**: 1211–1224.

Kwon, O. H., Kikuchi, A., Yamato, M., Sakurai, Y., and Okano, T. (2000). Rapid cell sheet detachment from poly(N-isopropylacrylamide)-grafted porous cell culture membranes. *J. Biomed. Mater. Res.* **50**: 82–89.

Lee, S. –W., and Laibinis, P. E. (1998). Protein-resistant coatings for glass and metal oxide surfaces derived from oligo(ethylene glycol)-terminated alkyltrichlorosilane. *Biomaterials* **19**: 1669–1675.

Lewandowska, K., Balachander, N., Sukenik, C. N., and Culp, L. A. (1989). Modulation of fibronectin adhesive functions for fibroblasts and neural cells by chemically derivatized substrata. *J. Cell. Physiol.* **141**: 334–345.

Loeb, G. E., Bak, M. J., Salcman, M., and Schmidt, E. M. (1977a). Parylene as a chronically stable, reproducible microelectrode insulator. *IEEE Trans. Biomed. Eng.* **BME-24**: 121–128.

Loeb, G. E., Walker, A. E., Uematsu, S., and Konigsmark, B. W. (1977b). Histological reaction to various conductive and dielectric films chronically implanted in the subdural space. *J. Biomed. Mater. Res.* **11**: 195–210.

Lopez, G. P., and Ratner, B. D. (1991). Substrate temperature effects of film chemistry in plasma deposition of organics. I. Nonpolymerizable precursors. *Langmuir* **7**: 766–773.

Lopez, G. P., Ratner, B. D., Rapoza, R. J., and Horbett, T. A. (1993). Plasma deposition of ultrathin films of poly(2-hydroxyethyl methacrylate): Surface analysis and protein adsorption measurements. *Macromolecules* **26**: 3247–3253.

Mao, G., Castner, D. G., and Grainger, D. W. (1997). Polymer immobilization to alkylchlorosilane organic monolayer films using sequential derivatization reactions. *Chem. Mater.* **9**(8): 1741–1750.

Maoz, R., Netzer, L., Gun, J., and Sagiv, J. (1988). Self-assembling monolayers in the construction of planned supramolecular structures and as modifiers of surface properties. *J. Chim. Phys.* **85**: 1059–1064.

Massines, F., Gherardi, N., and Sommer, F. (2000). Silane-based coatings of polypropylene, deposited by atmospheric pressure glow discharge plasmas. *Plasmas Polymers* **5**: 151–172.

Matsuda, T., and Inoue, K. (1990). Novel photoreactive surface modification technology for fabricated devices. *Trans. Am. Soc. Artif. Internal Organs* **36**: M161–M164.

Matsuzawa, M., Umemura, K., Beyer, D., Sugioka, K., and Knoll, W. (1997). Micropatterning of neurons using organic substrates in culture. *Thin Solid Films* **305**: 74–79.

Meller, P., Peters, R., and Ringsdorf, H. (1989). Microstructure and lateral diffusion in monolayers of polymerizable amphiphiles. *Colloid Polymer Sci.* **267**: 97–107.

Mrksich, M., and Whitesides, G. M. (1996). Using self-assembled monolayers to understand the interactions of manmade surfaces with proteins and cells. *Annu. Rev. Biophys. Biomol. Struct.* **25**: 55–78.

Nastasi, M., Mayer, J., and Hirvonen, J. K. (1996). *Ion–solid interactions: fundamentals and applications.* Cambridge Univ. Press, Cambridge, UK.

Nguyen, S.V. (1986). Plasma assisted chemical vapor deposited thin films for microelectronic applications. *J. Vac. Sci. Technol. B* **4**: 1159–1167.

Nichols, M. F., Hahn, A. W., James, W. J., Sharma, A. K., and Yasuda, H. K. (1984). Evaluating the adhesion characteristics of glow-discharge plasma-polymerized films by a novel voltage cycling technique. *J. Appl. Polymer Sci. Appl. Polymer Symp.* **38**: 21–33.

Panchalingam, V., Poon, B., Huo, H. H., Savage, C. R., Timmons, R. B., and Eberhart, R. C. (1993). Molecular surface tailoring of biomaterials via pulsed RF plasma discharges. *J. Biomater. Sci. Polymer. Ed.* 5(1/2): 131–145.

Picraux, S. T., and Pope, L. E. (1984). Tailored surface modification by ion implantation and laser treatment. *Science* **226**: 615–622.

Plueddemann, E. P. (1980). Chemistry of silane coupling agents. in *Silylated Surfaces*, D. E. Leyden, ed. Gordon and Breach Science Publishers, New York, pp. 31–53.

Pomerantz, M., Segmuller, A., Netzer, L., and Sagiv, J. (1985). Coverage of Si substrates by self-assembling monolayers and multilayers as measured by IR, wettability and x-ray diffraction. *Thin Solid Films* **132**: 153–162.

Ratner, B. D. (1980). Characterization of graft polymers for biomedical applications. *J. Biomed. Mater. Res.* **14**: 665–687.

Ratner, B. D. (1983). ESCA studies of extracted polyurethanes and polyurethane extracts: Biomedical implications. in *Physicochemical Aspects of Polymer Surfaces*, K. L. Mittal, ed. Plenum Publishing, New York, pp. 969–983.

Ratner, B. D., and Hoffman, A. S. (1980). Surface grafted polymers for biomedical applications. in *Synthetic Biomedical Polymers. Concepts and Applications*, M. Szycher and W. J. Robinson, eds. Technomic Publishing, Westport, CT, pp. 133–151.

Ratner, B. D., and Yoon, S. C. (1988). Polyurethane surfaces: solvent and temperature induced structural rearrangements. in *Polymer Surface Dynamics*, J. D. Andrade, ed. Plenum Press, New York, pp. 137–152.

Ratner, B. D., Horbett, T. A., Hoffman, A. S., and Hauschka, S. D. (1975). Cell adhesion to polymeric materials: Implications with respect to biocompatibility. *J. Biomed. Mater. Res.* **9**: 407–422.

Ratner, B. D., Hoffman, A. S., Hanson, S. R., Harker, L. A., and Whiffen, J. D. (1979). Blood compatibility–water content relationships for radiation grafted hydrogels. *J. Polymer Sci. Polymer Symp.* **66**: 363–375.

Ratner, B. D., Chilkoti, A., and Lopez, G. P. (1990). Plasma deposition and treatment for biomaterial applications. in *Plasma Deposition, Treatment and Etching of Polymers*, R. D'Agostino, ed. Academic Press, San Diego, pp. 463–516.

Rye, R. R., Nelson, G. C., and Dugger, M. T. (1997). Mechanistic aspects of alkylchlorosilane coupling reactions. *Langmuir* **13**(11): 2965–2972.

Safrany, A. (1997). Radiation processing: synthesis and modification of biomaterials for medical use. *Nucl. Instrum. Methods Phys. Res., Sect. B* **131**: 376–381.

Sara, M., and Sleytr, U. B. (1996). Crystalline bacterial cell surface layers (S-layers): from cell structure to biomimetics. *Prog. Biophys. Mol. Biol.* **65**(1/2): 83–111.

Sawin, H. H., and Reif, R. (1983). A course on plasma processing in integrated circuit fabrication. *Chem. Eng. Ed.* **17**: 148–152.

Sheu, M.-S., Hoffman, A. S., and Feijen, J. (1992). A glow discharge process to immobilize PEO/PPO surfactants for wettable and nonfouling biomaterials. *J. Adhes. Sci. Technol.* **6**: 995–1101.

Singh, A., and Silverman, J., editors. (1992). *Radiation Processing of Polymers.* Oxford Univ. Press, New York.

Sioshansi, P. (1987). Surface modification of industrial components by ion implantation. *Mater. Sci. Eng.* **90**: 373–383.

Somorjai, G. A. (1990). Modern concepts in surface science and heterogeneous catalysis. *J. Phys. Chem.* **94**: 1013–1023.

Somorjai, G. A. (1991). The flexible surface. Correlation between reactivity and restructuring ability. *Langmuir* **7**: 3176–3182.

Spivack, M. A., and Ferrante, G. (1969). Determination of the water vapor permeability and continuity of ultrathin parylene membranes. *J. Electrochem. Soc.* **116**: 1592–1594.

Srinivasan, R., and Lazare, S. (1985). Modification of polymer surfaces by far-ultraviolet radiation of low and high (laser) intensities. *Polymer* **26**: 1297–1300.

Stannett, V. T. (1990). Radiation grafting—state-of-the-art. *Radiat. Phys. Chem.* **35**: 82–87.

Swanson, M. J. (1996). A unique photochemical approach for polymer surface modification. in *Polymer Surfaces and Interfaces: Characterization, Modification and Application*, K. L. Mittal and K. W. Lee eds. VSP, The Netherlands.

Takahara, A., Ge, S., Kojio, K., and Kajiyama, T. (2000). In situ atomic force mircroscopic observation of albumin adsorption onto phase-separated organosilane monolayer surface. *J. Biomater. Sci. Polymer. Ed.* **11**(1): 111–120.

Tanaka, T., Atsuta, M., Nakabayashi, N., and Masuhara, E. (1988). Surface treatment of gold alloys for adhesion. *J. Prosthet. Dent.* **60**: 271–279.

Tsuji, H., Satoh, H., Ikeda, S., Ikemoto, N., Gotoh, Y., and Ishikawa, J. (1998). Surface modification by silver-negative-ion implantation for controlling cell-adhesion properties of polystyrene. *Surf. Coat. Technol.* **103–104**: 124–128.

Tyler, B. J., Ratner, B. D., Castner, D. G., and Briggs, D. (1992). Variations between Biomer lots. 1. Significant differences in the surface chemistry of two lots of a commercial polyetherurethane. *J. Biomed. Mater. Res.* **26**: 273–289.

Ulman, A. (1990). Self-assembled monolayers of alkyltrichlorosilanes: Building blocks for future organic materials. *Adv. Mater.* **2**: 573–582.

Ulman, A. (1991). *An Introduction to Ultrathin Organic Films.* Academic Press, Boston.

von Recum, A. F. (1986). *Handbook of Biomaterials Evaluation*, 1st ed. Macmillan, New York.

Ward, R. S. (1989). Surface modifying additives for biomedical polymers. *IEEE Eng. Med. Biol. Mag.* June, pp. 22–25.

Wasserman, S. R., Tao, Y. –T., and Whitesides, G. M. (1989). Structure and reactivity of alkylsiloxane monolayers formed by reaction of alkyltrichlorosilanes on silicon substrates. *Langmuir* **5**: 1074–1087.

Wen, J. M., Gabor, S., Lim, F., and Ward, R. (1997). XPS study of surface composition of a segmented polyurethane block copolymer modified by PDMS end groups and its blends with phenoxy. *Macromolecules* **30**: 7206–7213.

Whitesides, G. M., Mathias, J. P., and Seto, C. T. (1991). Molecular self-assembly and nanochemistry: a chemical strategy for the synthesis of nanostructures. *Science* **254**: 1312–1319.

Wilson, R. J., and Chiang, S. (1987). Surface modifications induced by adsorbates at low coverage: A scanning-tunneling-microscopy study of the Ni/Si(111) square-root-19 surface. *Phys. Rev. Lett.* **58**: 2575–2578.

Wu, S. (1982). *Polymer Interface and Adhesion.* Marcel Dekker, New York.

Xiao, S. J., Textor, M., Spencer, N. D., Wieland, M., Keller, B., and Sigrist, H. (1997). Immobilization of the cell-adhesive peptide arg-gly-asp-cys (RGDC) on titanium surfaces by covalent chemical attachment. *J. Mater. Sci.: Mater. Med.* **8:** 867–872.

Yasuda, H. K. (1979). Competitive ablation and polymerization (CAP) mechanisms of glow discharge polymerization. in *Plasma Polymerization,* ACS Symposium Series 108, M. Shen and A. T. Bell, eds. American Chemical Society, Washington, D.C., pp. 37–52.

Yasuda, H. K., and Gazicki, M. (1982). Biomedical applications of plasma polymerization and plasma treatment of polymer surfaces. *Biomaterials* **3:** 68–77.

2.15 TEXTURED AND POROUS MATERIALS

John A. Jansen and Andreas F. von Recum

INTRODUCTION

Surface irregularities on medical devices, such as grooves/ridges, hills, pores, and pillars, are expected to guide many types of cells (including immunological, epithelial, connective-tissue, neural, and muscle cells) and to aid tissue repair after injury. With the growing interest in tissue engineering, porous scaffold reactions *in vitro* and *in vivo* are assuming increasing importance (see Chapter 8.4). The final response to rough or porous materials is reflected in the organization of the cytoskeleton, the orientation of extracellular matrix (ECM) components, the amount of produced ECM, and angiogenesis. Although significant progress has been made, the exact cellular and molecular events underlying cellular and matrix orientation are not yet completely understood.

This chapter will provide information about how surface roughness is defined, prepared, and measured. In addition, it will cover the biological effects of surface irregularities on cells.

DEFINITION OF SURFACE IRREGULARITIES

Surface irregularities can be considered as deviations from a geometrically ideal (flat) surface. They can be created accidentally by the production process or engineered for specific purposes. Surface irregularities can be classified according to their dimensions and the way they are achieved. In view of this, surface irregularities can be classified into six classes (Sander, 1991). The main distinctive characteristic is their horizontal pattern. Thus, Class 1 irregularities are associated with form errors of the substrate surface such as straightness, flatness, roundness, and cylindricity. Class 2 surface features deal with so-called waviness deviations. Waviness is considered to occur if the wave spacing is larger than the wave depth. Class 3, 4, and 5 irregularities all refer to surface roughness. Roughness is assumed if the space between two hills is about 5 to 100 times larger than the depth. Depending on the manufacturing process used, roughness can be periodic or random. A periodic surface roughness is also referred to as surface texture

and represents a regular surface topography with well-defined dimensions and surface distribution. Further, distinction has to be made among macro, micro, and nano surface roughness. Microroughness deals with surface features sized in cellular and subcellular dimensions. Considering their appearance and morphological structure, class 3 surface roughness has a groove-type appearance; class 4 roughness deals with score marks, flakes, and protuberances, for example created by grit-blasting procedures; and class 5 surface roughness is the result of the crystal structure of a material.

POROSITY

Besides the surface irregularities as mentioned earlier, porosity can also be considered as surface irregularity. Porosity can occur only at the substrate surface or can completely penetrate throughout a bulk material. It consists of individual openings and spacings or interconnecting pores. Porosity can be created intentionally by a specific production process, such as sintering of beads, leaching of salt, sugar, or starch crystals, or knitting and weaving of fibers. On the other hand, porosity can also arise as a manufacturing artifact, for example, in casting procedures.

For many biomedical applications, there is a need for porous implant materials. They can be used for artificial blood vessels, artificial skin, drug delivery, bone and cartilage reconstruction, periodontal repair, and tissue engineering (Lanza *et al.,* 1997). For each application, the porous materials have to fulfil a number of specific requirements. For example, for bone ingrowth the optimum pore size is in the range of 75–250 μm (Pilliar, 1987). On the other hand, for ingrowth of fibrocartilagenous tissue the recommended pore size ranges from 200 to 300 μm (Elema *et al.,* 1990). Besides pore size, other parameters play a role, such as compressibility, pore interconnectivity, pore interconnection throat size, and possibly degradability of the porous material (de Groot *et al.,* 1990).

Although porosity can also be discerned as a different class of surface irregularity, the following sections will consider porosity as microtexture, much like other surface features. This choice is based on the many reports that emphasize the importance of this type of surface morphology for cell and tissue response.

PREPARATION OF SURFACE MICROTEXTURE

For the production of microtextured implant surfaces, numerous techniques are available ranging from simple manual scratching to more controlled fabrication methods. For example, from semiconductor technology, photolithographic techniques used in conjunction with reactive plasma and ion-etching, LIGA and electroforming, have become available. Deep reactive ion etching (DRIE) enhances the depth of surface etched features and gives parallel sidewalls—it is especially well suited for microelectromechanical systems (MEMS) fabrication. Microcontact printing (μCP) allows patterns to be transferred to biomaterial surfaces by a rubber stamp.

Because these techniques are relatively fast and cheap, and also allow the texturing of surfaces of reasonable size, they appear to be promising for biomedical research and applications. Other methods that offer the ability to texture and pattern surfaces include UV laser machining, electron-beam etching, and ion-beam etching.

Reactive Plasma and Ion Etching

For this method the material, usually silicon, is first cleaned and dried with filtered air (den Braber *et al.*, 1998a; Hoch *et al.*, 1996; Jansen *et al.*, 1996). Then it is coated with a primer and photoresist (PR) material. Photolithography is used to create a micropattern in the photoresist layer. Masks with predetermined dimensions are exposed with either UV light or electron beams depending on the size of the required surface configuration. Subsequently, the exposed resist is developed and rinsed off. Finally, this lithographically defined photoresist pattern is transferred into the underlying material by etching. This etching can be performed under wet or dry conditions. In the first situation, materials are placed in chemicals. Etch direction is along the crystal planes of the material. In the second situation, dry etching is performed using directed ions from a plasma or ion beam as etchants. This technique of physical etching allows a higher resolution than the wet technique. It is also applicable in noncrystalline materials because of the etch directionality without using crystal orientation. Finally, after the etching process, the remaining resist is removed. If a substrate is formed with microgrooves, the dimensions of the texture are usually described in pitch (or spacing), ridge width, and groove width (von Recum *et al.*, 1995).

Plasma and ion etching techniques can be used to create micropatterns in a wide variety of biopolymers. The micropatterns can be prepared directly in the polymer surface or transferred into the polymer surface via solvent-casting or injection-molding methods, whereby a micropatterned silicon wafer is used as a template (Fig. 1).

FIG. 1. Scanning electron micrograph of a micropatterned silicon wafer, which can be used as a template in a solvent-casting replication process.

LIGA

Another technology suitable for creating surface microtextures is the so-called LIGA process (Rogner *et al.*, 1992). LIGA refers to the German "Lithographie, Galvanoformung, Abformung" (lithography, electroplating, molding). The LIGA technique differs completely from that described in the preceding section, since it is not based on etching. In the LIGA process a thick X-ray-resistant layer is exposed to synchrotron radiation using a special X-ray mask membrane. Subsequently, the exposed layer is developed, which results in the desired resist structure. Then, metal is deposited onto the remaining resist structure by galvanization. After removal of the remaining resist either a metal structure or mold for subsequent cost-effective replication processes is achieved.

Microcontact Printing

The microcontact printing (μCP) method, developed in the laboratory of George Whitesides, provides a simple method to create patterns over large surface areas at the micro and even nanoscale (Kumar *et al.*, 1994). A master silicon template or mold is formed by conventional photolithographic and etching methods generating the micron-scale pattern of interest. Onto that template, a curable silicone elastomer is poured. When the silicone polymer cures, it is peeled off and then serves as a rubber stamp. The stamp can be "inked" in thiols, silanes, proteins or other polymers (see Chapter 2.14). Flat and curved surfaces can be patterned with these μCP stamps.

PARAMETERS FOR THE ASSESSMENT OF SURFACE MICROTEXTURE

Since the final biological performance of a microtextured surface is determined by the size and dimensions of the surface features, specific surface parameters have to be provided to describe and define the surface structure.

The definition of surface parameters is mostly based on a two-dimensional profile section. Occasionally, three-dimensional profiles are created (see the next two sections).

In general, for the quantitative description of surface microtexture, three parameters can be used:

1. Amplitude parameters, to obtain information about height variations
2. Spacing parameters, to describe the spacing between features
3. Hybrid parameters, a combination of height and spacing parameters

These parameters are presented as Ra, Rq, Rt, Rz, Rsk, Rku (amplitude parameters), Scx, Scy, Sti (spacing parameters), and Δq and λq (hybrid parameters). The R-parameters are denominations for a two-dimensional description. The S-parameters stand for a three-dimensional evaluation. These S-denominations are generally accepted since the work of

TABLE 1   Definition of Surface Parameters

Parameter	Definition
lm	Evaluation length = the horizontal limitation for the assessment of surface parameters
lv	Pre-travel length = the distance traversed by the tracing system over the sample before the tracing (lt) starts
ln	Over-travel length = the distance traversed or area scanned by the tracing system over the sample after the tracing (lt)
lt	Tracing length = the distance traversed by the tracing system when taking a measurement. It comprises the pre- and overtravel, and the evaluation length
le	Sampling length = a standardized number of evaluation lengths/areas as required to obtain a proper surface characterization
Ra/Sa	Arithmetical mean roughness = the arithmetical average value of all vertical departures of the profile or surface from the mean line throughout the sampling length/area
Rq/Sq	Root-mean square roughness = the root-mean square value of the profile or surface departures within the sampling length/area
Rt/St	Maximum roughness depth = the distance between the highest and lowest points of the profile or surface within the evaluation length/area
Rz/Sz	Mean peak-to-valley height = the average of the single peak-to valley heights of five adjoining sampling lengths/areas
Rsk/Ssk	Skewness = measure of the symmetry of the amplitude density function (ADF)
ADF	Amplitude density function = the graphical representation of the material distribution within the evaluation length/area
Rku/Sku	Kurtosis = fourth central moment of the profile or surface amplitude density with the evaluation length/area. Kurtosis is the measure of the sharpness of the profile or surface
Rcx/Rcy Scx/Scy	= mean spacing between surface peaks of the surface/area profile along the X or Y direction
Sti	= surface texture index, i.e. min. (Rq/Sq divided by max. Rq/Sq + min. Rsk/Ssk divided by max Rsk/Ssk + min.(q divided by max.)q + min (Rc/Sc divided max. Rc/Sc) divided by 4
Δq	= the root mean square slope of the rough profile throughout the evaluation length/area
λq	= the root mean square of the spacings between local peaks and valleys, taking into account their relative amplitudes and individual spatial frequencies

Stout *et al.* (1993). For a detailed description of available surface parameters, reference can be made to Sander (1991) and Wennerberg *et al.* (1992). A brief summary is given in Table 1.

Further, it has to be emphasized that for a correct assessment of surface parameters various requirements have to be met. A first condition is the provision of a reference line to which measurements can be related. Also, surface parameters have to be determined with a clear separation between roughness and waviness components. This separation has to be achieved by an electronic filtering procedure. In view of this, perhaps the most important measurement requirements are the parameters measuring length over the substrate surface and cutoff wavelength of the filter used. Measuring or tracing length has to be described in terms of real evaluation length (lm) and pre- and overtravel (lv resp. ln). The function of the electronic filter is to eliminate waviness and roughness frequencies out of the surface profile. As surface features differ in both their wavelength and surface profile depths, various filters are available. The filter type to be selected for a specific surface profile is defined in DIN standards. Use of the wrong filter will result in incorrect measurements (Sander, 1991).

CHARACTERIZATION OF SURFACE TOPOGRAPHY

Various methods are available to describe surface features. Scanning electron microscopy can be used to obtain a qualitative image of the surface geometry. Contact and noncontact profilometry are methods to quantify the surface roughness.

Contact Profilometry

The principle of contact profilometry is that a finely pointed stylus moves over the detected area. The vertical movements of the stylus are switched into numerical information. This method results in a two-dimensional description of the surface. The advantage of contact profilometry is that the method is inexpensive, direct, and reproducible. Contact profilometry can be applied on a wide variety of materials. The major disadvantage is that the diameter of the pointed stylus limits its use to surface features larger than the stylus point diameter. Another problem is that, because of the physical contact between the stylus and substrate surface, distortion of the surface profile can occur.

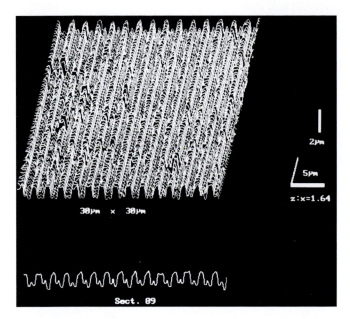

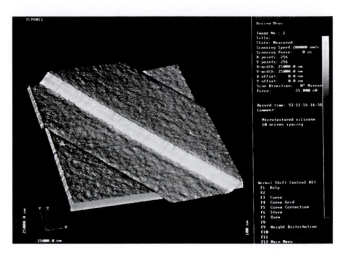

FIG. 3. Three-dimensional representation of an AFM measurement of a silicon wafer provided with 10-μm-wide and 0.5-μm-deep microgrooves. The raised wall of the edge shows a small inclination. This is a distortion due to the size and movement of the tip over the silicon surface.

FIG. 2. Results of a confocal laser scanning microscope (CLSM) surface analysis of a microgrooved substratum. CLSM has to be considered as a noncontact technique. A three- and two-dimensional surface representation is obtained, composed from 256 optical Z sections. To the right of the 3D surface profile, the size of the scanned area (30 μm²) and the difference in X versus Z enlargement can be found (Scale 1 : 1.64).

Noncontact Profilometry

In this method, the pointed stylus is replaced by a light or laser spot. This spot never touches the substrate surface. The light or laser beam is focused on the surface and the light is reflected and finally converted to an electrical signal. In this way both two- and three-dimensional surface profiles can be created (Fig. 2). Occasionally, techniques are used in which the reflected light is not directly translated to an electrical signal. In these so-called interferometers a surface profile is created by combining light reflecting off the surface with light reflecting off a reference substrate. When those two light bundles combine, the light waves interfere to produce a pattern of fringes, which are used to determine surface height differences.

The resolution of noncontact methods can be in the nanometer range. The limiting factor is the spot size. Several scans have to be taken to obtain a representative surface area. Occasionally, this is impossible or too laborious. In light beam interferometry, an additional disadvantage is that the substrate surface has to provide at least some reflectivity.

Atomic Force Microscopy

Atomic force microscopy (AFM) is a direct method for determining high-resolution surface patterns (Binnig *et al.*, 1986; van der Werf *et al.*, 1993) (also see Chapters 1.4 and 5.6). In AFM the substrate surface is brought close to a tip on a small cantilever which is attached to a piezo tube. The deflection of the cantilever, generated by interaction forces between tip and substrate surface, is detected and used as an input signal for a measuring system. AFM is frequently used as a contact method. However, noncontact and transient contact modes of analysis are also available. The advantage of AFM above other contact techniques is that AFM is generally not as destructive. Considering resolution, a limiting factor in AFM is again the size of the used tip (Fig. 3). Still, a significantly smaller tip diameter is used compared with conventional contact methods such as profilometry.

BIOLOGICAL EFFECTS OF SURFACE MICROTEXTURE

The role of standardized surface texture in inducing a specific cellular response is a field of active research. For example, various reports have suggested that a regular surface microtexture can benefit the clinical success of skin penetrating devices by preventing epithelial downgrowth (Brunette *et al.*, 1983; Chehroudi *et al.*, 1988) and reduce the inflammatory response (Campbell *et al.*, 1989) and fibrous encapsulation (Chehroudi *et al.*, 1991) of subcutaneous implants. Closely related to these studies, certain porosities have led to an increase of the vascularity of the healing response and a reduction of collagenous capsule density (Brauker *et al.*, 1995; Sharkawy *et al.*, 1998). The literature on the effect of surface texture on the healing of silicone breast implants is extensive (for example, see Pollock, 1992). Therefore, much current research has been focused on the effect of standardized surface roughness on the soft tissue reaction. Excellent reviews on the effect of surface microtexturing on cellular growth, migration, and attachment have been written by Singhvi *et al.* (1994), von Recum and van Kooten (1995), Brunette (1996), Curtis and Wilkinson (1997), and Folch and Toner (2000).

Hypotheses on Contact Guidance

Contact guidance is the phenomenon that cells adapt and orient to the substrate surface microtopography (Harrison, 1912). Early studies on contact guidance describe the alignment of cells and focal adhesions to microgrooves with dimensions 1.65–8.96 μm in width and 0.69 μm in depth. This cellular behavior was suggested to be due to the mechanical properties of the cytoskeleton (Dunn, 1982; Dunn and Brown, 1986). The relative inflexibility of cytoskeletal components was considered to prevent bending of cell protrusions over surface configurations with too large an angle.

Later studies and hypotheses focused on the relationships among cell contact site, deposited extracellular matrix, surface microtexture, and cell response. For example, a microtextured surface was supposed to possess local differences in surface free energy resulting in a specific deposition pattern of the substratum bound attachment proteins (Brunette, 1996; Maroudas, 1972; von Recum and van Kooten, 1995). The spatial arrangement of the adsorbed proteins and their conformational state were hypothesized to be affected. In addition to wettability properties, the specific geometric dimensions of the cell adhesion sites were suggested to induce a cell orientational effect (Dunn, 1982; Dunn and Brown, 1986; Ohara et al., 1979). A recent hypothesis suggests that contact guidance on microtextured surfaces is a part of the cellular efforts to achieve a biomechanical equilibrium condition with a resulting minimal net sum of forces. The signficance of this theory has been described extensively by Ingber (1993, 1994) in his tensegrity models. According to this model, the anisotropic geometry of substratum surface features establishes stress- and shear-free planes that influence the direction of cytokeletal elements in order to create a force economic situation (Oakley and Brunette, 1993, 1995; O'Neill et al., 1990).

The in Vitro Effect of Surface Microtexturing

A considerable number of in vitro studies have been performed to determine which of the hypotheses mentioned in the preceding section can be experimentally supported. Up to this point, we have to emphasize that comparison of the obtained data is difficult because most of the studies had differences in the surface textures of the materials explored. In addition, different bulk materials were also applied. Modern surface feature fabrication methods have allowed more precise surfaces to be fabricated so studies from different groups might be compared.

In the experiments performed by Curtis et al. (Clark et al., 1987, 1990, 1991; Curtis and Wilkinson, 1997) with fibroblasts and macrophages cultured on microgrooved glass substrates, groove depth was observed to be more important than groove width in the establishment of contact guidance. Therefore, these experiments believe that cytoskeletal flexibility and the possibility of making cellular protrusions are the determining cellular characteristics for contact guidance. As a consequence of these studies, other reseaches further explored the involvement of cytoskeletal elements in cell orientation processes. Also, the possibility of a relationship between cytoskeletal organization and cell–substrate contact sites was investigated (den Braber et al., 1995, 1996, 1998b; Meyle et al.,

1991, 1993; Oakley et al., and Brunette, 1993, 1995; Oakley 1997; Walboomers et al., 1998a, 1999). Although these studies varied in cell type used, substrate surface feature dimensions and substrate bulk chemical composition, the results clearly confirmed that very fine microgrooves (≤ 2 μm) have an orientational effect on both cell body and cytoskeletal elements. Transmission electron microscopy observations showed that cells were only able to penetrate into very shallow (≤ 1 μm) or wide (≥ 5 μm) microgrooves. Cells were also observed to possess cell adhesion structures that were wrapped around the edge of a ridge or attached to the wall of the ridge. On the basis of these findings, these investigators suggested that the mechanical properties of cellular structures can never be the only determining factor in contact guidance.

Further, a mechanical model to explain contact guidance suggests that the "surface feature stimulus" is transduced to the cytoskeleton via cell contact sites and cell surface receptors. In this model, the cytoskeleton is considered as a static structure. This is incorrect. The cytoskeleton is a highly dynamic system (Lackie, 1986), which is constantly broken down and elongated in living cells. Consequently, if the mechanical theory is still true, the fundamentals should be derived from other processes than just the remodeling of the cytoskeleton (Walboomers et al., 1998a). Studies on cell nuclear connections to the cytoskeleton may offer insights into the relationships between surface features and cell behavior (Maniotis et al., 1997).

Apart from changes in cell size, shape, and orientation, surface microtopography has been reported to influence other cell processes. For example, several studies described changes in cellular differentiation, DNA/RNA transcription, cellular metabolism, and cellular protein production of cells cultured on microtextured surfaces (Chou et al., 1995; Hong and Brunette 1987; Matsuzaka et al., 1999; von Recum and van Kooten, 1995; Singhvi et al., 1997; Wójciak-Stothard et al., 1995). A study using μCP surfaces with square cell adhesive and nonadhesive domains has shown that where surface adhesive domains are small (< 75 μm), apoptosis levels in endothelial cells is high (particularly so for 5 μm $\times$ 5 μm domains) and when cells are placed on larger domains, cell spreading and growth occurs (Chen et al., 1997). Whether these additional effects have to be considered as independent phenomena is still a topic of discussion. According to Hong and Brunette (1987), the good news was that surface microtopography can enhance the production of specific, perhaps favorable proteins. On the other hand, the production or secretion of less favorable metabolic products can also be enhanced. If this occurs, this might have a deleterious effect on the overall cell response. For example, a rise in the production or release of proteinases may not be beneficial for connective tissue cell response. This example shows that, at least at the molecular level, the regulation of cell function by substrate surface microtexture may be a complex affair.

The in Vivo Effect of Surface Microtexturing

Based upon interesting results from in vitro experiments, in vivo studies with microtextured implants have been performed. Unfortunately, the results from the various studies are not consistent. For example, in some animal experiments it was demonstrated that silicone-coated filters and bulk silicone rubber implants provided with surface features of 1–3 μm showed a minimal inflammatory response with direct fibroblast attachment and a very reduced connective capsule (Campbell and von Recum, 1989; Schmidt and von Recum, 1991, 1992). In contrast, other animal studies suggested that implant surface microgrooves were unable to influence the wound healing process at all (den Braber *et al.*, 1997; Walboomers *et al.*, 1998b). These differing results may hint at multiple surface-texture-related factors that are not yet identified and controlled.

Besides the effect on wound healing, microtextured implants have also been used to inhibit epidermal downgrowth along skin penetrating devices (Chehroudi *et al.*, 1989, 1990, 1992). This downgrowth is considered as a major failure mode for this type of implant. Indeed, the experiments suggested that epidermal downgrowth can be prevented or delayed by percutaneous devices provided with surface microgrooves.

DIRECTIONS FOR FURTHER DEVELOPMENTS

Considering the *in vitro* experiments, none of the earlier mentioned hypotheses to explain contact guidance has been fully supported. Therefore, based on various findings we suggest a new theory that is a refinement of the "mechanical" theory discussed earlier. The breakdown and formation of fibrous cellular components, especially in the filopodium, is influenced by the microgrooves. These microgrooves create a pattern of mechanical stress, which affects cell spreading and causes the alignment of cells. On the other hand, we must also notice that the ECM possesses mechanical properties. The ECM is not a rigid structure, but a dynamic mass of molecules. Many *in vitro* studies have already indicated that cell-generated forces of tension and traction can reorganize the ECM into structures that direct the behavior of single cells (Erickson, 1994; Choquet *et al.*, 1997; Janmey and Chaponnier, 1995; Janmey, 1998). As cells cannot penetrate very shallow or small grooves, we suppose that on those surfaces the forces as exerted by the cells will result in an enhanced reorganization of the deposited ECM proteins. Consequently, contact guidance and other cell behaviors are induced. No doubt, cell surface receptors and inside–outside cell signaling phenomena play an important role in this process. As far as *in vivo* applications of surface microtexturing, more research has to be done to learn and understand the full impact of surface microtexturing for medical devices. A first step is the development of techniques that enable the production of standardized microstructures on nonplanar surfaces. Evidently, this development will benefit not only biomaterial research, but also the production of microelectronic, mechanical, and optical devices and subsytems. As a second step, the relationship between the surface topographical design of an implant and histocompatibility has to be further documented. These studies must focus not only on the soft tissue response; they must also involve bone tissue behavior.

Bibliography

Binnig, G., Quate, C. F., and Gerber, C. (1986). Atomic force microscopy. *Phys. Rev. Lett.* **56:** 930–933.

Brauker, J. H., Carr-Brendel, V. E., Martinson, L. A., Crudele, J., Johnston, W. D., and Johnson, R. C. (1995). Neovascularization of synthetic membranes directed by membrane microarchitecture. *J. Biomed. Mater. Res.* **29:** 1517–1524.

Brunette, D. M. (1996). Effects of surface topography of implant materials on cell behavior in vitro and in vivo. in *Nanofabrication and Biosystems*, H. C. Hoch, L. W. Jelinski, and H. G. Craighead, eds. Cambridge University Press, Cambridge, UK, pp. 335–355.

Brunette, D. M., Kenner, G. S., and Gould, T. R. L. (1983). Grooved titanium surfaces orient growth and migration of cells from human gingival explants. *J. Dent. Res.* **62:** 1045–1048.

Campbell, C. E., and von Recum, A. F. (1989). Microtopography and soft tissue response. *J. Invest. Surg.* **2:** 51–74.

Chehroudi, B., Gould, T. R., and Brunette, D. M. (1988). Effects of a grooved epoxy substratum on epithelial cell behavior in vitro and in vivo. *J. Biomed. Mater. Res.* **22:** 459–473.

Chehroudi, B., Gould, T. R. L., and Brunette, D. M. (1989). Effects of a grooved titanium-coated implant surface on epithelial cell behavior *in vitro* and *in vivo*. *J. Biomed. Mater. Res.* **23:** 1067–1085.

Chehroudi, B., Gould, T. R., and Brunette, D. M. (1991). A light and electron microscope study of the effects of surface topography on the behavior of cells attached to titanium-coated percutaneous implants. *J. Biomed. Mater. Res.* **25:** 387–405.

Chehroudi, B., Gould, T. R. L., and Brunette, D. M. (1990). Titanium coated micromachined grooves of different dimensions affect epithelial and connective tissue cells differently *in vivo*. *J. Biomed. Mater. Res.* **24:** 1203–1219.

Chehroudi, B., Gould, T. R. L., and Brunette, D. M. (1992). The role of connective tissue in inhibiting epithelial downgrowth on titanium-coated percutaneous implants. *J. Biomed. Mater. Res.* **26:** 493–515.

Chen, C. S., Mrksich, M., Huang, S., Whitesides, G. M., and Ingber, D. E. (1997). Geometric control of cell life and death. *Science* **276:** 1425–1428.

Choquet, D., Felsenfeld D. P., and Sheetz, M. P. (1997). Extracellular matrix rigidity causes strengthening of integrin-cytoskeleton linkages. *Cell* **88:** 39–48.

Chou, L. S., Firth, J. D., Uitto, V. J., and Brunette, D. M. (1995). Substratum surface topography alters cell shape and regulates fibronectin mRNA level, mRNA stability, secretion and assembly in human fibroblasts. *J. Cell Sci.* **108:** 1563–1573.

Clark, P., Connoly, P., and Curtis, A. S. G. (1987). Topographical control of cell behavior I: simple step clues. *Development* **99:** 439–448.

Clark, P., Connoly, P., and Curtis, A. S. G. (1990). Topographical control of cell behavior II: multiple grooved substrata. *Development* **108:** 635–644.

Clark, P., Connoly, P., and Curtis, A. S. G. (1991). Cell guidance by ultrafine topography *in vitro*. *J. Cell Res.* **86:** 9–24.

Curtis, A. S. G., and Wilkinson, C. (1997). Topographical control of cells. *Biomaterials* **18:** 1573–1583.

den Braber, E. T., Ruijter, J. E., de Smits, H. T. J., Ginsel, L. A., Recum, A. F., and von Jansen, J. A. (1995). Effects of parallel surface microgrooves and surface energy on cell growth. *J. Biomed. Mat. Res.* **29:** 511–518.

den Braber, E. T., Ruijter, J. E., de Smits, H. T. J., Ginsel, L. A., Recum, A. F., and von Jansen, J. A. (1996). Quantitative analysis of cell proliferation and orientation on substrata with uniform parallel surface micro grooves. *Biomaterials* **17:** 1093–1099.

den Braber, E. T., Ruijter, J. E., and Jansen, J. A. (1997). The effect of a subcutaneous silicone rubber implant with shallow surface micro grooves on the surrounding tissue in rabbits. *J. Biomed. Mater. Res.* 37: 539–547.

den Braber, E. T., Jansen, H. V., de Boer, M. J., Croes, H. J. E., Elwenspoek, M., Ginsel, L. A., and Jansen, J. A. (1998a). Scanning electron microscopic, transmission electron microscopic, and confocal laser scanning microscopic observation of fibroblasts cultured on microgrooved surfaces of bulk titanium substrata. *J. Biomed. Mater. Res.* 40: 425–433.

den Braber, E. T., Ruijter, J. E., Ginsel, L. A., von Recum, A. F., and Jansen, J. A. (1998b). Orientation of ECM protein deposition, fibroblast cytoskeleton, and attachment complex components on silicone microgrooved surfaces. *J. Biomed. Mater. Res.* **40**: 291–300.

Dunn, G. A. (1982). Contact guidance of cultured tissue cells: a survey of potentially relevant properties of the substratum. in *Cell Behavior*, R. Bellairs, A. S. G. Curtis and G. A. Dunn, eds. Cambridge University Press, Cambridge, UK, pp. 247–280.

Dunn, G. A., and Brown, A. F. (1986). Alignment of fibroblasts on grooved surfaces described by a simple geometric transformation. *J. Cell Sci.* **83**: 313–340.

Elema. H., Groot, J. H., de Nijenhuis, A. J., Pennings, A. J., Veth, R. P. H., Klompmaker, J., and Jansen, H. W. B. (1990). Biological evaluation of porous biodegradable polymer implants in menisci. *Colloid Polymer Sci.* **268**: 1082–1088.

Erickson, H. P. (1994). Reversible unfolding of fibronectin type III and immunoglobulin domains provides the structural basis for stretch and elasticity of titin and fibronectin. *Proc. Natl. Acad. Sci. USA* **91**: 10114–10118.

Folch, A., and Toner, M. (2000). Microengineering of cellular interactions. *Annu. Rev. Biomed. Eng.* **2**: 227–256.

Groot, J. H., de Nijenhuis, A. J., Bruin, P., Pennings, A. J., Veth, R. P. H., Klompmaker, J., and Jansen, H. W. B. (1990). Preparation of porous biodegradable polyurethanes for the reconstruction of meniscal lesions. *Colloid Polymer Sci.* **268**: 1073–1081.

Harrison, R. G. (1912). The cultivation of tissues in extraneous media as a method of morphogenetic study. *Anat. Rec.* **6**: 181–193.

Hoch, H. C., Jelinski, L. W., and Craighead, H. G. (1996). *Nanofabrication and Biosystems.* Cambridge University Press, Cambridge, UK.

Hong, H. L., and Brunette, D. M. (1987). Effect of cell shape on proteinase secretion. *J. Cell Sci.* **87**: 259–267.

Ingber, D. E. (1993). Cellular tensegrity; defining new rules of biological design that govern the cytoskeleton. *J. Cell Sci.* **104**: 613–627.

Ingber, D. E. (1994). Cellular tensegrity and mechanochemical transduction. *in Cell Mechanics and Cellular Engineering*, V. C. Mow, F. Guilak, R. Tran-Son-Tay, and R. M. Hochmuth, eds. Springer-Verlag, New York, pp. 329–342.

Janmey, P. A., and Chaponnier, C. (1995). Medical aspects of the actin cytoskeleton. *Curr. Opin. Cell Biol.* **7**: 111–117.

Janmey, P. A. (1998). The cytoskeleton and cell signaling: component localization and mechanical coupling. *Physiol. Rev.* **78**: 763–781.

Jansen, H. V., Gardeniers, J. G. E., de Boer, M. J., Elwenspoek, M. E., and Fluitman, J. H. J. (1996). A survey on the reactive ion etching of silicon in microtechnology. *J. Micromech. Microeng.* **6**: 14–28.

Kumar, A., Biebuyck, H. A., and Whitesides, G. M. (1994). Patterning self-assembled monolayers: applications in materials science. *Langmuir* 10(5): 1498–1511.

Lackie, J. M. (1986). *Cell Movement and Cell Behaviour.* Allen & Unwin, London.

Lanza, R. P., Langer, R., and Chick, W. L. (1997). *Principles of Tissue Engineering.* Academic Press, San Diego.

Maniotis, A. J., Chen, C. S., and Ingber, D. E. (1997). Demonstration of mechanical connections between integrins, cytoskeletal filaments, and nucleoplasm that stablize nuclear structure. *Proc. Natl. Acad. Sci. USA* **94**: 849–854.

Maroudas, N. G. (1972). Anchorage dependence: correlation between amount of growth and diameter of bead, for single cells grown on individual glass beads. *Exp. Cell Res.* **74**: 337–342.

Matsuzaka, K., Walboomers, X. F., de Ruijter, J. E., and Jansen, J. A. (1999). The effect of poly-L-lactic acid with parallel surface micro groove on osteoblast-like cells *in vitro. Biomaterials* 20(14): 1293–1301.

Meyle, J., von Recum, A. F., and Gibbesch, B. (1991). Fibroblast shape conformation to surface micromorphology. *J. Appl. Biomat.* **2**: 273–276.

Meyle, J., Gültig, K., Wolburg, H., and von Recum, A. F. (1993). Fibroblast anchorage to microtextured surfaces. *J. Biomed. Mater. Res.* **27**: 1553–1557.

Oakley, C., and Brunette, D. M. (1993). The sequence of alignment of microtubules, focal contacts and actin filaments in fibroblasts spreading on smooth and grooved titanium substrata. *J. Cell Sci.* **106**: 343–354.

Oakley, C., and Brunette, D. M. (1995). Topographic compensation: guidance and directed locomotion of fibroblasts on grooved micromachined substrata in the absence of microtubules. *Cell Motil. Cytoskeleton* **31**: 45–58.

Oakley, C., Jaeger, N. A., and Brunette, D. M. (1997). Sensitivity of fibroblasts and their cytoskeletons to substratum topographies: topographic guidance and topographic compensation by micromachined grooves of different dimensions. *Exp. Cell Res.* **234**: 413–424.

Ohara, P. T., and Buck, R. C. (1979). Contact guidance *in vitro*. A light, transmission, and scanning electron microscopic study. *Expl. Cell Res.* **121**: 235–249.

O'Neill, C., Jordan, P., and Riddle, P. (1990). Narrow linear strips of adhesive substratum are powerful inducers of both growth and total focal contact area. *J. Cell Sci.* **95**: 577–586.

Pilliar, R. M. (1987). Porous-surfaced metallic implants for orthopaedic applications. *J. Biomed. Mater. Res.* **21**: 1–33.

Pollock, H. (1992): Breast capsular contracture: A retrospective study of textured versus smooth silicone implants. *Plast. Reconstr. Surg.* 91(3): 404–407.

Rogner, A., Eichner, J., Münchmeyer, D., Peters, R.-P., and Mohr, J. (1992). The LIGA technique—what are the opportunities? *J. Micromech. Microeng.* **2**: 133–140.

Sander, M. (1991). *A Practical Guide to the Assessment of Surface Texture.* Feinprüf Perthen Gmbh, Göttingen.

Schmidt, J. A., and von Recum, A. F. (1991). Texturing of polymer surfaces at the cellular level. *Biomaterials* **12**: 385–389.

Schmidt, J. A., and von Recum, A. F. (1992). Macrophage response to microtextured silicone. *Biomaterials* **13**: 1059–1069.

Sharkawy, A., Klitzman, B., Truskey, G. A., and Reichert, W. M. (1998). Engineering the tissue which encapsulates subcutaneous implants. II. Plasma–tissue exchange properties. *J. Biomed. Mater. Res.* **40**: 586–597

Singhvi, R., Stephanopoulos, G., and Wang, D. I. C. (1994). Review: effects of substratum morphology on cell physiology. *Biotechnol. Bioeng.* **43**: 764–771.

Stout K.-J., Sullivan, P. J., Dong, W. P., Mainsah, E., Luo, N., Mathia, T., and Zahouni, H. (1993). The devlopment of methods for the characterization of roughness in three dimensions. EUR 15178 EN of Commission of the European Communities, University of Birmingham, Birmingham, UK.

von Recum, A. F., and van Kooten, T. G. (1995). The influence of microtopography on cellular response and the implications for silicone implants. *J. Biomater. Sci. Polymer Ed.* 7: 181–198.

Walboomers, X. F., Croes, H. J. E., Ginsel, L. A., and Jansen, J. A. (1998a). Growth behavior of fibroblasts on microgrooved polystyrene. *Biomaterials* 19: 1861–1868.

Walboomers, X. F., Croes, H. J. E., Ginsel, L. A., and Jansen, J. A. (1998b). Microgrooved subcutaneous implants in the goat. *J. Biomed. Mater. Res.* 42: 634–641.

Walboomers, X. F., Monaghan, W., Curtis, A. S. G., and Jansen, J. A. (1999). Attachment of fibroblasts on smooth and microgrooved polystyrene. *J. Biomed. Mater. Res.* 46(2): 212–220.

Wennerberg, A., Albrektsson, T., Ulrich, H., and Krol, J. (1992). An optical three-dimensional technique for topographical descriptions of surgical implants. *J. Biomed. Eng.* 14: 412–418.

Werf, K. O., van der, Putman, C. A. J., de Grooth, B. G., Segerink, F. B., Schipper, E. H., van Hulst, N. F., and Greve, J. (1993). Compact stand-alone atomic force microscope. *Rev. Sci. Instrum.* 64: 2892–2897.

Wójciak-Stothard, B., Madeja, Z., Korohoda, W., Curtis, A., and Wilkinson, C. (1995). Activation of macrophage-like cells by multiple grooved substrata: topographical control of cell behaviour. *Cell Biol. Int.* 19: 485–490.

2.16 SURFACE-IMMOBILIZED BIOMOLECULES

Allan S. Hoffman and Jeffrey A. Hubbell

Biomolecules such as enzymes, antibodies, affinity proteins, cell receptor ligands, and drugs of all kinds have been chemically or physically immobilized on and within biomaterial supports for a wide range of therapeutic, diagnostic, separation, and bioprocess applications. Immobilization of heparin on polymer surfaces is one of the earliest examples of a biologically functional biomaterial. Living cells may also be combined with biomaterials, and the fields of cell culture, artificial organs, and tissue engineering are additional, important examples. These "hybrid" combinations of natural and synthetic materials confer "biological functionality" to the synthetic biomaterial. Since many sections and chapters in this text cover many aspects of this topic, including adsorption of proteins and adhesion of cells and bacteria on biomaterial surfaces, nonfouling surfaces, cell culture, tissue engineering, artificial organs, drug delivery, and others, this chapter will focus on the methodology involving physical adsorption and chemical immobilization of biomolecules on biomaterial surfaces, especially for applications requiring bioactivity of the immobilized biomolecule.

Among the different classes of biomaterials that could be biologically modified, polymers are especially interesting because their surfaces may contain reactive groups *de novo*, or they may be readily derivatized with reactive groups that can be used to covalently link biomolecules. Another advantage of polymers as supports for biomolecules is that the polymers may be fabricated in many forms, including films, membranes, tubes, fibers, fabrics, particles, capsules, and porous structures. Furthermore, polymer compositions vary widely, and molecular structures include homopolymers, and random, alternating, block, and graft copolymers. Living anionic polymerization

techniques, along with newer methods of living free radical polymerizations, now provide fine control of molecular weights with narrow distributions. The molecular forms of solid polymers include un-cross-linked chains that are insoluble at physiologic conditions, cross-linked networks, physical blends, and interpenetrating networks (IPNs) (e.g., Piskin and Hoffman, 1986; see also Chapter 2.2). When surfaces of metals or inorganic glasses or ceramics are involved, biological functionality can sometimes be added via a chemically immobilized or physisorbed polymeric or surfactant adlayer, or by use of techniques such as plasma gas discharge to deposit polymer compositions having functional groups (see also Chapter 2.14).

Patterned Surfaces

Biomaterial surfaces may be functionalized uniformly or in geometric patterns (Bernard *et al.*, 1998; Blawas and Reichert, 1998; James *et al.*, 1998; Kane *et al.*, 1999; Ito, 1999; Folch and Toner, 2000). Sometimes the patterned surfaces will have regions that repel proteins ("nonfouling" compositions) while others may contain covalently-linked cell receptor ligands (Neff *et al.*, 1999; Alsberg *et al.*, 2002; Csucs *et al.*, 2003; VandeVondele *et al.*, 2003), or may have physically adsorbed cell adhesion proteins (McDevitt *et al.*, 2002; Ostuni *et al.*, 2003). There has also evolved a huge industry based on "biochips" that contain microarrays of immobilized, single-stranded DNA (for genomic assays) or peptides or proteins (for proteomic assays) (Housman and Mrksich, 2002; Lee and Mrksich, 2002). The majority of these microarrays utilize inorganic silica chips rather than polymer substrates directly, but it is possible to incorporate functionality through chemical modification with silane chemistries (Puleo, 1997) or adsorption of a polymeric adlayer (Scotchford *et al.*, 2003; Winkelmann *et al.*, 2003). A variety of methods have been used for the production of these patterned biochips, including photocontrolled synthesis (Ellman and Gallop, 1998; Folch and Toner, 2000), microfluidic fluid exposure (Ismagilov *et al.*, 2001), and protection with adhesive organic protecting layers that are lifted off after exposure to the biomolecular treatment (Jackman *et al.*, 1999).

Immobilized Biomolecules and Their Uses

Many different biologically functional molecules can be chemically or physically immobilized on polymeric supports (Table 1) (Laskin, 1985; Tomlinson and Davis, 1986). When some of these solids are water-swollen they become hydrogels, and biomolecules may be immobilized on the outer gel surface as well as within the swollen polymer gel network. Examples of applications of these immobilized biological species are listed in Table 2. It can be seen that there are many diverse uses of such biofunctional systems in both the medical and biotechnology fields. For example, a number of immobilized enzyme supports and reactor systems (Table 3) have been developed for therapeutic uses in the clinic (Table 4) (De Myttenaere *et al.*, 1967; Kolff, 1979; Sparks *et al.*, 1969; Chang, 1972; Nose *et al.*, 1983, 1984; Schmer *et al.*, 1981; Callegaro and Denti, 1983; Lavin *et al.*, 1985; Sung *et al.*, 1986).

TABLE 1   Examples of Biologically Active Molecules that May Be Immobilized on or within Polymeric Biomaterials

Proteins/peptides
  Enzymes
  Antibodies
  Antigens
  Cell adhesion molecules
  "Blocking" proteins

Saccharides
  Sugars
  Oligosaccharides
  Polysaccharides

Lipids
  Fatty acids
  Phospholipids
  Glycolipids

Other
  Conjugates or mixtures of the above

Drugs
  Antithrombogenic agents
  Anticancer agents
  Antibiotics
  Contraceptives
  Drug antagonists
  Peptide, protein drugs

Ligands
  Hormone receptors
  Cell surface receptors (peptides, saccharides)
  Avidin, biotin

Nucleic acids, nucleotides
  Single or double-stranded DNA, RNA (e.g., antisense oliogonucleotides)

TABLE 2   Application of Immobilized Biomolecules and Cells

Enzymes	Bioreactors (industrial, biomedical)
	Bioseparations
	Biosensors
	Diagnostic assays
	Biocompatible surfaces
Antibodies, peptides, and other affinity molecules	Biosensors
	Diagnostic assays
	Affinity separations
	Targeted drug delivery
	Cell culture
Drugs	Thrombo-resistant surfaces
	Drug delivery systems
Lipids	Thrombo-resistant surfaces
	Albuminated surfaces
Nucleic acid derivatives and nucleotides	DNA probes
	Gene therapy
Cells	Bioreactors (industrial)
	Bioartificial organs
	Biosensors

TABLE 3   Bioreactors Supports and Designs

"Artificial cell" suspensions
  (microcapsules, RBC ghosts, liposomes, reverse micelles [w/o] microspheres)

Biologic supports
  (membranes and tubes of collagen, fibrin ± glycosaminoglycans)

Synthetic supports
  (porous or asymmetric hollow fibres, particulates, parallel plate devices)

TABLE 4   Examples of Immobilized Enzymes in Therapeutic Bioreactors

Medical application	Substrate	Substrate action
Cancer treatment		
L-Asparaginase	Asparagine	Cancer cell nutrient
L-Glutaminase	Glutamine	Cancer cell nutrient
L-Arginase	Arginine	Cancer cell nutrient
L-Phenylalanine lyase	Phenylalanine	Toxin
Indole-3-alkane α hydroxylase	Tryptophan	Cancer cell nutrient
Cytosine deaminase	5-Fluorocytosine	Toxin
Liver failure (detoxification)		
Bilirubin oxidase	Bilirubin	Toxin
UDP-Gluceronyl transferase	Phenolics	Toxin
Other		
Heparinase	Heparin	Anticoagulant
Urease	Urea	Toxin

Immobilized Cell Ligands

Cell interactions with foreign materials are usually mediated by a biological intermediate, such as adsorbed proteins, as described in Chapter 3.2. An approach using biologically functional materials can be much more direct, by adsorbing or covalently grafting ligands for cell-surface adhesion receptors to the material surface. This has been accomplished with peptides grafted randomly over a substrate (Massia and Hubbell, 1991) as well as with peptides presented in a pre-clustered manner (Irvine et al., 2001). The latter has important advantages: Cells normally cluster their adhesion receptors into assemblies referred to as focal contacts, and preassembly confers benefits in terms of both adhesion strength (Ward and Hammer, 1993) and cell signaling (Maheshwari et al., 2000). In addition to peptides, saccharides have also been grafted to polymer surfaces to confer biological functionality (Griffith and Lopina, 1998; Chang and Hammer, 2000).

Specific biomolecules can be immobilized in order to control cellular interactions; one important example is the polypeptide growth factor. Such molecules can be immobilized and retain their ability to provide biological cues that signal specific cellular behavior, such as support of liver-specific function in hepatocytes (Kuhl and Griffith-Cima, 1996), induction of neurite extension in neurons (Sakiyama-Elbert et al., 2001), induction of angiogenesis (Zisch et al., 2001), or the differentiation of mesenchymal stem cells into bone-forming osteoblasts (Lutolf et al., 2003b). Other molecules may be immobilized that can partake in enzymatic reactions at the surface. McClung et al. (2001, 2003) have immobilized lysines, whose ε-amino groups may interact with pre-adsorbed tissue plasminogen activator (tPA) during coagulation, to enhance fibrin clot dissolution at that surface.

TABLE 5 Some Advantages and Disadvantages of
Immobilized Enzymes

Advantages
  Enhanced stability
  Can modify enzyme microenvironment
  Can separate and reuse enzyme
  Enzyme-free product
  Lower cost, higher purity product
  No immunogenic response (therapeutics)
Disadvantages
  Difficult to sterilize
  Fouling by other biomolecules
  Mass transfer resistances (substrate in and product out)
  Adverse biological responses of enzyme support surfaces (*in vivo*
    or *ex vivo*)
  Greater potential for product inhibition

Some of the advantages and disadvantages of immobilized biomolecules are listed in Table 5, using enzymes as an example.

IMMOBILIZATION METHODS

There are three major methods for immobilizing biomolecules (Table 6) (Stark, 1971; Zaborsky, 1973; Dunlap, 1974). It can be seen that two of them are physically based, while the third is based on covalent or "chemical" attachment to the support molecules. Thus, it is important to note that the term "immobilization" can refer either to a transient or to a long-term localization of the biomolecule on or within a support. In the case of a drug delivery system, the immobilized drug is supposed to be released from the support, while an immobilized enzyme or adhesion-promoting peptide in an artificial organ is designed to remain attached to or entrapped within the support over the duration of use. Either physical or chemical immobilization can lead to "permanent" or long-term retention on or within a solid support, the former being due to the large size of the biomolecule. If the polymer support is biodegradable, then

TABLE 6 Biomolecule Immobilization Methods

Physical adsorption
  van der Waals
  Electrostatic
  Affinity
  Adsorbed and cross-linked

Physical "entrapment"
  Barrier systems
  Hydrogels
  Dispersed (matrix) systems

Covalent attachment
  Soluble polymer conjugates
  Solid surfaces
  Hydrogels

the chemically immobilized biomolecule may be released as the matrix erodes or degrades away. The immobilized biomolecule may also be susceptible to enzymatic degradation *in vivo*, and this remains an interesting aspect that has received relatively little attention.

A large and diverse group of methods have been developed for covalent binding of biomolecules to soluble or solid polymeric supports (Weetall, 1975; Carr and Bowers, 1980; Dean *et al.*, 1985; Shoemaker *et al.*, 1987; Yang *et al.*, 1990; Park and Hoffman, 1990; Gombotz and Hoffman, 1986; Schense *et al.*, 1999; Lutolf *et al.*, 2003a). Many of these methods are schematically illustrated in Fig 1. The same biomolecule may be immobilized by many different methods; specific examples of the most common chemical reactions utilized are shown in Fig. 2.

For covalent binding to an inert solid polymer surface, the surface must first be chemically modified to provide reactive groups (e.g., $-OH$, $-NH_2$, $-COOH$, $-SH$, or $-CH=CH_2$) for the subsequent immobilization step. If the polymer support does not contain such groups, then it is necessary to modify it in order to permit covalent immobilization of biomolecules to the surface. A wide number of solid surface modification techniques have been used, including ionizing radiation graft copolymerization, plasma gas discharge, photochemical grafting, chemical modification (e.g., ozone grafting), and chemical derivatization (Hoffman *et al.*, 1972, 1986; Hoffman, 1987, 1988; Gombotz and Hoffman, 1986, 1987). (See also Chapter 2.14.)

A chemically immobilized biomolecule may also be attached via a spacer group, sometimes called an "arm" or a "tether" (Cuatrecasas and Anfinsen, 1971; Hoffman *et al.*, 1972; Hoffman, 1987). One of the most popular tethers is PEG that has been derivatized with different reactive end groups (Kim and Feijen, 1985), and some companies offer a variety of chemistries of heterobifunctional linkers having activated coupling end groups such as *N*-hydroxysuccinimide (NHS), maleimide, pyridyl disulfide, and vinyl sulfone. Such spacer groups can provide greater steric freedom and thus greater specific activity for the immobilized biomolecule, especially in the case of smaller biomolecules. The spacer arm may also be either hydrolytically or enzymatically degradable, and therefore will release the immobilized biomolecule as it degrades (Kopecek, 1977; Hern and Hubbell, 1998).

Inert surfaces, whether polymeric, metal, or ceramic, can also be functionalized through modification of an polymeric adlayer. Such physisorbed or chemisorbed polymers can be bound to the surface via electrostatic interactions (VandeVondele *et al.*, 2003), hydrophobic interactions (Neff *et al.*, 1999), or specific chemical interactions, such as that between gold and sulfur atoms (Harder *et al.*, 1998; Bearinger *et al.*, 2003). Metal or ceramic surfaces may also be derivatized with functional groups using silane chemistry, such as with functionalized triethoxysilanes (Massia and Hubbell, 1991; Puleo, 1997). Plasma gas discharge has been used to deposit polymeric amino groups for conjugation of hyaluronic acid to a metal surface (Verheye *et al.*, 2000).

As noted earlier, hydrophobic interactions have been used to functionalize surfaces, utilizing ligands attached to hydrophobic sequences (e.g., Ista *et al.*, 1999; Nath and Chilkoti, 2003).

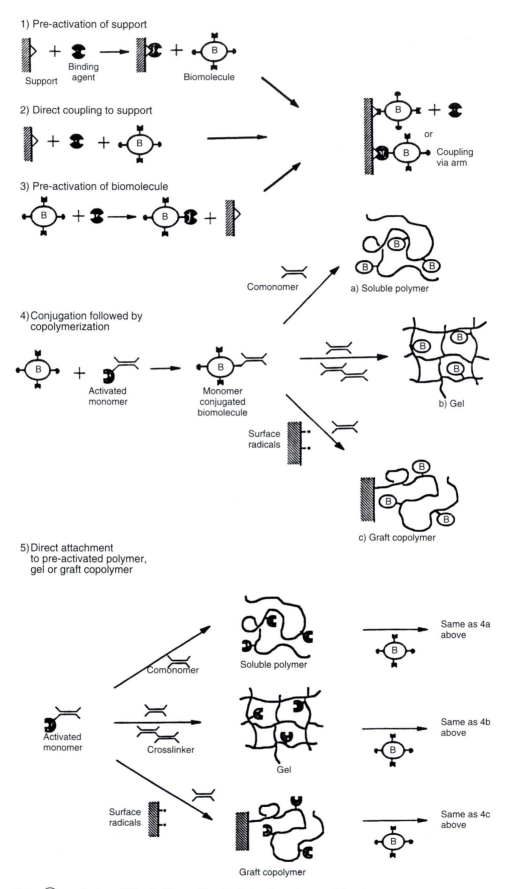

Note : Ⓑ may be immobilized with or without a "tether" arm in any of the above

FIG. 1.   Schematic cartoons showing various methods for covalent biomolecule immobilization.

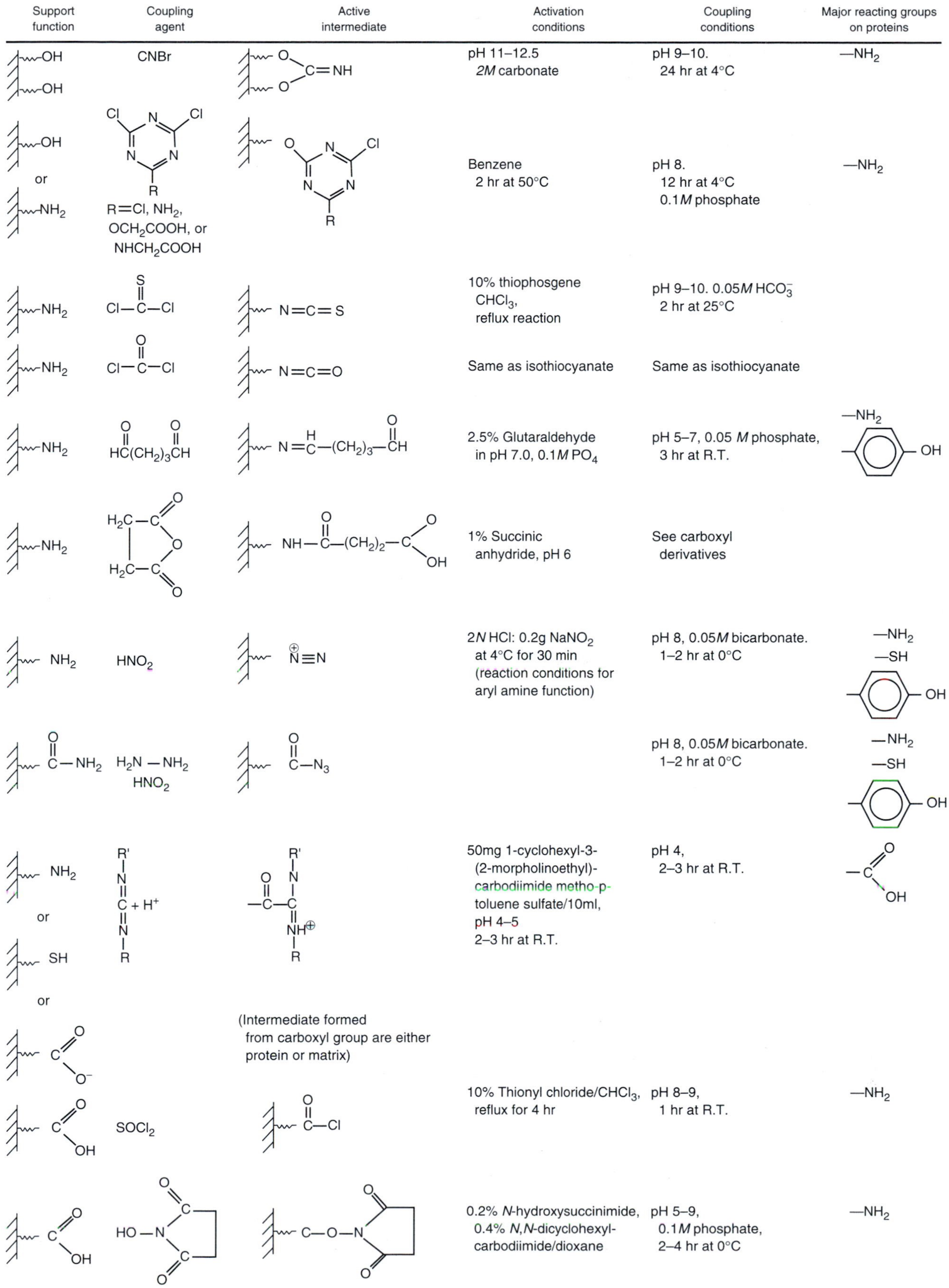

Support function	Coupling agent	Active intermediate	Activation conditions	Coupling conditions	Major reacting groups on proteins
~OH, ~OH	CNBr	~O-C=NH (cyclic)	pH 11–12.5 2M carbonate	pH 9–10. 24 hr at 4°C	—NH₂
~OH or ~NH₂	triazine, R=Cl, NH₂, OCH₂COOH, or NHCH₂COOH	triazine with O-support, Cl, R	Benzene 2 hr at 50°C	pH 8. 12 hr at 4°C 0.1M phosphate	—NH₂
~NH₂	Cl—C(=S)—Cl	~N=C=S	10% thiophosgene CHCl₃, reflux reaction	pH 9–10. 0.05M HCO₃⁻ 2 hr at 25°C	
~NH₂	Cl—C(=O)—Cl	~N=C=O	Same as isothiocyanate	Same as isothiocyanate	
~NH₂	HC(CH₂)₃CH (di-aldehyde)	~N=C—(CH₂)₃—CH(=O)	2.5% Glutaraldehyde in pH 7.0, 0.1M PO₄	pH 5–7, 0.05 M phosphate, 3 hr at R.T.	—NH₂ / —OH (aromatic)
~NH₂	succinic anhydride (H₂C–C(=O), H₂C–C(=O), O)	~NH—C(=O)—(CH₂)₂—C(=O)OH	1% Succinic anhydride, pH 6	See carboxyl derivatives	
~ NH₂	HNO₂	~ N⁺≡N	2N HCl: 0.2g NaNO₂ at 4°C for 30 min (reaction conditions for aryl amine function)	pH 8, 0.05M bicarbonate. 1–2 hr at 0°C	—NH₂ —SH / —OH (aromatic)
~C(=O)—NH₂	H₂N—NH₂ HNO₂	~C(=O)—N₃		pH 8, 0.05M bicarbonate. 1–2 hr at 0°C	—NH₂ —SH / —OH (aromatic)
~ NH₂ or ~ SH	R'—N=C=N—R + H⁺ (carbodiimide)	R'—N, —C(=O)—C=NH⁺—R	50mg 1-cyclohexyl-3-(2-morpholinoethyl)-carbodiimide metho-p-toluene sulfate/10ml, pH 4–5 2–3 hr at R.T.	pH 4, 2–3 hr at R.T.	—C(=O)OH
~C(=O)—O⁻		(Intermediate formed from carboxyl group are either protein or matrix)			
~C(=O)OH	SOCl₂	~C(=O)—Cl	10% Thionyl chloride/CHCl₃, reflux for 4 hr	pH 8–9, 1 hr at R.T.	—NH₂
~C(=O)OH	HO—N (succinimide)	~C(=O)—O—N (succinimide)	0.2% N-hydroxysuccinimide, 0.4% N,N-dicyclohexyl-carbodiimide/dioxane	pH 5–9, 0.1M phosphate, 2–4 hr at 0°C	—NH₂

FIG. 2. Examples of various chemical methods used to bond biomolecules directly to reactive supports (Carr and Bowers, 1980).

TABLE 7 Biomolecule Immobilization Methods

Method:	Physical and electrostatic adsorption	Cross-linking (after physical adsorption)	Entrapment	Covalent binding
Ease:	High	Moderate	Moderate to low	Low
Loading level possible:	Low (unless high S/V)	Low (unless high S/V)	High	(depends on S/V and site density)
Leakage (loss):	Relatively high (sens. to ΔpH salts)	Relatively low	Low to none[a]	Low to none
Cost:	Low	Low to moderate	Moderate	High

[a] Except for drug delivery systems.

Surfaces with hydrophobic gradients have also been prepared for this purpose (Detrait *et al.*, 1999). An interesting surface active product was developed several years ago that was designed to convert a hydrophobic surface to a cell adhesion surface by hydrophobic adsorption; it had an RGD cell adhesion peptide coupled at one end to a hydrophobic peptide sequence.

Sometimes more than one biomolecule may be immobilized to the same support. For example, a soluble polymer designed to "target" a drug molecule may have separately conjugated to it a targeting moiety such as an antibody, along with the drug molecule, which may be attached to the polymer backbone via a biodegradable spacer group (Ringsdorf, 1975; Kopecek, 1977; Goldberg, 1983). In another example, the wells in an immunodiagnostic microtiter plate usually will be coated first with an antibody and then with albumin or casein, each physically adsorbed to it, the latter acting to reduce nonspecific adsorption during the assay. In the case of affinity chromatography supports, the affinity ligand may be covalently coupled to the solid packing, and in some cases a "blocking" protein such as albumin or casein is then added to block nonspecific adsorption to the support.

It is evident that there are many different ways in which the same biomolecule can be immobilized to a polymeric support. Heparin and albumin are two common biomolecules that have been immobilized by a number of widely differing methods. These are illustrated schematically in Figs. 3 and 4.

Some of the major features of the different immobilization techniques are compared and contrasted in Table 7. The important molecular criteria for successful immobilization of a biomolecule are that a large fraction of the available biomolecules should be immobilized, and a large fraction of those immobilized biomolecules should retain an acceptable level of bioactivity over an economically and/or clinically appropriate time period.

CONCLUSIONS

It can be seen that there is a wide and diverse range of materials and methods available for immobilization of biomolecules and cells on or within biomaterial supports. Combined with the great variety of possible biomedical and biotechnological applications, this represents a very exciting and fertile field for applied research in biomaterials.

Bibliography

Alsberg, E., Anderson, K. W., Albeiruti, A., Rowley, J. A., and Mooney, D. J. (2002). Engineering growing tissues. *Proc. Natl. Acad. Sci.* **99:** 12025.

Bearinger, J. P., Terrettaz, S., Michel, R., Tirelli, N., Vogel, H., Textor, M., and Hubbell, J. A. (2003). Chemisorbed poly(propylene sulphide)-based copolymers resist biomolecular interactions. *Nat. Mater.* **2:** 259–264.

Bernard, A., Delamarche, E., Schmid, H., Michel, B., Bosshard, H. R., and Biebuyck, H. (1998). Printing patterns of proteins. *Langmuir*, **14:** 2225–2229.

Blawas, A. S., and Reichert, W. M. (1998). Protein patterning. *Biomaterials* **19:** 595–609.

Callegaro, L., and Denri, E. (1983). Applications of bioreactors in medicine. *Int. J. Artif. Organs* **6**(Suppl 1): 107.

Carr, P. W., and Bowers, L. D. (1980). *Immobilized Enzymes in Analytical and Clinical Chemistry: Fundamentals and Applications.* Wiley, New York.

Chang, K. C., and Hammer, D. A. (2000). Adhesive dynamics simulations of sialyl-Lewis(x)/E-selectin-mediated rolling in a cell-free system. *Biophys. J.* **79:** 1891–1902.

Chang, T. M. S. (1972). *Artificial Cells.* C.C. Thomas, Springfield, IL.

Csucs, G., Michel, R., Lussi, J. W., Textor, M., and Danuser, G. (2003). Microcontact printing of novel co-polymers in combination with proteins for cell-biological applications. *Biomaterials* **24:** 1713–1720.

Cuatrecasas, P., and Anfinsen, C. B. (1971). Affinity chromatography. *Ann. Rev. Biochem.* **40:** 259.

Dean, P. D. G., Johnson, W. S., and Middle, F. A., eds. (1985). *Affinity Chromatography.* IRL Press, Oxford.

De Myttenaere, M. H., Maher, J., and Schreiner, G. (1967). Hemoperfusion through a charcoal column for glutethimide poisoning. *Trans. ASAIO* **13:** 190.

Detrait, E., Lhoest, J. B., Bertrand, P., and de Aguilar, V. B. (1999). Fibronectin–pluronic coadsorption on a polystyrene surface with increasing hydrophobicity: relationship to cell adhesion. *J. Biomed. Mater. Res.* **45:** 404–413.

Dunlap, B. R. ed. (1974). *Immobilized Biochemicals and Affinity Chromatography.* Plenum, New York.

Ellman, J. A., and Gallop, M. A. (1998). Combinatorial chemistry. *Curr. Opin. Chem. Biol.* **2:** 17–319.

Folch, A., and Toner, M. (2000). Microengineering of cellular interactions. *Annu. Rev. Biomed. Eng.* **2:** 227–256.

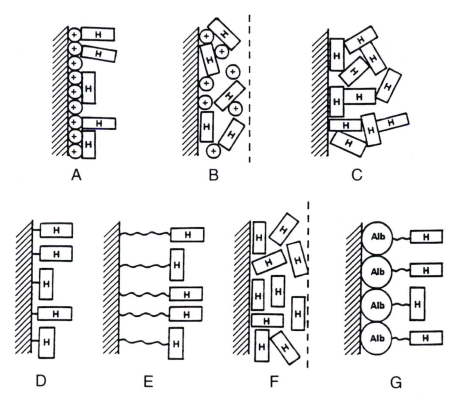

FIG. 3. Various methods for heparinization of surfaces: (A) heparin bound ionically on a positively charged surface; (B) heparin ionically complexed to a cationic polymer, physically coated on a surface; (C) heparin physically coated and self-cross-linked on a surface; (D) heparin covalently linked to a surface; (E) heparin covalently immobilized via spacer arms; (F) heparin dispersed into a hydrophobic polymer; (G) heparin–albumin conjugate immobilized on a surface (Kim and Feijen, 1985).

Goldberg, E., ed. (1983). *Targeted Drugs*. Wiley-Interscience, New York.

Gombotz, W. R., and Hoffman, A. S. (1986). Immobilization of biomolecules and cells on and within synthetic polymeric hydrogels. in *Hydrogels in Medicine and Pharmacy*, Vol. 1, N. A. Peppas, ed. CRC Press, Boca Raton, FL, pp. 95–126.

Gombotz, W. R., and Hoffman, A. S. (1987). Gas discharge techniques for modification of biomaterials. in *Critical Reviews in Biocompatibility*, Vol. 4, D. Williams, ed. CRC Press, Boca Raton, FL, pp. 1–42.

Griffith, L. G., and Lopina, S. (1998). Microdistribution of substratum-bound ligands affects cell function: Hepatocyte spreading on PEO-tethered galactose. *Biomaterials* **19**: 979–986.

Harder, P., Grunze, M., Dahint, R., Whitesides, G. M., and Laibinis, P. E. (1998). Molecular conformation and defect density in oligo (ethylene glycol)-terminated self-assembled monolayers on gold and silver surfaces determine their ability to resist protein adsoption. *J. Phys. Chem. B*. **102**: 426–436.

Hern, D. L., and Hubbell, J. A. (1998). Incorporation of adhesion peptides into nonadhesive hydrogels useful for tissue resurfacing. *J. Biomed. Mater. Res.* **39**: 266–276.

Hoffman, A. S. (1987). Modification of material surfaces to affect how they interact with blood. in *Blood in Contact with Natural and Artificial Surfaces*, E. Leonard, L. Vroman and V. Turitto, eds., *Ann. N.Y. Acad. Sci.* **516**: 96–101.

Hoffman, A. S. (1988). Applications of plasma gas discharge treatments for modification of biomaterial surfaces. *J. Appl. Polymer Sci. Symp.*, H. Yasuda and P. Kramer, eds. **42**: 251.

Hoffman, A. S., Schmer, G., Harris, C., and Kraft, W. G. (1972). Covalent binding of biomolecules to radiation-grafted hydrogels on inert polymer surfaces. *Trans. Am. Soc. Artif. Intenal. Organs* **18**: 10.

Hoffman, A. S., Gombotz, W. R., Uenoyama, S., Dong, L. C., and Schmer, G. (1986). Immobilization of enzymes and antibodies to radiation grafted polymers for therapeutic and diagnostic applications. *Radiat. Phys. Chem.* **27**: 265–273.

Houseman, B. T., and Mrksich, M. (2002). Towards quantitative assays with peptide chips: A surface engineering approach. *Trends Biotechnol.* **20**: 279–281.

Irvine, D. J., Mayes, A. M., and Griffith, L. G. (2001). Nanoscale clustering of RGD peptides at surfaces using comb polymers. 1. Synthesis and characterization of comb thin films. *Biomacromolecules* **2**: 85–94.

Ismagilov, R. F., Ng, J. M. K., Kenis, P. J. A., and Whitesides, G. M. (2001). Microfluidic arrays of fluid-fluid diffusional contacts as detection elements and combinatorial tools. *Anal. Chem.* **73**: 5207–5213.

Ista, L. K., Pérez-Luna, V. H., and López, G. P. (1999) Surface-grafted, environmentally sensitive polymers for biofilm release. *Appl. Environ. Microbiol.* **65**: 1603–1609.

Ito, Y. (1999). Surface micropatterning to regulate cell functions. *Biomaterials* **20**: 2333–2342.

Jackman, R. J., Duffy, D. C., Cherniavskaya, O., and Whitesides, G. M. (1999). Using elastomeric membranes as dry resists and for dry lift-off. *Langmuir* **15**: 2973–2984.

James, C. D., Davis, R. C., Kam, L., Craighead, H. G., Isaacson, M., Turner, J. N., and Shain, W. (1998). Patterned protein layers on

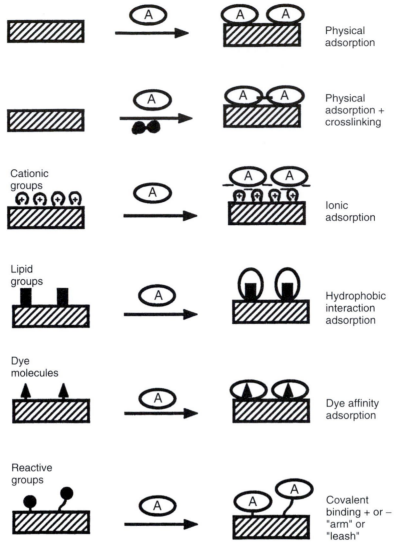

FIG. 4. Schematic of various ways that albumin may be immobilized on a surface. Albumin is often used as a "passivating" protein, to minimize adsorption of other proteins to a surface.

solid substrates by thin stamp microcontact printing. *Langmuir* **14:** 741–744.

Kane, R. S., Takayama, S., Ostuni, E., Ingber, D. E., and Whitesides, G. M. (1999). Patterning proteins and cells using soft lithography. *Biomaterials* **20:** 2363–2376.

Kim, S. W., and Feijen, J. (1985). Methods for immobilization of Heparin. in *Critical Reviews in Biocompatibility*. D. Williams, ed. CRC Press, Boca Raton, FL, pp. 229–260.

Kolff, W. J. (1979). Artificial organs in the seventies. *Trans. ASAIO* **16:** 534.

Kopecek, J. (1977). Soluble biomedical polymers. *Polymer Med.* **7:**191.

Kuhl, P. R., and Griffith-Cima, L. G. (1996). Tethered epidermal growth factor as a paradigm for growth factor-induced stimulation from the solid phase. *Nat. Med.* **2:** 1022–1027.

Laskin, A. I., ed. (1985). *Enzymes and Immobilized Cells in Biotechnology*. Benjamin/Cummings, Menlo Park, CA.

Lavin, A., Sung, C., Klibanov, A. M., and Langer, R. (1985). Enzymatic removal of bilirubin from blood: A potential treatment for neonatal jaundice. *Science* **230:** 543.

Lee, Y. S., and Mrksich, M. (2002). Protein chips: From concept to practice. *Trends Biotechnol.* **20:** S14–S18.

Lutolf, M. P., Raeber, G. P., Zisch, A. H., Tirelli, N., and Hubbell, J. A. (2003a). Cell-responsive synthetic hydrogels. *Adv. Mater.* **15:** 888–892.

Lutolf, M. R., Weber, F. E., Schmoekel, H. G., Schense, J. C., Kohler, T., Muller, R., and Hubbell, J. A. (2003b). Repair of bone defects using synthetic mimetics of collagenous extracellular matrices. *Nat. Biotechnol.* **21:** 513–518.

McClung, W. G., Clapper, D. L., Hu, S. P., and Brash, J. L. (2001). Lysine-derivatized polyurethane as a clot lysing surface: Conversion of plasminogen to plasmin and clot lysis in vitro. *Biomaterials* **22:** 1919–1924.

McClung, W. G., Clapper, D. L., Anderson, A. B., Babcock, D. E., and Brash, J. L. (2003). Interactions of fibrinolytic system proteins with lysine-containing surfaces. *J. Biomed. Mater. Res.* **66A:** 795–801.

McDevitt, T. C., Angelo, J. C., Whitney, M. L., Reinecke, H., Hauschka, S. D., Murry, C. E., and Stayton, P. S. (2002). In vitro generation of differentiated cardiac myofibers on micropatterned laminin surfaces. *J. Biomed. Mater. Res.* **60:** 472–479.

Maheshwari, G., Brown, G., Lauffenburger, D. A., Wells, A., and Griffith, L. G. (2000). Cell adhesion and motility depend on nanoscale RGD clustering. *J. Cell Sci.* **113:** 1677–1686.

Massia, S. P., and Hubbell, J. A. (1991). An RGD spacing of 440 nm is sufficient for integrin $\alpha_v\beta_3$-mediated fibroblast spreading and 140 nm for focal contact and stress fiber formation. *J. Cell Biol.* **114:** 1089–1100.

Nath, N., and Chilkoti, A. (2003). Fabrication of reversible functional arrays of proteins directly from cells using a stimuli responsive polypeptide. *Anal. Chem.* **75:** 709–715.

Neff, J. A., Tresco, P. A., and Caldwell, K. D. (1999). Surface modification for controlled studies of cell–ligand interactions. *Biomaterials* **20:** 2377–2393.

Nose, Y., Malchesky, P. S., and Smith, J. W., eds. (1983). *Plasmapheresis: New Trends in Therapeutic Applications.* ISAO Press, Cleveland, OH.

Nose, Y., Malchesky, P. S., and Smith, J. W., eds. (1984). *Therapeutic Apheresis: A Critical Look.* ISAO Press, Cleveland, OH.

Ostuni, E., Grzybowski, B. A., Mrksich, M., Roberts, C. S., and Whitesides, G. M. (2003). Adsorption of proteins to hydrophobic sites on mixed self-assembled monolayers. *Langmuir* **19:** 1861–1872.

Park, T. G., and Hoffman, A. S., eds. (1990). Immobilizaiton of *Arthrobacter simplex* in a thermally reversible hydrogel: effect of temperature cycling on steroid conversion. *Biotech. Bioeng.* **35:** 152–159.

Piskin, E., and Hoffman, A.S., eds. (1986). *Polymeric Biomaterials.* M. Nijhoff, Dordrecht, The Netherlands.

Puleo, D. A. (1997). Retention of enzymatic activity immobilized on silanized Co–Cr–Mo and Ti-6Al-4V. *J. Biomed. Mater. Res.* **37:** 222–228.

Ringsdorf, H. (1975). Structure and properties of pharmacologically active polymers. *J. Polymer Sci.* **51:** 135.

Sakiyama-Elbert, S. E., Panitch, A., and Hubbell, J. A. (2001). Development of growth factor fusion proteins for cell-triggered drug delivery. *FASEB J.* **15:** 1300–1302.

Schense, J. C., and Hubbell, J. A. (1999). Cross-linking exogenous bifunctional peptides into fibrin gels with factor XIIIa. *Bioconjugate Chem.* **10:** 75–81.

Schmer, G., Rastelli, L., Newman, M. O., Dennis, M. B., and Holcenberg, J. S. (1981) The bioartificial organ: review and progress report. *Internat. J. Artif. Organs* **4:** 96.

Scotchford, C. A., Ball, M., Winkelmann, M., Voros, J., Csucs, C., Brunette, D. M., Danuser, G., and Textor, M. (2003). Chemically patterned, metal-oxide-based surfaces produced by photolithographic techniques for studying protein- and cell-interactions. II: Protein adsorption and early cell interactions. *Biomaterials* **24:** 1147–1158.

Shoemaker, S., Hoffman, A. S., and Priest, J. H. (1987). Synthesis and properties of vinyl monomer/enzyme conjugates: Conjugation of L-asparaginase with N-succinimidyl acrylate. *Appl. Biochem. Biotechnol.* **15:** 11.

Sparks, R. E., Solemme, R. M., Meier, P. M., Litt, M. H., and Lindan, O. (1969). Removal of waste metabolites in uremia by microencapsulated reactants. *Trans. ASAIO* **15:** 353.

Stark, G. R., ed. (1971). *Biochemical Aspects of Reactions on Solid Supports.* Academic Press, New York.

Sung, C., Lavin, A., Klibanov, A., and Langer, R. (1986). An immobilized enzyme reactor for the detoxification of bilirubin. *Biotech. Bioeng.* **28:** 1531.

Tomlinson, E., and Davis, S. S. (1986). *Site-Specific Drug Delivery: Cell Biology, Medical and Pharmaceutical Aspects.* Wiley, New York.

VandeVondele, S., Voros, J., and Hubbell, J. A. (2003). RGD-grafted poly-L-lysine-*graft*-(polyethylene glycol) copolymers block non-specific protein adsorption while promoting cell adhesion. *Biotechnol. Bioeng.* **82:** 784–790.

Verheye, S., Markou, C. P., Salame, M. Y., Wan, B., King III, S. B., Robinson, K. A., Chronos, N. A. F., and Hanson, S. R. (2000). Reduced thrombus formation by hyaluronic acid coating of endovascular devices. *Arterioscler. Thromb. Vasc. Biol.* **20:** 1168–1172.

Ward, M. D., and Hammer, D. A. (1993). A theoretical-analysis for the effect of focal contact formation on cell–substrate attachment strength. *Biophys. J.* **64:** 936–959.

Weetall, H. H., ed. (1975). *Immobilized Enzymes, Antigens, Antibodies, and Peptides: Preparation and Characterization.* Dekker, New York.

Winkelmann, M., Gold, J., Hauert, R., Kasemo, B., Spencer, N. D., Brunette, D. M., and Textor, M. (2003). Chemically patterned, metal oxide based surfaces produced by photolithographic techniques for studying protein– and cell–surface interactions I: Microfabrication and surface characterization. *Biomaterials* **24:** 1133–1145.

Yang, H. J., Cole, C. A., Monji, N., and Hoffman, A. S. (1990). Preparation of a thermally phase-separating copolymer, poly(N-isopropylacrylamide-*co*-N-acryloxysuccinimide) with a controlled number of active esters per polymer chain. *J. Polymer Sci. A. Polymer Chem.* **28:** 219–220.

Zaborsky, O. (1973). *Immobilized Enzymes.* CRC Press, Cleveland, OH.

Zisch, A. H., Schenk, U., Schense, J. C., Sakiyama-Elbert, S. E., and Hubbell, J. A. (2001). Covalently conjugated VEGF-fibrin matrices for endothelialization. *J. Controlled Release* **72:** 101–113.

II

Biology, Biochemistry, and Medicine

3

Some Background Concepts

SUZANNE G. ESKIN, THOMAS A. HORBETT, LARRY V. MCINTIRE, RICHARD N. MITCHELL,
BUDDY D. RATNER, FREDERICK J. SCHOEN, AND ANDREW YEE

3.1 BACKGROUND CONCEPTS

Buddy D. Ratner

Much of the richness of biomaterials science lies in its inter-disciplinary nature. The two pillars of fundamental knowledge that support the structure that is biomaterials science are *materials science*, introduced in Part I, and the *biological-medical sciences*, introduced here. Complete introductory texts and a large body of specialized knowledge dealing with each of the chapters in this section, are available. However, these four chapters present sufficient background material so that a reader might reasonably follow the arguments presented later in this volume on biological interactions, biocompatibility, material performance and clinical performance.

In as short a time as can be measured after implantation in a living system (< 1 second), proteins are already observed on biomaterial surfaces. In seconds to minutes, a monolayer of protein adsorbs to most surfaces. The protein adsorption event occurs well before cells arrive at the surface. Therefore, cells see primarily a protein layer, rather than the actual surface of the biomaterial. Since cells respond specifically to proteins, this interfacial protein film may be the event that controls subsequent bioreaction to implants. Protein adsorption is also of concern for biosensors, immunoassays, array diagnostics, marine fouling and a host of other phenomena. Protein adsorption concepts are introduced in Chapter 3.2.

After proteins adsorb, cells arrive at an implant surface propelled by diffusive, convective or active (locomotion) mechanisms. The cells can adhere, release active compounds, recruit other cells, grow in size, replicate and die. These processes often occur in response to the proteins on the surface. Cell processes lead to responses (some desirable and some undesirable) that physicians and patients observe with implants. Cell processes at artificial surfaces are also integral to the unwanted buildup of marine organisms on ships, bacterial biofilms on implants and the useful growth of cells in bioreactors used to manufacture biochemicals. Cells at surfaces are discussed in Chapter 3.3.

After cells arrive and interact at implant surfaces, they may differentiate, multiply, communicate with other cell types and organize themselves in into tissues comprised of one or more cell types. Cells secrete extracellular matrix (ECM) molecules that fill the spaces between cells and serve as attachment structures for proteins and cells. The processes of angiogenesis (small blood vessel formation) and vasculogenesis (formation of larger blood vessels) are critical to provide this new tissue with nutrition and remove wastes. Finally, these tissues have distinctive reactions to irritation and injury. The development, organization and response to injury of tissues must be understood to appreciate the interplay between synthetic materials and tissues. Tissue structure and organization is reviewed in Chapter 3.4.

Finally, cells and tissues respond to mechanical forces. Two samples made of the same material, one a triangle shape and the other a disk, implanted in soft tissue, will show different healing reactions with considerably more fibrous reaction at the asperities of the triangle than along the circumference of the circle. Blood cells and the endothelial lining of the blood stream show distinctly different behaviors depending upon whether they are exposed to high or low shear forces. In recent years, much has been learned about the way external mechanical forces are transduced at cell surfaces into chemical signals that, in turn, direct cytoskeleton formation (or dissolution) and influence the nucleus of the cell to up- and down-regulate genes and messenger RNAs. Chapter 3.5 overviews mechanical effects on blood cells.

3.2 THE ROLE OF ADSORBED PROTEINS IN TISSUE RESPONSE TO BIOMATERIALS

Thomas A. Horbett

INTRODUCTION

The replacement of injured or diseased tissues with devices made from materials that are not of biologic origin is the central approach in current biomaterials science and clinical practice. The prevalence of this approach is due largely to the fact that these materials are not attacked by the immune system, unlike donor tissues or organs. This fundamental difference arises

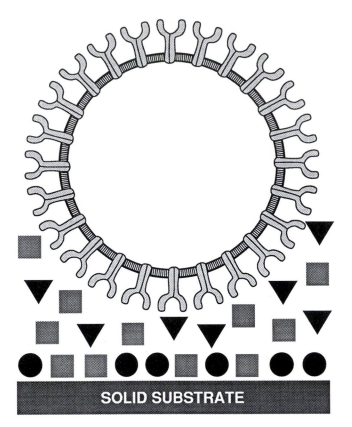

SOLID SUBSTRATE

FIG. 1. Cell interactions with foreign surfaces are mediated by integrin receptors with adsorbed adhesion proteins that sometimes change their biological activity when they adsorb. The cell is shown as a circular space with a bilayer membrane in which the adhesion receptor protein molecules (the slingshot-shaped objects) are partly embedded. The proteins in the extracellular fluid are represented by circles, squares, and triangles. The receptor proteins recognize and cause the cell to adhere only to the surface-bound form of one protein, the one represented by a solid circle. The bulk phase of this same adhesion protein is represented by a triangle, indicating that the solution and solid phase forms of this same protein have a different biological activity. The figure is schematic and not to scale. (From Horbett, 1996.)

protein adsorption are illustrated and discussed, including rapid kinetics, monolayer adsorption, and competitive adsorption. Molecular spreading events related to the conformational stability of the protein are presented at some length as background for a section on how the biological activity of adsorbed adhesion proteins is affected by the substrate. The final section summarizes the principles underlying the role of adsorbed proteins in mediating platelet response to biomaterials as an illustrative case study representative of many other types of cellular responses.

This chapter includes material that is discussed in greater detail in several previous review articles by the author (Horbett, 1993, 1994, 1999; Horbett and Brash, 1995). Those articles also give citations to all the work discussed here.

EXAMPLES OF THE EFFECTS OF ADHESION PROTEINS ON CELLULAR INTERACTIONS WITH MATERIALS

Protein adsorption to materials can be performed with a single protein, typically in a buffer solution, or from complex, multiprotein solutions such as blood plasma that can contain hundreds of proteins. Single proteins in buffer can be used to model fundamental aspects of protein adsorption or to study biological reactions to one protein. Adsorption from complex media approximates the reaction observed for implanted biomaterials. Examples of both types of adsorption are presented.

The Effects of Preadsorption with Purified Adhesion Proteins

Preadsorption of certain kinds of proteins onto a solid substrate such as tissue culture polystyrene greatly increases its adhesiveness to many kinds of cells, and such proteins are called adhesion proteins. The increased adhesiveness is because many cells have receptors on their cell membrane that bind specifically to the these specialized proteins. The adhesion receptors are called integrins. For example, fibronectin preadsorption greatly increases adhesion of fibroblasts, whereas albumin preadsorption prevents it. Experiments of this type have been done with a wide variety of cells and adhesion proteins, with basically similar results. Adhesion proteins also promote the flattening out or spreading of the cell onto the surface. A specific example is provided by measuring the percentage of attached cells that spread on surfaces pretreated with increasing concentrations of fibronectin. Spreading is only about 5% in the absence of fibronectin, but increases to nearly 100% as the fibronectin concentrations in the preadsorption solution are increased from 0.03 to 3 μg/ml.

Another example of the effect of fibronectin is shown in Fig. 2, which also contrasts it with the effects of the nonadhesive protein immunoglobulin G. As shown in the figure, the adhesion of the fibroblast-like 3T3 cells to a series of polymers and copolymers of 2-hydroxyethyl methacrylate (HEMA) and ethyl methacrylate (EMA) varies, being much less on the hydrophilic polyHEMA-rich surfaces than on the hydrophobic polyEMA-rich surfaces. The preadsorption of the surfaces

from the presence of immunologically recognizable biologic motifs on donor tissue and their absence on synthetic materials.

Nonetheless, there are other types of biological responses to implanted biomaterials that often impair their usefulness, including the clotting of blood and the foreign body reaction. Clearly, the body does recognize and respond to biomaterials. The basis for these reactions is the adsorption of adhesion proteins to the surface of the biomaterials that are recognized by the integrin receptors present on most cells. The adsorption of adhesion proteins to the biomaterial converts it into a biologically recognizable material, as illustrated in Fig. 1.

The interaction of adhesion receptors with adhesion proteins thus constitutes a major cellular recognition system for biomaterials. Therefore, the role of adsorbed adhesion proteins in mediating cellular interactions with biomaterials will be the primary focus of this chapter. Examples illustrating the ability of adsorbed adhesion proteins to influence cellular interactions with foreign materials are presented first. Then, some of the major physicochemical characteristics of

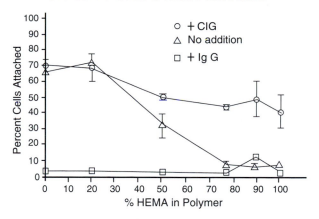

FIG. 2. 3T3 cell adhesion to HEMA-EMA copolymers: effect of preadsorption with fibronectin (designated CIG in the figure) or immunoglobulin G. Unpublished data from the author's laboratory.

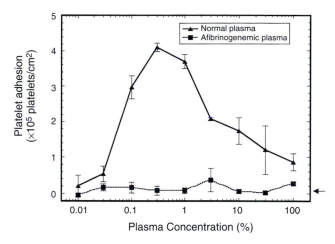

FIG. 3. Platelet adhesion to Immulon preadsorbed with normal and afibrinogenemic plasma. Platelet adhesion to Immulon I preadsorbed with normal (triangles) or afibrinogenemic plasma (squares). The solid line represents the platelet adhesion to Immulon I preadsorbed with a series of dilutions of normal plasma, whereas the dotted line represents the platelet adhesion to Immulon I preadsorbed with a series dilutions of afibrinogenemic plasma. The arrow at lower right corner indicates platelet adhesion to Immulon I preadsorbed with 2% BSA only. Source: Fig. 4 in Tsai and Horbett (1999).

with immunoglobulin G greatly reduces the adhesion of the cells to all the surfaces. In contrast, surfaces preadsorbed with fibronectin (designated CIG in the figure) are all fairly adhesive, much more so than the IgG preadsorbed surfaces or to the HEMA-rich surfaces not preadsorbed with fibronectin.

Preadsorption of adhesion proteins also affects cell interactions with surfaces studied under *in vivo* conditions. For example, platelet deposition onto polymeric arteriovenous shunts in dogs is greatly increased when fibrinogen or fibronectin are preadsorbed to the surfaces.

Depletion Studies

Although adsorption of purified adhesion proteins to a surface is one way to see their effect on cell adhesion, as presented in the preceding section, it does not mimic very well what occurs with implanted biomaterials. This is because implants are exposed to complex mixtures of proteins such as plasma or serum, so the adhesion protein must compete with many others for adsorption to the biomaterial. In that condition, despite its presence in the bulk phase, a given adhesion protein may really play little or no role. It is possible that very little of the adhesion protein may adsorb to the surface, as it is outcompeted by other proteins for the limited surface sites. Thus, a more biologically relevant way to understand the role of an adhesion protein in reactions to implants is to study the effect of their selective depletion from the complex mixture. The observations presented in this section and the articles they are based on are presented in greater detail in a review by the author (Horbett, 1999).

Selective depletion means that only one of the proteins is removed from the mixture at a time, by immunoadsorption chromatography, by use of plasma from mutant individuals lacking the adhesion protein of interest, or by selective enzymatic degradation. Thus, the more important role of adsorbed vitronectin as opposed to fibronectin in mediating attachment and spreading of cells on many surfaces has emerged from immunoadsorption studies. Several studies illustrate the important role that adsorbed fibrinogen plays in the adhesion of platelets, neutrophils, and macrophages.

The effects of removal of fibronectin or vitronectin or both proteins from serum on the adhesion of endothelial cells depends on surface chemistry. On tissue culture polystyrene (TCPS), fibronectin removal has little effect, whereas vitronectin removal greatly reduces adhesion. The results clearly show the primary role of adsorbed vitronectin in supporting endothelial cell adhesion to TCPS. In contrast, adhesion to a surface modified by ammonia in a glow discharge requires fibronectin, since removal of that protein greatly reduces adhesion to this surface, while vitronectin removal has little effect. However, the results for TCPS are more typical, i.e., on most surfaces studied by this method it appears that vitronectin, not fibronectin, is the primary agent responsible for cell adhesion.

Platelet adhesion to surfaces preadsorbed with plasma deficient in fibrinogen is much less than to the same surface preadsorbed with normal plasma, as illustrated in Fig. 3. Most of the adhesion supporting activity can be restored to fibrinogen deficient plasma by addition of normal levels of exogenous fibrinogen. In contrast, removal of fibronectin or vitronectin or von Willebrand's factor from plasma has little effect on platelet adhesion (not shown), even though these other plasma proteins act as adhesion proteins when adsorbed as single proteins to surfaces. It appears that too little of these other proteins adsorbs from plasma to make much difference, i.e., competition from fibrinogen and other proteins keeps their surface concentration too low and so their removal has no effect.

When mice are depleted of fibrinogen by treatment with an enzyme that degrades it, the adhesion of neutrophils and macrophages to a polymer implanted in their peritoneal cavity is greatly reduced. The fibrinogen depleted animals exhibited near normal neutrophil and macrophage adhesion to the implants if the implants are preadsorbed with fibrinogen. These studies illustrate the power of the depletion method very well. Previously, it had been thought that either complement or IgG would be the main adhesion proteins for neutrophils and monocytes because of the presence of receptors on these cells that bind these proteins. Instead, it appears that an integrin receptor for fibrinogen (CD11b/CD18, also known as Mac-1) plays a major role, at least during the initial or acute phase of the foreign body response in the mouse peritoneal cavity.

Inhibition of Receptor Activity with Antibodies

Another way to show the role of adhesion proteins in cell interactions with biomaterials is to add specific inhibitors of their function. Adhesion proteins cause cell adhesion by binding to integrin receptors that specifically recognize the adhesion protein. One way to inhibit this reaction is to add an antibody that binds to the receptor, blocking access to the adhesion protein. Examples of this approach are now presented.

Platelet-receptor-mediated interactions appear to be the primary mechanism of platelet interaction *in vivo* with certain vascular grafts because platelet deposition is largely inhibited by antibodies to the glycoprotein III/IIIa receptor. *In vitro* platelet adhesion to surfaces preadsorbed with blood plasma is also inhibited by anti-glycoprotein IIb/IIIa in a dose-dependent manner, as illustrated in Fig. 4. In this study, samples of the polyurethane Biomer were preadsorbed with plasma and then exposed to platelets in an albumin containing buffered saline suspension that had increasing amounts of the antibody. As shown in the figure, adhesion declined to very low values when high concentrations of the anti-integrin antibody were present.

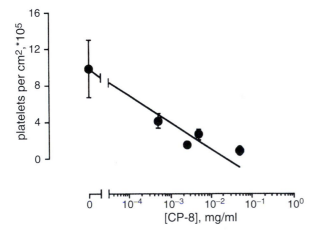

FIG. 4. Effect of anti-IIb/IIIa antibody on platelet adhesion to Biomer preadsorbed with plasma. Adhesion of platelets incubated in monoclonal antibody LJ-cp8 (monovalent Fab' fragment) directed against the glycoprotein (GP) IIb/IIIa complex to Biomer. Substrates were contacted with 1.0% plasma for 2 hr, then with washed, antibody-treated platelets for 2 hr. From Chinn, Horbett, and Ratner (1991).

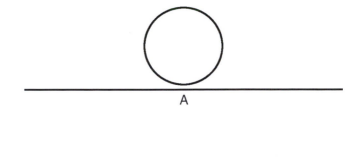

FIG. 5. The conversion of nonwettable polystyrene surface (top panel) into one completely wettable by water (bottom panel) is due to the adsorption of proteins.

THE ADSORPTION BEHAVIOR OF PROTEINS AT SOLID/LIQUID INTERFACES

Adsorption Transforms the Interface

Figure 5 illustrates an experiment that is performed by the author to demonstrate the adsorption of proteins to surfaces. As illustrated in part A, a water droplet placed on the surface of an unused polystyrene cell culture dish is easily visible because it beads up, i.e., the contact angle between the droplet and the polystyrene surface is very high because of the water-repellant, hydrophobic nature of polystyrene. If a cell were placed on a polystyrene dish instead of the water droplet, it also would encounter a very nonwettable surface. Part B of the figure illustrates the results of placing a water droplet on the surface of a polystyrene dish that had first been exposed to a protein solution for a short time and then rinsed extensively with water. As illustrated, no water droplet can be seen on this surface, reflecting the fact that in this case the added drop of water completely spread out over the surface of the preadsorbed dish. This happens because the water in part B was not able to interact with the polystyrene surface, because the surface had become coated with a layer of the hydrophilic protein adsorbate. Similarly, cells that come into contact with surfaces adsorbed with proteins do not directly "see" the substrate, but instead they interact with the intervening protein adsorbate.

Rapid Adsorption Kinetics and Irreversibility

The kinetics of adsorption of proteins to solid surfaces typically consists of a very rapid initial phase, followed by a slower phase upon approach to the steady-state value. Initially, proteins adsorb as quickly as they arrive at the largely empty surface. In this phase, adsorption is linear when plotted against time$^{1/2}$, characteristic of a diffusion-controlled process. In the later, slower phase, it is more difficult for the arriving proteins to find and fit into an empty spot on the surface.

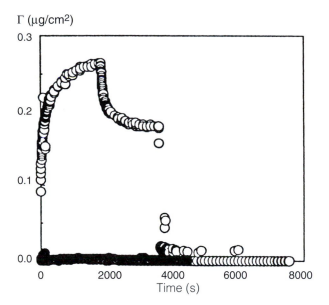

FIG. 6. The adsorption kinetics of lysozyme to a silica surface as studied with ellipsometry. The adsorbed amount versus time for adsorption of lysozyme to silica followed by buffer rinsing after 1800 sec, addition of surfactant (SDS) after 3600 sec, and a final rinse with buffer after 5400 sec (open circles). Adsorption from a mixture of the protein and surfactant for 1800 sec followed by rinsing is also included (closed circles). The experiments were carried out at 25°C in 0.01 M phosphate buffer, 0.15 M NaCl, pH 7. From Arnebrandt and Wahlgen (1995).

Figure 6 shows the time course of adsorption of lysozyme on silica measured with a high-speed, automated ellipsometer capable of very rapid measurements. At the earliest measurement time, less than a second into the study, the adsorption has reached almost half of the steady-state value. At 2000 sec, the protein solution was replaced with a buffer, resulting in some removal of loosely bound protein, but the adsorption stabilizes and would have remained at this value for much longer than shown, due to the tight, irreversible binding. At 3600 sec, a solution of the detergent sodium dodecyl sulfate (SDS) was infused, leading to complete removal of the protein. Thus, this experiment illustrates the rapid adsorption of proteins. It also illustrates that most of the adsorbed protein is irreversibly bound, as indicated by the fact that washing the surface with buffer does not remove the protein. The adsorbed protein is only removed when a strong surfactant (SDS in this example) is used. All these features are characteristic of protein adsorption to solid surfaces.

The Monolayer Model

The existence of a close packed monolayer of adsorbed protein is suggested by studies with single protein solutions in which a saturation effect can often be observed in the adsorption isotherm (Fig. 7). Adsorption to surfaces exposed to different concentrations of protein until steady state adsorption is achieved (2 hours or more) increases steeply at low bulk-phase concentrations but typically reaches a plateau or saturation value at higher bulk concentrations. Usually, the plateau value falls within the range expected for a close-packed monolayer of protein (about 0.1 to 0.5 µg/cm², depending on the diameter and orientation assumed for the protein).

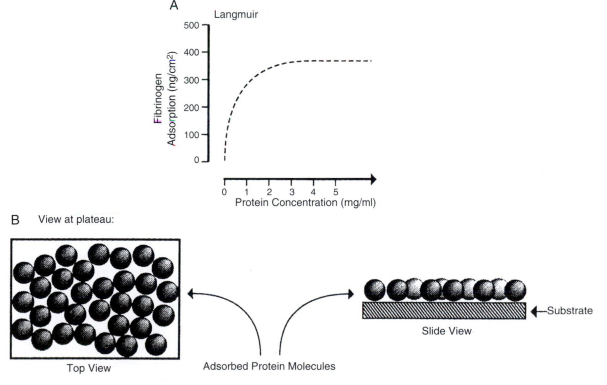

FIG. 7. Adsorption isotherms and the monolayer concept. From Horbett (1996).

The adsorption values from complex protein mixtures also typically are in the monolayer range. For example, the sum total of the amount of adsorption of the three major proteins in plasma (albumin, IgG, and fibrinogen) on the HEMA-EMA series of surfaces is also in the range of 0.1 to 0.5 $\mu g/cm^2$. In addition, the fact that competition exists between proteins for adsorption to a surface (see next section) also indicates the existence of limited sites. Thus, when a monolayer is the limit, there must be some selection for which proteins are present in the adsorbed film. It should be noted that well-defined plateaus are not always observed, but instead adsorption rises much more slowly at higher bulk-phase concentrations than at low concentrations, i.e., Freundlich isotherms do occur.

Competitive Adsorption of Proteins to Surfaces from Protein Mixtures

Adsorption from mixtures of proteins is selective, leading to enrichment of the surface phase in certain proteins. In this context, enrichment means that the fraction of the total mass of the adsorbed protein layer corresponding to a given protein is often much higher than the fraction of this protein in the bulk phase mixture from which it adsorbed. Since the solid can accommodate only a small fraction of the total protein present in the bulk phase, and proteins vary greatly in their affinity for surfaces, some adsorbed proteins are present in greater amounts than others. Studies of surfaces exposed to plasma have shown that many different proteins are present in the adsorbed film.

The competitive phenomena underlying differential enrichment from multi protein mixtures are most clearly illustrated in binary mixtures of proteins. Figure 8 has three separate curves in it, which overlap at the high and low ends. These curves represent the typical outcome of binary-mixture studies, but for three different conditions. For example, when a radiolabeled protein such as fibrinogen ("A" in the figure) is mixed with various amounts of an unlabeled protein such as albumin ("B" in the figure), the adsorption of fibrinogen ("A") always declines when sufficiently high amounts of albumin ("B") are present. However, the amount of competing protein needed to inhibit the adsorption of the labeled protein is different in each curve. This is meant to illustrate that, for a given pair of competing proteins, the competition curves will be different if

the surfaces they are competing for are different. In addition, if the surfaces are kept the same, but the competition of different pairs of proteins are studied, the curves will differ because the ability of proteins to compete for surface sites is quite different for different proteins.

For example, inhibition of fibrinogen adsorption to polyethylene requires a roughly 10-fold excess by weight of lower-affinity competing proteins such as albumin, but is effectively inhibited by the higher-affinity protein hemoglobin even when hemoglobin is present at only 10% of the mass of fibrinogen. However, the amounts needed for this inhibition will vary with the surfaces. Affinity variation is thus a major principle determining the outcome of competitive adsorption processes.

Examples of surface enrichment from complex protein mixtures are readily available. Although fibrinogen is only the third most concentrated protein in plasma, after IgG and albumin, biomaterials exposed to plasma are enriched in fibrinogen in the adsorbed phase. Hemoglobin is present in very low concentrations in plasma (0.01 mg/ml or less), but it is still adsorbed in amounts similar to the more predominant proteins because of its high surface affinity. Albumin, a lower-affinity protein, presents a counterexample. Albumin concentration in plasma is much higher than fibrinogen, yet the surface concentration of albumin adsorbed from plasma is typically about the same as fibrinogen. The high concentration of albumin in the plasma drives it onto the surface according to the law of mass action. Similarly, fibrinogen adsorption is higher from plasmas that contain higher concentrations of fibrinogen. Thus, mass concentration in the bulk phase is the second major factor determining competitive adsorption behavior.

The adsorption of fibrinogen from plasma exhibits some unusual behavior. On some surfaces, fibrinogen adsorption is maximal at intermediate dilutions of plasma (see example in Fig. 9A). In addition, fibrinogen adsorption from full-strength or moderately diluted plasma is higher at very early adsorption times than at later times (example shown in Fig. 9B). These are examples of the Vroman effect for fibrinogen. This phenomenon is a clear example of the unique effects of competitive adsorption on both the steady-state and the transient composition of the adsorbed layer that forms from plasma. The Vroman effect appears to involve displacement of initially adsorbed fibrinogen by later-arriving, more surface-active plasma proteins, especially high-molecular-weight kininogen, and transitions in the adsorbed fibrinogen that make it less displaceable with adsorption time (reviewed in Slack and Horbett, 1995).

MOLECULAR SPREADING EVENTS: CONFORMATIONAL AND BIOLOGICAL CHANGES IN ADSORBED PROTEINS

Proteins that adsorb to solid surfaces can undergo conformational changes because of the relatively low structural stability of proteins and the tendency to unfold to allow further bond formation with the surface. Conformational changes can be detected with many types of physicochemical methods and also by measuring changes in the biological activity of the adsorbed proteins.

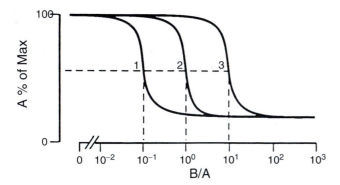

FIG. 8. Competitive adsorption of two proteins from a mixture. From Horbett (1993).

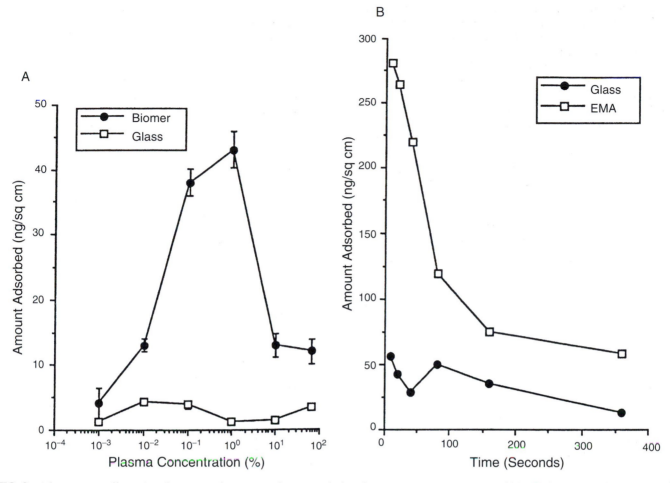

FIG. 9.   The Vroman effect. (A) Fibrinogen adsorption to biomer and glass from various concentrations of blood plasma. (B) Time course of fibrinogen adsorption to glass and poly(ethyl methacrylate) (PEMA) from undiluted blood plasma. Source: Fig. 1 in Slack and Horbett, (1995).

Physicochemical Studies of Conformational Changes

"Soft" proteins are found to adsorb more readily and more tenaciously than "hard" proteins. In this context, a "soft" protein is one with a low thermodynamic stability, whereas a "hard" protein is more stable to unfolding in solution in response to denaturing conditions such as elevated temperature. This concept and the articles supporting the following discussion are presented in detail elsewhere (Horbett and Brash, 1995).

Comparison of the adsorptive behavior of different proteins to their molecular properties indicates that less stable proteins are more adsorptive. The important role of structural stability in adsorption is also supported by recent studies with engineered mutant proteins with single amino acid substitutions that vary in stability. Lysozyme adsorption at the solid/liquid interface and tryptophan synthase occupation of the air/water interface are greater for less stable mutants.

Several studies with differential scanning calorimetry (DSC) methods seemed to indicate that adsorbed proteins may lose much of their structure, depending on how "soft" they are. Heat is taken up at a certain temperature for proteins in

solution due to unfolding of the native protein at the transition temperature. An absence or reduction of this effect for an adsorbed protein suggests that the adsorbed protein has already undergone the transition, i.e., that it has already unfolded upon adsorption. The transition enthalpy of lysozyme adsorbed to negatively charged polystyrene was much less than for the protein in solution (0–170 kJ/mol for the adsorbed protein versus about 600 kJ/mol for the native protein, depending on the pH). However, for lysozyme adsorbed on hematite, the unfolding enthalpy is only about 20% less than for the native protein, indicating that changes in the enthalpy of unfolding depend on the adsorbing surface. Furthermore, for lactalbumin the heat released is nearly zero when adsorbed on either the polystyrene or the hematite surface, suggesting complete unfolding of lactalbumin on both surfaces. These observations are consistent with the lower stability of lactalbumin in comparison to lysozyme. Several proteins adsorbed to pyrolytic carbon do not show any release of heat at the expected transition temperature, suggesting that pyrolytic carbon induces complete unfolding, a result that is consistent with the tenacious binding of proteins to this surface. It has also been shown that albumin and lysozyme adsorbed to polystyrene exhibit no unfolding enthalpy, whereas lysozyme adsorbed to a hydrophilic contact lens still exhibits about 50% of the heat released by the native protein. Streptavidin adsorbed to polystyrene displays an unfolding enthalpy that is very similar to that for the native protein in solution, probably because of the greater stability of streptavidin in comparison to lysozyme or albumin.

However, more recent studies of adsorbed proteins by the DSC method in conjunction with other, more direct conformational measurements such as circular dichroism (CD) show that at least some adsorbed proteins that appear to be completely denatured as judged by DSC still have considerable amounts of their native structure as measured by CD. It thus appears that some proteins become somewhat more stable after adsorption, and thus do not show heat release at the normal melting temperature.

The concept of molecular spreading of the adsorbed protein suggested by these observations has been used to explain differences in the amount of IgG adsorbed during stepwise adsorption. When the final concentration of bulk protein is achieved in a series of smaller concentration steps, as opposed to bringing the bulk concentration to its final value in one step, adsorption is smaller. Conformational changes upon adsorption of fibronectin to polystyrene beads and Cytodex microcarrier beads have been detected using electron spin resonance spectroscopy. Many other physicochemical studies are consistent with partial unfolding of the adsorbed proteins (Andrade, 1985; Horbett and Brash, 1995; Lundstrom, 1985).

Biological Changes in Adsorbed Proteins

Although physicochemical studies sometimes suggest complete denaturation of adsorbed proteins, most probes for biological activity suggest the changes are more limited (reviewed with citations in Horbett, 1993). Thus, enzymes retain at least some of their activity in the adsorbed state, especially when the surface are more fully loaded with enzyme. Measurements of enzyme activity or retention times during passage over hydrophobic chromatography matrices have shown that the degree of denaturation is highly dependent on the protein, the length of time the protein has spent on the matrix, the solvent, and other conditions, and is not necessarily complete.

Changes in the binding of a monoclonal antibody to fragment D of fibrinogen upon fibrinogen adsorption to polystyrene have been attributed to changes in the conformation of fibrinogen after adsorption. Thus, solution-phase fibrinogen does not bind the antibody raised to fragment D, but the surface-adsorbed fibrinogen does, and furthermore, bulk fibrinogen does not compete for the binding of the antibody to the surface-adsorbed fibrinogen. The RIBS (receptor-induced binding site) antibodies are similar: They bind to fibrinogen only after the fibrinogen has bound to either a solid surface or to the platelet IIb/IIIa receptor. The binding of the RIBS antibodies and others that bind to the platelet-binding regions of fibrinogen varies with the length of time after adsorption of the fibrinogen. Platelet adhesion to polymethacrylates has been correlated with the amount of antifibrinogen binding,

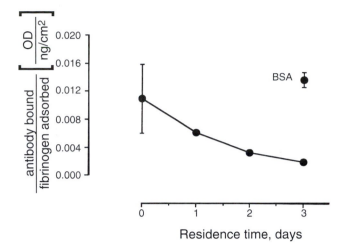

FIG. 10. Transitions in adsorbed fibrinogen. The effect of three-day residence in buffer or buffered albumin solution upon anti-fibrinogen binding to fibrinogen adsorbed from dilute plasma to biomer is shown. From Fig. 3A in Chinn *et al.* (1992).

suggesting that the adsorbed fibrinogen is in different conformations on the various polymethacrylates.

Fibrinogen undergoes a time-dependent transition after its adsorption to a surface that results in reduced platelet and antibody binding to the adsorbed fibrinogen as well as reduction in the sodium dodecyl sulfate and plasma displaceability, and changes in amide II frequency of the adsorbed fibrinogen. The losses in platelet binding, antibody binding, and SDS elutability are prevented if albumin is included in the storage buffer. An example showing time-dependent losses in antibody binding to fibrinogen and its prevention by albumin is shown in Fig. 10 (from Chinn *et al.*, 1992). Vitronectin also appears to undergo conformational changes upon adsorption that affect its ability to bind heparin and its infrared spectra.

Modulation of the biologic activity of fibronectin has been shown in several studies in which the ability of fibronectin adsorbed to various surfaces to support cell attachment or spreading was found to differ. For example, fibronectin adsorbed to tissue-culture-grade polystyrene was able to support cell attachment and spreading, whereas fibronectin adsorbed to ordinary polystyrene does not support spreading very well unless some albumin is added to the fibronectin solution. Fibronectin adsorbed to self-assembled monolayer films (SAMs) with various functional end groups also varies. On hydrophobic SAMs there is poor cell spreading unless albumin is coadsorbed (albumin "rescuing"). The albumin "rescuing" phenomenon observed for SAMs is similar to the albumin effect on fibronectin's ability to promote cell spreading on polystyrene, and to the effect of albumin addition in preventing losses in platelet adhesion to fibrinogen adsorbed surfaces discussed above. The ability of fibronectin adsorbed to a series of polymers to support the outgrowth of corneal epithelial cells has also been found to vary a great deal, despite the presence of similar amounts of fibronectin on the surfaces.

The effect of albumin addition in enhancing the adhesivity of fibronectin-coated surfaces is opposite of what one

TABLE 1   Principles Underlying the Influence of Adsorbed Plasma Proteins on Platelet Interactions with Biomaterials

1. Synthetic foreign materials acquire bioreactivity only after first interacting with dissolved proteins. The principal means by which the transformation from an inert, nonthrombogenic polymer to a biologically active surface takes place is the interaction of the proteins with the surface that then mediates cell adhesion.
2. Platelets are a major example of why and how adsorbed proteins are influential in cell–biomaterials interactions.
3. Sensitivity of platelets to adsorbed proteins is due to:
   a. Some proteins in plasma are strongly adhesive for platelets: fibrinogen, fibronectin, vitronectin, and von Willebrand factor.
   b. Concentrating, localizing, immobilizing effects of adsorbing the proteins at the interface accentuates the receptor-adhesion protein interaction.
   c. Platelets have receptors (IIb/IIIa and Ib/IX) that bind specifically to a few of the plasma proteins, mediating adhesion.
4. Principles of protein adsorption to biomaterials
   a. Monolayer adsorption and consequent competition for available adsorption sites means that not all proteins in the plasma phase can be equally represented on the surface.
   b. Driving forces for adsorption are the intrinsic surface activity and bulk phase concentration of the proteins.
   c. Surfaces vary in selectivity of adsorption.
   d. Biological activity of the adsorbed protein varies on different surfaces.

might expect, because the added albumin should reduce the amount of adsorbed fibronectin, as albumin competes for sites on the surfaces. The explanation for the albumin effect is thought to be that by adsorbing along with the fibronectin to the surface, the albumin molecules occupy surface sites near the fibronectin molecules. The adsorbed albumin molecules thus keep the adsorbed fibronectin molecules from undergoing structural changes that they would otherwise do in trying to spread into formerly empty surface sites but cannot do so if albumin molecules fill those sites.

The studies with platelets, fibroblasts, and epithelial cells show that substrate properties somehow modulate the ability of adsorbed proteins to interact with cells. These differences may arise at least in part from differences in the availability of epitopes on adhesive proteins for the cell surface receptor. That is, both the amount of the adsorbed adhesive protein and its "bioreactivity" are actively influenced by the properties of the surface to which it is adsorbed.

THE IMPORTANCE OF ADSORBED PROTEINS IN BIOMATERIALS

Table 1 summarizes the principles underlying the influence of adsorbed proteins in biomaterials used in contact with the blood. All of the principles listed also apply in other environments such as the extravascular spaces, albeit with other proteins and other cell types (e.g., macrophages in the peritoneum adhere via other receptors and other adhesion proteins). The platelets therefore provide a "case study," and we close this chapter by considering this case.

The sensitivity of platelet/surface interactions to adsorbed proteins is fundamentally due to the presence of adhesion receptors in the platelet membrane that bind to certain plasma proteins. There are only a few types of proteins in plasma that are bound by the adhesion receptors. The selective adsorption of these proteins to synthetic surfaces, in competition with

the many nonadhesive proteins that also tend to adsorb, is thought to mediate platelet adhesion to these surfaces. However, since the dissolved, plasma-phase adhesion proteins do not bind to adhesion receptors unless the platelets are appropriately stimulated, whereas unstimulated platelets can adhere to adsorbed adhesion proteins, it appears that adsorption of proteins to surfaces accentuates and modulates the adhesion receptor/adhesion protein interaction. The type of surface to which the adhesion protein is adsorbed affects the ability of the protein to support platelet adhesion (Horbett, 1993). The principles that determine protein adsorption to biomaterials include monolayer adsorption, the intrinsic surface activity and bulk concentration of the protein, and the effect of different surfaces on the selectivity of adsorption and biologic activity of the adsorbed protein.

More generally, all proteins are known to have an inherent tendency to deposit very rapidly on surfaces as a tightly bound adsorbate that strongly influences subsequent interactions of many different types of cells with the surfaces. It is therefore thought that the particular properties of surfaces, as well as the specific properties of individual proteins, together determine the organization of the adsorbed protein layer, and that the nature of this layer in turn determines the cellular response to the adsorbed surfaces.

Bibliography

Andrade, J. D. (1985). Principles of protein adsorption. in *Surface and Interfacial Aspects of Biomedical Polymers*, J. D. Andrade, ed. Plenum Press, New York, pp. 1–80.

Arnebrandt, T., and Wahlgen, M. (1995). Protein–surfactant interactions at solid surfaces in *Proteins at Interfaces II: Fundamentals and Applications*, ACS Symposium Series 602, T. A. Horbett and J. Brash, eds. American Chemical Society, Washington, D.C., pp. 239–254.

Chinn, J. A., Horbett, T. A., and Ratner, B. D. (1991). Baboon fibrinogen adsorption and platelet adhesion to polymeric materials. *Thromb. Haemostas.* **65:** 608–617.

Chinn, J. A., Posso, S. E., Horbett, T. A. and Ratner, B. D. (1992). Post-adsorptive transitions in fibrinogen adsorbed to polyurethanes: changes in antibody binding and sodium dodecylsulfate elutability. *J. Biomed. Mater. Res.* **26**: 757–778.

Horbett, T. A. (1993). Principles underlying the role of adsorbed plasma proteins in blood interactions with foreign materials. *Cardiovasc. Pathol.* **2**: 137S–148S.

Horbett, T. A. (1994). The role of adsorbed proteins in animal cell adhesion. *Colloid Surf. B: Biointerfaces* **2**: 225–240.

Horbett, T. A. (1996). Proteins: structure, properties, and adsorption to surfaces. in *Biomaterials Science,* B. D. Ratner, A. S. Hoffman, F. Schoen, and J. E. Lemons, eds. Academic Press, San Diego, pp. 133–141.

Horbett, T. A. (1999). The role of adsorbed adhesion proteins in cellular recognition of biomaterials. *BMES Bull.* **23**: 5–9.

Horbett, T. A., and Brash, J. L. (1995). Proteins at interfaces: an overview. in *Proteins at Interfaces II: Fundamentals and Applications,* ACS Symposium Series 602, T. A. Horbett and J. Brash, eds. American Chemical Society, Washington, D.C., pp. 1–25.

Lundstrom, I. (1985). Models of protein adsorption on solid surfaces. *Prog. Colloid Polymer Sci.* **70**: 76–82.

Slack, S. M., and Horbett, T.A. (1995). The Vroman effect: a critical review. in *Proteins at Interfaces II: Fundamentals and Applications,* ACS Symposium Series 602, T. A. Horbett and J. Brash, eds. American Chemical Society, Washington, D.C., pp. 112–128.

Tsai, W.-B., and Horbett, T.A. (1999). The role of fibronectin in platelet adhesion to plasma preadsorbed polystyrene. *J. Biomater. Sci. Polymer Ed.* **10**: 163–181.

3.3 CELLS AND CELL INJURY

Richard N. Mitchell and Frederick J. Schoen

Composed of nucleic acids, proteins, and other large and small molecules, cells constitute the basic structural building blocks of all living matter. They are held together by cell-to-cell junctions to form tissues comprising four general types: epithelium, connective tissue, muscle, and nerve. Organs are assembled from these basic tissues, "glued" together by a largely proteinaceous extracellular matrix (ECM) synthesized by the individual cells. The organs, in turn, perform the various functions required by the intact living organism, including circulation, respiration, digestion, excretion, movement, and reproduction.

A major goal in this and the following chapter is to describe how biological structure is adapted to perform physiologic function. This general and overarching concept extends from cells (and their subcellular constituents) to the organization of tissues and of organs. Beginning with the smallest subunits of cellular organization we will build to progressively more complex systems. In the following chapter, we will extend the concepts of structure–function correlation beyond cells to include extracellular matrix and complex tissues, and will also describe the technologies by which histologists and pathologists examine normal and abnormal tissues. In these chapters, we also provide an introduction to the physiologic responses to environmental stimuli, the mechanisms of cell injury, cell–materials interactions, and the methodologies by which these are all studied.

In this chapter on cells and cell injury and in the following chapter on tissues and the extracellular matrix, we will highlight the following fundamental concepts:

Cells and cell injury:

- General characteristics and functions of cells
- Compartmentalization of regionally specialized function by membranes
- Cellular specialization to facilitate unique functions
- Regulation and coordination of cell function
- The response of cells to injury, including mechanisms of cell death

Tissues and the extracellular matrix:

- The structure and function of the extracellular matrix
- Grouping of cells into tissues
- Integration of tissues into organs
- Remodeling of the extracellular matrix
- The interaction of cells, tissues, and foreign materials
- Basic methods used to study cells and tissues

NORMAL CELL HOUSEKEEPING

In very broad strokes, we will first outline the general organization of a prototypical cell, using it to identify the functional considerations required for maintaining a living cell. We will then revisit each of these structural features to illustrate important concepts in cellular maintenance and response to environment.

Conceptually, cells may be viewed as independent collections of self-replicating enzymes and structural proteins that carry out certain general functions. The most essential cell attributes are:

- Self-replication
- Protection from the environment
- Acquisition of nutrients
- Movement
- Communication
- Catabolism of extrinsic molecules
- Degradation and renewal of senescent intrinsic molecules
- Energy generation

Intracellular constituents exist in a microcosm of water, ions, sugars, and small-molecular-weight molecules called the cytosol or cytoplasm. Within the cytosol is also a source of energy, typically adenosine triphosphate (ATP). Although long conceptualized as a randomly diffusing bag of soluble molecules, the cell is, in fact, a structurally highly ordered and functionally integrated assembly of organelles, cytoskeletal elements, and enzymes.

The cytosol is delimited and protected from the environment by a phospholipid bilayer, the plasma membrane, which permits the cell to maintain cytosolic constituents at concentrations different from those in the surrounding environment. Because of its hydrophobic inner core, the plasma membrane is impermeant to charged and/or large polar molecules; however, it is rendered selectively permeable (i.e., permits

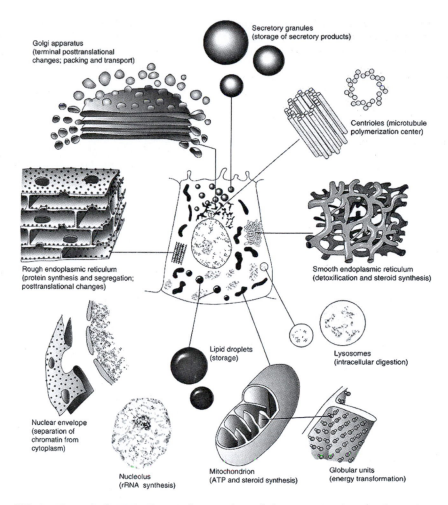

Golgi apparatus
(terminal posttranslational
changes; packing and transport)

Secretory granules
(storage of secretory products)

Centrioles (microtubule
polymerization center)

Rough endoplasmic reticulum
(protein synthesis and segregation;
posttranslational changes)

Smooth endoplasmic reticulum
(detoxification and steroid synthesis)

Nuclear envelope
(separation of
chromatin from
cytoplasm)

Lipid droplets
(storage)

Lysosomes
(intracellular digestion)

Nucleolus
(rRNA synthesis)

Mitochondrion
(ATP and steroid synthesis)

Globular units
(energy transformation)

FIG. 1.   General schematic of a typical mammalian cell, demonstrating the general organization and major organelles. Note that each compartment has distinct functions made possible
by selectively permeable membranes. (Reproduced by permission from Bergman, A. R., Afifi,
A. K., and Heidger, P. M., eds., 1996. *Histology.* Saunders, Philadelphia.)

specific passage) to incoming or outgoing material (ions, amino acids, etc.) by channel or transport proteins inserted through it. Most nutrient acquisition is thereby accomplished by the movement of desired substrates either through pores or by energy-driven transport. Cells also have the capacity to internalize material from the outside environment by capturing bits of the extracellular environment in invaginated folds of the plasma membrane called vesicles. Depending on the volume and size of the ingested material, the process may be called phagocytosis ("cell eating") or pinocytosis ("cell drinking"). Transcytosis is the movement of vesicles from one side of a cell to another, and it may play an important role in mediating the increased vascular permeability ("leaky vessels") that occurs around tumors or at sites of inflammation. The plasma membrane may also express a variety of specific surface molecules that facilitate interactions with other cells, soluble ligands (e.g., insulin), and/or with the extracellular matrix (communication).

Many of a cell's normal "housekeeping" functions are compartmentalized within membrane-bounded intracellular organelles (Fig. 1) thus permitting adjacent regions of the cell

to have vastly different chemistries. By isolating certain cellular functions within distinct compartments, potentially injurious degradative enzymes or toxic metabolites can be kept at usefully high concentrations locally without causing damage to more delicate intracellular constituents. Moreover, compartmentalization also allows the creation of unique intracellular environments (e.g., low pH, high calcium, or high concentration of a potent enzyme) that permit more efficient functioning of certain chemical processes, enzymes, or metabolic pathways.

The enzymes and structural proteins of the cell are constantly being renewed by ongoing synthesis tightly balanced with intracellular degradation. Oversight for the new synthesis of macromolecules, including deoxyribonucleic acid (DNA) and ribonucleic acid (RNA), is provided by the nucleus. New proteins destined for the plasma membrane or for secretion into the extracellular environment are synthesized and packaged in the rough endoplasmic reticulum (RER) and Golgi apparatus; proteins intended for remaining in the cytosol are synthesized on free ribosomes. Smooth endoplasmic reticulum (SER) may be abundant in certain cell types where it is used for steroid hormone and lipoprotein synthesis, as well as for the modification of hydrophobic compounds into water-soluble molecules for export. Degradation of internalized molecules or senescent self-molecules into their constituent amino acids, sugars, and lipids (catabolism) is the primary responsibility of the lysosomes and proteasomes. Peroxisomes play a specialized role in the breakdown of fatty acids, generating hydrogen peroxide in the process. Intracellular vesicles busily shuttle internalized material to appropriate intracellular site(s) for catabolism, or direct newly synthesized materials to the plasma membrane or relevant target organelle. The architecture of the cell is maintained by a scaffolding of intracellular proteins collectively called the cytoskeleton, analogous in some ways to the support provided by bones of our bodies.

Cell movement, including both movement of organelles and proteins within the cell, as well as movement of the cell in its environment, is accomplished through rearrangement of the cytoskeleton. These structural proteins also provide basic cellular shape and intracellular organization, which are necessary for the maintenance of cell polarity (differences in cell structure and function at the top of a cell versus its side or base). For example, in many cell types, and particularly in epithelial tissues, it is critical for cells to distinguish—and keep separated—the top (apical) versus the bottom and side (basolateral) surfaces. The major energy source for macromolecular synthesis, metabolite degradation, and intracellular transport is the mitochondrion, using oxidative phopsphorylation to generate ATP. Finally, all of these organelles must be replicated (organellar biogenesis) and correctly apportioned in daughter cells following mitosis.

The specific function(s) of a given cell are reflected by the relative amount and types of organelles it contains. The relative predominance of specific types of organelles can be inferred by examination of tissue sections prepared by standard histologic techniques and can be confirmed by transmission electron microscopy. For example, cells with high energy requirements can be expected to have a significantly greater capacity to generate that energy. Thus, kidney tubular epithelial cells (which reabsorb sodium and chloride against concentration gradients), and cardiac myocytes (which rhythmically contract 50–100 times per minute) have a generous complement of mitochondria. Conversely, cells specifically adapted to synthesize and export selected proteins (e.g., insulin in a pancreatic islet cell, or antibody produced by a plasma cell) have a well-developed rough endoplasmic reticulum.

THE PLASMA MEMBRANE: PROTECTION, NUTRIENT ACQUISITION, AND COMMUNICATION

Plasma membranes, as well as all other organellar membranes, are dynamic, fluid, and inhomogeneous lipid bilayers containing a variety of embedded proteins. Biologic membranes are composed of phospholipids that are amphipathic; that is, they have a polar head group that prefers to interact with water (hydrophilic), and nonpolar fatty acid chains that resist interacting with water (hydrophobic). These phospholipids will spontaneously assemble to form two-dimensional

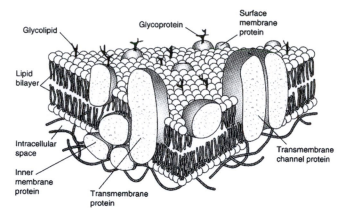

FIG. 2. Model of a prototypical cell membrane. Note the lipid bilayer with outer, hydrophilic (exposed to an aqueous environment) and inner, hydrophobic (maintains a barrier to solute movement) domains. Inserted through the membrane or in either inner or outer planes are various proteins that permit transport, cell–cell and signal–molecule interactions, and linkage of the membrane to the intracellular cytoskeleton. (Reproduced by permission from Bergman, A. R., Afifi, A. K., and Heidger, P. M., eds., 1996. *Histology.* Saunders, Philadelphia.)

sheets with their hydrophilic head groups facing toward the aqueous cytoplasm or extracellular fluid, and their hydrophobic lipid tails interacting to form a central core that largely excludes water. Since the lipid core of plasma membrane is intrinsically resistant to the movement of large or polar molecules, continued cell function requires that specific protein channels or transport mechanisms be in place to facilitate uptake of ions and metabolites. In addition, the plasma membrane is also the interface between the extracellular environment and the inner cellular domains; interactions of proteins at the cell surface with extracellular molecules or even other cells can trigger a cascade of intracellular signaling events (Fig. 2).

The lipid confers structural integrity and barrier function to the membrane and the inserted proteins provide specialized membrane functions (see later discussion). In addition, the specific composition of the lipid bilayer, e.g., relative amounts of various phospholipids, cholesterol, and glycolipids, will alter the physicochemical properties of the membrane, as well as the functioning of associated proteins. Although the various lipid and protein constituents can move about easily in the plane of the membrane, certain components have a predilection for each other, and different domains with distinct lipid compositions are thereby created. Since inserted membrane proteins have different solubilities in these various lipid domains, the membrane lipid inhomogeneities result in functionally distinct islands and patches of protein molecules. This has significance in terms of cell–cell and cell–matrix interactions, as well as in intracellular signaling. The lipid component of the plasma membrane is also asymmetric, that is, there are different general compositions of the inner and outer leaflets. The asymmetry has functional significance in that gangliosides, conferring a net negative charge, and glycolipids—both of which are on the outer face of the bilayer—are important for cell–cell and cell–matrix interactions, local electrostatic effects, and creation of barriers to infection. Inositol phospholipids, on the inner face, are important for intracellular signaling.

How proteins associate with membranes reflects their function. Most proteins inserted into membranes are integral or transmembrane proteins, having one or more relatively hydrophobic segments that traverse the lipid bilayer and securely anchor the protein. Occasionally, proteins are attached to the membrane via weaker lipid–protein or protein–protein interactions. This has functional significance in that nonintegral proteins can potentially interact with a variety of membrane molecules and can be used to transduce transient signals. Proteins involved in forming pores or transporting other molecules will typically be transmembrane, whereas those involved in signaling or cytoskeletal interactions will not.

Plasma membrane proteins frequently function together as large complexes. These complexes may be primarily assembled as the proteins are synthesized in the RER, or may arise by lateral diffusion in the plasma membrane. Such de novo complex formation, followed by signal cascade inside the cell, is a typical paradigm employed to translate ligand–surface receptor binding into an intracellular response. Large complexes also form the basis for intercellular connections such as tight junctions (see later discussion), and interactions between the same proteins on adjacent cells (homotypic interactions) create a zone that separates the apical and basolateral aspects of cells in epithelial layers, thereby establishing cell polarity.

The hydrophobic lipid core of plasma membranes is an effective barrier to the passage of most polar molecules. Small, nonpolar molecules such as O_2 and CO_2 readily dissolve in lipid bilayers, and therefore rapidly diffuse across them. Large hydrophobic molecules such as steroid hormones (testosterone and estrogen, for example) also readily cross lipid bilayers. Even polar molecules, if sufficiently small (e.g., water, ethanol, and urea at molecular weights of 18, 46, and 60 daltons, respectively) rapidly cross membranes. In contrast, glucose, at a molecular weight of only 180 daltons, is effectively excluded, and lipid bilayers are completely impermeant to ions, regardless of size, because of their charge and high degree of hydration. Therefore, to facilitate the entrance or disposal of various molecules, specific transport proteins are required. For small-molecular-weight molecules (ions, glucose, nucleotides, and amino acids up to approximately 1000 daltons) there are two main categories, carrier proteins and channel proteins. For larger molecules or even particles, uptake is mediated by specific receptors, internalized via endocytosis. Large molecules destined for export are packaged in secretory vacuoles that fuse with the plasma membrane and expel their contents in a process called exocytosis.

Each type of transported molecule (ion, sugar, nucleotide, etc.) requires a unique membrane transport protein to facilitate passage. These transporters typically exhibit strong specificity. Thus, a particular transporter will move glucose but not galactose; another transporter will move potassium but not sodium. Carrier proteins bind their specific ligand and undergo a series of conformational changes to transfer it across the membrane; their transport is relatively slow. In comparison, channel proteins create hydrophilic pores; when open, these permit rapid movement of selected solutes. In most cases, a concentration and/or electrical gradient between the inside and outside of the cell drives solute movement by passive transport (virtually all plasma membranes have an electrical potential difference across them, with the inside of a cell negative relative to the outside). In some cases, active transport against a concentration gradient is accomplished by carrier molecules (never channels) and requires energy expenditure (provided by the breakdown of ATP). Transporters also include the multidrug resistance (MDR) protein, which pumps polar compounds (for example, chemotherapeutic drugs) out of cells and may render cancer cells more resistant to treatment. Similar transport mechanisms also regulate intracellular and intra-organellar pH; human beings and most of their cytosolic enzymes prefer to work at pH 7.2, whereas lysosomes function best at pH 5 or less.

Because the plasma membrane is freely permeable to water, water will move into or out of cells along its concentration gradient (by osmosis). Extracellular salt in excess of that seen in the cytoplasm (hypertonicity) will cause a net movement of water out of cells; conversely, hypotonicity will cause a net movement of water into cells. Since the intracellular environment is rich in charged molecules that attract a large number of charged counterions (tending to increase osmolarity), cells need to constantly actively regulate their intracellular osmolarity by pumping out small inorganic ions (typically sodium and chloride). Loss of the ability to generate energy (e.g., in a cell injured by toxins or lacking oxygen) therefore results in an osmotically swollen, and eventually ruptured, cell.

Uptake and metabolism of large extracellular molecules requires vesicle targeting and membrane recycling. Proteins, large carbohydrates, and other macromolecules cannot enter cells by either channels or carriers. Instead, they are internalized by endocytosis (see Fig. 3). Endocytosis begins at a specialized region of the plasma membrane called the clathrin-coated pit, which rapidly invaginates and pinches off to form a clathrin-coated vesicle (about 2500 per minute in a typical fibroblast). Clathrin is a hexamer of proteins that spontaneously assemble into a basket-like lattice to drive the budding process of endocytosis. Trapped within the vesicle will be a minute gulp of the extracellular milieu, as well as any molecules specifically bound to receptors on the internalized bit of plasma membrane (receptor-mediated endocytosis). This is the pathway, for example, by which cells internalize iron from the circulation: ionized iron, bound to a protein called transferrin, interacts with cell surface transferrin receptors, which are then internalized via receptor-mediated endocytosis.

The vesicles rapidly lose their clathrin coat and then fuse with an acidic intracellular structure called the endosome, where they discharge their contents for digestion and further passage to the lysosome (Fig. 4 and later discussion). After release of bound ligand, receptors can either return to the plasma membrane for another cycle (e.g., the transferrin receptor) or may be degraded [e.g., the low-density lipoprotein (LDL) receptor]. Degradation of a receptor after internalization (receptor down-regulation) provides an important control for receptor expression and receptor-mediated signaling. Endocytosis can also deliver material completely across a cell, i.e., from the apical surface to the basolateral face (transcytosis), forming the basis for transport of nutrients from the gastrointestinal tract to the blood stream. Endocytosis is an

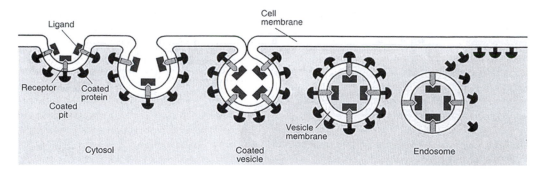

FIG. 3.   Schematic of receptor-mediated endocytosis, enabling large molecules (e.g., transferrin with bound iron) to enter the cell. After binding to specific receptors, proteins will be internalized on clathrin-coated pits, which pinch off to form clathrin-coated vesicles. The clathrin coat is then removed and the vesicle fuses with endosomes delivering its bound contents. After fusion, vesicles may pinch off from the endosome and recycle to the plasma membrane (exocytosis, not shown). (Reproduced by permission from Bergman, A. R., Afifi, A. K., and Heidger, P. M., eds., 1996. *Histology*. Saunders, Philadelphia.)

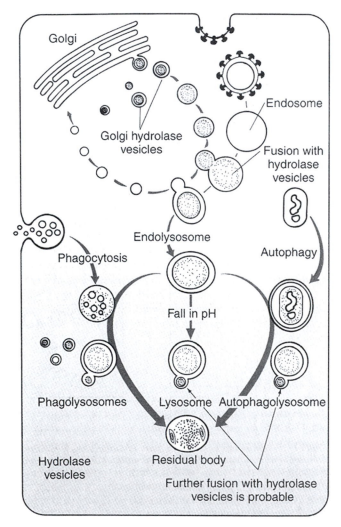

FIG. 4.   Schematic demonstrating the pathways by which internalized material is degraded. There is convergence of endocytic, phagocytic, and autophagic vesicles as they fuse with newly synthesized hydrolase-containing vesicles or with preexisting lysosomes. The contents are nearly completely degraded to constituent amino acids, sugars, and lipids; nondegradable material will accumulate in residual bodies. (Reproduced by permission from Bergman, A. R., Afifi, A. K., and Heidger, P. M., eds., 1996. *Histology*. Saunders, Philadelphia.)

ongoing process, with constant recycling of vesicles back to the plasma membrane (exocytosis). Endocytosis and exocytosis must be tightly coupled since a cell will internalize 10–20% of its own cell volume each hour, or about 1–2% of its plasma membrane each minute.

Cell communication is critical in multicellular organisms. At the most basic level, extracellular signals may determine whether a cell lives or dies, whether it remains quiescent, proliferates, migrates, or otherwise becomes active to perform its specific function(s). Intercellular signaling is critical in the developing embryo in order that that cells appear in the correct quantity and location, and in maintaining tissue organization. Intercellular signaling is also important in the intact organism, ensuring that all tissues act in appropriate concert in response to stimuli as divergent as food or threat to life. Loss of communication may be reflected in a congenital structural defect in the first instance, or in unregulated cell growth (cancer) or an ineffective response to an extrinsic stress in the second.

Cells communicate over short or long distances. Adjacent cells may communicate via gap junctions, which are narrow, hydrophilic channels that effectively connect the two cells' cytoplasm. The channels permit movement of small ions, various metabolites, and potential second messenger molecules, but not larger macromolecules. Extracellular signaling via soluble mediators occurs in three different forms:

1. Paracrine, meaning that it affects cells only in the immediate vicinity. To accomplish this, there can be only minimal diffusion, with the signal rapidly degraded, taken up by other cells, or trapped in the ECM.
2. Synaptic, where activated neural tissue secretes neurotransmitters at a specialized cell junction (synapse) onto target cells.
3. Endocrine, where a regulatory substance, such as a hormone, is released into the blood stream and acts on target cells at a distance.

Since most signaling molecules are present at extremely low concentrations ($\leq 10^{-8}\ M$), binding to the appropriate target cell receptor is typically a high-affinity and exquisitely specific interaction. Receptor proteins may be on the cell surface or they may be intracellular; in the latter case, ligands must be sufficiently hydrophobic to enter the cell (e.g., vitamin D, or steroid and thyroid hormones). For intracellular receptors, ligand binding leads to formation of a receptor–ligand complex that directly associates with nuclear DNA and subsequently either activates or turns off gene transcription. For cell-surface receptors, the binding of ligand can (1) open ion channels (e.g., at the synapse between electrically excitable cells), (2) activate an associated GTP-binding regulatory protein (G protein), or (3) activate an associated enzyme. Secondary intracellular downstream events frequently involve phosphorylation or dephosphorylation of target molecules, with subsequent changes in enzymatic activity.

An individual cell is exposed to a remarkable cacophony of signals, which it must sort through and integrate into a rational response. One ligand may induce a given cell type to differentiate, another may signal proliferation, and yet another may direct the cell to perform a specialized function. Multiple ligands at once, in a certain ratio, may signal yet another unique response. Many cells require certain signals just to continue living; in the absence of appropriate exogenous ligand, they may undergo a form of cellular suicide called apoptosis, or programmed cell death.

THE CYTOSKELETON: CELLULAR INTEGRITY AND MOVEMENT

The ability of cells to adopt a particular shape, maintain cell polarity, organize the relationship of intracellular organelles, and move depends on the intracellular scaffolding of proteins called the cytoskeleton. In eukaryotic cells there are three major classes of cytoskeletal proteins: 6- to 8-nm-diameter actin microfilaments, 10-nm-diameter intermediate filaments, and 25-nm-diameter microtubules. Although these proteins impart some structure (especially the intermediate filaments), it should also be emphasized that they are all dynamic. In particular, actin and microtubules are also used by cells to achieve movement or cellular contraction.

The globular protein actin (G-actin, 43,000 daltons) is the major subunit of microfilaments and is the most abundant cytosolic protein in cells. The monomers polymerize into long double-stranded helical filaments (F-actin), with a defined polarity (one end is stable, the other end grows or shrinks). In muscle cells, the filamentous protein myosin binds to actin and moves along it, driven by ATP hydrolysis forming the basis of muscle contraction. In nonmuscle cells, F-actin and an assortment of actin-binding proteins form well-organized bundles and networks that control cell shape and surface movements.

Intermediate filaments comprise a large and heterogeneous family of closely related structural proteins. Although they generally perform similar functions, each member of the family has a relatively restricted expression in specific cell types. These ropelike fibers are found predominantly in a stable polymerized form within cells; they are not usually actively reorganizing like actin and microtubules. Intermediate filaments impart strength and carry mechanical stress, e.g., in epithelia where they connect spot desmosomes (see below). They also form the major structural proteins of skin and hair (i.e., keratin).

Microtubules consist of polymerized dimers of α- and β-tubulin arrayed in constantly elongating or shrinking hollow tubes. These have a defined polarity with ends designated "+" or "−." In most cells, the "−" end is embedded in a microtubule organizing center (MTOC or centrosome) that lies near the nucleus; the "+" end elongates or recedes in response to various stimuli by the addition or subtraction of tubulin dimers. Within cells, microtubules may serve as mooring lines for protein "molecular motors" that hydrolyze ATP to move vesicles, organelles, or other molecules around cells; the polarity of the microtubules allows cells to direct whether attached structures are "coming" or "going" relative to the nucleus. In neurons, microtubules are critical for the delivery of molecules synthesized in the nuclear area to the far-flung reaches of the cytosol of the axon, which may be as far away as 10,000 times the width of the cell body (indeed, the nuclei of some motor neurons in the spinal cord have axons extending to the muscles of the big toe over 3 feet away!). Microtubules are also used to move chromosomes apart during mitosis, and thus play a basic role in cellular proliferation. Finally, in certain cells, microtubules and their associated molecular motors have been harnessed to facilitate cellular motility—they form cilia to move mucus and dust out of the airways and flagella to propel sperm.

Maintaining cellular and tissue integrity requires cell–cell and cell–ECM interactions, mediated through the plasma membrane and translated into the cytoskeleton. The organization of tissues requires attaching cells together and to the underlying ECM scaffolding. These surface attachments are connected via transmembrane proteins to the cytoskeletal elements. Thus, extracellular perturbations in a tissue may be translated into intracellular events.

As discussed earlier, the external face of a cell membrane is diffusely studded with carbohydrate-modified (glycosylated) proteins and lipids. This cell coat (or glycocalyx) functions primarily as a chemical and mechanical barrier, but also serves an important role in cell–cell and cell–matrix interactions, including sperm–egg attachment, blood clotting, lymphocyte recirculation, and inflammatory responses. For example, neutrophils (cells involved in acute inflammation) are recruited to sites of infection by local vascular wall cell expression of lectin-like molecules (selectins) that bind specific sugars expressed on the circulating cells.

Cell–cell connections include occluding junctions (tight junctions) and anchoring junctions (desmosomes) (Fig. 5) [gap junctions (described earlier) function primarily in cell-to-cell communication and do not materially contribute to cellular adhesiveness]. Tight junctions seal cells together in a continuous sheet, preventing even small molecules (but not water) from leaking from one side to the other. These junctions are the basis of the high electrical resistance of many epithelia, as well as the ability to segregate apical and basolateral spaces. It is important to remember, however, that these junctions (as well as the desmosomes) are also dynamic structures that intermittently dissociate and reform. This permits processes as diverse as healing of an epithelial wound, or allowing passage of inflammatory cells across vascular endothelium to sites of infection. Intracellular actin microfilaments are connected to these tight junctions and span from side to side across the cell. This creates a circumferential band (adhesion belt) of cytoskeleton that provides structural integrity and shear strength to the sheets of interconnected cells.

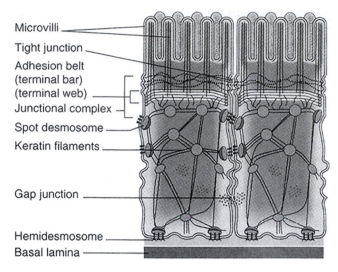

FIG. 5. Diagram of two prototypical epithelial cells (imagine these as part of a larger planar sheet) demonstrating attachments to the underlying basal lamina (hemidesmosomes) and to each other (tight junction, adhesion belt, and spot desmosomes). Note the cytoskeletal elements underlying each attachment point giving structural integrity to the individual cells as well as to the larger epithelial structure. Gap junctions do not confer intercellular adhesion, but are responsible for intracellular signaling. (Reproduced by permission from Bergman, A. R., Afifi, A. K., and Heidger, P. M., eds., 1996. *Histology.* Saunders, Philadelphia.)

Desmosomes mechanically attach cells (and their cytoskeletons) to other cells or the ECM. When they occur in broad belts or bands between cells, they are referred to as belt desmosomes; when they are small and rivet-like, they are denoted as spot desmosomes. Desmosomes are formed by a homotypic association (two proteins of the same type) of transmembrane glycoproteins called cadherins. The cytosolic ends of cadherins are associated with cytoskeletal actin microfilaments and intermediate filaments. Focal adhesion complexes or hemidesmosomes (literally, half a desmosome) are "spot welds" that connect cells to the extracellular matrix; in the case of epithelial tissues, the connections are to a dense ECM meshwork called the basal lamina or basement membrane (see Chapter 3.4). The plasma membrane proteins forming the basis of these interactions are called integrins; like cadherins, they attach to intracellular intermediate filaments. These focal adhesion complexes, connecting cells to the ECM, also act to generate intracellular signals when cells are subjected to increased shear stress (such as endothelium in a turbulent area of the blood stream).

THE NUCLEUS: CENTRAL CONTROL

With the exception of the terminally differentiated hematopoietic cells (erythrocytes and platelets), every human cell has a central regulatory nucleus containing nucleic acids (DNA and RNA) and proteins that determine the sequence and rate of macromolecular synthesis. The full complement of DNA in a cell is called its genome.

The nucleus is not a uniform static repository of molecules. At any given time, and depending on the functional state of

the cell, there will be active replication (duplication of DNA) or transcription (conversion of DNA into messenger RNA) of selected subsets of the genome. Clearly, proliferating cells will be busily generating an entire copy of all nuclear material so that daughter cells can be each afforded a complete set of genetic material. It is also intuitive that cells actively synthesizing and exporting specific proteins will have frenzied activity surrounding those relevant genes. However, even apparently quiescent cells have a constant turnover of proteins and organelles, requiring a tightly orchestrated on-and-off switching of the correct genes in the correct sequence.

Nuclear proteins and nucleic acids are organized into clumps and clusters called chromatin. Two basic forms of chromatin are recognizable by light microscopy and routine staining correlating with the activity of the gene. When genes are transcriptionally inactive, they are tightly coiled in compact aggregates wrapped around protein histones and are not accessible to transcription machinery. This results in a cytochemically dense appearance called heterochromatin. In actively transcribing genes, the nuclear material uncoils into a more extended linear form, which is cytochemically disperse and called euchromatin. The degree of cellular activation or gene transcription may thus be inferred from the general staining characteristics of nuclei (Fig. 6). The nucleolus is a subcompartment of the nucleus dedicated to ribosomal RNA

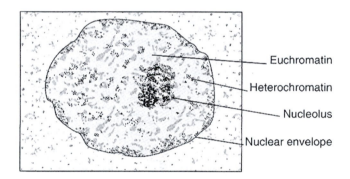

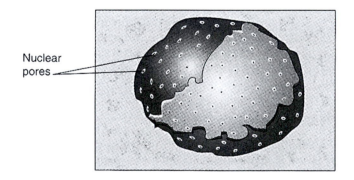

FIG. 6. Diagram of the nucleus and nuclear pore system. (*Top*). Illustration of heterochromatin (clumped nontranscribing nuclear material), euchromatin (dispersed, transcribing nuclear material), and nucleolus (site of ribosome synthesis). (*Bottom*) Schematic representation of nuclear pores; note that they are not passive openings in the nuclear envelope, but rather are selective transporters. (Reproduced by permission from Bergman, A. R., Afifi, A. K., and Heidger, P. M., eds., 1996. *Histology.* Saunders, Philadelphia.)

synthesis and ribosome assembly. Its size may also reflect the translational activity of the cell; the greater the protein synthesis, the larger the nucleolus.

Movement of molecules into and out the nucleus is restricted and tightly regulated. The nucleus is surrounded by a nuclear envelope formed by two concentric membranes supported by networks of intermediate filaments. The outer membrane is continuous with the endoplasmic reticulum and is joined to the inner membrane at numerous nuclear pores that punctuate the envelope. The pores are elaborate, gated structures that permit the active transport of molecules to and from the cytosol. Since the pores are freely permeable only to globular proteins ≤ 50,000 daltons, the process of moving large molecules and complexes (e.g., ribosomes out of the nucleus, or histones and polymerase complexes into the nucleus) is accomplished by specific receptor proteins. Macromolecules destined for the nucleus are identified by specific nuclear localization signals (typically certain amino acid sequences) that permit binding to the pore receptor proteins. Of note, these localization signals may be cryptic, as in some inactive steroid receptor proteins. Subsequent binding of the steroid ligand causes a conformational change that uncovers the localization signal; the receptor can then be translocated to the nucleus.

ROUGH AND SMOOTH ENDOPLASMIC RETICULUM, AND GOLGI APPARATUS: BIOSYNTHETIC MACHINERY

The endoplasmic reticulum (ER) is the site for the synthesis of all the transmembrane proteins and lipids for most of a cell's organelles, including the ER itself. It is also the initial site for the synthesis of the majority of molecules destined for residence in the inside of ER, Golgi, and lysosomes, or for export out of the cell. The ER is organized into a mesh-like maze of interconnecting branching tubes and flattened vesicles (see Fig. 1) forming a continuous sheet around a single highly convoluted space topologically equivalent to the extracellular environment. The ER is composed of contiguous but distinct domains, distinguished by the presence (rough ER or RER) or absence (smooth ER or SER) of ribosomes.

Membrane-bound ribosomes on the cytosolic face of RER are actively translating mRNA into proteins, which are folded and edited in the lumen of the ER (Fig. 7). This process, called translation, is directed by amino acid signal peptides generally present on the N-termini of nascent proteins; if a new protein with a signal peptide is produced on a free ribosome, the signal peptide directs the entire complex to attach to the ER membrane. Proteins synthesized in this way are directly inserted into the ER as they are being made. Within the ER lumen, the proteins fold and form multisubunit complexes under the scrutiny of ER chaperone molecules. These chaperones interact with a variety of proteins within the ER ensuring that they are completely assembled and functional. Failure to appropriately fold results in retention and eventually degradation within the ER; thus the ER has an editing function for safeguarding the fidelity of the transcriptional apparatus.

For proteins lacking a signal sequence, the translation apparatus remains on free ribosomes in the cytosol, frequently

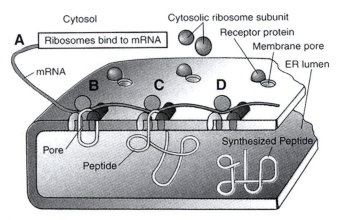

FIG. 7. Schematic demonstrating the general steps in the synthesis of proteins on rough endoplasmic reticulum (RER). (A) Peptide synthesis from mRNA begins on free ribosomes in the cytosol. (B) Signal peptide sequences on the nascent proteins direct the entire complex to attach to the ER membrane with insertion of the synthesizing protein into the RER lumen. (C) The signal sequence is cleaved and the protein is completely synthesized, eventually detaching from the ribosome. (D) The protein is folded and assembled (if necessary) into multisubunit complexes; the ribosome detaches from the ER surface and returns to the cytoplasm. (Reproduced by permission from Bergman, A. R., Afifi, A. K., and Heidger, P. M., eds., 1996. *Histology*. Saunders, Philadelphia.)

forming polyribosomes as multiple translation units attach to the mRNA; the resulting protein remains within the cytoplasm and is not exported from the cell.

After leaving the RER, proteins and lipids are modified in the Golgi apparatus and sorted for intracellular delivery (Fig. 8). In the Golgi proteins are glycosylated—modified by

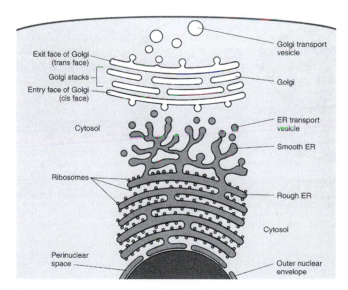

FIG. 8. Functional, schematic diagram of the relationship among the rough endoplasmic reticulum (RER), smooth endoplasmic reticulum (SER), and the Golgi apparatus. Proteins destined for export or for other intracellular organelles pass from the RER to the SER where they form vesicles for transport to the Golgi; there they are progressively modified and sorted, eventually leaving in transport vesicles to their appropriate final destination. (Reproduced by permission from Bergman, A. R., Afifi, A. K., and Heidger, P. M., eds., 1996. *Histology*. Saunders, Philadelphia.)

a stepwise addition of various sugars. These modifications are important in subsequent sorting of molecules to various intracellular sites, and also because glycosylation is critical for surface molecules involved in cell–cell or cell–matrix interactions. In addition to the stepwise glycosylation of lipids and proteins, the trans Golgi network sorts molecules for dispatch to other organelles (including the plasma membrane) or secretory vacuoles destined for extracellular release. The Golgi complex is especially prominent in cells specialized for secretion, including goblet cells of the intestine or bronchial epithelium (making large amounts of polysaccharide-rich mucus), and plasma cells (secreting antibodies).

The SER has a role in steroid hormone synthesis, modification of certain metabolites, and intracellular calcium regulation. The SER in most cells is relatively sparse. However, in cells that synthesize steroid hormones (e.g., the adrenal cortex or gonads) or that catabolize lipid-soluble molecules (e.g., liver cells make certain drugs more water-soluble so that they may be excreted), the SER may be particularly conspicuous. The SER is also responsible for sequestering intracellular Ca^{2+}; release of Ca^{2+} from the SER is a mechanism by which cells can rapidly respond to extracellular signals. Finally, in muscle cells, specialized SER called sarcoplasmic reticulum regulates the successive cycles of myofiber contraction (Ca^{2+} released into the cytosol) and relaxation (Ca^{2+} pumped back into the SER).

LYSOSOMES AND PROTEASOMES: WASTE DISPOSAL

To digest internalized macromolecules or senescent organelles, cells primarily rely upon lysosomes. Lysosomes are membrane-bounded organelles containing a large assortment (>40) of acid hydrolase enzymes including proteases, nucleases, lipases, glycosidases, phosphatases, and sulfatases. Each has an optimal activity at pH 5, which is a protective feature since these enzymes will do less damage should they leak into the pH 7.2 cytosol. Materials destined for catabolism arrive in the lysosomes by one of three pathways:

1. Internalized by fluid-phase or receptor-mediated endocytosis. Material passes from plasma membrane to early endosome to late endosome, and ultimately into the lysosome (see Fig. 4).
2. Obsolete organelles within cells (the average mitochondrion, for example, only lives 10 days) are shuttled into lysosomes by a process called autophagy. The resultant autophagosome then fuses with lysosomes and the organelle is catabolized.
3. Phagocytosis of microorganisms or large fragments of matrix or debris occurs primarily in professional phagocytes (macrophages or neutrophils). The material is engulfed to form a phagosome that subsequently fuses with a lysosome.

Proteasomes degrade cytosolic molecules that are senescent or require constant turnover to regulate their activity. The cytosol also needs to have a mechanism to degrade misfolded proteins (like the ER chaperone function) and to regulate the longevity of certain other proteins that turn over at discrete rates. To do this, many proteins destined for destruction are marked by covalently binding one or more 76-amino-acid proteins called ubiquitin; ubiquitin-tagged molecules are then fed into a large polymeric protein complex called the proteasome. Each proteasome is a cylinder composed of several different proteases, each with its active site pointed at the hollow core; proteins are degraded into small (6–12 amino acids) fragments. This degradative mechanism is thought to be a holdover from prokaryotes (i.e., primitive cellular organisms) that lack lysosomes.

MITOCHONDRIA: ENERGY GENERATION

Energy to run intracellular processes is extracted from adenosine triphosphate generated by mitochondria. Mitochondria accomplish ATP production by utilizing by-products of carbohydrate oxidation (such as glucose) to CO_2 and water through the glycolytic pathway that breaks down glucose to pyruvate or lactate (in the cell cytosol), the metabolism of pyruvate to carbon dioxide and water through the Krebs cycle, and oxidative phosphorylation. (The latter two occur in the mitochondria.) These reactions, and particularly oxidative phosphorylation (the process of generating ATP from substrate oxidation), are critically dependent on the availability of oxygen. When oxygen is present, 38 ATP are generated from metabolism of one glucose molecule; in the absence of oxygen, only two molecules of ATP are generated. ATP hydrolysis, the chemical reaction that removes a terminal phosphate from ATP, is accompanied by the release of a large amount of energy. Breakdown of fats and proteins also contributes to ATP production. The energy derived from the hydrolysis of ATP is used for cell needs such as active transport of ions and molecules across membranes, synthesis of molecules for cell housekeeping and for export, and specialized cell functions such as contraction of muscle.

Each mitochondrion has two separate and specialized membranes. There is a core matrix space (containing most of the enzymes for breaking down glucose and its primary metabolites) surrounded by an inner membrane (containing the enzymes to transfer electrons to oxygen) folded into cristae; these constitute the major working parts of the organelle. The inner membrane is enclosed by the intermembrane space (site of ADP to ATP phosphorylation), which is in turn surrounded by the outer membrane; the latter is studded with a transport protein called porin, which forms aqueous channels permeable to low-molecular-weight substrates (Fig. 9). Mitochondria are also central to the pathways leading to apoptosis (see later discussion).

Mitochondria probably evolved from ancestral prokaryotes engulfed by primitive eukaryotes about 1–2 billion years ago. That explains why mitochondria contain their own DNA (circularized, about 1% of the total cellular DNA) encoding for approximately one-fifth of the proteins involved in oxidative phosphorylation. Although the mitochondrial DNA codes for only a very small number of proteins, mitochondria have the machinery necessary to carry out all the steps of DNA replication, transcription, and translation. Consistent with its evolutionary origin, mitochondrial translational machinery is

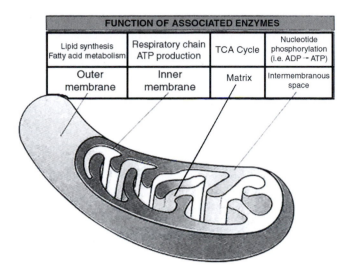

FUNCTION OF ASSOCIATED ENZYMES			
Lipid synthesis Fatty acid metabolism	Respiratory chain ATP production	TCA Cycle	Nucleotide phosphorylation (i.e. ADP → ATP)
Outer membrane	Inner membrane	Matrix	Intermembranous space

FIG. 9.   Schematic diagram of the mitochondrion demonstrating the functional segregation of the enzymatic machinery required to generate ATP. (Reproduced by permission from Bergman, A. R., Afifi, A. K., and Heidger, P. M., eds., 1996. *Histology.* Saunders, Philadelphia.)

similar to present-day bacteria. For example, protein synthesis is initiated with the same modified amino acid as bacteria (*N*-formylmethionine) and is sensitive to the same antibiotics. It is noteworthy that new mitochondria can only derive from preexisting mitochondria. Thus, since the ovum contributes the vast majority of cytoplasmic organelles in the fertilized zygote, mitochondrial DNA is maternally inherited.

CELL SPECIALIZATION AND DIFFERENTIATION

As discussed earlier, basic functional attributes of cells include nutrient absorption and assimilation, respiration, synthesis of macromolecules, growth, and reproduction. Without these basic activities, cells cannot live. However, most cells also exhibit specialization, that is, they have additional capabilities, such as irritability, conductivity, absorption, or secretion of molecules (for use by other cells). Multicellular organisms are thus composed of individual cells with marked specialization of structure and function. These differentiated cells allow a division of labor in the performance and coordination of complex functions carried out in architecturally distinct and organized tissues.

Differentiated cells have developed well-defined structural and/or functional characteristics associated with increasing specialization. For example, striated muscle cells have well-organized actin and myosin filaments that slide over one another, facilitating cellular contraction. Gastric (stomach) epithelial cells have large numbers of mitochondria to generate the ATP necessary to pump hydrogen ions out of the cell against a concentration gradient and acidify the stomach contents. Skin keratinocytes function as a protective barrier by losing their organelles and becoming scale-like structures filled with durable, nonliving keratin (an intermediate filament). Specialized phagocytic cells of the immune system must detect infectious microorganisms (e.g., bacteria,

parasites, and viruses), actively migrate to them, and then ingest and destroy them. Polymorphonuclear leukocytes (also called PMNs or neutrophils) are particularly active against bacteria, and macrophages react to other types of organisms and foreign material. B lymphocytes are not phagocytic but contribute to immunity by producing antibodies that can bind and neutralize infectious agents.

The structural and functional changes that occur during cellular differentiation are usually irreversible. Moreover, increasing specialization results in a loss of cell potentiality, as well as a loss in the capacity for cell division. For example, the newly fertilized ovum is absolutely undifferentiated and has the capacity to divide extensively, ultimately giving rise to progeny that make up all the cells of the body. These initially undifferentiated cells are said to be totipotential or pluripotential. As cells differentiate into particular tissue pathways, they lose the ability to interconvert and develop into all cell types; with further differentiation, cells may lose the ability to replicate at all. Thus, cells capable of dividing to produce several (but not all) types of cells are multipotential. Cells capable of both dividing and yielding differentiated cells of one or more types are called stem cells. Conversely, nerve cells and heart muscle cells are considered to be examples of highly specialized cells that, according to traditional teaching, can neither differentiate into other tissue types nor reproduce. Obviously, this has clinical significance when these terminally differentiated cells are injured (e.g., by stroke or heart attack). Nevertheless, recent evidence suggests that cells of end-stage, specialized and highly differentiated tissues can, under certain conditions such as following injury, dedifferentiate into multipotent forms or serve as stem cells capable of generating more specialized cells (Lee *et al.*, 2003; Nadal-Ginard *et al.*, 2003; Zheng *et al.*, 2003).

Cellular differentiation involves an alteration in gene expression. Every cell in the body has the same complement of genes (called the genotype). With progressive differentiation, selected subsets of genes are preferentially expressed, yielding a distinct biological profile (called the phenotype). As cells progressively specialize, more and more of the "unnecessary" genes in the differentiating cell are irreversibly turned off. Some genes are active at all times (constitutively expressed); others may be selectively activated or modulated depending on external influences (e.g., injury).

CELL INJURY AND REGENERATION

Cells constantly adjust their structure and function to accommodate alterations in their environment, particularly responding to chemical and mechanical stressors (Fig. 10). Cells attempt to maintain their intracellular milieu within relatively narrow physiologic parameters—i.e., they maintain normal homeostasis. Consequently, as they encounter physiologic stresses or pathologic stimuli, cells and tissues can adapt, achieving a new steady state and preserving viability. The principal adaptive responses are:

Hypertrophy: an increase in size of individual cells
Hyperplasia: an increase in cell number

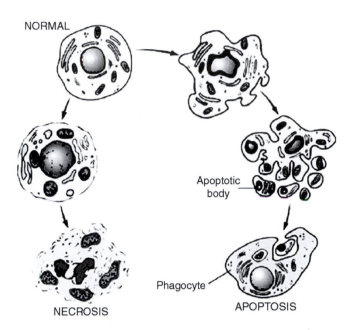

FIG. 10. The sequential cellular structural changes seen in necrosis (left) versus apoptosis (right). In necrosis, there is chromatin clumping, organelle swelling, and eventually membrane damage; dead cells typically must be degraded and digested by an influx of inflammatory cells. In apoptosis, the initial changes consist of nuclear chromatin condensation and fragmentation, followed by cytoplasmic budding of apoptotic bodies, and eventual phagocytosis by adjacent cells of the extruded cell fragments. (Reproduced by permission from Cotran, R. S., Kumar, V., and Collins, T., 1999. *Robbins Pathologic Basis of Disease*, 6th ed. Saunders, Philadelphia.)

Atrophy: a decrease in size, without appreciable change in cell number

Metaplasia: transformation from one mature cell type to another

There may also be more subtle changes in the expression of selected genes that are functionally beneficial but are not necessarily reflected in alterations of morphology.

Usually, if the extracellular stressors recede, the cells and tissues will revert to their prestressed state. However, if the stressors persist and a cell's adaptive capability is exceeded, cell injury develops. Up to a point, cell injury itself is reversible, and with normalization of the stimulus, the cell returns to its baseline state, usually no worse for wear. However, with severe or persistent stress, the cell suffers irreversible injury and dies.

For example, when heart muscle cells are subjected to persistent increased load (e.g., high blood pressure), the cells adapt by undergoing hypertrophy (enlargement of the individual myocytes, and eventually the entire heart) to compensate for the higher pressures they must pump against. Conversely, in periods of prolonged starvation (as can happen in prolonged illness or with malignant tumors), all myocytes (and thus the heart) will undergo atrophy. The same myocytes, subjected to an imbalance between blood supply and energy demand due to an occluded coronary artery (ischemia), may be reversibly injured if the occlusion is incomplete or sufficiently brief; alternatively, they may undergo irreversible injury (i.e., cell death,

as in myocardial infarction) following complete or prolonged occlusion.

Moreover, certain genetic abnormalities and/or environmental stimuli may trigger abnormal tissue growth that is uncoordinated relative to normal tissues, has lost its responsiveness to normal growth controls, and persists after cessation of the stimuli that initiated it. This condition is called "neoplasia"; in its malignant form, this is more commonly called cancer (see Chapter 4.7).

CAUSES OF CELL INJURY

Hypoxia and ischemia. Hypoxia is decreased O_2 supply relative to the needs of a particular tissue. Anoxia is the complete absence of oxygen. Hypoxia due to diminished blood flow is called ischemia; irreversible tissue injury (necrosis) due to ischemia is called infarction. Note that although diminished blood flow will invariably lead to hypoxia, oxygen deficiency can occur in the setting of adequate tissue perfusion.

Chemical injury. Chemical agents include components of food, naturally occurring toxins, hormones, neurotransmitters, synthetic drugs, environmental pollutants, poisons, ethanol, tobacco, even toxic biomaterials. Chemicals induce cell injury by one of two general mechanisms:

- By combining directly with a critical molecular component or cellular organelle and thereby inhibiting its normal activity. Chemotherapeutic drugs generally fall into this category.
- Chemicals that are not intrinsically biologically active may be converted to toxic metabolites during normal physiologic breakdown. Such modification is usually accomplished by the P-450 mixed function oxidases in the smooth endoplasmic reticulum (SER) of the liver, and the most important mechanism of cell injury is by formation of free radicals (see below). Acetaminophen belongs to this category of compounds.

Biologic agents. Infectious agents run the gamut from virus and bacteria to fungi, protozoans, and helminths. There is generally a preferred cell or tissue of invasion (called a tropism), and therefore each agent tends to have a defined spectrum of potential injury. Viruses multiply intracellularly, appropriating host biosynthetic machinery in the process; cell lysis may occur directly, or as a result of the immune system's recognition and destruction of infected cells. More insidiously, viruses may compromise the ability of a cell or tissue to perform its normal functions; worse still, viruses may play a role in transformation to malignant neoplasms. Bacteria have toxic cell wall constituents (e.g., endotoxin) and can release a variety of exotoxins. Moreover, the very process of eradicating infections can also cause injury.

Physical injury. Injury can result by direct mechanical force (trauma, pressure), temperature extremes (burn, frostbite), electric shock, or ionizing radiation.

Genetic defects. Mutations in a variety of cellular proteins can lead to cellular dysfunction and eventually irreparable cellular injury. Congenital defects generally manifest as progressive disorders; examples include lysosomal storage diseases where progressive accumulation of certain nondegradable metabolites eventually causes cell rupture, disorders of muscle (myopathies) due to defective energy synthesis by mitochondria, and sickle-cell anemia caused by a mutated hemoglobin that results in stiff, nondeformable red blood cells.

PATHOGENESIS OF CELL INJURY

There are two basic mechanisms of cell injury due to any cause:

Oxygen and oxygen-derived free radicals. Lack of oxygen obviously underlies the pathogenesis of cell injury in ischemia, but in addition partially reduced, activated oxygen species are important mediators of cell death. Free radicals are chemical species with a single unpaired electron in an outer orbital; they are extremely unstable and readily react with inorganic or organic chemicals. The most important ones in biological systems are oxygen-derived and include hydroxyl ($OH^\bullet$) radicals (from the hydrolysis of water, e.g., by ionizing radiation), superoxide radicals ($O_2^{-\bullet}$), and nitric oxide radicals ($NO^\bullet$). Free radicals initiate autocatalytic reactions; molecules that react with free radicals are in turn converted into free radicals, further propagating the chain of damage. When generated in cells, they cause single-strand breaks in DNA, fragment lipids in membranes via lipid peroxidation, and fragment or cross-link proteins leading to accelerated degradation or loss of enzymatic activity. Free-radical damage is a pathogenic mechanism in such varied processes as chemical and radiation injury, oxygen and other gaseous toxicity, cellular aging, microbial killing by phagocytic cells, inflammatory damage, and tumor destruction by macrophages, among others.

It is important to note that besides being a consequence of chemical and radiation injury, free-radical generation is also a normal part of respiration and other routine cellular activities, including microbial defense. It therefore makes sense that cells and tissues have developed mechanisms to degrade free radicals and thereby minimize any injury. Fortunately, free radicals are inherently unstable and generally decay spontaneously; superoxide, for example, rapidly breaks down in the presence of water into oxygen and hydrogen peroxide. The rate of such decay is significantly increased by the action of superoxide dismutases (SODs) found in many cell types. Other enzymes, e.g., glutathione (GSH) peroxidase, also protect against injury by catalyzing free radical breakdown, and catalase directs the degradation of hydrogen peroxide. In addition, endogenous or exogenous antioxidants, e.g., vitamin E, may either block free radical formation or scavenge them once they have formed.

Failure of intracellular ion homeostatic mechanisms is also important. Cytosolic free calcium is normally maintained by ATP-dependent calcium transporters at extremely low concentrations (less than $0.1\ \mu M$); this is in the face of sequestered mitochondrial and endoplasmic reticulum calcium stores, and an extracellular calcium typically at 1.3 mM (or about a 10^4-fold gradient). Ischemic or toxin-induced injuries allow a net influx of extracellular calcium across the plasma membrane, followed by release of calcium from the intracellular stores. Increased cytosolic calcium in turn activates a variety of phospholipases (promoting membrane damage), proteases (catabolizing structural and membrane proteins), ATPases (accelerating ATP depletion), and endonucleases (fragmenting genetic material).

RESPONSES TO CELL INJURY

Whether a specific form of stress induces adaptation or causes reversible or irreversible injury depends not only on the nature and severity of the stress, but also on several other cell-specific variables including vulnerability, differentiation, blood supply, nutrition, and previous state of the cell.

- Cellular response to injurious stimuli depends on the type of injury, its duration, and its severity. Thus, low doses of toxins or a brief period of low blood flow (ischemia) may lead only to reversible cell injury, whereas larger toxin doses or longer ischemic intervals may result in irreversible injury and cell death.

- Consequences of an injurious stimulus are also dependent on the type of cell being injured, its current status (nutritional, hormonal, etc.), and its adaptability. For example, striated skeletal muscle in the leg can tolerate complete ischemia for 2–3 hours without suffering irreversible injury, whereas cardiac muscle will die after only 20–30 minutes, and CNS neurons are dead after 2–3 minutes. Similarly, a well-nourished liver can withstand an ischemic or anaerobic challenge far better than a liver without any energy reserve.

  - Four intracellular systems are particularly vulnerable to injury:

    (i) Aerobic respiration, important in generating the adenosine triphosphate (ATP) energy stores that maintain the intracellular ion gradients (by active pumping) and synthetic pathways
    (ii) Cell membrane integrity, critical to cellular ionic and osmotic homeostasis
    (iii) Protein synthesis
    (iv) Integrity of the genetic apparatus

Most injury alters (in one form or another) the ability of cells to generate energy (make ATP) to run the various intracellular housekeeping chores. Hypoxia and ischemia are the most common ways that energy production is abated in the human body. The first effect of hypoxia is on the cell's aerobic respiration, i.e., oxidative phosphorylation by mitochondria;

as a consequence of reduced oxygen tension, the intracellular generation of ATP is markedly reduced. In quick succession:

1. The activity of the plasma-membrane ATP-driven sodium pump (Na^+/K^+-ATPase) declines with accumulation of intracellular sodium and diffusion of potassium out of the cell. The net gain of sodium solute is accompanied by an isosmotic gain of water, producing acute cellular swelling.
2. This is further exacerbated by the increased osmotic load from the accumulation of other metabolites, including inorganic phosphates, lactic acid, and purine nucleosides, as the cell struggles to generate ATP via anaerobic pathways. The cytoplasm also becomes acidic.
3. Ribosomes begin to detach from the rough endoplasmic reticulum and polysomes dissociate into monosomes, with consequent reduction in protein synthesis.
4. Worsening mitochondrial function and increasing membrane permeability cause further morphologic deterioration with dispersion of the cytoskeleton and formation of cell surface "blebs." Organelles, and indeed whole cells, appear swollen because of loss of osmotic regulation. Precipitation of intracellular proteins and organelles in conjunction with the cellular edema leads to the microscopic appearance of "cloudy swelling."

NECROSIS

If the injurious stimulus is removed, all the above disturbances are potentially reversible; if it persists, cell death may follow. For example, in schemic injury, if oxygen is restored, all of the above disturbances are potentially reversible. However, if ischemia persists, irreversible injury follows; the cells and tissue become necrotic (undergo necrosis). Two phenomena consistently characterize irreversibility. The first is the inability to reverse mitochondrial dysfunction (lack of oxidative phosphorylation and ATP generation) even upon restoration of oxygen; the second is the development of profound disturbances in membrane function. Massive calcium influx into the cell occurs, particularly if ischemic tissue is reperfused after the point of irreversible injury, with broad activation of calcium-dependent catabolic enzymes. Precipitation of calcium salts in cells (calcification) is discussed in detail in Chapter 6.4. Proteins, essential coenzymes, and ribonucleic acids seep out through the newly permeable membranes, and the cells also lose metabolites vital for the reconstitution of ATP. Injury to the lysosomal membranes results in leakage of their enzymes into the cytoplasm; the acid hydrolases are activated in the reduced intracellular pH of the ischemic cell and will further degrade cytoplasmic and nuclear components.

APOPTOSIS

Apoptosis (from root words meaning "a falling away from") has in the past decade been appreciated as a relatively distinctive and important mode of cell death that should be differentiated from standard-variety necrosis (Fig. 10). Apoptosis is responsible for the programmed cell death (or cellular "suicide") in several important physiologic (as well as pathologic) processes:

- The programmed destruction of cells during embryogenesis, including implantation, organogenesis, and developmental involution
- Hormone-dependent physiologic involution, such as the endometrium during the menstrual cycle, or the lactating breast after weaning; or pathologic atrophy, as in the prostate after castration
- Cell deletion in proliferating populations such as intestinal crypt epithelium, or cell death in tumors
- Deletion of autoreactive T cells in the thymus, cell death of cytokine-starved lymphocytes, or cell death induced by cytotoxic T cells

Indeed, failure of cells to undergo physiologic apoptosis may result in unimpeded tumor proliferation, autoimmune diseases, or aberrant development.

Apoptosis usually involves single cells or clusters of cells with condensed nuclear chromatin or chromatin fragments. The cells rapidly shrink, form cytoplasmic buds, and fragment into apoptotic bodies composed of membrane-bound vesicles of cytosol and organelles (see Fig. 10). These fragments are quickly extruded, phagocytosed, or degraded by neighboring cells and do not elicit an inflammatory response. The nuclear changes are due to fragmentation of DNA into histone-sized pieces, presumably through the activation of endonucleases.

The mechanisms underlying apoptosis are the subject of extensive and evolving investigation (Fig. 11). The process may be triggered by granzymes, degradative enzymes released by cytotoxic T cells, by activation of intrinsic pathways, e.g., in embryogenesis or direct radiation injury, or by interaction of a number of related plasma membrane receptors [e.g., Fas, or the tumor necrosis factor (TNF) receptor]. The plasma membrane receptors share an intracellular "death domain" protein sequence that, when multimerized, leads to a cascade of enzyme activation culminating in cell death.

Current data suggest that the various activators of the apoptosis pathway are ultimately funneled through the synthesis and/or activation of a number of cytosolic proteases. These proteases are termed caspases because they have an active-site cysteine and cleave after aspartic acid residues. In experimental systems, overexpression of any of the caspases will result in cellular apoptosis, suggesting that under normal circumstances, they must be tightly controlled. Increased cytosolic calcium can directly activate some intracellular proteases; in addition, increased intracellular calcium induces a "permeability transition" in mitochondria, resulting in caspase-3 activation. Initial activation of one or more such enzymes with broad specificity putatively leads to a cascade of activation of other proteases, inexorably culminating in cell suicide. For example, downstream endonuclease activation results in the characteristic DNA fragmentation, whereas cell volume and shape changes are caused by breakdown of the cytoskeleton.

The general framework of cell injury is summarized in Fig. 12. In the following chapter, we will extend the concepts

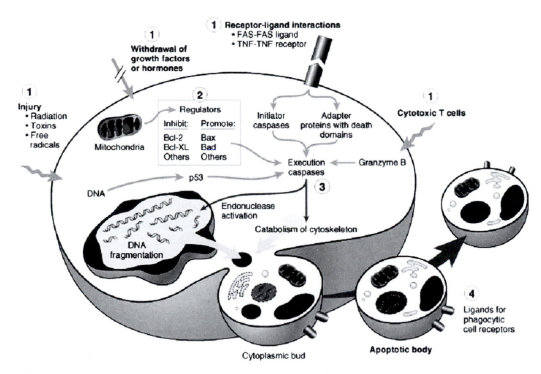

FIG. 11. Schematic of events occurring in apoptosis. Items labeled (1) are various stimuli for apoptosis; some involve direct activation of caspases (cytotoxic T cells) while others act via adaptor proteins (e.g., surface receptors such as FAS), or via mitochondrial release of cytochrome c. Items labeled (2) are an expanding set of inhibitors or promoters that fine-tune the death pathways leading to activation of the caspase mediators. Execution caspases (3) activate endonucleases that degrade nuclear chromatin, and intracellular proteases which degrade the cytoskeleton. The end result (4) is extruded apoptotic bodies containing various organelles and cytosolic components and that express a new surface molecule that induces their uptake by adjacent phagocytic cells. (Reproduced by permission from Cotran, R. S., Kumar, V., and Collins, T., 1999. *Robbins Pathologic Basis of Disease,* 6th ed. Saunders, Philadelphia.)

of structure–function correlation beyond cells to include the extracellular matrix and complex tissues, examine what happens following cell and tissue injury, and describe how histologists and pathologists examine normal and abnormal tissues.

Bibliography

Cell Biology:

Alberts, B., Johnson, A., Lewis, J., Raff, M., Roberts, K., and Walter, P. (2002). *Molecular Biology of the Cell,* 4th ed. Garland Publishing, New York.

Cooper, G. M. (2000). *The Cell: A Molecular Approach,* 2nd ed. Sinauer Associates, Sunderland, MA.

Lodish, H., Berk, A., Zipursky, S.L., Matsudaira, P., Baltimore, D., and Darnell, J. (2000). *Molecular Cell Biology,* 4th ed. W. H. Freeman, New York.

Cell Injury:

Cotran, R. S., Kumar, V., and Collins, T. (1999). *Robbins Pathologic Basis of Disease,* 6th ed. W. B. Saunders, Philadelphia.

Farber, J. L. (1994). Mechanisms of cell injury by activated oxygen species. *Environ. Health Perspect.* **102** (Suppl 10): 17–24.

Granville, D. J., Carthy, C. M., Hunt, D. W. C., and McManus, B. M. (1998). Apoptosis: molecular aspects of cell death and disease. *Lab. Invest.* **78**: 893–913.

Guo, M., and Hay, B. A. (1999). Cell proliferation and apoptosis. *Curr. Opin. Cell Biol.* **11**: 745–752.

Knight, J. A. (1995). Diseases related to oxygen-derived free radicals. *Ann. Clin. Lab. Sci.* **25**: 111–121.

Kroemer, G., Zamzami, N., and Susin, S. A. (1997). Mitochondrial control of apoptosis. *Immunol. Today* **18**: 44–51.

Lee, J. M., Zipfel, G. J., and Choi, D. W. (1999). The changing landscape of ischaemic brain injury mechanisms. *Nature* **399**: A7.

Lo, E. H., Dalkara, T., and Moskowitz, M. A. (2003). Mechanisms, challenges and opportunities in stroke. *Nat. Rev. Neurosci.* **4**: 399–415.

Majno, G., and Joris, I. (1995). Apoptosis, oncosis, and necrosis; an overview of cell death. *Am. J. Pathol.* **146**: 3–15.

Nadal-Ginard, B., Kajstura, J., Leri, A., and Anversa, P. (2003). Myocyte death, growth, and regeneration in cardiac hypertrophy and failure. *Circ. Res.* **92**: 139.

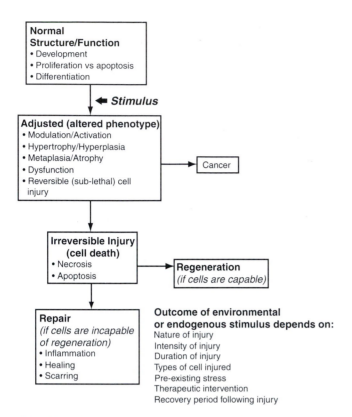

FIG. 12. Cellular mechanisms of disease, emphasizing the general concepts of activation and other phenotypic alterations, reversible and irreversible cell injury, and the possible outcomes of cell injury.

Nagata, S. (1997). Apoptosis by death factor. *Cell* **88**: 355–365.

Raff, M. (1998). Cell suicide for beginners. *Nature* **396**: 119–122.

Richter, C., Gogvadze, V., Laffranchi, R., Schlapbach, R., Schweitzer, M., Suter, M., Walter, P., and Yaffee, M. (1995). Oxidants in mitochondria: from physiology to diseases. *Biochim. Biophys. Acta* **1271**: 67–74.

Riley, P. A. (1994). Free radicals in biology: oxidative stress and the effects of ionizing radiation. *Int. J. Radiation Biol.* **65**: 27–33.

Taubes, G. (1996). Misfolding the way to disease. *Science* **271**: 1493–1495.

Trump, B. F., and Berezesky, I. K. (1995). Calcium-mediated cell injury and cell death. *FASEB J.* **9**: 219–228.

Zheng, Z., Lee, J. E., and Yanari, M. A. (2003). Stroke: Molecular mechanisms and potential targets for treatment. *Curr. Mol. Med.* **3**: 361–372.

3.4 TISSUES, THE EXTRACELLULAR MATRIX, AND CELL–BIOMATERIAL INTERACTIONS

Frederick J. Schoen and Richard N. Mitchell

This chapter will extend the concepts discussed previously in Cells and Cell Injury to describe how cells are organized to form specialized tissues and organs. Particularly important will be understanding the role of cell interactions with extracellular matrix, the material that surrounds cells and support tissues. The chapter will also show how environmental stimuli and injury affect tissue structure and function, and how tissues respond to various insults, including those related to the insertion of biomaterials. Four general areas will be covered:

1. Key principles governing the structure and function of normal tissues and organs
2. Basic processes leading to and resulting from abnormal (injured or diseased) tissues and organs
3. Key concepts in cell–biomaterials interactions
4. Approaches to studying the structure and function of tissues

Germane to this discussion are two basic definitions: Histology is the microscopic study of tissue structure; pathology is the study of the molecular, biochemical, and structural alterations and their consequences in diseased tissues and organs, and the underlying mechanisms that cause these changes. Some excellent general references are available (Fawcett, 1986; Cormack, 1987; Cotran *et al.*, 1999; Lodish *et al.*, 1999; Alberts *et al.*, 2002).

STRUCTURE AND FUNCTION OF NORMAL TISSUES

Biologic tissue is composed of three basic components: cells, intercellular (interstitial) substances, especially extracellular matrix, and various body fluids. Cells, the living component of the body, were discussed in detail in the previous chapter. They are surrounded by and obtain their nutrients and oxygen, including the body fluids blood, tissue fluid (known as extracellular fluid), and lymph. Blood consists of blood cells suspended in a slightly viscous fluid called plasma. Capillaries exude a clear watery liquid called tissue fluid that permeates the amorphous intercellular substances lying between capillaries and cells. More tissue fluid is produced than can be absorbed back into the capillaries; the excess is carried away as lymph by a series of vessels called lymphatics, which ultimately empty the lymph into the bloodstream.

In all tissues, cells are assembled during embryonic development into coherent groupings by virtue of specific cell–cell and cell–matrix interactions. Each type of tissue has a distinctive pattern of structural organization adapted to its particular function, which is strongly influenced by metabolic (Carmeliet, 2000) and/or mechanical factors (Ingber, 2002; Carter *et al.*, 1996).

The Need for Tissue Perfusion

Since all mammalian cells require perfusion (i.e., blood flow bringing oxygen and nutrients and carrying away wastes) for their survival, most tissues have a rich vascular network and the circulatory system is a key feature of tissue and organ structure and function (Fig. 1). Perfusion is provided by the cardiovascular system, composed of a pump (the heart), a series of distributing and collecting tubes (arteries and veins),

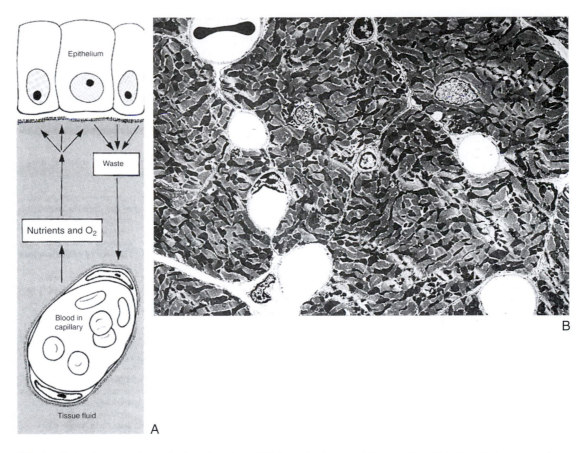

FIG. 1.   Role of the vasculature in tissue function. (A) Schematic diagram of the route by which the cells in a tissue obtain their nutrients and oxygen from underlying capillaries. Metabolic waste products pass in the reverse direction and are carried away in the blood stream. In each case, diffusion occurs through the tissue fluid that permeates the amorphous intercellular substances lying between the capillaries and the tissue cells. (B) Myocardium, a highly metabolic tissue, has a rich vascular/capillary network as demonstrated by transmission electron microscopy. Capillaries are visualized as six open spaces. A red blood cell is noted in the capillary at upper left. This photo contains nearly four complete myocyte profiles and parts of two others at the lower left and right corners. (A) Reproduced by permission from Cormack, D. H., 1987. *Ham's Histology*, 9th ed. Lippincott, Philadelphia.)

and an extensive system of thin-walled vessels (capillaries) that permit exchange of substances between the blood and the tissues. Circulation of blood transports and distributes essential substances to the tissues and removes by-products of metabolism. Implicit in these functions are the intrinsic capabilities of the cardiovascular system to buffer pulsatile flow in order to ensure steady flow in the capillaries, regulate blood pressure and volume at all levels of the vasculature, maintain circulatory continuity while permitting free exchange between capillaries/venules and the extravascular compartments, and control hemostasis (managing hemorrhage by a coordinated response of vasoconstriction and plugging of vascular defects by coagulation and platelet clumps). Other functions of the circulation include such homeostatic (control) mechanisms as regulation of body temperature, and distribution of various regulating substances (e.g., hormones, inflammatory mediators, growth factors). Moreover, the circulatory system distributes immune and inflammatory cells to their sites of action and the endothelium itself has important immunological and inflammatory functions (Schoen and Cotran, 1999).

Although the cardiac output is intermittent owing to the cyclical nature of the pumping of the heart, continuous flow to the periphery occurs by virtue of distention of the aorta and its branches during ventricular contraction (systole), and elastic recoil of the walls of the large arteries with forward propulsion of the blood during ventricular relaxation (diastole). Blood moves rapidly through the aorta and its arterial branches, which become narrower and whose walls become thinner and change histologically toward the peripheral tissues. By adjusting the degree of contraction of their circular muscle coats, the distal arteries (arterioles) control the distribution of tissue blood flow among the various capillary beds and permit regulation of blood pressure. Blood returns to the heart from the capillaries, the smallest and thinnest-walled vessels, by passing through venules and then through veins of increasing size.

Blood entering the right ventricle of the heart via the right atrium is pumped through the pulmonary arterial system at mean pressures about one-sixth of those developed in the systemic arteries. The blood then passes through the lung capillaries in the alveolar walls, where carbon dioxide is released to and oxygen taken up from the alveoli. The oxygen-rich blood returns through the pulmonary veins to the left atrium and ventricle to complete the cycle.

Large-diameter blood vessels are effective in delivering blood. Small vessels are most effective in diffusional transport to and from the surrounding tissues. Thus, owing to the very thin walls and slow velocity of blood in the capillaries, which falls to approximately 0.1 cm/sec from 50 cm/sec in the aorta, most exchange of oxygen, nutrients, and cellular wastes takes place through capillaries. Owing to the diffusion limit of oxygen of 100 to 200 μm in most highly metabolic tissues, cells are generally located no more than that distance from capillaries (recall Fig. 1). Thus, three-dimensional tissue formation and growth requires the formation of new blood vessels, a process called angiogenesis (Carmeliet, 2003). It also follows that tissues that require less nutrition and those that are relatively thin (e.g., heart valve leaflets) may either require a sparse vascular network or none at all.

In the following sections we will cover the general functional principles of tissue organization and response to various types of injury, highlighted by specific examples where illustrative.

Extracellular Matrix

Extracellular matrix (ECM) comprises the biological material produced by, residing in between, and supporting cells. ECM, cells, and capillaries are physically integrated in functional tissues (Fig. 2; see also Fig. 5). The ECM holds cells together by providing physical support and a matrix to which cells can adhere, signal each other, and interact. During normal development and as a component of the response of tissues to injury, adhesive interactions coordinate interactions with cell-surface receptors and subsequently, the cytoskeleton and the nucleus (Bokel and Brown, 2002). The resultant intracellular signaling affects a variety of events including gene expression and cell proliferation, mobility, and differentiation.

ECM consists of large molecules synthesized by cells, exported to the intercellular space and linked together into a structurally supportive composite. ECM is composed of (1) fibers (collagen and elastin) and (2) a largely amorphous interfibrillary matrix (mainly proteoglycans, noncollagenous cell-binding adhesive glycoproteins, solutes, and water). The principal functions of the ECM are:

- Mechanical support for cell anchorage
- Determination of cell orientation
- Control of cell growth
- Maintenance of cell differentiation
- Scaffolding for orderly tissue renewal
- Establishment of tissue microenvironment
- Sequestration, storage, and presentation of soluble regulatory molecules

Some extracellular matrices are specialized for a particular function, such as strength (tendon), filtration (the basement membranes in the kidney glomerulus), or adhesion (basement membranes supporting most epithelia). To produce additional mechanical strength in bones and teeth, the ECM is calcified.

Even in a tissue as "simple" as a heart valve leaflet, the coordinated interplay of several ECM components is critical to function (Schoen, 1997).

ECM components are synthesized, secreted, and remodeled by cells in response to appropriate stimuli. Virtually all cells secrete and degrade ECM to some extent. Certain cell types (e.g., fibroblasts and smooth muscle cells), are particularly active in production of interstitial ECM (i.e., the ECM between cells). Epithelial cells also synthesize the ECM of their basement membranes (see Fig. 5D).

Matrix components and the mechanical forces that cells experience markedly influence the maintenance of cellular phenotypes and affect cell shape, polarity, and differentiated function though receptors for specific ECM molecules on cell surfaces (such as integrins). The resultant changes in cytoskeletal organization and in production of second messengers can modify gene expression. ECM plays a critical role in cytodifferentiation and organogenesis, and as a scaffold allowing orderly repair following injury. The reciprocal instructions between cells and ECM are termed dynamic reciprocity.

In most tissues, the ECM is constantly turning over and being remodeled. Although matrix turnover is generally quite low in normal mature tissues, rapid and extensive proteolytic destruction characterizes various adaptive and pathologic states and accommodates changes in tissue form during embryogenesis. Regulated remodeling of ECM occurs during wound repair, tumor cell invasion, and metastasis, and at specific sites of embryonic tissue morphogenesis. The processes of directed cell motility and invasion require highly localized proteolytic reactions to carve through dense, structural areas of tissue. Two main classes of enzyme that have been implicated in the degradation of the protein components of the ECM are the plasminogen activator/plasmin family and the metalloproteinase family (Mutsaers et al., 1997; Shapiro, 1998).

ECM consists of large molecules interlinked to form a reticulum that schematically resembles a fiber-reinforced composite; in reality, ECM forms an expansile glycoprotein–water gel held in dynamic equilibrium by fibrillar proteins. Present to some degree in all tissues and particularly abundant as an intercellular substance in connective tissues, ECM has both fibrous and amorphous components. The key constituents of ECM include fibrillar proteins such as collagen and elastin, amorphous matrix components exemplified by glycosaminoglycans (GAGs) and proteoglycans, and adhesive proteins such as fibronectin and laminin (Fig. 3).

Collagens and Elastin

Collagen comprises a family of closely related but genetically, biochemically, and functionally distinct molecules, which are responsible for tissue tensile strength. The most common protein in the animal world, collagen provides the extracellular framework for all multicellular organisms. The collagens are composed of a triple helix of three polypeptide α chains; about 30 different α chains form at nearly 20 distinct collagen types. Types I, II, and III are the interstitial or fibrillar collagens and are the most abundant. Types IV, V, and VI are nonfibrillar

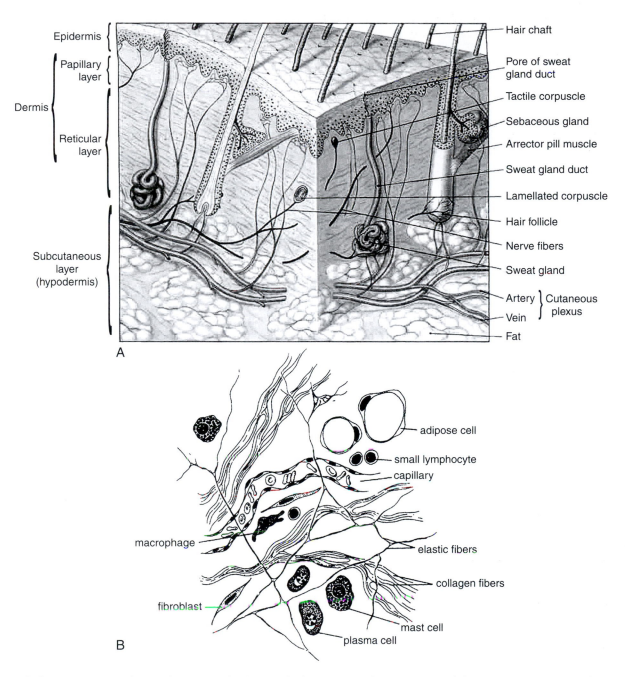

FIG. 2. Organization of tissue demonstrated at low and high power. (A) The components of the integumentary system (skin) demonstrating relationships among the major components. The entire diagram represents approximately 1 mm. (B) Diagrammatic representation of the cells and fibers of loose connective tissue. These microscopically identifiable components lie embedded in amorphous ground substance and are continuously bathed with tissue fluid produced by capillaries. The entire diagram represents approximately 100 μm. (A, Reproduced by permission from Martini, F. H., 2001. *Fundamentals of Anatomy and Physiology*. Prentice Hall, Upper Saddle River, NJ. B, Reproduced by permission from Cormack, D. H., 1987. *Ham's Histology*, 9th ed. Lippincott, Philadelphia.)

(or amorphous) and are present in interstitial tissue and basement membranes.

Since some individual structural components of the ECM such as collagen are substantially larger than the cells that produce them, they must be synthesized in discrete protein subunits that are secreted into the extracellular environment and are self-assembled there. For example, collagens are synthesized as soluble procollagen precursors, which are secreted and proteolytically processed to mature insoluble collagen molecules in the extracellular space. The main steps in collagen synthesis are

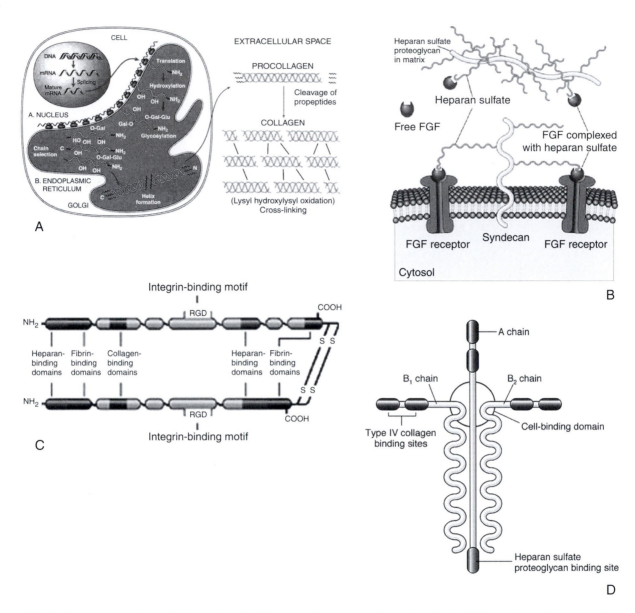

FIG. 3. Key concepts of extracellular matrix. (A) Collagen synthesis (see text for explanation). (B) Proteoglycans. Heparan sulfate proteoglycan in matrix and syndecan, cell surface proteoglycan. Its core protein spans the plasma membrane and can modulate the activity of fibroblast growth factor (FGF). The fibronectin molecule (C) consists of a dimmer held together by S–S bonds. Note the various domains that bind to extracellular matrix and the cell-binding domain containing an arginine-glycine-aspartic acid (RGD) sequence. The cross-shaped laminin (D) molecule spans basement membranes and has extracellular matrix (ECM)- and cell-binding domains. (Reproduced by permission from Cotran, R. S., Kumar, V., and Collins, T., 1999. *Robbins Pathologic Basis of Disease*, 6th ed. Saunders, Philadelphia.)

shown in Fig. 3A. After synthesis on ribosomes, the α chains are subjected to a number of enzymatic modifications, including hydroxylation of proline and lysine residues, providing collagen with a high content of hydroxyproline (10%). Vitamin C is required for hydroxylation of the collagen propeptide, a requirement that explains inadequate wound healing in vitamin C deficiency (scurvy). After the modifications, the procollagen chains align and form the triple helix. At this stage, the procollagen molecule is still soluble and contains N-terminal and C-terminal propeptides. During or shortly after

secretion from the cell, procollagen peptidases clip the terminal propeptide chains, promoting formation of fibrils, often called tropocollagen, and oxidation of specific lysine and hydroxylysine residues occurs by the extracellular enzyme lysyl oxidase. This results in cross-linkages between α chains of adjacent molecules stabilizing the array that is characteristic of collagen.

Elastic fibers confer passive recoil to various tissues; they are critical components of heart valves and the aorta, where repeated pulsatile flow would cause unacceptable shears on noncompliant tissue, and of intervertebral disks, where the repetitive forces of ambulation along the spine are dissipated. Elastin also forms layers called laminae in the walls of arteries. Unlike collagen, elastin can be stretched. The stretching of an artery every time the heart pumps blood into an artery is followed by the recoil of elastin, which restores the artery's former diameter between heartbeats.

Amorphous Matrix: Glycosaminoglycans (GAGs), Proteoglycans and Hyaluronan

Amorphous intercellular substances contain carbohydrate bound to protein. The carbohydrate is in the form of long-chained polysaccharides called glycosaminoglycans (GAGs). When GAGs are covalently bound to proteins, the molecules are called proteoglycans. GAGs are highly charged (usually sulfated) polysaccharide chains up to 200 sugars long, composed of repeating unbranched disaccharide units (one of which is always an amino sugar—hence the name glycosaminoglycan). GAGs are divided into four major groups on the basis of their sugar residues:

- Hyaluronic acid: a component of loose connective tissue and of joint fluid, where it acts as a lubricant
- Chondroitin sulfate and dermatan sulfate
- Heparan sulfate and heparin
- Keratin sulfate

With the exception of hyaluronic acid (which is unique among the GAGs because it is not sulfated), all GAGs are covalently attached to a protein backbone to form proteoglycans, with a structure that schematically resembles a bottle brush (Fig. 3B). Proteoglycans are remarkable in their diversity, owing to different core proteins, and different glycosaminoglycans. Proteoglycans are named according to the structure of their principal repeating disaccharide. Some of the most common are heparan sulfate, chondroitin sulfate, and dermatan sulfate. Proteoglycans can also be integral membrane proteins and are thus modulators of cell growth and differentiation. The syndecan family has a core protein that spans the plasma membrane and contains a short cytosolic domain as well as a long external domain to which a small number of heparan sulfate chains are attached. Syndecan binds collagen, fibronectin, and thrombospondin in the ECM and can modulate the activity of growth factors. Hyaluronan consists of many repeats of a simple disaccharide stretched end-to-end and binds a large amount of water, forming a viscous hydrate gel, which gives connective tissue turgor pressure and an ability to resist compression factors. This ECM component helps provide resilience as well as a lubricating feature to many types of connective tissue, notably that found in the cartilage in joints.

Adaptor/Adhesive Molecules

Adhesive proteins, including fibronectin, laminin, and entactin permit the attachment to and movement of cells within the ECM.

- Fibronectin is a ubiquitous, multidomain glycoprotein possessing binding sites for a wide variety of other ECM components, including collagen, heparins A and B, fibrin, and chondroitin sulfate (Fig. 3C). It is synthesized by many different cell types, with the circulating form produced mainly by hepatocytes. Fibronectin is important for linking cells to the ECM via cell surface integrins; the adaptor molecules. Fibronectin's adhesive character also makes it a crucial component of blood clots and of pathways followed by migrating cells. Thus, fibronectin-rich pathways guide and promote the migration of many kinds of cells during embryonic development and wound healing.
- Laminin is an extremely abundant component of the basal lamina, a tough, thin, sheetlike substratum on which cells sit and important for cell differentiation and tissue remodeling. The basal lamina or basement membrane contains a meshlike type IV collagen framework, laminin, and heparan sulfate proteoglycan; laminin facilitates cell binding to the basal lamina. Laminin polypeptides are arranged in the form of an elongated cross, with individual chains held together by disulfide bonds (Fig. 3D). Like fibronectin, laminin has a distinct domain structure; different regions of the molecule bind to type IV collagen, heparin sulfate, entactin (a short protein that cross-links each laminin molecule to type IV collagen), and cell surface integrins.

Cell–Matrix Interactions

Like cell–cell interactions, cell–matrix interactions have a high degree of specificity, requiring initial recognition, physical adhesion, electrical and chemical communication, cytoskeletal reorganization, and/or cell migration. Moreover, adhesion receptors may also act as transmembrane signaling molecules that transmit information about the environment to the inside of cells and mediate the effects of signals initiated by growth factors or compounds controlling tissue differentiation (Fig. 4). Moreover, the components of the extracellular matrix (ligands) with which cells interact are immobilized and not in solution. However, soluble (secreted) factors also modulate cell–cell communication in the normal and pathologic regulation of tissue growth and maturation. Cell surface adhesion molecules that interact with ECM include the integrin adhesion receptors, and the vascular selectins.

The integrins comprise a family of cell receptors with diverse specificity that bind ECM proteins, other cell surface proteins and plasma proteins, and control cell growth, differentiation, gene expression, and motility (Bokel and Brown, 2002). Some integrins bind only a single component of the ECM, e.g., fibronectin, collagen, or laminin (see above). Other integrins can interact with several of these polypeptides. In contrast to hormone receptors, which have high affinity and low abundance, the integrins exhibit low affinity and high abundance, so that they can bind weakly to several different but related matrix molecules. This property allows the integrins to promote cell–cell interactions as well as cell–matrix binding.

Basic Tissues

Humans have more than 100 distinctly different types of cells variously allocated to four types of basic tissues (Table 1

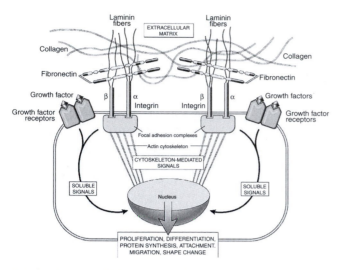

FIG. 4. Integrin ECM interaction. Schematic showing the mechanisms by which ECM (e.g., fibronectin and laminin) and growth factors can influence cell growth, motility, differentiation, and protein synthesis. Integrins bind ECM and interact with the cytoskeleton at focal adhesion complexes (protein aggregates that include vinculin, α-actinin, and talin). This can initiate the production of intracellular messengers, or can directly mediate nuclear signals. Cell surface receptors for growth factors also initiate second signals. Collectively, these are integrated by the cell to yield various responses, including changes in cell growth, locomotion, and differentiation. (Reproduced by permission from Cotran, R. S., Kumar, V., and Collins, T., 1999. *Robbins Pathologic Basis of Disease,* 6th ed. Saunders, Philadelphia.)

TABLE 1 The Basic Tissues: Classification and Examples

Basic Tissues	Examples
Epithelial tissue	
Surface	Skin epidermis, gut mucosa
Glandular	Thyroid follicles, pancreatic acini
Special	Retinal or olfactory epithelium
Connective tissue	
Connective tissue proper	
Loose	Skin dermis
Dense (regular, irregular)	Pericardium, tendon
Special	Adipose tissue
Hemopoietic tissue, blood and lymph	Bone marrow, blood cells
Supportive tissue	Cartilage, bone
Muscle tissue	
Smooth	Arterial or gut smooth muscle
Skeletal	Limb musculature, diaphragm
Cardiac muscle	Heart
Nerve tissue	Brain cells, peripheral nerve

and Fig. 5): (1) epithelium, (2) connective tissue, (3) muscle, and (4) nervous tissue. The basic tissues play specific functional roles and have distinctive microscopic appearances. They have their origins in embryological development; early events include the formation of a tube with three layers in its wall: (1) an outer layer of ectoderm, (2) an inner layer of endoderm, and (3) a middle layer of mesoderm (Fig. 6).

Epithelium covers the internal and external body surfaces. It provides a protective barrier (e.g., skin epidermis), on an absorptive surface (e.g., gut lining), and can generate internal and external secretions (e.g., endocrine and sweat glands, respectively). Epithelium derives mostly from ectoderm and endoderm, but also from mesoderm.

Epithelia accommodate diverse functions. An epithelial surface can be (1) a protective dry, cutaneous membrane (as in skin); (2) a moist, mucous membrane, lubricated by glandular secretions (digestive and respiratory tracts); (3) a moist, membrane lined by mesothelium, lubricated by fluid that derives from blood plasma (peritoneum, pleura, pericardium); and (4) the inner lining of the circulatory system, called endothelium. Epithelial cells play fundamental roles in the directional movement of ions, water, and macromolecules between biological compartments, including absorption, secretion, and exchange. Therefore the architectural and functional organization of epithelial cells includes structurally, biochemically, and physiologically distinct plasma membrane domains that contain region-specific ion channels, transport proteins, enzymes and lipids, and cell–cell junctional complexes. These integrate multiple cells to form an interface between biological compartments in organs. Subcellular epithelial specializations are not apparent to the naked eye—or even necessary light microscopy; they are perhaps best studied by transmission electron microscopy (TEM) and by functional assays (e.g., assessing synthetic products, permeability and transport).

Supporting the other tissues of the body, connective tissue arises from mesenchyme, a derivative of mesoderm. Connective tissue also serves as a scaffold for the nerves and blood vessels that support the various epithelial tissues. Other types of tissue with varying functions are also of mesenchymal origin. These include dense connective tissue, adipose (fat) tissue, cartilage and bone, and circulating cells (blood cells and their precursors in bone marrow), as well as inflammatory cells that defend the body against infectious organisms and other foreign agents.

Muscle cells develop from mesoderm and are highly specialized for contraction. They have the contractile proteins actin and myosin in varying amounts and configuration, depending on cell function. Muscle cells are of three types: smooth muscle, skeletal muscle, and cardiac muscle. The latter two have a striated microscopic appearance, owing to their discrete bundles of actin and myosin organized into sarcomeres. Smooth muscle cells, which have a less compact arrangement of myofilaments, are prevalent in the walls of blood vessels and the gastrointestinal tract. Their slow, nonvoluntary contraction regulates blood vessel caliber and proper movement of food and solid waste, respectively.

Nerve tissue, which derives from ectoderm, is highly specialized with respect to irritability and conduction. Nerve cells not only have cell membranes that generate electrical signals called action potentials, but also secrete neurotransmitters, molecules that trigger adjacent nerve or muscle cells to either transmit an impulse or to contract.

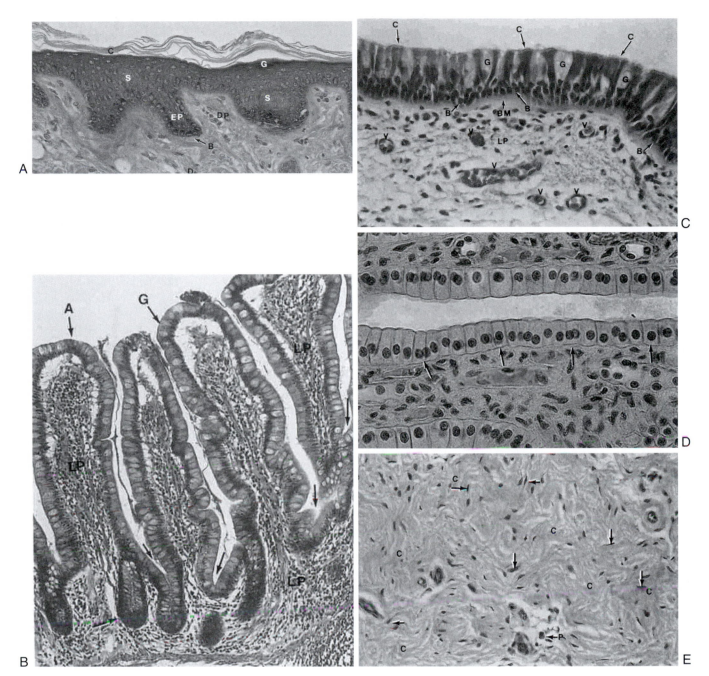

FIG. 5.   Photomicrographs of basic tissues, emphasizing key structural features. (A–D) Epithelium; (E, F) connective tissue; (G) muscle; and (H) nervous tissue. (A) Skin. Note the thin stratum corneum (c) and stratum granulosum (g). Also shown are the stratum spinosum (s), stratum basale (b), epidermal pegs (ep), dermal papilla (dp), and dermis (d). (B) Trachea, showing goblet cells (g), ciliated columnar cells (C), and basal cells (B). Note the thick basement membrane (bm) and numerous blood vessels (v) in the lamina propria (lp). (C) Mucosa of the small intestine (ileum). Note the goblet (g) and columnar absorbing (a) cells, the lamina propria (lp), muscularis mucosae (mm), and crypts (arrows). (D) Epithelium of a kidney collecting duct resting on a thin basement membrane (arrows). (E) Dense irregular connective tissue. Note the wavy unorientated collagen bundles (c) and fibroblasts (arrows). p, plasma cells. (F) Cancellous bone clearly illustrating the morphologic difference between inactive bone lining (endosteal, osteoprogenitor) cells (bl) and active osteoblasts (ob). The clear area between the osteoblasts and calcified bone represents unmineralized matrix or osteoid. cl, cement lines; o, osteocycles. (G) Myocardium (cardiac muscle). The key features are the centrally placed nuclei of the cardiac myocytes, intercalated discs (representing specialized end-to-end junctions of adjoining cells) and the sarcomeric structure visible as cross-striations. (H) Small nerve fascicles (n) with perineurium (p) separating it from two other fascicles (n). (A–F and H reproduced by permission from Berman, I., 1993. *Color Atlas of Basic Histology.* Appleton and Lange, 1993. G reproduced by permission from Schoen F. J., The heart. in *Robbins Pathologic Basis of Disease*, 7th ed., V. Kumar, N. Fausto, and A. Abbas, eds. Saunders, Philadelphia, in press.) (See color plate)

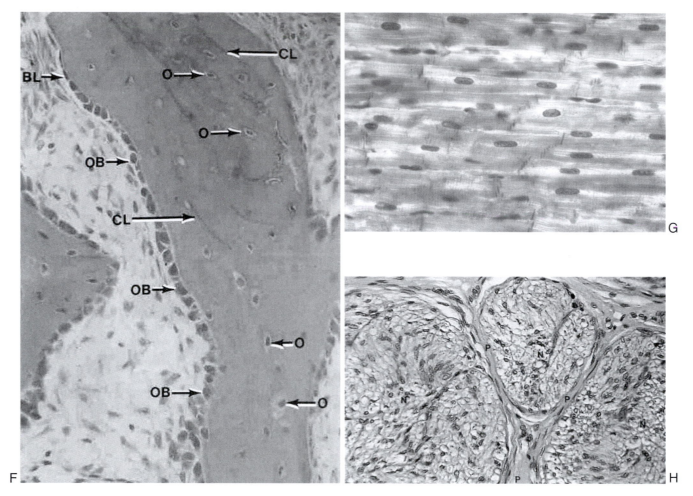

FIG. 5. *Continued*

Organs

Several different types of tissues arranged into a functional unit constitute an organ. These have a composite structure in which epithelial cells typically perform the specialized work of the organ, while connective tissue and blood vessels support and provide nourishment to the epithelium. There are two basic organ patterns: tubular (or hollow) and compact (or solid) organs. Tubular organs include the blood vessels and the digestive, urinary–genital, and respiratory tracts; they have similar architectures in that each is composed of layers of tissue arranged in a specific sequence. For example, each has an inner coat consisting of a lining of epithelium, a middle coat consisting of layers of muscle (usually smooth muscle) and connective tissue, and an external coat consisting of connective tissue and often covered by epithelium for example the intestines or vascular walls (Fig. 7). Specific variations reflect organ-specific functional requirements. Whereas the outer coat of an organ that blends into surrounding structures is called the adventitia, the outside epithelial lining of an organ suspended in a body cavity is called a serosa.

The histologic composition and organization, as well as the thickness, of these three layers vary characteristically with the physiologic functions performed by specific segments of the cardiovascular system and are particularly well exemplified in the circulation (Fig. 8). Blood vessels have three layers: an intima (primarily endothelium), a media (primarily smooth muscle and elastin), and an adventitia (primarily collagen). The amounts and relative proportions of layers of the blood vessels are influenced by mechanical factors (especially blood pressure, which determines the amount and arrangement of muscular tissue) and metabolic factors (reflecting the local nutritional needs of the tissues). Three features will illustrate the variation in site-specific structure–function correlations:

- As discussed earlier, capillaries have a structural reduction of the vascular wall to only endothelium and minimal supporting structures to facilitate exchange. Thus, capillaries are part of the tissues they supply and, unlike larger vessels, do not appear as a separate anatomic unit.
- Arteries and veins have distinctive structures. The arterial wall is generally thicker than the venous wall, in order to withstand the higher blood pressures that prevail within arteries compared with veins. Thickness of the arterial wall gradually diminishes as the vessels themselves become smaller, but the wall-to-lumen ratio becomes

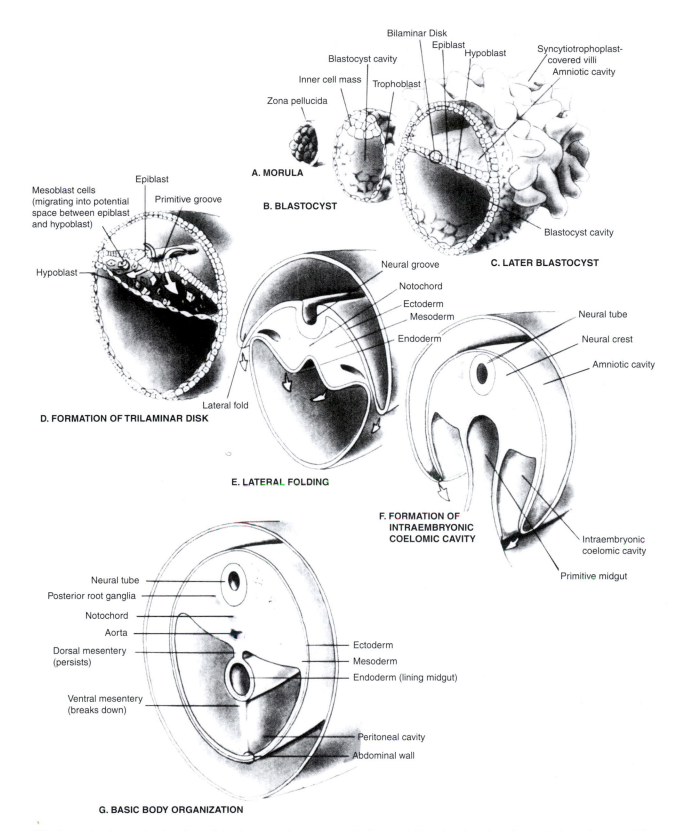

FIG. 6. Early phases of embryological development, demonstrating both essential layering that gives rise to the basic tissues and the early derivation of tubular structures. (Reproduced by permission from Cormack, D. H., 1987. *Ham's Histology*, 9th ed. Lippincott.)

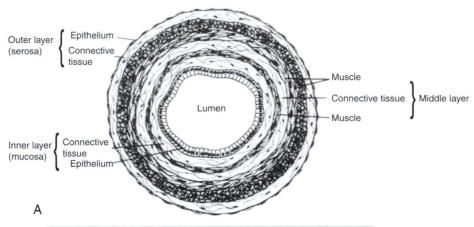

Outer layer (serosa) { Epithelium / Connective tissue

Muscle

Connective tissue } Middle layer

Muscle

Lumen

Inner layer (mucosa) { Connective tissue / Epithelium

A

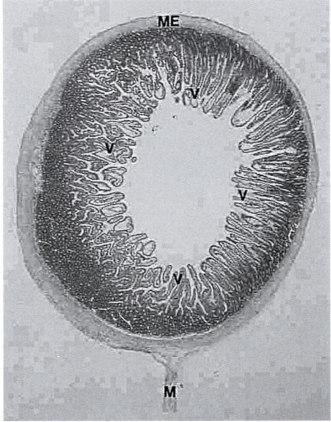

B

FIG. 7. (A) Organization of tissue layers in the digestive tract (e.g., stomach or intestines). (B) Photomicrograph of the dog jejunum illustrating villi (v), the muscularis external (me), and mesentery (m). In this organ the epithelium is organized into folds (the villi) in order to increase the surface area for absorption. (A, Reproduced by permission from Borysenko, M., and Beringer, T., *Functional Histology*, 3rd ed. Copyright 1989 Little, Brown, and Co. B, Reproduced by permission from Berman, I., 1993. *Color Atlas of Basic Histology*, Appleton and Lange.) (See color plate)

greater in the periphery. Veins have a larger overall diameter, a larger lumen, and a narrower wall than corresponding arteries with which they course (Fig. 8B).

• In essence, the heart is a blood vessel specialized for rhythmic contraction; its media is the myocardium, containing muscle cells (cardiac myocytes).

The blood supply of an organ comes from its outer aspect. In tubular organs, large vessels penetrate the outer coat, perpendicular to it, and give off branches that run parallel to the tissue layers (Fig. 9). These vessels divide yet again to give off penetrating branches that course through the muscular layer, and branch again in the connective tissue parallel to the layers.

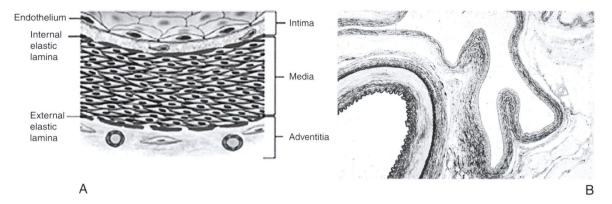

FIG. 8. The vascular wall. (A) Graphic representation of a small muscular artery (e.g., renal or coronary artery). (B) Photomicrograph of histologic section containing artery (A) and adjacent vein (V). Elastic membranes are stained black (internal elastic membrane of artery highlighted by arrow). Exposed to higher pressures, the artery has a thicker wall that maintains a circular lumen, even when blood is absent. Moreover, the elastin of the artery is more organized than in the corresponding vein. In contrast, the vein has a larger, but collapsed lumen and the elastin in its wall is diffusely distributed. (A, Reproduced by permission from Cotran, R. S., Kumar, V., and Collins, T., 1999. *Robbins Pathologic Basis of Disease*, 6th ed. Saunders, Philadelphia. B, Courtesy of Mark Flomenbaum, M.D., Ph.D., Office of the Chief Medical Examiner, New York City; reproduced by permission from Schoen, F. J., The heart, in *Robbins Pathologic Basis of Disease*, 7th ed., V. Kumar, N. Fausto, and A. Abbas, eds. Saunders, Philadelphia.)

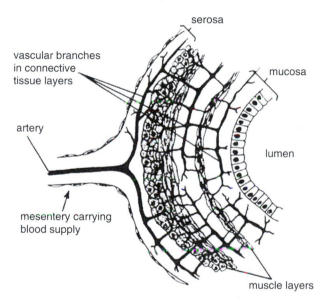

FIG. 9. Vascularization of hollow organs. (Reproduced by permission from M. Borysenko and T. Beringer, *Functional Histology*, 3rd ed. Copyright 1989, Little, Brown, and Co.)

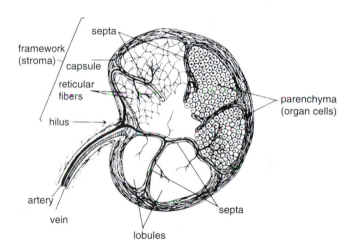

FIG. 10. Organization of compact organs. (Reproduced by permission from M. Borysenko and T. Beringer, *Functional Histology*, 3rd ed. Copyright 1989, Little, Brown, and Co.)

The small blood vessels have junctions (anastomoses) with one another in the connective tissue. These junctions may provide collateral pathways that can allow blood to bypass obstructions. Compact, solid organs have an extensive connective tissue framework, surrounded by a dense, connective tissue capsule (Fig. 10). Such organs have a hilus or area of thicker connective tissue where blood vessels and other conduits (e.g., airways in the lungs) enter the organ. From the hilus, strands of connective tissue extend into the organ and may divide it into lobules. The remainder of the organ has a delicate structural framework, including supporting cells, extracellular matrix, and vasculature (essentially the "maintenance" or "service core"), which constitutes the stroma.

The dominant cells in specialized tissues comprise the parenchyma (e.g., thyroglobulin-hormone-producing epithelial cells in the thyroid, or cardiac muscle cells in the heart). Parenchyma occurs in masses (e.g., endocrine glands), cords (e.g., liver), or tubules (e.g., kidney). Parenchymal cells can be arranged uniformly in an organ, or they may be segregated into a subcapsular region (cortex) and a deeper region (medulla), each performing a distinct functional role. In compact organs, the blood supply enters the hilus and then branches repeatedly to small arteries and ultimately capillaries in the parenchyma.

In both tubular and compact organs, veins and nerves generally follow the course of the arteries.

Parenchymal cells are generally more sensitive to chemical, physical, or ischemic (i.e., low blood flow) injury than is stroma. Moreover, when an organ is injured, orderly repair and regrowth of parenchymal cells requires an intact underlying stroma.

TISSUE RESPONSE TO INJURY

Inflammation and Repair

Inflammation and repair follow cell and tissue injury induced by various exogenous and endogenous stimuli. Inflammation is a protective response that eliminates (i.e., dilutes, destroys, or isolates) the cause of the injury (e.g., microbes or toxins) and disposes of both the necrotic cells and tissues that occur as a result of the injury. In doing so, the inflammatory response initiates the process that heals and reconstitutes the normal tissue. During the reparative phase, the injured tissue is replaced by native parenchymal cells, or by filling up the defect with fibroblastic scar tissue, or both. The outcome depends primarily on the tissue type and the extent and persistence of the injury (Fig. 11). When (1) tissue injury is transient or short-lived, (2) tissue destruction is small, and (3) the tissue is capable of regeneration, the outcome is restoration of normal structure and function; however, when the injury is extensive or occurs in tissues that do not regenerate, scarring results (Singer and Clark, 1999). An abscess is the outcome when an infection cannot be eliminated (i.e., localized collection of acute inflammation and infectious organisms); the body "controls" the infection by creating a wall around it.

Inflammation and repair constitutes an overlapping sequence of several processes (Fig. 12):

- *Acute inflammation.* The immediate and early response to injury, of relatively short duration, is characterized by fluid and plasma protein exudation into the tissue, and by accumulation of neutrophils (polymorphonuclear leukocytes).
- *Chronic inflammation.* This phase is manifested histologically as lymphocytes and macrophages, often with concurrent tissue destruction, and can evolve into repair involving fibrosis and new blood vessel proliferation. A special type of inflammation characterized by activated macrophages and often multinucleated giant cells is called a granuloma or granulomatous inflammation. The pattern occurs in pathologic states where the inciting agent is not removable, including the foreign body reaction, a characteristic inflammatory reaction to the implantation of a biomaterial.
- *Scarring.* In situations where repair cannot be accomplished by regeneration, scarring occurs as a composite of three sequential processes: (1) formation of new blood vessels (angiogenesis), (2) deposition of collagen (fibrosis), and (3) maturation and remodeling of the scar (remodeling). The early healing tissue rich in new capillaries and proliferation of fibroblasts is called granulation tissue. The essential features of the healing process are usually advanced by 4–6 weeks although full scar remodeling may require much longer.

Inflammation is also associated with the release of chemical mediators from plasma, cells, or extracellular matrix, which regulate the subsequent vascular and cellular events

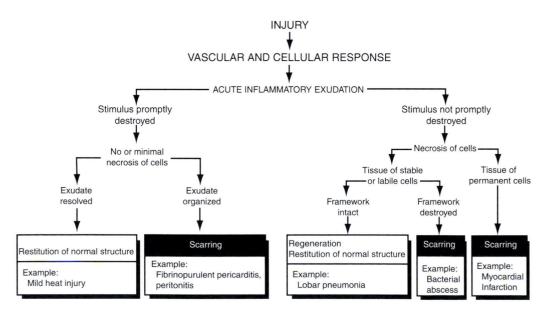

FIG. 11. Pathways of reparative responses after acute inflammatory injury. (Reproduced by permission from Cotran, R. S., Kumar, V., and Collins, T., 1999. *Robbins Pathologic Basis of Disease*, 6th ed. Saunders, Philadelphia.)

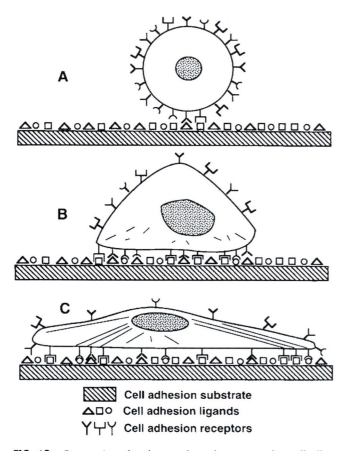

▨	Cell adhesion substrate
△□○	Cell adhesion ligands
⋎⋔⋎	Cell adhesion receptors

FIG. 12. Progression of anchorage-dependent mammalian cell adhesion. (A) Initial contact of cell with solid substrate. (B) Formation of bonds between cell surface receptors and cell adhesion ligands. (C) Cytoskeletal reorganization with progressive spreading of the cell on the substrate for increased attachment strength. (Reproduced by permission from Massia, S. P., 1999. Cell–extracellular matrix interactions relevant to vascular tissue engineering. in *Tissue Engineering of Prosthetic Vascular Grafts*, P. Zilla and H. P. Greisler, eds., RG Landes Co.)

and may modify their evolution. The chemical mediators of inflammation include the vasoactive amines (e.g., histamine), plasma proteases (of the coagulation, fibrinolytic, kinin, and complement systems), arachidonic acid metabolites (eicosinoids) produced in the cyclooxygenase pathway (the prostaglandins) and the lipoxygenase pathway (the leukotrienes), platelet-activating factor, cytokines [e.g., interleukin1 (IL-1), tumor necrosis factor-α (TNF-α), and interferon-γ (IFN-γ)], nitric oxide and oxygen-derived free radicals, and various intracellular constituents, particularly the lysosomal granules of inflammatory cells. Polypeptide growth factors also influence repair and healing by affecting cell growth, locomotion, contractility, and differentiation. Growth factors may act by endocrine (systemic), paracrine (stimulating adjacent cells) or autocrine (same cell carrying receptors for their own endogenously produced factors) mechanisms. Growth factors involved in mediating angiogenesis, fibroblast migration, proliferation, and collagen deposition in wounds include epidermal growth factor (EGF, important in proliferation of epithelial cells and fibroblasts), platelet-derived growth factor (PDGF, involved in fibroblast and smooth muscle cell migration), fibroblast growth factors (FGFs), transforming growth factor-β (TGF-β, with a central role in fibrosis), and vascular endothelial growth factor (VEGF, with a central role in angiogenesis).

Cell Regenerative Capacity

Most types of cell populations can undergo turnover, but the process is highly regulated, and the production of cells of a particular kind generally ceases until some are damaged or another need arises. Rates of proliferation are different among various cell populations and are frequently divided into three categories: (1) renewing (also called labile) cells have continuous turnover, with proliferation balancing cell loss that accrues by death or physiological depletion; (2) expanding (also called stable) cells, normally having a low rate of death and replication, retain the capacity to divide following stimulation; and (3) static (also called permanent) cells not only have no normal proliferation, but have lost their capacity to divide. The relative proliferative (and regenerative) capacity of various cells is summarized in Table 2.

In renewing (labile) cell populations (e.g., skin, intestinal epithelium, bone marrow), stem cells proliferate to form daughter cells that can become differentiated and repopulate the damaged cells. A particular stem cell produces many such daughter cells, and, in some cases, several different kinds of cells can arise from a common multipotential ancestor cell (e.g., bone marrow multipotential cells lead to several different types of blood cells). In epithelia, the stem cells are at the base of the tissue layer, away from the surface; differentiation and maturation occur as the cells move toward the surface. In expanding (stable) populations, cells can increase their rate of replication in response to suitable stimuli. Stable cell populations include glandular epithelial cells, liver, fibroblasts, vascular smooth muscle cells, osteoblasts, and endothelial cells. In contrast, permanent (static) cells have minimal if any normal mitotic capacity and, in general, cannot be induced to regenerate. In labile or stable populations, cells that die are generally replaced by new ones of the same kind, but more specialized (i.e., permanent) cells are generally replaced by scar. The inability to regenerate certain tissue types results in a clinically important deficit, since the function of the damaged tissue is irretrievably lost. For example, an area of heart muscle that is damaged by ischemic injury (myocardial infarction) cannot be effectively replaced by viable cells; the necrotic area is repaired by scar, which itself has no contractile potential. Therefore, the remainder of the heart muscle must assume the workload of the lost tissue.

Although the classical concepts just enumerated continue to hold true from a practical standpoint, recent evidence suggests that some regeneration of neural tissue and heart muscle cells can occur under certain circumstances following injury. Both the extent to which this can occur and strategies to harness this potential and exciting source of new tissue are yet unknown (Nadal-Ginard *et al.*, 2003; Nadareishvili and Hallenbech, 2003; Kokaia and Lindvall, 2003).

TABLE 2 Regenerative Capacity of Cells Following Injury

Category	Normal rate of replication	Response to stimulus/injury	Examples
Renewing/ labile	High	Modest increase	Skin, intestinal mucosa, bone marrow
Expanding/ stable	Low	Marked increase	Endothelium, fibroblasts, liver cells
Static/ permanent	None	No replication; replacement by scar	Heart muscle cells, nerves

Extracellular Matrix Remodeling

The maintenance of the extracellular matrix requires constant collagen remodeling, itself dependent on continued collagen synthesis and collagen catabolism. Turnover of the extracellular matrix is a unique biological problem because of the high collagen content of most extracellular matrix structures and the resistance of these triple helical molecules to the action of most proteases. Connective tissue remodeling, either physiological or pathological, is in most cases a highly organized process that involves the selective action of a group of related proteases that collectively can degrade most, if not all, components of the extracellular matrix. These proteases are known as the matrix metalloproteinases (MMPs). Subclasses include the interstitial collagenases, stromelysins, and gelatinases.

Enzymes that degrade collagen are synthesized by macrophages, fibroblasts, and epithelial cells. Collagenases are specific for particular types of collagens, and many cells contain two or more different such enzymes. For example, fibroblasts synthesize a host of matrix components, as well as enzymes involved in matrix degradation, such as MMPs and serine proteases. Particularly important in tissue remodeling are myofibroblasts, a particular phenotype of cells that show both features of smooth muscle cells (contractile proteins such as α-actin) and features of fibroblasts (rough endoplasmic reticulum in which proteins are synthesized). These cells may also be responsible for the production of (and likely respond to) tissue forces during remodeling, thereby regulating the evolution of tissue structure according to mechanical requirements.

Evidence suggests that growth factors and hormones (autocrine, paracrine, and endocrine) are pivotal in orchestrating both synthesis and degradation of ECM components. Cytokines such as TGF-β, PDGF, and IL-1 clearly play an important role in the modulation of collagenase and TIMP expression. MMP enzymatic activities are regulated by tissue inhibitors of metalloproteinases (TIMPs), which are especially important during wound repair. These natural inhibitors of the MMPs are multifunctional proteins with both MMP inhibitor activity and cell growth modulating properties. Turnover of the extracellular matrix is mediated by an excess of MMP over TIMPs activity. Distortion of the balance between matrix synthesis and turnover may result in altered matrix composition and amounts.

CELL/TISSUE-BIOMATERIALS INTERACTIONS

For most applications, biomaterials are in contact with cells and tissues (hard tissue, including bone; soft tissue, including cardiovascular tissues; and blood in the case of cardiovascular implants or extracorporeal devices), often for prolonged periods. Thus, rational and sophisticated use of biomaterials and design of medical devices requires some knowledge of the general concepts concerning the interaction of cells with nonphysiological surfaces. This discussion complements that described in Chapter 9.3.

Cell interactions with the external environment are mediated by receptors in the cell membrane, which interact with proteins and other ligands that adsorb to the material surface from the surrounding plasma and other fluids (Lauffenburger and Griffith, 2001). Cell adhesion triggers multiple functional biochemical signaling pathways within the cell.

Most tissue-derived cells require attachment to a solid surface for viability, growth, migration, and differentiation. The nature of that attachment is an important regulator of those functions. Moreover, the behavior and function of adherent cells (e.g., shape, proliferation, synthetic function) depend on the characteristics of the substrate, particularly its adhesiveness.

Following contact with tissue or blood, a bare surface of a biomaterial is covered rapidly (usually in seconds) with proteins that are adsorbed from the surrounding body fluids. The chemistry of the underlying substrate (particularly as it affects wettability and surface charge) controls the nature of the adherent protein layer. For example, macrophage fusion and platelet adhesion/aggregation are strongly dependent on surface chemistry. Moreover, although cells are able to adhere, spread, and grow on bare biomaterials surfaces *in vitro*, proteins absorbed from the adjacent tissue environment or blood and/or secreted by the adherent cells themselves markedly enhance cell attachment, migration, and growth. Cell adhesion to biomaterials is mediated by cytoskeletally associated receptors in the cell membrane, which interact with cell adhesion proteins that adsorb to the material surface from the surrounding plasma and other fluids (Fig. 12) (Saltzman, 2000).

Cell binding to the extracellular matrix through specific cell–substratum contacts is critical to cell-growth control through mechanical forces mediated through associated changes in cell shape and cytoskeletal tension (Ingber, 2002). Focal adhesions are considered to represent the strongest such interactions. They comprise a complex assembly of intra- and extracellular proteins, coupled to each other through transmembrane integrins. Cell-surface integrin receptors promote cell attachment to substrates, and especially those covered with the extracellular proteins fibronectin and vibronectin. These receptors transduce biochemical signals to the nucleus by activating the same intracellular signaling pathways that are used by growth factor receptors. The more cells spread, the higher their rate of proliferation. The importance of cell spreading on their proliferation has been emphasized by experiments that used endothelial cells cultured on microfabricated substrates containing fibronectin-coated islands of various defined shapes and sizes of a micrometer scale (Fig. 13) (Chen *et al.*, 1997).

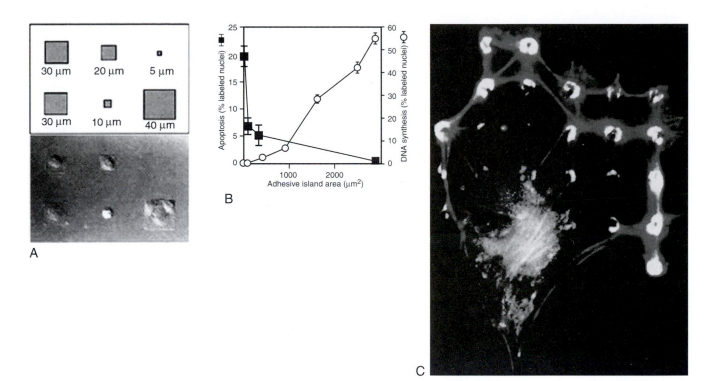

FIG. 13. Effect of spreading on cell growth and apoptosis. (A) Schematic diagram showing the initial pattern design containing different-sized square adhesive islands and Nomarski views of the final shapes of bovine adrenal capillary endothelial cells adherent to the fabricated substrate. Distances indicate lengths of the squares' sides. (B) Apoptotic index (percentage of cells exhibiting positive TUNEL staining) and DNA synthesis index (percentage of nuclei labeled with 5-bromodeoxyuridine) after 24 hours, plotted as a function of the projected cell area. Data were obtained only from islands that contained single adherent cells; similar results were obtained with circular or square islands and with human or bovine endothelial cells. (C) Fluorescence micrograph of an endothelial cell spread over a substrate containing a regular array of small (5-μm-diameter) circular ECM islands separated by nonadhesive regions created with a microcontact printing technique. Yellow rings and crescents indicate colocalization of vinculin (green) and F-actin (red) within focal adhesions that form only on the regulatory spaced circular ECM islands. (A, B, Reproduced by permission from Chen, C. S., *et al.*, 1997. Geometric control of cell life and death. *Science* **276**: 1425. C, Reproduced by permission from Ingber, D. E., 2003. Mechanosensation through integrins: Cells act locally but think globally. *Proc. Natl. Acad. Sci. USA* **100**: 1472.) (See color plate)

Cells spread to the limits of the islands containing a fibronectin substrate; cells on circular islands were circular while cells on square islands became square in shape and had 90° corners. When the spreading of the cells was restricted by small adhesive islands (10–30 μm), proliferation was arrested, whereas larger islands (80 μm) permitted proliferation. When the cells were grown on micropatterned substrates with 3- to 5-μm dots forming multiple adhesive islands that permitted the cells to extend over multiple islands while maintaining a total ECM contact similar to that of one small island (that was associated with inhibited growth), they proliferated. This confirmed that the ability to proliferate depended directly on the degree to which the cells were allowed to distend physically, and not on the actual surface area of substrate binding. Thus, cell distortion is a critical determinant of cell behavior.

Interactions of cells with ECM differ from those with soluble regulatory factors owing to the reciprocal interactions between the ECM and the cell's actin cytoskeleton (Ingber, 2003). For example, rigid substrates promote cell spreading and growth in the presence of soluble mitogens; in contrast, flexible scaffolds that cannot resist cytoskeletal forces promote cell retraction, inhibit growth, and promote differentiation. Thus, the properties of the nature and configuration of the surface-bound ECM on a substrate and the properties of the substrate itself can regulate cell–biomaterials interactions. The key concept is that a biomaterial surface can contain specific chemical and structural information that controls tissue formation, in a manner analogous to cell–cell communication and patterning during embryological development.

The exciting potential of this strategy is exemplified by tissue engineering approaches that employ biomaterials with surfaces designed to stimulate highly precise reactions with proteins and cells at the molecular level. Such materials provide the scientific foundation for molecular design of scaffolds that could be seeded with cells in vitro for subsequent implantation or specifically attract endogenous functional cells *in vivo*.

The binding domains of the extracellular matrix (ECM) environment can be mimicked by a multifunctional cell-adhesive surface created by specific proteins, peptides, and other biomolecules immobilized onto a material. For example, molecular modifications of resorbable polymer systems drive specific interactions with cell integrins (an important class of adhesion receptors that bind to ECM) and thereby direct cell proliferation, differentiation, and ECM production. The prototypical binding site present in the adhesive proteins fibronectin and vitronectin is the three amino acid sequence arginine-glycine-aspartic acid (RGD) which binds to a specific type ($\alpha_4\beta_1$) of integrin receptors on the cell surface (see Fig. 3C). Proteins (such as those with RGD sequences) that induce desirable cell behaviors (e.g., cell adhesion, spreading, and other functions) have been incorporated into biomaterials to control tissue reactions. This sequence supports the adhesion and spreading of human endothelial cells but not smooth muscle cells, fibroblasts, or blood platelets (Hubbell, 1999). Moreover, cellular responses induced can vary with the surface density of RGD peptides immobilized (Koo *et al.*, 2002). Through judicious selection of ligands, surfaces can be designed to reduce protein and cell adhesion, to prevent coagulation, to encourage endothelial cell attachment and retention, to promote capillary infiltration, and to prevent excessive smooth-muscle proliferation and collagen production. This manipulation of cell–integrin interactions with engineered ligands on synthetic biomaterials could improve function in existing applications such as the healing of vascular grafts.

A particularly exciting and active area is the use of chemically patterned surfaces to control cell behavior by creating adhesive and non-adhesive regions and perhaps even chemical gradients. By varying the size and chemistry of the various regions, and thereby the type, architecture, directional migration, and function of cells, a sort of two-dimensional organ can be grown. With photolithography, self-assembly, and other new technologies for micropatterning, the opportunity to truly engineer biological responses is emerging. Thus, the possibility emerges to design devices with surfaces that have selective cell adhesion and potentially can sort and organize a complex mixture of cells that form a specialized tissue (Fig. 14). This strategy has been used to "engineer" constructs of hepatic tissue in which hepatocytes and endothelial cells self-sort to form endothelium-lined liver cell plates (Kim *et al.*, 1998). A key challenge in tissue engineering is to understand quantitatively how cells respond to molecular signals and integrate multiple inputs to generate a given response, and to control nonspecific interactions between cells and a biomaterial, so that cell responses specifically follow desired receptor–ligand interactions. Exquisite control of scaffold architecture and overall and regional surface chemistry is now also possible; these features (and potentially the mechanical properties of the substrate) may precisely regulate cell behavior (Makohliso *et al.*, 1998; Bhatia *et al.*, 1999; Huang and Ingber, 2000).

Hepatocyte/Endothelial Cell Sorting

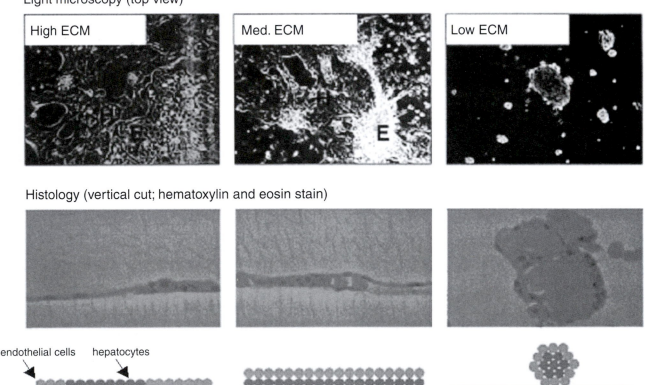

FIG. 14. Different levels of type 1 collagen coating on a culture dish result in different organization of endothelial cells and hepatocytes. High collagen levels cause both cell types to spread across the substratum (left). On intermediate collagen levels, endothelial cells form a layer on the substratum whereas hepatocytes form a layer on top of the endothelial cells (center). Low levels of collagen result in an inner layer of hepatocyte aggregate surrounded by endothelial cells (right). (Reproduced by permission from Lauffenburger, D. A., *et al.*, 2001. Who's got pull around here? Cell organization in development and tissue engineering. *Proc. Natl. Acad. Sci. USA* **98:** 4282.) (See color plate)

In addition, special relationships may accrue for biodegradable polymers, especially in tissue engineering applications, since the polymer disappears as functional tissue regenerates. Thus, polymer degradation may yield a dynamic surface whose chemistry might be unpredictable, but could possibly be manipulated to provide an additional level of control over cell interactions. Moreover, covalently immobilized growth factor, for example epidermal growth factor, can retain its biological activity, and potentially DNA delivered upon a biomaterial surface can be efficiently taken up by cells and the encoded gene expressed in a wound-healing environment (Swindle *et al.*, 2001; Richardson *et al.*, 2001).

Topography has also been studied for its effect on cell behavior, including depth and width of groove, and roughness (Von Recum and van Kooten, 1995). Surface texture influences cell behavior, including adhesion and movement attachment, spreading area, proliferation, orientation of cells to the topography, biochemical activity, and neurite (nerve) growth. Moreover, fibroblasts, neurons, and other cells will orient along fibers, ridges, and grooves with potential therapeutic application in nerve regeneration, and texture has been shown to influence macrophage spreading and fibroblast growth. This is particularly demonstrated by the correlations of tumorigenicity with substrate structural features, including roughness, perforations, and more complex surface features (see Chapter 4.7).

TECHNIQUES FOR ANALYSIS OF CELLS AND TISSUES

A number of techniques are available to observe living cells directly in culture systems; these are extremely useful in investigating the structure and functions of isolated cell types. Cells in culture (*in vitro*) often continue to perform many of the normal functions they have in the body (*in vivo*). Through measurement of changes in secreted products under different conditions, for example, culture methods can be used to study how cells respond to certain stimuli. However, since cells in culture do not have the usual intercellular organizational environment, normal physiological function may not always be present.

Techniques commonly used to study the structure of either normal or abnormal tissues, and the purpose of each mode of analysis, are summarized in Table 3 and Fig. 15. The most widely used technique for examining tissues is light microscopy, described below. Details of other useful procedures are available (Schoen, 1989).

Light Microscopy

The conventional light microscopy technique involves obtaining the tissue sample, followed by fixation, paraffin embedding, sectioning, mounting on a glass slide, staining, and examination. Photographs of conventional tissue sections taken through a light microscope (photomicrographs) were illustrated in Fig. 5. Photographs of a tissue sample, paraffin block, and resulting tissue section on a glass slide are shown

TABLE 3 Techniques for Studying Cells and Tissues[a]

Technique	Purpose
Gross examination	Overall specimen configuration; many diseases and processes can be diagnosed at this level
Light microscopy (LM)	Study overall microscopic tissue architecture and cellular structure; special stains for collagen, mucin, elastin, organisms, etc. are available.
Transmission electron microscopy (TEM)	Study ultrastructure (fine structure) and identify cells and their organelles and environment
Scanning electron microscopy (SEM)	Study topography and structure of surfaces
Enzyme histochemistry	Demonstrate the presence and location of enzymes in gross or microscopic tissue sections
Immunohistochemistry	Identify and locate specific molecules, usually proteins, for which a specific antibody is available
In situ hybridization	Localizes specific DNA or RNA in tissues to assess tissue identity or recognize a cell gene product
Microbiologic cultures	Diagnose the presence of infectious organisms
Morphometric studies (at gross, LM or TEM levels)	Quantitate the amounts, configuration, and distribution of specific structures
Chemical, biochemical, and spectroscopic analysis	Assess concentration of molecular or elemental constituents
Energy-dispersive X-ray analysis (EDXA)	Perform site-specific elemental analysis on surfaces
Autoradiography (at LM or TEM levels)	Locate the distribution of radioactive material in sections

[a]Modified by permission from F. J. Schoen, *Interventional and Surgical Cardiovascular Pathology: Clinical Correlations and Basic Principles*, W. B. Saunders, 1989.

in Fig. 16. The key processing steps are summarized in the following paragraphs.

Tissue Sample

The tissue is obtained by surgical excision (removal), biopsy (sampling), or autopsy (postmortem examination). A sharp instrument is used to remove and dissect the tissue to avoid distortion from crushing. Specimens should be placed in fixatives as soon as possible after removal.

Fixation

To preserve the structural relationships among cells, their environment, and subcellular structures in tissues, it is necessary to cross-link and preserve (i.e., fix) the tissue in a permanent state. Fixative solutions prevent degradation of the tissue

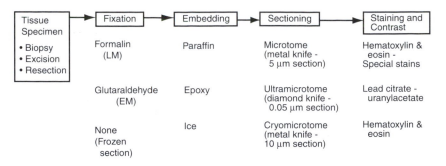

FIG. 15. Key features of tissue processing for examination by light and electron microscopy.

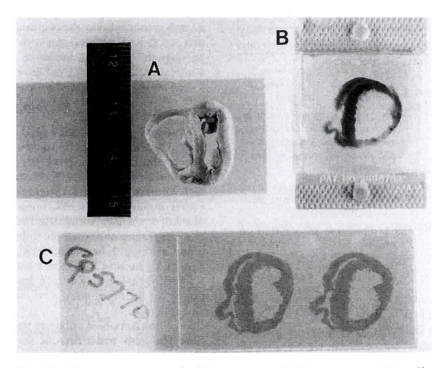

FIG. 16. Tissue processing steps for light microscopy. (A) Tissue section. (B) Paraffin block. (C) Resulting histologic section.

when it is separated from its source of oxygen and nutrition (i.e., autolysis) by coagulating (i.e., cross-linking, denaturing, and precipitating) proteins. This prevents cellular hydrolytic enzymes, which are released when cells die, from degrading tissue components and spoiling tissues for microscopic analysis. Fixation also immobilizes fats and carbohydrates, reduces or eliminates enzymic and immunological reactivity, and kills microorganisms present in tissues.

A 37% solution of formaldehyde is called formalin; thus, 10% formalin is approximately 4% formaldehyde. This solution is the routine fixative in pathology for light microscopy. For TEM and scanning electron microscopy (SEM), glutaraldehyde preserves structural elements better than formalin. Adequate fixation in formalin and/or glutaraldehyde requires tissue samples less than 1.0 and 0.1 cm, respectively, in

largest dimension. For adequate fixation, the volume of fixative into which a tissue sample is placed should generally be at least 5 to10 times the tissue volume.

Dehydration and Embedding

In order to support the specimen during sectioning, specimen water (approximately 70% of tissue mass) must be replaced by paraffin wax or other embedding medium, such as glycol methacrylate. This is done through several steps, beginning with dehydration of the specimen through increasing concentrations of ethanol (eventually to absolute). However, since alcohol is not miscible with paraffin (the final embedding medium), xylol (an organic solvent) is used as an intermediate solution.

Following dehydration, the specimen is soaked in molten paraffin and placed in a mold larger than the specimen, so that tissue spaces originally containing water, as well as a surrounding cube, are filled with wax. The mold is cooled, and the resultant solid block containing the specimen can then be easily handled.

Sectioning

Tissue specimens are sectioned on a microtome, which has a blade similar to a single-edged razor blade and is progressively advanced through the specimen block. The shavings are picked up on glass slides. Sections for light microscopic analysis must be thin enough to both transmit light and avoid superimposition of various tissue components. Typically sections are approximately 5 μm thick—slightly thicker than a human hair, but thinner than the diameter of most cells. If thinner sections are required (e.g., approximately 0.06-μm-thick ultrathin sections are necessary) for TEM analysis, a harder supporting (embedding) medium (usually epoxy plastic) and a correspondingly harder knife (usually diamond) are used. Sections for TEM analysis are cut on an ultramicrotome. Because the conventional paraffin technique requires overnight processing, frozen sections can be used to render an immediate diagnosis (e.g., during a surgical procedure that might be modified according to the diagnosis). In this method, the specimen itself is frozen, so that the solidified internal water acts as a support medium, and sections are then cut in a cryostat (i.e., a microtome in a cold chamber). Although frozen sections are extremely useful for immediate tissue examination, the quality of the appearance is inferior to that obtained by conventional fixation and embedding methods.

Staining

Tissue components have no intrinsic contrast and are of fairly uniform optical density. Therefore, in order for tissue to be visible by light microscopy, tissue elements must be distinguished by selective adsorption of dyes (Luna, 1968). Since most stains are aqueous solutions of dyes, staining requires that the paraffin in the tissue section be removed and replaced by water (rehydration). The stain used routinely in histology involves sequential incubation in the dyes hematoxylin and eosin (H&E). Hematoxylin has an alkaline (basic) pH that stains blue-purple; substances stained with hematoxylin typically have a net negative charge and are said to be "basophilic" (e.g., cell nuclei containing DNA). In contrast, substances that stain with eosin, an acidic pigment that colors positively charged tissue components pink-red, are said to be "acidophilic" or "eosinophilic" (e.g., cell cytoplasm, collagen). The tissue sections shown in Fig. 5 were stained with hematoxylin and eosin.

Special Staining

There are special staining methods for highlighting components that do not stain well with routine stains (e.g., microorganisms) or for indicating the chemical nature or the location of a specific tissue component (e.g., collagen, elastin; Table 4). There are also special techniques for demonstrating the specific chemical activity of a compound in tissues (e.g.,

TABLE 4 Stains for Light Microscopic Histology[a]

To demonstrate	Stain
Overall morphology	Hematoxylin and eosin (H & E)
Collagen	Masson's trichrome
Elastin	Verhoeff-van Gieson
Glycosoaminoglycans (GAGs)	Alcian blue
Collagen–elastic–GAGs	Movat
Bacteria	Gram
Fungi	Methenamine silver or periodic acid-Schiff (PAS)
Iron	Prussian blue
Calcium phosphates (or calcium)	von Kossa (or alizarin red)
Fibrin	Lendrum or phosphotungstic acid hematoxylin (PTAH)
Amyloid	Congo red
Inflammatory cell types	Esterases (e.g., chloroacetate esterase for neutrophils, nonspecific esterase for macrophages)

[a]Reproduced by permission from F. J. Schoen, *Interventional and Surgical Cardiovascular Pathology: Clinical Correlations and Basic Principles*, Saunders, 1989.

enzyme histochemistry). In this case, the specific substrate for the enzyme of interest is reacted with the tissue; a colored product precipitates in the tissue section at the site of the enzyme. In contrast, immunohistochemical staining takes advantage of the immunological properties (antigenicity) of a tissue component to demonstrate its nature and location by identifying sites of antibody binding. Antibodies to the particular tissue constituent are attached to a dye, usually a compound activated by a peroxidase enzyme, and reacted with a tissue section (immunoperoxidase technique), or the antibody is attached to a compound that is excited by a specific wavelength of light (immunofluorescence). Although some antigens and enzymatic activity can survive the conventional histological processing technique, both enzyme activity and immunological reactivity are often largely eliminated by routine fixation and embedding. Therefore, histochemistry and immunohistochemistry are frequently done on frozen sections; special preservation and embedding techniques now available often allow immunological methods to be carried out on carefully preserved tissue.

Electron Microscopy

Contrast in the electron microscope depends on relative electron densities of tissue components. Sections are stained with salts of heavy metals (osmium, lead, and uranium), which react differentially with different structures, creating patterns of electron density that reflect tissue and cellular architecture. An example of an electron photomicrograph is shown in Fig. 1B.

It is often possible to derive quantitative information from routine tissue sections using various manual or computer-aided methods. Morphometric or stereologic methodology, as these techniques are called, can be extremely useful in providing an objective basis for otherwise subjective measurements (Loud and Anversa, 1984).

Three-Dimensional Interpretation

Interpretation of tissue sections depends on the reconstruction of three-dimensional information from two-dimensional observations on tissue sections that are usually thinner than a single cell. Therefore, a single section may yield an unrepresentative view of the whole. A particular structure (even a very simple one) can look very different, depending on the plane of section. Figure 17 shows how multiple sections must be

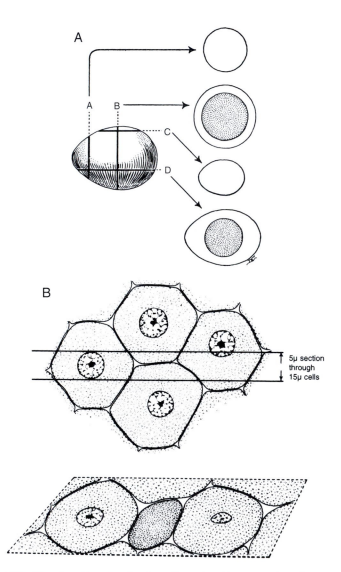

FIG. 17. Considerations for three-dimensional interpretation of two-dimensional information. Sections through a subject in different levels and orientations can give different impressions about its structure, here illustrated for a hard-boiled egg. (Modified by permission from D. H. Cormack: *Essential Histology*, Copyright 1993, Lippincott.)

examined to appreciate the actual configuration of an object or a collection of cells.

Artifacts

Artifacts are unwanted or confusing features in tissue sections that result from errors or technical difficulties in obtaining, processing, sectioning, or staining the specimen. Recognition of artifacts avoids misinterpretation. The most frequent and important artifacts are autolysis, tissue shrinkage, separation of adjacent structures, precipitates formed by poor buffering or by degradation of fixatives or stains, folds or wrinkles in the tissue sections, knife nicks, or rough handling (e.g., crushing) of the specimen.

Identification, Genotyping, and Functional Assessment of Cells, Including Synthetic Products, in Cells or Tissue Sections

It is frequently necessary to accurately ascertain or verify the identity of a cell, or to determine some aspect of its function, including the production of synthetic molecules. For such an assay, either isolated cells or whole tissues are used, depending on the objective of the study. Isolated cells or minced tissue have the advantage of allowing molecular and/or biochemical analyses on the cells and/or products, and often allow the acquisition of quantitative data. Nevertheless, the major advantage of whole-tissue preparations is the ability to spatially localize molecules of interest in the context of architectural features of the tissue.

Cellular apoptosis and proliferation can be quantified (Watanabe *et al.*, 2002). Immunohistochemical markers allow detection of proteins that are highly expressed in a tissue section. However, the relevant antibodies to proteins expressed in high concentration must be available and the expense of such studies limits their usefulness. *In situ* hybridization permits similar investigation of gene expression but, as with immunohistochemistry, only a discrete panel of previously predicted genes can be probed.

Several very exciting new and evolving techniques are available. Gene expression profiling shows the complete array of genes expressed in cells or tissues; the technology may identify pathogenetically distinct subtypes of any lesion and search for fundamental mechanisms even when candidate genes are unknown (Todd *et al.*, 2002; Bertucci *et al.*, 2003). Confocal microscopy helps localize a particular component in a living cell by observing a series of optical sections (planes) that are reconstructed into a three-dimensional image (Howell *et al.*, 2002). Tissue microassays permit the comparative examination of potentially hundreds of individual specimens in a single paraffin block. In addition, laser-assisted microdissection techniques permit isolation of individual or a homogenous population of cells on selected cell populations under direct visualization from a routine histological section of complex, heterogeneous tissue (Eltoum *et al.*, 2002). Very exciting new imaging technology, termed molecular imaging, may permit analysis of viable and *in vivo* tissues (Stephens and Allan, 2003; Weissleder and Ntziachristos, 2003; Webb *et al.*, 2000).

Bibliography

Alberts, B., Johnson, A., Lewis, J., Raff, M., Roberts, K., and Walters, P. (2002). *Molecular Biology of the Cell*, 4th ed. Garland Publishers.

Bertucci, F., Viens, P., Tagett, R., Nguyen, C., Houlgatte, R., and Birnbaum, D. (2003). DNA arrays in clinical oncology: promises and challenges. *Lab Invest.* **83:** 305–316.

Bhatia, S. N., Balis, U. J., Yarmush, M. L., and Toner, M. (1999). Effect of cell–cell interactions in preservation of cellular phenotype cocultivation of hepatocytes and nonparenchymal cells. *FASEB J.* **13:** 1883–1900.

Bokel, C., and Brown, N. H. (2002). Integrins in development: Moving on, responding to, and sticking to the extracellular matrix. *Dev. Cell* **3:** 311–321.

Carmeliet, P. (2000). Mechanisms of angiogenesis and arteriogenesis. *Nat. Med.* **6:** 389–395.

Carmeliet, P. (2003). Angiogenesis in health and disease. *Nat. Med.* **9:** 653–660.

Carter, D. R., van der Meulen, M. C. H., and Beaupre, G. S., (1996). Mechanical factors in bone growth and development. *Bone* **18:** 5S–10S.

Chen, C. S., Mrksich, M., Huang, S., Whitesides, G., and Ingber, D. E. (1997). Geometric control of cell life and death. *Science* **276:** 1425–1428.

Cormack, D. H. (1987). *Ham's Histology*, 9th ed. Lippincott, Philadelphia.

Cotran, R. S., Kumar, V., and Collins, T. (1999). *Robbins Pathologic Basis of Disease*, 6th ed. W. B. Saunders, Philadelphia.

Eltoum, I. A., Siegal, G. P., and Frost, A. R. (2002). Microdissection of histologic sections: past, present and future. *Adv. Anat. Pathol.* **9:** 316–322.

Fawcett, D. W. (1986). *Bloom and Fawcett's: A Textbook of Histology*. Saunders, Philadelphia.

Howell, K., Hopkins, N., and McLoughlin, P. (2002). Combined confocal microscopy and stereology: a highly efficient and unbiased approach to quantitative structural measurement of tissues. *Exp. Physiol.* **87:** 747–756.

Huang, S., and Ingber, D. E. (2000). Shape-dependent control of cell growth, differentiation, and apoptosis: switching between attractors in cell regulatory networks. *Exp. Cell. Res.* **261:** 91–103.

Hubbell, J. A. (1999). Bioactive biomaterials. *Curr. Opin. Biotechnol.* **10:** 123–129.

Ingber, D. E. (2002). Mechanical signaling and the cellular response to extracellular matrix in angiogenesis and cardiovascular physiology. *Circ. Res.* **91:** 877–887.

Ingber, D. E. (2003). Mechanosensation through integrins: cells act locally but think globally. *Proc. Natl. Acad. Sci. USA* **100:** 1472–1474.

Kim, S. S., Utsunomiya, H., Koski, J. A., Wu, B. M., Cima, M. J., Sohn, J., Mukai, K., Griffith, L., and Vacanti, J. P. (1998). Survival and function of hepatocytes on a novel three-dimensional synthetic biodegradable polymer scaffold with an intrinsic network of channels. *Ann. Surg.* **228:** 8–13.

Kokaia, Z., and Lindvall, O. (2003). Neurogenesis after ischaemic brain insults. *Curr. Opin. Neurobiol.* **13:** 127–132.

Koo, L. Y., Irvine, D. J., Mayes, A. M., Lauffenburger, D. A., and Griffith, L. G. (2002). Co-regulation of cell adhesion by nanoscale RGD organization and mechanical stimulus. *J. Cell Sci.* **115:** 1423–1433.

Lauffenburger, D. A., and Griffith, L. G. (2001). Who's got pull around here? Cell organization in development and tissue engineering. *Proc. Natl. Acad. Sci. USA* **98:** 4282–4284.

Lodish, H., Berk, A., Zipursky, S. L., Matsudaira, P., Baltimore, D., Darnell, J., and Zipursky, L. (1999). *Molecular Cell Biology*, 4th ed. W. H. Freeman and Co.

Loud, A. V., and Anversa, P. (1984). Morphometric analysis of biological processes. *Lab. Invest.* **50:** 250–261.

Luna, M. G. (1968). *Manual of Histologic Staining Methods of the Armed Forces Institute of Pathology*, 3rd ed. McGraw-Hill, New York.

Makohliso, S. A., Giovangrandi, L., Leonard, D., Mathieu, H. J., Llegems, M., and Aebischer, P. (1998). Application of Teflon-AF thin films for bio-patterning of neural adhesion. *Biosens. Bioelectron.* **13:** 1227–1235.

Mutsaers, S. E., Bishop, J. E., McGrouther, G., and Laurent, G. J. (1997). Mechanisms of tissue repair: from wound healing to fibrosis. *Int. J. Biochem. Cell Biol.* **29:** 5–17.

Nadal-Ginard, B., Kajstura, J., Leri, A., and Anversa, P. (2003). Myocyte death, growth, and regeneration in cardiac hypertrophy and failure. *Circ. Res.* **92:** 139–150.

Nadareishvili, Z., and Hallenbech, J. (2003). Neuronal regeneration after stroke. *N. Engl. J. Med.* **348:** 2355–2356.

Palsson, B. O., and Bhatia, S. (2003). *Tissue Engineering*. Academic Press, Boston.

Richardson, T. P., Murphy, W. L., and Mooney, D. J. (2001). Polymeric delivery of proteins and plasmid DNA for tissue engineering and gene therapy. *Crit. Rev. Eukaryot. Gene Expr.* **11:** 47–58.

Saltzman, W. M. (2000). Cell interactions with polymers. in *Principles of Tissue Engineering*, 2nd ed., R. P. Lanza, R. Langer, and J. Vacanti, eds. Academic Press, New York, pp. 221–235.

Schoen, F. J. (1989). *Interventional and Surgical Cardiovascular Pathology: Clinical Correlations and Basic Principles*. Saunders, Philadelphia.

Schoen, F. J. (1997). Aortic valve structure-function correlations: role of elastic fibers no longer a stretch of the imagination. *J. Heart Valve Dis.* **6:** 1–6.

Schoen, F. J., and Cotran, R. S. (1999). Blood vessels. in *Robbins Pathologic Basis of Disease*, 6th ed., R. S. Cotran, V. Kumar, and T. Collins, eds. W.B. Saunders, Philadelphia. pp. 493–541.

Shapiro, S. D. (1998). Matrix metalloproteinase degradation of extracellular matrix: biological consequences. *Curr. Opin. Cell Biol.* **10:** 602–608.

Singer, A. J., and Clark, R. A. (1999). Cutaneous wound healing. *N. Engl. J. Med.* **341:** 738–746.

Stephens, D. J., and Allan, V. J. (2003). Light microscopy techniques for live cell imaging. *Science* **300:** 82–86.

Swindle, G. S., Tran, K. T., Johnson, T. D., Banerjee, P., Mayes, A. M., Griffith, L., and Wells, A. (2001). Epidermal growth factor (EGF)-like repeats of human tenascin-C as ligands for EGF receptor. *J. Cell Biol.* **154:** 459–468.

Todd, R., Lingen, M. W., and Kuo, W. P. (2002). Gene expression profiling using laser capture microdissection. *Expert Rev. Mol. Diagn.* **2:** 497–507.

Von Recum, A. F., and van Kooten. (1995). The influence of microtopography on cellular response and the implication for silicone implants. *J. Biomater. Sci. Polymer Ed.* **7:** 181–198.

Watanabe, M., Hitomi, M., van der Wee, K., Rothenberg, F., Fisher, S. A., Zucl, R., Svobada K. K., Goldsmith, E. C., Heiskanen, K. M., and Nieminen, A. L. (2002). The pros and cons of apoptosis assays for use in the study of cells, tissues and organs. *Microsc. Microanal.* **8:** 375–391.

Webb, K., Hlady, V., and Tresco, P. A. (2000). Relationships among cell attachment, spreading, cytoskeletal organization, and migration rate for anchorage-dependent cells on model surfaces. *J. Biomed. Mater. Res.* **49:** 362–368.

Weissleder, R., and Ntziachristos, V. (2003). Shedding light onto live molecular targets. *Nat. Med.* **9:** 123–128.

3.5 MECHANICAL FORCES ON CELLS

Larry V. McIntire, Suzanne G. Eskin, and Andrew Yee

INTRODUCTION

Mechanical properties of materials have already been covered in Chapter 1.2 on bulk properties of materials. In this chapter, we will cover the effects of mechanical forces on cells within tissues, on the surfaces of biomaterials, or within polymer scaffolds. Because host cells interact with implanted materials, or because cells are implanted as therapeutic entities in themselves, often within a biomaterial scaffold, the response of cells to mechanical forces is important to consider in order to predict the success of an implant. Cells that are particularly adapted for functioning in concert with physical forces are those of the cardiovascular and musculoskeletal systems. In fact, the local mechanical environment may be crucial for maintenance of proper cell phenotype. Cells that make up the blood vessels are constantly subjected to blood flow and pulsatile pressure. Skeletal support tissues (bone, cartilage, tendon, and ligament) are made up of cells that withstand gravitational forces and directional stresses developed by muscle contraction.

VASCULAR CELL RESPONSES TO MECHANICAL FORCES

Mechanical forces resulting from blood flow directly affect cellular functions and thus, the physiology of the cardiovascular system. Providing insight into the cellular reactions to mechanical forces, the "response to injury" hypothesis describes how atherogenesis results from wound repair (Ross and Glomset, 1976). To repair the localized injury of the arterial wall, underlying smooth muscle cells proliferate, eventually constricting the vessel and disrupting the blood flow pattern; the resulting complex flow ultimately accelerates the disease progression, leading to potentially lethal conditions. Localization of atherosclerotic plaques in humans at bifurcations and areas in the vasculature characterized by low shear stress and often by complex or recirculating flow patterns support this hypothesis (Glagov et al., 1988). Constricted flow further affects other areas of the cardiovascular system, leading to vascular remodeling. Occlusions of supply arteries may lead to arteriogenesis, a condition in which preexisting collateral arterioles enlarge in response to increased blood flow (Carmeliet, 2000). For example, an *in vivo* study by van Gieson et al. illustrates the dilation of mesenteric microvessels in response to a rise in pressure and circumferential strain (van Gieson et al., 2003). Ligation of mesenteric arteries and veins adjacent to an implanted observation window redirected blood flow to the "collateral zone" without a change in the measured shear rate of the observed microvascular network. Accompanying the enlargement of the microvasculature, immature smooth-muscle cells surrounding the microvessels differentiate, exhibiting a quiescent, contractile phenotype. The mechanism(s) leading to this phenomenon remain to be determined.

The principles behind the effects of mechanical forces on cells also apply to vascular grafts. Regardless of the material used (autologous vein or synthetic graft), small-diameter arterial grafts (less than 6 mm) often fail within the first 5 years (Greenwald and Berry, 2000). Long-term failure of these conduits results from stress localization and flow disturbance at the anastomosis and from compliance mismatch. At anastomotic sites, flow separation, vortex formation, flow stagnation, and circumferential tension gradients occur. Moreover, intimal hyperplasia correlates with anastomotic sites, suggesting that these forces alone or in concert contribute to the eventual stenosis of vascular substitutes (Remuzzi et al., 2003). However, as observed by Mattson et al., increasing the flow rate in vascular grafts while maintaining laminar flow prevents neointimal thickening in the central region of the graft where flow is not complex (Fig. 1). Presumably, the increased shear stress signals inhibition and regression of intimal hyperplasia through the endothelium (Mattson et al., 1997).

Endothelial cells form the endothelium that lines the entire lumen of the vasculature. The forces imparted on this lining by blood flow include shear stress, circumferential strain, and normal stress (Fig. 2). The flow of blood over the endothelium generates viscous drag forces in the direction of flow. The resulting tangential force exerted per unit area of vessel surface at the blood–endothelium surface defines shear stress. Mathematically, the product between the viscosity and the velocity gradient at the wall, also known as the shear rate, equates to shear stress. With ventricular contraction, momentum propagates as waves down the aorta, but diminishes in amplitude down the arterial side of the circulation, giving rise to pulsatile arterial flow and thus pulsatile shear stress. Typical mean arterial values of shear stress range from 6 to 40 dyn/cm^2 but can vary from 0 to well over 100 dyn/cm^2 elsewhere in the vasculature (Patrick and McIntire, 1995). While pulsing down the arterial tree, blood flow remains mostly laminar; however, it often becomes complex and/or disturbed (reversing and/or recirculating) at areas of arterial branching, triggering spatial and/or temporal gradients in shear stress.

Along with momentum, pressure propagates as waves down the arterial tree, leading to a periodic normal stress across the vessel lumen. Since the arterial wall is compliant, this periodic pressure difference gives rise to a cyclic wall strain. Since native blood vessels and synthetic substrates on which cells are cultured are nearly elastic, cyclic strain can be measured as the percent change in diameter between the systolic and diastolic pressures. In normal circulation, the internal diameter, and thus, circumference of large mammalian arteries increases cyclically between 2% and 18% over the cardiac cycle at a frequency of approximately 1 Hz (60 cycles/min) (Dobrin, 1978). The normal stress is measured as the blood pressure within the particular branch of the arterial tree; typical systole/diastole values in normal physiology of large human arteries range from 90/50 mm Hg to 120/80 mm Hg (Mills et al., 1970).

Extensive studies that apply shear stress and cyclic strain *in vitro* confirm that endothelial cells actively participate in vascular physiology (Papadaki and Eskin, 1997; Vouyouka et al., 1998). Although only one cell thick, the endothelium provides a permeability barrier; controls thrombosis and hemostasis by maintaining an active thromboresistant surface, which no biomaterial has yet been able to match; and acts as a mechanosensor for the underlying tissue. Lying beneath the

Normal Blood Flow

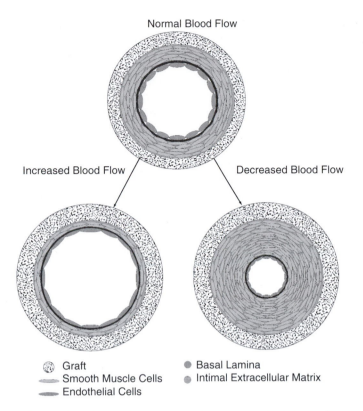

Increased Blood Flow   Decreased Blood Flow

◎ Graft   ● Basal Lamina
⬭ Smooth Muscle Cells   ◑ Intimal Extracellular Matrix
▬ Endothelial Cells

FIG. 1.   Intimal hyperplasia in synthetic arterial grafts. Polytetrafluorethylene (PTFE) grafts endothelialized by transmural capillary ingrowth
under normal blood flow result in neointimal thickening from proliferation of smooth muscle cells, which narrows the lumen. Ligating adjacent,
native vessels diverts all blood flow through the graft, increasing the shear
stress over the endothelium, which signals the underlying smooth muscle
cells to atrophy since vasodilation is impossible because of the rigidity of
PTFE. This response demonstrates that the endothelium regulates shear
stress by decreasing the shear rate.

basal lamina in the medial layer of the tissue, smooth muscle
cells contract, relax, proliferate, or migrate in response to shear
stress, cyclic strain, and paracrine factors (biochemical signals)
from endothelial cells. Vessels denuded of the endothelium
expose the smooth muscle cells to blood flow, resulting in
binding of platelets and subsequently, thrombosis.

As a mechanosensor, the endothelium, in response to changing shear stress, biochemically signals (via the production of
compounds such as nitric oxide and prostacyclin; Frangos
et al., 1985) the smooth muscle cells to dilate or contract.
Vasomotor control over lumen diameter primarily resides with
smooth muscle cells. Responding to changes in blood flow
and pressure, smooth muscle cells contract/relax and secrete a
matrix of collagen and elastin, which confer strength and elasticity, respectively, to modulate the circumferential tension. By
increasing the flow rate while holding the viscosity constant
or increasing the viscosity but sustaining a constant flow rate
through an excised cremaster arteriole, Koller *et al.* demonstrated that vasodilation does not depend on the mass transport
of vasoactive agents but on the response of the endothelium to
shear stress (Koller *et al.*, 1993). Treating the arterioles with
indomethacin abrogated the dilative response of the vessel,

implicating prostaglandin E_2 and I_2 as primary mediators. Furthermore, denudation of the arterioles elicited similar responses
to those treated with indomethacin, confirming the role of
the endothelial cells in modulating vascular tone in response
to changing shear stress (Du *et al.*, 1995). However, smooth
muscle cells also react to shear stress (from interstitial fluid
flow) and to cyclic strain (from pulse pressures of blood flow).
Excessive cyclic strain may be a more injurious force than
shear stress, since it affects the whole thickness of the artery
wall, including smooth muscle cells, fibroblasts (connective tissue cells in the adventitial layer), and the extracellular matrix
(Carosi *et al.*, 1994).

To determine the effects of mechanical forces on vascular
cells, investigators have isolated different cell types in culture
and individually subjected them to specific forces. *In vitro* studies of steady, laminar shear stress, although more prevalent in
capillaries and veins *in vivo* than in large arteries, provide the
necessary baseline for interpreting the effects of more complex
flow regimes. The parallel-plate flow chamber and the cone
and plate viscometer are normally used to study shear-stress
effects on endothelial and smooth muscle cells cultured on
rigid surfaces, usually glass slides, coverslips, or tissue culture

A

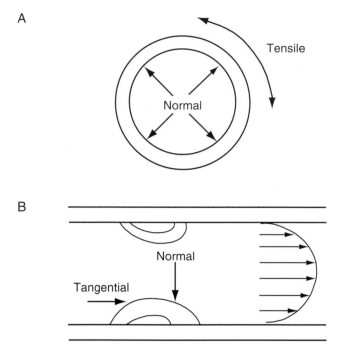

B

FIG. 2. Diagram of forces acting on blood vessel wall. (A) Cross-sectional view of cylindrical vessel showing normal and tensile forces due to hydrostatic pressure and circumferential deformation, respectively. (B) The forces on the endothelial lining include normal and tangential forces. The tangential force due to fluid flow is shown by the horizontal arrow. A tensile force on the endothelium would result from the normal force causing expansion in the circumferential direction. The resulting tensile force would be perpendicular to both normal and tangential forces.

dishes coated with extracellular matrix protein(s). For the parallel-plate flow chamber, a continuously filled supply reservoir elevated above a receiving reservoir creates a pressure drop proportional to the difference in height between the two reservoirs (Fig. 3). Since the two reservoirs are held at constant but different heights, gravity drives flow, and shear stress is steady. A small channel height allows viscous forces to dominate over momentum forces for a wide range of flow rates, making flow laminar (Frangos *et al.*, 1998). The Reynolds number quantifies the ratio between these two forces:

$$Re = \frac{Uh\rho}{\mu} \qquad (1)$$

where U = bulk velocity, h = channel height, ρ = density, and μ = viscosity. For a Newtonian fluid flowing through a parallel-plate flow chamber with a rectangular geometry, the steady, laminar shear stress at the wall is:

$$\tau_w = \frac{6\mu Q}{bh^2} \qquad (2)$$

where τ_w = wall shear stress, Q = flow rate, b = channel width, and h = channel height. By varying the chamber geometry, the flow rate, or the viscosity, the entire physiological range of wall shear stresses can be investigated.

For the cone and plate viscometer, cells are similarly cultured on a protein-coated rigid surface but recessed flush into the plate (Bussolari and Dewey, 1982). For small cone angles, the shear rate (and therefore the shear stress) is essentially constant throughout the flow field.

Recently, methods to investigate shear stresses with a more physiological waveform have been developed for both of these devices (Yee *et al.*, unpublished). The flow or shear rate measured from an artery through magnetic resonance imaging or ultrasound serves as input for the controller. Instead of gravity-driven flow, a computer-controlled pump regulates the instantaneous flow rate through a parallel-plate flow chamber. Similarly, a stepper motor can control the rotational speed of the cone of the viscometer (Blackman *et al.*, 2000), giving rise to pulsatile shear stress on the endothelial cells cultured on the plate.

If the cell property studied responds to shear stress in a dose-dependent fashion (i.e., varies directly with the level of shear stress imposed), it is referred to as shear stress dependent. If the property under study responds to changes in shear rate, but not directly with the level of shear stress imposed, it is referred to as flow dependent. The difference arises from inclusion of mass-transport phenomena (convection and diffusion) in flow-dependent processes, whereas the term "shear-stress dependent" specifies the mechanical force only. Differences between effects due to shear stress alone and those due to flow (mass and momentum transfer) can be examined by circulating media of different viscosities (with the addition of high molecular weight dextrans).

To study cyclic strain effects, the cells must be cultured on deformable substrates, usually silicone rubber or segmented polyurethane coated with an extracellular matrix proteins. In general, two kinds of devices have been used to impose cyclic strain on cells—(1) uniaxial or biaxial strain devices driven mechanically by an eccentric cam (Fig. 4), which imposes a uniform strain across the substrates, and (2) the Flexercell Strain Unit (Flexcell International, Corp.), driven by vacuum pressure pulled beneath the substrate on which cells are cultured, thus deforming substrate and cells (Haseneen *et al.*, 2003). The Flexercell unit applies a maximum strain of 25% around the edges of the substrate, decreasing to a minimum of less than 3% at the center (Gilbert *et al.*, 1994). However, it should be noted that not only do the mechanical forces that impinge on a population of cultured cells alter their function, but also the substrate on which the cells are cultured may affect cell response.

Modulation of vascular cell function by mechanical forces takes place at several levels, with the most central being gene regulation in the nucleus. Specific DNA sequences (genes) encode for synthesis of specific proteins (via mRNA), which have different roles in cell function. To affect the cell at the gene level, the stimulus (mechanical force) must be (1) perceived at the cell membrane (e.g., by receptors, ion channels, or integrins), (2) transmitted through the cytoplasm, perhaps physically (mechanotransduction through the cytoskeleton), biochemically, or both, and (3) stimulate (or inhibit) transcription in the nucleus (Fig. 5).

Significant progress has been made in the past decade in understanding many of the signal transduction pathways involved in mammalian cell sensing of mechanical forces (Hojo *et al.*, 2002; Shyy and Chien, 2002; Li *et al.*, 1997; Tai *et al.*, 2002; Wittstein *et al.*, 2000). The relative

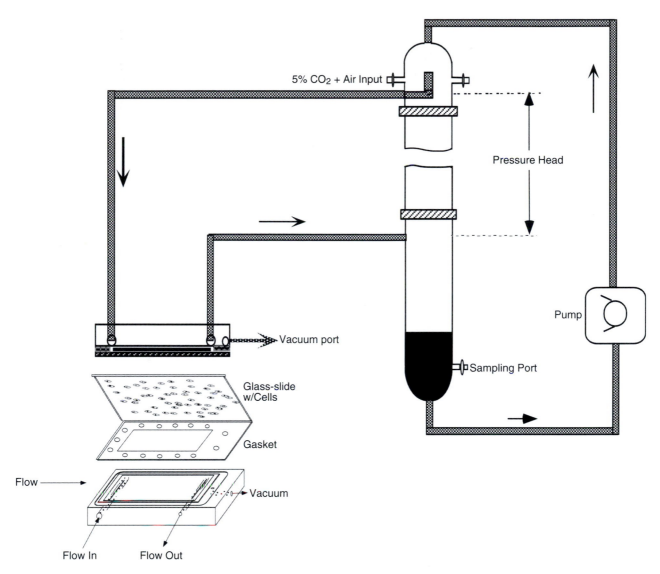

FIG. 3. Schematic diagram of the flow loop with parallel-plate flow chamber, showing side view of chamber assembled with loop components, and exploded view on lower left. The exploded view includes the polycarbonate base with two slits through which medium enters and exits the channel, the Silastic gasket that determines the channel height, and the glass slide on which the cells are cultured. The flow rate is controlled by the relative distance between the two reservoirs. The medium is recirculated from the lower to the upper reservoir by a roller pump. The channel depth is normally 220 μm; the area of cells exposed to flow is approximately 16 cm². The entire circuit is gassed with 5% CO_2 in humidified air and run at 37°C. Medium samples can be removed from the lower reservoir without disturbing the flow field. The medium volume is about 20 ml and the residence time in the flow chamber and its tubing is about 10 sec (modified from Frangos *et al.,* 1988).

contribution of diffusable second messengers (Ca^{2+}, pH, etc.) and direct mechanical transmission via cytoskeletal elements (tensegrity theory) is still a matter of some debate (Gudi *et al.,* 2003; Ingber, 2003).

To elucidate potential signal transduction elements and pathways, specific inhibitors such as calcium chelators (BAPTA-AM for intracellular Ca^{2+} and EGTA for extracellular Ca^{2+}), G-protein inhibitors (pertussis toxin and cholera toxin), ion-channel inhibitors, and kinase inhibitors are typically employed while changes in cellular activities such as

transcription factor regulation, intracellular pH, phosphorylation events, membrane perturbation, and cytoskeletal remodeling are monitored. However, to determine the effects of deformation and displacement of subcellular structures on cellular properties and activities, new optical methods such as atomic force microscopy, optical sectioning, and high resolution, four-dimensional (4D) fluorescence microscopy have been developed (Barbee, 2002; Helmke and Davies, 2002; Stamatas and McIntire, 2001). These new techniques have revealed the heterogeneity in the mechanical properties of the

A

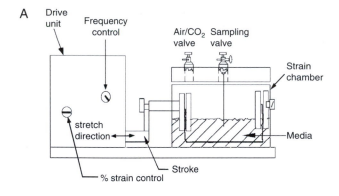

B

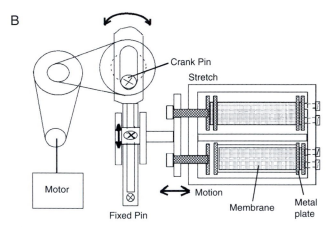

FIG. 4. Schematic diagram of cyclic strain unit: (A) Side view of poly-carbonate unit connected to motorgear assembly. Placement of the adjustable offset pin determines the level of elastic membrane strain. Frequency is set by controlling current to direct current motor with gear box. (B) Top view of unit showing two cell culture chambers; the top chamber shows the membrane clamped at both ends to impose cyclic strain. In the bottom chamber the membrane is fixed to the polycarbonate and clamped at only one end to provide a fluid motion control. Both chambers are maintained at 37°C and gassed with 5% CO_2 in humidifed air.

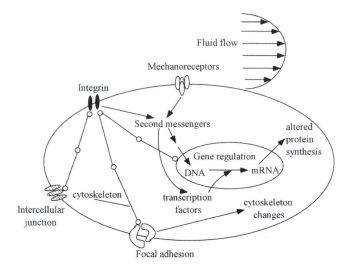

FIG. 5. Some of the possible mechanotransduction pathways in vascular cells. Putative mechanoreceptors on the luminal cell membrane may alter their conformation in response to mechanical stimuli, activating a signal transduction cascade involving diffusible second messengers. This cascade of chemical reactions leads to activation of gene transcription in the nucleus. Alternatively, mechanotransduction may be direct via elements of the cytoskeleton. Forces may be transmitted through these networks from the luminal cell membrane to focal adhesion sites, intracellular junctions or directly to the nucleus. Translation of the mechanical force to a chemical signal at these sites would then lead to alterations of cell function and gene regulation.

cytoskeletal proteins and their putative roles in mechanotransduction, perhaps explaining the spatial dependence on signal transduction at the intracellular level (Butler *et al.*, 2001).

Following mechanotransduction of shear stress, cyclic strain, or normal stress, vascular cells may alter the production levels of many substances at the genetic level. To claim gene regulation of a protein requires the following: (1) the corresponding mRNA level must change, (2) the change in mRNA must not be the result of stabilization or destabilization of the mRNA, and (3) the protein production must also change in response to the stimulus. Although less than 2% of the DNA functionally encodes for proteins, analysis of the effects of mechanical stress on the 30,000 to 40,000 protein-encoding human genes remains a daunting task (Consortium IHGS, 2001; Venter, 2001). Before DNA microarray technology, the choice of which gene to study was based on the suspected physiological significance of the corresponding protein in vasodilation or vasocontriction (endothelin-1, or nitric oxide synthase), in thrombosis (tPA and PAI-1, thrombomodulin), in proliferation (growth factors), in permeability

(cadherins, connexins), or in adhesion of blood borne substances to the endothelium (VCAM-1, MCP-1). Functional assays verify the expected protein function(s) and can further illustrate protein quantity alterations, caused by the mechanical stimulus, at the transcriptional level. Factors that directly influence DNA transcription, such as changes in transcription factors in response to shear stress on endothelial cells, have been reported (Chien *et al.*, 1998; Nagel *et al.*, 1999; Bao *et al.*, 2000). Some of the base sequences in the promoter regions of the DNA that are required for the gene specific shear stress response have been identified (Korenaga *et al.*, 1997; Resnick *et al.*, 1993). Tables 1 and 2 list some responses to shear stress of endothelial and smooth muscle cells, respectively. Table 3 shows some effects of cyclic strain on endothelial cells and smooth muscle cells.

To interpret and understand the effects of mechanical forces *in vivo*, *in vitro* shear stress studies can be compared to *in vitro* cyclic strain studies of the same cell type. Typically, cells are subjected to one type of force at a time, but *in vitro* studies with both forces acting in concert are beginning to appear as more suitable substrates are developed. Two ways to approach such a study are (1) to choose a cell product that varies in a dose-dependent manner throughout the physiological range of shear stresses, has a large amplitude variation, and responds differently under cyclic strain than under shear stress, then proceed to study how it is controlled by other physiologically relevant inputs (e.g., growth factors, cytokines, cytoskeletal elements); (2) to compare many genes in parallel, using DNA microarray techniques or differential display, and

TABLE 1 Shear Stress Mediated Endothelial Cell Responses

Response	Effect
General functions	
Morphology	Alignment and elongation
Actin stress fibers	Formation and alignment
Pinoctosis	Increase
LDL uptake	Slight increase
Cation channel	Activation
DNA synthesis	No effect (if laminar flow)
DNA synthesis	Stimulated (if turbulent flow)
G proteins	Activation
Apoptosis	Decrease
Protein synthesis and secretion	
Tissue plasminogen activator	Increase
Endothelin-1	Decrease
Plasminogen activator inhibitor	No effect
Fibronectin	Slight decrease
VCAM-1	Decrease
ICAM-1	Transient increase
Protease activated receptor-1	Decrease
Nitric oxide synthase	Increase
mRNA levels	
Tissue plasminogen activator	Increase
Endothelin-1	Decrease
Fibroblast growth factor-2	No effect
Glyceraldehyde-3-phosphate dehydrogenase	No effect
VCAM-1	Decrease
ICAM-1	Increase
Protease activated receptor-1	Decrease
Monocyte chemoattractant protein-1	Transient increase then decrease
Arachidonic acid metabolism	
PGI_2 synthesis	Large increase
Arachidonate uptake	Increase for phosphatidylinositol
Second messengers	
Inositol trisphosphate	Increase
Intracellular Ca^{2+}	Increase (ATP mediated)
cAMP	Increase mediated (PGI_2 mediated)
Diacylglycerol	Increase
Nitric oxide (NO)	Increase
pH	Decrease (HUVEC)

TABLE 2 Shear Stress Mediated Smooth Muscle Cell Responses

Response	Effect
Morphology	No alignment and orientation
Proliferation	Decrease
Tissue plasminogen activator	Increase
Protease activated receptor-1	Decrease
Nitric oxide	Increase
Prostaglandins E_2, I_2	Increase
Intracellular pH	Increase
Intracellular Ca^{2+}	No change

TABLE 3 Cyclic Strain Mediated Endothelial and Smooth Muscle Cell Responses

Response	Effect
Endothelial cells	
Endothelin 1	Increase
Tissue plasminogen activator	No change
Plasminogen activator inhibitor-1	Increase
Prostaglandin I_2	Increase
Platelet-derived growth factor-B	No change
c-fos	Increase
c-jun	Increase
Nuclear factor kappa-B	Increase
Nitric Oxide	Increase
MCP-1	Increase
VCAM-1	Decrease
Smooth muscle cells	
Proliferation	Increase
Platelet-derived growth factor	Increase
Fibroblast growth factor-2	Release
Protease activated receptor-1	Increase
Matrix metalloprotease-1	Decrease

focus on genes that yield the greatest differences by these techniques. Recently, with rapid progress in the genome sequencing projects, molecular biological techniques have been developed that allow investigators to examine a large number of genes in populations of cells subjected to shear stress or cyclic strain. Sequences from up to several thousand known genes can be arranged on a single chip or microtiter plate.

An example of the first approach is a study of the control of thrombin receptor, or protease activated receptor-1 (PAR-1), in vascular smooth-muscle cells by shear stress and cyclic strain (Papadaki, *et al.*, 1998). Thrombin receptor *in vivo* is upregulated in injured arteries, presumably in order to bind thrombin for cleavage of fibrinogen, leading to thrombus formation. Thus up-regulation of PAR-1 can be considered a thrombogenic response. However, activation of PAR-1 by thrombin is also a potent mitogenic stimulus for vascular smooth-muscle cells. PAR-1 protein decreases by two-thirds in smooth muscle cells subjected to 25 dyn/cm^2 shear stress for 24 hours. Under cyclic strain (20%, 1 Hz) PAR-1 protein increases 2.5-fold after 24 hours (Nguyen *et al.*, 2001). Thus smooth muscle cells under cyclic strain express a thrombogenic product, whereas arterial levels of fluid-induced shear stress down-regulate this protein.

In the second approach, the gene expression profiles are initially unknown, and genes of interest are selected after comparing treated cells to untreated cells at the genomic (1000–10,000 genes) or pathway-specific (~100 genes) level. For example, differences in gene expression between endothelial cells subjected to shear stress and endothelial cells kept under static culture are detected with microarrays (Chen *et al.*, 2001; Garcia-Cardena *et al.*, 2001; McCormick *et al.*, 2001). In the McCormick *et al.* study, shear stress of 25 dyn/cm^2 for 24 or 6 hours altered the expression of 52 genes (32 up-regulated, 20 down-regulated). Genes were considered up-regulated or down-regulated if the expression increased or decreased two-fold, respectively. Changes in the most notably altered genes—CYP1A1, CYP1B1, CTGF, ET-1, PGHS-2, NADH dehydrogenase—were verified by Northern blot densitometry. Particularly, up-regulation of CYP1A1 and CYP1B1 and down-regulation of CTGF, ET-1, and MCP-1 support the theory that laminar shear stress protects the endothelium from fibrotic and atherosclerotic disease. Data analysis of the huge data set is a challenge in these studies. Statistical methods for microarray data analysis continue to evolve into significance tests that leave fewer genes suspected of altered expression to chance.

An example of the second approach is the differential display study of Topper *et al.* (1996), in which mRNAs from shear stressed versus control endothelial cells were examined for differences in expression using oligonucleotide primer sets, and then amplifying the mRNAs that differ most. As in DNA microarray studies, the genes of interest are not chosen by the investigators. Three "atheroprotective genes," manganese superoxide dismutase, cyclooxygenase-2, and endothelial nitric oxide synthase, were up-regulated by steady laminar shear stress (10 dyn/cm^2 for 1 or 6 hours), but not by turbulent shear stress for the same exposure times at the same average wall shear stress.

SKELETAL CELL RESPONSES TO MECHANICAL FORCES

Bone, cartilage, ligament and tendon are all skeletal tissues derived from mesenchymal stem cells. In the specific tissue environments, musculoskeletal cells differ in the composition of the matrix they secrete, which results in different tissue mechanical properties. It has been known for more than 100 years that bone density patterns are a function of load (Wolff's law) (Stoltz *et al.*, 2003). Bone remodels in response to stress and loses density in unstressed regions. Studies under the aegis of NASA have dramatically enhanced appreciation of the effects of weightlessness in promoting bone loss in astronauts. Bone loss from microgravity produced by space flight has provided conclusive *in vivo* evidence for the positive effects of physiological loading of skeletal tissues (Duncan and Turner, 1995).

Exercise-stimulated bone remodeling may be the response of osteoblasts to interstitial fluid flow (IFF) (Hillsley and Frangos, 1994; Reich *et al.*, 1991). Furthermore, temporal gradients in IFF, imposed by pulsatile shear stress, stimulate osteoblast proliferation *in vitro* (Jiang *et al.*, 2002), whereas steady shear stress does not stimulate osteoblast signal transduction or proliferation. Continually changing flow regimens (such as pulsatility) *in vivo* simulate the increase in temporal gradients in blood flow that occur with increased physical activity.

Bone tissue comprises bone cells (osteocytes), osteoblasts (responsible for bone formation), osteoclasts (responsible for bone resorption), and an extracellular matrix composed of hydroxyapatite (a complex molecule of hydrated calcium phosphate) and collagen, which becomes calcified or ossified.

Mechanical loading generates fluid shear sress, hydrostatic compression, uniaxial stretch, or biaxial stretch on bone cells.

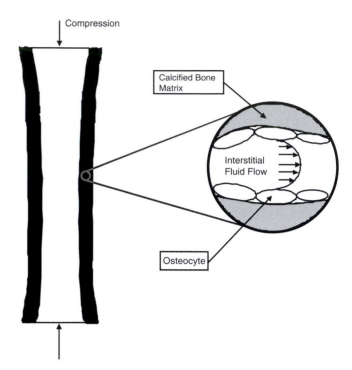

FIG. 6. Compressive force on long bone due to gravity and/or locomotion gives rise to shear stress on the bone cells due to interstitial fluid flow. Loading of porous bone or cartilage structure (or porous polymer scaffolds in load-bearing tissue-engineering applications) leads to interstitial fluid motion through small channels, leading to generation of significant shear stress on cells attached to the polymer matrix.

Cortical (long) bone tissue is porous, and interstitial fluid flow due to mechanical loading is thought to be the primary mechanism for action of mechanical forces on bone (Fig. 6). Devices for studying osteoblast response to these forces have been reviewed (Basso and Heersche, 2002). In addition, there is evidence that low-level vibrations are key to determining the structure of bone (Rubin *et al.*, 2001). Rats with depressed bone formation due to suspension of their hind legs had dramatically improved bone formation after 10 min per day vibration therapy (0.25 g, 90 Hz) for 28 days.

Cartilage, located on the articulating surfaces of joints, provides low friction for freely moving joints. In the growing embryo, cartilage is the precursor to bone. Cartilage cells (chondrocytes) secrete a matrix of collagen fibers embedded in mucopolysaccharide (e.g., chondroitin sulfate). Cartilage responds to mechanical loading similarly to bone. Increased load leads to increased matrix production, resulting in stronger tissue. Cartilage must withstand tensile and shear forces in addition to compression (Kim *et al.*, 1994). Most of the work on cartilage loading focuses on compression and hydrostatic pressure. Cyclic compression of explants (0.1 Hz, 2–3% compression) stimulated matrix synthesis by chondrocytes (Shieh and Athanasiou, 2003).

Tendons and ligaments join muscle to bone and bone to bone, respectively. These tissues withstand unidirectional tension. Tendons are primarily composed of collagen (86%

dry weight), secreted by fibroblasts and aligned unidirectionally. Tendon has one of the highest tensile strengths of any tissue (50–150 MPa). When resected, the tendon retracts and must be mechanically loaded for healing to occur (although this is not well studied as to timing or load). Ligaments, like tendons, are primarily composed of collagen fibers, but the fibers are less densely packed (70% dry weight) and woven, unlike the parallel arrangement in tendon. Ligaments contain more elastin, and thus are more extensible. Both are viscoelastic, and both dissipate energy and show hysteresis in the stress–strain curves.

Biomaterials problems related to skeletal tissues are mainly of two types: joint replacements and bone defect or loss replacement. By far most joint replacements are hip or knee arthroplasties (covered in Chapter 7.7 on orthopedic applications). These prostheses must replace the functions of bone, cartilage, and tendon. Most implants consist of metal alloys, materials with long fatigue life and high corrosion resistance. Articulating surfaces have more recently been coated with materials of low coefficient of friction and low wear rate. Joint replacement requires the presence of both load-bearing compressive forces on the bones and periodic flexing and extending of the bone for functional healing of bone, cartilage, and tendon to occur. To replace bone defects (e.g., as the result of tumor removal) presents a different challenge and has previously required autologous bone. As in the case of vascular cells, applied forces are transduced across the cell membranes of the skeletal cells leading to cell responses, most notably tissue remodeling.

Bone, which appears inert macroscopically, is a very dynamic tissue with high cell turnover rates, much higher than those of noninjured vascular cells. In addition, there are important differences between bone and cartilage tissue structure. Bone is highly vascularized, cellular, and innervated, whereas cartilage has low cell density, little blood supply, and no innervation. The effects of mechanical forces on bone and cartilage cells have been studied *in vitro* using devices similar to those employed for vascular cells. The relative quantities of collagen and glycosaminoglycans (GAGs) and cell density in tissue-engineered cartilage (chondrocytes seeded on polyglycolic acid scaffolds) are functions of the mechanical environment in which they are cultured (Gooch *et al.*, 1998). Table 4 gives a summary of mechanical force effects on skeletal tissue.

Bone cement, poly(methyl methacrylate), has been used for years to fill defects in bone. However, it is not biodegradable, and since the cement bears the load (stress shielding), the healing bone does not regenerate and the implant ultimately fails. However, resorting to degradable polymers seeded with cells requires considering the cell type, the mechanical properties of the polymer, and the strategy to maintain cell survivability. Some recent studies have begun to study mechanical loading of musculoskeletal cells grown in three-dimensional polymer scaffolds. This would give an environment much closer to that seen by the cells *in vivo*, which is inherently three-dimensional for these tissues. There are, however, significant problems in knowing the actual forces and strains the cells experience in these constructs and also what fluid motion is induced by the external loading. Thus separating mechanical strain and fluid shear stress effects on cell metabolism and tissue growth is difficult.

TABLE 4 Effect of Mechanical Forces on Skeletal Cell Responses

Cell/tissue	Force	Effect/responses
Cultured osteoblasts	Shear stress	Increase cAMP Increase PGE_2
Cultured osteocytes	Cyclic strain	Increase IP_3 Increase c-fos Increase COX-2 Increase osteopontin Activates focal adhesion kinase
Cortical bone	Bending load	Increase osteopontin Decrease myeloperoxidase
Cartilage disks	Static compression Cyclic strain	Decrease glycosaminoglycan Increase glycosaminoglycan
Chondrocytes	Cyclic strain	Increase collagen type II Increase aggrecan No change in B_1 integrin

In addition, specifying the macroscopic strain at the boundaries does not specify the local strain seen by the cells in these porous three-dimensional scaffolds. The mechanical properties of the polymer and those of the cells are quite different.

Bibliography

Bao, X., Clark, C. B., and Frangos, J. A. (2000). Temporal gradient in shear-induced signaling pathway involvement of MAP kinase, c-fos, and connexin 43. *Am. J. Physiol.* Heart and Circulatory Physiology. 2000; 278: H1598–H1605.

Barbee, K. A. (2002). Role of subcellular shear-stress distributions in endothelial cell mechanotransduction. *Ann. Biomed. Eng.* 30: 472–482.

Basso, N., and Heersche, J. N. M. (2002). Characteristics of *in vitro* osteoblastic cell loading models. *Bone.* 30: 347–351.

Blackman, B. R., Barbee, K. A., and Thibault, L. E. (2000). *In vitro* cell shearing device to investigate the dynamic response of cells in a controlled hydrodynamic environment. *Ann. Biomed. Eng.* 28: 363–372.

Bussolari, S. R., and Dewey, C. F. J. (1982). Apparatus for subjecting living cells to fluid shear stress. *Rev. Sci. Instruments* 53: 1851–1854.

Butler, P. J., Tsou, T-J., Li, J. Y-S., Usami, S., and Chien, S. (2002). Rate sensitivity of shear-induced changes in the lateral diffusion of endothelial cell membrane lipids: a role for membrane perturbation in shear-induced MAPK activation. *FASEB J.* 16: 216–218.

Carmeliet, P. (2000). Mechanisms of angiogenesis and arteriogenesis. *Nat. Med.* 6: 389–395.

Carosi, J. A., McIntire, L. V., and Eskin, S. G. (1994). Modulation of secretion of vasoactive materials from human and bovine endothelial cells by cyclic strain. *Biotechnol. Bioeng.* 43: 615–621.

Chen, B. P. C., Li, Y-S., Zhao, Y., Chen, K-D., Li, S., Lao, J., Yuan, S., Shyy, JY-J., and Chien, S. (2001). DNA microarray analysis of gene expression endothelial cells in response to 24-h shear stress. *Physiol. Genomics* 7: 55–63.

Chien, S., Li, S., and Shyy, Y. J. (1998). Effects of mechanical forces on signal transduction and gene expression in endothelial cells. *Hypertension* 31(1 Pt 2): 162–169.

Consortium IHGS. (2001). Initial sequencing and analysis of the human genome. *Nature* **409**: 860–921.

Dobrin P. B. (1978). Mechanical properties of arteries. *Physiol. Rev.* **58**: 397–460.

Du, W., Mills, I., and Sumpio, B. E. (1995). Cyclic strain causes heterogeneous induction of transcription factors, AP-1, CRE binding protein and NF-kB, in endothelial cells: species and vascular bed diversity. *J. Biomechan.* **28**: 1485–1491.

Duncan, R. L., and Turner, C. A. (1995). Mechanotransductions and the functional response of bone to mechanical strain. *Calcified Tissue Int.* **57**: 344–358.

Frangos, J. A., Eskin, S. G., McIntire, L. V., and Ives, C. L. (1985). Flow effects on prostacyclin production by cultured human endothelial cells. *Science.* **227**: 1477–1479.

Frangos, J. A., McIntire, L. V., and Eskin, S. G. (1988). Shear stress induced stimulation of mammalian cell metabolism. *Biotechnol. Bioeng.* **32**: 1053–1060.

Garcia-Cardena, G., Comander, J., Anderson, KR., Blackman, B. R., and Gimbrone, M. A, Jr. (2001). Biomechanical activation of vascular endotohelium as a determinant of its functional phenotype. *Proc. Natl. Acad. Sci. USA* **98**: 4478–4485.

Gilbert, J. A, Weinhold, P. S, Banes, A. J, Link, G. W, and Jones, G. L. (1994). Strain profiles for circular cell culture plates containing flexible surfaces employed to mechanically deform cells *in vitro*. *J. Biomechan.* **9**: 1169–1177.

Glagov, S., Zarins, C., Giddens, D. P., and Ku, D. N. (1988). Hemodynamics and atherosclerosis. *Arch. Pathol. Lab. Med.* **112**: 1018–1031.

Gooch, K. J., Blunk, T., Vunjak-Hovakovic, G., Langer, R., Freed, L. E., and Tennant, C. J. (1998). Mechanical forces and growth factors. in *Frontiers in Tissue Engineering*, C. W. J. Patrick, A. G. Mikos, and L. V. McIntyre, eds. Pergamon-Elsevier Science, New York, pp. 61–82.

Greenwald S. W., and Berry C. L. (2000). Improving vascular grafts: the importance of mechanical and hemodynamic properties. *J. Pathol.* **190**: 292–299.

Gudi, S., Huvar, I., White, C. R., McKnight, N. L., Dusserre, N., Boss, G. R., and Frangos, J. A. (2003). Rapid activation of ras by fluid flow is mediated by Gaq and Gbg subunits of heterotrimeric G proteins in human endothelial cells. *Arterioscler. Thrombosis Vas. Biol.* **23**: 994–1000.

Haseneen, N. A., Vaday, G. G., Zucker, S., and Foda, H. D. (2003). Mechanical stretch induces MMP-2 release and activation in lung endothelium: role of EMMPRIN. *Am. J. Physiol. Lung Cell. Mol. Physiol.* **284**.

Helmke, B. P., Davies, P. F. (2002). The cytoskeleton under external fluid mechanical forces: hemodynamic forces acting on the endothelium. *Ann. Biomed. Eng.* **30**: 2848.

Hillsley, M. V, and Frangos, J. A. (1994). Review: Bone tissue engineering: the role of interstitial fluid flow. *Biotechnol. Bioeng.* **43**: 573–581.

Hojo, Y., Saito, Y., Tanimoto, T., Hoefen, R. J., Baines, C. P., Yamamoto, K., Haendeler, J., Asmis, R., and Berk, B. C. (2002). Fluid shear stress attenuates hydrogen peroxide–induced c-Jun NH2-terminal kinase activation via a glutathione reductase–mediated mechanism. *Circ. Res.* **91**: 712–718.

Ingber, D. E. (2003). Tensegrity II. How structural networks influence cellular information processing networks. *J. Cell Sci.* **116**: 1397–1408.

Jiang, G-L., White, C. R., Stevens, H. Y., and Frangos, J. A. (2002). Temporal gradients in shear stimulate osteoblastic proliferation via ERK1/2 and retinoblastoma protein. *Am. J. Physiol. Endocrinol. Metab.* **283**: E383–E389.

Kim, Y-J., Sah, R. L. Y., Grodzinsky, A. J., Plaas, A. H. K., and Sandy, J. D. (1994). Mechanical regulation of cartilage biosynthetic behavior: physical stimuli. *Arch. Biochem. Biophys.* **311**: 1–12.

Koller, A., Sun, D., and Kaley, G. (1993). Role of shear stress and endothelial prostaglandins in flow- and viscosity-induced dilation of arterioles *in vitro*. *Circ. Res.* **72**: 1276–1284.

Korenaga, R., Ando, J., Kosaki, K., Isshiki, M., Takada, Y., and Kamiya, A. (1997). Negative transcriptional regulation of the VCAM-1 gene by fluid shear stress in murine endothelial cells. *Am. J. Physiol.* **273**: C1506–C1515.

Li, S., Kim, M., Hu, Y-L., Jalai, S., Schlaepfer, D. D., Hunter, T., Chien, S., and Shyy, JY-J. (1997). Fluid shear stress activation of focal adhesion kinase. *J. Biol. Chem.* **272**: 30455–30462.

Mattson, E. J. R., Kohler, T., Vergel, S. M., and Clowes, A. W. (1997). Increased blood flow induces regression of intimal hyperplasia. *Arterioscler. Thrombosis Vasc. Biol.* **17**: 2245–2249.

McCormick, S. M., Eskin, S. G., McIntire, L. V., Teng, C. L., Lu, C-M., Russell, C. G., and Chittur K. K. (2001). DNA microarray reveals changes in gene expression of shear stressed human umbilical vein endothelial cells. *Proc. Natl. Acad. Sci. USA* **98**: 8955–8960.

Mills, C. J., Gabe, I. T., Gault, J. H., Mason, D. T., Ross, J. J., Braunwald, E., and Shillingford, J. P. (1970). Pressure-flow relationships and vascular impedance in man. *Cardiovasc. Res.* **4**: 405–417.

Nagel, T., Resnick, N., Dewey. C. F. J., and Gimbrone, M. A. J. (1999). Vascular endothelial cells respond to spatial gradients in fluid shear stress by enhanced activation of transcription factors. *Arterioscler. Thrombosis Vasc. Biol.* **19**: 1825–1834.

Nguyen, K. T., Frye, S. R., Eskin, S. G., Patterson, C., Runge, M. S., and McIntire, L. V. (2001). Cyclic strain increases protease-activated receptor-1 expression in vascular smooth muscle cells. *Hypertension.* **38**: 1038–1043.

Papadaki, M., and Eskin, S. G. (1997). Effects of fluid shear stress on gene regulation of vascular cells. *Biotechnol Prog.* **13**: 209–221.

Papadaki, M., Ruef, J., Nguyen, K. T., Li, F., Patterson, C., Eskin, S. G., McIntire, L. V., and Runge, M.S. (1998). Differential regulation of protease activated receptor-1 and tissue plasminogen activator expression of shear stress in vascular smooth muscle cells. *Circ. Res.* **83**: 1027–1034.

Patrick, C. W. J., and McIntire, L. V. (1995). Shear stress and cyclic strin modulation of gene expression in vascular endothelial cells. *Blood Purif.* **13**: 112–124.

Reich, K. M., Gay, C. V., and Frangos, J. A. (1991). Fluid shear stress as a mediator of osteoblast cyclic adenosine monophosphate production. *J. Cell Physiol.* **261**: C428–C432.

Remuzzi, A., Ene-Iordacche, B., Mosconi, L., Bruno, S., Anghileri, A., Antiga, L., and Remuzzi, G. (2003). Radial artery wall shear stress evaluation in patients with arteriovenous fistula for hemodialysis access. *Biorheology.* **40**: 423–430.

Resnick, N., Collins, T., Atkinson, W., Bothron, D. T., Dewey, C. F. J., and Gimbrone, M. A. J. (1993). Platelet-derived growth factor B chain promoter contains a cis-acting shear-stress-responsive element. *Proc. Natl. Acad. Sci. USA.* **90**: 4591–4595.

Ross, R., and Glomset, J. A. (1976). The pathogenesis of atherosclerosis. *N. Eng. J. Med.* **295**: 369–420.

Rubin, C., Gang, X., and Judex, S. (2001). The anabolic activity of bone tissue, suppressed by disuse, is normalized by brief exposure to extremely low-magnitude mechanical stimuli. *FASEB J.* **15**: 2225–2229.

Shieh, A. C., and Athanasiou, K. A. (2003). Principles of cell mechanics for cartilage tissue engineering. *Ann. Biomed. Eng.* **31**: 1–11.

Shyy, J. Y., and Chien, S. (2002). Role of integrins in endothelial mechanosensing of shear stress. *Circ. Res.* **91:** 769–775.

Stamatas, G. N., and McIntire, L. V. (2001). Rapid flow-induced responses in endothelial cells. *Biotechnol. Prog.* **17:** 383–402.

Stoltz, J. F., Dumas, D., Wang, X., Payan, E., Mainard, D., Paulus, F., Maurice, G., Netter, P., and Muller, S. (2003). Influence of mechanical forces on cells and tissues. *Biorheology* **37:** 3–14.

Tai, L-K., Okuda, M., Abe, J-I., Chen, Y., and Berk, B. C. (2002). Fluid shear stress activates proline-rich tyrosine kinase via reactive oxygen species-dependent pathway. *Arterioscler. Thrombosis Vasc. Biol.* **22:** 1790–1796.

Topper, J. N., Cai, J., Falb, D., and Gimbrone, M. A. J. (1996). Identification of vascular endothelial cells differentially responsive to fluid mechanical stimuli: cyclooxygenase-2, manganese superoxide dismutase, and endothelial cell nitric oxide synthase are selectively up-regulated by steady laminar shear stress. *Proc. Natl. Acad. Sci. USA.* **93:** 10417–10422.

van Gieson, E. J., Murfee, W. L., Skalak, T. C., and Price, R. J. (2003). Enhanced smooth muscle cell coverage of microvessels exposed to increased hemodynamic stresses *in vivo*. *Circ. Res.* **92:** 929–936.

Venter, J. G. (2001). The sequence of the human genome. *Science* **291:** 1304–1351.

Vouyouka, A. G., Powell, R. J., Ricotta, J., Chen, H., Dudrick, D. J., Sawmiller, C. J., Dudrick, S. J., and Sumpio, B. E. (1998). Ambient pulsatile pressure modulates endothelial cell proliferation. *J. Mol. Cell. Cardiol.* **30:** 609–615.

Wittstein, I. S., Qiu, W., Ziegelstein, R. C., Hu, Q., and Kass, D. A. (2000). Opposite effects of pressurized steady versus pulsatile perfusion on vascular endothelial cell cytosolic pH. *Circ. Res.* **86:** 1230–1236.

4

Host Reactions to Biomaterials and Their Evaluation

JAMES M. ANDERSON, GUY COOK, BILL COSTERTON, STEPHEN R. HANSON,
ARNE HENSTEN-PETTERSEN, NILS JACOBSEN, RICHARD J. JOHNSON,
RICHARD N. MITCHELL, MARK PASMORE, FREDERICK J. SCHOEN,
MARK SHIRTLIFF, AND PAUL STOODLEY

4.1 INTRODUCTION

Frederick J. Schoen

Biomaterials and medical devices are now commonly used as prostheses in cardiovascular, orthopedic, dental, ophthalmological, and reconstructive surgery, in interventions such as angioplasty (stents) and hemodialysis (membranes), in surgical sutures or bioadhesives, and as controlled drug release devices. Most implants serve their recipients well for extended periods by alleviating the conditions for which they were implanted. However, some implants and extracorporeal devices ultimately develop complications—adverse interactions of the patient with the device, or vice versa—which constitute device failure and thereby may cause harm to or death of the patient. Complications result largely as a consequence of biomaterial–tissue interactions, which all implants have with the environment into which they are placed. Effects of both the implant on the host tissues and the host on the implant are important in mediating complications and device failure (Fig. 1).

Chapter 4 contains overview discussions of the most important host reactions to biomaterials and their evaluation, including nonspecific inflammation and specific immunological reactions, systemic effects, blood–materials interactions, tumor formation, and infection. To a great extent, these interactions arise from alterations of physiological (normal) processes (e.g., immunity, inflammation, blood coagulation) comprising host defense mechanisms that function to protect an organism from the deleterious external threats (such as bacteria and other microbiologic organisms, injury and foreign materials). Chapter 6 addresses degradation mechanisms in biomaterials (i.e., the effect of the host on biomaterials). Several key concepts of biomaterials–tissue interactions are emphasized here in an effort to guide the reader and facilitate the use of this chapter.

THE INFLAMMATORY REACTION TO BIOMATERIALS

In their respective chapters, Anderson, Mitchell, and Johnson describe the inflammatory and potential immunological interactions that occur with biomaterials. In contrast to living organ transplants, biomaterials are not generally "rejected." The process of organ rejection denotes an inflammatory process that results from a specific immune response and which causes tissue death, which synthetic biomaterials typically do not generate. The usual response to biomaterials comprises nonspecific inflammation (see Chapters 4.2, 4.3, and 4.4).

However, as summarized by Mitchell (2001), tissue-derived biomaterials (such as bioprosthetic heart valves) may express foreign histocompatibility antigens and be antigenic and capable of eliciting an immune response, including antibodies and antigen-specific T cells. Nevertheless, it is important to understand the following:

1. Tissue immunogenicity does not necessarily induce immunologically mediated device dysfunction.
2. Specific immunological responses can be not only a cause of but can result from device failure.
3. Although mononuclear inflammatory cell infiltrates (containing macrophages and lymphocytes) are characteristically associated with organ/tissue rejection on histological examination, mononuclear inflammatory infiltrates are themselves nonspecific and comprise a largely stereotyped and generic response to tissue injury. Therefore, the presence of mononuclear cells does not necessarily denote a rejection pathogenesis.

In order to invoke an immunological reaction to a biomaterial as the cause of a device failure, an immunological variant of the classical Koch's postulates, which are the objective criteria for concluding that a disease is infectious and caused by a specific microbiologic agent, would be appropriate. The classic Koch's postulates state that:

1. A suspected infectious agent should be recoverable from the pathologic lesions of the human host.
2. The agent should cause the pathologic lesions when inoculated into an animal host.
3. The agent should then be recoverable from the pathologic lesions in the animal.

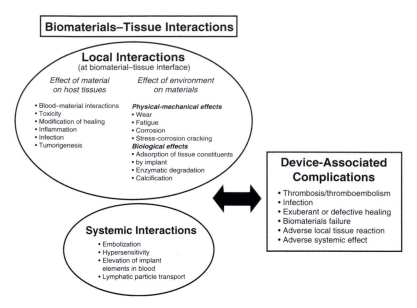

FIG. 1. Biomaterials–tissue interactions (reproduced from Schoen FJ). In: *Advances in Cardiovascular Medicine* (Harvey 1602–2002 Symposium, on the 4th Centenary of William Harvey's Graduation at the University of Pauda), Thiene G, Pessina AC (eds.), Universita degli Studi di Padova, 2002; 289–307.

Mitchell describes an immunological variant of Koch's postulates to test an immunological hypothesis for calcific and noncalcific bioprosthetic valve failure (Schoen, 1999) as follows:

1. Antigen-specific elements (antibodies or cells) should be directly associated with failing valves. Moreover, control experiments should be performed to demonstrate that any antibodies or cells on implanted valves are not simply present because of surgical manipulation or aberrant flow conditions.
2. Antibodies or cells from experimental animals that have dysfunctional implanted valves transferred into an appropriate second host (immunologically matched to the original valve donor) should cause valve failure.
3. Adoptively transferred cells or antibodies should be detectable on a failed valve in the second recipient.

Although some evidence for these criteria can be obtained in humans, carefully designed animal investigations provide the only rigorous way to satisfy them. With respect to tissue heart valves, although some investigators have demonstrated that such tissues can be immunogenic, there exists no evidence that valve destruction or loss of function is mediated by immune elements, or that blockade of immune mechanisms by immunosuppression prevents that outcome.

Most biomaterials of potential clinical interest typically elicit the foreign body reaction (FBR), a special form of nonspecific inflammation. The most prominent cells in the FBR are macrophages, which attempt to phagocytose the material and are variably successful, but complete engulfment and degradation are often difficult. The macrophages, activated in the process of interacting with a biomaterial, may elaborate cytokines that stimulate inflammation or fibrosis. Multinucleated giant cells in the vicinity of a foreign body are generally considered evidence of a more severe FBR. The

more "biocompatible" the implant, the more quiescent (less inflammation in) the ultimate response. When the implant is a source of particles not easily controlled, such as wear debris from articulating joint surfaces (Jacobs *et al.*, 2001), the inability of inflammatory cells to adhere to but not phagocytose particles larger than a critical size ("frustrated phagocytosis") can lead to release of enzymes (exocytosis) and cytokines and other chemical mediators (e.g., prostaglanding, tumor necrosis factor-alpha, and interleukin-1) and cause harm to the extracellular environment. Thus, inflammatory cell products that are critical in killing microorganisms in typical inflammation can damage tissue adjacent to foreign bodies.

The nature of the reaction is largely dependent on the chemical and physical characteristics of the implant. For most inert biomaterials, the late tissue reaction is encapsulation by a relatively thin fibrous tissue capsule (composed of collagen and fibroblasts). Tissue interactions can be modified by changing the chemistry of the surface (e.g., by adding specific chemical groupings to stimulate adhesion or bone formation in orthopedic implants), inducing roughness or porosity to enhance physical binding to the surrounding tissues, incorporating a surface-active agent to chemically bond the tissue, or using a bioresorbable component to allow slow replacement by tissue to simulate natural healing properties.

SYSTEMIC AND REMOTE EFFECTS

Hensten-Pettersen and Jacobsen summarize biomaterials-related systemic toxicity and hypersensitivity reaction (through lymphatics and the bloodstream) in animals and patients with either stainless steel or cobalt-base orthopedic total joint replacement components, elevations of metallic components occur in tissue (at both local and remote sites) and in serum and urine. Transport of particulates over large distances by macrophages to regional lymph nodes and to the lungs has been considered a systemic and remote effect. As a consequence of silicone migrated through lymphatic vessels to lymph nodes, an enlarged, hard axillary lymph node in a woman who received a silicone-gel breast prosthesis for reconstruction following mastectomy for a carcinoma can be misdiagnosed as tumor.

"Metal allergy" is well-recognized and is frequently associated in women with the use of cheap, high-nickel-alloy costume jewelry or earrings and can occur in association with metallic implants (Hallab *et al.,* 2001). By themselves, metal ions lack the structural complexity required to challenge the immune system. However, when combined with proteins, such as those available in the skin, connective tissues, and blood, a wide variety of metals induce immune responses and this can have clinical effects. Cobalt, chromium, and nickel are included in this category, with nickel perhaps the most potent; at least 10% of a normal population will be sensitive by skin test to one or more of these metals, at some threshold level.

THROMBOEMBOLIC COMPLICATIONS

Hanson emphasizes that exposure of blood to an artificial surface can induce thrombosis, embolization, and consumption of platelets and plasma coagulation factors, as well as the systemic effects of activated coagulation and complement products, and platelet activation. It is clear that no synthetic or modified biological surface generated by man is as resistant to thrombosis (thromboresistant) as normal unperturbed endothelium (the cellular lining of the circulatory system). However, it is important to understand that under some circumstances endothelial cells can be "dysfunctional" and although physically intact can express prothrombotic molecules that can induce thrombosis (Bonetti *et al.,* 2003).

Thromboembolic complications are a major cause of mortality and morbidity with cardiovascular devices. Both fibrin (red) thrombus and platelet (white) thrombus form in association with valves and other cardiovascular devices. As in the cardiovascular system in general, Virchow's triad (i.e., the conditions of surface thrombogenicity, hypercoagulability, and locally static blood flow) largely predicts the relative propensity toward thrombus formation and often the location of thrombotic deposits with cardiovascular prostheses (Anderson and Schoen, 1992). However, despite over a quarter century of intense research, the physical and chemical characteristics of materials that control the outcome of blood–surface interaction are incompletely understood.

When non-physiologic surfaces contact blood, three events comprise thrombotic interactions: 1) plasma protein deposition, 2) adhesion of platelets and leukocytes, and 3) bulk fibrin formation (blood coagulation). All foreign materials exposed to blood spontaneously and rapidly (seconds) absorb a film of plasma protein, largely fibrinogen. This is followed by cellular thrombogenesis (beginning with platelet adhesion to the first adsorbed plasma proteins). If conditions of relatively static flow are present, the fiber-forming steps of the coagulation process occur, and macroscopic thrombus ensues.

Considerable evidence implicates a primary regulatory role for blood platelets in the thrombogenic response to artificial surfaces. Platelet adhesion to artificial surfaces strongly resembles that of adhesion to the vascular subendothelium that has been exposed by injury. Nevertheless, the major clinical approach to controlling thrombosis in cardiovascular devices is the use of systemic anticoagulants, particularly coumadin (warfarin), which inhibits thrombin and fibrin formation but does not inhibit platelet-mediated thrombosis.

TUMORIGENESIS

Schoen emphasizes that although animals frequently have sarcomas at the site of an experimental biomaterial implant, neoplasms in humans occurring at the site of implanted medical devices are rare, despite the large numbers of implants used clinically over an extended duration. Moreover, the presence of a neoplasm at an implant site does not prove that the implant had a causal role. Cancers associated with foreign bodies can appear at any postoperative interval but tend to occur many years postoperatively. The pathogenesis of implant-induced tumors is not well understood; most experimental data indicate that the physical rather than chemical characteristics of the foreign body primarily determine tumorigenicity.

INFECTION

Infection occurs in as many as 5 to 10% of patients with implanted prosthetic devices and is a major source of morbidity and mortality (Jansen and Peters, 1993; Klug *et al.,* 1997; Kunin *et al.,* 1988; Mulcahy, 1999; Schierholz and Beuth, 2001; Tanner *et al.,* 1997; Vlessis *et al.,* 1997). Infections associated with medical devices are often resistant to antibiotics and host defenses, often persisting until the devices are removed. Early implant infections (less than approximately 1 to 2 months postoperatively) are most likely due to intraoperative contamination from airborne sources or nonsterile surgical technique, or to early postoperative complications such as wound infection. In contrast, late infections likely occur by a hematogenous (blood-borne) route and are often initiated by bacteremia induced by therapeutic dental or genitourinary procedures. Perioperative prophylactic antibiotics and periodic antibiotic prophylaxis given shortly before diagnostic and therapeutic procedures protect against implant infection. Infections associated with foreign bodies are characterized microbiologically by a high prevalence of *Staphylococcus epidermidis* and other staphylococci, especially *S. aureus*. Ordinarily, *S. epidermidis* is an organism with low virulence and thus an infrequent cause of non-prosthesis-associated deep infections. This emphasizes the unique environment in the vicinity of a foreign body.

The presence of a foreign body per se potentiates infection. A classic experiment indicated that the staphylococcal bacterial inoculum required to cause infection in the presence of a foreign implant was 10,000 less than that when no foreign body was present (Elek and Conen, 1957). Devices could facilitate infection in several ways. Microorganisms are provided access to the circulation and to deeper tissue by damage to natural barriers against infection during implantation or subsequent function of a prosthetic device. Moreover, an implanted foreign body could (1) limit phagocyte migration into infected tissue or (2) interfere with inflammatory cell phagocytic mechanisms, through release of soluble implant components or surface-mediated interactions, thus allowing bacteria to survive adjacent to the implant. As Costerton *et al.* emphasize, adhesion of bacteria to the prosthetic surface and the formation of microcolonies within an adherent biofilm are fundamental steps in the pathogenesis of clinical and experimental infections associated with foreign bodies.

Bibliography

Anderson, J. M., and Schoen, F. J. (1992). Interactions of blood with artificial surfaces. in *Current Issues in Heart Valve Disease: Thrombosis, Embolism and Bleeding*, E. G. Butchart and E. Bodnar, eds. ICR Publishers, pp. 160–171.

Bonetti, P. O., Lerman, L. O., and Lerman, A. (2003). Endothelial dysfunction: a marker of atherosclerotic risk. *Arterioscler. Thromb. Vasc. Biol.* 23: 168–175.

Elek, S. D., and Conen, P. E. (1957). The virulence of *Staphylococcus pyogenes* for man: a study of the problems of wound infection. *Br. J. Exp. Pathol.* 38: 573–586.

Hallab, N., Merritt, K., and Jacobs, J. J. (2001). Metal sensitivity in patients with orthopaedic implants. *J. Bone Joint Surg.* 83: 428–436.

Jacobs, J. J., Roebuck, K. A., Archibeck, M., Hallab, N. J., and Glant, T. T. (2001). Osteolysis: basic science. *Clin. Orthop.* 393: 71–77.

Jansen, B., and Peters, G. (1993). Foreign body associated infection. *J. Antimicrob. Chemother.* 32: A69–A75.

Klug, D., Lacroix, D., Savoye, C., Goullard, L., Grandmougin, D., Hennequin, J. L., Kacet, S., and Lekieffre, J. (1997). Systemic infection related to endocarditis on pacemaker leads: clinical presentation and management. *Circulation* 95: 2098–2107.

Kunin, C. M., Dobbins, J. J., Melo, J. C., Levinson, M. M., Love, K., Joyce, L. D., and DeVries, W. (1988). Infectious complications in four long-term recipients of the Jarvik-7 artificial heart. *JAMA* 259: 860–864.

Mitchell, R. N. (2001). Don't blame the lymphocyte: immunologic processes are NOT important in tissue valve failure. *J. Heart Valve Dis.* 10: 467–470.

Mulcahy, J. J. (1999). Management of the infected penile implant—concepts of salvage techniques. *Int. J. Impot. Res.* 11: S58–S59.

Schierholz, J. M., and Beuth, J. (2001). Implant infections: a haven for opportunistic bacteria. *J. Hosp. Infect.* 49: 87–93.

Schoen, F., and Levey, R. J. (1999). Tissue heart valves: Currrent challenges and future research perspectives. *J. Biomed. Mater. Res.* 47: 439–465.

Tanner, A., Maiden, M. F., Lee, K., Shulman, L. B., and Weber, H. P. (1997). Dental implant infections. *Clin. Infect. Dis.* 25: S213–S217.

Vlessis, A. A., Khaki, A., Grunkemeier, G. L., Li, H. H. and Starr, A. (1997). Risk, diagnosis and management of prosthetic valve endocarditis: a review. *J. Heart Valve Dis.* 6: 443–465.

4.2 INFLAMMATION, WOUND HEALING, AND THE FOREIGN-BODY RESPONSE

James M. Anderson

Inflammation, wound healing, and foreign body reaction are generally considered as parts of the tissue or cellular host

TABLE 1   Sequence/Continuum of Host Reactions Following Implantation of Medical Devices

Injury

Blood–material interactions

Provisional matrix formation

Acute inflammation

Chronic inflammation

Granulation tissue

Foreign-body reaction

Fibrosis/fibrous capsule development

responses to injury. Table 1 lists the sequence/continuum of these events following injury. Overlap and simultaneous occurrence of these events should be considered; e.g., the foreign body reaction at the implant interface may be initiated with the onset of acute and chronic inflammation. From a biomaterials perspective, placing a biomaterial in the *in vivo* environment requires injection, insertion, or surgical implantation, all of which injure the tissues or organs involved.

The placement procedure initiates a response to injury by the tissue, organ, or body and mechanisms are activated to maintain homeostasis. The degrees to which the homeostatic mechanisms are perturbed and the extent to which pathophysiologic conditions are created and undergo resolution are a measure of the host reactions to the biomaterial and may ultimately determine its biocompatibility. Although it is convenient to separate homeostatic mechanisms into blood–material or tissue–material interactions, it must be remembered that the various components or mechanisms involved in homeostasis are present in both blood and tissue and are a part of the physiologic continuum. Furthermore, it must be noted that host reactions may be tissue-dependent, organ-dependent, and species-dependent. Obviously, the extent of injury varies with the implantation procedure.

OVERVIEW

Inflammation is generally defined as the reaction of vascularized living tissue to local injury. Inflammation serves to contain, neutralize, dilute, or wall off the injurious agent or process. In addition, it sets into motion a series of events that may heal and reconstitute the implant site through replacement of the injured tissue by regeneration of native parenchymal cells, formation of fibroblastic scar tissue, or a combination of these two processes.

Immediately following injury, there are changes in vascular flow, caliber, and permeability. Fluid, proteins, and blood cells escape from the vascular system into the injured tissue in a process called exudation. Following changes in the vascular system, which also include changes induced in blood and its components, cellular events occur and characterize the inflammatory response. The effect of the injury and/or biomaterial *in situ* on plasma or cells can produce chemical factors

TABLE 2 Cells and Components of Vascularized Connective Tissue

Intravascular (blood) cells
 Erythrocytes (RBC)
 Neutrophils (PMNs, polymorphonuclear leukocytes)
 Monocytes
 Eosinophils
 Lymphocytes
 Plasma cells
 Basophils
 Platelets

Connective tissue cells
 Mast cells
 Fibroblasts
 Macrophages
 Lymphocytes

Extracellular matrix components
 Collagens
 Elastin
 Proteoglycans
 Fibronectin
 Laminin

that mediate many of the vascular and cellular responses of inflammation.

Blood–material interactions and the inflammatory response are intimately linked, and in fact, early responses to injury involve mainly blood and vasculature. Regardless of the tissue or organ into which a biomaterial is implanted, the initial inflammatory response is activated by injury to vascularized connective tissue (Table 2). Since blood and its components are involved in the initial inflammatory responses, blood clot formation and/or thrombosis also occur. Blood coagulation and thrombosis are generally considered humoral responses and may be influenced by other homeostatic mechanisms such as the extrinsic and intrinsic coagulation systems, the complement system, the fibrinolytic system, the kinin-generating system, and platelets. Thrombus or blood clot formation on the surface of a biomaterial is related to the well-known Vroman effect (see Chapter 3.2), in which a hierarchical and dynamic series of collision, absorption, and exchange processes, determined by protein mobility and concentration, regulate early time-dependent changes in blood protein adsorption. From a wound-healing perspective, blood protein deposition on a biomaterial surface is described as provisional matrix formation. Blood interactions with biomaterials are generally considered under the category of hematocompatibility and are discussed elsewhere in this book.

Injury to vascularized tissue in the implantation procedure leads to immediate development of the provisional matrix at the implant site. This provisional matrix consists of fibrin, produced by activation of the coagulation and thrombosis systems, and inflammatory products released by the complement system, activated platelets, inflammatory cells, and endothelial cells. These events occur early, within minutes to hours following implantation of a medical device. Components within

or released from the provisional matrix, i.e., fibrin network (thrombosis or clot), initiate the resolution, reorganization, and repair processes such as inflammatory cell and fibroblast recruitment. The provisional matrix appears to provide both structural and biochemical components to the process of wound healing. The complex three-dimensional structure of the fibrin network with attached adhesive proteins provides a substrate for cell adhesion and migration. The presence of mitogens, chemoattractants, cytokines, and growth factors within the provisional matrix provides for a rich milieu of activating and inhibiting substances for various cellular proliferative and synthetic processes. The provisional matrix may be viewed as a naturally derived, biodegradable, sustained release system in which mitogens, chemoattractants, cytokines, and growth factors are released to control subsequent wound-healing processes. In spite of the increase in our knowledge of the provisional matrix and its capabilities, our knowledge of the control of the formation of the provisional matrix and its effect on subsequent wound healing events is poor. In part, this lack of knowledge is due to the fact that much of our knowledge regarding the provisional matrix has been derived from *in vitro* studies, and there is a paucity of *in vivo* studies that provide for a more complex perspective. Little is known regarding the provisional matrix which forms at biomaterial and medical device interfaces *in vivo*, although attractive hypotheses have been presented regarding the presumed ability of materials and protein adsorbed materials to modulate cellular interactions through their interactions with adhesive molecules and cells.

The predominant cell type present in the inflammatory response varies with the age of the inflammatory injury (Fig. 1). In general, neutrophils predominate during the first several days following injury and then are replaced by monocytes as the predominant cell type. Three factors account for this change in cell type: neutrophils are short lived and disintegrate and disappear after 24–48 hour; neutrophil emigration from the vasculature to the tissues is of short duration; and chemotactic factors for neutrophil migration are activated early in the inflammatory response. Following emigration from the vasculature, monocytes differentiate into macrophages and these cells are very long-lived (up to months). Monocyte emigration

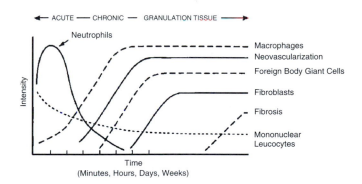

FIG. 1. The temporal variation in the acute inflammatory response, chronic inflammatory response, granulation tissue development, and foreign-body reaction to implanted biomaterials. The intensity and time variables are dependent upon the extent of injury created in the implantation and the size, shape, topography, and chemical and physical properties of the biomaterial.

may continue for day to weeks, depending on the injury and implanted biomaterial, and chemotactic factors for monocytes are activated over longer periods of time.

The temporal sequence of events following implantation of a biomaterial is illustrated in Fig. 1. The size, shape, and chemical and physical properties of the biomaterial may be responsible for variations in the intensity and duration of the inflammatory or wound-healing process. Thus, intensity and/or time duration of the inflammatory reaction may characterize the biocompatibility of a biomaterial.

While injury initiates the inflammatory response, the chemicals released from plasma, cells, or injured tissue mediate the inflammatory response. Important classes of chemical mediators of inflammation are presented in Table 3. Several points must be noted in order to understand the inflammatory response and how it relates to biomaterials. First, although chemical mediators are classified on a structural or functional basis, different mediator systems interact and provide a system of checks and balances regarding their respective activities and functions. Second, chemical mediators are quickly inactivated or destroyed, suggesting that their action is predominantly local (i.e., at the implant site). Third, generally the lysosomal proteases and the oxygen-derived free radicals produce the most significant damage or injury. These chemical mediators are also important in the degradation of biomaterials.

TABLE 3 Important Chemical Mediators of Inflammation Derived from Plasma, Cells, or Injured Tissue

Mediators	Examples
Vasoactive agents	Histamines, serotonin, adenosine, endothelial-derived relaxing factor (EDRF), prostacyclin, endothelin, thromboxane α_2
Plasma proteases	
Kinin system	Bradykinin, kallikrein
Complement system	C3a, C5a, C3b, C5b-C9
Coagulation/fibrinolytic system	Fibrin degradation products, activated Hageman factor (FXIIA), tissue plasminogen activator (tPA)
Leukotrienes	Leukotriene B$_4$ (LTB$_4$), hydroxyeicosatetranoic acid (HETE)
Lysosomal proteases	Collagenase, elastase
Oxygen-derived free radicals	H$_2$O$_2$, superoxide anion
Platelet activating factors	Cell membrane lipids
Cytokines	Interleukin 1 (IL-1), tumor necrosis factor (TNF)
Growth factors	Platelet-derived growth factor (PDGF), fibroblast growth Factor (FGF), transforming growth factor TGF-α or (TGF-β), epithelial growth factor (EGF)

ACUTE INFLAMMATION

Acute inflammation is of relatively short duration, lasting for minutes to hours to days, depending on the extent of injury. Its main characteristics are the exudation of fluid and plasma proteins (edema) and the emigration of leukocytes (predominantly neutrophils). Neutrophils (polymorphonuclear leukocytes, PMNs) and other motile white cells emigrate or move from the blood vessels to the perivascular tissues and the injury (implant) site. Leukocyte emigration is assisted by "adhesion molecules" present on leukocyte and endothelial surfaces. The surface expression of these adhesion molecules can be induced, enhanced, or altered by inflammatory agents and chemical mediators. White cell emigration is controlled, in part, by chemotaxis, which is the unidirectional migration of cells along a chemical gradient. A wide variety of exogenous and endogenous substances have been identified as chemotactic agents. Specific receptors for chemotactic agents on the cell membranes of leukocytes are important in the emigration or movement of leukocytes. These and other receptors also play a role in the transmigration of white cells across the endothelial lining of vessels and activation of leukocytes. Following localization of leukocytes at the injury (implant) site, phagocytosis and the release of enzymes occur following activation of neutrophils and macrophages. The major role of the neutrophil in acute inflammation is to phagocytose microorganisms and foreign materials. Phagocytosis is seen as a three-step process in which the injurious agent undergoes recognition and neutrophil attachment, engulfment, and killing or degradation. In regard to biomaterials, engulfment and degradation may or may not occur, depending on the properties of the biomaterial.

Although biomaterials are not generally phagocytosed by neutrophils or macrophages because of the disparity in size (i.e., the surface of the biomaterial is greater than the size of the cell), certain events in phagocytosis may occur. The process of recognition and attachment is expedited when the injurious agent is coated by naturally occurring serum factors called "opsonins." The two major opsonins are immunoglobulin G (IgG) and the complement-activated fragment, C3b. Both of these plasma-derived proteins are known to adsorb to biomaterials, and neutrophils and macrophages have corresponding cell-membrane receptors for these opsonization proteins. These receptors may also play a role in the activation of the attached neutrophil or macrophage. Other blood proteins such as fibrinogen, fibronectin, and vitronectin may also facilitate cell adhesion to biomaterial surfaces. Owing to the disparity in size between the biomaterial surface and the attached cell, frustrated phagocytosis may occur. This process does not involve engulfment of the biomaterial but does cause the extracellular release of leukocyte products in an attempt to degrade the biomaterial.

Henson has shown that neutrophils adherent to complement-coated and immunoglobulin-coated nonphagocytosable surfaces may release enzymes by direct extrusion or exocytosis from the cell. The amount of enzyme released during this process depends on the size of the polymer particle, with larger particles inducing greater amounts of enzyme release. This suggests that the specific mode of cell activation in the inflammatory response in tissue depends upon the size of the implant

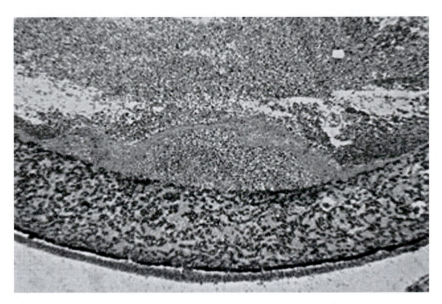

FIG. 2. Acute inflammation, secondary to infection, of an ePTFE vascular graft. A focal zone of polymorphonuclear leukocytes is present at the lumenal surface of the vascular graft, surrounded by a fibrin cap, on the blood-contacting surface of the ePTFE vascular graft. Hematoxylin and eosin stain. Original magnification 4×. (See color plate)

and that a material in a phagocytosable form (i.e., powder or particulate) may provoke a different degree of inflammatory response than the same material in a nonphagocytosable form (i.e., film).

Acute inflammation normally resolves quickly, usually less than 1 week, depending on the extent of injury at the implant site. However, the presence of acute inflammation (i.e., PMNs) at the tissue/implant interface at time periods beyond 1 week (i.e., weeks, months, or years) suggests the presence of an infection (Fig. 2).

CHRONIC INFLAMMATION

Chronic inflammation is less uniform histologically than acute inflammation. In general, chronic inflammation is characterized by the presence of macrophages, monocytes, and lymphocytes, with the proliferation of blood vessels and connective tissue. Many factors can modify the course and histologic appearance of chronic inflammation.

Persistent inflammatory stimuli lead to chronic inflammation. While the chemical and physical properties of the biomaterial in themselves may lead to chronic inflammation, motion in the implant site by the biomaterial or infection may also produce chronic inflammation. The chronic inflammatory response to biomaterials is usually of short duration and is confined to the implant site. The presence of mononuclear cells, including lymphocytes and plasma cells, is considered chronic inflammation, whereas the foreign-body reaction with the development of granulation tissue is considered the normal wound healing response to implanted biomaterials (i.e., the normal foreign-body reaction). Chronic inflammation with the presence of collections of lymphocytes and monocytes

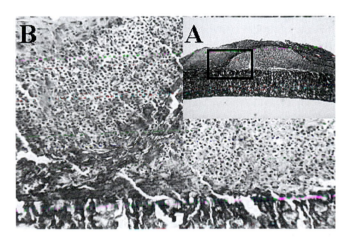

FIG. 3. Chronic inflammation, secondary to infection, of an ePTFE arteriovenous shunt for renal dialysis. (A) Low-magnification view of a focal zone of chronic inflammation. (B) High-magnification view of the outer surface with the presence of monocytes and lymphocytes at an area where the outer PTFE wrap had peeled away from the vascular graft. Hematoxylin and eosin stain. Original magnification (A) 4×, (B) 20×. (See color plate)

at extended implant times (weeks, months, years) also may suggest the presence of a long-standing infection (Fig. 3A, B).

Lymphocytes and plasma cells are involved principally in immune reactions and are key mediators of antibody production and delayed hypersensitivity responses. Although they may be present in nonimmunologic injuries and inflammation their roles in such circumstances are largely unknown. Little is known regarding humoral immune responses and cell-mediated immunity to synthetic biomaterials. The role of macrophages must be considered in the possible development

TABLE 4 Tissues and Cells of MPS and RES

Tissues	Cells
Implant sites	Inflammatory macrophages
Liver	Kupffer cells
Lung	Alveolar macrophages
Connective tissue	Histiocytes
Bone marrow	Macrophages
Spleen and lymph nodes	Fixed and free macrophages
Serous cavities	Pleural and peritoneal macrophages
Nervous system	Microglial cells
Bone	Osteoclasts
Skin	Langerhans' cells
Lymphoid tissue	Dendritic cells

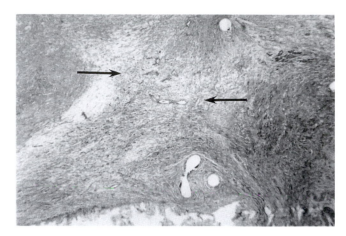

FIG. 4. Granulation tissue in the anastomotic hyperplasia at the anastomosis of an ePTFE vascular graft. Capillary development (red slits) and fibroblast infiltration with collagen deposition (blue) from the artery form the granulation tissue (arrows). Masson's Trichrome stain. Original magnification 4×. (See color plate)

of immune responses to synthetic biomaterials. Macrophages process and present the antigen to immunocompetent cells and thus are key mediators in the development of immune reactions.

Monocytes and macrophages belong to the mononuclear phagocytic system (MPS), also known as the reticuloendothelial system (RES). These systems consist of cells in the bone marrow, peripheral blood, and specialized tissues. Table 4 lists the tissues that contain cells belonging to the MPS or RES. The specialized cells in these tissues may be responsible for systemic effects in organs or tissues secondary to the release of components or products from implants through various tissue–material interactions (e.g., corrosion products, wear debris, degradation products) or the presence of implants (e.g., microcapsule or nanoparticle drug-delivery systems).

The macrophage is probably the most important cell in chronic inflammation because of the great number of biologically active products it can produce. Important classes of products produced and secreted by macrophages include neutral proteases, chemotactic factors, arachidonic acid metabolites, reactive oxygen metabolites, complement components, coagulation factors, growth-promoting factors, and cytokines.

Growth factors such as platelet-derived growth factor (PDGF), fibroblast growth factor (FGF), transforming growth factor-β (TGF-β), TGF-α/epidermal growth factor (EGF), and interleukin-1 (IL-1) or tumor necrosis factor (TNF-α) are important to the growth of fibroblasts and blood vessels and the regeneration of epithelial cells. Growth factors released by activated cells can stimulate production of a wide variety of cells; initiate cell migration, differentiation, and tissue remodeling; and may be involved in various stages of wound healing.

GRANULATION TISSUE

Within 1 day following implantation of a biomaterial (i.e., injury), the healing response is initiated by the action of monocytes and macrophages. Fibroblasts and vascular endothelial cells in the implant site proliferate and begin to form granulation tissue, which is the specialized type of tissue that is the hallmark of healing inflammation. Granulation tissue derives its name from the pink, soft granular appearance on the surface of healing wounds, and its characteristic histologic features include the proliferation of new small blood vessels and fibroblasts (Fig. 4). Depending on the extent of injury, granulation tissue may be seen as early as 3–5 days following implantation of a biomaterial.

The new small blood vessels are formed by budding or sprouting of preexisting vessels in a process known as neovascularization or angiogenesis. This process involves proliferation, maturation, and organization of endothelial cells into capillary vessels. Fibroblasts also proliferate in developing granulation tissue and are active in synthesizing collagen and proteoglycans. In the early stages of granulation tissue development, proteoglycans predominate but later collagen, especially type III collagen, predominates and forms the fibrous capsule. Some fibroblasts in developing granulation tissue may have the features of smooth muscle cells, i.e., actin microfilaments. These cells are called myofibroblasts and are considered to be responsible for the wound contraction seen during the development of granulation tissue. Macrophages are almost always present in granulation tissue. Other cells may also be present if chemotactic stimuli are generated.

The wound-healing response is generally dependent on the extent or degree of injury or defect created by the implantation procedure. Wound healing by primary union or first intention is the healing of clean, surgical incisions in which the wound edges have been approximated by surgical sutures. Healing under these conditions occurs without significant bacterial contamination and with a minimal loss of tissue. Wound healing by secondary union or second intention occurs when there is a large tissue defect that must be filled or there is extensive loss of cells and tissue. In wound healing by secondary intention, regeneration of parenchymal cells cannot completely reconstitute the original architecture and much

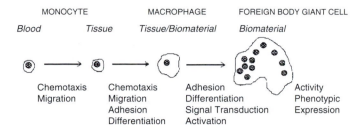

FIG. 5. *In vivo* transition from blood-borne monocyte to biomaterial adherent monocyte/macrophage to foreign-body giant cell at the tissue–biomaterial interface. Little is known regarding the indicated biological responses, which are considered to play important roles in the transition to FBGC development.

larger amounts of granulation tissue are formed that result in larger areas of fibrosis or scar formation. Under these conditions, different regions of tissue may show different stages of the wound-healing process simultaneously.

Granulation tissue is distinctly different from granulomas, which are small collections of modified macrophages called epithelioid cells. Langhans' or foreign-body-type giant cells may surround nonphagocytosable particulate materials in granulomas. Foreign-body giant cells are formed by the fusion of monocytes and macrophages in an attempt to phagocytose the material (Fig. 5).

FOREIGN-BODY REACTION

The foreign-body reaction to biomaterials is composed of foreign-body giant cells (FBGCs) and the components of granulation tissue (e.g., macrophages, fibroblasts, and capillaries in varying amounts, depending upon the form and topography of the implanted material; (Fig. 6). Relatively flat and smooth surfaces such as that found on breast prostheses have a foreign-body reaction that is composed of a layer of macrophages one

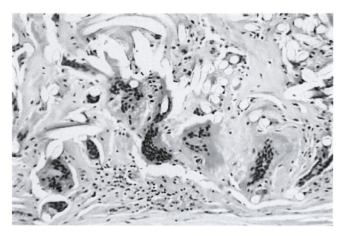

FIG. 7. Foreign-body reaction with multinucleated foreign body giant cells and macrophages at the periadventitial (outer) surface of a Dacron vascular graft. Fibers from the Dacron vascular graft are identified as clear oval voids. Hematoxylin and eosin stain. Original magnification 20×. (See color plate)

to two cells in thickness. Relatively rough surfaces such as those found on the outer surfaces of expanded poly tetrafluoroethylene (ePTFE) or Dacron vascular prostheses have a foreign-body reaction composed of macrophages and foreign-body giant cells at the surface. Fabric materials generally have a surface response composed of macrophages and foreign body giant cells, with varying degrees of granulation tissue subjacent to the surface response (Fig. 7).

As previously discussed, the form and topography of the surface of the biomaterial determine the composition of the foreign-body reaction. With biocompatible materials, the composition of the foreign-body reaction in the implant site may be controlled by the surface properties of the biomaterial, the form of the implant, and the relationship between the surface area of the biomaterial and the volume of the implant. For example, high-surface-to-volume implants such as fabrics, porous

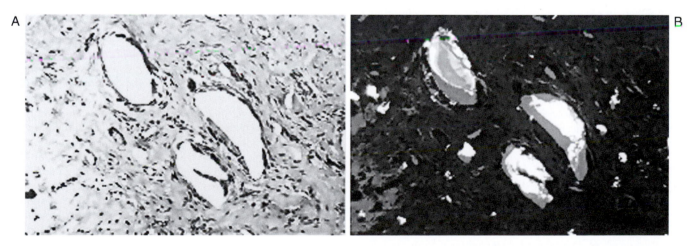

FIG. 6. (A) Focal foreign-body reaction to polyethylene wear particulate from a total knee prosthesis. Macrophages and foreign-body giant cells are identified within the tissue and lining the apparent void spaces indicative of polyethylene particulate. Hematoxylin and eosin stain. Original magnification 20×. (B) Partial polarized light view. Polyethylene particulate is identified within the void spaces commonly seen under normal light microscopy. Hematoxylin and eosin stain. Original magnification 20×. (See color plate)

materials, particulate, or microspheres will have higher ratios of macrophages and foreign-body giant cells in the implant site than smooth-surface implants, which will have fibrosis as a significant component of the implant site.

The foreign-body reaction consisting mainly of macrophages and/or foreign-body giant cells may persist at the tissue–implant interface for the lifetime of the implant (Fig. 1). Generally, fibrosis (i.e., fibrous encapsulation) surrounds the biomaterial or implant with its interfacial foreign-body reaction, isolating the implant and foreign-body reaction from the local tissue environment. Early in the inflammatory and wound-healing response, the macrophages are activated upon adherence to the material surface.

Although it is generally considered that the chemical and physical properties of the biomaterial are responsible for macrophage activation, the subsequent events regarding the activity of macrophages at the surface are not clear. Tissue macrophages, derived from circulating blood monocytes, may coalesce to form multinucleated foreign-body giant cells. It is not uncommon to see very large foreign-body giant cells containing large numbers of nuclei on the surface of biomaterials. While these foreign-body giant cells may persist for the lifetime of the implant, it is not known if they remain activated, releasing their lysosomal constituents, or become quiescent.

Figure 5 demonstrates the progression from circulating blood monocyte to tissue macrophage to foreign-body giant cell development that is most commonly observed. Indicated in the figure are important biological responses that are considered to play an important role in FBGC development. Material surface chemistry may control adherent macrophage apoptosis (i.e., programmed cell death) (see Chapter 3.3) that renders potentially harmful macrophages nonfunctional, while the surrounding environment of the implant remains unaffected. The level of adherent macrophage apoptosis appears to be inversely related to the surface's ability to promote diffusion of macrophages into FBGCs, suggesting a mechanism for macrophages to escape apoptosis.

Figure 8 demonstrates the sequence of events involved in inflammation and wound healing when medical devices are implanted. In general, the PMN predominant acute inflammatory response and the lymphocyte/monocyte predominant chronic inflammatory response resolve quickly (i.e., within 2 weeks) depending on the type and location of the implant. Studies using IL-4 or IL-13, respectively, demonstrate the role for Th2 helper lymphocytes in the development of the foreign body reaction at the tissue/material interface. Th2 helper lymphocytes have been described as "antiinflammatory" based on their cytokine profile, of which IL-4 is a significant component.

FIBROSIS/FIBROUS ENCAPSULATION

The end-stage healing response to biomaterials is generally fibrosis or fibrous encapsulation (Fig. 9). However, there may be exceptions to this general statement (e.g., porous materials inoculated with parenchymal cells or porous materials implanted into bone) (Fig. 10). As previously stated, the tissue response to implants is in part dependent upon the extent of injury or defect created in the implantation procedure and the amount of provisional matrix.

Repair of implant sites can involve two distinct processes: regeneration, which is the replacement of injured tissue by parenchymal cells of the same type, or replacement by connective tissue that constitutes the fibrous capsule. These processes are generally controlled by either (1) the proliferative capacity of the cells in the tissue or organ receiving the implant and the extent of injury as it relates to the destruction, or (2) persistence of the tissue framework of the implant site. See Chapter 3.4 for a more complete discussion of the types of cells present in the organ parenchyma and stroma, respectively.

The regenerative capacity of cells allows them to be classified into three groups: labile, stable (or expanding), and permanent (or static) cells (see Chapter 3.3). Labile cells continue to proliferate throughout life; stable cells retain this capacity but do not normally replicate; and permanent cells cannot reproduce themselves after birth. Perfect repair with restitution of normal structure can theoretically occur only in tissues consisting of stable and labile cells, whereas all injuries to tissues composed of permanent cells may give rise to fibrosis and fibrous capsule formation with very little restitution of the normal tissue or organ structure. Tissues composed of permanent cells (e.g., nerve cells and cardiac muscle cells) most commonly undergo an organization of the inflammatory exudate, leading to fibrosis. Tissues of stable cells (e.g., parenchymal cells of the liver, kidney, and pancreas); mesenchymal cells (e.g., fibroblasts, smooth muscle cells, osteoblasts, and chondroblasts); and vascular endothelial and labile cells (e.g., epithelial cells and lymphoid and hematopoietic cells) may also follow this pathway to fibrosis or may undergo resolution of the inflammatory exudate, leading to restitution of the normal tissue structure.

The condition of the underlying framework or supporting stroma of the parenchymal cells following an injury plays an important role in the restoration of normal tissue structure. Retention of the framework with injury may lead to restitution of the normal tissue structure, whereas destruction of the framework most commonly leads to fibrosis. It is important to consider the species-dependent nature of the regenerative capacity of cells. For example, cells from the same organ or tissue but from different species may exhibit different regenerative capacities and/or connective tissue repair.

Following injury, cells may undergo adaptations of growth and differentiation. Important cellular adaptations are atrophy (decrease in cell size or function), hypertrophy (increase in cell size), hyperplasia (increase in cell number), and metaplasia (change in cell type). Other adaptations include a change by cells from producing one family of proteins to another (phenotypic change), or marked overproduction of protein. This may be the case in cells producing various types of collagens and extracellular matrix proteins in chronic inflammation and fibrosis. Causes of atrophy may include decreased workload (e.g., stress-shielding by implants), and diminished blood supply and inadequate nutrition (e.g., fibrous capsules surrounding implants).

Local and systemic factors may play a role in the wound-healing response to biomaterials or implants. Local factors include the site (tissue or organ) of implantation, the adequacy

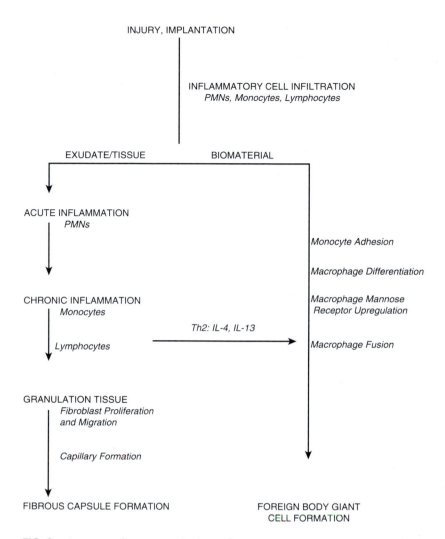

FIG. 8. Sequence of events involved in inflammatory and wound-healing responses leading to foreign-body giant cell formation. This shows the important of Th2 lymphocytes in the transient chronic inflammatory phase with the production of IL-4 and IL-13, which can induce monocyte/macrophage fusion to form foreign-body giant cells.

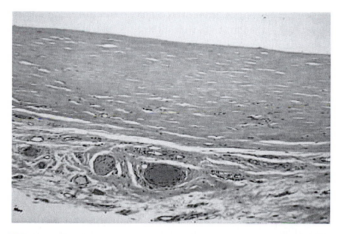

FIG. 9. Fibrous capsule composed of dense, compacted collagen. This fibrous capsule had formed around a Mediport catheter reservoir. Loose connective tissue with small arteries, veins, and a nerve is identified below the acellular fibrous capsule. (See color plate)

of blood supply, and the potential for infection. Systemic factors may include nutrition, hematologic derangements, glucocortical steroids, and preexisting diseases such as atherosclerosis, diabetes, and infection.

Finally, the implantation of biomaterials or medical devices may be best viewed at present from the perspective that the implant provides an impediment or hindrance to appropriate tissue or organ regeneration and healing. Given our current inability to control the sequence of events following injury in the implantation procedure, restitution of normal tissue structures with function is rare. Current studies directed toward developing a better understanding of the modification of the inflammatory response, stimuli providing for appropriate proliferation of permanent and stable cells, and the appropriate application of growth factors may provide keys to the control of inflammation, wound healing, and fibrous encapsulation of biomaterials.

FIG. 10. Fibrous capsule with a focal foreign-body reaction to silicone gel from a silicone gel–filled silicone-rubber breast prosthesis. The breast prosthesis–tissue interface is at the top of the photomicrograph. Oval void spaces lined by macrophages and a few giant cells are identified and a focal area of foamy macrophages (arrows) indicating macrophage phagocytosis of silicone gel is identified. Hematoxylin and eosin stain. Original magnification 10×. (See color plate)

Bibliography

Anderson, J. M. (2000). Multinucleated giant cells. *Curr. Opin. Hematol.* **7**: 40–47.

Anderson, J. M. (2001). Biological responses to materials. *Ann. Rev. Mater. Res.* **31**: 81–110.

Brodbeck, W. G., Shive, M. S., Colton, E., Nakayama, Y., Matsuda, T., and Anderson, J. M. (2001). Influence of biomaterials surface chemistry on apoptosis of adherent cells. *J. Biomed. Mater. Res.* **55**: 661–668.

Brodbeck, W. G., Voskerician, G., Ziats, N. P., Nakayama, Y., Matsuda T., and Anderson, J. M. (2001). In vivo leukocyte cytokine mRNA responses to biomaterials is dependent on surface chemistry. *J. Biomed. Mater. Res.* **64A**: 320–329.

Browder, T., Folkman, J., and Pirie-Shepherd, S. (2000). The hemostatic system as a regulator of angiogenesis. *J. Biol. Chem.* **275**: 1521–1524.

Clark, R. A. F., ed. (1996). *The Molecular and Cellular Biology of Wound Repair.* Plenum Publishing, New York.

Cotran, R. Z., Kumar, V., and Robbins, S. L., eds. (1999). Inflammation and repair. in *Pathologic Basis of Disease*, 6th ed. WB Saunders, Philadelphia, pp. 50–112.

Gallin, J. I., Snyderman, R., eds. (1999). *Inflammation: Basic Principles and Clinical Correlates*, 3rd ed. Raven Press, New York.

Henson, P. M. (1971). The immunologic release of constituents from neutrophil leukocytes: II. Mechanisms of release during phagocytosis, and adherence to nonphagocytosable surfaces. *J. Immunol.* **107**: 1547.

Hunt, T. K., Heppenstall, R. B., Pines, E., and Rovee, D., eds. (1984). *Soft and Hard Tissue Repair.* Praeger Scientific, New York.

Hynes, R. O. (2002). Integrins: bidirectional, allosteric signaling machines. *Cell* **110**: 673–687.

Hynes, R. O., and Zhao, Q. (2000). The evolution of cell adhesion. *J. Cell. Biol.* **150**: F89–96.

McNally, A. K., and Anderson, J. M. (1995). Interleukin-4 induces foreign body giant cells from human monocytes/macrophages. Differential lymphokine regulation of macrophage fusion leads to morphological variants of multinucleated giant cells. *Am. J. Pathol.* **147**: 1487–1499.

McNally, A. K., and Anderson, J. M. (2002). Beta1 and beta2 integrins mediate adhesion during macrophage fusion and multinucleated foreign body giant cell formation. *Am. J. Pathol.* **160**: 621–630.

Nguyen, L. L., and D'Amore, P. A. (2001). Cellular interactions in vascular growth and differentiation. *Int. Rev. Cytol.* **204**: 1–48.

Pierce, G. F. (2001). Inflammation in nonhealing diabetic wounds: the space-time continuum does matter. *Am. J. Pathol.* **159**(2): 399–403.

4.3 INNATE AND ADAPTIVE IMMUNITY: THE IMMUNE RESPONSE TO FOREIGN MATERIALS

Richard N. Mitchell

This is a fairly extensive topic, typically encompassing an entire course (with its own introductory text) called "Immunology." Thus, an overview chapter can only begin to acquaint the reader with the complexities of innate and adaptive immunity. Nevertheless, the goal here is to understand the broad organization of the immune system (specific and nonspecific components), how the different elements recognize perceived "invaders," and what effector responses are elicited. The end result is to understand the response of the body to the insertion of a foreign device, and to predict the potential outcome. For more extensive discussion of some aspect of the immune system, you are encouraged to refer to any of a number of excellent basic immunology texts (Abbas and Lichtman, 2003; Benjamini *et al.*, 2000; Janeway, 2001).

OVERVIEW

The function of the immune system is ultimately to defend the host against infectious organisms. The immune system is triggered into action whenever the host perceives tissue injury, anticipating that with that injury, microbial agents either have been causal or will become secondarily involved. The immune system accomplishes its protective role by stratifying the plethora of molecules it may ultimately contact as either "self" or "non-self." In general, the immune system does not react to self molecules or injure host tissues. However, when a particular molecule is perceived as nonself, the full gamut of immune responses are brought to bear in an attempt to remove or isolate it. In most cases, the immune response is so exquisitely specific and well-regulated that host tissues are not significantly affected. However, severe infections, persistent injury, or autoimmunity (inappropriate immune response to self) can lead to substantial tissue injury directly attributable to the host immune system. Thus, although the system evolved primarily to identify and eliminate infectious agents, noninfectious foreign materials also elicit immune responses, occasionally culminating (even if not infected) in severe tissue injury. Consequently, a more inclusive definition of immunity is a reaction to any foreign substance (microbes, proteins, polysaccharides, Silastic implants, etc.) regardless of the pathologic consequences. In order to understand the basics of the immune response, we will initially focus in this chapter on the physiologic pathways of immune responses to infectious agents. Once we understand those pathways, the

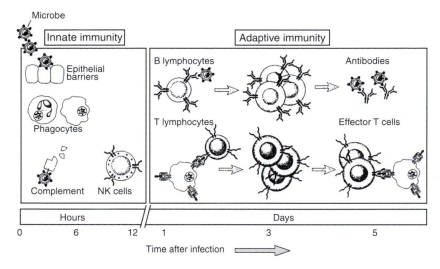

FIG. 1. Innate and adaptive immunity. Although with a limited ability to recognize invading organisms, the relatively primitive (evolutionarily speaking) components of innate immunity provide the first line of defense against microbial infections. Adaptive immunity, with exquisite specificity to any particular infectious agent, develops sometime later after innate components have responded. Notably, the elements of innate immunity not only respond first, but direct subsequent adaptive immunity; in turn, elements of the adaptive immune responses orchestrate a more efficient and vigorous response by the components of innate immunity. The specific kinetics of the responses shown are approximations and may vary depending on the inciting agent. Figure reprinted with permission from Abbas and Lichtman (2003).

mechanisms underlying a response to a foreign body will be briefly examined.

INNATE AND ADAPTIVE IMMUNITY

Defense against microbes is a two-stage process, beginning with a relatively nonspecific innate response to "injury," followed by a targeted adaptive response more uniquely focused on the specific causal agent (Fig. 1, Table 1).

INNATE IMMUNITY

Rapid (hours) response to infection is accomplished by the components of innate immunity (also called "natural" or

TABLE 1 Components of Innate versus Adaptive Immunity

	Innate	Adaptive
Physical/chemical barriers	Skin, mucosal epithelium, antimicrobial proteins	Lymphocytes in epithelia, secreted antibodies
Blood proteins	Complement	Antibodies
Cells	Phagocytes (macrophages, neutrophils), natural killer cells	Lymphocytes

Adapted from Abbas and Lichtman (2003).

"native" immunity) (Medzhitov and Janeway, 2000). Innate immunity is an evolutionarily primitive system found even in invertebrates and to some extent in plants. Cells and proteins of this system constitute the first line of defense, and in many cases, can also quite capably eliminate infections on their own. The components of innate immunity are also critical in mobilizing all subsequent effectors—including elements of adaptive immunity—to clear invading microorganisms (Fig. 2). Innate immunity is triggered by molecular structures that are common to groups of related microbes (Fig. 3) (Janeway and Medzhitov, 2002). The receptors for recognition therefore have a fairly limited diversity (fewer than 20 different types of molecules) and have no capacity to make fine distinctions between different substances; each cell of the innate system also expresses essentially the same cohort of receptors. The components of innate immunity react in essentially the same way each time they encounter the same infectious agent, so that there is no functional memory to allow more rapid or specific responses upon second encounter with the same agent. The principal components of innate immunity are:

- Physical and chemical barriers such as epithelia and antimicrobial proteins (e.g., defensins)
- Phagocytic cells (neutrophils and macrophages) that ingest (phagocytize) and destroy microbes (Fig. 4) (Underhill and Ozinsky, 2002)
- Natural killer (NK) cells that kill non-self targets
- Circulating proteins (complement, coagulation factors, C-reactive protein, etc.) that either directly insert pore-forming proteins in microbes that lead to cell death

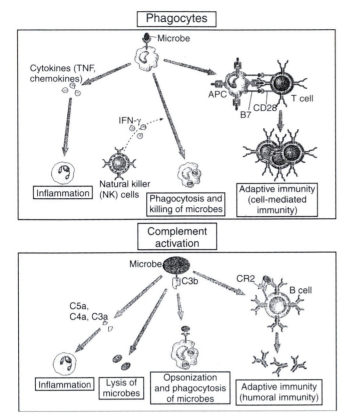

FIG. 2. Basic mechanisms of innate immunity. The two principal components of innate immunity in defense against microbial infection are phagocytes (cells) and complement (a collection of proteins, see also Chapter 4.3). Phagocytes (primarily the macrophage cell type shown here) will directly bind, ingest, and intracellularly degrade various microbes; in addition, macrophages can secrete cytokines to recruit and activate additional inflammatory cell types (e.g., neutrophils) to sites of infection and will help drive the activation of the T lymphocytes of the adaptive immune response. Note that macrophages may also require the production of cytokines by other cell types (in the figure, "IFN-γ" is interferon-γ, a cytokine with a variety of activities) in order to be most active. Complement proteins form pores in the membranes of microbes to cause direct cell lysis; in addition, complement components will incite inflammatory cell recruitment, augment phagocytosis (opsonize microbes), and participate in the activation of B lymphocytes in the adaptive immune response. Figure reprinted with permission from Abbas *et al.* (2000).

(e.g., complement, see Chapter 4.3), or that opsonize microbes (rendering them more "attractive" and readily phagocytized) (Fig. 5) (Barrington *et al.*, 2001; Walport, 2001a, b).

- Cytokines: proteins secreted by cells of innate or adaptive immunity that regulate and coordinate the cellular response (Seder and Gazzinelli, 1998).

ADAPTIVE IMMUNITY

Cellular and circulating protein (also called humoral) mediators that temporally follow innate immunity in recognizing and resolving infections (also called "specific" or "acquired"

immunity) are termed adaptive immunity. Adaptive immunity is more evolutionarily advanced, first seen in phylogenic development with the jawed vertebrates. It has the cardinal features of exquisite specificity for distinct macromolecules and memory, the latter being the ability to respond more vigorously to subsequent exposures of the same microorganism. Adaptive immunity also has a virtually limitless diversity, with the capacity to recognize 10^9–10^{11} distinct antigenic determinants (Fig. 3). Each cell of the adaptive immune system can recognize and respond to only a single antigenic determinant, so that the immense diversity of the system requires an equally large number of different cells. Foreign substances that induce these specific immune responses are called antigens and each antigen will typically activate only one (or a small set) of cells. The principal components of adaptive immunity (see also later discussion) are:

- T lymphocytes (also known as T cells), functionally divided into helper T cells (Th cells), which provide signals and soluble protein mediators (cytokines) to orchestrate the activity of other cell types, and cytotoxic T cells (Tc cells) which kill selected target cells
- B lymphocytes (also known as B cells), responsible for making antibodies
- Antibodies: proteins secreted by B lymphocytes with specificity for a specific antigen
- Cytokines: proteins secreted by cells of innate or adaptive immunity that regulate and coordinate the cellular response.

Innate and adaptive immunity are integrated components in the host defense response; the cells and proteins of each system function cooperatively. Thus, the initial innate response to microbes stimulates adaptive immune responses and influences how the adaptive immune system will respond (e.g., antibodies versus cellular mediators). Moreover, adaptive immunity directs and utilizes the effector mechanisms of innate immunity to clear infectious agents.

RECOGNITION AND EFFECTOR PATHWAYS IN INNATE IMMUNITY

The components of the innate system recognize structures that are characteristic of microbial pathogens and are not present on mammalian tissues; thus recognition via this pathway can distinguish self and non-self (Fig. 3) (Janeway and Medzhitov, 2002). Because the microbial products that are recognized are usually essential for survival of the microorganism, they cannot be discarded or mutated. These structures may be recognized by the cells or by humoral elements of the innate system and include:

- Double-stranded RNA found in cells containing replicating viruses. This induces cytokine production by infected cells leading to destruction of the intracellular virus.
- Unmethylated CpG DNA sequences characteristic of bacterial infections. These induce autocrine macrophage activation and more effective intracellular killing of phagocytosed organisms.

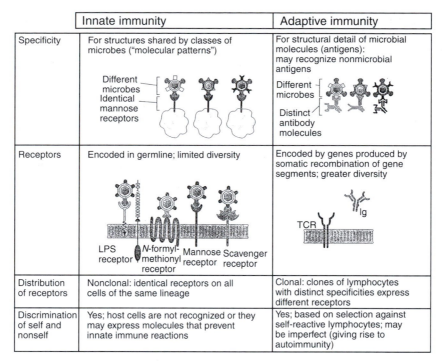

	Innate immunity	Adaptive immunity
Specificity	For structures shared by classes of microbes ("molecular patterns")	For structural detail of microbial molecules (antigens): may recognize nonmicrobial antigens
Receptors	Encoded in germline; limited diversity	Encoded by genes produced by somatic recombination of gene segments; greater diversity
Distribution of receptors	Nonclonal: identical receptors on all cells of the same lineage	Clonal: clones of lymphocytes with distinct specificities express different receptors
Discrimination of self and nonself	Yes; host cells are not recognized or they may express molecules that prevent innate immune reactions	Yes; based on selection against self-reactive lymphocytes; may be imperfect (giving rise to autoimmunity)

FIG. 3. Characteristics of innate and adaptive immunity. Cells of innate immunity have a limited number of receptors for foreign molecular structures; the same receptors are present on all cells, and because the number of different receptors is relatively small, they are all encoded in the germline. In comparison, the recognition of specific antigens by the adaptive immune system is specific and unique for each potential antigen. In order to encode such a huge diversity (10^9–10^{11} variations), antibodies and T cell receptors (TCR) are formed by somatic recombination of different gene segments. Moreover, each T or B cell will express only a single receptor type and can therefore recognize only one antigen. Lack of response to self is controlled by a number of mechanisms (not always perfect, hence the development of autoimmune disease), including destruction of self-reactive clones, or induction of specific "unresponsiveness." Figure reprinted with permission from Abbas and Lichtman (2003).

- N-Formylmethionine peptides from bacterial protein synthesis. Binding to receptors on neutrophils and macrophages causes chemotaxis (movement up a concentration gradient) and activation. Similar chemotaxis can be engendered by protein fragments released during complement activation, lipid mediators of inflammation, and chemokine proteins released by "stressed" cells.
- Mannose-rich oligosaccharides from bacterial or fungal cell walls. Engagement of receptors on macrophages induce phagocytosis; soluble mannose-binding protein in the plasma opsonizes or enhances phagocytosis of microbes bearing mannose.
- Bacterial or fungal wall oligosaccharides directly activate complement and induce either direct microbial lysis or microbial coating with complement that markedly enhance phagocytosis.
- Phosphorylcholine in bacterial cell walls binds to circulating C-reactive protein (CRP; also called pentraxin); CRP induces opsonization and also activates complement.
- Lipopolysaccharide (LPS) from certain (gram-negative) bacteria binds to circulating LPS-binding protein which in turn binds to CD14 surface molecules on macrophages. This binding elicits a wide range of cytokine responses

from the macrophages including tumor necrosis factor (TNF) and interleukin-12, which recruit and activate neutrophils and NK cells, respectively. By the same pathways, LPS induces severe systemic responses that culminate in septic shock (Hack et al., 1997).
- Teichoic acid from gram-positive bacteria elicits responses comparable to LPS.

Components of the innate system also recognize sites of injury, anticipating that these either may be primarily caused by infection or may be subject to subsequent infection. Thus, components of the coagulation cascade or denatured connective tissue elements (such as might occur at sites of trauma) may bind to macrophage cell surface receptors and induce activation. These become especially important in the context of the implantation of foreign bodies where otherwise minor trauma and the presence of denatured ECM bound on "inert" surfaces lead to macrophage activation (Tang and Eaton, 1999).

The function of all these recruiting and activating factors is to attract phagocytes to ingest and destroy microbes. The primary responding cell type in the earliest stages of injury or infection is the neutrophil, a short-lived (hours) phagocytic cell capable of ingesting and destroying microbes, as well

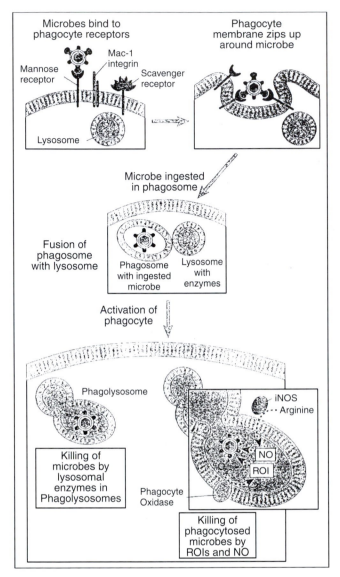

FIG. 4. Phagocytosis and intracellular destruction of microorganisms. Surface receptors on phagocytes either can bind microbes directly or may bind opsonized microbes (for example, Mac-1 integrin binds microorganisms after they have been coated with complement proteins). After binding to one (or more) of the variety of surface receptors, microbes are internalized into phagosomes, which subsequently fuse with intracellular lysosomes to form phagolysosomes. Fusion results in generation of reactive oxygen intermediates (ROI) and nitric oxide (NO) which kill the microbes largely via free radical injury; fusion with lysosomes also results in release of lysosomal enzymes that will also digest the microbes. Figure reprinted with permission from Abbas and Lichtman (2003).

as releasing a variety of potent proteases. Macrophages are secondarily recruited to sites of injury but are much longer-lived (they can last the lifetime of the host!) and persist at sites of inflammation, making them the dominant effector cell type in late-stage innate immunity. These phagocytes are recruited to sites of injury by changes in adhesion molecule expression on endothelial cells in the vicinity, and by chemotactic signals (acting, e.g., through G-protein-coupled receptors

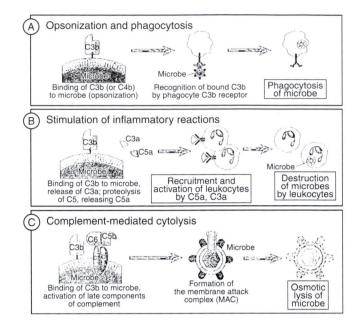

FIG. 5. Functions of complement. (A) Complement components will bind to microbe surfaces and render them more readily internalized by phagocytes (opsonization). (B) Fragments resulting from proteolytic activation of complement will recruit and activate inflammatory cells (shown here are neutrophils). (C) Complement also forms pores in the microbial membrane (so-called membrane attack complex or MAC) that result in osmotic rupture of microorganisms. Figure reprinted with permission from Abbas and Lichtman (2003).

on neutrophils) delivered by injured cells (i.e., chemokines), by complement components (generated during complement activation), and by microorganisms themselves (see the preceding list) (Fig. 6). The phagocytic cells clear opsonized microorganisms, kill them with reactive oxygen intermediates (superoxide, oxyhalide molecules, nitric oxide, and hydrogen peroxide), and degrade them with proteases (Fig. 4). However, release of such cytotoxic and degradative molecules into the neighboring environment can also cause local tissue injury. Severe local injury due to excessive neutrophil activation results in an abscess with total destruction of parenchyma and stroma.

In addition, activated macrophages release a variety of cytokines and other factors that can have both local and systemic effects (Fig. 6) (Seder and Gazzinelli, 1998):

- Tumor necrosis factor (TNF) recruits and activates neutrophils
- Interleukin-12 (IL-12) activates T cells and NK cells
- Coagulation pathways (tissue factor elaboration)
- Angiogenic factors (new blood vessel formation)
- Fibroblast activating factors (e.g., platelet-derived growth factor; PDGF) that induce fibroblast proliferation
- Transforming growth factor-β (TGF-β) expression increasing extracellular matrix (ECM) synthesis
- Matrix metalloproteinases that remodel the ECM

Thus, in the setting of prolonged activation, macrophages will ultimately mediate tissue fibrosis and scarring (Fig. 6).

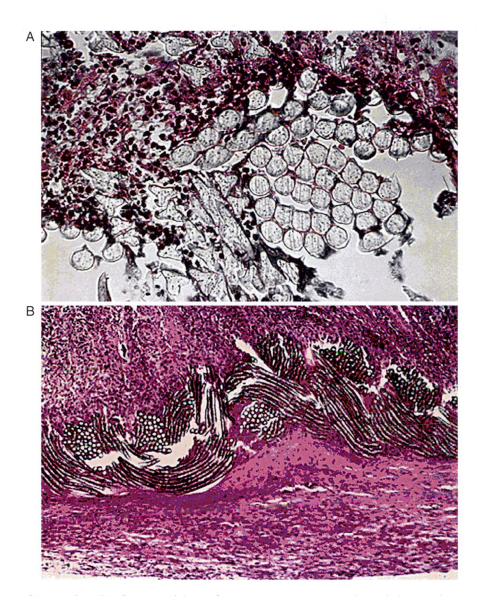

Chapter 2.4, Fig. 9 (A) Weft knit inflammatory response at 4 weeks (Golaski Microkit); (B) Warp knit inflammatory response at 3 days (Microvel).

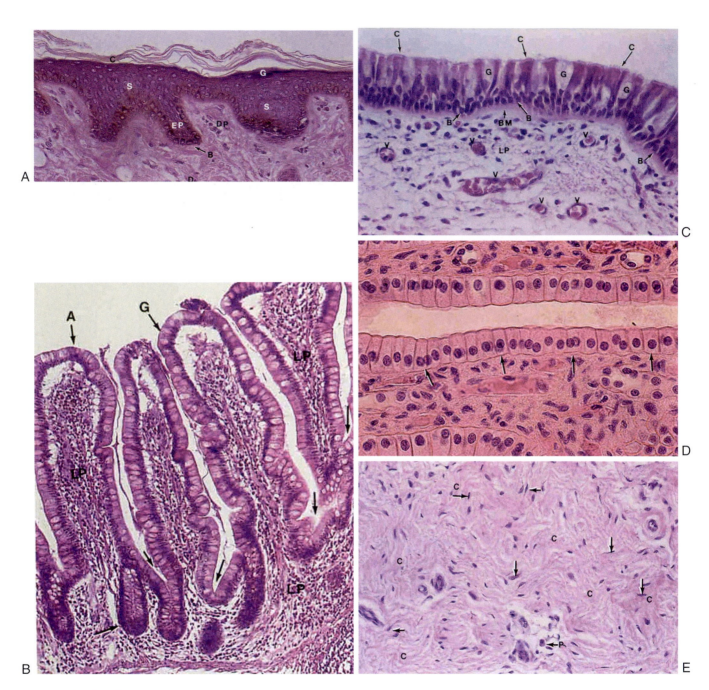

Chapter 3.4, Fig. 5 Photomicrographs of basic tissues, emphasizing key structural features. (A–D) Epithelium; (E, F) connective tissue; (G) muscle; and (H) nervous tissue. (A) Skin. Note the thin stratum corneum (c) and stratum granulosum (g). Also shown are the stratum spinosum (s), stratum basale (b), epidermal pegs (ep), dermal papilla (dp), and dermis (d). (B) Trachea, showing goblet cells (g), ciliated columnar cells (C), and basal cells (B). Note the thick basement membrane (bm) and numerous blood vessels (v) in the lamina propria (lp). (C) Mucosa of the small intestine (ileum). Note the goblet (g) and columnar absorbing (a) cells, the lamina propria (lp), muscularis mucosae (mm), and crypts (arrows). (D) Epithelium of a kidney collecting duct resting on a thin basement membrane (arrows). (E) Dense irregular connective tissue. Note the wavy unorientated collagen bundles (c) and fibroblasts (arrows). p, plasma cells. (F) Cancellous bone clearly illustrating the morphologic difference between inactive bone lining (endosteal, osteoprogenitor) cells (bl) and active osteoblasts (ob). The clear area between the osteoblasts and calcified bone represents unmineralized matrix or osteoid. cl, cement lines; o, osteocycles. H) Small nerve fascicles (n) with perineurium (p) separating it from two other fascicles (n). (A–F and H reproduced by permission from Berman, I., 1993. *Color Atlas of Basic Histology.* Appleton and Lange, 1993. G reproduced by permission from Schoen F. J., The heart. in *Robbins Pathologic Basis of Disease,* 7th ed., R. S. Cotran, V. Kumar and T. Collins, eds. Saunders, Philadelphia, in press.)

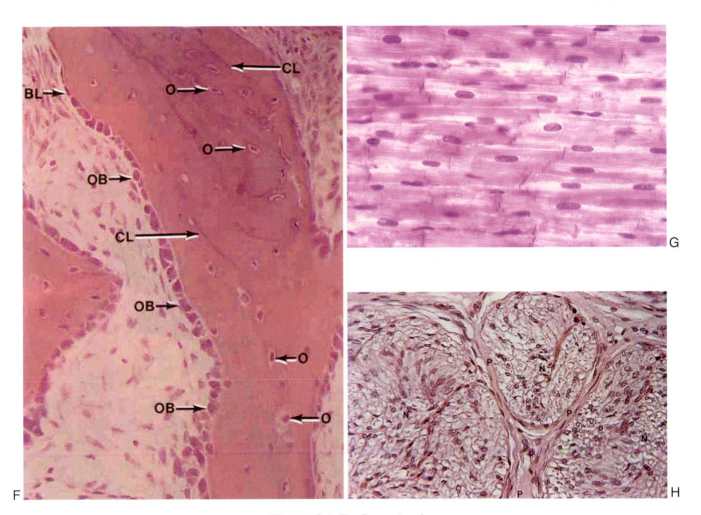

Chapter 3.4, Fig. 5—continued

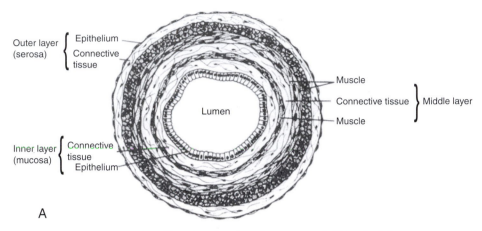

Outer layer (serosa) { Epithelium / Connective tissue

Muscle

Connective tissue } Middle layer

Muscle

Lumen

Inner layer (mucosa) { Connective tissue / Epithelium

A

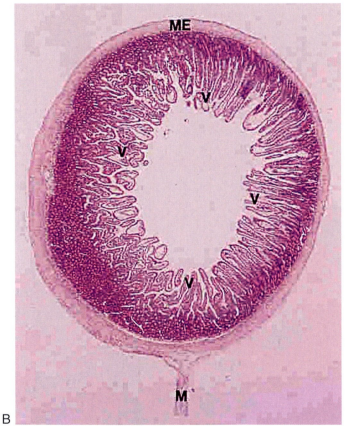

ME

V

V

V

V

M

B

Chapter 3.4, Fig. 7 (A) Organization of tissue layers in the digestive tract (e.g., stomach or intestines). (B) Photomicrograph of the dog jejunum illustrating villi (v), the muscularis external (me), and mesentery (m). In this organ the epithelium is organized into folds (the villi) in order to increase the surface area for absorption. (A, Reproduced by permission from Borysenko, M., and Beringer, T., *Functional Histology*, 3rd ed. Copyright 1989 Little, Brown, and Co. B, Reproduced by permission from Berman, I., 1993. *Color Atlas of Basic Histology*, Appleton and Lange.)

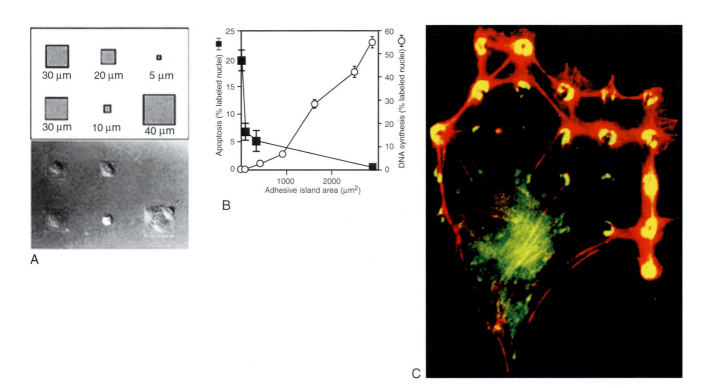

Chapter 3.4, Fig. 13 Effect of spreading on cell growth and apoptosis. (A) Schematic diagram showing the initial pattern design containing different-sized square adhesive islands and Nomarski views of the final shapes of bovine adrenal capillary endothelial cells adherent to the fabricated substrate. Distances indicate lengths of the squares' sides. (B) Apoptotic index (percentage of cells exhibiting positive TUNEL staining) and DNA synthesis index (percentage of nuclei labeled with 5-bromodeoxyuridine) after 24 hours, plotted as a function of the projected cell area. Data were obtained only from islands that contained single adherent cells; similar results were obtained with circular or square islands and with human or bovine endothelial cells. (C) Fluorescence micrograph of an endothelial cell spread over a substrate containing a regular array of small (5-μm-diameter) circular ECM islands separated by nonadhesive regions created with a microcontact printing technique. Yellow rings and crescents indicate colocalization of vinculin (green) and F-actin (red) within focal adhesions that form only on the regulatory spaced circular ECM islands. (A, B, Reproduced by permission from Chen, C. S., *et al.*, 1997. Geometric control of cell life and death. *Science* **276**: 1425. C, Reproduced by permission from Ingber, D. E., 2003. Mechanosensation through integrins: Cells act locally but think globally. *Proc. Natl. Acad. Sci. USA* **100**: 1472.)

Hepatocyte/Endothelial Cell Sorting

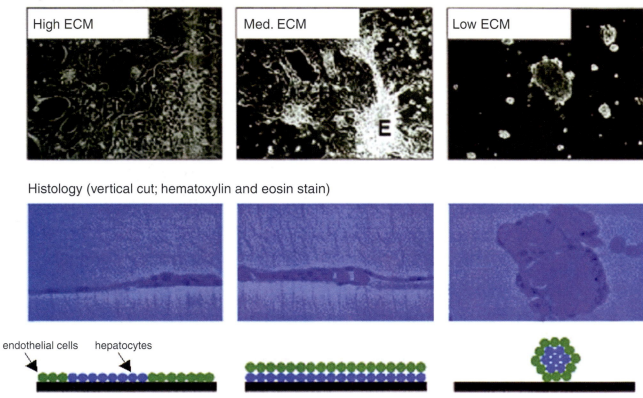

Chapter 3.4, Fig. 14 Different levels of type 1 collagen coating on a culture dish result in different organization of endothelial cells and hepatocytes. High collagen levels cause both cell types to spread across the substratum (left). On intermediate collagen levels, endothelial cells form a layer on the substratum whereas hepatocytes form a layer on top of the endothelial cells (center). Low levels of collagen result in an inner layer of hepatocyte aggregate surrounded by endothelial cells (right). (Reproduced by permission from Lauffenburger, D. A., *et al.*, 2001. Who's got pull around here? Cell organization in development and tissue engineering. *Proc. Natl. Acad. Sci. USA* **98**: 4282.)

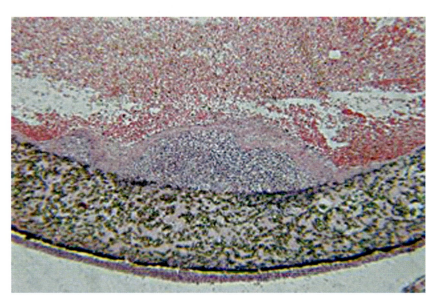

Chapter 4.2, Fig. 2 Acute inflammation, secondary to infection, of an ePTFE vascular graft. A focal zone of polymorphonuclear leukocytes is present at the lumenal surface of the vascular graft, surrounded by a fibrin cap, on the blood-contacting surface of the ePTFE vascular graft. Hematoxylin and eosin stain. Original magnification 4×.

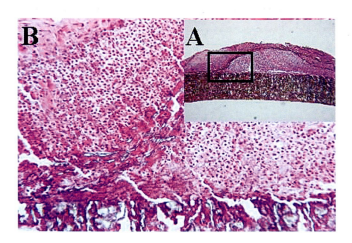

Chapter 4.2, Fig. 3 Chronic inflammation, secondary to infection, of an ePTFE arteriovenous shunt for renal dialysis. (A) Low-magnification view of a focal zone of chronic inflammation. (B) High-magnification view of the outer surface with the presence of monocytes and lymphocytes at an area where the outer PTFE wrap had peeled away from the vascular graft. Hematoxylin and eosin stain. Original magnification (A) 4×, (B) 20×.

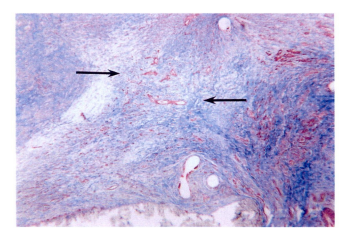

Chapter 4.2, Fig. 4 Granulation tissue in the anastomotic hyperplasia at the anastomosis of an ePTFE vascular graft. Capillary development (red slits) and fibroblast infiltration with collagen deposition (blue) from the artery form the granulation tissue (arrows). Masson's Trichrome stain. Original magnification 4×.

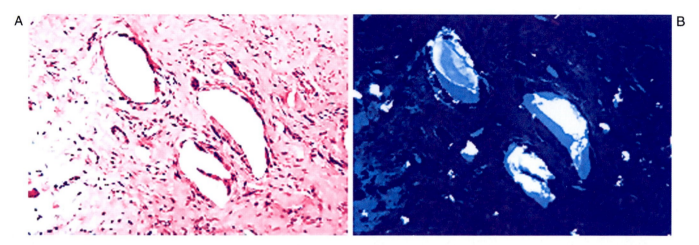

Chapter 4.2, Fig. 6 (A) Focal foreign-body reaction to polyethylene wear particulate from a total knee prosthesis. Macrophages and foreign-body giant cells are identified within the tissue and lining the apparent void spaces indicative of polyethylene particulate. Hematoxylin and eosin stain. Original magnification 20×. (B) Partial polarized light view. Polyethylene particulate is identified within the void spaces commonly seen under normal light microscopy. Hematoxylin and eosin stain. Original magnification 20×.

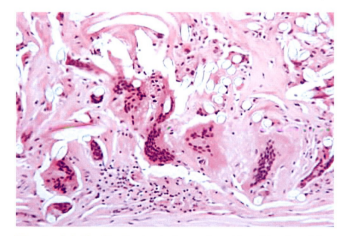

Chapter 4.2, Fig. 7 Foreign-body reaction with multinucleated foreign body giant cells and macrophages at the periadventitial (outer) surface of a Dacron vascular graft. Fibers from the Dacron vascular graft are identified as clear oval voids. Hematoxylin and eosin stain. Original magnification 20×.

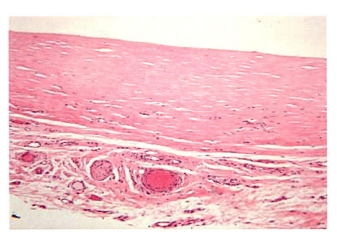

Chapter 4.2, Fig. 9 Fibrous capsule composed of dense, compacted collagen. This fibrous capsule had formed around a Mediport catheter reservoir. Loose connective tissue with small arteries, veins, and a nerve is identified below the acellular fibrous capsule.

Chapter 4.2, Fig. 10 Fibrous capsule with a focal foreign-body reaction to silicone gel from a silicone gel–filled silicone-rubber breast prosthesis. The breast prosthesis–tissue interface is at the top of the photomicrograph. Oval void spaces lined by macrophages and a few giant cells are identified and a focal area of foamy macrophages (arrows) indicating macrophage phagocytosis of silicone gel is identified. Hematoxylin and eosin stain. Original magnification 10×.

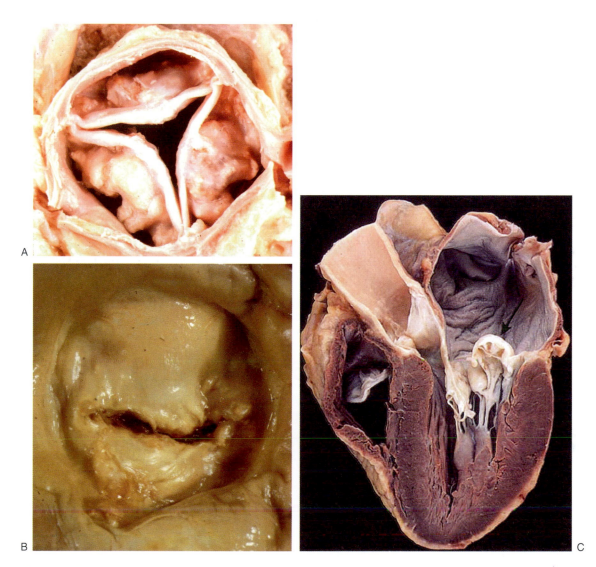

Chapter 7.3, Fig. 2 (A) Severe degenerative calcification of a previously anatomically normal tricuspid aortic valve, the predominant cause of aortic stenosis. (B) Chronic rheumatic heart disease, manifest as mitral stenosis, viewed from the left atrium. (C) Myxomatous degeneration of the mitral valve, demonstrating hooding with prolapse of the posterior mitral leaflet into the left atrium (*arrow*). A, B: Reproduced by permission from Schoen, F. J., and Edwards, W. D. (2001). Valvular heart disease: General priciples and stenosis. in *Cardiovascular Pathology*, 3rd ed. Silver, M. D., Gotlieb, A. I., and Schoen, F. J., eds. Churchill Livingstone, New York. C: Reproduced by permission from Schoen, F. J. (1999). The heart. in *Robbins Pathologic Basis of Disease*, 6th ed., R. S. Cotran, V. Kumar, T. Collins, eds. W.B. Saunders, Philadelphia.

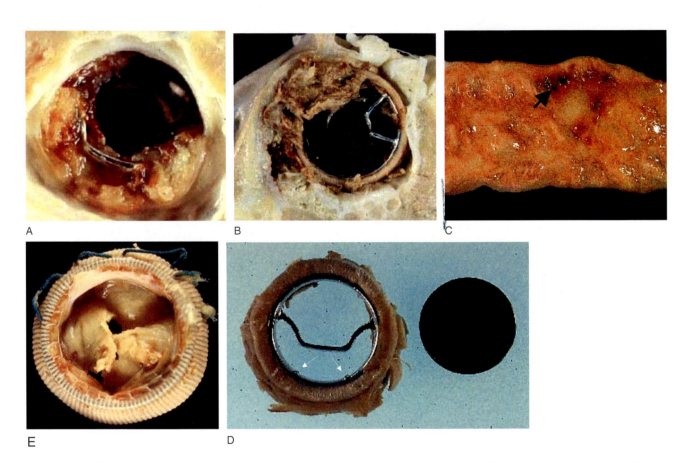

Chapter 7.3, Fig. 6   Prosthetic valve complications. (A) Thrombosis on a Bjork–Shiley tilting disk aortic valve prosthesis, localized to outflow strut near minor orifice, a point of flow stasis. (B) Thromboembolic infarct of the small bowel (arrow) secondary to embolus from valve prosthesis. (C) Prosthetic valve endocarditis with large ring abscess, viewed from the ventricular aspect of an aortic Bjork–Shiley tilting disk aortic valve. (D) Strut fracture of Bjork–Shiley valve, showing valve housing with single remaining strut and adjacent disk. (E) Structural valve dysfunction (manifest as calcific degeneration with tear) of porcine valve. B: Reproduced by permission from Schoen, F. J. (2001). Pathology of heart valve substitution with mechanical and tissue prostheses. in *Cardiovascular Pathology*, 3rd ed. M. D. Silver, A. I. Gotlieb, and F. J. Schoen, eds. Churchill Livingstone, New York. C: Reproduced by permission from Schoen, F. J. (1987). Cardiac valve prostheses: pathological and bioengineering considerations. *J. Card. Surg.* **2**: 65. A and D: Reproduced by permission from Schoen, F. J., Levy, R. J., and Piehler, H. R. (1992). Pathological considerations in replacement cardiac valves. *Cardiovasc. Pathol.* **1**: 29.

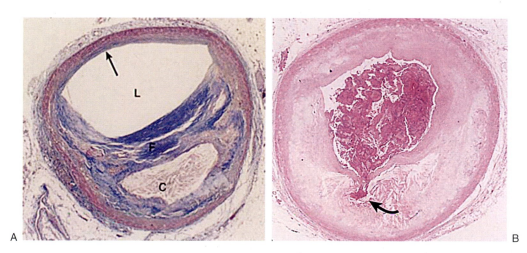

Chapter 7.3, Fig. 7 Atherosclerotic plaque in the coronary artery. (A) Overall architecture demonstrating a fibrous cap (F) and a central lipid core (C) with typical cholesterol clefts. The lumen (L) has been moderately narrowed. Note the plaque-free segment of the wall (*arrow*). (B) Coronary thrombosis superimposed on an atherosclerotic plaque with focal disruption of the fibrous cap (*arrow*), triggering fatal myocardial infarction. A: Reproduced by permission from Schoen, F. J., and Cotran, R. S. (1999). Blood vessels. in *Robbins Pathologic Basis of Disease*, 6th ed., R. S. Cotran, V. Kumar, and T. Collins, eds. W.B. Saunders, Philadelphia. B: Reproduced by permission from Schoen, F. J. (1989). *Interventional and Surgical Cardiovascular Pathology: Clinical Correlations and Basic Principles*. W.B. Saunders, Philadelphia.

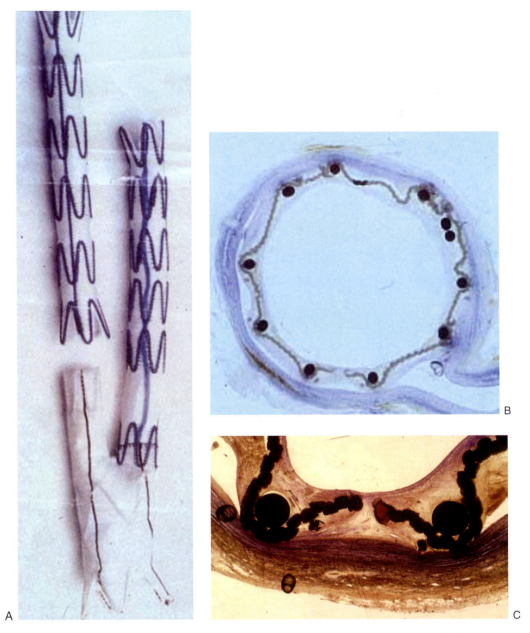

Chapter 7.3, Fig. 10 Stent grafts. (A) Configuration of device showing composite metal and fabric portions. (B) Low-power photomicrograph of well-healed experimental device explanted from a dog aorta. The lumen is widely patent and the fabric and metal components are visible. (C) High-power photomicrograph of stent graft interaction with the vascular wall, demonstrating mild intimal thickening. B and C: courtesy Jagdish Butany, MD, University of Toronto.

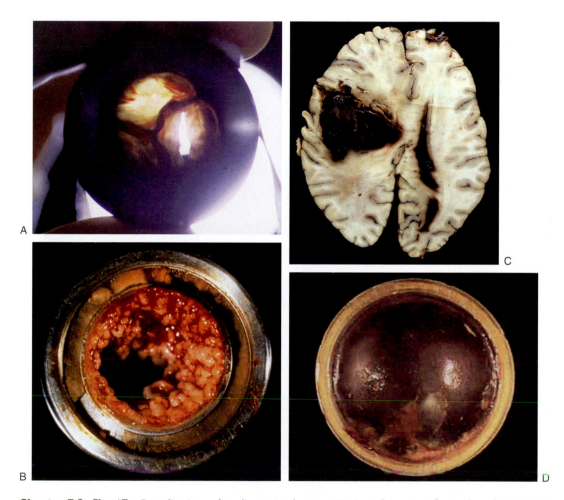

Chapter 7.3, Fig. 17 Complications of cardiac assist devices. (A) Cuspal tear in inflow valve of LVAD. (B) Hemorrhage into the brain in a patient with an LVAD. (C) Fungal infection in LVAD outflow graft. (D) Thrombosis on pumping bladder. A: Reproduced by permission from Schoen, F. J., and Padera, R. F. (2003). Pathologic considerations in the surgery of adult heart disease. in *Cardiac Surgery in the Adult*, 2nd ed., L. H. Cohn, ed. McGraw-Hill, New York. C: Reproduced by permission from Schoen, F. J., and Edwards, W. D. (2001). Pathology of cardiovascular interventions. in *Cardiovascular Pathology*, 3rd ed., M. D. Silver, A. I. Gotlieb, and F. J. Schoen, eds. Churchill Livingstone, New York. D: Reproduced by permission from Fyfe, B., and Schoen, F. J. (1993). Pathologic analysis of 34 explanted Symbion ventricular assist devices and 10 explanted Jarvik-7 total artificial hearts. *Cardiovasc. Pathol.* **2**: 187–197.

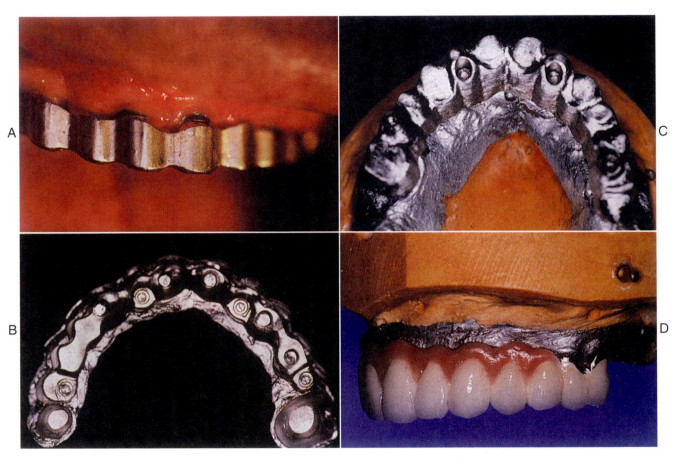

Chapter 7.8, Fig. 21 (A) Electron discharge machining or spark erosion techniques were used to mill this mesostructure bar shown attached to implants with fixation screws. (B) The armature or skeleton to which the porcelain is baked and the denture teeth are processed is milled intimately to fit by a precise frictional relationship to the mesostructure bar. (C) The external surface of this structure is prepared to receive the processed prosthesis. (D) The completed spark erosion fabricated mesostructure bar, with a totally porcelain baked superstructure, may be maintained by a frictional relationship or, if additional retention is desired, by the use of strategically related latches, such as Ceka-like attachments.

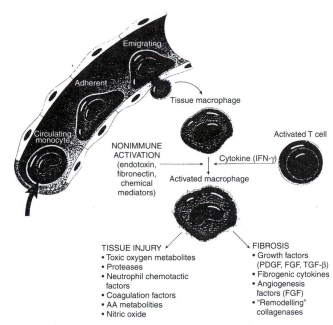

FIG. 6. Macrophage recruitment and local tissue effects after activation. Circulating monocytes are recruited to sites of tissue injury by changes in adhesion molecule expression on endothelial cells in the vicinity, and by chemotactic signals (chemokines) delivered by injured cells or neutrophils, by complement components (generated during complement activation), and by microorganisms themselves. Once these monocytes emigrate into the tissues, they become macrophages and may be activated by IFN-γ from various sources (including activated NK cells or T cells) or by nonimmunologic stimuli such as endotoxin. Activated macrophages will ingest microorganisms and necrotic debris, but will also make a number of eicosanoids (arachadonic acid or AA metabolites), reactive oxygen intermediates, and cytokine mediators that will affect the local tissue environment. Figure reprinted with permission from Kumar *et al.* (2003).

TYPES OF ADAPTIVE IMMUNITY

Adaptive immune responses adopt two basic (and interrelated) forms, humoral and cell-mediated immunity. These are accomplished by different components of the immune system and function to eliminate different types of microorganisms (Fig. 7).

- *Humoral immunity* is mediated by proteins called antibodies that are produced by B lymphocytes. Antibodies bind to unique microbial (or any molecular) antigens with exquisite specificity and target bound molecules or microbes for elimination by phagocytosis and digestion (e.g., via neutrophils and macrophages), direct killing (via NK cells), or lysis (via complement). Humoral immunity is the principal adaptive defense response against extracellular microorganisms (or their toxins), since antibodies can bind to them and assist in their clearance. Antibodies come in different types, e.g., IgA, IgG, IgM, IgE; IgG is the most common, although it is further subdivided into several different subtypes with different functionalities. The different antibody types are specialized to activate specific effector mechanisms (e.g., phagocytosis,

complement activation, or release of mediators from mast cells); the details are beyond the scope of this discussion.
- *Cellular immunity* is mediated by T lymphocytes, this form of immune response can participate in the elimination of extracellular microbes (Fig. 7). However, it is also the main mechanism by which intracellular pathogens (e.g., viruses and certain bacteria) that are not accessible to circulating antibodies can be targeted (Fig. 8).

T cells have surface receptors that cannot "see" intact foreign antigen, but rather recognize digested antigen fragments ("processed antigen") presented on the surface of certain host cell types in association with major histocompatibility complex (MHC) molecules (see later discussion). For helper T cells, these accessory or antigen-presenting cells include macrophages, one of the major cell types of the innate immune response. Thus, the innate system directs the response of the adaptive immune system. In return, recognition of foreign peptides leads to T-cell activation. Helper T cells (identified by their expression of the CD4 surface marker) assist in B-cell activation, as well as in the recruitment and activation of macrophages and neutrophils of the innate immune system. Helper T cells can also participate in the activation of NK cells, as well as cytototoxic or killer T cells (identified by their expression of the CD8 surface marker); in this manner infected cells containing intracellular pathogens may be recognized and deleted (Fig. 8).

Not all the possible responses are elicited at the same time in response to a particular pathogen. In some cases, it may be more advantageous to induce primarily a B-cell antibody-mediated response; in other circumstances, a cytotoxic T-cell response may be most warranted. Moreover, the adaptive immune response needs to be tightly regulated to prevent ongoing tissue injury, and therefore a negative-regulatory feedback must exist. The central regulation of these potential outcomes derives from the helper T cells, and more specifically the nature of the cytokines that they produce. Two basic types of helper T cells are currently recognized, called Th1 and Th2, each secreting fairly distinct subsets of cytokines (other T cell subsets are increasingly being identified, but the basic Th1 vs Th2 paradigm is sufficient for this discussion) (Abbas *et al.*, 1996). Thus, whether helper T cells induce or inhibit macrophage activation (for example) is largely a function of their differentiation and their ultimate cytokine repertoires (Fig. 9). The regulatory pathways that determine helper T-cell differentiation are an extremely active area of investigation.

RECOGNITION AND EFFECTOR PATHWAYS IN ADAPTIVE IMMUNITY

Any given T or B cell can only recognize one antigen; we are therefore able to respond to the wide diversity of foreign molecules because of an enormous repertoire of cells arising as a consequence of somatic recombination (see earlier discussion), each with different antigen specificity. Although antibodies and B cells bind to intact foreign molecules, most B-cell responses also require interactions with helper T cells. It is important to reiterate that T cells cannot recognize

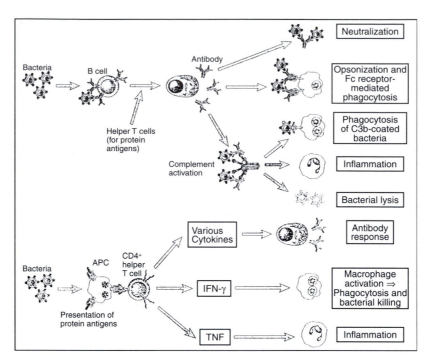

FIG. 7. T- and B-cell adaptive immune responses to extracellular microbes. Adaptive immune responses to extracellular microorganisms (and their toxins) include B-cell responses to generate antibody, and helper T-cell responses that can direct both B-cell antibody production and secondary cellular activation of macrophages and other inflammatory cells. Binding of antibodies can prevent microbes from entering host tissues (neutralization), can opsonize them for phagocytes, or can help activate complement more efficiently to increase inflammatory responses or induce microbial lysis. T-cell activation requires that antigen presenting cells (APC, such as macrophages) degrade the microbe first and present peptide fragments. After helper T-cell activation, and depending on the nature of the cytokines that are produced, B-cell responses can be augmented, or macrophages and other inflammatory cells may be activated. IFN-γ, interferon-γ; TNF, tumor necrosis factor. Figure reprinted with permission from Abbas and Lichtman (2003).

proteins until they have been degraded into smaller fragments and been bound to self histocompatibility molecules on antigen-presenting cells. Thus, most of the adaptive immune response, involving both B and T cells, is dependent on recognition of processed antigen fragments in the context of self histocompatibility proteins (Fig. 10).

In all mammals, histocompatibility molecules are grouped together on chromosomes into clusters generically called major histocompatibility complexes or MHCs. Proteins of this complex are denoted as "histocompatibility" molecules because they were first recognized as the major determining element in tissue ("histo") compatibility in organ transplantation. When inbred strains of animals shared the same MHC determinants, tissue grafts could be transplanted with relative impunity; if the donor and host were MHC-disparate, grafts were said to be histo-incompatible and the organs ultimately failed by a process called rejection (see discussion at the end of the chapter).

In humans, this MHC cluster occurs on chromosome 6, and the molecules are called human leukocyte antigens or HLA. There are two general categories of MHC molecules, called class I and class II. In humans, MHC class I (MHC I) molecules are called HLA-A, -B, and -C; MHC class II (MHC II)

molecules are called HLA-DP, -DQ, and -DR. The MHC class I molecules present peptide fragments derived from the antigens of intracellular pathogens to CD8+ cytotoxic T cells; MHC class II molecules present peptide fragments from the antigens of extracellular pathogens to CD4+ helper T cells (Fig. 10) (Klein and Sato, 2000a, b).

There is a basic dichotomy of responses depending on the original source of a particular antigen. Thus, proteins that come from the inside of cell (for example, viruses) associate with MHC I molecules and are recognized selectively by cytotoxic T cells (also called CD8+ T cells). Proteins that come from the outside of cells (for example, bacteria) associate with MHC II molecules and are selectively recognized by helper T cells (also called CD4+ T cells) (Fig. 10) (Germain, 1994).

When cytotoxic T cells encounter their specific antigen, their response is to kill the target cell bearing that antigen. When helper T cells encounter their specific antigen, their response is to make stimulatory molecules (cytokines) that cause the proliferation and activation of other cells; the major cytokine resulting in lymphocyte proliferation is interleukin-2 (IL-2). Besides increasing the numbers of T and B cells in the area of the immune response, helper T cells can also (a) activate B cells to secrete antibody; (b) activate macrophages and neutrophils

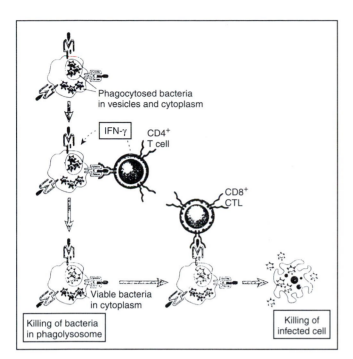

FIG. 8. Helper (CD4$^+$) and cytotoxic (CD8$^+$) T-cell collaboration in defense against intracellular microbes. Intracellular bacteria are partially degraded within the phagolysosomes of APC such as macrophages; the resulting peptide fragments are presented in the context of MHC molecules to activate helper and/or killer T cells. Cytokines elaborated by activated helper T cells can participate in turning on cytotoxic T cells, as well as in the activation of the original APC. In this manner, either cytotoxic T cells will directly kill the infected cell, or the additional booster activation of the APC by helper cytokines will enable them to completely destroy the microbe. Similar pathways exist to allow activated cytotoxic T cells or NK cells to kill cells infected with viruses. Figure reprinted with permission from Abbas and Lichtman (2003).

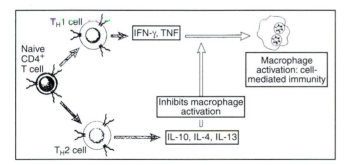

FIG. 9. Role of helper T-cell cytokines in determining immune responses. Naïve CD4$^+$ helper T cells can differentiate into either Th1 or Th2 type cells, each with distinct cytokine profiles and with distinct functions in immune regulation. In the example shown, interferon-γ (IFN-γ) and TNF secreted by Th1-type helper T cells drive macrophage activation, whereas interleukins-4, -10, and -13 (IL-4, IL-10, and IL-13) made by Th2 helper T cells inhibit macrophage activation. A similar dichotomy exists for the activation of the other elements of both the innate and adaptive immune response. Figure reprinted with permission from Abbas and Lichtman (2003).

to help clear infectious agents; (c) activate natural killer cells to be more cytotoxic; and (d) activate endothelium lining blood vessels to recruit even more inflammatory cells (Figs. 7 and 8).

T-cell recognition of antigen fragments bound to MHC molecules results in T-cell activation (Garcia *et al.*, 1999). This recognition step is accomplished by T cell receptors (TCR) on the surface of T lymphocytes; the TCR interact with a group of molecules (collectively called the CD3 complex) and send a signal to the nucleus resulting in cellular stimulation. Complete activation of T cells also requires additional interplay between other molecules (called costimulator molecules) on the surface of T cells and antigen-presenting cells of the innate immune system. Incomplete activation of T cells (i.e., without the costimulators) may result in anergy (no response) to the antigen (Fig. 11).

It bears repeating that although many aspects of immunity involve exquisitely sensitive responses to only selected foreign molecules (antigens), the immune response also involves cells (macrophages, neutrophils, and natural killer cells) and proteins (complement and cytokines) which are antigen nonspecific. Antigen-specific and nonspecific pathways interact with each other.

PATHOLOGY ASSOCIATED WITH IMMUNE RESPONSES

The innate and adaptive immune system exists primarily to defend us against infection (immune surveillance to neoplasm was a later evolutionary adaptation). Unfortunately, immune activation leads not only to the activation of host defenses and production of protective immunoglobulins and T-cells, but also occasionally to the development of responses that may potentially damage host tissues.

Both innate and adaptive immune responses may be implicated in causing disease states. As highlighted earlier, in the setting of prolonged activation, macrophages of the innate immune response will ultimately mediate tissue fibrosis and scarring. Indeed, the response to foreign materials—causing much of the local pathology associated with implants—is attributable to such persistent macrophage activation. Moreover, certain bacterial toxins (LPS) nonspecifically stimulate macrophages (as well as other cell types) and result in systemic pathology from excessive cytokine elaboration.

By having increased specificity, adaptive immunity might be expected to lead overall to less secondary damage. Normally, an exquisite system of checks and balances optimizes the antigen-specific eradication of infecting organisms with only trivial innocent by-stander injury. However, certain types of infection (e.g., virus) may require destroying host tissues to eliminate the disease (see Fig. 10B). Still other types of infections (e.g., tuberculosis) may only be controlled by a cellular response that walls off the offending agent with activated macrophages and scar, often at the expense of adjacent normal parenchyma (similar to foreign-body responses). Even when the host response to an infectious agent is specific antibody, the antibody occasionally cross-reacts with self-antigens (e.g., anti-cardiac antibodies following certain streptococcal infections, causing rheumatic heart disease). Immune complexes

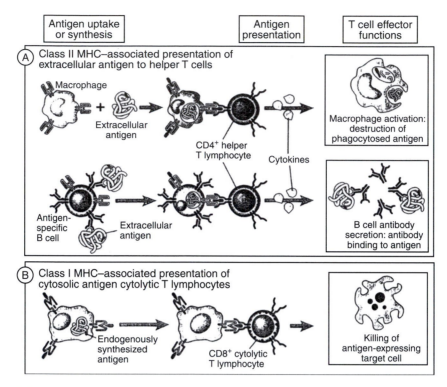

FIG. 10.   Presentation of extracellular versus intracellular antigens to cytotoxic versus helper T cells. (A) Extracellular antigens (e.g., from extracellular bacteria) are ingested and degraded by macrophages or other APC (such as B cells), and are then presented in association with MHC II surface molecules to CD4$^+$ helper T cells. Helper T cells activated in this manner lead to macrophage and/or B-cell activation that will eliminate the extracellular microbe antigens. (B) Intracellular antigens (e.g., from intracellular viruses) are degraded and presented in association with MHC I surface molecules to CD8$^+$ cytotoxic T cells. Killer T cells activated in this manner then lyse (kill) the cell that originally harbored the intracellular pathogen. Figure reprinted with permission from Abbas and Lichtman (2003).

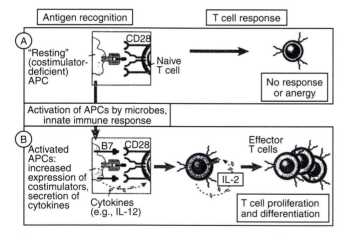

FIG. 11.   Role of costimulation in T-cell activation. (A) Antigen-presenting cells (APC) that are not activated will express few or no costimulator molecules. In that setting, even though the APC display processed antigen in the appropriate MHC context, the T cells will fail to respond. Indeed, such costimulator-poor APC presentation may result in a long-term anergy (inability to respond) to particular antigens. (B) Microbes and cytokines produced during innate immune responses activate APC to make costimulator molecules (such as B7, shown here) that will result in "complete" activation of the T cells. Activated APC also produce additional cytokines such as interleukin-12 (IL-12) that also participate in stimulating T-cell activation and differentiation. Figure reprinted with permission from Abbas and Lichtman (2003).

composed of specific antibody and circulating antigens may precipitate at inappropriate sites (see later discussion) and cause injury by activation of the complement cascade, or by facilitating binding of neutrophils and macrophages (e.g., poststreptococcal glomerulonephritis). If the antibody made in response to a particular antigen is IgE, any subsequent response to that antigen will be immediate hypersensitivity (allergy), potentially culminating in anaphylaxis. Finally, not all antigens that attract the attention of lymphocytes are exogenous. The immune system occasionally (but fortunately, rarely) loses tolerance for endogenous antigens, which results in autoimmune disease.

All of these forms of immune-mediated injury are collectively denoted as hypersensitivity. As discussed below and in Chapter 4.5, they are traditionally subdivided into four types; three are variations on antibody (immunoglobulin or Ig)-mediated injury, while the fourth is cell-mediated:

- IgE-mediated "immediate hypersensitivity"; allergy and anaphylaxis
- Mediated by antibody against fixed or circulating tissue antigens
- Immune complex (antigen–antibody)-mediated
- Immune cell–mediated

Pathogenesis of Antibody-Mediated Disease

Antibodies involved in immune-mediated diseases may bind to antigenic determinants that are intrinsic to (synthesized by) a particular tissue or cell, or that are exogenous and have been passively adsorbed (e.g., certain antibiotics or foreign proteins). Regardless of what they recognize, or how they got there, antibodies bound to the surfaces of cells or to extracellular matrix cause injury by certain basic mechanisms.

IgE-Mediated (Immediate Hypersensitivity)

Mast cells and basophils express surface Fc-receptors that can bind the Fc constant region of immunoglobulin E (IgE), one of the five basic immunoglobulin isotypes (Kay, 2001a, b). When circulating IgE's bind to the Fc-receptors and are subsequently cross-linked by specific allergen (antigen), they induce mast cell or basophil degranulation with release of preformed mediators, as well as synthesis of other potent effectors (Fig. 12):

- Preformed mediators: amines such as histamine and serotonin (cause vasodilation and increased vascular permeability)
- Mediators synthesized de novo:

    Prostaglandins (e.g., PGD_2) that can affect vessel and airway contraction and vascular permeability
    Leukotrienes (e.g., LTC_4, LTD_4, and LTE_4) that are exceptional vasoconstrictors and bronchoconstrictors previously identified as "slow-reacting substance(s) of anaphylaxis" (SRS-A)
    Platelet activating factor (PAF), a rapidly catabolized phospholipid derivative that increases vascular permeability and diminishes vascular smooth muscle tone; it also causes bronchoconstriction
    Cytokines, in particular TNF (recruits sequential waves of neutrophils and monocytes), and IL-4 (interleukin 4, induces local epithelial and macrophage expression of chemokines such as eotaxin, and also induces endothelial adhesion molecule expression: the combined effect will be to recruit eosinophils).

In most vascular beds, the overall result is vasodilation and increased vascular permeability, with a variable infiltrate classically predominated by eosinophils. Eosinophils are an inflammatory cell type classically associated with parasitic infections, as well as with allergies; they contain specific granules with potent cytotoxic activity for a variety of cell types. In the respiratory tree, the net result of an allergic stimulus is increased mucus secretion and bronchoconstriction.

The nature of the symptoms in any particular instance will depend on the portal of antigen entry, e.g., cutaneous (hives and rash, although these can also occur with inhaled or ingested allergen), inhaled (wheezing, airway congestion), ingested (diarrhea, cramping), or systemic (hypotension). The associated diseases range from the merely annoying (seasonal rhinitis or "hay fever") to debilitating (asthma) to life-threatening (anaphylaxis).

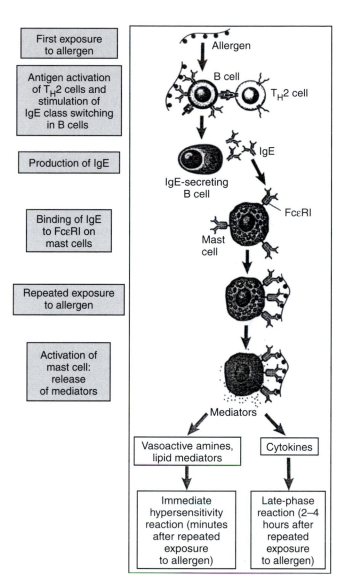

FIG. 12. Events in immediate-type hypersensitivity (allergy). Immediate hypersensitivity is initiated following contact with a specific allergen (an antigen that induces an IgE response). For unclear reasons, allergens induce in a susceptible host a predominant Th2 response that ultimately promotes an IgE antibody response. IgE then binds to mast cells in tissues (and basophils in the circulation, not shown) via specific IgE Fc receptors. Subsequent encounter with the relevant allergen results in IgE-Fc receptor cross-linking which activates the mast cells and basophils. Once activated, the cells secrete preformed mediators causing the characteristic immediate response (vasodilation and increased vascular permeability; may also cause bronchoconstriction). Over the next few hours (up to 24 hours), these activated cells will also synthesize and release additional mediators (prostaglandins, leukotrienes, PAF, and cytokines; see text). Figure reprinted with permission from Abbas and Lichtman (2003).

Antibody Bound to Cell Surfaces or Fixed Tissue Antigens

Antibodies bound to either intrinsic or extrinsic tissue antigens can induce tissue injury by promoting complement activation, inducing opsonization, or by interacting with important cell-surface molecules (Fig. 13).

Recall that complement may induce injury either by direct cytolysis via the C5b-9 membrane-attack complex (MAC)

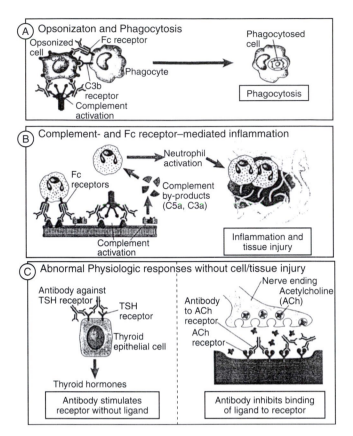

FIG. 13. Effector mechanisms in antibody-mediated disease. (A) Antibodies, with or without complement activation, will opsonize cells leading to phagocytosis and destruction. (B) Antibodies and secondarily generated complement fragments bound to large non-phagocytosable cells or tissues will recruit inflammatory cells such as neutrophils and macrophages. If these inflammatory cells cannot complete ingest the target, frustrated phagocytosis will result in the release of lysosomal contents and reactive oxygen intermediates into the tissues with subsequent extracellular damage. (C) Antibodies can also elicit pathology without causing tissue damage. In the panel on the left, antibodies to the thyroid stimulating hormone (TSH) receptor will mimic authentic TSH and will cause hyperstimulation of the thyroid (Graves' disease). In the panel on the right, antibodies to the acetylcholine (ACh) receptor at the neuromuscular junction will block normal ACh stimulation of muscle contraction leading to weakness (myasthenia gravis). Figure reprinted with permission from Abbas and Lichtman (2003).

punching holes in a cell's plasma membrane, or by opsonization (via the C3b fragment), enhancing phagocytosis by macrophages and neutrophils. In addition to direct cell killing, local activation of the complement cascade will result in the generation of complement fragments such as C3a and C5a (Fig. 5 and Chapter 4.3) (Barrington *et al.*, 2001; Walport, 2001a, b).

- C3a and C5a (so-called anaphylotoxins) mediate increased vascular permeability and smooth muscle relaxation (vasodilation), mainly via releasing histamine from mast cells
- C5a also activates the lipoxygenase pathway in arachidonic acid catabolism, resulting in increased leukotriene synthesis

- C5a-mediated chemotaxis of PMN and monocytes
- On circulating blood cells, bound complement may directly mediate cell lysis; in addition, bound antibody and opsonizing complement fragments induce efficient uptake and destruction by cells of the splenic and hepatic mononuclear phagocyte system

Antibody binding in conjunction with C3b opsonization may also lead indirectly to tissue injury. Large, nonphagocytosable cells or tissue may promote "frustrated phagocytosis" by neutrophils or macrophages; the attempted intracellular lysis results instead in the extracellular release of proteases and toxic oxygen metabolites (Fig. 13B).

Instead of fixing complement, target cells coated with low concentrations of antibody can also attract a variety of non-sensitized cells of innate immunity with Fc-receptors, most importantly the natural killer (NK) cells. These bind to the exposed Fc portion of the bound immunoglobulin and induce cell lysis without phagocytosis.

Binding of antibodies to certain receptors can induce pathology even without causing tissue injury. For example, in the case of Graves' disease, antibodies bind to the thyroid stimulating hormone (TSH) receptor on thyroid epithelial cells and mimic authentic TSH ligand interaction; the result is autonomous stimulation of the gland with hyperthyroidism. Alternatively, antibodies that cross-react with the acetylcholine receptor at the nerve–muscle synapse can block binding of acetylcholine and result in the weakness seen in the disease myasthenia gravis (Fig. 13).

Immune Complex (IC)-Mediated Injury

In many circumstances, circulating antigen and antibody combine to form insoluble aggregates called immune complexes (IC). These are usually efficiently cleared by macrophages in the spleen and liver, but occasionally deposit in certain vascular beds. Once ICs are deposited, the mechanism of injury is basically the same regardless of where or for what reason ICs have accumulated; the major sources of pathology are complement activation (see above) and neutrophil and/or macrophage injury (Fig. 14).

Pathogenesis of Cell-Mediated Disease

T-cell-mediated responses are of two general types (Fig. 15):

- *T cell-mediated cytolysis (caused by antigen-specific CD8+ cytotoxic T lymphocytes or CTL).* In CTL-mediated reactions, cytotoxic lymphocytes recognize specific antigen in association with class I MHC and induce direct cytolysis. It is important to emphasize that CTL-mediated cytolysis is highly specific, without significant "innocent bystander" injury.
- *Delayed-type hypersensitivity (mediated by cytokines and antigen nonspecific effector cells).* In the case of cell-mediated immunity, CD4+ helper T-cells recognize specific antigen in the context of class II MHC, and respond by producing a host of soluble antigen-nonspecific cytokines. These soluble mediators induce further T-lymphocyte recruitment and proliferation, and attract

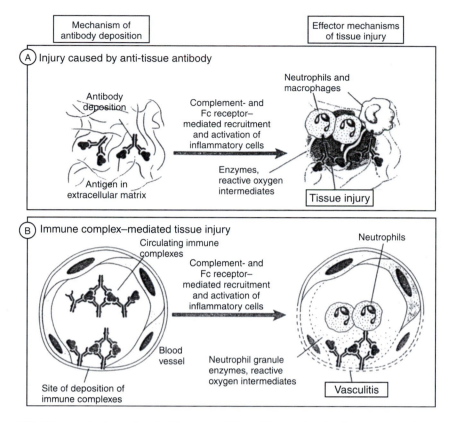

FIG. 14. Antibody-mediated pathology. (A) Direct binding of antibodies to tissue antigens will cause tissue injury by recruiting inflammatory cells and activating complement. (B) Circulating antigen–antibody complexes (also called immune complexes) can deposit in vessels and tissues also leading to inflammatory cell recruitment and complement activation. Figure reprinted with permission from Abbas and Lichtman (2003).

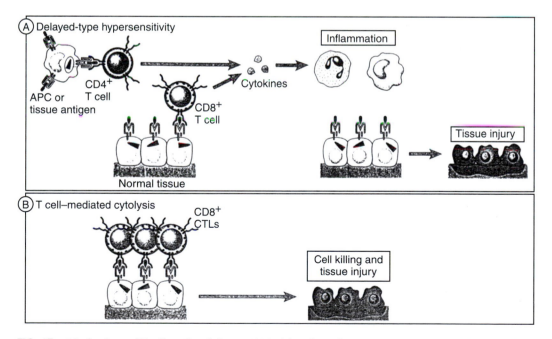

FIG. 15. Mechanisms of T-cell-mediated disease. (A) In delayed-type hypersensitivity responses, T cells (typically CD4+ helper T cells) respond to tissue or cellular antigens by secreting cytokines that stimulate inflammation, and ultimately promote tissue injury (APC, antigen-presenting cell). (B) In some diseases, CD8+ cytotoxic T cells directly kill tissue cells. Figure reprinted with permission from Abbas and Lichtman (2003).

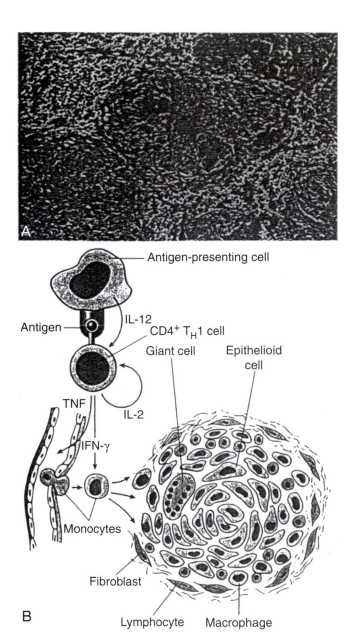

FIG. 16. Granulomatous inflammation. (A) A histologic section of a lymph node showing numerous granulomas, in this case, in response to tuberculosis. Granulomas are aggregates of activated macrophages, surrounded by activated lymphocytes. Note the presence of numerous multinucleated forms of the macrophages, so-called giant cells, which result from cell–cell fusion of macrophages under the influence of certain T-cell cytokines. (B) Schematic illustration of the events that lead to granuloma formation in response to persistent antigens. Antigen-presenting cells (APC) of the innate immune system process antigen and subsequently present it to CD4+ helper T cells; the APC also provide interleukin-12 (IL-12) and other cytokines to drive T-cell activation. Activated T cells, in turn, elaborate cytokines such as tumor necrosis factor (TNF) that will recruit inflammatory cells, and interferon-γ (IFN-γ) that will induce the activation of the recruited cells, in particular macrophages. These cytokines can also induce macrophage fusion to generate giant cells. If the antigen is not effectively eliminated, the constant cycle of T-cell and macrophage activation leads to the accumulation of an aggregate of activated cells. Activated macrophages will also elaborate mediators that result in tissue injury, as well as cytokines resulting in tissue fibrosis (see also Fig. 6). The end result

and activate antigen nonspecific macrophages; at the site of a CD4+ T-cell mediated response, the vast majority (greater than 90%), of newly recruited cells are not specific for the original inciting antigen. Cytokine-mediated CMI is critical in clearing intracellular infections not accessible to antibodies or CTL (e.g., tuberculosis, leishmania, histoplasmosis), as well as a variety of large infectious agents not well controlled by antibodies alone (e.g., fungi, protozoans, parasites). Although tightly regulated, the relatively nonspecific effector components of cell-mediated immunity (cytokines and activated macrophages) are largely responsible for the injury seen in delayed-type hypersensitivity (DTH).

In comparison to CTL, cytokine-mediated immunity may ultimately develop an antigen-nonspecific component; that is, after the initial antigen-specific T-cell response, the recruited antigen-nonspecific T-cells and macrophages can cause significant bystander injury. Macrophages in particular are an important component of the recruited inflammatory cells in DTH and mediate much of the subsequent immune effector responses. By virtue of the release of reactive oxygen intermediates, prostaglandins, lysosomal enzymes, and cytokines such as TNF (which, in turn, have potent effects, e.g., on the synthetic function of fibroblasts, lymphocytes, and endothelium), activated macrophages can potentially wreak significant havoc.

An important variant of DTH with a prominent localized component of activated macrophages is called granulomatous inflammation (Fig. 16). Granulomas (the designation of a nodule of granulomatous inflammation is a granuloma) are the characteristic response of the immune system to foreign objects (such as implanted devices), and are thus important elements in most tissue–materials interactions. Granulomas can be mediated by the same basic DTH pathways (antigen-specific T-cells and recruited nonspecific macrophages) in the setting of persistent antigenic stimuli (such as tuberculosis bacteria that may be difficult to eradicate). With persistent antigen, chronic macrophage activation results in cytokine elaboration culminating in a surrounding fibrosis. Presumably by organizing a local accumulation of activated macrophages, granulomas serve to eradicate, or at least wall off, infectious organisms that would otherwise be difficult to contain.

Granulomas also occur in the setting of large, inert, or indigestible substances (see list below); in that case, direct macrophage activation occurs by binding to denatured or modified host proteins that have adsorbed on the surfaces of the foreign materials via the receptors used for innate immunity (Tang and Eaton, 1993, 1999; Tang *et al.*, 1996). A diagnosis of granuloma suggests only a limited number of disease entities; clinically, the most common are foreign body, tuberculosis, and sarcoidosis (see also Table 2). Final confirmation of the

FIG. 16. is loss of tissue function and scar formation. In the case of "inert" foreign bodies, adsorption of host proteins onto the foreign-body surface with subsequent denaturation and modification can lead to direct macrophage activation via the receptors involved in innate immunity. Figure reprinted with permission from Kumar *et al.* (2003).

TABLE 2   Examples of Granulomatous Inflammation

Direct macrophage activation
  Dusts, e.g., beryllium, silica
  Foreign body, e.g., surgical suture, breast implant
  Gout (urate crystals)

T-cell-mediated macrophage activation
  Infections (TB, leprosy, syphilis, cat-scratch
    disease, schistosomiasis, fungus)
  Necrotizing vasculitis with granulomas (Wegener's
    granulomatosis, temporal arteritis)
  "Autoimmune" disorders with granulomas
    (Crohn's disease, de Quervain's thyroiditis)
  Sarcoidosis (inciting agent unknown)

particular inciting agent requires cultures, serologies, or special stains, or may be a diagnosis of exclusion (e.g., sarcoidosis).

Injury associated with granulomas may be due to displacement, compression, and necrosis of adjacent healthy tissue, or may be a consequence of the persistent chronic inflammation that led to the granuloma in the first place (e.g., berylliosis). Granulomas associated with a variety of "autoimmune disorders," such as temporal arteritis, Crohn's disease, and Wegener's granulomatosis, presumably reflect diseases with persistent antigen stimulation, or a heightened DTH response to specific self antigens.

SIMILARITIES AND DIFFERENCES BETWEEN ORGAN REJECTION AND THE RESPONSE TO SYNTHETIC MATERIALS OR TISSUE-DERIVED BIOMATERIALS

When foreign cells or organs are transplanted into a new host, the histocompatibility proteins on the cell surfaces of the graft are recognized by the components of adaptive immunity as being non-self. Note that except for minor genetic polymorphisms, most of the structural proteins and other molecular components in a graft are nearly identical to those that the host will also express (e.g., the contractile proteins in heart muscle, the collagenous extracellular matrix, the usual housekeeping proteins). The MHC molecules, however, are distinctly different between most humans (except identical twins!) and will elicit helper and cytotoxic T-cell activation, as well as B-cell antibody production. Clearly, once these pathways have been activated, the usual physiologic effector mechanisms (direct cell killing, complement activation, phagocytosis, cytokine elaboration, etc.) will be brought to bear on the graft and will in most cases effect its destruction. Again, although components of innate immunity are recruited and activated in the process of graft damage, the initial recognition step and the driving force for transplant rejection is via the cells of adaptive immunity (you are also referred to the basic immunology texts for excellent overviews of the rejection phenomenon; see Abbas and Lichtman, 2003; Benjamini *et al.*, 2000; Janeway, 2001). To prevent or reverse such rejection requires a whole armamentarium of immunosuppressive agents (e.g., cytotoxic drugs or agents such as cyclosporine, which put the recipient at risk of serious infections and certain tumors).

The point is emphasized here because in the immunologic sense, the synthetic materials that make up implanted devices are not rejected. In addition, tissue- or collagen-based biomaterials (e.g., a biological heart valve substitute or a processed collagen) derived from the same species (or sufficiently related species so that there are not major antigenic differences in, e.g., collagen proteins) are also not rejected. Such materials/devices do not elicit specific (adaptive) immune responses, and therefore will not have antibodies or lymphocytes that recognize the materials and cannot therefore drive the overall response. Moreover, although tissue-derived biomaterials derived from non-self [e.g., heart valve from another person (homograft) or an animal (porcine aortic valve or bovine pericardial bioprostheses)] may express foreign histocompatibility antigens, be antigenic, and be capable of eliciting adaptive immune responses (including antibodies and antigen-specific T cells), any failure of the device does not necessarily equate to immune-mediated device dysfunction. Stated another way, even immunogenetic tissue does not necessarily progress to device failure. Moreover, specific immunological responses can even be secondarily induced by device failure, but have nothing to do causally with the actual failure of the device. As a corollary statement, simply finding inflammatory cells (and even T cells and antibodies) does not in any way prove that the response is "rejection"; such elements will accrue at any site of injury in a nonspecific way (recall that some 90% of T cells in a DTH response are not antigen-specific, but are nonspecifically recruited to the site of injury). This is much more than a semantic point, in that synthetic or natural biomaterial device functions or longevity are not likely to benefit from specific immunosuppression. Of course, if a device incorporates viable cells in its manufacture (e.g., endothelial cells lining a vascular conduit), those cells will express MHC proteins and will elicit adaptive immune responses that materially contribute to device failure. In that instance, it will be necessary in the long term either to engineer such devices using cells derived from the individual who will eventually receive the implant, or to rely on long-term immunosuppression much as is done for organ transplants.

It should also be emphasized that although synthetic and biomaterials are not rejected in the immunologic sense, components of the immune system (particularly innate immunity) can contribute to device dysfunction and failure. In particular, and as described above, nonspecific activation of macrophages and complement will lead to local tissue damage via proteolysis, accumulation of other inflammatory cells, and/or cytokine elaboration; in most cases, with an ongoing, persistent innate response to a device that cannot be eliminated, fibrous scar tissue will also develop.

Thus, under certain circumstances, synthetic materials and biomaterials can have failure modes that are attributable to activation of the immune system (particularly innate immunity). An "inert" Silastic-clad breast prosthesis, for example, can accumulate dense scar tissue around it (secondary to persistent macrophage activation) that is not aesthetically ideal. Similarly, a metal hip prosthesis can induce ongoing macrophage activation that in the bone will lead to cytokine production that ultimately drives bone resorption and prosthesis loosening. Although administration of steroids in these settings may have some beneficial effect by limiting macrophage activation, it may also induce complications since steroids (among other side effects) also inhibit healing and increase susceptibility to infections.

Bibliography

Abbas, A., and Lichtman, A. (2003). *Cellular and Molecular Immunology*, 5th ed. W.B. Saunders, Philadelphia.

Abbas, A., Lichtman, A., and Pober, J. (2000). *Cellular and Molecular Immunology*, 4th ed. W.B. Saunders, Philadelphia.

Abbas, A., Murphy, K., and Sher, A. (1996). Functional diversity of helper T lymphocytes. *Nature* **383**: 787–793.

Barrington, R., Zhang, M., Fischer, M., and Carroll, M. C. (2001). The role of complement in inflammation and adaptive immunity. *Immunol. Rev.* **180**: 5–15.

Benjamini, E., Coico, R., and Sunshine, G. (2000). *Immunology: A Short Course*, 4th ed. Wiley-Liss, New York.

Garcia, K., Teyton, L., and Wilson, I. (1999). Structural basis of T cell recognition. *Ann. Rev. Immunol.* **17**: 369–397.

Germain, R. (1994). MHC-dependent antigen processing and peptide presentation: providing ligands for T lymphocyte activation. *Cell* **76**: 287–299.

Hack, C., Aarden, L., and Thijs, L. (1997). Role of cytokines in sepsis. *Adv. Immunol.* **66**: 101–195.

Janeway, C. (2001). *Immunobiology*, 5th ed. Garland, New York.

Janeway, C., and Medzhitov, R. (2002). Innate immune recognition. *Ann. Rev. Immunol.* **20**: 197–216.

Kay, A. (2001a). Allergy and allergic diseases. *N. Eng. J. Med.* **344**: 30–37.

Kay, A. (2001b). Allergy and allergic diseases. *N. Eng. J. Med.* **344**: 109–113.

Klein, J., and Sato, A. (2000a). The HLA system. *N. Eng. J. Med.* **343**: 782–786.

Klein, J., and Sato, A. (2000b). The HLA system. *N. Eng. J. Med.* **343**: 702–709.

Kumar, V., Cotran, R., and Robbins, S. (2003). Basic Pathology. W.B. Saunders, Philadelphia.

Medzhitov, R., and Janeway, C. (2000). Innate immunity. *N. Eng. J. Med.* **343**: 338–344.

Seder, R., and Gazzinelli, R. (1998). Cytokines are critical in linking the innate and adaptive immune responses to bacterial, fungal, and parasitic infection. *Adv. Int. Med.* **44**: 144–179.

Tang, L., and Eaton, J. (1993). Fibrin(ogen) mediates acute inflammatory responses to biomaterials. *J. Exp. Med.* **178**: 2147–2156.

Tang, L., and Eaton, J. (1999). Natural responses to unnatural materials: a molecular mechanism for foreign body reactions. *Mol. Med.* **5**: 351–358.

Tang, L., Ugarora, T. P., Plow E. F., and Eaton, J. W. (1996). Molecular determinants of acute inflammatory responses to biomaterials. *J. Clin. Invest.* **97**: 1329–1334.

Underhill, D., and Ozinsky, A. (2002). Phagocytosis of microbes: complexity in action. *Ann. Rev. Immunol.* **20**: 825–852.

Walport, M. (2001a). Complement. *N. Eng. J. Med.* **344**: 1058–1066.

Walport, M. (2001b). Complement. *N. Eng. J. Med.* **344**: 1141–1144.

4.4 THE COMPLEMENT SYSTEM

Richard J. Johnson

As discussed in the previous chapter, the immune system acts to protect each of us from the constant exposure to pathogenic agents such as bacteria, fungi, viruses, and cancerous cells that pose a threat to our lives. The shear multitude of structures that the immune system must recognize, differentiate from "self," and mount an effective response against has driven the evolution of this system into a complex network of proteins, cells, and distinct organs. An immune response to any foreign element involves all of these components, acting in concert, to defend the host from intrusion. Historically, the immune system has been viewed from two perspectives: cellular or humoral. This is a somewhat subjective distinction, since most humoral components (such as antibodies, complement components, and cytokines) are made by cells of the immune system and, in turn, often function to regulate the activity of these same cells. The focus of this chapter will be on the basic biochemistry and pathobiology of the complement system, a critical element of the innate immune response, and its relevance to biomaterials research and development.

INTRODUCTION

Complement is a term devised by Paul Ehrlich to refer to plasma components that were known to be necessary for antibody-mediated bactericidal activity. We now know that complement is composed of more than 30 distinct plasma and membrane bound proteins involving three separate pathways: classical, alternative, and the more recently described lectin pathway. The complement system directly and indirectly contributes both to innate inflammatory reactions and to cellular (i.e., adaptive) immune responses. This array of effector functions is due to the activity of a number of complement components and their receptors on various cells. These activities are summarized in Table 1, along with the responsible complement protein(s). One of the principal functions of complement is to serve as a primitive self–nonself discriminatory defense system. This is accomplished by coating a foreign material with complement fragments and recruiting phagocytic cells that attempt to destroy and digest the "intruder." Although the system evolved to protect the host from the invasion of adventitious pathogens, the nonspecific and spontaneous nature of the alternative pathway permits activation by various biomaterial surfaces. Because complement activation can follow three distinct but interacting pathways, the various ways of activating the cascade will be outlined separately below.

TABLE 1   Complement Activities

Activity	Complement protein
Identification/opsonization of pathogens	C3, C4
Recruitment/activation of inflammatory cells	C3a, C5a
Lysis of pathogens/cytotoxicity	C5b-9 (MAC)
Clearing immune complexes and apoptotic cells	C1q, C3b, C4b
Augment cellular immune responses (T and B cells)	C3, C4, C3a, C5a

CLASSICAL PATHWAY

The classical pathway (CP) is activated primarily by immune complexes (ICs) composed of antigen and specific antibody, although other proteins such as C-reactive protein, serum amyloid protein, and amyloid fibrils as well as apoptotic bodies can also activate the CP (Cooper, 1985). The proteins of this pathway are C1, C2, C4, C1 inhibitor (C1-Inh), and C4 binding protein (C4bp). Some of their basic biochemical characteristics are summarized in Table 2.

Complement activation by the CP is illustrated in Fig. 1. This system is an example of an enzyme cascade in which each step in the series, from initiation to the final product, involves

TABLE 2 Proteins of the Classical Pathway of Complement

Protein	Molecular weight	Subunits	Plasma concentration (μg/ml)
C1q	410,000	6A, 6B, 6C	70
C1r	85,000	1	35
C1s	85,000	1	35
C2	102,000	1	25
C4	200,000	$\alpha\beta\gamma$	600
C1-Inh	104,000	1	200
C4bp	570,000	8	230

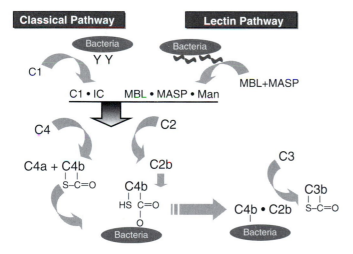

FIG. 1. Complement activation by the classical pathway (CP). Upon binding to the Fc region of an immune complex, C1 is activated and cleaves C4, exposing its thioester, which permits covalent attachment of C4b to the activating surface. C2 is cleaved, producing C2b, which binds to C4b to form the CP C3 convertase. C2b is a serine protease that specifically acts on C3 to generate C3b and C3a. The lectin pathway is also illustrated. MBL recognizes certain sugar residues (mannose, *N*-acetylglucosamine) on the surface of an activator (bacteria). MASP-1 appears to activate MASP-2, which then cleaves both C4 and C2 of the CP, generating the CP C3 convertase.

enzymatic reactions (in this case, proteolytic cleavage reactions) that result in some degree of amplification. Recent work with knockout mice (mice deficient in C1q, C2, C4, or IgG) has shown that the CP is in a state of continuous low-level activation, essentially primed to react vigorously in the presence of an IC. When an IC forms, the cascade is initiated when C1 binds to the Fc portion of an antigen–antibody complex. The C1 protein is composed of three different types of subunits called C1q, C1r, and C1s (Fig. 2). One end of C1q binds to an IC formed between an antigen and one molecule of (pentameric) immunoglobulin (Ig) M or several closely spaced IgG molecules. This interaction is believed to produce a conformational change in the C1q that results in activation (i.e., autocatalytic proteolysis) of the two C1r and then the two C1s subunits, bound to the other end of the C1q protein. Both C1r and C1s are zymogen serine proteases that are bound to the C1q in a calcium-dependent manner that is inhibited by calcium chelators such as citrate or EDTA. The proteolysis of C1s completes the activation of C1, which then proceeds to act on the next proteins in the cascade, C4 and C2.

C4 is composed of three separate chains, α, β and γ (Fig. 2), bound together by disulfide bonds. Activated C1s cleaves C4 near the amino-terminus of the α chain, yielding a 77-amino acid polypeptide called C4a and a much larger (190,000 Da) C4b fragment. The C4 protein contains a unique structural element called a thioester (Fig. 2). Thioesters have only been detected in two other plasma proteins, α2-macroglobulin and C3. Upon cleavage of C4, the buried thioester becomes exposed and available to react with a surface containing amino or hydroxyl moieties. About 5% of the C4b molecules produced react through the thioester and become covalently attached to the surface. This represents the first amplification step in the pathway since each molecule of C1 produces a number of surface-bound C4b sites.

The C4b protein, attached to the surface, acts as a receptor for C2. After binding to C4b, C2 becomes a substrate for C1s. Cleavage of C2 yields two fragments: A smaller C2a portion diffuses into the plasma, while the larger C2b remains bound to the C4b. The C2b protein is another serine protease that, in association with C4b, represents the classical pathway C3/C5 convertase.

As the name implies, the function of the C4b·C2b complex is to bind and cleave C3. The C3 protein sits at the juncture of the classical and alternative pathways and represents one of the critical control points. Cleavage of C3 by C2b yields a 9000-Da C3a fragment and a 175,000-Da C3b fragment that is very similar to C4b in both structure and function. Cleavage of C3 produces a conformational change in the C3b protein that results in exposure of its thioester group (Fig. 2). Condensation with water or surface carbohydrates results in covalent attachment of 10–15% of the C3b to the surface of the activator. This is the second amplification step in the sequence since as many as 200 molecules of C3b can become attached to the surface surrounding every C4b·C2b complex. Eventually one of the C3b molecules reacts with a site on the C4b protein, creating a C3b–C4b·C2b complex that acts as a C5 convertase (Fig. 4).

In contrast to C3, which can be cleaved in the fluid phase (see later discussion), proteolytic activation of C5 occurs only after it is bound to the C3b portion of the C5 convertase on the

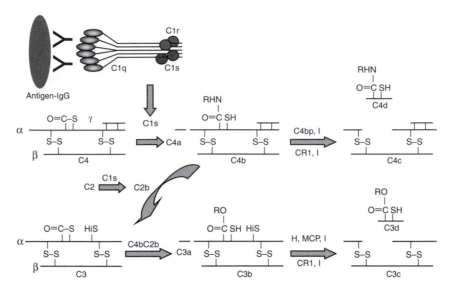

FIG. 2. Schematic illustration of C4 and C3 protein structures. O=C–S represents the reactive thioester bond that permits covalent attachment to surface nucleophiles (hydroxyl or amino groups). The pattern of proteolytic degradation and the resulting fragments are also shown. Although factor I is the relevant *in vivo* protease, some of these same fragments can be generated with trypsin, plasmin, and thrombin.

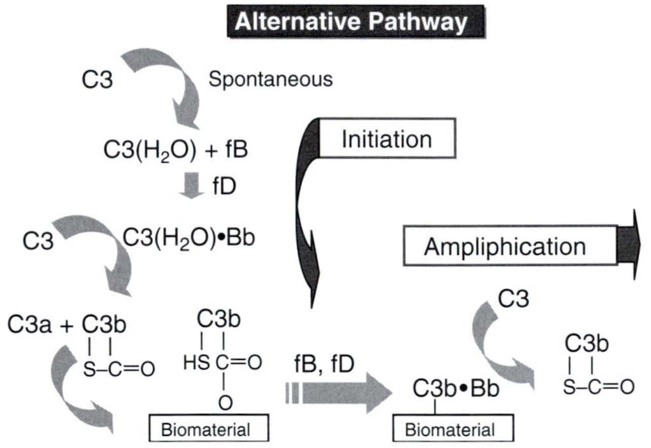

Alternative Pathway

FIG. 3. Complement activation by the alternative pathway (AP). The spontaneous conversion of C3 to C3(H$_2$O) permits the continuous production of C3b from C3, a process called C3 tickover. In the presence of an activating surface, the C3b is covalently bound and becomes the focal point for subsequent interactions. The bound C3b is recognized by factor B, which is then cleaved by factor D to produce a surface-bound C3 convertase (C3b·Bb). This results in amplification of the original signal to produce more convertase.

surface of an activator (e.g., the immune complex). Like C3, C5 is also cleaved by C2b to produce fragments designated C5a (16,000 Da) and C5b (170,000 Da). The C5b molecule combines with the proteins of the terminal components to form the membrane attack complex described later. C5a is a potent inflammatory mediator and is responsible for many of the adverse reactions normally attributed to complement activation in various clinical settings.

LECTIN PATHWAY

In the 1990s, investigators working with a protein called mannose binding protein (or mannan binding lectin, MBL) discovered of a third pathway that leads to complement activation (Matsushita, 1996). This scheme is called the lectin pathway and is composed of lectins such as MBL and two MBL-associated serine proteases or MASPs (Table 3). MBL is an acute phase protein, so its concentration in plasma increases substantially during an infection. MBL binds to terminal mannose, N-acetylglucosamine, and N-acetylmannosamine residues in complex carbohydrate structures. MBL has long been recognized as an opsonin, i.e., a protein that facilitates phagocytosis of bacteria. Low concentrations of MBL in children are associated with recurrent bacterial infections. MBL is similar in structure to C1q, having an amino-terminal domain with a collagen-like structure that binds the MASP proteins, followed by a globular carbohydrate recognition domain (CDR) that binds to sugar residues. There are two MASP proteins, called MASP-1 and MASP-2, that are very similar in structure to C1r and C1s (Wong *et al.*, 1999). Upon activation of MBL·MASP-1·MASP-2, the MASP protease components cleave C4 and C2, forming a CP C3 convertase (Fig. 1).

ALTERNATIVE PATHWAY

The alternative pathway (AP) was originally discovered in the early 1950s by Pillemer *et al.* (1954). Pillemer's group studied the ability of a yeast cell wall preparation, called zymosan, to consume C3 without affecting the amount of C1, C2, or C4. A new protein, called properdin, was isolated and implicated

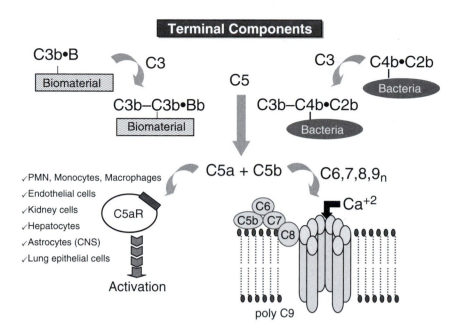

FIG. 4. Conversion of C5 produces C5a and leads to formation of the membrane attack complex (MAC). C5a binds to receptors on a variety of cells and results in numerous activities. C5b, formed by the CP, lectin, or the AP, binds C6 and C7 to form a complex that associates with the plasma membrane. This C5b67 multimer then binds C8, which results in the formation of a small hole in the lipid bilayer that allows small molecules to pass through. Association of multiple C9 proteins enlarges the pore, leading to loss of membrane integrity and cell death.

TABLE 3 Proteins of the Lectin Pathway of Complement

Protein	Molecular weight	Subunits	Plasma concentration (μg/ml)
MBL	270–650,000	18	1–3
MASP-1	93,000	2 (H,L)	6
MASP-2	76,000	2 (H,L)	

in initiating C3 activation independent of the CP. This new scheme was called the properdin pathway. However, this work fell into disrepute when it was realized that plasma contains natural antibodies against zymosan, which implied CP involvement in Pillemer's experiments. The pathway was rediscovered in the late 1960s with the study of complement activation by bacterial lipopolysaccharide and with the discovery of a C4-deficient guinea pig. The 1970s witnessed the isolation and characterization of each of the proteins of this pathway until it was possible to completely reconstruct the entire AP by recombining each of the purified proteins (Schreiber *et al.*, 1978). Most biomaterials activate complement through the AP, although there is evidence that the CP can also contribute (presumably subsequent to IgG binding).

The proteins of this pathway are described in Table 4. Their actions can be conceptually divided into three phases: initiation, amplification, and regulation (Figs. 3 and 5). Initiation is

a spontaneous process that is responsible for the nonselective nature of complement. During this stage, a small portion of the C3 molecules in plasma undergo a conformational change that results in hydrolysis of the thioester group, producing an activated form of C3 called $C3(H_2O)$ ("C3-water"), that will bind to factor B. The $C3(H_2O)\cdot B$ complex is a substrate for factor D (another serine protease), which cleaves the B protein to form a solution-phase alternative pathway C3 convertase: $C3(H_2O)\cdot Bb$. Analogous to C2b in the CP, Bb is a serine protease that (in association with $C3(H_2O)$) will cleave more C3 to form C3b. Under normal physiological conditions, most of the C3b produced is hydrolyzed and inactivated, a process that has been termed "C3 tickover." C3 tickover is a continuous process that ensures a constant supply of reactive C3b molecules to deposit on foreign surfaces, such as cellulosic- or nylon-based biomaterials. Recognition of the C3b by factor B, cleavage by factor D, and generation of more C3 convertase leads to the amplification phase (Fig. 3). During this stage, many more C3b molecules are produced, bind to the surface, and in turn lead to additional C3b·Bb sites. Eventually, a C3b molecule attaches to one of the C3 convertase sites by direct attachment to the C3b protein component of the enzyme. This C3b–C3b·Bb complex is the alternative pathway C5 convertase and, in a manner reminiscent of the CP C5 convertase, converts C5 to C5b and C5a (Fig. 4).

Recent work with purified proteins and techniques to measure direct interactions with polymer surfaces has revealed an additional potential mechanism for alternative pathway activation (Andersson *et al.*, 2002). Both C3b and C3 will adsorb

TABLE 4   Proteins of the Alternative Pathway of Complement

Protein	Molecular weight	Subunits	Plasma concentration (μg/ml)
C3	185,000	α β	1300
B	93,000	1	210
D	24,000	1	1
H	150,000	1	500
I	88,000	α β	34
P	106–212,000	2–4	20

TABLE 5   Proteins of the Membrane Attack Complex

Protein	Molecular weight	Subunits	Plasma concentration (μg/ml)
C5	190,000	α β	70
C6	120,000	1	60
C7	105,000	1	60
C8	150,000	α β	55
C9	75,000	1	55
S-protein	80,000	1	500

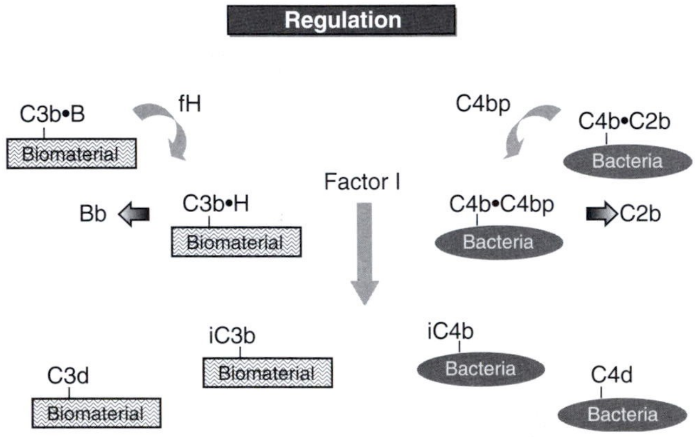

FIG. 5.   Control of complement activation by factors H, I, and C4 binding protein. The extent to which complement activation occurs on different surfaces is dependent on the ability of fH or C4BP to recognize C3b or C4b on the surface. Degradation by factor I results in irreversible inactivation and the production of C3 and C4 fragments recognized by various complement receptors on WBC.

to (not react with) polystyrene. A portion (about 10%) of the bound C3 or C3b binds factor B. This complex is recognized by factor D, which then catalyzes the formation of an AP C3 convertase. This process is facilitated by properdin, which increases the amount of convertase formed under these conditions. Interestingly, while the C3b·Bb convertase is controlled by factors H and I (see later discussion), the surface-bound C3·Bb convertase is not. The adsorption of C3 does not occur if the polystyrene surface is precoated with fibrinogen, so the extent to which this occurs in whole blood, where many other proteins can compete with C3 for binding to a biomaterial surface, has not been demonstrated.

MEMBRANE ATTACK COMPLEX

All three pathways lead to a common point: cleavage of C5 to produce C5b and C5a. C5a is a potent inflammatory mediator and is discussed later in the context of receptor-mediated white-blood-cell activation. The production of C5b initiates the formation of a macromolecular complex of proteins called the membrane attack complex (MAC) that disrupts the cellular

lipid bilayer, leading to cell death (Table 5). The sequence of events in MAC formation is outlined in Fig. 4.

Following cleavage of C5 by C5 convertase, the C5b remains weakly bound to C3b in an activated state in which it can bind C6 to form a stable complex called C5b6. This complex binds to C7 to form C5b67, which has ampiphilic properties that allow it to bind to, and partially insert into, lipid bilayers. The C5b67 complex then binds C8 and inserts itself into the lipid bilayer. The C5b678 complex disrupts the plasma membrane and produces small pores ($r \sim 1$ nm) that permit leakage of small molecules. The final step occurs when multiple copies of C9 bind to the C5b678 complex and insert into the membrane. This enlarges the pore to about 10 nm and can lead to lysis and cell death. Even at sublytic levels, formation of MAC on host cells results in a number of activation responses (elevated Ca^{2+}, arachidonic acid metabolism, cytokine production).

In addition to the usual means of generating C5b (i.e. through C5 convertase activity), several groups have shown that C5 can be modified by a variety of oxidizing agents (H_2O_2, superoxide anion, and others) to convert C5 into a C5b-like structure that will bind C6. The oxidized C5·C6 complex can bind C7, C8, and C9 to form lytic MAC. This mechanism of MAC formation may be relevant at sites where neutrophils and macrophages attempt to phagocytize a biomaterial, producing a variety of reactive oxygen species, or in hypoxia/reperfusion settings (angioplasty, cardiopulmonary bypass [CPB]).

CONTROL MECHANISMS

Various types of control mechanisms have evolved to regulate the activity of the complement system at numerous points in the cascade (Liszewski *et al.*, 1996). These mechanisms are shown in Fig. 6 and include (1) decay (dissociation) of convertase complexes, (2) proteolytic degradation of active components that is facilitated by several cofactors, (3) protease inhibitors, and (4) association of control proteins with terminal components that interfere with MAC formation. Without these important control elements, unregulated activation of the cascade results in overt inflammatory damage to various tissues and has been demonstrated to contribute to the pathology of many diseases (discussed later).

Control of Complement Activation

Decay Acceleration	Cofactor Activity	Protease/ Inhibitor	MAC Inhibitor
CR1	CR1	Factor I	CD59
Factor H	Factor H	C1Inh	S Protein
DAF/CD55	MCP/CD46	sCPN	Clusterin
C4BP	C4BP		

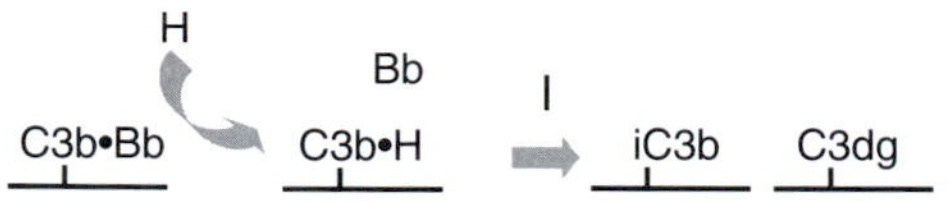

$$\text{C3b·Bb} \xrightarrow{\text{H}} \text{C3b·H} \xrightarrow{\text{I}} \text{iC3b} \quad \text{C3dg}$$

FIG. 6. Control of complement activation occurs by various mechanisms and is facilitated by a number of different proteins in the plasma (fH, fI, C1 Inh, C4BP, sCPN, S-protein, and clusterin) or on cell surfaces (CR1, DAF, MCP and CD59). Decay acceleration refers to the increased rate of displacement of either C2b or fBb from CP or AP convertases. Cofactor activity refers to the increase rate of factor I–mediated proteolysis facilitated by some proteins.

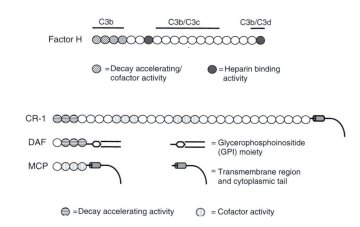

FIG. 7. Structure–activity relationships for complement control proteins. Each circle represents one short consensus repeat (SCR) domain, made up of about 60 amino acids. These SCR domains are strung together to create the different structures shown.

Starting at the top of the cascade, control of the classical and lectin pathway activation is mediated by a protein called C1 esterase inhibitor (C1-Inh). C1-Inh acts by binding to activated C1r and C1s subunits in C1 as well as MASP proteases bound to MBL. C1-Inh actually forms a covalent bond with these proteases, thus irreversibly inactivating these proteins. The effectiveness of this interaction is illustrated by the short half-life of C1s under physiological conditions (13 sec). The classical/lectin pathway C3/C5 convertase (C4b·C2b complex) spontaneously "decays" by dissociation of the C2b catalytic subunit. The rate of dissociation is increased by C4 binding protein (C4bp), which competes with C2 for a binding site on C4b. C4bp also acts as a cofactor for another control protein called factor I, which destroys the C4b by proteolytic degradation (Figs. 2 and 6).

The alternative pathway is also highly regulated by mechanisms that are very similar to the CP. The intrinsic instability of the C3b thioester bond (half-life $= 60$ μsec) ensures that most of the C3b (80–90%) is inactivated in the fluid phase. Once formed, the C3 convertase (C3b·Bb complex) also spontaneously dissociates and the rate of "decay" is increased by factor H. Aside from accelerating the decay of C3 convertase activity, factor H also promotes the proteolytic degradation of C3b by factor I (Figs. 2 and 6). Factors H and I also combine to limit the amount of active C3(H$_2$O) produced in the fluid phase.

In addition to factor H, there are several cell-membrane-bound proteins that have similar activities and structures (Fig. 7). These proteins act to limit complement-mediated damage to autologous, bystander cells. Decay-accelerating factor, or DAF, displaces Bb from the C3 convertase and thus destroys the enzyme activity. DAF is found on all cells in the blood (bound to the plasma membrane through a unique lipid group) but is absent in a disease called proximal nocturnal hemoglobinuria (PNH), which manifests a high spontaneous rate of red blood cell lysis. In addition to DAF, there are two other cell-bound control proteins: membrane cofactor protein (MCP) and CR1 (complement receptor 1, see later discussion).

MCP is found on all blood cells except erythrocytes, while CR1 is expressed on most blood cells as well as cells in tissues such as the kidney. Both of these proteins display cofactor activity for the factor I–mediated cleavage of C3b. CR1 also acts like factor H and DAF to displace Bb from the C3 convertase. A soluble recombinant form of CR1 (sCR1) was originally described by Weisman et al. (1990) and later produced commercially (Avant Immunotherapeutics). A number of investigations have used sCR1 to limit complement activation in various disease models (Larsson et al., 1997; Couser et al., 1995).

In contrast to the inhibitory proteins discussed above, properdin, the protein originally discovered by Pillemer et al., functions by binding to surface-bound C3b and stabilizing the C3 and C5 convertase enzymes. Although properdin is not necessary for activation of the alternative pathway, a genetic deficiency of this protein has been associated with an increased susceptibility to meningococcal infections.

As with the other stages of the cascade, there are several control mechanisms that operate to limit MAC formation and the potential for random lysis of "bystander" cells. The short half-life of the activated C5b (2 min) and the propensity of the C5b67 complex to self-aggregate into a nonlytic form help limit MAC formation. In addition, a MAC inhibitor, originally called S protein and recently shown to be identical to vitronectin, binds to C5b67 (also C5b678 and C5b6789) and prevents cell lysis. Recently another group of control proteins called homologous restriction factors (HRFs) have been discovered. They are called HRFs because they control assembly of the MAC on autologous cells (i.e., human MAC on human cells) but do not stop heterologous interactions (e.g., guinea-pig MAC on sheep red blood cells). One well-characterized member of this group is called CD59. It is widely distributed, found on erythrocytes, white blood cells, endothelial cells, epithelial cells, and hepatocytes. CD59 functions by interacting with C8 and C9, preventing functional expression of MAC complexes on autologous cells.

TABLE 6 Receptors for Complement Proteins

Receptors name/ligand	Structure	Cellular distribution/response
CR1/C3b, C4b	200,000-Da single chain	RBC, PMN, monocytes, B and T cells/clearance of immune complexes, phagocytosis, facilitates cleavage of C3b to C3dg by Factor I
CR2/C3dg	140,000-Da single chain	B cells/regulate B-cell proliferation
CR4	150,000-Da α chain 95,000-Da β chain	PMN, platelets, B cells/leukocyte–endothelial cell interaction
CR3/iC3b, ICAM-1, β glucan fibrinogen, factor Xa	185,000-Da α chain 95,000-Da β chain	PMN, monocyte/phagocytosis of microorganisms; respiratory burst activity
C5a/C5a		PMN, monocytes, T cells, epithelial cells, endothelial cells, hepatocytes, CNS, fibroblasts/chemotaxis, degranulation, hyperadherence, respiratory burst, cytokine production (IL-6, IL-8)
C3a/C3a	65,000 Da	Mast cells, eosinophils (various tissues)/histamine release, IL-6 production
C1q/C1q	70,000 Da	PMN, monocytes, B cells/respiratory burst activity
H/H	50,000 Da (three chains)	B cells, monocytes/secretion of factor I, respiratory burst activity

COMPLEMENT RECEPTORS

Except for the cytotoxic action of the MAC, most of the biological responses elicited by complement proteins result from ligand-receptor-mediated cellular activation (Sengelov, 1995). These ligands are listed in Table 6 and are discussed briefly here.

The ability of complement to function in the opsonization of foreign elements is accomplished in large part by a set of receptors that recognize various C3 and C4 fragments bound to these foreign surfaces. These receptors are called complement receptor 1, 2, 3, or 4 (CR1, CR2, etc). CR1 is found on a variety of cells including RBCs, neutrophils, monocytes, B cells, and some T cells and recognizes a site within the C3c region of C3b (Fig. 2). On neutrophils and monocytes, activated CR1 will facilitate the phagocytosis of C3b- and C4b-coated particles. On RBCs, CR1 acts to transport C3b–immune complexes to the liver for metabolism. As discussed above, CR1 is also a complement regulatory protein. CR2 is structurally similar to CR1 (with 16 SCR domains; see Fig. 7), but recognizes the C3d fragment of C3b that is bound to antigen. CR2 is expressed on antigen-presenting cells such as follicular dendritic cells and B cells where it facilitates the process of antigen–immune complex-driven B-cell proliferation, providing a link between innate and adaptive immunity. CR3 represents another complement receptor that binds to iC3b, and β-glucan structures found on zymosan (yeast cell wall). Also, on activated monocytes, CR3 has been shown to bind fibrinogen and factor Xa (of the coagulation cascade). CR3 is a member of the β2-integrin family of cell adhesion proteins that includes leukocyte functional antigen-1 (LFA-1) and CR4. LFA-1, CR3, and CR4 are routinely referred to as CD11a, CD11b, and CD11c, respectively. Each of these proteins associates with a molecule of CD18 to form a α–β heterodimer that is then transported and expressed on the cell surface. These proteins help mediate the cell–cell interactions necessary for such activities as chemotaxis and cytotoxic killing. A genetic deficiency in CR3/LFA proteins

leads to recurrent life-threatening infections. CR4 is found on neutrophils and platelets and binds C3d and iC3b. CR4 may facilitate the accumulation of neutrophils and platelets at sites of immune complex deposition.

In contrast to the ligands discussed earlier, which remain attached to activating surfaces, C3a, C4a, and C5a are small cationic polypeptides that diffuse into the surrounding medium to activate specific cells. These peptides are called anaphylatoxins because they stimulate histamine release from mast cells and cause smooth muscle contraction, which can produce increased vascular permeability and lead to a fatal form of shock called anaphylactic shock. These activities are lost when the peptides are converted to their des Arg analogs (i.e., with the loss of their carboxyl terminal arginine residue, referred to as C3a $_{des Arg}$, C5a $_{des Arg}$, etc.). This occurs rapidly *in vivo* and is catalyzed by serum carboxypeptidase N.

In addition to its anaphylatoxic properties, C5a and C5a $_{des Arg}$ bind to specific receptors originally found on neutrophils and monocytes. Recently the receptors for both C5a and C3a have been cloned and sequenced. The C5aR (CD88) has been shown to be expressed on endothelial cells (EC), hepatocytes, epithelial cells (lung and kidney tubules), T cells, cells in the CNS as well as on the myeloid cell lines. In addition, expression levels of C5aRs are increased on EC and hepatocytes by exposure to LPS and IL-6. In myeloid cells (neutrophils and monocytes), the C5a-receptor interaction leads to a variety of responses, including chemotaxis of these cells into an inflammatory locus; activation of the cells to release the contents of several types of secretory vesicles and produce reactive oxygen species that mediate cell killing; increased expression of CR1, CR3, and LFA-1, resulting in cellular hyperadherence; and the production of other mediators such as various arachidonic acid metabolites and cytokines, e.g., IL-1, -6, and -8. Many of the adverse reactions seen during extracorporeal therapies, such as hemodialysis, are directly attributable to C5a production. C3aRs are expressed on a variety of cell types

TABLE 7 Clinical Settings Involving Complement

Hemodialysis and cardiopulmonary bypass

Kidney disease (especially glomerulonephritis)

Ischemia/reperfusion injury (e.g., angioplasty following heart attack)

Sepsis and adult respiratory distress syndrome

Recurrent infections

Transplantation

Rheumatoid arthritis

Systemic lupus erythematosus

Asthma

Alzheimer's disease

Hereditary angioedema

TABLE 8 Types of Studies Demonstrating a Role for Complement in Kidney Disease

Deficiency or loss of complement regulatory activity results in tissue damage

Mechanistic and knockout studies implicate complement and C5 in particular

Ongoing glomerular disease is associated with various indices of complement activation

Inhibition of complement activation attenuates tissue damage in model systems

TABLE 9 Clinical Symptoms Associated with Cuprophan-Induced Biocompatibility Reactions

Cardiopulmonary:	Pulmonary hypertension
	Hypoxemia
	Respiratory distress (dyspnea)
	Neutropenia (pulmonary leukosequestration)
	Tachycardia
	Angina pectoris
	Cardiac arrest
Other:	Nausea, vomiting, diarrhea
	Fever, chills, malaise
	Urticaria, pruritus
	Headache

including eosinophils, neutrophils, monocytes, mast cells, and astrocytes (in the CNS), as well as γ-IFN-activated T cells. In eosinophils, C3a elicits responses similar to C5a, including intracellular calcium elevation, increases endothelial cell adhesion, and the generation of reactive oxygen intermediates.

CLINICAL CORRELATES

The normal function of complement is to mediate a localized inflammatory response to a foreign material. The complement system can become clinically relevant in situations where it fails to function or where it is activated inappropriately; some of these settings are shown in Table 7 (Lambris and Holers, 2000). In the first instance, a lack of activity due to a genetic deficiency in one or more complement proteins has been associated with increased incidence of recurrent infections (MBL deficiency in children), autoimmune disease (over 90% of C1-deficient patients develop SLE), and other pathologies (for example, a deficiency of C1 inhibitor is known to result in hereditary angioedema, where various soft tissues become extremely swollen because of overproduction of various vasoactive mediators). The second instance, inappropriate activation, also occurs in a variety of circumstances. It is now recognized that endothelial cells exposed to hypoxic conditions (ischemia due to angioplasty or a blocked artery due to atherosclerosis) activate complement following reperfusion of the blocked vessel. This results in further damage to the vessel wall and eventually to the surrounding tissue. Activation of the classical pathway by immune complexes occurs in various autoimmune diseases such as systemic lupus erythematosus and rheumatoid arthritis. Glomerular deposition of immune complexes results in local inflammation that can contribute to a type of kidney damage called glomerulonephritis (GN). Quite a number of experimental and clinical data have been accumulated demonstrating that complement directly contributes to the initiation and/or progression of GN (Table 8), resulting in the development of end-stage renal disease and the necessity of hemodialysis therapy.

One of the major settings where complement has been implicated in adverse clinical reactions is during extracorporeal therapies such as hemodialysis, cardiopulmonary bypass, and apheresis therapy. The same nonspecific mechanism that permits the alternative pathway to recognize microbes results in complement activation by the various biomaterials found in different medical devices. The following discussion summarizes the clinical experience with hemodialysis and cardiopulmonary bypass, but many of the observations concerning complement activation and WBC activation are relevant to other medical biomaterial applications.

One of the most investigated materials (from the perspective of complement activation) is the cellulosic Cuprophan membrane used extensively for hemodialysis. Some of the adverse reactions that occur during clinical use of a Cuprophan dialyzer are listed in Table 9. In 1977, Craddock *et al.* showed that some of these same manifestations (neutropenia, leukosequestration, and pulmonary hypertension) could be reproduced in rabbits and sheep when the animals were infused with autologous plasma that had been incubated *in vitro* with either Cuprophan or zymosan. This effect could be abrogated by treatment of the plasma to inhibit complement activation (heating to 56°C or addition of EDTA), thus linking these effects with complement. The development and use of specific radioimmunoassays (RIAs) to measure C3a and C5a by Dennis Chenoweth (1984) led to the identification of these complement components in the plasma of patients during dialysis therapy. A typical patient response to a Cuprophan

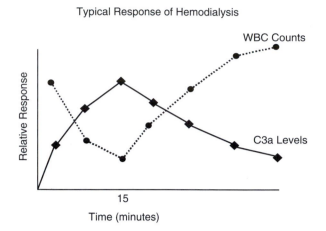

Typical Response of Hemodialysis

FIG. 8. A typical response pattern to dialysis with a complement-activating hemodialysis membrane. Many investigators have noted that the extent of C3a production is directly proportional to the degree of neutropenia at the same time points.

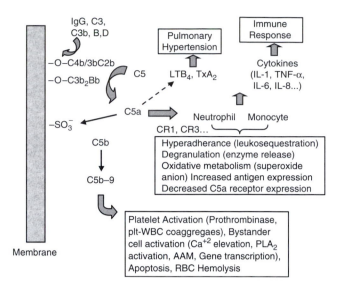

FIG. 9. The biochemical basis for complement-mediated adverse reactions during extracorporeal therapy. Production of C5a leads to receptor-dependent white blood cell activation. This results in profound neutropenia, increased concentrations of degradative enzymes, and reactive oxygen species that ultimately may lead to tissue damage and dysfunction of these important immune cells. Generation of secondary mediators, such as arachidonic acid metabolites (TxA2, LTB4) and cytokines, can have profound consequences on whole organ systems. Finally, formation of the MAC (C5b-9) has been linked with increased hemolysis during cardiopulmonary bypass and formation of microparticles and shown to increase platelet prothrombinase activity *in vitro*. This last observation suggests that surfaces that activate complement aggressively may be more thrombogenic.

membrane is shown schematically in Fig. 8. The C3a (and C5a) levels rise during the first 5–15 min, peaking between 10 and 20 min. For a Cuprophan membrane, typical peak C3a levels range from 4000 to 6000 ng/ml. During this period the white blood cells become hyperadherent and are trapped in the lung, resulting in a peripheral loss of these cells (neutropenia). As complement activation is controlled (e.g., by factor H), the C3a and C5a levels decrease to baseline levels and the WBCs return to the peripheral circulation, now in a more activated (primed) state. This is a very consistent response and many authors have noted a direct correlation between the extent of complement activation and the degree of neutropenia (as well as other responses such as CR3 expression) seen with various dialysis membranes.

Based on our understanding of the biochemistry of complement and its biological actions, the following scenario can be drawn (Fig. 9). Blood contact with the membrane results in initial protein deposition, including IgG, C3, and especially C3b, eventually leading to the formation of C3 and C5 convertase enzymes. Conversion of C5 results in C5a production, which leads to receptor-mediated neutrophil and monocyte activation. Production of C5b leads to MAC formation, which binds to bystander cells and results in subsequent activation of these cells through calcium-dependent mechanisms. Recognition of biomaterial-bound C3 and C4 fragments by WBC results in cell adherence and further activation of these cells. These various responses accounts for much of the pathophysiology seen clinically. The critical role of C5a in mediating many of these adverse reactions has been confirmed in experiments employing purified sheep C5a. Infusion of this isolated peptide into sheep, in a manner that would simulate exposure to this molecule during hemodialysis, produced dose-dependent responses identical to that seen when the sheep are subjected to dialysis (Johnson *et al.*, 1996). In addition, numerous *in vivo* and *in vitro* studies have documented the relationship between complement activation by biomaterials, the extent of WBC activation and the resulting inflammatory response illustrated

in Fig. 9 (Tang *et al.*, 1998; Gemmell *et al.*, 1996; Larsson *et al.*, 1997; Lewis and Van Epps, 1987).

In the same time frame in which clinicians were linking complement with leukopenia in the hemodialysis setting, a number of cardiovascular scientists were demonstrating complement activation by the materials used to make cardiopulmonary bypass circuits. Typical levels of C3a produced in these procedures ranged from 300 to 2400 ng/ml. These investigations soon associated C3a and C5a production with a group of symptoms known as "postperfusion" or "postpump" syndrome (Table 10). Further analysis showed that complement was activated by the materials in the circuit (such as the polypropylene membranes and the nylon filters) but was also activated by during neutralization of the heparin anticoagulant with the protamine sulfate that was given to each patient at the end of the operation. This was further exacerbated by complement activation that occurred in the ischemic vascular bed upon reperfusion of the tissue that also occurred at the end of the procedure. The importance of complement activation, and C5 conversion in particular, to the clinical outcome of CPB patients was clearly demonstrated in a study by Fitch *et al.* (1999). Using a single chain anti-C5 antibody fragment that inhibited C5a and MAC generation during the procedure, these investigators showed that this antibody fragment lowered WBC activation, blood loss, cognitive deficits and myocardial injury. These results are consistent with other studies (Velthuis *et al.*, 1996; Hsu, 2001) using heparin-coated CPB circuits

TABLE 10 Postperfusion or Postpump Syndrome

Increased capillary permeability with accumulation of interstitial fluid

Blood loss requiring transfusions

Fever

Leukocytosis (increased WBC counts)

Organ dysfunction: heart, liver, kidney, brain and GI tract

that demonstrate lower inflammatory indices (complement, cytokine, and elastase levels) that are associated with improved clinical outcomes (decreased blood loss, length of ICU stays, and morbidity).

The CPB experience with heparin-coated devices demonstrates that modification of device materials (or the blood-contacting surfaces of those materials) can dramatically limit complement activation and the subsequent inflammatory response. Based in part on this and similar observations, hemodialyzer/membrane manufacturers began developing new membranes to produce more biocompatible (i.e., less complement-activating) devices. These new membranes tend to fall into two groups: moderately activating modified cellulosics [such as cellulose acetate (CA), hemophane, and cellulose triacetate (CT)] and low activating synthetics [such as polyacrylonitrile (AN69), poly(methyl methacrylate) (PMMA), and polysulfone (PS)]. Moderately activating modified cellulosics produce C3a levels and neutropenic responses that are about 50% of Cuprophan levels, while the synthetic materials display 0–20% activation compared to Cuprophan. Based on the known properties of complement and the structures of these membranes, the reasons for the improved biocompatibility can be rationalized as follows. Most of these materials have a diminished level of surface nucleophiles. In theory, this should result in lower deposition of C3b, and in fact this has been verified experimentally. The diminished capacity to bind C3b results in lower levels of C3 and C5 convertase activity and consequently an abated production of C3a and C5a. Patient exposure to C5a is reduced even further by materials that allow for transport through the membrane to the dialysate (for example, high-flux membranes such as polysulfone will do this) or by absorbing the peptide back onto the surface (the negatively charged AN69 has been shown to have a high capacity for binding cationic C5a). Thus, limiting C3b deposition and C5a exposure are two proven mechanisms of avoiding the clinical consequences of complement activation.

The same result can be also accomplished by facilitating the normal control of C3 convertase by factor H. Kazatchkine *et al.* (1979) have shown that heparin coupled to either zymosan or Sepharose limits the normal complement activation that occurs on these surfaces by augmenting C3b inactivation through factors H and I. Presumably, this accounts for the improved biocompatibility of heparin-coated circuits used in CPB described above. Mauzac *et al.* (1985) have prepared heparin-like dextran derivatives that are extensively modified with carboxymethyl and benzylamine sulfonate groups. These researchers have shown that these modifications diminish complement activation by the dextran substrate. A simple

modification of cellulose membranes (Cuprophan) with maleic anhydride has been shown to limit the complement-activating potential of these materials by over 90% (Johnson *et al.*, 1990). Again, increased binding of factor H to surface-bound C3b appears to account for the improved biocompatibility of maleated cellulose. Thus materials that limit complement activation through normal regulatory mechanisms are on hand and may prove to be the next generation of complement-compatible materials. In addition, as the studies of Fitch *et al.* have shown, pharmaceutical control of complement is possible with agents that are now in clinical development.

SUMMARY AND FUTURE DIRECTIONS

The immune response to a biomaterial involves both humoral and cellular components. Activation of the complement cascade by classical, lectin, or alternative pathways leads to the deposition of C4b and C3b proteins. Recognition of these molecules by receptors on granulocytes can cause activation of these cells, leading to the production of degradative enzymes and destructive oxygen metabolites. Recognition of C4b or C3b by other proteins in the cascade leads to enzyme formation (C3 and C5 convertases), which amplifies the response and can lead to the production of a potent inflammatory mediator, C5a. C5a binds to specific receptors found on PMNs and monocytes. The interaction of C5a with these cells elicits a variety of responses including hyperadherence, degranulation, superoxide production, chemotaxis, and cytokine production. Systemic exposure to C5a during extracorporeal therapies has been associated with neutropenia and cardiopulmonary manifestations (Tables 9 and 10) that can have pathologic consequences. The other portion of the C5 protein, C5b, leads to formation of a membrane attack complex that activates cells at sublytic levels and has cytotoxic potential if produced in large amounts. The control of these processes is understood well enough to begin designing materials that are more biocompatible. Limiting C3b deposition (nucleophilicity), adsorbing C5a to negatively charged surface groups, and facilitating the role of factors H and I are three approaches that have been shown to be effective. Translating the last mechanism into commercial materials is one of the major challenges facing the development of truly complement-compatible membranes.

Bibliography

Anderson, J., Ekdahl, K. N., Larson, R., Nilsson, U. R., and Nilsson B. (2002). C3 absorbed to a polymer surface can form initiating alternative pathway convertase. *J. Immunol.* **168**: 5786–5791.

Chenoweth, D. E. (1984). Complement activation during hemodialysis: clinical observations, proposed mechanisms and theoretical implications. *Artificial Organs.* **8**: 231–287.

Cooper, N. R. (1985). The classical pathway of complement: activation and regulation of the first component. *Adv. Immunol.* **61**: 201–283.

Couser, W. G., Johnson, R. J., Young, B. A., Yeh, C. G., Toth C. A., and Rudolph, A. R. (1995). The effects of soluble complement receptor 1 on complement-dependent glomerulonephritis. *J. Am. Soc. Nephrol.* **5**: 1888–1894.

Craddock, P. R., Fehr, J., Brigham, K. L., Kronenberg, R. S., and Jacob, H. S. (1977). Complement and leukocyte-mediated pulmonary dysfunction in hemodialysis. *N. Eng. J. Med.* **296**: 769–774.

Fitch, J. C. K., Rollins, S., Matis, L., Alford, B., Aranki, S., Collard, C., Dewar, M., Elefteriades, J., Hines, R., and Kopf, G. (1999). Pharmacology and biological efficacy of a recombinant, humanized, single-chain antibody C5 complement inhibitor in patients undergoing coronary artery bypass graft surgery with cardiopulmonary bypass. *Circulation* **100**: 2499–2506.

Gemmell, C. H., Black, J. P., Yeo, E. L., and Sefton, M. V. (1996). Material-induced up-regulation of leukocyte CD11b during whole blood contact: material differences and a role for complement. *J. Biomed. Mater. Res.* **32**: 29–35.

Hsu, L-C. (2001). Heparin-coated cardiopulmonary bypass circuits: current status. *Perfusion* **16**: 417–428.

Johnson, R. J., Lelah, M. D., Sutliff, T. M., and Boggs, D. R. (1990). A modification of cellulose that facilitates the control of complement activation. *Blood Purif.* **8**: 318–328.

Johnson, R. J., Burhop, K. E., and Van Epps, D. E. (1996). Infusion of ovine C5a into sheep mimics the inflammatory response of hemodialysis. *J. Lab. Clin. Med.* **127**: 456–469.

Kazatchkine, M., Fearon, D. T., Silbert, J. E., and Austen, K. F. (1979). Surface-associated heparin inhibits zymosan included activation of the human alternative complement pathway by augmenting the regulatory action of control proteins. *J. Exp. Med.* **150**: 1202–1215.

Lambris, J. D., and Holers, V. M., eds. (2000). *Therapeutic Interventions in the Complement System.* Humana Press, Totowa, NJ.

Larsson, R., Elgue, G., Larsson, A., Nilsson Ekdahl, K., Nilsson, U. R., and Nilsson, B. (1997). Inhibition of complement activation by soluble recombinant CR1 under conditions resembling those in a cardiopulmonary circuit: upregulation of CD11b and complete abrogation of binding of PMN to the biomaterial surface. *Immunopharmacology* **38**: 119–127.

Lewis, S. L., and Van Epps, D. E. (1987). Neutrophil and monocyte alterations in chronic dialysis patients. *Am. J. Kidney Dis.* **9**: 381–395.

Liszewski, M. K., Farries, T. C., Lubin, D. M., Rooney, I. A., and Atkinson, J. P. (1996). Control of the complement system. *Adv. Immunol.* **61**: 201–282.

Matsushita, M. (1996). The lectin pathway of the complement system. *Microbiol. Immunol.* **40**: 887–893.

Mauzac, M., Maillet, F., Jozefonvicz, J., and Kazatchkine, M. (1985). Anticomplementary activity of dextran derivatives. *Biomaterials* **6**: 61–63.

Pillemer, L., Blum, L., Lepow, I. H., Ross, O. A., Todd, E. W., and Wardlaw, A. C. (1954). The properdin system and immunity. I. Demonstration and isolation of a new serum protein, properdin, and its role in immune phenomena. *Science* **120**: 279–285.

Ross, G. D. (1986). *Immunobiology of the Complement System.* Academic Press, New York.

Schreiber, R. D., Pangburn, M. K., Lesaure, P. H., and Muller-Eberhard, H. J. (1978). Initiation of the alternative pathway of complement: recognition of activators by bound C3b and assembly of the entire pathway from six isolated proteins. *Proc. Natl. Acad. Sci. USA* **75**: 3948–3952.

Sengelov, H. (1995). Complement receptors in neutrophils. *Crit. Rev. Immunol.* **15**: 107–131.

Tang, L., Liu, L., and Elwing, H. B. (1998). Complement activation and inflammation triggered by model biomaterial surfaces. *J. Biomed. Mater. Res.* **41**: 333–340.

Velthuis, H., Jansen, P. G. M., Hack, E., Eijsman, L., and Wildevuur, C. R. H. (1996). Specific complement inhibition with heparin-coated extracorporeal circuits. *Ann. Thorac. Surg.* **61**: 1153–1157.

Weisman, H. F., Bartow, T., Leppo, M. K., Marsch, H. C., Jr., Carson, G. R., Concino, M. F., Boyle, M. P., Roux, K. H., Weisfeldt, M. L.,

and Fearon, D. T. (1990). Soluble human complement receptor type 1: *In vivo* inhibitor or complement suppressing postischemic myocardial inflammation and necrosis. *Science* **249**: 146–151.

Wong, N. K. H., Kojima, M., Dobo, J., Ambrus, G., and Sim, R. B. (1999). Activities of the MBL-associated serine proteases (MASPS) and their regulation by natural inhibitors. *Mol. Immunol.* **36**: 853–861.

4.5 SYSTEMIC TOXICITY AND HYPERSENSITIVITY

Arne Hensten-Pettersen and Nils Jacobsen

Artificial implant devices comprise a variety of metallic alloys, polymers, ceramics, hydrogels, or composites for a large number of purposes and with widely different properties. With the exception of drug delivery systems, sutures, and other degradable biomaterial systems (Chapter 2.7), the implant devices are intended to resist chemical and biochemical degradation and to have minimal leaching of structural components or additives. However, synthetic devices are influenced by chemical and in some cases enzymatic processes resulting in the release of biomaterials-associated components. Since there is no natural repair mechanisms parallel to natural tissues, degradation (biodegradation) is a "one-way" process that brings about microscopic and macroscopic surface and bulk changes of the devices, sometimes enhanced by the biomechanical and bioelectrical conditions that the devices are intended to resist. With the exception of pathologic calcification of certain polymer implants, the surface changes may not be significant for the mechanical strength of the implant, whereas in contrast the released substances very often have biological effects on the surrounding tissues or, possibly, at other remote locations. Inflammatory, foreign body, or other local host reactions and tumorigenesis are discussed in Chapters 4.2, 4.3, and 4.7. The following discussion is concerned with the possibility of systemic toxic reactions and/or hypersensitive reactions caused by biomaterials-derived xenobiotics.

KINETICS AND NATURE OF BIOMATERIALS COMPONENTS

Xenobiotic components derived from *in vivo* medical devices have parenteral contact with connective tissue or other specialized tissues such as bone, dentin, and vascular or ocular tissue, whereas leachables from skin- and mucosa-contacting devices have to pass the epithelial lining of the oral mucosa, the skin, the gastrointestinal tract, or—for volatiles—the lung alveoli to get "inside" the body. In either case, further distribution of foreign substances to other tissues and organs is dependent on membrane diffusion into blood capillaries and lymph vessels. The transport may be facilitated by reversible binding to plasma proteins, globulins (metal, metal compounds), and chylomicrons (lipophilic substances). Storage—and later release—may take place for certain components in tissues such as fat and bone.

In addition to particulate matter the released components consist of chemical substances of different atomic and molecular size, solubility, and other chemical characteristics depending on the mother material. Examples are metal ions such as cobalt, chromium, nickel, molybdenum, and titanium from metallic orthopedic implants or prosthodontic materials, or residual monomers, chemical initiators, inhibitors, plasticizers, antioxidants, etc., from polymer implants and dental materials. Other degradation products from inorganic, organic, and composite devices also "rub off" to the surrounding tissues. The kinetic mechanisms for biomaterials components are in part the same as those of xenobiotics introduced by food or environmental exposure, i.e., the released components are subject to oxidation, reduction, and hydrolysis followed by conjugation mechanisms. All metabolic changes are in their nature intended to eliminate them by way of the urine, bile, lungs, and to a certain degree in salivary, sweat, and mammary glands and hair (deBruin, 1981).

A key question is, do the released components or their metabolites have any systemic toxic effect on the host and/or could they induce unwanted immunological reactions?

TOXICODYNAMIC CONSIDERATIONS

Systemic toxicity depends on toxic substances hitting a target organ with high sensitivity to a specific toxicant. Target organs are the central nervous system, the hematopoietic system, the circulatory system, and visceral organs such as liver, kidney, and lungs, in that order. The toxicity is based on interference with key cell functions and depends on the dose and the duration of the exposure. Serious effects may be incompatible with continued life, but most effects are local and reversible cell damage. However, some sublethal effects may include somatic cell mutation expressed as carcinogenesis, or germinal cell mutation, resulting in reproductive toxicity.

The key word in the evaluation of general toxicity is the dose, defined as the amount of a substance an organism is exposed to, usually expressed as mg per kg body weight. Adverse effects of foreign substances are often the result of repeated, chronic exposure to small doses that over a prolonged period of time may have deleterious effects similar to one large, short time exposure, provided that the repeated doses exceed a certain threshold level. This level is determined by the capacity of metabolism and elimination. Another important factor is the possibility of synergistic potentiating effects when several toxicants are present simultaneously. Whatever mechanism is involved, the principle of systemic toxicity presupposes a dose-dependent reaction that may be measured and described, and that may be explained by specific reactions at distinct molecular sites (Eaton and Klaassen, 1996).

The components derived from biomaterials represent a large series of widely different foreign substances with few characteristics in common and with a largely unknown concentration. Most of them have to be characterized as toxic per se, with large variations regarding their place on a ranking list of potential toxicity. Metal ions and salts derived from biomaterials devices, such as mercury, nickel, and chromium, are classified as toxicants. A similar statement could be made for components associated with polymeric materials. However, clinically relevant data on the concentration of degradation products are scarce, e.g., phthalate additives and degradation products from chemical additives derived from poly(methyl methacrylate) dental prostheses have been quantified in saliva (Lygre et al., 1993). In vitro experiments have shown that chromium and nickel are released from base metal orthodontic appliances, although the amounts are not comparable with the amounts calculated in food intake (Park and Shearer, 1983). In addition, the proportion of uptake by mucosa is unknown. The presence of leachable substances has also been demonstrated in the surrounding tissues of implants, but quantification is difficult. Information is available on the release and uptake of mercury derived from dental amalgam. For example a series of studies has shown the presence mercury in plasma and urine after inhalation of metallic mercury released from dental amalgam (Mackert and Berglund, 1997). Accumulation of mercury in tissues belonging to the central nervous system has been shown after occupational exposure (Nylander et al., 1989). Reproductive toxicity has been of specific concern. However, similar to other metals such as chromium and nickel, mercury exposure also takes place through food and through respiratory air. Careful scrutiny of the large number of partly controversial data by national and international scientific committees has not resulted in a consensus conclusion that the application of mercury amalgam should be discontinued as a dental biomaterial, although mercury is a significant environmental concern (The European Commission, 1998).

When occupational exposure is disregarded, the possibility of systemic toxicity or reproductive toxicity has not been seriously considered for other biomaterial components or metabolites, because of their low concentration as compared with their toxic potential. A fair conclusion at this point would be that there are no data indicating any systemic toxicity caused by biomaterials-derived xenobiotics. However, this field of interest is characterized by the increasing number of synthetic biomaterials on the market. Despite the premarketing testing programs it is difficult to predict single or synergistic toxic effects of leachable components and degradation products in the future.

ADVERSE EFFECTS OF DEFENSE MECHANISMS

The low probability of direct systemic adverse effects on target organs caused by biomaterial products does not rule out deleterious effects by other, dose-independent mechanisms. All substances not recognized as natural components of the tissues are subject to possible clearance by several mechanisms, e.g., phagocytic cells such as polymorphonuclear leukocytes, macrophages, and monocytes attempt to degrade and export the components. Larger foreign components are subject to more aggressive reactions by giant cells causing an inflammatory foreign-body reaction. Enzymes and other bioactive molecules associated with the phagocytosis and foreign-body reaction may cause severe local tissue damage. In addition, phagocytic cell contact and the contact with the circulatory system of lymph and blood opens up another way of neutralizing foreign substances by way of the immune system, introducing a biologic memory of previously encountered foreign substances and an enhancement system for their neutralization.

HYPERSENSITIVITY AND IMMUNOTOXICITY

The immune system is an indispensable biologic mechanism to fight potentially adverse invaders, most commonly of microbial origin. However, the immune system occasionally strikes invading molecules—adverse or not—with an intensity that stands in contrast to the sometimes minute amounts of foreign substances, and with the ability to cause host tissue damage. This phenomenon is called hypersensitivity. The resulting injury is part of a group of adverse reactions classified as immunotoxic.

In principle, immunologic hypersensitivity comprises two different mechanisms: allergy and intolerance. Allergy is a acquired condition resulting in an overreaction upon contact with a foreign substance, given a genetic disposition and previous exposure to the substance. Allergic reactions may include asthma, rhinitis, urticaria, intraoral and systemic symptoms, and eczema. Intolerance is an inherited reaction that resembles allergy and has common mediators and potentiating factors, such as complement activation, and histamine release, but is not dependent on a previous sensitization process. The intolerance reactions have been associated with drugs such as acetylsalicylic acid, whereas intolerance to leachable biomaterial components such as benzoic acid is conceivable but not known.

ALLERGY AND BIOMATERIALS

A foreign substance able to induce an allergic reaction is called an allergen. There is no acceptable way of predicting whether a substance or a compound is potentially allergenic only on the basis of its chemical composition and/or structure. However, experimental evidence and years of empirical results after testing substances causing allergic reactions have given some leads, e.g., large foreign molecules such as proteins and nucleoproteins are strong allergens, whereas lipids are not. However, the strongest chemical allergens associated with biomaterials are often chemically active substances of low molecular weight, often less than 500 Da, such as lipid-soluble organic substances derived from polymer materials or metal ions and metal salts. These are called haptens, i.e., they become full allergens only after reaction or combination with proteins that may be present in macrophages and Langerhans cells of the host.

TYPES OF ALLERGIES

The allergies are most often categorized into four main groups (type I–IV) according to the reaction mechanisms. The types I to III are associated with humoral antibodies initiated by B-lymphocytes that develop to immunoglobulin-producing plasma cells. The immunoglobulins are classified into five different classes, Ig E, A, D, G, and M, according to their basic structure and size. A variable portion of the immmunoglobulin is specific for the antigen that induced its production (Roitt et al., 1997). The type IV reaction is a cell-mediated reaction caused by T-lymphocytes. These interactions are also discussed in Chapter 4.3.

The types II and III allergies comprise antigen/antibody encounters including complement activation, cell lysis, release of vasoactive substances, inflammatory reaction, and tissue damage. Necrosis of periimplant tissue with histologic appearance and serum complement analyses consistent with Type III hypersensitivity has been observed in cases of atypical loosening of total hip prostheses (Hensten-Pettersen, 1993). However, an FDA document (Immunotoxicity Testing Guidance, 1999) omits the type II and III reactions for reasons of being "relatively rare and less likely to occur with medical devices/materials" leaving the types I and IV as relevant in the present context.

Type I Hypersensitivity

The type I reaction is based on an interaction between an intruding allergen and IgE immunoglobulins located in mast cells, basophils, eosinophils, and platelets, resulting in release of active mediators such as histamine and other vasoactive substances. The results are local or systemic reactions seen within a short time (minutes). The symptoms depend on the tissue or organ subject to sensitization, e.g., (1) inhaled allergens such as pollen or residual proteins associated with surgical latex gloves or other natural latex products that may result in asthmatic seizures, swelling of the mucosa of the throat, or worse; or (2) decreased blood pressure and anaphylactic shock. Food allergies may also give systemic symptoms. This type of host reaction is usually associated with full antigens. Since the potential allergens associated with biomaterials are small molecular haptens, the probablility of IgE-based allergic reactions is low, although IgE antibodies to chromium and nickel have been reported (Hensten-Pettersen, 1993). Reports on adverse reactions to orthopedic devices describe patients with urticarial reactions. Contact urticaria is a wheal and flare response to compounds applied on intact skin. The role of immunological contact urticaria in relation to medical devices is not clear.

Type IV Hypersensitivity

The cell-mediated hypersensitivity is referred to as "delayed" because it takes more than 12 hours to develop, often 24–72 hours. Prolonged challenges of macrophage-resistant allergens, usually of microbial origin, may result in persistant immunological granuloma formation. The T-lymphocytes producing the response have been sensitized by a previous encounter with an allergen and act in concert with other lymphocytes and mononuclear phagocytes to create four histologically different types characterized by skin-related tissue reactions. The reactions are elicited by interaction of cells and mediators that comprise (1) swelling (the

Jones–Mote type); (2) induration (the granulomatous type); (3) swelling and induration and possibly fever (the tuberculin type); and (4) eczema (the contact type) (Roitt *et al.*, 1997). The latter form of delayed hypersensitivity has been of specific importance in relation to biomaterials. Most information on this reaction has been obtained by studying the reaction patterns following exposure to external environmental and occupation-related chemicals.

Allergic contact dermatitis is acquired through previous sensitization with a foreign substance. The hapten is absorbed by the skin or mucosa and binds to certain proteins associated with the Langerhans cells, forming a complete antigen. The antigen is brought in contact with the regional lymph nodes, resulting in the formation of activated, specialized T cells that are brought into circulation. Upon new exposure, the allergen may again be transported from the site of entrance. The new contact between the allergen and the activated, specialized T cells releases inflammatory mediators, resulting in further production and attraction of T cells causing tissue damage. The reactions are not necessarily limited to the exposure site.

The presence of allergic contact dermatitis is evaluated by allergologists or dermatologists by applying the suspected haptens using epidermal or intradermal skin tests and reading the dermal or epidermal reaction after specified amounts of time. Commecial test kits for epidermal testing are available for a series of chemical substances related to different occupations. A vast amount of information on the allergenic characteristics of biomaterials-related substances has been obtained in this way, especially as regards dental materials (Kanerva *et al.*, 1995).

Many biomaterials employed in dentistry such as metal alloys and resin-based materials have medical counterparts, and both categories of biomaterials have materials counterparts met with in everyday life. The sensitization process therefore often has taken place before the biomaterialsc contact.

ATOPY

Atopic individuals have a constitutional predisposition for IgE-based hypersensitive reactions caused by environmental and food allergens. The reactions include histamine-mediated hay fever, asthma, gastrointestinal symptoms, or skin rashes and are more pronounced at an early age. Atopics have an increased risk of acquiring irritant contact dermatitis to external biomaterial devices such as orthodontic appliances. The relation to allergic contact dermatitis is unclear (Lindsten and Kurol, 1997); so also is the relationship between atopy and allergens or haptens from biomaterials exposed parenterally.

IMMUNOLOGIC TOXICITY OF MEDICAL DEVICES

Immunologic toxicity to surface medical devices and external communicating devices (dialyzers, laparoscopes, etc.) may represent mechanisms of sensitization and hypersensitive reactions similar to those of orally exposed biomaterials. Hypersensitivity reactions to implants in clinically inobservable locations are difficult to recognize unless they have dermal or systemic expressions. In addition, such reactions may be part of local toxic and/or mechanically induced inflammatory reactions using similar mediators for tissue response. For lack of more distinct discriptions, such reactions have been referred to as "deep tissue" reactions of type IV hypersensitivity.

A vast battery of *in vitro* and *in vivo* experimental studies have been performed to study potential adverse effects of biomaterial devices such as artificial joints, heart valves, and breast prostheses (Rodgers *et al.*, 1997). Aseptic loosening of metallic hip prostheses have been associated with "biologic" causes in addition to biomechanical factors and wear debris. However, it is currently unclear whether metal sensitivity is a contributing factor to implant failure (Hallab *et al.*, 2001). In fact, it is argued that the loosening process enhances the immunological sensitization, indicating that the cause/effect relation may be reversed (Milavec-Puretic *et al.*, 1998). What is clear is that local and general eczematous reactions have been observed following the insertion of metallic implants in patients subsequently shown to be allergic to cobalt, chromium, and nickel. Many case reports also describe the immediate healing of dermal reactions associated with metal implants (Al-Saffar and Revell, 1999). Metal allergy has also been discussed as a possible contributing factor in the development of in-stent coronary restenosis, although there is little evidence for this effect (Hillen *et al.*, 2002). However, established metal allergy in a patient does not as a rule seem be accompanied by clinical reactions to implant alloys containing the metal. If this statement is true, it is in line with clinical observations made in surveys on the use of metallic alloys in prosthodontics and orthodontics. Inhomogeneiety or mixture of alloys appear to determine the efflux of potentially hypersensitive metal ions, and hence increase the possibility of eliciting hypersensitive reactions (Grimsdottir *et al.*, 1992).

Methyl methacrylate bone cement is another potential allergenic factor in orthopedic surgery parallell with reactions in dentistry and cosmetics (Kaplan *et al.*, 2002), and immune-mediated disease and silicone based implants has been a matter of discussion for some time. However, a scientifically valid cause and effect relationship between immune based disease and silicone based implants has not been established (Rodgers *et al.*, 1997).

An extensive literature reflects clinical surveys and research activities related to natural latex used as barrier material by the health professions. It is accepted that residual latex proteins and chemicals associated with the production process may cause immediate and delayed reactions in patients and health personnel (Turjanmaa *et al.*, 1996).

OTHER INTERACTIONS

The FDA testing guidance referred to above also lists other interactions of medical devices, extracts of medical devices, or adjuvants with the immune system such as impairment of the normal immunologic protective mechanisms (immunosuppression), and long-term immmunological activity (immunostimulation) that may lead to harmful autoimmune responses. The autoimmune reaction is explained by the biomaterials-associated agent acting as an adjuvant that is stimulating to antibody/complement-based tissue damage by cross reactions with human protein. Chronic inflammatory, immune-related granuloma may take part in the development of autoimmune reactions.

CONCLUDING REMARKS

Biocompatibility issues related to medical devices form a multidimensional crossroad of technology and biology. One dimension is the various classes of biomaterials, such as plastics and other polymers, metals, ceramics, and glasses, depending on expert design to obtain maximal mechanical properties and minimal chemical dissolution. Another is the mode of application, ranging from skin and mucosal contact to totally submerged implants, with external communicating devices in between. A third dimension is the duration of contact, ranging from minutes to the expected lifetime, and the fourth, and decisive, is the biological reactions that can be expected. These circumstances prevent general statements on biomaterials. The present overview is aimed at students and limited to focus on collective mechanisms determining systemic toxicity and discuss hypersensitivity reactions documented by clinical reports.

Bibliography

Al-Saffar, N., and Revell, P. A. (1999). Pathology of the bone–implant interfaces. *J. Long-Term Effects Med. Implants* 9: 319–347.

deBruin, A. (1981). The metabolic fate of foreign compounds. in *Fundamental Aspects of Biocompatibility*, D. F. Williams, ed. CRC Press, Boca Raton, Fl, Vol. II, pp. 3–43.

Eaton, D. L., and Klaassen, C. D. (1996). Principles of toxicology. in *Casaretts and Doulls Toxicology*, C. D. Klaassen, ed. McGraw Hill, New York, pp.13–33.

European Commission (1998). Dental Amalgam. A report with reference to The Medical Device Directive 93/42/EEC from an ad hoc Working Group mandated by DG III.

Food and Drug Administration (1999). Immunotoxicity Testing Guidance, Document issued May 6. U.S. Department of Health and Human Services, pp. 1–15.

Grimsdottir, M. R., Gjerdet, N. R., and Hensten-Pettersen, A. (1992). Composition and in vitro corrosion of orthodontic appliances. *Am. J. Dentofac. Orthop.* 101: 23–30.

Hallab, N., Merrit, K., and Jacobs, J. J. (2001). Metal sensitivity in patients with orthopedic implants. *J. Bone Joint Surg.* 83: 428–436.

Hensten-Pettersen, A. (1993). Allergy and hypersensitivity. in *Biological, Material, and Mechanical Considerations of Joint Replacement*, B. F. Morrey, ed. Raven Press, New York, pp. 353–361.

Hillen, U., Haude, M., Erbel, R., and Goos, M. (2002). Evaluation of metal allergies in patients with coronary stents. *Contact Dermatitis.* 47: 353–356.

Kanerva, L., Estlander, T., and Jolanki, R. (1995). Dental problems. in *Practical contact Dermatitis,* J. D. Guin, ed. McGraw-Hill Health Profession Division, pp. 397–432.

Kaplan, K., Della Valle, C. J., Haines, H., and Zuckerman, J. D. (2002). Preoperative identification of a bone-cement allergy in a patient undergoing total knee arthroplasty. *J. Arthoplasty* 17: 788–791.

Lindsten, R., and Kurol, J. (1997). Orthodontic appliances in relation to nickel hypersensitivity. *J. Orofac. Orthop/Fortschr. Kieferorthop.* 58: 100–108.

Lygre, H., Klepp, K. N., Solheim, E., and Gjerdet, N. R. (1993). Leaching of additives and degradation products from cold-cured orthodontic resins. *Acta Odontol. Scand.* 52: 150–156.

Mackert, J. R., and Berglund, A. (1997). Mercury exposure from dental amalgam fillings: absorbed dose and the potential for adverse health effects. in *Dental Amalgam and Alternative Direct Restorative Materials,* I. A. Mjör, and G. N. Pakhomov, eds. Oral Health Division of Noncommunicable Diseases, WHO, Geneva, pp. 47–60.

Milavec-Puretic, V., Orlic, D., and Marusic, A. (1998). Sensitivity to metals in 40 patients with failed hip endoprosthesis. *Arch. Trauma Surg.* 117: 383–386.

Nylander, M., Friberg, L., Eggleston, R., and Björkmann, L. (1989). Mercury accumulation in tissues from dental staff and controls in relation to exposure. *Swed. Dent. J.* 13: 225–245.

Park, H. Y., and Shearer, P. D. (1983). In vitro release of nickel and chromium from simulated orthodontic aplliances. *Am. J. Orthod.* 84: 156–159.

Rodgers, K., Klykken, P., Jacobs, J., Frondoza, C., Tomazic, V., and Zelikoff, D. (1997). Immunotoxicity of medical devices. Symposium overview. *Fund. Appl. Toxicol.* 36: 1–14.

Roitt, I., Brostoff, J., and Male, D. (1997). *Immunology.* Churchill Livingstone, Edinburgh; Gower Medical Publishing, London, 2nd ed. Chapters 19 and 22.

Turjanmaa, K., Alenius, H., Mäkinen-Kiljunen, S., Reunala, T., and Palosuo, T. (1996). Natural rubber latex allergy (review). *Allergy* 51: 593–602.

4.6 BLOOD COAGULATION AND BLOOD–MATERIALS INTERACTIONS

Stephen R. Hanson

The hemostatic mechanism is designed to arrest bleeding from injured blood vessels. The same process may produce adverse consequences when artificial surfaces are placed in contact with blood. These events involve a complex set of interdependent reactions between (1) the surface, (2) platelets, and (3) coagulation proteins, resulting in the formation of a clot or thrombus that may subsequently undergo removal by (4) fibrinolysis. The process is localized at the surface by opposing activation and inhibition systems, which ensure that the fluidity of blood in the circulation is maintained. In this chapter, a brief overview of the hemostatic mechanism is presented. Although a great deal is known about blood responses to injured arteries and blood-contacting devices, important relationships remain to be defined in many instances. More detailed discussions of hemostasis and thrombosis have been provided elsewhere (Colman *et al.,* 2001; Esmon, 2003; Forbes and Courtney, 1987; Gresle *et al.,* 2002; Stamatoyannopoulos *et al.,* 1994).

PLATELETS

Platelets ("little plates") are nonnucleated, disk-shaped cells having a diameter of 3–4 µm and an average volume of 10×10^{-9} mm^3. Platelets are produced in the bone marrow, circulate at an average concentration of about 250,000 cells per microliter of whole blood, and occupy approximately 0.3% of

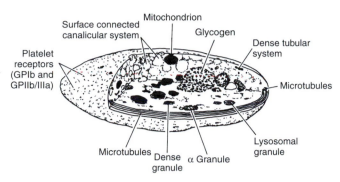

FIG. 1. Platelet structure.

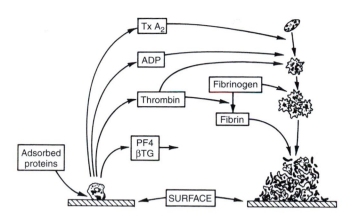

FIG. 2. Platelet reactions to artificial surfaces. Following protein adsorption to surfaces, platelets adhere and release α-granule contents, including platelet factor 4 (PF4) and β-thromboglobulin (β-TG), and dense granule contents, including ADP. Thrombin is generated locally through coagulation reactions catalyzed by procoagulant platelet surface phospholipids. Thromboxane A_2 (TxA$_2$) is synthesized. ADP, TxA$_2$, and thrombin recruit additional circulating platelets into an enlarging platelet aggregate. Thrombin-generated fibrin stabilizes the platelet mass.

the total blood volume. In contrast, red cells typically circulate at 5×10^6 cells per microliter and may make up 40–50% of the total blood volume. As discussed later, platelet functions are designed to (1) initially arrest bleeding through formation of platelet plugs, and (2) stabilize the initial platelet plugs by catalyzing coagulation reactions leading to the formation of fibrin.

Platelet structure provides a basis for understanding platelet function. In the normal (nonstimulated) state, the platelet discoid shape is maintained by a circumferential bundle (cytoskeleton) of microtubules (Fig. 1). The external surface coat of the platelet contains membrane-bound receptors (e.g., glycoproteins Ib and IIb/IIIa) that mediate the contact reactions of adhesion (platelet–surface interactions) and aggregation (platelet–platelet interactions). The membrane also provides a phospholipid surface that accelerates important coagulation reactions (see below), and forms a spongy, canal-like (canalicular) open network that represents an expanded reactive surface to which plasma factors are selectively adsorbed. Platelets contain substantial quantities of muscle protein (e.g., actin, myosin) that allow for internal contraction when platelets are activated. Platelets also contain three types of cytoplasmic storage granules: (1) α-granules, which are numerous and contain the platelet-specific proteins platelet factor 4 (PF-4) and β-thromboglobulin (β-TG), and proteins found in plasma (including fibrinogen, albumin, fibronectin, and coagulation factors V and VIII); (2) dense granules that contain adenosine diphosphate (ADP), calcium ions (Ca^{2+}), and serotonin; and (3) lysosomal granules containing enzymes (acid hydrolases).

Platelets are extremely sensitive cells that may respond to minimal stimulation. Activation causes platelets to become sticky and change in shape to irregular spheres with spiny pseudopods. Activation is accompanied by internal contraction and extrusion of the storage granule contents into the extracellular environment. Secreted platelet products such as ADP stimulate other platelets, leading to irreversible platelet aggregation and the formation of a fused platelet thrombus (Fig. 2).

Platelet Adhesion

Platelets adhere to artificial surfaces and injured blood vessels. At sites of vessel injury, the adhesion process involves the interaction of platelet glycoprotein Ib (GP Ib) and connective tissue elements that become exposed (e.g., collagen) and requires plasma von Willebrand factor (vWF) as an

essential cofactor. GP Ib (about 25,000 molecules per platelet) acts as the surface receptor for vWF (Colman *et al.*, 2001). The hereditary absence of GP Ib or vWF results in defective platelet adhesion and serious abnormal bleeding.

Platelet adhesion to artificial surfaces may also be mediated through platelet glycoprotein IIb/IIIa (integrin $\alpha_{IIb}\beta_3$) as well as through the GP Ib-vWF interaction. GP IIb/IIIa (about 80,000 copies per resting platelet) is the platelet receptor for adhesive plasma proteins that support cell attachment, including fibrinogen, vWF, fibronectin, and vitronectin (Gresle *et al.*, 2002). Resting platelets do not bind these adhesive glycoproteins, events which normally occur only after platelet activation causes a conformational change in GP IIb/IIIa. Platelets that have become activated near artificial surfaces (for example, by exposure to factors released from already adherent cells) could adhere directly to surfaces through this mechanism (e.g., via GP IIb/IIIa binding to surface-adsorbed fibrinogen). Also, normally unactivated GP IIb/IIIa receptors could react with surface proteins that have undergone conformational changes as a result of the adsorption process (Chapter 3.2). The enhanced adhesiveness of platelets toward surfaces preadsorbed with fibrinogen supports this view. Following adhesion, activation, and release reactions, the expression of functionally competent GP IIb/IIIa receptors may also support tight binding and platelet spreading through multiple focal contacts with fibrinogen and other surface-adsorbed adhesive proteins.

Platelet Aggregation

Following platelet adhesion, a complex series of reactions is initiated involving (1) the release of dense granule ADP, (2) the formation of small amounts of thrombin (see later discussion), and (3) the activation of platelet biochemical processes leading to the generation of thromboxane A_2. The release of ADP, thrombin formation, and generation of thromboxanes all act in concert to recruit platelets into the growing platelet aggregate (Fig. 2). Platelet stimulation by these agonists causes the expression on the platelet surface of activated GP IIb/IIIa, which then binds plasma proteins that support platelet aggregation. In normal blood, fibrinogen, owing to its relatively high concentration (Table 1), is the most important protein supporting platelet aggregation. The platelet–platelet interaction involves Ca^{2+}-dependent bridging of adjacent platelets by fibrinogen molecules (platelets will not aggregate in the absence of fibrinogen, GP IIb/IIIa, or Ca^{2+}). Thrombin binds directly to platelet thrombin receptors and plays a key role in platelet aggregate formation by (1) activating platelets, which then catalyze the production of more thrombin; (2) stimulating ADP release and thromboxane A_2 formation; and (3) stimulating the formation of fibrin, which stabilizes the platelet thrombus.

Platelet Release Reaction

The release reaction is the secretory process by which substances stored in platelet granules are extruded from the platelet. ADP, collagen, epinephrine, and thrombin are physiologically important release-inducing agents and interact with the platelet through specific receptors on the platelet surface. Alpha-granule contents (PF-4, β-TG, and other proteins) are readily released by relatively weak agonists such as ADP. Release of the dense granule contents (ADP, Ca^{2+}, and serotonin) requires platelet stimulation by a stronger agonist such as thrombin. Agonist binding to platelets also initiates the formation of intermediates that cause activation of the contractile–secretory apparatus, production of thromboxane A_2, and mobilization of calcium from intracellular storage sites. Elevated cytoplasmic calcium is probably the final mediator of platelet aggregation and release. As noted, substances that are released (ADP), synthesized (TxA_2), and generated (thrombin) as a result of platelet stimulation and release affect other platelets and actively promote their incorporation into growing platelet aggregates. *In vivo*, measurements of plasma levels of platelet-specific proteins (PF-4, β-TG) have been widely used as indirect measures of platelet activation and release.

Platelet Coagulant Activity

When platelets aggregate, platelet coagulant activity is produced, including expression of negatively charged membrane phospholipids (phosphatidylserine) that accelerate two critical steps of the blood coagulation sequence: factor X activation and the conversion of prothrombin to thrombin (see below). Platelets may also promote the proteolytic activation of factors XII and XI. The surface of the aggregated platelet mass thus serves as a site where thrombin can form rapidly in excess of the neutralizing capacity of blood anticoagulant mechanisms. Thrombin also activates platelets directly and generates polymerizing fibrin, which adheres to the surface of the platelet thrombus.

Platelet Consumption

In man, platelets labeled with radioisotopes are cleared from circulating blood in an approximately linear fashion over time with an apparent life span of approximately 10 days. Platelet life span in experimental animals may be somewhat shorter. With ongoing or chronic thrombosis that may be produced by cardiovascular devices, platelets may be removed from circulating blood at a more rapid rate. Thus steady-state elevations in the rate of platelet destruction, as reflected in a shortening of platelet life span, have been used as a measure of the thrombogenicity of artificial surfaces and prosthetic devices (Hanson *et al.*, 1980, 1990).

COAGULATION

In the test tube, at least 12 plasma proteins interact in a series of reactions leading to blood clotting. Their designation as Roman numerals was made in order of discovery, often before their role in the clotting scheme was fully appreciated. Their biochemical properties are summarized in Table 1. Initiation of clotting occurs either intrinsically by surface-mediated reactions, or extrinsically through factors derived from tissues. The two systems converge upon a final common pathway that leads to the formation of thrombin, and an insoluble fibrin gel when thrombin acts on fibrinogen.

Coagulation proceeds through a "cascade" of reactions by which normally inactive factors (e.g., factor XII) become enzymatically active following surface contact, or after proteolytic cleavage by other enzymes (e.g., surface contact activates factor XII to factor XIIa). The newly activated enzymes in turn activate other normally inactive precursor molecules (e.g., factor XIIa converts factor XI to factor XIa). Because this sequence involves a series of steps, and because one enzyme molecule can activate many substrate molecules, the reactions are quickly amplified so that significant amounts of thrombin are produced, resulting in platelet activation, fibrin formation, and arrest of bleeding. The process is localized (i.e., widespread clotting does not occur) owing to dilution of activated factors by blood flow, the actions of inhibitors that are present or are generated in clotting blood, and because several reaction steps proceed at an effective rate only when catalyzed on the surface of activated platelets or at sites of tissue injury.

Figure 3 presents a scheme of the clotting factor interactions involved in both the intrinsic and extrinsic systems and their common path. Except for the contact phase, calcium is required for most reactions and is the reason why chelators of calcium (e.g., citrate) are effective anticoagulants. It is also clear that the *in vitro* interactions of clotting factors, i.e., clotting, is not identical with coagulation *in vivo*, which may be triggered by artificial surfaces and by exposure of the cell-associated protein, tissue factor. There are also interrelationships between the intrinsic and extrinsic systems, such that under some conditions "crossover" or reciprocal activation reactions may be important (Colman *et al.*, 2001; Bennett *et al.*, 1987).

MECHANISMS OF COAGULATION

In the intrinsic clotting system, contact activation refers to reactions following adsorption of contact factors onto a negatively charged surface. Involved are factors XII, XI,

TABLE 1   Properties of Human Clotting Factors

Clotting factor	Apparent molecular weight (number of chains)	Approximate normal plasma concentration (μg/ml)	Active Form
Intrinsic clotting system			
Prekallikrein	100,000 (1)	50	Serine protease
High molecular weight kininogen	120,000 (1)	80	Cofactor
Factor XII	80,000 (1)	30	Serine protease
Factor XI	143,000 (2)	3–6	Serine protease
Factor IX	57,000 (1)	3–5	Serine protease
Factor VIII[a]	330,000 (1)	0.2	Cofactor
Von Willebrand factor[a]	250,000 (1)	10	Cofactor for platelet adhesion
Extrinsic clotting system			
Tissue factor	44,000 (1)	0[b]	Cofactor
Factor VII	50,000 (1)	1	Serine protease
Common pathway			
Factor X	59,000 (2)	5	Serine protease
Factor V	330,000 (1)	5–12	Cofactor
Prothrombin	72,000 (1)	140	Serine protease
Fibrinogen	340,000 (6)	2500	Fibrin polymer
Factor XIII	320,000 (4)	10	Transglutaminase

[a] In plasma, factor VIII is complexed with von Willebrand factor which circulates as a series of multimers ranging in molecular weight from about 600,000 to 2×10^6.

[b] The tissue factor concentration in cell free plasma is low since tissue factor is an integral cell membrane–associated protein expressed by vascular and inflammatory cells, although a role in coagulation and thrombosis for a circulating form of soluble tissue factor has recently been postulated.

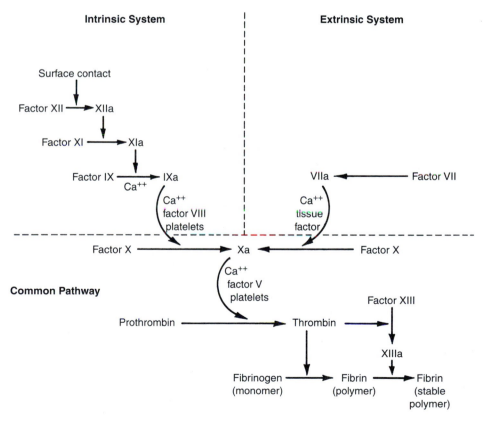

FIG. 3.   Mechanisms of clotting factor interactions. Clotting is initiated by either an intrinsic or extrinsic pathway with subsequent factor interactions that converge upon a final, common path.

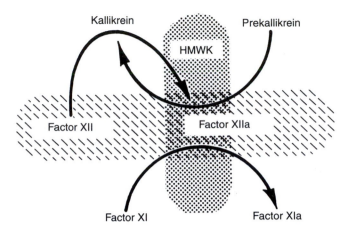

FIG. 4. Contact activation. The initial event *in vitro* is the adsorption of factor XII to a negatively charged surface (hatched, horizontal ovoid) where it is activated to form factor XIIa. Factor XIIa converts prekallikrein to kallikrein. Additional factor XIIa and kallikrein are then generated by reciprocal activation. Factor XIIa also activates factor XIa. Both prekallikrein and factor XI bind to a cofactor, high-molecular-weight kininogen (HMWK; dotted, vertical ovoid), which anchors them to the charged surface.

prekallikrein, and high-molecular-weight kininogen (HMWK) (Fig. 4). All contact reactions take place in the absence of calcium. Kallikrein also participates in fibrinolytic system reactions and inflammation (Bennett *et al.,* 1987). Although these reactions are well understood *in vitro,* their pathologic significance remains uncertain. For exampa, in hereditary disorders, factor XII deficiency is not associated with an increased bleeding tendency, and only a marked deficiency of factor XI produces abnormal bleeding.

A middle phase of intrinsic clotting begins with the first calcium-dependent step, the activation of factor IX by factor XIa. Factor IXa subsequently activates factor X. Factor VIII is an essential cofactor in the intrinsic activation of factor X, and factor VIII first requires modification by an enzyme, such as thrombin, to exert its cofactor activity. In the presence of calcium, factors IXa and VIIIa form a complex (the "tenase" complex) on phospholipid surfaces (expressed on the surface of activated platelets) to activate factor X. This reaction proceeds slowly in the absence of an appropriate phospholipid surface and serves to localize the clotting reactions to the surface (versus bulk fluid) phase. The extrinsic system is initiated by the activation of factor VII. When factor VII interacts with tissue factor, a cell membrane protein that may also circulate in a soluble form, factor VIIa becomes an active enzyme which is the extrinsic factor X activator. Tissue factor is present in many body tissues; is expressed by stimulated white cells and endothelial cells; and becomes available when underlying vascular structures are exposed to flowing blood upon vessel injury.

The common path begins when factor X is activated by either factor VIIa–tissue factor or by the factor IXa–VIIIa complex. After formation of factor Xa, the next step involves factor V, a cofactor, which (like factor VIII) has activity after modification by another enzyme such as thrombin. Factor Xa–Va, in the presence of calcium and platelet phospholipids,

forms a complex ("prothrombinase" complex) that converts prothrombin (factor II) to thrombin. Like the conversion of factor X, prothrombin activation is effectively surface catalyzed. The higher plasma concentration of prothrombin (Table 1), as well as the biologic amplification of the clotting system, allows a few molecules of activated initiator to generate a large burst of thrombin activity. Thrombin, in addition to its ability to modify factors V and VIII and activate platelets, acts on two substrates: fibrinogen and factor XIII. The action of thrombin on fibrinogen releases small peptides from fibrinogen (e.g., fibrinopeptide A) that can be assayed in plasma as evidence of thrombin activity. The fibrin monomers so formed polymerize to become a gel. Factor XIII is either trapped within the clot or provided by platelets and is activated directly by thrombin. A tough, insoluble fibrin polymer is formed by interaction of the fibrin polymer with factor XIIIa.

CONTROL MECHANISMS

Obviously, the blood and vasculature must have mechanisms for avoiding massive thrombus formation once coagulation is initiated. At least four types of mechanisms may be considered. First, blood flow may reduce the localized concentration of precursors and remove activated materials by dilution into a larger volume, with subsequent removal from the circulation following passage through the liver. Second, the rate of several clotting reactions is fast only when the reaction is catalyzed by a surface. These reactions include the contact reactions, the activation of factor X by factor VII–tissue factor at sites of tissue injury, and reactions that are accelerated by locally deposited platelet masses (activation of factor X and prothrombin). Third, there are naturally occurring inhibitors of coagulation enzymes, such as antithrombin III, which are potent inhibitors of thrombin and other coagulation enzymes (plasma levels of thrombin–antithrombin III complex can also be assayed as a measure of thrombin production *in vivo*). Another example of a naturally occuring inhibitor is tissue factor pathway inhibitor (TFPI), a protein that in association with factor Xa inhibits the tissue factor/factor VII complex. Fourth, during the process of coagulation, enzymes are generated that not only activate coagulation factors, but also degrade cofactors. For example, the fibrinolytic enzyme plasmin (see below) degrades fibrinogen and fibrin monomers and can inactivate cofactors V and VIII. Thrombin is also removed when it binds to thrombomodulin, a protein found on the surface of blood vessel endothelial cells. The thrombin–thrombomodulin complex then converts another plasma protein, protein C, to an active form that can also degrade factors V and VIII. *In vivo,* the protein C pathway is a key physiologic anticoagulant mechanism (Colman *et al.,* 2001; Esmon, 2003).

In summary, the platelet, coagulation, and endothelial systems interact in a number of ways that promote localized hemostasis while preventing generalized thrombosis. Figure 5 depicts some of the relationships and inhibitory pathways that apply to blood reactions following contact with both natural and artificial surfaces.

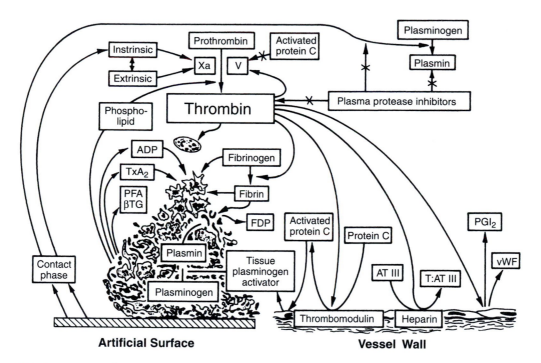

FIG. 5. Integrated hemostatic reactions between a foreign surface and platelets, coagulation factors, the vessel endothelium, and the fibrinolytic system.

Fibrinolysis

The fibrinolytic system removes unwanted fibrin deposits to improve blood flow following thrombus formation, and to facilitate the healing process after injury and inflammation. It is a multicomponent system composed of precursors, activators, cofactors and inhibitors, and has been studied extensively (Colman *et al.*, 2001; Forbes and Courtney, 1987). The fibrinolytic system also interacts with the coagulation system at the level of contact activation (Bennett *et al.*, 1987). A simplified scheme of the fibrinolytic pathway is shown in Fig. 6.

The most well-studied fibrinolytic enzyme is plasmin, which circulates in an inactive form as the protein plasminogen. Plasminogen adheres to a fibrin clot, being incorporated into the mesh during polymerization. Plasminogen is activated to plasmin by the actions of plasminogen activators that may be

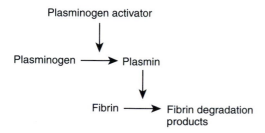

FIG. 6. Fibrinolytic sequence. Plasminogen activators, such as tissue plasminogen activator (tPA) or urokinase, activate plasminogen to form plasmin. Plasmin enzymatically cleaves insoluble fibrin polymers into soluble degradation products (FDPs), thereby effecting the removal of unnecessary fibrin clot.

present in blood or released from tissues, or that may be administered therapeutically. Important plasminogen activators occurring naturally in man include tissue plasminogen activator (tPA) and urokinase. Following activation, plasmin digests the fibrin clot, releasing soluble fibrin–fibrinogen digestion products (FDP) into circulating blood, which may be assayed as markers of *in vivo* fibrinolysis (e.g., the fibrin D-D dimer fragment). Fibrinolysis is inhibited by plasminogen activator inhibitors (PAIs), and by a thrombin-activated fibrinolysis inhibitor (TAFI) that promotes the stabilization of fibrin and fibrin clots (Colman *et al.*, 2001).

Complement

As detailed in Chapter 4.4, the complement system is primarily designed to effect a biologic response to antigen–antibody reactions. Like the coagulation and fibrinolytic systems, complement proteins are activated enzymatically through a complex series of reaction steps (Bennett *et al.*, 1987). Several proteins in the complement cascade function as inflammatory mediators. The end result of these activation steps is the generation of an enzymatic complex that causes irreversible damage (by lytic mechanisms) to the membrane of the antigen-carrying cell (e.g., bacteria).

Since there are a number of interactions between the complement, coagulation, and fibrinolytic systems, there has been considerable interest in the problem of complement activation by artificial surfaces, prompted in part by observations that devices having large surface areas (e.g., hemodialyzers) may cause (1) reciprocal activation reactions between complement enzymes and white cells, and (2) complement activation that may mediate both white-cell and platelet adhesion to artificial surfaces. Further observations regarding the complement activation pathways involved in blood–materials interactions are likely to be of interest.

Red Cells

Red cells are usually considered as passive participants in processes of hemostasis and thrombosis, although under some conditions (low shear or venous flows) red cells may form a large proportion of total thrombus mass. The concentration and motions of red cells have important mechanical effects on the diffusive transport of blood elements. For example, in flowing blood, red-cell motions may increase the effective dissusivity of platelets by several orders of magnitude. Under some conditions, red cells may also contribute chemical factors that influence platelet reactivity (Turitto and Weiss, 1980). The process of direct attachment of red cells to artificial surfaces has been considered to be of minor importance and has therefore received little attention in studies of blood–materials interactions.

White Cells

The various classes of white cells perform many functions in inflammation, infection, wound healing, and the blood response to foreign materials. White-cell interactions with artificial surfaces may proceed through as-yet poorly defined mechanisms related to activation of the complement, coagulation, fibrinolytic, and other enzyme systems, resulting in the expression by white cells of procoagulant, fibrinolytic, and inflammatory activities. For example, stimulated monocytes express tissue factor, which can initiate extrinsic coagulation. Neutrophils may contribute to clot dissolution by releasing potent fibrinolytic enzymes (e.g., neutrophil elastase). White cell interactions with devices having large surface areas may be extensive (especially with surfaces that activate complement), resulting in their marked depletion from circulating blood. Activated white cells, through their enzymatic and other activities, may produce organ dysfunction in other parts of the body. In general, the role of white-cell mechanisms of thrombosis and thrombolysis, in relation to other pathways, remains an area of considerable interest.

CONCLUSIONS

Interrelated blood systems respond to tissue injury in order to quickly minimize blood loss, and later to remove unneeded deposits after healing has occurred. When artificial surfaces are exposed, an imbalance between the processes of activation and inhibition of these systems can lead to excessive thrombus formation and an exaggerated inflammatory response. Whereas many of the key blood cells, proteins, and reaction steps have been identified, their reactions in association with artificial surfaces have not been well defined in many instances. Therefore, blood reactions that might cause thrombosis continue to limit the potential usefulness of many cardiovascular devices for

applications in man. Consequently, these devices commonly require the use of systemic anticoagulants, which present an inherent bleeding risk.

Acknowledgments

This work was supported by research grant HL-31469 from the National Institutes of Health, U.S. Public Health Service, and by the ERC Program of the National Science Foundation under Award EEC-9731643.

Bibliography

Bennett, B., Booth, N. A., and Ogston, D. (1987). Potential interactions between complement, coagulation, fibrinolysis, kinin-forming, and other enzyme systems. in *Haemostasis and Thrombosis*, 2nd ed., A. L. Bloom and D. P. Thomas, eds. Churchill Livingstone, New York, pp. 267–282.

Colman, R. W., Hirsh, J., Marder, V. J., Clowes, A. W., and George, J. N., eds. (2001). *Hemostasis and Thrombosis*, 4th ed. Lippincott, New York.

Esmon, C. T. (2003). The protein C pathway. *Chest* **124**(3 Suppl): 26S–32S.

Forbes, C. D., and Courtney, J. M. (1987). Thrombosis and artificial surfaces. in: *Haemostasis and Thrombosis*, 2nd ed., A. L. Bloom and D. P. Thomas, eds. Churchill Livingstone, New York, pp. 902–921.

Gresle, P., Page, C. P., Fuster, F., and Vermylen, J. (2002). *Platelets in Thrombotic and Non-thrombotic Disorders*, 1st ed. Cambridge Univ. Press, Cambridge.

Hanson, S. R., Harker, L. A., Ratner, B. D., and Hoffman, A. S. (1980). In vivo evaluation of artificial surfaces using a nonhuman primate model of arterial thrombosis. *J. Lab. Clin. Med.* **95**: 289–304.

Hanson, S. R., Kotze, H. F., Pieters, H., and Heyns, A. du P. (1990). Analysis of 111-indium platelet kinetics and imaging in patients with aortic aneurysms and abdominal aortic grafts. *Arteriosclerosis* **10**: 1037–1044.

Stamatoyannopoulos, G., Nienhuis, A. W., Majerus, P. W., and Varmus, H. (1994). *The Molecular Basis of Blood Diseases*, 2nd ed. W.B. Saunders, Philadelphia.

Turitto, V. T., and Weiss, H. J. (1980). Red cells: their dual role in thrombus formation. *Science* **207**: 541–544.

4.7 TUMORIGENESIS AND BIOMATERIALS

Frederick J. Schoen

The possibility that implant materials could cause tumors or promote tumor growth has long been a concern of surgeons and biomaterials researchers. This chapter describes general concepts in neoplasia, the association of tumors with implants in human and animals, and the pathobiology of tumor formation adjacent to biomaterials.

GENERAL CONCEPTS

Neoplasia, which literally means "new growth," is the process of excessive and uncontrolled cell proliferation (Cotran *et al.*, 1999; Kumar *et al.*, 1997). The new growth is called

TABLE 1 Characteristics of Benign and Malignant Tumors

Characteristics	Benign	Malignant
Differentiation	Well defined; structure may be typical of tissue of origin	Less differentiated with bizarre (anaplastic) cells; often atypical structure
Rate of growth	Usually progressive and slow; may come to a standstill or regress; cells in mitosis are rare	Erratic and may be slow to rapid; mitoses may be absent to numerous and abnormal
Local invasion	Usually cohesive, expansile, well-demarcated masses that neither invade nor infiltrate the surrounding normal tissues	Locally invasive, infiltrating adjacent normal tissues
Metastasis	Absent	Frequently present; larger and more undifferentiated primary tumors are more likely to metastasize

a neoplasm or tumor (i.e., a swelling, since most neoplasms are expansile, solid masses of abnormal tissue). Tumors are either benign (when their pathologic characteristics and clinical behavior are relatively innocent) or malignant (harmful, often deadly). Malignant tumors are collectively referred to as cancers (derived from the Latin word for crab, to emphasize their obstinate ability to adhere to adjacent structures and spread in many directions simultaneously). The characteristics of benign and malignant tumors are summarized in Table 1. Benign tumors do not penetrate (invade) adjacent tissues, nor do they spread to distant sites. They remain localized and surgical excision can be curative in many cases. In contrast, malignant tumors have a propensity to invade contiguous tissues. Moreover, owing to their ability to gain entrance into blood and lymph vessels, cells from a malignant neoplasm can be transported to distant sites, where subpopulations of malignant cells take up residence, grow, and again invade as satellite tumors (called metastases).

The primary descriptor of any tumor is its cell or tissue of origin. Benign tumors are identified by the suffix "oma," which is preceded by reference to the cell or tissue of origin (e.g., adenoma—from an endocrine gland; chondroma—from cartilage). The malignant counterparts of benign tumors carry similar names, except that the suffix "carcinoma" is applied to cancers derived from epithelium (e.g., squamous- or adeno-carcinoma, from protective and glandular epithelia, respectively) and "sarcoma" (e.g., osteo- or chondro-sarcoma, producing bone and cartilage, respective) to those of mesenchymal origin. Malignant neoplasms of the hematopoietic system, in which the cancerous cells circulate in blood, are called leukemias; solid tumors of lymphoid tissue are

called lymphomas. The major classes of malignant tumors are illustrated in Fig. 1.

Cancer cells express varying degrees of resemblance to the normal precursor cells from which they derive. Thus, neoplastic growth entails both abnormal cellular proliferation and modification of the structural and functional characteristics of the cell types involved. Malignant cells are generally less differentiated than normal cells. The structural similarity of cancer cells to those of the tissue of origin enables specific diagnosis (source organ and cell type); moreover, the degree of resemblance usually predicts prognosis of the patient (i.e., expected outcome based on biologic behavior of the cancer). Therefore, poorly differentiated tumors generally are more aggressive (i.e., display more malignant behavior) than those that are better differentiated. The degree to which a tumor mimics a normal cell or tissue type is called its grade of differentiation. The extent of spread and other effects on the patient determine its stage.

Neoplastic growth is unregulated. Neoplastic cell proliferation is therefore unrelated to the physiological requirements of the tissue and is unaffected by removal of the stimulus which initially caused it. These characteristics differentiate neoplasms from (1) normal proliferations of cells during fetal development or postnatal growth, (2) normal wound healing following an injury, and (3) hyperplastic growth that adapts to a physiological need, but that ceases when the stimulus is removed.

All tumors, benign and malignant, have two basic components: (1) proliferating neoplastic cells that constitute their parenchyma, and (2) supportive stroma made up of connective tissue and blood vessels. Although the parenchyma of neoplasms is characteristic of the specific cells of origin, the growth and evolution of neoplasms are critically dependent on the nonspecific stroma, usually composed of blood vessels, connective tissue, and inflammatory cells.

ASSOCIATION OF IMPLANTS WITH HUMAN AND ANIMAL TUMORS

Neoplasms occurring at the site of implanted medical devices are unusual, despite the large numbers of implants used clinically over an extended period of time. Nevertheless, cases of both human and veterinary implant-related tumors have been reported (Black, 1988; Jennings et al., 1988; Pedley et al., 1981; Schoen, 1987). In all, more than 50 cases of tumors associated with foreign material have been reported, of which approximately half were adjacent to therapeutic implants. The remainder include tumors related to bullets, shrapnel, other metal fragments, sutures, bone wax, and surgical sponge. Implant-related tumors have been reported both short and long term following implantation. More than 25% of tumors associated with foreign bodies have developed within 15 years, and more than 50% within 25 years (Brand and Brand, 1980).

The vast majority of malignant neoplasms associated with clinical fracture fixation devices, total joint replacements, mechanical heart valves, and vascular grafts and experimental foreign bodies in both animals and humans are sarcomas. They comprise various histologic subtypes, including fibrosarcoma,

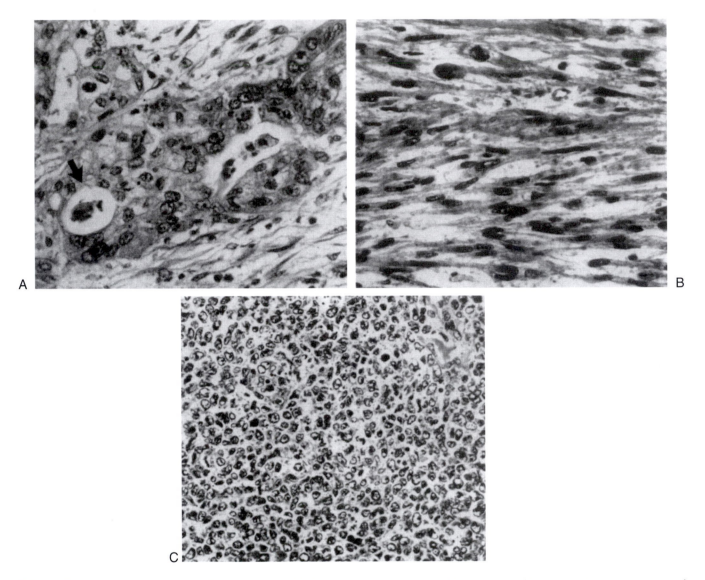

FIG. 1.   Types of malignant tumors. (A) Carcinoma, exemplified by an adenocarcinoma (gland formation noted by arrow). (B) Sarcoma (composed of spindle cells). (C) Lymphoma (composed of malignant lymphocytes). All stained with hematoxylin and eosin; all × 310.

osteosarcoma (osteogenic sarcoma), chondrosarcoma, and angiosarcoma, and are characterized by rapid and locally infiltrative growth. Carcinomas, reported far less frequently, have usually been restricted to situations where an implant has been placed in the lumen of an epithelium-lined organ. Illustrative reported cases are noted in Table 2; descriptions of others are available (Goodfellow, 1992; Jacobs *et al.*, 1992; Jennings *et al.*, 1988). Lymphomas have been reported in association with the capsules surrounding breast implants (Gaudet *et al.*, 2002; Keech and Creech, 1997; Sahoo *et al.*, 2003). A tumor forming adjacent to a clinical vascular graft is illustrated in Fig. 2. A non-implant-related primary tumor (gastric cancer) with a metastasis to a total knee replacement has also been reported (Kolstad and Hgstorp, 1990).

Whether there is a causal role for implanted medical devices in local or distant malignancy remains controversial. In an individual case, caution is necessary in implicating the implant in the formation of a neoplasm; demonstration of a tumor occurring adjacent to an implant does not necessarily prove that the implant caused the tumor (Morgan and Elcock, 1995). Large-scale epidemiological studies and reviews of available data have concluded that there is no evidence in humans for tumorgenecity of non-metallic and metallic surgical implants (McGregor *et al.*, 2000). Indeed, the risk in populations must be low, as exemplified by recent cohorts of patients with both total hip replacement and breast implants who show no detectable increases in tumors at the implant site (Berkel *et al.*, 1992; Deapen and Brody, 1991; Mathiesen *et al.*, 1995; Brinton and Brown, 1997). A clinical and experimental study even suggested that the evidence of breast carcinoma may be decreased in women with breast implants (Su *et al.*, 1995). However, one study suggested a small increase in the number of lung and vulvar cancers in patients with breast implants (Deapen and Brody, 1991). Importantly, the presence of an implant does

TABLE 2 Tumors Associated with Implant Sites in Humans—Representative Reports

Device (adjacent material)[a]	Tumor[b]	References	Postimplantation (years)
Fracture fixation			
Intramedullary rod (V)	L	McDonald (1980)	17
Smith–Petersen (V)	OS	Ward et al. (1987)	9
Total hip			
Charnley–Mueller			2
(UHMWPE, PMMA)	MFH	Bago-Granell et al. (1984)	1+
Mittlemeier (Al$_2$O$_3$)	STS	Ryu et al. (1987)	
Charnley–Mueller			
(UHMWPE)	OS	Martin et al. (1988)	10
Charnley–Mueller			
(SS, PMMA)	SS	Lamovec et al. (1988)	12
Unknown	OS	Adams et al. (2003)	3
(porous Ti–cobalt alloy)			
Total knee			
Unknown (V)	ES	Weber (1986)	4
Vascular graft			
Abdominal aortic graft (D)	MFH	Weinberg and Maini (1980)	1+
Abdominal aortic graft (D)	AS	Fehrenbacker et al. (1981)	12
Heart valve prosthesis			
St. Jude Medical (Carbon,			
Sizone-coated Dacron sewing			
cuff)	RS	Grubitzsch et al. (2001)	<1

[a] Materials: D, Dacron; PMMA, poly(methyl methacrylate) bone cement; SS, stainless steel; Ti, titanium; UHMWPE, ultrahigh-molecular-weight polyethylene; V, Vitallium.

[b] Tumor types: AS, angiosarcoma; ES, epithelioid sarcoma; L, lymphoma; MFH, malignant fibrous histiocytoma; OS, osteosarcoma; RS, rhabdomyosarcoma; SS, synovial sarcoma; STS, soft tissue sarcoma.

not impair the diagnosis of breast cancer (Brinton and Brown, 1997).

Moreover, neoplasms are common in both humans and animals and can occur naturally at the sites at which biomaterials are implanted. Most clinical veterinary cases have been observed in dogs, a species with a relatively high natural frequency of osteosarcoma and other tumors at sites where orthopedic devices are implanted. Moreover, spontaneous human musculoskeletal tumors are not unusual. However, since sarcomas arising in the aorta and other large arteries are rare, the association of primary vascular malignancies with clinical polymeric grafts may be stronger than that with orthopedic devices.

Clinically benign but exuberant foreign-body reactions may simulate neoplasms. For example, fibrohistiocytic lesions resembling malignant tumors may occur as a reaction to silica, previously injected as a soft-tissue sclerosing agent (Weiss et al., 1978). Moreover, regional lymphadenopathy (i.e., enlargement of lymph nodes) may result from an exuberant foreign-body reaction to material that has migrated from a prosthesis. This has been documented in cases of silicone emanating from both finger joints (Christie et al., 1977) and breast prostheses (Hausner et al., 1978), as well as in association with polymeric replacements of the temporomandibular joint, and with conventional metallic, ceramic, and polymeric total replacements of large joints (Jacobs et al., 1995). A mass lesion caused by foreign-body granuloma in a lymph node can masquerade as a neoplasm on physical examination (sometimes called a pseudotumor). Potentially, it could evolve into a lymphoma owing to chronic stimulation of the immune system.

PATHOBIOLOGY OF FOREIGN BODY TUMORIGENESIS

Considerable progress has been made over the past several decades in the understanding of the molecular basis of cancer (Cotran et al., 1999; Kumar et al., 1997). Four principles are fundamental and well accepted: (1) Neoplasia is associated with and often results from nonlethal genetic damage (or mutation), either inherited or acquired by the action of environmental agents such as physical effects (e.g., radiation, fibers or foreign bodies; Fry, 1989), chemicals or viruses. (2) The principal targets of the genetic damage are cellular regulatory genes (normally present and necessary for physiologic cell function, inducing cellular replication, growth and repair of damaged DNA). (3) The tumor mass evolves from the clonal expansion of a single progenitor cell that has incurred the genetic damage. (4) Tumorigenesis is a multistep process, generally owing to accumulation of successive genetic lesions. After a tumor has been initiated, the most important factors in its growth are the kinetics (i.e., balance of replication or loss) of cell number change and its blood supply. The formation of new vessels (angiogenesis) is essential for enlargement of tumors and for their access to the vasculature and, hence, metastasis (Carmeliet, 2003).

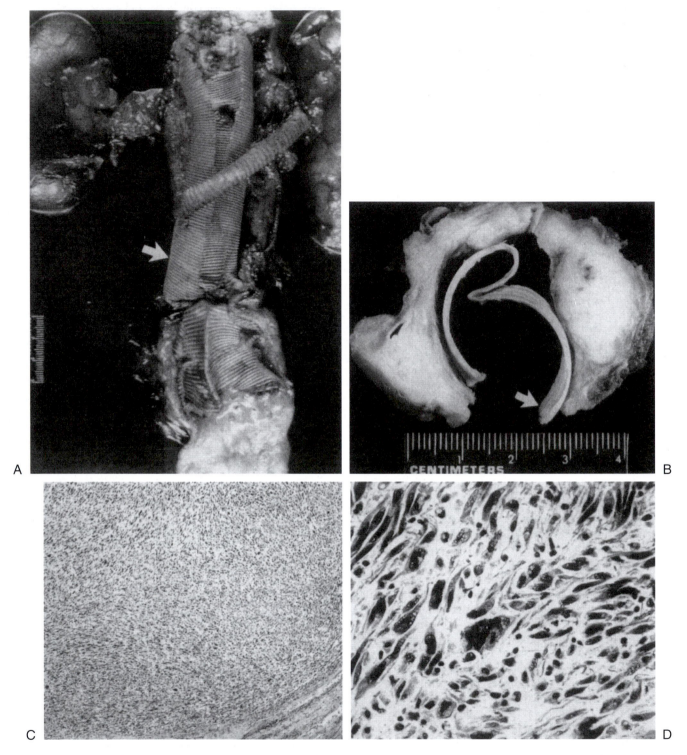

FIG. 2. Sarcoma arising 1 year following and in association with Dacron graft repair of abdominal aortic aneurysm. (A) and (B) Gross photographs (graft designated by arrow, surrounded by tumor mass). (C) and (D) Histologic appearance of tumor. (C) and (D) Stained with hematoxylin and eosin. (C) × 49, (D) × 300. [(A), (C), and (D) reproduced by permission from D. S. Weinberg and B. S. Maini, *Cancer* **46**: 398–402, 1980.]

The pathogenesis of implant-induced tumors is not well understood, yet most experimental data indicate that physical effects rather than the chemical characteristics of the foreign body are the principal determinants of tumorigenicity (Brand *et al.*, 1975). Tumors are induced experimentally by a wide array of materials of diverse composition, including those that could be considered essentially nonreactive, such as certain glasses, gold or platinum, and other relatively pure metals and polymers. Indeed, one surgeon performed a much-maligned experiment in which dimes inserted in rats yielded a rate of 60% sarcomas in 16 months (prompting the suggestion that dimes and probably all metallic coins were carcinogenic and should be discontinued!) (Moore and Palmer, 1997). Solid materials implanted in a configuration with high surface area are most tumorigenic. Materials lose their tumorigenicity when implanted in pulverized, finely shredded or woven form, or when surface continuity is interrupted by multiple perforations. This trend is often called the Oppenheimer effect. Thus, foreign-body neoplasia is generally considered to be a transformation process mediated by the physical state of implants; it is largely independent of the composition of the materials, unless specific carcinogens are present.

Solid-state tumorigenesis depends on the development of a fibrous capsule around the implant. Tumorigenicity corresponds directly to the extent and maturity of tissue encapsulation of a foreign body and inversely with the degree of active cellular inflammation. Thus, an active, persistent inflammatory response inhibits tumor formation in experimental systems. Host (especially genetic) factors also affect the propensity to form tumors as a response to foreign bodies. Humans are less susceptible to foreign-body tumorigenesis than are rodents, the usual experimental model. In rodent systems, tumor frequency and latency depend on species, strain, sex, and age. Concern has recently been raised over the possibility that foreign-body neoplasia can be induced by the release of wear debris or needlelike elements from composites in a mechanism that is analogous to that of asbestos-related mesothelioma (Brown *et al.*, 1990; Jaurand, 1991). However, animal experiments suggest only particles with very high length-to-diameter ratios (>100) produce this effect. Particles with this high aspect ratio are highly unlikely to arise as wear debris from orthopedic implants.

Nevertheless, cancer at foreign-body sites may be mechanistically related to that which occurs in diseases in which tissue fibrosis is a prominent characteristic, including asbestosis (i.e., lung damage caused by chronic inhalation of asbestos), lung or liver scarring, or chronic bone infections (Brand, 1982). However, in contrast to the mesenchymal origin of most implant-related tumors, other cancers associated with scarring are generally derived from adjacent epithelial structures (e.g., mesothelioma with asbestos).

Chemical induction effects are also possible. With orthopedic implants, the stimulus for tumorigenesis could be metal particulates released by wear of the implant (Harris, 1994). Indeed, implants of chromium, nickel, cobalt, and some of their compounds, either as foils or debris, are carcinogenic in rodents (Swierenza *et al.*, 1987), and the clearly demonstrated widespread dissemination of metal debris from implants (to

lymph nodes, bone marrow, liver, and spleen, particularly in subjects with loose, worn joint prostheses) not only could cause damage to distant organs, but also could be associated with the induction of neoplasia (Case *et al.*, 1994). Although unequivocal cases of metal particles or elemental metals provoking the formation of malignant tumors are not available, continued vigilance and further study of the problem in animal models is warranted (Lewis and Sunderman, 1996).

"Nonbiodegradable" and "inert" implants have been shown to contain and/or release trace amounts of substances such as remnant monomers, catalysts, plasticizers, and antioxidants. Nevertheless, such substances injected in experimental animals at appropriate test sites (without implants), in quantities comparable to those found adjacent to implants, are generally not tumorigenic. Moreover, chemical carcinogens such as nitrosamines or those contained in tobacco smoke may potentiate scar-associated cancers.

A chemical effect has been considered in the potential carcinogenicity of polyurethane biomaterials (Pinchuk, 1994). Under certain conditions (i.e., high temperatures in the presence of strong bases), diamines called 2,4-toluene diamine (TDA) and 4,4'-methylene dianiline (MDA) can be produced from the aromatic isocyanates used in the synthesis of polyurethanes. TDA and MDA are carcinogenic in rodents. However, it is uncertain whether (1) TDA and MDA are formed *in vivo*, and (2) these compounds are indeed carcinogenic in humans, especially in the low dose rate provided by medical devices. Although attention has been focused on polyurethane foam-coated silicone gel-filled breast implants, one type of which contained the precursor to TDA, the risk is considered zero to negligible (Expert Panel, 1991).

Foreign-body tumorigenesis is characterized by a long latent period, during which the presence of the implant is required for tumor formation. Available data suggest the following sequence of essential developmental stages in foreign-body tumorigenesis (summarized in Table 3): (1) cellular proliferation in conjunction with tissue inflammation associated with the foreign-body reaction (specific susceptible preneoplastic cells may be present at this stage); (2) progressive formation of a well-demarcated fibrotic tissue capsule surrounding the implant; (3) quiescence of the tissue reaction (i.e., dormancy and phagocytic inactivity of macrophages attached to the foreign body), but direct contact of clonal preneoplastic cells with the foreign body surface; (4) final maturation of preneoplastic cells; and (5) sarcomatous proliferation. Support for this

TABLE 3 Steps in Implant-Associated Tumorigenesis: A Hypothesis[a]

1. Cellular foreign-body reaction
2. Fibrous capsule formation
3. Preneoplastic cells contact implant surface during quiescent tissue reaction
4. Preneoplastic cell maturation and proliferation
5. Tumor growth

[a]Following K. G. Brand and colleagues.

multistep hypothesis for foreign body tumorigenesis comes from an experimental study by Kirkpatrick (2000) in which premalignant lesions were frequently found in implant capsules. A spectrum of lesions was observed, from proliferative lesions without atypical calls to atypical proliferation to incipient sarcoma.

The essential hypothesis is that initial acquisition of neoplastic potential and the determination of specific tumor characteristics does not depend on direct physical or chemical interaction between susceptible cells and the foreign body, and, thus, the foreign body per se probably does not initiate the tumor. However, although the critical initial event occurs early during the foreign-body reaction, the final step to neoplastic autonomy (presumably a genetic event) is accomplished only when preneoplastic cells attach themselves to the foreign-body surface. Subsequently, there is proliferation of abnormal mesenchymal cells in this relatively quiescent microenvironment, a situation not permitted with the prolonged active inflammation associated with less inert implants.

Thus, the critical factors in sarcomas induced by foreign bodies include implant configuration, fibrous capsule development and remodeling, and a period of latency long enough to allow progression to neoplasia in a susceptible host. The major role of the foreign body itself seems to be that of stimulating the formation of a fibrous capsule conducive to neoplastic cell maturation and proliferation. The rarity of human foreign body–associated tumors suggests that cancer-prone cells are infrequent in the foreign-body reactions to implanted human medical devices.

CONCLUSIONS

Neoplasms associated with therapeutic clinical implants in humans are rare; causality is difficult to demonstrate in any individual case. Experimental implant-related tumors are induced by a large spectrum of materials and biomaterials, dependent primarily on the physical and not the chemical configuration of the implant. The mechanism of experimental tumor formation, as yet incompletely understood, appears related to the implant fibrous capsule.

Bibliography

Adams, J. E., Jaffe, K. A., Lemons, J. E., and Siegal, G. P. (2003). Prosthetic implant associated sarcomas: a case report emphasizing surface evaluation and spectroscopic trace metal analysis. *Ann. Diagn. Pathol.* **7**: 35–46.

Bago-Granell, J., Aguirre-Canyadell, M., Nardi, J., and Tallada, N. (1984). Malignant fibrous histiocytoma at the site of a total hip arthroplasty. *J. Bone Joint Surg.* **66B**: 38–40.

Berkel, H., Birdsell, D. C., and Jenkins, H. (1992). Breast augmentation: a risk factor for breast cancer? *N. Engl. J. Med.* **326**: 1649–1653.

Black, J. (1988). *Orthopedic Biomaterials in Research and Practice.* Churchill-Livingstone, New York.

Brand, K. G. (1982). Cancer associated with asbestosis, schistosomiasis, foreign bodies and scars. in *Cancer: A Comprehensive Treatise,* 2nd ed., F. F. Becker, ed. Plenum, New York, Vol. I, pp. 661–692.

Brand, K. G., and Brand, I. (1980). Risk assessment of carcinogenesis at implantation sites. *Plast. Reconstr. Surg.* **66**: 591–595.

Brand, K. G., Buoen, L. C., Johnson, K. H., and Brand, I. (1975). Etiological factors, stages, and the role of the foreign body in foreign body tumorigenesis: a review. *Cancer Res.* **35**: 279–286.

Brinton, L. A., and Brown, S. L. (1997). Breast implants and cancer. *J. Natl. Cancer Inst.* **89**: 1341–1349.

Brown, R. C., Hoskins, J. A., Miller, K., and Mossman, B. T. (1990). Pathogenetic mechanisms of asbestos and other mineral fibres. *Mol. Aspects Med.* **11**: 325–349.

Carmeliet, P. (2003). Angiogenesis in health and disease. *Nat. Med.* **9**: 653–660.

Case, C. P., Langkamer, V. G., James, C., Palmer, M. R., Kemp, A. J., Heap, R. F., and Solomon, L. (1994). Widespread dissemination of metal debris from implants. *J. Bone Joint Surg. [Br].* **76-B**: 701–712.

Christie, A. J., Weinberger, K. A., and Dietrich, M. (1977). Silicone lymphadenopathy and synovitis. Complications of silicone elastomer finger joint prostheses. *JAMA* **237**: 1463–1464.

Cotran, R. S., Kumar, V., and Collins, T. (1999). *Robbins Pathologic Basis of Disease,* 6th ed. W.B. Saunders, Philadelphia, pp. 260–327.

Deapen, D. M., and Brody, G. S. (1991). Augmentation mammaplasty and breast cancer: A 5-year update of the Los Angeles study. *Mammaplast Breast Cancer* **89**: 660–665.

Expert Panel on the Safety of Polyurethane-covered Breast Implants (1991). Safety of polyurethane-covered breast implants. *Can. Med. Assoc. J.* **145**: 1125–1128.

Fehrenbacker, J. W., Bowers, W., Strate, R., and Pittman, J. (1981). Angiosarcoma of the aorta associated with a Dacron graft. *Ann. Thorac. Surg.* **32**: 297–301.

Fry, R. J. M. (1989). Principles of carcinogenesis: Physical. in *Cancer. Principles and Practice of Oncology,* 3rd ed. V. DeVita, ed. Lippincott, Philadelphia, pp. 136–148.

Gaudet, G., Friedberg, J. W., Weng, A., Pinkus, G. S., and Freedman, A. S. (2002). Breast lymphoma associated with breast implants: two case-reports and a review of the literature. *Leuk. Lymphoma* **43**: 115–119.

Goodfellow, J. (1992). Malignancy and joint replacement. *J. Bone Joint. Surg.* **74B**: 645.

Grubitzsch, H., Wollert, H. G., and Eckel, L. (2001). Sarcoma associated with silver coated mechanical heart valve prosthesis. *Ann. Thorac. Surg.* **72**: 1730–1740.

Harris, W. H. (1994). Osteolysis and particle disease in hip replacement. *Acta Orthop. Scand.* **65**: 113–123.

Hausner, R. J., Schoen, F. J., and Pierson, K. K. (1978). Foreign body reaction to silicone in axillary lymph nodes after prosthetic augmentation mammoplasty. *Plast. Reconst. Surg.* **62**: 381–384.

Jacobs, J. J., Rosenbaum, D. H., Hay, R. M., Gitelis, S., and Black, J. (1992). Early sarcomatous degeneration near a cementless hip replacement. *J. Bone Joint. Surg. Br.*. **74B**: 740–744.

Jacobs, J. J., Urban, R. M., Wall, J., Black, J., Reid, J. D., and Veneman, L. (1995). Unusual foreign-body reaction to a failed total knee replacement: Simulation of a sarcoma clinically and a sarcoid histologically. *J. Bone Joint Surg.* **77**: 444–451.

Jaurand, M. C. (1991). Observations on the carcinogenicity of asbestos fibers. *Ann. N.Y. Acad. Sci.* **643**: 258–270.

Jennings, T. A., Peterson, L., Axiotis, C. A., Freidlander, G. E., Cooke, R. A., and Rosai, J. (1988). Angiosarcoma associated with foreign body material. A report of three cases. *Cancer* **62**: 2436–2444.

Keech, J. A., Jr., and Creech, B. J. (1997). Anaplastic T-cell lymphoma in proximity to a saline-filled breast implant. *Plast. Reconstr. Surg.* **100**: 554–555.

Kirkpatrick, C. J., Alves, A., Kohler, H., Kriegsmann, J., Bittinger, F., Otto, M., Williams, D. F., and Eloy, R. (2000). Biomaterial-induced sarcoma. *Am. J. Pathol.* **156**: 1455–1467.

Kolstad, K., and Hgstorp, H. (1990). Gastric carcinoma metastasis to a knee with a newly inserted prosthesis. *Acta Orthop. Scand.* **61**: 369–370.

Kumar, V., Cotran, R. S., and Robbins, S. L. (1997). *Basic Pathology*, 6th ed. W.B. Saunders, Philadelphia, pp. 132–174.

Lamovec, J., Zidar, A., and Cucek-Plenicar, M. (1988). Synovial sarcoma associated with total hip replacement. *J. Bone Joint Surg.* **70A**: 1558–1560.

Lewis, C. G., and Sunderman, F. W., Jr. (1996). Metal carcinogenesis in total joint arthroplasty. *Clin. Orthop. Related Res.* **329S**: S264–S268.

Martin, A., Bauer, T. W., Manley, M. T., and Marks, K. H. (1988). Osteosarcoma at the site of total hip replacement. *J. Bone Joint Surg.* **70A**: 1561–1567.

Mathiesen, E. B., Ahlbom, A., Bermann, G., and Lindsgren, J. U. (1995). Total hip replacement and cancer. A cohort study. *J. Bone Joint Surg. Br.* **77B**: 345–350.

McDonald, W. (1980). Malignant lymphoma associated with internal fixation of a fractured tibia. *Cancer* **48**: 1009–1011.

McGregor, D. B., Baan, R. A., Partensky, C., Rice, J. M., and Wilbourn, J. D. (2000). Evaluation of the carcinogenic risks to humans associated with surgical implants and other foreign bodies — a report of an IARC Monographs Programme Meeting. *Eur. J. Cancer* **36**: 307–313.

Moore, G. E., and Palmer, Q. N. (1977). Money causes cancer. Ban it. *JAMA* **238**: 397.

Morgan, R. W., and Elcock, M. (1995). Artificial implants and soft tissue sarcomas. *J. Clin. Epidemiol.* **48**: 545–549.

Pedley, R. B., Meachim, G., and Williams, D. F. (1981). Tumor induction by implant materials. in *Fundamental Aspects of Biocompatibility*, D. F. Williams, ed. CRC Press, Boca Raton, FL, Vol. II, pp. 175–202.

Pinchuk, L. (1994). A review of the biostability and carcinogenicity of polyurethanes in medicine and the new generation of "biostable" polyurethanes. *J. Biomater. Sci. Polymer Ed.* **6**: 225–267.

Ryu, R. K. N., Bovill, E. G., Jr., Skinner, H. B., and Murray, W. R. (1987). Soft tissue sarcoma associated with aluminum oxide ceramic total hip arthroplasty. A case report. *Clin. Orthop. Rel. Res.* **216**: 207–212.

Sahoo, S., Rosen, P. P., Feddersen, R. M., Viswanatha, D. S., and Clark, D. A. (2003). Anaplastic large cell lymphoma arising in a silicone breast implant capsule: Case report and review of the literature. *Arch. Pathol. Lab. Med.* **127**: e115–e118.

Schoen, F. J. (1987). Biomaterials-associated infection, tumorigenesis and calcification. *Trans. Am. Soc. Artif. Int. Organs* **33**: 8–18.

Su, C. W., Dreyfuss, D. A., Krizek, T. J., and Leoni, K. J. (1995). Silicone implants and the inhibition of cancer. *Plast. Reconstr. Surg.* **96**: 513–520.

Swierenza, S. H. H., Gilman, J. P. W., and McLean, J. R. (1987). Cancer risk from inorganics. *Cancer Metas. Rev.* **6**: 113–154.

Ward, J. J., Dunham, W. K., Thornbury, D. D., and Lemons, J. E. (1987). Metal-induced sarcoma. *Trans. Soc. Biomater.* **10**: 106.

Weber, P. C. (1986). Epithelioid sarcoma in association with total knee replacement. A case report. *J. Bone Joint Surg.* **68B**: 824–826.

Weinberg, D. S., and Maini, B. S. (1980). Primary sarcoma of the aorta associated with a vascular prosthesis. A case report. *Cancer* **46**: 398–402.

Weiss, S. W., Enzinger, F. M., and Johnson, F. B. (1978). Silica reaction simulating fibrous histiocytoma. *Cancer* **42**: 2738–2743.

4.8 BIOFILMS, BIOMATERIALS, AND DEVICE-RELATED INFECTIONS

Bill Costerton, Guy Cook, Mark Shirtliff, Paul Stoodley, and Mark Pasmore

INTRODUCTION

Tens of millions of medical devices are used each year and, in spite of many advances in biomaterials, a significant proportion of each type of device becomes colonized by bacteria and becomes the focus of a device-related infection. Topical devices (e.g., contact lenses) are colonized as soon as they are placed on tissue surfaces, transcutaneous devices (e.g., vascular catheters) are progressively colonized by skin organisms, and even surgically implanted devices regularly become foci of infection. Implanted devices may be colonized by bacteria at the time of surgery, or they may be colonized by organisms that gain access to their surfaces by a hematogenous route, from a distant source. The most significant factor in the development of device-related infections appears to be the skill of the surgical team, with prosthetic hips being infected in less than 0.2% of cases in large specialized clinics, and as many as 4% in less proficient facilities. Generally, large and complex medical devices that require long and complicated surgery for their placement are at high risk of bacterial infection, and transcutaneous devices in this category (e.g., the Jarvik heart) automatically become infected. In many areas of medicine, the risk of infection limits the use of devices that constitute the epitome of the engineer's skill and imagination and incorporate the finest and most sophisticated biomaterials available in this fast-moving field.

As medical devices came into more regular use, the surgeons who placed them used their well-developed observation skills to define the "classic" device-related infection. These infections were often very slow to develop, with overt symptoms sometimes being seen almost immediately and sometimes being seen months or even years after the device was installed. Inflammation and pus formation were often local, especially in transcutaneous devices, but a certain proportion of patients with device-related infections suddenly developed acute disseminated infections caused by the same species that had colonized the device. These acute exacerbations of device-related infections responded well to antibiotic therapy. However, this treatment almost never reversed the local symptoms, and colonized devices often gave rise to a predictable series of acute exacerbations, so that good medical management usually dictated their removal. The bacteria that caused device-related infections were common skin biota (e.g., *Staphylococcus epidermidis*) and common environmental organisms (e.g., *Pseudomonas aeruginosa*), and certain species predominated in infections of certain devices. Because the infecting bacteria, and occasional fungi (e.g., *Candida albicans*), were so ubiquitous in the modern human environment, device recipients always had good immunity against these low-level pathogens, but these antibodies failed to prevent infection. It was the "front-line" medical specialists (e.g., orthopedic surgeons) who gradually persuaded medical microbiologists and infectious disease specialists that device-related infections differed from acute bacterial infections in several important respects.

The biofilm concept was developed and articulated (Costerton et al., 1978) in environmental microbiology, and it was introduced into medical microbiology when Tom Marrie et al. (1982) examined the surfaces of devices that had failed because of bacterial infection. This concept states that bacteria, in all but the most nutrient-deprived ecosystems, grow preferentially in matrix-enclosed communities attached to surfaces (Costerton et al., 1987). Electron microscopy proved to be useful in the examination of the surfaces of failed medical devices, because both scanning (SEM) and transmission (TEM) electron microscopy involve dozens of washing steps that remove floating or loosely adherent bacteria. Therefore any bacterial or fungal cells that remained on the surfaces of the device, after processing, were *bona fide* biofilm organisms. With medical colleagues leading the search (Khoury et al., 1992; Marrie and Costerton, 1984; Nickel et al., 1985), our morphological team examined hundreds of types of failed medical devices and found biofilms on all of their surfaces. Biofilms were seen on the surfaces of contact lenses that had been worn by volunteers (McLaughlin-Borlace et al., 1998), and very extensive sessile communities were seen on the surfaces of lenses that had been stored overnight in storage cases (Gray et al., 1995; McLaughlin-Borlace et al., 1998). Some of the most extensive biofilms we ever saw on a medical device were found on the surfaces of intrauterine contraceptive devices (Marrie et al., 1982), and teeth and dental devices were equally heavily colonized. It was in this area of topical medical devices that the distinction was made between colonization, which is the simple presence of microbial biofilms on a surface, and the infections that occur when this presence of a biofilm elicits a pathogenic response.

The surface of skin is colonized by a wide variety of bacteria and fungi, most of which are removed or killed by surgical preparations, but the deeper layers are also colonized by bacteria (mostly *S. epidermidis*) that escape skin sterilants. This cutaneous biota rapidly colonizes the surfaces of any transcutaneous device, and the biofilm moves along any device that is placed in a subcutaneous "tunnel," until the entire surface of the device is colonized. In this manner a microbial biofilm is introduced into the normally sterile environment of the peritoneum, by the Techkhoff catheter (Dasgupta et al., 1987), or into the normally sterile environment of the heart, by devices like the Hickman (Tenney et al., 1986) and the Swan-Ganz (Mermel et al., 1991) catheters. The inevitable colonization of transcutaneous devices, which is usually complete in 3–4 weeks, does not automatically lead to infection. All of the Hickman catheters in our National Cancer Institute study (Tenney et al., 1986) were seen to be colonized, and one was even partially blocked by a very exuberant biofilm, but only four of the 81 patients experienced overt infection and bacteremia. Chronic ambulatory peritoneal dialysis (CAPD) patients all have well-developed biofilms on their Tenckhoff catheters, but many do not develop peritonitis if their humoral and cellular immune mechanisms can "keep up" (Dasgupta et al., 1990) with the planktonic (floating) cells that are released from these sessile communities.

When implanted medical devices become colonized, the presence of these microbial biofilms always triggers pathogenic changes in the surrounding tissues, but symptoms are often slow to develop. Mechanical heart valves and vascular grafts can fail because biofilms on stitches that hold them in place cause inflammation, weaken the tissues involved, and lead to their detachment and displacement (Hyde et al., 1998). Orthopedic devices may develop "aseptic loosening" in that the device is loosened by bone dissolution, but there are no signs of inflammation. The biofilms of the causative pathogens are so coherent that routine cultures of the device and the tissues are almost always negative. Biofilms elicit few symptoms, because their matrix-enclosed cells produce few toxins and stimulate only cursory immune responses and inflammation, but local symptoms will be produced when planktonic cells are released from these sessile communities.

The examination of failed medical devices frequently reveal microbial biofilms. Therefore, the unique characteristics of device-related infections can be explained in terms of the characteristics of biofilms (Costerton et al., 1999). The slow development and asymptomatic nature of many device-related infections can be explained by the observation that biofilm bacteria produce few toxins and elicit little inflammatory response. Many device-related infections are negative in routine microbiological cultures because biofilms release a limited number of planktonic cells, large biofilm fragments grow up as a single colony on plates, and sessile cells do not grow well on agar surfaces. Common bacterial species predominate in device-related infections because they form biofilms very effectively in their natural environments (e.g., skin), and this biofilm mode of growth protects them from the immune responses that occur in all potential hosts. The biofilm mode of growth protects the causative agents of device-related infections from both humoral and cell-mediated immunity (see Chapter 4.3) (Leid et al., 2002), so these infections occur in healthy individuals, and they are never resolved by even the most active host defense mechanisms. Exacerbations of device-related infections are caused by the release of planktonic cells, and antibiotics can kill these floating cells and reverse the symptoms of acute infection, but the infection persists because the causative biofilm is resistant to these antibacterial agents. Most, if not all, of the characteristics of device-related infections can be explained in terms of the characteristics of biofilms, so it may be useful to examine the burgeoning field of biofilm microbiology, as an early step in the search for new biomaterials that will control these infections.

BIOFILM MICROBIOLOGY

Many of the concepts and techniques that have served microbiologists well, in the virtual conquest of epidemic bacterial diseases caused by planktonic organisms, now serve us only poorly in the study of device-related and other chronic bacterial diseases. This section on biofilm microbiology will focus on the central fact that biofilm bacteria differ from their planktonic counterparts in so many ways that they are as different as spores are from vegetative bacteria, and it is imperative that special biofilm methods be used in studies of the bacterial colonization of biomaterials.

Bacterial Adhesion to Surfaces

Often, the DLVO theory is applied to the study of bacterial adhesion to surfaces (van Loosdrecht *et al.*, 1990). This classic concept of colloid behavior visualizes a planktonic bacterial cell as a smooth colloid particle that interacts with the surface in a manner based on the charges on both surfaces, which overcome the basic repulsion of individual particles. Examinations of the surfaces of planktonic bacteria, using special preparations and electron microscopy, have clearly shown that these surfaces are not smooth. In addition to proteinaceous appendages (flagella and pili) that project 2–6 μm from the cell, the entire surfaces of planktonic cells of natural strains of bacteria are covered by a matrix of hydrophobic exopolysaccharide (EPS) fibers, and sometimes by a highly structured protein "coat." The external EPS layer of planktonic cells is anchored to the polysaccharide O antigen fibers that project from the lipopolysaccharide (LPS) of the outer membrane of gram-negative cells, and to the polysaccharide teichoic acid fibers that project from the cell walls of gram-positive cells. Elegant freeze-substitution microscopy preparations have shown that the actual surface of planktonic bacterial cells that would be capable of interacting with the surface to be colonized is a 0.2 to 0.4-μm-thick forest of protein and polysaccharide fibers. The planktonic bacterial cell is not a smooth-surfaced colloid particle, and the actual interaction of these cells with surfaces is based on the bridging of bacterial fibers with fibers adsorbed to the surface being colonized. Thus, DLVO theory is of limited application in the study of bacterial adhesion.

Another conventional microbiology method, the reliance on pure cultures of bacteria isolated from the system of interest, but subcultured hundreds of times in rich media, also does not serve us well in biofilm studies relevant to medical devices. This method, which dates from Robert Koch in the 1850s, produces lab-adapted strains of bacteria that are selected in favor of planktonic growth, because the simple act of subculturing leaves adherent cells behind in the old culture and transfers only free-floating cells. These lab-adapted strains lack many surface structures that would be necessary for their survival in a hostile "wild" environment, but they are not challenged by antibacterial agents, so they survive in the test tube, but perish if they are released into natural ecosystems. When these lab-adapted strains are used in studies of bacterial adhesion to biomaterials, they come close to the smooth-surfaced colloidal particles visualized in the DLVO theory, and data that are misleading for the understanding of medical-device-centered infection are generated. Several companies have spent millions of dollars on novel biomaterials to which lab-adapted strains of bacteria would adhere to only very poorly, only to have these biomaterials heavily colonized by "wild" natural bacteria, and to find that they performed unsatisfactorily in clinical tests. Most microbiologists who focus on biofilms never do adhesion experiments on strains that are more than one transfer from an infected patient, if their objective is to assess the propensity of a biomaterial for colonization by a putative pathogen. Scientists

at the FDA and EPA are aware of this necessity to use "wild" bacteria in biomaterials testing.

When planktonic cells adhere to a surface, which they do with considerable avidity, they exhibit behaviors that have been divided into "reversible" and "irreversible" patterns (Marshall *et al.*, 1971). The most actively motile organisms (e.g., *P. aeruginosa*) may use their flagella as landing mechanisms, and then may use their type IV pili to produce a twitching motility that allows them to pile up into elaborate structures, some of which resemble the fruiting bodies of the myxobacteria. Other less mobile organisms produce "windrows" of cells (Korber *et al.*, 1995) following adhesion, while cells that have neither flagella nor pili simply stay in place if the location is favorable, and detach if it is not. Movies showing these postadhesion behaviors of bacteria are available on the Center for Biofilm Engineering (CBE) Web site (www.erc.montana.edu). Biofilm engineers have generated surprising data (Stoodley *et al.*, 2001a, b) showing that many cells that adhere to surfaces also detach and leave the area, before they make the genetic switch to attach irreversibly and initiate the process of biofilm formation. Many people in the biomaterials field have speculated, intuitively, that key surface characteristics must favor (or inhibit) bacterial adhesion, and almost every possible combination of these characteristics has been tried in the search for colonization-resistant biomaterials. Wild bacteria adhere equally well to very hydrophobic (e.g., Teflon) and to very hydrophilic (e.g., PVC) surfaces, they colonize smooth surfaces as well as they adhere to rough surfaces (Marrie and Costerton, 1984; Sottile *et al.*, 1986), and they colonize smooth surfaces in very high shear flow systems (Characklis, 2003). Thus, we have no perfect biomaterial surface that resists bacterial colonization by virtue of its inherent surface properties, but nonfouling surfaces show limited potential for this application (Chapter 2.13).

Biofilm Formation on Surfaces

When a bacterial cell has "made the decision" to colonize a surface it sets in motion a pattern of gene expression that profoundly alters its previous planktonic phenotype, to produce a unique biofilm phenotype that may differ by as much as 70% in the proteins expressed (Sauer and Camper, 2001). Some of the first genes that are up-regulated in adherent cells are those involved in the production of the EPS material that will form the matrix of the biofilm and will also anchor the cell irreversibly to the surface. In *P. aeruginosa* the up-regulation of *algC*, which is a part of the alginate synthesis pathway, occurs within 18 minutes of initial cell adhesion (Davies and Geesey, 1995), and we see the secretion of matrix material by these cells within 30 minutes of adhesion. The genes that are up-regulated in the biofilm phenotype of many bacterial species are being analyzed by proteomics (Miller and Diaz-Torres, 1999; Oosthuizen *et al.*, 2002; Sauer and Camper, 2001; Sauer *et al.*, 2002; Svensater *et al.*, 2001; Tremoulet *et al.*, 2002a, b) and by microarray analysis (Schembri *et al.*, 2003; Schoolnik *et al.*, 2001; Stoodley *et al.*, 2002; Wagner *et al.*, 2003; Whiteley *et al.*, 2001), and individual genes involved in this profound phenotype shift are being identified daily. Sauer and her colleagues have reported that the phenotype of planktonic cells of both *P. aeruginosa* and *P. putida* differ from that of their biofilm counterparts more than they differ from that of planktonic cells of other species in the same genus. The inherent resistance of biofilm bacteria to antibiotics, all of which were selected on the basis of their ability to kill planktonic cells, is now largely attributed to the altered gene expression pattern of the biofilm phenotype. Scientists at Microbia Ltd (Boston) have identified one specific gene (fmt C) that is responsible for this inherent antibiotic resistance in biofilms formed by all staphylococcal species, and the deletion or blockage of this gene produces biofilms that are susceptible to conventional antibiotics.

Once attached cells have triggered the conversion to the biofilm phenotype, the multicellular community on the colonized surface begins to accrete larger numbers of cells by binary fission and by further recruitment of planktonic cells from the bulk water phase. As they increase in numbers and produce large amounts of EPS matrix material, the attached cells form microcolonies in which they constitute approximately 15% of the volume and the matrix occupies approximately 85% of the volume. The microcolonies assume tower-like and mushroom-like shapes (Fig. 1) in most natural and cultured biofilms, but many other morphologies may be dictated by species characteristics and by nutrient availability. The microcolonies may occupy the colonized surface, as discrete entities separated by open water channels (Fig. 1), or they may pile up in several layers to form thick sessile communities, but they always maintain their structural integrity and move independently under shear stress. As the biofilm matures and undergoes more phenotypic changes (Stoodley *et al.*, 2002), the processes of cell division and recruitment come into balance with programmed detachment of planktonic cells from the sessile community and sloughing. Most natural biofilms reach a mature thickness and

a stable community structure within a week or two of their initiation of colonization, and many remain relatively stable for years. Stoodley *et al.* (2002) have concluded that the biofilm phenotype would favor bacterial survival in the harsh environment of the primitive earth, and they have suggested that the planktonic phenotype may have developed considerably later, as a mechanism for dissemination.

Natural Control of Biofilm Formation on Surfaces

The complex structure of biofilm communities (Fig. 1) stimulated a lively discussion of what signals (e.g., hormones or pheromones) must be operative to allow the development of shaped microcolonies and sustained water channels. In 1998 the first demonstration that biofilm development in *P. aeruginosa* is controlled by an acyl homoserine lactone (AHL) quorum-sensing signal was published (Davies *et al.*, 1998), and it has subsequently been reported that agr, sar, and RAP (Balaban *et al.*, 1998) signals control this same process in gram-positive organisms. Additionally, it was shown that the autoinducer II signal (furanone) controls biofilm formation, and many other processes, in virtually all bacterial species (Schauder *et al.*, 2001; Xavier and Bassler, 2003). Most biofilm specialists agree that these signals are simply the tip of the iceberg, that many more signals will be discovered, and that specific blockage of many of these simple chemical signals offers a practical way to control virtually any bacterial "behavior." It has already been shown that specific signal inhibitors can block toxin production in *S. aureus*, and can even render specific bacteria essentially nonpathogenic in animal models (Balaban *et al.*, 2000).

The manipulation of biofilm formation is a very attractive target for new agents to prevent device-related infections. If bacteria that make contact with biomaterials were "locked" in the planktonic phenotype and were unable to assume the protected biofilm phenotype, they would be readily killed by host defense mechanisms and/or by antibiotic therapy. Several chemical analogs that block signal activity by interfering with the binding of the signal to its cognate receptor have been shown to be effective in inhibiting biofilm formation in specific pathogens (Balaban *et al.*, 1998). One such analog (RIP) prevents the binding of a biofilm control signal (RAP) to its receptor (TRAP). This signal blocker has been shown to prevent biofilm formation in an animal model of a device-related infection, and to allow complete eradication of the bacteria with conventional antibiotic therapy (Balaban *et al.*, 2003a, b). The researchers involved in the search for biofilm-control signal inhibitors are acutely aware of the subtle nature of the signal network in bacterial cells. It is highly unlikely that we will find a single ON/OFF switch for biofilm formation, and blockers that prevent biofilm formation may up-regulate invidious bacterial behaviors (e.g., toxin production), but we are encouraged by several observations made in natural ecosystems. Marine plants and animals control biofilm formation on their surfaces, presumably because biofilm/silt accretion would bury them and preclude photosynthesis in the plants, and at least one of the compounds that they use for this purpose is a signal inhibitor (de Nys *et al.*, 1995). In these natural systems, plant and animal surfaces are ideal locations for biofilm formation and growth,

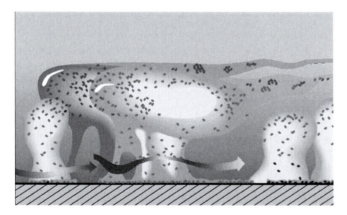

FIG. 1. Diagrammatic representation of the tower-like and mushroom-like microcolonies that are the basic structural units of the biofilms on colonized surfaces. Note that the matrix material occupies ±85% of the mass of these structures (while the cells comprise ±15%), that the microcolonies can be deformed into oscillating streamers by shear forces, and that water moves through these complex communities in a convective pattern. It was the complex differentiated structure of these microcolonies, and the maintenance of the open water channels, that stimulated speculation that some form of "hormone-like" cell–cell signaling must be involved in the formation of microbial biofilms.

but no bacterial mutant capable of thwarting the action of these natural biofilm control agents has emerged in millions of years of evolution.

Novel Engineering Approaches to Biofilm Control

The current therapy for device-related infections consists of trying to kill a biological entity (the bacteria) with a chemical (the antibiotic), with the only variable parameters being concentration and time of contact. Engineers have suggested that a number of physical forces could be harnessed to deliver higher concentrations of the antibiotic to the infecting organisms, or to compromise the bacteria in ways that make them more susceptible to the agents concerned. Two technologies that offer considerable promise involve the use of ultrasonic energy (Nelson *et al.*, 2002; Rediske *et al.*, 1998), and the imposition of a very weak sustained DC field (Costerton *et al.*, 1994) across the biofilm, and both have been shown to render sessile microbial populations susceptible to conventional antibiotics. Practical research is currently underway in the modification of biomaterials, and of device design, to harness the potential of these physical biofilm control technologies in our general strategies for the control of device-related infections.

BIOFILM-RESISTANT BIOMATERIALS

Biofilm-related complications have cost many lives in clinical settings. This unfortunate outcome may be reduced if the concepts and methods of modern biofilm microbiology can be inculcated into the development process for antibiofilm biomaterials. We will discuss some of the new agents that may give us effective control over the colonization of biomaterials and the incidence of device-related infections, and then we will discuss new methods for the release/delivery of these agents at the surfaces of biomaterials.

Testing for Antibacterial and Antibiofilm Properties of Biomaterials

There are serious concerns with the utility and information content of some of the methods used to assess the biofilm resistance of biomaterials. If a biomaterial gives a positive zone of inhibition test, what does it mean? It means that the biomaterial contained an antibacterial agent, which it released in the moist environment of the surface of an agar plate, and the agent killed the planktonic bacteria that had been deposited on this same surface to produce a "lawn." The major parameter operative in the test is the diffusion of the antibacterial agent through the agar, or in the fluid on the agar surface, more than the effectiveness of the agent. A very effective agent would have a very small zone, if it moves slowly through agar, and a weak agent would have an enormous zone if it diffused well through agar, or if it diffused well through water and the plate was wet. The release kinetics of the agent from the biomaterial are those of a biomaterial on a moist agar surface, which is not a common use target for biomaterials. Flask tests, in which candidate biomaterials are suspended in a medium

that is simultaneously inoculated with planktonic bacteria, are equally naive. If the biomaterials release enough of an antibacterial agent in the first few minutes of the test, all of the planktonic cells will be killed, the medium will be sterile, and there will be no organisms to colonize the biomaterial. So an antibiotic-releasing biomaterial that "dumps" all of its antibacterial agents in a few minutes will emerge from this test with flying colors! If the bacteria used in these relatively crude tests are lab-adapted by repeated subculturing, and thus defective in both antibacterial resistance and adhesion to surfaces, the biomaterials will be seen as promising. Yet both these tests are inappropriate and tend to lead to biomaterials that fail in biofilm resistance in animal and clinical trials.

The most appropriate tests are ones that mimic the conditions in the systems in which the biomaterial is targeted for use. If the biomaterial will be subjected to flow, or even to fluid exchange, the test should include these parameters. If the biomaterial will be used in a body fluid, such as blood or urine, an accurate simulation of that fluid should be used in the test, and the bacteria supplying the challenge should be adapted to the fluid. Bacteria used to challenge the putative biomaterial should be "wild" strains, recently isolated from clinical sources, and the challenge should come from fast-growing exponential-phase cells supplied by a chemostat, and not from variable cells from a "batch" culture. All of these parameters are best delivered using a flow cell, fed by a chemostat, and one of the most popular designs for such a system is given in Stoodley *et al.* (2001a). The flow cell also allows direct observation of large areas of the surface of the biomaterial, especially if the flow cell is mounted on the stage of a confocal scanning laser microscope (CSLM), and surface colonization can be monitored continuously (Cook *et al.*, 2000). Because the confocal microscope can resolve bacteria on opaque surfaces, and because this microscope allows us to examine living hydrated preparations, we can actually see the first microbial cells that adhere to biomaterial surface (Fig. 2A). If the adherent cells survive, they will initiate biofilm formation, and the adherent cells will gradually form matrix-enclosed communities (Fig. 2B) within which the cells will be separated by 3–5 μm of slime-filled space. The formation of biofilms requires that the adherent cells must be alive, so the observation of structured biofilms (Fig. 2C) on a surface that makes antibacterial and antibiofilm claims could have unfortunate clinical consequences (Cook *et al.*, 2000).

The observation of cells on the surface of a biomaterial is not necessarily negative data, especially if the cells are not very numerous and have not formed biofilms, because some antibacterial agents kill "incoming" planktonic cells and the dead cells remain on the surface (Fig. 2D). Even though we prefer biomaterials whose active agents kill "incoming" bacteria and do not retain these dead cells on the surface, agents that kill and retain bacterial cells are of some interest. For this reason, one of several available live/dead probes to ascertain the viability of adherent bacterial cells on biomaterials can be used. All of these methods give "snapshot" data, in that they necessitate the termination of colonization and the removal of biomaterial from the test system, but they yield accurate and useful data. The BacLite Live/Dead probe (Cook *et al.*, 2000) distinguishes live cells on the basis of membrane integrity, and live cells stain green while dead cells stain red (Fig. 2B). Living cells can also be distinguished from dead cells on the basis of their respiration using tetrazolium salts that produce an orange color when they are reduced by metabolically active organisms. In very practical terms, biomaterials set up in flow cells can be exposed to realistic flowing fluid containing active cells of a "wild" strain of potential pathogen, and the resultant colonization of its surface can be monitored by CSLM. When adherent bacterial cells are few and intermittent, live/dead data are not germane and the test can run for days without interruption. When the biomaterial becomes heavily colonized, by biofilm-forming bacteria that stain as living cells in the live/dead assay, the material is designated as having exceeded its period of colonization resistance. Because a layer of surviving cells provides

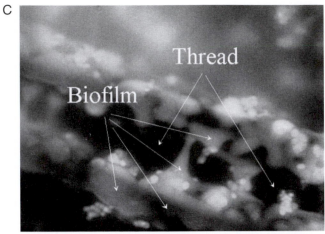

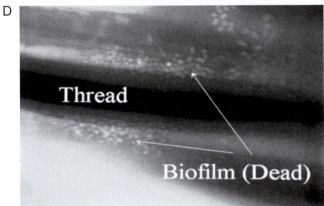

FIG. 2.—continued

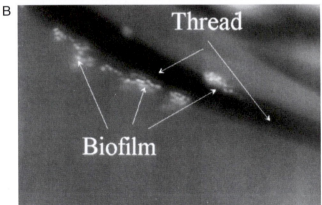

FIG. 2. These micrographs are confocal scanning laser microscope (CSLM) images of living unfixed biofilms formed on individual fibers of the clothlike material used to form the sewing cuffs of mechanical heart valves. (A–C) Images of biofilms formed when fibers of a silver-coated medical device were exposed to planktonic cells of *Staphylococcus epidermidis*, in a flow cell, as described in Cook *et al.* (2000). The biofilm seen in (A) was formed on the silver-coated fiber after 24 hours exposure, and special staining with the Live/Dead BacLite probe (B) showed that live cells (green) outnumbered dead cells (red) by a wide margin. When these same fibers were exposed to this planktonic challenge for 48 hours, as seen in (C), a mature biofilm had formed and individual bacterial cells were seen to be buried in large aggregates of matrix material. (D) Large numbers of dead cells (red) as seen in a Live/Dead stain of an effective antibacterial biomaterial that killed sessile cells of *S. epidermidis* as they attempted to colonize its surface. This putative biomaterial was, unfortunately, too toxic for clinical use.

an optimum surface for further bacterial colonization, biomaterials that have exceeded their colonization-resistant period tend to accrete biofilms and fail very rapidly, as in the case of the materials seen in (Fig. 2A–C).

Although direct observations are obviously the "gold standard" in tests of the resistance of biomaterials to bacterial colonization, CSLM microscopes are complex and relatively expensive. Many groups, including the CBE, have struggled with the inherent difficulties of removing sessile biofilm bacteria from colonized surfaces, and of enumerating these cells by standard microbiological techniques. The technique usually used is called "scraping and plating," and it involves breaking up the clumps of bacteria in the biofilm fragments and spreading these dispersed cells on the surface of an agar plate so that each cell gives rise to one colony when the plate is incubated. Many difficulties can contribute to the inaccuracy of this method, but all can be resolved or rationalized if all of the steps are monitored by microscopy. First, some cells may be left on the surface by the scraping method, which must be calibrated by microscopic examination of the scraped surface to see how many cells remain. Second, the scraped material must be homogenized or sonicated to break up clumps of bacteria in the scraped biofilm fragments to ensure that each living bacterium gives rise to one colony on the agar plate. Sonication may kill some cells, so it is important to calibrate the sonication time for each type of biofilm, until microscopy shows that the resultant suspension is mostly single cells, and most of these cells are alive. Cells that have assumed the biofilm phenotype may not grow well on the surfaces of agar plates, when they have been removed from the sessile community and suspended in an unfamiliar milieu, so that "committed" biofilm cells may not grow well on plates. When scraping and plating are used without the calibrations discussed above, this method can yield data that are 4 log values lower than the bacterial numbers seen by direct microscopy. However, the scrape and plate method can yield accurate and consistent data when it is properly calibrated by microscopy, and the first biofilm method using this enumeration technique has now been accepted as ASTM Method E 2196-02.

Potential Agents for the Control of Microbial Colonization of Biomaterials

The continued search for biomaterials that resist microbial colonization by virtue of their inherent surface properties may still give us valuable information on minimizing adhesion, but we should use modern biofilm methods to conduct this research. The search may be somewhat quixotic, because the large sums of money spent to date have yielded only a handful of materials of questionable utility. Thus, candidate materials should be subjected to *sine qua non* testing in realistic systems early in their development cycles.

The strategy most commonly used in current antibacterial biomaterials is the incorporation of conventional antibiotics into the material, with the objective of killing incoming planktonic cells, before they can adhere and initiate biofilm formation. Although there are some successful applications of this basically sound approach, the problem lies in the typical release kinetics of such materials. Most of the agent is liberated in the first short time period, while the remainder is made available slowly and over a long period of time, thus exposing the bacteria to sublethal antibiotic concentrations that may stimulate the development of resistant strains. Antibiotics with very specific targets, such as penicillin (penicillin binding protein) and ciprofloxacin (DNA gyrase), may induce bacterial mutations that render the mutants dramatically more resistant to the agent in question. For this reason many biomaterial designers have chosen to use multitarget antimicrobial agents (e.g., chlorhexidine), because mutants are only marginally more resistant, but most of these nonspecific compounds are not approved for systemic use in humans.

This quandary of balancing efficacy against the danger of acquired bacterial resistance does not affect the large cohort of bacterial manipulation molecules that is currently moving briskly toward the biomaterials market. Some of these biofilm control molecules are specifically targeted on quorum-sensing mechanisms, such as RIP on the TRAP two-component system in gram-positives and the brominated furanones on the AHL systems of gram-negatives, but others are simply known to affect biofilm formation. It is now obvious that signal control of bacterial behavior is a subtle process, in which many factors interact to control a network of activities, so we do not expect to find a simple ON/OFF switch that controls biofilm development. Nonetheless, we have found several signal blockers that inhibit biofilm formation and sharply reduce pathogenicity in animal models. The pivotal concept is that bacteria in contact with a biomaterial would be prevented from forming a biofilm on its surface, would be "locked" in the planktonic phenotype, and would be killed by host defenses (antibodies and activated leukocytes) and any antibiotics that might be present. Balaban *et al.* (2003b) have shown that the RIP analog of the RAP signal, which controls biofilm formation in all species of the *Staphylococcus* genus, prevents biofilm formation by these organisms on subcutaneous Dacron implants in rats. When specific antibiotics were administered to these test animals, while the challenging bacteria were locked in the planktonic phenotype, no live cells could be recovered from the biomaterial surfaces of the surrounding tissues. This approach to the control of device-related infections is rational and much more focused than conventional antibiotic therapy, and its proponents visualize a whole new series of species- and genus-specific agents that will control both biofilm formation and toxin production. Biofilm specialists take comfort from new observations that plants protect themselves from pathogenic biofilm colonization by the use of similar signals and signal blockers, and millions of years of coevolution have not produced resistant bacterial strains (Stoodley *et al.*, 2002).

Delivery of Biofilm Control Agents at Biomaterial Surfaces

The more candid among the surgeons who install medical devices have confided that some operations proceed perfectly, and the device slides into place rapidly and smoothly, while others take longer and "just don't feel right." It is these later cases that sometimes develop biofilms and device-related infections because the skin and environmental bacteria present near the biomaterial surfaces will have time to adhere and to initiate biofilm formation. Killing the planktonic bacteria before they have time to initiate biofilm formation is the objective of many programs in this area, and this can be accomplished by three general strategies:

1. Systemic antibiotic therapy that produces bactericidal concentrations in the body fluids in the operative field
2. Release of antibiotics and other bacterial manipulation molecules from the biomaterials to produce high and sustained concentrations of the agent in the immediate vicinity of the device
3. Irrigation and other techniques that deliver antibiotics to the biomaterial surface after the device is installed, before the operative wound is closed

Most surgeons use systemic antibiotic therapy in the perioperative time fame, and most also use this strategy in subsequent operations (including dental procedures) if a device has recently been installed and might not be fully epithelialized. The simplest manifestation of the antibiotic-releasing biomaterial strategy is a class of materials that can be "loaded," like a sponge, by soaking them in a solution of the antibiotic in question immediately before the device is installed. This tactic has backfired in many cases in which bacteria resistant to the antibiotic have started to grow in the fluids of the vessel in which the device is being soaked, have formed a biofilm on its surface, and have caused serious infections. We must be fastidious in the installation of medical devices, in that no preformed biofilms must ever be implanted, because preformed biofilms automatically give rise to biofilm infections (Ward *et al.*, 1992). It is equally important that the surfaces of biomaterials be absolutely clean, because any residue of dead biofilm or other organic accretion radically increases the colonization of that surface by planktonic bacteria and increases the chance of a biofilm infection. Also, some gram-negative bacterial cell-wall residues (endotoxin) can lead to inflammatory reactions.

The most commonly employed strategy in infection prevention is the impregnation of biomaterials with recognized antibacterial ions or molecules, with the intent of killing planktonic bacteria before they can colonize the material concerned. Whenever an ion or molecule is loaded into or onto a polymer, Fick's laws dictate that large amounts will be released in the early time frame, and that the release will taper off during the long period in which the concentration in this reservoir is being depleted (see Chapter 7.14). Many biomaterial designers have become adept at manipulating the initial concentration of the agent and the release rate, but we are always left with certain "spectra" of concentrations and polymer configurations that require choices of the "Hobson's" variety (that is, no real choice). If we put a large amount of ionic silver on a surface and release it quickly, we are flirting with silver toxicity. If we put a very stable form of silver on a surface and silver ions are released very slowly, bacteria will grow all over the silver coating (Fig. 2C) just as they colonize metallic copper (McLean *et al.*, 1993) if few copper ions are present. Westaim Biomedical, Inc., has introduced an exciting new silver coating for burn bandages that uses a galvanic combination of silver and copper and releases silver and copper ions at a steady rate that control bacterial colonization for a useful period of time. The galvanic potentials set up by these side-by-side "lakes" of copper and silver may also have an inhibitory effect on bacterial adhesion, biofilm formation, and the inherent resistance of biofilm bacteria to antibacterial agents.

Because many modern antibiotics are much less toxic than metal ions, the release patterns of these agents from biomaterials pose a different problem. We can obtain high and very effective concentrations of antibiotics, in the immediate vicinity of devices, for lengths of time that have already been found to be effective in certain clinical situations. These biomaterials are useful, but we cannot control the low-level release of the agent for months or years after this effective time frame. This produces a prolonged period in which the agent is present at a sublethal concentration, near the device and sometimes in the whole body, and raises the specter of the development of acquired antibiotic resistance in many potentially dangerous bacterial species. A new development at the University of Washington Engineered Biomaterials (UWEB) Engineering Research Center has addressed this problem. Biomaterials can be coated with a molecular "skin," a self-assembled surface layer, that completely contains molecules loaded into an

underlying plastic and can be temporarily deranged (by ultrasonic energy) to yield a controlled release of the molecule in question (Kwok *et al.*, 2001). This coating has been used to deliver insulin, in controlled pulses, and the UWEB and the CBE are currently adapting this ultrasonic-sensitive coating for the controlled release antibacterial agents (and bacterial manipulation agents) from implanted biomaterials. High concentrations of these agents could be released at the surfaces of medical devices, perioperatively or at any preliminary signs of device-related infection, and no further release would occur if the coating was not stimulated ultrasonically.

Many surgeons have expressed an interest in being able to sterilize a medical device *in situ*, after it has been installed and before the operative wound is closed. This strategy is rational, because the device is accessible, and any planktonic bacteria present in the operative field will initiate colonization of the surface of the device if they are not killed or manipulated to preclude biofilm formation. Irrigation with antibiotics is presently used, biofilm-inhibiting signals and signal blockers are being developed, and this *in situ* procedure may be the perfect opportunity for the use of ultrasonic energy and/or DC electric fields to enhance the killing of bacteria in nascent biofilms. We can readily contaminate sham animal operations with bacteria and determine the efficacy of several possible procedures for *in situ* sterilization by using the live/dead probe to examine the surfaces of devices recovered at intervals after the procedure. The final proof of the efficacy of the *in situ* sterilization of medical devices would be obtained in clinical tests, in which significant reductions in device-related infections could be documented.

SUMMARY

The concept that has been distilled from decades of clinical experience with device-related infections has now been fused with the biofilm concept, which states that bacteria live predominantly in matrix-enclosed protected communities. This fusion is reassuring, and intellectually satisfying, because it asserts that bacteria employ the same strategy for survival in the human body that they use in all other ecosystems. The mental image that is invoked is one in which a biofilm forms on the surface of a biomaterial, and that it has all of the properties of the sessile communities that predominate in industrial and environmental systems. Its cells express the distinct biofilm phenotype: They are resistant to antibacterial agents and to uptake by phagocytes, most of them grow slowly and adopt many different metabolic strategies, and they detach planktonic cells and biofilm fragments in a programmed manner. The clinical consequences of this mode of bacterial growth are that antibiotics are useful in treating acute planktonic exacerbations, but that these agents cannot clear the biofilm reservoir on the biomaterial, and the device must usually be removed to resolve the infection.

As we fuse the device-related infection concepts with biofilm concepts, we can discard several older methods that have been used to assess the efficacy of putative antibacterial biomaterials. New biofilm methods allow us to visualize bacteria on opaque surfaces, to determine the viability of these organisms, and even to identify the cells by genus and species. We realize that freshly isolated "wild" bacteria adhere avidly to plastic and metal surfaces that have been "conditioned" by exposure to body fluids, and we are more sanguine about claims that biomaterials can resist colonization by virtue of their surface properties alone. New technologies that deliver antibiotics in controlled and effective doses, at the surfaces of novel biomaterials, offer a solution to the problem of bacterial resistance induced by sublethal concentrations of these agents from "exhausted" biomaterials. The discovery that cells in biofilms communicate with each other by means of chemical signals brings to the medical biomaterials area a whole set of new molecules that can manipulate bacterial behaviors, such as toxin production and biofilm formation. These bacterial manipulation agents, many of which control biofilms in natural environments, have already been shown to "lock" targeted bacteria in the planktonic phenotype and to make them susceptible to conventional antibiotics and host defense factors. Physical treatments (e.g., ultrasonic radiation and DC electric fields) that make biofilm cells susceptible to antibacterial agents are also made available for use in medical systems, because of the fusion of the biofilm field with the study of device-related infections. This synthesis of concepts may accelerate the development of biomaterials that truly resist bacterial colonization, and these materials may allow us to build medical devices that will be substantially less susceptible to device-related infections.

Bibliography

Balaban, N., Collins, L. V., Cullor, J. S., Hume, E. B., Medina-Acosta, E., Vieira, D. M., O'Callaghan, R., Rossitto, P. V., Shirtliff, M. E., Serafim, D. S., Tarkowski, A., and Torres, J. V. (2000). Prevention of diseases caused by *Staphylococcus aureus* using the peptide RIP. *Peptides* 21: 1301–1311.

Balaban, N., Giacometti, A., Cirioni, O., Gov, Y., Ghiselli, R., Mocchegiani, F., Viticchi, C., Del Prete, M. S., Saba, V., Scalise, G., and Dell'Acqua, G. (2003a). Use of the quorum-sensing inhibitor RNAIII-inhibiting peptide to prevent biofilm formation in vivo by drug-resistant *Staphylococcus epidermidis*. *J. Infect. Dis.* 187: 625–630.

Balaban, N., Goldkorn, T., Nhan, R. T., Dang, L. B., Scott, S., Ridgley, R. M., Rasooly, A., Wright, S. C., Larrick, J. W., Rasooly, R., and Carlson, J. R. (1998). Autoinducer of virulence as a target for vaccine and therapy against *Staphylococcus aureus*. *Science* 280: 438–440.

Balaban, N., Gov, Y., Bitler, A., and Boelaert, J. R. (2003b). Prevention of *Staphylococcus aureus* biofilm on dialysis catheters and adherence to human cells. *Kidney Int.* 63: 340–345.

Characklis, W.G. (2003). Biofilm processes. in *Biofilms*, Characklis, W. G. and K. C. Marshall, eds. John Wiley & Sons, New York, pp. 195–231.

Cook, G., Costerton, J. W., and Darouiche, R. O. (2000). Direct confocal microscopy studies of the bacterial colonization in vitro of a silver-coated heart valve sewing cuff. *Int. J. Antimicrob. Agents* 13: 169–173.

Costerton, J. W., Cheng, K. J., Geesey, G. G., Ladd, T. I., Nickel, J. C., Dasgupta, M., and Marrie, T. J. (1987). Bacterial biofilms in nature and disease. *Ann. Rev. Microbiol.* 41: 435–464.

Costerton, J. W., Ellis, B., Lam, K., Johnson, F., and Khoury, A. E. (1994). Mechanism of electrical enhancement of efficacy of antibiotics in killing biofilm bacteria. *Antimicrob. Agents Chemother.* 38: 2803–2809.

Costerton, J. W., Geesey, G. G., and Cheng, K. J. (1978). How bacteria stick. *Sci. Am.* 238: 86–95.

Costerton, J. W., Stewart, P. S., and Greenberg, E. P. (1999). Bacterial biofilms: a common cause of persistent infections. *Science* 284: 1318–1322.

Dasgupta, M. K., Bettcher, K. B., Ulan, R. A., Burns, V., Lam, K., Dossetor, J.B., and Costerton, J.W. (1987). Relationship of adherent bacterial biofilms to peritonitis in chronic ambulatory peritoneal dialysis. *Peritoneal Dialysis Bull.* 7: 168–173.

Dasgupta, M. K., Larabie, M., Lam, K., Bettcher, K. B., Tyrrell, D. L., and Costerton, J. W. (1990). Growth of bacterial biofilms on Tenckhoff catheter discs in vitro after simulated touch contamination of the Y-connecting set in continuous ambulatory peritoneal dialysis. *Am. J. Nephrol.* 10: 353–358.

Davies, D. G., and Geesey, G. G. (1995). Regulation of the alginate biosynthesis gene *algC* in *Pseudomonas aeruginosa* during biofilm development in continuous culture. *Appl. Environ. Microbiol* 61: 860–867.

Davies, D. G., Parsek, M. R., Pearson, J. P., Iglewski, B. H., Costerton, J. W., and Greenberg, E. P. (1998). The involvement of cell-to-cell signals in the development of a bacterial biofilm. *Science* 280: 295–298.

de Nys, R., Steinberg, P., Willemsen, P., Dworjanyn, S. A., Gabelish, C. L., and King, R. J. (1995). Broad spectrum efffects of secondary metabolites from the red alga *Delisea puchra* in antifouling assays. *Biofouling* 8: 259–271.

Gray, T. B., Cursons, R. T., Sherwan, J. F., and Rose, P. R. (1995). Acanthamoeba, bacterial, and fungal contamination of contact lens storage cases. *Br. J. Ophthalmol.* 79: 601–605.

Hyde, J. A., Darouiche, R. O., and Costerton, J. W. (1998). Strategies for prophylaxis against prosthetic valve endocarditis: a review article. *J. Heart Valve. Dis.* 7: 316–326.

Khoury, A. E., Lam, K., Ellis, B., and Costerton, J. W. (1992). Prevention and control of bacterial infections associated with medical devices. *ASAIO J.* 38: M174–M178.

Korber, D. R., Lawrence, J. R., Lappin-Scott, H. M., and Costerton, J. W. (1995). Growth of microorganisms on surfaces. in *Microbial Biofilms*. H. M. Lappin-Scott and J. W. Costerton, ed. Cambridge University Press, Cambridge, U.K., pp. 15–45.

Kwok, C. S., Mourad, P. D., Crum, L. A., and Ratner, B. D. (2001). Self-assembled molecular structures as ultrasonically-responsive barrier membranes for pulsatile drug delivery. *J. Biomed. Mater. Res.* 57: 151–164.

Leid, J. G., Shirtliff, M. E., Costerton, J. W. and Stoodley, A. P. (2002). Human leukocytes adhere to, penetrate, and respond to *Staphylococcus aureus* biofilms. *Infect. Immun.* 70: 6339–6345.

Marrie, T. J., and Costerton, J. W. (1984). Scanning and transmission electron microscopy of *in situ* bacterial colonization of intravenous and intraarterial catheters. *J. Clin. Microbiol.* 19: 687–693.

Marrie, T. J., Nelligan, J., and Costerton, J. W. (1982). A scanning and transmission electron microscopic study of an infected endocardial pacemaker lead. *Circulation* 66: 1339–1341.

Marshall, K. C., Stout, R., and Mitchell, R. (1971). Selective sorption of bacteria from seawater. *Can. J. Microbiol* 17: 1413–1416.

McLaughlin-Borlace, L., Stapleton, F., Matheson, M., and Dart, J. K. (1998). Bacterial biofilm on contact lenses and lens storage cases in wearers with microbial keratitis. *J. Appl. Microbiol.* 84: 827–838.

McLean, R. J., Hussain, A. A., Sayer, M., Vincent, P. J., Hughes, D. J., and Smith, T. J. (1993). Antibacterial activity of multilayer silver–copper surface films on catheter material. *Can. J. Microbiol* 39: 895–899.

Mermel, L. A., McCormick, R. D., Springman, S. R., and Maki, D. G. (1991). The pathogenesis and epidemiology of catheter-related infection with pulmonary artery Swan-Ganz catheters: a prospective study utilizing molecular subtyping. *Am. J. Med.* **91**: 197S–205S.

Miller, B. S., and Diaz-Torres, M. R. (1999). Proteome analysis of biofilms: growth of *Bacillus subtilis* on solid medium as model. *Methods Enzymol.* **310**: 433–441.

Nelson, J. L., Roeder, B. L., Carmen, J. C., Roloff, F., and Pitt, W. G. (2002). Ultrasonically activated chemotherapeutic drug delivery in a rat model. *Cancer Res.* **62**: 7280–7283.

Nickel, J. C., Gristina, A. G., and Costerton, J. W. (1985). Electron microscopic study of an infected Foley catheter. *Can. J. Surg.* **28**: 50–54.

Oosthuizen, M. C., Steyn, B., Theron, J., Cosette, P., Lindsay, D., Von Holy, A., and Brozel, V. S. (2002). Proteomic analysis reveals differential protein expression by *Bacillus cereus* during biofilm formation. *Appl. Environ. Microbiol.* **68**: 2770–2780.

Rediske, A. M., Hymas, W. C., Wilkinson, R., and Pitt, W. G. (1998). Ultrasonic enhancement of antibiotic action on several species of bacteria. *J. Gen. Appl. Microbiol* **44**: 283–288.

Sauer, K., and Camper, A. K. (2001). Characterization of phenotypic changes in *Pseudomonas putida* in response to surface-associated growth. *J. Bacteriol.* **183**: 6579–6589.

Sauer, K., Camper, A. K., Ehrlich, G. D., Costerton, J. W., and Davies, D. G. (2002). *Pseudomonas aeruginosa* displays multiple phenotypes during development as a biofilm. *J. Bacteriol.* **184**: 1140–1154.

Schauder, S., Shokat, K., Surette, M. G., and Bassler, B. L. (2001). The LuxS family of bacterial autoinducers: biosynthesis of a novel quorum-sensing signal molecule. *Mol. Microbiol.* **41**: 463–476.

Schembri, M. A., Kjaergaard, K., and Klemm, P. (2003). Global gene expression in *Escherichia coli* biofilms. *Mol. Microbiol.* **48**: 253–267.

Schoolnik, G. K., Voskuil, M. I., Schnappinger, D., Yildiz, F. H., Meibom, K., Dolganov, N. A., Wilson, M. A., and Chong, K. H. (2001). Whole genome DNA microarray expression analysis of biofilm development by *Vibrio cholerae* O1 E1 Tor. *Methods Enzymol.* **336**: 3–18.

Sottile, F. D., Marrie, T. J., Prough, D. S., Hobgood, C. D., Gower, D. J., Webb, L. X., Costerton, J. W., and Gristina, A. G. (1986). Nosocomial pulmonary infection: possible etiologic significance of bacterial adhesion to endotracheal tubes. *Crit. Care Med.* **14**: 265–270.

Stoodley, P., Hall-Stoodley, L., and Lappin-Scott, H. M. (2001a). Detachment, surface migration, and other dynamic behavior in bacterial biofilms revealed by digital time-lapse imaging. *Methods Enzymol.* **337**: 306–319.

Stoodley, P., Sauer, K., Davies, D. G., and Costerton, J. W., (2002). Biofilms as complex differentiated communities. *Ann. Rev. Microbiol.* **56**: 187–209.

Stoodley, P., Wilson, S., Hall-Stoodley, L., Boyle, J. D., Lappin-Scott, H. M., and Costerton, J. W. (2001b). Growth and detachment of cell clusters from mature mixed-species biofilms. *Appl. Environ. Microbiol.* **67**: 5608–5613.

Svensater, G., Welin, J., Wilkins, J. C., Beighton, D., and Hamilton, I. R. (2001). Protein expression by planktonic and biofilm cells of *Streptococcus mutans. FEMS Microbiol. Lett.* **205**: 139–146.

Tenney, J. H., Moody, M. R., Newman, K. A., Schimpff, S. C., Wade, J. C., Costerton, J. W., and Reed, W. P. (1986). Adherent microorganisms on lumenal surfaces of long-term intravenous catheters. Importance of *Staphylococcus epidermidis* in patients with cancer. *Arch. Intern. Med.* **146**: 1949–1954.

Tremoulet, F., Duche, O., Namane, A., Martinie, B., and Labadie, J. C. (2002a). A proteomic study of *Escherichia coli* O157:H7 NCTC 12900 cultivated in biofilm or in planktonic growth mode. *FEMS Microbiol. Lett.* **215**: 7–14.

Tremoulet, F., Duche, O., Namane, A., Martinie, B., and Labadie, J. C. (2002b). Comparison of protein patterns of *Listeria monocytogenes* grown in biofilm or in planktonic mode by proteomic analysis. *FEMS Microbiol. Lett.* **210**: 25–31.

van Loosdrecht, M. C., Norde, W., and Zehnder, A. J. (1990). Physical chemical description of bacterial adhesion. *J. Biomater. Appl.* **5**: 91–106.

Wagner, V. E., Bushnell, D., Passador, L., Brooks, A. I., and Iglewski, B. H. (2003). Microarray analysis of *Pseudomonas aeruginosa* quorum-sensing regulons: effects of growth phase and environment. *J. Bacteriol.* **185**: 2080-2095.

Ward, K. H., Olson, M. E., Lam, K., and Costerton, J. W. (1992). Mechanism of persistent infection associated with peritoneal implants. *J. Med. Microbiol.* **36**: 406–413.

Whiteley, M., Bangera, M. G., Bumgarner, R. E., Parsek, M. R., Teitzel, G. M., Lory, S., and Greenberg, E. P. (2001). Gene expression in *Pseudomonas aeruginosa* biofilms. *Nature* **413**: 860–864.

Xavier, K. B., and Bassler, B. L. (2003). LuxS quorum sensing: more than just a numbers game. *Curr. Opin. Microbiol.* **6**: 191–197.

5

Biological Testing of Biomaterials

JAMES M. ANDERSON, RICHARD W. BIANCO, JOHN F. GREHAN, BRIAN C. GRUBBS,
STEPHEN R. HANSON, KIP D. HAUCH, MATT LAHTI, JOHN P. MRACHEK,
SHARON J. NORTHUP, BUDDY D. RATNER, FREDERICK J. SCHOEN,
ERIK L. SCHROEDER, CLARK W. SCHUMACHER, AND CHARLES A. SVENDSEN

5.1 INTRODUCTION TO TESTING BIOMATERIALS

Buddy D. Ratner

How can biomaterials be evaluated to determine if they are biocompatible and will function in a biologically appropriate manner in the *in vivo* environment? Meaningful testing procedures are overviewed in the five chapters in this section. Chapter 9.4 offers further insights in correlating physical measurements with biological performance. This introduction is an aid in coalescing themes that are common to all biomaterials biological testing.

Evaluation under *in vitro* (literally "in glass") conditions can provide rapid and inexpensive data on biological interaction (Chapter 5.2). However, the question must always be raised—will the *in vitro* test measure parameters relevant to what will occur in the much more complex *in vivo* environment? For example, tissue culture polystyrene, a surface modified polymer, will readily attach and grow most cells in culture. Untreated polystyrene will neither attach nor grow mammalian cells. Yet when implanted *in vivo*, both materials heal almost indistinguishably with a thin foreign body capsule. Thus, the results of the *in vitro* test do not provide information relevant to the implant situation. *In vitro* tests minimize the use of animals in research, a desirable goal. Also, *in vitro* testing is required by most regulatory agencies in the device approval process for clinical application. When appropriately used, *in vitro* testing provides useful insights that can dictate whether a device need be further evaluated in expensive *in vivo* experimental models.

Animals are used for testing biomaterials to model the environment that might be encountered in humans (Chapter 5.3). However, there is great range in animal anatomy, physiology, and biochemistry. Will the animal model provide data useful for predicting how a device performs in humans? Without validation to human clinical studies, it is often difficult to draw strong conclusions from performance in animals. The first step in designing animal testing procedures is to choose an animal model that offers a reasonable parallel anatomically or biochemically to the situation in humans. Experiments designed to minimize the number of animals needed, ensure that the

animals are treated humanely (e.g., NIH guidelines for the use of laboratory animals), and maximize the relevant information generated by the testing procedure are essential.

Some biomaterials fulfill their intended function in seconds. Others are implanted for a lifetime (10 years? 70 years?). Are 6-month implantation times useful to learn about a device intended for 3-minute insertion? Will six months implantation in a test model provide adequate information to draw conclusions about the performance of a device intended for lifetime implantation? These are not easy questions. However, they must be addressed and carefully considered in designing a biomaterials testing protocol.

This is a textbook on biomaterials with the special focus being materials. There are obviously important differences between implanting a sheet of cellulose (a material) in an animal and evaluating the performance and biological response of the same sheet of cellulose used as a dialysis membrane in an artificial kidney. The pros and cons of material testing (a relatively low cost procedure providing opportunities for carefully controlled experiments) versus evaluation in a device configuration (an expensive and difficult to control, but completely relevant situation) must always be weighed.

Experimental variability in the testing data is expected, particularly in tests in living systems. The more complex the system (e.g., a human in contrast to cells in culture), the larger the variability that might be expected. In order to draw defensible conclusions from expensive testing protocols, statistical methods assist us in ensuring that the results are meaningful within some probability range. Statistics should be used at two stages in testing biomaterials. Before an experiment is performed, statistical experimental design will indicate the minimum number of samples that must be evaluated to yield meaningful results. After the experiment is completed, statistics will help to extract the maximum amount of useful information from experiments.

Assistance in the design of many biomaterials tests is available through national and international standards-organizations. Thus, the American Society for Testing Materials (ASTM) and the International Standards Organization (ISO) can often provide detailed protocols for widely accepted, carefully thought out testing procedures (Chapter 10.2). Other testing protocols are available through government agencies (e.g., the FDA) and through commercial testing laboratories.

5.2 *In Vitro* Assessment of Tissue Compatibility

Sharon J. Northup

The term "cytotoxicity" means to cause toxic effects (death, alterations in cellular membrane permeability, enzymatic inhibition, etc.) at the cellular level. It is distinctly different from physical factors that affect cellular adhesion (surface charge of a material, hydrophobicity, hydrophilicity, etc.). This chapter reviews the evaluation of biomaterials by methods that use isolated, adherent cells in culture to measure cytotoxicity and biological compatibility.

HISTORICAL OVERVIEW

Cell culture methods have been used to evaluate the biological compatibility of materials for more than two decades (Northup, 1986). Most often today the cells used for culture are from established cell lines purchased from biological suppliers or cell banks (e.g., the American Type Culture Collection, Manassas, VA). Primary cells (with the exception of erythrocytes in hemolysis assays) are seldom used because they have less assay repeatability, reproducibility, efficiency, and, in some cases, availability. Several methods have been validated for repeatability (comparable data within a given laboratory) and reproducibility (comparable data among laboratories). These methods have been incorporated into national and international standards used in the commercial development of new products. In addition, there are a wide variety of methods in the research literature that have been used in specialized applications and that are on the leading edge of scientific development. These are not discussed in this chapter. As the science of biomaterials evolves, some of these research methods may become incorporated into routine products.

BACKGROUND CONCEPTS

Toxicity

A toxic material is defined as a material that releases a chemical in sufficient quantities to kill cells either directly or indirectly through inhibition of key metabolic pathways. The number of cells that are affected is an indication of the dose and potency of the chemical. Although a variety of factors affect the toxicity of a chemical (e.g., compound, temperature, test system), the most important is the dose or amount of chemical delivered to the individual cell.

Delivered and Exposure Doses

The concept of delivered dose refers to the dose that is actually absorbed by the cell. It differs from the concept of exposure dose, which is the amount applied to a test system. For example, if an animal is exposed to an atmosphere containing a noxious substance (exposure dose), only a small portion of the inhaled substance will be absorbed and delivered to the internal organs and cells (delivered dose). Because different cells have differing susceptibilities to the toxic effects of xenobiotics (foreign substances), the cells that are most sensitive are referred to as the target cells. Taken together, these two concepts mean that cell culture methods evaluate target cell toxicity by using delivered doses of the test substance. This distinguishes cell culture methods from whole-animal studies, which evaluate the exposure dose and do not determine the target cell dose of the test substance. This difference in dosage at the cellular level accounts for a significant portion of the difference in sensitivity (i.e., quantitation range) of cell culture methods compared with whole animal toxicity data. To properly compare the sensitivity of cell culture methods with *in vivo* studies, data from local toxicity models such as dermal irritation, implantation, and direct tissue exposure should be compared. These models reduce the uncertainties of delivered dose associated with absorption, distribution, and metabolism that are inherent in systemic exposure test models.

Safety Factors

A highly sensitive test system is desirable for evaluating the potential hazards of biomaterials because the inherent characteristics of the materials often do not allow the dose to be exaggerated. There is a great deal of uncertainty in extrapolating from one test system to another, such as from animals to humans. To allow for this, toxicologists have used the concept of safety factors to take into account intra- and interspecies variation. This practice requires being able to exaggerate the anticipated human clinical dosage in the nonhuman test system. In a local toxicity model in animals, there is ample opportunity for reducing the target cell dose by distribution, diffusion, metabolism, and changes in the number of exposed cells (because of the inflammatory response). On the other hand, in cell culture models, in which the variables of metabolism, distribution, and absorption are minimized, the dosage per cell is maximized to produce a highly sensitive test system.

Solubility Characteristics

The principal components in medical devices are water-insoluble materials (polymers, metals, and ceramics), meaning that less than one part of the material is soluble in 10,000 parts of water. Other components may be incorporated into the final product to obtain the desired physical, functional, manufacturing, and sterility properties. For example, plastics may contain plasticizers, slip agents, antioxidants, fillers, mold release agents, or other additives, either as components of the formulation or trace additives from the manufacturing process. The soluble components may be differentially extracted from the insoluble material. Till *et al.* (1982) have shown that the migration of chemicals from a solid plastic material into liquid solvents is controlled by diffusional resistance within the solid, chemical concentration, time, temperature, mass transfer resistance on the solvent side, fluid turbulence at the solid–solvent interface, and the partition coefficient of the chemical in the solvent. Because of these variables, the conditions for preparing extractions of biomaterials have been carefully standardized to improve the reproducibility of the data.

Complete dissolution of biomaterials is an alternative approach for *in vitro* testing. Its main limitation is that it does not simulate the intended clinical application or may create degradation products that do not occur in the clinical application. Therefore, the actual clinical dosage or agent exposed to the cells in pharmacokinetic terms may be exaggerated because the rate of diffusion from the intact material or device may be very slow or different than that for complete dissolution.

ASSAY METHODS

Three primary cell culture assays are used for evaluating biocompatibility: direct contact, agar diffusion, and elution (also known as extract dilution). These are morphological assays, meaning that the outcome is measured by observations of changes in the morphology of the cells. The three assays differ in the manner in which the test material is exposed to the cells. As indicated by the nomenclature, the test material may be placed directly on the cells or extracted in an appropriate solution that is subsequently placed on the cells. The choice of method varies with the characteristics of the test material, the rationale for doing the test, and the application of the data for evaluating biocompatibility.

To standardize the methods and compare the results of these assays, the variables of number of cells, growth phase of the cells (period of frequent cell replication), cell type, duration of exposure, test sample size (e.g., geometry, density, shape, thickness), and surface area of test sample must be carefully controlled. This is particularly true when the amount of toxic extractables is at the threshold of detection where, for example, a small increase in sample size could change the outcome from nontoxic to moderate or severe toxicity. Below the threshold of detection, differences in these variables are not observable. Within the quantitation range of these assays, varying slopes of the dose-response curve or exposure–effect relationship (Klaassen, 1986) will occur with different toxic agents in a manner similar to that in animal bioassays.

In general, cell lines that have been developed for growth *in vitro* are preferred to primary cells that are freshly harvested from live organisms because the cell lines improve the reproducibility of the assays and reduce the variability among laboratories. That is, a cell line is the *in vitro* counterpart of inbred animal strains used for *in vivo* studies. Cell lines maintain their genetic and morphological characteristics throughout a long (sometimes called infinite) life span. This provides comparative data with the same cell line for the establishment of a database. The L-929 mouse fibroblast cell has been used most extensively for testing biomaterials. Initially, L-929 cells were selected because they were easy to maintain in culture and produced results that had a high correlation with specific animal bioassays (Northup, 1986). In addition, the fibroblast was specifically chosen for these assays because it is one of the early cells to populate a healing wound and is often the major cell in the tissues that attach to implanted medical devices. Cell lines from other tissues or species may also be used. Selection of a cell line is based upon the type of assay, the investigator's

experience, measurement endpoints (viability, enzymatic activity, species specific receptors, etc.), and various other factors. It is not necessary to use human cell lines for this testing because, by definition, these cells have undergone some dedifferentiation and lost receptors and metabolic pathways in the process of becoming cell lines.

Positive and negative controls are often included in the assays to ensure the operation and suitability of the test system. The negative control of choice is a high-density polyethylene material. Certified samples may be obtained from the U.S. Pharmacopeial Convention, Inc., Rockville, MD. Several materials have been proposed as candidates for positive controls. These are low-molecular-weight organotin-stabilized poly(vinyl chloride), gum rubber, and dilute solutions of toxic chemicals, such as phenol and benzalkonium chloride. All of the positive controls are commercially available except for the organotin-stabilized poly(vinyl chloride).

The methodologies for the three primary cell culture assays are described in the U.S. Pharmacopeia, and in standards published by the American Society for Testing and Materials (ASTM), the British Standards Institute (BSI), and the International Standards Organization (ISO). There are minor variations in the methods among these sources because of the evolving changes in methodology, the time when the standards were developed, and the individual experiences of those participating in standards development. Pharmacopeial assays are legally required by the respective ministries of health in the United States (Food and Drug Administration), Europe, Japan, Australia, and other countries. It is expected that the ISO methods will replace the individual national regulations in Europe, whereas the ASTM and BSI standards are voluntary, consensus standards. The basic methodologies, as described in the U.S. Pharmacopeia (2004), are described in the following paragraph.

Direct Contact Test

A near-confluent monolayer of L-929 mammalian fibroblast cells is prepared in a 35-mm-diameter cell culture plate. The culture medium is removed and replaced with 0.8 ml of fresh culture medium. Specimens of negative or positive controls and the test article are carefully placed in individually prepared cultures and incubated for 24 hr at $37 \pm 1°C$ in a humidified incubator. The culture medium and specimens are removed and the cells are fixed and stained with a cytochemical stain such as hematoxylin blue. Dead cells lose their adherence to the culture plate and are lost during the fixation process. Live cells adhere to the culture plate and are stained by the cytochemical stain. Toxicity is evaluated by the absence of stained cells under and around the periphery of the specimen.

At the interface between the living and dead cells, microscopic evaluation will show an intermediate zone of damaged cells. The latter will have a morphological appearance that is abnormal. The change from normalcy will vary with the toxicant and may be evidenced as increased vacuolization, rounding due to decreased adherence to the culture plate, crenation, swelling, etc. For example, dying cells may round up and detach from the culture plate before they disintegrate. Crenation and swelling are often related to osmotic or oncotic pressures. Vacuolization frequently occurs with basic substances and is due to lysosomal uptake of the toxicant and fluids. This interface area should be included in determining the toxicity rating.

Agar Diffusion Test

A near-confluent monolayer of L-929 is prepared in a 60-mm-diameter plate. The culture medium is removed and replaced with a culture medium containing 2% agar. After the agar has solidified, specimens of negative and positive controls and the test article are placed on the surface of the same prepared plate and the cultures incubated for at least 24 hours at $37 \pm 1°C$ in a humidified incubator. This assay often includes neutral red vital stain in the agar mixture, which allows ready visualization of live cells. Vital stains, such as neutral red, are taken up and retained by healthy, viable cells. Dead or injured cells do not retain neutral red and remain colorless. Toxicity is evaluated by the loss of the vital stain under and around the periphery of the specimens. The interface area should be evaluated as described previously.

Selection of a proper agar for use in this assay continues to be a major problem. Agar is a generic name for a particular colloidal polymer derived from a red alga. There are many different grades of agar that are distinguished by their molecular weight and extent of cross-linking of the colloid. The mammalian tissue culture product called agar agar and agarose seem to work best. Agarose is a chemical derivative of agar that has a lower gelling temperature and is less likely to cause thermal shock. The thickness of the agar should be constant because the diffusion distance affects the cellular concentration of a toxicant. From a theoretical viewpoint, it could be expected that different chemicals will diffuse through the agar at different rates. This is true from a broad perspective, but because most toxicants are low molecular weight (less than 100 Da), the diffusion rate will not be sufficiently dissimilar within the 24-hour assay period.

Elution Test

An extract of the material is prepared by using (1) 0.9% sodium chloride or serum-free culture medium and a specified surface area of material per milliliter of extractant and (2) extraction conditions that are appropriate for the application and physical characteristics of the material. Alternatively, serum-containing culture media may be used with an extraction temperature of $37 \pm 1°C$. The choice of extractant sets an upper limit on the quantitation range of the assay in that, without added nutrients, 0.9% sodium chloride will itself be toxic to the cells after a short incubation period. The extract is placed on a prepared, near-confluent monolayer of L-929 fibroblast cells and the toxicity is evaluated after 48 hours of incubation at $37 \pm 1°C$ in a humidified incubator. Live or dead cells may be distinguished by the use of histochemical or vital stains as described earlier.

Interpretation of Results

Each assay is interpreted roughly on the basis of quartiles of affected cells. This corresponds to the customary morphological and clinical rating scales of no, slight, moderate, and severe response grades. The terms used to describe these grades refer to the characteristics of the assays. In the direct-contact and agar-diffusion assays, one expects a concentration gradient of toxic chemicals, with the greatest amount appearing under the specimen and then diffusing outward in more or less concentric areas. Physical trauma from movement of the specimen in the direct-contact assay is evident by patches of missing cells interspersed with normal healthy cells. This is not a concern with the agar-diffusion assay because the agar cushions the cells from physical trauma. Interpretation of the elution test is based upon what happens to the total population of cells in the culture plate. That is, any toxic agent is evenly distributed in the culture plate and toxicity is evaluated on the basis of the percent of affected cells in the population. Generally, more experience in cell culture morphology is required to appropriately evaluate the elution test than is required for the other two techniques.

Table 1 lists the advantages and disadvantages of the three assays. The chief concern in each of the assays is the transfer or diffusion of some chemical(s) X from the test sample to the cells. This involves the total available amount of X in the material, the solubility limit of X in the solution phase, the equilibrium partitioning of X between the material surface and the solution, and the rate of migration of X through the bulk phase of the material to the material surface. If sufficient analytical data are available to verify that there is one and only one leachable chemical from a given material, then empirical toxicity testing in vitro or in vivo could be replaced with literature reviews and physiologically based pharmacokinetic modeling of hazard potential. Usually a mixture of chemicals migrate from materials and therefore, empirical testing of the biological effects of the mixture is necessary.

The direct-contact assay mimics the clinical use of a device in a fluid path, e.g., blood path, in which the material is placed directly in the culture medium and extraction occurs in the presence of serum-containing culture media at physiological temperatures. The presence of serum presumably aids in the solubilization of leachable substances through protein binding, the in vivo mechanism for transporting water-insoluble substances in the blood path. The direct contact assay may be used for testing samples with a specific geometry (for example 1 × 1-cm^2 squares using extruded sheeting or molded plaques of material) or with indeterminate geometries (molded parts). The major difficulty with this assay is the risk of physical trauma to the cells from either movement of the sample or crushing by the weight of a high-density sample. In most direct contact assays, there will be a zone of affected cells around the periphery of a toxic sample because of a slow leaching rate from the surface and bulk matrix of the material being tested. However, if the toxicant is water soluble, the rate of leaching may be sufficient to cause a decrease in the entire cell population in the culture plate rather than only those cells closest to the sample.

The disadvantages of the direct contact assay can be avoided by using the agar diffusion assay. The layer of agar between the test sample and the cells functions as a diffusion barrier to enhance the concentration gradient of leachable toxicants while also protecting the cells from physical trauma. The test sample itself may also be tested as a diffusion barrier to the migration of inks or labeling materials through the material

TABLE 1   Advantages and Disadvantages of Cell Culture Methods

	Direct contact	Agar diffusion	Elution
Advantages	Eliminate extraction preparation	Eliminate extraction preparation	Separate extraction from testing
	Zone of diffusion	Zone of diffusion	Dose response effect
	Target cell contact with material	Better concentration gradient of toxicant	Extend exposure time
	Mimic physiological conditions	Can test one side of a material	Choice of extract conditions
	Standardize amount of test material or test indeterminate shapes	Independent of material density	Choice of solvents
	Can extend exposure time by adding fresh media	Use filter paper disk to test liquids or extracts	
Disadvantages	Cellular trauma if material moves	Requires flat surface	Additional time and steps
	Cellular trauma with high density materials	Solubility of toxicant in agar	
	Decreased cell population with highly soluble toxicants	Risk of thermal shock when preparing agar overlay	
		Limited exposure time	
		Risk of absorbing water from agar	

matrix to the cellular side of the sample. Even contact between the test sample and the agar ensures diffusion from the material surface into the agar and cell layers. That is, diffusion at the material–solution interface is much greater than that at the material–air interface. Absorbant test samples, which could remove water from the agar layer, causing dehydration of the cells below, should be hydrated prior to testing in this assay.

The elution assay separates the extraction and biological testing phases into two separate processes. This could exploit the extraction to the extent of releasing the total available pool of chemical X from the material, especially if the extraction is done at elevated temperatures that presumably enhance the rate of migration and solubility limit of chemical X in a given solvent. However, when the extractant cools to room temperature, chemical X may precipitate out of solution or partition to the material surface. In addition, elevated extraction temperatures may foster chemical reactions and create leachable chemicals that would not occur in the absence of excessive heating. For example, the polymeric backbone of polyamides and polyurethanes may be hydrolyzed when these polymers are heated in aqueous solutions. Basically, these arguments lead back to a standardized choice of solvents and extraction conditions for all samples rather than optimized solvents for each material.

As with any biological or chemical assay, these assays occasionally are affected by interferences, false negatives, and false positives. For example, a fixative chemical such as formaldehyde or glutaraldehyde will give a false negative in the direct contact but not the agar diffusion assay, which uses a vital stain. A highly absorbent material could give a false positive in the agar diffusion assay because of dehydration of the agar. Severe changes in onconicity, osmolarity, or pH can also interfere with the assays. Likewise, a chelating agent that makes an essential element such as calcium unavailable to the cells could appear as a false positive result. Thus, a judicious evaluation of the test material and assay conditions is required for an appropriate interpretation of the results.

CLINICAL USE

The *in vitro* cytotoxicity assays are the primary biocompatibility screening tests for a wide variety of elastomeric, polymeric, and other materials used in medical devices. After the cytotoxicity profile of a material has been determined, then more application-specific tests are performed to assess the biocompatibility of the material. For example, a product which will be used for *in vitro* fertilization procedures would be tested initially for cytotoxicity and then application-specific tests for adverse effects on a very low cell population density would be evaluated. Similarly, a new material for use in culturing cells would be initially tested for cytotoxicity, followed by specific assays comparing the growth rates of cells in contact with the new material with those of currently marketed materials.

Current experience indicates that a material that is judged to be nontoxic *in vitro* will be nontoxic in *in vivo* assays. This does not necessarily mean that materials that are toxic *in vitro* could not be used in a given clinical application. The clinical acceptability of a material depends on many different factors, of which target cell toxicity is but one. For example, glutaraldehyde-fixed porcine valves produce adverse effects *in vitro* owing to low residues of glutaraldehyde; however, this material has the greatest clinical efficacy for its unique application.

In vitro assays are often criticized because they do not use cells with significant metabolic activity such as the P-450 drug-metabolizing enzymes. That is, the assays can only evaluate the innate toxicity of a chemical and do not test metabolic products that may have greater or lesser toxicity potential. In reality, biological effects of the actual leachable chemicals are the most relevant clinically because most medical devices are in contact with tissues having very low metabolic activity (e.g., skin, muscle, subcutaneous or epithelial tissues) or none. Metabolic products do not form at the implantation site, but rather, require transport of the leachable chemical to distant tissues which are metabolically active. In the process, there is significant dilution of concentration in the blood, tissues, and total body water to the extent that the concentration falls below the threshold of biological activity.

NEW RESEARCH DIRECTIONS

The current interest in developing alternatives to animal testing has resulted in the development and refinement of a wide variety of *in vitro* assays. Cell cultures have been used for several decades for screening anticancer drugs and evaluating genotoxicity (irreversible interaction with the nucleic acids). Babich and Borenfreund (1987) have modified the elution assay for use with microtiter plates to evaluate the dose-response cytotoxicity potential of alcohols, phenolic derivatives, and chlorinated toluenes. This system has also been modified to include a microsomal (S-9) activating system to permit drug metabolism *in vitro* when evaluating pure chemicals such as chemotherapeutic and bacteriostatic agents (Borenfreund and Puerner, 1987). The microtiter methods are likely to have increased application because they provide reproducible, semiautomatic, quantitative, spectrophotometric analyses. The major hurdle will be in identifying the appropriate benchmark or quantitation range for interpreting the data for clinical risk assessment. In earlier quantitative methods for *in vitro* biocompatibility assays, statistical differences in biocompatibility, which are attainable with quantitative assays, were not found to be biologically different (Johnson *et al.,* 1985). That is, the objective data were biologically different only when they were separated into quartiles of response similar to the subjective data. Thus, the major direction of new research will be in defining the benchmarks for application of quantitative methodology.

Bibliography

ASTM (1995a). Practice for direct contact cell culture evaluation of materials for medical devices. Annual Book of ASTM Standards, 13.01, **F813**: 233–236.

ASTM (1995b). Standard test method for agar diffusion cell culture screening for cytotoxicity. Annual Book of ASTM Standards 13.01, **F895**: 247–250.

Babich, H., and Borenfreund, E. (1987). Structure-activity relationship (SAR) models established *in vitro* with the neutral red cytotoxicity assay. *Toxicol. in Vitro* **1**: 3–9.

Borenfreund, E., and Puerner, J. A. (1987). Short-term quantitative *in vitro* cytotoxicity assay involving an S-9 activating system. *Cancer Lett.* **34**: 243–248.

ISO (1992). "In vitro" method of test for cytotoxicity of medical and dental materials and devices. International Standards Organization, Pforzheim, W. Germany. ISO/10993-5.

Johnson, H. J., Northup, S. J., Seagraves, P. A., Atallah, M., Garvin, P. J., Lin, L., and Darby, T. D. (1985). Biocompatibility test procedures for materials evaluation *in vitro*. II. Objective methods of toxicity assessment. *J. Biomed. Mater. Res.* **19**: 489–508.

Klaassen, C. D. (1986). Principles of toxicology. in *Casarett and Douell's Toxicology*, 3rd ed. C. D. Klaassen, M. O. Amdur, and J. Doull, eds. Macmillan, New York, pp. 11–32.

Northup, S. J. (1986). Mammalian cell culture models. in *Handbook of Biomaterials Evaluation: Scientific, Technical and Clinical Testing of Implant Materials,* A. F. von Recum, ed. Macmillan, New York, pp. 209–225.

Till, D. E., Reid, R. C., Schwartz, P. S., Sidman, K. R., Valentine, J. R., and Whelan, R. H. (1982). Plasticizer migration from polyvinyl chloride film to solvents and foods. *Food Chem. Toxicol.* **20**: 95–104.

U.S. Pharmacopeia (2004). Biological reactivity tests in-vitro. in *U.S. Pharmacopeia 23*. United States Pharmacopeial Convention, Inc., Rockville, MD. Vol. 27, pp. 2173–2175.

5.3 *In Vivo* Assessment of Tissue Compatibility

James M. Anderson and Frederick J. Schoen

INTRODUCTION

The goal of *in vivo* assessment of tissue compatibility of a biomaterial, prosthesis, or medical device is to determine the biocompatibility or safety of the biomaterial, prosthesis, or medical device in a biological environment. Biocompatibility has been defined as the ability of a medical device to perform with an appropriate host response in a specific application, and biocompatibility assessment is considered to be a measurement of the magnitude and duration of the adverse alterations in homeostatic mechanisms that determine the host response. In this chapter, the term "medical device" will be used to describe biomaterials, prostheses, artificial organs, and other medical devices, and the terms "tissue compatibility assessment," "biocompatibility assessment," and "safety assessment" will be considered to be synonymous.

From a practical perspective, the *in vivo* assessment of tissue compatibility of medical devices is carried out to determine that the device performs as intended and presents no significant harm to the patient or user. Thus, the goal of the *in vivo* assessment of tissue compatibility is to predict whether a medical device presents potential harm to the patient or user by evaluations under conditions simulating clinical use.

Recently, extensive efforts have been made by government agencies, i.e., FDA, and regulatory bodies, i.e., ASTM, ISO, and USP, to provide procedures, protocols, guidelines, and standards that may be used in the *in vivo* assessment of the tissue compatibility of medical devices. This chapter draws heavily on the ISO 10993 standard, Biological Evaluation of Medical Devices, in presenting a systematic approach to the *in vivo* assessment of tissue compatibility of medical devices.

In the selection of biomaterials to be used in device design and manufacture, the first consideration should be fitness for purpose with regard to characteristics and properties of the biomaterial(s). These include chemical, toxicological, physical, electrical, morphological, and mechanical properties. Relevant to the overall *in vivo* assessment of tissue compatibility of a biomaterial or device is knowledge of the chemical composition of the materials, including the conditions of tissue exposure as well as the nature, degree, frequency, and duration of exposure of the device and its constituents to the intended tissues

TABLE 1   Biomaterials and Components Relevant to *In Vivo* Assessment of Tissue Compatibility

The material(s) of manufacture
Intended additives, process contaminants, and residues
Leachable substances
Degradation products
Other components and their interactions in the final product
The properties and characteristics of the final product

TABLE 2   Medical Device Categorization by Tissue Contact and Contact Duration

Tissue contact	
Surface devices	Skin
	Mucosal membranes
	Breached or compromised surfaces
External communicating devices	Blood path, indirect
	Tissue/bone/dentin communicating
	Circulating blood
Implant devices	Tissue/bone
	Blood
Contact duration	Limited, ≤ 24 hours
	Prolonged, > 24 hours and < 30 days
	Permanent, > 30 days

in which it will be utilized. Table 1 presents a list of biomaterial components and characteristics that may affect the overall biological responses of the medical device. Knowledge of these components in the medical device, i.e., final product, is necessary. The range of potential biological hazards is broad and may include short-term effects, long-term effects, or specific toxic effects, which should be considered for every material and medical device. However, this does not imply that testing for all potential hazards will be necessary or practical.

SELECTION OF *IN VIVO* TESTS ACCORDING TO INTENDED USE

In vivo tests for assessment of tissue compatibility are chosen to simulate end-use applications. To facilitate the selection of appropriate tests, medical devices with their component biomaterials can be categorized by the nature of body contact of the medical device and by the duration of contact of the medical device. Table 2 presents medical device categorization by body contact and contact duration. The tissue contact categories and subcategories as well as the contact duration categories have been derived from standards, protocols and guidelines utilized in the past for safety evaluation of medical devices. Certain devices may fall into more than one category, in which case testing appropriate to each category should be considered.

The ISO 10993 standard and the FDA guidance document (FDA blue book memorandum #G95-1) present a structured program for biocompatibility evaluation in which matrices are presented which indicate required tests according to specific types of tissue contact and contact duration. These matrices are not presented here but the *in vivo* tests are indicated in Table 3.

BIOMATERIAL AND DEVICE PERSPECTIVES IN *IN VIVO* TESTING

Two perspectives may be considered in the *in vivo* assessment of tissue compatibility of biomaterials and medical devices. The first perspective involves the utilization of *in vivo* tests to determine the general biocompatibility of newly developed biomaterials for which some knowledge of the tissue compatibility is necessary for further research and development. In this type of situation, manufacturing and other processes necessary to the development of a final product, i.e., the medical

TABLE 3   *In Vivo* Tests for Tissue Compatibility

Sensitization
Irritation
Intracutaneous reactivity
Systemic toxicity (acute toxicity)
Subchronic toxicity (subacute toxicity)
Genotoxicity
Implantation
Hemocompatibility
Chronic toxicity
Carcinogenicity
Reproductive and developmental toxicity
Biodegradation
Immune responses

device, have not been carried out. However, the *in vivo* assessment of tissue compatibility at this early stage of development can be used to evaluate the general tissue responses to the biomaterial as well as provide additional information relating to the proposed design criteria in the production of a medical device. While it is generally recommended that the identification and quantification of extractable chemical entities of a medical device should precede biological evaluation, it is quite common to carry out preliminary *in vivo* assessments to determine if there may be unknown chemical entities that produce adverse biological reactions. Utilized in this fashion, early *in vivo* assessment of the tissue compatibility of a biomaterial may provide insight into the biocompatibility of a material and may permit its further development into a medical device. Obviously, problems observed at this stage of development would require further efforts to improve the biocompatibility of the biomaterial and identify the agents responsible for the adverse reactions. As the *in vivo* assessment of tissue compatibility of a biomaterial or medical device is focused on the end-use application, it must be appreciated that a biomaterial considered compatible for one application may not be compatible for another application.

The second perspective regarding the *in vivo* assessment of tissue compatibility of medical devices focuses on the biocompatibility of the final product, that is, the medical device in the condition in which it is to be implanted. Although medical devices in their final form and condition are commonly implanted in carefully selected animal models to determine function as well as biocompatibility, it may be not appropriate to carry out all of the recommended tests necessary for regulatory approval on the final device. In these situations, some tests may be carried out on biomaterial components of devices that have been prepared under the manufacturing and sterilization conditions and other processes utilized in the final product.

SPECIFIC BIOLOGICAL PROPERTIES ASSESSED BY IN VIVO TESTS

In this section, brief perspectives on the general types of *in vivo* tests are presented. Details regarding these tests are found in the references. ISO 10993 standards advise that the biological evaluation of all medical device materials include testing for cytotoxicity, sensitization, and irritation. (Cytotoxicity tests are *in vitro*.) Beyond these fundamentals, the selection of further tests for *in vivo* biocompatibility assessment is based on the characteristics and end-use application of the device or biomaterial under consideration.

Sensitization, Irritation, and Intracutaneous (Intradermal) Reactivity

Exposure to or contact with even minute amounts of potential leachables from medical devices or biomaterials can result in allergic or sensitization reactions. Sensitization tests estimate the potential for contact sensitization to medical devices, materials, and/or their extracts. Symptoms of sensitization are often seen in skin and tests are often carried out topically in guinea pigs. Test design should reflect the intended route (skin, eye, mucosa) and nature, degree, frequency, duration, and conditions of exposure of the biomaterial in its intended clinical use. While sensitization reactions are immune-system responses to contact with chemical substances, ISO guidelines suggest irritation to be a local tissue inflammation response to chemicals, without a systemic immunological component. The most severely irritating chemical leachables may be discovered prior to *in vivo* studies by careful material characterization and *in vitro* cytotoxicity tests. Irritant tests emphasize utilization of extracts of the biomaterials to determine the irritant effects of potential leachables. Intracutaneous (intradermal) reactivity tests determine the localized reaction of tissue to intracutaneous injection of extracts of medical devices, biomaterials, or prostheses in the final product form. Intracutaneous tests may be applicable where determination of irritation by dermal or mucosal tests are not appropriate. Albino rabbits are most commonly used.

Since these tests focus on determining the biological response of leachable constituents of biomaterials, their extracts in various solvents are utilized to prepare the injection solutions. Critical to the conduct of these tests is the preparation of the test material and/or extract solution and the choice of solvents which must have physiological relevance. Solvents should be chosen to include testing for both water-soluble and fat-soluble leachables.

Systemic Toxicity Acute, Subacute, and Subchronic Toxicity

Systemic toxicity tests estimate the potential harmful effects *in vivo* on target tissues and organs away from the point of contact with either single or multiple exposure to medical devices, biomaterials, and/or their extracts. These tests evaluate the systemic toxicity potential of medical devices that release constituents into the body. These tests also include pyrogenicity testing, which assesses the induction of a systemic inflammatory response often measured as fever.

In tests using extracts, the form and area of the material, the thickness, and the surface area to extraction vehicle volume are critical considerations in the testing protocol. Appropriate extraction vehicles, i.e., solvents, should be chosen to yield a maximum extraction of leachable materials for use in the testing. Mice, rats, or rabbits are the usual animals of choice for the conduct of these tests and oral, dermal, inhalation, intravenous, intraperitoneal, or subcutaneous application of the test substance may be used, depending on the intended application of the biomaterial. Acute toxicity is considered to be the adverse effects that occur after administration of a single dose or multiple doses of a test sample given within 24 hours. Subacute toxicity (repeat-dose toxicity) focuses on adverse effects occurring after administration of a single dose or multiple doses of a test sample per day during a period of from 14 to 28 days. Subchronic toxicity is considered to be the adverse effects occurring after administration of a single dose or multiple doses of a test sample per day given during a part of the life span, usually 90 days but not exceeding 10% of the life span of the animal.

Pyrogenicity tests are also included in the systemic toxicity category to detect material-mediated fever-causing reactions to extracts of medical devices or materials. Although the rabbit pyrogen test has been the standard, the Limulus amebocyte lysate (LAL) reagent test has been used increasingly in recent years. It is noteworthy that no single test can differentiate pyrogenic reactions that are material-mediated from those due to endotoxin contamination.

Genotoxicity

In vivo genotoxicity tests are carried out if indicated by the chemistry and/or composition of the biomaterial (see Table 1) or if *in vitro* test results indicate potential genotoxicity [changes in deoxyribonucleic acid (DNA)]. Initially, at least three *in vitro* assays should be used and two of these assays should utilize mammalian cells. The initial *in vitro* assays should cover the three levels of genotoxic effects: DNA destruction, gene mutations, and chromosomal aberrations (as assessed by cytogenetic analysis). *In vivo* genotoxicity tests include the micronucleus test, the *in vivo* mammalian bone marrow cytogenetic tests—chromosomal analysis, the rodent dominant lethal tests, the mammalian germ cell cytogenetic assay, the mouse spot test, and the mouse heritable translocation assay. Not all of the *in vivo* genotoxicity tests need be performed and the most common test is the rodent micronucleus test. Genotoxicity tests are performed with appropriate extracts or dissolved materials using appropriate media as suggested by the known composition of the biomaterial.

Implantation

Implantation tests assess the local pathological effects on the structure and function of living tissue induced by a sample of a material or final product at the site where it is surgically implanted or placed into an implant site or tissue appropriate to the intended application of the biomaterial or medical device. The most basic evaluation of the local pathological effects is carried out at both the gross level and the microscopic level. Histological (microscopic) evaluation is used to characterize various biological response parameters (Table 4). To address specific questions, more sophisticated studies may need to be done. Examples include immunohistochemical staining of histological sections to determine the types of cells present, and studies of collagen formation and destruction. For short-term implantation evaluation out to 12 weeks, mice, rats, guinea pigs, or rabbits are the usual animals utilized in these studies. For longer-term testing in subcutaneous tissue, muscle, or bone, animals such as rats, guinea pigs, rabbits, dogs, sheep, goats, pigs, and other animals with relatively long life expectancy are suitable. If a complete medical device is to be evaluated, larger species may be utilized so that human-sized devices may be used in the site of intended application. For example, substitute heart valves are usually tested as heart valve replacements in sheep, whereas calves are usually the animal of choice for ventricular assist devices and total artificial hearts.

Hemocompatibility

Hemocompatibility tests evaluate effects on blood and/or blood components by blood-contacting medical devices or materials. *In vivo* hemocompatibility tests are usually designed to simulate the geometry, contact conditions, and flow dynamics of the device or material in its clinical application. From the ISO standards perspective, five test categories are indicated for hemocompatibility evaluation: thrombosis, coagulation, platelets, hematology, and immunology (complement and leukocytes). Two levels of evaluation are indicated: Level 1

TABLE 4 Biological Response Parameters as Determined by Histological Assessment of Implants

Number and distribution of inflammatory cells as a function of distance from the material/tissue interface
Thickness and vascularity of fibrous capsule
Quality and quantity of tissue ingrowth (for porous materials)
Degeneration as determined by changes in tissue morphology
Presence of necrosis
Other parameters such as material debris, fatty infiltration, granuloma, dystrophic, calcification, apoptosis, proliferation rate, biodegradation, thrombus formation, endothelialization, migration of biomaterials or degradation products

(required), and Level 2 (optional). Regardless of blood contact duration, hemocompatibility testing is indicated for external communicating devices—blood path indirect; external communicating devices—circulating blood; and blood-contacting implant devices. Chapter 5.4 gives further details on the testing of blood–material interactions.

Several issues are important in the selection of tests for hemocompatibility of medical devices or biomaterials. *In vivo* testing in animals may be convenient, but species' differences in blood reactivity must be considered and these may limit the predictability of any given test in the human clinical situation. While blood values and reactivity between humans and nonhuman primates are very similar, European community law prohibits the use of nonhuman primates for blood compatibility and medical device testing. Hemocompatibility evaluation in animals is complicated by the lack of appropriate and adequate test materials, for example, appropriate antibodies for immunoassays. Use of human blood in hemocompatibility evaluation implies *in vitro* testing, which usually requires the use of anticoagulants that are not usually present with the device in the clinical situation, except for perhaps the earliest implantation period. Although species differences may complicate hemocompatibility evaluation, the utilization of animals in short- and long-term testing is considered to be appropriate for evaluating thrombosis and tissue interaction.

Chronic Toxicity

Chronic toxicity tests determine the effects of either single or multiple exposures to medical devices, materials, and/or their extracts during a period of at least 10% of the lifespan of the test animal, e.g. over 90 days in rats. Chronic toxicity tests may be considered an extension of subchronic (subacute) toxicity testing and both may be evaluated in an appropriate experimental protocol or study.

Carcinogenicity

Carcinogenicity tests determine the tumorigenic potential of medical devices, materials, and/or their extracts from either single or multiple exposures or contacts over a period of the major portion of the lifespan of the test animal. Since tumors associated with medical devices have been rare (see Chapter 4.7) carcinogenicity tests should be conducted only if data from other sources suggest a tendency for tumor induction. In addition, both carcinogenicity (tumorigenicity) and chronic toxicity may be studied in a single experimental study. With biomaterials, these studies focus on the potential for solid-state carcinogenicity, i.e., the Oppenheimer effect (see Chapter 4.7). In carcinogenicity testing, controls of a comparable form and shape should be included; polyethylene implants are a commonly used control material. The use of appropriate controls is imperative as animals may spontaneously develop tumors and statistical comparison between the test biomaterial/device and the controls is necessary. To facilitate and reduce the time period for carcinogenicity testing of biomaterial, the FDA is exploring the use of transgenic mice carrying the human prototype c-Ha-ras gene as a bioassay mode for rapid carcinogenicity testing.

Reproductive and Developmental Toxicity

These tests evaluate the potential effects of medical devices, materials, and/or their extracts on reproductive function, embryonic development (teratogenicity), and prenatal and early postnatal development. The application site of the device must be considered and tests and/or bioassays should only be conducted when the device has a potential impact on the reproductive potential of the subject.

Biodegradation

Biodegradation tests determine the effects of a biodegradable material and its biodegradation products on the tissue response. They focus on the amount of degradation during a given period of time (the kinetics of biodegradation), the nature of the degradation products, the origin of the degradation products (e.g., impurities, additives, corrosion products, bulk polymer), and the qualitative and quantitative assessment of degradation products and leachables in adjacent tissues and in distant organs. The biodegradation of biomaterials may occur through a wide variety of mechanisms, which in part are biomaterial dependent, and all pertinent mechanisms related to the device and the end-use application of the device must be considered. Test materials comparable to degradation products may be prepared and studied to determine the biological response of degradation products anticipated in long-term implants. An example of this approach is the study of metallic and polymeric wear particles that may be present with long-term orthopedic joint prostheses. Further insights on biodegradation are available in Chapters 6.2 and 6.3.

Immune Responses

Immune response evaluation is not a component of the standards currently available for *in vivo* tissue compatibility assessment. However, ASTM, ISO, and the FDA currently have working groups developing guidance documents for immune response evaluation where pertinent. Synthetic materials are not generally immunotoxic. However, immune response evaluation is necessary with modified natural tissue implants such as collagen, which has been utilized in a number of different types of implants and may elicit immunological responses. The Center for Devices and Radiological Health of the FDA has released a draft immunotoxicity testing guidance document whose purpose is to provide a systematic approach for evaluating potential adverse immunological effects of medical devices and constituent materials. Immunotoxicity is any adverse effect on the function or structure of the immune system or other systems as a result of an immune system dysfunction. Adverse or immunotoxic effects occur when humoral or cellular immunity needed by the host to defend itself against infections or neoplastic disease (immunosuppression) or unnecessary tissue damage (chronic inflammation, hypersensitivity, or autoimmunity) is compromised. Potential immunological effects and responses

TABLE 5 Potential Immunological Effects and Responses

Effects
  Hypersensitivity
      Type I—anaphylactic
      Type II—cytotoxic
      Type III—immune complex
      Type IV—cell-mediated (delayed)
  Chronic inflammation
  Immunosuppression
  Immunostimulation
  Autoimmunity
Responses
  Histopathological changes
  Humoral responses
  Host resistance
  Clinical symptoms
  Cellular responses
    T cells
    Natural killer cells
    Macrophages
    Granulocytes

that may be associated with one or more of these effects are presented in Table 5.

Representative tests for the evaluation of immune responses are given in Table 6. Table 6 is not all-inclusive and other tests that specifically consider possible immunotoxic effects potentially generated by a given device or its components may be applicable. Examples presented in Table 6 are only representative of the large number of tests that are currently available. However, direct and indirect markers of immune responses may be validated and their predictive value documented, thus providing new tests for immunotoxicity in the future. Direct measures of immune system activity by functional assays are the most important types of tests for immunotoxicity. Functional assays are generally more important than tests for soluble mediators, which are more important than phenotyping. Signs of illness may be important in *in vivo* experiments but symptoms may also have a significant role in studies of immune function in clinical trials and postmarket studies.

SELECTION OF ANIMAL MODELS FOR *IN VIVO* TESTS

Animal models are used to predict the clinical behavior, safety, and biocompatibility of medical devices in humans (Table 7). The selection of animal models for the *in vivo* assessment of tissue compatibility must consider the advantages and disadvantages of the animal model for human clinical application. Several examples follow, which exemplify the advantages and disadvantages of animal models in predicting clinical behavior in humans (also see Chapter 5.5).

As described earlier, sheep are commonly used for the evaluation of heart valves. This is based on size considerations and also the propensity to calcify tissue components of bioprosthetic heart valves and thereby be a sensitive model for this complication. Thus, the choice of this animal model for bioprosthetic heart valve evaluation is made on the basis of

TABLE 6   Representative Tests for the Evaluation of Immune Responses

Functional Assays	Phenotyping	Soluble Mediators	Signs of Illness
Skin testing	Cell surface markers	Antibodies	Allergy
Immunoassays (e.g., ELISA)	MHC markers	Complement	Skin rash
Lymphocyte proliferation		Immune complexes	Urticaria
Plaque-forming cells		Cytokine patterns (T-cell subsets)	Edema
Local lymph node assay		Cytokines (IL-1, IL-1ra, TNFα, IL-6, TGF-β, IL-4, IL-13)	Lymphadenopathy
Mixed lymphocyte reaction		Chemokines	
Tumor cytotoxicity		Basoactive amines	
Antigen presentation			
Phagocytosis			

ELISA, Enzyme-linked immunosorbent assay; IL, Interleukin; TNF, Tumor necrosis factor; TGF, Transforming growth factor; MHC, Major histocompatibility complex.

TABLE 7   Animal Models for the *In Vivo* Assessment of Medical Devices

Device Classification	Animal
Cardiovascular	
Heart valves	Sheep
Vascular grafts	Dog, pig
Stents	Pig, dog
Ventricular assist devices	Calf
Artificial hearts	Calf
Ex-vivo shunts	Baboon, dog
Orthopedic/bone	
Bone regeneration/substitutes	Rabbit, dog, pig, mouse, rat
Total joints—hips, knees	Dog, goat, nonhuman primate
Vertebral implants	Sheep, goat, baboon
Craniofacial implants	Rabbit, pig, dog, nonhuman primate
Cartilage	Rabbit, dog
Tendon and ligament substitutes	Dog, sheep
Neurological	
Peripheral nerve regeneration	Rat, cat, nonhuman primate
Electrical stimulation	Rat, cat, nonhuman primate
Ophthalmological	
Contact lens	Rabbit
Intraocular lens	Rabbit, monkey

accelerated calcification in rapidly growing animals, which has its clinical correlation in young and adolescent humans. Nevertheless, normal sheep may not provide a sensitive assessment of the propensity of a valve to thrombosis, which may be potentiated by the reduced flow seen in abnormal subjects but diminished by the specific coagulation profile of sheep.

The *in vivo* assessment of tissue responses to vascular graft materials is an example in which animal models present a particularly misleading picture of what generally occurs in humans. Virtually all animal models, including nonhuman primates, heal rapidly and completely with an endothelial blood-contacting surface. Humans, on the other hand, do not show extensive endothelialization of vascular graft materials and the resultant pseudointima from the healing response in humans

has potential thrombogenicity. Consequently, despite favorable results in animals, small-diameter vascular grafts (less than 4 mm in internal diameter) yield early thrombosis in humans, the major mechanism of failure, which is secondary to the lack of endothelialization in the luminal surface healing response.

The use of appropriate animal models is an important consideration in the safety evaluation of medical devices that may contain potential immunoreactive materials. The *in vivo* evaluation of recombinant human growth hormone in poly(lactic-*co*-glycolic acid)(PLGA) microspheres demonstrates the appropriate use of various animal models to evaluate biological responses and the potential for immunotoxicity. Utilizing biodegradable PLGA microspheres containing recombinant human growth hormone (rhGH), Cleland *et al.* used Rhesus monkeys, transgenic mice expressing hGH, and normal control (Balb/C) mice in their *in vivo* evaluation studies. Rhesus monkeys were utilized for serum assays in the pharmacokinetic studies of rhGH release as well as tissue responses to the injected microcapsule formulation. Placebo injection sites were also utilized and a comparison of the injection sites from rhGH PLGA microspheres and placebo PLGA microspheres demonstrated a normal inflammatory and wound-healing response with a normal focal foreign-body reaction. To further examine the tissue response, transgenic mice were utilized to assess the immunogenicity of the rhGH PLGA formulation. Transgenic mice expressing a heterologous protein have been previously used for assessing the immunogenicity of structural mutant proteins. With the transgenic animals, no detectable antibody response to rhGH was found. In contrast, the Balb/C control mice had a rapid onset of high-titer antibody response to the rhGH PLGA formulation. This study points out the appropriate utilization of animal models to not only evaluate biological responses but also one type of immunotoxicity (immunogenicity).

FUTURE PERSPECTIVES ON *IN VIVO* MEDICAL DEVICE TESTING

As presented earlier in this chapter, the *in vivo* assessment of tissue compatibility of biomaterials and medical devices is dependent on the end-use application of the biomaterial or medical device. In this sense, the development and utilization of new biomaterials and medical devices will dictate the development of new test protocols and procedures for evaluating these new products. Furthermore, it must be understood that the *in vivo* assessment of tissue compatibility of biomaterials and medical devices is open-ended and new end-use applications will require new tests.

Over the past half-century, medical devices and biomaterials have generally been "passive" in their tissue interactions. That is, a mechanistic approach to biomaterials/tissue interactions has rarely been used in the development of biomaterials or medical devices. Heparinized biomaterials are an exception to this statement but considering the five subcategories of hemocompatibility, these approaches have minimal impact on the development of blood-compatible materials.

In the past decade, increased emphasis has been placed on bioactivity and tissue engineering in the development of biomaterials and medical devices for potential clinical application. Rather than a "passive" approach to tissue interactions, bioactive and tissue-engineered devices have focused on an "active" approach in which biological or tissue components, i.e., growth factors, cytokines, drugs, enzymes, proteins, extracellular matrix components, and cells that may or may not be genetically modified, are used in combinations with synthetic, i.e., passive, materials to produce devices that control or modulate a desired tissue response. Obviously, *in vivo* assessment of the targeted biological response of a tissue-engineered device will play a significant role in the research and development of that device as well as in its the safety assessment. It is clear that scientists working on the development of tissue-engineered devices will contribute significantly to the development of *in vivo* tests for biocompatibility assessment as these tests will also be utilized to study the targeted biological responses in the research phase of the device development.

Regarding tissue-engineered devices, it must be appreciated that biological components may induce varied effects upon tissue in the *in vivo* setting. For example, a simplistic view of the potentially complex problems that might result from a device releasing a growth factor to enhance cell proliferation is presented. Cell types in the implant site may react differently to the presence of an extrinsic growth factor. Autocrine, paracrine, and endocrine signaling may be different between the same cell types and different cell types in the implant site. Signal transduction systems may be variable depending on the different cells that are present within the implant site. The presence of a growth factor may result in markedly different cell proliferation, differentiation, protein synthesis, attachment, migration, shape change, etc., which would be cell type dependent. Thus, different cell type–dependent responses in an implant site, reacting to the presence of a single exogenous growth factor, may result in inappropriate, inadequate, or adverse tissue responses. These perspectives must be integrated into the planned program for *in vivo* assessment of tissue compatibility of tissue-engineered devices. Finally, a major challenge to the *in vivo* assessment of tissue compatibility of tissue-engineered devices is the use of animal tissue components in the early phase of device development, whereas the ultimate goal is the utilization of human tissue components in the final device

for end-use application. Novel and innovative approaches to the *in vivo* tissue compatibility of tissue-engineered devices must be developed to address these significant issues. Importantly, the development of clinically useful tissue-engineered devices will require enhanced understanding of the influence of the patient and biomechanical factors on the structure and function of healed and remodeled tissues. It will also require methodology that permits assessment of the dynamic progression of remodeling *in vivo*, perhaps through imaging of cellular gene expression and extracellular matrix remodeling noninvasively (Rabkin and Schoen, 2002). Careful studies of retrieved implants to establish biomarkers and mechanisms of structural evolution will be critical (see Chapter 9.5).

Bibliography

An, Y. H., and Friedman, R. J., editors (1999). *Animal Models in Orthopaedic Research.* CRC Press, Boca Raton, FL.

Association for the Advancement of Medical Instrumentation (1998). AAMI Standards and Recommended Practices, Vol. 4, Biological Evaluation of Medical Devices, 1997; Vol. 4S, Supplement, 1998.

Chapekar, M. S. (1996). Regulatory concerns in the development of biologic–biomaterial combinations. *J. Biomed. Mater. Res. Appl. Biomat.* **33:** 199–203.

Cleland, J. L., Duenas, E., Daugherty, A., Marian, M., Yang, J., Wilson, M., Celniker, A. C., Shahzamani, A., Quarmby, V., Chu, H., Mukku, V., Mac, A., Roussakis, M., Gillette, N., Boyd, B., Yeung, D., Brooks, D., Maa, Y.-F., Hsu, Ch., and Jones, A. J. S. (1997). Recombinant human growth hormone poly(lactic-co-glycolic acid) (PLGA) microspheres provide a long lasting effect. *J. Control. Release* **49:** 193–205.

FDA (1995). Blue Book Memorandum G95-1: FDA-modified version of ISO 10,993-Part 1, Biological Evaluation of Medical Devices— Part 1: Evaluation and Testing.

Langone, J. J. (1998). Immunotoxicity Testing Guidance. Draft Document, Office of Science and Technology, Center for Devices and Radiological Health, Food and Drug Administration.

Rabkin, E., and Schoen, F. J. (2002). Cardiovascular tissue engineering. *Cardiovasc. Pathol.* **11:** 305.

Yamamoto, S., Urano, K., Koizumi, H., Wakana, S., Hioki, K., Mitsumori, K., Kurokawa, Y., Hayashi, Y., and Nomura, T. (1998). Validation of transgenic mice carrying the human prototype c-Ha-ras gene as a bioassay model for rapid carcinogenicity testing. *Environ. Health Perspect.* **106**(Suppl. 1): 57–69.

Standards

ISO 10,993, Biological Evaluation of Medical Devices, International Standards Organization, Geneva, Switzerland:

ISO 10,993-1.	Evaluation and testing
ISO 10,993-2.	Animal welfare requirements
ISO 10,993-3.	Tests for genotoxicity, carcinogenicity, and reproductive toxicity
ISO 10,993-4.	Selection of tests for interactions with blood
ISO 10,993-5.	Tests for cytotoxicity: *In vitro* methods
ISO 10,993-6.	Tests for local effects after implantation
ISO 10,993-7.	Ethylene oxide sterilization residuals
ISO 10,993-9.	Framework for the identification and quantification of potential degradation products

ISO 10,993-10.   Tests for irritation and sensitization

ISO 10,993-11.   Tests for systemic toxicity

ISO 10,993-12.   Sample preparation and reference
materials

ISO 10,993-13.   Identification and quantification of
degradation products from polymers

ISO 10,993-14.   Identification and quantification of
degradation products from ceramics

ISO 10,993-15.   Identification and quantification of
degradation products from metals
and alloys

ISO 10,993-16.   Toxicokinetic study design for
degradation products and leachables

ASTM, American Society for Testing and Materials, Annual Book of
ASTM Standards, 1999:

ASTM F-619-97   Practice for Extraction of Medical
Plastics

ASTM F-720-96   Practice for Testing Guinea Pigs for
Contact Allergens: Guinea Pig
Maximization Test

ASTM F-748-95   Practice for Selecting Generic
Biological Test Methods for
Materials and Devices

ASTM F-749-98   Practice for Evaluating Material
Extracts by Intracutaneous
Injection in the Rabbit

ASTM F-981-93   Practice for Assessment of
Compatibility of Biomaterials
(Nonporous) for Surgical Implants
with Respect to Effect of Materials
on Muscle and Bone

ASTM F-1439-96   Guide for the Performance of Lifetime
Bioassay for the Tumorigenic
Potential of Implant Materials

ASTM F-763-93   Practice for Short-Term Screening of
Implant Materials

5.4 EVALUATION OF BLOOD–MATERIALS INTERACTIONS

Stephen R. Hanson and Buddy D. Ratner

Every day, thousands of devices made from synthetic materials or processed natural materials are interfaced with blood (see Chapters 7.2, 7.3, and 7.5). How can the biomaterials engineer know which materials might be best used in the fabrication of a blood-contacting device? This chapter will outline some methods and concerns in evaluating the blood compatibility of biomaterials, and the blood compatibility of medical devices. It does not automatically follow that if the materials comprising a device are blood compatible, a device fabricated from those materials will also be blood compatible. This important point should be clear upon completion of this chapter. Before considering the evaluation of materials and devices, the reader should be familiar with the protein and cellular reactions of blood coagulation, platelet responses, and fibrinolysis as discussed in Chapter 4.6. The history of methods

to assess blood compatibility is briefly addressed in Ratner (2000).

WHAT IS BLOOD COMPATIBILITY?

A discussion of the nature of blood compatibility would be straightforward if, following the introductory paragraph, there were a list of standard tests that might be performed to evaluate blood compatibility. By simply performing the tests outlined in such a list, a material could be rated "blood compatible" or "not blood compatible." Unfortunately, no widely recognized, standard list of blood compatibility tests exists. Because of the complexity of blood–materials interactions (BMIs), there is a basic body of ideas that must be mastered in order to appreciate what blood interaction tests actually measure. This section introduces the rationale for BMI testing and addresses a few important measurement schemes.

"Blood compatibility" can be defined as the property of a material or device that permits it to function in contact with blood without inducing adverse reactions. Unfortunately, this simple definition offers little insight into what a blood-compatible material is. More useful definitions become increasingly complex. This is because there are many mechanisms that the body has available to respond to material intrusions into the blood. A material that will not trigger one response mechanism may be highly active in triggering another mechanism. The mechanisms by which blood responds to materials have been discussed in Chapter 4.6. A more recent definition of blood compatibility integrates a multiparameter assessment of BMI with some of the parameters defined quantitatively (Sefton *et al.*, 2000). This textbook chapter will discuss how one measures the blood compatibility of materials in light of these response mechanisms and definitions.

We can also view blood compatibility from a different perspective by considering a material that is *not* blood compatible, i.e., a thrombogenic material. Such a material would produce specific adverse reactions when placed in contact with blood: formation of clot or thrombus composed of various blood elements; shedding or nucleation of emboli (detached thrombus); the destruction of circulating blood components; and activation of the complement system and other immunologic pathways (Salzman and Merrill, 1987). Most often, in designing blood-contacting materials and devices our aim is to minimize these generally undesirable blood reactions. However, consider the case where our aim is to develop a hemostatic device to promote the rapid induction of clotting.

WHY MEASURE BLOOD COMPATIBILITY?

Many devices and materials are presently used in humans to treat, or to facilitate treatment of, various disease states. Such devices include the extracorporeal pump-oxygenator (heart–lung machine) used in many surgical procedures, hollow fiber hemodialyzers for treatment of kidney failure, catheters for blood access and blood vessel manipulation (e.g., angioplasty), heart assist devices, stents for luminally supporting blood vessels, and devices for the permanent replacement of diseased heart valves (prosthetic heart valves) and arteries (vascular grafts). Since these and other blood-contacting devices have been successfully used in patients for many years and are judged to be therapeutically beneficial, it is reasonable to ask: (1) is there a continued need for assessing BMI?, and (2) are there important problems that remain to be addressed? The answer in both cases is clearly "yes." For example, many existing devices are frequently modified by incorporation of new design features or synthetic materials primarily intended to improve durability, physical, and mechanical characteristics, i.e., devices may be modified to improve characteristics other than BMI. However, since these changes may also affect blood responses, and since BMIs are not entirely predictable based on knowledge of device composition and configuration, blood compatibility testing is nearly always required to document safety.

The performance of many existing devices is also less than optimal (Salzman and Merrill, 1987; Williams, 1987; McIntire *et al.*, 1985; Ratner, 2000). For example, prolonged cardiopulmonary bypass and membrane oxygenation can produce a severe bleeding tendency. Mechanical heart valves occasionally shed emboli to the brain producing stroke. Angiographic catheters can lead to strokes. Synthetic vascular grafts perform less well than grafts derived from natural arteries or veins; graft failure due to thrombosis can lead to ischemia (lack of oxygen) and death of downstream tissue beds; small-diameter vascular grafts (< 4 mm i.d.) cannot be made. Thus, while performance characteristics have been judged to be acceptable in many instances (i.e., the benefit/risk ratio is high), certain existing devices could be improved to extend their period of safe operation (e.g., oxygenators), and to reduce adverse BMI long-term (e.g., heart valves). Further, many devices are only "safe" when anticoagulating drugs are used (e.g., oxygenators, mechanical heart valves, hemodialyzers). Device improvements that would reduce adverse BMI and thereby eliminate the need for anticoagulant therapy would have important implications both for health (fewer bleeding complications due to drug effects) and cost (complications can be expensive to treat). The reusability of devices that can undergo repeated blood exposure in individual patients (e.g., dialyzers) is also an important economic consideration.

For certain applications there are no devices presently available that perform adequately (due to adverse BMIs) even when antithrombotic drugs are used. Thus, there is a need for devices that could provide long-term oxygenation for respiratory failure, cardiac support (total artificial heart), and intravascular measurement of physiologic parameters (O_2, CO_2, pH), as well as for small-diameter vascular grafts (<5 mm internal diameter) and other conduits (e.g., stents) for reconstruction of diseased arteries and veins. Overall, there is a compelling need for continued and improved methods for evaluating BMI.

WHAT IS THROMBOGENICITY?

The thrombogenic responses induced by a material or device can be phenomenologically categorized into two groups.

First, as the term implies, a thrombogenic device may cause the accumulation of various blood elements that are preferentially concentrated locally relative to their concentrations in circulating blood (thrombus formation). Cardiovascular devices may also exhibit regions of disturbed flow or stasis that lead to the formation of blood clots (coagulated whole blood). These *local* effects may compromise device functions such as the delivery of blood through artificial blood vessels, the mechanical motions of heart valves, gas exchange through oxygenators, and the removal of metabolic waste products through dialysis membranes. These local blood reactions may also produce effects in other parts of the host organism, i.e., *systemic* effects. Thus thrombi may detach from a surface (embolize) and be carried downstream, eventually occluding a blood vessel of comparable size and impairing blood flow distal to the site of occlusion. Chronic devices may produce steady-state destruction or "consumption" of circulating blood elements, thereby lowering their concentration in blood (e.g., mechanical destruction of red cells by heart prostheses producing anemia, or removal of platelets due to ongoing thrombus formation), with a concomitant rise in plasma levels of factors released from those blood elements (e.g., plasma hemoglobin, platelet factor 4). Mediators of inflammatory responses and vessel tone may also be produced or released from cells (e.g., platelets, white cells, the complement pathway) following blood–surface interactions that can affect hemodynamics and organ functions at other sites. Thus "thrombogenicity" may be broadly defined as the extent to which a device, when employed in its intended use configuration, induces the adverse responses outlined above. Although all artificial surfaces interact with blood, an acceptably nonthrombogenic device can be defined as one that would produce neither local nor systemic effects of health consequence to the host organism.

With "thrombogenicity" now defined in a global manner in terms of adverse outcome events associated with device usage, the obvious goal is to design and improve devices using materials that are blood compatible (nonthrombogenic) for specific applications. Ideally, the biomaterials engineer would like to consult a handbook for a list of materials useful in the fabrication of a device. Unfortunately, there is little consensus as to what materials are blood compatible. Because of this lack of consensus, there is no "official" list of blood-compatible materials. As a result, an individual interested in learning which materials might be suitable for construction of a new blood-contacting device generally consults published studies, or directly performs studies on candidate materials.

Despite intensive efforts, the blood compatibility of specific materials for particular device applications is not well established because (1) the types of devices used are numerous, (2) they may exhibit complex flow geometries, (3) they are continuously being improved, (4) the possible blood responses are numerous, complex, dynamic, and not fully understood, and (5) it is difficult and expensive to measure device thrombogenicity (clinically significant local thrombosis or systemic effects) in a systematic way in either experimental animals or humans (Williams, 1987). Most tests purported to measure blood

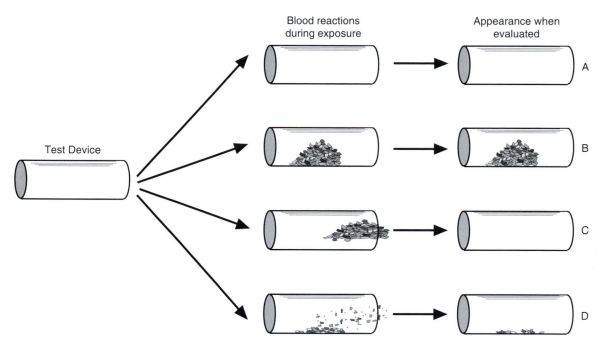

FIG. 1. Possible scenarios for blood–materials interactions, and limitations of evaluating only local thrombus formation at fixed time points. (A) Device remains free of thrombus. (B) Large thrombus forms and remains attached. (C) Large thrombus forms but detaches (embolizes). (D) Surface is highly reactive toward blood but deposited material is quickly removed through microembolism and or lysis. Inspection of devices C and D could lead to the incorrect conclusion that these surfaces are blood compatible.

compatibility really evaluate certain blood–material interactions, which are the events that occur (and observations that are made) when blood contacts a material. Figure 1 illustrates how alternate interpretations can be applied to data from "blood compatibility" tests. This concept is further expanded upon in Fig. 2. These alternative interpretations often invalidate or modify conclusions one would like to draw from such tests. For accuracy, the term "BMI assessment" will be used for the remainder of this chapter instead of "blood compatibility test." Based upon the characteristics of the evaluation method (i.e., what is really being measured) the biomaterials scientist must relate the significance of the events being observed (the BMIs) to the blood compatibility of the material or device. A solid understanding of the physical and biological mechanisms of blood–materials interactions is required to make this connection in a rational way.

In more specific terms, BMIs are the interactions (reversible and irreversible) between surfaces and blood solutes, proteins, and cells (adsorption, absorption, adhesion, denaturation, activation, spreading) that occur under defined conditions of exposure time, blood composition, and blood flow. Since each of these variables influence BMIs, we generally cannot (1) extrapolate results obtained under one set of test conditions to another set of conditions, (2) use short-term testing to predict long-term results, and (3) predict *in vivo* device performance based on BMIs testing of materials per se in idealized flow geometries. Nonetheless, such tests have provided important insights into the mechanisms of thrombus formation in general,

and the relationships between BMIs and blood compatibility. These studies also permit some general guidelines for device construction and, to a limited extent, may allow prediction of device performance in humans. These points are addressed in subsequent sections of this chapter.

The foregoing considerations suggest that no material may be simply "blood compatible" or "nonthrombogenic" since this assessment will depend strongly on details of the test system or usage configuration. In fact, under conditions of sluggish (low shear) blood flow or stasis, most if not all polymeric materials may become associated with localized blood clotting and thus be considered "thrombogenic." This is because synthetic materials, unlike the vascular endothelium (the perfect "blood-compatible material" that lines all blood vessels), cannot actively inhibit thrombosis and clotting by directly producing and releasing inhibitors or by inactivating procoagulant substances. The possibility that there may be no "biomaterials solution" to certain problems, or that device performance could be improved by emulating strategies found in nature, has led some investigators to consider coating devices with endothelial cells, antithrombotic drugs, or anticoagulating enzymes. Although there is no evidence that these methods have solved the problem of biomaterial thrombogenicity for any device, the approaches appear promising and are being widely explored. As for conventional synthetic materials and devices, establishing the usefulness of biologic surfaces and drug delivery devices requires appropriate methods for evaluating their blood interactions.

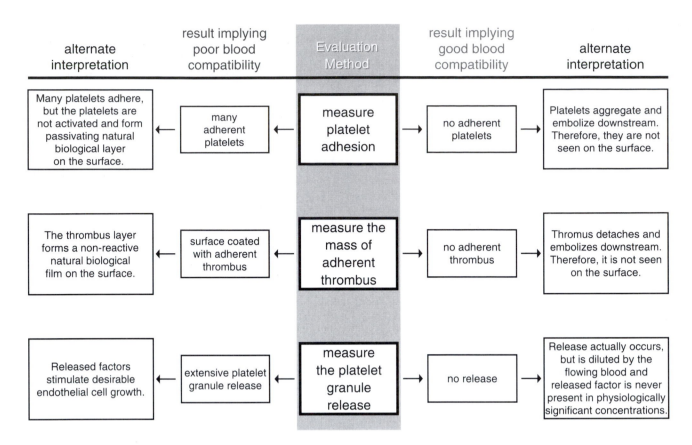

alternate interpretation	result implying poor blood compatibility	Evaluation Method	result implying good blood compatibility	alternate interpretation
Many platelets adhere, but the platelets are not activated and form passivating natural biological layer on the surface.	many adherent platelets	measure platelet adhesion	no adherent platelets	Platelets aggregate and embolize downstream. Therefore, they are not seen on the surface.
The thrombus layer forms a non-reactive natural biological film on the surface.	surface coated with adherent thrombus	measure the mass of adherent thrombus	no adherent thrombus	Thromus detaches and embolizes downstream. Therefore, it is not seen on the surface.
Released factors stimulate desirable endothelial cell growth.	extensive platelet granule release	measure the platelet granule release	no release	Release actually occurs, but is diluted by the flowing blood and released factor is never present in physiologically significant concentrations.

FIG. 2. Alternative scenarios that can be applied for interpreting results of blood–material interaction assays.

KEY CONSIDERATIONS FOR BMI ASSESSMENT

In 1856 Rudolph Virchow proposed that the three factors that contribute to the coagulation of blood are the blood chemistry, the blood-contacting surface, and the flow regime (commonly referred to as Virchow's triad).

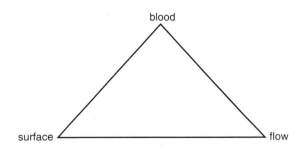

This assessment is still valid and provides a framework for more formally introducing the variables important in any system intended to evaluate BMIs. Also, the interaction time of blood with materials (ranging from seconds to years) has an influence on the three components of the triad. As described later, these variables may each profoundly influence the results and interpretation of BMI testing, and the ranking of blood compatibility of materials according to that test. It is assumed that the reader has reviewed the mechanisms of blood response to artificial surfaces as outlined in Chapter 4.6.

Blood: Factors Affecting Its Properties

The source and methods for handling blood can have important effects on BMI. Blood obtained from humans and various animal species has been employed *in vitro* and *in vivo*, in both the presence and absence of anticoagulants. Blood reactivity is also influenced by the extent and period of manipulation *in vitro*, the surface-to-volume ratio of blood placed in extracorporeal circuits, and the use of pumps for blood recirculation. These aspects are discussed next and summarized in Table 1.

The blood chemistry of each animal species is different. In particular, blood may vary with respect to the concentrations and functions of blood proteins and cells that participate in coagulation, thrombosis, and fibrinolysis (Chapter 4.6). The size of blood-formed elements may also differ. A comparison of blood chemistry between man and the more commonly used animal species has been published (McIntire et al., 1985). While human blood is obviously preferable for BMI, it is often impossible to use human blood in certain experiments. In addition there are significant health concerns in experimenting with human blood, and animals are commonly used for both *in vitro* and *in vivo* studies. Unfortunately, most investigations have employed a single animal species or blood source. There

TABLE 1 Factors Important in the Acquisition and Handling of Blood for BMI Experiments

Species of the blood donor
Health, gender, and age of the blood donor
Blood reactivity of the donor (individual physiological differences)
Time interval between blood draw and BMI experiment
Care with which the puncture for the blood draw was made
Temperature (for blood storage and testing)
Anticoagulation
Drugs and anesthetics present in the blood
Blood damage due to centrifugation and separation operations
Blood damage due to contact with foreign surfaces prior to the
  BMI experiment
(syringe, needle, blood bag, bottles, tubing, etc.)
Blood damage due to the air–blood interface
Blood damage due to pumping and recirculation

have been few comparisons between human and animal blood responses for evaluating BMI in any particular test situation (also see Chapter 5.5). In many instances differences between human and animal blood responses are likely to be large and must be borne in mind when interpreting experimental results. For example, the initial adhesiveness of blood platelets for artificial surfaces appears to be low in man and some primates, and high in the dog, rat, and rabbit (Grabaowski *et al.*, 1976). Following the implantation of chronic blood-contacting devices (e.g., vascular grafts) there may be large differences between man and all other animal species in terms of device healing and incorporation of surrounding tissues, which will be reflected as differences in the *time course* of BMIs. Further, although laboratory animals serving as blood donors for BMI experiments may represent a relatively homogeneous population in terms of age, health status, blood responses, etc., the human recipients of blood-contacting devices may vary considerably in terms of these parameters. Thus the results obtained with any animal species must be viewed with great caution if conclusions are to be drawn as to the significance of the results for humans.

Despite these limitations, animal testing has been particularly helpful in defining *mechanisms* of BMI and thrombus formation, and the interdependence of blood biochemical pathways, the nature of the surface, and the blood flow regime. In addition, while results of animal testing may not quantitatively predict results in man, in many cases results are likely to be qualitatively similar. These aspects are discussed further hereafter. In general, studies in lower animal species such as the rabbit, rat, and guinea pig may be useful to screen for profound differences between materials, for example, by incorporation of an antithrombotic drug delivery system into an otherwise thrombogenic device. Short-term screening to identify markedly reactive materials, and longer-term studies to evaluate the effects of healing phenomena on BMI, can also be performed in other species such as the dog and pig. When differences in BMI are likely to be modest (for example, as a consequence of subtle changes in surface chemistry or device configuration) the ranking of materials based on tests with lower animal species may be unrelated to results that would be obtained in man; studies with primates, which are

hematologically similar to man, are more likely to provide results that are clinically relevant. However, the relationships between human and primate BMI are also not well established in many models and applications, and should therefore be interpreted with caution.

In vitro testing generally requires anticoagulation of the blood that can have a profound effect on BMI. *In vivo* testing and the use of extracorporeal circuits are also commonly performed with anticoagulants. Two anticoagulants are most frequently used: sodium citrate, a chelator for calcium ions that are required for certain reactions of platelets and coagulation proteins, and heparin, a natural polysaccharide used to block the action of the coagulation protease thrombin (Chapter 4.6). Both can markedly affect BMI. In particular, the removal of calcium ions may profoundly depress platelet–surface reactions and the capacity of platelets to form aggregates and thrombi. Thus the relationship between BMI in the presence of citrate anticoagulant, and "thrombogenicity" in the absence of anticoagulant, is questionable. Similar concerns apply to heparin anticoagulation. Although this agent is less likely to interfere with the earliest platelet reactions, platelet thrombus formation may be impaired by inhibition of thrombin activity. The use of heparin is appropriate for evaluation of devices whose use normally requires heparinization *in vivo* (e.g., oxygenators, dialyzers). In general, results with anticoagulated systems cannot be used to predict performance in the absence of anticoagulants. A discussion of anticoagulants in the context of BMI has been presented (McIntire *et al.*, 1985).

Blood is a fragile tissue that begins to change from the moment it is removed from the body. It may become more active (activated) or less active (refractory) in several ways, even with effective anticoagulation. Thus, BMI evaluations with blood aged more than a few hours are questionable. If purified blood components or cells are used (e.g., platelets, fibrinogen), studies must be performed to insure that they remain functionally normal. In most cases the volume of blood used, relative to test surface area, should be large. Similarly, the area of non-test surfaces, including exposure to air interfaces, should be minimized. Changes in blood or test surface temperature, or exposure to intense light sources (Haycox *et al.*, 1991), can also produce artifactual results. When blood is pumped, the recirculation rate (fraction of total blood volume pumped per unit time) should be minimized since blood pumping alone can induce platelet and red cell damage, platelet release reactions, and platelet refractoriness (Haycox and Ratner, 1993).

Flow: How It Affects Blood Interactions

Blood flow controls the rate of transport (by diffusion and convection) of cells and proteins in the vicinity of artificial surfaces and thrombi. This subject has been reviewed (Leonard, 1987; Turitto and Baumgartner, 1987; Chapter 3.5). Although physiologically encountered blood shear forces probably do not damage or activate platelets directly, such forces can dislodge platelet aggregates and thrombi that may attach farther downstream or are carried away (embolize) to distal circulatory beds. Platelet diffusion in flowing blood, and early platelet attachment to surfaces, may be increased 50- to 100-fold by the presence of red blood cells that greatly enhance the movement of platelets across parallel streamlines. At higher shear forces, red cells may also contribute chemical factors that enhance platelet reactivity (Turitto and Baumgartner, 1987).

A number of studies using well-characterized flow geometries have suggested that the initial attachment of platelets to artificial surfaces increases with increasing blood flow, or, more specifically, with increasing wall shear rate (the slope of the velocity profile at the surface). Under conditions of low wall shear rate flow (less than $\sim 1000 \ \text{sec}^{-1}$) early platelet adhesion (over the first minutes of exposure) may depend more upon the platelet arrival rate (i.e., platelet availability) than on substrate surface properties (Friedman *et al.*, 1970). Under these conditions the platelet-surface reaction rate is said to be *diffusion controlled*. At higher shear rates, platelet adhesion may depend upon both the rate of platelet transport as well as surface properties (Schaub *et al.*, 2000); thus, studies designed to assess the role of surface properties are best performed under flow conditions where platelet transport is not limiting. Following initial platelet adhesion, subsequent processes of platelet aggregation and *in vivo* thrombus formation (over minutes to hours) may be partly *reaction controlled*. For example, platelet accumulation on highly thrombogenic artificial surfaces (e.g., fabric vascular grafts) or biologic surfaces (e.g., collagen) may be quite rapid and depends on both the substrate reactivity and factors influencing platelet availability (shear rate, hematocrit, and the platelet content of blood) (Harker *et al.*, 1991). Under other circumstances, the rate of platelet–surface interactions may be almost entirely reaction controlled. For example, with smooth-walled artificial surfaces that cause repeated embolization of small platelet aggregates continuously over days, the overall rate of platelet destruction depends strongly on material properties but not on blood flow rate or circulating platelet numbers over wide ranges of these variables (Hanson *et al.*, 1980). These concepts of diffusion and reaction control are further explained in Fig. 3.

It has been observed that under arterial flow conditions (high wall shear rate), thrombus that forms *in vivo* may be largely composed of platelets ("white thrombus"), whereas thrombus that forms under venous flow conditions (low shear rate) may contain mostly red cells entrapped in a fibrin mesh ("red thrombus"). The process of platelet thrombus formation may not be affected by administration of heparin in normal anticoagulating amounts (i.e., arterial thrombosis may be heparin-resistant), while venous thrombosis is effectively treated with heparin. This lack of effect against platelet reactions is somewhat surprising since the procoagulant enzyme thrombin, one of the most potent activators of platelets, is strongly inhibited by heparin. These observations have been incorrectly interpreted to mean that arterial and venous thrombosis are separable processes, with the former depending only on platelet reactions and the latter depending only on coagulation-related events. However, while platelet-dependent (arterial) thrombosis may be little affected by heparin, it is blocked quite effectively by other inhibitors of thrombin (Hanson and Harker, 1988; Wagner and Hubbell, 1990), indicating that heparin is limited in its capacity to block the thrombin enzyme when thrombin is produced locally in high concentrations through reactions that may be catalyzed on the platelet

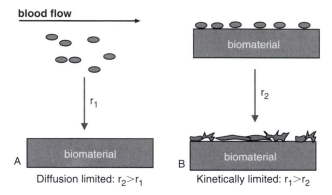

FIG. 3.   The effects of flow and material surface properties on platelet transport to, and reaction with, surfaces. One can define two rate processes, r_1, the rate of platelet transport from blood to the surface (higher in rapidly flowing blood), and r_2, the rate of reaction of a platelet with the surface. (A) In low wall shear rate blood flow, platelets can be slow to reach the surface and r_1 dominates the kinetics of the reaction—for reactive surfaces, the surface can be "starved" of platelets for reaction. (B) In high wall shear rate blood flow, platelets are transported to the surface much more rapidly than they can react with the surface, and the intrinsic reactivity of the surface, r_2, can be observed.

surface (Chapter 4.6). The formation of fibrin, due to the action of thrombin on fibrinogen, is also important for thrombus formation and stabilization since (1) fibrinolytic enzymes can reduce platelet thrombus formation, and (2) arterial thrombi are often composed of alternating layers of platelets and fibrin. Thus, in most circumstances, thrombin is a key promotor of *local* platelet and fibrin accumulation (on surfaces), under both high-shear and low-shear conditions. Thrombi may differ in appearance because under high-flow conditions, thrombin and precursor procoagulant enzymes (e.g., factor Xa) may be diluted sufficiently to prevent *bulk phase* clotting and trapping of red cells. These ideas have been addressed quantitatively (Basmadjian *et al.*, 1997). Also, flow effects on surface-induced thrombosis have been reviewed and the importance of pharmacological interventions on modulating this process have been described (Hanson and Sakariassen, 1998). Thus, the shear dependency of surface thrombosis depends on the surfaces' chemical composition. For example, anticoagulation effects may be more pronounced on a tissue factor–rich surface than on a collagen surface, particularly at low wall shear rates. Further, some platelet inhibitors such as aspirin are shear dependent while others seem not to be.

In summary, thrombus formation requires the transport by flow of platelets and coagulation proteins to surfaces. Fibrin polymerization, as well as local platelet activation and recruitment into growing thrombi, requires conversion of prothrombin to thrombin, the end product of a sequential series of coagulation reactions that are also catalyzed by platelets, and may be amplified or inhibited by various feedback mechanisms (Chapter 4.6). Blood flow regulates each reaction step such that under low (venous) flow conditions fibrin formation is abundant; thrombi may resemble coagulated whole blood with many entrapped red cells. Under high (arterial) flows platelets, stabilized by much smaller amounts of fibrin, may comprise the greater proportion of total thrombus mass.

Surfaces: The Properties of Biomaterials and Devices and Their Relationship to Blood Interactions

Many different artificial surfaces, in various device applications, are used in contact with blood. As discussed subsequently, tests designed to assess certain blood–materials interactions have shown that the surface physicochemical properties of materials and devices can have important effects on early events, for example, on protein adsorption and platelet adhesion, yet how these effects relate to subsequent thrombus formation remains uncertain.

When placed in contact with blood, most, if not all, artificial surfaces first acquire a layer of adsorbed blood proteins whose composition and mass may vary with time in a complex manner depending on substrate surface type (Chapter 3.2). This layer mediates the subsequent attachment of platelets and other blood cells that can lead to the development of platelet aggregates and thrombi. The relationship between material properties, the protein layer, and the propensity of a material or device to accumulate thrombus is not well understood because (1) protein–surface reactions involve complex, dynamic processes of competitive adsorption, denaturation, and activation; (2) cell–surface interactions may modify the protein layer, i.e., cells may deposit lipid and protein "footprints" derived from the cell membrane; (3) the importance of specific adsorbed proteins for subsequent cell interactions is not well defined; and (4) there have been few relevant tests in which both protein adsorption and later thrombus formation have been assessed. Under conditions of low blood shear, the capacity of negatively charged surfaces (such as glass) to activate intrinsic coagulation (via factor XII) can lead to thrombin production with subsequent platelet deposition and fibrin clot formation. Under other circumstances the availability on surfaces of adhesive plasma proteins, such as fibrinogen, may be important for regulating cell attachment (Horbett *et al.*, 1986).

With anticoagulated blood, initial platelet attachment to a variety of surfaces may be comparable and limited to a partial platelet monolayer, suggesting that surface properties may be "inconsequential" for early platelet adhesion, especially where platelet transport to the surface may be rate-limiting (Friedman *et al.*, 1970). In the absence of anticoagulants, initial platelet attachment may vary, but no general relationship to substrate surface properties has been demonstrated. In attempts to establish such relationships, thrombus formation has been studied using devices implanted in animals and composed of various materials including polymers, metals, carbons, charged surfaces, and hydrogels. Correlations have been sought between the blood response and surface properties such as charge (anionic-cationic), hydrophilicity, hydrophobicity, polarity, contact angle, wettability, and critical surface tension (Salzman and Merrill, 1987; Williams, 1987; McIntire *et al.*, 1985). These parameters have not proven satisfactory for predicting device performance even in idealized test situations, reflecting the complexity of the phenomena being investigated, the limitations of animal modeling (e.g., Fig. 1), and, in some cases, inadequate characterization of material surface properties (see Chapters 1.4 and 9.4 and Ratner, 1993a).

In many cases, material properties are constrained by the intended device application. For example, vascular grafts and the sewing rings of prosthetic heart valves are composed of fabric or porous materials to permit healing and tissue anchoring. Other materials must be permeable to blood solutes and gases (dialysis and oxygenator membranes) or distensible (pump ventricles, balloon catheters) and may necessarily exhibit complex flow geometries. In general, devices with flow geometries that cause regions of flow recirculation and stasis tend to produce localized clotting in the absence of heparin anticoagulation. On a microscopic scale, surface imperfections, cracks, and trapped air bubbles may serve as a focus to initiate thrombus formation. Although surface smoothness is usually desirable, many devices having a fabric or microporous surface (e.g., vascular grafts) function well if the layer of thrombus that forms is not thick enough to interfere with device function (Salzman and Merrill, 1987).

The complexity of understanding blood interactions with materials, as illustrated by the lack of consensus on what materials are or are not blood compatible, was pointed out many years ago (Ratner, 1984). The difficulty in assigning a label of "blood compatible" or ranking materials as to their suitability for use in blood still persists as is illustrated in two studies. One study, using a number of different screening assays, questioned the effects of surface properties on blood interactions since trends from each of the assays could not be correlated with each other (Sefton *et al.*, 2001). Another study, using fluorescent fiber optic microscopy to observe platelet deposition, ranked materials as to the magnitude of their interaction with blood (Schaub *et al.*, 2000). Both studies were of high quality and offered insights into testing blood interactions and measuring BMI. Yet, taken as a set, they offer little insight that will allow us to conclude exactly which surface properties will yield the most blood-compatible material.

Blood Interaction Times with Materials and Devices

Different events may occur at short and long BMI times. A test performed where blood contacts a device for seconds or minutes may yield a result that will have no meaning for devices used for hours or days, or which may be implanted chronically. Thus, measurements of protein adsorption may not predict levels of platelet adhesion. Platelet adhesion alone is not an adequate measure of thrombogenicity and does not predict local or systemic thrombogenic effects that could be harmful to the host organism. However, several studies indicate that an early maximum in platelet thrombus accumulation may be seen within hours of device exposure and this can be sufficient to produce device failure (e.g., small-diameter vascular graft occlusion) (Harker *et al.*, 1991). Therefore, short-term testing (over hours) may be appropriate for predicting the clinical usefulness of devices that can produce an acute, severe thrombotic response. In general, the nature and extent of BMIs may change continuously over the entire period of device exposure. An exception to this rule may be chronic implants that do not undergo tissue coverage (e.g., heart valve struts, arteriovenous shunts) and that may interact with blood elements at a constant rate as shown, for example, by steady-state increases in rates of platelet consumption (Hanson *et al.*, 1980).

EVALUATION OF BMIs

In this section we provide a general summary and interpretation of more commonly used *in vitro* and *in vivo* animal testing procedures to evaluate BMIs. It is emphasized that a thorough characterization of material properties is critical for the interpretation of these tests (Chapters 1.2 and 1.4 and Ratner, 1993a). A few specific tests that have been historically influential, and tests that are used today are summarized in Table 2.

BMIs can be evaluated *in vitro* and *in vivo*. Also, we can look at the BMIs of either biomaterials in a test configuration or the blood interactions of real devices. There are typically unique systems, apparatus and geometries for *in vitro*, *in vivo*, and device-based assays. However, there are commonalities in the measurements of blood parameters. For example, blood can be circulated through a tube *in vitro*, circulated through a biomaterial shunt *in vivo*, or circulated through a hemodialyzer (*in vivo*, or *in vitro*). In all cases, the blood emerging from the test system might be assessed by a partial thromboplastin time (PPT) test, a laser scattering assay of emboli produced, or a flow cytometric analysis of activated platelets. Thus, the methods for contacting blood with biomaterials and the methods for assessing blood change can often be considered separately.

In Vitro *Tests of BMIs*

In vitro BMI tests involve placing blood or plasma in a container composed of a test material, or recirculating blood through a flow chamber in which test materials contact blood under well-defined flow regimes that simulate physiologic flow conditions. Many flow geometries have been studied including tubes, parallel plates, packed beds, annular flows, rotating probes, and spinning disks. The older literature on these test methods has been reviewed (McIntyre *et al.*, 1985; Turitto and Baumgartner, 1987), and has yielded considerable insight into how proteins and platelets are transported to, and react with, artificial surfaces. Such studies provided a wealth of morphologic information at the cellular level regarding details of platelet–surface and platelet–platelet interactions (Sakariassen *et al.*, 1989). However, as discussed earlier, these tests are usually of short duration and are strongly influenced by the

TABLE 2 A Few Tests for Assessing Blood–Materials Interactions

Test name	Parameter(s) measured	Literature reference
Historical		
Vena cava ring test	Thrombus in a test ring in the vena cava	*Bull. N. Y. Acad. Med.* **48**(2): 482 (1972).
Renal embolus test	Thrombus in a test ring in the renal artery; embolization from the ring	*Bull. N.Y. Acad. Med.* **48**(2): 468 (1972).
Modified Lee White test	Activation of the intrinsic clotting system (whole blood clotting time)	*Ann. N. Y. Acad. Sci.* **146**: 11–20 (1968).
Baboon A-V shunt test	Platelet consumption, blood elements on shunt	*J. Clin. Invest.* **64**: 559 (1979).
Chandler loop	Blood cell deposition, platelet release	*Lab. Invest.* **7**: 110 (1958).
Currently used test systems		
Flow cells	Adherent blood elements, activation, cell release	*J. Mater. Sci. Mat. Med.* **8**: 119 (1997); *JBMR* **49**: 460 (2000).
Recirculating flow tubes	Adherent blood elements, activation, cell release, emboli	*J. Biomater. Sci. Polymer Ed.* **11**(11): 1147 (2000). *J. Biomed. Mater. Res.* **27**: 1181 (1993).
Cardiopulmonary bypass model	Thrombin–antithrombin III (TAT) complex, polymorphonuclear (PMN) elastase, complement C3a formation, Blood cell count analysis	*J. Biomater. Sci. Polymer Ed.* **11**(11): 1147 (2000).
Ventricular assist device model	(See Table 4)	*J. Biomater. Sci. Polymer Ed.* **11**(11): 1239 (2000).
Blood assessment methods		
Light-scattering embolus measurement	Embolization	*J. Biomed. Optics* **8**(1): 70–79 (2003).
Doppler ultrasound measurement	Embolization	*J. Vasc. Surg.* **25**: 179 (1997).
Flow Cytometry	Platelet P-selectin, Mac-1 receptor, platelet activation by annexin V, platelet microparticles	*J. Biomater. Sci. Polymer Ed.* **11**(11): 1197 (2000). *J. Biomater. Sci. Polymer Ed.* **11**(11): 1239 (2000).
Gamma Camera imaging	Thrombus accumulation, embolization	*Arteriosclerosis* **10**: 1037 (1990).
Activated Clotting Time measurement	Partial thromboplastin time (PTT) Recalcified activated clotting time	*Blood* **49**(2): 171 (1997).
Fibrinopeptide A measurement	Fibrin production	*J. Biomed. Mater. Res.* **31**(1): 145 (1996).
Complement activation assays	C3a, C5b-9	*J. Biomater. Sci. Polymer Ed.* **11**(11): 1239 (2000).
Light microscopic assessment	Cell attachment, activation, embolization	*J. Biomed. Mater. Res.* **49**: 460 (2000).

blood source, handling methods, and the use of anticoagulants. Thus, *in vitro* test results generally cannot be used to predict longer-term BMI and *in vivo* outcomes and can provide only the most general guidelines for the selection of materials for particular devices.

However, *in vitro* tests may be useful for screening materials that are highly reactive toward blood. Tests of the whole-blood clotting time and variations thereof involve placing non-anticoagulated whole blood (or blood anticoagulated with sodium citrate that is then recalcified) into containers of test material and measuring the time for a visible clot to form. Materials that quickly activate intrinsic coagulation and cause blood to clot within a few minutes (such as negatively charged glass) are probably unsuitable for use in devices with low shear blood flow, or in the absence of anticoagulants.

Recirculation of heparinized blood or citrated blood through tubular devices and materials may lead to platelet deposition onto highly thrombogenic materials with the appearance in plasma of proteins released from platelets (Kottke-Marchant *et al.*, 1985; Haycox and Ratner, 1993). Thus, these and similar methods may identify materials that might cause rapid platelet accumulation *in vivo* over short time periods, and therefore be unsuitable for certain applications such as small diameter vascular grafts or blood conduits. Both recirculation tests and *in vitro* clotting assays can be considered for preliminary screening and identification of materials that could be highly thrombogenic. Most artificial surfaces in common use would probably "pass" these tests. Since small differences in test results are likely to be meaningless for predicting material performance in actual use applications, these tests are not appropriate for optimizing or refining material properties. *In vivo* testing is therefore required.

Contemporary methods for evaluating damage to blood in *in vitro* test systems may expand the significance of these methods. For example, platelet-derived microparticles in blood have been used as a marker of blood damage induced by biomaterial surfaces (Gemmell, 2000).

In Vivo Tests of BMIs

Many studies have been performed in which test materials, in the form of rings, tubes, and patches, are inserted for short or long time periods into the arteries or veins of experimental animals (Salzman and Merrill, 1987; Williams, 1987; McIntire *et al.*, 1985). For the following reasons, most of these tests are of questionable relevance to the use of biomaterials in man: (1) The timing and type of measurements may be such that important blood responses are unrecognized. In particular, the measurement of gross thrombus formation at a single point in time may lead to incorrect conclusions about local thrombus formation (e.g., Fig. 1) and does not provide assessment of systemic effects of thrombosis such as embolization and blood element consumption. (2) With more commonly used animal species (e.g., dogs), blood responses may differ from those in humans both quantitatively and qualitatively. (3) The hemodynamics (blood flow conditions) of the model may not be controlled or measured. (4) There may be variable blood vessel trauma and tissue injury that can cause local thrombus

formation through the extrinsic pathway of blood coagulation (Chapter 4.6). Thus, *in vivo* testing of materials in idealized flow geometries (rather than actual device configurations) may provide few insights into selection of materials for use in man.

Evaluations of BMI may be performed in animals having arteriovenous (A-V) or arterioarterial (A-A) shunts, i.e., tubular blood conduits placed between an artery and vein, or between an artery and artery. A-V shunts have been studied in a variety of animals including baboons, dogs, pigs, and rabbits (McIntyre *et al.*, 1985). Qualitatively similar results have been obtained with shunts in dogs and baboons (Sefton *et al.*, 2000). An A-V shunt system is illustrated in Fig. 4. Once established, shunts may remain patent (not occluded) for long periods of time (months) without the use of anticoagulants. Test materials or devices are simply inserted as extension segments or between inlet and outlet portions of the chronic shunt. These systems have the advantages that (1) blood flow is easily controlled and measured, (2) native or anticoagulated blood can be employed, (3) the animal's physiology removes damaged blood elements and makes new blood with each circulation through the body, and (4) both short-term and long-term BMIs, including both local and systemic effects, can be evaluated. The downsides of these tests are complex surgery, the high expense, and ethical issues associated with working with larger animals.

As an example, consider the A-V shunt model in baboons—baboons are used because they are hematologically similar to man. The blood responses to tubular biomaterials and vascular grafts have been quantitatively compared with respect to (1) localized thrombus accumulation, (2) consumption of circulating platelets and fibrinogen, (3) plasma levels of factors released by platelets and coagulation proteins during thrombosis, and (4) embolization of microthrombi to downstream circulatory beds (Harker *et al.*, 1991). These studies in primates are consistent with observations in man that certain commonly used polymers [e.g., polytetrafluoroethylene,

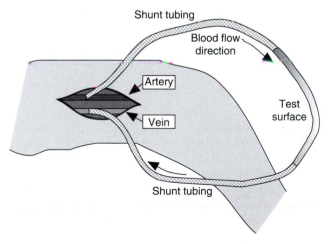

FIG. 4. Illustration of an arteriovenous (A-V) shunt placed between the femoral artery and vein (in the leg) of an experimental animal. Materials to be tested (in this case a tubular device) are interposed between inlet and outlet segments of the shunt.

TABLE 3 Blood–Materials Responses and Their Evaluation

Blood Components	Blood Response	Assessment[a]
Thrombus	Clot	Direct visual and histologic evaluation; noninvasive imaging (angiography, ultrasound, radioisotope, magnetic resonance); evidence of device dysfunction.
	Thromboembolism	Emboli detection (ultrasound, laser); evidence of organ/limb ischemia, stroke.
Platelets	Consumption	Increased removal of radioisotopically labeled cells; reduced blood platelet count.
	Dysfunction[b]	Reduced platelet aggregation in vitro; prolonged bleeding time.
	Activation	Increased plasma levels of platelet factor 4 and B-thromboglobulin; platelet membrane alterations (e.g., by flow cytometry).
Red cells[b]	Destruction	Decreased red cell count; increased plasma hemoglobin.
White cells[b]	Consumption/activation	Decreased counts of white cell populations; increased white cell plasma enzymes (e.g., neutrophil elastase).
Coagulation factors	Consumption[b]	Reduced plasma fibrinogen, factor V, factor VIII.
	Thrombin generation	Increased plasma levels of prothrombin fragment 1.2 and thrombin : antithrombin III complex.
	Fibrin formation	Increased plasma level of fibrinopeptide A.
	Dysfunction[b]	Prolonged plasma clotting times.
Fibrinolytic proteins	Consumption[b]	Reduced plasma plasminogen level.
	Plasmin generation	Increased plasma level of plasmin : antiplasmin complex.
	Fibrinolysis	Increased plasma level of fibrin D-dimer fragment.
Complement proteins[b]	Activation	Increased plasma levels of complement proteins C3a, C3b, C5a, C5b-9.

[a] Radioimmunoassays (RIA) and enzyme-linked immunoassays (ELISA) may not be available for detection of nonhuman proteins.
[b] Tests which may be particularly important with long-term and/or large surface area devices.

polyethylene, plasticized poly(vinyl chloride), silicone rubbers] and some vascular grafts (e.g., polytetrafluoroethylene) are relatively nonthrombogenic in extracorporeal circuits and arteries. Thus, results with shunt models, particularly in higher animal species, may predict BMI in humans when employed under comparable flow conditions (laminar unidirectional flow with arterial shear rates). Since extracorporeal shunts exclude modulating effects of blood vessel cells and tissue injury, results with these models may be less relevant to the behavior of devices that are placed surgically or whose responses may be mediated by interactions with the vessel wall as well as the blood (e.g., heart valves, grafts, indwelling catheters, and sensors).

In Vivo Evaluation of Devices

Since the blood response to devices is complex and not well predicted by testing of materials per se in idealized configurations, animal testing, and ultimately clinical testing, of functioning devices is required to establish safety and efficacy. Broad guidelines, based on the type of device being considered, are given next and apply to both animal and clinical testing. A summary of in vivo blood responses to devices, and of commonly used methods which have proven useful for evaluating those responses, is given in Table 3. Table 4, based upon one paper, lists the wide range of screens that have been used in one group over a decade to understand blood interaction with devices, specifically ventricular assist devices (Wagner et al., 2000).

TABLE 4 Parameters Measured in the Assessment of Blood Interactions with Ventricular Assist Devices (VADs)

Flow assessment
Laser Doppler anemometry, Fluorescent image tracking velocimetry

Coagulation
prothrombin fragment F1.2, thrombin–antithrombin (TAT)

Emboli
Transcranial Doppler ultrasound, Flow cytometric assays for the quantification of circulating platelet-containing microaggregates

Fibrinolysis
D-dimer

Platelet activation and deposition
platelet factor 4 (PF4), beta thromboglobulin (BTG), flow cytometric detection of p-selectin expression, platelet deposition

Complement
C3a, C5b-9

Leukocyte activation
flow cytometric detection of monocyte tissue factor expression, monocyte–platelet microaggregates, granulocyte–platelet microaggregates

[based upon Wagner, et al. (2000)]

Devices that have relatively small surface areas and are exposed for short periods of time (hours to days) include catheters, guidewires, sensors, and some components of extracorporeal circuits. With these devices the primary concern is the formation of significant thrombus that could interfere with device function (e.g., cause blood sensor malfunction), produce vessel occlusion (catheters), and either embolize spontaneously or be stripped from the device surface when it is removed from the body (e.g., during catheter withdrawal through a vessel insertion site), producing occlusion of distal vessels and tissue ischemia. Devices exposed for short periods that have large surface areas (dialyzers) and complex circuitries (pump-oxygenators) may, in addition, produce (1) a marked depletion of circulating blood cells and proteins (e.g., platelets and coagulation factors), (2) an immune/inflammatory response through activation of complement proteins and white cells, and (3) organ dysfunction mediated by hemodynamic, hematologic, and inflammatory reactions. Mechanical devices that are used for long periods of time (heart assist devices, extracorporeal membrane oxygenators) may produce profound systemic effects and organ dysfunction such that their use in man remains problematic. Other long-term implants (grafts, heart valves, stents) might be further improved by extending their period of safe operation and patency, and by reducing the frequency of embolic phenomena (e.g., stroke) and requirements for the concurrent use of antithrombotic drugs.

With both long-term and short-term device applications, thrombus formation can be assessed directly and indirectly. Important indirect assessments include depletion of circulating blood cells and proteins consumed in the process of thrombus formation, and the appearance in plasma of proteins generated in the process of thrombus formation (e.g., fibrinopeptide A, platelet factor 4). Direct assessment of blood flow rate, flow geometry, and extent of flow channel occlusion can in many cases be achieved using sophisticated methods including angiography, ultrasound imaging, and magnetic resonance imaging. Devices that are removed from the circulation should be visually inspected to assess whether thrombus has formed at particular sites or on certain materials. Emboli in flowing blood may be detected using ultrasound and laser-based techniques, although these methods are not used widely at present. Thrombus formation and rates of platelet destruction by both acutely placed and chronically implanted devices can be determined quantitatively by measurements of platelet lifespan and scintillation camera imaging of radioisotopically labeled blood elements (McIntire *et al.*, 1985; Hanson *et al.*, 1990).

In one study, a circulating loop system to assess materials under well controlled conditions was directly compared to results obtained on an actual device in flowing blood (Münch *et al.*, 2000). A Chandler loop was used as the well-defined test system. A cardiopulmonary bypass (CPB) device was used to observe blood reaction in a device. Both the recirculating loop system and the CPB device were surface treated with two types of heparin coatings. The parameters measured after blood flow were thrombin–antithrombin III (TAT) complex, platelet count, red blood cell count, white blood cell count, polymorphonuclear leukocyte elastase, and complement C3 activation. There were many observations reported including some tests that showed no significant difference between the two surfaces in the loop model and statistically significant

differences in the CPB model, and vice versa. The authors concluded:

> In the more complex and realistic simulated CPB model, experimental design and cost factors prevented easy/optimum manipulation of critical variables such as blood donor (use of paired samples) and heparin level. Testing in the simpler loop model, on the other hand, readily offered manipulation of these variables, and produced endings which overlapped with observations from the more complex CPB model. Thus, the models described here complimented.

Finally, it is important to emphasize that thrombosis occurs dynamically, such that thrombi continuously undergo processes of both formation and dissolution. Device failure represents the imbalance of these processes. Older thrombi may also be reorganized considerably by the enzymatic and lytic mechanisms of white cells. While the initial consequences of surgical device placement include tissue injury, thrombosis due to tissue injury, and foreign-body reactions, the flow surface of long-term implants may become covered with a stable lining of cells (e.g., ingrowth of vascular wall endothelial and smooth muscle cells onto and into vascular grafts) or blood-derived materials (e.g., compacted fibrin). Certain reactions of blood elements (e.g., platelets, thrombin) may also stimulate the healing response. Ultimately, long-term devices such as the small-caliber graft may fail because of excessive tissue ingrowth that could be largely unrelated to biomaterials properties, or may be stimulated by the biomaterials.

We summarize this section on device testing recognizing that many device applications described earlier, as well as laboratory and clinical methods for evaluating their biologic responses, will be unfamiliar to the bioengineer. However, it is important to appreciate that (1) each device may elicit a unique set of blood responses, both short-term and long-term, (2) methods are available to assess systemic changes in the blood and host organism that indirectly reflect thrombus formation, and (3) localized thrombus formation can usually be measured directly and quantitatively. Whenever possible, serial and dynamic studies should be performed to establish the time course of ongoing thrombus formation and dissolution. These measurements will ultimately predict device performance and allow for the rational selection of biomaterials that will minimize adverse blood-device interactions.

CONTEMPORARY CONCEPTS IN BMI EVALUATION

A few trends are noted in the recent literature that describe powerful methods to assess BMI. Two of these trends will be highlighted here: flow cytometry and emboli detection.

The use of flow cytometry to analyze activation of blood elements can be an effective method of measuring blood cell reaction and isolating specific pools of cells that have or have not undergone reaction. The basic technology behind flow cytometry (sometimes called fluorescence activated cell sorting, FACS) involves the laser separation of fluorescently labeled cells from a narrow flow stream of cells. The fluorescently labeled cells can be diverted from the flow stream to another flow channel where they are counted and harvested. One of the earliest studies using flow cytometry for blood compatibility studies showed that in contact with blood-activating synthetic materials, thrombotic membrane fragments called microparticles are released (Gemmell *et al.*, 1995). Since that important study, flow cytometry has been used to look at the up-regulation of platelet P-selectin (CD62), a consequence of α-granule release, monocyte and neutrophil CD11b (Mac-1 receptor) up-regulation (Gemmell, 2000), platelet activation by annexin V binding to the negative phospholipid found on activated platelets (Wagner *et al.*, 2000) and a number of other biospecific factors relating to platelet activation, white cell activation and platelet aggregate formation. The power of the flow cytometry method resides in its ability to pinpoint an event in a large pool of cells that can be identified by a fluorescent tag. The method is applicable to blood that has been contacted with synthetic materials *in vitro* or *in vivo*.

The ability to go beyond measurements of adhesive phenomena in blood (accumulating thrombus or platelets) and to expand studies to nonadhesive encounters leading to blood activation and damage represents an important growth area for BMI evaluation. Cytometric studies allow us to do systemic measurements rather than just local assessment. Methods to study embolus production in real time permit a key, clinically relevant parameter to be measured and quantified. Doppler ultrasound seems to have potential to do this, though the instrumentation is expensive. Other alternatives to study microemboli in blood include use of the Coulter counter on sampled blood and filtration pressure methods. However, a method that seems to have excellent potential to advance our understanding of how surfaces induce embolus formation is laser light scattering. Initial developments making this method suitable for embolus analysis in whole blood were published over 20 years ago (Reynolds and Simon, 1980) and then used for the assessment of BMI in conjunction with a baboon shunt model (Garfinkle *et al.*, 1984). New developments in light scattering instrumentation and data analysis should make laser microembolus detection a key tool for studying biomaterial reactions with blood (Solen *et al.*, 2003).

CONCLUSIONS

The most blood compatible material known is the natural, healthy, living lining of our blood vessels. This "material" functions well by a combination of appropriate surface chemistries, good blood flow characteristics, and active biochemical processes involving removal of prothrombotic substances and secretion of natural anticoagulants. It seems unlikely that we will ever match this performance in a synthetic material or device, although attempts to imitate aspects of the natural system represent a promising strategy for developing a new generation of blood-compatible devices (Chapters 2.14, 2.16, 3.4, and 7.18). At present, however, synthetic materials that perform less well than the vessel wall, but still satisfactorily, will be needed. This chapter provides only a brief outline of the issues involved in evaluating materials and devices to find those that are minimally damaging or activating towards blood. The subject of blood compatibility testing is complex, and advanced study of the subject is required before considering experiments

intended to elucidate basic mechanisms or improve human health. Further discussion elaborating upon the complexity of the issues involved in BMI testing can be found in Ratner, 1984, 1993b, 2000. Detailed discussion on the characterization of materials for biomaterials application and on BMI testing can also be found in a recent publication coordinated by the Device and Technology Branch of the National Heart, Lung and Blood Institute, NIH (Harker *et al.*, 1993).

Bibliography

Basmadjian, D., Sefton, M. V., and Baldwin, S. A. (1997). Coagulation on biomaterials in flowing blood: some theoretical considerations. *Biomaterials* **18**(23): 1511–1522.

Friedman, L. I., Liem, H., Grabowski, E. F., Leonard, E. F., and McCord, C. W. *et al.* (1970). Inconsequentiality of surface properties for initial platelet adhesion. *Trans. Am. Soc. Artif. Intern. Organs* **16**: 63–70.

Garfinkle, A. M., Hoffman, A. S., Ratner, B. D., Reynolds, L. O., and Hanson, S. R. (1984). Effects of a tetrafluoroethylene glow discharge on patency of small diameter Dacron vascular grafts. *Trans. Am. Soc. Artif. Int. Organs* **30**: 432–439.

Gemmell, C. H. (2000). Flow cytometric evaluation of material-induced platelet and complement activation. *J. Biomater. Sci. Polymer Ed.* **11**(11): 1197–1210.

Gemmell, C. H., Yeo, E. L., and Sefton, M. V. (1997). Flow cytometric analysis of material-induced platelet activation in a canine model: elevated microparticle levels and reduced platelet lifespan. *J. Biomed. Mater. Res.* **37**: 176–181.

Gemmell, C. H., Ramirez, S. M., Yeo, E. L., and Sefton, M. V. (1995). Platelet activation in whole blood by artificial surfaces: identification of platelet-derived microparticles and activated platelet binding to leukocytes as material induced activation events. *J. Lab. Clin. Med.* **125**: 276.

Grabowski, E. F., Herther, K. K., and Didisheim, P. (1976). Human versus dog platelet adhesion to cuprophane under controlled conditions of whole blood flow. *J. Lab. Clin. Med.* **88**: 368–373.

Hanson, S. R., and Harker, L. A. (1988). Interruption of acute platelet-dependent thrombosis by the synthetic antithrombin D-phenylalanyl-L-prolyl-L-arginyl chloromethylketone. *Proc. Natl. Acad. Sci. USA* **85**: 3184–3188.

Hanson, S. R., and Sakariassen, K. S. (1998). Blood flow and antithrombotic drug effects. *Am. Heart J.* **135**: S132–S145.

Hanson, S. R., Harker, L. A., Ratner, B. D., and Hoffman, A. S. (1980). In vivo evaluation of artificial surfaces using a nonhuman primate model of arterial thrombosis. *J. Lab. Clin. Med.* **95**: 289–304.

Hanson, S. R., Kotze, H. F., Pieters, H., and Heyns, A. du P. (1990). Analysis of 111-indium platelet kinetics and imaging in patients with aortic aneurysms and abdominal aortic grafts. *Arteriosclerosis* **10**: 1037–1044.

Harker, L. A., Kelly, A. B., and Hanson, S. R. (1991). Experimental arterial thrombosis in non-human primates. *Circulation* **83**(6): IV-41–IV-55.

Harker, L. A., Ratner, B. D., and Didisheim, P. (1993). Cardiovascular biomaterials and biocompatibility: a guide to the study of blood–material interaction. *Cardiovasc. Pathol.* **2** Suppl.(3).

Haycox, C. L., and Ratner, B. D. (1993). In vitro platelet interactions in whole human blood exposed to biomaterial surfaces: insights on blood compatibility. *J. Biomed. Mater. Res.* **27**: 1181–1193.

Haycox, C. L., Ratner, B. D., and Horbett, T. A. (1991). Photo-enhancement of platelet adhesion to biomaterial surfaces observed with epi-fluorescent video microscopy (EVM). *J. Biomed. Mater. Res.* **25**: 1317–1320.

Horbett, T. A., Cheng, C. M., Ratner, B. D., Hoffman, A. S., Hanson, S. R. (1986). The kinetics of baboon fibrinogen absorption to polymers; in vitro and in vivo studies. *J. Biomed. Mater. Res.* **20**: 739–772.

Kottke-Marchant, K., Anderson, J. M., Rabinowitch, A., Huskey, R. A., and Herzig, R. (1985). The effect of heparin vs. citrate on the interaction of platelets with vascular graft materials. *Thromb. Haemost.* **54**: 842–849.

Leonard, E. F. (1987). Rheology of thrombosis. in: *Hemostasis and Thrombosis,* 2nd ed., R. W. Colman, J. Hirsh, V. J. Marder, and E. W. Salzman (eds.). J. B. Lippincott, Philadelphia, pp. 1111–1122.

McIntire, L. V., Addonizio, V. P., Coleman, D. L., Eskin, S. G., Harker, L. A., Kardos, J. L., Ratner, B. D., Schoen, F. J., Sefton, M. V., and Pitlick, F. A. (1985). Guidelines for Blood–Material Interactions—Devices and Technology Branch, Division of Heart and Vascular Diseases, National Heart, Lung and Blood Institute. NIH Publication No. 85-2185, revised July 1985. U. S. Department of Health and Human Services, Washington, DC.

Münch, K., Wolf, M. F., Gruffaz, P., Ottenwaelter, C., Bergan, M., Schroeder, P., and Fogt, E. J. (2000). Use of simple and complex *in vitro* models for multiparameter characterization of human blood–material/device interactions. *J. Biomater. Sci. Polymer Ed.* **11** (11): 1147–1163.

Ratner, B. D. (1984). Evaluation of the blood compatibility of synthetic polymers: consensus and significance. in *Contemporary Biomaterials: Materials and Host Response, Clinical Applications, New Technology and Legal Aspects,* J. W. Boretos and M. Eden, eds. Noyes Publications, Park Ridge, N. J. pp. 193–204.

Ratner, B. D. (1993a). Characterization of biomaterial surfaces. *Cardiovasc. Pathol.* **2** Suppl.(3): 87S–100S.

Ratner, B. D. (1993b). The blood compatibility catastrophe. *J. Biomed. Mater. Res.* **27**: 283–287.

Ratner, B. D. (2000). Blood compatibility—a perspective. *J. Biomater. Sci. Polymer Ed.* **11** (11): 1107–1119.

Reynolds, L. O., and Simon, T. L. (1980). Size distribution measurements of microaggregates in stored blood. *Transfusion* **20**(6): 669–678.

Sakariassen, K. S., Muggli, R., and Baumgartner, H. R. (1989). Measurements of platelet interaction with components of the vessel wall in flowing blood. *Methods Enzymol.* **169**: 37–70.

Salzman, E. W., and Merrill, E. D. (1987). Interaction of blood with artificial surfaces. in: *Hemostasis and Thrombosis,* R. W. Colman, J. Hirsh, V. J. Marder, and E. W. Salzman, eds. 2nd ed., J. B. Lippincott, Philadelphia, pp. 1335–1347.

Schaub, R. D., Kameneva, M. V., Borovetz, H. S., and Wagner, W. R. (2000). Assessing acute platelet adhesion on opaque metallic and polymeric biomaterials with fiber optic microscopy. *J. Biomed. Mater. Res.* **49**: 460–468.

Sefton, M. V., Gemmell, C. H., and Gorbett, M. B. (2000). What really is blood compatibility? *J. Biomater. Sci. Polymer Ed.* **11** (11): 1165–1182.

Sefton, M. V., Sawyer, A., Gorbet, M., Black, J. P., Cheng, E., Gemmell, C., and Pottinger-Cooper, E. (2001). Does surface chemistry affect thrombogenicity of surface modified polymers? *J. Biomed. Mater. Res.* **55**(4): 447–459.

Solen, K., Sukavaneshvar, S., Zheng, Y., Hanrahan, B., Hall, M., Goodman, P., Goodman, B., and Mohammad, F. (2003). Light-scattering instrument to detect thromboemboli in blood. *J. Biomed. Optics* **8**(1): 70–79.

Turitto, V. T. and Baumgartner, H. R. (1987). Platelet–surface interactions. in: *Hemostasis and Thrombosis,* 2nd ed., R. W. Colman, J. Hirsh, V. J. Marder, and E. W. Salzman (eds.). J. B. Lippincott, Philadelphia, pp. 555–571.

Wagner, W. R., and Hubbell, J. A. (1990). Local thrombin synthesis and fibrin formation in an in vitro thrombosis model result in platelet recruitment and thrombus stabilization on collagen in heparinized blood. *J. Lab. Clin. Med.* **116**: 636–650.

Wagner, W. R., Schaub, R. D., Sorensen, E. N., Snyder, T. A., Wilhelm, C. R., Winowich, S., Borovetz, H. S., and Kormos, R. L. (2000). Blood biocompatibility analysis in the setting of ventricular assist devices. *J. Biomater. Sci. Polymer Ed.* **11** (11): 1239–1259.

Williams, D. ed. (1987). *Blood Compatibility.* CRC Press, Boca Raton, FL.

5.5 Large Animal Models in Cardiac and Vascular Biomaterials Research and Testing

Richard W. Bianco, John F. Grehan, Brian C. Grubbs, John P. Mrachek, Erik L. Schroeder, Clark W. Schumacher, Charles A. Svendsen, and Matt Lahti

INTRODUCTION

The purpose of using animal models in biomaterials research is to examine biomechanical systems as they exist in one species (the model) and apply that knowledge to human physiologic systems. Such an approach has long been employed and has resulted in improved health and increased longevity for humankind. The importance of this approach has been realized by many great investigators including Charles Darwin, who poignantly describes this link and its importance in a letter to a Swedish professor of physiology in 1881: "I know that physiology cannot possibly progress except by means of experiments on living animals, and I feel the deepest conviction that he who retards the progress of physiology commits a crime against mankind."

The development of new technologies, including *in vitro* methods and computer modeling, that has occurred in the last few decades has enhanced the approach to the study of physiology. In spite of all this progress, however, the use of *in vivo* models still remains a necessity in investigations into certain physiologic phenomena, especially those related to the pathophysiologic mechanisms of disease. Traditional mammalian models in biomedical research, such as rats, mice, guinea pigs, hamsters, and rabbits, have provided the scientific community with voluminous amounts of information about human disease and basic biological processes. For the evaluation of cardiac and vascular biomaterials, however, investigations using larger domestic animals have proven to be more valuable because these models portray anatomical and physiological characteristics that more closely resemble those observed in humans. This similarity allows creation of models that are capable of providing more accurate predictions of a bioprosthetic device's or a biomaterial's future clinical performance.

This chapter will focus on three commonly used animal models for *in vivo* biomaterials research and testing in cardiac and vascular surgery: dogs, pigs, and sheep. The chapter will begin with a short discussion on responsible use of animals in biomaterials research and testing, followed by specific considerations and examples of existing animal models for each species. Finally, current standards and testing requirements for *in vivo* preclinical evaluation of cardiac and vascular devices, as well as proposed future requirements, will be discussed.

RESPONSIBLE USE OF ANIMALS

Introduction

There are two basic principles governing the use of animals in research, education, and testing: (1) scientific reliance on the use of live animals should be minimized; (2) pain, distress, and other harm to laboratory animals should be reduced to the minimum necessary to obtain valid scientific data. Strict adherence to these basic ethical principles will not only enhance the quality of each *in vivo* preclinical evaluation but will also ensure that current and future generations of scientists will be able to employ all valid tools available to predict clinical safety of new or modified medical devices and other technology designed to improve human health.

Investigator and Institutional Responsibilities

When considering animal models for biomaterials research the investigator and sponsoring institution should first assess their responsibilities with respect to federal, state, and local laws and regulations. Additionally, given the growing public and political sensitivity to the use of research animals, all laboratories using animals must ensure that their programs recognize and comply with current definitions of humane animal care and use. The importance of this issue is reflected in a resolution adopted by the American Association for the Advancement of Science, February 19, 1990. This section briefly presents legal obligations and provides practical guidelines for responsible use of laboratory animals.

The Animal Welfare Act (Public Law 89-544, as amended) provides federal regulations that cover laboratory animal care and use. The regulations [Title 9 of the Code of Federal Regulations (CFR), Chapter 1, subchapter A—Animal Welfare, Parts 1, 2, 3] are enforced by the Animal and Plant Health Inspection Service (APHIS), U.S. Department of Agriculture. The Act covers dogs, cats, nonhuman primates, guinea pigs, hamsters, rabbits, and "any other warm-blooded animal, which is used or is intended for use in research, teaching, testing, experimentation . . . " (9 CFR subchapter A, Part 1, Section 1.1). Species specifically exempt from the Act include birds, rats, and mice bred for use in research, and horses and livestock species used in agricultural research. Institutions using those species covered by the Act must be registered by APHIS. Continued registration is dependent on submission of annual reports to APHIS by the institution, as well as satisfactory inspection of the institution's animal facility during unannounced site visits by APHIS inspectors. The Act provides specifications for animal procurement (i.e., from licensed suppliers), husbandry, and veterinary care that are used to determine compliance with the Act. An annual report must be supplied to APHIS and should contain a list of all species and the numbers of animals used by the institution in the previous year. The animals listed in the annual report must be categorized by the level of discomfort or pain they were thought to experience in the course of the research, education, or testing process they were used for.

In 1985 amendments were made to the Act (7 U.S.C. 2131, et seq.) and were implemented on October 30, 1989, August 15, 1990, and March 18, 1991. These amendments extended the Act to cover the institution's administrative review and control of its animal research program (Federal Register, Vol. 54, No. 168, pp. 36112–36163; Vol. 55, No. 36, pp. 28879–28881; Vol. 56, No. 32, pp. 6426–6505). Specifically, the new regulation requires that all animal research protocols be reviewed and approved by an Institutional Animal Care and Use Committee (IACUC) before they are initiated. Furthermore, the submitted protocols must state in writing that less harmful alternatives were considered but are not available, and that the proposed research is not unnecessarily duplicative. Additional requirements increase the scope of husbandry requirements for laboratory dogs and nonhuman primates.

The other relevant federal body is the National Institutes of Health (NIH). The Health Research Extension Act of 1985 (Public Law 99-158) required the director of NIH to establish guidelines for the proper care of laboratory animals and IACUC oversight of that care. Broad policy is described in the Public Health Service Policy on Humane Care and Use of Laboratory Animals (NIH, 1986), which identifies the Guide for the Care and Use of Laboratory Animals (NIH, 1985a) as the reference document for compliance. This policy applies to all activities involving animals either conducted by or supported by the U.S. Public Health Service (PHS). Before an institution can receive animal research funding from NIH or any other PHS agency, that institution must file a statement of assurance with the PHS Office for Protection from Research Risks (OPRR) that it is complying with these guidelines.

As required by the Animal Welfare Act, the PHS policy mandates that an annual update on animal research use and IACUC review of animal protocols and facilities be submitted. Unlike the Animal Welfare Act, however, the PHS policy covers all vertebrate species, does not include an enforcement arm for routine inspections (but does provide for inspection in cases of alleged misconduct), and penalizes noncompliant institutions by withdrawing funding support.

State and local laws and regulations may also affect animal research programs. These mainly involve restrictions on the availability of cats and dogs from municipal shelters. Within the past decade several states have also enacted registration and inspection statutes similar to the Animal Welfare Act. In several states, court rulings have required IACUC reports and deliberations at state-supported institutions to be conducted in public (e.g., Florida, Massachusetts, North Carolina, Washington).

Housing and Handling

The physical and psychological well-being of laboratory animals is determined in large part by their environment. Thus, careful attention must be paid to both the housing and handling of any investigational animal as improper handling or care may result in undue stress. The Institute of Laboratory

Animal Resources of the National Academy of Sciences publishes a detailed guide with specific recommendations for the housing of many laboratory animal species (Guide for the Care and Use of Laboratory Animals, 1996). Adequate assessment of available facilities and their compliance with these regulations must be considered conducting any animal investigation. This assessment should be done not only for regulatory affairs compliance, but also for scientific accuracy as data obtained from subjects that are not cared for properly may be inaccurate.

Euthanasia

In addition to housing and handling issues, investigators must become familiar with the issues pertaining to the appropriate practice of euthanasia. Study protocol completion or undue suffering of an individual study subject are both inviolable indications for euthanasia. Cachexia, anorexia, and a moribund state are also indications for euthanasia. In this phase of the investigation, as in all other phases of an animal investigation, great care should be taken to avoid any undue stress to the animal. To facilitate this, euthanasia techniques used should result in rapid unconsciousness followed by cardiac and respiratory arrest. The 1993 Report of the American Veterinary Medical Association (AVMA, 1993) provides multiple acceptable techniques for euthanasia and is an invaluable resource for investigators in the process of designing a study.

ANIMAL MODELS AND SPECIES CONSIDERATION

Canine

Introduction

The history of the use of canines in experimental surgery is extensive. Early experimental surgery required an animal model that was inexpensive, readily available, and easily managed. The solution to this problem came in the form of mongrel dogs. These animals could be obtained from pounds inexpensively and were often familiar with humans. Today, the emotional content of using canines for research has prohibitively increased the cost of these animals, and society no longer deems them appropriate for chronic studies. Although some acute studies may involve the use of pound animals, the majority of biomaterials research and testing involving canines requires the additional expense of using purpose-bred animals and has forced many researchers to look for alternative animal models. Despite this controversy, the canine model has been and still is considered the gold standard for many investigations.

Comparative Anatomy and Physiology

The canine cardiovascular physiology is similar to that of humans (Table 1). This homology, along with ease of vascular access and low body fat, make canines an ideal model for cardiac and peripheral vascular studies. A thoracotomy performed through the left fourth intercostal space will generally provide adequate access to the canine cardiovascular and pulmonary systems. There are only three significant anatomic variants to be aware of in canine surgery. First, the long thoracic nerve must be preserved because it innervates the serratus ventralis muscle, which is extremely important for ambulation in quadrupeds. As its course takes it along the dorsal aspect of the more cranial ribs, it can be easily transected while making the thoracotomy incision. The second important feature of canine anatomy is the drainage pattern of the right azygous vein. This vein empties into the superior vena cava (SVC) just proximal to where the SVC enters the right atrium. Care must be taken when encircling the SVC to avoid disruption of the azygous drainage. Last, the canine coronary vasculature is a highly collateralized system, which makes the canine model less desirable for regional ischemia/reperfusion investigations.

Canines show remarkable hematologic and metabolic similarity to humans as well (Table 2). Notable differences include a higher hemoglobin (range 14–19 g/dl) and white blood cell (WBC) numbers (range 6–16 thousand per cubic millimeter). The WBC differential and platelet counts are nearly identical to those in humans. Serum chemistries and liver functions also show remarkable similarities with the exception of serum amylase levels, which can be up to five times those in humans. Blood typing within the canine model is of arguable importance. Numerous canine blood groups have been described but only eight dog erythrocyte antigen (DEA) systems are currently recognized by international standards (Hale, 1995). Universal donor animals can be used if necessary; however, transfusion of typed and cross-matched blood is recommended as it reduces the risk of a transfusion reaction.

Perioperative Care

Anesthesia Dogs should be fasted for 12 hours prior to their operation to reduce the risk of aspiration during anesthetic induction and endotracheal intubation. Preparation of the animal involves administration of a mild intramuscular (IM) sedative such as acepromazine, a phenothiazine tranquillizer. Peripheral intravenous (IV) accesses can then be established. Prior to intubation, 0.05 mg/kg of atropine, an anticholinergic, may be administered to reduce the amount of oral secretions and ease endotracheal (ET) intubation. Once a secure airway has been obtained, anesthesia can be fully induced with a barbiturate such as thiopental sodium (12.5 mg/kg IV). Table 3 lists several of the recommended drugs and doses for sedation and anesthesia in dogs.

General anesthesia is maintained intraoperatively with an inhalation anesthetic such as isoflurane, with close monitoring of the physiological states of relaxation, oxygenation, and circulation. Periodic arterial blood gas analysis allows the adjustment of mechanical ventilation. Noninvasive blood pressure analysis may be sufficient for shorter peripheral vascular cases; however, invasive arterial blood pressure analysis should be instituted for all cases involving cardiopulmonary bypass. Peripheral arteries, such as the femoral or carotid arteries, or a more central artery such as the internal mammary artery, are equally sufficient for the placement of pressure monitoring cannulae. Core temperature analysis and continuous EKG monitoring are also essential elements of any operative procedure.

TABLE 1   Hemodynamic Values of Species Used to Evaluate Cardiac and Vascular Bioprostheses

Hemodynamic values	Dog	Pig	Sheep	Human
Heart rate (beats/min)	98 ± 17.9	105 ± 10	95 ± 24.2	70 ± 14
Mean arterial pressure (mm Hg)	96 ± 6.7	102 ± 9	70 ± 24	95 ± 11
Systolic pressure (mm Hg)	126 ± 23	127 ± 8	85 ± 3	126 ± 14
Diastolic pressure (mm Hg)	90 ± 20	86 ± 7	58 ± 15	79 ± 10
Stroke volume (ml/beat/kg)	2.0 ± 0.4	1.34 ± 0.26	1.03 ± 0.78	1.14 ± 0.31
Cardiac index (ml/min/kg)	88 ± 19.2	99 ± 19.9	115 ± 30.8	93 ± 20
Systemic vascular resistance (dyn sec cm^{-5})	3272 ± 66	2759 ± 70	1463 ± 183	1200 ± 600
Pulmonary vascular resistance (dyn s cm^{-5})	179 ± 57	441 ± 62	180 ± 53	120 ± 30
Resting total O_2 consumption (O_2/min/kg)	19.4 ± 3.7	6.6 ± 1.27	6.7 ± 3.0	3.9 ± 1.9

Values represent mean ± standard deviation.

Table created from: McKenzie JE (1996). Swine as a Model in Cardiovascular Research. *Advances in Swine in Biomedical Research*. Plenum Press, New York, pp. 7–17; Gross DR (1994). *Animal Models in Cardiovascular Research*, 2nd ed. Kluwer Academic Publishers. Dordrecht, The Netherlands; Kaneko JJ, Harvey JW, Bruss ML (1997). *Clinical Biochemistry of Domestic Animals*, 5th ed. Academic Press, San Diego; Fauci AS, Braunwald E, Isselbacher KJ, Wilson JD, Martin JB, Kasper DI, Hauser SL, Longo DL (1998). *Harrison's Principles of Internal Medicine*. McGraw-Hill, New York.

TABLE 2   Metabolic and Hematologic Values of Species Used to Evaluate Cardiac and Vascular Biomaterials

Metabolic/Hematologic Values	Dog	Pig	Sheep	Human
Arterial pH	7.37 ± 0.05	7.48 ± 0.03	7.45 ± 0.06	7.41 ± 0.03
Arterial pCO$_2$ (mm Hg)	38 ± 5.5	40 ± 2.3	38 ± 8.5	40 ± 5
Arterial pO$_2$ (mm Hg)	90 ± 5.0	82 ± 4.2	88 ± 7.5	90 ± 10.0
Arterial plasma HCO$_3^-$ (mEq/L)	22 ± 7.4	29 ± 2.2	26.6 ± 5.5	25.5 ± 4.5
Sodium (mEq/L)	146 ± 7.0	138 ± 3.5	152 ± 12.0	140.5 ± 4.5
Potassium (mEq/L)	4.4 ± 0.8	4.4 ± 0.4	5.5 ± 1.3	4.3 ± 0.8
Chloride (mEq/L)	112 ± 9.0	106 ± 7.8	117 ± 4.0	102 ± 4.0
Bicarbonate (mEq/L)	22 ± 7.4	29 ± 2.2	27 ± 6.0	26 ± 5.0
Urea (mM/L)	2.5 ± 0.8	3.2 ± 1.2	5.0 ± 2.1	5.4 ± 1.8
Creatinine (mM/L)	88.4 ± 44.2	89 ± 19.5	137.0 ± 31.0	<133
Calcium (mg/dl)	10.15 ± 1.15	4.80 ± 0.29	12.15 ± 0.65	5.05 ± 0.55
Magnesium (mg/dl)	2.2 ± 0.7	1.4 ± 0.2	2.1 ± 0.3	2.0 ± 0.6
Phosphate (mg/dl)	4.4 ± 1.8	4.0 ± 0.6	6.2 ± 1.2	3.8 ± 0.8
Albumin (g/dl)	3.1 ± 1.0	7.6 ± 0.8	3.6 ± 0.9	4.5 ± 1.0
Globulin (mEq/L)	7.3 ± 2.4	6.1 ± 0.9	7.4 ± 1.7	6.8 ± 1.3
Glucose (mM/L)	5.1 ± 1.5	4.6 ± 0.7	3.6 ± 0.8	5.3 ± 1.1
Lactate (mM/L)	0.8 ± 0.6	1.0 ± 0.3	1.2 ± 0.2	1.2 ± 0.6
WBC (cells $\times$ 10^3/ml)	11.5 ± 5.5	14.8	8 ± 4	7.4 ± 3.4
Hemoglobin (g/dl)	15 ± 3	12	12 ± 3	15 ± 3
Hematocrit (%)	45 ± 10	41	36 ± 9	44.5 ± 7.5
Platelets (Platelets $\times$ 10^9/L)	350 ± 150	350 ± 150	550 ± 250	265 ± 135

Values represent mean ± standard deviation.

Table created from: Hannon JP, Bossone CA, Wade CE (1990). Normal Physiological Values for Conscious Pigs Used in Biomedical Research. *Laboratory Animal Science*. 40(3):293–298; McKenzie JE (1996). Swine as a Model in Cardiovascular Research. *Advances in Swine in Biomedical Research*. Plenum Press, New York, pp. 7–17; Fauci AS, Braunwald E, Isselbacher KJ, Wilson JD, Martin JB, Kasper Dl, Hauser SL, Longo DL (1998). *Harrison's Principles of Internal Medicine*. McGraw-Hill, New York; Kaneko JJ, Harvey JW, Bruss ML (1997). *Clinical Biochemistry of Domestic Animals*, 5th ed. Academic Press, San Diego; Gross DR (1994). *Animal Models in Cardiovascular Research*, 2nd ed. Kluwer Academic Publishers. Dordrecht, The Netherlands.

Analgesia   Postoperatively, assessment of pain and discomfort may be based on a number of criteria. Canines will often display an abnormal posture, due to muscle flaccidity or rigidity when experiencing discomfort. Additionally, when approached, they may exhibit markedly different responses to their caretaker than observed preoperatively such as abnormal aggressiveness, withdrawal to touch, or increased vocalizations. As with all pain management, less medication is required if it is administered in a scheduled manner. This laboratory has had good results using a narcotic agonist/antagonist combination (e.g., buprenorphine) delivered through either a subcutaneous injection or intravenously.

TABLE 3　Anesthetics and Sedatives Acceptable for Use in Species Used to Evaluate Cardiac and Vascular Biomaterials

Drug name	Dose/kg			Route of delivery				
	Dog	Pig	Sheep	PO	SQ	IM	IP	IV
Preanesthetic anticholinergics								
Atropine	0.05 mg	0.05–0.5 mg	0.05 mg		X	X		X
Glycopyrrolate	0.011 mg	0.011 mg				X		
Phenothiazine/buterophenone sedatives								
Acepromazine or chlorpromazine	0.05–0.2 mg	0.05–0.2 mg	0.05–0.2 mg		X	X		X
	0.5–2 mg	0.5–2 mg		X				
Azaperone (*Stresnil*)		2.2 mg				X		
Benzodiazapene sedatives/anxiolytics								
Diazepam (*Valium*) — Class IV Controlled	0.25–0.5 mg	0.25–0.5 mg	0.25–0.5 mg					X
	2.5–5 mg	2.5–5 mg	2.5–5 mg			X	X	
Midazolam (*Versed*) — Class IV Controlled		0.1–1 mg	0.5 mg			X		X
Zolazepam (see *Telazol*)								
Thiazine sedatives								
Xylazine	0.5–1 mg	2–4 mg	0.05–0.2 mg			X		X
	0.5–1 mg				X			
Medetomidine		0.25 mg				X		
Narcotics/opiates								
Innovar-Vet (Fentanyl/droperidol) — Class II	0.13 ml	0.1 ml				X		
Butorphanol (*Torbugesic*)	0.05–0.5 mg	0.05–0.5 mg	0.01–0.05 mg		X	X		X
	0.05–0.5 mg	0.05–0.5 mg					X	
Morphine — Class II Controlled	0.25–2 mg	0.5–1 mg			X	X		
Oxymorphone — Class II Controlled	0.1–0.2 mg	0.15 mg			X	X		X
Barbituate anesthetics								
Phentobarbitol — Calss II Controlled	25–30 mg	25–30 mg	15–30 mg					X
Thiamylal/thiopental — Class III Controlled	10–20 mg	6–10 mg	2–5 mg					X
Dissociative anesthetics/cataleptics								
Ketamine	10–30 mg	10–30 mg	10–30 mg			X		X
Telazol (Tiletamine/Zolazepam) — Class III	5–10 mg	5–10 mg	2.2 mg			X		X
Other anesthetics/hypnotics								
Propofol	7.5–15 mg							X
Alpha chloralose — Class IV Controlled	50–100 mg	55–86 mg	35–62 mg					X

* PO, oral; SQ, subcutaneous; IM, intramuscular; IP, intraperitoneal; IV, intraveneous.
Table adapted from the University of Minnesota Animal Care and Use Manual 1997 reprint.

Table 4 lists several commonly used analgesics and their doses for use in dogs.

Existing Models

Cardiac Devices　Availability, familiarity, ease of housing, and anatomical similarities made canines the natural choice for early cardiac studies. Cardiopulmonary bypass, coronary artery bypass, and early valve studies were all developed primarily with canine models. An experimental protocol for coronary artery bypass grafting in canines was first documented in 1968 (Green *et al.*, 1968). Wakabayashi (Wakabayashi and Connolly, 1968) developed the technique in 1967 in an animal model and the Cleveland Clinic Group (Favalaro, 1969) independently developed and performed the technique in humans at approximately the same time. The concurrent development and improvement of cardiopulmonary bypass machines along with a safe technique for aortic cross-clamping (Wakabayashi *et al.*, 1971) made surgical revascularization possible. Further refinements such as internal mammary artery

use and radial artery grafts were also developed using the canine model (Rossiter *et al.*, 1974).

The canine model is suitable for testing both synthetic and biosynthetic vascular grafts used for coronary artery bypass, and still represents the gold standard for animal models of coronary artery bypass. This laboratory and others (Tomizawa *et al.*, 1994) have used the canine model for biosynthetic graft evaluations. In this model, a coronary artery bypass is performed using aortic cross-clamping and standard cardiopulmonary bypass with cooling. The left anterior descending artery is dissected free, ligated proximally, and looped just distal to the circumflex bifurcation. The distal anastomosis is sewn just past the origin of the first diagonal coronary artery. The aortic cross-clamp is then removed to reestablish distal blood flow. The proximal anastomosis is sewn into the aorta, with the aid of a partially occluding clamp for hemostasis. Important technical considerations for this model include perioperative antibiotics and anticoagulation with aspirin (325 mg/day) postoperatively.

The canine model offers several advantages for these studies. There is a large historical database for comparison, canines

TABLE 4 Analgesics Commonly Used when Testing Biomaterials in Large Animal Models

Drug name	Dose			Frequency[a]	Route of delivery[b]			
	Dog	Pig	Sheep		PO	SQ	IM	IV
Nonsteroidal antiinflammatory drugs								
Acetaminophen	15 mg/kg			TID	X			
Aspirin	25–50 mg/kg	25–50 mg/kg		BID	X			
Flunixin meglumine (*Banamine*)	1.1 mg/kg	1.1 mg/kg	1.1 mg/kg	QD			X	X
Ketoprofen	1–2 mg/kg	1–2 mg/kg	1–2 mg/kg	TID	X			X
Ketorolac (*Toradol*)	0.3–0.7 mg/kg	0.3–0.7 mg/kg	0.3–0.7 mg/kg	TID	X	X	X	X
Narcotic agonist/antagonists								
Buprenorphine (*Buprenex*)	0.005–0.02 mg/kg	0.1 mg/kg	0.005–0.02 mg/kg	BID-TID			X	
Butorphanol (*Torbugesic*)	0.2–0.4 mg/kg	0.2–0.4 mg/kg	0.2–0.4 mg/kg	q4h			X	
Nalbuphine (*Nubain*)	0.75–3 mg/kg	0.75–3 mg/kg	0.75–3 mg/kg	q4h			X	
Naloxone	0.04 mg/kg	0.04 mg/kg	0.04 mg/kg	q4h			X	
Narcotic agonist								
Morphine (15 mg/ml)	0.25–2 mg/kg	0.25–2 mg/kg	0.25–2 mg/kg	QID			X	X
Topical/regional anesthetics								
Lidocaine (*xylocaine*)	These are applied topically to painfull wounds			TID				

[a] QD, once daily; BID, twice daily; TID, three times daily; QID, four times daily; q4h, every 4 hours.
[b] PO, oral; SQ, subcutaneous; IM, intramuscular; IV, intravenous.
Table adapted from the University of Minnesota Animal Care and Use Manual 1997 reprint.

tolerate bypass well, and the availability of autologous saphenous vein for grafting allows researchers to perform positive controls. Additionally, canines have multiple sites for vascular access, which facilitates repeated angiographic evaluation of the conduit being studied. Although dogs tend to be more hypercoagulable than humans, patency in a canine model portends long-term patency in humans. Perhaps the most important aspect of this model is that it evaluates the new device/biomaterial in its intended clinical position.

Coronary artery stent placement is one of the newer techniques available to interventional cardiologists. Because of the recent shift away from the canine as a model for chronic device evaluation, only a few of the early studies (Schatz *et al.*, 1987; Roubin *et al.*, 1990) were performed in a canine model; the current standard uses a swine model (see later discussion). Nevertheless, a few recent reports have been published in which a canine model is used to study stent thrombogenesis (Santos *et al.*, 1998; Ueda *et al.*, 1999).

The evaluation of artificial heart valves began in the 1950s, using a canine model and valves created from materials such as polyurethane (Akutsu *et al.*, 1959; Doumanian and Ellis, 1961). Because early studies with artificial valves were hampered by short survival times secondary to thrombosis, anticoagulation was introduced in 1961 (Doumanian and Ellis, 1961) and led to a significant increase in survival. This advancement was followed by the rapid introduction of prosthetic cardiac valves into clinical use (Starr and Edwards, 1961). It was noted that although anticoagulation was used, humans tended to be less thrombogenic than dogs, a fact that would hinder more extensive use of the canine model for years to come. A relationship was also found to exist between bacteremia and postoperative thrombus formation. The combination of preoperative antibiotics, short bypass times, postoperative anticoagulation, and aseptic blood sampling techniques was eventually found to have a significant impact on survival of canines with artificial valves (Bianco *et al.*, 1986). Table 5 lists various antibiotics used in cardiovascular animal research models and their recommended dosing regimens.

The use of preoperative antibiotics, short cardiopulmonary bypass runs, and postoperative anticoagulation also validated the canine model for chronic studies involving artificial valves. At the time, a canine model represented a cost-effective alternative to other more expensive large animal models. Later studies, however, questioned the accuracy of biosynthetic valve calcification data derived from the canine model (Chanda *et al.*, 1997). Indeed, variation in the rate of calcification was recognized and eventually the FDA endorsed the juvenile sheep model for chronic bioprosthetic valve studies. The juvenile sheep model remains the model of choice for chronic evaluation of bioprosthetic valve; however, the canine remains an excellent model for chronic or acute hemodynamic testing.

Vascular Devices The canine model also has a long history in the research and testing of peripheral vascular biomaterials and devices. Early studies utilized large vessels such as the abdominal or thoracic aorta. However, expansion into other vascular beds began in the late 1960s and early 1970s with the use of the carotid artery to test small vascular prosthetics (Sharp, 1970). Ease of access to the carotid artery, along with its high flow rate and high pressure, led to investigations with multiple graft materials. In these investigations it was quickly discovered that graft diameter had a significant impact on thrombogenesis. Polytetrafluoroethylene (PTFE) grafts (Kelly and Eisemen, 1982) were developed with the help of a canine model, and again, it was recognized that this model represents a more thrombogenic environment than the human.

TABLE 5 Antibiotics Acceptable for Use in Large Animal Models

Drug name	Dose/kg Dog	Dose/kg Pig	Dose/kg Sheep	Frequency[a]	Route of delivery[b] PO	SQ	IM	IV
Pencillins								
Amoxicillin	11–22 mg	11–22 mg	11–22 mg	BID-TID	X		X	
Amoxicillin/clavulanate	14 mg	14 mg	14 mg	BID-TID	X			
Ampicillin	11–22 mg	11–22 mg	11–22 mg	BID-TID	X		X	X
Penicillin benzathine	44,000 IU	44,000 IU	44,000 IU	q48h		X	X	
Penicillin procaine	44,000 IU	44,000 IU	44,000 IU	QD-BID		X	X	
Ticarcillin	50–300 mg	50–300 mg	50–300 mg	TID-QID				X
Cephalosporins								
Cephadroxil	22 mg	22 mg	22 mg	BID	X			
Ceftiofur	3–5 mg	3–5 mg	3–5 mg	QD		X	X	X
Cephalexin	10–30 mg	10–30 mg	10–30 mg	BID	X			
Cephalothin	20–35 mg	20–35 mg	20–35 mg	TID			X	X
Cephazolin	10–25 mg	10–25 mg	10–25 mg	TID			X	X
Aminoglycosides								
Amikacin	5 mg	5 mg	5 mg	TID			X	X
Gentamicin	2–3 mg	2–3 mg	2–3 mg	BID-TID			X	X
Neomycin	7–15 mg	7–15 mg	7–15 mg	QD	X			
Tetracyclines								
Chlortetracycline	10–20 mg	10–20 mg	10–20 mg	TID	X			
Oxytetracycline	20 mg	20 mg	20 mg	q48h			X	
Tetracycline	10–20 mg	10–20 mg	10–20 mg	TID	X			X
Chloramphenicol	45–80 mg	45–80 mg	45–80 mg	TID	X		X	X
Macrolides								
Erythromycin	10–15 mg	10–15 mg	10–15 mg	TID	X			
	2–8 mg	2–8 mg	2–8 mg	QD			X	
Tylosin	7–12 mg	7–12 mg	7–12 mg	QD-TID	X		X	
Lincosamides								
Clindamycin	5–40 mg	5–40 mg	5–40 mg	BID	X		X	X
Lincomycin	10–25 mg	10–25 mg	10–25 mg	BID	X		X	
Sulfonamides								
Ormetoprin-sulfadimethoxine	15–30 mg	15–30 mg	15–30 mg	QD	X			
Sulfadimethoxine	25–50 mg	25–50 mg	25–50 mg	QD	X			
Trimethoprim-sulfamethoxazole	15–30 mg	15–30 mg	15–30 mg	BID	X	X		X
Fluoroquinolones								
Ciprofloxacin	2.5–7.5 mg	2.5–7.5 mg	2.5–7.5 mg	BID	X			
Enrofloxacin	2.5–5 mg	2.5–5 mg	2.5–5 mg	BID	X		X	X
Metronidazole	50–60 mg	50–60 mg	50–60 mg	QD-BID	X			X

[a] QD, once daily; BID, twice daily; TID, three times daily; QID, four times daily; hqx, every x hours.
[b] PO, oral; SQ, subcutaneous; IM, intramuscular; IV, intravenous.
Table adapted from the University of Minnesota Animal Care and Use Manual 1997 reprint.

Intimal hyperplasia of PTFE grafts was noted in these early canine studies and subsequently led to further research into this important complication. As in humans, long vascular grafts were shown to develop intimal hyperplasia at the anastomotic sites but poor endothelialization of the midportion of the grafts (Brophy *et al.*, 1991). More recently, however, swine have replaced the canine model for studies of these phenomena. Interestingly, the observations made in these early canine studies using PTFE grafts led investigators to develop clinical protocols to improve their long-term patency. In fact, one of these canine protocols, systemic aspirin therapy, is now recognized as the therapy of choice in improving graft patency (Curl *et al.*, 1986). Aspirin administration reduces thromboxane B2 levels and prevents platelet aggregation. In this laboratory,

aspirin (325 mg/day) is used routinely as postoperative therapy (beginning the first post-operative day) in investigations involving peripheral vascular prostheses.

Despite the cost, the canine model has several advantages that still make it the model of choice for many peripheral vascular studies. Ease of repeated vascular access, good long-term patency, and physiologic conditions similar to humans have all been cited for promulgation of the model. The increased thrombogenicity of smaller grafts can also be thought of as an advantage in that long-term patency represents success under rigorous conditions, thereby implying an increased potential for success in clinical trials. Physical durability, patency, and biomaterial–host interactions can all be tested (at the site of intended clinical use) within the same model.

Overall, the canine animal model has provided a long history of significant contributions to our understanding of cardiovascular physiology and to the development and testing of cardiovascular biomaterials and devices. In more recent times, there has been a shift away from the dog for many studies because of public concerns and rising animal cost. As a result other large domestic animals have become useful models for ongoing biomaterials research.

Swine

Introduction

It is believed that the majority of the breeds of swine we now know are descended from the Eurasian wild boar (*Sus scrofa*). Archaeological evidence from the Middle East indicates domestication of the pig occurred as early as 9000 years ago, with some evidence for domestication even earlier in China. From here the pig spread across Asia, Europe, Africa, and ultimately North America (Towne and Wentworth, 1950; Mellen, 1952; Clutton-Brock, 1999b). Today, swine serve many purposes in society from food source and pets to biomedical research subjects. In this latter role, swine have proven useful during the past four decades in studying a variety of human ailments including cardiovascular (Ramo *et al.*, 1970; Cevallos *et al.*, 1979), gastrointestinal (Kerzner *et al.*, 1977; Leary and Lecce, 1978; Pinches *et al.*, 1993), and hepatic (Mersmann *et al.*, 1972; Soini *et al.*, 1992; Sielaff *et al.*, 1997) diseases. Studies in swine have also provided extensive information in the areas of organ transplantation (Calne *et al.*, 1974; Marino and De Luca, 1985; Grant *et al.*, 1988; Al-Dossari *et al.*, 1994; Granger *et al.*, 1994).

Multiple authors have reported advantages to using swine as models in biomedical research including, but not limited to, anatomic and physiologic similarities to humans, low cost, availability, and the ability of swine to produce large litters of hearty newborns (Stanton and Mersmann, 1986; Swindle, 1992; Tumbleson and Schook, 1996). In addition, naturally occurring models of human disease (e.g., atherosclerosis) can be found in various strains of swine, making them ideal candidates for the study of these disease processes. Also helpful to investigators using swine is a large body of agricultural literature in the areas of swine nutrition, reproduction, and behavior that may have application in biomedical research. For all these reasons swine have become increasingly popular in the biomedical research laboratory.

Comparative Anatomy and Physiology

Swine are the one readily available species in which cardiovascular anatomy and physiology most closely resembles those in humans, likely because of a similar phylogenetic development that led to an omnivorous species with a cardiovascular system that has accommodated to a relative lack of exercise (McKenzie, 1996). The heart of an adult swine weighs 250–500 grams and comprises approximately 0.25% of the body weight (Ghoshal and Nanda, 1975), similar to that of humans. The coronary vasculature of a pig heart is nearly identical to that of humans in anatomic distribution, reactivity, and paucity of collateral flow (Hughes, 1986). The right and left coronary arteries are similar in size to each other. The left coronary supplies the greater part of the wall of the left ventricle and auricle, including the interventricular septum, by means of its circumflex and paraconal interventricular branches. The right coronary supplies the wall of the right ventricle and the auricle via its circumflex and subsinuosal branches (Ghoshal and Nanda, 1975). These similarities make swine ideal for the study of ischemia/reperfusion and coronary stent technology.

There are certain cardiovascular anatomic differences between swine and humans that must be considered when using swine as investigational subjects. First, the left azygous vein drains directly into the right atrium in swine compared to drainage into the brachiocephalic vein in humans (Ghoshal and Nanda, 1975). Second, there are only two arch vessels coming off of the aorta in swine. The most proximal of these is the brachiocephalic trunk, which gives rise to the right subclavian and a common carotid trunk that bifurcates into right and left carotid arteries. The second arch vessel is the left subclavian artery (Ghoshal and Nanda, 1975).

Electrophysiologic parameters of pigs more closely resemble those of man than any nonprimate animal. These parameters include similar P and R-wave amplitudes, P-R intervals of 70–113 msec (longer in older animals), mean QRS duration of 39 msec, a Q-T interval of 148–257 (again longer in older animals), as well as ventriculoatrial (V-A) conduction when the right ventricular endocardium is stimulated faster than normal sinus rate (Hughes, 1986). These similarities make the pig a good model for studying pacemakers. Some investigators have found swine to be highly susceptible to ventricular arrhythmias, often requiring prophylactic pharmacologic protection (i.e., lidocaine, flecainide, or bretylium). Alternatively, other investigators have exploited this ventricular instability to study "induced" tachyarrhythmias, and the pharmacologic agents and/or electrical interventions to manage them (Hughes, 1986).

Physiologically, swine have hemodynamic and metabolic values that are similar to those of humans (Tables 1 and 2). They also share similar pulmonary function parameters such as respiratory rate and tidal volumes (Willette *et al.*, 1996). Although they have lower levels of hemoglobin, circulating red cell volume, arterial oxygen saturation, and markedly lower venous oxygen saturations, they have similar platelet function, and lipoprotein patterns. Other physiologic characteristics that may play a role in experimentation include a higher core body temperature, a contractile spleen that sequesters 20–25% of the total red cell mass, very fragile arteries and veins, a higher plasma pH, and a higher plasma bicarbonate level. Swine also have 50% more functional extracellular space than humans and a slightly increased volume of total body water (Hannon *et al.*, 1990; Swindle *et al.*, 1986).

Perioperative Care

Anesthesia   Swine should be fasted at least 12 hours prior to the induction of anesthesia, longer if one desires the large bowel to be empty prior to surgery. Unless gastric surgery is to be performed, the animals may be given water without restriction. It is common to administer an IM preanesthetic agent followed by an IV anesthetic agent after IV access has been obtained. IM injections can safely be administered in the neck just behind the ear, in the triceps muscle of the forelimb, or in the semimembranosus–semitendinosus muscles of the hindlimb. IV access is most commonly obtained in the lateral auricular veins (Riebold and Thurmon, 1985; Swindle and Smith, 1994; Swindle, 1994). The preanesthetic/anesthetic protocol used in this laboratory involves the administration of an IM injection of a combination drug (Telazole) consisting of a dissociative anesthetic and a tranquilizer (tiletamine and zolazepam) at a dose of 2–4 mg/kg for sedation. Following this IV thiopental sodium (6–25 mg/kg) is administered if necessary. ET intubation follows and anesthesia is maintained via inhaled isoflurane (1–2 vol%). Commonly used anesthetics, doses, and routes of delivery are given in Table 3.

Analgesia Postoperative analgesia is a very important component of any animal experiment. Appropriate use of postoperative analgesia involves rapid recognition of discomfort and appropriate therapy. When in pain or distress swine will demonstrate changes in social behavior, gait, posture, and a lack of bed-making. Most swine will become very reluctant to move when in pain, and if forced to move will vocalize with even greater enthusiasm than they generally display (Gross, 1994). Buprenorphine is currently the analgesic of choice used in swine (Swindle and Smith, 1994; Swindle, 1994); we have found it to be very effective, and it can be administered at 0.05–0.1 mg/kg every 8–12 hours. Standard narcotics such as morphine and fentanyl are effective pain relievers but have a very short half-life in swine. Additional analgesic agents are listed in Table 4.

Existing Models

Cardiac Devices Orthotopic valve replacement in large animals is an important component of preclinical prosthetic valve assessment for the initial evaluation of surgical handling characteristics, hemodynamic performance, and valve-related pathology. Early investigators using swine for cardiovascular research described difficulties with venous access (Swan and Piermattei, 1971), anesthesia (Piermattei and Swan, 1970), cardiopulmonary bypass (Swan and Meagher, 1971), and several anatomic peculiarities (Swan and Piermattei, 1971). Subsequently, cardiovascular surgical procedures using swine have become technically feasible, and several investigators have conducted acute and chronic studies for the assessment of hemodynamic profiles of cardiac prostheses (Hasenkam et al., 1988a,b, 1989; Hazekamp et al., 1993). However, only a few long-term swine studies have been performed to examine the potential of prosthetic heart valve implants for thrombogenicity (Gross et al., 1997; Henneman et al., 1998; Grehan et al., 2000).

It has been reported that heart valve replacement in swine requires certain measures not needed with other species. These include the use of a crystalloid prime without plasma volume expanders (especially starch based), prophylactic administration of pharmacological protection against ventricular arrhythmias, insurance of adequate hypothermic cardioprotection during the time of cross-clamp, administration of "shock" doses of corticosteroid just prior to reperfusion, and the use of inotropic support during bypass weaning (Gross et al., 1997).

In our experience, (see later discussion) however, we have found some of these precautions unnecessary for the successful use of the pig in the evaluation of cardiac prosthesis.

In our investigation, 22 swine underwent mitral valve replacement with no operative deaths. All animals were weaned from cardiopulmonary bypass without inotropic assistance. No cardioplegic protection was used, and only mild total body hypothermia was instituted. Pathologic analysis of valves explanted from animals surviving more than 30 days demonstrated extensive fibrous sheath formation leading to valve orifice obstruction and restriction of leaflet motion in a significant number of animals. The extensive fibrous sheath formation (and subsequent valvular dysfunction) represents a chronic tissue response observed in many species following mitral valve replacement. This chronic tissue response, however, was noted to have developed sooner in swine than in our previous studies involving prosthetic valve implantation into other species (i.e., sheep). Additionally, we did not find this model to be useful in predicting device-related thrombogenicity, as there were no clinical or pathologic differences observed among the valve designs studied (Grehan et al., 2000). The utility of swine in predicting the thrombogenic potential of cardiac valvular prostheses is therefore limited.

Improved biotechnology has brought left ventricular assist devices (LVAD) into the clinical arena as bridge to transplantation therapy for patients with end-stage cardiac disease or those suffering from acute myocardial infarction (Frazier et al., 1992; Oz et al., 1997; Park et al., 2000). Clinical observations of patients who have received LVAD have revealed that these patients have a propensity for developing right ventricular dysfunction. To study this phenomenon, a model of iatrogenic congestive heart failure (CHF) using 7 days of rapid pacing to induce cardiomyopathy has been developed in swine (Chow and Farrar, 1992). This model has been used to investigate the hemodynamics and septal positioning of the assist devices that may contribute to right ventricular dysfunction (Chow and Farrar, 1992; Hendry et al., 1994).

Vascular Devices Initially, the pig was developed as a model to study solid organ transplantation. However, more recently its use has made an impact on the study of vascular disease. The pig is currently the species of choice for in vivo evaluation of vascular stents, restenosis biology, and balloon injury. The advantages of using swine include (1) easy access to coronary arteries with present catheterization and angioplasty techniques; (2) coronary arteries of sufficient size for catheters used in adult humans; (3) balanced coronary circulation that is anatomically similar to humans; (4) spontaneous development of atherosclerosis; and (5) comparable coagulation and fibrinolytic systems and lipid metabolism to humans (Fritz et al., 1980; White et al., 1989; Karas et al., 1992; Willette et al., 1996).

Numerous in vitro studies indicate that the coagulation and fibrinolytic systems in swine closely resemble those of humans (Wilbourn et al., 1993; Karges et al., 1994; Reverdiau-Moalic et al., 1996; Gross, 1997), which would suggest that this may be the ideal species for the preclinical evaluation of the thrombogenic potential of a prosthetic device. in vivo studies specifically looking at the thrombogenicity of implantable devices, both cardiac and noncardiac, have already been performed in swine (Rodgers *et al.*, 1990; Walpoth *et al.*, 1993; Scott *et al.*, 1995). Similar investigations have shown that stent thrombosis may occur early in this model (within 6 hours of implantation), suggesting that if a pig survives the first 12 hours (coronary stent remains patent), it is highly likely that it will not suffer a stent occlusion later (Schwartz and Holmes, 1994). Observations of such rapid restenosis led Schwartz *et al.* to suggest that the pig should be used as an accelerated thrombotic model capable of predicting stent thrombosis within hours of implant (Schwartz and Holmes, 1994). Stenting or balloon inflation in the swine model leads to intimal smooth muscle cell proliferation that closely resembles the cell size, density, and histopathological appearance of that seen in human restenosis (Karas *et al.*, 1992; Bonan *et al.*, 1996; Willette *et al.*, 1996). Additionally, as in humans, there is a predictable relationship between vascular injury and restenosis (i.e., more neointima is formed with deeper lesions) (Willette *et al.*, 1996; Jordan *et al.*, 1998).

Although investigation of stents and balloon injury has mainly been done on coronary arteries (Karas *et al.*, 1992; Bonan *et al.*, 1996; Willette *et al.*, 1996), studies on other arteries such as the iliacs have also been performed (White *et al.*, 1989). Swine have also been used to study the feasibility of percutaneous repair of abdominal aortic aneurysms (AAA). Jordan *et al.* (1998) developed an AAA model using rectus abdominus fascia and demonstrated prolonged survival in stented AAA in comparison to unstented controls. Unfortunately, this AAA model did not accurately mimic aneurysms seen in humans, as the AAA created was saccular and did not reproduce the geometric configuration or collateral circulation and back bleeding observed in human disease (Jordan *et al.*, 1998). Whitbread *et al.* (1996) created an AAA model in swine by interposing fusiform segments of glutaraldehyde-treated bovine internal jugular vein into the infrarenal aorta. This resulted in a pulsatile, nonthrombogenic AAA that approximates human dimensions (20 mm) and geometry (fusiform) (Whitbread *et al.*, 1996). Both AAA models have been used to develop stents and stent placement technology with the goal of reducing the pulsatility and diameter of the aneurysm while maintaining location of the stent without occluding nearby vessels.

Swine have become a useful and popular model to study biomaterials and the devices manufactured from these materials. Anatomic and physiologic similarities to humans have made swine the species of choice to study many cardiac and vascular prostheses, especially in acute studies looking at hemodynamic profiles of different devices. However, difficulties with husbandry and species temperament and a rapid rate of somatic growth have made swine less desirable as models for the chronic evaluation of devices in our experience.

Sheep

Introduction

As a gentle, docile animal, sheep were among the first species to be domesticated approximately 10,000 years ago, as evidenced by pictures and statuettes depicting woolen sheep in the Middle Eastern region. They remain popular to this day as a source of food and goods. The prevailing theory for the origin of the domestic sheep is that they originated from the wild sheep known as mouflon (*Ovis musimon*). Currently, mouflon can be found in two separate geographic locales: Asiatic mouflon in Asia Minor and southern Iran; and European mouflon on the islands of Sardinia and Corsica (Clutton-Brock, 1999a). The rapid expansion of agriculture and human consumption has been both the cause and effect of routine crossbreeding of sheep. Currently more than 200 different breeds exist worldwide. Common breeds used in surgical research include smaller breeds commonly used for meat commercially, such as the Hampshire and Dorset varieties, or the larger sheep, commonly used for wool commercially, such as the Merino and Rambouillet breeds.

Comparative Anatomy and Physiology

Specific organ differences between humans and sheep are notable (e.g., ruminant gut anatomy). However, in many instances the similarity in organ size of adult sheep to their human counterparts allows adequate approximation for the study of mechanical and bioprosthetic biomaterials and devices prior to human implantation. The ovine thorax has an exaggerated conical shape with a narrow spax (cranial aperture or thoracic inlet) and a caudal thoracic aperture that is six times wider. Cardiopulmonary anatomy is generally similar to that of humans, making sheep ideal models for cardiac research; however, a few anatomic differences require consideration (Hecker, 1974): for example, the single brachiocephalic artery arising from the aortic arch. In addition, tracheal anatomy varies slightly from humans, with the cranial lobe of the right lung arising directly from the trachea rather than the right mainstem bronchus. This variation can pose a difficulty during endotracheal intubation on induction of anesthesia and can result in nonventilation of a significant pulmonary segment (Carroll and Hartsfield, 1996).

Physiologic parameters (heart rate, blood pressure, cardiac index, and intracardiac pressures) have been established through numerous studies in both anesthetized and conscious sheep, including the percentage of blood flow to each major organ system (Matalon *et al.*, 1982; Nesarajah *et al.*, 1983; Newman *et al.*, 1983). In most instances, the hemodynamic and metabolic values are similar to those of other large mammals including humans (Tables 1 and 2).

In cardiovascular bioprosthesis studies, the similarities between human and animal hematologic systems is important. Adequate blood group typing should take place in the event transfusion is necessary. Coagulation parameters and the hematologic profile between sheep and humans reveal some differences, which become important in assessing the thrombogenicity of implanted cardiovascular prostheses. Notably, sheep possess decreased fibrinolytic activity and increased platelet number and adhesiveness (Gajewski and Povar, 1971; Tillman *et al.*, 1981; Karges *et al.*, 1994).

Perioperative Care

Anesthesia   Sheep should be fasted preoperatively to decrease the likelihood of regurgitation on induction of anesthesia and ET intubation. The reticulum, the most prox- imal rumen, has no sphincter mechanism at its oral end and often needs intubation and suction decompression to prevent regurgitation and aspiration (Holmberg and Olsen, 1987; Car- roll and Hartsfield, 1996). Orogastric intubation is especially important if preoperative fasting does not occur. Typically, a sedative is administered IM to allow easy and safe placement of IV catheters for controlled administration of medication. Sev- eral sedatives are useful in the pre- and postoperative periods to aid in animal handling, and for performing minor proce- dures as shown in Table 3. Large named veins such as the internal or external jugular or femoral veins can be used to secure IV access for fluid administration and frequent blood draws. Specialized catheters such as pulmonary artery catheters or long-term tunneled venous catheters (Hickman) may also be placed in these easily accessible, large veins (Tobin and Hunt, 1996). Following placement of a secure IV catheter and the administration of short-acting relaxing agents, ET intuba- tion can safely occur using standard cuffed ET tubes 5–10 mm in internal diameter. Anticholinergics or parasympatholytics, especially atropine, have been used on induction of anesthe- sia to decrease salivary and respiratory tract secretions. Use of these agents, however, is controversial in ruminants such as sheep, goats, and cattle because they inhibit gastrointestinal smooth muscle activity, which can cause rumen stasis in sheep (Carroll and Hartsfield, 1996). Table 3 lists the recommended dosing of anesthetic agents.

Analgesia Scheduled medication administration and periodic evaluation of behavior can help to ensure adequate pain relief. Sheep should be assessed for changes from their pre- operative temperament. Sheep tend to react stoically to pain, and their refusal to move can often indicate even minor dis- comfort (Carroll and Hartsfield, 1996). Other common clinical signs of postprocedural pain include tachycardia, tachypnea, grunting, grinding teeth, decreased appetite, vocalization on movement, and guarding of the operative site. Commonly used analgesics include narcotics administered via a subcutaneous, IV, or epidural route. Nonsteroidal antiinflammatory agents can be used in divided doses for additional analgesia. Please refer to Table 4 for recommended dosing of analgesics.

Existing Models

Cardiac Devices Because of the similar anatomic and physiologic characteristics of sheep and humans discussed ear- lier, sheep are often used in cardiovascular biomaterial research and several models have been developed for studying com- patibility and function of biomaterials. A total artificial heart model was developed in sheep at the University of Utah's Arti- ficial Heart Laboratory in the 1980s (Holmberg and Olsen, 1987). Other studies examined the potential use of synthetic or xenograft pericardial substitutes to reduce adhesion for- mation following coronary artery bypass or valve replacement (Gabbay et al., 1989; Bunton et al., 1990).

Many investigators have employed sheep in the preclinical testing of mechanical valves in both the aortic and mitral posi- tions (Barak et al., 1989; Vallana et al., 1992; Irwin et al., 1993; Cremer et al., 1995; Bhuvaneshwar et al., 1996; Okazaki et al.,

1996). Our experience reveals juvenile sheep (younger than 20 weeks) to be an excellent long-term model for mechanical prosthetic valve evaluation. Careful standardization to pre- vent infectious complications, short cardiopulmonary bypass time, whole sheep blood transfusion to minimize anemia, and rumen decompression with an orogastric tube to prevent vena caval compression were found to be useful in the successful implementation of this model (Irwin et al., 1993).

Acute and chronic models of stented bioprosthetic valves are well established through several studies in both the mitral and aortic position in sheep (Gott et al., 1992; Liao et al., 1993; Bianco et al., 1996; Vyavahare et al., 1997; Northrup et al., 1998; Ouyang et al., 1998; Salerno et al., 1998; Spyt et al., 1998). Sheep grow at a rate that allows valves to be subjected to changing physical environment. In contrast, larger ruminants such as cattle and large varieties of swine experience rapid and prolonged growth, until reaching a large adult size (Bianco et al., 1996; Grehan et al., 2000). Such rapid growth can cause difficulties in size matching between prosthetic valves and the native valve annulus, resulting in paravalvular leaks and functional valve stenosis (Braunwald and Bonchek, 1967; Gallo and Frater, 1983; Spyt et al., 1998). Premature leaflet cal- cification can occur in stented bioprosthetic valves because of abnormal mechanical loading characteristics. This is most con- sistently reproduced in juvenile sheep (younger than 6 months). We and others currently employ a model in sheep younger than 20 weeks to study both the dynamic environment related to their growth and the chronic effect of calcification in stented porcine bioprosthetic valves in the aortic position (Gott et al., 1992; Liao et al., 1993; Bianco et al., 1996; Ouyang et al., 1998).

Sheep also provide an important model for the study of low- profile stentless aortic valves (David et al., 1988; Brown et al., 1991; Hazekamp et al., 1993; Schoen et al., 1994; Salerno et al., 1998). In this regard we found that, with careful surgical tech- nique including a two-thirds transverse aortotomy for superior exposure and meticulous avoidance of coronary ostial obstruc- tion by the scalloped valve edges, sheep provide an excellent *in vivo* preclinical model to test these aortic valves (Salerno et al., 1998).

In addition to the subcoronary implantation of these stent- less devices, this laboratory has also developed a juvenile sheep model of aortic root replacement to allow for the pre-clinical evaluation of new or modified stentless devices using implan- tation techniques that parallel those seen in clinical practice. Again, careful attention to surgical technique proved to be very important in the success of this model including creation of very generous-sized coronary buttons, inclusion of a posterior "rim" of native aorta in the proximal anastamosis, and the cre- ation of a tension-free distal anastamosis by removing little if any native aorta (Grehan et al., 2001).

Vascular Devices In addition to cardiac devices, researchers have studied the interface between blood and sev- eral biomaterial surfaces in the evaluation of biosynthetic vascular grafts placed in sheep. Several authors have noted the reduction in thrombogenicity provided by endothelial seed- ing of small-diameter synthetic vascular grafts constructed of expanded PTFE or woven Dacron prior to placement in sheep

(James *et al.*, 1992; Taylor *et al.*, 1995; Jensen *et al.*, 1997; Dunn *et al.*, 1996; Poole-Warren *et al.*, 1996). Placement of vascular grafts composed of several engineered surface textures into sheep has also been performed in order to assess characteristics of the pseudointima layer (Fujisawa *et al.*, 1999). Additionally, the important problem of intimal hyperplasia in the venous anastomosis of high-flow arteriovenous fistulas has recently been evaluated in a sheep model (Kohler and Kirkman, 1999).

Recent progress in the development of large-vessel endovascular stents has occurred in part due to research with specific vascular disease models in adult sheep and has been spurred by interest in minimally invasive techniques. As in graft studies, there is significant interest in both the short- and long-term patency of endovascular stents. Stent incorporation into native tissue and the role of foreign-body reaction in normal arterial systems with respect to thrombogenicity and intimal hyperplasia has been evaluated in sheep aortic and iliac arteries (Rousseau *et al.*, 1987; Neville *et al.*, 1994; Schurmann *et al.*, 1998; White *et al.*, 1998). Current efforts have been directed at determining the role of endovascular stenting in the prevention of rupture of abdominal aortic aneurysms and the prevention of complications such as spinal cord ischemia through the use of these stents (Boudghene *et al.*, 1998; Beyguii *et al.*, 1999).

Current cardiovascular research has embraced the humane use of sheep for the testing of a variety of cardiac and vascular devices composed of many different biomaterials. In all of these investigations, sheep have proved a useful and reliable model, as evidenced by this laboratory's continued success using both juvenile and adult sheep for a variety of investigations. Current research supports the broader application of emerging endovascular technologies. With many anatomic and physiologic similarities to humans and a temperament that facilitates handling, sheep should continue to serve an important role in future investigations into biomaterials testing and development.

TESTING HIERARCHIES

Introduction

The ultimate fate for many new or modified biomaterials is to become a cardiac or vascular device intended for clinical use. Before this can occur, however, extensive testing must be performed and specific criteria must be met. This section will focus on preclinical *in vivo* evaluation standards that new or modified cardiac and vascular devices must meet before clinical evaluation can be considered. These criteria have been established by the International Organization for Standardization (ISO) and can be found in document CV5840-1996 (Cardiovascular implants—Cardiac valve prostheses) and CV7198-1994 (Cardiovascular implants—Tubular vascular prostheses). These documents contain information on all aspects of testing from the initial *in vitro* assessment to final packaging requirements. We will limit our discussion to the established standards for *in vivo* preclinical evaluation of cardiac valvular and tubular vascular prostheses. The standards that apply to the experimental methodology used in the investigation and the specific

study criteria that must be met, as well as required results documentation, will be delineated. Following this, a short discussion on the future directions of these standards and the predicted impact these changes may have on the study of these types of devices will be presented.

Current Recommendations

Cardiac Devices

Rationale The objective of preclinical *in vivo* testing is to evaluate the performance characteristics of cardiac devices such as surgical handling characteristics, hemodynamic performance, and the development of valve-related pathology. Although it is not feasible to assess the long-term durability of a cardiac device using an animal model, valuable data on hemodynamic performance and biological compatibility can be obtained. Additionally, during *in vivo* testing unanticipated side effects of the device can be discovered. In order to ensure that new or modified cardiac valvular prostheses are assessed in a uniform manner, the ISO has established a set of guidelines to direct both designers and investigators.

Methods Before any new or modified cardiac valve is allowed to undergo clinical evaluation the ISO requires that it be tested in an *in vivo* animal model in at least one of its intended anatomic positions. The device tested must be of clinical quality and of the same design and size of those intended for clinical use. Additionally, ISO requires that implantation be carried out using the same surgical technique for all of the implants being evaluated (e.g., suture technique, anatomic location, and valve orientation). The devices should be evaluated in at least six animals of the same species (preferably the same sex and age as well) for a duration of 20 weeks. At least two of the animals should be implanted with a reference (clinically approved) device to serve as concurrent controls.

Further specifications require each animal in which a cardiac device has been implanted to undergo a post-mortem examination. This ensures that data will be obtained from all animals; whether or not they survive the 20-week study period. An evaluation of hemodynamic performance of the device during or after the 20-week implantation period must be performed. This assessment should include measurement of the pressure gradient across the implanted device at a cardiac index of approximately 3 L/min/m^2. Additionally, an assessment of resting cardiac output and prosthesis regurgitation should be performed. A pathologic evaluation of the study animals should also be performed and should include an assessment of the major organs for valve-related pathology; an evaluation for any device-related hematological consequences; and an assessment of any structural changes of the heart valve substitute.

Reporting Results from these evaluations should be documented in an official report that describes the findings of the investigation in detail. This report should include the name of the institution(s) and investigator(s) involved in the study. A detailed description of the animal model used in the investigation and the rationale for its use should also be included. A pretest health assessment, including study animal age at the time of implantation, should also be included for each animal used in the study. A description of the operative procedure, including the suture technique, the orientation of the heart valve substitute, and any operative complications, should also be detailed. Descriptions of any adjuvant procedures the animals underwent during the study period (e.g., phlebotomy, angiography) should also be provided. The results of all blood tests, including a statement of the time elapsed between implantation and procurement of the sample should be provided. Blood tests should include an evaluation of the hematologic profile (with emphasis on hemolysis), and blood chemistries. Additionally, the report should include a subjective assessment of specific surgical handling characteristics of the device and any accessories. This assessment should include a discussion of unusual and/or unique attributes the device might possess. The report must contain a complete list of medications administered to the animals during the study period. Finally, the report should contain the results obtained from the hemodynamic studies to assist in the assessment of hemodynamic performance.

Pathologic documentation should include a gross and microscopic report on each animal in which a heart valve substitute was implanted, including any animal that did not survive for the minimum postimplantation period. Documentation should include visual records of the heart valve substitute *in situ* and evidence of any thromboembolic events occurring in the major organs. The cause of death of any animal that was not euthanized should be investigated and reported. Detailed examination of the explant must be performed, with specific attention to any structural changes in the device. If appropriate, further *in vitro* functional studies of the heart valve substitute should be undertaken (i.e., hydrodynamic testing).

Vascular Devices

Rationale As with cardiac devices, the purpose of *in vivo* preclinical testing of vascular prostheses is to evaluate the characteristics of a prosthesis that are difficult, if not impossible, to obtain using *in vitro* testing models. The capacity of the prosthesis to maintain physiologic function when used in the circulatory system, the biologic compatibility of the prosthetic, and the surgical handling characteristics of the prothesis are all important features that can be evaluated using an *in vivo* model and cannot be obtained with *in vitro* testing methods. This testing is not intended to demonstrate the long-term performance of the prosthesis, but rather the short-term (less than 20 weeks) response and patency of the prosthesis being investigated.

Methods In designing an experimental protocol that will ensure clinical relevance, researchers should consider the intended application of the device being studied as well as the necessary diameter and length of the prosthesis. Additionally, investigators should reflect on any specific biological characteristic of the chosen animal model and the impact of those characteristics on their conclusions. Current ISO requirements state that a rationale for the use of a particular species, the site of implantation, the selection of a control prosthesis, the method and interval of patency observations, and the number of animals used in each group being studied be provided. Each

type of prosthesis shall have been tested by implantation at the intended, or at an analogous, vascular site in not fewer than six animals for not less than 20 weeks in each animal unless a justification for a shorter-term study can be provided. A prosthesis shall not be tested in a species from which it was derived, unless appropriate justification can be provided. Angiography or Doppler should be used to monitor the duration of patency for each prosthesis and the results recorded. Each investigation should include a control group in which a clinically approved vascular prosthesis is implanted and studied in the same fashion as the experimental prosthesis.

Preoperative data should include the sex and weight of the animal, documentation of satisfactory preoperative health status, and any preoperative medications. Operative data must include the name of the implanting surgeon and a detailed description of the surgical procedure, to include the type and technique used in performing the proximal and distal anastomoses. The *in situ* length and diameter of the prosthesis in addition to any adverse intraoperative events (e.g., transmural blood leakage) must also be documented. Postoperative medications, patency assessments (method and interval from implantation), and any adverse event or deviation from protocol must be noted. Loss of patency before the intended study duration does not necessarily exclude the animal from the study population used to assess prosthetic function and host tissue response. All animals implanted with either test or control prostheses, including those excluded from the final analysis, shall be recorded and reported. Termination data should include an assessment of prosthesis patency (documenting the method used) and an assessment of prosthesis explant pathology.

Reporting At the conclusion of the study, a detailed report should be compiled in which the study protocol is delineated. The report should include the rationale for selection of animal species; implantation site; control prosthesis; method of patency assessment; and the intervals of observation. Additionally, documentation of sample size and pertinent operative, perioperative, and termination data must be presented. The results should include the patency rates and any adverse event that was encountered during the study interval and should be compared between the study and control groups. All animals entered into the study must be accounted for in the report. If data are excluded a rationale must be provided. A summary of pathology including photographs of the prosthesis *in situ* and micrographs should also be provided. In addition to these objective results, a subjective statement of the investigator's opinion of the device must be included. Finally, conclusions drawn from the study and a summary of the data auditing procedures should be documented.

Future Directions

The use of animals to predict clinical safety is an essential (and required) phase in the development of new or modified medical device technology. It is extremely unlikely, however, that preclinical use of animal models to assess new technology will ever evolve to the point that absolute statements can be offered regarding future clinical performance. Therefore, the enhancement of the predictive qualities of human safety and efficacy based on preclinical *in vivo* evaluations should be our priority.

Enhanced prediction of human safety and efficacy can be achieved by developing increasingly more relevant models of human disease and developing models in which new or modified devices are implanted orthotopically in each intended clinical location (site specific testing). Additionally, each preclinical study design should include an appropriate control "arm." This "arm" of the study should include animals implanted with a device that is currently approved and being used clinically. The control device should be of similar design (or preservation technique) as the device undergoing investigation and should be implanted using the same techniques. The use of concurrent control implants offers several advantages. First, it provides a direct clinical bridge to assess performance of a new or modified device. Second, it provides an assessment of the model utilized in the investigation relative to a device with a known clinical history. Finally, it can reduce the number of implants required in each investigation by providing a correlation to historical animal and clinical data.

The use of animals to predict human safety will and should remain a crucial part of the regulatory requirements governing medical device development and modification. In order to enhance the predictive quality of preclinical *in vivo* evaluations current models need to be refined and new models need to be developed that more closely mimic both human disease and the intended clinical use of the device under investigation. As these models are developed and perfected they should become part of the current standards to be met by investigators working in prostheses development. In addition, each preclinical evaluation should include the concurrent implantation of appropriate control devices so that comparisons of the investigational data obtained can be made to that of concurrent and historic animal and clinical control data. This will allow more accurate assessments of both the investigational device and the model to be made.

Acknowledgments

The authors thank Agustin P. Dalmasso for his editorial assistance and the staff and student employees of the University of Minnesota Department of Surgery's Division of Experimental Surgical Services for their commitment to achieving scientific accuracy and providing humane care to the investigational animals.

Bibliography

Adzick, N. S., and Longaker, M. T. (1991). Animal models for the study of fetal tissue repair. *J. Surg Res.* 51: 216–222.

Akutsu, T., Dreyer, B., and Kolff, W. J. (1974). Polyurethane artificial heart valves in animals. in *Canine Surgery*, J. Archibald and J. V. LaCroix eds. American Veterinary Publications, Inc, Santa Barbara, CA.

Al-Dossari, G. A., Kshettry, V. R., Jessurun, J., Bolman, R. M. III. (1994). Experimental large-animal model of obliterative bronchiolitis after lung transplantation. *Ann. Thorac. Surg.* 58(1): 34–39.

American Veterinary Medical Association (1993). 1993 Report of the AVMA Panel on Euthanasia. *JAVMA* 202: 229–249.

Barak, J., Einav, S., Tadmor, A., Vidne, B., and Austen, W. G. (1989). The effect of colloid osmotic pressure on the survival of sheep following cardiac surgery. *Int. J. Artif. Organs* 12: 47–50.

Beygui, R. E., Kinney, E. V., Pelc, L. R., Krievins, D., Whittmore, J., Fogarty, T. J., and Zarins, C. K. (1999). Prevention of spinal cord ischemia in an ovine model of abdominal aortic aneurysm treated with a self-expanding stent-graft. *J. Endovasc. Surg.* 6(3): 287–284.

Bhuvaneshwar, G. S., Muraleedharan, C. V., Vijayan, G. A., Kumar, R. S., and Valiathan, M. S. (1996). Development of the Chitra tilting heart valve prosthesis. *J. Heart Valve Dis.* 5: 448–458.

Bianco, R. W., Phillips, R., Mrachek, J., and Witson, J. (1996). Feasibility evaluation of a new pericardial bioprosthesis with dye mediated photo-oxidized bovine pericardial tissue. *J. Heart. Valve Dis.* 5: 317–322.

Bianco, R. W., St. Cyr, J. A., Schneider, J. R., Rasmussen, T. M., Clack, R. M., Shim, H. S., Sandstad, J., Rysavy, J., and Foker, J. E. (1986). Canine model for long-term evaluation of prosthetic mitral valves. *J. Surg. Res.* 41: 134–140.

Bonan, R., Paiement, P., and Leung, T. K. (1996). Swine model of coronary restenosis: effect of a second injury. *Cathet. Cardiovasc. Diagn.* 38: 44–49.

Boudghene, F. P., Sapoval, M. P., Bonneau, M., LeBlanche, A. F., Lavaste, F. C., and Michel, J. B. (1998). Abdominal aortic aneurysms in sheep: prevention of rupture with endoluminal stent-grafts. *Radiology* 206(2): 447–454.

Braunwald, N. S., and Bonchek, L. I. (1967). Prevention of thrombus formation on rigid prosthetic heart valves by the ingrowth of autogenous tissue. *J. Thorac. Cardiovasc. Surg.* 54: 630–638.

Brophy, C. M., Ito, R. K., Quist, W. C., Rosenblatt, M. S., Contreras, M., Tsoukas, A., and LoGerfo, F. W. (1991). A new canine model for evaluating blood prosthetic arterial graft interactions. *J. Biomed. Mater. Res.* 25: 1031–1038.

Brown, W. M. 3rd, Jay, J. L., Gott, J. P., Pan-Chih, Dorsey, L. M., Churchwell, A., and Guyton, R. A. (1991). Placement of aortic valve bioprostheses in sheep via a left thoracotomy: implantation of stentless porcine heterografts. *Trans. Am. Soc. Artif. Intern. Organs* 37: M445–M446.

Bunton, R. W., Xabregas, A. A., and Miller, A. P. (1990). Pericardial closure after cardiac operations. An animal study to assess currently available materials with particular reference to their suitability for use after coronary artery bypass grafting. *J. Thorac. Cardiovasc. Surg.* 100(1): 99–107.

Calne, R. Y., Bitter-Suermann, H., Davis, D. R., Dunn, D. C., Herbertson, B. M., Reiter, F. H., Sampson, D., Smith, D. P., and Webster, I. M. (1974). Orthotopic heart transplantation in the pig. *Nature* 247(437): 140–142.

Carroll, G. L., and Hartsfield, S. M. (1996). General anesthetic techniques in ruminants. *Vet. Clin. North Am. Food Anim. Pract.* 12(3): 627–661.

Cevallos, W. H., Holmes, W. L., Nyers, R. N., and Smink, R. D. (1979). Swine in atherosclerosis research—development of an experimental animal model and study of the effect of dietary fats on cholesterol metabolism. *Atherosclerosis* 34(3): 303–317.

Chanda, J., Kuribayashi, R., Abe, T., Sekine, S., Shibata, Y., and Yamagishi, I. (1997). Is the dog a useful model for accelerated calcification study of cardiovascular bioprostheses? *Artif. Organs* 21(5): 391–395.

Chow, E., and Farrar, D. J. (1992). Right heart function during prosthetic left ventricular assistance in a porcine model of congestive heart failure. *J. Thorac. Cardiovasc. Surg.* 104: 569–578.

Clutton-Brock, J. (1999a). Sheep and goats. in *A Natural History of Domesticated Mammals*, 2nd ed. Cambridge University Press, Cambridge, UK, pp. 69–80.

Clutton-Brock, J. (1999b). Pigs. in *A Natural History of Domesticated Mammals,* 2nd ed. Cambridge University Press, Cambridge UK, pp. 91–99.

Commission on Life Sciences (1996). Animal environment, housing, and management. in *Guide for the Care and Use of Laboratory Animals.* National Academy Press, pp. 21–55.

Cremer, J., Boetel, C., Fredow, G., Gebureck, P., Haverich, A. (1995). Radiographic assessment of structural defects in Bjork-Shiley convexo-concave prostheses. *Eur. J. Cardiothorac. Surg.* 9: 373–377.

Curl, G. R., Jakubowski, J. A., Deykin, D., and Bush, H. L. (1986). Beneficial effect of aspirin in maintaining the patency of small-caliber prosthetic grafts after thrombolysis with urokinase or tissue-type plasminogen activator. *Circulation* 74(Suppl I): 1–21.

Darwin, C. (1959). *The Life and Letters of Charles Darwin,* F. Darwin ed. Basic Books, New York, pp. 382–383.

David, T. E., Ropchan, G. C., and Butany, J. W. (1988). Aortic valve replacement with stentless porcine bioprostheses. *J. Card. Surg.* 3: 501–505.

Doumanian, A. V., and Ellis, F. H. (1961). Prolonged survival after total replacement of the mitral valve in dogs. *J. Thorac. Cardiovasc. Surg.* 42(5): 683–695.

Dunn, P. F., Newman, K. D., Jones, M., Yamada, I., Shayani, V., Virmani, R., and Dichek, D. A. (1996). Seeding of vascular grafts with genetically modified endothelial cells. Secretion of recombinant TPA results in decreased seeded cell retention in vitro and in vivo. *Circulation* 93(7): 1439–1446.

Favalaro, R. G. (1969). Saphenous vein graft in the surgical treatment of coronary artery disease: Operative technique. *J. Thorac. Cardiovasc. Surg.* 58: 178–185.

Frazier, O. H., Rose, E. A., Macmanus, Q., Burton, N. A., Lefrak, E. A., Poirier, V. L., and Dasse, K. A. (1992). Multicenter clinical evaluation of the HeartMate 1000 IP left ventricular assist device. *Ann. Thorac. Surg.* 53: 1080–1090.

Fritz, K. E., Daoud, A. S., Augustyn, J. M., and Jarmolych, J. (1980). Morphological and biochemical differences among grossly-defined types of swine aortic atherosclerosis induced by a combination of injury and atherogenic diet. *Exp. Mol. Pathol.* 32: 61–72.

Fujisawa, N., Poole-Warren, L. A., Woodard, J. C., Bertram, C. D., and Schindhelm, K. (1999). A novel textured surface for blood-contact. *Biomaterials* 20(10): 955–962.

Gabbay, S., Guindy, A. M., Andrews, J. F., Amato, J. J., Seaver, P., and Khan, M. Y. (1989). New outlook on pericardial substitution after open heart operations. *Ann. Thorac. Surg.* 48(6): 803–812.

Gajewski, J., and Povar, M. L. (1971). Blood coagulation values in sheep. *Am. J. Vet. Res.* 32: 405–409.

Gallo, I., and Frater, R. W. M. (1983). Experimental atrioventricular bioprosthetic valve insertion: a simple and successful technique. *Thorac. Cardiovasc. Surg.* 31: 288–290.

Ghoshal, N. G., and Nanda, B. S. (1975). Porcine heart and arteries. in *Sisson and Grossman's The Anatomy of the Domestic Animals,* 5th ed., R. Getty, ed. W.B. Saunders, Philadelphia.

Gott, J. P., Pan-Chih, Dorsey, L. M., Jay, J. L., Jett, G. K., Schoen, F. J., Girardot, J. M., and Guyton, R. A. (1992). Calcification of porcine valves: a successful new method of anti-mineralization. *Ann. Thorac. Surg.* 53: 207–216.

Granger, D. K., Matas, A. J., Jenkins, M. K., Moss, A. A., Chen, S. C., and Almond, P. S. (1994). Prolonged survival without posttransplant immunosuppression in a large animal model. *Surgery* 116(2): 236–241.

Grant, D., Duff, J., Zhong, R., Garcia, B., Lipohar, C., Keown, P., and Striller, C. (1988). Successful intestinal transplantation in pigs treated with cyclosporine. *Transplantation* 45(2): 279–284.

Green, G. E., Stertzer, S. H., and Reppert, E. H. (1968). Coronary artery bypass grafts. *Ann. Thorac. Surg.* 5(5): 443–450.

Grehan, J. F., Hilbert, S. L., Ferrans, V. J., Droel, J. S., Salerno, C. T., and Bianco, R. W. (2000). Development and evaluation of a swine model to assess the preclinical safety of mechanical heart valves. *J. Heart Valve Dis.* 9: 710–720.

Grehan, J. F., Casagrande, I., Oliveira, E. L., Santos, P. C., Pessa, C. J., Gerola, L. R., Buffolo, E., Mrachek, J. P., Norris, M. E., Lahti, M. T., and Bianco, R. W. (2001). A juvenile sheep model for the long-term evaluation of stentless bioprostheses implanted as aortic root replacements. *J. Heart Valve Dis.* 10(4): 505–512.

Gross, D. R. (1994). Recognition of pain and the use of analgesics. in *Animal Models in Cardiovascular Research,* 2nd revision ed., *Developments in Cardiovascular Medicine,* Vol. 153. Kluwer Academic Publishers, Dordrecht, The Netherlands, pp. 10–11.

Gross, D. R. (1997). Thromboembolic phenomena and the use of the pig as an appropriate animal model for research on cardiovascular devices. *Int. J. Artif. Organs* 20: 195–203.

Gross, D. R., Dewanjee, M. K., Zhai, P., Lanzo, S., and Wu, S. M. (1997). Successful prosthetic mitral valve implantation in pigs. *ASAIO J.* 43: M382–M386.

Hale, A. S. (1995). Canine blood groups and their importance in veterinary transfusion medicine. *Vet. Clin. North. Am. Small Anim. Prac.* 25(6): 1323–1332.

Hannon, J. P., Bossone, C. A., and Wade, C. E. (1990). Normal physiologic values for conscious pigs used in biomedical research. *Lab. Anim. Sci.* 40(3): 293–298.

Hasenkam, J. M., Ostergaard, J. H., Pederson, E. M., Paulsen, P. K., Nygaard, H., Schurizek, B. A., and Johannsen, G. (1988a). A model for acute haemodynamic studies in the ascending aorta in pigs. *Cardiovasc. Res.* 22: 464–471.

Hasenkam, J. M., Pedersen, E. M., Ostergaard, J. H., Nygaard, H., Paulsen, P. K., Johannsen, G., and Schurizek, B. A. (1988b). Velocity fields and turbulent stress downstream of a biological and mechanical aortic valve prostheses implanted in pigs. *Cardiovasc. Res.* 22: 472–483.

Hasenkam, J. M., Nygaard, H., Pedersen, E. M., Ostergaard, J. H., Paulsen, P. K., and Johannsen, G. (1989). Turbulent stresses downstream of porcine and pericardial aortic valves implanted in pigs. *J. Cardiac Surg.* 4: 74–78.

Hazekamp, M. G., Goffin, Y. A., and Huysmans, H. A. (1993). The value of the stentless biovalve prothesis: an experimental study. *Eur. J. Cardio-Thoracic Surg.* 7: 514–519.

Hecker, J. F. (1974). *Experimental Surgery on Small Ruminants.* Butterworths, London.

Hendry, P. J., Ascah, K. J., Rajagopalan, K., and Calvin, J. E. (1994). Does septal position affect right ventricular function during left ventricular assist in an experimental porcine model? *Circulation* Suppl 90(2): II353–II358.

Henneman, O. D., Van Rijk-Zwikker, Bruggemans, E. F., Rosendal, F. R., Delemarre, B. J., and Huysmans, H. A. (1998). The pig as an in vivo model for the evaluation of the thrombogenicity of mechanical heart valves. in *Workshop on Prosthetic Heart Valves: Future Directions,* February 18–22, Hilton Head, SC, p. 9.

Holmberg, D. L., and Olsen, D. B. (1987). Anesthesia and cardiopulmonary bypass technique in calves and sheep. *Vet. Surg.* 16(6): 463–465.

Hughes, H. C. (1986). Swine in cardiovascular research. *Lab. Anim. Sci.* 36(4): 348–350.

International Organization for Standardization Document 5840 (1996). Cardiovascular implants—cardiac valve prostheses.

International Organization for Standardization Document 7198-1 and 2 (1994). Cardiovascular implants—tubular vascular prostheses.

Irwin, E., Lang, G., Clack, R., St. Cyr, J., Runge, W., Foker, J., and Bianco, R. (1993). Long-term evaluation of prosthetic mitral valves in sheep. *J. Invest. Surg.* **6**: 133–141.

James, N. L., Schindhelm, K., Slowiaczek, B. K., Milthorpe, B., Graham, A. R., Munro, V. F., Johnson, G., and Steele, J. G. (1992). In vivo patency of endothelial cell-lined expanded polytetrafluroethylene prostheses in an ovine model. *Artif Organs* **16**: 346–353.

Jensen, N., Lindblad, B., Ljungberg, J., Leide, S., and Bergqvist, D. (1997). Early attachment of leukocytes, platelets and fibrinogen in endothelial cell-seeded Dacron venous conduits. *B. J. Surg.* **84**(1): 52–57.

Jordan, W. D., Sampson, L. K., Iyer, S., Anderson, P. G., Lyle, K., Brown, R. J., Luo, J., and Roubin, G. S. (1998). Abdominal aortic aneurysm repair via percutaneous endovascular stenting in the swine model. *Am. Surg.* **64**: 1070–1073.

Karas, S. P., Gravanis, M. B., Santoian, E. C., Robinson, K. A., Anderberg, K. A., and King, S. B., 3rd (1992). Coronary intimal proliferation after balloon-injury and stenting in swine: An animal model of restenosis. *J. Am. Coll. Cardiol.* **20**: 467–474.

Karges, H. E., Funk, K. A., and Ronneberger, H. (1994). Activity of coagulation and fibrinolyis parameters in animals. *Arzneim.-Forsch./Drug Res.* **44**: 793–797.

Katayama, H., Krzeski, R., Frantz, E. G., Ferreiro, J. I., Lucas, C. L., Ha, B., and Henry, G. W. (1993). Induction of right ventricular hypertrophy with obstructing balloon catheter. Nonsurgical ventricular preparation for the arterial switch operation in simple transposition. *Circulation* **89**(4): 1911–1912.

Kelly, G. L., and Eisemen, B. (1982). Development of a new vascular prosthetic. *Arch. Surg.* **117**: 1367–1370.

Kerzner, B., Kelly, M. H., Gall, D. G., Butler, D. G., and Hamilton, J. R. (1977). Transmissible gastroenteritis: sodium transport and the intestinal epithelium during the course of viral enteritis. *Gastroenterology* **72**(3): 457–461.

Kitchen, H. (1977). Sheep as animal models in biomedical research. *JAVMA* **170**(6): 615–9.

Kohler, T. R., and Kirkman, T. R. (1999). Dialysis access failure: A sheep model of rapid stenosis. *J. Vasc. Surg.* **30**(4): 744–751.

Leary, H. L. Jr., and Lecce, J. G. (1978). Effect of feeding on the cessation of transport of macromolecules by enterocytes of neonatal piglet intestine. *Biol. Neonate* **34**(3), 174–176.

Liao, K., Gong, G., and Hoffman, D. (1993). Spontaneous host endothelial growth on bioprosthetic valves and its relation to calcification. *Eur. J. Cardiothorac. Surg.* **7**: 591–596.

Lipsett, J., Cool, J. C., Runciman, S. I., Ford, W. D., Kennedy, J. D., and Martin, A. J. (1998). Effect of antenatal tracheal occlusion on lung development in the sheep model of congenital diaphragmatic hernia: a morphometric analysis of pulmonary structure and maturity. *Pediatr. Pulmonol.* **25**(4): 257–269.

Lumley, J. S. P., Green, C. J., Lear, P., and Angell-James, J. E. (1990). *Essentials of Experimental Surgery*. Butterworth & Co., London.

Marino, I. R., and De Luca, G. (1985). Orthotopic liver transplantation in pigs. An evaluation of different methods of avoiding the revascularization syndrome. *Transplantation* **40**(5): 494–498.

Marrie, T. J. (1995). Coxiella burnettii (Q fever) pneumonia. *Clin. Infect. Dis.* **21**(Suppl 3): S253–S264.

Matalon, S., Nesarajah, M. S., Krasney, J. A., and Farhi, L. E. (1982). Pulmonary and circulatory changes in conscious sheep exposed to 100% O2 at 1 ATM. *J. Appl. Physiol.* **53**(1): 110–116.

McKenzie, J. E. (1986). Swine as a model in cardiovascular research. in *Swine in Biomedical Research*, M. E. Tumbleson, and L. B. Schook, eds. Plenum Press, New York, pp. 7–17.

McKenzie, J. E. (1996). Swine as a model in cardiovascular research. in *Advances in Swine in Biomedical Research*, Vol. I, M. E. Tumbleson, and L. B. Schook, eds. Plenum Press, New York, pp. 3–17.

Mellen, I. M. (1952). *The Natural History of the Pig*. Exposition Press, New York.

Mendenhall, H. V. (1999). Surgical procedures. in *Handbook of Biomaterials Evaluation*, 2nd ed., A. F. von Recum, ed. Taylor and Francis, Philadelphia, pp. 481–493.

Mersmann, H. J., Goodman, J., Houk, J. M., and Anderson, S. (1972). Studies on the biochemistry of mitochondria and cell morphology in the neonatal swine hepatocyte. *J. Cell Biol.* **53**(2): 335–347.

National Academy Press Web site: http://www.nap.edu/catalog/5140.html

Nesarajah, M. S., Matalon, S., Krasney, J. A., and Fahri, L. E. (1983). Cardiac output and regional oxygen transport in the acutely hypoxic conscious sheep. *Respir. Physiol.* **53**(2): 161–172.

Neville, R. F. Jr., Bartorelli, A. L., Sidawy, A. N., and Leon, M. B. (1994). Vascular stent deployment in vein bypass grafts: observations in an animal model. *Surgery* **116**(1): 55–61.

Newman, J. H., Loyd, J. E., English, D. K., Ogletree, M. L., Fulkerson, W. J., and Brigham, K. L. (1983). Effects of 100% oxygen on lung vascular function in awake sheep. *J. Appl. Physiol.* **54**(3): 1379–1386.

Northrup, W. F. III, Mrachek, J. P., McClay, C., and Feeny, D. A. (1998). A novel annuloplasty system with rigid and flexible elements: initial experimental results in sheep and the case for untreated autologous pericardium. *J. Heart Valve Dis.* **7**(1): 62–71.

Okazaki, Y., Wika, K. E., Matasuyoshi, T., Fukamachi, K., Kunitoma, R., Tweden, K. S., and Harasaki, H. (1996). Platelets are deposited early post-operatively on the leaflet of a mechanical heart valve in sheep without post-operative antiplatelet agents. *Am. Soc. Artif. Organs J.* **42**: M750–M754.

Ouyang, D. W., Salerno, C. T., Pederson, T. S., Bolman, R. M., III, and Bianco, R. W. (1998). Long term evaluation of orthotopically implanted stentless bioprosthetic aortic valves in juvenile sheep. *J. Invest. Surg.* **11**(3): 175–183.

Oz, M. C., Argenziano, M., Catanese, K. A., Gardocki, M. T., Goldstein, D. J., Ashton, R. C., Gelijns, A. C., Rose, E. A., and Levin, H. R. (1997). Bridge experience with long-term implantable left ventricular assist devices. Are they an alternative to transplantation? *Circulation* **95**: 1844–1852.

Papadakis, K., Luks, F. I., Deprest, J. A., Evrard, V. E., Flageole, H., Miserez, M., and Lerut, T. E. (1998). Single-port tracheoscopic surgery in the fetal lamb. *J. Pediatr. Surg.* **33**(6): 918–920.

Park, S. J., Nguyen, D. Q., Bank, A. J., Ormaza, S., and Bolman, RM III. (2000). Left ventricular assist device bridge therapy for acute myocardial infarction. *Ann. Thorac. Surg.* **69**(4): 1146–1151.

Piermattei, D. L., and Swan, H. (1970). Techniques for general anesthesia in miniature pigs. *J. Surg. Res.* **10**(12): 587–592.

Pinches, S. A., Gribble, S. M., Beechey, R. B., Ellis, A., Shaw, J. M., and Shirazi-Beechey, S. P. (1993). Preparation and characterization of basolateral membrane vesicles from pig and human colonocytes: the mechanism of glucose transport. *Biochem. J.* **294**(Pt 2): 529–534.

Poole, T. B, and Robinson, R. (eds). (1987). *The UFAW Handbook on the Care and Management of Laboratory Animals*, 6th ed. Churchill Livingstone, New York.

Poole-Warren, L. A., Schindhelm, K., Graham, A. R., Slowiaczek, P. R., and Noble, K. R. (1996). Performance of small diameter synthetic vascular prostheses with confluent autologous endothelial cell linings. *J. Biomed. Mater. Res.* **30**: 221–229.

Ramo, B. W., Peter, R. H., Ratliff, N., Kong, Y., McIntosh, H. D., and Morris, J. J., Jr. (1970). The natural history of right coronary arterial occlusion in the pig. Comparison with left anterior descending arterial occlusion. *Am. J. Cardiol.* **26**(2): 156–161.

Reverdiau-Moalic, P., Watier, H., Vallee, I., Lebranchu, Y., Bardos, P., and Gruel, Y. (1996). Comparative study of porcine and human blood coagulation systems: possible relevance in xenotransplantation. *Transplant Proc.* 28: 643–644.

Riebold, T. W., and Thurmon, J. C. (1985). Anesthesia in swine. in *Swine in Biomedical Research*, M. E. Tumbleson, ed. Plenum Press, New York, pp. 243–254.

Risdahl, J. University of Minnesota Research Animal Resources Web site: http://www.ahc.umn.edu/rar/restraint/pigrestr.html

Rodgers, G. P., Minor, S. T., Robinson, K., Cromeens, D., Wollbert, S. C., Stephens, L. C., Guyton, J. R., Wright, K., Roubin, G. S., and Raizner, A. E. (1990). Adjuvant therapy for intracoronary stents. Investigations in atherosclerotic swine. *Circulation* 82: 560–569.

Rossiter, S. J., Brody, W. R., Kosek, J. C., Lipton, M. J., and Angell, W. W. (1974). Internal mammary artery versus autogenous vein for coronary artery bypass graft. *Circulation* 50: 1236–1243.

Roubin, G. S., King, S. B., III, Douglas, J. S., Jr., Lembo, N. J., and Robinson, K. A. (1990). Intracoronary stenting during percutaneous transluminal coronary angioplasty. *Circulation* 81(3Suppl): IV92–100.

Rousseau, H., Puel, J., Joffre, F., Sigwart, U., Duboucher, C., Imbert, C., Knight, C., Kropf, L., and Wallsten, H. (1987). Self-expanding endovascular prosthesis: an experimental study. *Radiology* 164(3): 709–714.

Salerno, C. T., Droel, J., and Bianco, R. W. (1998). Current state of in vivo preclinical heart valve evaluation. *J. Heart Valve Dis.* 7(2): 158–162.

Santos, R. M., Tnaguay, J. F., Crowley, J. J., Kruse, K. R., Sanders-Millare, D., Zidar, J. P., Phillips, H. R., Merhi, Y., Garcia-Cantu, E., Bonan, R., Cote, G., and Stack, R. S. (1998). Local administration of L-703,081 using a composite polymeric stent reduces platelet deposition in canine coronary arteries. *Am. J. Cardiol.* 82(5): 673–675.

Schatz, R. A., Palmaz, J. C., Tio, F. O., Garcia, F., Garcia, O., and Reuter, S. R. (1987). Balloon-expandable intracoronary stents in the adult dog. *Circulation* 76(2): 450–457.

Schoen, F. J., Hirsch, D., Bianco, R. W., and Levy, R. J. (1994). Onset and progression of calcification in porcine aortic bioprosthetic valves implanted as orthotopic mitral valve replacements in juvenile sheep. *J. Thorac. Cardiovasc. Surg.* 108: 880–887.

Schurmann, K., Vorwerk, D., Bucker, A., Grosskortenhaus, S., and Gunther, R. W. (1998). Single and tandem stents in sheep iliac arteries: is there a difference in patency? *Cardiovasc. Intervent. Radiol.* 21(5): 411–418.

Schwartz, R. S., and Holmes, D. R. (1994). Pigs, dogs, baboons, and man: lessons for stenting from animal studies. *J. Interven. Cardiol.* 7: 355–368.

Scott, N. A., Robinson, K. A., Nunes, G. L., Thomas, C. N., Viel, K., King, S. B. III, Harker, L. A., Rowland, S. M., Juman, I., and Cipolla, G. D. (1995). Comparison of the thrombogenicity of stainless steel and tantalum coronary stents. *Am. Heart. J.* 129: 866–872.

Sharp W. V. (1970). The carotid artery—a test site for the small vessel prosthetics. *J. Surg. Res.* 10(1): 41–46.

Sielaff, T. D., Nyberg, S. L., Rollins, M. D., Hu, M. Y., Amiot, B., Lee, A., Wu, W. S., and Cerra, F. B. (1997). Characterization of the three-compartment gel-entrapment porcine hepatocyte bioartificial liver. *Cell Biol. Toxicol.* 13(4-5): 357–364.

Soini, H. O., Takala, J., Nordin, A. J., Makisalo, H. J., and Hockerstedt, K. A. (1992). Peripheral and liver tissue oxygen tensions in hemorrhagic shock. *Crit. Care Med.* 20(9): 1330–1334.

Spyt, T. J., Fisher, J., Reid, J., Anderson, J. D., and Wheatley, D. J. (1998). Animal evaluation of a new pericardial bioprosthetic heart valve. *Artif. Organs* 12(4): 328–336.

Stanton, H. C., and Mersmann, H. J. (1986). Preface. in *Swine in Cardiovascular Research*, H. C. Stanton, and H. J. Mersmann, eds. Vol. I. CRC Press, Boca Raton, FL.

Starr, A., and Edwards, L. (1961). Mitral replacement: clinical experience with a ball-valve prosthesis. *Ann. Surg.* 154(4): 726–740.

Swan, H., and Meagher, M. (1971). Total body bypass in miniature pigs. *J. Thorac. Cardiovasc. Surg.* 61: 956–967.

Swan, H., and Piermattei, D. L. (1971). Technical aspects of cardiac transplantation in the pig. *J. Thorac. Cardiovasc. Surg.* 61: 710–723.

Swindle, M. M. (1992). Preface. in *Swine as Models in Biomedical Research*, M. M. Swindle, ed. Iowa State University Press, Ames, IA, pp. ix–x.

Swindle, M. M. (1994). Anesthetic and perioperative techniques in swine: an update. in *Charles River Laboratories Technical Bulletin*, Charles River Laboratories, Wilmington, DE, pp. 1–3.

Swindle, M. M., and Smith A. C. (1994). Swine: anesthesia and analgesia. in *Research Animal Anesthesia, Analgesia and Surgery*, A. C. Smith, and M. M. Swindle, eds. Scientists Center for Animal Welfare, Greenbelt, MD, pp. 107–110.

Swindle, M. M., Horneffer, P. J., Gardner, T. J., Gott V. L., Hall, T. S., Stuart R. S., Baumgartner, W. A., Borkon, A. M., Galloway, E., and Reitz, B. A. (1986). Anatomic and anesthetic considerations in experimental cardiopulmonary surgery in swine. *Lab. Anim. Sci.* 36(4): 357–361.

Taylor, A., Ao, P., and Fletcher, J. (1995). Inhibition of intimal hyperplasia and occlusion in Dacron graft with heparin and low molecular weight heparin. *Int. Angio.* 14(4): 375–380.

Tillman, P., Carson, S. N., and Talken, L. (1981). Platelet function and coagulation parameters in sheep during experimental vascular surgery. *Lab. Anim. Sci.* 31: 262–267.

Tobin, E., and Hunt, E. (1996). Supplies and technical considerations for ruminant and swine anesthesia. *Vet. Clin. North Am. Food Anim. Pract.* 12(3): 531–547.

Tomizawa, Y., Moon, M. R., DeAnda, A., Castro, L. J., Kosek, J., and Miller, D. C. (1994). Coronary bypass grafting with biological grafts in a canine model. *Circulation* 90(part 2): II-160–II-166.

Towne, C. W., and Wentworth, E. H. (1950). *Pigs from Cave to Corn Belt*. University of Oklahoma Press, Norman, OK.

Tumbleson, M. E., and Schook, L. B. (1996). Advances in swine in biomedical research. in *Advances in Swine in Biomedical Research*, M. E. Tumbleson, and L. B. Schook, eds. Vol. I. Plenum Press, New York, pp. 1–4.

Ueda, Y., Ditakaze, M., Imakita, M., Ishibashi-Ueda, H., Minamino, T., Asanuma, H., Ozaki, T., Imamura, E., Kuzuya, T., and Hori, M. (1999). Glycoprotein IIb/IIIa antagonist FK633 could not prevent neointimal thickening in stent implantation model of canine coronary artery. *Arterioscler. Thromb. Vasc. Biol.* 19(2): 343–347.

University of Minnesota Research Animal Resources Web site: http://www.ahc.umn.edu/rar/housing.html#farm

Vallana, F., Rinaldi, S., Galletti, P. M., Nguyen, A., and Piwnica, A. (1992). Pivot design in bileaflet valves. *ASAIO J.* 38: M600–M606.

Vyavahare, N., Hirsch, D., Lerner, E., Baskin, J. Z., Schoen, F. J., Bianco, R. W., Kruth, H. S., Zand, R., and Levy, R. J. (1997). Prevention of bioprosthetic heart valve calcification by ethanol preincubation: efficacy and mechanisms. *Circulation* 95: 479–458.

Wakabayashi, A., and Connolly, J. E. (1968). Comparative flow studies of myocardial revascularization grafts. *J. Thorac. Cardiovasc. Surg.* 56: 633–642.

Wakabayashi, A., Connolly, J. E., Adachi, R. T., Black, R. M., Stemmer, E. A., and Eisenman, J. J. (1971). Experimental development of the ascending aorta-coronary artery bypass. *Arch. Surg.* 103: 36–40.

Walpoth, B. H., Ammon, A., Galdikas, S., Ris, H. B., Schaffner, T., Hoflin, F., Schilt, W., Nettler, D., Nachnbur, B., and Althaus, U. (1993). Experimental assessment of thrombogenicity in vascular prostheses before and during prostaglandin E treatment. *Eur. J. Vasc. Surg.* 7(5): 493–499.

Whitbread, T., Birch, P., Rogers, S., Majeed, A., Rochester, J., Beard, J. D., and Gaines, P. (1996). A new animal model for abdominal aortic aneurysms: initial results using a multiple-wire stent. *Eur. J. Vasc. Endovasc. Surg.* 11: 90–97.

White, C. J., Ramee, S. R., Banks, A. K., Wiktor, D., and Price, H. L. (1989). The Yucatan miniature swine: an atherogenic model to assess the early patency rates of an endovascular stent. in *Swine as Models in Biomedical Research*, M. M. Swindle, ed. Iowa State University Press, Ames, IA, pp. 185–196.

White, J. G., Mulligan, N. J., Gorin, D. R., D'Agostino, R., Yucel, E. K., and Menzoian, J. O. (1998). Response of normal aorta to endovascular grafting: a serial histopathological study. *Arch. Surg.* 133: 246–249.

Wilbourn, B., Harrison, P., Lawvie, A., Savariau, E., Savidge, G., and Cramer, E. M. (1993). Porcine platelets contain an increased quantity of ultra-high molecular weight von Willebrand factor and numerous alpha-granular tubular structures. *Br. J. Haematol.* 83: 608–615.

Willette, R. N., Zhang, H., Louden, C., and Jackson, R. K. (1996). Comparing porcine models of coronary resetnosis. in *Swine in Biomedical Research*, M. E. Tumbleson, and L. B. Schook, eds. Plenum Press, New York, pp. 595–606.

5.6 MICROSCOPY FOR BIOMATERIALS SCIENCE

Kip D. Hauch

Along with test tubes, beakers, and a white lab coat, the microscope is an icon of the professional scientist's toolchest. Of all the senses, the scientist relies most heavily on vision; thus the advent of the light microscope is among the earliest and arguably most powerful tools in the history of science. Today, microscopy in its various forms still remains as the preeminent tool at the forefront of scientific exploration. Thus the biomaterials scientist makes extensive use of both light and electron microscopes to help fabricate and characterize new materials, coatings, and devices, and to study the behavior of cells and tissues at the biomaterials interface. This chapter will familiarize the reader with the more common microscopy tools used in current biomaterials research. The key concepts of magnification, resolution, and contrast are first introduced and then their meaning explored in the context of light microscopy. Digital Imaging is very briefly addressed. Attention then moves to the electron microscopies, specifically SEM. Fortunately, the ubiquitous importance of microscopy in the practice of science means that there exist numerous excellent resources to consult on these topics. A brief guide to some useful resources is found at the end of the chapter.

MAGNIFICATION, RESOLUTION, AND CONTRAST

The art of Scientific Instrumentation is to devise tools to observe natural phenomena and present them appropriately to our senses for interpretation (e.g., to our eyes as an image). Three concepts apply to this process: magnification, resolution, and contrast. The three concepts are interdependent, and frequently limitations in one aspect or the others lead to trade-offs.

Magnification is the appropriate scaling of the phenomena to our detectors (ultimately, for our purposes here, the human visual system). In a standard light microscope this is accomplished through the millennia old traditions of refracting light through curved glass lenses (e.g., an objective lens, or an ocular). In the electron microscope, as well as the other scanning microscopies (e.g., laser scanning confocal, multiphoton, scanning tunneling, or atomic force microscopies), photons or electrons are used to probe the sample discretely at a single point. The location of this sampling point is then rastered across the sample. Magnification is achieved by displaying the combined results from this "probing" on a more appropriate scale.

Resolution is dependent on the information present in the collected data sufficient to allow one to achieve perception. In microscopy we are usually interested in spatial resolution—the ability to resolve one location from another in an image. Resolution is often limited by the physical nature of the probe and/or the measurement of the signal thereof.

Contrast is the ability to detect specific differences in the signal. More than just signal/noise, contrast implies some aspect of specificity—the ability to tell one particular part of an image or signal from another. Resolution and contrast are often related—one is often gained at the expense of the other. With respect to light microscopes, the fundamental limits of magnification and resolution have generally been reached long ago. New and varied methods for the generation of contrast are what lead to the many different modalities of light microscopy—and as will be shown here, continue to be a source of technological development.

CONFIGURATIONS

A specimen can be imaged by passing illuminating photons or electrons through the sample (transmission microscopy, diascopic illumination) or by reflecting the illumination off the sample (reflected microscopy, episcopic illumination) to a detector.

In transmission microscopy, the sample must be thin enough to allow the probe to pass through the sample and reach the detector. For example, in transmitted light microscopy (probe = photons), samples may be up to 100 μm thick (although for typical high-resolution histology work, sections of tissue are prepared that are nominally only 5 μm thick.) In transmission electron microscopy (TEM, probe = electrons), samples are prepared that are only 50 nm thick.

Reflected techniques are used when the sample is opaque. The sample must reflect the illumination or otherwise generate a signal that is directed from the surface back to a detector. Reflected light is how we are used to viewing the everyday world around us. Reflected light techniques include simple magnifying hand lenses, stereoscopes and dissecting/operating room microscopes, and compound microscopes commonly used in the material sciences or manufacturing sector. As will be discussed, fluorescence-based techniques are almost always performed with episcopic illumination. Scanning electron microscopy (SEM) can be considered in this category as well, although the signal generated is somewhat more complex than simply the detection of reflected electrons.

LIGHT MICROSCOPY

The biomaterials scientist relies daily on a variety of light microscopes for tasks ranging from routine materials inspection to detailed complex biologic analyses. The interests of the biomaterials engineer are often focused on the biologic responses at the cell–biomaterial interface. These are studied *in vitro*, e.g., using cell culturing approaches, or studied *in vivo* through the examination of a tissue–implant interface. In this regard, the microscopy needs of the biomaterials scientist overlap significantly with those of the molecular cell biologist or pathologist. The presence of a biomaterial in the sample is often just another (sometimes complicating) factor to be dealt with.

Platforms

Light microscopes can be configured on several different platforms—for different purposes and with varying levels of customization. Hand lenses, loupes, stereoscopes, and dissecting microscopes are essential for fabrication and inspection of small biomedical devices and coatings. Likewise, operating microscopes are often helpful for implant work in small animal models. An upright microscope is one in which the sample is observed from above by an objective lens. Upright microscopes are used most often for histology work and are equipped with high-resolution but short-working-distance optics. A typical sample might be a slice of tissue mounted between a microscope slide and a coverslip. An inverted microscope is the mainstay of the tissue culture laboratory, where specimens are viewed through objective lenses positioned underneath the specimen stage. This allows for the observation of live cells in culture, e.g., through the transparent bottoms of tissue culture flasks or multiwell plates. The large open stage of the inverted instrument allows for more experimental versatility, e.g., the use of complicated flow cells, micromanipulators, or environmental enclosures. An inverted instrument is often equipped with slightly lower resolution, but longer working-distance optics. Either platform may be augmented for the specialized techniques to be discussed later, namely phase contrast, differential interference contrast, and fluorescence microscopy. With respect to research-grade instruments, there are more similarities than differences in the capabilities of the upright and inverted platforms. However, it is crucial that the choice of instrument platform and the experimental design receive mutual consideration early in the experimental planning stages.

Magnification in Light Microscopy

Magnification refers to the ratio of a feature dimension in the presented image to the corresponding feature dimension in the original specimen. In common practice, the range of magnification available in light microscopy reaches from $5\times$ to about $1000\times$ (a $100\times$ objective lens and a $10\times$ ocular). These values describe the magnification of the sample at the plane of the primary detector—our eyes at the ocular. Herein lies danger for the modern microscopist, since invariably it is not our eyes, but rather a digital camera that acquires an image for dissemination to our colleagues. The camera does not use the $10\times$ ocular; rather it uses a combination of extension tubes, transfer optics, intermediate magnifiers or reducers, camera coupling lenses, etc. Further, the acquired digital data are then displayed on computer monitors directly as images or embedded in mixed content files (.pdf) at different "zoom levels"; printed on paper for reports, reduced in journal pages, or plastered on poster boards for conferences; perhaps even displayed using data projectors onto huge screens in an auditorium. What then does magnification mean?

The days of conveying the level of magnification in an image by reporting the magnification of the objective lens alone, or using vague terms such as "low power" or "high power," are long gone. In the digital age it is far preferable to include a true size scale bar as part of every published image. This is conveniently done by separately acquiring an image containing objects of known size, most commonly a stage micrometer available through any microscopy supply house. With an image containing an object of known dimension, it is then straightforward to establish the relationship between distance (μm) in the specimen and pixels, the individual element of the acquired digital image. A scale bar can then be embedded in the images used for presentation, and measurements of feature size reported directly in appropriate units. Further—no matter how the image is then later displayed, the viewer will have a direct feel for the size of the features in the image. Modern software makes these operations relatively painless, and the side-by-side presentation of images taken at macroscopic, microscopic, and nanoscopic scales makes this practice indispensable to the scientific audience.

Resolution in Light Microscopy

The Rayleigh criterion is often used to describe the resolving power in light microscopy.

$$d = 1.22\lambda/(NA_{condenser} + NA_{obj}) \text{ for brightfield,}$$

and

$$d = 0.61\lambda/NA_{obj} \text{ for fluorescence,}$$

where λ is the wavelength of light. NA is the numerical aperture of the condenser or objective lens and is defined as

$$NA = n \sin\Theta$$

where n is the index of refraction of the medium between the lens and the sample (1.0 for air, 1.515 for oil) and Θ is the half-angle of the cone of light collected by the lens (Fig. 1). Similarly, the depth of focus, which is the thickness in the sample within which features will all appear in focus, is given by

$$Z = n\lambda/NA_{obj}^2$$

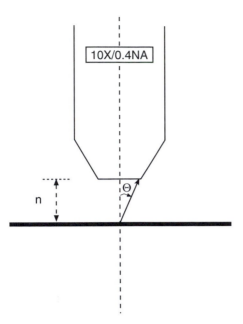

10X/0.4NA

FIG. 1. Numerical aperture: Θ = half the acceptance angle of the lens, n = the refractive index of the medium between the sample and the objective, $n = 1.00$ for air, $n = 1.515$ for oil.

These equations are idealized and are based on the first overlap of the diffraction patterns of two infinitely small point sources in the specimen. For monochromatic light at 546 nm and a high-quality oil immersion lens with a NA of 1.3, this equates to a resolving power of about 0.26 μm, easily adequate to resolve many subcellular organelles. Most modern research-grade microscopes are quite capable of realizing this resolution limit. However, the use of polychromatic white light may increase this value twofold. Further, the large values of NA_{cond}, which lead to optimal resolution, actually reduce the contrast available in the image. Thus, we may achieve superb resolution but not be able to distinguish the features which we seek to resolve! Here we find one of the classic trade-offs—between resolution and contrast.

It is important to understand the resolution available in an image when considering the tools for digital image acquisition. Ideally, the physical size of the individual sensing elements (e.g., the pixel on a digital imager) will be less than half the smallest resolvable feature size, after magnification by the objective and any intermediate magnifiers between the microscope and the imaging plane. The appropriate matching of physical pixel size in the detector to the resolving power of the optics is crucial, and if overlooked can result in degraded system performance. One finds that our eyes, and high-quality film, are well suited to the task. Lower power but relatively higher NA objectives (e.g. $10\times/0.4$, $20\times/0.75$) are the most demanding, i.e., they require the smallest pixels, and hence largest number of pixels in the detector.

Contrast and the Various Modes of Light Microscopy

The interaction of light with matter is what generates the contrast that we detect in a microscope image. Light can

be characterized by its amplitude, wavelength, phase, and polarization. Unfortunately, our human eyes, and most other electronic detectors, are sensitive only to amplitude and wavelength! Thus a light microscope must convert these many interactions of light and matter into differences in amplitude or wavelength that can be detected in the image. The various ways in which contrast is generated lead to the various light microscopy techniques, or modalities. These are described here in quite general terms—the reader is encouraged to explore the many excellent introductory texts on optical microscopy for more thorough coverage.

Transmitted Light Microscopy

The simplest way to view a thin specimen is to shine light through the specimen and view the magnified image directly. In this method, brightfield, we rely on the sample to provide its own contrast. Different features in the specimen may absorb some of the light and/or do so unequally at various wavelengths, thus leading to amplitude or color contrast. In many instances this may work—but consider a layer of unadulterated cultured cells on a glass slide. These cells, if well spread, may only be a few microns thick; made mostly of water, they absorb very little light and provide almost no contrast. Indeed it may be very difficult to see the unstained cells at all!

Another technique is darkfield microscopy. In darkfield the sample is illuminated obliquely, at such glancing angles that the illumination light, if undisturbed by the specimen, does not even enter the objective lens. Thus the empty areas (the background areas or "surround") in the image are completely black. Only where elements in the sample scatter light into the objective lens is light collected. Thus a darkfield image appears as pinpoints of light on a black background. Darkfield is particularly effective if the specimen contains small particulates.

Adding Contrast Agents: Chromophores and Fluorophores

How then can one generate more contrast in a specimen? The simplest answer is to cheat! Exogenous contrast agents can be added to the specimen. These are strongly absorbing or fluorescing large molecules or nanoparticles that are placed into the sample and ideally reside only in certain structures or features of the investigator's choosing. This, of course, is what is done in the staining of cells or tissue sections for histology. Many of these chromophores or stains date to ancient times when they were used to dye biologic materials, such as plant and animal tissues, skins, or textile fibers. Perhaps the most common histologic stains are hematoxylin and eosin (H and E). Hematoxylin (which appears blue) is termed a basophilic dye and binds to basic components such as those found in the cell nucleus. Eosin (which appears pink and is also quite fluorescent) is acidophilic and generally stains cytoplasmic proteins. A plethora of histologic stains is available to stain a myriad of cells, or structures within cells and tissues. Other common examples include Masson's trichrome, which is used to stain connective tissue or fibrosis around implants (collagen is green); alizarin red, for calcific deposits; ver Hoeff's for elastin; Geimsa and Wright's stain for blood cells; and the Gram stain for bacteria.

It is easy to see the beauty in chromophores (both literally and figuratively) since not only do they generate contrast in the image, but they do so with specificity. The investigator not only can see the cell, but can now see which part is a nucleus, determine the specific location of structural proteins, etc. This power of specificity can be enhanced by directly (covalently) linking chromophores or fluorophores to specific proteins or other molecules of interest, and then identifying and following the location of these reporter-linked conjugates on surfaces, inside cells, and in tissues.

Further, chromogenic substrates have been synthesized for certain key enzymes. These substrates are converted from colorless to colored (or otherwise undergo a color change) upon cleavage by the enzyme. Examples are the many substrates available for horseradish peroxidase, alkaline phosphatase, and β-galactosidase. These substrates can be used to detect enzymatic activity in certain cells and tissues directly.

Reaching yet another level of specificity, these enzymes can be conjugated to antibodies to impart molecular-level specificity in their binding. Thus the location of specific molecular antigens can be elucidated by allowing these antibody conjugates to bind, and then revealing the location of the antibodies by using the chromogenic substrates. This process (immunohistochemistry) is directly analogous to forms of the enzyme-linked immunosorbent assays (ELISA) used on the benchtop. Alternatively, the antibodies may be labeled directly with a fluorophore (direct immunofluorescence), or their locations revealed by the binding of a second labeled antibody that recognizes the first (secondary or indirect immunofluorescence). Finally, similarly labeled nucleic acid strands can be used to probe for and reveal the presence of certain sequences of genomic material in cells (in situ hybridization) or on surfaces (so-called microarray analysis) and thus give a glimpse into the genomic profile of cells and tissues.

The immense power and versatility afforded by the use of chromophores and fluorophores is now made clear. Beyond simply generating contrast, they provide molecular-level specificity. This enables the investigator not only to simply visualize cells and structures, but to begin to ask specific cell-physiology and bioengineering questions: is a particular species (a molecule, protein, receptor, enzyme) present? If so where is it, when does it appear, and in what quantities? Here in this generation of contrast lies the power of modern microscopy and its contribution to the cell biology and pathology fields so integral to the study of the biomaterials–biology interface. Because fluorophores play such a prominent role in these techniques, and since their use requires specialized technical approaches, fluorescence microscopy will be addressed in more detail in a separate section.

Optical Tricks: More Contrast—but No Cheating

So far, to generate contrast we have relied on the inherent contrast in the specimen (often inadequate) or the addition of contrast agents. Many of these contrast agents are toxic to cells. Histologic stains are typically used on tissue sections where cells have been fixed, embedded in a support matrix, and sliced open. Probes that bind to intracellular components have to cross the cell membrane; thus cells are permeabilized to allow access. Although there are "vital dyes" that can be used on living cells, many of the approaches described above interfere with the normal physiologic functioning of the cell. Thus, they are best suited for endpoint analysis. How then can one more safely generate contrast in living cells? How can one generate contrast without the addition of these exogenous agents? How can one generate contrast without "cheating?"

Fortunately the answer lies in several types of optical "tricks." Cells have limited ability to cause absorbance or color changes. However, there is plenty of diffraction (bending of light around organelles), refraction (bending of light at interfaces of structures with different indices of refraction, e.g., membranes), and other optical phenomena that can be used to our advantage. Unfortunately, our eyes are insensitive to all but changes in light intensity and color. The tricks, therefore, will be used to convert these other phenomena into differences in intensity or color, which we then can detect. These techniques all involve optical elements placed in the light path that modulate the incoming light, as well as elements that modulate the light collected by the objective. As such the sample usually resides in the middle of these two elements, and thus these can be thought of as the "sandwich techniques" (Foster, 1997).

Polarization

Molecules and materials that exhibit a strong regular repeat in their electronic structures are likely to induce orientationally specific changes in polarization during their interactions with light. Examples of such polarizing (birefringent) materials include metals, inorganic and organic crystals, collagen, certain organelles, muscle sarcomeres, and many others. Recall that light polarized by a polarizing filter will not be transmitted through a second polarizing filter oriented at 90° to the first (extinction). If the first filter is placed in the illumination path before reaching the sample, and the second filter is placed in the imaging path after sample, extinction will take place. No light will get through, provided that the sample does not impart any additional polarization to the transmitted light. However, if the sample does impart additional polarization, the conditions for extinction will be unsatisfied and the location of the polarizing materials immediately revealed. The degree of this additional polarization induced by the sample can even be measured with yet a third polarizing element (a compensator). The directionality of the polarization can be determined by rotating the specimen stage.

Synthetic polymers, of course, can also contain strong regular repeats in their electronic structure. This can be very useful in locating and identifying remnants of implanted polymeric materials in explanted tissue. The polarization induced by polymers can also be a disadvantage, however, as the plastics used in common tissue-culture ware make polarization microscopy nearly impossible using these substrates.

Phase Contrast and Hoffman Modulation Contrast

While there may be little amplitude change when light traverses a cell, there is plenty of diffraction of light by organelles, structures, and membranes. These rays that have interacted with the specimen experience an (unnoticeable) change in phase, having experienced an optical path-length difference $\Delta = (n_2 - n_1)(\text{thickness})$. Typically, the light is retarded in phase by an amount approximately $-\lambda/4$. In the 1930s, Dutch physicist Fritz Zernicke invented a method to convert these phase changes into changes in intensity. The result was phase contrast microscopy and the method helped foster the major advances in cell biology developed through the study of living cells in culture. For his work, Zernicke received the Nobel prize in 1953.

The key to phase contrast is to somehow separate those rays of light that pass through the specimen undisturbed (undiffracted) and thus contribute to the background (called the surround, S), from those rays that have been diffracted by interaction with the specimen (D). This can be accomplished spatially at a special plane in the optical path, the diffraction plane. At this plane an image is not formed; rather one can envision that rays of light are separated spatially by their degree of diffraction. Here each can be manipulated or treated selectively. In phase contrast, an annulus is placed in the illumination light path in a plane conjugate to this diffraction plane (Fig. 2). This results in a ring or cone of illumination approaching the specimen. Next, a special plate is placed at the diffraction plane, beyond the sample (actually located within the objective lens). This plate is constructed such that its thickness is less in a ring shaped region that matches exactly that of the ring in the annulus plate. Thus rays of light that are not diffracted (S) will pass through this "phase ring," while diffracted rays (D) will pass through the remaining portions of the phase plate. This phase ring results in a relative $+\lambda/4$ advancement in the phase of light that is undiffracted. Further this ring is made semitransparent, so that the light transmitted (i.e., the light that will contribute to the background in the final image) is reduced in intensity and thus appears gray. The diffracted rays have already experienced a retardation in their phase caused by their interactions with the specimen ($-\lambda/4$). When the surround (S) and the diffracted rays (D) are recombined in the real image plane, the resulting phase difference is ($\lambda/2$) and destructive interference takes place. Now, the changes in phase are converted to a reduction in intensity—which is made manifest in the recombined image as areas of dark (black).

We see that phase contrast is again a "sandwich technique" utilizing a phase ring in the illumination path, and a matching ring in the image-forming path. The sample is in the middle. This latter ring is actually physically built into the objective lens, and thus phase contrast microscopy requires a special phase contrast objective and the use of a matching and carefully aligned phase annulus for each objective. Fortunately phase contrast objectives can also be used for routine brightfield observation and most fluorescence applications, so for many labs they are a good value.

Phase contrast is ideal for thinly spread cells such as those in cell culture. They appear in high contrast, their organelles

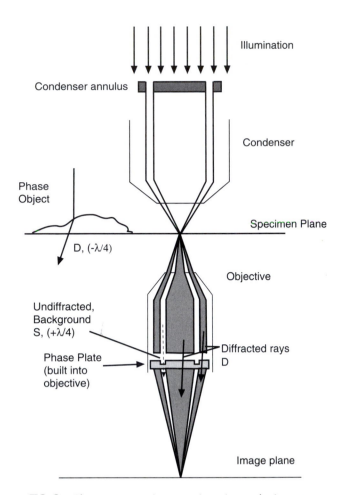

FIG. 2.   Phase contrast microscopy in an inverted microscope.

and pseudopodia revealed in areas of bright and dark on a gray background. Rounded cells, however, are surrounded by a strong halo of bright light. Simple phase contrast systems (often with fixed and prealigned annuli) are commonplace on inexpensive inverted microscopes found in tissue culture suites.

In Hoffman modulation contrast (no relation to this text's editor!) use is made again of the spatial separation of diffracted and undiffracted rays at the diffraction plane (Fig. 3). This time, instead of a cone of illumination an asymmetric slit is placed in the illumination path. Again a modulation plate is built into the objective placed at the diffraction plane. Also again, a semitransparent region is constructed to catch the rays of light emanating from the illumination slit and transversing the specimen undisturbed, thus resulting in a gray background. To one side of this surround region in the modulation plate, the plate is made fully transparent, while to the other side the plate is made opaque. The result then is that rays that are diffracted slightly in one direction appear quite bright, while rays diffracted in the opposite direction appear dark. In the recombined real image these variations in intensity appear as shades from light to dark across structural boundaries in the specimen. Our eyes conveniently interpret these gradations as shadows! The specimen takes on a distinct and crisp three-dimensional appearance. Hoffman modulation contrast also

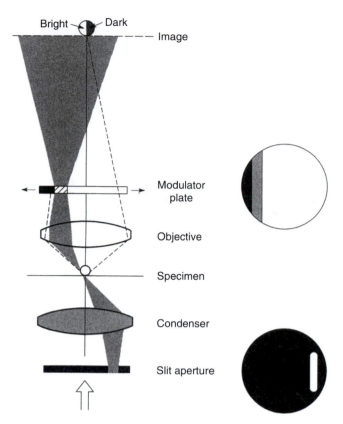

FIG. 3. Hoffman modulation contrast. Oblique illumination is provided by an off-axis slit in the condenser aperture. A modulator plate with matching complementary slit is placed in the objective back aperture and differentially blocks one sideband of diffracted light. From Murphy (2001).

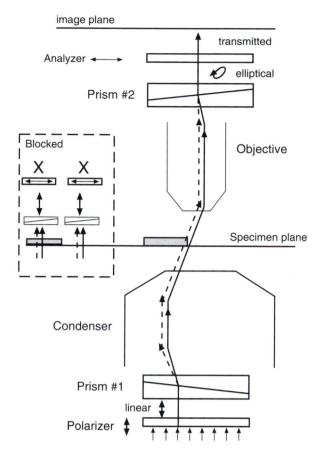

FIG. 4. Differential interference contrast on an upright microscope. Light is first linearly polarized, then split by the condenser prism (#1) into two spatially separate beams (shown as solid and dashed). If the two beams do not experience an optical path difference (inset), then they are recombined by the objective prism (#2) into a single ray that is again linearly polarized. This ray is then rejected by the second polarizer (the analyzer). If the two beams do experience an optical path difference in the sample, then the recombined ray is elliptically polarized and is partially transmitted by the analyzer. In this way spatial difference or gradients of refractive index in the sample are converted to differences in intensity (After Murphy, 2001).

requires the use of special objectives and illumination systems. Its use is compatible with tissue culture plasticware, unlike the other, more familiar shadow-like technique, DIC, which is discussed next.

Differential Interference Contrast Microscopy

The techniques discussed so far have converted differences in the optical path length through the specimen into variations in image intensity by utilizing oblique illumination and tricks at the diffraction plane. In differential interference contrast (DIC) microscopy; also called Nomarski, gradients in optical path length are sensed in a slightly different way—by splitting the incoming light into two parallel rays and allowing each ray to sample nearby areas of the sample, before being recombined to form an image. DIC is akin to an imaging version of a dual-beam interferometer.

DIC starts with linearly polarized light; thus a polarizer is again the first extra element in the illumination path (Fig. 4). Next, mounted near the condenser is a modified Wollaston prism that separates the incoming light into two parallel rays. These two rays of light are physically separated by a small fraction of a micron—indeed, a distance less than the resolved distance of the microscope. These two rays traverse

the specimen, and should they experience the same optical path through the specimen (i.e., no optical path difference), they are then recombined by a second prism located beyond the objective, again at the diffraction plane. This recombined light is again linearly polarized and can be selectively rejected by the use of a second polarizer (termed an analyzer) oriented at 90° to the first. In the presence of an optical path difference (e.g., due to a gradient in the index of refraction in the sample), the two rays are recombined and result in an elliptically polarized beam, which is now only partially attenuated by the analyzer. Thus, gradients in refractive index present in the sample are converted into gradients in intensity in the image. Our eyes perceive these as shadows and the sample takes on the striking appearance of three-dimensional relief. In practice, the objective prism is typically offset slightly so that the surround is not purely attenuated (black), but rather appears in shades of gray along the diagonal of the image.

Since DIC utilizes polarization in generating the contrast, it is not suitable for use with tissue culture plasticware, which will induce extraneous polarization. DIC produces crisp clear images in unstained specimens, with extraordinarily shallow depth of focus. This makes it ideal for optical sectioning tasks. DIC components are expensive and must be made from strain-free glasses. DIC is particularly amenable to combination with fluorescence techniques—where the fluorescence labeling provides the spatial localization of specific intracellular features (e.g., a nucleus or cytoskeleton) and DIC provides clarity to the body of the cell and other organelles.

FLUORESCENCE MICROSCOPY

Like chromophores, fluorophores are (usually) exogenous large molecules whose unique spectral properties make them readily identifiable under the microscope and thus generate contrast. Fluorescence techniques have had a wide-ranging and tremendous impact in the fields of molecular and cell biology, genomics, biophysics, bioengineering, and many others.

In the process of fluorescence, the fluorophore absorbs energy of a particular wavelength. While a small amount of this absorbed energy is lost to heat, the remainder is very quickly given up by the fluorophore as it reemits the energy as light of a slightly longer (less energetic) wavelength. The difference between these excitation and emission wavelengths is called the Stokes shift and provides the basis upon which the incoming and outgoing (emitted) light can be spectrally separated.

Fluorescein (known, together with its covalent linkage group isothiocyanate, as FITC) is perhaps the best known fluorophore. Today, there exist hundreds of new fluorophores, with various chemistries and spectral properties ranging from those excited by light in the UV range to the far red. The synthesis of new fluorophores with improved properties (e.g., increased quantum efficiency, resistance to photobleaching, narrower spectral characteristics) is an active enterprise. Excellent resources are available from vendors to help guide the choice of fluorophores for a myriad of applications (e.g., Molecular Probes, Inc.).

The applications of fluorescence are indeed numerous. As discussed earlier, fluorophores can be conjugated to a variety of biologically important molecules (lipids, membrane components, proteins, enzymes, antibodies, nucleotides, specific organelles) to serve a variety of reporting functions. Just a few common examples are the fluorophore DAPI to label cell nuclei and observe nuclear fragmentation during apoptosis; fluorescent conjugates of the mushroom toxin phalloidin which bind to f-actin in the cytoskeleton; and DiI for long-term cell tracking. The enzymatic activity of cells can be assessed by fluorescence, e.g., in the Live/Dead cell assay (Molecular Probes), where a masked nonfluorescent form of the green fluorophore calcein is cleaved by intracellular esterases in living cells, while dead cells fail to exclude the nuclear binding dye ethidium homodimer from their nuclei, which are then stained red. In this and many other systems, simultaneous labeling and imaging of multiple fluorophores is the norm.

Fluorophores can also be used to sense microenvironmental parameters, e.g., pH or the concentration of various ions such as Na^+, K^+, Cl^-, Ca^{2+}. The technique of FRAP, fluorescence recovery after photobleaching, can be used to study the intracellular dynamics or mobility of labeled macromolecules. In the technique named FRET, Förster resonance energy transfer, the proximity of two labeled molecules (or even two labeled domains on the same macromolecule) can be detected, as a donor fluorophore directly excites an acceptor fluorophore only when the two species are within a molecular distance of each other. This is truly an intermolecular ruler! Finally, the gene for the photoprotein from the jellyfish *Aqueorea victoria*, green fluorescent protein (GFP), has been cloned. This has been used to create numerous fluorescent chimeric proteins that can be directly expressed in living cells. A number of mutations have yielded blue, cyan, and yellow variants. Another class of photoproteins, dsRed, from Anemone is also in use. These advances give biologists unprecedented power to observe the processes of protein expression in cell populations.

Instrumentation for Fluorescence

Because of the small number of fluorophores often present and the potentially low quantum efficiency, fluorescence techniques require special, strong illumination sources. These are typically mercury or xenon arc lamps, or lasers. Fluorescence benefits from the use of very efficient collection optics, i.e., objective lenses with the highest possible NA. Detection is typically achieved with sensitive imagers, e.g., cooled CCD cameras or photomultipler tubes (PMTs).

Fluorescence microscopy can be performed on upright or inverted microscopes equipped with an arc lamp, and appropriate filters and mirrors. This is known as wide-field fluorescence microscopy. (An alternative technique, laser scanning confocal, and multiphoton instruments will be discussed in the next section.) In wide-field epifluorescence microscopy, illumination light from a Hg arc lamp is first passed through a band-pass filter that rejects all light save those wavelengths required to excite the fluorophore (the excitation band pass filter) (Fig. 5). Next, this excitation light hits a special mirror, the dichroic mirror, located directly behind the objective lens. The purpose of the dichroic mirror is to reflect shorter wavelength excitation light, but allow longer wavelength emission light to pass. The shorter wavelength excitation light is reflected by the dichroic mirror into the objective and delivered to the specimen. There the light is absorbed by the fluorophore and reemitted at a longer wavelength. This light is emitted in all directions; however, a portion of the emitted light is collected again by the objective lens to form the image. This light passes back through the objective and again hits the dichroic mirror. Now the longer wavelength emitted light is passed by the dichroic mirror (not reflected) and the beam travels on to one last filter. This last filter, the emission filter or barrier filter, rejects all but the emission wavelengths, thus leaving a clean signal that is composed only of light from the excited fluorophore. This illumination is apiscopic, a reflected light technique. The substrate can be opaque as long as the fluorophore is accessible at the surface.

The three filter elements, the excitation filter, the dichroic mirror and the emission filter, together form a filter set. Each filter set is specific to the spectral characteristics of one

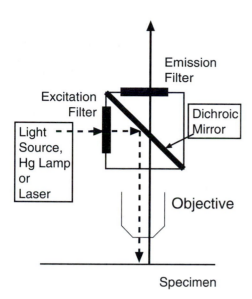

FIG. 5. Filter set. Dashed line represents shorter wavelength excitation light; solid line is longer wavelength emission light.

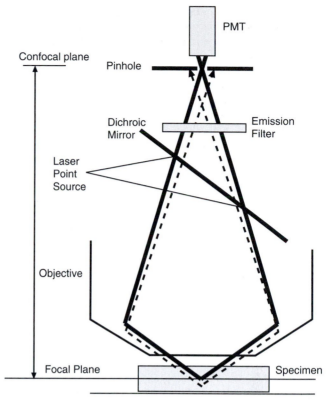

FIG. 6. Laser scanning confocal microscopy. The sample is illuminated by a laser that is reflected into the objective via dichroic mirror. This beam forms a diffraction-limited spot in the specimen at the focal plane. This beam is rastered across the specimen by scanning mirrors (not shown). Fluorescence from the focal plane is captured by the objective, passes through the dichroic mirror and emission filters, and arrives precisely at the opening of the pinhole aperture, which is found directly in front of the detector (a photomultiplier tube). Fluorescence from an out-of-focus plane (a below-focus plane is shown, dotted) arrives to either side of the pinhole and is rejected. (After Murphy, 2001).

particular fluorophore. More complex filter sets that allow for the simultaneous use of two or more fluorophores are available, but it is usually more convenient to capture separate images, each comprising the signal from an individual fluorophore, and then overlay these images later.

Confocal Microscopy: A Special Tool for Fluorescence Microscopy

Excited fluorophores in the sample send their emitted light in all directions. In wide-field fluorescence microscopy it is inevitable that, in addition to the fluorescence emanating from the fluorophores residing in the in-focus plane, some fluorescence from slightly out-of-focus planes (above and below the desired focal plane) will also be collected and contribute to the image. The result can be a muddled or hazy image, especially in thick or highly scattering specimens. A solution is the confocal microscope.

In a laser scanning confocal microscope (LSCM), a physical barrier (a pinhole) is placed at a confocal location in the imaging path between the objective and the detector (Fig. 6). This pinhole allows only those rays of light that emanate from precisely the focal plane of the objective lens to pass through to the detector. The result is a very clean, if very weak, signal. Further, instead of flooding the field of view with illumination from an arc lamp, a laser is used to excite the fluorophore. The laser beam forms a small diffraction-limited spot in the plane of the specimen. This spot is then rastered across the specimen using scanning mirrors and optics. Instead of imaging the entire field of view at once with an imager, a simpler more sensitive PMT can be used in synchrony with the laser to simply count the photons arriving as the result of the laser illumination at each location in the specimen. The completed image is then composed by digitally presenting the measured intensities at

each location (pixel) in the specimen. The diffraction-limited illumination spot provides for superb lateral resolution, and the confocal pinhole rejects all out-of-focus fluorescence contributions. The result is a crisp clean image that represents a thin optical plane virtually sliced through the specimen. By repeating the process at different focal planes (with the help of nanometer-resolution stepper motors attached to the focus drive) one can collect images from hundreds of planes in thick specimens (up to 100–200 μm deep) and composite these into three-dimensional fluorescent representations of the sample.

Laser scanning confocal microscopes were once considered to require a dedicated operator to care for complex lasers, complicated electronics, and fragile detectors. Presently, the advent of solid-state lasers and simpler, more robust electronics, scanners, and detectors has led to new generations of confocal instruments that promise improved affordability and ease of use. Like wide-field fluorescence, laser scanning confocal microscopy can be implemented on either upright or inverted platforms.

Two new advances in confocal microscopy are worthy of note. The first is the use of multispectral array detectors and spectral unmixing to provide for the simultaneous use of up to eight fluorophores. This technology is featured in an instrument produced by Zeiss. The second is the advent of the multiphoton microscope.

A fluorophore can be excited equally well by one properly tuned excitation photon; or by two excitation photons, each half the nominal excitation energy (twice the wavelength); or by three excitation photons, each a third the nominal excitation energy, etc. These multiple excitation photons must all reach the fluorophore molecule at precisely the same time, however—at least within the fluorescence decay lifetime of the fluorophore, which is measured in nanoseconds. This requires a femtosecond pulsed laser excitation source at long wavelengths (almost infrared). There are several advantages to this mode of illumination. First, longer wavelengths can penetrate deeper into cells and tissue, and do so with less biologic damage. Second, the pulsed nature of the excitation laser means that lower energy doses are delivered to the specimen. The effective excitation volume, the volume at the focal point of the objective where the spatial and temporal density of excitation photons is actually great enough to trigger a fluorescence emission, is exceedingly small—on the order of femtoliters. Thus, the emission signal produced and detected is always of very high spatial resolution, in both lateral as well as the axial dimensions.

DIGITAL IMAGING

It is not enough that we build instruments to produce magnified and resolved images of the microscopic world. We must be able to detect these images, to visualize the information, record and archive images, and make useful quantitative measurements of our observations. This is the science of imaging.

Detectors: Eyes, Film, Digital Imagers

Our human eyes are by far the most commonly used and relied upon detectors in light microscopy. Together with the ocular lenses, the lens of the eye serves as part of the microscope system and the final image is focused on the retina, the detector. The human eye is a remarkable, yet limited, living detection instrument. Obviously, our eyes are acutely tuned to the visible portion of the light spectrum, with the greatest sensitivity found in the mid-green. The eye can be very sensitive (as few as 100 photons in the dark-adapted eye) and can adapt to a wide range of signal intensity from very dim to very bright. However, this adaptation takes time; and the intra-scene dynamic range, the range of dim to bright within one field of view, is only about 1000-fold. The individual sensing elements in the eye, the rods and cones, have an effective size of about $2\,\mu m$, which can be compared to the pixel size in other detectors. The signal response in the eye is nonlinear, and the signals from our eyes receive considerable postprocessing by the visual systems in the brain. This is particularly true with regard to edges, shadows, and objects in motion. We are all familiar with many "optical illusions" that prey upon some of the foibles of the human visual system. As discussed earlier, we

make use of some of these illusions in light microscopy—e.g., the gradations of intensity in DIC and Hoffman are interpreted as shadows, when really there is little topographic information in the image.

Unfortunately, scientists with photographic memory are rare; thus the storage and archival capability of the human visual system is nil. The quantitative abilities of the eye are also limited. Although we can distinguish thousands of colors, we can barely discern between roughly 64–100 different gradations of gray (intensity)! Thus data that come from human observations are qualitative, or semiquantitative at best. The scientist must rely on careful subjective measurements, ideally with standardized scoring schemes, copious controls, blinded observers, and comparisons between multiple independent observers.

For more than a century the standard media for imaging was photographic film. Film can be made to detect broader spectral ranges than the eye, can be very sensitive, and has a dynamic range of about 2.5 log units. The grain size (individual sensing element or pixel) in film is typically $30\,\mu m$. The archival capability of film is excellent, and when properly used, the signal response can be made to be linear and the data analyzed quantitatively. However, the cost, time and chemical reagents required for film development are significant. The replacement of film with digital imagers in modern scientific imaging is essentially complete.

Digital Imaging Devices

Solid-state imaging devices are now a common part of our everyday world. Digital imaging devices are found in photocopiers, scanners, professional and consumer-grade video cameras and still cameras, surveillance cameras, toys, personal computers, and cell phones. The medical and scientific imaging fields have benefited greatly from this explosion of technology. The gaps in performance and sensitivity between consumer-grade and research-grade imagers are quickly closing.

The most common digital imaging device is the charge-coupled device (CCD). A CCD is a silicon chip that comprises a physical array of photodetectors, which correspond to the picture elements (pixels) in the image. As photons arrive at each photodiode they are converted to electrons, and the charges are accumulated and held locally on the chip in a potential well. Later, these groups of charges are transferred off the chip in a sequential manner, and the signal from each well is amplified and converted to a voltage. At this point the signal is still analog. An analog-to-digital converter is used to convert the steady stream of voltages into an array of numeric values that represent the relative intensity of photons at each photodiode in the detector array. It is this array of numbers that is the digital image.

Charge-coupled devices are available in several different architectures. In a full-frame camera, every spot on the chip is exposed to light. A physical shutter is used to control the exposure. After the exposure, the shutter is closed, and the charges are transferred and read. The operation is much like that found in a standard film camera. In a frame transfer camera, the CCD array is divided into two separate adjacent fields. The first is used for sensing light, while the second area on the chip is protected from light by an opaque mask and is used only to transfer and hold charges. The frame transfer camera has the advantage of speed. Thus in a frame transfer camera, one image is acquired in the sensing region, then the charges are transferred quickly to the masked area. While the charges in the masked area are being read out, a new image can be acquired in the sensing region. Finally, many modern cameras take this concept one step further by simply masking every other line of photodiodes on a chip, a so-called interline camera. This allows for imaging at video frame rates; built-in microlenses can be used to funnel photons away from the masked pixels and to the sensing pixels.

The performance of a scientific imaging device depends on several important parameters. The resolution is determined by the number of sensing elements (photodiodes, i.e., pixels) as well as the physical size of each photodiode. The size of each element also determines the capacity of each well, i.e., the number of charges that can be held before reaching saturation. This "full-well capacity" determines the dynamic range, which is the range between the brightest and the darkest intensities possible in the image. The sensitivity is determined by the signal/noise ratio. The signal is affected by the quantum yield, which is the ratio of charges created to photons hitting the detector. The quantum yield can vary significantly with wavelength. The noise in the signal can arise from several factors. At very low light levels, the stochastic nature of the incoming photons themselves contributes some variability, the "shot noise." Thermal events also generate electrons in the chip. Thus very sensitive detectors are often cooled with thermoelectric cooling devices or even liquid nitrogen to reduce this "thermal noise." Finally, noise can be generated in the process of transferring, reading, and amplifying the signals, and this "read noise" often scales with the speed of the reading process. Thus, very different and specialized imaging systems are used for various imaging tasks: e.g., very low-light fluorescence imaging; video-rate imaging; or very high-speed imaging for following millisecond or microsecond events.

Although there is some variation in the quantum efficiency with wavelength, a CCD camera is basically a monochrome imaging detector. Often this is adequate. For example, in fluorescence imaging the signal is ideally composed of only single-emission-wavelength photons—fluorescence microscopy is actually a monochrome technique! But as discussed earlier in this section, we often use multiple colored stains to create complex contrast in an image, and this requires a color imaging device.

A color CCD camera achieves color sensitivity by using filters to separate the image into its red, green, and blue components. This can be accomplished in several different ways. In the first example, a separate tunable color filter is placed in front of a monochrome camera. The filter is sequentially set for red, green, or blue, and three separate images are acquired. The images can then be compiled computationally to represent the color digital image. An advantage is that the full resolution of the imaging camera is maintained; however, the time required for the acquisition of three images means that the sample must hold still for a while (no sample motion!). In the second approach, a prism is used to split the incoming image into three separate images, representing red, green, and blue.

Three separate imaging chips are placed after the prism to collect the three separate images simultaneously. However, the third and most common approach is to use microprisms or colored filters mounted as a mask directly on the chip itself. In this arrangement, some photodiodes will be assigned to sense red, some green, and some blue. A computer algorithm keeps track of which pixels are which, and then combines the signals from adjacent R, G, and B sensing pixels and reports the result as one "virtual" pixel with R, G, and B values. Since the values from three (or more typically four) photodiodes are used to create one pixel, some resolution is sacrificed. Like all other areas of computing and electronics, imaging technology is advancing at a very rapid pace. Thus the reader is urged to consult current literature and vendor specifications for the latest advancements.

Digital Images, Image Processing, and Image Output

A full treatment of image processing and image analysis is beyond the scope of this chapter. However, a few thoughts are worthwhile for any scientist dealing with image data.

A digital image is ultimately an array of numbers where the value at each location in the array represents the intensity of one pixel in the image. A color image is represented by three values at each pixel. This array of numbers may be stored electronically in any of a myriad of proprietary file formats or in cross-platform formats such as bitmap (.bmp) or TIFF (.tif). In any event, the data should always be safely stored and archived in their raw, unadulterated form, before any image-processing steps or image-compression methods are applied. Subsequent processing steps should always be applied to a copy of the original data. Most vendors of imaging devices also offer software that controls image acquisition and offers some image-processing capabilities. Common image-processing packages include the NIH IMAGE software available from the U.S. National Institutes of Health, Research Services Branch (http://rsb.info.nih.gov/nih-image), as well as programs primarily meant for the professional graphic arts world such as Adobe Photoshop. There is often uncertainty about the scientific propriety of digital image manipulation. In general the least amount of processing necessary is preferable, and no processing should be applied if quantitative data are being extracted from the images. Certain modest operations such as histogram stretching (which is scaling the intensity values in an image so that they span a more full range of output values) or gamma correction (a nonlinear display of data to accentuate dim features while maintaining bright features) are commonly applied for aesthetic images where quantitative analysis is not employed. Scientific guidelines are beginning to be formulated and a good reference is the Microscopy Society of America's policy statement (www.microscopy.org). As with any treatment of scientific data, the methods of image manipulation must be documented and communicated such that others are capable of reproducing the findings from the raw data set, or through independent experimental work.

Of course, the road to imaging productivity is not complete until the image is communicated to others in the scientific community. Image output must be considered as part of the scientific process. Digital images are embedded in other media files and viewed using a variety of means. Typical CRT computer monitors or LCD displays unfortunately have fairly limited resolution, displaying only about a million pixels at roughly 70 dots per inch. Scientific journals typically print at about 300 dpi. Standard data projectors also only display about a million pixels; the fact that the image is hugely magnified on a screen does not "create" added resolution! Likewise, images overly enlarged for printing on scientific posters can suffer if they do not contain sufficient resolution (pixels). If color information is important, then special care must be taken to ensure color fidelity throughout the image acquisition, display and output processes. The responsibility for clear unbiased image presentation ultimately lies with the scientist. The image must be presented to the audience with adequate magnification, resolution, and contrast such that an independent observer can identify the features that are important and independently reach the conclusions the author wishes to convey.

ELECTRON MICROSCOPY

The resolution of any probing technique is ultimately dependent upon the size of the probe and the nature of the probe–specimen interaction. The Rayleigh criterion above describes how resolving power is related to wavelength in light microscopy. How then can we achieve much greater resolution? Although we can certainly use shorter wavelengths, the visible light spectrum is actually quite limited. To make really large strides in the achievement of resolution, the answer is to reduce the size of the probe drastically, and use electrons. Electrons indeed provide extraordinary resolution. However, in addition to interacting with the samples, electrons are also absorbed and scattered by even the small molecules in the atmosphere! Thus electron microscopy techniques must be conducted under high-vacuum conditions such as those described in Chapter 1.4.

Electron microscopy can be performed in either transmission or reflected modes. In transmission electron microscopy (TEM), images are formed by diffraction or phase contrast mechanisms, much as has been described earlier for transmitted light microscopy. The detectors are phosphor screens coupled with digital imagers. Samples must be very, very thin (50 nm). This is achieved by embedding samples in hard plastic resins and using very sharp diamond knife ultramicrotomes. TEM provides extraordinary resolution of organelle and membrane structures in single cells. For biomaterials, there is perhaps no other instrument that provides a higher resolution image of the intimate cell membrane–biomaterial interface. Unfortunately the challenging nature and limitations of TEM sample preparation mean that this technique is underutilized in biomaterials. By far the most common electron microscopy tool is scanning electron microscopy (SEM).

Scanning Electron Microscopy

The scanning electron microscope was developed in the 1960s and 1970s, and has had a tremendous impact on the materials sciences and engineering fields, as well as the biologic community. SEM provides 3D-like topographical information of bulk specimens at nanometer resolutions. It is remarkable that, while complex to generate, SEM images are easily accessible for even a child to interpret. The SEM has been truly instrumental in opening doors to the micro- and nano-world for the general public! In biomaterials the SEM is on an equal footing with Light Microscopy, as it is used to study the micro and nanoscale structure of materials and devices, and cell–biomaterial interactions.

The scanning electron microscope (Fig. 7) is somewhat analogous to the scanning confocal microscope discussed earlier. Instead of a diffraction limited laser beam, a beam of electrons is focused and swept across the surface of the specimen. The interaction of the electrons with the sample is localized and results in the generation of a signal, in this case other electrons escaping the specimen. These electrons are collected and counted by a detector, just as the photons are collected and counted by a photomultiplier tube (PMT) in LSCM. The counts collected at each raster point are then compiled into a digital image. The LSCM has very shallow depth of focus, and three-dimensional information has to be reconstructed by taking many images at various focus planes. A key advantage of the SEM is its extraordinarily large depth of focus, allowing for the direct imaging of complex, even somewhat tortuous, microscale structures.

To start with, in SEM, we wish a plentiful (bright) and narrow beam of electrons for illumination. Conventional SEMs generate electrons by heating a hairpin-shaped filament of

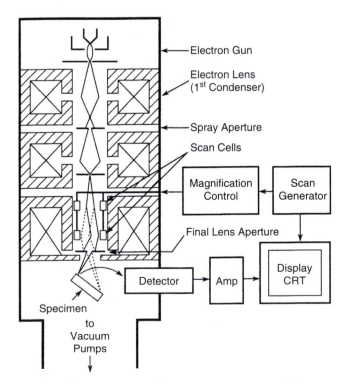

FIG. 7. Schematic drawing of the electron column in an SEM showing the electron gun, lenses, the scanning system and the electron detector. From Goldstein *et al.* (2003).

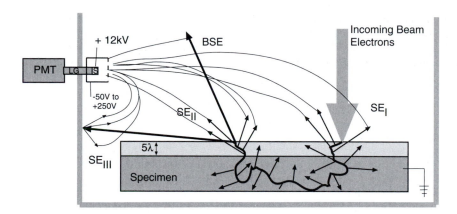

FIG. 8. Signal generation in SEM. Incident electrons enter the sample, collide with sample atoms, and may eventually exit the sample as backscattered electrons (BSE). Secondary electrons (SE$_I$, SE$_{II}$) are generated and can escape the sample if they are generated close enough to the surface (typically within five mean free path lengths). BSEs can also collide with the chamber walls or other components to generate even more secondary electrons (SE$_{III}$). The electrons are drawn to the E–T detector by the positively charged Faraday cage and the positively charged coating on the scintillator (S). The electrons are converted to photons and carried by the light guide (LG) out of the chamber, where the photons are counted by a photomultiplier tube (PMT). After Goldstein *et al.* (2003).

tungsten. The emitted electrons are quickly focused to a point by the Wehnelt cap and then accelerated toward an anode at high voltage. Alternatively a crystal of LaB$_6$ can be used as a source—this provides 5–10× greater brightness (more electrons) but requires a higher vacuum (10^{-4} Pa) and is much more expensive. Many modern instruments now utilize an alternative source, the field emission (FE) gun. The element is again tungsten, but the filament is now shaped into a very sharp wire tip (<100 nm radius). Electrons are pulled from the tip by a very high electric field, either without (cold field emission) or with (Schottky or thermal FE) the aid of heating. An FE source can provide a very narrow beam that is 1000× brighter than the conventional thermal source. However, the vacuum required at the tip is much higher, on the order of 10^{-9} Pa.

Focusing the illuminating beam of electrons onto the sample is accomplished by lenses; however, these lenses are electromagnetic, not glass. Typically one or two condenser lenses begin to narrow the beam down and a final "objective" lens focuses the beam onto the sample. Scanning coils are used to deflect the beam slightly from side to side to accomplish the rastering of the beam across the specimen. Unlike glass lenses, the focal length of magnetic lenses can be changed simply by changing the magnetic field. A beam-limiting aperture is placed in the path and used to limit the diameter of the probe and ensure a narrow beam angle. A standard conically shaped objective lens can accommodate almost any size sample capable of fitting into the chamber. The lens may reside 1–20 mm away from the sample (the working distance). If the sample is small enough, it actually may be placed inside the objective lens field (an immersion lens instrument) with the advantages of lower aberrations, smaller probe size, and greater resolution.

Operation of the instrument requires balancing several operational parameters and making trade-offs. The diameter of the probe, the number of electrons in the probe (probe current),

and the aperture angle are all crucial and quite interrelated parameters. Resolution is dependent upon small probe size, but there must be sufficient electrons to generate contrast (probe current): the familiar resolution/contrast trade-off again. Aperture angle also affects these, as well as determining the depth of focus. Thus operational parameters such as source current, condenser strength, working distance, aperture size, and acceleration voltage are all adjusted depending upon the particular needs of the situation: e.g., high-resolution imaging, large depth of focus, a need for high current, or surface sensitivity (low voltage).

The probe diameter of the incident electron beam may be as small as a few nanometers as it arrives at the surface. Unfortunately, the signal that is measured arises from interactions of this beam from within a much larger volume in the sample, the interaction volume. The incoming electrons may result in several measurable signals upon hitting the specimen (Fig. 8). First, the very energetic electrons may bounce off, or far more likely, bounce through the upper layers of the specimen, before getting turned around and again escaping the sample and returning to the chamber. These elastically scattered electrons are detected as backscattered electrons (BSEs). Depending on their energies, and the nature of the specimen (e.g., the atomic number), these backscattered electrons might travel as deep as 5–10 μm into the sample and reemerge as far as a micrometer or two away from their point of entry! Next, as these BSEs are bouncing through the specimen, much weaker secondary electrons (SEs) are generated. If these secondary electrons are generated near enough to the sample surface (say within five mean free path lengths), then they too will escape and can be detected as a secondary electron signal. Not only can secondary electrons be generated and escape near the point of entry of the incident beam, but secondary electrons can also be generated near the point of exit of a BSE, and these also escape and count toward the measured SE signal. Finally, escaped BSEs that hit the walls of the chamber can also generate SEs, and these too are counted. Most commonly, the signal used to form an image is the combination of counts from BSEs and all the SEs mentioned. Understanding the size and depth of the interaction volume in the specimen is crucial to interpreting the resolution and surface sensitivity in SEM imaging.

The detector is most commonly an Everhart–Thornly (E-T) detector located somewhere on the wall of the chamber. The E-T detector works by attracting escaped electrons (BSEs and all the SEs) with a positively biased grid placed directly in front of a positively charged scintillator. Electrons hitting the scintillator generate photons of light that are carried out of the chamber through a light guide to a PMT where they are counted. A negative bias on the grid will result in only the BSEs being detected.

All the incident electrons used for illumination have to be drained somewhere. Thus samples need to be conductive (or at least semiconductive) and grounded to the instrument. When current is unable to be discharged adequately it builds up on the surface and is manifest as "charging" in the image that degrades system performance. For this reason, many nonconducting samples (including biologic samples and polymers) are often coated with a very thin layer of metal, typically Au/Pd. The energetic incident electrons can also be damaging to some beam-sensitive samples, including polymers and delicate biologic structures.

Elemental Analysis in the SEM

In addition to the BSE and SE discussed above, incident electrons may also cause the emission of X-rays from the sample. Measurement of the energy and intensity of these X-rays can provide for elemental composition mapping across the specimen. This technique, energy dispersive X-ray (EDX) analysis, is conceptually the reverse of ESCA or XPS discussed in the chapter on surface analysis. This technique is electrons in, X-rays out.

Low-Voltage Imaging

Another key advantage of instruments designed around FE sources is that they can be operated at very low accelerating voltages, as low as 100 eV as compared to the typical 5–30 KeV in conventional instruments. This results in a much smaller interaction volume, and thus more sensitivity to surface topography. The samples receive less beam damage, and in fact many samples may be imaged without a metalized coating.

Variable Pressure and ESEM

A major drawback to SEM is the requirement of operating under high vacuum, 10^{-3}–10^{-6} Pa or even lower pressures. Biologic samples must be fixed and dehydrated with hopes that the structure will be preserved somehow under vacuum. Polymers too can undergo significant structural rearrangement when under high-vacuum conditions (e.g., hydrogels). Porous structures such as bone can be difficult to pump down to such vacuum levels.

Solutions are now appearing in the form of the Environmental SEM (ESEM) and numerous variable-pressure (VP) SEM instruments. The approach in both instruments is the same: basically, to make a "leaky" SEM. These instruments are differentially pumped, meaning that the portions of the instrument housing the electron gun and optics are kept under high vacuum, while the sample chamber is allowed to reach some larger (but not really atmospheric!) pressure. The opening between the optics column and the specimen chamber (the leak) is kept as small as possible, and the sample is kept as close to this opening as possible so that the incident electrons have only a short distance to travel through the scattering gas atmosphere. The ESEM, which has several fairly elaborate differential pumping systems, can reach chamber pressures as high as 2700 Pa (20 torr) and utilizes a special SE detector capable of operating in the presence of such gas pressures. This pressure is the saturated vapor pressure of water at room temperature, and thus fully hydrated specimens can be viewed directly. All SEM vendors now offer some form of these variable-pressure systems with chamber pressures reaching 250 Pa (2 torr) and utilizing a variety of detectors. This pressure is not adequate to keep samples fully hydrated, but samples can be observed "partially hydrated."

The advantages are obvious. Cells and tissues have been imaged under fully hydrated conditions. Colloids, emulsions, and hydrated polymers are no longer excluded from SEM analysis. The resolution in this operating mode, however, does not equal that under high vacuum, and scattering of the incident beam from column to sample is the culprit. For this reason, standard thermal electron guns and high accelerating voltages are typically used. Fortunately, the presence of the gas near the sample means that charging is much reduced. The flexibility that this mode of operation offers is tremendously attractive, and "VP-capable" SEMs have become big sellers.

Focused Ion Beam Instruments

Instead of an incident beam of electrons, a focused beam of positively charged ions can be delivered to the specimen, e.g., gallium ions from a liquid metal ion gun. This beam can also be used to generate images of very high resolution, but the technique is destructive. However, this can be an advantage, in that the focused ion beam (FIB) can be used as an extremely precise machining tool at the microscopic and, indeed, nanoscopic scale. Several new instruments now combine a FIB column with a SEM column to create the ultimate fabrication and imaging instrument for nanoscale engineering.

CONCLUSIONS

Although many of the techniques discussed here can trace their roots back for centuries, advances in modern microscopic imaging continue to abound. This section has not addressed scanning tunneling microscopy (STM) and atomic force imaging (AFM), important newer molecular imaging tools, which are covered elsewhere in this text. Surface analytical tools, such as ESCA and SIMS, are now being adapted to provide imaging capability at the microscopic scale. Conventional medical imaging tools, such as X-ray computed tomography (CT) and magnetic resonance imaging (MRI), are now being practiced at nearly microscopic resolutions. Electron tomography applied in high-voltage TEM instruments is revealing extraordinary three-dimensional representations of intracellular organelles and other structures. Modern computing power is reinvigorating the serial sectioning and reconstruction approaches of the past. The detection of intrinsic fluorescence via second harmonic generation in the multiphoton microscope is opening new doors to the imaging of biologic structures deeper in tissues. Optical coherence tomography promises important advances in tissue imaging for both clinical and research use. New nanoparticle fluorophores (quantum dots) provide extraordinarily bright emissions with narrow spectral signatures, enabling even further advancements in multilabel fluorescence imaging, biophotonics, and biosensors.

At first blush, microscopy may seem little more than a suite of tools for making little things appear larger so they can be examined. This is an accurate, but very incomplete, description of how microscopy is used in modern science. The wealth of microscopy tools available to the modern scientist allows one to move beyond simple visual exploration, and to begin to ask tailored mechanistic questions, pose hypotheses, and prove them with solid quantitative data. This is essential for scientific publications, as well as for the documentation required by regulatory agencies. The detailed description of structure; the localization of species in two and three dimensions with molecular specificity; the ability to follow and visualize molecular biological process such as gene expression over time—these are all powerful tools in the biomaterial scientist's arsenal. To be successful, the scientist must include microscopy and imaging considerations from the very onset of any experimental planning.

Bibliography

Web Resources

The Histochemical Society: http://www.histochemicalsociety.org

Hundreds of general-purpose microscopy Web sites exist, including independent sites and those offered by all major instrument and equipment vendors. Here are a few to get started:
http://www.microscopy.fsu.edu/primer/index.html
http://www.microscopyu.com/index.html
http://swehsc.pharmacy.arizona.edu/exppath/micro.html

The Microscopy Society of America: contains links to the society's journals and publications, tutorials, and a forum for asking questions: http://www.microscopy.org

Molecular Probes, Inc.—the leading vendor of fluorophores and home of the Molecular Probes Handbook, available online www.probes.com

The National Society for Histotechnology USA: http://www.nsh.org

The Royal Microscopy Society UK: http://www.rms.org.uk

Books

Foster, B. (1997). *Optimizing Light Microscopy for Biological and Clinical Laboratories*. Kendall/Hunt Publishing Company, Dubuque, IA.

Goldstein, J., Newbury, D., Joy, D., Lyman, C., Echlin, P., Lifshin, E., Sawyer, L., and Michael, J., (eds.) (2003). *Scanning Electron Microscopy and X-Ray Microanalysis*. Kluwer Academic/Plenum Publishers, New York.

Inoué, S., and Spring, K. R. (1997). *Video Microscopy: The Fundamentals,* 2nd ed. Plenum Press, New York.

Mason, W. T. (ed.) (1993). Fluorescence and Luminescence Probes for Biological Activity. Academic Press, New York.

Murphy, D. B. (2001). *Fundamentals of Light Microscopy and Electronic Imaging*. Wiley-Liss, New York.

Pawley, J. B. (ed.) (1995). *Handbook of Biological Confocal Microscopy,* 2nd ed. Plenum Press, New York.

Russ, J. C. (ed.) (2002). *The Image Processing Handbook,* 4th ed., CRC Press, Boca Raton, FL.

6

Degradation of Materials in the Biological Environment

ARTHUR J. COURY, ROBERT J. LEVY, BUDDY D. RATNER, FREDERICK J. SCHOEN,
DAVID F. WILLIAMS, AND RACHEL L. WILLIAMS

6.1 INTRODUCTION: DEGRADATION OF MATERIALS IN THE BIOLOGICAL ENVIRONMENT
Buddy D. Ratner

The biological environment is surprisingly harsh and can lead to rapid or gradual breakdown of many materials. Superficially, one might think that the neutral pH, low salt content, and modest temperature of the body would constitute a mild environment. However, many specialized mechanisms are brought to bear on implants to break them down. These are mechanisms that have evolved over millennia specifically to rid the living organism of invading foreign substances and they now attack our contemporary biomaterials. First, consider that, along with the continuous or cyclic stress many biomaterials are exposed to, abrasion and flexure may also take place. This occurs in an aqueous, ionic environment that can be electrochemically active to metals and plasticizing (softening) to polymers. Then, specific biological mechanisms are invoked. Proteins adsorb to the material and can enhance the corrosion rate of metals. Cells secrete powerful oxidizing agents and enzymes that are directed at digesting the material. The potent degradative agents are concentrated between the cell and the material where they act undiluted by the surrounding aqueous.

To understand the biological degradation of implant materials, synergistic pathways should be considered. For example, the cracks associated with stress crazing open up fresh surface area to reaction. Swelling and water uptake can similarly increase the number of site for reaction. Degradation products can alter the local pH, stimulating further reaction. Hydrolysis of polymers can generate more hydrophilic species, leading to polymer swelling and entry of degrading species into the bulk of the polymer. Cracks might also serve as sites initiating calcification.

Biodegradation is a term that is used in many contexts. It can be used for a reaction that occurs over minutes or over years. It can be engineered to happen at a specific time after implantation, or it can be an unexpected long-term consequence of the severity of the biological environment. Implant materials can solubilize, crumble, become rubbery, or become rigid with time. The products of degradation may be toxic to the body, or they may be designed to perform a pharmacologic function. Degradation is seen with metals, polymers, ceramics, and composites. Thus, biodegradation as a subject is broad in scope and rightfully should command considerable attention for the biomaterials scientist. This chapter, in three sections, introduces biodegradation issues for a number of classes of materials and provides a basis for further study on this complex but critical subject.

6.2 CHEMICAL AND BIOCHEMICAL DEGRADATION OF POLYMERS
Arthur J. Coury

Biodegradation is the chemical breakdown of materials by the action of living organisms which leads to changes in physical properties. It is a concept of vast scope, ranging from decomposition of environmental waste involving microorganisms to host-induced deterioration of biomaterials in implanted medical devices. Yet it is a precise term, implying that specific biological processes are required to effect such changes (Williams, 1989). This chapter, while grounded in biodegradation, addresses other processes that contribute to the often complex mechanisms of polymer degradation. Its focus is the unintended chemical breakdown in the body of synthetic solid-phase polymers. (See Chapter 2.7 for a description of systems engineered to break down in the body.)

POLYMER DEGRADATION PROCESSES

Polymeric components of implantable devices are generally reliable for their intended lifetimes. Careful selection and extensive preclinical testing of the compositions, fabricated components, and devices usually establish functionality and durability. However, with chronic, indwelling devices, it is infeasible during qualification to match all implant conditions in real time for years or decades of use. The accelerated aging, animal implants, and statistical projections employed cannot expose all of the variables that may cause premature deterioration of performance. The ultimate measure of the acceptability of a material for a medical device is its functionality for the device's intended lifetime as ascertained in human postimplant surveillance (Coury, 1999).

No polymer is totally impervious to the chemical processes and mechanical action of the body. Generally, polymeric biomaterials degrade because body constituents attack the biomaterials directly or through other device components, sometimes with the intervention of external factors.

Numerous operations are performed on a polymer from the time of its synthesis to its use in the body (see, e.g., Table 1). Table 2 lists mechanisms of physical and chemical deterioration, which may occur alone or in concert at various stages of a polymer's history. Moreover, a material's treatment prior

TABLE 1   Typical Operations on an Injection-Moldable Polymer Biomaterial

---

Polymer: Synthesis, extrusion, pelletizing

Pellets: Packaging, storage, transfer, drying

Components: Injection molding, post-mold finishing, cleaning, inspecting, packaging storage

Device: Fabrication storage (presterilization) cleaning, inspecting, packaging, storage (packaged), sterilization, storage (sterile), shipment, storage (preimplant), implantation, operation is body

---

TABLE 2   Mechanisms Leading to Degradation of Polymer Properties[a]

Physical	Chemical
Sorption	Thermolysis
Swelling	Radical scission
Softening	Depolymerization
Dissolution	Oxidation
Mineralization	Chemical
Extraction	Thermooxidative
Crystallization	Solvolysis
Decrystallization	Hydrolysis
Stress cracking	Alcoholysis
Fatigue fracture	Aminolysis, etc.
Impact fracture	Photolysis
	Visible
	Ultraviolet
	Radiolysis
	Gamma rays
	X-rays
	Electron beam
	Fracture-induced radical reactions

[a]Some degradation processes may involve combinations of two or more individual mechanisms.

to implant may predispose it to stable or unstable end-use behavior (Brauman et al., 1981; Greisser et al., 1994; Ling et al., 1998). A prominent example of biomaterial degradation caused by preimplant processing is the gamma irradiation sterilization of ultra-high molecular weight polyethylene used in total joint prostheses. The process generates free radicals within the material which react with oxygen to produce undesirable oxidation products. Chain oxidation and scission can occur for periods of months to years, causing loss of strength and embrittlement with limited shelf life (McKellop et al., 1995; Furman and Li, 1995; Weaver et al., 1995; Daly et al., 1998; Blanchet and Burroughs, 2001). Polypropylene and polytetrafluoroethylene are also notable as polymers that are generally chemically stable in the body but can be severely degraded during processing by sterilization with ionizing radiation (Williams, 1982; Portnoy, 1997). Gamma irradiation may also cause optical changes such as darkening of poly(methyl methacrylate) intraocular lenses (Hoffman, 1999). It is crucially important, therefore, that appropriate and rigorous processing and characterization protocols be followed for all operations (Coury et al., 1988; Shen et al., 1999).

After a device has been implanted, adsorption and absorption processes occur. Polymeric surfaces in contact with body fluids immediately adsorb proteinaceous components, and the bulk begins to absorb soluble components such as water, ions, proteins, and lipids. Cellular elements subsequently attach to the surfaces and initiate chemical processes. With biostable components, this complex interplay of factors is of little functional consequence. At equilibrium fluid absorption, there may be some polymer plasticization, causing dimensional and mechanical property changes (Coury et al., 1988). On the surface, a powerful acute attack by cells and many chemical agents, including oxidants and enzymes, will have been substantially withstood. With the resolution of this acute inflammatory phase, a fibrous capsule will likely have formed over the device, and the rate of release of powerful chemicals from activated cells will have markedly decreased.

For those polymers subjected to chemical degradation in vivo, few if any reports have comprehensively described the multistep processes and interactions that comprise each mechanism. Rather, explant analysis and occasionally metabolite evaluation are used to infer reaction pathways. The analysis of chemically degraded polymers has almost always implicated either hydrolysis or oxidation as an essential component of the process.

HYDROLYTIC BIODEGRADATION

Structures of Hydrolyzable Polymers

Hydrolysis is the scission of susceptible molecular functional groups by reaction with water. It may be catalyzed by acids, bases, salts, or enzymes. It is a single-step process in which the rate of chain scission is directly proportional to the rate of initiation of the reaction (Schnabel, 1981). A polymer's susceptibility to hydrolysis is the result of its chemical structure, its morphology, its dimensions, and the body's environment.

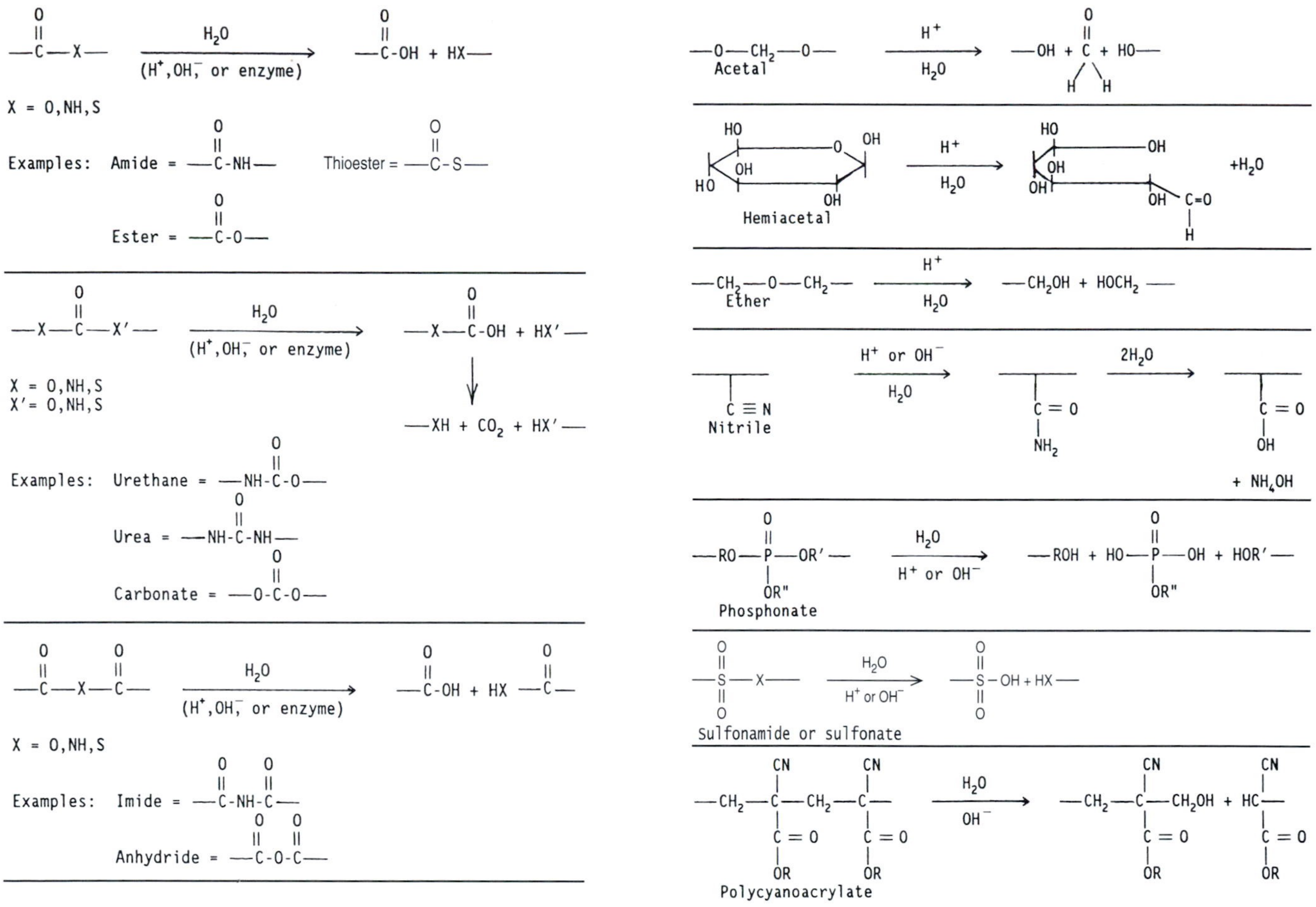

FIG. 1. Hydrolyzable groups in polymer biomaterials.

In a commonly used category of hydrolyzable polymeric biomaterials, functional groups consist of carbonyls bonded to heterochain elements (O, N, S). Examples include esters, amides, urethanes, and carbonates, and anhydrides. (Fig. 1). Other polymers containing groups such as ether, acetal, nitrile, phosphonate, sulfonate, sulfonamide or active methylenes hydrolyze under certain conditions (Fig. 1). Hydrolytically susceptible groups exhibit differing rates of degradation, which are dependent on the intrinsic properties of the functional group and on other molecular and morphological characteristics. Among carbonyl polymers, anhydrides display the highest hydrolysis rates followed, in order, by esters and carbonates. Polymers containing such groups, in fact, comprise many of the resorbable devices (Chapter 2.7). Other carbonyl groups such as urethane, imide, amide and urea can demonstrate long-term stability *in vivo* if contained in a hydrophobic backbone or highly crystalline morphologic structure. Groups that are normally very stable to hydrolysis are indicated in Fig. 2.

The rate of hydrolysis tends to increase with a high proportion of hydrolyzable groups in the main or side chain, other polar groups which enhance hydrophilicity, low crystallinity, low or negligible cross-link density, a high ratio of exposed surface area to volume, and, very likely, mechanical stress. Porous hydrolyzable structures undergo especially rapid property loss because of their large surface area. Factors that tend to suppress hydrolysis rate include hydrophobic moieties (e.g., hydrocarbon or fluorocarbon), cross-linking, high crystallinity due to chain order, thermal annealing or orientation, low stress, and compact shape. While the molecular weight of linear polymers *per se* may not have a great effect on degradation rate, physical property losses may be retarded for a given number of chain cleavage events with relatively high-molecular-weight polymers. Property loss caused by chain cleavage is more pronounced in polymers with weak intermolecular bonding forces.

Host-Induced Hydrolytic Processes

The body is normally a highly controlled reaction medium. Through homeostasis, the normal environment of most implants is maintained at isothermal (37°C), neutral (pH 7.4), aseptic, and photoprotected aqueous steady state. By *in vitro* standards, these conditions may appear mild. However, complex interactions of humoral and cellular components of body fluids involving activators, receptors, inhibitors, etc., produce aggressive responses to any foreign bodies through the processes of adhesion, chemical reaction, and particulate transport.

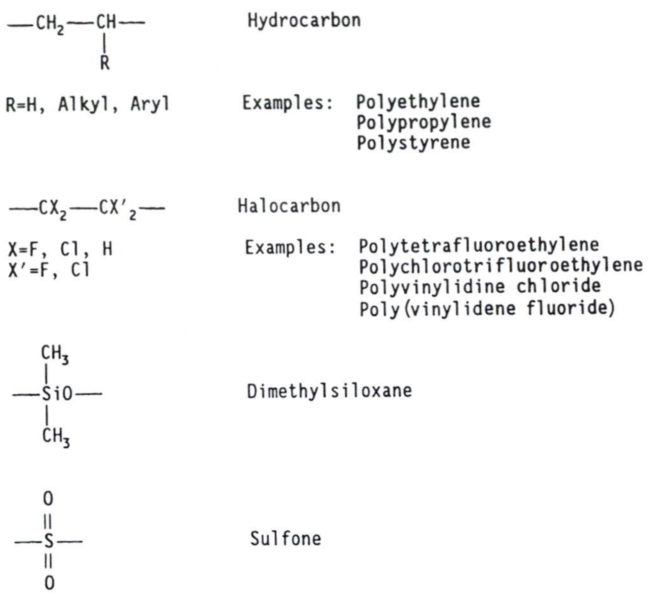

—CH₂—CH—   Hydrocarbon
         |
         R

R=H, Alkyl, Aryl   Examples: Polyethylene
                               Polypropylene
                               Polystyrene

—CX₂—CX′₂—   Halocarbon

X=F, Cl, H   Examples: Polytetrafluoroethylene
X′=F, Cl   Polychlorotrifluoroethylene
                         Polyvinylidine chloride
                         Poly(vinylidene fluoride)

         CH₃
          |
—SiO—   Dimethylsiloxane
          |
         CH₃

         O
         ‖
—S—   Sulfone
         ‖
         O

FIG. 2. Groups highly stable to hydrolysis.

Several scenarios leading to hydrolysis in the host can be considered. Whatever the scenario, hydrolysis can only occur at a site other than the surface of a polymer mass after water of permeation reaches that site. First, essentially neutral water is capable of hydrolyzing certain polymers (e.g., polyglycolic acid) at a significant rate (Chapter 2.7 and Zaikov, 1985). However, this simple mechanism is unlikely to be significant in polymer compositions selected for long-term *in vivo* biostability.

Next, ion-catalyzed hydrolysis offers a likely scenario in body fluids. Extracellular fluids contain ions such as: H^+, OH^-, Na^+, Cl^-, HCO_3^-, PO_4^{3-}, K^+ Mg^{2+}, Ca^{2+}, and SO_4^{2-}. Organic acids, proteins, lipids, lipoproteins, etc. also circulate as soluble or colloidal components. It has been shown that certain ions (e.g., PO_4^{3-}) are effective hydrolysis catalysts, enhancing, for example, reaction rates of polyesters by several orders of magnitude (Zaikov, 1985). Ion catalysis may be a surface effect or a combined surface-bulk effect, depending on the hydrophilicity of the polymer. Very hydrophobic polymers (e.g., those containing <2% water of saturation) absorb negligible concentrations of ions. Hydrogels, on the other hand, which can absorb large amounts of water (>15% by weight) are essentially "sieves," allowing significant levels of ions to be absorbed with consequent bulk hydrolysis via acid, base, or salt catalysis.

Localized pH changes in the vicinity of the implanted device, which usually occur during acute inflammation or infection, can cause catalytic rate enhancement of hydrolysis (Zaikov, 1985). Organic components, such as lipoproteins, circulating in the bloodstream or in extracellular fluid, appear to be capable of transporting catalytic inorganic ions into the polymer bulk by poorly defined mechanisms.

Enzymes generally serve a classic catalytic function, altering reaction rate (via ion or charge transfer) without being consumed by modifying activation energy but not thermodynamic equilibrium. While enzymes function in extracellular fluids, they are most effectively transferred onto target substrates by direct cell contact (e.g., during phagocytosis). Hydrolytic enzymes or hydrolases (e.g., proteases, esterases, lipases, glycosidases) are named for the molecular structures they affect. They are cell-derived proteins which act as highly specific catalysts for the scission of water-labile functional groups.

Enzymes contain molecular chain structures and develop conformations that allow "recognition" of chain sequences (receptors) on biopolymers. Complexes form between segments of the enzyme and the biopolymer substrate which result in enhanced bond cleavage rates. Lacking the recognition sequences of susceptible natural polymers, most synthetic polymers are more resistant to enzymatic degradation. Nevertheless, comparative studies have shown some enhancement of hydrolysis rates by enzymes, particularly with synthetic polyesters and polyamides (Zaikov, 1985; Smith *et al.*, 1987; Kopecek *et al.*, 1983). Apparently the enzymes can recognize and interact with structural segments of the polymers, or more accurately, of the polymers coated with serum proteins, to initiate their catalytic action *in vivo* (Pitt, 1992).

Enzymes with demonstrated effects on hydrolysis rates can be quite selective in the presence of several hydrolyzable functional groups. For example, polyether urethane ureas and polyester urethane ureas exposed to hydrolytic enzymes (an esterase, cholesterol esterase, and a protease, elastase) were observed for rate and site of hydrolysis. Enzyme catalysis was clearly observed for the ester groups while the hydrolytically susceptible urea, urethane, and ether groups did not show significant hydrolysis as indicated by release of radiolabeled degradation products (Santerre *et al.*, 1994; Labow *et al.*, 1995).

Many enzymes exert predominantly a surface effect because of their great molecular size, which prevents absorption. Even hydrogels [e.g., poly(acrylamide)], which are capable of absorbing certain proteins, have molecular weight cutoffs for absorption well below those of such enzymes. However, as the degrading surface becomes roughened or fragmented, enzymatic action may be enhanced as a result of increased surface area if the substrates remain accessible to phagocytic cells that contain the active enzymes. Implanted devices that are in continuous motion relative to neighboring tissue can provoke inflammation, stimulating enzyme release.

Hydrolysis: Preclinical and Clinical Experience

A discussion of *in vivo* responses of several prominent polymer compositions known to be susceptible to hydrolysis follows. The structures of these polymers are described in Chapter 2.2.

Polyesters

Poly(ethylene terephthalate) (PET), in woven, velour, or knitted fiber configurations, remains a primary choice of cardiovascular surgeons for large-diameter vascular prostheses, arterial patches, valve sewing rings, etc. It is a strong, flexible polymer, stabilized by high crystallinity as a result of chain rigidity and orientation and is often considered to be biostable.

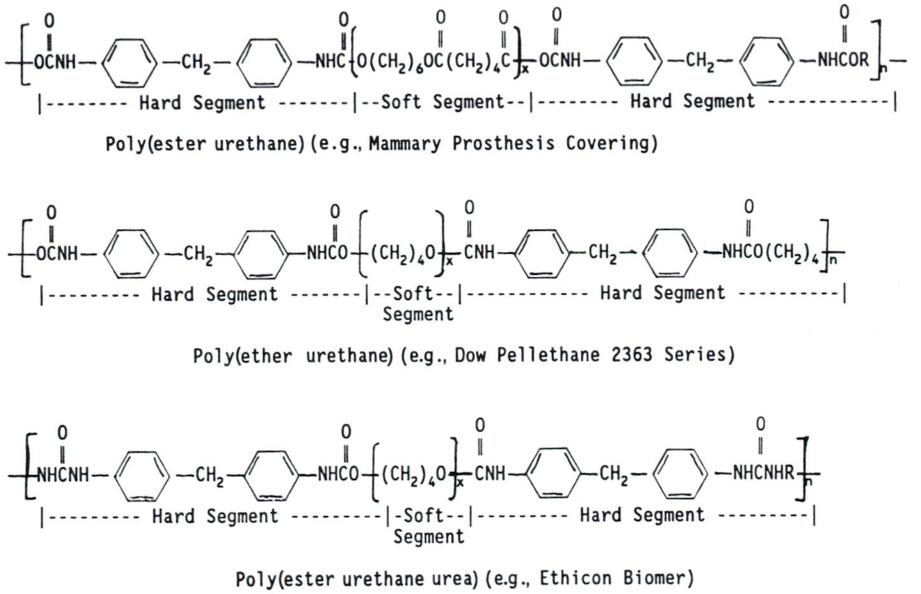

FIG. 3. Structure of implantable poly(ester urethane), poly(ether urethane) and poly(ether urethane urea).

Yet, over several decades, there have been numerous reports of long-term degeneration of devices *in vivo*, owing to breakage of fibers and device dilation. Proposed causes have been structural defects, processing techniques, handling procedures, and hydrolytic degradation (Cardia *et al.*, 1989).

Systematic studies of PET implants in healthy dogs have shown slow degradation rates, which were estimated to be equivalent to those in humans. For woven patches implanted subcutaneously, a mean total absorption time by the body of 30 ± 7 years, with 50% deterioration of fiber strength in 10 ± 2 years was projected. In infected dogs, however, where localized pH dropped to as low as 4.8, degradation was enhanced exponentially, with complete loss of properties within a few months (Zaikov, 1985). Human implant retrieval studies have shown significant evidence of graft infection (Vinard *et al.*, 1991). Besides the obvious pathological consequences of infection, the enhanced risk of polymer degradation is a cause for concern.

Aliphatic polyesters are most often intended for use as biodegradable polymers, with poly(caprolactone), for example, undergoing a significant decrease in molecular weight as indicated by a drop of 80–90% in relative viscosity within 120 weeks of implant (Kopecek *et al.*, 1983).

Poly(ester urethanes)

The earliest reported implants of polyurethanes, dating back to the 1950s, were cross-linked, aromatic poly(ester urethane) foam compositions (Blais, 1990; Bloch *et al.*, 1972). Their use in plastic and orthopedic reconstructive surgery initially yielded promising results. Acute inflammation was low. Tissue ingrowth promoted thin fibrous capsules. However, within months they were degraded and fragmented, producing untoward chronic effects (Bloch *et al.*, 1972). Foci of initial degradation are generally considered to be the polyadipate ester

soft segments which undergo hydrolysis (Fig. 3). By comparison, corresponding poly(ether urethanes) are very resistant to hydrolysis, although more susceptible to oxidation (see the section on oxidative biodegradation). Whether such hydrolytically degraded poly(ester urethanes) subsequently produce meaningful levels of aromatic amines (suspected carcinogens) by hydrolysis of urethane functions *in vivo* is currently an unresolved subject of considerable debate (Szycher *et al.*, 1991; Blais, 1990).

It is noteworthy that poly(ester urethane) foam-coated silicone mammary implants have survived as commercial products for decades (Blais, 1990), despite their known tendency to degrade. Apparently, the type of fibrous capsules formed on devices containing degradable foam were favored by some clinicians over those caused by smooth-walled silicone implants. In large device, unstabilized by tissue ingrowth, the frictional effects of sliding may cause increased capsule thickness and contraction (Snow *et al.*, 1981) along with extensive chronic inflammation.

Polyamides

Nylon 6 (polycaproamide) and nylon 6,6 [poly(hexamethylene adipamide)] contain a hydrolyzable amide connecting group, as do proteins. These synthetic polymers can absorb 9–11% water, by weight, at saturation. It is predictable, then, that they degrade by ion-catalyzed surface and bulk hydrolysis (Fig. 1). In addition, hydrolysis due to enzymatic catalysis leads to surface erosion (Zaikov, 1985). Quantitatively, nylon 6,6 lost 25% of its tensile strength after 89 days, and 83% after 726 days in dogs (Kopecek, 1983). An example of polyamide degradation of particular consequence involved the *in vivo* fragmentation of the nylon 6 tail string of an intrauterine contraceptive device. This string consisted of a nylon 6-sheath around nylon 6 multifilaments. The combination of fluid absorption (>10%) and hydrolysis was claimed to produce environmental stress cracking. The cracked coating allegedly provided a pathway for bacteria to travel from the vagina into the uterus, resulting in significant pelvic inflammatory disease (Hudson and Crugnola, 1987).

Degradation of a poly(arylamide) intended for orthopedic use (the fiber-reinforced polyamide from *m*-xylylene diamine and adipic acid) was also shown in a rabbit implant study. [Although the material provoked a foreign-body reaction comparable to a polyethylene control, surface pitting associated with resolving macrophages was noted at 4 weeks and became more pronounced by 12 weeks. This result was not predicted since polyarylamides are very resistant to solvents and heat (Finck *et al.*, 1995).]

Polyamides with long aliphatic hydrocarbon chain segments [e.g., poly(dodecanamide)] are more hydrolytically stable than shorter chain nylons and correspondingly degrade slower *in vivo*.

Poly(alkyl cyanoacrylates)

This class of polymers used as tissue adhesives is noteworthy as a rare case in which carbon–carbon bonds are cleaved by hydrolysis (Fig. 1). This occurs because the methylene (–CH_2–) hydrogen in the polymer is highly activated inductively by electron-withdrawing neighboring groups. Formation of the polymer adhesive from monomers is base catalyzed, with adsorbed water on the adherend being basic enough to initiate the reaction.

Catalysts for equilibrium reactions affect the reverse, as well as the forward reaction. Therefore, water associated with tissue can induce polycyanoacrylate hydrolysis by a "reverse Knoevenagel" reaction (Fig. 1). More basic conditions and (as suggested by *in vitro* cell culture or implant studies) enzymatic processes are much more effective. In chick embryo liver culture (a rich source of a variety of enzymes), methyl cyanoacrylate degraded much faster than in cell culture medium alone. In animal implants, poly(methyl cyanoacrylate) was extensively degraded within 4–6 months (Kopecek, 1983). Higher alkyl (e.g., butyl) homologs degraded slower than the methyl homolog and were less cytotoxic (Hegyeli, 1973; Vauthier *et al.*, 2003).

Polymers Containing Hydrolyzable Pendant Groups

Certain polymers intended for long-term implantation consist of biostable main-chain sequences and hydrolyzable pendant groups. Poly(methyl methacrylate) (PMMA) used in bone cements and intraocular lenses is an example of a hydrophobic polymer with a stable hydrocarbon main chain and hydrolyzable ester side groups. It has been proven, over decades of use, to provide reliable, stable service.

Another polymer system with a hydrocarbon backbone, poly(methyl acrylate-*co*-2-hydroxyethyl acrylate) also contains hydrolyzable ester side groups. This polymer, which forms hydrogels in an aqueous environment, has been used as a "scleral buckling" device for retinal detachment surgery. Basically, the dry polymer, shaped as a band or ring, placed as a "belt" around the sclera, expands through hydration to create an indentation in the zone of the retinal detachment to reestablish retinal contact. The device is left in place as a permanent implant (or "exoplant" as it is sometimes called because it is external to the sclera) (Braunstein and Winnick, 2002).

This hydrogel device, introduced into clinical practice in the 1980s (Refojo and Leong, 1981; Colthurst *et al.*, 2000), apparently performed satisfactorily for years as an approved product. However, in the 1990s, reports of long-term complications of these hydrogel scleral buckles began to surface (Hwang and Lim, 1997; Roldan-Pallares *et al.*, 1999). The hydrogel structures resumed swelling, sometimes with fragmentation, after maintaining stable dimensions for years. One report described a difficult explanation 13 years after implantation (Braunstein and Winnick, 2002). Pressures applied to the eye by this swelling have led to blindness and loss of the eyeball. Hydrogel scleral buckles are no longer used in retinal surgery (Watt, 2001).

Very little speculation has been provided in published articles about the mechanism of failure of acrylate scleral buckling devices other than that chemical degradation has occurred (Roldan-Pallares, *et al.*, 1999). I suggest that a likely mechanism involves hydrolysis of the ester side groups enhanced by the hydrophilic nature of the polymer (as contrasted to hydrophobic polymers such as PMMA). Hydrolysis of either of the two acrylate esters in the polymer chain provides an acrylic acid moiety. Linear poly(acrylic acid) is fully water soluble and each hydrolytic event renders the polymer more hydrophilic and subject to enhanced swelling. This process is slow but inexorable in the case of the scleral buckling device. The valuable lesson is that devices with intrinsically susceptible groups can eventually degrade by predictable mechanisms. This may take longer than is required for pivotal preclinical qualification studies (typically 2-year animal implants). If late degradation is suspected, therefore, accelerated aging studies should be performed *in vitro* with correlations made to *in vivo* studies. Such efforts give valuable, if not completely trustworthy, information. (See this chapter, "**Polymer Degradation Processes**" section.)

OXIDATIVE BIODEGRADATION

Oxidation Reaction Mechanisms and Polymer Structures

While much is known about the structures and reaction products of polymers susceptible to oxidative biodegradation, confirmation of the individual reaction steps has not yet been demonstrated analytically. Still, mechanistic inferences are possible from extensive knowledge of physiological oxidation processes and polymer oxidation *in vitro*.

The polymer oxidation processes to be discussed may be consistent with a homolytic chain reaction or a heterolytic mechanism. Species such as carbonyl, hydroxyl, and chain scission products are detectable. Classic initiation, propagation, and termination events for homolysis and ionic heterolytic processes are detailed in Fig. 4.

Except for the nature of susceptible functional groups, the principles of polymer degradation resistance stated in the

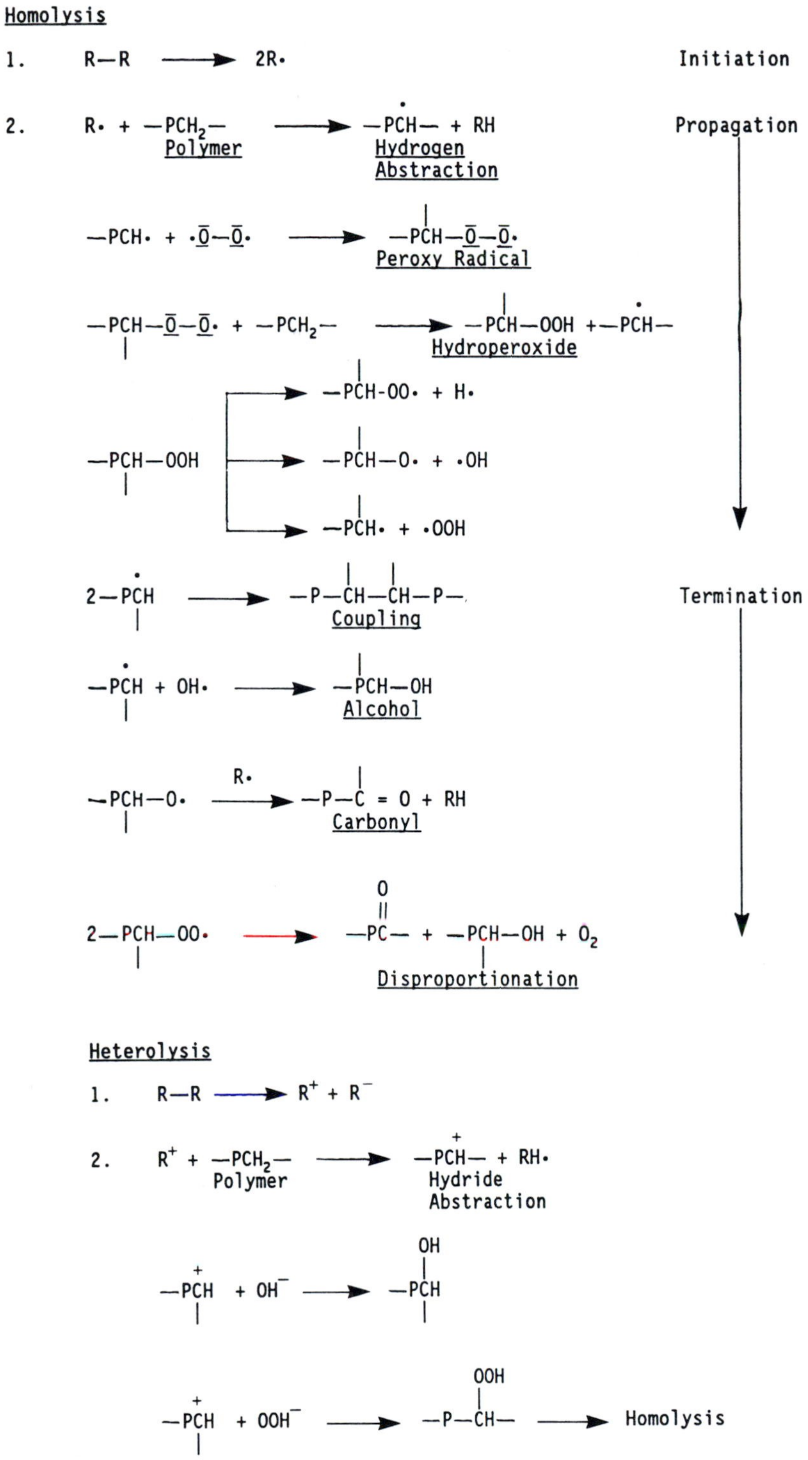

Homolysis

1. $R\!-\!R \longrightarrow 2R\cdot$ Initiation

2. $R\cdot + -PCH_2-\ \longrightarrow\ -P\overset{\cdot}{C}H- + RH$ Propagation
          Polymer   Hydrogen
                                     Abstraction

$-PCH\cdot + \cdot\bar{O}\!-\!\bar{O}\cdot\ \longrightarrow\ -PCH-\bar{O}\!-\!\bar{O}\cdot$
                                                  Peroxy Radical

$-PCH-\bar{O}\!-\!\bar{O}\cdot + -PCH_2-\ \longrightarrow\ -PCH-OOH + -\overset{\cdot}{P}CH-$
       | Hydroperoxide

$-PCH-OOH\ \Big\{\ \begin{array}{l} \longrightarrow -PCH\text{-}OO\cdot + H\cdot \\ \longrightarrow -PCH-O\cdot + \cdot OH \\ \longrightarrow -PCH\cdot + \cdot OOH \end{array}$

$2-\overset{\cdot}{P}CH\ \longrightarrow\ -P\overset{|}{C}H-\overset{|}{C}H-P-$ Termination
         | Coupling

$-\overset{\cdot}{P}CH + OH\cdot\ \longrightarrow\ -PCH-OH$
        | Alcohol

$-PCH-O\cdot\ \xrightarrow{\ R\cdot\ }\ -P-\overset{|}{C}=O + RH$
       | Carbonyl

$2-PCH-OO\cdot\ \longrightarrow\ -P\overset{\overset{\displaystyle O}{\|}}{C}- + -PCH-OH + O_2$
        | Disproportionation

Heterolysis

1. $R\!-\!R \longrightarrow R^+ + R^-$

2. $R^+ + -PCH_2-\ \longrightarrow\ -P\overset{+}{C}H- + RH$
          Polymer   Hydride
                                     Abstraction

$-P\overset{+}{C}H + OH^-\ \longrightarrow\ -P\overset{\overset{\displaystyle OH}{|}}{C}H$

$-P\overset{+}{C}H + OOH^-\ \longrightarrow\ -P\overset{\overset{\displaystyle OOH}{|}}{C}H-\ \longrightarrow\ \text{Homolysis}$

FIG. 4.   Proposed homolytic chain reaction and heterolytic oxidation mechanisms.

section on the structures of hydrolyzable polymers are valid for predicting relative oxidation resistance of polymers. Sites favored for initial oxidative attack, consistent with a homolytic or heterolytic pathway, are those that allow abstraction of an atom or ion and provide resonance stabilization of the resultant radical or ion. Figure 5 provides a selection of readily oxidized groups and the atom at which initial attack occurs. In Fig. 6, examples of radical and ion stabilization by resonance in ether and branched hydrocarbon structures are provided. Peroxy, carbonyl, and other radical intermediates are stabilized by

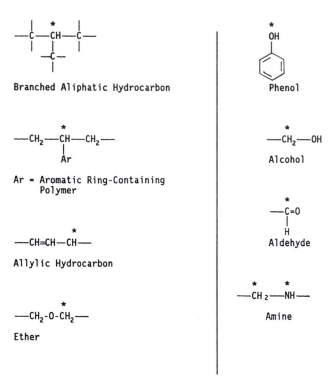

FIG. 5.   Readily oxidizable functional groups (* is site of homolysis or heterolysis).

similar resonance delocalization of electrons from the elements C, O, H, or N.

Two general categories of oxidative biodegradation, based on the source of initiation of the process, are direct oxidation by the host and device or external environment-mediated oxidation.

Direct Oxidation by Host

In these circumstances, host-generated molecular species effect or potentiate oxidative processes directly on the polymer. Current thinking, based on solid analytical evidence, is that such reactive molecues are derived from activated phagocytic cells responding to the injury and the properties of the foreign body at the implant site (Zhao *et al.*, 1991). These cells, which originate in the bone marrow and populate the circulatory system and connective tissues, are manifest as two types, the neutrophils (polymorphonuclear leukocytes, PMNs) and the monocytes. The latter can differentiate into macrophage and foreign body giant cell (FBGC) phenotypes.

Much work is under way to elucidate the sequence of events leading to phagocytic oxidation of biomaterials. Certain important processes of wound healing in the presence of biologically derived foreign bodies such as bacteria and parasites, are showing some relevance to biomaterial implants (Northup, 1987).

Neutrophils, responding to chemical mediators at the wound site, mount a powerful but transient chemical attack within the first few days of injury (Northup, 1987; Test

and Weiss, 1986). Chemically susceptible biomaterials may be affected if they are in close apposition to the wound site (Sutherland *et al.*, 1993). Activated macrophages subsequently multiply and subside within days at a benign wound site or in weeks if stimulants such as toxins or particulates are released at the site. Their fusion products, foreign body giant cells, can survive for months to years on the implant surface. Macrophages also remain resident in collagenous capsules for extended periods.

While we recognize that the mechanism of cellular attack and oxidation of biomaterials is as yet unconfirmed, the following discussion attempts to provide logical biological pathways to powerful oxidants capable of producing known degradation products.

Both PMNs and macrophages metabolize oxygen to form a superoxide anion (O_2^-). This intermediate can undergo transformation to more powerful oxidants on conceivably on can initiate homolytic reactions on the polymer. Superoxide dismutase (SOD), a ubiquitous peroxidase enzyme, can catalyze the conversion of superoxide to hydrogen peroxide, which in the presence of myeloperoxidase (MPO) derived from PMNs, is converted to hypochlorous acid (HOCl). A potent biomaterial oxidant in its own right (Coury *et al.*, 1987), hypochlorite can oxidize free amine functionality (e.g., in proteins) to chloramines that can perform as long-lived sources of chlorine oxidant (Test and Weiss, 1986, Figs. 7, 8). Hypochlorite can oxidize other substituted nitrogen functional groups (amides, ureas, urethanes, etc.) with potential chain cleavage of these groups.

The following paragraphs describe potential cooperative reactions involving acquired peroxidase and free ferrous ions. Macrophages contain essentially no MPO, so their hydrogen peroxide is not normally converted to HOCl. However, PMN-derived MPO can bind securely to foreign body surfaces (Locksley *et al.*, 1982), and serve as a catalyst reservoir for macrophage- or FBGC-derived HOCl production. If free ferrous ion, which is normally present in negligible quantities in the host, is released to the implant site by hemolysis or other injury, it can catalyze the formation of the powerfully oxidizing hydroxyl radical via the Haber–Weiss cycle (Klebanoff, 1982; Fig. 7).

Figure 8 shows radical and ionic intermediates of HOCl that may initiate biomaterial oxidation. Figure 9 is a diagram showing a leukocyte phagocytic process that employs endogenous MPO catalysis of HOCl formation. In a more general sense, the MPO may come from within or outside of the cell.

The foregoing discussion of sources of direct oxidation focused primarily on acute implant periods in which bursts of PMN activity followed by macrophage activity normally resolve within weeks. However, since the foreign body subsequently remains implanted, a sustained if futile attempt to phagocytose an implanted device provides a prolonged release of chemicals onto the biomaterial. This phenomenon, called exocytosis, occurs over months to possibly years (Zhao *et al.*, 1990) and results primarily from the macrophage-FBGC line. It can contribute to long-term chemical degradation of the polymer.

The oxidation processes induced by phagocytes are the result of oxidants produced by general foreign body responses,

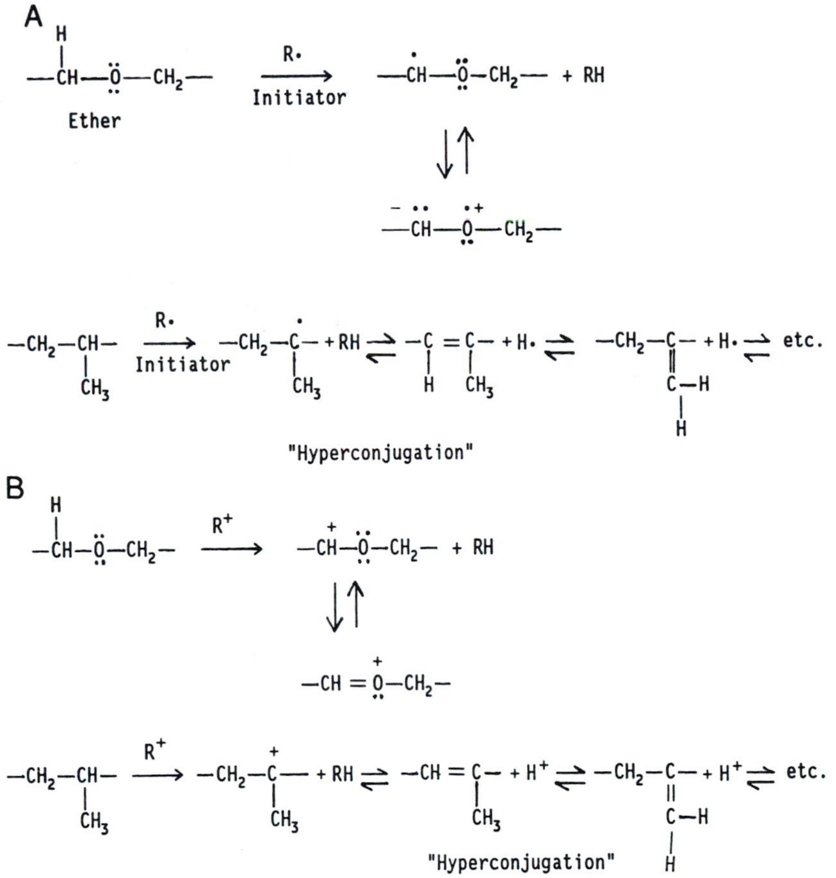

A

$$\text{Ether} \quad -\overset{\overset{H}{|}}{CH}-\overset{..}{\overset{..}{O}}-CH_2- \quad \xrightarrow[\text{Initiator}]{R\cdot} \quad -\overset{\cdot}{CH}-\overset{..}{\overset{..}{O}}-CH_2- \; + \; RH$$

$$\Big\downarrow\Big\uparrow$$

$$-\overset{-}{CH}-\overset{..}{\overset{..}{O}}\overset{\cdot +}{}-CH_2-$$

$$-CH_2-\underset{\underset{CH_3}{|}}{CH}- \quad \xrightarrow[\text{Initiator}]{R\cdot} \quad -CH_2-\underset{\underset{CH_3}{|}}{\overset{\cdot}{C}}- \; + \; RH \rightleftharpoons -\underset{\underset{H}{|}}{C}=\underset{\underset{CH_3}{|}}{C}- \; + \; H\cdot \rightleftharpoons -CH_2-\underset{\underset{\underset{H}{|}}{C-H}}{\overset{}{C}}- \; + \; H\cdot \rightleftharpoons \; etc.$$

"Hyperconjugation"

B

$$-\overset{\overset{H}{|}}{CH}-\overset{..}{\overset{..}{O}}-CH_2- \quad \xrightarrow{R^+} \quad -\overset{+}{CH}-\overset{..}{\overset{..}{O}}-CH_2- \; + \; RH$$

$$\Big\downarrow\Big\uparrow$$

$$-CH=\overset{+}{\overset{..}{O}}-CH_2-$$

$$-CH_2-\underset{\underset{CH_3}{|}}{CH}- \quad \xrightarrow{R^+} \quad -CH_2-\underset{\underset{CH_3}{|}}{\overset{+}{C}}- \; + \; RH \rightleftharpoons -CH=\underset{\underset{CH_3}{|}}{C}- \; + \; H^+ \rightleftharpoons -CH_2-\underset{\underset{\underset{H}{|}}{C-H}}{\overset{}{C}}- \; + \; H^+ \rightleftharpoons \; etc.$$

"Hyperconjugation"

FIG. 6. (A) Resonance stabilization of ether and hydrocarbon radicals. (B) Resonance stabilization of ether and hydrocarbon cations.

not direct receptor–ligand catalysis by oxidase enzymes. Attempts to degrade oxidatively susceptible polymers by direct contact with oxidase enzymes have produced short-range or limited effects (Santerre *et al.*, 1994; Sutherland *et al.*, 1993).

Macrophages mediate other processes, such as fibrous capsule formation around the device. Their release of cellular regulatory factors stimulates fibroblasts to populate the implant site and produce the collagenous sheath. Any knowledge of the effects of factors such as fibroblasts or fibrous capsules on rates and mechanisms of polymer degradation is, at this time, extremely rudimentary.

Stress Cracking

An important category of host-induced biodegradation with an oxidative component is stress cracking as manifest in poly(ether urethane) elastomers. It differs from classic environmental stress cracking (ESC), which involves a susceptible material at a critical level of stress in a medium which may permeat but does not dissolve the polymer. Classic ESC is not accompanied by significant chemical degradation (Stroke, 1988). Instead, stress cracking of polyurethanes is characterized by surface attack of the polymer and by chemical changes induced by relatively specific *in vivo* or *in vitro* oxidizing

$$\underset{\substack{\text{Neutrophil}\\\text{or}\\\text{Macrophage}}}{} + O_2 + e^- \quad \xrightarrow{\text{Activating Factors}} \quad \underset{\text{Superoxide Anion}}{O_2^-}$$

$$2O_2^- + 2H^+ \quad \xrightarrow{\text{SOD}} \quad O_2 + \underset{\substack{\text{Hydrogen}\\\text{Peroxide}}}{H_2O_2}$$

$$O_2^- + Fe^{3+} \quad \xrightarrow{} \quad O_2 + Fe^{2+}$$
Ferric Ion

$$O_2^- + H^+ \quad \xrightarrow{} \quad HO_2\cdot$$

$$H_2O_2 + Fe^{2+} \quad \xrightarrow{} \quad Fe^{3+} + OH\cdot + OH^-$$
Hydroxyl Radical

$$H_2O_2 + Cl^- + H^+ \quad \xrightarrow{\text{MPO}} \quad \underset{\substack{\text{Hypochlorous}\\\text{Acid}}}{HOCl} + H_2O$$

$$HOCl + R_2NH \quad \underset{}{\overset{}{\rightleftharpoons}} \quad \underset{\text{Chloramine}}{R_2N\text{-}Cl} + H_2O$$

$$NO\cdot + O_2^- \quad \xrightarrow{} \quad ONOO^-$$
Nitric Oxide   Peroxynitrite Anion

FIG. 7. Generation of potential oxidants by phagocytic processes.

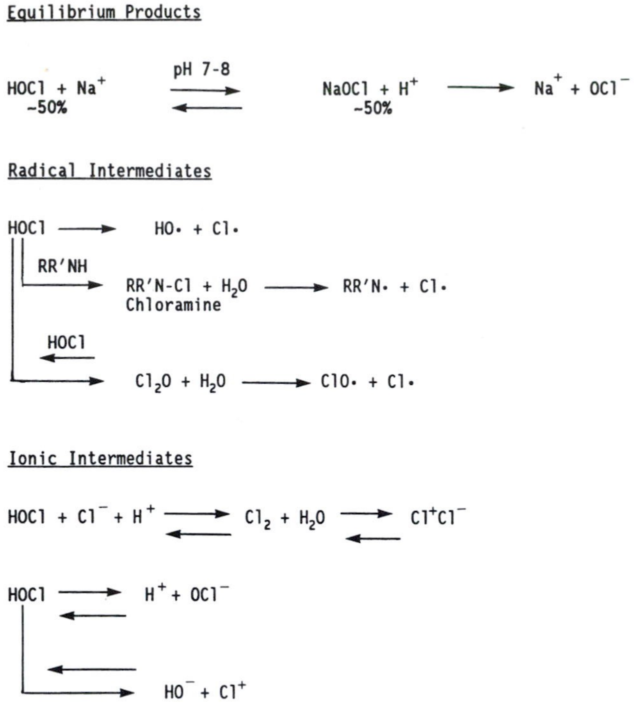

Equilibrium Products

$$HOCl + Na^+ \xrightleftharpoons[\sim 50\%]{pH\ 7-8} NaOCl + H^+ \longrightarrow Na^+ + OCl^-$$
$$\sim 50\%$$

Radical Intermediates

$$HOCl \longrightarrow HO\cdot + Cl\cdot$$

$$\downarrow RR'NH$$

$$RR'N-Cl + H_2O \longrightarrow RR'N\cdot + Cl\cdot$$
Chloramine

$$\downarrow HOCl$$

$$Cl_2O + H_2O \longrightarrow ClO\cdot + Cl\cdot$$

Ionic Intermediates

$$HOCl + Cl^- + H^+ \longrightarrow Cl_2 + H_2O \longrightarrow Cl^+Cl^-$$

$$HOCl \longrightarrow H^+ + OCl^-$$

$$\longrightarrow HO^- + Cl^+$$

FIG. 8. Hypochlorous acid: formation and potential reaction intermediates.

conditions. Conditions relevant to stress cracking of certain poly(ether urethane) compositions are stated in Table 3.

Recent information on the stress cracking of poly(ether urethanes) and poly(ether urethane ureas) (e.g., Fig. 3) has provided insights which may be valid for these and other compositions that can be oxidized (e.g., polypropylene; Altman *et al.*, 1986; polyethylenes, Wasserbauer *et al.*, 1990; Zhao *et al.*, 1995).

Poly(ether urethanes), which are resistant to hydrolysis *in vivo*, are used as connectors, insulators, tines, and adhesives for cardiac pacemakers and neurological stimulators (Fig. 10). They have performed with high reliability in chronic clinical applications since 1975. Certain poly(ether urethane) pacing leads have displayed surface cracks in their insulation after residence times *in vivo* of months to years. These cracks are directly related in frequency and depth to the amount of residual stress (Figs. 11, 12) and the ether (soft segment) content of the polyurethane (Coury *et al.*, 1987; Martin *et al.*, 2001).

Morphologically, the cracks display regular patterns predominately normal to the force vectors with very rough walls, occasionally with "tie fibers" bridging the gaps, indicative of ductile rather than brittle fracture (Figs. 13, 14). Infrared analysis indicates that oxidation does not take place detectably in the bulk, but only on the surface where extensive loss of ether functionality ($1110\,\text{cm}^{-1}$) and enhanced absorption in the hydroxyl and carbonyl regions are observed (Stokes *et al.*, 1987). Possible mechanisms for the oxidative degradation of ethers are presented in Fig. 15. The participation of molecular oxygen in the degradation mechanism is supported by studies which showed that poly(ether urethane urea) degradation *in vitro* correlated with oxygen diffusion into the polymer bulk after surface oxidation was initiated by hydrogen peroxide/cobalt chloride (Schubert *et al.*, 1997).

In a seminal study, Zhao *et al.* (1990) placed polyurethane tubing in cages permeable to fluids and cells under strain (therefore under high initial stress, which was subject to subsequent stress relaxation) and implanted them in rats. In certain cases, antiinflammatory steroids or cytotoxic polymers were

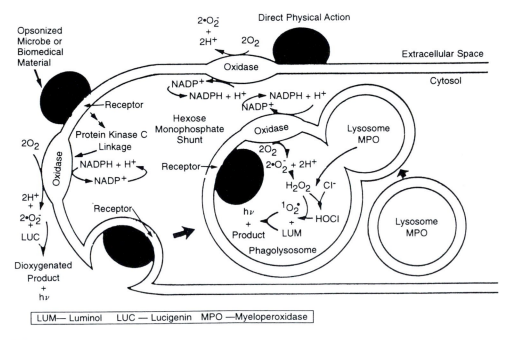

FIG. 9. Activation of phagocyte redox metabolism: chemiluminigenic probing with luminol and lucigenin. From R. C. Allen, personal communication.

TABLE 3 Characteristics of Poly(ether urethanes) That Cracked *In Vivo*

Components contained residual processing and/or applied mechanical stresses/strains

Components were exposed to a medium of viable cellular and extracellular body constituents

Polymers had oxidatively susceptible (aliphatic ether) groups

Analysis of polymers showed surface oxidation products

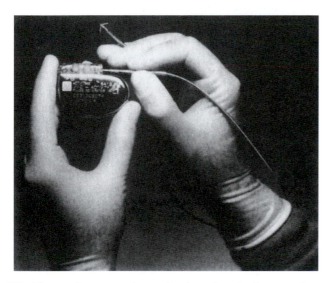

FIG. 10. Cardiac pacemaker with polyurethane lead, tine, and connector. Courtesy of Medtronic, Inc.

coimplanted in the cages. Implants of up to 15 weeks were retrieved. The only prestressed samples to crack were those that did not reside in the cages with the coimplants. The authors concluded that adherent cells caused the stress cracking, and cell necrosis or deactivation inhibited crack induction.

Subsequently, viable phagocytic cells were implicated as a cause of crack initiation *in vivo* (Zhao *et al.*, 1991). By removing adherent foreign body giant cells after implantation of a curved poly(ether urethane urea) film in a wire cage for up to 10 weeks, exposed "footprints" showed localized surface cracking on the order of several microns deep and wide. Adjacent areas of polymer which were devoid of attached cells were not cracked. Owing to relatively low stresses in the implanted film, deep crack propagation was not observed.

In vitro studies of strained (Stokes, 1988) and unstrained poly(ether urethane) films (Phua *et al.*, 1987; Bouvier *et al.*, 1991; Ratner *et al.*, 1988; Wiggins *et al.*, 2003) using oxidants, enzymes, etc., have sought to duplicate *in vivo* stress cracking. Although some surface chemical degradation with products similar to those seen *in vivo* was demonstrated, stress crack morphology was not closely matched *in vitro* until recently, in two studies. A test which involves immersing stressed poly(ether urethane) tubing in a medium of glass wool, hydrogen peroxide, and cobalt chloride produces cracks which duplicate those produced *in vivo* but with rate acceleration

of up to seven times (Zhao *et al.*, 1995). These investigators also showed that human plasma proteins, particularly alpha, 2-macroglobulin and ceruloplasmin, enhance *in vitro* stress cracking by oxidants in patterns morphologically similar to those observed *in vivo* (Zhao *et al.*, 1993). The potential of macrophages to contribute to stress cracking of poly(ether urethanes) was verified in a recent *in vitro* study which succeeded in potentiating macrophage oxidative effects with ferrous chloride and inhibiting them with the antiinflammatory steroid dexamethasone (Casas *et al.*, 1999). In another study, comparable crack patterns were produced when specimens of stressed tubing in rats were compared with those incubated with PMNs in culture (Sutherland *et al.*, 1993). Moreover, this study revealed a difference in chemical degradation products with time of implant which correlated with products from oxidants generated primarily by PMNs (HOCl) and macrophages (ONOO⁻). Early implant times, activated PMNs, and HOCl caused preferential decrease in the urethane oxygen stretch peak while longer implant times and ONOO⁻ caused selective loss of the aliphatic ether stretch peak (by infrared spectroscopy).

Taken together, the foregoing observations are consistent with a two-step mechanism for stress cracking *in vivo*. This hypothesis, as yet unproven, is under investigation. In the first step, surface oxidation induces very shallow, brittle microcracks. The second step involves propagation of the cracks in which specific body fluid components act on the formed cracks to enhance their depth and width without inducing major detectable bulk chemical reactions. Should this hypothesis prove correct, the term "oxidation-initiated stress cracking" would be reasonably descriptive.

The above description of stress cracking has generally considered static stress such as that formed during the cooling of molten parts or the assembly of components. Dynamic stresses and strains such as those occurring during the operation of diaphragm or bladder heart pumps or artificial joints can cause related cracking in areas of high flex. The cracking has been purported to increase with time of device operation but to display only minor surface chemical changes (Wu *et al.*, 1999; Tomita *et al.*, 1999).

This type of stress cracking has been controlled by reducing residual stress, isolating the polymer from cell contact (Tang *et al.*, 1994), protecting the polymer from stress cracking media, or using stress crack-resistant polymers (e.g., in the case of urethanes, ether-free) (Takahara *et al.*, 1994; Coury *et al.*, 1990; Tanzi *et al.*, 1997) and use of antioxidants such as hindred phenols (e.g., vitamin E, Monsanto Santowhite powder) (Schubert *et al.*, 1997). Stress cracking is next compared with another type of degradation, metal ion-induced oxidation.

Device- or Environment-Mediated Oxidation

Metal Ion–Induced Oxidation

A process of oxidative degradation that has, thus far, only been reported clinically for poly(ether urethane) pacemaker leads, requires, as does stress cracking, a very specific set of conditions. The enabling variables and fracture morphology are quite different from stress cracking, although oxidative degradation products are similar. Biodegradation of implanted

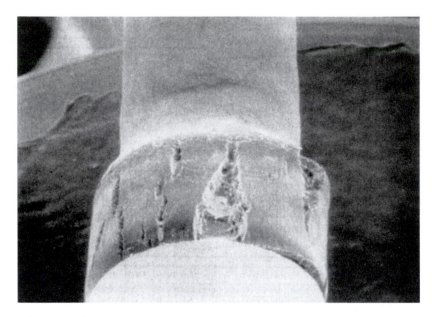

FIG. 11. Pellethane 2463-80A pacemaker lead tubing with high applied radial stress showing total breach.

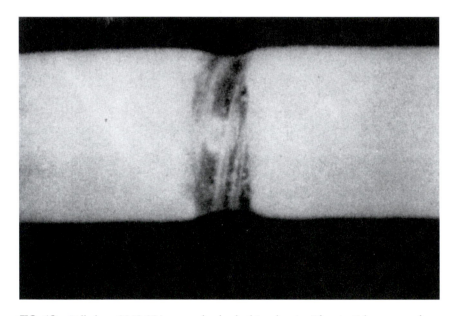

FIG. 12. Pellethane 2363-80A pacemaker lead tubing showing "frosting" due to stress from tight ligature.

devices through stress cracking always occurs on polymer surfaces exposed to cells and provides characteristic rough-walled fissures (indicative of ductile fracture) oriented perpendicular to the stress vector (Figs. 11–14). Metal ion–induced oxidation initiates on the enclosed inner surfaces of pacing lead insulation near corroded metallic components and their entrapped corrosion products. Smooth crack walls and microscopically random crack orientation is indicative of brittle fracture (Figs. 16, 17). Macroscopically, crack patterns that track metal component configurations may be present (Fig. 18). Degradation

products which may be found deeper in the bulk than with stress cracking are again indicative of brittle fracture.

This phenomenon called metal ion–induced oxidation has been confirmed by *in vitro* studies in which polyether urethanes were aged in metal ion solutions of different standard oxidation potentials. Above an oxidation potential of about +0.77, chemical degradation was severe. Below that oxidation potential, changes in the polymer that are characteristic of simple plasticization were seen (Coury *et al.*, 1987; Table 4). This technique also showed that metal ion–induced oxidation was

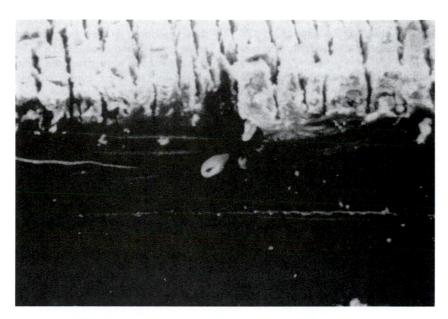

FIG. 13. Stress crack pattern (frosting) near tight ligature. × 14.

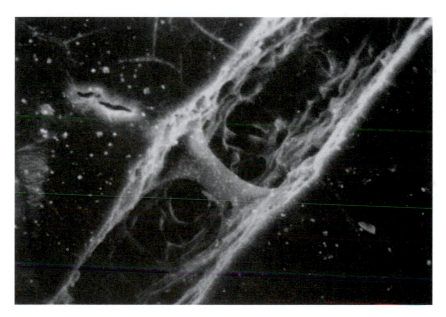

FIG. 14. Single stress crack in pacemaker lead tubing with rough walls and "tie fibers" indicative of ductile fracture. × 700.

proportional to the ether content of the polyurethane (Coury *et al.*, 1987; Table 5).

The effect of various metals on oxidation *in vitro* and *in vivo* HAS also been studied. Different metallic components of pacing lead conductors were sealed in poly(ether urethane) (Dow Pellethane 2363-80A) lead tubing and immersed in 3% hydrogen peroxide at 37°C for up to 6 months (Stokes *et al.*, 1987) or implanted in rabbits for up to 2 years (Stokes *et al.*, 1990). Both techniques resulted in corroded metals and degraded tubing lumen surfaces under certain conditions within 30 days.

In particular, the *in vivo* interaction of body fluids with cobalt and its alloys resulted in oxidative cracking of the polymer.

The metal ion–induced oxidation process clearly involves corrosion of metallic elements to their ions and subsequent oxidation of the polymer. In operating devices, the metal ion may be formed by solvation, galvanic or electrolytic corrosion, or chemical or biochemical oxidation (Fig. 19). In turn, these metal ions develop oxidation potentials that may well be enhanced in body fluids over their standard half-cell potentials. As strong oxidants, they produce intermediates or attack the

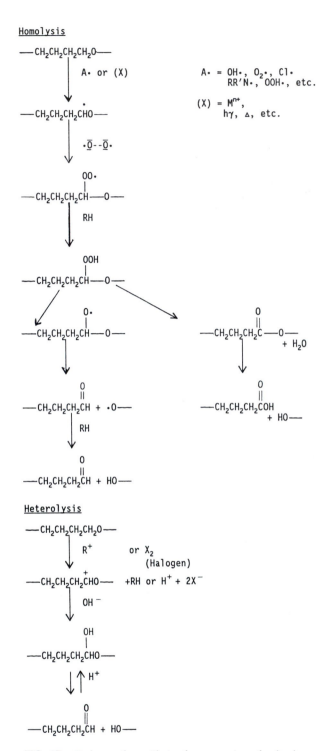

FIG. 15. Pathways for oxidative fragmentation of polyethers.

polymer to initiate the chain reaction (Fig. 20). Metal ion–induced oxidation is therefore the result of a highly complex interaction of the device, the polymer, and the body.

Should metal ion–induced oxidation be a possibility in an implanted device, several approaches are available to control this problem. They are not universally applicable, however,

and should be incorporated only if functionality and biocompatibility are retained. Potentially useful techniques include using corrosion-resistant metals, "flushing" corrosive ions away from the susceptible polymer, isolating the metals and polymer from electrolyte solutions, incorporating appropriate antioxidants, and using oxidation-resistant polymers if available.

Recently, polyurethane elastomers with enhanced oxidation stability have been developed. They are segmented, ether- and ester-free polymers with unconventional soft segments, including, for example, hydrogenated polybutadiene, polydimethylsiloxane, polycarbonate, and dimerized fat acid derivatives (Takahara *et al.*, 1991, 1994; Coury *et al.*, 1990; Pinchuk *et al.*, 1991; Kato *et al.*, 1995; Mathur *et al.*, 1997). In implant tests, they have shown reduced tendency to stress crack, and some of them have shown high resistance to metal ion oxidants *in vitro*. Early attempts to stabilize polyurethanes by laminating more biostable polymers (such as silicone rubbers) to tissue-facing surfaces have met with limited success (Pinchuk, 1992) in dynamic applications due to delamination tendencies.

More recent approaches to stabilizing polyurethanes to oxidative attack *in situ* have involved the use of surface modifying macromolecules (SMMs) (Santerre *et al.*, 2000) and surface modifying end groups (SMEs) (Ward, *et al.*, 1995, 1998). SMMs, typically fluorocarbon-based polymers, are blended with the polyurethanes during processing and migrate to the surface prior to implantation. SMEs are moieties (typically polysiloxane) bonded to polyurethane as end groups. The covalently modified polyurethanes may be used in bulk or as additives to conventional polyurethanes. Both approaches have provided enhanced *in vivo* stability for polyurethane implants; however, the long-term effects of these treatments are not, as yet, known. SMMs have been covalently modified with bioactive agents such as antioxidants to provide further degradation resistance (Ernsting *et al.*, 2002).

All of the "barrier" strategies to protecting polyurethanes described above appear to have validity, at least for protecting polyurethanes in the short term. The long-term (multiyear) benefits of these approaches remain to be seen in light of issues such as surface dynamics, interfacial interactions, and coating durability.

A caveat relevant to the opening paragraphs of this chapter is that all of these polyurethane modifications, while potentially providing enhanced resistance to biodegradation, still allow susceptibility to attack by biological components, often at slow rates. With poly(carbonate urethanes), for example, superior oxidation resistance has been observed in several studies (Tanzi *et al.*, 1997; Mathur *et al.*, 1997). However, in aqueous media *in vitro* and *in vivo*, slow degradation attributable to simple hydrolysis was also detected (Zhang *et al.*, 1997). The body fluid environment provides a relatively stable long-term hydrolytic medium, generally less subject to "respiratory bursts" that strongly enhance oxidative processes. Although phagocytic processes may also produce hydrolytic enzymes, their effects on synthetic polymers are specific and limited (Labow *et al.*, 2002). Hydrolysis, therefore, may be expected to take place continuously with poly(carbonate urethane) integrity susceptible to a combination of mechanical stress and vigorous oxidizing conditions (Fare *et al.*, 1999;

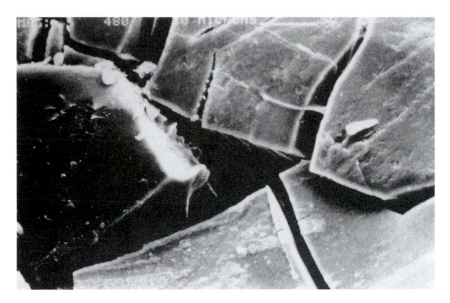

FIG. 16. Random crack pattern of Pellethane 2363-80A lead insulation caused by metal ion–induced oxidation. ×480.

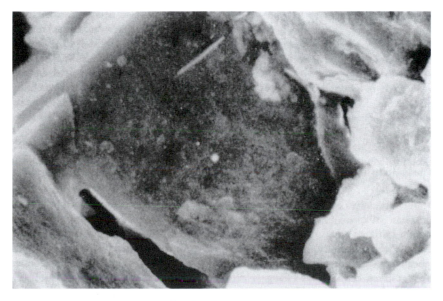

FIG. 17. Smooth crack wall indicative of brittle fracture caused by metal ion–induced oxidation. ×830.

Labow *et al.*, 2002). Studies of up to 3 years' implantation have indicated detectable hydrolysis (Seifalian *et al.*, 2003). Only long-term implant studies (e.g., 5 years or greater) would confirm the acceptability of poly(carbonate urethanes) or, for that matter, other new polymers having potentially susceptible groups.

Oxidative Degradation Induced by External Environment

Under very limited circumstances, the body can transmit electromagnetic radiation that may affect the integrity of implanted polymers. For example, the cornea and vitreous humor of the eye as well as superficial skin layers allow the passage of long-wave (320–400 nanometer) "ultraviolet A" radiation. Absorption of ultraviolet radiation causes electron excitation that can lead to photo-oxidative degradation. This process has been suggested in the breakdown of polypropylene components of intraocular lenses (Altman *et al.*, 1986).

In maxillofacial exo- and very likely endoprostheses, elastomers may undergo undesirable changes in color and physical properties as a consequence of exposure to natural sunlight-frequency radiation (Craig *et al.*, 1980). Photo-oxidation

FIG. 18. Crack pattern on inner lumen of polyether urethane lead insulation tracking coil indicative of metal ion-induced oxidation. ×100.

TABLE 4 Effect of Metal Ion Oxidation Potential on Properties of Poly(ether urethane) (Pellethane 2363-80A)[a]

Aqueous solution	Standard oxidation potential	Change in tensile strength (%)	Change in elongation (%)
PtCl₂	Ca + 1.2	− 87	− 77
AgNO₃	+ 0.799	− 54	− 42
FeCl₃	+ 0.771	− 79	− 10
Cu₂Cl₂	+ 0.521	− 6	+ 11
Cu₂(OAc)₂	+ 0.153	− 11	+ 22
Ni(OAc)₂	− 0.250	− 5	+ 13
Co(OAc)₂	− 0.277	+ 1	+ 13

[a] Conditions: 0.1 M solutions/90°C/35 days vs controls aged in deionized water; ASTM (D-1708) microtensile specimens; specimens were tested wet.

TABLE 5 Effect of Ether Content of Poly(ether urethane) on Susceptibility to Metal Ion–Induced Oxidation[a]

Polyetherurethane	Polyether content	Change in tensile strength (%)	Change in elongativon (%)
Pellethane 2363-80A	High	− 54	− 42
Pellethane 2363-55D	Low	− 23	− 10
Model segmented polyurethane	None	+ 9	+ 3

[a] Conditions: 0.1 M AgNO₃/90°C/35 days vs controls aged in deionized water; ASTM (D-1708) microtensile specimens.

Electrolysis

$$M° \xrightarrow{\quad\quad} M^{+n} + ne^-$$

$$2M° + 2H^+ \xrightarrow{\quad\quad} 2M^+ + H_2$$

$$4M° + O_2 + 2H_2O \xrightarrow{\quad\quad} 4M^+ + 4OH^-$$

$$M° + HOOH \xrightarrow{\quad\quad} M^+ + HO^- + HO\cdot$$

$$M° + HOCl \xrightarrow{\quad\quad} M^+ + HO\cdot + Cl^-$$

etc.

FIG. 19. Formation of metal ion from metal.

$$M^{+n} + H_2O \xrightarrow{\quad\quad} M^{+(n-1)} + HO\cdot + H^+$$

$$M^{+n} + -PH- \xrightarrow{\quad\quad} M^{+(n-1)} + -P\cdot + H^+$$
$$\text{(PH = Polymer)} \qquad\qquad\qquad\qquad |$$

$$M^{+n} + O_2 \xrightarrow{\quad\quad} M^{+(n+1)} + O_2^-$$

$$M^{+n} + HOCl \xrightarrow{\quad\quad} M^{+(n+1)} + HO\cdot + Cl^-$$

$$M^{+n} + -PH- \xrightarrow{\quad\quad} M^{+(n-1)}H + -P^+-$$

$$M^{+n} + H_2O_2 \xrightarrow{\quad\quad} M^{+(n+1)} + OH^- + HO\cdot$$

$$M^{+n} + H_2O_2 \xrightarrow{\quad\quad} M^{+(n-1)} + HO_2\cdot + H^+$$

FIG. 20. Initiation of oxidation pathways by metal ions.

mechanisms involving the urethane function of aromatic poly(ether- or poly(ester urethanes) are shown in Fig. 21. Antioxidants and ultraviolet absorbers provide limited protection for these materials.

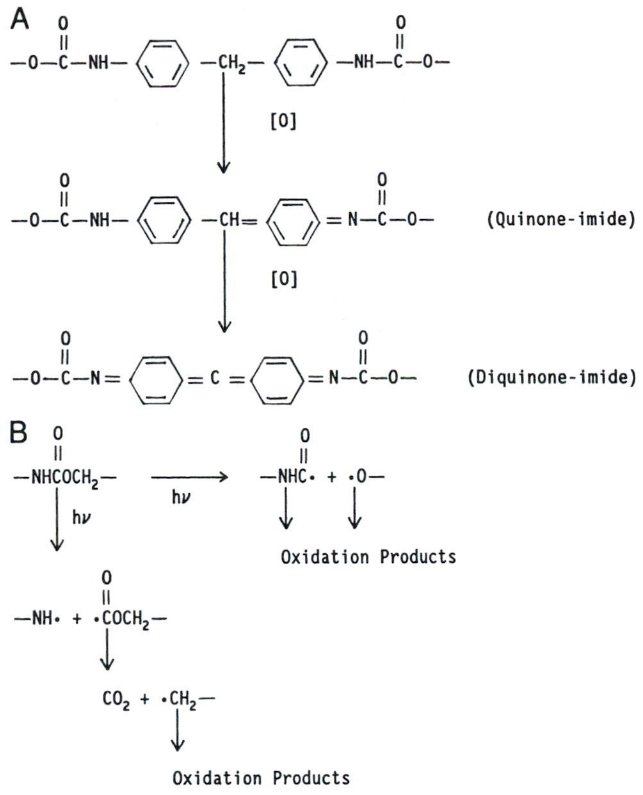

FIG. 21. Photo-oxidative reactions of aromatic polyurethanes. (A) Formation of quinone-imide from aromatic polyurethane. (From A. J. Coury *et al.*, *J. Biomater. Appl.* **3**, 1988.) (B) Photolytic cleavage of urethane link. (From S. K. Brauman *et al.*, *Ann. Biomed. Eng.* **9**, 1981.)

CONCLUSION

Polymers that are carefully chosen for use in implanted devices generally serve effectively for their intended lifetimes if they are properly processed and device–material–host interactions are adequately addressed. In certain limited circumstances, unintended hydrolytic or oxidative biodegradation occurs. This may be induced by direct attack by the host or via the intermediacy of the device or the outside environment. With susceptible polymers, protective measures can be taken to ensure extended efficacy, although new, biodegradation resistant polymer which are on the horizon may require less protection. Knowledge of biodegradation mechanisms and the employment of appropriate countermeasures will promote the continued growth in compositions and uses of polymers as implantable biomaterials.

Acknowledgments

The author is very grateful to Dr. R. C. Allen for providing the drawing on activated phagocyte redox metabolism. For their technical, advice and contributions, I sincerely thank Drs. James Anderson, Jonathan Sears, John Eaton, Allan Woffman, John Mahoney, Maurice Kreevoy, Grace Picciolo, and Buddy Ratner. For the preparation of this manuscript, I am deeply indebted to my "computer wizard," Mrs. Jayne McCaughey. For help in updating literature sources, I thank Ms. Mari Ferentinos.

QUESTIONS

1. The two major mechanisms of chemical degradation of polymers *in vivo* are hydrolysis and oxidation. Given the following polymers, indicate whether they are susceptible to hydrolysis, oxidation or both processes. If they would be highly resistant to both processes, so indicate.

   - Poly(carbonate urethane)
   - Poly(ether urethane)
   - Poly(ester urethane)
   - Aromatic polyester, poly(ethylene terephthalate)
   - Polypropylene
   - Polyethylene (linear)
   - Polytetrafluoroethylene
   - Polydimethylsiloxane

2. What are some common polymer functional groups susceptible to hydrolysis?

3. In the past, investigators have fabricated heart valves from "aromatic polyurethanes" which contain polyether, urethane, and urea functional groups. These devices

were intended to last for several years in use, but have generally failed to perform for those periods.

- What physical and chemical forces are acting on the heart valves *in vivo*?
- What are the most likely mechanisms (physical and chemical) of degradation leading to failure of these devices? State at least three mechanisms.

4. As a materials scientist, you have experience with polyurethanes as biomaterials. Among the readily available commercial elastomers, they demonstrate the best combination of physical properties, but are susceptible to biodegradation, mostly through their polyether or polyester soft segments. You are charged with designing a biostable elastomer (polyurethane or otherwise).

- Choose an approach to produce a chemically stable elastomer that retains reasonable physical properties for at least 3 years.
- Describe 5 tests/analyses (*in vitro/in vivo*) which may be used to characterize this elastomer and confirm its potential stability.

Bibliography

Allen, R. C. (1991). Activation of phagocyte redox metabolism: chemiluminigenic probing with luminol and lucigenin, Drawing Provided.

Altman, J. J., Gorn, R. A., Craft, J., and Albert, D. M. (1986). The break-down of polypropylene in the human eye: Is it clinically significant? *Ann. Ophthalmol.* 18: 182–185.

Blais, P. (1990). Letter to the editor. *J. Appl. Biometer.* 1: 197.

Blanchet, T. A., and Burroughs, B. R. (2001). Numerical oxidation model for gamma radiation — sterilized UHMWPE: consideration of dose-depth profile. *J. Biomed. Mater. Res.* 58(6): 684–693.

Bloch, B., and Hastings, G. (1972). *Plastics Materials in Surgery*, 2nd ed. Charles C. Thomas, Springfield, IL, pp. 97–98.

Booth, A. E. (1995). Industrial sterilization technologies: New and old trends shape manufacturer choices. *Med. Dev. Diagn. Industry* February: 64–72.

Bouvier, M. Chawla, A. S., and Hinberg, L., (1991). *In vitro* degradation of a poly (ether urethane) by trypsin. *J. Biomed. Mater. Res.* 25: 773–789.

Brauman, S. K., Mayorga, G. D., and Heller, J. (1981). Light stability and discoloration of segmented polyether urethanes. *Ann. Biomed. Eng.* 9: 45–58.

Braunstein, R. A., and Winnick, M. (2002). Complications of Miragel: Pseudotumor. *Arch. Ophthalmol.* 120: 228–229.

Cardia, G., and Regina, G. (1989). Degenerative Dacron graft changes: Is there a biological component in this textile defect?—A case report. *Vasc. Surg.* 23(3): 245–247.

Casas, J., Donovan, M., Schroeder, P., Stokes, K., and Untereker, D. (1999). *In vitro* modulation of macrophage phenotype and inhibition of polymer degradation by dexamethasone in a human macrophage/Fe/stress system. *J. Biomed. Mater. Res.* 46: 475–484.

Colthurst, M. J., Williams, R. L., Hiscott, P. S., and Grierson, I. (2000). Biomaterials used in the posterior segment of the eye. *Biomaterials* 21: 649–665.

Coury, A. J. (1999). Biostable polymers as durable scaffolds for tissue engineered vascular prostheses, in *Tissue Engineering of Vascular Prosthetic Grafts*, P. Zilla and H. Greisler (Eds.). R. G. Landes company, Austin, TX. Vol. 43, 469–480.

Coury, A. J., Slaikeu, P. C., Cahalan, P. T., and Stokes, K. B. (1987). Medical applications of implantable polyurethanes: Current issues. *Prog. Rubber Plastics Tech.* 3(4): 24–37.

Coury, A. J., Slaikeu, P. C., Cahalan, P. T., Stokes, K, B., and Hobot, C. M. (1988). Factors and interactions affecting the performance of polyrethane elastomers in medical devices. *J. Biomater. Appl.* 3: 130–179.

Coury, A. J., Hobot, C. M., Slaikeu, P. C., Stokes, K. B., and Cahalan, P. T., (1990). A new family of implantable biostable polyurethanes. *Trans. 16th Annual Meeting Soc. for Biomater.* May 20–23, 158.

Craig, R. G., Koran, A., and Yus, R. (1980). Elastomers for maxillofacial applications. *Biomaterials* 1(Apr.): 112–117.

Daly, B. M., and Yin, J. (1998). Subsurface oxidation of polyethylene. *J. Biomed. Mater. Res.* 42: 523–529.

Ernsting, M. J., Santerre, J. P., and Labow, R. S. (2002). Surface modification of a polycarbonate-urethane using a Vitamin E derivatized fluoroalkyl surface modifier. *Trans. 28th Annual Meeting Soc. Biomater.*, April 24–27, p. 16.

Faré S., Petrini, P., Motta, A., Cigada, A., and Tanzi, M. C. (1999). Synergistic efforts of oxidative environments and mechanical stress on *in vitro* stability of polyetherurethanes and polycarbonateurethanes. *J. Biomed. Mater. Res.* 45: 62–74.

Finck, K. M., Grosse-Siestrup, C., Bisson, S., Rinck, M., and Gross, U. (1994). Experimental *in vivo* degradation of polyarylamide. *Trans. 20th Annual Meeting Soc. for Biomater.* April 5–9, p. 210.

Furman, B., and Li, S. (1995). The effect of long-term shelf life aging of ultra high molecular weight polyethylene. *Trans. 21st Annual Meeting Soc. for Biomater.* March 18–22, p. 114.

Greisser, H. J., Gengenbach, T. R., and Chatelier, R. C. (1994). Long-term changes in the surface composition of polymers intended for biomedical applications. *Trans. 20th Annual Meeting Soc. for Biomater.* April 5–9, p. 19.

Hegyeli, A. (1973). Use of organ cultures to evaluate biodegradation of polymer implant materials. *J. Biomed. Mater. Res.* 7: 205–214.

Hoffman, A. (1999) Personal Communication.

Hudson, J., and Crugnola, A. (1987). The *in vivo* biodegradation of nylon 6 utilized in a particular IUD. *J. Biomater. Appl.* 1: 487–501.

Hwang, K. I., and Lim, J. I. (1997). Hydrogel explant fragmentation 10 years after scleral buckling surgery. *Arch. Ophthalmol.* 115: 1205–1206.

Kato, Y. P., Dereume, J. P., Kontges, H., Frid, N., Martin, J. B., MacGregor, D. C., and Pinchuk, L. (1995). Preliminary mechanical evaluation of a novel endoluminal graft. *Trans. 21st Annual Meeting Soc. for Biomater.* March 18–22, p. 81.

Klebanoff, S. (1982). Iodination catalyzed by the xanthine oxidase system: Role of hydroxyl radicals. *Biochemistry* 21: 4110–4116.

Kopecek, J., and Ulbrich, K. (1983). Biodegradation of biomedical polymers. *Prog. Polym. Sci.* 9: 1–58.

Labow, R. S., Erfle, D. J., and Santerre, J. P. (1995). Neutrophil-mediated degradation of segmented polyurethanes. *Biomaterials* 16: 51–59.

Labow, R. S., Tang, Y., McCloskey, C. B., and Santerre, J. P. (2002a). The effect of oxidation on the enzyme-catalyzed hydrolytic degradation of polyurethanes. *Can. J. Biomater. Sci., Polymer Ed.* 13(6): 651–665.

Labow, R. S., Meek, E., Matherson, L. A., and Santerre, J. P. (2002b). Human macrophage-mediated biodegradation of polyurethanes: assessment of candidate enzyme activities. *Biomaterials* 23(19): 3969–3975.

Ling, M. T. K., Westphal, S. P., Qin, S., Ding, S., and Woo, L. (1998). Medical plastics failures from heterogeneous contamination. *Med. Plast. Biomater.* 5(2): 45–49.

Ling, M. T. K., Westphal, S. P., Qin, S. Ding, S., and Woo, L. (1998). Medical plastics failures from heterogeneous contamination. *Med. Plast. Biomater.* 5(2): 45–49.

Locksley, R., Wilson, C., and Klebanoff, S. (1982). Role of endogenous and acquired peroxidase in the toxoplasmacidal activity of murine and human mononuclear phagocytes. *J. Clin. Invest.* 69(May): 1099–1111.

Martin, D. J., Poole Warren, L. A., Gunatillake, P. A., McCarthy, S. J., Meijs, G. F., and Schindhelm, K. (2001). New methods for the assessment of *in vitro* and *in vivo* stress cracking in biomedical polyurethanes. *Biomaterials* 22(9): 973–978.

Mathur, A. B., Collier, T. O., Kao, W. J., Wiggins, M., Schubert, M. A., Hiltner, A., and Anderson, J. M. (1997). in vivo biocompatibility and biostability of modified polyurethanes. *J. Biomed. Mater. Res.* 36: 246–257.

McKellop, H., Yeom, B., Campbell, P., and Salovey, R. (1995). Radiation induced oxidation of machined or molded UHMWPE after seventeen years. *Trans. 21st Annual Meeting Soc. for Biomater.* March 18–22, p. 54.

Northup, S. (1987). Strategies for biological testing of biomaterials. *J. Biomater. Appl.* 2: 132–147.

Phua, S. K., Castillo, E., Anderson, J. M., and Hiltner, A. (1987). Biodegradation of a polyurethane *in vitro*. *J. Biomed. Mater. Res.* 21: 231–246.

Pinchuk, L. (1992). Adhesive less bonding of silicone rubber to polyurethanes and the use of bonded materials. U.S. Patent 5,147,725, September 15.

Pinchuk, L., Esquivel, M. C., Martin, J. B., and Wilson, G. J. (1991). Corethane: A new replacement for polyether urethanes for long-term implant applications. *Trans. 17th Annual Meeting of the Soc. for Biomater.*, May 1–5, p. 98.

Pitt, C. G. (1992). Non-microbial degradation of polyesters: Mechanisms and modifications. in *Biodegradable Polymers and Plastics*, M. Vert, J. Feijin, A. Albertson, G. Scott, and E. Chiellini, eds. R. Soc. Chem., Cambridge, UK, pp. 1–19.

Portnoy, R. (1997). Clear, radiation-tolerant autoclavable polypropylene. *Med. Plast. Biomater.* 4(1): 40–48.

Ratner, B. D., Gladhill, K. W., and Horbett, T. A. (1988). Analysis of *in vitro* enzymatic and oxidative degradation of polyurethanes. *J. Biomed. Mater. Res.* 22: 509–527.

Refojo, M. F., and Leong, F. L. (1981). Poly(methylacrylate–co-hydroxyethyl acrylate) hydrogel implant material of strength and softness. *J. Biomed. Mater. Res.* 15: 497–509.

Roldan-Pallares, M., del Castillo, J. L., Awad-El Susi, S., and Refojo, M. F. (1999). Long-term complications of silicone and hydrogel explants in retinal reattachment surgery. *Arch. Ophthalmol.* 177: 197–201.

Santerre, J. P., Labow, R. S., Duguay, D. G., Erfle, D., and Adams, G. A. (1994). Biodegradation evaluation of polyether- and polyester-urethanes with oxidative and hydrolytic enzymes. *J. Biomed Mater. Res.* 28: 1187–1199.

Santerre, J. P., Meek, E., Tang, Y. W., and Labow, R. S. (2000). Use of fluorinated surface modifying macromolecules to inhibit the degradation of polycarbonate-urethanes by human macrophages. Trans. 6th World Biomaterials Congress: 77.

Schnabel, W. (1981). *Polymer Degradation Principles and Practical Applications*, Macmillan, New York, pp. 15–17, 179–185.

Schubert, M. A., Wiggins, M. J., Anderson, J. M., and Hiltner, A. (1997). Comparison of two antioxidants for poly(etherurethane urea) in an accelerated *in vitro* biodegradation system. *J. Biomed. Mater. Res.* 34: 493–505.

Schubert, M. A., Wiggins, M. J., Anderson, J. M., and Hiltner, A. (1997). Role of oxygen in biodegradation of poly(etherurethane urea) elastomers. *J. Biomed. Mater. Res.* 34: 519–530.

Seifalian, A. M., Salacinski, H. J., Tiwari, A., Edwards, E., Bowald, S., and Hamilton, G. (2003). in vivo biostability of a poly(carbonate-urea) urethane graft. *Biomaterials* 24(14): 2549–2557.

Shen, F. W., Yu, Y. J., and McKellop, H. (1999). Potential errors in FTIR measurement of oxidation in ultra-high molecular weight polyethylene implants. *J. Biomed. Mater. Res. (App. Biometer.)* 48: 203–210.

Smith, R., Oliver, C., and Williams, D. F. (1987). The enzymatic degradation of polymers *in vitro*. *J. Biomed. Mater. Res.* 21: 991–1003.

Snow, J., Harasaki, H., Kasick, J., Whalen, R., Kiraly, R., and Nosè, Y. (1981). Promising results with a new textured surface intrathoracic variable volume device for LVAS. *Trans. Am. Soc. Artif. Intern. Organs* XXVII: 485–489.

Stokes, K. (1988). Polyether polyurethanes: Biostable or not? *J. Biomater. Appl.* 3(Oct.): 228–259.

Stokes, K., Coury, A., and Urbanski, P. (1987). Autooxidative degradation of implanted polyether polyurethane devices. *J. Biomater. Appl.* 1(Apr.): 412–448.

Stokes, K., Urbanski, P., and Upton, J., (1990). The *in vivo* autooxidation of polyether polyurethane by metal ions. *J. Biomater. Sci. Polymer Edn.* 1(3): 207–230.

Sutherland, K., Mahoney, J. R., II, Coury, A. J., and Eaton, J. W. (1993). Degradation of biomaterials by phagocyte-derived oxidants. *J. Clin. Invest.* 92: 2360–2367.

Szycher, M., and Siciliano, A. (1991). An assessment of 2,4-TDA formation from Surgitek polyurethane foam under stimulated physiological conditions. *J. Biomater. Appl.* 5: 323–336.

Takahara, A., Coury, A. J., Hergenrother, R. W., and Cooper, S. L. (1991). Effect of soft segment chemistry on the biostability of segmented polyurethanes. I. *in vitro* oxidation. *J. Biomed. Mater. Res.* 25: 341–356.

Takahara, A., Coury, A. J., and Cooper, S. L. (1994). Molecular design of biologically stable polyurethanes. *Trans. 20th Annual Meeting Soc. for Biomater.* April 5–9, p. 44.

Tang, W. W., Santerre, J. P., Labow, R. S., Waghray, G., and Taylor, D. (1994). The use of surface modifying macromolecules to inhibit biodegradation of segmented polyurethanes. *Trans. 20th Annual Meeting Soc. for Biomater.* April 5–9, p. 62.

Tanzi, M. C., Mantovani, D., Petrini, P., Guidoin, R., and Laroche, G. (1997). Chemical stability of polyether urethanes versus polycarbonate urethanes. *J. Biomed. Mater. Res.* 36: 550–559.

Test, S., and Weiss, S. (1986). The generation of utilization of chlorinated oxidants by human neutrophils. *Adv. Free Radical Biol. Med.* 2: 91–116.

Tomita, N., Kitakura, T., Onmori, N., Ikada, Y., and Aoyama, E. (1999). Prevention of fatigue cracks in ultra-high molecular weight polyethylene joint components by the addition of vitamin E. *J. Biomed. Mater. Res. (App. Biomater.)* 48: 474–478.

Vauthier, C., Dubernet, C., Fattal, E., Pinto-Alphandary, H., and Couvreur, P. (2003). Poly (alkylcyanoacrylates) as biodegradable materials for biomedical applications. *Advanced Drug Delivery Rev.* 55(4): 519–548.

Vinard, E., Eloy, R., Descotes, J., Brudon, J. R., Giudicelli, H., Patra, P., Streichenberger, R., and David, M. (1991). Human vascular graft failure and frequency of infection. *J. Biomed. Mater. Res.* 25: 499–513.

Ward, R. S., White, K. A., Gill, R. S., and Wolcott, C. A. (1995). Development of biostable thermoplastic polyurethanes with oligomeric polydimethylsiloxane end groups. *Trans. 21st Annual Meeting Soc. for Biomater.* March 18–22, p. 268.

Ward, R. S., Tian, Y., and White, K. A. (1998). Improved polymer biostability via oligomeric end groups incorporated during synthesis. *Polymeric Mater. Sci. Eng.* 79: 526–527.

Wasserbauer, R., Beranova, M., Vancurova, D., and Dolezel, B. (1990). Biodegradation of polyethylene foils by bacterial and liver homogenates. *Biomaterials* 11(Jan.): 36–40.

Watt, D. R. (2001), Miragel sponge complications, www. retinadoc. com/scripts/retina.pl?function=viewquestions&forum=retina, 11/07/01.

Weaver, K. D., Sauer, W. L., and Beals, N. B. (1995). Sterilization induced effects on UHMWPE oxidation and fatigue strength. *Trans. 21st Annual Meeting Soc. for Biomater.* March 18–22, p. 114.

Wiggins, M. J., Anderson, J. M., and Hiltner, A. (2003). Biodegradation of polyurethane under fatigue loading. *J. Biomed. Mater. Res. Part A.* 65A(4): 524–535.

Williams, D. F. (1982). Review: Biodegradation of sugical polymers. *J. Mater. Sci.* 17: 1233–1237.

Williams, D. F. (1989). *Definitions in Biomaterials.* Elsevier, Amsterdam.

Wu, L., Weisberg, D. M., Runt, J., Felder III, G., Snyder, A. J., and Rosenberg, G. (1999). An investigation of the in vivo stability of poly(ether urethaneurea) blood sacs. *J. Biomed. Mater. Res.* 44: 371–380.

Zaikov, G. E. (1985). Quantitative aspects of polymer degradation in the living body. *JMS-Rev. Macromol. Chem. Phys.* C25(4): 551–597.

Zhang, Z., Marois, Y., Guidoin, R., Bull, P., Marois, M., How, T., Laroche, G., and King, M. (1997). Vascugraft® polyurethane arterial prosthesis as femoro-popliteal and femoro-peroneal bypass in humans: Pathological, structural and chemical analyses of four excised grafts. *Biomaterials* 18: 113–124.

Zhao, Q., Agger, M., Fitzpatrick, M., Anderson, J., Hiltner, A., Stokes, P., and Urbanski, P. (1990). Cellular interactions with biomaterials: *In vivo* cracking of pre-stressed pellethane 2363-80A. *J. Biomed. Mater. Res.* 24: 621–637.

Zhao, Q., Topham, N., Anderson, J. M., Hiltner, A., Lodoen, G., and Payet, C. R. (1991). Foreign-body giant cells and polyurethane biostability: *In vivo* correlation of cell adhesion and surface cracking. *J. Biomed. Mater. Res.* 25: 177–183.

Zhao, Q. H., McNally, A. K., Rubin, K. R., Renier, M., Wu, Z., Rose-Capara, V., Anderson, J. M., Hiltner, A., Urbanski, P., and Stokes, K. (1993). Human plasma α_2-macroglobulin promotes *in vitro* oxidative stress cracking of Pellethane 2363-80A. *Biomed. Mater. Res.* 27: 379–389.

Zhao, Q., Casas-Bejar, C., Urbanski, P., and Stokes, K. (1995). Glass wool–H_2O_2/$COCl_2$ for *in vitro* evaluation of biodegradative stress cracking in polyurethane elastomers. *J. Biomed. Mater. Res.* 29: 467–475.

Ziats, N., Miller, K., and Anderson, J. (1988). *In vitro* and *in vivo* interactions of cells with Biomaterials. *Biomaterials* 9(Jan.): 5–13.

6.3 DEGRADATIVE EFFECTS OF THE BIOLOGICAL ENVIRONMENT ON METALS AND CERAMICS

David F. Williams and Rachel L. Williams

The environment to which biomaterials are exposed during prolonged use (i.e., the internal milieu of the body) can be described as an aqueous medium containing various anions, cations, organic substances, and dissolved oxygen. The anions are mainly chloride, phosphate, and bicarbonate ions. The principal cations are Na^+, K^+, Ca^{2+}, and Mg^{2+}, but with

TABLE 1 Ionic Concentrations in Blood Plasma and Extracellular Fluid

Anion, cation	Blood plasma (mM)	Extracellular fluid (mM)
Cl^-	96–111	112–120
HCO_3^-	16–31	25.3–29.7
HPO_4^{2-}	1–1.5	193–102
SO_4^{2-}	0.35–1	0.4
$H_2PO_4^-$	2	—
Na^+	131–155	141–15
Mg^{2+}	0.7–1.9	1.3
Ca^{2+}	1.9–3	1.4–1.55
K^+	35–5.6	3.5–4

(Bundy, 1994)

TABLE 2 Major Proteins and Other Organic Constituents of Blood Plasma

Major proteins and organic molecules in blood plasma (gl^{-1} unless stated otherwise)	
Albumin	30–55
α-Globulins	5–10
β-Globulins	6–12
γ-Globulins	6.6–15
α-Lipoproteins	3.5–4.5
Fibrinogen	1.7–4.3
Total cholesterol	1.2–2.5
Fatty acids	1.9–4.5
Glucose	0.65–1.1
Lactate	0.5–2.2 mM
Urea	3–7 mM

(Bundy, 1994)

smaller amounts of many others. Table 1 presents the range of values for the anion and cation concentrations in blood plasma and extracellular fluid (Bundy, 1994). This represents an environment with a chloride concentration of approximately a third of that of sea water (Hanawa, 2002). The concentration of dissolved oxygen also influences the aggressive nature of the environment and in venous blood is approximately a quarter of that in air. The organic substances include low-molecular-weight species as well as relatively high-molecular-weight proteins and lipids. Table 2 gives examples of the concentration of various organic components of blood plasma. The protein content of the environment is known to have a significant influence on the corrosive nature of body fluids (Williams, 1985; Khan *et al.*, 1999a). The pH in this well-buffered system is around 7.4, although because of inflammation it may change for short periods following surgery to as low as 4 or 5 (Bundy, 1994). The temperature remains constant around 37°C.

On the basis of existing knowledge of the stability of materials in various environments, we should predict that metals, as a generic group, should be relatively susceptible to corrosion in this biological environment, whereas ceramics should display a varying susceptibility, depending on solubility. This

TABLE 3   Chemical Composition of Implant Alloys

Implant alloys	Composition
316L stainless steel	Cr, Mo, Ni, Mn, C, S, Si, P, Fe
Co–Cr based alloys	Cast Co–Cr–Mo
	Wrought Co–Cr–Mo
	Wrought Co–Cr–W–Ni
	Wrought Co–Cr–Mo–Ni
	Wrought Co–Cr–Mo–Ni–Fe
	Wrought Co–Cr–Mo–Ni–W–Fe
Cp titanium	Ti + traces of O, N, C, H, Fe
Titanium alloys	Ti-6Al-4V
	Ti-6Al-7Nb
	Ti-15Mo
	Ti-12Mo-6Zr-2Fe
	Ti-13Nb-13Zr
	Ti-15Mo-2.8Nb-0.2Si-0.26O
	Ti-16Nb-10Hf
	Ti-15Mo-5Zr-3Al
	Ti–Ni

TABLE 4   Chemical Composition of Common Implant Ceramics

Implant ceramics	Composition
Alumina	$-Al_2O_3$ + <0.3 wt% MgO
Zirconia	Yttria stabilized tetragonal zirconia
	ZrO_2 + 2–3 mol% Y_2O_3
Calcium phosphates	$Ca_3(PO_4)_2$–α- or β-tricalcium phosphate
	$Ca_{10}(PO_4)_6(OH)_2$—hydroxyapatite

correlates fairly well with experimental observations and clinical experience, since it is well known that all but the most corrosion-resistant metals will suffer significant and destructive attack upon prolonged implantation. Also, even the most noble of metals and those that are most strongly passivated (i.e., naturally protected by their own oxide layer) will still show some degree of interaction. The important passivating implant alloys and their compositions are presented in Table 3.

There are some ceramics that have a combination of very strong partially ionic, partially covalent bonds that are sufficiently stable to resist breakdown within this environment, such as the pure simple oxide ceramics, and others in which certain of the bonds are readily destroyed in an aqueous medium so that the material essentially dissolves, for example certain calcium phosphates. Typical implant ceramics and their compositions are presented in Table 4.

With these general statements in mind, we have to consider the following questions in relation to the corrosion and degradation of metals and ceramics:

1. Within these groups, how does the susceptibility to corrosion and degradation vary; by what precise mechanisms do the interfacial reactions take place; and how

is material selection (and treatment) governed by this knowledge?

2. Are there variables within this biological environment other than those described above that can influence these processes?

3. What are the consequences of such corrosion and degradation phenomena?

We review each of these questions in this chapter. It is particularly important to bear in mind some general points as these questions are discussed.

1. Material selection cannot be governed solely by considerations of stability, and mechanical and physical properties especially may be of considerable importance. Since corrosion is a surface phenomenon, however, it may be possible to optimize corrosion resistance by attention to or treatment of the surface rather than by manipulation of the bulk chemistry (Trepanier *et al.*, 1999). This offers the possibility of developing sufficient corrosion resistance in materials of excellent bulk mechanical and physical properties. Thus, noble metals such as gold and platinum are rarely used for structural applications (apart from dental restoration) because of their inferior mechanical properties, even though they have excellent corrosion resistance; instead, base metal alloys with passivated or protected surfaces offer better all-around properties.

2. Medical devices are not necessarily used in mechanically stress-free conditions and indeed the vast majority of those using metals or ceramics are structurally loaded. It is well known that mechanical stress plays a very important role in the corrosion of metals (Gilbert *et al.*, 1993; Jacobs *et al.*, 1998) and the degradation process in ceramics (Piconi and Maccauro, 1999), both potentiating existing effects and initiating others. This has to be taken into consideration.

3. We cannot expect the biological environment to be constant. Within the overall characteristics described earlier, there are variations (with time, location, activity, health status, etc.) in, for example, oxygen levels, availability of free radicals, and cellular activity, all of which may cause variations in the corrosive nature of the environment (Fonseca and Barbosa, 2001; Tengvall *et al.*, 1989). Most important, corrosion is not necessarily a progressive homogeneous reaction with zero-order kinetics. Corrosion processes can be quiescent but then become activated, or they can be active but then become passivated and localized, with transient fluctuations in conditions playing a part in these variations.

4. The effects of corrosion or degradation may be twofold. First, and in the conventional metallurgical sense, the most obvious, the problem can lead to loss of structural integrity of the material and function. This may be undesirable, as in the case of many long-term prostheses, or desirable, as in devices intended for short-term function (e.g., ceramics for drug delivery systems) or where the material is replaced by tissue during the degradation process, as with ceramic bone substitution. In addition to this, however, and usually of much greater

significance with biomaterials, when released into the tissue, the corrosion or degradation products can have a significant and controlling effect on that tissue (Jacobs *et al.*, 1998; Bravo *et al.*, 1990). Indeed, it is likely that the corrosion process is the most important mediator of the tissue response to metallic materials. It is therefore important that we know both the nature of the reaction products and their rate of generation. In this respect it is important to recognize that a very small release of certain metallic ions that cause adverse biological reactions may be more significant than a larger amount of a less stimulating by product of corrosion or degradation.

METALLIC CORROSION

Basic Principles

The most pertinent form of corrosion related to metallic biomaterials is aqueous corrosion. This occurs when electrochemical reactions take place on a metallic surface in an aqueous electrolyte. There are always two reactions that occur: the anodic reaction, which yields metallic ions, for example, involving the oxidation of the metal to its salt:

$$M \rightarrow M^{(n+)} + n(\text{electrons}) \tag{1}$$

and the cathodic reaction, in which the electrons so generated are consumed. The precise cathodic reaction will depend on the nature of the electrolyte, but two of the most important in aqueous environments are the reduction of hydrogen:

$$2H^+ + 2e^- \rightarrow H_2 \tag{2}$$

and the reduction of dissolved oxygen:

$$O_2 + 4H^+ + 4e^- \rightarrow 2H_2O \tag{3}$$

in acidic solutions or:

$$O_2 + 2H_2O + 4e^- \rightarrow 4OH^- \tag{4}$$

in neutral or basic solutions.

In all corrosion processes, the rate of the anodic or oxidation reaction must equal the rate of the cathodic or reduction reaction. This is a basic principle of electrochemically based metallic corrosion. It also explains how variations in the local environment can affect the overall rate of corrosion by influencing either the anodic or cathodic reactions. The whole corrosion process can be arrested by preventing either of these reactions.

From a thermodynamic point of view, first consider the anodic dissolution of a pure metal isolated in a solution of its salt. The metal consists of positive ions closely surrounded by free electrons. When the metal is placed in a solution, there will be a net dissolution of metal ions since the Gibbs free energy (ΔG) for the dissolution reaction is less than for the reverse reaction. This leaves the metal with a net negative charge, thus making it harder for the positive ions to leave the surface and increasing the ΔG for the dissolution reaction. There will come a point when the ΔG for the dissolution reaction will equal the ΔG for the reverse reaction. At this point, a dynamic equilibrium is reached and a potential difference will be set up across

TABLE 5   Electrochemical Series

Metal	Potential (V)
Gold	1.43
Platinum	1.20
Mercury	0.80
Silver	0.79
Copper	0.34
Hydrogen	0
Lead	−0.13
Tin	−0.14
Molybdenum	−0.20
Nickel	−0.25
Cobalt	−0.28
Cadmium	−0.40
Iron	−0.44
Chromium	−0.73
Zinc	−0.76
Aluminum	−1.33
Titanium	−1.63
Magnesium	−2.03
Sodium	−2.71
Lithium	−3.05

the charged double layer surrounding the metal. The potential difference will be characteristic of the metal and can be measured against a standard reference electrode. When this is done against a standard hydrogen electrode in a 1 N solution of its salt at 25°C, it is defined as the standard electrode potential for that metal (Table 5). The position of a metal in the electrochemical series primarily indicates the order with which metals displace each other from compounds, but it also gives a general guide to reactivity in aqueous solutions. Those at the top are the noble, relatively unreactive metals, whereas those at the bottom are the more reactive. This is the first guide to corrosion resistance, but, as we shall see, there are major difficulties related to the use and interpretation of reactions from this simple analysis.

Now consider a system in which the metal is in an aqueous solution that does not contain its ions. In this situation, the electrode potential at equilibrium (i.e., when the rate of the anodic reaction equals the rate of the cathodic reaction) will be shifted from the standard electrode potential and can be defined by the Nernst equation:

$$E = E_0 + (RT/nF \ln(a_{\text{anod}}/a_{\text{cath}})$$

where E_0 is the standard electrode potential, RT/F is a constant, n is the number of electrons transferred, and a is the activity of the anodic and cathodic reactants. At low concentrations, the activity can be approximated to the concentration. In this situation, there is a net dissolution of the metal and a current will flow. At equilibrium, the rate of the metal dissolution is equal to the rate of the cathodic reaction, and the rate of the reaction is directly proportional to the current density by Faraday's law; therefore:

$$i_{\text{anodic}} = i_{\text{cathodic}} = i_{\text{corrosion}}$$

and the Nernst equation can be rewritten:

$$E - E_0 = \pm\beta \ln(i_{corr}/i_0)$$

where β is a constant and i_0 is the exchange current density, which is defined as the anodic (or cathodic) current density at the standard electrode potential. Current density is the current, measured in amperes, normalized to the surface area of the metal.

These conditions represent convenient models for the basic mechanisms of corrosion, but they are hardly realistic. Indeed, in this situation of a homogeneous pure metal existing within an unchanging environment, an equilibrium is reached in which no further net movement of ions takes place. In other words, the corrosion process takes place only transiently, but is effectively stopped once this equilibrium is reached.

In reality, we usually have neither entirely homogeneous surfaces or solutions, nor complete isolation of the metal from other parts of the environment, and this equilibrium is easily upset. If the conditions are such that the equilibrium is displaced, the metal is said to be polarized. There are several ways in which this can happen. Two main factors control the behavior of metals in this respect and determine the extent of corrosion in practice. The first concerns the driving force for continued corrosion (i.e., the reasons why the equilibrium is upset and the nature of the polarization), and the second concerns the ability of the metal to respond to this driving force.

It is self-evident that if either the accumulating positive metal ions in the surrounding media or the accumulating electrons in the metal are removed, the net balance between the dissolution and the replacement of the ions will be disturbed. This will occur in the biological environment surrounding implanted alloys due to the interaction of the proteins with the metal ions. Metal ions can form complexes with proteins (Steinemann, 1996; Jacobs et al., 1998; Büdinger and Hertl, 2000) and these complexes can be transported away from the immediate vicinity. This removes the metal ions from the charged double layer at the interface allowing further release of metal ions to reestablish the equilibrium. Similarly, relative movement between the implant and the tissue, for example, at a bearing surface or on a cyclically loaded implant, will cause mixing at the interface and will modify the composition of the electrolyte and may modify the surface of the alloy (Khan et al., 1999b). The equilibrium is established precisely because of the imbalance of charge, so that if the charge balance is disturbed, further corrosion will occur to attempt to reestablish the balance. The result will be continued dissolution as the system attempts to achieve this equilibrium, in other words, sustained corrosion. An environment that allows the removal of electrons in contact with the metal or stirring of the electrolyte will achieve this.

The process of galvanic corrosion may be used to demonstrate this effect. Consider a single homogeneous pure metal, A, existing within an electrolyte (Fig. 1). The metal will develop its own potential, V_A, with respect to the electrolyte. If a different metal electrode, B, is placed into the same electrolyte, but without contacting A, it will develop its own potential V_B. If V_A is not equal to V_B, there will be a difference in the numbers of excess free electrons in each. This is of no consequence if A and B are isolated from each other, but should they be placed

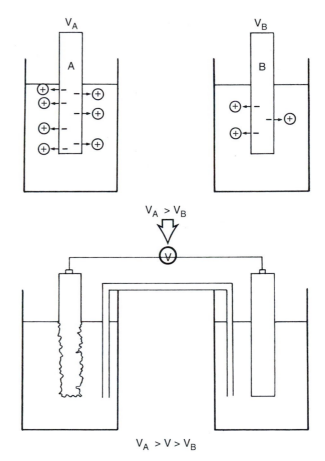

FIG. 1.   When electrical contact is made between electrodes A and B, electrode B acts as an electron sink, thus upsetting the equilibrium and causing continued dissolution of A.

in electrical contact, electrons will flow from that metal with the greater potential in an attempt to make the two electrodes equipotential. This upsets the equilibrium and causes continued and accelerated corrosion of the more active metal (anodic dissolution) and protects the less active (cathodic protection).

Galvanic corrosion may be seen whenever two different metals are placed in contact in an electrolyte. It has been frequently observed with complex, multicomponent surgical implants such as modular total joint designs consisting of titanium alloy femoral stems and cobalt alloy femoral heads (Jacobs et al., 1998). It is not necessary for the components to be macroscopic, monolithic electrodes for this to happen, and the same effect can be seen when there are different microstructural features within one alloy, such as the multiphase microstructure evident in implants of sensitized stainless steel where the grain boundaries become depleted in chromium and corrode preferentially to the remaining surface (Disegi and Eschbach, 2000). In practice, it is the regional variations in electrode potential over an alloy surface that are responsible for much of the generalized surface corrosion that takes place in metallic components.

Many of the commonly used surgical alloys contain highly reactive metals (i.e., with high negative electrode potentials), such as titanium, aluminum, and chromium. Because of this high reactivity, they will react with oxygen upon initial exposure to the atmosphere. This initial oxidation leaves an impervious oxide layer firmly adherent to the metal surface; thus all other forms of corrosion may be significantly reduced because the oxide layer acts as a protective barrier, passivating the metal. The manufacturing process for implant alloys may include a passivating step to enhance the oxide layer prior to implantation, for example nitric acid treatment of 316L stainless steel (Fraker and Griffith, 1985).

In summary, the basic principles of corrosion determine that:

1. In theory, corrosion resistance can be predicted from standard electrode potentials. This explains the nobility of some metals and the considerable reactivity of others, but is not useful for predicting the occurrence of corrosion of most alloy systems in practice.

2. Irrespective of standard electrode potentials, the corrosion resistance of many materials is determined by their ability to become passivated by an oxide layer that protects the underlying metal.

3. Corrosion processes in practice are influenced by variations in surface microstructural features and in the environment that disrupt the charge transfer equilibrium.

INFLUENCE OF THE BIOLOGICAL ENVIRONMENT

It is reasonable to assume that the presence of biological macromolecules will not cause a completely new corrosion mechanism. However, they can influence the rate of corrosion by interfering in some way with the anodic or cathodic reactions discussed earlier. Four ways in which this could occur are discussed next:

1. The biological molecules could upset the equilibrium of the corrosion reactions by consuming one or other of the products of the anodic or cathodic reaction. For example, proteins can bind to metal ions and transport them away from the implant surface. This will upset the equilibrium across the charged double layer and allow further dissolution of the metal; in other words, it will decrease ΔG for the dissolution reaction.

2. The stability of the oxide layer depends on the electrode potential and the pH of the solution. Proteins and cells can be electrically active and interact with the charges formed at the interface and thus affect the electrode potential (Bundy, 1994). Bacteria (Laurent et al., 2001) and inflammatory cells (Hanawa, 2002; Fonseca and Barbosa, 2001) can alter the pH of the local environment through the generation of acidic metabolic products that can shift the equilibrium.

3. The stability of the oxide layer is also dependent on the availability of oxygen. The adsorption of proteins and cells onto the surface of materials could limit the diffusion of oxygen to certain regions of the surface. This could cause preferential corrosion of the

oxygen-deficient regions and lead to the breakdown of the passive layer. Alternatively the biomolecule adsorption layer could act as a capacitor preventing the diffusion of molecules from the surface (Hiromoto et al., 2002).

4. The cathodic reaction often results in the formation of hydrogen, as shown earlier. In a confined locality, the buildup of hydrogen tends to inhibit the cathodic reaction and thus restricts the corrosion process. If the hydrogen can be eliminated, then the active corrosion can proceed. It is possible that bacteria in the vicinity of an implant could utilize the hydrogen and thus play a crucial role in the corrosion process.

There is sufficient evidence to support the premise that the presence of proteins and cells can influence the rate of corrosion of some metals (Williams, 1985; Khan et al., 1999a, b; Hanawa, 1999, 2002). Studies have examined these interactions electrochemically and have found very few differences in many of the parameters measured (e.g., electrode potential, polarization behavior, and current density at a fixed potential). However, analysis of the amount of corrosion through weight loss or chemical analysis of the electrolyte has shown significant effects from the presence of relatively low concentrations of proteins. These effects have varied from severalfold increases for some metals under certain conditions, to slight decreases under other conditions.

It has been shown that proteins adsorb onto metal surfaces and that the amount adsorbed appears to be different on a range of metals (Williams and Williams, 1988; Wälivaara et al., 1992). Similarly, proteins have been shown to bind to metal ions and it is suggested that they are transported away from the local site as a protein–metal complex and distributed systemically in the body (Jacobs et al., 1998). It is therefore likely that proteins will influence the corrosion reactions that occur when a metal is implanted, although there is no direct evidence to explain the mechanism of the interaction at this time.

CORROSION AND CORROSION CONTROL IN THE BIOLOGICAL ENVIRONMENT

The need to ensure minimal corrosion has been the major determining factor in the selection of metals and alloys for use in the body. Two broad approaches have been adopted. The first has involved the use of noble metals, that is, those metals and their alloys for which the electrochemical series indicates excellent corrosion resistance. Examples are gold, silver, and the platinum group of metals. Because of cost and relatively poor mechanical properties, these are not used for major structural applications, although it should be noted that gold and its alloys are extensively used in dentistry; silver is sometimes used for its antibacterial activity; and platinum-group metals (Pt, Pd, Ir, Rh) are used in electrodes.

The second approach involves the use of the passivated metals. Of the three elements that are strongly passivated (i.e., aluminum, chromium, and titanium), aluminum cannot be used on its own for biomedical purposes because of toxicity problems; however, it has an important role in several Ti alloys.

Chromium is very effectively protected but cannot be used in bulk. It is, however, widely used in alloys, especially in stainless steels and in the cobalt–chromium-based alloys, where it is normally considered that a level of above 12% gives good corrosion resistance and about 18% provides excellent resistance. Titanium is the best in this respect and is used as a pure metal or as the major constituent of alloys (Long and Rack, 1998). In alloys the passivating layer promoting the corrosion resistance is predominantly composed of one of these metal oxides. For example, chromium oxide passivates 316L stainless steel and Co–Cr-based alloys and Ti oxide in Ti alloys. The other alloying elements may be present in the surface oxide and this can influence the passivity of the layer (Sittig *et al.*, 1999). Careful pretreatment of the alloys can be used to control the passivity of these alloys (Shih *et al.*, 2000; Trepanier *et al.*, 1998). In particular, production procedures need to be controlled because of their influence on the surface oxides, for example, the cleaning (Aronsson *et al.*, 1997) and sterilization (Thierry *et al.*, 2000) procedures.

Although these metals and alloys have been selected for their corrosion resistance, corrosion will still take place when they are implanted in the body. Two important points have to be remembered. First, whether noble or passivated, all metals will suffer a slow removal of ions from the surface, largely because of local and temporal variations in microstructure and environment. This need not necessarily be continuous and the rate may either increase or decrease with time, but metal ions will be released into that environment. This is particularly important with biomaterials, since it is the effect of these potentially toxic or irritant ions that is the most important consequence of their use. Even with a strongly passivated metal, there will be a finite rate of diffusion of ions through the oxide layer, and possibly a dissolution of the layer itself. It is well known that titanium is steadily released into the tissue from titanium implants (Jacobs *et al.*, 1998; Hanawa *et al.*, 1998). Second,

some specific mechanisms of corrosion may be superimposed on this general behavior; some examples are given in the next section.

Pitting Corrosion

The stainless steels used in implantable devices are passivated by the chromium oxide that forms on the surface. It has been shown, however, that in a physiological saline environment, the driving force for repassivation of the surface is not high (Seah *et al.*, 1998). Thus, if the passive layer is broken down, it may not repassivate and active corrosion can occur.

Localized corrosion can occur as a result of imperfections in the oxide layer, producing small areas in which the protective surface is removed (Rondelli and Vicentini, 1999). These localized spots will actively corrode and pits will form in the surface of the material. This can result in a large degree of localized damage because the small areas of active corrosion become the anode and the entire remaining surface becomes the cathode. Since the rate of the anodic and cathodic reactions must be equal, it follows that a relatively large amount of metal dissolution will be initiated by a small area of the surface, and large pits may form (Fig. 2).

Fretting Corrosion

The passive layer may be removed by a mechanical process (Khan *et al.*, 1999b; Okazaki, 2002). This can be a scratch that does not repassivate, resulting in the formation of a pit, or a continuous cyclic process in which any reformed passive layer is removed. This is known as fretting corrosion, and it is suggested that this can contribute to the corrosion observed between a fracture fixation plate and the bone screws attaching the plate to the bone. There are three reasons why fretting can

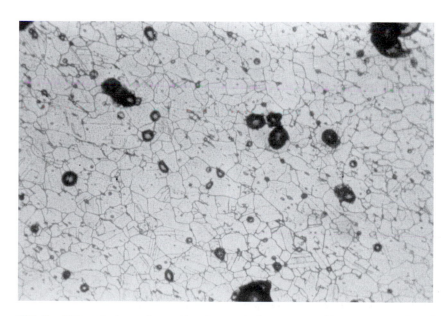

FIG. 2. This etched metallographic micrograph demonstrates the pitting corrosion of stainless steel.

affect the corrosion rate. The first is due to the removal of the oxide film as just discussed. The second is due to plastic deformation of the contact area; this can subject the area to high strain fatigue and may cause fatigue corrosion. The third is due to stirring of the electrolyte, which can increase the limited current density of the cathodic reaction.

Crevice Corrosion

The area between the head of the bone screw and countersink on the fracture fixation plate can also be influenced by the crevice conditions that the geometry creates (Fig. 3) (Cook *et al.,* 1987). Porous coated implants may also demonstrate crevice corrosion (Seah *et al.,* 1998). Accelerated corrosion can be initiated in a crevice by restricted diffusion of oxygen into the crevice. Initially, the anodic and cathodic reactions occur uniformly over the surface, including within the crevice. As the crevice becomes depleted of oxygen, the reaction is limited to metal oxidation balanced by the cathodic reaction on the remainder of the surface. In an aqueous sodium chloride solution, the buildup of metal ions within the crevice causes the influx of chloride ions to balance the charge by forming the metal chloride. In the presence of water, the chloride will dissociate to its insoluble hydroxide and acid. This is a rapidly accelerating process since the decrease in pH causes further metal oxidation.

Intergranular Corrosion

As mentioned earlier, stainless steels rely on the formation of chromium oxides to passivate the surface. If some areas of the alloy become depleted in chromium, as can happen if carbides are formed at the grain boundaries, the regions adjacent to the grain boundaries become depleted in chromium. The passivity of the surface in these regions is therefore affected and preferential corrosion can occur (Fig. 4). Although this problem can easily be overcome by heat treating the alloys (Disegi and Eschbach, 2000), it has been observed on retrieved implants (Walczak *et al.,* 1998) and can cause severe problems since once initiated it will proceed rapidly and may well cause fracture of the implant and the release of large quantities of corrosion products into the tissue.

Stress Corrosion Cracking

Stress corrosion cracking is an insidious form of corrosion since an applied stress and a corrosive environment can work together and cause complete failure of a component, when neither the stress nor the environment would be a problem on their own. The stress level may be very low, possibly only residual, and the corrosion may be initiated at a microscopic crack tip that does not repassivate rapidly. Incremental crack growth may then occur, resulting in fracture of the implant. Industrial uses of stainless steels in saline environments have shown susceptibility to stress corrosion cracking and therefore it is a potential source of failure for implanted devices.

Galvanic Corrosion

If two metals are independently placed within the same solution, each will establish its own electrode potential with respect to the solution. If these two metals are placed in electrical contact, then a potential difference will be established between them, electrons passing from the more anodic to the more cathodic metal. Thus equilibrium is upset and a continuous process of dissolution from the more anodic metal will take place. This accelerated corrosion process is galvanic corrosion. It is important if two different alloys are used in an implantable device when the more reactive may corrode freely.

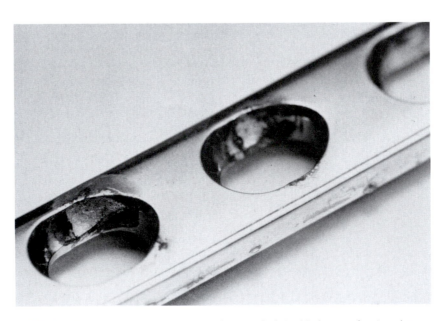

FIG. 3.   Crevice corrosion is evident in the screw hole in this fracture fixation plate.

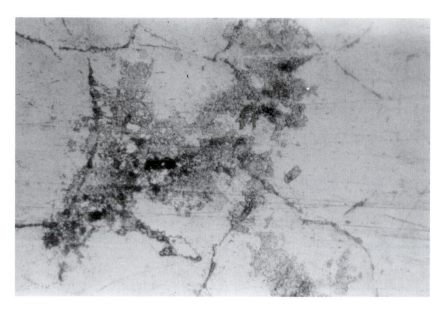

FIG. 4. Intergranular corrosion is demonstrated on this etched stainless steel specimen.

FIG. 5. Extensive corrosion on the titanium stem of a modular hip prosthesis.

Whenever stainless steel is joined to another implant alloy, it will suffer from galvanic corrosion. If both alloys remain within their passive region when coupled in this way, the additional corrosion may be minimal. Some modular orthopedic systems are made of titanium alloys and cobalt-based alloys on the basis that both should remain passive, but evidence of corrosion has been reported (Gilbert *et al.*, 1993). Certainly, as shown in Fig. 5, titanium stems of modular prostheses can exhibit extensive corrosion. Galvanic corrosion may also take place on a microscopic scale in multiphase alloys where phases are of considerably different electronegativity. In dentistry,

some amalgams may show extensive corrosion because of this mechanism.

CERAMIC DEGRADATION

The rate of degradation of ceramics within the body can vary considerably from that of metals in that they can be either highly corrosion resistant or highly soluble. As a general rule, we should expect to see a very significant resistance to degradation with ceramics and glasses. Since the corrosion process in metals is one of a conversion of a metal to ceramic structure (i.e., metal to a metal oxide, hydroxide, chloride, etc.) we must intuitively conclude that the ceramic structure represents a lower energy state in which there would be less driving force for further structural degradation. The interatomic bonds in a ceramic, being largely ionic but partly covalent, are strong directional bonds and large amounts of energy are required for their disruption. As extraction metallurgists know, it takes a great deal of energy to extract aluminum metal from the ore aluminum oxide, but as we have seen, the reverse process takes place readily by surface oxidation. Thus, we should expect ceramics such as Al_2O_3, ZrO_2, TiO_2, SiO_2, and TiN to be stable under normal conditions (Dalgleish and Rawlings, 1981). This is what is observed in clinical practice. There is limited evidence to show that some of these ceramics (e.g., polycrystalline Al_2O_3 and ZrO_2) do show "aging" phenomena (Marti, 2000; Piconi and Maccauro, 1999), with reductions in some mechanical properties, but the significance of this is unclear.

Alternatively, there will be many ceramic structures that, although stable in the air, will dissolve in aqueous environments. Consideration of the classic fully ionic ceramic structure NaCl and its dissolution in water demonstrates this point. It is possible, therefore, on the basis of the chemical structure, to identify ceramics that will dissolve or degrade in the body, and the opportunity exists for the production of structural materials with controlled degradation.

Since any material that degrades in the body will release its constituents into the tissue, it is necessary to select anions and cations that are readily and harmlessly incorporated into metabolic processes and utilized or eliminated. For this reason, it is compounds of sodium, and especially calcium, including calcium phosphates and calcium carbonates, that are primarily used.

The degradation of such compounds will depend on chemical composition and microstructure (Bohner, 2000). For example, tricalcium phosphate $[Ca_3(PO_4)_2]$ is degraded fairly rapidly while calcium hydroxyapatite $[Ca_{10}(PO_4)_6(OH)_2]$ is relatively stable. Within this general behavior, however, porosity will influence the rates so that a fully dense material will degrade slowly, while a microporous material will be susceptible to more rapid degradation.

In general, dissolution rates of these ceramics *in vivo* can be predicted from behavior in simple aqueous solution. However, there will be some differences in detail within the body, especially with variations in degradation rate seen with different implantation sites. It is possible that cellular activity, either by phagocytosis or the release of free radicals, could be responsible for such variations.

In between the extremes of stability and intentional degradability lie a small group of materials in which there may be limited activity. This is particularly seen with a number of glasses and glass ceramics, based on Ca, Si, Na, P, and O, in which there is selective dissolution on the surface involving the release of Ca and P, but in which the reaction then ceases because of the stable SiO_2-rich layer that remains on the surface. This is of considerable interest because of the ability of such surfaces to bond to bone, and this subject is dealt with elsewhere in this book.

On the basis of this behavior, bioceramics are normally classified under three headings:

Inert, or "nearly inert" ceramics
Resorbable ceramics
Ceramics of controlled surface reactivity

This area is discussed in detail in other chapters within this book.

SUMMARY

This chapter has attempted to demonstrate that metals are inherently susceptible to corrosion and that the greatest care is needed in using them within the human body. In general, ceramics have much less tendency to degrade, but care still has to be taken over aging phenomena. The human body is very aggressive toward all of these materials.

Questions

1. (a) What are the three most common implant alloys used in structural applications? For each one, state which element in the alloy is chosen to enhance corrosion resistance, and how do they do it.
    (b) Describe the mechanisms of intergranular corrosion and fretting corrosion.
    (c) If you had an orthodontic appliance where Ni–Ti wire was placed in the groove of a stainless steel bracket, what corrosion problems might you encounter?
2. Consider a situation in which a 316L stainless steel fracture fixation plate has been used to treat a tibial fracture. Discuss the possible mechanisms of corrosion of the device with reference to the alloy composition, the geometry of the device, and the mechanical environment. Include a discussion on the fate of any corrosion products and their possible effect on the patient.
3. Discuss the potential disadvantages of using two different alloys for the components of modular orthopedic prostheses.

Bibliography

Aronsson, B.-O., Lausmaa, J., and Kasemo, B. (1997). Glow discharge plasma treatment for surface cleaning and modification of metallic biomaterials. *J. Biomed. Mater. Res.* **35:** 49–73.

Bohner, M. (2000). Calcium orthophosphates in medicine: from ceramics to calcium phosphate cements. *Injury* **31(S-4):** 37–47.

Bravo, I., Carvalho, G., Barbosa, M., and de Sousa, M. (1990). Differential effects of eight metal ions on lymphocyte differentiation antigens in vitro. *J. Biomed. Mater. Res.* **24:** 1059–1068.

Büdinger, L., and Hertl, M. (2000). Immunological mechanisms in hypersensitivity reactions to metal ions: An overview. *Allergy* **55:** 108–115.

Bundy K. J. (1994). Corrosion and other electrochemical aspects of biomaterials. *Crit. Rev. Biomed. Eng.* **22(3/4):** 139–251.

Cook, S. D., Tomas, K. A., Harding, A. F., Collins, C. L., Haddad, R. J., Milicic, M., and Fischer, W. L. (1987). The in vivo performance of 250 internal fixation devices; a follow up study. *Biomaterials* **8**: 177–184.

Dalgleish, B. J., and Rawlings, R. D. (1981). A comparison of the mechanical behaviour of aluminas in air and simulated body environments. *J. Biomed. Mater. Res.* **15**: 527–542.

Disegi, J. A., and Eschbach, L. (2000). Stainless steel in bone surgery. *Injury* **31**(suppl 4): 2–6.

Fonseca, C., and Barbosa, M. A. (2001). Corrosion behaviour of titanium in biofluids containing H_2O_2 studied by electrochemical impedance spectroscopy. *Corr. Sci.* **43**: 547–559.

Fraker, A. C., and Griffith, C. D., eds. (1985). *Corrosion and Degradation of Implant Materials*. ASTM S.T.P. No. 859, American Society for Testing and Materials, Philadelphia.

Gilbert, J. L., Buckley, C. A., and Jacobs, J. J. (1993). in vivo corrosion of modular hip prosthesis components in mixed and similar metal combinations. The effect of crevice, stress, motion and alloy coupling. *J. Biomed. Mater. Res.* **27**: 1533–1544.

Hanawa, T. (1999). in vivo metallic biomaterials and surface modification. *Mater. Sci. Eng.* **A267**: 260–266.

Hanawa, T. (2002). Evaluation techniques of metallic biomaterials in vitro. *Sci. Technol. Adv. Mater.* **3**: 289–295.

Hanawa, T., Asami, K., and Asaoka, K. (1998). Repassvation of titanium and surface film regeneration in simulated bioliquid. *J. Biomed. Mater. Res.* **40**: 530–538.

Hiromoto, S., Noda, K., and Hanawa, T. (2002). Development of electrolytic cell with cell-culture for metallic biomaterials. *Corr. Sci.* **44**: 955–965.

Jacobs, J. J., Gilbert, J. L., and Urban, R. M. (1998). Current concepts review corrosion of metal orthopaedic implants. *J. Bone Joint Surg.* **80-A**: 268–282.

Khan, M. A., Williams, R. L., and Williams, D. F. (1999a). The corrosion behaviour of Ti-6Al-4V, Ti-6Al-7Nb and Ti-13Nb-13Zr in protein solutions. *Biomaterials* **20**(7): 631–637.

Khan, M. A., Williams, R. L., and Williams, D. F. (1999b). Conjoint corrosion and wear in titanium alloys. *Biomaterials* **20**(8): 765–772.

Laurent, F., Grosgogeat, B., Reclaru, L., Dalard, F., and Lissac, M. (2001). Comparison of corrosion behaviour in presence of oral bacteria. *Biomaterials* **22**: 2273–2282.

Long, M., and Rack, H. J. (1998). Titanium alloys in total joint replacement—a materials science perspective. *Biomaterials* **19**: 1621–1639.

Marti, A. (2000). Inert bioceramics (Al_2O_3, ZrO_2) for medical applications. *Injury* **31**(**5-4**): 33–36.

Okazaki, Y. (2002). Effect of friction on anodic polarization properties of metallic biomaterials. *Biomaterials* **23**: 2071–2077.

Piconi, C., and Maccauro, G. (1999). Zirconia as a ceramic biomaterial. *Biomaterials* **20**: 1–25.

Rondelli, G., and Vicentini, B. (1999). Localized corrosion behaviour in simulated human body fluids of commercial Ni–Ti orthodontic wires. *Biomaterials* **20**: 785–792.

Seah, K. H. W., Thampuran, R., and Teoh, S. H. (1998). The influence of pore morphology on corrosion. *Corr. Sci.* **40**: 547–556.

Shih, C.-C., Lin, S.-J., Chung, K.-H., Chen, Y.-L., and Su, Y.-Y. (2000). Increased corrosion resistance of stent materials by converting current surface film of polycrystalline oxide into amorphous oxide. *J. Biomed. Mater. Res.* **52**: 323–332.

Sittig, C., Textor, M., Spencer, N. D., Wieland, M., and Vallotton, P.-H. (1999). Surface characterization of implant materials c.p.Ti, Ti-6Al-7Nb and Ti-6Al-4V with different pretreatments. *J. Mater. Sci.: Mater. Med.* **10**: 35–46.

Steinemann, S. G. (1996). Metal implants and surface reactions. *Injury* **27**(**S-3**): 16–22.

Tengvall, P., Lundström, I., Sjökvist, L. Elwing, H., and Bjursten, L. (1989). Titanium–hydrogen peroxide interactions: model studies of the influence of the inflammatory response on titanium implants. *Biomaterials* **10**: 166–175.

Thierry, B., Tabrizian, M., Savadogo, O., and Yahia, L'H. (2000). Effects of sterilization processes on NiTi alloy: Surface characterization. *J. Biomed. Mater. Res.* **49**: 88–98.

Trepanier, C., Tabrizian, M., Yahia, L'H., Bilodeau, L., and Piron, D. (1998). Effect of modification of oxide layer on NiTi stent corrosion resistance. *J. Biomed. Mater. Res.* **43**: 433–440.

Trepanier, C., Leung, T. K., Tabrizian, M., Yahia, L. H., Bienvenu, J.-G., Tanguay, J.-F., Piron, D. L., and Bildeau, L. (1999). Preliminary investigation of the effects of surface treatments on biological response to shape memory NiTi stents. *J. Biomed. Mater. Res. (Appl. Biomater.)* **48**: 165–171.

Walczak, J., Shahgaldi, F., and Heatley, F. (1998). in vivo corrosion of 316L stainless-steel hip implants: morphology and elemental compositions of corrosion products. *Biomaterials* **19**: 229–237.

Wälivaara, B., Askendal, A., Elwing, H., Lundström, I., and Tengvall, P. (1992). Antisera binding onto metals immersed in human plasma in vitro. *J. Biomed. Mater. Res.* **26**: 1205–1216.

Williams, D. F. (1985). Physiological and microbiological corrosion. *Crit. Rev. Biocompat.* **1**(1): 1–24.

Williams, R. L., and Williams, D. F. (1988). The characteristics of albumin adsorption on metal surfaces. *Biomaterials* **9**(3): 206–212.

6.4 PATHOLOGICAL CALCIFICATION OF BIOMATERIALS

Frederick J. Schoen and Robert J. Levy

Biomaterials and prosthetic devices, particularly those used in the circulatory system, but also at other sites, may be affected by the formation of nodular deposits of calcium phosphate or other calcium-containing compounds, a process known as calcification or mineralization (Table 1). In many cases, this causes device failure. Calcification has been encountered in association with both synthetic and biologically derived biomaterials in various clinical and experimental settings, including bioprosthetic or homograft cardiac valve substitutes and vascular replacements, blood pumps used as cardiac assist devices, breast implants, intrauterine contraceptive devices, urinary prostheses, and soft contact lenses.

Deposition of mineral salts of calcium occurs as a normal process in bones and teeth (physiologic mineralization). Moreover, it is desirable that some implant biomaterials calcify, e.g., osteoinductive materials used for orthopedic or dental applications (Begley *et al.*, 1995). However, nonskeletal tissues and the biomaterials that comprise other medical devices are not intended to calcify (e.g., heart valves, breast implants), since mineral deposits can interfere with their function. Therefore, calcification of these tissues or biomaterials is abnormal or pathologic. The mature mineral phase of biomaterial-related and other forms of pathologic calcifications is a poorly crystalline calcium phosphate known as apatite. It closely resembles calcium hydroxyapatite, the mineral that provides the structural rigidity of bone and has the chemical formula $Ca_{10}(PO_4)_6(HO)_2$. Indeed, we will see later that many features are shared between biomaterials calcification, other pathologic

TABLE 1 Prostheses and Devices Affected by Calcification of Biomaterials

Configuration	Biomaterial	Clinical Consequence
Cardiac valve prostheses	Glutaraldehyde-pretreated porcine aortic valve or bovine pericardium, and allograft aortic/pulmonary valves	Valve obstruction or incompetency
Cardiac ventricular assist bladders	Polyurethane	Dysfunction by stiffening or cracking
Vascular grafts	Dacron grafts and aortic allografts	Graft obstruction or stiffening
Soft contact lens	Hydrogels	Opacification
Intrauterine contraceptive devices	Silicone rubber, polyurethane or copper	Birth control failure by dysfunction or expulsion
Urinary prostheses	Silicone rubber or polyurethane	Incontinence and/or infection

calcifications on the one hand and normal bone mineralization on the other. Pathologic calcification is also common in native arteries and heart valves, where it occurs as an important feature of the serious diseases atherosclerosis and degenerative aortic stenosis, respectively (Schoen, 1999).

Pathologic calcification is further classified as either dystrophic or metastatic, depending on its setting. Dystrophic calcification is the deposition of calcium salts (usually calcium phosphates) in damaged or diseased tissues or biomaterials in individuals with normal calcium metabolism. In contrast, metastatic calcification is the deposition of calcium salts in previously normal tissues in individuals with deranged mineral metabolism (for example, with elevated blood calcium levels). The conditions favoring dystrophic and metastatic calcification can act synergistically; thus, in the presence of abnormal mineral metabolism, calcification associated with biomaterials or injured tissues is enhanced. Moreover, the ability to form bone is physiologically regulated through adjustment of enhancing and inhibiting substances, many of which circulate in the blood. In young individuals the balance appropriately favors bone formation. However, this same chemical environment favors enhanced calcification of biomaterials in the young.

The cells and extracellular matrix of dead tissues are the principal sites of pathologic calcification. Calcification of an implant biomaterial can occur deep within the tissue (intrinsic calcification) or at the surface, associated with attached cells and proteins (extrinsic calcification). An important instance of extrinsic calcification is that associated with tissue heart valve infection (prosthetic valve endocarditis) (Schoen and Hobson, 1985).

THE SPECTRUM OF PATHOLOGIC BIOMATERIALS AND MEDICAL DEVICE CALCIFICATION

Heart Valves and Vascular Replacements

Calcific degeneration of glutaraldehyde-pretreated porcine bioprosthetic heart valves (Fig. 1) is the most clinically significant dysfunction of a medical device due to biomaterials calcification (Levy et al., 1986a; Schoen and Levy, 1984; Schoen et al., 1988; Schoen and Levy, 1999). The predominant pathologic process is intrinsic calcification of the valve cusps,

largely initiated in the deeply seated cells and the tissue from which the valve was fabricated and often involving collagen. Calcification leads to failure most commonly by causing cuspal tears, less frequently by cuspal stiffening, and rarely by inducing distant emboli. Overall, more than half of porcine bioprostheses fail within 12–15 years. Calcification is more rapid and aggressive in the young; for example, the rate of failure of bioprostheses is approximately 10% in 10 years in elderly recipients, but is nearly uniform in less than 4 years in most adolescent and preadolescent children (Grunkemeier et al., 1994). Calcification has also complicated the clinical use and experimental investigation of heart valves composed of other tissues (e.g., bovine pericardium) (Schoen et al., 1987; McGonagle-Wolff and Schoen, 1992) and polymers (e.g., polyurethane) (Hilbert et al., 1987; Schoen et al., 1992c).

In some young individuals with congenital cardiac defects or acquired aortic valve disease, human allograft/homograft aortic (or pulmonary) valves surrounded by a sleeve of aorta (or pulmonary artery) are used. Allograft valves are valves that are removed from a person who has died and transplanted to another individual; the tissue is usually cryopreserved but not chemically cross-linked. Allograft vascular segments (without a valve) can be used to replace a large blood vessel. Whether containing an aortic valve or nonvalved, allograft vascular tissue can undergo severe calcification, particularly in the wall; calcification can lead to allograft valve dysfunction or deterioration (Mitchell et al., 1998). Synthetic vascular replacements composed of Dacron or expanded polytetrafluoroethylene (e-PTFE) also calcify in some patients.

Polymeric Bladders in Blood Pumps

Deposition of calcific crystals on flexing bladder surfaces (which are usually composed of polyurethane) occurs in and may limit the functional longevity of blood pumps used as ventricular assist systems or total artificial hearts. Massive deposition of mineral leading to failure has been noted in experimental animals, and a lesser degree of calcification has been encountered following extended human implantation (Fig. 2). Mineral deposits can result in deterioration of pump or valve performance through loss of pliability or the initiation of tears. Blood pump calcification, regardless of the type of polyurethane used, generally predominates along the flexing margins of the diaphragm, emphasizing the

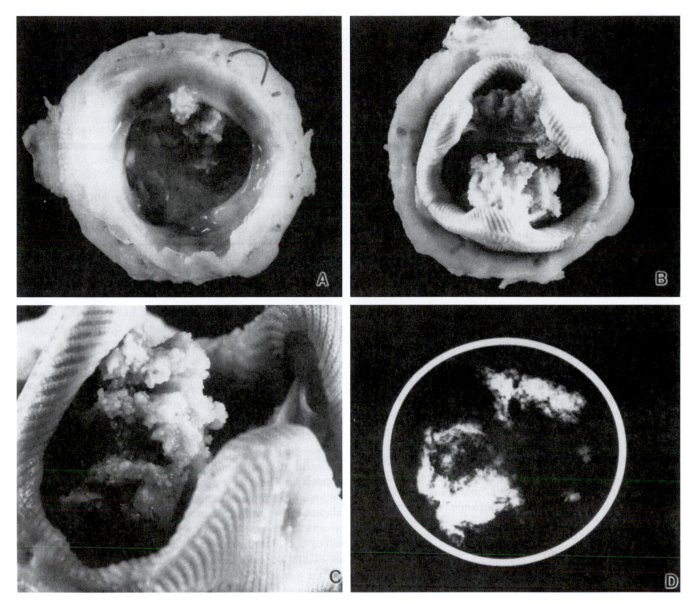

FIG. 1. Calcified clinical porcine bioprosthetic valve removed because of extensive calcification, causing stenosis. (A) Inflow surface of valve. (B) Outflow surface of valve. (C) Closeup of large calcific nodule ulcerated through cuspal surface. (D) Radiograph of valve illustrating radio-opaque, dense calcific deposits.

important potentiating role of mechanical factors in this system (Coleman *et al.*, 1981; Harasaki *et al.*, 1987).

Calcific deposits associated with blood pump components can occur either within the adherent layer of deposited proteins and cells (pseudointima) on the blood-contacting surface (extrinsic mineralization) or below the surface (intrinsic calcification) (Joshi *et al.*, 1996). In some cases, calcific deposits are associated with microscopic surface defects, either originating during bladder fabrication or resulting from cracking during function.

Breast Implants

Calcification of silicone-gel breast implant capsules occurs as discrete calcified plaques at the interface of the inner fibrous capsule with the implant surface. Capsular calcification has also been encountered with breast implants in patients with silicone envelopes filled with saline. Calcification could interfere with effective tumor detection and diagnosis, which could potentially delay treatment, particularly in patients who have breast implants following reconstructive surgery for breast cancer. In a study of breast implants removed predominantly for capsular contraction, 16% overall demonstrated calcific deposits, including 26% of implants inserted for 11–20 years and all those > 23 years (Peters and Smith, 1995).

Capsular mineralization has also been associated with the Dacron patches used on silicone-gel implants in the 1960s and early 1970s to anchor implants to the chest wall in an attempt to prevent implant migration and sagging. Ivalon (polyvinyl

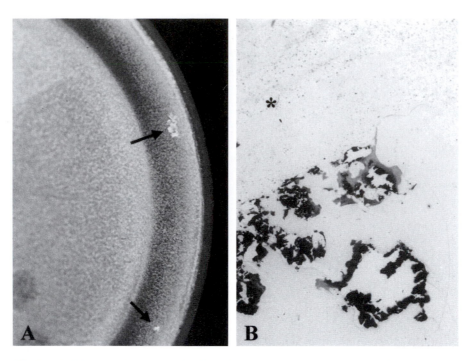

FIG. 2. Calcification of the flexing bladder of a ventricular assist pump removed from a person after 257 days. A) Gross photograph. Calcific masses are noted by arrows. B) Photomicrograph demonstrating calcific nodule (black) at blood-contacting surface of polyurethane membrane (asterisk). B) von Kossa stain (calcium phosphates black), 100X.

alcohol) sponge prostheses, used quite extensively during the 1950s, were also frequently associated with calcification. In Japan, where augmentation mammoplasty was frequently performed using injection of foreign material (liquid paraffin from approximately 1950 until 1964, and primarily liquid-silicone injections thereafter), the incidence of calcification has been much higher. One study showed calcification in 45% of breast augmentations which were done by injection (Koide and Katayama, 1979).

Intrauterine Contraceptive Devices

Intrauterine contraceptive devices (IUDs) are composed of plastic or metal and placed in a woman's uterus chronically to prevent implantation of a fertilized egg. Device dysfunction due to calcific deposits can be manifested as contraceptive failure or device expulsion. For example, accumulation of calcific plaque could prevent the release of the active contraception-preventing agent—either ionic copper from copper-containing IUDs or an active agent from hormone-releasing IUD systems. Studies of explanted IUDs using transmission and electron microscopy coupled with X-ray microprobe analysis have shown that surface calcium deposition is ubiquitous but variable among patients (Khan and Wilkinson, 1985).

Urinary Prostheses

Mineral crusts form on the surfaces of polymeric prostheses to alleviate urinary obstruction or incontinence (Goldfarb *et al.*, 1989). Observed in male and female urethral implants and artificial ureters, this problem can lead to obstruction and

device failure. The mineral crust consists of either hydroxyapatite or struvite, an ammonium- and magnesium-containing phosphate mineral derived from urine. There is some evidence that encrustation may both result from and predispose to bacterial infection.

Soft Contact Lenses

Calcium phosphate deposits can opacify soft contact lenses, typically composed of poly(2-hydroxyethyl methacrylate) (HEMA). Growing progressively larger with time, they are virtually impossible to remove without destroying the lens (Bucher *et al.*, 1995). Calcium from tear fluid is considered to be the source of the deposits found on HEMA contact lenses, and calcification may be potentiated in patients with systemic and ocular conditions associated with elevated tear calcium levels (Klintworth *et al.*, 1977).

ASSESSMENT OF BIOMATERIALS CALCIFICATION

Calcific deposits are investigated using morphologic and chemical techniques (Table 2). Morphologic techniques facilitate detection and characterization of the microscopic and ultrastructural sites and distribution of the calcific deposits and their relationship or tissue or biomaterials structural details. Such analyses yield important qualitative (but not quantitative) information. In contrast, chemical techniques, which require destruction of the tissue specimen, permit both identification and quantitation of bulk elemental composition and determination of crystalline mineral phases. However, such techniques

TABLE 2 Methods for Assessing Calcification

Technique	Sample preparation	Analytical results
Morphologic procedures		
Gross examination	Gross specimen	Overall morphology
Radiographs	Gross specimen	Calcific distribution
Light microscopy—von Kossa or alizarin red	Formalin or glutaraldehyde fixed	Microscopic phosphate or calcium distribution, respectively
Transmission electron microscopy	Glutaraldehyde fixed	Mineral ultrastructure
Scanning electron microscopy with electron microprobe	Glutaraldehyde fixed	Element localization/quantitation
Electron energy loss spectroscopy	Glutaraldehyde fixed or rapidly frozen	Elemental localization/quantitation (highest sensitivity)
Chemical procedures		
Atomic absorption	Ash or acid hydrolyzate	Bulk calcium
Colorimetric phosphate analysis	Ash or acid hydrolyzate	Bulk phosphorus
X-ray diffraction	Powder	Nature of crystal phase
Infrared spectroscopy	Powder	Carbonate mineral phase

generally cannot relate the location of the mineral to the details of the underlying tissue structure. The most comprehensive studies characterize both morphologic and chemical aspects of calcification.

Morphologic Evaluation

Morphologic assessment of calcification is done by means of several readily available and well-established techniques that range from macroscopic (gross) examination and radiographs (X-rays) of explanted prostheses to sophisticated electron energy loss spectroscopy. Each technique has advantages and limitations; several techniques are often used in combination to obtain an understanding of the structure, composition, and mechanism of each type of calcification. Careful visual examination of the specimen, often under a dissecting (low power) microscope, and radiography assess distribution of mineral in explanted bioprosthetic heart valves and ventricular assist systems. Specimen radiography typically involves placing the explanted prosthesis on an X-ray film plate and exposing to an X-ray beam in a special device used for small samples (e.g., we use the Faxitron, Hewlett-Packard, McMinnville, CA, with an energy level of 35 keV for 1 min for valves). Deposits of mineral appear as bright densities that have locally blocked the beam from exposing the film (see Fig. 1D).

Light microscopy of calcified tissues is widely used. Identification of mineral is facilitated through the use of either calcium- or phosphorus-specific stains, such as alizarin red (which stains calcium) or von Kossa (which stains phosphates) (see Fig. 2B and Fig. 3). These histologic stains are readily available, can be easily applied to tissue sections embedded in either paraffin or plastic, and are most useful for confirming and characterizing suspected calcified areas which have been noted by routine hematoxylin and eosin staining techniques. Sectioning of calcified tissue that has been embedded in paraffin often leads to considerable artifact due to fragmentation; embedding of tissue with calcific deposits in glycolmethacrylate polymer yields superior section quality.

Electron microscopic techniques, which involve the bombardment of the specimen with a highly focused electron beam in a vacuum, have much to offer in the determination of early sites of calcific deposits. In transmission electron microscopy (TEM), the beam traverses an ultra-thin section ($0.05\,\mu m$) (Fig. 4); observation of the ultrastructure (submicron tissue features) of calcification by TEM facilitates the understanding of the mechanisms by which calcific crystals form. Scanning electron microscopy (SEM) images the specimen surface, and can be coupled with elemental localization by energy-dispersive X-ray analysis (EDXA), allowing a semiquantitative evaluation of the local progression of calcium and phosphate deposition in a site-specific manner. Electron energy loss spectroscopy (EELS) couples transmission electron microscopy with highly sensitive elemental analyses to provide a most powerful localization of incipient nucleation sites and early mineralization (Webb *et al.*, 1991). In general, the more highly sensitive and sophisticated morphologic techniques require more demanding and expensive preparation of specimens to avoid unwanted artifacts. Forethought about and careful planning of specimen handling optimizes the yield provided by the array of available techniques, and allows multiple approaches to be used on a single specimen.

Chemical Assessment

Quantitation of calcium and phosphorus in biomaterial calcifications permits characterization of the progression of deposition, comparison of severity of deposition among specimens and determination of the effectiveness of preventive measures (Levy *et al.*, 1983a, 1985a; Schoen *et al.*, 1985, 1986, 1987; Schoen and Levy, 1999). However, such techniques destroy the configuration of the specimen during preparation. Calcium has been quantitated by atomic absorption spectroscopy of acid-hydrolyzed or ashed samples. Recently, highly sensitive multielement integrated coupled plasma (ICP) instrumentation has become available. This permits high-resolution quantitation of not only calcium, but other relevant elements,

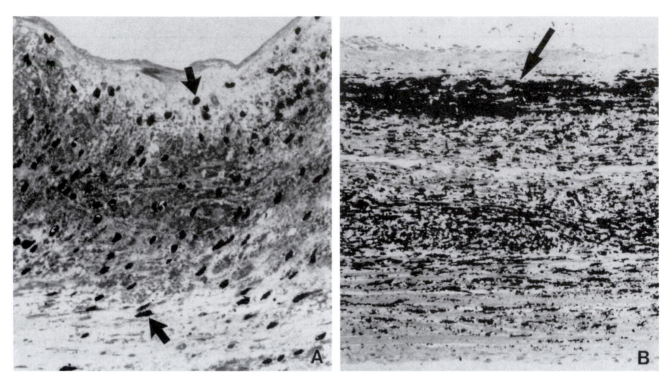

FIG. 3. Light microscopic appearance of progressive calcification of experimental porcine aortic heart valve tissue implanted subcutaneously in 3-week-old rats, demonstrated by specific staining. (A) 72-hr implantation illustrating initial discrete deposits (arrows). (B) 21-day implant demonstrating early nodule formation (arrow). Both stained with von Kossa stain (calcium phosphates black). (A) ×356; (B) ×190. (Reproduced with permission from F. J. Schoen *et al.*, *Lab. Invest.* **52:** 526, 1985.)

such as aluminum and ferric ion in the same sample. Phosphorus is usually quantitated as phosphate, using a molybdate complexation technique with spectrophotometric detection. The crystalline form of calcium phosphate (mineral phase) can be determined by X-ray diffraction. Carbonate-containing mineral phases may also be analyzed by infrared spectroscopy.

PATHOPHYSIOLOGY

General Considerations

The determinants of biomaterial mineralization include factors related to (1) host metabolism, (2) implant structure and chemistry, and (3) mechanical factors (Fig. 5). Natural cofactors and inhibitors may also play a role (see below). The most important host metabolic factor is related to young age, with more rapid calcification taking place in immature patients or experimental animals (Levy *et al.*, 1983a). Although the relationship is well established, the mechanisms accounting for this effect are uncertain. The structural elements of the biomaterial and their modification by processing may be important implant factors for bioprosthetic tissue is the pretreatment with glutaraldehyde, done to preserve the tissue (Golomb *et al.*, 1987; Grabenwoger *et al.*, 1996). It has been hypothesized that the cross-linking agent glutaraldehyde stabilizes and perhaps modifies phosphorus-rich calcifiable structures in the bioprosthetic tissue. These sites seem to be capable of mineralization upon

implantation when exposed to the comparatively high calcium levels of extracellular fluid. Calcification of the two principal types of biomaterials used in bioprostheses—glutaraldehyde-pretreated porcine aortic valves or glutaraldehyde-pretreated bovine pericardium—is similar in extent, morphology, and mechanisms. Furthermore, both intrinsic and extrinsic mineralization of a biomaterial is generally enhanced at the sites of intense mechanical deformations generated by motion, such as the points of flexion in heart valves or circulatory assist devices. In both physiologic and pathologic calcification, nucleation of apatite crystals is more difficult than subsequent growth, which occurs relatively easily since the concentrations of both calcium and phosphorus in blood and extracellular fluid are near saturation.

Regulation of Pathologic Calcification

Calcification has typically been considered a passive, unregulated, and degenerative process. However, the observations of matrix vesicles, hydroxyapatite mineral, and bone-related morphogenetic and noncollagenous proteins in pathological calcifications have suggested that the mechanisms responsible for pathologic calcification may be regulated, similarly to normal mineralization of bone and other hard tissues (Giachelli, 1999; Speer and Giachelli, 2004). In normal blood vessels and valves, inhibitory mechanisms outweigh procalcification inductive mechanisms; in contrast, in bone and pathologic tissues, inductive mechanisms dominate. In the process of

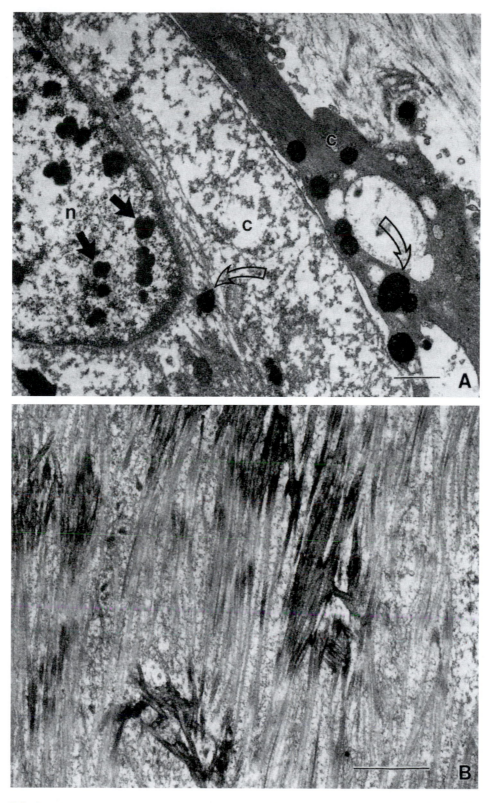

FIG. 4. Transmission of electron microscopy of calcification of experimental porcine aortic heart valve implanted subcutaneously in 3-week-old rats. (A) 48-hr implant demonstrating focal calcific deposits in nucleus of one cell (closed arrows) and cytoplasm of two cells (open arrows), n, nucleus; c, cytoplasm. (B) 21-day implant demonstrating collagen calcification. Bar = 2 μm. Ultrathin sections stained with uranyl acetate and lead citrate. (Figure 4A reproduced with permission from F. J. Schoen *et al.*, *Lab Invest.* **52**: 521, 1985.)

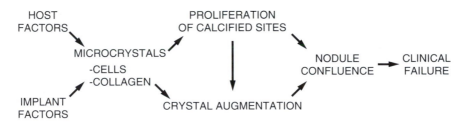

FIG. 5. Hypothesis for calcification of clinical bioprosthetic heart valves emphasizing relationships among host and implant factors, nucleation and growth of calcific nodules, and clinical failure of the device. (Reproduced with permission from F. J. Schoen *et al.*, *Lab. Invest.* **52:** 531, 1985.)

normal bone calcification, the growth of apatite crystals is regulated by several noncollagenous matrix proteins including: (1) osteopontin, an acidic calcium-binding phosphoprotein with high affinity to hydroxyapatite that is abundant in foci of dystrophic calcification; (2) osteonectin, and (3) osteocalcin, and other γ-carboxyglutamic acid (GLA)-containing proteins, such as matrix GLA protein (MGP). Naturally occurring inhibitors to crystal nucleation and growth may also play a role in biomaterial and other cardiovascular calcification (Schinke *et al.*, 1999). Specific inhibitors in this context include osteopontin (Steitz *et al.*, 2002) and high-density liproprotein (HDL, the "good" cholesterol) (Parhami *et al.*, 2002). An active area of research is the role in pathological mineralization of naturally-occurring mineralization cofactors, such as inorganic phosphate (Jono *et al.*, 2000) bone morphogenetic protein (Bostrom *et al.*, 1993) and proinflammatory lipids (Demer, 2002) and other substances (e.g., cytokines) as well as inhibitors. The noncollagenous proteins osteopontin, TGF-beta1, and tenascin-C involved in bone matrix formation and tissue remodeling have been demonstrated in clinical calcified bioprosthetic heart valves, natural valves, and atherosclerosis, suggesting that they play a regulatory role in these forms of pathologic calcification in humans (Srivasta *et al.*, 1997; Bini *et al.*, 1999; Jian *et al.*, 2001; Li QY *et al.*, 2002, Jian *et al.*, 2003).

Evidence for the active regulation of cardiovascular calcification also derives from tissue culture models of vascular cell calcification, which mimic pathologic vascular calcification *in vivo*, and genetic studies in mice. For example, osteopontin inhibits and proinflammatory lipids and cytokines enhance the mineralization of smooth muscle cell cultures (Wada *et al.*, 1999, Parhami *et al.*, 2002). In transgenic mouse models, in which the gene for the matrix GLA protein (MGP) was knocked out (Luo *et al.*, 1997) or the osteopontin gene was inactivated (Speer *et al.*, 2002), severe calcification of blood vessels resulted. Moreover, inhibition of matrix remodeling metalloproteinases inhibits calcification of elastin implanted subcutaneously in rats (Vyavahare *et al.*, 2000).

Experimental Models for Biomaterials Calcification

Animal models have been developed for the investigation of the calcification of bioprosthetic heart valves, aortic homografts, cardiac assist devices, and trileaflet polymeric valves (Table 3). Experimental models used to investigate the pathophysiology of bioprosthetic tissue calcification and as a preclinical screen of new or modified materials and design

TABLE 3 Experimental Models of Calcification

Type	System	Typical Duration
Calcification of bioprosthetic or other tissue heart valve	In-vitro incubation of tissue fragment or flexing valves	Days to weeks
	Rat subdermal implant of tissue fragment	3 weeks
	Calf or sheep orthotopic valve replacement	3–5 months
	Rat or sheep descending aorta	1–5 months
Calcification of polyurethane	Rat subdermal implant of material sample	1–2 months
	Calf or sheep artificial heart implant	5 months
	Trileaflet polymeric valve implant in calf or sheep	5 months
Calcification of hydrogel	Rat subdermal implant of material sample	3 weeks
Calcification of collagen	Rat subdermal implant of material sample	3 weeks
Urinary encrustation	*In vitro* incubation	Hours to days
	In vivo bladder implants (rats and rabbits)	10 weeks

configurations include tricuspid or mitral replacements or conduit-mounted valves in sheep or calves, and isolated tissue (i.e., not in a valve) samples implanted in and around the heart and subcutaneously in mice, rabbits, or rats (Levy *et al.*, 1983a; Schoen *et al.*, 1985, 1986). In both circulatory and noncirculatory models, bioprosthetic tissue calcifies progressively with a morphology similar to that observed in clinical specimens, but with markedly accelerated kinetics. Static *in vitro* models of biomaterials calcification have been investigated but have generally not been useful (Schoen *et al.*, 1992a; Mako and Vesely, 1997). However, several groups have used flexing valve models for bioprosthetic and polymeric valve calcification, in which the morphology of the resulting mineralization seems more representative of pathologic calcification that occurs *in vivo* (Bernacca *et al.*, 1992, 1994, 1997; Deiwick *et al.*, 2001; Pettenazzo *et al.*, 2001).

Compared with the several years normally required for calcification of clinical bioprostheses, valve replacements in sheep or calves calcify extensively in 3 to 6 months (Schoen *et al.*, 1985, 1994). However, expense, technical complexity, and stringent housing and management procedures pose important limitations to all the circulatory models using large animals. In addition, implantation in the heart requires the use of complex procedures such as cardiopulmonary bypass as well as a high level of surgical expertise and postoperative care. These limitations stimulated the development of subdermal (synonym subcutaneous—under the skin) implant models. In subdermal bioprosthetic implants in rats, rabbits, and mice, (1) calcification occurs at a markedly accelerated rate in a morphology comparable to that seen in circulatory explants; (2) the model is economical so that many specimens can be studied with a given set of experimental conditions, thereby allowing quantitative characterization and statistical comparisons; and (3) specimens are rapidly retrieved from the experimental animals, facilitating the careful manipulation and rapid processing required for detailed and high-resolution analyses (Levy *et al.*, 1983a; Schoen *et al.*, 1985, 1986). The subcutaneous model is a technically convenient and economically advantageous vehicle for investigating host and implant determinants and mechanisms of mineralization, as well as for screening potential strategies for its inhibition (anticalcification). Promising approaches may be investigated further in a large-animal valve implant model. Large-animal implants as valve replacements are also used (1) to elucidate further the processes accounting for clinical failures, (2) to evaluate the performance of design and biomaterials modifications in valve development studies, (3) to assess the importance of blood/surface interactions, and (4) to provide data required for approval by regulatory agencies (Schoen, 1992b). Polyurethane calcification has also been studied with subdermal implants in rats (Joshi *et al.*, 1996). Subcutaneous implants may also be used to investigate calcification of biomaterials intended for clinical use in other anatomic sites, for example, polyhydroxyethymethacrylate hydrogels used in soft contact lenses.

Pathophysiology of Bioprosthetic Heart Valve Calcification

Data from valve explants from patients and subdermal and circulatory experiments in animal models using bioprosthetic heart valve tissue have elucidated the pathophysiology of this important clinical problem and enhanced our understanding of pathologic calcification in general (Fig. 6). The similarities of calcification in the different experimental models and clinical bioprostheses suggest a common pathophysiology, independent of implant site. Calcification appears to depend on exposure of a susceptible substrate (often containing phosphorus) to extracellular fluid containing calcium; both mechanical factors and local implant-related or circulating substances may play regulatory roles. However, since the morphology and extent of calcification in subcutaneous implants is analogous to that observed in clinical and experimental circulatory implants, despite the lack of the dynamic mechanical activity that occurs in the circulatory environment, it is clear that dynamic stress promotes but is not prerequisite for calcification of bioprosthetic tissue. Interestingly, in the subcutaneous model, calcification is enhanced in areas of

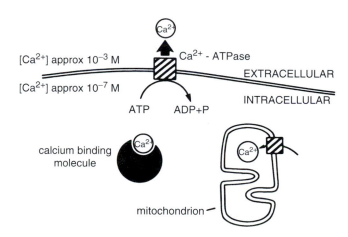

FIG. 6. There is a substantial physiologic (normal) gradient of free calcium across the cell membrane (10^{-3} M outside, 10^{-7} M inside) which is maintained as an energy-dependent process. With cell death or membrane dysfunction, calcium phosphate formation can be initiated at the membranous cellular structures. Reproduced by permission from Schoen, F. J., et al. (1988). Biomaterials-associated calcification: pathology, mechanisms, and strategies for prevention. *J. Applied. Biomater.* 22: 11–36.

tissue folds, bends, and areas of shear, suggesting that static mechanical deformation also potentiates mineralization (Levy *et al.*, 1983a, and unpublished results). Although these data suggest that local tissue disruption mediates the mechanical effect, the precise mechanisms by which mechanical factors influence calcification are uncertain.

Moreover, no definite role has been demonstrated for circulating macromolecules or cells and many lines of evidence suggest that neither nonspecific inflammation nor specific immunologic responses appear to favor bioprosthetic tissue calcification. Nevertheless, a potential role for inflammatory and immune processes has been postulated by some investigators (Love, 1993; Human and Zilla 2001a, b). Proponents of an immunological mechanism for failure cite the evidence that (1) experimental animals can be sensitized to both fresh and cross-linked bioprosthetic valve tissues, (2) antibodies to valve components can be detected in some patients following valve dysfunction, and (3) failed tissue valves often have brisk mononuclear inflammation; no causal immunologic basis has been demonstrated for bioprosthetic valve calcification. Nevertheless, in experiments in which valve cusps were enclosed in filter chambers that prevent host cell contact with tissue but allow free diffusion of extracellular fluid and implantation of valve tissue in congenitally athymic ("nude") mice, who have essentially no T-cell function, calcification morphology and extent are unchanged (Levy *et al.*, 1983b). Clinical and experimental data detecting antibodies to valve tissue after failure probably reflect a secondary response to valve damage rather than a cause of failure. The conditions that must be satisfied to prove a cause immunological mechanism are summarized by Mitchell in Chapter 4.3.

The initial calcification sites in bioprosthetic tissue are predominantly dead cells and cell membrane fragments (Schoen and Levy, 1999) (Fig. 4A). This occurs because the normal handling of calcium ions is disrupted in cells which have been rendered nonviable by glutaraldehyde fixation. Normally, plasma calcium concentration is 1 mg/ml (approximately $10^{-3} M$); since the membranes of healthy cells pump calcium out, the concentration of calcium in the cytoplasm is 1000–10,000 times lower (approximately $10^{-7} M$). Cell membranes and intracellular organelles are high in phosphorus (as phospholipids, especially phosphatidylserine, which can bind calcium); they can serve as nucleators of calcific crystals. Mitochondria are also enriched in calcium. Other initiators under various circumstances include collagen and elastic fibers of the extracellular matrix, denatured proteins, phosphoproteins, fatty acids, blood platelets, and bacteria. We have hypothesized that cells calcify after glutaraldehyde pretreatment because this cross-linking agent stabilizes all the phosphorus stores, but the normal mechanisms for elimination of calcium from the cells are not available in glutaraldehyde-pretreated tissue (Schoen et al., 1986). Once initial calcification deposits form, they can enlarge and coalesce, resulting in grossly mineralized nodules that can cause a prosthesis to malfunction.

In addition to the calcification of valve cusps, calcification of the adjacent aortic wall portion of glutaraldehyde-pretreated porcine aortic valves and valvular allografts and vascular segments is also observed clinically and experimentally. Mineral deposition occurs throughout the vascular cross section but is accentuated in the dense bands at the inner and outer media, and cells and elastin (which itself not generally a prominent site of mineralization in cusps) are the major sites. In nonstented porcine aortic valves that have greater portions of aortic wall exposed to blood than in the currently used stented valves, calcification of the aortic wall is potentially deleterious. It could stiffen the root, altering hemodynamic efficiency, cause nodular calcific obstruction, potentiate wall rupture, or provide a nidus for emboli. Moreover, some anticalcification agents (see later) including 2-amino-oleic acid (AOA) and ethanol prevent experimental cuspal but not aortic wall calcification (Chen et al., 1994b).

Calcification of Collagen and Elastin

Calcification of the extracellular matrix structural proteins collagen and elastin has been observed in clinical and experimental implants of bioprosthetic and homograft valvular and vascular tissue and has been studied using a rat subdermal model. Collagen-containing implants are widely used in various surgical applications, such as tendon prostheses and surgical absorptive sponges, but their usefulness is compromised owing to calcium phosphate deposits and the resultant stiffening. Cross-linking by either glutaraldehyde or formaldehyde promotes the calcification of collagen sponge implants made of purified collagen but the extent of calcification does not correlate with the degree of cross-linking (Levy et al., 1986b). In contrast, the calcification of elastin appears independent of pretreatment (Vyavahare et al., 1999).

PREVENTION OF CALCIFICATION

Three general strategies have been investigated for preventing calcification of biomaterial implants: (1) systemic therapy with anticalcification agents; (2) local therapy with implantable drug delivery devices; and (3) biomaterial modifications, whether by removal of a calcifiable component, addition of an exogenous agent, or chemical alteration.

Investigations of an anticalcification strategy must demonstrate not only the effectiveness of the therapy but also the absence of adverse effects (Schoen et al., 1992b). Adverse effects in this setting could include systemic or local toxicity, tendency toward thrombosis on infection, induction of immunological effects or structural degradation, with either immediate loss of mechanical properties or premature deterioration and failure. Indeed, there are several examples whereby an antimineralization treatment contributed to unacceptable degradation of the tissue (Jones et al., 1989; Gott et al., 1992; Schoen, 1998). The treatment should not impede normal valve performance, such as hemodynamics and durability. As summarized in more detail in Table 4, a rational approach for preventing bioprosthetic calcification must integrate safety and efficacy considerations with the scientific basis for inhibition of calcium phosphate crystal formation. This will of necessity involve the steps summarized in Table 5, before appropriate clinical trials can be done (Schoen et al., 1992b; Vyavahare et al., 1997a).

Experimental studies using bioprosthetic tissue implanted subcutaneously in rats have clearly demonstrated that adequate doses of systemic agents used to treat clinical metabolic bone disease can prevent its calcification (Levy et al., 1987). However, because these agents may interfere with calcium metabolism or growth of calcific deposits, systemic drugs are associated with many side effects, including interruption of physiologic calcification (i.e., bone growth), and animals receiving doses sufficient to prevent bioprosthetic tissue calcification suffer growth retardation. Thus, the principal disadvantage of the systemic use of anticalcification agents for preventing pathologic calcification relates to side effects on bone. This difficulty can be avoided by localized drug release using coimplants of a drug delivery system adjacent to the prosthesis, in which the effective drug concentration is confined to the site at which it is needed (i.e., near the implant) and systemic

TABLE 4   Criteria for Efficacy and Safety of Antimineralization Treatments

Efficacy
Effective and sustained calcification inhibition

Safety
Adequate performance (i.e., unimpaired hemodynamics and durability)
Does not cause adverse blood-surface interactions (e.g., hemolysis, platelet adhesion, coagulation protein activation, complement activation, inflammatory cell activation, binding of vital serum factors)
Does not enhance local or systemic inflammation (e.g., foreign body reaction, immunologic reactivity, hypersensitivity)
Does not cause local or systemic toxicity
Does not potentiate infection

(Modified from Schoen FJ et al. Antimineralization treatments for bioprosthetic heart valves. Assessment of efficacy and safety. J Thorac Cardiovasc Surg 1992; 104:1285–1288.)

TABLE 5 Preclinical Efficacy and Safety Testing of Antimineralization Treatments

Type of study	Information derived
Subcutaneous implantation in rats	Initial efficacy screen Mechanisms Dose-response Toxicity
Biomechanical evaluation	Hemodynamics Accelerated wear
Morphologic studies of unimplanted valves	Structural degradation assessed by light and transmission electron microscopy Scanning electron microscopy
Circulatory implants in large animals	Device configuration, surgical technique, *in vivo* hemodynamics, explant pathology Durability, thrombi, thromboembolism, hemolysis, cardiac and systematic pathology

side effects would be prevented (Levy *et al.*, 1985b). Studies incorporating EHBP (see below) in nondegradable polymers, such as ethylene–vinyl acetate (EVA), polydimethylsiloxane (silicone), and polyurethanes, have shown the effectiveness of this strategy in animal models. This approach, however, has been difficult to implement in a clinically useful manner.

The approach that most likely to yield an improved clinical valve involves modification of the substrate, either by removing or altering a calcifiable component or binding an inhibitor. Forefront strategies should also consider (1) a possible synergism provided by multiple anticalcification agents and approaches used simultaneously; (2) new materials, and (3) the possibility of tissue-engineered heart valve replacements (Rabkin and Schoen, 2002). The agents most widely studied, for efficacy, mechanisms, lack of adverse effects, and potential clinical utility, are summarized hereafter and in Table 6. Combination therapies using multiple agents may potentially provide synergy of beneficial effects to permit simultaneous prevention of calcification in both cusps and aortic wall, particularly beneficial in stentless aortic valves (Levy *et al.*, 2003).

Inhibitors of Hydroxyapatite Formation

Bisphosphonates

Ethane hydroxybisphosphonate (EHBP) has been approved by the FDA for human use to inhibit pathologic calcification and to treat hypercalcemia of malignancy. Compounds of this type probably inhibit calcification by poisoning the growth of calcific crystals. Either cuspal pretreatment or systemic or local therapy of the host with diphosphonate compounds inhibits experimental bioprosthetic valve calcification (Levy *et al.*, 1985b, 1987; Johnston *et al.*, 1993). Controlled clinical trials have orally administered bisphosphonates have demonstrated the ability to stabilize osteoporosis. These agents, such

as Alindronate (Phosphomax, Merck, Inc.), are hypothesized to act by stabilizing bone mineral. However, the effects of such agents on bioprosthetic valve or other pathologic biomaterial calcification are not yet known.

Trivalent Metal Ions

Pretreatment of bioprosthetic tissue with iron and aluminum (e.g., $FeCl_3$ and $AlCl_3$) inhibits calcification of subdermal implants with glutaraldehyde-pretreated porcine cusps or pericardium (Webb *et al.*, 1991). Such compounds are hypothesized to act through complexation of the cation (Fe or Al) with phosphate, thereby preventing calcium phosphate crystal formation and growth. Both ferric ion and the trivalent aluminum ion inhibit alkaline phosphatase, an important enzyme used in bone formation, and this may also be a component of the mechanisms by which they prevent initiation of calcification. Furthermore, recent research from our laboratories has demonstrated that aluminum chloride prevents elastin calcification through a permanent structural alteration of the elastin molecule. Iron and aluminum may also active when released from polymeric controlled-release implants.

Calcium Diffusion Inhibitor

Amino-oleic Acid

2-α-Amino-oleic acid (AOA, Biomedical Design, Inc., Atlanta, GA) bonds covalently to bioprosthetic tissue through an amino linkage to residual aldehyde functions and inhibits calcium flux through bioprosthetic cusps (Chen *et al.*, 1994a, b). AOA is effective in mitigating cusp but not aortic wall calcification in rat subdermal and cardiovascular implants. This compound is used in an FDA-approved porcine aortic valve (Fyfe and Schoen, 1999).

Removal/Modification of Calcifiable Material

Surfactants

Incubation of bioprosthetic tissue with sodium dodecyl sulfate (SDS) and other detergents extracts the majority of acidic phospholipids (Hirsch *et al.*, 1993); this is associated with reduced mineralization, probably resulting from suppression of the initial cell-membrane oriented calcification (Fig. 7). This compound is used in an FDA-approved porcine valve (David *et al.*, 1998).

Ethanol

Ethanol preincubation of glutaraldehyde-cross-linked porcine aortic valve bioprostheses prevents calcification of the valve cusps in both rat subdermal implants and sheep mitral valve replacements (Vyavahare *et al.*, 1997a, 1998). Pretreatment with 80% ethanol (1) extracts almost all phospholipids and cholesterol from glutaraldehyde-cross-linked cusps, (2) causes a permanent alteration in collagen conformation as assessed by attenuated total reflectance–Fourier transform infrared spectroscopy (ATR-FTIR), (3) affects cuspal interactions with water and lipids, and (4) enhances cuspal resistance to collagenase. Ethanol is in clinical use as a porcine valve cuspal

TABLE 6 Prototypical Agents for Mechanism-Based Prevention of Calcification

Mechanisms	Strategy/Agent
Inhibition of hydroxyapatite formation	Ethane hydroxybisphosphonate (EHBP)
Inhibition of calcium uptake	Alpha-amino-oleic acid (AOA)[TM]
Inhibition of Ca-P crystal growth; inhibition of alkaline phosphate; chemical modification of elastin	Ferric/aluminum chloride exposure
Phospholipid extraction	Sodium dodecyl sulfate (SDS)
Phospholipid extraction and collagen conformation modification	Ethanol exposure
Eliminate GA potentiation of calcification	Modification of (alternatives to) glutaraldehyde fixation
• amino acid neutralization of glutaraldehyde residues	
• polyepoxide (polyglycidal ether), acyl azide, carbodiimide, cyanimide and glycerol crosslinking	
• dye-mediated photooxidation	

AOA[TM] (α-amino-oleic acid) is a trademark of Biomedical Designs, Inc., of Atlanta GA.

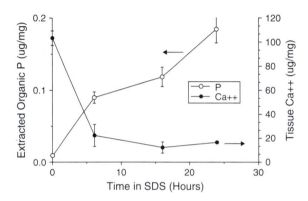

FIG. 7.   Reduction of calcification of bioprosthetic tissue by preincubation in 1% SDS demonstrated in a rat subcutaneous model of glutaraldehyde cross-linked porcine aortic valve. These results support the concept that phospholipid extraction is an important but perhaps not the only mechanism of SDS efficacy. Reproduced by permission from Schoen, F. J., Levy, R. J., and Piehler, H. R. (1992). Pathological considerations in replacement cardiac valves. *Cardiovasc. Pathol.* **1:** 29–52.

pretreatment in Europe, and its use in combination with aluminum treatment of the aortic wall of a stentless valve is under consideration.

Decellularization

Since the initial mineralization sites are devitalized connective cells of bioprosthetic tissue, some investigators have removed these cells from the tissue, with the intent of making the bioprosthetic matrix less prone to calcification (Wilson *et al.*, 1995; Courtman *et al.*, 1994).

Use of Tissue Fixatives Other Than Glutaraldehyde and Modification of Glutaraldehyde Fixation

Since previous studies have demonstrated that conventional glutaraldehyde fixation is conducive to calcification of bioprosthetic tissue, several studies have investigated modifications of and alternatives to conventional glutaraldehyde pretreatment. Paradoxically, fixation of bioprosthetic tissue

by extraordinarily high concentrations of glutaraldehyde (5–10× those normally used) appear to inhibit calcification (Zilla *et al.*, 1997, 2000). Residual glutaraldehyde residues in bioprosthetic tissue can be neutralized (detoxified) by treatment with lysine or diamine; this inhibits calcification of subdermal implants (Grabenwoger *et al.*, 1992; Trantina-Yates *et al.*, 2003). Non-glutaraldehyde cross-linking of bioprosthetic tissue with epoxides, carbodiimides, acylazides, and other compounds reduces their calcification in rat subdermal implant studies (Myers *et al.*, 1995; Xi *et al.*, 1992). Photooxidative preservation inhibits experimental calcification, possibly owing to the formation of unique calcification-resistant cross-links (Moore and Phillips, 1997).

Alternative Materials

Polyurethane trileaflet valves have been fabricated and investigated as a possible alternative to bioprostheses or mechanical valve prostheses. Despite versatile properties, such as superior abrasion resistance, hydrolytic stability, high flexural endurance, excellent physical strength, and acceptable blood compatibility, the use of polyurethane has been hampered by calcification, thrombosis, tearing, and biodegradation. Although the exact mechanism of polyurethane calcification is as yet unclear, it is believed that several physical, chemical, and biologic factors (directly or indirectly) play an important role in initiating this pathologic disease process (Schoen *et al.*, 1992c; Thoma and Phillips, 1995; Joshi *et al.*, 1996).

Tissue Engineered Heart Valve Replacements

In the approach called tissue engineering, an anatomically appropriate construct containing cells seeded on a resorbable scaffold is fabricated *in vitro* in a bioreactor, then implanted (Langer and Vacanti, 1993; Mayer *et al.*, 1997; Rabkin and Schoen, 2002). Progressive tissue remodeling *in vivo* is intended to ultimately recapitulate normal functional architecture. Autologous tissue-engineered valve cusps have been implanted in the pulmonary valve position in lambs, demonstrating the initial feasibility of the concept of a tissue-engineered heart valve leaflet (Shinoka *et al.*, 1996). With this concept, the cells comprising the prosthesis are intended to be viable and capable of renewal, thus theoretically inhibiting calcification. Heart valves utilizing this strategy have been implanted in growing sheep for extended periods (to 20 weeks) without calcification (Stock *et al.*, 1999; Hoerstrup *et al.*, 2000).

CONCLUSIONS

Calcification of biomaterial implants is an important pathologic process affecting a variety of tissue-derived biomaterials as well as synthetic polymers in various functional configurations. The pathophysiology has been partially characterized with a number of useful animal models; a key common feature is the involvement of devitalized cells and cellular debris. Although no clinically useful preventive approach has been proven to be safe and effective, several strategies based on either modifying biomaterials or local drug administration appear to be promising in some contexts.

Bibliography

Anderson, H. C. (1988). Mechanisms of pathologic calcification. *Rheum. Dis. Clin. N. Am.* **14**: 303–319.

Anderson, H. C. (1989). Mechanism of mineral formation in bone. *Lab. Invest.* **60**: 320–330.

Begley, C. T., Doherty, M. J., Mollan, R. A., and Wilson, D. J. (1995). Comparative study of the osteoinductive properties of bioceramic, coral and processed bone graft substitutes. *Biomaterials* **16**: 1181–1185.

Bernacca, G. M., Mackay, T. G., and Wheatley, D. J. (1992). in vitro calcification of bioprosthetic heart valves: report of a novel method and review of the biochemical factors involved. *J. Heart Valve Dis.* **1**: 115–130.

Bernacca, G. M., Tobasnick, G., and Wheatley, D. J. (1994). Dynamic in vitro calcification of porcine aortic valves. *J. Heart Valve Dis.* **3**:684–687.

Bernacca, G.M., Mackay, T.G., Wilkinson, R., and Wheatley, D. J. (1997). Polyurethane heart valves: fatigue failure, calcification, and polyurethane structure. *J. Biomed. Mater. Res.* **34**: 371–379.

Bini, A., Mann, K. G., Kudryk, B. J., and Schoen, F. J. (1999). Non-collagenous bone proteins, calcification and thrombosis in carcinoid artery atherosclerosis. *Arterio. Thromb. Vasc. Biol.* **19**: 1852–1861.

Bonucci, E. (1987). Is there a calcification factor common to all calcifying matrices? *Scanning Microsc.* **1**: 1089–1102.

Bostrom, K., Watson, K. E., Horn, S., Wortham, C., Herman, I. M., and Demer, L. L. (1993). Bone morphogenetic protein expression in human atherosclerotic lesions. *J. Clin. Invest.* **91**: 1800–1809.

Bucher, P. J., Buchi, E. R., and Daicker, B. C. (1995). Dystrophic calcification of an implanted hydroxyethylmethacrylate intraocular lens. *Arch. Opthalmol.* **113**: 1431–1435.

Chen, W., Kim, J. D., Schoen, F. J., and Levy, R. J. (1994a). Effect of 2-amino oleic acid exposure conditions on the inhibition of calcification of glutaraldehyde crosslinked porcine aortic valves. *J. Biomed. Mater. Res.* **28**: 1485–1495.

Chen, W., Schoen, F. J., and Levy, R. J. (1994b). Mechanism of efficacy of 2-amino oleic acid for inhibition of calcification of glutaraldehyde-pretreated porcine bioprosthetic valves. *Circulation* **90**: 323–329.

Cheng, P. T. (1988). Pathologic calcium phosphate deposition in model systems. *Rheum. Dis. Clin. N. Am.* **14**: 341–351.

Coleman, D., Lim, D., Kessler, T., *et al.* (1981). Calcification of non-textured implantable blood pumps. *Trans. Am. Soc. Artif. Intern. Organs* **27**: 97–103.

Courtman, D. W., Pereira, C. A., Kashef, V., McComb, D., Lee, J. M., and Wilson, G. J. (1994). Development of a pericardial acellular matrix biomaterial: biochemical and mechanical effects of cell extraction. *J. Biomed. Mater. Res.* **28**: 655–666.

Courtman, D. W., Pereira, C. A., Omar, S., Langdon, S. E., Lee, J. M., and Wilson, G. J. (1995). Biomechanical and ultrastructural comparison of cryopreservation and a novel cellular extraction of porcine aortic valve leaflets. *J. Biomed. Mater. Res.* **29**: 1507–1516.

David, T. E., Armstrong, S., and Sun, Z. (1998). The Hancock II bioprosthesis at 12 years. *Ann. Thorac. Surg.* **66**: S95–S98.

Deiwick, M., Glassmacher, B., Pettenazzo, E., Hammel, D., Castellon, W., Thiene, G., Reul, H., Berendes, E., and Scheld, H. H. (2001). Primary tissue failure of bioprostheses: new evidence from in vitro tests. *Thorac. Cardiovasc. Surg.* **49**: 78–83.

Demer, L. L. (2002). Vascular calcification and osteoporosis: Inflammatory responses to oxidized lipids. *Int. J. Epidemiol.* **31**: 737–741.

Fyfe, B., and Schoen, F. J. (1999). Pathologic analysis of removed non-stented Medtronic Freestyle aortic root bioprostheses treated with amino oleic acid (AOA). *Semin. Thorac. Cardiovasc. Surg.* **11**(Suppl 1): 151–156.

Giachelli, C. M. (1999). Ectopic calcification: gathering hard facts about soft tissue mineralization. *Am. J. Pathol.* **154**: 671–675.

Goldfarb, R. A., Neerhut, G. J., and Lederer, E. (1989). Management of acute hydronephrosis of pregnancy by urethral stenting: risk of stone formation. *J. Urol.* **141**(4): 921–922.

Golomb, G., Schoen, F. J., Smith, M. S., Linden, J., Dixon, M., and Levy, R. J. (1987). The role of glutaraldehyde-induced crosslinks in calcification of bovine pericardium used in cardiac valve bioprostheses. *Am. J. Pathol.* **127**: 122–130.

Gott, J. P., Chih, P., Dorsey, L. M. A., *et al.* (1992). Calcification of porcine valves: a successful new method of antimineralization. *Ann. Thorac. Surg.* **53**: 207–216.

Grabenwoger, M., Sider, J., Fitzal, F., *et al.* (1996). Impact of glutaraldehyde on calcification of pericardial bioprosthetic heart valve material. *Ann. Thorac. Surg.* **62**: 772–777.

Grabenwoger, M., Grimm, M., Ebyl, E., *et al.* (1992). Decreased tissue reaction to bioprosthetic heart valve material after L-glutamic acid treatment. A morphological study. *J. Biomed. Mater. Res.* **26**: 1231–1240.

Grunkemeier, G. L., Jamieson, W. R., Miller, D. C., and Starr, A. (1994). Actuarial versus actual risk of porcine structural valve deterioration. *J. Thorac. Cardiovasc. Surg.* **108**: 709–718.

Harasaki, H., Moritz, A., Uchida, N., *et al.* (1987). Initiation and growth of calcification in a polyurethane coated blood pump. *Trans. Am. Soc. Artif. Intern. Organs* **33**: 643–649.

Hilbert, S. L., Ferrans, V. J., Tomita, Y., Eidbo, E. E., and Jones, M. (1987). Evaluation of explanted polyurethane trileaflet cardiac valve prostheses. *J. Thorac. Cardiovasc. Surg.* **94**: 419–429.

Hirsch, D., Drader, J., Thomas, T. J., Schoen, F. J., Levy, J. T., and Levy, R. J. (1993). Inhibition of calcification of glutaraldehyde pretreated porcine aortic valve cusps with sodium dodecyl sulfate: preincubation and controlled release studies. *J. Biomed. Mater. Res.* **27**: 1477–1484.

Hoerstrup, S. P., Sodian, R., Daebritz, S., Wang, J., Bacha, E. A., Martin, D. P., Moran, A. M., Gulesarian, K. J., Sperling, J. S., Kaushal, S., Vacanti, J. P., Schoen, F. J., and Mayer, J. E. (2000).

Functional living trileaflet heart valves grown in vitro. *Circulation* 102: III44–III49.

Human, P., and Zilla, P. (2001a). Inflammatory and immune processes: The neglected villain of bioprosthetic degeneration? *J. Long Term Eff. Med. Implants* 11: 199–220.

Human, P., and Zilla, P. (2001b). The possible role of immune responses in bioprosthetic heart valve failure. *J. Heart Valve Dis.* 10: 460–466.

Jian, B., Jones, P. L., Li, Q., Mohler, E. R. 3rd, Schoen, F. J., and Levy, R. J. (2001). Matrix metalloproteinase-2 is associated with tenascin-C in calcific aortic stenosis. *Am. J. Pathol.* 159: 321–327.

Jian, B., Narula, N., Li, Q. Y., Mohler, E. R., 3rd, and Levy, R. J. (2003). Progression of aortic valve stenosis: TGF-beta 1 is present in calcified aortic valve cusps and promotes aortic valve interstitial cell calcification via apoptosis. *Ann. Thorac. Surg.* 75: 457–465.

Johnston, T. P., Webb, C. L., Schoen, F. J., and Levy, R. J. (1992). Assessment of the in-vitro transport parameters for ethanehydroxy diphosphonate through a polyurethane membrane. A potential refillable reservoir drug delivery device. *ASAIO* 38: M611–M616.

Johnston, T. P., Webb, C. L., Schoen, F. J., and Levy, R. J. (1993). Site-specific delivery of ethanehydroxy diphosphonate from refillable polyurethane reservoirs to inhibit bioprosthetic tissue calcification. *J. Control. Rel.* 25: 227–240.

Jones, M., Eidbo, E. E., Hilbert, S. L., *et al.* (1989). Anticalcification treatments of bioprosthetic heart valves: in-vivo studies in sheep. *J. Cardiac Surg.* 4: 69–73.

Jono, S., McKee, M. D., Murry, C. E., Shiroi, A., Nishizawa, Y., Mori, K., Morii, H., and Giachelli, C. M. (2000). Phosphate regulation of vascular smooth muscle cell calcification. *Circ. Res.* 87: e10–e17.

Joshi, R. R., Underwood, T., Frautschi, J. R., Phillips, R. E., Jr., Schoen, F. J., and Levy, R. J. (1996). Calcification of polyurethanes implanted subdermally in rats is enhanced by calciphylaxis. *J. Biomed. Mater. Res.* 31: 201–207.

Khan, S. R., and Wilkinson, E. J. (1985). Scanning electron microscopy, x-ray diffraction, and electron microprobe analysis of calcific deposits on intrauterine contraceptive devices. *Hum. Pathol.* 16: 732–738.

Klintworth, G. K., Reed, J. W., Hawkins, H. K., and Ingram, P. (1977). Calcification of soft contact lenses in patient with dry eye and elevated calcium concentration in tears. *Invest. Ophthalmol. Vis. Sci.* 16: 158–161.

Koide, T., and Katayama, H. (1979). Calcification in augmentation mammoplasty. *Radiology* 130: 337–338.

Langer, R., and Vacanti, J. P. (1993). Tissue engineering. *Science* 260: 920–926.

Lentz, D. L., Pollock, E. M., Olsen, D. B., and Andrews, E. J. (1982). Prevention of intrinsic calcification in porcine and bovine xenograft materials. *Trans. Am. Soc. Artif. Intern. Organs* 28: 494–497.

Levy, R. J., Schoen, R. J., Levy, J. T., Nelson, A. C., Howard, S. L., and Oshry, L. J. (1983a). Biologic determinants of dystrophic calcification and osteocalcin deposition in glutaraldehyde-reserved porcine aortic valve leaflets implanted subcutaneously in rats. *Am. J. Pathol.* 113: 142–155.

Levy, R. J., Schoen, F. J., and Howard, S. L. (1983b). Mechanism of calcification of porcine aortic valve cusps: Role of T-lymphocytes. *Am. J. Cardiol.* 52: 629–631.

Levy, R. J., Hawley, M. A., Schoen, F. J., Lund, S. A., and Liu, P. Y. (1985a). Inhibition by diphosphonate compounds of calcification of porcine bioprosthetic heart valve cusps implanted subcutaneously in rats. *Circulation* 71: 349–356.

Levy, R. J., Wolfrum, J., Schoen, F. J., Hawley, M. A., Lund, S. A., and Langer, R. (1985b). Inhibition of calcification of bioprosthetic heart valves by local controlled-released diphosphonate. *Science* 229: 190–192.

Levy, R. J., Schoen, F. J., and Golomb, G. (1986a). Bioprosthetic heart valve calcification: Clinical features, pathobiology and prospects of prevention. *CRC Crit. Rev. Biocompatibil.* 2: 147–187.

Levy, R. J., Schoen, F. J., Sherman, F. S., Nichols, J., Hawley, M. A., and Lund, S. A. (1986b). Calcification of subcutaneously implanted type I collagen sponges: effects of glutaraldehyde and formaldehyde pretreatments. *Am. J. Pathol.* 122: 71–82.

Levy, R. J., Schoen, F. J., Lund, S. A., and, Smith M. S. (1987). Prevention of leaflet calcification of bioprosthetic heart valves with diphosphonate injection therapy. Experimental studies of optimal dosages and therapeutic durations. *J. Thorac. Cardiovasc. Surg.* 94: 551–557.

Levy, R. J., Vyavahare, N., Ogle, M., Ashworth, P., Bianco, R., and Schoen, F. J. (2003). Inhibition of cusp and aortic wall calcification in ethanol- and aluminum-treated bioprosthetic heart valves in sheep: Background, mechanisms, and synergism. *J. Heart Valve Dis.* 12: 209–216.

Li, Q. Y., Jones, P. L., Lafferty, R. P., Safer, D., and Levy, R. J. (2002). Thymosin beta4 regulation, expression and function in aortic valve interstitial cells. *J. Heart Valve Dis.* 11: 726–735.

Love, J. W. (1993). *Autologous Tissue Heart Valves.* R.G. Landes, Texas.

Luo *et al.* (1997). Spontaneous calcification of arteries and cartilage in mice lacking matrix GLA protein. *Nature* 386: 78–81.

Mako, W. J., and Vesely, I. (1997). In-vivo and in-vitro models of calcification in porcine aortic valve cusps. *J. Heart Valve Dis.* 6: 316–323.

Mayer, J. E. Jr., Shin'oka, T., and Shum-Tim, D. (1997). Tissue engineering of cardiovascular structures. *Curr. Opin. Cardiol.* 12: 528–532.

McGonagle-Wolff, K., and Schoen, F. J. (1992). Morphologic findings in explanted Mitroflow pericardial bioprosthetic valves. *Am. J. Cardiol.* 70: 263–264.

Mitchell, R. N., Jonas, R. A., and Schoen, F. J. (1998). Pathology of explanted cryopreserved allograft heart valves: comparison with aortic valves from orthotopic heart transplants. *J. Thorac. Cardiovasc. Surg.* 115: 118–127.

Moore, M. A., and Phillips, R. E. (1997). Biocompatibility and immunologic properties of pericardial tissue stabilized by dye-mediated photooxidation. *J. Heart Valve Dis.* 6:307–315.

Myers, D. J., Nakaya, G., Girardot, G. M., and Christie, G. W. (1995). A comparison between glutaraldehyde and diepoxide-fixed stentless porcine aortic valves: biochemical and mechanical characterization and resistance to mineralization. *J. Heart Valve Dis.* 4: S98–S101.

Parhami, F., Basseri, B., Hwang, J., Tintut, Y., and Demer, L. L. (2002). High-density lipoprotein regulates calcification of vascular cells. *Circ. Res.* 91: 570–576.

Peters, W., Smith, D., Lugowski, S., McHugh, A., Keresteci, A., and Baines, C. (1995). Analysis of silicon levels in capsules of gel and saline breast implants and of penile prostheses. *Ann. Plast. Surg.* 34: 578–584.

Pettenazzo, E., Deiwick, M., Thiene, G., Molin, G., Glasmacher, B., Martignag, F., Bottio, T., Reul, H., and Valente, M. (2001). Dynamic in vitro calcification of bioprosthetic porcine valves: evidence of apatite crystallization. *J. Thorac. Cardiovasc. Surg.* 121: 428–430.

Rabkin, E., and Schoen, F. J. (2002). Cardiovascular tissue engineering. *Cardiovasc. Pathol.* 11: 305–317.

Schinke, T., McKee, M. D., and Karsenty, G. (1999). Extracellular matrix calcification: Where is the action? *Nat. Genet.* 21: 150–151.

Schoen, F. J., and Levy, R. J. (1984). Bioprosthetic heart valve failure: pathology and pathogenesis. *Cardiol. Clin.* 2: 717–739.

Schoen, F. J., Levy, R. J., Nelson, A. C., Bernhard, W. F., Nashef, A., and Hawley, M. (1985). Onset and progression of experimental bioprosthetic heart valve calcification. *Lab. Invest.* **52**: 523–532.

Schoen, F. J., and Hobson, E. (1985). Anatomic analysis of removed prosthetic heart valves: causes of failure of 33 mechanical valves and 58 bioprostheses, 1980 to 1983. *Hum. Pathol.* **16**: 549–559.

Schoen, F. J., Tsao, J. W., and Levy, R. J. (1986). Calcification of bovine pericardium used in cardiac valve bioprostheses. Implications for mechanisms of bioprosthetic tissue mineralization. *Am. J. Pathol.* **123**: 143–154.

Schoen, F. J., Kujovich, J. L., Webb, C. L., and Levy, R. J. (1987). Chemically determined mineral content of explanted porcine aortic valve bioprostheses: correlation with radiographic assessment of calcification and clinical data. *Circulation* **76**: 1061–1066.

Schoen, F. J., Harasaki, H., Kim, K. H., Anderson, H. C., and Levy, R. J. (1988). Biomaterials associated calcification: pathology, mechanisms, and strategies for prevention. *J. Biomed. Mater. Res.: Appl. Biomater.* **22A1**: 11–36.

Schoen, F. J., Golomb, G., and Levy, R. J. (1992a). Calcification of bioprosthetic heart valves: a perspective on models. *J. Heart Valve Dis.* **1**: 110–114.

Schoen, F. J., Levy, R. J., Hillbert, S. L., and Bianco, R. W. (1992b). Antimineralization treatments for bioprosthetic heart valves. Assessment of efficacy and safety. *J. Thorac. Cardiovasc. Surg.* **104**: 1285–1288.

Schoen, F. J. (1999). The heart. in *Robbins Pathologic Basis of Disease*, 6th ed., R. S. Cotran, V. Kumar, T. Collins eds. W.B. Saunders, Philadephia, pp. 543–599.

Schoen, F. J., Levy, R. J., and Piehler, H. R. (1992c). Pathological considerations in replacement cardiac valves. *Cardiovasc. Pathol.* **1**: 29–52.

Schoen, F. J., Hirsch, D., Bianco, R. W., and Levy, R. J. (1994). Onset and progression of calcification in porcine aortic bioprosthetic valves implanted as orthotopic mitral valve replacements in juvenile sheep. *J. Thorac. Cardiovasc. Surg.* **108**: 880–887.

Schoen, F. J. (1998). Pathologic findings in explanted clinical bioprosthetic valves fabricated from photo oxidized bovine pericardium. *J. Heart Valve Dis.* **7**: 174–179.

Schoen, F. J., and Levy, R. J. (1999). Tissue heart valves: current challenges and future research perspectives. *J. Biomed. Mater. Res.* **47**: 439–465.

Shinoka, T., Ma, P. X., Shum-Tim, D., *et al.* (1996). Tissue-engineered heart valves. Autologous valve leaflet replacement study in a lamb model. *Circulation* **94**: II164–II168.

Speer, M. Y., and Giachelli, C. M. (2004). Regulation of vascular calcification. *Cardiovasc. Pathol.* In press.

Speer, M. Y., McKee, M. D., Guldberg, R. E., Liaw, L., Yang, H.Y., Tung, E., Karsenty, G., Giachelli, C. M. (2002). Inactivation of the osteopontin gene enhances vascular calcification of matrix Gla protein-dificient mice: evidence for osteopontin as an inducible inhibitor of vascular calcification in-vivo. *J. Exp. Med.* **196**: 1047–1055.

Srivasta, S. S., Maercklein, P. B., Veinot, J., Edwards, W. D., Johnson, C. M., Fitzpatrick, L. A. (1997). Increased cellular expression of matrix proteins that regulate mineralization is associated with calcification of native human and porcine xenograft bioprosthetic heart valves. *J. Clin. Invest.* **5**: 996–1009.

Steitz, S. A., Speer, M. Y., McKee, M. D., Liaw, L., Almeida, M., Yang, H., and Giachelli, C. M. (2002). Osteopontin inhibits mineral deposition and promotes regression of ectopic calcification. *Am. J. Pathol.* **161**: 2035–2046.

Stock. U. A., Nagashima, M., Khalil, P. N., *et al.* (1999). Tissue-engineered valved conduits in the pulmonary circulation. *J. Thorac. Cardiovasc. Surg.* **119**: 732–740.

Thoma, R. J., and Phillips, R. E. (1995). The role of material surface chemistry in implant device calcification: a hypothesis. *J. Heart Valve Dis.* **4**: 214–221.

Trantina-Yates, A. E., Human, P., and Zilla, P. (2003). Detoxification on top of enhanced, diamine-extended glutaraldehyde fixation significantly reduces bioprosthetic root calcification in the sheep model. *J. Heart Valve Dis.* **12**: 93–100.

Vyavahare, N. R., Chen, W., Joshi, R., *et al.* (1997a). Current progress in anticalcification for bioprosthetic and polymeric heart valves. *Cardiovasc. Pathol.* **6**: 219–229.

Vyavahare, N., Hirsch, D., Lerner, E., *et al.* (1997b). Prevention of bioprosthetic heart valve calcification by ethanol preincubation. Efficacy and mechanism. *Circulation* **95**: 479–488.

Vyavahare, N. R., Hrisch, D., Lerner, E., *et al.* (1998). Prevention of calcification of glutaraldehyde-crosslinked porcine aortic cusps by ethanol preincubation: mechanistic studies of protein structure and water-biomaterial relationships. *J. Biomed. Mater. Res.* **40**: 577–585.

Vyavahare, N. R., Ogle, M., Schoen, F. J., and Levy, R. J. (1999). Mechanisms of elastin calcification and its prevention with A13C13. *Am. J. Pathol.* **155**: 973–982.

Vyavahare, N., Jones, P. L., Tallapragada, S., and Levy, R. J. (2000). Inhibition of matrix metallopriteinase activity attenuates tenascin-C production and calcification of implanted purified elastin in rats. *Am. J. Pathol.* **157**: 885–893.

Wada, T., McKee, M. D., Steitz, S., and Giachelli, C. M. (1999). Calcification of vascular smooth muscle cell cultures. Inhibition by osteopontin. *Circ. Res.* **84**: 166–178.

Webb, C. L., Schoen, F. J., Flowers, W. E., Alfrey, A. C., Horton, C., and Levy, R. J. (1991). Inhibition of mineralization of glutaraldehyde-pretreated bovine pericardium by AlCl$_3$. Mechanisms and comparisons with FeCl$_3$ LaCl$_3$ and Ga(NO$_3$)$_3$ in rat subdermal model studies. *Am. J. Pathol.* **138**: 971–981.

Wilson, G. J., Courtman, D. W., Klement, P., Lee, J. M., and Yeger, H. (1995). Acellular matrix: a biomaterials approach for coronary artery and heart valve replacement. *Ann. Thorac. Surg.* **60**: S353–S358.

Xi, T., Ma, J., Tian, W., Lei, X., Long, S., and Xi, B. (1992). Prevention of tissue calcification on bioprosthetic heart valve by using epoxy compounds: a study of calcification tests in-vitro and in-vivo. *J. Biomed. Mater. Res.* **26**: 1241–1251.

Zilla, P., Weissenstein, C., Bracher, M., Zhang, Y., Koen, W., Human, P., and von Uppel, U. (1997). High glutaraldehyde concentrations reduce rather than increase the calcification of aortic wall tissue. *J. Heart Valve Dis.* **6**: 490–491.

Zilla, P., Weissenstein, C., Human, P., Dower, T., and von Oppell, U. O. (2000). High glutaraldehyde concentrations mitigate bioprosthetic root calcification in the sheep model. *Ann. Thorac. Surg.* **70**: 2091–2095.

Application of Materials in Medicine, Biology, and Artificial Organs

HARVEY S. BOROVETZ, JOHN F. BURKE, THOMAS MING SWI CHANG, ANDRÉ COLAS,
A. NORMAN CRANIN, JIM CURTIS, CYNTHIA H. GEMMELL, BARTLEY P. GRIFFITH,
NADIM JAMES HALLAB, JORGE HELLER, ALLAN S. HOFFMAN, JOSHUA J. JACOBS,
RAY IDEKER, J. LAWRENCE KATZ, JACK KENNEDY, JACK E. LEMONS, PAUL S. MALCHESKY,
JEFFERY R. MORGAN, ROBERT E. PADERA, JR., ANIL S. PATEL, MIGUEL F. REFOJO,
MARK S. ROBY, THOMAS E. ROHR, FREDERICK J. SCHOEN, MICHAEL V. SEFTON,
ROBERT L. SHERIDAN, DENNIS C. SMITH, FRANCIS A. SPELMAN, PETER J. TARCHA,
RONALD G. TOMPKINS, RAMAKRISHNA VENUGOPALAN, WILLIAM R. WAGNER,
PAUL YAGER, AND MARTIN L. YARMUSH

7.1 INTRODUCTION

Jack E. Lemons and Frederick J. Schoen

Synthetic biomaterials have been evaluated and used for a wide range of medical and dental applications. From the earliest uses (~1000 B.C.) of gold strands as soft tissue sutures for hernia repairs, silver and gold as artificial crowns, and gemstones as tooth replacements (inserted into bone and extending into the oral cavity), biomaterials have evolved to standardized formulations. Since the late 1930s, high-technology polymeric, metallic, and ceramic substrates have played a central role in expanding the application of biomaterial devices.

Most students enter the biomaterials discipline with a strong interest in applications. Critical to understanding these applications is the degree of success and failure and, most important, what can be learned from a careful evaluation of past successes and failures. The following chapters present topics across the spectrum of applications, ranging from cardiovascular, orthopedic, ophthalmological, and dental therapeutic devices to skin substitutes, drug delivery systems, and sensors for diagnostic purposes. A central emphasis is the correlation of application limitations with the basic properties of the various biomaterials and devices and the biological interactions of the recipient with the biomaterials and strategies to extend and improve existing applications. Although the range of devices that constitute artificial organs is at present limited in clinical use, considerable research and development has involved devices that have active mechanical, biologic, or mass-exchange functions.

It is estimated that approximately 20 million individuals have an implanted medical device. Costs associated with prostheses and organ replacement therapies exceed $300 billion U.S. dollars per year and comprise nearly 8% of total healthcare spending worldwide (Lysaght and O'Loughlin, 2000).

Most implants serve their recipients well for extended periods by alleviating the conditions for which they were implanted. Considerable effort is expended in understanding biomaterials–tissue interactions and eliminating patient–device complications (the clinically important manifestations of biomaterials–tissue interactions). Moreover, many patients receive substantial and extended benefit despite complications. For example, heart valve disease is a serious medical problem. Patients with diseased aortic heart valves have a 50% chance of dying within approximately 3 years without surgery. Surgical replacement of a diseased valve leads to an expected survival of 70% at 10 years, a substantial improvement over the natural course. However, of these patients whose longevity and quality of life have clearly been enhanced, approximately 60% will suffer a serious valve-related complication within 10 years after the operation. Thus, long-term failure of biomaterials leading to a clinically significant event does not preclude clinical success overall.

The range of tolerable risk of adverse effects varies directly with the medical benefit obtained by the therapy. Benefit and risk go hand-in-hand and clinical decisions are made to maximize the ratio of benefit to risk. The tolerable benefit–risk ratio may depend on the type of implant and the medical problem it is used to correct. Thus, more risk can be tolerated in a heart assist device (a life-sustaining implant) than in a prosthetic hip joint (an implant that relieves pain and disability, while enhancing function) or, further along the spectrum, than a breast implant (an implant with predominantly cosmetic benefit).

Considering other examples, NIH sponsored consensus conferences (1990s) have provided documentation that total hip arthroplasties demonstrate that 90% of those placed will be in place and function after a decade for individuals over 65 years old. More recently, as described in the section on orthopaedics, concerns have been raised about longer-term influences of dilute concentrations of elements transferred from devices. As with other disciplines, a confirmed need exists for longer-term controlled clinical trials based on prospective protocols.

This chapter explores the most widely used therapeutic approaches and applications of materials in medicine, biology and artificial organs. The progress made in many of these areas is substantial. In most cases, the sections describe a device category from the perspective of the clinical need, the armamentarium of devices available to the practitioner, the results and complications, and the challenges to the field that limit success.

Bibliography

Lysaght, M. J., and O'Loughlin, J. A. (2000). Demographic scope and economic magnitude of contemporary organ replacement therapies. *ASAIO J.* **46**: 515–521.

7.2 NONTHROMBOGENIC TREATMENTS AND STRATEGIES

Michael V. Sefton and Cynthia H. Gemmell

In 1963, Dr. Vincent Gott at the Johns Hopkins University changed the field of biomaterials by failing to reproduce an earlier experiment. He was trying to show that an applied electric field could minimize thrombogenesis on a metal surface. He obtained this result, but was somewhat mystified when he discovered that the wire leading to his negative graphite electrode was broken. He soon realized the importance of rinsing his electrode with a common disinfectant (benzalkonium chloride) and heparin prior to implantation. Thus, the first heparinized material was born. Bob Lehninger (Leininger *et al.*, 1966) at Battelle Memorial Institute in Columbus, Ohio, followed up with better quaternary ammonium compounds and soon afterwards a host of chemical derivatization methods were devised to adapt the original GBH (graphite benzalkonium heparin) method to plastics and other materials. The principles underlying these strategies and others for lowering the thrombogenicity of materials are detailed here with a few examples. For a comprehensive review of nonthrombogenic treatments and strategies the reader is referred to a number of reviews (Sefton *et al.*, 1987; Engbers and Feijen, 1991; Amiji and Park, 1993; Ratner, 1995; Kim and Jaeobs, 1996).

CRITERIA FOR NONTHROMBOGENICITY

Thrombogenicity is defined (Williams, 1987) as the ability of a material to induce or promote the formation of thromboemboli. Here we are concerned with strategies to lower thrombogenicity, if not actually reduce it to zero, "nonthrombogenicity." Thrombogenicity should be thought of as a rate parameter, since low rates of thrombus or emboli formation are probably tolerable since the fibrinolytic or other clearance systems exist to remove "background" levels of thromboemboli. We are principally concerned with rates of thrombus formation that are sufficient to occlude flowpaths in medical devices (e.g.,

block the lumen of catheters) or rates of embolus formation that cause downstream problems such as myocardial infarction or transient ischemic attacks. The mechanism of thrombogenicity is described in Chapter 4.6, while the effects of fluid flow on thrombus development and embolization are described in Chapter 3.5.

Thrombi are produced through aggregation of activated platelets and/or the thrombin-dependent polymerization of fibrinogen into fibrin. Thrombin is directly responsible for fibrin formation but it is also an important agonist of platelet activation. A simple model of thrombin generation is illustrated in Fig. 1A. The variety of mechanisms that lead to thrombin generation are lumped into a single parameter, k_p (cm/sec), a rate constant that relates the rate of production of thrombin (per unit area), R_p (g/cm^{2}sec), to the thrombin concentration at the surface of a material C_w (g/ml):

$$R_p = k_p C_w \tag{1}$$

k_p includes both the procoagulant effect of the material (via clotting factors and platelets) less any coagulation inhibition processes. A material balance (Basmadjian, 1990; Rollason and Sefton, 1992) equating the rate of production at the surface to the rate of transport away from the surface $k_L(C_w - C_b)$ for tubes greater than about 0.1 mm in diameter (Leveque region) gives:

$$\frac{C_w}{C_b^0} = \frac{1}{1 - k_p/k_L(x)} \tag{2}$$

where C_b^0 is the concentration of thrombin at the inlet to a tube. $k_L(x)$ is the local mass transfer coefficient which is infinite at the tube inlet and decreases as one proceeds down the tube. Hence, C_w/C_b^0 increases progressively down the tube (Fig. 1B). When $k_L = k_p$, C_w becomes infinite and a thrombus is expected. For a simple tube in laminar flow, k_L is on the order of 10^{-3} cm/sec and so k_p must be less than this to avoid a thrombus. Experimental results suggest that k_p is on the order of 10^{-3} for simple materials such as polyethylene but $<10^{-4}$ for heparinized materials (Rollason and Sefton, 1992). According to this model, only such low-k_p materials can be expected to minimize thrombin production. This is one of the reasons heparinization and other active methods of inhibiting thrombin formation are so popular as strategies for imparting low thrombogenicity. There are, however, other criteria that must also be met.

It is also a requirement that platelet interactions with the surface do not lead to thrombosis. To some extent platelet adhesion is an inevitability. Once adherent, platelets change shape and release their granule contents, which can activate bulk platelets. Although it is intuitive to suggest that a nonthrombogenic surface should not support platelet adhesion it has not, unfortunately, been that simple. While most studies focus on evaluating the platelet compatibility of surfaces by measuring *in vitro* platelet adhesion, some *ex vivo* studies (Hanson *et al.*, 1980; Cholakis *et al.*, 1989) have demonstrated that even in the absence of adhesion, platelets can be "consumed." That is, the platelets are activated by the material, leading to their premature removal from the circulation. This becomes apparent as a significant shortening of platelet life span, if not also a decrease in systemic platelet count. Whether this process is initiated by nonadhesive direct contact with the

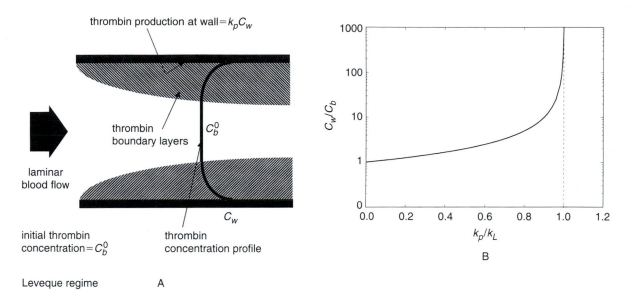

FIG. 1. (A) Model illustrating thrombin production at the surface of a tube as a balance between autoaccelerative production at the surface and mass transfer away from the wall. (B) Schematic illustration of Eq. 2 showing the dramatic increase in wall to bulk concentration ratio as the mass transfer coefficient k_L becomes equal to the first-order autocatalytic production constant (k_p). Since k_L decreases with increasing axial position down a tube, increasing k_p/k_L corresponds to increasing axial distance x. Adapted from Rollason and Sefton (1992).

material or the result of adherent platelet release or even the effect of complement activation is currently unknown. Nevertheless, such an observation suggests that low platelet adhesion is not a sufficient criterion of *in vivo* platelet compatibility. Rather, low thrombogenicity is characterized by both low platelet adhesion and low platelet activation (see Chapter 4.6 for definition of activation); the latter may even be more important than the former. Leukocyte activation (and complement activation) may also be key components in thrombogenicity (see Chapters 4.4 and 4.6); in the future, these too may become critical parameters defining the thrombogenicity of the surface. In an attempt to evaluate more than just platelet adhesion on biomaterials, flow cytometric techniques have been utilized to measure bulk (or circulating) activated platelets, platelet microaggregates and activated leukocytes (Baker *et al.*, 1998; Gemmell *et al.*, 1995). In future, it is hoped that such assays could provide better correlation between *in vitro* and *ex vivo* performances.

INERT (BIOTOLERANT) MATERIALS

Much of the effort in biomaterials research over the past 30 years has been directed toward the development of inert materials that do not react with platelets and coagulation factors. As outlined in Fig. 2, a number of approaches, often conflicting, have been developed. For example, there is still no consensus as to whether a surface should be hydrophilic or hydrophobic. The lack of agreement is largely due to our incomplete understanding of the biological pathways to materials failure and our inability to fully evaluate blood material responses. In developing materials with lower thrombogenicity researchers have primarily focused their efforts on modifying the surfaces of existing polymeric materials,

such as polyurethanes, silicone rubber, and polyethylene. This approach is reasonable since it is only the surface chemistry of a material that should dictate its biological responses; the mechanical characteristics of the material are primarily

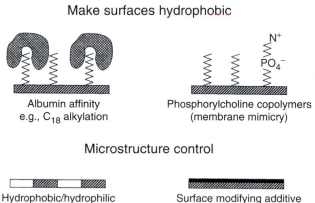

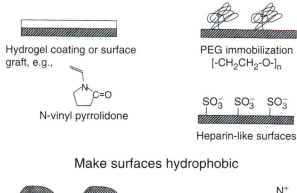

FIG. 2. Potential strategies for preparing inert surfaces with low thrombogenicity.

dictated by bulk chemistry. As illustrated by the following discussion the full range of surface modification strategies outlined in Chapter 2.14 have been used, albeit with limited success. Despite the successes in reducing protein and cellular deposits on some materials, a truly nonthrombogenic surface does not exist.

Hydrogels

A popular method to improve the blood compatibility of biomaterials is to increase surface hydrophilicity by incorporating a hydrogel at the surface. All the commonly used hydrogels (see Chapter 2.5) that can be cast, chemically cross-linked, or surface grafted have been used. By definition, hydrogels permit the retention of large amounts of water without dissolution of the polymer itself. This makes them very similar to biological tissues in that they are permeable to small molecules and possess low interfacial tension. It was Andrade who first postulated that since the interfacial free energy between blood and vascular endothelium was near zero, material surfaces that tend to have an interfacial energy of zero should have minimal thrombogenicity (Andrade *et al.*, 1973). Today, considerable experimental evidence supports the claim that materials with minimal interfacial energy, like hydrogels, do not strongly support cell and/or thrombus adhesion. Unfortunately, such generalizations are always flawed by exceptions. A hydrogel (cellulose) used as a dialysis membrane material (Cellophane) is well recognized as being thrombogenic, possibly because it is also a strong complement activator (see Chapter 4.4). Furthermore low-adhesion (or low protein adsorption) is not the same as low thrombogenicity.

In the early 1980s, a number of polymers were grafted with hydrogel surfaces in an effort to decrease their thrombogenicity. For example, in a large *ex vivo* study (Hanson *et al.*, 1980), a variety of surface-grafted copolymers were prepared and evaluated for platelet consumption in a baboon shunt model of arterial thrombogenesis. Although few platelets were found adherent to the graft surface, the higher the water content, the greater the rate at which the graft tubing caused the destruction of the circulating platelets (Fig. 3). Even though this study concluded that hydrogels do not possess low thrombogenicity, utilization of hydrogels remains a popular approach to lower thrombogenicity. It is likely that platelet consumption will be of concern only in applications with large surface areas. On the other hand, platelet consumption is evidence of platelet activation and the local (as opposed to systemic) consequences of such activation have yet to be defined.

Rather than radiation grafting of hydrogels onto materials, surfaces can also be simply coated with hydrophilic polymers, such as poly(vinyl pyrrolidone), PVP. While coating with PVP is intended to increase lubricity and ease catheter insertion, beneficial effects on thrombogenicity (and bacterial adhesion) have been noted (Francois *et al.*, 1996).

Poly(ethylene glycol) (PEG) Immobilization

Immobilization of the water-soluble synthetic polymer poly(ethylene glycol) (PEG, $-CH_2CH_2O-$) is a very popular

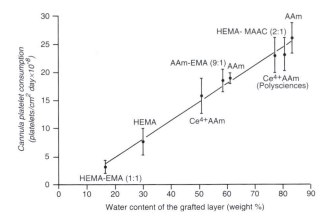

FIG. 3. Rate of cannula platelet consumption per unit area (in baboons) is directly related to the graft water content of shunts grafted with eight acrylic and methacrylic polymers and copolymers. HEMA, hydroxyethyl methacrylate; EMA, ethyl methacrylate; AAm, acrylamide; MAAC, methacrylic acid. Mean values ± 1 S.E. (from Hanson *et al.*, 1980, with permission).

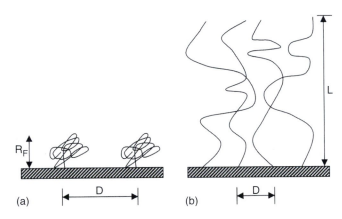

FIG. 4. Structure of polyethylene glycol (PEG) modified surfaces. (a) At low surface coverage ($D \gg R_F$), molecules assume conformation and size similar to random coil. (b) At high surface coverage ($D \ll R_F$), molecules are more extended chain-like and form a dense brush. Adapted from deGennes (1980).

means of making a biomaterial surface more protein and cell resistant.[1] It also makes the surface hydrogel-like. This approach was promoted by Merrill *et al.*, who recognized the lack of protein binding (hydrogen donor) sites on the PEO molecule (Merrill and Salzman, 1983). A widely recognized theory that helps explain the nonthrombogenicity of PEG containing surfaces was presented by Nagaoka *et al.* (1984). They reasoned that the presence of diffuse hydrophilic polymers, covering a significant portion of a biomaterial surface, would exert a steric repulsion effect toward blood proteins and cells (Fig. 4). The dominance of steric repulsion over the van der Waals attractive forces was hypothesized to be dependent on the extension and flexibility of the polymer chain in the bulk solution. The resulting excluded volume effect results from a loss in configurational entropy of the PEG that in turn results

---

[1] PEG, >10 kDa, is called PEO [poly(ethylene oxide)] reflecting the different monomer and polymerization process used.

from the rise in the local osmotic pressure occurring when PEG chains are compressed when blood elements approach the surface. This effect is dependent on both the chain length (N, monomers/chain) and the surface density of chains (σ, number of chains per unit area). A simple scaling relationship relates these parameters (and a, the monomer size) to the thickness, L, of the polymer layer at the surface, for the case of a good solvent (specifically an athermal solvent) and when the chain density is high (distance between chains, D) is less than the Flory radius, R_F (deGennes, 1980):

$$L \cong N a \sigma^{1/3} \qquad (3)$$

Jeon et al. (Jeon and Andrade, 1991; Jeon et al., 1991) among others, have theoretically modeled protein–surface interactions in the presence of PEG and concluded that steric repulsion by surface-bound PEG chains was largely responsible for the prevention of protein adsorption on PEG-rich surfaces.

As shown in Fig. 5, a number of approaches have been used to enrich surfaces with PEG. For example, it has been grafted to surfaces via a backbone hydrogel polymer, such as in the preparation of methoxypolyethylene glycol monomethacrylate copolymers (Nagaoka et al., 1984). It has also been covalently bonded directly to substrates via derivatization of its hydroxyl end groups with an active coupling agent or, alternatively, the hydroxyl end groups have been reacted with active coupling agents introduced onto the surface (Tseng and Park, 1992; Desai and Hubbell, 1991a; Chaikof et al., 1992). The availability of a large number of reactive PEG molecules (for example, amino-PEG, tresyl-PEG, N-hydroxysuccinimidyl-PEG, e.g., from Shearwater Polymers, Inc, now Nektar Therapeutics) has greatly facilitated the use of covalent immobilization strategies. Unfortunately, it is difficult to achieve the needed high surface coverages by immobilization since the first molecules immobilized sterically repel later molecules that are attached, unless thermodynamically poor solvents are used. The surface fraction may then be too low to completely "mask" the other functional groups that may be present. Pure monolayers of star

PEO have been grafted to surfaces in an attempt to increase surface coverage (Sofia and Merrill, 1998). Other investigators have used block copolymers (e.g., Pluronic) of PEO and PPO [poly(propylene oxide)] by adsorption, by gamma irradiation, or as an additive (McPherson et al., 1997); some have combined PEO with other strategies such as phosphoryl choline (Kim et al., 2000).

PEO has also been incorporated, by both ends, into polyetherpolyurethanes (Merrill et al., 1982; Liu et al., 1989; Okkema et al., 1989). Unfortunately, the results depend on a combined effect of surface microphase separation and the hydrogel effect of the PEO side chains. While some have noted lower thrombogenicity, others have not. For example, Okkema et al. (1989) synthesized a series of poly(ether urethanes) based on PEO and poly(tetramethylene oxide)(PTMO) soft segments and noted that the higher PEO-containing polymers were more thrombogenic in a canine ex vivo shunt model. Since PEO-containing polyurethanes have a considerable non-PEO phase Chaikof et al. (1989) prepared a cross-linked network of PEO chains using only small polysiloxane units.

A number of investigators have noted that the beneficial effect of PEG is molecular weight dependent. To some extent, however, the benefit of high molecular weight PEO is compromised by the crystallizability of long-chain PEO or the benefit may reflect particular process advantages of longer PEO chains. Chaikof et al. (1992) with end-linked PEO and Desai and Hubbell (1991b) with physically entrapped PEO found the lowest protein and platelet or cell deposition with high-molecular-weight PEO (>18,000 Da). Nagaoka et al. (1984) were among the first to demonstrate that increasing the PEG chain length of hydrogels containing methoxy poly(ethylene glycol) monomethacrylates led to reductions in protein and platelet adhesion (Fig. 6). It is clear that incorporation of PEG results in reduced levels of cell (including platelet) adhesion and protein adsorption, when compared to unmodified and typically hydrophobic substrates. It is far less clear whether the reduced adhesion or adsorption translates to lower material thrombogenicity (Llanos and Sefton, 1993a,b). Further it is not clear whether reduced adhesion/adsorption is due specifically to the thermodynamic effects of poly(ethylene oxide) or to the increase in surface hydrophilicity after its immobilization. While the in vitro results have looked very promising, the lack of correlation between the few in vitro and ex vivo studies is of concern. More recent efforts with plasma-deposited tetraglyme (Shen et al., 2001) have led to surfaces with ultralow adsorbed fibrinogen, suggesting that previous attempts at using PEG modification have not been succesful because of the inability to achieve the desired ultralow (<5 ng/cm^2) levels of adsorbed protein.

Albumin Coating and Alkylation

The early observation that surfaces coated with albumin did not support protein adsorption and platelet adhesion (reviewed by Andrade and Hlady, 1986) led many investigators to lower material thrombogenicity by either albumin coating or enhancing the affinity of albumin for surfaces via alkylation. Albumin is an abundant acidic globular plasma protein with

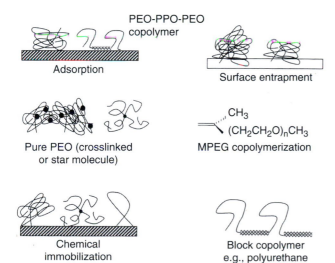

FIG. 5. Methods for incorporating poly(ethylene glycol) (PEG) onto the surfaces of materials.

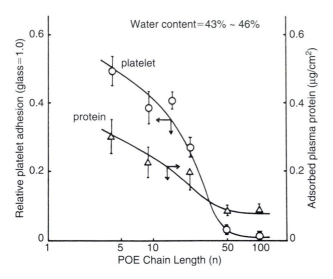

FIG. 6. Effect of PEG (or "POE") chain length (n) on the adhesion of platelets (from activated platelet-rich plasma) and adsorption of plasma proteins (from plasma) onto poly(methyl methacrylate-*co*-methoxypolyethylene glycol methacrylate) gels. Mean values and standard deviation are shown. Both protein adsorption and platelet adhesion were measured by total protein analysis. A relative platelet adhesion value of 0.3 corresponded to an SEM platelet adhesion density of ~80 platelets/1000 μm² (from Nagaoka *et al.*, 1984, with permission).

high aqueous solubility and stability. It is widely stated that albumin adsorption lowers material thrombogenicity since it does not possess the peptide sequences to enable interaction with cells (platelets and leukocytes) or the enzyme receptors in the coagulation cascade. It adsorbs relatively tightly onto hydrophobic surfaces while having a poor affinity for hydrophilic surfaces. Consequently, utilization of this strategy entails increasing the hydrophobicity of the surface, which is the opposite of the strategies discussed earlier whereby the hydrophilicity of the surface was increased. Albumin coating is consistent with another hypothesis, developed early in the 1970s, that suggested that hydrophobic surfaces with critical surface tensions (conceptually related to surface free energy) around 20–30 dyn/cm would be have a lower thrombogenicity (Baier *et al.*, 1970).

A limitation of relying on albumin coating is that other proteins will adsorb to the surface displacing the albumin thereby limiting the long term effectiveness of this strategy (see Chapter 3.2). Hence an example of coating a surface with albumin involved immersing a Dacron arterial prosthesis in albumin/glutaraldehdye solution followed by quenching of the free aldehyde groups (Kottke-Marchant *et al.*, 1989). The use of glutaraldehyde was expected to reduce the extent of exchange with proteins with higher affinity for the surface than albumin. Covalent immobilization of albumin has also been used for the same purpose (Hoffman *et al.*, 1972; Matsuda and Inoue, 1990). Unfortunately, these strategies do not address the inevitable denaturing of albumin with time.

Since albumin has binding pockets for long alkyl chains, Munro *et al.* (1981) hypothesized and demonstrated that surfaces with long carbon chains (C-16 or C-18) would have

a high affinity for albumin and would provide for a dynamically renewable natural albumin layer. Cellulose membranes were alkylated by grafting 4-vinylpyridine to the surface of Cuprophan and subsequently alkylating with non-fatty-acid-like C10 and fatty-acid-like C16 aliphatic chains (Frautschi *et al.*, 1995). Albumin adsorption (from dilute plasma) was enhanced on the alkylated surfaces. Strzinar and Sefton (1992) reacted a C18 and C4 isocyanate with poly(vinyl alcohol) hydrogel in order to improve the platelet reactivity of the hydrogel. Butylation of PVA dramatically reduced platelet reactivity in a dog shunt suggesting that the reduced reactivity might be independent of chain length and hence more related to the hydrophobicity of the substrate and not to albumin affinity per se. Unfortunately, the incorporation of C18 on the hydrophilic PVA surface was complicated by a nonuniform surface incorporation and this was a limitation of this strategy.

Phospholipid-Mimicking Surfaces

A number of investigators have hypothesized that a surface similar to the external phospholipid membrane of cells should be nonthrombogenic. Since phosphorylcholine, the major lipid head group on the external surface of blood cells, is inert in coagulation assays it has been the choice of investigators for incorporation into surfaces. Upon cellular activation, the negatively charged phospholipids that are preferentially localized on the cytoplasmic face of the plasma membrane flip to the outer membrane, which accelerates blood clotting by enabling assembly of the prothrombinase and tenase complexes.

Various approaches have been used to incorporate the electrically neutral zwitterionic head group into surfaces. Durrani *et al.* (1986) prepared a series of reactive derivatives of phosphorylcholine that were designed to react with surface hydroxyl groups and surface acid chlorides on various materials. Another approach involved coating materials (Campbell *et al.*, 1994; Lewis *et al.*, 2000) or blending a polyurethane (Ishihara *et al.*, 1995, 1996, 1999) with a methacryloylphosphorylcholine (MPC, Fig. 7)/polyacrylate copolymer. Platelet adhesion was significantly reduced on phosphorylcholine-coated expanded polytetrafluoroethylene (ePTFE) grafts at 90 min in dogs, and anastomotic neointimal hyperplasia and neointimal cell proliferation were also reduced (Chen *et al.*, 1998). The modified material's strong affinity for phospholipids due to the phosphorylcholine group is thought to impart

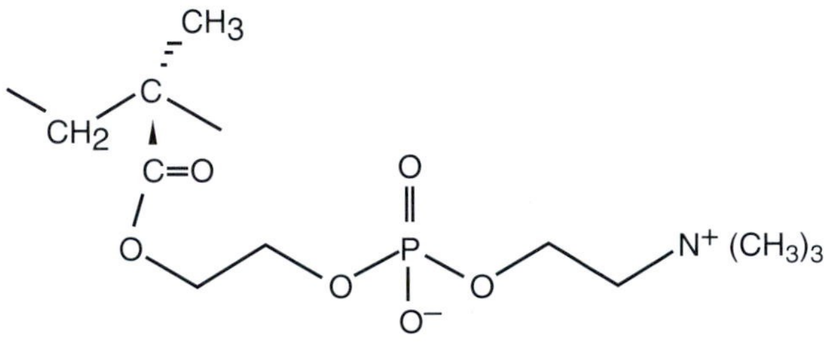

FIG. 7. Chemical structure of poly(2-methacryloyloxyethyl phosphorylcholine).

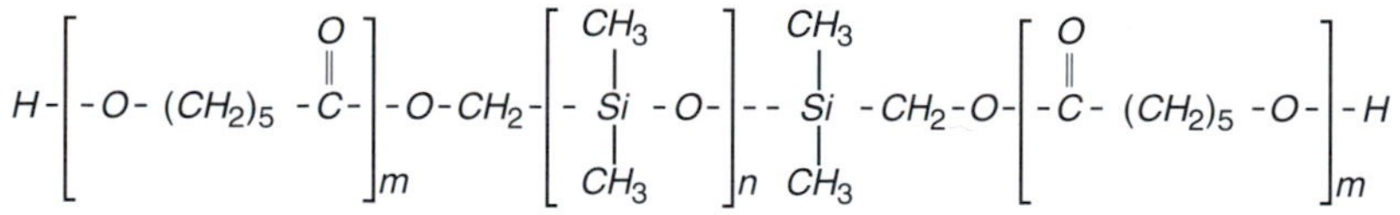

FIG. 8. General formulation of surface-modifying additive (SMA) block copolymer based on polycaprolactone (PCC) and poly(dimethyl siloxane) (PDMS). Adapted from Tsai *et al.* (1994).

the observed low thrombogenicity, although Ishihara *et al.* (1998) have attributed the low protein adsorption of their phospholipid polymers to the high free-water fraction. The mechanism of action of such materials remains unclear.

Surface-Modifying Additives

The blending of a copolymer, composed of polar and non-polar blocks, to a base polymer appears to be a successful means of lowering material thrombogenicity. The strategy, originally developed by Thoratec Laboratories Corporation, is a way to alter the surface properties of materials without affecting bulk properties (Ward *et al.*, 1984). The copolymers, added in low concentration, migrate to the base polymer surface during and after fabrication and dramatically change the outermost surface molecular layers that form the region that determines biocompatibility (Tsai *et al.*, 1994). The copolymers have a structure which is amphipathic, that is, certain groups or segments will have an attraction for the continuous phase (major polymer component of the blend), while other portions of the molecule will have little attraction for the base polymer and will be of lower polarity (Ward *et al.*, 1984). The generic formulation of a polycaprolactone–polysiloxane block copolymer that has been tested for biocompatibility is shown in Fig. 8. They have been used to lower the thrombogenicity of cardiopulmonary bypass and hemodialysis components by using SMA blended polymers or SMA coated surfaces. A recent clinical evaluation of the effects of SMA on cardiopulmonary bypass circuits demonstrated a reduction in platelet interactions with no effect on complement activation (Gu *et al.*, 1998).

Fluorination

The incorporation of fluorine into materials is also a strategy to lower thrombogenicity. It is believed that the fluorine group's low surface energy inhibits protein adsorption and platelet adhesion/activation. Preparing surfaces with fluorine is often facilitated by the tendency of fluorine-containing functional groups to concentrate on the surface of the polymer during preparation. Fluoroalkyl groups as chain extenders have been used to prepare nonthrombogenic polyurethanes (Kashiwagi *et al.*, 1993). Kiaei *et al.* (1988) found a strong effect of a fluoropolymer plasma on thrombus formation. A number of new fluorinated surface-modifying additives have also been developed (Tang *et al.*, 1996).

Plasma Treatments

Another surface modification technique resulting in a covalently attached coating is plasma deposition or glow discharge deposition (described in Chapter 2.14). This technique enables the rapid polymerization and deposition of organic compounds that cannot be easily polymerized by conventional polymer mechanisms. Active species, generated by energy absorption, react with and covalently bind to the substrate surface. Dacron vascular grafts were treated with tetrafluoroethylene in a radio frequency glow discharge (RFGD) apparatus by Bohnert *et al.* (1990). They demonstrated that while the deposition of fluorocarbon polymers did not lower the adsorption levels of fibrinogen and albumin it did substantially reduce the ability of detergent to elute the adsorbed proteins. The significance of this in terms of thrombogenicity is not clear. A study (Sefton *et al.*, 2001) comparing many such plasma-modified surfaces did not identify a modification chemistry that was superior to the base material in terms of platelet or leukocyte activation.

Heparin-Like Materials

A number of synthetic polymers have been synthesized or modified in order to prepare polymers with chemical similarity to heparin and thus possess heparin-like activities. For example, Fougnot *et al.* (1983) synthesized sulfonate/amino acid sulfamide polystyrene derivatives in order to create insoluble heparin-like materials. Some investigators (e.g., Grasel and Cooper, 1989) only incorporate sulfonate groups in an effort to lower thrombogenicity by surface thrombin inhibition. Enhanced binding and inactivation of thrombin was found suggesting that these weak heparin-like molecules are sufficiently dense so as to lower thrombogenicity. A new approach combines sulfonation and PEG-like materials in the use of sulfonated cyclodextrin polymers (Park *et al.*, 2002).

Self-Assembled Surface Layers

Self-assembled surface layers (Whitesides *et al.*, 1991) have been envisioned as useful templates to nucleate or organize ordered, designed biomaterials (Ratner, 1995). Molecular self-assembly is the spontaneous association of molecules under equilibrium conditions into stable, structurally well-defined aggregates joined by noncovalent bonds (Whitesides *et al.*, 1991). The advantages of self-assembly films are their high order and orientation and well-defined head-group geometry that exposes only one functional group to the outside (Ratner, 1995). This approach remains to be exploited for the preparation of practical low-thrombogenicity materials. So far, self-assembled monolayers of alkylsilanes supported on oxidized poly(dimethyl siloxane) (PDMS) rubber have been used as a model system (Silver *et al.*, 1999). The authors reported that surfaces grafted with hydrophobic head groups as –CH_3 and CF_3 had significantly lower platelet and fibrinogen deposition than the surfaces composed of hydrophilic head groups in a canine *ex vivo* arteriovenous series shunt model.

ACTIVE (BIOACTIVE) MATERIALS

The limited successes of the various surface treatments for lowering thrombogenicity has encouraged researchers to pursue other strategies. The most popular and the one that started the field is the heparinization of surfaces, although the incorporation of other antithrombotic and antiplatelet agents into materials is gaining popularity (Fig. 9). Currently, the antithrombotic agents utilized have mainly been against thrombin (hirudin, PPACK, [D-phenylalanyl-L-propyl-L-arginine chloromethyl ketone], thrombomodulin). In the future, it is likely that agents against factor Xa (low-molecular-weight heparins, LMWH) or recently developed inhibitors of tissue

factor expression (extrinsic pathway) could also impart a lower degree of thrombogenicity for some devices.

The incoporation of antiplatelet agents into materials has been limited by the instability and complex mechanism of action of many of these agents. The new generation of platelet GPIIb/IIIa antagonists (e.g., ReoPro, Centocor) will likely be more useful given their greater stability and potency. The future use of pharmacological agents could also include agents to inhibit complement. Complement activation is known to occur on biomaterials although the availability of inhibitory agents is limited. A clearer understanding of how activated complement proteins mediate cellular reactivity could provide the impetus for incorporating such agents into materials (Gemmell, 1998).

Heparinization

Heparinization of surfaces continues to be the most popular technique for lowering the thrombogenicity of materials. The heparin-like activity of the microvascular endothelium provides logic to such an approach but it should be appreciated that prostacyclin, a potent platelet inhibitor, also plays a role in maintaining the thromboresistance of the natural endothelium. As illustrated in Fig. 10, heparin is a linear, acidic carbohydrate composed of repeating disaccharide units that are O- and N-sulfated. The molecular weights of heparin chains range from 1200 to 40,000 Da with a mean molecular weight of approximately 10,000. Depending on its molecular weight and structure, heparin is able to inhibit, in association with its cofactor (antithrombin III), the serine proteases: thrombin and factor X. Since thrombin is also a potent platelet activator heparin's presence on a surface should help minimize platelet activation. Investigators have been able to covalently, ionically, and physically attach heparin to various substrates utilizing a number of chemistries given its remarkable stability. The main concern has been that once immobilized, the heparin should be able to assume its native conformation and be able to interact with antithrombin III.

The effectiveness of heparin as an agent capable of increasing synthetic venous graft patencies and reducing downstream anastomotic neointimal hyperplasia and cell proliferation has been demonstrated. Utilizing a novel polytetrafluoroethylene-based local drug delivery device heparin was infused within those blood layers adjacent to the graft wall and at downstream anastomotic sites for 14 days and demonstrated to be highly effective (Chen *et al.*, 1995). Such results continue to motivate researchers to heparinize surfaces.

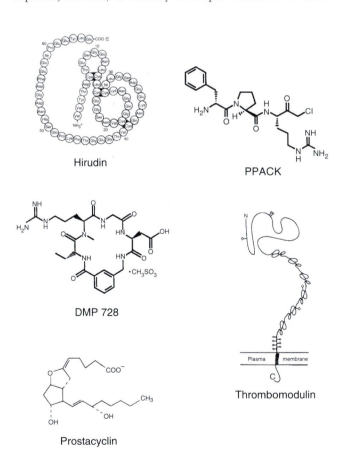

FIG. 9.   Agents, other than heparin, that have been (or could be) incorporated into biomaterials to actively control thrombogenicity. Hirudin and PPACK are thrombin inhibitors. Prostacyclin (PGI_2) is a natural inhibitor of platelet function while DMP 728 is a small-molecule GPIIb/IIIa antagonist. Thrombomodulin inhibits thrombin by activating protein C.

Ionically Bound Heparin and Controlled Release Systems

If heparin is bound ionically to the surface then heparin will be slowly released over time as a result of exchange with the ionic components of blood. A similar effect is obtained if heparin is dispersed within a hydrophobic polymer. Selected techniques to produce materials that release heparin at biologically significant rates have been discussed in depth by a number of investigators (Kim and Jacobs, 1996; Sefton *et al.*, 1987; Amiji and Park, 1993). Ionic approaches involve binding

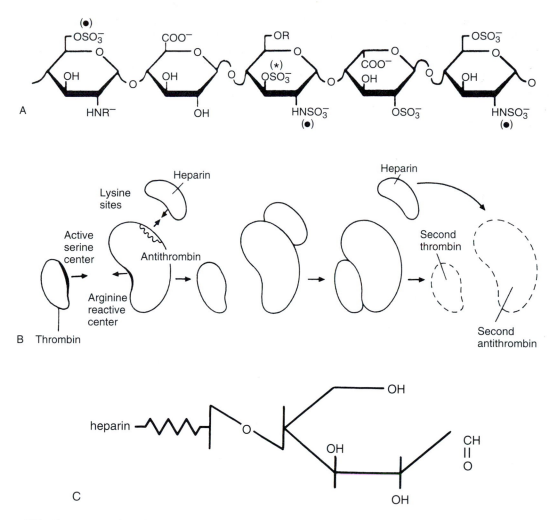

FIG. 10. (A) Antithrombin III–binding pentasaccharide of heparin. The pentasaccharide sequence is composed of three glucosamine (units 1, 3, and 5), one glucuronic acid (unit 2), and one iduronic acid (unit 4) units. Structural variants are indicated by –R' (–H or –SO$_3^-$) or –R'' (–COCH$_3$ or –SO$_3^-$). The 2-O-sulfate groups (asterisk), a marker component for the anti-thrombin-binding region, and sulfate groups indicated by (●) are essential for high-affinity binding to antithrombin. (B) The mechanism of action of heparin and antithrombin III. (C) The reducing end of heparin formed by treatment with nitrous acid. (A from Fiore and Deykin, 1995; B from Bauer and Rosenberg, 1995, with permission.)

the highly negatively charged heparin onto a cationic surface through ionic binding. The limitation is that the leaching of heparin will eventually leave the surface unprotected. Certain nonionic approaches are also characterized by high release rates (due to relatively unstable bonds), so that determining heparin release rate (in addition to amount bound) is a primary means of characterizing these surfaces. Limitations of the use of dispersed systems include the loading capacity of heparin, which prevents long-term usage, and the heterogeneity of heparin, leading to the early release of lower molecular weight chains.

The thromboresistance of heparinized materials based on controlled release appears to be due to a microenvironment of heparin in solution at the blood material interface. The relationship between release rate (N, g/cm^{2}s) and surface concentration (C_s, g/cm^3), for heparin or any other agent released from the inside wall of a tube of radius r_0 is given by (Basmadjian and Sefton, 1983):

$$\frac{C_s}{Nr_0/D} = 1.22 \left[\frac{x/r_0}{\text{ReSc}} \right]^{1/3} \quad (4)$$

where D = diffusivity (cm^2/sec), r_o = tube radius, x = axial position, Re = Reynolds number = $2r_0 v \rho/\mu$, Sc = Schmidt number = $\mu/\rho D$, v = average velocity, ρ = density, μ = viscosity. The diffusivity of heparin is ~7.5 × 10^{-7} cm^2/sec and the critical C_s (minimum therapeutic level) for heparin is ~0.5 μg/ml.

Use of tridodecyl methylammonium chloride (TDMAC, Lehninger *et al.*, 1966), a lipophilic hydrocarbon, eliminated the need for the graphite coating that was part of the original graphite–benzalkonium–heparin (GBH) method (Gott *et al.*, 1963). TDMAC and the many other quaternary ammonium compounds enabled heparin to be ionically bound to a wide range of biomaterials. Unfortunately, such compounds are surfactants with potentially toxic consequences and are known to be eluted from the surfaces within a week.

Quaternizable amino groups have also been incorporated directly into polymers to improve the stablility of the ionically bound heparin. Tanzawa *et al.* (1973) synthesized a graft copolymer with a dimethylaminoethyl group by UV-initiated copolymerization of *N,N*-dimethylaminoethyl methacrylate (DMAEM) and methoxypolyethylene glycol methacrylate (MPEG). Use of this technique (Angiocath, Toray Industries) to coat polyurethane catheters indicated that a minimal heparin elution rate of 0.04 μg/cm^2 min was needed to render the catheter thrombus free *in vivo* (Idezuki *et al.*, 1975). This value is consistent with Eq. 4.

A commercial procedure (Baxter Bentley Healthcare Systems, Irvine, CA) to ionically bind heparin (Duraflo II) has been used to coat cardiopulmonary bypass circuits and other devices. While heparin coating of extracorporeal circuits is designed to reduce surface thrombus formation, during clinical trials its effect on complement, contact activation, and inflammation has been unclear and often contradictory (Wildevuur *et al.*, 1997; Fosse *et al.*, 1997).

Covalently Bound Heparin

To impart a degree of activity longer than that possible with ionic linkages, heparin has been covalently immobilized to material surfaces. It is now recognized that the conformation of the attached heparin and the point of attachment (end

point versus multipoint) are critical factors determining the catalytic efficiency of the immobilized heparin. Lindhout *et al.* (1995) studied the antithrombin activity of surface-bound heparin under flow conditions. They demonstrated that the rate of thrombin inactivation of the antithrombin-heparin surface equals the maximal rate of transport of thrombin toward the surface when the surface coverage of antithrombin exceeds 10 pmol/cm^2, thus indicating that a higher intrinsic catalytic efficiency of a surface does not necessarily result in a higher antithrombin activity.

Many of the coupling methods listed in Chapter 2.16 have been used for heparin; see also Fig. 3 in Chapter 2.16. For example, Heyman *et al.* (1985) attached 1-ethyl-3-(3-dimethylaminopropyl)carbodiimide-activated heparin covalently onto chemically modified poly(ether urethane) catheters through a diaminoalkane spacer. The immobilized heparin retained its ability to bind and inactivate thrombin and factor Xa. Unfortunately, carbodiimides (Fig. 11) are recognized as less than ideal activating agents for immobilizing heparin since the acidic conditions used can result in a loss of anticoagulant activity.

Larm *et al.* (1983) developed a technique by which heparin can be covalently end-point-attached (commercialized as the Carmeda Bioactive Surface, Stockholm, Sweden) to the surface of plastics as well as glass and steel. Heparin is first partially depolymerized by deaminative cleavage using nitrous acid to produce heparin fragments terminating in an aldehyde group (Fig. 10C). The heparin is then covalently linked to the primary amino groups of polyethyleneimine (PEI) (Fig. 12). This technique results in a highly stable, low-thrombogenicity coating that has been demonstrated to retain its efficacy *in vivo* for 4 months in dogs (Arnander *et al.*, 1987) and during patient treatment with an artificial lung (Bindslev *et al.*, 1987).

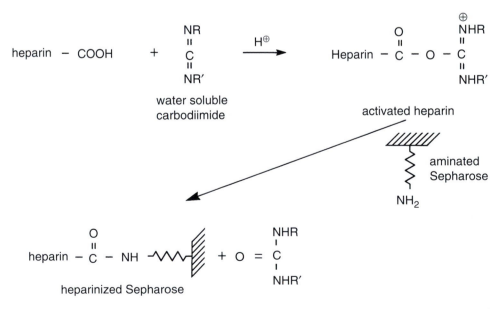

FIG. 11. One step carbodiimide activation of carboxyl groups of heparin for binding to aminated Sepharose. For EDC, R = (CH$_2$)$_3$N(CH$_3$)$_2$. For a carboxylated substrate (e.g., hydrolyzed polymethyl acrylate) its carboxyl groups may be activated by carbodiimide in a separate step for subsequent reaction with the free amine groups of heparin.

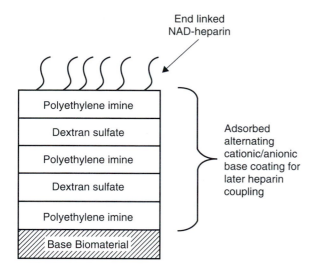

FIG. 12. CARMEDA method for immobilizing nitrous acid treated heparin (end-linked heparin) to dextran sulfate and polyethyleneimine-treated surface.

There was also a beneficial effect on *in vivo* bacterial colonization of treated polyurethane central venous catheters (Appelgren *et al.*, 1996).

Heparin has also been covalently immobilized via its terminal serine at the end of the protein–carbohydrate linkage region to poly(vinyl alcohol) hydrogel. The heparin is bound during the Lewis acid–catalyzed cross-linking of poly(vinyl alcohol) chains with glutaraldehyde. The heparinized poly(vinyl alcohol) hydrogel possessed significant anticoagulant activity although the platelet incompatibility of the hydrogel substrate led to significant platelet consumption in an *ex vivo* canine shunt model (Cholakis *et al.*, 1989).

Park *et al.* (1988) have covalently bound heparin to a polyurethane using PEO spacer groups. The increasingly mobile nature of the longer hydrophilic spacer chains were considered to have increased the observed bioactivity of immobilized heparin by providing a more bulk-like environment for the heparin. An intravascular oxygenator and carbon dioxide removal device (IVOX; CardioPulmonics, Salt Lake City, UT) also uses covalent heparin bound to a PEG spacer. The PEG is grafted onto a plasma-polymerized polysiloxane on a base material after plasma amination of the surface.

Thrombin Inhibition without Heparin

Recognizing the importance of inhibiting material-induced thrombin generation, agents other than heparin have been used to inhibit thrombin and thereby lower the thrombogenicity of the material. For example, hirudin, a potent thrombin inhibitor, has been covalently bound to biomaterials. Compared with heparin, hirudin is advantageous because it has no influence on platelet function, has no immune-mediated platelet-activating activity, and, most importantly, does not require the presence of endogenous cofactors such as antithrombin III. Thus the polypeptide hirudin is thought able to block clot-bound thrombin that is inaccessible to inhibition by a heparin/antithrombin III complex. On the other hand, heparin functions as a catalyst. That is immobilized heparin, if not blocked by adsorbed protein, can continually accelerate the inactivation of thrombin by antithombin III. It is unclear how long a covalently immobilized hirudin surface would remain effective and even whether an immobilized hirudin could gain access to the clot-bound thrombin. Nevertheless Ito *et al.* (1991) used heterobifunctional cross-linking reagents to derivatize hirudin to form covalent cross-links between hirudin and albumin, producing active conjugates for immobilization to surfaces (Ito *et al.*, 1991), and others have immobilized hirudin to a polylactide–glycolide copolymer (Seifert *et al.*, 1997). Another group has exploited controlled release of hirudin for this purpose (Kim *et al.*, 1998). Small-molecule thrombin inhibitors such as PPACK and others under development may prove useful as alternatives.

Another approach involves immobilizing thrombomodulin, an endothelial cell–associated protein that inhibits thrombin by activating protein C. Using a small-scale dialyzer it was demonstrated that immobilized human thrombomodulin (on cellulose) still had coenzyme activity for activation of protein C and anticoagulant activity (Kishida *et al.*, 1995). An amino-terminated silane was used to couple thrombomodulin to glass, and the authors reported both anticoagulant activity and reduced platelet adhesion (Han *et al.*, 2001). Further work is needed to appreciate the potential of such surfaces.

Immobilization of Anti-Platelet Agents

There has been considerable interest in incorporating anti-platelet agents into materials to lower material thrombogenicity, especially given the inevitable presence of platelets on biomaterials. The fact that the natural vessel wall helps prevent thrombus formation through a potent yet unstable anti-platelet agent, prostacyclin (to elevate platelet cyclic AMP levels, Fig. 9), supports such an approach. Efforts to maintain the biological activity of the agent during immobilization coupled with concern for the ability of bulk platelets to be affected by the immobilized agent has limited the enthusiasm for this approach. A simpler approach for a number of these agents would be to incorporate them into materials for release. Unfortunately, the inability to load sufficient drug into materials for release will limit the materials' life span. It remains to be seen whether suppression of initial blood/material interactions is sufficient for sustained biocompatibility.

Despite the challenges, anti-platelet agents have been incorporated into surfaces. For example, Ebert *et al.* (1982) immobilized prostaglandin $F_{2\alpha}$, which was subsequently converted to the unstable prostacyclin using a diaminododecane spacer arm. The benefits of spacer arms are described elsewhere (Chapter 2.16). A related approach is to add an anti-platelet agent such as PGE_1, in addition to heparin, into a polymer. A PGE_1–heparin compound was synthesized and incorporated into a polyurethane (Jacobs *et al.*, 1985). Also, prostacylin has been incorporated into polymer matrices for controlled release (McRea and Kim, 1978). Aspirin, capable of inhibiting the generation of the platelet activator thromboxane A2, has also been incorporated into poly(vinyl alcohol) membranes used in hemodialysis (Paul and Sharma, and 1997), and a dipyrimadole (Persantin) derivative was photoimmobilized on a polyurethane (Aldenhoff et al., 1997). A novel approach was to exploit endogenous S-nitrosoproteins in plasma to produce NO from immobilized cysteine to minimize platelet adhesion on a polyurethane and a poly(ethylene terepthalate) (Duan and Lewis, 2002).

An exciting new generation of anti-platelet agents based on inhibiting fibrinogen binding to activated platelet GPIIb/IIIa receptors offers a fresh approach to lowering the thrombogenicity of surfaces via drug release. These agents, from blocking monoclonal antibodies to small peptides and compounds (e.g., DMP728, Fig. 9), not only block platelet aggregation but will also likely block platelet adhesion to artificial surfaces. Some have already been incorporated into stents for release (see Section IV).

Immobilization of Fibrinolytic Agents

Some investigators have sought to promote fibrinolysis on artificial surfaces by promoting the surface generation of plasmin. Clot lysis is achieved by the action of plasmin on fibrin, after the former's formation from plasminogen. Whether the action of such surfaces is "too little too late" is unknown. Sugitachi and Takagi (1978) immobilized urokinase, a fibrinolytic enzyme that acts on plasminogen, on various materials. The long-term enzymatic stability of such agents (typically large proteins) after immobilization is of concern. A slightly different approach, recently pursued, is to immobilize plasminogen and then convert it to plasmin so as to impart fibrinolytic activity to a surface (Marconi et al., 1996). A disadvantage of utilizing streptokinase and urokinase to cleave plasminogen to plasmin is that these agents activate both circulating and fibrin-bound plasminogen. This contrast with the action of tissue plasminogen activator (tPA), an endogenous serine protease that converts only fibrin-bound plasminogen to plasmin.

A nonpharmacological approach to developing fibrinolytic surfaces for blood-contacting applications was the preparation of lysine-derivatized polyurethane surfaces (Woodhouse and Brash, 1992). The expectation is that these surfaces exhibit fibrinolytic activity because the lysine residues promote the selective adsorption of plasminogen from plasma through the lysine-binding sites in the plasminogen molecule. In the presence of tissue plasminogen activator, adsorbed plasminogen was cleaved twice as fast on the lysinized material as opposed to the precursor sulfonated material (Woodhouse et al., 1996).

USE OF ENDOTHELIAL CELLS AND RGD PEPTIDES

It is intuitive to believe that the ideal nonthrombogenic surface for vascular grafts/artificial hearts will consist of an intact luminal endothelial cell layer. Herring et al. (1984) seeded Dacron and e-PTFE grafts with endothelial cells in a preliminary clotting step with blood and endothelial cells.

Encouraging results have always been obtained in dogs although there are a number of issues that exist regarding the use of endothelial cells for vascular grafting. A full discussion of this subject is presented elsewhere (Zilla et al., 1999). Some attempt has been made to modify materials to give them a higher affinity for endothelial cell attachment. Sipehia (1990) used an anhydrous ammonia gaseous plasma technique to modify polystyrene and PTFE surfaces. The modified surfaces contained amide and amine groups and it was concluded that they facilitated the attachment of endothelial cells. Another approach is to precoat them with substances that promote endothelial cell adhesion (Bos et al., 1999). Van Wachem et al. (1987) noted that surfaces precoated with fibrinogen or fibronectin resulted in higher numbers of adhering cells. The most elegant approach has been that of Hubbell (Massia and Hubbell, 1990), who has utilized immobilized RGD peptides (see Chapter 2.16) to encourage endothelial cell attachment. This approach has been adopted by many others, although there is a concern that bare spots would lead to enhanced platelet and thrombus deposition.

STRATEGIES TO LOWER THE THROMBOGENICITY OF METALS

The recent deployment of stents (Chapter 7.3) for the purpose of maintaining patency in coronary arteries after angioplasty introduces metallic surfaces to the circulatory system. The minimal surface area of the device coupled with the concern over smooth-muscle cell proliferation led many investigators to hypothesize that thrombus formation would not be of great concern. On the other hand, anti-platelet therapy has led to reduced restenosis rates and lower platelet adhesion values and for a time there was great interest in heparin-coated stents. This early work has now been supplanted by rapamycin-eluting stents, which have the dramatic effect of eliminating in-stent restenosis for at least 6 months after deployment (Morice et al., 2002). Rapamycin (Sirolimus) is a macrolide antibiotic that was previously used as an immunosupressant The drug was blended in a proprietary mixture of nonerodible polymers and coated onto the surface of a stainless steel balloon-expandable stent. The use of this and other therapeutic agents (e.g., taxol) in polymer-coated stents is rapidly becoming standard in interventional cardiology.

Prior to the success with rapamycin, other strategies to lower stent thrombogenicity had been reported. For example, tantalum stents were polymer coated with poly(ether urethane) and parylene to reduce platelet adhesion density (Fontaine et al., 1996) and N-vinylpyrrolidone (NVP) has been gamma radiation grafted onto plasma-treated stainless steel stents (Seeger et al., 1995). Hyaluronic acid coating also reduced platelet thrombus formation on stainless steel stents and tubes in a primate thrombosis model (Verheye et al., 2000). In another study, metal/polymer composite stents loaded (40% by weight, >90% elution in 89 hours) with a potent anti-platelet agent (GPIIb/IIIa antagonist) were demonstrated to reduce, by almost a factor of 2, platelet adhesion in dogs 2 hours after stent deployment (Santos et al., 1998).

SUMMARY

It is an axiom that the interactions between materials and blood are complex. Hence it is no surprise that developing low-thrombogenicity materials (let alone materials with zero thrombogenicity) is very difficult. Medical device manufacturers relying on elegant device designs and systemic pharmacological agents have done wonders with existing materials. Adverse effects are minimized and existing devices, if not risk free, provide sufficient benefit to outweigh the risks. The focus of research in biomaterials is to make better materials that have fewer risks and greater benefits. Stents that "actively" prevent restenosis are a great example of how modifying a material can have a dramatic clinical effect. Using a less thrombogenic material so that catheters that did not occlude due to thrombosis would be highly desirable as well.

Inert materials, such as those with immobilized PEG, can resist protein and platelet deposition, but these may only be at best surrogate markers for thromboembolic phenomena. On the other hand, incorporating anticoagulants such as heparin can be an effective means of reducing thrombin production rates below critical values (e.g., $k_p < 10^{-4}$ cm/sec), but this may not be sufficient to prevent platelet activation and consumption. Many strategies for lowering thrombogenicity have been identified, and they all show a beneficial effect in at least one assay of thrombogenicity. However, few if any have made the transition from a one parameter benefit to multiple benefits or from *in vitro* to *in vivo*. These issues are discussed more fully in Chapter 4.6.

Which approach will ultimately be successful is impossible to predict. Certainly there is much activity in biomembrane mimicry and immobilizing PEG. Creating stable self-assembled monolayers may enable more sophisticated designer surfaces. There are many new anticoagulants and antithrombotics under development and few have yet to be incorporated into material surfaces. Combining approaches to address thrombin production and platelet activation simultaneously may lead to new opportunities. Finally, as new hypotheses are developed to understand cardiovascular material failure, new approaches will be identified for inhibiting particular pathways of failure. Perhaps the failure to produce the ideal nonthrombogenic material, despite 30 years of research, has merely reflected our limited understanding of blood–materials interaction. Perhaps, the right strategy for producing a nonthrombogenic material will have little to do with controlling platelets or thrombin, but will be directed toward leukocytes or complement (Gemmell, 1998; Wetterö *et al.*, 2003). Further research throughout the world is expected to improve the blood interactions of materials used in medicine.

Bibliography

Aldenhoff, Y. B., Blezer, R., Lindhout, T., and Koole, L. H. (1997). Photo-immobilization of dipyridamole (Persantin) at the surface of polyurethane biomaterials: reduction of in-vitro thrombogenicity. *Biomaterials* **18**: 167–172.

Amiji, M., and Park, K. (1993). Surface modification of polymeric biomaterials with poly(ethylene oxide), albumin, and heparin for reduced thrombogenicity. *J. Biomaterials Sci., Polymer Ed.* **4**: 217–234.

Andrade, J. D., and Hlady, V. (1986). Protein adsorption and materials biocompatibility: a tutorial review and suggested hypotheses. *Adv. Polymer Sci.* **79**: 1–63.

Andrade, J. D., Lee, H. B., John, M. S., Kim, S. W., and Hibbs, J. B., Jr. (1973). Water as a biomaterial. *Trans. Am. Soc. Artif. Internal Organs* **19**: 1.

Appelgren, P., Ransjo, U., Bindslev, L., Espersen, F., and Larm, O. (1996). Surface heparinization of central venous catheters reduces microbial colonization *in vitro* and *in vivo*: results from a prospective, randomized trial. *Crit. Care Med.* **24**: 1482–1489.

Arnander, C., Bagger-Sjoback, D., Frebelius, S., Larsson, R., and Swedenborg, J. (1987). Long-term stability *in vivo* of a thromboresistant heparinized surface. *Biomaterials* **8**: 496–499.

Baier, R. E., Gott, V. L., and Furuse, A. (1970). Surface chemical evaluation of thromboresistant materials before and after venous implantation. *Trans. Am. Soc. Artificial Organs* **16**: 50–57.

Baker, L. C., Davis, W. C., Autieri, J., Watach, M. J., Yamazaki, K., Litwak, P., and Wagner, W. R. (1998). Flow cytometric assays to detect platelet activation and aggregation in device-implanted calves. *J. Biomed. Mater. Res.* **41**: 312–321.

Basmadjian, D. (1990). The effect of flow and mass transport in thrombogenesis. *Ann. Biomed. Eng.* **18**: 685–709.

Basmadjian, D., and Sefton, M. V. (1983). Relationship between release rate and surface concentration for heparinized materials. *J. Biomed. Mater. Res.* **17**: 509–518.

Bauer, K. A., and Rosenberg, R. D. (1995). Control of coagulation reactions. in *Williams Hematology*, 5th ed. (E. Beutler, M. A. Lichtman, B. S. Collen, and T. J. Kipps, eds.) McGraw-Hill, New York, p. 1241.

Bindslev, L., Eklund, J., Norlander, O. Swedenborg, J., Olsson, P., Nilsson, E., Larm, O., Gouda, I., Malmberg, A., and Scholander, E. (1987). Treatment of acute respiratory failure by extracorporeal carbon dioxide elimination performed with a surface heparinized artificial lung. *Anesthesiology* **67**: 117–120.

Bohnert, J. L., Fowler, B. C., Horbett, T. A., and Hoffman, A. S. (1990). Plasma gas discharge deposited fluorocarbon polymers exhibit reduced elutability of adsorbed albumin and fibrinogen. *J. Biomater. Sci., Polymer Ed.* **1**: 279–297.

Bos, G. W., Scharenborg, N. M., Poot, A. A. Engbers, G. H. Beugeling, T., van Aken, W. G., and Feijen, J. (1999). Blood compatibility of surfaces with immobilized albumin–heparin conjugate and effect of endothelial cell seeding on platelet adhesion. *J. Biomed. Mater. Res.* **47**: 279–291.

Campbell, E. J., O'Byrne, V., Stratford, P. W., Quirk, I., Vick, T. A., Wiles, M. C., and Yianni, Y. P. (1994). Biocompatible surfaces using methacryloylphosphorylcholine laurylmethacrylate copolymer. *ASAIO J.* **40**: M853–M857.

Chaikof, E. L., Coleman, J. E., Ramberg, K. Connolly, R. J., Merrill, E. W., and Callow, A. D. (1989). Development and evaluation of a new polymeric material for small calibre vascular prostheses. *J. Surg. Res.* **47**: 193–199.

Chaikof, E. L., Merrill, E. W., Callow, A. D., Connolly, R. J., Verdon, S. L., and Ramberg, K. (1992). PEO enhancement of platelet deposition, fibrinogen deposition and complement activation. *J. Biomed. Mater. Res.* **26**: 1163–1168.

Chen, C., Hanson, S. R., and Lumsden, A. B. (1995). Boundary layer infusion of heparin prevents thrombosis and reduces neointimal hyperplasia in venous polytetrafluoroethylene grafts without systemic anticoagulation. *J. Vasc. Surg.* **22**: 237–245.

Chen, C., Ofenloch, J. C., Yianni, Y. P., Hanson, S. R., and Lumsden, A. B. (1998). Phosphorylcholine coating of ePTFE reduces platelet deposition and neointimal hyperplasia in arteriovenous grafts. *J. Surg. Res.* **77**: 119–125.

Cholakis, C. H., Zingg, W., and Sefton, M. V. (1989). Effect of heparin-PVA hydrogel on platelets in a chronic arterial venous shunt. *J. Biomed. Mater. Res.* **23**: 417–441.

DeGennes, P. G. (1980). Conformation of polymers attached to an interface. *Macromolecules* **13**: 1069–1075.

Desai, N. P., and Hubbell, J. A. (1991a). Biological responses to polyethylene oxide modified polyethylene terephthalate surfaces. *J. Biomed. Mater. Res.* **25**: 829–843.

Desai, N. P., and Hubbell, J. A. (1991b). Solution technique to incorporate polyethylene oxide and other water-soluble polymers into surfaces of polymeric biomaterials. *Biomaterials* **12**: 144–153.

Duan, X., and Lewis R. S. (2002). Improved haemocompatibility of cysteine-modified polymers via endogenous nitric oxide. *Biomaterials* **23**: 1197–1203.

Durrani, A. A., Hayward, J. A., and Chapman, D. (1986). Biomembranes as models for polymer surfaces. II. The synthesis of reactive species for covalent coupling of phosphorylcholine to polymer surfaces. *Biomaterials* **7**: 121–125.

Ebert, C. D., Lee, E. S., and Kim, S. W. (1982). The antiplatelet activity of immobilized prostacyclin. *J. Biomed. Mater. Res.* **16**: 629–638.

Engbers, G. H., and Feijen, J. (1991). Current techniques to improve the blood compatibility of biomaterial surfaces. *Int. J. Artif. Organs* **14**: 199–215.

Fiore, L., and Deykin, D. (1995). Anticoagulant therapy. in *Williams Hematology*, 5th ed. (E. Beutler, M. A. Lichtman, B. S. Collen, and T. J. Kipps, eds.) McGraw-Hill, New York, p. 1563.

Fontaine, A. B., Koelling, K., Passos, S. A., Cearlock, J., Hoffman, R., and Spigos, D. G. (1996). Polymeric surface modification of tantalum stents. *J. Endovas. Surg.* **3**: 276–283.

Fosse, E., Thelin, S., Svennevig, J. L., Jansen, P., Mollnes, T. E., Hack, E., Venge, P., Moen, O., Brockmeier, V., Dregelid, E., Halden, E., Hagman, L., Videm, V., Pedersen, T., and Mohr, B. (1997). Duraflo II coating of cardiopulmonary bypass circuits reduces complement activation but does not affect the release of granulocyte enzymes: a European multicentre study. *Euro. J. CardioThorac. Surg.* **11**: 320–327.

Fougnot, C., Dupiller, M. P., and Jozefowicz, M. (1983). Anticoagulant activity of amino acid modified polystyrene resins: influence of the carboxylic acid function. *Biomaterials* **4**: 101–104.

Francois, P., Vaudaux, P., Nurdin, N., Mathieu, H. J., Descouts, P., and Lew, D. P. (1996). Physical and biological effects of a surface coating procedure on polyurethane catheters. *Biomaterials* **17**: 667–678.

Frautschi, J. R., Eberhart, R. C., and Hubbell, J. A. (1995). Alkylated cellulosic membranes with enhanced albumin affinity: influence of competing proteins. *J. Biomater. Sci., Polymer Ed.* **7**: 563–575.

Gemmell, C. H. (1998). Platelet adhesion onto artificial surfaces: inhibition by benzamidine, pentamidine and pyridoxal-5-phosphate as demonstrated by flow cytometric quantification of platelet adhesion to microspheres. *J. Lab. Clin. Med.* **131**: 84–92.

Gemmell, C. H., Ramirez, S. M., Yeo, E. L., and Sefton, M. V. (1995). Platelet activation in whole blood by artificial surfaces: identification of platelet-derived microparticles and activated platelet binding to leukocytes as material-induced activation events. *J. Lab. Clin. Med.* **125**: 276–287.

Gott, V. L., Whiffen, J. D., and Dutton, R. C. (1963). Heparin bonding on colloidal graphite surfaces. *Science* **142**: 1297.

Grasel, T. G., and Cooper, S. L. (1989). Properties and biological interactions of polyurethane anionomers: effect of sulfonate incoporation. *J. Biomed. Mater. Res.* **23**: 311–338.

Gu, Y. J., Boonstra, P. W., Rijnsburger, A. A., Haan, J., and van Oeveren, W. (1998). Cardiopulmonary bypass circuit treated with surface-modifying additives: a clinical evaluation of blood compatibility. *Ann. Thorac. Surg.* **65**: 1342–1347.

Han, H. S., Yang, S. L., Yeh, H. Y., Lin, J. C., Wu, H. L., and Shi, G. Y. (2001). Studies of a novel human thrombomodulin immobilized substrate: surface characterization and anticoagulation activity evaluation. *J. Biomater. Sci., Polymer Ed.* **12**: 1075–1089.

Hanson, S. R., Harker, L. A., Ratner, B. D., and Hoffman, A. S. (1980). In vivo evaluation of artifcial surfaces with a nonhuman primate model of arterial thrombosis. *J. Lab. Clin. Med.* **95**: 289–304.

Herring, M. B., Baughman, S., Glover, J., Kesler, K., Jesseph, J., Dilley, R., Evan, A., and Gardner, A. (1984). Endothelial seeding of Dacron and polytetrafluorethylene grafts: the cellular events of healing. *Surgery* **96**: 745–754.

Heyman, P. W., Cho, C. S., McRea, J. C., Olsen, D. B., and Kim, S. W. (1985). Heparinized polyurethanes: *in vitro* and *in vivo* studies. *J. Biomed. Mater. Res.* **19**: 419–436.

Hoffman, A. S., Schmer, G., Harris, C., and Kraft, W. G. (1972). Covalent binding of biomolecules to radiation-grafted hydrogels on inert polymer surfaces. *Trans. Am. Soc. Artif. Internal Organs* **18**: 10–17.

Idezuki, Y., Watanabe, H., Hagiwara, M. Kanasugi, K., and Mori, Y. (1975). Mechanism of antithrombogenicity of a new heparinized hydrophilic polymer: chronic *in vivo* and clinical application. *Trans. Am. Soc. Artif. Internal Organs* **21**: 436–448.

Ishihara, K., Hanyuda, H., and Nakabayashi, N. (1995). Synthesis of phospholipid polymers having a urethane bond in the side chain as coating material on segmented polyurethane and their platelet adhesion-resistant properties. *Biomaterials* **16**: 873–879.

Ishihara, K., Tanaka, S., Furukawa, N., Kurita, K., and Nakabayashi, N. (1996). Improved blood compatibility of segmented polyrethanes by polymeric additives having phospholipid polar groups. I. Molecular design of polymeric additves and their functions. *J. Biomed. Mater. Res.* **32**: 391–399.

Ishihara, K., Nomura, H., Mihara, T., Kurita, K., Iwasaki, Y., and Nakabayashi, N. (1998). Why do phospholipid polymers reduce protein adsorption? *J. Biomed. Mater. Res.* **39**: 323–330.

Ishihara, K., Fukumoto, K., Iwasaki, Y., and Nakabayashi, N. (1999). Modification of polysulfone with phospholipid polymer for improvement of the blood compatibility. Part 1. Surface characterization. *Biomaterials* **201**: 545–551.

Ito, R. K., Phaneuf, M. D., and LoGerfo, F. W. (1991). Thrombin inhibition by covalently bound hirudin. *Blood Coag. Fibrinol.* **2**: 77–81.

Jacobs, H., Okano, T., Lin, J. Y., and Kim, S. W. (1985). PGE1-heparin conjugate releasing polymers. *J. Controlled Release* **2**: 313–319.

Jeon, S. I., and Andrade, J. D. (1991). Protein–surface interactions in the presence of polyethylene oxide. II. Effect of protein size. *J. Colloid Interf. Sci.* **142**: 159–166.

Jeon, S. I., Lee, J. H., Andrade, J. D., and DeGennes, P. G. (1991). Protein–surface interactions in the presence of polyethylene oxide. I. Simplified theory. *J. Colloid Interf. Sci.* **142**: 149–158.

Kashiwagi, T., Ito, Y., and Imanishi, Y. (1993). Synthesis of nonthrombogenicity of fluorolkyl polyetherurethanes *J. Biomater. Sci., Polymer Ed.* **5**: 157–166.

Kiaei, D., Hoffman, A. S., Ratner, B. D., and Horbett, T. A. (1988). Interaction of blood with gas discharge treated vascular grafts. *J. Appl. Polymer Sci. Appl. Polymer Symp.* **42**: 269–283.

Kim, D. D., Takeno, M. M., Ratner, B. D., and Horbett, T. A. (1998). Glow discharge plasma deposition (GDPD) technique for the local controlled delivery of hirudin from biomaterials. *Pharmaceut. Res.* **15**: 783–786.

Kim, K., Kim, C., and Byun, Y. (2000). Preparation of a PEG-grafted phospholipid Langmuir–Blodgett monolayer for blood-compatible material. *J. Biomed. Mater. Res.* **52**: 836–840.

Kim, S. W., and Jacobs, H. (1996). Design of nonthrombogenic polymer surfaces for blood-contacting medical devices. *Blood Purif.* **14:** 357–372.

Kishida, A., Akatsuka, Y., Yanagi, M., Aikou, T., Maruyama, I., and Akashi, M. (1995). *In vivo* and *ex vivo* evaluation of the antithrombogenicity of human thrombomodulin immobilized biomaterials. *ASAIO J.* **41:** M369–M374.

Kottke-Marchant, K., Anderson, J. M., Umemura, Y., and Marchant, R. E. (1989). Effect of albumin coating on the *in vitro* blood compatibility of Dacron arterial prostheses. *Biomaterials* **10:** 147–155.

Larm, O., Larsson, R., and Olsson, P. (1983). A new nonthrombogenic surface prepared by selective covalent binding of heparin via a modified reducing terminal residue. *Biomater. Med. Dev. Artif. Organs* **11:** 161–173.

Leininger, R. I., Cooper, C. W., Falb, R. D. and Grode, G. A. (1966). Nonthrombogenic plastic surfaces. *Science* **152:** 1625–1626.

Lewis, A. L., Hughes, P. D., Kirkwood, L. C., Leppard, S. W., Redman, R. P., Tolhurst, L. A., and Stratford, P. W. (2000). Synthesis and characterisation of phosphorylcholine-based polymers useful for coating blood filtration devices. *Biomaterials* **21:** 1847–1859.

Lindhout, T., Blezer, R., Schoen, P., Willems, G. M., Fouache, B., Verhoeven, M., Hendriks, M., Cahalan, L., and Cahalan, P. T. (1995). Antithrombin activity of surface-bound heparin studied under flow conditions. *J. Biomed. Mater. Res.* **29:** 1255–1266.

Liu, S. Q., Ito, Y., and Imanishi, Y. (1989). Synthesis and nonthrombogencity of polyurethanes with poly(oxyethylene) side chains in soft segment regions. *J. Biomater. Sci., Polymer Ed.* **1:** 111–122.

Llanos, G. R., and Sefton, M. V. (1993a). Does polyethylene oxide possess a low thrombogenicity? *J. Biomater. Sci., Polymer Ed.* **4:** 381–400.

Llanos, G. R., and Sefton, M. V. (1993b). Immobilization of poly(ethylene glycol) onto a poly(vinyl alcohol) hydrogel: 2. Evaluation of thrombogenicity. *J. Biomed. Mater. Res.* **27:** 1383–1391.

Marconi, W., Piozzi, A., and Romoli, D. (1996). Preparation and evaluation of polyurethane surfaces containing plasminogen. *J. Biomater. Sci., Polymer Ed.* **8:** 237–249.

Massia, S. P., and Hubbell, J. A. (1990). Covalently attached GRGD on polymer surfaces promotes biospecific adhesion of mammalian cells. *Ann. N.Y. Acad. Sci.* **589:** 261–270.

Matsuda, T., and Inoue, K. (1990). New photoreactive surface modification technology for fabricated devices. *Trans. Am. Soc. Artif. Internal Organs* **36:** M161–M164.

McPherson, T. B., Shim, H. S., and Park, K. (1997). Grafting of PEO to glass, nitinol, and pyrolytic carbon surfaces by gamma irradiation. *J. Biomed. Mater. Res.* **38:** 289–302.

McRea, J. C., and Kim, S. W. (1978). Characterization of controlled release of prostaglandin from polymer matrices for thrombus prevention. *Trans. Am. Soc. Artif. Internal Organs* **24:** 746–750.

Merrill, E. W., and Salzman, E. W. (1983). Polyethylene oxide as a biomaterial. *Am. Soc. Artif. Internal Organs J.* **6:** 60–64.

Merrill, E. W., Salzman, E. W., Wan, S., Mahmud, N., Kushner, L., Lindon, J. N., and Curme, L. (1982). *Trans. Am. Soc. Artif. Internal Organs* **28:** 482–487.

Morice, M. C., Serruys, P. W., Sousa, J. E., Fajadet, J., Ban Hayashi, E., Perin, M., Colombo, A., Schuler, G., Barragan, P., Guagliumi, G., Molnar, F., and Falotico, R., RAVEL Study Group (2002). Randomized study with the Sirolimus-coated Bx velocity balloon-expandable stent in the treatment of patients with de novo native coronary artery lesions. A randomized comparison of a sirolimus-eluting stent with a standard stent for coronary revascularization. *N. Engl. J. Med.* **346:** 1773–1780.

Munro, M. S., Quattrone, A. J., Ellsworth, S. R., Kulkari, P., and Eberhart, R. C. (1981). Alkyl substituted polymers with enhanced albumin affinity. *Trans. Am. Soc. Artif. Internal Organs* **27:** 499–503.

Nagaoka, S., Mori, Y., Takiuchi, H., Yokota, K., Tanzawa, H., and Nishiumi, S. (1984). Interaction between blood components and hydrogels with poly(oxyethylene) chains. in *Polymers as Biomaterials*, (S. Shalaby, A. S. Hoffman, B. D. Ratner, and T. A. Horbett, eds.) Plenum Press, New York, pp. 361–371.

Okkema, A. Z., Grasel, T. G., Zdrahala, R. J., Solomon, D. D., and Cooper, S. L. (1989). Bulk, surface, and blood-contacting properties of polyetherurethanes modified with polyethylene oxide. *J. Biomater. Sci., Polymer Ed.* **1:** 43–62.

Park, H. D., Lee, W. K., Ooya, T., Park, K. D., Kim, Y. H., and Yui, N. (2002). Anticoagulant activity of sulfonated polyrotaxanes as blood-compatible materials. *J. Biomed. Mater. Res.* **60:** 186–190.

Park, K. D., Okano, T., Nojiri, C., and Kim, S. W. (1988). Heparin immobilized onto segmented polyurethaneureas surfaces: effects of hydrophilic spacers. *J. Biomed. Mater. Res.* **22:** 977.

Paul, W., and Sharma, C. P. (1997). Acetylslicylic acid loaded poly(vinyl alcohol) hemodialysis membranes: effect of drug release on blood compatibility and permeability. *J. Biomater. Sci., Polymer Ed.* **8:** 755–764.

Ratner, B. D. (1995). Surface modification of polymers: chemical, biological and surface analytical challenges. *Biosensors Bioelectron.* **10:** 797–804.

Rollason, G., and Sefton, M. V. (1992). Measurement of the rate of thrombin production in human plasma in contact with different materials. *J. Biomed. Mater. Res.* **26:** 675–693.

Santos, R. M., Tanguay, J.-F., Crowley, J. J., Kruse, K. R., Sanders-Millare, D., Zidar, J. P., Phillips, H. R., Merhi, Y., Garcia-Cantu, E., Bonan, R., Cote, G., and Stack, R. S. (1998). Local Administration of L-703,081 using a composite polymeric stent reduces platelet deposition in canine coronary arteries. *Am. J. Cardiol.* **82:** 673–675.

Seeger, J. M., Ingegno, M. D., Bigatan, E., Klingman, N., Amery, D., Widenhouse, C., and Goldberg, E. P. (1995). Hydrophilic surface modification of metallic endoluminal stents. *J. Vasc. Surg.* **22:** 327–335.

Sefton, M. V., Cholakis, C. H., and Llanos, G. (1987). Preparation of nonthrombogenic materials by chemical modification. in *Blood Compatibility*, Vol. 1, D. F. Williams, ed. CRC Press, Boca Raton, FL, pp. 151–198.

Sefton, M. V., Sawyer, A., Gorbet, M., Black, J. P., Cheng, E., Gemmell, C., and Pottinger-Cooper, E. (2001). Does surface chemistry affect thrombogenicity of surface modified polymers? *J. Biomed. Mater. Res.* **55:** 447–459.

Seifert, B., Romaniuk, P., and Groth, T. (1997). Covalent immobilization of hirudin improves the haemocompatibility of polylactide-polyglycolide in vitro. *Biomaterials* **8:** 1495–1502.

Shen, M., Pan, Y. V., Wagner, M. S., Hauch, K. D., Castner, D. G., Ratner, B. D., and Horbett, T. A. (2001). Inhibition of monocyte adhesion and fibrinogen adsorption on glow discharge plasma deposited tetraethylene glycol dimethyl ether. *J. Biomater. Sci., Polymer Ed.* **12:** 961–978.

Silver, J. H., Lin, J. C., Lim, F., Tegoulia, V. A., Chaudhury, M. K., and Cooper, S. L. (1999). Surface properties and hemocompatibility of alkyl-siloxane monolayers supported on silicone rubber: effect of alkyl chain length and ionic functionality. *Biomaterials* **20:** 1533–1543.

Sipehia, R. (1990). The enhanced attachment and growth of endothelial cells on anhydrous ammonia gaseous plasma modified surfaces of polystyrene and poly(tetrafluoroethylene). *Biomater. Artif. Cells Artif. Organs* **18:** 437–446.

Sofia, S. J., and Merrill, E. W. (1998). Grafting of PEO to polymer surfaces using electron beam irradiation. *J. Biomed. Mater. Res.* **40**: 153–163.

Strzinar, I., and Sefton, M. V. (1992). Preparation and thrombogenicity of alkylated polyvinyl alcohol coated tubing. *J. Biomed. Mater. Res.* **26**: 577–592.

Sugitachi, A., and Takagi, K. (1978). Antithrombogenicity of immobilized urokinase—clinical application. *Int. J. Artif. Organs* **1**: 88–92.

Tang, Y. W., Santerre, J. P., Labow, R. S., and Taylor, D. G. (1996). Synthesis of surface-modifying macromolecules for use in segmented polyurethanes. *J. Appl. Polymer Sci.* **62**: 1133–1145.

Tanzawa, H., Mori, Y., Harumiya, N., Miyama, H., Hori, M., Ohshima, N., and Idezuki, Y. (1973). Preparation and evaluation of a new antithrombogenic heparinized hydrophillic polymer for use in cardiovascular systems. *Trans. ASAIO* **19**: 188.

Tsai, C.-C., Deppisch, R. M., Forrestal, L. J., Ritzau, G. H., Oram, A. D., Gohn, H. J., and Voorhees, M. E. (1994). Surface modifying additives for improved device–blood compatibility. *ASAIO J.* **40**: M619–M624.

Tseng, Y. C., and Park, K. (1992). Synthesis of photoreactive poly(ethylene glycol) and its application to the prevention of surface-induced platelet activation. *J. Biomed. Mater. Res.* **26**: 373–391.

Van Wachem, P. B., Vreriks, C. N., Beugeling, T., Feijen, J., Bantjes, A., Detmers, J. P., and van Aken, W. G. (1987). The influence of protein adsorption on interactions of cultured human endothelial cells with polymers. *J. Biomed. Mater. Res.* **21**: 701–718.

Verheye, S., Markou, C. P., Salame, M. Y., Wan, B., King III, S. B., Robinson, K. A., Chronos, N. A., and Hanson, S. R. (2000). Reduced thrombus formation by hyaluronic acid coating of endovascular devices. *Arteriosclerosis, Thrombosis & Vascular Biology.* **20**: 1168–1172.

Ward, R., White, K., and Hu, C. (1984). Use of surface modifying additives in the development of a new biomedical polyurethane-urea. in *Polyurethanes in Biomedical Engineering.* (H. Plank, G. Egbers, and I. Syre, eds.) New York, Elsevier. pp. 181–200.

Wetterö, J., Askendal, A., Tengval, P., and Bengtsson, T. (2003). Interactions between surface-bound actin and complement, platelets, and neutrophils. *J. Biomed. Mater. Res.* **66**: 162–175.

Whitesides, G. M., Mathias, J. P., and Seto, C. T. (1991). Molecular self-assembly and nano-chemistry: a chemical strategy for the synthesis of nanostructures. *Science* **254**: 1312–1319.

Wildevuur, C. R., Jansen, P. G., Bezemer, P. D., Kuik, D. J., Eijsman, L., Bruins, P., DeJong, A. P., Van Hardevelt, F. W., Biervliet, J. D., Hasenkam, J. M., Kure, H. H., Knudesen, L., Bellaiche, L., Ahulburg, P., Loisance, D. Y., Baufreton, C., Le Besnerais, P., Bajan, G., Matta, A., Van Dyck, M., Renotte, M. T., Ponlot, L. A., Baele, P., McGovern, E. A., and Ahlvin, E. (1997). Clinical evaluation of Duraflo II heparin treated extracorporeal circulation ciruits (2nd version). The European Working Group on heparin coated extracorporeal circulation circuits. *Eur. J. Cardio-Thorac. Surg.* **11**: 616–623.

Williams, D. F., ed. (1987). *Definitions in Biomaterials: Proceedings of a Consensus Conference of the European Society for Biomaterials, Chester, England, March 3–5, 1986.* Elsevier, Amsterdam.

Woodhouse, K. A., and Brash, J. L. (1992). Adsorption of plasminogen from plasma to lysine-derivatized polyurethane surfaces. *Biomaterials* **13**: 1103–1108.

Woodhouse, K. A., Weitz, J. I., and Brash, J. L. (1996). Lysis of surface-localized fibrin clots by adsorbed plasminogen in the presence of tissue plasminogen activator. *Biomaterials* **17**: 75–77.

Zilla, P., Deutsch, M., and Meinhart, J. (1999). Endothelial cell transplantation. *Semin. Vasc. Surg.* **12**: 52–63.

7.3 CARDIOVASCULAR MEDICAL DEVICES

Robert F. Padera, Jr., and Frederick J. Schoen

INTRODUCTION

In no area of medicine have biomaterials played a more critical role in the life-saving treatment of patients than in the cardiovascular system. Blood oxygenators used in cardiopulmonary bypass have made possible open heart surgeries such as coronary artery bypass surgery, valve replacement, and repair of congenital or acquired structural cardiac defects. Heart-valve prostheses, both mechanical and bioprosthetic, are used to replace dysfunctional natural valves with substantial enhancement of both survival and quality of life. In this situation, the benefit is substantial and has been well documented. Specifically, the mortality of unrepaired critical aortic stenosis is 50% at 2–3 years, a natural history more severe than many cancers. In contrast, survival following valve replacement is 50–70% at 10–15 years (Rahimtoolla, 2003). Although this intervention represents a tremendous improvement, patients with artificial valves still do not fare as well as similarly aged individuals without valve disease; complications related to the device are a major reason (Fig. 1). Metallic cylindrical mesh stents, inserted via catheters and without surgery during percutaneous transluminal coronary angioplasty (also known as PTCA, in which a balloon is threaded into a diseased vessel and inflated, thereby deforming the atherosclerotic plaque and partially relieving the obstruction to blood flow), have revolutionized the treatment of coronary artery disease and myocardial infarction. These interventions have markedly increased the longevity of hundreds of thousands of patients suffering from atherosclerotic vascular disease, the major cause of mortality in the developed world (ACC/AHA, 2001; Schoen, 1999a; Al Suwaidi *et al.*, 2000). More than one million PTCA procedures are performed annually worldwide, the majority of which employ intracoronary stenting. Synthetic vascular grafts used to repair weakened vessels or bypass blockages primarily in the abdomen and lower extremities have saved countless individuals from massive bleeding from ruptured degenerated aortas and have resulted in enhanced blood flow to and salvage of severely ischemic (i.e., blood-starved) organs and limbs, and are also used to obtain vascular access for hemodialysis treatment of patients with chronic renal failure. Devices to aid or replace the pumping function of the heart include intraaortic balloon pumps, ventricular assist devices, and total implantable artificial hearts. Pacemakers and automatic internal cardioverter defibrillators (AICDs) are used widely to override or correct aberrant life-threatening cardiac arrhythmias.

Most of these devices either alleviate the conditions for which they were implanted or provide otherwise enhanced function and serve the patients who receive them well and for extended periods. Nevertheless, device failure and/or other tissue–biomaterials interactions frequently cause clinically observable complications and necessitate reoperation or cause death. These deleterious outcomes may follow many years of uneventful benefit to the patient. Thus, precipitous or progressive "failure" can follow long-term "success."

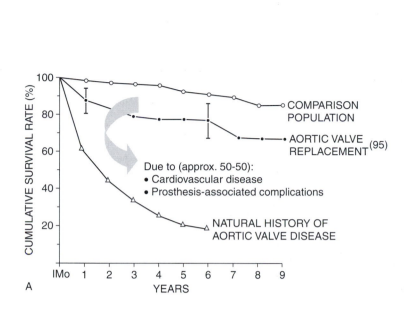

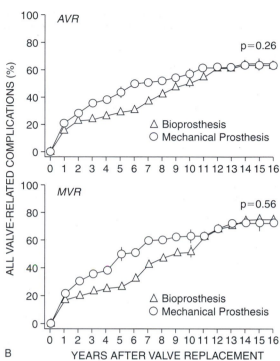

FIG. 1. Outcome following cardiac valve replacement. (A) Survival curves for patients with untreated aortic valve stenosis (natural history of valve disease) and aortic valve stenosis corrected by valve replacement, as compared with an age-matched control population without a history of aortic valve stenosis. The numbers presented in this figure for survival following valve replacement nearly 4 decades ago remain accurate today. This reflects the fact that improvements in valve substitutes and patient management have been balanced by a progressive trend toward operations on older and sicker patients with associated medical illnesses. Modified by permission from Roberts, L., *et al.* (1976). Long-term survival following aortic valve replacement. *Am. Heart J.* **91**: 311–317. (B) Frequency of valve-related complications for mechanical and tissue valves following mitral valve replacement (MVR) and aortic valve replacement (AVR). Reproduced by permission from Hammermeister, K., *et al.* (2000). Outcomes 15 years after valve replacement with a mechanical versus a bioprosthetic valve: final report of the Veterans Affairs Randomized Trial. *J. Am. Coll. Cardiol.* **36**: 1152–1158.

TABLE 1 Complications of Cardiovascular Devices

Heart valve prostheses	Vascular grafts	Circulatory assist devices
Thrombosis/thromboembolism	Thrombosis/thromboembolism	Thrombosis/thromboembolism
Anticoagulant-related hemorrhage	Infection	Endocarditis
Prosthetic valve endocarditis	Erosion into adjacent structures	Extraluminal infection
Intrinsic structural deterioration	Perigraft seroma	System component fractures
(wear, fracture, poppet escape,	(Anastomotic) false aneurysm	Bladder/valve calcification
cuspal tear, calcification)	(Anastomotic) intimal fibrous hyperplasia	Hemolysis
Nonstructural dysfunction	Mechanical failure	Mechanical failure
(pannus overgrowth, tissue or		
suture entrapment, paravalvular		
leak, inappropriate sizing,		
hemolytic anemia, noise)		

Some important mechanisms of tissue–biomaterials interaction are similar across device types, and several generic types of device-related complications can occur in recipients of nearly all cardiovascular implants. These complications, which will be discussed in detail later, include thromboembolic complications, infection, dysfunction owing to materials degeneration, and abnormal healing, either too much or too little. The clinical manifestations and relative frequencies of these prosthesis-associated problems vary among different device types; additionally some problems are unique to specific applications and models (Table 1).

This chapter will discuss the most widely used cardiovascular medical devices from three perspectives: descriptions of the devices, the diverse pathologies for which they are indicated, and complications that may arise from their use. Cardiovascular devices that are described in detail elsewhere in this volume (see Chapters 7.4 and 7.6), will only be briefly covered here.

SUBSTITUTE HEART VALVES (FOR VALVULAR HEART DISEASE)

The four valves in the human heart play a critical role in assuring the forward blood flow that is critical to proper cardiac function. The tricuspid valve allows flow from the right atrium to the right ventricle, the pulmonary valve from the right ventricle to the pulmonary artery, the mitral valve from the left atrium to the left ventricle, and the aortic from the left ventricle to the aorta. Disorders of these valves can cause stenosis (i.e., obstruction to flow), regurgitation (i.e., reverse flow across the valve), or a combination of both stenosis and regurgitation. Some disease processes such as infective endocarditis (i.e., infection of a heart valve) can cause rapid destruction of the affected valve and can lead to abrupt heart failure and death; others such as degenerative calcific aortic stenosis can take many decades to develop (during which the disease is inapparent) before clinical manifestations appear.

There are several major forms of valvular heart disease (Fig. 2). The most common indication for valve replacement overall is calcific aortic stenosis—obstruction at the aortic valve secondary to wear-and-tear induced calcific degeneration of the cusps of a previously normal tricuspid (i.e., with

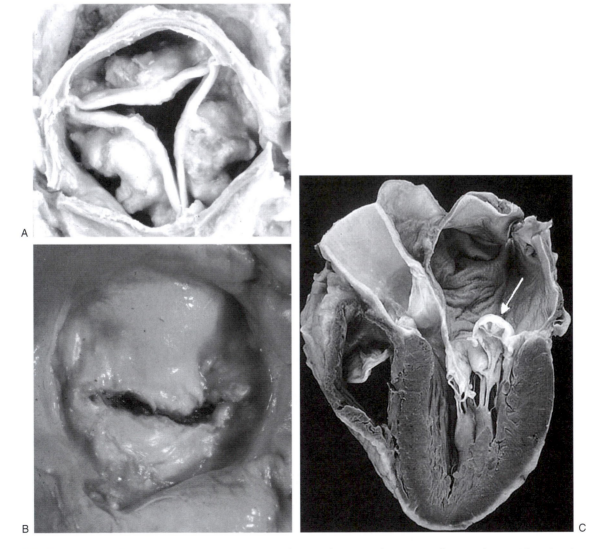

FIG. 2. Heart valve disease. (A) Severe degenerative calcification of a previously anatomically normal tricuspid aortic valve, the predominant cause of aortic stenosis. (B) Chronic rheumatic heart disease, manifest as mitral stenosis, viewed from the left atrium. (C) Myxomatous degeneration of the mitral valve, demonstrating hooding with prolapse of the posterior mitral leaflet into the left atrium (*arrow*). A, B: Reproduced by permission from Schoen, F. J., and Edwards, W. D. (2001). Valvular heart disease: General priciples and stenosis. in *Cardiovascular Pathology*, 3rd ed. Silver, M. D., Gotlieb, A. I., and Schoen, F. J., eds. Churchill Livingstone, New York. C: Reproduced by permission from Schoen, F. J. (1999). The heart. in *Robbins Pathologic Basis of Disease*, 6th ed., R. S. Cotran, V. Kumar, T. Collins, eds. W.B. Saunders, Philadelphia. (See color plate)

three cusps) aortic valve (Fig. 2A). This condition typically produces symptoms in the eighth decade of life. Calcification of the valve cusps does not allow them to fully open, causing pressure overload and resultant hypertrophy (enlargement) of the mass of the left ventricle. Patients who have congenitally (i.e., are born with) abnormal valves develop valve dysfunction and thereby symptoms at younger ages—for example approximately 15 years earlier when they are among the 1–2% of all individuals who are born with a bicuspid (i.e., with two cusps) aortic valve. Aortic regurgitation (also known as insufficiency) is most often caused by dilation of the aortic root, preventing closing of the cusps and allowing backflow across the valve. This leads to volume overload of the left ventricle. Mitral stenosis (Fig. 2B) is most often caused by chronic rheumatic heart disease that leads to scarring and stiffening of the mitral leaflets, usually many years following a bout of acute rheumatic fever suffered in childhood. Rheumatic fever is a complication of streptococcal pharyngitis (a common form of childhood throat infection) in a small percentage of individuals. Mitral regurgitation results from many different conditions; the most frequent include myxomatous degeneration (also known as floppy mitral valve, in which the strength of the mitral valve tissue is deficient and causing the leaflets to deform excessively) (Fig. 2C), conditions in which the left ventricle is abnormal and consequently the valve is not supported properly, and infective endocarditis. Diseases of the tricuspid and pulmonic valves are much less common and often do not require surgical intervention. The major complication of valvular heart disease is cardiac failure secondary to changes in the myocardium induced by pressure or volume overload of the chambers upstream or downstream of the diseased valve.

The surgical treatments available for valvular heart disease include repair and valve replacement. Reconstructive procedures to eliminate mitral insufficiency of various etiologies and to minimize the severity of rheumatic mitral stenosis are now highly effective and commonplace, accounting presently for over 70% of mitral valve operations (Bolling, 2001). Repairs are generally preferable, if they can be done. The advantages of repair over replacement relate to the elimination of both the risk of prosthesis-related complications and the need for chronic anticoagulation (which will be discussed later) that is required in many patients with substitute valves, and to a lower rate of postoperative valve-related infection (infective endocarditis). In conjunction with valve repair, the annulus (valve ring) is stabilized with or without a prosthetic annuloplasty ring.

In the many cases of valve disease in which repair cannot be done, severe symptomatic valvular heart disease is treated by excision of part or all of the diseased valve and replacement by a functional substitute (Isom, 2002). Since the early 1960s, following the first aortic valve replacements by Dwight Harken and mitral valve replacements by Albert Starr with caged ball valves, nearly 100 models of prosthetic heart valves have been developed and used (Harken *et al.*, 1960; Starr and Edwards, 1961). Today, more than 80,000 valve replacement procedures are performed each year in the United States and more than 275,000 per year worldwide. From a design standpoint, the ideal replacement valve would be nonthrombogenic, nonhemolytic, infection resistant, chemically inert, durable, and easily inserted. It would open fully and close quickly and completely, would heal appropriately in place, and would not be noticed by the patient (noise-free) (Harken *et al.*, 1962; Sapirstein and Smith, 2001). Cardiac valvular substitutes are of two generic types, mechanical and biological tissue (Vongpatanasin *et al.*, 1996; Schoen, 1995a; Korossis *et al.*, 2000). It is estimated that slightly more than half of all valves implanted in the present era are mechanical, the remainder are tissue.

Mechanical valves (Fig. 3) are composed of nonphysiologic biomaterials that employ a rigid, mobile occluder (usually a pyrolytic carbon disk) in a metallic cage (cobalt-chrome or titanium alloy) as in the Bjork-Shiley, Hall-Medtronic, and OmniScience valves, or two carbon hemidisks in a carbon housing as in the St. Jude Medical, or CarboMedics CPHV, the Medical Carbon Research Institute or On-X prostheses. Pyrolytic carbon is a material that has high strength, fatigue and wear resistance, and exceptional biocompatibility including thromboresistence (Cao, 1995). The sewing cuff, which anchors the valve into the native orifice, is composed of expanded polytetrafluoroethylene (ePTFE), Dacron, or other

FIG. 3. Mechanical prosthetic heart valves. (*Left*) Starr–Edwards caged-ball valve. (*Middle*) Bjork–Shiley tilting disk valve. (*Right*) St. Jude Medical bileaflet tilting disk heart valve. Reproduced by permission from Schoen, F. J. (2001). Pathology of heart valve substitution with mechanical and tissue prostheses. in *Cardiovascular Pathology*, 3rd ed., M. D. Silver, A. I. Gotlieb, and F. J. Schoen, eds. Churchill Livingstone, New York.

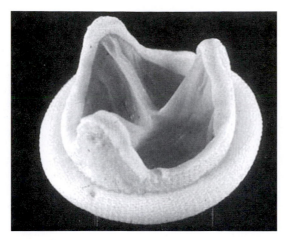
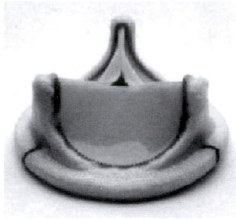

FIG. 4. Tissue heart valves. (*Left*) Hancock porcine valve. (*Right*) Carpentier–Edwards bovine pericardial valve. Reproduced by permission from Schoen, F. J. (2001). Pathology of heart valve substitution with mechanical and tissue prostheses. in *Cardiovascular Pathology*, 3rd ed. M. D. Silver, A. I. Gotlieb, and F. J. Schoen, eds. Churchill Livingstone, New York.

fabric to allow suturing and subsequently tissue integration into the host tissue. The opening and closing of the valve is purely a passive phenomenon, with the moving parts [occluder or disk(s)] responding to changes in pressure and blood flow within the chambers of the heart and great vessels. Patients receiving mechanical valves must be treated with lifelong anticoagulation to reduce the risk of thrombosis and thromboembolic events.

Tissue valves (Fig. 4) are anatomically more similar to natural valves than are mechanical prostheses. Most tissue valves are composed of three cusps of tissue derived from animals—most frequently either porcine (pig) aortic valve or bovine (cow) pericardium—treated with glutaraldehyde. This fixation preserves the tissue, kills the cells within the valve, and decreases the immunological reactivity of the tissue, so that no immunosuppression is required for these xenografts as is required for kidney or heart transplants. However, since these valves no longer contain viable cells, the cusps themselves cannot remodel or respond to injury as does normal tissue. These cusps are mounted on a metal or plastic stent with three posts (or struts) to simulate the geometry of a native valve. The base ring is covered by a Dacron- or ePTFE-covered sewing cuff to facilitate surgical implantation and healing. The most commonly used bioprosthetic valves are the Hancock porcine and Carpentier–Edwards porcine and Carpentier–Edwards pericardial tissue valves. The major advantages of tissue valves compared to mechanical prostheses are their pseudoanatomic central flow and relative nonthrombogenicity; patients with tissue valves usually do not require anticoagulant therapy. As reflected in overall heart valve substitution industry data, innovations in tissue valve technologies and design have stimulated this segment of the market to grow disproportionately in the past decade by expanding indications for tissue valve use (Fig. 5) (Rahimtoolla, 2003; Fann and Burdon, 2001).

Tissue valves derived from human cadavaric aortic or pulmonary valves (allografts) with or without the associated vascular conduit have exceptionally good hemodynamic profiles,

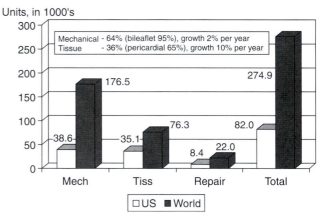

FIG. 5. Comparison of total number of mechanical and tissue valve replacements (and repairs) in the United States and the world in the year 2000. Reproduced by permission from Schoen, F. J., and Padera, R. F. (2003). Cardiac surgical pathology. in *Cardiac Surgery in the Adult*, 2nd ed., L. H. Cohn, and L. H. Edmunds, Jr., eds. McGraw-Hill, New sYork.

a low incidence of thromboembolic complications without chronic anticoagulation, and a low reinfection rate following valve replacement for endocarditis (O'Brien *et al.*, 2001). Early allografts sterilized and/or preserved with chemicals or irradiation suffered a high rate of leaflet calcification and rupture. Nevertheless, subsequent technical developments have led to cryopreserved allografts, in which freezing is performed with protection from crystallization by dimethylsulfoxide; storage until valve use is carried out at −196°C in liquid nitrogen. Contemporary allograft valves yield freedom from degeneration and failure equal to or better than those of conventional porcine bioprosthetic valves, but are limited by availability, difficulty in obtaining the proper size, and a more complex surgical procedure.

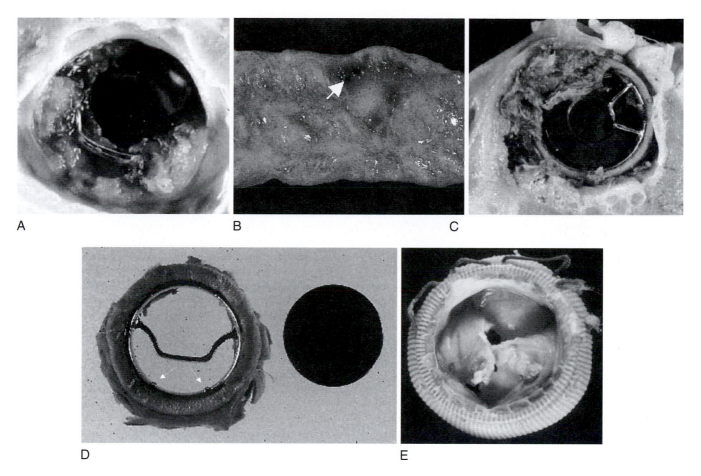

FIG. 6. Prosthetic valve complications. (A) Thrombosis on a Bjork–Shiley tilting disk aortic valve prosthesis, localized to outflow strut near minor orifice, a point of flow stasis. (B) Thromboembolic infarct of the small bowel (arrow) secondary to embolus from valve prosthesis. (C) Prosthetic valve endocarditis with large ring abscess, viewed from the ventricular aspect of an aortic Bjork–Shiley tilting disk aortic valve. (D) Strut fracture of Bjork–Shiley valve, showing valve housing with single remaining strut and adjacent disk. (E) Structural valve dysfunction (manifest as calcific degeneration with tear) of porcine valve. B: Reproduced by permission from Schoen, F. J. (2001). Pathology of heart valve substitution with mechanical and tissue prostheses. in *Cardiovascular Pathology*, 3rd ed. M. D. Silver, A. I. Gotlieb, and F. J. Schoen, eds. Churchill Livingstone, New York. C: Reproduced by permission from Schoen, F. J. (1987). Cardiac valve prostheses: pathological and bioengineering considerations. *J. Card. Surg.* **2:** 65. A and D: Reproduced by permission from Schoen, F. J., Levy, R. J., and Piehler, H. R. (1992). Pathological considerations in replacement cardiac valves. *Cardiovasc. Pathol.* **1:** 29. (See color plate)

Valve prosthesis reliability and host–tissue interactions play a major role in patient outcome. Four categories of valve-related complications (Fig. 6) are most important: thrombosis and thromboembolism, infection, structural dysfunction (i.e., failure or degeneration of the biomaterials making up a prosthesis), and nonstructural dysfunction (i.e., miscellaneous complications and modes of failure not encompassed in the previous groups). (Rahimtoolla, 2003; Vongpatanasin *et al.*, 1996; Schoen, 1995b; Schoen and Levy, 1999).

Thromboembolic complications are the major cause of mortality and morbidity after cardiac valve replacement with mechanical valves, and patients with them require chronic therapeutic anticoagulation with warfarin derivatives (Height and Smith, 1999). Thrombotic deposits that form on valve prostheses can immobilize the occluder or shed emboli (Fig. 6A,B). Tissue valves are less thrombogenic than mechanical valves, with most patients not requiring long-term anticoagulation unless they have atrial fibrillation or another specific propensity

to thrombose the valve. Nevertheless, the rate of thromboembolism in patients with mechanical valves on anticoagulation is not widely different from that in patients with bioprosthetic valves without anticoagulation (2–4% per year). Chronic oral anticoagulation also increases the risk of hemorrhage.

Prosthetic valve infection (endocarditis) occurs in 3–6% of recipients of substitute valves (Fig. 6C) and often involves the prosthesis–tissue junction at the sewing ring with accompanying tissue destruction in this area (Piper *et al.*, 2001; Mylonakis and Calderwood, 2001). This complication can occur at any time following valve implantation. The microbial etiology of early (less than 60 days postoperatively) prosthetic valve endocarditis is dominated by the staphylococcal species *S. epidermidis* and *S. aureus*, even though prophylactic antibiotic regimens used today are targeted against these microorganisms. The clinical course of early prosthetic valve endocarditis tends to be fulminant. In late endocarditis, a probable source of infection can be found in 25–80% of patients, the most frequent initiators being dental procedures, urologic infections and interventions, and indwelling catheters. The most common organisms in these late infections are *S. epidermidis, S. aureus, Streptococcus viridans,* and enterococci. Surgical reintervention is often required. Rates of infection of bioprostheses and mechanical valves are similar, but previous endocarditis markedly increases the risk.

Prosthetic valve dysfunction owing to materials degradation can necessitate reoperation or cause prosthesis-associated death. Durability considerations vary widely for mechanical valves and bioprostheses, for specific types of each, for different models of a particular prosthesis utilizing different materials or having different design features (e.g., different generations of a valve model such as Starr–Edwards caged ball valve (Schoen, 1995b; Schoen, 2001) or Bjork–Shiley tilting disk valves (Schoen, 1995b; Schoen, 2001)), and even for the same model prosthesis placed in the aortic rather than the mitral site (e.g., Braunwald–Cutter valve in the aortic but not the mitral site failing frequently (Schoen, 1995b; Schoen, 2001)). Fractures of metallic or carbon valve components occur rarely, but are catastrophic and life threatening (Fig. 6D) (Watarida *et al.,* 2001). In contrast, structural dysfunction is the major cause of failure of the most widely used bioprostheses (Fig. 6E), resulting in slowly progressive symptomatic deterioration and usually requiring reoperation (Sacks, 2001; Butany and Leask, 2001; Schoen and Levy, 1999; Schoen, 1999b). Within 15 years following implantation, 30–50% of porcine aortic valves implanted as either mitral or aortic valve replacements require replacement because of primary tissue failure. Cuspal mineralization is the major responsible pathologic process with regurgitation through tears the most frequent failure mode in porcine valves. Calcification is markedly accelerated in younger patients, with children and adolescents having an especially accelerated course. Bovine pericardial valves also suffer primarily design-related tearing, with calcification frequent but less limiting.

Paravalvular defects usually caused by inadequate healing may be clinically inconsequential or may aggravate hemolysis (destruction of red blood cells) or cause heart failure through regurgitation. Hemolysis owing to turbulent flow and blood–material surface interactions is an ever-present risk. Although severe hemolytic anemia is unusual with contemporary valves, paravalvular leaks or dysfunction owing to materials degeneration may induce clinically important hemolysis.

Methods are being actively studied and some are being used clinically to prevent calcification in bioprosthetic valves (Vyahavare *et al.,* 2000; Schoen and Levy, 1999; Levy *et al.,* 2003). Other approaches to provide improved valves include modifications of bioprosthetic valve stent design and tissue mounting techniques to reduce cuspal stresses, tissue treatment modifying or alternative to conventional glutaraldehyde pretreatment to enhance durability and postimplantation biocompatibility, nonstented porcine valves, minimally cross-linked autologous pericardial valves, flexible trileaflet polymeric (polyurethane) prostheses, and mechanical and tissue valves with novel design features to improve hemodynamics, enhance durability, and reduce thromboembolism. Some investigators are designing valves that could potentially be securely and safely inserted by a catheter rather than a major surgical

procedure (Cribier *et al.,* 2002). Tissue engineering, in which a functional, viable valve replacement is grown *in vitro* prior to implantation, is an important emerging field that may also lead to improved outcomes for patients with valvular disease (Rabkin and Schoen, 2002).

STENTS AND GRAFTS (FOR ATHEROSCLEROTIC VASCULAR DISEASE)

Atherosclerosis is a chronic, progressive, multifocal disease of the vessel wall intima of which the atheromatous plaque is the characteristic lesion. Atherosclerosis primarily affects the large elastic arteries and large and medium-sized muscular arteries of the systemic circulation, particularly at points of branches, sharp curvatures, and bifurcations. Atherosclerosis of native coronary arteries generally is limited to the large superficial epicardial vessels before they give off branches that bring blood to the heart muscle. Mature atherosclerotic plaques consist of a central core of lipid and cholesterol crystals and cells such as macrophages, smooth muscle cells and foam cells along with necrotic debris, proteins and degenerating blood elements (Fig. 7A) (Schoen and Cotran, 1999; Libby, 2000; Ross, 1999). This core is separated from the lumen by a fibrous cap rich in collagen. The major complications of atherosclerosis result from progressive obstruction of a vascular lumen, disruption of a plaque followed by thrombus formation (Fig. 7B), or destruction of the underlying vascular wall. The most important complication is myocardial infarction, which is permanent injury to heart muscle initiated by complete thrombotic occlusion following rupture of an atherosclerotic plaque that previously was only partially obstructive. The natural history of atheromatous plaque and the efficacy and safety of interventional therapies depend in part on relative plaque composition, the spatial distribution of the constituents and the integrity of the fibrous cap, which largely determines plaque stability (Kolodgie *et al.,* 2001; Huang *et al.,* 2001). Plaque mechanical properties can determine the propensity to complications as well as influence the success rate of interventions such as percutaneous transluminal coronary angioplasty (PTCA). Risk factors for atherosclerosis and coronary artery disease include diabetes, systemic arterial hypertension, hypercholesterolemia, and smoking.

Coronary Artery Stents

PTCA is used in patients with stable angina, unstable angina, or acute myocardial infarction to restore blood flow through a diseased portion of the coronary circulation obstructed by atherosclerotic plaque and/or thrombotic deposits (Fig. 8A) (Landau *et al.,* 1994). In this procedure developed and implemented first by Andreas Gruntzig in the late 1970s, a long catheter is passed retrograde from the femoral artery up the aorta to the openings of the coronary arteries that arise from the aorta immediately distal to the aortic valve cusps. Using radioopaque dye and fluoroscopy, areas of stenosis can be identified. A deflated balloon is passed over a guidewire to a site of stenosis, where the balloon is inflated using progressive and substantial expansile force ($\sim$10 atm). Enlargement of the

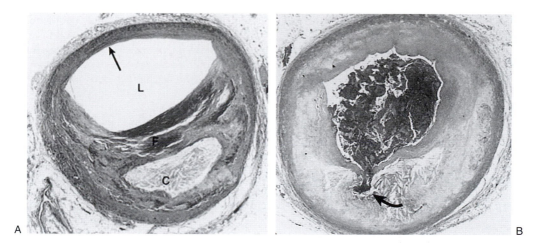

FIG. 7. Atherosclerotic plaque in the coronary artery. (A) Overall architecture demonstrating a fibrous cap (F) and a central lipid core (C) with typical cholesterol clefts. The lumen (L) has been moderately narrowed. Note the plaque-free segment of the wall (*arrow*). (B) Coronary thrombosis superimposed on an atherosclerotic plaque with focal disruption of the fibrous cap (*arrow*), triggering fatal myocardial infarction. A: Reproduced by permission from Schoen, F. J., and Cotran, R. S. (1999). Blood vessels. in *Robbins Pathologic Basis of Disease*, 6th ed., R. S. Cotran, V. Kumar, and T. Collins, eds. W.B. Saunders, Philadelphia. B: Reproduced by permission from Schoen, F. J. (1989). *Interventional and Surgical Cardiovascular Pathology: Clinical Correlations and Basic Principles*. W.B. Saunders, Philadelphia. (See color plate)

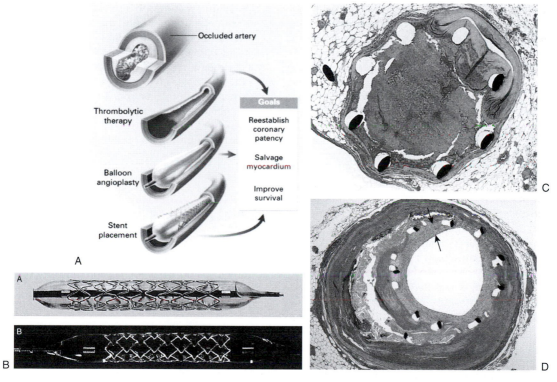

FIG. 8. Intravascular stents. (A) Catheter-based interventions for opening occluded coronary arteries, including thrombolysis, PTCA, and stenting. (B) Metallic stents. (C) Early thrombosis associated with a metallic coronary artery stent. (D) Mild restenosis in stent implanted for 1 month; arrows represent thickness of proliferative restenosis. A: Reproduced by permission from Lange, R. A., and Hillis, L. D. (2002). Methods of reperfusion in acute myocardial infarction. *N. Engl. J. Med.* **346**: 955. B: Reproduced by permission from Al Suwaidi, J., Berger, P. B., and Holmes, D. R. (2000). Coronary artery stents. *JAMA* **284**: 1828–1836. C: Reproduced by permission from Schoen, F. J., and Edwards, W. D. (2001). Pathology of cardiovascular interventions. in *Cardiovascular Pathology*, 3rd ed., M. D. Silver, A. I. Gotlieb, and F. J. Schoen, eds. Churchill Livingstone, New York.

lumen and increased blood flow occurs by plaque reduction via compression, embolization, or redistribution of the plaque contents and by overall mechanical expansion of the vessel wall (Virmani *et al.*, 1994).

Short-term failure of this procedure (i.e., closure of the treated vessel within hours to days) can occur via several mechanisms, including elastic recoil of the vessel wall, acute thrombosis at the site of angioplasty, and acute dissection (i.e., blood within the wall itself) of the vessel beyond the area of angioplasty. The major problem is that the long-term success of PTCA is limited by the development of progressive, proliferative restenosis, which occurs in 30–40% of patients, most frequently within the first 4–6 months (Haudenschild, 1993). The usual process causing restenosis after PTCA is fibrous tissue formation in the lumen, owing to excessive medial smooth muscle proliferation as an exaggerated response to angioplasty-induced injury, similar to features of atherosclerosis itself and vascular graft healing (see later discussion). Locally delivered pharmacologic and molecular therapies have not effectively mitigated restenosis after PTCA (Riessen and Isner, 1994; Kibbe *et al.*, 2000).

Stents (Fig. 8B) are expandable tubes of metallic mesh that have been developed to address these negative sequelae of balloon angioplasty. Stents have been used in patients since the late 1980s; today, the majority of patients undergoing PTCA will also receive a stent. Stents preserve luminal patency and provide a larger and more regular lumen by acting as a scaffold to support the disrupted vascular wall and minimize thrombus formation following PTCA and thereby reduce the impact of postangioplasty restenosis (Serruys *et al.*, 1994).

Stent technologies have undergone a rapid evolution. The majority of stents in use today are composed of balloon-expandable 316L stainless steel or nitinol mesh tubes that range from 8 to 38 mm and from 2.5 to 4.0 mm in diameter. Development has focused on permitting stents to become more flexible and more easily delivered and deployed, allowing the treatment of a greater number and variety of lesions. The choice of stent is based on several factors, including the characteristics of a given plaque, such as its diameter, length, and location within the coronary anatomy, and the experience of the interventional cardiologist with a particular stent.

Stenting has been shown to be superior to angioplasty alone in several lesions and situations, including in vessels greater than 3 mm in diameter, in chronic total occlusions, in stenotic vein grafts, in restenotic lesions after angioplasty alone, and in patients with myocardial infarction (Stone *et al.*, 2002). The early complications of stenting generally involve subacute stent thrombosis that occurs in 1 to 3% of patients within 7 to 10 days of the procedure (Fig. 8C). This complication has largely been overcome by aggressive multidrug treatment with antiplatelet agents such as clopidogrel, aspirin, and glycoprotein IIb/IIIa inhibitors. The major long-term complication of stenting is in-stent restenosis, which occurs within 6 months in 50% of patients who are stented (Fig. 8D). Tissue interactions with an implanted stent are complex (Welt and Rogers, 2002). There is early damage to the endothelial lining and stretching of the vessel wall, stimulating adherence and accumulation of platelets and leukocytes. Covered initially by a variable platelet–fibrin coating, stent wires may eventually become completely covered by a endothelium-lined neointima, with the wires embedded in a layer of intimal thickening consisting of smooth muscle cells in a collagen matrix (Farb *et al.*, 1999). This tissue may thicken secondary to the release of growth factors, chemotactic factors, and inflammatory mediators from platelets and other inflammatory cells that result in increased migration and proliferation of smooth muscle cells, and increased production of extracellular matrix molecules, narrowing the lumen and resulting in restenosis (Virmani and Farb, 1999).

Many approaches have been used in an attempt to reduce in-stent restenosis. Intracoronary radiotherapy is a procedure in which a beta or gamma source is brought into close proximity to the stent to deliver local radiation. This is thought to block cell proliferation, induce cell death, and inhibit migration of smooth muscle cells in the area of the stent to reduce neointimal accumulation. Several studies have shown promising initial results in reducing the restenosis rate, but have also shown some long-term complications including late thrombosis, increased restenosis at the edge of the treated field, and damage to the wall (Salame *et al.*, 2001).

The most promising results have been attained with polymer-coated drug-eluting stents (Fig. 9) (Fattori and Piva, 2003; Sousa *et al.*, 2003a,b). Two of the drugs currently in clinical trials are rapamycin (sirolimus) (Sousa *et al.*, 2003c) and paclitaxel (Park *et al.*, 2003). Rapamycin, a drug used for immunosuppression in solid-organ transplant recipients, also inhibits proliferation, migration, and growth of smooth muscle cells and extracellular matrix synthesis. Along with its anti-inflammatory properties, this drug targets the major mechanisms of restenosis discussed above. Paclitaxel, a drug used in the chemotherapeutic regimens for several types of cancer, also has similar anti–smooth muscle cell activities. These drugs are embedded in a polymer matrix (such as a copolymer of poly-*n*-butyl methacrylate and polyethylene–vinyl acetate or a gelatin–chondroitin sulfate coacervate film) that is coated onto the stent. The drug is released by diffusion and/or polymer

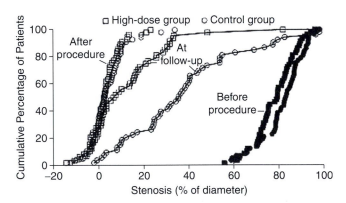

FIG. 9. Cumulative distribution of the percentage of stenosis in the high-dose and control groups. The distributions were similar at baseline (about 80%) and immediately after stent placement (about 0%). At 6 months follow-up, the distribution in the high-dose group remained similar to the distribution immediately after stenting, whereas the control group suffered greater restenosis (about 40%). Reproduced by permission from Park, S. J., *et al.* (2003). A paclitaxel-eluting stent for the prevention of coronary restenosis. *N. Engl. J. Med.* **348:** 1537–1545.

degradation over varying periods of time that can be engineered by the specifics of the polymer–drug system. These coated stents have had excellent initial success, virtually eliminating restenosis over time periods of 2 years and longer, and are felt to represent a major breakthrough in the treatment of coronary artery disease. Drug eluting stents are commercially available and most PTCA procedures will likely employ them (Poon *et al.*, 2002).

Peripheral Stents and Stent Grafts

Peripheral vascular disease results mainly from narrowing of the aorta and its branches secondary to atherosclerosis, the same accumulation of plaque within the arterial wall described above in coronary artery disease (Schoen, 1999b). As the degree of stenosis increases, blood flow to the distal tissues is impeded, causing ischemia in the tissues served by the diseased artery. Occlusions often occur in the abdominal aorta or the iliac arteries, which are the arteries serving the legs. When this occurs, pain is felt in the legs (especially the calf), buttocks, or hips during times of exertion; the pain usually diminishes with rest. In severe cases of vascular compromise, healing of even minor injuries can be inefficient leading to gangrene and requiring amputation of the affected limb. The treatment for many patients with peripheral vascular disease involves using a vascular graft to perform a bypass around the area of blockage to restore ample blood flow. Aneurysms (i.e., ballooning of the vessel due to the weakness of its wall), especially of the abdominal aorta, may result from atherosclerosis and are at risk of rupture once they reach a certain size. Aneurysms are also repaired with synthetic vascular grafts.

Surgical procedures such as open abdominal aortic aneurysm repair and aorto-femoral bypass grafting can have significant associated complications in certain patient populations. A minimally invasive approach such as that afforded by PTA with or without stenting is appropriate in these settings. Stents and stent grafts can be employed in the peripheral circulation to increase lumen size in a similar fashion to their use in the coronary circulation (Faries *et al.*, 2002; Ramaswami and Marin, 1999).

Stents for treatment of peripheral vascular disease are generally constructed of stainless steel or nitinol and may be coated with compounds such as ePTFE. Stent grafts (Fig. 10) are composed of a metallic frame covered by a fabric tube and combine the features of stents and vascular grafts; they can be deployed endovascularly. Stent grafts are used to treat aortic aneurysms, where the aortic wall has been weakened and threatens to rupture, as well as stenosis of other arterial sites. The graft portion, usually composed of polyester or ePTFE, can sit on either the luminal or abluminal (outside) aspect of the metallic stent and is intended to provide a mechanical barrier to prevent intravascular pressure from being transmitted to the weakened wall of the aneurysm, thus excluding the aneurysm from the flow of blood. These stents and stent grafts are deployed in a similar manner to those in the coronary circulation, either as self-expanding units or over an inflatable balloon. The stent used for a given application is selected by diameter, length, and geometry of the lesion and location of side braches or branch points.

Stent and stent grafts have been especially successful in treating subtotally occluded short (5–10 cm) segments of the iliac artery that can cause significant chronic lower extremity ischemia, and in the treatment of stenosis of the renal arteries and the smaller arteries of the lower extremity (femoral, popliteal, and tibial arteries), the carotid arteries, the celiac artery, and the superior mesenteric artery. Stents and stent grafts in these sites suffer mechanical failure at a greater rate than those in the coronary circulation (Jacobs *et al.*, 2003). Fracture of the struts that make up the stent, fabric erosion, and device fatigue are modes of failure of these devices.

Vascular Grafts

The concept of using synthetic material as a conduit in the vascular system dates back to the early 1900s when animal experiments were carried out using aluminum, silver, glass, and Lucite tubes as vascular replacements. Fabrics such as Vinyon N, a cloth used in parachutes, were employed in the mid-1900s as vascular conduits that could be fashioned from commercially available textiles. Current synthetic vascular grafts are typically fabricated from poly(ethylene terephthalate) (Dacron) or expanded polytetrafluoroethylene (ePTFE), with the Dacron grafts being used for larger vessel applications and the ePTFE to bypass smaller vessels. These grafts can be made porous to enhance healing but they are then impregnated with connective tissue proteins to aid clotting, reduce the blood loss through the pores of the graft upon implantation, and stimulate tissue ingrowth, and with antibiotics to reduce the risk of infection of the graft. Loosely woven or porous synthetic grafts that are not impregnated need to be preclotted with the patient's own blood for this same purpose.

Synthetic grafts (Fig. 11A) perform well in large-diameter, high-flow, low-resistance locations such as the aorta and the iliac and proximal femoral arteries, with grafts used for aortofemoral bypass having 5- to 10-year patency rates of 90% (Clagett, 2002). In contrast, synthetic small-diameter vascular grafts (<6 to 8 mm in diameter) generally perform less well with 5-year patency less than 50%. In general, the longer the interposition or bypass graft, and the smaller the recipient vessel, with a corresponding increase in resistance to flow, the less favorable are both short- and long-term patency rates. For this reason, a major superficial vein in the leg, the saphenous vein, is typically removed surgically and moved to the site needed to bypass blockages in the coronary circulation (for coronary artery bypass grafting) and in the distal extremities (e.g., for femoropopliteal bypass grafting) when an adequate length of disease-free vein segment can be harvested for the given application (Fig. 11B). Other natural vessels such as the internal mammary artery in the coronary circulation may be used.

When a synthetic graft is implanted, the luminal surface of the graft becomes coated with a film of plasma proteins, primarily fibrinogen. This layer develops over time into a platelet–fibrin aggregate termed a pseudointima. When endothelial cells cover this layer simulating the inner layer of a native blood vessel and serving as a nonthrombogenic surface, it is termed a neointima. Humans have a limited ability to endothelialize vascular grafts, resulting in confluent

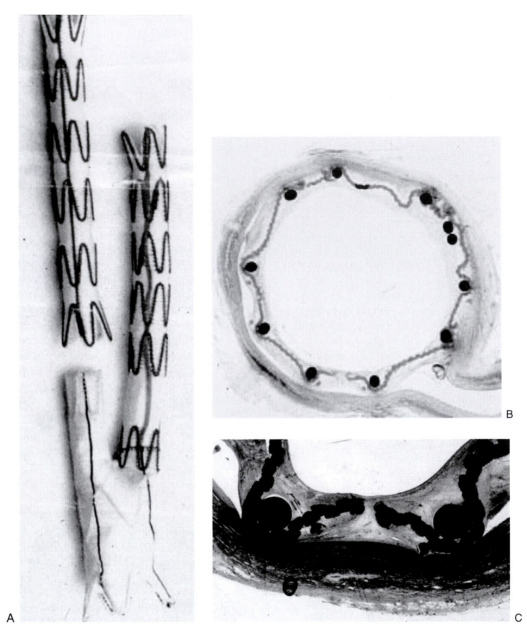

FIG. 10. Stent grafts. (A) Configuration of device showing composite metal and fabric portions. (B) Low-power photomicrograph of well-healed experimental device explanted from a dog aorta. The lumen is widely patent and the fabric and metal components are visible. (C) High-power photomicrograph of stent graft interaction with the vascular wall, demonstrating mild intimal thickening. B and C: courtesy Jagdish Butany, MD, University of Toronto. (See color plate)

endothelium covering only a 10- to 15-mm zone adjacent to the anastomosis. Thus, except adjacent to an anastomosis (the sutured connection of the graft to the native artery), a compacted platelet–fibrin pseudointima comprises the inner lining of clinical fabric grafts, even after long-term implantation (Fig. 12A). Because firm adherence of such linings to the underlying graft may be impossible, dislodgment of the lining leading to distal embolization or formation of a flap-valve can occur and cause acute obstruction.

Tissue lining the inner wall of a vascular graft has three possible sources: (1) overgrowth from the host vessel across anastomotic sites, (2) tissue ingrowth through fabric interstices,

and (3) deposition of functional endothelial cells and/or multipotential stem cells from the circulating blood (Fig. 12B). In a graft with interstices large enough to permit ingrowth of fibrovascular elements, endothelial cells can arise from capillaries extending from outside to inside the graft and migrate to the luminal surface at a large distance from the anastomosis. However, since most clinical vascular grafts are impervious in order to obviate hemorrhage, existing grafts and other fabrics used as cardiovascular implants heal primarily by ingrowth of endothelium and smooth muscle cells from the cut edges of the adjacent artery or other tissue. Although there is considerable current interest in harnessing the potential of

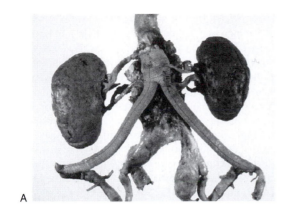

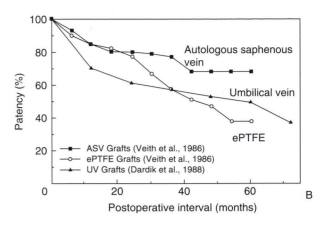

FIG. 11. (A) Dacron aortobifemoral bypass graft *in situ*. (B) Comparison of patency rates of autologous saphenous vein grafts, umbilical vein grafts, and ePTFE small-diameter vascular grafts. A: Courtesy of Dr. Bruce McManus, St. Paul's Hospital, Vancouver, Canada. Schoen, F. J. (2001). Prosthetics. in B. M. McManus, and E. Braunwald, eds. *Atlas of Cardiovascular Pathology*, Current Medicine, Inc., p. 216. B: Data from Veith, F. J., Gupta, S. K., Ascer, E., White-Flores, S., Samson, R. H., Scher, L. A., Towne, J. B., Bernhard, V. M., Bonier, P., Flinn, W. R., et al. (1986). Six-year prospective multicenter randomized comparison of autologous saphenous vein and expanded polytetrafluoroethylene grafts in infrainguinal arterial reconstructions. *J. Vasc. Surg.* **3**: 104–114. Dardik, H., Miller, N., Dardik, A., Ibrahim, I., Sussman, B., Berry, S. M., Wolodiger, F., Kahn, M., and Dardik, I. (1988). A decade of experience with the glutaraldehyde-tanned human umbilical cord vein graft for revascularization of the lower limb. *J. Vasc. Surg.* **7**: 336–346.

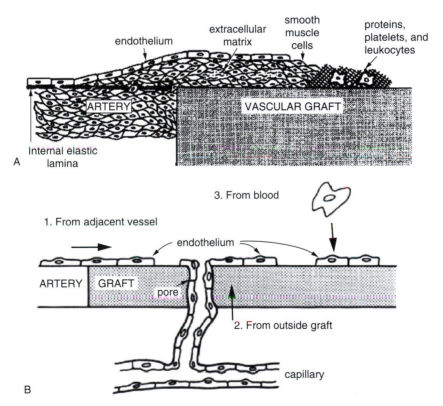

FIG. 12. Vascular graft healing. (A) Schematic diagram of pannus formation, the major mode of graft healing with currently available vascular grafts. Smooth muscle cells migrate from the media to the intima of the adjacent artery and extend over and proliferate on the graft surface; this smooth muscle cell layer is covered by a proliferating layer of endothelial cells. (B) Possible sources of endothelium on the blood-contacting surface of the vascular graft. Reproduced by permission from Schoen, F. J. (1989). *Interventional and Surgical Cardiovascular Pathology: Clinical Correlations and Basic Principles*. W.B. Saunders, Philadelphia.

circulating endothelial cell precursors to cover grafts, this has not yet been realized (Rafii, 2000).

An implanted graft becomes encapsulated in the surrounding connective tissue and elicits a typical foreign-body reaction. The tissue covering the graft on its exterior surface, separating it from normal tissue, consists of a foreign-body inflammatory reaction containing giant cells adjacent to the material, covered by collagen, fibroblasts, blood vessels, and other cellular and extracellular connective tissue elements. This foreign-body capsule extends from the outside graft surface to the surrounding undisturbed body tissues. Graft interstices may be filled with fibrin or connective tissue elements, including cells and extracellular matrix and, where ingrowth has occurred, blood vessels.

The major complications of vascular grafts are thrombosis/thromboembolism, infection, periprosthetic fluid collection, pseudoaneurysm (i.e., an extravascular hematoma that communicates with the intravascular space), intimal hyperplasia, and structural degeneration. Failure of small-diameter vascular prostheses is most frequently due to occlusion by thrombus formation or generalized or anastomotic fibrous hyperplasia. As in any cardiovascular site, "Virchow's triad" of surface thrombogenicity, hypercoagulability, and locally static or low blood flow largely predicts the propensity toward thrombus formation. This predicts thrombosis in small-diameter synthetic grafts where there can be low flow states due to poor flow out of the end of the graft (termed runoff) from atherosclerosis beyond the anastomosis. In larger vessels with higher blood flow such as the aorta, the surface thrombogenicity of the graft is overcome by brisk flow.

Prophylactic perioperative systemic antibiotics limit infection of implanted vascular prostheses to 6% of patients or less. Early infections typically are related to the surgical procedure or to perioperative complications such as wound infection. Late infections usually occur secondary to seeding from the blood of the synthetic material in patients with low-grade bacteremia, often secondary to dental or gastrointestinal procedures. Since the anastomotic suture line is usually involved, an infected vascular graft usually has a partially disrupted connection to the natural artery; thus, rupture with hemorrhage at the graft site may bring the patient to clinical attention. Surgical removal of an infected graft is usually necessary to cure the infection.

The healing of synthetic vascular grafts can yield exuberant fibrous tissue at the anastomotic site as an overactive physiologic repair response (Fig. 13). Late failure of clinical and experimental vascular grafts, especially in those less than 8 mm in diameter, is frequently due to diffuse fibrous thickening of the inner capsule. This intimal hyperplasia results primarily from smooth muscle cell migration, proliferation, and extracellular matrix elaboration following and possibly mediated by acute or ongoing endothelial cell injury. Contributing factors include (1) surface thrombogenesis, (2) delayed or incomplete endothelialization of the fabric, (3) disturbed flow across the anastomosis, and (4) mechanical factors at the junction of implant and host tissues. In vein grafts, intimal hyperplasia is often diffuse, leading to progressive luminal reduction of the entire graft, but focal lesions can cause isolated stenoses at anastomoses. In contrast, synthetic vascular prostheses tend to develop intimal hyperplasia predominantly at or near anastomoses, particularly at the distal site.

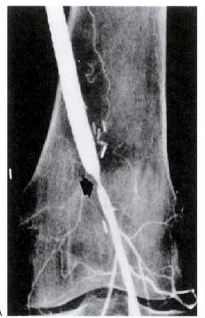
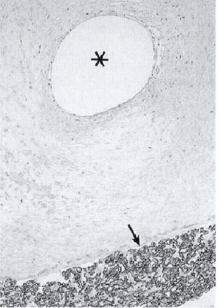

A B

FIG. 13. Anastomotic hyperplasia of the distal anastamosis of a synthetic femoropopliteal graft. (A) Angiogram demonstrating constriction. (B) Photomicrograph demonstrating ePTFE graft (arrow) with prominent intimal proliferation and very small residual lumen (*). Reproduced by permission from Schoen, F. J., and Cotran, R. S. (1999). Blood vessels. in *Robbins Pathologic Basis of Disease*, 6th ed., R. S. Cotran, V. Kumar, and T. Collins, eds. W.B. Saunders, Philadephia.

Progressive deterioration of a synthetic vascular graft can cause mechanical failure at the anastomotic site or in the body of the prosthesis leading to aneurysm formation or the formation of an external collection of blood due to rupture (false aneurysm). The incidence of long-term graft deterioration is unknown. The causes of delayed failure of a synthetic prosthesis include chemical, thermal, or mechanical damage to polymeric yarn materials during manufacture, fabric defects induced during manufacture (e.g., dropped stitches) or during insertion (even "atraumatic" vascular clamps can damage yarn fibers), and postoperative biodegradation of graft material.

In light of the complications associated with vascular grafts, current research has focused on improvement of synthetic vascular grafts and on alternatives such as tissue-engineered blood vessels. Attempts have been made to covalently modify the luminal surface of the grafts in order to (1) prevent coagulation, (2) prevent platelet adhesion/aggregation, (3) promote fibrinolysis, (4) inhibit smooth muscle cell adhesion/proliferation, and (5) promote endothelial cell adhesion and proliferation. Endothelialization of the entire graft would improve the thrombogenicity of the graft and may also help prevent bacterial attachment to the graft and subsequent infection. Attempts to seed the luminal surface with unmodified or genetically modified endothelial cells are also underway. Engineering an artery using biodegradable polymer matrices seeded with smooth muscle and endothelial cells is a promising approach to solving the problem of developing an adequate small-diameter vascular graft (Consigny, 2000; Seifalian *et al.*, 2001).

PACEMAKERS AND ICDS (FOR CARDIAC ARRHYTHMIAS)

The normal cardiac electrical cycle (Fig. 14A) begins with an impulse initiated by the sinoatrial (SA) node, which is located in the right atrium near its junction with the superior vena cava. The impulse spreads through both left and right atria, causing depolarization of the cardiac myocytes resulting in atrial contraction. The impulse arrives at the atrioventricular (AV) node, located in the posterior right atrium near the intraatrial septum, the ostium of the coronary sinus, and the tricuspid valve. After a short delay, the impulse passes through the AV node to the bundle of His and into the left and right bundle branches, located in the intraventricular septum. The impulse spreads through the right and left ventricles causing myocyte depolarization and coordinated ventricular contraction.

Cardiac arrhythmias reflect disturbances of either impulse initiation or impulse conduction (Huikuri *et al.*, 2001). Abnormal foci of impulse-generating (automatic) cells outside the SA node, called ectopic foci, may initiate propagated cardiac impulses that generate suboptimal ventricular contractions. Intrinsic SA node dysfunction also can account for disturbances of impulse initiation. Disturbances of impulse conduction mainly consist of conduction blocks or reentry. Conduction blocks constitute a failure of propagation of the usual impulse as a result of a disease process (such as ischemia or inflammation) or a drug. Blocks can be complete (no impulse propagation) or incomplete (impulse propagates

more slowly than normal or intermittently) and can be permanent or transient. Reentry is said to occur when a cardiac impulse traverses a loop of cardiac fibers and reexcites previously excited tissue without a second impulse from the SA node. For patients in whom these cardiac arrhythmias cannot be controlled by pharmacological therapy (i.e., antiarrhythmic drugs), two therapeutic options are available: (1) electrical therapy to control the cardiac rhythm (i.e., direct current cardioversion, implantable electrical devices such as pacemakers and implantable cardioverter-defibrillators), and (2) interventional/surgical therapy to remove the affected tissue or interrupt the abnormal pathway (i.e., endocardial resection, cryoablation, and other ablative techniques).

Cardiac Pacemakers

The implantation of the first cardiac pacemaker was performed in 1958 and since then, cardiac pacing has become a well-established therapeutic tool (Atlee and Bernstein, 2001). More than 500,000 patients in the United States currently have pacemakers (Fig. 14B,C) and more than 150,000 new permanent pacemakers are implanted each year; pacemaker placement, revision, or removal is a commonly performed procedure. Permanent cardiac pacing is used for various types of conduction block. These commonly lead to bradycardia (abnormally low heart rate) resulting in inadequate cardiac output. Most cardiac pacemakers are implanted in patients older than 60 years but they are also used in children, including infants, when necessary.

Modern cardiac pacing (Kusumoto and Goldschlager, 2002) is achieved by a system of interconnected components consisting of (1) a pulse generator which includes a power source and electric circuitry to initiate the electric stimulus and to sense normal activity; (2) one or more electrically insulated conductors leading from the pulse generator to the heart muscle, with a bipolar electrode at the distal end of each; and (3) a tissue, or blood and tissue, interface between electrode and adjacent stimulatable myocardial cells.

Pacemakers are powered by lithium-iodide batteries and have an expected service life of 5–12 years. The first pacemakers were large (40–200 cm^3) by today's standards (9–45 cm^3) and contained few of the features that characterize current devices. The pacemaker delivers a small current (2–4 mA) to the myocardium via the electrodes, resulting in depolarization and contraction of the heart. A single-chamber pacemaker delivers a stimulus based on a programmed timing interval. The pacemaker also senses intrinsic cardiac activity and can be inhibited from providing unnecessary or inappropriate stimuli. This "demand" pacing is valuable in a patient whose problem is intermittent. A dual-chamber pacemaker with electrodes in both the atrium and ventricle delivers the sequential atrial and ventricular signals to approximate the timing of the normal heartbeat. This device also senses intrinsic atrial and ventricular depolarizations and delivers stimuli at the appropriate time to maintain proper synchrony of the chambers.

Cardiac pacing may be either permanent or temporary. Permanent cardiac pacing involves long-term implantation of both pulse generator and electrode leads. The generator is

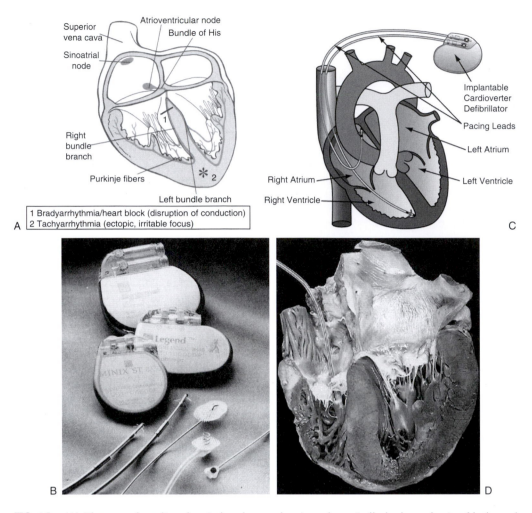

FIG. 14. (A) The normal cardiac electrical pathways showing schematically both conduction blocks and ectopic foci of impulse generation. (B) Two bipolar pulse generators: the dual chamber Synergist II, the rate-responsive Legend ST miniature ventricular device. Left to right are two pipolar urethane-insulated tined endocardial leads and three myocardial leads; a stab-in, urethane-insulated lead with silicone sewing pad and Dacron mesh disk; a screw-in lead, insulated with silicone; and an epicardial urethane-insulated lead with a silicone sewing pad. (C) Schematic demonstrating ICD lead placement in right ventricle. (D) Transvenous pacing lead placed in right ventricle demonstrating fibrosis of the distal portion of the lead (arrow). B: Courtesy of Arthur J. Coury, Medtronic, Inc., Minneapolis, MN. (now Genzyme, Inc., Cambridge, MA) D: Reproduced by permission from Schoen, F. J., and Cotran, R. S. (1999). Blood vessels. in *Robbins Pathologic Basis of Disease*, 6th ed., R. S. Cotran, V. Kumar, and T. Collins, eds. W.B. Saunders, Philadelphia.

placed in a tissue pocket beneath the skin on the anterior chest with the leads advanced transvenously through the left subclavian vein (which runs under the left clavicle) with the electrodes terminating at the endocardial surface of the heart. The tips of the electrodes are typically placed within the right atrium and/or right ventricle depending on the pacing modality.

Temporary pacing is most frequently used for the patients with acute myocardial infarction complicated by cardiac conduction system disturbances that could progress to complete heart block. Leads for temporary cardiac pacing are generally directed transvenously into the apex of the right ventricle and the pulse generator is located outside the body. In the context of cardiac surgery when the surface of the heart (epicardium) is already exposed, temporary pacing is achieved

by placing insulated wires with bare ends to the epicardial surfaces of the atria or ventricles with the leads emerging transthoracically from the anterior chest to permit easy withdrawal. Ultimately, the temporary pacemaker is either replaced by a permanent device or discontinued.

The interface between the electrode and depolarizable myocardial tissue is of critical importance in the proper functioning of the pacemaker (Fig. 14D). Typically, a layer of nonexcitable fibrous tissue induced by the electrode forms around the tip of the electrode. This nonexcitable tissue is undesirable as it increases the strength of the threshold pacing stimulus required to initiate myocyte depolarization. Fibrosis between the lead and the heart muscle can be reduced with improved lead designs and the use of slow, local release of corticosteroid drugs (Cornacchia *et al.*, 2000). The practical point is that, if pulse generator output is not set sufficiently high in the early postimplantation phase, loss of pacing with potentially fatal consequences can result. By contrast, maintaining output at such high levels once thresholds have stabilized greatly shortens battery life. Thus, pacemakers with adjustable variations in output have been developed.

An ideal endocardial pacing lead should provide stable fixation immediately from the time of implantation, achieve and maintain a minimal threshold for stimulation, maximize sensing, and function reliably for many years. Electrode fixation to the inside heart surface (endocardium) may be active or passive. In active fixation, the electrode is designed to grasp the endocardial surface to achieve immediate fixation at implantation. A very effective aid to passive fixation is the addition of projecting tines, or fins, in the region of the electrode tip. A different approach to improving fixation has been the development of electrodes with porous metal surfaces. An endocardial pacemaker lead may require a special design if it is implanted at a particular site. One example is the J-shaped atrial lead, which is curved to facilitate placing the electrode tip in the right atrial appendage, inherently the most stable site for fixation.

The complications of pacemaker leads and electrodes include infection, electrode dislodgment, lead fracture, thrombosis and thromboembolism, myocardial penetration or perforation, electrode corrosion and insulation failure, and an unduly high pacing threshold. Complications related to pacing leads may be related to the body of the lead, as distinct from the lead–pacemaker pack interface or the electrodes. Pacing lead improvements over the years have included helical coil and multifilament designs to decrease electrical resistance and enhance flexibility and durability.

Infection is a dreaded complication of the pacemaker leads and pulse generator (Voet *et al.*, 1999). The infection may originate in the pacemaker pocket and track along the lead, which acts as a contaminated foreign body. Alternatively, it may occur by implantation of bacteria on traumatized endocardium or thrombus contiguous with the lead. The most common organism cultured is *Staphylococcus epidermidis*. Septicemia may develop and septic pulmonary emboli have occurred. The fundamental therapeutic principle in pacemaker-related endocarditis is removal of at least the lead and, when the pacemaker pocket is involved, the entire pacing system. Most leads can be removed by prolonged gentle traction, but recourse to cardiotomy with cardiopulmonary bypass may be needed if the lead is incarcerated in fibrous tissue.

The most common complications related to the pulse generator are erosion due to pressure necrosis of the skin overlying the pulse generator, infection of the pacemaker pocket, and migration or rotation of the pacemaker pack. Under extreme circumstances, the entire pacemaker pack may extrude from the wound, causing a loss of pacing. Perhaps the most obvious complication related to the pacemaker is loss of the stimulus for pacing attributable either to failure of an electronic component or, more commonly, to battery depletion. In the past, many reports appeared on interference with pacemaker function by devices ranging from electric razors, toothbrushes, and microwave ovens at home to electrosurgical and diathermy

apparatus in the hospital. Fortunately, recent generations of cardiac pacemakers have been greatly improved with regard to their resistance to electromagnetic interference.

Implantable Cardioverter-Defibrillator (ICD)

The first implantable cardioverter-defibrillator was placed in 1980 (Jacobs *et al.*, 2003); currently more than 30,000 ICDs are implanted annually in the United States. The goal of ICDs is to prevent sudden death in patients with certain life-threatening arrhythmias. ICDs have been shown to revert sustained ventricular tachycardia (abnormally high ventricular rate, which often initiates sudden death) and ventricular fibrillation (uncoordinated electrical/myocardial activity) in multiple prospective clinical trials. Benefit in overall mortality has been documented (AVID, 1997).

ICDs consist of similar components to a pacemaker, namely a pulse generator and leads for tachyarrhythmia detection and therapy. The pulse generator is a self-powered, self-contained computer with one or two 3.2-V lithium–silver vanadium oxide batteries used to power all components of the system including aluminum electrolytic storage capacitors. The devices have a service life of 5–8 years. The lead is generally placed in the right ventricle through a transvenous approach. The ICD constantly monitors the ventricular rate and, when the rate exceeds a certain value, provides therapy. Current devices will initially provide a short burst of rapid ventricular pacing that terminates some types of ventricular tachyarrhythmias without providing a large shock. This approach can terminate up to 96% of episodes of ventricular tachycardia without the need for a shock. If this pacing fails to break the arrhythmia, the ICD delivers a shock of 10–30 joules between the electrode in the right ventricle and the surface of the pulse generator to terminate the dysrhythmic episode. These devices also keep a running record of arrhythmias and treatment results. ICDs are indicated in patients at high risk for ventricular arrhythmias (primary prevention) and in patients who have already had an episode of sudden cardiac death from which they have been resuscitated (secondary prevention).

Implantable cardioverter–defibrillators are subject to many of the same complications described earlier for implantable pacing systems. However, the more extensive hardware of the ICD system may contribute an increased frequency of complications relative to pacemakers, such as infection and late accumulation of pleural fluid. Several additional considerations are specific to ICDs. The consequences of repeated defibrillations can cause the following effects: (1) direct effect of repeated discharges on the myocardium and vascular structures, and (2) possible thrombogenic potential of the indwelling intravascular electrodes. Another major complication of ICDs from the standpoint of the patient, other than the inability to sense or terminate an arrhythmia leading to sudden death, is an inappropriate shock. In addition to being startling and quite painful at the time of the shock, patients receiving multiple inappropriate shocks have been known to develop posttraumatic stress disorder symptoms.

CARDIAC ASSIST AND REPLACEMENT DEVICES (FOR HEART FAILURE)

Congestive heart failure constitutes a deficiency of the pump function of the cardiac muscle, the myocardium, and is an extremely common condition (Jessup and Brozena, 2003). Each year in the United States, congestive heart failure is the principal cause of death in 40,000 individuals, a contributing factor in more than 200,000 deaths, diagnosed in 400,000 individuals, and the primary diagnosis in more than 900,000 hospitalizations. Cardiac transplantation cannot be the solution in most such individuals. The increasing discrepancy between the number of available donor hearts (only 3000 per year) and the large number of patients who might benefit from cardiac transplantation (estimated at 30,000–100,000 per year) has prompted efforts toward the development of mechanical devices to augment or replace cardiac function (Hunt and Frazier, 1998; Goldstein *et al.*, 1998; Jessup, 2001).

Congestive heart failure is the final common pathway of many cardiac conditions, including valvular heart disease, coronary artery atherosclerosis with resultant ischemic heart disease, and diseases that affect the cardiac muscle directly (often called cardiomyopathy). Heart failure can occur precipitously, as in myocardial infarction or myocarditis induced by viruses, or it can be a slow, progressive worsening of exercise tolerance and shortness of breath over many months or years owing to slow, progressive deterioration of the heart muscle. It can manifest itself in the postoperative period after both cardiac surgery (e.g., valve replacement, cardiac transplantation) and noncardiac surgery (e.g., abdominal aortic aneurysm repair).

As one might expect, therefore, the natural history of heart failure depends on the cause and progression of the underlying disease process. For example, patients with heart failure after cardiac surgery (called postcardiotomy shock) often recover the vast majority of their cardiac function after a short period of time if they are otherwise supported by mechanical support. In contrast, patients with a chronic cardiomyopathy, one of the most common reasons for cardiac transplantation, often need long-term mechanical support; studies have shown that at least 50% of such individuals would die in 3–5 years from their disease without it. One must take these clinical considerations into account when designing mechanical support systems, as different devices may best serve patients with different problems (DiGiorgi *et al.*, 2003).

Cardiopulmonary Bypass

First used in 1953 by John Gibbon, cardiopulmonary bypass devices pump blood external to the body and thereby permit complex cardiac surgical procedures to be done safely and effectively. Bypass machines are useful in extracorporeal membrane oxygenation (ECMO) to assist in the transport of oxygen and carbon dioxide for patients (especially neonates and infants) with pulmonary diseases such as respiratory distress syndrome (Alpard and Zwischenberger, 2002).

The basic operating principles of the current heart–lung machines are quite straightforward and have changed little in the past half century. Deoxygenated blood returning from the systemic circulation into the right atrium is withdrawn by gravity siphon into a cardiotomy reservoir and is then pumped into an oxygenator. The most common type of oxygenator is a membrane oxygenator, where oxygen is passed through the tube side of a shell-and-tube type device while the blood passes through the shell side. Oxygen and carbon dioxide are exchanged via diffusion through synthetic membranes (usually polypropylene or silicone) with high permeablity to these respiratory gases. The oxygenated blood is then passed through a heat exchanger to adjust the temperature of the blood and the blood is returned to the systemic circulation via the aorta. At the beginning of the procedure, the patient is anticoagulated with heparin to reduce the risk of thrombosis within the device; as the patient is weaned from bypass, the anticoagulation can be quickly reversed by the use of a drug called protamine. The heat exchanger lowers the temperature of the blood and therefore the core body temperature during the operation, decreasing the metabolic requirements of the body and protecting the organs (including the heart) against ischemic damage. At the end of the operation, the blood can be warmed to normal physiologic temperature as the patient is weaned from the bypass machine. During the surgical procedure, a specially trained perfusionist controls the operation of the heart–lung machine, allowing the surgeon and anesthesiologist to concentrate on their respective tasks. This device, therefore, provides the function of both the heart (maintaining systemic blood flow and pressure) and the lungs (oxygenating blood and removing carbon dioxide), allowing the heart to be effectively stopped for delicate surgical procedures that would be next-to-impossible to perform on a beating, moving heart.

Many improvements to the original design of cardiopulmonary bypass machines have been made since their inception. One of the problems with the original heart–lung machines was the trauma that they would cause to the blood, including red blood cells (hemolysis) leading to functional anemia and thrombocytopenia (low numbers of or dysfunctional platelets), leading to bleeding problems. This problem has been largely overcome with advanced pump designs and the use of the membrane oxygenators. Roller pumps and centrifugal pumps are commonly used because they cause a lesser degree of hemolysis and shear forces; it is important in the design of these pumps to determine the optimum balance between pumping function and hemolysis/shear stress to the formed blood elements. Bubble oxygenators, which directly pass bubbles of oxygen gas through the blood, cause more hemolysis, protein denaturation, and platelet dysfunction than membrane oxygenators and are currently less frequently used. In addition, newer devices allow blood that has escaped the circulation within the sterile operating field around the heart to be processed and returned to the patient, reducing the need for blood transfusion during the procedure.

Cardiopulmonary bypass can result in many pathophysiologic changes, including complement activation from the prolonged interaction of blood with synthetic surfaces (see Chapter 4.4), platelet and neutrophil activation and aggregation, changes in systemic vascular resistance, and expression of other proinflammatory mediators (Levy and Tanaka, 2003). When these changes are severe, the use of the heart–lung machine can result in confusion, renal insufficiency, pulmonary dysfunction, low-grade hepatic dysfunction, and increased susceptibility to infection. Together, these manifestations are termed the postperfusion syndrome. Current research is aimed at reducing these changes through the use of novel materials with improved biocompatibility, shortened and "off pump" surgeries, and certain therapeutic strategies to reduce the systemic effects of the proinflammatory mediators.

Intraaortic Balloon Pump

Since the original use of the intraaortic balloon pump (IABP) in 1968 by Adrian Kantrowitz, the basic design and function of the current device has remained relatively similar during the ensuing decades, but the indications for IABP have increased. IABPs (Fig. 15) are catheter-based polyethylene or polyurethane balloons with volumes of 30–50 ml, although smaller devices are used in the pediatric population. Helium is used as the inflating gas; its low viscosity allows for rapid inflation and deflation and it is rapidly dissolved in the bloodstream in the event of inadvertent balloon rupture.

IABPs (Baskett *et al.*, 2002; Nanas and Moulopoulos, 1994) are generally positioned under fluoroscopic guidance in the descending thoracic aorta after percutaneous insertion via the femoral artery. They are timed to inflate during diastole (ventricular relaxation) and deflate during systole (ventricular contraction) using the patient's electrocardiogram or arterial pressure curve for synchronization. This is counterpulsation, which is out of phase with the patient's heartbeat and causes volume displacement of blood proximally and distally within the aorta. Several beneficial effects serve to improve cardiac function. Coronary blood flow (the majority of which occurs in diastole) is increased by the rise in diastolic pressure, delivering more oxygenated blood to the myocardium. In addition, left ventricular afterload (the pressure the myocardium must attain to pump blood into the aorta) is decreased, reducing the workload and therefore the oxygen requirement of the myocardium. The combination of these two hemodynamic factors therefore improves the balance between myocardial oxygen supply and demand and results in improved cardiac performance. The device also directly improves systemic circulation, but only to a modest degree (approximately 10%), in contrast to other cardiac assist modalities.

IABP (Torchiana *et al.*, 1997) therapy permits the heart to rest and recover enough function to support adequate circulation after the device has been removed, usually after only a few days. The major contraindications for IABP use include aortic valve regurgitation and aortic dissection. Complications, which occur in 11–33% of patients with IABPs, include limb ischemia from problems at the insertion site requiring early balloon removal or vascular surgery, aortic dissection, balloon rupture, hematoma, and sepsis.

Ventricular Assist Devices

Ventricular assist devices and artificial hearts have been used primarily in two settings: (1) for potentially reversible heart

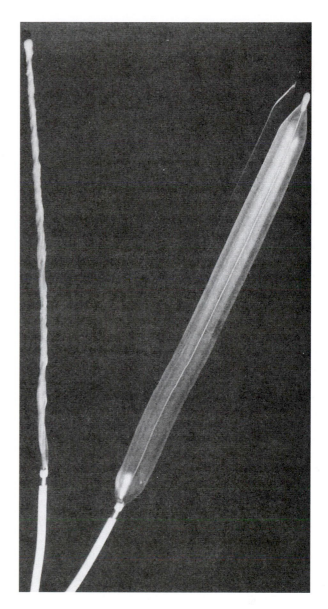

FIG. 15. Percutaneous intra-aortic balloon pump. Left, balloon deflated for insertion. Right, balloon inflated. (Courtesy S. Volvek, Datascope Corp., Oakland, NJ.)

failure, in which cardiac function is likely to recover with cardiac rest (e.g., postcardiotomy shock); and (2) for end-stage cardiac failure not likely to recover and where mechanical support will provide a bridge to transplantation. Left ventricular assist devices (LVADs) also provide long-term cardiac support with survival and quality-of-life improvements over optimal medical therapy for patients with end-stage congestive heart failure who are not transplant candidates (Rose *et al.*, 2001). These devices represent life-sustaining "destination therapy" for heart failure patients with contraindications to transplantation such as metastatic cancer. LVADs are also being investigated as a "bridge-to-recovery" in patients with congestive heart failure to induce ventricular changes that might improve cardiac function and eventually allow device

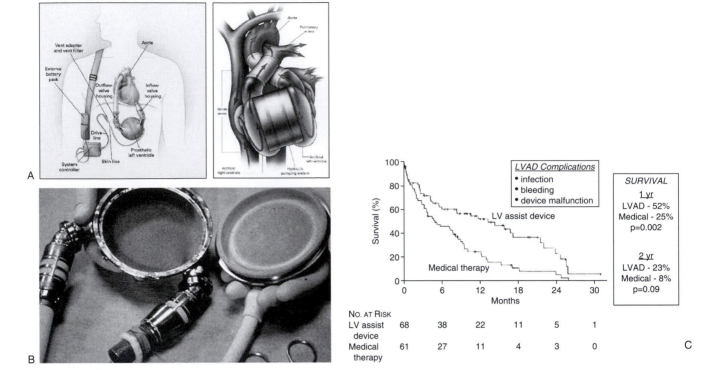

FIG. 16. Left ventricular assist device (LVAD) and total artificial heart. (A) Schematic representation of Thoratec HeartMate LVAD and Abiocor Total Implantable Artificial Heart. See also www.heartpioneers.com/abiocorimages.html. (B) HeartMate LVAD with textured internal surfaces. (C) Overall survival data for patients with end-stage heart failure treated with LVAD or with optimal medical treatment. A and C: Reproduced by permission from Rose, E. A., *et al.* (2001). Long-term use of a left ventricular assist device for patients with end-stage heart failure. *N. Engl. J. Med.* **345:** 1435–1443. B: Reproduced by permission from Schoen, F. J., and Edwards, W. D. (2001). Pathology of cardiovascular interventions. in *Cardiovascular Pathology*, 3rd ed., M. D. Silver, A. I. Gotlieb, and F. J. Schoen, eds. Churchill Livingstone, New York.

removal. Research in this area focuses on the mechanisms of cardiac recovery, identification of patients who could achieve recovery, and specifics such as the timing and duration of therapy (Kumpati *et al.*, 2001; Young, 2001).

Ventricular assist devices, first successfully employed by Michael DeBakey in 1963, can replace ventricular function for extended periods, in contrast to the short-term duration of cardiopulmonary bypass or IABP use. The heart remains in place with the VAD inflow connected to the atrium or ventricle and the VAD outflow connected to the aorta or pulmonary artery (Fig. 16A). In contrast, an artificial heart is composed of two pumping chambers that together replace the heart, analogous to heart transplantation. FDA-approved VADs include univentricular (used as a left or right ventricular assist device) and biventricular extracorporeal and implantable pulsatile devices. These pumps have mechanical or bioprosthetic valves to ensure unidirectional flow, fabric and metal conduits, and a blood-containing chamber with an elastomeric moving bladder/membrane in a rigid housing. Some devices are pneumatically driven by pulses of air generated by an external pneumatic unit and transmitted through tubes to the pumping unit, whereas others are electrically driven. The pumps themselves can be extracorporeal or implanted in the abdominal cavity. Some component of the system traverses the skin in most types of devices; the blood-containing conduits traverse the skin as they pass between the heart/aorta and the device in extracorporeal systems, while the tubing containing the pneumatic driveline or the electrical cord and air vent passes through the skin in implantable devices. The goal is to eliminate all through-the-skin connections and to transmit electrical energy via radio frequency across intact skin.

The Abiomed BVS 5000 is an extracorporeal, uni- or biventricular VAD that is one of the most commonly used devices for short-term mechanical support. The Thoratec VAD is an extracorporeal uni- or biventricular VAD that uses a pneumatic pusher-plate pump and is often used to provide long-term biventricular support as a bridge-to-transplant. The Novacor N1000PC is an LVAD system with an implantable pump and externalized vent tube, controller, and batteries that employs smooth polyurethane blood-contacting surfaces. These three devices require the patient to be on long-term anticoagulation to lower the risk of thrombosis and thromboembolic events. The Thoratec HeartMate system has both pneumatic and electrical models and consists of an implantable, long-term univentricular support system that was the first VAD to be approved as a bridge-to-transplant. This device (Fig. 16B) has a textured polyurethane-coated diaphragm surface along with a metallic surface covered with titanium beads, both of which encourage the deposition of circulating cells and coagulation proteins to form a stable pseudointima that minimizes additional thrombus formation and thromboemboli even without long-term anticoagulation. A recently published clinical trial (REMATCH) indicated both the efficacy of chronic VAD support in enhancing the survival of patients with otherwise refractory heart failure, and the safety of the pump surface design in preventing thromboembolism (Fig. 16C). Three classes of advanced devices are under development, including (1) pneumatic devices that are wrapped around the heart and thus do not directly contact blood, (2) compact intracardiac axial-flow pumps (Goldstein, 2003), and (3) a total implantable artificial heart.

Total Implantable Artificial Heart

The Jarvik-7 heart has served as a long-term mechanical cardiac replacement in several individuals since the first implantation of this device in 1982; this and several other types of replacement devices have been used as a bridge to transplantation. The first total artificial heart implants in the United States since the mid-1980s are currently being evaluated (AbioCor, Abiomed Corp.) (Marshalls, 2002).

The Jarvik-7 total artificial heart consisted of two pneumatically powered polyurethane sac-type blood pumps, each with either a 70 or 100 ml stroke volume and powered by separate power units, which provided both systemic and pulmonary circulations. Tilting-disk-type prosthetic inflow and outflow valves provided unidirectional blood flow in the two ventricles. Air tubes from the external pneumatic power unit traveled to the device through the skin via infection-resistant skin buttons. Air pulses entered the space between the cases to compress and expand the sac, providing cardiac output of 4–8 liters per minute. Each ventricle was attached to the native remnant atrium with an atrial suture cuff, while the outlet port was attached to the aorta or pulmonary artery by a conventional vascular graft. Clinical studies showed that the available artificial hearts would fit in the chest of an appropriately sized patient, would provide an adequate cardiac output, and could maintain cardiovascular pressures within a physiologic range. Infection and thromboembolism were frequent complications in these already critically ill patients, and ultimately resulted in withdrawal of this device from the commercial market. Implant retrieval and evaluation studies of these devices demonstrated that transitions of materials and geometries within the pumps were the most vulnerable sites for thrombus formation (Fyfe and Schoen, 1993).

The AbioCor implantable replacement heart (Fig. 16A) is designed to fully sustain the circulatory system and to mimic the function of the human heart while being a completely implantable system with no wires or ports traversing the skin. The internal thoracic unit, weighing 1 kilogram and about the size of a grapefruit, consists of two ventricles with artificial valves and a motorized hydraulic pumping system. In addition, there is an internal rechargeable battery, a miniaturized electronics package, and an external battery pack. The electronics package allows the control and monitoring of the heart rate based on the physiologic requirements of the patient at any given time. The external battery pack continually transmits energy to the internal battery transcutaneously, eliminating the problem of device infection via percutaneous lines. The internal lithium battery can hold enough power for about 30 minutes of use, allowing the patient to be free of the external pack for short periods of time. The internal moving parts of the device are fabricated from proprietary polyurethane (AngioFlex) that is designed to be a smooth biocompatible blood-contacting surface and to have flexibility and durability to withstand many years of continuous use. To implant the device, most of the native heart is removed; the left and right ventricular inflows are attached to the remnants of the left and right atria, with the outflows attached to the remnant aorta and pulmonary artery, respectively.

Complications of Cardiac Assist Devices

The major complications of cardiac assist devices are hemorrhage, thrombosis/thromboembolism, and infection. Additional common complications include hemolysis, pannus formation around anastomotic sites, and device failure. Device failure can occur secondary to fracture or tear of one of the prosthetic valves within the device conduits as this application provides a particularly severe test of valve durability (Fig. 17A), or can occur secondary to damage to the pumping bladder.

Hemorrhage (Fig. 17B) continues to be a problem in device recipients, although the risk of major hemorrhage has been decreasing with improved therapies and methods. Many factors predispose to perioperative hemorrhage, including (1) coagulopathy secondary to liver dysfunction, poor nutritional status, and anticoagulation therapy, (2) platelet dysfunction secondary to contact of blood with synthetic surfaces, and (3) the extensive nature of the required surgery. Hemorrhage is also a complication of anticoagulation in general, even in the absence of a medical device.

Nonthrombogenic blood-contacting surfaces and appropriate blood flow characteristics are essential for a clinically useful cardiac assist device or artificial heart. Thrombi form primarily in association with crevices and voids, especially in areas of disturbed blood flow such as connections of conduits and other components to each other and to the natural heart. Many pumps have used smooth polymeric pumping bladders; these are frequently associated with thromboembolic complications. One approach discussed earlier (Thoratec HeartMate) is the use of textured surfaces, which accumulate a controlled layer of thrombotic material that seems to be resistant to both further thrombus and thromboemboli (Long, 2001).

Infectious complications (Fig. 17C) have been a major limiting factor in the prolonged use of cardiac assist devices. Infection can occur within the device but may also be associated with the percutaneous pneumatic or electrical lines. Susceptibility to infection is potentiated not only by the usual prosthesis-associated factors, but also by the multisystem organ damage from the underlying disease, by the periprosthetic culture medium provided by postoperative hemorrhage, and by prolonged hospitalization with the associated risk of nosocomial infections. These infections are a significant cause of morbidity and mortality and are often resistant to antibiotic therapy and host defenses.

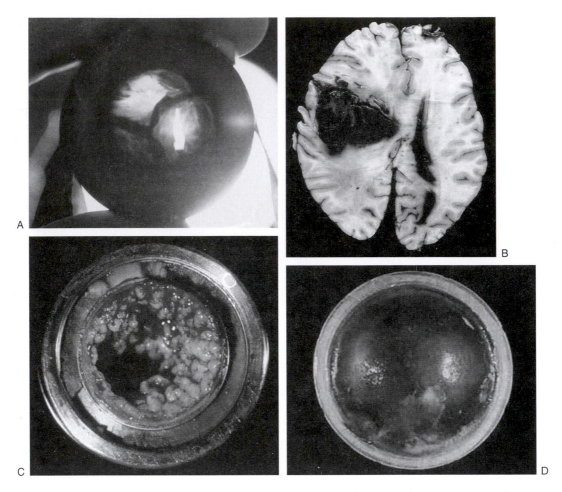

FIG. 17. Complications of cardiac assist devices. (A) Cuspal tear in inflow valve of LVAD. (B) Hemorrhage into the brain in a patient with an LVAD. (C) Fungal infection in LVAD outflow graft. (D) Thrombosis on pumping bladder. A: Reproduced by permission from Schoen, F. J., and Padera, R. F. (2003). Pathologic considerations in the surgery of adult heart disease. in *Cardiac Surgery in the Adult*, 2nd ed., L. H. Cohn, ed. McGraw-Hill, New York. C: Reproduced by permission from Schoen, F. J., and Edwards, W. D. (2001). Pathology of cardiovascular interventions. in *Cardiovascular Pathology*, 3rd ed., M. D. Silver, A. I. Gotlieb, and F. J. Schoen, eds. Churchill Livingstone, New York. D: Reproduced by permission from Fyfe, B., and Schoen, F. J. (1993). Pathologic analysis of 34 explanted Symbion ventricular assist devices and 10 explanted Jarvik-7 total artificial hearts. *Cardiovasc. Pathol.* 2: 187–197. (See color plate)

MISCELLANEOUS CARDIOVASCULAR DEVICES

Closure Devices: Atrial Septal Defect and Patent Ductus Arteriosus

In prenatal life, the circulation is different than it is in postnatal life (Schoen, 1999a). The lungs of the fetus are not expanding and oxygenation of fetal blood is provided via the placenta and maternal circulation. This requires two important shunts to be present that close immediately after birth. The foramen ovale, a hole in the fetal intraatrial septum, allows oxygenated blood returning to the right atrium from the placenta to preferentially pass into the left atrium. This blood passes through the mitral valve into the left ventricle and is pumped out through the aorta into the systemic circulation. The ductus arteriosus, present between the pulmonary artery and aorta, allows deoxygenated blood pumped from

the right ventricle to bypass the lungs and directly reenter the systemic circulation, as the prenatal pulmonary circulation has a high vascular resistance (owing to the nonexpanded lungs). After birth, these functional shunts should close to completely separate the right and left circulations; failure to do so results in an atrial septal defect (ASD) or patent ductus arteriosus (PDA) that can allow inappropriate shunting of blood in the postnatal circulation. ASDs or ventricular septal defects (VSDs) can also result from abnormal formation of the muscle of the atrial septum or ventricular septum.

Although these defects can be closed via an open surgical procedure (sutures and/or fabric patches for ASD or VSD, simple ligation for PDA), efforts have been made to allow closure of these defects using a minimally invasive approach (Hornung *et al.*, 2002). The decision to close an ASD, VSD, or PDA depends on the size of the shunt and the symptoms, if any, of the patient. The first catheter-based closure of a PDA was performed in 1967 by Porstmann using an Ivalon plug placed in the PDA to occlude flow. Of the many PDA closure devices that have been developed over the years, most are metal-based devices that work by causing thrombosis of the PDA with subsequent organization and fibrosis, permanently preventing flow through the residual ductus arteriosus. Mills and King reported the first transcatheter closure of an ASD in 1976 using a double umbrella device that covered the opening from both the right and left atrial sides. Their occlusion device consisted of a skeleton of ePTFE-coated wire supporting an occluder of Dacron fabric delivered through a catheter. Improvements over the years include better device fixation methods and smaller caliber introducers. Several ASD closure devices are being evaluated in clinical trials. Advantages of nonsurgical closure devices include shorter hospital stay, more rapid recovery, and no residual thoracotomy scar. As experience is gained in the transcatheter closure of PDAs and ASDs, this technology is being extended to the closure of VSDs, particularly in high-risk patients thought to be poor operative risks.

Inferior Vena Cava Filters: Deep Venous Thrombosis and Pulmonary Embolus

Venous thromboembolic disease is a significant cause of morbidity and mortality, largely due to the complication of pulmonary embolism (PE). The most common scenario is for a thrombus to form in the deep venous system of the lower extremities (so-called deep venous thrombus or DVT), become detached from the wall of the vein, travel through the inferior vena cava to the right side of the heart, and lodge as an embolus in one of the large branches of the pulmonary artery. Therapy for patients with DVT or PE usually involves anticoagulation with warfarin and/or low-molecular-weight heparin. However, when anticoagulation is contraindicated because of active or threatened bleeding, or when there is recurrent DVT/PE despite adequate anticoagulation, placement of an inferior vena cava (IVC) filter is indicated.

The concept of placing a barrier to catch DVTs destined for the pulmonary circulation dates back to Trousseau in the 1860s, but such filters did not become clinical reality until the late 1960s (Greenfield and Michna, 1988). Prior to the development of these devices, surgical ligation of the IVC during an open abdominal operation was performed. Intervential radiologists began to place filters using a minimally invasive percutaneous approach in the 1980s. There are five types of filters commonly used in current clinical practice whose common design elements include metallic wires in a configuration to catch emboli in the bloodstream, a mechanism to anchor the device securely to the wall of the IVC, and an ability to be deployed through a catheter percutaneously. These devices (Streiff, 2000; Greenfield and Proctor, 2000) (Fig. 18) include the two Greenfield-style filters (one titanium and one stainless steel), the bird's-nest filter (stainless steel), the Simon nitinol filter (nickel–titanium alloy with thermal memory properties), and the Vena Tech filter (alloy of cobalt, chromium, iron, nickel, molybdenum, magnesium, carbon, and beryllium). Complications of these devices include thrombosis at the insertion site, thrombi forming on the filter itself (Fig. 18B), thrombosis and obstruction of the IVC, migration or tilting of the device, and penetration of the IVC wall. Many of these complications are rare and/or without clinical consequence. As these devices have become safer and more effective, the indications for their use have expanded. Current research involves creating devices with a lower profile to make insertion easier and reduce insertion site thrombosis, designing removable devices, and continuing to evolve the devices with the goal of catching potentially lethal emboli while maintaining adequate vena caval blood flow.

CONCLUSION

Cardiovascular medical devices have been used for more than a half-century for myriad applications that save, prolong, and enhance the quality of life for countless individuals. Still, complications from these devices can cause significant morbidity and mortality for individual patients, even years after implantation. Ongoing research aims to improve existing devices, reduce the frequency and severity of complications and develop novel approaches to the treatment of cardiovascular disease.

Bibliography

ACC/AHA (2001). Guidelines for Percutaneous Coronary Intervention (Revision of the 1993 PTCA Guidelines) Executive Summary. *Circulation* **103**: 3019–3041.

Alpard, S. K., and Zwischenberger, J. B. (2002). Extracorporeal membrane oxygenation for severe respiratory failure. *Chest Surg. Clin. N. Am.* **12**: 355–378.

Al Suwaidi, J., Berger, P. B., Holmes, D. R. Jr. (2000). Coronary artery stents. *J. Am. Med. Assoc.* **284**: 1828–1836.

Antiarrhythmics versus Implantable Defibrillators (AVID) Investigators (1997). A comparison of antiarrhythmic-drug therapy with implantable defibrillators in patients resuscitated from near-fatal ventricular arrhythmias. *N. Engl. J. Med.* **337**: 1576–1583.

Atlee, J. L., and Bernstein, A. D. (2001). Cardiac rhythm management devices (part I): indications, device selection, and function. *Anesthesiology* **95**: 1265–1280.

Baskett, R. J. F., Ghali, W. A., Maitland, A., and Hirsch, G. M. (2002). The intraaortic balloon pump in cardiac surgery. *Ann. Thorac. Surg.* **74**: 1276–1287.

Bolling, S. F. (2001). Mitral valve reconstruction in the patient with heart failure. *Heart Fail. Rev.* **6**: 177–185.

Butany, J., and Leask, R. (2001). The failure modes of biological prosthetic heart valves. *J. Long Term Eff. Med. Implants* **11**: 115–135.

Cao, H. (1995). Mechanical performance of pyrolytic carbon in prosthetic heart valve applications. *J. Heart Valve Dis.* **5**: S32–S49.

Clagett, G. P. (2002). What's new in vascular surgery? *J. Am. Coll. Surg.* **194**: 165–201.

Consigny, P. M. (2000). Endothelial cell seeding on prosthetic surfaces. *J. Long Term Eff. Med. Implants* **10**: 79–95.

Cornacchia, D., Fabbri, M., Puglisi, A., Moracchini, P., Bernasconi, M., Nastasi, M., Menozzi, C., Mascioli, G., Marotta, T., and de Seta, F. (2000). Latest generation of unipolar and bipolar steroid eluting leads: long-term comparison of electrical performance in atrium and ventricles. *Europace* **2**: 240–244.

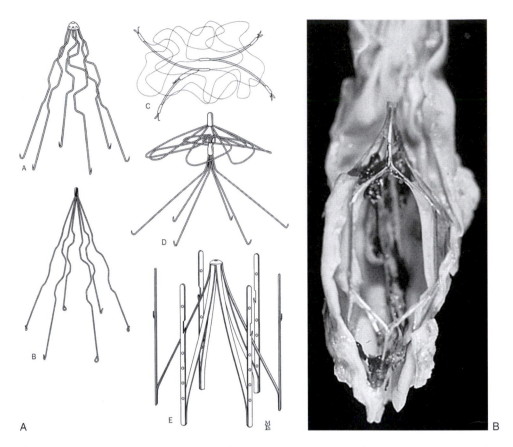

FIG. 18. Vena caval filters. (A) Diagram of the permanent vena caval filter models. A, The stainless steel Greenfield filter; B, the modified-hook titanium Greenfield filter; C, the bird's-nest filter; D, the Simon nitinol filter; E, the Vena Tech filter. (B) Photo of vena caval filter at autopsy demonstrating some thrombus at the filter site. A: Reproduced by permission from Streiff, M. B. (2000). Vena caval filters: a comprehensive review. *Blood* **95**: 3669–3677.

Cribier, A., Eltchaninoff, H., Borenstein, N., Tron, C., Baver, F., Derumeaux, G., Anselma, F., Laborda, F., Lean, M. B., and Bash, A. (2002). Percutaneous transcatheter implantation of an aortic valve prosthesis for calcific aortic stenosis. *Circulation* **106**: 3006–3008.

DiGiorgi, P. L., Rao, V., Naka, Y., and Oz, M. C. (2003). Which patient, which pump? *J. Heart Lung Transplant* **22**: 221–235.

Fann, J. I., and Burdon, T. A. (2001). Are the indications for tissue valves different in 2001 and how do we communicate these changes to our cardiology colleagues? *Curr. Opin. Cardiol.* **16**: 126–135.

Farb, A., Sangiorgi, G., Carter, A. J., Walley, V. M., Edwards, W. D., Schwartz, R. S., and Virmani, R. (1999). Pathology of acute and chronic coronary stenting in humans. *Circulation* **99**: 44–52.

Faries, P., Morrissey, N. J., Teodorescu, V., Gravereaux, E. C., Burks, Jr., J. A., Carroccio, A., Kent, K. C., Hollier, L. H., and Marin, M. L. (2002). Recent advances in peripheral angioplasty and stenting. *Angiology* **53**: 617–626.

Fattori, R., and Piva, T. (2003). Drug-eluting stents in vascular intervention. *Lancet* **361**: 247–249.

Fyfe, B., and Schoen, F. J. (1993). Pathologic analysis of 34 explanted Symbion ventricular assist devices and 10 explanted Jarvik-7 total artificial hearts. *Cardiovasc. Pathol.* **2**: 187–197.

Goldstein, D. J., Oz, M. C., and Rose, E. A. (1998). Implantable left ventricular assist devices. *N. Engl. J. Med.* **339**: 1522–1533.

Goldstein, D. J. (2003). Worldwide experience with the MicroMed DeBakey Ventricular Assist Device® as a bridge to transplantation. *Circulation* **108**[suppl 2]: II-272–II-277.

Greenfield, L. J., and Michna, B. A. (1988). Twelve-year clinical experience with the Greenfield vena cava filter. *Surgery* **104**: 706–712.

Greenfield, L. J., and Proctor, M. C. (2000). Current status of inferior vena cava filters. *Ann. Vasc. Surg.* **14**: 525–528.

Harken, D. E., Soroff, H. S., Taylor, W. J., LeFemine, A. A., Gupta, S. K., and Lunzer, S. (1960). Partial and complete prostheses in aortic insufficiency. *J. Thorac. Cardiovasc. Surg.* **40**: 744–762.

Harken, D. E., Taylor, W. J., LeFemine, A. A., Lunzer, S., Low, H. B., Cohen, W. L., and Jacobey, J. A. (1962). Aortic valve replacement with a caged ball valve. *Am. J. Cardiol.* **9**: 292–299.

Haudenschild, C. C. (1993). Pathobiology of restenosis after angioplasty. *Am. J. Med.* **94**(Suppl): 40–44.

Height, S. E., and Smith, M. P. (1999). Strategems for anticoagulant therapy following mechanical heart valve replacement. *J. Heart Valve Dis.* **8**: 662–664.

Hornung, T. S., Benson, L. N., and McLaughlin, P. R. (2002). Catheter interventions in adult patients with congenital heart disease. *Curr. Cardiol. Rep.* **4**: 54–62.

Huang, H., Virmani, R., Younis, H., Burke, A. P., Kamm, R. A., and Lea, R. T. (2001). The impact of calcification on the biomechanical stability of atherosclerotic plaques. *Circulation* **103**: 1051–1056.

Huikuri, H. V., Castellanos, A., and Myerburg, R. J. (2001). Sudden death due to cardiac arrhythmias. *N. Engl. J. Med.* **345**: 1473–1482.

Hunt, S. A., and Frazier, O. H. (1998). Mechanical circulatory support and cardiac transplantation. *Circulation* 97: 2079–2090.

Isom, O. W. (2002). Mitral commissurotomy and valve replacement for mitral stenosis: observations on selection of surgical procedures. *Adv. Cardiol.* 39: 114–121.

Jacobs, T. S., Won, J., Gravereaux, E. C., Favres, P. L., Morrissey, N., Teodorescu, V. J., Hollier, L. H., and Marin, M. L. (2003). Mechanical failure of prosthetic human implants: a 10-year experience with aortic stent graft devices. *J. Vasc. Surg.* 37: 16–26.

Jessup, M. (2001). Mechanical cardiac-support devices—dreams and devilish details. *N. Engl. J. Med.* 345: 1490–1493.

Jessup, M., and Brozena, S. (2003). Heart failure. *N. Engl. J. Med.* 348: 2007–2018.

Kibbe, M. R., Billiar, T. R., and Tzeng, E. (2000). Gene therapy for restenosis. *Circ. Res.* 86: 829–833.

Kolodgie, F. D., Burke, A. P., and Farb, A. (2001). The thin-cap fibroatheroma: A type of vulnerable plaque: the major precursor lesion to acute coronary syndromes. *Curr. Opin. Cardiol.* 16: 285–292.

Korossis, S. A., Fischer, J., and Ingham, E. (2000). Cardiac valve replacement: a bioengineering approach. *Biomed. Mater. Eng.* 10: 83–124.

Kumpati, G. S., McCarthy, P. M., and Hoercher, K. J. (2001). Left ventricular assist device bridge to recovery: a review of the current status. *Ann. Thorac. Surg.* 71: S103–S108.

Kusumoto, F. M., and Goldschlager, N. (2002). Device therapy for cardiac arrhythmias. *JAMA* 287: 1848–1852.

Landau, C., Lange, R. A., and Hillis, L. D. (1994). Percutaneous transluminal coronary angioplasty. *N. Engl. J. Med.* 330: 981–993.

Levy, J. H., and Tanaka, K. A. (2003). Inflammatory response to cardiopulmonary bypass. *Ann. Thorac. Surg.* 75: S715–S720.

Levy, R. J., Vyavahave, N., Ogle, M., Ashworth, P., Bianco, R., and Schoen, F. J. (2003). Inhibition of cusp and aortic wall calcification in ethanol- and aluminum-treated bioprosthetic heart valves in sheep: background, mechanisms and synergism. *J. Heart Valve Dis.* 12: 209–216.

Libby, P. (2000). Changing concepts in atherogenesis. *J. Intern. Med.* 247: 349–358.

Long, J. W. (2001). Advanced mechanical circulatory support with the HeartMate left ventricular assist device in the year 2000. *Ann. Thorac. Surg.* 71: S176–S182.

Marshall, E. (2002). A space age vision advances in the clinic. *Science* 295: 1000–1001.

Mylonakis, E., and Calderwood, S. B. (2001). Infective endocarditis in adults. *N. Engl. J. Med.* 345: 740–746.

Nanas, J. N., and Moulopoulos, S. D. (1994). Counterpulsation: historical background, technical improvements; hemodynamic and metabolic effects. *Cardiology* 84: 156–167.

O'Brien, M. F., Harrocks, S., Stafford, E. G., Gardner, M. A., Pohlner, P. G., Tesar, P. J., and Stephens, F. (2001). The homograft aortic valve: a 29-year, 99.3% follow up of 1,022 valve replacements. *J. Heart Valve Dis.* 10: 334–344.

Park, S. J., Shim, W. H., Ho, D. S., Raizner, A. E., Park, S. W., Hong, M. K., Lee, C. W., Choi, D., Jang, Y., Lam, R., Weissman, N. J., and Mintz, G. S. (2003). A paclitaxel-eluting stent for the prevention of coronary restenosis. *N. Engl. J. Med.* 348: 1537–1545.

Piper, C., Korfer, R., and Horstkotte, D. (2001). Prosthetic valve endocarditis. *Heart* 85: 590–593.

Poon, M., Badimon, J. J., and Fuster, V. (2002). Overcoming restenosis with sirolimus: from alphabet soup to clinical reality. *Lancet* 359: 619–622.

Rabkin, E., and Schoen, F. J. (2002). Cardiovascular tissue engineering. *Cardiovasc. Pathol.* 11: 305–317.

Rafii, S. (2000). Circulating endothelial precursors: mystery, reality, and promise. *J. Clin. Invest.* 105: 17–19.

Rahimtoolla, S. H. (2003). Choice of prosthetic heart valve for adult patients. *J. Am. Coll. Cardiol.* 41: 893–904.

Ramaswami, G., and Marin, M. L. (1999). Stent grafts in occlusive arterial disease. *Surg. Clin. North Am.* 79: 597–609.

Riessen, R., and Isner, J. M. (1994). Prospects for site-specific delivery of pharmacologic and molecular therapies. *J. Am. Coll. Cardiol.* 23: 1234–1244.

Rose, E. A., Gelijns, A. C., Moskowitz, A. J. *et al.* (2001). Long-term use of a left ventricular assist device for end-stage heart failure. *N. Engl. J. Med.* 345: 1435–1443.

Ross, R. (1999). Atherosclerosis. An inflammatory disease. *N. Engl. J. Med.* 340: 115–126.

Sacks, M. S. (2001). The biomechanical effects of fatigue on the porcine bioprosthetic heart valve. *J. Long Term Eff. Med. Implants* 11: 231–247.

Salame, M. Y., Verheye, S., Crocker, I. R., *et al.* (2001). Intracoronary radiation therapy. *Eur. Heart. J.* 22: 629–647.

Sapirstein, J. S., Smith, P. K. (2001). The "ideal" replacement heart valve. *Am. Heart. J.* 141: 856–860.

Schoen, F. J. (1995a). Approach to the analysis of cardiac valve prostheses as surgical pathology or autopsy specimens. *Cardiovasc. Pathol.* 4: 241–255.

Schoen, F. J. (1995b). Pathologic considerations in replacement heart valves and other cardiovascular prosthetic devices. in: *Cardiovascular Pathology: Clinicopathologic Correlations and Pathogenetic Mechanisms*. F. J. Schoen, and M. A. Gimbrone, eds. Williams and Wilkins, Baltimore, p. 194.

Schoen, F. J. (1999a). The heart. in: *Robbins Pathologic Basis of Disease*, 6th ed., R. S. Cotran, V. Kumar, and T. Collins, eds. W.B. Saunders, Philadephia, pp. 543–599.

Schoen, F. J. (1999b). Future directions in tissue heart valves: impact of recent insights from biology and pathology. *J. Heart Valve Dis.* 8: 350–358.

Schoen, F. J. (2001). Pathology of heart valve substitution with mechanical and tissue prostheses. in: *Cardiovascular Pathology*, 3rd ed., M. D. Silver, A. I. Gotlieb, and F. J. Schoen, eds. W.B. Saunders, pp. 629–677.

Schoen, F. J., and Cotran, R. S. (1999). Blood vessels. in: *Robbins Pathologic Basis of Disease*, 6th ed., R. S. Cotran, V. Kumar, and T. Collins, eds. W.B. Saunders, Philadephia, pp. 493–541.

Schoen, F. J., and Levy, R. J. (1994). Pathology of substitute heart valves: new concepts and developments. *J. Cardiac Surg.* 9(Suppl): 222–227.

Schoen, F. J., and Levy, R. J. (1999). Tissue heart valves: current challenges and future research perspectives. *J. Biomed. Mater. Res.* 47: 439–465.

Seifalian, A. M., Tiwari, A., Hamilton, G., and Salacinski, H. J. (2001). Improving the clinical patency of prosthetic vascular and coronary artery bypass grafts: the role of seeding and tissue engineering. *Artif. Organs* 26: 307–320.

Serruys, P. W., de Jaegere, P., Kiemeneij, F., Macaya, C., Rutsch, W., Heyndrick, G., Emanuelsson, H., Marco, J., Legrand, V., Materne, P., Belardi, J., Sigwart, U., Colombo, A., Goy, J. J., van der Heuvel, P., Delcan, J., and Murel, M-A. (1994). A comparison of balloon-expandable-stent implantation with balloon angioplasty in patients with coronary artery disease. *N. Engl. J. Med.* 331: 489–495.

Sousa, J. E., Serruys, P. W., and Costa, M. A. (2003a). New frontiers in cardiology: drug-eluting stents—pt I. *Circulation* 107: 2274–2279.

Sousa, J. E., Serruys, P. W., and Costa, M. A. (2003b). New frontiers in cardiology: drug-eluting stents—pt II. *Circulation* 107: 2283–2289.

Sousa, J. E., Costa, M. A., Sousa, A. G., Abizaid, A. C., Seixou, A. C., Abizaid, A. S., Feres, F., Mattos, L. A., Falotico, R., Jaeger, J., Pompa, J. J., and Serruys, P.W. (2003c). Two-year angiographic and intravascular utrasound follow-up after implantation of sirolimus-eluting stents in human coronary arteries. *Circulation* **107**: 381–383.

Starr, A., and Edwards, M. L. (1961). Mitral replacement: clinical experience with a ball valve prosthesis. *Ann. Surg.* **154**: 726–740.

Stone, G. W., Grines, C. L., Cox, D. A., Garcia, E., Tcheng, J. E., Griffin, J. J., Guagliumi, G., Stuckey, T., Turco, M., Carroll, J. D., Rutherford, B. D., and Lanksy, A. J. (2002). Comparison of angioplasty with stenting, with or without abciximab, in acute myocardial infarction. *N. Engl. J. Med.* **346**: 957–966.

Streiff, M. B. (2000). Vena caval filters: a comprehensive review. *Blood* **95**: 3669–3677.

Torchiana, D. F., Hirsch, G., Buckley, M. J., *et al.* (1997). Intraaortic balloon pumping for cardiac support: trends in practice and outcome, 1968 to 1995. *J. Thorac. Cardiovasc. Surg.* **113**: 758–769.

Virmani, R., and Farb, A. (1999). Pathology of in-stent restenosis. *Curr. Opin. Lipid.* **10**: 499–506.

Virmani, R., Farb, A., and Burke, A. P. (1994). Coronary angioplasty from the perspective of atherosclerotic plaque: morphologic predictors of immediate success and restenosis. *Am. Heart J.* **127**: 163–179.

Voet, J. G., Vandekerckhove, Y. R., Muyldermans, L. L., Missault, L. H., and Matthys, L. J. (1999). Pacemaker lead infection: report of three cases and review of the literature. *Heart* **81**: 88–91.

Vongpatanasin, W., Hillis, L. D., and Lange, R. A. (1996). Prosthetic heart valves. *N. Engl. J. Med.* **335**: 407–416.

Vyavahare, N., Jones, P. L., Hirsch, D., Schoen, F. J., and Lavy, R. J. (2000). Prevention of glutaraldehyde-fixed bioprosthetic heart valve calcification by alcohol pretreatment: further mechanistic studies. *J. Heart Valve Dis.* **9**: 561–566.

Watarida, S., Shiraishi, S., Nishi, T., Imura, M., Yamamoto, Y., Hirokawa, R., and Fujita, M. (2001). Strut fracture of Bjork–Shiley convexo-concave valve in Japan—risk of small valve size. *Ann. Thorac. Cardiovasc. Surg.* **7**: 246–249.

Welt, F. G., and Rogers, C. (2002). Inflammation and restenosis in the stent era. *Arterioscler. Thromb. Vasc. Biol.* **22**: 1769–1776.

Young, J. B. (2001). Healing the heart with ventricular assist device therapy: mechanisms of cardiac recovery. *Ann. Thorac. Surg.* **71**: S210–S219.

7.4 IMPLANTABLE CARDIAC ASSIST DEVICES

William R. Wagner, Harvey S. Borovetz, and Bartley P. Griffith

CLINICAL NEED AND APPLICATIONS

Heart failure results in more than 43,000 deaths per year in the United States and contributes to the death of another 220,000 individuals. There are approximately 400,000 new cases diagnosed annually and those diagnosed with this condition have a mortality rate of about 50% at 5 years (American Heart Association, 1999). When the severity of heart failure is graded, those in the sickest class (New York Heart Association class IV) have an even poorer prognosis, with a survival rate of only about 25% at 2 years (Schocken *et al.*, 1992). For these end-stage patients heart transplantation has become an effective treatment over the past 20 years, largely as a result of the introduction of successful immune suppression drugs such as cyclosporine. Actuarial survival rates for heart transplant patients are currently 75% at 3 years (United Network for Organ Sharing, 1999).

The success of heart transplantation is limited, however, by an inadequate and stagnant donor supply. Although 35,000–64,000 patients could potentially benefit from heart transplantation annually (Funk, 1991), in 2003, only 2055 cardiac transplants were performed in the United States. While waiting for scarce donor organs to become available, approximately 20% of individuals listed for heart transplantation die annually (United Network for Organ Sharing, 1999). To address the need to support the circulation in patients with end-stage heart failure a wide variety of mechanical devices have been developed over the past several decades. Current devices that are widely used for the purpose of bridge-to-cardiac transplantation will be discussed in this chapter, as will circulatory support devices that may enter the market in the near future.

Roles for Ventricular Assist Devices

Three devices are currently approved by the FDA for the provision of circulatory support while patients await a donor heart. These devices do not replace the patient's heart (as would a total artificial heart), but rather work as ventricular assist devices (VADs). The general concept behind a VAD can be seen in Fig. 1. The inflow conduit of the device is connected to the apex of the left ventricle. Blood from the patient's ventricle enters the VAD conduit and flows toward a unidirectional valve that directs flow into a pumping sac. The pumping sac

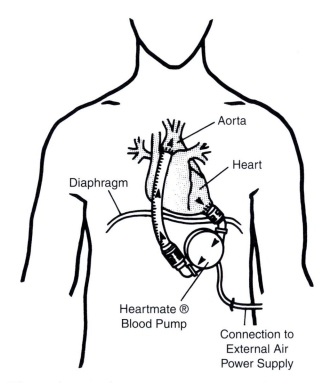

FIG. 1. ThermoCardiosystems HeartMate VAD demonstrating anatomical placement and blood flow. (Reprinted from F. A. Arabia, R. G. Smith, D. S. Rose, D. A. Arzouman, G. K. Sethi, and J. G. Copeland, *ASAIO J.* **42**: M542–M546, 1996.)

can be triggered to compress by several methods: sensing pump filling, compressing at a constant frequency, or triggering off the patient's electrocardiogram. Upon compression the inflow valve is closed and the outflow valve opens, permitting blood flow through the outflow conduit toward the patient's ascending aorta. The VAD thus "unloads" the left ventricle of the heart and provides the necessary work to periodically propel blood into the arterial tree.

As a bridge-to-cardiac transplantation, VADs have provided circulatory support to end-stage heart failure patients who otherwise would have a low likelihood of surviving until a donor organ became available. Nearly 70% of patients undergoing VAD support have survived until heart transplantation (Mehta *et al.*, 1995). Patients undergoing VAD "bridging" generally do not undergo further deterioration in their condition and, in fact, demonstrate markedly improved end-organ function and an improvement in their health status due to improved perfusion (Frazier *et al.*, 1994). Ironically, the success of VAD bridging has been implicated in intensifying the donor shortage by including recipients who would otherwise have not survived until transplantation (Massad *et al.*, 1996). A second result of this trend is the need for increasingly extended periods of VAD support prior to organ availability.

The improvement in end organ function generally associated with extended circulatory support has led investigators to evaluate whether the heart muscle itself may undergo recovery from the disease process of heart failure during the support period. A variety of recent studies in VAD-supported patients have shown a reduction in the inflammatory mediators associated with heart failure as well as decreased myocyte necrosis and apoptosis, improved myocyte contractility, and indications of improved function in the left ventricle (Mann and Willerson, 1998). These reports have lead to the investigation of VAD support as a "bridge to recovery" where patients are weaned after extended periods of circulatory support. The attractiveness in bridging to recovery lies in freedom from transplantation and immunosuppressive therapy for the patient and an increased donor organ supply for the community. Early clinical experience with this procedure indicates that, although success can be achieved, patients must be carefully screened for this procedure and the majority of VAD patients are unlikely to meet selection criteria (Loebe *et al.*, 1999).

Another result of the increasingly positive experience of supporting patients until transplantation with VADs is the concept of using VADs as a permanent source of circulatory support or as a "destination therapy." A number of permanent VAD implants have occurred in Europe in the past several years and a randomized trial comparing VAD implantation with medical therapy in patients who require, but are ineligible for heart transplantation, has been initiated in the United States with support from the National Heart, Lung, and Blood Institute (Rose *et al.*, 1999). A number of factors have led to the consideration and investigation of permanent support: the limited donor organ supply, a significant patient population that fails to meet age and medical eligibility requirements for transplant listing, and improved quality of life with recent portable VAD designs. Major concerns regarding this option focus primarily on the complications that remain associated with VAD support. As one might expect, biomaterial performance and

device biocompatibility are central to the perceived limits of utilizing VADs as alternatives to transplant in end-stage heart failure patients.

In looking back on the history of mechanical circulatory support, it is interesting that the field has returned to the challenge that was taken up in the early to mid-1980s with the first implants of the Jarvik-7 total artificial heart (TAH) intended for permanent support (DeVries *et al.*, 1984). Medical complications such as infection and thromboembolism, as well as manufacturing concerns, led to the discontinuation of TAH use as permanent support. Also at this time the success of immunosuppressive therapies was being reported, raising heart transplantation to the role of preferred therapy for end-stage heart failure. The TAH was subsequently transferred to a different corporate sponsor (CardioWest, Tucson, AZ) and utilized in the bridge to transplant role (Guy, 1998). In the meantime VADs were being used as investigational devices for bridging patients to transplantation with demonstrated success. In the 1990s several VADs received FDA approval for use as a bridge to cardiac transplantation. Supported primarily by the National Institutes of Health, research and development of TAHs is ongoing; however, clinical circulatory support is accomplished almost exclusively by VADs. This is not to say, however, that the TAH is not currently used in the bridge-to-cardiac-transplantation role. The CardioWest TAH has been implanted in 150 patients since 1993 and has a success rate of bridging patients to transplantation comparable to that of the more widely utilized VADs (Arabia *et al.*, 1999; Copeland *et al.*, 1997).

This chapter will focus only on current VAD designs widely utilized in the bridge-to-cardiac-transplantation role when discussing biomaterials and biocompatibility considerations.

VENTRICULAR ASSIST DEVICE DESIGN AND BLOOD CONTACTING MATERIALS

Thoratec

The Thoratec VAD (Thoratec Laboratories, Pleasanton, CA) is based upon the Pierce–Donachy VAD originally developed in the mid-1970s at Pennsylvania State University and currently has been implanted in over 1000 patients. In Fig. 2 the interface of this VAD with the circulatory system is presented. Of particular note is the paracorporeal nature of this device. Both inflow and outflow conduits cross the skin and the pumping sac rests on the lower abdomen. One advantage of this type of design is the potential to implant smaller patients. While the other commonly utilized VADs are often incompatible in terms of size with smaller, often female patients, the Thoratec device has been implanted in patients ranging in weight from 37 to 317 pounds. Also of note is the potential for this device to be used in a biventricular support mode with two devices implanted in tandem as shown in Fig. 2. Although it has been found that the majority of patients experiencing end-stage heart failure require only left ventricular assistance (Kormos *et al.*, 1996), for those who require support of the right ventricle, the Thoratec pump can be placed in conjunction with another Thoratec pump or other VAD supporting the left ventricle.

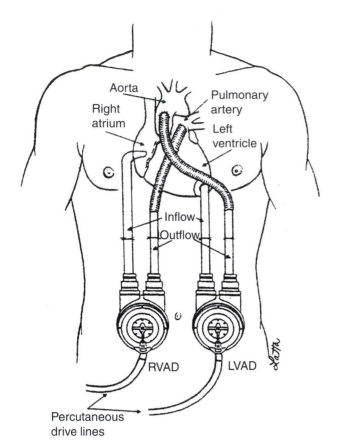

FIG. 2. Thoratec VADs utilized for biventricular support. The right ventricular assist device (RVAD) draws blood from the right atrium, while the left ventricular assist device (LVAD) is connected to the apex of the left ventricle. Both pumping sacs rest outside of the body with inflow and outflow conduits crossing the skin. (Reprinted from S. A. Hunt and O. H. Frazier, *Circulation* 97: 2079–2090, 1998.)

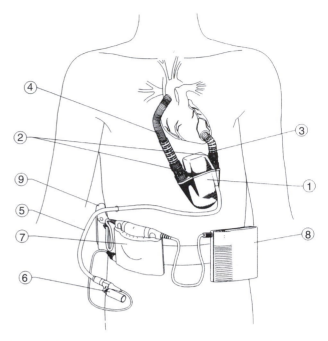

FIG. 3. Novacor VAD with wearable power pack and controller. 1, Pump/drive unit; 2, valved conduits; 3, inflow conduit; 4, outflow conduit; 5, percutaneous line; 6, filter; 7, compact controller; 8, primary power pack; 9, reserve power pack. (Reprinted from A. El-Banayosy, M. Deng, D. Y. Loisance, H. Vetter, E. Gronda, M. Loebe, and M. Vigano, *Eur. J. Cardiothorac. Surg.* 15: 835–841, 1999.)

The Thoratec VAD has the potential to draw blood from cannulas designed to interface with the ventricle or atrium, with the former being more commonly utilized for extended left ventricular support. The inflow cannula and the pumping sac are composed of Thoralon, a proprietary polyurethane elastomer blended with a surface-modifying agent. The surface-modifying agent is designed to increase the potential for surface molecular rearrangement and reduce thrombogenicity (Farrar *et al.*, 1988). The outflow cannula is composed proximally of Thoralon polyurethane that fuses distally with a Dacron graft, to allow suturing to the ascending aorta. The Dacron graft is preclotted with patient blood and thrombin or cryoprecipitate and thrombin prior to implantation to seal the graft pores. Where the cannulas cross the skin, polyester velour is present on external surfaces of the Thoralon to encourage tissue integration. The opposing valves controlling flow into the pumping sac are tilting disk mechanical valves (Bjork–Shiley Monostrut). Compression of the full pumping sac produces a stroke volume of approximately 65 ml and is accomplished pneumatically, requiring connection to a controller providing positive and negative air pressure. Although a portable pneumatic system is in clinical trials, many current Thoratec patients are tethered to a rather large drive console that limits mobility and the potential for discharge to home.

Novacor

The Novacor Left Ventricular Assist System shown in Fig. 3 is a VAD designed exclusively for left-sided support of the heart. Developed by Peer Portner and colleagues in the 1970s (Portner *et al.*, 1978) the Novacor system has undergone design refinement over the past two decades and has now been implanted in almost 1000 patients worldwide. Unlike the Thoratec device, the Novacor VAD rests predominantly within the body cavity with only a single percutaneous line to transmit power, provide pump control, and vent air displaced by pumping sac compression and relaxation. Compression of the pumping sac is accomplished electrically with a pulsed-solenoid energy converter coupled to pusher plates on opposing sides of the sac by a flat spring mechanism. The controller and power pack have been reduced in size such that patients can wear these units on a belt. The reduced size of the complete system allows great patient mobility and, in many instances, discharge to home while the patient awaits word of donor organ availability. The impact of such discharge on the economics of supporting patients with end-stage heart failure is great (i.e., intensive care unit daily charges versus negligible costs for patients at home). Such portable systems also offer quality of life improvements that have lead to the implantation of the Novacor system as an alternative to transplantion in Europe (Dew *et al.*, 1999; El-Banayosy *et al.*, 1999a).

The inflow and outflow conduits of the Novacor VAD are 22-mm-diameter low porosity, gelatin-sealed woven polyester grafts (Sulzer Vascutek). Since the grafts are sealed, no pre-clotting is required in the operating room. The inflow conduit is nonconvoluted, while the outflow is crimped. To protect against conduit kinking, external polypropylene supports are present on the external surfaces of the conduits. A sewing ring at the proximal end of the inflow conduit allows suturing to the apex of the left ventricle. The inflow conduit connects distally to a valved Dacron conduit housing a 22-mm porcine xenograft valve. This valved conduit connects to the smooth Biomer polyurethane pumping sac that has a maximum stroke volume of 70 ml. The outflow valved conduit and outflow conduit are similar to the inflow conduits with the exception that the distal section of the outflow conduit is not protected with polypropylene support. This allows sizing of the outflow graft to the patient.

The percutaneous lead of the Novacor, also known as the driveline, has a Dacron coating that begins at the pump and extends 34 cm distally. This Dacron coating is intended to encourage tissue integration, particularly at the skin exit site. At approximately 1 week after implantation granulation tissue is expected to adhere to the driveline to reduce the potential for bacterial entry along this biomaterial surface. Driveline infections have the potential to become life threatening if bacteria migrate up the line in the patient to colonize the exterior surfaces of VAD. This potential complication is common to all circulatory support devices that must physically communicate across the patient's skin.

HeartMate

The HeartMate VAD (developed by ThermoCardiosystems [Woburn, MA] and now produced by Thoratec) is similar to the Novacor in that the inflow and outflow conduits as well as the pumping sac reside entirely within the patient as seen in Fig. 1. Two designs of this VAD have been widely applied clinically. The older pneumatic model utilizes an external air pump to compress the pumping sac in a fashion akin to that of the Thoratec device, while the more recent vented electric design uses an integral electric motor powered by an external battery pack. The pneumatic model provides for some patient mobility with a portable 8.5-kg pneumatic console, but patient mobility is much improved in the electric model with wearable battery pack. From a blood-contacting materials perspective, both models are identical. The HeartMate VAD is the most widely utilized VAD design to date with more than 2000 patients having been implanted worldwide.

From a materials perspective, the HeartMate design departs substantially from the Thoratec and Novacor VADs. The inflow from the apex of the left ventricle occurs through a titanium alloy (Ti-6Al-4V) cannula attached to a low-porosity Dacron graft. The blood and tissue contacting surfaces of the titanium cannula are surface coated with sintered titanium microspheres 50–75 μm in diameter, as seen in Fig. 4A. The Dacron portion of the inflow conduit houses a 25-mm porcine xenograft valve (Medtronic). The outflow conduit is also a valved Dacron conduit with an identical opposing valve at the exit of the pumping sac. The pumping sac is composed

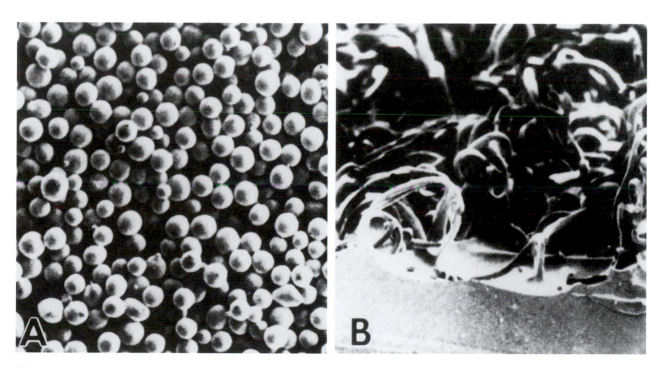

FIG. 4. (A) Blood-contacting surface of the rigid housing in the HeartMate VAD with sintered titanium microspheres (spheres are 50–75 μm in diameter). (B) Integrally textured polyurethane surface of the flexible diaphragm (fibrils are approximately 18 μm in diameter). (Reprinted from M. J. Menconi, S. Pockwinse, T. A. Owen, K. A. Dasse, G. S. Stein, and J. B. Lian, *J. Cell. Biochem.* **57**: 557–573, 1995.)

on one side of a titanium alloy surface with sintered titanium microspheres. On the opposing side the flexible pusher plate diaphragm is made of Biomer polyurethane that has been textured on its blood-contacting surface with surface integral fibrils approximately 18 μm in diameter and 300 μm in length, as seen in Fig. 4B (Menconi *et al.*, 1995). The maximum stroke volume of the pumping sac is 83 ml.

The implication of having textured surfaces contacting the blood in the HeartMate VAD is that these surfaces rapidly clot upon device placement. While this may seem like a negative scenario for the interior surfaces of a VAD, the concept behind the design is that while clots rapidly form on these surfaces, these thrombi are densely adherent and do not appear to embolize into the bloodstream in a clinically relevant manner. Over time additional blood cell interaction occurs, similar to an inflammatory reaction (see Chapter 4.2). A heterogeneous surface containing platelets, monocytes, macrophages, foreign-body giant cells, lymphocytes, and pluripotent hematopoietic cells is deposited. It is postulated that pluripotent hematopoietic cells differentiate into fibroblasts, myofibroblasts, and in some cases endothelial cells, that are reported to populate the surface (Rafii *et al.*, 1995; Spanier *et al.*, 1999; Frazier *et al.*, 1993). The fibroblastic cells may then secrete extracellular matrix components such as collagen, which is routinely found on textured VAD surfaces following extended implants. Unlike the inflammatory response associated with porous cardiovascular biomaterial surfaces, the cells populating the interior Heart-Mate VAD surface are likely derived entirely from the passing blood. Cellular migration from tissue onto these surfaces in the time frame that the cellular deposits are observed is considered unlikely. The blood-contacting biological interface that develops from the initial surface coagulum after approximately 1 week is referred to as the pseudointimal layer (Fig. 5) and

has been shown not to grow in thickness exceeding 150 μm over periods of implantation on the order of 1 year (Menconi *et al.*, 1995).

COMPLICATIONS AND VAD BIOCOMPATIBILITY ISSUES

Complications associated with the implantation and operation of VADs include those that may be related to biomaterial-centered processes discussed earlier in this text, including thromboembolism, infection, and bleeding. It is worth noting, however, that the mechanisms behind these complications vary among patients and are multifactorial. Many of the processes contributing to the complication may be independent of the biomaterial implant. Of particular relevance here is the health status of the patient at the time of device implantation. Patients in end-stage heart failure suffer from a variety of conditions secondary to the poor perfusion of their organs. They undergo a variety of invasive treatments and monitoring that precede VAD insertion and have often spent extended periods sedentary in the hospital environment.

Independent of VAD implantation end-stage heart failure patients are at risk for infection due to poor perfusion, compromised immune function, extended exposure to nosocomial infections, and invasive monitoring devices. Thromboembolism can occur in this patient group as a result of ventricular arrhythmia and ventricular mural thrombi, poor distal hemodynamics, and systemic vascular disease. Bleeding risks are elevated as a result of compromised liver function secondary to poor organ perfusion, and invasive monitoring and therapy short of VAD implantation. With these considerations in mind, there is certainly a relationship between the

FIG. 5.   Pseudointima developed on the interior textured surfaces of the HeartMate VAD after an implantation period of 243 days. On the left is the titanium surface with sintered microspheres and on the right is the textured polyurethane surface. (Reprinted from V. L. Poirier, *Thorac. Cardiovasc. Surg.* 47(suppl): 316–320, 1999.)

extensive biomaterial surface implantation of a VAD and the above complications. The blood-contacting surface areas of the VADs discussed above are in the 400–500 cm² range— a substantial implantation. These major complications will be discussed next.

Thromboembolism

Thromboembolic rates for VAD support vary widely among the specific devices studied and the reporting centers. Rates can vary simply due to the average implant period of the VAD, the health of the patients selected for support, and the patient medical management routine. Thromboembolism can be defined in a conservative manner to include only obvious neurologic events (strokes) lasting for more than 24 hours that are coupled with imaging evidence of a recent cerebral infarction and that cannot be explained as originating from a source other than the VAD. Alternatively, the definition can be liberalized to include suspected neurologic events such as blurred vision, weakness on one side, compromised field of view, and other events that are transient in nature, that may or may not be followed up with diagnostic imaging, and that may potentially be explained by some other risk factor. It is also possible to investigate peripheral embolization with imaging techniques that would not likely be employed in the majority of clinical centers (Schmid *et al.*, 1998). Finally, although not generally included in reported thromboembolic rates, tissue infarcts attributable to VAD support can be quantified at the time of autopsy in patients who die while on the device. With these limitations to reported thromboembolic rates in mind, the rates generally tend to be in the 10–30% range for the Thoratec and Novacor VADs and considerably less at approximately 5% for the HeartMate VAD (Mehta *et al.*, 1995; El-Banayosy *et al.*, 1999a; DeRose *et al.*, 1997).

To explain the principal mechanisms of thromboembolism in VAD patients one needs to recall the mechanisms of thrombosis on artificial surfaces discussed in Chapter 4.6 and consider the role of flow in this thrombotic process as discussed in Chapter 5.4. Low pump volumetric outputs have been directly associated with thrombotic deposition on the valves of an earlier design of the Novacor VAD (Wagner *et al.*, 1993). This design was subsequently altered to reduce the potential for low blood velocities to develop within the pumping sac and valve system. With all VAD designs, excessively low volumetric flow is considered to be a risk factor for thromboembolism.

The hemostatic system is affected with the implantation of all types of VADs. With the HeartMate device it has been shown that platelets, coagulation, and fibrinolysis are all activated early in the implant period at 2 hours after the operation (Livingston *et al.*, 1996). This activation is in excess of the marked elevations resulting from the extensive biomaterial contact involved in cardiopulmonary bypass for the implant surgery. The textured, prothrombotic surfaces of this VAD likely contribute to this effect. Longer-term elevations in hemostatic markers have also been reported for all VAD designs. Ongoing elevations in platelet activation, coagulation, and fibrinolysis have been reported for all three VADs discussed, suggesting that the device surfaces remain active during the

implant period in bridge-to-transplantation patients (Dewald *et al.*, 1997; Wilhelm *et al.*, 1999; Spanier *et al.*, 1996).

Attempting to relate activation of coagulation, platelets, or fibrinolysis with the risk for thromboembolism remains difficult for VAD patients. Although thromboembolic events are uncomfortably common, most centers do not have the numbers of patients to collect the number of observations necessary to statistically relate specific hemostatic alterations to thromboembolic events. As a result, markers of thromboembolism that do not result in an obvious clinical event (e.g., stroke) have been utilized to relate to measurable thrombotic markers in the blood (as discussed in Chapter 5.4). Microembolic signals detected in the cerebral arteries of VAD patients using ultrasound techniques have been explored as a potential indicator of stroke risk in VAD patients. Some reports have suggested that microembolic signals increase on days when patients experience a clinically obvious thromboembolic event (Schmid *et al.*, 1998). Others have shown that thrombin generation increases on days when large numbers of microemboli are detected in the cerebral arteries (Wilhelm *et al.*, 1999).

As with all cardiovascular devices, anticoagulation and antiplatelet therapy are options to control the reactions at the device–blood interface. A variety of specific anticoagulation and antiplatelet drug and monitoring regimens have been developed for VAD patients. Most commonly patients with Novacor and Thoratec devices are managed acutely with heparin and chronically with oral anticoagulants (i.e., warfarin) and antiplatelet agents. Anticoagulation is monitored with measurement of the activated partial thromboplastin time during heparin therapy and with the prothrombin time or international normalized ratio during oral anticoagulation. For the HeartMate VAD patients generally are not given chronic anticoagulation, and in only a fraction of the patient population are antiplatelet agents administered.

The fact that HeartMate VAD patients do not require chronic anticoagulation, yet still appear to have a markedly lower rate for thrombembolic complications merits discussion and has important ramifications for our definition of material biocompatibility in this setting. In much of the clinical literature the lower rate for thromboembolic complications in HeartMate VAD patients is attributed to the textured surfaces discussed earlier which lead to the formation of a biological pseudointima. This concept of encouraging well-anchored clot formation as a means of ultimately encouraging a biological lining goes back to the mid-1960s and earlier when Sharp, Hall, and other investigators researched this concept (Sharp *et al.*, 1964; Hall *et al.*, 1967).

There are few reports to date regarding the mechanisms through which the HeartMate pseudointima might act to reduce thromboembolism. Recent investigations have shown the mature surface to be dominated by inflammatory and fibroblastic cell lineages expressing tissue factor, inflammatory cytokines, and laying down matrices that include type I collagen (Spanier *et al.*, 1999; Menconi *et al.*, 1995). Such a surface would not appear to have antithrombotic properties. Indeed, Spanier *et al.* (1999) have suggested that such a surface may work by creating a sustained prothrombotic and potentially proinflammatory environment that triggers ongoing coagulation through the tissue factor pathway. The

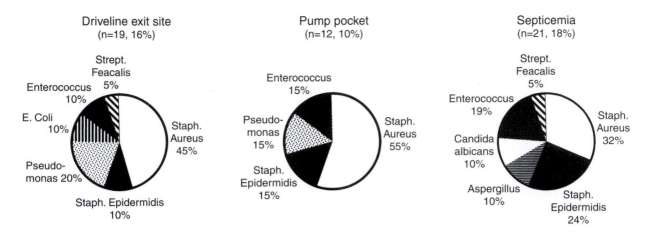

FIG. 6. Incidence of infectious organisms grouped by site. Cultures were performed of driveline exit site infection, pump pocket infection, and blood (septicemia) in a group of 118 Novacor VAD patients. The number of patients with infection for each site are presented above each pie chart as is the percentage of total patients experiencing infection at this site. (Reprinted from A. El-Banayosy, M. Deng, D. Y. Loisance, H. Vetter, E. Gronda, M. Loebe, and M. Vigano, *Eur. J. Cardiothorac. Surg.* **15**: 835–841, 1999.)

body's extended fibrinolytic response to the ongoing coagulation on this surface may serve as autoanticoagulation to prevent large potential thromboemboli from developing. Alternative explanations for the HeartMate VAD's impressive clinical results should also be considered. The development of small thrombi on the surface of the pump may occur with regularity throughout the implant period, but these thrombi may have a much lower likelihood of embolizing because of stronger adhesion to the extracellular matrix present on the surface. Deposition of the pseudointima may also act to "smooth" discontinuities in pump surfaces that otherwise would serve as a nidus for thrombi formation. The macroscopic flow patterns within the pumping chamber should also be considered. This VAD has been designed to have a "wandering vortex" that may act to minimize the size of thromboemboli generated from pumping-chamber surfaces (Slater *et al.*, 1996).

Infection

Infection is considered by many physicians and investigators to be the most serious complication facing VAD patients and a major obstacle to the implementation of current VAD designs as permanent implants to treat end-stage heart failure. As implant periods for bridge-to-transplant patients increase, the primary cause of death becomes infection (El-Banayosy *et al.*, 1999a). As with thromboembolism, the reported infection rates in VAD patients vary widely between centers because of the criteria selected to classify this complication as well as patient selection and medical management differences. Infection rates can be reported in a very general sense to include all patient infections, usually defined as clinical evidence of infection reflected by elevated leukocyte count, fever, or both in the presence of a positive culture. With this general definition, infection rates generally fall near 50% for all three aforementioned types of VADs (Sun *et al.*, 1999; McBride *et al.*, 1999; El-Banayosy *et al.*, 1999a). Alternatively, only device-related

infections may be reported. These infections can be classified as positive cultures from percutaneous drivelines and cannulas, and from the VAD pocket or mediastinum. With this definition the infection rates tend to fall in the 20–30% range.

Infection is directly responsible for the death of 10–15% of bridge-to-transplant VAD patients (Holman *et al.*, 1999; El-Banayosy *et al.*, 1999a). The seriousness of the infection is largely related to its location. Infections of the blood-contacting interior surfaces of the VAD, confirmed at device explant, are associated with high rates of mortality (~50%). Infection may also contribute to embolic damage and death by the generation of septic emboli, or through enhancement of the thromboembolic process. Patient infections not directly related to the device are not as serious, with the exception of positive blood cultures, which may indicate interior VAD surface vegetations. As discussed in Chapter 4.8, there are distinct bacterial and fungal species that tend to infect biomaterial surfaces and often these infections include multiple organisms. In Fig. 6 the composition of infections in a recent multicenter study are presented in terms of the infected site. Common skin bacteria that are traditionally associated with biomaterial infections dominate the VAD-infecting strains.

Management of infection in patients can be accomplished to a reasonable degree with antibiotics and antifungal agents. It is fortunate that the patient population remains dominated by individuals awaiting device removal at the time of transplantation. Surprisingly, a history of infection during VAD support is not a counterindication for heart transplantation, and studies have shown that, despite the use of antirejection drugs in the posttransplant period, these patients do well (Argenziano *et al.*, 1997). For those patients with suspected infection of the interior VAD surfaces, it has been suggested that device replacement may be an effective method of treatment if transplant is not a short-term option (Argenziano *et al.*, 1997). In Fig. 7 a bioprosthetic inflow valve removed from a Novacor VAD patient suffering from a *Candida albicans* infection is seen. With the extent of this growth, alteration in pump function

FIG. 7. *Candida albicans* vegetations partially obstructing the inflow valve of a Novacor VAD patient at the time of valve replacement. (Kindly provided by R. L. Kormos, M. D., Director, and S. Winowich of the University of Pittsburgh Medical Center Mechanical Circulatory Support Program.)

was apparent and a procedure was performed to replace the infected valves. The patient recovered from this procedure and was successfully transplanted.

Prevention of infection is a primary goal during device implantation and later patient medical management. Prophylactic antibiotics and, in some instances, antifungal agents are utilized in the perioperative period. Careful cleaning and attention to the percutaneous driveline and cannula sites are also of primary importance. From a biomaterials perspective two important design issues arise. First, it is critical for the exterior surfaces of the drivelines and cannulas to rapidly encourage tissue healing and to discourage bacterial colonization. Second, the mechanical properties of the percutaneous lines are of relevance in that large mismatches in stiffness between the skin and lines places stress at the biomaterial–tissue interface. The resulting abrasion may prevent or retard adequate wound healing. The VAD design improvement most likely to affect infection rates will be the introduction of transcutaneous power transmission and control systems, and elimination of the percutaneous line. Systems involving this technology are currently undergoing animal testing (Weiss *et al.*, 1999) and, as of this writing, one device (the LionHeart VAD produced by Arrow International) has been implanted clinically for the first time in two patients

in Germany. Such devices should greatly reduce VAD-related infections and improve quality of life. The totally implantable VAD systems will be of particular interest in the permanent VAD setting.

Bleeding

Bleeding remains a potentially lethal complication associated with VAD implantation, but one that generally occurs in the early postoperative period, when numerous wound sites exist. Reported bleeding rates fall in the 30–35% range (McBride *et al.*, 1999; El-Banayosy *et al.*, 1999b). Bleeding can be defined in terms of the administration of blood products, chest tube drainage after surgery, or surgical interventions to treat hemorrhage. Subsequent trips to the operating room to address bleeding place patients at risk for bacterial and fungal colonization, linking this complication to infection risk. Also of concern in this patient population is the exposure and sensitization of transplant candidates to human antigens associated with blood products (Massad *et al.*, 1997). The use of leukocyte-filtered products has been utilized to address this latter concern.

A number of mechanisms contribute to hemorrhage in VAD patients. As mentioned previously, VAD implantation surgery is associated with consumption and activation of patients' platelets and coagulation system, as well as activation of the fibrinolytic pathway (Livingston *et al.*, 1996). End-stage heart failure patients undergoing VAD implantation have deficiencies in coagulation factors even in the absence of obvious liver dysfunction (Wang *et al.*, 1995). After postoperative stabilization, Thoratec and Novacor patients are placed on chronic anticoagulation therapy, which carries an ongoing risk for over-anticoagulation (and possible bleeding). Of particular concern in the extended support setting are cerebral bleeds leading to neurologic complications. Anticoagulation in these VAD patients is thus carefully monitored throughout the implant period. To reduce the incidence of bleeding in the perioperative period, antifibrinolytic drugs, such as aprotinin, are now routinely utilized and have had marked effects in reducing blood loss and perioperative mortality (Goldstein *et al.*, 1995).

Mechanical Failure

In the extensive clinical experience with VADs over the past decade there have been very few reported incidents of mechanical failure related to material issues. This is likely due to the long development phases through which all three aforementioned FDA approved devices have passed. The pulsatile VAD systems subject internal diaphragms, cams, bearings, and springs to cyclic loads that would likely lead to more wear-associated failure were these devices asked to perform beyond their design specifications. In bridge-to-transplant implantations, bearing wear has been reported but has not led to catastrophic failure (Sun *et al.*, 1999). Other complications reported involve tearing of Dacron inflow conduits in the HeartMate device as seen in Fig. 8 (Scheld *et al.*, 1997). This complication was likely due to wear from the metal support cage rubbing on the graft surface, and the design has since been altered to constrain cage motion.

As pulsatile VADs move into the area of permanent support, it will be important to develop diagnostic techniques to noninvasively assess the wear process in implanted electrical and mechanical components. Acoustic and power consumption monitoring have been reported as innovative methods for evaluating wear in a patient with a permanent Novacor VAD on device support for more than 1000 days (Dohmen *et al.*, 1999). Durability and reliability testing *in vitro* will continue to be essential for discovering the typical time course of electromechanical failure modes and to provide the opportunity to develop such diagnostic methods. As an example of this, a report on *in vitro* reliability and durability testing with the Novacor VAD has demonstrated a single failure mode in test periods up to approximately 5 years (Lee *et al.*, 1999). This failure can be readily detected and may thus prevent unnecessary replacement of devices at fixed implant periods, while allowing detection of failure and remediation in those devices that may develop early signs of this wear. With knowledge of the failure mode engineers are provided a focal point to work on redesign of the device, leading to further improvements in long-term durability and reliability.

ROTARY BLOOD PUMPS FOR CHRONIC CIRCULATORY SUPPORT

In 1994 the National Heart, Lung and Blood Institute (NHLBI) issued a request for proposals (RFP) entitled, "Innovative Ventricular Assist Systems—IVAS" (NHLBI-HV-94-25). Some of the features desired in these innovative mechanical circulatory support devices were:

- Five year operation with 90% reliability
- Totally implantable with no external venting
- Size and weight characteristics suitable for implantation in both males and females between 18 and 70 years of age
- Demonstrated biocompatibility including avoidance of clinically significant thromboembolism, bleeding, tissue overgrowth, infection, tissue heating, and leakage of device fluids into surrounding tissues

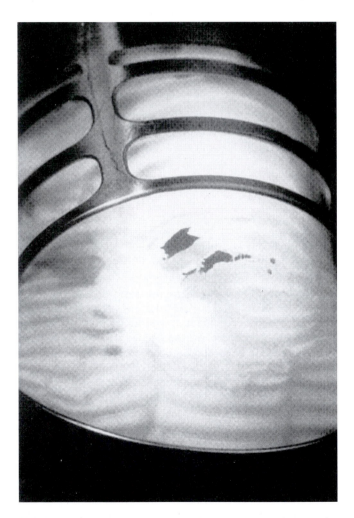

FIG. 8.   Ruptured Dacron inflow conduit of a HeartMate VAD that lead to sudden blood loss after 22 days of pump implantation. (Reprinted from H. H. Scheld, R. Soeparwata, C. Schmid, M. Loick, M. Weyand, and D. Hammel, *J. Thorac. Cardiovasc. Surg.* **114:** 287–289, 1997.)

- Acceptably low levels of audible noise and mechanical vibration
- Flow capacity of 3–7 L/min against a mean arterial pressure of 90 mm Hg (when operating as a VAD in support of the left ventricle)
- Capability for transmission of energy from outside to inside the body, and diagnostics into and out of the implantable VAD

Of the six proposals funded by the National Institutes of Health in response to this RFP, three involved rotary pump designs. As discussed next, these pumps are fundamentally different from the VADs described earlier in this chapter.

Rotary blood pumps have a number of advantages over the VADs currently used clinically, which generate pulsatile flow. A continuous-flow pump requires no bladder to create a stroke volume comparable to the native ventricle; rather, flow is accomplished with an impeller, in either an axial-flow (straight through) or centrifugal (center to tangential edge) configuration. The rotary pumps are significantly smaller, have fewer moving parts, and are free from valves and cyclic actuators. This simplified design should lead to an increase in overall reliability and durability. Lower overall volumes of implanted materials will make these systems applicable to patients generally excluded from current implantable VADs, e.g., smaller women and children. Also, with the smaller pump size, less of a pocket needs to be created in the body for implantation and this may reduce bleeding and infection rates. Without valves and the compression of a pumping sac, the rotary pumps are also much quieter than current pulsatile-flow generating devices.

As noted above, for permanent circulatory support a fully implantable system is desirable to reduce infection risk and improve quality of life. Both pulsatile-flow VADs and rotary blood pumps can avoid the need for percutaneous power and control wires by utilizing transcutaneous energy transmission systems. With pulsatile bladder pumps such as the three VADs described earlier, another constraint arises regarding complete implantability. The air displaced in the compression and expansion of the pumping sac must shuttle to another location within the body. Current VADs address this need by venting air to and from the device through the percutaneous lines that also serve to supply power and control VAD operation. Completely implantable systems must utilize a compliance chamber that expands and contracts counter to the pumping sac. A number of concerns arise with the implantation of a compliance chamber intended for extended VAD support. First, the foreign-body response (as described in Chapter 4.2) ultimately leads to fibrotic tissue deposition around these sacs and inhibits elastic deformation. This elasticity change alters the forces involved in pump filling and emptying. Second, it is difficult to create polymers that simultaneously exhibit low gas permeability and high elasticity. Current compliance-chamber designs leak gas slowly across the compliance-chamber wall, which if not accounted for, would ultimately lead to altered blood filling and pumping in the VAD. To address this problem the compliance chamber is periodically refilled with gas every 4–6 weeks through a subcutaneous infusion port (Weiss et al., 1999).

By their mode of operation, rotary blood pumps do not require a compliance chamber.

Although free from some of the shortcomings of VADs that generate pulsatile flow, rotary pumps introduce some unique biocompatibility concerns. The high revolutions per minute (RPM) required of rotors to generate flows in these pumps (~5000–10,000 RPM) subjects blood to high shear stresses not encountered physiologically. Although these stresses are of very short duration, concerns arise regarding hemolysis and platelet activation due to this exposure. Biocompatibility testing with rotary blood pumps invariably involves assessment of hemolytic indices, evaluation of thrombotic deposition, and, in animal studies, assessment of potential end-organ embolic damage. To minimize shear-related effects fluid flow over turbine-blade designs is often visualized or modeled in an effort to minimize regions of flow turbulence or stagnation (Kerrigan et al., 1996; Burgreen et al., 1999). A second flow-related concern regards the reduced pulsatility of blood flow that occurs when a continuous-flow, rotary pump is connected to empty the left ventricle and pump blood to the aorta. The chronic effects of diminished pulsatility on long-term patient health are not known, although the use of continuous-flow pumps for cardiopulmonary bypass and acute circulatory support have not led to major physiologic complications, and large-animal implants appear to tolerate this type of circulation for periods beyond 6 months. The current data on this topic have been reviewed by Jett (1999).

The mechanical requirements inherent in small rotary blood pumps have resulted in the selection of metals, primarily titanium alloys, as the material of choice for impellers, flow straighteners, and pump housings. In some current rotary systems inflow and outflow cannulas do not differ remarkably in design and materials selection from those used with current VADs, whereas in others the departure is significant, illustrated by designs wherein the pump is moved into the ventricle (Westaby et al., 1998; Wampler et al., 1999). Design and control concerns arise in association with the inflow orifice of rotary pumps. When ventricular pressures are reduced, the ventricular septum has the potential to be drawn into or around the inflow orifice, obstucting flow and potentially damaging the ventricular wall (Amin et al., 1998). Such concerns do not arise with the passive-filling, pulsatile-flow-generating VADs currently used clinically.

The design of inflow and outflow pump bearings (which support the spinning rotor and resist axial and radial thrust loads) represents a major challenge for rotary blood-pump designers. The bearings used in this setting must effectively dissipate heat, exhibit acceptably low wear, support radial and axial loads, and be readily machined to very high tolerances. Bearing designs currently under investigation utilize magnetic suspension, blood lubrication, or lubrication provided by a purge fluid. Since sliding contact may occur during transients in pressure and flow, starts and stops, and unexpected patient acceleration and deceleration, it is important that bearing materials be wear-resistant. With blood or plasma contact, it is also important that these bearings exhibit appropriate blood biocompatibility. Current reported materials that have been evaluated for rotary pump bearings include zirconium–niobium alloy (Zr-2.5Nb) and titanium–zirconium–niobium

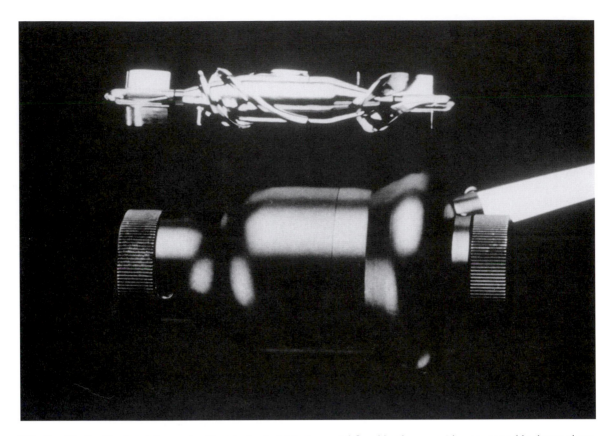

FIG. 9. Nimbus-Pittsburgh Innovative Ventricular Assist System axial flow blood pump with rotor assembly shown above. (Kindly provided by K. C. Butler, President, Nimbus, Inc.).

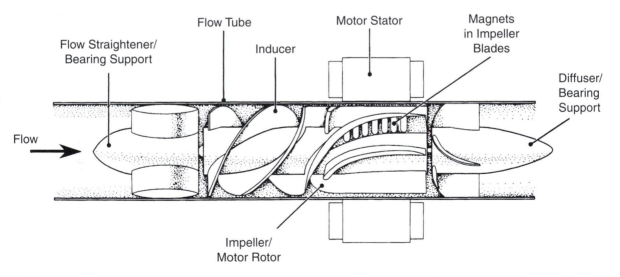

FIG. 10. Cross-section of the DeBakey VAD. (Reprinted from M. E. DeBakey, *Ann. Thorac. Surg.* **68**: 637–640, 1999.)

alloy (Ti-13Zr-13Nb), as well as carbon coating techniques (Golding *et al.*, 1998).

Rotary blood pumps currently in development include those described by Butler *et al.* (1997) seen in Fig. 9, DeBakey (1999) illustrated in cross-section in Fig. 10, Golding *et al.*

(1998), Wampler *et al.* (1999), and Westaby *et al.* (1998). One pump, the DeBakey VAD produced by MicroMed Technology (Houston, TX), has entered Phase I clinical trials in Europe. At the time of this writing the DeBakey VAD had been implanted in 18 patients with at least six recipients supported for over 60 days. This VAD prototype is currently not completely implantable, having a percutaneous line for power transmission, pump monitoring, and speed (RPM) control. Implants to date have been for bridge to cardiac transplantation.

CONCLUSIONS

Major advances have occurred in mechanical circulatory support technology over the past three decades. Bridging patients to cardiac transplantation has now become routine with three devices currently approved by the FDA to fill this role in the United States. This success has set the stage for the application of VADs as alternatives to cardiac transplantation in the transplant-ineligible population. As one would expect with an extensive blood-contacting biomaterial implantation accompanied by percutaneous drivelines, thromboembolism, infection, and bleeding remain complications of primary concern in VAD patients. Completely implantable VADs and innovative rotary blood pumps may serve to reduce these complications and improve device durability and reliability. In the next decade the challenges associated with supporting patients for years, as opposed to months, will be faced. Biomaterial and biocompatibility issues will likely be central to the problems and solutions that arise as mechanical circulatory support devices continue to evolve.

Acknowledgments

The authors thank Dr. Robert L. Kormos, Director of the University of Pittsburgh Medical Center Mechanical Circulatory Support Program, for his expert guidance and innumerable insights on circulatory support issues, as well as the engineers and staff of this program for their assistance on this manuscript. Special thanks must be extended to our VAD patients and their loved ones who have always encouraged us to better understand the challenges and opportunities of implantable circulatory assist devices.

Bibliography

American Heart Association (1999). "1999 Heart and Stroke Statistical Update." American Heart Association, Dallas, TX.

Amin, D. V., Antaki, J. F., Litwak, P., Thomas, D., Wu, Z. J., and Watach, M. (1998). Induction of ventricular collapse by an axial flow blood pump. *ASAIO J.* 44: M685–M690.

Arabia, F. A., Copeland, J. G., Smith, R. G., Banchy, M., Foy, B., Kormos, R., Tector, A., Long, J., Dembitsky, W., Carrier, M., Keon, W., Pavie, A., and Duveau, D. (1999). CardioWest total artificial heart: a retrospective controlled study. *Artif. Organs* 23: 204–207.

Argenziano, M., Catanese, K. A., Moazami, N., Gardocki, M. T., Weinberg, A. D., Clavenna, M. W., Rose, E. A., Scully, B. E., Levin, H. R., and Oz, M. C. (1997). The influence of infection on survival and successful transplantation in patients with left ventricular assist devices. *J. Heart Lung Transplant* 16: 822–831.

Burgreen, G. W., Antaki, J. F., Wu, J., le Blanc, P., and Butler, K. C. (1999). A computational and experimental comparison of two outlet stators for the Nimbus LVAD. *ASAIO J.* 45: 328–333.

Butler, K., Thomas, D., Antaki, J., Borovetz, H., Griffith, B., Kameneva, M., Kormos, R., and Litwak, P. (1997). Development of the Nimbus/Pittsburgh axial flow left ventricular assist system. *Artif. Organs* 21: 602–610.

Copeland, J. G., Arabia, F. A., Smith, R. G., Sethi, G. K., Nolan, P. E., and Banchy, M. E. (1997). Arizona experience with CardioWest Total Artificial Heart bridge to transplantation. *Ann. Thorac. Surg.* 68: 756–760.

DeBakey, M. E. (1999). A miniature implantable axial flow ventricular assist device. *Ann. Thorac. Surg.* 68: 637–640.

DeRose, J. J. Jr., Argenziano, M., Sun, B. C., Reemtsma, K., Oz, M. C., and Rose, E. A. (1997). Implantable left ventricular assist devices: an evolving long-term cardiac replacement therapy. *Ann. Surg.* 226: 461–470.

DeVries, W. C., Anderson, J. L., Joyce, L. D., Anderson, F. L., Hammond, E. H., Jarvik, R. K., and Kolff, W. J. (1984). Clinical use of the total artificial heart. *N. Engl. J. Med.* 310: 273–278.

Dew, M. A., Kormos, R. L., Winowich, S., Nastala, C. J., Borovetz, H. S., Roth, L. H., Sanchez, J., and Griffith, B. P. (1999). Quality of life outcomes in left ventricular assist system inpatients and outpatients. *ASAIO J.* 45: 218–225.

Dewald, O., Fischlein, T., Vetter, H. O., Schmitz, C., Godje, O., Gohring, P., and Reichart, B. (1997). Platelet morphology in patients with mechanical circulatory support. *Eur. J. Cardiothorac. Surg.* 12: 634–641.

Dohmen, P. M., Laube, H., de Jonge, K., Baumann, G., and Konertz, W. (1999). Mechanical circulatory support for one thousand days or more with the Novacor N100 left ventricular assist device. *J. Thorac. Cardiovasc. Surg.* 117: 1029–1030.

El-Banayosy, A., Deng, M., Loisance, D. Y., Vetter, H., Gronda, E., Loebe, M., and Vigano, M. (1999a). The European experience of Novacor left ventricular assist (LVAS) therapy as a bridge to transplant: a retrospective multi-centre study. *Eur. J. Cardiothorac. Surg.* 15: 835–841.

El-Banayosy, A., Korfer, R., Arusoglu, L., Minami, K., Kizner, L., Fey, O., Schutt, U., and Morshuis, M. (1999b). Bridging to cardiac transplantation with the Thoratec ventricular assist device. *Thorac. Cardiovasc. Surg.* 47(suppl): 307–310.

Farrar, D. J., Litwak, P., Lawson, J. H., Ward, R. S., White, K. A., Robinson, A. J., Rodvien, R., and Hill, J. D. (1988). In vivo evaluations of a new thromboresistant polyurethane for artificial heart blood pumps. *J. Thorac. Cardiovasc. Surg.* 95: 191–200.

Frazier, O. H., Baldwin, R. T., Eskin, S. G., and Duncan, J. M. (1993). Immunochemical identification of human endothelial cells on the lining of a ventricular assist device. *Tex. Heart Inst. J.* 20: 78–82.

Frazier, O. H., Macris, M. P., Myers, T. J., Duncan, J. M., Radovancevic, B., Parnis, S. M., and Cooley, D. A. (1994). Improved survival after extended bridge to cardiac transplantation. *Ann. Thorac. Surg.* 57: 1416–1422.

Funk, M. (1991). Epidemiology of end-stage heart disease. in *The Artificial Heart: Prototypes, Policies, and Patients*, J. R. Hogness, and M. VanAntwerp, eds. National Academy Press, Washington, D.C., pp. 251–261.

Golding, L., Medvedev, A., Massiello, A., Smith, W., Horvath, D., and Kasper, R. (1998). Cleveland Clinic continuous flow blood pump: progress in development. *Artif. Organs* 22: 447–450.

Goldstein, D. J., Seldomridge, J. A., Chen, J. M., Catanese, K. A., DeRosa, C. M., Weinberg, A. D., Smith, C. R., Rose, E. A., Levin, H. R., and Oz, M. C. (1995). Use of aprotinin in LVAD recipients reduces blood loss, blood use, and perioperative mortality. *Ann. Thorac. Surg.* 59: 1063–1067.

Guy, T. S. (1998). Evolution and current status of the total artificial heart: the search continues. *ASAIO J.* 44: 28–33.

Hall, C. W., Liotta, D., Ghidoni, J. J., DeBakey, M. E., and Dressler, D. P. (1967). Velour fabrics applied to medicine. *J. Biomed. Mater. Res.* **1**: 179–196.

Holman, W. L., Skinner, J. L., Waites, K. B., Benza, R. L., McGiffin, D. C., and Kirklin, J. K. (1999). Infection during circulatory support with ventricular assist devices. *Ann. Thorac. Surg.* **68**: 711–716.

Hosenpud, J. D., Bennett, L. E., Keck, B. M., Fiol, B., Boucek, M. M., and Novick, R. J. (1998). The registry of the International Society for Heart and Lung Transplantation: fifteenth official report—1998. *J. Heart Lung Transplant.* **17**: 656–668.

Jett, G. K. (1999). Physiology of nonpulsatile circulation: acute versus chronic support. *ASAIO J.* **45**: 119–122.

Kerrigan, J. P., Yamazaki, K., Meyer, R. K., Mori, T., Otake, Y., Outa, E., Umezu, M., Borovetz, H. S., Kormos, R. L., Griffith, B. P., Koyanagi, H., and Antaki, J. F. (1996). High-resolution fluorescent particle-tracking flow visualization within an intraventricular axial flow left ventricular assist device. *Artif. Organs* **20**: 534–540.

Kormos, R. L., Gasior, T. A., Kawai, A., Pham, S. M., Murali, S., Hattler, B. G., and Griffith, B. P. (1996). Transplant candidate's clinical status rather than right ventricular function defines need for univentricular versus biventricular support. *J. Thorac. Cardiovasc. Surg.* **111**: 773–782.

Lee, J., Miller, P. J., Chen, H., Conley, M. G., Carpenter, J. L., Wihera, J. C., Jassawalla, J. S., and Portner, P. M. (1999). Reliability model from the in vitro durability tests of a left ventricular assist system. *ASAIO J.* **45**: 595–601.

Livingston, E. R., Fisher, C. A., Bibidakis, E. J., Pathak, A. S., Todd, B. A., Furukawa, S., McClurken, J. B., Addonizio, V. P., and Jeevanandam, V. (1996). Increased activation of the coagulation and fibrinolytic systems leads to hemorrhagic complications during ventricular assist implantation. *Circulation* **94**(suppl II): II-227–II-234.

Loebe, M., Muller, J., and Hetzer, R. (1999). Ventricular assistance for recovery of cardiac failure. *Curr. Opin. Cardiol.* **14**: 234–248.

Mann, D. L., and Willerson, J. T. (1998). Left ventricular assist devices and the failing heart: a bridge to recovery, a permanent assist device, or a bridge too far? *Circulation* **98**: 2367–2369.

Massad, M. G., Cook, D. J., Schmitt, S. K., Smedira, N. G., McCarthy, J. F., Vargo, R. L., and McCarthy, P. M. (1997). Factors influencing HLA sensitization in implantable LVAD recipients. *Ann. Thorac. Surg.* **64**: 1120–1125.

Massad, M. G., McCarthy, P. M., Smedira, N. G., Cook, D. J., Ratliff, N. B., Goormastic, M., Vargo, R. L., Navia, J., Young, J. B., and Stewart, R. W. (1996). Does successful bridging with the implantable left ventricular assist device affect cardiac transplantation outcome? *J. Thorac. Cardiovasc. Surg.* **112**: 1275–1283.

McBride, L. R., Naunheim, K. S., Fiore, A. C., Moroney, D. A., and Swartz, M. T. (1999). Clinical experience with 111 Thoratec ventricular assist devices. *Ann. Thorac. Surg.* **67**: 1233–1239.

Mehta, S. M., Aufiero, T. X., Pae, W. E. Jr., Miller, C. A., and Pierce, W. S. (1995). Combined registry for the clinical use of mechanical ventricular assist pumps and the total artificial heart in conjunction with heart transplantation: sixth official report—1994. *J. Heart Lung Transplant.* **14**: 585–593.

Menconi, M. J., Pockwinse, S., Owen, T. A., Dasse, K. A., Stein, G. S., and Lian, J. B. (1995). Properties of blood-contacting surfaces of clinically implanted cardiac assist devices: gene expression, matrix composition, and ultrastructural characterization of cellular linings. *J. Cell. Biochem.* **57**: 557–573.

Portner, P. M., Oyer, P. E., Jassawalla, J. S., Miller, P. J., Chen, H., LaForge, D. H., and Skytte K. W. (1978). An implantable permanent left ventricular assist system for man. *Trans. Am. Soc. Artif. Intern. Organs* **24**: 99–103.

Rafii, S., Oz, M. C., Seldomridge, J. A., Ferris, B., Asch, A. S., Nachman, R. L., Shapiro, F., Rose, E. A., and Levin, H. R. (1995). Characterization of hematopoetic cells arising on the textured surface of left ventricular assist devices. *Ann. Thorac. Surg.* **60**: 1627–1632.

Rose, E. A., Moskowitz, A. J., Packer, M., Sollano, J. A., Williams, D. L., Tierney, A. R., Heitjan, D. F., Meier, P., Ascheim, D. D., Levitan, R. G., Weinberg, A. D., Stevenson, L. W., Shapiro, P. A., Lazar, R. M., Watson, J. T., Goldstein, D. J., and Gelijns, A. C. (1999). The REMATCH trial: rationale, design, and end points. *Ann. Thorac. Surg.* **67**: 723–730.

Scheld, H. H., Soeparwata, R., Schmid, C., Loick, M., Weyand, M., and Hammel, D. (1997). Rupture of inflow conduits in the TCI-HeartMate system. *J. Thorac. Cardiovasc. Surg.* **114**: 287–289.

Schmid, C., Weyand, M., Nabavi, D. G., Hammel, D., Deng, M. C., Ringelstein, E. B., and Scheld, H. H. (1998). Cerebral and systemic embolization during left ventricular support with the Novacor N100 device. *Ann. Thorac. Surg.* **65**: 1703–1710.

Schocken, D. D., Arrieta, M. I., Leaverton, P. E., and Ross, E. A. (1992). Prevalence and mortality rate of congestive heart failure in the United States. *J. Am. Coll. Cardiol.* **20**: 301–306.

Sharp, W. V., Finelli, A. F., Falor, W. H., and Ferraro, J. W. Jr. (1964). Latex vascular prosthesis: patency rate and neointimization related to prosthesis lining and electrical conductivity. *Circulation* **29**(suppl): 165–170.

Slater, J. P., Rose, E. A., Levin, H. R., Frazier, O. H., Roberts, J. K., Weinberg, A. D., and Oz, M. C. (1996). Low thromboembolic risk without using advanced-design left ventricular assist devices. *Ann. Thorac. Surg.* **62**: 1321–1328.

Spanier, T., Oz, M., Levin, H., Weinberg, A., Stamatis, K., Stern, D., Rose, E., and Schmidt, A. M. (1996). Activation of coagulation and fibrinolytic pathways in patients with left ventricular assist devices. *J. Thorac. Cardiovasc. Surg.* **112**: 1090–1097.

Spanier, T. B., Chen, J. M., Oz, M. C., Stern, D. M., Rose, E. A., and Schmidt, A. M. (1999). Time-dependent cellular population of textured-surface left ventricular assist devices contributes to the development of a biphasic systemic procoagulant response. *J. Thorac. Cardiovasc. Surg.* **118**: 404–413.

Sun, B. C., Catanese, K. A., Spanier, T. B., Flannery, M. R., Gardocki, M. T., Marcus, L. S., Levin, H. R., Rose, E. A., and Oz, M. C. (1999). 100 long-term implantable left ventricular assist devices: the Columbia Presbyterian interim experience. *Ann. Thorac. Surg.* **68**: 688–694.

United Network for Organ Sharing (1999). Transplant Patient Data Source.

Wagner, W. R., Johnson, P. C., Kormos, R. L., and Griffith, B. P. (1993). Evaluation of bioprosthetic valve-associated thrombus in ventricular assist device patients. *Circulation* **88**: 2023–2029.

Wampler, R., Lancisi, D., Indravudh, V., Gauthier, R., and Fine, R. (1999). A sealless centrifugal blood pump with passive magnetic and hydrodynamic bearings. *Artif. Organs* **23**: 780–784.

Wang, I. W., Kottke-Marchant, K., Vargo, R. L., and McCarthy, P. M. (1995). Hemostatic profiles of HeartMate ventricular assist device recipients. *ASAIO J.* **41**: M782–M787.

Weiss, W. J., Rosenberg, G., Snyder, A. J., Pierce, W. S., Pae, W. E., Kuroda, H., Rawhouser, M. A., Felder, G., Reibson, J. D., Cleary, T. J., Ford, S. K., Marlotte, J. A., Nazarian, R. A., and Hicks, D. L. (1999). Steady state hemodynamic and energetic characterization of the Penn State/3M Health Care Total Artificial Heart. *ASAIO J.* **45**: 189–193.

Westaby, S., Katsumata, T., Houel, R., Evans, R., Pigott, D., Frazier, O. H., and Jarvik, R. (1998). Jarvik 2000 Heart: potential for bridge to myocyte recovery. *Circulation* **98**: 1568–1574.

Wilhelm, C. R., Ristich, J., Knepper, L. E., Holubkov, R., Wisniewski, S. R., Kormos, R. L., and Wagner, W. R. (1999). Measurement of hemostatic indices in conjunction with transcranial doppler sonography in ventricular assist device patients. *Stroke* 30: 2554–2561.

7.5 ARTIFICIAL RED BLOOD CELL SUBSTITUTES

Thomas Ming Swi Chang

INTRODUCTION

The potential presence of HIV in donor blood in the 1980s was the main stimulus leading to serious development of the idea of modified hemoglobin started many years ago (Chang, 1957, 1964, 1965; Bunn and Jandl, 1968). Much progress has been made in the past 10 years leading to the final phase of phase III clinical trials before routine clinical application. The present modified hemoglobin can replace the plasma component and red blood cell component of blood. However, it cannot replace the platelets or the leukocytes of blood. During this period, much experience has been gained and much has been learned (Chang, 1997a). This chapter is a brief overview of this very large area. The readers are referred to much detailed description in the references (Chang, 1997a, 2000, 2002), which is updated on the Web site (www.artcell.mcgill.ca).

HEMOGLOBIN

Hemoglobin is the protein in red blood cells that is responsible for the transport of oxygen from the lung to the other tissues. When hemoglobin is extracted from red blood cells, it can be ultrafiltered and sterilized to remove and inactivate bacteria, viruses and other microorganisms. Thus unlike red blood cells, potential problems related to HIV, hepatitis viruses, parasites, and others could be eliminated. However, hemoglobin extracted from red blood cells cannot be used as a blood substitute for the following reasons. Each hemoglobin molecule is a tetramer consisting of four subunits—two alpha subunits and two beta subunits (Perutz, 1980) (Fig. 1). There is a requirement for 2,3-diphosphoglycerate (2,3-DPG) to be present in the 2,3-DPG pocket of the hemoglobin molecule. This is to decrease the affinity of hemoglobin for oxygen, thus allowing it to release oxygen properly. When infused into the body, there is little or no 2,3-DPG in the plasma to bind to the 2,3-DPG pocket. This results in hemoglobin with high affinity for oxygen, preventing oxygen from being easily released as required to the tissue. What is even more of a problem is that when free in the circulating blood, each hemoglobin molecule breaks down into potentially toxic half-molecules, dimers. The dimers are highly toxic especially their renal toxicity(Savitsky *et al.*, 1978). There are also other problems related to hemoglobin in free solution. The challenge is therefore to modify the basic biomaterial, hemoglobin, before it can be used as a blood substitute.

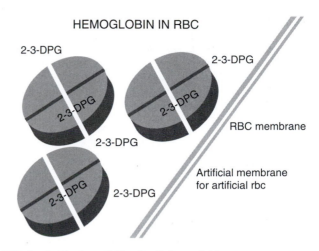

FIG. 1. Inside the red blood cell, hemoglobin stays as a tetramer. The red blood cell membrane retains 2,3DPG in the cell to bind to hemoglobin. This allows hemoglobin to release oxygen as needed by the tissues with P_{50} of about 28 mm Hg. In 1957 Chang prepared artificial red blood cells by removing the red blood cell membrane and replacing it with artificial membrane. Reprinted with permission from Chang, T. M. S. (1992). *Biomaterials, Artificial Cells and Immobilization Biotechnology* 20: 154–174, courtesy of Marcel Dekker, Inc.

MODIFIED HEMOGLOBIN

Polyhemoglobin

The first modified hemoglobin was formed by removing the biological cell membranes of red blood cells and replacing them with artificial membranes (Fig. 1) (Chang, 1957, 1964, 1965). These early artificial red blood cells can readily bind and release oxygen since 2,3-DPG is retained inside the artificial cells (Chang, 1957). However, although they are about the same size as red blood cells, they are not sufficiently flexible to allow them to survive for a long time after infusion. A number of approaches were used to change the surface properties of these artificial red blood cells (Chang, 1964, 1965, 1972). This includes surface modification to result in changes in surface charge, surface composition and membrane composition. Since hemoglobin has many surface and internal amino groups, Chang (1964, 1965, 1972) studied the use of a bifunctional agent, sebacyl chloride, to cross-link hemoglobin molecules into a cross-linked hemoglobin artificial red blood cell membrane. By decreasing the diameter, all the hemoglobin in the artificial red blood cells could be cross-linked into polyhemoglobin (Fig. 2). The reaction is as follows:

$$Cl-CO-(CH_2)_8-CO-Cl \ + \ HB-NH_2$$

Sebacyl chloride   Hemoglobin

$$= HB-NH-CO-(CH_2)_8-CO-NH-HB$$

Cross-linked hemoglobin

In 1971, Chang reported the use of another bifunctional agent, glutaraldehyde, to cross-link hemoglobin and a red blood cell

CROSSLINKED HB: POLYHEMOGLOBIN

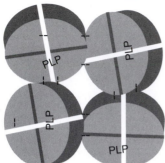

Sebacyl chloride (Chang, 1964)
Glutaraldehye (Chang, 1971)
Pyridoxalation (Benesch, 1975)

FIG. 2. Cross-linking the hemoglobin molecules together presents their breakdown into dimers. Following the use of sebacyl chloride (Chang, 1964, 1965) the same author reported (Chang, 1971) the use of glutaraldehyde as another cross-linker. Since then, many other investigators have extended and improved this, resulting in glutaraldehyde cross-linked polyhemoglobin, now in phase III clinical trial. Benesch uses pyrodoxal phosphate to replace the 2-3-DPG required for human polyhemoglobin. Reprinted with permission from Chang, T. M. S. (1992). *Biomaterials, Artificial Cells and Immobilization Biotechnology* **20:** 154–174, courtesy of Marcel Dekker, Inc.

enzyme, catalase, into soluble polyhemoglobin (Fig. 2). The reaction is as follows:

$$H-CO-(CH_2)_3-CO-H \ + \ HB-NH_2$$
Glutaraldehyde   Hemoglobin

$$= HB-NH-CO-(CH_2)_3-CO-NH-HB$$
Cross-linked hemoglobin

Cross-linking hemoglobin into soluble hemoglobin prevents the breakdown of hemoglobin into dimers. Furthermore, when infusing modified hemoglobin, their colloid osmotic pressure should normally be about the same as that of plasma proteins. This means that only about 7 g/dl of hemoglobin can be infused. Now, colloid osmotic pressure depends on the number of solute particles more than on the total amount of protein in solution. Polyhemoglobin is formed from the cross-linking of an average of four or five hemoglobin molecules together into a soluble macromolecular complex. This way, 15 g/dl of polyhemoglobin would have about the same colloid osmotic pressure as 7 g/dl of hemoglobin (Chang, 1997a). This means that a much larger amount of hemoglobin can be used when in the form of polyhemoglobin. However, there is no 2,3-DPG in the polyhemoglobin to allow them to have lower affinity for oxygen so that they can release oxygen to the tissue as required. Benesch *et al.* in 1975 made the important finding that the addition of a 2,3DPG analog, pyridoxal phosphate, to hemoglobin can also decrease oxygen affinity (Fig. 2). This is therefore added to cross-linked hemoglobin hemoglobin. The 1971 glutaraldehyde approach of Chang has been combined with Benesch's 1975 use of pyridoxal phosphate and developed into pyridoxalated human polyhemoglobin blood substitutes by Dudziak and Bonhard in 1976; Moss' group (Sehgal *et al.*) in 1980; DeVenuto and Zegna in 1982; and Chang's group (Keipert *et al.*) in 1982. The Northfield group of Gould *et al.* (1998) has developed this very extensively to its

present refined state. At present, Gould's group has completed phase II clinical trial and showed in control studies that this is effective in replacing blood loss in surgical patients. They are now in phase III clinical trial in trauma surgery using replacement of up to 10,000 ml which is equal to twice the total blood volume of a patient (Gould *et al.*, 1998). This approach has also been developed extensively by the Biopure group (Pearce and Gawryl, 1998) using bovine hemoglobin to form bovine polyhemoglobin. Since hemoglobin from cows does not need 2,3-DPG to decrease the affinity for oxygen, pyridoxal phosphate is not required. They have completed their phase II clinical trial in patients and have started phase III clinical trial using as much as 11,000 ml in individual patients'. Biopure's veterinary product has been approved by the FDA for routine use in canine veterinary medicine. Clinical trials on the Biopure Polyhemoglobin (Sprung *et al.*, 2002; LaMuraglia *et al.*, 2000) have led to its being approved for routine clinical uses in South Africa (Lok, 2001) and awaiting approval in the United States. Hsia (1991) originated a dialdehyde prepared from open ring sugars to form polyhemoglobin. This cross-linker also modified the 2,3-DPG pocket so that the resulting polyhemoglobin does not require 2,3-DPG or pyridoxal phosphate. This has been developed and extended by Hemosol (Adamson and Moore, 1998).

Other Types of Soluble Modified Hemoglobin

Bunn and Jandl (1968) cross-linked hemoglobin intramolecularly and found that this prevents the loss of hemoglobin through the kidney and serves to decrease renal toxicity. Walder *et al.* in 1979 reported the use of a 2,3-DPG pocket modifier, bis (3,5-dibromosalicyl) fumarate (DBBF) to intramolecularly cross-link the two α subunits of the hemoglobin molecule forming tetrameric hemoglobin. This prevents dimer formation and improves P_{50}. P_{50} is the ease with which oxyhemoglobin releases its oxygen and is measured by the oxygen tension at which 50% of the oxygen is released from oxyhemoglobin. Unlike polyhemoglobin, cross-linked tetrameric hemoglobin is a single hemoglobin molecule. Although its molecular weight is comparable to that of albumin, hemoglobin has much less surface negative charge. As a result, it can cross the intercellular junction of the endothelial cells lining the vasculature much more readily (Fig. 3). This may have resulted in the removal of nitric oxide. Nitric oxide decreases the contraction of smooth muscles including those in vasculature. This may explain the increase in esophageal and intestinal contraction and increase in blood pressure when using these types of cross-linked tetrameric hemoglobin. Baxter has extensively developed this (Nelson, 1998), but the results of their clinical trials have led them to discontinue this approach and to look at another approach.

Another very exciting area is the use of recombinant technology to produce recombinant human hemoglobin from *Escherichia coli* (Hoffman *et al.*, 1990). This results in the fusing of the alpha subunits of the hemoglobin molecule and preventing the breakdown of hemoglobin into dimers. Furthermore, this also resulted in the modification of the 2,3-DPG

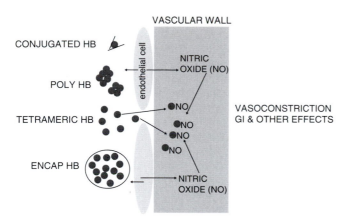

FIG. 3.   Different types of modified hemoglobin (Hb). Tetrameric Hb being smaller can move across the intercellular junction of the endothelial cells. This Hb binds nitric oxide (NO) in the interstitial space and lowers the NO concentration. This results in vasoconstriction. Other types of modified hemoglobin contain a varying amount of tetrameric hemoglobin that can also act similarly. Removal of these smaller tetrameric hemoglobin from polyhemoglobin, conjugated hemoglobin and microencapsulated hemoglobin would prevent this. Reprinted with permission from Chang, T. M. S. (1997). *Artificial Cells, Blood Substitutes and Immobilization Biotechnology* **25**: 1–24. Courtesy of Marcel Dekker, Inc.

pocket in such a way that there is no requirement for 2,3-DPG or pyridoxal phosphate. This has been developed by Somatogen and used in a phase II clinical trial (Freytag and Templeton, 1997). Being a single tetrameric hemoglobin, it can also cause the same effect on smooth muscle as described above for cross-linked tetrameric hemoglobin. They have now prepared a new generation of recombinant human hemoglobin in which they have blocked the binding site for nitric oxide (Doherty *et al.*, 1998). The resulting recombinant human hemoglobin no longer causes vasoconstriction. This is being developed by Baxter toward a clinical trial.

Chang (1964, 1965, 1972) has used interfacial polymerization to form a conjugated membrane consisting of polyamides and hemoglobin around hemoglobin. By decreasing the diameter of the conjugated membrane, all the hemoglobin can be cross-linked with polymer into conjugated hemoglobin. Wong (1988) cross-linked hemoglobin to dextran to form soluble conjugated hemoglobin. A PEG-conjugated bovine hemoglobin is now in phase II clinical trial (Shorr *et al.*, 1996). A conjugated human hemoglobin (Iwashita *et al.*, 1996) is also in phase II clinical trial.

Molecular Weight Distribution

Molecular weight (MW) distribution varies widely among the different types of cross-linked hemoglobin blood substitutes (Chang, 1997a). For intramolecularly cross-linked hemoglobin or recombinant hemoglobin, they are in the tetrameric form with a molecular weight of about 68,000. In polyhemoglobin the molecular weight distribution can vary widely. In polyhemoglobin, the molecular weights follow a distribution curve from tetrameric to much larger molecular weight. Some groups use a larger mean molecular weight. Other groups prefer

using a smaller mean molecular weight. Those with lower mean molecular weight also have more tetrameric hemoglobin. Those with higher mean molecular weight have less tetrameric hemoglobin because of the higher degree of polymerization. The pyridoxalated polyhemoglobin prepared from our laboratory procedure can be used as an example. The molecular weight distribution can be varied at will by changing the reaction time, ratio of glutaraldehyde to hemoglobin, and other reaction parameters. To measure the MW distribution, pyridoxalated polyhemoglobin (PP-PolyHb) was run on Sephadex G-200 (1.6 × 70 cm column) in 0.1 M Tris-HCl (pH 7.5) at 12 ml/hr. The fractions of polymerized hemoglobin were in the molecular weight ranges of 750,000 (10%); 470,000 (23%); 260,000 (39%); 130,000 (21%); and 66,000 (7%). Thus, 60% falls in the 130,000–350,000 range and 33% ranges in molecular weight from 350,000 to 750,000. Only 7% was in the tetrameric form. Gould's group (1998) has emphasized the need to remove as much tetrameric hemoglobin as possible. After their preparation of polyhemoglobin, they carried out another step to remove much of the tetrameric hemoglobin resulting in a preparation with less than 1% hemoglobin in the 68,000 MW range. Biopure has also removed much of the tetrameric hemoglobin resulting in preparation with less than 5% 68,000 MW hemoglobin (Pearce and Gawryl, 1998). These two preparations do not cause vasoconstriction or smooth muscle contraction because insignificant amounts of tetrameric hemoglobin crossing the endothelial cell junctions (Fig. 3).

In Vitro *Biocompatibility Screening Test Using Human Plasma/Blood*

Detailed *in vitro* and *in vivo* tests are required by the FDA before approval of using the blood substitutes for clinical trials (Fratantoni, 1991). However, even when the blood substitute is shown safe in animals, the response in animals is not always the same as for humans. This is especially true in tests for hypersensitivity, complement activation, and immunology. How do we bridge the gap between animal safety studies and use in humans? Complement activation is important in many adverse reactions of humans to modified hemoglobin (Chang and Lister, 1990, 1993a, b, 1994). Modified hemoglobin may be contaminated with trace amounts of blood-group antigen that can form antigen–antibody complexes. Other materials can potentially cause complement activation. These include endotoxin, microorganisms, insoluble immune complexes, chemicals, polymers, organic solvents, and others.

We have devised an *in vitro* test tube screening test using human plasma or blood (Chang and Lister, 1990, 1993a, b, 1994). The use of human plasma or blood gives the closest response to human testing next to actual injection. If we want to be more specific, we can use the plasma of the same patient who is to receive the blood substitute. Many components of human blood or plasma can be used for this *in vitro* screening test. If one were to select only one test, perhaps the most useful one would be the effect of modified hemoglobin on complement activation (C3a) when added to human plasma or blood. This simple test consists of adding 0.1 ml of modified

TABLE 1   Screening Test Based on Complement Activation (Chang and Lister, 1990, 1993a, b, 1994)

1 Immediately before use, the plasma sample is thawed.

2 400 μl of the plasma is pipetted into 4-ml sterile polypropylene tubes.

3 100 μl of pyrogen-free saline (or Ringer's lactate) for injection is added to the 400 μl of human plasma as control; 100 μl of hemoglobin or modified hemoglobin is added to each of the other tubes containing 400 μl of human plasma.

4 The reaction mixtures are incubated at 37°C at 60 RPM for 1 hour in a Lab-Line Orbit Environ Shaker (Fisher Scientific, Montreal, Canada).

5 After 1 hour the reaction is quenched by adding 0.4 ml of this solution into a 2 ml EDTA sterile tube containing 1.6 ml of sterile saline.

6 The samples are immediately stored at −70°C until analysis.

7 The analytical kit for human complement C3a is purchased from Amersham Canada. The method of analysis is the same as the instructions in the kit with two minor modifications. Centrifugation is carried out at 10,000 g for 20 min. After the final step of inversion, the inside wall of the tubes is carefully blotted with Q-Tips.

hemoglobin to a test tube containing 0.4 ml of human plasma or blood. Then, the plasma is analyzed for complement activation after incubation for 1 hour. Blood preparation is described in Table 1.

Blood is obtained by clean venous puncture from human volunteers into 50-ml polypropylene (Sastedt) heparinized tubes (10 IU heparin/ml of blood). Immediately separate plasma by centrifugation at 5500 G at 2°C for 20 min and freeze the plasma in separate portions at −70°C. Do not use serum because coagulation initiates complement activation. EDTA should not be used as an anticoagulant, because it interferes with complement activation.

The exact procedure and precise timing described in Table 1 are important in obtaining reproducible results. The baseline control level of C3a complement activation will vary with the source and procedure for obtaining the plasma. Therefore, a control baseline level must be used for each analysis. Furthermore, all control and test studies should be carried out in triplicate. Much practice is needed to establish this procedure when used for the first time. Reproducibility must be established before this test can be used.

Instead of using plasma and the need to withdraw blood with a syringe, a simpler procedure involves obtaining blood from finger pricks (Chang and Lister, 1993a). Sterile methods are used to prick a finger. Blood is collected in heparinized microhematocrit tubes. The tubes are kept at 4°C, then used immediately. Each blood sample used in testing contains 80 μl of whole blood and 20 μl of saline. Each test solution is added to a blood sample. Test solutions include saline (negative control); Zymosan (positive control); or hemoglobin. This is incubated at 37°C at 60 rpm. After 1 hour of incubation, EDTA solution is added to the sample to stop the reaction. The analysis for C3a is then carried out as described. The test kit is based on the ELISA C3a enzyme immunoassay (Quidel Co., San Diego).

In research, development, or industrial production of blood substitute, different chemicals, reagents, and organic solvents are used. This includes cross-linkers, lipids, solvents, chemicals, polymers, and other materials. Some of these can potentially result in complement activation and other reactions in humans. Other potential sources of problems include trace contaminants from ultrafilters, dialyzers, and chromatography. The test described in Table 1 is a simple *in vitro* test based on using human plasma or blood. For example, this screening test can detect several problems related to potential hypersensitivity reactions, and anaphylactic reactions, effects due to antibody–antigen complexes, and others. These potential problems may be related to contamination with trace blood group antigens, polymeric extracts, organic solvents, emulsifiers, and others. This is useful for preclinical trial studies and for screening before clinical use. This is also useful for screening batches of modified hemoglobin blood substitutes for industrial production to rule out potential problems. It is also useful in research and development. What is very exciting is that this approach may be the basis of a large-scale *in vitro* "clinical trial" in a large number of patients without infusing the product. By doing this, we can analyze the percentage of patients who may have adverse reactions without having to introduce the product into patients. Of course, ultimately, human clinical trials are required.

Second-Generation Modified Hemoglobin Blood Substitutes

The present, first-generation blood substitutes described above are effective for short-term uses especially in elective surgery (Winslow, 1996; Chang, 1997a). On the other hand, these first-generation blood substitutes can stay in the body's circulation only for about 12–24 hours (a typical red blood cell lasts about 100 days). Thus, their present role is restricted to short-term applications. For example, substitutes are being tested in humans for replacing blood lost during some cardiac, cancer, orthopedic, and trauma surgeries. Furthermore, new generations of blood substitutes may be needed for other conditions such as sustained severe hemorrhagic shock, stroke, or other conditions with sustained ischemia (low oxygen). When oxygen from blood substitute is reperfused to ischemic organs or tissues, oxygen radicals will be produced (Fig. 4). The present generation of blood substitutes do not have the enzymes needed to protect the body against oxidants such as oxygen radicals. Unchecked, oxygen radicals may cause reperfusion injuries and other problems (Alayash *et al.*, 1998). Furthermore, peroxide formed can break down hemoglobin resulting in further increase in oxygen radicals (Fenton reaction). Researchers are studying ways to solve this problem, including cross-linking the required enzymes to the hemoglobin or further modifying the molecular structure of the hemoglobin. For example, we have reported the cross-linking of superoxide dismutase (SOD) and catalase to hemoglobin to result in polyhemoglobin–catalase–SOD (Fig. 4) (D'Agnillo and Chang, 1998a,b; Chang *et al.*, 1998). In a rat intestinal ischemia-reperfusion model, we showed that polyhemoglobin–catalase-SOD can markedly decrease the level

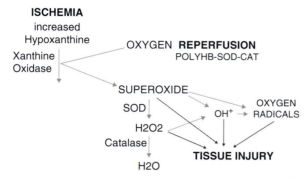

ISCHEMIA
increased
Hypoxanthine

Xanthine
Oxidase

OXYGEN **REPERFUSION**
POLYHB-SOD-CAT

SUPEROXIDE

SOD

OH⁺

OXYGEN
RADICALS

H_2O_2

Catalase

TISSUE INJURY

H_2O

FIG. 4. Ischemic reperfusion injuries. Ischemia leads to accumulation of hypoxanthine and activation of xanthine oxidase. Reperfusion bringing oxygen results in superoxide formation. This and other resulting oxidants and oxygen radicals can cause tissue injury. First-generation modified hemoglobin is prepared from ultrapure hemoglobin and therefore does not contain the required deoxidant enzymes. We showed that cross-linking polyhemoglobin with superoxide dismutase and catalase can significantly decrease the formation of oxygen radicals (D'Agnillo and Chang, 1998). Reprinted with permission from Chang, T. M. S. (1997). *Artificial Cells, Blood Substitutes and Immobilization Biotechnology* 25: 1–24. Courtesy of Marcel Dekker, Inc.

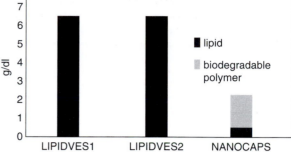

MEMBRANE MATERIAL IN 100 ml SUSPENSION

g/dl

lipid

biodegradable
polymer

LIPIDVES1 LIPIDVES2 NANOCAPS

FIG. 5. Amount of membrane materials in each 100-ml suspension. Hemoglobin lipid vesicles(LIPIDVES) compared to that of biodegradable polymeric hemoglobin nanocapsules (NANOCAPS). Reprinted with permission from Chang, T. M. S. (1997). *Artificial Cells, Blood Substitutes and Immobilization Biotechnology* 25: 1–24. Courtesy of Marcel Dekker, Inc.

PERFLUOROCHEMICALS

Introduction

Silicone and fluorocarbon are known for their ability to carry oxygen. Thus, Clark and Gollan (1966) showed that mice immersed in oxygenated silicone oil or liquid fluorocarbon could breathe in the liquid. In the same year, Chang (1966) showed that artificial cells formed from a hybrid of silicone rubber and hemolysate were very efficient in carrying and releasing oxygen. However, these solid elastic silicone-rubber artificial cells were removed rapidly from the circulation. Sloviter and Kamimoto (1967) showed that perfusion using finely emulsified fluorocarbon could maintain rat brain function for several hours. Geyer *et al.* (1968) showed that finely emulsified fluorocarbon could replace essentially all the blood of rats with the rats surviving and recovering. This was extended and developed in Japan to produce Fluosol-DA 20 suitable for clinical testing (Naito and Yokoyama, 1976; Mitsuno and Naito, 1979).

Earlier Perfluorocarbons: Fluosol-DA

Fluosol-DA is a 20% (w/v) mixture of seven parts of perfluorodecalin and three parts perfluorotripropylamine, with 2.7% Pluronic F-68 as an emulsifier and 0.4% egg yolk phospholipids to form a membrane coating on the emulsion. The average particle size of the emulsion is 0.118 μm. Because of the viscosity of the fluorocarbon emulsion at high concentrations, the maximum amount used is only 20%. Patients have to breathe 70% to 90% oxygen for this amount of fluosol-DA to carry enough oxygen. The amount of oxygen dissolved in fluorocarbon determines the amount of available oxygen. When oxygen is increased to 70–90% from the atmospheric oxygen of 20%, more oxygen is dissolved in the fluorocarbon. CO_2 also dissolves in the fluorocarbon, which carries it to the lung for elimination.

Mitsuno and Ohyanagi (1985) reviewed the Japanese experience with 401 patients who received Fluosol-DA (20%).

of oxygen radicals when using polyhemoglobin with the cross-linked enzymes (Razack *et al.*, 1997). Unlike polyhemoglobin, cross-linked polyhemoglobin–superoxide dismutase–catalase supplies oxygen without causing blood–brain barrier disruption or brain edema in a rat model of transient global brain ischemia–reperfusion (Powanda and Chang, 2002).

Third-Generation Blood Substitutes

Second-generation substitutes, like the first-generation blood substitutes, only circulate with a half-time of 12–24 hours. They are therefore more suitable for shorter term applications. Furthermore, the hemoglobin molecules will not be protected by a red blood cell membrane and highly purified hemoglobin is needed. Thus, researchers are working on more complicated, third-generation blood substitutes that will encapsulate hemoglobin and the required enzymes inside artificial red blood cells. Encapsulated hemoglobin is the first modified hemoglobin studied (Fig. 1) (Chang, 1957, 1964, 1972), but being more complicated, it has only been seriously developed more recently. One method is to encapsulate hemoglobin inside lipid vesicles about 0.2 μm in diameter (Djordjevich and Miller, 1980; Rudolph *et al.*, 1997; Tsuchida, 1995). This technique also increases the circulation time. A more recent approach is to use nanotechnology methods to encapsulate hemoglobin and enzymes inside biodegradable polylactic acid membrane nanocapsules some 0.15 μm in diameter (Yu and Chang, 1996; Chang and Yu, 1997, 1998) (Fig. 5). Nano-dimension artificial red blood cells with ulthrathin membranes of PEG-PLA [poly(ethylene glycol)–polylactide] copolymer stay in the circulation twice as long as polyhemoglobin (Chang and Yu, 2001).

Complement activation was the major clinical problem. Adverse effects were observed in some patients due to complement activation caused by the Pluronic surfactant used in Fluosol. The FDA approved the clinical use of Fluosol-DA only for coronary artery balloon angioplasties in 1989.

Modern Perfluorochemicals

Modern perfluorochemicals been discussed in detail (Riess, 1998; Keipert, 1998). Two new types of preparations have been developed. One type is based on perfluoroctyl bromide ($C_8F_{17}Br$) and perfluorodichoroctane ($C_8F_{16}Cl_2$). Both types allow the use of higher concentrations of PFC. Oxygent from Alliance Pharmaceutical Corp., San Diego, is prepared from perfluoroctyl bromide ($C_8F_{17}Br$) with egg yolk lecithin as the surfactant. The use of egg yolk lecithin instead of Pluronic surfactant has solved the problem of complement activation. Another approach, Oxyfluor from HemoGen, St. Louis, is based on the use of perfluorodichloroctane ($C_8F_{16}Cl_2$) with triglyceride and egg yolk lecithin.

Safety clinical trials using Oxygent shows that doses up to 1.8 g PFC/kg could be given without side effects (Keipert, 1998). Further increase in dosages may result in transient mild febrile response (38–39°C), a transient decrease in platelets. Oxygent is being used in phase II clinical trials in surgical patients. The use of 0.9 g/kg of Oxygent can avoid the need for 2 units of blood. This is also being combined with autologous blood transfusion in which about 1000 ml of each patient's blood is removed and stored before surgery. During surgery, Oxygent is given as needed. After surgery, the patient's blood is reinfused. There are several other potential applications for perfluorochemicals. One of the most important potential use is in patients who because of religious beliefs cannot use red blood cells or hemoglobin from animal sources.

GENERAL DISCUSSION

The basic ideas of cross-linked hemoglobin and encapsulated hemoglobin date back to before the 1960s. Concentrated efforts to develop blood substitutes for human use only seriously started after 1986 because of public concerns regarding HIV in donor blood. Unfortunately, a product must undergo years of research and development followed by clinical trials before it is ready for use in patients. It will take at least another one to two years for the first generation blood substitutes to be available for routine use. Had there been a serious development effort in the 1960s, blood substitutes would have already been available in 1986. As it is, the public has continued to be exposed to the potential, though extremely rare, hazard of HIV in donor blood. It is hoped that investigators will have the chance to develop new generations of blood substitutes discussed here long before another urgent need arrives. The principle of artificial cells has also been developed for other areas of medical applications including enzyme therapy, cell therapy, and other applications (Chang, 1997b; Chang and Prakash, 1998, 2001) (www.artcell.mcgill.ca).

Bibliography

Adamson, J. G., and Moore, C. (1998). Hemolink, an o-Raffinose crosslinked hemoglobin-based oxygen carrier. in *Blood Substitutes: Principles, Methods, Products and Clinical Trials*, T. M. S. Chang, ed., Vol. 2. Basel, Karger, pp. 62–79.

Alayash, A. I., Brockner-Ryan, B. A., McLeod, L. L., Goldman, D. W., and Cashon, R. E. (1998). Cell-free hemoglobin and tissue oxidants: probing the mechanisms of hemoglobin cytotoxicity. in *Blood Substitutes: Principles, Methods, Products and Clinical Trials*, Vol. 2, T. M. S. Chang, ed. Basel, Karger, pp. 157–174.

Benesch, R., Benesch, R. E., Yung, S., and Edalji, R. (1975). Hemoglobin covalently bridged across the polyphosphate binding site. *Biochem. Biophys. Res. Commun.* 63: 1123.

Bunn, H. F., and Jandl, J. H. (1968). The renal handling of hemoglobin. *Trans. Assoc. Am. Physicians* 81: 147.

Chang, T. M. S. (1957). Hemoglobin corpuscles. Report of a research, McGill University, 1–25, 1957. Medical Library, McIntyre Building, McGill University, 1957 (reprinted in 30th anniversary in artificial red blood cell research, *Biomater. Artifi. Cells Artif. Organs* 16: 1–9, 1988).

Chang, T. M. S. (1964). Semipermeable microcapsules. *Science* 146(3643): 524–525.

Chang, T. M. S. (1965). Ph.D. Thesis. McGill University Medical Library.

Chang, T. M. S. (1966). Semipermeable aqueous microcapsules ("artificial cells"): with emphasis on experiments in an extracorporeal shunt system. *Trans. Am. Soc. Artif. Int. Organs* 12: 13–19.

Chang, T. M. S. (1971). Stabilization of enzyme by microencapsulation with a concentrated protein solution or by crosslinking with glutaraldehyde. *Biochem. Biophys. Res. Commun.* 44: 1531–1533.

Chang, T. M. S. (1972). *Artificial Cells*. Charles C. Thomas, Springfield, IL.

Chang, T. M. S. (1997a). *Blood Substitutes: Principles, Methods, Products and Clinical Trials*, Vol. I. Karger, Basel.

Chang, T. M. S. (1997b). Artificial cells. in *Encyclopedia of Human Biology*, 2nd ed. Academic Press, San Diego, pp. 457–463.

Chang, T. M. S. (2000). Red blood cell substitutes. *Best Pract. Res. Clin. Haematol.* 13(4): 651–668.

Chang, T. M. S. (2002). Oxygen carriers. *Curr. Opin. Invest. Drugs* 3(8): 1187–1190.

Chang, T. M. S., and Lister, C. (1990). A screening test of modified hemoglobin blood substitute before clinical use in patients—based on complement activation of human plasma. *Biomater. Artif. Cells Artif. Organs* 18(5): 693–702.

Chang, T. M. S., and Lister, C. W. (1993a). Screening test for modified hemoglobin blood substitute before use in human. U.S. Patent No. 5,200,323, April 6, 1993.

Chang, T. M. S., and Lister, C. W. (1993b). Use of finger-prick human blood samples as a more convenient way for *in vitro* screening of modified hemoglobin blood substitutes for complement activation: a preliminary report. *Biomater. Artif. Cells Immobilization Biotechnol.* 21: 685–690, 1993.

Chang, T. M. S., and Lister, C. (1994). Assessment of blood substitutes: II. In vitro complement activation of human plasma and blood for safety studies in research, development, industrial production and preclinical analysis. *Artif. Cells Blood Subs. Immobilization Biotechnol.* 22: 159–170.

Chang, T. M. S., and Prakash, S. (1998). Microencapsulated genetically engineered cells: comparison with other strategies and recent progress. *Mol. Med. Today* 4: 221–227.

Chang, T. M. S., and Prakash, S. (2001). Procedure for microencapsulation of enzymes, cells and genetically engineered microorganisms. *Mol. Biotechnol.* 17: 249–260.

Chang, T. M. S., and Yu, W. P. (1997). Biodegradable polymer membrane containing hemoglobin for blood substitutes. U.S. Patent No. 5,670,173, September 23, 1997.

Chang, T. M. S., and Yu, W. P. (1998). Nanoencapsulation of hemoglobin and red blood cell enzymes based on nanotechnology and biodegradable polymer. in *Blood Substitutes: Principles, Methods, Products and Clinical Trials*, Vol. 2, T. M. S. Chang, ed. Basel, Karger, pp. 216–231.

Chang, T. M. S., and Yu, W. P. (2001). Biodegradable polymeric nanocapsules and uses thereof. U.S. Provisional Patent Application No. 60/316,001 (August 31, 2001).

Chang, T. M. S., D'Agnillo, F., and Razack, S. (1998). A second generation hemoglobin based blood substitute with antioxidant activities. in *Blood Substitutes: Principles, Methods, Products and Clinical Trials*, Vol. 2, T. M. S. Chang, ed. Karger, Basel, pp. 178–186.

Clark, L. C., Jr., and Gollan, F. (1966). Survival of mammals breathing organic liquids equilibrated with oxygen at atmospheric pressure. *Science* **152**: 1755.

D'Agnillo, F., and Chang, T. M. S. (1997). Modified hemoglobin blood substitute from Cross linked hemoglobin–superoxide dismutase–catalase. U.S. Patent No. 5,606,025, February, 1997.

D'Agnillo, F., and Chang, T. M. S. (1998a). Polyhemoglobin–superoxide dismutase–catalase as a blood substitute with antioxidant properties. *Nat. Biotechnol.* **16**(7): 667–671.

D'Agnillo, F., and Chang, T. M. S. (1998b). Absence of hemoprotein-associated free radical events following oxidant challenge of cross-linked hemoglobin–superoxide dismutase–catalase. *Free Radical Biol. Med.* **24**(6): 906–912.

DeVenuto, F., and Zegna, A. I. (1982). Blood exchange with pyridoxalated-polymerized hemoglobin. *Surg. Gynecol. Obstet.* **155**: 342–346.

Djordjevich, L., and Miller, I. F. (1980). Synthetic erythrocytes from lipid encapsulated hemoglobin. *Exp. Hematol.* **8**: 584.

Doherty, D. H., Doyle, M. P., Curry, S. R., Vali, R. J., Fattor, T. J., Olson, J. S., and Lemon, D. D. (1998). Rate of reaction with nitric oxide determines the hypertensive effect of cell-free hemoglobin. *Nat. Biotechnol.* **16**: 672–676.

Dudziak, R., and Bonhard, K. (1980). The development of hemoglobin preparations for various indications. *Anesthesist* **29**: 181–187.

Frantantoni, J. C. (1991). Points to consider in the safety evaluation of hemoglobin based oxygen carriers. *Transfusion* **31**(4): 369–371.

Freytag, J. W., and Templeton, D. (1997). Optro (Recombinant Human Hemoglobin): a therapeutic for the delivery of oxygen and the restoration of blood volume in the treatment of acute blood loss in trauma and surgery. in *Red Cell Substitutes; Basic Principles and Clinical Application*, A. S. Rudolph, R. Rabinovici, and G. Z. Feuerstein, eds. Marcel Dekker, New York, pp. 325–334.

Geyer, R. P., Monroe, R. G., and Taylor, K. (1968). Survival of rats totally perfused with a fluorocarbon-detergent preparation. in *Organ Perfusion and Preservation*, J. C. Norman, J. Folkman, W. G. Hardison, L. E. Rudolf, and F. J. Veith, eds. Appleton Century Crofts, New York, pp. 85–96.

Gould, S. A., Sehgal, L. R., Sehgal, H. L., DeWoskin, R., and Moss, G. S. (1998). The clinical development of human polymerized hemoglobin. in *Blood Substitutes: Principles, Methods, Products and Clinical Trials*, Vol. 2, T. M. S. Chang, ed., Vol. 2, Karger, Basel, pp. 12–28.

Hoffman, S. J., Looker, D. L., and Roehrich, J. M. (1990). Expression of fully functional tetrameric human hemoglobin in *Escherichia coli*. *Proc. Natl. Acad. Sci. USA* **87**: 8521–8525.

Hsia, J. C. (1991). o-Raffinose polymerized hemoglobin as red blood cell substitute. *Biomater. Artif. Cells Immobilization Biotechnol.* **19**: 402.

Iwashita, Y., Yabuki, A., Yamaji, K., Iwasaki, K., Okami, T., Hirati, C., and Kosaka, K. (1996). A new resuscitation fluid "stabilized hemoglobin" preparation and characteristics. *Biomater. Artif. Cells Artif. Organs* **16**: 271–280.

Keipert, P., Minkowitz, J., and Chang, T. M. S. (1982). Cross-linked stroma-free polyhemoglobin as a potential blood substitute. *Int. J. Artif. Organs* **5**: 383–385.

Keipert, P. E. (1998). Perfluorochemical emulsions: future alternatives to transfusion, development, pool. in *Blood Substitutes: Principles, Methods, Products and Clinical Trials*, Vol. 2, T. M. S. Chang, ed. Karger/Landes, Basel, Georgetown, Texas, pp. 101–121.

LaMuraglia, G. M., O'Hara, P. J., Baker, W. H., Naslund, T. C., Norris, M. W., Li, J., O'Hara, P. J., and Vandemeersch, E. (2000). The reduction of the allogenic transfusion requirement in aortic surgery with a hemoglobin-based solution. *J. Vasc. Surg.* **31**: 299–308.

Lok, C. (2001). Blood product from cattle wins approval for use in humans. *Nature* **410**: 855.

Mitsuno, T., and Naito, R. (eds.) (1979). *Perfluorochemical Blood Substitutes*. Excerpta Medica, Amsterdam.

Mitsuno, T., and Ohyanagi, H. (1985). Present status of clinical studies of fluosol-DA (20%) in Japan. in *Perfluorochemical Oxygen Transport*, K. K. Tremper, ed. Little Brown & Co, Boston, pp. 169–184.

Naito, R., and Yokoyama, K. (1978). An improved perfluorodecalin emulsion. in *Blood Substitutes and Plasma Expanders*, G. A. Jamieson and T. J. Greenwalt, eds. Alan R. Liss, New York, p. 81.

Nelson, D. J. (1998). Blood and HemAssist (DCLHb): potentially a complementary therapeutic team. in *Blood Substitutes: Principles, Methods, Products and Clinical Trials*, Vol. 2, T. M. S. Chang, ed. Karger, Basel, pp. 39–57.

Pearce, L. B., and Gawryl, M. S. (1998). Overview of preclinical and clinical efficacy of Biopure's HBOCs. in *Blood Substitutes: Principles, Methods, Products and Clinical Trials*, Vol. 2, T. M. S. Chang, ed. Karger, Basel, pp. 82–98.

Perutz, M. F. (1980). Stereochemical mechanism of oxygen transport by hemoglobin. *Proc. R. Soc. Lond. B* **208**: 135–147.

Powanda, D., and Chang, T. M. S. (2002). Cross-linked polyhemoglobin–superoxide dismutase–catalase supplies oxygen without causing blood brain barrier disruption or brain edema in a rat model of transient global brain ischemia-reperfusion. *Artif. Cells Blood Subst. Immob. Biotechnol.* **30**: 25–42.

Razack, S., D'Agnillo, F., and Chang, T. M. S. (1997). Effects of polyhemoglobin–catalase–superoxide dismutase on oxygen radicals in an ischemia–reperfusion rat intestinal model. *Artif. Cells Blood Subst. Immobilization Biotechnol.* **25**: 181–192.

Riess, J. G. (1998). Fluorocarbon-based oxygen-delivery: basic principles and product development, pool. in *Blood Substitutes: Principles, Methods, Products and Clinical Trials*, Vol. 2, T. M. S. Chang, ed. Karger, Basel, pp. 101–121.

Rudolph, A. S., Rabinovici, R., and Feuerstein, G. Z. (eds.) (1997). *Red Blood Cell Substitutes*. Marcel Dekker, New York.

Savitsky, J. P., Doozi, J., Black, J., Arnold, J. D. (1978). A clinical safety trial of stroma free hemoglobin. *Clin. Pharm. Ther.* **23**: 73.

Sehgal, L. R., Rosen, A. L., Gould, S. A., Sehgal, H. L., Dalton, L., Mayoral, J., and Moss, G. S. (1980). *In vitro* and *in vivo* characteristics of polymerized pyridoxalated hemoglobin solution. *Fed. Proc.* **39**: 2383.

Shorr, R. G., Viau, A. T., and Abuchowski, A. (1996). Phase 1B safety evaluation of PEG hemoglobin as an adjuvant to radiation therapy in human cancer patients. *Artif. Cells Blood Subst. Immobilization Biotechnol.* **24**(abstracts issue): 407.

Sloviter, H., and Kamimoto, T. (1967). Erythrocyte substitute for perfusion of brain. *Nature* **216**: 458.

Sprung, J., Kindscher, J. D., Wahr, J. A., Levy, J. H., Monk, T. G., Moritz, M. W., and O'Hara, P. J. (2002). The use of bovine hemoglobin glutamer-250 (Hemopure) in surgical patients: results of a multicenter, randomized, single-blinded trial. *Anesth. Analg.* **94:** 799–808.

Tsuchida, E. (ed.) (1995). *Artificial Red Cells.* John Wiley & Sons, New York.

Walder, J. A., Zaugg, R. H., Walder, R. Y., Steele, J. M., and Klotz, I. M. (1979). Diaspirins that crosslink alpha chains of hemoglobin: bis(3,5-dibromosalicyl)succinate and bis(3,5-dibromosalicyl)fumarate. *Biochemistry* **18:** 4265–4270.

Winslow, R. M. (1996). *Blood Substitutes in Development.* Ashley Publications, pp. 27–37.

Wong, J. T. (1988). Rightshifted dextran hemoglobin as blood substitute. *Biomater. Artif. Cells Artif. Organs* **16:** 237–245.

Yu, W. P., and Chang, T. M. S. (1996). Submicron polymer membrane hemoglobin nanocapsules as potential blood substitutes: preparation and characterization. *Artif. Cells Blood Subst. Immobilization Biotechnol.* **24:** 169–184.

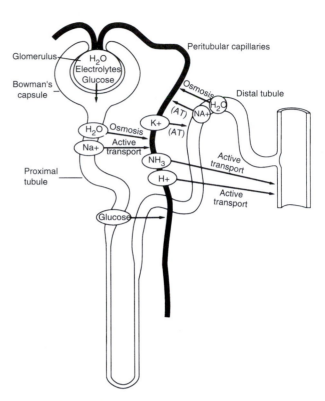

FIG. 1. Schematic representation of the nephron.

7.6 EXTRACORPOREAL ARTIFICIAL ORGANS

Paul S. Malchesky

Historically, the term "extracorporeal artificial organs" has been reserved for life-support techniques requiring the on-line processing of blood outside the patient's body. The substitution, support, or replacement of organ functions is performed when the need is only temporary or intermittent support may be sufficient. The category of extracorporeal artificial organs does not include various other techniques that may justifiably be considered as such, such as infusion pumps or dermal patches for drug delivery, artificial hearts used extracorporeally, eyeglasses and contact lenses for vision, and orthotic devices and manipulators operated by neural signals to control motion.

This chapter focuses on artificial organ technologies that perform mass-transfer operations to support failing or impaired organ systems. The discussion begins with the oldest and most widely employed kidney substitute, hemodialysis, and outlines other renal assist systems such as hemofiltration for the treatment of chronic renal failure and fluid overload and peritoneal dialysis. The blood treatment process of hemoperfusion, and apheresis technologies, which include plasma exchange, plasma treatment, and cytapheresis—used to treat metabolic and immunologic diseases—are also discussed. In addition, blood-gas exchangers, as required for heart–lung bypass procedures, and bioartificial devices that employ living tissue in an extracorporeal circuit are addressed. Significant concerns and associated technological considerations regarding these technologies, including blood access, anticoagulation, the effects of the extracorporeal circulation, including blood cell and humoral changes, and the biomodulation effects of the procedure and materials of blood contact are also briefly discussed.

KIDNEY ASSIST

The kidney's function is to maintain the chemical and water balance of the body by removing waste materials from the blood. The kidneys do this by sophisticated mechanisms of filtration and active and passive transport that take place within the nephron, the single major functioning unit in the kidney, of which there are about 1 million per kidney (Fig. 1). In the glomerulus, the blood entry portion of the nephron, a blood filtration process occurs in which solutes up to 60,000 Da are filtered. As this filtrate passes through the nephron's tubule system, its composition is adjusted to the exact chemical requirements of the individual's body to produce urine. In addition to these functions, the kidneys support various other physiological processes by performing secretory functions that aid red blood cell production, bone metabolism, and blood-pressure control. Renal failure occurs when the kidneys are damaged to the extent that they can no longer function to detoxify the body. The failure may be acute or chronic. In chronic renal failure or in the absence of a successful transplant, the patient must be maintained on dialysis.

Dialysis

Dialysis is the process of separating substances in solution by means of their unequal diffusion through a semipermeable membrane. The essentials for dialysis are (1) a solution containing the substance to be removed (blood); (2) a semipermeable membrane permeable to the substances to be removed and impermeable to substances to be retained (synthetic membrane in hemodialysis or the peritoneal membrane in peritoneal dialysis); and (3) the solution to which the permeable substances are to be transferred (dialysate).

The device containing the semipermeable membrane is called a dialyzer. Hemodialysis is performed on the majority (over 85%) of the patients while peritoneal dialysis is used for the rest.

Hemodialysis

Hemodialysis is the dialysis of blood. More than 300,000 patients in the United States and more than one million worldwide undergo hemodialysis typically three times a week, totaling more than 150 million treatments per year. The patient's blood is typically anticoagulated with heparin throughout the extracorporeal treatment, although other drugs may be used. Blood access is usually from a permanent fistula (a surgical connection between artery and vein) made in the patient's forearm. The blood is drawn from the body at a flow rate of 50 to about 500 ml/min, depending on patient conditions and procedural requirements, and then pumped with a roller pump through the extracorporeal circuit that includes the dialyzer. The transport process of water, electrolytes, and simple metabolites between the blood and the dialysate takes place in the dialyzer (Fig. 2A). The dialysate is a water solution of electrolytes of about the same concentration as normal plasma.

The membranes used, which initially were primarily cellulosic, are now various synthetic membranes, including polyacrylonitrile, poly(methyl methacrylate), ethyl vinyl alcohol, and polysulfone. The membranes permit the low-molecular-weight solutes (typically less than about 5000 Da) such as

urea, creatinine, uric acid, electrolytes, and water to pass freely, but prevent the passage of high-molecular-weight proteins and blood cellular elements. However, some membranes utilized will pass appreciable amounts of solutes of a greater molecular weight (below about 20,000 Da) and with high rates of water transport. Such membranes are called high-flux membranes.

The stimulus for the introduction of synthetic membranes has primarily come from the need to improve biocompatibility. The term "biocompatibility" is loosely defined and generally refers to the response and effects of blood contact with the material. However, the influence of the flow, the properties of the blood, the choice of anticoagulant, and the dialyzer design and its method of sterilization may also be important. Major indices of biocompatibility studied include variations in blood coagulation, blood cell changes, variations in platelet and leukocyte activation, and complement system activation. Different membranes elicit different responses. The degree of the response is related also to the surface area of blood contact.

Most dialyzers in use are of the hollow-fiber design, employing hollow-fiber membranes. Also available are parallel-plate designs employing sheet membrane films. Many variations of these basic designs are commercially available. Generally, blood and dialysate flow in opposite directions (countercurrent).

Water is removed in hemodialysis primarily by ultrafiltration. Convective flux of water and associated solutes takes place under a hydraulic pressure gradient between the blood and the dialysate compartments. The functions of the dialysate delivery system are to prepare and deliver dialysate of the required chemical makeup for use in the hemodialyzer. Monitoring and control equipment is included as part of the system to ensure that the dialysate composition is correct and ready for use by the hemodialyzer and that the procedure is carried out in a safe manner.

Methods of preparing and delivering dialysate are either batch or continuous. In a continuous dialysate-supply system, concentrate and processed tap water (filtered to remove particulates, ions, and organic matter) are continuously mixed during the course of the dialysis and delivered to the dialyzer. This type of system eliminates the space required for mixing an entire batch of dialysate. However, to be effective, the system must be closely controlled, since any malfunction can result in an improperly mixed dialysate. Dialysate can be recirculated in the dialyzer, although fresh dialysate is most effective because it produces a high concentration gradient with the blood and therefore a high removal rate for solutes. In a single-pass system, the flow of dialysate is usually kept low (about 500 ml/min) to limit the amount required.

Dialysate delivery systems also include monitoring, control, and safety equipment. These systems range from the use of simple components to automated systems capable of operating without an attendant. Equipment includes flow meters, temperature and pressure monitors, dialysate-conductivity probes, and display meters. Control equipment includes thermostat-controlled heaters and mixing valves for regulating dialysate temperatures and composition, valves for regulating flow rates and patient fluid loss, and adjustable high and low limits on various safety and monitoring devices. Water conditioning and treatment equipment is usually available separately

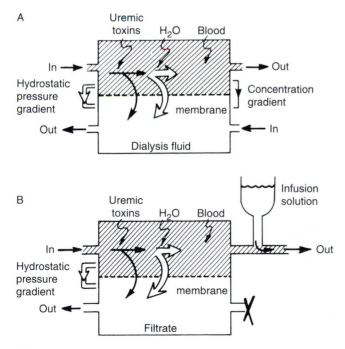

FIG. 2. Comparative operation principles in (A) hemodialysis and (B) hemofiltration with postfiltration addition of fluid.

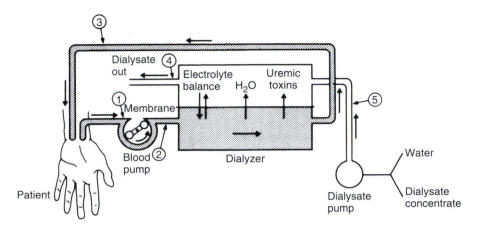

FIG. 3. Schematic of hemodialysis circuitry with on-line dialysate preparation. Numbered locations refer to instrumentation listed in Table 1.

from the dialysate delivery system. Safety equipment includes devices designed to indicate or correct any factor in dialysis that exceeds the established limits for safe operation. This equipment includes audio and/or visual alarms and failsafe shutdown sequences that would be employed during the course of dialysis. Present-day equipment allows one nurse to oversee several dialysis stations at one time.

Hemodialysis is generally performed for periods of 3–5 hours, three times a week. Hemodialysis may be performed in the hospital, in a dialysis center, or at home. Equipment requirements vary, depending upon where and by whom the dialysis is performed. Figure 3 schematically shows the circuitry with on-line dialysate preparation, and Table 1 outlines the most common factors monitored during hemodialysis with reference to the site of monitoring shown in Fig. 3. Portable or wearable systems have also been developed.

Peritoneal Dialysis

Peritoneal dialysis is carried out in the peritoneal cavity of the patient. The peritoneum is a thin membrane lining the abdominal cavity and covering the abdominal organs. It forms a closed sac. Through a cannula placed through the skin or a catheter permanently implanted, dialysate solution (about 2 liters in an adult) is infused, allowed to dwell for a designated time period, and drained. This process is repeated according to the needs of the patient. This semipermeable membrane permits transfer of solutes from the blood to the dialysate.

The efficiency of the process is strongly dependent upon the blood flow through the peritoneal membrane; the permeability of the peritoneal membrane; and the dialysate conditions of flow, volume, temperature, and net concentration gradient. Peritoneal dialysis may be performed intermittently or continuously.

The low volume of dialysate, the low degree of agitation of the dialysate in the peritoneal cavity, and blood flow dynamics in the peritoneum contribute to a lower efficiency for small solute transfer than hemodialysis. Thus, peritoneal dialysis requires longer treatment times than does hemodialysis.

In intermittent peritoneal dialysis, a single cycle for dialysate solution infusion, dwell, and drainage is generally accomplished in less than 30 min and the procedure continued for 8 to 12 hours per treatment (Fig. 4). Dialysate is available in commercially prepared bags or bottles or can be made on site, as in hemodialysis, from dialysate concentrate and water. Additional precautions must be taken with its preparation in comparison with the dialysate used in hemodialysis because of the sterility requirement for infusion of the dialyzing solution into the body. Monitoring and control equipment for the preparation and delivery of the dialyzing fluid is generally automated and capable of operating without an attendant: it includes automatic timers, pumps, electrically operated valves, thermostat-controlled heaters, conductivity meter, and alarms.

In the more popular form of peritoneal dialysis referred to as continuous ambulatory peritoneal dialysis (CAPD), the infused dialysate is allowed to reside for only a few hours. Generally, five to six exchanges are made per day. The difference between this technique and that of intermittent peritoneal dialysis and hemodialysis is that body chemistries are more stable and not fluctuating between the extremes of the pre- and postdialysis periods.

In peritoneal dialysis, the hydraulic pressure difference between the blood and dialysate is insufficient to cause any appreciable water removal so osmotic forces are utilized. The necessary osmotic gradient between the blood and dialysate is achieved by utilizing high concentrations of dextrose in the dialysate. Typically, solutions containing 1.5 to 4.5% dextrose are used.

An advantage of peritoneal dialysis over hemodialysis is that direct blood contact with foreign surfaces is not required, eliminating the need for anticoagulation. A disadvantage includes the increased risk of peritonitis. Less than 15% of the U.S. patients with chronic renal failure are supported by peritoneal dialysis.

Hemofiltration

Hemofiltration refers to the removal of fluid from whole blood. Solute and water are removed strictly by convective flux. The standard hemodialysis membrane, however, is a poor reproduction of the filtration properties of the nephron's

TABLE 1 Factors Most Commonly Monitored during Hemodialysis

Factor	Equipment position (see Fig. 3)	Operation	Remarks
Extracorporeal blood pressure	1, 2, and/ or 3	Measure pressure in drip chambers by mechanical or electronic manometer; abnormal pressure indicates any one of several malfunctions (increased line resistance, clotting, blood leak) and has high and low alarm limits	Installation in location 3 is considered mandatory and provides most meaningful information with respect to changes in flow (such as due to clots) or bloodline leak
Blood-leak detector	4	Photoelectric pickups in effluent dialysate line detect optical transmission changes due to presence of blood	Detection threshold is adjustable with alarm circuit to shut off blood pump or bypass the dialysate flow
Dialysate pressure	5 and/or 4	Measures pressure of dialysate inlet and/or outlet by mechanical or electronic manometer; usually has high and low alarm limit; abnormal operation can result in membrane rupture or improper ultrafiltration	Transmembrane pressure may be displayed with possible control of blood-side pressure
Dialysate temperature	5	Thermostatic measurement generally used to control electric heaters; dial thermometer or thermocouple gauge readout; out-of-range operation can result in patient discomfort or fatal blood damage	In central dialysate delivery system, the dialysate is centrally heated with trimming at individual stations
Dialysate flow rate	5	Measures and displays flow in a rotameter; unless extremely low, cannot result in undue harm to patient	Dialysate is normally used at a rate of 500 ml/min
Dialysate concentration	4 and 5	Flow differential is used to determine patient fluid loss and ultrafiltration rate	Ultrafiltration control is particularly important with the use of high-flux dialyzers
	5	Measures and displays electric conductivity of dialysate; improper concentration can result in blood cell and central nervous system damage	Continuous concentration measurement is a necessity in delivery systems using continuous proportioning of dialysate
Air trap and air bubble detector	3	Collects air prior to blood return to patient; if air is detected, clamp is activated to stop flow to patient	Mandatory

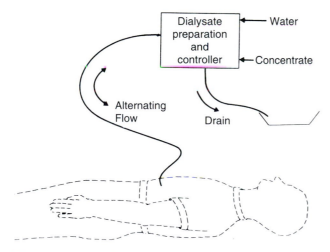

FIG. 4. Schematic of intermittent peritoneal dialysis with on-site dialysate preparation.

glomerulus. The standard hemodialysis membrane can separate from the blood solutes that are typically less than about 5000 Da. In an effort to make the process more like that of the natural glomerulus and to duplicate the process in the natural kidney, membranes of higher permeability (complete passage of solutes less than about 50,000 Da and more typically less than 20,000 Da) have been developed. In addition to their applicability for dialysis under controlled ultrafiltration conditions (referred to as high flux dialysis), such membranes have also found use as ultrafilters of fluid from blood. Membranes used for hemofiltration therapy are made of polysulfone, polyacrylonitrile, poly(methyl methacrylate), poly(ether sulfone), polyamide, and the cellulosics.

Hemofiltration was originally designed for treating chronic renal failure by removing a substantial fraction of fluid and reconstituting blood by adding fluid either before or after filtration (Fig. 2B). Owing to the high physiological sensitivity to changes in circulatory volume, very sensitive monitoring and control equipment for fluid withdrawal and infusion must be

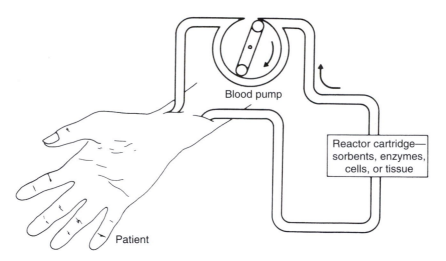

FIG. 5. Circuitry for sorbents or reactor (enzymatic, cellular, or tissue) hemoperfusion.

used to maintain the fluid flow balance accurately, as required for net removal of fluid from the patients. Only a few centers in the world are investigating this technique for treating chronic renal failure.

Modes of hemofiltration applications include continuous arteriovenous hemofiltration (CAVH), continuous venovenous hemofiltration (CVVH), or slow continuous ultrafiltration (SCUF). In these extracorporeal procedures, the patient's blood, under the driving force of arterial pressure (CAVH), or by an extracorporeal pump (CVVH), passes through a low-resistance device (typically of hollow-fiber design) and then is returned to the patient. Under force of the hydraulic resistance and aided by applied vacuum pressure if necessary on the filtrate side, fluid (ultrafiltrate) is removed from the blood. CAVH, CVVH, or SCUF is used to treat acute renal failure and fluid overload, particularly in critically ill patients who are in renal failure and require fluids, as in the form of parenteral nutrition. To enhance the solute transfer, the circuit may be modified to include the pumping of dialysate. This technique is referred to as hemodiafiltration. A further application of hemofiltration is the on-line removal of excess fluid during extracorporeal heart–lung bypass.

Hemoperfusion

Hemoperfusion refers to the direct perfusion of blood. In general, it refers to the perfusion of whole blood over a sorbent or reactor and therefore is different in this respect from hemodialysis (Fig. 5). In principle, the sorbent or reactor may consist of a wide variety of agents such as activated charcoal, nonionic or ionic resins, immunosorbents, enzymes, cells, or tissue. The purpose of biologically active agents may be to remove specific toxins from the blood or to carry out specific biochemical reactions. For practical reasons, such as biocompatibility concerns, hemoperfusion has been limited to the use of a small number of sorbents such as activated charcoal or resins in the treatment of drug intoxication or hepatic support.

The sorption of solute from the blood is based primarily upon its chemical affinity for the sorbents and less on its molecular size. The surface area for sorption may be as high as hundreds of square meters per gram of sorbent.

The use of hemoperfusion is limited primarily by biocompatibility concerns, including particulate release from the sorbent or reactor system. Sorbent and reactor systems are also applied to plasma perfusion (see the next section).

APHERESIS

Apheresis[1] refers to a procedure of separating and removing one or more of the various components of blood. This procedure can be used either therapeutically to remove cells or solutes that are considered harmful, or to obtain plasma or blood cells from normal donors. The procedure is referred to as plasmapheresis when plasma is separated and removed from the whole blood. If the plasma is discarded and replaced with donor plasma or a protein solution, the process is referred to as plasma exchange. If the separated plasma is processed and returned to the patient, the process is referred to as plasma treatment. Cytapheresis refers to the removal of cells and includes various types: leukocytapheresis, which involves the separation and removal of white blood cells; lymphocytapheresis, in which the lymphocytes are removed; erythrocytapheresis, in which the red blood cells are removed; and thrombocytapheresis, in which the platelets are removed.

Plasma Exchange

Plasma is routinely separated from whole blood in the medical laboratory and clinics. Millions of liters of plasma are collected annually in the United States for transfusion

---

[1]Definitions taken from Report of the Nomenclature Committee in Plasma Separation and Plasma Fractionation, M. J. Lysaght and H. J. Gurland, eds., Karger, Basel, pp. 331–334.

and the production of plasma products. Therapeutic plasma replacement or on-line plasma treatment has been applied in a number of disease states, including renal, hematological, neurological, autoimmune, rheumatological, and hepatic. Plasma exchange is now a recognized therapy in selected disease states and is being investigated in many others. Traditionally, plasma separation has been carried out by centrifugal techniques. However, within the past decades, membrane devices have been used increasingly in therapeutic and bulk plasma collection. In centrifugal separation, the difference in density between blood cells and plasma is exploited. In membrane separation, the plasma is separated from the blood cells by the application of a low (typically less than 50 mm Hg) transmembrane pressure gradient across a microporous membrane permeable to all plasma components and impermeable to blood cells.

A number of membrane materials in the flat-film and hollow-fiber configuration have been utilized, including polyethylene, polypropylene, cellulose diacetate, poly(methyl methacrylate), polysulfone, poly(vinyl chloride), poly(vinyl alcohol), and polycarbonate. The membranes are of the tortuous path type and differ significantly in their physical properties such as porosity, mean pore size and its distribution, and pore number. Differences in the chemical nature of the membranes affect the amount and types of proteins that are deposited on the membrane surface. Such differences in protein adsorption can affect the filtration and biocompatibility properties of the membranes. Extensive filtration studies with bloods of various species and varying compositions show that membranes of different polymer types and varying microstructures have different plasma separation rates. Plasma separation efficiency for all membrane or module designs correlates with blood shear rate. Biocompatibility considerations similar to those used in assessing hemodialysis apply.

In therapeutic plasmapheresis, the separated plasma is discarded and replaced with an oncotic plasma substitute such as albumin solution or fresh-frozen plasma. Plasma exchange has several limitations, including the reliance on plasma products, reactions or contaminations (particularly viral) caused by plasma product infusion, and loss of essential plasma components.

Centrifugal Plasma Separation

The centrifugal separation of plasma from blood is schematically depicted in Figs. 6 and 7. As the blood volume is rotated, the contents of the fluid exert a centrifugal force outward to the walls. Since the density of the blood particulate components, the cells, is higher than that of the plasma medium, sedimentation of the cells occurs outwardly through the plasma. The centrifugal acceleration exceeds that of gravity by a factor of a thousand.

For particle settling in the Stokes' law range (Reynolds number is less than 1 for most blood centrifugal processes) the terminal velocity, V_t, at a radial distance, r, from the center of rotation is:

$$V_t = \frac{\omega^2 r D_p^2 (\rho_p - \rho)}{18\mu},$$

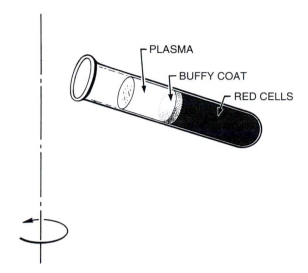

FIG. 6. Plasma separation in a test tube placed in a centrifugal field. The centrifugal separation of plasma from whole blood results in generally three distinct layers: the plasma, the buffy coat consisting of platelets and white blood cells, and the packed red cell layer.

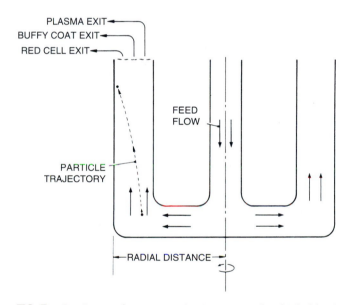

FIG. 7. Continuous plasma separation in a rotating bowl. The blood components are continuously or discontinuously ported off. Several commercial systems are available for clinical use.

where ω, is the angular velocity, D_p is the particle diameter, ρ_p the particle density, ρ the liquid density, and μ the liquid viscosity. This basic formula serves as the theoretical foundation for investigations of the phenomena of the centrifugal separation of particulate matter from the suspending fluid. Corrections to this equation for the nonspherical nature of blood cells and the effects of cell concentration have been proposed by various authors and serve as a basis of continued investigations of this process.

Blood is a complex mixture consisting of plasma, plasma solutes in a broad range of sizes, and cells of varying densities

TABLE 2 Comparison of Cell Properties

	Normal concentration in blood	Cell diameter (μm)	Mean density (g/ml)	Sedimentation coefficient ($S \times 10^7$)
Red blood cell	4.2–6.2×10^{12}/liter	8	1.098 (1.095–1.101)	12.0
White blood cell	4.0–11.0×10^9/liter			
Lymphocyte	1.5–3.5×10^9/liter	7–18	1.072 (1.055–1.095)	1.2
Granulocyte	2.5–8.0×10^9/liter	10–15	1.082	
Monocyte	0.2–0.8×10^9/liter	12–20	1.062	
Platelet	150–400×10^9/liter	2–4	1.058	0.032
Plasma	—	—	1.027 (1.024–1.030)	

TABLE 3 Plasma Separation Considerations
in Centrifugation

Operational and system-related parameters
  Whole blood and component flow rates
  Dimensions of centrifugal apparatus
  Angular velocity
  Radius of cell–plasma interface

Blood properties
  Cell diameter
  Cell concentration
  Specific weight of cell
  Specific weight of blood
  Viscosity of blood and plasma

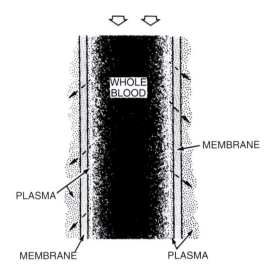

FIG. 8. Plasma separation from blood by a membrane. Based upon the membrane pore structure and operating conditions, the plasma is separated free of cells. Membranes of differing polymer types have been developed for clinical applications.

and sizes (Table 2). When blood is allowed to sit in a tube or is centrifuged, the higher density red blood cells will settle to the bottom of the tube first, followed by the white blood cells and platelets (which are of intermediate density between the red cells and the plasma). Based on the sedimentation coefficients, the order of separation is red cell > white cell > platelet. For centrifugal speeds that are nondestructive to the cells, no appreciable separation of the individual plasma constituents occurs (proteins, amino acids, electrolytes). The high sedimentation rate of red cells is related to their ability to aggregate, forming rouleaux. With rouleaux formation, the effective cell diameter increases, augmenting the settling rate (see equation).

Table 3 lists the operational parameters and blood properties that affect the sedimentation rate of blood cells and that therefore dictate the plasma separation rate. From an operational point of view, the force produced during centrifugation is proportional to the radial distance of blood in the centrifugal field and the square of the angular velocity, or revolutionary rate, of the centrifugal apparatus. Whole-blood flow rates in and component flow rates out, which will determine cell residence times (and therefore particle radial travel), establish the degree of separation between the various cell groups in a given centrifugal design.

In clinical operation, visual inspection or optical sensors are employed to select component pumping rates. For specific collection purposes, two or more separation steps may be employed. For example, the buffy coat is drawn from the whole blood and subjected to a second centrifugal separation or another process that concentrates specific white blood-cell types. In the preparation of blood cells for transfusion, further processing by filtration means may be necessary to purify the cell preparations (see Cytapheresis).

Owing to the varying densities and sedimentation velocities of the blood cells in routine operation of clinical centrifuges, a distinct separation of the cell types from plasma is not achieved. Thus in routine plasma collection, the plasma contains cells, most typically platelets.

Membrane Plasma Separation

Membrane separation of plasma is schematically shown in Fig. 8. Under low hydraulic transmembrane pressures, the plasma is separated. With the fine pore-size membranes employed, the cells do not pass; thus the separated plasma is devoid of any cells. Plasma solute removal, as assessed by its sieving coefficient (ratio of the concentration in the separated plasma to the concentration in the plasma entering the separator), has been shown to be near complete for all macromolecular weight solutes (albumin, immunoglobulins, lipids) of clinical interest. It has been shown that exceeding the safe limits of blood and plasma flow rates and transmembrane pressure leads to deterioration of plasma flux and solute sieving and may cause hemolysis. In comparison with other membrane filtration processes such as reverse osmosis, dialysis, and ultrafiltration, the separation fluid flux (ml/min-m^{-2}) and solute sieving rates are high for membrane plasma separation. This is due to the membrane styles employed. Such microporous membranes have relatively high porosities and high mean membrane pore diameters. Studies have also shown that the cellular and macromolecule concentrations of the blood will to a significant extent dictate operational limits of the membrane plasma separation process. Thus in the control of the process, membrane and module properties, blood properties, and operating conditions should be considered (Table 4).

Conventional filtration theories have been applied to the use of membranes to separate plasma from blood with only limited success. With increasing amounts of data collected on modules and membranes of varying properties, the theories have had to be modified to improve experimental correlations.

Operation of a hollow-fiber membrane plasma separation system is shown in Fig. 9. This schematic may be used to depict the control necessary for operation of membrane systems. For a given membrane or module design, and operation with a blood of given specification (cell and solute concentrations), it has been determined from studies on multiple types of membrane units that a unique relationship exists among the plasma flow and the transmembrane pressure for a given membrane type. Exceeding specified plasma flows causes the transmembrane pressure to rise unsteadily and the plasma

flux to decline (Fig. 10). The plasma flux–transmembrane pressure relationship has been found to be highly dependent on module design and membrane properties, blood flow and membrane properties, blood flow conditions, and blood properties. It is recognized that membrane polymer type and module sterilization method influence the degree of complement activation, coagulation, and solute–material interactions (such as adsorption) that occur at the membrane–plasma interface and that can influence the net pore size of the membrane during use. However, there is no evidence to suggest that the membrane chemical composition directly influences filtration.

Plasma separation is controlled through the blood and plasma flow rates (Fig. 9). The maximum value of the transmembrane pressure ($P_i - P_F$) is an indicator of the stability of the plasma separation process, and the feedback control for the blood and plasma pumps. The value of ($P_i - P_F$) is generally

TABLE 4 Plasma Separation Considerations in Membrane Systems

Operational and module-related parameters
Blood flow rate (shear rate)
Plasma flow rate (filtration velocity)
Transmembrane pressure
Membrane structure (pore number and pore size)
Filtration area
Number of fluid paths
Blood channel dimensions
Blood properties
Cell properties
Cell concentration
Macromolecule concentration
Viscosity of blood and plasma

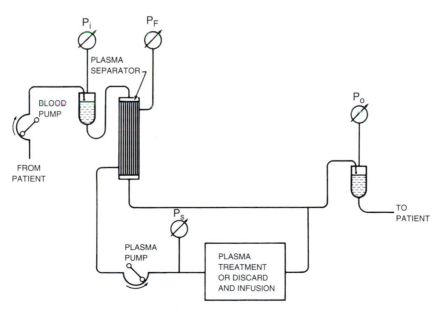

FIG. 9. Scheme of operation with a hollow-fiber membrane plasma separator.

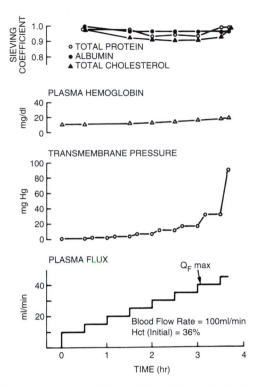

FIG. 10. *In vitro* test with bovine blood of a poly(vinyl alcohol) hollow-fiber plasma separator. Note that sieving for macromolecules is close to unity and that plasma flow below a given value (in this case, 40 ml/min) shows stability with respect to the transmembrane pressure. Unstable plasma separation is readily noted by the rising transmembrane pressure when the plasma flow is increased above 40 ml/min.

a more sensitive indicator than the mean transmembrane pressure $(P_i + P_o)/2 - P_F)$ because P_o may be monitored at a point downstream that will cause a significant difference between the mean and maximum values. As shown in Fig. 10, when the maximum allowable plasma flow rate is exceeded, the transmembrane pressure will rise and not stabilize. This has been shown to be the case for all membrane modules evaluated.

Increases in transmembrane pressure with increases in plasma flow in the stable transmembrane pressure region have been shown to depend on the membrane microstructures (pore size and pore number). An unstable operation is recognizable when the maximum transmembrane pressure does not stabilize with time for a given constant plasma flow. The plasma separation rate can be controlled manually or automatically through the transmembrane pressure change, or by preselecting a maximum operating transmembrane pressure.

The pore properties of the membranes employed for plasma separation are too small to pass even the smallest of the blood cells, the platelet, at operating conditions for which the plasma separation rate is stable. A plasma product devoid of particulate is deemed more desirable when on-line treatment of the plasma is carried out since the particulate may be interfering with the on-line treatment processes. Furthermore, the separation of cellular elements with plasma may result in biological systemic reactions. On-line treatment systems being

employed at present include membrane filtration, sorption, and processing by enzyme and biological reactors.

Plasma Treatment

The rationale for therapeutically applying plasmapheresis is that the removal from plasma of macromolecules accumulated in a disease will correct the abnormality. For plasmapheresis techniques to be effective, the macromolecules must be accessible through the circulation and exhibit favorable kinetics with respect to their rates of production and intercompartmental transfer (tissue, or interstitial, to plasma). In theory, plasma exchange can remove all solutes and achieve a postexchange solute concentration that is dependent on the initial solute concentration, the patient's plasma volume, and rate of plasma exchange as described by the equation:

$$C_f = C_i \exp\left(-Q_p t / V\right),$$

where C_f is the postexchange solute concentration, C_i is the initial plasma solute concentration, Q_p is the mean plasma exchange flow rate for the treatment time period t, or $Q_p t$ is the volume exchange per treatment period, and V is the patient's plasma volume (one-compartment model).

Plasma exchange is carried out when the specific factors in the plasma to be removed are not known or a specific technique for their selective removal is not available, or a group of solutes are to be removed and exchange is the simplest technique to be employed. However, plasma exchange has certain limitations. Since a large volume of plasma is discarded, many plasma constituents that are important to maintain homeostasis are discarded. This necessitates reliance on plasma products to replace the discarded plasma. Typically, albumin is used to replace the discarded plasma but in some cases whole plasma or other plasma fractions such as immunoglobulins are required. Annually, 12–14 million liters of plasma are collected in the United States, yet in recent years there has been a critical shortage of albumin, thus placing limits on its use. In addition to the limited supply of plasma product, the products are a potential source of contamination.

Schemes of plasma treatments that would remove specific plasma factors, minimize total volume and mass of plasma solutes removed, and eliminate or minimize the need for replacement protein solution could be particularly beneficial and more cost effective. A number of systems for on-line treatment have been developed. These are generally classified into their broad categories: membrane plasma fractionation, plasma sorption, and other physicochemical methods.

Membrane Plasma Filtration

Membrane filtration is the technique by which plasma is filtered through single or multiple membranes following its separation from the blood. The principle of macromolecule solute removal from plasma by membrane filtration is based on solute size differences. Molecules that are smaller than the pores of the membrane pass through, whereas molecules larger than the size of the pores are retained by the membrane.

The membranes used for plasma filtration are of the microporous types, with permeability and pore sizes typically much larger than those of the standard hemodialysis membranes, and smaller than those of the plasma separation membranes. Most membranes available have a broad range of pore sizes and therefore produce plasma solute separation efficiencies that are a function of the membrane media and the conditions of operation. Separation efficiency may be defined in terms of sieving coefficient (SC) or rejection coefficient (R) where $R = 1 - SC$. Sieving coefficient is calculated as the ratio of the concentration of the solute in filtrate to that in the incoming plasma. A sieving coefficient of 1 means the solute passes completely; a sieving coefficient of 0 means the solute is completely rejected and therefore retained by the membrane. Most macromolecular-weight plasma solutes have sieving coefficients between these extremes.

On-line membrane plasma fractionation techniques can be classified according to the temperature range of their separation, namely: (1) cascade (double) filtration (filtration at ambient temperature); (2) cryofiltration (filtration below physiological temperature); and (3) thermofiltration (filtration at or above physiological temperature). These techniques have been developed and used in clinical plasma therapy. The temperature and membrane chosen for various procedures are dependent on the targeted macromolecules to be removed. For instance, cryofiltration is used in treating disease states associated with cold-aggregating serum solutes. Thermofiltration has been used to selectively remove low-density lipoproteins (LDLs) in hyperlipidemic patients.

Sorption Plasma Fractionation

Sorption methods have been developed and used in plasma therapy to selectively remove pathological solutes to eliminate the need for substitution fluids. As in membrane plasma fractionation, the procedure involves the separation of plasma followed by perfusion through a sorbent column. Following treatment, the plasma is mixed with the concentrated cell fraction from the plasma separator and returned to the patient. Over the past two decades the development and use of sorbents in hemoperfusion or on-line plasma perfusion has been applied in treating drug overdose, uremia, liver insufficiency, autoimmune disorders, and familial hypercholesterolemia. Table 5 outlines the sorbents evaluated. Only a limited number of sorbents have reached clinical use.

Other Physicochemical Methods

Plasma fractionation may also be carried out by other methods. For example, on-line precipitation of LDL or the globulin fraction of plasma has been studied. Convective electrophoresis has also been investigated for protein separation.

Cytapheresis

Cytapheresis is most commonly performed by centrifugation (see the section on centrifugal plasma separation). The

TABLE 5 Sorbents[a]

Ligand or material for adsorption	Agent sorbed
Polylysin methylated albumin	T4 phage DNA
Anion-exchange resin Polyanion	bilirubin[b]
Dextran sulfate	LDL[b]
Heparin: heparin agarose	LDL
Tryptophan IM-TR	anti-acetycholine receptor Ab,[b] IC, RF[b]
Phenylalanine IM-PH	anti-MBP Ab,[b] IC, RF
Modified PVA gel I-02	RF, IC anti-DNA Ab anti-RNP Ab anti-Sm Ab
Oligosaccharide	anti-blood type AB
Charcoal sorbent	bilirubin, creatinine, urea, potassium
DNA	anti-DNA Ab[b]
Ag blood-type Ag	anti-blood type Ab[b]
Insulin	anti-insulin Ab
Factor VIII	anti-factor VIII Ab[b]
Factor IX	anti-factor FIX Ab
Anti-LDL Ab	LDL[b]
Ab Anti-α- Feto Ab	α-fetoprotien
Anti-HBs Ab	HBs
Anti-IgE Ab	IgE
Clq	IC
Protein A	IC, IgG,[b] C₁

[a] PVA, polyvinyl alcohol; RF, rheumatoid factor; IC, immune complexes.
[b] Clinical application stage.

most common application of cytapheresis has been in the treatment of leukemia by leukocytapheresis. Lymphocytapheresis has been investigated in a number of autoimmune disorders, including rheumatoid arthritis and renal allograft rejection. Erythrocytapheresis has been employed in sickle-cell anemia and severe parasitemia such as malaria. Thrombocytapheresis has been used to treat patients whose platelet concentrations are greater than 1×10^{12}/liter.

In addition to the centrifugal technique, filters have been developed specifically for the removal of leukocytes and lymphocytes. It has been shown that granulocytes adhere to various kinds of fibers such as nylon or cotton. Acrylic and polyester fibers have been found to bind granulocytes and lymphocytes. In such procedures, the blood is generally perfused on-line through a device containing the fibers. Several liters of blood may be processed over a couple of hours. The adhesion of cells to the fiber materials is related to the diameter of the fiber. Fiber devices are also used in blood collections to reduce certain cell populations. Specific adsorbent materials and magnetic separation techniques are also being developed to more specifically separate and collect cells.

A unique white blood cell treatment scheme referred to as photopheresis is now clinically available. In photopheresis, patients are orally administered the drug methoxypsoralen. The drug enters the nucleus of the white blood cells and weakly binds to the DNA. Blood is drawn, centrifugation is carried out, and the separated plasma and the lymphocytes are combined and exposed to ultraviolet A light. The photoactivated drug is locked across the DNA helix and blocks its replication. After the irradiation, this fraction is combined with the remainder of the blood and reinfused. The technique is used to treat skin manifestations of cutaneous T-cell lymphoma.

LUNG SUBSTITUTES AND ASSIST

The lungs are the body's interface with the air. They provide the means of getting oxygen into the blood and carbon dioxide out. The lungs perform their transport processes via a membrane that separates the air from the blood. As the blood cells deform to pass through the intricate and very fine capillary network, oxygen and carbon dioxide transport occur through the membrane. Failure of the lungs or their inability to carry out these functions requires artificial respiratory support in the form of mechanical assistance or substitution.

A ventilator provides mechanical assistance in breathing. It can be of two types, volume or pressure controlled. The cycling of the ventilator may be controlled and in most cases ventilators are set to operate continuously and independently of the patient's inspiratory efforts.

Artificial lungs are of two general types: direct blood contacting and membrane. Direct blood contacting devices may be of two types: bubble or film. In bubble-type oxygenators, oxygen gas is bubbled through the blood. The large surface area of the bubbles and their intimate contact with blood promote high gas-transport rates. Since the blood cells can be damaged by the mixing action, this type is restricted to short-term use for routine cardiac surgery. In film-type oxygenators, the gas contacts the blood, which is spread or distributed in films. These devices generally incorporate a mechanical mixing device so that the blood film is continually renewed. Since such complex devices are not made in disposable form, they are no longer employed clinically.

In membrane oxygenators, the blood and gas are separated by a membrane. The membrane may be of two types, diffusion (as silicone rubber) or microporous (as polypropylene), and may be in the form of film or hollow fibers. Combination-type membranes (thinly coated microporous membrane) are also available. Such membranes are designed to reduce water vapor transfer and improve biocompatibility. In hollow-fiber devices, the blood flow may be on the inside or outside of the fiber. In membrane oxygenators, the gas transport is similar to that in the natural lungs where a thin tissue layer separates the blood and gas. Damage to blood in membrane oxygenators is generally considered to be less than in direct-contacting devices. This allows them to be used for extended periods of time, even up to several weeks as in extracorporeal membrane oxygenation (ECMO) used in lung disorders, such as acute respiratory distress syndrome (ARDS), carbon dioxide retention, and in particular, ARDS caused by immature lung development in the newborn. Such extended uses are rare at present.

More than 500,000 open heart surgery procedures are performed each year worldwide and these procedures require a blood-gas exchanger or so-called artificial lung or oxygenator. Membrane oxygenators are used predominantly for routine cardiopulmonary bypass. In open-heart surgery, the oxygenator is used in conjunction with a pump ("heart–lung machine") as shown in Fig. 11. Membrane oxygenators are also used in organ preservation to provided oxygen and remove carbon dioxide from the perfusate circulated through the organ.

Most recent has been the investigation of membrane oxygenators as intravascular devices. When placed in the vascular system, such devices can provide partial support.

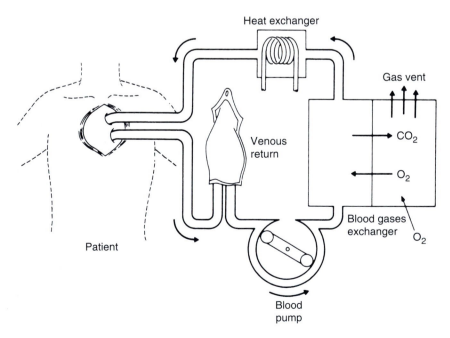

FIG. 11. Circuitry for heart–lung bypass.

BIOARTIFICIAL DEVICES

Because of the limitations of mechanical and passive mass transport devices, there is a growing interest in the development of hybrid artificial organ technologies. Hybrid artificial organs combine some form of biological material and nonbiological material to make a substitution or assist an organ system. The biological system may consist of enzymes or cells that carry out the biological and chemical functions absent or destroyed in the diseased organ system. The nonbiological material is used to encapsulate or enable the utilization of the biological material. Such applications typically involve the use of membranes. Two areas of investigation involve hepatic (liver) and pancreatic assist.

The liver is a complex organ, often referred to as the body's chemical factory. Its many functions can be classified as secretion of bile; metabolism of proteins, fats, and carbohydrates; detoxification of drugs and metabolites; and storage of vital substances such as iron and vitamins. Under certain clinical conditions the liver may regenerate itself in a short span of time, thereby making assist systems practical. However, substitution for the required metabolic support is not possible with present knowledge. The culturing of hepatic cells or the use of liver cells or tissue in membrane modules that permit extracorporeal perfusion but prevent direct blood-to-hepatic-cell contact is being experimentally investigated. The membrane may provide a means for containing the cells; ensure the transport of oxygen, carbon dioxide, nutrients, and cell products; and provide a barrier against immunologic interactions between the cells or tissue and the patient. Because of this last concern, membranes with a molecular weight below about 70,000 Da and typically below 50,000 Da are employed.

A major function of the pancreas is the secretion of insulin by its beta cells in the islet tissue in response to glucose increases in the blood. The lack of such control leads to diabetes.

In addition to mechanical and chemical modes of insulin delivery, efforts are directed at transplanting islet cells. Because of the immunologic-related problems discussed earlier, macroencapsulation (as in semipermeable hollow-fiber membranes) or microencapsulation of islets within thin semipermeable films is being investigated. Such systems may be implanted or used outside the body.

IMPORTANT AREAS OF CONCERN IN DEVELOPMENT OF EXTRACORPOREAL ARTIFICIAL ORGANS

Significant progress has been made and applications not considered possible have become reality; however, the field of extracorporeal artificial organs is still considered to be in its infancy. For example, chronic hemodialysis has only been possible since 1960. At present, all extracorporeal applications require the use of anticoagulants. Although efforts are being expanded to develop materials and coatings that will not require anticoagulation, few practical applications are possible at present. By design, extracorporeal artificial organs are invasive; blood access in many cases limits carrying out the procedures safely and effectively.

Extracorporeal artificial organs require the use of varied types of materials in contact with whole blood or plasma cells. Blood–materials interactions, through generally of only intermittent and short duration, can greatly affect various physiological systems, including the complement and coagulation cascades and the humoral and cellular immunological systems. While investigations point to changes and disturbances in such systems, the short- and long-term impacts are not known. The effects of a procedure may extend beyond the treatment times.

Since the systems are applied in a variety of disease states, the effects are also clouded by the variables related to the clinical situation (concomitant drug therapy, stage of the disease, type of disease, and frequency and duration of the treatment). An extracorporeal artificial organ should not be viewed in engineering analysis as a black box but must be viewed as actively interacting with the physiological system. Therapeutic applications tend not to be tailored to the needs of the patients. Hemodialysis is commonly prescribed without quantitation of the procedure. The question of adequacy is still debated. The prescription requirements for other extracorporeal therapies are even more vague and there are no markers or guidelines to assess adequacy and efficacy. Even so, dialysis for renal assist and some of the other technologies discussed are viewed as mature developments and generally taken for granted.

Transplantation benefits only a small percentage of those patients with chronic renal failure. Dialysis continues to support the large majority of patients, yet research on this technique or on improved technologies is poorly funded. Optimization of the devices used for extracorporeal therapies and processes is rarely achieved. These problems point to the need for a better understanding of the machine–human interface.

Regulatory requirements for new devices have considerably increased the time from concept to clinical application through marketing. Reimbursement issues on new technologies also greatly affect the research and development cycles. Concerns about animal research and the relationship of animal testing to the clinical situation also affect the development cycle.

The move to the development of hybrid organs emphasizes the lack of knowledge on organ system design as well as the lack of technology to substitute for the organ functions, and the need to make technologies more user friendly. Studies on organ substitution have been made possible by progress in such technologies as biomaterials and separation science. Developments in such fields will advance the science of and applications of extracorporeal artificial organs. This field is challenging. With a better understanding of the functions of the body, it holds promise for providing organ substitutes not only in end organ failure or as a bridge to transplantation, but also prophylactically.

Bibliography

Akizawa, T. (1998). Adsorbent: a determinant for the future development of therapeutic apheresis. *Therapeutic Apheresis* 2: 1.

Buckwald, H. (1987). Insulin replacement: bionic or natural. *Trans. Am. Soc. Artif. Intern. Organs* 33: 806.

Galletti, P. M. (1992). Bioartificial organs. *Artif. Organs* 16: 55.

Galletti, P. M., and Brecher, G. A. (1962). *Heart-Lung Bypass: Principles and Techniques of Extracorporeal Circulation*. Grune & Stratton, New York.

Gurland, H. J., Dau, P. C., Lysaght, M. J., Nosé, Y., Pussey, C. D., and Siafaca, K. (1983). Clinical plasmapheresis: who needs it? *Trans. Am. Soc. Artif. Intern. Organs* 29: 774.

Henderson, L. W. (1980). Historical overview and principles of hemofiltration. *Dialysis Transpl.* 9: 220.

Heshmati, F., Tavakoli R., Michel, A., Achkar, A., Guillemain, R., Couetil, J. P., and Andreu, G. (1997). Extracorporeal photo-chemotherapy: a treatment for organ graft rejection. *Therapeutic Apheresis* 1: 121.

HEW (1977). Evaluation of Hemodialyzers, and Dialysis Membranes. DHEW Publication No. (NIH) 77-1294, U.S. Department of Health, Education and Welfare.

Hori, H. (1986). Artificial liver: present and future. *Artif. Organs* 10: 211.

Isbister, J. P. (1997). Cytapheresis: the first 25 years. *Therapeutic Apheresis* 1: 17.

Jauregui, H. (1997). Spin doctors: new innovations for centrifugal apheresis. *Therapeutic Apheresis* 1: 284.

Kambic, H. E., and Nosé Y. (1997). Spin doctors: new innovations for centrifugal apheresis. *Therapeutic Apheresis* 1: 284.

Kasai, S., Sawa, M., and Mito, M (1994). Is the biological artificial liver clinically applicable? A historic review of biological artificial liver support systems. *Artif. Organs* 18: 348.

Ledebo, I. (1998). Principles and practices of hemofiltration and hemodiafiltration. *Artif. Organs* 22: 20.

Leypoldt, J. K., and Cheung, A. K. (1996). Characterization of molecular transport in artificial kidneys. *Artif. Organs* 20: 381.

Malchesky, P. S. (1986). Immunomodulation: bioengineering aspects. *Artif. Organs* 10: 128–134.

Malchesky, P. S. (1994). Nonbiological liver support: historic overview. *Artif. Organs* 18: 324.

Malchesky P. S. (2001). Membrane processes for plasma separation and plasma filtration: guiding principles for clinical use. *Therapeutic Apheresis* 5: 270–282.

Malchesky, P. S., and Nosé, Y. (1987). Control in plasmapheresis. in *Control Aspects of Biomedical Engineering*, M. Nalecz, ed. Int. Fed. Automatic Control, Pergamon Press, Oxford, pp. 111–122.

Maher, J. F. (1980). Pharmacology of peritoneal dialysis and permeability of the membrane. *Dialysis Transpl.* 9: 197.

Martis, L., Chen, C., and Moberly, J. B. (1998). Peritoneal dialysis solutions for the 21st century. *Artif. Organs* 20: 13.

Matsushita, M. and Nosé, Y. (1986). Artificial liver. *Artif. Organs* 10: 378.

Michaels, A. S. (1966). Operating parameters and performance criteria for hemodialyzers and other membrane separation devices. *Trans. Am. Soc. Artif. Intern. Organs* 12: 387.

Mito, M. (1986). Hepatic assist: present and future. *Artif. Organs* 10: 378.

Mortensen, J. D. (1992). Intravascular oxygenator: a new alternative method for augmenting blood gas transfer in patients with acute respiratory failure. *Artif. Organs* 16: 75.

Nolph, K. D., Ghods, A. J., Brown, P., *et al.* (1977). Factors affecting peritoneal dialysis efficiency. *Dialysis Transpl.* 6: 52.

Nosé, Y. (1969). *The Artificial Kidney*. Mosby, St. Louis.

Nosé, Y. (1973). *The Oxygenator*. Mosby, St. Louis.

Rozga, J., and Demetriou, A. A. (1995). Artificial liver: evolution and future perspectives. *ASAIO J.* 44: 831.

Sawada, K., Malchesky P. S., and Nosé, Y. (1990). Available removal systems: State of the art, in *Therapeutic Hemapheresis in the 1990s*, U. E. Nydegger, ed. S. Karger, Basel, pp. 51–113.

Scheen, A. J. (1992). Devices for the treatment of diabetes; today. *Artif. Organs* 16: 163.

Selam, J. L. (1997). Management of diabetes with glucose sensors and implantable insulin pumps: from the dream of the 60s to the realities of the 90s. *ASAIO J.* 43: 137.

Sueoka, A. (1998). Applications of membrane technologies for therapeutic apheresis. *Therapeutic Apheresis* 2: 252.

Woffindin, C., and Hoenich, N. A. (1995). Hemodialyzer performance: a review of the trends over the past two decades. *Artif. Organs* 19: 1113.

Zwischenberger, J. B., Tao, W., and Bidani, A. (1999). Intravascular membrane oxygenator and carbon dioxide removal devices: a review of performance and improvements. *ASAIO J.* 45: 41.

7.7 ORTHOPEDIC APPLICATIONS

Nadim James Hallab, Joshua J. Jacobs, and J. Lawrence Katz

Orthopedic biomaterials are enormously successful in restoring mobility and quality of life to millions of individuals each year. Orthopedic implants include reconstructive implants, fracture management products, spinal products, rehabilitation products, arthroscopy products, electrical stimulation products, and casting products. These products are generally used for either fracture fixation or joint replacement. More specific orthopedic applications within these two categories are listed below.

1. Fracture fixation devices

   - Spinal fixation devices
   - Fracture plates
   - Wires, pins and screws
   - Intramedullary devices
   - Artificial ligaments

2. Joint replacement (Fig. 1)

   - Hip arthroplasty
   - Knee arthroplasty
   - Ankle arthroplasty
   - Shoulder arthroplasty
   - Elbow arthroplasty
   - Wrist arthroplasty
   - Finger arthroplasty

Orthopedic Biomaterials Market

The overwhelming success of orthopedic biomaterials is perhaps best exemplified by their worldwide market, which dominated biomaterial sales at approximately $14 billion in 2002 with an expected growth rate of 7% to 9% annually. Global sales of fracture management products only totaled approximately $1.5 billion in 2000, whereas the other approximately $12 billion was spent on joint replacements (Fig. 1). Global sales of knee implant products equaled approximately $2.5 billion in 2002, representing approximately 700,000 knee replacement surgeries, which include first-time joint replacement procedures and revision procedures for replacement, repair, or enhancement of an implant product or component from a previous procedure. Revision procedures are growing at

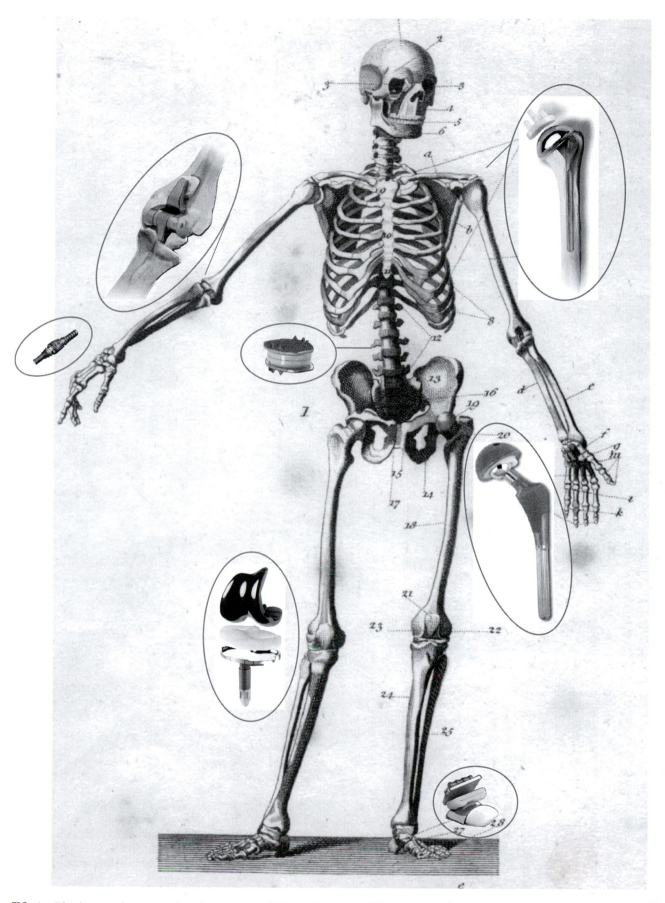

FIG. 1. The diagram shows seven locations where total joint arthroplasties (TJAs) are currently used to replace poorly functioning joints.

an accelerated rate of approximately 60% in the United States. Global sales of hip implant products equaled approximately $2.5 billion in 2002, representing approximately 700,000 hip replacement surgeries. Because of the clinical success of total joint replacement procedures for knees and hips, demand for total joint replacement of other joints, such as the shoulder and elbow, continues to grow. Approximately 55,000 shoulder and elbow implant procedures were conducted in 2000.

Orthopedic Biomaterials

Orthopedic biomaterials are generally limited to those materials that withstand cyclic load-bearing applications. Although metals, polymers, and ceramics are used in orthopedics, it is metals that have, over the years, uniquely provided the appropriate material properties such as high strength, ductility, fracture toughness, hardness, corrosion resistance, formability, and biocompatibility that are necessary for most load-bearing roles required in fracture fixation and total joint arthroplasty (TJA). The use of orthopedic biomaterials generally falls into one of three surgical specialty categories: upper extremity, spine, or lower extremity. Each specialty is typically divided into three general categories: pediatric, trauma, and reconstruction. Despite these numerous specialties and the hundreds of orthopedic applications, there are only a few orthopedic metals, ceramics, and polymers that dominate. Knowing the general properties, uses, and limitations of the "primary" orthopedic biomaterials is requisite to understanding what is required to improve the performance of current implant materials and why only a few dominate the industry. A summary of seven more prevalent orthopedic biomaterials and their primary use(s) are listed in Table 1.

Orthopedic Biomaterials Design

The developer of new biomaterials for orthopedic purposes faces the same duality of concerns present in all other implant use: (1) the material must not adversely affect its biological environment, and (2) in return the material must not be adversely affected by the surrounding host tissues and fluids. And it must do this in a way that exceeds that of present materials used for the same application, if any exist. In order for new "improved" orthopedic biomaterials to address this issue, there should be some understanding of the interrelationship between the structure and properties of the natural tissues that are being replaced. An appreciation of the "form–function" relationship in calcified tissues will help provide insight into factors determining implant design as well as deciding which are the materials of choice to meet a specific orthopedic need.

STRUCTURE AND PROPERTIES OF CALCIFIED TISSUES

There are several different calcified tissues in the human body and several different ways of categorizing them. All calcified tissues have one thing in common: in addition to the principal protein component, collagen, and small amounts of other organic phases, they all have an inorganic component hydroxyapatite [abbreviated OHAp, HA or $Ca_{10}(PO_4)_6(OH)_2$]. In the case of long bones such as the tibia or femur, an understanding of the organization of these two principal components is the beginning phase of the characterization of bone structure according to scale (i.e., the level of the observation technique). It has been convenient to treat the structure of compact cortical bone (e.g., the dense bone tissue found in the shafts of long bones) on four levels of organization.

The first level or molecular level of organization is the collagen triple helical structure (tropocollagen) and OHAp crystallography. It forms a hexagonal unit cell with space group symmetry $P6_3/m$ and lattice constants $a = 9.880$ Å and $c = 6.418$ Å, containing two molecular units, $Ca_5(PO_4)_3OH$, per unit cell. How cells produce this mineral phase and whether it is the first calcium phosphate laid down are subjects of considerable research at present. Because of its small crystallite size in bone (approximately $2 \times 20 \times 40$ nm), the X-ray diffraction pattern of bone exhibits considerable line broadening, compounding the difficulty of identifying additional phases. A Ca-bearing inorganic compound in one of the components of calcified tissues has led to the development of a whole class of ceramic and glass-ceramic materials that are osteophilic within the body (i.e., they present surfaces that bone chemically attaches to).

As yet, we do not know fully how the two components, collagen and OHAp, are arranged or what forms hold them

TABLE 1   Most Common Orthopedic Biomaterials

Material	Primary use(s)
Metals	
Ti alloy (Ti-6%Al –4%V)	Plates, screws, TJA components (nonbearing surface)
Co–Cr–Mo alloy	TJA components
Stainless steel	TJA components, screws, plates, cabling
Polymers	
Poly(methyl methacrylate) (PMMA)	Bone cement
Ultrahigh-molecular-weight polyethylene (UHMWPE)	Low-friction inserts for bearing surfaces in TJA
Ceramics	
Alumina (Al_2O_3)	Bearing-surface TJA components
Zirconia (ZrO_2)	Bearing-surface TJA components

together at this molecular level. Whatever the arrangement, when it is interfered with (as is apparently the case in certain bone pathologies in which the collagen structure is altered during formation), the result is a bone that is formed with seriously compromised physical properties.

The second or ultrastructural level may be loosely defined as the structural level observed with transmission electron microscopy (TEM) or high-magnification scanning electron microscopy (SEM). Here too, we have not yet achieved a full understanding of the collagen-OHAp organization. It appears that the OHAp can be found both inter- and intrafibrillarly within the collagen. As we shall see later, at this level, it appears that we can model the elastic properties of this essentially two-component system by resorting to some sort of linear superposition of the elastic moduli of each component, weighted by the percent volume concentration of each.

The third or microstructural level of organization is where these fibrillar composites form larger structures, fibers, and fiber bundles, which then pack into lamellar-type units that can be observed with both SEM and optical microscopy. The straight lamellar units forming the plexiform (lamellar) bone are found generally in young quadruped animals the size of cats and larger. This is the structural level that is described when the term "bone tissue" is used or when histology is generally being discussed. At this level, composite analysis can also be used to model the elastic properties of the tissue, thus providing an understanding of the macroscopic properties of bone (i.e., those associated with the behavior of the whole bone, or fourth level of structure). Unfortunately, this modeling is much very complex and a complete description lies beyond the scope of this chapter. Interested readers are referred to some of the original sources. (Katz, 1980a, b).

Since a significant portion of bone is composed of collagen, it is not surprising to find that in addition to being anisotropic and inhomogeneous, bone is also viscoelastic like all other biological tissues. Duplicating such properties with a long-lasting synthetic biomaterial remains an unrealized goal of orthopedic biomaterials where the history of implant development has been characterized by the elimination of available candidate materials based on their poor performance rather than production of biocompatible synthetic bone-mimetic materials.

Perhaps the best example of how advances in biomaterials over the past century have directed implant design improvements and resulted in widespread popularity is embodied in that of the total joint replacement (TJR), the earliest and most popular of which is the total hip replacement (THR). Many if not all of the biomaterial related issues (both mechanical and biological) that affect the performance of the THRs embody those applicable to other orthopedic implants. Therefore this chapter will detail the history of the total hip arthroplasty, current orthopedic materials technology, and future developments to highlight clinical concerns of orthopedic biomaterial development and current technology. The history of total hip arthroplasty is particularly pertinent to biomaterials science because it is one of the best illustrations of how an implant first used over a century ago has evolved to the highly successful status it has, primarily because of advances in biomaterials. Current clinical issues associated with orthopedic biomaterials biocompatibility focus on the degradation product and can be broken down into four basic questions: (1) how much material is released from the implant; (2) where the material is transported and in what quantity; (3) what the chemical form of the released degradation products is (e.g., inorganic precipitate versus soluble organometallic complex); and (4) what the pathophysiological interactions and consequences of such degradation are.

HISTORY OF TOTAL HIP ARTHROPLASTY

The earliest attempts to restore mobility to painful and deformed hip joints took place in the 1820s (White, 1822; Barton, 1827) and centered on simply removing the affected femoral and acetabular bone involved. This evolved in the 1830–1880s into ghastly attempts to restore mobility using interpositional membranes between the femoral head and acetabulum, where such materials as wooden blocks and animal (e.g., pig) soft tissue were tried. The first prosthetic hip replacement is dated to 1890 where Gluck published a description of a carved ivory femoral head replacement using bone-cement-like materials such as pumice and plaster of Paris to secure the implants in place (Walker, 1978; Stillwell, 1987).

The interpositional membrane strategy continued from the 19th into the 20th century where the use of new implant materials in the early 1900s (1900–1920) included organic materials (e.g., pig bladders and periimplant soft tissues) and inorganic materials such as gold foil. The use of the individual's own soft tissues was the most popular method of interpositional membrane hip surgery. The limited success of this procedure prevented widespread use and thus the treatment of painful, disfigured, and "frozen" (ankylosed) hip joints remained commonplace into the 1920s.

Mold Arthroplasty

It was not until 1923 when Marius Smith-Peterson was credited with ushering in the modern era of total joint replacement with his development of the "mold" arthroplasty (Fig. 2). Made of glass, it was inspired by a shard of glass found in a patient's back with a benign synovial-like membrane around it. This mold or cup arthroplasty was designed as a cup that fit in between the femoral head and the acetabular cup and articulated on both surfaces prompting a "tissue engineered" synovial/cartilage-like layer. This was the first widespread attempt to develop a better interpositional membrane, a technique that had been in practice for the previous 100 years. The efforts of Smith-Peterson and his colleagues over the years from 1923 to 1938 was spent improving the fracture resistance of the glass mold arthroplasty cup design using materials such as early polymers (e.g., celluloid or phenol–formeldehyde Bakelite or Formica) and improved glass, e.g., Pyrex. But it was not until 1939 when the first metal, a cobalt alloy termed Vitallium, was available and used by Venable, Stuck, and Beach that the corrosion resistance of the hip arthroplasty provided sufficient biocompatibility and performance (Fig. 2).

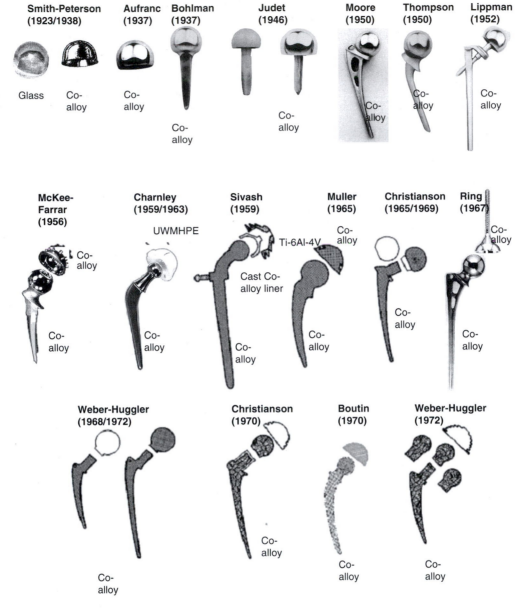

FIG. 2. The history of total hip arthroplasty is particularly pertinent to biomaterials science because it is one of the best illustrations of how an implant first used over a century ago has evolved to the highly successful status it has, primarily because of advances in biomaterials.

In 1937 Venable, Stuck, and Beach published a landmark article that was the first to analyze in a systematic fashion the electrolytic effects of various metals on bone and tissues (e.g., aluminum, copper, iron, nickel, lead, gold, magnesium, silver, stainless steel, and other alloys) and arrived at the conclusion that Vitallium (a cobalt–chromium alloy) was superior to the other metals in corrosion resistance and mechanical properties required for an implant (Venable *et al.*, 1937; Charnley, 1979). By observing the effects of corrosion and proposing guidelines for performance Venable, Stuck, and Beach set the standard by which future metallic alloys were selected for use in hip and other types of implants.

The superior material properties of the Vitallium alloy facilitated further design modifications of the mold arthroplasty by Otto E. Aufranc (Fig. 2) where the rim of the Smith–Peterson mold was removed (which often was the cause of adhesions and cup "freezing," and subsequent pain and immobility) and matching curves on the inner and outer surface were machined to meet at the rounded outer edge. Despite the high short-term success rates (<4 years) reported by Aufranc (>82%), the overall failure rate remained high (>50%). Another design modification of the mold arthroplasty in the 1940–1950s was the fixation of the mold to the acetabulum rim with screws, by such physicians as Albee-Pearson and Gaenslen. Although used in only four cases, Gaeslen reported using a cobalt alloy mold fixed to the acetabulum and another fixed to the femoral head creating a metal-on-metal total hip replacement. The popularity of mold arthroplasties endured into the 1970s when they remained touted as the treatment of choice for traumatic arthritis of the hip by leading orthopedic surgeons (Harris, 1969). However, back in the 1930s the natural progress in THA development was the progression from mold arthroplasty to short-stem prosthesis.

Femoral Head Prostheses/Short-Stem Prostheses

Femoral-head prosthetics were first made of such materials as ivory (Gluck, 1890) and rubber (Delbet, 1919) and were cemented (using a plaster-like cement) for stability (Walker, 1978; Stillwell, 1987). At about the same time these replacement heads were first fitted with a short stem by Earnest Hey Groves, who used an ivory nail to replace the articular surface of the femur. These types of implants were rare and remained unpopular compared to mold arthroplasties until 1937 when Harold Bohlman, using the work of Venable and Stuck, designed a corrosion resistant cobalt–chromium alloy femoral-head replacement with a short stem. This design was popularized by the Judet brothers in Paris in 1946, who used poly(methyl methacrylate) (PMMA), which was presumed biologically inert *in vivo*, to manufacture short-stemmed prostheses (Fig. 2). Initial good results were soon replaced with problems of implant fracture and excessive wear debris and by the early 1950s these implants were losing favor and being removed by surgeons. Vitallium (cobalt–chromium alloy) eventually replaced acrylic in several other short-stem designs. However, there were sound short stem designs as early as 1938 where Wiles introduced the cobalt alloy femoral shell attached to the femur with a central nail. This design was later popularized by Peterson in 1950, where he used a similar Vitallium shell design with a central nail and a plate attached to the nail for added stability. Others adopted and adapted the Judet brothers design using Vitallium, such as J. Thompson (1951) and Rossignal (1950). Rossignal designed large threads onto the stem to aid in fixation. These short stem designs were subject to what was deemed high shear stress and resulted in early loosening and failure in some patients. Short-stem designs were gradually replaced by longer stem designs that provided less stress concentration.

Long-Stem Prostheses

Long-stem prostheses continued the trend established by short-stemmed prostheses, that is, more and more weight-bearing forces were transferred to the femur though an intramedullary stem. The pattern for a long-stem prosthesis was established in 1940 by Bohlman in collaboration with Austin T. Moore in which they implanted a 12-inch Vitallium prosthesis that replaced the femoral head and had long supports that were screwed into the outside of the femoral shaft (Moore, 1943). And although there were innovations in long-stem design in the 1940s, such as the doorknob design of Earl McBride, where a threaded stem was screwed into the

intramedullary canal of the femur for fixation and load transferal, these designs were not popular. It was not until 1950 with the designs of Frederick R. Thompson and Austin T. Moore that long-stemmed prostheses became popular (Fig. 2). These designs were cast in Vitallium (cobalt–chrome alloy) and required the removal of the femoral head but only part of the neck. The design of Moore differed from that of Thompson in that it had fenestrations through the implant to allow bone growth and it had a rear vane to enhance rotational stability. Initially these implants were used without bone cement. Evidence for the successful designs of the Thompson and Moore prostheses is proved by their continued use with only slight variations from the original. Despite the excellent design of these early long-stemmed prostheses they were primarily successful when used in place of diseased femoral heads and did not work well when acetabular reaming was required. Therefore, this inadequacy prompted the development of the total hip replacement arthroplasty.

Total Hip Replacement Arthroplasty

Philip Wiles is credited with first total hip arthroplasty in 1938 when he used a stainless steel ball secured to the femur with a bolt and a stainless acetabular liner secured with screws (Wiles, 1953). The results of this design were disappointing because of the poor corrosion resistance of early stainless steel *in vivo* and the high stress concentrations of short-stemmed prostheses. An adaptation of this design that proved successful was developed by G. K. McKee and J. Watson-Farrar in 1951. They used a stainless steel cup and long-stemmed prosthesis (Thompson stem), which failed rapidly because of the poor corrosion resistance of the stainless steel, and was then changed to cobalt–chromium alloy with greater success. The McKee–Farrar prosthesis evolved quickly to incorporate a true spherical femoral head that was undercut at the neck to reduce the impingement of the head on the rim of the acetabular prosthesis and provide for a greater range of mobility (Fig. 2) (McKee and Watson-Farrar, 1943).

The next milestone in the evolution of modern total hip arthroplasty was the advent/popularization of acrylic dental bone cement, first used by Sven Kiar in 1950 to attach a plastic prosthesis to bone (Charnley, 1964). Later that year the Hospital for Joint Diseases in New York used poly(methyl methacrylate) (acrylic) bone cement as a means of fixation in total hip arthroplasties (Charnley, 1960; Wilson and Scales, 1970). The development of acrylic bone cement dramatically reduced the rates of loosening associated with metal–metal total hip arthroplasty. The Stanmore metal–metal design, which used a horseshoe-shaped cup, was popular but led to excessive wear and was replaced by a complete cup. McKee and Watson-Farrar adapted their design to facilitate bone cement with a land-mine-like studded acetabular cup intended to maximize mechanical fixation.

The 1950s marked the introduction and popularization of the total hip arthroplasty where it became simple and reliable enough to be practiced on a wide scale by the average orthopedic surgeon. However, the squeaking reported to occur in Judet and some later metal prostheses was identified by

Charnley to be a result of the relatively high frictional forces in the joint. These high torque and frictional forces resulted in the generation of significant metallic debris which purportedly resulted in early loosening. In 1960 Charnley developed a "low friction arthroplasty" device using shells of polytetrafluorothylene, PTFE (commonly called Teflon in related publications) on the femoral and acetabular sides, which resulted in early/immediate failures because of the massive debonding and wear debris. This was quickly followed by a thick-walled Teflon acetabular component articulating on a small head designed to reduce the shearing forces and torque. However, this design also generated excessive wear debris that produced immediate and severe inflammation and failure of the prosthesis. Charnley then replaced the Teflon with high-density polyethylene, which was not as friction-free as Teflon but was 1000 times more wear resistant. This prototype of total hip arthroplasty developed in 1962 was the basis for future designs, which remain the most popular form of total hip arthroplasty performed today (Fig. 2).

The basic design of Charnley was modified by Muller with variable neck sizes and larger heads. At the same period metal-on-metal designs by Smith, Ring and others (Ring, 1968) were unsuccessful challengers to the basic Charnley metal-on-polymer design. Other currently adopted design modifications were developed by Ling, Aufranc, Turner, Amstutz, Harris, and Galante, which include such innovations as femoral prosthesis geometrical modification for increases in stability and mobility, modular components for increased customization, porous coatings, and surface texturing/coating to increase fixation and bone ingrowth. Charnley is often deified in orthopedic literature as the metaphorical spark that lit the flames of innovation in prosthetic design. This is a typical surgeon-centered overglorification. For one thing other implant designs that predate Charnley such as the all-metal McKee–Farrar THA implant have enjoyed similar success rates to those reported by Charnley. More importantly, total hip arthroplasty is, perhaps, the best example of how orthopedic biomaterials and implant success have evolved over the past century through the innovation and hard work of many scientists and physicians and advances in areas of materials technology, biomechanics, biochemistry, immunology, infectious diseases, thrombosis, and pharmacology, to name a few.

TOTAL HIP ARTHROPLASTY: CURRENT TECHNOLOGY

Today the archetype of the total hip implant remains much as it was in the 1970s, albeit with a greater variety of implant materials and geometries (Fig. 3). Current THA is typically constructed of a titanium or cobalt–chromium alloy femoral stem [cemented with poly(methyl methacrylate), PMMA, or press fit into place, Fig. 4], connected to a "modular" cobalt–chromium alloy or ceramic head that articulates on a ultrahigh-molecular-weight polyethylene (UHMWPE) or ceramic acetabular cup fitted into a titanium or cobalt–chromium cup liner that is cemented, screwed, or press-fit into place (Fig. 4). Despite this simple archetype of the total hip replacement there exist hundreds of variations on this theme offered to today's orthopedic surgeons, with little in terms of absolute guidelines as

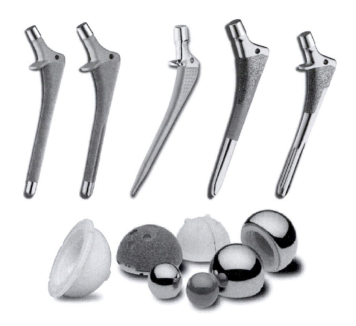

FIG. 3. Examples of a typical current total hip arthroplasty (THA) components available from a single manufacturer, titanium alloy stem, with a cobalt-base alloy (ASTM F-75) modular head bearing on an ultrahigh-molecular-weight polyethylene (UHMWPE) liner within a titanium alloy cup. Also shown is a ceramic head and three acetabular sockets with various surfaces for both cemented and cementless fixation. From left to right the stems are all components of the Versys Hip System (Zimmer Inc., Warsaw, IN) and from left to right are designated Beaded Fullcoat, Beaded Fullcoat with distal flutes, Cemented, Fiber Metal Taper, and Fiber Metal Midcoat. (Photographs courtesy of Zimmer Inc.)

to which type of implant is (or which of the more than 10 major manufacturers has) the best for well-defined orthopedic disease states. However, there are some general guidelines. Typically implants in older individuals (>80 years of age) are cemented into place with PMMA bone cement, because the chance for revision is minimal when compared to younger individuals (<60 years) and removing bone cement is both technically challenging and may compromise the availability of bone stock. Generally there are choices of surface roughness, coatings, geometry, material composition, etc., and each manufacturer claims that its product is superior to the rest. This, in combination with little to no publicly available information tracking the performance of each type of implant in patients, precludes accurate scientific analysis of which implant materials and design perform best. Additionally, stiff competition between manufacturers and the requisite attention to marketing required to compete in the marketplace has resulted in a dizzying array of new implants released each year claiming to be improved over last year's model. These claims are suspect, because the typical total hip replacement enjoys a success rate of over 90% at 7 years; therefore in most cases a minimum of 7 to 10 years must pass before such claims can be substantiated, and even then proof of superior performance is compromised by a myriad external factors such as surgeon, region of the country, and average activity of patient populations. This conflict between science and marketing and market share may (in

Press-Fit   Cemented

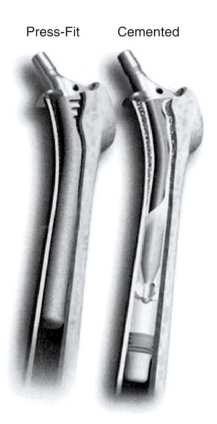

FIG. 4. Examples of cemented and noncemented stems showing the plug and centralizer used with a cemented stem (picture courtesy of Zimmer Inc.).

the opinion of this author) represent the single biggest obstacle to the scientific determination of superior implant design and progress. The unenviable responsibility rests with the FDA to prevent the zeal of economic pressure from undermining implant design in a regressive fashion. Today we see the result of implant technology that has excelled to the point where very real and hard decisions of implant selection are based on material alone (e.g., material couple selection in THA; Fig. 5).

Polymers

Polymers are most commonly used in orthopedics as articulating bearing surfaces of joint replacements (Fig. 3) and as an interpositional cementing material between the implant surface and bone (Fig. 4). Polymers used as articulating surfaces must have low coefficients of friction and low wear rates when in articulating contact with the opposing surface, which is usually made of metal. Initially, John Charnley used Teflon (PTFE) for the acetabular component of his total hip arthroplasty (Fig. 2). However, its accelerated creep and poor stress corrosion (for the material he used) caused it to fail *in vivo*, requiring replacement with his ultimate choice, ultrahigh-molecular-weight polyethylene (UHMWPE). Thus, the twofold demands of "not doing damage" and "not being damaged" were preeminent once again. Polymers used for fixation as a structural interface between the implant component and bone tissue require appropriate mechanical properties of a polymer,

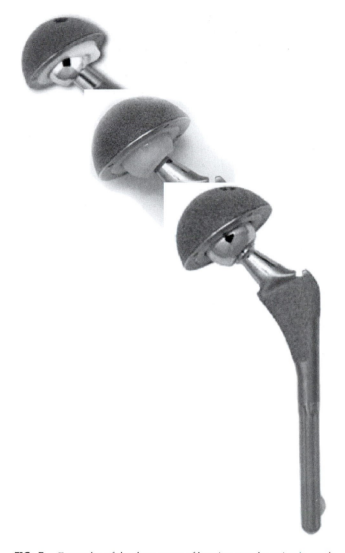

FIG. 5. Examples of the three types of bearing couples using in modern TJA. From top to bottom: metal-on-polymer, ceramic-on-ceramic, and metal-on-metal (Lineage line from Wright Medical Technology, Inc., Arlington TN).

which can be molded into shape and cured *in vivo*. The first type to be used, PMMA, was again popularized by Charnley, who borrowed from the field of dentistry. He adapted PMMA as a grouting material to fix both the stem of the femoral component and the acetabular component in place, and thus distribute the loads more uniformly from the implants to the bone. Since high interfacial stresses result from the accommodation of a high-modulus prosthesis within the much lower modulus bone, the use of a lower modulus interpositional material has been a goal of alternatives seeking to improve upon PMMA fixation. Thus, polymers such as polysulfone have been tried as porous coatings on the implant's metallic core to permit mechanical interlocking through bone and/or soft tissue ingrowth into the pores. However to date PMMA remains the method of choice for orthopedic surgeons. This requires that polymers have surfaces that resist creep under the stresses found in clinical situations and have high enough yield strengths to minimize plastic deformation. As indicated earlier, the important mechanical properties of orthopedic polymers are yield stress, creep resistance, and wear rate. These factors are controlled by such parameters as molecular chain structure, molecular weight, and degree of branching or (conversely) of chain linearity.

One of the more prevalent polymerics used in orthopedics today is a highly cross-linked ultrahigh-molecular-weight polyethylene, which is typically used in total joint arthroplasty as a load-bearing articulating surface, designed to provide low-friction load-bearing articulation. Polyethylene is available commercially in three different grades: low-density, high-density, and UHMWPE. The better packing of linear chains within UHMWPE results in increased crystallinity and provides improved mechanical properties required for orthopedic use even though there is a decrease in both ductility and fracture toughness. In total hip arthroplasty applications an acetabular cup of UHMWPE typically articulates against a femoral ball of cobalt–chromium alloy. The predominant problem presented by these metal–polymer articulating surfaces is the production of wear particles, i.e., polymer debris. The resultant wear of the polyethylene bearing purportedly produces billions of submicron-sized wear particles annually, in the <1–10 μm range. Producing greater cross-linking of polyethylene, using chemical and radiation techniques, has only recently improved its wear resistance in orthopedic applications. Wear tests have shown that the wear resistance of UHMWPE is improved by cross-linking with gamma irradiation at 2.5–5.0 Mrad and below as evidenced by simulator studies; however, this can negatively affect such physical properties as tensile strength (McKellop *et al.*, 2000). Therefore, care must be taken to minimize any negative oxidative effects while preserving high wear characteristics. Although newer, more highly cross-linked polyethylene has generally been accepted as superior to previous implant UHMWPE, there remains incomplete data regarding its ultimate long-term performance. In order to maximize the performance characteristics of polyethylene it is cross-linked prior to fabrication into its final form, e.g., an acetabular cup. Typically, an extruded bar of polyethylene is cross-linked using conventional gamma irradiation and then heat treated to reduce residual free radicals.

Ceramics

In recent years, ceramics and glass-ceramics have played an increasingly important role in implants. Although used in Europe for over a quarter century, only recently (Feb. 3, 2003) has the FDA approved the first ceramic-on-ceramic bearing hip implant to be used in total hip replacement procedures (Fig. 5). The primary reason for the introduction of this alternative bearing surface is the superior wear resistance of ceramics when compared to metal–metal or metal–polymer bearing surfaces. This and other improved properties such as resistance to further oxidation (implying inertness within the body), high stiffness, and low friction requires the use of full-density, controlled, small, uniform grain size (usually less than 5 μm) ceramic materials. The small grain size and full density are important since these are the two principal bulk parameters controlling the ceramic's mechanical properties. Any voids within the ceramic's body will increase stress, degrading the mechanical properties. Grain size controls the magnitude of the internal stresses produced by thermal contraction during cooling. In ceramics, such thermal contraction stresses are critical because they cannot be dissipated as they can in ductile materials via plastic deformation.

Alumina (Al_2O_3) and zirconia (ZrO_2) ceramics have been used in orthopedic THA for the past 30 years. The first ceramic couple (alumina/alumina) was implanted in 1970 by Pierre Boutin. From the outset the theoretical advantage of hard-on-hard articulating surfaces was low wear. Ceramics, because of their ionic bonds and chemical stability, are also relatively biocompatible. Initial concerns about fracture toughness and wear have been addressed by lowering grain size, increasing purity, lowering porosity, and improving manufacturing techniques (e.g., hot isostatic pressing, HIP). Early failures of these couples were plagued with both material-related and surgical errors. The very low wear rates combined with steadily decreasing rates of fracture (now estimated to occur 1/2000 over 10 years) have resulted in the growing popularity of all-ceramic bearings.

Zirconia was introduced in 1985, as a material alternative to Al_2O_3 for ceramic femoral heads, and has been gaining market share because of its demonstrable enhanced mechanical properties in the laboratory when compared to alumina. Femoral heads of zirconia can typically withstand 250 kN (or 25 tons), a value generally exceeding that possible with alumina or metal femoral heads. However, mechanical integrity of all ceramic components is extremely dependent on manufacturing quality controls, as evidenced by the recall of thousands of zirconia ceramic femoral heads by their manufacturer, St. Gobain Desmarquest, on August 14, 2001. This was because of *in vivo* fracture of some components due to a slight unintended variation in the manufacturing sintering process caused when the company bought a newer high-throughput assembly-line-type oven. In general ceramic particulate debris are chemically stable and biocompatible and cause untoward biologic responses at high concentrations.

There have been attempts to take advantage of the osteophilic surface of certain ceramics and glass ceramics. These materials provide an interface of such biological compatibility with osteoblasts (bone-forming cells) that these cells lay down bone in direct apposition to the material in some form of direct chemicophysical bond. Special compositions of glass-ceramics, termed bioglasses, are used for implant applications in orthopedics. The model proposed for the "chemical" bond formed between glass and bone is that the former undergoes a controlled surface degradation, producing an SiO-rich layer and a Ca,P-rich layer at the interface. Originally amorphous, the Ca,P-rich layer eventually crystallizes as a mixed hydroxycarbonate apatite structurally integrated with collagen, which permits subsequent bonding by newly formed mineralized tissues. There is still an entirely different series of inorganic compounds that also have been shown to be osteophilic. These include OHAp, which is the form of the naturally occurring inorganic component of calcified tissues, and calcite, CaCO, and its Mg analog, dolomite, among others being studied. The most extensive applications in both orthopedics and dentistry

TABLE 2 Mechanical Properties of Dominant Orthopedic Biomaterials

Orthopedic biomaterial	ASTM designation	Trade name and company (examples)	Elastic modulus (Young's modulus) (GPa)	Yield strength (elastic limit) (MPa)	Ultimate strength (MPa)	Fatigue strength (endurance limit) (MPa)	Hardness HVN	Elongation at fracture (%)
Cortical bone[a]								
Low strain			15.2	114t	150c/90t	30–45	—	—
High strain			40.8	—	400c–270t	—	—	—
Polymers								
UHMWPE			0.5–1.3	20–30	30–40t	13–20	60–90 (MPa)	130–500
PMMA			1.8–3.3	35–70	38–80t	19–39	100–200 (MPa)	2.5–6
Ceramics								
Al_2O_3			366	—	3790c 310t		20–30 (GPa)	—
ZrO_2			201	—	7500c 420t	—	12 (GPa)	—
Metals								
Stainless steels Co–Cr alloys	ASTM F138	Protusul S30-Sulzer	190	792	930t	241–820	130–180	43–45
	ASTM F75	Alivium-Biomet CoCrMo-Biomet Endocast SlL-Krupp Francobal-Benoist Girard Orthochrome- DePuy Protosul 2-Sultzer Vinertia-Deloro Vitallium C-Howmedica VitalliumFHS- Howmedica Zimaloy-Zimmer Zimalloy Micrograin	210–253	448–841	655–1277t	207–950	300–400	4–14
	ASTM F90	Vitallium W-Howdmedica	210	448–1606	1896t	586–1220	300–400	10–22
	ASTM F562	HS25l-Haynes Stellite MP35N-Std Pressed Steel Corp.	200–230	300–2000	800–2068t	340–520	8–50 (RC)	10–40
	ASTM 1537	TJA 1537-Allvac Metasul - Sulzer	200–300	960	1300t	200–300	41 (RC)	20
Ti alloys								
CPTi	ASTM F67	CSTi-Sulzer	110	485	760t	300	120–200	14–18
Ti-6Al-4V	ASTM 136	Isotan-Aesculap Werke Protosul 64WF-Sulzer Tilastan-Waldemar Link Tivaloy 12-Biomet Tivanium-Zimmer	116	897–1034	965–1103t	620–689	310	8

[a] Cortical bone is both anisotropic and viscoelastic thus properties listed are generalized.

c, compression; t, tension; *, no current ASTM standard; RC, Rockwell Hardness Scale; HVN, Vickers Hardness Number, kg/mm.

have involved OHAp. This has been used as a cladding for metal prostheses for the former, and in dense, particulate form for the latter. The elastic properties of OHAp and related compounds are compared with those of bone, dentin, and enamel in Table 2. The use of both OHAp and the glass ceramics as claddings on the metallic stems of hip prostheses is still another method of providing fixation instead of using PMMA. In these cases, the fixation is via the direct bonding of bone to the cladding surface.

Metals

Since the principal function of the long bones of the lower body is to act as load-bearing members, it was reasonable that the initial materials introduced to replace joints, such as artificial hips, were metals. Both stainless steel, such as 316L, and Co–Cr alloys became the early materials of choice, because of their relatively good corrosion resistance and reasonable fatigue life within the human body. Of course, their stiffness, rigidity, and strength considerably exceeded those of bone. However, in certain applications, owing to size restrictions and design limitations (e.g., in rods used to straighten the spine in scoliosis), fatigue failures did occur. Metals remain the central material component of state of the art total hip arthroplasties. Metals provide appropriate material properties such as high strength, ductility, fracture toughness, hardness, corrosion resistance, formability, and biocompatibility necessary for use in load-bearing roles required in fracture fixation and total joint arthroplasty (TJA). Implant alloys were originally developed for maritime and aviation uses where mechanical properties such as high strength and corrosion resistance are paramount. There are three principal metal alloys used in orthopedics and particularly in total joint replacement: (1) titanium-based alloys, (2) cobalt-based alloys, and (3) stainless steel alloys. Alloy specific differences in strength, ductility, and hardness generally determine which of these three alloys is used for a particular application or implant component. However, it is the high corrosion resistance of all three alloys, more than anything, which has led to their widespread use as load-bearing implant materials. These material properties of metals (Table 2) are due to the miraculous nature of the metallic bond, molecular microstructure, and elemental composition of metals.

Stainless Steel Alloys

Stainless steels were the first metals to be used in orthopedics in 1926. However, it was not until 1943, when ASTM 304 was recommended as a standard implant alloy material, that steels were reliable as an implant alloy. All steels are composed of iron and carbon and may typically contain chromium, nickel, and molybdenum. Trace elements such as manganese, phosphorus, sulfur, and silicon are also present. Carbon and the other alloy elements affect the mechanical properties of steel through alteration of its microstructure.

The form of stainless steel most commonly used in orthopedic practice is designated 316LV (American Society for Testing and Materials F138, ASTM F138; others include F139, F899, F1586, F621, etc.). "316" classifies the material as austenitic, and the "L" denotes the low carbon content and "V" the vacuum under which it is formed. The carbon content must be kept at a low level to prevent carbide (chromium–carbon) accumulation at the grain boundaries.

Although the mechanical properties of stainless steels are generally less desirable than those of the other implant alloys (lower strength and corrosion resistance), stainless steels do possess greater ductility indicated quantitatively by a threefold greater "percentage of elongation at fracture" when compared to other implant metals (Table 2). This aspect of stainless steel has allowed it to remain popular as a material for cable fixation components in total knee arthroplasty, and a low-cost alternative to Ti and Co alloys.

Cobalt–Chromium Alloys

Of the many Co–Cr alloys available, there are currently only two predominantly used as implant alloys (Table 3). These two are (1) cobalt–chromium–molybdenum (CoCrMo), which is designated ASTM F-75 and F-76, and (2) cobalt–nickel–chromium–molybdenum (CoNiCrMo), designated as ASTM F-562. Other Co alloys approved for implant use include one that incorporates tungsten (W) (CoCrNiW, ASTM F-90) and another with iron (CoNiCrMoWFe, ASTM F-563). Co–Ni–Cr–Mo alloys that contain large percentages of Ni (25–37%) promise increased corrosion resistance, yet raise concerns of possible toxicity and/or immunogenic reactivity (discussed later) from released Ni. The biologic reactivity of released Ni from Co–Ni–Cr alloys is cause for concern under static conditions, and because of their poor frictional (wear) properties Co–Ni–Cr alloys are also inappropriate for use in articulating components. Therefore the dominant implant alloy used for total joint components is CoCrMo (ASTM F-75).

Cobalt alloys are generally cast into their final shape because they are susceptible to work-hardening at room temperatures. That is, the improvements in strength and hardness gained by cold working are not worth the loss in fracture toughness. Thus Co–Cr–Mo alloy hip implant components

TABLE 3 Approximate Weight Percent of Different Metals within Popular Orthopedic Alloys

Alloy	Ni	N	Co	Cr	Ti	Mo	Al	Fe	Mn	Cu	W	C	Si	V
Stainless steel (ASTM F138)	10–15.5	<0.5	*	17–19	*	2–4	*	61–68	*	<0.5	<2.0	<0.06	<1.0	*
CoCrMo alloys														
(ASTM F75)	<2.0	*	61–66	27–30	*	4.5–7.0	*	<1.5	<1.0	*	*	<0.35	<1.0	*
(ASTM F90)	9–11	*	46–51	19–20	*	*	*	<3.0	<2.5	*	14–16	<0.15	<1.0	*
(ASTM F562)	33–37	*	35	19–21	<1	9.0–11	*	<1	<0.15	*	*	*	<0.15	*
Ti alloys														
CPTi (ASTM F67)	*	*	*	*	99	*	*	0.2–0.5	*	*	*	<0.1	*	*
Ti-6Al-4V (ASTM F136)	*	*	*	*	89–91	*	5.5–6.5	*	*	*	*	<0.08	*	3.5–4.5
45TiNi	55	*	*	*	45	*	*	*	*	*	*	*	*	*
Zr alloy														
(95% Zr, 5% Nb)	*	*	*	*	*	*	*	*	*	*	*	*	*	*

*Indicates less than 0.05%.

Note: Alloy compositions are standardized by the American Society for Testing and Materials (ASTM vol. 13.01).

are predominantly manufactured using lost wax (investment) casting methods.

Although Co–Cr–Mo alloys are the strongest, hardest, and most fatigue resistant of the metals used for joint replacement components, care must be taken to maintain these properties because the use of finishing treatments can also function to reduce these same properties (Table 2). For example, sintering of porous coatings onto femoral or tibal TJA Co–Cr–Mo stems can decrease the fatigue strength of the alloy from 200–250 MPa to 150 MPa after heating (annealing) the implant at 1225°C.

Titanium Alloys

Titanium alloys were developed in the mid-1940s for the aviation industry and were first used in orthopedics around the same time. Two post–World War II alloys, commercially pure titanium (CPTi) and Ti-6Al-4V, remain the two dominant titanium alloys used in implants. Commercially pure titanium (CPTi, ASTM F67) is 98–99.6% pure titanium. Although CPTi is most commonly used in dental applications, the stability of the oxide layer formed on CPTi (and consequently its high corrosion resistance, Table 4) and its relatively higher ductility (i.e., the ability to be cold worked), compared to Ti-6Al-4V, has led to the use of CPTi in porous coatings (e.g., fiber metal) of TJA components. Generally, joint replacement components (i.e., TJA stems) are made of Ti-6Al-4V (ASTM F-136) rather than CPTi, because of its superior mechanical properties (Table 2).

Titanium alloys are particularly good for THA components because of their high corrosion resistance compared with stainless steel and Co–Cr–Mo alloys. A passive oxide film (primarily of TiO_2) protects both Ti-6Al-4V and CPTi alloys. Generally Ti-6Al-4V has mechanical properties that exceed those of stainless steel, with a flexural rigidity less than stainless steel and Co–Cr–Mo alloys. The torsional and the axial stiffness (moduli) of Ti alloys are closer to those of bone and theoretically provide less stress shielding than do Co alloys and stainless steel. However, titanium alloys are particularly sensitive to geometrical factors, in particular notch sensitivity. This reduces the effective strength of a component by increasing the material's susceptibility to crack propagation through the component. Therefore care is taken both in the design geometry and in the fabrication of Ti alloy components. Perhaps the greatest drawback to Ti alloys is their relative softness compared to Co–Cr–Mo alloys (Table 2) and their relatively poor wear and frictional properties. Ti-6Al-4V is >15% softer than Co–Cr–Mo alloys and also results in significantly more wear than Co–Cr–Mo when used in applications requiring articulation, e.g., TKA or THA femoral heads. Thus Ti alloys are seldom used as materials where hardness or resistance to wear is the primary concern.

New Alloys and Surface Coatings

The quest for new THA metal alloys with improved biocompatibility and mechanical properties remains an ongoing one. The use of Ti alloys, Co–Cr–Mo alloys, or stainless steels in a specific application generally involves trade-offs of one desirable property for another. Some examples of this are the sacrifice of chemical inertness for hardness (wear resistance), as is the case with Ti alloy for Co–Cr–Mo in TJA bearing surfaces, and the compromise of strength for ductility when using stainless steel instead of Ti and Co–Cr–Mo alloys for bone fixation cables. Although new alloys claim to be just that, "new," they are often merely variations of the three categories of implant metals previously described (which are already approved for use as implant materials). These improved alloys usually contain only the minor addition of new elements to protect assertions of substantial equivalence to existing ASTM- and FDA-approved alloys, therefore easing the burden of regulatory approval. These new alloys generally fall under one of four categories: (1) titanium alloys, (2) cobalt alloys, (3) stainless steels, and, less approved, (4) refractory metals.

New Zirconium and Tantalum Alloys

Zirconium (Zr) and tantalum (Ta) are characterized as refractory metals (others include molybdenum and tungsten) because of their relative chemical stability (passive oxide layer)

TABLE 4 Electrochemical Properties of Implant Metals (Corrosion Resistance) in 0.1 M NaCl at pH 7

Alloy	ASTM designation	Density (g/cm³)	Corrosion potential[a] (vs calomel) (mV)	Passive current density (mA/cm²)	Breakdown potential (mV)
Stainless steel	ASTM F138	8.0	−400	0.56	200–770
Co–Cr–Mo Alloys	ASTM F75	8.3	−390	1.36	420
Ti alloys					
CPTi	ASTM F67	4.5	−90 to −630	0.72–9.0	>2000
Ti-6Al-4V	ASTM 136	4.43	−180 to −510	0.9–2.0	>1500
Ti5Al2.5Fe	**	4.45	−530	0.68	>1500
Ni45Ti	**	6.4–6.5	−430	0.44	890

**, No current ASTM standard.

[a]The corrosion potential represents the open circuit potential (OCP) between the metal and a calomel electrode. The more negative the OCP, the more chemically reactive and thus the less corrosion resistance. Generally low current density indicates greater corrosion resistance. The higher the breakdown potential the better, (i.e. the more elevated the breakdown potential, the more stable the protective layer).

FIG. 6.   Examples of new THA and TKA oxidized zirconium components currently gaining popularity because of enhanced mechanical and biocompatibility properties (Oxinium, Smith and Nephew Inc., Memphis, TN) (photographs courtesy of Smith and Nephew Inc.).

and high melting points. Because of their high strength, chemical stability, and resistance to wear, alloys such as Zr (e.g., Oxinium) are likely to gain popularity as orthopedic biomaterials. Because of their surface oxide layer's stability, Zr and Ta are highly corrosion resistant. Corrosion resistance generally correlates with biocompatibility (although not always) because more stable metal oxides tend to be less chemically active and/or biologically available and are thus less participatory in biologic processes. This enhanced biocompatibility is produced by the relatively thick surface oxide layer (approximately 5 μm) and the ability to extend ceramic-like material properties (i.e., hardness) into the material through techniques such as oxygen enrichment has resulted in the production of new implant components using these alloys (Fig. 6) (e.g., Oxidized Zirconium TKA femoral components, Smith and Nephew Inc.). Although new Zr alloys such as Oxinium generally possess high levels of hardness (12 GPa) and wear resistance (approximately 10-fold that of Co and Ti alloys, using abrasion testing), which makes them well suited for bearing-surface applications, they are costly to manufacture and currently are sought after in special circumstances where issues such as a metal allergy (or more accurately metal hypersensitivity) require particular attention to biocompatibility. As difficulties associated with the cost of forming and machining

these metals are overcome the use of these materials is expected to grow (Black, 1992).

New Titanium Alloys

One new group of Ti alloys put forward for orthopedic component uses molybdenum at concentrations greater than 10%. The addition of Mo acts to stabilize the BCC (beta) phase at room temperature; thus these alloys are referred to as beta Ti alloys. These beta Ti alloys promise 20% lower moduli, which are closer to bone and thus provide better formability with maintenance of other mechanical properties typical of Ti-6Al-4V.

Other attempts at improving traditional Ti-6Al-4V alloys seek to improve biocompatibility and mechanical properties by the substitution of V (a relatively toxic metal) with other less toxic metals. Two such Ti alloys include Ti5Al2.5Fe and Ti6Al17Nb which substitute Fe and Nb for V, respectively. These alloys have similar properties to traditional Ti-6Al-4V, yet they claim higher fatigue strength and a lower modulus, thus enhancing bone-to-implant load transfer.

New Cobalt Alloys

Some "new" cobalt alloys are identical in composition to traditional alloys, but use novel processing techniques to

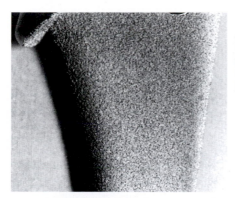

Co-Cr alloy THA femoral stem with a
Co-Cr beaded surface

Pure-titanium fiber metal coating on a
Ti-alloy THA stem

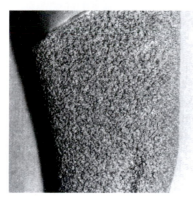

Plasma sprayed titanium surface on a
Ti-alloy stem

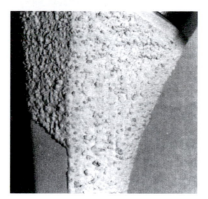

Hydroxyapatite coating on a roughened
titanium alloy THA stem

FIG. 7. Examples of currently used surface coatings on stems of THA to enhance both short- and long-term fixation.

manipulate the microstructure of the implant materials to improve their mechanical properties. One such example recently patented, TJA-1537, although compositionally identical to ASTM F-75, claims enhanced wear resistance and fatigue strength through elimination of carbide, nitride, and second-phase particles (Allegheny Technologies). These particles normally form at the grain boundaries within a standard F75 CoCrMo alloy and act to decrease wear and fatigue resistance. Other new Co alloys under development for use in orthopedics seek to improve biocompatibility by eliminating Ni and improve mechanical properties by reducing the carbon content, thus avoiding carbide precipitation at grain boundaries.

New Stainless Steels

The relatively poor corrosion resistance and biocompatibility of stainless steels when compared to Ti and Co–Cr–Mo alloys provide incentive for development of improved stainless steels. New alloys such as BioDur 108 (Carpenter Technology Corp.) attempt to solve the problem of corrosion with an essentially nickel-free austenitic stainless alloy. This steel contains a high nitrogen content to maintain its austenitic structure and boasts improved levels of tensile yield strength, fatigue strength, and improved resistance to pitting corrosion

and crevice corrosion as compared to nickel-containing alloys such as Type 316L (ASTM F138).

Surfaces and Coatings

A variety of surface coatings are currently used to enhance the short- and long-term performance of implants by encouraging bone ingrowth and providing enhanced fixation. These different surfaces include roughened titanium, porous coatings made of cobalt chromium or titanium beads, titanium wire mesh (fiber mesh), plasma-sprayed titanium, and bioactive nonmetallic materials such as hydroxyapatite or other calcium phosphate compositions (Fig. 7). Currently osteoconductive and osteoinductive growth factors such as transforming growth factor beta (TGF beta) are being developed for use as osteogenic surface coating treatments to enhance orthopedic implant fixation.

ORTHOPEDIC BIOMATERIALS: CLINICAL CONCERNS

The millions of total joint replacements implanted over the years have restored quality of life and longevity to millions of people worldwide and an aging population demographic in the United States. Current implant designs represent the cumulative efforts of scientists, engineers, and physicians and generally last 10–20 years with an approximate revision rate of 7% after 10 years of service. The benefits provided to patients by orthopedic implants in terms of pain, mobility, and quality of life are immeasurable. Therefore the following sections, which focus on the problems associated with implants, seek to provide the student of biomaterials a foundation for understanding the relevant issues for orthopedic research and is not intended to serve as an indictment of orthopedic materials.

Despite their overwhelming success over the long term (>7 years), orthopedic biomaterials have been associated with adverse local and remote tissue responses. It is generally the degradation products of orthopedic biomaterials (generated by wear and electrochemical corrosion) that mediate these adverse effects. This debris may be present as particulate wear, colloidal nanometer-size complexes (specifically or nonspecifically bound by protein), free metallic ions, or inorganic metal salts/oxides, or in an organic storage form such as hemosiderin. Clinical aspects of biocompatibility regarding polymer and metal release from orthopedic prosthetic devices have taken on an increasing sense of urgency because of the escalating rates of people receiving implants and the recognition of extensive implant debris within local and remote tissues. Particulate debris have enormous specific surface areas available for interaction with the surroundings and chronic elevations in serum metal content. Clinical issues associated with biomaterial degradation can be broken down into four basic questions: (1) How much material is released from the implant? (2) Where is the material transported and in what quantity? (3) What is the chemical form of the released degradation products (e.g., inorganic precipitate versus soluble organometallic complex)? (4) What are the pathophysiological interactions and consequences of such degradation? The answers to these questions, over the long term, remain largely unknown. There is a growing body of literature addressing the issues associated with the first two questions. However, little is currently known with regard to the latter two questions. The remainder of this chapter will focus on that which is known (and of orthopedic clinical concern) regarding biomaterial degradation (through wear and electrochemical corrosion), dissemination of debris, and consequent local/systemic effects.

Orthopedic Biomaterial Wear

The generation of wear debris, and the subsequent tissue reaction to such debris, is central to the longevity of total joint replacements. In fact, particulate debris are currently extolled as the primary factor affecting the long-term performance of joint replacement prostheses, and the primary source of orthopedic biomaterial degradation (based on overall implant mass or volume lost). Particulate debris generated by wear, fretting, or fragmentation induces the formation of an inflammatory reaction, which at a certain point promotes a foreign-body granulation tissue response that has the ability to invade the bone–implant interface. This commonly results in progressive, local bone loss that threatens the fixation of both cemented and cementless devices alike (Jacobs *et al.*, 1994b; Jacobs, 1995; Jacobs *et al.*, 2001).

Mechanisms of Wear Debris Generation

Wear involves the loss of material in particulate form as a consequence of relative motion between two surfaces. Two materials placed together under load will only contact over a small area of the higher peaks or asperities. Electrorepulsive and atomic binding interactions occur at the individual contacts and, when the two surfaces slide relative to one another, these interactions are disrupted. This results in the release of material in the form of particles (wear debris). The particles may be lost from the system, transferred to the counterface, or remain between the sliding surfaces. There are primarily three processes which can cause wear: (1) abrasion—by which a harder surface "plows" grooves in the softer material; (2) adhesion—by which a softer material is smeared onto a harder countersurface, forming a transfer film; and (3) fatigue—by which alternating episodes of loading and unloading result in the formation of subsurface cracks which propagate to form particles that are shed from the surface.

Wear Rates

During an initial "wearing in" period, the relative motion of surfaces causes a large number of asperities to break resulting in a high wear rate. After this initial period, the actual contact area increases and the two surfaces can be said to have adapted to one other. Over time, the wear rates decrease and eventually become linearly dependent on the contact force and sliding distance represented by the steady-state wear equation:

$$V = KFx \qquad (1)$$

where V is volumetric wear (mm^3/year), K is a material constant of the material couple, F is the contact force (N), and x is the distance of relative travel (mm).

In general, the softer of two bearing materials will wear more rapidly. In the most popular joint replacement couple, a metal-on-polymer pair, the polymer wears almost exclusively, whereas in the case of a metal-on-ceramic pair (not commercially available), the metal wears to a greater extent. The *in vitro* wear rates for the socket (in hip-joint simulation studies) range from 0 to 3000 mm^3/year depending on the type of couple employed and the environment (e.g., lubricant). There is a great deal of variability associated with *in vivo* wear rates of orthopedic biomaterials, which are generally measured by radiographic follow-up studies. Radiographic wear measurements are expressed as linear wear rates whereas *in vitro* studies generally report volumetric wear. Volumetric wear can be directly related to the number of wear particles released into periprosthetic fluids (typically on the order of 1×10^9 of particles per year). The most common wear couple for hip and knee arthroplasty currently in use in the United States is a cobalt-base alloy head (most commonly a Co–Cr–Mo alloy ASTM F-75,) bearing on an ultrahigh-molecular-weight polyethylene (UHMWPE) cup or liner. The wear rates of this couple are generally on the order of 0.1 mm/year, with particulate generation as high as 1×10^6 particles per step or per cycle. Clinically, implant wear rates have been found to increase with the following: (1) physical activity, (2) weight of the patient, (3) size of the femoral head (32 versus 28 mm), (4) roughness of the metallic counterface, and (5) oxidation of the polyethylene (Jacobs *et al.*, 1994b; Jacobs, 1995; Jacobs *et al.*, 2001).

Orthopedic Biomaterial Corrosion

Electrochemical corrosion occurs to some extent on all metallic surfaces including implants. This is undesirable for two primary reasons: (1) the degradative process may reduce structural integrity of the implant, and (2) the release of products of degradation is potentially toxic to the host. Metallic biomaterial degradation may result from electrochemical dissolution phenomena or wear, but most commonly occurs through a synergistic combination of the two. Electrochemical processes include generalized corrosion uniformly affecting an entire surface, and localized corrosion affecting either areas of a device relatively shielded from the environment (crevice corrosion), or seemingly random sites on the surface (pitting corrosion). Additionally, these electrochemical and other mechanical processes interact, potentially causing premature structural failure and/or accelerated metal release (e.g., stress corrosion cracking, corrosion fatigue, and fretting corrosion; Brown and Merritt, 1981; Cook *et al.*, 1983; Bundy *et al.*, 1991; Brown *et al.*, 1992; Collier *et al.*, 1992b; Gilbert and Jacobs, 1997).

Corrosion Mechanisms

Corrosion of orthopedic biomaterials is a multifactorial phenomenon and is dependent on five primary factors: (1) geometric variables (e.g., taper crevices in modular component hip prostheses), (2) metallurgical variables (e.g., surface microstructure, oxide structure and composition), (3) mechanical variables (e.g., stress and/or relative motion), (4) solution variables (e.g., pH, solution proteins, enzymes) and (5) the mechanical loading environment (e.g., degree of movement, contact forces). Current investigational efforts to minimize the corrosion of orthopedic biomaterials deal directly with the complex interactions of these factors.

There are two essential features associated with how and why a metal corrodes. The first has to do with thermodynamic driving forces, which cause corrosion (oxidation/reduction) reactions. In general, whether or not corrosion will take place under the conditions of interest depends on the chemical driving force (ΔG) and the charge separation. This separation contributes to what is known as the electrical double layer (Fig. 8), which creates an electrical potential across the metal–solution interface (much like a capacitor):

$$\Delta G = -nF\Delta E \qquad (2)$$

where n is the valence of the ion, F is the Faraday constant (95,000 coulombs/mole electrons), and E is the voltage across the metal–solution interface. This potential is a measure of the reactivity of the metals or the driving force for metal oxidation. Therefore the more negative the potential of a metal in solution, the more reactive it will tend to be (i.e., the greater is ΔG for reduction).

The second factor governing the corrosion process of metallic biomaterials is the kinetic barrier to corrosion (e.g., surface

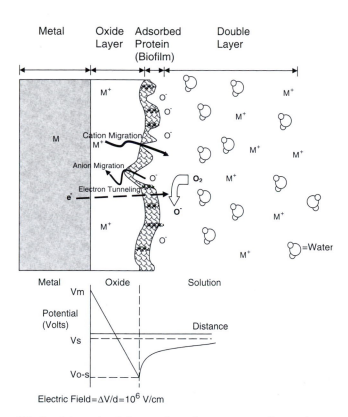

FIG. 8. Schematic of the interface of a passivating alloy surface in contact with a biological environment.

oxide layer). Kinetic barriers prevent corrosion not by energetic considerations but by physically limiting the rate at which oxidation or reduction processes can take place. The well-known process of passivation or the formation of a metal-oxide passive film on a metal surface is one example of a kinetic limitation to corrosion. In general, kinetic barriers to corrosion prevent the migration of metallic ions from the metal to the solution, the migration of anions from solution to metal, or the migration of electrons across the metal–solution interface. Passive oxide films are the most well known forms of kinetic barriers in corrosion, but other kinetic barriers exist, including polymeric coatings (Gilbert and Jacobs, 1997; Jacobs *et al.*, 1998a).

Passivating Oxide Films

Most alloys used in orthopedic appliances rely on the formation of passive films to prevent significant oxidation from taking place. These films consist of metal oxides, which form spontaneously on the surface of the metal in such a way that they prevent further transport of metallic ions and/or electrons across the film. Passive films must have certain characteristics to be able to limit further oxidation. The films must be compact and fully cover the metal surface, they must have an atomic structure that limits the migration of ions and/or electrons across the metal oxide–solution interface, and they must be able to remain on the surface of these alloys even with the mechanical stressing and abrasion that can be expected with orthopedic devices.

Passivating oxide films, which spontaneously grow on the surface of metals, have five primary structural and physical characteristics, which are particularly relevant to implant degradation processes:

1. First, these oxide films are very thin, typically on the order of 5 to 70 Å, which depends on the potential across the interface as well as solution variables (e.g., pH). Furthermore, the oxide structure may be amorphous or crystalline. Since the potential across the metal solution interface for these reactive metals is typically 1 to 2 volts, the electric field across the oxide is very high, on the order of 10^6–10^7 V/cm. One of the more widely accepted models is based on the theory of Mott and Cabrera, which states that oxide film growth depends on the electric field across the oxide. If the potential across the metal oxide–solution interface is decreased (i.e., made closer to the electrochemical series potential), then the film thickness will decrease by reductive dissolution processes at the oxide. If the interfacial potential is made sufficiently negative or the pH of the solution is made low enough, then these oxide films will no longer be thermodynamically stable and will undergo reductive dissolution, without which corrosion will increase (Gilbert and Jacobs, 1997; Jacobs *et al.*, 1998a).

2. Second, oxide films have the characteristics of semiconductors with an atomic defect structure, which determines the ability for ionic and electronic transport across films. Metal cations and oxygen anions require the presence of cationic or anionic vacancies (respectively) in the oxide for transport of these species across the film. If there is a deficit of metal ions in the oxide film (i.e., there are cationic vacancies), for example, then metal ion transport is possible and these oxides are known as p-type semiconductors. Chromium oxide (Cr_2O_3) is such a metal-deficit oxide. On the other hand, if there is an excess of metal ions in the oxide (or a deficit of anions) then cation transport is limited but anion transport can occur. These oxides will also have excess electrons and are known as n-type semiconductors. TiO_2 spontaneously formed on titanium alloy implant (Ti-6Al-4V) surfaces is one such n-type semiconductor. The greater the number of defects (vacancies or other valence species) the less able is the oxide film to prevent migration of ionic species and the lower is the kinetic barrier to corrosion. TiO_2 is very close to being stoichiometric (chemically homogeneous) and hence does not have many ionic defects, resulting in an increased resistance to ionic transport. Other defects may be present in these passive oxide films that may alter their ability to limit corrosion. For instance, the addition of other metal ions with valence states, which are different from the native metal ions, can alter both the electronic and ionic transport of charge across the interface. These additions may enhance or degrade the ability of the oxide to prevent corrosion depending on the nature of the oxide. One example of improved corrosion resistance from mixed oxides comes from what is known as a spinel. Spinels are typically mixed oxides of the form (A_2O_3)BO, where A and B are +3 and +2 valence metal ions. In Co–Cr alloys, for instance, a spinel of (Cr_2O_3)CoO can form on the surface. Spinels are typically known to have higher strengths and better resistance to diffusion of ions compared to single metal ion oxides. Therefore a high concentration of spinels in the oxide layer will act to resist dissolution

of a metal implant (Gilbert and Jacobs, 1997; Jacobs *et al.*, 1998a).

3. Third, the ratio of the "oxide specific volume" to metal alloy specific volume (i.e., Pilling Bedworth ratio) will determine if the oxide will adhere to the metal or not. If there is too great a mismatch between the metal and oxide lattice parameters then consequential stresses will be generated between the two. The magnitude of the internal stress will vary with the thickness of the oxide. Too great an oxide thickness will thus result in spontaneous fracture or spalling of the oxide, lowering the kinetic barrier effect of the oxide to corrosion (Gilbert and Jacobs, 1997; Jacobs *et al.*, 1998a).

4. Fourth, the morphology of these oxide films is not one of a smooth, flat, continuous sheet of adherent oxide covering the metal. Transmission electron microscopy (TEM) and atomic force microscopy (AFM) techniques have shown that oxides of titanium, for instance, consist of needle or dome shapes. The size and shape of these oxide domes change with applied potential when immersed in oxalic and other acids (Gilbert and Jacobs, 1997; Jacobs *et al.*, 1998a).

5. Finally, mechanical factors such as fretting, micromotion, or applied stresses may abrade or fracture oxide films. When an oxide film is ruptured from the metal substrate, fresh unoxidized metal is exposed to solution. When these films reform or repassivate, the magnitude of the repassivation currents that are subsequently generated may be great. This is because large driving forces exist for oxidation and when the kinetic barrier is removed these large driving forces can operate to cause oxidation. However, the extent and duration of the oxidation currents will depend on the repassivation kinetics for oxide film formation. Hence, the mechanical stability of the oxide films, as well as the nature of their repassivation process, is central to the performance of oxide films in orthopedic applications (Gilbert and Jacobs, 1997; Jacobs *et al.*, 1998a).

Corrosion at Modular Interfaces of Joint Replacements

One issue associated with orthopedic alloys is the corrosion observed in the taper connections of retrieved modular joint replacement components (Fig. 9). With the growing number of total joint designs that use metal on metal conical tapers as modular connectors between components, the effects of crevices, stress, and motion take on increasing importance. Severe corrosion attack can take place in the crevices formed by these tapers *in vivo*. Gilbert *et al.* have shown that of 148 retrieved implants, approximately 35% showed signs of moderate to severe corrosion attack in the head–neck taper connections of total hip prostheses. This attack was observed in components that consisted of Ti-6Al-4V alloy stems and Co–Cr heads as well as Co–Cr stems on Co–Cr heads. This corrosion process is likely the result of a combination of stress and motion at the taper connection and the crevice geometry of the taper. The stresses resulting from use cause fracturing and abrasion of the oxide film covering these passive metal surfaces. This, in turn, causes significant changes in the metal surface potential (makes it more negative) and in the crevice solution chemistry as the oxides continuously fracture and repassivate. These changes may result in deaeration (loss of O_2) of the crevice solution and a lowering of the pH in the crevice as

FIG. 9. Modular junction taper connection of a total hip arthroplasty showing corrosion of the taper connections. Macrograph of deposits of $CrPO_4$ corrosion particle products on the rim of a modular cobalt–chrome femoral head.

is expected in crevice corrosion attack. The ultimate result of this process is a loss of the oxide film and its kinetic barrier effect and an increase in the rate of corrosive attack in the taper region. The corrosion processes in the Co–Cr alloys have been observed to consist of intergranular corrosion, etching, selective dissolution of cobalt, and the formation of Cr-rich particles. In isolated cases, this occurs to such an extent that intergranular corrosion caused fatigue failure in the neck of a Co–Cr stem. Corrosion attack of titanium alloy stems has also been observed in some cases.

Very little is known about the mechanical stability of passive oxide films and the electrochemical reactions (e.g., ion and particle release) that occur when the oxide film is fractured. What is known is that when the oxide films of these orthopedic alloys are abraded or removed from the surface by rubbing, the open circuit potential can decrease to as low as −500 mV (versus standard calomel electrode). These voltage potentials may be significant and prolonged enough to cause changes in the oxide structure and stability by bringing the interface potential into the active corrosion range of the alloy, thereby dramatically accelerating the corrosion rate and decreasing implant performance (Brown *et al.*, 1992; Collier *et al.*, 1992a; Gilbert *et al.*, 1993; Bobyn *et al.*, 1994; Jacobs *et al.*, 1995; Brown *et al.*, 1995; Urban *et al.*, 1997b).

Local Tissue Effects of Wear and Corrosion

Normal bone maintenance relies on the balance of bone formation and bone resorption, which mainly involves the coordinated function of osteoblasts and osteoclasts. Thus, either a decrease in osteoblastic bone formation or an increase in osteoclastic bone resorption can result in net bone loss and osteolysis. Bone loss (i.e., osteolysis) around an implant is the primary concern associated with the local effects of orthopedic implant degradation. This osteolysis causing implant debris occurs through both wear and corrosion mechanisms. Osteolysis is observed either as diffuse cortical thinning or as a focal cyst-like lesion. It was initially thought that reaction to particulate poly(methyl methacrylate) bone cement produced osteolytic lesions based on histological studies demonstrating cement debris associated with macrophages, giant

cells, and a vascular granulation tissue. More recently, however, osteolysis has been recognized in association with loose and well-fixed uncemented implants, demonstrating that the absence of acrylic cement does not preclude the occurrence of osteolysis (Jacobs *et al.*, 2001; Vermes *et al.*, 2001a).

There are numerous secretory products created by the cells around implants that can negatively affect bone turnover (Fig. 10). These include the pro-inflammatory cytokines IL-1, IL-6, and TNF-α, to name a few. In addition, antiinflammatory cytokines such as IL-10 may modulate this process. Other factors involved with bone resorption include the enzymes responsible for catabolism of the organic component of bone. These include matrix metalloproteinases collagenase and stromelysin. Prostaglandins, in particular PGE_2, also are known to be important intercellular messengers in the osteolytic cascade produced by implant debris. More recently, several mediators known to be involved in stimulation or inhibition of osteoclast differentiation and maturation, such as RANKL (also referred to as osteoclast differentiation factor) and osteoprotegerin, respectively, have been suggested as key factors in the development and progression of bone loss (osteolytic lesions) produced from implant debris.

Goldring *et al.* (1983) were among the first to describe the synovial-like character of the bone implant interface in patients with loose total hip replacements and determine that the cells within the membrane have the capacity to produce large amounts of bone-resorbing factors PGE_2 and collagenase. However, since studies typically can only document the end stage of the loosening process, rather than those initiating processes, pharmacologic interventions have been limited.

Osteolysis associated with total knee arthroplasty has been reported less frequently than that associated with total hip arthroplasty. It is unclear why. However, in addition to obvious factors such as implant/bone mechanical loading environments, other more subtle differential mechanisms of hip and knee wear, and differences in interfacial barriers to migration of debris have been postulated to account for this apparent disparity.

Although polyethylene particles are generally recognized as the most prevalent particles in the periprosthetic milieu, metallic and ceramic particulate species are also present in

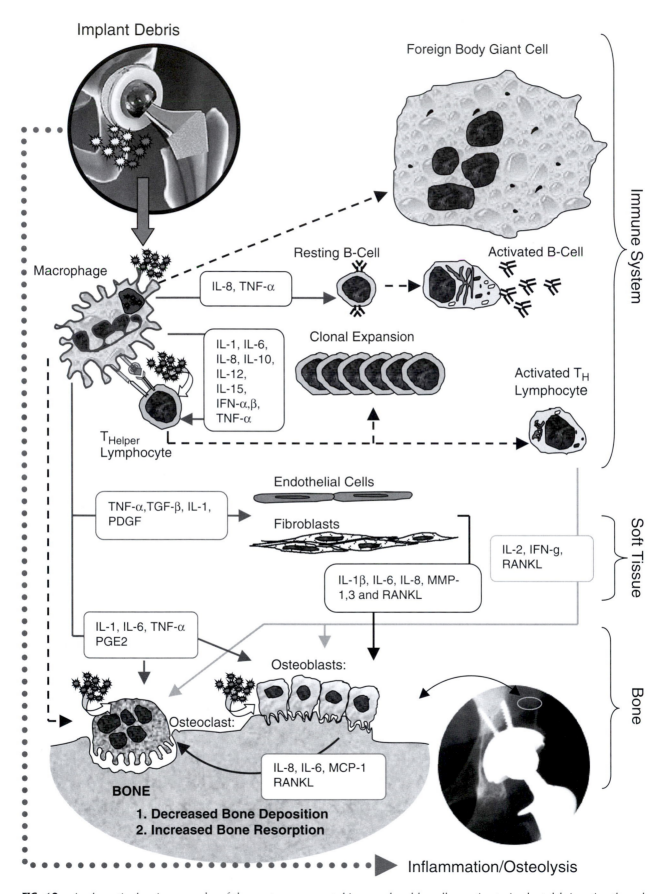

FIG. 10. A schematic showing examples of the most common cytokines produced by cells reacting to implant debris acting through a variety of pathways to negatively affect bone turnover.

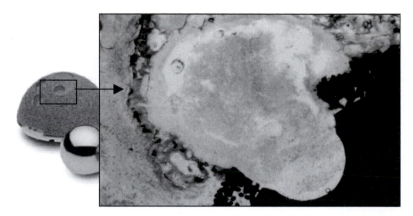

FIG. 11. Photomicrograph (5×) of a section through an acetabular section of a femoral stem retrieved at autopsy, 89 months after implantation. Note that the periprosthetic cavity surrounded development of a granuloma emanating from an unfilled screw hole.

variable amounts and may have important repercussions. The bulk of this debris originates from the articular surface and has easy access to local bone. When present in sufficient amounts, particulates generated by wear, corrosion, or a combination of these processes can induce the formation of an inflammatory, foreign-body granulation tissue with the ability to invade the bone–implant interface (Fig. 11). Localized osteolytic lesions in these areas are common, but their clinical significance is limited unless large granulomatous lesions develop.

The common observance of particle-induced osteolysis remote from the articulation surfaces has shown there is substantial particle migration between the joint space and the distal regions of the THA implant space. Autopsy specimens of retrieved implants have demonstrated the presence of connective tissue macrophages (histiocytes) in cavities surrounding regions of the femoral component. While the overall incidence of femoral osteolysis associated with THA tends to be proximal in the initial stages, over time it tends to progress distally. The volume of debris generated from THA polyethylene is related to a number of variables, including the smoothness of the concave metallic surface of the acetabular component, the tolerance between polyethylene and metal shell, and the relative stability of the insert (LaBerge, 1998; Shanbhag *et al.*, 1998; Wimmer *et al.*, 1998).

The relative contribution of each of the particulate species to the overall process of periprosthetic bone loss remains unkown. *In vitro* cell culture studies have demonstrated that the macrophage and fibroblast response to particulate debris is a function of particle size, composition, and dose. However, particles of different compositions may exhibit differential cytotoxicity. Polyethylene particles are believed to be the most biologically active by virtue of their greater number and small size, giving rise to an enormous surface area for interaction with the surrounding tissues (Jacobs *et al.*, 2001).

Histological Features

The tissues surrounding modern implants may include areas of osseointegration, fibrous encapsulation, and a variable

presence of the foreign-body response to polyethylene and cement debris in joint replacement devices. Absent is any specific histologic evidence of the slow release of metallic species that is known to occur with all metallic implants. However, accelerated corrosion and a tissue response that can be directly related to identifiable corrosion products can be demonstrated in the tissues surrounding multi-part devices (Urban *et al.*, 1997b, 1998, 2000).

Stainless Steels Histological sections of the tissues surrounding stainless steel internal fixation devices generally show encapsulation by a fibrous membrane with little or no inflammation over most of the device. At screw-plate junctions, however, the membranes often contain macrophages, foreign-body giant cells, and a variable number of lymphocytes in association with two types of corrosion products. The first consists of iron-containing granules. The second, termed microplates, consists of relatively larger particles of a chromium compound. Microplates are found within the tissues as closely packed, platelike particle aggregates ranging in size from 0.5 mm to 5.0 mm.

Hemosiderin-like granules often surround the collections of microplates, but the granules are found alone as well. The granules are yellow-brown, mainly spherical, and 0.1 to 3 or more micrometers in diameter. They are predominantly intracellular, most often in macrophages, but may also be found in other periprosthetic cells (e.g., fibroblasts). X-ray diffraction has indicated that the granules consist of a mixture of two or more of the iron oxides, αFe_2O_3 and σFe_2O_3, and the hydrated iron oxides, $\alpha Fe_2O_3 \cdot H_2O$ and $\sigma Fe_2O_3 \cdot H_2O$.

Cobalt-Base Alloys The nature of corrosion at modular connection products is similar whether modular heads are mated with cobalt–chromium alloy or Ti-6Al-4V alloy femoral stems. The principal corrosion product identified by electron microprobe energy-dispersive X-ray analysis and Fourier-transform infrared microprobe spectroscopy is a chromium phosphate $(Cr(PO_4)4H_2O)$ hydrate rich material termed "orthophosphate." This corrosion product can be found at the

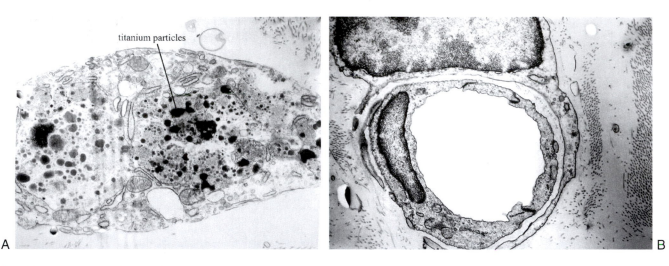

FIG. 12. Transmission electron photomicrographs. (A) Macrophage containing phagocytized titanium particles. (B) Endothelial cell lining with embedded titanium debris. These specimens were obtained from a tissue sample overlying the posterolateral fusion mass (16-week autograft + titanium) (original TEM magnification 20,000×).

modular head–neck junction and as particles within the joint capsules, at the bone–implant interfacial membranes, and at sites of femoral osteolytic lesions. Particles of the orthophosphate material have been found at the bearing surface of the UHMWPE acetabular liners, suggesting their participation in three-body wear and an increased production of polyethylene debris. Particles of the chromium orthophosphate hydrate-rich corrosion product found in the tissues ranged in size from submicron to aggregates of particles up to 500 micrometers.

Titanium-Base Alloys The degradation products observed in histologic sections of tissues adjacent to titanium base alloys are of a different nature than the precipitates associated with stainless steel and cobalt base alloys. Despite the remarkable corrosion resistance of titanium base alloys, there have been persistent reports of tissue discoloration due to metallic debris in the periprosthetic tissues. These particulates observed in local tissues surrounding titanium alloy implants have the same elemental composition as the parent alloy, as opposed to the precipitated corrosion products that occur with stainless steel and cobalt–chromium alloys (Fig. 12). However, wear debris present an enormous surface area for electrochemical dissolution, which in all likelihood is a major factor contributing to observed systemic elevations in titanium of patients with titanium implants (Urban *et al.*, 1996, 1997a, 1997b).

Remote and Systemic Effects of Wear and Corrosion

Implant surfaces and wear debris generated from the implant may release chemically active metal ions into the surrounding tissues. While these ions may stay bound to local tissues, there is an increasing recognition that released metal products bind to specific protein moieties and are transported in the bloodstream and/or lymphatics to remote organs. The concern about the release and distribution of

metallic degradation products is justified by the known potential toxicities of the elements used in modern orthopedic implant alloys: titanium, aluminum, vanadium, cobalt, chromium, and nickel. In general terms, metal toxicity may occur through (1) metabolic alterations, (2) alterations in host/parasite interactions, (3) immunologic interactions of metal moieties by virtue of their ability to act as haptens (specific immunological activation) or antichemotactic agents (nonspecific immunological suppression), and (4) by chemical carcinogenesis (Luckey and Venugopal, 1979; Beyersmann, 1994; Goering and Klaasen, 1995; Britton, 1996; Hartwig, 1998).

Cobalt, chromium, and possibly nickel and vanadium are essential trace metals in that they are required for certain enzymatic reactions. In excessive amounts, however, these elements may be toxic. Excessive cobalt may lead to polycythemia, hypothyroidism, cardiomyopathy, and carcinogenesis. Excessive chromium can lead to nephropathy, hypersensitivity, and carcinogenesis. Nickel can lead to eczematous dermatitis, hypersensitivity, and carcinogenesis. Vanadium can lead to cardiac and renal dysfunction and has been associated with hypertension and depressive psychosis.

Biologically nonessential metallic elements also possess specific toxicities. Titanium, although generally regarded as inert, has been associated with pulmonary disease in patients with occupational exposure and with platelet suppression in animal models. Aluminum toxicity is well documented in renal failure and has been associated with anemia, osteomalacia, and neurological dysfunction, possibly including Alzheimer's disease. However, when considering the variety of documented toxicities of these elements, it is important to keep in mind that the toxicities generally apply to soluble forms of these elements and may not apply to the chemical species that result from prosthetic implant degradation.

At this time, the association of metal release from orthopedic implants with any metabolic, bacteriologic, immunologic, or carcinogenic toxicity is conjectural since cause and effect have not been well established in human subjects. However, this is due in large part to the difficulty of observation in that most symptoms attributable to systemic and remote toxicity can be expected to occur in any population of orthopedic patients (Jacobs *et al.*, 1999a).

Metal Ion Release

In the long clinical experience of permanent and temporary metallic implants, there has always been concern with local tissue reactions. Implants, or wear debris generated from implants, may release chemically active metal ions into the surrounding tissues. While these ions may stay bound to local tissues, metal ions may also bind to protein moieties that are then transported in the bloodstream and/or lymphatics to remote organs (Jacobs *et al.*, 2001).

There is a considerable publication concerning serum and urine chromium (Cr), cobalt (Co), and nickel (Ni) levels following total joint replacement, but relatively fewer studies examining titanium (Ti), aluminum (Al), and vanadium (V) levels. Many investigations have been hampered by technical limitations of the analytical instrumentation. Normal human serum levels of prominent implant metals are approximately 1–10 ng/ml Al, 0.15 ng/ml Cr, <0.01 ng/ml V, 0.1–0.2 ng/ml Co, and <4.1 ng/ml Ti. Following total joint arthroplasty, levels of circulating metal (Al, Cr, Co, Ni, Ti, and V) have been shown to increase (Table 5).

Multiple studies have demonstrated chronic elevations in serum and urine Co and Cr following successful primary total joint replacement. In addition, transient elevations of urine and serum Ni have been noted immediately following surgery. This hypernickelemia/hypernickeluria may be unrelated to the implant itself since there is such a small percentage of Ni within these implant alloys. Rather, this may be related to the use of stainless steel surgical instruments or the metabolic changes associated with the surgery itself.

Chronic elevations in serum Ti and Cr concentrations are found in subjects with well-functioning Ti- and/or Cr-containing THR components without measurable differences in urine and serum Al concentrations. Vanadium concentrations have not been found greatly elevated in patients with TJA (Table 5) (Dorr *et al.*, 1990; Michel *et al.*, 1991; Stulberg *et al.*, 1994; Jacobs *et al.*, 1994c, 1998a).

Metal ion levels within serum and urine of TJA patients can be affected by a variety of factors. For example, patients with total knee replacement components containing Ti-base alloy and carbon fiber reinforced polyethylene wear couples demonstrated tenfold elevations in serum Ti concentrations at an average of 4 years after implantation. Up to a hundred times higher than normal control values of serum Ti elevations have also been reported in patients with failed metal-backed patellar components where unintended metal/metal articulation was possible. However, even among these TJA patients, there was no elevation in serum or urine Al, serum or urine V levels, or urine Ti levels. Mechanically assisted crevice corrosion in patients with modular femoral stems from total hip arthroplasty has been associated with elevations in serum Co and urine Cr. It has been previously assumed that extensively porous coated cementless stems would give rise to higher serum and urine chromium concentrations due to the larger surface area available for passive dissolution. More recent studies suggest that disseminated Cr can predominantly come from fretting corrosion of the modular head/neck junction.

TABLE 5 Approximate Average Concentrations (ng/ml or ppb) of Metal in Human Body Fluids with and without Total Joint Replacements (Dorr *et al.*, 1990; Michel *et al.*, 1991; Stulberg *et al.*, 1994; Jacobs *et al.*, 1994c, 1998a)

		Ti	Al	V	Co	Cr	Mo	Ni
Serum	Normal	2.7	2.2	<0.8	0.18	0.05–0.15	*	0.4–3.6
	THA	4.4	2.4	1.7	0.2–0.6	0.3	*	<9.1
	THA-F	8.1	2.2	1.3	*	0.2	*	*
	TKA	3.2	1.9	<0.8	*	*	*	*
	TKA-F	135.6	3.7	0.9	*	*	*	*
Urine	Normal	<1.9	6.4	0.5	*	0.06	*	*
	TJA	3.55	6.53	<0.4	*	0.45	*	*
Synovial fluid	Normal	13	109	5	5	3	21	5
	TJA	556	654	62	588	385	58	32
Joint capsule	Normal	723	951	122	25	133	17	3996
	TJA	1540	2053	288	1203	651	109	2317
	TJA-F	19,173	1277	1514	821	3329	447	5789
Whole blood	Normal	17	13	6	0.1–0.1.2	2.0–4.0	0.5–1.8	2.9–7.0
	TJA	67	218	23	20	110	10	29

Normal: Subjects without any metallic prosthesis (not including dental).
THA: Subjects with well-functioning total hip arthroplasty.
THA-F: Subjects with poorly functioning total hip arthroplasty (needing surgical revision).
TKA: Subjects with well-functioning total knee arthroplasty.
TKA-F: Subjects with poorly functioning total knee arthroplasty (needing surgical revision).
TJA: Subjects with well-functioning total joint arthroplasty.
TJA-F: Subjects with poorly functioning total joint arthroplasty (needing surgical revision).
*Not tested.

TABLE 6   Concentrations of Metal in Body Tissue of Humans with and without Total Joint Replacements (μg/g)

		Cr	Co	Ti	Al	V
Skeletal muscle	Normal	<12	<12	*	*	*
	TJA	570	160	*	*	*
Liver	Normal	<14	120	100	890	14
	TJA	1130	15,200	560	680	22
Lung	Normal	*	*	710	9830	26
	TJA	*	*	980	8740	23
Spleen	Normal	10	30	70	800	<9
	TJA	180	1600	1280	1070	12
Psuedocapsule	Normal	150	50	<65	120	<9
	TJA	3820	5490	39,400	460	121
Kidney	Normal	<40	30	*	*	*
	TJA	<40	60	*	*	*
Lymphatic tissue	Normal	690	10	*	*	*
	TJA	690	390	*	*	*
Heart	Normal	30	30	*	*	*
	TJA	90	280	*	*	*

TJA, Subjects with a well-functioning total joint arthroplasty.
* Not tested.

However, wear of the articulating surface remains the purported predominant source of metallic implant debris (Jacobs *et al.*, 1998a, b, 1999b).

Homogenates of remote organs and tissue obtained postmortem from subjects with Co base alloy total joint replacement components have indicated that significant increases in Co and Cr concentrations occur in the heart, liver, kidney, spleen, and lymphatic tissue (Table 6). Similarly, patients with Ti base alloy implants demonstrated elevated Ti, Al, and V levels in joint pseudocapsules (with up to 200 ppm of Ti six orders of magnitude greater than that of controls, 880 ppb of Al, and 250 ppb of V). Spleen Al levels and liver Ti concentrations can also be markedly elevated in patients with failed titanium-alloy implants (Jacobs *et al.*, 1994a).

Particle Distribution

Variables influencing accumulation of wear debris in remote organs are not clearly identified. When the magnitude of particulate debris generated by a prosthetic device is increased, it seems likely that a corresponding elevation in both the local and systemic burden of particles may be expected. Thus, component loosening, duration of implantation, and the modular designs of contemporary hip and knee replacement prostheses provide the potential for increased generation of metallic and polymeric debris Fig. 12. Wear particles found disseminated beyond the periprosthetic tissue are primarily in the submicron size range. Numerous case reports document the presence of metallic, ceramic, or polymeric wear debris from hip and knee prostheses in regional and pelvic lymph nodes (Fig. 13) along with the findings of lymphadenopathy, gross pigmentation due to metallic debris, fibrosis (buildup of fibrous tissue), lymph-node necrosis, and histiocytosis (abnormal function of tissue macrophages), including complete effacement of nodal architecture. The inflammatory response to metallic and polymeric debris in lymph nodes has been demonstrated to include

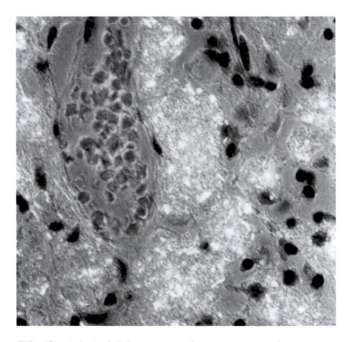

FIG. 13. Polarized light micrograph (orig. 190×) of parraaortic lymph node demonstrates the abundance and morphology of birefringent particles within macrophages. The larger filamentous particles were identified by infrared spectroscopy to be polyethylene.

immune activation of macrophages and associated production of cytokines. Metallic wear particles have been detected in the paraaortic lymph nodes in up to 70% of patients with total joint replacement components.

Lymphatic transport is thought to be a major route for dissemination of wear debris. Wear particles may migrate via perivascular lymph channels as free or phagocytosed particles within macrophages. Within the abdominal paraaortic lymph

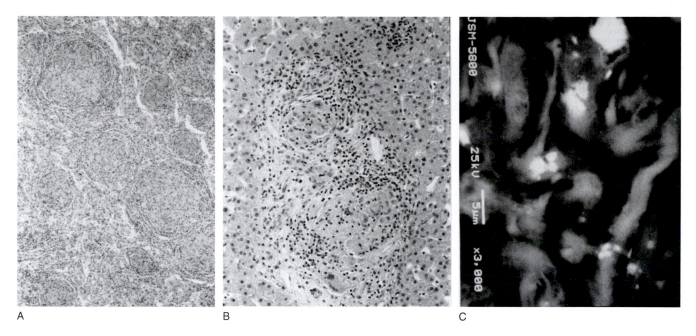

FIG. 14. Epithelioid granulomas (A) within the portal tract of the liver (orig. 40×) and (B) within the splenic parenchyma (orig. 15×) in a patient with a failed titanium-alloy total hip replacement and symptomatic hepatitis. (C) Backscattered SEM of a granuloma in the spleen (orig. 3000×) demonstrating titanium alloy particles.

nodes the majority of disseminated particles are submicron in size; however, metallic particles as large as 50 μm and polyethylene particles as large as 30 μm have also been identified. These particles may further disseminate to the liver or spleen where they are found within macrophages or, in some cases, as epithelioid granulomas throughout the organs. Within liver and spleen, the maximum size of metallic wear particles is nearly an order of magnitude less than that in lymph nodes, indicating there may be additional stages of filtration proceeding the lymphatic system or alternate routes of particle migration. In the liver and spleen, as in the lymph nodes, cells of the mononuclear–phagocyte system may accumulate small amounts of a variety of foreign materials without apparent clinical significance. However, accumulation of exogenous particles can induce granulomas or granulomatoid lesions in the liver and spleen (Fig. 14). It is likely that the inflammatory reaction to particles in the liver, spleen and lymph nodes is modulated, as it is in other tissues, by (1) material composition, (2) the number of particles, (3) their rate of accumulation, (4) the duration for which they are present, and (5) the biologic reactivity of cells to these particles. Metallic particles in the liver or spleen have been more prevalent in subjects with previously failed arthroplasties when compared with cases of well-functioning primary joint replacements. Metal particles, unlike polyethylene debris, can be characterized using an electron microprobe, which allows identification of individual, submicron metallic wear particles against a background of particulates from environmental or sources other than the prosthetic components. Overall, the smallest disseminated particles identifiable, using the microprobe are approximately 0.1 μm in diameter. However, metallic wear debris may extend into the nanometer range, suggesting that additional methods of specimen preparation and analytic instrumentation may be required to more fully define the high burden of metallic wear particles in remote tissues (Urban et al., 1995, 2000).

Polyethylene particles comprise a substantial fraction of the disseminated wear particles in subjects with revision and those with primary TJAs. While the presence of these polyethylene particles in lymph nodes can be confirmed by Fourier transform infrared spectroscopy microanalyses, polyethylene particulates in liver and spleen have so far precluded unequivocal identification. In these sites, the size of wear particles may be much smaller than 0.1 μm, making differentiation impossible by polarized light microscopy or infrared spectroscopy.

Diseases that cause obstruction of lymph flow through lymph nodes, such as metastatic tumor, or that cause generalized disturbances of circulation, such as chronic heart disease or diabetes, may be expected to decrease particle migration to remote organs. Other diseases, such as acute or chronic-active inflammation in the periprosthetic tissues, may increase particle migration (Jacobs et al., 1999a, 2001; Vermes et al., 2001b).

Hypersensitivity

Some adverse responses to orthopedic biomaterials are subtle and continue to foster debate and investigation. One of these responses is "metal allergy" or hypersensitivity to metallic biomaterials. Released ions, although not sensitizers in on their own, can activate the immune system by forming complexes with native proteins. These metal–protein complexes are considered to be candidate antigens (or allergens) in human clinical applications. Polymeric wear debris are not easily chemically degraded *in vivo* and have not been implicated as sources of allergic-type immune responses. This is presumably due to the relatively large degradation products associated with the mechanical wear of polymers *in vivo*, which may be large enough to prevent the formation of polymer–protein haptenic complexes with human antibodies (Hallab *et al.*, 2000a, b, c, 2001a, b).

Metal hypersensitivity is a well-established phenomenon. Moreover, dermal hypersensitivity to metal is common, affecting about 10–15% of the population. Dermal contact and ingestion of metals have been reported to cause immune reactions that most typically manifest as skin hives, eczema, redness, and itching. Although little is known about the short- and long-term pharmacodynamics and bioavailability of circulating metal degradation products *in vivo*, there have been many reports of immunologic type responses temporally associated with implantation of metal components. Individual case reports link hypersensitivity immune reactions with adverse performance of metallic clinical cardiovascular, orthopedic and plastic surgical and dental implants. These phenomenon were discussed in Chapter 4.5.

Metals accepted as sensitizers (haptenic moieties in antigens) include beryllium, nickel, cobalt, and chromium, while occasional responses have been reported to tantalum, titanium, and vanadium. Nickel is the most common metal sensitizer in humans followed by cobalt and chromium. Cross sensitivity reactions between metals are common. Nickel and cobalt are, reportedly, the most frequently cross reactive. The amounts of these metals found in medical-grade alloys are shown in Table 3.

Type IV delayed type hypersensitivity (DTH) is a cell-mediated type of response with which orthopedic implant–associated hypersensitivity reactions (metal sensitivity or metal allergy) are generally associated. Metal-antigen sensitized T-DTH lymphocytes release various cytokines that result in the accumulation and activation of macrophages. The majority of DTH participating cells are macrophages. Only 5% of the participating cells are antigen-specific T lymphocytes (T-DTH cells), within a fully developed DTH response. The effector phase of a DTH response is initiated by contact of sensitized T cells with antigen. In this phase T cells, which are antigen-activated, are termed T-DTH cells and secrete a variety of cytokines that recruit and activate macrophages, monocytes, neutrophils, and other inflammatory cells. These released cytokines include IL-3 and GM-CF, which promote hematopoiesis of granulocytes; monocyte chemotactic activating factor (MCAF) which promotes chemotaxis of monocytes toward areas of DTH activation; INF-□ and TNF-□, which produce a number of effects on local endothelial cells facilitating infiltration; and migration inhibitory factor (MIF), which inhibits the migration of macrophages away from the site of a DTH reaction. Activation, infiltration, and eventual migration inhibition of macrophages is the final phase of the DTH response. Activated macrophages, because of their increased ability to present class II MHC and IL-2, can trigger the activation of more T-DTH cells, which in turn activates more macrophages, which activates more T-DTH cells, and so on. This DTH self-perpetuating response can create extensive tissue damage.

The first apparent correlation of eczematous dermatitis to metallic orthopedic implants was reported in 1966 by Foussereau and Lauggier, where nickel was associated with hypersensitivity responses. Over the past 20 years growing numbers of case reports link immunogenic reactions with adverse performance of metallic cardiovascular, orthopedic and plastic surgical and dental implants. In some instances clinical immunological symptoms have led directly to device removal. In these cases reactions of severe dermatitis (inflammation of the skin), urticaria (intensely sensitive and itching red round wheals on the skin), and/or vasculitis (patch inflammation of the walls of small blood vessels) have been linked with the relatively more general phenomena of metallosis (metallic staining of the surrounding tissue), excessive periprosthetic fibrosis, and muscular necrosis. The temporal and physical evidence leaves little doubt that the phenomenon of hypersensitivity to metal released from orthopedic implants does occur in some patients. These cases of severe metal sensitivity raise the greatest concern.

Incidence of Hypersensitivity Responses among Patients with Metal Implants

Some studies have shown that the incidence of metal sensitivity among patients with both well and poorly functioning implants is roughly twice as high as that of the general population, approximately 25% (Fig. 15). Furthermore, the average incidence of metal sensitivity among patients with a "failed" implant (in need of revision surgery) is approximately 50–60% (Fig. 15). This is greater than five times the incidence of metal sensitivity found in the general population and two to three times that of patients with metal implants. This increased prevalence of metal sensitivity among patients with a loose prosthesis has prompted the speculation that immunological processes may be a factor in implant loosening.

Specific types of implants with greater propensity to release metal *in vivo* may be more prone to induce metal sensitivity. Failures of total hip prostheses with metal-on-metal bearing surfaces were associated with greater incidence of metal allergy than similar designs with metal-on-ultrahigh molecular weight polyethylene bearing surfaces. Alternatively, several published reports have indicated that after total joint replacement with metallic components some patients show an induction of metal tolerance; that is, previously diagnosed metal sensitivity abated after implantation of a metallic prosthetic.

Additionally confounding to any clear connection between metal sensitivity and implant failure is the lack of any reported correlation between incidence of metal sensitivity and implant residence time, infection, reason for removal, or pain. This lack of causal evidence implicating cell-mediated immune responses has prompted some to conclude that implantation of cemented metal-to-plastic joint prosthesis is safe, even in the case of a preexisting metal allergy. However, this is not a consensus opinion. At this time, however, it is unclear whether metal sensitivity causes implant loosening or whether implant loosening results in the development of metal sensitivity.

The majority of investigations conclude that metal sensitivity can be a contributing factor to implant failure. Such cases include instances in which clinical immunological symptoms lead directly to the need for device removal. In these

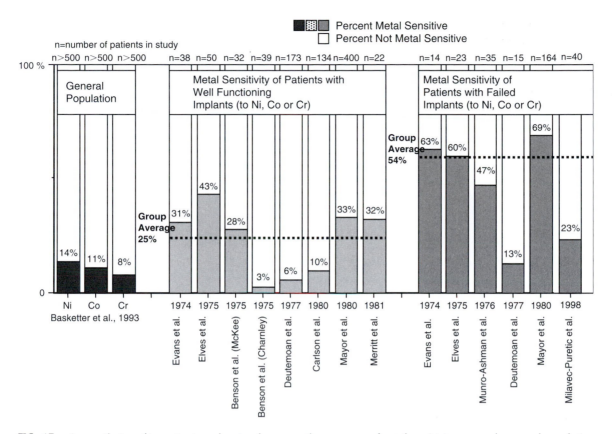

FIG. 15. A compilation of investigations showing the averaged percentages of metal sensitivity among the general population for nickel, cobalt, and chromium, among patients after receiving a metal containing implant, and among patient populations with failed implants. All subjects were tested by means of a patch or LIF test.

or similar cases there have been reported reactions of severe dermatitis, urticaria, and/or vasculitis all presumably linked to what has been reported as metallosis, excessive periprosthetic fibrosis, and muscular necrosis. The clinical observation of apparent immune sensitivity to metallic implants is not limited to orthopedic surgery. Some case reports suggest metal sensitivity to pacemakers, heart valves, and reconstructive, dental, and general-surgical devices. The temporal and physical evidence associated with such cases leaves little doubt that the phenomenon of metal-induced hypersensitivity does occur in some cases, currently accepted within the orthopedic community to be <1% of patients. However, it is currently unclear whether metal sensitivity exists only as an unusual complication in a few susceptible patients, or is more subtle and common phenomenon, which over time plays a significant role in implant failure. It is likely that cases involving implant-related metal sensitivity have been underreported because alternate causes were attributed to failure of the implant. Mechanisms by which *in vivo* metal sensitivity occurs have not been well characterized. Thus, the degree to which a precondition of metal hypersensitivity may elicit an overaggressive immune response in a patient receiving an implant remains unpredictable. Continuing improvements in immunologic testing methods will likely enhance future assessment of patients

susceptible to hypersensitivity responses (Hallab *et al.*, 2000a, b, c, 2001a, b).

Carcinogenesis

The carcinogenic potential of the metallic elements used in TJA remains an area of on-going research. Animal studies have documented the carcinogenic potential of orthopedic implant materials. Small increases in rat sarcomas were noted to correlate with high serum cobalt, chromium, or nickel content from metal implants. Furthermore, lymphomas with bone involvement were also more common in rats with metallic implants. Implant site tumors in dogs and cats—primarily osteosarcoma and fibrosarcoma—have been associated with stainless steel internal fixation devices.

Initially, epidemiological studies implicated cancer incidence in the first and second decades following total hip replacement. However, larger, more recent studies have found no significant increase in leukemia or lymphoma; although these studies did not include as large a proportion of subjects with a metal-on-metal prosthesis. There are constitutive differences in the populations with and without implants that are independent of the implant itself, which confound the interpretation of epidemiological investigations.

The association of metal release from orthopedic implants with carcinogenesis remains conjectural since causality has not been definitely established in human subjects. The identification of such an association depends both on the availability of comparative epidemiology and on the ability to perform tests on the patient before and after device removal. The actual number of cases of tumors associated with orthopedic implants is likely underreported. However, with respect to the number of devices implanted on a yearly basis the incidence of cancer at the site of implantation is relatively rare. Continued surveillance and longer-term epidemiological studies are required to fully address these issues (Gillespie *et al.*, 1988; Visuri and Koskenvuo, 1991; Matheisen *et al.*, 1995; Nyren *et al.*, 1995).

Preventive Strategies and Future Directions

Current strategies designed to address the problem of biomaterial-related implant failure are primarily aimed at decreasing the amount of periprosthetic particulate burden and any subsequent effects. Recently there has been a great deal of innovation regarding stronger, more wear-resistant polyethylene. These more highly cross-linked UHMWPE polymers are currently in various phases of clinical trial. However, initial results show demonstrable decrease in polyethylene wear with potential for less particulate induced bioreactivity/osteolysis and therefore greater implant performance. In the same vein, femoral heads with diameters of 32-mm have been associated with increased volumetric polyethylene wear; to combat this smaller 28-mm heads are currently extolled as more biocompatible. Manufacturing flaws such as fusion defects and foreign-body inclusions have also been suggested to contribute to adverse polyethylene wear properties. The elimination of polyethylene is another approach being investigated clinically in various centers. With the realization that early problems may have been related to the design and not the articulation, there has been a renewed interest in the application of metal–metal and ceramic–ceramic bearings. Future designs that attempt to reduce wear include: improved tolerances between polyethylene inserts and their metal backing, improved surface finish on the metallic concave surfaces, secure locking mechanisms, and the avoidance of holes on the convex portion of the acetabular prosthesis.

Metallic wear is also being addressed through techniques such as nitriding and nitrogen ion implantation to decrease the potential for abrasive wear and fretting in titanium alloy and cobalt alloy stems. Fabrication of metallic bearing surfaces with extremely low roughness can be expected to decrease articular wear rates. A polished metal head can be made as smooth as a ceramic head. Polishing of the stem will remove surface asperities and decrease particle generation from stem/bone fretting. In addition, enhanced polishing could minimize any secondary surface contamination.

New metallic biomaterials are being proposed that attempt to improve load transfer to the bone and reduce the incidence of loosening and thigh pain. Currently used alloys (Co–Cr–Mo alloy, E = 227 GPa and Ti-6Al-4V alloy, E = 115 GPa) have relatively high elastic moduli that limit smooth transfer of load to the surrounding bone in THA. Designs to improve load transfer can use a reduced cross-sectional area to increase flexibility, but at the expense of adequate stability of the implant within the bone. Additionally, the stresses may exceed the relatively low fatigue strength of Co–Cr–Mo implant alloy. Lower-modulus, more corrosion-resistant implant alloys are being developed. A Ti-13Nb-13Zr (E = 79 GPa) alloy is one such alloy that contains fewer elements of questionable cell response (i.e., Co, Cr, Mo, Ni, Fe, Al, V), and that possesses comparable strength and toughness to existing Ti-6Al-4V implant alloy. The Nb and Zr constituents seek to improve bone biocompatibility and corrosion resistance. Additionally, novel surface treatments on implant alloys such as the diffusion hardening (DH) treatment proposed for the Ti-13Nb-13Zr alloy can produce a hardened surface with wear resistance superior to that of Co–Cr–Mo alloy, currently the industry leader. These enhanced surface properties may lead to an improvement in the resistance to microfretting occuring within femoral head–neck taper regions and modular interfaces of current implant designs.

Electrochemical corrosion of orthopedic implants remains a significant clinical concern. Although the freely corroding implants used in the past have been replaced with modern corrosion-resistant "super alloys," deleterious corrosion processes have been observed in certain clinical settings. Attention to metallurgical processing variables, tolerances of modular connections, surface processing modalities, and appropriate material selection can diminish corrosion rates and minimize the potential for adverse clinical outcomes. For example, nitriding can reportedly significantly reduce the magnitude of fretting corrosion of Ti-6Al-4V devices. A need to further investigate the mechanical–electrochemical interactions of metal oxide surfaces in implants persists. Characterization of the stresses and motion needed to fracture passivating oxide films as well as the effects of repeated oxide abrasion on the electrochemical behavior of the interface and ultimately the implant remain avenues of active investigation.

The clinical significance of elevated metal content in body fluids and remote organs of patients with metallic implants needs to be further elucidated. Considerably more work will be required to discern the specific chemical forms and distribution of metal degradation products associated with the various forms of implant degradation. Additionally uncharacterized is how these degradation products interact with proteins *in vivo* in terms of (1) metal-ion/protein complexes, (2) nanometer-particle/protein complexes (ion-like particles) and (3) particle/protein-biofilm complexes. Although much has been revealed regarding the deleterious end effects of particulate debris (e.g., osteolysis) little is known of how in terms of cellular mechanics this phenomenon occurs. Therefore, an understanding of both the constituents of orthopedic implant degradation and their biological ligands is necessary to investigate and ultimately characterize their associated toxicity and bioreactivity. The importance of this line of investigation in the evaluation of orthopedic biomaterial performance is growing as the use of present implant is increasing, as new orthopedic implants are being developed (Fig. 16) and as expectations of implant durability and performance increase (Jacobs *et al.*, 1996); (Black, 1996).

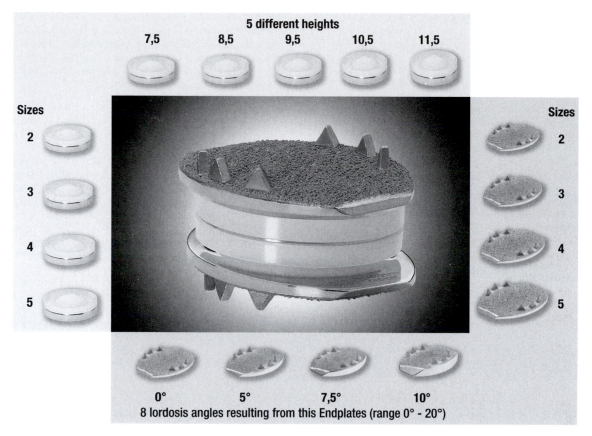

5 different heights

7,5   8,5   9,5   10,5   11,5

Sizes

2

3

4

5

Sizes

2

3

4

5

0°   5°   7,5°   10°
8 lordosis angles resulting from this Endplates (range 0° - 20°)

FIG. 16. The LINK SB Charité III artificial disk showing the range of standard sizes available. This design consists of an UHMWPE sliding core, which articulates unconstrained between two highly polished metal endplates, simulating the natural movement of spine (i.e., the nucleus within its annular containment).

QUESTIONS

1. What are the general qualifications for orthopedic bio-materials and what are the two major types of orthopedic implants?
   ANS: Generally orthopedic implants must be capable of load bearing and are used in either in bone fixation or total joint arthroplasty.

2. (a) What is the primary problem associated with the longevity of current total joint replacement implants and why? (b) How could this be solved by improved biomaterials?
   ANS: (a) Particle and wear debris generation at the articulating surfaces of total joint replacements causes aseptic loosening of the implants through the induction of inflammation around the implant (macrophage-particle overload) which causes a decreased production of new bone and increased bone resorption. (b) More wear-resistant materials at the articulating surface while maintaining current available levels of other mechanical properties.

3. What are the primary cytokines associated with particle-induced osteolysis, and what is the primary cell type involved in this process?

ANS: The primary cytokines associated with osteolysis are TNF-alpha, IL-6 IL-1, prostaglandin E_2, etc., produced by macrophage interaction with particulate debris from implants.

4. What are the seven primary orthopedic implant bioma-terials?
   ANS: (a) Co-base alloys, Ti-base alloys, stainless steel, PMMA, UHMWPE, alumina, and zirconia.

5. Name one new orthopedic biomaterial and why it was introduced (i.e., what specific problem/deficit it is attempting to solve).
   ANS: Zirconium alloy was introduced to decrease the amount of metallic wear debris being generated at the bearing surface and as an alternative metal less likely to induce hypersensitivity reactions.

6. What is one way in which current designs of orthopedic implants are creating more challenging conditions for biomaterials to overcome and why?
   ANS: Increasing popularity and degree of orthopedic implant component modularity is creating more interfacial surfaces from which implant debris can be generated through fretting corrosion and interact both locally and systemically.

Bibliography

Beyersmann, D. (1994). Interactions in metal carcinogenicity. *Toxicol. Lett.* **72**: 333–338.

Black, J. (1992). *Biomaterials*. Marcel Dekker, Inc., New York.

Black, J. (1996). *Prosthetic Materials*, VCH Publishers, Inc., New York, pp. 141–162.

Bobyn, J. D., Tanzer, M., Krygier, J. J., Dujovne, A. R., and Brooks, C. E. (1994). Concerns with modularity in total hip arthroplasty. *Clin. Orthop.* **298**: 27–36.

Britton, R. S. (1996). Metal-induced Hepatoxicity. *Semin. Liver Dis.* **16**: 3–12.

Brown, S. A. and Merritt, K. (1981). Fretting corrosion in saline and serum. *J. Biomed. Mater. Res.* **15**: 479–488.

Brown, S. A., Flemming, C. A. C., Kawalc, J. S., Vassaux, C. J., Payer, J. H., Kraay, M. J., and Merritt, K. (1992). Fretting acclereated crevice corrosion of modular hips. *Trans Soc. Biomater. Implant Retrieval Symp.* **15**: 59.

Brown, S. A., Flemming, C. A., Kawalec, J. S., Placko, H. E., Vassaux, C., Merritt, K., Payer, J. H., and Kraay, M. J. (1995). Fretting corrosion accelerates crevice corrosion of modular hip tapers. *J. Appl. Biomater.* **6**: 19–26.

Bundy, K. J., Williams, C. J., and Luedemann, R. E. (1991). Stress-enhanced ion release—the effect of static loading. *Biomaterials* **12**: 627–639.

Charnley, J. (1960). Anchorage of the femoral head prosthesis to the shaft of the femur. *J. Bone Joint Surg. [Br]* **42**: 28.

Charnley, J. (1964). The bonding of prosthesis to bone by cement. *J. Bone Joint Surg. [Br]* **46**: 518.

Charnley, J. (1979). *Low Friction Arthroplasty of the Hip, Theory and Practice*, Springer-Verlag, Berlin.

Collier, J. P., Mayor, M. B., Jensen, R. E., Surprenant, V. A., Surprenant, H. P., McNamar, J. L., and Belec, L. (1992a). Mechanisms of failure of modular prostheses. *Clin. Orthop.* **285**: 129–139.

Collier, J. P., Surprenant, V. A., Jensen, R. E., Mayor, M. B., and Surprenant, H. P. (1992b). Corrosion between the components of modular femoral hip prostheses. *J. Bone Joint Surg. [Am]* **74-B**: 511–517.

Cook, S. D., Gianoli, G. J., Clemow, A. J., and Haddad, R. J. J. (1983). Fretting corrosion in orthopedic alloys. *Biomater. Med. Devices Artif. Organs* **11**: 281–292.

Dorr, L. D., Bloebaum, R., Emmanual, J., and Meldrum, R. (1990). Histologic, biochemical and ion analysis of tissue and fluids retrieved during total hip arthroplasty. *Clin. Orthop. Rel. Res.* **261**: 82–95.

Foussereau, J., and Laugier (1966). Allergic eczemas from metallic foriegn bodies. *Trans. St.John's Hosp. Derm. Soc.* **52**: 220–225.

Gilbert, J. L. and Jacobs, J. (1997). The mechanical and electrochemical processes associated with taper fretting crevice corrosion: a review. in: *ASTM STP 1301 Modularity of Orthopedic Implants*. ASTM, Philadelphia, pp. 45–59.

Gilbert, J. L., Buckley, C. A., and Jacobs, J. J. (1993). In vivo corrosion of modular hip prosthesis components in mixed and similar metal combinations. The effect of crevice, stress, motion, and alloy coupling. *J. Biomed. Mater. Res.* **27**: 1533–1544.

Gillespie, W. J., Frampton, C. M., Henderson, R. J., and Ryan, P. M. (1988). The incidence of cancer following total hip replacement. *J. Bone Joint Surg. [Br]* **70**: 539–542.

Goering, P. L., and Klaasen, C. D. (1995). *Hepatoxicity of Metals*, Academic Press, New York, pp. 339–388.

Goldring, S. R., Schiller, A. L., Roelke, M., Rourke, C. M., O'Neill, D. A., and Harris, W. H. (1983). The synovial-like membrane at the bone-cement interface in loose total hip replacements and its proposed role in bone lysis. *J. Bone Joint Surg.* **65A**: 575–584.

Hallab, N., Jacobs, J. J., and Black, J. (2000a). Hypersensitivity to metallic biomaterials: a review of leukocyte migration inhibition assays [In Process Citation]. *Biomaterials* **21**: 1301–1314.

Hallab, N. J., Jacobs, J. J., Skipor, A., Black, J., Mikecz, K., and Galante, J. O. (2000b). Systemic metal–protein binding associated with total joint replacement arthroplasty. *J. Biomed. Mater. Res.* **49**: 353–361.

Hallab, N. J., Mikecz, K., and Jacobs, J. J. (2000c). A triple assay technique for the evaluation of metal-induced, delayed-type hypersensitivity responses in patients with or receiving total joint arthroplasty. *J. Biomed. Mater. Res.* **53**: 480–489.

Hallab, N., Merritt, K., and Jacobs, J. J. (2001a). Metal sensitivity in patients with orthopedic implants. *J. Bone Joint Surg. Am.* **83-A**: 428–436.

Hallab, N. J., Mikecz, K., Vermes, C., Skipor, A., and Jacobs, J. J. (2001b). Differential lymphocyte reactivity to serum-derived metal–protein complexes produced from cobalt-based and titanium-based implant alloy degradation. *J. Biomed. Mater. Res.* **56**: 427–436.

Harris, W. H. (1969). Traumatic arthritis of the hip after dislocation and acetabular fractures: treatment by mold arthroplasty. An end-result study using a new method of result evaluation. *J. Bone Joint Surg. Am.* **51**: 737–755.

Hartwig, A. (1998). Carcinogenicity of metal compounds: possible role of DNA repair inhibition. *Toxicol. Lett.* **102–103**: 235–239.

Jacobs, J. J. (1995). Particulate wear. *JAMA* **273**: 1950–1956.

Jacobs, J. J., Gilbert, J. L., and Urban, R. M. (1994a). Corrosion of metallic implants. in *Advances in Orthopedic Surgery*, Vol. 2, R. N. Stauffer, ed. Mosby, St. Louis, pp. 279–319.

Jacobs, J. J., Shanbhag, A., Glant, T. T., Black, J., and Galante, J. O. (1994b). Wear debris in total joints. *J. Am. Acad. Orthop. Surg.* **2**: 212–220.

Jacobs, J. J., Skipor, A. K., Urban, R. M., Black, J., Manion, L. M., Starr, A., Talbert, L. F., and Galante, J. O. (1994c). Systemic distribution of metal degradation products from titanium alloy total hip replacements: an autopsy study. *Trans. Orthop. Res. Soc. New Orleans*, 838.

Jacobs, J., Urban, R. M., Gilbert, J. L., Skipor, A., Black, J., Jasty, M. J., and Galante, J. O. (1995). Local and distant products from modularity. *Clin. Orthop.* **319**: 94–105.

Jacobs, J. J., Skipor, A. K., Doorn, P. F., Campbell, P., Schmalzried, T. P., Black, J., and Amstutz, H. C. (1996). Cobalt and chromium concentrations in patients with metal on metal total hip replacements. *Clin. Orthop. Rel. Res.* **S329**: S256–S263.

Jacobs, J. J., Gilbert, J. L., and Urban, R. M. (1998a). Corrosion of metal orthopedic implants. *J. Bone Joint Surg. [Am]* **80**: 268–282.

Jacobs, J. J., Skipor, A. K., Patterson, L. M., Hallab, N. J., Paprosky, W. G., Black, J., and Galante, J. O. (1998b). Metal release in patients who have had a primary total hip arthroplasty. A prospective, controlled, longitudinal study. *J. Bone Joint Surg. [Am]* **80**: 1447–1458.

Jacobs, J., Goodman, S., Sumner, D. R., and Hallab, N. (1999a). Biologic response to orthopedic implants. in *Orthopedic Basic Science*. American Academy of Orthopedic Surgeons, Chicago, pp. 402–426.

Jacobs, J. J., Silverton, C., Hallab, N. J., Skipor, A. K., Patterson, L., Black, J., and Galante, J. O. (1999b). Metal release and excretion from cementless titanium alloy total knee replacements. *Clin. Orthop.* **358**: 173–180.

Jacobs, J. J., Roebuck, K. A., Archibeck, M., Hallab, N. J., and Glant, T. T. (2001). Osteolysis: basic science. *Clin. Orthop.* **393**: 71–77.

Katz, J. L. (1980a). Anisotropy of Young's modulus of bone. *Nature* **283**: 106–107.

Katz, J. L. (1980b). The structure and biomechanics of bone. *Symp. Soc. Exp. Biol.* **34**: 137–168.

LaBerge, M. (1998). Wear. in *Biomaterial Properties*, J. Black and M. C. Hastings, eds. Chapman and Hall, London, pp. 364–405.

Luckey, T. D. and Venugopal, B. (1979). *Metal Toxicity in Mammals.* Plenum, New York.

Matheisen, E. B., Ahlbom, A., Bermann, G., and Lindgren, J. U. (1995). Total hip replacement and cancer. *J. Bone Joint Surg. [Br]* **77-B**: 345–350.

McKee, G. K., and Watson-Farrar, J. (1943). Replacement of the arthritic hips to the McKee-Farrar replacement. *J. Bone Joint Surg. [Br]* **48**: 245.

McKellop, H., Shen, F. W., Lu, B., Campbell, P., and Salovey, R. (2000). Effect of sterilization method and other modifications on the wear resistance of acetabular cups made of ultra-high molecular weight polyethylene. A hip-simulator study. *J. Bone Joint Surg. Am.* **82-A**: 1708–1725.

Michel, R., Nolte, M., Reich, M., and Loer, F. (1991). Systemic effects of implanted prostheses made of cobalt–chromium alloys. *Archi. Orthop. Trauma Surg.* **110**: 61–74.

Moore A. T. (1943). Metal hip joint: a case report. *J. Bone Joint Surg. [Am]* **25**: 688.

Nyren, O., Mclaughlin, J. K., Anders-Ekbom, G. G., Johnell, O., Fraumeni, and Adami, H. (1995). Cancer risk after hip replacement with metal implants: a population-based cohort study in Sweden. *J. Nat. Cancer Inst.* **87**: 28–33.

Ring, P. A. (1968). Complete replacement arthroplasty of the hip by the Ring prosthesis. *J. Bone Joint Surg. [Br]* **50**: 720.

Shanbhag, A. S., Hasselman, C. T., Jacobs, J. J., and Rubash, H. E. (1998). Biological response to wear debris. in *The Adult Hip*, J. J. Callaghan, A. G. Rosenberg, and H. Rubash, eds. Lippincott-Raven Publishers, Philadelphia, pp. 279–288.

Stillwell, W. T. (1987). *The Art of Total Hip Arthroplasty.* Grune & Stratton, Orlando, FL.

Stulberg, B. N., Merritt, K., and Bauer, T. (1994). Metallic wear debris in metal-backed patellar failure. *J. Biomed. Mat. Res. Appl. Biomater.* **5**: 9–16.

Urban, R. M., Jacobs, J. J., Tomlinson, M. J., Gavrilovic, J., and Andersen, M. (1995). *Migration of Corrosion Products from the Modular Head Junction to the Polyethylene Bearing Surface and Interface Membranes of Hip Prostheses.* Raven Press, New York.

Urban, R. M., Jacobs, J. J., Tomlinson, M. J., Black, J., Turner, T. M., Sauer, P. A., and Galante, J. O. (1996). Particles of metal alloys and their corrosion products in the liver, spleen and para-aortic lymph nodes of patients with total hip replacement prosthesis. *Orthop. Trans.* **19**: 1107–1108.

Urban, R. B., Jacobs, J., Gilbert, J. L., Rice, S. B., Jasty, M., Bragdon, C. R., and Galante, G. O. (1997a). Characterization of solid products of corrosion generated by modular-head femoral stems of different designs and materials. in *STP 1301 Modularity of Orthopedic Implants*, D. E. Marlowe, J. E. Parr, and M. B. Mayor, eds. ASTM, Philadelphia, pp. 33–44.

Urban, R. L., Jacobs, J. J., Gilbert, J. L., Rice, S. B., Jasty, M., Bragdon, C. R., and Galante, J. O. (1997b). Characterization of solid products of corrosion generated by modular-head femoral stems of different designs and materials. in: *Modularity of Orthopedic Implants*, D. E. Marlow, J. E. Parr, and M. B. Mayor, eds. ASTM, Philadelphia, PA, pp. 33–44.

Urban, R. L., Hall, D. J., Sapienza, C. I., Jacobs, J. J., Sumner, D. R., Rosenberg, A. G., and Galante, J. O. (1998). A comparative study of interface tissues in cemented vs. cementless total knee replacement tibial components retrieved at autopsy. *Trans. SFB* **21**.

Urban, R. M., Jacobs, J. J., Tomlinson, M. J., Gavrilovic, J., Black, J., and Peoc'h, M. (2000). Dissemination of wear particles to the liver, spleen, and abdominal lymph nodes of patients with hip or knee replacement. *J. Bone and Joint Surg. [Am]* **82**: 457–476.

Venable C. S., Stuck, W. G., and Beach A. (1937). The effects on bone of the presence of metals; based upon electrolysis. An experimental study. *Ann. Surg.* **105**: 917.

Vermes, C., Chandrasekaran, R., Jacobs, J. J., Galante, J. O., Roebuck, K. A., and Glant, T. T. (2001a). The effects of particulate wear debris, cytokines, and growth factors on the functions of MG-63 osteoblasts. *J. Bone Joint Surg. Am.* **83**: 201–211.

Vermes, C., Glant, T. T., Hallab, N. J., Fritz, E. A., Roebuck, K. A., and Jacobs, J. J. (2001b) The potential role of the osteoblast in the development of periprosthetic osteolysis: review of *in vitro* osteoblast responses to wear debris, corrosion products, and cytokines and growth factors. *J. Arthroplasty* **16**: 95–100.

Visuri, T., and Koskenvuo, M. (1991). Cancer risk after McKee–Farrar total hip replacement. *Orthopedics* **14**: 137–142.

Walker, P. S. (1978). *Human Joints and Their Artificial Replacements.* Charles C. Thomas, Springfield, IL.

Wiles, P. (1953). The surgery of the osteoarthritic hip. *Br. J. Surg.* **45**: 488.

Wilson, J. N. and Scales, J. T. (1970). Loosening of total hip replacements with cement fixation. Clinical findings and laboratory studies. *Clin. Orthop. Rel. Res.* **72**: 145–160.

Wimmer, M., Berzins, A., Kuhn, H., Bluhm, A., Nassutt, R., Schneider, E., and Galante, J. O. (1998). Presence of multiple wear directions in autopsy retrieved acetabular components. *Trans. ORS* **23**.

7.8 DENTAL IMPLANTATION

A. Norman Cranin and Jack E. Lemons

The most frequent surgical procedure performed upon humans is dental extraction. Without question, the most frequently employed prosthesis is the complete or partial denture (Fig. 1). These are available as both fixed and removable devices. Most often, they are supplied in the latter form owing to their relative ease of production as well as economy. There are numerous instances where, because of a paucity of natural teeth, a lack of posterior teeth (required as abutments), or in the presence of anatomic aberration resulting from traumatic, congenital, or metabolic causes, the most desirable type of prosthesis, the fixed or nonremovable type, cannot be used. In such circumstances, surgical implants may be placed to anchor prosthetic denture superstructures. Depending on the areas requiring such devices and the purposes which they must serve, dental implants are available in a variety of materials and an abundant number of designs (Cranin *et al.*, 1987; Worthington, 1988).

HISTORY

The Egyptians were known to have secured teeth to the jawbone with gold wire, and through the centuries, surgeons have sought techniques for implanting tooth substitutes, with little reported success. In the days of the American Revolution, seamen were attacked in darkened waterfront areas and their teeth extracted, to be implanted into the mouths of the wealthy and more fortunate gentry. Those biologic specimens were

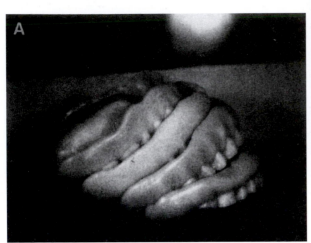

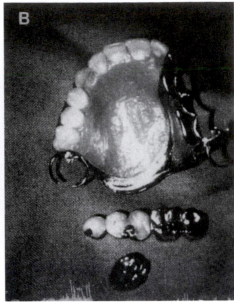

FIG. 1. (A) Complete, removable dentures are the prosthetic replacement used by most edentulous people. As can be seen in this picture, some denture wearers find it difficult to accommodate to conventional prostheses. They represent one segment of the population who can be helped with the use of dental implants. (B) Removable partial dentures, fixed partial dentures (bridges), and crowns are widely accepted replacements for missing teeth.

lost quickly owing to the immune responses of the recipients (Guerini, 1969; Cranin, 1970; Driskell, 1987).

In the early 20th century, Greenfield reported on the implantation of circular gold and platinum cribs as artificial dental roots, to which porcelain teeth were attached by a slotted coupling device. Later, the Strock brothers, using a Vitallium cast in the shape of self-tapping wood screws (after the work of Venable and Stuck), implanted more than dozen cast alloy screws as free-standing dental implants. One was reported to have survived for over 20 years (Strock, 1939; Strock and Strock, 1949). The most consistent dental implant reported in the first half of this century, and which initiated the current era of enthusiasm, was the subperiosteal device. It was introduced by Gershkoff and Goldberg (1949). Enthusiasts flourished, and reports of long-term success came from all parts of the world. These castings, made from nonferrous, cobalt–chromium-based alloys, were employed at first only for completely edentulous mandibles, and in so doing satisfied a significant and popular need. The individual who required single or unilateral implants, however, was not served well by adaptations of these early subperiosteal implants (which, for the first few years were not fabricated from direct bone impressions, but rather, by using a "guesswork" X-ray or intraoral replica template measuring system) (Weinberg, 1950). In 1951, Berman introduced the direct bone impression, which created opportunities for more accurate, longer lasting, and more predictable subperiosteal implants (Berman, 1951) (Fig. 2).

In 1947, Formaggini discovered a technique for placing the first of the modern-day, endosteal root-form implants (Formaggini, 1947). His method was improved by Rafael Chercheve, whose system was heralded universally (Chercheve,

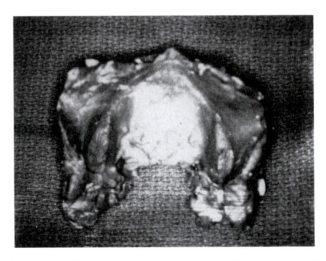

FIG. 2. This direct bone impression of a surgically exposed maxilla using a polysulfide material shows the accurate anatomic reproduction produced. The model produced from this impression will be used to fabricate a cast cobalt-based alloy subperiosteal implant (see Fig. 6).

1956) (Fig. 3). Those root forms, however, failed to serve all patients who were in need because their large dimensions could not be accommodated by the commonly found posterior alveolar ridge, which was thin because of atrophy after tooth loss.

Roberts in 1967 made a laminar-shaped stainless steel device that he called a blade implant. It was placed in a narrow osteotomy that all but the thinnest ridges could accept and showed considerable promise (Roberts and Roberts, 1970).

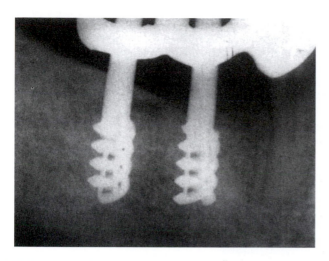

FIG. 3. These Chercheve implants are being used to help support a fixed (nonremovable) bridge in conjunction with the natural teeth (1974).

FIG. 5. The NobelBiocare implant (designed by Branemark) is completely buried under the gum with its superior border level with or slightly below the crest of bone. These submergible implants are left buried for a healing period of 3–6 months, at which time they are uncovered and coupled to permucosal abutments that will be used a foundations for dental prostheses.

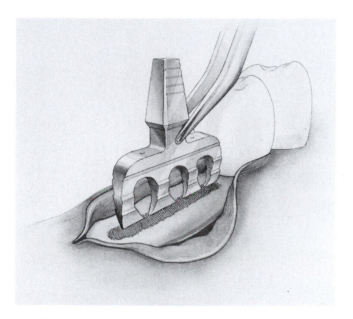

FIG. 4. The one-piece blade implant made of commercially pure titanium is a laminar appliance designed to fit into jawbone ridges too thin to accommodate the three-dimensional root-form types. The insertion operation requires the cutting of a longitudinal bone groove into the alveolar crest.

Linkow developed this device into a globally acceptable system by creating a wide variety of sizes and shapes designed to fit almost any clinical requirement and by changing the biomaterial to unalloyed titanium (Linkow, 1968). The 1970s were the era of popularity for the blade endosteal implant (Fig. 4).

The concept of osseointegration was developed and the term coined by Professor Per-Invar Branemark at the University of Gotegurg, Sweden. Osseointegration has been defined as a direct structural and functional connection between ordered, living bone and the surface of a load-carrying implant (Branemark, 1983). The initial concept stemmed from vital microscopic studies of bone and other tissues, which involved the use of an implanted titanium chamber containing an optical system for transillumination. The investigators observed that the screw-shaped titanium chambers became incorporated within the bone tissue, which actually grew onto the irregularities of the titanium surface (Branemark *et al.*, 1969). These findings led to research exploring the possibilities of artificial root replacement. The experimental studies that followed involved extraction of teeth in dogs and their replacement by osseointegrated screw-shaped implants. Fixed prostheses were connected after an initial healing period of 3 to 4 months during which they bore minimum load (Fig. 5). Radiologic and histologic studies showed that integration could be maintained for 10 years in dogs, without inflammatory reactions. Further, the load-bearing capacity of the individual implants was shown to be extremely high. Based on these findings, the Foundation for osseointegration and the Branemark Implant System were established in 1962. Basic research continued in the years that followed and the first edentulous patient was treated in 1965 (Branemark, 1983).

Clinical data on osseointegrated implants were presented to North America at a scientific meeting held at Toronto in 1982 (Zarb, 1983). Success rates of 90–100%, initially presented by the Swedish group, were confirmed by longitudinal follow-up studies from over 50 osseointegrated implant centers around the world (Adell, 1983; Laney *et al.*, 1986; Albrektsson *et al.*, 1988).

Today, osseointegrated implants are used to treat both partially edentulous and completely edentulous ridges. Severely atrophied ridges, however, required several surgical procedures involving bone augmentation to prepare the host site for acceptance of endosteal devices. A staunch group of implantologists continues to utilize the subperiosteal implant for those patients who are considered untreatable by endosteal root-form or blade-type devices.

The main disadvantage of the subperiosteal implant continued to be the necessity of performing a two-stage operation; the first, a direct bone impression in order to make the implant casting, and the second, 24 hours to 6 weeks later, in order to insert the casting surgically. Using computer-based design and machining technology (CAD/CAM), Truitt and James, in 1982, introduced a method by which computerized axial tomography (CT) scans of facial skeletons could be employed to generate relatively accurate models of the jawbones. These models then were used to develop subperiosteal implant castings (Truitt *et al.*, 1986). Such castings have served successfully and have eliminated the need for first-stage, bone-impression surgery (Fig. 6). More recently, because of studies performed by Cranin (1998), which demonstrated inaccuracies of CAD-CAM technology,

newer techniques for replication, such as stereolithography, were introduced. The 1980s also witnessed a wide range of root-form, two-stage endosteal implants following the work of Branemark, using numerous adaptations of his original threaded device. These have incorporated materials and designs that include press-fit (nonthreaded) configurations and coatings of alumina, hydroxylapatite, and titanium oxide plasma spray, as well as roughened surfaces created by blasting or etching techniques (Lemons, J. E. 1977. Surface conditions for surgical implants and biocompatibility. *J. Oral Imsl.* 7(3): 362–374. Niznick, 1982; Kay *et al.*, 1986; Kirsch, 1986)

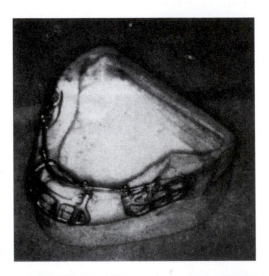

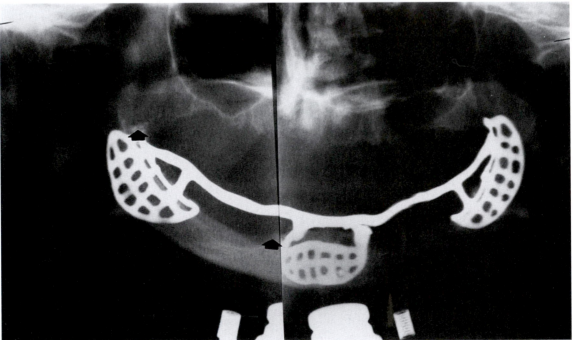

FIG. 6. CAD/CAM-generated models of the mandible or maxilla eliminate first-stage bone impression surgical procedures (see Fig. 2) for subperiosteal implants. Seen here is the mandibular tripodal subperiosteal implant that fits on the bone beneath the gum. Only the post and bars protrude to serve as retainers for dental prostheses. (B) A panoramic radiograph shows the aggressive extensions of newer designs.

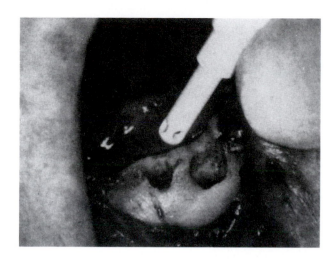

FIG. 7. Exact size and shape holes (osteotomies) are made to accommodate (endosteal) implants placed within the bone. This hydroxylapatite (Integral) coated press-fit implant is stabilized initially by a precise frictional grip.

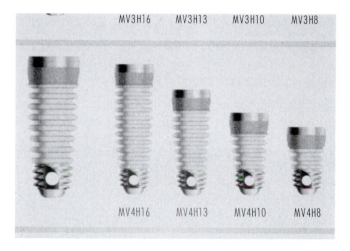

FIG. 8. These threaded, root-form implants are supplied in a wide variety of sizes and lengths. Each is supplied with a number of abutments designed to accommodate many prosthetic situations.

(Figs. 7 and 8). Self-tapping implants have also broadened the field to a level that now offers the clinician a large and often confusing spectrum of implant varieties, the benefits of some remain unclear, and the selection of which is dependent upon a number of poorly defined criteria and prejudices (English, 1988). Of more recent interest are the one-stage root form implant, first introduced by ITI, and the immediate-load philosophy (Schnitman, 1988), which can be practiced with all forms of threaded root-form implants, both one- and two-stage. The latter is accomplished by simply attaching the abutment to the implant body at the time of its insertion. The obvious advantages of this technique are the saving of from 3 to 6 months of time awaiting osseointegration, the elimination of a second surgical procedure, and the availability of abutments for immediate use.

CURRENTLY USED IMPLANT MODALITIES

Designs and Methods

Dental implant designs can be separated into two categories called endosteal (endosseous), which extend into the bone tissue, or subperiosteal systems, which contact the exterior bone surfaces (Lemons, 1988). These are shown schematically in Figs. 5 and 6. The endosteal forms, such as root forms (cylinders, screws), blades (plates), transosseous or staples, ramus frame, and endodontic stabilizers, are placed into the bone as shown. In contrast, the subperiosteal devices are fitted to the bone surface as customized shapes while bone plates are placed into the bone under the periosteum and fixed with endosteal screws.

Synthetic materials for root form devices were originally fabricated from precious metals such as gold, platinum, iridium, and palladium. Cost and strength considerations soon resulted in the development and use of alloys and reactive group metals such as tantalum, titanium, and zirconium, and the inert cast and wrought cobalt- (stellites) and iron (stainless steel)-based alloy systems. Since the 1950s, materials formerly used for general industrial applications (automotive, aircraft, aerospace, etc.) have been reconstituted to include a wide range of metals and alloys, ceramics and carbons, polymers, combinations and composites (ASTM, 2002). Some commonly used biomaterials for early dental implants are listed in Table 1.

The more commonly used biomaterials for endosteal implants are titanium and titanium alloy, aluminum oxide, and surface coatings of hydroxyapatite. The cobalt based casting alloys are most often used for the subperiosteal implant devices (McKinney, 1991). Since the properties of these biomaterials are quite different from one another, specific criteria must be applied to each device design and clinical application. These details are discussed in the following sections.

Subperiosteal

Subperiosteal dental devices may be used for partially and completely edentulous jaws and are the implants of choice for those regions that contain insufficient bone to accommodate endosteal implants of either the blade- or root-form varieties (Schnitman, 1987). They consist of a mesh-type infrastructure cast of a surgical-grade cobalt–chromium–molybdenum alloy to which are attached from four to six permucosal abutments. Atop these protrusions into the oral cavity may be prosthetic abutments to serve as retainers for fixed bridge prostheses or retentive bars, which connect the abutments together into a single structure. These may be used to gain the retention of overdentures (Yurkstas, 1967; Dalise, 1979; Cranin et al., 1978) (Figs. 9 and 10). Subperiosteal implants are cast to models of maxillae or mandibles made either by direct bone impressions using polysulfide or poly(vinyl siloxane) elastomeric impression materials or by CAD/CAM-generated models. They are designed to rest on the cortical bone and are entrapped and fixed by a reattachment of periosteal fibers through the numerous interstices incorporated into their infrastructural designs. When successful, they are purely incidental to local tissue physiology (James, 1983).

TABLE 1 Examples of Early (Prior) Endosteal Implants

Name	Material	Description	Stages	Surface	Primary retention	Length (mm)	Diameter (mm)
Bränemark (Nobelpharma)	Titanium, commercially pure (CP)	Threaded screw	2	Machined	Threading (bone tapping, self tapping)	7, 10, 13, 15, 18, 20	3.75
						7, 10, 13, 15, 18	4.0
Core-vent	Titanium alloy	Threaded hollow basket	2	Sand blasted	Threading, core (self-tapping)	8, 10.5, 13, 15	4.3, 5.3, 6.3
TPS screw	Titanium (CP)	Threaded screw	1	Plasma spray	Threading	8, 11, 14, 17, 20	3.5, 4.0
IMZ (Interpore)	Titanium (CP)	Bullet shaped	2	Plasma spray, optional hydroxylapatite	Press fit	8, 10, 13, 15	3.3
						8, 11, 13, 15	4.0
Integral (Calcitek)	Titanium (CP)	Bullet shaped	2	Hydroxylapatite	Pres fit	8, 10, 13, 15	3.25, 4.0
Kyocera (Bioceram)	Single crystal sapphire	Threaded screw	1	Smooth, porous	Threading	E: 19.5, 22, 25	3.3, 4.2
						S: 19.5, 23	3.0, 4.0
						20.5, 24	4.2, 4.8
Tubingen	Aluminum oxide polycrystal porous	Stepped	1	Aluminum oxide stippled	Press fit	21, modifiable	4.0, 5.0 6.0, 7.0
Stryker	Titanium alloy	Finned	2	Machined, optional hydroxylapatite	Press fit	8, 11, 14	3.5, 4.0 5.0
Bausch and Lomb (Steri-Oss)	Titanium (CP)	Threaded screw, bullet shaped	2	Machined, optional hydroxylapatite	Threaded screw, Press fit	12, 16, 20	3.5
						8, 10, 12, 14, 16	3.8
						12, 16, 21	4.0

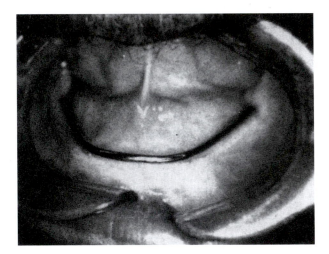

FIG. 9. The Brookdale bar is attached to an underlying subperiosteal infrastructure (see Fig. 6) by four to six permucosal posts. A denture prosthesis is designed with clips that attach to the bar for stabilization.

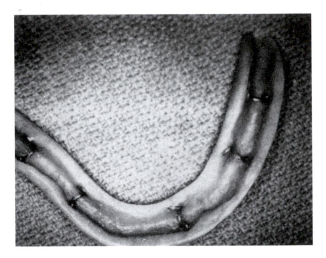

FIG. 10. The undersurface of a denture designed to clip on to a subperiosteal implant bar.

Relatively high levels of success have been attributed to these subperiosteal implants, particularly the newer mandibular tripodal configurations as well as the pterygohamular maxillary designs (Linkow, 1986; Cranin *et al.*, 1985) (Fig. 11). When CAD/CAM-generated castings were used, slight inaccuracies were sometimes experienced which were treated with particulate bone substitute materials such as hydroxylapatite, which were used as local fillers (Fig. 12).

Endosteal

Blade Type The blade- or plate-type designs are available in one-piece (with attached abutment) or submergible (with separate to-be-attached abutment) configurations

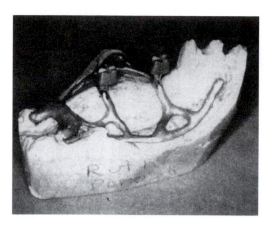

FIG. 11. The maxillary unilateral pterygohamular subperiosteal implant is used to provide posterior abutments for nonremovable (fixed) bridges in the maxilla. It is designed to fit against the maxillary, palatal, zygomatic, and pterygoid bones, which are reliable sources of cortical support.

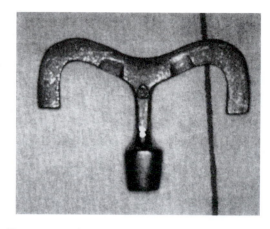

FIG. 13. One-piece blade implants such as the Cranin anchor design have been used successfully as distal abutments or pier abutments for fixed (nonremovable) bridges. The entire structure, with the exception of the protruding dental abutment, is placed within the confines of the jawbone. The abutment protrudes to serve as a prosthetic retainer.

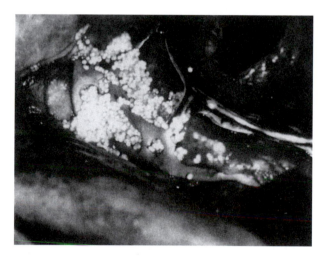

FIG. 12. Minor discrepancies that may occur between the subperiosteal implant fabricated on a CAD/CAM model and the jaw upon which it rests are grafted with hydroxylapatite granules, a synthetic bone replacement material. Materials like this (tricalcium phosphate, pmma/pHema) are osteoconductive materials that serve as reliable but nonosseous fillers.

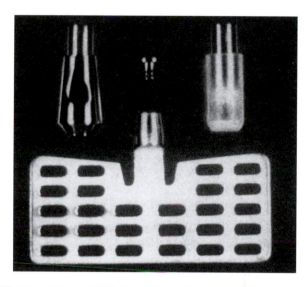

FIG. 14. Submergible blade implants with prosthetic heads that are attached to the implant at the date subsequent to surgery allow for a stress-free healing period.

(Figs. 13 and 14). They are fabricated in a great variety of designs of relatively pure surgical-grade titanium or titanium–aluminum–vanadium alloy. With the aid of clear plastic templates, these may be employed in an almost limitless number of host sites. In addition, several manufacturers supply the surgeon with one- and two-piece titanium blade-form blanks. These may be fashioned and cut to almost any shape or size that the potential surgical site demands. If the implantologist cannot find an implant that will satisfy a need within the broad spectrum of manufactured designs, or believes in the anchor ("shoulderless") philosophy (Fig. 13), a custom casting may be made by using distortion-free dental and CT scan radiographs to incorporate dimensional accuracy (Cranin, 1980;

Baumhammers, 1972). Cobalt alloy is the biomaterial most often used for such castings.

True osseointegration (a direct bony interface) does not routinely occur when intraosseous slot osteotomies are created for blade implant placement by using high-speed fissure-bur operative techniques. There is a question as to whether two-stage submergible blades fare better than the one-piece implants (particularly since the buried components permit the protrusion of a threaded cervix through the mucosa). Nonetheless, reports from the literature support blade-form endosteal implants with success for periods exceeding 10 years. James and others reported that the mechanism of support is fibro-osseous suspension (integration). They indicated that a sling of connective tissue fibers suspends the implant within its bony crypt, and

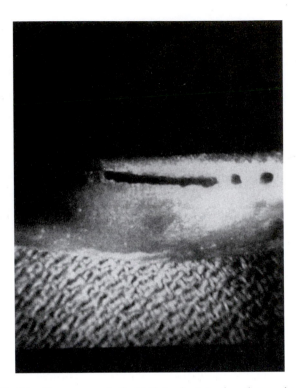

FIG. 15. The receptor site for blade implants (such as the one shown in Figs. 13 and 14) is made freehand with a saline-cooled high-speed bur (250,000 rpm). First, a dotted line following the length and position of the implant is made. Then the dots are connected to form the groove receptor site. (See Fig. 4.)

that upon biomechanical stimulation of these fibers caused by mastication, an environment of osteogenesis occurs which creates continual bone anabolism. (James, 1988; Weiss, 1986). These implants, which require a reasonable level of surgical skill to insert, demand the use of freehand, high-speed, saline-cooled capabilities (Fig. 15). They may be used as unilateral posterior abutments for fixed bridge prostheses or in a wide variety of prosthetic functions, such as supports for full coping bar/overdentures or even as single tooth supports. Their prognosis is judged by some to be less predictable because of the support mechanism afforded them—a fibrous sling or hammock that permits micromovement and is accompanied by a slow exteriorization with epithelial downgrowth from the point of permucosal penetration.

Root-Form Types Long-term clinical experience has proven osseointegration to be a reliable modality for treating edentulism, making root-form implants the most popular and compelling implant design. Currently more than 80 companies are manufacturing implants and implant-related products. Most of the root-form implants available today follow the basic principles of design for successful osseointegration, which were outlined by the Swedish group at Goteburg (Albrektsson, 1983) (Fig. 8) (Table 1).

The first factor is the materials and biomaterials used in manufacturing such implants. The Branemark (NobelBiocare) and many other implants are made from commercially pure

(CP) titanium. Upon exposure to aid or water, it quickly forms an oxide layer of 8–10 nm at room temperature. This oxide layer makes titanium extremely corrosion resistant, particularly in the biologic environment. The alloyed form of titanium (Ti6Al4V) is also widely used to manufacture these type devices. Titanium alloy has been shown to be as biocompatible as commercially pure titanium and to have superior mechanical properties (Williams, 1977).

The second factor is the design of the implant. It must be constructed to encourage initial stability in its osteotomy. It has been shown that micromovement or displacement of the implant relative to its host site during initial healing can destroy the network of immature preosseous collagenous tissue, which serves as the scaffold for bone development. Such "toggling" will lead to fibrous repair instead of bony regeneration (Brunski, 1979). In the root-form type systems initial stability is achieved by the helical shape of the implant, which is inserted in a threaded osteotomy (in cases of less dense bone, the implant itself may be used for threading during the insertion procedure). Other endosteal designs include cylindrical or press-fit configurations, spiral shapes, finned cylinders, sintered balls, and partially hollow cylinders. Cylindrical implants are sometimes coated with either titanium plasma spray or calcium phosphate compounds (hydroxylapatite). These surface treatments offer more surface area and initial retention, which allow for precision of fit of the implants in the osteotomies, prevention of micro motion during the initial healing period and subsequent improved prognoses, particularly in less dense bone. Hydroxylapatite, because of its chemical similarity to osseous tissues, forms an attachment with bone, independent of any mechanical interlocking mechanism. This concept has been described by the term "biointegration" (Kay, 1986; Meffert, 1988; Niznick, 1982; Kirsch, 1986).

The third factor necessitates the use of a surgical technique, which prevents excessive heat generation during preparation of the host site. It has been shown that the heating of bone to a temperature of 47°C or above for over 30 sec causes vascular damage and alters its vitality (Ericsson et al., 1983). To minimize mechanical and thermal trauma, the surgical protocol involves the use of specially designed spiral, helical, or spade drills of graduated lengths and diameters (Cranin, 1991) (Fig. 16). Most bone preparations are carried out at a maximum rotational speed of 2000 rpm and under profuse internal and external irrigation through hollow-core drills.

Finally, another factor for achieving a direct bone-to-implant interface is the suggested technique of maintaining newly placed implants within bone spared from mechanical forces or loads. It has been shown that for successful osseointegration the individual implant should not be loaded for a period of at least 3 months in the mandible and 6 months in the maxilla. Most currently used systems achieve this by producing two-piece designs. As such, these submerged implant systems require two surgical procedures. At the first-stage surgery, the implants (submergible portions) are inserted into the bone and the overlying mucosa is closed, protecting the implants from mechanical loads and oral microbiota. After a 3- to 6-month (or longer in cases of bone grafted sites) period of healing, the second surgical stage of abutment connection is carried out. The bone remodeling process continues after the implant

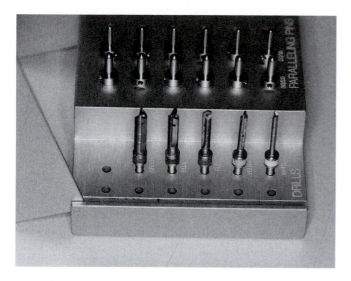

FIG. 16. A variety of instruments is used to create precise holes within the bone for the root-form implants. Twist drills, bi- and tri-spade drills (two cutting edges and three cutting edges, respectively), as well as bone taps to thread the bone all help to make an exacting atraumatic procedure. This set of color-coded drills designed by Cranin and produced by the Brasseler Corp. is available in its own sterilizable carrier. Each drill is no more than 500 μm greater in diameter than the next in the series. These small increments ensure greater and less traumatic drilling accuracy. The final osteotomies are threaded with the specific bone taps made by the company that supplies the root form implants. They are used at speeds as low as 6 rpm and have hollow cores that allow passage of a bone-cooling irrigant.

is loaded with dental prostheses for an additional period of months to years (Fig. 17). More recently, the concept of immediate loading has gained more popularity. There are a growing number of one-piece, nonsubmerged implant systems available, although the ITI System championed the technique in the 1980s. These are placed in a single surgical procedure and are permitted to protrude into the oral cavity immediately after insertion. Some practitioners follow this philosophy, which permits the implant to be placed into immediate service as a functioning abutment. Clinical longitudinal studies have demonstrated successful osseointegration using such one-stage implantation techniques (Buser *et al.*, 1990; Misch, 1999) (Fig. 18). Among the benefits offered by the one-stage method is a great telescoping of treatment time. In addition, Hruska (Cranin, 1987) introduced an intraoral welding machine that provided the prosthodontist with an instant post-surgical connecting bar permitting the immediate provision of an internally clipped overdenture (Figs. 9 and 10).

In the past 5 years, with the levels of root form implant success becoming so predictable, attention of surgeons have turned to more sophisticated problems such as esthetics. Great attention has been given to the appearance of the completed prosthesis at the gum margin. Whereas pink acrylic had been the medium used to solve the unattractive dark shadows caused by metal crown margins, recent skills have been devoted to the surgical creation and/or preservation of natural gingival papillae and margins with great and rewarding success (Tarnow, 2000).

One of the significant problems encountered by planning teams is a paucity of bone required for the placement of root-form implants, both in height and in width. Advanced techniques in optimizing potential host sites by bone grafting, both autogenous and allogeneic, in granular and block forms have been introduced in the past 15 years (Cranin, 1991, 1999). Significant assistance is offered to the surgeon with use of CT scans taken of their patients with radiopaque markers placed in their mouths. The resultant images delineate with great accuracy the potential host site locations, dimensions, and bone quality (Fig. 19). Recently, surgeons that are more aggressive have discovered the benefits of the zygomatic buttresses as sources for implant retention. The 30+-mm-long implants utilized for these sites are inserted via a palatal-oblique approach and require significant skill in order to be successful (S. Perel, 2000).

Root-form implants are used as abutments for single crown, fixed-cementable bridges, fixed-detachable bridges (anchored by screws), and overdentures. The latter may be retained to the underlying mesostructure bars using Lew attachments (locking bars) (Lew, 1973) (Fig. 20), Hader and similar clips (Fig. 10), "O" rings, and other similar, simple devices, but also may employ the extremely complex technique of retention by spark erosion (Fig. 21).

There are a wide variety of other endosteal implant designs that employ bone anchorage to stabilize loose teeth (endodontic implants), to serve as prosthetic abutments (transosteal implants), and to anchor long-span bridges (with a paucity of natural teeth) to the bone (C-M pins) (Cranin, 1999) (Fig. 22).

Endodontic Stabilizer Designs

Endodontic stabilizers are long pins or screws that are of a small enough diameter to be passed through a tooth root canal. Examples of endodontic stabilizers and pins are shown in Fig. 23. These designs of implants are sometimes placed into teeth that have minimal bone and periodontal ligament support. The stabilizer is intended to improve the mechanical force transfer from the tooth and implant to the bone, to increase the tooth stability, and thereby to enhance long-term usefulness. These devices are used with endodontically treated teeth, and a relatively stable seal at the tooth root apex area is critical to long-term stability.

Stabilizers are fabricated with a mechanically rolled or swaged surface configuration to enhance mechanical strength. These are usually titanium, although some designs utilize titanium and cobalt alloy. Smooth surface designs are also available in single-crystal sapphire (aluminum oxide) ceramic.

Transosseous and Staple Designs

Early transosseous screw or post designs were fabricated from cast cobalt alloy, as shown in Fig. 24. These were placed through the anterior region of the mandible and protruded through the gingival into the oral cavity (Cranin, 1970). These protrusions or abutments (usually bilateral at the location) were used to stabilize standard type, soft tissue supported lower dentures. An extension of this clinical concept the endosteal staple (Small, 1980) incorporated an inferior burden plate that connected the two transosteal threaded pins. Additional pins

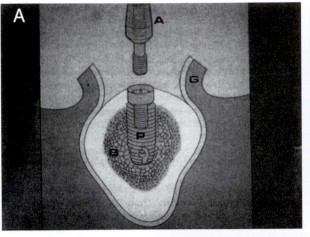

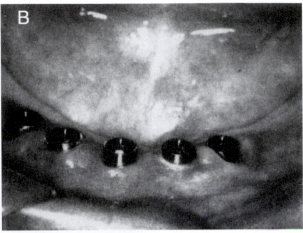

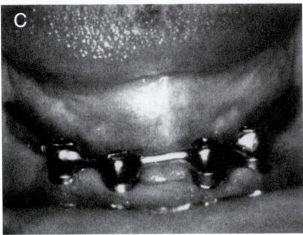

FIG. 17. (A) An abutment is connected to the implant after a 3- to 6-month healing period. The implant (P) rests between the gum (G) while the bone (B) grows into and around the metal structure, granting the firm anchorage known as osseointegration. This cross section shows the gum being opened surgically for placement of the abutment (A). (B) The abutment attaches to the implant, which is buried within the bone. The abutment passes through the gum into the oral cavity. (C) A final anchoring prosthesis such as this bar with four copings is screwed into the implant abutments, which have internally threaded holes. This permits the practitioner to remove this mesostructure bar for evaluation, hygiene, and when necessary, for repair.

and subsequently screws were placed along the plate to penetrate the inferior border of the mandible and thereby provide endosteal support. These designs were fabricated from titanium alloy. Some designs were also fabricated from gold alloy. More recent versions now include bone screw stabilizers and hydroxyapatite coatings to enhance mechanical stability within the bone.

Ramus Frames

The ramus frame implants are also intended for treatment of the edentulous mandibular arch. Several commercially marketed ramus frame designs are available in Fig. 25. The anterior endosteal portion of these devices is placed into the central symphysis of the mandible, while the posterior extensions or blades are implanted into the ascending ramus or distal mandibular regions. The extensions along the posterior (distal) regions, and additional surface tabs were added as design changes in an attempt to enhance resistance to bone resorption and to

progressive settling into the bone of the mandible during long-term use. These devices are utilized as one-stage restorative systems in patient treatments. Many of these devices are placed in patients who are completely edentulous and the patient starts to use the denture the first day after surgery. Thus, the frame-supported lower denture, stabilized by the bone insertions, articulates with a tissue-supported upper denture.

Plates, Screws, and Wires

Bone plates, screws, and wires used in oral and maxillofacial surgery are very similar to those used in orthopedic surgery. On a relative basis, more of the oral surgical implants are fabricated from titanium as compared with orthopedic devices, and, in general, they are smaller in dimension. A wide range of dental design configurations exists that is intended to stabilize both simple and complex craniofacial bone lesions. All commonly used plates, screws, and wires are constructed from metals and alloys (Fig. 26).

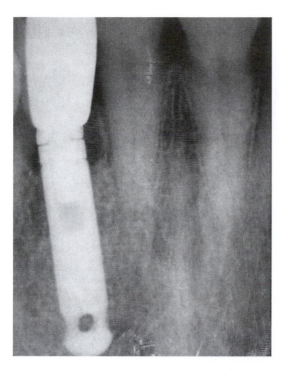

FIG. 18. This X-ray shows an implant that was loaded immediately after its insertion. Three days later a prosthetic crown was cemented to it.

FIG. 20. The Lew attachment is a stainless steel, two-piece device (Park Dental Research Corp.) that is embedded into the flange of an overdenture. It serves when in closed position (C) to lock the prosthesis beneath an implant bar, and when the button is pulled out by the patient (O), it serves to release it.

Intramucosal Inserts

One design listed as a dental implant is the mucosal implant series. These are small, mushroom-shaped button that are fabricated into the denture base and protrude from the denture surface (Fig. 27). The buttons are located using a surgical procedure so that they penetrate into the adjacent mucosa. Epithelial cells line the crypt around the buttons during healing; therefore

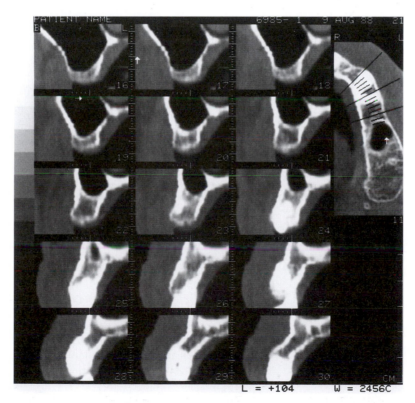

FIG. 19. These cross-sectional images, derived from a CT scan (DentaScan), show a series of cuts through the canin/premolar area of the maxilla. Depth, width, and location of the maxillary sinus are in evidence. In addition, the radiopaque objects in the lower sections indicate the planned location of an implant as delineated by a barium-filled tooth placed in the radiographic template.

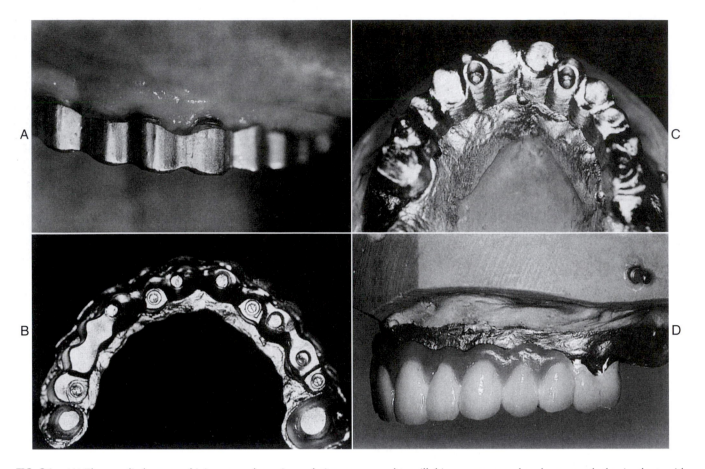

FIG. 21. (A) Electron discharge machining or spark erosion techniques were used to mill this mesostructure bar shown attached to implants with fixation screws. (B) The armature or skeleton to which the porcelain is baked and the denture teeth are processed is milled intimately to fit by a precise frictional relationship to the mesostructure bar. (C) The external surface of this structure is prepared to receive the processed prosthesis. (D) The completed spark erosion fabricated mesostructure bar, with a totally porcelain baked superstructure, may be maintained by a frictional relationship or, if additional retention is desired, by the use of strategically related latches, such as Ceka-like attachments. (See color plate)

these systems are not true implants. However, since they are often placed into freshly prepared soft tissue lesions, implant quality materials and surface conditions are required (e.g., most are passivated stainless steel or titanium). As mentioned most are fabricated from type 316L stainless steel or titanium, with some made from aluminum oxide ceramics. The denture base is constructed from poly(methyl methacrylate), which also is used as an implant biomaterial for orthopedic (bone cement) and dental (bone substitute) applications.

TISSUE INTERFACES

Occlusal forces from the mastication of food are transferred from the intraoral prosthesis throughout the implant abutment and neck (connector) and into the implant-to-tissue interface region. These forces are dissipated through the associated tissues, and the quality of functional stability can be correlated with the relative interfacial stability (micromotion)

over time (Brunski, 1988). Two types of interfacial conditions have been described for functional dental implants: fibrous tissue integration (called pseudo-ligament fibro-osteal integration) (Weiss, 1986) and bone tissue integration (called osseo- or osteointegration) (Branemark *et al.*, 1977) (Figs. 24A, B).

Osseointegration

The concept of osseo- or osteointegration was initially associated with root-form implant designs, two-stage restorative treatments, and a titanium oxide implant to bone interfacial condition (Branemark *et al.*, 1977). This is described as a direct bone-to-biomaterial interface (without fibrous tissue) for a functioning implant at the optical microscopy limits of resolution (0.5 μm). A wide range of implant biomaterials and designs have been shown to exhibit this type of interface for functional dental systems (Rizzo, 1988). Proponents support the concepts that (1) greater force can be transferred along

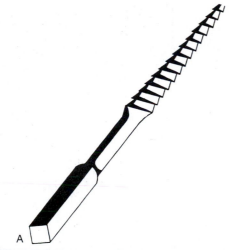

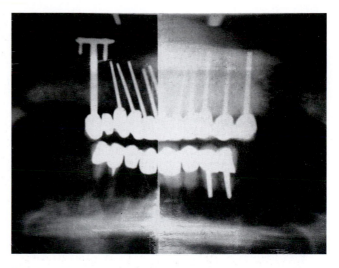

FIG. 23. This X-ray shows a transmandibular implant that is anchored by screws to the cortex at the inferior border of the mandible. Adjacent to it are nine endodontic implant stabilizers. These chrome alloy pins are placed through the root canals of teeth with poor bone support and their ends, after passing through the root apices, extend into the surrounding bone, which lends additional retention. They are most successful when cemented into multiple adjacent teeth.

outcomes for different systems, and reports at 10 years show very promising outcomes for multiple design and surface types.

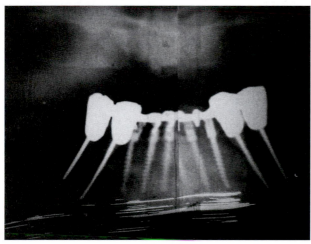

FIG. 22. This thin ridge implant, CM (crete mince) is manufactured in Switzerland. Its maximum width is 2 mm and it is supplied in lengths of up to 20 mm (A). To be effective they must be used in multiples (B).

CLINICAL ENVIRONMENT

Repair

As might be expected, bone loss will sometimes be seen around the cervices of root form implants. This phenomenon is known as saucerization (Cranin, 1970). Meffert (1992) established standards for the surgical repair of ailing, failing, and failed implants. The last group was defined as demonstrating mobility and mandated removal. The other two groups, all of which were found to be stable, were essentially salvageable if the prescribed techniques were followed.

bone interfaces; (2) direct association or attachment eliminates or minimizes interfacial movement (slip); and (3) periodontal soft-tissue regions are more biomechanically stable.

Interestingly, both osseous and fibrous tissue interfaces have provided functional situations for dental implant devices over 15 or more years (Rizzo, 1988). Clinical trials that have been developed using prospective analysis protocols have provided opinions about the questions related to "which is the best interface." However, in general, the clinical use of endosteal implants, in terms of sales, demonstrates a strong preference for root-form design titanium and hydroxyapatite type implant systems.

Hydroxyapatite, glass ceramics, porous layers added to metals, and other biomaterial surface modifications have been introduced and developed to enhance bone adaptation to endosteal and subperiosteal implant surfaces (Ducheyne and Lemons, 1988). These surface modifications could result in a more stable implant-to-tissue interface. Clinical data continue to be developed to evaluate relative

Biomechanics

The body of the dental implant contacts the bone and soft tissue interfaces within the submucosal regions. These contact zones within or along the bone surface provides the areas for mechanical force transfer (Bidez *et al.*, 1986). Therefore, the implant-to-tissue interface becomes a critical area for force transfer and thereby the focal point for quality and stability of intraoral function. The implant body is directly connected to the transmucosal (or transgingival) post, which provides the base for the dental bridge abutment. The intraoral prosthesis (crown, bridge, or denture) is attached to the abutment using cement, screws or clips. This arrangement was schematically shown in Fig. 1B.

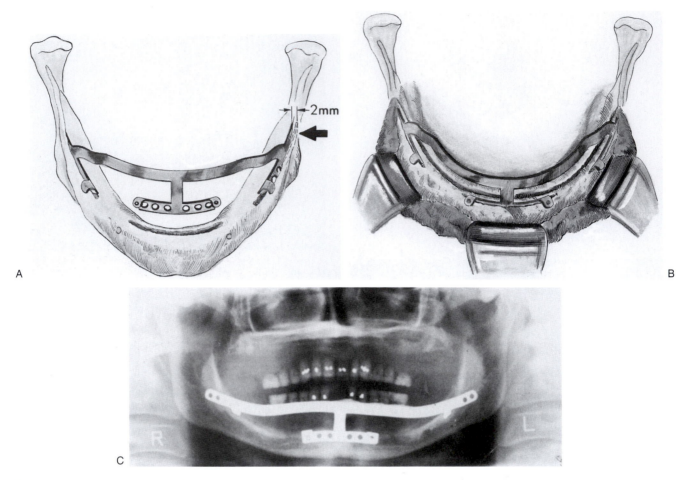

FIG. 24. (A) The ramus frame implant (Roberts and Roberts, 1970) is placed into three blade-type osteotomies. (see Fig. 4.) (B) It is tapped into the precut bone slots until the seating tabs become flush with the surrounding bone. (C) When successful, these implants will survive as retentive devices for overdentures, for many years.

The dental implant device provides a percutaneous connection from the submucosal to the oral environment resulting in a very complex set of conditions (Meffert, 1988). The oral cavity represents a multivariate external environment with a wide range of circumstances (e.g., foods and abrasion, pH, temperatures from 5 to 55°C, high magnitude forces, bacteria). Stability requires some type of seal between the external and internal environments, which of course depends upon both the device and the patient. The most significant reason for this is discovery by Robert James (1974) of hemidesmasomes, i.e., microscopic suction cups that were shown to be responsible for the adhesion of epithelial cells to polished implant cervices, thereby sealing their infrastructures from the onslaught of the challenging oral environment.

Fortunately, the oral environment and the associated tissues are known to be very tolerant and resistant to external challenges.

Crown, Bridge, and Denture Restorations

Intraoral prostheses involve a wide range of metals, ceramics, polymers, and composites. Since many of these intraoral devices are mechanically or chemically (cemented) attached to the dental implants, biomaterial interactions become a part of the overall restorative treatment. Therefore, each implant modality must be evaluated to ensure that adverse biomaterial of biomechanical conditions are not introduced through incorrect selection (Lemons, 1988). For detailed information, the reader is referred to dental material textbooks (Craig, 1985; Leinfelder and Lemons, 1988; Phillips, 1991; Cranin, 1999; Lemons and Cranin-Prior edition of book).

Fixed Restorations

Most root-form and blade (plate) systems use restorative treatments in which the implant is rigidly attached to the intraoral prosthesis. Some systems provide polyethylene or polyoxymethylene spacers within the core of the implant to enhance mobility through the interposition of lower modulus materials, while others require that occlusal surfaces of the teeth be fabricated from polymer-based restorative materials (Kirsch, 1986; Branemark *et al.*, 1977). These variations result in a wide range of treatments being utilized clinically. In this regard, device retrieval analyses have shown some situations where galvanic coupling of dissimilar alloys (see Chapter 6.3)

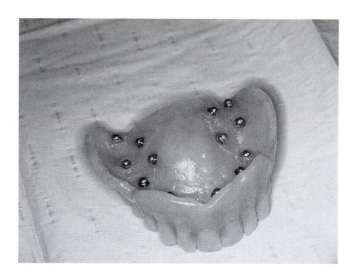

FIG. 25. Intramucosal inserts (Dahl, 1944; Jermyn, 1953) are stainless steel or CP titanium, arrow-head-shaped buttons (Park Dental Research Corp.) that are processed into the tissue-borne surfaces of nonretentive maxillary dentures. The mucosal receptor sites that they lock into offer significant anchorage to them.

or inadequate quality castings that contained porosities and high caron content have influenced device longevities (Lemons, 1988). At this time, most implant device-related treatment includes a prescription or recommendation for the design and biomaterial of the intraoral prosthesis.

Removable Dentures

The subperiosteal, transosseous, staple, ramus-frame, and mucosal-insert implant designs are associated with removable denture prostheses. The dentures are fastened to the various intraoral connectors or abutments shown in previous figures. The dentures can be attached by a variety of connector types that range from magnets to simple recesses into one of the components. Dentists and dental laboratory personnel have been ingenious in developing these connector systems. In some situations, interlocking devices have evolved to provide what is called "fixed-removable" prostheses.

TRENDS IN RESEARCH AND DEVELOPMENT

The dental implant field has provided a significant contribution to our overall understanding of basic biomaterial and biomechanical properties. The oral cavity provides 28 to 32 candidate implant sites, a necessity for crossing the epithelium to provide functional support, a wide range of anatomical shapes, and an even wider range of environmental conditions. The oral cavity and associated tissues provide a very severe environment for implant biomaterials and designs, and it is interesting to note the trends in research and development within this field.

A

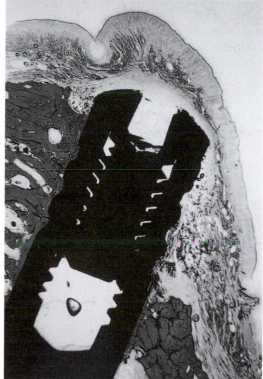

B

FIG. 26. These two photomicrographs (original magnification 8×) show the appearance of the bone, connective tissues, and epithelium supporting successful and failing root form, endosteal implants. The histology shows dramatic, stark differences (Cranin *et al.*, 1999).

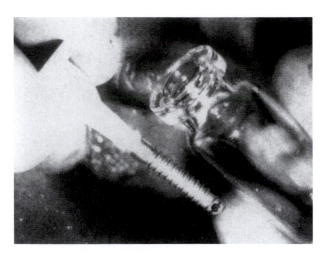

FIG. 27. Root-form implants are packaged within glass or plastic vials sterile and ready for use. Many companies supply the implants attached to their vial covers, which offer easy handling, a primary device for torquing the implant into place, and a method that prevents metallic contamination when other transport instruments might be used.

Conservative Treatment Modalities

In the 1970s, dental implants were judged to be in a longitudinal research and development phase. Typically, they were only used in clinical situations where all other treatments had failed and where bone and soft-tissue health was not compromised for implant-based restorations.

As a result of the reported clinical longevities and the functionalities of several dental implant designs and biomaterials, opinions now reflect the use of implants on a routine basis where they are indicated for dental restoration of normal function. Dental implant–based treatments have moved from the longitudinal research phase in 1970 to a recognized conservative modality in 2003. Important to this change is the quality and quantity of bone and soft tissue available at the time of implant placement. Advances in hard and soft tissue grafting have allowed many patients, who were previously considered poor candidates for dental implants, to undergo oral rehabilitation with the use of dental implants. This trend is expected to continue to provide ever-improved device function and longevity.

GENERAL ASPECTS OF PACKAGING AND PREPARATION

Of the three basic implant types: blades, root forms, and subperiosteals, the first two are usually prefabricated and the last is custom cast. Blades are supplied in titanium or titanium alloy directly from the manufacturer either in sterile packages or in plastic envelopes. Handling them is difficult, because even if they are prepared properly by the maker or user, the benefits of presurgical treatment such as radiofrequency glow discharge, sterility, and freedom from surface lipids or contaminants (e.g., talc from rubber gloves) will be lost owing to the

frequent and sometimes aggressive manipulations necessitated during implantation procedures.

In most instances, root forms are successfully placed on their first insertion, so innovative sterile packaging in small glass or plastic vials is the current method of presentation. These implants need not be touched by the surgeon's fingers and are transferred from package to host site either by the vial covers into which they are inserted and which serve as handles, or in the titanium-tipped forceps, mounting platforms and fixture mounts (Fig. 27).

Because subperiosteal implants are individually cast of cobalt–chromium–molybdenum alloy, the laboratory and the surgical team must treat them individually. They require passivation with dilute acids, defatting using acetone or other organic solvents, ultrasonic cleaning, and, finally, sterilization by autoclaving at 270°F for 20 min (Baier *et al.*, 1986). During surgical insertion, no other metals (stainless steel retractors, suction tips, or seating instruments) should be permitted to touch their surfaces. Baier and co-workers contend that such implants may achieve a more compatible surface condition with the use of radio-frequency glow discharge (RFGD), which sterilizes and creates a surface of maximum wettability. In 1993, Linkow began to use the Ohara technique for casting subperiosteal implants of titanium. They were light, strong, and highly compatible. The problem presented by the technique was that even with a specialized casting machine and method, the metal could not be successfully utilized for complete subperiosteal implants, therefore limiting its benefits.

The implant surgeon, the manufacturer, and the biomaterials scientist, in addition to their basic skills, must acquire a thorough understanding of metallurgy, biocompatibility, and the benefits of proper cleanliness and sterilization so that an appropriate host-site environment will be created to ensure the long-term success of implants.

QUESTIONS

1. What are the most popular metals in use today in dental implantology?
2. How well does bone tolerate invasive procedures? Is heat generated? Are there safe limits to induced temperatures? How can the health of host bone be assured?
3. What are some of the presurgical diagnostic techniques used to assist in planning?
4. Name some of the less frequently used dental implant modalities.

Bibliography

Adell, R. (1983). Clinical results of osseointegrated implants supporting fixed prosthesis in edentulous jaws. *J. Prosthetic. Dent.* 50: 251.

Adell, R., Lekholm, U., Rockler, B., and Branemark, P. I. (1981). A 15-year study of osseointegrated implants in the treatment of the edentulous jaw. *Int. J. Oral Surg.* 10: 387.

Albrektsson, T. (1983). Direct bone anchorage of dental implants. *J. Prosthetic Dent.* 50: 255–261.

Albrektsson, T., Dahl, E., and Endom, L. (1988). Osseointegrated oral implants. A Swedish multicenter study of 8,139 consecutively inserted Nobelpharma implants. *J. Periodontol.* **59**: 287.

Albrektsson, T., Jansson, T., and Lekholm, U. (1986). Osseointegrated dental implants. *Dent. Clin. North Am.* **30**: 151.

ASTM (2003). Annual Book of ASTM Standards. Vol 13.01, ASTM Press, W. Conshohocken, PA.

Baier, R. E., Natiella, J. R., Meyer, A. E., and Carter, J. M. (1986). Importance of implant surface preparation for biomaterials with different intrinsic properties in *D. van Series 29*. Excerpta Medica, Amsterdam, pp. 13–40.

Baumhammers, A. (1972). Custom modifications and specifications for blade vent implant designs to increase their biologic compatibility. *Oral Implantol.* **2**: 276.

Berman, N. (1951). An implant technique for full lower denture. *Dent. Digest* **57**: 438.

Bidez, M. W., Lemons, J. E., and Isenburg, B. P. (1986). Displacements of precious and nonprecious dental bridges utilizing endosseous implants as distal abutments. *J. Biomed. Mater. Res.* **20**: 785–797.

Branemark, P. I. (1983). Osseointegration and its experimental background. *J. Prosthet. Dent.* **50**: 399.

Branemark, P. I., Breine, U., Adell, R., Hansson, B. O., Lindstorm, J., and Ohlsson, A. (1969). Intra-osseous anchorage of dental prostheses. I. Experimental studies. *Scand. J. Plast. Reconstr. Surg.* **16**: 17.

Branemark, P. I., Hanssen, B. O., Adell, R., Brien, U., Lindstrom, J., Hallen, O., and Ohman, A. (1977). Osseointegrated implants in the treatment of the edentulous jaw. *Scand. J. Plast. Reconstr. Surg.* *Suppl.* **16**.

Brunski, J. B., Moccia, A. F., Pollack, S. R., Korostoff, E., and Trachtenberg, D. I. (1969). The influence of functional use of endosseous implants on the tissue-implant interface: histological aspects. *J. Dent. Res.* **58**: 1953.

Buser, D., Weber, H. P., and Lang, N. P. (1990). Tissue integration of non-submerged implants: 1-year results of a longitudinal study with ITI hollow-screw and hollow-cylinder implants. *Clin. Oral. Implantol. Res.* **1**: 33.

Cherchève, R. (1956). Nouveaux apercus sure le probleme des implants dentaires chez l'edente comple. Implants et tuteurs. *Rev. Fr. Odont. Stomatol.*, July.

Cranin, A. N. (1970). Some philosophic comments on the endosteal implant. *Dental Clinics of NA.* **14**: 173–175.

Cranin, A. N. (1991). Endosteal implants in patients with corticosteroid dependence. *J. Oral Impl.* **17**: 414–417.

Cranin, A. N., Klein, M., and Simons, A. (1991). *Atlas of Oral Implantology*, 1st Ed. Theime, Stuttgart, 194–208.

Cranin, A. N., and Sirakian, A. (1993). Preparing host sites for generic root form implants. *New York State Dent. J.* **59**: 12–17.

Cranin, A. N., Klein, M., Ley, J. P., Andrews, J., and DiGregorio, R. (1998). An in vitro comparison of the computerized tomography/CAD-CAM and direct bone impression techniques for subperiosteal implant model generation. *J. Oral Impl.* **24**: 74–79.

Cranin, A. N., Sirakian, A., Russell, D., and Klein, M. (1998). The role of incision design and location in the healing processes of alveolar ridges and host sites. *Intl. Jour. Oral Maxillofacial. Implants.* **13**: 483–491.

Cranin, A. N., Klein, M., and Simons, A. (1999). Atlas of Oral Implantology 2nd Ed. Mosby, St. Louis, 109–170.

Cranin, A. N. (1970). *Oral Implantology*. Thomas, Springfield, IL.

Cranin, A. N. (1980). The anchor endosteal implant. *Dent. Clin. North Am.* **24**: 505.

Cranin, A. N., Schnitman, P., and Rabkin, M. (1978). The Brookdale bar subperiosteal implant. *Trans. Soc. Biomater.* **2**: 331.

Cranin, A. N., Satler, N., and Shpuntoff, R. (1985). The unilateral pterygohamular subperiosteal implant: evolution of a technique. *J. Am. Dent. Assoc.* **110**: 496.

Cranin, A. N., Gelbman, J., and Dibling, J. (1987). Evolution of dental implants in the twentieth century. *Alpha Omegan* **80**: 24–31.

Dahl, G. S. (1958). Some aspects of the button technique. *J. Impl. Dent.* **5**: 49–53.

Dalise, D. (1979). The micro-ring for full subperiosteal implant and prosthesis construction. *J. Prosthet. Dent.* **42**: 211.

Doundoulakis, J. H. (1987). Surface analysis of titanium after sterilization: role in implant tissue interface and bioadhesion. *J. Prosthet. Dent.* **58**: 471.

Driskell, T. D. (1987). History of implants. *CDIA* **15**: 10.

Ducheyne, P., and Lemons, J. E. (1998). Bioceramics: Material characteristics versus *in vivo* behavior. *Ann. New York Acad. Sci.* **523**: 000–000.

English, C. (1988). Cylindrical implants. *J. Calif. Dent. Assoc.* **1**: 18.

Ericsson, R. A., and Albrektsson, T. (1983). Temperature threshold levels for heat-induced bone tissue injury. *Prosth. Dent.* **50**: 101–107.

Formaggini, M. (1947). Prostesi dentaria a mezzo di infibulazione directa endoalveolare. *Rev. Ital. Stomatol.*, March.

Gershkoff, A., and Goldberg, N. I. (1949). The implant lower denture. *Dent. Dig.* **55**: 490.

Guerini, V. A. (1969). *History of Dentistry*. Milford House, New York.

Hruska, A. R. (1987). Intraoral welding of pure titanium. *Quintessence. Intl.* **10**: 683–685.

James, R. A. (1983). Subperiosteal implant designs based on periimplant tissue behavior. *N. Y. J. Dent.* **53**: 407.

James, R. A. (1988). Connective tissue dental implant interface. *J. Oral Implantol.* **13**: 607.

James, R. A., and Keller, E. E. (1974). A histopathological report on the nature of the epithelium and underlying connective tissue which surrounds oral implants. *J. Biomed. Mater. Res.* **8**: 373–383.

Jermyn, A. C. (1955). Mucosal Inserts. *J. Impl. Dent.* **2**: 29–35.

Kay, J. M., Jarco, M., Logan, G., and Liu, S. T. (1986). The structure and properties of HA coatings on metal in *Trans. Soc. for Biomat.* **8**: 13.

Kirsch, A. (1986). Plasma-sprayed titanium IMZ implants. *J. Oral Implantol.* **12**: 494.

Laney, W. R., Tolman, D. E., Keller, E. E., Desjardins, R. P., Van Roekel, N. B., and Branemark, P. I. (1986). Dental implants: Tissue integrated prostheses utilizing the osseointegraion concept. *Mayo Clin. Proc.*, 362.

Lemons, J. E. (1988). Dental implant retrieval analyses. *J. Dent. Ed.* **52**: 748–757.

Lew, I. Greene, B. D., and Maresca, M. J. (1977). The Lew passive retainer for overlay and partial prostheses. *J. Oral Impl.* **7**: 124–137.

Linkow, L. I. (1968). Prefabricated endosseous implant prostheses. *Dent. Concepts* **11**: 3.

Linkow, L. I. (1986). Tripodal subperiosteal implants. *J. Oral Implantol.* **12**: 228.

Meffert, R. M. (1988). The soft tissue interface in dental implantology. *J. Dent. Ed.* **52**: 810–812.

Meffert, R. M., Langer, B., and Fritz, M. E. (1992). Dental implants: A review. *J. Periodental* **63**: 859–870.

Meffert, R. and Cranin, A. N., et al. (1999). *Atlas of Oral Implantology*, 2nd Ed. Mosby, St. Louis, 425–432.

Misch, C. E. (1999). *Contemporary Implant Dentistry*, 2nd Ed. Mosby, St. Louis, 94–106.

Niznick, G. A. (1982). The Core-vent implant system. *Oral Implantol.* **10**: 379–418.

Perel, S. (2000). The zygomatic extension implant. Lecture presented to the NYU MaxiCourse®.

Rizzo, A. A. (ed.) (1988). Proceedings of the consesnsus development conference on dental implants. *J. Dent. Ed.* 52: 678–827.

Roberts, H. D., and Roberts, R.A. (1970). The ramus endosseous implant. *J. S. Calif. Dent. Assoc.* 38: 571.

Schnitman, P. A. (1987). Diagnosis, treatment planning, and the sequencing of treatment for implant reconstructive procedures. *Alpha Omegan* 80: 32.

Small, I. A. (1980). Benefit and risk of mandibular staple boneplates. In Dental Implants: Benefit and Risk, PHS Pub. 81-1531, 139–152, U.S. Public Health Service, Washington, D.C.

Strock, A. E. (1939). Experimental work on direct implantation in the alveolus. *Am. J. Orthol. Oral Surg.* 25: 5.

Strock, A. E., and Strock, M. S. (1949). Further studies on inert metal implantation for dental replacement. *Alpha Omegan* 25: 467–472.

Tarnow, D. P. Cho, S-C., and Wallace, S. (2000). Th effect of the interdental distance on the height of the interdental bone crest. *J. Periodontology* 71: 546–549.

Truitt, H. P., James, R. A., and Lozado, J. (1986). Noninvasive technique for mandibular subperiosteal implants: a preliminary report. *J. Prosthet. Dent.* 55: 494.

Weinberg, D. D. (1950). Subperiosteal implantation of Bitallium (cobalt–chromium alloy), artificial abutment. *J. Am. Dent. Assoc.* 40: 549.

Weiss, C. M. (1986). Tissue integration of dental endosseous implants: Description and comparative analysis of fibro-osseous integration and osseous integration systems. *J. Oral. Implantol.* 12: 169.

Williams, D. F. (1977). Titanium as a metal for implantation. Part I. Physical properties. *J. Med. Eng. Technol.* 6: 195–198, 202–203.

Worthington, P. (1988). Current implant usage. in *Proc. NIH Consensus Development Conference on Dental Implants*, June 13–15. NIDR, Bathesda, MD.

Yurkstas, A. A. (1967). The current status of implant prosthodontics. *Newslett. AAID* 16:

Zarb, G. A. (1983). Introduction to osseointegration in clinical dentistry. *J. Prosthet. Dent.* 49: 824.

Zarb, G. A., and Symington, J. M. (1983). Osseointegrated dental implants: Preliminary report on a replication study. *J. Prosthet. Dent.* 50: 271.

7.9 ADHESIVES AND SEALANTS

Dennis C. Smith

According to a definition of the American Society for Testing and Materials, an adhesive is a substance capable of holding materials together by surface attachment. Inherent in the concept of adhesion is the fact that a bond that resists separation is formed between the substrates or surfaces (adherends) comprising the joint and work is required to separate them.

"Adhesive" is a general term that covers designations such as cement, glue, paste, fixative, and bonding agent used in various areas of adhesive technology. Adhesive systems may comprise one- or two-part organic and/or inorganic formulations that set or harden by several mechanisms.

Commercial adhesive systems are often designed to result in only a thin layer of adhesive for efficient bonding of the two surfaces since thick layers may contain weakening defects such as air voids or contaminants. Such systems may be low-viscosity liquids. In other situations where, for example, the surfaces to be joined are irregular, gap-filling qualities are required of the bonding agent. These systems may be solid–liquid (filled) adhesives or viscous liquids and are usually referred to as cements, glues, or sealants. Thus, the term "sealant" implies not only that good bonding and gap-filling characteristics are present in the material, but also that the bonded joint is impervious, for example, to penetration by water or air. Since most adhesives, including biomaterials, are used to joint dissimilar materials that are subjected to a variety of physical, mechanical, and chemical stresses, good resistance to environmental degradative processes, including biodegradation, is essential.

The applications of adhesive biomaterials range from soft (connective) tissue adhesives, used both externally to temporarily fix adjunct devices such as colostomy bags and internally for wound closure and sealing, to hard (calcified) tissue adhesives used to bond prosthetic materials to teeth and bone on a more permanent basis. All of these biological environments are hostile, and a major problem in the formulation of medical and dental adhesives is to develop a material that will be easy to manipulate, interact intimately with the tissue to form a strong bond, and also be biocompatible. Over the past two decades, more success at a clinical level has been achieved in bonding to hard tissues than to soft tissues.

The opposing concept to adhesion is *abhesion*, i.e., the deliberate prevention of bonding between surfaces. Such situations may arise in prevention of staining or fouling of surfaces or of adhesions between parenchymatous and surrounding tissue after surgery.

More details on the background to adhesion and adhesives can be found in recent texts (Mittal and Pizzi, 1994; Comyn, 1997; Pocius, 1997; Mittal and Pizzi, 1999). General information on wound closure and surgical adhesives is given in Sierra and Salz (1996) and Chu *et al.* (1997). This chapter does not cover mucosal adhesives such as Polycarbophil or Carbopol.

HISTORICAL OVERVIEW

Wound closure by means of sutures extends back many centuries. The idea of using an adhesive is more recent but dates back to at least 1787 when it was noted "that many workmen glue their wounds with solid glue dissolved in water" (Haring, 1972). Hide glue is similar to gelatin, which itself derives from collagen. Other biological adhesives, such as blood and egg white, have also been known for centuries; however, first attempts to develop adhesives with specific chemical structures began in the late 1940s and 1950s.

Cyanoacrylates

Natural materials such as cross-linked gelatin and thrombin–plasma were investigated, but a major stimulus was provided by the discovery in 1951 of methyl 2-cyanoacrylate by Cooper *et al.* (1972). This clear liquid monomer and its higher homologs (ethyl, butyl, octyl, etc.) were found to polymerize rapidly in the presence of moisture or blood, giving rapid hemostasis and highly adherent films. Extensive clinical and laboratory investigations on the cyanoacrylates took place in the 1960s and 1970s (Matsumoto, 1972), but problems of manipulation and biocompatibility, including reports of cancer in laboratory animals, have limited their current use to surface applications on oral mucosa and life-threatening arteriovenous situations.

Protein Glues

The discovery of the adhesive properties of the cyanoacrylates prompted numerous studies on synthetic adhesive systems designed to interact with tissue protein side-chain groups to achieve chemical bonding (Cooper *et al.*, 1972). Few systems were found to possess the requisite combination of biocompatibility, ease of manipulation, and effectiveness. As a result of this experience and the more strictly controlled regulatory situation of today, only limited new research is being done on novel tissue adhesives. Work has been reported on synthetic polymerizable systems containing the reactive cyanoacrylate or isocyanate groups (Ikada, 1997; Flagle *et al.*, 1999), but attention has been more focused clinically on materials based on natural origins. Some studies still continue on the gelatin–resorcinol–formaldehyde (GRF) combination (Ikada, 1997) but the main emphasis has been on fibrin glues derived from a fibrinogen-thrombin combination. Other protein- and peptide-based materials have been investigated including fibrin–collagen (Prior *et al.*, 1999), cross-linked albumin (Feldman *et al.*, 1999), and derivatized collagen (DeVore, 1999).

Hydrogels

Light-cured poly(ethylene glycol)-based monomers have been developed to form a series of aqueous hydrogels. These materials can also be used as drug delivery matrices (Coury *et al.*, 1999).

Tooth and Bone Cements

As with soft tissues, interest in adhesive bonding to skeletal or hard tissues as a replacement for, or supplementation to, gross mechanical fixation such as screws has developed mainly in this century and particularly in the past 40 years. Fixation of orthopedic joint components by a cement dates back at least to Gluck (1891) and retention of metal or ceramic inlays and crowns on teeth by dental cements to about 1880. The development of acrylic room-temperature polymerizing (cold-curing) systems for dental filling applications in the 1950s led to their use as dental cements and later to their application for fixation of hip joint components by Charnley and Smith (Charnley, 1960, 1970; Smith, 1971). These situations involved bonding by mechanical interlocking into surface irregularities.

Tooth Adhesives

In the case of tooth restorations, bonding of similar acrylic filling systems using mechanical interlocking resulted in leakage along the bonded interface. This so-called microleakage initiated an intensive effort over the past 40 years to develop adhesive dental cements and filling (restorative) materials (Phillips and Ryge, 1961; Smith, 1991, 1998). This led to the development of poly(acrylic acid)-based adhesive cements, first the zinc polycarboxylates (Smith, 1967) and then, in 1972, the glass ionomer cements (Smith, 1998), and, to resin bonding systems.

Bonding materials and techniques are now a major component of clinical dentistry (Degrange and Roulet, 2000). Effective clinical bonding of polymerizable fluid dimethacrylate monomers and composite formulations to dental enamel, the most highly calcified (98%) tissue in the body, has been achieved by using phosphoric acid etching of the surface (the "acid-etch" technique). Bonding to tooth dentin is currently achieved by using acidic primer monomeric systems containing functional groups such as polycarboxylate or polyphosphate and hydrophilic monomers such as hydroxyethyl methacrylate (Roulet and Degrange, 2000). Similar materials have been investigated for adhesion to bone that is compositionally similar to dentin (Lee and Brauer, 1989).

Table 1 gives a summary of the compositions of the various materials in current clinical use.

BACKGROUND CONCEPTS

As indicated previously, significant advances in adhesive biomaterials have occurred over the 30–40 years as real progress has taken place in the science and technology of adhesion and adhesives. This development is continuing since the fundamental aspects of the formation of adhesive bonds at interfaces are not yet fully understood even though successful application of adhesives in technically demanding situations has been achieved.

Experience and, to some extent, theory have shown that severe hostile environments such as biological milieu may require specific surface pretreatments for the surfaces being joined in addition to selection of an adhesive with appropriate characteristics (Comyn, 1997). Such surface pretreatments may involve cleaning or etching processes designed to remove contaminants and expose wettable surfaces, and may require the application of primers to achieve specific chemical reactivity at the surface (Mittal and Pizzi, 1999). These procedures are a reflection of the need for intimate interfacial contact between the bonding agent and the adherends in order to form adhesive bonds across the interface. These adhesive forces must hold the materials together throughout the required service life of the joint. However, it must also be appreciated that the factors of the design of an adhesive joint, the applied loads, and the service environment it must withstand will all affect its mechanical performance and life expectancy (Kinloch, 1987; Comyn, 1997; Mittal and Pizzi, 1999).

The establishment of intimate molecular contact between the adhesive and adherend requires, ideally, the adhesive and/or primer to (1) exhibit a zero or near-zero contact angle when liquid, (2) have a low viscosity during bonding, and (3) be able to displace air and contaminants during application. As discussed elsewhere in this volume (Chapters 1.4 and 9.4), surface wetting to achieve these requirements involves an understanding of wetting kinetics and equilibria on clean, high-energy surfaces, the kinetics of spreading of the adhesive including the effects of microroughness, and the minimization of surface contaminants, including moisture, during the bonding process.

TABLE 1   Classification and Composition of Tissue Adhesives

Type	Components	Setting mechanism
Soft–tissue adhesives		
Cyanoacrylate	Butyl or isobutyl cyanoacrylate	Addition polymerization
Fibrin sealants	A. Fibrinogen,	Fibrin clot formation
Factor XIII	B. Thrombin, $CaCl_2$	
GRF glue.	Gelatin, resorcinol, formaldehyde	Condensation
		Polymerization
Hydrogel	Block copolymers of PEG, poly(lactic acid) and acrylate esters	Photoinitiated addition polymerization
Hard–tissue adhesives		
Bone		
Acrylic bone cement	Methyl methacrylate and poly (methyl methacrylate)	Pertoxide–amine initiated addition polymerization
Teeth		
Dental cements	Zinc oxide powder, phosphoric acid liquid	Acid–base reaction,
Zinc phosphate		Zn complexation
Zinc polycarboxylate	Zinc oxide powder, aqueous poly (acrylic acid)	Acid–base reaction,
		Zn complexation
Glass ionomer (polyalkenoate)	Ca, Sr, Al silicate glass powder aqueous poly (acrylic-itatomic acid) or (acrylic-maleic acid)	Acid–base reaction, metal ion complexation
Resin-based	Aromatic or urethane dimethacrylate monomers, silicate or other glass fillers aqueous poly(acrylic acid–itaconic acid–methacrylate) comonomers	Peroxide–amine or photoinitated polymerization / Photoinitiated addition polymerization
Resin-modified glass ionomer	Hydroxyethyl methacrylate aromatic or urethane diamethacrylates, Ca, Sr, Al glass powder	
Dentin adhesive	Etchant: phosphoric acid primer: carboxylate or phosphate	Photoinitiated addition polymerization
	Monomers hydroxyethyl methacrylate/water/solvent	
	Bonding agent: urethane or aromatic dimethacrylate monomers	

Four main mechanisms of adhesion at the molecular level have been proposed (Comyn, 1997; Mittal and Pizzi, 1999): (1) mechanical interlocking, (2) adsorption (including chemical bonding), (3) diffusion theory, and (4) electronic theory. More complex interpretations have been proposed (Schulz and Nardin, 1999) but the validity of each theory is influenced by the system under consideration.

Mechanical Interlocking

This adhesion involves the penetration of the bonding agent into surface irregularities or porosity in the substrate surface. Gross examples of this mechanism include the retention of (a) dental filling materials in mechanically prepared tooth cavities, (b) crowns by dental cements on teeth (Fig. 1), and (c) the fixation of artificial joint components by acrylic bone cement (Fig. 2). Even apparently smooth surfaces are pitted and rough at the microscopic level, and strong bonding can arise with an adhesive that can penetrate at this level. The use of primers (chemical pretreatments) can create surface irregularities or porosities at the microscopic level, or can form porous layers that similarly provide effective micromechanical interlocking. Examples include the etching of dental enamel by 35–40% phosphoric acid (Fig. 3) and primer treatment of tooth dentin with acidic agents (Fig. 4). In each case the unpolymerized bonding agent penetrates 5–50 μm into the surface, creating numerous resin "tags" that provide a strong bond. A further

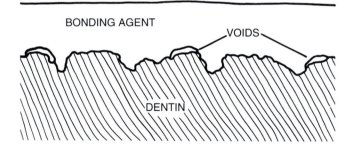

FIG. 1. Diagrammatic representation of mechanical interlocking by a cement to tooth dentin. Note voids at interface due to imperfect adaption.

effect of surface roughness may be an increase in the interfacial energy dissipated during joint separation.

Adsorption Theory

This theory postulates that if intimate interfacial molecular contact is achieved, interatomic and intermolecular forces will establish a strong joint. Such forces include van der Waals and hydrogen bonds, donor–acceptor bonds involving acid–base interactions, and primary bond (ionic, covalent, metallic) formation (chemisorption). Numerous studies have suggested that

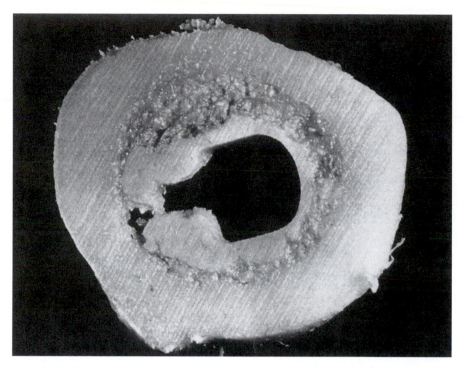

FIG. 2. Section through femur after removal of stem of hip prosthesis showing mechanical interlocking by bone cement into cancellous bone. (After J. Charnley, personal communication.)

FIG. 3. Dental enamel etched by 30-sec treatment with 35% phosphoric acid showing prismatic structure. Prisms are about 5 μm in diameter.

FIG. 4. Treatment of dentin surface by acidic primer showing demineralised collagen fibers in surface zone.

secondary bonds (van der Waals and hydrogen bonds) alone are sufficient to establish strong bonding. However, where environmental attack is severe (e.g., by water as in biological systems), the formation of primary bonds across the interface eems to be essential (Kinloch, 1987).

Evidence of primary bond formation has been found for commercial adhesives, particularly as a result of the use of chemical primers (e.g., silane coupling agents) on ceramics (Kinloch, 1987; Schulz and Nardin, 1999). In the biomedical field, the poly(acrylic acid)-based dental cements (zinc polycarboxylate and glass ionomer cements) (Fig. 5) have been shown to undergo carboxylate bonding with Ca in enamel and dentin (Smith, 1991, 1998). Silane primers are also used in the bond formation between dental resin adhesives and dental ceramics. Formation of covalent bonds by reaction of acid chloride, aldehyde, or other reactive groups with hydroxyl or

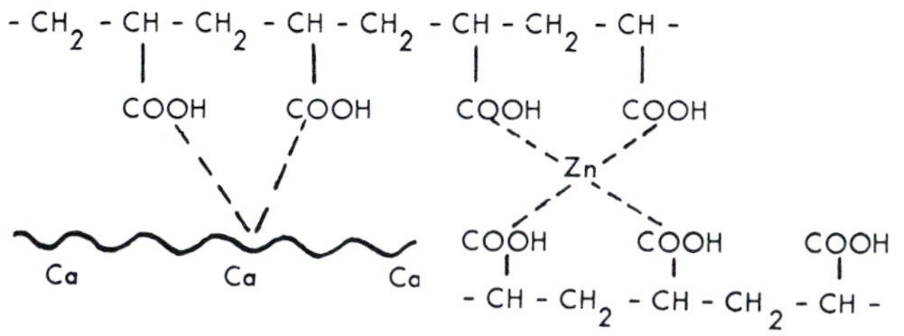

amino groups in dentin collagen and/or complexation reactions between dentin Ca and phosphate or carboxyl groups have also been postulated in several reactive polymerizable dentin bonding systems, but no unequivocal evidence for this has yet been presented (Eliades, 1993; Smith, 1998).

Diffusion Theory

This theory states that the intrinsic adhesion of polymers to substrates and each other involves mutual interdiffusion of polymer molecules or segments across the interface. This can occur only when sufficient chain mobility is present. The application of this theory is limited to specific situations. Diffusion of polymers into intimate contact with metallic or ceramic surfaces may in fact result in enhanced adsorption or even micromechanical interlocking as a source of improved bonding. This concept (and others) have led to the idea of an "interphase" that is formed between adhesive and substrate and influences bonding behavior (see below).

Electronic Theory

Electronic theory postulates that electronic transfer between adhesive and adherend may lead to electrostatic forces that result in high intrinsic adhesion. Such interactions may arise in certain specialized situations, but for typical adhesive–substrate interfaces, any electrical double layer generated does not contribute significantly to the observed adhesion (Kinloch, 1987; Schulz and Nardin, 1999).

Weak Boundary Layer

A theory that has an impact on the foregoing adhesion mechanisms is the concept of weak boundary layers at the surface of materials to be bonded, either intrinsic or formed by contamination of the adherends. Such layers may result in the formation of an interphase between adhesive and substrate that may exhibit reduced or altered bonding behavior. Such behavior has led to extensive investigation of cleaning procedures and adhesion primers to optimize the bond (Mittal and Pizzi, 1999).

The evidence presently available suggests that for most biological adhesives, adsorption phenomena or micromechanical interlocking account for the bond formation and behavior observed. Since few practical surfaces, especially tissues, are completely smooth and nonporous, it is likely that both mechanisms exist in practical clinical situations, with one or the other predominating according to the type of adhesive system, surface preparation technique, and bonding environment.

COMPOSITION AND CHARACTERISTICS OF ADHESIVE BIOMATERIALS

Soft-Tissue Adhesives

Most soft-tissue adhesives are intended to be temporary. That is, they are removed or degrade when wound healing is sufficiently advanced for the tissue to maintain its integrity.

FIG. 5. Diagramatic representation of setting of zinc polyacrylate and bonding to calcific surface.

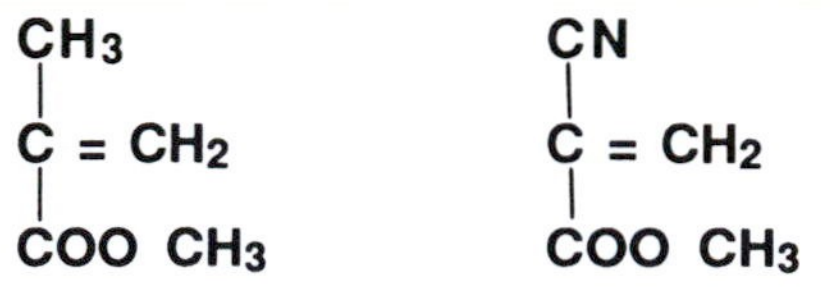

CH₃   CN
|
C = CH₂   C = CH₂
|
COO CH₃   COO CH₃

METHYL   **METHYL**
METHACRYLATE   **CYANOACRYLATE**

FIG. 6. Structures of methyl cyanoacrylate and methyl methacrylate.

Effective adhesion can be obtained on dry skin or wound surfaces by using wound dressing strips with acrylate-based adhesives. However, on wound surfaces that are wet with tissue fluid or blood, the adhesive must be able to be spread easily on such a surface, provide adequate working time, develop and maintain adhesion, desirably provide hemostasis, facilitate wound healing, and maintain biocompatibility. Positive antimicrobial action would be an additional advantage (Ikada, 1997).

Few, if any, systems comply with all these requirements. Currently, there are two principal systems in widespread clinical use—cyanoacrylate esters and fibrin tissue adhesives. Another glue based on a gelatine–resorcinol–formaldehyde combination still receives limited use. An interesting but still experimental system based on polypeptides from marine organisms (mussel adhesive) does not seem to have developed into practical use. These materials have been reviewed (Sierra and Salz, 1996; Chu *et al.*, 1997). More recently a family of synthetic implantable resorbable hydrogel has been developed (A. Coury, 1999, personal communication) that show promise as sealants for the lung and other sites.

Cyanoacrylate Esters

These esters are fluid, water-white monomers that polymerize rapidly by an anionic mechanism in the presence of weak bases such as water or NH₂ groups. Initially, methyl cyanoacrylate (Fig. 6) was used but in the past two decades isobutyl and *n*-butyl cyanoacrylate have been found more acceptable. The higher cyanoacrylates spread more rapidly on wound surfaces and polymerize more rapidly in the presence of blood. Furthermore, they degrade more slowly over several weeks in contrast to the methyl ester, which hydrolyzes rapidly yielding formaldehyde that results in an acute inflammatory response.

These materials achieve rapid hemostasis as well as a strong bond to tissue. However, the polymer film is somewhat brittle and can be dislodged on mobile tissue and the materials can be difficult to apply on large wounds. Because of adverse tissue response and production of tumors in laboratory animals, cyanoacrylates are not approved for routine clinical use in the United States, although a commercial material based on *n*-butyl cyanoacrylate is approved by several other countries.

The current uses are as a surface wound dressing in dental surgery, especially in periodontics, and in life-threatening applications such as brain arteriovenous malformations. Reports of sarcomas in laboratory animals (Reiter, 1987), late complications after dura surgery (Chilla, 1987), evidence of

in vitro cytotoxicity (Ciapetti *et al.*, 1994), and lack of regulatory approval have restricted their further use in spite of work on synthesis of new types of cyanoacrylate.

Fibrin Sealants

Fibrin sealants involve the production of a synthetic fibrin clot as an adhesive and wound-covering agent. The concept of using fibrin dates back to 1909 but was placed on a specific basis by Matras *et al.* in 1972. The commercial materials first available (Tisseel, Tissucol, Fibrin-Kleber) consisted of two solutions that are mixed immediately before application to provide a controlled fibrin deposition. Later a "ready-to-use" formulation (Tisseel Duo) was introduced (Schlag and Redl, 1987).

The essential components of these solutions are as follows:

Solution A:	Solution B:
Fibrinogen	Thrombin
Factor XIII	CaCl₂

The fibrinogen is at a much higher concentration (~70 mg/ml) than that in human plasma. On mixing the two solutions using a device such as a twin syringe with a mixing nozzle a reaction similar to that of the final stages of blood clotting occurs in that polymerization of the fibrinogen to fibrin monomers and a white fibrin clot is initiated under the action of thrombin and CaCl₂. Aprotinin, an inhibitor of fibrinolysis, may also be included in solution A. The composition may be adjusted to promote hemostasis, for example, or to minimize persistence of the clot to avoid fibrosis.

There are now several commercial products available in Europe, Japan, and Canada; none is presently approved for use in the United States (MacPhee, 1996). Fibrinogen for these commercial materials is manufactured from the pooled plasma of selected donors using processes such as cryoprecipitation. The material is subjected to in-process virus activation and routinely screened for hepatitis and HIV. To minimize these risks recent processes produce the fibrinogen in a "closed" (single-donor) blood bank or utilize the patient's own blood. This autologous fibrin glue is preferred but its quality is partly determined by the fibrinogen level in the donor plasma (Ikada, 1997). Other complications include formation of antibodies and thrombin inhibitors as well as potential risks of BSE (bovine spongiform encephalopathy) if bovine thrombin is used. More recently, human-derived thrombin has been employed.

Fibrin sealant has four main advantages: (1) it is hemostatic, (2) it adheres to connective tissue, (3) it promotes wound healing, and (4) it is biodegradable with excellent tissue tolerance (Sierra and Salz, 1996; Ikada, 1997; Scardino *et al.*, 1999). The adhesive strength is not as high as that of cyanoacrylates but is adequate for many clinical situations. Thorough mixing of the ingredients and application techniques or devices that allow uniform spreading are essential to success.

The material has been used in a wide variety of surgical techniques for hemostasis and sealing involving thoracic–cardiovascular, neurologic, plastic, and ophthalmic surgery and as a biodegradable adhesive scaffold for meshed skin grafts in burn patients (Sierra and Salz, 1996; Ikada, 1997; Feldman *et al.*, 1999). A symposium has reviewed current clinical uses (Spotnitz, 1998). An auxiliary aspect of fibrin sealants is their use as a delivery vehicle at local sites for antibiotics and growth factors (Sierra and Salz, 1996).

Attempts to modify fibrin sealant have been directed toward (a) improvements in ease of application and control of setting by use of a one-component light-activated product (Scardino *et al.*, 1999), (b) improvements in strength and performance by addition of fibrillar collagen (Sierra and Salz, 1996), and (c) development of a formulation containing gelatin (Ikada, 1997).

Gelatin–Resorcinol–Formaldehyde Glue

This glue was developed in the 1960s by Falb and co-workers (Falb and Cooper, 1966; Cooper *et al.*, 1972) as a less toxic material than methyl cyanoacrylate. The material is fabricated by warming a 3:1 mixture of gelatin and resorcinal and adding an 18% formaldehyde solution. Cross-linking of the gelatin and resorcinol by the formaldehyde takes place in about 30 seconds.

The material was used in a variety of soft-tissue applications but technical problems and toxicity have limited its application in recent years to aortic dissection (Ikada, 1997). In attempts to overcome the toxicity and potential mutagenecity/carcinogenicity of the formaldehyde component, modified formulations have been developed in which other aldehydes such as glutaraldehyde and glyoxal (Ennker *et al.*, 1994a, b) are substituted for the formaldehyde. Favorable results with this material (GR-DIAL) have been reported (Ennker *et al.*, 1994a, b). Concerns with toxicity remain, however, and this material has not received FDA approval for commercial use. Less toxic gelatin cross-linking agents have been investigated *in vitro* without substantial improvement.

The three foregoing types of adhesives have received extensive clinical trial in Europe and Japan. Fibrin sealant is the most widely used material currently. However, all of these systems have significant deficiencies. This is illustrated by their relative characteristics listed in Table 2 (Ikada, 1997).

Bioadhesives

Bioadhesives are involved in cell-to-cell adhesion, adhesion between living and nonliving parts of an organism, and adhesion between an organism and foreign surfaces. Adhesives produced by marine organisms such as the barnacle and the mussel have been extensively investigated over the past 30 years because of their apparent stable adhesion to a variety of surfaces under adverse aqueous conditions. These studies have shown that these organisms secrete a liquid acidic protein adhesive that is cross-linked by a simultaneously secreted enzyme system. The bonding probably involves hydrogen bonding and ionic bonding from the acidic groups (Waite, 1989).

The adhesive from the mussel has been identified as a polyphenolic protein, molecular weight about 130,000, which is cross-linked by a catechol oxidase system in about 3 minutes. A limiting factor in the practical use of this material is the difficulty of extraction from the natural source. The basic unit of the polyphenolic protein has been identified as a specific decapeptide (Waite, 1989; Green, 1996). Recombinant DNA technology and peptide synthesis have been used in attempts to produce an affordable adhesive with superior properties. Little information has been reported on the performance (including biocompatibility) of these materials. Green (1996) investigated their potential in ocular sites.

Hydrogel Sealants

A new approach is the development of synthetic sealants based on poly(ethylene glycol) (PEG) hydrogels that are derived from original work of Sawhney, Pathak, and Hubbell (1993). A family of fully synthetic implantable, resorbable hydrogels intended for use as surgical sealants has been developed (Ranger *et al.*, 1997; Alleyne *et al.*, 1998; Tanaka *et al.*, 1999; Moody *et al.*, 1996), for barrier coatings (West *et al.*, 1996) and for drug delivery matrices (Lovich *et al.*, 1998). The hydrogels (Focal Inc.) are formed by *in situ* deposition of aqueous formulations based on specialized macromers followed by photopolymerization to highly cross-linked structures.

The macromers are reactive block copolymers consisting of a water-soluble core such as polyoxyethylene flanking biodegradable oligomers such as poly(lactic acid) or poly(trimethylene carbonate) and polymerizable end caps such as acrylate esters (Sawhney *et al.*, 1993). Control of the physical properties and degradation rates is achieved by specifying the molecular structures and concentration of the reactants in the formulation (Goodrich *et al.*, 1997). Photopolymerization can be effected with ultraviolet or visible light using appropriate photoinitiators. Typically, the initiator system eosin Y–triethanolamine is employed with visible illumination in the 450–550 nm range.

In most applications for these hydrogels, strong bonding to tissue is required. This is achieved by use of a two-part sealant system consisting of primer and topcoat. Strong, durable bonding to a wide variety of internal tissues has been demonstrated (Coury *et al.*, 1999).

Published reports to date have shown these hydrogels to be effective as sealants for the lung (Ranger *et al.*, 1997), dura mater (Alleyne *et al.*, 1998), and blood vessels (Tanaka *et al.*, 1999; Moody *et al.*, 1996) and as local drug-delivery depots (Lovich *et al.*, 1998). Commercialization of the formulations

TABLE 2 Characteristics of Currently Available Tissue Adhesives (after Ikada, 1997)

	Fibrin glue	Cyanoacrylate	GRF
Handling	Excellent[a]	Poor[b]	Poor
Set time	Medium	Short	Medium
Tissue bonding	Poor	Good	Excellent
Pliability	Excellent	Poor	Poor
Toxicity	Low[c]	Medium[d]	High
Resorbability	Good	Poor	Poor
Cell infiltration	Excellent	Poor	Poor

[a] Spray type.
[b] Low viscous type.
[c] Not autologous.
[d] Long alkyl chains.

and associated delivery devices for sealant applications has commenced.

Hard-Tissue Adhesives

Bone Cement

As previously noted, attachment of prostheses to calcified tissues (bone, tooth enamel, dentin) can be achieved by gross mechanical interlocking to machined surfaces. Thus, room temperature–polymerizing methyl methacrylate (Fig. 7) systems are used for fixation of orthopedic implants (e.g., acrylic bone cement; see Chapter 7.7) and for dental restorations. The former, in a closed system, have been relatively successful. However, conditions are much more stringent in the mouth because of the changing environment, thermomechanical stresses on the bond, and the presence of oral bacteria that result in marginal leakage at the tooth–filling interface and renewed tooth decay. Thus, considerable development of new dental cements and adhesive systems has occurred in recent years in the attempts to provide a leak-proof bond to attach fillings, crowns, and veneers to the tooth (Fig. 8).

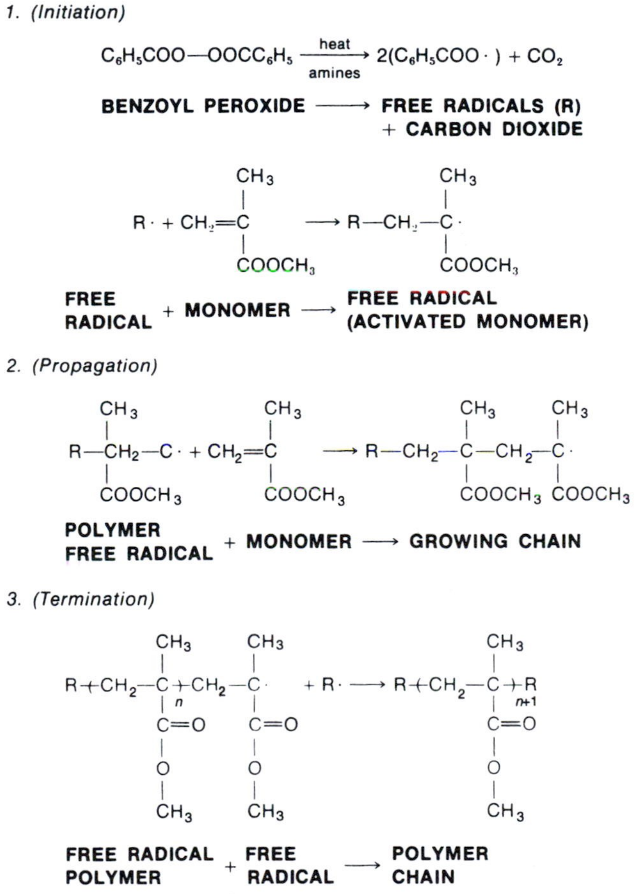

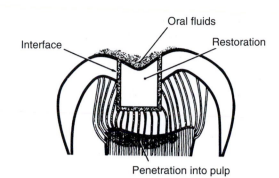

FIG. 8. Leakage of oral fluids and bacteria around dental filling material in tooth crown. (From R. W. Phillips, *Science of Dental Materials*, 9th ed., p. 62. W. B. Saunders, Philadelphia, with permission.)

Dental Cements

Dental cements are, traditionally, fast-setting pastes obtained by mixing solid and liquid components. Most of these materials set by an acid–base reaction and more recent resin cements harden by polymerization (Smith, 1988, 1991, 1998). The compositions of these materials are given in Table 1. The classes and typical mechanical properties are given in Table 3.

Zinc phosphate cement is the traditional standard. This material is composed primarily of zinc oxide powder and a 50% phosphoric acid solution containing Al and Zn. The mixed material sets to a hard, rigid cement (Table 1) by formation of an amorphous zinc phosphate binder. Although the cement is gradually soluble in oral fluids and can cause pulpal irritation, it is clinically effective over 10 to 20 year periods. The bonding arises entirely from penetration into mechanically produced irregularities on the surface of the prepared tooth and the fabricated restorative material. Some interfacial leakage occurs because of cement porosity and imperfect adaptation (Fig. 1), but this is usually acceptable since the film thickness is generally below 100 μm.

Poly(carboxylic acid) cements were developed in 1968 (Smith, 1967, 1998) to provide materials with properties comparable to those of phosphate cements but with adhesive properties to calcified tissues. Zinc polyacrylate (polycarboxylate) cements are formed from zinc oxide and a polyacrylic acid solution. The metal ion cross-links the polymer structure via carboxyl groups, and other carboxyl groups complex to Ca ions in the surface of the tissue (Fig. 5). The zinc polycarboxylate cements have adequate physical properties, excellent biocompatibility in the tooth, and proven adhesion to enamel and dentin (Smith, 1991) but are opaque. The need for a translucent material led to the development of the glass-ionomer cements (Smith, 1998).

The glass ionomer cements are also based on polyacrylic acid or its copolymers with itaconic or maleic acids, but utilize a calcium aluminosilicate glass powder instead of zinc oxide (Smith, 1998). In this case, the cements set by cross-linking of the polyacid with Ca and Al ions from the glass together with formations of a silicate gel structure. The set structure and the residual glass particles yield a stronger, more rigid cement (Table 1) but with similar adhesive properties to the zinc polyacrylate cements. Both cements are widely used clinically.

1. (Initiation)

$$C_6H_5COO\text{—}OOCC_6H_5 \xrightarrow[\text{amines}]{\text{heat}} 2(C_6H_5COO\cdot) + CO_2$$

BENZOYL PEROXIDE $\longrightarrow$ **FREE RADICALS (R) + CARBON DIOXIDE**

$$R\cdot + CH_2=\underset{\underset{COOCH_3}{|}}{\overset{\overset{CH_3}{|}}{C}} \longrightarrow R\text{—}CH_2\text{—}\underset{\underset{COOCH_3}{|}}{\overset{\overset{CH_3}{|}}{C}}\cdot$$

FREE RADICAL + **MONOMER** $\longrightarrow$ **FREE RADICAL (ACTIVATED MONOMER)**

2. (Propagation)

$$R\text{—}CH_2\text{—}\underset{\underset{COOCH_3}{|}}{\overset{\overset{CH_3}{|}}{C}}\cdot + CH_2=\underset{\underset{COOCH_3}{|}}{\overset{\overset{CH_3}{|}}{C}} \longrightarrow R\text{—}CH_2\text{—}\underset{\underset{COOCH_3}{|}}{\overset{\overset{CH_3}{|}}{C}}\text{—}CH_2\text{—}\underset{\underset{COOCH_3}{|}}{\overset{\overset{CH_3}{|}}{C}}\cdot$$

POLYMER FREE RADICAL + **MONOMER** $\longrightarrow$ **GROWING CHAIN**

3. (Termination)

$$R\text{—}(CH_2\text{—}\underset{\underset{\underset{\underset{CH_3}{|}}{O}}{\overset{\overset{C=O}{|}}{C}}\text{—})_n CH_2\text{—}\underset{\underset{\underset{\underset{CH_3}{|}}{O}}{\overset{\overset{C=O}{|}}{C}}\cdot + R\cdot \longrightarrow R\text{—}(CH_2\text{—}\underset{\underset{\underset{\underset{CH_3}{|}}{O}}{\overset{\overset{C=O}{|}}{C}}\text{—})_{n+1} R$$

FREE RADICAL POLYMER + **FREE RADICAL** $\longrightarrow$ **POLYMER CHAIN**

FIG. 7. Autopolymerizing methyl methacrylate systems as used in dental resins and acrylic bone cement. (From R. Roydhouse, in *Dental Materials: Properties and Selection*, W. J. O'Brien, ed. Quintessence Books, Chicago, with permission.)

TABLE 3   Properties of Dental Cements and Sealants

Material	Strength (MPa)			
	Compressive (MPa)	Tensile (MPa)	Modulus of elasticity (GPa)	Fracture toughness $K_1C\ MN^{-3/2}$
Zinc phosphate	80–100	5–7	13	~0.2
Zinc polycarboxylate	55–85	8–12	5–6	0.4–0.5
Glass ionomer	70–200	6–7	7–8	0.3–0.4
Resin sealant unfilled	90–100	20–25	2	0.3–0.4
Resin sealant filled	150	30	5	
Resin cement	100–20	30–40	4–6	
Composite resin filling material	350–40	45–70	15–20	1.6

In recent materials, the polyacid molecule contains both ionic carboxylate and polymerizable methacrylate groups and is induced to set, by both an acid–base reaction and visible light polymerization. These dual-cure cements are widely used clinically. Adhesive bonding but not complete sealing is obtained because of imperfect adaptation to the bonded surfaces under practical conditions. A further development involves glass-filled photopolymerized carboxylated methacrylate polymers (resin-modified materials) that are closer to composite resin materials (Smith, 1998).

Resin cements are fluid or paste-like monomer systems based on aromatic or urethane dimethacrylates (Fig. 9). Silanated ceramic fillers may be present to yield a composite composition. The two-component materials polymerize on mixing through a two-part organic peroxide–tertiary amine initiator–activator system in about 3 minutes. More recent are one-component materials containing diketone polymerization initiators that achieve polymerization in about 30 seconds by exposure to visible (blue) light energy. Dual-cure systems are also available. These set materials are strong, hard, rigid, insoluble, cross-linked polymers (Table 1). Bonding is achieved by mechanical interlocking to surface roughness. In recent materials, reactive adhesive monomers may also be present (see below), also conferring a presumed chemisorption mechanism.

Enamel and Dentin Adhesives

Enamel and dentin bonding systems are similar polymer–ceramic composite formulations to resin cements but are more complex systems containing reactive monomers. Their use usually involves first an acidic pretreatment of the tooth surface, then an unfilled bonding composition (primer resin) to achieve good wetting of the tooth surface, followed by a filled bonding agent for the bulk of the bond. These materials are used to attach composite resin restorative materials, ceramic veneers, and orthodontic metal and ceramic brackets to enamel and dentin surfaces. Because of the different composition and physical properties of enamel and dentin (see Chapter 3.4) more complex and greater demands are placed on multi-purpose adhesive systems intended for both tissues. Clinical application of these systems are reviewed by Roulet and Degrange (2000).

Enamel Bonding

Bonding to enamel is achieved by pretreating the surface with 35–50% phosphoric acid for 30–60 seconds as described earlier (Fig. 3). This resulting washed and dried surface is readily wettable and penetratable by resin cements and bonding agents. The resulting resin tags (5–50 μm long) in the surface of the tissue result in efficient micromechanical interlocking with a potential tensile bond strength of about 20 MPa, equivalent to cohesive failure in the resin or in the enamel.

Dentin Bonding

Bonding to dentin currently involves pretreatment of the prepared (machined) surface with acidic solutions (phosphoric, nitric, maleic acids) or ethylenediaminetetraacetic acid (EDTA) to remove cutting debris (the smear layer). This procedure opens the orifices of the cut dentinal tubules and creates microporosity in the surface (Fig. 4). A primer treatment is then applied that comprises a reactive monomer system (Fig. 10) containing a carboxylate function or a polyphosphate function, depending on the type of product. These primers also contain hydrophilic monomers such as hydroxyethyl methacrylate (HEMA) and may also contain water.

The function of the primer is to penetrate the demineralized dentin surface and facilitate wetting by an unfilled dimethacrylate bonding resin that is subsequently applied. Polymerization of this treatment layer by visible-light activation results in micromechanical bond formation by penetration into the dentin and surface tubules, so forming the so-called hybrid layer (Nakabayashi et al., 1991) or resin-interdiffusion zone (Van Meerbeck et al., 1992). Chemical interaction with the hydroxyapatite and/or proteinaceous phases of the dentin surface may also occur. However, direct chemical evidence has not yet been provided for the postulated interactions (Eliades, 1993; Degrange and Roulet, 2000; Smith, 1998). Under the best conditions, initial tensile bond strengths of 15–25 MPa can be obtained depending on test conditions. The long-term durability of these bonds under oral conditions is being investigated (Roulet and Degrange, 2000).

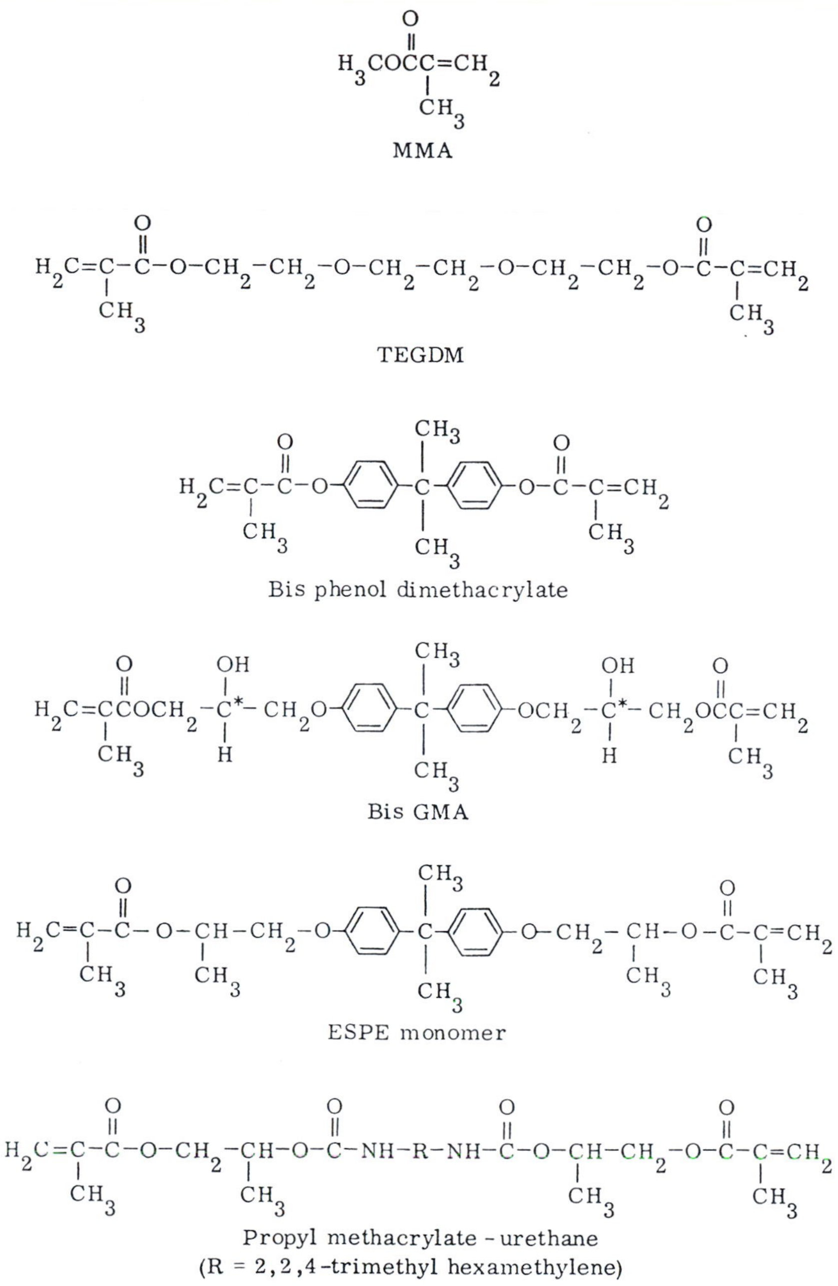

$$H_3COCC=CH_2$$

MMA

TEGDM

Bis phenol dimethacrylate

Bis GMA

ESPE monomer

Propyl methacrylate - urethane
(R = 2,2,4-trimethyl hexamethylene)

FIG. 9. Structures of some dimethacrylate monomers used in dental composite filling and bonding systems.

NEW RESEARCH DIRECTIONS

As a result of the experience of the past two decades, the problems involved in developing an adhesive system for both soft and hard tissues have been addressed and identified. A practical limitation in many systems remains ease of manipulation and application. For example, as noted earlier, the effectiveness of the fibrin sealant is critically dependent on proper mixing of the ingredients and uniform application. It has proved difficult to reconcile short- and long-term biocompatibility needs with chemical adhesion mechanisms that use reactive monomer system.

Where relatively temporary (less than 30 days) adhesion is required, as in wound healing, systems based on natural models that allow biodegradation of the adhesive and interface and subsequent normal tissue remodeling appear to merit further development. For longer term (years) durability in both soft and hard tissues, hydrophilic monomers and polymers of low toxicity that can both diffuse into the tissue surface and form ionic bonds across the interface seem to be the most promising approaches. Evidence has been obtained of the need for

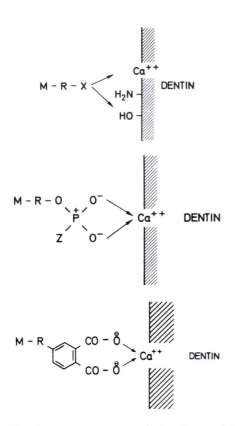

FIG. 10. Reactive monomer structures for bonding to calcific tissues. (M-R = monomer portion of molecule. (After Asmussen, E., Aranjo, P. A., and Pentsfeld, A., 1989, *Trans. Acod. Dent. Mater.* **2**: 59.)

hydrophobic–hydrophilic balance in adhesive monomer systems (Nakabayashi *et al.*, 1991), and the use of hydrophilic monomers such as hydroxyethyl methacrylate in commercial materials has facilitated surface penetration.

Recent development trends in soft-tissue adhesives appear to reflect these approaches. Newer materials (Chu, 1997) comprise cross-linked collagen materials (Colcys), light-cured polymerizable and biodegradable polyethylene glycols (FocaSeal), as noted earlier, photopolymerized derivatized collagen (DeVore, 1999), a bioresorbable hemostatic collagen-derived matrix with thrombin (Floseal), and serum albumin cross-linked with a derivatized poly(ethylene glycol) sealant. The development of synthetic peptides and materials based on human recombinant components are being investigated.

On calcified surfaces, the use of hydrophilic electolytes such as the polycarboxylates has demonstrated that proven ionic bonding *in vitro* can also be achieved *in vivo* (Smith, 1998). An advantage of such systems is that surface molecular reorientations can improve bonding with time (Peters *et al.*, 1974). Encouraging preliminary results have been obtained with new glass ionomer hybrid systems and there is considerable scope for the future developments of modified polyelectrolyte cements. Analogous materials based on polyphosphonates are being developed.

The development of more efficient adhesives and sealants that, in addition to enhancing the durability of current applications, would permit new applications such as osteogenic bone space fillers, percutaneous and permucosal seals, and functional attachment of prostheses is still a challenging problem for the future.

Bibliography

Alleyne, C. H. Jr., Cawley, C. M., Barrow, D. L., Poff, B. C., Powell, M. D., Sawhncy, A. S., and Dellehay, D. L. (1998). Efficacy and biocompatibility of a photopolymerized synthetic, absorbable hydrogel as a dural sealant in a canine craniotomy model. *J. Neursurg.* **88**: 308–313.

Charnley, J. (1960). Anchorage of the femoral head prosthesis to the shaft of the femur. *J. Bone Joint Surg. (B)* **42**: 28–30.

Charnley, J. (1970). *Acrylic cement in orthopaedic surgery.* E. S. Livingstone, Edinburgh.

Chilla, R. (1987). Histoacryl-induzierte Spätkomplikationen nach Duraplastiken an der Fronto- und Otobasis. *HNO* **35**: 250–251.

Chu, C. C. (1997). New emerging materials for wound closure. in *Wound Closure Biomaterials and Devices*, C. C. Chu, J. A. von Fraunhofer, and H. P. Greisler, eds. CRC Press, Boca Raton, FL, pp. 347–384.

Chu, C. C., von Fraunhofer, J. A., and Greisler, H. P. (1997). *Wound Closure Biomaterials and Devices.* CRC Press, Boca Raton, FL.

Ciapetti, G., Stea, S., Cenni, E., Sudanese, A., Marraro, D., Toni, A., and Pizzoferrato, A. (1994). Toxicity of cyanoacrylates in vitro using extract dilution assay on cell cultures. *Biomaterials* **15**: 92–96.

Comyn, J. (1997). *Adhesion Science.* Royal Society of Chemistry, London.

Cooper, C. W., Grode, G. A., and Falb, R. D. (1972). The chemistry of cyanoacrylate adhesives. in *Tissue Adhesives in Surgery*, T. Matsumoto, ed. Medical Examination Publishing Company, New York.

Coury, A., Hebida, P., Mao, J., Medalie, D., Barman, S., Doherty, E., Hoodrich, S., Poff, B., Fajaratnam, J., Terrazzano, J., Ban Lue, S., and Warnock, D. (1999). In-vivo bonding efficacy of PEG-based hydrogels to a variety of internal tissues. *Trans. Soc. Biomater.* **25**: 45.

DeVore, D. P. (1999). Photopolymerized collagen-based adhesives and surgical sealants. *Trans. Soc. Biomater.* **25**: 161.

Eliades, G. C. (1993). Dentin bonding systems. in *State of the Art on Direct Posterior Filling Materials and Dentine Bonding*, G. Vanherle, M. Degrange, G. Willems, eds. Cavex Holland BV, Haarlem, pp. 49–74.

Ennker, J., Ennker, I. C., Schoon, D., Schoon, H. A., Dorge, S., Messler, M., Rimpler, M., and Hetzer, R. (1994a). The impact of gelatine–resocinal–formaldehyde glue on aortic tissue: a histomorphologic examination. *J. Vasc. Surg.* **20**: 34–43.

Ennker, I. C., Ennker, J., Schoon, D., Schoon, H. A., Rimpler, M., and Hetzer, R. (1994b). Formaldehyde free collagen glue in experimental lung gluing. *Ann. Thorac. Surg.* **57**: 1622–1627.

Falb, R. D., and Cooper, C. W. (1966). Adhesives in surgery. *New Scientist*, 308–309.

Feldman, D. S., Barker, T. H., Blum, B. E., Kilpadi, D. V., and Reddon, R. A. (1999). Fibrin as a tissue adhesive and scaffold for meshed skin grafts in burn patients. *Trans. Soc. Biomater.* **25**: 42.

Flagle, J., Allan, J. M., Dovley, R. L., and Shalaby, S. W. (1999). Absorbable tissue adhesives in skin wound repair. *Trans. Soc. Biomater.* **25**: 373.

Gluck, T. (1891). Referat über die durch das moderne chirurgische Experiment gewonenen positiven Resultate, betroffend die Naht und den Ersatz von Defecten höherer Gewebe sowie über die Verwerthung resorbirbarer und lebendiger Tampons in der Chirurgie. *Langenbecks Archiv fur Klinische Chirugie* **41**: 187–239.

Goodrich, S. D., Barman, S. P., Coury, A. J., Metts, H. A., Nason, W. C., Weaver, D. J., and Yao, F. (1997). *Trans. Soc. Biomater.* **23**: 249.

Green, K. (1996). Mussel adhesive protein. in *Surgical Adhesives and Sealants*, D. H. Sierra and R. Salz, eds. Technomic Publishing, Lancaster, pp. 19–28.

Haring, R. (1972). Current status of tissue adhesives in Germany. in *Tissue Adhesives in Surgery*, T. Matsumoto, ed. Medical Examination, New York, p. 430.

Ikada, Y. (1997). Tissue adhesives. in *Wound Closure Biomaterials and Devices*, C. C. Chu, J. A. von Fraunhofer, and H. P. Greisler, eds. CRC Press, Boca Raton, FL, pp. 317–346.

Kinloch, A. J. (1987). *Adhesion and Adhesives.* Chapman and Hall, London.

Lovich, M. A., Philbrook, M., Sawyer, S., Weselcouch, E., and Edelman, E. R. (1998). Aterial heparin deposition; role of diffusion, convection and extravascular space. *Am. J. Physiol.* **275** (Heart Circ. Physiol. 44): H2236–H2242.

MacPhee, M. J. (1996). Commercial pooled-source fibrin sealant. in *Surgical Adhesives and Sealants*, D. H. Sierra and R. Salz, eds. Technomic Publishing, Lancaster, pp. 13–18.

Matras, H. (1972). Suture-free interfascicular nerve transplantation in animal experiments. *Wiener Medizinische Wochenschrift* **122**: 517–523.

Matsumoto, T. (ed.) (1972). *Tissue Adhesives in Surgery.* Medical Examination Publishing Company Inc., New York, p. 430.

Mittal, K. L., and Pizzi, A. (eds.) (1999). *Adhesion Promotion Techniques.* Marcel Dekker, New York.

Moody, E. W., Levine, M. A., Rodowsky, R., and Sawhney, A. (1996). A synthetic photopolymerized biodegradable hydrogel for sealing arterial leaks. 5th World Biomater Cong., 587.

Nakabayashi, N., Nakaumura, M., Yasuda, N. (1991). Hybrid layer as a dentin bonding mechanism. *J. Esthet. Dent.* **3**: 133–135.

Peters, W. J., Jackson, R. W., and Smith, D. C. (1974). Studies of the stability and toxidity of zine polyacralyte (polycarboxylate) cements (PAZ). *J. Biomed. Mater. Res.* **8**: 53.

Phillips, R. W., and Ryge, G. (eds.) (1961). *Adhesive Restorative Dental Materials.* National Institutes of Health, U.S. Public Health Service.

Pocius, A. V. (1997). *Adhesion and Adhesives Technology: An Introduction.* Hanser Publishers, Munich.

Prior, J. J., Wallace, D. G., Harner, A., and Powers, N. (1999). Performance of a sprayable hemostat containing fibrillar collagen, bovine thrombin and autologous plasma. *Trans. Soc. Biomater.* **25**: 159.

Ranger, W. R., Halpin, D., Sawhney, A. S., Lyman, M., and Locicero, J. (1997). Pneumostasis of experimental air leaks with a new photopolymerized synthetic tissue sealant. *Am. Surg.* **63**: 788–795.

Reiter, A. (1987). Induction of sarcomas by the tissue-binding substance Histoacryl-blau in the rat. *Z. für Exper. Chirg. Transpl. Kunsliche Organe* **20**: 55–60.

Roulet, J-F., and Degrange, M. (eds.) (2000). *Adhesion: The Silent Revolution in Dentistry.* Quintessence Publishing Company, Inc., Chicago.

Sawhney, A. S., Pathak, C. P., and Hubbell, J. A. (1993). Biorodible hydrogels based on photopolymerized poly(ethylene glycol)-co-poly(x-hydroxy acid) diacrylate macromers. *Macromolecules* **26**: 581–587.

Scardino, M. S., Swain, S. F., Morse, G. S., Sartin, E. A., Wright, J. C., and Hoffman, C. E. (1999). Evaluation of fibrin sealants in cutaneous wound closure. *J. Biomed. Mater. Res. (Appl. Biomater.)* **48**: 315–321.

Schlag, G., and Redl, H. (1987). Fibrin Sealant in Operative Medicine, Vol. 4. *Plastic Surgery, Maxillofacial and Dental Surgery.* Springer-Verlag, Berlin.

Schulz, J., and Nardin, M.(1999). Theories and mechanisms of adhesion. in *Adhesion Promotion Techniques*, K. L. Mittal and A. Pizzi, eds. Marcel Dekker, New York, pp. 1–26.

Sierra, D. H., and Salz, R. (eds.) (1996). *Surgical Adhesives and Sealants.* Technomic Publishing, Lancaster, PA.

Smith, D. C. (1967). A new dental cement. *Brit. Dent. J.* **123**: 540–541.

Smith, D. C. (1971). Medical and dental applications of cements. *J. Biomed. Mater. Res. Symp.* **1**: 189–205.

Smith, D. C. (1991). Dental cements. *Curr. Opin. Dent.* **1**: 228–234.

Smith, D. C. (1998). Development of glass-ionomer cement systems. *Biomaterials* **19**: 467–478.

Spotnitz, W. D. (ed.) (1998). Topical issue on clinical use of fibrin sealants and other tissue adhesives. *J. Long Term Effects Med. Implants* **8**: 81–174.

Tanaka, K., Tadamoto, S., Ohtsuda, T., and Kotsuda, Y. (1999). Advaseal for acute aortic dissection: experimental study. *Eur. J. Cardio-thoracic Surg.* **15**: 114–115.

Van Meerbeek, B., Inokoshi, S., Braem, M., Lambrechts, P., and Vanherle, G. (1992). Morphological aspects of resin–dentin inter-diffusion zone with different dentin adhesive systems. *J. Dent. Res.* **71**: 1530–1540.

Waite, J. H. (1989). The glue protein of ribbed mussels (Genkenska denissa): a natural adhesive with some features of collagen. *J. Comp. Physiol. [B]* **159**(5): 517–525.

West, J. L., Chowdhury, S. M., Sawhney, A. S., Pathak, C. P., Dunn, R. C., and Hubbell, J. A. (1996). Efficacy of adhesion barriers. *J. Reproduct. Med.* **1**: 149–154.

7.10 OPHTHALMOLOGICAL APPLICATIONS

Miguel F. Refojo

Light that penetrates into the eye is partially refracted in the cornea, passes through the aqueous humor and the pupil (the opening in the center of the iris), is further refracted in the crystalline lens, passes through the vitreous humor, and converges on the retina (Fig. 1). Diverse polymeric devices, such as spectacles, contact lenses, and intraocular implants, are used to correct the optical function of the eye. The materials used in spectacle lenses are outside the scope of this chapter. Contact lenses, however, being in intimate contact with the tissues of the eye, are subject to the same regulations that govern the use of implant materials, and they are included in this chapter with other biomaterials used to preserve and to restore vision such as intraocular implants.

CONTACT LENSES

General Properties

Contact lenses are optical devices that must have good transmission of visible light. Pigments and dyes are added to some contact lenses for cosmetic effect. Contact lenses also may have ultraviolet (UV) light-absorbing additives, usually copolymerized in the contact lens material, to protect the eye from the harmful effects of UV light. UV light not absorbed by the normal crystalline lens is harmful to the retina and may contribute to the clouding of the lens (cataract) (Miller, 1987).

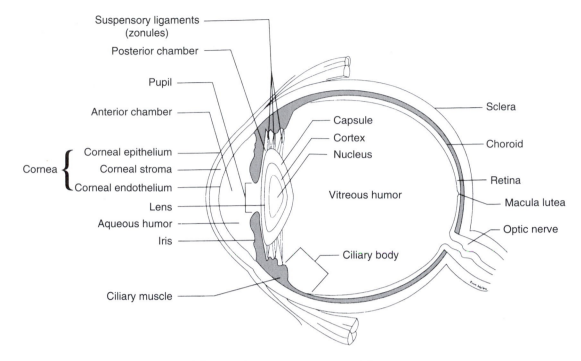

FIG. 1. Schematic representation of the eye.

The principal properties sought in contact lens materials, in addition to the required optical properties, chemical stability, and amenability to manufacture at reasonable cost, are high oxygen transmissibility (to meet the metabolic requirements of the cornea), tear-film wettability (for comfort), and resistance to accumulation on the lens surfaces of mucus/protein/lipid deposits from the tear film and other external sources. Contact lenses also must be easy to clean and disinfect (Kastl, 1995).

Most of the available contact lenses were developed with the most important property of oxygen permeability in mind. The oxygen permeability coefficient, P, is a property characteristic of a material. [$P = Dk$, where D is the diffusivity, in cm^2/sec, and k is the Henry's law solubility coefficient, in $cm^3(O_2 STP)/[cm^3(polymer)$ mm HG] For a given contact lens, its oxygen transmissibility (Dk/L) is more important than the Dk of the material; oxygen transmissibility is defined as the oxygen permeability coefficient of the material divided by the average thickness of the lens (L, in cm) (Holden *et al.*, 1990).

For oxygen permeability, the ideal contact lens would be made of polydimethylsiloxane. For better mechanical properties and manufacture, most silicone elastomeric lenses have been made of diverse poly(methyl phenyl vinyl siloxanes). Because of its hydrophobic character, to be useful, a silicone rubber lens must be treated in an RF-plasma reactor or other suitable procedure to make its surface hydrophilic and tolerated on the eye. (see Chapter 2.4, "Thin Films, Grafts, and Coatings"). Nevertheless, the silicone rubber lenses have not been successful for general cosmetic use, not only because of surface problems and comfort, but principally because they have a strong tendency to adhere to the cornea.

There are currently a wide variety of contact lens materials with diverse physical properties that determine the fitting characteristics of the lens on the eye (Lai *et al.*, 1993).

Soft Hydrogel Contact Lenses

There are essentially two kinds of hydrogel contact lenses (Refojo, 1996). The most common are the original homogeneous hydrogels (see Chapter 2.5), and the newest kind that are made of heterogeneous hydrogels that consist of a water-poor hydrophobic phase of siloxane moieties dispersed within a water-rich hydrophilic phase similar to the standard hydrogels. These heterogeneous hydrogels have highly enhanced oxygen transmissibility compared with the standard hydrogels of similar hydration (Alvord *et al.*, 1998).

Standard Hydrogel Contact Lenses

The standard hydrogel lenses are made of slightly cross-linked hydrophilic polymers and copolymers. The original hydrogel contact lens material was poly(2-hydroxyethyl methacrylate) (PHEMA) (Wichterle and Lim, 1960); at equilibrium swelling in physiological saline solution it contains about 40% water of hydration. (Hydration of hydrogel contact lenses is customarily given as a percentage of water by weight, on a wet basis.)

The soft hydrogel contact lenses (SCL) are supple and fit snugly on the corneal surface. Because there is little tear exchange under these lenses, most of the oxygen that reaches the cornea must permeate through the lens. The oxygen permeability coefficient of the standard hydrogel materials increases

TABLE 1 Chemical Composition of some Hydrogel
Contact Lenses

Polymer	USAN[a]	%H$_2$O
2-Hydroxyethyl methacrylate (HEMA) with ethylene glycol dimethacrylate (EGDM)	Polymacon	38
HEMA with methacrylic acid (MAA) and EGDM	Ocufilcon A	44
	Ocufilcon C	55
HEMA with sodium methacrylate and 2-ethyl-2-(hydroxymethyl)-1,3-propanediol trimethacrylate	Etafilcon A	58
HEMA with divinylbenzene, methyl methacrylate (MMA), and 1-vinyl-2-pyrrolidone (VP)	Tetrafilcon A	43
HEMA with VP and MAA	Perfilcon A	71
HEMA with N-(1,1-dimethyl-3-oxobutyl)acrylamide and 2-ethyl-2(hydroxymethyl)-1,3-propanediol trimethacrylate	Bufilcon A	45
	Bufilcon B	55
2,3-Dihydroxypropyl methacrylate with MMA	Crofilcon A	39
VP with MMA, allyl methacrylate, and EGDM	Lidofilcon A	70
	Lidofilcon B	79
MAA with HEMA, VP, and EGDM	Vifilcon A	55

[a]United States Adopted Name.

exponentially with the water content. These hydrogel materials have a hypothetical limit in oxygen permeability that approaches the oxygen permeability of pure water.

The oxygen transmissibility of the rather thick original PHEMA hydrogel contact lenses was soon found to be insufficient for normal corneal metabolism. Therefore, new hydrogel contact lenses were developed with higher water content or with a water content similar to that of PHEMA but more amenable to fabrication in an ultrathin modality. This takes advantage of the law of diffusion which, applied to contact lenses, will guarantee that for any lens type under the same conditions of wear the oxygen flux through the lens will double when the thickness is halved.

Other conventional hydrogel contact lens materials include HEMA copolymers with other monomers such as methacrylic acid, acetone acrylamide, and N-vinylpyrrolidone. Commonly used materials are also copolymers of N-vinylpyrrolidone and of glyceryl methacrylate with methyl methacrylate. Newer hydrogel contact lenses are made of poly(vinyl alcohol) and a copolymer of HEMA with a methacrylate of phosphorylcholine (Formula 1). The poly(vinyl alcohol) lens (CIBA) is a daily disposable lens. The other lens (omafilcon A, Proclear, Biocompatibles Inc.) has 59% hydration and apparently better water retention than other conventional hydrogel lenses.

$$\underset{\substack{|\\O}}{\overset{\substack{CH_3\\|}}{CH_2{=}C{-}C{-}O{-}CH_2{-}CH_2{-}O}}\underset{\substack{|\\O^-}}{\overset{\substack{O^-\\|}}{P}}{-}O{-}CH_2{-}CH_2{-}\underset{\substack{|\\CH_3}}{\overset{\substack{CH_3\\|}}{N^+}}{-}CH_3$$

(1)

A variety of other monomers and cross-linking agents are used as minor ingredients in many commercial standard hydrogel contact lenses (Lai *et al.*, 1993) (Table 1).

Hydrogel lenses have been classified by the Food and Drug Administration (FDA) in four general groups: I, low water (<50% H$_2$O), nonionic; II, high water (>50% H$_2$O), nonionic; III, low water, ionic; and IV, high water, ionic. The ionic character is usually due to the presence of methacrylic acid, which is highly hydrophilic and enhances hydration but is also responsible for higher protein binding to the contact lenses. High water of hydration is a desirable property for good oxygen permeability, but it carries some disadvantages such as friability and protein penetration into the polymer network. Physiologically and optical, untrathin low-water-content contact lenses can perform very well as daily-wear lenses.

As a result of temperature changes and water evaporation, all hydrogel contact lenses dehydrate to some degree on the eye. Higher-water-content lenses dehydrate more than low-water-content lenses, and thin lenses dehydrate more easily than thick lenses (Refojo, 1991). An important advantage of the use of high-water-content, thin hydrogel contact lenses is their relatively high oxygen transmissibility. However, a drawback of these lenses is that as they dehydrate on the eye, they induce corneal epithelium injuries by a mechanism still unclear.

The transparent cornea does not contain blood vessels. Therefore, the cornea, to maintain its metabolism and transparency, uses oxygen from the ambient air when the eye is open (partial pressure O$_2$ about 160 mm Hg at sea level), and when the eye is closed from the blood vessels of the palpebral conjunctiva (partial pressure O$_2$ 55 mm Hg). The ideal contact lens should not be a barrier to the passage of oxygen to the cornea. The standard hydrogel contact lenses are used preferably for daily wear (open eye), but some with higher oxygen transmissibility are used for up to 1 week extended wear (day and night) (Holden *et al.*, 1984).

Siloxane-Hydrogel Contact Lenses

Striving to enhance the oxygen transmissibility of the soft contact lenses, while maintaining the comfort of the standard hydrogel lenses, the contact lens industry developed a new kind of siloxane-hydrogel contact lenses with sufficiently high oxygen transmissibility to allow their use for up to 1 month extended wear. These lenses are made essentially by copolymerizing hydrophilic monomers, similar to those used in the standard hydrogel lenses, with macromers of methacrylate, or vinyl, end-capped polydimethylsiloxanes (Formulas 2 and 3 (Künzler, 1996; Refojo, 1996), and/or with methacrylates of tris(trimethylsiloxy)silane (TRIS-like) (see Formula 5). The silicone macromers and the TRIS-like monomers are functionalized with hydrophilic groups to make them compatible with the hydrophilic components of the hydrogel. In these hydrogels, the oxygen permeation takes place mainly through the siloxane-rich phase and to a relatively lower degree though the water of hydration of the hydrogel. Because the relatively high content of siloxane moieties on the surface of these materials, the lenses must be treated in an RF-plasma reactor or other suitable procedure to make their surface hydrophilic and tolerated on the eye, as was mentioned above for the silicone elastomeric lenses. The two commercial siloxane-hydrogel contact lenses are PureVision (balafilcon A, Bausch & Lomb), which at 36% water has O_2 permeability about of 110×10^{-11} cm² (STP) cm²/(cm³ sec mm Hg), and Focus Night & Day (lotrafilcon A, Ciba Vision Corp.) that at 24% water has oxygen transmissibility of about 175×10^{-11}. Both of these lenses compare very favorably with the oxygen permeability of standard hydrogel contact lenses (Young and Benjamin, 2003). The lotrafilcon polymer contains a macromer similar to that given in Formula 4, but with a silicone block inserted at the midpoint of the perfluoropolyether block. On the other hand, the balafilcon polymer contains a siloxanyl methacrylate similar to that in Formula 5, but with a hydrophilic urethane group inserted between the alkyl and methacrylate moieties.

Rigid Contact Lenses

The rigid contact lenses fit loosely on the cornea and move with the blink more or less freely over the tear film that separates the lens from the corneal surface. The mechanical properties of rigid contact lenses must be such that any flex on the lens provoked by the blink must recover instantaneously at the end of the blink.

The original rigid contact lenses were made of poly(methyl methacrylate) (PMMA), which is an excellent material in almost all respects expect for lack of permeability to oxygen. Several materials that were specially developed for the manufacture of rigid gas-permeable (RGP) contact lenses are copolymers of methyl methacrylate with a siloxanylalkyl methacrylates, most often methacryloxypropyltris

$$CH_2{=}C{-}CO{-}(CH_2)_2NHCO(CH_2)_3{-}Si{-}O(Si{-}O)_n{-}Si(CH_2)_3OCNH(CH_2)_2{-}OC{-}C{=}CH_2 \qquad (2)$$

$$CH_2{=}C{-}C{-}O{-}(CH_2)_n{-}Si{-}O(Si{-}O)_x{-}(Si{-}O)_y{-}Si{-}(CH_2)_n{-}O{-}C{-}C{=}CH_2 \qquad (3)$$

(trimethylsiloxy)silane (Formula 5). To compensate for the hydrophobic character imparted to the polymer by the high siloxane content of these copolymers (required for oxygen permeability), the copolymer also contains some hydrophilic comonomers. The most commonly used hydrophilic comonomer in rigid lenses is methacrylic acid. There are also minor ingredients and cross-linking agents.

$$CH_2{=}C{-}C{-}O{-}CH_2{-}CH_2{-}CH_2{-}Si{-}O{-}Si{-}CH_3 \qquad (5)$$

Flexible Perfluoropolyether Lenses

The only perfluoropolyether lens that reached the market, but which is not longer available, was made from a copolymer of a telechelic perfluoropolyether (Formula 4) (which imparts high oxygen permeability) with vinylpyrrolidone (which imparts wettability) and methyl methacrylate (which imparts rigidity). This flexible, nonhydrated contact lens, made by the molding procedure, had a high oxygen permeablility and, because of its high fluorine content, was claimed to be more resistant to coating by tear proteins than are other contact lens materials.

This methacrylate end-capped perfluoropolyether can be also used to make highly oxygen-permeable hydrogels.

$$CH_2{=}C{-}CO(CH_2)_2NHCOCH_2CF_2O(CF_2CF_2O)_m(CF_2O)_nCF_2CH_2OCNH(CH_2)_2OC{-}C{=}CH_2 \qquad (4)$$

TABLE 2 Composition of Some Rigid Gas-Permeable Contact Lenses

Polymer	USAN
Cellulose acetate dibutyrate	Porofocon
	Cabufocon
3-[3,3,5,5,5-Pentamethyl-1,1-bis[(pentamethyldisiloxanyl)oxy]trisiloxanyl]propyl methacrylate, with methyl methacrylate (MMA), methacrylic acid (MAA), and tetraethylene glycol dimethacrylate (TEGDMA)	Silafocon
MMA with MAA, EGDMA, 3-[3,3,3,-trimethyl-1,1-bis(trimethylsiloxy)disiloxanyl] propyl methacrylate (TRIS), and N-(1,1-dimethyl-3-oxybutyl)acrylamide.	Nefocon
VP with HEMA, TRIS, allyl methacrylate, and α-methacrylonyl-ω-(methacryloxy)poly(oxyethylene)-co-oxy(dimethylsilylene)-co-oxyethylene.	Mesifilcon
TRIS with MMA, dimethyl itaconate, MAA, and TEGDMA.	Itafocon
TRIS with 2,2,2,-trifluoro-1-(trifluoromethyl)ethyl methacrylate, 1-vinyl-2-pyrrolidone (VP), MAA, and ethyleneglycol dimethacrylate (EGDMA).	Melafocon
TRIS with 2,2,2-trifluoroethyl methacrylate, MAA, MMA, VP with EGDMA.	Paflufocon

A diversity of RGP contact lenses, consisting of different but closely related comonomers used in a variety of proportions to obtain the most desirable properties, are commercially available (Table 2). However, any subtle change in the chemistry of a contact lens material might strongly affect its clinical performance. As a general rule, the oxygen permeability coefficient of these original RGP siloxanylalkyl methacrylate contact lens materials is inversely proportional to the density (Refojo, 1994).

The development of the fluorine-containing RGP contact lenses was the realization that the fluoroderivatives may improve oxygen permeability and resistance to deposit formation on the lenses. Thereafter, the contact lens chemists included fluoroalkyl methacrylates or a similar fluorine-content monomer as an additional ingredient in the siloxanylalkyl methacrylate-co-methyl methacrylate RGP contact lens materials. These perfluoroalkyl–siloxanylalkyl–methyl methacrylate contact lenses have high oxygen permeability and, supposedly, better surface properties than the non-fluorine-containing rigid contact lenses, but their Dk is not directly related to their density like that of the original RGP contact lenses. Some RGP contact lenses made of siloxanyl alkylmethacrylates with styrene derivatives (Menicon) have high enough oxygen transmissibility to be used for extended wear (Benjamin and Cappelli, 2002).

Cellulose acetate butyrate (CAB) was also used as a rigid oxygen-permeable contact lens material. However, CAB has not only relatively low oxygen permeability compared with the siloxanyl–alkyl methacrylate copolymers but also low scratch resistance, and it tends to wrap with humidity changes.

Other copolymers have been used as contact lens materials are isobutyl and isopropyl styrene, with hydrophilic comonomers of the type of HEMA or vinylpyrrolidone.

CORNEAL IMPLANTS

The cornea is an avascular tissue that consists of three principal layers (Fig. 1). The outermost layer, which itself consists of about five cellular layers, is the epithelium. The central and main portion of the cornea is the stroma, a collagenous connective tissue that is 78% hydrated in its normal state. Normal corneal hydration is disrupted by hypoxia and injury to the limiting epithelial and endothelial membranes. The endothelium is the innermost single layer of cells, which by means of a "pump-leak" mechanism is mostly responsible for maintaining normal corneal hydration. Swelling, tissue proliferation, and vascularization may compromise the transparency of the cornea. There are several types of corneal implants that replace all or part of the cornea (Refojo, 1986a; Abel, 1988).

Epikeratphakia and Artificial Epithelium

To correct the optics of the eye after cataract extraction, the surgeon may perform an epikerotophakia procedure which consists of transplanting a slice of donor cornea. The transplanted tissue heals into a groove carved into the recipient corneal surface and is reepithelialized with the recipient corneal epithelium (Werblin *et al.*, 1987) (Fig. 2). A modification of this technique attempts to obtain similar results with an artificial material that would heal into the donor cornea and be able to grow the epithelium of the donor cornea on its surface (Evans *et al.*, 2003).

An epithelium that has become irregular through swelling and proliferation has been replaced by the artificial epithelium made of a hard PMMA contact lens glued with a cyanoacrylate

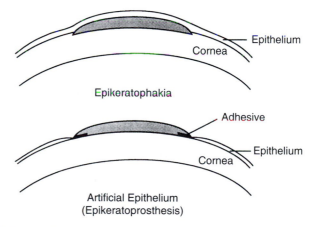

FIG. 2. Schematic representation of superficial corneal implants. (Top) In the epikeratophakia procedure the corneal epithelium is removed before the implant is placed on the stromal surface and epithelium grows over the implant. (Bottom) The artificial epithelium or epikeratoprosthesis is a contact lens glued to a deepithelialized cornea. Ideally, the epithelium should not grow over or under the glued-on lens.

adhesive to the corneal stroma (Fig. 2). This procedure has not been successful mainly because of failure of the glue to maintain a tight attachment of the prosthesis to the corneal stroma and also because of epithelium penetration between the prosthesis and the cornea.

Plastic Corneas

Corneal transplants from donor eyes are usually highly successful. In the instance of transplant failure, an opaque cornea can be replaced with an artificial cornea (keratoprosthesis) (Barber, 1988). These are usually through-and-through corneal implants, consisting of a central optical portion and some modality of skirt that fixes the prosthesis to the recipient cornea (Fig. 3). Another plastic cornea model is the "collarbutton" type that holds the optical cylinder of the prosthesis to the recipient cornea by means of front and back plastic plates (Nouri *et al.*, 2001). Many materials have been tested to make all or part of these devices, but most artificial corneas implanted in patients have been made of PMMA. The main problem with through-and-through keratoprostheses is common to all kinds of implants that are not fully buried in the recipient tissue: faulty tissue–prosthesis interface, epithelium downgrowth, retrocorneal membranes, and tissue ulceration and infection around the prosthesis. The most feasible solution to these problems would be the development of a material for the optical portion of the keratoprosthesis that would accept growth and attachment of transparent epithelium on its surface. Also needed are biomaterials that would heal into the recipient corneal tissue (Chirila *et al.*, 1998a).

Artificial Endothelium

The corneal endothelium has been replaced, but not very successfully in the long term, by a silicone-rubber membrane that passively controls corneal hydration (Fig. 3). Unfortunately, this procedure was unsuccessful because the membrane served as a barrier not only to water but also to the nutrients that the cornea normally receives from the aqueous humor.

Intracorneal Implants

Ophthalmic surgeons may use diverse polymeric devices to correct the optical function of the eye. Thus, intracorneal implants can be used instead of spectacles or contact lenses to correct nearsightedness and farsightedness (Fig. 4). Some intracorneal implants are made of hydrogel materials tailored to have high permeability to metabolites and able to correct severe myopia (McCarey *et al.*, 1989). The stromal cells (keratocytes) and the epithelium receive their nutrients from the aqueous humor and also release waste products in the same direction. Therefore, some previously used intrastromal lens implants that were made of PMMA and polysulfone, which are impermeable to metabolites, resulted in the ulceration and vascularization of the overlying stroma. A newer development is the intrastromal ring of PMMA or silicone rubber that obviates the corneal nutritional problem of the lens implants. More recently the intracorneal implant consists of two 180° narrow segments of PMMA that change the corneal curvature and, hence, the eye's optical power. These implants can make the corneal curvature steeper, increasing the refractive power, or flatter, decreasing the refractive power (Fink *et al.*, 1999).

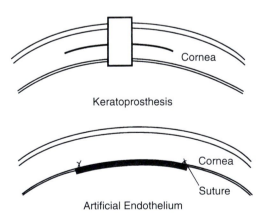

FIG. 3. (Top) Schematic representation of a through-and-through artificial cornea (keratoprosthesis), that consists of an optical cylinder that penetrates the opaque tissue. The prosthesis has an intrastromal rim that holds the prosthesis in the cornea. (Bottom) Schematic representation of the artificial corneal endothelium that consists of a transparent membrane sutured to the posterior part of the a cornea denuded of its endothelium. This membrane acts as a barrier to the inflow of aqueous fluid into the corneal stroma.

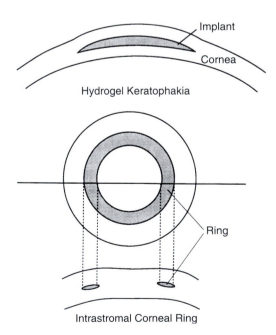

FIG. 4. Schematic representation intracorneal implants used to change the curvature of the cornea in refractive keratoplasty. (Top) An intrastromal hydrogel intracorneal implant. (Bottom) An intrastromal corneal ring, that in a newer version is implanted as two 180° PMMA segments.

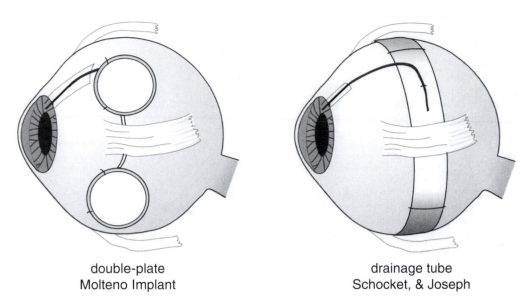

double-plate
Molteno Implant

drainage tube
Schocket, & Joseph

FIG. 5. Schematic representation of two typical models of aqueous humor drainage implants for glaucoma. (Yamanaka and Refojo, 1990).

IMPLANTS FOR GLAUCOMA

Polymeric devices are used to control abnormally high intraocular pressure in otherwise intractable glaucoma (Krupin et al., 1988). These devices consist essentially of tiny silicone tubes that transport the aqueous humor—which normally maintains the physiological intraocular pressure and flows in and out of the eye in a well-regulated manner—from the anterior chamber to some artificially created space between the sclera and the other tissues that surround the eyeball. This artificial space is maintained by one or more end plates connected to the anterior chamber by the silicone tubing (Fig. 5). The artificial space fills with the aqueous humor and forms a bleb from which the aqueous humor is absorbed into the blood circulation. The most common materials used in the endplates are silicone rubber and polypropylene, although other materials have also been investigated (Ayyala et al., 1999). Tissue proliferation, or capsule formation, takes place around all implanted biomaterials and may retard or even stop the outflow of aqueous humor from the bleb.

INTRAOCULAR LENS IMPLANTS

Intraocular lenses (IOLs) are used after cataract extraction to replace the opaque crystalline lens of the eye (Apple et al., 1984), and are dealt with in Patel's chapter (7.11).

IMPLANTS FOR RETINAL DETACHMENT SURGERY

A retina detached from its source of nutrition in the choroidal circulation ceases to be sensitive to light. The choroid is the vascular layer between the retina and the sclera (Fig. 1).

In cases of retinal detachment, the surgeon must reattach the retina to restore vision. Retinal detachment could result from traction of a retracting vitreous humor or from seepage of liquefied vitreous through a retinal hole between the retina and the choroid. Retina surgeons can often restore vision to these eyes with vitreous implants and scleral buckling materials (Refojo, 1986b) (Fig. 6).

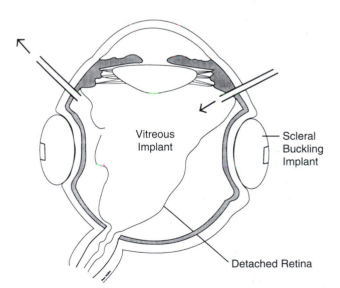

Vitreous Implant

Scleral Buckling Implant

Detached Retina

FIG. 6. Schematic representation of an eye with a detached retina. The retina can be pushed back into its normal place by injecting a fluid in the vitreous cavity (inside arrow) while the subretinal fluid is drained (outside arrow). A scleral buckling implant (the drawing represents is an encircling implant) is placed over retinal tears to counteract the traction on the retina of a shrinking vitreous and to reapproximate the retina to the underlying tissues.

Vitreous Implants

Vitreous implants are desirable in certain difficult cases of retinal detachment surgery (Refojo, 1986b). Physiological saline solution, air, sulfur hexafluoride, perfluorocarbon gases, and low-molecular-weight perfluorocarbon liquids frequently are injected into the vitreous cavity during vitreoretinal surgery. These fluids may perform well only as short-term retinal tamponade agents. Hyaluronic acid solutions and many other hydrophilic polymers have been investigated for long-term vitreous replacement without much success (Chirila *et al.*, 1998b). Therefore, for long-term vitreous replacement the only substance used at this time, as last resort and with variable results, in silicone oils of high viscosity (1000 to 5000 centistokes). (Giordano and Refojo, 1998). Retinal toxicity, oil emulsification, glaucoma, and corneal clouding are some of the complications of intraocular silicone oils. These complications may be avoided by removing the oil after choroidal–retinal adhesion has been achieved, but oil removal involves further surgery, is difficult to achieve completely, and carries the risk of recurrence of retinal detachment.

Scleral Buckling Materials

The scleral implants are used for the external treatment of retinal detachment. By indentation of the sclera these implants bring together the detached retina and the underlying choroidal layer (Schepens and Acosta, 1991). Scleral buckling materials for retinal detachment surgery must be soft elastic. Solid silicone rubber and silicone sponge have been used successfully for many years. An acrylic hydrogel made of a copolymer of 2-hydroxyethyl acrylate with methyl acrylate has also been used for scleral bucking and improves the rate of infection resulting from the use of the sponge and the potential for long-term pressure necrosis of the more rigid solid silicone-rubber implants (Refojo, 1986b). However, long-term biodegradation of the acrylic hydrogel resulted in complications due to implant swelling (Roldán-Pallarés *et al.*, 1999).

SURGICAL ADHESIVES

As in most applications of polymers as biomedical implants, in ophthalmology any polymeric device must be as free as possible of residual monomer. However, in the unique case of the cyanoacrylate surgical adhesives, the monomers are applied directly to the tissues and almost instantaneously polymerize and adhere tenaciously to the tissues. (See Chapter 7.9, "Adhesives and Sealants.") The cyanoacrylate adhesives have been used in many diverse applications in the eye but have been particularly useful in corneal perforation and ulcer, as well as in gluing artificial epithelium to the corneal surface and repairing retinal detachments (Refojo *et al.*, 1971).

Bibliography

Abel, R., Jr. (1988). Development of an artificial cornea: I. History and materials. in *The Cornea: Transations of the World Congress on the Cornea III*, H. D. Cavanagh, ed. Raven Press, New York, pp. 225–230.

Alvord, L., Court, J., Davis, T., Morgan, C. F., Schindhelm, K., Vogt, J., and Winterton, L. (1998). Oxygen permeability of a new type of high *Dk* soft contact lens material. *Optom. Vis. Sci.* **75**: 30–36.

Apple, D. J., Loftfield, K., Mamalis, N., Normal D. K.-V., Brady, S. E., and Olson, R. J. (1984). Biocompatibility of implant materials: a review and scanning electron microscopic study. *Am. Intraocular Implant Soc. J.* **10**: 53–66.

Ayyala, R. S., Harman, L. E., Michilini-Norris, B., Ondrovic, L. E., Heller, E., Margo, C. E., and Stevens, S. X. (1999). Comparison of different biomaterials for glaucoma drainage devices. *Arch. Ophthalmol.* **117**: 233–236.

Barber, J. C. (1988). Keratoprostheses: past and present. *Int. Ophthalmol. Clin.* **28**: 103–109.

Benjamin, W. J., and Cappelli, Q. A. (2002). Oxygen permeability (*Dk*) of thirty-seven rigid contact lens materials. *Optom. Vis. Sci.* **79**: 103–111.

Chirila, T. V., Hicks, C. R., Dalton, P. D., Vijayasekaran, S., Lou, X., Clayton, A. B., Ziegelaar, B. W., Fitton, J. H., Platten, S., Crawford, G. J., and Constable, I. J. (1998a). Artificial cornea. *Prog. Polymer Sci.* **23**: 447–473.

Chirila, T. V., Hong, Y., Dalton, P. D., Constable, I. J., and Refojo, M. F. (1998b). The use of hydrophilic polymers as artificial vitreous. *Prog. Polymer Sci.* **23**: 475–508.

Evans, M. D. M., Taylor, S., Dalton, B. A., and Lohmann, D. (2003). Polymer design for corneal epithelial tissue adhesion: Pore density. *J. Biomed. Mater. Res.* **64A**: 357–363.

Fink, A. M., Gore, C., and Rosen, E. S. (1999). Corneal changes associated with intrastromal corneal ring segments *Arch. Ophthalmol.* **117**: 282.

Giordano, G. G., and Refojo, M. F. (1998). Silicone oils and vitreous substitutes. *Prog. Polymer Sci.* **23**: 509–532.

Grobe, G. L., Künzler, J., Seelye, D., and Salomone, J. C. (1999). Silicone hydrogels for contact lens application. *Polymer Mater. Sci. Eng.* **80**: 108–109.

Holden, B. A., Newton-Homes, J., Winterton, L., Fatt, I., Hamano, H., La Hood, D., Brennan, N. A., and Efron, N. (1990). The *Dk* project: an interlaboratory comparison of *Dk/L* measurements. *Optom. Vis. Sci.* **67**: 476–481.

Holden, B. A., Sweeney, D. F., and Sanderson, G. (1984). The minimum precorneal oxygen tension to avoid corneal edema. *Invest. Ophthalmol. Vis. Sci.* **25**: 476–480.

Kastl, P. R. (ed.) (1995). *Contact Lenses: The CLAO Guide to Basic Science and Clinical Practice.* Kendall/Hunt, Dubuque, IA.

Krupin, T., Ritch, R., Camras, C. B., Brucker, A. J., Muldoon, T. O., Serle, J., Podos, S. M., and Sinclair, S. H. (1988). A long Krupin-Denver valve implant attached to a 180° scleral explant for glaucoma surgery. *Ophthalmology* **95**: 1174–1180.

Künzler, J. F. (1996) Contact lenses, gas permeable. in *Polymeric Materials Encyclopedia*, J. C. Salmone, ed. CRC Press, Boca Raton, FL, pp. 1497–1503.

Lai, Y. C., Wilson, A. C., and Zantos S. G. (1993). Contact Lenses. in *Kirk-Othmer: Encyclopedia of Chemical Technology*, 4th ed. Wiley, New York, Vol. 7, pp. 192–218.

McCarey, B. E., McDonald, M. B., van Rij, G., Salmeron, B., Pettit, D. K., and Knight, P. M. (1989). Refractive results of hyperopic hydrogel intracorneal lenses in primate eyes. *Arch. Ophthalmol.* **107**: 724–730.

Miller, D. (ed.) (1987). *Clinical Light Damage to the Eye.* Springer-Verlag, New York.

Nouri, M., Terada, H., Alfonso, E. C., Foster, S., Durand, M. L., and Dohlman, C. H. (2001). Endophthalmitis after keratoprosthesis. *Arch. Ophthalmol.* **119**: 484–489.

Refojo, M. F. (1986a). Current status of biomaterials in ophthalmology. in *Biological and Biomechanical Performance of Biomaterials*, P. Christel, A. Meunier, and A. J. C. Lee, eds. Elsevier, Amsterdam, pp. 159–170.

Refojo, M. F. (1986b). Biomedical materials to repair retinal detachments. in *Biomedical Materials*, J. M. Williams, M. F., Nichols, and W. Zingg, eds. Materials, Research Society Symosia Proceedings, Materials Research society, Pittsburgh, Vol. 55, pp. 55–61.

Refojo, M. F. (1991). Tear evaporation considerations and contact lens wear. in *Contact Lens Use under Adverse Conditions*, P. E. Flattu, ed. National Research Council, Washington, DC, pp. 38–43.

Refojo, M. F. (1994). Chemical composition and properties, in *Contact Lens Practice*, M. Ruben and M. Guillon, eds. Chapman & Hall Medical, London, pp. 27–41.

Refojo, M. F. (1996). Contact lenses, hydrogels. in *Polymeric Materials Encyclopedia*. J. C. Salomane, ed. CRC Press, Boca Raton, FL, pp. 1504–1509.

Refojo, M. F., Dohlman, C. H., and Koliopoulos, J. (1971). Adhesives in ophthalmology: a review. *Surv. Ophthalmol.* **15**: 217–236.

Roldán-Pallarés, M., Castillo-Sanz, J. L., Awad-El Susi, S., and Refojo, M. F. (1999). Long-term complications of silicone and hydrogel explants in retinal reattachment surgery. *Arch. Ophthalmol.* **117**: 197–201.

Schepens, C. L., and Acosta, F. (1991). Scleral implants: an historical perspective. *Surv. Ophthalmol.* **35**: 447–453.

Werblin, T. P., Peiffer, R. L., and Patel, A. S. (1987). Synthetic keratophakia for the correction of aphakia. *Ophthalmology* **94**: 926–934.

Wichterle, O., and Lim, D. (1960). Hydrophilic gels for biological use. *Nature* **185**: 117–118.

Yamanaka, A., and Refojo, M. F. (1990). Biomaterials in ophthalmology. *J. Jpn. Soc. Ophthalm. Surgeons* **3**: 493–503.

Young, M. D., and Benjamin, W. D. (2003). Oxygen permeability of the hypertransmissible contact lenses. *Eye Contact Lenses* **29**: S17–S21.

7.11 INTRAOCULAR LENS IMPLANTS: A SCIENTIFIC PERSPECTIVE

Anil S. Patel

INTRODUCTION TO OPTICS OF THE EYE, CATARACTS, AND INTRAOCULAR LENS IMPLANTS

Vision is considered to be our dominant sense. We see in the brain through optic nerves, which carry information of the optical image formed in the eye. The physiological optics of the eye for formation of the image, the photochemical transduction in the neurosensory retina at the optical image plane, and the subsequent neural processing in the retina and the brain are quite complex and beyond the scope of this chapter (LeGrand and El Hage, 1980). As an introduction, a brief description of optical imaging of the eye is necessary.

The eye is schematically illustrated in Figure 1 of Section 7.10. The average healthy adult eye is about 24.2 mm long axially from anterior surface of the cornea to the retina. It can be considered as a camera with the cornea and the natural crystalline lens as its two lenses, while the iris forming the pupil is an aperture between them, and the retina is the imaging plane. The image of an outside object of interest is formed on the retinal photoreceptors by the fixed refractive power of the cornea in combination with the variable refractive power of the natural lens, permitting imaging of any object at various far and near distances from the eye. The fixed refractive power of the cornea is about 43 diopters, which is about 70% of the total refractive power of the eye. The refractive power of the natural lens is about 20 diopters for seeing far objects. For seeing near objects, it increases as required to an upper limit, which gradually diminishes with age. When young, the refractive power of the natural lens can increase by up to 14 diopters, but by about the age of 42, this capability is less than the 3 diopters required for near reading distance. The physiology of variable refractive power of the natural lens is as follows.

As shown in Figure 1 of Section 7.10, the natural lens has a capsular bag containing a nucleus surrounded by the cortex. It is attached by the suspensory ligaments called zonules to the ciliary muscle. For an average healthy eye, the refractive power of the cornea and its natural lens are, in combination, adequate to form a focused image on the retina for any object at 20 feet or greater distance when the ciliary muscle is relaxed. In this physiologically relaxed condition of the eye, the zonules are in tension, and the natural lens is also in tension, providing about 20 diopters of required refractive power for distant vision. For seeing an object at any nearer distance, the ciliary muscle appropriately contracts radially and moves anteriorly, thereby relaxing the zonules, which in turn relaxes the natural lens. This results in the needed change in the shape of the lens, and thus increases its refractive power as required for near vision. This complex mechanism, which results in the ability to increase the refractive power of the natural lens for near vision, is called accommodation (Kaufman, 1992). This ability gradually diminishes with age as the lens grows and eventually become less than 3 diopters required for near reading distance, and the eye is considered presbyopic. This presbyopia of the eye is treated with eyeglasses, contact lenses, or evolving refractive surgery in order to restore the ability to see near objects.

Further aging of the lens results in loss of its clarity as opacities begin to appear in its structures, and the optical image formation deteriorates. Eventually, such a compromised natural lens is unacceptable and is called cataractous since it is transformed to an unacceptable structure for the primary function—to allow formation of a clear image on the retina (U.S. Department of Health and Human Services, 1983). This normal aging process related to cataract formation eventually leads to blindness and is the leading cause of blindness in the world. In the developed world, depending upon the need for vision, cataract surgery is carried out much more promptly before the total blindness stage when the degradation in the vision by the cataractous lens reaches an unacceptable level (AAO, 1989). During a relatively safe and short-duration surgery, not only is the cataractous lens removed, but in its place a biomaterial medical device called an intraocular lens (IOL) is implanted to provide the clear optical imaging function of the normal lens (Alpar and Fechner, 1986). While currently, worldwide, an estimated 10 million IOLs are implanted, primarily in the developed world and emerging countries, the lack of resources and rapid increase in life expectancy accounts for estimated 20 million cataract-related blind people in the underdeveloped world. Cataract blindness is preventable by a simple surgical procedure using the IOL implant, and the number of such procedures is expected to grow. Even then, cataracts are expected to remain the leading cause of blindness with an estimated 40 million cases by the year 2020 (Brian and Taylor, 2001).

WHY ARE IOLS SUCCESSFUL?

Within a span of only 30 years, the number of cataract surgeries and IOL implantations has rapidly expanded for sound reasons, both in terms of improvements in surgical procedure and visual outcome from the patient's perspective, as summarized below.

Prior to IOL implantation, thick spectacles of about 10 diopters refractive power were needed to attempt to restore the function of the natural lens after cataract surgery. Imaging by such external spectacle lenses and the cornea creates a 25% magnification of the resultant image on the retina compared with imaging by the natural lens and the cornea. Thus, when a cataract develops in one eye of the patient, its functional restoration by cataract surgery and a spectacle lens is unacceptable to the brain since, for fusion of images of the two eyes, the brain requires them to be no more than 6% different in size. Thus, restoration of vision by spectacles after cataract removal required waiting for maturation of cataracts in both eyes before surgeries were carried out to achieve equal image sizes for both eyes. This resulted in significantly poor vision in the unoperated cataractous eye for a prolonged period before cataract surgery was done. Additionally, cataract surgery 30 years ago was associated with other risks and poor visual outcome (as compared in Table 1), and was thus relatively discouraged until too late an age, when the compromised vision was totally unacceptable in a relatively advanced state of dysfunction, compared with today's practice in the developed world. Binocular vision with spectacles also had a significantly reduced field of vision and severe astigmatism on account of the required large size of the incision for cataract surgery.

The size of the optic of the IOL and its intraocular location eliminated image size magnification and reduced visual field problems, which were present for the spectacles. Foldable IOLs encouraged development of small-incision, safer, cataract surgery techniques without surgically induced astigmatism, and thus achieved quicker and better vision after cataract surgery (Apple et al., 2000a). Table 1 captures all the reasons for the overwhelming success of IOLs, where needed resources are available.

EMERGING FUNCTIONAL VARIATIONS OF IOLS

As a replacement of the cataractous natural lens, the majority of currently used IOLs are monofocal IOLs of fixed refractive spherical diopteric power, and thus cannot correct existing astigmatism as needed for some eyes, or restore accommodation as was provided by the pre-presbyopic natural lens. Toric IOLs, multifocal IOLs, and emerging accommodative

IOLs are relatively recent developments, and they will be discussed later as IOLs with emerging variations of the optical function (Sanders et al., 1992; Maxwell and Nordan, 1990; Samalonis, 2002).

The eye with its natural lens is called the phakic eye. The eye after removal of the natural lens, either clear or cataractous, is called an aphakic eye. The eye with replacement of the natural lens with an IOL is called a pseudophakic eye. An ophthalmic implant placed in a phakic eye without removal of its natural lens, for example, an additional IOL in order to correct refractive error, is called a phakic IOL (Obstbaum, 2003). Surgical implantation of a phakic IOL is a form of refractive surgery that provides needed refractive correction in a myopic or hyperopic eye as an option to eyeglasses or contact lenses or corneal refractive surgery such as LASIK. The current status of phakic IOLs will also be discussed later as a variation of optical function.

Thus, the IOL as an ophthalmic intraocular implant is now evolving to provide a variety of optical functions. The biomaterials for such variety of IOLs have been many, and new materials are emerging in order to meet the requirements of their various optical functions and designs.

BIOMATERIALS FOR IOLS

Biomaterial selection for IOLs is based on specialized requirements to maintain a stable, clear path for optical imaging as well as long-term safe acceptance as a permanent implant by the unique biology of the eye (Ratner, 1998; Mamalis, 2002). Additionally, the selected biomaterial is required to meet IOL design objectives such as targeted incision size for insertion and needed mechanical fixation, typically via anchoring components called loops or haptics. The haptics permit the IOL to achieve stable location within the eye and also permit the targeted optics design to achieve the functional goal for the IOL. Figure 1a–d illustrates various lens haptics.

Historical Beginning of Poly(methyl methacrylate) (PMMA) as an IOL Material

Harold Ridley of London, in 1949, was the first to implant an IOL made from a PMMA sheet originally formulated for canopies of the British Royal Air Force airplanes (Ridley, 1951). His selection of the material was based on an accidental implantation of pieces of the canopy in pilot's eyes, which were observed by him to be quietly tolerated without noticeable inflammatory reaction or other unacceptable biological consequence. Also, since it remained optically transparent, he chose this Imperial Chemical Industry's PMMA formulation, known as Perspex CQ (Clinical Quality), to create a biconvex 10.5-mm diameter disk-shaped IOL, a historical first. This Ridley IOL was too big and bulky, weighing about 110 mg in air, compared with current IOLs, weighing about 20 mg on average (Patel et al., 1999). It was difficult to mechanically fixate such a bulky IOL in the eye, and serious surgical and postoperative complications led to discontinuation of its use.

TABLE 1 Why Are IOLs Successful? Comparison of Cataract Surgery in the Developed World for Restoration of Vision 30 Years Ago with Spectacles and Today with Foldable IOLs

30 years ago	Foldable IOL today	Improvements (2000+)
• Ten days of hospitalization • Use of general anesthesia or retrobulbar injection in the back of the eye with long needle	• Same-day outpatient surgery in a few hours • Use of topical drops and intracameral anesthetic or peribulbar injection if needed	• Convenience and cost • Lowered risk and complications
• Large >12-mm incision requiring significant stitching and resultant astigmatism	• Small 3-mm incision requiring no stitching	• No stitching-related surgically induced astigmatism
• Use of alpha-chymotrypsin enzyme to weaken zonules and cryoextraction of the entire cataractous lens through the large incision	• Use of ultrasonic phacoemulsifier and irrigating/aspirating instruments to break up and remove the nucleus and cortex, leaving behind the capsular bag with an anterior opening for insertion of IOL in the bag	• Significantly reduced complications of retinal detachment, excessive inflammation, and infection
• No protection of cornea and intraocular tissues during surgery	• Use of a viscoelastic biomaterial to protect the corneal endothelium and other intraocular tissues	• Significantly reduced risk of losing corneal clarity and inflammation of the uveal tissues
• Thick spectacles to restore function of the natural lens resulting in a 25% image magnification, which is incompatible with monocular cataract surgery; also resultant reduced field of vision	• No restriction on monocular cataract surgery	• No delay in cataract surgery in the first eye since IOL provides natural unchanged size of image; also unchanged field of vision
• Overall unsatisfying vision achieved several months after surgery; anxiety and fear of cataract surgery	• Satisfying vision next day after surgery; minimal anxiety and fear of cataract surgery and IOL implantation	• Minimal duration of compromised vision and interruption from occupation; prompt restoration of excellent vision

This concept of the IOL implantation is recognized as a pioneering breakthrough. Also, PMMA remained the dominant IOL material for the next 40 years until the 1990s when foldable IOLs of other materials emerged.

Evaluation of Monofocal IOL Designs and Biomaterials

Although the Ridley IOL was unacceptable, it stimulated many other surgeons, who created several generations of monofocal IOL designs with a much smaller optic made of PMMA and a variety of fixation means and materials (Apple et al., 1989).

The anatomical confines of the eye's anterior chamber (AC) were selected for fixation of IOL. Several AC-IOL designs emerged between 1952 and 1962 with a relatively heavy footplate extension of the optic, which rested in the angle formed by the peripheral iris structure and the collagen fibers of the scleral tissue at the junction with the cornea (Choyce, 1958; Strampelli, 1961). While fixation of the IOL was ensured by these AC-IOL designs, the unique biocompatibility needs of the nearby cornea and iris uveal tissue were revealed. Contact of the IOL with the cornea, for example by rubbing, damaged the corneal endothelium and resulted in loss of corneal clarity. The uveal tissue responded with chronic inflammation. Since exact anatomical sizing was difficult, mis-sized AC-IOLs aggravated the foregoing complications by protrusive erosion or loose rotation within the anterior chamber (Ellingson, 1978). These AC-IOL designs were abandoned, and iris-fixated or -supported designs

emerged between 1959 and 1973 (Binkhorst, 1959). These iris-supported IOLs introduced fixation using several biomaterials attached to the PMMA optic. Fixation structures were shaped as round loops of various materials including nylon 6, nylon 66, and polypropylene suture materials. Even platinum–iridium loops and titanium clips were used. Nylon suture degenerated in the eye (Kronenthal, 1977). Metal loops or clips were heavy, causing injury to the iris and hemorrhage within the anterior chamber (Shephard, 1977). Only iris-supported IOL designs with polypropylene loops created by Cornelius Binkhorst seemed viable. His two-loop iridocapsular IOL was implanted with the loops supported by the iris while the optic was captured by the capsular opening (Alpar and Fechner, 1986). The nucleus and the cortex of the cataractous lens were removed earlier during the surgery through this opening. This method of cataract surgery is called extracapsular cataract extraction (ECCE). Prior to this surgical technique, removal of the entire cataractous lens using an enzyme to weaken the zonules and a cryoextractor was more prevalent. This method of cataract surgery is called intracapsular cataract surgery (ICCE). Iridocapsular supported Binkhorst two-loop IOL implantation after ECCE gave encouraging results, but with the passage of 6 to 10 years, late decompensation of cornea occurred, often requiring corneal transplantation. Also, erosion of the iris, IOL dislocation with dilation of the pupil and difficulty of implantation discouraged its use (Nicholson, 1982; Obstbaum, 1984). The Binkhorst two-loop iridocapsular IOL with ECCE is considered the forerunner of current capsular-fixated posterior chamber IOL (PC-IOL) with ECCE using phacoemulsification.

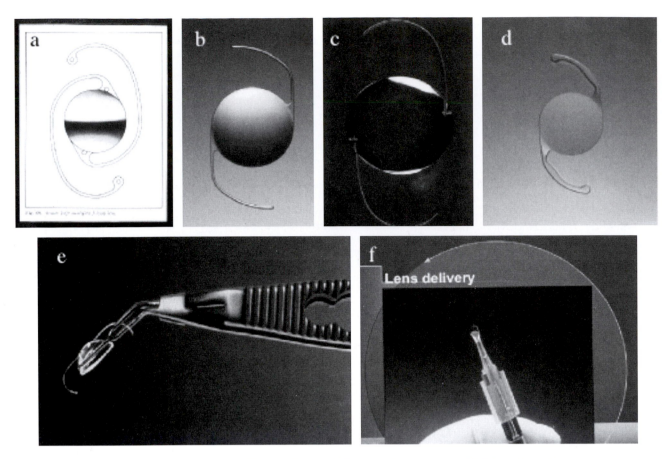

FIG. 1.   IOL transition occurring in developed world is best illustrated here: (a) Past: three-piece PMMA IOL; (b) Current: single-piece PMMA IOL; (c) Past: single-piece foldable IOL; (d) Current: single-piece foldable IOL; (e) Past: insertion forceps held folded IOL; (f) Current: folded IOL emerging from an injector tip cartridge.

Failure of iris-supported IOL led to revisiting AC-IOLs in the 1970s and 1980s with a thinner footplate and the evolution of haptics, much thinner and somewhat flexible extensions of the optic, for fixation. These haptics were an integral part of the IOL made from the same PMMA material, or the haptic could be formed using other biomaterials. Polypropylene suture material, extruded monofilament from other PMMA formulations, and polyamide materials were used for haptics. Even these newer AC-IOL designs gave unsatisfactory long-term outcomes with undesirable corneal decompensation, chronic uveal inflammation, and eventual glaucoma (Apple *et al.*, 1987). Many AC-IOLs with finely looped haptics were associated with micro movement and rubbing with sensitive tissues of the anterior chamber, resulting in loss of corneal endothelial cells, iris chafing, and adhesions around the haptics. While the latest AC-IOL design with a one-piece PMMA optic with integral open haptics, which were wider and yet somewhat flexible, improved the results by avoiding the complications just discussed, the search intensified for fixating the IOL in the posterior chamber where the natural lens is located.

Finally, the search for a PC-IOL led to an IOL by Stevens Shearing in 1979 with a PMMA optic and two J-shaped polypropylene haptics designed for placement behind the iris in the anatomical space called the ciliary sulcus, in front of the posterior capsule (Shearing, 1979; similar to Fig. 1c). Soon, a variety of shapes of haptics emerged with a different degree of flexibility and contact area for ease of implantation and stable fixation. Also, PMMA monofilament as a haptic material replaced polypropylene after some concern for its degradation was raised (Apple *et al.*, 1984). The PC-IOL designs achieved such success that manufacturing processes even created one-piece PMMA IOLs with integral flexible haptics (Fig. 1a,b).

PC-IOL implantation required the ECCE cataract surgery method, and, while large numbers of IOLs were implanted with this method from 1977 to 1992, several refinements emerged to address complications of the sulcus-fixated PC-IOL. Inflammation on account of iris chafing, decentration of the IOL when one haptic was fixated in the sulcus and the other unintentionally went into the bag, and posterior capsule opacification emerged as unacceptable complications of the sulcus-fixated IOL. Placement of the entire IOL (both haptics and the optic) within the capsular bag avoided iris chafing, reduced inflammation, and improved central fixation. C-shaped haptics and the surgical technique of opening the capsular bag by manually controlled tearing called anterior curvilinear capsulorrhexis provided reliable implantation of the PC-IOL in the capsular bag with good centration (Grimbel and Neuhann, 1990). As a result, a large number of surgeons with varying skill levels began to implant PC-IOLs.

By the early 1980s, Perspex CQ PMMA was replaced by most IOL manufacturers with differing PMMA formulations with UV chromophore additives in order to protect the retina from damage by UV radiation.

Viscoelastic materials, such as sodium hyaluronate, were introduced and used to protect the corneal endothelium and other intraocular tissues (Miller and Stegmann, 1982). Even coatings for IOLs and surface modifications for IOLs were investigated in order to minimize damage to corneal endothelium due to accidental touch during insertion into the eye, as well as to minimize foreign-body reaction on the surface of the IOL. Heparin surface-modified PMMA IOLs were introduced, but since they did not solve the posterior capsular opacification complication and also required a longer incision than needed for foldable IOLs, their use was minimal and short-lived (Phillipson *et al.*, 1990; Winther-Nielsen *et al.*, 1998).

The ECCE method of cataract removal, which is required for PC-IOL implantation, retains the entire capsular bag except the capsulorrhexis-excised anterior capsule face. There is a region near the equator of the lens capsule where the residual lens epithelial cells (LEC) with potency for mitosis and migration remain. These LEC proliferate and migrate and spread onto the anterior capsule as well as onto the posterior capsule and even potentially on the IOL surface. These LEC cells undergo morphological changes reducing their transparency resulting in anterior capsule opacification (ACO) and posterior capsular opacification (PCO), as well as contraction of the capsular bag (Kruger *et al.*, 2001; Amon, 2001). PCO is the complication that reduces clarity of the optical path and degradation of the quality of image formation to the extent that it is also called "secondary cataract" formation. While making an adequate opening in the cloudy posterior capsule by a Nd-YAG laser beam without surgically opening the eye restored the clarity of the optical path and the image quality on the retina, there are several disadvantages of this surgical procedure. They include the cost of the procedure, the degraded vision prior to it, the possibility of damaging the lens, and the potential for inducing retinal detachment because of the shock wave created by the focused laser. These disadvantages stimulated approaches to prevent PCO after PC-IOL implantation in the capsular bag. The shape of the optic was changed from convexo-plano to biconvex, and haptics were angulated to achieve complete contact of the posterior surface of the PC-IOL to the posterior capsule. Ridges were also added on the posterior surface to stop migration of LEC without much success. Thus, Nd-YAG laser-created posterior capsulotomy remained a necessary procedure for "secondary cataracts" due to PCO. The rate of need for this posterior capsulotomy depends on the type of IOL material and design, success in surgical removal of much of the cortical contents of the lens, age-related vigor of remaining LEC, and other inflammatory stimulating factors. With some foldable IOL materials, as discussed later, and using the latest surgical techniques for ECCE

with phacoemulsification, the PCO formation rate and the posterior capsulotomy as a necessary second surgical procedure have been significantly reduced.

Rationale and Emergence of Soft Biomaterials for Foldable IOLs

While PC-IOLs of PMMA material were implanted in very large numbers in the early 1980s, another ECCE surgical method drew attention, A probe with a titanium needle that is axially oscillated by an ultrasonic instrument achieved removal of the cataractous nucleus by fragmenting and aspirating it. Cortical material was aspirated out by another probe as a component of the instrument system known as an ultrasonic phacoemulsifier. It also provided necessary simultaneous irrigation into the eye to prevent collapse of the anterior chamber and damage to the cornea. Advancement in this technology with further refinement in the surgical techniques resulted in eyes with less inflammation postoperatively. Also, the incision length required to enter and remove the cataract by this technology is only 3 mm compared with an approximately 9-mm incision required for the manual ECCE extraction method for removal of the nucleus of the cataract. This reduction in incision length resulted in significant reduction in surgically induced astigmatism. Additionally, ECCE using the phacoemulsifier spared mechanical trauma to intraocular tissue encountered during manual ECCE extraction method, thus minimizing tissue irritation and inflammation.

Postoperative large astigmatism induced by the relatively large 9-mm incision required for manual ECCE was degrading the optical image formation and resultant vision (Luntz and Livingston, 1977). While it was managed by surgical and suturing techniques and by appropriate eyeglass correction, it was recognized that the surgically induced astigmatism by a 3-mm incision to enter the eye for ECCE by phacoemulsification was negligible, but subsequent enlargement to insert the PMMA IOL with the preferred 6-mm diameter optic size increased the surgically induced astigmatism (Kohnen *et al.*, 1995; Olson and Crandall, 1998). A soft biomaterial for designing foldable IOLs for insertion through a 3-mm incision was needed to eliminate this surgically induced astigmatism. Additionally, foldable IOL insertion through a 3-mm incision after ECCE with phacoemulsification promised elimination of the need for stitching, and retention of the minimal post surgical inflammation provided by this method of cataract surgery (Gills and Sanders, 1991).

In the early 1980s, a transparent polysiloxane, a formulation originally created for other industrial applications, was used as a biomaterial for foldable IOLs by Starr, Inc. (Mazzocco *et al.*, 1986). In the subsequent 15 years, several formulations of polysiloxane with different refractive indexes, mechanical properties, and UV transmission characteristics emerged from other companies. PolyHEMA, a hydrogel material for contact lens, was also used as a biomaterial for foldable IOLs, but fixation was difficult and it produced unacceptable postoperative complications after Nd-YAG posterior capsulotomy.

In 1994, the first new soft biomaterial for IOLs was introduced by Alcon Laboratories Inc.; as AcrySof IOL. This

TABLE 2 Types of Biomaterials for IOLs in the Market or Clinical Investigation by Leading and Newly Emerging Manufacturers

Manufacturer	Type of IOL	Biomaterials

Advanced Medical Optics	Monofocal IOLs
	Foldable IOLs:
	  Monofocal
	  Multifocal
		PMMA with UV absorber
		Proprietary polysiloxane with
		UV absorber
		Proprietary copolymer of acrylates with UV absorber
		(hydrophobic), Sensar IOL, PMMA monofilament haptics

ALCON	Monofocal IOLs
	
	Foldable IOLs:
	
	  Monofocal
	Foldable IOLs:
	  Toric
	  Multifocal
	  Phakic
	Foldable:
	  Blue Blocking Monofocal IOLs
		PMMA with UV absorber
		PMMA monofilament haptics for three-piece IOLs
		Proprietary copolymer of acrylates with UV absorber
		(hydrophobic) ACRYSOF® IOL, PMMA monofilament
		haptics for three-piece IOLs
		
		Same as above
		
		Same as above with yellow bondable dye

Bausch and Lomb	Monofocal IOLs Foldable monofocal IOLs Phakic AC IOL	PMMA with UV absorber Proprietary polysiloxane with UV absorber Proprietary copolymer of acrylates with UV absorber (hydrophilic). Hydroview® IOL PMMA with UV absorber
C&C Vision	Foldable accommodative IOL—movement based	Proprietary polysiloxane with UV absorber as optic and polyamide haptics, Crytalens® AT-45
Calhoun Vision	Foldable monofocal light Adjustable power IOL	Proprietary photopolymerizable polysiloxane macromers dispersed within polymerized polysiloxane matrix
CIBAvision	Prefolded monofocal IOL Phakic AC-IOL Phakic PC-IOL	Proprietary copolymer of acrylates with UV absorber (hydrophilic), Memory® lens that unfolds upon hydration and warming in the eye Same as above with PMMA footplate covered by the copolymer, Vivarte® IOL Proprietary polysiloxane with UV absorber, PRL, IOL
CORNEAL Laboratories	Foldable monofocal IOLs	PMMA with UV absorber Proprietary polysiloxane with UV absorber Proprietary hydrogel
Human Optics, AG	Foldable accommodative IOL—movement based	Proprietary copolymer of acrylates with UV absorber (hydrophilic)
IOL TECH Laboratories	Foldable monofocal IOL Foldable phakic AC-IOL	Proprietary hydrogel with UV absorber (28% water content) Same as above
Medenium	Foldable monofocal IOL Foldable phakic PC-IOL	Proprietary copolymer of acrylates with UV absorber (hydrophobic), Matrix® IOL Same as given above as CIBAvision PRL® IOL
Ophtec	Phakic AC-IOL Toric phakic IOL	PMMA with UV absorber Iris-fixated Artisan® IOL
Pfizer (Pharmaclia)	Monofocal IOL Foldable monofocal IOLs Multifocal IOL	Heparin surface-modified PMMA with UV absorber Proprietary polysiloxane with UV absorber PMMA with UV absorber
Rayner IOL Ltd., (Manufacturer of Ridley IOL)	Monofocal IOL Foldable monofocal IOLs	PMMA with UV absorber Proprietary copolymer of acrylates with UV absorber (hydrophilic)
STARR Surgical	Foldable monofocal IOL Foldable toric IOL Foldable phakic PC-IOL	Proprietary polysiloxane with UV absorber, also proprietary hydrogel, Collamer® IOL Same as that for Collamer®, known as Implantable Contact Lens (ICL)

biomaterial was specifically created to meet the desired optical, mechanical, and biological properties for foldable IOLs (Anderson *et al.*, 1993). This copolymer of phenylethyl acrylate and phenylethyl methacrylate with a cross-linking agent and a bonded UV absorbing chromophore provided a higher refractive index of 1.55. This allowed the design of a thinner foldable IOL from this hydrophobic soft acrylate material. Its tailored mechanical properties resulted in slow unfolding of the IOL, unlike the rapid, uncontrolled, springlike opening of early silicone IOL with plate haptics. Early three-piece AcrySof IOLs with PMMA monofilament C-shaped haptics were folded and inserted using forceps, similar to three-piece silicone IOL designs (Fig. 1e). Subsequently IOL injectors were developed and are currently used for most of the foldable IOLs (Fig. 1f). Also, AcrySof IOL is now available as a single-piece foldable IOL with integral haptics of the same biomaterial as optic.

The AcrySof IOL soon became the most widely used foldable IOL, and millions of these IOLs had been implanted globally with overall satisfactory outcomes. Besides being a small-incision IOL, it unfolds gently in the eye, remains well centered, and postoperatively results in "quieter" eyes without significant inflammation. Some glistening, which is judged cosmetic, and edge-related optical photic phenomena are reported and now have been minimized through improvements. The newer generation of silicone foldable IOLs and other hydrophobic as well as hydrophilic soft acrylate foldable IOLs also emerged (see Table 2) (Apple *et al.*, 2000a). This success for soft, foldable IOLs, and in particular, AcrySof, is attributed not only to the above-discussed good short-term postoperative results, but also, most importantly, to long-term reduction in PCO formation and the needed Nd-YAG posterior capsulotomy (Hollick *et al.*, 1999). The Nd-YAG capsulotomy rate for AcrySof IOL is <5%, the lowest thus far reported for all IOLs (Apple *et al.*, 2001). Scientific investigations have explained this, based on both the properties of its biomaterial and its design. These investigations are still ongoing with other foldable IOL materials and designs. The surface property of the material and the square-edge design of the optic are attributed as factors for reduction of PCO (Linnola, 1997; Linnola *et al.*, 2000).

In addition to several proprietary hydrophobic soft acrylates and polysiloxanes, several hydrophilic soft acrylates and newer hydrogels have also been introduced as foldable IOLs. But their disadvantages such as greater tendency for fibrotic membrane formation over the anterior surface of the IOL, ACO, and postoperative surface or deep calcification, higher PCO rates, and lack of proven advantage over hydrophobic soft acrylates or polysiloxane IOLs have reduced their acceptance (Koch *et al.*, 1999; Apple *et al.*, 2000b; Werner *et al.*, 2001; Izak *et al.*, 2003).

Currently, hydrophobic acrylates and polysiloxane foldable IOLs are predominantly used in developed countries. Special insertion tools have been designed and used to implant them through a 3-mm incision, the same as required for ECCE cataract extraction by phacoemulsifier.

Though the AcrySof IOL is widely used, many surgeons find silicone IOLs also as acceptable for most cataract patients without other ocular pathology, especially when the cost of the IOL is a consideration. Silicone IOLs are made by several manufacturers. Concern has been expressed about optical complications associated with silicone oil adhesion on silicone IOLs if the use of silicone oil is required in the future for vitreo-retinal surgery (Kaushik *et al.*, 2001; Khawly *et al.*, 1998). IOLs fabricated from other soft hydrophobic acrylates are also now emerging in the market.

Monofocal IOLs with fixed spherical refractive power, discussed earlier, remain the dominant IOL implant for replacement of the cataractous lens. Transition from hard PMMA IOLs to soft foldable IOLs for small-incision cataract surgery has mostly occurred in the developed world. This was primarily driven by improved postoperative results for the patients. Therefore this transition is also expected in the developing world when the necessary resources and skills are available.

IOLS WITH VARIATIONS OF THE OPTICAL FUNCTION

Monofocal Toric IOLs

Since monofocal IOLs with spherical refractive power do not correct preexisting corneal astigmatism, it is corrected either surgically by certain techniques during the cataract surgery or postoperatively using sphere-cylindrical power spectacles. With the same biomaterials as used for monofocal IOLs, toric IOLs with toric optics have recently emerged as another alternative for the more precise correction of significant preexisting corneal astigmatism. The axis of the cylinder for the spherocylindrical optic of the monofocal toric IOL is marked on the IOL for necessary proper alignment during the cataract surgery. Postoperative rotation of a toric silicone IOL with plate haptics (from Staar Surgical) has been widely reported since it is difficult to size it for needed fixation in the capsular bag (Leyland *et al.*, 2001). Such rotation significantly compromises the optical objective of the toric IOL. Another toric IOL design with flexible haptics (from Alcon Inc.), that can provide improved stable fixation has been clinically investigated with much better results without significant rotation (Lane, 2003).

Multifocal IOLs

Since monofocal IOLs optically focus on an object at only one chosen distance with a relatively narrow depth of focus, postoperatively, patients are required to wear spectacles for other distances. Even though almost all elderly cataract patients are accustomed to wearing spectacles because of their presbyopia prior to cataract, the investigation of the opportunity to eliminate spectacles using multifocal IOLs began in the 1980s. With the same biomaterials as for monofocal IOLs, several multifocal optical designs using refractive optical zones or diffractive optical structures have been investigated, implementing simultaneous optical imaging for objects located at a distance as well as near (Maxwell and Nordan, 1990). This optical design principle is called simultaneous vision and has the potential for halo and other unwanted optical consequences depending on the optical design. Multifocal IOLs with refractive optical zones are sensitive to pupil size in terms of their optical performance. The FDA-approved Array Multifocal IOL by AMO, Inc., has a radial array of alternating variable optical power to provide both distant and near optics for any pupil size (Steinert *et al.*, 1999). Although for pupils of 3-mm diameter or larger such a design does provide multifocal vision, the near vision is reported to be suboptimal for a smaller pupil size. Also, halos appear with nighttime driving when the pupil is larger, and these remain a potentially unacceptable complication for some patients (Haring *et al.*, 2001).

Earlier, the 3M Corporation, and, more recently, Pharmacia Corporation (now merged with Pfizer, Inc.), introduced a diffractive optic design that provides pupil-size-independent distant and near optics, but still retains potentially unacceptable halos with nighttime driving. With better understanding of the cause of such halos, a unique optical design that combines a limited central diffractive optical zone with refractive optics beyond it was introduced by Alcon, Inc., as the AcrySof Multifocal IOL to minimize nighttime halo and retain pupil-independent optics for daytime (Dublineau, 2003).

The simultaneous vision principle based multifocal IOLs demands a more accurate selection of the necessary IOL power in order to achieve spectacle-free vision postoperatively. Also, since the vision achieved is not an actual accommodation, it is called pseudoaccommodation.

Applications of multifocal IOLs for treatment of presbyopia, hyperopia, and myopia are currently being clinically investigated in certain suitable patients. For treatment of any one of these refractive errors of the eye, its noncataractous, clear natural lens is surgically removed and is replaced by a multifocal IOL (Packer *et al.*, 2002). This lensectomy is carried out by the same surgical procedure as is required for cataract surgery. Unpredictability of achieving necessary targeted refraction and acceptance of unavoidable optical effects of the simultaneous vision optics of multifocal IOLs remain key hurdles for its widespread use for treatment of refractive errors.

Phakic IOLs

Currently, for correction of refractive errors of myopia or hyperopia, treatment alternatives include spectacles, contact lenses, and a variety of corneal refractive surgery procedures including LASIK. As another modality of treatment, the phakic IOL, which is an IOL to correct the refractive error of the eye, is placed in the phakic eye in addition to the natural crystalline lens. Phakic IOLs have been clinically investigated for many years with continuous improvement to achieve long-term safety and, it is hoped, widespread use. Unique designs with newer biomaterials are currently being investigated for achieving this goal.

For correction of myopia, the angle-supported phakic AC-IOL designs of PMMA, similar to earlier AC-IOLs for cataract surgery, resulted in complications of damaged corneal endothelial cells and pupil ovalization. More flexible, soft and foldable phakic AC-IOL designs of hydrophobic soft acrylate material (Alcon's AcrySof Phakic AC-IOL) and of hydrogel material (Cibavision's Vivarte Phakic AC-IOL), when properly sized, are reported to have improved acceptable results in ongoing long-term clinical investigations (Knorz, 2003; Henahan,

2001). Iris-fixated phakic AC-IOLs of PMMA (Ophtec's Artisan Phakic AC-IOL), when properly implanted, also have been reported to have acceptable results, but the surgery requires special surgical techniques (Maloney *et al.*, 2002). Additionally, long-term safety data for the cornea and the iris are not yet available with well-controlled studies. The anterior chamber of the hyperopic eye is anatomically relatively small and thus less suitable for a phakic AC-IOL.

Phakic posterior chamber IOL designs using silicone and hydrogel materials are placed between the iris and the natural lens to correct myopic and hyperopic refractive errors. They have been clinically investigated over the past few years. Their anatomical placement has potential for iris chafing and related complication of pigmentary glaucoma. Also, cataractogenesis of the natural lens by either metabolic disturbances or accidental touch of a surgical instrument is reported, especially with silicone phakic PC-IOLs (Sanchez-Galeana *et al.*, 2003). A hydrogel phakic PC-IOL (Starr's ICL), after several design iterations in its final design, is reported to have fewer complications, and U.S. FDA approval is expected (Sanders *et al.*, 2003).

Accommodative IOL

In order to provide true accommodation after cataract surgery, as well as to treat refractive errors of myopia, hyperopia, and presbyopia, a safe and effective accommodative IOL with 3 or more diopters of accommodation ability to replace the cataract or the natural lens has been long sought after.

Designs implementing the concept of forward movement of the optic of the IOL to achieve accommodation were pursued by C&C Vision, Inc., of the United States and Human Optics of Germany (Samalo, 2002). The Crystalens AT-45 of C&C Vision is a silicone IOL with polyamide haptics and two hinges that permit axial forward movement of the optic for accommodation and subsequent backward movement for disaccommodation (Cumming *et al.*, 2001). Although only 1 to 2 diopters of such movement-based accommodation is reported, the results remain variable and without scientific understanding of factors responsible for such movement. Since there is no other significant safety issue, even this limited success is encouraging.

In order to achieve a greater magnitude of accommodation, a mechanically coupled pair of optic-based accommodative IOL was investigated in an animal model by Hara in Japan, and then by Sarfarazi in United States (Hara *et al.*, 1992). Recent results in monkey eyes by Sarfarazi are encouraging for this approach, which does not require any new materials for the design (Sarfarazi, 2003).

Another concept under investigation in animal models requires removal of the contents of the natural cataractous or clear lens through the smallest possible anterior capsulotomy and subsequently filling it by appropriate silicone polymer to achieve the needed geometrical shape and optical power (Haefliger *et al.*, 1987). Other gel biomaterials have also been investigated, but a host of surgical, biological, and optical issues remain as hurdles to this approach to restore accommodation using the natural capsular bag, zonules, and ciliary muscle for a physiology-based truly accommodative IOL. In another variation of this approach, Nishi of Japan has reported implantation of a silicone-filled balloon as an IOL into the mostly intact capsular bag to achieve 4 to 8 diopters of accommodation in monkey eyes (Nishi *et al.*, 1993). Thus, a search for truly accommodative IOL may challenge interdisciplinary resources, including newer biomaterials for some of the above concepts.

Adjustable Power IOLs

For achieving the targeted postoperative refraction for any pseudophakic eye, careful measurement of the axial length of the eye and refractive power of its cornea are required before IOL implantation surgery. Subsequently, using one of many available IOL power selection mathematical formulas, the necessary IOL power is selected by the surgeon for implantation. Errors in this procedure and deviation of the actual axial position of the IOL from that anticipated by the formula result in postoperative residual refractive error. Most of the time, postoperative spectacle correction to address this error is acceptable, but occasionally excessive error requires explantation of the IOL to address unacceptable imbalance of refraction between the two eyes. Also, to achieve the goal of spectacle-free vision with multifocal IOLs or phakic IOLs, achievement of targeted refraction is more critical and necessary. Thus, for many years, concepts for a postoperative adjustable power IOL have been investigated. Several concepts requiring surgical reentry into of the eye were investigated, but did not result in products because of concern for safety and cost. A new, unique biomaterial-based light adjustable IOL by Calhoun Vision, Inc., is being pursued clinically in the human eye (Schwierling *et al.*, 2002). This approach is attractive since it does not require surgical reentering of the eye for adjustment of the refractive power of the already implanted IOL. This new biomaterial technology consists of an IOL, which is implanted as a polymerized silicone matrix containing nonpolymerized silicone macromers with an attached photoinitiator dispersed within it for subsequent staged photopolymerization (Maloney, 2003). The IOL is implanted as per current practice for IOL power selection for the targeted refraction. Postoperatively, the achieved refraction is measured. Depending upon sign and magnitude of the refractive error, for necessary adjustment of power of the IOL, the IOL is judiciously exposed to long UV radiation from a slit-lamp based light delivery instrument leading to partial photopolymerization. Spatial intensity distribution is used for either exposing the central optical zone, to increase the power of the IOL, or exposing the peripheral optical zone, to decrease the power of the IOL. After the required duration for redistribution of the remaining unpolymerized macromers in the IOL, the patient is reexamined for the consequent change in the shape, and, hence, power of the IOL toward the targeted correction of the refractive error. If the IOL power adjustment is not achieved, the procedure of partial polymerization is repeated with needed adjustment until targeted refraction is achieved. At this stage, as a final step, the entire optic is exposed with UV radiation for complete "lock-in" of the shape of the IOL. After the safety of this IOL was established in rabbit eyes, currently it is under clinical investigation required to establish safety and efficacy in terms of accuracy, precision, reproducibility, and acceptance of resultant optical shapes.

Yellow-Tinted Blue Blocking IOL

Menicon and Hoya Corporations of Japan, for their PMMA IOLs, and recently Alcon, Inc., for its AcrySof IOL, have each incorporated a yellow dye into their IOL materials in order to block visible blue light in addition to complete blocking of UV radiation. The AcrySof Natural IOL mimics the UV-visible transmission of a 53-year-old human natural lens (Cionni, 2003). The scientific rationale for such a yellow-tinted IOL is to prevent unnatural excessive blue color perception, and to protect the more susceptible aging retina after cataract surgery. Blue light can be toxic to the retina and is considered one of the potential risk factors of age-related macular degeneration, a serious sight-threatening disease.

OVERALL SUMMARY AND FUTURE OF IOLs

In a relatively brief span of a few decades, IOLs have emerged as one of the most successful biomaterial-based implants. They are considered a safe and effective means to prevent the leading cause of age-related degradation of vision, cataracts, that eventually leads to functional blindness. Currently, an estimated 10 million IOLs are implanted annually for cataract replacement, mostly in the developed world, and in some regions of the developing world where resources are available. If the prevailing rate of cataract surgery in the United States is projected for the approximately 6 billion population of the world, then it is estimated that 45 million IOLs are needed annually! Few ophthalmologists and little money have led to delayed (or no) cataract treatment in the developing world, resulting in a tragic backlog of 20 million cataract-related blind people. Thus, with expected continuous global development and simultaneous increases in population as well as life expectancy, and with increasing demand for good vision, the future of IOLs for cataract surgery is assured.

The successful developments of IOLs were outcomes of interdisciplinary team efforts by creative surgeons, scientists, and engineers in both academia and industry. Starting from a biomaterial selection based on accidental implantation of pieces of PMMA from the canopies of fighter airplanes, a variety of newer biomaterials have been specifically created to achieve the various designs and functional goals of IOLs, as outlined in Table 2. Creative interactions and an understanding of the opportunities and issues of biomaterials, optical and mechanical designs, and evolving surgical techniques, along with improved understanding of the unique biology of the eye, have spearheaded the sophistication we see in modern IOLs. These efforts are expected to continue to achieve development of truly accommodative IOLs with the capability of mimicking the physiological function of the natural lens. Multidisciplinary efforts will also continue to achieve improved phakic IOLs to correct refractive error of presbyopia, myopia, and hyperopia. When long-term safety and efficacy are demonstrated, phakic IOLs or truly accommodative IOLs have the potential to replace the current alternatives of spectacles, contact lenses, and a variety of corneal refractive surgery procedures, including LASIK. Thus, globally, the number of IOLs required for all functional goals could be manyfold greater than the given projection for IOLs needed for cataract replacement.

Further opportunities for improved optical designs of IOLs for correction of corneal aberration have just started with the Pharmacia Tecnis IOL (now Pfizer) (Mester *et al.*, 2003). With further understanding of the optical needs for achieving optimum vision, eventually a custom IOL may evolve for each individual eye for all of the projected functions of IOLs for aphakic and phakic eyes (Kohnen, 2003). Achievement of these goals will require advances in biomaterials for IOLs and biomaterials scientists working closely with physicists, chemists, engineers, surgeons, regulatory affairs experts, and business people. The success of IOLs in improving vision worldwide genuinely showcases the strength of the biomaterials endeavor and the highly multidisciplinary nature of this field.

QUESTIONS

1. What are the advantages of IOL implantation over spectacles for correction of vision after removal of cataract? What other improvements have been achieved in cataract surgery over the past 30 years?
2. Which was the first biomaterial selected for IOL implant? What was the basis for this selection?
3. What is the rationale for foldable IOLs? Which foldable soft hydrophobic materials are preferred over soft hydrophilic materials and why?

Bibliography

Alpar, J. J., and Fechner, P. U. (1986). *Fechner's Intraocular Lenses*. Thieme.

American Academy of Ophthalmology (1989). *Cataract in the Otherwise Healthy Adult Eye*. AAO, San Francisco.

Amon, M. (2001). Biocompatibility of intraocular lenses. *J. Cataract Refract. Surg.* 27: 178–179.

Anderson, C., Koch, D. D., Green, G., Patel, A., and Vannoy, S. (1993). Alcon AcrySof Acrylic Intraocular Lens. in *Foldable Intraocular Lenses*, R. G. Martin, J. P. Gills, D. R. Sanders, eds. Slack, Thorofare, NJ, pp. 161–177.

Apple, D. J., Mamlis, N., Bray, S. F., *et al.* (1984). Biocompatibility of implant materials: a review and scanning electron microscopic study. *J. Am. Intraocular Implant Soc.* I: pp. 53–66.

Apple, D. J., Brems, R. N., Park, R. B., *et al.* (1987). Anterior chamber lenses. Part I: Complications and pathology and a review of design. *J. Cataract Refractive Surgery* 13: 157–174.

Apple D. J., Kincaid, Inc., Mamlis, N., and Olson R. J. (1989). *Intraocular Lenses: Evolution, Designs, Complications and Pathology*. Williams and Wilkins, Baltimore.

Apple, D. J., Auffarth, G. U., Peng, Q., and Visessook. N. (2000a). *Foldable Intraocular Lenses: Evolution, Clinicopathologic, Correlations and Complications*. Slack, Thorofare, NJ.

Apple, D. J., Werner, L., Escobar-Gomez, M., and Pandey, S. K. (2000b). Deposits on the optical surfaces of Hydroview intraocular lenses (letter). *J. Cataract Refract. Surg.* 26: 1773–1777.

Apple, D. J. *et al.* (2001). Eradication of posterior capsule opacification: documentation of a marked decrease in Nd-YAG laser posterior capsulotomy rates noted in analysis of 5,416 pseudophakic human eyes obtained post-mortem. *Ophthalmology* 108: 505–518.

Binkhorst, C. D. (1959). Iris-supported artificial pseudophakia: a new development in intraocular artificial lens surgery (iris clip lens). *Trans. Ophthalmol. Soc. UK* 79: 859–584.

Brian, G., and Taylor, H. (2001). Cataract blindness: challenges for the 21st century. *Bull. World Health Org.*, 249–256.

Choyce, D. P. (1958). Correction of uni-ocular aphakia by means of anterior chamber acrylic implants. *Trans. Ophthalmol. Soc. UK* 78: 459–470.

Cionni, R. J. (2003). Clinical study results of the AcrySof Natural IOL. in *Abstracts, Symposium on Cataract, IOL, and Refractive Surgery*. ASCRS, Fairfax, VA, p. 11.

Cumming, J. S., Slade, S. G., and Chayet, A. (2001). Clinical evaluation of the model AT-45 silicone accommodative intraocular lens; results of feasibility and the initial phase of a Food and Drug Administration clinical trial: the AT-45 Study Group. *Ophthalmology* 108: 2005–2009; discussion by Werblin, T. P., p. 2010.

Dublineau, P. (2003). Experience with the AcrySof RESTOR IOL. in *Abstracts, Symposium on Cataract, IOL, and Refractive Surgery*. ASCRS, Fairfax, VA, p. 57.

Ellingson, F. T. (1978). The uveitis–glaucoma–hyphema syndrome associated with the Mark VIII Anterior Chamber Lens Implant. *Am. Intraocular Implant Soc. J.* 4: 50–53.

Gills, J. P., and Sanders, D. R. (1991). Use of small incision to control induced astigmatism and inflammation following cataract surgery. *J. Cataract Refract. Surg.* 17(supplement): 740–744.

Grimbel, H. V., and Neuhann, T. (1990). Development, advantages, and methods of the continuous circular capsulorrhexis technique; *J. Cataract Refract. Surg.* 16: 31–37.

Haefliger, E., Parel, J. M., Fantes, F., *et al.* (1987). Accommodation of an endocapsular silicone lens (phaco-ersatz) in the non-human primate; *Ophthalmology* 94: 471–477.

Hara, T., Hara T. Yasuda, A., *et al.* (1992). Accommodative intraocular lens with spring action—part 2. Fixation in the living rabbit. *Ophthalm. Surg.* 23: 632–635.

Haring, G., Dick, H. B., Krummenaur, F., Weissmantel, U., and Uroncke, W. (2001). Subjective photic phenomena with refractive multifocal and monofocal intraocular lenses; results of a multicenter questionnaire. *J. Cataract Refract. Surg.* 27: 245–249.

Henahan, J. F. (2001). A first look at the presbyobic phakic IOL. *Eyeworld News*. ASCRS, Fairfax, VA 6: 76.

Hollick, E. J., Spalton, D. J., and Ursell, P. G. (1999). The effect of polymethyl methacrylate, silicone, and polyacrylic intraocular lenses on posterior capsule opacification three years after cataract surgery. *Ophthalmology* 106: 49–55.

Izak, A. M., Werner, L., Pardey, S. K., and Apple, D. J. (2003). Calcification of modern foldable hydrogel intraocular lens designs. *Eye* 17: 393–406.

Kaufman, P. L. (1992). Accommodation and presbyopia: neuromuscular and biophysical aspects. in *Adler's Physiology of the Eye: Clinical Application*, W. M. Hart, Jr., ed. 9th edition St. Louis, MO, pp. 391–411.

Kaushik, R. J., Brar, G. S., and Gupta, A. (2001). Neodymium-YAG capsulotomy rates following phacoemulsification with implantation of PMMA, silicone, and acrylic intraocular lenses. *Ophthalmic Surg. Lasers* 32: 375–382.

Khawly, J. A., Lambert, R. J., and Jaffe, J. G. (1998). Intraocular lens changes after short- and long-term exposure to intraocular silicone oil; an *in vivo* study. *Ophthalmology* 105: 1227–1233.

Knorz, M. (2003). Three-year phase I clinical results of the AcrySof phakic ACL. in *Abstracts, Symposium on Cataract, IOL and Refractive Surgery.* ASCRS, Fairfax, VA, p. 25.

Koch, M. U., Kalicharan, D., and Vanderwant, J. J. L. (1999). Lens epithelial cell formation related to hydrogel foldable intraocular lenses. *J. Cataract Refract. Surg.* 25: 1637–1640.

Kohnen, T., Dick, B., and Jacobi, K. W. (1995). Comparison of the induced astigmatism after temporal clear corneal tunnel incision of different size. *J. Cataract Refract. Surg.* 21: 417–424.

Kohnen, T. (2003). Aberration-correcting intraocular lenses. *J. Cataract Refract. Surg.* 29: 627–628.

Kronenthal, R. L. (1977). Intraocular degradation of nonabsorbable suture. *Am. Intraocular Implant Soc. J.* 3: 222–228.

Kruger, A. J., Amon, M., Schauersberger, J., *et al.* (2001). Anterior capsule opacification and lens epithelial outgrowth on the intraocular lens surface after curettage. *J. Cataract Refract. Surg.* 127: 1987–1991.

Lane, S. S. (2003). Rotational stability of the AcrySof Toric single-piece IOL. *Abstracts, Symposium on Cataract, IOL, and Refractive Surgery.* ASCRS, Fairfax, VA, p. 9.

LeGrand, Y., and El Hage, S. G. (1980). *Physiological Optics*, Springer Verlag Series in Optical Sciences, Vol. 13. New York.

Leyland, M., Zinicola, E., Bloom, P., and Lee, N. (2001). Prospective evaluation of a plate haptic toxic intraocular lens. *Eye* 15(2): 202–205.

Linnola, R. J., (1997). The sandwich theory: a bioactivity-based explanation for posterior capsule opacification after cataract surgery. *J. Cataract Refract. Surg.* 23: 1539–1542.

Linnola, R. J., Werner, L., Pandey, S. K. *et al.* (2000). Adhesions of fibronectin, vitronectin, laminin and collagen-type IV to intraocular lens materials in pseudophakic human autopsy eyes, part II: explanted IOLS. *J. Cataract Refract. Surg.* 26: 1807–1818.

Luntz, M. H., and Livingston, D. G. (1977). Astigmatism in cataract surgery. *Br. J. Ophthalmol.* 61: 360–365.

Maloney, R. (2003). The changing shape of customized IOLs. *Rev. Ophthalmol.* 10: 01.

Maloney, R. K., Nguyen, L. H., and John, M. E. (2002). Artisan phakic intraocular lens for myopia: short-term results of a prospective, multicenter study. *Ophthalmology* 109: 1631–1641.

Mamalis, N., (editorial) (2002). IOL biocompatibility. *J. Cataract Refract. Surg.* 28: 1–2.

Maxwell, W. A., Nordan, L. T. (eds.) (1990). *Current Concepts of Multifocal Intraocular Lenses.* Slack, Thorofare, NJ.

Mazzoco, T. R., Rajacich, G. M., and Epstein, E. C. (1986). *Soft Implant Lenses in Cataract Surgery.* Slack, Thorofare, NJ.

Mester, V., Dillinger, P., and Anterist, N. (2003). Impact of a modified optic design on visual function: clinical comparative study. *J. Cataract Refract. Surg.* 29: 652–660.

Miller, D., and Stegmann, R. (1982). The use of Healon in intraocular lens implantation. *Int. Ophthalmol. Clin.* 22: 177–187.

Nicholson, D. H. (1982). Occult iris erosion: a treatable case of recurrent hyphema in iris-supported intraocular lenses. *Ophthalmology* 84: 113–120.

Nishi, O., Nakai, Y., Yamada, Y., and Mizymoto, Y. (1993). Amplitudes of accommodation of primate lenses refilled with two types of inflatable endocapsular balloons. *Arch. Ophthalmol.* 111: 1677–1684.

Obstbaum, S. A. (1984). Complications of intraocular lenses. Membranes, discolorations, inflammation, and management of the posterior capsule. in *Cataract Surgery: Current Options and Problems*, J. M. Engelstein, ed. Grune and Stratton; Orlando, FL; pp. 509–533.

Obstbaum, S. A. (editorial) (2003). Emergence of the role of cataract and IOL surgery in the correction of refractive errors. *J. Cataract Refract. Surg.* 29: 857.

Olson, R. J., and Crandall, A. S. (1998). Prospective randomized comparison of phacoemulsification cataract surgery with a 3.2-mm versus a 5.5-mm sutureless incision. *Am. J. Ophthalmol.* 125: 612–620.

Packer, M., Fine, I. H., and Hoffman, R. S. (2002). Refractive lens exchange with the Array multifocal intraocular lens. *J. Cataract Refract. Surg.* 26: 421–422.

Patel, A. S., Carson, D. R., and Patel, P. H. (1999). Evaluation of an unused 1952 Ridley intraocular lens. *J. Cataract Refract. Surg.* 25: 1535–1539.

Phillipson, B., Fagerholm, P., Calel, B., *et al.* (1990). Heparin surface modified intraocular lenses: a one-year followup of a safety study. *Aceta. Ophthalmol. (Copenhagen)* 68: 601–603.

Ratner, B. D. (guest editorial) (1998). Ophthalmologic Biocompatibility: Anachronism or oxymoron? *J. Cataract Refract. Surg.* 24: 288–290.

Ridley, H. (1951). Intra-ocular acrylic lenses. *Trans. Ophthalmol. Soc. UK* 71: 617–621.

Sarfarazi, F. M. (2003). Optical and mechanical design for human implantation of the Sarfarazi elliptical accommodating IOL. *Abstracts, Symposium on Cataract, IOL, and Refractive Surgery.* ASCRS, Fairfax, VA, p. 189.

Samalonis, L. B. (ed.) (2002). The 21st century IOL. *Eye World News*: ASCRS, Fairfax, VA 7: 30.

Samalonis, L. B. (ed.) (2002). Accommodative IOLs coming a long way. *Eyeworld*, March. American Society of Cataract and Refractive Surgery, Fairfax, VA.

Sanchez-Galeama, C. A., Smith, R. J., Sanders, D. R., *et al.* (2003). Lens opacities after posterior chamber phakic intraocular lens implantation. *Ophthalmology* 110: 781–785.

Sanders, D. R., Grabow, H.B., and Shepard, J. (1992). The toric IOL. in *Sutureless Cataract Surgery: An Evolution Towards Minimally Invasive Technique*, J. P. Gills, R. G. Martin, and D. R. Sanders, eds. Slack, Thorofare, NJ; pp. 183–197.

Sanders, D. R., Vukich, J. A., Doney, K., and Gaston, M. (2003). U. S. Food and Drug Administration clinical trial of the implantable contact lens for moderate to high myopia. *Ophthalmology* 110: 225–266.

Schwierling, J. T., Schwartz, D. M., Sandsedt, C. A., and Jethmalani, J. (2002). Light-adjustable intraocular lenses. *Rev. Refract. Surg.* 3: 01.

Shearing, S. P. (1979). Mechanism of fixation of the Shearings posterior chamber intra-ocular lens. *Contact Intraocular Lens Med. J.* 5: 74–77.

Shepard, D. D. (1977). The dangers of metal-looped intraocular lenses. *Am. Intraocular Implant Soc. J.* 3: 42.

Steinert, R. F., Aker, B. L., Trentacost, D. L., Smith, P. J., and Tarantino, N. (1999). A prospective comparative study of the AMO ARRAY zonal-progressive multifocal silicone intraocular lens and a monofocal intraocular lens. *Ophthalmology* 106: 1243–1255.

Strampelli, B. (1961). Anterior chamber lenses: present technique. *Arch. Ophthalmol.* 66: 12–17.

U.S. Department of Health and Human Services (1983). Cataract in Adults: Management of Functional Impairment, Clinical Practice; Guideline No. 4; Washington, DC.

Werner, L. Apple, D. J., Kaskaloglu, M., and Pardey, S. K. (2001). Dense opacification of the optical component of a hydrophilic acrylic intraocular lens: a clinicopathologic analysis of 9 explanted lenses. *J. Cataract Refract. Surg.* 27: 1485–1492.

Winther-Nielsen, A., Johansen, J., Pederson, G. K., and Corydon, L. (1998). Posterior capsular opacification and neodymium-YAG capsulotomy with heparin surface modified intraocular lenses. *J. Cataract Refract. Surg.* 24: 940–944.

7.12 BURN DRESSINGS AND SKIN SUBSTITUTES

Jeffrey R. Morgan, Robert L. Sheridan, Ronald G. Tompkins, Martin L. Yarmush, and John F. Burke

Every year approximately 12,000 people die from severe burns and thermal injury. Most of these deaths are due to the catastrophic problems that ensue when the skin's integrity is disrupted. The major lethal problems are massive fluid losses and microbial invasion. Prompt replacement of the integrity of the skin is a cornerstone of therapy for these patients, but lack of available natural skin makes this an almost impossible task for those patients with very large burns.

The outlook for those who have suffered serious burns has improved dramatically over the past 20 years, in terms of both survival and postinjury quality of life. Twenty years ago a 30% burn was a serious threat to life and smaller burns resulted in protracted hospitalization and disability. Today such injuries are routinely managed with hospital stays measured in days, followed by prompt return to work or school. These improvements arose after the value of early excision and immediate autograft closure of full-thickness wounds was recognized and widely adopted. However, in those with large injuries, these favorable outcomes have not been realized because of the inability to accomplish immediate definitive wound closure. Until a durable and reliable skin replacement has been developed, these patients will continue to require protracted acute hospitalizations and find their reconstructive options limited by a shortage of autograft. The modern search for a suitable skin substitute has been under way since the 1940s, with steady progress as the principles of wound healing and the functions and properties of the skin became better understood. This chapter reviews those principles and examines their application as they relate to the search for the ideal skin substitute for patients with massive burns.

BRIEF HISTORY OF ADVANCES IN BURN TREATMENT

Burn patients die for two reasons: burn shock during the first few postinjury days and, in those who survive burn shock, wound sepsis during the first few postinjury weeks. Burn shock resulting from the diffuse capillary leak triggered by the burn wound, which results in inadequate circulating volume and progressive hypotension and eventual shock. The necrotic skin and subcutaneous tissue predictably become infected over the first postburn week, with resulting systemic inflammation and progressive sepsis. Recognition of these processes and the development of effective interventions have had a profound impact on the natural history of the injuries.

The exaggerated fluid requirements of burn patients were first understood in the 1930s. Various fluid resuscitation formulas were developed and refined over the ensuing 30 years. Although there is still room to improve fluid resuscitation techniques and to modify the physiology behind the high volume requirements, burn shock has been almost eliminated as a cause of death, except in those with cardiovascular disease or massive injuries.

Traditional burn management was expectant. Even in the 1950s and 1960s, when fluid resuscitation was becoming very effective, burn wounds were treated with various salves while the eschar (necrotic skin and fat) became colonized with bacteria, liquefied, and separated from the underlying viable tissue. This resulted in the regular occurrence of wound sepsis and nearly universal death in those with larger wounds. At this time there was therefore no need for a skin substitute. The odor from the liquefying wounds led to the separation of burn patients from other parts of the hospital. In the 1970s, the favorable impact of early identification and excision of full-thickness wounds with immediate skin graft placement on underlying viable tissue was recognized (Burke *et al.*, 1974). These procedures are not simple and were initially practiced only in those with small injuries, where truncated hospital stays and enhanced functional outcomes were noted. Those with large injuries continued to die; there was still no need for a skin substitute. With better availability of blood banks and intensive care units, burn excision is now practiced in those with large injuries where enhanced survival and shortened stays have been reported. However, closure of the massive wounds thus generated has become a very difficult clinical problem.

PRINCIPLES OF WOUND COVERAGE AND HEALING

Hospitalization of the massively burned patient is divided into four general phases: initial evaluation and resuscitation, initial wound excision and biologic closure, definitive wound closure, and rehabilitation and reconstruction. Currently, even those patients who suffer very large injuries can be expected to recover and become reintegrated with family and community. Length of hospital stay can be roughly estimated as 8 days + 0.8 days per percent burn and, in those with large injuries, is primarily limited by the absence of a reliable permanent skin substitute.

To understand the requirements necessary to replace lost skin, we must first understand the general mechanism of wound healing. Burn injuries result in loss of skin structure to varying degrees (Tompkins and Burke, 1989). Usually burns are first degree (the loss of the epidermal layer), second degree (the loss of the epidermal layer and a portion of the dermis), or third degree [the loss of tissue through the dermis, including the hair follicles and sweat glands, and extending into the hypodermis (subcutaneous) layers] (Fig. 1). Occasionally, as often seen in electrical injuries, a deep, full-thickness wound is referred to as a fourth-degree burn, which is defined as extending downward through the subcutaneous tissues to involve tendon, bone, muscle, and other deep structures.

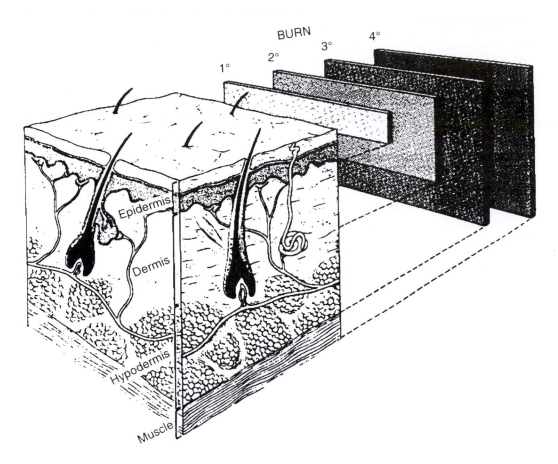

FIG. 1.   First-, second-, and third-degree burns and the depth of a split-thickness skin graft.

Second-degree burns can generally be classified as either superficial or deep. The interface between the epidermis and dermis is not linear, but consists of many papilla formed by the rete pegs of the epidermis. Also the skin varies in thickness at different locations on the body. In superficial injury, enough of the deep epidermal or superficial dermal layers may remain to allow spontaneous healing of the wound by reepithelialization. Other sources of epidermal cells are the epidermal appendages, including the hair follicles, sweat glands, and sebaceous glands. These wounds can be treated by simple dressings without topical antibiotics and will heal spontaneously within 10 to 14 days. Deep second-degree burns have completely destroyed the epidermis and extend further into the dermis, with large amounts of necrotic tissue being present. Both the fluid and bacterial barriers are severely compromised, putting the patient at a much higher risk. These wounds, if allowed to heal on their own (because the dermis has been grossly distorted or destroyed), result in hypertrophic scarring, with nonoptimal functional and cosmetic results. These types of wounds should be treated as if they were third-degree burns to allow for faster and better healing.

A third-degree burn is defined as having damaged the skin all of the way through the dermis and into the subcutaneous tissue. The wound is freely permeable to fluids, proteins, and bacteria. The constant proteinaceous exudate

from these wounds, combined with an abundance of necrotic skin above, make an ideal medium for bacterial growth. In these injuries, all the dermis and all of the epidermal cells within the wound have been destroyed, including those in the epidermal appendages. Demis cannot regenerate, so it must be replaced, whereas epidermis can regenerate; however, without replacement, epidermal cells must migrate from the distant wound edges. Therefore, prompt debridement of the wound followed by autografting or other methods of wound coverage is the treatment of choice. Small third-degree wounds (those of less than 2 cm) may be allowed to heal spontaneously by ingrowth from the wound edges, but this process requires at least 6 weeks; the wound may never heal completely and always results in a generous scar.

When a patient suffers a third-degree burn, the necrotic layer of tissue that was once viable skin is referred to as the "eschar." This layer is removed surgically and any of the underlying dermis and hypodermis is then covered. If the freshly excised open wound is not immediately closed either with skin or skin replacement, then "granulation tissue" is formed by the local invasion of small blood vessels and fibroblasts from beneath the wound's surface. Granulating wounds become reddened with these new vascular structures, and if no grafting is performed and the wound is simply allowed to spontaneously heal, the continued invasion of fibroblastic tissue

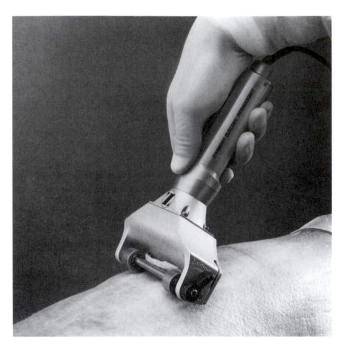

FIG. 2. A dermatome used to harvest split-thickness skin grafts.

FIG. 3. A skin meshing device.

will eventually form a hypertrophic scar through the unorganized deposition of collagen. New epidermis seen around the edges of the wound is usually inadequate for reepithelialization in all but the smallest injuries. Myofibroblasts, specialized fibroblastic cells with the contractile properties of smooth muscle, invade the wound and begin to pull the edges inward and result in contractures and restriction in movement of the surrounding skin and its structures. This process can result in severe deformation of the surrounding features and is particularly troublesome in places such as the face, neck, and limbs. In some cases of massive wounds, even contraction of the wound will fail to bring the edges of the wound together, and the center of the wound remains open permanently.

Clearly the best coverage for the wound is natural skin taken from the individual himself (an autograft) to avoid specific immunological incompatibility. If the burn injury is anywhere from 35 to 50% of the total surface area of the body, it is frequently possible to transplant partial-thickness skin grafts from other noninjured areas of the patient. These grafts are usually about 0.3 to 0.5 mm thick and include the epidermal layer and a thin portion of the underlying dermis. They are harvested from the donor site using a reciprocating blade such as an electric dermatome or a hand-held knife (Fig. 2). The graft is placed on the freshly excised wounds and survives by simple diffusion of nutrients for the first 72 hours until neovascularization of the graft occurs. The graft must be placed on a wound bed free of dead tissue, infection, hemorrhage, or significant exudate or it will not survive. Wounds that have been freshly excised of all dead tissue offer the best short- and long-term results in all instances because granulation tissue prompts an inflammatory process that eventually results in exacerbation

of scar formation. The epidermis at the donor site where the graft was originally harvested regenerates in 2–3 weeks from the deep epidermal elements (hair follicles and sweat glands) which were left behind, leaving a thinner, less functional dermis because the dermis removed by the graft will not regenerate. Even in small burns there is a considerable price to the patient when a split-thickness graft is harvested. The threat of infection and hypertrophic scar formation is always present and the harvesting of a split-thickness skin graft always leaves a permanent scar and often unwanted pigment changes resulting in permanent cosmetic deformity.

Many times the graft is "meshed," which involves passing the graft through a mesher which makes many small linear incisions in the graft (Fig. 3). These incisions allow the graft to be gently spread and expanded in size. By increasing the edge area of the epithelium, meshing allows the graft to cover from 1.5 to 9 times its initial area, although 1.5 to 3 expansions are most common. The mesh also allows fluids (tissue exudates and blood) to drain from the bed, which helps to increase the likelihood of graft survival. Unfortunately the mesh pattern is usually visible for extended periods of time after healing because the dermis does not properly regenerate within the interstices and therefore meshing makes the cosmetic results less desirable than with nonmeshed grafts.

WOUND COVERAGE AND SKIN SUBSTITUTES

Once an assessment has been made of the severity of the wound, decisions can be made as to the course of treatment and the appropriate types of materials that can be used to achieve closure of the wound. For first-degree and superficial second-degree burns the choice of medication or wound coverage is a never-ending source of discussion. However, as long as wounds are carefully monitored for infection and kept clean and moist, most medications and membranes will perform well. Topical medications range from petrolatum through antibiotic- or antiseptic-containing aqueous or ointment preparations

TABLE 1 Partial Listing of Some Commonly Used Topical Wound Medications and Their Principal Characteristics

Medication	Selected characteristics
Silver sulfadiazene	Broad spectrum, painless
Aqueous silver nitrate	Broad spectrum including fungi, electrolyte issues
Mafenide acetate	Broad spectrum, penetrates eschar
Petrolatum	Bland and nontoxic
Various debriding enzymes	Selected utility
Various antibiotic ointments	Selected utility

through debriding enzymes. A partial listing of available topical medications and their characteristics is given in Table 1. When properly used by knowledgeable persons in a program of burn care that includes regular wound evaluation, cleansing, and monitoring, all can be effective.

Wound membranes and substitutes either provide transient physiologic wound closure or can contribute to definitive wound closure (Fig. 4). Physiologic wound closure implies down-regulation of the intense inflammatory reaction characteristic of an open wound as well as a degree of protection from mechanical trauma, vapor transmission characteristics similar to skin, and a physical barrier to bacteria. Temporary membranes can contribute to the creation of a moist wound environment with a low bacterial density and have been used as a dressing on clean superficial wounds while they await

epithelialization. All synthetic membranes are more or less occlusive. If a membrane dressing is placed over devitalized tissue, submembrane purulence can occur in the closed space with potentially disastrous results. There are a large number of such membranes in common use, some of which are listed in Table 2.

TEMPORARY SKIN SUBSTITUTES

There are four common uses for temporary skin substitutes: (1) as a dressing on donor sites to facilitate pain control and epithelialization from skin appendages, (2) as a dressing on clean superficial wounds for the same reasons, (3) to provide temporary physiologic closure of deep dermal and full-thickness wounds after excision while awaiting autograft availability or healing of underlying widely meshed autografts, and (4) as a "test" graft in questionable wound beds. There are many temporary substitutes and membranes in common use (Pruitt and Levine, 1984), classes of which will be described next.

Allografts from Cadavers

Split-thickness human allograft, procured from organ and tissue donors, remains the standard by which other temporary skin covers are judged. This tissue is procured, processed, stored, distributed, and tracked by skin banks. Although commonly used after refrigeration for 7 days or less, it is quite

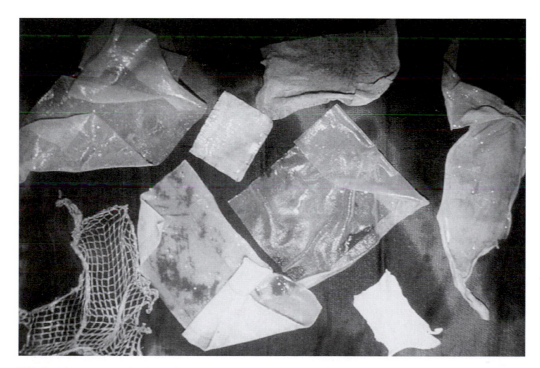

FIG. 4. Various wound covers, clockwise starting at the bottom left-hand corner: meshed split-thickness autograft, TransCyte, Epicel, cryopreserved cadaver allograft, Biobrane, split-thickness autograft, EZ Derm, and Integra Dermal Regeneration Template.

TABLE 2 Partial Listing of Some Commonly Used Wound Membranes and Their Principal Characteristics

Membrane	Selected characteristics
Temporary	
Porcine xenograft	Adheres to coagulum, excellent pain control
Biobrane[a]	Bilaminate, fibrovascular ingrowth into inner layer
Split-thickness allograft	Vascularizes and provides durable temporary closure
Various semipermeable membranes	Provides vapor and bacterial barrier
Various hydrocolloid dressings	Provides vapor and bacterial barrier, absorbs exudate
Various impregnated gauzes	Provides barrier while allowing drainage
Allogeneic dressings	Provides temporary cover while supplementing growth factors
Permanent	
Epicel[b]	Provides autologous epithelial layer
Integra[c]	Provides scaffold for neodermis, requires delayed thin autograft grafting
AlloDerm[d]	Consists of cell-free human dermal scaffold, requires immediate thin autograft

[a] Mylan Laboratories, Inc.
[b] Genzyme Biosurgery, Inc., Cambridge, MA.
[c] Integra LifeSciences Corporation, Plainsboro, NJ.
[d] LifeCell, Inc., Branchburg, NJ.

effective when cryopreserved according to well-established protocols designed to maximize cell viability. Cryopreserved skin can be made widely available through express mailing. When viable split-thickness allograft skin is applied to a clean excised wound, the tissue vascularizes and provides durable biologic cover until it is recognized as foreign by the host. This process generally results in loss of the allogeneic epithelium approximately 3 weeks after application in most burn patients. An added benefit to allografts that undergo revascularization is that, like autografts that revascularize, bacterial counts in the wound drop dramatically. Allografts at least provide temporary physiologic wound closure and allow time for reepithelialization of a donor site, but all allografts must eventually be replaced by autograft. When donor sites for autografts have reepithelialized it is possible to reharvest autograft from the same site; however, there is a limit to the number of times that a donor site can be reharvested because the dermis cannot regenerate and repeated removal of even thin layers of dermis will leave a donor area with unsatisfactory function. When modern screening techniques are followed, the risk of disease transmission by cadaver allografts is very low.

Allografts of Cultured Cells and Collagen

Using technology for the culture of skin cells that is described in detail later in this chapter, it is possible to culture large numbers of cells and to incorporate these cells into a composite skin substitute. Typically, discarded tissues, such as foreskins, are used as the starting source of cells. Cells from a 1-cm^2 piece of skin can be expanded to prepare approximately 1600 m^2 of certain types of cultured skin grafts, like the one described here. Since these cells did not originate from the patients to whom they will ultimately be transplanted, they are also allografts.

A cultured allogeneic composite graft with both an epidermal layer of cultured epidermal keratinocytes and a dermal analog of cultured dermal fibroblasts in a collagen gel has been developed based on work by Eugene Bell and colleagues (Bell *et al.*, 1981). To prepare these grafts, cultured allogeneic fibroblasts of the dermis are mixed with a solution of purified bovine collagen type 1. This solution is allowed to gel and the surface of this gel is seeded with cultured allogeneic epidermal keratinocytes. When the surface of this bilayered skin equivalent is exposed to the air, the epidermal keratinocytes stratify, differentiate, and form a cornified layer much like the epidermis of normal skin. This cultured allograft (Apligraf, Organogenesis, Inc.) is approved for use in the treatment of diabetic foot ulcers and venous leg ulcers (Fig. 5). The allograft helps provide temporary wound closure and seems to stimulate local healing events that lead to closure of the ulcer.

A slightly different composite graft using cultured allogeneic skin cells is also commercially available (OrCel, Ortec International, Inc.). Fibroblasts and keratinocytes isolated from foreskins are grown *in vitro* and seeded on opposite sides of a bilaminate matrix of bovine collagen. The epidermal surface of the bilaminate matrix is a nonporous collagen gel, so keratinocytes seeded on the surface are able to form an epidermal cover. The opposite side or dermal side of the bilaminate matrix is a porous collagen sponge, so fibroblasts seeded on this surface are able to disperse and penetrate the sponge. These allogeneic composite grafts are matured *in vitro* (10–14 days). This product is approved for use on fresh, clean split-thickness donor sites in burn patients.

Xenografts

The most common source of xenograft (i.e., a graft from a different species) for temporary wound closure is the domestic swine. Typically high-quality frozen hides of 1 to 2-year-old animals are obtained from a reliable slaughterhouse. Split-thickness grafts (unmeshed and meshed) of desired sizes are

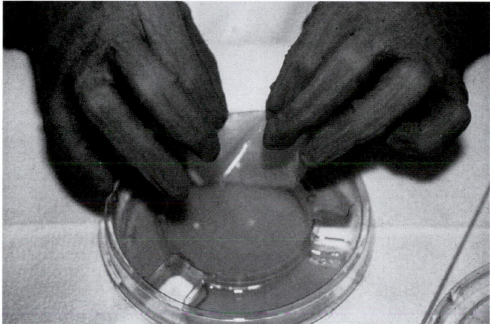

FIG. 5. The manufacturing process and close-up of Apligraf.

harvested from these hides. The final grafts, which have non-viable epidermal and dermal layers, are sterilized and stored frozen (Mediskin I, Brennan Medical, Inc.). Alternatively, the epidermis of the grafts is removed and the porcine dermis is subjected to aldehyde cross-linking to increase strength and durability. These grafts can be stored at room temperature (EZ Derm).

Porcine xenograft is used for the temporary coverage of clean wounds such as superficial second-degree burns and donor sites. Its use has been favorably reported in patients with toxic epidermal necrolysis, and it has been combined with silver to suppress microbial colonization of wounds. Porcine xenografts, at best, only adhere to the wound bed with a fibrinous interaction, but this can temporarily close a wound. The grafting success rate is lower than that with fresh cadaver allograft. The porcine grafts must be immobilized for several days to allow bonding of the graft to the wound bed by growth of fibroblasts into the dermal layer of the graft. As with any

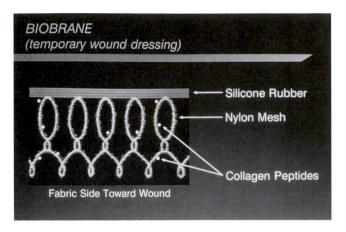

FIG. 6. Diagram of Biobrane.

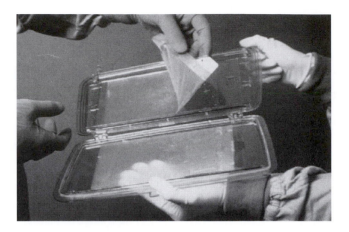

FIG. 7. Photograph of TransCyte.

skin replacement procedure, the wound bed must be debrided thoroughly and be free of significant bacterial contamination. Although porcine xenograft does not vascularize, it will adhere to a clean superficial wound and can provide excellent pain control while the underlying wound epithelializes. Because xenografts will ultimately separate from the wound bed, they must be removed and replaced with autograft material.

Synthetic Membranes

There are a number of monolayer semipermeable membrane dressings that are designed to provide vapor and bacterial barriers and to control pain while the underlying superficial wound or donor site heals. These can be used with good effect on clean superficial wounds as well as split-thickness donor sites. There is one bilayer synthetic membrane in general use at present. This product (Biobrane, Mylan Laboratories, Inc.) is a biosynthetic wound dressing constructed of a silicon film that forms a barrier to water loss and microbial invasion (Fig. 6). Partially embedded within the silicon film is a nylon fabric that has been covalently cross-linked with hydrophilic type I porcine collagen. The wound bed comes into contact with this complex three-dimensional fabric, which helps promote clotting in the nylon matrix and adherence of the membrane to the wound bed. The membrane stays attached to the wound until reepithelialization occurs. Biobrane is flexible enough to deform, maintaining adequate contact with the contours of the wound while still having the mechanical stability necessary for ease of handling and durability as a dressing. Biobrane is not biodegradable and serves at best only as a temporary closure material, lasting up to a month. This and other membranes like it have been used to good effect in clean superficial burns and donor sites.

Hydrocolloid Dressings

Hydrocolloid dressings are semiocclusive dressings that combine a synthetic membrane or foam with a hydrocolloid. Elastomers and adhesives such as polyurethane provide a membrane that is conformable and lightly adhesive to the wound while providing a barrier to moisture loss and microbial invasion. To capture the exudate produced by many wounds, the polyurethane is combined with a hydrocolloid such as carboxymethycellulose that forms a gel when it absorbs liquids. The gel helps to provide a moist environment that may enhance wound healing, and the semiocclusive nature of the membrane helps to alleviate pain. Since the moisture-capturing capacity of the hydrocolloid can be saturated, hydrocolloid dressings are changed every 3–5 days. Although, like monolayer membranes, they are often associated with submembrane fluid collections, they can be used to good effect in selected superficial wounds.

Cells Combined with Synthetic Membranes

A hybrid product of cultured cells and synthetic polymers has also been produced as a temporary substitute. This product (TransCyte, Smith & Nephew) is commercially available for the treatment of burns (Fig. 7). A nylon mesh, coated with porcine dermal collagen, is bonded to a thin silicon membrane. These membranes are inserted into a cassette-like bioreactor containing culture medium. The nylon matrix is seeded with allogeneic fibroblasts that grow within the three-dimensional nylon mesh. During this growth, the cells deposit additional collagen, extracellular matrix proteins, and growth factors. When growth is complete, the culture medium is removed from the bioreactor, replaced with a freezing solution, and the entire cassette is frozen. The product is shipped and kept frozen. For use, the product is thawed, the freezing solution decanted, and the membrane removed and placed on the wound so that the fibroblast side contacts the wound. After the freeze and thaw process, the allogeneic fibroblasts are no longer viable. This product is used as a temporary cover for surgically excised full-thickness and deep partial-thickness as well as clean superficial burn wounds. The intended use of this product is as an alternative to cadaver allografts. The silicon layer and nylon matrix are translucent, making it possible to visually monitor the wound bed for the possibility of microbial infections. Since the nylon mesh is not biodegradable, this product must eventually be removed.

Areas of Research

One promising area of research is the identification of growth factors that promote wound healing by stimulating cell proliferation and/or cell chemotaxis. The genes encoding many of the growth-factor proteins that are secreted by cells in the wound environment have been identified and large amounts of these proteins have been produced by recombinant DNA technology. Several of these growth-factor proteins are being tested as ointments or incorporated into biomaterials for sustained delivery to wounds. A gene therapy approach is also being investigated as another means to achieve sustained delivery of growth factors. Cultured keratinocytes have been genetically modified to secrete high levels of wound-healing growth factors and are being tested in preclinical studies (Eming *et al.*, 1998).

PERMANENT SKIN SUBSTITUTES

Although no ideal substitute exists at present, there are a number of skin substitutes that may contribute to the permanent reconstitution of the dermis, epidermis, or both. This is an area of active investigation that is of particular importance for the massively burned patient.

Autografts of Sheets of Cultured Keratinocytes

Using methods pioneered by Howard Green and colleagues, keratinocytes, the principal cell of the epidermis, can be grown in large numbers from a small piece of freshly harvested skin (~25 cm^2) (Green *et al.*, 1979). This full-thickness biopsy of skin is procured from an unburned area and sent to a company (Epicel, Genzyme BioSurgery, Inc.) that specializes in cultured skin. The biopsy is minced, and the pieces treated with specific enzymes to disaggregate the tissue into single cells. The resulting cell suspension is plated in culture flasks along with a "feeder layer" of murine fibroblasts that have been treated with a nonlethal dose of radiation that prevents them from multiplying. The murine fibroblasts provide insoluble matrix proteins that facilitate the clonal growth of keratinocytes as well as other growth factors. The flasks also contain a rich culture medium that contains numerous factors which stimulate the growth of keratinocytes, such as fetal calf serum, epidermal growth factor, insulin, hydrocortisone, and cholera toxin. Within a few days after plating the cell suspension, small colonies of two to four cells form and eventually expand into larger multicell layered colonies of keratinocytes (Fig. 8). To expand cell numbers even further, these primary cultures of keratinocytes can be treated with trypsin and the resulting cell suspension passed to more culture flasks for secondary cultures. To ready the cells for transplantation, the keratinocyte colonies are grown to confluence by letting the colonies merge into one continuous sheet of cells that covers the entire surface of the flask. This also results in dislodgement of the murine fibroblasts. The resulting sheets of keratinocytes are removed from the dishes as intact sheets by treatment with the enzyme dispase. The thin sheets of cells, which are 30 cm^2 in size and approximately two to eight cells thick, are attached to petrolatum gauze for ease of handling and shipped back to the hospital. The company specializing in this procedure can now deliver 192 sheets of cultured epithelium within 21 days of receiving the biopsy. Depending on the size of the patient, this is equivalent to 45% of the total body surface area.

Since the patient's own skin is used in the process, the cultured cells are autologous and the resulting epithelial sheet

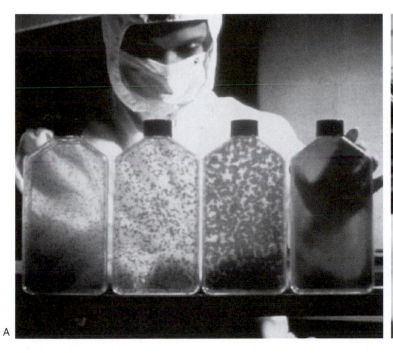

FIG. 8. Stained colonies of cultured keratinocytes at various stages of growth and close-up of Epicel, a sheet of cultured keratinocytes.

is an autograft that is capable of permanent engraftment. Cultured epidermal sheets have been used widely for the treatment of very large burns with mixed results and, at times, suboptimal rates of engraftment. When compared to the performance of standard split-thickness autografts, the gold standard of graft performance, autografts of cultured epithelial sheets have several deficiencies. The cell sheets are thin and fragile and thus are sensitive to mechanical dislodgement, dehydration, and microbial destruction. Moreover, since the cell sheets lack a dermis, cosmetic, functional and durable properties of engrafted epithelial sheets is not ideal.

Analog of the Dermis as Part of a Bilayered Skin Replacement

Ioannis Yannas and John Burke developed a bilaminar skin replacement with an outer layer of Silastic and inner layer of collagen sponge (Burke *et al.*, 1981; Yannas *et al.*, 1982a). They performed extensive design and optimization steps to produce a skin replacement that had a synthetic epidermal layer with performance characteristic similar to native epidermis and a dermal layer that acts as a biodegradable scaffold to facilitate vascularization and population by normal healthy host fibroblasts. This material is approved by the FDA for burn treatment and is commercially available (Integra Dermal Regeneration Template, Integra LifeSciences, Inc.; Figs. 9, 10). The outer Silastic layer is 0.1 mm thick and provides the bacterial barrier needed while maintaining the proper water flux through the membrane. In order to prevent the swelling of the wound bed and the accumulation of fluid between the graft and its bed, all grafts must be permeable to moisture.

The exact permeability must be controlled, as too large a water flux of moisture will result in dehydration of the wound surface and disruption of the wound–graft interface, whereas too small a water flux will cause the accumulation of exudative fluid beneath the graft and lift it from the wound bed. Flux through the Silastic membrane is 1 to 10 mg/cm^2/hr, which is similar to that of normal epidermal tissue. The Silastic layer also provides the mechanical rigidity needed to suture the graft in place, preventing movement of the material during wound healing.

A principal component of the dermal analog of this material is collagen. Collagen is an excellent material owing to many of its unique properties. Collagen is a polypeptide compound, hydrophilic in nature, and subject to degradation by distinct classes of extracellular enzymes. The substance is very well studied, allowing for the control of many key physical parameters, including the small strain viscoelastic properties and the rate of degradation (which can be controlled either by cross-linking the collagen fibers or by incorporation of other materials in the suspension). A fairly pure form can be extracted in large amounts from many commercial sources. Collagen is inherently low in antigenicity and exerts a hemostatic effect on the fragile and vascular wounds. Collagen can be formulated in many different ways from gels to films, depending on the properties desired and methods for application. Cross-linking the collagen fibers increases the tensile strength, but unfortunately it also tends to make the fibers stiff and brittle. The permeability of collagen materials may be controlled by their thickness, but mechanical properties must be considered if the material becomes too thick. By incorporating dilutants, the Young's modulus of collagen can be varied over a range

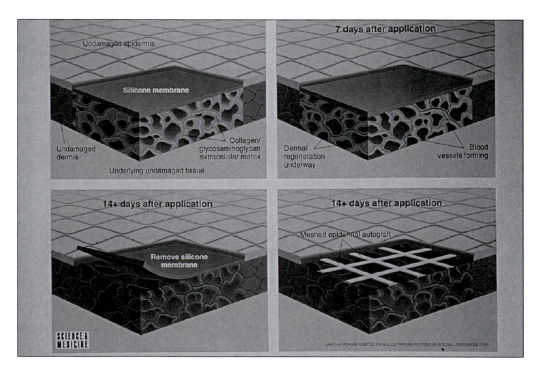

FIG. 9.   Schematic of Integra Dermal Regeneration Template placed in an excised burn wound.

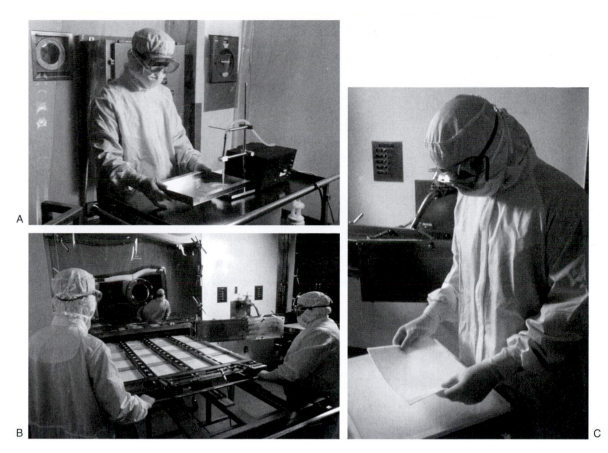

FIG. 10. Manufacturing process of Integra Dermal Regeneration Template. Bovine tendons are cleaned, frozen, processed into a slurry with glycosaminoglycans, and freeze dried into sheets.

of 10^5. A problem noted with collagen was that by increasing the cross-linking (in order to add strength) the compound became stiff and brittle. This could be overcome by incorporating a second macromolecule, a glycosaminoglycan (GAG) into the membrane (discussed next).

Another property of collagen is that it naturally adheres to the wound initially because of its binding with fibrin. The collagen–chondroitin 6-sulfate adheres to the wound within minutes (in contrast to collagen sponge), and neovascularization can be observed within 3–5 days. The strong attachment of the collagen–GAG to the wound is evident by measuring the peel strength of the graft, which is 9 N/m at 24 hours and increases to 45 N/m by 10 days. Successful adherence of the artificial skin to the wound results in physiologic wound closure and its beneficial effects.

The dermal analog of the Integra Artificial Skin is actually a mixture of fibers of collagen (isolated from bovine tendon) and chondroitin 6-sulfate, a type of glycosaminoglycan (GAG) isolated from shark cartilage. During the manufacturing process, the collagen fibers and GAGs are coprecipitated and this slurry is poured into metal pans, frozen using a controlled-rate freezer and then freeze dried (Fig. 10). This process creates a collagen/GAG sponge with controlled pore size (Chamberlain and Yannas, 1999). Proper pore size facilitates

optimal ingrowth of fibrovascular cells. An impermeable graft without pores would not provide any means for such migration. Since most of the epithelial and mesenchymal cells found in the wound bed are on the order of 10 μm in diameter, pores consistent with that order of magnitude would allow free access into the graft material. Pore size of the membrane was shown to be exquisitely sensitive to the method of drying. Dagalakis *et al.* (1980) showed that instantaneous freezing followed by slow sublimation at constant low temperatures was necessary to ensure the maximum mean pore size.

Also influencing final pore size was the inclusion of GAGs. Glycosaminoglycans are found in significant quantity in the ground substance of extracellular matrix and may contribute up to 30% by dry weight of this intercellular region. GAGs consist of a family of macromolecules (molecular weight from 10^5 to 10^6) that share the characteristic of having multiple long polysaccharide chains covalently bound to a core protein by a glycosidic bond. In contrast to glycoproteins, these compounds have a significant quantity of carbohydrate and exhibit behaviors closer to those of a polysaccharide than to those of protein. Each of the various GAGs has its own unique repeating disaccharide that forms the long carbohydrate chains. In the case of chondroitin 6-sulfate, the sugar is *N*-acetylgalactosamine (Fig. 11). The molecules are highly charged and polyanionic

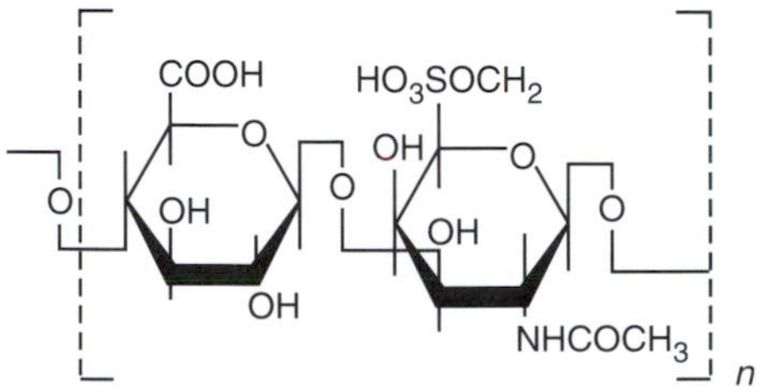

FIG. 11. Repeating disaccharide of chondroitin 6-sulfate.

in nature. They have multiple properties, including the ability to change the mechanical properties of their environment. Hyaluronic acid, which is found in the synovial fluid of joints, contributes to lubrication. For the artificial skin or skin replacement developed by Yannas and Burke, the GAG chondroitin 6-phosphate was chosen because of its important properties: (1) low antigenicity; (2) nontoxic breakdown products; (3) ability to decrease the rate of collagen breakdown; and (4) ability to increase the strength of collagens while making them more elastic. When examined by electron microscopy, the pore structure of the collagen–GAG sponges was significantly more open than that of GAG-free collagen and one could control the pore size by adjusting the collagen–GAG mixture. Bonding of chondroitin 6-sulfate to the collagen is required because at neutral pH the GAG will dissociate from collagen and elute from the material in the absence of cross-linking.

Cross-linking of the sponge was shown to contribute to pore size as well as strength, elasticity and importantly, the rate of degradation. In their design, Yannas and Burke sought to balance the rate of degradation with the time required for cellular regeneration. To balance these terms, they estimated that the time constant for biodegradation (t_b) should be on the same order of magnitude as the time required for the wound to heal (t_h), approximately 25 days, or

$$(t_b)/(t_h) \sim 1$$

Since t_h is constant for any patient, the design of the graft was controlled for t_b. They reasoned that if the graft degraded too quickly, the wound would effectively be left uncovered. On the other hand, if the graft remained intact for long periods of time after the wound was healed, then scar formation and incorporation of the device into the wound would become problems.

Cross-linking of the collagen and GAG was performed using glutaraldehyde. The coherent membrane was removed from the glutaraldehyde and placed in distilled water to remove any unreacted aldehydes. If the membrane was freeze dried, it was first necessary to partially cross-link the collagen in a 105°C vacuum oven. After this step, it was possible to immerse the membrane in a glutaraldehyde solution without it collapsing; the remainder of the cross-linking was then carried out in the standard manner. The amount of cross-linking, which in turn affects the strength, elasticity, and pore size of the membrane, was controlled by adjusting several parameters. The time the membrane remained in the vacuum oven determined the initial retention of GAG by the membrane;

the second step of cross-linking by glutaraldehyde was carried out under acidic conditions (pH 3.2), although they slow the cross-linking process because the GAG forms an unstable ionic complex with collagen at neutral pH. By adjusting the concentration of glutaraldehyde in the mixture, it was possible to tightly control the molecular weight between cross-links (M_c, an inverse measure of cross-link density) between a very loose network ($M_c = 30,000$) and a very tight one ($M_c = 10,000$). This processing resulted in a collagen–GAG complex with $8.2 \pm 0.8\%$ GAG by weight, a pore volume fraction of $96 \pm 2\%$, and a mean pore size of $50 \, \mu m \pm 20 \, \mu m$.

The final step in the production is the addition of the Silastic layer. The Silastic layer was applied as a liquid monomer, and curing takes place at room temperature on the surface of the collagen after exposure to ambient moisture. The final bilayer copolymer composite is then stored at room temperature either in 70% alcohol or after being freeze dried. Integra LifeSciences provides this material as 10×25 cm aseptic sheets packaged in 70% isopropyl alcohol, which are stored at 2–8°C prior to use.

This off-the-shelf product is used for the coverage of freshly excised full-thickness burns. In 2 to 3 weeks when the dermal analog has revascularized, the outer silicone membrane is removed and replaced with an ultrathin epithelial autograft. Procuring thin autografts has several advantages because donor sites heal faster with less morbidity and repeat harvesting of these sites can be performed sooner. Initial reports of this material were successful, and postmarketing trials are in progress. Histological cross sections of wounds closed with this artificial skin show complete replacement of bovine collagen at 7 weeks with remodeled human dermis. Minimal evidence of hypertrophic scar formation has been noted, and the grafted areas are more supple than those areas simply covered with meshed autograft. This artificial skin is the only long-term dermal replacement with a large experience with burn patients (Heimbach et al., 1988). Although submembrane purulence must be watched for and promptly treated, the material is likely to play an important role in the management of serious burns.

Acellular Allogeneic Dermis

Cadaver skin has been processed to yield an acellular product that is used as an analog of the dermis. Split-thickness cadaver skin is treated with a high-salt solution and the allogeneic epidermis is removed and discarded. This deepidermalized dermis is further treated with a solution containing a detergent to eliminate dermal cells and to inactivate any possible pathogens such as viruses. The resulting acellular dermis has no cells but has bundles of collagen and elastin fibers organized as both reticular and papillary dermis. The undulating papillary surface has many of the proteins of the basement membrane such as laminin, and collagen types IV and VII that are important for the attachment and migration of epidermal keratinocytes. The acellular dermis also has basement membrane proteins that line the channels previously occupied by dermal blood vessels. The material is further washed in a freezing medium and freeze-dried under controlled conditions to avoid disruption of the matrix proteins. This acellular dermis, which is commercially available (AlloDerm, LifeCell, Inc.) is supplied as a freeze-dried sheet (meshed or unmeshed) approximately 200 cm². Prior to use, acellular dermis is rehydrated in normal saline, becomes soft and pliable, and can be secured with sutures if necessary.

Acellular dermis has been widely used as an implant in oral and reconstructive surgery, whereas for burns, there has been limited use. Acellular dermis has been used for the treatment of full-thickness wounds by applying it to a clean vascularized wound bed and overlaying the acellular dermis with a thin (0.003 to 0.006 inch) split-thickness autograft. Thin autografts, with minimal dermis, are recommended since excessive dermal tissue can interfere with graft take. The grafts are dressed to prevent infection and dehydration. Since the acellular dermis has had its cells removed, the primary targets of immune rejection are lost and acellular dermis is not rejected like conventional cadaver allografts. Rather, the acellular dermis acts as a scaffold for the ingrowth of fibrovascular cells and the thin autograft reepithelializes the surface. Clinical experience with this material in acute and reconstructive burn wounds is still early and limited, but appears favorable.

Permanent Skin Substitutes in Development

Most investigators have come to the conclusion that the ideal skin substitute for the patient suffering from large burns will provide for immediate replacement of both lost dermis and lost epidermis. Both elements are required for optimal function and skin appearance; however, a reliably successful composite substitute that performs as well as split-thickness autograft has not yet been developed. Much exciting work is underway and such a device might very well revolutionize burn care.

Yannas and Burke have extended their studies by seeding the dermal analog of their bilayer skin replacement with autologous keratinocytes obtained from the patient. A small sample of skin is removed from the patient and the top cornified layer of the epidermis is discarded. The basal proliferative cells are then dissociated from one another with trypsin, and suspended in growth medium. These cells are introduced into the dermal analog, either by direct injection or by centrifugal force (this entire procedure can be accomplished in the guinea pig model in under 4 hours). The keratinocytes proliferate and form sheets of keratinized cells. Functional skin replacement has been achieved in guinea pigs in under 4 weeks (Yannas *et al.*, 1982a).

Cultured cells have also been combined with the collagen–GAG dermal analog and tested in burn patients with some success. The surface of the dermal analog has been laminated with a thin layer of collagen and this surface seeded with cultured autologous epidermal keratinocytes (Boyce and Hansbrough, 1987). In addition, the porous underside of the dermal analog is seeded with cultured autologous dermal fibroblasts. When grown at the air/liquid interface, the epidermal keratinocytes stratify and form a cornified layer. This composite skin graft of biomaterials and autologous cells has been tested in a limited clinical trial for the permanent closure of full-thickness burns with some success, and further investigations of this potentially exciting technology continue.

CONCLUSIONS

There are two major categories of skin substitutes and wound coverings: temporary and permanent. If a wound is superficial and will heal in reasonable time and produce acceptable cosmetic results, then this wound may benefit from coverage with a temporary skin substitute. As a group, temporary skin substitutes all provide a barrier to microbial invasion, help prevent fluid loss, serve to control pain, and provide coverage while healing is taking place under the substitute. There are a variety of temporary substitutes available that are widely used with good success. Depending on their specific use, these temporary substitutes are fabricated from a variety of synthetic materials, purified naturally occurring biomaterials, and processed tissues as well as cultured cells of the skin. These are often combined to produce hybrid devices that have two or all of these components. Since these are temporary skin replacements, allogeneic cells from unrelated donors are used to allow for the large-scale production of these devices prior to use.

Deep third-degree burns will not heal on their own and so require permanent skin closure. The gold standard for the permanent closure of these wounds is the split-thickness autograft obtained from an unburned area of the same patient. However, for the massively burned patient, these donor sites are limited. These patients have a desperate need for a permanent skin replacement. Temporary skin substitutes used to close wounds on these patients can help to provide additional time for the healing of donor sites that can be reharvested. However, hospitalization is prolonged and cosmetic and functional outcomes are not ideal. Permanent skin replacements are being developed from synthetic materials, purified naturally occurring biomaterials, and processed tissues as well as cultured cells of the skin. Since these are permanent skin replacements, it is important that autologous cells from the patient be used with one or more of these components. Promising permanent skin replacements are presently being developed and evaluated; however, the ideal skin replacement for this group of patients does not yet exist.

Bibliography

Bell, E., Ehrlich, H. P., Buttle, D. J., and Nakatsuji, T. (1981). Living tissue formed in vitro and accepted as skin-equivalent tissue of full thickness. *Science* **211**: 1052–1054.

Boyce, S. T., and Hansbrough, J. F. (1987). Biological attachment, growth, and differentiation of cultured human epidermal keratinocytes on a graftable collagen and chondroitin-6-sulfate substrate. *Surgery* **103**: 422–431.

Burke, J. F., Bondoc, C. C., and Quinby, W. C. (1974). Primary burn excision and immediate grafting: a method of shortening illness. *J. Trauma* **14**: 389–395.

Burke, J. F., Yannas, I. V., Quinby, W. C., Bondoc, C. C., and Jung, W. K. (1981). Successful use of a physiologically acceptable artificial skin in the treatment of extensive burn injury. *Ann. Surg.* **194**: 413–428.

Chamberlain, L. J., and Yannas, I. V. (1999). Preparation of collagen–glycosaminoglycan copolymers for tissue regeneration. in *Methods in Tissue Engineering*, J. R. Morgan, and M. L. Yarmush eds. Humana Press, Totowa, NJ, pp. 3–17.

Dagalakis, N., Flink, J., Stasikelis, P., Burke, J. F., and Yannas, I. V. (1980). Design of an artificial skin. Part III. Control of pore structure. *J. Biomed. Mater. Res.* **14**: 511–528.

Eming, S. A., Yarmush, M. L., and Morgan, J. R. (1998). Genetically modified keratinocytes expressing PDGF-A enhance the performance of a composite skin graft. *Hum. Gene Ther.* **9**: 529–539.

Green, H., Kehinde, O., and Thomas, J. (1979). Growth of cultured human epidermal cells into multiple epithelia suitable for grafting. *Proc. Natl. Acad. Sci. USA* **76**: 5665–5668.

Heimbach, D., Luterman, A., Burke, J., Cram, A., Herndon, D., Hunt, J., Jordan, M., McManus, W., Solem, L., Warden, G., and Zawacki, B. (1988). Artificial dermis for major burns, a multi-center randomized clinical trial. *Ann. Surg.* **208**: 313–320.

Pruit, B. A., and Levine, N. S. (1984). Characteristics and uses of biological dressings and skin substitutes. *Arch. Surg.* **119**: 312–322.

Tompkins, R. G., and Burke, J. F. (1989). Burn wound. in *Current Surgical Therapy*, J. L. Cameron, ed. B. C. Decker, Philadelphia, pp. 695–702.

Yannas, I. V., and Burke, J. F. (1980). Design of an artificial skin. I. Basic design principles. *J. Biomed. Mater. Res.* **14**: 65–81.

Yannas, I. V., Burke, J. F., Orgill, D. P., and Skrabut (1982a). Wound tissue can utilize a polymeric template to synthesize a functional extension of skin. *Science* **215**: 174–176.

Yannas, I. V., Burke, J. F., Warpehoski, M., Stasikelis, P., Skrabut, E. M., and Orgill, D. P. (1982b). Design principles and preliminary clinical performance of an artificial skin. in *Biomaterials: Interfacial Phenomena and Applications*, S. L. Cooper and N. A. Peppas, eds. American Chemical Society, New York, pp. 476–481.

7.13 SUTURES

Mark S. Roby and Jack Kennedy

INTRODUCTION

A suture is a strand of material that is used to approximate tissues or to ligate blood vessels during the wound-healing period. The use of linen suture as early as 4000 years ago makes these biomaterials one of the earliest recorded medical devices (Chu, 1991). Over the years, a variety of other materials have been used with catgut and silk being the most prevalent until the development of synthetic fibers during the 1950s. It is worth noting that catgut-based sutures used by Galen to treat wounded gladiators around 150 A.D. are still in use today (Barrows, 1986).

One class of sutures, the synthetic absorbable sutures, comprise the fastest-growing segment of the suture market and represent an area of considerable development. For these reasons, the following discussion will focus more on this class of sutures.

CLASSIFICATION

Sutures can be classified according to three characteristics—origin, absorption profile and fiber construction (Table 1).

TABLE 1 Classification of Sutures

Characteristic	Categories
Origin	Natural
	Synthetic
Absorption	Absorbable
	Nonabsorbable
Fiber construction	Multifilament Braid
	Monofilament

Table 2 summarizes the major types of sutures, general applications, and representative products and manufacturers.

Origin

Natural sutures are based on fibers derived from animals or plants. The two most important natural sutures are catgut based on animal collagen and silk spun by the silkworm. Synthetic sutures are based on manmade polymers or metals. Manmade polymers include many common textile fibers such as PET, nylon, polypropylene, and PTFE as well as those based on α-hydroxy acids.

Absorption

Originally, an absorbable suture was one that lost a significant portion of its mechanical strength over a period of 2 months while a nonabsorbable suture was one that maintained a significant portion of its strength longer than 2 months (Swanson and Tromovitch, 1982). However, this period has been extended to three months for certain synthetic absorbable monofilaments and up to a year for a new braided suture developed since that time (Chu, 1997a; Guttman and Guttman, 1994), further blurring the line between absorbable and nonabsorbable materials.

In the case of absorbable sutures, it is important to distinguish between the loss of mechanical strength and the absorption and elimination of the material from the body. A suture may lose its tensile strength over a relatively short period of time, for example weeks, but require months or even years to absorb completely and be eliminated from the body. The tensile strength loss profile is an important parameter in suture selection since mechanical support is necessary during the critical wound-healing period.

The primary mode of degradation for natural materials is enzymolysis, whereas for synthetic absorbable materials it is hydrolysis (Capperauld and Bucknall, 1984; Chu, 1985; Salthouse and Matlaga, 1976; Singhal *et al.*, 1988; Zhong *et al.*, 1993). However, there are several reports indicating that enzymes may play a role in the degradation of synthetic absorbable materials (Chu and Williams, 1983; Williams, 1982). Most likely enzymes play a greater role in the degradation and elimination of the hydrolysis by-products from synthetic absorbable sutures (Salthouse and Matlaga, 1976; Singhal *et al.*, 1988).

TABLE 2 Types of Sutures

Suture type	Generic structure	Major clinical application[a]	Type[b]	Representative product	Representative manufacturer
Natural materials					
Catgut	Protein	Plain: Subcutaneous, rapid-healing tissues, ophthalmic	T	Surgical Gut	Ethicon
		Chromic: Slower-healing tissues	T	Chromic, Plain Gut	Syneture
			T	Mild Chromic Gut	Syneture
Silk	Protein	General suturing, ligation	B	Perma-Hand	Ethicon
			B	Softsilk	Syneture
Synthetic nonabsorbable materials					
Polyester	PET	Heart valves, vascular prostheses, general	B	Ethibond Excel	Ethicon
			B	Surgidac	Syneture
			B	Ti-Cron	Syneture
			B	Tevdek	Teleflex
	Polybutester	Plastic, cuticular	M	Novafil	Syneture
		Cardiovascular	M	Vascufil	Syneture
Polypropylene PP		General, vascular anastomosis	M	Prolene	Ethicon
			M	Surgipro	Syneture
			M	Surgipro II	Syneture
			M	Deklene II	Teleflex
Polyamide	Nylon-6, -6,6	Skin, microsurgery, tendon	M	Ethilon	Ethicon
			M	Monsof	Syneture
			M	Dermalon	Syneture
			B	Nurolon	Ethicon
			B	Surgilon	Syneture
Stainless steel	Steel alloy	Abdominal and sternal closures, tendon	M, T	Ethisteel	Ethicon
			M, T	Steel	Syneture
			M, T	Flexon	Syneture
Fluoropolymers	ePTFE	General, vascular anastomosis	M	Gore-Tex	W.L. Gore
	PVF/PHFP		M	Pronova	Ethicon
Synthetic absorbable materials					
Braids	PGA/PLLA	Peritoneal, fascial, subcutaneous	B	Vicryl	Ethicon
	PGA/PLLA		B	Vicryl Rapide	Ethicon
	PGA/PLLA		B	Panacryl	Ethicon
	PGA/PLLA		B	Polysorb	Syneture
	PGA		B	Dexon	Syneture
	PGA		B	Bondek	Teleflex
Monofilaments	PDO	Application dependent on tensile strength loss	M	PDS II	Ethicon
	PGA/PCL		M	Monocryl	Ethicon
	PGA/PTMC/PDO	Profile required	M	Biosyn	Syneture
	PGA/PTMC		M	Maxon	Syneture
	PGA/PCL/ PTMC/PLLA		M	Caprosyn	Syneture

[a] Gupta (1996); Mathes and Moelleken (1995).
[b] T, twisted monofilament; M, monofilament; B, multifilament braid.

Fiber Construction

Sutures are fabricated into fibers in order to maximize the resulting strength. For materials that possess a high tensile modulus, yarns composed of low-denier filaments are fabricated into a multifilament braid in order to achieve adequate suppleness. Materials that possess a tensile modulus of about 500 kpsi or less can be fabricated into monofilaments with acceptable handling characteristics although the optimum range is about 350 kpsi or less. Catgut is a twisted multifilament with a polished outer surface that handles similarly to and is therefore often classified as a monofilament.

Because a suture is required to pass through tissues with minimal friction, coatings are normally applied to multifilament braids to increase the surface lubricity. Like the suture itself, coatings can be based on absorbable or nonabsorbable materials.

GENERAL CHARACTERISTICS

Sutures can be characterized by a number of different properties, all of which can be categorized into three groups—physical, handling, and biological (Table 3). There is no one

TABLE 3   Suture Properties and Definitions

Characteristic	Definition
Physical	
Dimensions	Size (diameter) and length
Tensile strength	Maximum tensile stress that can be applied to the strand
Knot-pull strength	Maximum tensile stress that can be applied to the ears of a knot
Needle attachment force	Force required to separate the needle and strand
Knot security	Force that a knot can withstand before slipping or untying
Coefficient of friction	Force required to slide a strand across itself
Stiffness	Force required to bend a strand to a predetermined angle
Memory	Tendency of a strand to return to its original shape
Creep	Deformation of a strand or knot over time when subjected to a constant load
Swelling	Tendency of a strand to swell in tissues
Capillarity	Degree to which a liquid is transported along a strand
Handling	
Knot tie down	Ease with which a knot can be slid down a strand
First throw hold	Force that a "surgeon's knot" (double turn) can withstand before slipping
Tissue drag	Force required to pass a strand through tissue
Package memory	Degree to which a suture retains its package configuration
Suppleness	The "feel" of a suture
Biological	
Tissue reaction	Cellular response to a strand
Tensile strength loss	The loss of tensile strength as a function of time
Absorption	Enzymatic and/or hydrolytic breakdown of a strand followed by elimination
Biocompatibility	Degree to which a strand and breakdown products affect surrounding tissue

ideal suture since the type of tissue, the patient condition, and surgeon preference all come into play when selecting an appropriate suture. The availability of several different types of suture materials offered by different manufacturers makes this selection process a difficult one without an understanding of the basic properties of each.

SUTURE MATERIALS

Natural Sutures

Catgut Suture

Catgut is generally derived from bovine intestinal serosa or from the submucosa of sheep or goat intestine. The primary constituent of catgut is collagen, a protein of three polypeptides arranged in a triple helix coil structure. Processing begins with the cleaning and splitting of the gut into longitudinal ribbons followed by additional chemical and mechanical cleaning. The resulting ribbons are then treated with dilute formaldehyde to increase strength and the resistance to enzymolysis by blocking the hydroxy and amino pendant groups on collagen (Chu, 1997b). The treated ribbons are then twisted and dried followed by mechanical grinding and polishing on a centerless grinder to achieve the desired diameter. This grinding process can result in the disruption of fibril integrity at the surface since the strand is composed of twisted ribbons. This leads to weak points that can contribute to inconsistency in strength and, more importantly, to fraying when a knot is run down the length of the suture (Benicewicz and Hopper, 1990; Chu, 1997b). Catgut that is treated in the manner just described is

referred to as plain gut. Tanning of catgut using chromium salts was first developed by Lister in 1840 and is referred to as chromic gut. Chromic gut slows the rate of absorption and results in lower tissue reaction (Chu, 1997b). Since catgut sutures become less pliable on drying, these materials are generally packaged in an alcohol solution.

Catgut has been used for centuries because it is a well known and inexpensive absorbable material. Disadvantages include inconsistent strength, inconsistent tissue absorption, fraying, and tissue reactivity.

Silk Suture

Silk is a fibrous protein derived from the silkworm, normally the *Bombyx mori* silkworm (Chu, 1997b). Processing involves the removal of natural waxes and gums followed by twisting or braiding of the fine filaments into a multifilament construction. To improve the smoothness, beeswax or a siloxane-based coating is applied to the braid. The level and the type of coating can have a significant effect on the knot rundown and the knot security. Silk slowly loses strength over a period of roughly 1 year and is absorbed after 2 or more years (Benewicz and Hopper, 1990; Stashak and Yturraspe, 1978; Van Winkle and Hastings, 1972). Silk is inexpensive and is considered to be the "gold standard" in suture handling characteristics. However, as a natural material, it can incite a tissue reaction.

Collagen Suture

Another type of collagen suture is reconstituted collagen derived from bovine flexor tendons by enzymatic or acidic digestion to a gel followed by extrusion into a coagulation bath. The resulting fibers are twisted and can be treated with chromium salts. Collagen sutures tend to be more consistent than catgut sutures and are simpler to process. The primary use for this suture is in microsurgery or ophthalmic surgery.

Cotton Suture

Cotton was introduced as a suture material in 1939 and gained wide use during World War II due to the scarcity of silk. It is based on cellulose derived from long-staple cotton fiber and is coated with wax. When implanted, it slowly loses tensile strength over a 2-year period (Van Winkle and Hastings, 1972). It possesses low strength that increases roughly 10% when the suture is wet (Stashak and Yturraspe, 1978; Herrmann, 1971). It elicits tissue reactivity similar to that of silk and possesses a high level of capillarity. Cotton suture material can also become electrostatically charged, causing it to cling to gloves and surgical linen (Bellenger, 1982). As with collagen the general use of cotton is limited.

Linen Suture

Linen is also based on cellulose and is derived from twisted long-staple flax. Like cotton, its tensile strength increases when wet. It possesses a higher tissue reactivity and greater capillarity than cotton and has found limited use (Stashak and Yturraspe, 1978).

Synthetic Nonabsorbable Sutures

Polyester Suture

Polyester-based sutures include those based on poly(ethylene terephthalate) (PET), poly(butylene terephthalate) (PBT), and polybutester, which is a polyether–ester derived from poly(tetramethylene glycol), 1,4-butanediol, and dimethyl terephthalate. All are produced by the condensation of the corresponding diol or glycol with terephthalic acid or an ester of terephthalic acid. PET is a hard, brittle material, while the longer segment in PBT renders it somewhat less brittle. Both PET and PBT are fabricated into multifilaments to obtain the desired handling characteristics. Coating with Teflon, siloxane-based materials, or poly(butylene adipate) is required to increase surface lubricity. PET suture, introduced in the 1950s is the polyester used most frequently, and it has exceptional strength and minimal tissue reactivity. PBT results in a less stiff fiber than PET. The longer soft segment with the ether linkage in the polybutester reduces the tensile modulus enough that it can be fabricated as a monofilament. With minimal tensile strength loss, the polyesters offer excellent long-term support.

Polyamide Suture

Both nylon-6,6 and nylon-6 are fabricated into sutures. Nylon-6,6 is derived from the condensation of hexamethylene diamine and adipic acid while nylon-6 is derived from the ring-opening polymerization of caprolactam. Both nylons can be fabricated into either monofilaments or multifilament braids,

although the monofilaments tend to be stiff, requiring additional throws to secure the knot (Stashak and Yturraspe, 1978; Bellenger, 1982; Guttman and Guttman, 1994). However, the addition of a pliabilizing fluid to the packaged monofilament suture can improve the pliability (Capperauld and Bucknall, 1984). The multifilament nylons are generally coated with poly(butylene adipate) or a siloxane-based material. Since the amide bond is susceptible to hydrolysis, nylon sutures lose tensile strength over time (Chu, 1997b).

Polypropylene Sutures

Polypropylene sutures are based on isotactic polypropylene, polymerized from propylene using a Ziegler–Natta catalyst. Introduced in 1970, polypropylene causes one of the lowest tissue reactions and does not lose strength after it is implanted. Because of its smooth surface, careful knot tying is required in order to secure knots (Bellenger, 1982; Guttman and Guttman, 1994; Stashak and Yturraspe, 1978). Differences in processing of polypropylene can lead to differences in suture toughness, surface smoothness, and surface hardness, characteristics that can contribute to the relative level of fraying and breaking when tying knots (Chu, 1997b).

Fluoropolymer Sutures

Polytetrafluoroethylene (PTFE) is derived from the free-radical polymerization of tetrafluoroethylene and is employed as an expanded monofilament. Incorporation of pores results in a microporous structure that has 50% or more air by volume (Chu, 1997b). Although this inclusion of pores leads to relatively lower strength, expanded PTFE fibers possess a needle-to-suture diameter ratio of nearly one with the unique advantage that blood leakage is reduced relative to other types of sutures that do not have the ability to be compressed. In addition, the surface characteristics of ePTFE suture results in very low tissue drag. Pronova, a new monofilament suture based on blends of poly(vinylidene fluoride) and poly(vinylidene fluoride-*co*-polyhexafluoropropylene), has been introduced for use in cardiovascular surgery. The homopolymer component provides strength while the copolymer provides suppleness. Suture sizes smaller than 4/0 are made from a blend high in the homopolymer, while the larger sizes are made from a 50/50 mixture. This material reportedly exhibits improved pliability and resistance to fraying and breaking when compared to polypropylene sutures (Ethicon, 1999). Monofilament sutures made from homopolymers of poly(vinylidene difluoride) (PVDF) are available in several markets outside the United States. They are used as a replacement for polypropylene, and have been shown to be more resistant to creep (Wada *et al.*, 2001).

Stainless Steel

A number of different types of metal have been used over the past several centuries including gold, silver, bronze, copper, nickel, tantalum, stainless steel, and other alloys. Austenitic stainless steel is the most suitable type for sutures and is composed of iron, chromium, nickel and molybdenum (Bellenger, 1982). While stainless steel possesses very high strength, is inert, and is noncapillary, it is difficult to handle

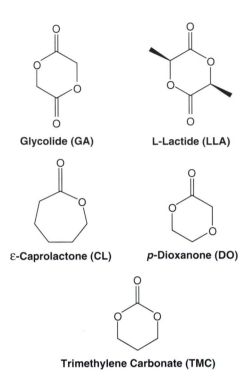

FIG. 1. Monomers used in synthetic absorbable sutures.

TABLE 4 Properties of Homopolymers Derived from Common Monomers

Polymer	Abbreviation	Strength	Flexibility	Rate of hydrolysis
Poly(glycolic acid)	PGA	Excellent	Poor	Fast
Poly-(L-lactic acid)	PLLA	Good	Poor	Slow
Polydioxanone	PDO	Fair	Good	Moderate
Poly(trimethylene carbonate)	PTMC	Poor	Excellent	Slow
Polycaprolactone	PCL	Fair	Excellent	Slow

TABLE 5 Block Construction Considerations[a]

Monomer	Polymerization rate	Residual monomer (%)	Glass transition (°C)	Peak melting (°C)
Glycolide	Fast	1–3	36	224–226
L-Lactide	Slow	2–7	65	185
p-Dioxanone	Moderate	4–15	−10–0	110–115
ε-Caprolactone	Slow	1–3	−60	70
Trimethylene carbonate	Moderate	1–5	−20	Amorphous

[a]Doddi *et al.* (1977); Frazza and Schmidt (1971); Roby *et al.* (1985); Shalaby and Johnson (1994).

because of kinking and possible puncturing of surgical gloves (Bellenger, 1982; Capperauld and Bucknall, 1984; Guttman and Guttman, 1994). In addition, the ends of stainless steel sutures can cause tissue irritation (Bellenger, 1982). Stainless steel is available either as a monofilament or as a multifilament braid.

Absorbable Synthetic Sutures

The need for an absorbable suture that was more consistent and more inert than catgut suture initiated the development of absorbable synthetic sutures. Today, synthetic absorbable sutures represent almost 42% of total suture usage worldwide.

Monomers

All of the major synthetic absorbable sutures are based on one or more of five different cyclic monomers—glycolide (GA), L-lactide (LLA), trimethylene carbonate (TMC), p-dioxanone, and ε-caprolactone (CL) (Fig. 1).

Because these monomers each produce polymers with different characteristics, absorbable polymers must be designed very carefully in order to achieve the desired properties (Tables 4 and 5). For example, construction of monofilaments requires the combination of flexible polymer segments such as that obtained with PTMC, PCL, or PDO with high-strength segments such as that obtained with PGA, PLLA, or PDO. When constructing copolymers in which the predominant block is PGA, the PGA block is prepared last since glycolide polymerizes rapidly with a melting point

above the polymerization temperature of the other common monomers.

Polymerization

The ring-opening polymerizations or copolymerizations of these monomers are accomplished using an organometallic catalyst such as stannous chloride or stannous octoate and a mono-or difunctional hydroxy initiator. Numerous catalysts have been investigated, but the acceptable biocompatibility of tin (II) makes this metal catalyst the one of choice in biomedical polymers (Zhang *et al.*, 1992). In addition, stannous octoate is compatible with glycolide/lactide mixtures ensuring adequate mixing of the catalyst (Wasserman and Versfelt, 1974). The mechanism is thought to involve the formation of an activated ester carbonyl/catalyst complex followed by ring-opening by the initiator (Frazza and Schmitt, 1971; Shalaby and Johnson, 1994; Zhang *et al.*, 1992).

There are several competing reactions in the polymerization of these monomers including depolymerization from polymer to monomer and ester interchange, as well as thermal and/or oxidative decomposition. All of these competing reactions are influenced by the type of monomer, the temperature and the catalyst (Nieuwenhuis, 1992; Zhang *et al.*, 1992). The presence of carboxyl end groups can also have a deleterious effect on the thermal stability both at elevated temperature and at ambient temperature (Brode and Koleske, 1972). For this reason the acid number is minimized by using hydroxyl initiators and by the appropriate purification and moisture protection procedures. Since the presence of impurities can have adverse

TABLE 6 Effect of Composition on Crystallinity[a]

Glycolide (mol%)	L-Lactide (mol%)	Morphology
0–25	75–100	Crystalline
25–65	35–75	Amorphous
65–100	0–35	Crystalline

[a] Gilding and Reed (1979).

effects on molecular weight, sophisticated handling, purification, and drying systems have been developed to ensure ultrahigh-quality monomer and to protect the entire monomer to finished suture process from excessive exposure to moisture and heat.

Synthetic Absorbable Multifilament Braids

Dexon suture introduced in 1971 by Davis & Geck (now Syneture) was the first successful commercial synthetic absorbable suture and is based on polyglycolic acid (Benicewicz and Hopper, 1990). Although PGA was first reported in 1893 by Bishoff and investigated by Carothers in 1932, it was not until the 1960s that Schmitt and co-workers were able to prepare PGA in the high molecular weights necessary for fiber applications such as sutures (Frazza and Schmitt, 1971).

One of the requirements for strength in a suture is a certain level of crystallinity to ensure adequate fiber properties. In the case of glycolide/lactide random copolymers, there are three general composition ranges possible as summarized in Table 6. Crystallinity reaches a maximum as the compositions approach 100% of either monomer and reaches a minimum for compositions comprising roughly equivalent concentrations of both.

Vicryl suture, based on a 90/10 PGA/PLLA copolymer, was introduced by Ethicon, in 1974 (Guttman and Guttman, 1994). Similarly, Polysorb suture was introduced by U.S. Surgical Corp. (now Syneture) in 1991, also based on a PGA/PLLA copolymer.

Although the homopolymer and copolymers exhibit similar properties, there are some differences, the most notable being the degradation profiles. Since the copolymers possess lower crystallinity than the PGA homopolymer, the resulting absorption time for the copolymers is somewhat less than that for the homopolymer (Capperauld and Bucknall, 1984; Chu, 1997c; Fredericks et al., 1984; Guttman and Guttman, 1994; Swanson and Tromovitch, 1982). Modifications to Vicryl suture resulted in the 1996 release of a product with a slightly longer strength loss profile from 40% to 50% of the original strength at week 3 (Ethicon, 1996).

Another distinguishing feature between the three synthetic absorbable braids just discussed is the coating applied to the braid for increased lubricity. Typically, a suture coating must meet three requirements. First, it must allow knots to be run down the length of the suture smoothly without wearing off. Second, the coating must enable the knot thus formed to remain secure. Finally, the coatings must absorb in a similar manner to the suture itself and not increase the tissue reactivity. The initial Dexon and Vicryl braids did not possess coatings and

as a consequence were difficult to use. Each of the three braid products has undergone a number of coating variations.

Multicomponent coatings composed of PGA/PLLA copolymers and water-insoluble fatty-acid salts were developed as one way of meeting these needs (Mattei, 1980). The Vicryl coating is a mixture of equal parts of a PGA/PLLA copolymer and calcium stearate while the Polysorb coating is a mixture of a PCL/PGA copolymer and calcium stearoyl lactylate (USS/DG, 1997; Ethicon, 1996). In each of these coatings, the calcium salt facilitates knot rundown and knot security while the copolymer serves as a film-forming matrix for the salt. Unlike Vicryl and Polysorb sutures, Dexon II suture is coated with polycaprolate (Davis and Geck, 1990). The coatings are formulated so as to exhibit similar absorption characteristics to the suture itself.

Vicryl Rapide suture was introduced by Ethicon as a faster-absorbing synthetic absorbable braid relative to Vicryl suture itself. It is obtained by reducing the molecular weight of Vicryl braid and thereby reducing the time frame for tensile-strength loss.

Ethicon developed a long-term absorbable braid called Panacryl suture. This braided suture is designed for orthopedic use and composed of a PGA/PLLA (3/97) copolymer with a PCL/PGA coating. The tensile loss is complete in 1 year, and mass loss is essentially complete in 1.5 to 2.5 years (Ethicon, 1999a).

The synthetic absorbable braids offer a more consistent strength loss profile than catgut with less tissue reactivity, with the result that the use of these materials has grown dramatically and they are now the most commonly used type of suture in surgery (Rodheaver et al., 1996).

Synthetic Absorbable Monofilaments

Much of the recent development in sutures has been in the area of synthetic absorbable monofilaments. As discussed previously, these materials offer the advantages of lower tissue drag, reduced capillarity, and, more recently, lower cost. To this end, five different sutures have been developed over the past 20 years. These include PDS II, Maxon, Monocryl, Biosyn, and Caprosyn sutures.

PDS suture is based on poly(p-dioxanone) and was introduced in 1983 by Ethicon, (Chu, 1997c). The appeal of this polymer system is the inherent simplicity (Doddi et al., 1977). It combines the strength and hydrolytic sensitivity of the glycolic ester linkage with the flexibility and hydrolytic stability of the ether linkage in an alternating structure based on only one monomer. Subsequent to the introduction of PDS suture, improvements resulted in PDS II suture with greater flexibility and better handling characteristics (Chu, 1997c). It has been reported that these improvements were a result of fiber processing modifications (Chu, 1997c; Broyer, 1994). This modification is achieved by heat treating the fiber surface for a short period of time at approximately 125°C to melt and thereby modify the surface morphology. With 25% of its original strength at 6 weeks, PDS II suture offers the longest support of any synthetic absorbable suture (Ethicon, 1989). This extended support allows it to be used in slower healing tissues that previously required a nonabsorbable material (Capperauld and Bucknall, 1984).

TABLE 7 Effect of Block structure on PGA/PTMC Copolymer Properties[a]

Block type	PGA content (wt %)	Peak melting (°C)	Tensile strength (kpsi)	Tensile modulus (kpsi)	21-day *in vivo* str. retention (% of original)	*In vivo* absorption (days)
Homopolymer	100	227	122	2100	20	<90
Random	88	209	65	970	3	<90
Block	65	216	42	610	53	<180
Modified block	65	213	61	260	69	<180

[a]Roby *et al.* (1985).

Maxon suture, based on a PGA/PTMC (67.5/32.5) triblock copolymer, was introduced in 1985 by Davis and Geck (Benicewicz and Hopper, 1991). The middle block is composed of a PTMC/PGA (85/15) random copolymer while the two end blocks are composed of predominantly PGA (Casey and Roby, 1984). The structure of this copolymer was the result of investigating several block structures and composition ratios of glycolide and TMC (Table 7). As can be observed, increasing the TMC content reduces the tensile modulus, extends the tensile strength loss profile, and increases the time required for absorption. Although the simplest to construct, the random copolymers begin to lose the required properties at a low level of PTMC incorporation. On the other hand, the diblock copolymers begin to approach the desirable properties but do not possess the required flexibility. However, incorporation of glycolide into the PTMC middle block achieves the desired characteristics of extended strength and flexibility. Here the PGA segments contribute high strength and hydrolytic sensitivity while the PTMC segments contribute flexibility and extend the tensile strength loss profile by virtue of its relatively lower hydrolytic sensitivity. It is important to note that the addition of small amounts of PGA to the predominantly PTMC middle block is essential to the absorption of the polymer since PTMC absorbs very slowly if at all. Maxon suture also offers longer support than the original synthetic absorbable braids, comparable to PDS II (USS/D&G and Ethicon IFU).

Unlike the first two monofilaments, Monocryl suture was developed as a shorter-term suture. Introduced by Ethicon, it was more directly positioned against catgut suture (Bezwada *et al.*, 1995). The structure of Monocryl suture is also the result of a two-step polymerization process based on a PGA/PCL (75/25) copolymer. It has been reported that the first block is a PCL/PGA (45/55) random copolymer while the end blocks are predominantly composed of PGA (Bezwada *et al.*, 1995). It was found that the solubility of the random PCL/PGA middle block in the glycolide added in the second step to form the end block was critical to obtaining a one-phase system with the desired properties. This solubility was dependent on the PCL/PGA middle block ratio with PCL contents in the range of 40–50% being preferred (Bezwada and Jamiolkowski, 1993).

Biosyn suture is a recent addition to the family of monofilaments, introduced in 1995 by U.S. Surgical Corp. Unlike other monofilaments that offer strength-loss profiles that are either shorter or longer than those of the absorbable braids, this product was positioned as a monofilament with a strength-loss profile similar to that of the absorbable braids. The most complex structure of all of the synthetic absorbable sutures, it is based on three different monomers—glycolide, TMC, and *p*-dioxanone in a 60/23/17 weight ratio (Roby *et al.*, 1997). Like Maxon and Monocryl copolymers, the Biosyn copolymer is based on a two-step process. The first block is a random PTMC/PDO (65/35 w/w) copolymer while the end blocks are random PGA/PDO (92/8 w/w) copolymers (Roby *et al.*, 1995). As was the case with the Maxon and Monocryl copolymers, the block structure and composition are critical to obtaining the desired properties (Table 8). It should be noted that small

TABLE 8 Effect of Composition and Block Structure on PGA/PTMC/PDO Copolymer Properties[a]

Total PGA content (wt %)	First block PDO (wt %)	Second block PDO (wt %)	Peak melting (°C)	Tensile modulus (kpsi)	Initial loop pull (kg)[b]	Week 3 loop pull (kg)[b]
72	36	4	194	920	8.7	0
69	49	4	206	700	9.1	0.7
69	65	7	194	610	7.5	0.2
67	69	7	181	290	7.3	0.2
64	52	5	184	300	7.9	0.4
64	35	4	190	190	7.0	1.3
60	34	8	186	140	6.1	2.2
60	53	9	186	135	6.4	0.7

[a]Roby *et al.* (1997).
[b]*In vivo.*

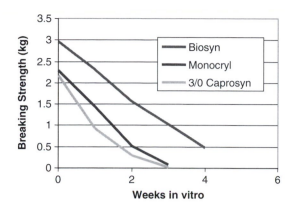

FIG. 2. *In vitro* strength loss for several synthetic absorbable monofilaments.

amounts of *p*-dioxanone added to the PGA end blocks is critical to achieving the desired results. Initial experience with this new suture indicates that it is an attractive alternative to braided synthetic absorbable sutures (Rodeheaver *et al.*, 1996).

The newest synthetic absorbable monofilament suture is Caprosyn, introduced by USS/DG in late 2002. This product was designed to meet the surgeon's needs for a supple, fast-absorbing monofilament for use in plastic surgery and to replace catgut. This material is a tetrapolymer, made by copolymerizing glycolide, ε-caprolactone, L-lactide, and TMC in a 68/17/7/7 weight ratio (Roby *et al.*, 2001). The single-step polymerization process gives rise to a random copolymer, where controlling ε-caprolactone conversion in a narrow window is key to obtaining the desired flexibility of the finished suture. Too much PCL in the final polymer and it will be very supple but weak; too little and it will be strong but stiff. Caprosyn loses all strength by 3 weeks and is absorbed by the body in 56 days, making it the fastest-absorbing synthetic monofilament. Figure 2 compares the *in vitro* tensile loss curves for the newest synthetic absorbable monofilaments.

Table 9 summarizes the chemical structures for all of the synthetic absorbable sutures discussed. As indicated previously, the glycolic ester unit is present in all of these polymers.

TABLE 9 Chemical Structure of Synthetic Absorbable Sutures

Suture	Block structure	Polymer composition (%)
Multifilament Braids		
Dexon	PGA homopolymer	—
Vicryl	PGA/PLLA random copolymer	90/10
Polysorb	PGA/PLLA random copolymer	90/10
Panacryl	PGA/PLLA random copolymer	3/97
Monofilaments		
PDS II	PDO homopolymer	—
Maxon	PGA–PTMC/PGA–PGA	100–85/15–100
Monocryl	PGA–PCL/PGA–PGA	100–45/55–100
Biosyn	PGA/PDO–PTMC/PDO–PGA/PDO	92/8–65/35–92/8
Caprosyn	PGA/PCL/PTMC/PLLA random copolymer	70/16/8/5

PROPERTY COMPARISONS

Few, if any, studies have been completed that compare properties listed in Table 3 for all sutures at one time. As a result, comparisons are generally limited to selected materials and for selected properties. Since conditions and/or testing procedures can vary, care must be taken when comparing the results of one investigation to that of another. A qualitative comparison of the major properties for the various types of sutures is summarized in Table 10.

Physical

Physical properties are fairly quantitative, lending themselves to repeatable test procedures. The United States Pharmacopeia (USP) is a set of standards that specify test procedures and product specifications. For sutures, these specifications includes length, diameter, tensile knot pull strength, and needle attachment force. Sutures that meet all of the USP specifications may be labeled as "USP." For sutures that meet all but the diameter specifications, the label may state "USP except for diameter" and must contain the diameter overage information in the package labeling. Sutures that do not meet the diameter specification require a larger-than-specified diameter in order to meet strength requirements. Diameter is important since a surgeon will employ the smallest diameter possible in order to minimize the amount of material that is drawn through the tissue and/or placed into the body. Meeting these USP specifications and stating the exceptions, if applicable, are required for regulatory approval by the FDA. A similar set of specifications exist in Europe known as the European Pharmacopoeia (EP).

Handling

Handling characteristics are the most difficult to describe, although attempts have been made to quantify these properties (Chu, 1997a). The tensile modulus of a material used as a monofilament and the characteristics of a coating used on a multifilament braid are responsible for a large share of a suture's handling characteristics. In addition, the application, type of tissue, patient condition, and surgeon preference also contribute to the overall assessment of a suture's handling characteristics. A suture that possesses excellent handling in one setting may possess only fair or even unacceptable handling in another. For this reason, the most reliable information regarding a suture's handling characteristics is generally obtained by performing evaluations in the appropriate *in vitro* or *in vivo* environment.

Biological

Assessment of tensile-strength loss and absorption are typically performed in an *in vivo* or *in vitro* setting. With respect to tensile strength loss, confusion can arise if the testing methods are not properly designated. Specifically, there are two important considerations when comparing the tensile-strength loss profiles for absorbable sutures. The first is the

TABLE 10 Suture Properties[a]

Suture type	Fiber type	Knot pull strength	Knot security	Handling	Tissue reactivity	In vivo strength loss
Natural materials						
Catgut, plain	T	Poor	Poor	Fair	High	7–10 days
Catgut, chromic					Mod.	—
	T		Fair	Fair	High	21–28 days
Silk	B	Fair	Good	Very good	High	1 year
Synthetic nonabsorbable materials						
Polyester	B	Good	Poor–Good	Good	Moderate	Indefinite
Polyester	M		Poor	Fair	Low	Indefinite
Polyamide	B	Fair	Fair	Good	Low	15–25% / year
Polyamide	M	Fair	Poor	Poor	Low	15–25% / year
Polypropylene	M	Fair	Poor	Poor	Low	Indefinite
PTFE	M	Poor	Very good	Very good	Low	Indefinite
Stainless steel	T,M	High	Good	Poor	Low	Indefinite
Synthetic absorbable materials						
Braids	B	Good	Fair–Good	Good	Low	10 days to 4 weeks, 1 year
Monofilaments	M	Fair	—			
		Good	Poor–Fair	Fair–Good	Low	10 days to 6 weeks

[a] Casey and Lewis (1986); Chu (1997a); Gupta (1996); Guttman and Guttman (1994); Herrmann (1971); Mathes and Moelleken (1995).

T, twisted monofilament; M, monofilament; B, braided multifilament; all braids are coated sutures.

type of tensile-strength test employed. The USP knot-pull test involves pulling the ears of the knot in opposite directions and measuring the force required to cause failure. The standard loop-pull test involves pulling the loop formed by the knot in opposite directions and measuring the force required to cause failure. The former test essentially measures the maximum force that can be applied by the surgeon, whereas the latter measures the maximum tissue support force (Rodeheaver *et al.*, 1996). Loop-pull forces are therefore higher than knot-pull forces because the loop measurement involves a double strand of the material, whereas the knot measurement involves the two ears or one strand of the material. Both yield valuable information.

The second consideration is whether the tensile strength is reported as an absolute force-to-break or as a percentage of the original tensile strength. Figures 3A and 3B illustrate the effects of this difference when Biosyn suture is compared to Vicryl suture. An *in vivo* tensile-strength loss comparison using the absolute loop-pull strength is summarized in Fig. 4 for synthetic absorbable sutures and indicates that a wide range of tensile-strength loss profiles is available from which to choose.

As mentioned previously, absorption of a suture can take place months after complete strength loss has occurred (Fig. 5). As expected, the order in which the synthetic absorbable sutures absorb is similar to the order in which the tensile strength is lost (Fig. 6). To support the tensile strength loss and absorption information collected from *in vivo* and *in vitro* studies such as those described above, pharmacokinetic studies are generally performed. Pharmacokinetics involve the implantation of ^{14}C-labeled polymer into animals and following the level of radioactivity for selected ^{14}C-labeled by-products. In addition, the tissues are analyzed to determine the extent of by-product deposition, if any, in

the various organs. These studies generally support the data summarized above (Table 11).

COLORANTS

Sutures can be provided dyed different colors or in their natural color. As sutures become wet with body fluids, they tend to blend into the surrounding tissue, making them hard to locate in the surgical field. Dyed sutures are preferred by

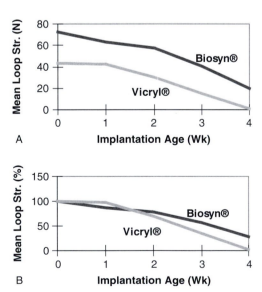

FIG. 3. Tensile strength retention plotted as load (N) and percentage of original strength. (Rodeheaver *et al.*, 1996.)

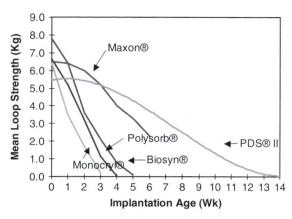

FIG. 4. *In vivo* tensile strength loss of synthetic absorbable sutures. (Roby, 1998.)

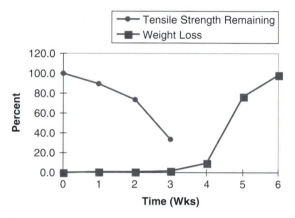

FIG. 5. *In vivo* tensile strength loss compared to mass loss for vicryl suture. (Fredericks *et al.*, 1984.)

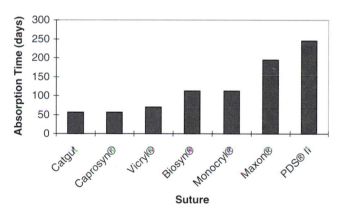

FIG. 6. Time required for essentially complete absorption. (Roby, 1998; USS/DG, 2002.)

TABLE 11 Absorption Times Using ^{14}C-Labeled Polymers

[14]C-Labeled polymer	Absorption time (days)	Reference
PDS[TM] II	180	Ray *et al.* (1997)
Maxon[TM]	182–213	Katz *et al.* (1985)
Monocryl[TM]	91–119	Bezwada *et al.* (1995)
Biosyn[TM]	105–126	Jiang *et al.* (1997)

TABLE 12 Popular Colorants Used in Sutures in the United States[a]

Suture material	Colorant allowed (U. S.)
Catgut	Pyrogallol
Silk	Logwood extract, D&C Blue No. 9
Polyester	D&C Green No. 6, D&C Blue No. 6
Nylon	Logwood extract, FD&C Blue No. 2, D&C Green No. 5
Polybutester	[Phthalocyaninato(2-)]Copper
Polypropylene	[Phthalocyaninato(2-)]Copper
PGA	D&C Green No. 6, D&C Violet No. 2
PDO, PGA Copolymers	D&C Violet No. 2
PGA/PTMC Copolymers	D&C Green No. 6

[a]Marmion (1991).

and toxicology data in order to receive approval, and limits are placed on the maximum amount of colorant allowed in the suture. Additionally, each batch of colorant is tested and certified by the FDA as meeting established purity criteria. A list of commonly used colorants in sutures can be found in Table 12.

NEEDLES

A needle enables the suture strand to be passed through the tissues that are to be approximated. The most important characteristics of a needle include sharpness, strength, profile, corrosion resistance, and durability. The variety of surgery and surgeons has created a demand for a great variety of needles. Different combinations of length, curvature, wire diameter, and point geometry have resulted in hundreds of types of surgical needles being available. Needles are generally classified by point geometry as shown in Table 13, but additional characteristics such as length, curvature, and wire diameter are considered by the surgeon during product selection.

The increase in sophistication of surgical technique over the years has given rise to a need for smaller and smaller wire diameter needles that will provide increased resistance to bending and breaking while penetrating tissue, as exemplified by the needle and suture shown in Fig. 7, which is typical of that used by CV surgeons during the coronary arterial bypass graft (CABG) procedure. This has been accomplished by the development of new stainless steel alloys that can offer the processability needed to drill, grind, and form a needle, combined with the stiffness and strength that the surgeon requires.

most surgeons because they are easily visualized in tissue. Undyed sutures are typically used by plastic surgeons in subcutaneous applications so that they will not be seen through the skin. The colorants used are approved by the FDA for use in specific suture materials through a Color Additive Petition (CAP). The CAP requires extensive biocompatibility, stability,

TABLE 13   Surgical Needle Types

Type	Description	Uses
Taper point	A circular cross section smoothly increasing in diameter from tip to body	Cardiovascular and general surgery
Cutting	A triangular cross section smoothly increasing in diameter from tip to body	Plastic surgery
Taper-cutting	A taper-point needle with small facets ground at the tip to provide some cutting effect	Cardiovascular and general surgery
Spatula or diamond	A sharp, flat tip tapering back to a triangular body	Ophthalmic

FIG. 7.   Surgipro needle and suture used in bypass surgery compared to 0.5-mm pencil lead.

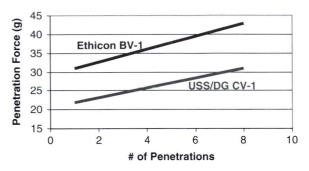

FIG. 8.   Needle penetration force as a function of the number of passes through a test medium. (USS/DG Sales Literature, 2002.)

The industry has moved the premium cardiovascular, ophthalmic, and plastic-surgery needles away from the 420 and 300 series stainless steel alloys that were prone to bending, to the precipitation-hardened martensitic 455-type alloys (Edlich *et al.*, 1993). The resulting materials are extremely hard, and lasers are typically used to drill the holes for attachment to the suture. These needles are marketed under the trade names of Ethalloy and Surgalloy in the United States.

All needles start with high-quality stainless steel wire. The first step is to drill a hole in the wire that will accommodate the suture to be attached to the needle. The holes are commonly drilled mechanically or with a laser. The grinding and polishing operations that form the proper tip and body geometry are performed, followed by coining or pressing of the sides to improve rigidity and stability in the needle drivers that surgeons use. If the needles are to be curved, that process is performed followed by cleaning and deburring. If a martensitic alloy is being used, aging at the optimal time and temperature under vacuum is performed in order to increase resistance to bending, which results in an increase of up to 50% in bend moment and tensile strength (Rizk *et al.*, 1995). The cleaning and vacuum steps are critical in order to avoid any discoloration during the heat treat process. Electropolishing may be done in order to improve the surface finish and sharpness of

cutting needles. Lastly, a polydimethylsiloxane (PDMS) based coating is applied and allowed to cure. The coating improves tissue penetration performance of the needle by up to 60% (Walther and Raddatz, 1999), and proper curing is essential to provide a durable coating that will not wear off after repeated penetrations through tissue. This is critical because when less force is required to pass the needle through tissue, there is less likelihood of bending or breaking the needle. A new coating for surgical needles has been developed by USS/D&G called NuCoat. It is based on a mixture of polydimethylsiloxanes with an improved curing cycle (Roby, 2003). This new coating process results in a needle that does not show increased penetration forces after repeated tissue passes when compared to standard commercial needles as shown in Fig. 8 (USS/D&G, 2002). Because most surgical procedures require surgeons to make multiple passes through tissue with the needles, it is an advantage to have a needle that maintains a consistently low penetration force.

The current generation of high-quality PDMS-coated 455 series stainless steel needles gives the surgeon a selection of extremely strong, durable needles in a wide range of sizes and styles that minimize trauma to tissue.

PACKAGING AND STERILIZATION

A suture package protects the suture from damage and biological contamination, allows for easy removal of the suture in the operating room, withstands the sterilization conditions used, and contains no contaminants that may transfer to the suture. Currently, a paper or plastic retainer is used to hold the suture and needle in place, with the needle being parked in foam or in a plastic slot or slit. The fiber can be straight, can be wrapped in a Fig. 8, or can follow an oval or spiral track. The materials used here are medical or food grades of paper, cardboard, and plastics (PVC, PETG, and PP) and must withstand the sterilization method without degradation. Biocompatibility testing is done on any component that may come into contact with the suture to ensure that there are no toxic materials present. If the suture is a synthetic absorbable, a foil moisture barrier must be used to protect the suture from degradation during shipping and storage. This is done by drying and sealing the sterile suture in a foil pouch, and if this is done properly the suture will show no degradation in the package for at least 5 years. Sutures based on synthetic absorbables, polypropylene, and sutures containing PTFE pledgets must be sterilized using ethylene oxide gas, because these materials will degrade when exposed to gamma irradiation or e-beam sterilization. Catgut must be sterilized using gamma sterilization, and the remaining sutures can be sterilized by either method.

REGULATORY CONSIDERATIONS

Sutures are regulated by the FDA as medical devices and are categorized into three different classes—Class I (General Controls), Class II (Performance Standards), and Class III (Premarket Approval).

The regulatory requirements, and therefore the development and commercialization time lines, are quite different for these three classes. Class I materials require certain controls such as those addressing possible adulteration and misbranding to ensure safety. Only stainless steel sutures fall into this class.

Class II materials require that the manufacturer apply the controls defined by Class I materials and demonstrate "substantial equivalence" to an existing approved Class II material, often referred to as the predicate device. Characterization of the new material and the predicate device includes physical and mechanical, biocompatibility, efficacy, and stability testing. This testing is submitted to the FDA in a filing that is referred to as a 510(k). Most suture materials were reclassified to Class II from Class III by the FDA in 1989 based on the large body of existing information and usage.

Class III materials require considerable characterization and testing to ensure safety and efficacy. This type of filing is referred to as a PMA (Premarket Approval). Extensive preclinical testing is required as well as multicenter clinical trials. Since the extent of characterization required and the approval timelines for a PMA submission are substantially greater than that required for a 510(k) submission, PMA materials require a longer commercialization time at a much higher cost. The use of a new monomer in significant quantities would probably require a PMA regulatory approval route and is therefore an important consideration in the design of new polymer materials for sutures.

MARKET TRENDS

Suture revenues in 2001 are estimated to have been $632 million (U.S.) and growing at an annual rate of roughly 2–3% (Frost and Sullivan, 2001). Roughly 13% of the usage was natural sutures, 41% synthetic nonabsorbable sutures, and 42% synthetic absorbable sutures. There has been a shift in the usage of the different materials for several reasons. First, there has been a continual decrease in the use of natural materials in favor of synthetic materials because of the inconsistent properties and the inflammatory tissue response that these materials can exhibit. In the case of catgut, the fear of transmission of TSE has resulted in the banning of the product in major international markets. Second, there is an increasing usage of absorbable synthetic materials over nonabsorbable materials in order to avoid a long-term foreign-body effect. Finally, cost containment in the health-care industry combined with the desire for a suture with less tissue drag and less potential to harbor bacteria has resulted in the development of and the increased usage of absorbable synthetic monofilaments over absorbable synthetic multifilament braids (Rodeheaver et al., 1996). In contrast to the suture market in general, these monofilament sutures are increasing in usage at an annual rate of roughly 15% (Roby, 1998).

Until recently the three major suppliers of sutures were Ethicon, Inc., Davis & Geck, and U.S. Surgical Corp. The last two suppliers were recently acquired by Tyco International and are now called Syneture, reducing the number of major suppliers to two. The required capital equipment, technology, and development time, combined with the near commodity nature of these products, makes a major entry into the suture business relatively unattractive. However, there are a number of smaller suture manufacturers successfully serving various niche markets within the business.

FUTURE DEVELOPMENT

Because of the nature of this mature cost-constrained market, future development will be limited to products that can demonstrate a performance benefit. In addition, it is unlikely that the use of new materials that require a PMA regulatory approval route will be developed unless this benefit is significant.

Suture-Related

Numerous synthetic absorbable suture products have been developed by the major suture manufacturers over the past 25 years (Table 14). These products cover a wide range of tensile strength loss profiles both in multifilament braid and

TABLE 14 Summary of Synthetic Absorbable Sutures

Suture	Short term (<3 weeks)	Medium term (3–4 weeks)	Long term (>4 weeks)
Multifilament	Vicryl Rapide	Vicryl Polysorb Dexon	Panacryl
Monofilament	Monocryl Caprosyn	Biosyn Monosyn	PDS II Maxon

in monofilament form. Competition will most likely result in additional products, and the development of additional multifilament braid coatings will continue in order to provide the best suture products possible.

As previously mentioned, absorbable sutures lose strength well before absorption is complete. New block structures and/or the incorporation of low levels of different monomers may be developed to decrease this lag time. Ideally, a suture should absorb rapidly following the loss of tensile strength.

Research is also focused on sutures that can do more than just provide mechanical support to the wound. Since the incidence of wound infection is roughly 8%, antibiotic and silver-based coatings or additives are being investigated as antimicrobial agents (Chu, 1997c). Ethicon has introduced a new absorbable braided suture, Vicryl Plus, containing the bacteristat triclosan (Ethicon, 2002), which has been shown to inhibit colonization of the suture by *Staph a.* and *e.* in *in vitro* studies. The addition of various agents to accelerate the wound-healing process are also being investigated.

Non-Suture-Related

Alternative methods of wound closure include staples, tapes, tissue sealants, and tissue adhesives. The first two methods provide rapid closure with lower infection rates. However, staples do not provide the same level of cosmetic closure as sutures, whereas tapes are generally limited to nonload-bearing topical areas (Quinn, 1998).

Most tissue sealants are based on fibrin glue and possess low strength, require long and sometimes complex preparation times, and require viral deactivation to eliminate the risk of blood-borne viral transmission. BioGlue (CryoLife) is a more recently approved material based on albumin and glutaraldehyde for use as an internal sealant in aortic dissections. Newer tissue sealants based on synthetic absorbable materials, such as CoSeal, FloSeal, and FocalSeal, circumvent some of the inherent weaknesses of fibrin-based sealants (Macchiarini *et al.*, 1999). Although these materials will probably not replace sutures, their hemostatic properties make this closure method a useful adjunct to sutures to reduce bleeding in vessel anastomosis or as an alternative closure method in friable tissues such as the lung.

Topical tissue adhesives based on cyanoacrylates such as Histoacryl, Indermil, and Dermabond adhesives react rapidly with moisture to form a strong bond to tissues. Although acute and inflammatory reactions were reported with the early short-chain derivatives in the 1950s, the development of longer-chain derivatives appears to have reduced the level of toxicity when adhesives are used in topical applications (Maw *et al.*, 1997; Quinn, 1998). Perhaps the largest threat to sutures is a tissue adhesive that is strong, nontoxic, bioabsorbable, and easy to use. This is currently an area of active research that could provide a quantum leap in wound-closure technology.

Bibliography

Barrows, T. H. (1986). Degradable implant materials: a review of synthetic absorbable polymers and their applications. *Clin. Mater.* **1**: 233–257.

Bellenger, C. R. (1982). Sutures. Part I: The purpose of sutures and available suture materials. *Comp. Ed. Pract. Vet.* **4**: 507–515.

Benicewicz, B. C. and Hopper, P. K. (1990). Polymers for absorbable surgical sutures—Part I. *J. Bioactive Compat. Polymers* **5**: 453–472.

Benicewicz, B. C., and Hopper, P. K. (1991). Polymers for absorbable surgical sutures—Part II. *J. Bioactive Compat. Polymers* **6**: 64–94.

Bezwada, R. S., and Jamiolkowski, D. D. (1993). Segmented copolymers of epsilon caprolactone and glycolide for new absorbable monofilament sutures. *Trans. 19th Annual Meeting Soc. for Biomater.*, April 28–May 2, 1993, p. 40.

Bezwada, R. S., Jamiolkowski, D. D., Lee, I. Y., Agarwal, V., Persivale, J., Trenka-Benthin, S., Erneta, M., Suryadevara, J., Yang, A., and Liu, S. (1995). Monocryl suture, a new ultra-pliable absorbable monofilament suture. *Biomaterials*, **16**(15): 1141–1148.

Brode, G. L., and Koleske J. V. (1972). Lactone polymerization and polymer properties. *J. Macromol. Sci. Chem.* **A6**(6): 1109–1144.

Broyer, E. (1994). Thermal treatment of theraplastic filaments for the preparation of surgical sutures, U.S. Patent No. 5,294,395 (to Ethicon, Inc.).

Capperauld, I., and Bucknall T. E. (1984). Sutures and dressings. in *Wound Healing for Surgeons*, T. E. Bucknall and H. Ellis, eds. Baillie're Tindall, Eastbourne, UK, pp. 75–93.

Casey, D. J., and Roby M. S. (1984). Synthetic copolymer surgical articles and method of manufacturing the same. U.S. Patent No. 4,429,080 (to American Cyanamid Co.).

Casey, D. J. and Lewis, O. G. (1986). Absorbable and nonabsorbable sutures. in *Handbook of Biomaterials Evaluation: Scientific, Technical, and Clinical Testing of Implant Materials*, A. von Recum, ed. Macmillan Co., New York, pp. 86–94.

Chu, C. C. (1985). The degradation and biocompatibility of suture materials. in *CRC Critical Reviews in Biocompatibility*, D. F. Williams, ed., Vol. 3. CRC Press, Boca Raton, FL, pp. 261–322.

Chu, C. C. (1991). Recent advancements in suture fibers for wound closure. *ACS Symp. Ser. High-Tech Fibrous Mater.* pp. 167–211.

Chu, C. C. (1997a). Classification and general characteristics of suture materials. Chapter 4 in *Wound Closure Biomaterials and Devices*, C. C. Chu, J. Von Fraunhofer, and H. P. Greisler, eds. CRC Press, Boca Raton, FL, pp. 39–63.

Chu, C. C. (1997b). Chemical structure and manufacturing processes. Chapter 5 in *Wound Closure Biomaterials and Devices*, C. C. Chu, J. Von Fraunhofer, and H. P. Greisler, eds. CRC Press, Boca Raton, FL, pp. 65–106.

Chu, C. C. (1997c). New emerging materials for wound closure. Chapter 12 in *Wound Closure Biomaterials and Devices*, C. C. Chu, J. Von Fraunhofer, and H. P. Greisler, eds. CRC Press, Boca Raton, FL, pp. 347–384.

Chu, C. C., and Williams, D. F. (1983). The effect of gamma irradiation on the enzymatic degradation of polyglycolic acid absorbable sutures. *J. Biomed. Mater. Res.* **17**: 1029–1040.

Davis+Geck (1990). Dexon II package insert card.

Davis+Geck (1995). Maxon package insert card.

Doddi, N., Versfelt, C. C., and Wasserman, D. (1977). Synthetic absorbable surgical devices of poly-dioxanone. U.S. Patent No. 4,052,988 (to Ethicon, Inc.).

Edlich, R., Thacker, J., McGregor, W., and Rodeheaver, G. (1993). Past, present, and future for surgical needles and needle holders. *Am. J. Surg.* **166**: 522–532.

Ethicon, Inc., (1989). PDS II package insert card.

Ethicon, Inc., (1996). Vicryl package insert card.

Ethicon, Inc., (1999a). Panacryl package insert card.

Ethicon, Inc., (1999b). Sales literature for pronova suture.

Ethicon, Inc., (2002). Vicryl Plus package insert card.

Frazza, E. J., and Schmitt, E. E. (1971). A new absorbable suture. *J. Biomed. Mater. Res. Symp.* **1**: 43–58.

Fredericks, R. J., Melveger, A. J., and Dolegiewitz, L. J. (1984). Morphological and structural changes in a copolymer of glycolide and lactide occurring as a result of hydroysis. *J. Polymer Sci.* **22** (*Poly. Phys. Ed.*): 57–66.

Frost and Sullivan (2001). Report A038-54, U.S. wound closure devices market.

Gilding, D. K., and Reed, A. M. (1979). Biodegradable polymers for use in surgery—polyglycolic/poly(lactic acid) homo- and copolymers: 1. *Polymer* **20**: 1459–1464.

Gupta, D. (1996). Sutures. Chapter 7.9 in *Biomaterials Science, an Introduction to Materials in Medicine*, B. D. Ratner, A. S. Hoffman, F. J. Schoen, and J. E. Lemons, eds. Academic Press, San Diego, pp. 356–360.

Guttman, B., and Guttman, H. (1994). Sutures: properties, uses and clinical investigation. in *Polymeric Biomaterials*, S. Dumitriu, ed. Marcel Dekker, New York, pp. 325–346.

Herrmann, J. B. (1971). Tensile strength and knot security of surgical suture materials. *Am. Surg.* **37**: 209–217.

Jiang, Y., Bobo, J., Bennett, S., and Roby, M. (1997). A new synthetic monofilament absorbable suture from block copolymers of trimethylene carbonate, dioxanone and glycolide. Part II. *In vivo* absorption, tissue distribution and elimination of ^{14}C labeled copolymers. *Trans. 23rd Annual Meeting Soc. for Biomater.*, April 30–May 4, 1997, p. 339.

Katz, A. R., Mukherjee, D. P., Kaganov, A. L., and Gordon, S. (1985). A new synthetic monofilament absorbable suture made from polytrimethylene carbonate. *Surg. Gyn. Obstet.* **161**: 213–222.

Kulkarni, R. K., Panl, K. C., Neuman, C., and Leonard, F. (1966). Polylactic acid for surgical implants. *Arch. Surg.* **93**: 839–843.

Macchiarini, P., Wain, J., Almy, S., and Dartevelle, P. (1999). Experimental and clinical evaluation of a new synthetic, absorbable sealant to reduce air leaks in thoracic operations. *J. Thorac. Cardiovasc. Surg.* **117**(4): 751–758.

Marmion, D. (1991). *Handbook of U. S. Colorants*, 3rd Ed. John Wiley & Sons, New York, pp. 24–31.

Mathes, S. J., and Moelleken, B. R. W. (1995). Wound healing. in *Surgery A Problem-Solving Approach*, J. H. Davis and G. F. Sheldon, eds. Mosby, St. Louis, pp. 415–455.

Mattei, F. V. (1980). Absorbable coating composition for sutures. U.S. Patent No. 4,201,216 (to Ethicon, Inc.).

Maw, J. L., Quinn, J. V., Wells, G. A., Ducic, Y., Odell, P. F., Lamothe, A., Brownrigg, P. J., and Sutcliffe, T. (1997). A prospective comparison of octylcyanoacrylate tissue adhesive and suture for the closure of head and neck incisions. *J. Otolaryngol.* **26**(1): 26–30.

Nieuwenhuis, J. (1992). Synthesis of polylactides, polyglycolides and their copolymers. *Clin. Mater.* **10**: 59–67.

Quinn, J. V. (1998). *Tissue Adhesives in Wound Care*. B. C. Decker, Inc., London.

Ray, J. A., Doddi, N., Regula, D., Williams, J. A., and Melveger, A. (1981). Polydioxanone (PDS), a novel monofilament synthetic absorbable suture. *Surg. Gyn. Obstet.* **153**: 497–507.

Rizk, S., Powers, W., and Samsel, S. (1995). Method of making heat treated surgical needles. U.S. Patent No. 5,411,613 (to United States Surgical Corp.).

Roby, M. S. (1998). Recent advances in absorbable sutures. International Conference on Advances in Biomaterials and Tissue Engineering, June 14–19, 1998, p. 46.

Roby, M. (2003). Plasma treated surgical needles and methods for their manufacturer. Publication #W003028770 (to Tyco Healthcare Group LP).

Roby, M., and Kennedy, J. (2003). A New Monofilament Synthetic Absorbable Suture, *Trans. 29th Annual Meeting Soc. for Biomater.*, April 30–May 4, 2003, p. 329.

Roby, M. S., Casey, D. J., and Cody, R. D. (1985). Absorbable sutures based on glycolide/trimethylene carbonate copolymers. *Trans. 11th Annual Meeting Soc. For Biomater.*, April 24–28, 1985, p. 216.

Roby, M. S., Bennett, S., Kokish, M., and Jiang, Y. (1995). Absorbable block copolymers and surgical articles fabricated therefrom. U.S. Patent No. 5,403,347 (to U.S. Surgical Corp.).

Roby, M. S., Bennett, S., Kokish, M., and Jiang, Y. (1997). A new synthetic monofilament absorbable suture from block coplymers of trimethylene carbonate, dioxanone and glycolide. Part I: Synthesis and processing. *Trans. 23rd Annual Meeting Soc. for Biomater.*, April 30–May 4, 1997, p. 338.

Roby, M., Kokish, L., Mehta, R., and Jonn, J. (2001). Absorbable polymers and surgical articles fabricated therefrom. U.S. Patent No. 6,235,869. (to U.S. Surgical Corp.).

Rodeheaver, G. T., Beltran, K. A., Green, C. W., Faulkner, B. C., Stiles, B. M., Stanamir, G. W., Traeland, H., Fried, G. M., Brown, H. C., and Edlich, R. F. (1996). Biomechanical and clinical performance of a new synthetic monofilament absorbable suture. *J. Long-Term Effects Med. Implants* **6**(3&4): 181–198.

Salthouse, T. N., and Matlaga, B. F. (1976). Polyglactin 910 suture absorption and the role of cellular enzymes. *Surg. Gyn. Obset.* **142**: 544–550.

Shalaby, W. S., and Johnson, R. A. (1994). Synthetic absorbable polyesters. in *Biomedical Polymers: Designed to Degrade Systems*, W. S. Shalaby, ed. Hanser Publishing, New York, pp. 1–33.

Singhal, J. P., Singh, H., and Ray, A. R. (1988). Absorbable suture materials: preparation and properties. *JMS Rev. Macromol. Chem. Phys.* **C28**(3,4): 475–502.

Stashak, E. S., and Yturraspe, D. J. (1978). Considerations for selection of suture materials. *J. Vet. Surg.* **7**: 48–55.

Swanson, N. A., and Tromovitch, T. A. (1982). Suture materials, 1980s: properties, uses and abuses. *Int. J. Dermatol.* **21**: 373–378.

Van Winkle, W., Jr., and Hastings, J. C. (1972). Considerations in the choice of suture material for various tissues. *Surg. Gyn. Obstet.* **135**: 113–126.

USS/DG (1997). Polysorb package insert card.

USS/DG (2002). Sales literature on CV needle penetration.

Wada, A., Kubota, H., Hatanaks, H., and Iwamoto, Y. (2001). Comparison of the mechanical properties of polyvinylidene fluoride and polypropylene monofilament sutures used for flexor tendon repair. 47th Annual Meeting, Orthopaedic Research Society, Feb. 25–28, 2001, San Francisco, CA.

Walther, C., and Raddatz, G. (1999). Process for coating surgical needles. U.S. Patent No. 5985355 (to Ethicon, Inc.).

Wasserman, D., and Versfelt, C. C. (1974). Use of stannous octoate catalyst in the manufacture of L(−)-lactide–glycolide copolymer sutures. U.S. Patent No. 3,839,297 (to Ethicon, Inc.).

Williams, D. F. (1982). Review: Biodegradation of surgical polymers. *J. Mater. Sci.* **17**: 1233–1246.

Zhang, X., Wyss, U. P., Pichora, D., and Goosen, F. A. (1992). An investigation of the synthesis and thermal stability of poly(D,L-lactide). *Polymer Bull.* **27**: 623–629.

Zhong, S. P., Doherty, P. J., and Williams, D. F. (1993). The degradation of glycolide acid/lactic acid copolymer *in vivo*. *Clin. Mater.* **14**: 145–153.

7.14 DRUG DELIVERY SYSTEMS

Jorge Heller and Allan S. Hoffman

A wide variety of polymeric biomaterials are components of the various formulations and devices that are routinely used for delivering drugs to the body. This chapter reviews the current state-of-the-art of drug delivery systems (DDS) that release drugs in a controlled manner. The major advantage of developing systems that release drugs in a controlled manner can be appreciated by examining Fig. 1, which illustrates changes in blood plasma levels following a single dose administration of a therapeutic agent. As shown, the blood plasma level rapidly rises and then exponentially decays as the drug is metabolized and/or eliminated from the body. The figure also shows drug concentrations above which the drug produces undesirable (e.g., toxic) side effects and below which it is not therapeutically effective. The difference between these two levels is known as the therapeutic index, which is usually based on a dose-response of the median 50% of a population.

Using only a single dose administration, the time during which the concentration of the drug is above the minimum effective level can be extended by increasing the size of the dose. However, when this done, blood plasma concentrations extend into the toxic response region, an undesirable situation. One can also administer safe doses at periodic intervals in order to maintain a desired drug concentration level; however, this is inconvenient, and patient compliance is often poor. For these reasons, there has been great interest in developing controlled-release formulations and devices that can maintain a desired blood plasma level for long periods of time without reaching a toxic level or dropping below the minimum effective level. These DDS are called "zero order" systems since they release drug at a constant rate.

Aside from the clear therapeutic advantage of controlled release products, there are also compelling business reasons for the development of such devices. Because of increasingly stringent FDA regulations, the cost for introducing new drug entities has escalated to more than $200 million U.S. dollars for each drug, and it is not uncommon for this development to require more than 10 years of research and development work. Thus, it is reasonable for pharmaceutical companies to attempt to maximize their financial return for each drug by research aimed at extending the patented lifetime of the drug, and one means of doing this is to patent a new controlled-release formulation with the same drug. However, the commercial feasibility of such a strategy is predicated on a demonstration that the controlled-release formulation is indeed superior in safety and efficacy to the single dose formulation, and most importantly that the cost of the controlled-release formulation is low enough to ensure a reasonable market penetration.

Successful efforts to produce pharmaceutical formulations that would prolong the action of therapeutic agents for any particular dose, thus minimizing the frequency between dosing, go back to the late 1940s and early 1950s with the introduction of the first commercial product, known as Spansules®. This product was designed to increase the duration of orally administered drugs and consisted of small drug spheres coated with a soluble coating. By using coatings with varying thickness, capsule dissolution times could be varied, thus prolonging the action of one dose of the therapeutic agent. Such formulations are now known as "sustained release" or "prolonged release products," and more recently they have been called "controlled-delivery" products.

However, the functionality of such products depends greatly on the local *in vivo* environment, and as such varies from patient to patient. For this reason, a major effort has been underway since the late 1960s focused on the development of products that are capable of releasing drugs by reproducible and predictable kinetics. Ideally, such products are not significantly affected by the local *in vivo* environment so that patient-to-patient variability is greatly reduced. Such devices are known as "controlled-release" products.

Controlled-release polymeric systems can be classified according to the mechanism that controls the release of the therapeutic agent as described in Table 1.

DIFFUSION-CONTROLLED DELIVERY SYSTEMS

There are two fundamentally different devices where the rate of drug release is controlled by drug diffusion. These are membrane-controlled devices and monolithic devices.

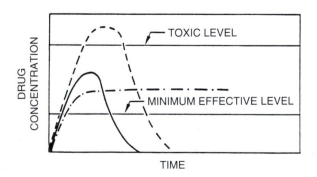

FIG. 1. Drug concentration following absorption of therapeutic agent as a function of time. (—) safe dose, (– – –) unsafe dose, (– · –) controlled release. [Reprinted from Roseman, T. J., and Yalkowsky, S. H. in *Controlled Release Polymeric Formulations*, D. R. Paul and F. W. Harris (eds.). ACS Symposium Series 33, American Chemical Society, Washington DC, pp. 33–52.]

TABLE 1   Controlled Release Drug Delivery Systems

- **Diffusion-Controlled DDS**
  Reservoir and Monolithic Systems
- **Water Penetration-Controlled DDS**
  Osmotic and Swelling-Controlled Systems
- **Chemically-Controlled DDS**
  Biodegradable Reservoir and Monolithic Systems
  Biodegradable Polymer Backbones with Pendant Drugs
- **Responsive DDS**
  Physically- and Chemically-Responsive Systems (T, solvents, pH, ions)
  Mechanical, Magnetic- or Ultrasound-Responsive Systems
  Biochemically-Responsive; Self-Regulated Systems
- **Particulate DDS**
  Microparticulates
  Polymer-Drug Conjugates
  Polymeric Micelle Systems
  Liposome Systems

Membrane-Controlled Reservoir Devices

In this particular delivery system the active agent is contained in a core or "reservoir" which is surrounded by a thin polymer membrane, and release to the surrounding environment occurs by diffusion through the rate-controlling membrane. This system is called a "reservoir device."

When the membrane is nonporous, diffusion can be described by Fick's first law:

$$J = -D\frac{dC_m}{dx} \tag{1}$$

where J is the flux per unit area, in g/cm²-sec, C_m is the concentration of the agent in the membrane in g/cm³, dC_m/dx is the concentration gradient, and D is the diffusion coefficient of the agent in the membrane in cm²/sec.

Because the concentration of the agent in the membrane cannot be readily determined, Eq. 1 can be rewritten using partition coefficients which describe the ratio of the concentration of the agent in the membrane in equilibrium with that in the surrounding medium.

$$J = \frac{DK\Delta C}{l} \tag{2}$$

where ΔC is the difference in concentration between the solutions on either side of the membrane, K is the partition coefficient, and l is the thickness of the membrane.

Reservoir devices can also be constructed with membranes that have well-defined pores connecting the two sides of the membrane. Diffusion in such microporous membranes occurs principally by diffusion through the liquid-filled pores, where the composition of the liquid will control the overall transport flux across the membrane. In this system, the flux is described by Eq. 3.

$$J = \frac{\varepsilon DK\Delta C}{\tau l} \tag{3}$$

where ε is the volume fraction porosity, which is normalized as fractional area (area of pores per unit area) of the membrane and τ is the tortuosity (average length of pore channel traversing the membrane).

One of the most important consequences of Eqs. 2 and 3 is that the flux J will remain constant provided that the membrane material does not change with time, i.e., provided that D, K, ε, and τ remain constant. However, another very important requirement for zero-order delivery is that ΔC also remain constant. The practical consequences of this requirement is that the concentration of the agent in the core cannot change with time and that the agent released from the device must be rapidly removed ("sink" conditions). Constant agent concentration in the core can be readily achieved by dispersing the agent in a medium in which it has a very low solubility, leading to a saturated drug concentration in the reservoir. However, ensuring that the concentration of an agent at the external surface of the device is essentially zero (i.e., does not build up around the device) is not always possible, particularly with agents having very low water solubility, and deviations from zero-order kinetics can be caused by such "boundary-layer effects."

Another factor that contributes to deviation from constant or zero-order drug release is migration of the agent from the core into the membrane on storage. Then, when the device is placed in a release medium, the initial drug release is rapid because the agent diffuses from the saturated membrane until the steady-state (zero-order) release rate is achieved. This initial nonlinear portion of the release is known as the "burst effect."

Monolithic Devices

In a monolithic device the therapeutic agent is uniformly dispersed or dissolved in a polymer matrix and its release is controlled by diffusion from the matrix. The matrix is assumed to be inert, i.e., non-swelling or fully swollen, and non-degrading. Mathematical treatment of diffusion depends on the solubility of the agent in the polymer, and two cases must be considered. In one case, the agent is present below its solubility limit and is dissolved in the polymer. In the other case, the agent is present well above its solubility limit and is dispersed in the polymer.

For an agent that is dissolved in the polymer, release can be calculated by two equations, known as early time approximation (Eq. 4) and late time approximation (Eq. 5) (Baker and Lonsdale, 1974).

$$\frac{dM_t}{dt} = 2M_x\left[\frac{D}{\pi l^2 t}\right]^{1/2} \tag{4}$$

$$\frac{dM_t}{dt} = \frac{8DM_x}{l^2}\exp\frac{\pi^2 Dt}{l^2} \tag{5}$$

These equations predict rate of release from a slab of thickness l, where D is the diffusion coefficient, M_x is the total amount of agent dissolved in the polymer, and M_t is the amount released at time t. According to Eq. 4, which is valid over about the first 60% of the release time, the rate decreases linearly with the square root of time. During the latter 40% of the release time the rate decays exponentially with time, as shown by Eq. 5. Plots of these two approximations are shown in Fig. 2.

When the agent is dispersed as particulates in the polymer, release kinetics can be calculated by the Higuchi equation

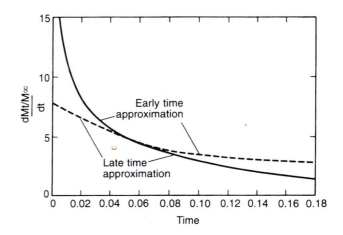

FIG. 2. Release rate of drug initially dissolved in a slab as a function of time. [Reprinted from Baker, R. W., and Lonsdale, H. K. (1974). in *Controlled Release of Biologically Active Agents*, A. C. Tankaray and R. E. Lacey (eds.). Plenum Press, New York, pp. 15–71.] 2: M = total drug released at $t = \bullet$.

(Higuchi, 1961). Equation 6 is for a planar slab model system.

$$\frac{dM_t}{dt} = \frac{A}{2}\left[\frac{2DC_sC_o}{t}\right]^{1/2} \quad (6)$$

where A is the slab area, C_s is the solubility of the agent in the matrix, and C_o is the total concentration of dissolved and dispersed agent in the matrix. Clearly, in this particular case, release rate decreases as the square root of time over the major portion of the delivery time and deviates only after the concentration of the active agent remaining in the matrix falls below the saturation value C_s.

Applications

Determining whether the reservoir or monolithic device is most appropriate for an intended application will depend upon the need for constant drug release rate, the manufacturing cost, and safety. Even though reservoir-type devices can yield zero order kinetics for long periods of time, the safety of this type of device may sometimes be of concern because rupture of the membrane can lead to sudden release of the drug within the core, especially when the core is either in a liquid state or is a fragile matrix. Thus, in such a situation, the amount of drug in the core and its toxicity must be considered.

Although long-term zero-order drug release achievable with reservoir devices is often highly desirable, manufacture of such devices is expensive and in many applications where device cost is an important factor, the less expensive matrix-type devices are used, even though release rate declines with time. The utilization of monolithic or matrix delivery devices is very common in the veterinary or agricultural fields, where low cost is of paramount importance.

Reservoir Devices: When cost is not an overriding consideration, as is the case in human therapeutics, then reservoir-type devices are an excellent choice and a number of such devices have been developed. The original reservoir DDS products were developed by Alza Corp. in the 1970s. One is the Ocusert®, which is inserted in the lower cul-de-sac and releases pilocarpine for control of glaucoma and is shown in Fig. 3.

Another reservoir device is the intrauterine contraceptive device Progestasert®, which is a T-shaped device capable of releasing progesterone from the vertical member of the device. A simpler, rod-shaped implant containing a different

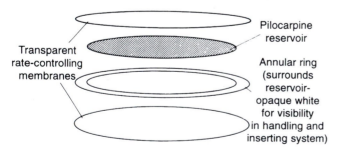

FIG. 3. Schematic diagram of Ocusert. [Reprinted from Chandrasekaran, S. K., Benson, H., and Urquhart, J. (1978). in *Sustained and Controlled Release Drug Delivery Systems*, J. R. Robinson (ed.). Marcel Dekker, New York, pp. 571–593.]

contraceptive steroid, is called the Norplant®, which was originally developed by the Population Council. The original Norplant® system consisted of six silicone tubes, 20×2 mm, and contained levonorgestrel in the core of the tubes, which were designed for implantation in the upper arm. They could maintain a therapeutically effective plasma concentration of levonorgestrel for as long as 5 years.

Drug-eluting stents: A recent example of a device coated with a drug-containing polymer is the Cypher® drug-eluting stent developed by the Cordis Corp. In this system, a 316L stainless steel stent is first gas-discharge coated with parylene C to prime the metal for the polymer–drug coating. It is then coated with a polymer matrix consisting of a blend of poly(ethylene-vinyl acetate) and poly(butyl methacrylate) plus the dispersed drug Sirolimus, also known as Rapamycin. The drug inhibits mitosis and proliferation of smooth muscle cells; such proliferation can cause restenosis and occlusion of the blood vessel. In some cases the Cypher® may be further coated with a pure polymer coating, (i.e., containing no drug) converting it from a monolithic matrix to a reservoir delivery system. The stent is implanted by balloon angioplasty into a partially or totally blocked small blood vessel (such as a coronary artery). The catheter is threaded to the site of the occlusion where a nylon balloon on the catheter is expanded, and this action expands the stent against the luminal wall of the vessel, pushing back the blocking tissue mass. Eighty percent of the drug is released over 30 days. (Information taken from Cordis Corp. web site.)

A different, competitive drug-coated stent is coming into clinical use as this chapter goes to press. This is the TAXUS Express2® Paclitaxel-eluting coronary stent system of Boston Scientific Corp. This stent is stainless steel, using the Express2® platform, and is coated with a monolithic matrix polymer-drug mixture. The coating is a dispersed mixture of Translute® polymer and Paclitaxel. Translute® is an exceptionally biostable, vascular compatible, hydrophobic triblock elastomeric copolymer. Paclitaxel is a multifunctional molecule acting on several pathways implicated in the restenotic cascade. Its primary mechanism of action in most cancer cell lines has been through the stablization of microtubule dynamics (Helmus, 2003). This stent may also be overcoated with a drug-free, rate-controlling polymer membrane, converting it to a reservoir system.

Even though reservoir-type devices can yield zero-order kinetics for long periods of time, the safety of this type of device may sometimes be of concern because rupture of the membrane can lead to sudden release of the drug within the core, especially when the core is either in a liquid state or is a fragile matrix. Thus, in such a situation, the amount of drug in the core and its toxicity must be considered.

Transdermal patches: To date, the most commercially successful use of drug diffusion-controlled systems is in transdermal applications. In these devices, a polymeric delivery system is held on the skin by an adhesive. The device contains the drug either in a reservoir with a rate-controlling membrane, or dispersed in a polymer matrix. A schematic of a membrane-controlled transdermal patch is shown in Fig. 4. The drug diffuses from the device through the skin and is taken up by the systemic circulation. Because the outer layer of the skin, called the stratum corneum, is highly impermeable to most

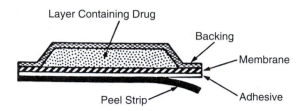

FIG. 4. Schematic of membrane-controlled transdermal device.

TABLE 2 Currently Marketed Transdermal Systems

Drug	Product name	Manufacturer
Testosterone	Testoderm®	Alza
Testosterone	Androderm®	SmithKline & Beecham
Estradiol	Alora®	P & G Pharm.
Estradiol	Climar®	Berlex Labs Pharma
Estradiol	FemPatch®	ParkeDavis
Estradiol	Menorest®	Ciba
Estradiol	Vivelle®	Ciba
Estradiol	Estraderm®	Ciba
Nicotine	Habitrol®	Ciba
Nicotine	Nicotrol®	McNeil
Nicotine	Prostep®	Elan
Nicotine	Nicoderm®	SmithKline & Beecham
Nitroglycerine	Deponit®	Schwarz
Nitroglycerine	Minitran®	3M
Nitroglycerine	Nitrodur®	Key
Nitroglycerine	Nitrodisc®	Searle
Nitroglycerine	Transderm Nitro®	Ciba
Clonidine	Catapres TTS®	Boehringer Ingelheim
Fentanyl	Duragesic®	Janssen
Scopolamine	Transderm Scop®	Ciba

drugs, either very potent drugs are used or flux of the drug through the skin may be augmented (a) by the use of penetration enhancers (such as alcohol or propylene glycol), (b) by electrical means such as iontophoresis, or (c) by transient physical disruption of the skin structure using ultrasound. Some of the current FDA-approved transdermal systems are shown in Table 2.

WATER PENETRATION-CONTROLLED DELIVERY SYSTEMS

In this type of delivery device, drug release rate is controlled by the rate of water diffusion into the device. Two general types of DDS are in use, osmotic and swelling-controlled systems.

Osmotically Controlled Delivery Devices

The operation of an osmotic device can be described by referring to the schematic representation in Fig. 5 (Theeuwes and Yum, 1976). In this device, an osmotic agent is contained within a rigid housing and is separated from the agent by a movable partition. One wall of the rigid housing is a semipermeable membrane, and when the device is placed in an aqueous environment, water is osmotically drawn across the semipermeable membrane. The resultant increase in volume within the osmotic compartment exerts pressure on the partition, which forces the agent out of the device through the delivery orifice.

The volume flux of water across the semipermeable membrane is given by:

$$\frac{dV}{dt} = \frac{A}{l} L_p \left[\sigma \Delta\pi - \Delta P \right] \tag{7}$$

where dV/dt is the volume flux, $\Delta\pi$ and ΔP are, respectively, the osmotic and hydrostatic pressure differences across the semipermeable membrane, L_p is the membrane mechanical permeability coefficient, σ is the reflection coefficient (usually ~ 1), and A and l are respectively the membrane area and thickness.

The mass rate of delivery dM/dt of the agent is then given simply by

$$\frac{dM}{dt} = \frac{dV}{dt} C \tag{8}$$

where C is the concentration of the agent in the solution that is pumped out of the orifice.

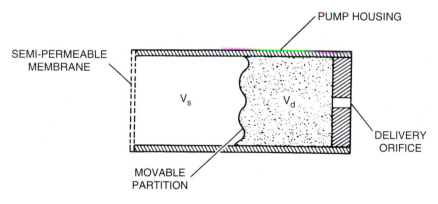

V_s VOLUME OF OSMOTIC AGENT COMPARTMENT
V_d VOLUME OF DRUG COMPARTMENT

FIG. 5. Schematic representation of an osmotic pump. [Reprinted from Theeuwes, F., and Yum, S. I. (1976). *Ann. Biomed. Eng.* **4:** 343–353.]

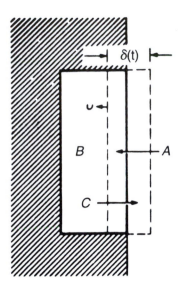

FIG. 6. Schematic representation of swelling-controlled release system. As penetrant A, water, enters the glassy polymer B, bioactive agent C is released through the gel phase of thickness $\delta(t)$. [Reprinted from Langer, R. S., and Peppas, N. A. (1983). in *Rev. Macromol. Chem. Phys.* **C23**: 61–126.]

Swelling-Controlled Devices

In this type of delivery system the agent is dispersed in a hydrophilic polymer that is stiff or glassy in the dehydrated state but when placed in an aqueous environment will swell. Because diffusion of molecules in a glassy matrix is extremely slow, no significant release occurs while the polymer is in the glassy state. However, when such a material is placed in an aqueous environment, water will penetrate the matrix and as a consequence of swelling, the glass transition temperature of the polymer is lowered below the ambient temperature, and diffusion of the drug from the polymer takes place.

The process, shown schematically in Fig. 6, is characterized by the movement of two fronts (Langer and Peppas, 1983). One front, the swelling interface, separates the glassy polymer from the swollen rubbery polymer and moves inward into the device at a velocity, v. The other front, the polymer interface, moves outward as it swells, and separates the swollen polymer from the pure dissolution medium. The polymer may go through a gel state as it swells. In systems where the glassy polymer is noncrystalline and linear, dissolution takes place. When the polymer is highly crystalline, or cross-linked, no dissolution takes place and the swollen polymer remains as a hydrogel.

Although many swellable drug formulations are available, the term "swelling-controlled" is only applicable to those formulations where the release is actually controlled by the swelling phenomenon just described.

Applications

Osmotic Devices: Two types of osmotic devices are in use. One device, known as Osmet®, is a capsule approximately 2.5 cm long and 0.6 cm in diameter and is intended as a research

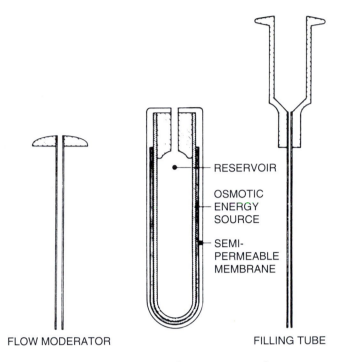

FLOW MODERATOR   FILLING TUBE

FIG. 7. Schematic representation of osmotic pump and components. [Reprinted from Theeuwes, F., and Yum, S. I. (1976). *Ann. Biomed. Eng.* **4**: 343–353.]

device that can be implanted in the tissues of experimental animals, where it delivers a therapeutic agent at a known constant rate. The agent is placed in an impermeable flexible rubber reservoir that is surrounded by an osmotic agent, which in turn is surrounded and sealed within a rigid cellulose acetate membrane (Theeuwes and Yum, 1976).

In an aqueous environment, water is osmotically drawn across the cellulose acetate membrane and the resultant pressure on the rubber reservoir forces the agent out of the orifice. The device is shown in Fig. 7. It is provided empty and can be filled with the desired therapeutic agent by the user. Because the driving force is osmotic transport of water across the cellulose acetate membrane, rate of release of the agent from the device is independent of the surrounding environment.

A second type of device, known as Oros®, is shown in Fig. 8A (Theeuwes, 1975). This device is intended for oral drug delivery applications and is manufactured by compressing an osmotically active agent plus the drug (which may be osmotically active itself) into a tablet, coating the tablet with a semipermeable membrane (usually cellulose acetate) and drilling a small hole through the coating with a laser. When the device is placed in an aqueous environment, water is drawn across the semipermeable membrane and a solution of the drug is pumped out of the orifice. A major advantage of this device is that a constant rate of release is achieved as it traverses the gastrointestinal tract.

However, this device works best only with water-soluble drugs that act as the osmotic agent. To be useful with water insoluble drugs, a different version, known as the push-pull Oros®, has been developed (Theeuwes, 1979). In this device,

A

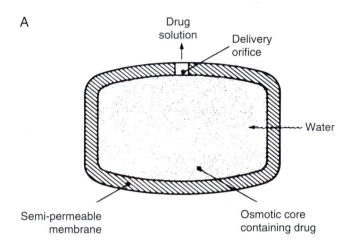

Semi-permeable membrane

Osmotic core containing drug

B

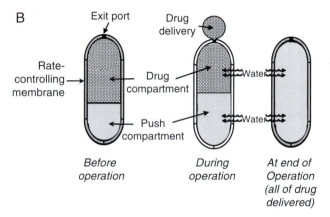

Before operation

During operation

At end of Operation (all of drug delivered)

FIG. 8. (A) Cross section of Oros. [Reprinted from Theeuwes, F. (1975). *J. Pharm. Sci.* **64**: 987–1991]. (B) Push-Pull Osmotic Pump. [Reprinted from Theeuwes, F., (1979). Novel drug delivery systems. in *Drug Absorption*, L. F. Prescott and W. S. Nimmo (eds.). ADIS Press, New York, pp. 157–176.]

two compartments are separated by a flexible partition. The top compartment contains drug and has a delivery orifice. The bottom "push" compartment contains a solid osmotic driving agent such as a water swelling or dissolving polymer. A semipermeable membrane surrounds both compartments and regulates water influx into each separately. The device is shown schematically in Fig. 8B. The "push" compartment is designed to deliver all of the drug out of the device.

An over-the-counter osmotic device is available and known as Acutrim®. It has been developed for use in appetite suppression.

Swelling-Controlled DDS: A swelling-controlled oral delivery device commercially available as Geomatrix® is shown schematically in Fig. 9. In this device, a drug is dispersed in a swellable polymer such as hydroxypropyl methyl cellulose, which is compressed into a tablet, and two sides are coated with a water-impermeable coating such a cellulose acetate propionate. This impermeable coating restricts the swelling of the matrix it to a "sideways" expansion, and as such, modifies diffusional release kinetics so that if the diffusion length increases

FIG. 9. Schematic representation of swelling-controlled oral system.

as the permeability, DK, increases, then reasonably constant release kinetics can be achieved (Colombo, 1993).

CHEMICALLY CONTROLLED DELIVERY SYSTEMS

There is some confusion in the use of the terms bioerodible, and biodegradable, and although there is no universally accepted definition, bioerodible refers to the solubilization of an initially water-insoluble material with or without changes in the chemical structure, while biodegradable refers to solubilization that occurs as a consequence of breakdown of main chain bonds. The prefix "bio" is used because the erosion or biodegradation takes place in a biological environment and carries no implication that erosion is assisted by enzymes or other active species in the biological milieu, although sometimes this is the case. Hydrolytic degradation is most common (Heller, 1984).

Degradable Reservoir Systems

This system contains the drug in the core of a solid implant covered by a biodegradable membrane. This delivery system is identical to the reservoir system already discussed, except that the membrane surrounding the drug core is bioerodible or biodegradable. Such systems combine the advantage of long-term zero-order drug release with bioerodibility or biodegradability.

Because constancy of drug release requires that the diffusion coefficient D of the agent in the membrane remain constant (Eq. 5), the bioerodible membrane must remain essentially unchanged over the duration of drug delivery. Typically, the bioerodible polymer and the drug are both hydrophobic. Further, the membrane must remain intact while there is still drug in the core to prevent its abrupt release. Thus, significant bioerosion cannot take place until drug delivery has been completed.

The only system that has used this approach is a delivery device for contraceptive steroids known as Capronor®, based on the biodegradable polymer polycaprolactone. The device is designed to release levonorgestrel at constant rates for 1 year and to completely bioerode in 3 years (Pitt *et al.*, 1980). Even though phase II clinical trials have been successfully completed, the device is no longer under development.

Drug Dispersed in a Biodegradable Matrix

This system is a matrix system and encompasses the majority of all work dealing with the development of bioerodible drug delivery systems. Four major classes of polymers dominate this type of DDS (See also Chapter 2.7).

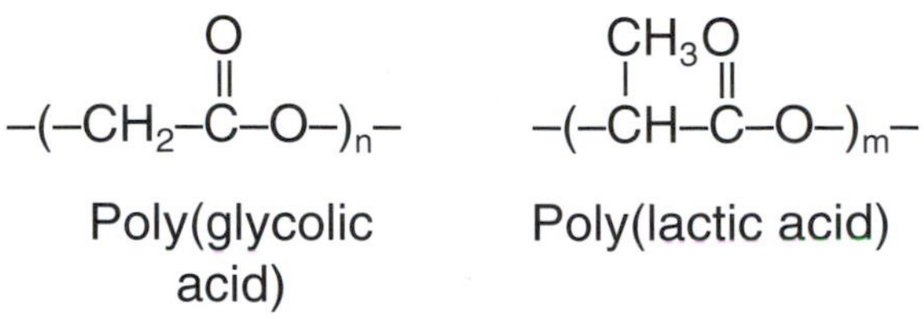

FIG. 10.   Structure of polymers based on glycolic acid and lactic acid.

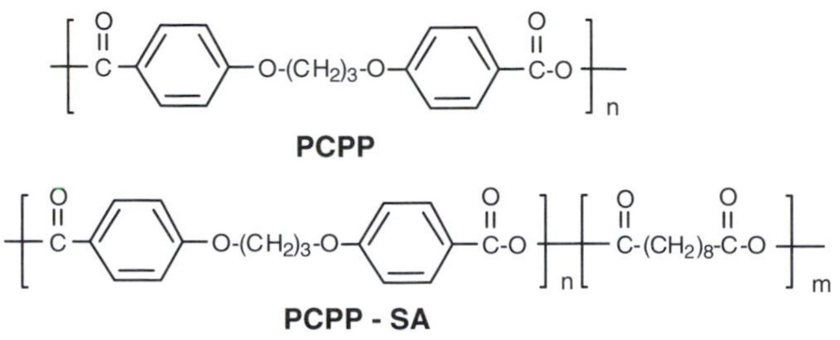

FIG. 11.   Structure of polymers based on bis(*p*-carboxyphenoxy) propane and sebacic acid.

Poly(lactic acid) and Poly(lactic-*co*-glycolic acid) Copolymers

These polymers were originally developed as biodegradable sutures and were approved because they degrade to the natural metabolites L-lactic acid and glycolic acid. To this day these polymers occupy a dominant place among biodegradable drug delivery systems (Heller, 1984). Their structure is shown in Fig. 10. Because lactic acid has a chiral center, poly(lactic acid) can exist in four stereoisomeric forms, poly(L-lactic acid), poly(D-lactic acid), *meso*-poly(D,L-lactic acid), and the racemic mixture of poly(L-lactic acid) and poly(D-lactic acid). The polymer is obtained by the cationic ring opening polymerization of the cyclic lactide, or a mixture of lactide and glycolide (Kulkarni *et al.*, 1971).

Erosion rate of poly(lactic-*co*-glycolic acid) copolymers can be controlled by varying the ratio of glycolic to lactic acid in the copolymer, and the higher the glycolic acid content, the faster the erosion rate (see Chapter 2.7). However, as the amount of glycolic acid in the copolymer is increased, crystalline domains of PGA may form, and solubility in toxicologically acceptable solvents decreases; thus the highest amount of PGA normally used is about 50 mol%. Such copolymers have an average lifetime of about 2–4 weeks.

Poly(L-lactic acid) has been shown to undergo an erosion process that begins at the outer perimeter of the device and moves gradually into the interior, followed by catastrophic disintegration (Li *et al.*, 1990a, b, c). The erosion process has been divided into four steps. In step 1, water diffuses into the polymer and ester bond cleavage starts. In step 2, differentiation between the surface and interior begins, with a drastic decrease in molecular weight in the inner part of the matrix, where the acidic environment accelerates the degradation. In step 3, low-molecular-weight oligomers begin to diffuse through the thinning outer layer, and when the molecular weight of these oligomers is low enough to allow solubilization in the medium, weight loss begins. In the final step 4, a polymer shell remains after the oligomers have solubilized and slow erosion of the shell takes place. Clearly, this complex degradation process does not yield a controlled release rate.

Such an erosion process can have a significant impact on the use of such polymers. When the polymer is used in orthopedic applications such as bone plates, screws, or intramedullar rods, the massive release of acidic products can exceed local tissue clearance capabilities, and it has been shown to produce serious inflammatory responses (Laurencin *et al.*, 1990). In addition to accelerating degradation, the low internal pH can also have adverse effects on acid-sensitive incorporated therapeutic agents, such as DNA, antigens, or some proteins.

Polyanhydrides

These materials were first prepared in 1909 (Bucher and Slade, 1909) and were subsequently investigated as potential textile fibers, but found unsuitable because of their hydrolytic instability. Although polyanhydrides based on poly[bis(*p*-carboxyphenoxy)alkane anhydrides] have significantly improved hydrolytic stability they retain sufficient hydrolytic instability to prevent commercialization, despite their good fiber-forming properties. However, it is this hydrolytic instability that makes them excellent candidates for the construction of bioerodible drug delivery systems. At this time, two families of polyanhydrides have been described.

Family I (PA I)   The first family of polyanhydrides is based on bis(*p*-carboxyphenoxy)propane and sebacic acid, and the use of this polymer as a bioerodible matrix for the controlled release of therapeutic agents was first reported in 1983 (Rosen *et al.*, 1983). Because aliphatic polyanhydrides hydrolyze very rapidly while aromatic polyanhydrides hydrolyze very slowly, good control over hydrolysis rate can be achieved by using copolymers of aliphatic and aromatic polyanhydrides. In this way, erosion rates from days to many months have been demonstrated (Leong *et al.*, 1985, 1986). The structure of a polymer based on bis(*p*-carboxyphenoxy)propane and sebacic acid is shown in Fig. 11.

Family II (PA II)   The second family of polyanhydrides is based on nonlinear dimers and trimers of erucic acid and sebacic acid (Domb and Maniar, 1993). These polyanhydrides had significantly improved mechanical properties relative to family I and by varying the ratio of erucic acid dimer to sebacic acid, materials having a wide range of mechanical properties could be prepared (Domb and Maniar, 1993). However, hydrolysis of the polymer produces sebacic acid and the dimer of erucic acid, which, being a 36-carbon material, is highly water insoluble. The structure of the polymer based on erucic acid dimer is shown in Fig. 12.

More recently, a copolymer based on ricinoleic acid and sebacic acid has been described (Teomin *et al.*, 1999). The polymer has been shown to be biocompatible, but no tissue-clearance data have been described. The structure of the copolymer based on ricinoleic acid and sebacic acid is shown in Fig. 13.

A phase II clinical trial of the erucic acid dimer/sebacic acid polyanhydride containing gentamycin for the treatment of osteomyelitis has been described (Li *et al.*, 2002).

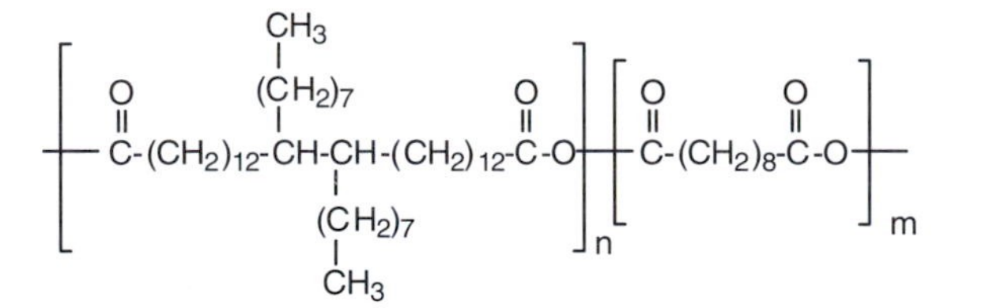

FIG. 12. Structure of copolymers based on erucic acid dimer and sebacic acid.

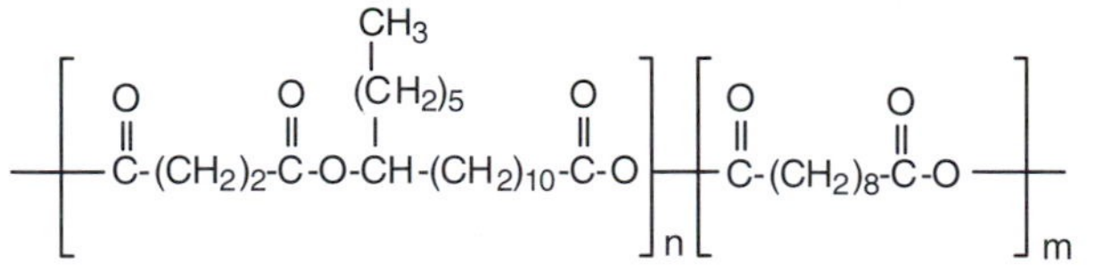

FIG. 13. Structure of copolymers based on ricinoleic acid and sebacic acid.

Poly(ortho esters)

The first poly(ortho ester) was described in the 1970s in a series of patents (Choi and Heller, 1978a, b, c, 1979) and represent the first example of a new synthetic polymer specifically designed for controlled drug delivery. To date, four families of such polymers have been described, and comprehensive reviews have been published (Heller, 1993; Heller *et al.*, 2002).

Family I-(POE I) This family of polymers was developed at the Alza Corporation and was known then as Alzamer®. Use of the polymer has been suggested for contraception (Benagiano *et al.*, 1979) and for narcotic addiction treatment (Capozza *et al.*, 1978). These polymers are no longer under development. The structure of POE I is shown in Fig. 14.

Family II-(POE II) This family of polymers was developed at the Stanford Research Institute (Heller, 1980) and has been extensively investigated in a number of applications. POE II represents a significant advance over POE I in both ease of synthesis and ability to manipulate both mechanical and thermal properties, as well as being able to control erosion times from days to many months. The polymer is highly hydrophobic (Nguyen *et al.*, 1985), so that despite the hydrolytic instability of ortho ester linkages, it is very stable because only limited amounts of water are able to penetrate the polymer. To control erosion rates, acidic excipients can be used to accelerate hydrolysis and basic excipients can be used to retard hydrolysis. Using the latter approach, materials that displayed surface

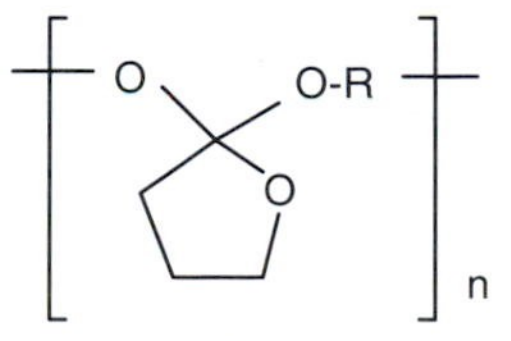

FIG. 14. Structure of poly(ortho ester) I.

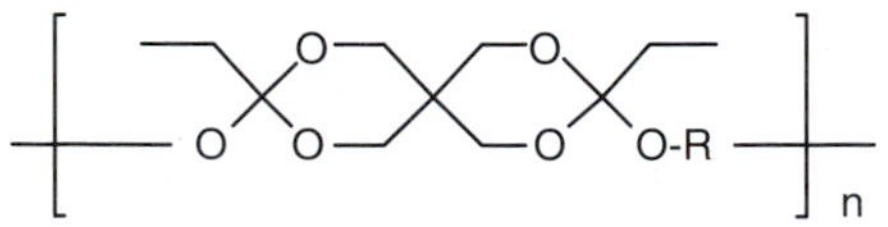

FIG. 15. Structure of poly(ortho ester) II.

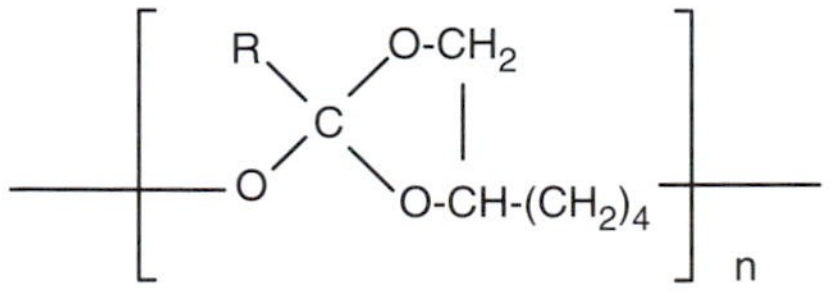

FIG. 16. Structure of poly(ortho ester) III.

erosion for over 1 year have been prepared (Heller, 1985). The structure of POE II is shown in Fig. 15.

Family III-(POE III) This family was also developed at the Stanford Research Institute (Heller, 1990a). Because this polymer has a highly flexible structure, it is a high viscosity, semisolid material at room temperature, even at fairly high molecular weights. The structure of POE III is shown in Fig. 16.

Family IV-(POE IV) This is the latest and most successful polymer and was developed at Advanced Polymer Systems, now A.P. Pharma (Ng *et al.*, 1997). The structure of this polymer is similar to that of POE II, but incorporates a short segment of lactic or glycolic acid in the polymer backbone. This segment acts as a latent acidic catalyst and eliminates the need to add acidic excipients to the polymer to control erosion. This represents a major improvement because the addition of acidic excipients makes formulation more difficult and leads to complications due to the diffusion of the acidic excipient from the polymer matrix. A detailed study of the hydrolysis mechanism has been published (Schwach-Abdellaoui *et al.*, 1999). The structure of POE IV is shown in Fig. 17.

Poly(phosphoesters) The general chemical structure of poly(phosphoesters) is shown in Fig. 18.

In the past, interest in phosphorus-containing polymers centered on their flame-retardant properties (Coover *et al.*, 1960). In view of their hydrolytic instability and relatively high cost, interest in these materials diminished. However, as with the polyanhydride system already discussed, it was their hydrolytic instability that renewed interest in such polymers for drug delivery applications (Richard *et al.*, 1991). Because the ultimate hydrolysis products are phosphate ions, alcohol, and a diol, the polymer has the potential to be nontoxic.

Poly(phosphoesters) are prepared by a dehydrochlorination reaction between a phosphorodichlorinate and a diol, as shown in Fig. 19.

These polymers exhibit considerable versatility in that mechanical and thermal properties can be controlled by varying the nature of the pendant R′ group and the group R in the polymer backbone.

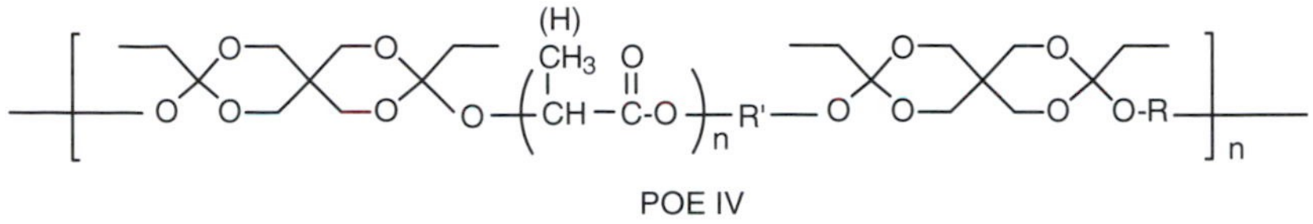

FIG. 17. Structure of poly(ortho ester) IV.

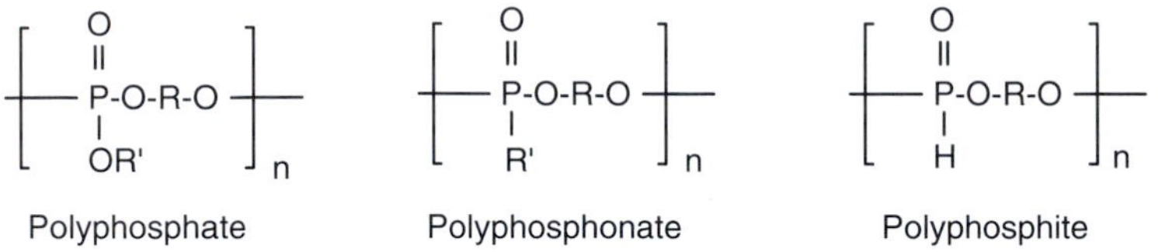

FIG. 18. Structure of various poly(phosphoesters).

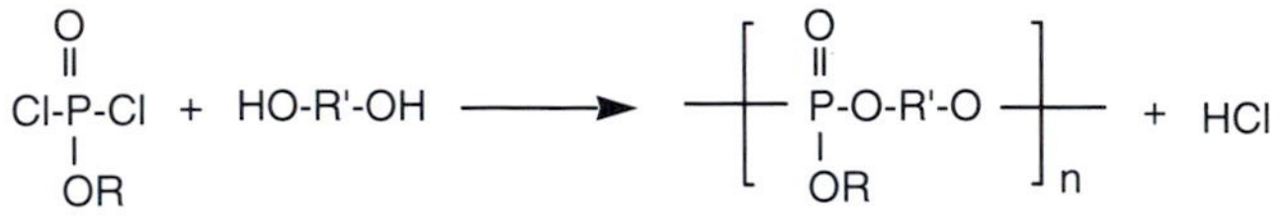

FIG. 19. Preparation of poly(phosphoesters).

Applications

Poly(lactic acid) and Poly(lactic-*co*-glycolic acid) Copolymers

Although the literature is full of reports using PLGA particulates for drug delivery, only a small number of drug delivery devices based on these copolymers have reached commercialization. They are shown in Table 3.

Of these, the ones that have realized the highest commercial success are devices that release a luteinizing hormone releasing hormone (LHRH) analog for the treatment of prostate cancer. This therapy is based on early work by Huggins *et al.*

(1941), who recognized the androgen dependency of prostatic adenocarcinomas, and one treatment modality is to suppress androgen levels by the systemic delivery of LHRH analogs. LHRH analogs are chemical modifications of the natural hormone that are many hundreds of times more potent and down-regulate its receptors with suppression of the production of testosterone, progesterone, and estrogen. Thus, repeated administration at 1-, or 3-month intervals is a common palliative treatment for prostate cancer in patients who are a poor surgical risk.

Another very important application is the incorporation of proteins into these polymers with full retention of biological activity. Although proteins are stable in organic solvents such as methylene chloride, conventional microencapsulation methods cannot be used because proteins will denature when exposed to an organic solvent/water interface. To circumvent this problem, Alkermes Corp. has marketed a proprietary process called Prolease® where micronized protein is dispersed

TABLE 3 Examples of Commercialized Devices Based on Poly(Lactic-*co*-Glycolic Acid) Copolymers

Trade name	Device form	Drug
Suprefact® Depot	Strand	Buserilin
Zoladex®	Strand	Goserilin
Zoladex LA®	Strand	Goserilin
Decapeptyl Depot®	Microspheres	Triptorelin
Enantone LP®	Microspheres	Leuprorelin
Somatulin LP®	Microspheres	Lanreotide
Parlodel LAR®	Microspheres	Bromocriptine
Sandostatin-LAR®	Microspheres	Ocreotide
Eligard 7.5®	Solution of Polymer and Drug in NMP	Leuprolide Acetate
Eligard 22.5®	Solution of Polymer and Drug in NMP	Leuprolide Acetate
Eligard 30.0®	Solution of Polymer and Drug in NMP	Leuprolide Acetate
Nutropin®	Microspheres	Recombinant Human Growth Hormone
Lupron®	Microspheres	Leuprolide Acetate

NMP is *N*-Methyl Pyrrolidone
Zoladex® – 1 month delivery
Zoladex LA® – 3 months delivery
Eligard 7.5® – 1 month delivery
Eligard 22.5® – 3 months delivery
Eligard 30.0® – 4 months delivery

in a lactide/glycolide solution in methylene chloride and then atomized into liquid nitrogen that contains a frozen layer of ethanol at the bottom of the vessel. The small particles of polymer solution with dispersed protein are instantly frozen and settle on top of the frozen ethanol layer. The liquid nitrogen is then evaporated, the ethanol melts, and the particles settle into the cold ethanol, where they harden as the methylene chloride is extracted, and they are then harvested (Khan *et al.*, 1992).

This cryogenic method was used to encapsulate recombinant human growth hormone (rhGH) using an 8-kDa, 50/50 copolymer of lactic and glycolic acid. A single injection of microspheres using a 23-gauge needle, suspended in 3% w/v low-viscosity carboxymethylcellulose, 1% polysorbate 20, and 0.9% NaCl in monkeys (*Macaca mulatta*) resulted in elevated blood plasma levels for more than 1 month. The protein was stabilized during the encapsulation process and during *in vivo* release by complexing with zinc (Johnson *et al.*, 1996). This product is now commercially available as Nutropin®.

Polyanhydrides

This family of polymers has been extensively investigated for the treatment of brain cancer, and particularly glioblastoma multiforme, in pivotal studies carried out by Brem and Langer (1996). A striking feature of malignant brain tumors is that they do not metastasize, but rather, recur within 2 cm of the original tumor. Thus, if a drug delivery device containing a suitable antineoplastic agent such as BCNU [1,3-bis(2-chloroethyl)-1-nitrosourea] is placed at the site of tumor resection, survival times should be improved because localized drug delivery would destroy malignant cells not removed by the tumor resection procedure.

A multicenter clinical trial has been carried out, and in a Cox regression model, taking into account the effect of age, Karnovsky score, and tumor type, the BCNU polymer treatment was significantly more effective than placebo polymer ($p = 0.01$). A device known as Glyadel® has been approved by the FDA as a secondary treatment for patients that have failed surgical treatment of glioblastoma multiforme. Glyadel® is a wafer approximately 1.45 cm in diameter and 1 mm thick. Each wafer contains 192.3 mg of a 80:20 poly[bis(*p*-carboxyphenoxy)propane and sebacic acid and 7.7 mg BCNU. A maximum of eight wafers per procedure are used. The wafers are quite brittle and unstable and must be stored at $-20°C$.

Poly(ortho esters)

While as yet there are no commercial applications for poly(ortho esters), a number of delivery devices based on POE IV are in advanced stages of development.

POE III One of the most significant aspects of POE III is its biocompatibility, which makes this system an excellent candidate for ocular applications. To achieve such biocompatibility it is necessary to use a very pure polymer fabricated aseptically to avoid the possible formation of acidic degradation products. In some applications it is also necessary to incorporate into

the polymer basic compounds such as $Mg(OH)_2$ that will neutralize acidic hydrolysis products. When such precautions are observed, devices containing 5-fluorouracil, (used as an adjunct drug for glaucoma-filtering surgery), or polymers designed for intravitreal injections can be developed. Animal studies are currently in advanced stages with excellent results (Einmahl *et al.*, 2003).

POE IV Because physical properties of this polymer system can be varied within very wide limits, materials having consistencies and physical properties varying from semisolids to polymers having relatively low softening temperatures can be prepared. The versatility of this polymer system can be illustrated with applications in the treatment of periodontal disease, postoperative pain control, ocular applications, and protein delivery.

Treatment of periodontal disease uses low-molecular-weight polymers based on decanediol and decanediol dilactide with incorporated tetracycline. Such materials are semisolids at room temperature and can be injected into the periodontal pocket with a blunt needle. Effectiveness in human clinical trials has been demonstrated (Heller *et al.*, 2002). In the treatment of postoperative pain, a semisolid formulation based on triethylene glycol and triethylene glycol glycolide is injected into the surgical incision. This treatment results in a high local concentration of an anaesthetic agent such as mepivacaine so that pain control can be achieved with low systemic concentration, thus minimizing undesirable side effects (Barr *et al.*, 2002). Phase II clinical trials of this delivery system are about to be initiated.

Recent work (Einmahl *et al.*, 2003) has shown that this family of poly(ortho esters) shows an ocular biocompatibility that is even better than that of POE III.

Materials suitable for protein delivery are constructed from diol and diol monoglycolide pairs chosen to permit the polymer to be extruded at temperatures low enough so that protein activity is not compromised. In this process, an intimate mixture of micronized protein and finely ground polymer is prepared and extruded into 1-mm-diameter strands, which are cut into suitable length and implanted using a trochar. Excellent release kinetics have been achieved (Rothen-Weinhold *et al.*, 2001).

Poly(phosphoesters)

A poly(phospho ester) with the trade-name of Polylactophate® has been investigated for the release of insulin (Kader *et al.*, 2000) and Paclitaxel. The Paclitaxel device has been trademarked as Paclimer® (Dang *et al.*, 1999). Polylactophate is a copolymer of ethyl phosphate and lactide and has the structure shown in Fig. 20.

Paclimer® was investigated for efficacy in a nude-mice human ovarian cancer cell line, OVCAR3. A significant survival time was achieved.

Pendant-Chain Systems

In one type of pendant-chain system a drug is covalently attached to a biodegradable polymer backbone. In early work

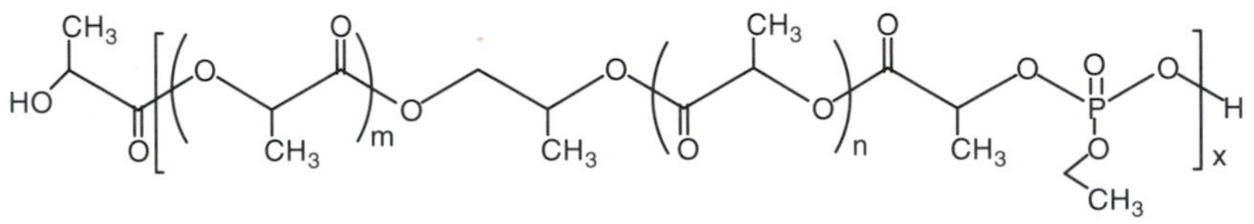

FIG. 20.   Structure of Polylactophate.

in this field, hydrophobic drugs were covalently attached via labile linkages to the water soluble poly(N^5-hydroxypropyl-L-glutamate) backbone. This drug polymer construct is water insoluble and as the drug is gradually released by cleavage of the labile linkages, the polymer again becomes water soluble. (Peterson *et al.*, 1980). Such systems are toxicologically complicated because the FDA requires unambiguous proof that the chemically attached drug is released in its native form and not as a chemical derivative. Pendant-chain systems are not currently under development with the exception of polymer conjugates used in tumor targeting, covered later in this chapter.

RESPONSIVE "SMART" SYSTEMS

Environmentally-Responsive Systems

Temperature-Responsive, Phase-Separating Polymer Systems

A well-known example of a polymer that undergoes a sharp, thermally induced phase separation is poly(N-isopropyl acrylamide) (Heskins and Guillet, 1968). Its transition at 32°C can be shifted to higher or lower temperatures by copolymerization with more hydrophilic or hydrophobic comonomers, respectively. Thus, this material could be of interest in drug delivery applications; however, it may not be suitable due to potential cell toxicity of the homopolymer and its copolymers, although this issue has not been thoroughly investigated.

Of more interest are triblock copolymers of poly(ethylene glycol) and poly(propylene glycol), PEO–PPO–PEO, known as Pluronics® or Poloxamer® polyols, some compositions of which have already been approved for use in the body. (Wang and Johnston, 1991). While many of these block copolymers have transition temperatures well above body temperature, they do exhibit transitions at body temperature in solutions at concentrations above 16 wt%. However, such materials are non-degradable and thus may not be suitable as injectable DDS. They have been used in ophthalmic applications, but discomfort due to vision blurring and crusting has been reported (Joshi *et al.*, 1993).

One of the more recent trends in thermogelling materials is the development of biodegradable systems suitable as injectable gelling systems. Such materials can be constructed from ABA block copolymers based on poly(ethylene glycol) and poly(lactic-*co*-glycolic acid) copolymers. The triblock may be designed as PLGA-PEO-PLGA or as PEO-PLGA-PEO, where the former will have a more pronounced thermal gelation capability. PLA has also been used as the hydrophobic, degradable block segment. When the block lengths and relative amounts are correctly chosen, a material is obtained that is soluble in water at room temperature and forms a firm gel at body temperature (Jeong *et al.*, 1999a). Such materials can be injected as a solution having a viscosity not much different from a saline solution. In addition, since only water is used with this material, there are no organic solvent/water interface problems and proteins can be incorporated without loss of activity. These materials are available as dry powders that are reconstituted prior to use. Another thermogelling material that is soluble in water at room temperature and forms a firm gel at body temperature has also been recently described (Jeong *et al.*, 2000). While the exact mechanism of thermogelling is not known with absolute certainty, it has been postulated that the hydrophobic blocks (PLGA or PLA) of the amphiphilic block copolymers aggregate together to form micelles that are in equilibrium with monomeric polymer chains (Jeong *et al.*, 1999b). Then, as the temperature is increased, the equilibrium shifts to micelle formation, and above the critical gelation temperature, the micelles pack together to occupy the entire volume, resulting in gel formation. Evidence for such a mechanism is provided by ^{13}C NMR studies, dye solubilization studies, and light-scattering studies, which all show an abrupt change in micellar diameter and aggregation at the critical gelation temperature (Jeong *et al.*, 1999b). It has also been postulated that the water content of the gel will determine its degradation rate and release kinetics of the incorporated therapeutic agent (Jeong *et al.*, 1999b).

Applications

A triblock copolymer based on poly(ethylene glycol) and poly(lactic-*co*-glycolic acid) copolymer is under development under the trade-name ReGel®. A formulation of paclitaxel in ReGel® is known as OncoGel® and a phase I clinical trial with a dose-escalating scheme to determine safety and efficacy in patients suffering from breast and neck tumors has been completed (Rohr *et al.*, 2002). All patients tolerated OncoGel® very well with no systemic side effects noted. At the highest dose, which was not specified, OncoGel® showed efficacy, and in one case, an 80% reduction in tumor volume was noted. The dose range for a phase II clinical trial has been identified.

Solvent-Responsive, Phase-Separating Polymer Systems

Another means of developing in situ-forming drug delivery systems is polymer precipitation in a poor solvent. Such an injectable delivery system is prepared from a water-insoluble biodegradable polymer dissolved in a water-miscible, biocompatible solvent. When this system is mixed with a drug and injected into the physiologic environment, the solvent diffuses out into the surrounding tissues while water diffuses in, and the polymer precipitates, forming a solid polymeric implant containing entrapped drug. This method has been used in human (Dunn *et al.*, 1990) and in veterinary applications (Dunn *et al.*, 1995).

The polymer most commonly used is a poly(lactic-co-glycolic acid) copolymer dissolved in *N*-methylpyrrolidone (NMP). This system is known as Atrigel®. (See below). There are a number of other systems, such as: (1) poly(lactic acid) dissolved in a mixture of benzyl benzoate and benzyl alcohol, known as PLAD (Duenas *et al.*, 2001); (2) poly(lactic-co-glycolic acid) copolymer dissolved in glycofurol (Eliaz and Kost 2001); (3) poly(lactic-co-glycolic acid) copolymer dissolved in benzyl benzoate known as Alzamer® and (4) sucrose acetate isobutyrate dissolved in NMP or Miglyol, known as SABER(Tipton and Holl, 1998).

Applications

Three formulations of PLGA dissolved in *N*-methyl-pyrrolidone and containing leuprolide acetate, have been used for the treatment of prostate cancer. These formulations are listed in Table 3.

Temperature- and pH-Responsive Hydrogel Systems

In this approach, responsive polymers and lightly crosslinked gels are incorporated into formulations that alter drug release in response to changes in temperature or pH. Usually these systems are hydrogels that contain absorbed drugs, and when a temperature or pH stimulus is applied, the gel shrinks and the drug is released. Such systems are not very effective, since the drug may gradually release by diffusion from the hydrogel before the stimulus causes it to release in a burst.

One example of polymers that can reversibly respond to temperature changes are hydrogels based on *N*-isopropyl-acrylamide (Hoffman *et al.*, 1986; Dong and Hoffman, 1987; Hoffman, 1987; Okano *et al.*, 1990). When this monomer is copolymerized with methylene bis-acrylamide the resulting crosslinked materials have been shown to shrink abruptly when heated just above 32°C. The collapse of the hydrogel occurs as a result of a phase transition of poly (*N*-isopropylacrylamide) which is soluble in water below its lower critical solution temperature (LCST) of 32°C but becomes insoluble when the temperature is increased above this temperature. Such materials can deliver biological molecules when triggered by temperature changes, but this is not very practical due to the isothermal condition in the body. However, one unique way to utilize such a thermally responsive hydrogel in the body was developed by Dong and Hoffman (1991). They prepared a combined pH- and temperature-sensitive hydrogel, based on a random copolymer of NIPAAm and AAc, and it was shown to release a model drug linearly over a four hour period as the pH went from those found in the gastric to the enteric regions of the GI tract. At 37°C body temperature the NIPAAm component was trying to maintain the gel in the collapsed state, while as the pH went from acidic (e.g., as it would in the gastric region) to neutral conditions (e.g., as it would in the enteric region),

the AAc component was becoming ionized, forcing the gel to swell and slowly release the drug.

Temperature-responsive smart polymer hydrogels can also be useful for topical delivery to open wounds, skin and mucosal surfaces such as the eye or nose. The temperatures of such surfaces may be a few degrees below 37°C but they are still well above ambient temperature, and that difference could be utilized to deliver a drug from a thermally sensitive polymer formulated with the drug.

Mucoadhesive Polymers

There are a number of mucoadhesive polymer carriers that have been investigated, but the most popular one is a lightly crosslinked poly(acrylic acid); two products are known commercially as Carbomer® and Polycarbophil®. These hydrogels may be loaded with a water-soluble drug at a slightly acidic pH where the polymeric –COOH groups are protonated. If this polymer/drug system is applied to a mucosal surface such as the eye, the neutralization of the –COOH groups to form the sodium salt can draw water into the formulation and cause the system to thicken. This viscosity increase, combined with the mucoadhesive property of the poly(acrylic acid), can lead to an enhanced residence time of the formulation on the mucosal surface, which enhances the efficacy of drug delivery. Poly(acrylic acid) is also an effective mucoadhesive polymer in the GI tract, especially at the low pH of the stomach. Mechanisms of mucoadhesion are largely attributed to a combination of H-bonding and molecular entanglements of the polymer with the glycoprotein, mucin. (Park and Robinson, 1985; Leussen *et al.*, 1997; Singla *et al.*, 2000, Robinson and Peppas, 2002).

Some researchers have used chitosan as a pH-sensitive mucoadhesive carrier (Henriksen, 1996; Leussen *et al.*, 1997) especially for nasal drug delivery. (Aspden *et al.*, 1997) In this case, the chitosan is not soluble at pH 7.4, but could swell and begin to dissolve at the slightly lower pH that exists in the nose. This could cause the chitosan to gel, adhere to the nasal mucosa, and maintain the drug in place within the nasal tissues. (Ilum, 2001)

Intracellular Drug Delivery

Drugs may interact with the body in three sites: (a) within the circulation or interstitial space, (b) at membrane receptors on cell surfaces, or (c) at various sites within the cell. When a drug or drug/carrier is taken up by a cell, the usual process is endocytosis, and the drug is localized within the endosome. The endosome has a proton pump in its membrane, causing the pH to drop from 7.4 to as low as 5.0; within one to a few hours the endosome is "trafficked" to the lysosome, which is also at a lowered pH. The lysosome contains enzymes that are active at the low pH of that vesicle. If the drug is enzyme-susceptible, its efficacy will be determined by how efficiently it can escape the endosome before it ends up in the lysosome, where it may be enzymatically degraded. Enzymatically susceptible drugs, such as DNA, RNA, antisense oligonucleotides, peptides or proteins must escape the endosome to be effective intracellularly.

pH- and Ion-Responsive DDS: Non-viral Intracellular Nucleic Acid Delivery

Gene therapy and antisense oligonucleotide (asODN) and silencing RNA delivery systems have become very actively researched in the past several years (Felgner *et al.*, 1996; Mahato *et al.*, 1997; Rolland, 1999; Hillery *et al.*, 2001). In particular, since DNA, RNA and asODNs are all highly negatively charged molecules, cationic polymers and cationic liposomes have been used to ionically complex and condense in size these nucleic acid drugs in the body. Plasmid DNA complexes with cationic polymers are called polyplexes, and complexes with cationic liposomes are called lipoplexes. Together these two methods are called non-viral delivery systems. They are much less efficient than viral carriers, but the latter may cause undesirable toxic and immunogenic responses in the body (although there is also some toxicity associated with polycations in the body). Some of the more common cationic polymers include poly(ethylene imine) (PEI), poly(L-lysine) (PLL), chitosan, poly(vinyl imidazole), poly(amidoamines) and acrylic copolymers of N,N dimethyl aminoethyl methacrylate (DMAEMA). The resultant polyplexes are typically slightly positively charged nanoparticles of ca. 50-200 nm in diameter. Once they are taken up into a cell by endocytosis, some of the amino groups on the polymers that have the appropriate pK have been postulated to buffer the acidic pH within the endosome by binding protons and thereby causing sodium ions to enter to maintain electrical neutrality. This leads to osmotic swelling and bursting of the endosome, releasing the polyplex (or lipoplex) to the cytosol. This is called the "proton sponge" effect (Behr, 1997), and it may be one step in the mechanism by which polyplexes (or lipoplexes) can transfect cells. It is not clear how and when the polycation actually releases the DNA. It is probably not an efficient process, since the polycation and the DNA must each be "reunited" with their counterions in that process. This may be one reason the non-viral systems are so inefficient. Some polyplexes and lipoplexes also tend to be unstable in serum, possibly due to disruption by proteins and lipoproteins in the circulation, and also potentially due to degradation of the DNA by nucleases. Another challenge to the stability of the polyplexes or lipoplexes may be heparan sulfate groups on cell surfaces that can compete with the DNA for the cationic groups on the polymer or liposome.

Viruses have evolved an efficient system for delivering DNA or RNA to the cytosol which utilizes the lowered pH within the endosome. Certain peptide sequences on the surface of the virus become hydrophobic when their acidic amino acids are protonated, causing those sequences to fuse with or cause pore formation in the endosomal membrane, allowing the virus to deliver its genomic contents to the cytosol. Based on this natural action of the virus, the Stayton/Hoffman research group has mimicked this viral action by designing and synthesizing pH-sensitive polymers, such as poly(propylacrylic acid), PPAA, that would similarly become hydrophobic at the endosomal pHs and disrupt the endosomal membrane. (Murthy *et al.*, 1999; Stayton *et al.*, 2000; Cheung *et al.*, 2001; Murthy *et al.*, 2001; Kyriakides *et al.*, 2002). A pendant glutathione-responsive disulfide group has been added to this acid-responsive polymer, such that when the polymer-drug complex or conjugate escapes the endosome, it releases the drug after reduction by glutathione in the cytosol (Bulmus *et al.*, 2003). Other approaches by these same researchers have been to design and synthesize membrane-disruptive polymers that are PEGylated via acid-degradable acetal bonds; these bonds degrade within the endosome, exposing the backbone which is membrane-disruptive, leading to the release of the drug to the cytosol (Murthy *et al.*, 2003).

pH-Responsive Enteric Coatings

An enteric coating is a polymer that is insoluble at a low pH and soluble at physiologic pH. The most common reasons for the use of enteric coatings is (a) to protect acid-sensitive drugs from gastric juices, (b) to prevent gastric distress due to irritation caused by certain drugs, (c) to prevent the absorption of certain drugs in the stomach, and (d) to deliver drugs to the small intestine if that is their prime absorption site. The USP test for enteric-coated tablets requires that the tablet withstand agitation in artificial gastric juices at 37°C for one hour and be disintegrated within two hours when agitated in artificial intestinal juices at 37°C.

Enteric coatings have been known for over 100 years and a great many materials have been investigated. A comprehensive tabulation of such materials has been made (Martin, 1965). However, modern enteric coatings are based on polycarboxylic acids that are hydrophobic and water-insoluble at a low pH because the carboxyl groups are unionized and become water soluble at a higher pH where the carboxyl groups ionize.

Examples of enteric coating compositions include partially esterified copolymers of methyl vinyl ether and maleic anhydride (Lappas and McKeehan, 1965) or copolymers of methacrylic acid and methacrylic acid esters. By varying the degree of esterification, or the ratio of methacrylic acid to methacrylic acid esters, the dissolution pH can be modified to any desired value. The copolymers of methacrylic acid and methacrylic acid esters are commercially available as Eudragit L and Eudragit S, and are widely used as enteric coatings. They dissolve at pH 6.0 and 7.0, respectively (Lehman, 1968).

Partially esterified copolymers of methyl vinyl ether and maleic anhydride have also been used as a matrix for the controlled-release of incorporated drugs. An erosion-controlled, zero-order drug release has been achieved (Heller *et al.*, 1978).

Another enteric coating composition of cellulose acetate phthalate has also been widely used. It is insoluble below a pH of about 6.0 and will dissolve at a pH higher than 6.0 (Malm *et al.*, 1951). However, cellulose acetate phthalate is hygroscopic so that proper storage conditions are important.

Mechanical- and Physically-Responsive Systems

Among the most advanced externally regulated devices are mechanical pumps which dispense drugs from a reservoir to the body by means of a catheter. Such pumps can be worn externally with a percutaneous delivery tube or needle penetrating the skin, or can be implanted in a suitable body site. A major application is control of diabetes by delivering insulin in response to blood glucose levels. A number of such pumps such as the CPI Lilly pump have sophisticated control mechanisms and microprocessors that allow a programmed insulin delivery (Brunetti *et al.*, 1991). They are also used for pain control of terminal cancer patients.

Another means of externally regulating drug delivery is by means of magnetism (Hsieh *et al.*, 1989). In this procedure, small magnetic spheres are embedded within a polymer matrix that contains a dispersed therapeutic agent. When an oscillating magnetic field is applied to the polymer, the normal diffusional release is significantly and reversibly increased.

Rate of drug release from some DDS can also be increased by the use of ultrasound (Kost *et al.*, 1989). This method has been applied to biodegradable polymers and it has been shown that rate of bioerosion and drug release can be significantly enhanced. Unlike magnetically enhanced devices, ultrasound can be used with biodegradable implants and represents a means by which enhanced insulin delivery could be realized. Ultrasound has also been found to enhance drug transport across the stratum corneum in transdermal drug delivery devices.

Self-Regulated, Feedback Systems

During the last three decades, controlled-release administration of therapeutic agents from various types of delivery systems has become an important area of research and significant advances in theories and methodologies have been made. However, even though devices that are capable of releasing therapeutic agents by well-defined kinetics are a significant improvement over conventional dosage forms, these devices do not yet represent the ultimate therapy because the agent is released without regard of the need of the recipient. Therefore, another very significant improvement could be realized if systems could be devised that are capable of adjusting drug output in response to a physiological need or signal. (Heller, 1988.) Such "self-regulated" devices should be capable of altering drug release with a "built-in" response to an environmental signal, such as an increase in insulin delivery rate in response to an increase in glucose concentration. This is different from systems described earlier, where an external stimulus, such as an increase in pumping rate delivers insulin from an implanated pump. The latter DDS are referred to as "externally regulated." Self-regulated devices based on enzyme reactions can be classified as "substrate-responsive" devices.

Self-Regulated Systems

Two different types of devices based on enzyme/substrate-responsive or substrate-responsive reactions have been described. In the first type, an enzyme-substrate reaction modulates drug release from the device while in the second type the substrate itself triggers drug release by an affinity recognition process.

Modulated Devices The first published example of a modulated device was based on a coated monolithic matrix system, where the coating was a hydrogel containing an entrapped enzyme, and the matrix was composed of a drug physically dispersed in a bioerodible polymer. (Heller and Trescony, 1979). When the enzyme reacted with its substrate, (e.g., an endogenous metabolite) the product caused a local pH change in the device that accelerated the erosion of the bioerodible polymer, increasing the rate of release of the drug. This particular system was based on the reaction between urea and the enzyme urease. The polymer used was a partially esterified copolymer of methyl vinyl ether and maleic anhydride containing a marker drug, hydrocortisone. This copolymer in the form of a disc was coated with a hydrogel containing urease. When this system was placed in solutions having differing concentrations of urea, the hydrocortisone release rate was proportional to the urea concentration, and the effect was "reversible" in the sense that when the urea concentration decreased, the drug release rate was similarly reduced. (Heller and Trescony, 1979).

Clearly, a system that releases hydrocortisone in response to external urea concentration is not therapeutically relevant, but it did demonstrate the feasibility of this concept. A more therapeutically relevant system utilizes pH changes resulting from the glucose oxidase conversion of glucose in the presence of oxygen to gluconic acid, and pH-sensitive polymers that can swell or collapse or degrade in response to that change. Such polymers can be in the form of:

(a) hydrogel membranes that reversibly alter their porosity by a protonation of tertiary amine functions in the polymer, causing swelling and enhanced permeation of insulin (Albin *et al.*, 1985, Ishihara and Matsui, 1986), or

(b) porous membranes, where polyacrylic acid (PAAc) has been grafted onto the pore walls, and where protonation of the carboxyl groups causes the PAAc coating to collapse against the pore walls, opening the pores to insulin permeation (Ito *et al.*, 1989, Iwata and Matsuda, 1988) or

(c) bioerodible polymers that alter their erosion rate in response to pH changes (Heller *et al.*, 1990b).

In a completely different, affinity-based system, insulin was glycosylated, and the glucose groups were complexed with the plant-derived carbohydrate-binding protein Concanavalin A (ConA) which has four binding sites for glucose and other sugars. In the presence of free glucose, the glycosylated insulin is displaced from the complex with ConA in amounts proportional to the external glucose concentration. When this system is entrapped within a glucose- and insulin-permeable, but ConA impermeable capsule, it can deliver insulin in direct response to blood glucose levels (Makino *et al.*, 1990). Nakamae, Hoffman and coworkers (Nakamae *et al.*, 1994) extended this concept to an affinity hydrogel based on poly(glucosylethyl methacrylate) which was crosslinked by affinity association with ConA. In the presence of free glucose, the ConA interactions with the pendant polymeric glucose groups were displaced, and the system swelled. Drug molecules such as insulin could then diffuse faster through the swollen hydrogel barrier system. ConA was also covalently linked into the polymer network, eliminating its potential diffusion out of the swollen gel. (Nakamae *et al.*, 1994; Lee and Park, 1996) (See also Chapter 2.6).

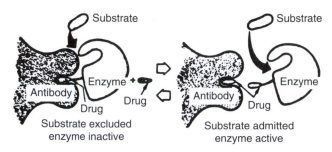

FIG. 21. Reversible enzyme inactivation by hapten-antibody interactions. In this particular case, drug is morphine and enzyme is lipase. [Reprinted from Schneider, R. S., Lindquist, P., Wong, E. T., Rubinstein, K. E., and Ullman, E. F. (1973). *Clin. Chem.* **19**: 821–825.]

Siegel and coworkers have used the glucose-stimulated swelling and collapse of pH-responsive hydrogels containing entrapped glucose oxidase to drive a hydrogel piston for release of insulin in a glucose-driven, feed-back manner. (Dhanarajan *et al.*, 2002).

Triggered Devices One important application is the development of a device capable of releasing the narcotic antagonist naltrexone in response to external morphine. Such a device would be useful in the treatment of narcotic addiction by blocking the opiate-induced euphoria by the morphine-triggered release of the antagonist. Development of this device utilizes the reversible inactivation of an enzyme by hapten-antibody interaction (Schneider *et al.*, 1973) as shown in Fig. 21.

In a device based on this principle, naltrexone is contained in a core surrounded by an enzymatically degradable coating that prevents naltrexone release. In the absence of external morphine the device is stable, but upon exposure to morphine, the reversibly inactivated enzyme is activated and degrades the protective coating, which results in naltrexone release. The actual device utilizes the enzyme lipase and a triglyceride protective coating (Roskos *et al.*, 1993).

PARTICULATE SYSTEMS

Microcapsules and Microspheres

A microcapsule includes a well-defined core containing the therapeutic agent, and a well-defined polymer membrane surrounding the core. In controlled drug delivery applications, such systems are of interest because drug release takes place by diffusion from the core through the rate-limiting membrane. Such systems fall under the category of membrane-controlled reservoir devices, already discussed. Consequently, microcapsules are capable of extended zero-order drug release provided that the conditions already discussed are met. Microcapsules may be prepared by a complex, three phase colloidal dispersion method. They may also be made by coating microspheres, which is a simpler process. An excellent review is available (Mathiowitz *et al.*, 1999) so only a few general comments will be made here. The reader is referred to the review article for additional details.

The attractiveness of achieving long-term zero-order drug release with microcapsules may be offset by (a) fairly expensive manufacturing methods that limit the use of microcapsules to human therapeutics that are not price-constrained and (b) the possibility of membrane rupture and abrupt release of the core material with the consequent overdosing potential.

On the other hand, microspheres are most often prepared by a simpler microemulsion polymerization technique. In microspheres, there is no outer membrane and such devices are in effect matrix devices with the drug more or less uniformly distributed in the polymer matrix. Release kinetics are typical of diffusion from a matrix system, as discussed above. Particle size distribution will be an important variable.

Microspheres and microcapsules are special categories of microparticulates. "Coarse" microparticles may be prepared by simple phase separation and lyophilization or solid diminution techniques, such as grinding and micronizing, using polymer/drug solutions, suspensions or solids. In these cases the particles are often not spherical, and size distributions will be broad, sometimes necessitating sieving.

Despite non-zero-order release kinetics, microsphere and microparticle delivery systems find wide application in both human and veterinary therapeutics because they can be inexpensive, are simple to manufacture and there is little danger of abrupt drug release. A number of particulate DDS have reached commercialization and they are included in Table 3.

Applications

There are numerous applications of macrocapsules and microspheres, and the reader is again referred to the review already mentioned (Mathiowitz and Kreitz, 1999). A number of applications are also listed in Table 3.

Polymer Therapeutics

The term "polymer therapeutics" has been coined to encompass polymeric drugs, polymer-protein conjugates, polymer-drug conjugates and complexes, and polymeric micelles with drugs covalently or physically incorporated in the core (Duncan *et al.*, 1996). An excellent review has recently been published (Duncan, 2003). Polymeric drug carriers are also discussed in other sections of this chapter.

The Enhanced Permeability and Retention Effect (EPR)

The principal rationale driving the development of polymer-drug conjugates, as well as micellar drug delivery systems is the enhanced permeability and retention (EPR) effect, first described by Maeda and coworkers (Matsumura and Maeda, 1986). This effect was attributed to the combination of two effects, (a) the disorganized pathology of tumor vasculature with a discontinuous endothelium which makes it hyperpermeable ("leaky") to circulating macromolecules and to particulates such as liposomes or micelles having an average diameter between 50 and 100 nanometers, plus (b) poor lymphatic drainage, resulting in retention of the delivery vehicle in the interstitial space of the tumor. Such materials, when administered intravenously will passively accumulate in tumors and can result in an intra-tumor concentration as much as 70-fold

TABLE 4 Polymer–Protein Conjugates on the Market or in Clinical Development[a]

Compound	Name	Year marketed	Indication
PEG–adenosine deaminase	Adagen®	1990	SCID syndrome
SMANCS	Zinostatin®, Stimalmer®	1993 (Japan)	Hepatocellular carcinoma
PEG–L-asparaginase	Oncaspar®	1994	Acute lymphoblastic leukemia
PEG–α-interferon 2b	PEG-INTRON®	2000	Hepatitis C
PEG–α-interferon 2b	PEG-INTRON®	Various clinical trials	Cancer, multiple sclerosis, HIV/AIDS
PEG–α-interferon 2a	PEGASYS®	2002	Hepatitis C
PEG–HGR	Pegvisomant®	2002 (approved EU)	Acromegaly
PEG–G-CSF	PEG-filgrastim®, Neulasta®	2002	Prevention of neutropenia associated with cancer chemotherapy
PEG–anti-TNF Fab	CD870	Phase II	Rheumatoid arthritis

[a]From Duncan (2003).

over that found systemically. Polymeric micelles and liposomes may similarly accumulate within the interstitial space of a tumor due to EPR (see below). This type of DDS has been called "passive targeting."

PEG-Protein Conjugates

Proteins are now recognized as an important therapeutic entity, but there are significant problems associated with their use. Dominant among these is antigenicity and proteolytic degradation, and both of these can result in very short plasma half-lives. These problems can, to some extent, be alleviated by conjugation with poly(ethylene glycol) (PEG), a process known as PEGylation, first described by Davis, Aubuchowski and colleagues (Davis, 2002). This concept is also discussed below as the "stealth" liposome DDS.

The exact mechanism by which PEG protects proteins from enzymatic degradation, reduces antigenicity and prolongs circulation time is not entirely understood. However, because the PEG chains bind and retain water (about 2–3 water molecules per ether oxygen group) it is likely that the PEG chains form a protective "cloud" at the protein surface that strongly retains hydration water and thus resists closer approach to the protein itself (e.g., Torchilin et al., 1995). This protective cloud then shields the protein from the action of degrading enzymes and reduces interaction with receptors, thus reducing immunogenicity. Further, it may also block recognition and capture by the reticuloendothelial system (RES) and further enhance circulation times (See also Chapter 2.13).

Since protein drugs have an active site, this protective cloud can also shield the active site and reduce biological activity. However, it has been suggested that the increased circulation time more than compensates for the reduced activity.

Applications

PEGylated protein drugs used clinically or in clinical development are shown in Table 4 (Duncan, 2003). A unique polymer-drug conjugate based on Styrene-Maleic Anhydride copolymer (SMA) and the drug NeoCardioStatin (NCS), called SMANCS, is also included in this table (Matsumura and Maeda, 1986).

Polymer Drug Conjugates

A schematic representation of a polymer-drug conjugate is shown in Fig. 22. This model was first proposed by Ringsdorf (Ringsdorf, 1975). The essential features of this model are a water-soluble backbone polymer with a covalently attached drug via a spacer. Optionally, a targeting residue can also be attached via a spacer. The water-soluble polymer backbone can be degradable, or non-degradable.

An effective polymer-drug conjugate is stable while in blood plasma, but then gradually releases the drug once it is localized by the EPR effect within a tumor, and the drug is then taken up by the tumor cells. One of the most extensively investigated polymers used in polymer-drug conjugates is poly[N-(2-hydroxypropyl) methacrylamide], known as HPMA (Kopecek and Bazilova, 1973; Duncan and Kopecek, 1984). In this particular polymer system, the drug is covalently attached to the water-soluble polymer backbone via a glycine-phenylalanine-leucine-glycine spacer. This spacer is stable in the circulation, but once it has been internalized by the tumor cells via endocytic uptake, the enzyme cathepsin B cleaves the peptide spacer within the lysosome, and the drug is liberated from the polymer conjugate and diffuses from the lysosome to the

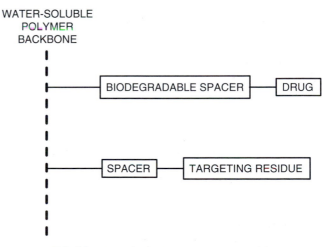

FIG. 22. Ringsdorf polymer conjugate model.

nucleus where it can kill the cell. (Duncan and Kopecek, 1984). This has been called "lysosomotropic" drug delivery.

However, HPMA has a non-degradable backbone so that the molecular weight of the polymer has to be kept below the renal threshold so that the polymer can be excreted. Unfortunately, this also reduces the time the conjugate is in the circulation. Since it has been shown that EPR-mediated passive targeting is directly proportional to the plasma concentration of the circulating conjugate (Seymour *et al.*, 1995) the construction of biodegradable, water-soluble polymer backbones where molecular weights are not limited by renal excretion values is clearly of interest.

Among the first biodegradable polymer-drug conjugates is poly(L-glutamic acid) containing bound paclitaxel, or campothecin. These conjugates are now in clinical trials (Schulz *et al.*, 2002, Shafter *et al.*, 2002). Recently, a polyacetal-doxorubicin conjugate designed to be relatively stable at pH 7.4 but to rapidly hydrolyze at pH 5.5 has been described (Tomlinson *et al.*, 2003).

Applications

Polymer-drug conjugates in clinical development are shown in Table 5 (Duncan, 2003).

Polymeric Micelles

Amphiphilic, A-B, block copolymers, i.e., block copolymers constructed from a hydrophilic and a hydrophobic segment, are known to assemble in an aqueous environment into polymeric micelles. This self-assembly is due to the large difference in solubility between the two segments. Polymeric micelles have a fairly narrow size distribution and have a unique core-shell structure where the hydrophobic segments form an inner core surrounded by the hydrophilic segments, known as the "corona." Polymeric micelles only exist above the critical micelle concentration (CMC). Below that concentration the micelles dissociate into the individual polymer molecules, known as unimers. Micelles can also be constructed from graft copolymers with a soluble main chain and insoluble grafts, or side chains. Kataoka, Okano and coworkers have published extensively on polymeric micelles, and have designed many interesting block copolymers for use in polymer micellar DDS. (For example, see Yokoyama *et al.*, 1990). An excellent recent review has also been published (Torchilin, 2001).

Polymeric micelles have excellent potential as DDS. Among these is their ability to solubilize poorly water-soluble drugs and thus increase drug bioavailability. They can also passively target tumors due to the EPR effect. Like liposomes, micelles can also be targeted by attachment of specific ligands at the ends of the hydrophilic block segment.

Although other hydrophilic polymers can be used to solubilize and stabilize hydrophobic drugs, (Torchilin *et al.*, 1994), poly(ethylene glycol) remains the polymer of choice. (See Chapter 2.13 on Nonfouling Surfaces). However, a variety of polymers have been used to form the hydrophobic block. These include poly(propylene oxide) (Kabanov *et al.*, 1989) poly(L-lysine) (Katayose *et al.*, 1998), poly(aspartic acid) (Yokoyama *et al.*, 1990) γ-benzoyl-L-aspartate (Kwon *et al.*, 1997) , γ-benzyl-L-glutamate (Jeong *et al.*, 1998), polycaprolactone (Kim *et al.*, 1998), poly(D,L-lactic acid) (Ramaswamy *et al.*, 1997) and poly(ortho esters) (Toncheva *et al.*, 2003).

Applications

Polymeric micelles have principally been used as drug carriers in tumor targeting applications. Micelles based on Pluronics [tri-block copolymers of poly(ethylene oxide) (PEO) and poly(propylene oxide) (PPO) with the general structure of PEO-PPO-PEO] with incorporated doxorubicin have been shown in a Phase II study to circumvent p-glycoprotein-mediated drug resistance (Batrakova *et al.*, 1996), a phenomenon where antineoplastic drugs are pumped out of the cell resulting in failures of chemotherapy regimes. Micelles based on a block copolymer of poly(ethylene glycol)-poly(aspartic acid), with a fraction of doxorubicin covalently bound and a fraction ionically complexed to the aspartic acid units, was investigated in a Phase I clinical trial as a formulation known as NK911 (Nakanishi *et al.*, 2001). While only the non-covalently bound doxorubicin was active, a three-to-four fold improvement in targeting was observed.

Liposomes

The first observation that ordered structures are obtained when water insoluble lipids, such as phospholipids, are mixed with an excess of water was made by Bangham and coworkers in the 1960's (e.g., see historical review by Bangham, 1983). These ordered structures eventually arrange into concentric, closed spherical membranes known as liposomes, or vesicles. Such liposomes can consist of one, or a multiplicity of bilayer membranes. When the diameter of such vesicles is about 25 to 100 nm, they are known as small unilamellar vesicles (SUV) and when the diameter is between about 100 nm to many microns,

TABLE 5 Polymer–Drug Conjugates in Clinical Trials as Anticancer Agents[a]

Compound	Name	Company	Linker	Development status
HPMA copolymer–doxorubicin	PK1; FCE28068	CRC/Pharmacia	Amide	Phase II
HPMA copolymer–doxorubicin–galactosamine	PK2; FCE28069	CRC/Pharmacia	Amide	Phase I/II
HPMA copolymer–paclitaxel	PNU-166945	Pharmacia	Ester	Phase I
HPMA copolymer–camphothecin	MAG-CPT PNU166148	Pharmacia	Ester	Phase I
HPMA copolymer–platinate	AP5280	Access Pharmaceuticals	Malonate	Phase I
Polyglutamate–paclitaxel	CT-2103 XYOTAX	Cell Therapeutics	Ester	Phase II/III
Polyglutamate–camphothecin	CT-2106	Cell Therapeutics	Ester	Phase I
PEG–camphothecin	PROTHECAN	Enzon	Ester	Phase II

[a] From Duncan (2003).

TABLE 6 Examples of Commercialized Liposome-Based Products[a]

Company	Product	Drug	Target disease
IGI	Newcastle disease virus	Killed Newcastle virus	Newcastle disease
IGI	Avian rheovirus vaccine	Killed avian rheovirus	Rheovirus infection in breeder chicken
Gilead	AmBisone®	Amphotericin B	Systemic fungal infections
Gilead	DaunoXome®	Daunorubicin	Cancer
Johnson & Johnson	Doxil®	Doxorubicin	Cancer
Johnson & Johnson	Amphocil®	Amphotericin B	Systemic fungal infections
Elan	ABLC®	Amphotericin B	Systemic fungal infections
Swiss Serum and Vaccine Institute	Epaxal-Berna Vaccine®	Inactivated hepatitis A virions	Hepatitis A
Swiss Serum and Vaccine Institute	Inflexal Vaccine®	Hemagglutinin and neraminidase from H1N1, H3N2, and B strains	Influenza

[a]From Gregoriades (2003).

they are known as multilamellar vesicles (MLV). A good review has recently been published (Gregoriades, 2003).

Liposomes are able to entrap water-soluble solutes in the aqueous inner core. Lipophilic drugs may be entrapped in the lipid bilayers by combining such drugs with the phospholipids used during their preparation.

Vesicles larger than about 200 nm are rapidly cleared from the blood and end up in the macrophages of the reticuloendothelial system (RES). Circulation times of liposomes can be augmented by coating with PEG or sialic acid surfactants. In the former case, PEG is conjugated to a lipid tail, which inserts into the liposome's lipid bilayer, "PEGylating" the liposome. These polymers form a water-retaining coating around the liposome and prevent recognition by the RES system (Gregoriades, 1995a). Such liposomes have been termed "stealth liposomes."

While passive targeting of liposomes to the RES is of some interest, a more interesting approach is one where liposomes with target recognition properties are constructed (Gregoriades, 1995b). Such "active targeting" requires the identification of suitable receptors on the surface of the target cells and appropriate ligands on the surface of the liposome that can recognize this target. When the liposome is PEGylated, the targeting ligand is conjugated to the outer end of the PEG molecules.

Applications

Examples of commercialized liposome-based products are listed in Table 6 (Gregoriades, 2003).

Bibliography

Albin, G., Horbett, T. A., and Ratner, B. D. (1985). Glucose sensitive membranes for controlled delivery of insulin: insulin transport studies. *J. Controlled Release* 2: 153–164.

Aspden, T. J., Mason, J. D. T., Jones, N. S., Lowe, J., Skaugrund, O., and Illum, L. (1997). Chitosan as a nasal delivery system. *J. Pharm. Sci.* 86: 509–513.

Baker, R. W., and Lonsdale, H. K. (1974). Controlled release: mechanisms and rates. in *Controlled Release of Biologically Active Agents*, A. C. Tanquary, and R. E. Lacey (eds.). Plenum Press, New York, pp. 15–71.

Bangham, A. D. (1983). Liposomes: a historical perspective, in *Liposomes*, M. J. Ostro (ed.). Marcel Dekker, New York.

Barr, J., Woodburn, K. W., Ng, S. Y., Shen, H.-R., and Heller, J. (2002). Post surgical pain management with poly(ortho esters). *Adv. Drug Deliv. Rev.* 54: 1041–1048.

Batrakova *et al.* (1996). Anthracycline antibiotics non-covalently incorporated into block copolymer micelles: in vivo evaluation of anticancer activity. *Br. J. Cancer* 74: 1545–1552.

Behr, J. P. (1997). The proton sponge: A trick to enter cells the viruses did not exploit. *Chimia* 51: 34–36.

Benagiano, G., Schmitt, E., Wise, D., and Goodman, M. (1979). Sustained release hormonal preparations for the delivery of fertility-regulating agents. *J. Polymer Sci. Polymer Symp.* 66: 129–148.

Brem, H., and Langer, R. (1996). Polymer-based drug delivery to the brain. *Sci. Med.* 3: 2–11.

Brunetti, P., Benedetti, M. M., Calabrese, G., and Reboldi, G. P. (1991). Closed loop delivery systems for insulin. *Int. J. Artif. Organs* 14: 216–226.

Bucher, J. E., and Slade, W. C. (1909). The anhydrides of isophthalic and terephthalic acids. *J. Am. Chem. Soc.* 31: 1319–1321.

Bulmus, V., Woodward, M., Lin, L., Stayton, P. S., and Hoffman, A. S. (2003). A new pH-responsive and glutathione-reactive, membrane-disruptive polymeric carrier for intracellular delivery of biomolecular drugs. *J. Controlled Release* 93: 105–120.

Capozza, R. C., Sendlebeck, L., and Balkenhol, J. (1978). Preparation and evaluation of a bioerodible naltrexone delivery system in: *Polymeric Delivery Systems*. R. J. Kostelnick (ed.). Gordon and Breach, New York, pp. 59–73.

Cheung, C. Y., Murthy, N., Stayton, P. S., and Hoffman, A. S. (2001). A pH-sensitive polymer that enhances cationic lipid-mediated gene transfer. *Bioconj. Chem.* 12: 906–910.

Choi, N. S., and Heller, J. (1978a). No. 4,093,038, March 14, 1978.

Choi, N. S., and Heller, J. (1978b). No. 4,093,709, June 6, 1978.

Choi, N. S., and Heller, J. (1978c). No. 4,131,648, December 26, 1978.

Choi, N. S., and Heller, J. (1979). No. 4,1238,344, February 6, 1979.

Colombo, P. (1993). Swelling-controlled release in hydrogel matrices for oral route. *Adv. Drug Deliv. Rev.* 11: 37–57.

Coover, H. W., McConnell, R., and McCall, M. (1960). Flame-resistant polymers: polyphosphates and poly(phosphonates) from dehydroxy aromatic compounds. *Ind. Eng. Chem.* 52: 409.

Dang, W., Dordunoo, S., Zhao, Z., Wang, H., Dhanesar, S., and Lapidus, R. (1999). Polyphosphoester paclitaxel microspheres (Paclimer) delivery system: in vitro and in vivo studies. *Proc. Int. Symp. Control. Rel. Biaoct. Mater.* 26: 513–514.

Davis, F. F. (2002). The origin of pegylation. *Adv. Drug Deliv. Rev.* 54: 457–458.

Dhanarajan, A. P., Misra, G. P., and Siegel, R. A. (2002). Autonomous chemomechanical oscillations in a hydrogel/enzyme system driven by glucose. *J. Phys. Chem.* **106:** 8835–8838.

Domb, A. J., and Maniar, M. (1993). Absorbable biopolymers derived from dimer fatty acid. *J. Polymer Sci. Part A: Polymer Chem.* **31:** 1275–1285.

Dong, L. C., and Hoffman, A. S. (1987). Thermally reversible hydrogels: swelling characteristics and activities of copoly(NIPAAm-AAm) gels containing immobilized asparaginase. in *ACS Symposium Series, 350, Reversible Polymeric Gels and Related Systems,* P. Russo (ed.). ACS, Washington, D.C., pp. 236–244.

Dong, L. C., and Hoffman, A. S. (1991). A novel approach for preparation of pH- and temperature-sensitive hydrogels for enteric drug delivery. *J. Contr. Release* **15:** 141–152.

Duenas, E., Okumu, F., Daugherty, A., and Cleland, J. (2001). Sustained delivery of rhVEGF from a novel injectable liquid. *Proc. Int. Symp. Control. Rel. Bioact. Mater.* **28:** 1014–1015.

Duncan, R. (2003). The dawning era of polymer therapeutics. *Nat. Rev.* **2:** 347–360.

Duncan, R., and Kopecek, J. (1984). Soluble synthetic polymers as potential drug carriers. *Adv. Polym. Sci.* **57:** 51–101.

Duncan, R., Dimitrijevic, S., and Evagorou, E. G. (1996). The role of polymer conjugates in the diagnosis and treatment of cancer. *S. T. P. Pharma Sci.* **6:** 237–263

Dunn, R. L., English, J. P., Cowsar, D. R., and Vanderbelt, D. P. (1990). U.S. Patent No. 4,938,763.

Dunn, R. L., Hardee, G., Polson, A., Bennett, Martin, S., Wardley, R., Moseley, W., Krinick, N., Foster, T., Frank, K., and Cox, S. (1995). In-situ forming biodegradable implants for controlled release veterinary applications. *Proc. Int. Symp. Control. Rel. Bioct. Mater.* **22:** 91–92.

Einmahl, S., Behar-Cohen, F., Tabatabay, C., Savoldelli, M., D'Hermies, F., Chauvaud, D., Heller, J., and Gurny, R. (2000). A viscous bioerodible poly(ortho ester) as a new biomaterial for intraocular application, *J. Biomed. Mater. Res.* **50:** 566–573.

Einmahl, S., Ponsart, S., Bejjani, R. A., D'Hermies, F. D., Savoldelli, M., Heller, J., Tabatabay, C., and Gurny, R. (2003). Ocular biocompatibility of an auto-catalyzed poly(ortho ester). *J. Biomed. Mater. Res.* **67A:** 44–53.

Eliaz, R., and Kost, J. (2001). U.S. Patent No. 6,206,921 B1.

Felgner, P. L., *et al.* (1996). *Artificial Self-Assembling Systems for Gene Delivery.* ACS Books, Washington, D.C.

Gregoriades, G. (1995a). Fate of liposomes in vivo: a historical perspective. in *Stealth Liposomes,* D. Lasic and F. Martin (eds.). CRC Press, Boca Raton, FL, pp. 7–12.

Gregoriades, G. (1995b). Engineering of targeted liposomes: progress and problems. *Trends Biotechnol.* **13:** 527–537.

Gregoriades, G. (2003). Liposomes in drug and vaccine delivery. *Drug Deliv. Syst. Sci.* **21:** 91–97.

Heller, J., and Trescony, P. V. (1979). Controlled drug release by polymer dissolution II. Enzyme-mediated delivery device. *J. Pharm. Sci.* **68:** 919–921.

Heller, J. (1980). Controlled release of biologically active compounds from bioerodible polymers. *Biomaterials* **1:** 51–57.

Heller, J. (1984). Biodegradable polymers in controlled drug delivery. *CRC Crit. Rev. Therap. Drug Carrier Syst.* **1:** 39–90.

Heller, J. (1985). Controlled drug release from poly (ortho esters): a surface eroding polymer. *J. Controlled Release* **2:** 167–177.

Heller, J. (1988). Chemically self-regulated drug delivery systems, *J. Controlled Release* **8:** 111–125.

Heller, J. (1993). Poly(ortho esters). *Adv. Polymer Sci.* **107:** 41–92.

Heller, J., Baker, R. W., Gale, R. M., and Rodin, J. O. (1978). Controlled drug release by polymer dissolution I. Partial esters of maleic anhydride copolymers. Properties and theory. *J. Appl. Polymer Sci.* **22:** 1991–2009.

Heller, J., Ng, S. Y., Fritzinger, B. K., and Roskos, K. V. (1990a). Controlled drug release from bioerodible hydrophobic ointments. *Biomaterials* **11:** 235–237.

Heller, J., Chang, A. C., Rodd, G., and Grodsky, G. M. (1990b). Release of insulin from a pH-sensitive poly(ortho ester). *J. Controlled Release* **14:** 295–304.

Heller, J., Barr, J., Ng, S. Y., Schwach-Abdelllauoi, K. and Gurny, R. (2002). Poly(ortho esters): synthesis, characterization, properties and uses. *Adv. Drug Deliv. Rev.* **54:** 1015–1039.

Helmus, M. (2003). Boston Scientific Corp., personal communication, September 2003.

Henriksen, I., *et al.* (1996). Bioadhesion of hydrated chitosans. *Int. J. Pharm.* **145:** 231–240.

Heskins, H., and Guillet, J. E. (1968). *J. Macromol. Sci. Chem.* A2 **6:** 1209.

Higuchi, T. (1961). Rates of release of medicaments from ointment bases containing drugs in suspension. *J. Pharm. Sci.* **50:** 874–875.

Hillery, A. M., Lloyd, A. W., and Swarbrick, J. (2001). *Drug Delivery and Targeting.* Taylor and Francis, London, pp. 371–398.

Hoffman, A. S. (1987). Applications of thermally reversible polymers and hydrogels in therapeutics and diagnostics, *J. Controlled Release* **6:** 297–305.

Hoffman, A. S., Afrassiabi, A., and Dong, L. C. (1986). Thermally reversible hydrogels: II. Delivery and selective removal of substances in aqueous solutions. *J. Controlled Release* **4:** 213–222.

Hsieh, D. S., Langer, R., and Folkman, J. (1981). Magnetic modulation of release of macromolecules from polymers. *Proc. Natl. Acad. Sci. USA* **78:** 1863–1867.

Huggins, C., Stevens, R. E., and Hodges, C. V. (1941). Studies on prostatic cancer. II. The effect of castration in advanced carcinoma of the prostate gland. *Arch. Surg.* **43:** 209–223.

Ilum, L. (2001). Oral presentation at U.S.–Japan Drug Delivery Meeting, Maui, Hawaii, Dec. 17–21.

Ishihara, K., and Matsui, K. (1986). Glucose-responsive insulin release from polymer capsule. *J. Polymer Sci., Polymer Lett. Ed.* **24:** 413–417.

Ito, Y., Casolaro, M., Kono, K., and Imanishi, Y. (1989). An insulin-releasing system that is responsive to glucose. *J. Controlled Release* **10:** 195–203.

Iwata, H., and Matsuda, T. (1988). Preparation and properties of novel environment-sensitive membranes prepared by graft polymerization onto a porous substrate. *J. Membrane Sci.* **38:** 185–199.

Jeong, B., Choi, Y. K., Bae, Y. H., Zentner, G., and Kim, S. W. (1999a). New biodegradable polymers for injectable drug delivery systems. *J. Controlled Release* **62:** 109–114.

Jeong, B., Bae, Y. H., and Kim, S. W. (1999b). Biodegradable thermosensitive micelles of PEG-PLGA-PEG triblock copolymers. *Colloids Surf. B: Interfaces* **16:** 185–193.

Jeong, B., Kibbey, M. R., Birnhaum, J. C., Won, Y.-Y., and Gutowska, A. (2000). Thermogelling biodegradable polymers with hydrophilic backbones: PEG-g-PLGA. *Macromolecules* **33:** 8317–8322.

Jeong, Y. I., Cheon, J. B., Kim, S. H., Nah, J. W., Lee, Y. M., Sung, Y. K., Akaike, T., and Cho, C. S. (1998). Clonazepam release from core-shell type nanoparticles in vitro. *J. Controlled Release* **51:** 169–178.

Johnson, O. F. L., Cleland, J. F., Lee, H. J., Charnis, M., Duenas, E., Jaworowicz, W., Shepard, D., Shazamani, A., Jones, A. J. S., and Putney, C. D. (1996). A month-long effect from a single injection of microencapsulated human growth hormone. *Nat. Med.* **2:** 795–799.

Joshi, A., Ding, S., and Himmelstein, K. J. (1993). U.S. Patent No. 5,252,318.

Kabanov, A. V., Chekhonin, V. P., Alakhov, V. Yu., Batrakova, E. V., Lebedev, A. S., Melik-Nubarov, N. S., Arzhakov, S. A., Levashov, A. V., Morozov, G. V., Severin, E. S., and Kabanov, V. A. (1989). The neuroleptic activity of haloperidol increases after its solubilization in surfactant micelles. *FEBS Lett.* **258**: 343–345.

Kader, A., Dordunoo, S., Zhao, Z., DePalma, P., Dhaneser, S., and Dang, W. (2000). Physicochemical characterizations of controlled release polilactofate polymeric microspheres with insulin. *Proc. Int. Symp. Controlled Release Bioact. Mater.* **27**: 1004–1005.

Katayose, A., and Kataoka, K. (1998). Remarkable increase in nuclease resistance of plasmid DNA through supramolecular assembly with poly(ethylene glycol)–poly(L-lysine) block copolymer. *J. Pharm Sci.* **87**: 160–163.

Khan, M. A., Healy, M. D. S., and Bernstein, H. (1992). Low temperature fabrication of protein loaded microspheres. *Proc. Int. Symp. Controlled. Release Bioact. Mater.* **19**: 518–519.

Kim, S. Y., Shin, I. G., Lee, Y. M., Cho, C. G., and Sung, Y. K. (1998). Methoxy poly(ethylene glycol) and ε-caprolactone amphiphilic block copolymers micelle containing indomethacin. II. Micelle formation and drug release behavior. *J. Controlled Release* **51**: 13–22.

Kopecek, J., and Bazilova, H. (1973). Poly(N-hydroxypropyl) methacrylamide. I. Radical polymerization and copolymerization. *Eur. Polymer J.* **9**: 7–14.

Kost, J., Leong, K., and Langer, R. (1989). Ultrasound-enhanced polymer degradation and release of incorporated substances. *Proc. Natl. Acad. Sci. USA* **86**: 7663–7666.

Kulkarni, R. K., Moore, E. G., Hegyeli, A. F., and Leonard, F. (1971). Biodegradable poly(lactic acid) polymers. *J. Biomed. Mater. Res.* **5**: 169–181.

Kwon, G. S., Naito, M., Yokoyama, M., Okano, T., Sakurai, Y., and Kataoka, K. (1997). Block copolymer micelles for drug delivery: loading and release of doxorubicin. *J. Controlled Release* **48**: 195–201.

Kyriakides, T. R., Cheung, C. Y., Murthy, N., Bornstein, P., Stayton, P. S., and Hoffman, A. S. (2001). pH-sensitive polymers that enhance intracellular drug delivery in vivo. *J. Controlled Release* **78**: 295–303.

Langer, R. S., and Peppas, N. A. (1983). Chemical and physical structure of polymers as carriers for controlled release of bioactive agents: a review. *Rev. Macromol. Chem. Phys.* **C23**: 61–126.

Lappas, L. C., and McKeehan (1965). Polymeric pharmaceutical materials. I. Preparation and properties. *J. Pharm. Sci.* **54**: 176–181.

Laurencin, C., Morris, C., Pierri-Jaques, H., Schwartz, E., and Zou, L. (1990). The development of bone-bioerodible polymer composites for skeletal tissue regeneration: studies of initial cell adhesion and spread. *Trans. Orthop. Res. Soc.* **36**: 183–190.

Lee, S. J., and Park, K. (1996). Synthesis and characterization of sol-gel phase-reversible hydrogels sensitive to glucose. *J. Molec. Recognition* **9**: 549–557.

Lehman, K. (1968). *Drugs made in Germany*, 11, 34.

Leong, K. W., Brott, B. C., and Langer, R. (1985). Bioerodible polyanhydrides as drug carrier matrices I: Characterization, degradation and release characteristics. *J. Biomed. Mater. Res.* **19**: 941–955.

Leong, K. W., D'Amore, P. D., Marletta, M., and Langer, R. (1986). Bioerodible polyanhydrides as drug carrier matrices II: biocompatibility and chemical reactivity. *J. Biomed. Mater. Res.* **20**: 51–64.

Leussen, H. L., *et al.* (1997). Mucoadhesive polymers in peroral peptide drug delivery. *J. Controlled Release* **45**: 15–23.

Li, L. C., Deng, J., Stephens, D. (2002). Polyanhydride implant for antibiotic delivery – from the bench to the clinic. *Adv. Drug. Deliv. Rev.* **54**: 963–986.

Li, S. W., Garreau, H., and Vert, M. (1990a). Structure–property relationships in the case of the degradation of massive aliphatic poly(α-hydroxy acids) in aqueous media, part 1: Poly(D,L-lactic acid). *J. Mater. Sci. Mater. Med.* **1**: 123–130.

Li, S. W., Garreau, H., and Vert, M. (1990b). Structure–property relationships in the case of the degradation of massive aliphatic poly(α-hydroxy acids) in aqueous media, part 2: Degradation of lactide–glycolide copolymers: PLA 37.5GA25 and PLA 75GA25. *J. Mater. Sci. Mater. Med.* **1**: 131–139.

Li, S. W., Garreau, H., and Vert, M. (1990c). Structure–property relationships in the case of the degradation of massive aliphatic poly(α-hydroxy acids) in aqueous media, part 3: Influence of the morphology of poly(L-lactic acid). *J. Mater. Sci. Mater. Med.* **1**: 198–206.

Mahato, R. I., *et al.* (1997). Non-viral vectors for in vivo gene therapy. *Crit. Rev. Ther. Drug Carrier Syst.*, **14**: 133–172.

Makino, K., Mack, E. J., Okano, T., and Kim, S. W. (1990). A microcapsule self-regulating delivery system for insulin. *J. Controlled Release* **12**: 235–239.

Malm, C. J., Emerson, J., and Hiatt, G. D. (1951). Cellulose acetate phthalate as an enteric coating. *J. Am. Pharm. Assn.* **40**: 520–525.

Martin, E. W. (1965). *Remington's Pharmaceutical Sciences*, 13th ed. Mack, Easton, PA, p. 604.

Mathiowitz, E., and Kreitz, M. R. (1999). Microencapulation. in *Encyclopedia of Controlled Drug Delivery*. Wiley and Sons, New York, Vol. 2, pp. 493–546.

Matsumura, Y., and Maeda, H. (1986). A new concept for macromolecular therapies in cancer chemotherapy: mechanism of tumouritropic accumulation of proteins and the antitumor agent SMANCS. *Cancer Res.* **6**: 67387–6392.

Murthy, N., Stayton, P. S., and Hoffman, A. S. (1999). The design and synthesis of polymers for eukaryotic membrane disruption. *J. Controlled Release* **61**: 137–143.

Murthy, N., Chang, I., Stayton, P. S., and Hoffman, A. S. (2001). pH-sensitive hemolysis by random copolymers of alkyl acrylates and acrylic acid. *Macromol. Symposia* **172**: 49–55.

Murthy, N., Campbell, J., Fausto, N., Hoffman, A. S., and Stayton, P. S. (2003). Cytoplasmic delivery from endosomes of drugs that are conjugated or complexed to membrane-disruptive polymers via pH-degradable linkages. *Bioconj. Chem.* **14**: 412–419.

Nakanishi *et al.* (2001). Development of the polymer micelle carrier system for doxorubicin. *J. Controlled Release* **74**: 295–302.

Ng, S. Y., Vandamme, T., Taylor, M. S., and Heller, J. (1997). Synthesis and erosion studies of self-catalyzed poly(ortho esters). *Macromolecules* **30**: 770–772.

Nguyen, T. H., Himmelstein, K. J., and Higichi, T. (1985). Some equilibrium and kinetics of water sorption in poly(ortho esters). *Int. J. Pharm.* **25**: 1–12.

Okano, T., Bae, Y. H., Jacobs, H. and Kim, S. W. (1990). Thermally on-off switching polymers for drug permeation and release. *J. Controlled Release* **11**: 255–265.

Park, H., and Robinson, J. (1985). Physico-chemical properties of water insoluble polymers important to mucin epithelial adhesion. *J. Controlled Release* **2**: 47–57.

Petersen, R. V., Anderson, R. G., Fang, S. M., Gregonis, D. E., Kim, S. W., Feijen, J., Anderson, J. M., and Mitra, S. (1980). Controlled release of progestins from poly(α-amino acid) carriers. in *Controlled Release of Bioactive Materials*, R. W. Baker (ed.). Academic Press, New York, pp. 45–60.

Pitt, C. G., Marks, T. A., and Schindler, A. (1980). Biodegradable delivery systems based on aliphatic polyesters: applications to contraceptives and narcotic antagonists. in: *Controlled Release of Bioactive Materials*, R. W. Baker (ed.). Academic Press, New York, pp. 19–43.

Ramaswamy, M., Zhang, X., Burt, H. M., and Wasan, K. M. (1997). Human plasma distribution of free paclitaxel and paclitaxel associated with diblock copolymers. *J. Pharm. Sci.* **86**: 460–464.

Rejmanova, P., Kopecek, J., Pohl, J., Baudys, M., Kostka, V. (1983). Polymers containing enzymatically degradable bonds. 8. Degradation of oligopeptide sequences in N-(2-hydroxypropyl) methacrylamide copolymers by bovine spleen cathepsin B. *Macromol. Chem.* **184**: 2009–2020.

Richard, M., Dahiyat, B. I., Arm, D. M., Lin, S., and Leong, K. W. (1991). Evaluation of polyphosphates and polyphosphonates as degradable biomaterials. *J. Biomed. Mater. Res.* **25**: 1151.

Ringsdorf, H. (1975). Structure and properties of pharmacologically active polymers. *J. Polymer Sci. Polymer Symp.* **51**: 135–153.

Robinson, D. N., and Peppas, N. A. (2002). Preparation and Characterization of pH-Responsive Poly(methacrylic acid-g-ethylene glycol) Nanospheres. *Macromol.* **35**: 3668–3674.

Rohr, U. D., Markmann, S., Oberhoff, C., Janat M. M., Schindler, A. E., McRea, J. C., and Zentner, G. (2002). Oncogel: A technological breakthrough in local treatment of paclitaxel sensitive cancers. *Proc. Int. Symp. Controlled Release Biaoct. Mater.* **29**: 33–334.

Rolland, A. (1999). *Advanced Gene Delivery: from Concepts to Pharmaceutical Products.* Harwood Press.

Rosen, H. B., Chang, J., Wnek, G. E., Linhardt, R. J., and Langer, R. (1983). Bioerodible polyanhydrides for controlled drug delivery. *Biomaterials* **4**: 131–133.

Roskos, K. V., Tefft, J. A., and Heller, J. (1993). A morphine-triggered delivery system useful in the treatment of heroin addiction. *Clin. Mater.*, **13**: 109–119.

Rothen-Weinhold, A., Schwach-Abdellaoui, K., Barr, J., Ng, S. Y., Shen, H.-R., Gurny, R., and Heller, J. (2001). Release of BSA from poly(ortho ester) extruded thin strands. *J. Controlled Release* **71**: 31–37.

Schneider, R. S., Lidquist, P., Wong, E. T., Rubenstein, K. E., and Ullman, E. F. (1973). Homogeneous enzyme immunoassay for opiates in urine. *Clin. Chem.* **19**: 821–825.

Schulz, J., *et al.* (2002). Phase II study of CT2103 in patients with colorectal cancer having recurrent disease after treatment with 5-fluorouracil-containimg regimen. *Proc. Am. Soc. Clin. Oncol.*, 2330.

Schwach-Abdellaoui, K., Heller, J., and Gurny, R. (1999). Hydrolysis and erosion studies of autocatalyzed poly(ortho esters) containing lactoyl–lactyl acid dimers. *Macromolecules* **32**: 301–307.

Seymour, L. W., Miyamoto, Y., Maeda, H., Brereton, M., Srohalm, J., Ulbrich, K., and Duncan, R. (1995). Influence of molecular weight on passive tumor accumulation of a soluble macromolecular drug carrier. *Eur. J. Cancer* **31**: 766–770.

Shafter, S. A., *et al.* (2002). Metabolism of poly-L-glutamic acid (PG)-paclitaxel (CT-2103); proteolysis by lysosomal cathepsin B and identification of intermediate metabolites. *Proc. Am. Assoc. Cancer Res.* **43**: 2067.

Siegel, R. A. (1990). pH-sensitive gels: swelling equilibria, kinetics and applications for drug delivery. in *Pulsed and Self-Regulated Drug Delivery*, J. Kost (ed.). CRC Press, Boca Raton, FL, pp. 129–157.

Singla, A. K. *et al.* (2000). Potential applications of carbomer in oral mucoadhesive controlled drug delivery systems: a review, *Drug Dev. Indust. Pharm.* **26**: 913–924.

Stayton, P. S., Hoffman, A. S., Murthy, N., Lackey, C., Cheung, C., Tan, P., Klumb, L. A., Chilkoti, A., Wilbur, F. S., and Press, O. W. (2000). Molecular engineering of proteins and polymers for targeting and intracellular delivery of therapeutics. *J. Controlled Release*, **65**: 203–220.

Teomin, D., Nyska, A., and Domb, A. J. (1999). Ricinoleic acid-based biopolymers. *J. Biomed. Mater. Res.* **45**: 258–267.

Theeuwes, F. (1975). Elementary osmotic pump. *J. Pharm. Sci.* **64**: 1987–1991.

Theeuwes, F. (1979). Novel drug delivery systems. in *Drug Absorption*, L. F. Prescott, and W. S. Nimmo, (eds.). ADIS Press, New York, pp. 157–176.

Theeuwes, F., and Yum, S. I. (1976). Principles of the design and operation of generic osmotic pumps for the delivery of semi-solid or liquid drug formulations. *Ann. Biomed. Eng.* **4**: 343–353.

Tipton, A. J., and Holl, R. J. (1998). No. 5,747,058.

Tomlinson, R., Heller, J., Brocchini, S., and Duncan, R. (2003). Polyacetal–doxorubicin conjugates designed for pH-dependent degradation. *Bioconj. Chem.* **14**: 1096–1106.

Toncheva, V., Schacht E., Ng, S. Y., Barr, J., and Heller J. Use of block copolymers of poly(ortho esters) and poly(ethylene glycol) as micellar drug carriers for tumor targeting applications. *J. Drug Targeting* **11**: 345–353.

Torchilin, V. P. (2001). Structure and design of polymeric surfactant-based drug delivery systems. *J. Controlled Release* **73**: 137–172.

Torchilin, V. P., and Papisov, M. I. (1994). Why do polyethylene glycol-coated liposomes circulate so long? *J. Liposome Res.* **4**: 725–739.

Torchilin, V. P., Trubetskoy, V. S., Whiteman, K. R., Caliceti, P., Ferruti, P., and Veronese, F. M. (1995). New synthetic amphiphilic polymers for steric protection of liposomes in vivo. *J. Pharm Sci.* **84**: 1049–1053.

Venkatraman, S., and Gale, R. (1998). Skin adhesives and skin adhesion 1. Transdermal drug delivery systems. *Biomaterials* **19**: 1110–1136.

Wang, P., and Johnston, T. P. (1991). Kinetics of sol-to-gel transition for Poloxamer polyols. *J. Appl. Polymer Sci.* **43**: 283–292.

Yang, H. J., Cole, C. A., Monji, N., and Hoffman, A. S. (1990). Preparation of thermally phase separating copolymer, poly(N-isopropylacrylamide-co-N-acryloxysuccinimide) with a controlled number of active esters per polymer chain. *J. Polymer Sci., Part A, Polymer Chem.* **28**: 219–226.

Yokoyama, M., Miyauchi, M., Yamada, N., Okano, T., Sakurai, K., Kataoka, K., and Inoue, S. (1990). Characterization and anti-cancer activity of the micelle-forming polymeric anticancer drug adriamycin-conjugated poly(ethylene glycol)–poly(aspartic acid) block copolymers. *Cancer Res.*, 1693–1700.

7.15 BIOELECTRODES

Ramakrishna Venugopalan and Ray Ideker

INTRODUCTION

Bioelectrodes are interdomain converters or intradomain modifiers and are used to transmit information into or out of the body. This chapter introduces the reader to the basic reactions that occur at an electrode–electrolyte interface, explains how these processes define the type of electrode, explores some simple equivalent circuit models for the electrode–electrolyte interface, reviews the factors influencing material selection for electrodes, and finally looks at some applications of such electrodes. Although the fundamentals discussed apply to both recording and stimulating electrodes, this chapter focuses on stimulating electrodes in the material selection and application sections because of their recent emergence as very effective treatment options.

Surface or transcutaneous electrodes used to monitor or measure electrical events that occur in the body are considered as monitoring or recording electrodes. Typical applications for recording electrodes include electrocardiography, electroencephalography, and electromyography. Electrodes used

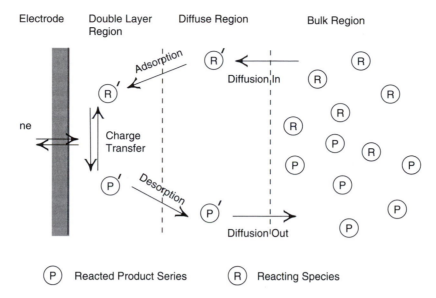

FIG. 1. Faradaic and nonfaradaic pathways of general electrode reactions. the charge transfer reaction at the surface is faradaic. The diffusion and the adsorption/desorption processes are nonfaradaic.

to transmit information into the body influence specific processes that occur in the body and are considered as stimulator electrodes. In all of these applications the electrodes are used to transmit voltage or current waveforms to specific target areas in the body for either direct control of a bioelectric event or indirect influence on the target area through a stimulated chemical change. Such stimulator electrodes are used in cardiac pacemakers and defibrillators to maintain or restore sinus rhythm, in transcutaneous electronic nerve stimulators for pain suppression, in neural stimulation systems for applications ranging from epilepsy control to auditory augmentation, in polarizing devices for intranscutaneous drug delivery, and in stimulators for tissue healing/regeneration.

ELECTRODE–ELECTROLYTE INTERFACE

In 1800 Volta demonstrated that the electrode–electrolyte interface was the source of electrical potential and initiated his research on direct current electricity. The nonlinearity of this interface with current lead to Ohm (1826) using a thermopile as a source of potential and resulted in the law that bears his name. However, the origin of the term "electrode" is attributed to one of the most prolific researchers in electrochemistry, Faraday (1834). In 1879 Helmholtz established the first model for the electrode–electrolyte interface. Several more complex models of this interface were proposed by Gouy (1910) and Stern (1924). This section will explain the basic reactions at the electrode–electrolyte interface, how they define a specific electrode type, and finally the make-up of this interface.

Faradaic and Nonfaradaic Processes

Bioelectrodes can interact with the body fluids (electrolytes) in two primary ways—faradaic and nonfaradaic (summarized in Fig. 1). An electrode may establish ohmic contact with

the surrounding environment thereby transferring electrons across the electrode/electrolyte interface via oxidation and/or reduction reactions. All such charge-transfer processes are governed by Faraday's law (i.e., the amount of chemical change occurring at an electrode–electrolyte interface is directly proportional to the current that flows through that interface) and hence are called faradaic processes. Electrodes at which faradaic processes occur are also known as charge-transfer electrodes. Under certain thermodynamically or kinetically unfavorable conditions an electrode–electrolyte interface may exhibit an absence of charge–transfer reactions. However, the electrode–electrolyte interface may change because of adsorption or desorption processes that occur at the interface, and transient external currents can flow due to changes in the environment. These processes at the electrode–electrolyte interface are known as nonfaradaic processes.

Polarizable and Nonpolarizable Electrodes

Electrodes in which charge transfer can occur in an unhindered manner represent truly faradaic or perfectly nonpolarizable electrodes. Conversely, electrodes in which no faradaic charge transfer can occur are called perfectly polarizable electrodes. In the latter case the half-cell potential of the electrode results in the formation of an electrical double layer of charge akin to two parallel plates of a capacitor, and charging or discharging of this capacitor monitors or influences a bioelectric event. Although an electrode can behave as either a polarizable or a nonpolarizable electrode under specific conditions at the interface, it will also exhibit a minimal amount of the other process (secondary) during clinical use.

Electrical Double Layer

At a given potential there will exist a charge on the metal electrode (q_M) and a charge in the solution (q_S). The potential

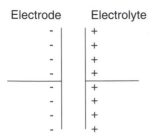

FIG. 2. Simplistic parallel-plate Helmholtz model for electrical double layer.

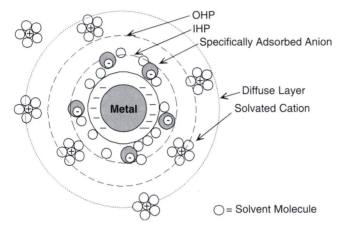

FIG. 3. Multilayer model for the electrode–electrode interface. IHP, inner Helmholz plane, OHP, outer Helmholz plane.

across the electrode–electrolyte interface and the composition of the electrolyte will determine if the charge on the electrode is positive or negative. The sum total of the charge at the interface will however be equal to zero at all times, i.e., $q_M = -q_S$. The charge on the metal is a result of an excess or a deficiency of electrons, and the charge in the electrolyte is due to an excess of anions or cations in close proximity to the electrode. Helmholtz (1879) established a model for this aligned array of charges referred to as the electrical double layer (EDL). This model is illustrated in Fig. 2.

Under realistic conditions this interface is more complex than the simplified parallel-plate model. The metal electrode still forms one charge array at the electrode–electrolyte interface. On the aqueous solution side of the interface, polar solvent molecules and specifically adsorbed ions or molecules make up the first layer. This layer, referred to as the compact, Stern, or Helmholtz layer, extends to the locus of the electrical centers of the specifically adsorbed ions (inner Helmholtz plane). Charged ions in the electrolyte attract their own sheath of solvent molecules and form what are called micelles or solvated ions. Because these ions are well insulated they interact with the conducting electrode surface primarily through long-range electrostatic forces. The size of these ions and their nonspecific adsorption mechanism results in the formation of a three-dimensional diffuse outer layer. This outer layer extends from the locus of the electrical centers of the solvated ions (outer Helmholtz plane) into the bulk of the solution. The resulting interfacial structure is illustrated in Fig. 3.

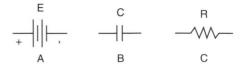

FIG. 4. Basic circuit elements. (A) The half-cell potential or battery circuit element. (B) Simplistic double-layer or parallel-plate capacitor circuit element. (C) Hindrance to faradaic transfer or the resistive circuit element.

EQUIVALENT CIRCUIT MODELS

Helmholtz (1879) not only described the nature of the electrode–electrolyte interface, but also proposed the first equivalent circuit model for it, i.e., an idealized parallel-plate capacitor. Warburg (1899), Fricke (1932), Randles (1947), and Sluyters-Rehbach (1970) contributed significantly to continued development of equivalent circuit models for specific conditions at the interface. However, their models did not account for the passage of direct current through this interface. In 1968 Geddes and Baker applied the concept of a shunting (parallel) faradaic leakage resistance across the non-faradaic components that resulted in two equivalent circuit models for the bioelectrode–electrolyte interface. These models accounted for the passage of direct current through this interface. Subsequent experimental investigations by Onaral and Schwan (1982) very strongly corroborated this concept of a shunting faradaic leakage resistance at low frequencies. Detailed analysis of different types of recording and stimulation electrodes and corresponding equivalent circuits are discussed by Geddes and Baker (1989) in their seminal textbook on the principles of biomedical instrumentation. A review article by Geddes (1997) provides a less rigorous historical perspective on the evolution of such circuits, their advantages, and their limitations.

Basic Circuit Elements

A voltage is developed across the electrical double layer and is referred to as the half-cell potential for that electrode under specific environmental factors. It is impossible to measure the potential developed at a single electrode. Hence an arbitrary standard electrode is used to close the circuit, and the total voltage is measured as a difference of the half-cell potentials of the two electrodes. This can be represented in terms of a voltage source element or battery represented in Fig. 4A. The electrical double layer in its most simplistic form is represented by two charged parallel plates with an insulated gap, i.e., a capacitor circuit element (Fig. 4B). The hindrance to the charge transfer that occurs at a faradaic electrode–electrolyte interface is represented by a resistive circuit element illustrated in Fig. 4C.

Simple Equivalent Circuits

The electrolyte/tissue surrounding an electrode will result in a resistive element referred to as the uncompensated or series resistance (R_u) that will be a contributing factor in all cases. An ideally polarizable electrode (nonfaradaic) will be represented

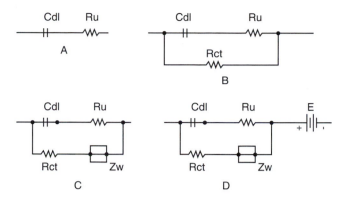

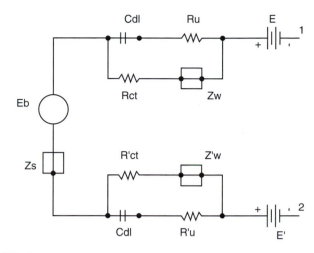

FIG. 5. Simple equivalent circuits. (A) Equivalent circuit for an ideally polarizable electrode. (B) Equivalent circuit element for a practical electrode. (C) A polarizable electrode subject to mass-transport limitations. (D) A polarizable circuit with mass-transport limitations and the half-cell potential of the electrode.

FIG. 6. The equivalent circuit model for a two-electrode scheme on a subject. E_b is the local biological event being monitored or initiated, and Z_S is the simplified black-box model used to represent the generic subject impedance.

TABLE 1 Factors Influencing Material Selection for Electrodes

Electrode	Surface area, geometry, and surface condition
Electrical	Potential, current, and quantity of charge
Environmental	Mass-transfer variables and solution variables
Engineering	Availability, cost, strength, and fabricability

by a series combination of the established double layer capacitance (C_{dl}) and R_u. This is illustrated in Fig. 5A. However, this model predicts infinite impedance for direct current (0 Hz). But, it is known that a direct current can be passed between an electrode–electrolyte interface. This disparity can be explained by placing a charge-transfer resistance (R_{ct}) or a faradaic leakage resistance across the model illustrated in Fig. 5A, as shown in Fig. 5B. Most physiological electrolytes diffuse in and out of the tissue surrounding an electrode. A polarizable electrode with such mass transport limitations (partial diffusion control) will have an impedance term called the Warburg impedance (Z_w) associated with it and will result in the model illustrated in Fig. 5C. Finally, a half-cell potential (E) is associated with every electrode placed in an electrolyte, and it is placed in series with the circuit in Fig. 5C, as shown in Fig. 5D.

Electrodes on a Subject

The equivalent circuit in Fig. 5D is a very good approximation of a single electrode–electrolyte interface. However, it is known that at least a two-electrode system is required either to make measurements of or to initiate a local biological event (E_b). The equivalent circuit models to represent the subject are quite complex because of the varying nature of tissue microstructure. Thus, a simplistic black box model (Z_s) is used to represent the generic subject impedance. The whole equivalent circuit comprising the subject impedance, the biological event, and the two electrodes is illustrated in Fig. 6.

FACTORS INFLUENCING MATERIAL SELECTION

While recording electrodes are not subject to active degradation because of the inherent nature of the application, they are subject to degradation by the environment. Stimulation electrodes, on the other hand, are subject to degradation by the very excitation applications they are used for. All the factors listed in Table 1 are of specific importance to material selection for electrodes (Fontana, 1986; Bard and Faulkner, 1980).

Stimulating electrodes work by producing transient electric fields in their vicinity. The uniformity and precision of

such electrical fields are dependent on the charge density distribution created by the stimulation electrode. Although ideally polarizable electrodes can create such fields with no chemical changes at the electrode–tissue interface, they are typically limited to less than 20 $\mu C/cm^2$ of true surface area for commonly used metallic electrodes (Robblee and Sweeney, 1996). Typical clinical applications require higher charge density distribution, and hence, most stimulation electrodes rely on predominantly faradaic charge transfer mechanisms for direct stimulation of a bioelectric event or a chemical reaction that would stimulate a bioelectric event. Some faradaic processes are reversible while other processes are not (Brummer and Turner, 1977). Irreversible reactions initiate the electrolysis of water (oxygen or hydrogen evolution), oxidation of chloride ions, and corrosion or dissolution of an electrode that produces soluble metal-ion complexes (Geddes and Baker, 1989). This is undesirable, as it can result in multiple local and system biological responses that may change the properties of surrounding tissue or the type of surrounding tissue itself, i.e., fibrous tissue encapsulation to isolate corroding nonbiocompatible object (Grill and Mortimer, 1994; Akers et al., 1977).

This change in property or type of surrounding tissue significantly influences the impedance at electrode–tissue interface (Danilovic and Ohm, 1998; Glikson et al., 1995; Beard et al., 1992). This has necessitated the estimation of a property called the "reversible charge injection limit." The charge injection limit is the maximum quantity of charge that can be injected or transferred before irreversible chemical changes occur. The charge injection limit of any bioelectrode will depend on the reversible processes available during stimulation (environmental variables), the shape and surface morphology of the electrode (electrode variables), and finally, the geometry and duration of the stimulation waveform. An excellent review article by Robblee and Rose (1990) discusses the relative importance of the aforementioned factors in achieving reversible charge injection.

Even if careful consideration is given to the reversible charge injection limit, any implanted device (foreign body) will produce local changes in the tissue as discussed before. Consequently, regulated-voltage stimulation (resultant current is dependent on tissue impedance) is susceptible to complex nonreproducible fluctuations in comparison to regulated-current stimulation (direct application of current/charge) for implanted electrode systems. The stimulation waveforms in their simplest configuration may be pulsatile or symmetric sinusoidal. The charge injected (q) by a pulsatile waveform (say a rectangular pulse) is proportional to the pulse current (i) and the time duration (t) for which it is applied, i.e., $q = it$. The charge can thus be controlled very precisely and is independent of the frequency with which it is delivered. In contrast, the charge delivered by a symmetric sinusoidal waveform is dependent on its frequency, i.e., current amplitudes that result in an acceptable charge density at high frequency may deliver an unacceptably high charge density at low frequency. But, symmetric sinusoidal waveforms possess no net dc component and thus have reduced residual effect on the periprosthetic tissue. The basics of simulation waveform geometry, duration, and frequency and how they relate to a specific application can be found in the book chapter on stimulation electrodes by Geddes and Baker (1989) in their textbook on biomedical instrumentation. The design of stimulation waveforms is also one of the strongest emerging research areas in interventional clinical research (Walcott *et al.*, 1998, 1995; Hillsley *et al.*, 1993).

ELECTRODE MATERIALS

5.1 Noble Metals

Platinum is the most popular electrode material because it is stable, inert, and corrosion resistant. However, pure platinum is a very soft metal and is typically alloyed with iridium to improve its mechanical properties, thereby facilitating construction or manufacture of small electrodes. The amount of iridium in a platinum–iridium alloy can vary from ~2% to ~30% without limiting the charge injection ability of the alloy (Robblee and Sweeney, 1996). Small-diameter intracortical stimulation needles that require flexural rigidity to achieve appropriate placement are examples of bioelectrodes that would use platinum–iridium alloys. Although platinum and platinum–iridium alloys exhibit extremely high corrosion resistance even at high injection current densities ($50–150\ \mu C/cm^2$) without gas evolution or electrolyte oxidation, metal dissolution is not avoidable and occurs at all charge densities (McHardy *et al.*, 1980; Robblee *et al.*, 1980).

Iridium metal can be used instead of platinum in applications that require superior mechanical properties. Although the iridium metal electrodes possess no advantage over the platinum electrodes in terms of charge injection ability, iridium oxide films exhibit significantly reduced interface impedance compared to bare metal (Glarum and Marshall, 1980). This can facilitate greater efficiency of charge flow, thus contributing to prolonged battery life of an implanted stimulation device. Multilayer oxides can be formed on iridium electrodes by cycling the electrochemical iridium potential between predetermined values versus a standard electrode. These multilayer oxides can transfer or inject significant charge densities by reversible transitions between two stable valence levels or oxides (Robblee *et al.*, 1983), but they are very susceptible to limiting charge injection densities due to the shape or type of stimulation waveform. For example, when subjected to a predominantly cathodal pulse of .2 msec the electrode can inject only $1000\ \mu C/cm^2$. But, the electrode can inject almost $3500\ \mu C/cm^2$ when the cathodal pulses are interspersed with an anodic bias voltage. The anodic bias voltage returns the electrode/oxide to its highest stable valence state between treatments, thereby increasing the charge injection ability of the electrode (Agnew *et al.*, 1986; Kelliher and Rose, 1989). Also, transport and conductivity limitations during the short treatment time frames restrict the utilization of the large charge injection capacity of such electrodes.

Non-noble Metals

Intramuscular electrodes that require high mechanical and fatigue strength may be made from non-noble metals/alloys such as 316LVM stainless steel or nickel–cobalt-based Elgiloy and MP35N. These materials rely on thin passive films for corrosion resistance and inject charge by reduction or oxidation of their passive films. Although large anodic pulses that push these electrodes into their transpassive behavior can lead to significant corrosion, it is important to note that corrosion can result at even relatively low cathodic pulses. These electrodes are capable of injecting ~$40\ \mu C/cm^2$ and are susceptible to corrosion-related failures even at that level (White and Gross, 1974).

Titanium and tantalum can form insulation oxide films during anodic polarization and can be used for manufacturing capacitive electrodes (electrodes that inject charge without faradaic reactions at the electrode–electrolyte interface). While the titanium oxide possesses a significantly higher dielectric strength compared to tantalum oxide, it is susceptible to significant DC leakage limiting its use in electrode applications. Capacitive electrodes based on tantalum–tantalum pentoxide have limited charge storage capacity ($100–150\ \mu C/cm^2$) and can be used only in applications above a particular geometric size ($0.05\ cm^2$) for neural prosthesis applications (Rose *et al.*, 1985).

Also, the translation of micro- or nanofabrication technology (Wise and Najafi, 1991) from the semiconductor to the biomedical community is likely to result in multielectrode arrays (Mastrototatro *et al.*, 1992) to either control or monitor different functions of a single event or multiple events.

APPLICATIONS

Cardiology

The pumping of the heart is controlled by regular electrical impulses generated by a group of special cells in the sinus node. The propagation of these impulses in a specific sequence causes the heart to contract, pump blood, and subsequently relax/expand once the signal passes. Any disorganization of this complex generation and propagation of electrical impulses (transmembrane potentials) or abnormal change of heart rate results in a condition called an arrhythmia (Janse, 1994).

When the arrhythmia results in slowing down of the heart, a device called a pacemaker is used to pace or restore the heart to its normal rate (Ellenbogen *et al.*, 1995). When the arrhythmias result in a very rapid heart rate with disorganization of propagation and asynchrony of contraction, a condition called atrial or ventricular fibrillation occurs (Epstein and Ideker, 1995). Ventricular fibrillation is known to be the cause of more than 250,000 deaths in the United States annually. This condition can be rectified using shock therapy, and the corrected rhythm can be maintained using a combination of medication and/or an electronic device called the implantable cardioverter/defibrillator (ICD).

For the first few years after the ICD became commercially available, the electrodes were placed directly on the heart during open-chest surgery (Ideker *et al.*, 1991). These electrodes were approximately 20–60 cm^2 and constructed out of titanium mesh. More recently, the ICD electrodes have been mounted on catheters that are inserted into the heart by way of the venous system so that major surgery to open the chest is no longer necessary (Singer *et al.*, 1998). These electrodes, typically 5–8 cm long, are frequently made of titanium wire wound around the catheter to form a tight coil with no space between the turns. The sealed, titanium-coated can housing the circuits and battery is used as one of the defibrillation electrodes in the latest ICDs (Fig. 7).

To effectively defibrillate, a certain minimum potential gradient or a certain minimum derivative of the potential gradient must be created by the shock field throughout all or most of the myocardium (Walcott *et al.*, 1997; Ideker *et al.*, 1990). Depending on the type of shock waveform, the minimum potential gradient needed to defibrillate is approximately 4–6 V/cm. The defibrillation electrode material, location, and dimensions affect the potential gradient field created by the shock, and hence, the shock strength needed to defibrillate (Ideker *et al.*, 1991). Other variables, such as orientation of the myocardial fibers, the shape and size of the heart, the presence of cardiac disease, and medications can also affect the shock strength needed for defibrillation (Baynham and Knisley, 1998).

Whereas a minimum shock strength is needed to defibrillate, too strong a shock can have detrimental effects including temporary electrical paralysis of the myocardium, the induction of new arrhythmias by shock itself, and damage or even death to the cardiac cells (Fotuhi *et al.*, 1999). These effects may be caused by electroporation of the cardiac cell membrane, which occurs when the potential difference created by the shock across the cell membrane is so strong that a

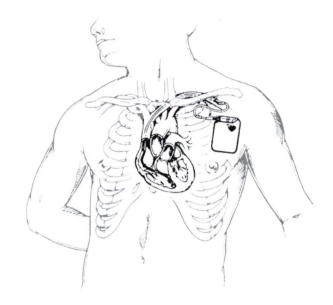

FIG. 7. Schematic illustrating the placement configuration of a implantable cardiovertor/defibrillator device.

dielectric breakdown forms holes in the membrane that allow indiscriminate passage of ions and other particles.

In contrast, an ablation electrode relies on causing irreversible changes in the target tissue (Iskos *et al.*, 1997). An ablation electrode is used to kill abnormal cells (Jais *et al.*, 1998; Kunze *et al.*, 1998) or normal cells in an abnormal location (Sebag *et al.*, 1997; Atie *et al.*, 1996; Haissaguerre *et al.*, 1994). In the former case, abnormal cells in the atria or the ventricles serve as rapidly firing foci of electrical activity that drive the remainder of the heart or allow the formation of a continuous reentrant loop pathway that excites the heart, initiating tachyarrhythmias. Ablation was originally performed by administering a large DC shock through a catheter electrode placed next to the target tissue and a large plate on the back of the patient. This shock damaged tissue by the direct effects of the large electric field and barotrauma caused by sudden gas bubble formation when arcing occurred between the shocking electrodes. Currently, a radio-frequency pulse applied through a platinum or platinum–iridium catheter electrode is used to kill the tissue by ablating or heating it during the application of the pulse. This technique is more localized and causes less damage to tissue surrounding the area of anomaly (Stancak *et al.*, 1996).

Neurology

Significant advances in imaging modalities, minimally invasive instrumentation, and robotic guidance/placement systems (Hefti *et al.*, 1998; Fenton *et al.*, 1997) are facilitating three-dimensional positioning of stimulation electrodes in the brain. Electrical stimulation of the brain can be used for a variety of applications ranging from pain management to treatment for Parkinson's disease. Stimulation within the brain must be executed without any charge transfer across the electrode-tissue interface to prevent permanent tissue damage. Capacitor electrodes are ideal for this application. The tissue acts as the dielectric between the capacitive plates and, since the surface of these electrodes are oxide coated, negligible charge transfer occurs (Rose *et al.*, 1985). Several clinical studies provide evidence that deep brain stimulation can be used to control medically refractory essential tremor (Ondo *et al.*, 1998; Lyons *et al.*, 1998) or untreatable Parkinsonian tremor (Ghika *et al.*, 1998; Krack *et al.*, 1998). Motor cortex stimulation has been shown to reduce poststroke pain in patients when muscle contraction was inducible (Katayama *et al.*, 1998).

The spinal cord can also be stimulated to control pain in various clinical conditions (Segal *et al.*, 1998). Percutaneous spinal electrodes are placed in patients who are not suitable candidates for coronary bypass grafting for treatment of ischaemic pain condition such as angina pectoris (Andersen, 1997). They may also be used to control chronic back pain if surgical outcomes are not successful (Rainov *et al.*, 1996). Transcutaneous electrical stimulation (TES) of peripheral nerves has been used to control chronic pain due to spinal cord injury or stoke damage (Taub *et al.*, 1997). Localized stimulation of nerve endings using retinal and cochlear implants has also been attempted to enhance the loss of vision or hearing, respectively (Heetderks and Hambrecht, 1988).

Functional neuromuscular simulation (FNS) refers to the electrical stimulation of skeletal muscle for artificially controlled exercise or restoration of motor function lost due spinal cord injury or disease (Sweeney, 1992). FNS systems are designed to provide upper-extremity control (Hoshimiya *et al.*, 1989), lower-extremity control (Graupe, 1989), or both. FNS takes advantage of the fact that most motor neurons below the level of injury remain intact. Thus electrical stimulation of the their axons can be used to cause muscle contraction. For example, the phrenic nerve can be stimulated to control respiration, or the sacral nerves can be stimulated to regulate the detrusor muscle in the bladder. Although FNS designers strive to mimic the body's control strategies, it is still extremely difficult to attain the fine gradation of force required to achieve adequate function. A review article by Heetderks and Hambrecht (1988) is an excellent source of information for understanding the potential advances in the clinical application of neural stimulation.

Other

Catheter ablation electrodes can be used via the urethra or perineal cavity to damage/kill benign or malignant hyperplasia in the prostate (Dixon, 1995). Ablation has also been used to reduce the size of mestastatic liver tumors and osteoid osteomas by damaging the irregular cells (Salbiati, 1997; Rosenthal *et al.*, 1995). Cryosurgery, i.e, ablation by extreme cooling, has been used to treat refractory hepatic neuroendocrine metastases (Bilchik *et al.*, 1997). Stimulating electrodes may also be used to enhance tooth movement in orthodontics or the growth of teeth. In this case, platinum and noble metals are not useful because they irritate the gingival tissue. Composite electrodes are used with biocompatible gels to keep the electrode separated from the moist gingival tissue (Beard *et al.*, 1992).

Future Trends—Healing and Regeneration

The U.S. Food and Drug Administration has approved a number of electrical bone-growth stimulating devices for treating nonunion fractures and for enhancing spinal fusion (Polk, 1995). These devices generally use one of three approaches to apply low-intensity fields to the tissues—direct current applied to the wound site through surgically implanted wire electrodes; high-frequency (>20 kHz) "capacitatively coupled" sine-wave signals applied through skin electrodes; and low-frequency pulsed electromagnetic fields (PEMFs) applied by Helmholtz-type coils strapped to the limb. The PEMF devices do not conduct current directly to the tissues and, in the strictest sense, may not be considered as electrodes.

The implanted electrodes usually consist of stainless steel or titanium wire, but may include other metals such as silver or platinum. The wires, serving as cathodes, are connected to small batteries supplying a continuous current of 2–20 μA. It is uncertain whether the resulting stimulation of bone growth is due to electrochemical changes in the tissue surrounding the cathodes or to the electric current itself (Black, 1987). Spadaro (1997) used a rabbit femur model to compare the bone growth stimulation by different metallic cathodes at the same current density. The observed differences in the degree of bone formation would seem to argue for an electrochemical, rather than an electrical, basis for the effects. Yet the same results might also be attributed to the mechanical properties of the wires; as the animals moved about, the stiffer wires would transmit more mechanical force to the bone and this micromotion might itself elicit an osteoinductive response.

The capacitatively coupled devices use 2 to 3-cm-diameter electrodes adjoined to the skin through a conductive gel. These are powered by a small battery pack and produce voltage gradients in bone estimated at 1–100 mV/cm at current densities in the μA/cm^2 range. As with the implanted electrodes, the precise basis for their effects are not completely understood but appear to involve the distribution of cations, particularly calcium, at the cell surface (Zhuang *et al.*, 1997).

The range of these applications should expand considerably when the molecular basis for electric field–tissue interactions is better understood. Research in animal models suggests that low-intensity fields may prove to be clinically useful for stimulating wound healing and nerve regeneration. For example, platinum cuff electrodes that surround the nerve or penetrating needle electrodes may be used to stimulate peripheral nerve regeneration (Heiduschka and Thanos, 1998). An electrode array with polymeric guidance tubules or a DC electrical field may also be applied to direct growth of new peripheral nerve axons. The nerves grow toward the cathode of the bioelectrode (Heiduschka and Thanos, 1998). It must be kept in mind, however, that this area of research remains controversial because of frequent difficulties in reproducing experimental results and the absence of a generally accepted model.

SUMMARY STATEMENT

Although significant progress has been made in our understanding of the bioelectrode–tissue interface, it is still not possible to predict all of its properties with certainty. New materials, micro- or nanofabricating capability, computer processing power, and a better understanding of biological processes over the past two decades have significantly improved our ability to make bioelectrodes that more than adequately monitor or influence a biological event. However, it is becoming apparent at this time that even the most advanced synthetic prosthesis cannot totally restore normal function. Hence, the human impact of interventional and potential regenerative procedures made possible due to current bioelectrodes will form the basis for further advances in this old, but still nascent, field.

QUESTIONS

1. What material would you choose for a deep brain stimulating electrode and why? Develop a simplistic equivalent circuit model for such an electrode–electrolyte interface. You do not have to model the tissue. Model only the interface.

2. Calculate the increase in charge injected by decreasing the frequency of a symmetric sinusoidal waveform by an order of magnitude assuming all other parameters remain constant.

3. Compare and contrast the region of oxide stability (potential range) for titanium alloy and 316L stainless steel electrodes. Assume that the oxides on titanium electrode and stainless steel electrodes are predominantly TiO_2 and Cr_2O_3, respectively. (Hint: Refer to the Pourbaix diagrams for titanium and chromium.)

4. Would a material used for a radio-frequency ablation electrode also be ideal for a cryoablation electrode (ablation due to freezing)?

5. Develop an equivalent circuit model for a capacitively coupled device used for healing nonunion fractures. You do not have to model the tissue. Model only the interface.

Bibliography

Agnew, W. F., Yuen, T. G. H., McCreedy, D. B., and Bullara, L. A. (1986). Histopathalogic evaluation of prolonged intracortical electrical stimulation. *Exper. Neurol.* 92: 162–185.

Akers, J. M., Peckman, P. H., Keith, M. W., and Merritt, K. (1997). Tissue response to chronically stimulated implanted epimysial and intramuscualr electrodes. *IEEE Trans. Rehab. Eng.* 5(2): 207–220.

Andersen, C. (1997). Complications in spinal cord stimulation for treatment of angina pectoris. Differences in unipolar and multipolar percutaneous inserted electrodes. *Acta Cardiologica* 52(4): 325–333.

Atie, J., Maciel, W., Pierobon, M. A., and Andrea, E. (1996). Radiofrequency ablation in patients with Wolff–Parkinson–White syndrome and other accessory pathways. *Arquivos Brasileiros de Cardiologia* 66 (Suppl 1): 29–37.

Bard, A. J., and Faulkner, L. R., eds. (1980). Introduction and overview of electrode processes. in *Electrochemical Methods.* John Wiley and Sons, New York, pp. 1–43.

Baynham, T. C., and Knisley, S. B. (1998). Combating heart disease with FEA. *Mech. Eng.*, Oct.: 70–72.

Beard, R. B., Hung, B. N., and Schmukler, R. (1992). Biocompatibility considerations at stimulation electrode interfaces. *Ann. Biomed. Eng.* 20: 395–410.

Bilchik, A. J., Sarantou, T., Foshag, L. J., Giuliano, A. E., and Ramming, K. P. (1997). Cryosurgical palliation of metastatic neuroendocrine tumors resistant to conventional therapy. *Surgery* 122(6): 1040–1047.

Black, J. (1987). *Electrical Stimulation. Its Role in Growth, Repair, and Remodeling of the Musculoskeletal System.* Praeger, New York, pp. 92–96.

Brummer, S. B., and Turner, M. J. (1977). Electrochemical considerations for safe electrical stimulation of nervous system with platinum electrodes. *IEEE Trans. Biomed. Eng.* **BME-24**: 59–63.

Danilovic, D., and Ohm, O. J. (1998). Pacing impedance variability in tined steriod eluting leads. *Pacing Clin. Electrophysiol.* 21(7): 1356–1363.

Dixon, C. M. (1995). Transurethral needle ablation for the treatment of benign prostatic hyperplasia. *Urol. Clin. N. Amer.* 22(2): 441–444.

Ellenbogen, K. A., Kay, G. N., and Wilkoff, B. L. (1995). *Clinical Cardiac Pacing.* W.B. Saunders, Philadelphia.

Epstein, A. E., and Ideker, R. E. (1995). Ventricular fibrillation. in *Cardiac Electrophysiology: From Cell to Bedside*, D. P. Zipes and J. Jalife (eds.) W. B. Saunders, Philadelphia, pp. 927–934.

Faraday, M. (1834). Experimental researches in electricity, 7th series. *Phil. Trans. R. Soc. Lond.* 124: 77–122.

Fenton, D. S., Geremia, G. K., Dowd, A. M., Papathanasiou, M. A., Greenlee, W. M., and Huckman, M. S. (1997). Precise placement of sphenoidal electrodes via fluoroscopic guidance. *Am. J. Neuroradiol.* 18(4): 776–778.

Fontana, M. G. (1986). Corrosion principles. in: *Corrosion Engineering*, 3rd ed. McGraw-Hill, New York, pp. 12–38.

Fotuhi, P. C., Epstein, A. E., and Ideker, R. E. (1999). Energy levels for defibrillation: what is of real clinical importance? *Am. J. Cardiol.* 83: 240–330.

Fricke, H. (1932). The theory of electrolytic polarization. *Phil. Mag.* 14: 310–318.

Geddes, L. A. (1997). Historical evolution of circuit models for the electrode–electrolyte interface. *Ann. Biomed. Eng.* 25: 1–14.

Geddes, L. A., and Baker, L. E., eds. (1989). Electrodes. in: *Principles of Applied Biomedical Instrumentation*, 3rd ed. Wiley-Interscience, New York, pp. 315–452.

Ghika, J., Villemure, J. G., Fankhauser, H., Favre, J., Assal, G., and Ghika-Schmid, F. (1998). Efficiency and safety of bilateral contemporaneous pallidal stimulation (deep brain stimulation) in levodopa-responsive patients with Parkinson's disease with severe motor fluctuations: a 2-year follow-up review. *J. Neurosurg.* 89(5): 713–718.

Glarum, S. H., and Marshall, J. H. (1980). The A-C response of iridium oxide films. *J. Electrochem. Soc.* 127: 1467–1474.

Glikson, M., von Feldt, L. K., Suman, V. J., and Hayes, D. L. (1995). Clinical surveillance of an active fixation, bipolar, polyurethane insulated pacing lead. Part II: The ventricular lead. *Pacing Clin. Electrophysiol.* 18(2): 374–425.

Gouy, M. (1910). Sur la constitution de la charge electricque a la surface d'un electrolyte. *J. Phys. (Paris)* 9: 457–468.

Graupe, D. (1989). EMG pattern analysis for patient-responsive control of FES in paraplegics for walker-supported walking. *IEEE Trans. Biomed. Eng.* 36(7): 711–719.

Grill, W. M., and Mortimer, J. T. (1994). Electrical properties of implant encapsulation tissue. *Ann. Biomed. Eng.* 22(1): 23–33.

Haissaguerre, M., Cauchemez, B., Marcus, F., Le Metayer, P., Lauribe, P., Poquet, F., Gencel, L., and Clementy, J. (1995). Characteristics of the ventricular insertion sites of accessory pathways with anterograde decremental conduction properties. *Circulation* 91(4): 1077–1085.

Heetderks, W. J., and Hambrecht, F. T. (1988). Applied neural control in the 90s. *Proc. IEEE* 76: 1115–1121.

Hefti, J. L., Epitaux, M., Glauser, D., and Frankhauser, H. (1998). Robotic three-dimensional positioning of electrode in the brain. *Comput. Aided Surg.* 3(1): 1–10.

Heiduschka, P., and Thanos, S. (1998). Implantable bioelectric interfaces for lost nerve functions. *Prog. Neurobiol.* 55(5): 433–461.

Helmholtz, H. (1879). Studien über electrische Grenzchichten. *Ann. Phys. Chem.* 7: 337–382.

Hillsley, R. E., Walker, R. G., Swanson, D. K., Rollins, D. L., Wolf, P. D., Smith, W. M., and Ideker, R. E. (1993). Is the second phase of a biphasic defibrillation waveform the defibrillating phase? *Pacing Clin. Electrophysiol.* 16(7 Pt 1): 1401–1411.

Hoshimiya, N., Naito, A., Yajima, M., and Handa, Y. (1989). A multichannel FES system for the restoration of motor functions in high spinal cord injury patients: a respiration-controlled system for multijoint upper extremity. *IEEE Trans. Biomed. Eng.* 36(7): 754–760.

Ideker, R. E., Chen, P. S., Zhou, X. H. (1990). Basic mechanisms of defibrillation. *J. Electrocardiol.* 23(Suppl): 36–38.

Ideker, R. E., Wolf, P. D., Alferness, C., Krassowska, W., and Smith, W. M. (1991). Current concepts for selecting the location, size and shape of defibrillation electrodes. *Pacing Clin. Electrophysiol.* 14(2 Pt 1): 227–240.

Iskos, D., Fahy, G. J., Lurie, K. G., Sakaguchi, S., Adkisson, W. O., and Benditt, D. G. (1997). Nonpharmacologic treatment of atrial fibrillation: current and evolving strategies. *Chest* 112(4): 1079–1090.

Jais, P., Haissaguerre, M., Shah, D. C., Takahashi, A., Hocini, M., Lavergne, T., Lafitte, S., Le Mouroux, A., Fischer, B., and Clementy, J. (1998). Successful irrigated-tip catheter ablation of atrial flutter resistant to conventional radiofrequency ablation. *Circulation* 98(9): 835–838.

James, M. J. (1994). *mechanisms of Arrhythmias.* Futura Publishing Company Inc., New York.

Katayama, Y., Fukaya, C., and Yamamoto, T. (1998). Poststroke pain control by chronic motor cortex stimulation: neurological characteristics predicting a favorable response. *J. Neurosurg.* 89(4): 585–591.

Kelliher, E. M., and Rose, T. L. (1989). Evaluation of charge injection properties of thin film redox materials for use as neural stimulation electrodes. *MRS Symp. Proc.* 110: 23–27.

Krack, P., Pollak, P., Limousin, P., Hoffmann, D., Benazzouz, A., and Benabid, A. L. (1998). Inhibition of levodopa effects by internal pallidal stimulation. *Movement Disorders* 13(4): 648–652.

Kunze, K. P., Hayen, B., and Geiger, M. (1998). Ambulatory catheter ablation. Indications, results and risks. *Herz* 23(2): 135–140.

Lyons, K. E., Pahwa, R., Busenbark, K. L., Troster, A. I., Wilkinson, S., and Koller, W. C. (1998). Improvements in daily functioning after deep brain stimulation of the thalamus. *Movement Disorders* 13(4): 690–692.

Mastrototaro, J. J., Massoud, H. Z., Pilkington, T. C., and Ideker, R. E. (1992). Rigid and flexible thin-film multielectrode arrays for transmural cardiac recording. *IEEE Trans. Biomed. Eng.* 39(3): 271–279.

McHardy, J., Robblee, L. S., Marston, J. M., and Brummer, S. B. (1980). Electrical stimulation with Pt. electrodes. IV. factors influencing Pt dissolution in inorganic saline. *Biomaterials* 1: 129–134.

Michaiel, J. J. (1993). *Mechanisms of Arrhythmias.* Futura Publishing Company, Armonk, NY.

Ohm, G. S. (1826). Bestimmung des Gesetzes nach welchem Metalle die Contaktelectricität leiten. *Schweiggers J. Chem. Phys.* 46: 137–166.

Onaral, B., and Schwan, H. P. (1982). Linear and nonlinear properties of platinum electrode polarization. *Med. Biol. Eng. Comput.* 20: 299–300.

Ondo, W., Jankovic, J., Schwartz, K., Almaguer, M., Simpson, R. K. (1998). Unilateral thalamic deep brain stimulation for refractory essential tremor and Parkinson's disease tremor. *Neurology* 51(4): 1063–1069.

Polk, C. (1995). Therapeutic applications of low-frequency sinusoidal and pulsed electric and magnetic fields. in *Handbook of Biomedical Engineering*, J. D. Bronzino, ed. CRC Press, Boca Ralon, FL, pp. 1404–1416.

Rainov, N. G., Heidecke, V., and Burkert, W. (1996). Short test-period spinal cord stimulation for failed back surgery syndrome. *Minimally Invasive Neurosurg.* 39(2): 41–44.

Randles, E. B. (1947). Rapid electrode reactions. *Discuss. Faraday Soc.* 1: 11–19.

Robblee, L. S., and Rose, T. L. (1990). Electrochemical guidelines for selection of protocols and electrode materials for neural stimulation. in *Neural Prostheses: Fundamental Studies*, W. F. Agnew and D. B. McCreedy, eds. Prentice Hall, Upper Saddle River, NJ, pp. 25–66.

Robblee, L. S., and Sweeney, J. D. (1996). Bioelectrodes. in *Biomaterials Science*, 1st ed., B. D. Ratner, A. S. Hoffman, F. J. Schoen, and J. E. Lemons (eds.). Academic Press, San Diego, pp. 371–375,

Robblee, L. S., Lefko, J. L., and Brummer, S. B. (1983). An electrode suitable for reverse charge injection in saline. *J. Electrochem. Soc.* 130: 731–733.

Robblee, L. S., McHardy, J., Marston, J. M., and Brummer, S. B. (1980). Electrical stimulation with Pt electrodes. V. The effects of protein on Pt dissolution. *Biomaterials* 1: 135–139.

Rose, T. L., Kelliher, E. M., and Robblee, L. S. (1985). Assessment of capacitor electrodes for intracortical neural stimulation. *J. Electrochem. Soc.* 130: 731–733.

Rosenthal, D. I., Springfield, D. S., Gebhardt, M. C., Rosenberg, A. E., and Mankin, H. J. (1995). Osteoid osteoma: percutaneous radio-frequency ablation. *Radiology* 197(2): 451–454.

Sebag, C., Lavergne, T., Motte, G., and Guize, L. (1997). Radiofrequency ablation of accessory atrioventricular pathways. *Archives des Maladies du Coeur et des Vaisseaux* 90(Spec No 1): 11–17.

Segal, R., Stacey, B. R., Rudy, T. E., Baser, S., and Markham, J. (1998). Spinal cord stimulation revisited. *Neurol. Res.* 20(5): 391–396.

Singer, I., Barold, S. S., and Camm, A. J. (1998). *Nonpharmacological Therapy of Arrhythmias for 21st century: The State of the Art.* Futura Publishing, Armonk, NY.

Sluyters-Rehbach, M., and Sluyters, J. H. (1970). Sine wave methods in the study of electrode processes. *Electroanal. Chem.* 4: 1–121.

Solbiati, L., Goldberg, S. N., Ierace, T., Livraghi, T., Meloni, F., Dellanoce, M., Sironi, S., and Gazelle, G. S. (1997). Hepatic metastases: percutaneous radio-frequency ablation with cooled-tip electrodes. *Radiology* 205(2): 367–373.

Spadaro, J. A. (1997) Mechanical and electrical interactions in bone remodeling. *Bioelectromagnetics* 18: 193–202.

Stancak, B., Pella, J., Resetar, J., Palinsky, M., and Bodnar, J. (1996). Ablation of supraventricular tachydysrhythmias with direct and radiofrequency current. *Vnitrni Lekarstvi* 42(11): 779–783.

Stern, O. (1924). Zur theory der electrolytischen Doppelschicht. *Z. Elektonchem.* 30: 508–516.

Sweeney, J. D. (1992). Skeletal mucsle responses to electrical stimulation. in *Electrical Stimulation and Electropathology*, J. P. Reilly (ed.). Cambridge University Press, Cambridge, UK, pp. 285–327.

Taub E., Munz, M., and Tasker, R. R. (1997). Chronic electrical stimulation of the gasserian ganglion for the relief of pain in a series of 3 and 4 patients. *J. Neurosurg.* 86(2): 197–202.

Volta, A. (1800). On the electricity excited by the mere contact of conducting substances of different kinds. A letter from Alexander

Volta, F.R.S. (Professor of Natural Philosophy, University of Pavia) to Rt. Hon. Sir Joseph Banks, K.B.P.R.S. *Phil. Trans. R. Soc. Lond.* 90: 744–746.

Walcott, G. P., Knisley, S. B., Zhou, X., Newton, J. C., and Ideker, R. E. (1997). On the mechanism of ventricular defibrillation. *Pacing Clin. Electrophysiol.* 20(2 Pt 2): 422–431.

Walcott, G. P., Melnick, S. B., Chapman, F. W., Jones, J. L., Smith, W. M., and Ideker, R. E. (1998). Relative efficacy of monophasic and biphasic waveforms for transthoracic defibrillation after short and long durations of ventricular fibrillation. *Circulation* 98(20): 2210–2215.

Walcott, G. P., Walker, R. G., Cates, A. W., Krassowska, W., Smith, W. M., and Ideker, R. E. (1995). Choosing the optimal monophasic and biphasic waveforms for ventricular defibrillation. *J. Cardiovasc. Electrophysiol.* 6(9): 737–750.

Warburg, E. (1899). Über das verhalten sogenanter unpolarsbarer Elektroden gegen Wechselstrons. *Ann. Phys. Chim.* 67: 493–499.

Warburg, E. (1901). Über die Polarizationscapacitat des Platins. *Ann. Phys.* 6: 125–135.

White, R. L., and Gross, T. J. (1974). An evaluation of the resistance to electrolysis of metals for use in biostimulation probes. *IEEE Trans. Biomed. Eng.* BME-21: 487–490.

Wise, K. D., and Najafi, K. (1991). Microfabrication techniques for integrated sensors and microsystems. *Science* 254: 1335–1342.

Zhuang, H., Wang, W., Seldes, R., Tahernia, A., Fan, H., and Brighton, C. (1997). Electrical stimulation induces the level of TGF-beta 1 mRNA in osteoblastic cells by a mechanism involving calcium/calmodulin pathway. *Biochim. Biophys. Res. Commun.* 237: 225–229.

7.16 COCHLEAR PROSTHESES

Francis A. Spelman

INTRODUCTION

Cochlear prostheses help the deaf to contact the auditory environment (NIH, 1995). The problem of sensorineural deafness is immense. The probable number of users of cochlear prostheses has been estimated by some as 900,000 in the United States alone (Levitt and Nye, 1980). Other estimates range as high as 2,000,000 sensorineural deaf in the United States. The treatment for sensorineural deaf patients is the cochlear implant (NIH, 1995). Sensorineural deaf subjects cannot perceive sound without extraordinary aid. That is so because the basic transduction system is lost either as a result of damage or destruction of the cochlea or because the auditory nerve is damaged (Levitt and Nye, 1980).

This chapter introduces cochlear prostheses and some of the materials problems that must be faced by their designers. The concentration of the chapter is on cochlear implants rather than on implants in the brainstem. Approximately 59,000 people use cochlear implants, whereas fewer than 100 have brainstem implants (NIDCD, 2003; Brackmann *et al.*, 1993). At present, cochlear implants are in clinical use, while brainstem implants are still experimental. There is no discussion of tactile prostheses for the deaf (Martin, 1985). This chapter introduces the physiology of the auditory system and describes cochlear prostheses and some of the issues related to biomaterials that bioengineers face in the design of such apparatus.

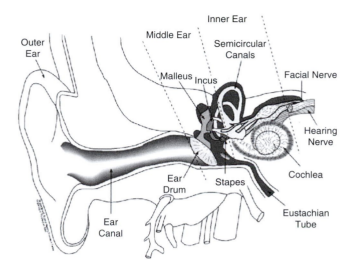

FIG. 1. Sketch of the auditory periphery, showing the outer, middle, and inner ear. Reproduced with permission from the V.M. Bloedel Hearing Research Center at the University of Washington.

OVERVIEW OF THE AUDITORY SYSTEM

The auditory system can be divided into its peripheral organs and the nuclei in the central nervous system that process the signals produced by the peripheral organs. This is a greatly abbreviated presentation. More details are found in texts like those of Geisler (1998) and Dallos *et al.* (1996). Geisler refers to several Web sites that demonstrate the behavior of the periphery, e.g., http://www.neurophys.wisc.edu/animations and the animations of the Association for Research in Otolaryngology, http://www.aro.org. The sketch of Fig. 1 shows the auditory periphery.

The Periphery

The auditory periphery consists of the outer ear, middle ear, and inner ear. The outer ear, the *pinna*, collects changes in pressure (condensations and rarefactions) produced by the auditory signals. Those signals are generated by sound sources in the environment of the listener. Acoustic signals are guided to the middle ear along the ear canal, an entry into the head of the subject that is lined by soft tissue. The ear canal is open at its peripheral end and bounded by the eardrum, the *tympanic membrane*, at its inner end. The length of the ear canal is about 3 cm in the human (Geisler, 1998).

The middle ear is bounded distally by the tympanic membrane and proximally by the *cochlea*, where the foot plate of the *stapes* contacts the oval window of the cochlea.[1] In the inner ear are three tiny bones that comprise the *ossicular chain*: the *malleus* (hammer), *incus* (anvil), and *stapes* (stirrup). The bones have flexible connections. They provide a mechanical advantage so that the eardrum can be driven by air and, in turn, can drive the dense fluids that are found in the cochlea. The motion of the foot plate of the stapes is about 75% of

---

[1] The terms *proximal* and *distal* refer to the brain in this case. Proximal is nearer to the brain and distal is further from the brain.

that of the tympanic membrane in the human (Geisler, 1998). Further mechanical advantage is provided by the relative areas of the tympanic membrane and the foot plate of the stapes: the tympanic membrane has about 20 times the area of that of the stapes (Geisler, 1998). Small pressure changes in air are transformed into larger changes in the fluids of the cochlea. Conversely, large displacements of the eardrum produce small displacements of the footplate of the stapes.

The *cochlea* is a snail-shaped organ that is located bilaterally in the temporal bones of the head. The cochlea is oriented such that its wide base faces in a medial and posterior direction, while the axis of its spiral points laterally and anteriorly (Fig. 1).

The cochlea contains three spiraling chambers or *scalae*: the *scala tympani*, *scala vestibuli*, and *scala media*. The *organ of Corti* lies within the *scala media* on the *basilar membrane*. The *hair cells* are located on the basilar membrane (Geisler, 1998; Dallos *et al.*, 1996). The basilar membrane is an elegant structure that acts as a mechanical Fourier analyzer: specific regions of the membrane vibrate maximally in response to the frequency of the sound waves that are imposed on the stapes. The membrane displacements produce maxima for high frequencies at the basal end and for low frequencies at the apical end of the cochlea. The hair cells residing on the membrane have *cilia* that are bent when the membrane vibrates. The hair cells synapse with the peripheral processes of the auditory nerve, the *hearing nerve* in Fig. 1. There are 25–30,000 afferent neurons that synapse with the hair cells (Geisler, 1998). The organization of the auditory nerve is by frequency. Indeed, the entire auditory system is organized by frequency (tonotopically) (Geisler, 1998; Popper and Fay, 1991).

Critical to the design of the cochlear prosthesis is the location of the peripheral processes of the auditory neurons, the cells of the auditory nerve.[2] The neurons are bipolar cells, and their peripheral processes are found in and under the *bony spiral lamina* (Geisler, 1998; Popper and Fay, 1991; Spelman and Voie, 1996), an osseous or bony structure that extends from the *modiolar* (medial) wall of the scala tympani in the cochlea. The cell bodies of the VIII nerve are located in Rosenthal's canal, a hollow structure in the modiolar bone.

The anatomy of the scala tympani has led to the design of the cochlear implant. The prosthesis is designed to stimulate the auditory neurons electrically. Placing the sites of electrodes near the neurons without violating the bony wall of the modiolus means that the electrode arrays of the implants are located in the scala tympani (see later discussion).

Highlights of the Central Auditory System

The tonotopic structure of the auditory system is found throughout the auditory system. The frequency-dependent structure of the auditory signal forms the responses of single neurons in the auditory nerve (Sachs and Young, 1979). As signals are produced binaurally, those signals travel from each cochlea to the *cochlear nucleus*, where the neurons send data to the *contralateral trapezoid body*, (Sachs and Young, 1979), the *inferior colliculus*, the *medial geniculate nucleus*, and the

auditory cortex (Rubel and Dobie, 1989). The properties of the auditory nerve are well understood and have been for more than a decade. The properties of the cochlear nucleus are under active investigation, as are those of the higher centers of the auditory system.

This chapter focuses on the auditory periphery. It is in the periphery where cochlear prostheses are most effective, although auditory prostheses have been used in the cochlear nucleus, notably for patients who suffer from damage to the VIII nerve (Shannon *et al.*, 1993; McCreery *et al.*, 1997).

Damage to the Periphery

Sensorineural deafness caused peripherally can result from serious damage to the hair cells or to the auditory nerve. Clearly, if the neurons of the hearing nerve are damaged, their peripheral processes cannot be driven and stimulation from sites in the cochlea will not work. In those cases, central prostheses have been used experimentally (Shannon *et al.*, 1993; McCreery *et al.*, 1997).

Damage to the hair cells can result from a number of causes. Pyman *et al.* cite 11 root causes in people over 6 years of age, and seven causes in people under 6 years of age (Pyman *et al.*, 1990). Their population was 65 people in the former case and 29 in the latter. Large numbers of subjects had unknown causes of deafness, but there were cases of meningitis, otosclerosis, and trauma that caused the problems (Pyman *et al.*, 1990). In another study, Hinojosa and Marion analyzed 65 ears and found six causes of congenital deafness in 19 subjects, and nine causes of acquired deafness in 46 subjects. In the latter population, otosclerosis caused the greatest damage, followed closely by bacterial infections (Hinojosa and Marion, 1983).

Damage to the hair cells from loud sounds requires special mention, since the popularity of painful audio systems in automobiles and as portable sources of entertainment appears to be increasing. Hair cells can be damaged by intense sounds (Popper and Fay, 1991); chronic exposure to loud sounds should be avoided despite the relatively small numbers cited by Hinojosa and Pyman (Pyman *et al.*, 1990; Hinojosa and Marion, 1983).

Neural Plasticity

The auditory system responds to stimulation, both anatomically and physiologically. The central nervous system can reorganize itself in response to auditory signals (Brugge, 1991). The plasticity of the system occurs more vigorously in children than in adults. That understanding has changed the application of cochlear prostheses from a focus only on adults to a large distribution of instruments for children as well (Clark, 1996). Deaf children are treated at 2 years of age and in some cases 12 months (Skinner, 2001; Osberger, 1997).

COCHLEAR PROSTHESES

Cochlear prostheses present one of the remarkable success stories of biomedical engineering. The idea of electrical

---

[2]The auditory nerve is also known as the VIII cranial nerve.

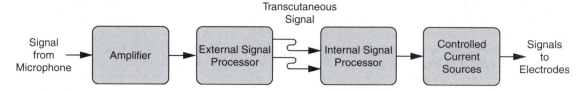

FIG. 2. Block diagram of a generic cochlear prosthesis. (After Spelman, 1999.)

stimulation of the peripheral auditory system is credited to Volta (1806) and cited in Simmons (1996). More modern approaches to solve the problem of deafness with a neural prosthesis were offered in the latter part of the 20th century (Simmons, 1996; Djourno and Eyries, 1957; House and Berliner, 1991).

Architecture of a Cochlear Prosthesis

The architecture of a cochlear prosthesis is shown in Fig. 2. The architecture shown here follows the textual architecture described on the NIDCD Web site. Here the receiver/stimulator described on that site is divided into a signal processor and controlled current sources (NIDCD, 2003). A microphone is the transducer that converts the auditory signal into an electrical signal. The microphone's signal is sent to an external signal processor. That signal processor decomposes the electrical information into amplitude and frequency data: the signal is filtered and analyzed to produce data about the envelope of the information within a particular frequency band (Loizou, 1999). Other processing techniques have been proposed, but have not been produced commercially (Clopton et al., 2002). Several bands are analyzed simultaneously to develop a vocoder model of the audio signal (Gold and Reder, 1967). The data are transferred across the skin as digital signals. An internal processor takes those data and converts them to current drive signals for the electrodes of the multichannel electrode array. In some implementations, single current sources are used and switched between contacts, while in others multiple sources can be driven simultaneously (Loizou, 1999). The currents that are sent to the electrodes can be either analog or pulsatile signals.

In the past 20 years, cochlear prostheses have become multichannel systems. The first implants employed single electrodes, whereas present devices use up to 22 contacts inside the scala tympani (Spelman, 1999; Loizou, 1999; House and Berliner, 1991). In most processing strategies, one electrode is driven at a time, although analog drives present currents to all electrodes simultaneously (Osberger and Fisher, 1999).

Commercially Available Systems

The following sections are brief descriptions of the strategies that are used by the three manufacturers of cochlear prostheses that are presently available to deaf patients. All of the manufacturers have Web sites that provide additional information.

The Nucleus Implant

Cochlear Pty., Limited, is the largest producer of cochlear prostheses in the world. For further information on the Nucleus prosthesis, visit www.cochlear.com. The Nucleus has an electrode array that uses 22 or 24 electrodes. Twenty-two electrodes are placed in the scala tympani. The 24-electrode array uses two external electrodes that can be used as distant return electrodes. They permit true monopolar excitation of the internal electrodes. The Nucleus Contour electrode array is curved in order to approximate the modiolar wall of the scala tympani, bringing the electrodes of the array into apposition with the cells of the auditory nerve. The external processor extracts frequency information from 20 filters, sampling the auditory signal to determine the frequency bands containing the maximum energy during a given 3.3-msec interval. The envelopes of the signals taken from the filters whose energy is greatest are sampled. The data are distributed to electrodes within the array, driving between five and 10 electrodes within a given sequence. One electrode is driven at a time. The electrodes are driven with biphasic rectangular pulses that repeat rapidly. The auditory signal's data are updated at rates that may be as high as 300 pulses per second (Vandali et al., 2000).

The silicone rubber electrode array, when straightened, approximates a truncated cone whose diameter varies from 0.4 mm at the apical end to 0.6 mm at the basal end. The array is composed of segments of platinum rings, designed to face the peripheral processes of the auditory nerve. Each ring is approximately 0.35 mm in width, and the edge-to-edge separation is about 0.4 mm. The insulated platinum/iridium wires that connect to each electrode are led back through the silicone rubber carrier of the electrodes (Patrick et al., 1990). During insertion, the array is straightened with an internal polymeric stiffener. The array is placed into the scala tympani via a small hole that is drilled through the temporal bone. The stiffener is withdrawn while the array is gently inserted by the surgeon. In the past, the implanted electrode array resided near the lateral wall of the scala tympani when it was implanted in the ear of a human subject (Skinner et al., 1994). The new array more nearly apposes the modiolar wall, although in two of 12 insertions into temporal bones the array pierced the basilar membrane (Tykocinski et al., 2001). Another study in human temporal bones showed that damage appeared to be a function of the technique of the implant surgeon during insertion (Rebscher et al., 2001).

The Clarion

The Clarion cochlear implant is the offering of the Advanced Bionics Corporation (Sylmar, CA). For the most recent information about the Clarion implant, visit www.cochlearimplant.com.

The Clarion Multi-Strategy device offers several processing strategies to deliver signals to patients (Kessler, 1999). Its processor can operate with several different processing strategies. The strategies include Simultaneous Analog Stimulation (SAS), the Paired Pulsatile Sampler (PPS), and the Continuous Interleaved Sampler (CIS). The microphone and amplifier drive an analog/digital converter signal-compressing software, most often logarithmic compression. The compressed signal is decomposed into a set of digitally filtered signals. The digitized and filtered outputs are transmitted across the skin with an RF link to an internal processor, employing a 49-MHz AM signal (Kessler, 1999). The data are demodulated and demultiplexed internally and delivered to current sources that drive electrodes directly with the compressed and filtered analog signals (Kessler, 1999). The sampling rate is 13,000 samples per second per channel, with an aggregate sampling rate of 104,000 samples per second. The SAS system operates similarly to the Compressed Analog (CA) systems that have been used earlier (Spelman, 1999; Loizou, 1999; Kessler, 1999). The SAS system drives several electrodes simultaneously. It still suffers from the interference that occurs as a result of field interactions among electrodes.

The Clarion processor provides other strategies as well. Continuous Interleaved Stimulation (CIS) was developed by Wilson and his colleagues to overcome the field interactions between channels in the Ineraid implant (Wilson *et al.*, 1991). Biphasic pulses are delivered one pair at a time to the electrodes of the implanted array in the cochlea. The pulses are interleaved (multiplexed) in time to eliminate interactions among the electric fields in the cochlea. The rates of excitation can be varied to optimize the signals delivered to specific patients. The acoustic signal is amplified, compressed, and band-pass filtered. The outputs of the filtered signals are rectified and low-pass filtered to obtain envelope information for each of the band-pass filters. The amplitudes of the pulse pairs are varied in proportion to the magnitudes of the envelopes of the signals in specific frequency bands at the time of analysis. The widths of the pulses delivered are constant. The Clarion processor updates the signals 833 times per second (Loizou, 1999). Eight electrodes are usually driven, although eight pairs of electrodes are provided via the Clarion electrode array.

The processor developed by Advanced Bionics allows the use of paired pulsatile stimulation (PPS) (Kessler, 1999). Paired Pulsatile Simulation (PPS) drives two electrodes simultaneously with paired biphasic pulses while maintaining the maximum physical separation between the driven electrodes. Physical separation limits field interactions between electrodes (Kessler, 1999; Zimmerman-Philips and Murad, 1999).

Osberger and Fisher (1999) reported on 71 patients who used the Clarion Multi-Strategy. The subjects improved their hearing performance when they were used a preferred mode of stimulation. The patients who preferred the SAS mode had been deaf for shorter times than those who preferred CIS (Osberger and Fisher, 1999).

Advanced Bionics has introduced a method to appose the electrodes of their array to the cells of the VIII cranial nerve in Rosenthal's canal, which lies behind the modiolar wall of the scala tympani. They produced a preshaped electrode that was used jointly with a polymeric Electrode Positioning System (EPS). The EPS applied pressure to the lateral wall of the cochlea, positioning the electrode array against the modiolus. The system was used for 3 years and then recalled in the summer of 2002. For further information, see the Web site of the U.S. FDA. The present offering of Advanced Bionics does not employ the EPS.

The Med-El Implant

The Combi 40 and the Combi 40+ implants are offered by Med-El Corporation (Innsbruck, Austria, www.medel.com). The Med-El electrode array is slim, having 24 contacts. Med-El provides data that indicate that their array can be placed within the entire length of the human scala tympani, 31 mm.

Med-El offers Continuous Interleaved Stimulation, and at this time they offer a high multiplexing rate of 18,000 pulses per second, extracting data with a Hilbert transform approach (Anonymous, 2003). The data can be updated at a rate of 1500 samples per second when 12 contacts are used, as they are in the Combi 40. Med-El employs another approach. They choose the frequencies in the auditory spectrum containing maximum energy, rectify and filter the signals, and then send amplitude-modulated pulses to the electrodes that correspond to the frequencies of maximum energy. The approach has been called an "n of m" strategy (www.cochlearimplants.com; www.medel.com) (Anonymous, 2003).

MATERIALS AND ELECTRODE ARRAYS

Materials issues are salient in the design and construction of electrode arrays. The polymers that are used to insulate the arrays must be compatible with the tissues of the scala tympani; the electrode sites must use metals that can deliver appropriate currents to excite neurons; and the mechanical characteristics of the arrays must allow safe and easy insertion into a volume that is both small and complex in shape. Rebscher and co-workers (1999) describe techniques and tests designed to ensure safe insertion.

Although other parts of the prosthesis are of concern, the concerns are small compared to those that surround the electrode arrays. In this chapter, I address issues of electrode design and choice of materials, the electrical and mechanical properties of insulating materials, and some of the issues that are related to the tissue sheaths that surround electrode arrays in the scala tympani of the cochlea.

Electrode Arrays

Electrode arrays change electrical currents into the ionic currents that can stimulate neurons. The number of electrode sites and the strategy used to drive the contacts determines the electric fields that are generated in the scala tympani, and, ultimately, the number of independent channels that can be driven simultaneously. Several people have suggested ways in which electrical currents could be combined to focus the neural stimuli. Suesserman and co-workers suggested an approach that was supported by a lumped-element, electrical model of the inner ear (Suesserman and Spelman, 1993). Later, Jolly and co-workers performed experiments to demonstrate that the predicted fields could be produced in the inner ear of the guinea pig (Jolly *et al.*, 1996) and that it might be possible to stimulate independent groups of neurons (Jolly *et al.*, 1997). Later, Middlebrooks and Bierer (2002) and Bierer and Middlebrooks (2002) showed clearly that focused fields produced focused excitation of neurons in the auditory system.

Contacts and Focusing

Simple models of point sources of current can be used to illustrate some of the effects of driving tissue with point sources of electric current. Consider a conductive medium that is semi-infinite in extent and bounded by a perfectly insulating surface. The current source lies on the insulating boundary at location x_0, y_0, z_0. The dimensions are given in meters. The method of images (Kong, 1986) can be used to show that the potential field produced by the source is:

$$V(x, y, z) = \frac{\rho I}{2\pi} \frac{1}{\left[(x - x_0)^2 + (y - y_0)^2 + (z - z_0)^2\right]^{0.5}}$$

where ρ is the resistivity of the medium, given in Ohm-m; I is the magnitude of current, given in amperes; and $V(x,y,z)$ is the electrical potential at location x, y, z, given in volts.

The preceding equation ignores the properties of the electrode, e.g., its polarization impedance, since they are overlooked entirely by the point source approximation (Macdonald, 1987).

Point sources cannot be produced. However, the equation just given describes a hemispherical source of radius r placed on the insulating surface with the source's center at the location of the point source, x_0, y_0, z_0. As long as the potential is computed for radial distances that are greater than r, the equation is valid and approximates the electrode in the absence of electrochemical effects.

A special case of the potential is illustrative. Let the location of the source be at the origin of a Cartesian coordinate system. Compute the field produced along the z-axis. Let the radius of the hemisphere be 50 μm. Consider a material, such as perilymph, the fluid that fills the scala tympani, whose resistivity is 0.63 ohm-m. Then,

$$V(z) = 0.1 \frac{I}{z} \text{ volts}, \quad z > 50 \text{ μm}$$

Let the current be 100 μA, a reasonable threshold for a 200-μsec pulse in an experimental animal. The potential decreases monotonically from 200 mV at the surface to 50 mV at a distance of 200 μm.

Problem 3 raises additional issues that can be addressed with this simple model: with a straightforward model of this sort, the engineer can gain a quick understanding of the benefits and limitations of using multiple sources to focus the electric fields in the cochlea. Combinations of positive and negative currents can change the widths of fields, but narrowing the fields will reduce peak potentials.

Thinking about dipole fields leads to that conclusion. A theoretical dipole consists of a pair of positive and negative point sources of current, placed at an infinitesmal distance from each other. The potential field decreases rapidly for distances that are much larger than the separation between the electrodes. The point of the thought experiment is to suggest that the smaller the distance between sources of opposite polarities, the smaller will be their peak potentials at a distance. Hence, sources and sinks of electric current must be used judiciously to focus potential fields.

The point source model of an electrode is suitable for spherical electrodes if their fields are modeled for radial distances greater than the radial dimension of the sphere. The same model works for a hemispherical electrode that is attached to a planar insulator. The electrodes that are used in cochlear prostheses are often more like finite, planar surfaces, and their solutions are not straightforward except in the simplest cases (Rubinstein *et al.*, 1987; Pearson, 1990). The potential field that is generated by a circular electrode has been understood since the time of Weber (Rubinstein *et al.*, 1987) and was described by Wiley and Webster more than 20 years ago (Wiley and Webster, 1982). The critical finding for the latter work is that the current density of such an electrode is nonuniform over its surface and becomes singular at the boundaries of the electrode. Current density is singular at the edges of many kinds of planar electrodes as was shown in an elegant demonstration by Rubinstein (1988). Recessing the electrode can eliminate the singularities of current density at its metallic surface, whereas singularities are found at the aperture of the recessed electrode (Rubinstein *et al.*, 1987). However, shaping the aperture can eliminate those singularities (Suesserman *et al.*, 1991).

Consideration of surface-mounted and recessed electrodes of finite size leads to an understanding of the potential for corrosion of such electrodes. Corrosion will be greater at the edges where the current and charge densities are high. Further, thinking about the boundary condition at the surface of a metallic electrode, that the potential is constant on the surface, suggests that the field produced by an electrode of finite size will differ from that produced by a point source. Figure 3 illustrates some of the points just made. The figure is a simple finite-element model of a strip line on the surface of an insulator. The strip line is 10% of the width of the insulator. The upper surface of the line and its insulating carrier bound a conductive space. The resistivity of the insulator is about 160 times that of the conductive space. The strip is a superconductor, and the bounding surfaces are held at zero potential. The potential across the strip is uniform. The arrows in the figure represent current density; the black contour lines are contours of equipotentials and the colors represent current densities. Note that the current densities are highest at the edges of the strip. Red color corresponds to high current density, while the white color shows current density that is nearly zero.

Finite electrodes produce a second effect, not seen for point sources. As the observer moves closer to the source, the field approaches a constant value over the surface of the source. That is a result of the boundary condition requiring that superconductors have uniform potentials on their surfaces. Of course, the potential decreases as the observer moves laterally away from the source. Figure 4 illustrates the condition. Two curves

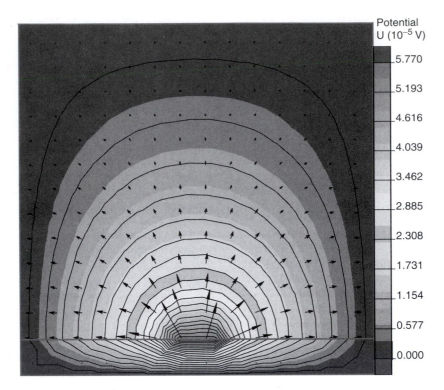

FIG. 3. The fields produced by current (0.1 mA) driven through a superconducting strip that is mounted on an insulator that faces a conductive, homogenous, isotropic medium. Done with the student version of QuickField.

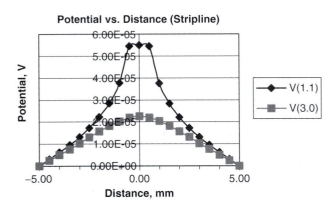

FIG. 4. Potential produced by the stripline of Fig. 3 in a conducting medium. The stripline lies on an insulator whose resistivity is 160 times that of the conducting medium and produces a current of 0.1 mA.

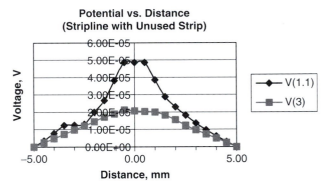

FIG. 5. Potential fields measured 0.1 mm and 3 mm above two parallel superconducting striplines, one of which measures current and one of which is passive.

of potential are shown, one calculated for a distance 0.1 mm above the plain containing the source, and the second calculated for a distance 3 mm above the source. The nearer contour shows clearly the uniform potential found when measurements are made near the source.

Another difference between point sources and finite sources is that unused point sources have no effect on the potential fields that are near them, whereas finite sources must sustain constant potentials at their surfaces. Peters did an analysis of that situation that showed clearly the effects that finite sources have on fields (Peters, 2000). The effect is shown in Fig. 5.

The stripline of Figs. 3 and 4 is driven with 0.1 mA. A second superconducting line lies near the source. When potential is plotted 0.1 mm above the 1-mm-wide line, the effect of the second conductor is obvious as a result of the plateau of potential that is computed over the second line. The effect is less obvious when observations are made 3 mm above the lines. Both curves show that the peak potentials decrease after the introduction of the second, passive line.

Contact Materials

What materials are suitable to make the electrode arrays that are used in cochlear implants? Noble metals, that is, gold, platinum, iridium, and some of their alloys, have been used extensively (Robblee and Rose, 1990). The arrays that are available clinically employ alloys of platinum and iridium, usually 90% Pt and 10% Ir. Those alloys have been studied and used for decades (Robblee and Rose, 1990; Brummer and Turner, 1977a; McCreery et al., 1992). The platinum provides a material that can carry charge efficiently, while the iridium provides structural strength.

More recently, the oxides of iridium have been investigated and have demonstrated characteristics that are markedly superior to those of platinum (Robblee and Rose, 1990; Meyer et al., 2001; Weiland et al., 2003). Electrode materials are studied with standard electrochemical techniques, notably cyclic voltammetry (Robblee and Rose, 1990) and electrochemical impedance spectroscopy (EIS) (Macdonald, 1987). The former technique provides direct evidence that shows the amount of charge density that a given material can support. EIS gives indirect evidence of the charge-carrying capacity, but provides direct information about the voltage that is required to drive a specific current.

Charge density is a critical variable to consider in the design of a cochlear electrode array. The charge density calculation is usually based on a "geometric" surface area, which assumes that the current density is uniform across the metallic surface. Although that assumption is not valid in most cases, there are few calculations of geometries that result in mixed boundary problems. For the purpose of this chapter, we will assume that geometric areas suffice, but warn the reader to be wary when calculating surface areas. That said, the measurement of the charge capacity of a particular electrode is certainly valid when it is done appropriately (Meyer et al., 2001). The charge density calculation may be in error unless great care is taken to ensure that the electrode's surface is smooth and that the current density profile is accounted for (Robblee and Rose, 1990; Meyer et al., 2001; Weiland et al., 2003).

Concern for the charge carrying capacity of electrode sites is critical for the design of cochlear electrode arrays, since it bears directly on their safety and longevity. Platinum–iridium arrays have followed the work of Brummer and Turner (1977a, b), usually applying a safety factor of at least 2. The charge densities are held below 150 $\mu C/cm^2$. Increasing the surface area of a platinum electrode with platinum black can increase the charge density by a factor of 30 (Jolly et al., 1996), but the approach has not been used outside of the laboratory.

Oxides of iridium, both activated and plated, offer real promise for cochlear electrode arrays. The charge densities measured range from 4 mC /cm^2 to 27 mC/cm^2 (Meyer et al., 2001; Weiland et al., 2003). Those charge densities are not used consistently, but a safety factor is applied for long-term operation. The charge density can be reduced by as much as a factor of 20 for long-term tests (Meyer et al., 2001). Even so, the promise of IrOx is great: the potential increase in charge density, and thus in maximum stimulus current is nearly a factor of 10. If Pt–Ir can tolerate stimuli whose charges lie below 150 $\mu C/cm^2$ and IrOx can withstand a long-term stimulus of 1.2 mC/cm^2. It is easy to compute maximum currents if we assume that we can deal with geometric surface areas. If two electrodes whose areas are 10^{-4} cm^2 are used, and they are driven with square pulses whose durations are 200 μsec, the maximum safe current is 75 μA for the Pt–Ir electrode and 600 μA for the IrOx electrode. The IrOx electrode can excite the auditory nerve over its full dynamic range, while the Pt–Ir electrode can barely meet a threshold value (Vollmer et al., 2001).

Another concern for cochlear implant designers is the geometry of their electrode arrays. The developers of thin-film arrays must be concerned with the traces that are used on the thin films. As the arrays get smaller, the traces must decrease in size as well. Although implant designers may think that the currents that they use are small (they are!), the traces are small as well. Temperature increase in the traces may well be a problem as the traces decrease in size (Brooks, 1998). UltraCAD Design, Inc. offers freeware to solve PCB trace problems with a regression relation (www.UltraCAD.com). For example, a copper trace that is 7.5 cm long, 12.5 μm wide, and 1 μm thick will have a temperature increase of nearly 1°C when it carries 800 μA. If several traces carry similar currents, temperature in the cochlea could increase artificially, possibly causing thermal damage to the tissue.

Material Properties

The choice of materials for cochlear prostheses extends beyond the selection of materials that will be used on the surfaces of electrode contacts. At issue are the compatibility of arrays with the tissues of the cochlea, and compatibility of the materials that cover the internal processors that are placed in the temporal bones of the recipients of implants. In this chapter, I do not consider the latter, since the ceramic and titanium cases and their silicone cases have not created measurable problems at the time of this writing, at least since some early problems with infection were taken care of in the late 1970s and early 1980s. The materials used as carriers for the electrode arrays have presented issues that will be discussed hereafter.

As was mentioned earlier, Advanced Bionics recalled its HiFocus electrode array. Some attributed that recall to potential concerns with the electrode positioning system (www.leifcabraser.com/cochlear.htm), a polymeric system that placed the array in contact with the modiolar wall.

Placement issues had been raised before about other implants (Skinner et al., 1994; Rebscher et al., 2001). Spiral tomographs of the implanted ear showed that the electrode arrays entered the scala vestibuli by way of the basilar membrane. The composition of the fluids in the scala tympani and the scala vestibuli are dramatically different. The perilymph in the scala tympani is rich in sodium ion, while the endolymph in the scala vestibuli is rich in potassium ion (Dallos et al., 1996). Mixing the two damages and can kill hair cells. Although that is not critical in the case of the sensorineural deaf patient, the electrode arrays are distant from the peripheral processes of the auditory neurons and the leakage of potassium-rich endolymph will damage those neurons as well.

In a study of the electrode arrays produced by Cochlear Corporation and Advanced Bionics corporation, four surgeons implanted both new and old designs of arrays into cadaveric human temporal bones. The investigators assessed the amount of trauma observed, the insertion depth, and proximity to the modiolar wall. The arrays showed no significant difference in trauma produced by the arrays of the two corporations; the greatest difference was found among the surgeons who did the implants (Rebscher et al., 2001). The Spiral could be inserted further than any of the other designs, and of the new designs, the Contour had the greatest insertion depth. Both the Contour and the HiFocus arrays apposed the modiolar wall (Rebscher et al., 2001).

The preceding paragraph raises the issue of insertion depth. The advantage of a cochlear electrode array that reaches the more apical turns of the cochlea is that it can reach the tonotopic locations that decode frequencies below 1 kHz. For example, insertion of 70% of the cochlear spiral reaches the 300-Hz region of the basilar membrane (Geisler, 1998). Since the range of frequencies necessary to decode speech is 300–3000 Hz (Levitt and Nye, 1980), it is desirable to reach the low-frequency regions of the cochlea with cochlear implants.

That problem and the need to place electrode contacts near to the modiolar wall of the cochlea has led designers to build arrays with mechanical characteristics that make the implants conform to the anatomy of the cochlea. Rebscher and his colleagues (1999) suggested a design that elegantly made use of the mechanical properties of the wires that interconnected the electrode sites and the drive electronics. They built an array that arranged the wires to form a central, vertical beam. The beam bent easily to conform to the spiral shape of the cochlea, and the pitch of the spiraling array could be controlled by the geometry of the beam in a basal–apical direction. Advanced Bionics has employed the design in two of their arrays. *In vitro* measures of the stiffness of Cochlear and Advanced Bionics arrays showed a clear anisotropy for the latter, but not for the former (Rebscher et al., 2001). Another approach suggested the use of shape-memory wire, e.g., Nitinol, within the array to match the cochlear spiral. The wire was shaped to fit the modiolar wall (Spelman et al., 1998), straightened for insertion into the cochlear electrode array, and electrically heated beyond its transition temperature after the array was inserted into the scala tympani. The system has not been commercialized.

The stiffness of a cochlear electrode array is important for ease of insertion, placement near the target neurons, and, possibly, for insertion trauma. The arrays are mechanical beams, and their mechanical properties can be investigated with classical methods in some cases, at least to obtain information about the relative stiffness of arrays that employ different materials (Boyd, 1935; Enderle et al., 2000). For a cantilevered beam of uniform cross section and homogeneous material, the maximum deflection that occurs when the beam is loaded at its free end is

$$y_{max} = \frac{Pl^3}{3EI}$$

where P is the load in newtons, l is the length in meters, E is the modulus of elasticity in pascals (N/m^2), and I is the cross-sectional area moment of inertia, m^4.

The preceding equation is a gross oversimplification of a complex problem. The assumptions that underlie the relationship are that the deflection produces small angles, that the cross section is uniform, and that the material is linear

and homogeneous. Cochlear electrode arrays are inhomogeneous: they contain both metallic conductors and polymeric substrates; they may be laminated. Cochlear electrode arrays are tapered. The number of conductors inside of them varies with length. The angles of bending can be large, particularly in the apical turns of the cochlea. The arrays are likely to be anisotropic. Thus, for a complete analysis, more sophisticated mathematics is necessary and may not be amenable to closed-form mathematical solutions. Numerical techniques, e.g., finite-element analysis, are likely to be required for better understanding.

Measurements

Cochlear prostheses are sophisticated instruments that require thorough measurement and analysis. Since they are Class III devices from the point of view of the FDA, they must be carefully tested and the tests documented before they are used.

Each component of a cochlear implant is tested before the assembly can be applied to a human subject. For example, the processing section of the prosthesis must be tested to ensure that it can parse the auditory signal and reassemble the decomposed aggregate into a new auditory signal that can be understood readily by hearing listeners. Although that is not a definitive test for its ultimate users, the deaf, it gives confidence that the processor is not introducing anomalous information. Further, such tests indicate the data rates that must be transmitted across the skin to an internal processor.

The data-transfer link must be tested to ensure that it can sustain the necessary data rates across an appropriate thickness of skin (about 1 cm in the adult human). Animal tests can provide confidence in this case.

The internal processor must be tested to learn whether it can select electrode sites unambiguously and reliably. The current drivers are tested to discover their linearity, repeatability, and voltage range over the full life of the battery. As batteries sag during use, does the transfer function of the current driver remain constant? Is the voltage range sufficient to drive the full dynamic range of every electrode?

The electrode array is tested carefully from the design through the manufacturing process. The tests include long-term soaking, electrochemical tests (Parker et al., 2000), field tests, and neurophysiological tests (Jolly et al., 1996, 1997). As an electrode design is introduced it must be placed in physiological saline solutions for several weeks to discover whether the insulating carriers promote leakage between the conductors of the arrays.

Electrochemical tests include tests of open-circuit potential, whose stability may indicate shorting across traces (Parker et al., 2001), as well as electrochemical impedance spectroscopy, which can also indicate open-circuited and short-circuited electrode sites as well as poorly plated electrode sites (Macdonald, 1987; Parker et al., 2001). (Note that Parker et al. is an abstract. A full manuscript can be downloaded from http://www.eng.monash.edu.au/ieee/ieeebio1999/p41.htm.) If more complete information about the characteristics of the electrode sites is needed, cyclic voltammetry is in order (Goodisman, 1987; Cogan, 2002).

Measurement of the electric fields produced by electrodes is useful during the initial design phase. It shows the characteristics of the potential fields that are produced by the excitation of single or multiple electrodes (Jolly *et al.*, 1996; Suesserman *et al.*, 1991). It also can show which electrodes have parasitic conductive paths that lie between them (Prochazka and Spelman, unpublished results).

When bench testing and *in vitro* testing are complete, when a cochlear prosthesis is ready for clinical trials, behavioral testing begins. That testing ranges from phsychophysical tests (Vandali *et al.*, 2000; Shannon, 1992; Pfingst *et al.*, 1997), to tests of monosyllabic words and simple sentences (Skinner, 2001; Osberger and Fisher, 1999), to investigations of the understanding of contextual information (Vandali *et al.*, 2000). Behavioral testing is the true "gold standard," since it determines the success or failure of the cochlear prosthesis.

Focusing Fields and the Interaction with Tissue

The cochlear prostheses that are in use today drive electrodes singly, with the exception of the Advanced Bionics SAS system. That is a result of the interference between the fields that are produced by the electrodes of the devices, fields that are defined by the size of the electrodes used, their distances from the target cells, and the magnitudes and phases of the signals that are applied to each contact (Spelman *et al.*, 1995). If fields are focused, it is likely that multiple groups of neurons can be driven simultaneously and independently (Jolly *et al.*, 1997). It is not clear whether such independence will be preferred by patients or that it will provide improved speech perception (Pfingst *et al.*, 1997). It seems logical from the operation of the normal auditory system (Sachs and Young, 1979), but Pfingst's work argues against it (Pfingst *et al.*, 1997).

Both modeling and experimental studies suggest clear benefits that can result from perimodiolar placement of electrode arrays. The modeling studies show that the fields are clearly more limited in extent when the electrode contacts are near to the target cells (Spelman *et al.*, 1995; Frijns *et al.*, 1996, 2001). Use of the simple model given in the earlier section on "Contacts and Focusing" can demonstrate the narrowing of the field that takes place when a point source is placed near its target.

Recall that finite-sized electrodes do not produce infinitesimally wide fields as they are approached; the fields approach the width of the electrode (potential curves in Fig. 4 illustrate that fact). However, a point can be made for either point sources or finite sources: the further from the source, the wider the field. That statement is clear for a single electrode, driven as a monopole; less clear for a tripolar configuration, driven as a quadrupole (Spelman *et al.*, 1995). In the latter case, the width of the potential field changes less with distance than for the former case. In both cases, the peak potential in the field is larger the closer the contact or contacts are to the target. Thus, intervening tissue between the electrodes and their target neurons is to be minimized or avoided entirely.

If tissue surrounds the implanted electrode array, then several problems can arise. First, the distance between the electrodes and their target cells will increase. Second, the sheaths that surround implanted electrodes have higher resistivities than do the solutions of the scala tympani and are anisotropic (Spelman, 2004). Third, tissue sheaths can cause an increase in the impedances of the electrodes and a concomitant increase in the voltage required to achieve excitation, with accompanying increases in power consumption.

The effect of increasing the distance becomes greater the closer is the electrode array to the neurons of the auditory nerve. Considering the simple case of a point source driven as a monopole, the electric field is proportional to the inverse of the distance between the source and the target. For a distance of 100 μm, an increase of 50 μm represents a 50% change in the distance between the cells and the target. If there is a distance of 500 μm initially, the change is 10%. In experimental animals sheaths of 50-μm thickness were found after a few months of implantation (Shepherd *et al.*, 1983; Leake *et al.*, 1990).

The introduction of connective tissue sheaths introduces multiple layers into the analysis of fields in the inner ear. Even a simple three-layered problem produces a complicated result that is not a closed-form solution and will not be addressed here (Spelman, 1989). It is enough to say that the signal produced by the stimulating current will be attenuated and, in the case when the sheath is anisotropic, will likely direct the excitation to undesirable locations (Spelman, 2004).

If cochlear electrode arrays could have their surfaces treated appropriately, they might resist the adhesion of cells and the growth of surrounding tissue sheaths (Dalsin *et al.*, 2003). That approach to combating the growth of cells that are produced by inflammatory responses may prove to be successful for future developments of electrode arrays. The approach can protect the surface of the substrate of the array, but not the metallic surfaces of the electrodes themselves.

Tissue on the surfaces of the electrodes must increase their impedances. Membranous tissues behave like leaky capacitors, introducing materials with specific capacitances of 33.8 μF/cm^2 (Junge, 1977). The introduction of such capacitances can affect the impedance of the electrodes of the array and their abilities to carry currents and stimulate cells. Thus, the materials to prevent cell adhesion to cochlear electrode arrays must be designed to maintain current-carrying capacity and low impedance while they prevent cell adhesion and the growth of undesirable tissue sheaths.

Costs and Benefits

That more than 50,000 people use cochlear prostheses is testimony to their effectiveness. The prostheses have improved monotonically over the past several decades, with substantial improvements in comprehension that occurred when multi-channel prostheses were developed, and more gradual improvements otherwise (Rubinstein and Miller, 1999).

At the same time, cochlear prostheses have been found to be cost effective as well. In careful cost analyses, Cheng and his co-workers investigated cochlear implants and found that they were competitive with other surgical interventions (Cheng and Niparko, 1999). The study accounted for the costs of surgery and rehabilitation, and considered change in the quality of life as an intangible.

DIRECTIONS FOR THE FUTURE

Clearly, multichannel cochlear prostheses will be produced and implanted in large numbers and will be used extensively worldwide. The three major producers of implants are likely to move to more cosmetic devices. All of them presently make behind-the-ear prostheses and are pursuing totally implantable devices. At this time, the greatest two stumbling blocks to the latter are battery and microphone technologies. Implantable microphones will require great attention to the issue of tissue encapsulation around the diaphragm of the microphone.

Another likely improvement will be the number of electrodes that are used in the array. Investigators have pursued a 72-contact array (Spelman *et al.*, 1998), a device that demands a quantum change in processor technology (Clopton *et al.*, 2002). High-density electrode arrays will require great care and attention paid to the issues of tissue growth and compatibility. If a high-density array and its processor are available, they could provide precise phase information that is unavailable now and that will likely make possible successful binaural cochlear prostheses (Clopton and Lineaweaver, 2001).

ACKNOWLEDGMENTS

This work was supported in part by Grants DC005531, NS37944, and DC04614 from the National Institutes of Health.

QUESTIONS

1. Sound at the level of the tympanic membrane demonstrates a pressure peak at a frequency of 2.5 kHz in the human. Consider the boundary conditions of the ear canal and explain why that resonance occurs.

2. Considering the areas of the stapes and the tympanic membrane, as well as the properties of the ossicular chain, compute the approximate ratio of pressures that is found when the tympanic membrane is driven at low frequency. Note that this is an approximate calculation that does not consider the dynamics of the system.

3. Focusing stimuli has been proposed by several people (Suesserman and Spelman, 1993; Jolly *et al.*, 1996) as a solution to the problem of field interference between monopole sources. Consider using dipole and quadrupole sources. Assume that the sources are 50-μm hemispheres located on the surface of an insulating boundary. Let the hemispheres be separated by 200 μm in both cases. Plot the potential fields along two lines: one that is 100 μm above and parallel to the sources, another that is 200 μm above and parallel to the sources. Describe the properties of the fields that are produced. Look particularly at the peak potentials and the half-amplitude widths of the fields. Hints: A dipole consists of two sources, one carrying current I and the other carrying current $-I$. A quadrupole is a special case of a tripole. Three sources are used. The central source carries current I, while the two flanking sources carry current $-I/2$. For ease of calculation, place one of the sources at the origin, and let the other sources lie on, e.g., the x-axis.

4. Wiley and Webster (1982) give an equation that describes the potential field produced by an electrode of radius a located at the origin of a cylindrical coordinate system. The electrode is placed on the surface of an insulating boundary.

$$V(r, z)$$

$$= \frac{2V_0}{\pi} \sin^{-1}\left\{\frac{2a}{\left[(r-a)^2 + z^2\right]^{0.5} + \left[(r+a)^2 + z^2\right]^{0.5}}\right\}$$

and

$$J_z(r, 0) = \frac{2V_0}{\rho\pi}\frac{1}{(a^2 - r^2)^{0.5}} = \frac{J_0}{2\left[1 - (r/a)^2\right]^{0.5}}$$

where

$$J_0 \equiv \frac{I_0}{\pi a^2}$$

I_0 is the total current flowing into the electrode, J_0 is a current density (defined earlier), ρ is the resistivity of the medium, and r is the radial distance $[x^2 + y^2]^{0.5}$. Place a point source at the origin and compare it to a circular, planar electrode with its center at the origin. Let the radius of the circular electrode be 100 μm. Compute the potential field for -1 mm $\leq x \leq 1$ mm at altitudes of 50 and 750 μm. What can you say about the shapes and the peak potentials?

5. A cochlear electrode array employs contacts that are hemispheres 100 μm in diameter. As the designer of the array, you compare Pt–Ir contacts with IrOx contacts. In one stimulus mode, you will use sinusoids of 100 and 1000 Hz. What is the maximum current that you can tolerate for each material at each frequency? Hint: compute charge per phase of the sinusoid.

6. A measurement of the IrOx electrode produces an impedance magnitude of 5 kilohms at 1000 Hz. If you drive a dipole pair with a 1-kHz sinusoid, what voltage range must the current source have if you drive the maximum allowable current that the electrode can tolerate? Assume that the tissue impedance of the cochlea is small compared to the impedance of the electrodes.

7. Two electrode array designs are considered. One uses a silicone substrate and the other a liquid crystal polymer substrate. If a circular cross section is used, with an outside diameter of 200 μm, find the force that would be exerted on a free end whose length is 1 mm and whose deflection is 20 μm. The modulus of elasticity of silicone is 2.76 MPa and that of liquid crystal polymer is 158 MPa. The polar moment of inertia is

$$I_p = \frac{\pi d^4}{32}$$

8. A simple model of electrode impedance is the Warburg model (Macdonald, 1987). This diffusion model can take the form of a parallel resistance/capacitance circuit in which the resistance and capacitance are both inversely proportional to the square root of frequency, i.e.,

$$R(f) = \frac{R_0}{\sqrt{f}}$$

$$C(f) = \frac{C_0}{\sqrt{f}}$$

Compute the magnitude and phase of the electrode's impedance if its impedance is 1 megohm at 10 Hz. Compute for frequencies from 10 Hz to 10 kHz. Assume that the electrode is circular in shape, and that its diameter is 100 μm. If it is covered with a membrane whose specific capacitance is that given in the text, how does the character of the impedance change? Plot the magnitudes and phases for both situations.

9. If the time constant of the membrane that covers the electrode is 125 msec, recompute the impedance for problem 8, accounting for the membrane resistance that parallels the membrane capacitance.

10. Consider the uncoated electrode of problem 8. If a controlled sinusoidal current were applied to the electrode, what would be the peak-to-peak voltage necessary to drive the current, undistorted, over the full frequency range? Assume that the electrode is driven as a monopole and that its counterelectrode has negligible impedance.

Bibliography

Anonymous, (2003). Med-El Web page, www.medel.com. Med-El, Innsbruck, Austria.

Bierer, J. A., and Middlebrooks, J. C. (2002). Auditory cortical images of cochlear-implant stimuli: dependence on electrode configuration. *J. Neurophysiol.* 87(1): 478–492.

Boyd, J. E., (1935). *Strength of Materials*, 4th ed., Vol. 1. McGraw-Hill, New York.

Brackmann, D. E., Hitselberger, W. E., Nelson, R. A., Moore, J., Waring, M. D., Portillo, F., Shannon, R. V., and Telischi, F. F. (1993). Auditory brainstem implant: I. Issues in surgical implantation. *Otolaryng. Head Neck Surg.*, 108(6): 624–633.

Brooks, D. (1998). Temperature Rise in PCB Traces. in *PCB Design Conference, West.* Miller Freeman, Inc.

Brugge, J. F. (1991). An overview of central auditory processing. in *The Mammalian Auditory Pathway: Neurophysiology*, A. N. Popper and R. R. Fay, (eds.). Springer-Verlag, New York, pp. 1–34.

Brummer, S. B., and Turner, M. J. (1977a). Electrical stimulation with Pt electrodes I: a method for determination of "real" electrode areas. *IEEE Trans. Biomed. Eng.* 24(5): 436–440.

Brummer, S. B., and Turner, M. J. (1977b). Electrical stimulation with Pt electrodes II: Estimation of maximum surface redox (theoretical non-gassing) limits. *IEEE Trans. Biomed. Eng.* 24(5): 440–444.

Cheng, A. K., and Niparko, J. K. (1999). Cost-utility of the cochlear implant in adults. *Arch. Otolaryngol. Head Neck Surg.* 125(11): 1214–1218.

Clark, G. M. (1996). Electrical stimulation of the auditory nerve: the coding of frequency, the perception of pitch and the development of speech processing strategies for profoundly deaf people. *Clin. Exp. Pharmacol. Physiol.* 23: 766–776.

Clopton, B., and Lineaweaver, S. (2001). The Importance of Phase Information for the Encoding of Speech. Internal report, Advanced Cochlear Systems.

Clopton, B. M., Lineaweaver, S. K. R., Corbett, III, S. S., Spelman, F. A., Method of processing auditory data. (2002). U.S. Patent No. 6,480,820. Advanced Cochlear Systems, USA.

Cogan, S. F., (2002). Stability of electro-active materials and coatings for charge-injection electrodes. in *Second Joint EMBS-BMES Conference 2002*, IEEE Press, Houston.

Dallos, P., Popper, A. N., and Fay, R. R. (eds.) (1996). *The Cochlea*. Springer Handbook of Auditory Research, Vol. 8. Springer, New York, p. 551.

Dalsin, J. L., Hu, B. H., Lee, B. P., and Messersmith, P. B. (2003). Mussel adhesive protein mimetic polymers for the preparation of nonfouling surfaces. *J. Am. Chem. Soc.* 125(14): 4253–4258.

Djourno, A., and Eyries, C. (1957). Prothese Autitive par excitation electrique a distance du nerf sensoriel a l'aide d'un bobinage inclus a demcure. *Presse Med.* 35: 14–17.

Enderle, J., Blanchard, S., and Bronzino, J. (2000). Introduction to Biomedical Engineering, Academic Press Series in Biomedical Engineering, J. Bronzino. (ed.), Vol. 1. Academic Press, San Diego.

Frijns, J. H. M., de Snoo, S. L., and ten Kate, J. H. (1996). Spatial selectivity in a rotationally symmetric model of the electrically stimulated cochlea. *Hear. Res.* 95: 33–48.

Frijns, J. H. M., Briaire, J. J., and Grote, J. J. (2001). The importance of human cochlear anatomy for the results of modiolus-hugging multichannel cochlear implants. *Otol. Neurol.* 22(3): 340–349.

Geisler, C. D. (1998). *From Sound to Synapse: Physiology of the Mammalian Ear*. Oxford University Press, New York, p. 381.

Gold, B., and Rader, C. M. (1967). The channel vocoder. *IEEE Trans. Audio Electroacoust.* AU-15: 148–161.

Goodisman, J. (1987). *Electrochemistry: Theoretical Foundations*. Wiley Interscience, New York.

Hinojosa, R., and Marion, M. (1983). Histopathology of profound sensorineural deafness. *Ann. N.Y. Acad. Sci.* 405: 458–484.

House, W. F., and Berliner, K. I. (1991). Cochlear implants: from idea to clinical practice. in *Cochlear Implants: A Practical Guide*, H. Cooper (ed.). Singular Publishing Group, San Diego, pp. 9–33.

Jolly, C. N., Spelman, F. A., and Clopton, B. M. (1996). Quadrupolar stimulation for cochlear prostheses: modeling and experimental data. *IEEE Trans. Biomed. Eng.* 43(8): 857–865.

Jolly, C. N., Clopton, B. M., Spelman, F. A., and Lineaweaver, S. K. (1997). Guinea pig auditory nerve response triggered by a high density electrode array. *Med. Prog. Technol.* 21(Suppl.): 13–23.

Junge, D. (1977). *Nerve and Muscle Excitation*, Sinauer and Associates, Sunderland, MA, p. 143.

Kessler, D. K. (1999). The CLARION Multi-Strategy Cochlear Implant. *Ann. Otol. Rhinol. Laryngol. Suppl.* 177: 8–16.

Kong, J. A., (1986). *Electromagnetic Wave Theory*, Vol. 1. Wiley, New York.

Leake, P. A., Kessler, D. K., and Merzenich, M. M. (1990). Application and safety of cochlear prostheses. in *Neural Prostheses: Fundamental Studies*, W. F. Agnew, and D. B. McCreery, (eds.) Prentice-Hall, Englewood Cliffs, NJ, pp. 253–296.

Levitt, H., and Nye, P. W. (1980). Sensory training aids for the hearing impaired (prologue). in *Sensory Aids for the Hearing Impaired*, H. Levitt, J. M. Pickett, and R. A. Houde (eds.). IEEE Press, New York, pp. 3–28.

Loizou, P. C. (1999). Signal-processing techniques for cochlear implants. *IEEE Eng. Med. Biol. Mag.* 18(3): 34–46.

Macdonald, J. R. (1987). *Impedance Spectroscopy*. Wiley, New York.

Martin, M. C. (1985). Alternatives to cochlear implants. in *Cochlear Implants*, R. A. Schindler and M. M. Merzenich (eds.). Raven Press, New York.

McCreery, D. B., *et al.* (1992). Stimulation with chronically implanted microelectrodes in the cochlear nucleus of the cat: histologic and physiologic effects. *Hear. Res.* 62(1): 42–56.

McCreery, D. B., Shannon, R. V., Moore, J. K., and Chatergee, M. (1997). *The Feasibility of a Cochlear Nucleus Auditory Prosthesis Based on Microstimulation*. Huntington Medical Research Institutes and House Ear Institute, Pasadena, CA.

Meyer, R. D., Cogan, S. F., Nguyen, T. H., and Rauh, R. D. (2001). Electrodeposited iridium oxide for neural stimulation and recording electrodes. *IEEE Trans. Neural Sys. Rehab. Eng.* 9(1): 2–11.

Middlebrooks, J. C., and Bierer, J. A. (2002). Auditory cortical images of cochlear-implant stimuli: coding of stimulus channel and current level. *J. Neurophysiol.* 87(1): 493–507.

NIDCD (2003). *Health Information: Cochlear Implants*. NIH (U.S. government): www.nidcd.nih.gov/health/hearing/coch.asp.

NIH (1995). NIH Consensus Statement: Cochlear Implants in Adults and Children. National Institutes of Health, Bethesda, MD.

Osberger, M. J. (1997). Current issues in cochlear implants in children. *Hearing Rev.* 4: 28–31.

Osberger, M. J., and Fisher, L. (1999). SAS-CIS preference study in postlingually deafened adults implanted with the Clarion cochlear implant. *Ann. Otol. Rhinol. Laryngol.* 108(4): 74–70.

Parker, J. R., *et al.* (2001). Testing of thin-film electrode arrays for cochlear implants of the future [abstract]. Asilomar, CA.

Patrick, J. F., Seligman, P. M., Money, D. K., and Kuzma, J. A. (1990). Engineering. in *Cochlear Prostheses*, G. M. Clark, Y. C. Tong, and J. F. Patrick (eds.). Churchill Livingstone, Edinburgh, pp. 99–125.

Pearson, C. E. (1990). *Handbook of Applied Mathematics*. 2nd ed., Vol. 1. Van Nostrand Reinhold, New York.

Peters, W. (2000). Near and far fields of electrodes of finite dimension in a conductive medium. Personal Communication.

Pfingst, B. E., Zwolan, T. A., and Holloway, L. A. (1997). Effects of stimulus configuration on psychophysical operating levels and on speech recognition with cochlear implants. *Hear. Res.* **112**: 247–260.

Popper, A. N., and Fay, R. R. (1991). *The Mammalian Auditory Pathway: Neurophysiology*. Springer Handbook of Auditory Research, Vol. 2. Springer-Verlag, New York, p. 431.

Pyman, B. C., Brown, A. M., Dowell, R. C., and Clark, G. M. (1990). Preoperative evaluation and selection of adults. in *Cochlear Prostheses*, G. M. Clark, Y. C. Tong, and J. F. Patrick (eds.). Churchill Livingstone, Melbourne, pp. 125–135.

Rebscher, S. J., Heilmann, M., Bruszewski, W., Talbot, N. H., Snyder, R. L., and Merzenich, M. M. (1999). Strategies to improve electrode positioning and safety in cochlear implants. *IEEE Trans. Biomed. Eng.* 46(3): 340–352.

Rebscher, S. J., Wardrop, P. J., Whinney, D., and Leake, P. A. (2001). Insertion trauma, mechanical performance and optimum dimensions: refining a second generation of cochlear implant electrodes. in *2001 Conference on Implantable Auditory Prostheses*, Pacific Grove, CA, USA.

Robblee, L. S., and Rose, T. L. (1990). Electrochemical guidelines for selection of protocols and electrode materials for neural stimulation. in *Neural Prostheses: Fundamental Studies*, W. F. Agnew and D. B. McCreery (eds.). Prentice Hall, Englewood Cliffs, NJ. pp. 25–66.

Rubel, E. W., and Dobie, R. A. (1989). The auditory system: central auditory pathways. in *Textbook of Physiology: Excitable Cells and Neurophysiology*, H. D. Patton *et al.* (eds.). W.B. Saunders, Philadelphia, pp. 386–411.

Rubinstein, J. T. (1988). Quasi-Static Analytical Models for Electrical Stimulation of the Auditory Nervous System. *Department of Bioengineering*, University of Washington, Seattle, WA, p. 96.

Rubinstein, J. T., and Miller, C. A. (1999). How do cochlear prostheses work? *Curr. Opin. Neurophysiol.* 9: 399–404.

Rubinstein, J. T., Spelman, F. A., Soma, M., and Suesserman, M. F. (1987). Current density profiles of surface mounted and recessed electrodes for neural prostheses. *IEEE Trans. Biomed. Eng.* BME-34(11): 864–875.

Sachs, M. B., and Young, E. D. (1979). Encoding of steady-state vowels in the auditory nerve: representation in terms of discharge rate. *J. Acoust. Soc. Am.* 667(2): 470–479.

Shannon, R. V. (1992). Temporal modulation transfer function in patients with cochlear implants. *J. Acoust. Soc. Am.* **91**: 2156–2164.

Shannon, R. V., Fayad, J., Moore, J., Lo, W. W., Otto, S., Nelson, R. A., and O'Leary, M. (1993). Auditory brainstem implant: II. Postsurgical issues and performance. *Otolaryng. Head Neck Surg.* **108**(6): 634–642.

Shepherd, R. K., Clark, G. M., Black, R. C., and Patrick, J. F. (1983). The histopathological effects of chronic electrical stimulation of the cat cochlea. *J. Laryngol. Otol.* 97(4): 333–341.

Simmons, F. B. (1966). Electrical stimulation of the auditory nerve in man. *Arch. Otolaryng.* 84: 24–76.

Skinner, M. W. (2001). Cochlear implants in children: What direction should future research take? in *2001 Conference on Implantable Auditory Prostheses*, Pacific Grove, CA, USA.

Skinner, M. W., Ketten, D. R., Vannier, M. W., Gates, G. A., Yoffe, R. T., and Kalender, W. A. (1994). Determination of the position of Nucleus cochlear implant electrodes in the inner ear. *Am. J. Otol.* 15(5): 644–651.

Spelman, F. A. (1989). Determination of tissue impedances of the inner ear: models and measurements. in *Cochlear Implants: Modeles of the Electrically Stimulated Ear*, J. M. Miller, and F. A. Spelman (eds.). Springer-Verlag, New York, p. 422.

Spelman, F. A. (1999). The past, present and future of cochlear prostheses. *IEEE Eng. Med. Biol. Mag.* 18(3): 27–33.

Spelman, F. A. (2004). Cochlear implants. in *Biomedical Instrumentation*, P.-Å. Öberg, T. Togawa, and F. A. Spelman (eds.). Wiley, Berlin, in press.

Spelman, F. A., and Voie, A. H. (1996). Fascicles of the auditory nerve in the human cochlea: Measurements in the region between the spiral ganglion and the osseous spiral lamina. in *Nineteenth Annual Midwinter Meeting of the Association for Research in Otolaryngology*. St. Petersburg, FL.

Spelman, F. A., Pfingst, B. E., Clopton, B. M., Jolly, C. N., and Rodenhiser, K. L. (1995). The effects of electrode configuration on potential fields in the electrically-stimulated cochlea: models and measurements. *Ann. Otol. Rhinol. Laryngol.* **104**(Suppl. 166): 131–136.

Spelman, F. A., Clopton, B. M., Voie, A. H., Jolly, C. N., Huynh, K., Boogaard, J., and Swanson, J. W. (1998). *Cochlear implant with shape memory material and method for implanting the same*. U.S. Pattent No. 5,800,500. Assigned to PI Medical (now MicroHelix) and University of Washington.

Suesserman, M. F., and Spelman, F. A. (1993). Lumped-parameter model for in vivo cochlear stimulation. *IEEE Trans. Biomed. Eng.* 40(3): 234–235.

Suesserman, M. F., Spelman, F. A., and Rubinstein, J. T. (1991). *In vitro* measurement and characterization of current density profiles produced by nonrecessed simple recessed and radially varying recessed stimulating electrodes. *IEEE Trans. Biomed. Eng.* 38(5): 401–408.

Tykocinski, M., Saunders, E., Cohen, L. T., Treba, C., Briggs, R. J. S., Gibson, P., Clark, G. M., and Cowan, R. S. C. (2001). The Contour electrode array: safety study and initial patient trials of a new perimodiolor design. *Otol. Neurol.* 22: 33–41.

Vandali, A. E., Whitford, L. A., Plant, K. L., and Clark, G. M. (2000). Speech perception as a function of electrical stimulation rate: using the Nucleus 24 cochlear implant system. *Ear Hear.* 21(6): 608–624.

Vollmer, M., Beitel, R. E., and Snyder, R. L. (2001). Auditory detection and discrimination in deaf cats: psychophysical and neural thresholds for intracochlear electrical signals. *J. Neurophysiol.* 86(5): 2330–2343.

Weiland, J. D., Anderson, D. J., and Humayun, M. S. (2003). *In vitro* electrical properties for iridium oxide versus titanium nitride stimulating electrodes. *IEEE Trans. Biomed. Eng.* 49(12): 1574–1579.

Wiley, J. D., and Webster, J. G. (1982). Analysis and control of the current distribution under circular dispersive electrodes. *IEEE Trans. Biomed. Eng.* BME-29(5): 381–385.

Wilson, B. S., Finley, C. C., Lawson, D. T., Wolford, R. D., Eddington, D. K., and Rabinowitz, W. M. (1991). Better speech recognition with cochlear implants. *Nature* 352: 236–238.

Zimmerman-Phillips, S., and Murad, C. (1999). Programming features of the CLARION Multi-Strategy Cochlear Implant. *Ann. Otol. Rhinol. Laryngol. Suppl.* 177: 17–21.

7.17 BIOMEDICAL SENSORS AND BIOSENSORS
Paul Yager

SENSORS IN MODERN MEDICINE

A convergence of factors is now resulting in the rapid development of sensors for biomedical use. These factors include:

A. Increasing knowledge of the physics, chemistry and biochemistry of physiology
B. Pressure for reduction in the cost of delivering medical care through more efficient treatment and shorter hospital stays
C. Steadily decreasing costs of microprocessor technology for data acquisition, analysis, and display
D. Reduction in the size of sensors due to silicon microfabrication and fiber optic technology
E. Rapidly advancing sensor technology in nonbiomedical fields and for *in vitro* use for clinical chemistry
F. Advances in biomaterials

Advances in computer technology have reached the point that the control of devices and processes is often limited only by the ability to provide reliable information to the computer. Furthermore, the technologies developed by the microprocessor and fiber-optics industries are now spawning a new generation of sensors. Areas benefiting now from these new sensors include the automotive and aerospace industries, chemical and biochemical processing, and environmental monitoring. While there has been ready adoption of new sensor technologies for *ex vivo* measurements, *in vivo* use of such sensors to clinical

TABLE 1 Chemical Indicators of Health

Small, simple	pH (acidity)
↓	Electrolytes (ions)
↓	Blood gases, including general anesthetics
↓	Drugs and neurotransmitters
↓	Hormones
↓	Proteins (antibodies and enzymes)
↓	Viruses
↓	Bacteria
↓	Parasites
Large, complex	Tumors

practice is lagging far behind. This is due to serious performance deficits of sensor operation *in vivo*, as well as the same regulatory factors that apply to any use of devices and materials *in vivo*. Sensors are as subject to the same problems of biocompatibility as are any other type of *in vivo* device; initial unbridled optimism on the part of analytical chemists for the potential applicability of their sensors to *in vivo* use has given way to a more sober appraisal of the potential of the field.

This chapter provides a brief overview of the current state of application of sensors to clinical medicine, with an emphasis on chemical sensors, biosensors, and the emerging field of microfabricated sensors. Several excellent reviews may be found in the literature (Rolfe, 1990; Collison and Meyerhoff, 1990; Turner *et al.*, 1987; Kohli-Seth *et al.*, 2000; West *et al.*, 2003; Nakamura *et al.*, 2003; Jain, 2003; Ziegler, 2003; Vo-Dinh *et al.*, 2000).

PHYSICAL VERSUS CHEMICAL SENSORS

Biomedical sensors fall into two general categories: physical and chemical. Physical parameters of biomedical importance include pressure, volume, flow, electrical potential, and temperature, of which pressure, temperature, and flow are generally the most clinically significant and lend themselves to the use of small *in vivo* sensors. Electrical potential measurement is covered in another chapter in this volume. Chemical sensing generally involves the determination of the concentration of a chemical species in a volume of gas, liquid, or tissue. The species can vary in size from the H^+ ion to a live pathogen (Table 1), and when the species is complex, an interaction with another biological entity is required to recognize it. When such an entity is employed the sensor is considered a biosensor. It is, in general, necessary to distinguish this chemical from a number of similar interferents, which can be technologically challenging, but this is an area in which biosensors excel. The clinical utility of monitoring compounds in the body has motivated great efforts to develop biosensors. The prime target is improvement upon current methods for determination of glucose concentration for treatment of diabetes. Because of the large numbers of diabetics worldwide and their requirement for frequent measurements of blood glucose, glucose monitoring continues to be the dominant market for enzyme-based biosensors.

INTERACTION OF THE SENSOR WITH ITS ENVIRONMENT

One helpful way to classify sensors is to consider the relationship between the sensor and the analyte, as shown schematically in Fig. 1. The more intimate the contact between the sensor and the analyte, the more complete is the information about the nature of the chemical species being measured. However, obtaining greater chemical information may involve some hazard to the physical condition of the system being studied. This is not a trivial problem when dealing with human subjects.

Noncontacting sensors produce only a minimal perturbation of the sample to be monitored. In general, such measurements are limited to the use of electromagnetic radiation such as light, or sampling the gas or liquid phase near a sample. It may be even necessary to add a probe molecule to the sample to make the determination. Two examples are monitoring the temperature of a sample by its infrared emission intensity, and spectroscopic monitoring of pH-dependence of the

optical absorption of a pH-sensitive dye. No damage is done to the sample, but limited types of information are available, and chemical selectivity is difficult to achieve *in vivo*.

Contacting sensors may be either noninvasive or invasive. Direct physical contact with a sample allows a rich exchange of chemical information; thus much effort has been expended to develop practical contacting sensors for biomedical purposes. A temperature probe can be in either category, but with the exception of removing or adding small quantities of heat, it does not change the environment of the sample. Few chemical sensors approach the nonperturbing nature of physical sensors. All invasive sensors damage the biological system to a certain extent, and physical damage invariably leads to at least localized chemical change. Tissue response can, in turn, lead to spurious sampling. Furthermore, interfacial phenomena and mass transport govern the function of sensors that require movement of chemical species into and out of the sensor. Restrictions on size of the invasive sensor allowable in the biological system can limit the types of measurements that can be made—even a 1-mm diameter pH electrode is of no use in measuring intracapillary pH values. Clearly, this is the most difficult type of sensor to perfect.

Most contacting sensors are derived from chemical assays first developed as sample removal sensors. Although it is certainly invasive and traumatic to remove blood or tissue from a live animal, removal of some fluids such as urine and saliva can be achieved without trauma. Once removed, a fluid can be pretreated to make it less likely to adversely affect the functioning of a sensor. For example, heparin can easily be added to blood to prevent clotting in an optical measurement cell. Cells that might interfere with optical assays can also be removed prior to measurement. Toxic reagents and probe molecules can be added at will, and samples can be fractionated to remove interfering species. The sensor and associated equipment can be of any size, be at any temperature, and use as much time as necessary to make an accurate measurement. Further, a sensor outside the body is much easier to calibrate. This approach to chemical measurement allows the greatest flexibility in sensor design and avoids many biocompatibility problems.

CONSUMING VERSUS NONCONSUMING SENSORS

There are at least two distinct ways in which a sensor can interact with its environment; these can be called consuming and nonconsuming (Fig. 2). A nonconsuming or equilibrium sensor is one that can give a stable reading with no net transport of matter or energy between the sensor and its environment. For example, while a thermometer is approaching equilibrium it takes up or releases heat, but when it has reached its ultimate temperature it no longer directly affects the temperature of its environment. Some chemical sensors such as ion-selective electrodes and some antibody-based sensors work similarly and have the advantage of being minimally perturbative of their environments. A nonconsuming sensor may become slower to respond after being coated by a biofilm, but may still provide the same ultimate reading.

The consuming or nonequilibrium sensors rely on constant unidirectional flux of energy or matter between the sensor and

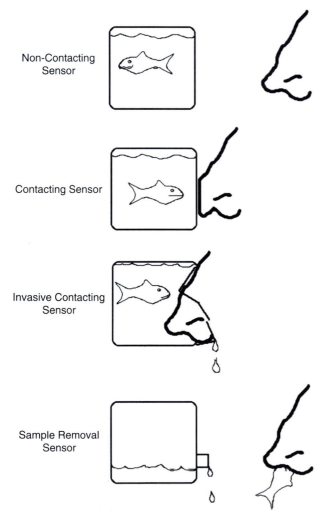

Non-Contacting Sensor

Contacting Sensor

Invasive Contacting Sensor

Sample Removal Sensor

FIG. 1. Different types of relationships may occur between a chemical sensor and the analyte to which it is sensitive.

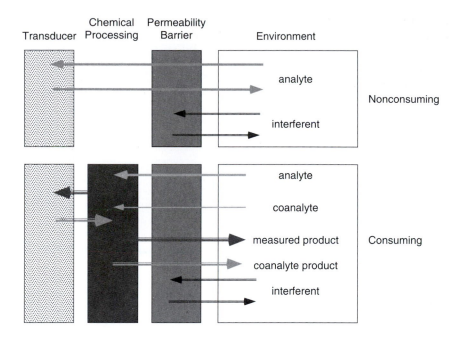

FIG. 2. A schematic diagram comparing consuming and nonconsuming sensors. The transducer is usually isolated from the environment by a permeability barrier to keep out interfering species, and in a consuming sensor, there is usually an intermediate layer in which active chemical processing occurs. In the consuming sensor there is a complex flux of analytes and products that makes it very sensitive to changes in permeability of the sensor–solution interface.

environment. The most common glucose sensor, for example (see later discussion), destroys glucose in the process of measuring it, thereby reducing its concentrations in the tissue in which it is measured, as well as reducing the pH and O_2 concentrations and generating toxic H_2O_2 as a by-product. This chemical measurement cannot be made *in situ* without affecting the system in which the measurement is made, and although it may be tolerable in flowing blood, it may not be acceptable in tissue over the long term. The greater the size of the sensor, the greater its sensitivity, but the more seriously it perturbs its environment. Many of the most fully developed specific *in vitro* analytical techniques for the determination of biochemicals involve irreversible consumption of the analyte, so sensors based on such techniques are difficult to implement *in vivo*.

SITE OF MEASUREMENT

Sample removal sensors are the mainstays of the clinical chemistry laboratory. For example, the first commercially manufactured biosensor—an electrode for measuring glucose—was made by Yellow Springs Instruments for a clinical chemistry analyzer. The major problem with the use of such devices is the time delay inherent in moving a sample to a distant location where the large, expensive instrument resides and waiting for delivery of the information derived from it, as well as the fact that useful instruments are often heavily utilized.

A major activity of modern bioanalytical chemistry is conversion of sample removal sensors into invasive contacting

sensors (Fig. 1). Not only can this approach improve care by reduction of the aforementioned delays, but it can also enable new types of procedures, such as automated feedback control of delivery of drugs to patients. However, development of such an approach often requires inventing new ways of making the measurements themselves. Chemical measurement at bedside or *in vivo* is technologically more challenging than in a prepared chemical laboratory with highly trained technicians. Clinical personnel must attend to the critical needs of the patient and have little time for fiddling with temperamental instrumentation. The instrumentation must therefore be made nearly foolproof, rugged, safe, reliable, and, if possible, self-calibrating.

DURATION OF USE

Two major questions in the design of any sensor are how often and for how long it is expected to be used. There are several factors to be considered:

Length of time for which monitoring is required. Determination of blood glucose levels must be made for the entire lifetime of diagnosed diabetics, whereas intraarterial blood pressure monitoring may only be needed during a few hours of surgery.

Frequency of measurements. Pregnancy testing may have to occur only once a month, whereas monitoring of blood pCO_2 during an operation may have to be made several times a minute.

Reusability of the sensor. Some chemical sensors contain reagents that are consumed in a single measurement. Such sensors are usually called "dip-stick sensors," such as are now found in pregnancy testing kits. High-affinity antibodies, for example, generally bind their antigens so tightly that they cannot be reused. On the other hand, most physical and chemical sensors are capable of measuring the concentration of their analyte on a continuous basis and are therefore inherently reusable.

Lifetime of the sensor. Chemical sensors all have limited lifetimes because of such unavoidable processes as oxidation, and although these may be extended through low-temperature storage, *in vivo* conditions are a threat to the activity of the most stable biochemical. Most sensors degrade with time, and the requirement for accuracy and precision usually limit their practical lifetime, particularly when recalibration is not possible. Mechanical properties can also limit lifetimes; although a thermocouple may have a shelf life of centuries, it can be broken on its first use by excessive flexing.

Appropriateness of repeated use. The need for sterility is the most important reason to avoid reuse of an otherwise reusable sensor. If it is not logistically possible or economically feasible to completely sterilize a used sensor, it will only be used on a single individual, and probably only once.

Biocompatibility. If the performance of a sensor is degraded by continuous contact with biological tissue (as discussed later), or if the risk to the health of the patient increases with the time in which a sensor is in place, the lifetime of the sensor may be much shorter *in vivo* than *in vitro*.

As a consequence of all of these factors in the design of an integrated sensing system, the probe—that part of the sensor that must be in contact with the tissue or blood—is often made disposable. Probes must therefore be as simple and as inexpensive to manufacture as possible, although it is often true that it is the sale of consumables such as probes that can be more profitable than the sale of the device itself.

BIOCOMPATIBILITY

The function of most chemical sensors is limited by the rate of diffusion through an unstirred layer of liquid at the interface, whether the alteration in rate controls response or merely response time. However, interfacial flux in biological media can be altered further by processes unique to living systems. Biocompatibility is an issue of importance for any material in contact with living systems, but is particularly so for chemical sensors. The properties of the interface are crucial to the ability of the sensor to make accurate, reproducible readings.

In Vitro *Use*

It is difficult to prevent biofilm formation on surfaces exposed to active media such as bacterial or eukaryotic cell cultures. The growth of a biofilm on a sensor degrades its performance in some way. In fact, in the most extreme case, the sensor can become sensitive only to conditions in the biofilm. Biofilms often consist of bacteria embedded in a secreted matrix of complex polysaccharides. Although the microenvironment of these films may be beneficial to the bacteria, it is detrimental to the function of the sensor for several reasons. First, the chemistry of the film may damage structural or active components of the sensor. Second, if the film completely encloses the sensor, only the film's microenvironment is sensed, rather than the solution that surrounds it. If the living components of the film metabolize the analyte to be sensed, it may never reach the sensor. Even a "dead" film may exclude certain analytes by charge or size and thereby lower the concentration available for sensing at the surface below the film. If a sensor used *in vitro* is fouled, it can often be removed, cleaned, and restored to its original activity. For example, carbon electrodes containing immobilized enzymes can be restored by simply polishing away the fouled surface (Wang and Varughese, 1990). When sensors are used *in vitro* for monitoring the chemistry of body fluids, preprocessing can be used to reduce the accretion of biofilms that might impede the function of the sensor.

In Vivo *Use*

Introduction of sensors into the body creates a complex set of problems. The sensor and the body act on each other in detrimental ways that must be minimized if not entirely avoided. Many of the problems (described later) worsen with time and may limit the utility of sensors to very short uses. In some cases, it has been found that the problems are almost immediate, producing spurious results from the outset. Subcutaneous glucose needle electrodes have been found to give accurate results *in vitro* before and after producing erroneous values *in vivo*. The biological environment may simply make it impossible to perform accurate chemical measurements with certain types of sensors.

Effects of Sensors on the Body

The introduction to the body of a sensor is a traumatic event, although the degree of trauma depends on the site of placement. The gastrointestinal tract can clearly be less traumatically accessed than the pulmonary artery. Critically ill patients often must have catheters placed into their circulatory systems for monitoring of blood pressure and administration of drugs, fluids, and food, so that no additional trauma is caused by including a small flexible sensor in that catheter. The size, shape, flexibility, and surface chemistry of the sensor probe are also important factors, although they are covered elsewhere in this volume. The outermost materials of the chemical sensor have at least one requirement not normally placed upon structural biomaterials: they must be permeable to the analyte in question. Thin films of polyurethanes are permeable to at least some analytes, and Nafion and porous Teflon work in other cases, but this issue is far from resolved. Some metals and graphite can be used directly as electrodes in the body.

When short-term implantation in tissue is possible, there is initial trauma at the site of insertion. Longer-term implantation increases the risk of infection along the surface of the implant. When the site of implantation of the probe is the circulatory system, the thrombogenicity of the probe is of paramount importance. Surface chemistry, shape, and placement within the vessel have all been shown to be of great importance in reducing the risk of embolism. Also, sensors based on chemical reactions often contain or produce toxic substances during the course of their operation, so great care must be taken to ensure that these are either not released or at low enough levels to avoid significant risk to the patient.

Effects of Surface Fouling on Sensor Function

The response of a consuming sensor depends on the rate of flux of the analytes and products across the interface. Indeed, most consuming sensors achieve linearity of response by making the diffusion into the sensor of analyte the rate-limiting step in the chemical reaction on which the sensor relies. Any new surface film over the interface will slow the diffusion and reduce the sensor output signal, producing an apparent reduction in the concentration of the analyte. Only equilibrium sensors avoid this problem, and even they experience a reduction in their response time.

These aforementioned problems of inert biofilms pale in comparison to the problems that arise when the film contains living cells, which is usually the case *in vivo*. As mentioned in other chapters, enzymatic and nonenzymatic degradation of polymers can be greatly accelerated, resulting in mechanical and eventual electrical failure of a sensor. Large changes in local pH, pO_2, and pCO_2 and concentration of other chemically active species such as superoxide can either chemically alter analytes or damage the function of the sensor itself. If, as often happens when a foreign body is implanted in soft tissue, a complete capsule of collagen and macrophages forms, the sensor within it may only be capable of sensing the microenvironment of activated macrophages, which is certainly not a normal sample of tissue.

Thrombus formation in blood is another common problem. Although such schemes as the use of polyurethane coatings and covalent grafting of heparin to that surface that work for vascular grafts are also helpful for sensors, thrombus is a complex, active substance that severely compromises the function of a sensor. In general, sensors in blood are used only for a few days at most and in the critical-care setting, during which time the patient may be undergoing anticoagulant therapy that reduces the problem of thrombus formation. As yet there is no device for permanent implantation in the bloodstream, despite great efforts in this direction.

Because of the trauma involved with insertion and removal of *in vivo* sensors, it is not generally feasible to remove a sensor to calibrate it. Consequently, any calibration must be done *in situ*, and since this would require the delivery of a calibration solution to the sensor, it is not generally done.

Successful *in vivo* sensors to date have been those that do not require recalibration during the lifetime of that sensor.

CLASSES OF SENSORS

New Technologies

As mentioned at the outset, the current rapid pace in development of biomedical sensors is fueled in part by a series of technological advances from other fields.

Fiber Optics

Advances in the field of optical communications have created a new technology for controlling light by using waveguides such as optical fibers. Whereas most of the technology for communication uses near-infrared light and most chemical measurements are made with ultraviolet and visible light, the fiber optic sensor field is bridging the gap by developing fibers that work well in the visible and chemical techniques that employ near-infrared light. Optical fibers have an advantage over wires in that they do not conduct current, so sensors made from such fibers (optrodes) are intrinsically safer than electrochemical sensors or even thermistors in that they reduce the risk of electrical shock to the patient.

Microprocessor-Controlled Devices

The recent availability of powerful, small, and relatively inexpensive computers has permitted designers of sensors to include sophisticated analytical procedures as part of the normal function of the sensors. Many previously manual operations such as calibration can now be completely automated. Rather than building a custom analog circuit to perform a particular function, one can now assemble stock electronic parts and customize only the software for the application. This has also relaxed the once-stringent requirement for linearity of response for the sensor itself—nearly any form of response can be programmed into a lookup table kept in memory. Home treatment of the elderly and chronically ill has been aided by the acquisition of data from sensors in the home. The data can be sent by radio or phone lines to central locations for more sophisticated analysis or routing of emergency services. Improvements in telemetry have led to the development of sensors that have no wires penetrating the body at all, which avoids the route for infection provided by continuous penetration of the body by catheters and electrical leads.

Microsensors and Microfabrication

Because of rapid progress in the semiconductor industry, micromachining of silicon into complex three-dimensional shapes with dimensions of less than 1 μm is now relatively commonplace. Devices can be electrical, such as electrodes, single transistors, and complex circuits; optical, such as photodiodes and optical waveguides; and mechanical, such as sensors, pumps, and microactuators. These diverse devices can be integrated into a single wafer, creating an entire "chemical laboratory on a chip." Furthermore, silicon microlithography is so successful in producing computers because it allows production of multiple copies of small and precisely manufactured devices, so the problems and costs inherent in manufacturing hand-made sensors can be avoided.

Physical Sensors

Temperature

The proper maintenance of a particular temperature, such as the 37°C of the human body core, is an indicator of health. Alteration of the temperature of the whole body or of a particular organ may be advantageous during certain medical procedures, such as surgery and preservation of organs, and this requires careful but minimally traumatic monitoring.

Thermometry began with development of devices based on calibrated changes in volume of liquids by Galileo in the 1600s and was perfected in the mid-18th century. Such thermometers are inexpensive and can be quite accurate, but are generally fragile, bulky, slow to respond, and require reading by eye, so they have been largely replaced in the clinical setting. If determination of the surface temperature of the skin is adequate, there are inexpensive techniques available relying on changes in the optical properties of a film of cholesteric liquid crystals, or expensive options such as the use of infrared radiometers. Modern methods of measuring the temperature of a bulk material such as blood or tissue are based on temperature-dependent electrical properties of matter. Devices include thermocouples, resistance temperature detector (RTD) sensors, thermistors, and silicon diodes, microcomputer-based applications of all of these. The sensor must be placed into tissue or blood, so to maintain accuracy there should be little transfer of heat into or out of the body along the leads to the thermometer. A small, self-contained telemetric temperature sensor with no external leads at all was developed by Human Technologies, Inc. It is capable of continuous readings of core temperature for the duration of the device's residence in the gastrointestinal tract.

Thermocouples

The Seebeck effect is responsible for temperature-dependent potentials (or current, in a closed circuit) across the junction between two different metals. Responses (Seebeck coefficients) of 10 to 80 μV/°C are the range for commonly used thermocouple pairs. In potential measurement circuits, the size of the contact region is immaterial, so very small thermocouples can be made with response times as short as milliseconds. The response is not linear, and either lookup tables or high-order polynomials are needed to linearize the responses. Precision better than 0.1°C is not generally practical. The thermocouple is self-powered and thus introduces no heat to the system being measured, but a reference junction at a known temperature is required, usually within the housing of the digital voltmeter (DVM) sensor electronics (Fig. 3). Thermocouples are cheap and as reproducible as the chemistry of the metals used. Because they can be made extremely small, they continue to be the sensor of choice in some applications. The Cardiovascular Devices Inc. (CDI) Systems 1000 fiber optic sensor for pH, pCO_2, and pO_2 (see later discussion), ironically, used a thermocouple for its required temperature reference at the tip of the optrode.

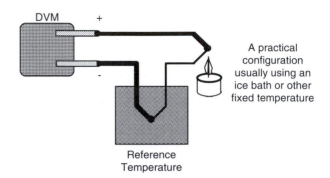

FIG. 3. A schematic representation of the manner in which thermocouples are used to measure temperature.

RTDs

The platinum RTD is based on the temperature dependence of the resistance of a metal. If care is taken to eliminate other sources of changes in resistance, chiefly mechanical strain, it is possible to measure temperature quite accurately and reproducibly. When well treated, RTDs are the most stable temperature measurement devices. The best RTDs are still hand-made coils of Pt wire, but these tend to be very expensive and bulky. Recently, deposition of Pt film on ceramics has been developed as a smaller, cheaper alternative with faster response times and nearly the same stability. Nonlinearities require adjustment in software for accurate readings, and because some current must be supplied to make a measurement, there is Joule heating of the device and the sample around it.

Thermistors

These are the most sensitive temperature measurement devices. They are generally made of semiconductive materials with negative temperature coefficients (decreasing R with increasing T). This effect is often quite large (several percent per °C), but also quite nonlinear. Very small devices can be fabricated with response times approaching those of thermocouples. Unfortunately, there are several drawbacks: the response of individual devices is highly dependent on processing conditions, the devices are fragile, and they self-heat. Nevertheless, because of their sensitivity they are the most common transducers for *in vivo* temperatures. For example, they are commonly used as the temperature sensors incorporated in the Swan-Ganz catheter used to determine cardiac output by thermodilution.

Optical Techniques

Optical techniques of temperature measurement have gained favor in recent years, in part because the use of optical fibers for measurements removes the necessity of passing metal wires into body that both allow a path for potentially lethal shock and can perturb electromagnetic fields such as those used in magnetic resonance imaging and other techniques requiring the use of microwaves. The Model 3000 Fluoroptic Thermometer (Luxtron Corp.) is such a device; it is based on the temperature dependence of the lifetime of phosphorescent emission from an inorganic material (magnesium fluorogermanate) placed at up to four different locations along four 250-μm plastic optical fibers. This allows four nearly independent temperature measurements at spacings as close as 3 mm, which is useful for monitoring microwave-induced hyperthermia. Accuracy to ± 0.2°C over a 40°C range is claimed. The expense of this type of device has initially limited its use to situations in which metal wires are not acceptable, but any of several recent temperature measurement schemes may prove substantially cheaper. Particularly when integrated with other sensors in a single multichannel device, optical thermometry may prove more popular than the use of thermistors.

Pressure

The most common sensor placed into the circulatory system of hospitalized patients is a blood pressure monitor. Both static and pulsatile blood pressure are key signs for monitoring the state of patients, particularly those with impaired cardiac function or undergoing trauma such as surgery. Although it is possible to measure the static blood pressure external to the body by using pressure cuffs and some acoustic or optical means of detection, such techniques are subject to a number of artifacts and do not work well in patients with impaired cardiac function. The site of measurement may be either an artery or vein, and the sensor may be implanted for short-term use during an operation, or over a longer term in an intensive-care-unit setting. It is even possible to monitor intraarterial pressure in ambulatory patients over long periods.

The pressure ranges generally seen in the circulatory system range from 0 to 130 mm Hg above the ambient 760 mm Hg. The most commonly available pressure transducers for accurate measurement in this range have been strain gauges, which are temperature-sensitive devices about an inch in diameter. The transducer itself is therefore placed outside the body, and pressure is transmitted to the transducer through a catheter. The transducer must be calibrated at least at turn on, and older models also required a two-point calibration against at least one calibrated pressure different from atmospheric and periodic rezeroing against ambient.

The mechanical properties of the catheter clearly can affect the accuracy of the waveform recorded by the transducer, as can the viscosity of the solution filling the catheter.

Blood would normally clot in the stagnant interior of the catheter, degrading the sensor response and causing a risk of embolism, so it is necessary to flush the catheter periodically with heparinized saline to keep both the lumen of the catheter and the artery patent. This requires the presence of a fairly complex set of sterile tubing and valves attached to a saline reservoir. Because flushing the catheter perturbs the pressure, it is not possible to obtain truly continuous measurements.

Most transducers have been fragile and expensive, and a factor that contributes to the cost of their use is the requirement that all materials in contact with blood must be sterilized. There is generally a diaphragm that separates the blood and saline from the mechanical transducer. This diaphragm and associated dome-shaped housing have in the past been a permanent part of the apparatus that required cleaning and sterilization between uses, but recent advances in manufacturing inexpensive, mechanically reproducible diaphragms have allowed this part also to be made disposable.

Other pressures of clinical importance are intracranial pressure and intrauterine pressure. Although in both these cases no direct invasion of the circulatory system is required, periodic flushing of all catheters is required. If, on the other hand, the sensor can be placed directly in the cavity in which the pressure is to be measured, flushing may not be necessary. Advances in silicon processing have allowed the manufacture of extremely small pressure sensors based on thin diaphragms suspended above evacuated cavities. The position of the diaphragm depends on the instantaneous pressure differential across it, so these devices can be used as both microphones and pressure monitors. The pressure can be monitored electrically, based on changes of capacitance between the diaphragm and the apposing wall of the cavity, or optically by monitoring changes in the reflectivity of the resonant optical cavity between the diaphragm and wall. This latter approach is the basis for fiber optic sensors introduced by FiberOptic Sensor Technologies, Inc., and MetriCor Corporation.

Chemical Sensors

A variety of chemical sensors employing different transduction mechanisms are currently employed for *in vivo* and *in vitro* measurements of biological parameters. These are summarized in Table 2.

TABLE 2 Transducers Used in Chemical Sensors

Transducer	Mode of measurement
Ion-selective electrode, concentration gas-selective electrode, FET	Potentiometry—determination of surface of charged species
Oxygen electrode, electrochemical electrode	Amperometry—monitoring available concentration of electrochemically active species
Low-impedance electrodes for monitoring conductance, impedance, admittance	Monitoring changes in bulk or surface electrical properties caused by altered molecular concentrations
Optical waveguides with detection of absorption, fluorescence, phosphorescence, chemiluminescence, surface plasmon resonance	Photometry
Thermistors, RTD, calorimeters	Monitoring temperature change induced by chemical reaction
Piezoelectric crystal SAW, BAW, etc., with chemically selective coating	Change in sound absorption or phase induced by binding to device

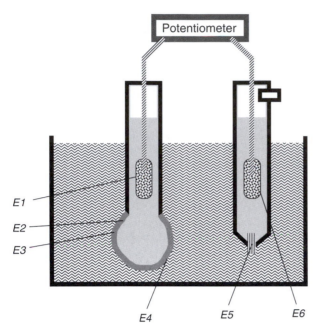

FIG. 4. Schematic of a pH meter and the potentials generated at various interfaces. Independent potentials generated in a pH measurement system: E1, measuring internal AgCl electrode potential; E2, internal reference solution–glass potential; E3, glass asymmetry potential; E4, analyte–glass potential; E5, reference liquid junction potential; E6, reference internal AgCl electrode potential. Any contamination or fouling at any of those interfaces causes degradation and drift. E4 is the potential that varies with solution pH, but a key source of error is E5, the liquid junction potential generated at the point where the reference electrolyte must leak slowly from the reference electrode body.

pH

The pH values of blood and tissue are normally maintained within narrow ranges; even slight deviations from theses values have great diagnostic value in critical care. Measurement of pH is made either by electrically monitoring the potential on the surface of pH-sensitive materials such as certain oxides or glasses (potentiometric sensor), or by optically monitoring the degree of protonation of a chemical indicator. The pH electrode has traditionally incorporated a thin glass membrane that encloses a reference solution (Eisenman, 1967) (Fig. 4), although of late some success in the use of pH-sensitive field effect transistors (pHFETs) has been reported and a solid-state pH electrode based on a pHFET is being marketed. Another device is the light-addressable potentiometric sensor now produced by Molecular Devices Inc., which has shown great promise for *in vitro* measurement for biosensing (Hafeman *et al.*, 1988). Electrical measurements require the use of a reference electrode that is often the source of problems and has a lifetime dependent on the volume of electrolyte that it contains. Consequently, miniaturization of pH electrodes for long-term *in vivo* use has been a difficult problem. Such sensors are nonconsuming, but surface fouling can influence the readings, and recalibration must be performed frequently.

Because of these drawbacks, there has been a shift toward optical indicator-based sensors using fiber-optic detection (Saari and Seitz, 1982; Jordan *et al.*, 1987; Jones and Porter,

1988; Benaim *et al.*, 1986), and at least two such sensors are now in advanced clinical trials. In these sensors, a small amount of dye with a pH-dependent fluorescence spectrum is immobilized in a polymer or hydrogel at the end of an optical fiber. Exciting light is sent down the fiber and the fluorescence emitted is returned to a photodetector along the same fiber. The detector can discriminate between the reflected exciting light and the probe's emission because they are at different frequencies. Such a sensor requires no reference and can function until the dye leaches from the probe or is bleached by the exciting light. However, both electrical and optical measurements of pH are dependent on the temperature, so pH measurements are always performed in conjunction with a temperature measurement as close to the site of the pH sensor as possible. Also, no pH sensor is absolutely specific, so in an uncontrolled environment there is always the risk of interferences. For example, the electrical sensors are subject to errors in the presence of biologically important metal ions, and fluorescence-based pH sensors are affected by fluorescence quenchers such as some inhalation general anesthetics.

Ions

Many simple ions such as K^+, Na^+, Cl^-, and Ca^{2+} are normally kept within a narrow range of concentrations, and the actual concentration must be monitored during critical care. Potentiometric sensors for ions (ion-selective electrodes or ISEs) operate similarly to pH electrodes; a membrane that is primarily semipermeable to one ionic species can be used to generate a voltage that obeys the Nernst (or more accurately the Nikolski) equation (Ammann, 1986):

$$E = \text{consant} + \frac{2.303\, RT}{zF} \log[a_i + k_{ij}(a_j)^{z/y}]$$

where E is the potential in response to an ion, i, of activity a_i, and charge z; k_{ij} is the selectivity coefficient; and j is any interfering ion of charge y and activity a_j.

Glasses exist that function as selective electrodes for many different monovalent and some divalent cations. Alternatively, a hydrophobic membrane can be made semipermeable if a hydrophobic molecule that selectively binds an ion (an ionophore) is dissolved in it. The selectivity of the membrane is determined by the structure of the ionophore. One can detect K^+, Mg^{2+}, Ca^{2+}, Cd^{2+}, Cu^{2+}, Ag^+, and NH_4^+ by using specific ionophores. Some ionophores are natural products, such as gramicidin, which is highly specific for K^+, whereas others such as crown ethers and cryptands are synthetic. S^{2-}, I^-, Br^-, Cl^-, CN^-, SCN^-, F^-, NO_3^-, ClO_4^-, and BF^- can be detected by using quaternary ammonium cationic surfactants as a lipid-soluble counterion. ISEs are generally sensitive in the 10^{-1} to 10^{-5} M range, but none is perfectly selective, so to unambiguously determine ionic concentrations it is necessary to use two or more ISEs with different selectivities. Also, ISEs require a reference electrode like that used in pH measurements. One can immobilize ionophore-containing membranes over planar potential-sensitive devices such as FETs, to create ion-sensitive FETs (ISFETs). As the potential does not depend on the area of the membrane, these work as well as larger bench-scale electrodes. An advantage of this approach is that a dense array of different ISFETs can be manufactured in a small area by using

TABLE 3 Electroactive Chemicals

Inorganic species
  Single-electron transfers:
    Solvated metal ions such as Fe^{2+}/Fe^{3+}
    All M^0/M^{n+} pairs
  Many species undergo multielectron transfer reactions:
    The oxygen–water series $O_2/H_2O_2/H_2O/OH/H^+$
Organic species
  Most aromatics, particularly generally nitrogen-containing aromatic
    heterocycles
  (Reactions usually involve changes in number of atoms attached to
    molecule, therefore are multistep, multielectron processes)
Metallo-organics:
  Ferrocene/ferrocinium
Biochemical species:
  Hemes, chlorophylls
  Quinones
  $NAD^+/NADH$ (not affected by O_2)
  $NADP^+/NADPH$ (not affected by O_2)
  FAD/FADH
  FMN/FMNH

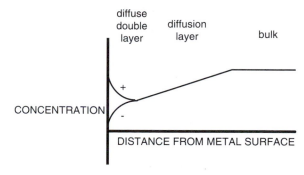

FIG. 5. A representation of the concentration of a charged analyte near the surface of an electrode (at left) at which it is being consumed. Note that there is a linear gradient of concentration in the unstirred diffusion layer and a further depletion or enrichment of the analyte when it is close enough to the electrode to sense the surface potential.

microfabrication techniques. Early problems with adhesion of the membranes to silicon have been largely solved by modifications of the design of the FETs themselves (Blockburn, 1987). The most typical membrane material used in ISEs is poly(vinyl chloride) plasticized with dialkyl sebacate or other hydrophobic chemicals. This membrane must be protected from fouling if an accurate measurement is to be made.

Electrochemically Active Molecules

If a chemical can be oxidized or reduced, there is a good chance that this process can occur at the surface of an electrode. Selectivity can be achieved because each compound has a unique potential below which it is not converted, so under favorable conditions a sweep of potential can allow identification and quantification of different species with a single electrode. This process is the basis for detection of a number of important biochemicals such as catecholamines (Table 3). Some species are determined directly at electrodes and others indirectly by interactions with mediator chemicals that are more easily detected at particular electrode surfaces.

Because detection involves conversion of one species to another, this is a consuming sensor, with all of the attendant problems. At least two electrodes are needed, and current must flow through the sample for a measurement to be made, although a precise reference electrode is not as necessary as in a potentiometric sensor. Near the electrode surface the concentration of either the oxidized or reduced species may differ greatly from the bulk concentration. This is partially because of depletion of the analyte near the surface (the diffusion layer), as well as attraction or repulsion of charged species from the charged electrode surface in the diffuse double layer (Fig. 5).

Since the current flow is the measured quantity and current is proportional to the number of molecules converted per unit time, the signal is controlled by mass transport, the electrostatics in the electric double layer, and specific chemical and physical interactions at the electrode surface. The great advantages of electrochemical detection are counterbalanced by its great sensitivity to surface fouling and any process that changes the resistance between the measuring electrodes. Also, there are interfering compounds present at high concentrations *in vivo* such as ascorbate that can swamp signals from more interesting but less concentrated analytes such as catecholamines. Tricks such as using selective membranes over the electrodes can solve some of these problems. For example, negatively charged Nafion allows passage of catecholamines but blocks access to the electrode by negatively charged ascorbate.

Blood Gases

Perhaps the most important physiological parameters after heart rate and blood pressure are the partial pressures of blood gases O_2 and CO_2 (pO_2 and pCO_2). It is also often useful to compare pressures of these gases in the arterial and venous circulation. The pulse oximeter, now manufactured by a number of vendors, allows noninvasive determination of the degree of saturation of hemoglobin in the arterial circulation, but does not give any information about the actual pO_2. It only works on the arterial circulation near the periphery and gives no information about pCO_2. An invasive fiber optic probe developed by Abbott Critical Care Systems (the Oximetrix 3 SvO_2) allows measurement of oxygen saturation directly in veins by measuring the reflected light at three wavelengths (Schweiss, 1983). Since the affinity of hemoglobin for O_2 depends on the pH, which in turn depends on pCO_2, it is necessary in many cases to measure the actual pO_2, pCO_2, and pH simultaneously. As all these sensors depend on temperature, a temperature probe is also required. The first successful fiber-optic measurement of *in vivo* pO_2 was reported by Peterson in 1984, and the principle employed has been used for most successful subsequent pO_2 sensors (Peterson *et al.*, 1984). The CDI System 1000 was a fiber optic sensor for pH, pCO_2, and pO_2, as shown in Fig. 6. It was the first complete blood gas sensor, and it combines the use of fiber optics with a smooth shape and size to avoid creating turbulence in the blood flow, and a covalent heparin coating to reduce thrombogenicity.

Two methods dominate for the measurement of pO_2. The first and most popular employs the amperometric Clark electrode, which consumes O_2 and generates H_2O_2 and OH^- as by-products. The internal electrolyte in the sensor is separated from the external medium by a Teflon or silicone-rubber membrane that readily passes O_2 but prevents both water and other electrochemically active species from passing. Severe fouling of the membrane can reduce the rate of O_2 diffusion and hence the response. The alternative optical approach relies on the efficient collisional quenching of most fluorophores by O_2. As the fluorescence intensity is inversely proportional to pO_2, the reduction in intensity can be used as the basis of a fiber-optic sensor. Because this method relies entirely on the intensity of the fluorescent signal, it is subject to drift and degradation from photobleaching and thus is not appropriate for long-term use.

The same hydrophobic membranes that are permeable to O_2 are also permeable to CO_2, so they may be placed over pH electrodes or pH-sensitive optical probes containing bicarbonate buffer for selective determination of pCO_2 (Zhujun and Seitz, 1984; Gehrich *et al.*, 1986).

BIOSENSORS

Definition and Classification

The repertoire of chemicals that can be determined by the sensors mentioned previously is relatively limited. To determine the presence or concentration of more complex biomolecules, viruses, bacteria, and parasites *in vivo*, it is necessary to borrow from nature (Fig. 7). Biosensors are sensors that use biological molecules, tissues, organisms, or principles. This definition is broad and by no means universally accepted, although it is more restrictive than the other common interpretation that would include all the sensors described in this chapter. The leading biological components of biosensors are summarized in Table 4. Enormous progress has been made in the development of biosensors in recent years, and this work has been recently and exhaustively reviewed (Turner *et al.*, 1987; Kohli-Seth *et al.*, 2000; West *et al.*, 2003; Nakamura *et al.*, 2003; Jain, 2003; Ziegler, 2003; Vo-Dinh *et al.*, 2000).

Most of the applications have been in the realm of analytical chemistry for use in chemical processing and fermentation, with the exception of development of enzyme-based glucose sensors, on which we will focus.

Currently, commercially available are biosensors for glucose (used first in an automated clinical chemistry analyzer and based on glucose oxidase), lactate, alcohol, sucrose, galactose, uric acid, alpha amylase, choline, and L-lysine. All are amperometric sensors based on O_2 consumption or H_2O_2 production in conjunction with the turnover of an enzyme in the presence of substrate. A urea sensor is based on urease immobilized on a pH glass electrode (Turner, 1989). Most of these sensors are macroscopic and are employed in the controlled environment of a clinical chemistry analyzer, but the ExacTech device, manufactured by Baxter since 1987, is a complete glucose sensor containing disposable glucose oxidase-based electrodes, power supply, electronics, and readout in a housing the size of a ball point pen. One places a drop of blood on the disposable electrode and a few seconds later a fairly accurate reading of blood glucose is obtained. It is widely believed that much more frequent measurement of blood glucose with correspondingly frequent adjustments of the dose of insulin delivered could significantly improve the long-term prognosis for insulin-dependent diabetics. Increasing the frequency of the current sampling method (i.e., puncturing the finger for drops of blood) is not acceptable. Much progress has been made toward the goal of producing a glucose sensor that could be implanted for a period of time in the tissue or blood (Thome-Duret *et al.*, 1996; Hu and Wilson, 1997; Reach and Wilson, 1992; Ishikawa *et al.*, 1998; Gerritsen *et al.*, 1998; Shichiri

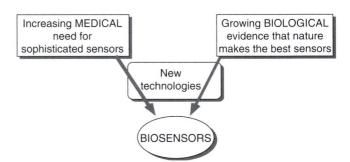

FIG. 7. The development of biosensors is driven by increased need for biochemical information in the medical community, and the knowledge that nature senses these chemicals best, combined through emerging technologies to interface the biochemicals with physical transducers.

TABLE 4 Biological Components of Biosensors

Binding	Catalysis
Antibodies	Enzymes
Nucleic acids	Organelles
Receptor proteins	Tissue slices
Small molecules	Whole organisms
Ionophores	

The two categories are not mutually exclusive, for example, some enzymes may be employed for binding alone.

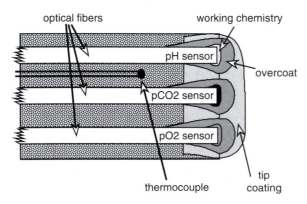

FIG. 6. A schematic drawing of the probe of the Cardiovascular Devices System 1000 fiber-optic blood gas sensor, based on three different combinations of selective membranes and fluorescent probes. Light enters and leaves from the left.

TABLE 5 Advantages and Disadvantages of Biochemicals for Chemical Detection

Advantages for binding:
  "Uniquely" high selectivity
  Possibility of raising antibodies to nearly all antigens
  Antibodies and biotin–avidin system allow selective attachment of markers and reporters of binding
  High binding constants possible
  Several possible detection modalities
  Ion flux through gated channels can provide gain
Advantages for catalysis:
  For every biochemical there is an enzyme that can be used to detect its presence
  High selectivity possible with some enzymes
  Several possible detection modalities
  Enzymatic cascades can provide gain
  Universality of redox coupling and pH effects permit common transduction schemes
Disadvantages of biosensors:
  Biomolecules generally have poor thermal and chemical stability compared to inorganic materials
  The function of the biological component usually dictates that they must have narrow operating ranges in temperature, pH, ionic strength
  Susceptibility to enzymatic degradation is universal
  Need for bacteriostatic techniques in their fabrication
  Time-dependent degradation of performance is guaranteed with the use of proteins
  Production and purification can be difficult and costly
  Immobilization can reduce apparent activity of enzymes or kill them outright
  Most live organisms need care and feeding

et al., 1998). However, the problem is a formidable one that epitomizes the attempts to develop biosensors for *in vivo* use.

Background

The Utility of Biochemical Approaches to Sensing

Some of the many advantages of using biochemicals for sensing are summarized in Table 5. The most important is that, despite the disadvantage of using chemically labile components in a sensor, they allow measurement of chemical species that cannot otherwise be sensed. Sensors have been fabricated that incorporate small biochemicals such as antibodies, enzymes and other proteins, ion channels, liposomes, whole bacteria and eukaryotic cells (both alive and dead), and even plant and animal tissue.

Immobilization

One of the key engineering problems in biosensors is the immobilization of the biochemistry used to the transducing device. Approaches range from simply trapping an enzyme solution between a semipermeable membrane and a metal electrode, to covalently cross-linking several enzymes to a porous hydrogel coated on a pH electrode, to covalently cross-linking a complete monolayer of antibody to the surface of an optical fiber. The immobilization of a layer of material

over the transducing device increases the response time, so for altered sensitivity and greatly enhanced selectivity, speed is often sacrificed. Monolayers do not contain much material, so to detect binding of so few molecules, it is generally necessary to employ some amplification scheme, such as attachment of an enzyme to an antibody that announces its presence by converting a subsequently added substrate to a large quantity of readily detected product. Such schemes add complexity and time to the detection. Immobilization also has unpredictable effects on the activity and stability of biochemicals.

Sensing Modalities

Potential-Based Sensors (pH and ISE)

Some of the first biosensors employed enzyme-catalyzed reactions (such as those of penicillinase, urease, and even glucose oxidase) that affect pH. By putting a pH electrode into the solution containing the enzyme it is possible to monitor the rate of enzymatic turnover. It is also possible to use pH to monitor the change in production of CO_2 by bacteria in the presence of substrates that they are capable of metabolizing (Simpson and Kobos, 1982). However, there is always a problem for *in vivo* use of pH-based sensors related to the fact that the external environment is capable of strong buffering of pH changes, and any change in pH in the immediate environment of the sensing surface is reduced toward the bulk pH by a degree that depends on the strength of that buffering.

Electrochemical Sensors

Many enzymes perform oxidation and reduction reactions and can be coupled, if indirectly, to electrodes. The electrochemically active species in enzymes is generally a cofactor (Table 3) that, when bound, is not accessible to the electrode surface at which the electron transfer must take place for detection. In the case of the glucose oxidase reaction, the normal biological reaction is:

$$\text{Glucose} + O_2 + H_2O \Leftrightarrow \text{Gluconic acid} + H_2O_2$$

The enzyme uses an FAD coenzyme to mediate the oxidation, and the resultant $FADH_2$ is directly oxidized by O_2 to return to FAD to prepare for the next catalytic reaction. Unlike NAD and NADP, FAD is tightly bound to the enzyme, so normally only a small diffusible molecule such as O_2 can gain access to it to alter its oxidation state. This means that under many circumstances, such as those present in tissue, the concentration of O_2 is rate limiting, so the sensor often measures not glucose but the rate at which O_2 can arrive at the enzyme to reoxidize its cofactor. There are two electrochemical ways to couple the reaction to electrodes: monitoring depletion of O_2 by reducing what is left at an electrode, or monitoring buildup of H_2O_2 by oxidizing it to O_2 and protons. The latter approach is generally used to avoid direct effects of O_2 variation on the electrode, but this does not completely solve the problem. The electrode reaction for peroxide oxidation is as follows:

$$H_2O_2 \Rightarrow O_2 + 2H^+ + 2e^-$$

The best solution to date to cast off the tyranny of the rate-limiting step of O_2 diffusion has been the use of electrochemical

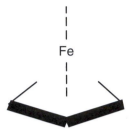

FIG. 8. The structure of the ferrocene–ferrocinium ion couple that allows one to overcome the dependence of the glucose oxidase reaction on pO_2. The two five-membered rings are cyclopentadienyl anions and the iron may be in either the Fe^{2+} or Fe^{3+} states, giving a total charge of 0 or +1.

mediators that are at a higher concentration than O_2 and can therefore shuttle back and forth between the protein and the electrode faster than the enzyme is reduced, so that the arrival of the substrate such as glucose is always rate limiting. A typical chemical that works in this way is ferrocene, a sandwich of an iron cation between two cyclopentadienyl anions (Fig. 8). It exists in neutral and +1 oxidation states that are readily interconvertible at metal or carbon electrodes. A proprietary modified ferrocene is used in the aforementioned ExacTech carbon electrode–based glucose sensor. Other glucose oxidase–based electrodes have been employed on catheters for *in vivo* determination of blood glucose, with varying degrees of success (Gough *et al.*, 1986). Thrombus formation is generally a problem, as is the possible alteration in localized glucose levels in tissue traumatized by insertion of probes, no matter how small. It may well be that use of the techniques employed in keeping pressure catheters clear will also work with biosensors such as this.

An elegant and oft-tried approach that avoids many of the problems inherent in implantable sensors is to extract fluid from the body and to measure the glucose concentration of that fluid. In the past few years, the Cygnus Corporation has developed the use of iontophoretic extraction of glucose through the skin for its GlucoWatch—a wristwatch-sized device that requires no puncturing of the skin. By passing a small current between two electrodes on the skin, ions and neutral species are made to flow from the extracellular fluid space to a gel pad under one electrode. Glucose oxidase in the gel pad reacts with the glucose, producing a measurable analyte (H_2O_2), which is subsequently detected electrochemically. The electrodes with their overlayer of gel-immobilized enzyme last 12 hours, and every few days the wearer has to move the watch to prevent skin irritation.

Optical Waveguide Sensors

Fiber optics can be used either as thin flexible pipes to transport light to and from a sensor at a remote site, or in a way that takes advantages of the unique properties of optical waveguides. The former mode still dominates, and the CDI blood gas sensor uses three fibers just to move photons to and from the small volumes of immobilized chemistry at the probe end. The Schultz fiber-optic glucose sensors involve a more sophisticated use of the light path exiting the optical fiber combined with clever use of lectin biochemistry. In principle, these sensors allow continuous measurement of blood glucose. There are at least two features specific to waveguides that have been used for sensors for *in vitro* measurement that may soon find themselves ready for *in vivo* use as well. In one, the ability of light sent down two fibers to interfere with itself on return to the source allows sensitive measurement of changes in the length or phase velocity of the fibers. This, in turn can be altered by enzymatically induced changes in the temperature of the fiber or its cladding in the volume surrounding the fiber. Another approach is to use the light in the evanescent wave that exists in the region just outside the waveguide to probe a small volume adjacent to the surface. If binding species such as antibodies are immobilized on the surface, it is possible to selectively excite and collect fluorescence from the surface layer even in the presence of high concentrations of fluorophore or other absorbers in the bulk solution. This technique has allowed the use of antibody-based detection of analytes such as theophylline in whole blood in a sensor designed by the ORD Corporation. These sensors are primarily for single use, and one fiber is used for each measurement. Nonspecific adsorption to the fiber surface, which is a serious interference in such sensors, can be reduced by using surface passivating films of proteins such as bovine serum album.

Complex fiber-optic sensors have been developed (Michael *et al.*, 1998). These use imaging fiber optic bundles (with as many as thousands of individual fibers) to create an array of multiple chemical sensors. By selectively illuminating the proximal end of single fibers one at a time (or in patterns), it is possible to photopolymerize selective chemistries onto the end of the illuminated fibers. This allows the creation of arrays of probes at the end of the bundle with as many different chemistries as there are polymerization steps. This approach has been applied to a variety of different chemical detection scenarios (While *et al.*, 1998; Healey *et al.*, 1997a, b).

Today the most rapidly growing type of waveguide sensor actually does not use a waveguide at all. In surface plasmon resonance (SPR), light is totally internally reflected from an interface that is coated with a very thin film metal such as gold. At a specific phase matching angle, light from the incoming beam is coupled strongly into plasmons in the metal surface film, where it propagates for a short distance. The advantage of this technique over conventional waveguide approaches is that a very large fraction of the input beam is coupled into a very thin layer near the surface. Since the phase matching condition is strongly dependent on the refractive index of medium adjacent to the metal film, but in the low-index layer, small changes in the protein content in that layer can be easily detected. Because the analyte can be measured with no requirement that it be modified by an optical or chemical label, this approach is inherently less expensive and faster than other techniques with comparable sensitivity. Commercial instruments in planar and fiber-optic configurations have become available, and SPR monitoring of proteins and other molecules is rapidly becoming a standard research laboratory technique. As yet there are no commercial instruments for *in vivo* use based on SPR.

Acoustic/Mechanical Sensors

Binding of material to surfaces changes its mass, which can change either the object's resonant frequency or the velocity of vibrations propagated through it. This has allowed development of sensors called surface acoustic wave (SAW) or bulk acoustic wave (BAW) detectors that are based on oscillating crystals. Sensitive detection of analytes is relatively easy in the gas phase, and while there have been reports of selective detection of analytes using immobilized antibodies, there is still controversy as to how or if the technique works when the oscillating detector is in contact with liquid. It is, however, unlikely that this technique will prove applicable to *in vivo* use, where some nonspecific adsorption of protein is almost unavoidable.

Thermal and Phase Transition Sensors

Chemical reactions can give up heat because they involve breaking and formation of chemical bonds, each of which has a characteristic enthalpy. There is also a strong effect of the heats of solution of the substrates and products, particularly charged species. Many enzymatic reactions release 25 to 100 kJ/mol, or 5 to 25 kcal/mol (Table 6). A 1-mM solution of substrate completely enzymatically converted to product with a 5-kcal/mol heat of reaction would increase in temperature by 0.005°C, which is readily measurable in laboratory conditions. Sensors based on this principle are in use as detectors in chromatography and in principle could be applied to almost any enzymatic reaction. Some reactions have little or no heat production (e.g., ester hydrolysis, such as the acetylcholinesterase reaction) but can be observed using "tricks" such as coupling the reaction to the heat of protonation of a buffer such as Tris:

$$\text{Acetylcholine} \rightarrow \text{H}_3\text{CCO}_2\text{H} + \text{choline} \qquad \Delta H \approx 0 \text{ kJ/mol}$$

$$\text{H}_3\text{CCO}_2\text{H} + \text{Tris}^- \rightarrow \text{H}_3\text{CCO}_2- + \text{TrisH} \quad \Delta H = -47 \text{ kJ/mol}$$

Alternatively, a sequence of enzymes such as glucose oxidase followed by catalase can be used, which converts the hydrogen peroxide produced by the oxidase to O_2 and water in another exothermic reaction (Danielsson and Mosbachs 1987). However, the technical difficulties in making such measurements in the thermally noisy environment of the human body have so far prevented application of this principle to development of *in vivo* sensors.

An alternative thermal approach is to use the depression in phase transition temperatures of pure compounds by dissolving in them dissimilar small molecules that prefer the fluid

TABLE 6 Enthalpies of Some Enzymatic Reactions

Enzyme	Substrate	$-\Delta H$ (kJ/mol)
Catalase	H_2O_2	100
Cholesterol oxidase	Cholesterol	53
Glucose oxidase	Glucose	80
Hexokinase	Glucose	28
Lactate dehydrogenase	Sodium pyruvate	62
NADH dehydrogenase	NADH	225
Penicillinase	Penicillin-G	67
Urease	Urea	61

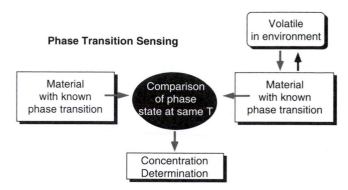

Phase Transition Sensing

FIG. 9. A schematic diagram of the process of phase transition sensing, which is an application of the well-known purity dependence of phase transition temperatures. Some diagnostic technique must be applied to allow a quantitative comparison of the phase states of two samples of the material, one of which is in equilibrium with small molecules in the environment, and another of which is at the same temperature, but chemically isolated.

phase over the crystalline phase. If the temperature is known, the concentration of the small molecule can be determined by the extent of the freezing-point depression (Fig. 9). This principle has been successfully applied to the detection of general anesthetics and is now being applied to the development of a fluorescence-based fiber-optic probe for *in vivo* use (Merlos, 1989; Merlo *et al.*, 1990).

Biomembrane-Based Sensors

A complex biological system that has been applied to the development of sensors is the biological membrane and the lipids and proteins that make it up. Numerous approaches have been made to apply lipid bilayers to chemical detection, including at least two sensors for general anesthetics (Merlo, 1989; Merlo *et al.*, 1990; Wolbeis and Posch, 1985). Membrane receptor proteins are responsible for transducing many important biological binding events and could be used to great advantage for monitoring such chemicals as hormones, neurotransmitters, and neuroactive drugs. Several schemes have been tried to this end, including immobilizing ligand-gated ion channel receptor proteins in fiber-optic devices and measuring the binding of fluorescently labeled ligands, and reconstituting them into defined lipid monolayers on solid electrodes (Eldefrawi, *et al.*, 1988) and across holes as bilayers (Ligler *et al.*, 1988). These electrical techniques promise the most sensitive detection, as a single channel opening can be monitored electrically, but also involve some of the most difficult technical challenges, including stabilizing of the normally fragile lipid bilayer. An ancillary benefit of the use of biomembranes is that phospholipids have been reported to enhance the biocompatibility of biomaterials, so they may have a dual role in bilayer-based sensors.

Microfabrication-Based Sensors

The use of microfabrication has become a central tool in the development of many types of sensors. The first attempt to meld semiconductors with chemical and biosensors was use of the chemical field-effect transistor, which has now become

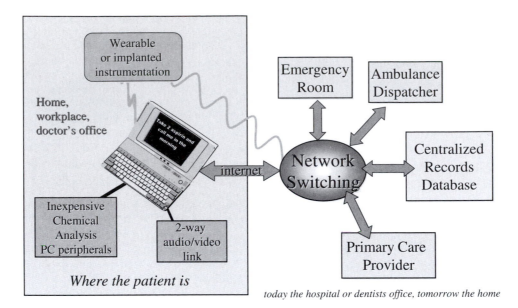

FIG. 10. A concept for a new type of doctor-patient interface as part of vision of distributed diagnosis and home healthcare (D2H2). By utilizing the existing infrastructure for wired and wireless data communication, as well as existing capabilities for storing large amounts of patient data, it will be possible in the near future to allow patients to maintain an up-to-date record of many different health parameters without frequent visits to hospitals and physicians. For biomaterials purposes, the most important data will concern the status of implanted devices and systems.

a commercial product, as least for the measurement of pH. Subsequent use of microfabrication has focused on fluidic channels, optical windows, and electrodes with dimensions ranging from millimeters to micrometers. Foremost among the applications of microfabrication has been the forming of small channels in insulating materials (such as glass or plastic) for capillary electrophoresis. Although strictly speaking a chromatographic technique and not a sensor technique, integrated systems that incorporate both microcapillary electrophoresis and optical or electrical detection have so many of the functions of a sensor that the differentiation is perhaps no longer meaningful. The strong interest in making highly parallel arrays of capillaries for high-throughput screening and DNA diagnostics has prompted intense development activity in microcapillary arrays both in academe and in industry.

The small dimensions of microchannels allow new types of devices to be designed that have no parallel in the macroscopic world. Several laboratories have lately exploited the properties unique to the low-Reynolds-number flows usually found in microdevices to create novel sensors and sensor systems. The primary use for such components has, so far, been in tabletop instruments. However, there is every reason to believe that such devices will soon work their way into *in vivo* systems as well.

MICROTOTAL ANALYTICAL SYSTEMS AND THE FUTURE

In the near future, several factors can be expected to influence the development of sensors. Once it has been demonstrated that a given type of sensor has practical value, the process of making it cheaply and reproducibly becomes supremely important in determining whether it sees the marketplace. Automation of the fabrication of small *in vivo* probes will be a high priority in the next few years. There will probably be a great increase in chemical sensors that are manufactured from the start with silicon microfabrication in mind, rather than the current practice, which is generally scaling beaker chemistry down to the size of microchips. The problem of ensuring that a sensor is as biocompatible as possible, while maintaining its function, will continue to be the most pressing problem for *in vivo* use for some time. Advances must continue to be made in materials, probe shape and size, site of use, and manufacturing. There will continue to be a strong emphasis on development of noninvasive techniques that will avoid the difficulties of biocompatibility. It may well be that near-infrared spectroscopy and magnetic resonance imaging techniques will be able to provide sufficient chemical information to diagnose some disease states. The biomolecules employed in biosensors have so far been restricted to natural enzymes and antibodies, but there is every reason to expect that as it becomes more common to tailor molecules for particular jobs, we will be able to improve on nature for transduction of chemical events.

One of the most exciting areas in chemical sensors research is microfabricated total analytical systems, or μTAS. Its great appeal is the potential to place complete miniaturized chemical sensing instruments in many more locations than is now possible. Current capabilities for both wired and wireless communication have advanced to the point that we can envision dense and extensive networks of sensors connected to central processing facilities for applications ranging from environmental monitoring to process control, agriculture, and biomedical diagnostics. It is also clear that medical care can be greatly improved by moving more biomedical diagnostics out of the centralized medical laboratory and into the operating room, the ambulance, by the patient's bedside (both in and out of the hospital), and into the medicine cabinet at home (see Fig. 10). With increasing reliance on complex courses of medication to improve and extend the lives of an aging population, there is a large potential market for medical chemical monitoring. Microfluidics will play a central role in enabling instruments capable of complex chemical measurements in instruments that will be small enough to be practical. Microfluidics can be viewed as an enabling technology in the decentralization of medicine.

Several near-term developments could be very advantageous for the development of chemical and biochemical sensors. These include the following:

- Development of microfluidic automation of most common sample preconditioning steps to eliminate the need for restricting chemical testing to use only by trained personnel
- Availability of a wide range of both semi-permanent and single-use disposable microfluidic chemical analytical systems for complex chemical analysis in packages the size of a cellular telephone
- Elimination of the need for large centralized laboratories and the associated long lag between sample collection and receipt by the user of interpreted chemical information
- Ready availability of inexpensive small sensors connected via wire and wireless links to networked two-way data transmission systems—"distributed chemical sensing networks"

The distribution of biomedical analysis of fluids to remote sites could not only improve medical care, but also empower the individual to be more active in the maintenance of his or her own health.

Clearly, this technology must be inexpensive, chemically versatile, and relatively accurate to be successful as a commercial product. The size of samples of blood will have to be on the order of a few drops. This immediately brings us into the realm of microfluidics. Practical implementations of microfluidics require dealing with samples far more complex than those regularly introduced into instrumentation with such narrow channels. Microfluidics systems are very vulnerable to problems inherent in unrefined samples to be encountered in the scenarios for which these instruments are being proposed. For example, many of the best sensing technologies in the analytical chemist's arsenal are not suited to complex mixtures of analytes with overlapping interfering signals. This is true for both optical and electrochemical detection methods. Surface fouling is a many-faceted problem in small channels, in that it can lead to converting every fluid transport channel into a chromatography column or, in the worst cases, can lead to loss of the entire sample to an irreversibly adsorbed layer upstream of the detector (see Chapter 2.13, "Nonfouling Surfaces"). Electrochemical sensors have always been vulnerable to fouling, but in microfluidic systems, the electrodes will usually not be removable for refurbishment by the user. Some types of ancillary techniques, such as electro-osmotic pumping, dependent as they are on the nature of the chemistry of the channel walls, are particularly susceptible to disruption by sample-to-sample variability.

Finally, colloidal suspensions, whether biological, organic, or inorganic in nature, are a threat to continuous operation. Depending on individual size, charge, and state of aggregation, they can clog small channels, remove coatings on channel surfaces, and interfere with optical measurements. The severity of the problems caused by fouling and clogging depends on whether the analytical instrument is planned for continuous multisample analysis or is designed around disposable single-use sample-contacting components. However, the presence of particles is a serious problem in both cases and has not been adequately dealt with in all but a few instrument designs. These instruments will be moving to nonlaboratory situations and environments in which users will have little or no training, so the instrument itself must perform nearly all sample preconditioning steps. The user will not perform anything more complicated than putting samples in the machine, and perhaps not even that!

To make point-of-care medical diagnostics work outside the hospital or doctor's office, the patient has to self-sample the blood. Current methods for removing a drop or two from the finger are too painful to be used frequently or casually. As a role model, blood-sucking insects have developed ways to remove small volumes of blood from the human body with little notice by the human. MEMS-based devices that could emulate the mosquito would enable the development of user-friendly monitors of blood chemistry.

Regardless of the nature of the sensor, it is clear that the continuing reduction in the size and cost of computers will be reflected in increasing use of sensors to provide crucial input in "smart" devices, be they for the control of drug delivery or of prosthetic limbs. Automated health-care delivery systems will reduce the need for reliance on the constant vigilance of overloaded hospital personnel and allow the chronically ill to be monitored and treated outside of a hospital setting.

SUMMARY

Physical and chemical sensors are already important in the diagnosis and treatment of the critically and chronically ill. New physical, chemical, biochemical, and biological sensing technologies are currently under development that could greatly augment our current *in vivo* capabilities. Biocompatibility remains the most important problem for all such sensors, particularly for biosensors and other chemical sensors in which transport of material is vital to function. The economic and human impact of improving the lot of diabetics is such that success of such a sensor will open the way for many other *in vivo* biosensors.

Acknowledgments

I thank the faculty and students of the Biomaterials group of the Center for Bioengineering for acquainting me with the state of the art in solutions to the biocompatibility problem, and my own students and postdocs for helping me continue to learn about sensing.

Bibliography

Ammann, D. (1986). *Ion Selective Microelectrodes; Principles, Design and Application.* Springer-Verlag, Berlin.

Benaim, N., Grattan, K. T. V., and Palmer, A. W. (1986). Simple fibre optic pH sensor for use in liquid titrations. *Analyst* **111**: 1095–1097.

Blackburn, G. F. (1987). Chemically sensitive field effect transistors. in *Biosensors*, A. P. F. Turner, I. Karube, and G. S. Wilson, eds. Oxford Science Publications, Oxford, pp. 481–530.

Collison, M. E., and Meyerhoff, M. E. (1990). Chemical sensors for bedside monitoring of critically ill patients. *Anal. Chem.* **62**: 425A–437A.

Danielsson, B., and Mosbach, K. (1987). Theory and application of calorimetric sensors. in *Biosensors*, A. P. F. Turner, I. Karube, and G. S. Wilson, eds. Oxford Science Publications, Oxford, pp. 575–595.

Eisenman, G. (1967). *Glass Electrodes for Hydrogen and Other Cations.* Marcel Dekker, New York.

Eldefrawi, M. E., Sherby, S. M., Andreou, A. G., Mansour, N. A., Annau, Z., Blum, N. A., and Valdes, J. J. (1988). Acetylcholine receptor-based biosensor. *Anal. Lett.* **21**: 1665–1680.

Gehrich, J. L., Lubbers, D. W., Opitz, N., Hansmann, D. R., Miller, W. W., Tusa, J. K., and Yaafuso, M. (1986). Optical fluorescence and its application to an intravascular blood gas monitoring system. *IEEE Trans. Biomed. Eng.* **33**: 117–131.

Gerritsen, M., Jansen, J. A., Kros, A., Nolte, R. J., and Lutterman, J. A. (1998). Performance of subcutaneously implanted glucose sensors: a review. *J. Invest. Surg.* **11**: 163–174.

Gough, D. A., Armour, J. C., Lucisano, J. Y., and McKean, B. D. (1986). Short-term *in vivo* operation of a glucose sensor. *Trans. Am. Soc. Artif. Intern. Organs.* **32**: 148–150.

Hafeman, D. G., Parce, J. W., and McConnell, H. M. (1988). Light-addressable potentiometric sensor for biochemical systems. *Science* **240**: 1182–1185.

Healey, B. G., Matson, R. S., and Walt, D. R. (1997a). Fiberoptic DNA sensor array capable of detecting point mutations. *Anal. Biochem.* **251**: 270–279.

Healey, B. G., Li, L., and Walt, D. R. (1997b). Multianalyte biosensors on optical imaging bundles. *Biosens. Bioelectron.* **12**: 521–529.

Hu, Y., and Wilson, G. S. (1997). A temporary local energy pool coupled to neuronal activity: fluctuations of extracellular lactate levels in rat brain monitored with rapid-response enzyme-based sensor. *J. Neurochem.* **69**: 1484–1490.

Ishikawa, M., Schmidtke, D. W., Raskin, P., and Quinn, C. A., (1998). Initial evaluation of a 290-micron diameter subcutaneous glucose sensor: glucose monitoring with a biocompatible, flexible-wire, enzyme- based amperometric microsensor in diabetic and nondiabetic humans. *J. Diabetes Complications* **12**: 295–301.

Jain, K. K. (2003). Current status of molecular biosensors. *Med. Device Technol.* **14**: 10–15.

Jones, T. P., and Porter, M. D. (1988). Optical pH sensor based on the chemical modification of a porous polymer film. *Anal. Chem.* **60**: 404.

Jordan, D. M., Walt, D. R., and Milanovich, F. P. (1987). Physiological pH fiber-optic chemical sensor based on energy transfer. *Anal. Chem.* **59**: 437.

Kohli-Seth, R., and Oropello, J. M. (2000). The future of bedside monitoring. *Crit. Care Clin.* **16**: 557–578.

Ligler, F. S., Fare, T. L., Seib, K. D., Smuda, J. W., Singh, A., Ahl, P., Ayers, M. E., Dalziel, A., and Yager, P. (1988). Fabrication of key components of a receptor-based biosensor. *Med. Instru.* **22**: 247–256.

Merlo, S. (1989). Development of a fluorescence-based fiber optic sensor for detection of general anesthetics. University of Washington, Seattle, WA.

Merlo, S., Yager, P., and Burgess, L. W. (1990). An optical method for detecting anesthetics and other lipid-soluble compounds. *Sensors Actuators* **A21–A23**: 1150–1154.

Michael, K. L., Taylor, L. C., Schultz, S. L., and Walt, D. R. (1998). Randomly ordered addressable high-density optical sensor arrays. *Anal. Chem.* **70**: 1242–1248.

Nakamura, H., and Karube, I. (2003). Current research activity in biosensors. *Anal. Bioanal. Chem.* **377**: 446–468.

Peterson, J. I., Fitzgerald, R. V., and Buckhold, D. K. (1984). Fiber-optic probe for *in vivo* measurement of oxygen partial pressure. *Anal. Chem.* **56**: 62–67.

Reach, G., and Wilson, G. S. (1992). Can continuous glucose monitoring be used for the treatment of diabetes? *Anal. Chem.* **64**: 381–387.

Rolfe, P. (1990). *In vivo* chemical sensors for intensive-care monitoring. *Med. Biol. Eng. Comput.* **28**: B34–B46.

Saari, L. A., and Seitz, W. R. (1982). pH sensor based on immobilized fluoresceinamine. *Anal. Chem.* **54**: 821–823.

Schweiss, J. F. (1983). Continuous measurement of blood oxygen saturation in the high risk patient. Abbot Critical Care Systems.

Shichiri, M., Sakakida, M., Nishida, K., and Shimoda, S. (1998). Enhanced, simplified glucose sensors: long-term clinical application of wearable artificial endocrine pancreas. *Artif. Organs* **22**: 32–42.

Simpson, D. L., and Kobos, R. K. (1982). Microbiological assay of tetracycline with a potentiometric CO_2 gas sensor. *Anal. Lett.* **15**: 1345–1359.

Thome-Duret, V., Reach, G., Gangnerau, M. N., Lemonnier, F., Klein, J. C., Zhang, Y., Hu, Y., and Wilson, G. S. (1996). Use of a subcutaneous glucose sensor to detect decreases in glucose concentration prior to observation in blood. *Anal. Chem.* **68**: 3822–3826.

Turner, A. P. F. (1989). Current trends in biosensor research and development. *Sensors Actuators* **17**: 433–450.

Turner, A. P. F., Karube, I., and Wilson G. S. E. (1987). *Biosensors: Fundamentals and Applications.* Oxford University Press, Oxford.

Vo-Dinh, T., and Cullum, B. (2000). Biosensors and biochips: advances in biological and medical diagnostics. *Fresenius J. Anal. Chem.* **366**: 540–551.

Wang, J., and Varughese, D. (1990). Polishable and robust biological electrode surfaces. *Anal. Chem.* **62**: 318–320.

West, J. L., and Halas, N. J. (2003). Engineered nanomaterials for biophotonics applications: improving sensing, imaging and therapeutics. *Annu. Biomed. Eng.* **9**: 1–149.

White, J., Dickinson, T. A., Walt, D. R., and Kauer, J. S. (1998). An olfactory neuronal network for vapor recognition in an artificial nose. *Biol. Cybern.* **78**: 245–251.

Wolfbeis, O. S., and Posch, H. E. (1985). Fiber optical fluorosensor for determination of halothane and/or oxygen. *Anal. Chem.* **57**: 2556–2561.

Zhujun, Z., and Seitz, W. R. (1984). A carbon dioxide sensor based on fluorescence. *Anal. Chim. Acta.* **160**: 305–309.

Ziegler, C. (2000). Cell-based biosensors. *Fresenius J. Anal. Chem.* **366**: 552–559.

7.18 DIAGNOSTICS AND BIOMATERIALS

Peter J. Tarcha and Thomas E. Rohr

INTRODUCTION

The ability to make quantitative measurements is critical to progress in any technical discipline. In medical diagnostics, very small quantities of analytes often need to be quickly and accurately measured in complex biological mixtures.

TABLE 1 Typical Analytes in Solid Phase Immunoassay

Category	Example
1. Clinical	
a. Therapeutic drugs	a. Digoxin
b. Hormones	b. Thyroid stimulating hormone
c. Pregnancy/fertility	c. Human chorionic gonadatrophin/ estradiol
d. Cardiac markers	d. Troponin I
e. Infectious disease	e. Hepatitis B surface antigen
f. Hemotological analytes	f. Ferritin
g. Cell surface markers	g. CD4/CD8 ratio
h. Cancer	h. Prostrate specific antigen
i. Allergy	i. Immunoglobulin E
j. Genetic testing	j. DNA sequences for paternity/ forensics/heredity
k. DNA probes	k. Hepatitis C
2. Other	
a. Agricultural	a. Aflatoxin
b. Environmental, e.g., pesticides	b. Diazinon
c. Veterinary	c. Canine heartworm

This is accomplished by using the ability of certain pairs of biomolecules to bind to one another at high affinities. Individual members of such a pair of molecules are generally referred to as ligands, and as members of a ligand binding pair. High-affinity binding confers high specificity, and the use of binding molecule pairs has allowed the design of assays which have revolutionized the field of medical diagnostics. Examples of some current diagnostic assays utilizing binding molecules are shown in Table 1.

The most important use of a biomaterial in a binding molecule-based clinical diagnostic assay is to serve as a compatible solid phase or as a support to which specific binding molecules, such as antibodies and antigens, will be attached. These immobilized binding molecules specifically capture analytes or other ancillary reagents from a test mixture. In the most common assay formats, the test mixture is then removed and the solid phase washed by addition and removal of a wash solution. Subsequent assay steps involve the addition and removal of required reagents, again followed by wash steps. In earlier formats, the inside of a polystyrene test tube or microtiter plate well often served as the solid phase. More recently, microparticles and nanoparticles have replaced coated vessels as the solid phase of choice. Particles provide a larger total surface area upon which to immobilize binding molecules and can be manufactured as liquid suspensions. A dispersed configuration also greatly reduces the diffusion distances of soluble reagents to the solid phase, shortening the incubation times required for recognition and binding. Furthermore, since the reaction vessels are not coated with binding molecules, they need no special handling, are much less expensive to produce, and are suitable for use with any of the assays of a product line. The use of microparticles can, however, complicate the separation of the solid-phase reagent from other components of the test mixture. Much of the current activity in the development of new ligand binding assay formats involves either automation of the steps of traditional assay schemes or design of homogeneous assay formats that do not require separation or wash steps. Four examples of traditional ligand binding immunoassay formats are illustrated in Fig. 1.

This chapter will not review the chemical properties, methods of characterization, or surface modifications of the many solid-phase biomaterials currently used in medical diagnostics. This subject has been reviewed previously (Tarcha, 1991) in a treatise on the theory and practice of solid-phase immunoassay (Butler, 1991). The current chapter is intended to present trends and new concepts in the use of biomaterials for medical diagnostics, many of which have not yet been commercialized.

NEW SOLID-PHASE MATERIALS FOR LIGAND BINDING ASSAYS

Particles

Some of the very attributes that make micro- and nanoparticulate solid-phase reagents attractive for use in ligand binding assays can also be detrimental. Unlike the case with coated reaction vessels, the separation of a microparticle solid phase reagent from other reaction components can be difficult. Particles can usually be separated from a reaction mixture by centrifugation; however, when there is no significant difference in density between the particle and the continuous phase, high g forces may be necessary. Centrifugal sedimentation of protein-coated microparticles can cause them to become aggregated, necessitating the use of energetic means (e.g., sonication) to redisperse them. In addition, assay formats using centrifugal force are not easily automated.

Microparticles can be separated from other reagents by capture upon appropriate filters or membranes, but subsequent recovery for further reaction steps can be difficult. Examples of commercially successful semiautomated clinical analysis systems using filter recovery of microparticles include the IMx and AxSym instruments marketed by Abbott Laboratories (North Chicago, IL). These instruments utilize microparticle capture enzyme immunoassay (MEIA) technology, which efficiently separates bound and unbound immunoreactants by the capture and washing of the microparticle solid-phase reagent on a glass fiber matrix. Further assay steps and signal development are performed on this matrix (Fiore *et al.*, 1988).

A unique reagent system has been described that used solution kinetics for the ligand binding reactions, followed by the formation of a microprecipate solid-phase *in situ* (Monji and Hoffman, 1987). The system took advantage of a water-soluble, thermally precipitating polymer, poly(N-isopropylacrylamide), that was conjugated to a monoclonal antibody. This polymer precipitated reversibly from water above a critical temperature of 31°C, enabling bound immune complexes to be precipitated, centrifuged, and washed repeatedly. Nonspecific binding, which normally produces backgrounds in conventional solid-phase systems, was low because of the hydrophilic nature of the polymer and resulting precipitate.

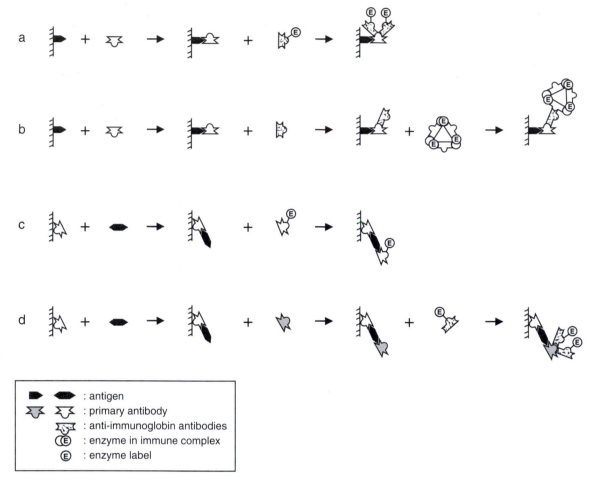

FIG. 1. Schematic outlines (a–d) for noncompetitive solid-phase enzyme immunoassay formats. The solid phase comprises antigen (formats a, b) or antibody (formats c, d) immobilized on a solid support. In the first incubation, these immobilized binding molecules capture the soluble analyte (antibody in formats a and b, antigen in formats c and d) from the sample. Different methods can be used to detect the captured analyte, indirectly with labeled antibody (formats a and c) by a bridge (format b), or by more elaborate indirect procedures (format d). (Reprinted with permission from Tijssen, 1985, copyright 1985, Elsevier Science.)

Intrinsically Colored Particles

In the past 7 or 8 years there has been an increasing use of rapid, "self-performing" assays utilizing chromatography strips (Pope *et al.*, 1996, 1997a, b; Tarcha *et al.*, 1991). In such a format, a binding molecule such as an antibody or antigen is immobilized in a "capture zone" midway along a porous strip. Colored colloidal particles coated with binding molecules are deposited between the capture zone and the site of test sample application. The particles become hydrated and are carried along with the test sample as it moves along the strip by capillary action from its site of application. Test samples can be applied by pipette or dropper, or by dipping one end of the strip in the test sample. In a direct "sandwich" assay, binding molecules on the colored particles capture analyte from the test sample as they travel along the strip. The particles become captured and produce a colored region when the immobilized binding molecules of the capture zone bind to analyte that has bound to the particles. The analyte becomes "sandwiched" between the binding molecules immobilized on the strip and those on the particles. In a "competitive" format, either the binding molecules on the particles or of the capture zone are analogs of the analyte. In the absence of free analyte, the binding molecules on the particles will bind directly to the binding molecules of the capture zone of the strip. The presence of analyte in the test sample inhibits binding of the colored particles at the capture zone as analyte molecules occupy the binding sites of one set of binding molecules. For qualitative assays these events can be observed visually without the aid of an instrument. Quantitative results can be obtained with the use of surface densitometers, colorimeters, or fluorometers.

Latex particles that have been imbibed with dyes are available commercially and have been used for tests of this sort. Higher apparent extinction coefficients can be gained through the use of particles that are intrinsically colored because of their chemical structure. For example, colloids made of selenium have been used as red labels in visual strip-based immunoassays (Yost *et al.*, 1990). Methods of producing selenium colloids for use in immunoassays and their surface properties have been described (Mees *et al.*, 1995).

Nanoparticles made of polypyrrole have also been used as intrinsically colored labels in immunoassays. The intense black color of polypyrrole (apparent molar absorption coefficient $\varepsilon_{max} \sim 10^5$ L mol^{-1} cm^{-1}) is presumably due to electronic transitions within the conjugated chain structure. These particles can be functionalized to permit covalent attachment of binding molecules (Tarcha *et al.*, 1991). More recent work (Bieniarz *et al.*, 1999) has indicated that intentional surface modification for covalent antibody immobilization may not be necessary because of the propensity of nucleophiles such as thiols and amines to add to the electrophilic sites of the polypyrrole. Composite nanoparticles of colloidal silica combined with polypyrrole were developed at the University of Sussex (Maeda and Armes, 1993, 1994; Maeda *et al.*, 1995) and have also been shown to function well in immunoassays (Pope *et al.*, 1996).

Magnetically Responsive Particles

Current Strategies The desire to automate microparticle-based ligand binding assays has led to a search for alternatives to centrifugation or filtration as a means of separating microparticles during reagent changes and wash steps. One approach is to use microparticles that will respond to a magnetic field. Application of an external magnetic field to a reaction vessel can cause magnetically responsive microparticles in the test mixture to migrate to the vessel wall, where they can be held by the field while reagent changes and wash steps are performed. Removal of the field releases the microparticles for reaction with the next assay reagent. For this assay format to be practical, the coated microparticles must have surface properties that make them resist irreversible aggregation during magnetic capture. Furthermore, although the particles must exhibit a strong degree of magnetization while in a magnetic field, their magnetization must be lost when the field is removed; otherwise they will continue to attract one another in the absence of the magnetic field and be difficult to disperse.

Materials displaying these magnetic properties are referred to as being *superparamagnetic*.

For a material to be superparamagnetic, the individual particles of magnetic material must be so small, typically 2–20 nm, that they constitute only a single magnetic domain, and they must be kept dispersed so no permanent long-range magnetic order can form in an applied magnetic field. Superparamagnetic microparticles can be produced by dispersing nanoparticles of magnetically responsive material within colloidal particles or by distributing them within the voids of porous microparticles. Superparamagnetism is also displayed by liquid dispersions of monodomain magnetic nanoparticles. Such dispersions are referred to as *ferrofluids*.

Superparamagnetic microparticles suitable for use in diagnostic ligand binding assays are available commercially from several suppliers (Table 2). For example, Bang's Labs distributes a superparamagnetic microparticle preparation produced by dispersing monodomain Fe_3O_4 nanoparticles in colloidal polystyrene. The result is polydisperse with an average particle diameter of 1 μm. This type of particle will have some of the Fe_3O_4 nanoparticles exposed on the microparticle surface, which can cause inactivation of some sensitive biological reagents, especially enzymes. For these cases, particles having a polystyrene coat applied over the Fe_3O_4-bearing particle are also offered. Polysciences, Inc., produces superparamagnetic "Biomag" microparticles by applying an aminosilane coating to magnetite nanoparticles. The resulting nanoparticle aggregates are somewhat polydisperse with diameters centered around 1.5 μm. Monodisperse superparamagnetic microparticles are produced by Spherotech, Inc., and Interfacial Dynamics Corp. A monodisperse polystyrene core particle is coated with a shell containing Fe_2O_3 nanoparticles, then with an outer shell of polystyrene. Because the magnetite is contained only in a shell region, these particles usually cannot display as strong a magnetic response as those having a uniform dispersion of magnetite throughout the particle. Superparamagnetic microparticles available from Dynal begin with a porous polyurethane latex core. Nanocrystals of Fe_2O_3 and Fe_3O_4 are precipitated within the pores of the core particle, then a

TABLE 2 Suppliers of Magnetically Responsive Particles Useful in Ligand Binding Assays

Supplier	Particle size	Composition	Comments
Bang's Labs, Indianapolis, IN	1 μm	Fe_3O_4 nanoparticles dispersed in colloidal polystyrene	Polydisperse
Interfacial Dynamics Corp., Portland, OR	2.8 μm	Colloidal polystyrene core coated with Fe_2O_3 shell, then polystyrene outer shell	Monodisperse
Polysciences, Inc., Warrington, PA	1.5 μm	Aggregates of aminosilane-coated magnetite nanoparticles	Polydisperse
Spherotech, Inc., Libertyville, IL	1.0, 2.5, 4.0, 7.0 μm	Colloidal polystyrene core coated with Fe_2O_3 shell, then polystyrene outer shell	Monodisperse
Dynal, Inc., Lake Success, NY	2.8, 4.5, 5.0 m	Fe_2O_3 and Fe_3O_4 precipitated inside porous microparticle	Monodisperse
CPG Corp., Lincoln Park, NJ	5 μm	Magnetite precipitated inside controlled-pore glass particles	Porous particles
Miltenyi Biotec Inc., Auburn, CA	30–70 nm	Magnetite core particle with dextran coat	Nanoparticles
Immunicon Corp., Huntindon Valley, PA	130–170 nm	Magnetite particles with protein overcoat	Nanoparticles

polystyrene coat is applied, resulting in a nonporous particle. CPG Corp. offers large, magnetically responsive porous glass particles in two size ranges of 37–74 and 74–125 μm diameter, with pore sizes ranging from 500 to 1000 Å.

Many of these microparticles are available with their surfaces modified with carboxyl, amino, or other chemical groups, which alter surface characteristics and allow the covalent attachment of binding molecules. Commercial suppliers also offer microparticles with various binding molecules already attached.

Another use of magnetically responsive reagents is in the sorting of particular types of living cells. Particles coated with binding molecules specific for unique surface ligands of certain types of cells will bind only to that cell type in a complex mixture of other cell types, such as whole blood. The cell–particle aggregate can then be separated from the mixture by the application of a magnetic field. For this application, aqueous-compatible ferrofluids are advantageous. Ferrofluids that can be dispersed in aqueous solutions are produced by dispersing nanocrystals of magnetite in very small (5–50 nm diameter) colloidal particles. Because of their small size, dispersed ferrofluids do not readily move in an applied magnetic field. The binding of multiple ferrofluid particles to a cell, however, can create an aggregate magnetic force on the cell–ferrofluid complex that is strong enough to achieve its separation. Use of ferrofluids has the advantage that unbound ferrofluid particles will not be captured. Immunicon and Miltenyi offer aqueous suspensions of somewhat larger nanoparticles with attached binding molecules for this purpose (Table 2).

The advantage of magnetically responsive reagents for automated immunodiagnostics is reflected in the fact that several of the commercial leaders in the field are using this technology in their current high-throughput immunoassay analyzers. Examples include Abbott Laboratories' Architect, Bayer's ACS 180 and Immuno 1, Beckman-Colter's ACCESS, Tosoh's AIA-1200 DX, and Johnson and Johnson's VITROS systems.

New Strategies Measurement of Magnetic Force   An alternative method to simplify medical diagnostic immunoassay formats is to use the solid phase itself as the signal generator, eliminating the need for additional reagents and steps. Measurement of the magnetic force exerted upon magnetically responsive reagents by a magnetic field can serve as the assay readout signal. In a "sandwich" format, superparamagnetic microparticles coated with ligand binding molecules specific for a particular analyte are mixed with the test mixture and a solid phase that has also been coated with ligand binding molecules specific for the analyte. Presence of the analyte in the test mixture will cause some of the microparticles to become bound to the solid phase. Unbound or weakly bound particles can be removed by application of a magnetic field. The reaction vessel can then moved near a magnet attached to a force-sensing device and the force exerted upon the magnet by the bound particles measured. The force exerted upon a typical superparamagnetic microparticle is more than nine orders of magnitude greater than that exerted by gravity upon a binding molecule the size of an antibody. This amplification allows assay sensitivities comparable to those of enzyme immunoassays to be achieved using conventional force measuring devices

such as electronic balances (Rohr, 1995). The resultant assay system contains only binding molecules and microparticles, greatly simplifying reagent production and stability. The binding of magnetically responsive microparticles to solid phases has also been measured using atomic force microscopes and micromachined cantilever devices (Baselt *et al.*, 1998).

Single-step assay formats are also possible using magnetically responsive microparticles. In Fig. 2A, a reaction vessel (RV) has a binding molecule (Ab1) specific for the analyte immobilized upon its bottom. A second binding molecule (Ab2) specific for a different site on the analyte is attached to a magnetically responsive microparticle to form the magnetic reagent (MR). The test sample containing the analyte of interest (A) and the magnetic reagent are introduced into the reaction vessel, which is fitted with a lid (L). The binding molecules on the vessel bottom and on the microparticles bind whatever analyte is present, resulting in some fraction of the microparticles becoming bound to the vessel bottom through the analyte. The reaction vessel is placed on an elevator (E) below a magnet (M) attached to a microbalance (B). The microbalance is zeroed; then the elevator (E) moves the reaction vessel close to the magnet (Fig. 2B). The increased intensity of the magnetic field causes the unbound microparticles to be pulled to the underside of the vessel lid. Because of their close proximity, the unbound

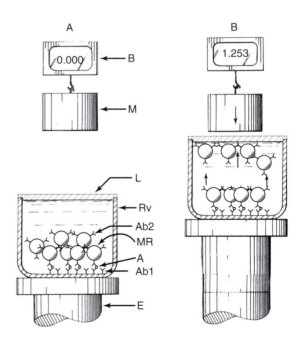

FIG. 2. (A) One-step sealed vessel magnetic force immunoassay. Antibodies (Ab1) specific for an epitope of the analyte of interest (A) are immobilized on the bottom of a reaction vessel (RV). The analyte and a magnetically responsive reagent (MR) are introduced into the reaction vessel, which is fitted with a lid (L). The magnetically responsive reagent consists of a superparamagnetic microparticle coated with a second antibody (Ab2) specific for a second epitope of the analyte. The result of incubation is the binding of the magnetically responsive reagent to the vessel bottom through the captured analyte. (B) An elevator (E) moves the reaction vessel close to a magnet (M), causing unbound magnetically responsive reagent to be pulled to the underside of the lid, where it exerts an increased force upon the magnet, increasing its apparent weight as measured by the attached balance (B) (Rohr, 1995).

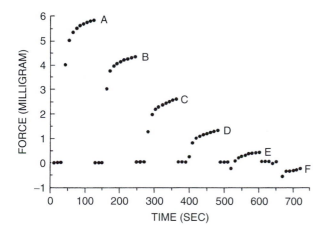

FIG. 3. Magnetic force immunoassay for alpha-fetoprotein (AFP). A test sample containing no analyte was mixed with the magnetically responsive reagent in the reaction vessel and incubated for 30 min at room temperature. The reaction vessel was then placed on the elevator in the remote position (Fig. 2A); the balance was zeroed and data collection begun. After 30 seconds, the elevator was raised to position the vessel lid close to the magnet (Fig. 2B). The apparent weight of the magnet rapidly increased as unbound magnetically responsive reagent was pulled to the underside of the vessel lid (response A), where a 6-mg weight change was observed after 90 sec. For measurements of test analyte solutions, the elevator was lowered and the reaction vessel replaced with one containing the magnetically responsive reagent and AFP at a concentration of 15 µg per ml. After 150 sec incubation, the elevator was again raised to bring the vessel close to the magnet (response B). The responses C, D, E, and F were obtained with test solutions containing concentrations of AFP of 50, 100, 200, and 350 µg/ml (Rohr, 1995).

microparticles collected under the lid exert a much stronger force upon the magnet than do those that remain bound to the well bottom. This force is displayed by the balance as an apparent change in the magnet's weight. By changing the distance between the reaction vessel and the magnet, the strength of the applied field can be adjusted to be just sufficient to pull nonspecifically bound particles off of the well bottom.

Results from an assay of this type developed for human alpha-fetal protein (AFP) are shown in Fig. 3. Assay of a solution containing no AFP resulted in virtually all of the magnetic reagent being pulled to the underside of the vessel lid, with an observed 6-mg force change (Fig. 3, trace A), or more than 10,000 times the rated sensitivity of the balance. Assay of samples containing 15, 50, 100, 200, and 350 ng of AFP resulted in decreasing force changes as more particles became bound to the well bottom through the captured AFP analyte (Fig. 3, traces B, C, D, E, and F, respectively). The sensitivity of this homogeneous immunoassay is comparable to that of current commercial immunoassays (Rohr, 1995).

Self-Assembled Monolayers

Unique possibilities for the micropatterning of metal surfaces containing specific biorecognition molecules have been shown by Whitesides and his group (Roberts *et al.*, 1998). They have engineered surfaces using self-assembled monolayers (SAMs) by designing a series of assembly molecules to promote specific binding while simultaneously inhibiting nonspecific binding. The basic structure of the assembly molecule is an alkane with a thiolate end group on one end and a poly(ethylene oxide) (PEO) spacer attached to the other. The free end of the poly(ethylene oxide) can be functionalized with a carboxylic acid group for activation and coupling with specific binding ligands. Thiols form bonds with metals such as gold and silver. When a surface of either of these metals is exposed to a thiolated SAM reagent, rapid *chemisorption* orients the carboxylated PEO outward from the surface for subsequent coupling to the specific binding molecule of interest.

Such reagents are ideally suited to analysis using a surface plasmon resonance (SPR)-based biosensor commercially available from BIAcore AB (Uppsala, Sweden). The sensor of this device can detect molecules binding to its surface and has the advantage of being able to monitor binding events in real time. SPR analysis is done by reflecting visible-to-near-infrared radiation off a textured surface that has been coated with a thin film of a coinage metal, typically silver or gold. The mobile electrons of the metal, combined with the topography of the textured surface define "surface plasmons," which are groups of electrons with inherent resonance energies. At a certain angle of incidence for a fixed wavelength of light, the oscillation frequency of the electric dipole of the incident photons will match the resonance frequency of the surface plasmons, resulting in absorption. The angle of incidence at which maximum energy absorption occurs is called the "plasmon notch" angle and can be precisely determined. Subsequent binding or absorption of anything to the surface of the sensor will cause the "plasmon notch" angle to shift (Johnsson *et al.*, 1991). The instrument is thus able to measure the extent of capture of an analyte by an immobilized antibody by measuring the resultant shift in the "plasmon notch" angle. Nonspecific binding of proteins or other substances at the interface will also give rise to a signal, which has limited the utility of the technique to some extent.

Using a BIAcore 1000 instrument, it was shown that applying mixed SAMs composed of thiol–alkane–PEO and thiol–alkane PEO–COOH significantly inhibited the nonspecific binding of the proteins lysozyme, ovalbumin, carbonic anhydrases, and fibrinogen. In contrast, nonspecific binding of the protein cytochrome *c*, which is known to interact with surface carboxylic groups, was more difficult to inhibit. The Whitesides group also described the immobilization of various other proteins to SAMs and demonstrated the measurement of biospecific binding for several analytes (Lahiri *et al.*, 1999; Rao *et al.*, 1999).

Molecularly Imprinted Surfaces

Molecular imprinting of polymeric surfaces is a process in which an analyte or other binding molecule serves as a template, around which a polymer is "molded." The template molecule is combined with one or more monomers that can be polymerized and cross-linked. Alternatively, the molecule of interest can be covalently modified through labile linkages with polymerizable groups. After polymerization, the template molecule is removed by extraction and/or chemical cleavage, leaving behind a void that has the size and shape of, and possibly electronic complimentarity to, the target analyte binding molecule (Mosbach and Ramstrom, 1996; Ramstrom *et al.*, 1996, and references contained within). Monomers are chosen such that the cross-linked imprint matrix is highly resistant to physical and chemical factors and can be reused repeatedly. Since some of the resultant binding constants for the target molecule approach those found in antibody–antigen interactions, these structures may find applicability in reusable biosensors and other diagnostic devices.

The concept of imprinting has been extended beyond the use of conventional synthetic monomers and polymers to natural polymers such as enzymes. In one example, an enzyme–inhibitor complex was precipitated in an organic solvent, locking in the secondary structure of the enzyme (Staahl *et al.*, 1991). After removal of the inhibitor, the enzyme was shown to be catalytically active in organic media.

Further extension of the concept of imprinting by taking advantage of the tremendous diversity of natural polymers as the matrix is an intriguing possibility. Work in this direction was done at the University of Washington using a radio-frequency glow discharge (RFGD) plasma deposition technique to form a polysaccharide-like surface with protein-imprinted "nanopits" (Shi *et al.*, 1999). Exposure to the templates of binary protein mixtures of bovine serum albumin and immunoglobulin G in a competitive manner revealed a significant preferential adsorption of these proteins to their respective imprints.

Surface-Enhanced Spectroscopies

Surface-Enhanced Raman Scattering

Another aspect of the surface plasmon resonance phenomenon (see "Self-Assembled Monolayers," above) is the dramatic increase in the intensity of Raman light scattering, fluorescence, and infrared absorption observed when some molecules are brought into close proximity to (but not necessarily in contact with) metal surfaces displaying surface plasmon activity. The surfaces need to be textured or coated with minute metal particles or have periodic structure such as that of an optical grating. Colloidal dispersions of certain metals, especially gold and silver, can also show these dramatic signal enhancements. In 1974 Dr. Richard P. Van Duyne first recognized this effect as a unique physical phenomenon for Raman scattering and coined the term "surface-enhanced Raman scattering" (SERS) (Jeanmarie and Van Duyne, 1974). The intensity of Raman scatter as the result of surface enhancement can be several million times greater than that observed with unenhanced solution-state spectroscopy.

In all solid-phase immunoassays up through the late 1980s, the solid phase was used to separate bound from unbound species. Tarcha and co-workers (Rohr *et al.*, 1989) realized that the surface enhancement effect could be used to design homogeneous, no-wash immunoassays. This generally involved the use of antibodies conjugated to Raman-active dyes. When these dye conjugates formed an immune complex with a binding molecule attached to a plasmon-active surface, the Raman scattering signal from the captured dye label

was enhanced because of the SERS effect. Uncaptured label was not enhanced, and hence not detected, thereby eliminating the need for separation steps. Other assay configurations that employed metal colloids to which dyes had been adsorbed were also demonstrated (Tarcha *et al.*, 1996). Using this approach, no-wash immunoassays for several large- and small-molecule analytes were performed [e.g., human chorionic gonadotropin (HCG) and theophylline, respectively].

Surface-Enhanced Sensors

Sensors based on the SERS effect have been described for the detection of gene probes (Isola *et al.*, 1998). Raman-active dye-labeled probes were shown to hybridize to amplified oligonucleotides, SERS signals being detected after deposition of a silver layer over the hybridized samples.

Sensors based on surface-enhanced infrared absorption have also been demonstrated using model systems consisting of *Salmonella* bacteria and the enzyme glucose oxidase and antibodies directed against them (Brown *et al.*, 1998). Characteristic infrared absorption bands were detected as binding molecules immobilized on a plasmon-active metal surface captured their complementary binding molecules. It was suggested that the fingerprints of the spectrum could be used for identification as well.

Reusable solid phases with potential for use in surface-enhanced fluorescence sensors have been described (Tarcha *et al.*, 1999). They were made by depositing well-defined layers of SiO_2 or SiO onto silver island films. The surface-enhancement factors observed for the fluorescence of dyes were only 10- or 20-fold, as is usually the case, but the study data were consistent with theory in that fluorescence from dyes having low quantum yields was more greatly enhanced than that from dyes with high quantum yields. In addition, the study showed that the SiO_2 coating allowed washing and reuse of the device without degradation of the metal surface's activity. The coating also helped reduce photodecomposition of adsorbates on the silver surface without substantial loss in the observed enhancement factor.

Issues related to the protein stability, biocompatibility, and reproducibility of SERS-active surfaces have been addressed by Natan and co-workers (Keating *et al.*, 1998). They made reproducible colloidal gold–protein conjugates and enhanced their SERS activity by forming aggregates of these conjugates with silver colloids. The stability of the model protein used in the conjugate, cytochrome *c*, was reported to be good.

As continued advances are made in the physical and chemical stability of the plasmon-active surfaces, in control of colloidal aggregate size, and in the biocompatibility of the surfaces with protein, the use of these surface-enhanced techniques will move beyond the research stage and into commercial instruments.

Other Biosensor Strategies

Many of the principles discussed in this chapter can be applied to the design of biosensors. A major goal of biosensor development is the production of a device that can be permanently implanted *in vivo*. Such a device could provide long-term feedback to an indwelling therapeutic pump as a closed-loop system. Insulin delivery to a diabetic patient is the most obvious application. Such sensor systems have not yet been successful commercially because of limited functional stability resulting from a loss of activity of the binding molecules. A second problem is fouling or encapsulation of the sensor, leading to decreased rates of analyte transport to the sensor surface. Implantation and reimplantation can be traumatic for the patient and always pose a risk of infection or other complications, so the device must be easily sterilized and able to function for an extended period of time before needing replacement.

Many reviews giving in-depth coverage to the current and future art of biosensor design are available (Wilkins and Atanasov, 1996; Pfeiffer, 1997; Bergveld, 1996; Wang, 1999). In addition, see chapters in this book on "Bioelectrodes" and "Biomedical Sensors and Biosensors." Here we will only mention two unique approaches to solving some classical biosensor problems. Researchers at the University of New Mexico made it possible to extend the life of an indwelling glucose sensor by recharging it with fresh immobilized enzyme (Xie and Wilkins, 1991). In the device enzyme immobilized on dispersed carbon powder was contained within a membrane that isolated it from the biological milieu. The enzyme-coated particles were replaced through recharge and discharge tubes attached to the device. Naturally, this is an invasive procedure and its commercial acceptance would be questionable if it had to be done frequently.

A second approach avoids the use of enzymes, providing for *in vivo* residence with no access port by using a noninvasive fluorescence readout through the skin, while addressing the problem of biocompatiblity and fouling (Russell *et al.*, 1999). This device uses biocompatible photopolymerized poly(ethylene glycol) hydrogel particles that contain fluorescein isothiocyanate–dextran (FITC-dextran) and tetramethylrhodamine isothiocyanate–labeled concanavalin A (TRITC-ConA). concanavalin A has a natural binding affinity for glucose and dextran, which is a polymer of glucose. In the absence of free glucose, the FITC-dextran binds to the TRITC-labeled concanavalin A, quenching the fluorescence of the fluorescein label through fluorescence energy transfer. Free glucose competes with the FITC-dextran for binding to the TRITC-labeled concanavalin A, thereby decreasing fluorescence quenching. As a result, observed fluorescence increases linearly over a glucose concentration range of 0 to 600 mg/dl. The authors demonstrate that it is possible to create a microparticle-based fluorescent glucose sensor suitable for subcutaneous implantation in a fashion similar to that of a tattoo.

LIGAND IMMOBILIZATION ON SOLID PHASES

Linker Arms

The sensitivity of a solid-phase-based immunoassay depends primarily on the quality of the antibody (high binding constant), quality of the immobilization procedure (minimal loss of binding activity after immobilization), and a low level of background signal, generally referred to as nonspecific binding.

In a sandwich-type immunoassay, the capture antibody is immobilized to the surface by adsorption or by random or specific chemical coupling. Loss of binding activity by the immobilized species can be minimized by reducing unproductive events such as denaturation or attachment in an unfavorable orientation. As shown in Fig. 1 for a sandwich immunoassay, the analyte of interest binds to the immobilized capture antibody, then a labeled second antibody binds to the bound analyte. The label on the second antibody allows detection of the binding event. Nonspecific binding of the labeled antibody to the solid phase must be minimized to avoid undesired background signal. In order to reduce nonspecific binding the solid phase is usually overcoated with one or more benign proteins or surface-active agents such as casein, bovine serum albumin, or a member of the Tween series of nonionic surfactants. As with biomaterials for *in vivo* implantation, the surface properties of biomaterials for medical diagnostics largely determine their performance, bulk properties being much less important. This subject is also covered in the chapters on "Surface Properties of Materials" and "Nonfouling Surfaces" found in this book.

Antibodies can be covalently linked to surfaces to improve their stability toward displacement and to orientate them optimally for binding to their complimentary ligand (Bieniarz, *et al.*, 1993; Husain and Bieniarz, 1994; Pope *et al.*, 1996). Further improvement may be obtained by using linking molecules that provide spacing from the surface. Linking molecules may be homo- or heterobifunctional with respect to their chemically reactive groups, heterobifunctionality providing for more versatile sequential conjugations. A representative review article (Wong and Wong, 1992) and a comprehensive monograph (Hermanson *et al.*, 1992) describe the general chemical characteristics and methods for the use of linker molecules in the immobilization of proteins. Linking molecules are commercially available from companies such as Pierce Chemical Company of Rockford, IL.

Linking molecules are also widely used to link, or *conjugate*, one molecule to another, rather than to a surface. For example, antibodies are often linked to an enzyme such as alkaline phosphatase to form conjugates, the enzyme serving as a reporter molecule or label. The performance of a linker molecule is influenced by its length, reactivity, and solution properties. An example of one linking strategy started with the commercially available heterobifunctional linker 1-[[4-[(2,5-dioxo-1-pyrrolidinyl)oxyl-carbonyl]cyclohexyl]methyl]-*H*-pyrrole 2,5-dione (SMCC). This linker results in a spacing of nine atoms between linked molecules. The SMCC linker was extended further by coupling a series of 6-aminocaproic acid units, forming linkers that provide spacings of 16, 23, and 30 atoms after conjugation (Fig. 4 and Bieniarz *et al.*, 1996). It was found that the activity of soluble antibody–enzyme conjugates improved with the length of the linker. The same series of extended linker arms were used to couple antibodies to nanoparticles for use in an immunoassay. Again, the apparent activity of the immobilized antibody increased with the length of the spacer.

In the immobilization of a biomolecule, the solubilization of the linker in contact with the aqueous medium may also play an important role in the resultant activity, perhaps because of the microenvironment it provides for the protein.

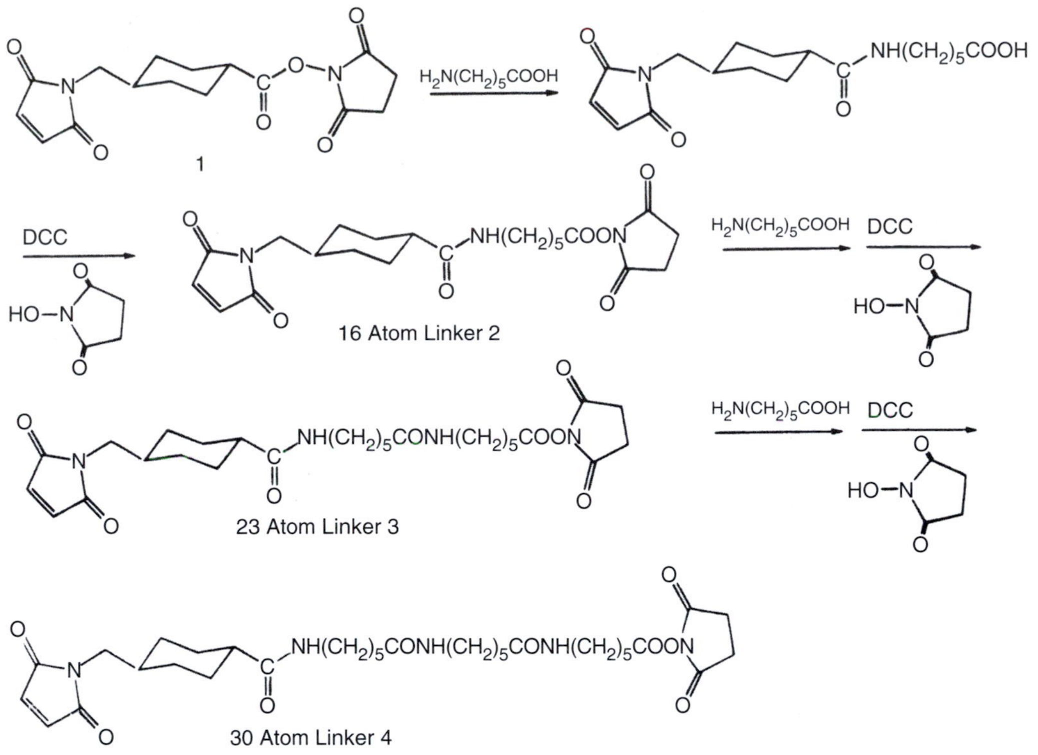

FIG. 4. Synthesis of extended heterobifunctional linkers. (Reprinted with permission from Bieniarz *et al.*, 1996, copyright 1996, American Chemical Society.)

For example, it was shown that two model enzymes, trypsin and alpha-chymotrypsin, could be made more stable toward thermal inactivation by modification of their accessible lysine groups with anhydrides or chloranhydrides of aromatic carboxylic acids (Mozhaev *et al.*, 1988). The authors attributed this result to a hydrophilization of the surface area of the protein globule. In earlier related work (Mozhaev *et al.*, 1983), the same enzymes were modified with acryloyl groups and copolymerized with acrylamide to form a hydrogel. In this structured, hydrophilic environment, the enzymes were found to be over 100 times more stable against irreversible thermal inactivation.

It should be anticipated that further improvements in binding activity and in reduction of nonspecific binding will occur through the use of specifically designed, soluble linker arms that provide a local environment benign to the immoblized or conjugated binding molecule. An additional benefit can be obtained if the linker is "protein-resistant" and repels adsorptive binding of the labeled second antibody (Lee *et al.*, 1989; Jeon *et al.*, 1991; Litauszki *et al.*, 1998). Linker molecules that are uncharged water-soluble oligomers or polymers can passivate a surface against the nonspecific binding of proteins through a combination of low interfacial free energy of the hydrated surface and high chain mobility.

The use of solubilizable linkers is taught in a patent describing the development of a waveguide-based apparatus for homogeneous fluorescence-based immunoassays (Herron *et al.*, 1997). In order to avoid wash steps, nonspecific binding must be minimized in this assay format. The authors describe the use of water-soluble poly(ethylene glycol) (PEG) spacers, both to covalently attach whole antibodies and Fab' fragments to waveguides and to simultaneously passivate the waveguide surface. In one technique a linker is formed by modifying both ends of PEG molecules of various lengths with ethylenediamine. The solid phase is pretreated with glutaraldehyde to generate an aldehyde surface. Excess linker is then added to the modified solid phase to generate a "PEGylated" surface that has amine end groups remaining for coupling of an antibody or antibody fragment.

A second technique, which appears to be more effective, is the use of a triblock copolymer composed of a hydrophobic block poly(propylene oxide) (PPO) flanked on either end by hydrophilic blocks of poly(ethylene oxide) (PEO). This class of copolymer surfactant, known in the literature under the tradename of Pluronic, is commercially available from BASF Corporation (Parsippany, NJ). In the method described, the surface was first coated with Pluronic PF108 or PF105. Next the free PEO chain ends of the copolymer were derivatized in a photoactivated coupling reaction with a bifunctional photoaffinity cross-linker. A logical option described is the use of a benzophenone having a maleimido group at one of the para positions. Upon irradiation, the photoactivatable group of the cross-linker covalently binds to the free PEO chains, leaving the reactive maleimido group of the linker available for coupling to a desired Fab' fragment. Using this procedure, the best data yielded ratios of nonspecific-to-specific binding approaching 0.003 on polystyrene supports.

Photolinking

In recent years there has been an increase in the use of controlled patterning in the immobilization of ligand binding molecules to accommodate various assay formats, such as immunoassays performed along membranes and microporous strips, and multiligand 2D matrices used for screening of various analytes. The same techniques are also applied to the patterning of molecules that promote the adhesion of cells to surfaces having potential as scaffolds for tissue engineering.

SurModics, Inc., of Eden Prairie, MN, has developed PhotoLink, a commercial process for the covalent immobilization of biomolecules onto the surface of any hydrocarbon-containing material. This process involves the application of photoactivatable reagents to the surface followed by exposure to light, to achieve covalent coupling to the surface through activation of the photogroups. The photoreagents are typically polymers that can serve as linkers or surface modifiers, or biologically active molecules to which photoactivatable groups have been attached. They can also be heterobifunctional coupling reagents having both photoactivatable and chemically reactive groups. In contrast to graft polymerization for surface modification, this approach does not involve photoinitiation of polymerization.

Biomolecules can be immobilized by either a one-step or a two-step method. In the one-step method, the biomolecule is prederivatized with a photoreactive moiety. This photoreagent is then applied to the surface and irradiated with light of suitable wavelength to effect the coupling. In the two-step method, the surface is prederivatized with a photoreactive coupling reagent. The coupling agent, either a photopolymer or a heterobifunctional photoreagent, has appropriate functional groups available to subsequently immobilize the molecule of interest by conventional coupling techniques. The building blocks of typical heterobifunctional photolinking reagents are shown in Fig. 5.

Aryl ketones, such as benzophenone derivatives, have most commonly been used as the photoactivatable species for such photoreagents, although aryl azides, such as azidonitrophenyl groups, have also been used. Benzophenone derivatives are typically attached to the polymers or other molecules to be immobilized onto the surface. For example, the technology can be applied to hydrophilic polymers, such as polyvinylpyrrolidone, polyacrylamide, and PEO, which are used in certain applications because of their unique protein-resistance properties. Though all of the mentioned polymers are effective, the last has been the most completely studied for protein resistance applications (Lee *et al.*, 1989; Jeon *et al.*, 1991; Litauszki *et al.*, 1998).

Upon photoactivation, aryl ketones abstract a hydrogen atom from the surface and form a new carbon–carbon bond. The ability of the activated intermediate to return to the ground state if no coupling occurs and the tendency of the benzophenone derivatives to associate with hydrophobic surfaces

Heterobifunctional Crosslinking Reagents

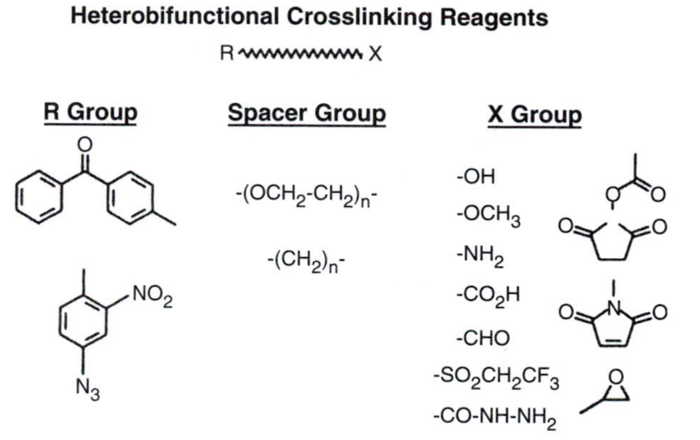

FIG. 5. Building blocks for typical heterobifunctional photocrosslinking reagents, where R is a photoreactive linker and X is a chemical linker. (Reprinted from Amos *et al.*, 1995, pp. 895–926, by courtesy of Marcel Dekker, Inc.)

both contribute to the efficiency of the photocoupling. The mechanism of aryl ketone photocoupling is shown in Fig. 6.

This basic photoimmobilization technology can be used in diagnostic applications to both prevent unwanted protein adsorption by immobilization of hydrophilic polymers and to covalently immobilize proteins of interest to the surface (Amos *et al.*, 1995). Figure 7 depicts an experiment where a variety of radiolabeled proteins with varying molecular weights were incubated in polyacrylamide-coated wells of polystyrene microtiter plates at a concentration of between 7 and 25 μg/ml. The plates had been previously photomodified with the hydrophilic polyacrylamide polymer to reduce nonspecific protein binding. After incubation with the protein, the surfaces were washed extensively using a buffer containing Tween-20 and the individual wells then separated and the quantity of bound protein determined by liquid scintillation. The proteins and their molecular weights included human gamma globulin, 150 kDa; thrombin, 36 kDa; chymotrypsinogen, 25 kDa; ribonuclease, 14 kDa; insulin B chain, 6 kDa; neurotensin, 1673 Da; and angiotensin, 1046 Da. This modification resulted in a 40–90% decrease in protein adsorption.

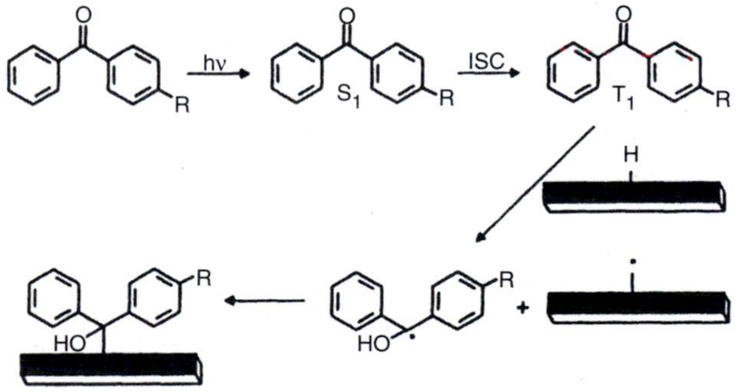

FIG. 6. Aryl ketone photocoupling mechanism to surfaces where S_1 = singlet excited state, T_1 = longer lived triplet state, and ISC = intersystem crossing. (Reprinted from Amos *et al.*, 1995, pp. 895–926, by courtesy of Marcel Dekker, Inc.)

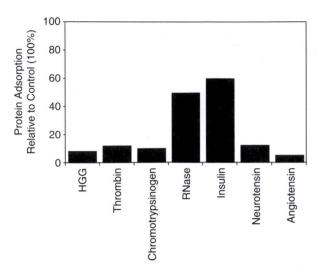

FIG. 7. Adsorption of various proteins onto modified polystyrene. Corning Costar Ultra Low Binding Lab Coat strip plates, modified with hydrophilic photopolymers, were compared with unmodified Costar strip plates. (Reprinted from Amos *et al.*, 1995, pp. 895–926, by courtesy of Marcel Dekker, Inc.)

Utilizing the same technology, biomolecules such as nucleic acids, antibodies, antigens, and enzymes can be covalently attached to surfaces for use in diagnostic kits. Covalent immobilization can be helpful for binding molecules that do not adsorb or adsorb very poorly. Covalent attachment can include site-specific immobilization of antibodies. For example, linkage of antibodies through the carbohydrate side chains usually found in the Fc region can increase sensitivity because of favorable surface orientation.

Enzyme immunoassays have been reported using visual detection on membranes having photoimmobilized antibodies (Gorovits *et al.*, 1991). These authors photochemically modified the surface of porous membranes using azido compounds having a second chemically reactive moiety for protein coupling. In the case of photolinking 4,4′-diazidostilbene-2-2′-disulfonate, proteins were subsequently immobilized by the activation of one or two of the reagent's sulfo groups with carbodiimide. When the diethylacetate *p*-azidobenzaldehyde linker was used, the protein was subsequently coupled to the immobilized aldehyde by reductive amination, involving Schiff base formation followed by reduction to the stable secondary amine with sodium borohydride. Regenerated cellulose proved to be the best support material in this work. Tests reported were for the analytes thyroxine, human immunoglobulin, human chorionic gonadotropin, and cells of the bacteria *Shigella sonnei*.

The surfaces of commercial polysulfone and polyethersulfone ultrafiltration membranes have been successfully modified using benzophenone or benzoylbenzoic acid to surface polymerize acrylic acid (Ulbricht *et al.*, 1996). The resulting acid groups were used for the covalent immobilization of various proteins. Because of the tendency for homolytic chain scissions at several locations with polymers of this structure, careful control of ultraviolet excitation energy was necessary.

These examples illustrate the growing use of photolinking rather than present an extensive review of the literature. Its use, combined with lithographic techniques, is bound to grow as the need for multiple analyte tests on smaller and smaller devices increases because of cost sensitivity and the need to reduce the volume of medical waste.

Steric Inhibition

Although immobilizing water-soluble polymers on the solid phase can suppress nonspecific binding of labeled proteins in many immunoassays, the presence of these polymers can have a detrimental effect in a chromatography-based assay format. This format requires the interaction of two solid-phases, both possessing attached binding molecules (see the section describing intrinsically colored particles earlier in this chapter). The detrimental effect of attached, water-soluble polymer acting as a steric stabilizer was illustrated in a competitive assay for the analyte estrone-3-glucuronide (E1G) (Pope *et al.*, 1997a, b). The assay used intrinsically black polypyrrole nanoparticles, made using poly(vinyl alcohol) as a colloidal stabilizer. As shown by X-ray photoelectron spectroscopy, even after extensive cleaning, the particles possessed a significant surface concentration of poly(vinyl alcohol). The presence of the polymer does not inhibit the immobilization of the antibodies to the particle surface, nor the ability of the immobilized antibody to bind soluble analyte; however, the ability of the particle-bound antibody to bind to an analyte immobilized on the capture zone of a chromatography strip was inhibited. This inhibition is likely due to steric stabilization of the particle against interaction with another hydrated surface. As a result, the binding molecules cannot approach close enough to one another to interact. Steric stabilization is distinguished from electrostatic stabilization in that in the former the stabilizing molecules are usually uncharged. In aqueous systems, these surface molecules are almost always water-soluble polymers. As such surfaces approach one another, the concentration of water-soluble polymer will increase in the region between the two surfaces. Coalescence is not favored in this case because of osmotic and entropic effects. The principles of steric stabilization are described in a seminal paper by Napper (1977).

Data confirming these proposed principles as they apply to chromatography-based assays was obtained (Pope *et al.*, 1997a, b). As shown in Fig. 8, selective degradation of the poly(vinyl alcohol) by cleavage of the polymer chain at vicinal diol sites [originating from infrequent head-to-head polymerization in the precursor poly(vinyl acetate)] increased the binding response of the immunoreagent 140-fold for the E1G immunoassay.

SPECIFIC ACTIVITY OF SOLID PHASE IMMUNOREAGENTS

In the development of biomaterials for implantation, *in vivo* biocompatibility is always of paramount concern. Biocompatibility can be somewhat predicted using *in vitro* cell-based assays, which monitor cell growth and spreading or the release

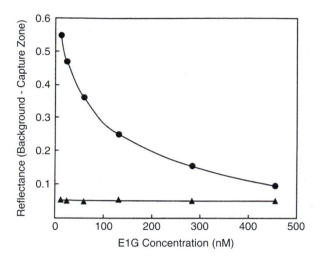

FIG. 8. Standard assay curves for a model competitive immunoassay for the hormone estrone-3-glucuronide (E1G) in a chromatography-based format. Polypyrrole latex immunoreagent reacted with periodic acid (•) and unreacted control latex immunoreagent (▲). (Reprinted with permission from Pope *et al.*, 1997a, copyright 1997, American Chemical Society.)

of cytokines. The most common methods of assessment are the retrieval of implanted material, the sectioning and histological analysis of encapsulation, and the measurement of cellular markers of inflammation. Similarly, in a solid-phase ligand binding assay, the biocompatibility of the surface of the materials with the binding molecules (e.g., antibodies, antibody fragments, antigens, oligonucleotide segments, and protein-conjugated haptens), is a critical element for a stable, reproducible, sensitive, and specific test.

The most important parameters to measure when evaluating a solid-phase ligand binding reagent are the total binding activity of the immobilized binding molecules and their binding affinity for the analyte of interest or other binding molecule. Total binding activity can be determined, for example, by measuring how many analyte molecules can be specifically bound by the binding molecules attached to the solid phase. If the number of binding molecules attached to the solid phase can be determined, the fraction of their original binding activity that was retained following the immobilization procedure can be calculated. Introducing a radioactive label into the binding molecules before the immobilization procedure is the most direct means of measuring the number of binding molecules on the solid-phase reagent. For example, a tritium label can usually be introduced into an antibody by oxidation of the carbohydrate side chains with periodate to produce aldehyde groups. These groups are then reduced with tritium-labeled sodium borohydride to produce tritium-labeled hydroxyls. Radioactive counting of the coated solid phase then directly yields the quantity of antibodies present. In practice, radiolabels are used less and less because of cost and regulatory safety issues. Alternatively, nonradioactive labels such as fluorescein can be used for this purpose, but the extent of surface quenching of the fluorescence signal cannot be accurately predicted.

A second method to determine the quantity of binding molecules attached to the solid phase is to measure their

depletion from solution during the immobilization procedure. A calculation of total solid-phase surface area and an assumption of an individual binding molecule's "footprint" area provides an estimate of the total surface concentration that should be present if monolayer coverage is complete.

Inevitably, because of denaturation or immobilization with improper orientation, some binding molecules will be inactivated during the coating process. To evaluate the efficiency of coating procedures, the retained specific binding activity of the immobilized binding molecules must be determined. Again, the most straightforward method is to directly measure the quantity of radiolabeled ligand captured by the solid-phase reagent using liquid scintillation counting. An alternative approach, which avoids radiolabels, is to force the immobilized binding molecules of the solid-phase reagent in question to compete for analyte binding with the immobilized binding molecules of a different solid phase. Microparticle solid phases especially lend themselves to this approach, an example of which is illustrated in Fig. 9. In format A, antibodies (1) specific for a particular analyte (A) are immobilized onto the bottom of a microtiter plate well. Analyte is captured by the antibodies on the well bottom, then a second antibody (2), conjugated to an enzyme (E) is captured by the captured analyte, i.e., the classic "sandwich" immunoassay format. The captured enzyme generates a signal in the presence of a substrate (S), the intensity of which is proportional to the quantity of analyte originally captured by the solid phase. In format B, capture of the analyte by the solid phase is inhibited by the presence of soluble capture antibody (1), which will occupy binding sites on the analyte, preventing its binding to the immobilized antibody (1) on the solid phase. As a result, less of the antibody–enzyme conjugate becomes bound to the solid phase and the intensity of the resulting signal is diminished. Usually, the concentration of soluble antibody necessary to achieve a 50% signal inhibition is determined. This quantity is a measure of the binding activity of the soluble antibody. In format C, microparticles (MP) coated with the same antibody (1) can be used to inhibit the assay in the same way. The number of antibody-coated microparticles necessary to achieve 50% signal inhibition is displaying the same binding activity as the quantity of soluble antibody necessary to achieve the same 50% signal inhibition. Comparison of this quantity of soluble antibody to that actually on the microparticles is a measure of the loss of binding activity resulting from the immobilization procedure. Even without knowledge of the actual quantity of antibody on the microparticles, such determinations can be used to compare the efficiency of different immobilization procedures. Using the methods just described, comparisons of the binding activity of two particulate immunoreagents were demonstrated (Pope *et al.*, 1996).

CONCLUSION

The medical diagnostic industry is under unprecedented pressure to reduce the cost and to maximize the throughput speed of clinical assays. The response has been the centralization of clinical testing and the concomitant development of highly automated, high-throughput clinical analyzers.

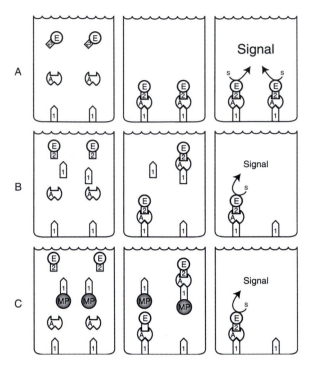

FIG. 9. Measurement of retained binding activity of antibodies immobilized on microparticles by inhibition. (Format A) A typical sandwich immunoassay in which immobilized antibodies (1) capture analyte (A), which in turn captures a second antibody (2) conjugated to an enzyme label (E) to form an immobilized immune complex. After removal of unbound antibody–label conjugate, conversion of the substrate (S) by the immobilized enzyme generates the signal. (Format B) Inhibition of immune complex capture by soluble antibody (competition assay). Soluble antibody (1) binds to the same site on the analyte as does the immobilized antibody (1), inhibiting its capture. As a result, less conjugated enzyme becomes immobilized and a less intense signal is generated. (Format C) Inhibition of immune complex capture by antibody-coated microparticles. Antibody (1) attached to microparticles (MP) binds to the same site on the analyte as does immobilized antibody (1), inhibiting its capture on the surface of the well. As a result, less conjugated enzyme is captured and a less intense signal is generated.

The development and application of specialized biomaterials has been instrumental in the successful design of these machines. Although much more efficient than their predecessors, these new designs primarily automate previous assay technology. To meet future demands in both the high-volume testing lab and low-volume point-of-care setting, the next generation of tests will have to depart from past designs.

Biologically derived binding molecules will remain the major source of the specificity needed to measure small quantities of analytes in complex biological mixtures, although synthesized polypeptides and polynucleic acid probes will find increasing use.

Technologies maturing in both related and distant fields, such as the lithographic layering techniques developed for the semiconductor industry, the automated optical scanning of miniaturized arrays used for drug discovery, and advances in the field of implantation biology, holographic optics, surface-enhanced spectroscopies, surface patterning using self-assembled monolayers, and pattern recognition, will

all provide new approaches to biomaterials development for clinical diagnostic assays.

Bibliography

Amos, R. A., Anderson, A. B., Clapper, D. L., Duquette, P. H., Duran, L. W., Hohle, S. G. Sogard, D. J., Swanson, M. J., and Guire, P. E. (1995). Biomaterial surface modification using photochemical coupling technology. in *Encyclopedic Handbook of Biomaterials and Bioengineering: Part A: Materials*, Vol. 1, Marcel Dekker, New York, pp. 895–926.

Baselt, D. R., Lee, G. U., Natesan, M., Metzger, S. W., Shehan, P. E., and Colton, R. J. (1998). A biosensor based on magnetoresistance technology. *Biosensors Bioelectron.* 13: 731–739.

Bergveld, P. (1996). The future of biosensors. *Sensors Actuators A* 56: 65–73.

Bieniarz, C., Husain, M., and Bond, H. E. (1993). Site specific labeling of immunoglobulins and detectable labels. U.S. Patent No. 5,191,066.

Bieniarz, C., Husain, M., Barnes, G., King, C. A., and Welch, C. J. (1996). Extended length heterobifunctional coupling agents for protein conjugations. *Bioconj. Chem.* 7: 88–95.

Bieniarz, C., Husain, M., and Tarcha, P. J. (1999). Evidence for the addition of nucleophiles to the surface of polypyrrole latex. *Macromolecules* 32: 792–795.

Brown, C. W., Li, Y., Seelenbinder, J. A., Pivarnik, P., Rand, A. G., Letcher, S. V., Gregory, O. J., and Platek, M. J. (1998). Immunoassays based on surface-enhanced absorption spectroscopy. *Anal. Chem.* 70: 2991–2996.

Butler, J. E. (1991). *Immunochemistry of Solid-Phase Immunoassay*. CRC Press, Boca Raton, FL.

Fiore, M., Mitchell, J., Doan, T., Nelson, R., Winter, G., Grandone, C., Zeng, K., Haraden, R., Smith, J., and Harris, K. (1988). The Abbott IMx automated benchtop immunochemistry analyzer system. *Clin. Chem.* 34: 1726–1732.

Gorovits, B. M., Osipov, A. P., Doseeva, V. V., and Egorov, A. M. (1991). New enzyme immunoassay with visual detection based on membrane photoimmobilized antibodies. *Anal. Lett.* 24: 1937–1966.

Hermanson, G. T., Mallia, A. K., and Smith, P. K. (1992). *Immobilized Affinity Ligand Techniques*. Academic Press, San Diego.

Herron, J. N., Christensen, D. A., Wang, H.-K., Caldwell, K. D., Janatova, V., and Huang, S.-C. (1997). Apparatus and methods for multi-analyte homogeneous fluoroimmunoassays. U.S. Patent No. 5,677,196.

Husain, M., and Bieniarz, C. (1994). Fc site-specific labeling of immunoglobulins with calf intestinal alkaline phosphatase. *Bioconj. Chem.* 5: 482–490.

Isola, N. R., Stokes, D. L., and Vo-Dinh, T. (1998). Surface-enhanced Raman gene probe for HIV detection. *Anal. Chem.* 70: 1352–1356.

Jeanmarie, D. L., and Van Duyne, R. P. (1977). Surface Raman spectroelectrochemistry. I. Heterocyclic aromatic and aliphatic amines adsorbed on the anodized silver electrode. *J. Electroanal. Chem.* 84: 1–20.

Jeon, S. I., Lee, J. H., Andrade, J. D., and DeGennes, P. G. (1991). Protein–surface interactions in the presence of polyethylene oxide. *J. Colloid Interface Sci.* 142: 149–166.

Johnsson, B., Lofas, S., and Lindquist, G. (1991). Immobilizaiton of proteins to a carboxymethyldextran-modified gold surface for biospecific interaction analysis in surface plasmon resonance sensors. *Anal. Biochem.* 2: 291–336.

Keating, C. D., Kovaleski, K. M., and Natan, M. J. (1998). Protein: colloid conjugates for surface enhanced Raman acattering:

stability and control of protein ortientation. *J. Phys. Chem. B* **102**: 9404–9413.

Lahiri, J., Lsaacs, L., Tien, J., and Whitesides, G. M. (1999). A strategy for the generation of surfaces presenting ligands for studies of binding based on an active ester as a common reactive intermediate: a surface plasmon resonance study. *Anal. Chem.* **71**: 777–790.

Lee, J. H., Kopecek, J., and Andrade, J. D. (1989). Protein-resistant surfaces prepared by PEO-containing block copolymer surfactants. *J. Biomed. Mater. Res.* **23**: 351–368.

Litauszki, L., Howard, L., Salvati, L., and Tarcha, P. J. (1998). Surfaces modified with PEO by the Williamson reaction and their affinity for proteins. *J. Biomed. Mater. Res.* **35**: 1–8.

Maeda, S., and Armes, S. P. (1993). Preparation of novel polypyrrole–silica colloidal nanocomposites. *J. Colloid Interface Sci.* **159**: 257–259.

Maeda, S., and Armes, S. P. (1994). Preparation and characterization of novel polypyrrole–silica colloidal nanocomposites. *J. Mater. Chem.* **4**: 935–942.

Maeda, S., Corradi, R., and Armes, S. P. (1995). Synthesis and characterization of carboxylic acid–functionalized polypyrole–silica microparticles. *Macromolecules* **28**: 2905–2911.

Mees, D. R., Pysto, W., and Tarcha, P. J. (1995). Formation of selenium colloids using sodium ascorbate as the reducing agent. *J. Colloid Interface Sci.* **170**: 254–260.

Monji, N., and Hoffman, A. S. (1987). A novel immunoassay system and bioseparation process based on thermal phase separating polymers. *Appl. Biochem. Biotechnol.* **14**: 107–120.

Mosbach, K., and Ramstrom, O. (1996). The emerging technique of molecular imprinting and its future impact on biotechnology. *Biotechnology* **14**: 163–170.

Mozhaev, V. V., Siksnis, V. A., Torchilin, V. P., and Martinek, K. (1983). Operational stability of copolymerized enzymes at elevated temperatures. *Biotechnol. Bioeng.* **25**: 1937–1945.

Mozhaev, V. V., Siksnis, V. A., Melik-Nubarov, N. S., Galkantaite, N. Z., Denis, G. J., Butkus, E. P., Zaslavsky, B. Y., Mestechkina, N. M., and Martinek, K. (1988). Protein stabilization via hydrophilization. *Eur. J. Biochem.* **173**: 147–154.

Napper, D. H. (1977). Steric stabilization. *J. Colloid Interface Sci.* **58**: 390–407.

Pfeiffer, D. (1997). Commercial biosensors for medical application. in *Frontiers in Biosensors II. Practical applications*, F. W. Scheller, F. Schubert and J. Fedrowitz (eds.). Birkhauser Verlag, Basel, pp. 149–160.

Pope, M. R., Armes, S. P., and Tarcha, P. J. (1996). Specific activity of polypyrole nanparticulate immunoreagents: Comparison of surface chemistry and immobilizaiton options. *Bioconj. Chem.* **7**: 436–444.

Pope, M. R., Putman, B., Mees, D. R., Joseph, M., Subotich, D., Pry, T., and Tarcha, P. J. (1997a). Effect of steric stabilizers on the performance of colloidal immunoreagents. *PMSE* **76**: 463–464.

Pope, M. R., Tarcha, P. J., Mees, D. R., Joseph, M. K., Pry, T. A., Putman, C. B., and Subotich, D. D. (1997b). Method for improving the performance of an immunoreagnet in an immunoassay. U.S. Patent No. 5,681,754.

Ramstrom, O., Yu, C., and Mosbach, K. (1996). Chiral recognition in adrenergic receptor binding mimics prepared by molecular imprinting. *J. Mol. Recog.* **9**: 691–696.

Rao, J., Yan, L., Xu, B., and Whitesides, G. M. (1999). Using surface plasmon resonance to study the binding of vancomycin and its dimer to self-assembled monolayers presenting D-Ala-D-Ala. *J. Am. Chem. Soc.* **121**: 2629–2630.

Roberts, C., Chen, C., Mrksich, M., Martichonok, V., Ingber, D. E., and Whitesides, G. M. (1998). Using mixed self-assembled monolayers presenting RDG and (EG)3OH groups to characterize long term attachment of bovine capillary endothelial cells to surfaces. *J. Am. Chem. Soc.* **120**: 6548–6555.

Rohr, T. E. (1995). Magnetically assisted binding assays using magnetically labeled binding members. U.S. Patent No. 5,445,970 and U.S. Patent No. 5,445,971.

Rohr, T. E., Cotton, T., Fan, N., and Tarcha, P. J. (1989). Immunoassay employing surface-enhanced Raman spectroscopy. *Anal. Biochem.* **182**: 388–398.

Russell, R. J., Pishko, M. V., Gefrides, C. C., McShane, M. J., and Cote, G. L. (1999). A fluorescence-based glucose biosensor using concanavalin A and dextran encapsulated in a poly(ethylene glycol) hydrogel. *Anal. Chem.* **71**: 3126–3132.

Shi, H., Tsai, W.-B., Garrison, M. D., Ferrari, S., and Ratner, B. D. (1999). Template-imprinted nanostructured surfaces for protein recognition. *Nature* **398**: 593–597.

Staahl, M., Mansson, M.-O., and Mosbach, K. (1990). The synthesis of a D-amino acid ester in an organic media with α-chymotrypsin modified by a bio-imprinting procedure. *Biotechnol. Lett.* **12**: 161–166.

Tarcha, P. J. 1991. *The Chemical Properties of Solid Phases and Their Interaction with Proteins, Ch. 2 in Immunochemistry of Solid-Phase Immunoassay*, J. E. Butler, ed. CRC Press, Boca Raton, FL.

Tarcha, P. J., Misun, D., Finley, D., Wong, M., and Donovan, J. J. (1991). Synthesis, analysis, and immunodiagnostic applications of polypyrrole latex and its derivatives. in *Polymer Latexes, Preparation, Characterization and Applications*, E. S. Daniels, E. D. Sudol, and M. S. El-Aasser (eds.). ACS Symposium Series 492. American Chemical Society, Washington, D.C., pp. 347–367.

Tarcha, P. J., Rohr, T. E., Markese, J. J., Cotton, T., and Rospendowski, B. N. (1996). Surface-enhanced Raman spectroscopy immunoassay method, composition and kit. U.S. Patent No. 5,567,628.

Tarcha, P. J. DeSaja-Gonzalez, J., Rodriguez-Llorente, S., and Aroca, R. (1999). Surface-enhanced fluorescence on SiO2-coated silver island films. *Appl. Spectrosc.* **53**: 43–48.

Tijssen, P. (1985). Practice and theory of enzyme immunoassays. in *Laboratory Techniques in Biochemistry and Molecular Biology*, Vol. 15, R. H. Burdon and P. H. van Knippenberg (eds.). Elsevier Science, Amsterdam, p. 15.

Ulbricht, M., Riedel, M., and Marx, U. (1996). Novel photochemical surface functionalization of polysulfone ultrafiltration membranes for covalent immobilization of biomolecules. *J. Membrane Sci.* **120**: 239–259.

Wang, J. (1999). Amperometric biosensors for clinical and therapeutic drug monitoring: a review. *J. Pharmaceut. Biomed. Anal.* **19**: 47–53.

Wilkins, E., and Atanasov, P. (1996). Glucose monitoring: state of the art and future possibilities. *Med. Eng. Phys.* **18**: 273–288.

Wong, S. S., and Wong, L.-J. (1992). Chemical crosslinking and the stabilization of proteins and enzymes. *Enzyme Microb. Technol.* **14**: 866–874.

Xie, S. L., and Wilkins, E. (1991). Rechargeable glucose electrodes for long-term implantation. *J. Biomed. Eng.* **13**: 375–378.

Yost, D. A., Russel, J. C., and Yang, H. (1990). Non-metal colloidal particle immunoasssay. U.S. Patent No. 4,954,452.

7.19 MEDICAL APPLICATIONS OF SILICONES

Jim Curtis and André Colas

MEDICAL APPLICATIONS

Silicones, with their unique material properties, have found widespread application in health care. Properties attributed to silicone include *biocompatibility* and *biodurability*, which can be expressed in terms of other material properties such as hydrophobicity, low surface tension, and chemical and thermal stability. These properties were the basis for silicone's initial use in the medical field. For example, their hydrophobic (water-repellent) character caused silicones to be considered for blood coagulation prevention in the mid-1940s. Researchers from the Universities of Toronto and Manitoba obtained a methylchlorosilane from the Canadian General Electric Company and coated syringes, needles, and vials with the material. When rinsed with distilled water, the silane hydrolyzed, forming a silicone coating on the substrate. The researchers published results from their clotting time study in 1946, finding that the silicone treatment "on glassware and needles gives a surface which preserves blood from clotting for many hours" (Jaques *et al.*, 1946). Researchers at the Mayo Clinic took notice of the work by their Canadian colleagues, indicating that silicone "was the most practical of any known [substance] for coating needle, syringe and tube" (Margulies and Barker, 1949). They also demonstrated that leaving blood in silicone-coated syringes had no significant effect on the blood as measured by coagulation time after being dispensed from the syringe. Soon the use of silicone precoating of needles, syringes, and blood collection vials became commonplace. In addition to the silicone's blood-preserving quality, it was soon discovered that silicone-coated needles were less painful. Today most hypodermic needles, syringes, and other blood-collecting apparatus are coated and/or lubricated with silicone.

Silicone's chemical stability and elastic nature are beneficial for many applications involving long-term implantation. The first published report of silicone elastomers being implanted in humans was in April 1946, when Dr. Frank H. Lahey told of his use of these materials for bile duct repair. He obtained the material, called "bouncing clay" at the time, from the experimental laboratory of the General Electric Company (GE). Citing its elastic properties, he reported, "It is flexible, it will stretch, it will bounce like rubber and it can be cast in any shape" (Lahey, 1946).

In 1948, Dr. DeNicola implanted an artificial urethra fashioned from the same type of GE silicone tubing used previously by Lahey. The first apparently successful replacement of the human male urethra by artificial means was conducted under general anesthesia. The 3¾ inch (9.5 cm) long silicone tube was threaded over a narrow catheter whose distal end was in the bladder. Fourteen months after implantation, the artificial urethra "had been retained with normal genito-urinary function. . . . There is no evidence at this time that the tube is acting as a foreign body irritant. . ." (DeNicola, 1950).

A particularly notable early silicone implant was the hydrocephalus shunt, which benefited from silicone's thermal stability. This application became quite celebrated when tenderly described in *Reader's Digest* (LaFay, 1957). Charles Case "Casey" Holter was born on the seventh of November 1955 with a neural tube defect called lumbo-sacral myelomeningocele. By December, the baby had contracted meningitis, and surgeons at the Children's Hospital of Philadelphia closed the defect. A few weeks later, hydrocephalus caused young Casey's head to swell as cerebrospinal fluid (CSF) collected in his brain.

At the time, there were few treatment options and this affliction was fatal for most children who contracted it. After infection concerns with the daily venting of CSF through the fontanels (spaces between cranial bones that have not completely fused), Dr. Eugene Spitz implanted a polyethylene shunt catheter in Casey to drain excess CSF from the brain into the atrium of the heart. A valve was needed to allow the CSF to drain when pressure would begin to build in the brain, but close to prevent backflow when the pressure equalized. Spitz had been a neurosurgery resident at the University of Pennsylvania in 1952 and had gained clinical experience with a ball and spring valve developed by the Johnson Foundation, an arm of Johnson & Johnson. So it was this valve that was first implanted in young Casey. Basically a scaled-down version of an automotive pressure relief valve, it frequently clogged with tissue. Casey's father John, a machinist by profession, asked Spitz about the valve, the CSF, and the pressures involved. Spitz confided in Holter that "a competent one-way valve" that would be stable in the human body was needed (Baru *et al.*, 2001).

It is said that necessity is the mother of invention, and this is a poignant example. A desperately concerned father went home to his garage workshop that evening and constructed a prototype valve from two rubber condoms and flexible tubing. However, autoclaving caused the material to shrink a bit and the valve to leak. Holter discussed the shrinkage problem with a local rubber company, where the head of research suggested he replace the natural rubber with a thermally stable material known as silicone. Holter obtained Silastic brand silicone elastomer and tubing free of charge from Dow Corning. In March of 1956, Holter believed the valve that would come to bear his name was ready. At the time, Holter's son was too ill to undergo the surgery; however, Spitz saw promise in the valve design and successfully implanted the ventriculo-atrial shunt in another hydrocephalic child (Baru *et al.*, 2001). The Holter valve was so successful that its production began that summer and the valve is still being made in almost unchanged form today (Aschoff *et al.*, 1999).

These early health-care applications resulted in substantial interest in the emerging silicone materials and their promising properties. The two leading silicone suppliers, General Electric and Dow Corning, began receiving inquiries from the medical field at unprecedented rates. By 1959, Dow Corning was so inundated with requests for materials and information that the Dow Corning Center for Aid to Medical Research was established to act as a clearing house for all information on medical uses of silicone, and to supply medical scientists with research quantities of various silicone materials, all without cost to the researcher. The Center corresponded with more than 35,000 physicians and researchers from all over the world and in numerous areas of health care (Braley, 1973).

The upsurge of interest in silicones for health-care applications continued in the early 1960s. Before the end of the decade, silicone materials were being employed or evaluated in numerous health-care applications—in orthopedics, catheters, drains and shunts of numerous descriptions, as components in kidney-dialysis, blood-oxygenator, and heart-bypass machines, heart valves, and aesthetic implants, to name just a few.

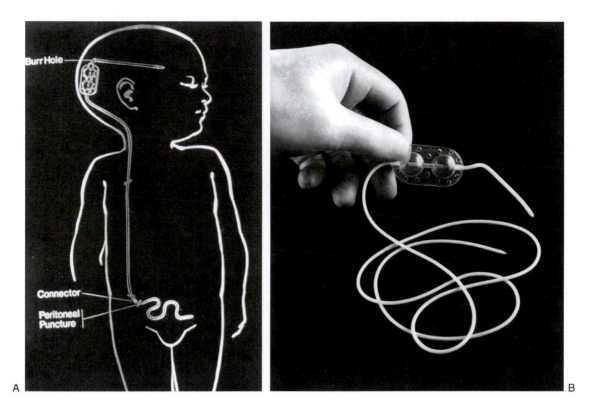

FIG. 1.   (A) Schematic and (B) photograph of silicone shunt for treatment of hydrocephalus.

Orthopedic Applications of Silicone

The most significant orthopedic applications of silicone are the hand and foot joint implants. Dr. Alfred Swanson, with assistance from Dow Corning, developed silicone finger joint implants such as those shown in Fig. 2 (Swanson, 1968). Similar implants were developed for the other small joints of the foot and hand. In addition to the double-stemmed finger joint implants in each of the metacarpophalangeal joints (large arrow), Fig. 2D also shows a single-stemmed Silastic ulnar head implant at the distal terminus of the ulna (small arrow).

Nearly four decades later, silicone remains the most prevalent type of small joint implant.

Another early orthopedic application of silicone was in 1969, when the French GUEPAR (Group d'Utilisation et d'Etude des Prothèses Articulaires) posterior-offset hinged total knee implant was introduced. This design was constructed of the metal Vitallium with a shock-absorbing silicone bumper that prevented impact of the anterior portions of the tibial and femoral components during knee extension (Mazas, 1973).

Catheters, Drains and Shunts

The properties of silicone elastomers have also found application in numerous catheters, shunts, drains, and the like (Fig. 3). These included devices fabricated with silicone extrusions, as well as devices with nonsilicone substrates that had been silicone-coated to provide less host reaction. For example,

although several all-silicone urology catheters have been utilized, the Silastic Foley is a latex catheter whose exterior and interior had been coated with silicone elastomer (Fig. 4).

Figure 5 shows the various components of the Cystocath suprapubic drainage system, which was used for bladder drainage after gynecological surgery that complicated or prevented normal urethral urination. The system included (a) the catheter, a silicone tube whose nonwetting surface minimized encrustation, (b) the body seal made of flexible silicone elastomer that conformed easily to the body contour and allowed patient freedom of movement, (c) pressure-sensitive silicone adhesive that adhered well to skin, and (d) the trocar needle used to pierce the bladder and overlying tissue. After vaginal surgery the bladder was inflated and located. The Silastic Medical Adhesive B was applied by brush to the clean abdomen over the bladder and the bottom of the body seal component. The pressure-sensitive adhesive—a reaction product of silicone polymers and silicone resin dispersed in a solvent to facilitate application and spreading—had excellent properties conducive to the application. It provided good adherence to dry or wet skin without causing irritation or sensitization, and good permeability to oxygen, carbon dioxide and moisture vapor; it also formed a waterproof and urine-proof seal. After a short wait for the solvent to evaporate allowing the adhesive to become tacky, the body seal was adhered to the abdomen. The trocar was advanced through the center of the body seal and pierced through the skin and bladder. The stylet was removed and the silicone catheter threaded through the needle

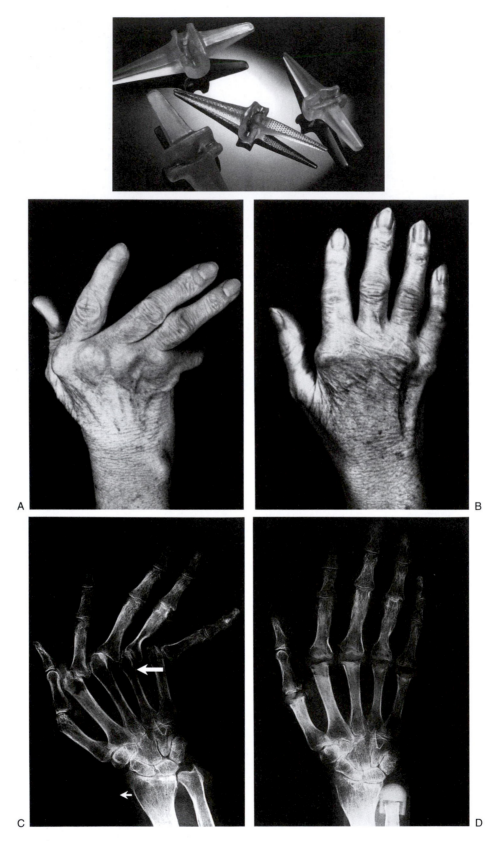

FIG. 2.   (A, C) Photograph and x-ray of arthritic right hand prior to restorative implantation surgery. (B, D) Postoperative photograph and x-ray view of the same hand.

FIG. 3. Examples of silicone in catheters, drains, tubes, and cannulas.

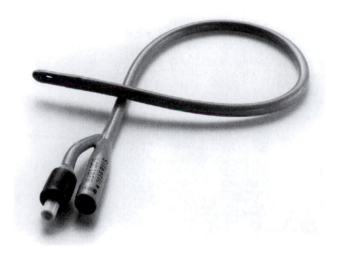

FIG. 4. Silastic Foley catheter.

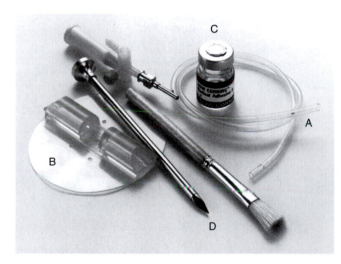

FIG. 5. Suprapubic drainage system.

and well into the bladder. The needle was withdrawn leaving the catheter in place. The silicone tube was secured in the retention groove and the distal end attached to a siphon drainage system.

Silicone pressure-sensitive adhesives (PSAs) based on this early material are in common use today, in transdermal drug delivery and other applications.

Extracorporeal Equipment

Silicone tubing and membranes found application in numerous extracorporeal machines, due in large part to their hemocompatibility and permeability properties. Silicone has been used in kidney dialysis, blood oxygenator, and heart bypass machines (Fig. 6). Blood compatibility was also a factor in silicone's application in several mechanical heart valves (Fig. 7).

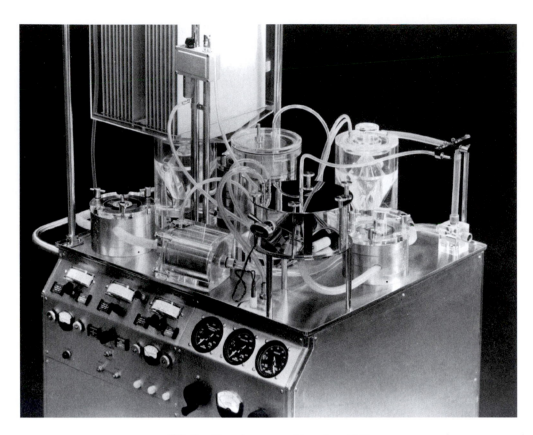

FIG. 6.   Heart bypass machine, circa 1964.

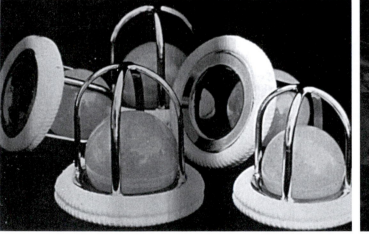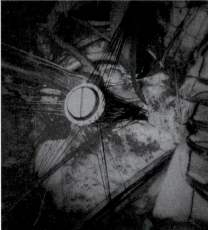

FIG. 7.   Examples of early heart valves containing silicone elastomer.

The use of silicone in extracorporeal applications continues today. Hemocompatibility testing has suggested that platinum-cured silicone tubing may be superior to PVC in several respects (Harmand and Briquet, 1999).

Aesthetic Implants

Silicones have been used extensively in aesthetic and reconstructive plastic surgery for over 40 years, and they continue to be so utilized. Silicone elastomer is used in implanted prosthetics of numerous descriptions. Silicone implants are widely used in the breast, scrotum, chin, nose, cheek, calf, and buttocks. Some of these devices may also employ a softer-feeling substance known as silicone gel. The gel is a lightly cross-linked silicone elastomer, without silica or other reinforcing filler, that is swollen with polydimethylsiloxane fluid. The gel is contained within an elastomer shell in breast, testicular, and chin implants. Surgeons implant these medical devices for

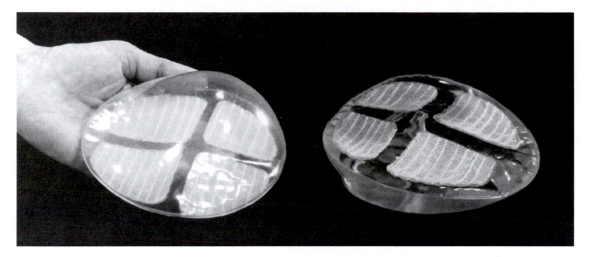

FIG. 8. Silastic mammary prosthesis, 1964.

aesthetic reasons, to correct congenital deformity, or during reconstructive surgery after trauma or cancer treatment.

The most prominent of these aesthetic implants is the silicone breast implant. Breast enlargement by artificial means has occurred for over a century. In 1895, Czerny reported transplanting a lipoma to a breast in order to correct a defect resulting from the removal of a fibroadenoma (Gerow, 1976). The insertion of glass balls into the breasts was described by Schwarzmann in 1930 and again by Thorek in 1942. The Ivalon sponge, introduced by Pangman in 1951, was the first augmentation prosthesis to be retained fairly consistently. This surgical sponge, formulated of poly(vinyl alcohol) cross-linked with formaldehyde, was at first hand-trimmed to the desired shape by the implanting surgeon and later preformed by Clay Adams, Inc. There was some early recognition of the tendency for tissue growth into the open-cell foam, and in 1958, Pangman patented the concept of encapsulating the foam with an alloplastic (manmade) envelope. His patent also contemplated the use of other fill materials in place of foam, materials such as silicones. The Polystan and Etheron polyurethane sponge implants began to be used as breast implants in 1959 and 1960, respectively. These sponge implants became popular in the early 1960s.

With silicone materials and prototypes supplied free of charge from Dow Corning, Doctors Cronin and Gerow developed and tested their silicone gel–filled breast implant in 1961. They implanted the first pair in a woman in 1962 (Cronin and Gerow, 1963). Word of their success and the superiority of these silicone implants to the existing foam type led to the popularity of the silicone gel breast implant.

Figure 8 shows the appearance of these implants in 1964. The shells of Cronin-type implants were vacuum-molded with anterior and posterior elastomer pieces sealed together creating a seam around the perimeter of the base. The posterior shell had exposed loops of Dacron mesh attached. The surgeons believed that prosthesis fixation to the chest wall was necessary to prevent implant migration.

Since this early 1964 design, Dow Corning and the numerous other companies that manufactured silicone breast implants made prosthesis design improvements, including elimination of the seam and realization that fixation is frequently not needed.

In the early 1990s, these popular devices became the subject of a torrent of contentious allegations regarding their safety. Although the legal controversy regarding silicone gel-filled implants continues in the United States, these medical devices are widely available worldwide and are available with some restriction in the United States. The controversy in the 1990s initially involved breast cancer, then evolved to autoimmune connective tissue disease, and continued to evolve to the frequency of local or surgical complications such as rupture, infection, or capsular contracture. Epidemiology studies have consistently found no association between breast implants and breast cancer (McLaughlin *et al.*, 1998; Brinton *et al.*, 2000; Mellemjkaer *et al.*, 2000; Park *et al.*, 1998). In fact, some studies suggest that women with implants may have decreased risk of breast cancer (Deapen *et al.*, 1997; Brinton *et al.*, 1996). Reports of cancer at sites other than the breast are inconsistent or attributed to lifestyle factors (Herdman and Fahey, 2001). The epidemiologic research on autoimmune or connective-tissue disease has also been remarkably uniform and concludes there is no causal association between breast implants and connective-tissue disease (Hennekens *et al.*, 1996; Sánchez-Guerrero *et al.*, 1995; Gabriel *et al.*, 1994; Nyrén *et al.*, 1998; Edworthy *et al.*, 1998; Kjoller *et al.*, 2001).

Largely without any specific safety concern or allegation critical of it, another silicone gel-filled implant was swept up in the breast implant controversy. At the time, most testicular implants were constructed of the same materials as silicone gel breast implants. Silicone artificial testicles had been used nearly as long as the breast implants. Dow Corning, one of several companies that manufactured these implants, was producing them as early as 1964. These devices served to ameliorate psychological stress associated with testicle loss due to cancer,

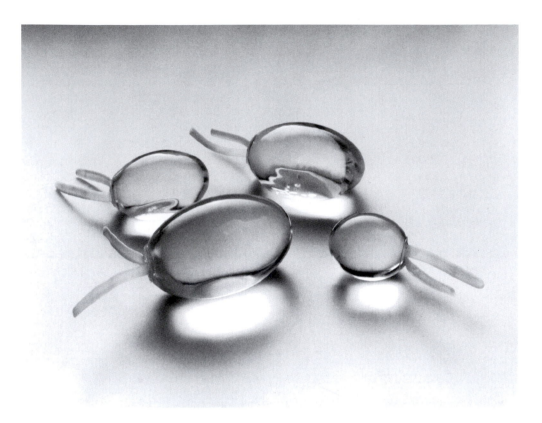

FIG. 9. Silicone testicular implants.

traumatic injury, or those absent at birth. The Teflon strips seen in Fig. 9 shield each implant shell during suturing through an elastomer loop at the superior pole to achieve fixation for proper anatomical orientation in the scrotum.

BIOCOMPATIBILITY

There has been much discussion regarding the various definitions of the term *biocompatibility*. We now take it to mean "the ability of a material to perform with an appropriate host response in a specific situation" (Black, 1992; Remes and Williams, 1992). Historically it has been tacitly understood that silicone materials are intrinsically biocompatible since they have been utilized successfully in so many health-care applications. However, given the modern definition of the term, no material can be assumed to be universally biocompatible, since such implies that it is suitable for every conceivable health-care application involving contact with the host patient.

Numerous silicone materials have undergone biocompatibility testing. Many of these materials have passed every bio-qualification test; however, others have not. There are several factors that can affect the results of such testing, including the material's composition. As described in Chapter 2.3, the basic polydimethylsiloxane (PDMS) polymer can be modified to replace methyl with other functional groups. In some cases, those groups may be responsible for untoward host response. There may be by-products from the preparation of silicone materials that might trigger tissue reaction. For example, these could come from the use of a peroxide initiator under inappropriate temperature and processing conditions.

Purity is another factor that can affect bio-test results. Medical silicone materials, including fluids, gels, elastomers, and adhesives, are manufactured by several companies today. Some of these firms manufacture these medical materials following GMP principles in dedicated, registered, and inspected facilities. Others sell materials generated on their industrial production line into the health-care market.

Selection of appropriate preclinical material bio-qualification tests for their application is the responsibility of the medical device or pharmaceutical manufacturer. Several national, international, and governmental agencies have provided guidance and/or regulation. Several silicone manufacturers offer special grades of materials that have met their specific requirements. The buyer should carefully investigate the supplier's definition since there are no universal special grade definitions. At Dow Corning, Silastic BioMedical Grade materials have been qualified to meet or exceed the requirements of ISO 10993-1, USP (United States Pharmacopeia) Class V Plastics tests (acute systemic toxicity and intracutaneous reactivity), hemolysis, cell culture, skin sensitization, mutagenicity, pyrogenicity, and 90-day implant testing. Other physiochemical qualification tests have been conducted, such as certain tests from the European Pharmacopoeia. Specific information

TABLE 1 Biodurability Studies of Silicone Elastomer and Medical Implants.

Year	Researcher	Synopsis
1960	Ames	Explant examination of a clinical silicone ventriculocisternostomy shunt used in the treatment of hydrocephalus showed "the silicone rubber tubing was unchanged by three years implantation in the tissues of the brain and in the cervical subarachnoid space." Similarly, after over 3 years implantation of silicone tubing in the peritoneal cavity of dogs, Ames wrote, "The physical properties of the tubing itself are apparently unchanged by prolonged contact with tissues."
1963	Sanislow and Zuidema	Silastic T-tubes were placed in the common ducts of dogs and explanted 9 months later. They were found to be free of bile-salt precipitation and completely patent. Four were tested for tensile strength and compared with a control sample from the same lot of elastomer. "These tests suggested that little physical change occurred in the Silastic as a result of prolonged contact with animal bile." The tensile strength after 9 months was reported as 1130 psi (7.8 MPa), the same value as reported for the unimplanted control.
1964	Leininger *et al.*	The Battelle Memorial Institute examined the biodurability of five plastics by implanting films in dogs for 17 months' duration. The materials tensile strength and elongation were measured and compared with unimplanted controls. Although large deterioration was seen in the tensile properties of polyethylene, Teflon, and nylon, the results for Mylar and Silastic remained essentially the same.
1974	Swanson and LeBeau	"Dog-bone"-shaped specimens of medical grade silicone rubber were implanted subcutaneously in dogs. Tensile properties and lipid content were measured at 6 months and 2 years postimplantation. A slight but statistically significant decrease in measured ultimate tensile strength and elongation were observed, as well as a small weight increase attributed to lipid absorption.
1976	Langley and Swanson	Mechanical test specimens were implanted in dogs for 2 years. Tensile strength, elongation, and tear resistance showed no statistically significant changes. Lipid absorption into the elastomer ranged from 1.4 to 2.6%.
2000	Curtis *et al.*	Six silicone breast implants surgically excised after 13.8 to 19.3 years and 10 similar nonimplanted units were tested to determine shell tensile properties and molecular weight of silicone gel extracts. The "study observed only minor changes (less than the explant or implant lot-to-lot variation range) in the tensile strength of Dow Corning silicone breast implants after nearly twenty years of human implantation." The gel extract molecular weight was either unchanged by implantation or increased slightly.
2003	Brandon *et al.*	In the most comprehensive study of breast implant biodurability heretofore published, the authors reported their results of tensile, cross-link density, and percent extractable measurements made on 42 explants and 51 controls. The study included some of the oldest explants, with human implantation durations up to 32 years. The researchers also performed a literature search and plotted all published explant tensile modulus data against implantation duration, finding no temporal relationship. Neither was a relationship with implant time seen for the cross-link density results, supporting the biodurability of the silicone elastomer utilized in the implant shells. The researchers concluded, "There was little or no degradation of the base polydimethylsiloxane during *in vivo* aging in any of the implants we examined."

regarding material biotesting can be found in other chapters of this text. Testing of the device in finished form should follow material bio-qualification tests such as those described above.

BIODURABILITY

Traditionally we have thought of *biocompatibility* as the situation in which the biomaterial has minimal adverse impact on the host. Conversely, *biodurability* is where the host has a minimal adverse effect on the biomaterial (see Chapter 6.2). Silicone's material properties, such as hydrophobicity, have been related to biocompatibility properties such as hemolytic potential, and its relative purity and high-molecular-weight polymeric nature and chemical structure provide a theoretical basis for its lack of toxicity. Silicone's biodurability in medical applications is probably related to its exceptional thermal and chemical stability properties.

Silicones are used in numerous applications requiring high temperature resistance (Noll, 1968; Stark *et al.*, 1982). During thermogravimetric analysis and in absence of impurities, polydimethylsiloxane degradation starts only at around 400°C. Thus, silicones remain essentially unaffected by repeated sterilization by autoclaving, and they can usually be dry-heat sterilized as well. Other sterilization methods can also be utilized, such as ethylene oxide exposure and gamma and e-beam irradiation—although care must be taken to ensure complete sterilant outgassing in the former and that dosage does not affect performance properties in the latter.

Although silicones can be chemically degraded, particularly at elevated temperatures, by substances capable of acting as depolymerization catalysts (Stark *et al.*, 1982), their hydrophobic nature limits the extent of their contact with many aqueous solutions. Typically the biologic milieu does not present a particularly hostile chemical environment for silicone. A notable exception, however, is the stomach, which excretes large amounts of hydrochloric acid, capable of attacking PDMS if it remains resident there too long.

Based on silicone elastomer performance in long-term implantation applications, its biodurability is generally considered excellent (Table 1).

The chemical stability associated with silicones became so well established that it has been formulated into other biomaterials, such as polyurethane, to enhance their biodurability (Pinchuk et al., 1988; Ward, 2000; Christenson et al., 2002).

Notwithstanding the chemical stability of silicone, certain factors have been shown to affect its durability in terms of long-term in vivo performance. The hydrophobic elastomer is somewhat lipophilic and can be swollen by lipids or other nonpolar agents. Early experience with in vivo failure of silicone-containing heart valves was traced to elastomer absorption of lipids from the blood that resulted in significant dimensional swelling (McHenry et al., 1970). In most cases the absorption was low and failures did not occur, but in a small percentage of cases, the silicone was absorbing quantities sufficient to render the valves variant. Researchers speculated that variations in silicone poppet manufacture, such as cure, might have been a factor (Carmen and Mutha, 1972). Absorption of lipids was a variable reported by Swanson and LeBeau (1974) and Langley and Swanson (1976). The work of Brandon et al. (2002, 2003) has shown that the shells of silicone gel-filled breast implants also absorb silicone fluid (from the gel) causing a minor diminution in mechanical properties, one that is reversed after extraction of the elastomer.

CONCLUSION

A variety of silicone materials have been prepared, many possessing excellent properties including chemical and thermal stability, low surface tension, hydrophobicity, and gas permeability. These characteristics were the origin of silicone's use in the medical field and are key to the materials' reported biocompatibility and biodurability. Since the 1960s silicones have enjoyed expanded medical application and today are one of the most thoroughly tested and important biomaterials.

Acknowledgments

The authors thank Doctors S. Hoshaw and P. Klein, both from Dow Corning, for their contribution regarding breast implant epidemiology.

Bibliography

Ames, R. H. (1960). Response to Silastic Tubing. *Bull. Dow Corning Center Aid Med. Res.* **2**(4): 1.

Aschoff, A., Kremer, P., Hashemi, B., and Kunze, S. (1999). The scientific history of hydrocephalus and its treatment. *Neurosurg. Rev.* **22**: 67.

Baru, J. S., Bloom, D. A., Muraszko, K., and Koop, C. E. (2001). John Holter's shunt. *J. Am. Coll. Surgeons* **192**: 79.

Black, J. (1992). *Biological Performance of Materials: Fundamentals of Biocompatibility*. Marcel Dekker, New York.

Braley, S. A. (1973). *Spare Parts for Your Body*. Dow Corning Center for Aid to Medical Research, Midland, MI.

Brandon, H. J., Jerina, K. L., Wolf, C. J., and Young, V. L. (2002). *In vivo* aging characteristics of silicone gel breast implants compared to lot-matched controls. *Plast. Reconstr. Surg.* **109**(6): 1927.

Brandon, H. J., Jerina, K. L., Wolf, C. J., and Young, V. L. (2003). Biodurability of retrieved silicone gel breast implants. *Plast. Reconstr. Surg.* **111**(7): 2295.

Brinton, L. A., Lubin, J. H., Burich, M. C., Colton, T., Brown, S. L., and Hoover, R. N. (2000). Breast cancer following augmentation mammaplasty (United States). *Cancer Causes Control* **11**: 819.

Brinton, L. A., Malone, K. E., Coates, R. J., Schoenberg, J. B., Swanson, C. A., Daling, J. R., and Stanford, J. L. (1996). Breast enlargement and reduction: results from a breast cancer case-control study. *Plast. Reconstr. Surg.* **97**(2): 269.

Carmen, R., and Mutha, S. C. (1972). Lipid absorption by silicone rubber heart valve poppets—*in-vivo* and *in-vitro* results. *J. Biomed. Mater. Res.* **6**: 327.

Christenson, E. M., Dadestan, M., Wiggins, M. J., Ebert, M., Ward, R., Hiltner, A., and Anderson, J. M. (2002). The effect of silicone on the biostability of poly(ether urethane). *Soc. Biomater. 28th Ann. Meeting Trans.* p. 111.

Cronin, T. D., and Gerow, F. J. (1963). Augmentation mammaplasty: a new "natural feel" prosthesis. Transactions of the Third International Congress of Plastic Surgery—*Excerp. Med. Intt. Cong.* **66**: 41.

Curtis, J. M., Peters, Y. A., Swarthout, D. E., Kennan, J. J., and VanDyke, M. E. (2000). Mechanical and chemical analysis of retrieved breast implants demonstrate material durability. Transactions of the Sixth World Biomaterials Congress, p. 346.

Deapen, D. M., Bernstein, L., and Brody, G. S. (1997). Are breast implants anticarcinogenic? A 14-year follow-up of the Los Angeles Study. *Plast. Reconstr. Surg.* **99**(5): 1346.

DeNicola, R. R. (1950). Permanent artificial (silicone)urethra. *J. Urol.* **63**(1): 168–172.

Edworthy, S. M., Martin, L., Barr, S. G., Birdsell, D. C., Brant, R. F., and Fritzler, M. J. (1998). A clinical study of the relationship between silicone breast implants and connective tissue disease. *J. Rheumatol.* **25**(2): 254.

Gabriel, S. E., O'Fallon, W. M., Kurland, L. T., Beard, C. M., Woods, J. E., and Melton, L. J. 3rd. (1994). Risk of connective-tissue diseases and other disorders after breast implantation. *New Engl. J. Med.* **330**(24): 1697.

Gerow, F. J. (1976). Breast implants. in *Reconstructive Breast Surgery*, N. G. Georgiade (ed.). Mosby, St. Louis.

Harmand, M. F., and Briquet, F. (1999). *In vitro* comparative evaluation under static conditions of the hemocompatibility of four types of tubing for cardiopulmonary bypass. *Biomaterials* **20**(17): 1561.

Hennekens, C. H., Lee, I. M., Cook, N. R., Hebert, P. R., Karlson, E. W., LaMotte, F., Manson J. E., and Buring, J. E. (1996). Self-reported breast implants and connective-tissue diseases in female health professionals. A retrospective cohort study. *JAMA* **275**(8): 616.

Herdman, R. C., and Fahey, T. J. (2001). Silicone breast implants and cancer. *Cancer Investi* **19**(8): 821.

Jaques, L. B., Fidlar, E., Feldsted, E. T., and MacDonald, A. G. (1946). Silicones and blood coagulation. *Can. Med. Assoc. J.* **55**: 26.

Kjoller, K., Friis, S., Mellemkjaer, L., McLaughlin, J. K., Winther, J. F., Lipworth, L., Blot, W. J., Fryzek, J., and Olsen, J. H. (2001). Connective tissue disease and other rheumatic conditions following cosmetic breast implantation in Denmark. *Arch. Int. Med.* **161**: 973.

LaFay, H. (1957). A father's last-chance invention saves his son. *Reader's Digest*, January, pp. 29–32.

Lahey, F. H. (1946). Comments made following the speech "Results from using Vitallium tubes in biliary surgery," read by Pearse HE before the American Surgical Association, Hot Springs, VA. *Ann. Surg.* 124: 1027.

Langley, N. R., and Swanson, J. W. (1976). *Effects of Subcutaneous Implantation, through Two Years, on the Physical Properties of Medical Grade Tough Rubber (MDF-0198)*. Internal Dow Corning report number 1976-I0030-4571, produced to FDA and in litigation discovery, MDL926 number P—000004408.

Leininger, R. I., Mirkovitch, V., Peters A., and Hawks, W. A. (1964). Change in properties of plastics during implantation. *Trans. Am. Soc. Artif. Internal Organs* **X**: 320.

Margulies, H., and Barker, N. W. (1973). The coagulation time of blood in silicone tubes. *Am. J. Med. Sci.* **218**: 42.

Mazas, F. B., and GUEPAR. (1973). GUEPAR Total Knee Prosthesis. *Clin. Orthop. Rel. Res.* **94**: 211.

McHenry, M. M., Smeloff, E. A., Fong, W. Y., Miller, G. E., and Ryan, P. M. (1970). Critical obstruction of prosthetic heart valves due to lipid absorption by Silastic. *J. Thorac. Cardiovasc. Surg.* **59**(3): 413.

McLaughlin, J. K., Nyrén, O., Blot, W. J., Yin, L., Josefsson, S., Fraumeni, J. F., and Adami, H. O. (1998). Cancer risk among women with cosmetic breast implants: a population-based cohort study in sweden. *J. Nat. Cancer Inst.* **90**(2): 156.

Mellemkjaer, L., Kjoller, K., Friis, S., McLaughlin, J. K., Hogsted, C., Winther, J. F., Breiting, V., Krag, C., Kruger Kjaer, S., Blot, W. J., and Olsen, J. H. (2000). Cancer occurrence after cosmetic breast implantation in Denmark. *Int. J. Cancer* **88**: 301.

Noll, W. (1968). *Chemistry and Technology of Silicones*. Academic Press, New York.

Nyrén, O., Yin, L., Josefsson, S., McLaughlin, J. K., Blot, W. J., Engqvist, M., Hakelius, L., Boice, J. D. Jr. and Adami, H. O. (1998). Risk of connective tissue disease and related disorders among women with breast implants: a nation-wide retrospective cohort study in Sweden. *Br. Med. J.* 316(7129): 417.

Pangman, W. J. (1958). U.S. Patent No. 2,842,775.

Park, A. J., Chetty, U., and Watson, A. C. H. (1998). Silicone breast implants and breast cancer. *Breast* 7(1): 22.

Pinchuk, L., Martin, J. B., Esquivel, M. C., and MacGregor, D. C. (1988). The use of silicone/polyurethane graft polymers as a means of eliminating surface cracking of polyurethane prostheses. *J. Biomater. Appl.* 3(2): 260.

Remes, A., and Williams, D. F. (1992). Immune response in biocompatibility. *Biomaterials* 13(11): 731.

Sánchez-Guerrero, J., Colditz, G. A., Karlson, E. W., Hunter, D. J., Speizer, F. E., and Liang, M. H. (1995). Silicon breast implants and the risk of connective-tissue diseases and symptoms. *New Engl. J. Med.* 332(25): 1666.

Sanislow, C. A., and Zuidema, G. D. (1963). The use of silicone T-tubes in reconstructive biliary surgery in dogs. *J. Sur. Res.* **III**(10): 497.

Schwartzmann, E. (1930). Die technik der mammaplastik. *Der. Chirurg.* 2(20): 932–945.

Stark, F. O., Falenda, J. R., and Wright, A. P. (1982). Silicones. in *Comprehensive Organometallic Chemistry*, Vol. 2. G. Wilkinson, F. G. A. Sone, and E. W. Ebel (eds.). Pergamon Press, Oxford, p. 305.

Swanson, A. B. (1968). Silicone rubber implants for replacement of arthritic or destroyed joints in the hand. *Surg. Clin. North Am.* **48**: 1113.

Swanson, J. W., and LeBeau, J. E. (1974). The effect of implantation on the physical properties of silicone rubber. *J. Biomed. Mater. Res.* 8: 357.

Thorek, M. (1942). Amastia, hypomastia and inequality of the breasts. in *Plastic Surgery of the Breast and Abdominal Wall*. Charles C. Thomas, Springfield, IL.

Ward, R. S. (2000). Thermoplastic silicone-urethane copolymers: a new class of biomedical elastomers. *Med. Dev. Diagnost. Ind.*, April.

CHAPTER

8

Tissue Engineering

SIMON P. HOERSTRUP, LICHUN LU, MICHAEL J. LYSAGHT, ANTONIOS G. MIKOS, DAVID REIN, FREDERICK J. SCHOEN, JOHNNA S. TEMENOFF, JOERG K. TESSMAR, AND JOSEPH P. VACANTI

8.1 INTRODUCTION

Frederick J. Schoen

Biomaterials investigation and development has been stimulated and informed by a logical evolution of cell and molecular biology, materials science, and engineering, and an understanding of the interactions of materials with the physiological environment. These developments have permitted the evolution of concepts of tissue–biomaterials interactions to evolve through three stages, overlapping over time, yet each with a distinctly different objective (Fig. 1) (Hench and Pollak, 2002). The logical and rapidly progressing state-of-the-art, called *tissue engineering*, is discussed in this chapter.

The goal of early biomaterials development and use in a wide variety of applications was to achieve a suitable combination of functional properties to adequately match those of the replaced tissue without deleterious response by the host. The "first generation" of modern biomaterials (beginning in the mid-20th century) used largely off-the-shelf, widely available, industrial materials that were not developed specifically for their intended medical use. They were selected because of a desirable combination of physical properties specific to the clinical use, and they were intended to be *bioinert* (i.e., they elicited minimal response from the host tissues). The widely used elastomeric polymer silicone rubber is prototypical (see Chapter 2.2). Pyrolytic carbon, originally developed in the 1960s as a coating material for nuclear fuel particles and now widely used in mechanical heart valve substitutes, exemplifies one of the first biomaterials whose formulation was studied, modified, and controlled according to engineering and biological principles specifically for medical application (Bokros, 1977).

Subsequently, technology enabled and certain applications benefited by "second-generation" biomaterials that were intended to elicit a nontrivial and controlled reaction with the tissues into which they were implanted, in order to induce a desired therapeutic advantage. In the 1980s, *bioactive* materials were in clinical use in orthopedic and dental surgery as various compositions of bioactive glasses and ceramics (Hench and Pollak, 2002), in controlled-localized drug release applications such as the Norplant hormone-loaded contraceptive formulation (Meckstroth, 2001), and in devices such

as the HeartMate® left ventricular assist device for patients with congestive heart failure, with an integrally-textured polyurethane surface that fosters a controlled thrombotic reaction to minimize the risk of thromboembolism (Rose *et al.*, 2001). Recently, drug-eluting endovascular stents have been shown to markedly limit in-stent proliferative restenosis following balloon angioplasty (Sousa *et al.*, 2003). The need for maximally effective dosing regimens, new protein– and nucleic acid–based drugs (which cannot be taken in classical pill form), and elimination of systemic toxicities have stimulated development of new implantable polymers and innovative systems for controlled drug delivery and gene therapy (LaVan *et al.*, 2002). Controlled drug delivery is now capable of providing a wide range of drugs that can be targeted (e.g., to a tumor, to a diseased blood vessel, to the pulmonary alveoli) on a one-time or sustained basis with highly regulated dosage and can regulate cell and tissue responses through delivery of growth factors and plasmid DNA containing genes that encode growth factors (Bonadio *et al.*, 1999; Richardson *et al.*, 2001).

The second generation of biomaterials also included the development of resorbable biomaterials with variable rates of degradation matched to the requirements of a desired application. Thus, the discrete interface between the implant site and the host tissue could be eliminated in the long term, because the foreign material would ultimately be degraded by the host and replaced by tissues. A biodegradable suture composed of poly(glycolic acid) (PGA) has been in clinical use since 1974. Many groups continue to search for biodegradable polymers with the combination of strength, flexibility, and a chemical composition conducive to tissue development (Hubbell, 1999; Griffith, 2000; Langer, 1999).

With engineered surfaces and bulk architectures tailored to specific applications, "third generation" biomaterials are intended to stimulate highly precise reactions with proteins and cells at the molecular level. Such materials provide the scientific foundation for molecular design of scaffolds that could be seeded with cells *in vitro* for subsequent implantation or specifically attract endogenous functional cells *in vivo*. A key concept is that a scaffold can contain specific chemical and structural information that controls tissue formation, in a manner analogous to cell–cell communication and patterning during embryological development. The transition from

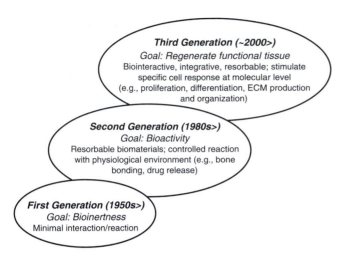

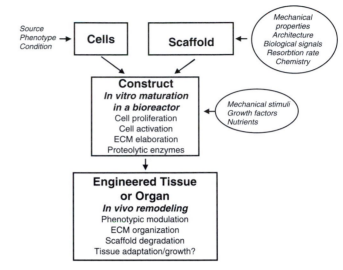

FIG. 1. Evolution of biomaterials science and technology. (From Rabkin, E., and Schoen, F. J. 2002. Cardiovascular tissue engineering. *Cardiovasc. Pathol.* **11**: 305.)

FIG. 2. Tissue engineering paradigm. In the first step of the typical tissue engineering approach, differentiated or undifferentiated *cells* are seeded on a bioresorbable *scaffold* and then the *construct* matured *in vitro* in a bioreactor. During maturation, the cells proliferate and elaborate extracellular matrix to form a "new" tissue. In the second step, the construct is implanted in the appropriate anatomical position, where remodeling *in vivo* is intended to recapitulate the normal tissue/organ structure and function. The key variables in the principal components—cells, scaffold, and bioreactor—are indicated. (From Rabkin, E., and Schoen, F. J., 2002, Cardiovascular tissue engineering. *Cardiovasc. Pathol.* **11**: 305.)

second- to third-generation biomaterials is exemplified by advances in controlled delivery of drugs or other biologically active molecules. Nanotechnology and the development of microelectromechanical systems (MEMS) have opened new possibilities for fine control of cell behavior through manipulation of surface chemistry and the mechanical environment (Chen *et al.*, 1997; Bhatia *et al.*, 1999; Huang and Ingber, 2000; Chiu *et al.*, 2000, 2003).

Tissue engineering is a broad term describing a set of tools at the interface of the biomedical and engineering sciences that use living cells or attract endogenous cells to aid tissue formation or regeneration, and thereby produce therapeutic or diagnostic benefit. In the most frequent paradigm, cells are seeded on a scaffold composed of synthetic polymer or natural material (collagen or chemically treated tissue), a tissue is matured *in vitro*, and the construct is implanted in the appropriate anatomic location as a prosthesis (Langer and Vacanti, 1993; Fuchs *et al.*, 2001; Griffith and Naughton, 2002; Rabkin and Schoen, 2002; Vacanti and Langer, 1999). A typical scaffold is a bioresorbable polymer in a porous configuration in the desired geometry for the engineered tissue, often modified to be adhesive for cells, in some cases selective for a specific circulating cell population.

The first phase is the *in vitro* formation of a tissue *construct* by placing the chosen cells and scaffold in a metabolically and mechanically supportive environment with growth media (in a *bioreactor*), in which the cells proliferate and elaborate extracellular matrix. In the second phase, the construct is implanted in the appropriate anatomic location, where remodeling *in vivo* is intended to recapitulate the normal functional architecture of an organ or tissue. The key processes occurring during the *in vitro* and *in vivo* phases of tissue formation and maturation are (1) cell proliferation, sorting, and differentiation, (2) extracellular matrix production and organization, (3) degradation of the scaffold, and (4) remodeling and potentially growth of the tissue. The general paradigm of tissue engineering is illustrated in Fig. 2. Biological and engineering challenges in tissue engineering are focused on the

TABLE 1 Control of Structure and Function of an Engineered Tissue

Cells	Biodegradable matrix/scaffold
Source	*Architecture/porosity/chemistry*
Allogenic	Composition/charge
Xenogenic	Homogeneity/isotropy
Autologous	Stability/resorption rate
	Bioactive molecules/ligands
Type/phenotype	Soluble factors
Single versus multiple types	
Differentiated cells from	*Mechanical properties*
primary or other tissue	Strength
Adult bone-marrow stem cells	Compliance
Pluripotent embryonic stem cells	Ease of manufacture
Density	
Viability	**Bioreactor conditions**
Gene expression	
Genetic manipulation	*Nutrients/oxygen*
	Growth factors
	Perfusion and flow conditions
	Mechanical factors
	Pulsatile
	Hemodynamic shear stresses
	Tension/compression

From Rabkin, E., and Schoen, F. J. (2002). Cardiovascular tissue engineering. *Cardiovasc. Pathol.* **11**: 305.

three principal components that comprise the "cell–scaffold–bioreactor system"; control of the various parameters in device fabrication (Table 1) may have major impact on the ultimate result. Exciting new possibilities are opened by advances in stem cell technology (Blau *et al.*, 2001; Bianco and Robey, 2001) and the recent evidence that some multipotential cells possibly capable of tissue regeneration are released by the bone marrow and circulating systemically (Hirschi *et al.*, 2002) while others may be resident in organs such as heart and the central nervous system formally not considered capable of regeneration (Hirschi and Goddell, 2002; Grounds *et al.*, 2002; Nadal-Ginard *et al.*, 2003; Johansson, 2003; Orlic *et al.*, 2003).

Tissue-engineered configurations for skin replacement have achieved clinical use. Further examples of previous and ongoing clinical tissue engineering approaches include cartilage regeneration using autologous chondrocyte transplantation (Brittberg, *et al.*, 1994) and a replacement thumb with bone composed of autologous periosteal cells and natural coral (hydroxyapatite) (Vacanti *et al.*, 2001). A key challenge in tissue engineering is to understand quantitatively how cells respond to molecular signals and integrate multiple inputs to generate a given response, and to control nonspecific interactions between cells and a biomaterial, so that cell responses specifically follow desired receptor–ligand interactions. Another approach uses biohybrid extracorporeal artificial organs using functional cells that are isolated from the recipient's blood or tissues by an impermeable membrane (Colton, 1995; Humes *et al.*, 1999; Strain and Neuberger, 2002). *Tissue engineering* also seeks to understand structure/function relationships in normal and pathological tissues (particularly those related to embryological development and healing) and to control cell and tissue responses to injury, physical stimuli, and biomaterials surfaces, through chemical, pharmacological, mechanical, and genetic manipulation. This is an immensely exciting field.

Bibliography

Bhatia, S. N., Balis, U. J., Yarmush, M. L., and Toner, M. (1999). Effect of cell–cell interactions in prevention of cellular phenotype: cocultivation of hepatocytes and nonparenchymal cells. *FASEB J.* **13**: 1883–1900.

Bianco, P., and Robey, P. G. (2001). Stem cells in tissue engineering. *Nature* **414**: 118–121.

Blau, H. M., Brazelton, T. R., and Weimann, J. M. (2001). The evolving concept of a stem cell: entity or function? *Cell* **105**: 829–841.

Bokros, J. C. (1977). Carbon biomedical devices. *Carbon,* **15**: 353–371.

Bonadio, J. E., Smiley, E., Patil, P., and Goldstein, S. (1999). Localized, direct plasmid gene delivery *in vivo*: prolonged therapy results in reproducible tissue regeneration. *Nat. Med.* **7**: 753–759.

Brittberg, M., Lindahl, A., Nilsson, A., Ohlsson, C., Isaksson, O., and Peterson, L. (1994). Treatment of deep cartilage defects in the knee with autologous chondrocyte transplantation. *N. Engl. J. Med.* **331**: 889–895.

Chen, C. S., Mrksich, M., Suang, S., Whitesides, G. M., and Ingber, D. E. (1997). Geometric control of cell life and death. *Science* **276**: 1425–1528.

Chiu, D. T., Jeon, N. L., Huang, S., Kane, R. S., Wargo, C. J., Choi, I. S., Ingber, D. E., and Whitesides, G. M. (2000). Patterned deposition of cells and proteins into surfaces by using three-dimensional microfluidic systems. *Proc. Natl. Acad. Sci. USA* **97**: 2408–2413.

Chiu, J-J., Chen, L-J., Lee, P-L., Lee, C-I., Lo, L-W., Usami, S., and Chien, S. (2003). Shear stress inhibits adhesion molecule expression in vascular endothelial cells induced by coculture with smooth muscle cells. *Blood* **101**: 2667–2674.

Colton, C. K. (1995). Implantable biohybrid artificial organs. *Cell Transplantation* **4**: 415–436.

Fleming, R. G., Murphy, C. J., Abrams, G. A., Goodman, S. L., and Nealey, P. F. (1999). Effects of synthetic micro- and nano-structured surfaces on cell behavior. *Biomaterials* **20**: 573–588.

Fuchs, J. R., Nasseri, B. A., and Vacanti, J. P. (2001). Tissue engineering: a 21st century solution to surgical reconstruction. *Ann. Thorac. Surg.* **72**: 577–590.

Griffith, L. G. (2000). Polymeric biomaterials. *Acta Mater.* **48**: 263–277.

Griffith, L. G., and Naughton, G. (2002). Tissue engineering—current challenges and expanding opportunities. *Science* **295**: 1009–1014.

Grounds, M. D., White, J. D., Rosenthal, N., and Bogoyevitch, M. A. (2002). The role of stem cells in skeletal and cardiac muscle repair. *J. Histochem. Cytochem.* **50**: 589–610.

Hench, L. L., and Pollak, J. M. (2002). Third-generation biomedical materials. *Science* **295**: 1014–1017.

Hirschi, K. K., and Goddell, M. A. (2002). Hematopoietic, vascular and cardiac fates of bone marrow–derived stem cells. *Gene Ther.* **9**: 648-652.

Huang, S., and Ingber, D. E. (2000). Shape-dependent control of cell growth, differentiation, and apoptosis: switching between attractors in cell regulatory networks. *Exp. Cell Res.* **261**: 91–103.

Hubbell, J. A. (1999). Bioactive biomaterials. *Curr. Opin. Biotechnol.* **10**: 123–129.

Humes, H. D., Buffington, D. A., MacKay, S. M., Funk, A., and Wetzel, W. E. (1999). Replacement of renal function in uremic animals with a tissue-engineered kidney. *Nat. Biotechnol.* **17**: 451–455.

Johansson, C. B. (2003). Mechanism of stem cells in the central nervous system. *J. Cell Physiol.* **196**: 409–418.

Langer, R. (1999). Selected advances in drug delivery and tissue engineering. *J. Controlled Release* **62**: 7–11.

Langer, R., and Vacanti, J. P. (1993). Tissue engineering. *Science* **260**: 920–926.

Lauffenburger, D. A., and Griffith, L. G. (2001). Who's got pull around here? Cell organization in development and tissue engineering. *Proc. Natl. Acad. Sci. USA* **98**: 4282–4284.

LaVan, D. A., Lynn, D. M., and Langer, R. (2002). Moving smaller in drug discovery and delivery. *Nat. Rev.* **1**: 77–84.

Meckstroth, K. R., and Darney, P. D. (2001). Implant contraception. *Semin. Reprod. Med.* **19**: 339–354.

Nadal-Ginard, B., Kajstura, J., Leri, A., and Anversa, P. (2003). Myocyte death, growth, and regeneration in cardiac hypertrophy and failure. *Circ. Res.* **92**: 139–150.

Orlic, D., Kajstura, J., Chimenti, S., Bodine, D. M., Leri, A., and Anversa, P. (2003). Bone marrow stem cells regenerate infarcted myocardium. *Pediatr. Transplant.* **7**(Suppl 3): 86–88.

Rabkin, E., and Schoen, F. J. (2002). Cardiovascular tissue engineering. *Cardiovasc. Pathol.* **11**: 305–317.

Richardson, T. P., *et al.* (2001). Polymeric delivery of proteins and plasmid DNA for tissue engineering and gene therapy. *Crit. Rev. Eukar. Gene Exp.* **11**: 47–58.

Rose, E. A., Gelijns, A. C., Muscowitz, A. J., Heitjan, D. F., Stevenson, L. W., Dembitsky, W., Long, J. W., Ascheim, D. D., Tierney, A. R., Levitan, R. G., Watson, J. T., Ronan, N. S., and Meier, P. (2001). Long-term mechanical left ventricular assist for end-stage heart failure. *N. Engl. J. Med.* **345**: 1435–1443.

Sousa, J. E., Serruys, P. W., and Costa, M. A. (2003). New frontiers in cardiology: drug-eluting stents—I and II. *Circulation* **107**: 2274–2279, 2383–2389.

Strain, A. J., and Neuberger, J. M. (2002). A bioartificial liver—state of the art. *Science* **295**: 1005–1009.

Vacanti, J. P., and Langer, R. (1999). Tissue engineering: the design and fabrication of living replacement devices for surgical reconstruction and transplantation. *Lancet* **354**: SI32–SI34.

Vacanti, C. A., Bonassar, L. J., Vacanti, M. P., and Shufflebarger, J. (2001). Replacement of an avulsed phalanx with tissue-engineered bone. *N. Engl. J. Med.* **344**: 1511–1514.

8.2 OVERVIEW OF TISSUE ENGINEERING

Simon P. Hoerstrup and Joseph P. Vacanti

INTRODUCTION

The loss or failure of an organ or tissue is a frequent, devastating, and costly problem in health care, occurring in millions of patients every year. In the United States, approximately 9 million surgical procedures are performed annually to treat these disorders, and 40 to 90 million hospital days are required. The total national health-care costs for these patients exceed $500 billion per year (Langer and Vacanti, 1993, 1999). Organ or tissue loss is currently treated by transplanting organs from one individual to another or performing surgical reconstruction by transferring tissue from one location in the human body to the diseased site. Furthermore, artificial devices made of plastic, metal, or fabrics are utilized. Mechanical devices such as dialysis machines or total joint replacement prostheses are used, and metabolic products of the lost tissue, such as insulin, are supplemented. Although these therapies have saved and improved millions of lives, they remain imperfect solutions.

Tissue engineering represents a new, emerging interdisciplinary field applying a set of tools at the interface of the biomedical and engineering sciences that use living cells or attract endogenous cells to aid tissue formation or regeneration (Rabkin *et al.*, 2002) to restore, maintain, or improve tissue function.

Engineered tissues using the patient's own (autologous) cells or immunologically inactive allogeneic or xenogeneic cells offer the potential to overcome the current problems of replacing lost tissue function and to provide new therapeutic options for diseases such as metabolic deficiencies.

CURRENT THERAPEUTIC APPROACHES FOR LOST TISSUE OR ORGAN FUNCTION

Transplantation

Organs or parts of organs are transplanted from a cadaveric or living-related donor into the patient suffering from lost organ function. Many innovative advances have been made in transplantation surgery during recent years and organ transplantation has been established as a curative treatment for end-stage diseases of liver, kidney, heart, lung, and pancreas (Starzl, 2001; Starzl *et al.*, 1989; Stratta *et al.*, 1994). Unfortunately, transplantation is substantially limited by a critical donor shortage. For example, fewer than 5200 liver donors are available annually for the approximately 18,500 who were on the waiting list in the year 2001 (http://www.ustransplant.org/annual.html). Besides the donor shortage, the other major problem of organ transplantation remains the necessity of lifelong immunosuppression therapy with a number of substantial and serious side effects (Keeffe, 2001).

Surgical Reconstruction

Organs or tissues are moved from their original location to replace lost organ function in a different location (e.g., saphenous vein as coronary bypass graft, colon to replace esophagus or bladder, myocutaneous flaps or freegrafts for plastic surgery). Nevertheless there are a number of problems associated with this method of therapy, since the replacement tissues, consisting of a different tissue type, cannot replace all of the functions of the original tissue. Moreover, long-term complications occur, such as the development of a malignant tumor in colon tissue replacing bladder function (Kato *et al.*, 1993; Kusama *et al.*, 1989) or calcification and resulting stenosis of vascular grafts (Kurbaan *et al.*, 1998). Finally, there is also the risk of complications and surgical morbidity at the donor site.

Artificial Prosthesis

The use of artificial, nonbiological materials in mechanical heart valves, blood vessels, joint replacement prostheses, eye lenses, or extracorporeal devices such as dialysis or plasmapheresis machines has improved and prolonged patients' lives dramatically. However, these methods are complicated by infection, limited durability of the material, lack of mechanism of biological repair and remodeling, chronic irritation, occlusion of vascular grafts, and the necessity of anticoagulation therapy and its side effects (Kudo *et al.*, 1999; Mow *et al.*, 1992). Regarding the pediatric patient population, not all artificial implants can provide a significant growth or remodeling potential, which often results in repeated operations associated with substantial morbidity and mortality (Mayer, 1995).

Supplementation of Metabolic Products of Diseased Tissues or Organs

In the case of the loss of endocrine tissue function, hormonal products such as insulin or thyroid, adrenal, or gonadal hormones can be successfully supplemented by oral or intravenous medication. In most cases, chronic supplementation is necessary. Unfortunately, supplementation therapy cannot replace natural feedback mechanisms, frequently resulting in dysregulation of hormone levels. As a consequence, clinical conditions such as hypo- or hyperglycemic crises or the long-term complications of chronic hormonal imbalances nephropathy or microvascular disease in patients with insulin-dependent diabetes mellitus continue to occur (Orchard *et al.*, 2002, 2003).

TISSUE ENGINEERING AS AN APPROACH TO REPLACE LOST TISSUE OR ORGAN FUNCTION

In the most frequent paradigm of tissue engineering, isolated living cells are used to develop biological substitutes for the restoration or replacement of tissue or organ function. Generally, cells are seeded on bioabsorbable scaffolds, a tissue is matured *in vitro*, and the construct is implanted in the appropriate anatomic location as a prosthesis. Cells used in tissue engineering may come from a variety of sources including application-specific differentiated cells from the patients themselves (autologous), human donors (allogeneic) or animal sources (xenogeneic), or undifferentiated cells comprising progenitor or stem cells. The use of isolated cells or cell aggregates enables manipulation prior to implantation, e.g., transfection of genetic material or modulation of the cell surface in order to prevent immunorecognition. Three general strategies have been adopted for the creation of new tissues including cell injection, closed, or flow-through systems, and tissue engineering using biodegradable scaffolds.

Cell Injection Method

The cell injection method avoids the complications of surgery by allowing the replacement of only those cells that supply the needed function. Isolated, dissociated cells are injected into the bloodstream or a specific organ of the recipient. The transplanted cells will use the vascular supply and the stroma provided by the host tissue as a matrix for attachment and reorganization (Matas *et al.*, 1976). This method offers opportunities for a number of applications in replacing metabolic functions as occurs in liver disease, for example (Grossman *et al.*, 1994). However, cell mass sufficient to replace lost metabolic functions is difficult to achieve and its application for replacing functions of structural tissues such as heart valves or cartilage is rather limited. Several cell types may be used for injection, such as bone marrow cells, blood-derived progenitor cells, and muscle satellite cells (see also the later subsection on tissue engineering of muscle).

Whole bone marrow contains multipotent mesenchymal stem cells (marrow stromal cells) that are derived from somatic mesoderm and are involved in the self-maintenance and repair of various mesenchymal tissues. These cells can be induced *in vitro* and *in vivo* to differentiate into cells of mesenchymal lineage, including fat, cartilage and bone, and cardiac and skeletal muscle. The first successful allogenic bone marrow transplant in a human was carried out in 1968. More than 40,000 transplants (from bone marrow, peripheral blood, or umbilical cord blood) were carried out worldwide in 2000 (http://www.ibmtr.org/newsletter/pdf/2002Feb.pdf). The most common indications for allotransplants are acute and chronic leukemias, myelodysplasia (MDS), and nonmalignant diseases (aplastic anemia, immune deficiencies, inherited metabolic disorders). Autotransplants are generally used for non-Hodgkin's lymphoma (NHL), multiple myeloma (MM), Hodgkin's lymphoma, and solid tumors.

Experimental studies suggested that bone marrow–derived or blood-derived progenitor cells may also contribute, e.g., to the regeneration of infarcted myocardium (Orlic *et al.*, 2001a).

Closed-System Method

Closed systems can be either implanted or used as extracorporeal devices. In this approach, cells are isolated from the body by a semipermeable membrane that allows diffusion of nutrients and the secreted cell products but prevents large entities such as antibodies, complement factors, or other immunocompetent cells from destroying the isolated cells. Protection is also provided to the recipient when potentially pathological (e.g., tumorigenic) cells are transplanted. Implantable systems (encapsulation systems) come in a variety of configurations, basically consisting of a matrix that cushions the cells and supports their survival and function and a surrounding porous membrane (Fig. 1). In vascular-type designs the transplanted secretory cells are housed in a chamber around a vascular conduit separated from the bloodstream by a semipermeable membrane. As blood flows through, it can absorb substances secreted by the therapeutic cells while the blood provides oxygen and nutrients to the cells. In macroencapsulation systems, a semipermeable membrane is used to encapsulate a relatively large (up to 50–100 million per unit) number of transplanted cells. Microcapsules are basically microdroplets of hydrogel with a diameter of less than 0.5 mm housing smaller numbers of cells. Macrocapsules are far more durable than microcapsule droplets and can be designed to be refillable in the body. Moreover, they can be retrieved, providing opportunities for

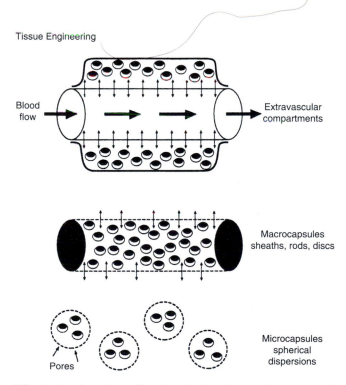

FIG. 1. Configurations of implantable closed-system devices for cell transplantation. (Reprinted with permission from Langer, R., and Vacanti, J. P., 1993. *Science* **260**: 920–926.)

more control than microcapsules. Their main limitation is the number of cells they can accommodate. In animal experiments, implantable closed-system configurations have been successfully used for the treatment of Parkinson´s disease as well as diabetes mellitus (Aebischer *et al.*, 1991, 1988; Kordower *et al.*, 1995; Date *et al.*, 2000). If islets of Langerhans are used, they will match the insulin released to the concentration of glucose in the blood. This has been successfully demonstrated in small and large animals with maintenance of normoglycemia even in long-term experiments (Kin *et al.*, 2002; Lacy *et al.*, 1991; Lanza *et al.*, 1999; Sullivan *et al.*, 1991). Major drawbacks of these systems are fibrous tissue overgrowth and resultant impaired diffusion of metabolic products, nutrients, and wastes, as well as the induction of a foreign-body reaction with macrophage activation resulting in destruction of the transplanted cells within the capsule (Wiegand *et al.*, 1993).

In extracorporeal systems (vascular or flow-through designs) cells are usually separated from the bloodstream. Great progress is being made in the development of extracorporeal liver assist devices for support of patients with acute liver failure. Currently four devices that rely on allogenic or xenogenic hepatocytes cultured in hollow-fiber membrane technology are in various stages of clinical evaluation (Patzer, 2001; Rozga *et al.*, 1994).

Tissue Engineering Using Scaffold Biomaterials

Open systems of cell transplantation with cells being in direct contact to the host organism aim to provide a permanent solution to the replacement of living tissue. The rationale behind the use of open systems is based on empirical observations: dissociated cells tend to reform their original structures when given the appropriate environmental conditions in cell culture. For example, capillary endothelial cells form tubular structures and mammary epithelial cells form acini that secrete milk on the proper substrata *in vitro* (Folkman and Haudenschild, 1980). Although isolated cells have the capacity to reform their respective tissue structure, they do so only to a limited degree since they have no intrinsic tissue organization and are hindered by the lack of a template to guide restructuring. Moreover, tissue cannot be transplanted in large volumes because diffusion limitations restrict interaction with the host environment for nutrients, gas exchange, and elimination of waste products. Therefore, the implanted cells will survive poorly more than a few hundred microns from the nearest capillary or other source of nourishment (Vacanti *et al.*, 1988). With these observations in mind, an approach has been developed to regenerate tissue by attaching isolated cells to biomaterials that serve as a guiding structures for initial tissue development. Ideally, these scaffold materials are biocompatible, biodegradable into nontoxic products, and manufacturable (Rabkin *et al.*, 2002). Natural materials used in this context are usually composed of extracellular matrix components (e.g., collagen, fibrin) or complete decellularized matrices (e.g., heart valves, small intestinal submucosa). Synthetic polymer materials are advantageous in that their chemistry and material properties (biodegradation profile, microstructure) can be well controlled. The majority of scaffold-based tissue

engineering concepts utilize synthetic polymers [e.g., poly (glycolic acid) (PGA), poly (lactid acid) (PLA), or poly (hydroxy alkanoate) (PHA)]. In general, these concepts involve harvesting of the appropriate cell types and expanding them *in vitro*, followed by seeding and culturing them on the polymer matrices. The polymer scaffolds are designed to guide cell organization and growth allowing diffusion of nutrient to the transplanted cells. Ideally, the cell–polymer matrix is prevascularized or would become vascularized as the cell mass expands after implantation. Vascularization could be a natural response to the implant or be artificially induced by sustained release of angiogenic factors from the polymer scaffold (Langer and Vacanti, 1999). Since the polymer scaffold is designed to be biodegradable, concerns regarding long-term biocompatability are obviated.

Cells used in tissue engineering may come from a variety of sources including cell lines from the patients themselves (autologous), human donors (allogeneic), or animal sources (xenogeneic). However, allogeneic and xenogeneic tissue may be subjected to immunorejection. Cell-surface modulation offers a possible solution to this problem by deleting immunogenic sites and therefore preventing immunorecognition. A bank of cryopreserved cells would then be possible and genetic engineering techniques could be used to insert genes (Raper and Wilson, 1993) to replace proteins, such as the LDL receptor (Chowdhury *et al.*, 1991) or factor IX (Armentano *et al.*, 1990).

APPLICATIONS OF TISSUE ENGINEERING:

Investigators have attempted to engineer virtually every mammalian tissue. In the following summary, we discuss replacement of ectodermal, endodermal, and mesodermal derived tissues.

Ectodermal Derived Tissue

Nervous System

Diseases of the central nervous system, such as a loss of dopamine production in Parkinsons's disease, represent an important target for tissue engineering. Transplantation of fetal dopamine-producing cells by stereotactically guided injection into the appropriate brain region has produced significant reversal of debilitating symptoms in humans (Lindvall *et al.*, 1990). Further benefit regarding survival, growth, and function has been demonstrated when implantation of dopamine-producing cells was combined with polymer-encapsulated cells continuously producing human glial cell line–derived growth factor (GDNF) (Sautter *et al.*, 1998). In the animal model PC12 cells, an immortalized cell line derived from rat pheochromocytoma, have been encapsulated in polymer membranes and implanted in the guinea pig striatum (Aebischer *et al.*, 1991) or primates (Date *et al.*, 2000; Kordower *et al.*, 1995), resulting in a dopamine release from the capsule detectable for many months. Similarly, encapsulated bovine adrenal chromaffin cells have been implanted into the subarachnoid space in rats, where through their continuous production of enkephalins and catecholamines they appeared to relieve chronic intractable pain (Sagen, 1992). Finally, investigations have been undertaken to achieve brain tissue by immobilization of neuronal and glia cells in N-methacrylamide polymer hydrogels. These cells have shown cell viability and maintained differentiation *in vitro* (Woerly *et al.*, 1996).

Nerve regeneration is another field of current investigations. When nerve injury results in gaps that are too wide for healing, autologous nerve grafts are used as a bridge. Several laboratories have shown in animal models that artificial guiding structures composed of natural polymers (laminin, collagen, chondroitin sulfate) or synthetic polymers can enhance nerve regeneration (Valentini *et al.*, 1992). Moreover, this process can be aided by placing Schwann cells seeded in polymer membranes (Guenard *et al.*, 1992). Polymers can also be designed so that they slowly release growth factors, possibly allowing regrowth of the damaged nerve over a greater distance (Aebischer *et al.*, 1989; Haller and Saltzman, 1998). In the case of neurodegenerative diseases such as amyotrophic lateral sclerosis (ALS), progression of the motor neuron disease could be successfully delayed in the animal model by polymer encapsulation of genetically modified cells to secrete neutrotrophic factors. This suggests that encapsulated cell delivery of neutrotrophic factors may provide a general method or effective administration of therapeutic proteins for the treatment of neurodegenerative diseases (Aebischer *et al.*, 1996a, b; Tan *et al.*, 1996).

Recently, a phase I/II clinical trial has been performed in 12 amyotrophic lateral sclerosis (ALS) patients to evaluate the safety and tolerability of intrathecal implants of encapsulated genetically engineered baby hamster kidney (BHK) cells releasing human ciliary neurotrophic factor (CNTF) (Aebischer *et al.*, 1996b, Zurn *et al.*, 2000). No adverse side effects have been observed in these patients in contrast to the systemic delivery of large amount of CNTF. However, antibodies against bovine fetuin have been detected because the capsules have been kept in a medium containing fetuin before transplantation.

Micorencapsulated cells may also be used for the treatment of malignant brain tumors (Thorsen *et al.*, 2000). Genetically modified cells secrete tumor controlling/suppressing substances such as the anti-angiogenic protein endostatin (Read *et al.*, 2001).

Cornea

The cornea is a transparent window covering the front surface of the eye that protects the intraocular contents and is the main optical element that focuses light onto the retina. Worldwide, millions of people suffer from bilateral corneal blindness. Transplant donors are limited and there is a risk of infectious agent transmission. Moreover, in the case that the limbal epithelial stem cells of the recipient are damaged (alkali burns, autoimmune conditions, or recurrent graft failures), the donor corneal epithelium desquamate and is replaced by conjuctivization and fibrovascular scarring in the recipient.

The cornea is avascular and immunologically privileged, making this tissue an excellent candidate for tissue engineering. Ideally, an artificial cornea would consist of materials that support adhesion and proliferation of corneal epithelial cells so that an intact continuous epithelial layer forms.

In addition, these materials should have appropriate nutrient and fluid permeability, light transparency, low light scattering, and no toxicity. Artificial cornea has been developed that consisted of a peripheral rim of biocolonizable microporous fluorocarbon polymer (polytetrafluoroethylene, PTFE) fused to an optical core made of polydimethylsiloxane coated with polyvinylpyrrolidone (Legeais and Renard, 1998). In contrast to this "hybrid" cornea, another group used poly-(2-hydroxyethyl methacrylate) (PHEMA) for both the porous skirt (opaque sponge, 10–30 μm) and the optical core (transparent gel) (Chirila, 2001; Crawford *et al.*, 2002). PHEMA is a biomaterial with a long record of ocular tolerance in applications such as contact lenses, intraocular lenses, and intracorneal inlays. The use of this material as porous sponge allowed cellular invasion, production of collagen, and vascularization, without the formation of a foreign-body capsule. Both devices have been tested preclinically.

Furthermore, tissue-engineered implantable contact lenses could obviate the need for surgery in patients who seek convenient, reversible correction of refractive error. An onlay involves debridement of the central corneal epithelium and placement of a synthetic lenticule on the exposed stromal surface, leaving Bowman's zone intact. The anterior surface of the lenticule is then covered by the recipient eye's corneal epithelium, incorporating the lenticule to achieve the desired refractive correction by altering the curvature of the anterior corneal surface. Porous collagen-coated perfluoropolyether (PFPE) was successfully tested in cats (Evans *et al.*, 2002) and Lidifilcon A, a copolymer of methyl methacrylate and N-vinyl-2-pyrrolidone (Allergan Medical Optics, Irvine, CA) was implanted in monkeys (McCarey *et al.*, 1989).

The multistep procedure of corneal reconstruction has been demonstrated using corneal cells from rabbit (Zieske *et al.*, 1994) and from fetal pig (Schneider *et al.*, 1999), human cells from donor corneas (Germain *et al.*, 1999), or immortalized cell lines from the main layers of the cornea (Griffith *et al.*, 1999). In these studies collagen matrices or collagen–chondritin sulfate substrates cross-linked with glutaraldehyde have been tested. More recently, carbodiimide cross-linking and composites using urethane/urea techniques have been evaluated for biocompatibility and epithelial ingrowth (Griffith *et al.*, 2002).

Skin

Several new types of tissue transplants are being studied for treatment of burns, skin ulcers, deep wounds, and other injuries. One approach to skin grafts involves the *in vitro* culture of epidermal cells (keratinocytes). Small skin biopsies are harvested from burn patients and expanded up to 10,000-fold. This expansion is achieved, e.g., by cultivating keratinocytes on a feeder layer of irradiated NIH 3T3 fibroblasts, which, in conjunction with certain added media components, stimulates rapid cell growth. This approach allows coverage of extremely large wounds. A disadvantage is the 3- to 4-week period required for cell expansion, which may be too long for a severely burned patient. Cryopreserved allografts may help to circumvent this problem (Nave, 1992). Another promising approach uses human neonatal dermal fibroblasts grown on poly(glycolic acid) mesh. In deep injuries involving all layers of skin the grafts are placed onto the wound bed and a skin graft is placed on top followed by vascularization of the graft. This results in formation of an organized tissue resembling dermis. Clinical trials have shown good graft acceptance without evidence of immune rejection (Hansbrough *et al.*, 1992). Fibroblasts have also been placed on hydrated collagen gel. Upon implantation, the cells migrate through the gel by enzymatic digestion of collagen, which results in reorganization of collagen fibrils (Bell *et al.*, 1979). ApliGraf, formerly known as Graftskin, is a commercially available two-layered tissue-engineered skin product composed of type I bovine collagen that contains living human dermal fibroblasts and an overlying cornified epidermal layer of living human keratinocytes. Both cell types are derived from neonatal foreskin and grow in a special mold that limits lateral contraction (Bell *et al.*, 1991a, b). ApliGraf has been investigated in a multicenter study after excisional surgery for skin cancer with good results (Eaglstein *et al.*, 1999).

The artificial skin developed by Burke and Yannas (Burke *et al.*, 1981), now called Integra, consists of collagen–chondritin 6-sulfate fibers obtained from bovine hide (collagen) and shark cartilage (chondritin 6-sulfate). It has been engineered into an open matrix of uniform porosity and thickness and covered with a uniformly thick (0.1-mm) silicone sheet. This artificial skin has been studied extensively in humans (Heimbach *et al.*, 1988; Sheridan *et al.*, 1994) and was approved for use in burn patients in 1997.

Besides clinical use of artificial skin, several companies have explored the possibilities of dermal substitutes for diagnostic purposes. There is particular interest in minimizing the use of animals for topological irritation, corrosivity, and other testing (Fentem *et al.*, 2001; Portes *et al.*, 2002). Gene therapy for genodermatoses (Spirito *et al.*, 2001), junctional epidermolysis bullosa (Robbins *et al.*, 2001), and ichthyosis (Jensen *et al.*, 1993) remains a topic of great interest using either transgenic fibroblasts or keratinocytes.

Endoderm

Liver

Liver transplantation is a routine treatment for end-stage liver disease, but donor organ shortage remains a serious problem. Many patients die while waiting for a transplant and those with chronic disease often deteriorate resulting in low survival rates after transplantation. Therefore a "bridging" device that would support liver function until a donor liver became available or the patient's liver recovered is of great interest. Most liver support systems remove toxins normally metabolized by the liver through dialysis, charcoal hemoperfusion, immobilized enzymes, or exchange transfusion. However, none of these systems can offer the full functional spectrum performed by a healthy liver. Hepatocyte systems aiming at replacement of liver function by transplantation of isolated cells are being studied for both extracorporeal and implantable applications. Extracorporeal systems can be used when the patient's own organ is recovering or as a bridge to transplantation. These systems provide a good control of the medium surrounding the cell

system and a minimized risk of immune rejection. Their design is primarily based on hollow-fiber, spouted-bed, or flat-bed devices (Bader *et al.*, 1995). Implantable hepatocyte systems, on the other hand, offer the possibility of permanent liver replacement (Yarmush *et al.*, 1992a). Successful hepatocyte transplantation depends on a number of critical steps. After cell harvest the hepatocytes must be cultured and expanded *in vitro* prior to transplantation. Hepatocyte morphology can be maintained by cultivating the cells on three-dimensional structures, such as sandwiching them between two hydrated collagen layers. Under these conditions the hepatocytes have been shown to secrete functional markers at physiological levels (Dunn *et al.*, 1991). Moreover the hepatocytes must be attached to the polymer substrata so that they maintain their differentiated function and viability. A sufficient mass of hepatocytes must become engrafted and remain functional to achieve metabolic replacement and vascularization, which is critical for graft survival (Yarmush *et al.*, 1992b). Finally, hepatocyte transplantation per se provides neither all cell types nor the delicate and complex structural features of the liver. Products normally excreted through bile may accumulate because of the difficulty in reconstructing the biliary tree solely from hepatocytes. However, hepatocytes placed on appropriate polymers can form tissues resembling those in the natural organ and have shown evidence of bile ducts and bilirubin removal (Uyama *et al.*, 1993). More recently, model systems in which the vascular architecture is mimicked in the device have been tested using three-dimensional printing, hepatocytes, and endothelial cells (Fig. 2; Kim *et al.*, 1998).

Four bioartificial liver devices have entered sustained clinical trials. The device rely all on hollow-fiber membranes to isolate hepatocytes from direct contact with patient fluids. They differ in source and treatment of hepatocytes prior to patient use and in the choice of perfusate: plasma or whole blood. Three devices are perfused with the patient's plasma. The HepatAssist is filled with freshly thawed cryopreserved primary porcine hepatocytes along with collagen-coated dextran beads for cell attachment (Chen *et al.*, 1997; Rozga *et al.*, 1993; Watanabe *et al.*, 1997). The ELAD system uses a HepG2 human hepatocyte cell line that has been grown to confluence in the extracellular space (Ellis *et al.*, 1996; Sussman *et al.*, 1994). The Gerlach BELS run either with human hepatocytes (if available) or with porcine primary hepatocytes embedded in a collagen matrix in the extraluminal space (Gerlach, 1997; Gerlach *et al.*, 1997). In contrast, the bioartificial liver support system (BLSS) is perfused with whole blood. This has the advantage that a greater rate of blood concentration reduction and lower endpoint blood concentration at equivalent perfusion times is achieved compared to systems using plasma perfusion. The detoxification is performed with primary porcine hepatocytes (Mazariegos *et al.*, 2001; Patzer *et al.*, 2002, 1999).

Pancreas

Each year more than 700,000 new cases of diabetes are diagnosed in the United States and approximately 150,000 patients die from the disease and its complications. Diabetes is characterized by pancreatic islet destruction leading to more

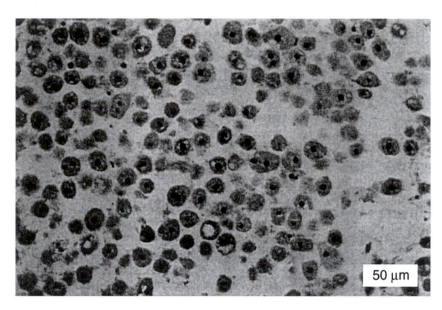

FIG. 2.   Histologic photomicrograph demonstrating viable hepatic cells after 2 days under flow conditions (hematoxylin and eosin; original magnification ×300). (Reprinted with permission from Kim, S. S., *et al.*, 1998. *Ann. Surg.* **228:** 8–13.)

or less complete loss of glucose control. Tissue engineering approaches to treatment have focused on transplanting functional pancreatic islets, usually encapsulated to avoid immune reaction. Three general approaches have been tested in animal experiments. In the first, a tubular membrane was coiled in a housing that contained islets. The membrane was connected to a polymer graft that in turn connected the device to blood vessels. This membrane had a 50-kDa molecular mass cutoff, thereby allowing free diffusion of glucose and insulin but blocking passage of antibodies and lymphocytes. In pancreatectomized dogs treated with this device, normoglycemia was maintained for more than 150 days (Sullivan *et al.*, 1991). In a second approach, hollow fibers containing rat islets were immobilized in polysaccharide alginate. When the device was placed intraperitoneally in diabetic mice, blood glucose levels were lowered for more than 60 days and good tissue biocompatability was observed (Lacy *et al.*, 1991). Finally, islets have been enclosed in microcapsules composed of alginate or polyacrylate. This method offers a number of distinct advantages over the use of other biohybrid devices, including greater surface-to-volume ratio and ease of implantation (simple injection) (Kin *et al.*, 2002; Lanza *et al.*, 1999, 1995). All of these transplantation strategies require a large, reliable source of donor islets. Porcine islets are used in many studies and genetically engineered cells that overproduce insulin are also under investigation (Efrat, 1999).

Tubular Structures

The current concept of using tubular structures of other organs for reconstruction of bladder, ureter and urethra, trachea, esophagus, intestine, and kidney represents a major

therapeutic improvement. A diseased esophagus, for example, can be treated clinically with autografts from the colon, stomach, skin, or jejunal segments. However, such procedures carry a substantial risk of graft necrosis, inadequate blood supply, infection, lack of peristaltic activity, and other complications. Copolymer tubes consisting of lactic and glycolic acid have been sutured into dogs after removal of esophageal segments, over time resulting in coverage of the polymer with connective tissue and epithelium (Grower *et al.*, 1989). Alternatively, elastin-based acellular aortic patches have been successfully used in experimental esophagus injury in the pig. While mucosal and submucosal coverage took place within 3 weeks, the majority of the elastin-based biomaterial degraded. However, the muscular layer did not regenerate (Kajitani *et al.*, 2001). In a similar approach fetal intestinal cells have been placed onto copolymer tubes and implanted in rats. Histological examination after several weeks revealed differentiated intestinal epithelium lining of the tubes and this epithelium appeared to secrete mucus (Vacanti *et al.*, 1988). Furthermore, intestinal epithelial organoid units transplanted on porous biodegradable polymer tubes have been shown to vascularize and to regenerate into complex tissue resembling small intestine (Kim *et al.*, 1999), and successful anastomosis between tissue-engineered intestine and native small bowel has been performed (Fig. 3; Kaihara *et al.*, 1999). Finally, Perez *et al.* demonstrated that tissue-engineered small intestine is capable of developing a mature immunocyte population and that mucosal exposure to luminal stimuli is critical to this development (Perez *et al.*, 2002). Despite these promising findings, the regeneration of the muscle layer seems to be a major problem. Autologous mesenchymal stem cells seeded onto a collagen sponge graft induced only a transient distribution

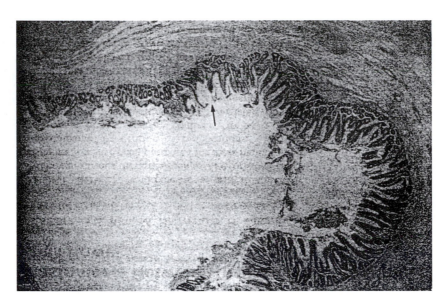

FIG. 3. Histology of a tissue-engineered intestine 10 weeks after implantation characterized by crypt villus structures. Arrow indicates anastomosis site; left site of the arrow is tissue-engineered intestine and right is native small bowel. (Reprinted with permission from Kaihara, S., *et al.*, 1999. *Transpl. Proc.* **31:** 661–662.)

of cells positive for α-smooth muscle actin (Hori *et al.*, 2002).

Tubular structures have also been used in kidney replacement. As a first step toward creating a bioartificial kidney, renal tubular cells have been grown on acrylonitrile–vinyl chloride copolymers or microporous cellulose nitrate membranes. *In vitro*, these cells transported glucose and tetraethylammonium cation in the presence of a hemofiltrate of uremic patients (Uludag *et al.*, 1990). In a further attempt to create bioartificial renal tubule, renal epithelial cells have been grown on hollow fibers and formed an intact monolayer exhibiting functional active transport capabilities (MacKay *et al.*, 1998). Finally, an extracorporeal device was developed using a standard hemofiltration cartridge containing renal tubule cells (Humes *et al.*, 1999; Nikolovski *et al.*, 1999). The pore size of the hollow fibers allows the membranes to act as scaffolds for the cells and as an immunoprotective barrier. *In vitro* and *in vivo* studies have shown that the cells keep differentiated active transport, differentiated metabolic transport, and important endocrine processes (Humes *et al.*, 2002, 2003).

For replacement of urether, urothelial cells were seeded onto degradable polyglycolic acid tubes and implanted in rats and rabbits resulting in two or three layers of urothelial cell lining (Atala *et al.*, 1992). More recently, an acellular collagen matrix from bladder submucosa seeded with cells from urethral tissue was also successfully used for tubularized replacement in the rabbit. In contrast, unseeded matrices lead to poor tissue development (de Filippo *et al.*, 2002). A neo-bladder has been created from urothelial and smooth muscle cells *in vitro* and after implantation in the animal, functional evaluation for up to 11 months has demonstrated a normal capacity to retain urine, normal elastic properties, and normal histologic architecture (Oberpenning *et al.*, 1999).

Mesoderm

Cartilage

More than 1 million surgical procedures in the United States each year involve cartilage replacement. Current therapies include cartilage transplantation and implantation of artificial polymer or metal prostheses. Unfortunately, donor tissue is limited and artificial implants may result in infection and adhesive breakdown at the host–prosthesis interface. Finally, a prosthesis cannot adapt in response to environmental stresses as does cartilage (Mow *et al.*, 1992). The need for improved treatments has motivated research aiming at creating new cartilage that is based on collagen–glycosaminglycan templates (Stone *et al.*, 1990), isolated chondrocytes (Grande *et al.*, 1989), and chondrocytes attached to natural or synthetic polymers (Cancedda *et al.*, 2003; Vacanti *et al.*, 1991; Wakitani *et al.*, 1989). It is critical that the cartilage transplant have an appropriate thickness and attachment to be mechanically functional. Chondrocytes grown in agarose gel culture have been shown to produce tissues with stiffness and compressibility comparable to those of articular cartilage (Freed *et al.*, 1993). The use of bioreactors for cultivating chondrocytes on polymer scaffolds *in vitro* enables nutrients to penetrate the center of this nonvascularized tissue, leading to relatively strong and thick (up to 0.5 cm) implants (Buschmann *et al.*, 1992). Moreover, it has been shown that the hydrodynamic conditions in tissue-culture bioreactors can modulate the composition, morphology, mechanical properties, and electromechanical function of engineered cartilage (Vunjak-Novakovic *et al.*, 1999). In other studies, chondrocytes were seeded onto PGA meshes and conditioned for several weeks on an orbital shaker. The functional cartilage was then combined with an osteoconductive support made of ceramic/collagen sponge. The composite was press-fitted in a large experimental osteochondral injury in a rabbit knee joint, where it showed good structural and functional properties (Schaefer *et al.*, 2002). With regard to the needs of reconstructive surgery, tissue-engineered autologous cartilage has been generated *in vitro* from tiny biopsies (Naumann *et al.*, 1998). Finally, some research has been undertaken to evaluate tissue engineering of cartilage even in space in order to elucidate the influence of micro/agravity on tissue formation (Freed *et al.*, 1997).

Bone

Current therapies of bone replacement include the use of autogenous or allogenic bone. Moreover, metals and ceramics are used in several forms: biotolerant (e.g., titanium), bioresorbable (e.g., tricalcium phosphate), porous (e.g., hydroxyapatite-coated metals), and bioactive (e.g., hydroxyapatite and glasses). Synthetic and natural polymers have been investigated for bone repair, but it has been difficult to create a polymer displaying optimal strength and degradation properties. Another approach involves implantation of demineralized bone powder (DBP), which is effective in stimulating bone growth. By inducing and augmenting formation of both cartilage and bone (including marrow), bone morphogenic proteins (BMP) or growth factors such as transforming growth factor-β (TGF-β) represent other promising strategies (Toriumi *et al.*, 1991; Yasko *et al.*, 1992). Bone growth can also be induced when cells are grown on synthetic polymers and ceramics. For example, when human marrow cells are grown on porous hydroxyapatite in mice, spongious bone formation was detectable inside the pores within 8 weeks (Casabona *et al.*, 1998). Femoral shaft reconstruction has been demonstrated using bioresorbable polymer constructs seeded with osteoblasts as bridges between the bone defect (Puelacher *et al.*, 1996), and similar experiences have been reported for craniofacial applications (Breitbart *et al.*, 1998). Formation of phalanges and small joints has been demonstrated with selective placement of periosteum, chondrocytes, and tenocytes into a biodegradable synthetic polymer scaffold (Isogai *et al.*, 1999). Large bone defects in tibia of sheep were successfully reconstructed using combinations of autologous marrow stromal cells and coral (Petite *et al.*, 2000). Similar results were obtained by Kadiyala *et al.*, who have treated experimentally induced nonunion defects in adult dog femora with autologous marrow-derived cells grown on a hydroxyapatite : beta tricalcium phosphate (65 : 35) scaffold (Kadiyala *et al.*, 1997). This approach was also successful in patients suffering from segmental bone defects. Abundant callus formation along the implants and good integration at the interface with the host bones was observed 2 months after surgery (Quarto *et al.*, 2001).

Muscle

The ability to generate muscle fibers has possible application regarding the treatment of muscle injury, cardiac disease, disorders involving smooth muscle of the intestine or urinary tract, and systemic muscular diseases such as Duchenne muscular dystrophy (DMD). Myoblasts from unaffected relatives have been transplanted into Duchenne patients and shown to produce dystrophin several months following the implantation.

Myoblasts can migrate from one healthy muscle fiber to another (Gussoni *et al.*, 1992); thus cell-based therapies may be useful in treating muscle atrophies. Creation of a whole hybrid muscular tissue was achieved by a sequential method of centrifugal cell packing and mechanical stress-loading resulting in tissue formation strongly resembling native muscle in terms of cell density, cell orientation, and incorporation of capillary networks (Okano *et al.*, 1998). Kim and Mooney demonstrated with regard to smooth muscle cells the importance of matching both the initial mechanical properties and the degradation rate of a predefined three-dimensional scaffold to the specific tissue that is being engineered (Kim and Mooney, 1998).

Loss of heart muscle tissue in the course of ischemic heart disease or cardiomyopathies is a major factor of morbidity and mortality in numerous patients. Once patients become symptomatic, their life expectancy is usually markedly shortened. This decline is mostly attributed to the inability of cardiomyocytes to regenerate after injury. Necrotic cells are replaced by fibroblasts leading to scar tissue formation and regional contractile dysfunction. In contrast, skeletal muscle has the capacity of tissue repair, presumably because of satellite cells that have regenerative capability. Satellite cells are undifferentiated skeletal myoblasts, which are located beneath the basal lamina in skeletal muscles. These cells have also been tested for myocardial repair (Chiu *et al.*, 1995; Menasche, 2003; Menasche *et al.*, 2001; Taylor *et al.*, 1998). In rats, myoblast grafts can survive for at least 1 year (Al Attar *et al.*, 2003). However, satellite cells transplanted into nonreperfused scar tissue do not transdifferentiate into cardiomyocytes but show a switch to slow-twitch fibers, which allow sustained improvement in cardiac function (Hagege *et al.*, 2003; Reinecke *et al.*, 2002).

Recent studies have suggested that bone marrow-derived or blood-derived progenitor cells contribute to the regeneration of infarcted myocardium and enhance neovascularization of ischemic myocardium (Kawamoto *et al.*, 2001; Orlic *et al.*, 2001a, b). In a pilot trial it was shown that also in patients with reperfused acute myocardial infarction, intracoronary infusion of autologous progenitor cells beneficially affected postinfarction left-ventricle remodeling processes (Assmus *et al.*, 2002).

An alternative approach to cell grafting techniques is the generation of cardiac tissue grafts *in vitro* and implanting them as spontaneously and coherently contracting tissues. As a model system, rat neonatal or embryonic chicken cardiomyocytes may be seeded on three-dimensional polymeric scaffolds (Carrier *et al.*, 1999) or collagen disks formed as a sponge (Radisic *et al.*, 2003) or by layering cell sheets three-dimensionally (Shimizu *et al.*, 2002). The latter two approaches are suitable for producing thicker cardiac tissue with more evenly distributed cells at a higher density. A principally different approach to generate engineered heart tissue was developed by Eschenhagen and colleagues. Neonatal or embryonic cardiomyocytes were mixed with freshly neutralized collagen I and cast into a cylindrical mold. After a few days the tissue patches were transferred to a stretching device, which induced hypertrophic growth and increased cell differentiation (Eschenhagen *et al.*, 2002b, 1997; Zimmermann *et al.*, 2000). Interestingly, the response to isoprenalin of stretched tissue was much more pronounced than in unstretched tissue (Eschenhagen *et al.*, 2002b; Zimmermann *et al.*, 2002).

Blood Vessels

Peripheral vascular disease represents a growing health and socioeconomic burden in most developed countries (Ounpuu *et al.*, 2000). Today, artificial prostheses made of expanded polytetrafluoroethylene (ePTFE) and poly(ethylene terephthalate) (PET, Dacron) are the most widely used synthetic materials. Although successful in large diameter (> 5-mm) high-flow vessels, in low-flow or smaller diameter sites they are compromised by thrombogenicity and compliance mismatch (Edelman, 1999). To circumvent these problems numerous modifications and techniques to enhance hematocompatibility and graft patency have been evaluated both *in vitro* and *in vivo*. These include chemical modifications, coatings (Gosselin *et al.*, 1996; Ye *et al.*, 2000), and surface seeding with endothelial cells (Pasic *et al.*, 1996; Zilla *et al.*, 1999). *In vitro* endothelialization of ePTFE grafts may result in patency rates comparable to state-of-the-art venous autografts (Meinhart *et al.*, 1997). Polymer surface modifications involving protein adsorption may also be desirable. Unfortunately, materials that promote endothelial cell attachment often simultaneously promote attachment of platelets and smooth muscle cells associated with the adverse side effects of clotting and pseudointimal thickening. A possible solution has been demonstrated with polymers containing adhesion molecules (ligands) specific for endothelial cells (Hubbell *et al.*, 1991).

To overcome the limitations just mentioned, tissue engineering procedures could lead to completely biological vascular grafts. In fact, there have already been case reports regarding first human pediatric applications of tissue-engineered large-diameter vascular grafts (Naito *et al.*, 2003; Shin'oka *et al.*, 2001). As to small-caliber grafts, there are three principal approaches involving (1) synthetic biodegradable scaffolds, (2) biological scaffolds, and (3) completely autologous methods.

1. Niklason *et al.* have shown in animal models that by utilizing flow bioreactors to condition biodegradable polymers loaded with vascular cells, it is possible to generate arbitrary lengths of functional vascular grafts with significant extracellular matrix production, contractile responses to pharmacological agents, and tolerance of supraphysiologic burst pressures (Mitchell and Niklason, 2003; Niklason *et al.*, 2001, 1999). Similar *in vitro* experiments based on human vascular-derived cells seeded on PGA/PHA copolymers demonstrated the feasibility of viable, surgically implantable human small-caliber vascular grafts and the important effect of a "biomimetic" *in vitro* environment on tissue maturation (Hoerstrup *et al.*, 2001).

2. A different approach to tissue engineering of vascular grafts comprises the use of decellularized natural matrices as initially introduced by Rosenberg *et al.* (1996). Histological examination of chemically decellularized carotid arteries revealed well-preserved structural matrix proteins. This provides an acellular scaffold that can be successfully repopulated *in vitro* prior to implantation (Teebken *et al.*, 2000). Such scaffolds have also been shown to be repopulated *in vivo* (Bader *et al.*, 2000). Recently, successful utilization of endothelial precursor cells for tissue engineering of vascular grafts based on decellularized matrices has been demonstrated (Kaushal *et al.*, 2001).

3. L'Heureux *et al.* cultured and conditioned sheets of vascular smooth muscle cells and their native extracellular matrix without any scaffold material in a flow system. Subsequently these sheets were placed around a tubular support device and after maturation the tubular support was removed and endothelial cells were seeded into the lumen. Thereby a complete scaffold-free vessel was created with a functional endothelial layer and a burst strength of more than 2000 mm Hg (L'Heureux *et al.*, 1998).

Angiogenesis (the formation of new blood vessels) is essential for growth, tissue repair, and wound healing. Therefore, many tissue-engineering concepts involve angiogenesis for the vascularization of the newly generated tissues. Unfortunately, so far advances have been compromised by the inability to vascularize thick, complex tissues, particularly those comprising large organs such as the liver, kidney, or heart. To overcome these limitations, several approaches have been investigated. Vacanti and co-workers used local delivery of basic fibroblast growth factor (bFGF) to increase angiogenesis and engraftment of hepatocytes in tissue-engineered polymer devices (Lee *et al.*, 2002). In another study sustained and localized delivery of vascular endothelial growth factor (VEGF) combined with the transplantation of human microvascular endothelial cells was used to engineer new vascular networks (Peters *et al.*, 2002). Using micromachining technologies on silicon, Kaihara *et al.* demonstrated *in vitro* generation of branched three-dimensional vascular networks formed by endothelial cells (Kaihara *et al.*, 2000).

Heart Valves

For treatment of heart-valve disease, mechanical or biological valves are currently in use. The drawbacks of mechanical valves include the need for lifelong anticoagulation, the risk of thromboembolic events, prosthesis failure, and the inability of the device to grow. Biological valves (homograft, xenograft, fixed by cryopreservation of chemical treatment) have a limited durability due to their immunogenic potential and the fact that they represent nonliving tissues without regeneration capacities. All types of contemporary valve prostheses basically consist of nonliving, foreign materials, posing specific problems to pediatric applications when devices with growth potential are required for optimal treatment.

The basic concept currently used for tissue engineering of heart valve structures is to transplant autologous cells onto a biodegradable scaffold, to grow and to condition the cell-seeded scaffold device *in vitro*, and finally to implant the tissue-like construct into the donor patient.

The heart-valve scaffold may be based on either biological or synthetic materials. Donor heart valves or animal-derived valves depleted of cellular antigens can be used as a scaffold material. Removing the cellular components results in a material composed of essentially extracellular matrix proteins that can serve as an intrinsic template for cell attachment (Samouillan *et al.*, 1999). In general, nonfixed acellularized valve leaflets have shown recellularization by the host, as demonstrated in dogs (Wilson *et al.*, 1995) and sheep (Elkins *et al.*, 2001; Goldstein *et al.*, 2000). However, first clinical applications of this concept in children resulted in rapid failure of the heart valves due to severe foreign-body-type reactions associated with a 75% mortality (Simon *et al.*, 2003). In a further approach, specific biological matrix constituents can be used as scaffold material. Collagen is one of the materials that show biodegradable properties and can be used as a foam (Rothenburger *et al.*, 2002), gel or sheet (Hutmacher *et al.*, 2001), or sponge (Taylor *et al.*, 2002), and even as a fiber-based scaffold (Rothenburger *et al.*, 2001). It has the disadvantage that it is difficult to obtain from the patient. Therefore, most of the collagen scaffolds are of animal origin. Another biological material displaying good controllable biodegradable properties is fibrin. Since fibrin gels can be produced from the patient's blood to serve as autologous scaffold, no toxic degradation or inflammatory reactions are expected (Lee and Mooney, 2001).

The use of synthetic materials as scaffolds has already been broadly demonstrated for cardiovascular tissue engineering. Initial attempts to create single heart-valve leaflets were based on synthetic scaffolds, such as polyglactin, PGA [poly(glycolic acid)], PLA [poly(lactic acid)], or PLGA (copolymer of PGA and PLA). To create complete trileaflet heart-valve conduits, PHA-based materials (polyhydroxyalkanoates) were used (Sodian *et al.*, 2000). These materials are thermoplastic and can therefore be easily molded into any desired three-dimensional shape. A combined polymer scaffold consisting of nonwoven PGA and P4HB (poly-4-hydroxybutyrate) has shown promising *in vivo* results (Hoerstrup *et al.*, 2000a).

In most cardiovascular tissue-engineering approaches cells are harvested from donor tissues, e.g., from peripheral arteries, and mixed vascular cell populations consisting of myofibroblasts and endothelial cells are obtained. Out of these, pure viable cell lines can be easily isolated by cell sorters (Hoerstrup *et al.*, 1998) and the subsequent seeding onto the biodegradable scaffold is undertaken in two steps. First, the myofibroblasts are seeded and grown *in vitro*. Second, the endothelial cells are seeded on top of the generated neotissue, leading to the formation of a native leaflet-analogous histological structure (Zund *et al.*, 1998).

Successful implantation of a single tissue-engineered valve leaflet has been demonstrated in the animal model (Shinoka *et al.*, 1996) and based on a novel *in vitro* conditioning protocol of the tissue-engineered valve constructs in bioreactor flow systems (pulse-duplicator) completely autologous, living heart-valves were generated (Fig. 4). Interestingly, these tissue-engineered valves showed good *in vivo* functionality and strongly resembled native heart valves with regard to biomechanical and morphological features (Hoerstrup *et al.*, 2000b; Rabkin *et al.*, 2002). With regard to clinical applications, several human cell sources have been investigated (Schnell *et al.*, 2001). Recently, cells derived from bone marrow or umbilical cord have been successfully utilized to generate heart valves and conduits *in vitro* (Hoerstrup *et al.*, 2002a, b). In contrast to vascular cells, these cells can be obtained without surgical

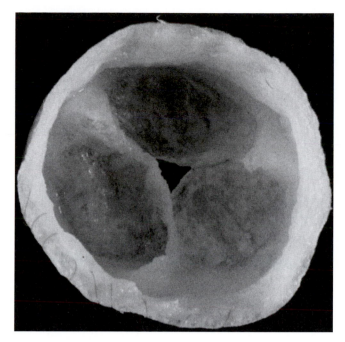

FIG. 4. Photograph of a living, tissue engineered heart valve after 14 days of biomimetic conditioning in a pulse–duplicator–bioreactor based on a rapidly biodegradable synthetic scaffold material. (Reprinted with permission from Hoerstrup *et al.*, 2000. *Circulation* **102**: III-44–III-49.)

interventions representing an easy-to-access cell source in a possible routine clinical scenario. Because of their good proliferation and progenitor potential, these cells are expected to be an attractive alternative for cardiovascular tissue-engineering applications.

Blood

There is a critical need for blood cell substitutes since donor blood suffers from problems such as donor shortage, requirements for typing and cross-matching, limited storage time, and, even more importantly in the era of AIDS, infectious disease transmission. Oxygen-containing fluids or materials as a substitute for red blood cells offer important applications in emergency resuscitation, shock, tumor therapy, and organ preservation. Several oxygen transporters are under investigation. Hemoglobin is a primary candidate, which not only serves as the natural oxygen transporter in blood but also functions in carbon dioxide transport, as a buffer, and in regulating osmotic pressure. Early clinical trials of cell-free hemoglobin were complicated by its lack of purity, instability, high oxygen affinity, and binding nitric oxide (NO), leading to cardiovascular side effects. These problems have been subsequently addressed by various chemical modifications such as intra- and intermolecular cross-linking using diacid, glutaraldehyde, or *o*-raffinose or conjugation to dextran or polyethylene glycol. Because of the limited hemoglobin availability, genetically engineered human hemoglobin or hemoglobin from bovine sources may represent a valid alternative. Several products are now in phase II/III clinical studies (Winslow, 2000). The latest developments include nanoencapsulated genetically engineered macromolecules of poly (hemoglobin–catalase–superoxide dismutase). Biodegradable polylactides and polyglycolides are used as carriers leading to artificial red blood cells containing hemoglobin and protective enzymes (Chang, 2003). Furthermore, perfluorocarbons (PFC) may be an alternative characterized by a high gas dissolving capacity (O_2, CO_2, and others), chemical and biological inertness, and low viscosity. However, hemoglobin binds significantly more oxygen at a given partial oxygen pressure than can be dissolved in PFC. Research to create functional substitutes for platelets by encapsulating platelet proteins in lipid vesicles has also been conducted (Baldassare *et al.*, 1985). Finally, stem cells have the potential to differentiate into the various cellular elements of blood (Thomson *et al.*, 1998).

FUTURE PERSPECTIVES

Current methods of transplantation and reconstruction are among the most time-consuming and costly therapies available today. Tissue-engineering offers future promise in the treatment of loss of tissue or organ function as well as for genetic disorders with metabolic deficiencies. Besides that, tissue engineering offers the possibility of substantial future savings by providing substitutes that are less expensive than donor organs and by providing a means of intervention before patients are critically ill. Few areas of technology will require more interdisciplinary research or have the potential to affect more positively the quality and length of life. Much must be learned from cell biology, especially with regard to what controls cellular differentiation and growth and how extracellular matrix components influence cell function. Immunology and molecular genetics will contribute to the design of cells or cell transplant systems that are not rejected by the immune system. With regard to the cell source, transplanted cells may come from cell lines or primary tissue, from the patients themselves, or from other human donors, animal tissue, or fetal tissue. In choosing the cell source a balance must be found between ethical issues, safety issues, and efficacy. These considerations are particularly important when introducing new techniques in the tissue-engineering field such as the generation of histocompatible tissue by cloning (nuclear transfer) (Lanza *et al.*, 2002) or by the creation of oocytes from embryonic stem cells (Hubner *et al.*, 2003).

The materials used in tissue engineering represent a major field of research regarding, e.g., polymer processing, development of controlled-release systems, surface modifications, and mathematical models possibly predicting *in vivo* cellular events.

Bibliography

Aebischer, P., Pochon, N. A., Heyd, B., Deglon, N., Joseph, J. M., Zurn, A. D., Baetge, E. E., Hammang, J. P., Goddard, M., Lysaght, M., Kaplan, F., Kato, A. C., Schluep, M., Hirt, L., Regli, F., Porchet, F., and De Tribolet, N. (1996a). Gene therapy for amyotrophic lateral sclerosis (ALS) using a polymer encapsulated xenogenic cell line engineered to secrete hCNTF. *Hum. Gene. Ther.* **7**: 851–860.

Aebischer, P., Salessiotis, A. N., and Winn, S. R. (1989). Basic fibroblast growth factor released from synthetic guidance channels facilitates peripheral nerve regeneration across long nerve gaps. *J. Neurosci. Res.* **23**: 282–289.

Aebischer, P., Schluep, M., Deglon, N., Joseph, J. M., Hirt, L., Heyd, B., Goddard, M., Hammang, J. P., Zurn, A. D., Kato, A. C., Regli, F., and Baetge, E. E. (1996b). Intrathecal delivery of CNTF using encapsulated genetically modified xenogeneic cells in amyotrophic lateral sclerosis patients. *Nat. Med.* **2**: 696–699.

Aebischer, P., Tresco, P. A., Winn, S. R., Greene, L. A., and Jaeger, C. B. (1991). Long-term cross-species brain transplantation of a polymer-encapsulated dopamine-secreting cell line. *Exp. Neurol.* **111**: 269–275.

Aebischer, P., Winn, S. R., and Galletti, P. M. (1988). Transplantation of neural tissue in polymer capsules. *Brain Res.* **448**: 364–368.

Al Attar, N., Carrion, C., Ghostine, S., Garcin, I., Vilquin, J. T., Hagege, A. A., and Menasche, P. (2003). Long-term (1 year) functional and histological results of autologous skeletal muscle cells transplantation in rat. *Cardiovasc. Res.* **58**: 142–148.

Armentano, D., Thompson, A. R., Darlington, G., and Woo, S. L. (1990). Expression of human factor IX in rabbit hepatocytes by retrovirus-mediated gene transfer: potential for gene therapy of hemophilia B. *Proc. Natl. Acad. Sci. USA* **87**: 6141–6145.

Assmus, B., Schachinger, V., Teupe, C., Britten, M., Lehmann, R., Dobert, N., Grunwald, F., Aicher, A., Urbich, C., Martin, H., Hoelzer, D., Dimmeler, S., and Zeiher, A. M. (2002). Transplantation of progenitor cells and regeneration enhancement in acute myocardial infarction (TOPCARE-AMI): *Circulation* **106**: 3009–3017.

Atala, A., Vacanti, J. P., Peters, C. A., Mandell, J., Retik, A. B., and Freeman, M. R. (1992). Formation of urothelial structures *in vivo* from dissociated cells attached to biodegradable polymer scaffolds *in vitro*. *J. Urol.* **148**: 658–662.

Bader, A., Knop, E., Fruhauf, N., Crome, O., Boker, K., Christians, U., Oldhafer, K., Ringe, B., Pichlmayr, R., and Sewing, K. F. (1995). Reconstruction of liver tissue *in vitro*: geometry of characteristic flat bed, hollow fiber, and spouted bed bioreactors with reference to the *in vivo* liver. *Artif. Organs* **19**: 941–950.

Bader, A., Steinhoff, G., Strobl, K., Schilling, T., Brandes, G., Mertsching, H., Tsikas, D., Froelich, J., and Haverich, A. (2000). Engineering of human vascular aortic tissue based on a xenogeneic starter matrix. *Transplantation* **70**: 7–14.

Baldassare, J. J., Kahn, R. A., Knipp, M. A., and Newman, P. J. (1985). Reconstruction of platelet proteins into phospholipid vesicles. Functional proteoliposomes. *J. Clin. Invest.* **75**: 35–39.

Bell, E., Ivarsson, B., and Merrill, C. (1979). Production of a tissue-like structure by contraction of collagen lattices by human fibroblasts of different proliferative potential *in vitro*. *Proc. Natl. Acad. Sci. USA* **76**: 1274–1278.

Bell, E., Parenteau, N., Gay, R., Nolte, C., Kemp, P., Bilbo, B., Ekstein, B., and Johnson, E. (1991a). The living skin equivalent: its manufacture, its organotypic properties and its responses to irritants. *Toxic. Vitro* **5**: 591–596.

Bell, E., Rosenberg, M., Kemp, P., Gay, R., Green, G. D., Muthukumaran, N., and Nolte, C. (1991b). Recipes for reconstituting skin. *J. Biomech. Eng.* **113**: 113–119.

Breitbart, A. S., Grande, D. A., Kessler, R., Ryaby, J. T., Fitzsimmons, R. J., and Grant, R. T. (1998). Tissue engineered bone repair of calvarial defects using cultured periosteal cells. *Plast. Reconstr. Surg.* **101**: 567–574; discussion 575–576.

Burke, J. F., Yannas, I. V., Quinby, W. C., Jr., Bondoc, C. C., and Jung, W. K. (1981). Successful use of a physiologically acceptable artificial skin in the treatment of extensive burn injury. *Ann. Surg.* **194**: 413–428.

Buschmann, M. D., Gluzband, Y. A., Grodzinsky, A. J., Kimura, J. H., and Hunziker, E. B. (1992). Chondrocytes in agarose culture synthesize a mechanically functional extracellular matrix. *J. Orthop. Res.* **10**: 745–758.

Cancedda, R., Dozin, B., Giannoni, P., and Quarto, R. (2003). Tissue engineering and cell therapy of cartilage and bone. *Matrix Biol.* **22**: 81–91.

Carrier, R. L., Papadaki, M., Rupnick, M., Schoen, F. J., Bursac, N., Langer, R., Freed, L. E., and Vunjak-Novakovic, G. (1999). Cardiac tissue engineering: cell seeding, cultivation parameters, and tissue construct characterization. *Biotechnol. Bioeng.* **64**: 580–589.

Casabona, F., Martin, I., Muraglia, A., Berrino, P., Santi, P., Cancedda, R., and Quarto, R. (1998). Prefabricated engineered bone flaps: an experimental model of tissue reconstruction in plastic surgery. *Plast. Reconstr. Surg.* **101**: 577–581.

Chang, T. M. (2003). Future generations of red blood cell substitutes. *J. Intern. Med.* **253**: 527–535.

Chen, S. C., Mullon, C., Kahaku, E., Watanabe, F., Hewitt, W., Eguchi, S., Middleton, Y., Arkadopoulos, N., Rozga, J., Solomon, B., and Demetriou, A. A. (1997). Treatment of severe liver failure with a bioartificial liver. *Ann. N. Y. Acad. Sci.* **831**: 350–360.

Chirila, T. V. (2001). An overview of the development of artificial corneas with porous skirts and the use of PHEMA for such an application. *Biomaterials* **22**: 3311–3317.

Chiu, R. C., Zibaitis, A., and Kao, R. L. (1995). Cellular cardiomyoplasty: myocardial regeneration with satellite cell implantation. *Ann. Thorac. Surg.* **60**: 12–18.

Chowdhury, J. R., Grossman, M., Gupta, S., Chowdhury, N. R., Baker, J. R., Jr., and Wilson, J. M. (1991). Long-term improvement of hypercholesterolemia after *ex vivo* gene therapy in LDLR-deficient rabbits. *Science* **254**: 1802–1805.

Crawford, G. J., Hicks, C. R., Lou, X., Vijayasekaran, S., Tan, D., Mulholland, B., Chirila, T. V., and Constable, I. J. (2002). The Chirila Keratoprosthesis: phase I human clinical trial. *Ophthalmology* **109**: 883–889.

Date, I., Shingo, T., Yoshida, H., Fujiwara, K., Kobayashi, K., and Ohmoto, T. (2000). Grafting of encapsulated dopamine-secreting cells in Parkinson's disease: long-term primate study. *Cell Transplant.* **9**: 705–709.

de Filippo, R. E., Yoo, J. J., and Atala, A. (2002). Urethral replacement using cell seeded tubularized collagen matrices. *J. Urol.* **168**: 1789–1792; discussion 1792–1793.

Dunn, J. C., Tompkins, R. G., and Yarmush, M. L. (1991). Long-term *in vitro* function of adult hepatocytes in a collagen sandwich configuration. *Biotechnol. Prog.* **7**: 237–245.

Eaglstein, W. H., Alvarez, O. M., Auletta, M., Leffel, D., Rogers, G. S., Zitelli, J. A., Norris, J. E., Thomas, I., Irondo, M., Fewkes, J., Hardin-Young, J., Duff, R. G., and Sabolinski, M. L. (1999). Acute excisional wounds treated with a tissue-engineered skin (Apligraf). *Dermatol. Surg.* **25**: 195–201.

Edelman, E. R. (1999). Vascular tissue engineering: designer arteries. *Circ. Res.* **85**: 1115–1117.

Efrat, S. (1999). Genetically engineered pancreatic beta-cell lines for cell therapy of diabetes. *Ann. N. Y. Acad. Sci.* **875**: 286–293.

Elkins, R. C., Dawson, P. E., Goldstein, S., Walsh, S. P., and Black, K. S. (2001). Decellularized human valve allografts. *Ann. Thorac. Surg.* **71**: S428–S432.

Ellis, A. J., Hughes, R. D., Wendon, J. A., Dunne, J., Langley, P. G., Kelly, J. H., Gislason, G. T., Sussman, N. L., and Williams, R. (1996). Pilot-controlled trial of the extracorporeal liver assist device in acute liver failure. *Hepatology* **24**: 1446–1451.

Eschenhagen, T., Didie, M., Heubach, J., Ravens, U., and Zimmermann, W. H. (2002a). Cardiac tissue engineering. *Transpl. Immunol.* **9**: 315–321.

Eschenhagen, T., Didie, M., Munzel, F., Schubert, P., Schneiderbanger, K., and Zimmermann, W. H. (2002b). 3D engineered heart tissue for replacement therapy. *Basic. Res. Cardiol.* **97**(Suppl 1): I146–I152.

Eschenhagen, T., Fink, C., Remmers, U., Scholz, H., Wattchow, J., Weil, J., Zimmermann, W., Dohmen, H. H., Schafer, H., Bishopric, N., Wakatsuki, T., and Elson, E. L. (1997). Three-dimensional reconstitution of embryonic cardiomyocytes in a collagen matrix: a new heart muscle model system. *FASEB. J.* **11**: 683–694.

Evans, M. D., Xie, R. Z., Fabbri, M., Bojarski, B., Chaouk, H., Wilkie, J. S., McLean, K. M., Cheng, H. Y., Vannas, A., and Sweeney, D. F. (2002). Progress in the development of a synthetic corneal onlay. *Invest. Ophthalmol. Vis. Sci.* **43**: 3196–3201.

Fentem, J. H., Briggs, D., Chesne, C., Elliott, G. R., Harbell, J. W., Heylings, J. R., Portes, P., Roguet, R., van de Sandt, J. J., and Botham, P. A. (2001). A prevalidation study on *in vitro* tests for acute skin irritation. Results and evaluation by the Management Team. *Toxicol. Vitro* **15**: 57–93.

Folkman, J., and Haudenschild, C. (1980). Angiogenesis *in vitro*. *Nature* **288**: 551–556.

Freed, L. E., Langer, R., Martin, I., Pellis, N. R., and Vunjak-Novakovic, G. (1997). Tissue engineering of cartilage in space. *Proc. Natl. Acad. Sci. USA* **94**: 13885–13890.

Freed, L. E., Marquis, J. C., Nohria, A., Emmanual, J., Mikos, A. G., and Langer, R. (1993). Neocartilage formation *in vitro* and *in vivo* using cells cultured on synthetic biodegradable polymers. *J. Biomed. Mater. Res.* **27**: 11–23.

Gerlach, J. C. (1997). Long-term liver cell cultures in bioreactors and possible application for liver support. *Cell Biol. Toxicol.* **13**: 349–355.

Gerlach, J. C., Lemmens, P., Schon, M., Janke, J., Rossaint, R., Busse, B., Puhl, G., and Neuhaus, P. (1997). Experimental evaluation of a hybrid liver support system. *Transplant Proc.* **29**: 852.

Germain, L., Auger, F. A., Grandbois, E., Guignard, R., Giasson, M., Boisjoly, H., and Guerin, S. L. (1999). Reconstructed human cornea produced *in vitro* by tissue engineering. *Pathobiology* **67**: 140–147.

Goldstein, S., Clarke, D. R., Walsh, S. P., Black, K. S., and O'Brien, M. F. (2000). Transspecies heart valve transplant: advanced studies of a bioengineered xeno-autograft. *Ann. Thorac. Surg.* **70**: 1962–1969.

Gosselin, C., Vorp, D. A., Warty, V., Severyn, D. A., Dick, E. K., Borovetz, H. S., and Greisler, H. P. (1996). ePTFE coating with fibrin glue, FGF-1, and heparin: effect on retention of seeded endothelial cells. *J. Surg. Res.* **60**: 327–332.

Grande, D. A., Pitman, M. I., Peterson, L., Menche, D., and Klein, M. (1989). The repair of experimentally produced defects in rabbit articular cartilage by autologous chondrocyte transplantation. *J. Orthop. Res.* **7**, 208–218.

Griffith, M., Hakim, M., Shimmura, S., Watsky, M. A., Li, F., Carlsson, D., Doillon, C. J., Nakamura, M., Suuronen, E., Shinozaki, N., Nakata, K., and Sheardown, H. (2002). Artificial human corneas: scaffolds for transplantation and host regeneration. *Cornea* **21**: S54–S61.

Griffith, M., Osborne, R., Munger, R., Xiong, X., Doillon, C. J., Laycock, N. L., Hakim, M., Song, Y., and Watsky, M. A. (1999). Functional human corneal equivalents constructed from cell lines. *Science* **286**: 2169–2172.

Grossman, M., Raper, S. E., Kozarsky, K., Stein, E. A., Engelhardt, J. F., Muller, D., Lupien, P. J., and Wilson, J. M. (1994). Successful *ex vivo* gene therapy directed to liver in a patient with familial hypercholesterolaemia. *Nat. Genet.* **6**: 335–341.

Grower, M. F., Russell, E. A., Jr., and Cutright, D. E. (1989). Segmental neogenesis of the dog esophagus utilizing a biodegradable polymer framework. *Biomater. Artif. Cells Artif. Organs.* **17**: 291–314.

Guenard, V., Kleitman, N., Morrissey, T. K., Bunge, R. P., and Aebischer, P. (1992). Syngeneic Schwann cells derived from adult nerves seeded in semipermeable guidance channels enhance peripheral nerve regeneration. *J. Neurosci.* **12**: 3310–1320.

Gussoni, E., Pavlath, G. K., Lanctot, A. M., Sharma, K. R., Miller, R. G., Steinman, L., and Blau, H. M. (1992). Normal dystrophin transcripts detected in Duchenne muscular dystrophy patients after myoblast transplantation. *Nature* **356**: 435–438.

Hagege, A. A., Carrion, C., Menasche, P., Vilquin, J. T., Duboc, D., Marolleau, J. P., Desnos, M., and Bruneval, P. (2003). Viability and differentiation of autologous skeletal myoblast grafts in ischaemic cardiomyopathy. *Lancet* **361**: 491–492.

Haller, M. F., and Saltzman, W. M. (1998). Nerve growth factor delivery systems. *J. Controlled Release* **53**: 1–6.

Hansbrough, J. F., Cooper, M. L., Cohen, R., Spielvogel, R., Greenleaf, G., Bartel, R. L., and Naughton, G. (1992). Evaluation of a biodegradable matrix containing cultured human fibroblasts as a dermal replacement beneath meshed skin grafts on athymic mice. *Surgery* **111**: 438–446.

Heimbach, D., Luterman, A., Burke, J., Cram, A., Herndon, D., Hunt, J., Jordan, M., McManus, W., Solem, L., Warden, G., and Zanvacki, B. (1988). Artificial dermis for major burns. A multi-center randomized clinical trial. *Ann. Surg.* **208**: 313–320.

Hoerstrup, S. P., Kadner, A., Breymann, C., Maurus, C. F., Guenter, C. I., Sodian, R., Visjager, J. F., Zund, G., and Turina, M. I. (2002a). Living, autologous pulmonary artery conduits tissue engineered from human umbilical cord cells. *Ann. Thorac. Surg.* **74**: 46–52; discussion.

Hoerstrup, S. P., Kadner, A., Melnitchouk, S., Trojan, A., Eid, K., Tracy, J., Sodian, R., Visjager, J. F., Kolb, S. A., Grunenfelder, J., Zund, G., and Turina, M. I. (2002b). Tissue engineering of functional trileaflet heart valves from human marrow stromal cells. *Circulation* **106**: I143–I150.

Hoerstrup, S. P., Sodian, R., Daebritz, S., Wang, J., Bacha, E. A., Martin, D. P., Moran, A. M., Guleserian, K. J., Sperling, J. S., Kaushal, S., Vacanti, J. P., Schoen, F. J., and Mayer, J. E., Jr. (2000a). Functional living trileaflet heart valves grown *in vitro*. *Circulation* **102**: III44–III49.

Hoerstrup, S. P., Sodian, R., Sperling, J. S., Vacanti, J. P., and Mayer, J. E., Jr. (2000b). New pulsatile bioreactor for *in vitro* formation of tissue engineered heart valves. *Tissue Eng.* **6**: 75–79.

Hoerstrup, S. P., Zund, G., Schoeberlein, A., Ye, Q., Vogt, P. R., and Turina, M. I. (1998). Fluorescence activated cell sorting: a reliable method in tissue engineering of a bioprosthetic heart valve. *Ann. Thorac. Surg.* **66**: 1653–1657.

Hoerstrup, S. P., Zund, G., Sodian, R., Schnell, A. M., Grunenfelder, J., and Turina, M. I. (2001). Tissue engineering of small caliber vascular grafts. *Eur. J. Cardiothorac. Surg.* **20**: 164–169.

Hori, Y., Nakamura, T., Kimura, D., Kaino, K., Kurokawa, Y., Satomi, S., and Shimizu, Y. (2002). Experimental study on tissue engineering of the small intestine by mesenchymal stem cell seeding. *J. Surg. Res.* **102**: 156–160.

Hubbell, J. A., Massia, S. P., Desai, N. P., and Drumheller, P. D. (1991). Endothelial cell-selective materials for tissue engineering in the vascular graft via a new receptor. *Biotechnology (N.Y.)* **9**: 568–572.

Hubner, K., Fuhrmann, G., Christenson, L. K., Kehler, J., Reinbold, R., De La Fuente, R., Wood, J., Strauss, J. F., 3rd, Boiani, M., and Scholer, H. R. (2003). Derivation of oocytes from mouse embryonic stem cells. *Science* **300**: 1251–1256.

Humes, H. D., Fissell, W. H., Weitzel, W. F., Buffington, D. A., Westover, A. J., MacKay, S. M., and Gutierrez, J. M. (2002). Metabolic replacement of kidney function in uremic animals with a bioartificial kidney containing human cells. *Am. J. Kidney Dis.* **39**: 1078–1087.

Humes, H. D., MacKay, S. M., Funke, A. J., and Buffington, D. A. (1999). Tissue engineering of a bioartificial renal tubule assist device: *in vitro* transport and metabolic characteristics. *Kidney Int.* **55**: 2502–2514.

Humes, H. D., Weitzel, W. F., Bartlett, R. H., Swaniker, F. C., and Paganini, E. P. (2003). Renal cell therapy is associated with dynamic and individualized responses in patients with acute renal failure. *Blood Purif.* **21**: 64–71.

Hutmacher, D. W., Goh, J. C., and Teoh, S. H. (2001). An introduction to biodegradable materials for tissue engineering applications. *Ann. Acad. Med. Singapore* **30**: 183–191.

Isogai, N., Landis, W., Kim, T. H., Gerstenfeld, L. C., Upton, J., and Vacanti, J. P. (1999). Formation of phalanges and small joints by tissue-engineering. *J. Bone. Joint. Surg. Am.* **81**: 306–316.

Jensen, T. G., Jensen, U. B., Jensen, P. K., Ibsen, H. H., Brandrup, F., Ballabio, A., and Bolund, L. (1993). Correction of steroid sulfatase deficiency by gene transfer into basal cells of tissue-cultured epidermis from patients with recessive X-linked ichthyosis. *Exp. Cell Res.* **209**: 392–397.

Kadiyala, S., Jaiswal, N., and Bruder, S. P. (1997). Culture expanded, bone marrow derived mesenchymal stem cells can regenerate a critical-sized segmental bone defect. *Tissue Eng.* **3**: 173–185.

Kaihara, S., Borenstein, J., Koka, R., Lalan, S., Ochoa, E. R., Ravens, M., Pien, H., Cunningham, B., and Vacanti, J. P. (2000). Silicon micromachining to tissue engineer branched vascular channels for liver fabrication. *Tissue Eng.* **6**: 105–117.

Kaihara, S., Kim, S. S., Benvenuto, M., Choi, R., Kim, B. S., Mooney, D., Tanaka, K., and Vacanti, J. P. (1999). Anastomosis between tissue-engineered intestine and native small bowel. *Transplant Proc.* **31**: 661–662.

Kajitani, M., Wadia, Y., Hinds, M. T., Teach, J., Swartz, K. R., and Gregory, K. W. (2001). Successful repair of esophageal injury using an elastin based biomaterial patch. *ASAIO J.* **47**: 342–345.

Kato, T., Sato, K., Miyazaki, H., Sasaki, S., Matsuo, S., and Moriyama, M. (1993). The uretero-ileoceco-proctostomy (ileocecal rectal bladder): early experiences in 18 patients. *J. Urol.* **150**: 326–331.

Kaushal, S., Amiel, G. E., Guleserian, K. J., Shapira, O. M., Perry, T., Sutherland, F. W., Rabkin, E., Moran, A. M., Schoen, F. J., Atala, A., Soker, S., Bischoff, J., and Mayer, J. E., Jr. (2001). Functional small-diameter neovessels created using endothelial progenitor cells expanded *ex vivo*. *Nat. Med.* **7**: 1035–1040.

Kawamoto, A., Gwon, H. C., Iwaguro, H., Yamaguchi, J. I., Uchida, S., Masuda, H., Silver, M., Ma, H., Kearney, M., Isner, J. M., and Asahara, T. (2001). Therapeutic potential of *ex vivo* expanded endothelial progenitor cells for myocardial ischemia. *Circulation* **103**: 634–637.

Keeffe, E. B. (2001). Liver transplantation: current status and novel approaches to liver replacement. *Gastroenterology* **120**: 749–762.

Kim, B.-S., and Mooney, D. J. (1998). Engineering smooth muscle tissue with a predefined structure. *J. Biomed. Mater. Res.* **41**: 322–332.

Kim, S. S., Kaihara, S., Benvenuto, M. S., Choi, R. S., Kim, B. S., Mooney, D. J., Taylor, G. A., and Vacanti, J. P. (1999). Regenerative signals for intestinal epithelial organoid units transplanted on biodegradable polymer scaffolds for tissue engineering of small intestine. *Transplantation* **67**: 227–233.

Kim, S. S., Utsunomiya, H., Koski, J. A., Wu, B. M., Cima, M. J., Sohn, J., Mukai, K., Griffith, L. G., and Vacanti, J. P. (1998). Survival and function of hepatocytes on a novel three-dimensional synthetic biodegradable polymer scaffold with an intrinsic network of channels. *Ann. Surg.* **228**: 8–13.

Kin, T., Iwata, H., Aomatsu, Y., Ohyama, T., Kanehiro, H., Hisanaga, M., and Nakajima, Y. (2002). Xenotransplantation of pig islets in diabetic dogs with use of a microcapsule composed of agarose and polystyrene sulfonic acid mixed gel. *Pancreas* **25**: 94–100.

Kordower, J. H., Liu, Y. T., Winn, S., and Emerich, D. F. (1995). Encapsulated PC12 cell transplants into hemiparkinsonian monkeys: a behavioral, neuroanatomical, and neurochemical analysis. *Cell. Transplant.* **4**: 155–171.

Kudo, T., Kawase, M., Kawada, S., Kurosawa, H., Koyanagi, H., Takeuchi, Y., Hosoda, Y., and Wanibuchi, Y. (1999). Anticoagulation after valve replacement: a multicenter retrospective study. *Artif. Organs* **23**: 199–203.

Kurbaan, A. S., Bowker, T. J., Ilsley, C. D., and Rickards, A. F. (1998). Impact of postangioplasty restenosis on comparisons of outcome between angioplasty and bypass grafting. Coronary Angioplasty versus Bypass Revascularisation Investigation (CABRI) Investigators. *Am. J. Cardiol.* **82**: 272–276.

Kusama, K., Donegan, W. L., and Samter, T. G. (1989). An investigation of colon cancer associated with urinary diversion. *Dis. Colon Rectum* **32**: 694–697.

Lacy, P. E., Hegre, O. D., Gerasimidi-Vazeou, A., Gentile, F. T., and Dionne, K. E. (1991). Maintenance of normoglycemia in diabetic mice by subcutaneous xenografts of encapsulated islets. *Science* **254**: 1782–1784.

Langer, R., and Vacanti, J. P. (1993). Tissue engineering. *Science* **260**: 920–926.

Langer, R. S., and Vacanti, J. P. (1999). Tissue engineering: the challenges ahead. *Sci. Am.* **280**: 86–89.

Lanza, R. P., Chung, H. Y., Yoo, J. J., Wettstein, P. J., Blackwell, C., Borson, N., Hofmeister, E., Schuch, G., Soker, S., Moraes, C. T., West, M. D., and Atala, A. (2002). Generation of histocompatible tissues using nuclear transplantation. *Nat. Biotechnol.* **20**: 689–696.

Lanza, R. P., Ecker, D. M., Kuhtreiber, W. M., Marsh, J. P., Ringeling, J., and Chick, W. L. (1999). Transplantation of islets using microencapsulation: studies in diabetic rodents and dogs. *J. Mol. Med.* **77**: 206–210.

Lanza, R. P., Kuhtreiber, W. M., Ecker, D., Staruk, J. E., and Chick, W. L. (1995). Xenotransplantation of porcine and bovine islets without immunosuppression using uncoated alginate microspheres. *Transplantation* **59**: 1377–1384.

Lee, H., Cusick, R. A., Browne, F., Ho Kim, T., Ma, P. X., Utsunomiya, H., Langer, R., and Vacanti, J. P. (2002). Local delivery of basic fibroblast growth factor increases both angiogenesis and engraftment of hepatocytes in tissue-engineered polymer devices. *Transplantation* **73**: 1589–1593.

Lee, K. Y., and Mooney, D. J. (2001). Hydrogels for tissue engineering. *Chem. Rev.* **101**: 1869–1879.

Legeais, J. M., and Renard, G. (1998). A second generation of artificial cornea (Biokpro II). *Biomaterials* **19**: 1517–1522.

L'Heureux, N., Paquet, S., Labbe, R., Germain, L., and Auger, F. A. (1998). A completely biological tissue-engineered human blood vessel. *FASEB. J.* **12**: 47–56.

Lindvall, O., Rehncrona, S., Brundin, P., Gustavii, B., Astedt, B., Widner, H., Lindholm, T., Bjorklund, A., Leenders, K. L., and Rothwell, J. C. (1990). Neural transplantation in Parkinson's disease: the Swedish experience. *Prog. Brain Res.* **82**: 729–736.

MacKay, S. M., Funke, A. J., Buffington, D. A., and Humes, H. D. (1998). Tissue engineering of a bioartificial renal tubule. *ASAIO J.* **44**: 179–183.

Matas, A. J., Sutherland, D. E., Steffes, M. W., Mauer, S. M., Sowe, A., Simmons, R. L., and Najarian, J. S. (1976).

Hepatocellular transplantation for metabolic deficiencies: decrease of plasma bilirubin in Gunn rats. *Science* **192**: 892–894.

Mayer, J. E., Jr. (1995). Uses of homograft conduits for right ventricle to pulmonary artery connections in the neonatal period. *Semin. Thorac. Cardiovasc. Surg.* **7**: 130–132.

Mazariegos, G. V., Kramer, D. J., Lopez, R. C., Shakil, A. O., Rosenbloom, A. J., DeVera, M., Giraldo, M., Grogan, T. A., Zhu, Y., Fulmer, M. L., Amiot, B. P., and Patzer, J. F. (2001). Safety observations in phase I clinical evaluation of the Excorp Medical Bioartificial Liver Support System after the first four patients. *ASAIO J.* **47**: 471–475.

McCarey, B. E., McDonald, M. B., van Rij, G., Salmeron, B., Pettit, D. K., and Knight, P. M. (1989). Refractive results of hyperopic hydrogel intracorneal lenses in primate eyes. *Arch. Ophthalmol.* **107**: 724–730.

Meinhart, J., Deutsch, M., and Zilla, P. (1997). Eight years of clinical endothelial cell transplantation. Closing the gap between prosthetic grafts and vein grafts. *ASAIO J.* **43**: M515–M521.

Menasche, P. (2003). Skeletal muscle satellite cell transplantation. *Cardiovasc. Res.* **58**: 351–357.

Menasche, P., Hagege, A. A., Scorsin, M., Pouzet, B., Desnos, M., Duboc, D., Schwartz, K., Vilquin, J. T., and Marolleau, J. P. (2001). Myoblast transplantation for heart failure. *Lancet* **357**: 279–280.

Mitchell, S. L., and Niklason, L. E. (2003). Requirements for growing tissue-engineered vascular grafts. *Cardiovasc. Pathol.* **12**: 59–64.

Mow, V. C., Ratcliffe, A., and Poole, A. R. (1992). Cartilage and diarthrodial joints as paradigms for hierarchical materials and structures. *Biomaterials* **13**: 67–97.

Naito, Y., Imai, Y., Shin'oka, T., Kashiwagi, J., Aoki, M., Watanabe, M., Matsumura, G., Kosaka, Y., Konuma, T., Hibino, N., Murata, A., Miyake, T., and Kurosawa, H. (2003). Successful clinical application of tissue-engineered graft for extracardiac Fontan operation. *J. Thorac. Cardiovasc. Surg.* **125**: 419–420.

Naumann, A., Rotter, N., Bujia, J., and Aigner, J. (1998). Tissue engineering of autologous cartilage transplants for rhinology. *Am. J. Rhinol.* **12**: 59–63.

Nave, M. (1992). Wound bed preparation: approaches to replacement of dermis. *J. Burn Care Rehabil.* **13**: 147–153.

Niklason, L. E., Abbott, W., Gao, J., Klagges, B., Hirschi, K. K., Ulubayram, K., Conroy, N., Jones, R., Vasanawala, A., Sanzgiri, S., and Langer, R. (2001). Morphologic and mechanical characteristics of engineered bovine arteries. *J. Vasc. Surg.* **33**: 628–638.

Niklason, L. E., Gao, J., Abbott, W. M., Hirschi, K. K., Houser, S., Marini, R., and Langer, R. (1999). Functional arteries grown *in vitro*. *Science* **284**: 489–493.

Nikolovski, J., Gulari, E., and Humes, H. D. (1999). Design engineering of a bioartificial renal tubule cell therapy device. *Cell. Transplant.* **8**: 351–364.

Oberpenning, F., Meng, J., Yoo, J. J., and Atala, A. (1999). De novo reconstitution of a functional mammalian urinary bladder by tissue engineering. *Nat. Biotechnol.* **17**: 149–155.

Okano, T., and Matsuda, T. (1998). Muscular tissue engineering: capillary-incorporated hybrid muscular tissues *in vivo* tissue culture. *Cell Transplant.* **7**: 435–442.

Orchard, T. J., Chang, Y. F., Ferrell, R. E., Petro, N., and Ellis, D. E. (2002). Nephropathy in type 1 diabetes: a manifestation of insulin resistance and multiple genetic susceptibilities? Further evidence from the Pittsburgh Epidemiology of Diabetes Complication Study. *Kidney Int.* **62**: 963–970.

Orchard, T. J., Olson, J. C., Erbey, J. R., Williams, K., Forrest, K. Y., Smithline Kinder, L., Ellis, D., and Becker, D. J. (2003). Insulin resistance-related factors, but not glycemia, predict coronary artery disease in type 1 diabetes: 10-year follow-up data from the Pittsburgh Epidemiology of Diabetes Complications Study. *Diabetes Care* 26: 1374–1379.

Orlic, D., Kajstura, J., Chimenti, S., Jakoniuk, I., Anderson, S. M., Li, B., Pickel, J., McKay, R., Nadal-Ginard, B., Bodine, D. M., Leri, A., and Anversa, P. (2001a). Bone marrow cells regenerate infarcted myocardium. *Nature* 410: 701–705.

Orlic, D., Kajstura, J., Chimenti, S., Limana, F., Jakoniuk, I., Quaini, F., Nadal-Ginard, B., Bodine, D. M., Leri, A., and Anversa, P. (2001b). Mobilized bone marrow cells repair the infarcted heart, improving function and survival. *Proc. Natl. Acad. Sci. USA* 98: 10344–10349.

Ounpuu, S., Anand, S., and Yusuf, S. (2000). The impending global epidemic of cardiovascular diseases. *Eur. Heart J.* 21: 880–883.

Pasic, M., Muller-Glauser, W., von Segesser, L., Odermatt, B., Lachat, M., and Turina, M. (1996). Endothelial cell seeding improves patency of synthetic vascular grafts: manual versus automatized method. *Eur. J. Cardiothorac. Surg.* 10: 372–379.

Patzer, J. F., 2nd (2001). Advances in bioartificial liver assist devices. *Ann. N. Y. Acad. Sci.* 944: 320–333.

Patzer, J. F., 2nd, Campbell, B., and Miller, R. (2002). Plasma versus whole blood perfusion in a bioartificial liver assist device. *ASAIO J.* 48: 226–233.

Patzer, J. F., 2nd, Mazariegos, G. V., Lopez, R., Molmenti, E., Gerber, D., Riddervold, F., Khanna, A., Yin, W. Y., Chen, Y., Scott, V. L., Aggarwal, S., Kramer, D. J., Wagner, R. A., Zhu, Y., Fulmer, M. L., Block, G. D., and Amiot, B. P. (1999). Novel bioartificial liver support system: preclinical evaluation. *Ann. N. Y. Acad. Sci.* 875: 340–352.

Perez, A., Grikscheit, T. C., Blumberg, R. S., Ashley, S. W., Vacanti, J. P., and Whang, E. E. (2002). Tissue-engineered small intestine: ontogeny of the immune system. *Transplantation* 74: 619–623.

Peters, M. C., Polverini, P. J., and Mooney, D. J. (2002). Engineering vascular networks in porous polymer matrices. *J. Biomed. Mater. Res.* 60: 668–678.

Petite, H., Viateau, V., Bensaid, W., Meunier, A., de Pollak, C., Bourguignon, M., Oudina, K., Sedel, L., and Guillemin, G. (2000). Tissue-engineered bone regeneration. *Nat. Biotechnol.* 18: 959–963.

Portes, P., Grandidier, M. H., Cohen, C., and Roguet, R. (2002). Refinement of the Episkin protocol for the assessment of acute skin irritation of chemicals: follow-up to the ECVAM prevalidation study. *Toxicol. Vitro* 16: 765–770.

Puelacher, W. C., Vacanti, J. P., Ferraro, N. F., Schloo, B., and Vacanti, C. A. (1996). Femoral shaft reconstruction using tissue-engineered growth of bone. *Int. J. Oral Maxillofac. Surg.* 25: 223–228.

Quarto, R., Mastrogiacomo, M., Cancedda, R., Kutepov, S. M., Mukhachev, V., Lavroukov, A., Kon, E., and Marcacci, M. (2001). Repair of large bone defects with the use of autologous bone marrow stromal cells. *N. Engl. J. Med.* 344: 385–386.

Rabkin, E., and Schoen, F. J. (2002). Cardiovascular tissue engineering. *Cardiovasc. Pathol.* 11: 305–317.

Rabkin, E., Hoerstrup, S. P., Aikawa, M., Mayer, J. E., Jr., and Schoen, F. J. (2002). Evolution of cell phenotype and extracellular matrix in tissue-engineered heart valves during *in-vitro* maturation and *in-vivo* remodeling. *J. Heart Valve Dis.* 11: 308–314; discussion 314.

Radisic, M., Euloth, M., Yang, L., Langer, R., Freed, L. E., and Vunjak-Novakovic, G. (2003). High-density seeding of myocyte cells for cardiac tissue engineering. *Biotechnol. Bioeng.* 82: 403–414.

Raper, S. E., and Wilson, J. M. (1993). Cell transplantation in liver-directed gene therapy. *Cell. Transplant.* 2: 381–400; discussion 407–410.

Read, T. A., Sorensen, D. R., Mahesparan, R., Enger, P. O., Timpl, R., Olsen, B. R., Hjelstuen, M. H., Haraldseth, O., and Bjerkvig, R.

(2001). Local endostatin treatment of gliomas administered by microencapsulated producer cells. *Nat. Biotechnol.* 19: 29–34.

Reinecke, H., Poppa, V., and Murry, C. E. (2002). Skeletal muscle stem cells do not transdifferentiate into cardiomyocytes after cardiac grafting. *J. Mol. Cell Cardiol.* 34: 241–249.

Robbins, P. B., Lin, Q., Goodnough, J. B., Tian, H., Chen, X., and Khavari, P. A. (2001). *In vivo* restoration of laminin 5 beta 3 expression and function in junctional epidermolysis bullosa. *Proc. Natl. Acad. Sci. USA* 98: 5193–5198.

Rosenberg, N., Martinez, A., Sawyer, P. N., Wesolowski, S. A., Postlethwait, R. W., and Dillon, M. L., Jr. (1966). Tanned collagen arterial prosthesis of bovine carotid origin in man. Preliminary studies of enzyme-treated heterografts. *Ann. Surg.* 164: 247–256.

Rothenburger, M., Vischer, P., Volker, W., Glasmacher, B., Berendes, E., Scheld, H. H., and Deiwick, M. (2001). *In vitro* modelling of tissue using isolated vascular cells on a synthetic collagen matrix as a substitute for heart valves. *Thorac. Cardiovasc. Surg.* 49: 204–209.

Rothenburger, M., Volker, W., Vischer, J. P., Berendes, E., Glasmacher, B., Scheld, H. H., and Deiwick, M. (2002). Tissue engineering of heart valves: formation of a three-dimensional tissue using porcine heart valve cells. *ASAIO J.* 48: 586–591.

Rozga, J., Podesta, L., LePage, E., Morsiani, E., Moscioni, A. D., Hoffman, A., Sher, L., Villamil, F., Woolf, G., McGrath, M., Kong, L., Rosen, H., Lanman, T., Vierling, J., Makowka, L., and Demetriou, A. A. (1994). A bioartificial liver to treat severe acute liver failure. *Ann. Surg.* 219: 538–544; discussion 544–546.

Rozga, J., Williams, F., Ro, M. S., Neuzil, D. F., Giorgio, T. D., Backfisch, G., Moscioni, A. D., Hakim, R., and Demetriou, A. A. (1993). Development of a bioartificial liver: properties and function of a hollow-fiber module inoculated with liver cells. *Hepatology* 17: 258–265.

Sagen, J. (1992). Chromaffin cell transplants for alleviation of chronic pain. *ASAIO J.* 38: 24–28.

Samouillan, V., Dandurand-Lods, J., Lamure, A., Maurel, E., Lacabanne, C., Gerosa, G., Venturini, A., Casarotto, D., Gherardini, L., and Spina, M. (1999). Thermal analysis characterization of aortic tissues for cardiac valve bioprostheses. *J. Biomed. Mater. Res.* 46: 531–538.

Sautter, J., Tseng, J. L., Braguglia, D., Aebischer, P., Spenger, C., Seiler, R. W., Widmer, H. R., and Zurn, A. D. (1998). Implants of polymer-encapsulated genetically modified cells releasing glial cell line-derived neurotrophic factor improve survival, growth, and function of fetal dopaminergic grafts. *Exp. Neurol.* 149: 230–236.

Schaefer, D., Martin, I., Jundt, G., Seidel, J., Heberer, M., Grodzinsky, A., Bergin, I., Vunjak-Novakovic, G., and Freed, L. E. (2002). Tissue-engineered composites for the repair of large osteochondral defects. *Arthritis. Rheum.* 46: 2524–2534.

Schneider, A. I., Maier-Reif, K., and Graeve, T. (1999). Constructing an *in vitro* cornea from cultures of the three specific corneal cell types. *In Vitro Cell Dev. Biol. Anim.* 35: 515–526.

Schnell, A. M., Hoerstrup, S. P., Zund, G., Kolb, S., Sodian, R., Visjager, J. F., Grunenfelder, J., Suter, A., and Turina, M. (2001). Optimal cell source for cardiovascular tissue engineering: venous vs. aortic human myofibroblasts. *Thorac. Cardiovasc. Surg.* 49: 221–225.

Sheridan, R. L., Tompkins, R. G., and Burke, J. F. (1994). Management of burn wounds with prompt excision and immediate closure. [see comments]. *J. Intensive Care Med.* 9: 6–17.

Shimizu, T., Yamato, M., Isoi, Y., Akutsu, T., Setomaru, T., Abe, K., Kikuchi, A., Umezu, M., and Okano, T. (2002). Fabrication of pulsatile cardiac tissue grafts using a novel 3-dimensional cell sheet manipulation technique and temperature-responsive cell culture surfaces. *Circ. Res.* **90**: e40.

Shinoka, T., Imai, Y., and Ikada, Y. (2001). Transplantation of a tissue-engineered pulmonary artery. *N. Engl. J. Med.* **344**: 532–533.

Shinoka, T., Ma, P. X., Shum-Tim, D., Breuer, C. K., Cusick, R. A., Zund, G., Langer, R., Vacanti, J. P., and Mayer, J. E., Jr. (1996). Tissue-engineered heart valves. Autologous valve leaflet replacement study in a lamb model. *Circulation* **94**: II164–II168.

Simon, P., Kasimir, M. T., Seebacher, G., Weigel, G., Ullrich, R., Salzer-Muhar, U., Rieder, E., and Wolner, E. (2003). Early failure of the tissue engineered porcine heart valve SYNERGRAFT in pediatric patients. *Eur. J. Cardiothorac. Surg.* **23**: 1002–1006.

Sodian, R., Hoerstrup, S. P., Sperling, J. S., Daebritz, S., Martin, D. P., Moran, A. M., Kim, B. S., Schoen, F. J., Vacanti, J. P., and Mayer, J. E., Jr. (2000). Early *in vivo* experience with tissue-engineered trileaflet heart valves. *Circulation* **102**: III22–III29.

Spirito, F., Meneguzzi, G., Danos, O., and Mezzina, M. (2001). Cutaneous gene transfer and therapy: the present and the future. *J. Gene Med.* **3**: 21–31.

Starzl, T. E. (2001). The birth of clinical organ transplantation. *J. Am. Coll. Surg.* **192**: 431–446.

Starzl, T. E., Demetris, A. J., and Van Thiel, D. (1989). Liver transplantation (1). *N. Engl. J. Med.* **321**: 1014–1022.

Stone, K. R., Rodkey, W. G., Webber, R. J., McKinney, L., and Steadman, J. R. (1990). Future directions. Collagen-based prostheses for meniscal regeneration. *Clin. Orthop.* 129–135.

Stratta, R. J., Taylor, R. J., Bynon, J. S., Lowell, J. A., Sindhi, R., Wahl, T. O., Knight, T. F., Weide, L. G., and Duckworth, W. C. (1994). Surgical treatment of diabetes mellitus with pancreas transplantation. *Ann. Surg.* **220**: 809–817.

Sullivan, S. J., Maki, T., Borland, K. M., Mahoney, M. D., Solomon, B. A., Muller, T. E., Monaco, A. P., and Chick, W. L. (1991). Biohybrid artificial pancreas: long-term implantation studies in diabetic, pancreatectomized dogs. *Science* **252**: 718–721.

Sussman, N. L., Gislason, G. T., and Kelly, J. H. (1994). Extracorporeal liver support. Application to fulminant hepatic failure. *J. Clin. Gastroenterol.* **18**: 320–324.

Tan, S. A., Deglon, N., Zurn, A. D., Baetge, E. E., Bamber, B., Kato, A. C., and Aebischer, P. (1996). Rescue of motoneurons from axotomy-induced cell death by polymer encapsulated cells genetically engineered to release CNTF. *Cell Transplant.* **5**: 577–587.

Taylor, D. A., Atkins, B. Z., Hungspreugs, P., Jones, T. R., Reedy, M. C., Hutcheson, K. A., Glower, D. D., and Kraus, W. E. (1998). Regenerating functional myocardium: improved performance after skeletal myoblast transplantation. *Nat. Med.* **4**: 929–933.

Taylor, P. M., Allen, S. P., Dreger, S. A., and Yacoub, M. H. (2002). Human cardiac valve interstitial cells in collagen sponge: a biological three-dimensional matrix for tissue engineering. *J. Heart Valve Dis.* **11**: 298–306.

Teebken, O. E., Bader, A., Steinhoff, G., and Haverich, A. (2000). Tissue engineering of vascular grafts: human cell seeding of decellularised porcine matrix. *Eur. J. Vasc. Endovasc. Surg.* **19**: 381–386.

Thomson, J. A., Itskovitz-Eldor, J., Shapiro, S. S., Waknitz, M. A., Swiergiel, J. J., Marshall, V. S., and Jones, J. M. (1998). Embryonic stem cell lines derived from human blastocysts. *Science* **282**: 1145–1147.

Thorsen, F., Read, T. A., Lund-Johansen, M., Tysnes, B. B., and Bjerkvig, R. (2000). Alginate-encapsulated producer cells: a potential new approach for the treatment of malignant brain tumors. *Cell Transplant.* **9**: 773–783.

Toriumi, D. M., Kotler, H. S., Luxenberg, D. P., Holtrop, M. E., and Wang, E. A. (1991). Mandibular reconstruction with a recombinant bone-inducing factor. Functional, histologic, and biomechanical evaluation. *Arch. Otolaryngol. Head Neck Surg.* **117**: 1101–1112.

Uludag, H., Ip, T. K., and Aebischer, P. (1990). Transport functions in a bioartificial kidney under uremic conditions. *Int. J. Artif. Organs* **13**: 93–97.

Uyama, S., Kaufmann, P. M., Takeda, T., and Vacanti, J. P. (1993). Delivery of whole liver-equivalent hepatocyte mass using polymer devices and hepatotrophic stimulation. *Transplantation* **55**: 932–935.

Vacanti, C. A., Langer, R., Schloo, B., and Vacanti, J. P. (1991). Synthetic polymers seeded with chondrocytes provide a template for new cartilage formation. *Plast. Reconstr. Surg.* **88**: 753–759.

Vacanti, J. P., Morse, M. A., Saltzman, W. M., Domb, A. J., Perez-Atayde, A., and Langer, R. (1988). Selective cell transplantation using bioabsorbable artificial polymers as matrices. *J. Pediatr. Surg.* **23**: 3–9.

Valentini, R. F., Vargo, T. G., Gardella, J. A., Jr., and Aebischer, P. (1992). Electrically charged polymeric substrates enhance nerve fibre outgrowth *in vitro*. *Biomaterials* **13**: 183–190.

Vunjak-Novakovic, G., Martin, I., Obradovic, B., Treppo, S., Grodzinsky, A. J., Langer, R., and Freed, L. E. (1999). Bioreactor cultivation conditions modulate the composition and mechanical properties of tissue-engineered cartilage. *J. Orthop. Res.* **17**: 130–138.

Wakitani, S., Kimura, T., Hirooka, A., Ochi, T., Yoneda, M., Yasui, N., Owaki, H., and Ono, K. (1989). Repair of rabbit articular surfaces with allograft chondrocytes embedded in collagen gel. *J. Bone Joint Surg. Br.* **71**: 74–80.

Watanabe, F. D., Mullon, C. J., Hewitt, W. R., Arkadopoulos, N., Kahaku, E., Eguchi, S., Khalili, T., Arnaout, W., Shackleton, C. R., Rozga, J., Solomon, B., and Demetriou, A. A. (1997). Clinical experience with a bioartificial liver in the treatment of severe liver failure. A phase I clinical trial. *Ann. Surg.* **225**: 484–491; discussion 491–494.

Wiegand, F., Kroncke, K. D., and Kolb-Bachofen, V. (1993). Macrophage-generated nitric oxide as cytotoxic factor in destruction of alginate-encapsulated islets. Protection by arginine analogs and/or coencapsulated erythrocytes. *Transplantation* **56**: 1206–1212.

Wilson, G. J., Courtman, D. W., Klement, P., Lee, J. M., and Yeger, H. (1995). Acellular matrix: a biomaterials approach for coronary artery bypass and heart valve replacement. *Ann. Thorac. Surg.* **60**: S353–S358.

Winslow, R. M. (2000). Blood substitutes. *Adv. Drug Deliv. Rev.* **40**: 131–142.

Woerly, S., Plant, G. W., and Harvey, A. R. (1996). Neural tissue engineering: from polymer to biohybrid organs. *Biomaterials* **17**: 301–310.

Yarmush, M. L., Dunn, J. C., and Tompkins, R. G. (1992a). Assessment of artificial liver support technology. *Cell Transplant.* **1**: 323–341.

Yarmush, M. L., Toner, M., Dunn, J. C., Rotem, A., Hubel, A., and Tompkins, R. G. (1992b). Hepatic tissue engineering. Development of critical technologies. *Ann. N. Y. Acad. Sci.* **665**: 238–252.

Yasko, A. W., Lane, J. M., Fellinger, E. J., Rosen, V., Wozney, J. M., and Wang, E. A. (1992). The healing of segmental bone defects, induced by recombinant human bone morphogenetic protein (rhBMP-2). A radiographic, histological, and biomechanical study in rats. *J. Bone Joint Surg. Am.* **74**: 659–670.

Ye, Q., Zund, G., Jockenhoevel, S., Schoeberlein, A., Hoerstrup, S. P., Grunenfelder, J., Benedikt, P., and Turina, M. (2000). Scaffold precoating with human autologous extracellular matrix for improved cell attachment in cardiovascular tissue engineering. *ASAIO. J.* **46**: 730–733.

Zieske, J. D., Mason, V. S., Wasson, M. E., Meunier, S. F., Nolte, C. J., Fukai, N., Olsen, B. R., and Parenteau, N. L. (1994). Basement membrane assembly and differentiation of cultured corneal cells: importance of culture environment and endothelial cell interaction. *Exp. Cell Res.* **214:** 621–633.

Zilla, P., Deutsch, M., and Meinhart, J. (1999). Endothelial cell transplantation. *Semin. Vasc. Surg.* **12:** 52–63.

Zimmermann, W. H., Fink, C., Kralisch, D., Remmers, U., Weil, J., and Eschenhagen, T. (2000). Three-dimensional engineered heart tissue from neonatal rat cardiac myocytes. *Biotechnol. Bioeng.* **68:** 106–114.

Zimmermann, W. H., Schneiderbanger, K., Schubert, P., Didie, M., Munzel, F., Heubach, J. F., Kostin, S., Neuhuber, W. L., and Eschenhagen, T. (2002). Tissue engineering of a differentiated cardiac muscle construct. *Circ. Res.* **90:** 223–230.

Zund, G., Hoerstrup, S. P., Schoeberlein, A., Lachat, M., Uhlschmid, G., Vogt, P. R., and Turina, M. (1998). Tissue engineering: a new approach in cardiovascular surgery: seeding of human fibroblasts followed by human endothelial cells on resorbable mesh. *Eur. J. Cardiothorac. Surg.* **13:** 160–164.

Zurn, A. D., Henry, H., Schluep, M., Aubert, V., Winkel, L., Eilers, B., Bachmann, C., and Aebischer, P. (2000). Evaluation of an intrathecal immune response in amyotrophic lateral sclerosis patients implanted with encapsulated genetically engineered xenogeneic cells. *Cell Transplant.* **9:** 471–484.

8.3 IMMUNOISOLATION

Michael J. Lysaght and David Rein

INTRODUCTION

In the context of tissue engineering and cellular medicine, the terms *immunoisolation* and *encapsulation* usually refer to devices and therapies in which living cells are separated from a host by a selective membrane barrier. This barrier permits bidirectional trafficking of small molecules between host and grafted cells, and protects foreign cells from effector agents of a host's immune system. In analogy with pharmacological immunosuppression, the degree of protection afforded by immunoisolatory barriers depends upon the circumstances of application and may be total or partial, long-term or short-term. Occasional reference to the concept, which is illustrated in Fig. 1, can be found as early as the late 1930s and appears sporadically in the literature of the 1950s and 1960s. The approach first received serious investigational attention in the mid-1970s. Interest has expanded considerably in the past two decades. Encapsulation currently encompasses a daunting array of therapy formats, device configurations, and biomaterials.

The first modern efforts involving cell encapsulation were directed at development of a long-term implant to replace the endocrine function of a diabetic pancreas. Other investigators quickly expanded this field of study to include short-term extracorporeal replacement of the failing liver. Later applications include the use of encapsulated cells for *in situ* synthesis and local delivery of naturally occurring and recombinant cell products for the treatment of chronic pain, Parkinson's disease, macular degeneration, and similar disorders. Devices employed for encapsulated cell therapy vary in size over several orders of magnitude from small spheres, with a volume of 10^{-5} cm^3, to large extracorporeal devices with a net volume of ~ 10 cm^3. Their anticipated service life ranges from a few hours in the case of the liver to several years for other therapeutic implants. In some cases, the immunoisolatory vehicles simply serve as constitutive sources of bioactive molecules; in other cases, regulated release is required; and for still others, host detoxification is the goal. Despite such a spectrum of application parameters, devices containing immunoisolated cells share many common features and design principles: (1) Cells are rarely deployed more than 500 μm (5×10^{-2} cm) from the host; cells much farther than this critical distance either undergo necrosis or cease to synthesize and release protein. (2) Cells generally are supported on a matrix or scaffold to provide some of the functions of normal extracellular matrix and to prevent the formation of large, unvascularized cellular aggregates. (3) Separative membranes are invariably self-supporting, thus requiring a design trade-off between transport characteristics and mechanical strength. (4) Both the membrane and the matrices usually are prepared from either hydrogels or reticulated foams, themselves chosen from relatively few among the many available candidates.

In the remainder of this overview, we will describe the immunological challenge of protecting cells with barrier materials; summarize critical components of immunoisolate devices, i.e., cells, membranes, and matrices; review the more common device configurations; and conclude with a short survey of the development status of principal applications.

THE CHALLENGE OF IMMUNOISOLATION

The immune system is a network that deploys a complex, redundant phalanx of pathways to distinguish self from non-self and to destroy non-self. Membrane barriers have proven remarkably effective in preventing immune destruction of *allogenic* cells, i.e., cells originating from the same species as the host. Protection of allogenic cells is possible because they normally will not be subjected to immune destruction in the absence of cell–cell contact between graft and host. Even protein-permeable membranes have been found to allow long-term function and survival of allogenic cells. This happy circumstance is marred by the reality that the supply of transplantable human cells is very limited, just as is the supply of human solid organs, and thus therapies based upon transplanted human cells are not likely to have much therapeutic impact. There has been some effort to create dividing cell lines from protein secreting human cells, or to genetically engineer naturally dividing human cells (e.g., fibroblasts) to produce useful proteins. In the future, stem cells may provide unconstrained supply of tissue. In the main, however, investigators have responded to the scarcity of human cells by turning to cells of animal origins. Such so-called *xenogenic* cells are far more difficult to encapsulate and success is constrained to special cases. [Examples are devices containing xenogenic cells implanted in certain immunoprivileged sites (spinal fluid, ventricles, eyes) where the avidity of the immune response is muted or placed in contact with acellular fluid (or flowing blood) to minimize

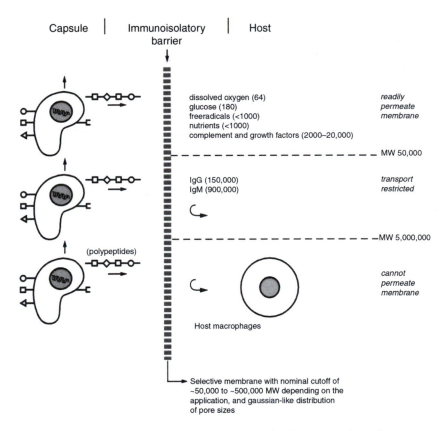

Capsule | Immunoisolatory barrier | Host

dissolved oxygen (64)
glucose (180)
freeradicals (<1000)
nutrients (<1000)
complement and growth factors (2000–20,000)

readily permeate membrane

—— MW 50,000

IgG (150,000)
IgM (900,000)

transport restricted

——MW 5,000,000

(polypeptides)

cannot permeate membrane

Host macrophages

Selective membrane with nominal cutoff of
~50,000 to ~500,000 MW depending on the
application, and gaussian-like distribution
of pore sizes

FIG. 1. The concept of encapsulation. A biocompatible selective membrane barrier surrounds naturally occurring or genetically modified cells. Nutrients, oxygen, and small bioactive materials freely transit the membrane but immunologically active species are too large to penetrate. Although such perfect selectivity is clearly an idealization, techniques have been successfully developed to the point of large-scale clinical evaluation.

the localized inflammatory reaction.] Or, in the case of the liver, xenogenic cells are utilized for periods of time that are much shorter than that required for the development of fulminant immune responses. All such strategies have proven successful, and there is abundant evidence of survival of xenogenic cells, in these limited circumstances, for 3 to 6 months or longer. In contrast, encapsulated xenogenic cells rarely survive much beyond 14 to 28 days when implanted subcutaneously or intraperitoneally in immunocompetent hosts.

Two mechanisms are invoked to explain the inability of membranes to universally protect xenogenic implants, and the different fates of encapsulated allogeneic and xenogeneic cells. First, membranes are not ideally semipermeable and thus allow passage of small quantities of large immunomolecules, including complement and both elicited and preformed immunoglobulins. Such agents are far more active against xenogenic cells than against allogenic grafts. A second problem is that soluble antigens "leak" from cell surfaces or are released upon cell necrosis. These protein constituents are not conserved between species and are, in varying degrees, immunogenic. Their gradual release results in a localized inflammatory response in the neighborhood of the membrane, readily visualized by histology. Inflammatory cells express a number of

low-molecular-weight toxins (including free radicals and cytotoxic cytokines) that pass through the membrane and attack the cells inside. Note that this mechanism requires a local nidus of inflammatory tissue and is thus unlikely to be encountered when implants are placed in cell-free fluid cavities. It is also not significant with allogenic cells whose only non-self proteins are confined to the MHC system.

Interestingly, pharmacological immunosuppression has rarely been used in conjunction with encapsulation, though on theoretical grounds the combination of mechanical and chemical agents would likely prove highly effective.

DEVICES FOR IMMUNOISOLATION

Depending on size and shape, implantable immunoisolation device designs can be categorized as either microcapsules or macrocapsules. Microcapsule beads are illustrated in Fig. 2 (upper right), along with other materials popular for immunoisolation. A current macrocapsule design is shown in Fig. 3. These different designs all share the common components of a permselective membrane, an internal matrix, and the living encapsulated tissue. Macrocapsules are small (100–600 μm,

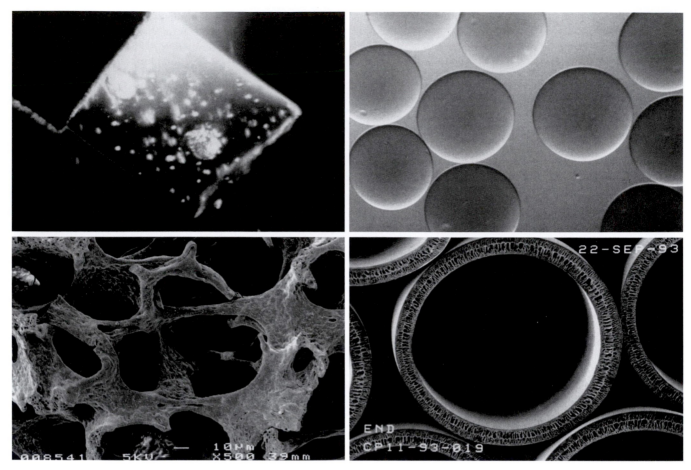

FIG. 2. Materials used as matrices or barrier materials. Micrographs or photomacrographs of hydrophilic materials in the form of matrices (top left) and microcapsules (top right) and of hydrophobic materials in the form of foams (bottom left) and membranes for use in macrocapsules (bottom right).

FIG. 3. Photograph of an implantable macrocapsule. The pencil and tweezers are included for scale.

0.01–0.06×10^{-4} cm in diameter) spherical beads containing up to several thousand cells. Typically, hundreds to thousands of microcapsules are implanted into the host to achieve a therapeutic dose. This design minimizes transport resistance, allows for easy implantation, and provides good dose control. However, microcapsules are difficult to explant and are usually quite fragile. Macrocapsules are much larger in size with the capacity to hold millions of cells, generally requiring a single device for a given therapy. These devices are implanted as tubular or flat sheet diffusion chambers with an inner diameter dimension of 0.5–2.0 mm and a length of 1–10 cm. Macrocapsules provide mechanical and chemical stability superior to those of microcapsules and are easily retrieved. A significant concern with this design is the geometric resistance to mass transport, which limits viability of encapsulated tissue. An alternative macrocapsule design involves connecting the device directly to the patient's circulatory system. The cells are contained in a chamber surrounding the macrocapsule, and the flowing blood can provide an efficient means of nutrient transport. A major challenge with this vascular design is maintaining shunt patency of the device.

Membranes

A wide variety of different materials have been used to formulate the permselective membranes for microcapsules and macrocapsules. In general, the membranes for macrocapsules have been engineered from synthetic thermoplastics, whereas those for microcapsules have been engineered using hydrogel-based materials. Table 1 and Fig. 2 illustrate the materials and appearance of hydrophilic and hydrophobic membrane materials used for immunoisolation.

The process to manufacture microcapsules typically starts with the creation of a slurry of the living cells in a dilute

TABLE 1   Materials Commonly Used in Encapsulation

Hydrophilic	Hydrophobic

Membranes

Alginate

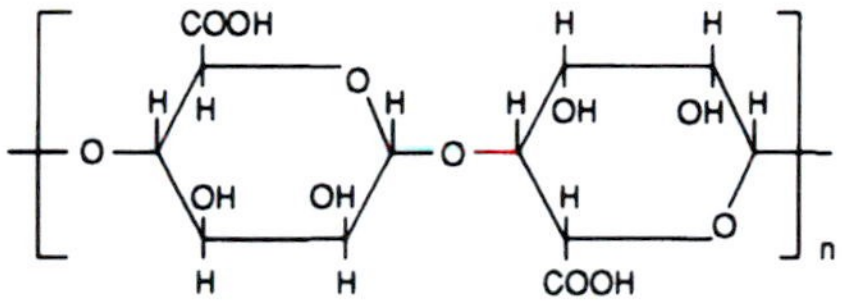

Polysulfone

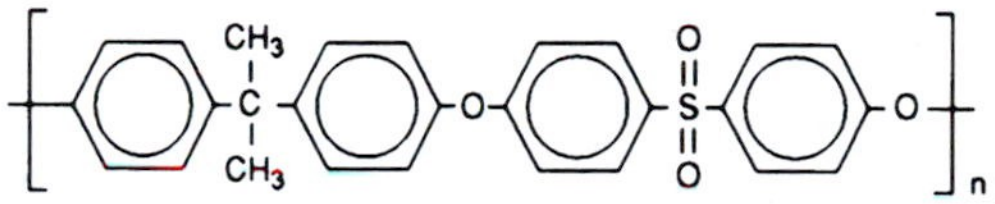

Polylysine

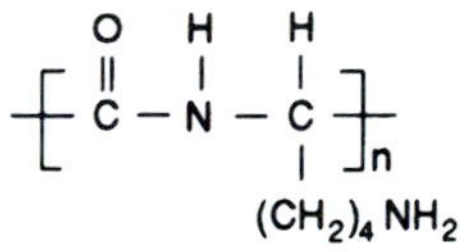

PAN-PVC

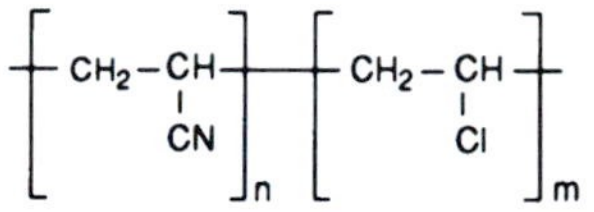

HEMA-MMA

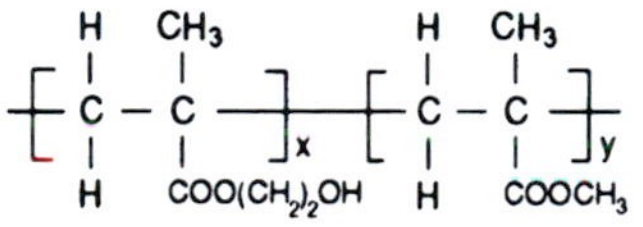

Matrices

Chitosan

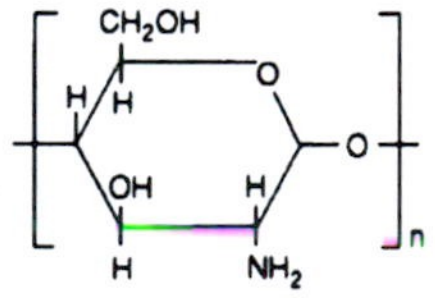

Dacron-polyethylene terephthalate

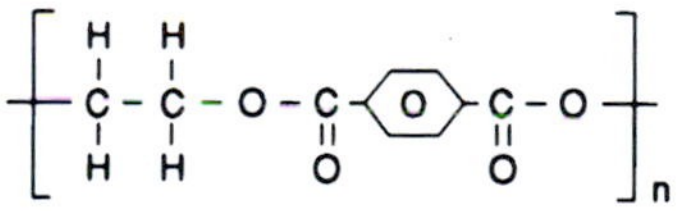

Collagen

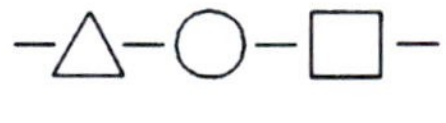

amino acid sequence

Polyvinyl Alcohol

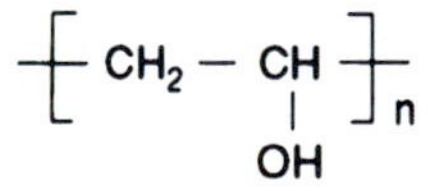

Alginate

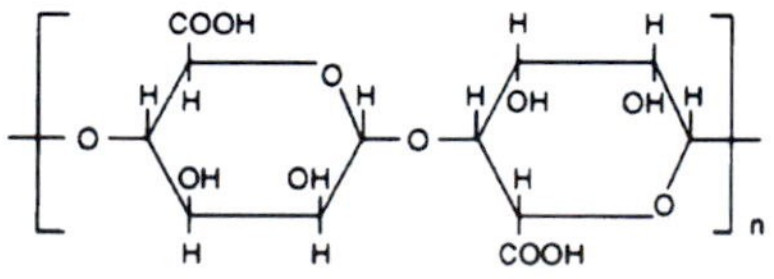

hydrogel solution. Next, small droplets are formed by extruding this mixture through an appropriate nozzle, followed by the cross-linking of the hydrogel to form the mechanically stable microcapsules with an immunoisolatory layer. In an alternative process, water-insoluble synthetic polymers are used in place of hydrogels to prepare the cell slurry. These microcapsules and the immunoisolatory layer are then formed upon interfacial precipitation of the polymer solution. The process to manufacture macrocapsules typically involves phase inversion of a thermoplastic polymer solution cast as a flat sheet or extruded as a hollow fiber. During phase inversion, the polymer solution is placed in controlled contact with miscible nonsolvent, resulting in the formation of the mechanically stable and immunoisolatory membrane. At a later stage, the living cells are aseptically introduced into the fiber or chamber, which is subsequently sealed.

The processes developed to manufacture macrocapsules and microcapsules are very versatile and allow for the formation of membranes with a wide variety of different transmembrane pore structures and outer surface microgeometries. Membrane selection has a strong influence on microcapsule or macrocapsule device performance and is characterized in terms of membrane chemistry, transport properties, outer surface morphology, and strength. Optimum parameters are dictated by the metabolic requirements of the encapsulated cells, the size of the therapeutic agent to be delivered, the required immunoprotection, and the desired biocompatibility. Membrane transport properties are chosen to maintain viability and functionality of the encapsulated cells and provide release of the therapeutic agent. This selection involves designing membranes that provide sufficient nutrient flux to meet the requirements of the encapsulated cells, while preventing flux of immunological species that would reject the tissue.

Biocompatibility is defined by the host reaction to the implant and has a significant impact upon device performance. Biocompatibility depends upon the nature of the encapsulated cell and both the transport properties and outer morphology of the membrane barrier.

Transport properties are routinely evaluated in combination with a physical characterization of the membrane to develop structure–property relations. Physical parameters such as inner diameter, wall thickness, wall morphology, and surface morphology can influence the transport behavior. Light micrometry is used to characterize membrane geometry, and scanning electron microscopy is used to analyze membrane morphology. The high-resolution techniques of atomic force microscopy and low-voltage scanning electron microscopy have been exploited to image the porosity and pore size of the permselective skin of ultrafiltration membranes. A wide range of membrane wall morphologies can be produced using the phase inversion process: most common are foamlike or trabecular structures. Outer surface morphology is generally characterized as rough (microgeometries > 2 μm) or smooth. Implanted into a host tissue site, rough surface will frequently evoke a significant host fibrotic reaction, whereas smooth surfaces will evoke a relatively mild reaction. In some cases, a vascularized host reaction can actually improve encapsulated device viability, by providing nutrients and oxygen to the perimembrane region.

Matrices

The second component of an immunoisolation device is the internal matrix. Hydrogels and solid scaffolds have been widely used and can be produced from synthetic or naturally derived materials (Table 1). Examples of hydrogel matrices include alginate, agarose, and poly(ethylene oxide), and examples of scaffold matrices include poly(ethylene terephthalate) yarn, poly(vinyl alcohol) foam, and cross-linked chitosan. This matrix serves two basic functions. The first is to provide mechanical support for the encapsulated cells in order to maintain a uniform distribution within the device. In the absence of this support, the cells often gravitate toward one region of the device and form a large necrotic cluster. The matrix also serves a biological function by stimulating the cells to secrete their own extracellular matrix, regulating cell proliferation, regulating secretory function, and maintaining the cells in a differentiated phenotype.

Selection of a matrix for a particular cell type involves several design considerations. Generally, suspension cell cultures prefer a hydrogel-based matrix, whereas anchorage-dependent cells prefer the attachment surfaces of a solid scaffold. The matrix must also be physically and chemically compatible with the permselective membrane. For example, scaffold matrices should not damage the integrity of the permselective membrane and soluble matrix components should not significantly affect the pore size. The stability of the matrix should also be considered and in general must at least match the lifetime of the device. Finally, the transport characteristics of the matrix candidates need to be considered. Certain matrices may exhibit significant resistance to the transport of small or large solutes, and thus affect overall performance.

Cells

The final component of the immunoisolation device is the encapsulated cells used to secrete the therapeutic molecules. These cells may be derived from "primary" cells, (i.e., postmitotic cells dividing very slowly if at all), continuously dividing cell lines, or genetically engineered tissue. All three cell types have been successfully encapsulated. Cell sourcing for a device begins with a definition of the desired secretory function of the implant. For example, chromaffin cells are a known source of the opoid peptide norepinephrine and have been used as a cellular delivery system to treat chronic pain. Such chromaffin cells are obtained as primary cultures from an enzymatic isolation of the bovine adrenal gland. Islets of Langerhans for the delivery of insulin to replace pancreatic function represent another widely investigated primary cell type. The PC12 rat pheochromocytoma line is an example of an immortalized cell line derived from a tumor that has been used for the delivery of L-dopa and dopamine in the treatment of Parkinson's disease. Cells engineered to secrete a variety of neurotrophic factors have been used in an encapsulated environment for the treatment of neurodegenerative diseases and include the Chinese hamster ovary (CHO) line, the Hs27 human foreskin fibroblast, and the baby hamster kidney (BHK) line.

Different cell types have different requirements for survival and function in a device and may result in a variety of levels of performance in any given implant site. These cell-specific considerations include metabolic requirements, proliferation rate within a device, and antigenicity and are assessed to ensure long-term device performance. For example, a highly antigenic encapsulated cell may be rapidly rejected in a nonimmunoprivileged site, such as the peritoneal cavity. This same encapsulated tissue may result in very satisfactory performance in an immunoprivileged site, such as the central nervous system. Similarly, a cell with a high nutrient requirement may provide superior performance in a nutrient-rich site, such as a subcutaneous pouch, and fail in a nutrient-poor site such as the cerebral spinal fluid.

Safety is another consideration in sourcing cells for eventual human implants. Grafts must be derived from healthy donors or from stable, contaminant-free cell lines. Before approving human clinical trials, regulatory bodies require testing for known transmittable diseases, mycoplasma, reverse transcriptase, cultivable viruses, and microbial contaminants.

APPLICATIONS

At this writing (mid-1999) several applications of immunoisolated cell therapy are in clinical trials but none have reached the stage of approval by regulatory agencies and routine clinical utilization. Table 2 summarizes the application status of the bioartificial liver, the bioartificial pancreas, and the delivery of cell and gene therapy. As in all areas of tissue engineering, technology is moving rapidly and Table 2 should be appreciated in historical rather than current context.

The bioartificial liver currently is being evaluated as a "bridge to transplant," i.e., to extend the lifetime of patients who are medically eligible for liver transplantation until a donor organ becomes available. Several designs and treatment protocols have been proposed; one appealing format is shown in Fig. 4. The extracorporeal circuit is broadly similar to

that used in dialysis for the treatment of kidney failure. Blood is continuously withdrawn from the patient's vasculature, at a rate of 200–300 ml/min, treated in a hollow-fiber bioartificial liver, and ultimately returned to the patient. A charcoal filter may be added to further detoxify the blood. Treatments are performed daily for 4 hours. Results in early human trials were quite encouraging and several cases of recovery without transplantation were observed. No results are yet available from the controlled and blinded trials.

The bioartificial pancreas has enjoyed very impressive success in rodent studies—so much so that no fewer than five reports on "proof of principle" experiments have appeared in the hallowed pages of *Science* magazine. Unfortunately, results from larger animal models and human studies have been disappointing. Investigators have not been able to reliably isolate the number of islets (500,000 to 1,000,000 or $\sim 2 \times 10^9$ cells) required for a large recipient. Moreover, species scaling of device design has proven difficult: formats that were suitable in rodents generally have been unsatisfactory in large animals. Some investigators believe that use of genetically engineered cells or their transgenic cohorts may solve the problem of cell source. Genetically engineered cells might be more productive than islets and ease some of the constraints on device design. Development of a clinically beneficial bioartificial pancreas remains an important challenge for biomedical engineering in the early 21st century.

Encapsulation is also being developed for the delivery of cell and gene therapy. Small quantities of cells producing a desired therapeutic molecule are placed inside the lumen of a sealed hollow fiber or encapsulated in microspheres. A therapeutic dose may involve a very manageable $1-10 \times 10^6$ cells (roughly two orders of magnitude fewer cells than would be required for a bioartificial pancreas). From one perspective, these devices represent a form of drug delivery providing point-source, time-constant, and site-specific delivery with the added benefit of a "regenerable" source of bioactive "drug." In another sense, when recombinant cells are involved, this approach can be considered a form of gene therapy in which the transplanted gene resides in cells housed in a capsule rather than directly in the cells of the recipient. The technical issues involved in this form of encapsulated cell therapy are largely resolved. Several successful preclinical and clinical trials have been reported. However, to date no blinded study with a control study has shown efficacy.

TABLE 2 Application Status of Immunoisolation (Late 1999)

Bioartificial liver	Several reports of clinical investigations in literature for bridge to transplant two successful phase I[a] trials. Two "pivotal[b] trials" underway.
Bioartificial pancreas	One case report of a single patient receiving a therapeutic dose of islets (and immunosuppression). Several reports of "survival studies" at smaller doses. Preclinical trials report outstanding success in rodents but not in dogs and nonhuman primates.
Delivery of cell and gene therapy	Pain: Successful phase I study completed; pivotal trial failed to show efficacy. ALS: Human clinical trials reported; Huntington trial is in progress. Numerous studies in primates, other large animals, and rodents.

[a]Phase I. Small trial to determine safety in ~10 patients.
[b]Pivotal. Large trial to determine efficacy. Includes control arm.

Bibliography

Aebischer, P., and Lysaght, M. J. (1995). Immunoisolation and cellular xenotransplantation. *Xeno* 3: 43–48.
Aebischer, P., Pochon, N. A., Heyd, B., Deglon, N., Joseph, J. M., Zurn, A. D., Baetge, E. E., Hammang, J. P., Goddard, M., Lysaght, M., Kaplan, F., Kato, A. C., Schluep, M., Hirt, L., Regli, F., Porchet, F., and DeTribolet, N. (1996a). Gene therapy for amyotrophic lateral sclerosis (ALS) using a polymer encapsulated xenogenic cell line engineered to secrete hCNTF. *Hum. Gene Ther.* 7: 851–860.
Aebischer, P., Schluep, M., Deglon, N., Joseph, J. M., Hirt, L., Heyd, B., Goddard, M., Hammang, J. P., Zurn, A. D., Kato, A. C., Regli, F., and Baetge, E. E. (1996b). Intrathecal delivery of CNTF using

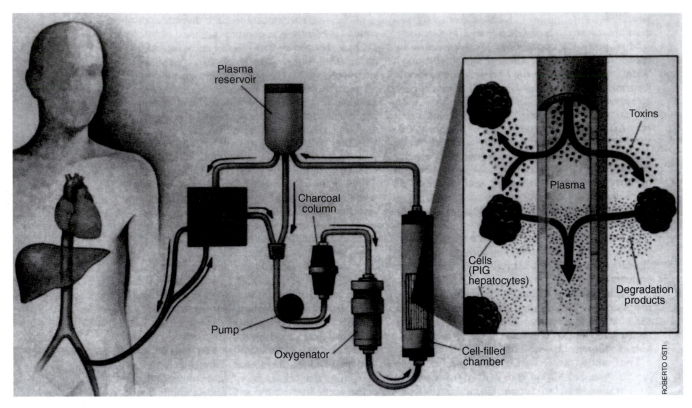

FIG. 4. The bioartificial liver. Several closely related versions of this intermittent extracorporeal therapy format have undergone preliminary clinical trials. Controlled evaluations are just beginning.

encapsulated genetically modified xenogeneic cells in amyotrophic lateral sclerosis. *Nat. Med.* **2**(6): 696–699.

Avgoustiniatos, E. S., and Colton, C. K. (1997). Effect of external oxygen mass transfer on viability of immunoisolated tissue. *Ann. N. Y. Acad. Sci.* **831**: 145–167.

Brauker, J., Carr-Brendel, V., Martinson, L., Crudele, J., Johnston, W., and Johnson, R. (1995). Neovascularization of synthetic membranes directed by membrane microarchitecture. *J. Biomed. Mater. Res.* **29**: 1517–1524.

Cabasso, I. (1980). Hollow fiber membranes. in *Kirk-Othmer Encyclopedia of Chemical Technology*, Vol. 12. Wiley, New York, pp. 492–517.

Chen, S. C., Mullon, C., Kahaku, E., Watanabe, F., Hewitt, W., Eguchi, S., Middleton, Y., Arkadopoulos, N., Rozga, J., Solomon, B., and Demetriou, A. A. (1997). Treatment of severe liver failure with a bioartificial liver. *Ann. N. Y. Acad. Sci.* **831**: 350–360.

Colton, C. K. (1995). Implantable biohybrid artificial organs. *Cell Transplant.* **4**(4): 415–436.

Dionne, K. E., Cain, B. M., Li, R. H., Bell, W. J., Doherty, W. J., Rein, D. H., Lysaght, M. J., and Gentile, F. T. (1996). Transport characterization of membranes for immunoisolation. *Biomaterials* **17**: 257–266.

Emerich, D. F., Winn, S. R., Hantraye, P. M., Peschanski, M., Chen, E. Y., Chu, Y., McDermott, P., Baetge, E. E., and Kordower, J. H. (1997). Protective effect of encapsulated cells producing neurotrophic factor CNTF in a monkey model of Huntington's disease. *Nature* **386**(6623): 395–399.

Emerich, D., Lidner, M., Winn, S. R., Chen, E., Frydel, B., Koedower, J. (1996). Implants of encapsulated human CNTF producing fibroblasts prevent behavioral deficits and striatal degeneration in a rodent model of Huntington's disease. *J. Neurosci.* **16**: 5168–5181.

Inoue, K., Fujisato, T., Gu, Y. J., Burczak, K., Sumi, S., Kogire, M., Tobe, T., Uchida, K., Nakai, I., Maetani, S., and Ikada, Y. (1992). Experimental hybrid islet transplantation: application of polyvinyl alcohol membrane for entrapment of islets. *Pancreas* **7**: 562–568.

Kordower, J. H., Liu, Y., Winn, S., and Emerich, D. (1995). Encapsulated PC12 cell transplants into hemiparkinsonian monkeys: a behavioral, neuroanatomical, and neurochemical analysis. *Cell Transplant.* **4**: 155–171.

Lanza, R. P., and Chick, W. L. (1997). Transplantation of pancreatic islets. *Ann. N. Y. Acad. Sci.* **831**: 323–331.

Lanza, R. P., Cooper, D. K. C., and Chick, W. L. (1997). Xenotransplantation. *Sci. Am.* **277**(1): 54–59.

Li, R. (1998). Materials for immunoisolated cell transplantation. *Adv. Drug Dev. Rev.* **33**(1–2): 87–109.

Lysaght, M. J., and Aebischer, P. A. (1999). Encapsulated cells as therapy. *Sci. Am.* Apr. 76–83.

Roberts, T., De Boni, U., and Sefton, M. V. (1996). Dopamine secretion by PC12 cells microencapsulated in a hydroxyethyl methacrylate–methyl methacrylate copolymer. *Biomaterials* **17**: 267–275.

Sagen, J., Wang, H., Tresco, P., and Aebischer, P. (1993). Transplants of immunologically isolated xenogeneic chromaffin cells provide a long-term source of pain neuroactive substances. *J. Neurosci.* **13**: 2415–2423.

Strathmann, H. (1985). Production of microporous media by phase inversion processes. In *Materials Science of Synthetic Membranes*, D. R. Lloyd (ed.). American Chemical Society, Washington, D. C.

Winn, S. R., and Tresco, P. A. (1994). Hydrogel applications for encapsulated cellular transplants. *Methods Neurosci.* **21**: 387–402.

8.4 SYNTHETIC BIORESORBABLE POLYMER SCAFFOLDS

Antonios G. Mikos, Lichun Lu, Johnna S. Temenoff, and Joerg K. Tessmar

SCAFFOLD DESIGN

Tissue engineering involves the development of functional substitutes to replace missing or malfunctioning human tissues and organs (Langer and Vacanti, 1993). Most strategies in tissue engineering have focused on using biomaterials as scaffolds to direct specific cell types to organize into three-dimensional (3D) structures and perform differentiated function of the targeted tissue. Synthetic bioresorbable polymers that are fully degradable into the body's natural metabolites by simple hydrolysis under physiological conditions are the most attractive scaffold materials. These scaffolds offer the possibility to create completely natural tissue or organ equivalents and thus overcome the problems such as infection and fibrous tissue formation associated with permanent implants.

These synthetic polymers must possess unique properties specific to the tissue of interest as well as satisfy some basic requirements in order to serve as an appropriate scaffold. One essential criterion is biocompatibility, i.e., the polymer scaffold should not invoke an adverse inflammatory or immune response once implanted (Babensee *et al.*, 1998). Some important factors that determine its biocompatibility, such as the chemistry, structure, and morphology, can be affected by polymer synthesis, scaffold processing, and sterilization. Toxic residual chemicals involved in these processes (e.g., monomers, stabilizers, initiators, cross-linking agents, emulsifiers, organic solvents) may be leached out from the scaffold with detrimental effects to the engineered and surrounding tissue.

The primary role of a scaffold is to provide a temporary substrate to which transplanted cells can adhere. Most organ cell types are anchorage-dependent and require the presence of a suitable substrate in order to survive and retain their ability to proliferate, migrate, and differentiate. Cell morphology correlates with cellular activities and function; strong cell adhesion and spreading often favor proliferation while a rounded cell shape is required for cell-specific function. For example, it has been demonstrated that the use of substrates with patterned surface morphologies or varied extracellular matrix (ECM) surface coatings can modulate cell shape and function (Chen *et al.*, 1998; Mooney *et al.*, 1992; Singhvi *et al.*, 1994).

For epithelial cells, cell polarity is essential for their function. Polarity refers to the distinctive arrangement, composition, and function of cell-surface and intracellular domains. This typically corresponds to a heterogeneous extracellular environment. For example, retinal pigment epithelium (RPE) cells have three major surface domains: the apical surface is covered with numerous microvilli; the lateral surface is joined with neighboring cells by junctional complexes; and the basal surface is convoluted into basal infoldings and connected to the basal lamina. The polymer scaffold for RPE transplantation should therefore provide proper surface chemistry and surface microstructure for optimal cell–substrate interaction and, along with appropriate culture conditions, be able to induce proper cell polarity (Lu *et al.*, 1998).

Besides cell morphology, the function of many organs is dependent on the 3D spatial relationship of cells with their ECM. The shape of a skeletal tissue is also critical to its function. Gene expression in cells is regulated differently by 2D versus 3D culture substrates. For instance, the differentiated phenotype of human epiphyseal chondrocytes is lost on 2D culture substrates but reexpressed when cultured in 3D agarose gels (Aulthouse *et al.*, 1989). A polymer scaffold should be easily and reproducibly processed into a desired shape that can be maintained after implantation so that it defines the ultimate shape of the regenerated tissue. A suitable scaffold should therefore act as a template to direct cell growth and ECM formation and facilitate the development of a 3D structure.

Porosity, pore size, and pore structure are important factors to be considered with respect to nutrient supply to transplanted and regenerated cells. To regenerate highly vascularized organs such as the liver, porous scaffolds with large void volume and large surface-area-to-volume ratio are desirable for maximal cell seeding, attachment, growth, ECM production, and vascularization. Small-diameter pores are preferable to yield high surface area per volume provided the pore size is greater than the diameter of a cell in suspension (typically 10 μm). However, topological constraints may require larger pores for cell growth. Previous experiments have demonstrated optimal pore sizes of 20 μm for fibroblast ingrowth, 20–125 μm for adult mammalian skin regeneration, and 200–400 μm for bone ingrowth (Boyan *et al.*, 1996; Whang *et al.*, 1995). The rate of tissue invasion into porous scaffold also depends on the pore size and polymer crystallinity (Mikos *et al.*, 1993c; Park and Cima, 1996; Wake *et al.*, 1994). Compared to isolated pore structure, an interconnected pore network enhances the diffusion rates to and from the center of the scaffold and facilitates vascularization, thus improving oxygen and nutrient supply and waste removal. The vascularization of an implant is a prerequisite for regeneration of most 3D tissues except for cartilage, which is avascular.

Mechanical properties of the polymer scaffold should be similar to the tissue or organ intended for regeneration. For load-bearing tissues such as bone, the scaffold should be strong enough to withstand physiological stresses to avoid collapse of the developing tissue. Also, transfer of load to the scaffold (stress shielding) after implantation may result in lack of sufficient mechanical stimulation to the ingrowing tissue. For the regeneration of soft tissues such as skin, the scaffolds are required to be pliable or elastic. The stiffness of the scaffold may affect the mechanical tension generated within the cell cytoskeleton, which is critical for the control of cell shape and function (Chicurel *et al.*, 1998). A more rigid surface may facilitate the assembly of stress fibers and enhance cell spreading and dividing. Scaffold compliance may also affect cell–cell contacts and aggregation (Moghe, 1996).

Understanding and controlling the degradation process of a scaffold and the effects of its degradation products on the body is crucial for long-term success of a tissue-engineered cell–polymer construct. The local drop in pH due to the release of acidic degradation products from some implants may cause tissue necrosis or inflammation. Polymer particles formed after long-term implantation of a scaffold or due to micromotion at the implantation site may elicit an inflammatory response. Microparticles of polymers have been shown to suppress initial rat-marrow stromal osteoblast proliferation in culture (Wake *et al.*, 1998). The mechanism by which the scaffold degrades should also be considered. For example the degradation products are released gradually by surface erosion, whereas during bulk degradation, the release of degradation products occurs only when the molecular weight of the polymer reaches a critical value. This late-stage burst effect may cause greater local pH drop.

The rate of scaffold degradation is tailored to allow cells to proliferate and secrete their own ECM while the polymer scaffold vanishes over a desired time period (from days to months) to leave enough space for new tissue growth. Since the mechanical strength of a scaffold usually decreases with degradation time, the degradation rate may be required to match the rate of tissue regeneration in order to maintain the structural integrity of the implant. The degradation rate of a scaffold can be affected by various factors listed in Table 1.

The design requirements of a tissue engineering scaffold are specific to the structure and function of the tissue to be regenerated. The polymer scaffold is typically engineered to mimic the natural ECM of the body. ECM proteins play crucial roles in the control of cell growth and function (Hay, 1993; Howe *et al.*, 1998). However, most synthetic polymer scaffolds do not possess the specific signals (ligands) that can be recognized by cell-surface receptors. It is therefore preferable that the polymer chain have chemically modifiable functional groups onto which sugars, proteins, or peptides can be attached. In addition, polymer–peptide hybrid molecules may be created or the ligand may be immobilized on the scaffold surface to

generate a biomimetic microenvironment (Shakesheff *et al.*, 1998; Shin *et al.*, 2003b).

SCAFFOLD MATERIALS

The range of physical, chemical, mechanical, and degradative properties that may be achieved using synthetic bioresorbable polymers renders them extremely versatile as scaffold materials. Their molecular weight and chemical composition may be precisely controlled during polymer synthesis. Copolymers, polymer blends, and composites with other materials may be manufactured to give rise to properties that are advantageous over homopolymers for certain applications. Moreover, many polymers can be functionalized by converting end groups or addition of side chains with various chemical groups to obtain polymers that can be self-cross-linked or cross-linked with proteins and other bioactive molecules (Behravesh *et al.*, 1999). By choosing an appropriate processing technique, scaffolds of specific architecture and structural characteristics may be fabricated.

Not all types of currently available synthetic bioresorbable polymers can be manufactured into 3D scaffolds because of their chemical and physical properties and processability (Table 2). The most widely utilized scaffold materials are poly(α-hydroxy esters) such as PGA, PLA, and PLGA. They have been fabricated into thin films, fibers, porous foams, and conduits and investigated as scaffolds for regeneration of several tissues. Furthermore, the lysine groups in poly(lactic acid-*co*-lysine) provide sites for addition of cell-adhesion sequences such as RGD peptides (Barrera *et al.*, 1995; Cook *et al.*, 1997).

Poly(propylene fumarate) (PPF), an unsaturated linear polyester that can be cross-linked through its fumarate double bonds, has been investigated as a bioresorbable bone cement (Peter *et al.*, 1997a). The cross-linking, mechanical, and degradative properties of an injectable composite scaffold made of PPF and β-tricalcium phosphate have been characterized (Peter *et al.*, 1999, 1998b, 1997b). The mechanical properties of PPF scaffolds can be further modified depending on the cross-linking parameters employed. An increase in compressive modulus was observed with the use of a cross-linking agent, PPF-diacrylate, and the choice of a photo-(light-based) initiator, rather than a thermally based initiator system (Timmer *et al.*, 2003).

Poly(ethylene glycol) (PEG), although nondegradable, is often used to fabricate copolymers or polymer blends to increase the hydrophilicity, biocompatibility, and/or softness of the scaffold. Poly(propylene fumarate-*co*-ethylene glycol) [P(PF-co-EG)] hydrogels have been developed for cardiovascular applications (Suggs *et al.*, 1998a, 1997, 1999). When used in combination with a pore-forming agent and modified with cell-adhesion ligands, P(PF-co-EG) has also been used in highly porous scaffolds for bone tissue engineering (Behravesh and Mikos, 2003). Results indicate that this hydrogel supported the proliferation, osteogenic differentiation and matrix production from seeded bone-marrow stromal cells during 28 days of *in vitro* culture (Behravesh and Mikos, 2003).

TABLE 1 Factors Affecting Scaffold Degradation

Polymer chemistry	Scaffold structure
Composition	Density
Structure	Shape
Configuration	Size
Morphology	Mass
Molecular weight	Surface texture
Molecular weight distribution	Porosity
	Pore size
Chain motility	Pore structure
Molecular orientation	Wettability
Surface-to-volume ratio	Processing method and conditions
Ionic groups	Sterilization
Impurities or additives	
In vitro	*In vivo*
Degradative medium	Implantation site
pH	Access to vasculature
Ionic strength	Mechanical loading
Temperature	Tissue growth
Mechanical loading	Metabolism of degradation products
Type and density of cultured cells	Enzymes

TABLE 2 Scaffold Materials and Their Applications[a]

Materials	Applications
Poly(α-hydroxy esters)	
Poly(L-lactic acid) (PLLA)	Bone, cartilage, nerve
Poly(glycolic acid) (PGA)	Cartilage, tendon, urothelium, intestine, liver, bone
Poly(D,L-lactic-co-glycolic acid) (PLGA)	Bone, cartilage, urothelium, nerve, RPE
PLLA-bonded PLGA fibers	Smooth muscle
PLLA coated with collagen or poly(vinyl alcohol) (PVA)	Liver
PLLA and poly(ethylene glycol) (PEG) block copolymer	Bone
PLGA and PEG blends	Soft tissue and tubular tissue
Poly(L-lactic acid-co-ε-caprolactone) (PLLACL)	Meniscal tissue, nerve
Poly(D,L-lactic acid-co-ε-caprolactone) (PDLLACL)	Vascular graft
Polyurethane/poly(L-lactic acid)	Small-caliber arteries
Poly(lysine-co-lactic acid)	Bone, cartilage, nerve
Poly(propylene fumarate) (PPF)	Bone
Poly(propylene fumarate-co-ethylene glycol) [P(PF-co-EG)]	Cardiovascular, bone
PPF/β-tricalcium phosphate (PPF/β-TCP)	Bone
Poly(ε-caprolactone)	Drug delivery
Polyhydroxyalkanoate (PHA)	Cardiovascular
Polydioxanone	Bone
Polyphosphates and polyphosphazenes	Skeletal tissue, nerve
Pseudo-poly(amino acids)	Bone
Tyrosine-derived polyiminocarbonates	
Tyrosine-derived polycarbonate	
Tyrosine-derived polyacrylates	

[a]Adapted from Babensee *et al.* (1998).

In addition, another material including PEG, oligo [poly(ethylene glycol) fumarate] (OPF), has been developed and characterized (Jo *et al.*, 2001). Because of the chemical structure of this oligomer, it can be used to form biodegradable hydrogels with a high degree of swelling (Jo *et al.*, 2001; Temenoff *et al.*, 2003). This material may find uses in guided tissue regeneration applications because it demonstrates relatively low general cell adhesion, but at the same time, possesses an ability to be modified with peptides that could encourage adhesion of specific cell types (Shin *et al.*, 2002; Temenoff *et al.*, 2003).

Poly(ε-caprolactone) (PCL) as well as blends and copolymers containing PCL have also been studied as scaffold materials (Suggs and Mikos, 1996). Polyphosphates and polyphosphazenes have been processed into scaffolds for bone tissue engineering (Behravesh *et al.*, 1999; Renier and Kohn, 1997). Pseudo-poly(amino acids), in which amino acids are linked by both amide and nonamide bonds (such as urethane, ester, iminocarbonate, and carbonate), are amorphous and soluble in organic solvents and thus processable into scaffolds. The most studied among these polymers are tyrosine-derived polycarbonates and polyacrylates (James and Kohn, 1996). By structural modifications of the backbone and pendant chains, polymer families with systematically and gradually varied properties can be created.

APPLICATIONS OF SCAFFOLDS

Tissue Induction

Tissue induction is the process by which ingrowth of surrounding tissue into a porous scaffold is effected (Fig. 1A). The scaffold provides a substrate for the migration and proliferation of the desired cell types. For example, an osteoinductive material can be used to selectively induce bone formation. This approach has been employed to regenerate several other tissues including skin and nerve.

Cell Transplantation

The concept is that cells obtained from patients can be expanded in culture, seeded onto an appropriate polymer scaffold, cultured, and then transplanted (Bancroft and Mikos, 2001) (Fig. 1B). The time at which transplantation takes place varies with a specific application. Usually the cells are allowed to attach to the scaffold, proliferate, and differentiate before implantation. A scaffold for bone cell transplantation should be osteoconductive, meaning that it has the capacity to direct the growth of osteoblasts *in vitro* and allow the integration of the transplant with the host bone. This strategy is the most

A

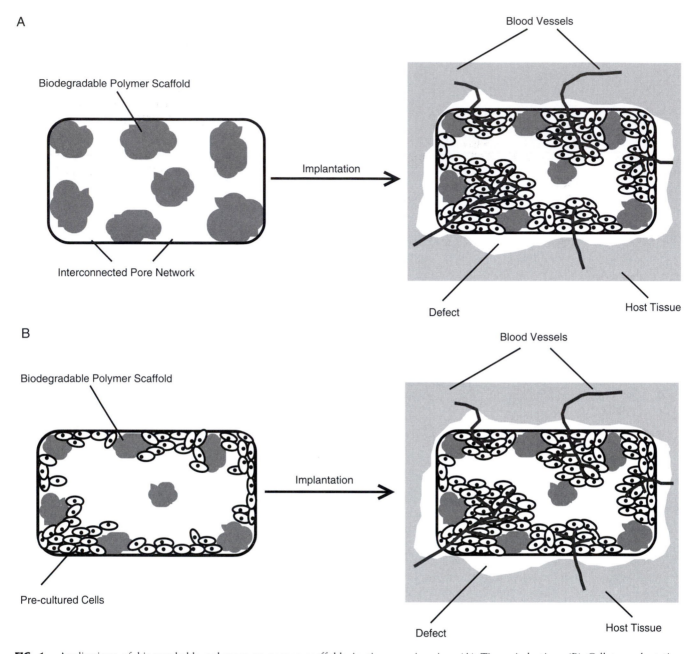

FIG. 1. Applications of bioresorbable polymers as porous scaffolds in tissue engineering. (A) Tissue induction. (B) Cell transplantation. (C) Prevascularization. (D) *In situ* polymerization. In all cases, the porous scaffolds allow tissue ingrowth as they degrade gradually.

widely used in tissue engineering and has been investigated for the transplantation of many cell types including osteoblasts, chondrocytes, hepatocytes, fibroblasts, smooth muscle cells, and RPE. This method also offers the possibility that genetically modified cells could be transplanted, thereby simultaneously presenting both the cells and the bioactive factors they produce to the site of interest, with the potential of further enhancing regeneration of the injured area (Blum *et al.*, 2003). This approach may be considered a combination of the applications

of cell transplantation and delivery of bioactive molecules, discussed later.

Prevascularization

The major obstacle in the development of large 3D transplants such as liver is nutrient diffusion limitation, because cells will not survive farther than a few hundred microns from

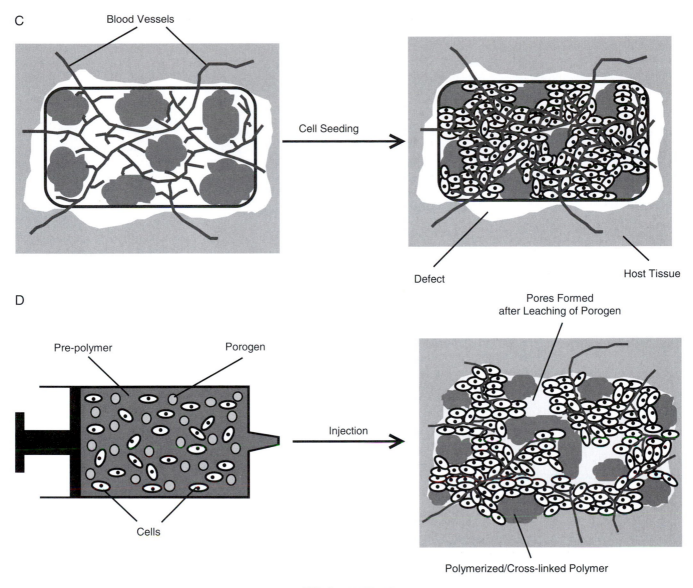

FIG. 1.—continued

the nutrient supply. Although the scaffold can be vascularized postimplantation, the rate of vascularization is usually insufficient to prevent cell death inside the scaffold. In this case, prevascularization of the scaffold may be necessary to allow the ingrowth of fibrovascular tissue or uncommitted vascular tissue such as periosteum (layer of connective tissue covering bone) before cell seeding by injection (Fig. 1C). The prevascularized scaffold will provide a substrate for cell attachment, growth, and function. The extent of prevascularization has to be optimized to allow sufficient nutrient diffusion as well as enough space for cell seeding and tissue growth (Mikos *et al.*, 1993c).

Some complex osseous defects created by bone tumor removal or extensive tissue damage exceed the critical size for normal healing and require a large transplant to restore function. A novel strategy is to prefabricate vascularized bone flaps by implanting a mold containing bioresorbable polymers with osteoinductive elements onto a periosteal site remote from the defect where prevascularization and ectopic bone formation can occur over a period of time as the scaffold degrades (Thomson *et al.*, 1999). The created autologous bone can then be transplanted to the defect site where vascular supply can be attached via microsurgery to existing vessels.

In Situ Polymerization

Injectable, in situ polymerizable, bioresorbable materials can be utilized to fill defects of any size and shape with minimal surgical intervention (Fig. 1D). For instance, PPF has been developed as an injectable bone cement that hardens within

10 to 15 minutes under physiological conditions. These materials do not require prefabrication but must meet additional requirements since polymerization or cross-linking reactions occur *in vivo*. All reagents and products must be biocompatible, and the reaction conditions such as temperature, pH, and heat release should not damage implanted cells or the surrounding tissue.

The hardened material (scaffold) must be highly porous and have interconnected pore structure in order to serve as a suitable template for guiding cell growth and differentiation. This can be achieved by combining a porogen such as sodium chloride crystals in the injectable paste that are eventually leached out, leaving a porous polymer matrix. Since the leaching step occurs *in vivo*, local high salt concentration may lead to high osmolarity and tissue damage. The amount of porogen incorporated has to be optimized to ensure biocompatibility, while enough porosity needs to be achieved to allow sufficient nutrient diffusion and vascularization.

PPF has also been developed for use in combination with cell transplantation applications through the incorporation of cells within the material during the cross-linking procedure. Because of the potentially non-cytocompatible conditions that may be present during the curing reaction, a composite material has been developed in which cells are first encapsulated in gelatin microspheres, and these are then included with the PPF during cross-linking (Payne *et al.*, 2002a, b). It has been shown that this encapsulation procedure enhances the viability, proliferation, and osteogenic differentiation of rat-marrow stromal cells (as compared to nonencapsulated cells) when cultured on cross-linking PPF *in vitro* (Payne *et al.*, 2002a, b).

Delivery of Bioactive Molecules

Cellular activities can be further modulated by various soluble bioactive molecules such as DNA, cytokines, growth factors, hormones, angiogenic factors, or immunosuppresant

drugs (Babensee *et al.*, 2000; Holland and Mikos, 2003; Kasper and Mikos, 2003). For instance, bone morphogenetic proteins (BMPs) have been identified as a family of growth factors that regulate differentiation of bone cells (Ripamonti and Duneas, 1996). Controlled local delivery of these tissue inductive factors to transplanted and regenerated cells is often desirable. This has led to the concept of incorporation of bioactive molecules into scaffolds for implantation. These factors can be bound into polymer matrix during scaffold processing (Behravesh *et al.*, 1999; Shin *et al.*, 2003a) (Fig. 2A). Alternatively, bioresorbable microparticles or nanoparticles loaded with these molecules can be impregnated into the substrates (Hedberg *et al.*, 2002; Holland *et al.*, 2003; Lu *et al.*, 2000) (Fig. 2B). By incorporating BMPs or other osteogenic molecules into the injectable paste, PPF can also serve as a delivery vehicle for bone growth factors to induce a bone-regeneration cascade (Hedberg *et al.*, 2002). The release of bioactive molecules *in vivo* is governed by both diffusion and polymer degradation (Hedberg *et al.*, 2002; Holland *et al.*, 2003). In addition, if the molecules are encapsulated within microparticles that are degraded through enzymatic actions, such as gelatin (Holland *et al.*, 2003), the concentration and activity of these enzymes may also affect the release profile of the factors from composite scaffolds.

SCAFFOLD PROCESSING TECHNIQUES

The technique used to manufacture synthetic bioresorbable polymers into suitable scaffolds for tissue regeneration depends on the properties of the polymer and its intended application (Table 3). Scaffold processing usually involves (1) heating the polymers above their glass transition or melting temperatures; (2) dissolving them in organic solvents; and/or (3) incorporating and leaching of porogens (gelatin microspheres, salt crystals, etc.) in water (Temenoff and Mikos, 2000).

A Bioactive Molecules

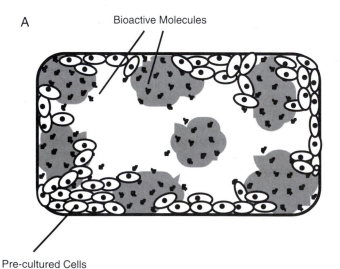

Pre-cultured Cells

B Microparticles

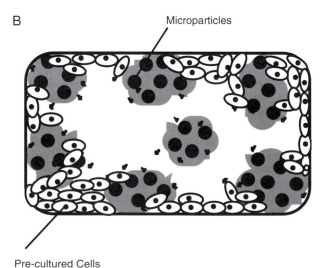

Pre-cultured Cells

FIG. 2. Localized delivery of bioactive molecules from scaffolds. (A) Release directly from the supporting matrix. (B) Microparticles or nanoparticles loaded with bioactive molecules are impregnated into scaffolds and serve as delivery vehicles.

TABLE 3 Examples of Scaffolds Processed by Various Techniques

Processing technique	Examples
Fiber bonding	PGA fibers; PLA-reinforced PGA fibers
Solvent casting and particulate leaching	PLA, PLGA, PPF foams
Superstructure engineering	PLA, PLGA membranes
Compression molding	PLA, PLGA foams
Extrusion	PLA, PLGA conduits
Freeze-drying	PLGA foams
Phase separation	PLA foams
High-pressure gas foaming	PLGA, P(PF-co-EG) scaffolds
Solid freeform fabrication	Complex 3D PLA, PLGA structures

These processes usually result in decrease in molecular weight and have profound effects on the biocompatibility, mechanical properties, and other characteristics of the formed scaffold. Incorporation of large bioactive molecules such as proteins into the scaffolds and retention of their activity have been a major challenge.

Fiber Bonding

Fibers provide a large surface-area-to-volume ratio and are therefore desirable as scaffold materials. PGA fibers in the form of tassels and felts have been utilized as scaffolds to demonstrate the feasibility of organ regeneration (Cima *et al.*, 1991; Vacanti *et al.*, 1991). However, these fibers lack the structural stability necessary for *in vivo* uses, which has led to the development of a fiber bonding technique (Mikos *et al.*, 1993a). With this method, PGA fibers are aligned in the shape of the desired scaffold and then embedded in a PLA/methylene chloride solution. After evaporation of the solvent, the PLA–PGA composite is heated above the melting temperatures of both polymers. PLA is removed by selective dissolution after cooling, leaving the PGA fibers physically joined at their cross-points without any surface or bulk modifications while maintaining their initial diameter. Stipulations concerning the choice of solvent, immiscibility of the two polymers, and their relative melting temperatures restrict the general application of this technique to other polymers.

An alternative method of fiber bonding has also been developed to prepare tubular scaffolds for the regeneration of intestine, blood vessels, and ureters (Mooney *et al.*, 1996b, 1994a). In this technique, a nonwoven mesh of PGA fibers is attached to a rotating Teflon cylinder. The scaffolds are reinforced by spray casting with solutions of PLA or PLGA, which results in a thin coat that bonds the cross-points of PGA fibers. The behavior of transplanted cells is therefore determined by the PLA or PLGA coating instead of the PGA mesh. The mechanical strength of the scaffold is provided by both

fibers and coating and is designed in such a way to withstand mechanical stresses or compromise degradation of PLA or PLGA. For example, PGA fiber-based matrices alone did not withstand contractile forces exerted by cultured smooth muscle cells, while scaffolds stabilized by spray-coating atomized PLA solution over the sides of the PGA matrices maintained their desired size and shape over 7 weeks in culture (Kim and Mooney, 1998). This method is very useful for fabrication of thin scaffolds; however, it does not allow the creation of complex 3D scaffolds since only a thin layer at the surface may be engineered by coating.

Solvent Casting and Particulate Leaching

In order to overcome some of the drawbacks associated with fiber bonding, a solvent-casting and particulate-leaching (SC/PL) technique has been developed to prepare porous scaffolds with controlled porosity, surface-area-to-volume ratio, pore size, and crystallinity for specific applications (Mikos *et al.*, 1994b). This method can be applied to PLA, PLGA, and any other polymers that are soluble in a solvent such as chloroform or methylene chloride. For example, sieved salt particles are dispersed in a PLA/chloroform solution that is used to cast a membrane onto glass petri dishes. After evaporating the solvent, the PLA/salt composite membranes are heated above the PLA melting temperature and then quenched or annealed by cooling at controlled rates to yield amorphous or semicrystalline foams with regulated crystallinity. The salt particles are eventually leached out by selective dissolution in water to produce a porous polymer matrix.

Highly porous PLA foams with porosities up to 93% and median pore diameters up to 500 μm have been prepared using the above technique (Mikos *et al.*, 1994b; Wake *et al.*, 1994). Porous PLGA foams fabricated by the same method have been shown to support osteoblasts growth both *in vitro* and *in vivo* (Ishaug *et al.*, 1997; Ishaug-Riley *et al.*, 1997, 1998). The porosity and pore size can be controlled independently by varying the amount and size of the salt particles, respectively. The surface-area-to-volume ratio depends on both initial salt weight fraction and particle size. In addition, the crystallinity, which affects both degradation and mechanical strength of the polymer, can be tailored to a particular application. A disadvantage of this method is that it can only be used to produce thin wafers or membranes with uniform pore morphology up to 3 mm thick (Wake *et al.*, 1996). The preparation of thicker membranes may result in the formation of a solid skin layer characteristic of asymmetric membranes. The two controlling phenomena are solvent evaporation of the surface and solvent diffusion in the bulk.

This method has been modified to fabricate tubular scaffolds (Mooney *et al.*, 1995a, 1994b). Porous PLGA membranes prepared using SC/PL are wrapped around Teflon cylinders, and the overlapping ends are fused together with chloroform. The Teflon core is then removed to leave a hollow tube. Because of the relatively brittle nature of the porous membranes used, this method is limited to tubular scaffolds with a low ratio of wall thickness to inner diameter.

To increase the pliability of the porous membranes, PEG has been blended with PLGA in the SC/PC process (Wake *et al.*, 1996). Micropores resulted from dissolution of PEG during leaching are believed to alter the structure of the pore walls and increase the pliability of the scaffold. These membranes can be rolled over into tubular scaffolds with a significantly higher ratio of wall thickness to inner diameter. The membranes fabricated from the polymer blend do not show any macroscopic damage during rolling as is observed for tubes made of PLGA alone.

Highly porous PPF scaffolds have also been formed using the SC/PL technique for both tissue induction and delivery of bioactive factors (Fisher *et al.*, 2003; Hedberg *et al.*, 2002). In this procedure, the PPF is cross-linked around the salt particles in molds of desired size. The samples are then removed from the molds and the salt is leached in water. Mechanical and degradation properties of the resulting scaffolds, with pore sizes ranging from 300 to 800 μm and porosities of 60–70%, have been characterized *in vitro* (Fisher *et al.*, 2003). These scaffolds were also found to induce a mild tissue response when implanted for up to 8 weeks either subcutaneously or in cranial defects in rabbits (Fisher *et al.*, 2002).

Superstructure Engineering

Polymer scaffolds with complex 3D architecture (superstructures) can be formed by superimposing defined structural elements such as pores, fibers, or membranes in order according to stochastic, fractal, or periodic principles (Wintermantel *et al.*, 1996). This approach may provide optimal spatial organization and nutritional conditions for cells. The coherence of structural elements determines the anisotropic structural behavior of the scaffold. The major concern in engineering superstructures is the spatial organization of the elements in order to obtain desired pore sizes and interconnected pore structure.

A simple example of this technique is membrane lamination to construct foams with precise anatomical shapes (Mikos *et al.*, 1993b). A contour plot of the particular 3D shape is first prepared. Highly porous PLA or PLGA membranes with the shapes of the contour are then manufactured using SC/PL. The adjacent membranes are bonded together by coating chloroform on their contacting surfaces. The final scaffold is thus formed by laminating the constituent membranes in the proper order. It has been shown that continuous pore structures are formed with no boundary between adjacent layers. In addition, the bulk properties of the 3D scaffold are identical to those of the individual membranes.

Compression Molding

Compression molding is an alternative technique of constructing 3D scaffolds. In this method, a mixture of fine PLGA powder and gelatin microspheres is loaded in a Teflon mold and then heated above the glass transition temperature of the polymer (Thomson *et al.*, 1995). The PLGA/gelatin composite is subsequently removed from the mold and gelatin microspheres are leached out. In this way, porous PLGA scaffolds with a geometry identical to the shape of the mold can be produced.

Polymer scaffolds of various shapes can be constructed by simply changing the mold geometry. This method also offers the independent control of porosity and pore size by varying the amount and size of porogen used, respectively. In addition, it is possible to incorporate bioactive molecules in either polymer or porogen for controlled delivery, because this process does not utilize organic solvents and is carried out at relatively low temperatures for amorphous PLGA scaffolds. This manufacturing technique may also be applied to PLA or PGA. However, higher temperatures are required (above the polymer melting temperatures) because these polymers are semicrystalline.

Compression molding can be combined with the SC/PL technique to form porous 3D foams. The dried PLGA/salt composites obtained by SC are broken into pieces of less than 5 mm in edge length and compression-molded into a desired 3D shape (Widmer *et al.*, 1998). The resulted composite material can then be cut into desired thickness. Subsequent leaching of the salt leaves an open-cell porous foam, with more uniform pore distribution than those obtained by SC/PL for increased thickness.

Highly porous poly(α-hydroxy ester) scaffolds, though desirable in many tissue engineering applications, may lack required mechanical strength for the replacement of load bearing tissues such as bone. Hydroxyapatite and β-tricalcium phosphate are biocompatible and osteoconductive materials and can be incorporated into these foams to improve their mechanical properties. Because the macroscopic mixing of three solid particulates (polymer powder, porogen, and ceramic) is difficult, a combined SC, compression-molding, and PL technique described earlier has been employed to fabricate an isotropic composite foam scaffold of PLGA reinforced with short hydroxyapatite fibers (15 μm in diameter and 45 μm in length) (Thomson *et al.*, 1998). Within certain range of fiber contents, these scaffolds have superior compressive strength compared to nonreinforced materials of the same porosity.

Extrusion

Various extrusion methods such as ram (piston-cylinder) extrusion, hydrostatic extrusion, or solid-state-extrusion (die drawing) have been applied to increase the orientation of polymer chains and thus produce high-strength, high-modulus materials (Ferguson *et al.*, 1996). More recently, an extrusion process has been successfully combined with the SC/PL technique to manufacture porous tubular scaffolds for guided tissue such as peripheral nerve regeneration (Widmer *et al.*, 1998). First the dry polymer/salt composite wafers obtained from SC are cut into pieces and placed in a customized piston extrusion tool (Fig. 3). The tool is then mounted into a hydraulic press and heated to the desired processing temperature. The temperature is allowed to equilibrate and the polymer/salt composite is then extruded by applying pressure. The extruded tubes are cut to appropriate lengths. Finally, the salt particles are leached out to yield highly porous conduits.

The pressure for extrusion at a constant rate is dependent on the extrusion temperature. High temperature may result in

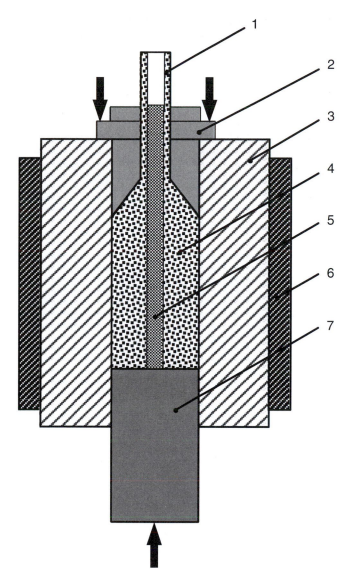

FIG. 3. Piston extrusion tool for the manufacture of tubular polymer/salt composite structures: 1, extruded polymer/salt construct; 2, nozzle defining the outer diameter of the tubular construct; 3, tool body; 4, melted polymer/salt mixture; 5, rod defining the inner diameter of the tubular construct; 6, heat band with temperature control; and 7, piston moving the melted polymer/salt mixture. The arrows indicate the attachment points for the forces involved in the extrusion process.

dissolved in a solvent such as glacial acetic acid or benzene to form a solution of desired concentration. The solution is then frozen and the solvent is removed by lyophilization under high vacuum. Several polymers including PLGA and PLGA/PPF have been prepared into porous foams with this method. The foams have either leaflet or capillary structures depending on the polymer and solvent used in fabrication. These foams are generally not suitable as scaffolds for cell transplantation. Subsequent compression of the foams by grinding and extrusion can generate matrices with varied densities. Foam density has been shown to determine the kinetics of drug release from these matrices.

An emulsion freeze-drying technique has also been developed to fabricate porous scaffolds (Whang *et al.*, 1995). In this technique, water is added to a PLGA/methylene chloride solution and the immiscible phases are homogenized. The created emulsion (water-in-oil) is then poured into a copper mold maintained in liquid nitrogen ($-196°C$). After quenching, the samples are freeze-dried to remove methylene chloride and water. Using this technique, PLGA foams with porosity in the range of 91–95% and median pore diameters of 13–35 μm with larger pores greater than 200 μm have been made by varying processing parameters such as water volume fraction, polymer weight fraction, and polymer molecular weight. Compared to the SC/PL technique, this method produces foams with smaller pore sizes but higher specific pore surface area and can produce thick (>1 cm) foams.

Phase Separation

The ability to deliver bioactive molecules from a degrading polymer scaffold is desirable for tissue regeneration. However, the activity of the molecule is often dramatically decreased because of the harsh chemical or thermal environments used in some polymer processing techniques. Using a novel phase separation technique, scaffolds loaded with small hydrophilic and hydrophobic bioactive molecules have been manufactured (Lo *et al.*, 1995). The polymer is dissolved in a solvent such as molten phenol or naphthalene, followed by dispersion of the bioactive molecule in this homogeneous solution. A liquid–liquid phase separation is induced by lowering the solution temperature. The resulting bicontinuous polymer and solvent phases are then quenched to create a two-phase solid. Subsequent removal of the solidified solvent by sublimation leaves a porous polymer scaffold loaded with bioactive molecules.

The fabricated PLA foams have pore sizes up to 500 μm with relatively uniform distributions. The properties of the foams depends on the polymer type, molecular weight, concentration, and solvent used. It has been shown that proteins such as alkaline phosphatase retain as much as 75% of their activity after scaffold fabrication with the naphthalene system, but the activity is completely lost in the phenol system. Although phenol has a lower melting temperature than naphthalene, it is a more polar solvent and can interact with proteins and weaken the hydrogen bonding within the protein structure, resulting in a loss of protein activity. The phenol system may be useful for the entrapment of small drugs or short peptides instead.

thermal degradation of the polymer. The porosity and pore size of the extruded conduits are determined by salt weight fraction, salt particle size, and processing temperatures. The fabricated conduits have an open-pore structure and are suitable for incorporation of cells or microparticles loaded with tissue inductive factors.

Freeze-Drying

Low-density polymer foams have been produced using a freeze-drying technique (Hsu *et al.*, 1997). Polymer is first

Gas Foaming

In one example of the gas foaming (GF) technique, solid disks of PLGA prepared by either compression molding or solvent casting are exposed to high-pressure CO_2 (5.5 MPa, 25°C) environment to allow saturation of CO_2 in the polymer (Mooney et al., 1996a). A thermodynamic instability is then created by reducing the CO_2 gas pressure to ambient level, which results in nucleation and expansion of dissolved CO_2 pores in the polymer particles. PLGA sponges with a porosity of up to 93% and a pore size of about 100 μm have been fabricated. The porosity and pore structure are dependent on the amount of CO_2 dissolved, the rate and type of gas nucleation, and the rate of gas diffusion to the pore nuclei.

The major advantage of this technique is that it involves no organic solvent or high temperature and therefore is promising for incorporating tissue induction factors in the polymer scaffolds. However, the effects of high pressure on the retention of activity of proteins still need to be assessed. In addition, this process yields mostly nonporous surfaces and a closed pore structure inside the polymer matrix, which is undesirable for cell transplantation. In an improved method, a porogen such as salt particles can be combined with the polymer to form composite disks before gas foaming (Harris et al., 1998). The expansion and fusion of the polymer particles lead to the formation of a continuous matrix with entrapped salt particles, which are subsequently leached out. The GF/PL process produces porous matrices with predominately interconnected macropores (created by leaching out salt) and smaller, closed pores (created by the nucleation and growth of gas pores in the polymer particles). The fabricated matrices have a more uniform pore structure and higher mechanical strength than those obtained with SC/PL.

For injectable scaffolds, a combination of ascorbic acid, ammonium persulfate, and sodium bicarbonate has been used at atmospheric conditions to form highly porous hydrogel materials for bone tissue engineering (Behravesh et al., 2002). In this case, as the hydrogel is cross-linked, carbon dioxide is produced, causing pore formation. The ratio of the three components just listed determines the final porosity (43–84%) and pore size (50–200 μm) of the scaffolds (Behravesh et al., 2002). As mentioned previously, these porous [P(PF-co-EG)] foams supported rat-marrow stromal cell differentiation and bone matrix production during in vitro culture (Behravesh and Mikos, 2003).

Solid Freeform Fabrication

Solid freeform fabrication (SFF) refers to computer-aided design, computer-aided manufacturing (CAD/CAM) methodologies such as stereolithography, selective laser sintering (SLS), ballistic particle manufacturing, and 3D printing (3DP) for the creation of complex shapes directly from CAD models. SFF techniques, although mainly investigated for industrial applications such as rapid prototyping, offer the possibility to fabricate polymer scaffolds with well-defined architecture because local composition, macrostructure, and microstructure can be specified and controlled at high resolution in the interior of the components. These methods build complex 3D objects by material addition and fusion of cross-sectional layers (2D slices decomposed from CAD models). In addition, they allow the formation of multimaterial structures by selective deposition. Prefabricated structures can also be embedded during material buildup. By carefully controlling the processing conditions, cells, bioactive molecules, or synthetic vasculature may be included directly into layers of polymer scaffolds during fabrication.

An example of the use of stereolithography is the development of a diethyl fumarate/PPF resin as a liquid base material for a custom-designed apparatus using a computer-controlled, ultraviolet laser and suitable photoinitiator (Cooke et al., 2002). In this case, the machine builds the desired structure from the bottom toward the top, with the resin allowed to wash over the sample after each layer is formed. This provides new base material to be photo-cross-linked in the desired geometry for the next layer using the computer-driven laser. The spatial resolution of such a system is 100 μm (Cooke et al., 2002).

In the SLS technique, a thin layer of evenly distributed fine powder is first laid down (Bartels et al., 1993; Berry et al., 1997). A computer-controlled scanning laser is then used to sinter the powder within a cross-sectional layer. The energy generated by the laser heats the powder into a glassy state and individual particles fuse into a solid. Once the laser has scanned the entire cross section, another layer of powder is laid on top and the whole process is repeated.

In the 3DP process, each layer is created by adding a layer of polymer powder on top of a piston and cylinder containing a powder bed and the part being fabricated. This layer is then selectively joined where the part is to be formed by ink-jet printing of a binder material such as an organic solvent. The printed droplet has a diameter of 50–80 μm. The printhead position and speed are controlled by computer. The piston, powder bed, and part are lowered and a new layer of polymer powder is laid on top of the already processed layer and selectively joined. The layered printing process is repeated until the desired part is completed.

The local microstructure within the component can be controlled by varying the printing conditions. The resolution of features currently attainable by 3DP for degradable polyesters is about 200 μm (Griffith et al., 1997). Using this technique, scaffolds with complex structures may be fabricated (Giordano et al., 1996). A model drug (dye) with a concentration profile specified by a CAD model has been successfully incorporated into a scaffold during the 3DP process, demonstrating the feasibility of producing complex release regimes using a single drug-delivery device (Wu et al., 1996). By mixing salt particles in the polymer powder and their subsequent leaching after 3DP process, porous PLGA scaffolds with an intrinsic network of interconnected branching channels have been fabricated for cell transplantation (Kim et al., 1998). This network of channels and micropores could provide a structural template to guide cellular organization, enhance neovascularization, and increase the capacity for mass transport. Furthermore, multiple printheads containing different binder materials can be used to modify local surface chemistry and structure. Patterned PLA substrates with selective cell-adhesion domains have been fabricated by 3DP (Park et al., 1998).

CHARACTERIZATION OF PROCESSED SCAFFOLDS

Various techniques are available to characterize the fabricated polymer scaffolds (Table 4). The molecular weight and polydispersity index of the polymer can be measured by gel permeation chromatography (GPC). Information on chemical composition and structure can be obtained by nuclear magnetic resonance (NMR) spectroscopy, X-ray diffraction, Fourier transform infrared (FTIR), and FT-Raman (FTR) spectroscopy. The thermal properties of the polymer such as glass transition temperature (T_g), melting temperature (T_m), and crystallinity (X_c) can be determined by differential scanning calorimeter (DSC). Porosity and pore size distribution of a porous scaffold are measured by mercury intrusion porosimetry. Scanning electron microscopy (SEM) is the most common method to view the pore structure and morphology. The 3D microstructure of porous PLGA matrices has been analyzed by confocal microscopy (Tjia and Moghe, 1998). Mechanical properties of the scaffolds such as tensile strength and modulus, compression strength and modulus, compliance/hardness, flexibility, elasticity, and stress and stain at yield can be measured using mechanical testing equipment. Some tests require the processing of scaffolds into a particular shape and dimensions specified by ASTM standards.

The *in vitro* degradation properties can be evaluated by placing the bioresorbable scaffolds in simulated body fluid, typically pH 7.4 phosphate-buffered saline (PBS). Changes in sample weight, molecular weight, morphology, and thermal and mechanical properties can then be measured at various time points until degradation process is completed. In addition, characterization of the chemical makeup of the degradation products may be possible through the use of GPC or high-performance liquid chromatography (HPLC) (Timmer *et al.*, 2002). However, such studies do not allow for the continuous

observation of changes within the scaffolds. An *in vivo* study is often necessary to predict the degradation behavior of the scaffolds for cell transplantation (Shin *et al.*, 2003b).

Material surfaces, which are usually different from the bulk, play a crucial role in regulating cell response. Electron spectroscopy for chemical analysis (ESCA) and static secondary ion mass spectrometry (SIMS) are the most powerful tools for analyzing surface chemistry and composition. Information on the orientation of chemical groups can be obtained by polarized IR and near-edge X-ray absorption fine structure (NEXAFS). Surface morphology can be characterized by SEM, scanning probe microscopy (SPM), and atomic force microscopy (AFM). Surface wettability and energy are assessed by contact-angle measurements.

CELL SEEDING AND CULTURE IN 3D SCAFFOLDS

The major obstacles to the *in vitro* development of 3D cell–polymer constructs for the regeneration of large organs or defects have been obtaining uniform cell seeding at high densities and maintaining nutrient transport to the cells inside the scaffolds. To achieve desired spatial and temporal distribution of cells and molecular cues affecting cellular function, cell culture conditions should provide control over hydrodynamic and biochemical factors in the cell environment.

Static Culture

The conventional static cell seeding technique involves the placement of the scaffold in a cell suspension to allow the absorption of cells. However, the resulting cell distribution in the scaffold is often not uniform, with the majority of the cells attached only to the outer surfaces (Wald *et al.*, 1993). Wetting hydrophobic polymer scaffolds with ethanol and water prior to cell seeding allows for displacement of air-filled pores with water and thus facilitates penetration of cell suspension into these pores (Mikos *et al.*, 1994a). Infiltration with hydrophilic polymers or surface hydrolysis of scaffolds has also been shown to increase the cell seeding density (Gao *et al.*, 1998; Mooney *et al.*, 1995b). Seeding cells by injection or applying vacuum to ensure penetration of the cell suspension through the 3D matrix could result in uniform cell seeding initially. However, the uniformity is lost under static culture conditions because of the nutrient and oxygen diffusion limitation within the scaffold.

Several dynamic cell seeding and culture techniques have been developed to ensure uniform cell distribution, which will lead to uniform tissue regeneration (Fig. 4). Compared to static culture conditions, mass transfer rates can be maintained at higher levels and cell growth is not restricted by the rate of nutrient supply under well-mixed culture conditions. These methods can be scaled up and are suitable for cell cultivation using multiple scaffolds.

Spinner Flask Culture

In a spinner flask, 3D polymer scaffolds are first fixed to needles attached to the cap of the flask, and then exposed to

TABLE 4 Characterization of Bioresorbable Polymer Scaffolds

Properties	Techniques
Bulk properties	
Molecular weight	GPC
Polydispersity index	GPC
Chemical composition, structure	NMR, X-ray diffraction, FTIR, FTR
Thermal properties (T_g, T_m, X_c, etc.)	DSC
Porosity, pore size	Mercury intrusion porosimetry
Morphology	SEM, confocal microscopy
Mechanical properties	Mechanical testing
Degradative properties	*In vitro, in vivo*
Surface properties	
Surface chemistry	ESCA, SIMS
Distribution of chemistry	Imaging methods (e.g., SIMS)
Orientation of groups	Polarized IR, NEXAFS
Texture	SEM, AFM, STM
Surface energy and wettability	Contact-angle measurement

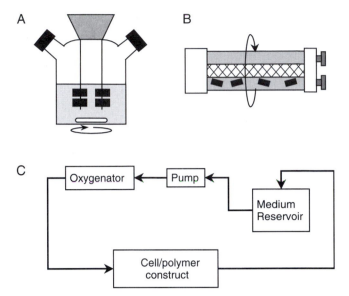

FIG. 4.   Dynamic cell seeding and culture techniques in 3D scaffolds. (A) Spinner flask. (B) Rotary vessel. (C) Perfusion system.

a uniform, well-mixed cell suspension (Fig. 4A) (Freed *et al.*, 1993). Using this method, porous PGA scaffolds have been uniformly seeded with chondrocytes at high yield and high kinetic rate (to minimize the time that cells stay in the suspension) (Vunjak-Novakovic *et al.*, 1998). Mixing has been found to promote the formation of cell aggregates with sizes of 20–32 μm. The spin rate, however, needs to be well adjusted to minimize cell damage under high shear stress. The spinner flask is also suitable for suspension culture of hepatocyte spheroids that exhibit enhanced liver function compared to monolayer culture in the long-term (Kamihira *et al.*, 1997).

Rotary Vessel Culture

The rotating-wall vessel (RWV) also allows enhanced mass transport and is useful for 3D cell culture (Fig. 4B). The polymer scaffolds are loaded into the vessel and a uniform cell suspension is added. Vessel rotation is initiated to allow dynamic cell seeding and increased to maintain a high rate of nutrient and oxygen diffusion. Alternately, the scaffolds can be preseeded with cells under static conditions before loading (Goldstein *et al.*, 1999). Several configurations of RWV have been used in microgravity tissue engineering (Freed and Vunjak-Novakovic, 1997).

Perfusion Culture

A flow perfusion culture system has been used for *in vitro* regeneration of large 3D tissues and organs (Fig. 4C) (Bancroft *et al.*, 2002; Glowacki *et al.*, 1998; Griffith *et al.*, 1997; Kim *et al.*, 1998). The cell–polymer constructs are maintained in a continuous-flow condition. The culture medium is pumped from a reservoir through an oxygenator and the cell–polymer constructs, and recirculated back to the reservoir. The flow rate

for cell survival is adjusted based on cell mass. The entire perfusion unit is maintained in normal sterile culture conditions. Compared to static culture, medium perfusion has been shown to significantly enhance cell viability and matrix production (Glowacki *et al.*, 1998). Additionally, the medium flow rate was found to influence ECM deposition and the timing of osteogenic differentiation when marrow stromal cells were cultured on three-dimensional scaffolds in a perfusion bioreactor (Bancroft *et al.*, 2002). These systems are useful for the development of complex tissue structures as well as the study of the effects of mechanical stimulation on cell viability, differentiation, and ECM production.

Other Culture Conditions

Ideally the culture conditions should provide all necessary signals that the cells normally experience *in vivo* for optimal tissue regeneration. For instance, mechanical stimulation plays an important role in the differentiation of mesenchymal tissues (Chiquet *et al.*, 1996; Goodman and Aspenberg, 1993). Application of well-controlled loads may stimulate bone growth into porous scaffolds. The degradation of the scaffolds can be affected by applied strain (Miller and Williams, 1984). Transwell culture systems that allow the use of different culture media for apical or basal sides are often employed to induce and maintain the polarity of epithelial cells. The growth and function of some retinal cells may be regulated by the light–dark cycle. In some cases, a gradient substrate with spatially controlled wettability or other properties may be desired (Ruardy *et al.*, 1995). Some cellular chemotactic responses may require the creation of concentration gradients of growth factors. Temporal presentation of signals is also important. For example, each phase of the differentiation of osteoblasts (proliferation, maturation of ECM, and mineralization) requires different signals (Lian and Stein, 1992; Peter *et al.*, 1998a). Coculture of several types may be preferred for *in vitro* organogenesis including angiogenesis.

CONCLUSIONS

Significant progress has been made to optimize the engineering of tissue and organ analogs. However, many challenges remain in the engineering of 3D tissues and organs for clinical use. Nevertheless, many advances have been made in synthetic polymer chemistry, scaffold processing methods, and tissue-culture techniques. These may eventually allow the generation of long-term functional complex cell–polymer constructs with precisely controlled local environment such as material microstructure, nutrient and growth factor concentration, and mechanical forces.

Bibliography

Aulthouse, A. L., Beck, M., Griffey, E., Sanford, J., Arden, K., Machado, M. A., and Horton, W. A. (1989). Expression of the human chondrocyte phenotype in vitro. *In Vitro Cell Dev. Biol.* **25**: 659–668.

Babensee, J. E., Anderson, J. M., McIntire, L. V., and Mikos, A. G. (1998). Host response to tissue engineered devices. *Adv. Drug Deliv. Rev.* **33**: 111–139.

Babensee, J. E., McIntire, L. A., and Mikos, A. G. (2000). Growth factor delivery for tissue engineering. *Pharm. Res.* **17**: 497–504.

Bancroft, G. N., and Mikos, A. G. (2001). Bone tissue engineering by cell transplantation. in *Tissue Engineering for Therapeutic Use*, Y. Ikada and N. Ohshima (eds.). Vol. 5. Elsevier Science, New York, pp. 151–163.

Bancroft, G. N., Sikavitsas, V. I., van den Dolder, J., Sheffield, T. L., Ambrose, C. G., Jansen, J. A., and Mikos, A. G. (2002). Fluid flow increases mineralized matrix deposition in 3D perfusion culture of marrow stromal osteoblasts in a dose-dependent manner. *Proc. Natl. Acad. Sci. USA* **99**: 12600–12605.

Barrera, D., Zylstra, E., Lansbury, P., and Langer, R. (1995). Copolymerization and degradation of poly(lactic-*co*-lysine). *Macromolecules* **28**: 425–432.

Bartels, K. A., Bovik, A. C., Crawford, R. C., Diller, K. R., and Aggarwal, S. J. (1993). Selective laser sintering for the creation of solid models from 3D microscopic images. *Biomed. Sci. Instrum.* **29**: 243–250.

Behravesh, E., and Mikos, A. G. (2003). Three-dimensional culture of marrow stromal osteoblasts in biomimetic poly(propylene fumarate-*co*-ethylene glycol)-based macroporous hydrogels. *J. Biomed. Mater. Res.* **66A**: 698–706.

Behravesh, E., Yasko, A. W., Engel, P. S., and Mikos, A. G. (1999). Synthetic biodegradable polymers for orthopaedic applications. *Clin. Orthop. Rel. Res.* **367**: S118–S129.

Behravesh, E., Jo, S., Zygourakis, K., and Mikos, A. G. (2002). Synthesis of in-situ cross-linkable macroporous biodegradable poly(propylene fumarate-*co*-ethylene glycol) hydrogels. *Biomacromolecules* **3**: 374–381.

Berry, E., Brown, J. M., Connell, M., Craven, C. M., Efford, N. D., Radjenovic, A., and Smith, M. A. (1997). Preliminary experience with medical applications of rapid prototyping by selective laser sintering. *Med. Eng. Phys.* **19**: 90–96.

Blum, J. S., Barry, M. A., and Mikos, A. G. (2003). Bone regeneration through transplantation of genetically modified cells. *Clin. Plast. Surg.* **30**: 611–620.

Boyan, B. D., Hummert, T. W., Dean, D. D., and Schwartz, Z. (1996). Role of material surfaces in regulating bone and cartilage cell response. *Biomaterials* **17**: 137–146.

Chen, C. S., Mrksich, M., Huang, S., Whitesides, G. M., and Ingber, D. E. (1998). Micropatterned surfaces for control of cell shape, position, and function. *Biotechnol. Prog.* **14**: 356–363.

Chicurel, M. E., Chen, C. S., and Ingber, D. E. (1998). Cellular control lies in the balance of forces. *Curr. Opin. Cell Biol.* **10**: 232–239.

Chiquet, M., Matthisson, M., Koch, M., Tannheimer, M., and Chiquet-Ehrismann, R. (1996). Regulation of extracellular matrix synthesis by mechanical stress. *Biochem. Cell Biol.* **74**: 737–744.

Cima, L. G., Vacanti, J. P., Vacanti, C., Ingber, D., Mooney, D., and Langer, R. (1991). Tissue engineering by cell transplantation using degradable polymer substrates. *J. Biomech. Eng.* **113**: 143–151.

Cook, A. D., Hrkach, J. S., Gao, N. N., Johnson, I. M., Pajvani, U. B., Cannizzaro, S. M., and Langer, R. (1997). Characterization and development of RGD-peptide-modified poly(lactic acid-*co*-lysine) as an interactive, resorbable biomaterial. *J. Biomed. Mater. Res.* **35**: 513–523.

Cooke, M. N., Fisher, J. P., Dean, D., Rimnac, C., and Mikos, A. G. (2002). Use of stereolithography to manufacture critical-sized 3D biodegradable scaffolds for bone ingrowth. *J. Biomed. Mater. Res. Part B: Appl. Biomater.* **64B**: 65–69.

Ferguson, S., Wahl, D., and Gogolewski, S. (1996). Enhancement of the mechanical properties of polylactides by solid-state extrusion.

II. Poly(L-lactide), poly(L/D-lactide), and poly(L/DL-lactide). *J. Biomed. Mater. Res.* **30**: 543–551.

Fisher, J. P., Vehof, J. W. M., Dean, D., van der Waerden, J. P. C. M., Holland, T. A., Mikos, A. G., and Jansen, J. A. (2002). Soft and hard tissue response to photocrosslinked poly(propylene fumarate) scaffolds in a rabbit model. *J. Biomed. Mater. Res.* **59**: 547–556.

Fisher, J. P., Holland, T. A., Dean, D., and Mikos, A. G. (2003). Photoinitiated cross-linking of the biodegradable polyester poly(propylene fumarate). Part II. *In vitro* degradation. *Biomacromolecules* **4**: 1335–1342.

Freed, L. E., and Vunjak-Novakovic, G. (1997). Microgravity tissue engineering. *In Vitro Cell. Dev. Biol.-Animal* **33**: 381–385.

Freed, L. E., Vunjak-Novakovic, G., and Langer, R. (1993). Cultivation of cell–polymer cartilage implants in bioreactors. *J. Cell. Biochem.* **51**: 257–264.

Gao, J., Niklason, L., and Langer, R. (1998). Surface hydrolysis of poly(glycolic acid) meshes increases the seeding density of vascular smooth muscle cells. *J. Biomed. Mater. Res.* **42**: 417–424.

Giordano, R. A., Wu, B. M., Borland, S. W., Cima, L. G., Sachs, E. M., and Cima, M. J. (1996). Mechanical properties of dense polylactic acid structures fabricated by three dimensional printing. *J. Biomater. Sci. Polymer Ed.* **8**: 63–75.

Glowacki, J., Mizuno, S., and Greenberger, J. S. (1998). Perfusion enhances functions of bone marrow stromal cells in three-dimensional culture. *Cell Transplant.* **7**: 319–326.

Goldstein, A. S., Zhu, G., Morris, G. E., Meszlenyi, R. K., and Mikos, A. G. (1999). Effect of osteoblastic culture conditions on the structure of poly(DL-lactic-*co*-glycolic acid) foam scaffolds. *Tissue Eng.* **5**: 421–433.

Goodman, S., and Aspenberg, P. (1993). Effects of mechanical stimulation on the differentiation of hard tissues. *Biomaterials* **14**: 563–569.

Griffith, L. G., Wu, B., Cima, M. J., Powers, M. J., Chaignaud, B., and Vacanti, J. P. (1997). *In vitro* organogenesis of liver tissue. *Ann. N. Y. Acad. Sci.* **831**: 382–397.

Harris, L. D., Kim, B.-S., and Mooney, D. J. (1998). Open pore biodegradable matrices formed with gas foaming. *J. Biomed. Mater. Res.* **42**: 396–402.

Hay, E. D. (1993). Extracellular matrix alters epithelial differentiation. *Curr. Opin. Cell Biol.* **5**: 1029–1035.

Hedberg, E. L., Tang, A., Crowther, R. S., Careny, D. H., and Mikos, A. G. (2002). Controlled release of an osteogenic peptide from injectable biodegradable polymeric composites. *J. Controlled Release* **84**: 137–150.

Holland, T. A., and Mikos, A. G. (2003). Review: advances in drug delivery for articular cartilage. *J. Controlled Release* **86**: 1–14.

Holland, T. A., Tabata, Y., and Mikos, A. G. (2003). *In vitro* release of transforming growth factor-β1 from gelatin microparticles encapsulated in biodegradable, injectable oligo(poly(ethylene glycol) fumarate) hydrogels. *J. Controlled Release* **91**: 299–313.

Howe, A., Aplin, A. E., Alahari, S. K., and Juliano, R. L. (1998). Integrin signaling and cell growth control. *Curr. Opin. Cell Biol.* **10**: 220–231.

Hsu, Y. Y., Gresser, J. D., Trantolo, D. J., Lyons, C. M., Gangadharam, P. R. J., and Wise, D. L. (1997). Effect of polymer foam morphology and density on kinetics of *in vitro* controlled release of isoniazid from compressed foam matrices. *J. Biomed. Mater. Res.* **35**: 107–116.

Ishaug, S. L., Crane, G. M., Miller, M. J., Yasko, A. W., Yaszemski, M. J., and Mikos, A. G. (1997). Bone formation by three-dimensional stromal osteoblast culture in biodegradable polymer scaffolds. *J. Biomed. Mater. Res.* **36**: 17–28.

Ishaug-Riley, S. L., Crane, G. M., Gurlek, A., Miller, M. J., Yasko, A. W., Yaszemski, M. J., and Mikos, A. G. (1997). Ectopic bone formation by marrow stromal osteoblast transplantation using poly(DL-lactic-*co*-glycolic acid) foams implanted into the rat mesentery. *J. Biomed. Mater. Res.* **36**: 1–8.

Ishaug-Riley, S. L., Crane-Kruger, G. M., Yaszemski, M. J., and Mikos, A. G. (1998). Three-dimensional culture of rat calvarial osteoblasts in porous biodegradable polymers. *Biomaterials* **19**: 1405–1412.

James, K., and Kohn, J. (1996). New biomaterials for tissue engineering. *Mater. Res. Soc. Bull.* **21**: 22–26.

Jo, S., Shin, H., Shung, A. K., Fisher, J. P., and Mikos, A. G. (2001). Synthesis and characterization of oligo(poly(ethylene glycol) fumarate) macromer. *Macromolecules* **34**: 2839–2844.

Kamihira, M., Yamada, K., Hamamoto, R., and Iijima, S. (1997). Spheroid formation of hepatocytes using synthetic polymer. *Ann. N. Y. Acad. Sci.* **831**: 398–407.

Kasper, F. K., and Mikos, A. G. (2003). Biomaterials and gene therapy. in *Molecular and Cellular Foundations of Biomaterials*, N. Peppas and M. V. Sefton (eds.). Academic Press, New York, pp. 131–163.

Kim, B.-S., and Mooney, D. J. (1998). Engineering smooth muscle tissue with a predefined structure. *J. Biomed. Mater. Res.* **41**: 322–332.

Kim, S. S., Utsunomiya, H., Koski, J. A., Wu, B. M., Cima, M. J., Sohn, J., Mukai, K., Griffith, L. G., and Vacanti, J. P. (1998). Survival and function of hepatocytes on a novel three-dimensional synthetic biodegradable polymer scaffold with an intrinsic network of channels. *Ann. Surg.* **228**: 8–13.

Langer, R., and Vacanti, J. P. (1993). Tissue engineering. *Science* **260**: 920–926.

Lian, J. B., and Stein, G. S. (1992). Concepts of osteoblast growth and differentiation: basis for modulation of bone cell development and tissue formation. *Crit. Rev. Oral. Biol. Med.* **3**: 269–305.

Lo, H., Ponticiello, M. S., and Leong, K. W. (1995). Fabrication of controlled release biodegradable foams by phase separation. *Tissue Eng.* **1**: 15–28.

Lu, L., Carcia, C. A., and Mikos, A. G. (1998). Retinal pigment epithelium cell culture on thin biodegradable poly(DL-lactic-*co*-glycolic acid) films. *J. Biomater. Sci. Polymer Ed.* **9**: 1187–1205.

Lu, L., Stamatas, G. N., and Mikos, A. G. (2000). Controlled release of transforming growth factor-β1 from biodegradable polymers. *J. Biomed. Mater. Res.* **50**: 440–451.

Mikos, A. G., Bao, Y., Cima, L. G., Ingber, D. E., Vacanti, J. P., and Langer, R. (1993a). Preparation of poly(glycolic acid) bonded fiber structures for cell attachment and transplantation. *J. Biomed. Mater. Res.* **27**: 183–189.

Mikos, A. G., Sarakinos, G., Leite, S. M., Vacanti, J. P., and Langer, R. (1993b). Laminated three-dimensional biodegradable foams for use in tissue engineering. *Biomaterials* **14**: 323–330.

Mikos, A. G., Sarakinos, G., Lyman, M. D., Ingber, D. E., Vacanti, J. P., and Langer, R. (1993c). Prevascularization of porous biodegradable polymers. *Biotechnol. Bioeng.* **42**: 716–723.

Mikos, A. G., Lyman, M. D., Freed, L. E., and Langer, R. (1994a). Wetting of poly(L-lactic acid) and poly(DL-lactic-*co*-glycolic acid) foams for tissue culture. *Biomaterials* **15**: 55–58.

Mikos, A. G., Thorsen, A. J., Czerwonka, L. A., Bao, Y., Langer, R., Winslow, D. N., and Vacanti, J. P. (1994b). Preparation and characterization of poly(L-lactic acid) foams. *Polymer* **35**: 1068–1077.

Miller, N. D., and Williams, D. F. (1984). The *in vivo* and *in vitro* degradation of poly(glycolic acid) suture material as a function of applied strain. *Biomaterials* **5**: 365–368.

Moghe, P. V. (1996). Soft-tissue analogue design and tissue engineering of liver. *Mater. Res. Soc. Bull.* **21**: 52–54.

Mooney, D., Hansen, L., Vacanti, J., Langer, R., Farmer, S., and Ingber, D. (1992). Switching from differentiation to growth in hepatocytes: control by extracellular matrix. *J. Cell. Physiol.* **151**: 497–505.

Mooney, D. J., Mazzoni, C. L., Organ, G. M., Puelacher, W. C., Vacanti, J. P., and Langer, R. (1994a). Stabilizing fiber-based cell delivery devices by physically bonding adjacent fibers. in *Biomaterials for Drug and Cell Delivery*, A. G. Mikos, R. M. Murphy, H. Bernstein, and N. A. Peppas (eds.). Vol. 331. Material Research Society, Pittsburgh, pp. 47–52.

Mooney, D. J., Organ, G., Vacanti, J. P., and Langer, R. (1994b). Design and fabrication of biodegradable polymer devices to engineer tubular tissues. *Cell Transplant.* **3**: 203–210.

Mooney, D. J., Breuer, C., McNamara, K., Vacanti, J. P., and Langer, R. (1995a). Fabricating tubular devices from polymers of lactic and glycolic acid for tissue engineering. *Tissue Eng.* **1**: 107–118.

Mooney, D. J., Park, S., Kaufmann, P. M., Sano, K., McNamara, K., Vacanti, J. P., and Langer, R. (1995b). Biodegradable sponges for hepatocyte transplantation. *J. Biomed. Mater. Res.* **29**: 959–965.

Mooney, D. J., Baldwin, D. F., Suh, N. P., Vacanti, J. P., and Langer, R. (1996a). Novel approach to fabricate porous sponges of poly(D,L-lactic-*co*-glycolic acid) without the use of organic solvents. *Biomaterials* **17**: 1417–1422.

Mooney, D. J., Mazzoni, C. L., Breuer, C., McNamara, K., Hern, D., Vacanti, J. P., and Langer, R. (1996b). Stabilized polyglycolic acid fibre-based tubes for tissue engineering. *Biomaterials* **17**: 115–124.

Park, A., and Cima, L. G. (1996). *In vitro* cell response to differences in poly-L-lactide crystallinity. *J. Biomed. Mater. Res.* **31**: 117–130.

Park, A., Wu, B., and Griffith, L. G. (1998). Integration of surface modification and 3D fabrication techniques to prepare patterned poly(L-lactide) substrates allowing regionally selective cell adhesion. *J. Biomater. Sci. Polymer Ed.* **9**: 89–110.

Payne, R. G., McGonigle, J. S., Yaszemski, M. J., Yasko, A. W., and Mikos, A. G. (2002a). Development of an injectable, *in situ* crosslinkable, degradable polymeric carrier for osteogenic cell populations. Part 2. Viability of encapsulated marrow stromal osteoblasts cultured on crosslinking poly(propylene fumarate). *Biomaterials* **23**: 4373–4380.

Payne, R. G., McGonigle, J. S., Yaszemski, M. J., Yasko, A. W., and Mikos, A. G. (2002b). Development of an injectable, *in situ* crosslinkable, degradable polymeric carrier for osteogenic cell populations. Part 3. Proliferation and differentiation of encapsulated marrow stromal osteoblasts cultured on crosslinking poly(propylene fumarate). *Biomaterials* **23**: 4381–4387.

Peter, S. J., Miller, M. J., Yaszemski, M. J., and Mikos, A. G. (1997a). Poly(propylene fumarate). in *Handbook of Biodegradable Polymers*, A. J. Domb, J. Kost and D. M. Wiseman (eds.). Harwood Academic Publishers, Amsterdam, pp. 87–97.

Peter, S. J., Nolley, J. A., Widmer, M. S., Merwin, J. E., Yaszemski, M. J., Yasko, A. W., Engel, P. S., and Mikos, A. G. (1997b). *In vitro* degradation of a poly(propylene fumarate)/β-tricalcium phosphate composite orthopaedic scaffold. *Tissue Eng.* **3**: 207–215.

Peter, S. J., Liang, C. R., Kim, D. J., Widmer, M. S., and Mikos, A. G. (1998a). Osteoblastic phenotype of rat marrow stromal cells cultured in the presence of dexamethasone, β-glycerolphosphate, and L-ascorbic acid. *J. Cell. Biochem.* **71**: 55–62.

Peter, S. J., Miller, S. T., Zhu, G., Yasko, A. W., and Mikos, A. G. (1998b). *In vivo* degradation of a poly(propylene fumarate)/β-tricalcium phosphate injectable composite scaffold. *J. Biomed. Mater. Res.* **41**: 1–7.

Peter, S. J., Kim, P., Yasko, A. W., Yaszemski, M. J., and Mikos, A. G. (1999). Crosslinking characteristics of an injectable poly(propylene fumarate)/β-tricalcium phosphate paste and mechanical properties of the cross-linked composite for use as a biodegradable bone cement. *J. Biomed. Mater. Res.* **44**: 314–321.

Renier, M. L., and Kohn, D. H. (1997). Development and characterization of a biodegradable polyphosphate. *J. Biomed. Mater. Res.* **34**: 95–104.

Ripamonti, U., and Duneas, N. (1996). Tissue engineering of bone by osteoinductive biomaterials. *Mater. Res. Soc. Bull.* **21:** 36–39.

Ruardy, T. G., Schakenraad, J. M., van der Mei, H. C., and Busscher, H. J. (1995). Adhesion and spreading of human skin fibroblasts on physicochemically characterized gradient surfaces. *J. Biomed. Mater. Res.* **29:** 1415–1423.

Shakesheff, K. M., Cannizzaro, S. M., and Langer, R. (1998). Creating biomimetic microenvironments with synthetic polymer-peptide hybrid molecules. *J. Biomater. Sci. Polymer Ed.* **9:** 507–518.

Shin, H., Jo, S., and Mikos, A. G. (2002). Modulation of marrow stromal osteoblast adhesion on biomimetic oligo(poly(ethylene glycol) fumarate) hydrogels modified with Arg-Gly-Asp peptides and a poly(ethylene glycol) spacer. *J. Biomed. Mater. Res.* **61:** 169–179.

Shin, H., Jo, S., and Mikos, A. G. (2003a). Review: biomimetic materials for tissue engineering. *Biomaterials* **24:** 4353–4364.

Shin, H., Ruhe, P. Q., and Mikos, A. G. (2003b). *In vivo* bone and soft tissue response to injectable, biodegradable oligo(poly(ethylene glycol) fumarate) hydrogels. *Biomaterials* **24:** 3201–3211.

Singhvi, R., Kumar, A., Lopez, G., Stephanopoulos, G. N., Wang, D. I. C., Whitesides, G. M., and Ingber, D. E. (1994). Engineering cell shape and function. *Science* **264:** 696–698.

Suggs, L. J., and Mikos, A. G. (1996). Synthetic biodegradable polymers for medical applications. in *Physical Properties of Polymers Handbook*, J. E. Mark (ed.). American Institute of Physics, Woodbury, NY, pp. 615–624.

Suggs, L. J., Payne, R. G., Yaszemski, M. J., Alemany, L. B., and Mikos, A. G. (1997). Synthesis and characterization of a block copolymer consisting of poly(propylene fumarate) and poly(ethylene glycol). *Macromolecules* **30:** 4318–4323.

Suggs, L. J., Kao, E. Y., Palombo, L. L., Krishnan, R. S., Widmer, M. S., and Mikos, A. G. (1998a). Preparation and characterization of poly(propylene fumarate-co-ethylene glycol) hydrogels. *J. Biomater. Sci. Polymer Ed.* **9:** 653–666.

Suggs, L. J., Krishnan, R. S., Garcia, C. A., Peter, S. J., Anderson, J. M., and Mikos, A. G. (1998b). *In vitro* and *in vivo* degradation of poly(propylene fumarate-co-ethylene glycol) hydrogels. **42:** 312–320.

Suggs, L. J., Shive, M. S., Garcia, C. A., Anderson, J. M., and Mikos, A. G. (1999). *In vitro* cytotoxicity and *in vivo* biocompatibility of poly(propylene fumarate-co-ethylene glycol) hydrogels. *J. Biomed. Mater. Res.* **46:** 22–32.

Temenoff, J. S., and Mikos, A. G. (2000). Formation of highly porous biodegradable scaffolds for tissue engineering. *Electr. J. Biotechnol.* **3:** http://www.ejb.org/content/vol3/issue2/full/5/index.html.

Temenoff, J. S., Steinbis, E. S., and Mikos, A. G. (2003). Effect of drying history on swelling properties and cell attachment to oligo(poly(ethylene glycol) fumarate) hydrogels for guided tissue regeneration applications. *J. Biomater. Sci. Polymer Ed.* **14:** 989–1004.

Thomson, R. C., Yaszemski, M. J., Powers, J. M., and Mikos, A. G. (1995). Fabrication of biodegradable polymer scaffolds to engineer trabecular bone. *J. Biomater. Sci. Polymer Ed.* **7:** 23–28.

Thomson, R. C., Yaszemski, M. J., Powers, J. M., and Mikos, A. G. (1998). Hydroxyapatite fiber reinforced poly(α-hydroxy ester) foams for bone regeneration. *Biomaterials* **19:** 1935–1943.

Thomson, R. C., Mikos, A. G., Beahm, E., Lemon, J. C., Satterfield, W. C., Aufdemorte, T. B., and Miller, M. J. (1999). Guided tissue fabrication from periosteum using preformed biodegradable polymer scaffolds. *Biomaterials* **20:** 2007–2018.

Timmer, M. D., Jo, S., Wang, C., Ambrose, C. G., and Mikos, A. G. (2002). Characterization of the cross-linked structure of fumarate-based degradable polymer networks. *Macromolecules* **35:** 4373–4379.

Timmer, M. D., Ambrose, C. G., and Mikos, A. G. (2003). Evaluation of thermal- and photo-crosslinked biodegradable poly(propylene fumarate)-based networks. *J. Biomed. Mater. Res.* **66A:** 811–818.

Tjia, J. S., and Moghe, P. V. (1998). Analysis of 3-D microstructure of porous poly(lactide–glycolide) matrices using confocal microscopy. *J. Biomed. Mater. Res.* **43:** 291–299.

Vacanti, C. A., Langer, R., Schloo, B., and Vacanti, J. P. (1991). Synthetic polymers seeded with chondrocytes provide a template for new cartilage formation. *Plast. Reconstr. Surg.* **88:** 753–759.

Vunjak-Novakovic, G., Obradovic, B., Martin, I., Bursac, P. M., Langer, R., and Freed, L. E. (1998). Dynamic cell seeding of polymer scaffolds for cartilage tissue engineering. *Biotechnol. Prog.* **14:** 193–202.

Wake, M. C., Patrick, C. W., Jr., and Mikos, A. G. (1994). Pore morphology effects on the fibrovascular tissue growth in porous polymer substrates. *Cell Transplant.* **3:** 339–343.

Wake, M. C., Gupta, P. K., and Mikos, A. G. (1996). Fabrication of pliable biodegradable polymer foams to engineer soft tissues. *Cell Transplant.* **5:** 465–473.

Wake, M. C., Gerecht, P. D., Lu, L., and Mikos, A. G. (1998). Effects of biodegradable polymer particles on rat marrow derived stromal osteoblasts *in vitro*. *Biomaterials* **19:** 1255–1268.

Wald, H. L., Sarakinos, G., Lyman, M. D., Mikos, A. G., Vacanti, J. P., and Langer, R. (1993). Cell seeding in porous transplantation devices. *Biomaterials* **14:** 270–278.

Whang, K., Thomas, C. H., Healy, K. E., and Nuber, G. (1995). A novel method to fabricate bioabsorbable scaffolds. *Polymer* **36:** 837–842.

Widmer, M. S., Gupta, P. K., Lu, L., Meszlenyi, R. K., Evans, G. R. D., Brandt, K., Savel, T., Gurlek, A., Patrick, C. W., Jr., and Mikos, A. G. (1998). Manufacture of porous biodegradable polymer conduits by an extrusion process for guided tissue regeneration. *Biomaterials* **19:** 1945–1955.

Wintermantel, E., Mayer, J., Blum, J., Eckert, K.-L., Luscher, P., and Mathey, M. (1996). Tissue engineering scaffolds using superstructures. *Biomaterials* **17:** 83–91.

Wu, B. M., Borland, S. W., Giordano, R. A., Cima, L. G., Sachs, E. M., and Cima, M. J. (1996). Solid free-from fabrication of drug delivery devices. *J. Controlled Release* **40:** 77–87.

III

Practical Aspects
of Biomaterials

Implants, Devices, and Biomaterials: Issues Unique to this Field

James M. Anderson, Stanley A. Brown, Allan S. Hoffman, John B. Kowalski,
Katharine Merritt, Robert F. Morrissey, Buddy D. Ratner,
and Frederick J. Schoen

9.1 INTRODUCTION

Frederick J. Schoen

Chapter 9 addresses some special concerns in the use of surgical implants and medical devices. The themes off this chapter are that implant retrieval and analysis contribute greatly to understanding and ultimately solving problems with implants, both individually and in cohorts, and that the characteristics of biomaterials' surfaces play a critical role in determining the tissue reactions to implants. Moreover, in this respect, sterilization of implants is not only critical to preventing infection, but also this procedure can alter the chemistry of biomaterials' surfaces (and indeed their bulk properties) and thereby inadvertently alter implant performance. A central concept of this chapter is that following contact of biomaterials with tissues, careful, informed and detailed analysis of the implants, the biomaterials that comprise them, and the surrounding tissues can be a powerful tool in understanding the mechanisms and causes of tissue-biomaterials interactions, both desired and adverse.

Implant retrieval and evaluation is an approach that contributes to ensuring the efficacy and safety of medical devices and understanding the mechanisms of interaction of the constituent biomaterials with the surrounding tissues. Implants can be retrieved at either reoperation or necropsy or autopsy of animals or humans, respectively. The literature contains numerous instances where problem-oriented medical implant research has yielded important insights into deficiencies and complications limiting the success of implants (Schoen, 1998). Many specific examples are discussed in Chapter 9.5. Implant research has guided development of new and modification of existing implant designs and materials, assisted in decisions of implant selection and management of patients and permitted *in vivo* study of the mechanisms of biomaterials-tissue interactions, both local and distant from the device.

Preclinical studies of modified designs and materials are crucial to developmental advances. These investigations usually include *in vitro* functional testing (such as fatigue studies at accelerated rates) and implantation of functional devices in the intended location in an appropriate animal model, followed by noninvasive and invasive monitoring, specimen explantation, and detailed pathological and material analysis. Relative to clinical studies, animal investigation may permit more detailed monitoring of device function and enhanced observation of morphologic detail as well as frequent assay of laboratory parameters, and allow in situ observation of fresh implants following elective sacrifice at desired intervals. In addition, specimens from experimental animals can minimize the artifacts that can occur inadvertently under clinical circumstances. Furthermore, advantageous technical adjuncts may be available in animal but not human investigations, such as injection of radiolabeled imaging markers or fixation by pressure perfusion that maintains tissues and cells in their physiological configuration following removal. Animal studies often facilitate observation of specific complications in an accelerated time frame, such as calcification of bioprosthetic valves, in which the equivalent of 5–10 yr in humans is simulated in 4–6 mo in juvenile sheep (Schoen *et al.*, 1985). Moreover, in preclinical animal studies, experimental conditions can generally be held constant among groups of subjects, including nutrition, activity levels, and treatment conditions, and concurrent control implants are often possible. This is clearly not possible in clinical studies.

Nevertheless, clinicopathologic analysis of cohorts of patients who have received a new or modified prosthesis type evaluates device safety and efficacy to an extent beyond that obtainable by either *in vitro* tests of durability and biocompatibility or preclinical investigations of implant configurations in large animals. Moreover, through analysis of rates and modes of failure as well as characterization of the morphology and mechanisms of specific failure modes, retrieval studies can contribute to the development of methods for enhanced clinical recognition of failures. The information gained should serve to guide both future development of improved prosthetic devices to eliminate complications and diagnostic and therapeutic management strategies to reduce the frequency and clinical impact of complications. For individual patients, demonstration of a propensity toward certain complications could impact greatly on further management. Moreover, history has shown that some medical devices demonstrate important complications only during clinical trials or postmarket surveillance (indeed, complications that were not predicted by animal investigations). Subsequent bench study of clinically important device failure mechanisms may yield prevention strategies; such strategies may then be screened *in-vivo* in animal studies; favorable therapies are then tested in clinical trials. Thus, an important future goal is the effective integrated use of data derived from implant retrieval investigations (along with other clinical and experimental data) to influence both regulatory decisions and device improvements in an ongoing, incremental and iterative fashion throughout the product life-cycle.

Clinical implant retrieval and evaluation can yield several additional benefits. Implant retrieval studies have demonstrated that success of a material or design feature in one application may not necessarily translate to another. Although analysis of implants and medical devices has traditionally concentrated on failed devices, important data can accrue from implants serving the patient well until death or removal for unrelated causes. Indeed, detailed analyses of implant structural features following implantation can yield an understanding of not only predisposition to specific failure modes but also structural correlates of favorable performance. Implant retrieval and evaluation may also provide specimens and data that can be used to educate patients, their families, physicians, residents, students, engineers, and biomaterials scientists, as well as the general public. As a basic research resource, the process of implant retrieval and evaluation yields data that can be used to develop and test hypotheses and to improve protocols and techniques.

For investigation of bioactive materials/devices and potentially tissue engineered medical devices, in which the interactions between the implant and the surrounding tissue are complex, research based on implant retrieval and evaluation continues to be critical. In such instances, novel and innovative approaches must be used in the investigation of *in vivo* tissue compatibility. In such implant types, the scope of the concept of "biocompatibility" is much broader and the approaches employed in implant retrieval and evaluation require identification of the phenotypes and functions of cells and the architecture and remodeling of extracellular matrix (Schwartz and Edelman, 2002; Rabkin *et al.*, 2002). These are circumstances in which individual patient characteristics (for example, genetic polymorphisms in molecules important in matrix remodeling) could have a profound influence on outcome in some patients (Ye, 2000; Jones *et al.*, 2003). This potentially yields a new area of study: "biomateriogenomics," conceptually analogous to the emerging area of pharmacogenomics (Weinshilboum, 2003). Indeed, individuals with genetic defects in coagulation proteins may be unusually susceptible to thrombosis of prosthetic heart valves (Gencbay *et al.*, 1998). Thus, a critical role of implant retrieval will be the identification of tissue characteristics (*biomarkers*) that will be predictive of (i.e., surrogates for) success and failure. A most exciting possibility is that such biomarkers may be used to non-invasively image/monitor the maturation/remodeling of tissue engineered devices *in vivo* in individual patients (Rabkin and Schoen, 2002; Sameni *et al.*, 2003).

Bibliography

Gencbay, M., Turan, F., Degertekin, M., Eksi, N., Mutlu, B., and Unalp, A. (1998). High prevalence of hypercoagulable states in patients with recurrent thrombosis of mechanical heart valves. *J. Heart Valve Dis.* 7: 601–609.

Jones, G. T., Phillips, V. L., Harris, E. L., Rossaak, J. I., and van Rij, A. M. (2003). Functional matrix metalloproteinases-9 polymorphism (C-1562T) associated with abdominal aortic aneurysm. *J. Vasc. Surg.* 38: 1363–1367.

Rabkin, E., Hoerstrup, S. P., Aikawa, M., Mayer, J. E., Jr., and Schoen, F. J. (2002). Evolution of cell phenotype and extracellular matrix in tissue-engineered heart valves during *in vitro* maturation and *in vivo* remodeling. *J. Heart Valve Dis.* 11: 308–314.

Rabkin, E., and Schoen, F. J. (2002). Cardiovascular tissue engineering. *Cardiovasc. Pathol.* 11: 305–317.

Sameni, M., Dosescu, J., Moin, K., and Sloane, B. F. (2003). Functional imaging of proteolysis: stromal and inflammatory cells increase tumor proteolysis. *Mol. Imaging* 2: 159–175.

Schoen, F. J. (1998). Role of Device Retrieval and Analysis in the Evaluation of Substitute Heart Valves, in Clinical Evaluation of Medical Devices: Principles and Case Studies, K. B., Witkin, (ed.) Humana Press, Inc., Totowa, N. J., pp. 209–231.

Schoen, F. J., Levy, R. J., Nelson, A. C., Bernhard, W. F., Nashef, A., and Hawley, M. (1985). Onset and progression of experimental bioprosthetic heart valve calcification. *Lab. Invest.* 52: 523–532.

Schwartz, R. S., and Edelman, E. R. (2002). Drug-eluting stents in preclinical studies. Recommended evaluation from a consensus group. *Circulation* 106: 1867–1873.

Weinshilboum, R. (2003). Inheritance and drug response. *N. Engl. J. Med.* 348: 529–537.

Ye, S. (2000). Polymorphism in matrix metalloproteinase gene promoters: implication in regulation of gene expression and susceptibility of various diseases. *Matrix Biol.* 19: 623–629.

9.2 STERILIZATION OF IMPLANTS AND DEVICES

John B. Kowalski and Robert F. Morrissey

Implants and devices (products) introduced transiently or permanently into the body of a human or an animal must be sterile to avoid subsequent infection that can lead to serious illness or death. In this chapter, we discuss the meaning of the term "sterile," give an overview of the development and validation of sterilization processes, and describe the various sterilization methods, including their advantages and disadvantages (Dempsey and Thirucote, 1989). Sterilization process development and validation are discussed from the point of view of compatibility of the implant/device and associated packaging with the sterilization process as well as meeting the requirements for sterility. It is recommended that the biomaterials specialist consider sterilization-related issues early in the developmental process so that the final product can be readily sterilized by the most cost-effective process.

STERILITY AS A CONCEPT

"Sterility" is defined as the absence of all living organisms. This especially includes the realm of microorganisms,

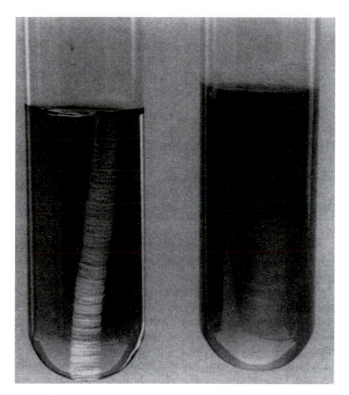

FIG. 1. Sterility tests of an experimental vascular graft illustrating negative (sterile, left) and positive (nonsterile, right) results.

such as bacteria, yeasts, molds, and viruses. The presence of even one viable bacterium on an implant renders it nonsterile. Sterility should not be confused with cleanliness. A polished stainless steel surface may easily be nonsterile (contaminated with numerous unseen microorganisms), whereas a rusty nail can be made sterile after exposure to a sterilization process that was properly developed and validated. It is true that implants and devices that are "microbiologically clean," having few viable microorganisms present (low bioburden), are more readily sterilized than those that are highly contaminated.

How then is sterility measured or proven? For relatively small numbers of product samples (assuming the product is not too large to test in its entirety), sterility can be determined by immersing each of the product samples into an individual container of sterile liquid microbiological culture medium and incubating the containers under the proper conditions. If the product is sterile, no microbial growth will occur; if it is nonsterile, the culture medium will become turbid as a result of microbial proliferation (Fig. 1). Testing small numbers of samples, however, does not give very meaningful information about the sterility of a large batch of products (often in the thousands of units) that have been processed in an industrial-scale sterilizer. Sterilization process development and validation studies are used to determine what is referred to as a sterility assurance level (SAL). The SAL is the probability that a product will be nonsterile after exposure to a specified sterilization process. The generally accepted maximum SAL for implants is 10^{-6} or a probability of no more than one in one million that the implant will be nonsterile. The process development and validation studies also document the compatibility

of the product and its package with the proposed sterilization process.

STERILIZATION PROCESS DEVELOPMENT AND VALIDATION

Product and Packaging Compatibility

The first concern in sterilization process development and validation is the demonstration that the product is compatible with the process; this also applies to the packaging, which maintains the integrity and sterility of the product on its journey to the medical practitioner and patient. The integrity of the product and the packaging system must be demonstrated shortly after sterilization and also after aging studies (often performed at elevated temperatures) to document the absence of delayed sterilization-related deleterious effects. Such delayed effects are most commonly encountered with radiation sterilization. During the product and packaging compatibility studies, "worst case" processing conditions must be used to ensure that the product and packaging are tested after exposure to the most rigorous conditions that may be encountered. For example, if the sterilization process specification allows a temperature range of 52 to 57°C, such tests must be conducted on samples exposed to a 57°C process. Also, the effect of multiple sterilization exposures must be considered in the event that a product must be exposed to a second sterilization process as a result of a condition that invalidates the initial sterilization such as a sterilizer malfunction.

When sterilizing with gaseous agents, the amount of residual sterilant and any by-product(s) in the medical device and associated packaging must be determined. If required, an aeration regimen must be developed to ensure that the residuals are below safe limits for patient use, and that manufacturing personnel are not exposed to the sterilant and/or by-product(s) that may be released into the workplace atmosphere.

Sterility Assurance Level

Sterilization validation studies must also document that the product attains the required SAL after exposure to the proposed process. By using the techniques of bioburden determination and fractional-run sterilization studies (see later discussion), the ability of the proposed process to consistently deliver an SAL of 10^{-6} or better must be conclusively documented. For these studies to be valid, the product samples must be produced under actual manufacturing conditions and be exposed to the process in their final packaging configuration. Also, the fractional-run sterilization studies must represent the least lethal conditions allowed by the process specification. In the example cited above, where the specified temperature range was 52 to 57°C, the fractional sterilization runs would be performed at the lowest temperature, 52°C, the temperature that would give the slowest rate of microorganism inactivation.

The determination of an SAL begins with the enumeration of the bioburden, the number of viable microorganisms on the product just prior to sterilization (Morrissey, 1981). Bioburden is usually determined on 10 to 30 product samples and involves washing, shaking, or sonicating the microorganisms from the item into a sterile recovery fluid such as a saline solution. By using conventional microbiological techniques, the number of microorganisms in the recovery fluid can be determined.

Once the bioburden is known, fractional-run sterilization studies can be performed to determine the microbial rate of kill or process lethality. In a fractional sterilization run, product samples (in packages) are exposed to a fraction of the desired sterilization process or dose. For example, if the proposed sterilization process has an exposure time of 2 hours, the fractional runs may have exposure times of 30, 40, and 50 min. Samples from these runs are tested for sterility and the results graphically analyzed to estimate the exposure time required to achieve a 10^{-6} SAL. The results from such a study are shown in Fig. 2. In this example, the average bioburden per sample was 240 colony-forming units. After the 30-, 40-, and 50-min fractional runs, there were, respectively, 28/50, 7/50, and 1/50 samples that remained nonsterile. The estimated time to achieve a 10^{-6} SAL for this hypothetical bioburden and sterilization process is approximately 100 minutes, which is within the proposed 120-min exposure time. Note that when there is less than one surviving organism per product unit, there is not actually 0.01 of an organism on each unit, but a probability of 1 in 100 that any given product unit is nonsterile.

In the example just given, 100 minutes of exposure to the hypothetical sterilization process was the *estimate* of the time required to attain an SAL of 10^{-6}. The actual exposure time used in the routine sterilization process will include a safety factor to ensure that natural variation in the number of microorganisms on product units (and/or differing resistance to sterilization) does not lead to a failure to attain the required 10^{-6} SAL. A good discussion of the various aspects of

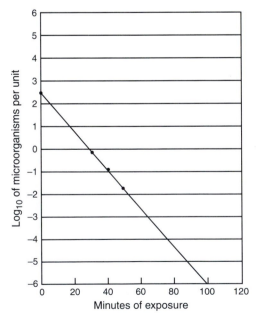

FIG. 2. Microbial kill curve based on data from a series of fractional sterilization runs. Time to achieve a 10^{-6} SAL is approximately 100 min.

sterilization process validation is contained in the specific International Standards Organization (ISO) sterilization documents (ISO 11134, 1994; ISO 11135, 1994; ISO 11137, 1995).

OVERVIEW OF STERILIZATION METHODS

The first sterilization method to be used for implants was moist heat or autoclaving, which involves exposure to saturated steam under pressure. Owing to the relatively high temperature of the process (121°C), most nonmetallic implants and packaging materials cannot be sterilized by this method. This limitation led to the development and use of ethylene oxide (EO) gas and ionizing radiation (gamma rays, accelerated electrons) to sterilize medical products (Association for the Advancement of Medical Instrumentation, 1999; Morrissey and Phillips, 1993; Block, 2000). The advances in complexity of medical products from stainless steel surgical instruments to drug–device combinations and biomaterials for tissue scaffolding has led to the further development of sterilization processes to include low-temperature gas plasma and new gaseous agents such as chlorine dioxide.

MOIST HEAT STERILIZATION

The first sterilization method applied to medical products was moist heat sterilization or autoclaving (ISO 11134, 1994).

Process and Mechanism of Action

With this method, sterilization is achieved by exposing the product to saturated steam, usually at 121°C to 125°C. The use of steam at this temperature requires a pressure-rated sterilization chamber; a typical industrial steam sterilizer is shown in Fig. 3. The design of the product must ensure that all surfaces are contacted by the steam, and the packaging must allow steam to penetrate freely. A typical moist heat sterilization process lasts 15 to 30 min *after all surfaces of the product reach a temperature of at least 121°C.*

Moist heat sterilization kills microorganisms by destroying metabolic and structural components essential to their replication. The coagulation of essential enzymes and the disruption of protein and lipid complexes are the main lethal events.

Applications—Advantages and Disadvantages

Currently, the main use of this method occurs in hospitals; it is the method of choice for the sterilization of metallic surgical instruments and heat-resistant surgical supplies (linen drapes, dressings). Hospitals also perform moist heat sterilization of metallic devices such as stainless steel sutures. A specialized form of moist heat sterilization is used for many intravenous solutions.

The advantages of moist heat sterilization are efficacy, speed, process simplicity, and lack of toxic residues. The high temperature and pressure limit the range of compatible products and packaging materials.

FIG. 3. An example of an industrial-scale steam sterilizer.

EO STERILIZATION

Sterilization with EO gas has been exploited as a low-temperature process that is compatible with a wide range of product and packaging materials (ISO 11135, 1994)

Process and Mechanism of Action

Below its boiling point of 11°C, EO is a clear, colorless liquid. It is toxic and considered a human carcinogen. Contact of liquid EO with the skin and eyes and inhalation of the gas should be avoided. EO may be used in the pure form or mixed with N_2, CO_2, or a non-ozone-depleting chlorofluorocarbon (CFC)-like compound. Pure EO and mixtures without a proven inerting compound are flammable and potentially explosive. Because of the negative effects of CFC compounds (CFC-12) on the earth's ozone layer, alternative inerting compounds have been developed.

For EO sterilization, products contained within gas-permeable packaging are loaded into a sterilization vessel, generally fabricated from stainless steel. The vessel is evacuated to remove air, at a rate and to a final pressure that is compatible with the product and packaging, and then moisture (from steam) is introduced to attain a relative humidity generally between 60 and 80%. The presence of moisture is required for sterilization efficacy with EO gas. The EO gas (or mixture) is then injected to a final concentration of ~600–800 mg/liter. The sterilizer is maintained at the desired gas concentration and temperature (typically 40 to 50°C) for a sufficient time to achieve the required SAL. The chamber is reevacuated to remove the EO, and "air flushes" are performed to reduce the EO levels to below acceptable limits. Often, further aeration outside of the sterilization chamber (in some instances, at elevated temperatures) is required to effectively remove residual EO (and by-products) from the product and packaging materials.

A vessel similar to that shown in Fig. 3 is employed for EO sterilization. It is somewhat more complex than its steam counterpart to allow for evacuation, moisture and gas addition and control, and air flushes. An EO sterilization process typically ranges from 2 to 16 hours in duration, depending on the time required for aeration inside the sterilization vessel. The lethal effect of EO on microorganisms is mainly due to alkylation of amine groups on nucleic acids.

Applications

EO is used to sterilize a wide range of medical products, including surgical sutures, intraocular lenses, ligament and tendon repair devices, absorbable and nonabsorbable meshes, neurosurgery devices, absorbable bone-repair devices, heart valves, vascular grafts, and stents coated with bioactive compounds.

EO Residuals Issues

Because of potential toxicity/carcinogenicity, residual EO and its by-product, ethylene chlorohydrin (EC) are of concern in medical products and packaging materials. Also, release of EO into the air in poststerilization manufacturing and storage areas is a concern due to the potential for personnel exposure. The maximum allowable limits for EO and EC are no longer expressed as parts-per-million (ppm) in a medical product but rather as a maximum allowable dose delivered to a patient (ISO 10993-7, 1995) (Table 1). Limits are given for devices categorized as "permanent contact," "prolonged exposure," "limited exposure," and "special situations." Current Occupational Health and Safety (OSHA) regulations dictate that a worker may not be exposed to more than 1 ppm of EO during an 8-hour time-weighted average work day.

Advantages and Disadvantages

The advantages of EO are its efficacy (even at low temperatures), high penetration ability, and compatibility with a wide range of materials. The main disadvantage centers on EO residuals with respect to both the implant and release into the environment.

The impact of CFC compounds on the earth's ozone layer forced facilities using the 12% CFC–12/88% EO mixture to switch to pure EO, a non-CFC gas mixture, or one of the alternative inerting compounds that have been developed. Pure EO and mixtures without an inerting compound require the use of costly explosion-proof equipment and damage-limiting building construction.

TABLE 1 Proposed EO Residue Limits on Medical Devices

Device type	Maximum ppm
Implant:	
Small (<10 g)	250
Medium (10–100 g)	100
Large (>100 g)	25
Intrauterine devices	5
Intraocular lenses	25
Devices contacting:	
Blood (*ex vivo*)	25
Mucosa	250
Skin	250
Surgical scrub sponges	25

RADIATION STERILIZATION

This method of sterilization utilizes ionizing radiation that involves either gamma rays from a ^{60}Co (cobalt-60) isotope source or machine-generated accelerated electrons. With this method, delivery of a sufficient radiation dose to the entire product will yield the required SAL (ISO 11137, 1995).

^{60}Co Sterilization

Of the radiation sterilization methods, exposure to ^{60}Co gamma rays is by far the most popular and widespread method. Gamma rays are highly penetrating, and the typical doses used for the sterilization of medical products are readily delivered and measured.

Process and Mechanism of Action

A schematic top view of a typical industrial ^{60}Co irradiator is shown in Fig. 4. The ^{60}Co isotope is contained in sealed stainless steel "pencils" ($\sim 1 \times 45$ cm) held in a planar array within a metal source rack. When the irradiator is not in use, the source rack is lowered into a water-filled pool (~ 25 feet deep).

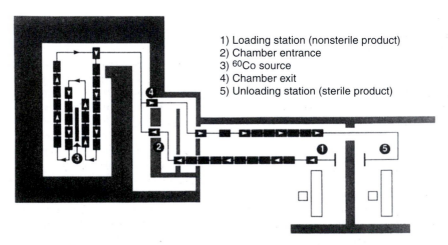

1) Loading station (nonsterile product)
2) Chamber entrance
3) ^{60}Co source
4) Chamber exit
5) Unloading station (sterile product)

FIG. 4. A schematic top view of a typical industrial ^{60}Co irradiator.

At this depth, the radiation cannot reach the surface, and it is then safe for personnel to enter the radiation cell. The outside walls and ceiling of the cell are constructed of thick, reinforced concrete for radiation shielding.

In use, materials to be sterilized are moved around the raised source rack by a conveyor system to ensure that the desired dose is uniformly delivered. Radiation measuring devices called dosimeters are placed along with the materials to be sterilized to document that the minimum dose required for sterilization was delivered and that the maximum dose for product and package integrity was not exceeded. The maximum dose divided by the minimum dose is referred to as the "overdose ratio." Irradiators are designed and product-loading patterns are configured to minimize this ratio. The most commonly validated dose used to sterilize medical products is 25 kGy.

The radioactive decay of ^{60}Co (5.3-year half-life) results in the formation of ^{60}Ni, the ejection of an electron, and the release of gamma rays. The gamma rays cause ionization of key cellular components, especially nucleic acids, which results in the death of microorganisms (Hutchinson, 1961). The ejected electron does not have sufficient energy to penetrate the wall of the pencil and therefore does not participate in the sterilization process.

Applications—Advantages and Disadvantages

^{60}Co radiation sterilization is widely used for medical products, such as surgical sutures and drapes, metallic bone implants, knee and hip prostheses, syringes, and neurosurgery devices. A wide range of materials are compatible with radiation sterilization, including polyethylene, polyesters, polystyrene, polysulfones, and polycarbonate. The fluoropolymer polytetrafluoroethylene (PTFE) is not compatible with this sterilization method because of its extreme radiation sensitivity. Undesirable materials effects with radiation sterilization are generally due to molecular-chain scission and/or cross-linking.

^{60}Co radiation sterilization approaches being the ideal sterilization method. It is a simple process that is rapid and effective, and it is readily measured and controlled through straightforward dosimetry methods. The main disadvantages are the very high capital costs associated with establishing an in-house sterilization operation and the incompatibility of some materials with this method. Another disadvantage is the continual decay of the isotope (even when the irradiator is idle), which results in longer processing times and ultimately the need for additional isotope to be added to the irradiator. Recently, a validation approach has been introduced for radiation and electron beam sterilization that is less labor intensive and requires considerably fewer product samples (AAMI TIR 27, 2001; Kowalski and Tallentire, 2003).

Electron Beam Sterilization

Medical products may also be sterilized with accelerated electrons (Cleland *et al.*, 1993). With this method, radioactive isotopes are not involved because the electron beam is machine-generated using an accelerator. The accelerator is also located within a concrete room to contain "stray electrons" but, in this case, when the accelerator is turned off, no radiation or radioactive material is present and therefore a water-filled pool is unnecessary.

Process and Mechanism of Action

With this method, articles to be sterilized are passed under the electron beam for a time that is sufficient to accumulate the desired dose (again, often 25 kGy). As with gamma rays, the lethality against microorganisms is related to ionization of key cellular components. In contrast to gamma rays, however, accelerated electrons have considerably less penetrating ability, making this method unsuitable for thick or densely packaged products.

Applications—Advantages and Disadvantages

Electron beam sterilization has the same potential range of applications and material compatibility characteristics as the ^{60}Co process. However, because of the issue of penetration distance, its use is much more limited; the availability of higher energy/higher power machines is lessening this limitation. A unique application for this method is the in-line sterilization of thin products immediately following primary packaging.

OTHER STERILIZATION PROCESSES

Traditional Methods

Because of the extreme temperatures involved ($>140°C$), dry heat sterilization is rarely if ever used for medical products. Occasionally, products are sterilized in hospitals by immersion in an aqueous glutaraldehyde solution (Block, 2000). This procedure is used only in special circumstances where the product is sensitive to heat and the aeration time after EO sterilization is not acceptable. Achieving an acceptable SAL with this method requires meticulous attention to detail and relatively long immersion times.

New Technologies

Several new technologies have emerged that have potential utility for the sterilization of medical products. These include low-temperature gas plasma, gaseous chlorine dioxide, ozone, vapor-phase hydrogen peroxide, and machine-generated X-rays (Block, 2000). The first four methods are being examined as potential alternatives to EO. Machine-generated X-rays have the advantage of machine generation (nonisotopic source) and penetrating power similar to gamma rays.

A hydrogen peroxide gas plasma system has been developed for the sterilization of heat- and moisture-sensitive devices (Favero, 2000). The process operates at $<50°C$ and is very rapid; the total process time is less than 75 min. This system was initially developed for hospital use with a key benefit being no requirement for poststerilization aeration. Instruments and devices processed in this system are ready for immediate use since aeration is not required to remove residual hydrogen peroxide in contrast to EO sterilization where the aeration period can be quite lengthy. The use of this technology has broadened to medical products manufacturers where a rapid process and the lack of a requirement for aeration offer operational advantages.

Gaseous chlorine dioxide has been shown to be effective for the sterilization of medical products (implants as well as other devices), pharmaceutical components, and barrier isolation systems (Kowalski *et al.*, 1988; Kowalski, 1998). The process is relatively rapid, generally 1.5 to 3 hours in duration, and there is little or no need for poststerilization aeration because of the low level of sterilant residuals with most materials. The design of the gas generation system allows for a wide range of sterilizer sizes and configurations. The green color of this gas allows for spectrophotometric measurement of gas concentration inside the sterilization chamber. The ability to readily monitor and control gas concentration is a significant benefit in sterilization process development and validation studies. Because chlorine dioxide (and also hydrogen peroxide) are oxidizing agents, it is important to carry out the product and package compatibility testing described earlier that must be a part of all sterilization process development and validation studies.

CHALLENGE TO THE BIOMATERIALS SPECIALIST

The challenge to the future biomaterials specialist will be to develop materials and products that have the desired biological and mechanical properties and that can be sterilized by cost-effective methods having minimum potential effects upon the patient, manufacturing personnel, and environment. Biomaterials specialists developing the next generation of implants and devices must consider sterilization issues and requirements at the earliest stages of the product development process. Failure to do so often results in a product "defaulting" to a less desirable sterilization process from a cost and/or fit into manufacturing point of view. Reliance upon EO will diminish somewhat in favor of radiation sterilization and some of the newly emerging technologies. The hydrogen peroxide gas plasma and gaseous chlorine dioxide sterilization processes offer operational benefits and these sterilants have a more favorable toxicity profile. New sterilization technologies that pose less risk to people and the environment should be exploited whenever possible (Jorkasky, 1987).

Bibliography

ISO 10993-7 (1995). Biological evaluation of medical devices—Part 7: Ethylene oxide sterilization residuals.

ISO 11134 (1994). Sterilization of health care products—Requirements for validation and routine control—Industrial moist heat sterilization.

ISO 11135 (1994). Medical devices—Validation and routine control of ethylene oxide sterilization.

ISO 11137 (1995). Sterilization of health care products—Requirements for validation and routine control—Radiation sterilization.

Association for the Advancement of Medical Instrumentation (1999). *AAMI Standards and Recommended Practices*, Vol. 1.3,

Sterilization, Part 3, *Industrial Process Control*. Association for the Advancement of Medical Instrumentation, Arlington, VA.

Association for the Advancement of Medical Instrumentation (2001). Technical Information Report 27. Sterilization of health care products—Radiation sterilization—Substantiation of 25 kGy as a sterilization dose—Method VD$_{max}$.

Block, S. S. (2000). *Disinfection, Sterilization, and Preservation*, 5th ed. Lippincott Williams & Wilkins, Philadelphia.

Cleland, M. R., O'Neill, M. T., and Thompson, C. C. (1993). Sterilization with accelerated electrons. in *Sterilization Technology*, R. F. Morrissey and G. B. Phillips (eds.). Van Nostrand Reinhold, New York.

Dempsey, D. J., and Thirucote, R. R. (1989). Sterilization of medical devices: a review. *J. Biomater. Appl.* 3: 454–523.

Favero, M. (2000). Hydrogen peroxide gas plasma low temperature sterilization. *Inf. Control Today* 4: 44–46.

Hutchinson, F. (1961). Molecular basis for action of ionizing radiation injury. *Science* 134: 533.

Jorkasky, J. F. (1987). Medical product sterilization changes and challenges. *Med. Device Diagn. Ind.* 9: 32–37.

Kowalski, J. B. (1998). Sterilization of medical devices, pharmaceutical components, and barrier isolation systems with gaseous chlorine dioxide. in *Sterilization of Medical Products, Vol. VII*, R. F. Morrissey and J. B. Kowalski (eds.). Polyscience Publications, Champlain, pp. 311–323.

Kowalski, J. B., and Tallentire, A. (2003). Aspects of putting into practice VD$_{max}$. *Radiat. Phys. Chem.* 67: 137–141.

Kowalski, J. B., Hollis, R. A., and Roman, C. A. (1988). Sterilization of overwrapped foil suture packages with gaseous chlorine dioxide. *Dev. Ind. Microbiol.* 29: 239–245.

Morrissey, R. F. (1981). Bioburden: a rational approach. in *Sterilization of Medical Products, Vol. II*, E. R. L. Gaughran and R. F. Morrissey (eds.). Multiscience Publications Limited, Montreal, pp. 11–24.

Morrissey, R. F., and Phillips, G. B., eds. (1993). *Sterilization Technology: A Practical Guide for Manufacturers and Users of Health Care Products*. Van Nostrand Reinhold, New York.

9.3 IMPLANT AND DEVICE FAILURE

Frederick J. Schoen and Allan S. Hoffman

Although the majority of clinical implants and other medical devices serve their patients well, often for extended intervals, some fail. Irrespective of implant site or desired function of the device, the overwhelming majority of clinical complications produced by medical devices fall into several well-defined categories that are summarized in Table 1. Determination of the cause(s) and contributory mechanism(s) of implant or other device failure is usually done by a process called implant retrieval and evaluation (see Chapter 9.5). The results of this analysis have several potential implications. They may, for example:

- Alter patient management, leading to a choice of a different type of prosthesis, and/or a change in the type or dose of drug therapy such as anticoagulation, and/or closer monitoring of a patient by a non-invasive therapy such as echocardiography for a heart valve or X-ray or bone scan to monitor a hip joint.
- Reveal a vulnerability of a specific prosthesis type to a particular mode or mechanism of failure, leading to

TABLE 1   Patient–Device Interactions Causing Clinical Complications of Medical Devices

Thrombosis
- Thrombotic occlusion
- Thromboembolism
- Anticoagulation-related hemorrhage (owing to the therapy to prevent thrombosis)

Infection

Inappropriate healing
- Too little/incomplete
- Normal/tissue overgrowth

Structural failure due to materials degeneration
- Wear
- Fracture
- Calcification
- Tearing

Adverse local tissue interactions
- Inflammation
- Toxicity
- Tumor formation

Migration
- Whole device
- Embolization or lymphatic spread of materials fragments

Systemic and miscellaneous effects
- Allergy
- Heart valve noise

TABLE 2   Potential Contributory Causes of Implant or Device Failure

Inadequate materials properties

Inadequate materials testing

Design flaw

Inadequate preclinical animal test models

Components missized/mismatched during fabrication

Defects introduced during manufacturing, packaging, shipping, and storage

Damage or contamination during sterilization

Damage or contamination during implantation

Technical error during implantation

Poor patient–prosthesis matching

Unavoidable, physiologic patient–prosthesis interactions

Inadequate patient management

Unusual/abnormal patient response

action by regulatory agencies such as withdrawal from use, and/or closer scrutiny of a group of patients with such a device and/or a change in design, materials selection, processing, or fabrication.
- Influence product liability litigation, either in an individual case or in a class-action situation involving multiple patients.

The purpose of this chapter is to define a conceptual approach to determination of the factors responsible for device or implant failure and to identify the variety of potential causes of failure of an extracorporeal device or an implant. Analysis relates to considerations that relate to either the development and testing process, and thereby potentially affect all devices in a cohort, or conditions germane primarily to the failed device and the patient who had it.

For example:

- Was there a design flaw, or a poor choice of materials using an appropriate design?
- Did failure arise because the preclinical testing of the device did not reveal some defect of design or materials that only became apparent following large-scale clinical utilization?
- Was the implant damaged during fabrication or implantation?
- Did the patient have an abnormal physiologic response to the implant, e.g., hypersensitivity or a tendency toward blood clotting?

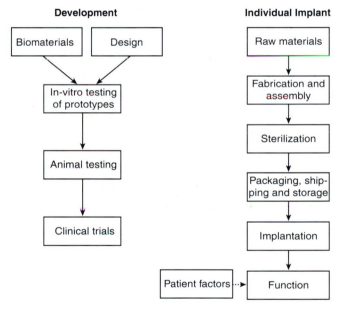

FIG. 1.   Potential causes of implant or device failure.

The array of potential causes is summarized in Table 2 and conceptualized in Fig. 1. As we will see, failure of a device in the clinical setting often involves multiple contributing factors. This chapter is not meant to describe the detailed mechanisms of either deterioration or biological responses of materials (see Chapter 6.2). Moreover, although this chapter focuses on selected examples (primarily cardiovascular), the approach and investigative pathways are broadly applicable to other systems (e.g., orthopaedic and dental) which is covered in other chapters (see Jacobs, Brown, Cranin, etc.)

BIOMATERIALS SELECTION

Materials are often implicated as the major cause of a device failure. The wide range of different material "breakdown" mechanisms that can cause an implant or device to fail is summarized in Table 3. If the wrong material is selected for a particular application, a device or implant can fail simply because the components do not have the requisite physical, chemical, or biological properties. For example, the poor resistance of polytetrafluoroethylene (PTFE or Teflon®-like) polymers to abrasive wear and its unsuitability for use in both hip-joint prostheses (Charnley et al., 1969) and artificial heart valves (Silver and Wilson, 1977) were appreciated following extensive clinical use. Similarly, the inadequate durability of a carbon-reinforced Teflon carbon-reinforced Teflon®-like composites in temporomandibular joint replacement was encountered in clinical usage (Trumpy et al., 1996). In addition, the earliest silicone ball poppets in the Starr–Edwards caged-ball heart valves frequently absorbed lipids from blood during function, and consequently swelled and became brittle and sometimes fractured (Chin et al., 1971). "Ball variance," as this complication was known, frequently led to abnormal poppet motion with slowly developing dysfunction or fracture-related catastrophic loss of parts of or the entire poppet from the cage. This sometimes caused blockage of blood vessels downstream from the heart due to the shedding of small fragments. This problem stimulated development of an improved curing protocol for medical silicone elastomers, and "ball variance" was thereby eliminated. Interestingly and conversely, a material used to form the femoral stem of a hip-joint prosthesis is too stiff, can cause "stress shielding," leading to structurally poor surrounding bone and consequent loosening (Huiskes, 1998).

TABLE 3   Some Mechanisms of Biomaterial "Breakdown"

Mechanical
- Creep
- Wear
- Stress cracking
- Fracture

Physiochemical
- Adsorption of biomolecules (such as proteins leading to irreversible fouling)
- Absorption of H_2O or lipids (leading to softening or hardening)
- Leaching of low-molecular-weight compounds (such as plasticizer) leading to weakening or embrittlement
- Dissolution (leading to disintegration)

Biochemical/chemical hydrolysis (such as amide or ester bonds, leading to bond scission)
- Oxidation or reduction (leading to bond scission or cross-linking)
- Mineralization/calcification

Electrochemical
- Corrosion

DESIGN

Device design is critical. Design of the femoral stems of hip-joint prostheses has evolved from clinical experience with fractured stems in human implants to computer-aided design and manufacturing techniques. In addition, the design of a heart valve affects the pattern of blood flow and associated platelet damage and/or presence of regions of blood stasis, both of which can lead to thrombosis (Yoganathan et al., 1978, 1981).

When a widely used tilting-disk heart valve was redesigned to allow more complete opening and thereby enhance its hemodynamic function as well as reduce the incidence of thrombosis, a large number of mechanical failures occurred (Walker et al., 1995). Nearly all had similar characteristics: they resulted from was metallic fatigue fracture of the welds anchoring metallic struts (which confine and guide the motion of the disk) to the housing, with consequent separation of the strut and escape of the disk. Pathology studies revealed that the underlying problem was metal fatigue resulting from design-related high-velocity overrotation of the disk during closure, excessively stressing the welds, potentially coupled with intrinsic flaws in the welded regions (Schoen et al., 1992). Interestingly, many such failures occurred in young patients during exercise, when cardiac forces and thereby valve closing velocities are enhanced. Another example was a specific design of a bileaflet tilting-disk heart valve which developed fractures that were initiated in various locations along the contacts of the disks with the housing. In this case the likely cause was a design flaw that led to cavitation bubbles as the disk moved away from the housing during the earliest phases of valve opening. The working hypothesis is that implosion of these bubbles initiated microscopic cracks in the pyrolytic carbon components that then precipitated gross fractures.

TESTING OF MATERIALS/DESIGN CONFIGURATIONS

In some cases, the *in vitro* or *in vivo* evaluation tests that were carried out on the material itself, or on the design and device prototype, did not adequately simulate conditions that are encountered in actual clinical use. In the case of a new bileaflet tilting-disk heart valve design that suffered clinical failures due to thrombosis initiated at the points where the hemidisks contacted and pivoted in the housing, the testing of implants in sheep failed to reveal a tendency toward thrombosis (Gross et al., 1996). Furthermore, appropriate tests are not available for some physiologic responses. For example, the potential for a significant immunological response, such as has been suggested but not proven in some patients with silicone gel breast implants, is not possible to evaluate in preclinical studies. It is also possible for available tests to have been overlooked because the specific problems that occurred later were not anticipated.

A particularly illustrative example that points out the need for improved preclinical testing regimes was the failure after 1 or more years of function of bioprosthetic heart valves fabricated from photooxidized bovine pericardium by abrasion-induced tearing of the pericardial tissue against the cloth that covered a portion of the stent (Schoen, 1998). The problem in this case was that photooxidized pericardium has a higher compliance than the glutaraldehyde-preserved pericardium used in an identical design that has been clinically successful. Higher compliance leads to greater excursion of the cuspal material during valve closing. The greater tissue movement of the more compliant material caused the tissue to contact and thereby abrade against the cloth. However, this problem occurred despite extensive preimplantation *in vitro* bench fatigue testing that did not reveal this problem, most likely owing to the markedly accelerated rate of *in vitro* testing in the system that limits tissue movements used (generally approximately 20 times or more actual). Moreover, animal testing done in sheep was not (and is typically not) extended more than approximately 5 months, except in a few specimens that in this case did not exhibit gross failure.

BIOLOGICAL TESTING OF IMPLANTS

After phototype devices have been designed and fabricated, biologic tests in animals are normally used to gain more information for eventual regulatory approval and introduction of such a device into the clinic. However, as with the *in vitro* tests on the materials or designs, *in vivo* tests on the final device or implant could be poorly chosen, key tests could be overlooked, and/or the animal model selected or permitted may not be the most appropriate one available. For example, there have been continuing arguments over the past 20 or more years among biomaterials scientists as to whether the dog is an appropriate animal model for evaluating blood–surface interactions, especially since dog platelets are much more adherent to foreign surfaces than are human platelets. Further, there is no adequate animal model to study the intense inflammatory reaction to wear debris and other particulates that can accumulate adjacent to a hip joint.

An interesting lesson was learned with respect to differences in healing in some animal models vs. humans with a caged-ball mechanical prosthetic heart valve type that had a ball fabricated from silicone and cloth-covered cage struts (Schoen *et al.*, 1984). This design was intended to encourage anchoring and organization of thrombus in the fabric and thereby decrease thromboembolism. Preclinical studies of this valve concept utilized mitral implants in pigs, sheep, and calves; in such models the cloth-covered struts were rapidly healed by endothelium-coated fibrous tissue. However, subsequent clinical implantation of these valves was complicated by cloth wear and embolization. This vivid "case history" illustrates and reinforces the concept that human implantation may reveal important problems not predicted by preclinical animal testing. The more vigorous healing that occurred in the preclinical implants in this case is typical in animals compared to humans; this sometimes makes prediction of such problems very difficult. This is the major rationale for closely monitored clinical trials during introduction of a new implant type, and continued surveillance following widespread general availability and implication thereafter.

Another area in which animal models must be used thoughtfully is in the study of calcification of tissue valves (Schoen and Levy, 1999). Two models are generally used: (a) subcutaneous implants of valve materials in weanling (approximately 3 weeks old) rats, which achieve clinical levels of calcification in 8 weeks or less, and (b) mitral valve replacements in juvenile sheep, which calcify extensively in 3–5 months. However, when older animals are used, calcification is vastly diminished. Use of inadequately severe models could lead to overestimation of the efficacy of an anticalcification strategy.

RAW MATERIALS, FABRICATION, AND STERILIZATION

In unusual circumstances, a batch of raw material is defective or damaged. The fabrication process used to manufacture a device can introduce defects or contaminants that can lead to failure of a device whose materials and design and their combination are well-matched and appropriate for the application, and otherwise intact at the start of device assembly. The weld fracture in the tilting-disk heart valve mentioned above illustrates well the introduction of defects during the fabrication process that can contribute to a tendency toward failure.

Ultra High Molecular Weight PolyEthylene (UHMW-PE) has been utilized for more than 40 years as a standard component for articulating load transferring prostheses. Much has been published about the influence(s) of processing on basic biomechanical and in vivo functional properties. More than two decades past, sterilization by low dose gamma radiation was tested and accepted as a regular production cycle. Within the past decade, retrieval and analysis studies showed alterations in structural and tribiology properties for some UHMW-PE components that were correlated with gamma radiation and shelf storage in air, near surface oxidation, structural and property changes and altered stabilities during *in vivo* function. The background and various changes introduced within the profession, related to this situation, have been described in detail within two books based on conferences and edited by Wright and Goodman (1996, 2001). Readers are referred to these books for additional information. Most interestingly, one result of "corrective measures" has been to introduce new generations of enhanced cross-linked and other polyethylene biomaterial products.

With some methods, incomplete sterilization or damage to the material by the sterilization process can occur. Certain materials place serious limitations on sterilization conditions. For example, sterilization of poly(methyl methacrylate) (PMMA) intraocular lenses may lead to discoloration. Moreover, such devices cannot be heat sterilized because the shape (and thereby optics) of the rigid PMMA lens would change above the glass transition temperature of the PMMA (100°C). Sterilization by ethylene oxide gas cannot be used with some biomaterials because of solubility within and/or reactivity with the biomaterial. These limitations are sometimes so severe that a sterilization protocol is used that is inadequate to sterilize the device.

Incomplete sterilization, of course, can lead to infection, as exemplified by a cohort of porcine tissue valves that were contaminated with *Myobacterium chelonei*, an organism related to that causing human tuberculosis (Rumisek *et al.*, 1985). It was later appreciated that the antibacterial and antifungal efficacy of low concentrations of glutaraldehyde (with which this tissue was treated) is poor. The use of combined sterilization agents (e.g., alcoholic glutaraldehyde solutions) solved this problem without causing damage to the valve tissue.

PACKAGING, SHIPPING, AND STORAGE

Contamination or degradation not only can occur during fabrication and sterilization of devices, but also can be introduced during packaging and shipping. In all medical devices and implants, inadequate packaging or improper storage can contribute to limited shelf life. Transdermal drug-delivery patches have to be carefully protected so that the drug does not leak out of the device. Tissue-derived heart valves and other devices may be degraded by excessive heat or freezing, and so are packaged with temperature indicators. Packaging and storage are also critical issues for condoms, usually made from natural rubber; this material is sensitive to oxidation, which can degrade the condom during storage on the shelf.

CLINICAL HANDLING AND SURGICAL PROCEDURE

Given an implant made of appropriate materials, properly designed, fabricated, sterilized, tested, manufactured, packaged, shipped, and stored, the "moment of truth" occurs at the instant when the package is opened and the device is handled and implanted or contacted extracorporeally with the patient's "biosystem." The most important issues here relate to the possibility of infection due to preimplantation contamination or mechanical damage caused by improper handling by surgical personnel and/or their instruments. Other examples are the exposure of devices to solutions/1ight-especially UV, contact with towels, starch on gloves, etc.

An example of a technical "error" is the implantation with kinking of a tubular vascular graft, thereby impeding its flow and possibly inducing thrombus.

THE PATIENT/USER

Every device or implant will have a certain "failure rate" based on normal population statistics, even if each one is identical to every other one. Nevertheless, such events can be very unfortunate for the rare patient who encounters such unexpected responses.

Factors related to the particular recipient may cause some failures. A patient can abuse or misuse an implant, or can exhibit an unexpected "abnormal" physiologic response. For example, a hip implant recipient who overexercises before adequate healing has occurred could cause implant loosening. Improper handling of a condom could precipitate a rip or pinhole.

Patients must be matched to their prostheses. For example, since young patients exhibit accelerated calcification of tissue heart valves, such devices are rarely used in them. Moreover, it would not be appropriate to use a mechanical valve (which requires anticoagulation therapy) in a patient with a known bleeding problem, such as a stomach ulcer.

Abnormal responses may be related to hypersensitivity or infection. For example, some patients are allergic (i.e., exhibit hypersensitivity) to nickel or other elements or compounds. Cases are reported in which patients with implants containing nickel had hypersensitive reactions that necessitated removal. Another example is that of implant recipients such as immuno-suppressed individuals with cancer or organ transplants who are particularly vulnerable to infection at any site; they may be at high risk to develop implant-related infections. Moreover, "abnormal" or "skewed" physiologic responses may also occur in any large population of patients. For example, some patients have genetic or acquired abnormalities of coagulation that render them particularly vulnerable to thrombosis. They are at heightened thrombotic risk for cardiovascular implant or device therapies (De Stefano, 1996; Girling and de Swiet, 1997).

Failure can also be related to use, independent of patient factors. Interposition grafts linking artery to vein in the arm are often used for the access to the vasculature required by hemodialysis. Use requires frequent puncture by needles. Repetitive puncture can lead to coalescent holes, graft fragmentation, and extensive local hemorrhage.

CONCLUSION

In this chapter, we have described the wide range and diversity of factors that can contribute to the failure of a medical device or implant. It is important to emphasize that inadequate properties or behavior of the biomaterials or designs are not always responsible for failure. "Unexpected," abnormal physiological responses of a patient due to an implant or treatment with a therapeutic device can be expected. It is up to the biomaterial scientists and engineers to alert and educate the public and their representatives to these possibilities. In this way, the materials scientist or engineer can play a major role in ensuring the success of medical devices and implants.

Bibliography

Charnley, J., Kamangar, A., and Longfield, M. D. (1969). The optimum size of prosthetic heads in relation to the wear of plastic sockets in total replacement of the hip. *Med. Biol. Eng.* 7: 31–39.

Chin, H. P., Harrison, E. C., Blankenhorn, D. H., and Moacanin, J. (1971). Lipids in silicone rubber valve prostheses after human implantation. *Circulation* 43: I-51–I-56.

De Stefano, V., Finazzi, G., and Mannucci, P. M. (1996). Inherited thrombophilia: pathogenesis, clinical syndromes, and management. *Blood* 87: 3531–3544.

Girling, J., and de Swiet, M. (1997). Acquired thrombophilia. *Bailleres Clin. Ob. Gyn.* 11: 447–462.

Gross, J. M., Shu, M. C. S., Dai, F. F., Ellis, J., and Yoganathan, A. P. (1996). A microstructural flow analysis within a bileaflet mechanical heart valve hinge. *J. Heart Valve Dis.* 5: 581–590.

Huiskes, R. (1998). The causes of failure of hip and knee arthroplasties. *Neder. Tijdsch. Voor Geneesk.* **142**: 2035–2040.

Rumisek, J. D., Albus, R. A., and Clarke, J. S. (1985). Late *Myobacterium chelonei* bioprosthetic valve endocarditis: activation of implanted contaminant? *Ann. Thorac. Surg.* **39**: 277–279.

Schoen, F. J. (1998). Pathologic findings in explanted clinical bioprosthetic valves fabricated from photooxidized bovine pericardium. *J. Heart Valve Dis.* **7**: 174–179.

Schoen, F. J., and Levy, R. J. (1999). Tissue heart valves: current challenges and future research perspectives. *J. Biomed. Mater. Res.* **47**: 439–465.

Schoen, F. J., Goodenough, S. H., Ionescu, M. I., and Braunwald, N.S. (1984). Implications of late morphology of Braunwald–Cutter mitral valve prostheses. *J. Thorac. Cardiovasc. Surg.* **88**: 208–216.

Schoen, F. J., Levy, R. J., and Piehler, H. R. (1992). Pathological considerations in replacement cardiac valves. *Cardiovasc. Pathol.* **1**: 29–52.

Silver, M. D., and Wilson, C. J. (1977). The pathology of wear in the Beall model 104 heart valve prosthesis. *Circulation* **56**: 617–622.

Trumpy, I. G., Roald, B., and Lyberg, T. (1996). Morphologic and immunohistochemical observation of explanted proplast–Teflon temporomandibular joint interpositional implants. *J. Oral Maxillofac. Surg.* **54**: 63–68.

Walker, A. M., Funch, D. P., Sulsky, S. I., and Dreyer, N. A. (1995). Patient factors associated with strut fracture in Bjork–Shiley 60° convexo-concave heart valves. *Circulation* **92**: 3235–3239.

Wright, T. M., and Goodman, S. B. eds. (2001). Implant Wear in Total Joint Replacement and (1996). Implant Wear: The Future of Total Joint Replacement Am Acad of Ortho Surgeons, 6300 N. River Road, Rosemont, IL 60018.

Yoganathan, A. P., Corcoran, W. H., Harrison, E. C., and Carl, J. R. (1978). The Bjork–Shiley aortic prosthesis: flow characteristics, thrombus formation and tissue overgrowth. *Circulation* **58**: 70–76.

Yoganathan, A. P., Reamer, H. H., Corcoran, W. H., Harrison, E. C., Shulman, I. A., and Parnassus, W. (1981). The Starr–Edwards aortic ball valve: flow characteristics, thrombus formation, and tissue overgrowth. *Artif. Organs* **5**: 6–17.

9.4 CORRELATION, SURFACES AND BIOMATERIALS SCIENCE

Buddy D. Ratner

INTRODUCTION

Physical or chemical measurements that can reliably predict *in vivo* biocompatibility are unavailable for most biomaterials. It would be ideal, when considering a new material for medical applications, to use a spectroscopic technique to measure the properties, and, from that physical measurement, to predict how well the material will work in a particular application. Animal experiments are expensive, are of questionable value for predicting performance in humans, and raise ethical issues (see Chapter 5.5). Human clinical trials are very expensive and also raise ethical issues (Chapter 10.4). Can we predict or prescreen *in vivo* or *in vitro* performance from measurements of surface and other properties? This chapter, addressing correlation, will examine this question, and offer suggestions for future exploration.

Correlations can take many forms. For biomaterials, three possibilities are:

1. Measure specific surface physicochemical properties (for example, contact angle) and correlate that with a biointeraction response such as protein adsorption or blood cell interaction
2. Measure protein adsorption or cell adhesion and predict *in vivo* performance
3. Measure an animal response to materials (for example, blood platelet consumption by the material) and predict clinical performance

This chapter will primarily focus on those correlations that fit within type (1), but the comments made will have relevance to many correlations that can be envisioned for biomaterials.

BIOCOMPATIBILITY AND MEDICAL DEVICE PERFORMANCE

A reexamination of the definition of biocompatibility, briefly presented in the Introduction chapter (i) of this textbook, is appropriate at this point.

> Biocompatibility is the ability of a material to perform with an appropriate host response in a specific application (Williams, 1987).

The "biocompatibility" of a medical device can be defined in terms of the success of that device in fulfilling its intended function. Thus, the blood filtration module of a hemodialysis system might be defined by its ability to appropriately fractionate soluble blood components, its robustness over its intended lifetime, and its nondamaging interaction with the patient's blood. Alternatively, we can define a "blood compatibility" for the membrane, a "blood compatibility" for the silicone rubber header, a "blood compatibility" for the tubing, and a "blood and soft-tissue compatibility" for the percutaneous connection between the apparatus and the patient's bloodstream. Similarly, for a hip-joint prosthesis, we can discuss the fatigue resistance of the device, the corrosion resistance of the device, the distribution of stresses transferred to the bone by the device, the solid angle of mobility provided, and the overall success of the device in restoring a patient to an ambulatory state. Again, the hip-joint prosthesis performance might also be couched in terms of the tissue reaction to the bone cement, the tissue reaction to an uncemented titanium prosthesis stem, and the tissue reaction to the acetabular cup. In both of these examples, two cases are offered: in the first case, a whole system (device) performance is assessed, and in the second case, the biological reaction to specific components of the device (the biomaterials) is examined. The difference between the consideration of the whole device and the materials that compose it is a critically important point. In certain contexts, only the performance of the complete device is defined as biocompatibility. This definition can be inferred from the U.S. Food and Drug Administration policy that only complete devices, and never materials, receive "approval." The performance of the individual materials is referred to sometimes as "biocompatibility" and sometimes as "bioreaction" or "bioresponse."

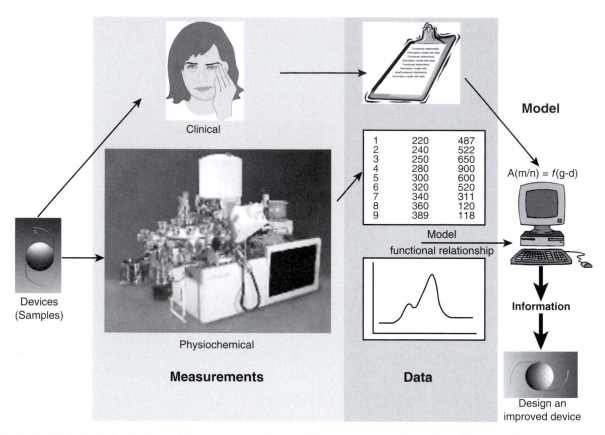

FIG. 1.   Clinical and physicochemical data are converted to useful information to optimize the performance of an implant device.

This is a textbook on biomaterials, so it would be appropriate to focus on the materials. "Bioreaction" is a much simpler word than "biocompatible," and this term will largely be used here. Bioreaction will be discussed here and defined by example.

Toxicology assays, often inappropriately referred to as biocompatibility assessment, were presented in Chapter 5.2. These assays deal with measurement of substances that leach from materials, most of which will induce some cell or tissue reaction. Such assays will be discussed here because the published correlations are clear, interpretable, and measurable. If we concentrate on bioreactions to implanted materials that do not intentionally leach substances (i.e., most biomaterials), the surface properties immediately assume a high profile as the prime candidate to control bioreaction. However, hardness, porosity, shape, movement, and specific implant site are also important. These issues will be addressed in working through important concepts in correlating material properties to bioreaction.

DATA, INFORMATION, AND STATISTICS

Laboratory experiments lead to the acquisition of data from the testing of the properties (biological and physical) of implant materials. Frequently we are presented with a column of numbers or a plot. Staring at these numbers or plots is often unhelpful in understanding the system under study. What we really desire is not *data*, but *information*, about our system.

This idea is illustrated in Fig. 1, proposing how clinical performance data and physicochemical measurement data might lead to the development of an improved implant device. Correlation is one way to process data so that it yields information.

Correlation is a tool in a class of methods called statistics that are useful for analyzing data so as to appreciate its variance, significance, and interrelationships. A general introduction to statistics is not presented here, but every student of biomaterials science should be versed in these mathematical tools. A few useful, general books on statistics as applied to scientific problems are Bevington and Robinson (2002), Mosteller and Tukey (1977), and Anderson and Sclove (1986). A book that can be useful in interpreting the meaning of statistics is Huck (2000). A search of Amazon.com found more than 4500 books on statistics and probability.

CORRELATION

Correlation is a relationship between two or more variables. Correlation does not necessarily mean cause and effect. For example, most people walking with open umbrellas will have wet feet. But, in this situation, it is obvious that umbrellas do not cause wet feet. We have a high correlation, but it misses the controlling factor (causative factor) in this example—the rain. Thus, we can propose that A (the umbrella) causes B (wet feet), B causes A, they cause each other, or they are both caused by C (the rain). We cannot prove causation, but it can be strongly suggested. Often, where correlations are observed, the causative factor is obscured, and so we have data but little useful information. Where relationships are established between a dependent variable and an independent (or explanatory) variable, this is referred to as regression analysis.

It may be more productive to look at this problem in terms of calibration and prediction. Calibration in a practical (e.g., analytical) sense has been defined by Martens and Naes as the use of empirical data and prior knowledge for determining how to predict unknown quantitative information Y from available measurements X, via some mathematical transfer function (Martens and Naes, 1989). This could be as simple as plotting Y versus X and using a least-squares fit to deal with modest levels of random noise, or as complex as a multivariate calibration model to accommodate noisy data, interferents, multiple causes, nonlinearities, and outliers. Multivariate methods will be elaborated upon toward the end of this chapter.

ASPECTS OF THE BIOREACTION TO BIOMATERIALS

Bioreaction, a process related to, but more general than, "biocompatibility," can have many manifestations. Some of these are listed in Table 1. A bioreaction is most simply defined as an observed response upon interaction of a material with a biological system or system containing biomolecules. Can simple measured physical properties of materials be correlated with bioreactions? There are many examples where this is indeed the case. Table 2 lists some of the surface physical measurements for materials that one might hypothesize as influencing bioreaction or biocompatibility.

THE CASE FOR CORRELATION—A BRIEF REVIEW OF THE LITERATURE

In the 1970s and 1980s there was great enthusiasm about using correlation to predict biomaterial performance. The lessons learned have provided insights and also tempered that

TABLE 1 Some Bioreactions

Protein adsorption	Macrophage adhesion
Protein retention	Neutrophil attachment
Lipid adsorption	Cell spreading
Bacterial adhesion	Phagocytosis
Platelet adhesion	Biodegradation
Hemolysis	Angiogenesis
Platelet activation	Apoptosis
Expression of genes	m-RNA production
Cytokine release	Fibrous encapsulation

Note: these reactions are commonly observed with implant materials. However, their relationship to "biocompatibility" is not direct.

TABLE 2 Physical Parameters of Biomaterial Surfaces That Might Correlate with Bioreactivity

Wettability	Subsurface features
Hydrophilic/hydrophobic	Distribution of functional groups
Polar/dispersive	Receptor sites
Surface chemistry	Modulus
Specific functional groups	Hydrogel (swelling) character
Surface electrical properties	Mobility
Roughness/porosity	Adventitious contamination
Domains (patterns) of chemistry	Trace levels of functional groups

enthusiasm. Still, papers continue to explore correlation and few modern efforts will be described here.

An early and influential paper demonstrating that physical measurements might be correlated with observed reactions to biomaterials concerned extractable from biomaterials (Homsy, 1970). Many biomaterials were examined in this study. Each was extracted in a pseudoextracellular fluid. The extract was examined by infrared absorbance spectroscopy for hydrocarbon bands that are indicative of organic compounds. A strong, positive correlation was observed between the strength of the IR absorbances and the reaction of the materials with a primary tissue culture of newborn mouse heart cells. This paper was important in scientifically justifying *in vitro* cell culture analysis for screening the toxicology of biomaterials, and for developing the principle that biomaterials should not unintentionally leach substances.

In the late 1960s, Robert Baier and colleagues offered an intriguing hypothesis concerning surface properties and bioreaction that continues to stimulate new experiments (Baier, 1972). This hypothesis is based upon interfacial energetics of surfaces as approximated from contact-angle measurements and suggests that materials with critical surface tensions (see Chapter 1.4) of approximately 22 dyn/cm would exhibit minimum bioreactivity (Fig. 2). Support for this hypothesis has been

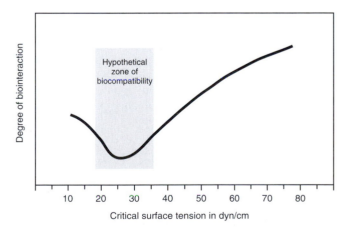

FIG. 2. A hypothesis for a minimum biointeraction for surfaces with critical surface tensions around 22 dyn/cm.

generated in a number of experiments spanning many different types of biointeractions (Baier *et al.*, 1985; Dexter, 1979). However, in a larger number of cases, this minimum has not been observed raising questions about the generality of this concept (Neumann *et al.*, 1979; Yasuda *et al.*, 1978; Mohandas *et al.*, 1974; Lyman *et al.*, 1970; Chang *et al.*, 1977).

Some of the clearest biointeraction correlations have been observed in simple, nonproteinaceous media. Linear trends of cell (mammalian and bacterial) adhesion versus various measures of surface energy have been noted (Chang *et al.*, 1977; Neumann *et al.*, 1979; Yasuda *et al.*, 1978; Mohandas *et al.*, 1974). For example, Chang and co-workers (1977) found that the adhesion of washed pig platelets to solid substrates increased with increasing water contact angle, a parameter that generally correlates well with solid surface tension. It is interesting that these simple linear trends often vanish or diminish if protein is present in the attachment medium (Neumann *et al.*, 1979; Chang *et al.*, 1977; van der Valk *et al.*, 1983). More complex surface energetic parameters have also been explored to correlate bioreaction to surface properties (Kaelble and Moacanin, 1977).

Correlations between material properties and long-term events upon implantation are less frequently seen in the literature. However, some important examples have been published. The baboon A-V shunt model of arterial thrombosis has yielded a number of intriguing correlations (Harker and Hanson, 1979). Using an *ex vivo* femoral–femoral shunt, this model measures a first-order rate constant of platelet destruction induced by the shunt material (the units are platelets destroyed/cm^2·day). The values for this surface reactivity parameter are independent of flow rate (after the flow rate is sufficiently high to ensure kinetically limited reaction), length of time that the reaction is observed, blood platelet count, and surface area of the material in contact with the blood. In one experiment, a series of hydrogels grafted to the luminal surfaces of 0.25-cm i.d. tubes was studied (Hanson *et al.*, 1980). The platelet consumption (see Chapter 5.4) was found to increase in a simple, linear fashion with the equilibrium water content of the hydrogels. This correlation, illustrated in Fig. 3, is particularly intriguing because the hydrogel materials studies were amide-, carboxylic acid-, and hydroxyl-based.

The only clear, correlating parameter was equilibrium water content. In another study, the platelet consumption of a series of polyurethanes was observed to decrease in a linear fashion as the fraction of the polyurethane C1s ESCA surface spectrum that was indicative of hydrocarbon moieties increased (Hanson *et al.*, 1982).

There have been many attempts to correlate specific protein adsorption with biological reaction. A 1975 study showed that the number of platelets adsorbed to polyurethanes correlated inversely with the amount of albumin adsorbed in competition with fibrinogen and IgG (Lyman *et al.*, 1975). For a series of polyurethanes, platelet attachment was shown to correlate with the amount of fibrinogen adsorbed (Chinn *et al.*, 1991). However, two surprising outlier points could not be explained—two materials that adsorbed high fibrinogen levels adhered low levels of platelets. Bailly *et al.* (1996) used a direct ELISA method to measure adsorbed fibrinogen on catheters and found that *in vitro* platelet adhesion and *in vivo* catheter thrombogenicity correlated with amount of adsorbed fibrinogen. Hu *et al.* (2001) have isolated specific domains within fibrinogen that help to explain their observation that the level of adsorbed fibrinogen correlates with the magnitude of the foreign body reaction at short (~3 day) implantation times. These studies all implicate levels of adsorbed fibrinogen with complex biological reactions.

A material parameter that lends itself to correlation is roughness or surface texture. Roughness is readily measured using a scanning electron microscope, a profilometer, or an atomic force microscope (see Chapter 1.4). Relationships between roughness and blood hemolysis (Wielogorski *et al.*, 1976) or thrombogenicity (Hecker and Scandrett, 1985) were reported. Textures and roughness are also extremely important to the fixation of materials into hard tissue and to the nature of the foreign-body response observed (Brauker *et al.*, 1992; Thomas and Cook, 1985; Schmidt and von Recum, 1991) (see Chapter 2.15). A complication in the use of the roughness parameters is differentiation between porosity and roughness, and also an appreciation of the difference between the average feature amplitude (often called R_a) and the nature of the roughness (e.g., are rolling hills and jagged rocks of the same height also the same roughness? See Fig. 4).

Clinical results correlated with material properties are rare, in part because materials used in clinical studies are generally not characterized to measure the parameters appropriate to make such correlations. However, a few such studies have been published. In the 1970s, a contact-angle measurement criterion was established for qualifying the clinical success of processed umbilical cord vascular grafts (Shapiro *et al.*, 1978). In another, example, rigid gas-permeable contact lens wettability was correlated with subject discomfort and a predictive trend was noted (Bourassa and Benjamin, 1989). Catheters that are used in humans were evaluated in a test system closely simulating clinical application (Wilner *et al.*, 1978). Catheters were classified into three groups related to their probable success, but clear relationships to surface properties were not discerned. Studies by Bailly *et al.* (1996) were more successful at establishing relationships between catheter properties and *in vivo* thrombogenicity. The complications in performing control studies, the difficulties in assembling a sufficiently large experimental

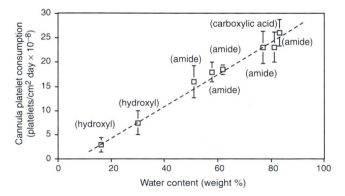

FIG. 3. A correlation between baboon platelet consumption as measured in an *ex vivo* shunt system and hydrogel water content. (From Hanson *et al.*, 1980.)

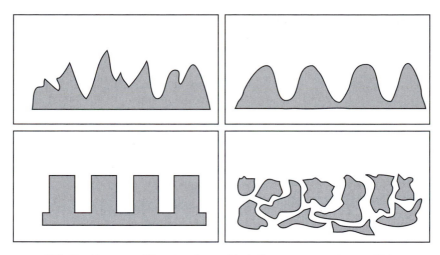

FIG. 4.   Four very different surfaces with similar average roughnesses (R_a).

population, and the complexity of the materials (devices) and the human biology make clinical correlation challenging.

ISSUES COMPLICATING SIMPLE CORRELATION

Correlations should allow us to take many readily measured physical properties and use that information in the design of improved biomaterials. Though many studies suggesting interesting possibilities are cited here, this textbook does not present these correlations as rules that can be used in biomaterials design. It should be clear by now that, although many correlations have been noted, there is considerable contradictory evidence about what the correlating factors are, and the nature of the correlations. Also, in many systems, no obvious correlations have been noted.

The most widely used correlating factor has been surface energetics, possibly because contact angles can be readily measured in any laboratory. Surface energetic parameters all relate back to the second law of thermodynamics and it is well established that the interactions of simple colloid particles can be modeled using thermodynamic and electrostatic arguments. If living cells are treated as simple colloid particles with fixed mass, charge density, polar forces, and hydrophobic interactions, thermodynamic (energetic) modeling may be appropriate (Gerson and Scheer, 1980; Fletcher and Pringle, 1985). However, living cells most often cannot be viewed as "hard, charged spheres." Living cells can change their surface characteristics in response to surfaces and other stimuli. Also, specific (e.g., receptor) interactions do not lend themselves to this simple thermodynamic modeling. For example, two surfaces with similar (but not the same) immobilized oligopeptides, and hence essentially the same surface energy, may interact very differently with cells, if one of the oligopeptides represents a minimal recognized sequence for the cell surface receptor. This was observed with fibroblast cell attachment where an immobilized peptide containing an RGD unit (arginine-glycine-aspartic acid) and a closely related immobilized peptide containing an RGE segment (where the E indicates

glutamic acid) were compared (Massia and Hubbell, 1990). The RGD peptide was highly active in inducing cell spreading while the RGE peptide was not (see Chapter 2.16 for details on this experiment). Finally, the nature of the correlations may be multivariate rather than univariate. This concept will now be discussed.

MULTIVARIATE CORRELATION

We are trained throughout our science education to appreciate cause-and-effect correlations. For example, if the temperature of a solution is increased, the reaction rate of two reactants in the solution will increase in some relatively simple manner. Unfortunately, many systems, particularly multicomponent systems such as are so often found in biomaterials science, have competing reactions that are dependent upon each other (e.g., the product of one reaction may influence the rate of another reaction). Thus, we do not see a simple relationship, but rather, many things changing simultaneously. Our eye cannot discriminate the key trend(s) in this "stew" of changing numeric values. Multivariate statistics is a class of statistical methods that looks for trends, patterns, and relationships among many variables. Also, where contemporary analytical instrumentation produces large amounts of complex (e.g. spectral) data, multivariate statistics can assist in examining the data for similarities, differences, and trends. Where large data sets overload our ability to discern relationships, multivariate methods thrive on large amounts of data and, in fact, become more accurate and useful. This class of statistical methods has come unto its own only with the introduction of powerful computers since the methods are computation-intensive. Many general introductions to multivariate statistics are available (Martens and Naes, 1989; Sharaf *et al.*, 1986; Brereton, 2003; Massart *et al.*, 1988, Wickens, 1995). Multivariate statistics applied to problems involving chemistry is often referred to as "chemometrics."

An important general principle in multivariate analysis is dimensionality reduction. A plot of x versus y requires us to think in two dimension. A "3D" plot of x, y, and z can still be easily visualized. Where we have w, x, y, and z as the axes, we lose the ability to absorb the information in graphical form and assign trends to the data. However, if we take our 3D example, we can visualize a projection (shadow) of the 3D data cluster in two dimensions. We have reduced the dimensionality from 3 to 2. Similarly, our 4D data set can be projected (by a computer) into a 3D space. Thus we have a data representation we can visualize in order to look for trends. This dimensionality reduction is readily performed by computers using standard linear algebra methods. The number of dimensions that the computer can work with is, for all practical purposes, unlimited.

There are many multivariate statistical algorithms useful for analyzing data. They are sometimes divided into classification methods (also called cluster analysis methods) and factor analysis methods (Mellinger, 1987). Classification methods find similarities in groups of data points and arrange them accordingly. Factor analysis methods take data and transform them into new "factors" that are linear combinations of the original data. In this way the dimensionality of the problem is reduced. Factor analysis methods useful for multivariate correlation with data sets such as are acquired in biomaterials research include principal component analysis (PCA)(Wold *et al.*, 1987) and partial least squares (PLS) regression (Geladi and Kowalski, 1986). Two important points about about these methods are that they do not require a hard model (rarely do we have such a quantitative model) and that they make use of all the data (i.e., we do not have to choose which data we want to put into the correlation model). Examples of the application of these methods to biomaterials research are increasing (Perez-Luna *et al.*, 1994; Wojciechowski and Brash, 1993). The power of these statistical methods is being recognized, and they will become standard data analysis tools. This is because they make efficient use of all data, thrive on large amounts of data produced by modern instruments, are objective in that we do not have to choose which data to use, are congruent with biomaterials studies that typically have many interrelated variables, and reduce the influence of noise and irrelevant variables thereby effectively increasing the signal-to-noise ratio.

Multivariate statistical methods can be a great boon to data analysis, but they will not solve all problems in biomaterials science. They should be considered as powerful hypothesis generators. The correlations and trends noted using such analysis represent a new view of the significance of data that we cannot appreciate by staring at spectra or tables. New hypotheses about the importance of materials variables can be formulated, and then they must be tested. Multivariate statistical methods also provide powerful tools for experimental design, but these will not be discussed here.

CONCLUSIONS

Perhaps the reader expected this chapter to provide instruction on the importance of wettability, or roughness, or carboxyl group concentration for biological reaction? Unfortunately, biomaterials science is not yet at the state where we can assemble a handbook of design data about the relationships between surface structures and biological reactions. We do not yet understand the controlling variables from the biology or even the materials science sufficiently well to generalize many of our observations. What this chapter does do is to suggest that such relationships probably exist, and to point out that there are powerful mathematical methods that have the potential to help us generalize our data into correlations and trends useful in biomaterials and medical device design.

Bibliography

Anderson, T. W., and Sclove, S. L. (1986). *The Statistical Analysis of Data*. The Scientific Press, Palo Alto, CA.

Baier, R. E. (1972). The role of surface energy in thrombogenesis. *Bull. N. Y. Acad. Med.* **48**: 257–272.

Baier, R. E., DePalma, V. A., Goupil, D. W., and Cohen, E (1985). Human platelet spreading on substrata of known surface chemistry. *J. Biomed. Mater. Res.* **19**: 1157–1167.

Bailly, A. L., Laurent, A., Lu, H., Elalami, I., Jacob, P., Mundler, O., Merland, J. J., Lautier, A., Soria, J., and Soria, C. (1996). Fibrinogen binding and platelet retention: relationship with the thrombogenicity of catheters. *J. Biomed. Mater. Res.* **30**: 101–108.

Bevington, P. R., and Robinson, D. K. (2002). *Data Reduction and Error Analysis for the Physical Sciences*, 3rd ed. McGraw-Hill, New York.

Bourassa, S., and Benjamin, W. J. (1989). Clinical findings correlated with contact angles on rigid gas permeable contact lens surfaces *in vivo. J. Am. Optom. Assoc.* **60**: 584–590.

Brauker, J., Martinson, L., Young, S., and Johnson, R. C. (1992). Neovascularization at a membrane-tissue interface is dependent on microarchitecture. *Transactions of the Fourth World Biomaterials Congress*, Berlin, April 24–28, 1992, p. 685.

Brereton, R. G. (2003). *Data Analysis for the Laboratory and Chemical Plant*. John Wiley & Sons, New York.

Chang, S. K., Hum, O. S., Moscarello, M. A., Neumann, A. W., Zingg, W., Leutheusser, M. J., and Ruegsegger, B. (1977). Platelet adhesion to solid surfaces. The effect of plasma proteins and substrate wettability. *Med. Progr. Technol.* **5**: 57–66.

Chang, S. K., Neumann, A. W., Moscarello, M. A., Zingg, W., and Hum, O. S. (1977). Substrate wettability and *in vitro* platelet adhesion. in *Abstracts of the 51st Colloid and Surface Science Symposium*, Grand Island, NY, June 19–22, 1977.

Chinn, J. A., Posso, S. E., Edelman, P. G., Ratner, B. D., and Horbett, T. A. (1991). Fibrinogen adsorption and platelet adhesion to novel polyurethanes. Abstracts, American Chemical Society Spring Meeting, April 14–19, 1991, Atlanta, GA.

Dexter, S. C. (1979). Influence of substratum critical surface tension on bacterial adhesion—*in situ* studies. *J. Colloid Interface Sci.* **70**: 346–354.

Fletcher, M., and Pringle, J. H. (1985). The effect of surface free energy and medium surface tension on bacterial attachment to solid surfaces. *J. Coll. Interf. Sci.* **104**: 5–14.

Geladi, P., and Kowalski, B. R. (1986). Partial least-squares regression: a tutorial. *Anal. Chim. Acta* **185**: 1–17.

Gerson, D. F., and Scheer, D. (1980). Cell surface energy, contact angles and phase partition III. Adhesion of bacterial cells to hydrophobic surfaces. *Biochim. Biophys. Acta* **602**: 506–510.

Hanson, S. R., Harker, L. A., Ratner, B. D., and Hoffman, A. S. (1980). *In vivo* evaluation of artificial surfaces with a nonhuman primate model of arterial thrombosis. *J. Lab. Clin. Med.* **95**: 289–304.

Hanson, S. R., Harker, L. A., Ratner, B. D., and Hoffman, A. S. (1982). Evaluation of artificial surfaces using baboon arteriovenous shunt model. in *Biomaterials 1980, Advances in Biomaterials* Vol. 3, G. D. Winter, D. F. Gibbons, and H. Plenk, Jr. (eds.). John Wiley and Sons, Chichester, UK, pp. 519–530.

Harker, L. A., and Hanson, S. R. (1979). Experimental arterial thromboembolism in baboons. Mechanism, quantitation, and pharmacologic prevention. *J. Clin. Invest.* **64**: 559–569.

Hecker, J. F., and Scandrett, L. A. (1985). Roughness and thrombogenicity of the outer surfaces of intravascular catheters. *J. Biomed. Mater. Res.* **19**: 381–395.

Homsy, C. A. (1970). Bio-compatibility in selection of materials for implantation. *J. Biomed. Mater. Res.* **4**: 341–356.

Hu, W. J., Eaton, J. W., and Tang, L (2001). Molecular basis of biomaterial-mediated foreign body reactions. *Blood* **98**: 1231–1238.

Huck, S. W. (2000). *Reading Statistics and Research*, 3rd ed. Addison-Wesley, Reading, MA.

Kaelble, D. H., and Moacanin, J. (1977). A surface energy analysis of bioadhesion. *Polymer* **18**: 475–482.

Lyman, D. J., Klein, K. G., Brash, J. L., and Fritzinger, B. K. (1970). The interaction of platelets with polymer surfaces. *Thrombos. Diathes. Haemorrh.* **23**: 120–128.

Lyman, D. J., Knutson, K., McNeill, B., and Shibatani, K. (1975). The effects of chemical structure and surface properties of synthetic polymers on the coagulation of blood. IV. The relation between polymer morphology and protein adsorption. *Trans. Am. Soc. Artif. Internal. Organs* **21**: 49–54.

Martens, H., and Naes, T. (1989). *Multivariate Calibration*. John Wiley and Sons, New York.

Massart, D. L., Vandeginste, B. G. M., Deming, S. N., Michotte, Y., Kaufman, L. (1988). *Chemometrics: A Textbook*. Elsevier Science, Amsterdam.

Massia, S. P., and Hubbell, J. A. (1990). Covalent surface immobilization of Arg-Gly-Asp- and Tyr-Ile-Gly-Ser-Arg-containing peptides to obtain well-defined cell-adhesive substrates. *Anal. Biochem.* **187**: 292–301.

Mellinger, M. (1987). Multivariate data analysis: its methods. *Chemometrics Intelligent Lab. Syst.* **2**: 29–36.

Mohandas, N., Hochmuth, R. M., and Spaeth, E. E. (1974). Adhesion of red cell to foreign surfaces in the presence of flow. *J. Biomed. Mater. Res.* **8**: 119–136.

Mosteller, F., and Tukey, J. W. (1977). *Data Analysis and Regression*. Addison-Wesley, Reading, MA.

Neumann, A. W., Moscarello, M. A., Zingg, W., Hum, O. S., and Chang, S. K. (1979). Platelet adhesion from human blood to bare and protein-coated polymer surfaces. *J. Polymer Sci., Polymer Symp.* **66**: 391–398.

Perez-Luna, V. H., Horbett, T. A., and Ratner, B. D. (1994). Developing correlations between fibrinogen adsorption and surface properties using multivariate statistics. *J. Biomed. Mater. Res.* **28**: 1111–1126.

Schmidt, J. A., and von Recum, A. F. (1991). Texturing of polymer surfaces at the cellular level. *Biomaterials* **12**: 385–389.

Shapiro, R. I., Cerra, F. B., Hoffman, J., and Baier, R. (1978). Surface chemical features and patency characteristics of chronic human umbilical vein arteriovenous fistulas. *Surg. Forum* **29**: 229–231.

Sharaf, M. A., Illman, D. L., and Kowalski, B. R. (eds.) (1986). *Chemometrics*. John Wiley & Sons, New York.

Thomas, K. A., and Cook, S. D. (1985). An evaluation of variables influencing implant fixation by direct bone apposition. *J. Biomed. Mater. Res.* **19**: 875–901.

van der Valk, P., van Pelt, A. W. J., Busscher, H. J., de Jong, H. P., Wildevuur, C. R. H., and Arends, J. (1983). Interaction of fibroblasts and polymer surfaces: relationship between surface free energy and fibroblast spreading. *J. Biomed. Mater. Res.* **17**: 807–817.

Wickens, T. D., (1995). *The Geometry of Multivariate Statistics*, Lawrence Earlbaum Associates, Publishers, Hillsdale, NJ.

Wielogorski, J. W., Davy, T. J., and Regan, R. J. (1976). The influence of surface rugosity on haemolysis occurring in tubing. *Biomed. Eng.* **11**: 91–94.

Williams, D. F. (1987). Definitions in biomaterials. in *Proceedings of a Consensus Conference of the European Society for Biomaterials, Chester, England, March 3–5, 1986*, Vol. 4, Elsevier, Amsterdam.

Wilner, G. D., Casarella, W. J., Baier, R., and Fenoglio, C. M. (1978). Thrombogenicity of angiographic catheters. *Circ. Res.* **43**: 424–428.

Wojciechowski, P. W., and Brash, J. L. (1993). Fibrinogen and albumin adsorption from human blood plasma and from buffer onto chemically functionalized silica substrates. *Coll. Surf. B: Biointerfaces* **1**: 107–117.

Wold, S., Esbensen, K., and Geladi, P. (1987). Principal component analysis. *Chemometrics Intelligent Lab. Syst.* **2**: 37–52.

Yasuda, H., Yamanashi, B. S., and Devito, D. P. (1978). The rate of adhesion of melanoma cells onto nonionic polymer surfaces. *J. Biomed. Mater. Res.* **12**: 701–706.

9.5 IMPLANT RETRIEVAL AND EVALUATION

James M. Anderson, Frederick J. Schoen, Stanley A. Brown, and Katharine Merritt

Implant retrieval and evaluation offers the opportunity to investigate and study the intended use of biomaterials *in vivo*. Implant retrieval and evaluation programs, in general, are designed to determine the efficacy and safety or biocompatibility of biomaterials, prostheses, and medical devices. The goal of safety testing is to determine if a biomaterial presents potential harm to the patient; it evaluates the interaction of the biomaterial with the *in vivo* environment and determines the effect of the host on the implant. Biocompatibility assessment is the determination of the ability of a biomaterial to perform with an appropriate host response in a specific application. In the *in vivo* environment, biocompatibility assessment is considered to be a measure of the degree and extent of adverse alteration(s) in homeostatic mechanism(s).

In this chapter, implant retrieval and evaluation are considered from the perspective of ultimate human clinical use. Therefore, this chapter draws upon many important perspectives presented in chapters dealing with materials science and engineering, biology, biochemistry, and medicine, host reactions and their evaluation, the testing of biomaterials, the degradation of materials in the biological environment, the application of materials in medicine and dentistry, the surgical perspective of implantation, the correlation of material surface properties with biological responses of biomaterials, and failure analysis. Although this chapter focuses on implant retrieval and evaluation in the clinical environment, many of the goals and perspectives presented are important to preclinical (i.e., animal) investigation. Emphasis will be on the rationale and overall contributions of hypothesis-driven explant analysis applicable to failed as well as nonfailed and, in some situations, unimplanted devices. In this chapter, implants are considered to be composed of one or more biomaterials that are part of

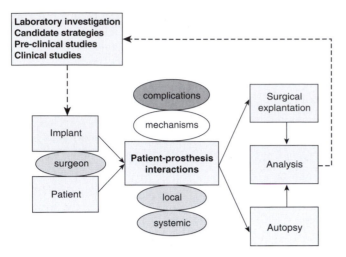

FIG. 1.   Role of implant retrieval and evaluation in the development and use of clinical devices.

TABLE 1   General Goals of Implant Retrieval and Evaluation

Determine rates, modes, and mechanisms of implant failure

Identify effects of patient and prosthesis factors on performance

Establish factors that promote implant success

Determine dynamics, temporal variations, and mechanisms of tissue–materials and blood–materials interactions

Develop design criteria for future implants

Determine adequacy and appropriateness of animal models

a specific design for a specific application. For these purposes, implants are prostheses, medical devices, or artificial organs.

GOALS

The general concept of implant retrieval that will guide our further discussion is presented in Fig. 1. An implant is placed in a patient by a surgeon. The potential impact of the surgeon (device selection, implantation technique, and implant manipulation, for example) on success or failure has major implications and constitutes a key difference between drugs and devices. All implants have interaction of the constituent biomaterials with the surrounding tissues. These interactions may be local at the site of implantation or distant from the device, and their mechanisms may be elucidated by careful and appropriate study. Clinically important deleterious biomaterials–tissue interactions manifest as *complications*. Implants are retrieved after successful function or failure by either surgical removal (at reoperation) or autopsy (by permission obtained after the the patient's death). The information derived from implant retrieval and evaluation is frequently used to guide development of new and modification of existing implant designs and materials, to assist in decisions of implant selection, and to otherwise alter the management of patients (such as anticoagulation drug regimens for prosthetic heart valves, or activity limitations for patients with prosthetic joints).

The general goals of retrieved implant evaluation are presented in Table 1. While many implant failures can be characterized as implant- or material-dependent or clinically or biologically dependent, many modes and mechanisms of failure are dependent on both implant and biological factors. To appropriately appreciate the dynamics and temporal variations of tissue–materials and blood–materials interactions of implants, a fundamental understanding of these interactions is important. An implant retrieval and evaluation program should elucidate materials, design, and biological factors in implant performance and enhance design criteria for future development. Finally, implant retrieval and evaluation should

offer the opportunity to determine the adequacy and appropriateness of animal models used in preclinical testing of implants and biomaterials. Thus, the strengths and weaknesses as well as the advantages and disadvantages of animal models can be evaluated.

The goals of routine hospital surgical pathology or autopsy examination of a prosthetic device are generally restricted to documentation of the removal of a specific device at reoperation or that the patient has died with, and diagnosis of a clinical abnormality that required therapeutic intervention. Detailed correlation of morphologic features with clinical signs, symptoms, and dysfunctional physiology is usually not performed. However, directed and informed pathological examination of prostheses retrieved during preclinical animal studies or at reoperation or autopsy of human patients can provide valuable additional information. First, preclinical studies of modified designs and materials are crucial to developmental advances. These investigations usually include implantation of functional devices in the intended location in an appropriate animal model, followed by noninvasive and invasive monitoring, followed by specimen explantation, and detailed pathological and material analysis. Second, for individual patients, demonstration of a propensity toward certain complications could have great impact on further management. Third, clinicopathologic analysis of cohorts of patients who have received a new or modified prosthesis type evaluates device safety and efficacy to an extent beyond that obtainable by either *in vitro* tests of durability and biocompatibility or preclinical investigations of implant configurations in large animals. Moreover, through analysis of rates and modes of failure as well as characterization of the morphology and mechanisms of specific failure modes in patients with implanted medical devices, retrieval studies can contribute to the development of methods for enhanced clinical recognition of failures. The information gained should serve to guide both future development of improved prosthetic devices to eliminate complications and diagnostic and therapeutic management strategies to reduce the frequency and clinical impact of complications. Emphasis is usually directed toward failed implants; however, careful and sophisticated analysis of removed prostheses that are functioning properly is also needed. Indeed, detailed analyses of implant structural features prior to and their evolution following implantation can yield an understanding of structural correlates of favorable performance and identify predisposition to specific failure modes, which can be extremely valuable.

Device retrieval analysis has an important regulatory role, as specified in the Safe Medical Devices Act of 1990 (PL101-629), the first major amendment to the Federal Food, Drug and Cosmetics Act since the Medical Device Amendments of 1976. The user requirements of the 1990 legislation require health-care personnel and hospitals to report (within 10 days) to the Food and Drug Administration (FDA) or manufacturers or both (depending on the nature of the occurrence) all prosthesis-associated complications that cause death, serious illness, or injury. Such incidents are often discovered during a pathologist's diagnostic evaluation of an implant in the autopsy suite or the surgical pathology laboratory.

Implant retrieval and evaluation programs may also provide specimens and data that serve as a teaching or research resource. The information gained can be used to educate patients, their families, physicians, residents, students, engineers, and materials scientists, as well as the general public. As a research resource, implant retrieval and evaluation yields data that can be used to develop and test hypotheses and to improve protocols and techniques.

COMPONENTS OF IMPLANT RETRIEVAL AND EVALUATION

Implant retrieval and evaluation is an interdisciplinary effort by scientists with expertise in materials science, materials engineering, biomechanics, biology, pathology, microbiology, radiology, medicine, and surgery. Table 2 lists the important components of such a program. Given these components, the importance of an interdisciplinary or team perspective is apparent.

The first component is the appropriate accessioning, cataloging, and identification of retrieved implants. Patient anonymity is required in any implant retrieval and evaluation program and this can be achieved through appropriate accessioning cataloging, coding of demographic information, and restriction of access. As is typical at many institutions, retrieval at University Hospitals of Cleveland/Case Western Reserve

TABLE 2 Important Components and Features of an Implant Retrieval and Evaluation Program

Entire activity is hypothesis-driven

Specimens are appropriately accessioned, cataloged, and identified

Known and potential failure modes are considered

Patient's medical history and laboratory results are reviewed

Data are collected on well-designed, study-specific forms

Careful gross examination, photography, and other basic analyses are always done

Advanced analytical techniques done by specialists are considered

Analytical protocols and techniques for assessing host and implant responses are rigorously followed

Correlations and cause-and-effect relationships among material, design, mechanical, manufacturing, clinical, and biological variables are sought

Quantitative data are collected wherever possible and appropriate

Statistical and multivariant analyses are used

University is carried out through the Surgical Pathology and Autopsy Divisions of the Institute of Pathology (see Table 3).

This program uses the general accessioning and cataloging scheme for surgical pathology specimens. Gross and microscopic diagnoses in a standardized format are provided for patients' charts, for the appropriate clinicians and surgeons, and for the database for Implant Retrieval and Evaluation Program. The Department of Surgery provides specimens to the Division of Surgical Pathology, with the patient's name and hospital identification number, clinical diagnoses, and notes on the patient and/or implant history.

As can be seen from Table 3, the number of types of implants retrieved is large and covers the orthopedic, cardiovascular, gynecological, and soft-tissue areas. Second, the number of

TABLE 3 Retrieved Implants, Division of Surgical Pathology, Institute of Pathology, University Hospitals of Cleveland

Category	Total	1999–2002	1994–1998	1993–1989	1988–1984	1983–1979
Hip	1624	314	383	497	311	119
Knee	808	178	203	250	108	69
Catheter	1135	135	243	427	199	131
Graft	870	179	195	231	166	99
IUD	210	38	30	55	60	27
General	1178	282	280	243	284	89
Breast	624	139	232	200	42	11
Finger	96	7	7	24	34	24
Orthopedic	3079	422	762	796	683	416
Heart valve	289	50	34	53	96	56
Total	9913	1744	2369	2776	1983	1041

TABLE 4   Patient Conditions and Other Factors Influencing Implant Failure or Success

Orthopedic	Cardiovascular
Polyarthritis syndromes	Atherosclerosis
Connective-tissue disorders	Diabetes
Osteoarthritis	Infection
Trauma	Hypertension
Infection	Ventricular hypertrophy
Metabolic disease	Arrhythmias
Endocrine disease	Coagulation abnormalities
Tumor	Cardiac function
Primary joint disease	Recipient activity level
Osteonecrosis	
Recipient activity level	

retrieved implants of each type has increased over the past decade. Several factors are responsible for this. They include increased recognition by surgeons of the importance of implant retrieval and evaluation as well as the increased numbers of implants.

An in-depth evaluation of a retrieved implant requires a review of the patient's medical history and radiographs, where pertinent. Table 4 provides a partial list of conditions that may influence the failure or success of orthopedic and cardiovascular implants. The identification of acute and chronic problems presented by the patient will provide guidance on how the evaluation of a specific implant should be carried out. For example, a clinical diagnosis of infection necessitates that certain techniques should be used in the evaluation. In addition, gross examination and photography play an important role and must be carried out before specific techniques are used to evaluate implants.

It is critical to use a strategy that will allow optimal information yield using appropriate analytical protocols and techniques to assess host and implant responses. This strategy is directed toward developing correlative and cause-and-effect relationships among material, design, mechanical, manufacturing, clinical, and biological variables. Finally, whenever possible, analytical protocols and techniques should produce quantitative information that can be analyzed statistically. These analyses may also include clinical information.

Procedures used to evaluate devices and prostheses after function in animals and humans are largely the same. However, subject to humane treatment considerations, enumerated in institutional and National Institutes of Health (NIH) guidelines that enforce the Federal Animal Welfare Act of 1992, animal studies permit more detailed monitoring of device function and enhanced observation of morphologic detail (including blood–tissue–biomaterials interaction), as well as frequent assay of laboratory parameters (such as indices of platelet function or coagulation), and allow *in situ* observation of fresh implants following elective sacrifice at desired intervals. In addition,

specimens from experimental animals can often be obtained rapidly, thereby minimizing the autolytic changes that occur when tissues are removed from their blood supply. Furthermore, advantageous technical adjuncts may be available in animal but not human investigations, including *in vivo* studies, such as injection of radiolabeled ligands for imaging platelet deposition, fixation by pressure perfusion that maintains tissues and cells in their physiological configuration following removal, and injection of various substances that serve as informative markers during analysis (such as indicators of endothelial barrier integrity). Animal studies often facilitate observation of specific complications in an accelerated time frame, such as calcification of bioprosthetic valves, in which the equivalent of 5–10 years in humans is simulated in 4–6 months in sheep. Moreover, in preclinical studies, experimental conditions can be held constant among groups of subjects, including nutrition, activity levels, and treatment conditions. Consequently, concurrent control implants, in which only a single critical parameter varies, are often available in animal but not human studies.

Most implant retrieval involves examination of failed implants that have been surgically removed from patients or have been encountered at autopsy. Nevertheless, critical information can accrue from examination of "successful" implants removed after either fulfilling their function (e.g., a bone fixation device) or death of the patient resulting from an unrelated cause (e.g., a heart valve prosthesis in a patient who dies of cancer). However, such studies may present difficulties, including but not limited to access to the anatomic site, timing of the explantation following excessive postmortem interval, and family permission (Jacobs *et al.*, 1999).

APPROACH TO ASSESSMENT OF HOST AND IMPLANT RESPONSES

Evaluation of the implant without attention to the tissue will produce an incomplete evaluation and no understanding of the host response. Table 5 provides a partial list of techniques for evaluating implants and tissues. We anticipate that novel implants and materials may dictate further techniques to be used in evaluating host–biomaterials interactions and new analytical techniques will likely be developed.

In general, analytical protocols and techniques for assessing host and implant responses can be divided into two categories: nondestructive and destructive testing procedures. Only after appropriate accessioning, cataloging and identification, and complete review of the patient's medical history and radiography can the analytical protocols and techniques for implant evaluation be specified.

It should be noted that the techniques for implant evaluation are most commonly destructive techniques, that is, the implant or portions of the implant must be destroyed or altered to obtain the desired information on the properties of the implant or material. The availability of the implant and tissue specimens will dictate the choice of technique.

Detailed analysis of the implant–tissue interface often necessitates that both a piece of tissue and a piece of the implant which it contacts be sectioned and examined in continuity (i.e., as one unit). This generally precludes any other analysis of that specimen. Similarly, chemical analysis of a piece of tissue for metal ion concentration requires acid digestion and thus precludes morphological analysis. Thus, consideration must be given to either selection of multiple specimens or division of a specimen before processing in order to permit part to be used for morphological and part for chemical analyses. Each of these portions may need to be further subdivided to be processed according to the requirements of specific tests. Moreover, since many of the procedures include cutting or sectioning a portion of the retrieved device, one must consider the legal aspects of device ownership and destruction of what may become evidence in a court of law. Strict institutional guidelines need to be adhered to, and permission for use of destructive methods may be necessary. For clinicopathologic correlation studies that involve utilization of both patient data and pathological findings, approval of the local institutional review board (IRB) is required. This is largely to ensure that the confidentiality and other interests of the patients from whom the implants were removed will be maintained, especially for purposes of publication or other dissemination of information.

Special Technical Problems of Bone and Calcified Tissue

Evaluation of an implant and the surrounding host tissue when the implant was placed into bone, bone has formed around it, local tissue has calcified, or teeth presently pose unique problems. Dental implants and orthopedic implants are usually associated with hard tissue, but other implants may have contact with calcified tissue. Material containing calcified tissue, bone, or teeth should be identified before the laboratory begins processing it or there may be harm to some of the equipment used for sectioning soft tissue. In addition some consensus standards, such as ASTM F561 and ISO/FDIA 12891-1 have been developed that address the special issues. Bone is a dynamic tissue and analysis of bone formation and bone resorption may be a critical part of the study. Selection of techniques for evaluation must be undertaken carefully. Consideration needs to be given to what questions are being addressed before anything is cut or any harsh solutions are used.

If the implant and tissue are handled correctly, the techniques listed in Table 5 are applicable to the study of hard tissue and associated implants. As stated earlier, some techniques are mutually exclusive. For example, histologic analysis of particle size and quantity and chemical digestion and analysis for metal ions cannot be done on the same sample.

Following routine fixation to preserve cellular detail, either of two generic approaches may be taken with implants adjacent to hard tissue: preparation of undecalcified sections or decalcification prior to embedding. Fully decalcified bone can be embedded and sectioned using methods common to soft-tissue pathology. To prepare undecalcified sections, hard tissue and the associated implant component or section are dehydrated and embedded in a hard plastic (usually methyl methacrylate)

TABLE 5 Techniques for Implant Evaluation

Implant	Tissue
Atomic absorption spectrophotometry	Atomic absorption spectrophotometry
ATR-FTIR	Autoradiography
Burst strength	Biochemical analysis
Compliance studies	Cell culture
Contact angle measurements	Chemical analysis
Digestability	Enzyme histochemistry
EDAX	Gel electrophoresis
ESCA	Histology
Extractability	Immunocytochemistry
Fatigue studies	Immunofluorescence
Fracture analysis	Immunohistochemistry
FTIR	Immunoperoxidase
Gel permeation chromatography	*In situ* hybridization
Glass transition temperature	Microbiologic cultures
Hardness studies	Molecular analysis for cellular gene expression
Light microscopy	
Macrophotography	Morphometry
Melt temperature	Radiography
Metallographic examination	Scanning electron microscopy
Particulate analysis	Tissue culture
Polarized light microscopy	Transmission electron microscopy
Porosity analysis	
Radiography	
Scanning electron microscopy	
Shrink temperature	
SIMS	
Stereomicroscopy	
Stress analysis	
Tensile studies	
Transmission electron microscopy	
Topography analysis	

and thick (approximately 500 μm) sections are cut with a diamond saw and ground down to form the final thin section (approximately 50 μm). The advantage of undecalcified sections is that areas of mineralization and newly formed bone can be clearly distinguished from soft tissue. Specimen X-rays can also be taken to identify sites and extent of mineralization. Assistance with techniques for evaluation of specimens containing bone is available from the Society of Histotechnology (Callis and Strerchi, 1998; McNeil *et al.*, 1997; Sanderson and Bachus, 1997).

Other Special Issues

Because implant retrieval programs may obtain data that will be used in regulatory or litigation procedures, it is important to document the disposition and appearance of the tissue and of the implant at each stage in the evaluation. Ample use of photography will greatly help in the description of the tissue and implant condition at each step of analysis. Careful labeling that is recorded during photography is helpful in later review of the evaluation procedure.

Concerns relative to the safety of laboratory personnel arise in the examination of retrieved materials. If the device and/or tissues arrive in the laboratory without having been adequately disinfected or treated, special precautions need to be observed with the use of barrier clothing and gloves and the use of biosafety cabinets. All safety regulations of the institution need to be known and followed. Procedures for disposal of hazardous chemicals also need to be carefully observed. If untreated devices or tissues are to be packaged and shipped from the clinical setting to the laboratory, care must be taken to follow the rules and regulations of the shipping agency and the Department of Transportation for transportation of biohazardous material.

An additional and very complex matter relates to implant ownership (Beyleveld *et al.*, 1995). Several parties may have (and claim) rights to possession of a removed device under various circumstances: the patient, the surgeon, the pathologist, the hospital, an entity such as an insurance company, and attorneys representing both the plaintiff and the defendant in product liability or malpractice litigation. Moreover, ownership may not automatically entail the right to test. These issues are both controversial and unresolved.

The Multilevel Strategy to Implant Evaluation

Implant retrieval and analysis protocols often utilize a multilevel approach along the spectrum of essential documentation to the use of sophisticated research tools. Some investigators have advocated a basic two-level approach (Table 6). Level I studies include routine evaluation modalities capable of being

TABLE 6   Study Prioritization

Level I studies
    Gross dissection
    Photographic documentation
    Microbiologic cultures
    Radiography
    Light microscopic histopathology

Level II studies
    Scanning electron microscopy
    Transmission electron microscopy
    Energy dispersive X-ray microanalysis (EDXA)
    Analysis of adsorbed and absorbed proteins
    Mechanical properties measurement
    Materials surface analysis
    Leukocyte immunophenotypic studies
    Molecular analyses for cellular gene expression

done in virtually any laboratory, and that characterize the overall safety and efficacy of a device. Level II studies comprise well-defined and meaningful test methods that are difficult, time-consuming, or expensive to perform, require special expertise, or yield more investigative or esoteric data. Since some level II analyses may be mutually exclusive, some material might routinely need to be accessioned, set aside (during level I analyses) and prepared for more specialized level II studies, in the event that they should be indicated later. Level II evaluation is usually undertaken with specific investigative objectives directed toward a focused research question or project. Prioritized, practical approaches using these guidelines have been described for heart valves, cardiac assist devices and artificial hearts, and other cardiovascular devices (Schoen *et al.*, 1990; Schoen, 1995; Borovetz, 1995; Schoen, 2001; Schwartz *et al.*, 2002).

Both ASTM and ISO implant retrieval standards (F561 and ISO 12891, respectively) have utilized a three-stage approach to implant analysis. Stage I analysis consists of routine device identification and description. Stage II is more detailed, time-consuming, and expensive and includes photography and nondestructive failure analysis. Since stage I and II protocols are the same for all material types, they are combined in F561. Stage III protocols include destructive analytical techniques, many of which are specific to particular material types, and thus separate guidelines are provided for metallic, polymeric, and ceramic materials. Combinations of these protocols provide guidelines for analysis of the various components of composites and tissue engineered materials.

THE ROLE OF IMPLANT RETRIEVAL IN DEVICE DEVELOPMENT

For implants placed into many anatomic sites, prosthesis-associated pathology is a major determinant of quality of life and prognosis of patients. Implant retrieval and evaluation plays a critical role in the evolution of medical devices through development and clinical use (Fig. 1). Implants are examined as (1) fabricated but unimplanted prototypes, to reveal changes in device components induced by the fabrication process that can predict (or lead to an understanding of) failure modes observed subsequently; (2) specimens that have been subjected to *in vitro* tests of biocompatibility or durability; (3) specimens removed from animal models following *in vivo* function, usually as actual devices; (4) specimens that accrue in carefully controlled clinical trials; and (5) failed or functioning specimens that are explanted in the course of ongoing surveillance of general clinical use of a device. In each case, implant retrieval is concerned with the documentation of device safety and efficacy and problems that dictate modification of design, materials, or use of the implants in patients (e.g., patient–device matching or patient management). Devices are explanted and analyzed with attention directed toward all complications, but especially those that may be considered special vulnerabilities engendered by the device type. Efforts are made to retrieve and evaluate as many such implants as possible, both failed and nonfailed. Evaluation of retrieved animal implants plays a special role in the documentation for the FDA-Investigational Device Exemption (IDE) required for a clinical trial and for the FDA-Premarket

Approval (PMA) required for general marketing. The literature contains numerous instances where problem-oriented implant evaluation studies have yielded important insights into the deficiencies and complications that have limited the success of various implants. Implant retrieval has also contributed to the assessment of the safety and efficacy of implant modifications intended to be improvements. The conceptual approach to the analysis of a failed implant is explored in Chapter 9.3.

WHAT CLINICALLY USEFUL INFORMATION HAS BEEN LEARNED FROM IMPLANT RETRIEVAL AND ANALYSIS?

Cardiovascular Implants

Cardiovascular implants commonly involve both blood and soft-tissue interactions with materials. The complications most commonly found with these implants, i.e., heart valve

TABLE 7 Complications of Heart Valve Substitutes, Vascular Grafts and Cardiac Assist/Replacement Devices

Heart valve prostheses	Vascular grafts	Cardiac assist/ replacement devices
Thrombosis	Thrombosis	Thrombosis
Embolism	Embolism	Embolism
Paravalvular leak	Infection	Endocarditis
Anticoagulation-related hemorrhage	Perigraft erosion	Extraluminal infection
Infective endocarditis	Perigraft seroma	Component fracture
Extrinsic dysfunction	False aneurysm	Hemolysis
Mechanical valves	Anastomotic hyperplasia	Calcification
Cloth wear	Disintegration/ degradation	
Incomplete valve closure		
Component fracture		
Tissue valves		
Cusp tearing		
Cusp calcification		
Hemolytic anemia		

TABLE 8 Complications of Vascular Stents

Thrombosis

Proliferative restenosis

Strut related inflammation

Foreign body reaction

Incomplete expansion

Overexpansion

Infection

prostheses, vascular grafts, cardiac assist devices/artificial hearts, and vascular stents are listed in Tables 7 and 8. Perspectives, approaches, and techniques for evaluating cardiovascular implants have been described (Schoen, 1990; Schoen, 1995; Borovetz, 1995; Schoen, 2001; Schwartz *et al.*, 2002). Specific examples from the field of cardiovascular prostheses are summarized next and in Table 9 to illustrate the scope and utility of this activity.

Experience with caged-ball mechanical prosthetic heart valves in the 1960s demonstrated both thrombus and degradation as complications. Degradation of silicone due to lipid absorption (called ball variance) was virtually eliminated by improved curing of silicone poppets (Hylen *et al.*, 1972). Subsequently, a cloth-covered caged-ball mechanical prosthetic heart valve that had a ball fabricated from silicone and a metallic cage covered by polypropylene mesh was introduced. The cloth-covered cage struts were designed to encourage tissue ingrowth and thereby decrease thromboembolism. Preclinical studies in pigs, sheep, and calves demonstrated rapid healing by fibrous tissue of the cloth-covered struts. However, subsequent clinical studies of these valves showed abundant cloth wear in some patients, sufficient to cause escape of the ball through the spaces between the cage struts, a circumstance that was rapidly fatal without emergency surgery (Schoen *et al.*, 1984). This analysis demonstrated the general concept that valve prostheses used in humans may have important complications that were not predicted by animal investigations, in this case because the vigorous healing that occurred in animals but not in humans obscured the potential problem.

Modification of a tilting-disk valve design intended to permit enhanced opening and thereby reduce thromboembolism also led to an unfortunate clinical problem. The design change necessitated a manufacturing change in the manner in which metallic struts were welded to the housing. An unusually large cluster of clinical implants of these valves failed by fracture of the welds anchoring one of the two metallic struts that guided and restricted disk motion (Walker *et al.*, 1995). This led to separation of a strut from the housing and consequent disk escape with acute valve failure, which was frequently fatal. Detailed materials analysis of retrieved implants that had failed revealed that the underlying problem was metal fatigue at the strut–housing junction resulting from overrotation of the disk during valve closure. This caused an abnormally hard strike of the disk on the strut, and consequent excessive bending stresses near the weld, which coupled with the intrinsic porosity and resultant vulnerability of welded joints induced fatigue failure (Schoen *et al.*, 1992).

Multiple coordinated studies of retrieved experimental and clinical implants have characterized and facilitated therapeutic approaches toward elimination of calcification-induced failure modes in bioprosthetic tissue heart-valve replacement (Schoen and Levy, 1999). Such studies have identified the causes and morphology of failure, patterns of mineral deposition, nature of the mineral phase, and early events in the mechanisms of calcification. Calcification of valve tissue in both circulatory and subcutaneous animal models exhibits morphological features similar to that observed in clinical specimens, but calcification is markedly accelerated in the experimental explants. Such studies have provided the means to test approaches to inhibit bioprosthetic valve calcification by modifying host, implant,

TABLE 9   Clinical Utility of Retrieval Studies—Heart Valve Substitutes

Implant type	Knowledge gained/lessons learned
Caged ball valves	• Poppets fabricated from industrial silicone absorbed blood lipids and became swollen and brittle • Fragments of degraded heart valve material may embolize to other organs • Thrombosis can occur at stasis points downstream of the ball
Cloth-covered caged ball valves	• Cloth wear can cause hemolysis and cloth emboli • Cloth wear accompanied by silicone poppet wear can precipitate poppet escape • Healing of fabric may be more vigorous in animals than in humans • Quantitation of data (e.g., polymeric poppet weight and dimensions) may facilitate the understanding of a failure mode
Caged disk valves	• Teflon has poor wear resistance as a valve occluder • Poor design features may potentiate thrombosis
Tilting-disk valves	• Thrombosis may initiate downstream to the edge of a partially open disk at a region of stasis • A "minor" change in valve design can result in a new propensity toward failure • Animal implants instrumented with strain gauges can be used to test a hypothetical mechanical failure mechanism • Understanding a failure mode can lead to both new methods for noninvasive diagnosis (e.g., X-ray and acoustic) and modified patient management strategies (e.g., drugs to depress cardiac contractility)
Bileaflet tilting-disk valves	• Cavitation may cause critical materials damage in some valve designs • Thrombosis may be initiated in regions of microstasis at component junctions • Microscopic areas of stasis may be predicted by computer-assisted computation • Animal implant models may fail to predict vulnerability to thrombosis in humans
Bioprosthetic heart valves	• Tissue calcification is a major failure mode • Calcification is most pronounced in areas of leaflet flexion, where deformations are maximal • Calcification is accelerated in young recipients, especially children • Heart valve calcification can be studied outside of the circulation (e.g., subcutaneous implants in rats) • Calcification is initiated principally at cell remnants deep in the tissue
Cryopreserved allograft valves	• These valves are not viable and cannot grow • Failure is not immunologically mediated and, therefore, immunosuppression is inappropriate
Substitution of new materials	• Pyrolytic carbon has favorable clinical durability • Detailed examination of functional (not failed) prostheses may yield worthwhile data

and mechanical influences. These studies serve to emphasize that biological as well as mechanical failure mechanisms can be understood using specifically chosen and controlled animal model investigations, guided by and correlated with the results of careful studies of retrieved clinical specimens. Furthermore, they show how implant retrieval studies can be used as a critical component of a program to assure efficacy and safety of potential therapeutic modifications.

An additional use of retrieval studies is exemplified by a case in which a change in biomaterials was shown to be beneficial. Implant retrieval studies of a low-profile disk valve composed of a disk and cage fabricated from Teflon demonstrated poor wear properties of the disk. After a new model of the valve with disk and struts fabricated from pyrolytic carbon was introduced, clinical experience suggested that the use of carbon contributed to a major advancement in the durability of prosthetic heart valves. Retrieval studies of carbon valves recovered at autopsy or surgery and analyzed by surface scanning electron microscopy and surface profilometry indicated that compared with valves composed of Teflon, carbon valves exhibited minimal abrasive wear (Schoen et al., 1982). This study revealed that although analysis of implants and medical devices has traditionally concentrated on those devices that

failed in service, important data can accrue from implants serving the patient until death or removal for unrelated causes. Thus, detailed examination of the properly functioning prostheses removed from patients after long duration of implantation may yield worthwhile data, provided that a focused question is asked of the material.

Other retrieval studies have revealed (1) the cause of excessive thrombosis of a new bileaflet tilting-disk design (Gross et al., 1996); (2) the cause of mechanical failure of a new pericardial bioprosthesis fabricated from photofixed bovine pericardium that had different mechanical properties than conventional glutaraldehyde-fixed tissue (Schoen, 1998a); and (3) the cause of failure and characteristic changes that occur in cryopreserved allograft aortic heart valves transplanted from one individual to another (Mitchell et al., 1998). Studies of ventricular assist devices have demonstrated the importance of valves or biomaterial junctions in initiating thrombosis (Fyfe and Schoen, 1993; Wagner et al., 1993), and several studies have described the failure modes of specific types of vascular grafts in different locations (Canizales et al., 1982; Downs et al., 1991; Guidoin et al., 1993) and the morphology of vascular stents (Anderson et al., 1992; van Beusekom et al., 1993; Farb et al., 1999).

Orthopedic and Dental Implants

Complications most commonly found with orthopedic and dental implants are presented in Table 10 and Table 11, respectively. The clinical utility of hard-tissue (orthopedic and dental) implant retrieval and evaluation is summarized in Table 12.

Early studies were concerned primarily with metallic devices used for internal fixation of fractures. Examination of some early designs showed failure at regions of poor metallurgy or weak areas due to poor design (Cahoon and Paxton, 1968). Other studies (Cook *et al.*, 1985; Wright *et al.*, 1982) reinforce the importance of material and design, such as the pitfall of stamping the manufacturer's name in the middle of the plate where stresses are high. Modern total hip replacements have also had fatigue fractures initiated at labels etched at high-stress regions (Woolson *et al.*, 1997). Galvanic corrosion due to mixed metal alloys was also demonstrated in early retrieval studies (Scales *et al.*, 1959). Analyses of fixation devices have shown a correlation between corrosion and biological reactions such as allergic reactions to metallic elements in the implants (Brown and Merritt, 1981; Cook *et al.*, 1988; Merritt and Rodrigo, 1996).

The development of the low-friction total hip arthroplasty by Charnley is a classical example of how implant retrieval can yield important clinical information which impacts implant materials and design (Charnley *et al.*, 1969). The initial choice for the low-friction polymeric cup (replacing the acetabulum) was a Teflon-like PTFE. As in the heart valves described above, this relatively soft material had a very high wear rate and within a few years, many hip replacements had been removed due to severe wear and intense inflammatory reactions. The analysis

TABLE 10 Complications of Orthopedic Implants

Bone resorption	Loosening
Corrosion	Mechanical mismatch
Fatigue	Fracture
Fibrosis	Motion and pain
Fixation failure	Particulate formation
Fracture	Loss
Incomplete osseous integration	Stress riser
Infection	Loss
Interface separation	Surface wear

TABLE 11 Complications of Dental Implants

Adverse foreign-body reaction	Loosening
Biocorrosion	Foreign body reaction
Electrochemical galvanic coupling	Corrosion
Fatigue	Particulate formation
Fixation failure	Loosening
Fracture	Loss
Infection	Loss
Interface separation	Wear
Loss of mechanical force transfer	Loosening

TABLE 12 Clinical Utility of Retrieval Studies: Hard-Tissue Implants

Implant type	Knowledge gained/lessons learned
Plates, screws, and rods used for fracture fixation	• Do not mix metal alloys in same device • Match the hardness (degree of cold work) and stiffness with the application • Metallic implant wear and corrosion may lead to problems associated with allergic reactions
Tooth root implants	• Careful surgical technique avoids thermal tissue necrosis of the implant–tissue interface • Implant surface finish is a critical determinant of outcome
Femoral stems of total hip replacements (fracture analysis)	• Cobalt alloys require high-quality casting • High-strength superalloys may be advantageous • Welded regions may fail in tensile-loaded locations • Part numbers should not be etched in tensile-loaded locations
Femoral stem, modular interface	• Corrosion at an interface is dependent on metallurgical and mechanical design factors
Polymeric component to total joint replacements (excluding one-piece flex hinge joints like finger)	• Teflon performs poorly in wear applications • Reinforcement of UHMWPE with chopped carbon fiber is ineffective • Laboratory simulation should be done using clinically realistic implant wear motion and loading patterns • Mechanical designs that produce high localized stresses may cause delamination • Radiation sterilization to produce cross-linking is useful but can lead to molecular-chain scission, oxidation, and aging
Analysis of tissues surrounding total joint replacements	• Wear particles derived from breakdown of the implant–bone interface or wear of the articulation (bearing) material cause inflammation and loosening.
Titanium and titanium-alloy implants	• A good material in one implant application may not necessarily be good in another • Commercially pure titanium is appropriate for tooth root implant or fracture fixation devices • Titanium alloys may yield severe wear in some total joint applications

of the retrievals has also demonstrated the relationships between large head diameter and the wear debris and excessive wear that have served as the basis for our understanding of joint bearing design. Clinical and retrieval studies are now examining the performance of metal-on-metal total hip bearing couples (Doorn *et al.*, 1996).

Teflon was replaced by ultrahigh-molecular-weight polyethylene (UHMWPE) as the acetabular cup bearing surfaces, and retrieval studies have continuously stimulated changes in the methods of processing UHMWPE. Retrieval studies have demonstrated that reinforcing UHMWPE with carbon fibers did not sufficiently improve its strength (Wright *et al.*, 1988). Heat pressing to improve surface hardness was not effective with some total knee designs (Bloebaum *et al.*, 1991; Wright *et al.*, 1992). Work was also directed toward assessing the effects of radiation sterilization on the polymer (Sutula *et al.*, 1995).

Implant retrieval and analysis has provided critical information toward understanding the anchoring of implants in bone. Branemark and associates demonstrated that direct bone–metal osseointegration of titanium tooth-root implants was possible with carefully controlled surgical technique and implant surface preparation (Albrektsson *et al.*, 1982). Porous coatings may be applied to implants to facilitate bony ingrowth; relationships between pore size and bony ingrowth have come from retrieval studies (Cook *et al.*, 1988).

Examination of the interface between the bone and the acrylic bone cement in long-term implant recipients examined at autopsy demonstrated the presence of inflammation with macrophages, thus providing the initial description of bone resorption due to particulate debris (Charnley, 1970). The relationship between the amount of particulate wear debris generated and bone resorption due to inflammation was graphically demonstrated by the retrieval and tissue studies of Willert and Semlitsch (1977). Extensive research has been conducted over the years looking at "particle disease" and characterization of the polymeric debris, as correlated with implant type and design (Schmalzied *et al.*, 1997). Studies on tissue reactions to particulates and wear debris require special histologic evaluation as described by Wright and Goodman (1996), in the various consensus standards, in other references cited earlier, and in many other studies readily available in the orthopedic or histotechnology literature.

Implant retrieval studies have demonstrated that success of a material in one application may not necessarily translate to another. For example, titanium does well as a bone–implant interface, but it may be subject to severe wear as a bearing surface (Agins *et al.*, 1988). An alternative approach in the hip was to use titanium alloy stems with a modular press-fit heads made of a more wear-resistant cobalt alloy. However, some of these designs demonstrated significant corrosion, which was first attributed to use of mixed metals (Collier *et al.*, 1991), and later to microgalvanic corrosion of the cast heads (Mathiasen *et al.*, 1991) and to micromotion and fretting corrosion between components (Brown *et al.*, 1995).

As with orthopedic devices, complications of dental implants may be related to the mechanical–biomechanical aspects of force transfer, or the chemical–biochemical aspects of elements transferred across biomaterial and tissue interfaces, or both. The complex synergism that exists between tissue and biomaterial responses presents a significant challenge in the identification of the failure mechanisms of dental implants. Since many dental devices are percutaneous, in that they contact bone but are also exposed to the oral cavity, the problems of infection are also significant (Moore, 1987; Sussman and Moss, 1993).

Lemons (1988) and others have provided appropriate perspectives to be taken in the evaluation of dental implants.

CONCLUSIONS

Although the focus of this chapter has been on retrieval and evaluation of implants in humans, the perspectives, approaches, techniques, and methods may also apply to evaluation of new material, preclinical testing for biocompatibility, and premarket clinical evaluation. Each type of *in vivo* or clinical setting has its unique implant–host interactions and therefore requires the development of a unique strategy for retrieval and evaluation.

Indeed, for investigation of bioactive materials and tissue-engineered medical devices, in which interactions between the implant and the surrounding tissue are complex, novel and innovative approaches must be used in the investigation of *in vivo* tissue compatibility. In such cases, the scope of the concept of "biocompatibility" is much broader and the approaches of implant retrieval and evaluation require measures of the phenotypes and functions of cells and the architecture and remodeling of extracellular matrix and the vasculature (Rabkin *et al.*, 2002; Rabkin and Schoen, 2002; Peters *et al.*, 2002; Mizuno *et al.*, 2002). Moreover, a critical role of implant retrieval will be the identification of tissue characteristics (biomarkers) that will be predictive of (i.e., surrogates for) success and failure. A most exciting possibility is that such biomarkers may be used to noninvasively image/monitor the maturation/remodeling of tissue-engineered devices *in vivo* in individual patients.

Bibliography

Agins, H. J., Alcock, N. W., Bansal, M., Salvati, E. A., Wilson, P. D., Pellicci, P. M., and Bullough, P. G. (1988). Metallic wear in failed titanium-alloy hip replacements. *J. Bone Joint Surg.* **70A**: 347–356.

Albrektsson, T., Branemark, P.-I., Hansson, H.-A., and Lindstrom, J. (1982). Osseo-integrated titanium implants. *Acta. Orthop. Scand.* **52**: 155–170.

Anderson, J. M. (1986). Procedures in the retrieval and evaluation of vascular grafts. in *Vascular Graft Update: Safety and Performance*, H. E. Kambic, A. Kantrowitz, and P. Sung (eds.). ASTM STP 898 American Society for Testing and Materials, Philadelphia, pp. 156–165.

Anderson, J. M. (1993). Cardiovascular device retrieval and evaluation. *Cardiovasc. Pathol.* **2**(3)(Suppl.): 199S–208S.

Anderson, P. G., Bajaj, R. K., Baxley, W. A., and Roubin, G. S. (1992). Vascular pathology of balloon-expandable flexible coil stent in humans. *J. Am. Coll. Cardiol.* **19**: 372–381.

ASTM F561–97 (1999). Practice for Retrieval and Analysis of Implanted Medical Devices, and Associated Tissues. *ASTM Annual Book of Standards*, Vol. 13.01.

Beyleveld, D., Howells, G. G., and Longley, D. (1995). Heart valve ownership: legal, ethical and policy issues. *J. Heart Valve Dis.* 4: S2–S5.

Bloebaum, R. D., Nelson, K., Dorr, L. D., Hoffman, A., and Lyman, D. J. (1991). Investigation of early surface delamination observed in retrieved hard-pressed tibial inserts. *Clin. Orthop.* 269: 120–127.

Borovetz, H. S., Ramasamy, N., Zerbe, T. R., and Portner, P. M. (1995). Evaluation of an implantable ventricular assist system for humans with chronic refractory heart failure. Device explant protocol. *ASAIO J.* 41: 42–48.

Brown, S. A., and Merritt, K. (1981). Metal allergy and metallurgy. in: *Implant Retrieval: Material and Biological Analysis*, NBS SP 601. Weinstein, Gibbons, Brown, Ruff (eds.), pp. 299–321.

Brown, S. A., Flemming, C. A. C., Kawalec, J. S., Placko, H. E., Vassaux, C., Merritt, K., Payer, J. H., and Kraay, M. J. (1995). Fretting accelerated crevice corrosion of modular hip tapers. *J. Appl. Biomater.* 6: 19–26.

Cahoon, J. R., and Paxton, H. W. (1968). Metallurgical analysis of failed orthopaedic implants. *J. Biomed. Mater. Res.* 2: 1–22.

Callis, G., and Strerchi, D. (1998). Decalcification of bone: literature review and practical study of various decalcifying agents. Methods, and their effects on bone histology. *J. Histotechnol.* 21: 49–58.

Canizales, S., Charara, J., Gill, F., Guidoin, R., Roy, P. E., Bonnaud, P., Laroche, G., Batt, M., Roy, P., Marois, M., *et al.* (1982). Expanded polytetrafluoroethylene prostheses as secondary blood access sites for hemodialysis: pathological findings in 29 excised grafts. *Can. J. Surg.* 154: 17–26.

Charnley, J. (1970). The reaction of bone to self-curing acrylic cement. A long-term histological study in man. *J. Bone. Joint Surg.* 53B: 340–353.

Charnley, J., Kamangar, A., and Longfield, M. D. (1969). The optimum size of prosthetic heads in relation to the wear of plastic sockets in total replacement of the hip. *Med. Biol. Eng.* 7: 31–39.

Collier, J. P., Suprenant, V. A., Jensen, R. E., and Mayor, M. B. (1991). Corrosion at the interface of cobalt-alloy heads on titanium alloy stems. *Clin. Orthop.* 271: 305–312.

Cook, S. D., Renz, E. A., Barrack, R. L., Thomas, K. A., Harding, A. F., Haddad, R. J., and Milicic, M. (1985). Clinical and metallurgical analysis of retrieved internal fixation devices. *Clin. Orthop.* 184: 236–247.

Cook, S. D., Barrack, R. L., Thomas, K. A., and Haddad, Jr., R. J. (1988a). Quantitative analysis of tissue growth into human porous total hip components. *J. Arthroplasty* 3: 249–262.

Cook, S. D., Thomas, K. A., and Haddad, R. J., Jr. (1988b). Histologic analysis of retrieved human porous-coated total joint components. *Clin. Orthop.* 234: 90–101.

Doorn, P. F., Campbell, P. A., and Amstutz, H. A. (1996). Metal versus polyethylene wear particles in total hip replacements. A review. *Clin. Orthop.* 329S: S206–S216.

Downs, A. R., Guzman, R., Formichi, M., Courbier, R., Jausseran, J. M., Branchereau, A., Juhan, C., Chakfe, N., King, M., and Guidoin, R. (1991). Etiology of prosthetic anastomotic false aneurysms: pathologic and structural evaluation in 26 cases. *Can. J. Surg.* 34: 53–58.

Farb, A., Sangiorgi, G., Carter, A. J., *et al.* (2002). Pathology of acute and chronic coronary stenting in humans. *Circulation* 105: 2932–2933.

Fyfe, B., and Schoen, F. J. (1993). Pathologic analysis of 34 explanted Symbion ventricular assist devices and 10 explanted Jarvik-7 total artificial hearts. *Cardiovasc. Pathol.* 2: 187–197.

Gross, J. M., Shu, M.C.S., Dai, F.F., Ellis, J., and Yoganathan, A. P. (1996). A microstructural flow analysis within a bileaflet mechanical heart valve hinge. *J. Heart Valve Dis.* 5: 581–590.

Guidoin, R., Chakfe, N., Maurel, S., How, T., Batt, M., Varois, M., and Gosselin, C. (1993). Expanded polytetrafluoroethylene arterial prostheses in humans: histopathological study of 298 surgically excised grafts. *Biomaterials* 14: 678–693.

Hylen, J. C., Hodam, R. P., and Kloste, F. E. (1972). Changes in the durability of silicone rubber in ball-valve prostheses. *Ann. Thorac. Surg.* 13: 324.

ISO 12891. Retrieval and Analysis of Implantable Medical Devices.

Jacobs, J. J., Patterson, L. M., Skipor, A. K., Urban, R. M., Black, J., and Galante, J. O. (1999). Postmortem retrieval of total joint replacement components. *J. Biomed. Mater. Res. (Appl. Biomater.)* 48: 385–391.

Lemons, J. E. (1988). Dental implant retrieval analyses. *J. Dent. Educ.* 52: 748–756.

Mathiesen, E. B., Lindgren, J. U., Biomgren, G. G. A., and Reinholt, F. P. (1991). Corrosion of modular hip prostheses. *J. Bone Joint Surg.* 73B: 569–575.

McNeil, P. J., Durbridge, T. C., Parkinson, I. H., and Moore, R. J. (1997). Simple method for the simultaneous determination of formation and resorption in undecalcified bone embedded in methyl methacrylate. *J. Histotechnol.* 20: 307–311.

Merritt, K., and Brown, S. A. (1981). Metal sensitivity reactions to orthopaedic implants. *Int. J. Dermatol.* 20: 89–94.

Merritt, K., and Rodrigo, J. J. (1996). Immune response to synthetic materials: sensitization of patients receiving orthopaedic implants. *Clin. Orthop.* 329S: S233–S243.

Mitchell, R. N., Jonas, R. A., and Schoen, F. J. (1998). Pathology of explanted cryopreserved allograft heart valves: comparison with aortic valves from orthotopic heart transplants. *J. Thorac. Cardiovasc. Surg.* 115: 118–127.

Mizuno, S. Tateishi, T., Ushida, T., and Glowacki, J. (2002). Hydrostatic fluid pressure enhances matrix synthesis and accumulation by bovine chondrocytes in three-dimensional culture. *J. Cell Biol.* 193: 319–327.

Moore, W. E. C. (1987). Microbiology of periodontal disease. *J. Periodontal Res.* 22: 335–341.

Peters, M. D., Polverini, P. J., and Mooney, D. J. (2002). Engineering vascular networks in porous polymer matrices. *J. Biomed. Mater. Res.* 60: 668–678.

Rabkin, E., and Schoen, F. J. (2002). Cardiovascular tissue engineering. *Cardiovasc. Pathol.* 11: 305.

Rabkin, E., Hoerstrup, S. P., Aikawa, M., Mayer, J. E. Jr., and Schoen, F. J. (2002). Evolution of cell phenotype and extracellular matrix in tissue-engineered heart valves during *in vitro* maturation and *in vivo* remodeling. *J. Heart Valve Dis.* 11: 308–314.

Sanderson, C., and Bachus, K. N. (1997). Staining technique to differentiate mineralized and demineralized bone in ground sections. *J. Histotechnol.* 20: 119–122.

Scales, J. T., Winter, G. D., and Shirley, H. T. (1959). Corrosion of orthopaedic implants. *J. Bone Joint Surg.* 41B: 810–820.

Schmalzied, T. P., Campbell, P., Schmitt, A. K., Brown, I. C., and Amstutz, H. C. (1997). Shapes and dimensional characteristics of polyethylene wear particles generated *in-vivo* by total knee replacements compared to total hip replacements. *J. Biomed Mater. Res. (Appl. Biomater.)* 38: 203–210.

Schoen, F. (1989). *Interventional and Surgical Cardiovascular Pathology*. W.B. Saunders, Philadelphia.

Schoen, F. J. (1995). Approach to the analysis of cardiac valve prostheses as surgical pathology or autopsy specimens. *Cardiovasc. Pathol.* 4: 241–255.

Schoen, F. J. (1998a). Pathologic findings in explanted clinical bioprosthetic valves fabricated from photooxidized bovine pericardium. *J. Heart Valve Dis.* 7: 174–179.

Schoen, F. J. (1998b). Role of device retrieval and analysis in the evaluation of substitute heart valves. in *Clinical Evaluation of Medical Devices: Principles and Case Studies*, K. B. Witkin (ed.). Humana Press, Totowa, NJ, pp. 209–231.

Schoen, F. J. (2001). Pathology of heart valve substitution with mechanical and tissue prostheses, in *Cardiovascular Pathology*, 3rd ed., M. D. Silver, A. Gotlieb, and F. J. Schoen (eds.). W.B. Saunders, Philadelphia, pp. 629–677.

Schoen, F. J., and Levy, R. J. (1999). Tissue heart valves: Current challenges and future research perspective. *J. Biomed Mater. Res.* 47: 439–465.

Schoen, F. J., Titus, J. L., and Lawrie, G. M. (1982). Durability of pyrolytic carbon-containing heart valve prostheses. *J. Biomed. Mater. Res.* 16: 559–570.

Schoen, F. J., Goodenough, S. H., Ionescu, M. I., and Braunwald, N. S. (1984). Implications of late morphology of Braunwald–Cutter mitral heart valve prostheses. *J. Thorac. Cardiovasc. Surg.* 88: 208–216.

Schoen, F. J., Anderson, J. M., Didisheim, P., Dobbins, J. J., Gristina, A. G., Harasaki, H., and Simmons, R. L. (1990). Ventricular assist device (VAD) pathology analyses: guidelines for clinical study. *J. Appl. Biomater.* 1: 49–56.

Schoen, F. J., Levy, R. J., and Piehler, H. R. (1992). Pathological considerations in replacement cardiac valves. *Cardiovasc. Pathol.* 1: 29–52.

Schwartz, R. S., and Edelman, E. R. (2002). Drug-eluting stents in preclinical studies. Recommended evaluation from a consensus group. *Circulation* 106: 1867–1873.

Sussman, H. I., and Moss, S. S. (1993). Localized osteomyelitis secondary to endodontic implant pathosis. A case report. *J. Periodontol.* 64: 306–310.

Sutula, L. C., Collier, J. P., Saum, K. A., *et al.* (1995). Impact of gamma sterilization on clinical performance of polyethylene in the hip. *Clin. Orthop.* 319: 28–40.

Van Beusekom, H. M. M., van der Giessen, W. J., van Suylen, R. J., Bos, E., Bosman, F. T., and Serruys, P. W. (1993). Histology after stenting of human saphenous vein bypass grafts: observations from surgically excised grafts 3 to 320 days after stent implantation. *J. Am. Coll. Cardiol.* 21: 45–54.

Wagner, W. R., Johnson, P. C., Kormos, R. L., and Griffith, B. P. (1993). Evaluation of bioprosthetic valve–associated thrombus in ventricular assist device patients. *Circulation* 88: 203–2029.

Walker, A. M., Funch, D. P., Sulsky, S. L., and Dreyer, N. A. (1995). Patient factors associated with strut fracture in Björk-Shiley 60° convexo-concave heart valves. *Circulation* 92: 3235–3239.

Weinstein, A., Gibbons, D., Brown, S., and Ruff, W. (eds.) (1981). *Implant Retrieval: Material and Biological Analysis.* U. S. Department of Commerce, National Bureau of Standards, Rockville, MD.

Willert, H. G., and Semlitsch, M. (1977). Reaction of articular capsule to wear products of artificial joint prostheses. *J. Biomed. Mater. Res.* 11: 157–164.

Woolson, S. T., Milbauer, J. P., Bobyn, J. D., Yue, S., and Maloney, W. J. (1997). Fatigue fracture of a forged cobalt–chromium–molybdenum femoral component inserted with cement. *J. Bone Joint Surg.* 79A: 1842–1848.

Wright, T. M., and Goodman, S. B. (eds.) (1996). *Implant Wear: The Future of Joint Replacement.* American Academy of Orthopedic Surgeons, Rosemont, IL.

Wright, T. M., Hood, R. W., and Burstein, A. H. (1982). Analysis of material failures. *Orthop. Clin. N. Am.* 13: 33–44.

Wright, T. M., Rimnac, C. M., Faris, P. M., and Bansal, M. (1988). Analysis of surface damage in retrieved carbon fiber–reinforced and plain polyethylene tibial components from posterior stabilized total knee replacements. *J. Bone Joint. Surg.* 70-A: 1312–1319.

Wright, T. M., Rimnac, C. M., Stulberg, S. D., Mintz, L., Tsao, A. K., Klein R. W., and McCrae, C. (1992). Wear of polyethylene in total joint replacements. Observations from retrieved PCA knee implants. *Clin. Orthop.* 276: 126–134.

USPHS (1985). U. S. Department of Health and Human Services, Public Health Service, National Institutes of Health. Guidelines for Blood–Material Interactions. NIH Publication No. 85–12185, National Institutes of Health, Bethesda, MD.

General References

Fraker, A. C., and Griffin, C. D. (eds.) (1985). *Corrosion and Degradation of Implant Materials: Second Symposium.* ASTM Special Publication, STP 859, Philadelphia.

Ratner, B. D. (ed.) (1988). *Surface Characterization of Biomaterials.* Elsevier, Amsterdam.

Transactions of the Society for Biomaterials Symposium on Retrieval and Analysis of Surgical Implants and Biomaterials (1988). Snowbird, UT, August 12–14, pp. 1–67.

von Recum, A. F. (1999). *Handbook of Biomaterials Evaluation: Scientific, Technical, and Clinical Testing of Implant Materials,* 2nd ed. Taylor & Francis, Washington, D.C.

Weinstein, A., Gibbons, D., Brown, S., and Ruff, W. (eds.) (1981). *Implant Retrieval: Material and Biological Analysis.* U. S. Dept of Commerce, NBS SP 601.

10

New Products and Standards

ELAINE DUNCAN, JACK E. LEMONS, JAY P. MAYESH,
PAMELA SAHA, SUBRATA SAHA, MARY F. SCRANTON

10.1 INTRODUCTION

Jack E. Lemons

A very dynamic situation continues to exist related to the development and introduction of new surgical implant devices. This, in part, relates to product and business competition. Significant amounts of proprietary research and development in conducted annually throughout the industrial sector, often first seen in product-related throughout or within regulatory submissions that become public domain.

Surgical implant biomaterial standards within the United States are based on consensus organizations with the American Society for Testing and Materials Committee F-4 (ASTM F-4) representing most implant devices. These standards are published annually in Volume 13.01 from the ASTM. Those interested in biomaterials and surgical implant devices should refer to the various national and international sources of product literature and standards as a key source of information. The section components of Chapter 10 will provide related references and sources for obtaining the industrial and standard related literature.

10.2 VOLUNTARY CONSENSUS STANDARDS

Jack E. Lemons

WHAT ARE STANDARDS?

Consensus standards are documents that have been developed by committees to represent a consensus opinion on test methods, materials, devices, or procedures. Most standards organizations review their documents within 5 years to ensure that they are up to date. The mechanisms by which they are developed are described in subsequent sections. Development of standards is an ongoing process and the latest publications should be consulted for new standards.

A test method standard describes the test specimen to be used, the conditions under which it is to be tested, how many specimens and what controls are to be tested, and how the data are to be analyzed. Many methods are validated by "round robin testing," meaning that several laboratories have followed the method and their results are analyzed to determine the degree to which they agree and to a specified degree of precision and accuracy. Once a test method has been standardized, it can be used in any other laboratory; the details are sufficient to ensure that different facilities will obtain similar results for the same samples. Stating that a test was "conducted in accordance with . . . " ensures that the results can be duplicated. Some representative test methods are listed in Table 1.

A material or specification standard describes the chemical, mechanical, physical and electrical properties of the material. Any test method standards cited are to be used to ensure that a significant sample meets the requirements of the standard. Some representative material standards are listed in Table 2.

For implant materials, there is also a requirement that the materials meet general biocompatibility test criteria. There are two formats for the biocompatibility language in the material standards of the American Society for Testing and Materials (ASTM). For materials that can be well characterized by chemical, mechanical, and physical tests and have demonstrated a well-characterized biological response, reference to the published biological testing data and clinical experience is often sufficient. For materials that are not well characterized, for example, the wide class of materials called epoxy resins, biological test methods are cited, and each particular formulation must be tested independently. This area is evolving for combination biological and synthetic (tissue-engineered) products.

A device standard describes the device and its laboratory-based performance. General design aspects, dimensions, and dimensional tolerances are given using schematic drawings. The materials to be used are described by reference to materials standards. Methods for testing the device are also cited. Since test methods only describe how to do a test, it is in the device-related standards that performance is addressed. For example, the fatigue life requirements or biocompatibility requirements of the device and its materials would be stated in a device standard. Some representative device standards are listed in Table 3.

TABLE 1   Some Representative ASTM Standard Test Methods

A. Mechanical testing standards
   ASTM D412 Test methods for rubber properties in tension
   ASTM D638 Test method for tensile properties of plastics
   ASTM D695 Test method for compressive properties of rigid plastics
   ASTM D790 Test methods for flexural properties of unreinforced and reinforced plastics and electrical insulating materials

B. Metallographic methods
   ASTM E3 Preparation of metallographic specimens
   ASTM E7 Terminology relating to metallography
   ASTM E45 Determining the inclusion content of steel
   ASTM E112 Determining the average grain size

C. Corrosion testing
   ASTM G3 Conventions applicable to electrochemical measurements in corrosion testing
   ASTM G5 Reference test method for making potentiostatic and potentiodynamic anodic polarization measurements
   ASTM G59 Conducting potentiodynamic polarization resistance measurements
   ASTM F746 Pitting and crevice corrosion of surgical alloys
   ASTM F897 Fretting corrosion of osteosynthesis plates and screws
   ASTM F1875 Practice for Fretting Corrosion Testing of Modular Implant Interfaces: Hip Femoral Head-Bore and Cone Taper Interface.
   F2129 Conducting Cyclic Potentiodynamic Polarization Measurements to determine the Corrosion susceptibility of small implant devices.

D. Polymer testing
   D2238 Test methods form absorbance of polyethylene due to methyl groups at 1378 cm-14
   D3124 Test method for vinylidene unsaturation in polyethylene by infrared spectrophotometry

TABLE 2   Some Typical ASTM Materials Standards (4)

ASTM F75 Cast cobalt–chromium–molybdenum alloy for surgical implant applications

ASTM F139 Stainless steel sheet and strip for surgical implants (special quality)

ASTM F451 Acrylic bone cements

ASTM F603 High-purity dense aluminum oxide for surgical implant applications

ASTM F604 Silicone elastomers used in medical applications

ASTM F641 Implantable epoxy electronic encapsulants

A procedure or guidance standard describes how to do something that would not be considered a test. Examples include standards for surface preparation and standardized procedures for sterilizing implants. Table 4 lists some typical procedure standards.

TABLE 3   Some Representative AAMI and ASTM Device Standards

AAMI CVP3 Cardiac valve prostheses

AAMI VP20 Vascular graft prostheses

AAMI RD17 Hemodialyzer blood tubing

AAMI ST8 Hospital steam sterilizers

E667 Clinical thermometers (maximum self-registering, mercury-in-glass)

ASTM F367 Holes and slots with spherical contour for metric cortical bone screws

ASTM F703 Implantable breast prostheses

ASTM F623 Foley catheters

TABLE 4   Some Representative AAMI (1) and ASTM (4) Procedure Standards

AAMI ROH-1986 Reuse of hemodialyzers

AAMI ST19 Biological indicators for saturated steam sterilization processes in health care facilities

AAMI ST21 Biological indicators for ethylene oxide sterilization processes in health care facilities

ASTM F86 Surface preparation and marking of metallic surgical implants

ASTM F561 Retrieval and analysis of implanted medical devices and associated tissues

ASTM F565 Care and handling of orthopedic implants and instruments

ASTM F983 Permanent marking of orthopedic implant components

WHO USES STANDARDS?

The term "voluntary standards" implies that the documents are not mandatory; anyone can use them. This terminology also refers to the way that standards are developed. Standards are used by manufacturers, users, test laboratories, and, in many instances, college professors and their students. One's use or compliance with a standard is voluntary. Using them is often to everyone's advantage. At this time, standards can also be utilized as a part of the regulatory (FDA) approval process (1997 FDA MA).

Manufacturers often use standards as guidelines in making and testing their materials and devices. Manufacturers also cite standards in their sales literature as a concise way of describing their product. Stating that a device is made from cast cobalt–chromium–molybdenum alloy in accordance with ASTM F75 tells the user precisely what the material is. Conformance to standards is also a way to expedite device review by FDA, as is described in Chapter 10.3.

On a more personal level, for example, after purchasing a piece of plastic pipe at the hardware store labeled with "ASTM D1784," one could go to ASTM Volume 8.02 and find that this is a specification for rigid poly(vinyl chloride) compounds. If you have "DIN" stamped on the bottom of your ski boots, you know they conform to the standards of the Deutsches Institut für Normung, and the ski shop will have standards for adjusting your bindings.

As an example of why one would want device standards used for medical devices, consider screws for fixing bone fractures. There are device standards for bone screws, plates, taps, and screwdrivers. The physician can purchase a screw and a screwdriver and be confident that the components will fit as intended. A surgeon about to remove a plate implanted at another hospital can evaluate radiographs and see that the device has 4.5-mm bone screws of a specific design. Knowing this, a standard 4.5-mm screwdriver can be used to remove the screws.

Standardized test methods should simplify life. For example, one of the authors taught two laboratory sessions as part of an undergraduate biomedical engineering course. One was on mechanical testing, the other on metallography and implant analysis. In the mechanical testing laboratory, several ASTM standard test methods for mechanical testing, such as D790, "Flexural properties of plastics and electrical insulating materials," were used. This method describes the samples, test apparatus, test speeds, and equations used to calculate the results. During the lab session the students followed the test directions. In writing the methods section of their reports, all they had to write is that "the test was done according to D790."

WHO WRITES STANDARDS?

In the United States, voluntary consensus standards are developed by a number of organizations. In the medical electronics, sterilization, vascular prosthesis, and cardiac valve areas, most standards are developed by committees within the Association for the Advancement of Medical Instrumentation (AAMI). In the implant materials and implants area, most standards are set by ASTM Committee F-04 on Medical and Surgical Materials and Devices. These documents may then be reviewed and accepted by the American National Standards Institute (ANSI). ANSI is the official U.S. organization that interacts with other national organizations in developing international standards within the International Standards Organization (ISO), such as TC 150 on Medical Materials and Devices and TC 194 on Biological Evaluation of Medical Devices. The USP provides information on minimal biocompatibility testing for materials intended to be used in medical devices.

Dental material standards are written by the American Dental Association (ADA). Similar committees exist in other countries: the Canadian Standards Association (CSA), the British Standards Institute (BSI), the Association Francaise de Normalisation (AFNOR) in France, and the Deutsches Institut für Normung (DIN) in Germany, which is a voluntary organization.

The initiation, development, and process for the completion of national ASTM F04 consensus standards for medical and surgical materials and devices has evolved significantly within recent years. In part, this has been a result of multiple interactions amongst those involved with the basic sciences, applied research and development, business marketing and sales, clinical applications, regulatory agencies, professional societies, legal and insurance professions, device recipients, and associated advocacy groups. To establish consensus opinions satisfactory to all of these interest groups is far from a simple process.

History and Current Structure of ASTM F04

The ASTM was organized in 1898, whereas Committee F04 on Medical Devices and Surgical Materials and Devices was founded in 1962. The committee has grown to include a current membership of approximately 500 individuals representing a variety of disciplines and interests. The ASTM F04 Committee has more than 100 active standards and is structured into more than 30 technical subcommittees. The overview structure includes five divisions divided according to responsibilities specific to organizational activities (process) or the development of specific types of standard document (e.g., full consensus standards include six types (1) classification, (2) guide, (3) practice, (4) specification, (5) terminology, and (6) test methods, plus a provisional status). The divisions are (I) resources, (II) orthopedic devices, (III) medical/surgical devices, (IV) tissue engineered products, and (V) Administration. Each division is subdivided into subcommittees and task groups according to areas of interest and activities. This structure is intended to be flexible and can be rearranged to suit new or more efficient operations at any scheduled meeting of the executive committee. The divisions, subcommittees, and task groups have an appointed chair and vice-chair, whereas the executive is nominated and elected on a 2-year cycle.

Standards Development Process

After a request for a new standard is received by a member or group (task group or subcommittee) and accepted by the executive committee, a task group activity is initiated. An appropriate chair is recommended and approved by the administrative committees and the process is started. Consensus standards developments follows a reference document for content, form, and structure. An assignment of an appropriate document number is made for record purposes and a staff manager confirms that the committee and subcommittee representation associated with this activity is classified and balanced with respect to producers and nonproducers. A first draft of the proposed document is reviewed (usually three to five times) before the task force reaches a consensus.

Critical to these proceedings is the necessity that adequate information is available within the public domain in order to substantiate the requirements listed within the final standard. If data are limited or unknown, round-robin tests or new test methods must be developed, and then confirmed to be applicable and valid. Sometimes, documents are held until basic (necessary) information for the standard is developed. Again, a standard must be based on known results and documents are not intended to represent areas of new research.

As a part of the process, once a draft document has been circulated within the task group and consensus is reached, the task group chair may recommend initial voting at the task group level. At this time, the opinions gained could lead to further improvements and no formal voting rules are required within the task group interactions. At the next stages, the sub and main committees of ASTM F04 must ensure a balance among the various voting interests with adequate representation from the general-interest, user, consumer, and producer segments of the membership. The total of the user, consumer, and general-interest votes must equal or exceed the number of producer votes. To prevent domination by any one interest group, only one vote per voting interest is permitted. The ASTM staff confirms the numerical status (adequate response and balance) for each formal vote and all members are permitted to vote on any ballot within their committee of membership. If approved or approved with editorial (no substantive) changes at the subcommittee level, the document proceeds to main committee ballot and society review. If, however, a negative vote is received during formal voting, the task group and subcommittee chairs must resolve the negative to the satisfaction of the negative voter or must provide a rationale and justification to find the negative voter nonpersuasive. This opinion must be accepted by the task force and approved at the subcommittee and main committee levels by a formal recorded vote based on written documentation and associated discussions. The staff manager works with the committee to confirm the validity of the vote and to document the action.

The general experience has been that it requires 3 to 5 years (six to ten meetings) to go from a first draft to a full standard accepted at the main committee level. Several procedural steps are also required in the process, including approval of a rationale statement, use of standardized units (SI) and terminology, and acceptance by the editorial and precision and bias subcommittees. At the stage of society acceptance, the final document is reviewed by the Committee on Standards prior to publication by ASTM.

After approval, a given standard may remain "active as published" for up to 5 years. At 5 years, that standard must be reaffirmed or revised to suit the information available at that time. For record purposes, the date of last formal approval is included as a part of the standards designation.

BIOCOMPATIBILITY STANDARDS

There is a wide range of tests that may be used to determine the biological response to materials. Short-term uses require only short-term tests. Long-term uses require tests applicable to the particular device and tissue type. Since not all tests are necessary for all applications, national and international standards organizations have developed matrix documents that indicate what methods are appropriate for specific applications. These documents can be used as guidelines in preparing a submission to the U.S. Food and Drug Administration (FDA), the

EC, and other national regulatory agencies for approval of a new material or device. Similar matrix documents have been standardized by the CSA, BSI, and ISO. Test method documents have also been developed by the National Institutes of Health (NIH) and the U.S. Pharmacopeia (USP). Guidelines for dental materials have also been developed by the ADA and ISO.

Much of the standards activity is now associated with the International Standards Organization (ISO) with biological evaluation of medical devices under the consideration of TC 194 and presented in the developing documents of ISO 10993. There are various parts to this document. Part 1 is definitions and the guidance on selection of evaluation test categories that should be done. The other parts of 10993 give more discussion and detail on the selection of individual tests that should be done for a particular biological interaction or biological effect (e.g., contact with blood, systemic toxicity, genotoxicity). Often, the details of test methods are not given in the ISO documents and reference is made to other documents such as ASTM and USP standards for procedures and methodology.

Material selection and evaluation of biological risk are integral components of the design process for medical devices being considered by TC 194. This evaluation is a component of the risk management plan in line with ISO/IEC 14971—Application of Risk Management to Medical Devices, encompassing identification of all hazards and the estimation of their associated risks. Criteria to define the acceptable biological (toxicological) risk must be established at the start of the risk assessment and design management processes. The biological safety evaluation must be designed and performed to demonstrate the achievement of the specified criteria for safety. Adequate risk assessment requires characterization of toxicological hazards and exposures. Following the risk management structure described in ISO 14971, a major component in hazard identification is material characterization.

In the following section we review some of the steps taken to establish the biocompatibility of a new material for a specific application, in this case a long-term orthopedic implant. We use ASTM standard F748 "Practice for Selecting Generic Biological Test Methods for Materials and Devices" as a guideline. The standard test methods described are those used within ASTM.

In Vitro *Tests*

F619. Practice for Extraction of Medical Plastics. A method for extraction of medical plastics in liquids that simulate body fluids. The extraction vehicle is then used for chemical or biological tests. Extraction fluids include saline, vegetable oil (sesame or cottonseed), and water.

F813. Practice for Direct Contact Cell Culture Evaluation of Materials for Medical Devices. A cell culture test using American Type Culture Collection (ATCC) L929 mouse connective tissue cells. This method or this type of cell culture method can be used as the first stage of biological testing. It is also used for quality control in a production setting. There are other ASTM standard cell culture methods, and others not standardized by ASTM that could also be used.

F756. Assessment of Hemolytic Properties of Materials. An *in vitro* test to evaluate the hemolytic properties of materials intended for use in contact with blood. Procedure A is static; procedure B is done under dynamic conditions.

Short-Term in Vivo Testing

F719. Testing Biomaterials in Rabbits for Primary Skin Irritation. A procedure to assess the irritancy of a bio-material in contact with intact or abraded skin. This test would be indicated for surgical glove material or skin dressings.

F720. Practice for Testing Guinea Pigs for Contact Allergens: Guinea Pig Maximization Test. A two-stage induction procedure employing Freund's complete adjuvant and sodium lauryl sulfate, followed 2 weeks later by a challenge with the extract material. Ten animals per test material.

F749. Practice for Evaluating Material Extracts by Intracutaneous Injection in the Rabbit. Extraction vehicles (as per F619) of saline and vegetable oil are injected intracutaneously and the skin reaction graded for erythema, edema, and necrosis. Two rabbits per extraction vehicle.

F750. Practice for Evaluating Material Extracts by Systemic Injection in the Mouse. Intravenous injection of saline extracts and intraperitoneal injection of oil extracts. Animals are observed for evidence of toxicity. Five mice per extract and five mice per extract vehicle controls.

F763. Practice for Short-Term Screening of Implant Materials. This method provides for several implant types and sites for short term screening *in vivo*. This method is essentially the first stage of animal testing of solid pieces of the implant material.

F1983. Assessment of Compatibility of Absorbable/Resorbable Biomaterials for Implant Applications. This material type presents unique features for tissue evaluation, in that the materials are not inert, and a chronic inflammatory reaction may be observed during the degradation period. The time periods at which reactions are examined are based on the anticipated rates of degradation of the test materials.

Additional tests for special issues are also included in ASTM standards such as examination and reactions to particles, immunotoxicity, and retrieval and analysis of implants and tissues.

There are additional *in vitro* tests that have not yet been standardized by ASTM:

Thrombogenicity. Tests for the propensity for materials to cause blood coagulation have not been standardized. Guidelines for such tests have been developed by the NIH Heart Lung and Blood Institute.

Mutagenicity. There are a number of *in vitro* and *in vivo* tests to determine if chemicals cause cell mutations.

Although not specifically developed for implants, guidelines do exist as part of the OECD guidelines for testing of chemicals, and within ASTM, e.g., E1262, Guide for the Performance of the Chinese Hamster Ovary Cell/Hypoxanthine Guanine Phosphoribosyl Transferase Gene Mutation Assay; E1263 Guide for Conduct of Micronucleus Assays in Mammalian Bone Marrow Erythrocytes; E1280, Guide for Performing the Mouse Lymphoma Assay for Mammalian Cell Mutagenicity; E1397, Practices for the *in Vitro* Rat Hepatocyte DNA Repair Assay; and E1398, Practices for the *in Vivo* Rat Hepatocyte DNA Repair Assay, which are in Vol 11.05 of the Annual book of Standards.

Pyrogenicity. A pyrogen is a chemical that causes fever. The USP rabbit test is a standard *in vivo* test. One can also test for bacterial endotoxins, which are pyrogenic, using the Limulus amebocyte lysate (LAL) test. The oxygen-carrying cell (amebocyte) of the horseshoe crab, *Limulus polyphemus*, lyses when exposed to endotoxin.

Long-Term Testing in Vivo

There are two aspects to the long-term testing issue. One is the response of tissue to the material; the other is the response of the material (degradation) to implantation.

F981. Practice for Assessment of Compatibility of Biomaterials for Surgical Implants with Respect to Effects of Materials on Muscle and Bone. Long-term implantation of test materials in the muscle and bone of rats, rabbits, and dogs. Two species are recommended. For rabbit muscle implants: the standard calls for four rabbits per sacrifice period, with one control and two test materials placed in the paravertebral muscles on each side of the spine. For bone implants in rabbits: the standard calls for three implants per femur.

A general necropsy is performed at the time of sacrifice. Muscle and bone implant sites are removed at sacrifice and the implants left *in situ* until the tissue has been fixed in formalin. Implants may be removed prior to embedding and sectioning.

No standards have been established for any long-term testing of devices. However, for a device intended for a particular application, it is essential to conduct a functional device test. For a fracture fixation plate, it could be proposed to use plates to fix femoral osteotomies in dogs. This study would consider the effects of the implant on the tissues, as well as the effect of implantation on the properties of the device, i.e., material degradation.

The methodology for long-term carcinogenicity testing of implants also has not yet been standardized by the ASTM, although F1439 (Standard Guide for Performance of Lifetime Bioassay for the Tumorigenic Potential of Implant Materials) does provide guidelines for test selection. This is normally a life survival and tumor production test, typically in rats. ISO 10993-3 provides considerations for genotoxicity, carcinogenicity, and reproductive toxicity testing with reference to some test methods.

SUMMARY

Standard test methods allow results to be reproduced or verified by other researchers. In the biomaterials field, standards can be used for physical, mechanical, chemical, electrical, and biological testing of materials. By using materials that conform to standards, a manufacturer can tell the user what is in the material and what to expect from it in terms of properties.

Bibliography

AAMI Resource Catalogue, Association for the Advancement of Medical Instrumentation, Arlington, VA.

ADA Document 41 for recommended standard practices for biological evaluation of dental materials (2001). Chicago.

AFNOR S90-700, Choice of tests enabling assessment of biocompatibility of medical materials and devices. Paris.

ASTM Annual Book of Standards. West Conshohocken, PA (published annually).

BSI BS5736, Evaluation of medical devices for biological hazards. British Standards Institution.

CSA CAN3-Z310.6-M84, Tests for biocompatibility. Canadian Standards Organization, Ontario.

ISO 7405, Preclinical evaluation of biocompatibility of medical devices used in dentistry-test methods. International Organization for Standardization, Geneva.

ISO 10993, Implants for surgery, Biocompatibility selection of biological test methods for materials and devices. International Organization for Standardization, Geneva.

NIH HLBI, Guidelines on characterization of biomaterials. (1984).

USP (2000). The U.S. Pharmacopeia 24. United States Pharmacopeial Convention, Rockville, MD.

10.3 DEVELOPMENT AND REGULATION OF MEDICAL PRODUCTS USING BIOMATERIALS

Elaine Duncan

INTRODUCTION

What we now call the "Food, Drug & Cosmetic Act" was amended in 1976 to broaden the powers of the U.S. Food and Drug Administration (FDA) over medical devices. Subsequent amendments have continued steadily to refine and broaden the powers of the agency to control veterinary drugs, biologics, and combination products. Narrowly crafted definitions in the 1976 amendment attempted to segregate the authority between the divisions of the agency, maintaining drugs under the Center for Drug Evaluation and Research (known as CDER). The portion of the agency now known as the Center for Devices and Radiological Health (CDRH) maintains authority over medical devices. Major amendments occurred again in 1990, 1997, and 2002. Regardless of these many amendments, the agency has never been granted the authority to specifically regulate biomaterials. Rather, the regulatory constraints on biomaterials are indirect and according to the intended use of the device, drug, biologic, or combination product.

Combination Products

Through the years, the distinctions among drugs, biologics, and devices have blurred. Combination products are products that are recognized to have device, biologics, or drugs in combination. The premarket review jurisdiction of such combination products is predetermined by interagency agreement. In October 2002, Congress passed the Medical Device User Fee and Modernization Act of 2002 that established the Office of Combination Products. The Office has not yet issued formal operational instructions to the medical device industry as this chapter goes to press, but it is expected that the mode of action of the product will continue to be the primary consideration for determination of review and compliance authority. However, historical precedence can be a factor.

Manufacturers may request FDA to make a determination of the review authority in advance if the manufacturers feel that the review authority is uncertain. FDA has pledged to make public certain determinations of authority, but for reasons of confidentiality the Agency cannot do this until the product under review has cleared the approval process. For combination products, joint review is commonly required and more than one type of premarket application may be necessary. As biomaterials become hybrids of or are developed in combination with tissue-derived molecules, polymers, and pharmaceutical agents, the medical device and biomaterials developer will increasingly encounter this type of regulatory environment.

Global Regulatory Strategy and Intended Use

Biomaterials may be incorporated into drug delivery systems, used as carriers within packages for biologics, and incorporated into key components of medical devices in broad applications from disposable tubing and syringes to implantable devices sustaining life or restoring organ or limb function. In all of these various applications, the biomaterials within or without are regulated indirectly by the intended use of the final product. Therefore, the regulation of the biomaterial depends primarily on the risk associated with the intended use of the product. Aside from drug delivery systems, most applications for conventional biomaterials are in the medical device field, so the focus of this section of the chapter will be on the regulatory constraints on product development of new devices using biomaterials.

The application of biomaterials to medical device technology in nearly every industrialized country is regulated according to the intended use of the product incorporating the material and the relative risk of the use of the materials. These regulations may be as a result of direct laws or regulations and usually take the form of requirements to comply with voluntary or mandatory certifications to recognized standards or norms. After a new product has cleared the requirements (which restrict marketing until complete), continuing compliance requirements attempt to ensure quality of the finished device.

Changes to an existing product (such as a change to a biomaterial) or a new product (with either a well-characterized material or a new biomaterial) may be subject to regulatory approval or review prior to market introduction, or companies may be required to document that the changes do not affect the safety and efficacy of the device. In general, all regulations that govern medical devices place the burden of proof upon the manufacturer to document the quality of the biomaterials from which the medical devices may be fabricated.

Typical biomaterial regulatory control follows one or more of the following types:

1. Guidance documents or device specific requirements
2. Adoption by reference to international standards for materials and test requirements
3. Requirements for validation and verification of material performance within the device
4. Manufacturing and purchasing controls to ensure continued quality and performance.

Regulations and requirements for biomaterials in medical devices typically focus on concerns for the safety and efficacy of biomaterials when the medical device is used as intended. The potential harm derived by the use of an unsafe material is viewed in the context of the device application. Therefore, more stringent testing of material safety and efficacy is required for devices with longer exposure, those used in life-supporting or life-sustaining devices, or where repeated exposure to the biomaterial component could engender greater risk.

Medical device manufacturers should also consider their global regulatory requirements prior to embarking on the development of a new product. These requirements should be a part of the tasks required in any initial design phase plan. Medical device manufacturers need to determine the compliance requirements on the products based on the market applications, historic precedence, and applicable standards. The U.S. FDA and European requirements, although setting the norm for many countries, are not necessarily the same as those that may be required in Brazil or in Japan. The Canadian and Australian systems have undergone significant changes in recent years and are similar but not identical to the European regulatory system. Most global regulatory strategies now follow the system of essential requirements. Harmonization committees and organizations are striving to make the safety and efficacy of medical devices a global requirement. A well-constructed technical file (a.k.a. Device Master Record or technical dossier) will go a long way in satisfying the regulatory requirements of most countries. Documentation of conformity to international standards and essential requirements will become an increasingly important function for all medical device developers. Harmonization efforts may soon do away with the historic "me too" practice of the 510(k) submission, which requires companies to compare the properties and performance of the product to a previously cleared device. Instead, an FDA submission for a low, or medium-risk device will certify conformity to standards or FDA published guidance documents and equivalence analysis may be less crucial.

Adoption of International Standards for Quality

In the years since the first edition of *Biomaterials Science: An Introduction to Materials in Medicine*, a shift in perspective

to a worldwide harmonized quality system has been adopted within the medical device community. This was less by design and more by rational mandate, as the European Union adopted the Medical Device Directives (MDD) and the Active Implantable Medical Device Directive (AIMDD). The general construction of these directives builds on recognized international standards and Essential Requirements. The Essential Requirements specifically require that the chemical, physical, and biological properties guarantee characteristics and performance of the product. Demonstration of conformance to the essential requirements is directly tied to demonstration of meeting a recognized European or international-level standard.

Beyond specific standards for biomaterial testing to demonstrate compliance (which are discussed later), additional requirements for total quality systems have become an international norm. The U.S. FDA Good Manufacturing Practices were modified to harmonize with globally recognized standards known as ISO 9000 (ISO is the abbreviation for International Standards Organization). The revised regulation, known as the Quality System Regulation (QSR), includes the general principles of Design Control and Review and Risk Analysis. The ISO 13485 standard for "Quality Management Systems—Medical Devices—System Requirements for Regulatory Purposes" is now recognized by Canada. Conformity is required in order to sell any new products in Canada; for those products sold prior to December 31, 2001, 1 year grace was allowed to establish conformity. ISO 13485–2003 has now issued as a "stand-alone" standard for medical device manufacturers, independent of ISO 9000–2000, and harmonizes with the FDA Quality System Regualtion.

Product-specific international standards for product quality and evaluation have become an integral part of many regulatory programs around the world. FDA has issued a list of standards for conformity acceptance, and even those not recognized officially by FDA can be useful for communicating safety and performance as part of the review dossier. Some of the more commonly applied standards are for electrical safety and the ISO 10993 series, discussed in detail on the following pages. Up-to-date lists of recognized standards are discussed in detail on page 790.

Design Control and Risk Analysis

Design control and review, coupled with risk analysis, form the foundation of the testing and development requirements for any Class II or Class III medical product. (For details on FDA product classifications, go to www.fda.gov/cdrh). Customer "inputs" and design outputs (specifications) are to be reviewed throughout the design plan. A design plan is reviewed at the beginning of the product development cycle, and again as the program evolves. Typical design stages that trigger design review are the concept phase (usually prior to initiation of major external resources), development phase (often includes animal and bench testing), clinical evaluation phase (often includes investigational trials on human subjects), and the market introduction phase (includes scale-up and transfer to manufacturing).

As they are enforced differently, regulatory requirements between countries can represent different degrees of difficulty for the new medical device. Most countries will require demonstration that the device has been shown to be safe and effective for the intended use and meets certain essential requirements for safety. Risk analysis represents one method medical device manufacturers have of communicating to regulators how the safety of the product has been assured. International standards have been recognized that provide guidance on conducting risk analysis (ISO 14971, EN 1441, IEC 1025). The risk analysis process has become a standardized procedure used to determine the type of controls and mitigations that are needed to ensure product safety and human factors for most new medical devices under development. Transferring design to manufacturing also uses risk analysis methods to assess the most critical process steps for quality performance.

Premarket Approval, Clearance, or "CE Mark"

For Europe, once a developer of a medical device has demonstrated conformance with the Essential Requirements of either the MDD or AIMDD, the company may file a Certificate of Conformance or request that a Notified Body conduct an examination to ensure conformance. The route to conformity assessment is spelled out in the directives and is determined by the classification of the device and the applicable rule, which is primarily a risk-based stratagem. The CE Mark must be applied, based upon assessment of conformance, prior to the sale of any medical device in a country within the European Union and in other countries that have voluntarily adopted the recognition of the CE Mark.

U.S. FDA requires that the manufacturer demonstrate safety and efficacy of the product (Pre-Market Approval Application, or PMA), unless the device can be shown to be substantially equivalent to a predicate product that is legally marketed in the United States [510(k)]. For most new products, those without predicates, and certain high-risk products (typically Class III), a PMA (or its equivalent Product Development Protocol, PDP) is required. Certain Class I and exempted Class II medical devices are permitted to be marketed without FDA's prior authorization, although manufacturing (Quality System Regulations) and/or documentation controls may still apply.

Premarket approval applications for medical devices containing biomaterials may take a variety of regulatory pathways, but for most new or innovative biomaterial-based, implantable devices, the formal Premarket Approval Application (PMA) is commonly required. [FDA has continued to permit certain Class II and Class III devices to enter the market after clearance of the 510(k)]. The specific regulations and classifications for medical products changes frequently, and the list is far too exhaustive to include in a textbook chapter. For the latest regulatory status of any type of medical product using biomaterials, contact the U.S. Food and Drug Administration at their Web site (www.fda.gov).

For companies undergoing the PMA process, a critical step is often the final review by the clinical advisory panel. This panel of medical experts makes recommendations to the FDA on the acceptance of a new medical device, but most particularly on whether, in their opinion, the clinical evaluation of the product has demonstrated efficacy. Proof of efficacy is highly dependent upon statistical analysis of prospective data collected under tightly controlled clinical evaluation protocols. The FDA has sought to take some of the long-range risk out of the approval process by offering manufacturers alternative approval routes. PMAs can now be reviewed in modules. The modular PMA is finding favor with medical device companies because they can submit sections of testing as completed. This does not, however, eliminate the risk for final rejection of the clinical testing conducted under Investigational Device Exemption regulations by the clinical panel.

The Product Development Protocol (PDP) offers an alternative route to market approval. The PDP allows the manufacturer to present a plan to the FDA and the clinical panel in advance of initiating the testing. At this time, FDA may restrict the application of any PDP to those devices for which FDA feels a model can be developed from prior regulatory experience.

Manufacturing Controls

In most regulatory schemes, manufacturing to acceptable practices is a standard requirement, although the specific obligations and criteria required are different from country to country. The U.S. FDA has developed a tiered system for continued inspection of manufacturing locations based on a risk-based strategy. There are typically limitations imposed on the variety and scope of changes that can be made by the manufacturer to a Class III product after market introduction without prior reapproval by FDA. The FDA has regulations concerning recalls, product removals, and corrections and heavily stresses the importance of corrective and preventive action (CAPA) programs for marketed products.

Compliance to regulations governing the sale and distribution of a medical device after the product has been approved typically requires that standard manufacturing practices be followed. Postapproval limitations to the distribution of the medical device typically include restrictions to be placed on the claims for the product and on the content of labeling. Some "off-label use" is tolerated, but promotion of new claims typically triggers a new application to support a different indication for use or claims of performance not previously reviewed by FDA.

Most regulatory schemes also require the reporting of adverse effects. The degree of hazard that must be reported to the governing agency will vary from country to country. Tracking of implantable or life-supporting devices is limited but still a common regulatory constraint for high-risk implantable devices, such as defibrillators.

Clinical Trials of Unapproved Devices

Basic requirements for the conduct of human clinical trials continue to be harmonized internationally, although significant differences remain. Barriers to initiating a clinical trial still differ between countries, but once the trial is underway, a similar level of patient rights and investigator obligations is expected. European concerns for patient privacy preceded U.S. adoption of strict patient privacy regulations, although privacy has always been a concern for the human subjects committees reviewing protocols.

Clinical evaluations for significant-risk devices containing biomaterials are typically cleared for study in human patients by FDA under an Investigational Device Exemption. A protocol or study plan is required, and with a few minor exceptions, informed consent from the patient must be documented prior to testing the device or material in an unproven device on a human subject. Investigators must be trained on the use of the product and sign an agreement in order to conduct the study. Investigators must now disclose certain financial aspects. Approval by an Institutional Review Board (IRB), also known as the human subjects committee or ethics committee, is required by the FDA prior to conducting studies funded by the National Institute of Health, and for most international regulatory agencies. Case report forms and data collection methods must be carefully crafted in order to document data required to support the claims for the product under study. All of these same requirements are stipulated in European medical device directives, but individual countries are responsible for their implementation and compliance.

Clinical trials for medical devices containing biomaterials must develop objective evidence of safety of the device, and indirectly the safety of the biomaterial as applied in the device. Any safety problem encountered with the use of a biomaterial during the study needs to be reported as a part of the adverse effects of the clinical trial. Any special performance claim attributed to the biomaterial, such as tissue ingrowth, needs to be evaluated both as a part of a preclinical animal study, and if possible, in the clinical study. In the United States, only claims for biomaterial performance supported with statistically valid evidence will be permitted in the device labeling. For this reason, many performance claims for biomaterials are embedded within the performance claims of the device, or never presented at all.

Often, the indirect or noninvasive measurements, such as X-rays, prove inadequate to demonstrate biomaterial performance criteria, so biomaterial performance claims may also depend upon high-quality animal studies to demonstrate interfacial characteristics, such as tissue ingrowth and biocompatibility. Animal studies must have some basis for prediction of performance in the human application, which can prove difficult given the differences in animal anatomy and healing characteristics when compared to humans. Increasingly these studies are conducted under the Good Laboratory Practices Regulations (CFR Part 58), which establish the quality standards for the research data that may be used in support of an FDA application.

ISO Standard for Biocompatibility Testing

Typical biocompatibility screens, such as those described by ISO 10993, are conducted prior to clinical evaluation of the device or clearance to market the product. Implementation of ISO 10993 varies, based on the risk associated with the use of

the product. The ISO 10993-1 references standardized biocompatibility screens, which are primarily focused on the chemical toxicity of leachable components or degradation by-products. Although ISO 10993-1 admonishes regulators to consider the material from a global suitability perspective, most commonly regulators use the table of tests for consideration as a requirements checklist. Application reviewers tend to give more credence to laboratory test reports. The exposure-based categories and duration segments have, through the years (the chart was adapted from the original Tripartite Agreement of 1987), been seen as a logical, although awkward, structure for determination of testing levels.

Fault does not lie in the standard as much as with its implementation. The standard, evolved from guidance, admonishes regulators that these tests do not substitute for an understanding of the biomaterial performance as a component of the device and the interaction of the body with the device and material. For example, these screens generally do not evaluate the long-term effects of the structural forces at work on the biomaterial when incorporated into an implant. Developers must recognize the need for animal studies designed to evaluate the function and potential failure of the implant in the intended use. Although beneficial as a standardized series of tests, the scope and intent of the ISO 10993 cannot incorporate all concerns for implant performance. Rather these test schemes should be viewed as a toxicological analysis with limited value in predicating the overall performance of a material in an implanted device.

Many of the biological screens that comprise the remaining parts of ISO 10993 are designed to evaluate specific aspects of biocompatibility, and therefore are not required for all medical devices. Part 3 briefly describes genotoxicity, carcinogenicity, and reproductive toxicity testing, but these descriptions do not constitute protocols. Detailed protocols must still be the role of the contract laboratory. Tests for blood interaction are listed in Part 4 with general requirements for the conduct of the tests, but again, specific protocols are beyond its scope. Cytotoxicity is described in Part 5 and evaluations for local effects after implantation are described in Part 6. Although this is meant to elaborate on evaluation of implanted materials, the scope of the evaluations listed specifically excludes implantation studies where the implant is subject to mechanical (functional) loading.

Part 7 describes the evaluation of medical devices for ethylene oxide residues and Part 9 is a technical report and considered still under development. It is intended to facilitate test procedures to evaluate biological responses to degradation products released from medical devices. Part 10 is a more fully developed standard providing specific protocol and acceptance criteria for the evaluation of irritation and sensitization. Sensitization testing has become one of the most expensive and time-consuming screens applied to even short-duration (less than 30 days) devices. Systemic toxicity is described in Part 11. Part 12 describes the preparation of samples and reference materials and provides recommendations for thickness or size of the sample proportional to the extraction ratio for testing. Part 13D could be viewed as a companion to Part 9, as Part 13 describes how to conduct an accelerated degradation test to determine if materials are prone to degradation and what the by-products are likely to be. Unfortunately, such accelerated procedures can introduce conditions within the materials that are not mimicked in actual application. Part 15 describes the identification and quantification of degradation products from metals and alloys and refers to induced electrochemical degradation that may or may not be appropriate for the application of the device. Part 16 describes toxicokinetic studies that should be considered if the device is designed to be bioresorbable or if leachables might be expected to migrate from the device. Since the standard states that leachables from metals, alloys, and ceramics are usually too low to justify such studies, it is apparent that the standard is meant to apply to polymeric biomaterials.

FDA and international regulatory agencies expect medical device manufacturers to provide a clear analysis of the risks associated with the use of the product. This frequently requires a full understanding of the hazards inherent with the biomaterials employed in the manufacture of the product. The basic performance requirements of these materials within the device must be fully characterized by the device manufacturer. Risk analysis forms the basis of the testing of new products through verification and validation and includes, where necessary, all *in vitro* and bench trials, animal studies, and clinical trials. It is apparent that risk-based rationale for device evaluation will continue to develop as the backbone for regulatory oversight. ISO 14538—Method for the Establishment of Allowable Limits for Residues in Medical Devices using Health Based Risk Assessment is yet another example of this trend. The FDA issued a guidance in May 1999 titled "Immunotoxicity Testing Guidance" that refers to the ISO 10993 but sets forth a flow chart and yet another table to serve as a guide for whether or not immunotoxity testing is required.

In conducting the risk assessment, regulatory authorities outside the United States recognize the "prior art" use of a medical material and give more credibility to literature studies in assessing risk. In the United States, despite statements that consideration will be given for prior use, the practical reality is that between manufacturers there is limited cooperation to share the information for prior use of a material in the same intended use. Even if this information can be determined from literature or predicate devices, it is seldom available at a level specific to formulation or sterilization method. As a result, for even mundane applications, the FDA may require extensive, costly, and time-consuming testing of medical polymers in acute toxicity screens despite years of empirical knowledge.

Where polymeric materials or metallic and ceramic materials are combined with growth factors (or with other biological agents) or pharmaceutical products, extensive testing is warranted. New concerns for the safety of biomaterials derived from human or animal products are increasing the regulations associated with medical products. The FDA has issued guidance for the use of animal materials used in medical devices. The guidance titled "Medical Devices Containing Materials Derived from Animal Sources (Except for *in Vitro* Diagnostic Devices")" has provided details on FDA's concerns and how the FDA expects manufacturers will control the sources of the materials.

Shelf-Life and Aging

Physical property characterization and durability testing of the biomaterial in its intended use are critical parameters to establish. If failures begin to occur in the field, differences from baseline parameters can be crucial evidence to assist in establishing the cause. Shelf-life aging of medical device materials has become a required practice. This area will take on increasing concern as medical device manufacturers incorporate drugs and biologics as surface treatments to devices. Sterilization of the product prior to use can frequently affect the performance of the biomaterial and some sterilization effects are not readily apparent. The FDA requires that performance testing for the materials, as well as safety testing of the device, be conducted after the same sterilization cycle that will be applied to the finished product.

Changes to Regulated Medical Devices

Once a medical device product is on the market, a change to a biomaterial can trigger a new 510(k) or a modification of a 510(k) application. The FDA now allows for abbreviated 510(k)s, special 510(k)s, and third-party reviews. Alternative 510(k) submissions are a relatively new process and have not received broad implementation to date. As medical device manufacturers become more accustom to the use of design control and review and the incorporation of international standards as a part of their input requirements to product design, it is expected that such abbreviated or special 510(k)s will be more commonly applied. For more information, the FDA has developed guidance: "Alternative Approaches to Demonstrating Substantial Equivalence in Premarket Notifications." The FDA has provided a checklist for the modification of 510(k) devices in the guidance: "Deciding When to Submit a 510(k) for a Change to an Existing Device." These helpful guides can assist manufacturers in determining whether or not a 510(k) would be expected by the FDA and the optimum way to present the information. In lieu of a 510(k) medical device manufacturers will need to document why the change to a biomaterial did not require a new 510(k).

Companies with Investigational Device Exemption (IDE) clinical studies, whether or not the studies are "significant risk" or non-significant risk, need to have a strategy for determining whether or not a change to a biomaterial triggers either a full IDE supplement, an IDE 5-Day Report, an Annual Report notation, or documentation to the file. In the circumstance of the "non-significant risk study," there is no IDE, so typically the FDA would not be notified. Nevertheless, it is prudent to alert the Institutional Review Board about any modification to a material, if there is any reasonable doubt of the safety and risk to the patient. (In most circumstances, a non-significant risk study does not involve a biomaterial that could put a patient at risk.) Changes to materials should be reviewed using a decision tree and according to standard operating procedures within the company. Hazard analysis and risk assessment should be reviewed again to determine whether the change in the biomaterial has affected the safety of the product. Any change to a biomaterial should go through the design review and documentation system set up by the company. When in doubt, the agency should be contacted, because the change to the material could require the study to be conducted under an IDE.

If changes are made to a marketed product as a result of a death or serious injury report or undertaken to prevent such injuries, the change could be subject to review by the U.S. Food and Drug Administration; particularly if existing products are removed from the market by way of "corrections or recalls." Consult a regulatory consultant or attorney before making changes under these circumstances.

SUMMARY

Medical device regulations and recognized practices for applying biomaterials to new medical products change faster than chapters can be updated. Although this is a broad outline of the current regulations, consult the FDA for the most up-to-date requirements. At the time of publication, the FDA Web site is www.fda.gov.

10.4 ETHICAL ISSUES IN THE DEVELOPMENT OF NEW BIOMATERIALS

Subrata Saha and Pamela Saha

This chapter presents a brief review of ethical questions raised by ethicists and scientists in the course of development of new biomaterials. Engineers, scientists, and physicians need to become familiar with these issues to take part in this debate and they should play an active role in directing the standards of professionalism in the field of biomaterials.

INTRODUCTION

The field of biomaterials science has changed dramatically since its beginning a half century ago. We are no longer confined to improvement of metal alloys, plastics, and ceramics for use in the human body. Actual grown tissue (tissue engineering), even the cloning of human cells and organs, is within our reach. As more is possible, new questions and ethical concerns are raised. Should science set limits on itself? For example, if human beings can one day be cloned, should this be considered off limits for ethical reasons? The environment in which research is being conducted is changing. We see this in the intense competitiveness for recognition. Other factors involve financial and economic opportunities, as well as conflicts of interest in conducting objective research. The fields of science and ethics are themselves changing with increased awareness of the need for appropriate treatment of animal and human subjects in research pitted against the demand for more rigorous unbiased test results. This chapter briefly examines concerns associated with clinical trials, animal experimentation, industrial support of research, authorship, regulation, and patents.

CLINICAL TRIALS

A biomaterials engineer faces significant challenges in planning a study involving the use of patients. This step takes the engineer out of the traditional laboratory more dramatically than any other previous stage in the development of a new implant. A human subjects review committee at all institutions performing experiments on human subjects closely oversees this planning process to ensure strict adherence to ethical standards. At this point we examine some contrasts between the practice of medicine and the engineering and science professions (Gifford, 1986; Kopelman, 1983). A timely overview of the surgeon's ethical considerations in clinical treatment and research has been published (Angelos *et al.*, 2003).

In engineering and science, there is the obligation to objectively test one's theories, applications, and products, and to reasonably anticipate all hazards and potential for misuse or accident. A confidence level, a statistically significant level of probability, must be demonstrated within strict standards of accepted research protocol. When engineering studies are submitted for review by peers, the methods, data, and conclusions are scrutinized for possible flaws that may cast doubt on the validity of the findings. This same level of scientific rigor is, of course, expected for biomedical engineers and biomaterials scientists. The fact that the engineering advances would be applied to patients intensified the rigor of scientific inspection.

The historical roots of research in the practice of medicine was based on methods other than randomization techniques. From the days of Hippocrates, physicians were guided by the experience and observations. Medical journals do publish anecdotal observations. Medical conferences, in the past, consisted in part of presentations of respected physicians relaying their personal opinions based on their individual experiences with patients. Questions from participants follow this exchange of personal epithets in a collegial manner, in contrast to the near-adversarial challenges of claims made by carefully conducted research at scientific conferences.

Whereas the methods of scientists and engineers emphasize the pursuit of knowledge and accuracy of claims, practicing physicians have an additional compelling concern—providing the best care to an individual patient. This is understandable since trust and confidence is placed in the physician who is expected to be the ally of the patient. More recently, evidence-based medicine is rapidly gaining acceptance as the appropriate and justified basis for therapeutic protocols. The debate over randomized double-blind clinical studies, and when and if studies should be discontinued when there appears to be evidence demonstrating substantial advantage or harm, is an attempt to resolve the often conflicting demands of scientific evidence and the obligations of the therapeutic alliance. How and when to obtain informed consent for a study, the design of a termination strategy into a clinical trial to protect the patients, the right of self-determination of individual patients in their medical care, and the need to have accurate assessment for safe and effective care of larger numbers of people are all important considerations in the design of a study (Gifford, 1986; Kopelman, 1983; Rosner, 1987; Saha and Saha, 1988). The evidence suggests that rigid guidelines cannot be broadly applied and that studies will need to be examined individually in terms of value of the knowledge, available avenues for obtaining it, risks involved to the patient subjects, informed consent problems and their impact on the study, and justifications for violating the therapeutic alliance. Most plans that involve randomization are likely to violate to some degree the therapeutic alliance with an individual patient, if this alliance is viewed in the extreme sense that only what is perceived best should be offered as treatment. However, if it is also true that treatment based on physician experience is not necessarily the optimal care then it can be argued that the traditional perception of a therapeutic alliance requires modification. Thus, some aspects of the practice of medicine and the scientific/engineering professions converge with the infusion of scientific values and acceptance of uncertainty. Also, engineers must appreciate the complex interaction between physician and patient, understand their own role in patient care, and come to share with the physician the obligations of the therapeutic alliance (Fielder, 1992).

ANIMAL EXPERIMENTATION

Another significant change over the post quarter of a century is our perception of animals and their place in research. That a research scientist has a moral obligation to provide a degree of humane treatment to animal subjects is now generally accepted without further debate. Guidelines as to how to comply with current standards for treatment of animals in research were first published in 1963 by the Institute of Laboratory Animal Resources, Commission on Life Sciences, National Research Council. This has been updated six times, most recently in 1996. A new biomaterials implant is likely to require testing in animals prior to testing in human subjects. Familiarity with ethical considerations, regulations, and review boards that oversee animal research is essential to successfully conducting such research.

Although there exists a general consensus in the scientific community that humane animal research is essential, there also is increased public questioning about the ethical appropriateness of all animal research. Militant protests including violence toward researchers and vandalism of research labs have escalated over the past 20 years (Saha and Saha, 1991, 1992). Despite these unacceptable methods of protest by some groups of animal rights activists, new concerns about animal research have entered the debate. A claim has been made that animals have rights and value equal to those of human beings. If this were to become generally accepted, it would certainly place near-insurmountable burdens on research in biomaterials. This would essentially be stating that the health and survival needs of human beings provide no moral justification for inflicting harm or risks on animals. This is the position of animal rights theorist Tom Regan (Whitbeck, 1998). This point of view leads to the inevitable conclusion that human beings are not justified in exploiting animals for human benefits. This would have to extend beyond research to the use of animals as sources of food, clothing, transportation, sport as in hunting and fishing, and beasts of burden. Even though animals can be said to have a moral standing with respect to human beings, perhaps even

a high moral standing, this need not be viewed as standing equal to that of human beings. Various criteria have been used to determine grounds for assigning degrees of moral standing, such as intelligence, ability to feel pain and pleasure, social interaction in a given species, resemblance to humans, position on the evolutionary ladder, and whether particular animals have been pets or even belong to a species commonly used as pets. All these factors present a case against which to weigh the benefits of a given study, and the possibility of achieving the same or acceptable goals by any other means. This has led to an increased interest in alternative models, i.e., computer models and the use of tissue culture. This may help reduce the use of animals but will not likely completely eliminate the need for some animal experimentation. Current knowledge of the biology of whole systems is far too limited to suggest that all pertinent factors could be programmed into a computer or that isolated cell colonies could provide an environment completely comparable to *in vivo* studies. While animal studies are essential for the advancement of the health and welfare of human beings as well as animals, and while exploitation of animals for this purpose may be justifiable, certainly every effort should be made to minimize the use of animal subjects, to provide for humane handling of animal subjects, and to restrict the use of mammals and other higher species.

INDUSTRIAL SUPPORT FOR RESEARCH

Not surprisingly, the successes stemming from biomedical research have brought about tremendous growth in the field. Demand and intensified activity have not only led to an increased need for funds for research, but an opening up of nontraditional sources of research support. As government funding can no longer meet the needs of academic researchers, industry has stepped in bringing with it ethical issues for the biomedical engineer (Saha and Saha, 1998). Justifiable concern over the influence of business interests and its impact on scientific integrity has been expressed. Controversy over conflict of interest issues raised by corporate sponsorship has recently led to editorial requirements to disclose funding sources on publications (Schulman *et al.*, 1994). This requirement has been extended by some journals to include revealing any other associated financial interests and additional bias, i.e., religion or even sexual orientation (Rothman, 1993).

This editorial position on disclosure of funding sources raises still other issues of fairness to authors as some authors protest that this disparages the research despite high quality content (Rothman, 1993). Still, useful guidelines must be developed for interactions between corporations and researchers that leave the integrity of the research intact.

The fact that business imposes problems as well as benefits for their scientist partners is exemplified in the story of Joan Sabaté, a Loma Linda University researcher who theorized that a diet high in nuts and low in fat could cut cholesterol (Stolberg, 1993). He felt that it would not be worthwhile to wade through the long, laborious and likely fruitless process of seeking government support. He therefore turned to industry and found support from the California Walnut Commission. The funding helped Sabaté obtain results that he published in

The New England Journal of Medicine. However, the study was only about walnuts, a modification of the original hypothesis that applied to all nuts. There is a clear suggestion here that the source of funding influenced the published outcome of the study, in spite of the undeniable value of the work.

More serious than the preceding example is the case of an engineer being asked by management to suppress data on a product that may delay or prevent FDA approval. Although the engineer has a duty to protect public safety that supersedes company loyalty, he should weigh and consider possible solutions to the problem that may best meet his responsibilities to all affected parties. Ethics is not just about judging behaviors as moral or immoral. It also has to do with determining the best course of action in a given case (Whitbeck, 1998). This implies that there is not just one possible right answer. There clearly are wrong answers, but the challenge is more than to simply reject these. In the case just offered, an engineer has been asked to do something that violates his responsibility to scientific integrity and to public safety. It is also illegal. What must the engineer do? One possible answer may be to inform the manager in question about the legal and ethical concerns raised by his order to suppress data. This would give the manager an opportunity to retract the order, thus possibly satisfying all allegiances. If the order was not retracted, there still could be other options before the engineer directly "blows the whistle" on the company: for example, going through the channels of the organization. The approach to ethical problem solving is to treat the situation as a practical matter that requires a solution. The goal would be to determine a plan that best addresses all conflicting loyalties and obligations. That is why discussions about these issues help honest and moral individuals learn new ways to solve ethical problems. Although there is little debate about judging certain specific acts of clearly unethical conduct such as fraud, the best right answer often takes serious contemplation and consideration.

AUTHORSHIP

Issues of credit, recognition, adequate as well as appropriate referencing, and plagiarism seem more pronounced today than they have in the past. This is due to several factors. There is what may be a disproportionate emphasis on quantity of publication for gaining tenure, for example. Despite the discussion of this topic in some recent publications, debate over how to assign credit and to protect claims to contributions in a field is an old matter. One need only read about the rivalry between Isaac Newton (1642–1727) and Leibniz (1646–1716) to understand the extent to which recognition is an important need among scientists and how such competition could lead to some destructive practices. One of the first efforts to address this problem among scientists was by the Royal Society in Great Britain, which established the convention of giving credit to the researcher who is first to publish or submit for publication a discovery (Whitbeck, 1998). Unfortunately little has been done today to establish conventions of assigning credit and recognition in a variety of contexts.

The one important question to answer about authorship is, who merits authorship of a particular work? After that,

how should the authors be listed? Conventions concerning authorship vary internationally as well as within disciplines and universities. Many of these are unspoken and efforts to discuss such matters openly may be met with some resistance. Caroline Whitbeck described such reluctance on the part of senior researchers and department heads: "One of the molecular biologists... said forcefully that he would refuse to discuss his criteria and practices with any student or post-doc, and tell that person if he wanted to work with him, he would just have to trust him" (Whitbeck, 1998). However, it is precisely this lack of open discussion that leads to misunderstanding and the distrust. As authorship concerns the whole community of scientists and is not confined to the needs of a single laboratory, professional organizations may be helpful in directing discussions on this matter to a fruitful end.

Assuming primary authorship over the work of those of lower rank by merit of position rather than contribution needs to be addressed. Department chairs, heads of laboratories, and research program heads may insist on authorship in exchange for financial support or use of a laboratory or equipment, or simply because he or she is the department head and has the power to insist on it. Standards of assigning authorship, referencing, and crediting others who have contributed in other ways need to be further discussed and examined (Armstrong, 1993). This is important to preserve a healthy environment for the production of meaningful biomedical research.

Authorship is not only about credit. It is also about responsibility. An author must be prepared to be accountable for a published work. That means that assignment of authorship can be harmful as well as helpful and so a scientist must consent to authorship. Some journals today require that all authors sign a submission form accepting authorship (Whitbeck, 1998). This practice was in response to some highly publicized cases of alleged fabrication of research that involved some scientists named on work for which they had not been responsible. Recently, *Science* has stated that it will publish short accounts written by each author describing their individual contributions to the published work.

Other problems of authorship include duplication of publications, gift authorship, stealing of ideas not yet published, and plagiarism by reviewers. These are not presented as issues in themselves as much as they point to a lack of redress or means for correction. A fair and effective mechanism for investigating charges of misconduct within the scientific community would be useful. This may help prevent excessive oversight by governmental agencies, which brings with it other problems such as sensationalistic coverage by the press.

REGULATION

Prior to 1976, the regulation of medical devices was limited to adequate labeling and prevention of marketing clearly fraudulent devices (Ratner *et al.*, 1996). When World War II came to an end, wartime advances in materials and engineering found promising use in medicine. Rapid progress in development of remarkable devices for use in the human body transformed the fields of cardiovascular and orthopedic surgery. Not surprisingly, there were failures of devices that caught public attention. In 1976, this led to the Medical Device Amendment to the 1938 Federal Food, Drug, and Cosmetic Act. This amendment defined a medical device and empowered the FDA to regulate the devices at all stages of their development and use (Ratner *et al.*, 1996). A classification system provided a means for the FDA to impose different controls on devices according to the amount of risk presented. Class I involved products of least risk such as adhesive bandages and manually adjustable hospital beds. Class II products such as oxygen masks, blood-pressure cuffs, and powered wheelchairs were of intermediate risk. Class III are devices of highest risk, artificial heart valves, for example. In 1990 these FDA regulatory controls were expanded still more by the Safe Medical Devices Act.

In addition to governmental regulation, there are other prominent organizations that have developed guidelines for the design and testing of biomaterials. These are the American Society for Testing and Materials (ASTM), the Association for the Advancement of Medical Instrumentation (AAMI), and the American National Standards Institute (ANSI). These organizations form what are referred to as voluntary consensus standards. Conformation to these standards is voluntary. However, this provides engineers, scientists, and manufacturers with direction and tools for the management of compulsory governmental standards. The advantages of self-regulation over that imposed by lawmakers outside the field are evident. This reduces the burden on Congress and the FDA as well. Appropriate use of the voluntary system of standards is important to maintain trust that the field can regulate itself. It would fall into disrepute if standards were made that gave unjustifiable advantages to one industry over another. Engineers serve an important role in the standards process and also advise federal agencies on this subject. Members of governmental bodies tasked with approval for new devices have the difficult tasks of encouraging advancement, setting aside possible loyalties, avoiding conflict of interests, and evaluating a new product for safety and effectiveness (Saha and Saha, 1987).

PATENTS

Issues of credit and intellectual property reach beyond the order of authorship on papers and extend to protection of one's ideas that have potential for earning profit. The law offers some protection through the use of copyright and patent. Copyright is the right to exclusive publication and sale of a work. This protects against another taking the same work and reproducing it without permission of the holder of the copyright. However, as copyright has only to do with the actual expression of a work and not the ideas expressed, this does not protect against another taking the ideas. Ideas themselves cannot be copyrighted. As soon as a work is in some concrete form, that work is considered protected by copyright until 50 years after the death of the author. Copyright can be reassigned or inherited.

A patent is an exclusive right to produce, use, and sell a particular creation and to prevent others from reproducing, using, or selling a design that one has invented. For useful devices the

time limit on patents in the United States is 20 years. Proving a claim is reinforced by notebook keeping and documentation on the progress of an idea. Not all inventions are patentable. To patent a device one must prove it useful and original. Establishing a patent makes the details of an invention plain and in order to be eligible for a patent it must be applied for within 1 year of public disclosure.

Another means of protecting intellectual property is a trade secret. A trade secret is some method or design that gives one an advantage over the competition as long as it remains secret. Courts will protect trade secrets as long as the holder of the secret takes sufficient precautions to keep the secret secure. Confidential information is legally protected from wrongful disclosure. However, reinventing a trade secret or learning it through reverse engineering of a purchased product is acceptable. Although a trade secret has no time limits as a patent does, it does not restrict others from use of an idea if obtained legitimately.

Claims on intellectual property are applied differently in an academic setting as compared to industry. Invention and new product development in industry are rewarded with promotions, bonuses, and salary raises as opposed to ownership or having one's name associated with the work. This is recognized by the National Society of Professional Engineers, which states in its code of ethics that "engineers' designs, data, records, and notes referring exclusively to an employer's work are the employer's property" (Whitbeck, 1998). However, if research is done at a university, there exist an expectation that the work should be publishable. The concepts of confidentiality and trade secrets have greater significance in industry in contrast to the tradition of freely sharing ideas in an academic setting. Profit motive and competitiveness that restrained sharing of discoveries in industry led to copyright and patent protection so that published discoveries could still have protection. However, the conflicts associated with intellectual property are becoming a problem in the academic setting as well. This can lead to secrecy within research groups and laboratories. This might also curtail the benefit of learning from the work of others or even improving upon the work of others. This raises the question as to whether the public has some claim in the discussion about intellectual property rights. When company A holds patent B, only company A can fully develop and market it. If company A fails to make use of the patent or improve upon it, that particular line of advancement is confined internally to company A. Suppose for example that company A is designing a new material that can be grafted onto and integrated into the human circulatory system and can act as lung tissue. Company A has the idea patented broadly in terms of applications so that the material cannot be used by others for any medical purpose. The discovery, if successful, could be a breakthrough for treatment of lung disease and should have other applications in health care. However, company A has limited resources and will have to put the development of its new idea on hold. Another example is more ominous. Suppose there is a test to determine which candidates will respond to a certain kind of treatment against AIDS; let us call it drug X. However, company Y that makes drug X has a patent on the test and has decided not to develop the test and make it available. Does intellectual property right carry with it some responsibility? Are there overriding moral concerns? While protecting ownership, are we losing sight of a greater duty to public health and welfare? Certainly a system that allows for the suppression of a new and better course of treatment in favor of a more profitable but less effective method of care needs reexamining. A healthy environment where respect for intellectual property can exist along with a sharing of ideas is best facilitated by the professionals who must deal with these issues daily. Openness in discussing these issues will help lead to a consensus among peers. Promotion of values by professional societies fosters the development of an ethically appropriate consensus.

CONCLUSION

We have introduced a few basic issues on the subject of ethics as it relates to biomaterials and medical devices. This is only a start. Change continues to transform our perceptions about what is possible, what we can do but perhaps what we should not do, about limits on research with humans and animals, and about the balance of objectivity and bias. Science and how it is conducted are changing. Medicine and the doctor–patient relationship are changing. The field of ethics has likewise grown. It has matured from what on the surface appeared to be a collection of conflicting principles into a useful basis for finding a resolution of potentially opposing considerations. One author has stated that ethical problems need to be approached in much the same manner as an engineer would approach a difficult problem in design (Whitbeck, 1998). This approach offers fresh insights useful for engineers, scientists, and physicians to address ethical problems.

Bibliography

Angelos, P., Lafreniere, R., Murphy, T. F., and Rosen, W. (2003). Ethical issues in surgical treatment and research. *Curr. Probl. Surg.* 40: 345–448.

Armstrong, J. D. II (1993). Plagiarism: what is it, whom does it offend, and how does one deal with it? *AJR* 161: 479–484.

Fielder, J. (1992). The bioengineer's obligations to patients. *J. Investi. Surg.* 5: 201–208.

Gifford, F. (1986). The conflict between randomized clinical trials and the therapeutic obligation. *J. Med. Phil.* 11: 347–366.

Institute of Laboratory Animal Resources Commission on Life Sciences, National Research Council (1996). *Guide for the Care and Use of Laboratory Animals*. National Academy Press, Washington, D.C.

Kopelman, L. (1983). Randomized clinical trials, consent and the therapeutic relationship. *Clin. Res.* 31: 1–11.

Ratner, B. D., Hofman, A. S., Schoen, F. J., and Lemons, J. E. (eds.) (1996). *Biomaterials Science: An Introduction to Materials in Medicine*. Academic Press, San Diego.

Rosner, F. (1987). The ethics of randomized clinical trials. *Am. J. Med.* 82: 283–290.

Rothman, K. J. (1993). Journal policies on conflict of interest. *Science* 261: 1661.

Saha, S., and Saha, P. (1987). Bioethics and applied biomaterials. *J. Biomed. Mater. Res. Appl. Biomater.* 21(A2): 181–190.

Saha, P. S., and Saha, S. (1988). Clinical trials of medical devices and implants: ethical concerns. *IEEE Eng. Med. Biol. Mag.* 85–87.

Saha, P. S., and Saha, S. (1991). Ethical issues on the use of animals in the testing of medical implants. *J. Long-Term Effects Med. Implants* 1(2): 127–134.

Saha, S., and Saha, P. (1992). Biomedical engineering and animal research. *BMES Bull.* 16: 2.

Saha, S., and Saha, P. S. (1998). Biomedical research: some ethical challenges. *Crit. Rev. Biomed. Eng.* 26(5, 6): 380.

Schulman, K., Sulmasy, P., and Roney, D. (1994). Ethics, economics, and the publication policies of major medical journals. *J. Am. Med. Assoc.* 272(2): 154–156.

Stolberg, S. (1993). Funding science–for a price," *Times*, June 8, 1993.

Whitbeck, C. (1998). *Ethics in Engineering Practice and Research*. Cambridge University Press, Cambridge, UK.

10.5 LEGAL ASPECTS OF BIOMATERIALS

Jay P. Mayesh and Mary F. Scranton

INTRODUCTION

Students of biomaterials engineering know that no product lasts forever and that implantable medical devices have unwanted side effects. In today's litigious society, these factors often transform patients and device manufacturers into warring parties in always unwelcome and sometimes financially disastrous products liability litigation over the safety of medical devices. Products liability law imposes legal responsibility on manufacturers of products (ladders, cars, and medical devices, to name a few) as well as other companies involved in the "stream of commerce," such as wholesalers, distributors, and retailers, for injuries incurred by the consumer.

Products liability plaintiffs typically rely on three theories of liability. First, they claim that the manufacturer was negligent, meaning that the manufacturer failed to use reasonable care in designing and manufacturing the product. Second, plaintiffs claim the manufacturer breached legally enforceable promises, called warranties, in that the product did not meet recognized performance expectations or have the expected qualities of products of its type. Finally, plaintiffs sue under strict liability, in which the manufacturer is held responsible for a product that was unreasonably dangerous to the consumer or carried inadequate warnings, regardless of the degree of care exercised, or any promises made, by the manufacturer. Strict liability rests on two assumptions about law and economics. The first of these is that imposing liability on a manufacturer, even without a showing of carelessness on the manufacturer's part, is fair because the manufacturer is best able to discover and correct defects in its products before they cause harm. The second assumption is that, in the event of personal injury attributable to a product, the manufacturer can afford to compensate the injured party, add the cost of the injury to the product, and, if necessary, raise the price to recover the cost. To avoid liability under these legal theories, medical device manufacturers must design, manufacture, and sell products that are reasonably safe, and they must disclose in written warnings to physicians (and sometimes to patients) all risks associated with the products.

Products liability falls within the area of civil law called *torts*. A tort is simply a wrongful act that may give rise to a lawsuit. The most common tort case is the auto negligence lawsuit. Professional malpractice claims against doctors and lawyers are also torts. If the plaintiff is successful, a tort lawsuit results in an award of money, called damages, against the defendant.

This chapter describes "mass tort" products liability litigation involving implantable medical devices—that is, personal injury litigation implicating thousands of individuals making a claim for injury as the result of failure of the same medical device. Although the individual injuries may vary, most if not all claims arise out of the same mode of failure or insufficient warning. Faced with numbers of cases in the tens of thousands and legal costs in the tens of millions of dollars, the parties often attempt to consolidate mass tort claims in order to streamline pretrial procedures and control costs, either by class action (Mayesh *et al.*, 1986) or by consolidated litigation, in which all cases are assigned to one judge. In the federal courts, consolidated litigation is termed "multidistrict" litigation.

Although medical device mass torts raise complicated scientific and medical issues, in the United States legal system, lay jurors, untrained in science, are the arbiters of reasonable safety, adequate disclosure, and causation. The vagaries of the jury system always create a degree of risk and uncertainty for litigants. That risk is compounded in medical device litigation when jurors are asked to discern which of many competing medical and scientific theories is more credible. Outcomes become even more unpredictable in cases involving unsettled or evolving areas of science.

To help illustrate the subtle and complex world of mass tort litigation, the five biggest subjects of medical device litigation are described below, followed by a discussion of some of the obstacles and advantages faced by litigants in prosecuting and defending these claims, including the problems posed by science in the courtroom.

INTRAUTERINE DEVICES

The first mass tort litigation involving a medical device arose out of injuries to women who received intrauterine devices (IUDs). In the early 1970s the IUD was presented to the public as a safe, effective alternative to oral contraceptives. In the following decades serious reproductive health risks associated with IUDs ensued, prompting women to sue the manufacturers.

A. H. Robins, the manufacturer of the Dalkon Shield, bore the brunt of IUD litigation. Because the Dalkon Shield predated the Medical Device Amendments of 1976 (see later discussion), no requirements for premarket testing were in place. As described in the case *Tetuan v. A. H. Robins Co.* (241 Kan. 441 [1987]), A. H. Robins was anxious to make inroads in the market and stave off competitors, and thus began marketing the device in 1971 having performed few safety and efficacy studies. While internal corporate documents reflected A. H. Robins' concerns with the paucity of information on the product, A. H. Robins nevertheless publicly touted the benefits of the Dalkon Shield, distributing product cards that claimed the Dalkon Shield was superior to other forms of contraception and placing advertisements in popular magazines. These efforts earned the Dalkon Shield a dominant position in the IUD market.

As further detailed in *Tetuan*, shortly after the device came onto the market, the manufacturer began to receive adverse incident reports from doctors about health problems believed to be induced by the Dalkon Shield, including septic abortions and an increased incidence of pelvic inflammatory disease (PID). The Dalkon Shield, like other IUDs, has a string that descends through the cervix from the uterus to allow the user to ensure the device is in place. Before marketing the Dalkon Shield, the manufacturer had knowledge that the string had a "wicking" tendency, meaning it could transport fluid by capillary action into the uterus and introduce bacteria into an otherwise sterile environment. This wicking was alleged to be the cause of the comparatively high rate of infections and injuries in women who used the Dalkon Shield.

By the end of 1975, the lawsuits against A. H. Robins were sufficiently numerous that they were consolidated before a district court in Kansas for pretrial purposes (*In re A. H. Robins Co., Inc. "Dalkon Shield" IUD Products Liability Litigation*, 406 F. Supp. 540 [J.P.M.L. 1975]), and the number of lawsuits continued to multiply. Although A. H. Robins was successful in early suits, once certain damaging corporate documents became public, the tide turned, and plaintiffs began to win large verdicts. In May of 1985, in the pivotal *Tetuan* case, a Kansas jury awarded $9.25 million in damages. Shortly after the *Tetuan* verdict, A. H. Robins filed for bankruptcy reorganization (Vairo, 1992). Under the reorganization plan, a trust fund of $2.475 billion was established to pay outstanding claims against the manufacturer. The bulk of the fund came from another health-care company that agreed to make the contribution in order to acquire A. H. Robins (Vairo, 1992).

IUD plaintiffs have claimed injuries such as uterine perforations, infections, ectopic pregnancies, spontaneous abortions, fetal injuries and birth defects, sterility, and hysterectomies (*In re Northern District of California Dalkon Shield IUD Products Liability Litigation*, 693 F.2d 847 [2d Cir. 1982]). The theories asserted include failure to warn, unsafe design, breach of warranty, and fraud. Plaintiffs prevailed at trial and were able to negotiate large settlements because the defenses that manufacturers typically rely on in medical device litigation were less likely to avail the IUD defendants. Sometimes medical device manufacturers argue that their device cannot be linked with the type of injuries suffered by plaintiffs. In IUD cases, causation has been a less defensible issue as compared to other medical device litigation, because medical experts generally agree on the type of injuries IUDs can produce (Vairo, 1992). Although an IUD manufacturer may successfully argue that the plaintiff's injury had another cause, such as a sexually transmitted disease, the general causative correlation between IUDs and certain injuries is not in question.

In light of evidence that they were not sufficiently candid about health problems associated with their devices, IUD manufacturers also had difficulty proving they satisfied their duty to warn physicians of the risks of IUD use.

For instance, in *Nelson v. Dalkon Shield Claimants Trust* (1994 WL 255392 [D.N.H. June 8, 1994]) the court refused to dismiss the plaintiff's claims because the evidence showed that A. H. Robins failed to issue public warnings about the risks associated with the Dalkon Shield's string. Statutes of limitation, which provide an injured party a fixed amount of time to file suit after an injury occurs, frequently have provided IUD manufacturers with their strongest defense. Injuries caused by IUDs may not manifest themselves for many years, and an even greater amount of time may pass before the user becomes aware that the IUD may have caused the injury.

ARTIFICIAL HEART VALVES

Artificial heart valves are composed primarily of metal or carbon alloys and are classified according to their structure as caged-ball, single tilting-disk, or bileaflet tilting-disk valves (Vongpatanasin *et al.*, 1996). All three types have been the subject of products liability litigation.

In the late 1970s, 15 suits were brought against a manufacturer of a caged-ball valve. Plaintiffs claimed that defects in the valve caused it to wear out prematurely, resulting in major embolic complications, premature open heart surgery to replace the valve, and catastrophic poppet-ball escape from the valve cage. The theories of liability included negligence in the design, manufacture, and testing of the valve; breach of express or implied warranties; and strict liability (*In re Cutter Labs, Inc. "Braunwald-Cutter" Aortic Valve Products Liability Litigation*, 465 F. Supp. 1295 [J.P.M.L. 1979]). The manufacturer incurred significant legal expenses yet was only partially successful in defending the lawsuits (*Lindsay v. Cutter Laboratories, Inc.*, 536 F. Supp. 799 [W.D. Wis. 1982]).

Beginning in the mid-1980s, a manufacturer of a tilting-disk valve, implanted in 50,000 to 100,000 patients between 1979 and 1986 (*Bowling v. Pfizer*, 143 F.R.D. 141 [S.D. Ohio 1992]), became the object of numerous lawsuits due to the valve's potential to fracture. The valve in question was developed around the time the Medical Device Amendments of 1976 were passed and was thus one of the first devices to undergo the premarket approval process that the statute established (see later discussion) (House, 1990). Although the valve was subjected to FDA scrutiny, it did not receive the comprehensive evaluation that would be made today, because the FDA had not yet finalized procedures for granting premarket approval (House, 1990).

Even before the valve was marketed, the first instance of valve failure due to "strut fracture" occurred in clinical trials. The valve consists of a disk located inside a metal ring covered by a Teflon sewing ring, which is sutured to the heart. The disk opens and closes rhythmically, allowing blood to pass through the heart. The disk is held in place by two wire holders, the inflow and outflow struts. When the outflow strut fractures, the disk escapes from the ring, causing uncontrolled blood flow through the heart, usually resulting in death. As of 1990, 389 fractures resulting in 248 deaths had been reported (House, 1990).

According to a detailed Congressional report on the valve, based in part on examination of the manufacturer's internal documents, strut fracture was most likely caused or exacerbated by deficiencies in quality control procedures in the manufacturing process. The report contained allegations that the welding process for the valve was "out of control," that inspections were inadequate, and that the manufacturer was not willing to make the financial commitment necessary to correct the lack of oversight in the manufacturing process (House, 1990). By putting adequate manufacturing controls in place, this manufacturer might have at least reduced the potential for strut fracture and also placed itself in a better position to survive public scrutiny if problems with its valve arose in the future.

In addition, the manufacturer was subjected to criticism for its less-than-forthcoming response to the strut fracture problem. The Congressional report alleged that the manufacturer repeatedly provided misleading information to the medical community and the FDA, minimizing the frequency of strut fracture. It further contended that the manufacturer provided unreliable assurances that the strut fracture problem had been eliminated (House, 1990). After the manufacturer became embroiled in litigation, it continued to cloak information about the strut fracture problem, settling every death suit that was brought as a result of fractured valves, and insisting that the settlements be held in confidence (*Bowling v. Pfizer*, 143 F.R.D. 141 [S.D. Ohio 1992]).

Still the number of lawsuits against the manufacturer mounted, and eventually the manufacturer entered a settlement with a class of plaintiffs implanted with its valve who had not yet experienced fracture. Under the terms of the settlement, the manufacturer agreed to establish a $75 million Patient Benefit Fund for research and heart valve replacement surgery and a Medical Compensation Fund of between $80 million and $130 million to provide cash payments to valve recipients. The settlement also guaranteed immediate cash payments in the event of a fracture and provided for contribution of another $10 million for spouses of class members (*Bowling v. Pfizer*, 143 F.R.D. 141 [S.D. Ohio 1992]). The manufacturer's overall clumsy handling of the fracture problem highlights the mistakes medical device manufacturers should avoid to avert litigation.

The issue of whether a plaintiff whose heart valve has not failed can make a legally valid claim against a manufacturer has been addressed repeatedly. Generally, courts have demonstrated little willingness to entertain lawsuits based on a heart valve recipient's fear of future valve failure because they find that either physical injury or an actual device failure is a prerequisite to imposing liability on a manufacturer. However, exceptions to the general trend can be found. For instance, in a California case, *Kahn v. Shiley, Inc.* (217 Cal. App.3d 848 [1990]), the court allowed a plaintiff to proceed with her fraud claim, because the court reasoned that fraud does not challenge the safety or efficacy of the medical device. Because fraud allegations focus exclusively on the defendant's conduct, the fact that the plaintiff's valve had not failed was immaterial. In another case, *Michael v. Shiley, Inc.* (46 F.3d 1316 [3d Cir. 1995]), a court allowed the plaintiff to proceed to trial because, unlike most plaintiffs asserting fear of future valve failure, she suffered a tangible injury because she underwent surgery to replace her heart valve.

PACEMAKERS

A pacemaker is a device that uses electrical impulses to reproduce or regulate the rhythms of the heart. Pacemakers have three components—a pulse generator, leads, and electrodes. The pulse generator produces and sends accurately timed electrical impulses to the heart. Then, the lead, an insulated wire, conducts the output of the generator to one or two electrodes located in the heart. Most currently implanted pacemakers are "bipolar," meaning they have two electrodes, whereas older leads are usually "unipolar" and have only one electrode (Cobbe and Morley-Daves, 1997).

Widely publicized instances of lead-wire fracture in certain pacemaker models sparked mass litigation. The type of lead wire at issue was encased in a polyurethane coating that tended to hold the lead in the proper shape. The lead wire had a potential to break, puncture its coating, and cause serious injury to the heart and blood vessels (Cohen, 1995).

Commonly, plaintiffs have alleged injuries resulting directly from the malfunction, as well as injuries incident to pacemaker replacement surgery. In *Rogerson v. Telectronics, Co.* (1998 WL 559788 [N.D. Ill. Aug. 25, 1998]), the plaintiff sued after she suffered heart failure when her pacemaker malfunctioned and she was subjected to two open-heart surgeries to remove and replace it. In other instances, as in *In re Cordis Corp. Pacemaker Product Liability Litigation* (1992 WL 754061 [S.D. Ohio Dec. 23, 1992]), plaintiffs have filed class actions seeking, among other relief, that manufacturers bear the cost of medical monitoring of patients implanted with a pacemaker model with a known potential to malfunction.

Pacemaker manufacturers have been largely successful in defending products liability suits, and, in several cases, they have succeeded in convincing courts to dismiss plaintiffs' claims before trial. For example, in one case, *Ellis v. Cardiac Pacemakers, Inc.* (1998 WL 401682 [W. D. N. Y. July 17, 1998]), the court dismissed plaintiff's manufacturing defect and design defect claims because the plaintiff failed to present evidence of any error in the manufacturing process or any evidence that a safer design was feasible. The failure-to-warn claim was dismissed under the "learned intermediary" doctrine, which provides that, in the case of most medical devices, a manufacturer has discharged its duty to warn once it has informed the medical community (not the patient) of the risks associated with a medical device.

Despite their record of success in litigation, pacemaker manufacturers in recent years have twice agreed to settlement of class actions brought against them. One manufacturer agreed to create a $60 million fund to settle 1500 claims brought by recipients of pacemakers with allegedly defective leads (Anonymous, 1996). Another manufacturer agreed to create a $57.2 million benefit fund for persons who received allegedly defective pacemaker lead models, although under the terms of the settlement, the manufacturer admitted no wrongdoing (Anonymous, 1998a).

PEDICLE SCREWS

Pedicle screws are bone screws that are implanted in the pedicles of the spine and are used to anchor a variety of stabilizing hardware. Litigation involving pedicle screws was sparked by a December 1993 exposé on the ABC news program, 20/20. The story publicized allegations that pedicle screws, which at that time had received FDA approval only for use in the long bones of arms and legs, were being implanted in the pedicles, a procedure associated with a high rate of complications (Brown and Sawicki, 1997). The story spawned a multitude of lawsuits, which were consolidated before a Pennsylvania district court for all pretrial procedures (*In re Orthopedic Bone Screw Products Liability Litigation*, 176 F.R.D. 158 [E.D. Pa. 1997]). Before that court relinquished control of the litigation, 2300 civil actions involving more than 5000 plaintiffs were before it (*In re Orthopedic Bone Screw Products Liability Litigation*, 1998 WL 118060[(E.D. Pa. Jan. 12, 1998]).

Plaintiffs alleged several variants of device failure. They complained that pedicle screws implanted in their spines broke, loosened, corroded, or were malpositioned, causing, among other injuries, pseudarthrosis, neurogenic bladder, and arachnoiditis. In addition to device failure, plaintiffs also alleged that manufacturers committed fraud on the FDA by seeking FDA approval for pedicle screw use only in the long bones of the arms and legs, but then promoting their use in spinal surgery. Plaintiffs have also attempted, unsuccessfully, to impose liability on doctors and medical associations, alleging they conspired with manufacturers to commit fraud by conducting medical seminars that promoted pedicle screws for unapproved uses.

Plaintiffs have argued that manufacturers should be liable for failing to warn of the risks inherent in pedicle screws. As with many pacemaker cases, failure-to-warn claims involving pedicle screws have frequently been dismissed by courts under the "learned intermediary" doctrine, particularly when the implanting surgeon testifies that he was aware of the risks through sources entirely independent of the manufacturer's product literature. In addition, plaintiffs have claimed, usually without success, that the pedicle screws were defective. In *Toll v. Smith & Nephew Richards, Inc.* (1998 WL 398062 [E.D. La. July 14, 1998]), for instance, the court dismissed the case because the plaintiff could not make the required showing that an alternative design capable of preventing the alleged injury existed. Courts have found that lack of FDA approval for use of pedicle screws in the spine does not strengthen plaintiffs' claims of defect, and that "off-label" use of a medical device (that is, use for a purpose other than that for which it is FDA-approved) is not prohibited. Finally, since spinal surgery can have complications regardless of instrumentation used, plaintiffs' claims have frequently failed because they could not show that their injuries were not complications of surgery unrelated to the use of the screws.

As the discussion indicates, many claims against pedicle screw manufacturers have been dismissed by courts prior to trial. However, a handful of cases have reached juries, and the results have been mixed. A Louisiana jury rendered a verdict against one manufacturer in the amount of $318,000 after finding the pedicle screw was unreasonably dangerous (Anonymous, 1995a), and a Texas jury awarded the plaintiff $451,000 against one manufacturer for misrepresenting the safe and effective use of its pedicle screws (Anonymous, 1998b). On the other hand, a Pennsylvania jury rendered a verdict for the defendant manufacturer (Anonymous, 1995b), and a Tennessee trial ended with a hung jury (Anonymous, 1998c).

The financial strain of the massive litigation prompted one manufacturer to settle and agree to contribute $100 million and the proceeds of its insurance policies to a settlement fund. In exchange, the manufacturer, as well as its distributors and physicians and hospitals who used its products, were released from suit under any products liability theory (*In re Orthopedic Bone Screw Products Liability Litigation*, 176 F.R.D. 158 [E.D. Pa. 1997]).

SILICONE BREAST IMPLANTS

Since the early 1960s, an estimated 2 million silicone breast implants have been implanted in women, for both breast augmentation and reconstruction. Silicone breast implants are made out of a polydimethylsiloxane (PDMS) elastomer shell, to which fumed amorphous silica is added, encasing PDMS gel or saline. Some implants have been manufactured with a thin layer of polyurethane foam covering the elastomer shell; this was thought to decrease the formation of capsular contracture (hardening of the scar capsule surrounding the implant with resultant disfigurement), one of the common local complications of breast implants.

Physicians have long recognized that silicone breast implants (including those filled with saline) occasionally cause local complications such as capsular contracture, and that implants can rupture, often necessitating surgery to remove and replace the implants. However, it was not until the early 1990s, spurred by a television exposé and the publication of several case reports of women with implants who developed autoimmune diseases, that breast implants exploded into mass tort litigation. The FDA, having classified silicone breast implants as Class III devices that "present a potential unreasonable risk of illness or injury" in 1988, enforced the requirement that manufacturers collect and provide safety data on silicone breast implants. At this point, no controlled epidemiological studies exploring the relationship between breast implants and systemic disease existed. Citing the lack of safety data, the FDA imposed a moratorium on use of silicone breast implants, except in clinical studies.

Meanwhile, tens of thousands of implant recipients sued the implant and raw material manufacturers, claiming that they had developed autoimmune diseases, such as lupus, scleroderma, and fibromyalgia. To establish that the implants caused illness, plaintiffs' lawyers relied on uncontrolled case reports of autoimmune disease in women with implants and on anecdotal testimony by treating physicians to persuade juries that the implants caused illness. They also relied on early case reports of autoimmune disease following direct injections of silicone liquid and paraffin into the breasts of Japanese women. Although none of these sources of evidence scientifically established

that breast implants cause autoimmune disease, several early plaintiffs won large verdicts.

After publication of the first well-controlled epidemiological studies refuting an association between implants and recognized autoimmune diseases, many plaintiffs modified their injury claims. They alleged that they had developed "atypical" autoimmune diseases with signs and symptoms that would not have been looked for in the studies that tracked the classic autoimmune diseases.

Plaintiffs premised liability on assertions that manufacturers defectively designed the implants and failed to warn physicians and patients of health risks. A central allegation in plaintiffs' lawsuits has been that silicone has myriad ill effects on the immune system. Plaintiffs claim that, as a result of rupture of the implants and "gel bleed" of low-molecular-weight PDMS through the elastomer shell, silicone microdroplets migrate to remote organs where the silicone causes a chronic inflammatory response and the development of silicone granulomas (Plaintiffs' Submission and Proposed Findings, 1997). Moreover, they claim that silicone is an antigen capable of eliciting an immunologic response in the body, and that silicone acts as an adjuvant, heightening the body's immunologic response to other substances. Plaintiffs claim that silicone, through these mechanisms, exacerbates existing autoimmune diseases and causes classic and atypical autoimmune diseases in exposed women.

In addition to citing gel bleed as a factor in causation, plaintiffs offered several theories that implicate biodegradation of silicone. They alleged that silicone, migrating throughout the body, is picked up by phagocytes and transformed into crystalline silica, which, they claim, is immunogenic and causes connective-tissue disorders. Moreover, they allege that the silicone biodegrades into silanols, relying on *in vivo* NMR spectroscopy studies that purported to identify silicone and its metabolic by-products in the blood and livers of exposed women. Other scientists have been unable to replicate these NMR findings (Macdonald *et al.*, 1995; Mayesh and O'Hea, 1997).

In a medical science atmosphere of case reports and elaborate scientific theories raising questions and uncertainties, even though no epidemiological studies supported causation, the industry was prepared to spend approximately $4 billion to settle claims globally. The plaintiffs demanded more money, and settlement fell apart. Faced with potential liability and enormous defense costs, Dow Corning, the largest manufacturer of breast implants, was forced into bankruptcy.

Over time, scientific evidence began to mount against causation. Each epidemiology study from leading institutions such as the Mayo Clinic, Harvard Medical School, Johns Hopkins University, the University of Michigan, and the University of California failed to show any association between silicone breast implants and any autoimmune disease or atypical disease. Nonetheless, occasionally manufacturers were still subjected to substantial jury verdicts.

In 1996, the Coordinating Judge before whom all lawsuits filed in federal court were consolidated for pretrial proceedings appointed a National Science Panel of four impartial scientific experts. An immunologist, epidemiologist, toxicologist, and rheumatologist were charged to evaluate the scientific data on silicone breast implants in relation to connective tissue diseases and immunologic dysfunction (National Science Panel, 1998). These experts were instructed to "review and critique the scientific literature pertaining to the possibility of a causal association between silicone breast implants and connective tissue diseases, related signs and symptoms, and immune system dysfunction."

On three occasions in 1996 and 1997, the panel heard testimony from medical and scientific experts chosen by the lawyers for plaintiffs and for the manufacturers. The panel received more than 2000 documents from counsel for plaintiffs and manufacturers, and the panel members performed their own literature searches. The panel concluded in November 1998 that available data do not support a connection between silicone breast implants and any of the defined connective tissue diseases or other autoimmune or rheumatic conditions (National Science Panel, 1998).

PREEMPTION

As discussed in the preceding section on regulation of medical devices, medical device manufacturers were not required to seek FDA approval for new medical devices prior to 1976. In that year, a new regulatory scheme was implemented under the Medical Device Amendments of 1976 (MDA). The MDA includes a "preemption" provision, Section 360k(a), which states that "no State . . . may establish . . . any requirement (1) which is different from, or in addition to, any requirement applicable under this chapter, and (2) which relates to the safety or effectiveness of the device. . . . " Manufacturers have frequently argued that the MDA's preemption provision precludes plaintiffs from bringing lawsuits under the tort law. The preemption defense succeeds if the court finds that Congress intended that the federal MDA scheme govern the question of device safety to the exclusion of the tort law.

In 1996, the Supreme Court addressed preemption in the context of medical device litigation in a case called *Medtronic Inc. v. Lohr* (518 U.S. 470 [1996]). The Court considered the issue of whether state tort law claims against the manufacturer of a pacemaker, approved by the FDA under the abbreviated premarket notification process, were preempted. The Court unanimously determined that the plaintiff's design defect claim was not preempted because the premarket notification process merely focuses on the "substantial equivalence" of a device, not its safety. The Court also unanimously agreed that the plaintiff's claims were not preempted to the extent that the plaintiff alleged that the pacemaker manufacturer violated FDA regulations. It found that the preemption provision, and its accompanying regulations, did not preclude claims that imposed requirements mirroring those imposed by federal law. Finally, the Court found that the plaintiff's manufacturing and labeling claims were not preempted.

Following the *Lohr* decision, the proper interpretation of the MDA's preemption provision has continued to divide courts. However, certain patterns are discernible. The extent to which tort law claims are preempted has usually been determined by the regulatory status of the medical device in question. When the device was marketed under the premarket notification process, courts, in accordance with *Lohr*, have found that tort law claims are not preempted. The impact of *Lohr* is highly significant in this respect. As the Supreme Court in *Lohr* pointed out, the FDA has had difficulty keeping up with the Pre Marketing Approval process, and as a result, most new medical devices obtain FDA approval under the "substantial equivalence" exception. Therefore, for manufacturers of the large number of devices marketed after obtaining FDA approval through the premarket notification process, the MDA's preemption provision will not provide a defense. In contrast, when a device was marketed only after securing FDA approval through the more comprehensive Pre Marketing Approval process, courts have frequently found that most tort law claims are preempted. Similarly, courts have found that claims against manufacturers of devices that have been granted an investigative device exemption are preempted.

It should be noted that no matter what the regulatory status of a particular medical device, under *Lohr* and subsequent judicial decisions, claims that are based on violations of FDA regulations are not preempted. Therefore, no manufacturer can assert preemption as a defense to all potential claims.

SCIENCE IN THE COURTROOM

When medical device lawsuits reach the trial phase, the expert witness assumes a critical role. Whether an allegedly defective medical device caused the plaintiff's injury is an issue deemed beyond the ken of the average juror, and courts therefore require the litigants to present scientific evidence in the form of expert opinions. The courtroom, however, is a forum ill-suited to discussion of scientific principles, and the scientific evidence communicated to jurors too often fails to meet the standards of reliability our judicial system envisions.

Laypersons who lack schooling in the fundamentals of scientific inquiry are prone to biases. For instance, a layperson would be more inclined to accept coincidence as proof. As Marcia Angell, editor of *The New England Journal of Medicine*, noted, many people might find reasonable the proposition that mere temporal relationship, i.e., that health complications followed breast implantation, is sufficient proof that the implants caused the injuries (Angell, 1996). A scientist, on the other hand, is trained to understand that association is not causation and anecdotal reports are no substitute for scientific data.

The gap between a layperson's and a scientist's understanding of cause and effect yawns even wider in the courtroom because both the goals and the methods of science and litigation are at odds (Mayesh and Ried, 1986). Litigation and science employ disparate standards of proof, as well as disparate measures of causation. Science examines causative correlations in the population at large, whereas litigation asks whether a particular device caused a particular plaintiff's injury. Moreover, litigation and scientific inquiry demand quantitatively different standards of proof. Scientific inquiry seeks to establish causal relationships to a 95% degree of certainty, whereas judicial inquiry requires only a 51% probability of correctness. Under these circumstances, it is easily imaginable that a jury could determine that a systemic disease was caused by exposure to an implanted device while a scientist would find the same evidence merely sufficient to suggest an interesting hypothesis (Mayesh and Ried, 1986b).

As a practical matter, the judicial system's goal of timely conflict resolution would not be well served by imposing the standards of the scientific community on the courtroom. Lawsuits must be resolved in a few years. Questions of science typically take decades of testing and data accumulation before repeatability can be achieved and the question thereby resolved. To resolve the tension, the judicial system has fashioned a compromise by allowing juries to hear only that scientific evidence deemed by the judge to be reliable and generally accepted within the scientific community.

The saga of DNA identification is an example of how judges serve as the gatekeepers of whether scientific evidence is allowed before the jury. Until DNA typing became generally accepted as reliable, it was not even allowed into the courtroom. Now that it has been established as accepted methodology, DNA proof may be put before the jury although, of course, subject to rigorous dispute over correct methodology. This model, however, does not always prove satisfactory.

Because scientific inquiry is typically characterized by a degree of uncertainty, deft litigants may succeed in casting doubt upon strong scientific proof. On the other side, experts for hire are sometimes allowed to spin unproven theories in order to strengthen the inclination of jurors to assume that anecdotes are the equal of data and association the correlate of cause. At the conclusion of the evidence, 12 laypersons, who have listened to experts express conflicting opinions, vote on which is more persuasive. The opinions themselves need not be held with more than a "reasonable degree" of scientific certainty, which the giver of the opinion may define as "more probable than not." Although the jurors need usually be unanimous, their measure of confidence need only reach "a fair preponderance of the credible evidence," meaning 51%. Little wonder, then, that there is a disconnect between real science and courtroom science.

BIOMATERIALS ACCESS ASSURANCE ACT

Plaintiffs suing the manufacturer of an allegedly defective device sometimes also join as defendants the suppliers of its raw materials and component parts. The incentive to sue a raw material supplier is particularly strong if the supplier has significant financial resources and the medical device manufacturer has limited assets. For instance, DuPont, a supplier of Teflon used in TMJ implants, was named in a succession of lawsuits after the small manufacturer of the implants went bankrupt (Birnbaum and Jackson, 1998; Harper, 1997). Although bulk suppliers have consistently succeeded in having suits against them dismissed, the cost of litigation can nonetheless be substantial. After weighing the risk of becoming embroiled in expensive litigation against the tiny profits derived from the medical device market, bulk suppliers began to deny manufacturers access to their products.

To address the dwindling supply of biomaterials essential for the manufacture of implantable devices, Congress passed the Biomaterials Access Assurance Act of 1998. The act shields suppliers of raw materials and component parts from liability unless the supplier is also the manufacturer or seller of the device, or furnished materials or components that did not comply with contractual requirements or certain other specifications. Under the act, a supplier named in a lawsuit is entitled to move for dismissal immediately. Once the motion to dismiss is filed, the supplier is excused from participation in any "discovery," the expensive and time-consuming process by which litigants seek information from each other before trial. Limited discovery is allowed only to determine whether a supplier failed to comply with contractual requirements or specifications.

After dismissal from the lawsuit, the supplier may still face liability in certain, limited circumstances. After a verdict against the device manufacturer, the plaintiff or the manufacturer can require the supplier to pay part of the judgment if the court determines that the supplier's negligent or intentional conduct was a cause of the plaintiff's injury. The plaintiff may only utilize this procedure if the full amount of damages cannot be recovered from the manufacturer.

LIABILITY OF THE DESIGN ENGINEER

For several reasons, an individual design engineer is an unlikely defendant in medical device cases. First, plaintiffs know that manufacturers are more likely to have the financial resources to compensate them for their injuries. Second, a plaintiff gains a "David and Goliath" tactical advantage by suing only the manufacturer, and not the design engineer. A jury will be more inclined to sympathize with a plaintiff if the case is viewed as a confrontation between a single individual and a corporate giant. However, if sued, a design engineer is theoretically not immune from liability.

The theories available to plaintiffs are less expansive when suing a design engineer as opposed to a manufacturer. Design engineers may be liable under a theory of professional malpractice, which asks whether the design engineer has exercised the degree of care reasonably expected in the profession. Liability will only be imposed if the design engineer is found to be at fault for failing to live up to that standard. In contrast, a medical device manufacturer may be found liable under a strict liability theory, which requires no showing of carelessness.

Courts have generally recognized that the policy objectives underlying strict liability would not be furthered by applying the theory of strict liability to design engineers. Courts have reasoned that design engineers provide a professional service and do not occupy the same superior position that allows manufacturers to discover defects and spread economic losses.

Although the law of strict liability is subject to some flux and uncertainty, the great weight of judicial precedent provides a good deal of assurance that a design engineer who develops a medical device will not be held liable. And, as a practical matter, a design engineer is rarely, if ever, named as a defendant.

DEFENSIVE MANUFACTURING AND MARKETING

Although little can be done to prevent a plaintiff from initiating a lawsuit, the best defense to a products liability suit is having manufactured a safe, well-designed product with legally adequate warnings (Mayesh and Rome, 1992). In virtually every lawsuit alleging injury caused by a medical device, the plaintiff—and eventually jurors—will have access to the manufacturer's internal corporate documents. These include lot histories, manufacturing specifications, results of toxicology and safety tests, quality assurance documents, FDA submissions and compliance reviews, adverse incident reports, and intracorporate memos discussing all of these, whether paper or e-mail. The majority of these documents will have been prepared many years prior to the lawsuit, often by personnel who are no longer employed by the manufacturer at the time of trial.

If the manufacturer makes the proper investment up front in designing, manufacturing, and selling the product, these documents can be the best proof that the manufacturer performed all necessary safety testing, that the product design conformed with all government and industry standards, that no manufacturing defects occurred, that warnings and instructions for use were legally appropriate, and that the manufacturer complied with all regulatory requirements for device approval. Moreover, the documents should establish that the manufacturer took appropriate postmarket actions. This includes documentation that it was responsive to adverse incident notices from doctors and patients, had a method of tracking such complaints as well as resolving them, and when appropriate, took postmarket action, such as product recall and issuing revised warnings.

Bibliography

Angell, M. (1996). Shattuck Lecture—Evaluating the Health Risks of Breast Implants: the Interplay of Medical Science, the Law, and Public Opinion. *N. Eng. J. Med.* **334**: 1513—1518.

Anonymous (1995a). Federal jury finds acromed device unreasonably dangerous, awards $318,000 in Reeves. *Mealey's Litig. Rep. Pedicle Screws* **1**(17).

Anonymous (1995b). Pedicle screw case ends in defense verdict. *Penn. Law Weekly*, 19 June.

Anonymous (1996). $60 million pacemaker lead class action settlement approved. *Mealey's Litig. Rep. Drugs Med. Dev.* **1**(1).

Anonymous (1998a). $57.2 million class settlement reached in telectronics pacemaker lead action. *Mealey's Litig. Rep. Drugs Med. Dev.* **3**(15).

Anonymous (1998b). Jury awards $451,000 in first pedicle screw lawsuit to reach a verdict. (1998). *Med.-Leg. Aspects Breast Implants* **6**(7): 1.

Anonymous (1998c). Tennessee pedicle trial ends with hung jury. (1998). *Mealey's Litig. Rep. Pedicle Screws* **4**(4).

Biomaterials Access Assurance Act (1998). *U.S. Code*, Vol. 21, secs. 1601–6.

Birnbaum, S. L., and Jackson, J. R. (1998). Products liability. *Nation. L. J.* 9 March.

Brown, P. L., and Sawicki, M. (1997). In the wake of the proposed pedicle screw settlement. *Med.-Leg. Aspects Breast Implant Litig.* **5**(5): 1.

Cobbe, S. M., and Morley-Davies, A. (1997). Cardiac pacing. *Lancet* **349**(9044): 41–46.

Cohen, N. (1995). Pacemaker wire recall requires thorough notification. *Leader's Prod. Liabil. L. Strat.* **8**(8): 5–6.

Forstadt, J., Mayesh, J. P., and Cogan, B. M. (1986). Product liability. in *Handbook of Management for Growing, Business*, Heyel and B. Mankus. Van Nostrand Reinhold, New York.

Harper, G. L. (1997). An analysis of the potential liabilities and defenses of bulk suppliers of titanium biomaterials. *Gonz. L. Rev.* **32**: 195–224.

House (1990). House Comm. on Energy and Commerce, 101st Cong., 2d Sess. The Bjork–Shiley Valve: Earn as you learn. (Comm. Print 1990).

Mayesh, J. P., and O'Hea, J. A. (1997). Can plaintiffs win second-generation claims? *Leader's Prod. Liabil. L. Strat.* **15**(7): 5, 6.

Mayesh, J. P., and Ried, W. M. (1986a). The problems caused by "junk science." *Leader's Prod. Liabil. Newslett.* **5**(3): 1, 6.

Mayesh, J. P., and Ried, W. M. (1986b). Junk experts cloud tort outlook. *Med. Malpract. L. Strat.* **4**(2): 4, 6.

Mayesh, J. P., and Rome, W. A. (1992). Setting the groundwork for a successful defense. *Leader's Prod. Liabil. L. Strat.* **10**(10): 8, 6.

Macdonald, Plavac, N., Peters, W., Lugowski, S., and Smith D. (1995). Failure of Si NMR to detect increased blood silicone levels in silicone gel breast implant recipients. *Anal. Chem.* **67**(20): 3799–3801.

Medical Device Amendments of 1976. (1976). *U.S. Code*, Vol. 21, secs. 360c–360k.

National Science Panel (1998). Silicone breast implants in relation to connective tissue diseases and immunologic dysfunction. Report to the Honorable Sam C. Pointer, Jr., Coordinating Judge for the Federal Breast Implant Multi-District Litigation (17 November 1998).

Plaintiff's Submission and Proposed Findings to the National Science Panel on Silicone Gel Breast Implants (6 October 1997).

Vairo, G. M. (1992). The Dalkon Shield claimants trust: paradigm lost (or found)? *Fordham L. Rev.* **61**: 617–660.

Vongpatanasin, W., Hillis, L. D., and Lange, R. A. (1996). Medical progress: prosthetic heart valves. *N. Eng. J. Med.* **335**: 407–416.

Perspectives and Possibilities in Biomaterials Science

BUDDY D. RATNER, FREDERICK J. SCHOEN, JACK E. LEMONS, AND ALLAN S. HOFFMAN

The field of biomaterials is a young one with perhaps 50 years of formal history. We have emerged from an era in which biomaterials research and development was driven by surgeon–visionary–entrepreneurs (the "surgeon hero"; see A History of Biomaterials) to the beginning of the 21st century where development activities have largely transferred to university, government and industry laboratories. Early 21st-century biomaterials science is characterized by the integration of exciting discoveries from molecular and cell biology, materials science, and clinical medicine. Experiments driven by an immediate need to address a problem with a patient (i.e., make a device and get it into the clinic as quickly as possible) are more and more being supplanted by systematic, hypothesis-driven investigation by teams of engineers, basic scientists and physicians. This new style of research invites the incorporation of recent scientific ideas into the development of biomaterials. In this epilogue to *Biomaterials Science: An Introduction to Materials in Medicine*, the editors offer to those interested in the future of biomaterials a potpourri of representative ideas and perspectives that have the potential to revolutionize the field. This chapter is intended to stimulate vision, dreaming, and planning. Some of the ideas presented in this section are predicated (1) upon molecular engineering (engineering from the molecules up rather than the bulk properties down), (2) on new ideas from physics, and (3) on advanced materials science. The growth in nanotechnology and engineered tissues also contributes to more biospecific and mechanically appropriate materials. Many of these basic ideas have been expanded upon elsewhere (Drexler, 1992; Lehn, 1988, 1995; McGee, 1991; Prime and Whitesides, 1991; Ratner, 1993; Tirrell *et al.*, 1991; Ulman, 1991; Ratner and Ratner, 2003; Fuchs *et al.*, 2001; Griffith, 2000; Griffith and Naughton, 2002; Hench and Pollak, 2002; Hubbell, 1999; Langer, 2000; Langer and Vacanti, 1993; LaVan *et al.*, 2002; Vacanti and Langer, 1999). Other ideas described in this chapter come from the contemporary medical and biomaterials literature.

Our choices for these "capsule reviews" of new ideas hardly present a complete list of exciting frontiers. There are many other new ideas, paradigms, and algorithms that might be equally influential. The important point is that biomaterial science can freely use many ideas—we, the biomaterials community, have long ago passed beyond the rigid barriers of a small collection of conventional academic disciplines, and we now concentrate on solving problems relating to human health and the interactions of biological systems with materials without concern for these disciplinary silos.

Here are some contemporary areas that offer the possibility to transform aspects of today's biomaterials science and lead to exciting advances in human health and the quality of life.

BIOMATERIALS AND GLUCOSE SENSING MEET PHOTONICS

The rainbow iridescence characteristic of opals is associated with multilayer diffraction from organized arrays of amorphous silica nanospheres embedded in an amorphous silica matrix. The colors in an opal are dictated by the size of spheres, their spacing, and the refractive index difference between spheres and matrix. Professor Sanford Asher of the University of Pittsburgh has brought this principle to glucose sensing (Alexeev *et al.*, 2003). Dr. Asher's group assembles monodisperse silica nanspheres into close-packed sphere arrays. In the interstitial space between the spheres he infiltrates a hydrogel monomer and then cross-link–polymerizes the system to a gel. The gel is then soaked in hydrofluoric acid to solubilize the silica spheres leaving hollow cells within the hydrogel (a template fabrication method). The material is seen to have an opal-like appearance. Interestingly, when the gel is swollen (in different solvents, for example), the color shifts distinctly (Fig. 1). By coupling a novel, reversible cross-linking reaction into the gel that causes the gel to swell proportionately to solution glucose concentration, he has created a system that changes colors profoundly in response to glucose concentration over the range from hyperglycemic to normal to hypoglycemic. A strip of this unique gel has been incorporated into a contact lens. Since tear glucose is in equilibrium with blood glucose,

Bragg Diffraction

$$m\lambda = 2nd \sin\theta$$

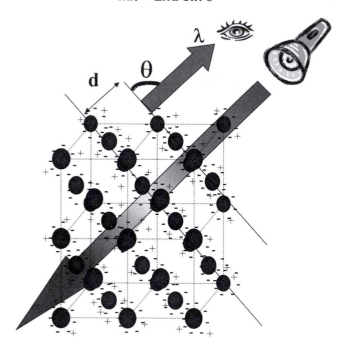

FIG. 1. The principle of a photonic crystal gel whose color changes with gel swelling and glucose concentration (courtesy of Professor Sanford Asher).

when a diabetic's blood sugar rises, the gel strip changes colors. A diabetic wearing this contact lens carries a small mirror with a built-in light source and a color code around the edge. By gently pressing down the lower eyelid to expose the lower portion of the eyelid, the diabetic can directly read his or her glucose level off the color chart. Importantly, tear proteins do not seem to damage or affect the glucose sensing.

This development is deserving of inclusion in this section on Perspectives and Possibilities because it brings the rapidly growing (and glowing) area of photonic materials to biomaterials. The template-synthesized glucose-sensitive gel also freely exploits ideas from materials science, physics, polymer chemistry, biochemistry, medicine, and optics and leads to a new concept that may dramatically improve the quality of life for diabetics, a far cry from the frequent needle sticks required at this time to draw blood.

LACK OF "PURE" AND "SAFE" NATURAL BIOMATERIALS

Nature offers us a wealth of materials that have shown potential for application in biomaterials. Some examples include alginate (from seaweed), chitin (from crab shells), silk (from silkworms) and nacre (from sea shells). Such materials are generally plentiful (hence, in principle, low cost) and can have remarkable functionality and material performance.

However, because of their origins in natural environments, these materials often have substantial contamination levels from bacteria, other living creatures, biosynthesized molecules, and pollutants (sometimes referred to as the "bioburden"). Furthermore, proteins and protein-based materials derived from sources such as cattle offer almost endless possibilities for controlled-release systems, therapeutics, and materials (for example, collagen or fibrin). Yet these proteins can harbor potentially deadly or undesirable agents such as viruses, prions, or bacteria. Furthermore, all these biomaterials and therapeutic agents are difficult (or impossible) to sterilize. There is a pressing need for methodologies to clean, purify, and sterilize biomaterials derived from natural sources. These methodologies must be scaleable to permit production at economically reasonable levels. Some progress has been made in removing endotoxin from protein preparations (Dudley *et al.*, 2003). Also, some manufacturers do sell these materials in highly pure forms, but costs are high. Beyond purity and sterilization a larger challenge remains—what remarkable biomaterials might be developed from materials from natural sources if safe, plentiful supplies of such materials were available?

SELF-ASSEMBLED MATERIALS

Physicist Richard Feynman, in his classic address that launched the nanotechnology revolution (Feynman, 1960), proposed the engineering materials from the bottom up—self-assembling the materials from atoms and molecules. Self-assembly is a central principle of nature and can be used for surface modification (Chapter 2.14). Self-assembly can also be used to make three-dimensional materials, probably the most impressive example being life, which self-assembles from molecules that are synthesized when the sperm meets the egg. Materials made by self-assembly have high degrees of order and orientation at various length scales including nanometer, micron, and millimeter. Numerous examples are available of self-assembled structures, typically made with specially synthesized designer molecules that have shown potential for biomaterials applications. Professor Sam Stupp uses an amphiphile approach to synthesize elegant structures (Stupp *et al.*, 1997). These structures have found application for bone repair and nerve regeneration. Professors Robert Langer and Gabor Somojai have used self-assembly to create smart surfaces that switch from hydrophobic to hydrophilic by turning on and off an electrical potential (Lahann, *et al.*, 2003b)(also described later in the sub-section on "novel elastic and 'smart' biopolymers.") DNA nucleotide bases have been noted to self-assemble (Boland and Ratner, 1994), and this assembly process has been implicated in the evolution of life on earth (Sowerby and Heckl, 1998). Nonadhesive, glycocalyx-like surfaces have been created using self-assembling oligosaccharide surfactant polymers (Holland *et al.*, 1998). Silk proteins assemble leading to the remarkable physical properties of these fibers (see later discussion). The examples are so numerous that the reader is referred to one of the many reviews and monographs on this subject. Importantly, this self-assembly design principle will lead to precision-engineered, readily manufacturable biomaterials of the future.

DNA TECHNOLOGIES

Nucleic acid polymers (DNA, RNA) are largely thought of as information coding systems. The reading of the 3 billion base pairs making up the human genome was a triumph of technology, biology, perseverance, and politics—there will inevitably be important advances in medicine and human health that will appear as this huge data set is translated into useful information (Lander *et al.*, 2001). However, the wider possibilities for nucleotides have been pointed out by Ronald Breaker in describing the nucleic acid space. With one monomer unit we might have four possible molecules (A, T, G, C). If we consider three monomer units, the possibilities go to 64 (codons in the genetic code). If $N = 85$ we could create 10,000 earth masses of DNA of unique molecules. Nature has not nearly exhausted the possibilities... we as engineers can explore this molecular space. What might be done with nucleotide polymers?

Interestingly, the molecular building blocks making up DNA and RNA have shown the potential for sophisticated, designed molecular-structure formation and specialized functionality. DNA is stable, is water soluble, and can be made in forms that are rigid or flexible. Endonucleases and restriction enzymes allow chains to cut in precise molecular positions while ligases permit them to be reconnected. This synthetic flexibility has triggered impressive creativity (Luo, 2003). For example, Professor Ned Seeman uses DNA chains to construct elegant two- and three-dimensional supramolecular structures (Seeman, 1997; see Fig. 2). The data storage abilities of DNA are not of special importance in this nanofabrication technology. The ability to create unique geometries and nanomachines dominates the thinking in creating these supramolecular structures.

SILK AS A BIOMATERIAL

Silk is a strong and tough natural biomaterial fiber spun by insects and spiders. Braided silk is used as a surgical suture

and in woven silk fabrics for consumer use. Spiders produce a variety of silks, and there is a clear link between protein sequence and structure–property relationships (Gosline *et al.*, 1999). There are up to seven different types of silk fiber and each silk is composed of peptide modules that confer distinct mechanical properties (Hinman *et al.*, 2000). An understanding of silk structure and an economical means of production would permit further exploitation of silk for a wide range of new uses, such as oxygen-permeable membranes and biocompatible materials (Altman *et al.*, 2003).

Recent research has led to an understanding of the detail of how silk proteins are assembled in nature (Lazaris, 2002). Biological synthesis and processing of polymers can be used as a route to gain insight into topics such as molecular recognition, self-assembly, and the formation of materials with well-defined architectures (Valluzzi *et al.*, 2002). The potential ability to mimic production of this natural material (a general process called *biomimetics*), the production of high-molecular-weight spider silk by analog proteins encoded by synthetic genes in several microbial systems (Fahnestock *et al.*, 2000), and the demonstration that appropriately transgenic goats can express spider-silk protein in their milk raise the possibility of large-scale biotechnological production of usable silk (Atkins, 2003).

NOVEL ELASTIC AND "SMART" BIOPOLYMERS

Novel polymers with unusual but highly desirable properties continue to be developed for medical devices. Recent examples include the synthesis and characterization of a biodegradable elastomer that has properties resembling those of the extracellular matrix protein elastin, and other polymers with properties that vary in a predictable, controllable manner with environmental conditions. Specifically, a tough, biodegradable elastomer with unique mechanical properties has been synthesized from sebacic acid and glycerol and formulated as a substrate for engineered tissue (Wang *et al.*, 2002b, 2003). Moreover, degradable, thermoplastic polymers that are

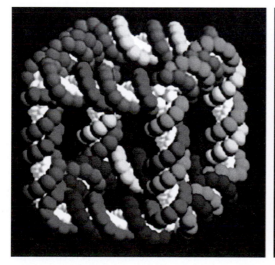

FIG. 2. Molecular models of DNA chains spliced into unique supramolecular geometries (images courtesy of Professor Nadrian Seeman).

able to change their shape predictably after an increase in temperature have been synthesized (Lenlein and Langer, 2002). In a particularly useful adjunct to minimally invasive surgical procedures, devices composed of these polymers could be compressed into a compact form that could be implanted into the body through small incisions and would expand at body temperature. Self-tying suture knots may be possible. Finally, another polymer that potentially rapidly and in a controlled manner drastically changes its surface groups and hence chemistry in controlled response to environmental change has been developed (Lahann et al., 2003b). This polymer exhibits dynamic changes in interfacial properties, such as wettability, in response to an electrical potential through conformational transitions between hydrophilic and hydrophobic states. Such a surface design enables translation of molecular-level conformational transitions to macroscopic changes in surface properties without altering the surface's chemical identity. Therapeutic medical devices could be fabricated so that their function could be uniquely and beneficially responsive to the chemical and physical environment. Moreover, surface-immobilized specifically functional molecules can support the self-assembly of cells, proteins, and antibodies to create cell-based bioassays (Lahann et al., 2003b).

NEURONAL ELECTRODE ARRAYS

Electrical stimulation of brain and neuronal tissue via electrodes has been explored since the invention of electricity. Clinically, stimulatory electrodes are used for pacemakers, Parkinson's disease, and the control of bowel function and for pacing numerous other organs (see chapter 7.15). More recently, new applications for arrays containing numerous electrodes have been demonstrated. The electrodes in the arrays are used for either stimulation or recording. These applications typically integrate biomaterials science, electrical engineering, computer processing, and medical science. The cochlear prosthesis is the first electrode array application to find widespread use. This implanted device permits the perception sound for the deaf (see Chapter 7.16). Many groups are working on a related technology for vision by either electrical stimulation of the retina (Humayun et al., 2003) or indirect stimulation near the visual centers of the brain. Recording electrode arrays have permitted mechanical devices to be controlled directly from the brain of an animal. For example, a monkey was able to control the cursor on a computer screen by just "thinking at it." Conversely, electrode arrays have been used to control the actions of an animal, an experiment that raised a number of ethical concerns. The possibilities in this area of research to offer sight to the blind or mobility to the paralyzed are tantalizing. The biocompatibility of the electrodes in contact with tissue is always a central issue in devices that depend on such arrays. Even a thin collagen encapsule can impede necessary electrical contact.

GENE EXPRESSION ANALYSIS

It is now possible to scan a large number of genes in a tissue sample at once and to determine which genes are active and used to make a protein (i.e., expressed). Gene expression analysis is done with postage-stamp-sized DNA chips (i.e.,

DNA microarrays)—DNA-covered silicon, glass, or plastic wafers capable of analyzing thousands of genes simultaneously to identify the ones that are active in a sample of cells (Lockhart and Winzelers, 2000; Stears et al., 2003). The genes expressed in the test sample are compared with those of a normal sample. A single chip containing all approximately 30,000 known human genes is available. In experimental work, this will permit identification and quantification of the complex cellular changes that occur in cells during pathologic processes, including biomaterials–tissue interactions, and may facilitate an understanding of the mechanisms of disease. Examples in biomaterials research that exemplify the power of this methodology are beginning to appear (Carinci et al., 2003; Ku et al., 2002; Pioletti et al., 2002; Risbud et al., 2001; Wang et al., 2002a). Clinically, this technology may make it easier to identify cancer-causing mutations and may permit the subclassification of tumors that have a similar appearance by conventional pathological analysis (Lakhani and Ashworth, 2001). Microarrays may provide a genetic profile of a cancerous cell that could reveal where the cancerous cell originated, how far it has progressed, and which therapies will work best to halt its further growth and spread (as in breast cancer; van de Vuver et al., 2002). It is expected that protein microarrays will also soon be widely available and useful (Banks, 2000; Liotta et al., 2001).

CIRCULATING ENDOTHELIAL PROGENITOR CELLS AND OTHER STEM CELLS

It has recently been appreciated that adult bone marrow is a rich source of a large spectrum of tissue-specific stem and progenitor cells that can populate functional organs with tissue-specific cells (Jiang et al., 2002). Evidence suggests that bone marrow-derived endothelial progenitor cells (EPCs) contribute to tissue vascularization (Rafii and Lyden, 2003). Several studies have shown that EPCs contribute to repair of vascular injury, endothelialization of vascular grafts, new blood-vessel formation during wound healing, postmyocardial infarction, and pathologic processes including atherosclerosis and tumor growth (Sata et al., 2002; Shimizu and Mitchell, 2003). The muscle of the heart, called myocardium and formerly considered a terminally differentiated organ incapable of endogenous regeneration, is also now believed to harbor and/or attract precursor cells (Nadal-Ginard et al., 2002). Research is needed to develop strategies to stimulate and regulate the production and release of functionally beneficial cells in the bone marrow, to cause them to target specific anatomic sites, and to induce them to adhere, proliferate, and appropriately differentiate. Moreover, it is possible that organs may themselves harbor endogenous EPC-like or other stem cells. Harnessing the potential of endogenous precursor and stem cells may permit the exciting possibility of directed organ regeneration following injury or repair.

MOLECULAR IMAGING

The noninvasive visualization of specific molecular targets, pathways, and physiological effects in vivo would facilitate study of cell phenotypes and function as well as extracellular synthesis and remodeling in the assessment of biomaterials–tissue interactions and evolution of engineered tissues. Macroscopic optical molecular imaging technologies using specific molecular biomarker targets deep inside living animals (research done primarily to date in mice) has become possible as a result of a number of advances, including design of biocompatible fluorescent probes and mathematical modeling of photon propagation in tissue (Weissleder and Mahmood, 2001; Weissleder and Ntziachristos, 2003). For example, the proteolytic activity in atherosclerotic plaque that potentially causes its rupture, resulting in myocardial infarction, has been visualized, raising the potential of catheter-based monitoring. Using this technology, cell structure and function may be visualized at high spatial and temporal resolution, thereby permitting unique visualization of molecular processes *in vivo*.

DENTAL MATERIALS: A PERSPECTIVE

One perspective related to the evolution and contribution of biomaterials science to the musculoskeletal surgical discipline is specific to dental restorative treatments based on synthetic (biomaterial) implants. Dental implant treatment outcomes have changed very significantly over the decades from 1960 to 2000. For example, some presentations at clinical and basic science meetings in the late 1960s and early 1970s presented success ratios as less than 60% within 5- to 10-year follow-up periods. This was summarized by a commissioned review paper by Natiella, *et al.* in 1972. They reviewed available literature and recommended that dental implant treatments remain in a "longitudinal research phase."

Results over the following decades were based on combined efforts among many, with major emphasis on the biomaterials and biomechanical sciences and associated clinical trials. The biomaterials research and development focused on bulk and surface properties of synthetics; the biomechanics on device design and functional force transfer; while the clinical studies presented results that included in-depth studies of surgical (procedures and placement), healing (passive versus active loading conditions), restoration (prosthodontics), and longer-term maintenance (hygiene and periodontal health). Associated with these studies were multiple changes in the constituting, manufacturing, sterilizing, and packaging processes. These changes resulted in subsequent descriptions and outcomes associated with "chemically and mechanically clean" devices.

It was realized that more controlled conditions for optimizing biomaterial properties, biomechanical designs, clinical placements, restorations, and maintenance could result in an implant surface-to-bone condition that was called "osseointegration." This biomaterial-to-bone interface condition was extended to many different device designs and thereby allowed treatments to be extended to a broader patient population, and other areas of musculoskeletal surgery.

The various studies and applications evolved rapidly during the 1980s and 1990s, clearly as a multidisciplinary coordination. Importantly to all involved and society at large, dental implant treatments now have realized success ratios greater than 95% at 5 years and 90–95% at 10 years. Some studies are now showing greater than 80% survival at 20 years post implant placement. At this time, most endosteal dental implant devices are fabricated from titanium or titanium alloy with or without calcium phosphate (called hydroxyapatite or HA) coatings on the implant body sections. Additionally, it now seems that many are considering the placement of dental implants after tooth loss as a conservative dental treatment. This trend was recognized by the American Dental Association sponsored report on advances in dentistry over the past 100 years, where dental implant treatments were included as an important change in the overall practice of dentistry.

Related to the current book, the success story of dental-implant-based clinical treatments is also a success story related to applications of biomaterials science and technology.

BIOCOMPATIBILITY

The concept of "biocompatibility" is central to biomaterials science and sets biomaterials apart from materials widely used in commerce and technology. In the introductory chapter to this textbook, the widely used definition of biocompatibility was offered: biocompatibility is the ability of a material to perform with an appropriate host response in a specific application. (Williams, 1987). This definition is accurate and serves us well when we consider the performance of implanted devices. However, for the response of the body to the materials themselves, contemporary thinking may call for a restructuring of this definition. Consider these points:

1. All biostable materials that pass routine toxicology and endotoxin assays heal almost indistinguishably when implanted as similar-sized specimens in soft-tissue sites. Why should polyethylene, Teflon, single-crystal alumina, gold, pyrolytic carbon, polyurethane, and a hydrogel all heal similarly, within a collagenous foreign-body capsule, although there are significant differences in surface properties of these materials?

2. Why does the body try to isolate itself from a "biocompatible" biomaterial, rather than integrate it?

3. The body has the ability to heal complex internal wounds with normal, vascularized reconstruction of the tissue. Why do "biocompatible" biomaterials turn off this normal wound-healing response?

4. Why are there activated macrophages around implanted biomaterials even years after implantation? Why is the inflammatory response never resolved?

5. Why do some unique porous structures heal in a more vascularized, less fibrotic manner (Brauker *et al.*, 1995)? Also, some processed natural tissues heal with no fibrotic foreign body reaction (Badylak...).

6. If a normal wound-healing reaction could happen around implanted biomaterials, biosensors, bioelectrodes, and drug delivery systems would not be impeded by the dense collagenous capsule. Also, devices such as breast implants, heart valve sewing rings, and intraocular lenses might not encapsulate.

Ideas from modern biology and new developments in materials science may eventually lead us to a new definition of biocompatible and a new generation of biomaterials that seamlessly integrate into the body (Ratner, 1993).

ETHICS

An ever-evolving frontier in biomaterials science is in the areas of ethics. Ethical questions are frequently raised and stimulate us to think about issues such as control and shaping of our own bodies; life and death; relationships among patient, doctor, and attorney; relationships with federal regulatory agencies; relationships between academic scientists and entrepreneurs; the cost of health care; and the use of animals in research. The ubiquity of these ethical issues in our ongoing biomaterials effort highlights a special strength and excitement in the field called biomaterials science: *we have direct impact on people*. The editors and authors of this volume hope that you become as excited about biomaterials—its intellectual challenges, humanitarian aspects, and rewards—as we are.

Bibliography

Alexeev, V. L., Sharma, A. C., Goponenko, A. V., Das, S., Lednev, I. K., Wilcox, C. S., Finegold, D. N., and Asher, S. A., (2003). High ionic strength glucose-sensing photonic crystal. *Anal. Chem.* 75(10): 2316–2323.

Altman, G. H., Diaz, F., Jakuba, C., Calabro, T., Horan, R. L., Chen, J., Lu, H., Richmond, J., and Kaplan, D. L. (2003). Silk-based biomaterials. *Biomaterials* 24: 401.

Atkins, E. (2003). Silk's secrets. Nature **424**: 1010.

Badylak, S. F. (2002). The extracellular matrix as a scaffold for tissue reconstruction. *Semin Cell Dev Biol.* 13(5): 377–383.

Banks, R. E., Dunn, M. J., Hochstrasser, D. F., Sanchez, J-C., Blackstock, W., Pappin, D. J., and Selby, P. (2000). Proteomics: new perspectives, new biomedical opportunities. *Lancet* **356**: 1749.

Boland, T., and Ratner, B. D. (1994). Two dimensional assembly of purines and pyrimidines on Au(111). Langmuir **10**: 3845–3852.

Brauker, J.H., Carr-Brendel, V. E., Martinson, L. A., Crudele, J., Johnston, W. D., and Johnson, R. C. (1995). Neovascularization of synthetic membranes directed by membrane microarchitecture. *J. Biomed. Mater. Res.* **29**: 1517–1524.

Bu, M., Melvin, T., Ensell G., Wilkinson, J. S., and Evans, A. G. R. (2003). Design and theoretical evaluation of a novel microfluidic device to be used for PCR. *J. Micromech. Microeng.* **13**: S125–S130.

Carinci, F., Volinia, S., Pezzetti, F., Francioso, F., Tosi, L., and Piatelli, A. (2003). Titanium-cell interaction: analysis of gene expression profiling. *J. Biomed. Mater. Res.* **66B**: 341.

Chen, J., Tung, C. H., Mahmood, U., Ntziachristos, V., Gyurko, R., Fishman, M. C., Huang, P. L., and Weissleder, R. (2002). *In vivo* imaging of proteolytic activity in atherosclerosis. *Circulation* **105**: 2766.

Drexler, K. E. (1992). *Nanosystems*. John Wiley & Sons, New York.

Dudley, A., McKinstry, W., Thomas, D. Best, J., and Jenkins, A. (2003). Removal of endotoxin by reverse phase HPLC abolishes anti-endothelial cell activity of ≠bacterially expressed plasminogen kringle 5. *Bio Techniques* **35**: 724–732.

Fahnestock, S. R., Yao, Z., and Bedzyk, L. A. (2000). Microbial production of spider silk proteins. *J. Biotechnol.* **74**: 105.

Feynman, R. (1960). There's plenty of room at the bottom. *Caltech Eng. Sci.*, Feb. (also see http://www.zyvex.com/nanotech/feynman.html).

Fire, A., Xu, S. Q., Montgomery, M. K., Kostas, S. A., Driver, S. E., and Mello, C. C. (1998). Potent and specific genetic interference by double-stranded RNA in *Caenorhabditis elegans*. *Nature* **391**: 806.

Fuchs, J. R., Nasseri, B. A., and Vacanti, J. P. (2001). Tissue engineering: 21st century solution to surgical reconstruction. *Ann. Thorac. Surg.* **72**: 577–590.

Gosline, J. M., Guerette, P. A., Ortlepp, C. S., and Savage, K. N. (1999). The mechanical design of spider silks: from fibroin sequence to mechanical function. *J. Exp. Biol.* **202**: 3295, 1999.

Griffith, L. G. (2000). Polymeric biomaterials. *Acta Mater.* **48**: 263.

Griffith, L. G., and Naughton, G. (2002). Tissue engineering—current challenges and expanding opportunities. *Science* **295**: 1009–1014.

Hench, L. L., and Pollak, J. M. (2002). Third-generation biomedical materials. *Science* **295**: 1014.

Hinman, M. B., Jones, J. A., and Lewis, R. V. (2000). Synthetic spider silk: a modular fiber. *Trends Biotechnol.* **18**: 374.

Holland, N. B., Qiu, Y., Ruegsegger, M., and Marchant, R. E. (1998). Biomimetic engineering of non-adhesive glycocalyx-like surfaces using oligosaccharide surfactant polymers. *Nature* **392**: 799–801.

Hubbell, J. A. (1999). Bioactive biomaterials. *Curr. Opin. Biotechnol.* **10**: 123.

Humayun, M. S., Weiland, J. D., Fujii, G. Y., Greenberg, R., Williamson, R., Little, J., Mech, B., Cimmarusti, V., Van Boemel, G., Dagnelie, G., and de Juan, E. (2003). Visual perception in a blind subject with a chronic microelectronic retinal prosthesis. *Vision Res.* 43(24): 2573–2581.

Jiang, Y., *et al.* (2002). Pluripotency of mesenchymal stem cells derived from adult marrow. *Nature* **418**: 41.

Jin, H.-J., and Kaplan, D. L. (2003). Mechanism of silk processing in insects and spiders. *Nature* **424**: 1057–1061.

Ku, C. H., Browne, M., Gregson, P. J., Corbeil, J., and Pioletti, D. P. (2002). Large-scale gene expression analysis of osteoblasts cultured on three different Ti-6A1-4V surface treatments. *Biomaterials* **23**: 4193.

Lahann, J., Balcells, M., Lu, H., Rodon, T., Jensen, K. F., and Langer, R. (2003a). Reactive polymer coatings: a first step toward surface engineering of microfluidic devices. *Anal. Chem.* **75**: 2117.

Lahann, J., Mitragotri, S., Tran, T. N., Kaido, H., Sundaram, J., Choi, I. S., Hoffer, S., Somorjai, G. A., and Langer R. (2003b). A reversibly switching surface. *Science* **299**: 371.

Lakhani, S. R., and Ashworth, A. (2001). Microarray and histopathological analysis of tumours: the future and the past? *Nat. Rev.* **1**: 151.

Lander, E. S., *et al.*, (2001). Initial sequencing and analysis of the human genome. *Nature* **409**: 860–921.

Langer, R. (2000). Biomaterials: status, challenges, and perspectives. *AIChE J.* **46**: 1286.

Langer, R., and Vacanti, J. P. (1993). Tissue engineering. *Science* **260**: 920.

LaVan, D. A., Lynn, D. M., and Langer, R. (2002). Moving smaller in drug discovery and delivery. *Nat. Rev.* **1**: 77.

Lazaris, A., *et al.* (2002). Spider silk fibers spun from soluble recombinant silk produced in mammalian cell. *Science* **295**: 472–476.

Lehn, J. M. (1988). Supramolecular chemistry—scope and perspectives: molecules, supermolecules, and molecular devices (Nobel lecture). *Angew. Chem. Int. Ed. Engl.* 27(1): 89–112.

Lehn, J.-M. (1995). *Supramolecular Chemistry: Concepts and Perspectives*. John Wiley & Sons, New York.

Lenlein, A., and Langer, R. (2002). Biodegradable, elastic shape-memory polymers for potential biomedical applications. *Science* **296**: 1673.

Liotta, L. A., *et al.* (2001). Clinical proteomics: personalized molecular medicine. *JAMA* **286**: 2211.

Lockhart, D. J., and Winzeler, E. A. (2000). Genomics, gene expression and DNA arrays. *Nature.* **405**: 827.

Luo, D. (2003). The road from materials to biology. *Mater. Today* **6**(11): 38–43.

McGee, H. A., Jr. (1991). *Molecular Engineering.* McGraw-Hill, New York.

Nadal-Ginard, B., Kajstura, J., Leri, A., and Anversa, P. (2003). Myocyte death, growth, and regeneration in cardiac hypertrophy and failure. *Circ. Res.* **92**: 139.

Natiella, J., *et al.* (1972). Current evaluation of dental implants. *J. Am. Dent. Assoc.* **84**: 1358.

Pioletti, D. P., Leoni, L., Genini, D., Takei, H., Du, P., and Corbeil, J. (2002). Gene expression analysis of ectoblastic cells contacted by orthopedic implant particles. *J. Biomed. Mater. Res.* **61**: 408.

Pirrung, M. C., Connors, R. V., Odenbaugh, A. L., Montague-Smith, M.P., Walcott, N.G., and Tollett, J.J. (2000). The arrayed primer extension method for DNA microcip analysis. Molecular computation of satisfaction problems. *J. Am. Chem. Soc.* **122**: 1873–1882.

Prime, K. L., and Whitesides, G. M. (1991). Self-assembled organic monolayers: model systems for studying adsorption of proteins at surfaces. *Science* **252**: 1164–1167.

Rafii, S., and Lyden, D. (2003). Therapeutic stem and progenitor cell transplantation for organ vascularization and regeneration. *Nat. Med.* **9**: 702.

Ratner, B. D. (1993). New ideas in biomaterials science—a path to engineered biomaterials. *J. Biomed. Mater. Res.* **27**: 837–850.

Ratner, M., and Ratner, D. (2003). Nanotechnology: a gentle introduction to the next big idea. Prentice Hall, Upper Saddle River, NJ.

Risbud, M., Ringe, J., Bhonde, R., and Sittinger, M. (2001). *In vitro* expression of cartilage-specific markers by chondrocytes on a biocompatible hydrogel: implications for engineering cardilage tissue. *Cell Transplant.* **10**: 755.

Saiki, R., Scharf, S., Faloona, F., Mullis, K. B., Horn, G. T., Erlich, H. A. and Arnheim, N. (1985). Enzymatic amplification of beta-globin genomic sequences and restriction site analysis for diagnosis of sickle cell anemia. *Science* **230**: 1350–1354.

Sata, M., *et al.* (2002). Hematopoietic stem cells differentiate into vascular cells that participate in the pathogenesis of atherosclerosis. *Nat. Med.* **8**: 403.

Seeman, N. C. (1997). DNA components for molecular architecture. *Acc. Chem. Res.* **30**(9): 357–363.

Shimizu, K., and Mitchell, R. N. (2003). Stem cell origins of intimal cells in graft arterial disease. *Curr. Atheroscler. Rep.* **5**: 230.

Soukup, G. A., and Breaker, R. R. (1999). Engineering precision RNA molecular switches. *Proc. Natl. Acad. Sci USA* **96**: 3584.

Sowerby, S. J., and Heckl, W. M. (1998). The role of self assembled monolayers of the purine and pyrimidine bases in the emergence of life. *Origins Life Evolut. Biosphere* **28**: 283–310.

Stears, R. L., Martinsky, T., and Schena, M. (2003). Trends in microarray analysis. *Nat Med* **9**(1): 140–145.

Stupp, S. I., LeBonheur, V., Walker, K., Li, L. S., Huggins, K. E., Keser, M., and Amstutz, A. (1997). Supramolecular materials: self-organized nanostructures. *Science* **276**: 384–389.

Tirrell, D. A., Fournier, M. J., and Mason, T. L. (1991). Genetic engineering of polymeric materials. *MRS Bull.* **16**(7): 23–28.

Ulman, A. (ed.) (1991). *An Introduction to Ultrathin Organic Films.* Academic Press, Boston.

Vacanti, J. P., and Langer, R. (1999). Tissue engineering: the design and fabrication of living replacement devices for surgical reconstruction and transplantation. *Lancet* **354**: SI32.

Valluzzi, R., Winkler, S., Wilson, D., and Kaplan, D. L. (2002). Silk: molecular organization and control of assembly. *Phil. Trans. R. Soc. Lond. B Biol. Sci.* **357**: 165.

van de Vuver, M. J., *et al.* (2002). A gene-expression signature as a predictor of survival in breast cancer. *N. Engl. J. Med.* **347**: 1999.

Wang, M. L., Nesti, L. J., Tuli, R., Lazatin, J., Danielson, K. G., Sharkey, P. F., and Tuan, R.S. (2002a). Titanium particles suppress expression of osteoblastic phenotype in human mesenchymal stem cells. *J. Orthop. Res.* **20**: 1175.

Wang, Y., Ameer, G. A., Sheppard, B. J., and Langer, R. (2002b). A tough biodegradable elastomer. *Nat. Biotechnol.* **20**: 602.

Wang, Y., Kim, Y. M., and Langer, R. (2003). *In vivo* degradation characteristics of poly(glycerol sebacate). *J. Biomed. Mater. Res.* **66A**: 192.

Weissleder, R., and Mahmood, U. (2001). Molecular imaging. *Radiology* **219**: 316.

Weissleder, R., and Ntziachristos, V. (2003). Shedding light onto live molecular targets. *Nat. Med.* **9**: 123.

White, T. J. (1996). The future of PCR technology: diversification of technologies and applications. *Trends Biotechnol.* **14**(12): 478–483.

Williams, D. F. (1987). in *Definitions in Biomaterials. Proceedings of a Consensus Conference of the European Society for Biomaterials*, Chester, England, March 3–5, 1986, Vol. 4. Elsevier, New York.

Yin, J. Q., and Wan, Y. (2002). RNA-mediated gene regulation system: now and the future [review]. *Int. J. Mol. Med.* **10**: 355–365.

Zhang, S. (2003). Fabrication of novel biomaterials through molecular self assembly. *Nat. Biotechnol.* **21**: 1171–1178.

A

Properties of Biological Fluids

Steven M. Slack

This appendix represents a compilation of information relevant to biomaterials scientists regarding the properties and composition of several body fluids, i.e., blood, plasma (serum), cerebrospinal fluid, synovial fluid, saliva, tear fluid, and lymph. Where possible, ranges of values are provided but the reader should recognize that significant variations are possible, particularly in states of disease. Further, the data reported here reflect adult measurements and may be substantially different in a pediatric population. Values for cerebrospinal fluid refer to the lumbar region, those for synovial fluid refer to the knee joint, and those for lymph refer to the thoracic duct, unless otherwise specified. Table A1 lists the physicochemical properties of these fluids. Table A2 provides the typical cellular composition of human blood. Table A3 shows the normal volumes of these fluids in males and females and presents equations

whereby such volumes can be estimated from the mass of the individual. Next, Table A4 lists the approximate concentrations of the major proteins present in various biological fluids. Tables A5 and A6 present the concentrations of inorganic and organic species, respectively, in these fluids. Table A7 provides data on the major plasma proteins, i.e., concentration, molecular weight, isoelectric point (pI), sedimentation coefficient (S), diffusion coefficient (D), extinction coefficient (E_{280}), partial specific volume (V_{20}), carbohydrate content (C), and half-life. Finally, Tables A8 and A9 present information on the proteins involved in the complement pathway and blood coagulation pathway, respectively. Some of the information contained in this appendix have been previously published in Black, J., and Hastings, G. (eds.) (1998). *Handbook of Biomaterial Properties*. Chapman & Hall, New York, pp. 114–124.

TABLE A1 Physicochemical Properties of Several Biological Fluids

Property	Whole blood	Plasma (serum)	Cerebrospinal fluid	Synovial fluid	Saliva	Tear fluid
Freezing-point depression[a]	0.557–0.577	0.512–0.568	0.540–0.603	—	0.07–0.34	0.572–0.642
Osmolality[b]	—	275–295	290–324	292–300	—	309–347
pH[c]	7.35–7.45	7.35–7.43	7.35–7.70	7.29–7.45	5.0–7.1	7.3–7.7
Refractive index[d]	16.2–18.5	1.3485–1.3513	1.3349–1.3351	—	—	1.3361–1.3379
Relative viscosity[e]	2.18–3.59	1.18–1.59	1.020–1.027	> 300	—	1.26–1.32
Specific conductivity[f]	—	0.0117–0.0123	0.0119	—	—	—
Specific gravity[g]	1.052–1.061	1.024–1.027	1.006–1.008	1.008–1.015	1.002–1.012	1.004–1.005
Specific heat[h]	0.87	0.94	—	—	—	—
Surface tension[i]	55.5–61.2	56.2	60.0–63.0	—	15.2–26.0	40–50

[a]Units are °C.
[b]Units are mosm/kg H_2O. Calculated from freezing-point depression.
[c]pH measured from arterial blood and plasma, and from cisternal portion of CSF.
[d]Measured at 20°C.
[e]Measured *in vitro* at 37°C for whole blood, plasma, and synovial fluid, and at 38°C for cerebrospinal fluid. The viscosity of serum is slightly less than plasma due to the absence of fibrinogen.
[f]Units are S/cm. Measured at 25°C for plasma, 18°C for CSF.
[g]Relative to water at 20°C.
[h]Units are cal/g °C.
[i]Units are dyn/cm. Measured at 20°C.

TABLE A2   Cellular Composition of Blood

Cell type	Cells/μl	Half-life in circulation
Erythrocytes	$4.6–6.2 \times 10^6$ (M) $4.2–5.2 \times 10^6$ (F)	25 ± 2 days
Leukocytes		
Neutrophils	3000–5800	6–8 hours
Eosinophils	50–250	8–12 hours
Basophils	15–50	?
Monocytes	300–500	1–3 days
Lymphocytes	1500–3000	Variable
Platelets	$1.5–3.5 \times 10^5$	3.2–5.2 days
Reticulocytes	$2.3–9.3 \times 10^4$	—

TABLE A3   Volumes of Various Biological Fluids[a]

Parameter	Whole blood	Erythrocytes	Plasma	CS fluid	Tear fluid
Volume (ml)	4490 (M) 3600 (F)	2030 (M) 1470 (F)	2460 (M) 2130 (F)	100–160	4.0–13

[a]The following equations can be used to estimate blood volume (BV), erythrocyte volume (EV), and plasma volume (PV) from the known body mass (b, kg) with a coefficient of variation of approximately 10%:

Males (M)	Females (F)
$BV = 41.0 \times b + 1530$	$BV = 47.16 \times b + 864$
$PV = 19.6 \times b + 1050$	$PV = 28.89 \times b + 455$
$EV = 21.4 \times b + 490$	$EV = 18.26 \times b + 409$

TABLE A4   Protein Concentrations (mg/dl) in Various Biological Fluids

Protein	Plasma (serum)	Cerebrospinal fluid	Synovial fluid	Saliva	Tear fluid	Lymph
Total	6000–8000	20–40	500–1800	140–640	430–1220	2910–7330
Albumin	4000–5500	11.5–19.5	400–1000	0.2–1.2	400	1500–2670
α_1-Acid glycoprotein	50–115	0.1–0.25	—	—	—	260
αA??????	—	—	—	6–70	—	—
α_1-Antitrypsin	85–185	0.4–1.0	45–110	—	1.5	—
Ceruloplasmin	15–60	0.07–1.0	1–7.5	—	4	—
Fibrinogen	200–400	0.065	—	—	—	—
Fibronectin (μg/ml)	150–300	1–3	150	< 1	3–9	—
Haptoglobin	70–140	0.075–0.4	—	9	—	—
Hemopexin	50–120	—	—	—	—	—
IgA	100–400	0.1–0.3	60–115	2.2–15	4–80	—
IgG	650–1600	0.7–2.0	150–46	0.3–1.8	4–60	780
IgM	30–120	—	9–20	0.1–1.2	trace	—
Lysozyme	0.3–0.8	—	—	13–66	100–280	—
α_2–Macroglobulin	150–450	0.3–0.65	10–50	—	—	—
Transferrin	200–320	0.5–1.2	—	—	10	—

TABLE A5 Concentrations of Major Inorganic Substances (mmol/L) in Various Biological Fluids

Electrolyte	Whole blood	Plasma (serum)	Cerebrospinal fluid	Synovial fluid	Saliva	Tear fluid	Lymph
Bicarbonate	19–23	21–30	21.3–25.9	—	2–13	20–40	—
Calcium	2.42	2.1–2.6	1.02–1.34	1.2–2.4	0.69–2.46	0.35–0.77	1.7–2.8
Chloride	77–86	98–109	122–132	87–138	6.5–42.9	110–135	87–103
Magnesium	1.48–1.85	0.80–1.05	0.55–1.23	< Serum	0.065–0.38	—	—
Total phosphorus	10.1–14.3	2.5–4.8	0.442–0.694	> Serum	3.9–9.3	0.11–10.3	2.0–3.6
Potassium	40–60	3.5–5.6	2.62–3.3	3.5–4.5	14–41	31–36	3.9–5.6
Sodium	79–91	125–145	137–153	133–139	5.2–24.4	126–166	118–132
Sulfate	0.1–0.2	0.31–0.58	—	Same as serum	—	—	—

TABLE A6 Concentrations of Organic Compounds (mg/dl) in Various Biological Fluids

Species	Whole blood	Plasma (serum)	Cerebrospinal fluid	Synovial fluid	Saliva	Tear fluid	Lymph
Amino acids	4.8–7.4	3.6–7.0	1.0–1.5	—	—	5.0	—
Bilirubin	0.3–1.1	0.2–0.8	< 0.01	—	—	—	0.8
Cholesterol	115–225	120–200	0.16–0.77	0.3–1.0	—	10.6–24.4	34–106
Creatine	2.9–4.9	0.13–0.77	0.46–1.9	—	—	—	—
Creatinine	1–2	0.6–1.2	0.65–1.05	—	0.5–2	—	0.8–8.9
Fatty acids	250–390	150–500	—	—	—	—	—
Glucose	80–100	85–110	50–80	—	10–30	10	140
Hyaluronic acid	—	—	—	250–365	—	—	—
Lipids, total	445–610	400–850	0.77–1.7	—	—	—	—
Phospholipid	225–285	150–300	0.2–0.8	13–15	—	—	—
Urea	20–40	20–30	13.8–36.4	—	14–75	20–30	—
Uric acid	0.6–4.9	2.0–6.0	0.5–2.6	7–8	0.5–2.9	—	1.7–10.8
Water (g)	81–86	93–95	94–96	97–99	99.4	98.2	81–86

TABLE A7 Properties of the Major Plasma Proteins

Protein	Plasma concentration (mg/ml)	Molecular weight (Da)	pI	S^a	D^b	E^c_{280}	V^d_{20}	C^e	Half-life (days)
Prealbumin	0.12–0.39	54,980	4.7	4.2	—	14.1	0.74	—	1.9
Albumin	40–55	66,500	4.9	4.6	6.1	5.8	0.733	0	17–23
α_1-Acid glycoprotein	0.5–1.15	44,000	2.7	3.1	5.3	8.9	0.675	41.4	5.2
α_1-Antitrypsin	0.85–1.85	54,000	4.0	3.5	5.2	5.3	0.646	12.2	3.9
C1q	0.05–0.1	459,000	—	11.1	—	6.82	—	8	—
C3	1.5–1.7	185,000	6.1–6.8	9.5	4.5	—	0.736	—	—
C4	0.3–0.6	200,000	—	10.0	—	—	—	—	—
Ceruloplasmin	0.15–0.60	160,000	4.4	7.08	3.76	14.9	0.713	8	4.3

continued

TABLE A7 Properties of the Major Plasma Proteins (*continued*)

Protein	Plasma concentration (mg/ml)	Molecular weight (Da)	pI	S^a	D^b	E_{280}^c	V_{20}^d	C^e	Half-life (days)
Fibrinogen	2.0–4.0	340,000	5.5	7.6	1.97	13.6	0.723	2.5	3.1–3.4
Fibronectin	0.15–0.2	450,000	—	13–13.6	2.1–2.3	13.5	0.72	4–9	0.33
α_2-Haptoglobin									
Type 1.1	1.0–2.2	100,000	4.1	4.4	4.7	12.0	0.766	19.3	2–4
Type 2.1	1.6–3.0	200,000	4.1	4.3–6.5	—	12.2	—	—	
Type 2.2	1.2–2.6	400,000	—	7.5	—	—	—	—	
Hemopexin	0.5–1.2	57,000	5.8	4.8	—	19.7	0.702	23.0	9.5
IgA (monomer)	1.0–4.0	162,000	—	7.0	3.4	13.4	0.725	7.5	5–6.5
IgG	6.5–16.5	150,000	6.3–7.3	6.5–7.0	4.0	13.8	0.739	2.9	20–21
IgM	0.3–1.2	950,000	—	18–20	2.6	13.3	0.724	12	5.1
Lysozyme	0.003–0.008	14,400	10.5	—	—	—	—	—	—
α_2 – Macroglobulin	1.5–4.5	725,000	5.4	19.6	2.4	8.1	0.735	8.4	7.8
Transferrin	2.0–3.2	76,500	5.9	5.5	5.0	11.2	0.758	5.9	7–10

[a] Sedimentation constant in water at 20°C, expressed in Svedberg units.
[b] Diffusion coefficient in water at 20°C, expressed in 10^{-7} cm^2/sec.
[c] Extinction coefficient for light of wavelength 280 nm traveling 1 cm through a 10 mg/ml protein solution.
[d] Partial specific volume of the protein at 20°C, expressed as ml g^{-1}.
[e] Carbohydrate content of the protein, expressed as the percentage by mass.

TABLE A8 Properties of Proteins Involved in the Complement System

Protein	Serum concentration (mg/L)	Relative molecular weight M_r (Da)	Sedimentation constant S_{20w} (10^{-13} sec)
C1q	50–100	459,000	11.1
C1r	35–40	83,000	7.5
C1s	32–40	83,000	4.5
C2	20–35	108,000	4.5
C3	1500–1700	185,000	9.5
C4	300–600	200,000	10.0
C5	120–180	185,000	8.7
C6	42–60	128,000	5.5
C7	4–60	121,000	6.0
C8	35–50	151,000	8.0
C9	45–70	71,000	4.5
Factor B	220–330	92,000	5–6
Factor D	Trace	24,000	3.0
Properdin	25–35	220,000	5.4
C1 inhibitor	145–170	100,000	—
Factor H	475–575	150,000	6.0
Factor I	30–45	88,000	5.5

TABLE A9 Properties of Proteins Involved in Blood Coagulation

Protein	Plasma concentration (μg/ml)	Relative molecular weight M_r (Da)	Biological half-life $t_{1/2}$ (hours)
Fibrinogen	2000–4000	340,000	72–120
Prothrombin	70–140	71,600	48–72
Factor III (tissue factor)	—	45,000	—
Factor V	4–14	330,000	12–15
Factor VII	Trace	50	2–5
Factor VIII	0.2	330,000	8–12
Factor IX	5.0	57,000	24
Factor X	12	58,800	24–40
Factor XI	2.0–7.0	160,000	48–84
Factor XII	15–47	80,000	50–60
Factor XIII	10	320,000	216–240
Protein C	4.0	62,000	10
Protein S	22	77,000	—
Protein Z	3.0	62,000	60
Prekallikrein	35–50	85,000	—
High-molecular-weight kininogen	70–90	120,000	—
Antithrombin III	210–250	58,000	67

B

Properties of Soft Materials

CHRISTINA L. MARTINS

TABLE B1 Some Mechanical and Physical Properties of the Most Common Polymers Used as Biomaterials and the Respective Application

Polymer	Tensile strength (MPa)	Tensile modulus (GPa)	Elongation (%)	T_g (°C)	T_m (°C)	Water absorption (%)	Water contact angle (°)	Biomedical applications
Polyethylene:								
Low-density polyethylene—LDPE[1]	4–16	0.1–0.3	90–800	−20	95–115	<0.01	93–95	Tubing[1,4,5]; shunts[1]; catheters[3–5]
High-density polyethylene—HDPE[1]	21–38	0.4–1.2	20–1000	−125	135–138	<0.01	91	Plastic surgery implants
Ultrahigh-molecular-weight polyethylene—UHMWPE[10]	22–42	0.65	—	—	125–135	<0.01	—	Acetabulum in total hip prostheses[4]; artificial knee prostheses[4]
Polypropylene[1]	30–38	1.1–1.6	200–700	0	165	<0.01	104	Heart valve structures[1]; oxygenator[3,4]; and plasmapheresis[1] membranes; unabsorbable sutures[1,3,4]; disposable syringes[1,3,4]
Poly(vinyl chloride)—PVC (rigid)[1]	35–62	2.4–4.1	2–40	87–90	212	0.1–0.4	80	Catheters[3]; maxillofacial prostheses[11]; blood bags[1,2,3,5]; tubings[1,2,5]; plasmapheresis membrane[1]
(plasticized)[1]	10–24	—	200–450					
Polytetrafluoroethylene—PTFE[1]	14–35	0.4	200–400	−10	327	0	110	Oxygenator membrane[1]; vascular graft[1–5]; catheter coating[1–5]; facial prosthesis[1]; heart valves structures[4]; stapes prosthesis (in tympanoplasty)[4,5]
Polydimethylsiloxane[1]	2–10	—	100–600	(−)120–(−)123	—	0.02	101–109	Oxygenator membrane[1,4]; tubing[1,2,5]; shunts[1,4]; breast, joint, tracheal, bladder, and maxillofacial prostheses[4]; heart pacemaker leads[4]; heart valve structures[4]; burn dressing[4,5]
Poly(ethylene terephthalate)—PET or Dacron[1]	59–72	2.8–4.1	50–300	69–82	265–270	0.1–0.2	73–78	Vascular grafts[1–4]; heart valve structures[3]; shunts[1]; laryngeal, esophageal, and bladder prostheses[5,4]; nonabsorbable sutures[3,4,11]; tendon reconstruction[11]

continued

TABLE B1 Some Mechanical and Physical Properties of the Most Common Polymers Used as Biomaterials and the Respective Application (*continued*)

Polymer	Tensile strength (MPa)	Tensile modulus (GPa)	Elongation (%)	T_g (°C)	T_m (°C)	Water absorption (%)	Water contact angle (°)	Biomedical applications
Polyamides (nylons)[1]	62–68	1.2–2.9	60–300	45–75	200–270	1.5	—	Hemodialysis membrane[1]; nonabsorbable sutures[1–3]
Poly(ether urethane) (e.g., Pellethane)[6]	35–48	< 0.01	350–600	(−)43–(−)60	188–204	—	62–107	Percutaneous leads[1]; catheters[1,2]; tubings[1,3]; intraaortic balloons[1]; wound dressing[1]
Poly(ether urethane urea) (e.g. Biomer)[6]	31–41	< 0.01	600–800	(−)53–(−)67	120–150	0.02	63–69	Artificial heart components[1]; heart valve[1]; tubing[4]; vascular graft prosthesis[4]
Polystyrene[1]	35–83	2.8–4.1	1–3	116	137	0.10	45[a]	Tissue culture dish[1,3]
Polycarbonate[1]	55–66	2.4	100–130	150	267	0.2	62[a]	Connectors[4]; oxygenator, hemodialyzer and plasmapheresis membranes[4]
Polysulfones[1]	70.3	2.5	50–100	190–285	—	0.22	—	Artificial heart components[1,2]; heart valve structures[1,2]; oxygenator, hemodialysis and plasmapheresis membrane[4]
Poly(methyl methacrylate)[1]	48–76	3.1	2–10	110	160	0.3–0.4	62	Dentures[1,4,11]; plasmapheresis[1] and hemodialysis[3,4] membranes; bone cement[1,3–5,11]; intraocular lens[1,3–5]; middle ear prosthesis[1,3]
Poly(2-hydroxy-ethylmethacrylate)—PHEMA[1]	0.3	0.8	50	115 (dry)	—	> 1	—	Catheter coating[1,2]; drug delivery device[1]; soft contact lens[1,11]; vascular prosthesis coating[1]; burn dressing[1,11]
Poly(vinyl alcohol)[4] (98–99% hydrolysed)	67–110		0–300	85	230	—	—	Drug delivery devices[4,11]
(87–89% hydrolysed)	24–79			58	180			
Poly(ε-caprolactone) (MW: 44,000)[7]	16	0.4	80	−62	57	< 0.2	—	Drug delivery devices[2]
Poly(glycolic acid) (MW: 50,000)[10]	647	6.5		36	187–222	—	—	Drug delivery devices[2,11]; absorbable sutures[2,11]
Poly(lactic acid)—PLA[7]								Drug delivery devices[11]
L-PLA (MW: 50,000–300,000)	28–48	1.2–3.0	6–2	54–59	170–178	—	—	
D,L-PLA (MW: 107,000–550,000)	29–35	1.9–2.4	6–5	51–53	—			
Poly(ortho ester)[7b] (MW: 99,700)	20	0.82	220	55	—	0.2–1	—	Drug delivery devices[1,2]
Cellulose acetate[4]	13.1–58.6	0.6–1.8	6–50	—	306	2.0–6.5	—	Hemodialysis membrane[2,11]
Collagen[8]	50–100	1	10	120[10]	230[10]	> 1	—	Hemostatic agent[8,11]; vascular prosthesis[8,11]; heart valves[11]; tendons and ligaments[8,11]; wound and burn dressings[8,11]; absorbable sutures[8,11]
Silk[9c]	610–690	0.015–0.017	4–16	—	—	0.2–1	—	Nonabsorbable sutures[2,11]

[a] Values are averages of advancing and receding contact angles.
[b] Transcyclohexane dimethanol (t-CDM): 1,6-hexanediol (1,6-HD) = 35 : 65.
[c] *Bombyx mori* silkworm silk.

Bibliography

Marchant, R. E., and Wang, I. (1994). Physical and chemical aspects of biomaterials used in humans. in *Implantation Biology—The Host Response and Biomedical Devices*, R. S. Greco (ed.). CRC Press, Boca Raton, FL, pp.13–38.

Helmus, M. N., and Hubbell, J. A., (1993). Materials selection. *Cardiovasc. Pathol.* **2**(suppl.): 53S–71S.

Lee, H. B., Kim, S. S., and Khang, G. (1995). Polymeric biomaterials. in *The Biomedical Engineering Handbook*, J. D. Bronzino (ed.). CRC Press, Boca Raton, FL, pp. 581–597.

Kroschwitz, J. I. (1990). *Concise Encyclopedia of Polymer Science and Engineering.* John Willey and Sons, New York.

Dumitriu, S., and Dimitriu-Medvichi, C. (1994). Hydrogel and general properties of biomaterials. in *Polymeric Biomaterials*, S. Dumitriu (ed.). Marcel Dekker, New York, pp. 3–97.

Lamba, N. M. K., Woodhouse, K. A., and Cooper, S. L. (1998). *Polyurethanes in Biomedical Applications.* CRC Press, Boca Raton, FL.

Engelberg, I., and Kohn, J. (1991). Physico-mechanical properties of degradable polymers used in medical applications: a comparative study. *Biomaterials* **12**: 292–304.

Li, S. T. (1995). Biological biomaterials: tissue-derived biomaterials (collagen). in *The Biomedical Engineering Handbook*, J. D. Bronzino (ed.). CRC Press, Boca Raton, FL, pp. 627–647.

Perez-Rigueiro, J., Viney, C., Llorca, J., and Elices, M. (2000). Mechanical properties of single-brin silkworm silk. *J. Appl. Polymer Sci.* **75**: 1270–1277.

Polymers: A. Property Database. http://polymersdatabase.com/

Williams, D. (1990). *Concise Encyclopedia of Medical Dental Materials.* Pergamon Press, UK.

C

Chemical Compositions of Metals Used for Implants

JOHN B. BRUNSKI

TABLE C1   Chemical Compositions of Metals Used for Implants

Material	ASTM designation	Common/trade names	Composition (wt.%)	Notes
Stainless steel	F55 (bar, wire) F56 (sheet, strip) F138 (bar, wire) F139 (sheet, strip)	AISI 316LVM 316L 316L 316L	60–65 Fe 17.00–20.00 Cr 12.00–14.00 Ni 2.00–3.00 Mo max 2.0 Mn max 0.5 Cu max 0.03 C max 0.1 N max 0.025 P max 0.75 Si max 0.01 S	F55, F56 specify 0.03 max for P, S. F138, F139 specify 0.025 max for P and 0.010 max for S. LVM = low vacuum melt.
Stainless steel	F745	Cast stainless steel cast 316L	60–69 Fe 17.00–20.00 Cr 11.00–14.00 Ni 2.00–3.00 Mo max 0.06 C max 2.0 Mn max 0.045 P max 1.00 Si max 0.030 S	
Co–Cr–Mo	F75	Vitallium Haynes-Stellite 21 Protasul-2 Micrograin-Zimaloy	58.9–69.5 Co 27.0–30.0 Cr 5.0–7.0 Mo max 1.0 Mn max 1.0 Si max 2.5 Ni max 0.75 Fe max 0.35 C	Vitallium is a trademark of Howmedica, Inc. Hayness-Stellite 21 (HS 21) is a trademark of Cabot Corp. Protasul-2 is a trademark of Sulzer AG, Switzerland. Zimaloy is a trademark of Zimmer USA.
Co–Cr–Mo	F799	Forged Co–Cr–Mo Thermomechanical Co–Cr–Mo FHS	58–59 Co 26.0–30.0 Cr 5.0–7.00 Mo max 1.00 Mn max 1.00 Si max 1.00 Ni max 1.5 Fe max 0.35 C max 0.25 N	FHS means, "forged high strength" and is a trademark of Howmedica, Inc.

continued

TABLE C1   Chemical Compositions of Metals Used for Implants (*continued*)

Material	ASTM designation	Common/trade names	Composition (wt.%)	Notes
Co–Cr–W–Ni	F90	Haynes-Stellite 25 Wrought Co–Cr	45.5–56.2 Co 19.0–21.0 Cr 14.0–16.0 W 9.0–11.0 Ni max 3.00 Fe 1.00–2.00 Mn 0.05–0.15 C max 0.04 P max 0.40 Si max 0.03 S	Haynes-Stellite 25 (HS25) is a trademark of Cabot Corp.
Co–Ni–Cr–Mo–Ti	F562	MP 35 N Biophase Protasul-10	29–38.8 Co 33.0–37.0 Ni 19.0–21.0 Cr 9.0–10.5 Mo max 1.0 Ti max 0.15 Si max 0.010 S max 1.0 Fe max 0.15 Mn	MP35 N is a trademark of SPS Technologies, Inc. Biophase is a trademark of Richards Medical Co. Protasul-10 is a trademark of Sulzer AG, Switzerland.
Pure Ti, grade 4	F67	CP Ti	Balance Ti max 0.10 C max 0.5 Fe max 0.0125–0.015 H max 0.05 N max 0.40 O	CP Ti comes in four grades according to oxygen content— Grade 1 has 0.18% max O Grade 2 has 0.25% max O Grade 3 has 0.35% max O Grade 4 has 0.40% max O
Ti–6Al–4V ELI	F136-79	Ti–6Al–4V	88.3–90.8 Ti 5.5–6.5 Al 3.5–4.5 V max 0.08 C max 0.0125 H max 0.25 Fe max 0.05 N max 0.13 O	

*A more recent specification can be found from ASTM, the American Society for Testing and Materials, under *F136-98e1 Standard Specification for Wrought Titanium-6 Aluminum-4 Vanadium ELI (Extra Low Interstitial) Alloy (R56401) for Surgical Implant Applications.*

D

The Biomaterials Literature

BOOKS

(This book list is updated and modified from a list posted on www.Biomat.net.)

Advanced Biomaterials—Characterization, Tissue Engineering, and Complexity (2002). S. C. Moss and L. K. Parker (eds.). Material Research Society, Massachusetts (Materials Research Society Symposia Proceedings, Vol. 711), ISBN 1558996478.

Advanced Biomaterials in Biomedical Engineering and Drug Delivery Systems (1996). O. Naoya, W. Kim Sung, and J. Feijen (eds.). Springer-Verlag, ISBN 443170163X.

Advances in Biomaterials 1 (1987). S. M. Lee (eds.). Technomic Publishing, ISBN 0877625042.

Antimicrobial/Anti-Infective Materials: Principles, Applications and Devices (1999). S. P. Sawan and G. Manivannan (eds.). Technomic Publishing Co. Inc., ISBN 1566767946.

Applications of Biomaterials in Facial Plastic Surgery (1991). A. I. Glasgold, and F. H. Silver (eds.). CRC Press, ISBN 0849352517.

Artificial Humans: Transplants and Bionics (1985). M. Thomas, H. Julian Menssner (eds.), ISBN 0671443674.

The Artificial Kidney: Physiological Modeling and Tissue Engineering (1999). J. K. Leypoldt (ed.). Landes Bioscience, ISBN 1570596026.

Artificial Organs (Lucent Overview Series) (1996). J. J. Presnall (ed.). Lucent Books, ISBN 1560062576.

Atlas of Oral Implantology (1999). A. N. Cranin, M. Klein, M. Simons, and A. Simons (eds.). Year Book Medical Pub, ISBN 155664552X.

Bioadhesive Drug Delivery Systems: Fundamentals, Novel Approaches and Developments (1999). E. Mathiowitz, D. Chickering III, and C.-M. Lehr (eds.). Marcel Dekker, ISBN 0911910131.

Bioartificial Organs II—Tissue Sourcing, Immunoisolation and Clinical Trials (2001). D. Hunkeler, A. Cherringto, A. Prokop, and R. Rajotte (eds.). New York Academy of Sciences, ISBN 1573313424.

Bioartificial Organs II: Technology, Medicine, and Materials (Annals of the New York Academy of Sciences, Vol. 875) (1999). D. Hunkeler, A. Prokop, and A. D. Cherrington (eds.). New York Academy of Sciences, ISBN 1573311952.

Bioartificial Organs: Science, Medicine, and Technology (Annals of the New York Academy of Sciences, Vol. 831) (1997). A. Prokop, D. Hunkeler, and A. Cherrington (eds.). New York Academy of Sciences, ISBN 1573310999.

Bioceramics (1999). *Advanced Ceramics*, Vol. 1. J. F. Shackelford (ed.). Gordon and Breach Science Publishers, ISBN 9056996126.

Bioceramics 16 (2004). M. A. Barbosa, F. J. Monteiro, R. Correia, and B. Leon (eds.). Proceedings of the 16th International Symposium on Ceramics in Medicine, Porto, Portugal, 6–9 November. Trans Tech Publications, ISBN 0878499326.

Bioceramics of Calcium Phosphate (1983). K. De Groot (ed.). CRC Press, ISBN 0849364566.

Bioceramics: Material Characteristics versus in Vivo Behaviour (Annals of the New York Academy of Sciences, Vol. 523) (1988). P. Ducheyne and J. E. Lemons (eds.). New York Academy of Sciences, ISBN 089766437X.

Bioceramics: Materials and Applications II (Ceramic Transactions, Vol. 63) (1996). R. P. Rusin and G. S. Fischman (eds.). American Ceramic Society, ISBN 1574980068.

Bioceramics: Materials and Applications III (Ceramic Transactions, Vol. 110) (2000). L. George, R. P. Rusin, G. S. Fischman, and V. Janas (eds.). American Ceramic Society, ISBN 1574981021.

Biocompatibility Assessment of Medical Devices and Materials (1997). J. H. Braybrook (eds.). John Wiley & Sons, ISBN 0471965979.

Biodegradable Hydrogels for Drug Delivery (1993). K. Park, W. S. W. Shalaby, and H. Park (eds.). Technomic Publishing, ISBN 1566760046.

Bio-Implant Interface: Improving Biomaterials and Tissue Reactions (2003). J. E. Ellingsen and S. P. Lyngstadaas (eds.). CRC Press, ISBN 0849314747.

Bioinstrumentation and Biosensors (1991). D. L. Wise (ed.). Marcel Dekker, ISBN 0824783379.

Biological Performance of Materials (1999). J. Black (ed.). Marcel Dekker, ISBN 0824771060.

Biology in Physics: Is Life Matter? (Polymers, Interfaces and Biomaterials Series) (2000). K. Y. Bogdanov (ed.). Academic Press, ISBN 0121098400.

Biomaterials, 2nd edition (1992). J. B. Park and R. S. Lakes (eds.). Kluwer Academic/Plenum Publishers, ISBN 0306439921.

Biomaterials (1991). D. Byrom (ed.). Macmillan Press, ISBN 0333514076.

Biomaterials and Bioengineering Handbook (2000). D. L. Wise (ed.). Marcel Dekker, ISBN 0824703189.

Biomaterials for Drug and Cell Delivery (1994). Materials Research Society Symposium, Boston, USA (1993). A. G. Mikos, R. M. Murphy, and H. Bernstein (eds.). Materials Research Society, ISBN 1558992308.

Biomaterials for Drug Delivery and Tissue Engineering (Materials Research Society Symposia Proceedings, Vol. 662, November 27–29, 2000. Boston, MA) (2001). S. Mallapragada, M. Tracy, B. Narasimhan, E. Mathiowitz, and R. Korsmeyer (eds.). Materials Research Society, ISBN 1558995722.

Biomaterials in Artificial Organs (1985). J. P. Paul, J. M. Courtney, J. D. S. Gaylor, and T. Gilchrist (eds.). Wiley-VCH, ISBN 3527152342.

Biomaterials in the Design and Reliability of Medical Devices (2002). M. N. Helmus (ed.). Kluwer Academic, ISBN 0306476908.

Biomaterials, Mechanical Properties (ASTM Special Technical Publication, Stp 1173) (1994). H. E. Kambic and A. T. Yokobori (eds.). American Society for Testing & Materials, ISBN 0803118945.

Biomaterials Regualting Cell Function and Tissue Development (*Materials Research Society Symposium Proceedings*, Vol. 530, April 13–14, 1998, San Francisco) (2000). R. C. Thomson, D. J. Mooney, K. E. Healy, Y. Ikada, and A. F. Mikos (eds.). Material Research Society, ISBN 155899436X.

Biomaterials Science and Biocompatibility (1999). F. H. Silver and D. L. Christiansen (eds.). Springer Verlag, ISBN 0387987118.

Biomaterials Science and Engineering (1984). J. B. Park (ed.). Kluwer Academic/Plenum Publishers, ISBN 0306416891.

Biomaterials: An Introduction (1992). J. B. Park and R. S. Lakes (eds.). Plenum Publishers, ISBN 0306439921.

Biomaterials: Interfacial Phenomena and Applications (1982). S. L. Cooper and N. A. Peppas (eds.). American Chemical Society, ISBN 0841206317.

Biomaterials: Principles and Applications (2002). J. B. Park and J. D. Bronzino (eds.). CRC Press, ISBN 0849314917.

Biomaterial–Tissue Interfaces (Proceedings of the 9th European Conference on Biomaterials, Chester, UK, 1991) (1992). P. J. Doherty, R. L. Williams, D. F. Williams, and A. J. C. Lee (eds.). Elsevier Science, ISBN 0444890653.

Biomechanics of Medical Devices (1981). D. Ghista (ed.). Marcel Dekker, ISBN 0824768485.

Biomedical Applications of Polyurethanes (2001). P. Vermette, G. Laroche, and H. J. Griesser (eds.). Landes Bioscience, ISBN 158706023X.

Biomedical Materials: Drug Delivery, Implants and Tissue Engineering (symposium held November 30–December 1, 1998, Boston) (1999). T. Neenan, M. Marcolongo and R. F. Valentini (eds.). Materials Research Society, ISBN 1558994564.

Biomedical Polymers (1993). D. Cohn and J. Kost (eds.). Elsevier Science, ISBN 1858610435.

Biomedical Polymers: Designed to Degrade Systems (1994). S. W. Shalaby (ed.). Rapra Technology Ltd.

Biomimetic Materials Chemistry (1995). S. Mann (ed.). Wiley-VCH, ISBN 1560816694.

Biomineralization (2000). E. Baeuerlein (ed.). Wiley, ISBN 3527299874.

The Bionic Patient: Health Promotion for People with Implanted Prosthetic Devices (2002). F. E. Johnson and K. S. Virgo (eds.). Humana Press, ISBN 0896039595.

Biopolymer Methods in Tissue Engineering (Methods in Molecular Biology Series) (2003). A. P. Hollander and P. V. Hatton (eds.). Humana Press, ISBN 0896039676.

Biopolymers at Interfaces (1998). M. Malmsten (ed.). Marcel Dekker.

Biorelated Polymers and Gels: Controlled Releases and Applications in Biomedical Engineering (Polymers, Interfaces, and Biomaterials) (1998). T. Okano (ed.). Academic Press, ISBN 0125250908.

Biosensors (1992). F. Scheller and F. Schubert (eds.). Elsevier Science, ISBN 0444987835.

Biosensors in the Body: Continuous in Vivo Monitoring (1997). D. M. Fraser (ed.). John Wiley & Sons, ISBN 0471967076.

Blood Substitutes: Principles, Methods, Products and Clinical Trials (Tissue Engineering, Basel, Switzerland) (1997). T. M. S. Chang (ed.). S. Karger Publishing, ISBN 3805565844.

Boing-Boing the Bionic Cat (2000). L. Hench (ed.). Amer. Cream. Pub. ISBN 1574981099.

Bone Cements: Up-to-Date Comparison of Physical and Chemical Properties of Commercial Materials (2000). K. D. Kuhn (ed.). Springer-Verlag, ISBN 3540672079.

Bone Engineering (2000). J. E. Davies (ed.). Em Squared, ISBN 096869800X.

Bone Graft Substitutes (2003). C. T. Laurecin (ed.). ASTM, ISBN 080313561.

Cardiovascular Biomaterials (1991). G. W. Hastings (ed.). Springer-Verlag, ISBN 3540196668.

Cell Encapsulation Technology and Therapeutics (1999). W. M. Kuhtreiber, R. P. Lanza, and W. L. Chick (eds.). Birkhauser, ISBN 081764010X.

Chemical Sensors & Biosensors for Medical & Biological Applications (1998). U. E. Spichiger-Keller (ed.). Wiley, ISBN 3527288554.

Chitin: Fulfilling a Biomaterials Promise (2001). E. Khor (ed.). Elsevier, ISBN 0080440185.

Clinical Applications of Biomaterials (2nd European Conference on Biomaterials, Gothenburg, Sweden, 1981) (1981). A. J. C. Lee (ed.). John Wiley & Sons, ISBN 0471104035.

Clinical Implant Materials (*Advances in Biomaterials*, No. 9) (1990). G. Heimke, U. Soltesz, and A. J. C. Lee (eds.). Elsevier Science, ISBN 044488226X.

Cobalt-Base Alloys for Biomedical Applications (1999). J. A. Disegi, R. L. Kennedy, and R. Pilliar (eds.). ASTM, ISBN 0803126085.

The Coming Biotech Age: The Business of Bio-materials (2000). R. W. Oliver (ed.). McGraw-Hill, ISBN 0071350209.

Commodity Thermoplastics for Medical Applications (1998). R. C. Portnoy (ed.). Technomic Publishing, ISBN 1566766966.

Composite Materials for Implant Applications in the Human Body (1993). R. D. Jamison and L. N. Gilbertson (eds.). ASTM, ISBN 080311852X.

Computer Technology in Biomaterials Science and Engineering (1999). J. V. Sloten (ed.). Wiley, ISBN 0471976024.

Concise Encyclopedia of Medical & Dental Materials (1990). D. Williams (ed.). Elsevier Science, ISBN 0080361943.

Controlled Release Dosage Form Design (1999). C. Kim (ed.). Technomic Publishing, ISBN 1566768101.

Definitions in Biomaterials (Proceedings of a Consensus Conference of the European Society for Biomaterials, March 3–5, 1986, Chester, UK) (1987). D. F. Williams (ed.). Elsevier Science, ISBN 0444428585.

Degradable Polymers: Principles & Applications (1995). G. Scott and D. Gilead (eds.). Chapman & Hall, ISBN 0412590107.

Degradation Phenomena on Polymeric Biomaterials (1992). H. Plank, M. Dauner, and M. Renardy (eds.). Springer-Verlag, ISBN 354055548X.

Dental Biomaterials (1999). E. Combe, W. H. Douglas, and T. Burke (eds.). Kluwer Academic, ISBN 0792385314.

Dental Implants: The Art and Science (2001). C. A. Babbush (ed.). W. B. Saunders Co, ISBN 0721677479.

Design and Applications of Hydrophilic Polyurethanes—Medical, Agricultural and Other Applications (2000). T. Thomson (ed.). Technomic Publishing, ISBN 1566768950.

Design Engineering of Biomaterials for Medical Devices (1997). D. Hill (ed.). John Wiley & Sons, ISBN 0471967084.

Design of Biomedical Devices and Systems (2003). P. H. King, R. C. Fires, and N. A. Donner (eds.). Marcel Dekker, ISBN 082470889X.

Drug Delivery: Engineering Principles for Drug Therapy (Topics in Chemical Engineering) (2001). W. M. Saltzman (ed.). Oxford University Press, ISBN 0195085892.

Electroforming in Restorative Dentistry: New Dimensions in Biologically Based Prostheses (2000). J. Wirz and A. Hoffman (eds.). Quintessence Publishing, ISBN 0867153768.

Encyclopedic Handbook of Biomaterials and Bioengineering (4 volumes) (1995). D. L. Wise, D. J. Trantolo, D. E. Altobelli, M. J. Yaszemski, J. D. Gresser, and E. R. Schwartz (eds.). Marcel Dekker, ISBN 0824796497.

Formulation and Delivery of Proteins and Peptides (ACS Symposium Series, No. 567) (1994). J. L. Cleland and R. Langer (eds.). American Chemical Society, ISBN 0841229597.

Frontiers in Biomedical Polymer Applications, Vol. 1 (1998). R. M. Ottenbrite (ed.). Technomic, ISBN 1566765773.

Frontiers in Biomedical Polymer Applications, Vol. 2 (1999). R. M. Ottenbrite (ed.). Technomic Publishing, ISBN 1566767148.

Frontiers in Tissue Engineering (1998). C. W. Patrick Jr, A. G. Mikos, and L. V. McIntire (eds.). Elsevier Science, ISBN 0080426891.

Functional Behavior of Orthopedic Biomaterials: Applications (1984). P. Ducheyne (ed.). CRC Press, ISBN 0849362660.

Functional Biomaterials (*Key Engineering Materials*, Vol 1 198–199) (2001). N. Katsube, W. O. Soboyejo, and M. Sacks (eds.). Trans Tech Publications, ISBN 0878498710.

Functional Tissue Engineering (2003). F. Guilak, D. L. Butler, S. A. Goldstein, and D. J. Mooney (eds.). Springer-Verlag, ISBN 0387955534.

Future Strategies for Tissue and Organ Replacement (2002). J. M. Polak, L. L. Hench, and P. Kemp (eds.). World Scientific, ISBN 1860943101.

Gels Handbook (2000). Y. Osada and K. Kajiwara (eds.). Academic Press, ISBN 0123946905.

Gene Therapy and Tissue Engineering in Orthopaedic and Sports Medicine (*Methods in Bioengineering*, Vol. XX) (2000). J. Huard and F. H. Fu (eds.). Birkhauser, ISBN 0817640711.

Handbook of Biomaterials Evaluation: Scientific, Technical, and Clinical Testing of Implant Materials (1999). A. V. Recum (ed.). Hemisphere Pub, ISBN 1560324791.

Handbook of Biomaterials Properties (1998). J. Black and G. Hastings (eds.). Chapman & Hall, ISBN 0412603306.

Handbook of Medical Device Design (2000). R. C. Fries (ed.). Marcel Dekker, ISBN 0824703995.

Handbook of Pharmaceutical Controlled Release Technology (2000). D. L. Wise (ed.). Marcel Dekker, ISBN 0824703693.

Human Biomaterials Applications (1999). M. J. Yaszemski, D. L. Wise, D. J. Trantolo, and D. E. Altobelli (eds.). Humana Press, ISBN 0896033376.

Hydrophilic Polymer Coatings for Medical Devices—Structure/ Properties, Development Manufacture and Applications (1997). R. J. Laporte (ed.). Technomic Publishing, ISBN 1566765048.

Hydroxyapatite and Related Materials (1994). P. W. Brown and B. Constantz (eds.). CRC Press, ISBN 0849347505.

Imaging Techniques in Biomaterials (1994). M. A. Barbosa and A. Campilho (eds.). North-Holland—Elsevier Science, ISBN 0444897747.

Implantation Biology: The Host Response and Biomedical Devices (1994). R. S. Greco (ed.). CRC Press, ISBN 0849344328.

Implants and Restorative Dentistry (2000). G. M. Scortecci, C. E. Misch, and K. U. Benner (eds.). Martin Dunitz Publishers, ISBN 1853177032.

Innovation and Invention in Medical Devices: Workshop Summary (2001). K. E. Hanna, F. J. Manning, P. Bouxsein, and A. Pope (eds.). National Academy Press, ISBN 0309082552.

Interfaces in Biomaterials Sciences (European Materials Research Society Symposium, 13) (1990). D. Muster (ed.). North-Holland, ISBN 0444888365.

Interventional and Surgical Cardiovascular Pathology: Clinical Correlations and Basic Principles (1989). F. J. Schoen and W. B. Saunders. ASIN 072162457X.

An Introduction to Bioceramics (1993). L. L. Hench and J. Wilson (eds.). American Ceramic Society, ISBN 9810214006.

An Introduction to Tissue–Biomaterials Interactions (2002). K. C. Dee, D. A. Puleo, and R. Bizios (eds.). Wiley-Liss, ISBN 0471253944.

In Vitro Testing of Biomaterials (1995). D. B. Jones (ed.). Kluwer Academic Publishers, ISBN 0412564807.

Macromolecular Biomaterials (1984). G. W. Hastings and P. Ducheyne (eds.). CRC Press, ISBN 0849362636.

Macromolecular Symposia 172: Polymers in Medicine (2001). J. Kahovec (ed.). Wiley-VCH, ISBN 3527303340.

Materials Science and Technology—A Comprehensive Treatment, Vol. 14, *Medical and Dental Materials* (1992). R. W. Cahn and E. J. Kramer (eds.). John Wiley & Sons.

Materials Technology Foresight in Biomaterials (1995). Materials Strategy Commission (eds.). Institute of Materials, ISBN 0901716871.

Mechanical Testing of Bone and the Bone Implant Interface (Stp, No. 1272) (1999). Y. H. An and R. A. Draughn (eds.). CRC Press, ISBN 0849302668.

Medical Applications of Titanium and Its Alloys: The Material and Biological Issues (1997). S. A. Brown and J. E. Lemons (eds.). American Society for Testing & Materials, ISBN 0803120109.

Medical Device Packaging Handbook, 2nd edition (1998). M. Sherman (eds.). Marcel Dekker.

Medical Device Sterilization (1999). A. Booth *et al.* (eds.). Technomic Publishing, ISBN 1566768497.

Medical Devices—International Perspectives on Health and Safety (1994). C. W. D. Van Gruting (ed.). Elsevier Science, ISBN 0444892427.

Medical Plastics Degradation Resistance and Failure Analysis (1998). R. C. Portnoy (ed.). Plastics Design Library/Society of Plastics Engineers.

Medical Textiles '96 (Proceedings, International Conference, 17–18 July 1996, Bolton, UK) (1997). Woodhead Publishing.

Medical Textiles for Implantation (Proceedings of the 3rd International ITV Conference on Biomaterials, Stuttgart, Germany, 1989) (1991). H. Planck, M. Dauner, and M. Renardy (eds.). Springer-Verlag, ISBN 0387527419.

Metals as Biomaterials (1998). J. A. Helsen and J. Breme (eds.). John Wiley & Sons, ISBN 0471969354.

Methods of Tissue Engineering (2001). A. Atala and R. P. Lanza (eds.). Academic Press, ISBN 0124366368.

Microfabrication in Tissue Engineering and Bioartificial Organs (Microsystems Series, 5) (1999). S. Bhatia (ed.). Kluwer Academic Pub, ISBN 0792385667.

Mineralization in Natural and Synthetic Biomaterials (*Materials Research Society Symposia Proceedings*, V. 599, November 29– December 1, 1999, Boston) (2000). P. Li (ed.). Material Research Society, ISBN 1558995072.

Modern Aspects of Protein Adsorption on Biomaterials (1991). Y. F. Missirlis and W. Lemm (eds.). Kluwer Academic Publishers, ISBN 0792309731.

Monitoring of Orthopedic Implants: A Biomaterials–Microelectronics Challenge (1993). F. Burny and R. Puers (eds.). North-Holland, ISBN 0444816208.

Natural and Living Biomaterials (1984). G. W. Hastings and P. Ducheyne (eds.). CRC Press, ISBN 0849362644.

Neural Prostheses for Restoration of Sensory and Motor Function (2000). J. K. Chapin and K. A. Moxon (eds.). CRC Press, ISBN 0849322251.

Novartis Foundation Symposium 249—Tissue Engineering of Cartilage and Bone (2003). Novartis (eds.). John Wiley & Sons, ISBN 0470844817.

Oral Implantology and Biomaterials (*Progress in Biomedical Engineering*, Vol. 7) (1989). H. Kawahara (ed.). Elsevier Science, ISBN 0444873473.

Physics of Biomaterials: Fluctuations, Self-Assembly and Evolution (1996). T. Riste (ed.). Kluwer Academic Publishers, ISBN 0792341317.

Piezoelectricity in Biomaterials and Biomedical Devices (1984). P. M. Galletti (ed.). Gordon & Breach Science Publishers, ISBN 0677404859.

Polymer Biomaterials in Solution, as Interfaces and as Solids (1995). S. L. Cooper, C. H. Bamford, and T. Tsuruta (eds.). VSP International Science Publishers, ISBN 9067641804.

Polymeric Biomaterials, 2nd edition (2001). S. Dumitriu (ed.). Marcel Dekker, ISBN 0824705696.

Polymeric Drugs and Drug Delivery Systems (2000). R. M. Ottenbrite and S. W. Kim (eds.). Technomic Publishing, ISBN 1566769566.

Polymers as Biomaterials (1985). S. W. Shalaby (ed.). Plenum, ISBN 030641886X.

Polymers for the Medical Industry (Conference Book of Papers, 29–30 November 1999, London) (1999). Rapra Technology, ISBN 1859572014.

Polymers for Tissue Engineering (1998). M. S. Shoichet and J. A. Hubbell (eds.). VSP ISBN 9067642894.

Polymers in Disposable Medical Devices—A European Perspective (Industry Analysis Report) (1999). C. R. Blass (ed.). Rapra Technology, ISBN 1859571824.

Polymers in Medical Applications (2001). B. J. Lambert, F.-W. Tang, and W. J. Rogers (eds.). Rapra Technology, ISBN 1859572596.

Polymers in Medicine: Biomedical and Pharmaceutical Applications (1992). R. M. Ottenbrite and E. Chiellini (eds.). Technomic Publishing, ISBN 0877629293.

Polymers: Biomaterials and Medical Applications (Encyclopedia Reprint Series) (1989). J. I. Kroschwitz (ed.). John Wiley & Sons, ISBN 0471512079.

Polyurethanes in Biomedical Applications (1998). N. M. K. Lamba, K. A. Woodhouse and S. L. Copper (eds.). CRC Press, ISBN 0849345170.

Porous Materials for Tissue Engineering (1997). D.-M. Liu and V. Dixit (eds.). Trans Tech Publications, ISBN 0878497730.

Principles of Tissue Engineering (2000). R. P. Lanza, R. Langer, and J. P. Vacanti (eds.). Academic Press, ISBN 0124366309.

Properties of Biomaterials in the Physiological Environment (1980). S. D. Bruck (ed.). CRC Press, ISBN 0849356857.

Protein-Based Materials (Bioengineering of Materials, Vol. XX) (1997). K. McGrath and D. Kaplan (eds.). Springer-Verlag, ISBN 0817638482.

Quantitative Characterization and Performance of Porous Implants for Hard Tissue Applications (Stp 953) (1998). J. E. Lemons (ed.). American Society for Testing & Materials, ISBN 0803109652.

Reconstructing the Body: Vol. I—*Implants in Surgery* and Vol. II—*Biomaterials and Tissue Engineering for the 21st Century* (2000). D. F. Williams (ed.). Liverpool University Press, ISBN 0853236658.

Reparative Medicine: Growing Tissues and Organs (Annals of the New York Academy of Medicine, Vol. 961) (2002). J. D. Sipe, C. A. Kelley and L. A. McNicol (eds.). New York Academy of Sciences, ISBN 1573313823.

The Role of Platelets in Blood–Biomaterial Interactions (1993). Y. F. Missirlis and J.-L. Wautier (eds.). Kluwer Academic Publishers, ISBN 0792321626.

The Role of Poly(vinyl chloride) in Healthcare (2001). C. R. Blass (eds.). Rapra Technology, ISBN 1859572588.

Safety Evaluation of Medical Devices (1997). S. C. Gad (ed.). Marcel Dekker, ISBN 0824798279.

Silicones for Pharmaceutical and Biomedical Applications (1998). A. L. Himstedt (ed.). Technomic Publishing, ISBN 1566766184.

Spare Parts: Organ Replacement in American Society (1992). R. C. Fox and J. P. Swazey (eds.). Oxford University Press, ISBN 0195076508.

Stainless Steels for Medical and Surgical Applications (2003). G. L. Winters and M. J. Nutt (eds.). ASTM, ISBN 0803134592.

Stem Cell Transplantation and Tissue Engineering (2002). A. Haverich and H. Graf (eds.). Springer-Verlag, ISBN 3540414959.

Structural Biomaterials (revised edition) (1990). J. F. V. Vincent (ed.). Princeton University Press, ISBN 0691025134.

Supramolecular and Colloidal Structures in Biomaterials and Biosubstrates (2000). M. Lal, P. J. Lillford, V. M. Naik, and V. Prakash (eds.). Imperial College Press, ISBN 1860942369.

Surface Analytical Techniques for Probing Biomaterial Processes (1996). J. Davies, A. De Millo, and L. R. Fisher (eds.). CRC Press, ISBN 0849383528.

Surface and Interfacial Aspects of Biomedical Polymers: Surface Chemistry and Physics (1985). J. D. Andrade (ed.). Plenum Press, ISBN 0306417413.

Surface Characterization of Biomaterials (Progress in Biomedical Engineering, Vol. 6) (1988). B. D. Ratner (ed.). Elsevier Science, ISBN 0444430164.

Surface Modification of Polymeric Biomaterials (Proceedings of the American Chemical Society International Symposium, Division of Polymer Chemistry, Anaheim, 1995) (1997). B. D. Ratner and D. G. Castner (eds.). Plenum Publishers, ISBN 0306455129.

Surface Properties of Biomaterials (1994). R. West and Batts (eds.). Butterworth–Heinemann, ISBN 0750622059.

Surgical Adhesives and Sealants: Current Technology and Applications (1996). D. Sierra and R. Saltz (eds.). Technomic Publishing, ISBN 1566763274.

Synthetic Bioabsorbable Polymers for Implants (2000). C. M. Agrawal, J. E. Parr, and S. T. Lin (eds.). ASTM, ISBN 0803128703.

Synthetic Biodegradable Polymer Scaffolds (Tissue Engineering, Vol. XX) (1997). A. Atala, D. J. Mooney, and A. Arbor (eds.). Birkhauser, ISBN 0817639195.

Synthetic Polymers for Biotechnology and Medicine (2003). R. Freitag (ed.). Landes Bioscience, ISBN 1587060272.

Szycher's Dictionary of Biomaterials & Medical Devices (1992). M. Szycher (ed.). Technomic Publishing, ISBN 0877628823.

Test Procedures for the Blood Compatibility of Biomaterials (1993). S. Dawids (ed.). Kluwer Academic Publishers, ISBN 0792321073.

Thin Films and Surfaces for Bioactivity and Biomedical Applications (1996). C. M. Cotell, A. E. Meyer, S. M. Gorbatkin, and G. L. Grobe III (eds.). Materials Research Society, ISBN 1558993177.

Tissue and Organ Regeneration in Adults (2001). I. V. Yannas (ed.). Springer, ISBN 0387952144.

Tissue Engineering (Principles and Applications in Engineering Series) (2003). B. Palsson, J. A. Hubbell, R. Plonsey, and J. D. Bronzino (eds.). Parthenon, ISBN 0849318122.

Tissue Engineering (2003). B. O. Palsson and S. N. Bhatia (eds.). Prentice Hall, ISBN 0130416967.

Tissue Engineering for Therapeutic Use 1 (Proceedings of the First International Symposium of Tissue Engineering for Therapeutic Use, Kyoto) (1998). International Symposium of Tissue Engineering for Therapeutic Use 1997 (eds.). Elsevier Science, ISBN 0444829938.

Tissue Engineering for Therapeutic Use 3 (1999). International Symposium of Tissue Engineering for Therapeutic Use 1998 (ed.). Pergamon Press, ISBN 0444500294.

Tissue Engineering for Therapeutic Use 4 (2000). Y. Ikada and Y. Shimizu (eds.). Excerpta Medica, ISBN 0444502939.

Tissue Engineering Methods and Protocols (Methods in Molecular Medicine, Vol. 18) (1998). J. R. Morgan, M. L. Yarmush, and J. R. Morgan (eds.). Humana Press, ISBN 0896035166.

Tissue Engineering of Vascular Prosthetic Grafts (1999). P. P. Zilla and H. P. Greisler (eds.). R. G. Landes, ISBN 1570595496.

Tissue Engineering: Current Perspectives (1994). E. Bell (ed.). Birkhauser, ISBN 0817636870.

Tissue Engineering: Dental Applications and Future Directions (1999). S. E. Lynch, R. Genco, and R. Marx (eds.). Quintessence Publishing, ISBN 0867153466.

Tissue-Inducing Biomaterials (Materials Research Society Symposium, Boston, 1991) (1992). L. G. Cima and E. S. Ron (eds.). Materials Research Society, ISBN 1558991468.

Titanium in Medicine: Material Science, Surface Science, Engineering, Biological Responses and Medical Applications (2001). D. M. Brunette, P. Tengvall, and M. Textor (eds.). Springer-Verlag, ISBN 3540669361.

Viscoelasticity of Biomaterials (1992). W. G. Glasser and H. Hatakeyama (eds.). American Chemical Society, ISBN 0841222215.

Water in Biomaterials Surface Science (2001). M. Morra (ed.). Nobil Bio recherche, ISBN 0471490415.

Williams Dictionary of Biomaterials (1999). D. F. Williams (eds.). Liverpool University Press, ISBN 0853239215.

Wound Closure Biomaterials and Devices (1996). C.-C. Chu and J. A. Von Fraunhofer (eds.). CRC Press, ISBN 0849349648.

BIOMATERIALS JOURNALS

Advanced Drug Delivery Reviews (Elsevier)

American Journal of Drug Delivery (Adis International)

American Society of Artificial Internal Organs Transactions

Annals of Biomedical Engineering (Blackwell—Official Publication of the Biomedical Engineering Society)

Annual Reviews of Biomedical Engineering

Artificial Organs (Raven Press)

Artificial Organs Today (ed. T. Agishi; VSP Publishers)

Biomacromolecules (American Chemical Society)

Biofouling (Harwood Academic Publishers)

Biomaterial–Living System Interactions (BioMir; Sevastianov, ed.)

Biomaterials (including *Clinical Materials*) (Elsevier)

Biomaterials, Artificial Cells and Artificial Organs (TMS Chang, ed.)

Biomaterials Forum (Society For Biomaterials)

Biomaterials: Processing, Testing and Manufacturing Technology (Butterworth)

Biomedical Engineering OnLine (electronic—http://www.biomedical-engineering-online.com/start.asp)

Biomedical Materials (Elsevier)

Bio-medical Materials and Engineering (ed. T. Yokobori, Pergamon Press)

Biomedical Microdevices (Kluwer)

Biosensors and Bioelectronics (Elsevier)

Cells and Materials (Scanning Microscopy International)

Cell Transplantation (Pergamon)

Clinical Biomechanics

Colloids and Surfaces B: Biointerfaces (Elsevier)

Dental Materials

Drug Delivery Systems & Sciences (Euromed Scientific)

Drug Targeting and Delivery (Academic Press)

Drug Delivery Technology

e-biomed: the Journal of Regenerative Medicine (http://rudolfo.ingentaselect.com/vl=25808836/cl=12/nw=1/rpsv/catchword/mal/15248909/contpl.htm)

European Cells and Materials (electronic—http://www.eurocellmat.org.uk)

Frontiers of Medical and Biological Engineering (Y. Sakurai, ed.; VSP Publishers)

IEEE Transactions on Biomedical Engineering

International Journal of Artificial Organs (Wichtig Editore)

Journal of Bioactive and Compatible Polymers (Technomics)

Journal of Biomaterials Applications (Technomics)

Journal of Biomaterials Science: Polymer Edition (VSP Publishers)

Journal of Biomedical Materials Research (Wiley; Official Publication of the Society for Biomaterials)

Journal of Biomedical Materials Research: Applied Biomaterials (Wiley)

Journal of Controlled Release (Elsevier)

Journal of Drug Targeting (Harwood Academic Publishers)

Journal of Engineering in Medicine (Institution of Mechanical Engineers)

Journal of Long Term Effects of Medical Implants (CRC Press)

Materials in Medicine (Chapman and Hall; Official Publication of the European Society for Biomaterials)

Medical Device and Diagnostics Industry (Canon Publications)

Medical Device Research Report (AAMI)

Medical Device Technology (Astor Publishing Corporation)

Medical Plastics and Biomaterials (Canon Communications, Inc.)

Nanobiology (Carfax Publishing Co.)

Nanobioscience

Nanobiotechnology

Nanotechnology (An Institute of Physics Journal)

Regenerative Medicine

Tissue Engineering (Mary Ann Liebert, Inc.)

Trends in Biomaterials & Artificial Organs (Society For Biomaterials And Artificial Organs—India)

SOME BIOMATERIALS BOOKS

Bibliography

Black, J. (1992). *Biological Performance of Materials: Fundamentals of Biocompatibility*, 2nd ed. Marcel Dekker, New York.

Boretos, J. W., and Eden, M. (eds.) (1984). *Contemporary Biomaterials: Material and Host Response, Clinical Applications, New Technology and Legal Aspects*. Noyes Publ., Park Ridge, NJ.

Glasgold, A. I., and Silver, F. H. (1991). *Applications of Biomaterials in Facial Plastic Surgery*, CRC Press, Boca Raton, FL.

Heimke, G. (1990). *Osseo-Integrated Implants*, CRC Press, Boca Raton, FL.

Hench, L. L., and Ethridge, E. C. (1982). *Biomaterials: An Interfacial Approach*, Academic Press, New York.

Park, J. B. (1979). *Biomaterials: An Introduction*, Plenum, New York.

Park, J. B. (ed.) (1984). *Biomaterials Science and Engineering*, Plenum, New York.

Schoen, F. J. (1989). *Interventional and Surgical Cardiovascular Pathology: Clinical Correlations and Basic Principles*, W. B. Saunders, Philadelphia.

Silver, F. H., and Doillon, C. (1989). *Biocompatibility: Interactions of Biological and Implanted Materials*, Vol. 1—Polymers, VCH, New York.

Von Recum, A. F. (ed.) (1986). *Handbook of Biomaterials Evaluation*, Macmillan, New York.

Williams, D. (ed.) (1990). *Concise Encyclopedia of Medical and Dental Materials*, Pergamon Press, Oxford.

Yamamuro, T., Hench, L. L., and Wilson, J. (1990). *CRC Handbook of Bioactive Ceramics*, CRC Press, Boca Raton, FL.

Index